Slope and Average Rate of Change

Slope of a line through (x_1, y_1) and (x_2, y_2): $m = \dfrac{\Delta y}{\Delta x} = \dfrac{y_2 - y_1}{x_2 - x_1}$

Average rate of change of $f(x)$ between (x_1, y_1) and (x_2, y_2): $\dfrac{f(x_2) - f(x_1)}{x_2 - x_1}$

Difference quotient: $\dfrac{f(x + h) - f(x)}{h}$

Graphs of Basic Functions

Constant Function

$f(x) = b$

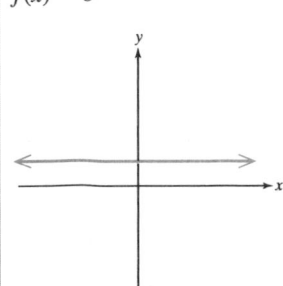

Linear Function

$f(x) = mx + b$

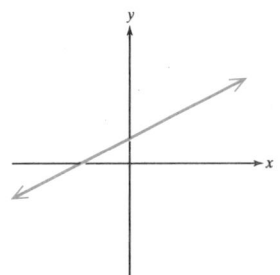

Identity Function

$f(x) = x$

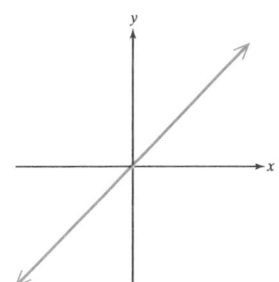

Quadratic Function

$f(x) = x^2$

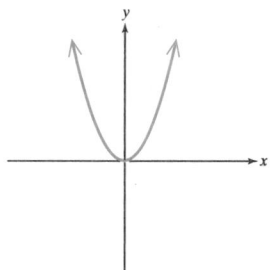

Cubic Function

$f(x) = x^3$

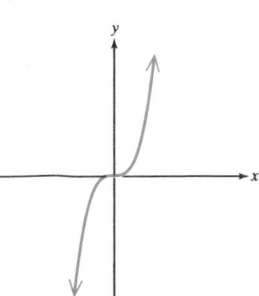

Absolute Value Function

$f(x) = |x|$

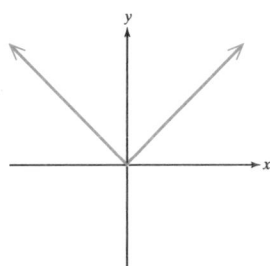

Square Root Function

$f(x) = \sqrt{x}$

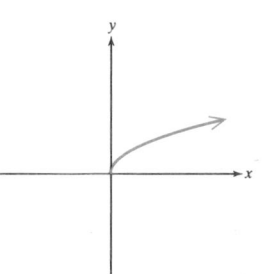

Cube Root Function

$f(x) = \sqrt[3]{x}$

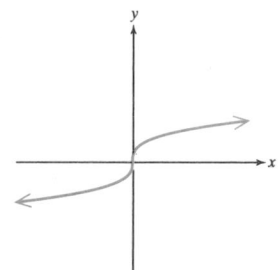

Reciprocal Function

$f(x) = \dfrac{1}{x}$

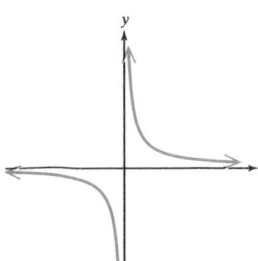

Greatest Integer Function

$f(x) = [x]$

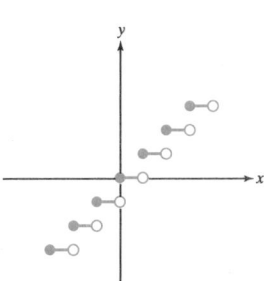

Exponential Function

$f(x) = b^x$, where $b > 0$ and $b \neq 1$

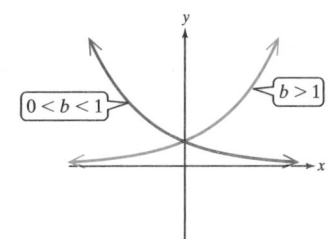

Logarithmic Function

$f(x) = \log_b x$, where $b > 0$ and $b \neq 1$

$y = \log_b x \Leftrightarrow b^y = x$

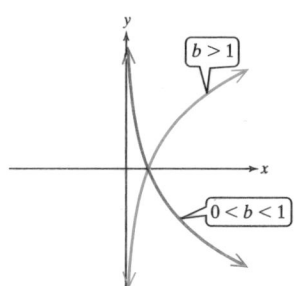

Transformations of Graphs

Given $c > 0$ and $h > 0$, the graph of the given function is related to the graph of $y = f(x)$ as follows:

$y = f(x) + c$ Shift the graph of $y = f(x)$ up c units.
$y = f(x) - c$ Shift the graph of $y = f(x)$ down c units.

$y = f(x - h)$ Shift the graph of $y = f(x)$ to the right h units.
$y = f(x + h)$ Shift the graph of $y = f(x)$ to the left h units.

$y = -f(x)$ Reflect the graph of $y = f(x)$ over the x-axis.
$y = f(-x)$ Reflect the graph of $y = f(x)$ over the y-axis.

$y = af(x)$ If $a > 1$, stretch the graph of $y = f(x)$ vertically by a factor of a.
   If $0 < a < 1$, shrink the graph of $y = f(x)$ vertically by a factor of a.

$y = f(ax)$ If $a > 1$, shrink the graph of $y = f(x)$ horizontally by a factor of $\frac{1}{a}$.
   If $0 < a < 1$, stretch the graph of $y = f(x)$ horizontally by a factor of $\frac{1}{a}$.

Tests for Symmetry

Consider the graph of an equation in x and y. The graph of the equation is

- Symmetric to the **y-axis** if substituting $-x$ for x results in an equivalent equation.

- Symmetric to the **x-axis** if substituting $-y$ for y results in an equivalent equation.

- Symmetric to the **origin** if substituting $-x$ for x and $-y$ for y results in an equivalent equation.

Even and Odd Functions

- f is an **even function** if $f(-x) = f(x)$ for all x in the domain of f.

- f is an **odd function** if $f(-x) = -f(x)$ for all x in the domain of f.

Properties of Logarithms

$\log_b 1 = 0$ $\log_b (xy) = \log_b x + \log_b y$

$\log_b b = 1$ $\log_b \left(\dfrac{x}{y}\right) = \log_b x - \log_b y$

$\log_b b^x = x$ $\log_b x^p = p \log_b x$

$b^{\log_b x} = x$

Change-of-base formula: $\log_b x = \dfrac{\log_a x}{\log_a b}$

$b^x = b^y$ implies that $x = y$.

$\log_b x = \log_b y$ implies that $x = y$.

Variation

y varies **directly** as x.
y is **directly** proportional to x. $\Big\} \; y = kx$

y varies **inversely** as x.
y is **inversely** proportional to x. $\Big\} \; y = \dfrac{k}{x}$

y varies **jointly** as w and x.
y is **jointly** proportional to w and x. $\Big\} \; y = kwx$

Precalculus

Julie Miller
Daytona State College

Donna Gerken
Miami-Dade College

Mc
Graw
Hill
Education

PRECALCULUS

Published by McGraw-Hill Education, 2 Penn Plaza, New York, NY 10121. Copyright © 2017 by McGraw-Hill Education. All rights reserved. Printed in the United States of America. No part of this publication may be reproduced or distributed in any form or by any means, or stored in a database or retrieval system, without the prior written consent of McGraw-Hill Education, including, but not limited to, in any network or other electronic storage or transmission, or broadcast for distance learning.

Some ancillaries, including electronic and print components, may not be available to customers outside the United States.

This book is printed on acid-free paper.

2 3 4 5 6 7 8 9 0 DOW/DOW 1 0 9 8 7 6

ISBN 978–0–07–803560–9
MHID 0-07-803560–0

ISBN 978–0–07–753818–7 (Annotated Instructor's Edition)
MHID 0-07-753818–8

Senior Vice President, Products & Markets: *Kurt L. Strand*
Vice President, General Manager, Products & Markets: *Marty Lange*
Vice President, Content Design & Delivery: *Kimberly Meriwether David*
Managing Director: *Ryan Blankenship*
Brand Manager: *Caroline Celano*
Director, Product Development: *Rose Koos*
Product Developer: *Emily Windelborn*
Director of Marketing: *Alex Gay*
Marketing Specialist: *Cherie Harshman*
Director of Digital Content Development: *Robert Brieler*
Digital Product Analyst: *Michael Lemke*
Associate Digital Product Analyst: *Adam Fischer*
Director, Content Design & Delivery: *Linda Avenarius*
Program Manager: *Lora Neyens*
Content Project Manager: *Peggy J. Selle*
Buyer: *Jennifer Pickel*
Design: *David W. Hash*
Content Licensing Specialist (Image): *Carrie Burger*
Cover Image: *©LewLong/CORBIS*
Compositor: *Aptara®, Inc.*
Typeface: *10.5/12 pt. Times LT Std.*
Printer: *R.R. Donnelley*

Library of Congress Cataloging-in-Publication Data

Names: Miller, Julie, 1962- | Gerken, Donna.
Title: Precalculus/Julie Miller, Donna Gerken.
Description: First edition. | New York, NY: McGraw-Hill Education, [2017] |

Includes index.
Identifiers: LCCN 2015040407 | ISBN 9780078035609 (alk. paper) | ISBN 0078035600 (alk. paper) |
ISBN 9780077538187 (alk. paper) | ISBN 0077538188 (alk. paper)
Subjects: LCSH: Precalculus—Textbooks. | Functions—Textbooks.
Classification: LCC QA303.2 .M534 2017 | DDC 510—dc23 LC record available at http://lccn.loc.gov/2015040407

The Internet addresses listed in the text were accurate at the time of publication. The inclusion of a website does not indicate an endorsement by the authors or McGraw-Hill Education, and McGraw-Hill Education does not guarantee the accuracy of the information presented at these sites.

About the Authors

Julie Miller is from Daytona State College where she has taught developmental and upper-level mathematics courses for 20 years. Prior to her work at DSC, she worked as a software engineer for General Electric in the area of flight and radar simulation. Julie earned a bachelor of science in applied mathematics from Union College in Schenectady, New York, and a master of science in mathematics from the University of Florida. In addition to this textbook, she has authored textbooks in developmental mathematics, trigonometry, and precalculus, as well as several short works of fiction and nonfiction for young readers.

"My father is a medical researcher, and I got hooked on math and science when I was young and would visit his laboratory. I remember doing simple calculations with him and using graph paper to plot data points for his experiments. He would then tell me what the peaks and features in the graph meant in the context of his experiment. I think that applications and hands-on experience made math come alive for me, and I'd like to see math come alive for my students."

Donna Gerken is a professor at Miami Dade College where she has taught developmental courses, honors classes, and upper-level mathematics classes for decades. Throughout her career she has been actively involved with many projects at Miami Dade including those on computer learning, curriculum design, and the use of technology in the classroom. Donna's bachelor of science in mathematics and master of science in mathematics are both from the University of Miami.

Letter from the Authors

Precalculus is a foundation course essential to success in Calculus and all subsequent courses in science, technology, engineering, and mathematics. To support student success in this course, we accommodate a variety of learning styles by bringing together a seamless integration of print and digital content. The clear, concise writing style and pedagogical features of our textbook continue throughout the online content in ConnectMath, throughout our instructional videos, and throughout the adaptive reading and learning experience of SmartBook. Furthermore, to enhance preparation for Calculus, we have dedicated exercise sets that specifically target skills needed in Calculus.

The main objectives of this Precalculus textbook and its print and digital content are threefold:

- To provide students with a clear and logical presentation of fundamental concepts that will prepare them for continued study in mathematics.
- To help students develop logical thinking and problem-solving skills that will benefit them in all aspects of life.
- To motivate students by demonstrating the significance of mathematics in their lives through practical applications.

Julie Miller julie.miller.math@gmail.com

donna gerken dgerken@mdc.edu

Dedications

To our students: Enjoy your journey with mathematics and the doors it opens to science, business, and the natural world around you. —Julie Miller

For Eddie and Aurora—true friendship needs no words, but it does require the ability to appear interested while someone talks about trigonometry applications during every shared holiday dinner. —Donna Gerken

Table of Contents

CHAPTER **3**   Exponential and Logarithmic Functions   355

CHAPTER **4**   Trigonometric Functions   445

CHAPTER **5**   Analytic Trigonometry   567

CHAPTER 6   Applications of Trigonometric Functions   629

CHAPTER 7   Trigonometry Applied to Polar Coordinate
Systems and Vectors   685

CHAPTER 8   Systems of Equations and Inequalities   765

Key Features

Clear, Precise Writing

Because a diverse group of students take this course, Julie Miller and Donna Gerken have written this manuscript to use simple and accessible language. Through their friendly and engaging writing style, students are able to understand the material easily.

Exercise Sets

The exercises at the end of each section are graded, varied, and carefully organized to maximize student learning:

- **Prerequisite Review Exercises** begin the section-level exercises and ensure that students have the foundational skills to complete the homework sets successfully.
- **Concept Connections** prompt students to review the vocabulary and key concepts presented in the section.
- **Core Exercises** are presented next and are grouped by objective. These exercises are linked to examples in the text and direct students to similar problems whose solutions have been stepped-out in detail.
- **Mixed Exercises** do *not* refer to specific examples so that students can dip into their mathematical toolkit and decide on the best technique to use.
- **Write About It** exercises are designed to emphasize mathematical language by asking students to explain important concepts.
- **Technology Connections** require the use of a graphing utility and are found at the end of exercise sets. They can be easily skipped for those who do not encourage the use of calculators.
- **Expanding Your Skills Exercises** challenge and broaden students' understanding of the material.

Problem Recognition Exercises

Problem Recognition Exercises appear in strategic locations in each chapter of the text. These exercises provide students with an opportunity to synthesize multiple concepts and decide which problem-solving technique to apply to a given problem.

Exercises for Calculus

- **Algebra for Calculus** exercises specifically target the skills students use in calculus such as the simplification of a limit, and the simplification after taking a derivative.
- **Equations and Inequalities for Calculus** is another exercise set that provide sample equations and inequalities seen in calculus. These include the important skills of finding the values that make an expression zero or undefined to find the critical numbers in calculus.

Examples

- The examples in the textbook are stepped-out in detail with thorough annotations at the right explaining each step.
- Following each example is a similar **Skill Practice** exercise to engage students by practicing what they have just learned.
- For the instructor, references to an even-numbered exercise are provided next to each example. These exercises are highlighted with blue circles in the exercise sets and mirror the related examples. With increased demands on faculty time, this has been a popular feature that helps faculty write their lectures and develop their presentation of material. If an instructor presents all of the highlighted exercises, then each objective of that section of text will be covered.

Modeling and Applications

One of the most important tools to motivate our students is to make the mathematics they learn meaningful in their lives. The textbook is filled with robust applications and numerous opportunities for mathematical modeling for those instructors looking to incorporate these features into their course.

Callouts

Throughout the text, popular tools are included to highlight important ideas. These consist of:

- **Tip** boxes that offer additional insight to a concept or procedure.
- **Avoiding Mistakes** boxes that fend off common mistakes.
- **Point of Interest** boxes that offer interesting and historical mathematical facts.
- **Instructor Notes** to assist with lecture preparation.

Graphing Calculator Coverage

Material is presented throughout the book illustrating how a graphing utility can be used to view a concept in a graphical manner. The goal of the calculator material is not to replace algebraic analysis, but rather, to enhance understanding with a visual approach. Graphing calculator examples are placed in self-contained boxes and may be skipped by instructors who choose not to implement the calculator. Similarly, the graphing calculator exercises are found at the end of the exercise sets and may also be easily skipped.

End-of-Chapter Materials

The textbook has the following end-of-chapter materials for students to review before test time:

- Brief summary with references to key concepts. A detailed summary is located at www.mhhe.com/millerprecalculus.
- Chapter review exercises.
- Chapter test.
- Cumulative review exercises. These exercises cover concepts in the current chapter as well as all preceding chapters.

Supplement Package

Supplements for the Instructor

Author-Created Digital Media

Digital assets were created exclusively by the author team to ensure that the author voice is present and consistent throughout the supplement package.

- Donna Gerken ensures that each algorithm in the online homework has a stepped-out solution that matches the textbook's writing style.
- Julie Miller created **video content** (lecture videos, exercise videos, graphing calculator videos, and Excel videos) to give students access to classroom-like instruction by the author.
- Julie Miller constructed over 50 **dynamic math animations** to accompany the text. The animations are diverse in scope and give students an interactive approach to conceptual learning. The animated content illustrates difficult concepts by leveraging the use of on-screen movement where static images in the text may fall short. They are organized in Connect hosted by ALEKS by chapter and section.

The *Instructor's Resource Manual* (**IRM**) is a printable electronic supplement put together by the author team. The IRM includes Guided Lecture Notes, Classroom Activities using Wolfram Alpha, and Group Activities.

- The Guided Lecture Notes are keyed to the objectives in each section of the text. The notes step through the material with a series of questions and exercises that can be used in conjunction with lecture.
- The Classroom Activities using Wolfram Alpha promote active learning in the classroom by using a powerful online resource.
- A Group Activity is available for each chapter of the book to promote classroom discussion and collaboration.

The *Instructor's Solution Manual* provides comprehensive, worked-out solutions to all exercises in the section exercises, review exercises, problem recognition exercises, chapter tests, and cumulative reviews. The steps shown in the solutions match the style and methodology of solved examples in the textbook.

TestGen is a computerized test bank utilizing algorithm-based testing software to create customized exams quickly. This user-friendly program enables instructors to search for questions by topic, format, or difficulty level; to edit existing questions, or to add new ones; and to scramble questions and answer keys for multiple versions of a single test. Hundreds of text-specific, open-ended, and multiple-choice questions are included in the question bank.

Annotated Instructor's Edition

- Answers to exercises appear adjacent to each exercise set, in a color used only for annotations.
- Instructors will find helpful notes within the margins to consider while teaching.
- References to even-numbered exercises appear in the margin next to each example for the instructor to use as Classroom Examples.

Power Points present key concepts and definitions with fully editable slides that follow the textbook. An instructor may project the slides in class or post to a website in an online course.

Supplements for the Student

Student Worksheets including guided lecture notes that step through the key objectives and Problem Recognition Exercise worksheets.

ALEKS® Prep for Precalculus

ALEKS Prep for Precalculus focuses on prerequisites and introductory material for Precalculus. These prep products can be used during the first 3 weeks of a course to prepare students for future success in the course and to increase retention and pass rates.

Connect Math® Hosted by ALEKS

Connect Math Hosted by ALEKS Corp. is an exciting, new assignment and assessment ehomework platform. Starting with an easily viewable, intuitive interface, students will be able to access key information, complete homework assignments, and utilize an integrated, media-rich eBook.

Smartbook® is the first and only adaptive reading experience available for the higher education market. Powered by the intelligent and adaptive LearnSmart engine, Smart-Book facilitates the reading process by identifying what content a student knows and doesn't know. As a student reads, the material continuously adapts to ensure the student is focused on the content he or she needs the most to close specific knowledge gaps.

Detailed Chapter Summaries are available at www.mhhe.com/millerprecalculus.

Efficient. Easy to Use. Effective. Engaging.
Gives Your Students the ALEKS Advantage

Are your students prepared for your course? Do they learn at different paces? Is there inconsistency between homework and test scores? ALEKS successfully addresses these core challenges and more.

With decades of scientific research behind its creation, ALEKS offers the most advanced adaptive learning technology that is proven to increase student success in math.

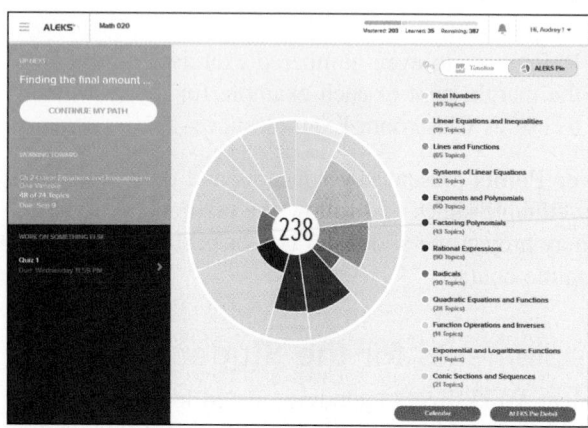

Student-Friendly Learning Experience

ALEKS is designed to meet the needs of today's students. A clean, modern, mobile-ready interface allows students to easily navigate their learning, track their progress and manage their assignments from anywhere.

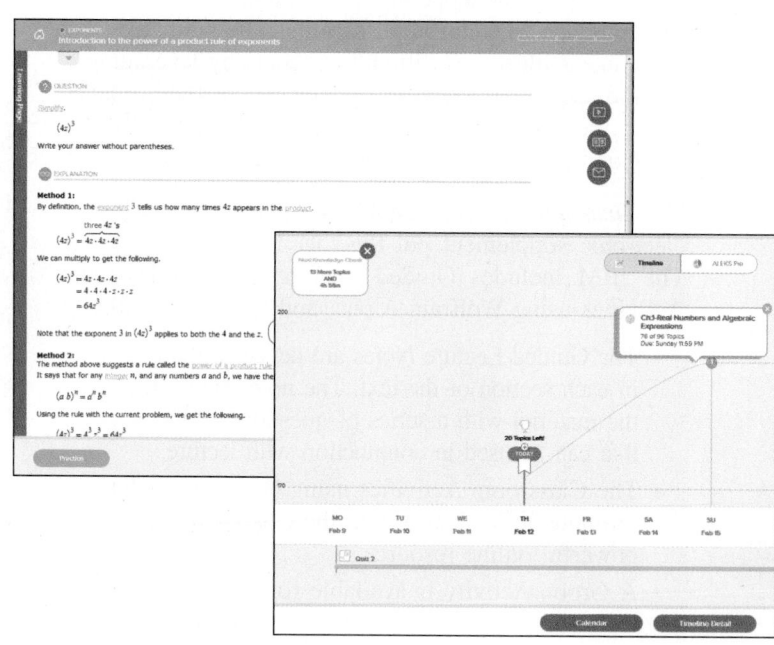

Optimized, Structured Learning

Using adaptive artificial intelligence, ALEKS identifies precisely what each student knows and doesn't know, and prescribes an individualized learning plan tailored to their unique strengths and weaknesses:

- Targets critical knowledge gaps
- Open-response environment
- Motivates student learning
- Presents only topics students are ready to learn
- Enhances learning with interactive resources
- Provides a structured learning path

> "I evaluated many different options, and ALEKS provided, by far, the best cycle of assessment and learning that allows for individualized instructional paths . . . no other program matches ALEKS."
> —Professor Eliza Gallagher, *Clemson University, SC*

Learn More: **Successinmath.com**

THE ALEKS Instructor Module includes intuitive customization and management features that help save you valuable time and effort. Easily manage your courses and track student progress, all through one simple interface.

- Create learning goals with due dates that align with your textbook/syllabi and to pace student learning
- Implementation services and training help make setup simple and timely
- 100% mobile-ready allows you to manage your classes from anywhere
- LMS integration offers single sign-on and gradebook sync capabilities

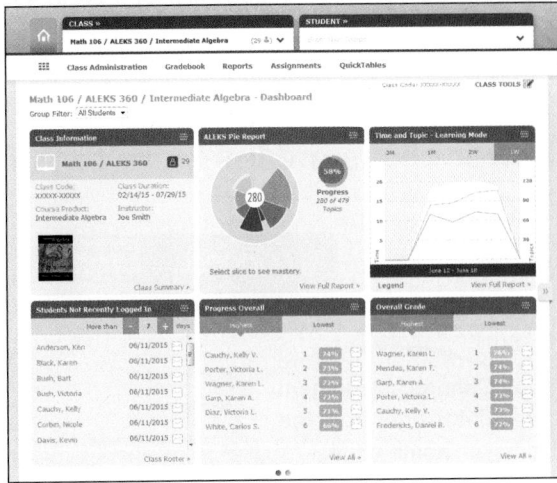

ALEKS Reporting provides detailed data on student progress, allowing you to quickly identify intervention needs and to better understand student learning behaviors. View data for individual students up to the multi-campus level.

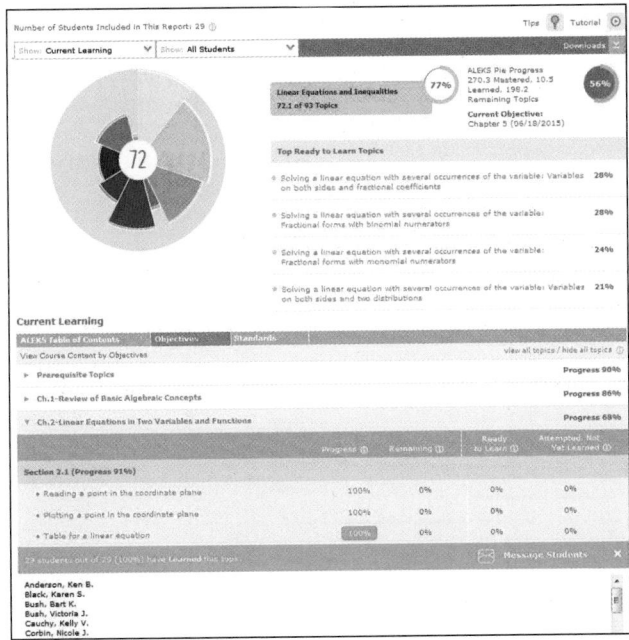

View student progress in a course area or at the individual topic level, including which topics students are struggling with the most.

Use this data to inform your teaching, group students based on similar knowledge levels, and shape a meaningful learning experience for your students.

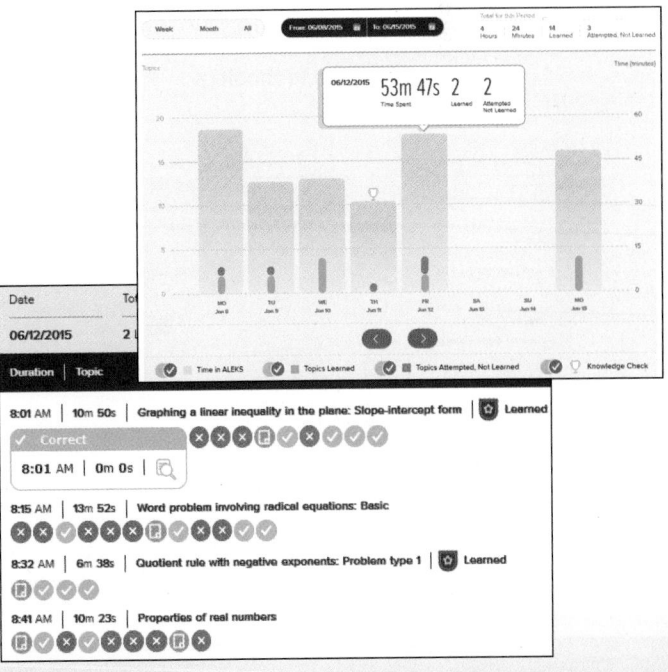

View a daily breakdown of how students learn in ALEKS, including the exact problems they attempt and their answers. This will help you to identify common mistakes and understand their behavior pattern, such as when they study and how frequently.

Learn More: **Successinmath.com**

Connect Math Hosted by ALEKS

Built By Today's Educators, For Today's Students

Fewer clicks means more time for you…

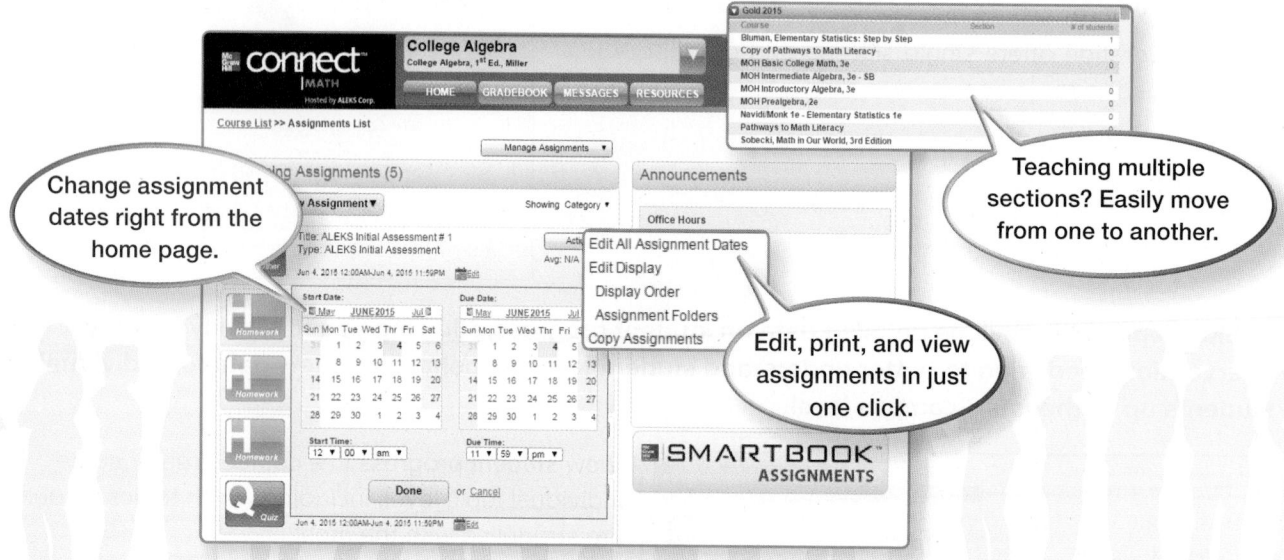

Change assignment dates right from the home page.

Teaching multiple sections? Easily move from one to another.

Edit, print, and view assignments in just one click.

…and your students.

Online Exercises were carefully selected and developed to provide a seamless transition from textbook to technology.

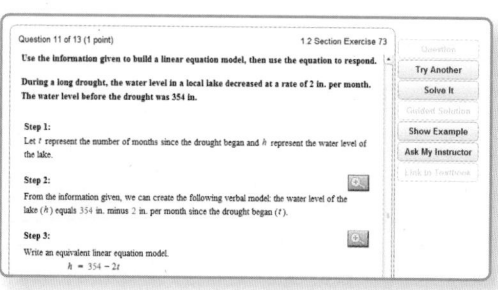

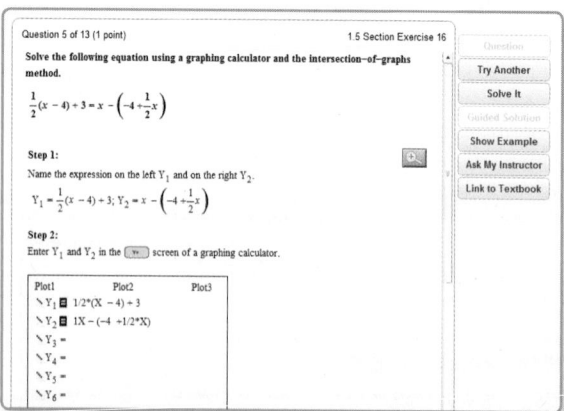

For consistency, the guided solutions match the style and voice of the original text as though the author is guiding the students through the problems.

successinmath.com

Quality Content For Today's Online Learners

Why SmartBook? Because it's more than just words on a page.

McGraw Hill Education | SMARTBOOK™

Students today thrive on efficiency, mobility, and motivation. And SmartBook delivers. SmartBook is the first adaptive reading experience designed to change the way students learn. Students and instructors can enjoy access to SmartBook anywhere, anytime (now available offline) with a new and improved mobile interface. If students still prefer holding a text as they study, they can **order a loose-leaf copy of their textbook for just $15.**

SmartBook breaks down the learning experience into four stages: *Preview, Read, Practice,* and *Recharge.* Each stage provides personalized guidance and just-in-time remediation to ensure students stay focused and learn as efficiently as possible. With LearnSmart technology, questions are designed to foster critical thinking and conceptual learning.

Reports are also available to both students and instructors that track progress and show each student's strengths and weaknesses. What does this mean for you? Teach a more informed classroom and provide more personalized guidance.

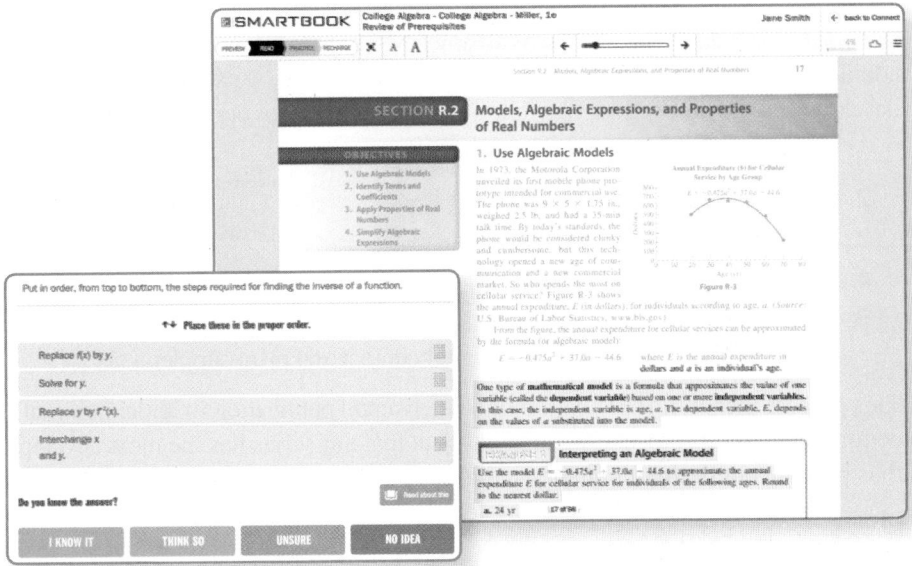

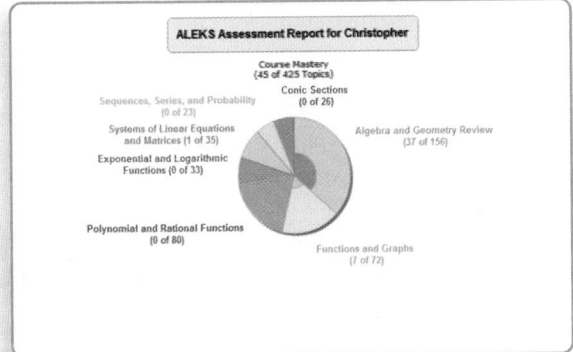

The ALEKS® Initial Assessment is an artificially intelligent (AI), diagnostic assessment that identifies precisely what a student knows. Instructors can then use this information to make more informed decisions on what topics to cover in more detail with the class.

successinmath.com

McGraw Hill Education connect® MATH

Hosted by **ALEKS Corp.**

Our Commitment to Market Development and Accuracy

McGraw-Hill's Development Process is an ongoing, never-ending, market-oriented approach to building accurate and innovative print and digital products. We begin developing a series by partnering with authors who desire to make an impact within their discipline to help students succeed. Next, we share these ideas and manuscript with instructors for review for feedback and to ensure that the authors' ideas represent the needs within that discipline. Throughout multiple drafts, we help our authors adapt to incorporate ideas and suggestions from reviewers to ensure that the series carries the same pulse as today's classrooms. With any new series, we commit to accuracy across the series and its supplements. In addition to involving instructors as we develop our content, we also utilize accuracy checks through our various stages of development and production. The following is a summary of our commitment to market development and accuracy:

1. 3 drafts of author manuscript
2. 2 rounds of manuscript review
3. 3 rounds of accuracy checking
4. 3 rounds of proofreading and copyediting
5. Toward the final stages of production, we incorporate additional rounds of quality assurance from instructors as they help contribute toward our digital content and print supplements

This process then will start again immediately upon publication in anticipation of the next edition. With our commitment to this process, we are confident that our series has the most developed content the industry has to offer, thus pushing our desire for quality and accurate content that meets the needs of today's students and instructors.

Acknowledgments:

Paramount to the development of *Precalculus* was the invaluable feedback provided by the instructors from around the country who reviewed the manuscript or attended a market development event over the course of the several years the text was in development.

A Special Thanks to All of the Event Attendees Who Helped Shape *Precalculus*.

Focus groups and symposia were conducted with instructors from around the country to provide feedback to editors and the authors and ensure the direction of the text was meeting the needs of students and instructors.

Halina Adamska, *Broward College–Central*
Mary Beth Angeline, *West Virginia University*
Colleen Beaudoin, *University of Tampa*
Rachel Black, *Central New Mexico Community College*
Tony Bower, *Saint Phillips College*
Bowen Brawner, *Tarleton State University*
Denise Brown, *Collin College*
Wyatt Bryant, *Tarleton State University*
Christine Bush, *Palm Beach State College*
Michelle Carmel, *Broward College–North*
Lydia Casas, *Saint Phillips College*
Carlos Corona, *San Antonio College*
Deric Davenport, *Pikes Peak Community College*
Alan Dinwiddie, *Front Range Commmunity College–Fort Collins*
Marion Foster, *Houston Community College*
Charles Gabi, *Houston Community College*

Jason Geary, *Harper College*
Steve Gonzales, *Northwest Vista College*
Jeffrey Guild, *Broward-Central Campus*
Craig Hardesty, *Hillsborough Community College–Southshore*
Lori Hodges, *University of New Orleans*
Carolyn Horseman, *Polk College*
Kimber Kaushik, *Houston Community College*
Lynette Kenyon, *Collin College–Plano*
Daniel Kernler, *Elgin Community College*
Sharon Kobrin, *Broward-Central Campus*
Daniel Kopsas, *Ozarks Technical Community College*
Danny Lau, *University of North Georgia*
Andreas Lazari, *Valdosta State University*
Joyce Lee, *Polk College*
Wayne Lee, *Saint Philips College*
Domingo Litong, *Houston Community College*

Tammy Louie, *Portland Community College*

Susan May, *Meridian Community College*

Michael McClendon, *University of Central Oklahoma*

Jerry McCormack, *Tyler Junior College*

Mikal McDowell, *Cedar Valley College*

Meagan McNamee, *Central Piedmont Community College*

Rebecca Muller, *Southeastern Louisiana University*

Denise Natasha, *Georgia Gwinnett College*

Lynne Nisbet, *St. Louis Community College*

Altay Ozgener, *State College of Florida–Manatee*

Denise Pendarvis, *Georgia Perimeter College*

Scott Peterson, *Oregon State University*

Davidson Pierre, *State College of Florida–Manatee*

Mihaela Poplicher, *University of Cincinnati*

Candace Rainer, *Meridian Community College*

Dee Dee Shaulis, *University of Colorado–Boulder*

Prem Singh, *Ohio University*

Rita Sowell, *Volunteer State Community College*

Pam Stogsdill, *Bossier Parish Community College*

Peter Surgent, *Community College of Baltimore*

Dustin Walsh, *Southeast Community College*

Kim Walters, *Mississippi State University*

Michael Warren, *Tarleton State University*

Jeff Weaver, *Baton Rouge Community College*

Carol Weideman, *Saint Petersburg College–Gibbs*

Benjamin Wescoatt, *Valdosta State University*

Sean Woodruff, *Saint Petersburg College–Tarpon Springs*

Manuscript Reviewers

Carol Abbott, *Ohio University*

Marylynne Abbott, *Ozarks Technical Community College*

Jay Abramson, *Arizona State University–Tempe*

Ryan Adams, *Northwest Florida State College*

Halina Adamska, *Broward College–Central*

Mark Ahrens, *Normandale Community College*

Douglas Aichele, *Oklahoma State University*

John Alford, *Sam Houston State University*

Kinnari Amin, *Georgia Perimeter College*

Patricia Anderson, *Arapahoe Community College*

Darry Andrews, *Ohio State University*

Raji Ariyaratna, *Houston Community College*

Peter Arvanites, *Rockland Community College*

Alvina Atkinson, *Georgia Gwinnett College*

Robin Ayers, *Western Kentucky University*

Mohamed Baghzali, *North Dakota State University*

Robert Banik, *Mississippi State University*

Terry Lee Barron, *Georgia Gwinnett College*

John Beatty, *Georgia Perimeter College*

Tim Bell, *San Jacint College*

Sergey Belyi, *Troy University*

Kenneth Bernard, *Virginia State University*

Jaime Bestard, *Miami Dade College*

Elizabeth Betzel, *Columbus State Community College*

Patricia Bezona, *Valdosta State University*

Nicholas Bianco, *Florida Gulf Coast University*

Susan Billimek, *Northwest Arkansas Community College*

Greg Bloxom, *Pensacola State College*

Laurie Boudreaux, *Nicholls State University*

Jeremy Bourgeois, *Nicholls State University*

Kristina Bowers, *Tallahassee Community College*

Troy Brachey, *Tennessee Technological University*

Stephanie Branham, *University of Tampa*

James Brawner, *Armstrong Atlantic State University*

James Brink, *Ohio University*

Jeffery Brown, *University of North Carolina*

Kristine Buddemeyer, *Seminole State College*

Annette Burden, *Youngstown State University*

Elsie Campbell, *Angelo State University*

Vollet Carly, *Portland Community College*

James Carolan, *Wharton County Junior College*

Lydia Casas, *Saint Philips College*

Jason Cates, *Brookhaven College*

Tim Chappell, *MCC Penn Valley Community College*

Jerry Chen, *Suffolk County Community College*

Lars Christensen, *Texas Tech University*

Elizabeth Chu, *Suffolk County Community College*

Ivette Chuca, *El Paso Community College*

Carl Clark, *Indian River State College*

Beth Clickner, *Hillsborough Community College*

Thomas Cooper, *University of North Georgia*

Christina Cornejo, *Erie Community College*

Jim Cotter, *Truckee Meadows Community College*

Cindy Cummins, *Ozarks Technical Community College*

Leslie Dalrymple, *Texas State University–San Marcos*

Hall Dave, *Portland Community College*

Nelson De La Rosa, *Miami Dade College*

Jerry Degroot, *Purdue University*

Noemi DeHerrera, *Houston Community College*

Reynolds Dennis, *Portland Community College*

Alan Dinwiddie, *Front Range Community College*

Christy Dittmar, *Austin Community College–Rio Grande*

Ginger Eaves, *Bossier Parish Community College*

Steven Edwards, *Southern Polytechnic State University*

Keith Erickson, *Georgia Gwinnett College*

Keith Erickson, *Georgia Gwinnett College*

Dihema Ferguson, *Georgia Perimeter College*

Barbara Finnegan, *Del Mar Community College*

Elise Fischer, *Johnson County Community College*

Stephanie Fitch, *Missouri University of Science and Technology*

Marion Foster, *Houston Community College*

Lana Fredrickson, *Front Range Community College*

John Fulk, *Georgia Perimeter College*

Darren Funk-Neubauer, *Colorado State University*

Ed Gallo, *Sinclair Community College*

Valdez Gant, *North Lake College*

David George, *Ohio State University*

Kevin Gibbs, *University of Toledo*

Vijaya Gompam, *Troy University*

Jennifer Gorman, *College of Southern Nevada*

William Griffiths, *Southern Polytechnic State University*

Barry Griffiths, *University of Central Florida*

Jeff Gutliph, *Georgia Perimeter College*

Sylvia Gutowska, *Community College of Baltimore County–Essex*

Ryan Hansen, *West Virginia University*

Debbie Hanus, *Brookhaven College*

Craig Hardesty, *Hillsborough Community College*

Jason Hasbrouck, *College of Lake County*

Tom Hayes, *Montana State University*

Mary Beth Headlee, *State College of Florida Manatee*

James Hearl, *Pasco-Hernando Community College*

Beata Hebda, *University of North Georgia*

Christy Hediger, *Lehigh Carbon Community College*

Matthew Henes, *Pasadena City College*

Jean Hindie, *Community College of Denver*

Gangadhar Hiremath, *University of North Carolina*

Linda Ho, *El Camino College*

Lori Hodges, *University of New Orleans*

Sarah Holliday, *Southern Polytechnic State University*

Heidi Howard, *Florida State College*

Keith Hubbard, *Stephen F. Austin State University*

Sharon Jackson, *Brookhaven College*

David James, *Howard University*

Erin Joseph, *Central New Mexico Community College*

Brian Kahl, *Community College of Baltimore County–Essex*

Chandra Karnati, *University of North Georgia*

Susan Keith, *Georgia Perimeter College*

Lynette Kenyon, *Collin College Plano*

Raja Khoury, *Collin College–Plano*

Minsu Kim, *University of North Georgia*

Christopher King, *College of Southern Nevada*

Sandra Kingan, *Brooklyn College*

Russell Kohl, *University of Maryland*

Daniel Kopsas, *Ozarks Technical Community College*

Elena Kravchuk, *University of Alabama*

Ramesh Krishnan, *South Plains College*

Bohdan Kunciw, *Salisbury University*

Weiling Landers, *Windward Community College*

Viktoriya Lanier, *Middle Georgia State College*

Rick Leborne, *Tennessee Technological University*

Sungwook Lee, *University of Southern Mississippi*

Deb Lehman, *State Fair Community College*

Xuhui Li, *California State University–Long Beach*

Folberg Lisa, *Portland Community College*

Domingo Litong, *Houston Community College*

Tammy Louie, *Portland Community College*

Rene Lumampao, *Austin Community College–Rio Grande*

Phil Maclean, *Columbus State Community College*

Edmund MacPherson, *Tyler Junior College*

Anna Madrid-Larranaga, *Central New Mexico Community College*

Crepin Mahop, *Howard University*

Jason Malozzi, *Lehigh Carbon Community College*

Cynthia Martinez, *Temple College*

Shawna Masters, *Collin College–Plano*

Ramon Mata-Toledo, *James Madison University*

Janet Mayeux, *SE Louisiana University*

Michael Mays, *West Virginia University*

Roderick McBane, *Houston Community College*

Cynthia McGinnis, *Northwest Florida State College*

Christine McKenna, *University of Nevada Las Vegas*

Tim McNicholl, *Iowa State University*

Mary Merchant, *Cedar Valley College*

Robert Mitchell, *Pennsylvania College of Technology*

Chris Mizell, *Northwest Florida State College*

Val Mohanakumar, *Hillsborough Community College*

Monica Montalvo, *Collin College–Plano*

Candy Moraczewski, *Delaware County Community College*

Malika T. Morris, *Houston Community College*

Dorothy Muhammad, *Houston Community College*

Linda Myers, *Harrisburg Community College*

Julie Nation, *Mississippi State University*

Chris Nazelli, *Wayne State University*

Mandri Obeyesekere, *Houston Community College*

Victor Obot, *Texas Southern University*

Charles Odion, *Houston Community College*

Chinenye Ofodile, *Howard University*

Donald Orr, *Miami Dade College*

Victor Pambuccian, *Arizona State University–West*

Stan Perrine, *Georgia Gwinnett College*

Haberman Peter, *Portland Community College*

Shawn Peterson, *Texas State University–San Marcos*

Gerri Petty, *Oklahoma State University*

David Platt, *Front Range Community College*

Wendy Pogoda, *Hillsborough Community College*

John Polhill, *Bloomsburg University of Pennsylvania*

Jonathan Poritz, *Colorado State University*

Didi Quesada, *Miami Dade College*

Brooke Quinlan, *Hillsborough Community College*

Mohammed Rajah, *Miracosta College*

Don Ransford, *Edison College–Fort Myers*

Carolynn Reed, *Austin Community College–Rio Grande*

Denise Reid, *Valdosta State University*

Nancy Resseguie, *Arapahoe Community College*

Shelia Rivera, *University of West Georgia*

Ken Roblee, *Troy University*

Michael Rosenthal, *Florida International University*

Bill Rosenthal, *Laguardia Community College*

Sean Rule, *Central Oregon Community College*

Haazim Sabree, *Georgia Perimeter College*

Fatemeh Salehibakhsh, *Houston Community College*

Fary Sami, *Harford Community College*

Jared Schlieper, *Armstrong Atlantic State University*

Julie Schmidt, *Central Oregon Community College*

Linda Schott, *Ozarks Technical Community College*

Mary Schuster, *University of Cinncinatti*

Mohammad Shakil, *Miami Dade College*

Lisa Shannon, *Meridian Community College*

Kenneth Shiskowski, *Eastern Michigan University*

Giorgi Shonia, *Ohio University*

Andrew Siefker, *Angelo State University*

Jennifer Siegel, *Broward College–Central*

Randell Simpson, *Temple College*

Premjit Singh, *Ohio University*

Sounny Slitine, *Saint Philips College*

David Slutzky, *University of North Georgia*

Shawn Smith, *Nicholls State University*

Cindy Soderstrom, *Salt Lake Community College*

Mary Ann Sojda, *Montana State University*

Shannon Solis, *San Jacint College*

Scott Sorrell, *University of Louisiana Lafayette*

Greg Stiffler, *Community College of Baltimore County–Essex*

Malgorzata Surowiec, *Texas Tech University*

Vic Swaim, *SE Louisiana University*

Paula Talley, *Temple College*

Mary Ann Teel, *University of North Texas*

Alain Togbe, *Purdue University*

Rae Tree, *Oklahoma State University*

Danielle Truszkowski, *Community College of Baltimore County–Essex*

Chris Turner, *Pensacola State College*

Joan Van Glabek, *Edison College–Fort Myers*

Phil Veer, *Johnson County Community College*

Sue Wall, *Wake Tech Community College*

Karen Walters, *Nova Community College*

James Wan, *Long Beach City College*

Walter Wang, *Baruch College*

Bingwu Wang, *Eastern Michigan University*

Jason Wetherington, *Pasco-Hernando Community College*

Emily Whaley, *Georgia Perimeter College*

Mark White, *University of North Carolina*

Dr. Bob Wieman, *Virginia State University*

Rebecca Wong, *West Valley College*

Jane-Marie Wright, *Suffolk County Community College*

Changyong Zhong, *Georgia State University*

Author Acknowledgments:

An editor once told us that publishing a book is like making a movie because there are so many people behind the scenes that make the final product a success. Words cannot begin to express our heartfelt thanks to all of you, but we'll do our best. First and foremost, we want to thank our editor Emily Windelborn who started with us on this project when it was just idea, and then lent her unwavering, day-to-day support through final publication. Without Emily, we'd still be on page 1. To Ryan Blankenship, Marty Lange, Kurt Strand, we are forever grateful for the amazing opportunities you and McGraw-Hill have given us.

To our brand manager, Caroline Celano, as the pilot on this long journey you set the standard for leadership. Each day, you lead by example to unlock our full potential and inspire our best work. You're strong but not rude; kind, but not weak; humble, but not timid. To the marketing team Michelle Greco, Leigh Jacka, Simon Wong, Megan Farber, Sara Swangard, Ashley Swafford, Jill Gordon, and Alex Gay: Your artistry and creative ideas for our project are 90% inspiration, 90% innovation, and 100% perspiration. Thank you for making us shine.

Our heartfelt gratitude goes to the content project manager Peggy Selle for steering the ship and keeping us all on task. To Laurie Janssen and David Hash, many thanks for a beautiful design, and to Carrie Berger for the beautiful photos and art. The book is gorgeous.

Special thanks to the digital team, Rob Brieler and Victor Pareja, for keeping the digital train on the tracks. Thank you for overseeing the enormous job of managing digital content and ensuring consistency of the author voice. Along these lines, we must express our utmost gratitude to digital authors Tim Chappell, Meghan Clovis, Vanessa Coffelt, Alina Coronel, Lizette Foley, Lance Gooden, Jeffrey Guild, Esmarie Kennedy, Michael Larkin,

Nathalie Vega-Rhodes, and Stephen Toner for their diligence working on digital content. All of you are amazing. To Jason Wetherington and Mary Beth Headlee thank you so much for your work on SmartBook and for the additional pairs of eyes on our manuscript. To Beth Clickner, many, many thanks for the beautiful and thorough PowerPoint presentations of our material. To our colleague and friend Kimberly Alacan, we're so grateful for your creativity in preparing the chapter openers and the group activities. To Nora Devlin, thank you for managing the solutions manual projects, lecture notes, and Internet Activities. No doubt, many instructors and students thank you as well.

To Patricia Steele, the best copy editor ever, thank you for mentoring us and for ensuring consistency throughout our work. To Elka Block, Jennifer Blue, and the team at DiacriTech, many thanks for doing multiple levels of accuracy checking. Your talents are absolutely amazing. Additional thanks to Julie Kennedy and Carey Lange for their tireless attention to detail proofreading pages.

Finally, to the dedicated people in the McGraw-Hill sales force, thank you so much for your continued confidence, encouragement, and support.

Most importantly, we want to give special thanks to all the students and instructors who use *Precalculus* in their classes.

—Julie Miller and Donna Gerken

Applications Index

R Review of Prerequisites

Chapter Outline

Athletes know that in order to optimize their performance they need to pace themselves and be mindful of their target heart rate. For example, a 25-year-old with a maximum heart rate of 195 beats per minute should strive for a target heart rate zone of between 98 and 166 beats per minute. This correlates to between 50% and 85% of the individual's maximum recommended heart rate (Source: American Heart Association, www. americanheart.org). The mathematics involved in finding maximum recommended heart rate and an individual's target heart rate zone use a linear model relating age and resting heart rate. An introduction to modeling is presented here in Chapter R along with the standard order of operations used to carry out these calculations.

Chapter R reviews skills and concepts required for success in college algebra. Just as an athlete must first learn the basics of a sport and build endurance and speed, a student studying mathematics must focus on necessary basic skills to prepare for the challenge ahead. Preparation for algebra is comparable to an athlete preparing for a sporting event. Putting the time and effort into the basics here in Chapter R will be your foundation for success in later chapters.

SECTION R.1 Sets and the Real Number Line

OBJECTIVES

1. **Identify Subsets of the Set of Real Numbers**
2. **Use Inequality Symbols and Interval Notation**
3. **Find the Union and Intersection of Sets**
4. **Evaluate Absolute Value Expressions**
5. **Use Absolute Value to Represent Distance**
6. **Apply the Order of Operations**
7. **Simplify Algebraic Expressions**
8. **Write Algebraic Models**

1. Identify Subsets of the Set of Real Numbers

A hybrid vehicle gets 48 mpg in city driving and 52 mpg on the highway. The formula $A = \frac{1}{48}c + \frac{1}{52}h$ gives the amount of gas A (in gal) for c miles of city driving and h miles of highway driving. In the formula, A, c, and h are called **variables** and these represent values that are subject to change. The values $\frac{1}{48}$ and $\frac{1}{52}$ are called **constants** because their values do not change in the formula.

For a trip from Houston, Texas, to Dallas, Texas, a motorist travels 36 mi of city driving and 91 mi of highway driving. The amount of fuel used by this hybrid vehicle is given by

$$A = \frac{1}{48}(36) + \frac{1}{52}(91)$$
$$= 2.5 \text{ gal}$$

The numbers used in day-to-day life such as those used to determine fuel consumption come from the set of real numbers, denoted by $\mathbb{R}$. A **set** is a collection of items called **elements.** The braces { and } are used to enclose the elements of a set. For example, {gold, silver, bronze} represents the set of medals awarded to the top three finishers in an Olympic event. A set that contains no elements is called the **empty set** (or **null set**) and is denoted by { } or $\varnothing$.

When referring to individual elements of a set, the symbol $\in$ means "is an element of," and the symbol $\notin$ means "is not an element of." For example,

$5 \in \{1, 3, 5, 7\}$ is read as "5 is an element of the set of elements 1, 3, 5, and 7."

$6 \notin \{1, 3, 5, 7\}$ is read as "6 is *not* an element of the set of elements 1, 3, 5, and 7."

A set can be defined in several ways. Listing the elements in a set within braces is called the **roster method.** Using the roster method, the set of the even numbers between 0 and 10 is represented by {2, 4, 6, 8}. Another method to define this set is by using **set-builder notation.** This uses a description of the elements of the set. For example,

$$\{x \mid x \text{ is an even number between 0 and 10}\}$$

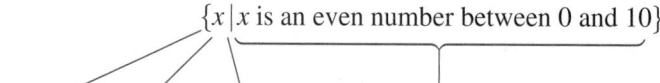

The set of   all x such that x is an even number between 0 and 10

In our study of college algebra, we will often refer to several important **subsets** (parts of) the set of real numbers (Table R-1).

Table R-1 Subsets of the Set of Real Numbers, $\mathbb{R}$

Set	Definition
Natural numbers, $\mathbb{N}$	$\{1, 2, 3, \ldots\}$
Whole numbers, $\mathbb{W}$	$\{0, 1, 2, 3, \ldots\}$
Integers, $\mathbb{Z}$	$\{\cdots, -3, -2, -1, 0, 1, 2, 3, \ldots\}$
Rational numbers, $\mathbb{Q}$	$\left\{\dfrac{p}{q} \middle\| p, q \in \mathbb{Z} \text{ and } q \neq 0\right\}$ • Rational numbers can be expressed as a ratio of integers where the denominator is not zero. <u>Examples:</u> $-\frac{6}{11}$ (ratio of -6 and 11) and 9 (ratio of 9 and 1). • All terminating and repeating decimals are rational numbers. <u>Examples:</u> 0.71 (ratio of 71 and 100), $0.\overline{6} = 0.666\ldots$ (ratio of 2 and 3).
Irrational numbers, $\mathbb{H}$	Irrational numbers are real numbers that cannot be expressed as a ratio of integers. The decimal form of an irrational number is nonterminating and nonrepeating. <u>Examples:</u> π and $\sqrt{2}$

TIP Notice that the first five letters of the word *rational* spell *ratio*. This will help you remember that a rational number is a *ratio* of integers.

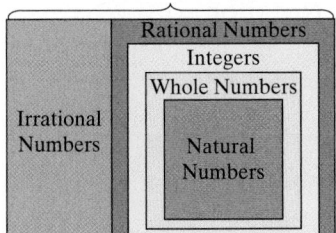

Real Numbers (R)

Figure R-1

The relationships among the subsets of real numbers defined in Table R-1 are shown in Figure R-1. In particular, notice that together the elements of the set of rational numbers and the set of irrational numbers make up the set of real numbers.

2. Use Inequality Symbols and Interval Notation

All real numbers can be located on the real number line. We say that a is less than b (written symbolically as $a < b$) if a lies to the left of b on the number line. This is equivalent to saying that b is greater than a (written symbolically as $b > a$) because b lies to the right of a.

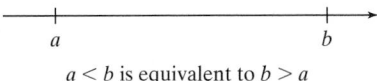

$a < b$ is equivalent to $b > a$

In Table R-2, we summarize other symbols used to compare two real numbers.

Table R-2   Summary of Inequality Symbols and Their Meanings

Inequality	Verbal Interpretation	Other Implied Meanings	Numerical Examples
$a < b$	a is less than b	b exceeds a b is greater than a	$5 < 7$
$a > b$	a is greater than b	a exceeds b b is less than a	$-3 > -6$
$a \leq b$	a is less than or equal to b	a is at most b a is no more than b	$4 \leq 5$ $5 \leq 5$
$a \geq b$	a is greater than or equal to b	a is no less than b a is at least b	$9 \geq 8$ $9 \geq 9$
$a = b$	a is equal to b		$-4.3 = -4.3$
$a \neq b$	a is not equal to b		$-6 \neq -7$
$a \approx b$	a is approximately equal to b		$-12.99 \approx -13$

An interval on the real number line can be represented in set-builder notation or in interval notation. In Table R-3, observe that a parenthesis) or (indicates that an endpoint is not included in an interval. A bracket] or [indicates that an endpoint *is* included in the interval. The real number line extends infinitely far to the left and right. We use the symbols $-\infty$ and ∞ to denote the unbounded behavior to the left and right, respectively.

Table R-3   Summary of Interval Notation and Set-Builder Notation

Let a, b, and x represent real numbers.

Set-Builder Notation	Verbal Interpretation	Graph	Interval Notation
$\{x \mid x > a\}$	the set of real numbers greater than a		(a, ∞)
$\{x \mid x \geq a\}$	the set of real numbers greater than or equal to a		$[a, \infty)$
$\{x \mid x < b\}$	the set of real numbers less than b		$(-\infty, b)$
$\{x \mid x \leq b\}$	the set of real numbers less than or equal to b		$(-\infty, b]$
$\{x \mid a < x < b\}$	the set of real numbers between a and b		(a, b)
$\{x \mid a \leq x < b\}$	the set of real numbers greater than or equal to a and less than b		$[a, b)$
$\{x \mid a < x \leq b\}$	the set of real numbers greater than a and less than or equal to b		$(a, b]$
$\{x \mid a \leq x \leq b\}$	the set of real numbers between a and b, inclusive		$[a, b]$
$\{x \mid x \text{ is a real number}\}$ $\mathbb{R}$	the set of all real numbers		$(-\infty, \infty)$

TIP As an alternative to using parentheses and brackets to represent the endpoints of an interval, an open dot or closed dot may be used. For example, $\{x \mid a \leq x < b\}$ would be represented as follows.

Classroom Examples: p. 12
Exercises 20, 26, 32

EXAMPLE 1   **Expressing Sets in Interval Notation and Set-Builder Notation**

Complete the table.

Graph	Interval Notation	Set-Builder Notation
–5 –4 –3 –2 –1 0 1 2 3 4 5		
	$\left(\frac{7}{2}, \infty\right)$	
		$\{y \mid -4 \le y < 2.3\}$

Solution:

Graph	Interval Notation	Set-Builder Notation	Comments
–5 –4 –3 –2 –1 0 1 2 3 4 5	$(-\infty, 2]$	$\{x \mid x \le 2\}$	The bracket at 2 indicates that 2 is included in the set.
–5 –4 –3 –2 –1 0 1 2 3 4 5	$\left(\frac{7}{2}, \infty\right)$	$\{x \mid x > \frac{7}{2}\}$	The parenthesis at $\frac{7}{2} = 3.5$ indicates that $\frac{7}{2}$ is *not* included in the set.
–5 –4 –3 –2 –1 0 1 2 3 4 5	$[-4, 2.3)$	$\{y \mid -4 \le y < 2.3\}$	The set includes the real numbers between -4 and 2.3, including the endpoint -4.

Skill Practice 1

a. Write the set represented by the graph in interval notation and set-builder notation.

–5 –4 –3 –2 –1 0 1 2 3 4 5

b. Given the interval, $\left(-\infty, -\frac{4}{3}\right]$, graph the set and write the set-builder notation.

c. Given the set, $\{x \mid 1.6 < x \le 5\}$, graph the set and write the interval notation.

3. Find the Union and Intersection of Sets

Two or more sets can be combined by the operations of union and intersection.

Union and Intersection of Sets

The **union** of sets A and B, denoted $A \cup B$, is the set of elements that belong to set A or to set B or to both sets A and B.

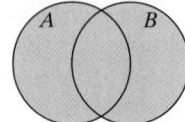

$A \cup B$
A union B
The elements in A or B or both

The **intersection** of sets A and B, denoted $A \cap B$, is the set of elements common to both set A and set B.

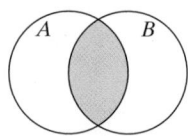

$A \cap B$
A intersection B
The elements common to A and B

In Examples 2 and 3, we practice finding the union and intersections of sets.

Answers

1. a. Interval notation: $(-4, \infty)$;
   Set-builder notation: $\{x \mid x > -4\}$
b.

–5 –4 –3 –2 –1 0 1 2 3 4 5 ;
Set-builder notation: $\{x \mid x \le -\frac{4}{3}\}$

c.

–5 –4 –3 –2 –1 0 1 2 3 4 5 ;
Interval notation: (1.6, 5]

Classroom Example: p. 12
Exercise 38

EXAMPLE 2   Finding the Union and Intersection of Sets

Find the union or intersection of sets as indicated.

$$A = \{-5, -3, -1, 1\} \qquad B = \{-5, 0, 5\} \qquad C = \{-4, -2, 0, 2, 4\}$$

a. $A \cap B$ **b.** $A \cup B$ **c.** $A \cap C$

Solution:

a. $A \cap B = \{-5\}$ The only element common to both A and B is -5. $A = \{-5, -3, -1, 1\}, B = \{-5, 0, 5\}$

b. $A \cup B = \{-5, -3, -1, 0, 1, 5\}$ The union of A and B consists of all elements from A along with all elements from B.

c. $A \cap C = \{\ \}$ Sets A and C have no common elements.

Skill Practice 2   From the sets A, B, and C defined in Example 2, find

    a. $B \cap C$ **b.** $B \cup C$ **c.** $A \cup C$

Classroom Example: p. 13
Exercise 42

EXAMPLE 3   Finding the Union and Intersection of Sets

Find the union or intersection as indicated.

$$D = \{x \mid x < 4\} \qquad E = \{x \mid x \geq -2\} \qquad F = \{x \mid x \leq -3\}$$

a. $D \cap E$ **b.** $D \cup E$ **c.** $D \cup F$ **d.** $E \cap F$

Solution:

Graph each individual set and take the union or intersection.

a.
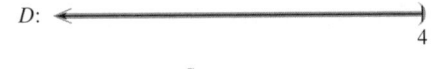

The intersection is the region of overlap.

Set notation: $D \cap E = \{x \mid -2 \leq x < 4\}$
Interval notation: $[-2, 4)$

b.
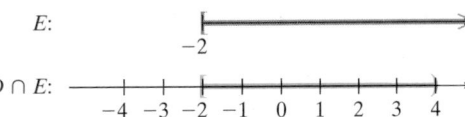

The union contains the elements from D along with those from E.

Set notation: $D \cup E = \mathbb{R}$
Interval notation: $(-\infty, \infty)$

c.
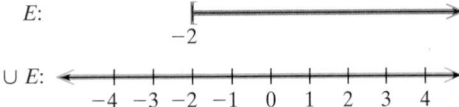

Set F is contained within set D. The union is set D itself.

Set notation: $D \cup F = \{x \mid x < 4\}$
Interval notation: $(-\infty, 4)$

Answers
2. a. $\{0\}$ **b.** $\{-5, -4, -2, 0, 2, 4, 5\}$
   c. $\{-5, -4, -3, -2, -1, 0, 1, 2, 4\}$

d. E:

There are no elements common to both sets E and F.

F:

$E \cap F$:

$$-4 \ -3 \ -2 \ -1 \ \ 0 \ \ 1 \ \ 2 \ \ 3 \ \ 4$$

Set notation: $E \cap F = \{\ \}$

Skill Practice 3   Given $X = \{x \,|\, x \geq -1\}$, $Y = \{x \,|\, x < 2\}$, and $Z = \{x \,|\, x < -4\}$, find the union or intersection of sets as indicated.

a. $X \cap Y$ **b.** $X \cup Y$ **c.** $Y \cap Z$ **d.** $X \cap Z$

---

4. Evaluate Absolute Value Expressions

Every real number x has an opposite denoted by $-x$. For example, $-(4)$ is the opposite of 4 and simplifies to -4. Likewise, $-(-2.1)$ is the opposite of -2.1 and simplifies to 2.1.

The **absolute value** of a real number x, denoted by $|x|$, is the distance between x and zero on the number line. For example:

$$|-5| = 5 \quad \text{because } -5 \text{ is 5 units from zero on the number line.}$$
$$|5| = 5 \quad \text{because } 5 \text{ is 5 units from zero on the number line.}$$

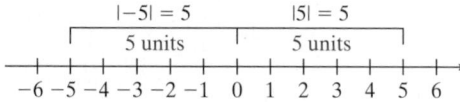

Notice that if a number is negative, its absolute value is the opposite of the number. If a number is positive, its absolute value is the number itself.

Definition of Absolute Value

Let x be a real number. Then $|x| = \begin{cases} x & \text{if } x \geq 0 \\ -x & \text{if } x < 0 \end{cases}$

Verbal Interpretation	Numerical Example
• If x is positive or zero, then $|x|$ is just x itself.	$|4| = 4$
	$|0| = 0$
• If x is negative, then $|x|$ is the opposite of x.	$|-4| = -(-4) = 4$

Classroom Examples: p. 13
Exercises 52, 58

EXAMPLE 4   Removing Absolute Value Symbols

Use the definition of absolute value to rewrite each expression without absolute value bars.

a. $|\sqrt{3} - 3|$ **b.** $|3 - \sqrt{3}|$ **c.** $\dfrac{|x - 4|}{x - 4}$ for $x < 4$

Answers

3. a. $\{x \,|\, -1 \leq x < 2\}$; $[-1, 2)$
   b. $\mathbb{R}$; $(-\infty, \infty)$
   c. $\{x \,|\, x < -4\}$; $(-\infty, -4)$
   d. $\{\ \}$

TIP Calculator
approximations can be
used to show that
$\sqrt{3} - 3 \approx -1.27$ is negative,
and $3 - \sqrt{3} \approx 1.27$ is
positive.

Solution:

a. $|\sqrt{3} - 3| = -(\sqrt{3} - 3)$
$\qquad\quad = -\sqrt{3} + 3 \quad$ or $\quad 3 - \sqrt{3}$

The value $\sqrt{3} \approx 1.73 < 3$, which implies that $\sqrt{3} - 3 < 0$. Since the expression inside the absolute value bars is negative, take the opposite.

b. $|3 - \sqrt{3}| = 3 - \sqrt{3}$

The value $\sqrt{3} \approx 1.73 < 3$, which implies that $3 - \sqrt{3} > 0$. Since the expression inside the absolute value bars is positive, the simplified form is the expression itself.

c. $\dfrac{|x - 4|}{x - 4}$ for $x < 4$

$\qquad = \dfrac{-(x - 4)}{x - 4}$

$\qquad = -1 \cdot \dfrac{x - 4}{x - 4}$

$\qquad = -1$

The condition $x < 4$, implies that $x - 4 < 0$. Since the expression inside the absolute value bars is negative, take the opposite.

Skill Practice 4   Use the definition of absolute value to rewrite each expression without absolute value bars.

a. $|5 - \sqrt{7}|$ **b.** $|\sqrt{7} - 5|$ **c.** $\dfrac{x + 6}{|x + 6|}$ for $x > -6$

5. Use Absolute Value to Represent Distance

Absolute value is also used to denote distance between two points on a number line.

Distance Between Two Points on a Number Line

The distance between two points a and b on a number line is given by

$$|a - b| \qquad \text{or} \qquad |b - a|$$

That is, the distance between two points on a number line is the absolute value of their difference.

Classroom Example: p. 13
Exercise 62

EXAMPLE 5   Determining the Distance Between Two Points

Write an absolute value expression that represents the distance between 4 and -1 on the number line. Then simplify.

Solution:

$|4 - (-1)| = |5| = 5$

$|-1 - 4| = |-5| = 5$

The distance between 4 and -1 is represented by $|4 - (-1)|$ or by $|-1 - 4|$.

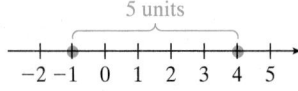

5 units

$-2\ -1\quad 0\quad 1\quad 2\quad 3\quad 4\quad 5$

Skill Practice 5   Write an absolute value expression that represents the distance between -9 and 2 on the number line. Then simplify.

Answers

4. a. $5 - \sqrt{7}$
   b. $5 - \sqrt{7}$
   c. 1
5. $|-9 - 2|$ or $|2 - (-9)|$;
    The distance is 11 units.

6. Apply the Order of Operations

TIP Note that the square of any real number is nonnegative. Therefore, there is no real number that is the square root of a negative number. For example, $\sqrt{-25}$ is not a real number because no real number when squared equals -25.

Note: The value $\sqrt{-25}$ is an imaginary number and will be discussed in Section R.3.

Repeated multiplication can be written by using exponential notation. For example, the product $5 \cdot 5 \cdot 5$ can be written as 5^3. In this case, 5 is called the **base** of the expression and 3 is the **exponent (or power).** The exponent indicates how many times the base is used as a factor.

To find a square root of a nonnegative real number, we reverse the process to square a number. For example, a square root of 25 is a number that when squared equals 25. Both 5 and -5 are square roots of 25, because $5^2 = 25$ and $(-5)^2 = 25$. A radical sign $\sqrt{}$ is used to denote the principal square root of a number. The **principal square root** of a nonnegative real number is the square root that is greater than or equal to zero. Therefore, the principal square root of 25, denoted by $\sqrt{25}$, equals 5.

$$\sqrt{25} = 5 \text{ because } 5 \geq 0 \text{ and } 5^2 = 25$$

The symbol $\sqrt[3]{}$ represents the cube root of a number. For example:

$$\sqrt[3]{64} = 4 \text{ because } 4^3 = 64$$

Many expressions involve multiple operations. In such a case, it is important to follow the order of operations.

Order of Operations

Step 1 Simplify expressions within parentheses and other grouping symbols. These include absolute value bars, fraction bars, and radicals. If nested grouping symbols are present, start with the innermost symbols.

Step 2 Evaluate expressions involving exponents.

Step 3 Perform multiplication or division in the order in which they occur from left to right.

Step 4 Perform addition or subtraction in the order in which they occur from left to right.

Classroom Example: p. 13
Exercise 72

EXAMPLE 6   Simplifying a Numerical Expression

Simplify. $7 - \{8 + 4[2 - (5 - \sqrt{64})^2]\}$

Solution:

$7 - \{8 + 4[2 - (5 - \sqrt{64})^2]\}$

$= 7 - \{8 + 4[2 - (5 - 8)^2]\}$ Simplify within inner parentheses, $\sqrt{64} = 8$. Subtract within the inner parentheses.

$= 7 - \{8 + 4[2 - (-3)^2]\}$

$= 7 - [8 + 4(2 - 9)]$ Continue simplifying within the inner parentheses. Simplify $(-3)^2$ to get 9.

$= 7 - [8 + 4(-7)]$ Simplify $(2 - 9)$ to get -7.

$= 7 - (8 - 28)$ Multiply before adding or subtracting.

$= 7 - (-20)$ Simplify within parentheses.

$= 27$

Skill Practice 6   Simplify. $50 - \{2 - [\sqrt{121} + 3(-1 - 3)^2]\}$

When simplifying an expression, particular care must be taken with expressions involving division and zero.

> ### Division Involving Zero
>
> To investigate division involving zero, consider the expressions $\frac{5}{0}$, $\frac{0}{5}$, and $\frac{0}{0}$ and their related multiplicative forms.
>
> 1. Division by zero is undefined.
>   <u>Example</u>: $\frac{5}{0} = n$ implies that $n \cdot 0 = 5$. No number, n, satisfies this requirement.
> 2. Zero divided by any nonzero number is zero.
>   <u>Example</u>: $\frac{0}{5} = 0$ implies that $0 \cdot 5 = 0$ which is a true statement.
> 3. We say that $\frac{0}{0}$ is **indeterminate** (cannot be uniquely determined). This concept is investigated in detail in a first course in calculus.
>   <u>Example</u>: $\frac{0}{0} = n$ implies that $n \cdot 0 = 0$. This is true for any number n. Therefore, the quotient cannot be uniquely determined.

7. Simplify Algebraic Expressions

An algebraic **term** is a product of factors that may include constants and variables. An algebraic **expression** is a single term or the sum of two or more terms. For example, the expression

$$-3xz^2 + \left(-\frac{4}{b} \right) + z\sqrt{x - y} + 5 \text{ has four terms.}$$

The terms $-3xz^2$, $-\frac{4}{b}$, and $z\sqrt{x - y}$ are **variable terms.** The term 5 is not subject to change and is called a **constant term.** The constant factor within each term is called the **coefficient** of the term. Writing the expression as $-3xz^2 + \left(-4 \cdot \frac{1}{b} \right) + 1z\sqrt{x - y} + 5$, we identify the coefficients of the four terms as -3, -4, 1, and 5, respectively.

The properties of real numbers summarized in Table R-4 are often helpful when working with algebraic expressions.

Table R-4   **Properties of Real Numbers**

Let a, b, and c represent real numbers or real-valued expressions.

Property	In Symbols and Words	Examples
Commutative property of addition	$a + b = b + a$ The order in which real numbers are added does not affect the sum.	<u>ex</u>: $4 + (-7) = -7 + 4$ <u>ex</u>: $6 + w = w + 6$
Commutative property of multiplication	$a \cdot b = b \cdot a$ The order in which real numbers are multiplied does not affect the product.	<u>ex</u>: $5 \cdot (-4) = -4 \cdot 5$ <u>ex</u>: $x \cdot 12 = 12x$
Associative property of addition	$(a + b) + c = a + (b + c)$ The order in which real numbers are grouped under addition does not affect the sum.	<u>ex</u>: $(3 + 5) + 2 = 3 + (5 + 2)$ <u>ex</u>: $-9 + (2 + t) = (-9 + 2) + t$ $= -7 + t$
Associative property of multiplication	$(a \cdot b) \cdot c = a \cdot (b \cdot c)$ The order in which real numbers are grouped under multiplication does not affect the product.	<u>ex</u>: $(6 \cdot 7) \cdot 3 = 6 \cdot (7 \cdot 3)$ <u>ex</u>: $8 \cdot \left(\frac{1}{8} \cdot y \right) = \left(8 \cdot \frac{1}{8} \right) \cdot y$ $= 1y$
Identity property of addition	$a + 0 = a$ and $0 + a = a$ The number 0 is called the **identity element of addition** because any number plus 0 is the number itself.	<u>ex</u>: $-5 + 0 = -5$ <u>ex</u>: $0 + \sqrt{z} = \sqrt{z}$
Identity property of multiplication	$a \cdot 1 = a$ and $1 \cdot a = a$ The number 1 is called the **identity element of multiplication** because any number times 1 is the number itself.	<u>ex</u>: $\sqrt{2} \cdot 1 = \sqrt{2}$ <u>ex</u>: $1 \cdot (2w + 3) = 2w + 3$

(Continued)

Property	In Symbols and Words	Examples
Inverse property of addition	$a + (-a) = 0$ and $(-a) + a = 0$ For any real number a, the value $-a$ is called the **additive inverse of a** (also called the **opposite of a**). The sum of any number and its additive inverse is the identity element for addition, 0.	ex: $4\pi + (-4\pi) = 0$ ex: $-ab^2 + ab^2 = 0$
Inverse property of multiplication	$a \cdot \frac{1}{a} = 1$ and $\frac{1}{a} \cdot a = 1$ where $a \neq 0$ For any nonzero real number a, the value $\frac{1}{a}$ is called the **multiplicative inverse of a** (also called the **reciprocal of a**). The product of any nonzero number and its multiplicative inverse is the identity element, 1.	ex: $-5 \cdot \left(-\frac{1}{5}\right) = 1$ ex: $\frac{1}{x^2} \cdot x^2 = 1$ for $x \neq 0$ *Note:* The number zero does not have a multiplicative inverse (reciprocal).
Distributive property of multiplication over addition	$a \cdot (b + c) = a \cdot b + a \cdot c$ The product of a number and a sum equals the sum of the products of the number and each term in the sum.	ex: $4 \cdot (5 + x) = 4 \cdot 5 + 4 \cdot x$ $= 20 + 4x$ ex: $2(x + \sqrt{3}) = 2x + 2\sqrt{3}$

The distributive property is used to "clear" parentheses when a factor outside parentheses is multiplied by multiple terms inside parentheses. To simplify expressions, we often use the distributive property to clear parentheses and combine like terms.

Two terms are **like terms** if they have the same variables and the corresponding variables are raised to the same power. We can combine like terms by using the distributive property. For example,

$8x^2 + 6x^2 - x^2$
$= 8x^2 + 6x^2 - 1x^2$ Note that the coefficient on the third term is -1.
$= (8 + 6 - 1)x^2$ Apply the distributive property.
$= 13x^2$ Simplify.

TIP Although the distributive property is used to add and subtract like terms, it is tedious to write each step. Adding or subtracting like terms can also be done by combining the coefficients and leaving the variable factor unchanged.

$8x^2 + 6x^2 - 1x^2 = 13x^2$ This method will be used throughout the text.

In Example 7, we simplify an expression by applying the distributive property to "clear" parentheses and combine like terms.

Classroom Example: p. 14
Exercise 94

EXAMPLE 7   **Clearing Parentheses and Combining Like Terms**

Simplify. $5 - 2(4c - 8d) + 3(1 - d) + c$

Solution:

$5 - 2(4c - 8d) + 3(1 - d) + c$ Apply the distributive property to clear parentheses.
$= 5 - 8c + 16d + 3 - 3d + c$ Combine like terms.
$= 8 - 7c + 13d$

TIP After applying the distributive property, the original parentheses are removed. For this reason, we often call this process "*clearing parentheses*."

Skill Practice 7   Simplify. $12 - 3(5x - 2y) + 5(3 - x) - y$

Answer
7. $-20x + 5y + 27$

8. Write Algebraic Models

An important skill in mathematics and science is to develop mathematical models. Example 8 offers practice writing algebraic expressions based on verbal statements.

Classroom Examples: p. 14
Exercises 100, 102

EXAMPLE 8   Writing an Algebraic Model

a. The maximum recommended heart rate M for adults is the difference of 220 and the person's age a. Write a model to represent an adult's maximum recommended heart rate in terms of age.

b. After eating at a restaurant, it is customary to leave a tip t for the server for at least 15% of the cost of the meal c. Write a model to represent the amount of the tip based on the cost of the meal.

Solution:

a. $M = 220 - a$ The word "difference" implies subtraction in the order given.

b. $t \geq 0.15c$ 15% of c implies multiplication.

Skill Practice 8

a. The sale price S on a lawn mower is the difference of the original price P and the amount of discount D. Write a model to represent the sale price.

b. The amount of simple interest I earned on a certain certificate of deposit is 4.65% of the amount of principal invested P. Write a model for the amount of interest.

Answers
8. a. $S = P - D$
b. $I = 0.0465P$

SECTION R.1   Practice Exercises

Concept Connections

1. A(n) _____set_____ is a collection of items called elements.

2. $\mathbb{Z} = \{..., -3, -2, -1, 0, 1, 2, 3, ...\}$ is called the set of ___integers___.

3. $\mathbb{R}$ is the notation used to denote the set of ____real____ ___numbers___.

4. A(n) __irrational__ number is a real number that cannot be expressed as a ratio of two integers.

5. Write an absolute value expression to represent the distance between a and b on the real number line. $|a - b|$ or $|b - a|$

6. The __commutative__ properties of addition and multiplication indicate that the order in which two real numbers are added or multiplied does not affect the sum or product.

7. The associative property of addition indicates that $a + (b + c) = $ _$(a + b) + c$_, and the associative property of multiplication indicates that $a(bc) = $ ___$(ab)c$___.

8. The statement $a(b + c) = ab + ac$ represents the __distributive__ property of multiplication over addition.

Objective 1: Identify Subsets of the Set of Real Numbers

For Exercises 9–10, write an English sentence to represent the algebraic statement.

9. a. $3 \in \mathbb{N}$ 3 is an element of the set of natural numbers.   **b.** $-3.1 \notin \mathbb{W}$ −3.1 is not an element of the set of whole numbers.

10. a. $\frac{2}{5} \in \mathbb{Q}$ $\frac{2}{5}$ is an element of the set of rational numbers.   **b.** $\pi \notin \mathbb{Q}$ π is not an element of the set of rational numbers.

For Exercises 11–12, determine whether the statement is true or false.

11. a. $-5 \in \mathbb{N}$ False   **b.** $-5 \in \mathbb{W}$ False   **12. a.** $\frac{1}{3} \in \mathbb{N}$ False   **b.** $\frac{1}{3} \in \mathbb{W}$ False

c. $-5 \in \mathbb{Z}$ True   **d.** $-5 \in \mathbb{Q}$ True   **c.** $\frac{1}{3} \in \mathbb{Z}$ False   **d.** $\frac{1}{3} \in \mathbb{Q}$ True

Objective 2: Use Inequality Symbols and Interval Notation

For Exercises 13–18, write each statement as an inequality.

13. a is at least 5. $a \geq 5$

14. b is at most -6. $b \leq -6$

15. $3c$ is no more than 9. $3c \leq 9$

16. $8d$ is no less than 16. $8d \geq 16$

17. The quantity $(m + 4)$ exceeds 70. $m + 4 > 70$

18. The quantity $(n - 7)$ is approximately equal to 4.
$n - 7 \approx 4$

For Exercises 19–24, write the interval notation and set-builder notation for each given graph. (See Example 1)

19. $(-7, \infty); \{x | x > -7\}$

20. $[2, \infty); \{x | x \geq 2\}$

21. $(-\infty, 4.1]; \{x | x \leq 4.1\}$

22. $(-\infty, -2.93); \{x | x < -2.93\}$

23. $[-6, 0); \{x | -6 \leq x < 0\}$

24. $(2, 8]; \{x | 2 < x \leq 8\}$

For Exercises 25–30, graph the given set and write the corresponding interval notation. (See Example 1)

25. $\{x | x \leq 6\}$

26. $\{x | x < -4\}$

27. $\left\{ x \left| -\dfrac{7}{6} < x \leq \dfrac{1}{3} \right. \right\}$

28. $\left\{ x \left| -\dfrac{4}{3} \leq x < \dfrac{7}{4} \right. \right\}$

29. $\{x | 4 < x\}$

30. $\{x | -3 \leq x\}$

For Exercises 31–36, interval notation is given for several sets of real numbers. Graph the set and write the corresponding set-builder notation. (See Example 1)

31. $(-3, 7]$

32. $[-4, -1)$

33. $(-\infty, 6.7]$

34. $(-\infty, -3.2)$

35. $\left[-\dfrac{3}{5}, \infty \right)$

36. $\left(\dfrac{7}{8}, \infty \right)$

Objective 3: Find the Union and Intersection of Sets

For Exercises 37–40, refer to sets A, B, C, X, Y, and Z and find the union or intersection of sets as indicated. (See Example 2)

$A = \{0, 4, 8, 12\},$ $B = \{0, 3, 6, 9, 12\},$ $C = \{-2, 4, 8\}$
$X = \{1, 2, 3, 4, 5\},$ $Y = \{1, 2, 3\},$ $Z = \{6, 7, 8\}$

37. a. $A \cup B$ $\{0, 3, 4, 6, 8, 9, 12\}$ **b.** $A \cap B$ $\{0, 12\}$ **c.** $A \cup C$ $\{-2, 0, 4, 8, 12\}$

d. $A \cap C$ $\{4, 8\}$ **e.** $B \cup C$ $\{-2, 0, 3, 4, 6, 8, 9, 12\}$ **f.** $B \cap C$ $\{\ \}$

38. a. $X \cup Z$ $\{1, 2, 3, 4, 5, 6, 7, 8\}$ **b.** $Y \cup Z$ $\{1, 2, 3, 6, 7, 8\}$ **c.** $Y \cap Z$ $\{\ \}$

d. $X \cup Y$ $\{1, 2, 3, 4, 5\}$ **e.** $X \cap Y$ $\{1, 2, 3\}$ **f.** $X \cap Z$ $\{\ \}$

39. a. $A \cup X$ $\{0, 1, 2, 3, 4, 5, 8, 12\}$ **b.** $A \cap Z$ $\{8\}$ **c.** $C \cap Y$ $\{\ \}$

d. $B \cap Y$ $\{3\}$ **e.** $C \cup Z$ $\{-2, 4, 6, 7, 8\}$ **f.** $A \cup Z$ $\{0, 4, 6, 7, 8, 12\}$

40. a. $A \cap X$ $\{4\}$ **b.** $B \cup Z$ $\{0, 3, 6, 7, 8, 9, 12\}$ **c.** $C \cup Y$ $\{-2, 1, 2, 3, 4, 8\}$

d. $B \cap Z$ $\{6\}$ **e.** $C \cap Z$ $\{8\}$ **f.** $A \cup Y$ $\{0, 1, 2, 3, 4, 8, 12\}$

41. Refer to sets C, D, and F and find the union or intersection of sets as indicated. Write the answers in set notation. (See Example 3)

$C = \{x | x < 9\},$ $D = \{x | x \geq -1\},$ $F = \{x | x < -8\}$

a. $C \cup D$ $\mathbb{R}$ **b.** $C \cap D$ $\{x | -1 \leq x < 9\}$ **c.** $C \cup F$ $\{x | x < 9\}$

d. $C \cap F$ $\{x | x < -8\}$ **e.** $D \cup F$ $\{x | x < -8 \text{ or } x \geq -1\}$ **f.** $D \cap F$ $\{\ \}$

42. Refer to sets M, N, and P and find the union or intersection of sets as indicated. Write the answers in set notation.

$M = \{y \,|\, y \geq -3\}$, $N = \{y \,|\, y \geq 5\}$, $P = \{y \,|\, y < 0\}$

a. $M \cup N$ $\{y \,|\, y \geq -3\}$ **b.** $M \cap N$ $\{y \,|\, y \geq 5\}$ **c.** $M \cup P$ $\mathbb{R}$

d. $M \cap P$ $\{y \,|\, -3 \leq y < 0\}$ **e.** $N \cup P$ $\{y \,|\, y < 0 \text{ or } y \geq 5\}$ **f.** $N \cap P$ $\{\ \}$

For Exercises 43–46, find the union or intersection of the given intervals. Write the answers in interval notation.

43. a. $(-\infty, 4) \cup (-2, 1]$ $(-\infty, 4)$ **b.** $(-\infty, 4) \cap (-2, 1]$ $(-2, 1]$

44. a. $[0, 5) \cup [-1, \infty)$ $[-1, \infty)$ **b.** $[0, 5) \cap [-1, \infty)$ $[0, 5)$

45. a. $(-\infty, 5) \cup [3, \infty)$ $(-\infty, \infty)$ **b.** $(-\infty, 5) \cap [3, \infty)$ $[3, 5)$

46. a. $(-\infty, -1] \cup [-4, \infty)$ $(-\infty, \infty)$ **b.** $(-\infty, -1] \cap [-4, \infty)$ $[-4, -1]$

Objective 4: Evaluate Absolute Value Expressions

For Exercises 47–58, simplify each expression by writing the expression without absolute value bars. (See Example 4)

47. $|-6|$ 6 **48.** $|-4|$ 4 **49.** $|0|$ 0

50. $|1|$ 1 **51.** $|\sqrt{17} - 5|$ $5 - \sqrt{17}$ **52.** $|\sqrt{6} - 6|$ $6 - \sqrt{6}$

53. a. $|\pi - 3|$ $\pi - 3$ **54. a.** $|m - 11|$ for $m \geq 11$ $m - 11$ **55. a.** $|x + 2|$ for $x \geq -2$ $x + 2$

    b. $|3 - \pi|$ $\pi - 3$ **b.** $|m - 11|$ for $m < 11$ $11 - m$ **b.** $|x + 2|$ for $x < -2$ $-x - 2$

56. a. $|t + 6|$ for $t < -6$ $-t - 6$ **57. a.** $\dfrac{|z - 5|}{z - 5}$ for $z > 5$ 1 **58. a.** $\dfrac{7 - x}{|7 - x|}$ for $x < 7$ 1

    b. $|t + 6|$ for $t \geq -6$ $t + 6$ **b.** $\dfrac{|z - 5|}{z - 5}$ for $z < 5$ -1 **b.** $\dfrac{7 - x}{|7 - x|}$ for $x > 7$ -1

Objective 5: Use Absolute Value to Represent Distance

For Exercises 59–64, write an absolute value expression to represent the distance between the two points on the number line. Then simplify without absolute value bars. (See Example 5)

59. 1 and 6 $|1 - 6|$ or $|6 - 1|$; 5 **60.** 2 and 9 $|2 - 9|$ or $|9 - 2|$; 7 **61.** 3 and -4 $|3 - (-4)|$ or $|-4 - 3|$; 7

62. -8 and 2 $|-8 - 2|$ or $|2 - (-8)|$; 10 **63.** 6 and 2π $|6 - 2\pi|$ or $|2\pi - 6|$; $2\pi - 6$ **64.** 3 and π $|3 - \pi|$ or $|\pi - 3|$; $\pi - 3$

Objective 6: Apply the Order of Operations

For Exercises 65–78, simplify the expressions. (See Example 6)

65. a. 4^2 16 **b.** $(-4)^2$ 16 **c.** -4^2 -16 **d.** $\sqrt{4}$ 2 **e.** $-\sqrt{4}$ -2 **f.** $\sqrt{-4}$ Not a real number

66. a. 9^2 81 **b.** $(-9)^2$ 81 **c.** -9^2 -81 **d.** $\sqrt{9}$ 3 **e.** $-\sqrt{9}$ -3 **f.** $\sqrt{-9}$ Not a real number

67. a. $\sqrt[3]{8}$ 2 **b.** $\sqrt[3]{-8}$ -2 **c.** $-\sqrt[3]{8}$ -2 **d.** $\sqrt{100}$ 10 **e.** $\sqrt{-100}$ Not a real number   **f.** $-\sqrt{100}$ -10

68. a. $\sqrt[3]{27}$ 3 **b.** $\sqrt[3]{-27}$ -3 **c.** $-\sqrt[3]{27}$ -3 **d.** $\sqrt{49}$ 7 **e.** $\sqrt{-49}$ Not a real number   **f.** $-\sqrt{49}$ -7

69. $20 - 12(36 \div 3^2 \div 2)$ -4 **70.** $200 - 2^2(6 \div \dfrac{1}{2} \cdot 4)$ 8 **71.** $6 - \{-12 + 3[(1 - 6)^2 - 18]\}$ -3

72. $-5 - \{4 - 6[(2 - 8)^2 - 31]\}$ 21 **73.** $-4 \cdot \left(\dfrac{2}{5} - \dfrac{7}{10}\right)^2$ $-\dfrac{9}{25}$ **74.** $6 \cdot \left[\left(\dfrac{1}{3}\right)^2 - \left(\dfrac{1}{2}\right)^2\right]$ $-\dfrac{5}{6}$

75. $9 - (6 + ||3 - 7| - 8|) \div \sqrt{25}$ 7 **76.** $8 - 2(4 + ||2 - 5| - 5|) \div \sqrt{9}$ 4 **77.** $\dfrac{|11 - 13| - 4 \cdot 2}{\sqrt{12^2 + 5^2} - 3 - 10}$ Undefined

78. $\dfrac{(4 - 9)^2 + 2^2 - 3^2}{|-7 + 4| + (-12) \div 4}$ Undefined

Objective 7: Simplify Algebraic Expressions

For Exercises 79–84, combine like terms.

79. $-14w^3 - 3w^3 + w^3$ $-16w^3$

80. $12t^5 - t^5 - 6t^5$ $5t^5$

81. $3.9x^3y - 2.2xy^3 + 5.1x^3y - 4.7xy^3$ $9x^3y - 6.9xy^3$

82. $0.004m^4n - 0.005m^3n^2 - 0.01m^4n + 0.007m^3n^2$
$-0.006m^4n + 0.002m^3n^2$

83. $\frac{1}{3}c^7d + \frac{1}{2}cd^7 - \frac{2}{5}c^7d - 2cd^7$ $-\frac{1}{15}c^7d - \frac{3}{2}cd^7$

84. $\frac{1}{10}yz^4 - \frac{3}{4}y^4z + yz^4 + \frac{3}{2}y^4z$ $\frac{11}{10}yz^4 + \frac{3}{4}y^4z$

For Exercises 85–90, apply the distributive property.

85. $-(4x - \pi)$ $-4x + \pi$

86. $-\left(\frac{1}{2}k - \sqrt{7}\right)$ $-\frac{1}{2}k + \sqrt{7}$

87. $-8(3x^2 + 2x - 1)$ $-24x^2 - 16x + 8$

88. $-6(-5y^2 - 3y + 4)$ $30y^2 + 18y - 24$

89. $\frac{2}{3}(-6x^2y - 18yz^2 + 2z^3)$ $-4x^2y - 12yz^2 + \frac{4}{3}z^3$

90. $\frac{3}{4}(12p^5q - 8p^4q^2 - 6p^3q)$ $9p^5q - 6p^4q^2 - \frac{9}{2}p^3q$

For Exercises 91–98, simplify each expression. (See Example 7)

91. $2(4w + 8) + 7(2w - 4) + 12$ $22w$

92. $3(2z - 4) + 8(z - 9) + 84$ $14z$

93. $-(4u - 8v) - 3(7u - 2v) + 2v$ $-25u + 16v$

94. $-(10x - z) - 2(8x - 4z) - 3x$ $-29x + 9z$

95. $12 - 4[(8 - 2v) + 5(-3w - 4v)] - w$ $88v + 59w - 20$

96. $6 - 2[(9z + 6y) - 8(y - z)] - 11$ $4y - 34z - 5$

97. $2y^2 - \left[13 - \frac{2}{3}(6y^2 - 9) - 10\right] + 9$ $6y^2$

98. $6 - \left[5t^2 - \frac{3}{4}(12 - 8t^2) + 5\right] + 11t^2$ 10

Objective 8: Write Algebraic Models

99. Jake is 1 yr younger than Charlotte.

  a. Write a model for Jake's age J in terms of Charlotte's age C. $J = C - 1$

  b. Write a model for Charlotte's age C in terms of Jake's age J. **(See Example 8)** $C = J + 1$

100. For a recent NFL season, Aaron Rodgers had 9 more touchdown passes than Joe Flacco.

  a. Write a model for the number of touchdown passes R thrown by Rodgers in terms of the number thrown by Flacco F. $R = F + 9$

  b. Write a model for the number of touchdown passes F thrown by Flacco in terms of the number thrown by Rodgers R. $F = R - 9$

101. At the end of the summer, a store discounts an outdoor grill for at least 25% of the original price. If the original price is P, write a model for the amount of the discount D. **(See Example 8)** $D \geq 0.25P$

102. When Ms. Celano has excellent service at a restaurant, the amount she leaves for a tip t is at least 20% of the cost of the meal c. Write a model representing the amount of the tip. $t \geq 0.20c$

103. Suppose that an object is dropped from a height h. Its velocity v at impact with the ground is given by the square root of twice the product of the acceleration due to gravity g and the height h. Write a model to represent the velocity of the object at impact.
$v = \sqrt{2gh}$

104. The height of a sunflower plant can be determined by the time t in weeks after the seed has germinated. Write a model to represent the height h if the height is given by the product of 8 and the square root of t. $h = 8\sqrt{t}$

105. A power company charges one household $0.12 per kilowatt-hour (kWh) and $14.89 in monthly taxes.

  a. Write a formula for the monthly charge C for this household if it uses k kilowatt-hours. $C = 0.12k + 14.89$

  b. Compute the monthly charge if the household uses 1200 kWh. $158.89

106. A utility company charges a base rate for water usage of $19.50 per month, plus $4.58 for every additional 1000 gal of water used over a 2000-gal base.

  a. Write a formula for the monthly charge C for a household that uses n thousand gallons over the 2000-gal base. $C = 4.58n + 19.50$

  b. Compute the cost for a family that uses a total of 6000 gal of water for a given month. $37.82

107. The cost C (in \$) to rent an apartment is \$640 per month, plus a \$500 nonrefundable security deposit, plus a \$200 nonrefundable deposit for each dog or cat.

    a. Write a formula for the total cost to rent an apartment for m months with n cats/dogs. $C = 640m + 200n + 500$

    b. Determine the cost to rent the apartment for 12 months, with 2 cats and 1 dog.   \$8780

108. For a certain college, the cost C (in \$) for taking classes the first semester is \$105 per credit-hour, \$35 for each lab, plus a one-time admissions fee of \$40.

    a. Write a formula for the total cost to take n credit-hours and L labs the first semester. $C = 105n + 35L + 40$

    b. Determine the cost for the first semester if a student takes 12 credit-hours with 2 labs.   \$1370

Mixed Exercises

For Exercises 109–114, evaluate each expression for the given values of the variables.

109. $\dfrac{-b}{2a}$ for $a = -1$, $b = -6$ -3

110. $\sqrt{b^2 - 4ac}$ for $a = 2$, $b = -6$, $c = 4$ 2

111. $\sqrt{(x_2 - x_1)^2 + (y_2 - y_1)^2}$ for $x_1 = 2$, $x_2 = -1$, $y_1 = -4$, $y_2 = 1$ $\sqrt{34}$

112. $\dfrac{y_2 - y_1}{x_2 - x_1}$ for $x_1 = -1.4$, $x_2 = 2$, $y_1 = 3.1$, $y_2 = -3.7$ -2

113. $\dfrac{1}{3}\pi r^2 h$ for $\pi \approx \dfrac{22}{7}$, $r = 7$, $h = 6$ 308

114. $\dfrac{4}{3}\pi r^3$ for $\pi \approx \dfrac{22}{7}$, $r = 3$ $\dfrac{792}{7}$

115. Under selected conditions, a sports car gets 15 mpg in city driving and 25 mpg for highway driving. The model $G = \frac{1}{15}c + \frac{1}{25}h$ represents the amount of gasoline used (in gal) for c miles driven in the city and h miles driven on the highway. Determine the amount of gas required to drive 240 mi in the city and 500 mi on the highway.   36 gal

116. Under selected conditions, a sedan gets 22 mpg in city driving and 32 mpg for highway driving. The model $G = \frac{1}{22}c + \frac{1}{32}h$ represents the amount of gasoline used (in gal) for c miles driven in the city and h miles driven on the highway. Determine the amount of gas required to drive 220 mi in the city and 512 mi on the highway.   26 gal

Write About It

117. When is a parenthesis used when writing interval notation?   A parenthesis is used if an endpoint to an interval is *not* included in the set.

118. When is a bracket used when writing interval notation?   A bracket is used if an endpoint to an interval *is* included in the set.

119. Explain the difference between the commutative property of addition and the associative property of addition.

120. Explain why 0 has no multiplicative inverse.   For any nonzero number a, the multiplicative inverse is defined as $\frac{1}{a}$. If $a = 0$, this would be $\frac{1}{0}$, which is undefined.

Expanding Your Skills

121. If $n > 0$, then $n - |n| = $ ____0____ .

122. If $n < 0$, then $n - |n| = $ ____$2n$____ .

123. If $n > 0$, then $n + |n| = $ ____$2n$____ .

124. If $n < 0$, then $n + |n| = $ ____0____ .

125. If $n > 0$, then $-|n| = $ ____$-n$____ .

126. If $n < 0$, then $-|n| = $ ____n____ .

For Exercises 127–130, determine the sign of the expression. Assume that a, b, and c are real numbers and $a < 0$, $b > 0$, and $c < 0$.

127. $\dfrac{ab^2}{c^3}$ positive

128. $\dfrac{a^2 c}{b^4}$ negative

129. $\dfrac{b(a + c)^3}{a^2}$ negative

130. $\dfrac{(a + b)^2 (b + c)^4}{b}$ positive or zero

For Exercises 131–134, write the set as a single interval.

131. $(-\infty, 2) \cap (-3, 4] \cap [1, 3]$ $[1, 2)$

132. $(-\infty, 5) \cap (-1, \infty) \cap [0, 3)$ $[0, 3)$

133. $[(-\infty, -2) \cup (4, \infty)] \cap [-5, 3)$ $[-5, -2)$

134. $[(-\infty, 6) \cup (10, \infty)] \cap [8, 12)$ $(10, 12)$

Technology Connections

For Exercises 135–138, use a calculator to approximate the expression to 2 decimal places.

135. $5000\left(1 + \dfrac{0.06}{12}\right)^{(12)(5)}$ 6744.25

136. $8500\left(1 + \dfrac{0.05}{4}\right)^{(4)(30)}$ 37,741.81

137. $\dfrac{-3 + 5\sqrt{2}}{7}$ 0.58

138. $\dfrac{6 - 3\sqrt{5}}{4}$ -0.18

SECTION R.2   Exponents and Radicals

1. Apply Properties of Exponents

In Section R.1, we learned that exponents are used to represent repeated multiplication. Applications of exponents appear in many fields of study, including computer science. Computer engineers define a *bit* as a fundamental unit of information having just two possible values. These values are represented by either 0 or 1. A *byte* is usually taken as 8 bits. Computer programmers know that there are 2^n possible values for an *n*-bit variable. So 1 byte has $2^8 = 256$ possible values.

Three bytes are often used to represent color on a computer screen. The intensity of each of the colors red, green, and blue ranges from 0 to 255 (a total of 256 possible values each). So the number of colors that can be represented by this system is

$$2^8 \cdot 2^8 \cdot 2^8 = (256)(256)(256) = 16,777,216$$

There are over 16 million possible colors available using this system. For example, the color given by red 137, green 21, blue 131, is a deep pink. See Figure R-2.

Figure R-2

The product $2^8 \cdot 2^8 \cdot 2^8 = (2 \cdot 2 \cdot 2 \cdot 2 \cdot 2 \cdot 2 \cdot 2 \cdot 2)(2 \cdot 2 \cdot 2 \cdot 2 \cdot 2 \cdot 2 \cdot 2 \cdot 2)(2 \cdot 2 \cdot 2 \cdot 2 \cdot 2 \cdot 2 \cdot 2 \cdot 2)$
$= 2^{8+8+8} = 2^{24} = 16,777,216.$

A similar approach can be used to show that $b^m b^n = b^{m+n}$ for natural numbers *m* and *n*.

We would like to extend this result to expressions where *m* and *n* are negative integers or zero. For example,

$$\underline{b^0} \cdot b^4 = b^{0+4} = b^4 \quad \text{For this to be true, the value } b^0 \text{ must be 1.}$$
$$\underline{1} \cdot b^4 = b^4$$

Now consider an example involving a negative exponent.

$$\underline{b^{-4}} \cdot b^4 = b^{-4+4} = b^0 = 1 \quad \text{For the product } b^{-4} \cdot b^4 \text{ to be equal to 1, it follows}$$
$$\underline{\frac{1}{b^4}} \cdot b^4 = 1 \qquad\qquad \text{that } b^{-4} \text{ must be the reciprocal of } b^4. \text{ That is, } b^{-4} = \frac{1}{b^4}.$$

These observations lead to two important definitions.

Definition of b^0 and b^{-n}	
If b is a nonzero real number and n is a positive integer, then	
$b^0 = 1$	Examples: $(5000)^0 = 1$ and $(-3x)^0 = 1$ for $x \neq 0$
$b^{-n} = \left(\dfrac{1}{b}\right)^n = \dfrac{1}{b^n}$	Examples: $4^{-1} = \left(\dfrac{1}{4}\right)^1 = \dfrac{1}{4}$ and $x^{-3} = \left(\dfrac{1}{x}\right)^3 = \dfrac{1}{x^3}$ for $x \neq 0$

The definitions given here have two important restrictions.

By definition, $b^0 = 1$ provided that $b \neq 0$. Therefore,

- The value of 0^0 is not defined here. The value of 0^0 is said to be indeterminate and is examined in calculus.

For a positive integer n, by definition, $b^{-n} = \dfrac{1}{b^n}$ provided that $b \neq 0$. Therefore,

- The value of 0^{-n} is not defined here.

All examples and exercises in the text will be given under the assumption that the variable expressions avoid these restrictions. For example, the expression x^0 will be stated with the implied restriction that $x \neq 0$.

Classroom Examples: p. 25
Exercises 10, 14

EXAMPLE 1 **Simplifying Expressions with a Zero or Negative Exponent**

Simplify.

a. $6y^0$ **b.** -5^0 **c.** 5^{-2} **d.** $\dfrac{1}{p^{-4}}$ **e.** $5x^{-8}y^2$

Solution:

a. $6y^0 = 6 \cdot y^0 = 6 \cdot 1 = 6$ The base for the exponent of 0 is y, not $6y$. The product of 6 and y^0 is $6 \cdot 1 = 6$.

b. $-5^0 = -1 \cdot 5^0 = -1 \cdot 1 = -1$ This is interpreted as the opposite of 5^0 or $-1 \cdot 5^0$.

c. $5^{-2} = \dfrac{1}{5^2}$ or $\dfrac{1}{25}$ Rewrite the expression using the reciprocal of 5 and change the exponent to positive 2.

d. $\dfrac{1}{p^{-4}} = \dfrac{1}{\dfrac{1}{p^4}} = 1 \cdot \dfrac{p^4}{1} = p^4$ Rewrite p^{-4} as $\dfrac{1}{p^4}$. Then divide fractions.

e. $5x^{-8}y^2 = 5 \cdot x^{-8} \cdot y^2$ The exponent of -8 applies to x only, not to the factors of 5 or y.

     $= 5 \cdot \dfrac{1}{x^8} \cdot y^2 = \dfrac{5y^2}{x^8}$

Skill Practice 1 Simplify.

a. $4z^0$ **b.** -4^0 **c.** 2^{-3} **d.** $\dfrac{1}{n^{-5}}$ **e.** $-3a^4b^{-9}$

The property $b^m b^n = b^{m+n}$ is one of several important properties of exponents that can be used to simplify an algebraic expression (Table R-5).

Table R-5 **Properties of Exponents**

Let a and b be real numbers and m and n be integers.*

Property	Example	Expanded Form
$b^m \cdot b^n = b^{m+n}$	$x^4 \cdot x^3 = x^{4+3} = x^7$	$x^4 \cdot x^3 = (x \cdot x \cdot x \cdot x)(x \cdot x \cdot x) = x^7$
$\dfrac{b^m}{b^n} = b^{m-n}$	$\dfrac{t^6}{t^2} = t^{6-2} = t^4$	$\dfrac{t^6}{t^2} = \dfrac{t \cdot t \cdot t \cdot t \cdot t \cdot t}{t \cdot t} = t^4$
$(b^m)^n = b^{m \cdot n}$	$(x^2)^3 = x^{2 \cdot 3} = x^6$	$(x^2)^3 = (x^2)(x^2)(x^2) = x^6$
$(ab)^m = a^m b^m$	$(4x)^3 = 4^3 x^3$	$(4x)^3 = (4x)(4x)(4x)$ $= (4 \cdot 4 \cdot 4)(x \cdot x \cdot x)$ $= 4^3 x^3$
$\left(\dfrac{a}{b}\right)^m = \dfrac{a^m}{b^m}$	$\left(\dfrac{2}{y}\right)^2 = \dfrac{2^2}{y^2}$	$\left(\dfrac{2}{y}\right)^2 = \left(\dfrac{2}{y}\right) \cdot \left(\dfrac{2}{y}\right) = \dfrac{2^2}{y^2}$

*The properties are stated under the assumption that the variables are restricted to avoid the expressions 0^0 and $\frac{1}{0}$. Expressions of the form 0^0 and $\frac{1}{0}$ are said to be indeterminate and are examined in calculus.

Answers

1. a. 4   **b.** -1 **c.** $\dfrac{1}{2^3}$ or $\dfrac{1}{8}$

   d. n^5 **e.** $\dfrac{-3a^4}{b^9}$

Classroom Examples: p. 25
Exercises 34, 36

EXAMPLE 2   Simplifying Expressions Containing Exponents

Simplify. Write the answers with positive exponents only.

a. $\left(\dfrac{14a^2b^7}{2a^5b}\right)^{-2}$ **b.** $(-3x)^{-4}(4x^{-2}y^3)^3$

Solution:

a. $\left(\dfrac{14a^2b^7}{2a^5b}\right)^{-2} = (7a^{2-5}b^{7-1})^{-2}$ Simplify within parentheses first.
    To divide common bases, subtract the exponents.

$= (7a^{-3}b^6)^{-2}$

$= (7)^{-2}(a^{-3})^{-2}(b^6)^{-2}$ Raise each factor within parentheses to the -2 power.

$= 7^{-2}a^6b^{-12}$

$= \dfrac{1}{7^2} \cdot a^6 \cdot \dfrac{1}{b^{12}}$ Simplify factors with negative exponents.

$= \dfrac{a^6}{49b^{12}}$ Simplify.

b. $(-3x)^{-4}(4x^{-2}y^3)^3$
$= (-3)^{-4}x^{-4} \cdot 4^3(x^{-2})^3(y^3)^3$ Raise each factor inside parentheses to the power outside parentheses.

$= (-3)^{-4} \cdot 4^3 \cdot x^{-4} \cdot x^{-6} \cdot y^9$ Regroup factors.

$= (-3)^{-4} \cdot 4^3 \cdot x^{-10} \cdot y^9$ To multiply the factors of x, add the exponents.

$= \dfrac{4^3y^9}{(-3)^4x^{10}}$ Simplify factors with negative exponents.

$= \dfrac{64y^9}{81x^{10}}$ Simplify.

Skill Practice 2   Simplify. Write the answers with positive exponents only.

a. $(3w)^{-3}(5wt^5)^2$ **b.** $\left(\dfrac{26c^5d^{-2}}{13cd^8}\right)^{-3}$

2. Apply Scientific Notation

In many applications of science, technology, and business we encounter very large or very small numbers. For example:

- eBay Inc. purchased the Internet communications company, Skype Technologies, for approximately $2,600,000,000. (*Source*: www.ebay.com)
- The diameter of a capillary is measured as 0.000 005 m.
- The mean surface temperature of the planet Saturn is $-300°F$.

Very large and very small numbers are sometimes cumbersome to write because they contain numerous zeros. Furthermore, it is difficult to determine the location of the decimal point when performing calculations with such numbers. For these reasons, scientists will often write numbers using scientific notation.

Answers

2. a. $\dfrac{25t^{10}}{27w}$ **b.** $\dfrac{d^{30}}{8c^{12}}$

Scientific Notation
A number expressed in the form $a \times 10^n$, where $1 \le |a| < 10$ and n is an integer is said to be in **scientific notation.**

Examples		
Skype Purchase	Capillary Size	Saturn Temp.
$2,600,000,000	0.000 005 m	$-300°F$
$= \$2.6 \times 1,000,000,000$	$= 5.0 \times 0.000001$ m	$= -3 \times 100°F$
$= \$2.6 \times 10^9$	$= 5.0 \times 10^{-6}$ m	$= -3 \times 10^2 \ °F$

To write a number in scientific notation, the number of positions that the decimal point must be moved determines the power of 10. Numbers 10 or greater require a positive exponent on 10. Numbers between 0 and 1 require a negative exponent on 10.

Classroom Examples: p. 26
Exercises 48, 54

EXAMPLE 3   Writing Numbers in Scientific Notation and Standard Decimal Notation

Write the numbers in scientific notation or standard decimal notation.
a. 0.0000002   **b.** 230 billion   **c.** 1.36×10^7 **d.** 3.9×10^{-3}

Solution:

a. $0.0000002 = 2.0 \times 10^{-7}$

7 place positions

> The number 0.0000002 is between 0 and 1. Use a negative power of 10.

b. 230 billion $= 230,000,000,000$

$230,000,000,000 = 2.3 \times 10^{11}$

11 place positions

> First write 230 billion in standard form.
>
> The number 230 billion is greater than 10. Use a positive power of 10.

c. $1.36 \times 10^7 = 13,600,000$

7 place positions

> 10^7 is greater than 10. Move the decimal point 7 places to the right. Insert zeros to the right as needed.

d. $3.9 \times 10^{-3} = 0.0039$

3 place positions

> 10^{-3} is between 0 and 1. Move the decimal point 3 places to the left. Insert zeros to the left as needed.

Skill Practice 3   Write the numbers in scientific notation or standard decimal notation.
a. 0.0000035   **b.** 49,500   **c.** 5.86×10^2 **d.** 5.0×10^{-2}

Example 4 demonstrates the process to multiply numbers written in scientific notation.

Classroom Example: p. 26
Exercise 62

EXAMPLE 4   Performing Calculations with Scientific Notation

A light-year is the distance that light travels in 1 yr. If light travels at a speed of 6.7×10^8 mph, how far will it travel in 1 yr (8.76×10^3 hr)?

Answers
3. a. 3.5×10^{-6} **b.** 4.95×10^4
    c. 586   **d.** 0.05

Solution:

Distance $=$ (Rate)(Time)

$= (6.7 \times 10^8 \text{ mph})(8.76 \times 10^3 \text{ hr})$ Regroup factors.
Multiply and add the
$= (6.7)(8.76) \times (10^8)(10^3) \text{ mi}$ powers of 10.

$= 58.692 \times 10^{11} \text{ mi}$ The number 58.692 is not
between 1 and 10. Rewrite
this as 5.8692×10^1.
$= (5.8692 \times 10^1) \times 10^{11} \text{ mi}$

$= 5.8692 \times 10^{12} \text{ mi}$ One light-year is approximately 5.87 trillion miles.

Skill Practice 4

A satellite travels 1.72×10^4 mph. How far does it travel in 24 hr
$(2.4 \times 10^1 \text{ hr})$?

---

3. Evaluate *n*th-Roots

We now want to extend the definition of b^n to expressions in which the exponent, n, is a rational number.

First we need to understand the relationship between *n*th-powers and *n*th-roots. From Section R.1, we know that for $a \geq 0$, $\sqrt{a} = b$ if $b^2 = a$ and $b \geq 0$. Square roots are a special case of *n*th-roots.

Definition of an *n*th-Root

For a positive integer $n > 1$, the principal *n*th-root of a, denoted by $\sqrt[n]{a}$, is a number b such that

$$\sqrt[n]{a} = b \text{ means that } b^n = a.$$

If n is even, then we require that $a \geq 0$ and $b \geq 0$.

For the expression $\sqrt[n]{a}$, the symbol $\sqrt[n]{}$ is called a **radical sign,** the value a is called the **radicand,** and n is called the **index.**

Classroom Examples: p. 26
Exercises 72, 76, 78

EXAMPLE 5   **Simplifying *n*th-Roots**

Simplify.

     a. $\sqrt[5]{32}$ **b.** $\sqrt{\dfrac{49}{64}}$ **c.** $\sqrt[3]{-0.008}$ **d.** $\sqrt[4]{-1}$ **e.** $-\sqrt[4]{1}$

Solution:

     a. $\sqrt[5]{32} = 2$ $\sqrt[5]{32} = 2$ because $2^5 = 32$.

     b. $\sqrt{\dfrac{49}{64}} = \dfrac{7}{8}$ $\sqrt{\dfrac{49}{64}} = \dfrac{7}{8}$ because $\dfrac{7}{8} \geq 0$ and $\left(\dfrac{7}{8}\right)^2 = \dfrac{49}{64}$.

     c. $\sqrt[3]{-0.008} = -0.2$ $\sqrt[3]{-0.008} = -0.2$ because $(-0.2)^3 = -0.008$.

     d. $\sqrt[4]{-1}$ is not a real number $\sqrt[4]{-1}$ is not a real number because no real number when raised to the fourth power equals -1.

     e. $-\sqrt[4]{1} = -1 \cdot \sqrt[4]{1}$ $-\sqrt[4]{1}$ is interpreted as $-1\sqrt[4]{1}$. The factor of -1 is outside the radical.
              $= -1 \cdot 1$
              $= -1$

Answer

4. 4.128×10^5 mi

Skill Practice 5   Simplify.

   a. $\sqrt[3]{-125}$ **b.** $\sqrt{\dfrac{144}{121}}$ **c.** $\sqrt[5]{0.00001}$ **d.** $\sqrt[6]{-64}$ **e.** $-\sqrt[6]{64}$

---

4. Simplify Expressions of the Forms $a^{1/n}$ and $a^{m/n}$

Next, we want to define an expression of the form a^n, where n is a rational number. Furthermore, we want a definition for which the properties of integer exponents can be extended to rational exponents. For example, we want

$$(25^{1/2})^2 = 25^{(1/2)\cdot 2} = 25^1 = 25$$

$25^{1/2}$ must be a square root of 25, because when squared, it equals 25.

Definition of an $a^{1/n}$ and $a^{m/n}$

Let m and n be integers such that m/n is a rational number in lowest terms and $n > 1$. Then,

$$a^{1/n} = \sqrt[n]{a} \qquad \text{and} \qquad a^{m/n} = \sqrt[n]{a^m} = \left(\sqrt[n]{a}\right)^m$$

If n is even, we require that $a \geq 0$.

The definition of $a^{m/n}$ indicates that $a^{m/n}$ can be written as a radical whose index is the denominator of the rational exponent. The order in which the nth-root and exponent m are performed within the radical does not affect the outcome. For example:

Take the 4th root first:	$16^{3/4} = \left(\sqrt[4]{16}\right)^3$ $= (2)^3$ $= 8$	or Cube 16 first:	$16^{3/4} = \sqrt[4]{16^3}$ $= \sqrt[4]{4096}$ $= 8$

Classroom Examples: p. 26–27
Exercises 80, 88

EXAMPLE 6   Simplifying Expressions of the Form $a^{1/n}$ and $a^{m/n}$

Write the expressions using radical notation and simplify if possible.

   a. $25^{1/2}$ **b.** $\left(\dfrac{64}{27}\right)^{1/3}$ **c.** $(-81)^{1/4}$ **d.** $32^{3/5}$ **e.** $81^{-3/4}$

Solution:

   a. $25^{1/2} = \sqrt{25} = 5$

   b. $\left(\dfrac{64}{27}\right)^{1/3} = \sqrt[3]{\dfrac{64}{27}} = \dfrac{4}{3}$

   c. $(-81)^{1/4}$ is undefined because $\sqrt[4]{-81}$ is not a real number.

   d. $32^{3/5} = \sqrt[5]{32^3} = \left(\sqrt[5]{32}\right)^3 = (2)^3 = 8$

   e. $81^{-3/4} = \dfrac{1}{81^{3/4}}$ Rewrite the expression to remove the negative exponent.

            $= \dfrac{1}{\left(\sqrt[4]{81}\right)^3} = \dfrac{1}{(3)^3} = \dfrac{1}{27}$ Rewrite $81^{3/4}$ using radical notation.

Answers

5. a. -5 **b.** $\dfrac{12}{11}$ **c.** 0.1
   d. Not a real number
   e. -2

Skill Practice 6   Simplify the expressions if possible.

a. $36^{1/2}$ **b.** $\left(\dfrac{1}{125}\right)^{1/3}$ **c.** $(-9)^{1/2}$ **d.** $(-1)^{4/3}$ **e.** $(-125)^{-2/3}$

5. Simplify Radicals

In Example 6, we simplified several expressions with rational exponents. Next, we want to simplify radical expressions. First consider expressions of the form $\sqrt[n]{a^n}$. The value of $\sqrt[n]{a^n}$ is not necessarily a. Since $\sqrt[n]{a}$ represents the principal nth-root of a, then $\sqrt[n]{a}$ must be nonnegative for even values of n. For example:

$$\sqrt{(5)^2} = 5 \quad \text{and} \quad \sqrt{(-5)^2} = |-5| = 5$$

> The absolute value is needed here to guarantee a nonnegative result.

$$\sqrt[4]{(2)^4} = 2 \quad \text{and} \quad \sqrt[4]{(-2)^4} = |-2| = 2$$

In Table R-6, we generalize this result and give three other important properties of radicals.

Table R-6   Properties of Radicals

Let $n > 1$ be an integer and a and b be real numbers. The following properties are true provided that the given radicals are real numbers.

Property	Examples	
1. If n is even, $\sqrt[n]{a^n} = \lvert a \rvert$.	$\sqrt{x^2} = \lvert x \rvert$	$\sqrt[4]{x^8} = \lvert x^2 \rvert = x^2$
2. If n is odd, $\sqrt[n]{a^n} = a$.	$\sqrt[3]{x^3} = x$	$\sqrt[5]{(y+4)^5} = y + 4$
3. Product Property $\sqrt[n]{a} \cdot \sqrt[n]{b} = \sqrt[n]{ab}$	$\sqrt[3]{7} \cdot \sqrt[3]{x} = \sqrt[3]{7x}$	$\sqrt{75} = \sqrt{25} \cdot \sqrt{3} = 5\sqrt{3}$
4. Quotient Property $\dfrac{\sqrt[n]{a}}{\sqrt[n]{b}} = \sqrt[n]{\dfrac{a}{b}}$	$\dfrac{\sqrt{125}}{\sqrt{5}} = \sqrt{\dfrac{125}{5}} = \sqrt{25} = 5$	
5. Nested Radical Property $\sqrt[m]{\sqrt[n]{a}} = \sqrt[mn]{a}$	$\sqrt[4]{\sqrt[3]{x}} = (x^{1/3})^{1/4} = x^{(1/3)(1/4)} = x^{1/12} = \sqrt[12]{x}$	

Properties 1–5 follow from the definition of $a^{1/n}$ and the properties of exponents. We use these properties to simplify radical expressions and must address four specific criteria.

Simplified Form of a Radical

Suppose that the radicand of a radical is written as a product of prime factors. Then the radical is simplified if all of the following conditions are met.

1. The radicand has no factor other than 1 that is a perfect nth-power. This means that all exponents in the radicand must be less than the index.
2. No fractions may appear in the radicand.
3. No denominator of a fraction may contain a radical.
4. The exponents in the radicand may not all share a common factor with the index.

In Example 7, we simplify expressions that fail condition 1. Also notice that the expressions in Example 7 are assumed to have positive radicands. This eliminates the need to insert absolute value bars around the simplified form of $\sqrt[n]{a^n}$.

Answers

6. a. 6   **b.** $\dfrac{1}{5}$

   c. Undefined (not a real number)

   d. 1   **e.** $\dfrac{1}{25}$

Classroom Examples: p. 27
Exercises 92, 96

EXAMPLE 7   Simplifying Radicals Using the Product Property

Simplify each expression. Assume that all variables represent positive real numbers.

a. $\sqrt[3]{c^5}$ b. $\sqrt{50}$ c. $\sqrt[4]{32x^9y^6}$

Solution:

a. $\sqrt[3]{c^5} = \sqrt[3]{c^3 \cdot c^2}$ Write the radicand as a product of a perfect cube and another factor.

$= \sqrt[3]{c^3} \cdot \sqrt[3]{c^2}$ Apply the product property of radicals.

$= c\sqrt[3]{c^2}$ Simplify $\sqrt[3]{c^3}$ as c.

Instructor Note:
Consider giving students a visual representation of simplifying a radical. For example:

$\sqrt{x^7}$
$= \sqrt{(x \cdot x) \cdot (x \cdot x) \cdot (x \cdot x) \cdot x}$
$= x^3\sqrt{x}$

b. $\sqrt{50} = \sqrt{5^2 \cdot 2}$ Factor the radicand. The radical is not simplified because the radicand has a perfect square.

$= \sqrt{5^2} \cdot \sqrt{2}$ Apply the product property of radicals.

$= 5\sqrt{2}$ Simplify.

c. $\sqrt[4]{32x^9y^6}$

$= \sqrt[4]{2^5x^9y^6}$ Factor the radicand.

$= \sqrt[4]{(2^4x^8y^4)(2xy^2)}$ Write the radicand as the product of a perfect 4th power and another factor.

$= \sqrt[4]{2^4x^8y^4} \cdot \sqrt[4]{2xy^2}$ Apply the product property of radicals.

$= 2x^2y\sqrt[4]{2xy^2}$ Simplify the first radical.

Skill Practice 7   Simplify each expression. Assume that all variables represent positive real numbers.

a. $\sqrt[4]{d^7}$ b. $\sqrt{45}$ c. $\sqrt[3]{54x^{13}y^8}$

6. Apply Operations on Radicals

In Example 8(a), we will use the product property of radicals to multiply two radical expressions. In Example 8(b), we will use the quotient property of radicals to divide radical expressions.

Classroom Examples: p. 27
Exercises 102, 112

EXAMPLE 8   Multiplying and Dividing Radical Expressions

Multiply or divide as indicated. Assume that x and y represent positive real numbers.

a. $\sqrt{6x} \cdot \sqrt{10x}$ b. $\dfrac{\sqrt[3]{3x^7y}}{\sqrt[3]{81xy^4}}$

Solution:

a. $\sqrt{6x} \cdot \sqrt{10x} = \sqrt{60x^2}$ The radicals have the same index. Apply the product property of radicals.

$= \sqrt{2^2 \cdot 3 \cdot 5 \cdot x^2}$ Factor the radicand.

$= \sqrt{(2x)^2 \cdot 3 \cdot 5}$ Write the radicand as the product of a perfect square and another factor.

$= \sqrt{(2x)^2} \cdot \sqrt{3 \cdot 5}$ Apply the product property of radicals.

$= 2x\sqrt{15}$ Simplify.

Answers
7. a. $d\sqrt[4]{d^3}$ b. $3\sqrt{5}$
   c. $3x^4y^2\sqrt[3]{2xy^2}$

TIP In Example 8(b), the purpose of writing the quotient of two radicals as a single radical is to simplify the resulting fraction in the radicand.

b. $\dfrac{\sqrt[3]{3x^7y}}{\sqrt[3]{81xy^4}} = \sqrt[3]{\dfrac{3x^7y}{81xy^4}}$ Apply the quotient property of radicals to write the expression as a single radical.

$= \sqrt[3]{\dfrac{x^6}{27y^3}}$ The numerator and denominator share common factors. Simplify the fraction.

$= \dfrac{x^2}{3y}$ Simplify.

Skill Practice 8 Multiply or divide as indicated. Assume that c, d, and y represent positive real numbers.

a. $\sqrt{15y} \cdot \sqrt{21y}$ **b.** $\dfrac{\sqrt[3]{625c^2d^{10}}}{\sqrt[3]{5c^5d}}$

In Example 8(b), we removed the radical from the denominator of the fraction. This is called **rationalizing the denominator.**

We can use the distributive property to add or subtract radical expressions. However, the radicals must be like radicals. This means that the radicands must be the same and the indices must be the same. For example:

$3\sqrt{2x}$ and $-5\sqrt{2x}$ are like radicals.

$3\sqrt{2x}$ and $-5\sqrt[3]{2x}$ are not like radicals because the indices are different.

$3\sqrt{2x}$ and $-5\sqrt{2y}$ are not like radicals because the radicands are different.

Classroom Example: p. 27 Exercise 118

EXAMPLE 9 **Adding and Subtracting Radicals**

Add or subtract as indicated. Assume that all variables represent positive real numbers.

a. $5\sqrt[3]{7t^2} - 2\sqrt[3]{7t^2} + \sqrt[3]{7t^2}$ **b.** $x\sqrt{98x^3y} + 5\sqrt{18x^5y}$

Solution:

a. $5\sqrt[3]{7t^2} - 2\sqrt[3]{7t^2} + 1\sqrt[3]{7t^2}$ The radicals are like radicals. They have the same radicand and same index.

$= (5 - 2 + 1)\sqrt[3]{7t^2}$ Apply the distributive property.

$= 4\sqrt[3]{7t^2}$ Simplify.

b. $x\sqrt{98x^3y} + 5\sqrt{18x^5y}$ Each radical can be simplified.
$x\sqrt{98x^3y} = x\sqrt{(7^2x^2) \cdot (2xy)} = 7x^2\sqrt{2xy}$
$5\sqrt{18x^5y} = 5\sqrt{(3^2x^4)(2xy)} = 15x^2\sqrt{2xy}$

$= 7x^2\sqrt{2xy} + 15x^2\sqrt{2xy}$ The terms are like terms.

$= (7 + 15)x^2\sqrt{2xy}$ Apply the distributive property.

$= 22x^2\sqrt{2xy}$ Simplify.

Skill Practice 9 Add or subtract as indicated. Assume that all variables represent positive real numbers.

a. $-4\sqrt[3]{5w} + 9\sqrt[3]{5w} - 11\sqrt[3]{5w}$ **b.** $\sqrt{75cd^4} + 6d\sqrt{27cd^2}$

Answers

8. a. $3y\sqrt{35}$ **b.** $\dfrac{5d^3}{c}$

9. a. $-6\sqrt[3]{5w}$ **b.** $23d^2\sqrt{3c}$

SECTION R.2   Practice Exercises

Concept Connections

1. For a nonzero real number b, the value of $b^0 =$ ___1___.

2. For a nonzero real number b, the value $b^{\boxed{-n}} = \dfrac{1}{b^n}$.

3. From the properties of exponents, $b^m b^n = b^{\boxed{m+n}}$.

4. If $b \neq 0$, then $\dfrac{b^m}{b^n} = b^{\boxed{m-n}}$.

5. From the properties of exponents, $(b^m)^n = b^{\boxed{m \cdot n}}$.

6. Given the expression $\sqrt[n]{a}$, the value a is called the ___radicand___ and n is called the ___index___.

7. The expression $a^{m/n}$ can be written in radical notation as ___$(\sqrt[n]{a})^m$ or $\sqrt[n]{a^m}$___, provided that $\sqrt[n]{a}$ is a real number.

8. The expression $a^{1/n}$ can be written in radical notation as ___$\sqrt[n]{a}$___, provided that $\sqrt[n]{a}$ is a real number.

Objective 1: Apply Properties of Exponents

For Exercises 9–14, simplify each expression. (See Example 1)

9. a. 8^0 1   **b.** -8^0 -1 **c.** $8x^0$ 8   **d.** $(8x)^0$ 1

10. a. 7^0 1   **b.** -7^0 -1 **c.** $7y^0$ 7   **d.** $(7y)^0$ 1

11. a. 8^{-2} $\dfrac{1}{64}$ **b.** $8x^{-2}$ $\dfrac{8}{x^2}$ **c.** $(8x)^{-2}$ $\dfrac{1}{64x^2}$ **d.** -8^{-2} $-\dfrac{1}{64}$

12. a. 7^{-2} $\dfrac{1}{49}$ **b.** $7y^{-2}$ $\dfrac{7}{y^2}$ **c.** $(7y)^{-2}$ $\dfrac{1}{49y^2}$ **d.** -7^{-2} $-\dfrac{1}{49}$

13. a. $\dfrac{1}{q^{-2}}$ q^2 **b.** q^{-2} $\dfrac{1}{q^2}$ **c.** $5p^3 q^{-2}$ $\dfrac{5p^3}{q^2}$ **d.** $5p^{-3}q^2$ $\dfrac{5q^2}{p^3}$

14. a. $\dfrac{1}{t^{-4}}$ t^4 **b.** t^{-4} $\dfrac{1}{t^4}$ **c.** $11t^{-4}u^2$ $\dfrac{11u^2}{t^4}$ **d.** $11t^4 u^{-2}$ $\dfrac{11t^4}{u^2}$

For Exercises 15–46, use the properties of exponents to simplify each expression. (See Example 2)

15. $x^7 \cdot x^6 \cdot x^{-2}$ x^{11}

16. $y^{-3} \cdot y^7 \cdot y^4$ y^8

17. $(-3c^2 d^7)(4c^{-5}d)$ $-\dfrac{12d^8}{c^3}$

18. $(-7m^{-3}n^{-8})(3m^{-5}n)$ $-\dfrac{21}{m^8 n^7}$

19. $\dfrac{y^{-3}y^6}{y^2}$ y

20. $\dfrac{z^{-8}z^{12}}{z^3}$ z

21. $\dfrac{18k^2 p^9}{27k^5 p^2}$ $\dfrac{2p^7}{3k^3}$

22. $\dfrac{10a^3 b^{11}}{25a^7 b^3}$ $\dfrac{2b^8}{5a^4}$

23. $\dfrac{2m^{-6}n^4}{6m^{-2}n^{-1}}$ $\dfrac{n^5}{3m^4}$

24. $\dfrac{4p^{-7}q^5}{2p^{-2}q^{-2}}$ $\dfrac{2q^7}{p^5}$

25. $(p^{-2})^7$ $\dfrac{1}{p^{14}}$

26. $(q^{-4})^2$ $\dfrac{1}{q^8}$

27. $\left(\dfrac{7a}{b}\right)^2$ $\dfrac{49a^2}{b^2}$

28. $\left(\dfrac{pq}{4}\right)^4$ $\dfrac{p^4 q^4}{256}$

29. $(4x^2 y^{-3})^2$ $\dfrac{16x^4}{y^6}$

30. $(-3w^{-3}z^5)^2$ $\dfrac{9z^{10}}{w^6}$

31. $\left(\dfrac{7k}{n^2}\right)^{-2}$ $\dfrac{n^4}{49k^2}$

32. $\left(\dfrac{5w^5}{v}\right)^{-3}$ $\dfrac{v^3}{125w^{15}}$

33. $\left(\dfrac{4x^3 z^{-5}}{12y^{-2}}\right)^{-3}$ $\dfrac{27z^{15}}{x^9 y^6}$

34. $\left(\dfrac{3z^2 w^{-1}}{15p^{-4}}\right)^{-3}$ $\dfrac{125w^3}{p^{12}z^6}$

35. $(-2y)^{-3}(6y^{-2}z^8)^2$ $-\dfrac{9z^{16}}{2y^7}$

36. $(-15z)^{-2}(5z^4 w^{-6})^3$ $\dfrac{5z^{10}}{9w^{18}}$

37. $\left(\dfrac{1}{2}\right)^{-3} - \left(\dfrac{1}{3}\right)^{-3} + \left(\dfrac{1}{6}\right)^{-3}$ 197

38. $\left(-\dfrac{1}{3}\right)^{-2} - \left(\dfrac{1}{4}\right)^{-2} + \left(\dfrac{1}{2}\right)^{-2}$ -3

39. $\left(\dfrac{1}{4x^3 y^{-5}}\right)^{-2}\left(\dfrac{8}{x^{-13}y^{14}}\right)^{-1}$ $\dfrac{2y^4}{x^7}$

40. $\left(\dfrac{1}{3a^2 b^{-4}}\right)^{-3}\left(\dfrac{3}{a^{-5}b^4}\right)^{-1}$ $\dfrac{9a}{b^8}$

41. $\dfrac{(4vw^{-3}x^2)^2}{(2v^2 w^3 x^{-2})^4} \cdot (-v^3 w^2 x^{-4})^{-5}$ $-\dfrac{x^{32}}{v^{21}w^{28}}$

42. $\dfrac{(14v^2 w^{-3}x^{-2})^3}{(21v^{-5}w^3 x^{-2})^{-1}} \cdot (-7v^4 w^{-1}x^{-4})^{-4}$ $\dfrac{24x^8}{v^{15}w^2}$

43. $(3x+5)^{14}(3x+5)^{-2}$ $(3x+5)^{12}$

44. $(2y-7z)^{-4}(2y-7z)^{13}$ $(2y-7z)^9$

45. $2^{-2} + 2^{-1} + 2^0 + 2^1 + 2^2$ $\dfrac{31}{4}$

46. $3^{-2} + 3^{-1} + 3^0 + 3^1 + 3^2$ $\dfrac{121}{9}$

Objective 2: Apply Scientific Notation

For Exercises 47–52, write the numbers in scientific notation. (See Example 3)

47. a. 350,000 3.5×10^5 **b.** 0.000035 3.5×10^{-5} **c.** 3.5 3.5×10^0

48. a. 2710 2.71×10^3 **b.** 0.00271 2.71×10^{-3} **c.** 2.71 2.71×10^0

49. The speed of light is approximately 29,980,000,000 cm/sec. 2.998×10^{10} cm/sec

50. The mean distance between the Earth and the Sun is approximately 149,000,000 km. 1.49×10^8 km

51. The size of an HIV particle is approximately 0.00001 cm. 1.0×10^{-5} cm

52. One picosecond is 0.000 000 000 001 sec. 1.0×10^{-12} sec

For Exercises 53–58, write the number in standard decimal notation. (See Example 3)

53. a. 2.61×10^{-6} 0.00000261   **b.** 2.61×10^6 2,610,000   **c.** 2.61×10^0 2.61

54. a. 3.52×10^{-2} 0.0352   **b.** 3.52×10^2 352   **c.** 3.52×10^0 3.52

55. A drop of water has approximately 1.67×10^{21} molecules of H_2O.   1,670,000,000,000,000,000,000 molecules

56. A computer with a 3-terabyte hard drive can store approximately 3.0×10^{12} bytes.   3,000,000,000,000 bytes

57. A typical red blood cell is 7.0×10^{-6} m.   0.000 007 m

58. The blue light used to read a laser disc has a wavelength of 4.7×10^{-7} m.   0.000 000 47 m

For Exercises 59–66, perform the indicated operation. Write the answer in scientific notation. (See Example 4)

59. $\dfrac{8.4 \times 10^{-6}}{2.1 \times 10^{-2}}$ 4×10^{-4}

60. $\dfrac{6.8 \times 10^{11}}{3.4 \times 10^3}$ 2×10^8

61. $(6.2 \times 10^{11})(3 \times 10^4)$ 1.86×10^{16}

62. $(8.1 \times 10^6)(2 \times 10^5)$ 1.62×10^{12}

63. $\dfrac{3.6 \times 10^{-14}}{5 \times 10^5}$ 7.2×10^{-20}

64. $\dfrac{3.68 \times 10^{-8}}{4 \times 10^2}$ 9.2×10^{-11}

65. $\dfrac{(6.2 \times 10^5)(4.4 \times 10^{22})}{2.2 \times 10^{17}}$ 1.24×10^{11}

66. $\dfrac{(3.8 \times 10^4)(4.8 \times 10^{-2})}{2.5 \times 10^{-5}}$ 7.296×10^7

67. Jonas has a personal music player with 80 gigabytes of memory (80 gigabytes is approximately 8×10^{10} bytes). If each song requires an average of 4 megabytes of memory (approximately 4×10^6 bytes), how many songs can Jonas store on the device? 2×10^4 songs

68. Joelle has a personal web page with 60 gigabytes of memory (approximately 6×10^{10} bytes). She stores math videos on the site for her students to watch outside of class. If each video requires an average of 5 megabytes of memory (approximately 5×10^6 bytes), how many videos can she store on her website? 1.2×10^4 videos

69. A typical adult human has approximately 5 L of blood in the body. If 1 μL (1 microliter) contains 5×10^6 red blood cells, how many red blood cells does a typical adult have? (*Hint*: $1 \text{ L} = 10^6$ μL.) 2.5×10^{13} red blood cells

70. The star Proxima Centauri is the closest star (other than the Sun) to the Earth. It is approximately 4.3 light-years away. If 1 light-year is approximately 5.9×10^{12} mi, how many miles is Proxima Centauri from the Earth? 2.537×10^{13} mi

Objective 3: Evaluate *n*th-Roots

For Exercises 71–78, simplify the expression. (See Example 5)

71. $\sqrt[4]{81}$ 3

72. $\sqrt[3]{125}$ 5

73. $\sqrt{\dfrac{4}{49}}$ $\dfrac{2}{7}$

74. $\sqrt{\dfrac{9}{121}}$ $\dfrac{3}{11}$

75. $\sqrt[4]{-81}$ Not a real number

76. $\sqrt[4]{-625}$ Not a real number

77. $\sqrt[3]{-\dfrac{1}{8}}$ $-\dfrac{1}{2}$

78. $\sqrt[3]{-\dfrac{64}{125}}$ $-\dfrac{4}{5}$

Objective 4: Simplify Expressions of the Forms $a^{1/n}$ and $a^{m/n}$

For Exercises 79–88, simplify each expression. (See Example 6)

79. a. $25^{1/2}$ 5   **b.** $(-25)^{1/2}$ Undefined   **c.** $-25^{1/2}$ -5

80. a. $36^{1/2}$ 6   **b.** $(-36)^{1/2}$ Undefined   **c.** $-36^{1/2}$ -6

81. a. $27^{1/3}$ 3   **b.** $(-27)^{1/3}$ -3 **c.** $-27^{1/3}$ -3

82. a. $125^{1/3}$ _5_   **b.** $(-125)^{1/3}$ _−5_   **c.** $-125^{1/3}$ _−5_

83. a. $\left(\dfrac{121}{169}\right)^{1/2}$ $\dfrac{11}{13}$ **b.** $\left(\dfrac{121}{169}\right)^{-1/2}$ $\dfrac{13}{11}$

84. a. $\left(\dfrac{49}{144}\right)^{1/2}$ $\dfrac{7}{12}$ **b.** $\left(\dfrac{49}{144}\right)^{-1/2}$ $\dfrac{12}{7}$

85. a. $16^{3/4}$ _8_   **b.** $16^{-3/4}$ $\dfrac{1}{8}$ **c.** $-16^{3/4}$ _−8_

   d. $-16^{-3/4}$ $-\dfrac{1}{8}$ **e.** $(-16)^{3/4}$ _Undefined_   **f.** $(-16)^{-3/4}$ _Undefined_

86. a. $81^{3/4}$ _27_   **b.** $81^{-3/4}$ $\dfrac{1}{27}$ **c.** $-81^{3/4}$ _−27_

   d. $-81^{-3/4}$ $-\dfrac{1}{27}$ **e.** $(-81)^{3/4}$ _Undefined_   **f.** $(-81)^{-3/4}$ _Undefined_

87. a. $64^{2/3}$ _16_   **b.** $64^{-2/3}$ $\dfrac{1}{16}$ **c.** $-64^{2/3}$ _−16_

   d. $-64^{-2/3}$ $-\dfrac{1}{16}$ **e.** $(-64)^{2/3}$ _16_   **f.** $(-64)^{-2/3}$ $\dfrac{1}{16}$

88. a. $8^{2/3}$ _4_   **b.** $8^{-2/3}$ $\dfrac{1}{4}$ **c.** $-8^{2/3}$ _−4_

   d. $-8^{-2/3}$ $-\dfrac{1}{4}$ **e.** $(-8)^{2/3}$ _4_   **f.** $(-8)^{-2/3}$ $\dfrac{1}{4}$

Objective 5: Simplify Radicals

89. a. For what values of t will the statement be true? $\sqrt{t^2} = t$ _t ≥ 0_

   b. For what value of t will the statement be true? $\sqrt{t^2} = |t|$ _All real numbers_

90. a. For what values of c will the statement be true? $\sqrt[4]{(c+8)^4} = c + 8$ _c ≥ −8_

   b. For what value of c will the statement be true? $\sqrt[4]{(c+8)^4} = |c + 8|$ _All real numbers_

For Exercises 91–100, simplify each expression. Assume that all variable expressions represent positive real numbers. (See Example 7)

91. a. $\sqrt{c^7}$ _$c^3\sqrt{c}$_   **b.** $\sqrt[3]{c^7}$ _$c^2\sqrt[3]{c}$_   **c.** $\sqrt[4]{c^7}$ _$c\sqrt[4]{c^3}$_   **d.** $\sqrt[9]{c^7}$ _$\sqrt[9]{c^7}$_

92. a. $\sqrt{d^{11}}$ _$d^5\sqrt{d}$_   **b.** $\sqrt[3]{d^{11}}$ _$d^3\sqrt[3]{d^2}$_   **c.** $\sqrt[4]{d^{11}}$ _$d^2\sqrt[4]{d^3}$_   **d.** $\sqrt[12]{d^{11}}$ _$\sqrt[12]{d^{11}}$_

93. a. $\sqrt{24}$ _$2\sqrt{6}$_   **b.** $\sqrt[3]{24}$ _$2\sqrt[3]{3}$_

94. a. $\sqrt{54}$ _$3\sqrt{6}$_   **b.** $\sqrt[3]{54}$ _$3\sqrt[3]{2}$_

95. $\sqrt[3]{250x^2y^6z^{11}}$ _$5y^2z^3\sqrt[3]{2x^2z^2}$_   **96.** $\sqrt[3]{40ab^{13}c^{17}}$ _$2b^4c^5\sqrt[3]{5abc^2}$_   **97.** $\sqrt[4]{96p^{14}q^7}$ _$2p^3q\sqrt[4]{6p^2q^3}$_   **98.** $\sqrt[4]{243m^{19}n^{10}}$ _$3m^4n^2\sqrt[4]{3m^3n^2}$_

99. $\sqrt{84(y-2)^3}$
   $2(y-2)\sqrt{21(y-2)}$ **100.** $\sqrt{18(w-6)^3}$ _$3(w-6)\sqrt{2(w-6)}$_

6. Apply Operations on Radicals

For Exercises 101–112, multiply or divide as indicated. Assume that all variable expressions represent positive real numbers. (See Example 8)

101. $\sqrt{10} \cdot \sqrt{14}$ _$2\sqrt{35}$_   **102.** $\sqrt{6} \cdot \sqrt{21}$ _$3\sqrt{14}$_   **103.** $\sqrt[3]{xy^2} \cdot \sqrt[3]{x^2y}$ _xy_

104. $\sqrt[4]{a^3b} \cdot \sqrt[4]{ab^3}$ _ab_   **105.** $(3\sqrt[4]{a^3})(-5\sqrt[4]{a^3})$ _$-15a\sqrt{a}$_   **106.** $(7\sqrt[6]{t^5})(-2\sqrt[6]{t^5})$ _$-14t\sqrt[3]{t^2}$_

107. $\sqrt{\dfrac{p^7}{36}}$ _$\dfrac{p^3\sqrt{p}}{6}$_   **108.** $\sqrt{\dfrac{q^{11}}{4}}$ _$\dfrac{q^5\sqrt{q}}{2}$_   **109.** $4\sqrt[3]{\dfrac{w^3z^5}{8}}$ _$2wz\sqrt[3]{z^2}$_

110. $8\sqrt[3]{\dfrac{c^6d^7}{64}}$ _$2c^2d^2\sqrt[3]{d}$_   **111.** $\dfrac{\sqrt[3]{5x^5y}}{\sqrt[3]{625x^2y^4}}$ _$\dfrac{x}{5y}$_   **112.** $\dfrac{\sqrt[3]{2m^2n^7}}{\sqrt[3]{16m^{14}n^4}}$ _$\dfrac{n}{2m^4}$_

For Exercises 113–120, add or subtract as indicated. Assume that all variables represent positive real numbers. (See Example 9)

113. $3\sqrt[3]{2y^2} - 9\sqrt[3]{2y^2} + \sqrt[3]{2y^2}$ _$-5\sqrt[3]{2y^2}$_   **114.** $8\sqrt[4]{3z^3} - \sqrt[4]{3z^3} + 2\sqrt[4]{3z^3}$ _$9\sqrt[4]{3z^3}$_

115. $\dfrac{1}{5}\sqrt{50} - \dfrac{7}{3}\sqrt{18} + \dfrac{5}{6}\sqrt{72}$ _$-\sqrt{2}$_   **116.** $\dfrac{2}{5}\sqrt{75} - \dfrac{2}{3}\sqrt{27} - \dfrac{1}{2}\sqrt{12}$ _$-\sqrt{3}$_

117. $-3x\sqrt[3]{16xy^4} + xy\sqrt[3]{54xy} - 5\sqrt[3]{250x^4y^4}$ _$-28xy\sqrt[3]{2xy}$_   **118.** $8\sqrt[4]{32a^5b^6} - 5b\sqrt[4]{2a^5b^2} - ab\sqrt[4]{162ab^2}$ _$8ab\sqrt[4]{2ab^2}$_

119. $12\sqrt{2y} + 5y\sqrt{2y}$ _$(12 + 5y)\sqrt{2y}$_   **120.** $-8\sqrt{3w} + 3w\sqrt{3w}$ _$(-8 + 3w)\sqrt{3w}$_

Mixed Exercises

For Exercises 121–122, use the Pythagorean theorem to determine the length of the missing side. Write the answer as a simplified radical.

121.

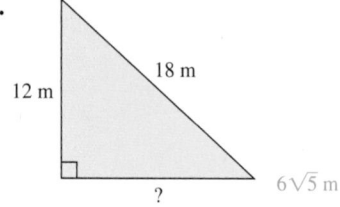

$6\sqrt{5}$ m

122.

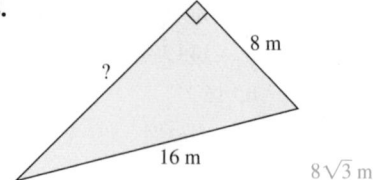

$8\sqrt{3}$ m

123. The slant length L for a right circular cone is given by $L = \sqrt{r^2 + h^2}$, where r and h are the radius and height of the cone. Find the slant length of a cone with radius 4 in. and height 10 in. Determine the exact value and a decimal approximation to the nearest tenth of an inch. $2\sqrt{29}$ in. ≈ 10.8 in.

$L = ?$ 10 in.   4 in.

124. The lateral surface area A of a right circular cone is given by $A = \pi r \sqrt{r^2 + h^2}$, where r and h are the radius and height of the cone. Determine the exact value (in terms of π) of the lateral surface area of a cone with radius 6 m and height 4 m. Then give a decimal approximation to the nearest square meter.
$12\pi\sqrt{13}$ m$^2 \approx 136$ m^2

4 m   6 m

125. The depreciation rate for a car is given by $r = 1 - \left(\frac{S}{C}\right)^{1/n}$, where S is the value of the car after n years, and C is the initial cost. Determine the depreciation rate for a car that originally cost \$22,990 and was valued at \$11,500 after 4 yr. Round to the nearest tenth of a percent.   15.9%

126. For a certain oven, the baking time t (in hr) for a turkey that weighs x pounds can be approximated by the model $t = 0.84x^{3/5}$. Determine the baking time for a 15-lb turkey. Round to 1 decimal place.   4.3 hr

Write About It

127. Explain the difference between the expressions $6x^0$ and $(6x)^0$.

128. Explain why scientific notation is used.

129. Explain the similarity in simplifying the given expressions.

    a. $2x + 3x$

    b. $2\sqrt{x} + 3\sqrt{x}$ In each case, add like terms or

    c. $2\sqrt[3]{x} + 3\sqrt[3]{x}$ like radicals by using the distributive property.

130. Explain why the given expressions cannot be simplified further.

    a. $2x + 3y$

    b. $2\sqrt{x} + 3\sqrt{y}$ The terms within each

    c. $2\sqrt[3]{x} + 3\sqrt{x}$ expression are not like terms or like radicals.

Expanding Your Skills

For Exercises 131–132, refer to the formula $F = \dfrac{Gm_1m_2}{d^2}$. This gives the gravitational force F (in Newtons, N) between two masses m_1 and m_2 (each measured in kg) that are a distance of d meters apart. In the formula, $G = 6.6726 \times 10^{-11}$ N-m^2/kg^2.

131. Determine the gravitational force between the Earth (mass $= 5.98 \times 10^{24}$ kg) and Jupiter (mass $= 1.901 \times 10^{27}$ kg) if at one point in their orbits, the distance between them is 7.0×10^{11} m.
1.55×10^{18} N

132. Determine the gravitational force between the Earth (mass $= 5.98 \times 10^{24}$ kg) and an 80-kg human standing at sea level. The mean radius of the Earth is approximately 6.371×10^6 m. 7.86×10^2 N

For Exercises 133–136, without the assistance of a calculator, fill in the blank with the appropriate symbol $<$, $>$, or $=$.

133. a. $5^{15} \underline{\ <\ } 5^{17}$ **b.** $5^{-15} \underline{\ >\ } 5^{-17}$

134. a. $\left(\frac{1}{5}\right)^{15} \underline{\ >\ } \left(\frac{1}{5}\right)^{17}$ **b.** $\left(\frac{1}{5}\right)^{-15} \underline{\ <\ } \left(\frac{1}{5}\right)^{-17}$

135. a. $(-1)^{86} \underline{\ >\ } (-1)^{87}$ **b.** $(1)^{86} \underline{\ =\ } (1)^{87}$

136. a. $(-1)^0 \underline{\ >\ } -1^{41}$ **b.** $(-1)^{42} \underline{\ =\ } (-1)^0$

For Exercises 137–142, write each expression as a single radical for positive values of the variable. (*Hint*: Write the radicals as expressions with rational exponents and simplify. Then convert back to radical form.)

137. $\sqrt[5]{x^3y^2} \cdot \sqrt[4]{x}$ $\sqrt[20]{x^{17}y^8}$

138. $\sqrt[4]{a^2b} \cdot \sqrt[3]{ab^2}$ $\sqrt[12]{a^{10}b^{11}}$

139. $\sqrt[6]{m}\,\sqrt[3]{m^2}$ $\sqrt[18]{m^5}$

140. $\sqrt[5]{y}\,\sqrt[4]{y^3}$ $\sqrt[20]{y^7}$

141. $\sqrt{x\sqrt{x\sqrt{x}}}$ $\sqrt[8]{x^7}$

142. $\sqrt[3]{y\,\sqrt[3]{y\,\sqrt[3]{y}}}$ $\sqrt[27]{y^{13}}$

For Exercises 143–144, evaluate the expression without the use of a calculator.

143. $\sqrt{\dfrac{8.0 \times 10^{12}}{2.0 \times 10^{4}}}$ 2.0×10^4

144. $\sqrt{\dfrac{1.44 \times 10^{16}}{9.0 \times 10^{10}}}$ 4.0×10^2

For Exercises 145–146, simplify the expression.

145. $\sqrt{\sqrt[3]{6} + \sqrt[4]{16} + \sqrt{\sqrt{25} + \sqrt{16} + \sqrt{9}}}$ $2\sqrt{2}$

146. $\sqrt{\sqrt[4]{11} + \sqrt[3]{125} + \sqrt{\sqrt{81} + \sqrt[3]{1000} + \sqrt{36} + \sqrt{25}}}$ $2\sqrt{3}$

SECTION R.3   Polynomials and Factoring

TIP The number 0 can be written in infinitely many ways: $0x$, $0x^2$, $0x^3$, and so on. For this reason, the degree of 0 is undefined.

TIP Some polynomials have more than one variable. In such a case, the degree of each term is the sum of the exponents on the variable factors. For example, the polynomial

$$-4x^4y^7 + 3xy^5$$

has two terms. The degree of the first term is 11 and the degree of the second term is 6.

1. Add and Subtract Polynomials

The Environmental Protection Agency (EPA) is responsible for providing fuel economy data (gas mileage information) that is posted on the window stickers of new vehicles. Many variables contribute to fuel consumption including the speed of the vehicle. For example, for one midsize sedan tested, the gas mileage G (in miles per gallon, mpg) can be approximated by

$$G = -0.008x^2 + 0.748x + 13.5,$$ where x is the speed of the vehicle in mph and $15 \le x \le 75$ mph.

The expression on the right side of this equation is called a polynomial. A **polynomial** in the variable x is a finite sum of terms of the form ax^n. In each term, the coefficient, a, is a real number, and the exponent, n, is a whole number. The degree of ax^n is n.

The terms of the polynomial $-0.008x^2 + 0.748x + 13.5$ are written in **descending order** by degree. The term with highest degree is written first. This is called the **leading term,** and its coefficient is called the **leading coefficient.** The **degree of the polynomial** is the same as the degree of the leading term. Therefore, the polynomial $-0.008x^2 + 0.748x + 13.5$ is a degree 2 polynomial.

The preceding discussion is meant as an informal introduction to polynomials and associated key vocabulary. However, as your level of mathematical sophistication increases, you should strive to understand definitions written in a more concise mathematical language. Take a minute to read the formal definition of a polynomial.

> **Definition of a Polynomial in x**
>
> A **polynomial in the variable x** is an expression of the form:
>
> $$a_nx^n + a_{n-1}x^{n-1} + a_{n-2}x^{n-2} + \cdots + a_1x + a_0$$
>
> The coefficients a_n, a_{n-1}, a_{n-2}, ... , a_0 are real numbers, where $a_n \ne 0$, and the exponents n, $n - 1$, $n - 2$, ... , 0 are whole numbers.
>
> The term a_nx^n is called the **leading term,** the coefficient a_n is the **leading coefficient,** and the exponent n is the **degree of the polynomial.**

In the preceding definition, subscript notation a_n (read as "a sub n"), a_{n-1} (read as "a sub $n-1$"), and so on, is used to denote the coefficients of the terms. Subscript notation is used rather than lettered variables such as a, b, c, and the like, when a large or undetermined number of terms is suggested.

A polynomial with one term is also called a **monomial.** A polynomial with two terms is called a **binomial,** and a polynomial with three terms is called a **trinomial.** To add or subtract polynomials, combine like terms. This is demonstrated in Example 1.

Classroom Examples: p. 40
Exercises 12, 14

EXAMPLE 1 Adding and Subtracting Polynomials

Add or subtract as indicated, and simplify.

a. $(-4w^3 - 5w^2 + 6w + 3) + (8w^2 - 4w + 2)$

b. $(6.1a^2b + 2.9ab - 4.5b^2) - (2.6a^2b - 4.1ab + 2.1b^2)$

Point of Interest

The sixteenth century brought the use of many modern symbols in algebra. Mathematicians Christoff Rudolff and Richard Stifel of Germany introduced the "+" and "−" signs, and Robert Record of England introduced the modern "=" sign.

Solution:

a. $(-4w^3 - 5w^2 + 6w + 3) + (8w^2 - 4w + 2)$

$= -4w^3 - 5w^2 + 8w^2 + 6w - 4w + 3 + 2$ Group like terms.

$= -4w^3 + 3w^2 + 2w + 5$ Combine like terms.

b. $(6.1a^2b + 2.9ab - 4.5b^2) - (2.6a^2b - 4.1ab + 2.1b^2)$

$= (6.1a^2b + 2.9ab - 4.5b^2) + (-2.6a^2b + 4.1ab - 2.1b^2)$

$= 6.1a^2b - 2.6a^2b + 2.9ab + 4.1ab - 4.5b^2 - 2.1b^2$ Group like terms.

$= 3.5a^2b + 7ab - 6.6b^2$ Combine like terms.

Skill Practice 1   Add or subtract as indicated, and simplify.

a. $(7t^5 - 3t^2 - 2t) + (2t^5 + 3t^2 - 5t - 4)$

b. $(0.08x^3y - 0.02x^2y - 0.1xy) - (0.05x^3y - 0.07x^2y + 0.02xy)$

TIP Addition and subtraction of polynomials can also be done by aligning like terms in columns. The difference of polynomials from Example 1(b) is shown here.

$$6.1a^2b + 2.9ab - 4.5b^2$$
$$-(2.6a^2b - 4.1ab + 2.1b^2) \xrightarrow{\text{Add the opposite.}} + (-2.6a^2b + 4.1ab - 2.1b^2)$$
$$\overline{3.5a^2b + 7ab - 6.6b^2}$$

2. Multiply Polynomials

To multiply polynomials we use the distributive property to multiply each term in the first polynomial by each term in the second polynomial.

Classroom Example: p. 40
Exercise 22

EXAMPLE 2 Multiplying Polynomials

Multiply and simplify.

$$(4x + 2)\left(x^2 - 6x + \frac{1}{2}\right)$$

Answers

1. a. $9t^5 - 7t - 4$

  b. $0.03x^3y + 0.05x^2y - 0.12xy$

TIP Multiplication of polynomials can also be performed vertically.

$$x^2 - 6x + \tfrac{1}{2}$$
$$\times \qquad 4x + 2$$
$$2x^2 - 12x + 1$$
$$4x^3 - 24x^2 + 2x$$
$$4x^3 - 22x^2 - 10x + 1$$

Solution:

$$(4x + 2)\left(x^2 - 6x + \frac{1}{2}\right)$$

$$= 4x(x^2) + 4x(-6x) + 4x\left(\frac{1}{2}\right) + 2(x^2) + 2(-6x) + 2\left(\frac{1}{2}\right)$$

Multiply each term in the first polynomial by each term in the second.

$$= 4x^3 - 24x^2 + 2x + 2x^2 - 12x + 1$$ Simplify.

$$= 4x^3 - 22x^2 - 10x + 1$$ Combine like terms.

Skill Practice 2   Multiply and simplify.

$$(3y - 6)\left(y^2 + 4y + \frac{2}{3}\right)$$

TIP The product of conjugates equals the square of the first term from the binomials, minus the square of the second term from the binomials.

TIP The square of a binomial results in the square of the first term in the binomial, plus twice the product of terms in the binomial, plus the square of the second term in the binomial.

Two expressions of the form $a - b$ and $a + b$ are called **conjugates.** The product of two conjugates results in a **difference of squares.**

$$(a - b)(a + b) = a^2 + ab - ab - b^2$$
$$= a^2 - b^2 \quad \text{(difference of squares)}$$

An expression of the form $(a + b)^2$ or $(a - b)^2$ is called a **square of a binomial.** In expanded form, the product is a **perfect square trinomial.**

$$(a + b)^2 = (a + b)(a + b) = a^2 + ab + ab + b^2$$
$$= a^2 + 2ab + b^2 \quad \text{(perfect square trinomial)}$$

$$(a - b)^2 = (a - b)(a - b) = a^2 - ab - ab + b^2$$
$$= a^2 - 2ab + b^2 \quad \text{(perfect square trinomial)}$$

Special Case Products

Product of Conjugates: $(a + b)(a - b) = a^2 - b^2$ (difference of squares)
Square of a Binomial: $(a + b)^2 = a^2 + 2ab + b^2$ (perfect square trinomial)
$(a - b)^2 = a^2 - 2ab + b^2$ (perfect square trinomial)

Classroom Examples: p. 40
Exercises 30, 34

EXAMPLE 3   **Applying Special Case Products**

a. Multiply. $(2t^2 - 9)(2t^2 + 9)$ **b.** Square the binomial. $(3x - 7)^2$

Solution:

a. $(2t^2 - 9)(2t^2 + 9)$ This is a product of conjugates: $(a - b)(a + b) = a^2 - b^2$.

$$= (2t^2)^2 - (9)^2$$ The product is a difference of squares.

$$= 4t^4 - 81$$ Simplify.

b. $(3x - 7)^2$ This is the square of a binomial, $(a - b)^2$, where $a = 3x$ and $b = 7$.

$$= (3x)^2 - 2(3x)(7) + (7)^2$$ The product is $a^2 - 2ab + b^2$.

$$= 9x^2 - 42x + 49$$ Simplify.

Answer
2. $3y^3 + 6y^2 - 22y - 4$

Classroom Examples: p. 41
Exercises 52, 56, 58

> **Skill Practice 3**
>
>   **a.** Multiply. $(5w^3 + 2y)(5w^3 - 2y)$ **b.** Square the binomial. $(6p + 5)^2$

3. Multiply Radical Expressions Involving Multiple Terms

The process used to multiply polynomials can be extended to algebraic expressions that are not polynomials. In Example 4 we multiply expressions containing two or more radical terms.

EXAMPLE 4   Multiplying Radical Expressions with Multiple Terms

Multiply and simplify. Assume that all variable expressions represent positive real numbers.

   a. $(3\sqrt{x} + 5)(2\sqrt{x} - 7)$ **b.** $(4\sqrt{5} + \sqrt{6})(4\sqrt{5} - \sqrt{6})$ **c.** $(3x + \sqrt{2})^2$

Solution:

a. $(3\sqrt{x} + 5)(2\sqrt{x} - 7)$

Apply the distributive property.

$= (3\sqrt{x})(2\sqrt{x}) + (3\sqrt{x})(-7) + (5)(2\sqrt{x}) + (5)(-7)$
$= 6\sqrt{x^2} - 21\sqrt{x} + 10\sqrt{x} - 35$
$= 6x - 11\sqrt{x} - 35$

Combine like terms.

b. $(4\sqrt{5} + \sqrt{6})(4\sqrt{5} - \sqrt{6})$

This is a product of conjugates: $(a + b)(a - b)$.

$= (4\sqrt{5})^2 - (\sqrt{6})^2$

The product is a difference of squares. Note that $(\sqrt{5})^2 = 5$ and $(\sqrt{6})^2 = 6$.

$= 16 \cdot 5 - 6$
$= 80 - 6$
$= 74$

c. $(3x + \sqrt{2})^2$

This is the square of a binomial, $(a + b)^2$, where $a = 3x$ and $b = \sqrt{2}$.

$= (3x)^2 + 2(3x)(\sqrt{2}) + (\sqrt{2})^2$ The product is $a^2 + 2ab + b^2$.
$= 9x^2 + 6\sqrt{2}x + 2$ Simplify.

> **Skill Practice 4**   Multiply and simplify. Assume that all variable expressions represent positive real numbers.
>
>   **a.** $(4\sqrt{t} + 10)(3\sqrt{t} - 8)$ **b.** $(3\sqrt{7} - \sqrt{5})(3\sqrt{7} + \sqrt{5})$ **c.** $(4y - \sqrt{3})^2$

Answers
3. a. $25w^6 - 4y^2$
   b. $36p^2 + 60p + 25$
4. a. $12t - 2\sqrt{t} - 80$ **b.** 58
   c. $16y^2 - 8\sqrt{3}y + 3$

4. Factoring by Grouping

We now look at the multiplication of polynomials in reverse. The goal is to decompose a polynomial into a product of factors. This process is called **factoring.**

There are many techniques used to factor a polynomial, but the first step is always to factor out the greatest common factor (GCF). For example, each term in the polynomial $12x^5 + 18x^4 - 24x^3$ shares a common factor of x^3 and a greatest common numerical factor of 6. To factor out $6x^3$ from the polynomial, we have

$$12x^5 + 18x^4 - 24x^3$$
The GCF is $6x^3$.

$$= 6x^3(2x^2) + 6x^3(3x) + 6x^3(-4)$$
Write each term as a product of the GCF and another factor.

$$= 6x^3(2x^2 + 3x - 4)$$
Apply the distributive property.

Sometimes it is preferable to factor out a negative factor from a polynomial. For example, consider the polynomial, $-4x^2 - 8x + 12$. If we factor out -4, then the leading coefficient of the remaining polynomial will be positive.

$$-4x^2 - 8x + 12$$

$$= (-4)(x^2) + (-4)(2x) + (-4)(-3)$$
Write each term as a product of -4 and another factor.

$$= -4(x^2 + 2x - 3)$$
Apply the distributive property.

To factor a polynomial containing four terms, we often try factoring by grouping. This is demonstrated in Example 5.

TIP The factored form of any polynomial can be checked by multiplication.

$$6x^3(2x^2 + 3x - 4)$$
$$= 12x^5 + 18x^4 - 24x^3 ✓$$

TIP When a negative factor is factored out of a polynomial, the remaining terms in parentheses will have signs opposite to those in the original polynomial.

Classroom Example: p. 41
Exercise 80

EXAMPLE 5   Factoring by Grouping

Factor by grouping.

$2ax - 6ay - 5x + 15y$

Solution:

$2ax - 6ay - 5x + 15y$
Consider the first pair of terms and second pair of terms separately.

$= 2a(x - 3y) - 5(x - 3y)$
Factor out the GCF from the first two terms and from the second two terms.

$= (x - 3y)(2a - 5)$
Factor out the common binomial factor $(x - 3y)$ from each term.

Check: $(x - 3y)(2a - 5)$
$= 2ax - 5x - 6ay + 15y$ ✓

Avoiding Mistakes

After factoring out the GCF from each pair of terms, the binomial factors must match to be able to complete the process of factoring by grouping.

Skill Practice 5   Factor by grouping. $7cd - c^2 + 14d - 2c$

5. Factor Quadratic Trinomials

Next we want to factor quadratic trinomials. These are trinomials of the form $ax^2 + bx + c$, where the coefficients a, b, and c are integers and $a \neq 0$. To understand the basis to factor a trinomial, first consider the product of two binomials.

product of $2x$ and x — product of 3 and 2

$$(2x + 3)(x + 2) = 2x^2 + \underline{4x + 3x} + 6$$

sum of products of inner terms and outer terms

To factor a trinomial, this process is reversed.

Answer
5. $(7d - c)(c + 2)$

EXAMPLE 6 **Factoring a Quadratic Trinomial with Leading Coefficient 1**

Factor. $x^2 - 8x + 12$

Solution:

factors of x^2

$x^2 - 8x + 12 = (\square x + \square)(\square x + \square)$ — First fill in the blanks so that the product of first terms in the binomials is x^2. In this case, we have $1x$ and $1x$.

factors of 12

$x^2 - 8x + 12 = (1x + \square)(1x + \square)$ — Fill in the remaining blanks with numbers whose product is 12 and whose sum is the middle term coefficient, -8. The numbers are -2 and -6.

$= (x - 2)(x - 6)$ — Check: $(x - 2)(x - 6) = x^2 - 6x - 2x + 12$
$= x^2 - 8x + 12$ ✓

Skill Practice 6 Factor. $y^2 - 9y + 14$

The trinomial in Example 6 has a leading coefficient of 1. This simplified the factorization process. For any trinomial with a leading coefficient of 1 ($x^2 + bx + c$), the constant terms in the binomial factors have a product of c and a sum equal to the middle term coefficient, b.

To factor a trinomial of the form $ax^2 + bx + c$, where $a \neq 1$, all possible factors of ax^2 must be tested with all factors of c until the correct middle term is found. This is demonstrated in Example 7.

EXAMPLE 7 **Factoring a Quadratic Trinomial**

Factor. $10x^3 + 105x^2y - 55xy^2$

Solution:

$10x^3 + 105x^2y - 55xy^2$
$= 5x(2x^2 + 21xy - 11y^2)$ — Factor out the GCF, $5x$.

factors of $2x^2$

$= 5x(2x^2 + 21xy - 11y^2) = 5x(\square x + \square y)(\square x + \square y)$ — Fill in the blanks so that the product of first terms in the binomials is $2x^2$. We have $2x$ and x.

factors of $-11y^2$

$= 5x(2x^2 + 21xy - 11y^2) = 5x(2x + \square)(x + \square)$ — Fill in the remaining blanks with factors of $-11y^2$.

$(2x + 11y)(x - 1y) = 2x^2 + 9xy - 11y^2$ — Incorrect.
$(2x + y)(x - 11y) = 2x^2 - 21xy - 11y^2$ — Incorrect.
$(2x - 11y)(x + y) = 2x^2 - 9xy - 11y^2$ — Incorrect.
$(2x - y)(x + 11y) = 2x^2 + 21xy - 11y^2$ — Correct! ✓

The trinomial $10x^3 + 105x^2y - 55xy^2$ factors as $5x(2x - y)(x + 11y)$.

Skill Practice 7 Factor. $30m^3 - 200m^2n - 70mn^2$

Answers
6. $(y - 7)(y - 2)$
7. $10m(3m + n)(m - 7n)$

The process to factor a trinomial can be challenging if the coefficients are large. In fact, we will learn other techniques that will come in handy in such cases. One technique we show here is to reverse the special case product formula for the square of a binomial. That is, given a perfect square trinomial, factor the trinomial as the square of a binomial.

Factored Form of a Perfect Square Trinomial

$$a^2 + 2ab + b^2 = (a + b)^2$$
$$a^2 - 2ab + b^2 = (a - b)^2$$

A perfect square trinomial factors as the square of a binomial.

Classroom Example: p. 41
Exercise 94

EXAMPLE 8   Factoring a Perfect Square Trinomial

Factor.

a. $4x^2 - 28x + 49$ **b.** $81c^4 + 90c^2d + 25d^2$

Solution:

a. $4x^2 - 28x + 49$
$= (2x)^2 - 2(2x)(7) + (7)^2$ — The trinomial fits the pattern $a^2 - 2ab + b^2$, where $a = 2x$ and $b = 7$.
$= (2x - 7)^2$ — Factor as $a^2 - 2ab + b^2 = (a - b)^2$.

b. $81c^4 + 90c^2d + 25d^2$
$= (9c^2)^2 + 2(9c^2)(5d) + (5d)^2$ — The trinomial fits the pattern $a^2 + 2ab + b^2$, where $a = 9c^2$ and $b = 5d$.
$= (9c^2 + 5d)^2$ — Factor as $a^2 + 2ab + b^2 = (a + b)^2$.

Skill Practice 8   Factor.

a. $64y^2 + 16y + 1$ **b.** $9m^4 - 30m^2n + 25n^2$

---

6. Factor Binomials

Recall that the product of conjugates results in a difference of squares. Therefore, the factored form of a difference of squares is a product of conjugates.

Factored Form of a Difference of Squares

$$a^2 - b^2 = (a + b)(a - b)$$

Note: If a and b share no common factors other than 1, then a sum of squares $a^2 + b^2$ is not factorable over the set of real numbers.

Classroom Examples: p. 41
Exercises 96, 100

EXAMPLE 9   Factoring a Difference of Squares

Factor completely.

a. $p^2 - 100$ **b.** $32y^4 - 162$

Solution:

a. $p^2 - 100$
$= (p)^2 - (10)^2$ — The binomial fits the pattern $a^2 - b^2$, where $a = p$ and $b = 10$.
$= (p + 10)(p - 10)$ — Factor as $a^2 - b^2 = (a + b)(a - b)$.

Answers
8. a. $(8y + 1)^2$ **b.** $(3m^2 - 5n)^2$

b. $32y^4 - 162$

$= 2(16y^4 - 81)$ Factor out the GCF, 2.

$= 2[(4y^2)^2 - (9)^2]$ The binomial fits the pattern $a^2 - b^2$, where $a = 4y^2$ and $b = 9$.

$= 2(4y^2 + 9)(4y^2 - 9)$ Factor as $a^2 - b^2 = (a + b)(a - b)$.

$= 2(4y^2 + 9)[(2y)^2 - (3)^2]$ The factor $(4y^2 - 9)$ is also a difference of squares, with $a = 2y$ and $b = 3$.

$= 2(4y^2 + 9)(2y + 3)(2y - 3)$ The polynomial is now factored completely.

Skill Practice 9   Factor completely.

     a. $t^2 - 121$ **b.** $625z^5 - z$

A binomial can also be factored if it fits the pattern of a difference of cubes or a sum of cubes.

TIP The factored form of a sum or difference of cubes is a binomial times a trinomial. To help remember the pattern for the signs, remember **SOAP**: **S**ame sign, **O**pposite signs, **A**lways **P**ositive.

Factored Form of a Sum and Difference of Cubes

Sum of cubes: $a^3 + b^3 = (a + b)(a^2 - ab + b^2)$

Difference of cubes: $a^3 - b^3 = (a - b)(a^2 + ab + b^2)$

                  Same sign   **A**lways **P**ositive

                      Opposite signs

Classroom Examples: p. 41
Exercises 102, 106

EXAMPLE 10   **Factoring a Sum and Difference of Cubes**

Factor completely.

     a. $x^3 + 125$ **b.** $8m^6 - 27n^3$

Solution:

a. $x^3 + 125$

$= (x)^3 + (5)^3$ The binomial fits the pattern of a sum of cubes $a^3 + b^3$, where $a = x$ and $b = 5$.

$= (x + 5)[(x)^2 - (x)(5) + (5)^2]$ Factor as $a^3 + b^3 = (a + b)(a^2 - ab + b^2)$.

$= (x + 5)(x^2 - 5x + 25)$ Simplify.

b. $8m^6 - 27n^3$

$= (2m^2)^3 - (3n)^3$ The binomial fits the pattern of a difference of cubes $a^3 - b^3$, where $a = 2m^2$ and $b = 3n$.

$= (2m^2 - 3n)[(2m^2)^2 + (2m^2)(3n) + (3n)^2]$ Factor as $a^3 - b^3 = (a - b)(a^2 + ab + b^2)$.

$= (2m^2 - 3n)(4m^4 + 6m^2n + 9n^2)$ Simplify.

Skill Practice 10   Factor completely.

     a. $u^3 + 27$ **b.** $64v^3 - 125z^6$

Answers

9. a. $(t + 11)(t - 11)$
     b. $z(25z^2 + 1)(5z + 1)(5z - 1)$
10. a. $(u + 3)(u^2 - 3u + 9)$
     b. $(4v - 5z^2)(16v^2 + 20vz^2 + 25z^4)$

7. Apply a General Strategy to Factor Polynomials

Factoring polynomials is a strategy game. The process requires that we identify the technique or techniques that best apply to a given polynomial. To do this we can follow the guidelines given in Table R-7.

Table R-7   Factoring Strategy

First Step	Number of Terms	Technique
Factor out the GCF	4 or more terms	Try factoring by grouping.
	3 terms	If possible write the trinomial in the form $ax^2 + bx + c$. • If the trinomial is a perfect square trinomial, factor as $$a^2 + 2ab + b^2 = (a + b)^2$$ $$a^2 - 2ab + b^2 = (a - b)^2$$ • Otherwise try factoring by the trial-and-error method.
	2 terms	• If the binomial is a difference of squares, factor as $$a^2 - b^2 = (a + b)(a - b)$$ • If the binomial is a sum of cubes, factor as $$a^3 + b^3 = (a + b)(a^2 - ab + b^2)$$ • If the binomial is a difference of cubes, factor as $$a^3 - b^3 = (a - b)(a^2 + ab + b^2)$$ *Note*: A sum of squares, $a^2 + b^2$, cannot be factored over the real numbers.

Classroom Example: p. 41
Exercise 108

EXAMPLE 11   Factoring a 4-Term Polynomial

Factor completely. $x^5 - 9x^3 + 8x^2 - 72$

Solution:

$x^5 - 9x^3 + 8x^2 - 72$ — Factor by grouping two terms with two terms.

$= x^3(x^2 - 9) + 8(x^2 - 9)$ — Factor out the GCF from the first two terms. Factor out the GCF from the second two terms.

$= (x^2 - 9)(x^3 + 8)$ — Factor the first binomial as a difference of squares and the second binomial as a sum of cubes.

$= (x + 3)(x - 3)(x + 2)(x^2 - 2x + 4)$

Skill Practice 11   Factor completely. $x^5 - x^3 - 64x^2 + 64$

Classroom Example: p. 41
Exercise 110

EXAMPLE 12   Factoring by Grouping 1 Term with 3 Terms

Factor completely. $25 - x^2 - 4xy - 4y^2$

Solution:

This polynomial has 4 terms. However, the standard grouping method does not work even if we try rearranging terms.

$25 - \underbrace{x^2 - 4xy - 4y^2}_{\text{perfect square trinomial}}$ — We might try grouping 1 term with 3 terms. The reason is that after factoring out -1 from the last 3 terms we have a perfect square trinomial.

$= 25 - (x^2 + 4xy + 4y^2)$

$= 25 - (x + 2y)^2$ — This results in a difference of squares $a^2 - b^2$ where $a = 5$ and $b = (x + 2y)$.

$= (5)^2 - (x + 2y)^2$

$= [5 + (x + 2y)][5 - (x + 2y)]$ — Factor as $a^2 - b^2 = (a + b)(a - b)$.

$= (5 + x + 2y)(5 - x - 2y)$ — Simplify.

Answer
11. $(x + 1)(x - 1)(x - 4)(x^2 + 4x + 16)$

Skill Practice 12   Factor completely. $36 - m^2 - 6mn - 9n^2$

In Example 12, it may be helpful to write the expression $25 - (x + 2y)^2$ by using a convenient substitution. If we let $u = x + 2y$, then the expression becomes $25 - u^2$. This is easier to recognize as a difference of squares.

$$25 - u^2 = (5 + u)(5 - u)$$
$$= [5 + (x + 2y)][5 - (x + 2y)] \quad \text{Back substitute.}$$
$$= (5 + x + 2y)(5 - x - 2y)$$

In Example 13, we practice making an appropriate substitution to convert a cumbersome expression into one that is more easily recognizable and factorable.

Classroom Example: p. 41
Exercise 112

EXAMPLE 13   Factoring a Trinomial by Using Substitution

Factor completely. $(x^2 - 5)^2 + 2(x^2 - 5) - 24$

Solution:

$(x^2 - 5)^2 + 2(x^2 - 5) - 24$	Notice that the first two terms in the polynomial share a common factor of $(x^2 - 5)$. Suppose that we let u represent this expression. That is, $u = x^2 - 5$.
$= u^2 + 2u - 24$	
$= (u + 6)(u - 4)$	Factor.
$= [(x^2 - 5) + 6][(x^2 - 5) - 4]$	Back substitute. Replace u by $(x^2 - 5)$.
$= (x^2 + 1)(x^2 - 9)$	Simplify.
$= (x^2 + 1)(x + 3)(x - 3)$	Factor $x^2 - 9$ as a difference of squares.

Skill Practice 13   Factor completely. $(x^2 - 2)^2 + 5(x^2 - 2) - 14$

Some polynomials cannot be factored with the techniques learned thus far. For example, no combination of binomial factors with integer coefficients will produce the trinomial $3x^2 + 9x + 5$.

$$(3x + 5)(x + 1) = 3x^2 + 3x + 5x + 5 \quad \text{Incorrect. Wrong middle term.}$$
$$= 3x^2 + 8x + 5$$
$$(3x + 1)(x + 5) = 3x^2 + 15x + x + 5 \quad \text{Incorrect. Wrong middle term.}$$
$$= 3x^2 + 16x + 5$$

In such a case we say that the polynomial is a **prime polynomial.** Its only factors are 1 and itself.

8. Factor Expressions Containing Negative and Rational Exponents

We now revisit the process to factor out the greatest common factor. In some applications, it is necessary to factor out a variable factor with a negative integer exponent or a rational exponent. Before we demonstrate this in Example 14, take a minute to review a similar example with positive integer exponents.

$2x^6 + 5x^5 + 7x^4$ The GCF is x^4. This is x raised to the *smallest exponent* to which it appears in any term.

$$= x^4(2x^2 + 5x^1 + 7x^0)$$

The powers on the factors of x within parentheses are found by subtracting 4 from the original exponents.

Answers
12. $(6 + m + 3n)(6 - m - 3n)$
13. $(x^2 + 5)(x + 2)(x - 2)$

Classroom Examples: p. 42
Exercises 132, 142

EXAMPLE 14 **Factoring Out Negative and Rational Exponents**

Factor. Write the answers with no negative exponents.

a. $2x^{-6} + 5x^{-5} + 7x^{-4}$ **b.** $x(2x + 5)^{-1/2} + (2x + 5)^{1/2}$

Solution:

a. $2x^{-6} + 5x^{-5} + 7x^{-4}$
$\boxed{-6 - (-6)}\ \boxed{-5 - (-6)}\ \boxed{-4 - (-6)}$

The smallest exponent on x is -6.
Factor out x^{-6}.

$= x^{-6}(2x^0 + 5x^1 + 7x^2)$

The powers on the factors of x within parentheses are found by subtracting -6 from the original exponents.

$= x^{-6}(2 + 5x + 7x^2)$

Simplify the negative exponent.

$= \dfrac{7x^2 + 5x + 2}{x^6}$

$b^{-n} = \dfrac{1}{b^n}$

b. $x(2x + 5)^{-1/2} + (2x + 5)^{1/2}$

The smallest exponent on $(2x + 5)$ is $-\frac{1}{2}$.

$\boxed{-1/2 - (-1/2)}\ \boxed{1/2 - (-1/2)}$ Factor out $(2x + 5)^{-1/2}$.

$= (2x + 5)^{-1/2}\,[x(2x + 5)^0 + (2x + 5)^1]$

The powers on the factors of $(2x + 5)$ within parentheses are found by subtracting $-\frac{1}{2}$ from the original exponents.

$= (2x + 5)^{-1/2}\,[x + (2x + 5)]$

The expression $(2x + 5)^0 = 1$, for $2x + 5 \neq 0$.

$= (2x + 5)^{-1/2}\,(3x + 5)$

Simplify.

$= \dfrac{3x + 5}{(2x + 5)^{1/2}}$

Simplify the negative exponent.

Answers

14. a. $\dfrac{2a^2 - 3a + 11}{a^5}$

b. $\dfrac{5x + 3}{(4x + 3)^{3/4}}$

Skill Practice 14 Factor completely and write the answer with no negative exponents.

a. $11a^{-5} - 3a^{-4} + 2a^{-3}$ **b.** $x(4x + 3)^{-3/4} + (4x + 3)^{1/4}$

SECTION R.3   Practice Exercises

Concept Connections

1. The ___leading___ term of a polynomial is the term of highest degree.

2. The leading ___coefficient___ of a polynomial is the numerical factor of the leading term.

3. A ___binomial___ is a polynomial that has two terms, and a ___trinomial___ is a polynomial with three terms.

4. The conjugate of $3 - \sqrt{x}$ is ___$3 + \sqrt{x}$___.

5. The binomial $a^3 + b^3$ is called a sum of ___cubes___ and factors as ___$(a + b)(a^2 - ab + b^2)$___.

6. The binomial $a^3 - b^3$ is called a ___difference___ of cubes and factors as ___$(a - b)(a^2 + ab + b^2)$___.

7. The trinomial $a^2 + 2ab + b^2$ is a ___perfect___ square trinomial. Its factored form is ___$(a + b)^2$___.

8. The binomial $a^2 - b^2$ is called a difference of ___squares___ and factors as ___$(a + b)(a - b)$___.

Objective 1: Add and Subtract Polynomials

For Exercises 9–10, determine if the expression is a polynomial.

9. a. $4a^2 + 7b - 3$ Yes   **b.** $\frac{3}{4}x^2y$ Yes   **c.** $6x + \frac{7}{x} + 5$ No   **d.** $\sqrt{p^2 + 2p - 5}$ No

10. a. $3x^5 - 9x^2 + \frac{2}{x^3}$ No   **b.** $\sqrt{2}ab^4$ Yes   **c.** $3|y| + 2$ No   **d.** $-7x^3 - 4x^2 + 2x - 5$ Yes

For Exercises 11–14, add or subtract as indicated and simplify. (See Example 1)

11. $(-8p^7 - 4p^4 + 2p - 5) + (2p^7 + 6p^4 + p^2)$ $-6p^7 + 2p^4 + p^2 + 2p - 5$

12. $(-7w^5 + 3w^3 - 6) + (9w^5 - 5w^3 + 4w - 3)$ $2w^5 - 2w^3 + 4w - 9$

13. $(0.05c^3b + 0.02c^2b^2 - 0.09cb^3) - (-0.03c^3b + 0.08c^2b^2 - 0.1cb^3)$ $0.08c^3b - 0.06c^2b^2 + 0.01cb^3$

14. $(0.004mn^5 - 0.001mn^4 + 0.05mn^3) - (0.003mn^5 + 0.007mn^4 - 0.07mn^3)$ $0.001mn^5 - 0.008mn^4 + 0.12mn^3$

Objective 2: Multiply Polynomials

For Exercises 15–22, multiply and simplify. (See Example 2)

15. $7m^2(2m^4 - 3m + 4)$ $14m^6 - 21m^3 + 28m^2$

16. $8p^5(-2p^2 - 5p - 1)$ $-16p^7 - 40p^6 - 8p^5$

17. $(2x - 5)(x + 4)$ $2x^2 + 3x - 20$

18. $(w + 7)(6w - 3)$ $6w^2 + 39w - 21$

19. $(4u^2 - 5v^2)(2u^2 + 3v^2)$ $8u^4 + 2u^2v^2 - 15v^4$

20. $(2z^3 + 5u^2)(7z^3 - u^2)$ $14z^6 + 33u^2z^3 - 5u^4$

21. $(3y + 6)\left(\frac{1}{3}y^2 - 5y - 4\right)$ $y^3 - 13y^2 - 42y - 24$

22. $(10v - 5)\left(\frac{1}{5}v^2 - 3v + 1\right)$ $2v^3 - 31v^2 + 25v - 5$

23. Write the expanded form for $(a + b)^2$. $a^2 + 2ab + b^2$ **24.** Write the expanded form for $(a + b)(a - b)$. $a^2 - b^2$

For Exercises 25–38, perform the indicated operations and simplify. (See Example 3)

25. $(4x - 5)(4x + 5)$ $16x^2 - 25$

26. $(3p - 2)(3p + 2)$ $9p^2 - 4$

27. $(3w^2 - 7z)(3w^2 + 7z)$ $9w^4 - 49z^2$

28. $(9v^3 + 2u)(9v^3 - 2u)$ $81v^6 - 4u^2$

29. $\left(\frac{1}{5}c - \frac{2}{3}d^3\right)\left(\frac{1}{5}c + \frac{2}{3}d^3\right)$ $\frac{1}{25}c^2 - \frac{4}{9}d^6$

30. $\left(\frac{1}{6}n - \frac{4}{5}p^4\right)\left(\frac{1}{6}n + \frac{4}{5}p^4\right)$ $\frac{1}{36}n^2 - \frac{16}{25}p^8$

31. $(5m - 3)^2$ $25m^2 - 30m + 9$

32. $(7v - 2)^2$ $49v^2 - 28v + 4$

33. $(4t^2 + 3p^3)^2$ $16t^4 + 24t^2p^3 + 9p^6$

34. $(2a^2 + 11b^3)^2$ $4a^4 + 44a^2b^3 + 121b^6$

35. $(w + 4)^3$ $w^3 + 12w^2 + 48w + 64$

36. $(p - 2)^3$ $p^3 - 6p^2 + 12p - 8$

37. $[(u + v) - w][(u + v) + w]$ $u^2 + 2uv + v^2 - w^2$

38. $[(c + d) - a][(c + d) + a]$ $c^2 + 2cd + d^2 - a^2$

39. Suppose that x represents the smaller of two consecutive integers.

  a. Write a polynomial that represents the larger integer. $x + 1$

  b. Write a polynomial that represents the sum of the two integers. Then simplify. $x + (x + 1); 2x + 1$

  c. Write a polynomial that represents the product of the two integers. Then simplify. $x(x + 1); x^2 + x$

  d. Write a polynomial that represents the sum of the squares of the two integers. Then simplify. $x^2 + (x + 1)^2; 2x^2 + 2x + 1$

40. Suppose that x represents the larger of two consecutive odd integers.

  a. Write a polynomial that represents the smaller integer. $x - 2$

  b. Write a polynomial that represents the sum of the two integers. Then simplify. $x + (x - 2); 2x - 2$

  c. Write a polynomial that represents the product of the two integers. Then simplify. $x(x - 2); x^2 - 2x$

  d. Write a polynomial that represents the difference of the squares of the two integers. Then simplify. $x^2 - (x - 2)^2; 4x - 4$

For Exercises 41–48, simplify each expression.

41. $(y + 7)^2 - 2(y - 3)^2$ $-y^2 + 26y + 31$

42. $(x - 4)^2 - 6(x + 1)^2$ $-5x^2 - 20x + 10$

43. $(x^n + 3)(x^n - 7)$ $x^{2n} - 4x^n - 21$

44. $(y^n + 4)(y^n - 5)$ $y^{2n} - y^n - 20$

45. $(z^n + w^m)^2$ $z^{2n} + 2w^mz^n + w^{2m}$

46. $(w^n - y^m)^2$ $w^{2n} - 2w^ny^m + y^{2m}$

47. $(6x + 5)(6x - 5) - (6x + 5)^2$ $-60x - 50$

48. $(2y - 7)(2y + 7) - (2y - 7)^2$ $28y - 98$

Objective 3: Multiply Radical Expressions Involving Multiple Terms

For Exercises 49–66, multiply and simplify. Assume that all variable expressions represent positive real numbers. (See Example 4)

49. $5\sqrt{2}(2\sqrt{2} + 6\sqrt{3}$ ...
$20 + 30\sqrt{6} - 20\sqrt{2}$

51. $(2\sqrt{y} - 3)(4\sqrt{y} + 5)$
$8y - 2\sqrt{y} - 15$

52. $(5\sqrt{z} - 4)(3\sqrt{z} +$ ...
$15z + 18\sqrt{z} - 24$

54. $(5\sqrt{2} - \sqrt{11})(5\sqrt{2} + \sqrt{11})$ 39

55. $(4x\sqrt{y} - 2y\sqrt{x})(4x$ ...
$16x^2y - 4xy^2$

57. $(6z - \sqrt{5})^2$ $36z^2 - 12\sqrt{5}z + 5$

58. $(2v - \sqrt{17})^2$ $4v^2$ ...

60. $(\sqrt{y + 2} - 4)(\sqrt{y + 2} + 4)$ $y - 14$

61. $(\sqrt{x + 1} - 5)^2$ x ...

63. $(\sqrt{5} + 2\sqrt{x})(\sqrt{5} - 2\sqrt{x})$
$\sqrt{25} - 4x$

64. $(\sqrt{7 + 3\sqrt{z}})(\sqrt{7}$ ...
$\sqrt{49} - 9z$

66. $(\sqrt{a + 2} - \sqrt{a - 2})^2$
$2a - 2\sqrt{a^2 - 4}$

Handwritten:

85. $2t^3 - 28t^3 + 80t$
$2t(t^2 - 14t + 40)$
$2t(t - 4)(t - 10)$

79. $12x^3 - 9x^2 - 40x + 30$
$3x^2(4x - 3) - 10(4x - 3)$
$(4x - 3)(3x^2 - 10)$

Objective 4: Factor by Grouping

For Exercises 67–72, factor out the greatest common factor.

67. $15c^5 - 30c^4 + 5c^3$ $5c^3(3c^2 - 6c + 1)$

68. $12m^4 + 15m^3 + 3m^2$ $3m^2(4m^2 + 5m + 1)$

69. $21a^2b^5 - 14a^3b^4 + 35a^4b$
$7a^2b(3b^4 - 2ab^3 + 5a^2)$

70. $36p^4q^7 + 18p^3q^5 - 27p^2q^6$
$9p^2q^5(4p^2q^2 + 2p - 3q)$

71. $5z(x - 6y) + 7(x - 6y)$ $(x - 6y)(5z + 7)$

72. $4t(u + 8v) - 3(u + 8v)$
$(u + 8v)(4t - 3)$

For Exercises 73–76, factor out the indicated common factor.

73. a. Factor out 3. $-6x^2 + 12x + 9$
$3(-2x^2 + 4x + 3)$
 b. Factor out -3. $-6x^2 + 12x + 9$
$-3(2x^2 - 4x - 3)$

74. a. Factor out 5. $-15y^2 - 10y + 25$ $5(-3y^2 - 2y + 5)$
 b. Factor out -5. $-15y^2 - 10y + 25$ $-5(3y^2 + 2y - 5)$

75. Factor out $-4x^2y$. $-12x^3y^2 - 8x^4y^3 + 4x^2y$
$-4x^2y(3xy + 2x^2y^2 - 1)$

76. Factor out $-7a^3b$. $-14a^4b^3 + 21a^3b^2 - 7a^3b$ $-7a^3b(2ab^2 - 3b + 1)$

For Exercises 77–82, factor by grouping. (See Example 5)

77. $8ax + 18a + 20x + 45$
$(2a + 5)(4x + 9)$

78. $6ty + 9y + 14t + 21$ $(3y + 7)(2t + 3)$

79. $12x^3 - 9x^2 - 40x + 30$
$(3x^2 - 10)(4x - 3)$

80. $30z^3 - 35z^2 - 24z + 28$
$(5z^2 - 4)(6z - 7)$

81. $cd - 8d + 4c - 2d^2$ $(c - 2d)(d + 4)$

82. $7t - 6v^2 + tv - 42v$
$(t - 6v)(v + 7)$

Objective 5: Factor Quadratic Trinomials

For Exercises 83–94, factor the trinomials. (See Examples 6–8)

83. $p^2 + 2p - 63$ $(p + 9)(p - 7)$

84. $w^2 + 5w - 66$ $(w - 6)(w + 11)$

85. $2t^3 - 28t^2 + 80t$ $2t(t - 4)(t - 10)$

86. $5u^4 - 40u^3 + 35u^2$ $5u^2(u - 1)(u - 7)$

87. $7y^3z - 40y^2z^2 - 12yz^3$
$yz(7y + 2z)(y - 6z)$

88. $11a^3b + 18a^2b^2 - 8ab^3$
$ab(11a - 4b)(a + 2b)$

89. $t^2 - 18t + 81$ $(t - 9)^2$

90. $p^2 + 8p + 16$ $(p + 4)^2$

91. $50x^3 + 160x^2y + 128xy^2$ $2x(5x + 8y)^2$

92. $48y^3 - 72y^2z + 27yz^2$ $3y(4y - 3z)^2$

93. $4c^4 - 20c^2d^3 + 25d^6$ $(2c^2 - 5d^3)^2$

94. $9m^4 + 42m^2n^4 + 49n^8$ $(3m^2 + 7n^4)^2$

Objective 6: Factor Binomials

For Exercises 95–106, factor the binomials. (See Examples 9–10)

95. $9w^2 - 64$
$(3w + 8)(3w - 8)$

96. $16t^2 - 49$
$(4t + 7)(4t - 7)$

97. $200u^4 - 18v^6$
$2(10u^2 + 3v^3)(10u^2 - 3v^3)$

98. $75m^6 - 27n^4$
$3(5m^3 + 3n^2)(5m^3 - 3n^2)$

99. $625p^4 - 16$
$(25p^2 + 4)(5p + 2)(5p - 2)$

100. $81z^4 - 1$
$(9z^2 + 1)(3z + 1)(3z - 1)$

101. $y^3 + 64$
$(y + 4)(y^2 - 4y + 16)$

102. $u^3 + 343$
$(u + 7)(u^2 - 7u + 49)$

103. $c^4 - 27c$
$c(c - 3)(c^2 + 3c + 9)$

104. $d^4 - 8d$
$d(d - 2)(d^2 + 2d + 4)$

105. $8a^6 - 125b^9$
$(2a^2 - 5b^3)(4a^4 + 10a^2b^3 + 25b^6)$

106. $27m^{12} - 64n^9$
$(3m^4 - 4n^3)(9m^8 + 12m^4n^3 + 16n^6)$

Objective 7: Apply a General Strategy to Factor Polynomials

For Exercises 107–130, factor completely. (See Examples 11–13)

107. $30x^4 + 70x^3 - 120x^2 - 280x$
$10x(3x + 7)(x + 2)(x - 2)$

108. $4y^4 - 10y^3 - 36y^2 + 90y$
$2y(2y - 5)(y + 3)(y - 3)$

109. $a^2 - y^2 + 10y - 25$
$(a + y - 5)(a - y + 5)$

110. $c^2 - z^2 + 8z - 16$
$(c + z - 4)(c - z + 4)$

111. $(x^2 - 2)^2 - 3(x^2 - 2) - 28$
$(x^2 + 2)(x + 3)(x - 3)$

112. $(y^2 + 2)^2 + 5(y^2 + 2) - 24$
$(y^2 + 10)(y + 1)(y - 1)$

113. $(x^3 + 12)^2 - 16$
$(x^3 + 16)(x + 2)(x^2 - 2x + 4)$

114. $(y^3 + 34)^2 - 49$
$(y^3 + 41)(y + 3)(y^2 - 3y + 9)$

115. $(x + y)^3 + z^3$
$(x + y + z)(x^2 + 2xy + y^2 - xz - yz + z^2)$

116. $(a + 5)^3 - b^3$
$(a + 5 - b)(a^2 + 10a + 25 + ab + 5b + b^2)$

117. $9m^2 + 42m(3n + 1) + 49(3n + 1)^2$
$(3m + 21n + 7)^2$

118. $4x^2 + 36x(7y - 1) + 81(7y - 1)^2$
$(2x + 63y - 9)^2$

119. $(c - 3)^2 - (2c - 5)^2$
$(3c - 8)(-c + 2)$ or $-(3c - 8)(c - 2)$

120. $(d + 6)^2 - (4d - 3)^2$
$3(5d + 3)(-d + 3)$ or $-3(5d + 3)(d - 3)$

121. $p^{11} - 64p^8 - p^3 + 64$
$(p - 4)(p^2 + 4p + 16)(p^4 + 1)(p^2 + 1)(p + 1)(p - 1)$

122. $t^7 + 27t^4 - t^3 - 27$
$(t + 3)(t^2 - 3t + 9)(t^2 + 1)(t - 1)(t + 1)$
123. $m^6 + 26m^3 - 27$
$(m + 3)(m^2 - 3m + 9)(m - 1)(m^2 + m + 1)$
124. $n^6 - 7n^3 - 8$
$(n - 2)(n^2 + 2n + 4)(n + 1)(n^2 - n + 1)$
125. $16x^6z + 38x^3z - 54z$
$2z(2x + 3)(4x^2 - 6x + 9)(x - 1)(x^2 + x + 1)$
126. $24y^7 + 21y^4 - 3y$
$3y(2y - 1)(4y^2 + 2y + 1)(y + 1)(y^2 - y + 1)$
127. $x^2 - y^2 - x + y$
$(x - y)(x + y - 1)$
128. $a^2 - b^2 - a - b$
$(a + b)(a - b - 1)$
129. $a^2 + ac - 2c^2 - c + a$
$(a - c)(a + 2c + 1)$
130. $x^2 + 2xy - 3y^2 - y + x$
$(x - y)(x + 3y + 1)$

Objective 8: Factor Expressions Containing Negative and Rational Exponents

For Exercises 131–142, factor completely. Write the answers with positive exponents only. (See Example 14)

131. $2x^{-4} - 7x^{-3} + x^{-2}$ $\dfrac{x^2 - 7x + 2}{x^4}$
132. $5t^{-7} + 2t^{-6} - t^{-5}$ $\dfrac{-t^2 + 2t + 5}{t^7}$
133. $y^{-2} - y^{-3} - 12y^{-4}$ $\dfrac{(y - 4)(y + 3)}{y^4}$

134. $w^{-3} + 10w^{-4} + 9w^{-5}$ $\dfrac{(w + 9)(w + 1)}{w^5}$
135. $2c^{7/4} + 4c^{3/4}$ $2c^{3/4}(c + 2)$
136. $10y^{9/5} - 15y^{4/5}$ $5y^{4/5}(2y - 3)$

137. $\dfrac{8}{3}x^{1/3} + \dfrac{5}{3}x^{-2/3}$ $\dfrac{8x + 5}{3x^{2/3}}$
138. $\dfrac{15}{2}x^{1/2} - \dfrac{3}{2}x^{-1/2}$ $\dfrac{3(5x - 1)}{2x^{1/2}}$
139. $5x(3x + 1)^{2/3} + (3x + 1)^{5/3}$ $(3x + 1)^{2/3}(8x + 1)$

140. $7t(4t + 1)^{3/4} + (4t + 1)^{7/4}$ $(4t + 1)^{3/4}(11t + 1)$
141. $x(3x + 2)^{-2/3} + (3x + 2)^{1/3}$ $\dfrac{2(2x + 1)}{(3x + 2)^{2/3}}$
142. $x(5x - 8)^{-4/5} + (5x - 8)^{1/5}$ $\dfrac{2(3x - 4)}{(5x - 8)^{4/5}}$

Write About It

143. Explain the similarity in simplifying the given expressions.
In each case, multiply by using the distributive property.
 a. $(3x + 2)(4x - 7)$
 b. $(3\sqrt{x} + \sqrt{2})(4\sqrt{x} - \sqrt{7})$

144. Explain the similarity in simplifying the given expressions.
Both expressions involve a product of conjugates $(a + b)(a - b)$. In each case, the product is $a^2 - b^2$.
 a. $(x + 3)(x - 3)$
 b. $(\sqrt{x} + \sqrt{3})(\sqrt{x} - \sqrt{3})$

145. Why is the sum of squares $a^2 + b^2$ not factorable over the real numbers?

146. Explain the similarity in the process to factor out the GCF in the following two expressions.
$$5x^4 + 4x^3 \quad \text{and} \quad 5x^{-4} + 4x^{-3}$$

Expanding Your Skills

We say that the expression $x^2 - 4$ is factorable over the integers as $(x + 2)(x - 2)$. Notice that the constant terms in the binomials are integers. The expression $x^2 - 3$ can be factored over the irrational numbers as $x^2 - 3 = (x + \sqrt{3})(x - \sqrt{3})$. For Exercises 147–152, factor each expression over the irrational numbers.

147. $x^2 - 5$ $(x + \sqrt{5})(x - \sqrt{5})$
148. $y^2 - 11$ $(y + \sqrt{11})(y - \sqrt{11})$
149. $z^4 - 36$ $(z^2 + 6)(z + \sqrt{6})(z - \sqrt{6})$

150. $w^4 - 49$ $(w^2 + 7)(w + \sqrt{7})(w - \sqrt{7})$
151. $x^2 - 2\sqrt{5}x + 5$ $(x - \sqrt{5})^2$
152. $c^2 - 2\sqrt{3}c + 3$ $(c - \sqrt{3})^2$

For Exercises 153–156, determine if the statement is true or false. If a statement is false, explain why.

153. The sum of two polynomials each of degree 5 will be degree 5.

154. The sum of two polynomials each of degree 5 will be less than or equal to degree 5. *True*

155. The product of two polynomials each of degree 4 will be degree 8. *True*

156. The product of two polynomials each of degree 4 will be less than degree 8.

PROBLEM RECOGNITION EXERCISES

Simplifying Algebraic Expressions

Many expressions in algebra look similar but the methods used to simplify the expressions may be different. For Exercises 1–14, simplify each expression. Assume that the variables are restricted so that each expression is defined.

1. a. $64^{1/2}$ 8
 b. $64^{1/3}$ 4
 c. $64^{2/3}$ 16
 d. 64^{-1} $\dfrac{1}{64}$
 e. $-64^{1/2}$ -8
 f. $(-64)^{1/2}$ *Undefined*
 g. $(-64)^{2/3}$ 16
 h. $(-64)^{-2/3}$ $\dfrac{1}{16}$

2. a. $(5ab^3)^2$ $25a^2b^6$
 b. $(5a + b^3)^2$ $25a^2 + 10ab^3 + b^6$
 c. $(5ab^3)^{-2}$ $\dfrac{1}{25a^2b^6}$
 d. $(5a + b^3)^{-2}$ $\dfrac{1}{25a^2 + 10ab^3 + b^6}$

3. a. $(2x^4y)^2$ $4x^8y^2$
 b. $(2x^4 - y)^2$ $4x^8 - 4x^4y + y^2$
 c. $(2x^4y)^{-2}$ $\dfrac{1}{4x^8y^2}$
 d. $(2x^4 - y)^{-2}$ $\dfrac{1}{4x^8 - 4x^4y + y^2}$

4. a. x^5x^3 x^8
 b. $\dfrac{x^5}{x^3}$ x^2
 c. $x^{-5}x^3$ $\dfrac{1}{x^2}$
 d. $(x^5)^{-3}$ $\dfrac{1}{x^{15}}$

5. a. $\sqrt{x^8}$ x^4
 b. $\sqrt[3]{x^8}$ $x^2\sqrt[3]{x^2}$
 c. $\sqrt[5]{x^8}$ $x\sqrt[5]{x^3}$
 d. $\sqrt[9]{x^8}$ $\sqrt[9]{x^8}$

6. a. $(3a + 4b^2) - (2a - b^2)$ $a + 5b^2$
   b. $(3a + 4b^2)(2a - b^2)$ $6a^2 + 5ab^2 - 4b^4$

7. a. $(a^2 - b^2) + (a^2 + b^2)$ $2a^2$
   b. $(a^2 - b^2)(a^2 + b^2)$ $a^4 - b^4$
   c. $(a - b)^2 - (a + b)^2$ $-4ab$
   d. $(a - b)^2(a + b)^2$ $a^4 - 2a^2b^2 + b^4$

8. a. $|a - b|$ for $a < b$ $b - a$
   b. $|a - b|$ for $a > b$ $a - b$

9. a. $|x + 2|$ for $x > -2$ $x + 2$
   b. $|x + 2|$ for $x < -2$ $-x - 2$

10. a. $\left(\sqrt{x}\right)^2 + \left(\sqrt{y}\right)^2$ $x + y$
   b. $\sqrt{x^2 + y^2}$ $\sqrt{x^2 + y^2}$
   c. $\left(\sqrt{x} + \sqrt{y}\right)^2$ $x + 2\sqrt{xy} + y$
   d. $\left(\sqrt{x} + \sqrt{y}\right)\left(\sqrt{x} - \sqrt{y}\right)$ $x - y$

11. a. $\sqrt[3]{2x} \cdot \sqrt[3]{2x}$ $\sqrt[3]{4x^2}$
   b. $\sqrt[3]{2x} + \sqrt[3]{2x}$ $2\sqrt[3]{2x}$

12. a. $\sqrt[4]{y} \cdot \sqrt[3]{y}$ $\sqrt[12]{y^7}$
   b. $\sqrt[4]{y} + \sqrt[3]{y}$ $\sqrt[4]{y} + \sqrt[3]{y}$

13. a. $36 \div 12 \div 6 \div 2$ $\frac{1}{4}$
   b. $36 \div (12 \div 6) \div 2$ 9
   c. $36 \div 12 \div (6 \div 2)$ 1
   d. $(36 \div 12) \div (6 \div 2)$ 1

14. a. $\sqrt{6^2 + 8^2}$ 10
   b. $\sqrt{6^2} + \sqrt{8^2}$ 14

SECTION R.4 Rational Expressions and More Operations on Radicals

OBJECTIVES

1. Determine Restricted Values for a Rational Expression
2. Simplify Rational Expressions
3. Multiply and Divide Rational Expressions
4. Add and Subtract Rational Expressions
5. Simplify Complex Fractions
6. Rationalize the Denominator of a Radical Expression

1. Determine Restricted Values for a Rational Expression

Suppose that an object that is originally 35°C is placed in a freezer. The temperature T (in °C) of the object t hours after being placed in the freezer can be approximated by the model

$$T = \frac{350}{t^2 + 3t + 10}$$ For example, 2 hr after being placed in the freezer the temperature of the object is

$$T = \frac{350}{(2)^2 + 3(2) + 10} = 17.5°C$$

The expression $\dfrac{350}{t^2 + 3t + 10}$ is called a rational expression. A **rational expression** is a ratio of two polynomials. Since a rational expression may have a variable in the denominator, we must be careful to exclude values of the variable that make the denominator zero.

Classroom Examples: p. 51
Exercises 10, 12, 14

EXAMPLE 1 **Determining Restricted Values for a Rational Expression**

Determine the restrictions on the variable for each rational expression.

a. $\dfrac{x - 3}{x + 2}$ **b.** $\dfrac{x}{x^2 - 49}$ **c.** $\dfrac{4}{5x^2y}$

Solution:

a. $\dfrac{x - 3}{x + 2}$
$\boxed{x \neq -2}$

Division by zero is undefined.
For this expression $x \neq -2$. If -2 were substituted for x, the denominator would be zero. $-2 + 2 = 0$

b. $\dfrac{x}{x^2 - 49} = \dfrac{x}{(x + 7)(x - 7)}$
$\boxed{x \neq -7}$ $\boxed{x \neq 7}$

For this expression, $x \neq -7$ and $x \neq 7$. If x were -7, then $-7 + 7 = 0$. If x were 7, then $7 - 7 = 0$. In either case, the denominator would be zero.

c. $\dfrac{4}{5x^2y}$
$\boxed{x \neq 0}$ $\boxed{y \neq 0}$

For this expression, $x \neq 0$ and $y \neq 0$. If either x or y were zero, then the product would be zero.

Skill Practice 1   Determine the restrictions on the variable.

a. $\dfrac{x+4}{x-3}$ **b.** $\dfrac{5}{c^2-16}$ **c.** $\dfrac{6}{7ab^3}$

2. Simplify Rational Expressions

A rational expression is simplified (in lowest terms) if the only factors shared by the numerator and denominator are 1 or -1. To simplify a rational expression, factor the numerator and denominator, then apply the property of equivalent algebraic fractions.

Equivalent Algebraic Fractions

If a, b, and c represent real-valued expressions, then

$$\frac{ac}{bc} = \frac{a}{b} \quad \text{for } b \neq 0 \text{ and } c \neq 0 \qquad \underline{\text{Example:}} \quad \frac{10x}{15x} = \frac{2 \cdot 5 \cdot \overset{1}{\cancel{x}}}{3 \cdot 5 \cdot \underset{1}{\cancel{x}}} = \frac{2}{3}$$

Classroom Examples: p. 51
Exercises 18, 22

EXAMPLE 2   Simplifying Rational Expressions

Simplify.

a. $\dfrac{x^2-16}{x^2-x-12}$ **b.** $\dfrac{8+2\sqrt{7}}{4}$

Solution:

a. $\dfrac{x^2-16}{x^2-x-12} = \dfrac{(x+4)(x-4)}{(x+3)(x-4)}$

Factor the numerator and denominator. We have the restrictions that $x \neq -3$, $x \neq 4$.

$\qquad = \dfrac{(x+4)(x-\overset{1}{\cancel{4}})}{(x+3)(x-\underset{1}{\cancel{4}})}$

Divide out common factors that form a ratio of $\frac{1}{1}$.

$\qquad = \dfrac{x+4}{x+3} \quad \text{for } x \neq -3, x \neq 4$

The same restrictions for the original expression also apply to the simplified expression.

b. $\dfrac{8+2\sqrt{7}}{4} = \dfrac{2(4+\sqrt{7})}{2 \cdot 2} = \dfrac{\overset{1}{\cancel{2}}(4+\sqrt{7})}{\underset{1}{\cancel{2}} \cdot 2} = \dfrac{4+\sqrt{7}}{2}$

Skill Practice 2   Simplify.

a. $\dfrac{x^2-8x}{x^2-7x-8}$ **b.** $\dfrac{3+9\sqrt{5}}{6}$

In Example 2(a), the expressions $\dfrac{x^2-16}{x^2-x-12}$ and $\dfrac{x+4}{x+3}$ are equal for all values of x for which *both* expressions are defined. This excludes the values $x = -3$ and $x = 4$.

Answers

1. a. $x \neq 3$ **b.** $c \neq -4$ and $c \neq 4$
   c. $a \neq 0$ and $b \neq 0$

2. a. $\dfrac{x}{x+1}$; $x \neq 8$, $x \neq -1$
   b. $\dfrac{1+3\sqrt{5}}{2}$

TIP The expressions $4 - x$ and $x - 4$ are opposite polynomials (the signs of their terms are opposites). The ratio of two opposite factors is -1.

The property of equivalent algebraic fractions tells us that we can divide out common factors that form a ratio of 1. We can also divide out factors that form a ratio of -1. For example:

$$\dfrac{4 - x}{x - 4} \xrightarrow{\text{Factor out } -1 \text{ from the numerator.}} \dfrac{-1(-4 + x)}{x - 4} = \dfrac{-1(x - 4)}{x - 4} = -1$$

Numerator and denominator are opposite polynomials. Their ratio is -1.

3. Multiply and Divide Rational Expressions

To multiply and divide rational expressions, use the following properties.

> **Multiplication and Division of Algebraic Fractions**
>
> Let a, b, c, and d be real-valued expressions.
>
> $$\dfrac{a}{b} \cdot \dfrac{c}{d} = \dfrac{ac}{bd} \quad \text{for } b \neq 0, d \neq 0$$
>
> $$\dfrac{a}{b} \div \dfrac{c}{d} = \dfrac{a}{b} \cdot \dfrac{d}{c} = \dfrac{ad}{bc} \quad \text{for } b \neq 0, c \neq 0, d \neq 0$$

To multiply or divide rational expressions, factor the numerator and denominator of each fraction completely. Then apply the multiplication and division properties for algebraic fractions.

Restriction Agreement

Operations on rational expressions are valid for all values of the variable for which the rational expressions are defined. In Examples 3–8 and in the exercises, we will perform operations on rational expressions without explicitly stating the restrictions. Instead, the restrictions on the variables will be implied.

Classroom Example: p. 51
Exercise 34

EXAMPLE 3   **Dividing Rational Expressions**

Divide. $\dfrac{x^3 - 8}{4 - x^2} \div \dfrac{3x^2 + 6x + 12}{x^2 - x - 6}$

Solution:

$$\dfrac{x^3 - 8}{4 - x^2} \div \dfrac{3x^2 + 6x + 12}{x^2 - x - 6}$$

$$= \dfrac{x^3 - 8}{4 - x^2} \cdot \dfrac{x^2 - x - 6}{3x^2 + 6x + 12}$$

Multiply the first fraction by the reciprocal of the second fraction.

$$= \dfrac{(x - 2)(x^2 + 2x + 4)}{(2 - x)(2 + x)} \cdot \dfrac{(x - 3)(x + 2)}{3(x^2 + 2x + 4)}$$

Factor. Note that $x^3 - 8$ is a difference of cubes.

$$= \dfrac{\overset{(-1)}{(x - 2)}\overset{1}{(x^2 + 2x + 4)} \cdot (x - 3)\overset{1}{(x + 2)}}{(2 - x)(2 + x) \cdot 3(x^2 + 2x + 4)}$$

Note that $(x - 2)$ and $(2 - x)$ are opposite polynomials, and their ratio is -1.

$$= -\dfrac{x - 3}{3}$$

Avoiding Mistakes

For the expression $-\dfrac{x - 3}{3}$ do not be tempted to "divide out" the 3 in the numerator with the 3 in the denominator. The 3's are *terms*, not factors. Only common *factors* can be divided out.

Answer

3. $-\dfrac{y + 5}{5}$

Skill Practice 3   Divide. $\dfrac{y^3 - 27}{9 - y^2} \div \dfrac{5y^2 + 15y + 45}{y^2 + 8y + 15}$

4. Add and Subtract Rational Expressions

Recall that fractions can be added or subtracted if they have a common denominator.

Addition and Subtraction of Algebraic Fractions

Let a, b, and c be real-valued expressions.

$$\frac{a}{b} + \frac{c}{b} = \frac{a+c}{b} \qquad \text{and} \qquad \frac{a}{b} - \frac{c}{b} = \frac{a-c}{b} \qquad \text{for } b \neq 0$$

If two rational expressions have different denominators, then it is necessary to convert the expressions to equivalent expressions with the same denominator. We do this by applying the property of equivalent fractions.

Property of Equivalent Fractions

Let a, b, and c be real-valued expressions.

$\dfrac{a}{b} = \dfrac{ac}{bc}$, where $b \neq 0$ and $c \neq 0$ <u>Example:</u> $\dfrac{5}{x} = \dfrac{5 \cdot xy}{x \cdot xy} = \dfrac{5xy}{x^2 y}$

When adding or subtracting numerical fractions or rational expressions, it is customary to use the least common denominator (LCD) of the original expressions and to follow these guidelines.

Adding and Subtracting Rational Expressions

Step 1 Factor the denominators and determine the LCD of all expressions. The LCD is the product of unique prime factors where each factor is raised to the greatest power to which it appears in any denominator.

Step 2 Write each expression as an equivalent expression with the LCD as its denominator.

Step 3 Add or subtract the numerators as indicated and write the result over the LCD.

Step 4 Simplify if possible.

Classroom Example: p. 52
Exercise 44

EXAMPLE 4   **Adding Rational Expressions**

Add the rational expressions and simplify the result. $\dfrac{7}{4a} + \dfrac{11}{10a^2}$

Instructor Note:
Point out to students that $\frac{5a}{5a}$ and $\frac{2}{2}$ are simply convenient forms of 1. We are multiplying the original fractions by 1.

Solution:

$$\frac{7}{4a} + \frac{11}{10a^2} = \frac{7}{2^2 a} + \frac{11}{2 \cdot 5a^2}$$

Step 1: Factor the denominators. The LCD is $2^2 \cdot 5a^2$ or $20a^2$.

$$= \frac{7 \cdot (5a)}{2^2 a \cdot (5a)} + \frac{11 \cdot (2)}{2 \cdot 5a^2 \cdot (2)}$$

Step 2: Multiply numerator and denominator of each expression by the factors missing from the denominators.

$$= \frac{35a}{20a^2} + \frac{22}{20a^2}$$

Step 3: Add the numerators and write the result over the common denominator.

$$= \frac{35a + 22}{20a^2}$$

Step 4: The expression is already simplified.

Skill Practice 4   Add the rational expressions and simplify the result.

$$\frac{8}{9y^2} + \frac{1}{15y}$$

Classroom Example: p. 52
Exercise 56

EXAMPLE 5   Subtracting Rational Expressions

Subtract the rational expressions and simplify the result.

$$\frac{3x + 5}{x^2 + 4x + 3} - \frac{x - 5}{x^2 + 2x - 3}$$

Solution:

$$\frac{3x + 5}{x^2 + 4x + 3} - \frac{x - 5}{x^2 + 2x - 3}$$

$$= \frac{3x + 5}{(x + 3)(x + 1)} - \frac{x - 5}{(x + 3)(x - 1)}$$

Factor the denominators. The LCD is $(x + 3)(x + 1)(x - 1)$.

$$= \frac{(3x + 5)(x - 1)}{(x + 3)(x + 1)(x - 1)} - \frac{(x - 5)(x + 1)}{(x + 3)(x - 1)(x + 1)}$$

$$= \frac{(3x + 5)(x - 1) - (x - 5)(x + 1)}{(x + 3)(x + 1)(x - 1)}$$

Subtract the numerators and write the result over the common denominator.

$$= \frac{3x^2 + 2x - 5 - (x^2 - 4x - 5)}{(x + 3)(x + 1)(x - 1)}$$

$$= \frac{3x^2 + 2x - 5 - x^2 + 4x + 5}{(x + 3)(x + 1)(x - 1)}$$

$$= \frac{2x^2 + 6x}{(x + 3)(x + 1)(x - 1)}$$

$$= \frac{2x(x + 3)}{(x + 3)(x + 1)(x - 1)}$$

Factor the numerator and denominator and simplify.

$$= \frac{2x}{(x + 1)(x - 1)}$$

Avoiding Mistakes

It is very important to use parentheses around the second trinomial in the numerator. This will ensure that all terms that follow will be subtracted.

Skill Practice 5   Subtract the rational expressions and simplify the result.

$$\frac{t}{t^2 + 5t + 6} - \frac{2}{t^2 + 3t + 2}$$

5. Simplify Complex Fractions

A **complex fraction** (also called a **compound fraction**) is an expression that contains one or more fractions in the numerator or denominator. We present two methods to simplify a complex fraction. Method I is an application of the order of operations.

Answers

4. $\dfrac{40 + 3y}{45y^2}$

5. $\dfrac{t - 3}{(t + 1)(t + 3)}$

> **Simplifying a Complex Fraction: Order of Operations (Method I)**
>
> **Step 1** Add or subtract the fractions in the numerator to form a single fraction. Add or subtract the fractions in the denominator to form a single fraction.
> **Step 2** Divide the resulting expressions.
> **Step 3** Simplify if possible.

Classroom Example: p. 52
Exercise 58

EXAMPLE 6   Simplifying a Complex Fraction (Method I)

Simplify. $\dfrac{\dfrac{x}{4} - \dfrac{4}{x}}{\dfrac{1}{4} + \dfrac{1}{x}}$

Solution:

$$\dfrac{\dfrac{x}{4} - \dfrac{4}{x}}{\dfrac{1}{4} + \dfrac{1}{x}} = \dfrac{\dfrac{x \cdot x}{4 \cdot x} - \dfrac{4 \cdot 4}{x \cdot 4}}{\dfrac{1 \cdot x}{4 \cdot x} + \dfrac{1 \cdot 4}{x \cdot 4}} = \dfrac{\dfrac{x^2 - 16}{4x}}{\dfrac{x + 4}{4x}}$$

Step 1: Subtract the fractions in the numerator. Add the fractions in the denominator.

$$= \dfrac{x^2 - 16}{4x} \cdot \dfrac{4x}{x + 4}$$

Step 2: Multiply the rational expression from the numerator by the reciprocal of the expression from the denominator.

$$= \dfrac{(x - 4)(x + 4)}{4x} \cdot \dfrac{4x}{x + 4}$$

Step 3: Simplify by factoring and dividing out common factors.

$$= x - 4$$

Skill Practice 6 Simplify. $\dfrac{\dfrac{1}{7} + \dfrac{1}{y}}{\dfrac{y}{7} - \dfrac{7}{y}}$

In Example 7 we demonstrate another method (Method II) to simplify a complex fraction.

> **Simplifying a Complex Fraction: Multiply by the LCD (Method II)**
>
> **Step 1** Multiply the numerator and denominator of the complex fraction by the LCD of all individual fractions.
> **Step 2** Apply the distributive property and simplify the numerator and denominator.
> **Step 3** Simplify the resulting expression if possible.

Classroom Example: p. 52
Exercise 62

EXAMPLE 7   Simplifying a Complex Fraction (Method II)

Simplify. $\dfrac{d^{-2} - c^{-2}}{d^{-1} - c^{-1}}$

Answer
6. $\dfrac{1}{y - 7}$

Solution:

$$\frac{d^{-2} - c^{-2}}{d^{-1} - c^{-1}} = \frac{\dfrac{1}{d^2} - \dfrac{1}{c^2}}{\dfrac{1}{d} - \dfrac{1}{c}}$$

First write the expression with positive exponents.

$$= \frac{c^2 d^2 \cdot \left(\dfrac{1}{d^2} - \dfrac{1}{c^2}\right)}{c^2 d^2 \cdot \left(\dfrac{1}{d} - \dfrac{1}{c}\right)}$$

Step 1: Multiply numerator and denominator by the LCD of all four individual fractions: $c^2 d^2$.

$$= \frac{\dfrac{c^2 d^2}{1} \cdot \dfrac{1}{d^2} - \dfrac{c^2 d^2}{1} \cdot \dfrac{1}{c^2}}{\dfrac{c^2 d^2}{1} \cdot \dfrac{1}{d} - \dfrac{c^2 d^2}{1} \cdot \dfrac{1}{c}}$$

Step 2: Apply the distributive property.

$$= \frac{c^2 - d^2}{c^2 d - c d^2}$$

$$= \frac{(c - d)(c + d)}{cd(c - d)} = \frac{c + d}{cd}$$

Step 3: Simplify by factoring and dividing out common factors.

Skill Practice 7   Simplify. $\dfrac{4 - 6x^{-1}}{2x^{-1} - 3x^{-2}}$

Classroom Example: p. 52
Exercise 66

EXAMPLE 8   **Simplifying a Complex Fraction (Method II)**

Simplify. $\dfrac{\dfrac{2}{1 + h} - 2}{h}$

Solution:

TIP The expression given in Example 8 is a pattern we see in a first-semester calculus course.

$$\frac{\dfrac{2}{1 + h} - 2}{h} = \frac{\dfrac{2}{1 + h} - \dfrac{2}{1}}{h} = \frac{(1 + h) \cdot \left(\dfrac{2}{1 + h} - \dfrac{2}{1}\right)}{(1 + h) \cdot (h)}$$

Step 1: Multiply numerator and denominator by the LCD, which is $(1 + h)$.

$$= \frac{\dfrac{(1 + h)}{1} \cdot \left(\dfrac{2}{1 + h}\right) - \dfrac{(1 + h)}{1} \cdot \left(\dfrac{2}{1}\right)}{(1 + h) \cdot (h)}$$

Step 2: Apply the distributive property.

$$= \frac{2 - 2(1 + h)}{h(1 + h)}$$

Step 3: Simplify.

$$= \frac{2 - 2 - 2h}{h(1 + h)} = \frac{-2h}{h(1 + h)} = \frac{-2}{1 + h} \quad \text{or} \quad -\frac{2}{1 + h}$$

Answers

7. $2x$

8. $-\dfrac{5}{1 + h}$

Skill Practice 8   Simplify. $\dfrac{\dfrac{5}{1 + h} - 5}{h}$

6. Rationalize the Denominator of a Radical Expression

The same principle that applies to simplifying rational expressions also applies to simplifying algebraic fractions. For example, $\frac{5}{\sqrt{x}}$ is an algebraic fraction, but not a rational expression because the denominator is not a polynomial.

From Section R.2, we outlined the criteria for a radical expression to be simplified. Conditions 2 and 3 are stated here.

- No fraction may appear in the radicand.
- No denominator of a fraction may contain a radical.

In Example 9, we use the property of equivalent fractions to remove a radical from the denominator of a fraction. This is called **rationalizing the denominator.**

Classroom Examples: p. 52
Exercises 70, 76

EXAMPLE 9   Rationalizing the Denominator

Simplify. Assume that x is a positive real number.

    a. $\dfrac{5}{\sqrt{x}}$ **b.** $\dfrac{4}{\sqrt{7} - \sqrt{5}}$

Solution:

a. $\dfrac{5}{\sqrt{x}} = \dfrac{5 \cdot \sqrt{x}}{\sqrt{x} \cdot \sqrt{x}}$ Multiply numerator and denominator by $\sqrt{x}$ so that the radicand in the denominator is a perfect square.

$\quad = \dfrac{5\sqrt{x}}{\sqrt{x^2}}$ Apply the product property of radicals.

$\quad = \dfrac{5\sqrt{x}}{x}$ Simplify the radical in the denominator.

b. $\dfrac{4}{\sqrt{7} - \sqrt{5}} = \dfrac{4 \cdot (\sqrt{7} + \sqrt{5})}{(\sqrt{7} - \sqrt{5}) \cdot (\sqrt{7} + \sqrt{5})}$ Multiply numerator and denominator by the conjugate of the denominator.

$\quad = \dfrac{4(\sqrt{7} + \sqrt{5})}{(\sqrt{7})^2 - (\sqrt{5})^2}$ Recall that $(a - b)(a + b) = a^2 - b^2$.

$\quad = \dfrac{4(\sqrt{7} + \sqrt{5})}{7 - 5}$ Simplify the radicals in the denominator.

$\quad = \dfrac{\overset{2}{4}(\sqrt{7} + \sqrt{5})}{2}$ Simplify the fraction.

$\quad = 2(\sqrt{7} + \sqrt{5})$ or $2\sqrt{7} + 2\sqrt{5}$

TIP Keep the numerator in factored form until the denominator is simplified completely. By so doing, it will be easier to identify common factors in the numerator and denominator.

Skill Practice 9   Simplify.

    a. $\dfrac{\sqrt{5}}{\sqrt{7}}$ **b.** $\dfrac{12}{\sqrt{13} - \sqrt{10}}$

In Example 9(b), we multiplied the numerator and denominator by the conjugate of the denominator. The rationale is that product $(a + b)(a - b)$ results in a difference of squares $a^2 - b^2$. If either a or b has a square root factor, then the product will simplify to an expression without square roots.

Answers

9. a. $\dfrac{\sqrt{35}}{7}$

  b. $4(\sqrt{13} + \sqrt{10})$ or $4\sqrt{13} + 4\sqrt{10}$

SECTION R.4   Practice Exercises

Concept Connections

1. A ___rational___ expression is a ratio of two polynomials.

2. The restricted values of the variable for a rational expression are those that make the denominator equal to ___zero___.

3. The expression $\dfrac{5(x+2)}{(x+2)(x-1)}$ equals $\dfrac{5}{x-1}$ provided that $x \neq$ ___−2___ and $x \neq$ ___1___.

4. The ratio of a polynomial and its opposite equals ___−1___.

5. A ___complex (or compound)___ fraction is an expression that contains one or more fractions in the numerator or denominator.

6. The process to remove a radical from the denominator of a fraction is called ___rationalizing___ the denominator.

Objective 1: Determine Restricted Values for a Rational Expression

For Exercises 7–14, determine the restrictions on the variable. (See Example 1)

7. $\dfrac{x-4}{x+7}$ $x \neq -7$

8. $\dfrac{y-1}{y+10}$ $y \neq -10$

9. $\dfrac{a}{a^2-81}$ $a \neq 9, a \neq -9$

10. $\dfrac{t}{t^2-16}$ $t \neq 4, t \neq -4$

11. $\dfrac{a}{a^2+81}$ No restricted values

12. $\dfrac{t}{t^2+16}$ No restricted values

13. $\dfrac{6c}{7a^3b^2}$ $a \neq 0, b \neq 0$

14. $\dfrac{11z}{8x^5y}$ $x \neq 0, y \neq 0$

Objective 2: Simplify Rational Expressions

15. Determine which expressions are equal to $-\dfrac{5}{x-3}$.

   a. $\dfrac{-5}{x-3}$ **b.** $\dfrac{5}{3-x}$

   c. $-\dfrac{5}{3-x}$ **d.** $\dfrac{-5}{3-x}$ a and b

16. Determine which expressions are equal to $\dfrac{-2}{a+b}$.

   a. $\dfrac{-2}{a-b}$ **b.** $-\dfrac{2}{a+b}$

   c. $\dfrac{2}{-a-b}$ **d.** $\dfrac{2}{a-b}$ b and c

For Exercises 17–26, simplify the expression and state the restrictions on the variable. (See Example 2)

17. $\dfrac{x^2-9}{x^2-4x-21}$ $\dfrac{x-3}{x-7}$; $x \neq -3, x \neq 7$

18. $\dfrac{y^2-64}{y^2-7y-8}$ $\dfrac{y+8}{y+1}$; $y \neq 8, y \neq -1$

19. $-\dfrac{12a^2bc}{3ab^5}$ $-\dfrac{4ac}{b^4}$; $a \neq 0, b \neq 0$

20. $-\dfrac{15tu^5v}{3t^3u}$ $-\dfrac{5u^4v}{t^2}$; $t \neq 0, u \neq 0$

21. $\dfrac{10-5\sqrt{6}}{15}$ $\dfrac{2-\sqrt{6}}{3}$

22. $\dfrac{12+4\sqrt{3}}{8}$ $\dfrac{3+\sqrt{3}}{2}$

23. $\dfrac{2y^2-16y}{64-y^2}$ $-\dfrac{2y}{8+y}$; $y \neq 8, y \neq -8$

24. $\dfrac{81-t^2}{7t^2-63t}$ $-\dfrac{9+t}{7t}$; $t \neq 9, t \neq 0$

25. $\dfrac{4b-4a}{ax-xb-2a+2b}$ $\dfrac{4}{x-2}$; $a \neq b; x \neq 2$

26. $\dfrac{2z-2y}{xy-xz+3y-3z}$ $-\dfrac{2}{x+3}$; $y \neq z, x \neq -3$

Objective 3: Multiply and Divide Rational Expressions

For Exercises 27–34, multiply or divide as indicated. The restrictions on the variables are implied. (See Example 3)

27. $\dfrac{3a^5b^7}{a-5b} \cdot \dfrac{2a-10b}{12a^4b^{10}}$ $\dfrac{a}{2b^3}$

28. $\dfrac{8x-3y}{x^3y^4} \cdot \dfrac{6xy^8}{24x-9y}$ $\dfrac{2y^4}{x^2}$

29. $\dfrac{c^2-d^2}{cd^{11}} \div \dfrac{8c^2+4cd-4d^2}{8c^4d^{10}}$ $\dfrac{2c^3(c-d)}{d(2c-d)}$

30. $\dfrac{m^{11}n^2}{m^2-n^2} \div \dfrac{18m^9n^5}{9m^2+6mn-15n^2}$ $\dfrac{m^2(3m+5n)}{6n^3(m+n)}$

31. $\dfrac{2a^2b-ab^2}{8b^2+ab} \cdot \dfrac{a^2+16ab+64b^2}{2a^2+15ab-8b^2}$ a

32. $\dfrac{2c^2-2cd}{3c^2d+2c^3} \cdot \dfrac{4c^2+12cd+9d^2}{2c^2+cd-3d^2}$ $\dfrac{2}{c}$

33. $\dfrac{x^3-64}{16x-x^3} \div \dfrac{2x^2+8x+32}{x^2+2x-8}$ $-\dfrac{x-2}{2x}$

34. $\dfrac{3y^2+21y+147}{25y-y^3} \div \dfrac{y^3-343}{y^2-12y+35}$ $\dfrac{3}{y(5+y)}$

Objective 4: Add and Subtract Rational Expressions

For Exercises 35–40, identify the least common denominator for each pair of expressions.

35. $\dfrac{7}{6x^5yz^4}$ and $\dfrac{3}{20xy^2z^3}$ $60x^5y^2z^4$

36. $\dfrac{12}{35b^4cd^3}$ and $\dfrac{8}{25b^2c^3d}$ $175b^4c^3d^3$

37. $\dfrac{2t+1}{(3t+4)^3(t-2)}$ and $\dfrac{4}{t(3t+4)^2(t-2)}$ $t(3t+4)^3(t-2)$

38. $\dfrac{5y-7}{y(2y-5)(y+6)^4}$ and $\dfrac{6}{(2y-5)^3(y+6)^2}$ $y(2y-5)^3(y+6)^4$

39. $\dfrac{x+3}{x^2+20x+100}$ and $\dfrac{3}{2x^2+20x}$ $2x(x+10)^2$

40. $\dfrac{z-4}{4z^2-20z+25}$ and $\dfrac{5}{12z^2-30z}$ $6z(2z-5)^2$

For Exercises 41–56, add or subtract as indicated. **(See Examples 4–5)**

41. $\dfrac{m^2}{m+3}+\dfrac{6m+9}{m+3}$ $m+3$

42. $\dfrac{n^2}{n+5}+\dfrac{7n+10}{n+5}$ $n+2$

43. $\dfrac{2}{9c}+\dfrac{7}{15c^3}$ $\dfrac{10c^2+21}{45c^3}$

44. $\dfrac{6}{25x}+\dfrac{7}{10x^4}$ $\dfrac{12x^3+35}{50x^4}$

45. $\dfrac{9}{2x^2y^4}-\dfrac{11}{xy^5}$ $\dfrac{9y-22x}{2x^2y^5}$

46. $\dfrac{-2}{3m^3n}-\dfrac{5}{m^2n^4}$ $\dfrac{2n^3+15m}{3m^3n^4}$

47. $\dfrac{2}{x+3}-\dfrac{7}{x}$ $\dfrac{5x+21}{x(x+3)}$

48. $\dfrac{4}{m-2}-\dfrac{3}{m}$ $\dfrac{m+6}{m(m-2)}$

49. $\dfrac{1}{x^2+xy}-\dfrac{2}{x^2-y^2}$ $\dfrac{1}{x(x-y)}$

50. $\dfrac{4}{4a^2-b^2}-\dfrac{1}{2a^2-ab}$ $\dfrac{1}{a(2a+b)}$

51. $\dfrac{5}{y}+\dfrac{2}{y+1}-\dfrac{6}{y^2}$ $\dfrac{7y^2-y-6}{y^2(y+1)}$

52. $\dfrac{5}{t^2}+\dfrac{4}{t+2}-\dfrac{3}{t}$ $\dfrac{t^2-t+10}{t^2(t+2)}$

53. $\dfrac{3w}{w-4}+\dfrac{2w+4}{4-w}$ 1

54. $\dfrac{2x-1}{x-7}+\dfrac{x+6}{7-x}$ 1

55. $\dfrac{4}{x^2+6x+5}-\dfrac{3}{x^2+7x+10}$ $\dfrac{1}{(x+1)(x+2)}$

56. $\dfrac{3}{x^2-4x-5}-\dfrac{2}{x^2-6x+5}$ $\dfrac{1}{(x+1)(x-1)}$

Objective 5: Simplify Complex Fractions

For Exercises 57–68, simplify the complex fraction. **(See Examples 6–8)**

57. $\dfrac{\dfrac{1}{27x}+\dfrac{1}{9}}{\dfrac{1}{3}+\dfrac{1}{9x}}$ $\dfrac{1}{3}$

58. $\dfrac{\dfrac{1}{8x}+\dfrac{1}{4}}{\dfrac{1}{2}+\dfrac{1}{4x}}$ $\dfrac{1}{2}$

59. $\dfrac{\dfrac{x}{6}-\dfrac{5x+14}{6x}}{\dfrac{1}{6}-\dfrac{7}{6x}}$ $x+2$

60. $\dfrac{\dfrac{x}{3}-\dfrac{2x+3}{3x}}{\dfrac{1}{3}+\dfrac{1}{3x}}$ $x-3$

61. $\dfrac{2a^{-1}-b^{-1}}{4a^{-2}-b^{-2}}$ $\dfrac{ab}{2b+a}$

62. $\dfrac{3u^{-1}-v^{-1}}{9u^{-2}-v^{-2}}$ $\dfrac{uv}{3v+u}$

63. $\dfrac{\dfrac{3}{1+h}-3}{h}$ $-\dfrac{3}{1+h}$

64. $\dfrac{\dfrac{4}{1+h}-4}{h}$ $-\dfrac{4}{1+h}$

65. $\dfrac{\dfrac{7}{x+h}-\dfrac{7}{x}}{h}$ $-\dfrac{7}{x(x+h)}$

66. $\dfrac{\dfrac{8}{x+h}-\dfrac{8}{x}}{h}$ $-\dfrac{8}{x(x+h)}$

67. $\dfrac{\dfrac{3}{x-1}-\dfrac{1}{x+1}}{\dfrac{6}{x^2-1}}$ $\dfrac{x+2}{3}$

68. $\dfrac{\dfrac{1}{x+1}}{\dfrac{-5}{x^2-3x-4}+\dfrac{1}{x-4}}$ 1

Objective 6: Rationalize the Denominator of a Radical Expression

For Exercises 69–84, simplify the expression. Assume that the variable expressions represent positive real numbers.
(See Example 9)

69. $\dfrac{4}{\sqrt{y}}$ $\dfrac{4\sqrt{y}}{y}$

70. $\dfrac{7}{\sqrt{z}}$ $\dfrac{7\sqrt{z}}{z}$

71. $\dfrac{4}{\sqrt[3]{y}}$ $\dfrac{4\sqrt[3]{y^2}}{y}$

72. $\dfrac{7}{\sqrt[4]{z}}$ $\dfrac{7\sqrt[4]{z^3}}{z}$

73. $\dfrac{\sqrt{12}}{\sqrt{x+1}}$ $\dfrac{2\sqrt{3x+3}}{x+1}$

74. $\dfrac{\sqrt{50}}{\sqrt{x-2}}$ $\dfrac{5\sqrt{2x-4}}{x-2}$

75. $\dfrac{8}{\sqrt{15}-\sqrt{11}}$ $2(\sqrt{15}+\sqrt{11})$

76. $\dfrac{12}{\sqrt{6}-\sqrt{2}}$ $3(\sqrt{6}+\sqrt{2})$

77. $\dfrac{x-5}{\sqrt{x}+\sqrt{5}}$ $\sqrt{x}-\sqrt{5}$ **78.** $\dfrac{y-3}{\sqrt{y}+\sqrt{3}}$ $\sqrt{y}-\sqrt{3}$ **79.** $\dfrac{2\sqrt{10}+3\sqrt{5}}{4\sqrt{10}+2\sqrt{5}}$ $\dfrac{5+4\sqrt{2}}{14}$ **80.** $\dfrac{3\sqrt{3}+\sqrt{6}}{5\sqrt{3}-2\sqrt{6}}$ $\dfrac{19+11\sqrt{2}}{17}$

81. $\dfrac{7}{\sqrt{3x}}+\dfrac{\sqrt{3x}}{x}$ $\dfrac{10\sqrt{3x}}{3x}$ **82.** $\dfrac{4}{\sqrt{11y}}+\dfrac{\sqrt{11y}}{y}$ $\dfrac{15\sqrt{11y}}{11y}$ **83.** $\dfrac{5}{w\sqrt{7}}-\dfrac{\sqrt{7}}{w}$ $-\dfrac{2\sqrt{7}}{7w}$ **84.** $\dfrac{13}{t\sqrt{2}}-\dfrac{\sqrt{2}}{t}$ $\dfrac{11\sqrt{2}}{2t}$

Mixed Exercises

85. The average round trip speed S (in mph) of a vehicle traveling a distance of d miles each way is given by

$$S = \dfrac{2d}{\dfrac{d}{r_1}+\dfrac{d}{r_2}}.$$

In this formula, r_1 is the average speed going one way, and r_2 is the average speed on the return trip.

a. Simplify the complex fraction. $S = \dfrac{2r_1 r_2}{r_1 + r_2}$

b. If a plane flies 400 mph from Orlando to Albuquerque and 460 mph on the way back, compute the average speed of the round trip. Round to 1 decimal place.   427.9 mph

86. The formula $R = \dfrac{1}{\dfrac{1}{R_1}+\dfrac{1}{R_2}}$ gives the total electrical

resistance R (in ohms, Ω) when two resistors of resistance R_1 and R_2 are connected in parallel.

a. Simplify the complex fraction. $R = \dfrac{R_1 R_2}{R_2 + R_1}$

b. Find the total resistance when $R_1 = 12\ \Omega$ and $R_2 = 20\ \Omega$.   7.5 Ω

87. The concentration C (in ng/mL) of a drug in the bloodstream t hours after ingestion is modeled by

$$C = \dfrac{600t}{t^3 + 125}.$$

a. Determine the concentration at 1 hr, 12 hr, 24 hr, and 48 hr. Round to 1 decimal place.

b. What appears to be the limiting concentration for large values of t?   0 ng/mL

88. An object that is originally 35°C is placed in a freezer. The temperature T (in °C) of the object can be

approximated by the model $T = \dfrac{350}{t^2 + 3t + 10}$,

where t is the time in hours after the object is placed in the freezer.

a. Determine the temperature at 2 hr, 4 hr, 12 hr, and 24 hr. Round to 1 decimal place.

b. What appears to be the limiting temperature for large values of t?   0°C

For Exercises 89–102, simplify the expression.

89. $\dfrac{2x^3 y}{x^2 y + 3xy} \cdot \dfrac{x^2 + 6x + 9}{2x + 6} \div 5xy^4$ $\dfrac{x}{5y^4}$

90. $\dfrac{2y^2 + 20y + 50}{12 - 4y} \cdot \dfrac{y-3}{y^2 + 12y + 35} \div (3y + 15)$ $-\dfrac{1}{6(y+7)}$

91. $\left(\dfrac{4}{2t+1} - \dfrac{t}{2t^2 + 17t + 8}\right)(t+8)$ $\dfrac{3t+32}{2t+1}$

92. $\left(\dfrac{2m}{6m+3} - \dfrac{1}{m+4}\right)(2m+1)$ $\dfrac{2m^2 + 2m - 3}{3(m+4)}$

93. $\dfrac{n-2}{n-4} + \dfrac{2n^2 - 15n + 12}{n^2 - 16} - \dfrac{2n-5}{n+4}$ 1

94. $\dfrac{c^2 + 13c + 18}{c^2 - 9} + \dfrac{c+1}{c+3} - \dfrac{c+8}{c-3}$ 1

95. $\dfrac{1 - a^{-1} - 6a^{-2}}{1 - 4a^{-1} + 3a^{-2}}$ $\dfrac{a+2}{a-1}$

96. $\dfrac{1 + t^{-1} - 12t^{-2}}{1 - 4t^{-1} + 3t^{-2}}$ $\dfrac{t+4}{t-1}$

97. $\dfrac{34}{2\sqrt{5} - \sqrt{3}}$ $4\sqrt{5} + 2\sqrt{3}$

98. $\dfrac{13}{2\sqrt{7} + \sqrt{2}}$ $2\sqrt{7} - \sqrt{2}$

99. $\dfrac{8 - \sqrt{48}}{6}$ $\dfrac{4 - 2\sqrt{3}}{3}$

100. $\dfrac{10 - \sqrt{50}}{15}$ $\dfrac{2 - \sqrt{2}}{3}$

101. $\dfrac{\dfrac{t+6}{1 + \dfrac{2}{t}} - t - 4}{\ }$ $-\dfrac{8}{t+2}$

102. $\dfrac{\dfrac{m-4}{1 - \dfrac{2}{m}} - m + 2}{\ }$ $-\dfrac{4}{m-2}$

For Exercises 103–110, write the expression as a single term, factored completely. Do not rationalize the denominator.

103. $1 - x^{-2} - 2x^{-3}$ $\dfrac{x^3 - x - 2}{x^3}$

104. $1 - 8x^{-5} + 30x^{-7}$ $\dfrac{x^7 - 8x^2 + 30}{x^7}$

105. $\dfrac{3}{2\sqrt{3x}} + \sqrt{3x}$ $\dfrac{3(2x+1)}{2\sqrt{3x}}$

106. $\dfrac{2}{\sqrt{x}} + \sqrt{x}$ $\dfrac{x+2}{\sqrt{x}}$

107. $\dfrac{\sqrt{x^2+1}+\dfrac{x^2}{\sqrt{x^2+1}}}{x^2+1}$ $\quad \dfrac{2x^2+1}{(x^2+1)\sqrt{x^2+1}}$

108. $\dfrac{\dfrac{x^2}{\sqrt{x^2+9}}-\sqrt{x^2+9}}{x^2}$ $\quad -\dfrac{9}{x^2\sqrt{x^2+9}}$

109. $2\sqrt{4x^2+9}+\dfrac{8x^2}{\sqrt{4x^2+9}}$ $\quad \dfrac{2(8x^2+9)}{\sqrt{4x^2+9}}$

110. $3\sqrt{9x^2+1}+\dfrac{27x}{\sqrt{9x^2+1}}$ $\quad \dfrac{3(9x^2+9x+1)}{\sqrt{9x^2+1}}$

Write About It

111. Explain why the expression $\dfrac{x}{x-y}$ is not defined for $x=y$.   *If $x=y$, then the denominator $x-y$ will equal zero. Division by zero is undefined.*

112. Is the statement $\dfrac{3(x-4)}{(x+2)(x-4)}=\dfrac{3}{x+2}$ true for all values of x? Explain why or why not.   *No. The expressions are equal for all real numbers except $x=-2$ and $x=4$.*

Expanding Your Skills

113. The numbers 1, 2, 4, 5, 10, and 20 are natural numbers that are factors of 20. There are other factors of 20 within the set of rational numbers and the set of irrational numbers. For example:

  a. Show that $\dfrac{14}{3}$ and $\dfrac{30}{7}$ are factors of 20 over the set of rational numbers. $\dfrac{14}{3}\cdot\dfrac{30}{7}=\dfrac{420}{21}=20$

  b. Show that $\left(5-\sqrt{5}\right)$ and $\left(5+\sqrt{5}\right)$ are factors of 20 over the set of irrational numbers.
  $\begin{aligned}(5-\sqrt{5})(5+\sqrt{5})&=(5)^2-(\sqrt{5})^2\\&=25-5\\&=20\end{aligned}$

114. a. Show that $\dfrac{15}{2}$ and $\dfrac{4}{5}$ are factors of 6 over the set of rational numbers. $\dfrac{15}{2}\cdot\dfrac{4}{5}=\dfrac{60}{10}=6$

  b. Show that $\left(3-\sqrt{3}\right)$ and $\left(3+\sqrt{3}\right)$ are factors of 6 over the set of irrational numbers.
  $\begin{aligned}(3-\sqrt{3})(3+\sqrt{3})&=(3)^2-(\sqrt{3})^2\\&=9-3\\&=6\end{aligned}$

For Exercises 115–120, simplify the expression.

115. $\dfrac{w^{3n+1}-w^{3n}z}{w^{n+2}-w^nz^2}$ $\quad \dfrac{w^{2n}}{w+z}$

116. $\dfrac{x^{2n+1}-x^{2n}y}{x^{n+3}-x^ny^3}$ $\quad \dfrac{x^n}{x^2+xy+y^2}$

117. $\dfrac{\sqrt{5}}{\sqrt[6]{2}}$ $\quad \dfrac{\sqrt[6]{5^32^4}}{2}$

118. $\dfrac{\sqrt{7}}{\sqrt[4]{3}}$ $\quad \dfrac{\sqrt[4]{7^23^3}}{3}$

119. $\dfrac{a-b}{\sqrt[3]{a}-\sqrt[3]{b}}$ (*Hint*: Factor the numerator as a difference of cubes over the set of irrational numbers.) $\sqrt[3]{a^2}+\sqrt[3]{ab}+\sqrt[3]{b^2}$

120. $\dfrac{x+y}{\sqrt[3]{x}+\sqrt[3]{y}}$ $\quad \sqrt[3]{x^2}-\sqrt[3]{xy}+\sqrt[3]{y^2}$

For Exercises 121–122, rationalize the numerator by multiplying numerator and denominator by the conjugate of the *numerator*.

121. $\dfrac{\sqrt{4+h}-2}{h}$ $\quad \dfrac{1}{\sqrt{4+h}+2}$

122. $\dfrac{\sqrt{x+h}-\sqrt{x}}{h}$ $\quad \dfrac{1}{\sqrt{x+h}+\sqrt{x}}$

SECTION R.5   Equations with Real Solutions

OBJECTIVES

1. Solve Linear Equations in One Variable
2. Solve Quadratic Equations with Real Solutions
3. Solve Rational Equations
4. Solve Absolute Value Equations
5. Solve Radical Equations
6. Solve Equations for a Specified Variable

1. Solve Linear Equations in One Variable

In this section we solve a variety of equations of different types beginning with linear equations in one variable.

> **Definition of a Linear Equation in One Variable**
>
> A **linear equation in one variable** is an equation that can be written in the form $ax+b=0$, where a and b are real numbers, $a\neq 0$, and x is the variable.

A linear equation in one variable is also called a **first-degree equation** because the degree of the variable term must be exactly one.

A **solution** to an equation is a value of the variable that makes the equation a true statement. The set of all solutions to an equation is called the **solution set** of the equation. **Equivalent equations** have the same solution set. To solve a linear equation in one variable, we form simpler, equivalent equations until we obtain an equation whose solution is obvious. The properties used to produce equivalent equations include the addition and multiplication properties of equality.

3. Solve Rational Equations

A rational equation is an equation in which each term is a rational expression. For example $\frac{12}{x} = \frac{6}{2x} + 3$ is a rational equation. One technique to solve a rational equation is to multiply both sides by the least common denominator of all the terms. This is demonstrated in Example 7.

Classroom Example: p. 69
Exercise 68

EXAMPLE 7   **Solving a Rational Equation**

Solve the equation and check the solution. $\frac{12}{x} = \frac{6}{2x} + 3$

Solution:

$$\frac{12}{x} = \frac{6}{2x} + 3$$ Restrict x so that $x \neq 0$.

$$2x\left(\frac{12}{x}\right) = 2x\left(\frac{6}{2x} + \frac{3}{1}\right)$$ Clear fractions by multiplying both sides by the LCD, $2x$. Since $x \neq 0$, this will produce an equivalent equation.

$$\frac{2x}{1}\left(\frac{12}{x}\right) = \frac{2x}{1}\left(\frac{6}{2x}\right) + \frac{2x}{1}\left(\frac{3}{1}\right)$$ Apply the distributive property.

$$24 = 6 + 6x$$ Simplify.

$$18 = 6x$$ Subtract 6 from both sides.

$$3 = x$$ Check: $\frac{12}{x} = \frac{6}{2x} + \frac{3}{1}$

The solution set is $\{3\}$.

$$\frac{12}{(3)} \stackrel{?}{=} \frac{6}{2(3)} + \frac{3}{1}$$

$$4 \stackrel{?}{=} 1 + 3 \checkmark \text{ true}$$

Skill Practice 7   Solve the equation and check the solution. $\frac{15}{y} = \frac{21}{3y} + 2$

In Example 7, notice that two of the expressions in the equation are undefined for $x = 0$ because the denominator would be zero. Thus, when we solve a rational equation, we must restrict the values of the variable to avoid division by zero. This is demonstrated in Example 8.

Classroom Example: p. 69
Exercise 80

EXAMPLE 8   **Solving a Rational Equation**

Solve. $\frac{2x}{x-4} - \frac{3}{x+2} = \frac{x^2 + 14}{x^2 - 2x - 8}$

Answer

7. $\{4\}$

Check: $\frac{15}{y} = \frac{21}{3y} + \frac{2}{1}$

$$\frac{15}{(4)} \stackrel{?}{=} \frac{21}{3(4)} + \frac{2}{1}$$

$$\frac{15}{4} \stackrel{?}{=} \frac{7}{4} + \frac{8}{4} \checkmark \text{ true}$$

Solution:

$$\frac{2x}{x-4} - \frac{3}{x+2} = \frac{x^2+14}{x^2-2x-8}$$

$$\frac{2x}{x-4} - \frac{3}{x+2} = \frac{x^2+14}{(x-4)(x+2)}$$

Factor the denominators.

The variable x has the restrictions that $x \neq 4$ and $x \neq -2$.

$$(x-4)(x+2)\left(\frac{2x}{x-4} - \frac{3}{x+2}\right) = (x-4)(x+2)\left[\frac{x^2+14}{(x-4)(x+2)}\right]$$

Multiply both sides by the LCD to clear fractions.

$$2x(x+2) - 3(x-4) = x^2 + 14$$

The resulting equation is quadratic.

$$2x^2 + 4x - 3x + 12 = x^2 + 14$$

$$x^2 + x - 2 = 0$$

$$(x+2)(x-1) = 0$$

$$\cancel{x = -2} \quad \text{or} \quad x = 1$$

Apply the zero product property.

The value −2 is not a solution because it is a restricted value. It does not check.

Check: $x = -2$

$$\frac{2(-2)}{(-2)-4} - \underbrace{\frac{3}{(-2)+2}}_{\text{undefined}} \overset{?}{=} \underbrace{\frac{(-2)^2+14}{(-2)^2-2(-2)-8}}_{\text{undefined}}$$

Check: $x = 1$

$$\frac{2(1)}{(1)-4} - \frac{3}{(1)+2} \overset{?}{=} \frac{(1)^2+14}{(1)^2-2(1)-8}$$

$$\frac{2}{-3} - 1 \overset{?}{=} \frac{15}{-9}$$

$$-\frac{5}{3} \overset{?}{=} -\frac{5}{3} \quad \checkmark \text{ true}$$

The solution set is {1}. The value −2 does not check.

Skill Practice 8   Solve. $\dfrac{3x}{x-5} = \dfrac{2}{x+1} + \dfrac{2x^2+40}{x^2-4x-5}$

4. Solve Absolute Value Equations

We now turn our attention to equations involving absolute value expressions. For example, given $|x| = 5$, the solution set is $\{-5, 5\}$. The equation can also be written as $|x - 0| = 5$, implying that the solutions are the values of x that are 5 units from 0 on the number line (Figure R-3).

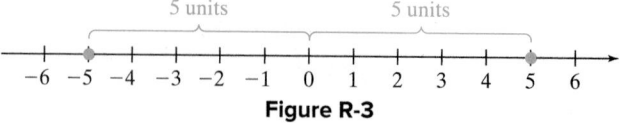

Figure R-3

Given a nonnegative real number k, the generic absolute value equation $|u| = k$ can be solved directly from the definition of absolute value. Recall that

$$|u| = \begin{cases} u \text{ if } u \geq 0 \\ -u \text{ if } u < 0 \end{cases}$$

Answer

8. {−6}; The value 5 does not check.

Thus, $|u| = k$ means that $u = k$ or $-u = k$. Solving for u, we have $u = k$ or $u = -k$.

This and three other properties summarized in Table R-8 follow directly from the definition of absolute value.

> **Table R-8   Absolute Value Equations**
>
> Let k represent a positive real number.
>
> **1.** $|u| = k$ is equivalent to $u = k$ or $u = -k$.
>
> **2.** $|u| = 0$ is equivalent to $u = 0$.
>
> **3.** $|u| = -k$ has no solution.
>
> **4.** $|u| = |w|$ is equivalent to $u = w$ or $u = -w$.

To solve an absolute value equation, first isolate the absolute value. Then solve the equation by rewriting the equation in its equivalent form given in Table R-8.

Classroom Examples: p. 70
Exercises 90, 94

EXAMPLE 9   Solving Absolute Value Equations

Solve the equations.   **a.** $2|3 - 2t| = 6$ **b.** $2 = |7w - 3| + 8$

Solution:

a.
$$2|3 - 2t| = 6$$
Divide by 2 to isolate the absolute value.

$$|3 - 2t| = 3$$
The equation is in the form $|u| = k$, where $u = 3 - 2t$ and $k > 0$.

$$3 - 2t = 3 \quad \text{or} \quad 3 - 2t = -3$$
Rewrite the equation in the form $u = k$ or $u = -k$.

$$-2t = 0 \quad \text{or} \quad -2t = -6$$

$$t = 0 \quad \text{or} \quad t = 3$$

$$\underline{\text{Check: } t = 0}$$
$$2|3 - 2(0)| \overset{?}{=} 6$$
$$2|3| = 6 \checkmark$$

$$\underline{\text{Check: } t = -3}$$
$$2|3 - 2(3)| \overset{?}{=} 6$$
$$2|3 - 6| \overset{?}{=} 6$$
$$2|-3| = 6 \checkmark$$

The solution set is $\{3, 0\}$.

b.
$$2 = |7w - 3| + 8$$
Subtract 8 from both sides to isolate the absolute value.

$$-6 = |7w - 3|$$
The equation is in the form $|u| = k$, where $u = 7w - 3$ and $k < 0$. By definition, an absolute value cannot be negative. There is no solution.

The solution set is $\{ \ \}$.

> **Skill Practice 9**   Solve the equations.
>
> **a.** $5|2 - 4t| = 50$ **b.** $5 = |6c - 7| + 9$

In Example 10, we solve equations involving two absolute values by writing $|u| = |w|$ in the equivalent form $u = w$ or $u = -w$.

Classroom Examples: p. 70
Exercises 96, 102

EXAMPLE 10   Solving Equations with Two Absolute Values

Solve the equations.   **a.** $|2x - 5| = |x + 1|$ **b.** $|6 - x| = |x - 6|$

Solution:

a. $|2x - 5| = |x + 1|$ The equation is in the form $|u| = |w|$.

$$2x - 5 = x + 1 \quad \text{or} \quad 2x - 5 = -(x + 1)$$
Rewrite the equation in the form $u = w$ or $u = -w$.

$$x - 5 = 1 \quad \text{or} \quad 2x - 5 = -x - 1$$
Solve each individual equation.

$$x = 6 \quad \text{or} \quad 3x = 4$$

$$x = \frac{4}{3}$$
Both solutions check in the original equation.

Answers
9. a. $\{-2, 3\}$ **b.** $\{\}$

The solution set is $\left\{ 6, \dfrac{4}{3} \right\}$.

b. $|6 - x| = |x - 6|$ The equation is in the form $|u| = |w|$.

$6 - x = x - 6$ or $6 - x = -(x - 6)$ Rewrite the equation in the form $u = w$ or $u = -w$.

$-2x = -12$ or $6 - x = -x + 6$ Solve each individual equation.

$x = 6$ or $6 = 6$ (identity)

> The solution to this equation is all real numbers (including 6).

The solution set is $\mathbb{R}$.

Skill Practice 10   Solve the equations.

  a. $|3x - 4| = |2x + 1|$ **b.** $|4 + x| = |-4 - x|$

---

5. Solve Radical Equations

An equation with one or more radicals containing a variable (such as $\sqrt[3]{x} = 5$) is called a **radical equation.** We can eliminate the radical by raising both sides of the equation to a power equal to the index of the radical.

$$\sqrt[3]{x} = 5$$
$$\left(\sqrt[3]{x}\right)^3 = (5)^3 \qquad \text{The index is 3. Therefore, raise both sides}$$
$$x = 125 \qquad \text{to the third power.}$$

By raising each side of a radical equation to a power equal to the index, a new equation is produced. However, some (or all) of the solutions to the new equation may *not* be solutions to the original equation. These are called **extraneous solutions.** For this reason, it is necessary to check all potential solutions in the original equation. For example, consider the equation $\sqrt{x} = -10$. By inspection, this equation has no solution because the principal square root of x must be nonnegative. However, if we square both sides of the equation, it appears as though a solution exists:

Square both sides.
$$\sqrt{x} = -10 \qquad \text{Solution set: \{ \}}$$
$$\left(\sqrt{x}\right)^2 = (-10)^2$$
$$x = 100 \qquad \text{The value 100 does not check in the original equation. Therefore, 100 is an extraneous solution.}$$

Solving a Radical Equation

Step 1   Isolate the radical. If an equation has more than one radical, choose one of the radicals to isolate.

Step 2   Raise each side of the equation to a power equal to the index of the radical.

Step 3   Solve the resulting equation. If the equation still has a radical, repeat steps 1 and 2.

*Step 4   Check the potential solutions in the original equation and write the solution set.

*In solving radical equations, extraneous solutions potentially arise when both sides of the equation are raised to an even power. Therefore, an equation with only odd-indexed roots will not have extraneous solutions. However, it is still recommended that all potential solutions be checked.

Answers

10. a. $\left\{\dfrac{3}{5}, 5\right\}$ **b.** $\mathbb{R}$

Classroom Example: p. 70
Exercise 106

EXAMPLE 11   Solving a Radical Equation

Solve. $\sqrt{x + 10} - 4 = x$

Avoiding Mistakes

When raising both sides of an equation to a power, be sure to enclose both sides of the equation in parentheses.

Solution:

$$\sqrt{x + 10} - 4 = x$$

$$\sqrt{x + 10} = x + 4 \qquad \text{Isolate the radical.}$$

$$\left(\sqrt{x + 10}\right)^2 = (x + 4)^2 \qquad \text{The index is 2. Therefore, raise both sides to the second power.}$$

$$x + 10 = x^2 + 8x + 16 \qquad \text{The resulting equation is quadratic.}$$

$$0 = x^2 + 7x + 6 \qquad \text{Set one side equal to zero.}$$

$$0 = (x + 6)(x + 1) \qquad \text{Factor.}$$

$$x = -6 \quad \text{or} \quad x = -1 \qquad \text{Apply the zero product rule.}$$

Both sides of the equation were raised to an even power. Therefore, it is necessary to check the potential solutions.

$$\underline{\text{Check: } x = -6}$$
$$\sqrt{x + 10} - 4 = x$$
$$\sqrt{(-6) + 10} - 4 \overset{?}{=} (-6)$$
$$\sqrt{4} - 4 \overset{?}{=} -6$$
$$2 - 4 \overset{?}{=} -6$$
$$-2 \overset{?}{=} -6 \text{ false}$$

$$\underline{\text{Check: } x = -1}$$
$$\sqrt{x + 10} - 4 = x$$
$$\sqrt{(-1) + 10} - 4 \overset{?}{=} (-1)$$
$$\sqrt{9} - 4 \overset{?}{=} -1$$
$$3 - 4 \overset{?}{=} -1$$
$$-1 \overset{?}{=} -1 \checkmark \text{ true}$$

The solution set is $\{-1\}$. The value -6 does not check.

Skill Practice 11   Solve the equation. $\sqrt{t + 7} = t - 5$

In Example 12, we solve the equation $\sqrt{m - 1} - \sqrt{3m + 1} = -2$. The first step is to isolate one of the radicals. However, the presence of the constant term, -2, makes it impossible to isolate both radicals simultaneously. As a result, it is necessary to square both sides of the equation twice.

Classroom Example: p. 70
Exercise 110

EXAMPLE 12   Solving an Equation Containing Two Radicals

Solve. $\sqrt{m - 1} - \sqrt{3m + 1} = -2$

Avoiding Mistakes

Exercise caution when squaring the two-term expression on the right.
$$\left(\sqrt{3m + 1} - 2\right)^2$$
$$= \left(\sqrt{3m + 1}\right)^2$$
$$\quad - 2\left(\sqrt{3m + 1}\right)(2) + (2)^2$$
$$= 3m + 1$$
$$\quad - 4\sqrt{3m + 1} + 4$$

Solution:

$$\sqrt{m - 1} - \sqrt{3m + 1} = -2$$

$$\sqrt{m - 1} = \sqrt{3m + 1} - 2 \qquad \text{Isolate one of the radicals.}$$

$$\left(\sqrt{m - 1}\right)^2 = \left(\sqrt{3m + 1} - 2\right)^2 \qquad \text{The index is 2. Therefore, raise both sides to the second power.}$$

$$m - 1 = 3m + 1 - 4\sqrt{3m + 1} + 4$$

$$m - 1 = 3m + 5 - 4\sqrt{3m + 1}$$

$$4\sqrt{3m + 1} = 2m + 6 \qquad \text{Isolate the remaining radical.}$$

$$2\sqrt{3m + 1} = m + 3 \qquad \text{Divide both sides by 2 to simplify.}$$

$$\left(2\sqrt{3m + 1}\right)^2 = (m + 3)^2 \qquad \text{The resulting equation has another radical. Isolate the radical, and square both sides again.}$$

$$4(3m + 1) = m^2 + 6m + 9$$

$$12m + 4 = m^2 + 6m + 9 \qquad \text{The resulting equation is quadratic.}$$

$$0 = m^2 - 6m + 5$$

$$0 = (m - 5)(m - 1) \qquad \text{Apply the zero product property.}$$

$$m = 5 \quad \text{or} \quad m = 1 \qquad \text{Both sides of the equation were raised to an even power. Check both potential solutions.}$$

Answer

11. $\{9\}$; The value 2 does not check.

Check: $m = 5$ | Check: $m = 1$

$$\sqrt{m - 1} - \sqrt{3m + 1} = -2$$
$$\sqrt{(5) - 1} - \sqrt{3(5) + 1} \stackrel{?}{=} -2$$
$$\sqrt{4} - \sqrt{16} \stackrel{?}{=} -2$$
$$2 - 4 \stackrel{?}{=} -2 \; \checkmark \; \text{true}$$

$$\sqrt{m - 1} - \sqrt{3m + 1} = -2$$
$$\sqrt{(1) - 1} - \sqrt{3(1) + 1} \stackrel{?}{=} -2$$
$$\sqrt{0} - \sqrt{4} \stackrel{?}{=} -2$$
$$0 - 2 \stackrel{?}{=} -2 \; \checkmark \; \text{true}$$

Both solutions check. The solution set is $\{1, 5\}$.

Skill Practice 12 Solve. $\quad 1 + \sqrt{n + 4} = \sqrt{3n + 1}$

In Example 13, we solve an equation containing rational exponents.

Classroom Example: p.70
Exercise 114

EXAMPLE 13 Solving an Equation Containing Rational Exponents

Solve the equation. $\quad 2(x + 1)^{2/3} = 8$

Solution:

$2(x + 1)^{2/3} = 8$

$2\sqrt[3]{(x + 1)^2} = 8$ Write the expression on the left in radical notation.

$\sqrt[3]{(x + 1)^2} = 4$ Divide by 2 to isolate the radical.

$\left[\sqrt[3]{(x + 1)^2}\right]^3 = (4)^3$ Raise both sides to a power equal to the index of the radical.

$(x + 1)^2 = 64$

$x + 1 = \pm\sqrt{64}$ Apply the square root property.

$x = -1 \pm 8$

$x = -1 - 8 = -9 \quad \text{or} \quad x = -1 + 8 = 7$

Check: $x = -9$ Check: $x = 7$

$(-9 + 1)^{2/3} \stackrel{?}{=} 4 \quad\quad (7 + 1)^{2/3} \stackrel{?}{=} 4$

The solution set is $\{-9, 7\}$.

$(-8)^{2/3} = 4 \; \checkmark \quad\quad (8)^{2/3} = 4 \; \checkmark$

Skill Practice 13 Solve the equation. $\quad 2(x - 4)^{3/4} = 54$

6. Solve Equations for a Specified Variable

The techniques presented in this section can also be used to manipulate an equation with multiple variables. In particular, we are often interested in solving an equation for a specified variable in terms of the other variables in the equation.

Classroom Example: p. 70
Exercise 140

EXAMPLE 14 Solving an Equation for a Specified Variable

Solve for r. $\quad V = \dfrac{1}{3}\pi r^2 h \quad (r > 0)$

Answers
12. $\{5\}$; The value 0 does not check.
13. $\{85\}$

Solution:

TIP The equation $V = \frac{1}{3}\pi r^2 h$ is linear in the variables V and h, and quadratic in the variable r.

$$V = \frac{1}{3}\pi r^2 h$$

This equation is quadratic in the variable r. The strategy in this example is to isolate r^2 and then apply the square root property.

$$3(V) = 3\left(\frac{1}{3}\pi r^2 h\right)$$

Multiply both sides by 3 to clear fractions.

$$3V = \pi r^2 h$$

$$\frac{3V}{\pi h} = \frac{\pi r^2 h}{\pi h}$$

Divide both sides by πh to isolate r^2.

$$\frac{3V}{\pi h} = r^2$$

TIP The formula $V = \frac{1}{3}\pi r^2 h$ gives the volume of a right circular cone with radius r. Therefore, $r > 0$.

$$r = \sqrt{\frac{3V}{\pi h}} \text{ or } r = \frac{\sqrt{3V\pi h}}{\pi h}$$

Apply the square root property. Since $r > 0$, we take the positive square root only.

Skill Practice 14   Solve for v. $E = \frac{1}{2}mv^2$ $(v > 0)$

Classroom Example: p. 70
Exercise 142

EXAMPLE 15   **Solving an Equation for a Specified Variable**

Solve for t. $mt^2 + nt = z$

Solution:

This equation is quadratic in the variable t. The strategy is to write the polynomial in descending order by powers of t. Then since there are two t terms with different exponents, we cannot isolate t directly. Instead we apply the quadratic formula.

Avoiding Mistakes

In the equation $mt^2 + nt - z = 0$, t is the variable, and m, n, and z are the coefficients.

$$mt^2 + nt = z$$
$$mt^2 + nt - z = 0$$

Write the polynomial in descending order by t.

$$a = m, b = n, c = -z$$

Identify the coefficients of each term.

$$t = \frac{-(n) \pm \sqrt{(n)^2 - 4(m)(-z)}}{2m}$$

Apply the quadratic formula.

$$t = \frac{-n \pm \sqrt{n^2 + 4mz}}{2m}$$

Simplify.

Skill Practice 15   Solve for p. $cp^2 - dp = k$

Answers

14. $v = \sqrt{\frac{2E}{m}}$ or $v = \frac{\sqrt{2Em}}{m}$

15. $p = \frac{d \pm \sqrt{d^2 + 4ck}}{2c}$

SECTION R.5 | Practice Exercises

Concept Connections

1. A _____quadratic_____ equation is a second-degree equation of the form $ax^2 + bx + c = 0$ where $a \neq 0$.

2. A _____linear_____ equation is a first-degree equation of the form $ax + b = 0$ where $a \neq 0$.

3. A _____conditional_____ equation is one that is true for some values of the variable and false for others.

4. An _____identity_____ is an equation that is true for all values of the variable for which the expressions in the equation are defined.

5. A _____contradiction_____ is an equation that is false for all values of the variable.

6. The square root property indicates that if $x^2 = k$, then $x = $ _____$\pm\sqrt{k}$_____.

7. Given $ax^2 + bx + c = 0$ ($a \neq 0$), write the quadratic formula. $x = \dfrac{-b \pm \sqrt{b^2 - 4ac}}{2a}$

8. A _____rational_____ equation is an equation in which each term contains a rational expression.

Objective 1: Solve Linear Equations in One Variable

For Exercises 9–20, solve the equation. (See Example 1)

9. $4 = 7 - 3(4t + 1)$ {0}

10. $11 = 7 - 2(5p - 2)$ {0}

11. $-6(v - 2) + 3 = 9 - (v + 4)$ {2}

12. $-5(u - 4) + 2 = 11 - (u - 3)$ {2}

13. $0.05y + 0.02(6000 - y) = 270$ {5000}

14. $0.06x + 0.04(10,000 - x) = 520$ {6000}

15. $2(5x - 6) = 4[x - 3(x - 10)]$ $\left\{\dfrac{22}{3}\right\}$

16. $4(y - 3) = 3[y + 2(y - 2)]$ {0}

17. $\dfrac{1}{2}w - \dfrac{3}{4} = \dfrac{2}{3}w + 2$ $\left\{-\dfrac{33}{2}\right\}$

18. $\dfrac{2}{5}p - \dfrac{3}{10} = \dfrac{7}{15}p - 1$ $\left\{\dfrac{21}{2}\right\}$

19. $\dfrac{n + 3}{4} - \dfrac{n - 2}{5} = \dfrac{n + 1}{10} - 1$ {41}

20. $\dfrac{t - 2}{3} - \dfrac{t + 7}{5} = \dfrac{t - 4}{10} + 2$ {110}

21. In the mid-nineteenth century, explorers used the boiling point of water to estimate altitude. The boiling temperature of water T (in °F) can be approximated by the model $T = -1.83a + 212$, where a is the altitude in thousands of feet.

   a. Determine the temperature at which water boils at an altitude of 4000 ft. Round to the nearest degree.   205°F

   b. Two campers hiking in Colorado boil water for tea. If the water boils at 193°F, approximate the altitude of the campers. Give the result to the nearest hundred feet.   10,400 ft

22. For a recent year, the cost C (in $) for tuition and fees for x credit-hours at a public college was given by $C = 167.95x + 94$.

   a. Determine the cost to take 9 credit-hours. $1605.55

   b. If Jenna spent $2445.30 for her classes, how many credit-hours did she take?   14 credit-hours

For Exercises 23–28, identify the equation as a conditional equation, a contradiction, or an identity. Then give the solution set. (See Example 2)

23. $2x - 3 = 4(x - 1) - 1 - 2x$ Contradiction; { }

24. $4(3 - 5n) + 1 = -4n - 8 - 16n$ Contradiction; { }

25. $-(6 - 2w) = 4(w + 1) - 2w - 10$ Identity; $\mathbb{R}$

26. $-5 + 3x = 3(x - 1) - 2$ Identity; $\mathbb{R}$

27. $\dfrac{1}{2}x + 3 = \dfrac{1}{4}x + 1$ Conditional equation; {−8}

28. $\dfrac{2}{3}y - 5 = \dfrac{1}{6}y - 4$ Conditional equation; {2}

Objective 2: Solve Quadratic Equations with Real Solutions

For Exercises 29–36, solve by applying the zero product property. (See Example 3)

29. $n^2 + 5n = 24$ {−8, 3}

30. $y^2 = 18 - 7y$ {−9, 2}

31. $8t(t + 3) = 2t - 5$ $\left\{-\dfrac{5}{2}, -\dfrac{1}{4}\right\}$

32. $6m(m + 4) = m - 15$ $\left\{-\dfrac{5}{6}, -3\right\}$

33. $3x^2 = 12x$ {0, 4}

34. $z^2 = 25z$ {0, 25}

35. $(m + 4)(m - 5) = -8$ {−3, 4}

36. $(n + 2)(n - 4) = 27$ {−5, 7}

For Exercises 37–42, solve by using the square root property. (See Example 4)

37. $x^2 = 81$ $\{9, -9\}$

38. $w^2 = 121$ $\{11, -11\}$

39. $5y^2 - 35 = 0$ $\{\sqrt{7}, -\sqrt{7}\}$

40. $6v^2 - 30 = 0$ $\{\sqrt{5}, -\sqrt{5}\}$

41. $(k + 2)^2 = 28$ $\{-2 \pm 2\sqrt{7}\}$

42. $3(z + 11)^2 - 10 = 110$ $\{-11 \pm 2\sqrt{10}\}$

For Exercises 43–48, determine the value of n that makes the polynomial a perfect square trinomial. Then factor as the square of a binomial.

43. $x^2 + 14x + n$ $n = 49; (x + 7)^2$

44. $y^2 + 22y + n$ $n = 121; (y + 11)^2$

45. $w^2 - 3w + n$ $n = \dfrac{9}{4}; \left(w - \dfrac{3}{2}\right)^2$

46. $v^2 - 11v + n$ $n = \dfrac{121}{4}; \left(v - \dfrac{11}{2}\right)^2$

47. $m^2 + \dfrac{2}{9}m + n$ $n = \dfrac{1}{81}; \left(m + \dfrac{1}{9}\right)^2$

48. $k^2 + \dfrac{2}{5}k + n$ $n = \dfrac{1}{25}; \left(k + \dfrac{1}{5}\right)^2$

For Exercises 49–54, solve by completing the square and applying the square root property. (See Example 5)

49. $y^2 + 22y - 4 = 0$ $\{-11 \pm 5\sqrt{5}\}$

50. $x^2 + 14x - 3 = 0$ $\{-7 \pm 2\sqrt{13}\}$

51. $2x(x - 3) = 4 + x$ $\left\{4, -\dfrac{1}{2}\right\}$

52. $5c(c - 2) = 6 + 3c$ $\left\{3, -\dfrac{2}{5}\right\}$

53. $-4y^2 - 12y + 5 = 0$ $\left\{-\dfrac{3}{2} \pm \dfrac{\sqrt{14}}{2}\right\}$

54. $-2x^2 - 14x + 5 = 0$ $\left\{-\dfrac{7}{2} \pm \dfrac{\sqrt{59}}{2}\right\}$

For Exercises 55–64, solve by using the quadratic formula. (See Example 6)

55. $x^2 - 3x - 7 = 0$ $\left\{\dfrac{3 \pm \sqrt{37}}{2}\right\}$

56. $x^2 - 5x - 9 = 0$ $\left\{\dfrac{5 \pm \sqrt{61}}{2}\right\}$

57. $(6x + 5)(x - 3) = -2x(7x + 5) + x - 12$ $\left\{\dfrac{1}{2}, -\dfrac{3}{10}\right\}$

58. $(5c + 7)(2c - 3) = -2c(c + 15) - 35$ $\left\{-\dfrac{2}{3}, -\dfrac{7}{4}\right\}$

59. $\dfrac{1}{2}x^2 - \dfrac{2}{7} = \dfrac{5}{14}x$ $\left\{\dfrac{5 \pm \sqrt{137}}{14}\right\}$

60. $\dfrac{1}{3}x^2 - \dfrac{7}{6} = \dfrac{3}{2}x$ $\left\{\dfrac{9 \pm \sqrt{137}}{4}\right\}$

61. $0.4y^2 = 2y - 2.5$ $\left\{\dfrac{5}{2}\right\}$

62. $0.09n^2 = 0.42n - 0.49$ $\left\{\dfrac{7}{3}\right\}$

63. $m^2 + 4m = -2$ $\{-2 \pm \sqrt{2}\}$

64. $n^2 + 8n = -3$ $\{-4 \pm \sqrt{13}\}$

Objective 3: Solve Rational Equations

For Exercises 65–66, determine the restrictions on x.

65. $\dfrac{3}{x - 5} + \dfrac{2}{x + 4} = \dfrac{5}{7}$ $x \neq 5, x \neq -4$

66. $\dfrac{2}{x + 1} - \dfrac{5}{x - 7} = \dfrac{2}{3}$ $x \neq -1, x \neq 7$

For Exercises 67–84, solve the equation. (See Examples 7–8)

67. $\dfrac{1}{2} - \dfrac{7}{2y} = \dfrac{5}{y}$ $\{17\}$

68. $\dfrac{1}{3} - \dfrac{4}{3t} = \dfrac{7}{t}$ $\{25\}$

69. $\dfrac{w + 3}{4w} + 1 = \dfrac{w - 5}{w}$ $\{-23\}$

70. $\dfrac{x + 2}{6x} + 1 = \dfrac{x - 7}{x}$ $\{-44\}$

71. $\dfrac{c}{c - 3} = \dfrac{3}{c - 3} - \dfrac{3}{4}$ $\{\ \}$; The value 3 does not check.

72. $\dfrac{7}{d - 7} - \dfrac{7}{8} = \dfrac{d}{d - 7}$ $\{\ \}$; The value 7 does not check.

73. $\dfrac{2}{x - 5} - \dfrac{1}{x + 5} = \dfrac{11}{x^2 - 25}$ $\{-4\}$

74. $\dfrac{2}{c + 3} - \dfrac{1}{c - 3} = \dfrac{10}{c^2 - 9}$ $\{19\}$

75. $\dfrac{-14}{x^2 - x - 12} - \dfrac{1}{x - 4} = \dfrac{2}{x + 3}$ $\{\ \}$; The value -3 does not check.

76. $\dfrac{2}{x^2 + 5x + 6} - \dfrac{2}{x + 2} = \dfrac{1}{x + 3}$ $\{\ \}$; The value -2 does not check.

77. $\dfrac{5}{x^2 - x - 2} - \dfrac{2}{x^2 - 4} = \dfrac{4}{x^2 + 3x + 2}$ $\{16\}$

78. $\dfrac{4}{x^2 - 2x - 8} - \dfrac{1}{x^2 - 16} = \dfrac{2}{x^2 + 6x + 8}$ $\{-22\}$

79. $\dfrac{3x}{x + 2} - \dfrac{5}{x - 4} = \dfrac{2x^2 - 14x}{x^2 - 2x - 8}$ $\{5\}$; The value -2 does not check.

80. $\dfrac{4c}{c - 5} - \dfrac{1}{c + 1} = \dfrac{3c^2 + 3}{c^2 - 4c - 5}$ $\{-2\}$; The value -1 does not check.

81. $\dfrac{m}{2m + 1} + 1 = \dfrac{2}{m - 3}$ $\left\{\dfrac{6 \pm \sqrt{51}}{3}\right\}$

82. $7 + \dfrac{20}{z} = \dfrac{3}{z^2}$ $\left\{\dfrac{1}{7}, -3\right\}$

83. $\dfrac{18}{m^2 - 3m} + 2 = \dfrac{6}{m - 3}$ $\{\ \}$; The value 3 does not check.

84. $\dfrac{48}{m^2 - 4m} + 3 = \dfrac{12}{m - 4}$ $\{\ \}$; The value 4 does not check.

Objective 4: Solve Absolute Value Equations

For Exercises 85–102, solve the equations. (See Examples 9 and 10)

85. a. $|p| = 6$ $\{6, -6\}$ **86. a.** $|w| = 2$ $\{2, -2\}$ **87. a.** $|x - 3| = 4$ $\{7, -1\}$ **88. a.** $|m + 1| = 5$ $\{4, -6\}$
b. $|p| = 0$ $\{0\}$ **b.** $|w| = 0$ $\{0\}$ **b.** $|x - 3| = 0$ $\{3\}$ **b.** $|m + 1| = 0$ $\{-1\}$
c. $|p| = -6$ $\{\}$ **c.** $|w| = -2$ $\{\}$ **c.** $|x - 3| = -7$ $\{\}$ **c.** $|m + 1| = -1$ $\{\}$

89. $2|3x - 4| + 7 = 9$ $\left\{\frac{5}{3}, 1\right\}$ **90.** $4|2t + 7| + 2 = 22$ $\{-1, -6\}$ **91.** $-3 = -|c - 7| + 1$ $\{11, 3\}$

92. $-4 = -|z + 8| - 3$ $\{-7, -9\}$ **93.** $2 = 8 + |11y + 4|$ $\{\}$ **94.** $6 = 7 + |9z - 3|$ $\{\}$

95. $|3y + 5| = |y + 1|$ $\left\{-2, -\frac{3}{2}\right\}$ **96.** $|2a - 3| = |a + 2|$ $\left\{5, \frac{1}{3}\right\}$ **97.** $\left|\frac{1}{4}w\right| = |4w|$ $\{0\}$

98. $|3z| = \left|\frac{1}{3}z\right|$ $\{0\}$ **99.** $|x + 4| = |x - 7|$ $\left\{\frac{3}{2}\right\}$ **100.** $|k - 3| = |k + 3|$ $\{0\}$

101. $|2p - 1| = |1 - 2p|$ $\mathbb{R}$ **102.** $|4d - 3| = |3 - 4d|$ $\mathbb{R}$

Objective 5: Solve Radical Equations

For Exercises 103–122, solve the equation. (See Examples 11–13)

103. $1 = -2 + \sqrt{2x + 7}$ $\{1\}$ **104.** $6 = 9 + \sqrt{5 - 3x}$ $\{\}$ **105.** $\sqrt{m + 18} + 2 = m$ $\{7\}$; The value -2 does not check.

106. $\sqrt{2n + 29} + 3 = n$ $\{10\}$; The value -2 does not check. **107.** $-4\sqrt[3]{2x - 5} + 6 = 10$ $\{2\}$ **108.** $-3\sqrt[5]{4x - 1} + 2 = 8$ $\left\{-\frac{31}{4}\right\}$

109. $\sqrt{8 - p} - \sqrt{p + 5} = 1$ $\{-1\}$; The value 4 does not check. **110.** $\sqrt{d + 4} - \sqrt{6 + 2d} = -1$ $\{5\}$; The value -3 does not check. **111.** $3 - \sqrt{y + 3} = \sqrt{2 - y}$ $\{-2, 1\}$

112. $\sqrt{k - 2} = \sqrt{2k + 3} - 2$ $\{3, 11\}$ **113.** $2(x + 5)^{2/3} = 18$ $\{-32, 22\}$ **114.** $3(x - 6)^{2/3} = 48$ $\{-58, 70\}$

115. $(3x + 1)^{3/2} + 2 = 66$ $\{5\}$ **116.** $(2x - 1)^{3/2} - 3 = 122$ $\{13\}$ **117.** $m^{3/4} = 5$ $\{5\sqrt[3]{5}\}$ or $\{5^{4/3}\}$

118. $n^{5/6} = 3$ $\{3\sqrt[5]{3}\}$ or $\{3^{6/5}\}$ **119.** $2p^{4/5} = \frac{1}{8}$ $\left\{\pm\frac{1}{32}\right\}$ **120.** $5t^{2/3} = \frac{1}{5}$ $\left\{\pm\frac{1}{125}\right\}$

121. $(2v + 7)^{1/3} - (v - 3)^{1/3} = 0$ $\{-10\}$ **122.** $(5u - 6)^{1/5} - (3u + 1)^{1/5} = 0$ $\left\{\frac{7}{2}\right\}$

Objective 6: Solve Equations for a Specified Variable

For Exercises 123–152, solve for the specified variable. (See Examples 14–15)

123. $A = lw$ for l $\quad l = \frac{A}{w}$ **124.** $E = IR$ for R $\quad R = \frac{E}{I}$ **125.** $P = a + b + c$ for c $\quad c = P - a - b$

126. $W = K - T$ for K $\quad K = W + T$ **127.** $7x + 2y = 8$ for y $\quad y = -\frac{7}{2}x + 4$ **128.** $3x + 5y = 15$ for y $\quad y = -\frac{3}{5}x + 3$

129. $5x - 4y = 2$ for y $\quad y = \frac{5}{4}x - \frac{1}{2}$ **130.** $7x - 2y = 5$ for y $\quad y = \frac{7}{2}x - \frac{5}{2}$ **131.** $S = \frac{n}{2}(a + d)$ for d

132. $S = \frac{n}{2}[2a + (n - 1)d]$ for a **133.** $6 = 4x + tx$ for x $\quad x = \frac{6}{4 + t}$ **134.** $8 = 3x + kx$ for x $\quad x = \frac{8}{3 + k}$

135. $6x + ay = bx + 5$ for x **136.** $3x + 2y = cx + d$ for x **137.** $A = \pi r^2$ for $r > 0$

138. $V = \pi r^2 h$ for $r > 0$ **139.** $a^2 + b^2 = c^2$ for $a > 0$ $\quad a = \sqrt{c^2 - b^2}$ **140.** $a^2 + b^2 + c^2 = d^2$ for $c > 0$ $\quad c = \sqrt{d^2 - a^2 - b^2}$

141. $kw^2 - cw = r$ for w **142.** $dy^2 + my = p$ for y **143.** $s = v_0 t + \frac{1}{2}at^2$ for t

144. $S = 2\pi rh + \pi r^2 h$ for r **145.** $\frac{1}{f} = \frac{1}{p} + \frac{1}{q}$ for p $\quad p = \frac{fq}{q - f}$ **146.** $\frac{1}{R} = \frac{1}{R_1} + \frac{1}{R_2} + \frac{1}{R_3}$ for R_3

147. $16 + \sqrt{x^2 - y^2} = z$ for x $\quad x = \pm\sqrt{(z - 16)^2 + y^2}$ **148.** $4 + \sqrt{x^2 + y^2} = z$ for y $\quad y = \pm\sqrt{(z - 4)^2 - x^2}$ **149.** $\frac{P_1 V_1}{T_1} = \frac{P_2 V_2}{T_2}$ for T_1 $\quad T_1 = \frac{P_1 V_1 T_2}{P_2 V_2}$

150. $\frac{t_1}{s_1 v_1} = \frac{t_2}{s_2 v_2}$ for v_2 $\quad v_2 = \frac{t_2 s_1 v_1}{s_2 t_1}$ **151.** $T = 2\pi\sqrt{\frac{L}{g}}$ for g $\quad g = \frac{4\pi^2 L}{T^2}$ **152.** $t = \sqrt{\frac{2s}{g}}$ for s $\quad s = \frac{t^2 g}{2}$

Mixed Exercises

For Exercises 153–156, solve the equation. (*Hint:* Use the zero product property.)

153. $-3x(2x - 1)(x + 6)^2 = 0$ $\left\{0, \frac{1}{2}, -6\right\}$

154. $5y(3 - y)(4y + 1)^2 = 0$ $\left\{0, 3, -\frac{1}{4}\right\}$

155. $98t^3 - 49t^2 - 8t + 4 = 0$ $\left\{\pm\frac{2}{7}, \frac{1}{2}\right\}$

156. $2m^3 + 3m^2 = 9(2m + 3)$ $\left\{-3, -\frac{3}{2}, 3\right\}$

Write About It

157. Explain why the value 5 is not a solution to the equation $\frac{x}{x - 5} + \frac{1}{5} = \frac{5}{x - 5}$.

158. Explain why the value 2 is not the only solution to the equation $2x + 4 = 2(x - 3) + 10$.

Expanding Your Skills

For Exercises 159–160, solve for the indicated variable.

159. $x^2 - xy - 2y^2 = 0$ for x $x = 2y$ or $x = -y$

160. $3a^2 + 2ab - b^2 = 0$ for a $a = \frac{b}{3}$ or $a = -b$

For Exercises 161–166, write an equation with integer coefficients and the variable x that has the given solution set.
[*Hint:* Apply the zero product property in reverse. For example, to build an equation whose solution set is $\left\{2, -\frac{5}{2}\right\}$ we have $(x - 2)(2x + 5) = 0$, or simply $2x^2 + x - 10 = 0$.]

161. $\{4, -2\}$ $x^2 - 2x - 8 = 0$ **162.** $\{7, -1\}$ $x^2 - 6x - 7 = 0$ **163.** $\left\{\frac{2}{3}, \frac{1}{4}\right\}$ $12x^2 - 11x + 2 = 0$ **164.** $\left\{\frac{3}{5}, \frac{1}{7}\right\}$ $35x^2 - 26x + 3 = 0$

165. $\left\{\sqrt{5}, -\sqrt{5}\right\}$ $x^2 - 5 = 0$ **166.** $\left\{\sqrt{2}, -\sqrt{2}\right\}$ $x^2 - 2 = 0$

Additional answers can be found in the Instructor Answer Appendix.

SECTION R.6   Complex Numbers and More Quadratic Equations

OBJECTIVES

1. Simplify Imaginary Numbers
2. Perform Operations on Complex Numbers
3. Solve Quadratic Equations over the Set of Complex Numbers
4. Use the Discriminant
5. Solve Equations in Quadratic Form

1. Simplify Imaginary Numbers

In our study of algebra thus far, we have worked exclusively with real numbers. However, as we encounter new types of equations, we need to look outside the set of real numbers to find solutions. For example, the equation $x^2 = 1$ has two solutions: 1 and -1. But what about the equation $x^2 = -1$? There is no real number x for which $x^2 = -1$. For this reason, mathematicians defined a new number i such that $i^2 = -1$. The number i is called an *imaginary number* and is used to represent $\sqrt{-1}$. Furthermore, the square root of any negative real number is an imaginary number that can be expressed in terms of i.

> **The Imaginary Number i**
> - $i = \sqrt{-1}$ and $i^2 = -1$
> - If b is a positive real number, then $\sqrt{-b} = i\sqrt{b}$.

Classroom Examples: p. 80
Exercises 8, 12, 16

EXAMPLE 1   **Writing Imaginary Numbers in Terms of i**

Write each expression in terms of i.

a. $\sqrt{-25}$ b. $\sqrt{-12}$ c. $\sqrt{-13}$

Solution:

a. $\sqrt{-25} = i\sqrt{25} = 5i$

b. $\sqrt{-12} = i\sqrt{12} = i \cdot 2\sqrt{3}$ The value $i \cdot 2\sqrt{3}$ can be written as $2i\sqrt{3}$ or as
   $\quad\quad = 2i\sqrt{3}$ or $2\sqrt{3}i$ $2\sqrt{3}i$. Note, however, that the factor i is written
                                                  outside the radical.

c. $\sqrt{-13} = i\sqrt{13}$ or $\sqrt{13}i$

Skill Practice 1   Write each expression in terms of i.

a. $\sqrt{-81}$ b. $\sqrt{-50}$ c. $\sqrt{-11}$

It is important to note that the multiplication and division properties of radicals can be used only if the radicals represent real-valued expressions.

$$\sqrt{a} \cdot \sqrt{b} = \sqrt{ab} \quad \text{provided that the roots represent real numbers.}$$

$$\frac{\sqrt{a}}{\sqrt{b}} = \sqrt{\frac{a}{b}} \quad \text{provided that the roots represent real numbers.}$$

For this reason, we write the radical expressions in terms of i first, before applying the multiplication or division property of radicals. For example,

$$\sqrt{-9} \cdot \sqrt{-25} = i\sqrt{9} \cdot i\sqrt{25} \qquad \text{Write each radical in terms of } i \text{ first,}$$
$$\qquad\qquad\qquad\qquad\qquad\qquad before \text{ multiplying.}$$
$$= 3i \cdot 5i \qquad\qquad \text{Simplify the radicals.}$$
$$= 15i^2 \qquad\qquad \text{Multiply.}$$
$$= 15(-1) \qquad\quad \text{By definition, } i^2 = -1.$$
$$= -15$$

$$\frac{\sqrt{-50}}{\sqrt{-2}} = \frac{i\sqrt{50}}{i\sqrt{2}} = \frac{\sqrt{50}}{\sqrt{2}} = \sqrt{\frac{50}{2}} = \sqrt{25} = 5$$

By definition, $i^2 = -1$, but what about other powers of i? Consider the following pattern.

TIP Notice that even powers of i simplify to 1 or −1.

- If the exponent is a multiple of 4, then the expression equals 1.
- If the exponent is even but *not* a multiple of 4, then the expression equals −1.

$i^1 = i$ $i^1 = i$
$i^2 = -1$ $i^2 = -1$
$i^3 = i^2 \cdot i = (-1)i = -i$ $i^3 = -i$ Pattern: $i, -1, -i, 1$
$i^4 = i^2 \cdot i^2 = (-1)(-1) = 1$ $i^4 = 1$
$i^5 = i^4 \cdot i = (1)i = i$ $i^5 = i$
$i^6 = i^4 \cdot i^2 = (1)(-1) = -1$ $i^6 = -1$ Pattern repeats: $i, -1, -i, 1, \ldots$
$i^7 = i^4 \cdot i^2 \cdot i = (1)(-1)i = -i$ $i^7 = -i$
$i^8 = i^4 \cdot i^4 = (1)(1) = 1$ $i^8 = 1$

Notice that the fourth powers of i (i^4, i^8, i^{12}, ...) equal the real number 1. For other powers of i, we can write the expression as a product of a fourth power of i and a factor of i, i^2, or i^3, which equals i, -1, or $-i$, respectively.

Answers

1. a. $9i$ b. $5i\sqrt{2}$ c. $i\sqrt{11}$

Classroom Example: p. 80
Exercise 28

EXAMPLE 2   **Simplifying Powers of *i***

Simplify.

   a. i^{48} **b.** i^{23} **c.** i^{50} **d.** i^{-19}

Solution:

   a. $i^{48} = (i^4)^{12} = (1)^{12} = 1$ The exponent 48 is a multiple of 4. Thus, i^{48} is equal to 1.

   b. $i^{23} = i^{20} \cdot i^3$ Write i^{23} as a product of the largest fourth power of *i* and a remaining factor.
$\quad\quad = (1) \cdot i^3 = -i$

   c. $i^{50} = i^{48} \cdot i^2$
$\quad\quad = (1)(-1) = -1$

   d. $i^{-19} = i^{-20} \cdot i^1$
$\quad\quad = (i^4)^{-5} \cdot i = (1) \cdot i = i$

Skill Practice 2   Simplify.

   a. i^{13} **b.** i^{103} **c.** i^{64} **d.** i^{-30}

TIP To simplify i^n, divide the exponent, *n*, by 4. The remainder is the exponent of the remaining factor of *i* once the fourth power of *i* has been extracted.

Example: $i^{50} = i^{48} \cdot i^2 = (1)i^2$

$$\begin{array}{r} 12 \\ 4\overline{)50} \\ \underline{48} \\ 2 \end{array}$$

So $i^{50} = (1) \cdot i^2 = -1$

2. Perform Operations on Complex Numbers

We now define a new set of numbers that includes the real numbers and imaginary numbers. This is called the set of complex numbers.

Complex Numbers

Given real numbers *a* and *b*, a number written in the form $a + bi$ is called a **complex number.** The value *a* is called the **real part** of the complex number and the value *b* is called the **imaginary part.**

$$\boxed{\text{Real part: 5}} \quad \boxed{\text{Imaginary part: } -7}$$
$$5 - 7i = 5 + (-7)i$$

Notes	Examples
• If $b = 0$, then $a + bi$ equals the real number *a*. This tells us that all real numbers are complex numbers.	$4 + 0i$ is generally written as the real number 4.
• If $a = 0$ and $b \neq 0$, then $a + bi$ equals bi, which we say is **pure imaginary.**	The number $0 + 8i$ is a pure imaginary number and is generally written as simply $8i$.

A complex number written in the form $a + bi$ is said to be in **standard form.** That being said, we sometimes write $a - bi$ in place of $a + (-b)i$. Furthermore, a number such as $5 + \sqrt{3}i$ is sometimes written as $5 + i\sqrt{3}$ to emphasize that the factor of *i* is not under the radical.

To add or subtract complex numbers, add or subtract their real parts, and add or subtract their imaginary parts. That is,

$$(a + bi) + (c + di) = (a + c) + (b + d)i$$
$$(a + bi) - (c + di) = (a - c) + (b - d)i$$

Answers

2. a. *i*   **b.** $-i$ **c.** 1   **d.** -1

Classroom Example: p. 80
Exercise 32

EXAMPLE 3   **Adding and Subtracting Complex Numbers**

Perform the indicated operations. Write the answer in the form $a + bi$.

$(-2 - 4i) + (5 + 2i) - (3 - 6i)$

Solution:

$(-2 - 4i) + (5 + 2i) - (3 - 6i)$ Combine the real parts and combine the

$= (-2 + 5 - 3) + [-4 + 2 - (-6)]i$ imaginary parts.

$= 0 + 4i$ Write the result in the form $a + bi$.

Skill Practice 3   Perform the indicated operations. Write the answer in the form $a + bi$.

$(8 - 3i) - (2 + 4i) + (5 + 7i)$

In Examples 4 and 5, we multiply complex numbers using a process similar to multiplying polynomials.

Classroom Example: p. 80
Exercise 40

EXAMPLE 4   **Multiplying Complex Numbers**

Multiply. Write the result in the form $a + bi$. $(-2 + 6i)(4 - 3i)$

Solution:

$(-2 + 6i)(4 - 3i)$ Apply the distributive property.

$= -2(4) + (-2)(-3i) + 6i(4) + 6i(-3i)$

$= -8 + 6i + 24i - 18i^2$

$= -8 + 30i - 18(-1)$ Recall that $i^2 = -1$.

$= -8 + 30i + 18$

$= 10 + 30i$ Write the result in the form $a + bi$.

Skill Practice 4   Multiply. Write the result in the form $a + bi$.

$(-5 + 4i)(3 - i)$

In Example 5, we make use of the special case products:

$$(a \pm b)^2 = a^2 \pm 2ab + b^2 \quad \text{and} \quad (a + b)(a - b) = a^2 - b^2$$

Classroom Examples: p. 80
Exercises 42, 48

EXAMPLE 5   **Evaluating Special Products with Complex Numbers**

Multiply. Write the results in the form $a + bi$.

   a. $(3 + 4i)^2$ **b.** $(5 + 2i)(5 - 2i)$

Answers

3. $11 + 0i$

4. $-11 + 17i$

Solution:

a. $(3 + 4i)^2 = (3)^2 + 2(3)(4i) + (4i)^2$ Apply the property
$(a + b)^2 = a^2 + 2ab + b^2$.

$= 9 + 24i + 16i^2$

$= 9 + 24i + 16(-1)$

$= 9 + 24i - 16$

$= -7 + 24i$ Write the result in the form $a + bi$.

b. $(5 + 2i)(5 - 2i) = (5)^2 - (2i)^2$ Apply the property
$(a + b)(a - b) = a^2 - b^2$.

$= 25 - 4i^2$

$= 25 - 4(-1)$

$= 25 + 4$

$= 29$ or $29 + 0i$ Write the result in the form $a + bi$.

Skill Practice 5   Multiply. Write the result in the form $a + bi$.

a. $(4 - 7i)^2$ **b.** $(10 - 3i)(10 + 3i)$

In Section R.3 we noted that the expressions $(a + b)$ and $(a - b)$ are conjugates. Similarly, the expressions $(a + bi)$ and $(a - bi)$ are called **complex conjugates.** Furthermore, as illustrated in Example 5(b), the product of complex conjugates is a real number.

$$(a + bi)(a - bi) = (a)^2 - (bi)^2$$
$$= a^2 - b^2i^2$$
$$= a^2 - b^2(-1)$$
$$= a^2 + b^2$$

Product of Complex Conjugates

If a and b are real numbers, then $(a + bi)(a - bi) = a^2 + b^2$.

In Example 6, we demonstrate division of complex numbers such as $\frac{8 + 2i}{3 - 5i}$. The goal is to make the denominator a real number so that the quotient can be written in standard form $a + bi$. This can be accomplished by multiplying the denominator by its complex conjugate. Of course, this means that we must also multiply the numerator by the same quantity.

Classroom Example: p. 80
Exercise 54

EXAMPLE 6   **Dividing Complex Numbers**

Divide. Write the result in the form $a + bi$. $\dfrac{8 + 2i}{3 - 5i}$

Solution:

$$\frac{8 + 2i}{3 - 5i} = \frac{(8 + 2i) \cdot (3 + 5i)}{(3 - 5i) \cdot (3 + 5i)}$$ Multiply numerator and denominator by the conjugate of the denominator.

$$= \frac{24 + 40i + 6i + 10i^2}{(3)^2 + (5)^2}$$ Apply the distributive property in the numerator.
Multiply conjugates in the denominator.

$$= \frac{24 + 46i + 10(-1)}{9 + 25}$$ Replace i^2 by -1.

$$= \frac{14 + 46i}{34}$$

$$= \frac{14}{34} + \frac{46}{34}i = \frac{7}{17} + \frac{23}{17}i$$ Write the result in the form $a + bi$.

Skill Practice 6   Divide. Write the result in the form $a + bi$.

$$\frac{5 + 6i}{2 - 7i}$$

3. Solve Quadratic Equations over the Set of Complex Numbers

In Examples 7–9, we solve quadratic equations over the set of complex numbers. This means that in addition to having real solutions to quadratic equations, we now consider solutions of the form $a + bi$.

Classroom Example: p. 81
Exercises 58

EXAMPLE 7   Solving a Quadratic Equation with Imaginary Solutions

Solve the equation by applying the square root property. $2y^2 + 36 = 0$

Solution:

$$2y^2 + 36 = 0$$

$$2y^2 = -36$$ Isolate the square term.

$$y^2 = -18$$ Write the equation in the form $y^2 = k$.

$$y = \pm\sqrt{-18}$$ Apply the square root property.

$$y = \pm 3i\sqrt{2}$$ Simplify the radical.

$$\sqrt{-18} = i\sqrt{18} = i\sqrt{3^2 \cdot 2} = 3i\sqrt{2}$$

The solution set is $\{\pm 3i\sqrt{2}\}$.   Both solutions check in the original equation.

TIP The solutions to the equation in Example 7 are written concisely as $\pm 3i\sqrt{2}$. Do not forget that this actually represents two solutions:

$$y = 3i\sqrt{2} \quad \text{and}$$
$$y = -3i\sqrt{2}$$

Skill Practice 7   Solve the equation by applying the square root property.

$$2c^2 + 80 = 0$$

Point of Interest

Unfortunate names? In the long history of mathematics, number systems have been expanded to accommodate meaningful solutions to equations. But the negative connotation of their names may suggest a reluctance by early mathematicians to accept these new concepts. Negative numbers for example are not unpleasant or disagreeable. Irrational numbers are not illogical or absurd, and imaginary numbers are not "fake." Instead these sets of numbers are necessary to render solutions to such equations as

$$2x + 10 = 0 \qquad x^2 - 5 = 0 \qquad x^2 + 4 = 0$$

Answers

6. $-\dfrac{32}{53} + \dfrac{47}{53}i$ **7.** $\{\pm 2i\sqrt{10}\}$

Classroom Example: p. 81
Exercise 70

EXAMPLE 8   **Using the Quadratic Formula**

Solve the equation by applying the quadratic formula. $\dfrac{3}{10}x^2 - \dfrac{2}{5}x + \dfrac{7}{10} = 0$

Solution:

$$\dfrac{3}{10}x^2 - \dfrac{2}{5}x + \dfrac{7}{10} = 0$$

The equation is in the form $ax^2 + bx + c = 0$.

$$10\left(\dfrac{3}{10}x^2 - \dfrac{2}{5}x + \dfrac{7}{10}\right) = 10(0)$$

Multiply by 10 to clear fractions.

$$3x^2 - 4x + 7 = 0$$

$$a = 3, \; b = -4, \; c = 7$$

Identify the values of a, b, and c.

$$x = \dfrac{-(-4) \pm \sqrt{(-4)^2 - 4(3)(7)}}{2(3)}$$

Apply the quadratic formula.
$$x = \dfrac{-b \pm \sqrt{b^2 - 4ac}}{2a}$$

$$= \dfrac{4 \pm \sqrt{-68}}{6}$$

Simplify.

$$= \dfrac{4 \pm 2i\sqrt{17}}{6}$$

Simplify the radical.

$$= \dfrac{2\left(2 \pm i\sqrt{17}\right)}{2 \cdot 3}$$

Factor the numerator and denominator.

$$= \dfrac{2 \pm i\sqrt{17}}{3}$$

Simplify the fraction.

$$= \dfrac{2}{3} \pm \dfrac{\sqrt{17}}{3}i$$

Write the solutions in standard form, $a + bi$.

The solution set is $\left\{\dfrac{2}{3} \pm \dfrac{\sqrt{17}}{3}i\right\}$.

The solutions both check in the original equation.

Skill Practice 8   Solve the equation by applying the quadratic formula.

$$\dfrac{5}{12}x^2 - \dfrac{1}{2}x + \dfrac{1}{4} = 0$$

4. Use the Discriminant

The solutions to a quadratic equation are given by $x = \dfrac{-b \pm \sqrt{b^2 - 4ac}}{2a}$. The radicand, $b^2 - 4ac$, is called the *discriminant*. The value of the discriminant tells us the number and type of solutions to the equation. We examine three different cases.

Answer

8. $\left\{\dfrac{3}{5} \pm \dfrac{\sqrt{6}}{5}i\right\}$

Using the Discriminant to Determine the Number and Type of Solutions to a Quadratic Equation

Given a quadratic equation $ax^2 + bx + c = 0$ $(a \neq 0)$, the quantity $b^2 - 4ac$ is called the **discriminant**.

Discriminant $b^2 - 4ac$	Number and Type of Solutions	Examples	Result of Quadratic Formula
$b^2 - 4ac < 0$	2 nonreal solutions	$2x^2 - 3x + 5 = 0$ $b^2 - 4ac = (-3)^2 - 4(2)(5)$ $= -31$	$x = \dfrac{3 \pm \sqrt{-31}}{4}$
$b^2 - 4ac = 0$	1 real solution	$x^2 + 6x + 9 = 0$ $b^2 - 4ac = (6)^2 - 4(1)(9)$ $= 0$	$x = \dfrac{-6 \pm \sqrt{0}}{2} = -3$
$b^2 - 4ac > 0$	2 real solutions	$2x^2 + 7x - 1 = 0$ $b^2 - 4ac = (7)^2 - 4(2)(-1)$ $= 57$	$x = \dfrac{-7 \pm \sqrt{57}}{4}$

Classroom Examples: p. 81
Exercises 74, 78, 80

EXAMPLE 9   Using the Discriminant

Use the discriminant to determine the number and type of solutions for each equation.

a. $5x^2 - 3x + 1 = 0$ **b.** $2x^2 = 3 - 6x$ **c.** $4x^2 + 12x = -9$

Solution:

Equation	$b^2 - 4ac$	Solution Type and Number
a. $5x^2 - 3x + 1 = 0$	$(-3)^2 - 4(5)(1)$ $= -11$	Because $-11 < 0$, there are two nonreal solutions.
b. $2x^2 = 3 - 6x$ $\quad 2x^2 + 6x - 3 = 0$	$(6)^2 - 4(2)(-3)$ $= 60$	Because $60 > 0$, there are two real solutions.
c. $4x^2 + 12x = -9$ $\quad 4x^2 + 12x + 9 = 0$	$(12)^2 - 4(4)(9)$ $= 0$	Because the discriminant is 0, there is one real solution.

Skill Practice 9   Use the discriminant to determine the number and type of solutions for each equation.

a. $2x^2 - 4x + 5 = 0$ **b.** $25x^2 = 10x - 1$ **c.** $x^2 + 10x = -9$

5. Solve Equations in Quadratic Form

We have learned to solve quadratic equations by applying the quadratic formula or by completing the square and applying the square root property. This is particularly important because many other equations are quadratic in form. That is, with a simple substitution, these equations can be expressed as quadratic equations in a new variable. For example:

Answers
9. a. Discriminant: -24
      (2 nonreal solutions)
  b. Discriminant: 0
      (1 real solution)
  c. Discriminant: 64
      (2 real solutions)

Classroom Example: p. 81
Exercise 86

TIP For an equation written in quadratic form, notice that the expression for u is taken to be the variable expression from the middle term.

Equation in Quadratic Form   **New Equation**

$$\left(2 + \frac{3}{x}\right)^2 - \left(2 + \frac{3}{x}\right) - 12 = 0 \quad \xrightarrow{\text{Let } u = 2 + \frac{3}{x}} \quad u^2 - u - 12 = 0$$

$$2w^{2/3} - 3w^{1/3} - 20 = 0 \quad \xrightarrow{\text{Let } u = w^{1/3}} \quad 2u^2 - 3u - 20 = 0$$

The equations on the right are quadratic and easily solved. Then using back substitution, we can solve for the original variable.

EXAMPLE 10   Solving an Equation in Quadratic Form

Solve. $(2x^2 - 3)^2 + 36(2x^2 - 3) + 35 = 0$

Solution:

$(2x^2 - 3)^2 + 36(2x^2 - 3) + 35 = 0$	The equation is in quadratic form.
$u^2 + 36u + 35 = 0$	Let $u = 2x^2 - 3$.
$(u + 35)(u + 1) = 0$	Set one side equal to zero and factor the other side.
$u = -35$ or $u = -1$	Apply the zero product property.
$2x^2 - 3 = -35$ or $2x^2 - 3 = -1$	Back substitute. Replace u by $2x^2 - 3$.
$2x^2 = -32$ or $2x^2 = 2$	
$x^2 = -16$ or $x^2 = 1$	Isolate the square term.
$x = \pm 4i$ or $x = \pm 1$	Apply the square root property.

The solution set is $\{\pm 4i, \pm 1\}$.   The solutions all check in the original equation.

Skill Practice 10   Solve. $(x^2 - 6)^2 + 33(x^2 - 6) + 62 = 0$

Classroom Example: p. 81
Exercise 92

EXAMPLE 11   Solving an Equation in Quadratic Form

Solve. $2w^{2/3} = 3w^{1/3} + 20$

Solution:

$2w^{2/3} = 3w^{1/3} + 20$	Set one side equal to zero, and write the expression on the left in descending order.
$2w^{2/3} - 3w^{1/3} - 20 = 0$	
$2(w^{1/3})^2 - 3(w^{1/3}) - 20 = 0$	The equation is in quadratic form.
$2u^2 - 3u - 20 = 0$	Let $u = w^{1/3}$.
$(2u + 5)(u - 4) = 0$	Factor.
$u = -\dfrac{5}{2}$ or $u = 4$	Apply the zero product property.
$w^{1/3} = -\dfrac{5}{2}$ or $w^{1/3} = 4$	Back substitute. Replace u by $w^{1/3}$.
$(w^{1/3})^3 = \left(-\dfrac{5}{2}\right)^3$ or $(w^{1/3})^3 = (4)^3$	Cube both sides.
$w = -\dfrac{125}{8}$ or $w = 64$	Both solutions check in the original equation.

The solution set is $\left\{-\dfrac{125}{8}, 64\right\}$.

TIP Consider the equation from Example 11:

$2w^{2/3} - 3w^{1/3} - 20 = 0$

As an alternative to using substitution, the expression on the left can be factored directly.

$(2w^{1/3} + 5)(w^{1/3} - 4) = 0$

Applying the zero product property results in the same solutions.

Answers

10. $\{\pm 5i, \pm 2\}$ **11.** $\left\{-125, \dfrac{27}{8}\right\}$

Skill Practice 11   Solve. $2t^{2/3} = 15 - 7t^{1/3}$

SECTION R.6   Practice Exercises

Concept Connections

1. The imaginary number i is defined so that $i = \sqrt{-1}$ and $i^2 =$ _____ -1 .

2. For a positive real number b, the value $\sqrt{-b} =$ _____ $i\sqrt{b}$.

3. Given a complex number $a + bi$, the value of a is called the _____ real _____ part and the value of b is called the _____ imaginary _____ part.

4. Given a complex number $a + bi$, the expression $a - bi$ is called the complex _____ conjugate _____ .

5. The equation $m^{2/3} + 10m^{1/3} + 9 = 0$ is said to be in _____ quadratic _____ form, because making the substitution $u =$ _____ $m^{1/3}$ _____ results in a new equation that is quadratic.

6. Consider the equation $(4x^2 + 1)^2 + 4(4x^2 + 1) + 4 = 0$. If the substitution $u =$ _____ $4x^2 + 1$ _____ is made, then the equation becomes $u^2 + 4u + 4 = 0$.

Objective 1: Simplify Imaginary Numbers

For Exercises 7–18, write each expression in terms of i and simplify. (See Example 1)

7. $\sqrt{-121}$ $11i$

8. $\sqrt{-100}$ $10i$

9. $\sqrt{-98}$ $7i\sqrt{2}$

10. $\sqrt{-63}$ $3i\sqrt{7}$

11. $\sqrt{-4}\sqrt{-9}$ -6

12. $\sqrt{-1}\sqrt{-36}$ -6

13. $\sqrt{-10}\sqrt{-5}$ $-5\sqrt{2}$

14. $\sqrt{-6}\sqrt{-15}$ $-3\sqrt{10}$

15. $\dfrac{\sqrt{-98}}{\sqrt{-2}}$ 7

16. $\dfrac{\sqrt{-45}}{\sqrt{-5}}$ 3

17. $\dfrac{\sqrt{-63}}{\sqrt{7}}$ $3i$

18. $\dfrac{\sqrt{-80}}{\sqrt{5}}$ $4i$

For Exercises 19–26, simplify each expression and write the result in standard form, $a + bi$.

19. $\dfrac{8 + 3i}{14}$ $\dfrac{4}{7} + \dfrac{3}{14}i$

20. $\dfrac{4 + 5i}{6}$ $\dfrac{2}{3} + \dfrac{5}{6}i$

21. $\dfrac{-4 - 6i}{-2}$ $2 + 3i$

22. $\dfrac{9 - 15i}{-3}$ $-3 + 5i$

23. $\dfrac{-18 + \sqrt{-48}}{4}$ $-\dfrac{9}{2} + \sqrt{3}i$ or $-\dfrac{9}{2} + i\sqrt{3}$

24. $\dfrac{-20 + \sqrt{-50}}{-10}$ $2 - \dfrac{\sqrt{2}}{2}i$

25. $\dfrac{14 - \sqrt{-98}}{-7}$ $-2 + \sqrt{2}i$

26. $\dfrac{-10 + \sqrt{-125}}{5}$ $-2 + \sqrt{5}i$ or $-2 + i\sqrt{5}$

Objective 2: Perform Operations on Complex Numbers

For Exercises 27–30, simplify the powers of i. (See Example 2)

27. a. i^{20} 1 b. i^{29} i c. i^{50} -1 d. i^{-41} $-i$

28. a. i^{32} 1 b. i^{47} $-i$ c. i^{66} -1 d. i^{-27} i

29. a. i^{37} i b. i^{-37} $-i$ c. i^{82} -1 d. i^{-82} -1

30. a. i^{103} $-i$ b. i^{-103} i c. i^{52} 1 d. i^{-52} 1

For Exercises 31–56, perform the indicated operations. Write the answers in standard form, $a + bi$. (See Examples 3–6)

31. $(2 - 7i) + (8 - 3i)$ $10 - 10i$

32. $(6 - 10i) + (8 + 4i)$ $14 - 6i$

33. $(15 + 21i) - (18 - 40i)$ $-3 + 61i$

34. $(250 + 100i) - (80 + 25i)$ $170 + 75i$

35. $(2.3 + 4i) - (8.1 - 2.7i) + (4.6 - 6.7i)$ $-1.2 + 0i$

36. $(0.05 - 0.03i) + (-0.12 + 0.08i) - (0.07 + 0.05i)$ $-0.14 + 0i$

37. $2i(5 + i)$ $-2 + 10i$

38. $4i(6 + 5i)$ $-20 + 24i$

39. $(3 - 6i)(10 + i)$ $36 - 57i$

40. $(2 - 5i)(8 + 2i)$ $26 - 36i$

41. $(3 - 7i)^2$ $-40 - 42i$

42. $(10 - 3i)^2$ $91 - 60i$

43. $(3 - \sqrt{-5})(4 + \sqrt{-5})$ $17 - i\sqrt{5}$

44. $(2 + \sqrt{-7})(10 + \sqrt{-7})$ $13 + 12i\sqrt{7}$

45. $(2 - i)^2 + (2 + i)^2$ 6

46. $(3 - 2i)^2 + (3 + 2i)^2$ 10

47. $(10 - 4i)(10 + 4i)$ 116

48. $(3 - 9i)(3 + 9i)$ 90

49. $(\sqrt{2} + \sqrt{3}i)(\sqrt{2} - \sqrt{3}i)$ 5

50. $(\sqrt{5} + \sqrt{7}i)(\sqrt{5} - \sqrt{7}i)$ 12

51. $\dfrac{6 + 2i}{3 - i}$ $\dfrac{8}{5} + \dfrac{6}{5}i$

52. $\dfrac{5 + i}{4 - i}$ $\dfrac{19}{17} + \dfrac{9}{17}i$

53. $\dfrac{8 - 5i}{13 + 2i}$ $\dfrac{94}{173} - \dfrac{81}{173}i$

54. $\dfrac{10 - 3i}{11 + 4i}$ $\dfrac{98}{137} - \dfrac{73}{137}i$

55. $\dfrac{5}{13i}$ $0 - \dfrac{5}{13}i$

56. $\dfrac{6}{7i}$ $0 - \dfrac{6}{7}i$

Objective 3: Solve Quadratic Equations over the Set of Complex Numbers

For Exercises 57–72, solve the equations. (See Examples 7–8)

57. $4u^2 + 64 = 0$ $\{4i, -4i\}$

58. $8p^2 + 72 = 0$ $\{3i, -3i\}$

59. $\left(t - \dfrac{1}{2}\right)^2 = -\dfrac{17}{4}$ $\left\{\dfrac{1}{2} \pm \dfrac{\sqrt{17}}{2}i\right\}$

60. $\left(a - \dfrac{1}{3}\right)^2 = -\dfrac{47}{9}$ $\left\{\dfrac{1}{3} \pm \dfrac{\sqrt{47}}{3}i\right\}$

61. $t^2 - 8t = -24$ $\{4 \pm 2i\sqrt{2}\}$

62. $p^2 - 24p = -156$ $\{12 \pm 2i\sqrt{3}\}$

63. $4z^2 + 24z = -160$ $\{-3 \pm i\sqrt{31}\}$

64. $2m^2 + 20m = -70$ $\{-5 \pm i\sqrt{10}\}$

65. $y^2 = -4y - 6$ $\{-2 \pm i\sqrt{2}\}$

66. $z^2 = -8z - 19$ $\{-4 \pm i\sqrt{3}\}$

67. $t(t - 6) = -10$ $\{3 \pm i\}$

68. $m(m + 10) = -34$ $\{-5 \pm 3i\}$

69. $-\dfrac{7}{6}c + \dfrac{1}{2} = -\dfrac{5}{6}c^2$ $\left\{\dfrac{7 \pm i\sqrt{11}}{10}\right\}$

70. $-\dfrac{5}{2}d + 1 = -3d^2$ $\left\{\dfrac{5 \pm i\sqrt{23}}{12}\right\}$

71. $9x^2 + 49 = 0$ $\left\{\pm\dfrac{7}{3}i\right\}$

72. $121x^2 + 4 = 0$ $\left\{\pm\dfrac{2}{11}i\right\}$

Objective 4: Use the Discriminant

For Exercises 73–80, (a) evaluate the discriminant and (b) determine the number and type of solutions to each equation. (See Example 9)

73. $3x^2 - 4x + 6 = 0$ a. -56
b. 2 nonreal solutions

74. $5x^2 - 2x + 4 = 0$ a. -76
b. 2 nonreal solutions

75. $-2w^2 + 8w = 3$ a. 40
b. 2 real solutions

76. $-6d^2 + 9d = 2$ a. 33
b. 2 real solutions

77. $3x(x - 4) = x - 4$ a. 121
b. 2 real solutions

78. $2x(x - 2) = x + 3$ a. 49
b. 2 real solutions

79. $-1.4m + 0.1 = -4.9m^2$ a. 0
b. 1 real solution

80. $3.6n + 0.4 = -8.1n^2$ a. 0
b. 1 real solution

Objective 5: Solve Equations in Quadratic Form

For Exercises 81–100, make an appropriate substitution and solve the equation. (See Examples 10–11)

81. $(2x + 5)^2 - 7(2x + 5) - 30 = 0$ $\left\{-4, \dfrac{5}{2}\right\}$

82. $(3x - 7)^2 - 6(3x - 7) - 16 = 0$ $\left\{\dfrac{5}{3}, 5\right\}$

83. $(x^2 + 2x)^2 - 18(x^2 + 2x) = -45$ $\{-5, -3, 1, 3\}$

84. $(x^2 + 3x)^2 - 14(x^2 + 3x) = -40$ $\{-5, -4, 1, 2\}$

85. $(x^2 + 2)^2 + (x^2 + 2) - 42 = 0$ $\{\pm 3i, \pm 2\}$

86. $(y^2 - 3)^2 - 9(y^2 - 3) - 52 = 0$ $\{\pm i, \pm 4\}$

87. $\left(m - \dfrac{10}{m}\right)^2 - 6\left(m - \dfrac{10}{m}\right) - 27 = 0$ $\{-5, -1, 2, 10\}$

88. $\left(x + \dfrac{6}{x}\right)^2 - 12\left(x + \dfrac{6}{x}\right) + 35 = 0$ $\{1, 2, 3, 6\}$

89. $\left(2 + \dfrac{3}{t}\right)^2 - \left(2 + \dfrac{3}{t}\right) = 12$ $\left\{\dfrac{3}{2}, -\dfrac{3}{5}\right\}$

90. $\left(\dfrac{5}{y} + 3\right)^2 + 6\left(\dfrac{5}{y} + 3\right) = -8$ $\left\{-1, -\dfrac{5}{7}\right\}$

91. $5c^{2/5} - 11c^{1/5} + 2 = 0$ $\left\{\dfrac{1}{3125}, 32\right\}$

92. $3d^{2/3} - d^{1/3} - 4 = 0$ $\left\{\dfrac{64}{27}, -1\right\}$

93. $y^{1/2} - y^{1/4} - 6 = 0$ $\{81\}$

94. $n^{1/2} + 6n^{1/4} - 16 = 0$ $\{16\}$

95. $9y^{-4} - 10y^{-2} + 1 = 0$ $\{-3, -1, 1, 3\}$

96. $100x^{-4} - 29x^{-2} + 1 = 0$ $\{-5, -2, 2, 5\}$

97. $4t - 25\sqrt{t} = 0$ $\left\{0, \dfrac{625}{16}\right\}$

98. $9m - 16\sqrt{m} = 0$ $\left\{0, \dfrac{256}{81}\right\}$

99. $30k^{-2} - 23k^{-1} + 2 = 0$ $\left\{\dfrac{3}{2}, 10\right\}$

100. $3q^{-2} + 16q^{-1} + 5 = 0$ $\left\{-\dfrac{1}{5}, -3\right\}$

Mixed Exercises

For Exercises 101–104, verify by substitution that the given values of x are solutions to the given equation.

101. $x^2 + 25 = 0$

    a. $x = 5i$ $(5i)^2 + 25 = 0$ ✓

    b. $x = -5i$ $(-5i)^2 + 25 = 0$ ✓

102. $x^2 + 49 = 0$

    a. $x = 7i$ $(7i)^2 + 49 = 0$ ✓

    b. $x = -7i$ $(-7i)^2 + 49 = 0$ ✓

103. $x^2 - 4x + 7 = 0$

    a. $x = 2 + i\sqrt{3}$ $(2 + i\sqrt{3})^2 - 4(2 + i\sqrt{3}) + 7 = 0$ ✓

    b. $x = 2 - i\sqrt{3}$ $(2 - i\sqrt{3})^2 - 4(2 - i\sqrt{3}) + 7 = 0$ ✓

104. $x^2 - 6x + 11 = 0$

    a. $x = 3 + i\sqrt{2}$ $(3 + i\sqrt{2})^2 - 6(3 + i\sqrt{2}) + 11 = 0$ ✓

    b. $x = 3 - i\sqrt{2}$ $(3 - i\sqrt{2})^2 - 6(3 - i\sqrt{2}) + 11 = 0$ ✓

105. Prove that $(a + bi)(c + di) = (ac - bd) + (ad + bc)i$.

106. Prove that $(a + bi)^2 = (a^2 - b^2) + (2ab)i$.

For Exercises 107–108, solve the equation in two ways.

a. Solve as a radical equation by first isolating the radical.

b. Solve by writing the equation in quadratic form and using an appropriate substitution.

107. $y + 4\sqrt{y} = 21$ $\{9\}$ **108.** $w - 3\sqrt{w} = 10$ $\{25\}$

Write About It

109. Explain the flaw in the following logic.
$$\sqrt{-9} \cdot \sqrt{-4} = \sqrt{(-9)(-4)} = \sqrt{36} = 6$$

110. Discuss the difference between the products $(a + b)(a - b)$ and $(a + bi)(a - bi)$.

111. Give an example of a complex number that is its own conjugate.

112. Give an example of two complex numbers that are not real numbers, but whose product is a real number.

113. Given a quadratic equation, what is the discriminant and what information does it provide about the given quadratic equation?
Given $ax^2 + bx + c = 0$, where $a \neq 0$, the discriminant is $b^2 - 4ac$. The discriminant indicates the number and type of solutions to the equation.

114. Explain how to determine if an equation is in quadratic form.   An equation is in quadratic form if, after a suitable substitution, the equation can be written in the form $au^2 + bu + c = 0$, where u is a variable expression.

Expanding Your Skills

For Exercises 115–120, factor the expressions over the set of complex numbers. For assistance, consider these examples.

- In Section R.3 we saw that some expressions factor over the set of integers. For example: $x^2 - 4 = (x + 2)(x - 2)$.
- Some expressions factor over the set of irrational numbers. For example: $x^2 - 5 = (x + \sqrt{5})(x - \sqrt{5})$.
- To factor an expression such as $x^2 + 4$, we need to factor over the set of complex numbers. For example, verify that $x^2 + 4 = (x + 2i)(x - 2i)$.

115. a. $x^2 - 9$ $(x + 3)(x - 3)$ **116. a.** $x^2 - 100$ $(x + 10)(x - 10)$ **117. a.** $x^2 - 64$ $(x + 8)(x - 8)$
  b. $x^2 + 9$ $(x + 3i)(x - 3i)$ **b.** $x^2 + 100$ $(x + 10i)(x - 10i)$ **b.** $x^2 + 64$ $(x + 8i)(x - 8i)$

118. a. $x^2 - 25$ $(x + 5)(x - 5)$ **119. a.** $x^2 - 3$ $(x + \sqrt{3})(x - \sqrt{3})$ **120. a.** $x^2 - 11$ $(x + \sqrt{11})(x - \sqrt{11})$
  b. $x^2 + 25$ $(x + 5i)(x - 5i)$ **b.** $x^2 + 3$ $(x + i\sqrt{3})(x - i\sqrt{3})$ **b.** $x^2 + 11$ $(x + i\sqrt{11})(x - i\sqrt{11})$

For Exercises 121–124, write an equation with integer coefficients and the variable x that has the given solution set.
[*Hint:* Apply the zero product property in reverse. For example, to build an equation whose solution set is $\{2, -\frac{5}{2}\}$ we have $(x - 2)(2x + 5) = 0$, or simply $2x^2 + x - 10 = 0$.]

121. $\{2i, -2i\}$ $x^2 + 4 = 0$ **122.** $\{9i, -9i\}$ $x^2 + 81 = 0$ **123.** $\{1 \pm 2i\}$ $x^2 - 2x + 5 = 0$ **124.** $\{2 \pm 9i\}$
  $x^2 - 4x + 85 = 0$

Additional answers can be found in the Instructor Answer Appendix.

SECTION R.7   Applications of Equations

OBJECTIVES

1. Solve Applications Involving Simple Interest
2. Solve Applications Involving Mixtures
3. Solve Applications Involving Uniform Motion
4. Solve Applications Involving Rate of Work Done
5. Solve Applications of Quadratic Equations

1. Solve Applications Involving Simple Interest

In Examples 1–6, we use linear and rational equations to model physical situations to solve applications. While there is no magic formula to apply to all word problems, we do offer the following guidelines to help you organize the given information and to form a useful model.

Problem-Solving Strategy

1. Read the problem carefully. Determine what the problem is asking for, and assign variables to the unknown quantities.
2. Make an appropriate figure or table if applicable. Label the given information and variables in the figure or table.
3. Write an equation that represents the verbal model. The equation may be a known formula or one that you create that is unique to the problem.
4. Solve the equation from step 3.
5. Interpret the solution to the equation and check that it is reasonable in the context of the problem.

Simple interest I for a loan or an investment is based on the principal P (amount of money invested or borrowed), the annual interest rate r, and the time of the loan t in years. The relationship among the variables is given by $I = Prt$.

For example, if $5000 is invested at 4% simple interest for 18 months (1.5 yr), then the amount of simple interest earned is

$$I = Prt$$
$$I = (\$5000)(0.04)(1.5)$$
$$= \$300$$

The formula for simple interest is used in Example 1.

Classroom Example: p. 89
Exercise 8

EXAMPLE 1   Solving an Application Involving Simple Interest

Kent invested a total of $8000. He invested part of the money for 2 yr in a stock fund that earned the equivalent of 6.5% simple interest. He put the remaining money in an 18-month certificate of deposit (CD) that earned 2.5% simple interest. If the total interest from both investments was $855, determine the amount invested in each account.

Solution:

We can assign a variable to *either* the amount invested in the stock fund or the amount invested in the CD.

Let x represent the principal invested in the stock fund.

Then, $(8000 - x)$ is the remaining amount in the CD.

The interest from each account is computed from the formula $I = Prt$. Consider organizing this information in a table.

	Stock Fund (6.5% yield)	CD (2.5% yield)	Total
Principal ($)	x	$8000 - x$	$8000
Interest ($)	$x(0.065)(2)$	$(8000 - x)(0.025)(1.5)$	$855

Avoiding Mistakes

The CD was invested for 18 months. Be sure to convert to years.

18 months = 1.5 yr

To build an equation, note that

$$\begin{pmatrix}\text{Interest from} \\ \text{stock fund}\end{pmatrix} + \begin{pmatrix}\text{Interest} \\ \text{from CD}\end{pmatrix} = \begin{pmatrix}\text{Total} \\ \text{interest}\end{pmatrix}$$

$x(0.065)(2) + (8000 - x)(0.025)(1.5) = 855$	Second row of table
$0.13x + 0.0375(8000 - x) = 855$	Simplify.
$0.13x + 300 - 0.0375x = 855$	Apply the distributive property.
$0.0925x + 300 = 855$	Combine like terms.
$0.0925x = 555$	Subtract 300 from both sides.
$x = 6000$	

Avoiding Mistakes

Check that the answer is reasonable.

Amount of interest:
($6000)(0.065)(2) = $780
($2000)(0.025)(1.5) = $75
Total: $780 + $75 = $855 ✓

The amount invested in the stock fund is x: $6000.
The amount invested in the CD is $8000 - x = \$8000 - \$6000 = \$2000$.

Skill Practice 1   Franz borrowed a total of $10,000. Part of the money was borrowed from a lending institution that charged 5.5% simple interest. The rest of the money was borrowed from a friend to whom Franz paid 2.5% simple interest. Franz paid his friend back after 9 months (0.75 yr) and paid the lending institution after 2 yr. If the total amount Franz paid in interest was $735, how much did he borrow from each source?

Answer

1. Franz borrowed $4000 from his friend and $6000 from the lending institution.

2. Solve Applications Involving Mixtures

In Example 1, we "mixed" money between two different investments. We had to find the correct distribution of principal between two accounts to produce the given amount of interest. Example 2 presents a similar type of application that involves mixing different concentrations of a bleach solution to produce a third mixture of a given concentration.

For example, household bleach contains 6% sodium hypochlorite (active ingredient). This means that the remaining 94% of liquid is some other mixing agent such as water. Therefore, given 200 cL of household bleach, 6% would be pure sodium hypochlorite, and 94% would be some other mixing agent.

$$\text{Pure sodium hypochlorite} = (0.06)(200 \text{ cL}) = 12 \text{ cL}$$
$$\text{Other mixing agent} = (0.94)(200 \text{ cL}) = 188 \text{ cL}$$

To find the amount of pure sodium hypochlorite, we multiplied the concentration rate by the amount of solution.

Classroom Example: p. 90
Exercise 10

EXAMPLE 2 Solving an Application Involving Mixtures

Household bleach contains 6% sodium hypochlorite. How much household bleach should be combined with 70 L of a weaker 1% hypochlorite solution to form a solution that is 2.5% sodium hypochlorite?

Solution:

Let x represent the amount of 6% sodium hypochlorite solution (in liters).
70 L is the amount of 1% sodium hypochlorite solution.
Therefore, $x + 70$ is the amount of the resulting mixture (2.5% solution).

The amount of pure sodium hypochlorite in each mixture is found by multiplying the concentration rate by the amount of solution.

	6% Solution	1% Solution	2.5% Solution
Amount of solution (L)	x	70	$x + 70$
Pure sodium hypochlorite (L)	$0.06x$	$0.01(70)$	$0.025(x + 70)$

To build an equation, note that

$$\begin{pmatrix} \text{Amount of sodium} \\ \text{hypochlorite in} \\ 6\% \text{ solution} \end{pmatrix} + \begin{pmatrix} \text{Amount of sodium} \\ \text{hypochlorite in} \\ 1\% \text{ solution} \end{pmatrix} = \begin{pmatrix} \text{Amount of sodium} \\ \text{hypochlorite in} \\ 2.5\% \text{ solution} \end{pmatrix}$$

$0.06x + 0.01(70) = 0.025(x + 70)$	Second row in the table
$0.06x + 0.7 = 0.025x + 1.75$	Solve the equation.
$0.035x = 1.05$	
$x = 30$	

Avoiding Mistakes

Check that the answer is reasonable. The total amount of the resulting solution is 30 L + 70 L, which is 100 L.

Amount of sodium hypochlorite:
$0.06(30 \text{ L}) = 1.8 \text{ L}$
$0.01(70 \text{ L}) = 0.7 \text{ L}$
$0.025(100 \text{ L}) = 2.5 \text{ L}$ ✓

The amount of household bleach (6% sodium hypochlorite solution) needed is 30 L.

Skill Practice 2 How much 4% acid solution should be mixed with 200 mL of a 12% acid solution to make a 9% acid solution?

Answer

2. 120 mL of the 4% acid solution should be used.

3. Solve Applications Involving Uniform Motion

Example 3 involves uniform motion. Recall that the distance that an object travels is given by

$$d = rt \quad \text{Distance} = (\text{Rate})(\text{Time})$$

It follows that
$$t = \frac{d}{r} \quad \text{Time} = \frac{\text{Distance}}{\text{Rate}}$$

Classroom Example: p. 90
Exercise 14

EXAMPLE 3   **Solving an Application Involving Uniform Motion**

Trent takes his boat 6 mi downstream with a 1.5-mph current. The return trip against the current takes 1 hr longer. Find the speed of the boat in still water (in the absence of current).

Solution:

Let b represent the speed of the boat in still water.

	Distance (mi)	Rate (mph)	Time (hr)
With current	6	$b + 1.5$	$\dfrac{6}{b + 1.5}$
Against current	6	$b - 1.5$	$\dfrac{6}{b - 1.5}$

Assign a variable to the unknown quantity.

Organize the given information in a figure or chart.

The return trip against the current takes longer. The difference in time between the return trip and the original trip is 1 hr.

$$\left(\begin{array}{c} \text{Time of trip} \\ \text{against current} \end{array} \right) - \left(\begin{array}{c} \text{Time of trip} \\ \text{with current} \end{array} \right) = 1$$

$$\frac{6}{b - 1.5} - \frac{6}{b + 1.5} = 1$$

The restrictions on b are $b \neq 1.5$ and $b \neq -1.5$.

$$(b - 1.5)(b + 1.5)\left(\frac{6}{b - 1.5} - \frac{6}{b + 1.5} \right) = (b - 1.5)(b + 1.5)(1)$$

$$6(b + 1.5) - 6(b - 1.5) = (b - 1.5)(b + 1.5) \qquad \text{Apply the distributive property.}$$

$$6b + 9 - 6b + 9 = b^2 - 2.25$$

$$20.25 = b^2$$

$$b = \pm\sqrt{20.25} \qquad \text{Apply the square root property.}$$

$$b = 4.5 \qquad \text{Reject the negative solution because } b \text{ represents the speed}$$

The speed of the boat in still water is 4.5 mph.   of the boat.

Skill Practice 3   A fishing boat can travel 60 km with a 2.5-km/hr current in 2 hr less time than it can travel 60 km against the current. Determine the speed of the fishing boat in still water.

Answer

3. The boat travels 12.5 km/hr in still water.

Point of Interest

The relationship $d = rt$ is a familiar formula indicating that distance equals rate times time, or equivalently that $t = \frac{d}{r}$. However, suppose that a spaceship travels to a distant planet and then returns to Earth. Einstein's theory of special relativity indicates that $t = \frac{d}{r}$ only represents the trip's duration for an observer on Earth. For a person on the spaceship, the time will be shorter by a factor of $\sqrt{1 - \frac{r^2}{c^2}}$, where r is the speed of the spaceship and c is the speed of light.

For example, suppose that a spaceship travels to a planet 10 light-years away (a light-year is the distance that light travels in 1 yr) and then returns. The round trip is 20 light-years. Further suppose that the spaceship travels at half the speed of light, that is, $r = 0.5c$.

To an observer on Earth, the elapsed time of travel (in yr) is

$$t_E = \frac{d}{r} = \frac{20}{0.5} = 40 \text{ yr}$$

To an observer on the spaceship, the elapsed time (in yr) is

$$t_S = \frac{d}{r}\sqrt{1 - \frac{r^2}{c^2}} = \frac{20}{0.5}\sqrt{1 - \frac{(0.5c)^2}{c^2}} \approx 34.6 \text{ yr}$$

The unit of measurement in each case is years. Thus, the observer on Earth perceives the time of travel to be 40 yr, whereas to an observer on the spaceship the time of travel is only 34.6 yr.

4. Solve Applications Involving Rate of Work Done

In Example 4, we investigate an application involving the combined rates at which two individuals work to complete a task. From this information, we can determine the amount of time required for the individuals to complete a task by working together.

Classroom Example: p. 90
Exercise 22

EXAMPLE 4   Solving an Application Involving "Work" Rates

At a mail-order company, Derrick can process 100 orders in 4 hr. Miguel can process 100 orders in 3 hr.

a. How long would it take them to process 100 orders if they work together?

b. How long would it take them to process 1400 orders if they work together?

Solution:

a. Let t represent the amount of time required to process 100 orders working together.

One method to approach this problem is to add the rates of speed at which each person works.

$$\left(\begin{array}{c}\text{Derrick's} \\ \text{speed}\end{array}\right) + \left(\begin{array}{c}\text{Miguel's} \\ \text{speed}\end{array}\right) = \left(\begin{array}{c}\text{Speed working} \\ \text{together}\end{array}\right)$$

$$\frac{1 \text{ job}}{4 \text{ hr}} + \frac{1 \text{ job}}{3 \text{ hr}} = \frac{1 \text{ job}}{t \text{ hr}} \qquad \text{1 job} = \text{100 orders.}$$

$$\frac{1}{4} + \frac{1}{3} = \frac{1}{t}$$

$$12t\left(\frac{1}{4} + \frac{1}{3}\right) = 12t\left(\frac{1}{t}\right) \qquad \text{Multiply both sides by the LCD, } 12t.$$

$$3t + 4t = 12 \qquad \text{Apply the distributive properly.}$$

$$7t = 12$$

$$t = \frac{12}{7} \quad \text{or} \quad 1\frac{5}{7}$$

Derrick and Miguel can process 100 orders in $1\frac{5}{7}$ hr working together.

b. The time required to process 1400 orders is 14 times as long as the time to process 100 orders. $\left(\frac{12}{7} \text{ hr}\right)(14) = 24 \text{ hr}.$

Skill Practice 4   Sheldon and Penny were awarded a contract to paint 16 offices in the new math building at a university. Once all the preparation work is complete, Sheldon can paint an office in 30 min and Penny can paint an office in 45 min.

   a. How long would it take them to paint one office working together?
   b. How long would it take them to paint all 16 offices?

5. Solve Applications of Quadratic Equations

In Example 5, we use the Pythagorean theorem and a quadratic equation to find the lengths of the sides of a right triangle.

Classroom Example: p. 91
Exercise 30

EXAMPLE 5   **Solving an Application Involving the Pythagorean Theorem**

A window is in the shape of a rectangle with an adjacent right triangle above (see figure). The length of one leg of the right triangle is 2 ft less than the length of the hypotenuse. The length of the other leg is 1 ft less than the length of the hypotenuse. Find the lengths of the sides.

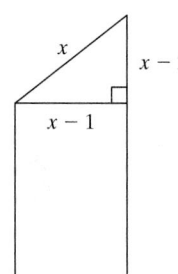

Solution:

Let x represent the length of the hypotenuse.
$x - 1$ represents the length of the longer leg.
$x - 2$ represents the length of the shorter leg.

Use the Pythagorean theorem to relate the lengths of the sides.

$a^2 + b^2 = c^2$	Pythagorean theorem
$(x - 1)^2 + (x - 2)^2 = (x)^2$	Substitute $x - 1$, $x - 2$, and x for the lengths of the sides.
$x^2 - 2x + 1 + x^2 - 4x + 4 = x^2$	
$x^2 - 6x + 5 = 0$	Set one side of the equation equal to zero.
$(x - 1)(x - 5) = 0$	Factor.
$x - 1 = 0$ or $x - 5 = 0$	Apply the zero product property.
$\cancel{x = 1}$ or $x = 5$	Reject $x = 1$ because if x were 1, the lengths of the legs would be 0 ft and -1 ft, which is impossible.

The hypotenuse is 5 ft.
The length of the longer leg is given by $x - 1$:   5 ft $-$ 1 ft = 4 ft.
The length of the shorter leg is given by $x - 2$:   5 ft $-$ 2 ft = 3 ft.

Avoiding Mistakes

In Example 5, be sure to square the binomials correctly. Recall that $(a - b)^2 = a^2 - 2ab + b^2$.

Therefore,
$(x - 1)^2 = x^2 - 2x + 1$
$(x - 2)^2 = x^2 - 4x + 4$

Skill Practice 5   A sail on a sailboat is in the shape of two adjacent right triangles. The hypotenuse of the lower triangle is 10 ft, and one leg is 2 ft shorter than the other leg. Find the lengths of the legs of the lower triangle.

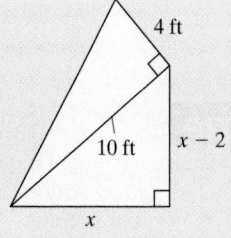

Answers
4. a. It would take 18 min to paint one office working together.
   b. It would take 288 min (4 hr 48 min) to paint 16 offices working together.
5. The longer leg is 8 ft and the shorter leg is 6 ft.

In the study of physical science, a common model used to represent the vertical position of an object moving vertically under the influence of gravity is given in Table R-9.

Table R-9   Vertical Position of an Object

Suppose that an object has an initial vertical position of s_0 and initial velocity v_0 straight upward. The vertical position s of the object is given by

$$s = -\frac{1}{2}gt^2 + v_0t + s_0, \text{ where}$$

g	is the acceleration due to gravity. On Earth, $g = 32$ ft/sec^2 or $g = 9.8$ m/sec^2.
t	is the time of travel.
v_0	is the initial velocity.
s_0	is the initial vertical position.
s	is the vertical position of the object at time t.

Instructor Note:
The value of g is actually the magnitude of the acceleration vector, and the value of v_0 is the magnitude of the velocity vector.

TIP The value of g is chosen to be consistent with the units for position and velocity. In this case, the initial height is given in ft. The initial velocity is given in ft/sec. Therefore, we choose g in ft/sec^2 rather than m/sec^2.

For example, suppose that a child tosses a ball straight upward from a height of 1.5 ft, with an initial velocity of 48 ft/sec.

The initial height is $s_0 = 1.5$ ft.

The initial velocity is $v_0 = 48$ ft/sec.

The acceleration due to gravity is $g = 32$ ft/sec^2.

The vertical position of the ball (in feet) is given by

$$s = -\frac{1}{2}gt^2 + v_0t + s_0$$

$$s = -\frac{1}{2}(32)t^2 + (48)t + (1.5)$$

$$= -16t^2 + 48t + 1.5$$

Classroom Example: p. 92
Exercise 36

EXAMPLE 6   Analyzing an Object Moving Vertically

A toy rocket is shot straight upward from a launch pad of 1 m above ground level with an initial velocity of 24 m/sec.

a. Write a model to express the height of the rocket s (in meters) above ground level.

b. Find the time(s) at which the rocket is at a height of 20 m. Round to 1 decimal place.

c. Find the time(s) at which the rocket is at a height of 40 m.

Solution:

a. $s = -\frac{1}{2}gt^2 + v_0t + s_0$

$s = -\frac{1}{2}(9.8)t^2 + (24)t + (1)$

$= -4.9t^2 + 24t + 1$

In this example,

$s_0 = 1$ m

$v_0 = 24$ m/sec

$g = 9.8$ m/sec^2

b. $20 = -4.9t^2 + 24t + 1$ Substitute 20 for s.

$4.9t^2 - 24t + 19 = 0$ Set one side equal to zero.

$t = \dfrac{-(-24) \pm \sqrt{(-24)^2 - 4(4.9)(19)}}{2(4.9)}$ Apply the quadratic formula.

TIP Choose $g = 9.8$ m/sec^2 because the height is given in meters and velocity is given in meters per second.

$$t = \frac{24 \pm \sqrt{203.6}}{9.8} \begin{cases} t = \dfrac{24 + \sqrt{203.6}}{9.8} \approx 3.9 \\[2ex] t = \dfrac{24 - \sqrt{203.6}}{9.8} \approx 1.0 \end{cases}$$

The rocket will be at a height of 20 m at 1 sec and 3.9 sec after launch.

c. $40 = -4.9t^2 + 24t + 1$ Substitute 40 for s.

$4.9t^2 - 24t + 39 = 0$

$t = \dfrac{-(-24) \pm \sqrt{(-24)^2 - 4(4.9)(39)}}{2(4.9)}$ Apply the quadratic formula.

$t = \dfrac{24 \pm \sqrt{-188.4}}{9.8}$ The solutions are not real numbers.

There is no real number t for which the height of the rocket will be 40 m. The rocket will not reach a height of 40 m.

Skill Practice 6   A fireworks mortar is launched straight upward from a pool deck 2 m off the ground at an initial velocity of 40 m/sec.

   a. Write a model to express the height of the mortar s (in meters) above ground level.

   b. Find the time(s) at which the mortar is at a height of 60 m. Round to 1 decimal place.

   c. Find the time(s) at which the rocket is at a height of 100 m.

Answers

6. a. $s = -4.9t^2 + 40t + 2$
  b. The mortar will be at a height of 60 m at 1.9 sec and 6.3 sec after launch.
  c. The mortar will not reach a height of 100 m.

| **SECTION R.7** | **Practice Exercises** |

Concept Connections

1. If $6000 is borrowed at 7.5% simple interest for 2 yr, then the amount of interest is _____$900_____.

2. Suppose that 8% of a solution is fertilizer by volume and the remaining 92% is water. How much fertilizer is in a 2-L bucket of solution?   0.16 L

3. If $d = rt$, then $t = \dfrac{d}{r}$

4. If $d = rt$, then $r = \dfrac{d}{t}$

Objective 1: Solve Applications Involving Simple Interest

5. Rocco borrowed a total of $5000 from two student loans. One loan charged 3% simple interest and the other charged 2.5% simple interest, both payable after graduation. If the interest he owed after 1 yr was $132.50, determine the amount of principal for each loan. **(See Example 1)**   Rocco borrowed $1500 at 3% and $3500 at 2.5%.

7. Fernando invested money in a 3-yr CD (certificate of deposit) that returned the equivalent of 4.4% simple interest. He invested $2000 less in an 18-month CD that had a 3% return. If the total amount of interest from these investments was $706.50, determine how much was invested in each CD.   Fernando invested $4500 in the 3-yr CD and $2500 in the 18-month CD.

6. Laura borrowed a total of $22,000 from two different banks to start a business. One bank charged the equivalent of 4% simple interest, and the other charged 5.5% interest. If the total interest after 1 yr was $910, determine the amount borrowed from each bank.   Laura borrowed $20,000 from the bank charging 4% interest, and $2000 from the bank charging 5.5% interest.

8. Ebony bought a 5-yr Treasury note that paid the equivalent of 2.8% simple interest. She invested $5000 more in a 10-yr bond earning 3.6% than she did in the Treasury note. If the total amount of interest from these investments was $5300, determine the amount of principal for each investment.   Ebony invested $7000 in the Treasury note and $12,000 in the bond.

Objective 2: Solve Applications Involving Mixtures

9. Ethanol fuel mixtures have "E" numbers that indicate the percentage of ethanol in the mixture by volume. For example, E10 is a mixture of 10% ethanol and 90% gasoline. How much E5 should be mixed with 5000 gal of E10 to make an E9 mixture? (**See Example 2**)   1250 gal of E5

10. A nurse mixes 60 cc of a 50% saline solution with a 10% saline solution to produce a 25% saline solution. How much of the 10% solution should he use?   100 cc

11. The density and strength of concrete are determined by the ratio of cement and aggregate (aggregate is sand, gravel, or crushed stone). Suppose that a contractor has 480 ft^3 of a dry concrete mixture that is 70% sand by volume. How much pure sand must be added to form a new mixture that is 75% sand by volume?   96 ft^3 of sand

12. Antifreeze is a compound added to water to reduce the freezing point of a mixture. In extreme cold (less than $-35°$F), one car manufacturer recommends that a mixture of 65% antifreeze be used. How much 50% antifreeze solution should be drained from a 4-gal tank and replaced with pure antifreeze to produce a 65% antifreeze mixture?   1.2 gal

Objective 3: Solve Applications Involving Uniform Motion

13. Jesse takes a 3-day kayak trip and travels 72 km south from Everglades City to a camp area in Everglades National Park. The trip to the camp area with a 2-km/hr current takes 9 hr less time than the return trip against the current. Find the speed that Jesse travels in still water. (**See Example 3**)   Jesse travels 6 km/hr in still water.

14. A plane travels 800 mi from Dallas, Texas, to Atlanta, Georgia, with a prevailing west wind of 40 mph. The return trip against the wind takes $\frac{1}{2}$ hr longer. Find the average speed of the plane in still air.   The plane travels 360 mph in still air.

15. Jean runs 6 mi and then rides 24 mi on her bicycle in a biathlon. She rides 8 mph faster than she runs. If the total time for her to complete the race is 2.25 hr, determine her average speed running and her average speed riding her bicycle.
Jean runs 8 mph and rides 16 mph.

16. Barbara drives between Miami, Florida, and West Palm Beach, Florida. She drives 50 mi in clear weather and then encounters a thunderstorm for the last 15 mi. She drives 20 mph slower through the thunderstorm than she does in clear weather. If the total time for the trip is 1.5 hr, determine her average speed in nice weather and her average speed driving in the thunderstorm.   Barbara drives 30 mph in the thunderstorm and 50 mph in nice weather.

17. Two passengers leave the airport at Kansas City, Missouri. One flies to Los Angeles, California, in 3.4 hr and the other flies in the opposite direction to New York City in 2.4 hr. With prevailing westerly winds, the speed of the plane to New York City is 60 mph faster than the speed of the plane to Los Angeles. If the total distance traveled by both planes is 2464 mi, determine the average speed of each plane.   The plane to Los Angeles travels 400 mph and the plane to New York City travels 460 mph.

18. Two planes leave from Atlanta, Georgia. One makes a 5.2-hr flight to Seattle, Washington, and the other makes a 2.5-hr flight to Boston, Massachusetts. The plane to Boston averages 44 mph slower than the plane to Seattle. If the total distance traveled by both planes is 3124 mi, determine the average speed of each plane.
The plane to Seattle travels 420 mph, and the plane to Boston travels 376 mph.

19. Darren drives to school in rush hour traffic and averages 32 mph. He returns home in mid-afternoon when there is less traffic and averages 48 mph. What is the distance between his home and school if the total traveling time is 1 hr 15 min?   The distance is 24 mi.

20. Peggy competes in a biathlon by running and bicycling around a large loop through a city. She runs the loop one time and bicycles the loop five times. She can run 8 mph and she can ride 16 mph. If the total time it takes her to complete the race is 1 hr 45 min, determine the distance of the loop.   The loop is 4 mi.

Objective 4: Solve Applications Involving Rate of Work Done

21. Joel can run around a $\frac{1}{4}$-mi track in 66 sec, and Jason can run around the track in 60 sec. If the runners start at the same point on the track and run in opposite directions, how long will it take the runners to cover $\frac{1}{4}$ mi? (**See Example 4**) $\frac{220}{7}$ sec or approximately 31.4 sec

22. Marta can vacuum the house in 40 min. It takes her daughter 1 hr to vacuum the house. How long would it take them if they worked together?   24 min

23. One pump can fill a pool in 10 hr. Working with a second slower pump, the two pumps together can fill the pool in 6 hr. How fast can the second pump fill the pool by itself?   15 hr

24. Brad and Angelina can mow their yard together with two lawn mowers in 30 min. When Brad works alone, it takes him 50 min. How long would it take Angelina to mow the lawn by herself?   75 min or 1 hr 15 min

Objective 5: Solve Applications of Quadratic Equations

25. A patio is configured from a rectangle with two right triangles of equal size attached at the two ends. The length of the rectangle is 20 ft. The base of the right triangle is 3 ft less than the height of the triangle. If the total area of the patio is 348 ft^2, determine the base and height of the triangular portions.   The base is 9 ft and the height is 12 ft.

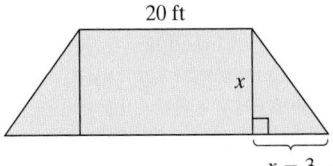

20 ft

x

$x - 3$

26. The front face of a house is in the shape of a rectangle with a Queen post roof truss above. The length of the rectangular region is 3 times the height of the truss. The height of the rectangle is 2 ft more than the height of the truss. If the total area of the front face of the house is 336 ft^2, determine the length and width of the rectangular region.   The length is 24 ft and the height is 10 ft.

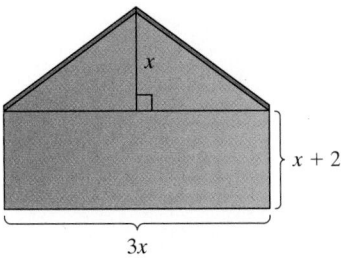

x

$x + 2$

$3x$

27. A baseball diamond is in the shape of a square with 90-ft sides. How far is it from home plate to second base? Give the exact value and give an approximation to the nearest tenth of a foot.   The distance is $90\sqrt{2}$ ft or approximately 127.3 ft.

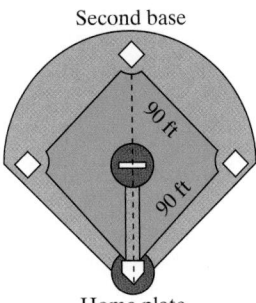

Second base

90 ft

90 ft

Home plate

28. The figure shown is a cube with 6-in. sides. Find the exact length of the diagonal through the interior of the cube d by following these steps.

a. Apply the Pythagorean theorem using the sides on the base of the cube to find the length of diagonal c. $c = 6\sqrt{2}$ in.

b. Apply the Pythagorean theorem using c and the height of the cube as the legs of the right triangle through the interior of the cube. $d = 6\sqrt{3}$ in.

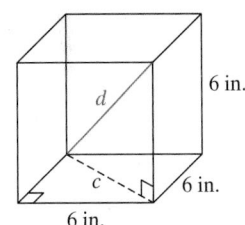

d

c

6 in.

6 in.

6 in.

29. The sail on a sailboat is in the shape of two adjacent right triangles. In the lower triangle, the shorter leg is 2 ft less than the longer leg. The hypotenuse is 2 ft more than the longer leg. **(See Example 5)**

a. Find the lengths of the sides of the lower triangle.   The lengths of the sides of the lower triangle are 6 ft, 8 ft, and 10 ft.

b. Find the total sail area.   The total area is 44 ft^2.

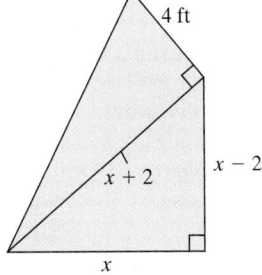

4 ft

$x + 2$

$x - 2$

x

30. A portion of a roof truss is given in the figure. The triangle on the left is configured such that the longer leg is 7 ft longer than the shorter leg, and the hypotenuse is 1 ft more than twice the shorter leg.

a. Find the lengths of the sides of the triangle on the left.   The lengths of the sides of the triangle on the left are 8 ft, 15 ft, and 17 ft.

b. Find the lengths of the sides of the triangle on the right.   The lengths of the sides of the triangle on the right are 9 ft, 12 ft, and 15 ft.

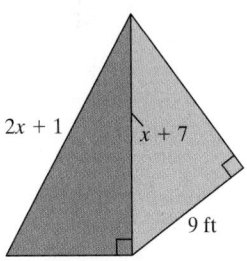

$2x + 1$

$x + 7$

9 ft

x

31. The display area on a cell phone has a 3.5-in. diagonal.

a. If the aspect ratio of length to width is 1.5 to 1, determine the length and width of the display area. Round the values to the nearest hundredth of an inch.   The length is approximately 2.91 in. and the width is approximately 1.94 in.

b. If the phone has 326 pixels per inch, approximate the dimensions in pixels.   Using the rounded values from part (a), the screen is approximately 949 pixels by 632 pixels.

32. The display area on a computer has a 15-in. diagonal. If the aspect ratio of length to width is 1.6 to 1, determine the length and width of the display area. Round the values to the nearest hundredth of an inch.   The length is approximately 12.72 in., and the width is approximately 7.95 in.

For Exercises 33–36, use the model $s = -\frac{1}{2}gt^2 + v_0t + s_0$ with $g = 32$ ft/sec² or $g = 9.8$ m/sec². (See Example 6)

33. NBA basketball legend Michael Jordan had a 48-in. vertical leap. Suppose that Michael jumped from ground level with an initial velocity of 16 ft/sec.

 a. Write a model to express Michael's height (in ft) above ground level t seconds after leaving the ground. $s = -16t^2 + 16t$

 b. Use the model from part (a) to determine how long it would take Michael to reach his maximum height of 48 in. (4 ft).   It would take Michael Jordan 0.5 sec to reach his maximum height of 4 ft.

34. At the time of this printing, the highest vertical leap on record is 60 in., held by Kadour Ziani. For this record-setting jump, Kadour left the ground with an initial velocity of $8\sqrt{5}$ ft/sec.

 a. Write a model to express Kadour's height (in ft) above ground level t seconds after leaving the ground. $s = -16t^2 + 8\sqrt{5}t$

 b. Use the model from part (a) to determine how long it would take Kadour to reach his maximum height of 60 in. (5 ft). Round to the nearest hundredth of a second.   It would take 0.56 sec to reach a height of 5 ft.

35. A bad punter on a football team kicks a football approximately straight upward with an initial velocity of 75 ft/sec.

 a. If the ball leaves his foot from a height of 4 ft, write an equation for the vertical height s (in ft) of the ball t seconds after being kicked. $s = -16t^2 + 75t + 4$

 b. Find the time(s) at which the ball is at a height of 80 ft. Round to 1 decimal place.   The ball will be at an 80-ft height 1.5 sec and 3.2 sec after being kicked.

36. In a classic *Seinfeld* episode, Jerry tosses a loaf of bread (a marble rye) straight upward to his friend George who is leaning out of a third-story window.

 a. If the loaf of bread leaves Jerry's hand at a height of 1 m with an initial velocity of 18 m/sec, write an equation for the vertical position of the bread s (in meters) t seconds after release. $s = -4.9t^2 + 18t + 1$

 b. How long will it take the bread to reach George if he catches the bread on the way up at a height of 16 m? Round to the nearest tenth of a second.   George will catch the bread 1.3 sec after release.

Mixed Exercises

37. Suppose that 40 deer are introduced in a protected wilderness area. The population of the herd P can be approximated by $P = \frac{40 + 20x}{1 + 0.05x}$, where x is the time in years since introducing the deer. Determine the time required for the deer population to reach 200.   16 yr

38. Starting from rest, an automobile's velocity v (in ft/sec) is given by $v = \frac{180t}{2t + 10}$, where t is the time in seconds after the car begins forward motion. Determine the time required for the car to reach a speed of 60 ft/sec (≈ 41 mph).   10 sec

39. Brianna's SUV gets 22 mpg in the city and 30 mpg on the highway. The amount of gas she uses A (in gal) is given by $A = \frac{1}{22}c + \frac{1}{30}h$, where c is the number of city miles driven and h is the number of highway miles driven. If Brianna drove 165 mi on the highway and used 7 gal of gas, how many city miles did she drive?   33 mi

40. Dexter's truck gets 32 mpg on the highway and 24 mpg in the city. The amount of gas he uses A (in gal) is given by $A = \frac{1}{24}c + \frac{1}{32}h$, where c is the number of city miles driven and h is the number of highway miles driven. If Dexter drove 60 mi in the city and used 9 gal of gas, how many highway miles did he drive?   208 mi

41. At a construction site, cement, sand, and gravel are mixed to make concrete. The ratio of cement to sand to gravel is 1 to 2.4 to 3.6. If a 150-lb bag of sand is used, how much cement and gravel must be used?   62.5 lb of cement and 225 lb of gravel

42. The property tax on a $180,000 house is $1296. At this rate, what is the property tax on a house that is $240,000? $1728

43. In addition to measuring a person's individual HDL and LDL cholesterol levels, doctors also compute the ratio of total cholesterol to HDL cholesterol. Doctors recommend that the ratio of total cholesterol to HDL cholesterol be kept under 4. Suppose that the ratio of a patient's total cholesterol to HDL is 3.4 and her HDL is 60 mg/dL. Determine the patient's LDL level and total cholesterol. (Assume that total cholesterol is the sum of the LDL and HDL levels.)   LDL is 144 mg/dL and the total cholesterol is 204 mg/dL.

44. For a recent Congress, there were 10 more Democrats than Republicans in the U.S. Senate. This resulted in a ratio of 11 Democrats to 9 Republicans. How many senators were Democrat and how many were Republican?   55 Democrat and 45 Republican

45. When studying wildlife populations, biologists sometimes use a technique called "mark-recapture." For example, a researcher captured and tagged 30 deer in a wildlife management area. Several months later, the researcher observed a new sample of 80 deer and determined that 5 were tagged. What is the total number of deer in the population?   480 deer

46. To estimate the number of bass in a lake, a biologist catches and tags 24 bass. Several weeks later, the biologist catches a new sample of 40 bass and finds that 4 are tagged. How many bass are in the lake?   240 bass

47. Seismographs can record two types of wave energy (P waves and S waves) that travel through the Earth after an earthquake. Traveling through granite, P waves travel approximately 5 km/sec and S waves travel approximately 3 km/sec. If a geologist working at a seismic station measures a time difference of 40 sec between an earthquake's P waves and S waves, how far from the epicenter of the earthquake is the station?   300 km

48. Suppose that a shallow earthquake occurs in which the P waves travel 8 km/sec and the S waves travel 4.8 km/sec. If a seismologist measures a time difference of 20 sec between the arrival of the P waves and the S waves, how far is the seismologist from the epicenter of the earthquake?   240 km

49. Henri needs to have a toilet repaired in his house. The cost of the new plumbing fixtures is $110 and labor is $60/hr.
   a. Write a model that represents the cost of the repair C (in $) in terms of the number of hours of labor x.
   $C = 110 + 60x$
   b. After how many hours of labor would the cost of the repair job equal the cost of a new toilet of $350?   4 hr

50. After a hurricane, repairs to a roof will cost $2400 for materials and $80/hr in labor.
   a. Write a model that represents the cost of the repair C (in $) in terms of the number of hours of labor x.
   $C = 2400 + 80x$
   b. If an estimate for a new roof is $5520, after how many hours of labor would the cost to repair the roof equal the cost of a new roof?   39 hr

51. On moving day, Guyton needs to rent a truck. The length of the cargo space is 12 ft, and the height is 1 ft less than the width. The brochure indicates that the truck can hold 504 ft³. What are the dimensions of the cargo space? Assume that the cargo space is in the shape of a rectangular solid. (**See Example 1**)   The dimensions of the cargo space are 6 ft by 7 ft by 12 ft.

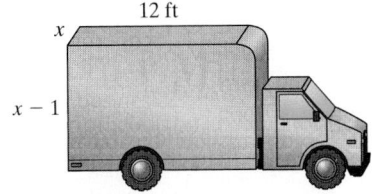
12 ft

52. Lorene plans to make several open-topped boxes in which to carry plants. She makes the boxes from rectangular sheets of cardboard from which she cuts out 6-in. squares from each corner. The length of the original piece of cardboard is 12 in. more than the width. If the volume of the box is 1728 in.³, determine the dimensions of the original piece of cardboard.   The dimensions of the cardboard sheet are 24 in. by 36 in.

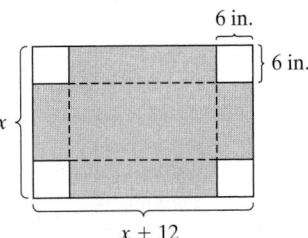

53. The population P of a culture of *Pseudomonas aeruginosa* bacteria is given by $P = -1718t^2 + 82{,}000t + 10{,}000$, where t is the time in hours since the culture was started. Determine the time(s) at which the population was 600,000. Round to the nearest hour.   There were 600,000 organisms approximately 9 hr and 39 hr after the culture was started.

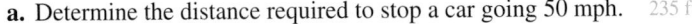

54. The distance d (in ft) required to stop a car that was traveling at speed v (in mph) before the brakes were applied depends on the amount of friction between the tires and the road and the driver's reaction time. After an accident, a legal team hired an engineering firm to collect data for the stretch of road where the accident occurred. Based on the data, the stopping distance is given by $d = 0.05v^2 + 2.2v$.
   a. Determine the distance required to stop a car going 50 mph.   235 ft
   b. Up to what speed (to the nearest mph) could a motorist be traveling and still have adequate stopping distance to avoid hitting a deer 330 ft away?   62 mph

Write About It

55. Is it possible for the measures of the angles in a triangle to be represented by three consecutive odd integers?
   Explain.   No. If x represents the measure of the smallest angle, then the equation $x + (x + 2) + (x + 4) = 180$ does not result in an odd integer value for x. Instead the measures of the angles would be even integers.

56. Bob wants to change a $100 bill into an equal number of $20 bills, $10 bills, and $5 bills. Is this possible?
   Explain.   No. If x represents the number of each type of bill, then the solution to the equation $20x + 10x + 5x = 100$ is not a whole number.

Expanding Your Skills

57. A **golden rectangle** is a rectangle in which the ratio of its length to its width is equal to the ratio of the sum of its length and width to its length: $\frac{L}{W} = \frac{L + W}{L}$ (values of L and W that meet this condition are said to be in the **golden ratio**).

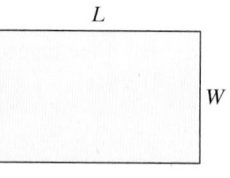 

L

W

    a. Suppose that a golden rectangle has a width of 1 unit. Solve the equation to find the exact value for the length. Then give a decimal approximation to 2 decimal places. $L = \frac{1 + \sqrt{5}}{2} \approx 1.62$

    b. To create a golden rectangle with a width of 9 ft, what should be the length? Round to 1 decimal place. 14.6 ft

58. An artist has been commissioned to make a stained glass window in the shape of a regular octagon. The octagon must fit inside an 18-in. square space. Determine the length of each side of the octagon. Round to the nearest hundredth of an inch. The sides are approximately 7.46 in.

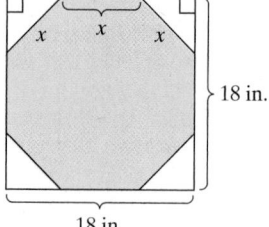

18 in.

18 in.

59. A farmer has 160 yd of fencing material and wants to enclose three rectangular pens. Suppose that x represents the length of each pen and y represents the width as shown in the figure.

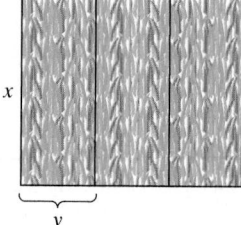

x

y

    a. Assuming that the farmer uses all 160 yd of fencing, write an expression for y in terms of x. $y = \dfrac{160 - 4x}{6}$ or $y = \dfrac{80 - 2x}{3}$

    b. Write an expression in terms of x for the area of one individual pen. $A = x\left(\dfrac{80 - 2x}{3}\right)$

    c. If the farmer wants to design the structure so that each pen encloses 250 yd^2, determine the dimensions of each pen. Each pen can be 25 yd by 10 yd, or it can be 15 yd by $\dfrac{50}{3}$ yd.

60. At noon, a ship leaves a harbor and sails south at 10 knots. Two hours later, a second ship leaves the harbor and sails east at 15 knots. When will the ships be 100 nautical miles apart? Round to the nearest minute. The ships will be 100 nautical miles apart at approximately 6:51 PM.

61. Pam is in a canoe on a lake 400 ft from the closest point on a straight shoreline. Her house is 800 ft up the road along the shoreline. She can row 2.5 ft/sec and she can walk 5 ft/sec. If the total time it takes for her to get home is 5 min (300 sec), determine the point along the shoreline at which she landed her canoe.

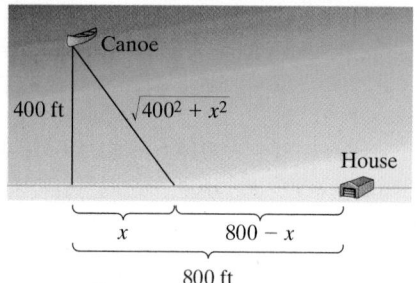

Canoe

400 ft $\sqrt{400^2 + x^2}$

House

x $800 - x$

800 ft

Pam can row to a point $166\frac{2}{3}$ ft down the beach or to a point 300 ft down the beach to be home in 5 min.

62. Debbie is in a boat in the ocean 48 mi from point A, the closest point along a straight shoreline. She needs to dock the boat at a marina x miles farther up the coast, and then drive along the coast to point B, 96 mi from point A. Her boat travels 20 mph, and she drives 60 mph. If the total trip took 4 hr, determine the distance x along the shoreline. The marina is 36 mi up the coast.

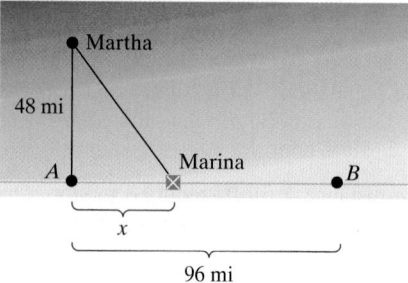

Martha

48 mi

A Marina B

x

96 mi

Linear, Compound, and Absolute Value Inequalities

OBJECTIVES

1. Solve Linear Inequalities in One Variable
2. Solve Compound Linear Inequalities
3. Solve Absolute Value Inequalities
4. Solve Applications of Inequalities

1. Solve Linear Inequalities in One Variable

Emily wants to earn an "A" in her College Algebra course and knows that the average of her tests and assignments must be at least 90. She has five test grades of 96, 84, 80, 98, and 88. She also has a score of 100 for online homework, and this carries the same weight as a test grade. She still needs to take the final exam and the final is weighted as two test grades. To determine the scores on the final exam that would result in an average of 90 or more, Emily would solve the following inequality (see Example 9):

$$\frac{96 + 84 + 80 + 98 + 88 + 100 + 2x}{8} \geq 90, \text{ where } x \text{ is Emily's score on the final.}$$

A linear equation in one variable is an equation that can be written as $ax + b = 0$, where a and b are real numbers and $a \neq 0$. A **linear inequality** is any relationship of the form $ax + b < 0$, $ax + b \leq 0$, $ax + b > 0$, or $ax + b \geq 0$. The solution set to a linear equation consists of a single element that can be represented by a point on the number line. The solution set to a linear inequality contains an infinite number of elements and can be expressed in set-builder notation or in interval notation.

Equation/ Inequality	Solution Set	Graph
$x + 4 = 0$	$\{-4\}$	
$x + 4 \geq 0$	$\{x \mid x \geq -4\}$ or $[-4, \infty)$	
$x + 4 < 0$	$\{x \mid x < -4\}$ or $(-\infty, -4)$	

To solve a linear inequality in one variable, we use the following properties of inequality.

Properties of Inequality

Let a, b, and c represent real numbers.

1. If $x < a$, then $a > x$.
2. If $a < b$ and $b < c$, then $a < c$.
3. If $a < b$ and $c < d$, then $a + c < b + d$.
4. If $a < b$, then $a + c < b + c$ and $a - c < b - c$.
5. If c is *positive* and $a < b$, then $ac < bc$ and $\dfrac{a}{c} < \dfrac{b}{c}$.
6. If c is *negative* and $a < b$, then $ac > bc$ and $\dfrac{a}{c} > \dfrac{b}{c}$.

These statements are also true expressed with the symbols $>$, $\leq$, and $\geq$.

Property 6 indicates that if both sides of an inequality are multiplied or divided by a negative number, then the direction of the inequality sign must be reversed.

EXAMPLE 1   Solving a Linear Inequality

Solve the inequality. Graph the solution set and write the solution set in set-builder notation and in interval notation.

$$-6x + 4 < 34$$

Solution:

$$-6x + 4 < 34$$
$$-6x + 4 - 4 < 34 - 4 \qquad \text{Subtract 4 from both sides.}$$
$$-6x < 30$$
$$\frac{-6x}{-6} > \frac{30}{-6} \qquad \text{Divide both sides by } -6. \text{ Reverse the inequality sign.}$$
$$x > -5$$

The solution set is $\{x \mid x > -5\}$.
Interval notation: $(-5, \infty)$

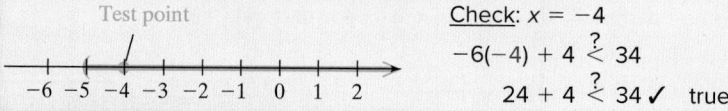

Skill Practice 1   Solve the inequality. Graph the solution set and write the solution set in set-builder notation and in interval notation. $-5t - 6 \geq 24$

TIP In Example 1, the solution set to the inequality $-6x + 4 < 34$ is $\{x \mid x > -5\}$. This means that all numbers greater than -5 make the inequality a true statement. You can check by taking an arbitrary test point from the interval $(-5, \infty)$. For example, the value $x = -4$ makes the original inequality true.

Test point

Check: $x = -4$
$$-6(-4) + 4 \overset{?}{<} 34$$
$$24 + 4 \overset{?}{<} 34 \checkmark \quad \text{true}$$

EXAMPLE 2   Solving a Linear Inequality Containing Fractions

Solve the inequality. Graph the solution set and write the solution set in set-builder notation and in interval notation.

$$\frac{x + 1}{3} - \frac{2x - 4}{6} \leq -\frac{x}{2}$$

Solution:

$$\frac{x + 1}{3} - \frac{2x - 4}{6} \leq -\frac{x}{2}$$
$$6\left(\frac{x + 1}{3} - \frac{2x - 4}{6}\right) \leq 6\left(-\frac{x}{2}\right) \qquad \text{Multiply both sides by the LCD of 6 to clear fractions.}$$
$$2(x + 1) - (2x - 4) \leq -3x$$
$$2x + 2 - 2x + 4 \leq -3x \qquad \text{Apply the distributive property.}$$
$$6 \leq -3x$$
$$\frac{6}{-3} \geq \frac{-3x}{-3} \qquad \text{Divide both sides by } -3. \text{ Since } -3 \text{ is a negative number, reverse the inequality sign.}$$
$$-2 \geq x \quad \text{or} \quad x \leq -2$$

The solution set is $\{x \mid x \leq -2\}$.
Interval notation: $(-\infty, -2]$

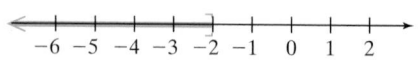

Skill Practice 2   Solve the inequality. Graph the solution set and write the solution set in set-builder notation and in interval notation.

$$\frac{m-4}{2} - \frac{3m+4}{10} > -\frac{3m}{5}$$

2. Solve Compound Linear Inequalities

In Examples 3–5, we solve **compound inequalities.** These are statements with two or more inequalities joined by the word "and" or the word "or." For example, suppose that x represents the glucose level measured from a fasting blood sugar test.

- The normal glucose range is given by $x \geq 70$ mg/dL and $x \leq 100$ mg/dL.
- An abnormal glucose level is given by $x < 70$ mg/dL or $x > 100$ mg/dL.

To find the solution sets for compound inequalities follow these guidelines.

Solving a Compound Inequality

Step 1   To solve a compound inequality, first solve the individual inequalities.

Step 2   • If two inequalities are joined by the word "and," the solutions are the values of the variable that simultaneously satisfy each inequality. That is, we take the *intersection* of the individual solution sets.

   • If two inequalities are joined by the word "or," the solutions are the values of the variable that satisfy either inequality. Therefore, we take the *union* of the individual solution sets.

Classroom Examples: p. 102
Exercises 30, 32

EXAMPLE 3   Solving a Compound Inequality "Or"

Solve. $x - 2 \leq 5$ or $\dfrac{1}{2}x > 6$

Solution:

$x - 2 \leq 5$ or $\dfrac{1}{2}x > 6$ First solve the individual inequalities.
Then take the *union* of the individual solution sets.

$x \leq 7$ or $x > 12$

The solution set is $\{x \mid x \leq 7$ or $x > 12\}$.

Interval notation: $(-\infty, 7] \cup (12, \infty)$

Skill Practice 3   Solve. $\dfrac{1}{4}y < -1$ or $3 + y \geq 5$

Answers

2.

$\{m \mid m > 3\}; (3, \infty)$

3. $\{y \mid y < -4$ or $y \geq 2\};$
$(-\infty, -4) \cup [2, \infty)$

In Example 4, we solve a compound inequality in which the individual inequalities are joined by the word "and." In this case, we take the intersection of the individual solution sets.

Classroom Examples: p. 102
Exercises 30, 32

EXAMPLE 4   Solving a Compound Inequality "And"

Solve. $-\dfrac{1}{4}t < 2$ and $0.52t \geq 1.3$

Solution:

$-\dfrac{1}{4}t < 2$ and $0.52t \geq 1.3$ First solve the individual inequalities. Multiply both sides of the first inequality by -4 (reverse the inequality sign). Divide the second inequality by 0.52.

$t > -8$ and $t \geq 2.5$ Take the *intersection* of the individual solution sets. The intervals overlap for values of t greater than or equal to 2.5.

$t > -8$

$t \geq 2.5$

$t > -8$ and $t \geq 2.5$

The solution set is $\{t \mid t \geq 2.5\}$. Interval notation: $[2.5, \infty)$

Skill Practice 4   Solve. $0.36w \leq 0.54$ and $-\dfrac{1}{2}w > 3$

Sometimes a compound inequality joined by the word "and" is written as a three-part inequality. For example:

$5 < -2x + 7$ and $-2x + 7 \leq 11$ In this example, two simultaneous conditions are imposed on the quantity $-2x + 7$.

$5 < -2x + 7 \leq 11$

To solve a three-part inequality, the goal is to isolate x in the middle region. This is demonstrated in Example 5.

Classroom Example: p. 103
Exercise 38

EXAMPLE 5   Solving a Three-Part Compound Inequality

Solve. $5 < -2x + 7 \leq 11$

Solution:

$5 < -2x + 7 \leq 11$

$5 - 7 < -2x + 7 - 7 \leq 11 - 7$ Subtract 7 from all three parts of the inequality.

$-2 < -2x \leq 4$

$\dfrac{-2}{-2} > \dfrac{-2x}{-2} \geq \dfrac{4}{-2}$ Divide all three parts by -2.

$1 > x \geq -2$ or equivalently $-2 \leq x < 1$.

The solution set is $\{x \mid -2 \leq x < 1\}$.
Interval notation: $[-2, 1)$

Instructor Note:
Remind students that a "three-part" inequality is used to imply that x is between two values. Show them why the following statements fail to do this.

$-5 > x < 4$
$-6 < x > 8$
$5 < x < -2$

Skill Practice 5   Solve. $-16 \leq -3y - 4 < 2$

3. Solve Absolute Value Inequalities

TIP The solution set to $|x| < 3$ is the set of real numbers within 3 units of zero on the number line.

The solution set to $|x| > 3$ is the set of real numbers more than 3 units from zero on the number line.

We now investigate the solutions to absolute value inequalities. For example:

Inequality	Graph	Solution Set
$|x| < 3$		$\{x \mid -3 < x < 3\}$
$|x| > 3$		$\{x \mid x < -3 \text{ or } x > 3\}$

We can generalize these observations with the following properties involving absolute value inequalities.

Properties Involving Absolute Value Inequalities

For a real number $k > 0$,

1. $|u| < k$ is equivalent to $-k < u < k$.   (1)
2. $|u| > k$ is equivalent to $u < -k$ or $u > k$.   (2)

Note: The statements also hold true for the inequality symbols $\leq$ and $\geq$, respectively.

Properties (1) and (2) follow directly from the definition: $|u| = \begin{cases} u \text{ if } u \geq 0 \\ -u \text{ if } u < 0 \end{cases}$

By definition, $|u| < k$ is equivalent to

$$u < k \quad \text{and} \quad -u < k$$
$$u < k \quad \text{and} \quad u > -k$$
$$-k < u < k \qquad (1)$$

By definition, $|u| > k$ is equivalent to

$$u > k \quad \text{or} \quad -u > k$$
$$u > k \quad \text{or} \quad u < -k$$
$$u < -k \quad \text{or} \quad u > k \qquad (2)$$

Classroom Example: p. 103
Exercise 48

EXAMPLE 6   Solving an Absolute Value Inequality

Solve the inequality and write the solution set in interval notation.

$$2|6 - m| - 3 < 7$$

Solution:

$2|6 - m| - 3 < 7$	First isolate the absolute value. Add 3 and divide by 2.
$|6 - m| < 5$	The inequality is in the form $|u| < k$, where $u = 6 - m$.
$-5 < 6 - m < 5$	Write the equivalent compound inequality $-k < u < k$.
$-11 < -m < -1$	Subtract 6 from all three parts.
$\dfrac{-11}{-1} > \dfrac{-m}{-1} > \dfrac{-1}{-1}$	Divide by -1 and reverse the inequality signs.

$11 > m > 1$ or equivalently $1 < m < 11$.

The solution set is $\{m \mid 1 < m < 11\}$.
Interval notation: $(1, 11)$

Skill Practice 6   Solve the inequality and write the solution set in interval notation. $3|5 - x| + 2 \leq 14$

Answer
6. $[1, 9]$

EXAMPLE 7 Solving an Absolute Value Inequality

Solve the inequality and write the solution set in interval notation.

$$-4 \geq -2|3x + 1|$$

Solution:

$-4 \geq -2\lvert 3x + 1\rvert$	First isolate the absolute value.
$\dfrac{-4}{-2} \leq \dfrac{-2\lvert 3x + 1\rvert}{-2}$	Divide both sides by -2 and reverse the inequality sign.
$2 \leq \lvert 3x + 1\rvert$ $\lvert 3x + 1\rvert \geq 2$	Write the absolute value on the left. Notice that the direction of the inequality sign is also changed. The inequality is now in the form $\lvert u\rvert \geq k$, where $u = 3x + 1$.
$3x + 1 \leq -2 \quad \text{or} \quad 3x + 1 \geq 2$	Write the equivalent form $u \leq -k$ or $u \geq k$.
$x \leq -1 \quad \text{or} \qquad x \geq \dfrac{1}{3}$	Take the union of the solution sets of the individual inequalities.

The solution set is $\left\{ x \,\middle|\, x \leq -1 \quad \text{or} \quad x \geq \dfrac{1}{3} \right\}$.

Interval notation: $(-\infty, -1] \cup \left[\dfrac{1}{3}, \infty\right)$

Skill Practice 7 Solve the inequality and write the solution set in interval notation. $\quad -18 > -3|2y - 4|$

An absolute value equation such as $|7x - 3| = -6$ has no solution because an absolute value cannot be equal to a negative number. We must also exercise caution when an absolute value is compared to a negative number or zero within an inequality. This is demonstrated in Example 8.

EXAMPLE 8 Solving Absolute Value Inequalities with Special Case Solution Sets

Solve the inequality and write the solution set in interval notation where appropriate.

a. $|x + 2| < -4$ **b.** $|x + 2| \geq -4$
c. $|x - 5| \leq 0$ **d.** $|x - 5| > 0$

Solution:

a. $|x + 2| < -4$

The solution set is $\{\ \}$.

By definition an absolute value is greater than or equal to zero. Therefore, the absolute value of an expression cannot be less than zero or any negative number. This inequality has no solution.

b. $|x + 2| \geq -4$

The solution set is $\mathbb{R}$.
Interval notation: $(-\infty, \infty)$

An absolute value of any real number is greater than or equal to zero. Therefore, it is also greater than every negative number. This inequality is true for all real numbers, x.

c. $|x - 5| \leq 0$
$|x - 5| = 0$
$x - 5 = 0$
$x = 5$
The solution set is $\{5\}$.

The absolute value of x minus 5 cannot be less than zero, but it can be *equal* to zero.

d. $|x - 5| > 0$

$|x - 5| > 0$ for all values of x except 5. When $x = 5$, we have $|5 - 5| = 0$, and this is not greater than zero. The solution set is all real numbers excluding 5.

The solution set is $\{x \mid x < 5 \text{ or } x > 5\}$.
Interval notation: $(-\infty, 5) \cup (5, \infty)$

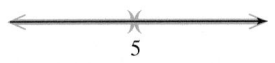

Skill Practice 8 Solve the inequality and write the solution set in interval notation where appropriate.

a. $|x - 3| < -2$ **b.** $|x - 3| > -2$
c. $|x + 1| \leq 0$ **d.** $|x + 1| > 0$

---

4. Solve Applications of Inequalities

In Example 9, we use a linear inequality to solve an application.

Classroom Example: p. 103
Exercise 70

EXAMPLE 9 **Using a Linear Inequality in an Application of Grades**

Emily has test scores of 96, 84, 80, 98, and 88. Her score for online homework is 100 and is weighted as one test grade. Emily still needs to take the final exam, which counts as two test grades. What score does she need on the final exam to have an average of at least 90? (This is the minimum average to earn an "A" in the class.)

Solution:

Let x represent the grade needed on the final exam.

$$\left(\begin{array}{c}\text{Average of}\\ \text{all scores}\end{array}\right) \geq 90$$

To earn an "A," Emily's average must be at least 90.

$$\frac{96 + 84 + 80 + 98 + 88 + 100 + 2x}{8} \geq 90$$

Take the sum of all grades. Divide by a total of eight grades.

$$\frac{546 + 2x}{8} \geq 90$$

$$8\left(\frac{546 + 2x}{8}\right) \geq 8\,(90)$$

Multiply by 8 to clear fractions.

$$546 + 2x \geq 720$$

Subtract 546 from both sides.

$$2x \geq 174$$

Divide by 2.

$$x \geq 87$$

Emily must earn a score of at least 87 to earn an "A" in the class.

Skill Practice 9 For a recent year, the monthly snowfall (in inches) for Chicago, Illinois, for November, December, January, and February was 2, 8.4, 11.2, and 7.9, respectively. How much snow would be necessary in March for Chicago to exceed its monthly average snowfall of 7.28 in. for these five months?

Answers

8. a. $\{ \ \}$ **b.** $\mathbb{R}; (-\infty, \infty)$
   c. $\{-1\}$ **d.** $(-\infty, -1) \cup (-1, \infty)$
9. Chicago would need more than 6.9 in. of snow in March.

SECTION R.8   Practice Exercises

Concept Connections

1. The multiplication and division properties of inequality indicate that if both sides of an inequality are multiplied or divided by a negative real number, the direction of the ___inequality___ sign must be reversed.

2. If a compound inequality consists of two inequalities joined by the word "and," the solution set is the ___intersection___ of the solution sets of the individual inequalities.

3. The compound inequality $a < x$ and $x < b$ can be written as the three-part inequality ___$a < x < b$___.

4. If a compound inequality consists of two inequalities joined by the word "or," the solution set is the ___union___ of the solution sets of the individual inequalities.

5. If k is a positive real number, then the inequality $|x| < k$ is equivalent to ___$-k$___ $< x <$ ___k___.

6. If k is a positive real number, then the inequality $|x| > k$ is equivalent to $x <$ ___$-k$___ or x ___$>$___ k.

7. If k is a positive real number, then the solution set to the inequality $|x| > -k$ is ___$\mathbb{R}$___.

8. If k is a positive real number, then the solution set to the inequality $|x| < -k$ is ___$\{\}$___.

Objective 1: Solve Linear Inequalities in One Variable

For Exercises 9–26, solve the inequality. Graph the solution set, and write the solution set in set-builder notation and interval notation. (See Examples 1–2)

9. $-2x - 5 > 17$

10. $-8t + 1 < 17$

11. $-3 \le -\frac{4}{3}w + 1$

12. $8 \ge -\frac{5}{2}y - 2$

13. $-1.2 + 0.6a \le 0.4a + 0.5$

14. $-0.7 + 0.3x \le 0.9x - 0.4$

15. $-5 > 6(c - 4) + 7$

16. $-14 < 3(m - 7) + 7$

17. $\frac{4 + x}{2} - \frac{x - 3}{5} < -\frac{x}{10}$

18. $\frac{y + 3}{4} - \frac{3y + 1}{6} > -\frac{1}{12}$

19. $\frac{1}{3}(x + 4) - \frac{5}{6}(x - 3) \ge \frac{1}{2}x + 1$

20. $\frac{1}{2}(t - 6) - \frac{4}{3}(t + 2) \ge -\frac{3}{4}t - 2$

21. $5(7 - x) + 2x < 6x - 2 - 9x$ $\{\}$

22. $2(3x + 1) - 4x > 2(x + 8) - 5$ $\{\}$

23. $5 - 3[2 - 4(x - 2)] \ge 6\{2 - [4 - (x - 3)]\}$

24. $8 - [6 - 10(x - 1)] \ge 2\{1 - 3[2 - (x + 4)]\}$

25. $4 - 3k > -2(k + 3) - k$

26. $2x - 9 < 6(x - 1) - 4x$

Objective 2: Solve Compound Linear Inequalities

For Exercises 27–34, solve the compound inequality. Graph the solution set, and write the solution set in interval notation. (See Examples 3–4)

27. **a.** $x < 4$ and $x \ge -2$
    b. $x < 4$ or $x \ge -2$

28. **a.** $y \le -2$ and $y > -5$
    b. $y \le -2$ or $y > -5$

29. **a.** $m + 1 \le 6$ or $\frac{1}{3}m < -2$
    b. $m + 1 \le 6$ and $\frac{1}{3}m < -2$

30. **a.** $n - 6 > 1$ or $\frac{3}{4}n \ge 6$
    b. $n - 6 > 1$ and $\frac{3}{4}n \ge 6$

31. **a.** $-\frac{2}{3}y > -12$ and $2.08 \ge 0.65y$
    b. $-\frac{2}{3}y > -12$ or $2.08 \ge 0.65y$

32. **a.** $-\frac{4}{5}m < 8$ and $0.85 \le 0.34m$
    b. $-\frac{4}{5}m < 8$ or $0.85 \le 0.34m$

33. **a.** $3(x - 2) + 2 \le x - 8$ or $4(x + 1) + 2 > -2x + 4$
    b. $3(x - 2) + 2 \le x - 8$ and $4(x + 1) + 2 > -2x + 4$

34. a. $5(t - 4) + 2 > 3(t + 1) - 3$ or $2t - 6 > 3(t - 4) - 2$

b. $5(t - 4) + 2 > 3(t + 1) - 3$ and $2t - 6 > 3(t - 4) - 2$

35. Write $-2.8 < y \le 15$ as two separate inequalities joined by "and." $-2.8 < y$ and $y \le 15$

36. Write $-\frac{1}{2} \le z < 2.4$ as two separate inequalities joined by "and." $-\frac{1}{2} \le z$ and $z < 2.4$

For Exercises 37–42, solve the compound inequality. Graph the solution set, and write the solution set in interval notation. (See Example 5)

37. $-3 < -2x + 1 \le 9$

38. $-6 \le -3x + 9 < 0$

39. $1 \le \dfrac{5x - 4}{2} < 3$

40. $-2 \le \dfrac{4x - 1}{3} \le 5$

41. $-2 \le \dfrac{-2x + 1}{-3} \le 4$

42. $-4 < \dfrac{-5x - 2}{-2} < 4$

Objective 3: Solve Absolute Value Inequalities

For Exercises 43–46, solve the equation or inequality. Write the solution set to each inequality in interval notation.

43. a. $|x| = 7$ $\{7, -7\}$

b. $|x| < 7$ $(-7, 7)$

c. $|x| > 7$ $(-\infty, -7) \cup (7, \infty)$

44. a. $|y| = 8$ $\{8, -8\}$

b. $|y| < 8$ $(-8, 8)$

c. $|y| > 8$ $(-\infty, -8) \cup (8, \infty)$

45. a. $|a + 9| + 2 = 6$ $\{-13, -5\}$

b. $|a + 9| + 2 \le 6$ $[-13, -5]$

c. $|a + 9| + 2 \ge 6$ $(-\infty, -13] \cup [-5, \infty)$

46. a. $|b + 1| - 4 = 1$ $\{-6, 4\}$

b. $|b + 1| - 4 \le 1$ $[-6, 4]$

c. $|b + 1| - 4 \ge 1$ $(-\infty, -6] \cup [4, \infty)$

For Exercises 47–60, solve the inequality, and write the solution set in interval notation if possible. (See Examples 6–7)

47. $3|4 - x| - 2 < 16$ $(-2, 10)$

48. $2|7 - y| + 1 < 17$ $(-1, 15)$

49. $2|x + 3| - 4 \ge 6$ $(-\infty, -8] \cup [2, \infty)$

50. $5|x + 1| - 9 \ge -4$ $(-\infty, -2] \cup [0, \infty)$

51. $|4w - 5| + 6 \le 2$ $\{\}$

52. $|2x + 7| + 5 < 1$ $\{\}$

53. $|5 - p| + 13 > 6$ $\mathbb{R}; (-\infty, \infty)$

54. $|12 - 7x| + 5 \ge 4$ $\mathbb{R}; (-\infty, \infty)$

55. $-11 \le 5 - |2p + 4|$ $[-10, 6]$

56. $-18 \le 6 - |3z + 3|$ $[-9, 7]$

57. $10 < |-5c - 4| + 2$

58. $15 < |-2d - 3| + 6$ $(-\infty, -6) \cup (3, \infty)$

59. $\left|\dfrac{y + 3}{6}\right| < 2$ $(-15, 9)$

60. $\left|\dfrac{m - 4}{2}\right| < 14$ $(-24, 32)$

For Exercises 61–68, write the solution set. (See Example 8)

61. a. $|x| = -9$ $\{\}$

b. $|x| < -9$ $\{\}$

c. $|x| > -9$ $\mathbb{R}; (-\infty, \infty)$

62. a. $|y| = -2$ $\{\}$

b. $|y| < -2$ $\{\}$

c. $|y| > -2$ $\mathbb{R}; (-\infty, \infty)$

63. a. $18 = 4 - |y + 7|$ $\{\}$

b. $18 \le 4 - |y + 7|$ $\{\}$

c. $18 \ge 4 - |y + 7|$ $\mathbb{R}; (-\infty, \infty)$

64. a. $15 = 2 - |p - 3|$ $\{\}$

b. $15 \le 2 - |p - 3|$ $\{\}$

c. $15 \ge 2 - |p - 3|$ $\mathbb{R}; (-\infty, \infty)$

65. a. $|z| = 0$ $\{0\}$

b. $|z| < 0$ $\{\}$

c. $|z| \le 0$ $\{0\}$

d. $|z| > 0$ $(-\infty, 0) \cup (0, \infty)$

e. $|z| \ge 0$ $\mathbb{R}; (-\infty, \infty)$

66. a. $|2w| = 0$ $\{0\}$

b. $|2w| < 0$ $\{\}$

c. $|2w| \le 0$ $\{0\}$

d. $|2w| > 0$ $(-\infty, 0) \cup (0, \infty)$

e. $|2w| \ge 0$ $\mathbb{R}; (-\infty, \infty)$

67. a. $|k + 4| = 0$ $\{-4\}$

b. $|k + 4| < 0$ $\{\}$

c. $|k + 4| \le 0$ $\{-4\}$

d. $|k + 4| > 0$ $(-\infty, -4) \cup (-4, \infty)$

e. $|k + 4| \ge 0$ $\mathbb{R}; (-\infty, \infty)$

68. a. $|c - 3| = 0$ $\{3\}$

b. $|c - 3| < 0$ $\{\}$

c. $|c - 3| \le 0$ $\{3\}$

d. $|c - 3| > 0$ $(-\infty, 3) \cup (3, \infty)$

e. $|c - 3| \ge 0$ $\mathbb{R}; (-\infty, \infty)$

Objective 4: Solve Applications of Inequalities

69. Marilee wants to earn an "A" in a class and needs an overall average of at least 92. Her test grades are 88, 92, 100, and 80. The average of her quizzes is 90 and counts as one test grade. The final exam counts as 2.5 test grades. What scores on the final exam would result in Marilee's overall average of 92 or greater? **(See Example 9)**   Marilee needs to score at least 96 on the final exam.

70. A 10-yr-old competes in gymnastics. For several competitions she received the following "All-Around" scores: 36, 36.9, 37.1, and 37.4. Her coach recommends that gymnasts whose "All-Around" scores average at least 37 move up to the next level. What "All-Around" scores in the next competition would result in the child being eligible to move up?   The child needs a score of at least 37.6.

71. Rita earns scores of 78, 82, 90, 80, and 75 on her five chapter tests for a certain class and a grade of 85 on the class project. The overall average for the course is computed as follows: the average of the five chapter tests makes up 60% of the course grade; the project accounts for 10% of the grade; and the final exam accounts for 30%. What scores can Rita earn on the final exam to earn a "B" in the course if the cut-off for a "B" is an overall score greater than or equal to 80, but less than 90? Assume that 100 is the highest score that can be earned on the final exam and that only whole-number scores are given. Rita needs to score between 77 and 100, inclusive.

73. A car travels 50 mph and passes a truck traveling 40 mph. How long will it take the car to be more than 16 mi ahead? It will take more than 1.6 hr or 1 hr 36 min.

75. For a certain bowling league, a beginning bowler computes her handicap by taking 90% of the difference between 220 and her average score in league play. Determine the average scores that would produce a handicap of 72 or less. Also assume that a negative handicap is not possible in this league.

77. Donovan has offers for two sales jobs. Job A pays a base salary of $25,000 plus a 10% commission on sales. Job B pays a base salary of $30,000 plus 8% commission on sales.

  a. How much would Donovan have to sell for the salary from Job A to exceed the salary from Job B? Donovan would need to sell more than $250,000 in merchandise.

  b. If Donovan routinely sells more than $500,000 in merchandise, which job would result in a higher salary? Job A

72. Trent earns scores of 66, 84, and 72 on three chapter tests for a certain class. His homework grade is 60 and his grade for a class project is 85. The overall average for the course is computed as follows: the average of the three chapter tests makes up 50% of the course grade; homework accounts for 20% of the grade; the project accounts for 10%; and the final exam accounts for 20%. What scores can Trent earn on the final exam to pass the course if he needs a "C" or better? A "C" or better requires an overall score of 70 or better, and 100 is the highest score that can be earned on the final exam. Assume that only whole-number scores are given. Trent needs to score between 63 and 100, inclusive.

74. A rectangular garden is to be constructed so that the width is 100 ft. What are the possible values for the length of the garden if at most 800 ft of fencing is to be used? The length must be 300 ft or less.

76. Body temperature is usually maintained between 36.5°C and 37.5°C, inclusive. Determine the corresponding range of temperatures in Fahrenheit. Use the relationship between degrees Celsius C and degrees Fahrenheit F: $C = \dfrac{5}{9}(F - 32)$. Normal body temperature is between 97.7°F and 99.5°F, inclusive.

78. Nancy wants to vacation in Austin, Texas. Hotel A charges $179 per night with a 14% nightly room tax and free parking. Hotel B charges $169 per night with an 18% nightly room tax plus a one-time $40 parking fee. After how many nights will Hotel B be less expensive? After 8 nights (9 or more), Hotel B will be less expensive.

Absolute value inequalities are often used to represent measurement error. For example, suppose that a machine is calibrated to dispense 8 fl oz of orange juice with a measurement error of no more than 0.05 fl oz. If x represents the actual amount of orange juice poured into the bottle, then x is a solution to the inequality $|x - 8| \le 0.05$. For Exercises 79–82,

a. Write an absolute value inequality to represent each statement.

b. Solve the inequality. Write the solution set in interval notation.

79. The variation between the measured value v and 16 oz is less than 0.01 oz.
  a. $|v - 16| < 0.01$ or equivalently $|16 - v| < 0.01$ **b.** (15.99, 16.01)

81. The value of x differs from 4 by more than 1 unit.
  a. $|x - 4| > 1$ or equivalently $|4 - x| > 1$ **b.** $(-\infty, 3) \cup (5, \infty)$

83. A refrigerator manufacturer recommends that the temperature t (in °F) inside a refrigerator be 36.5°F. If the thermostat has a margin of error of no more than 1.5°F,

  a. Write an absolute value inequality that represents an interval in which to estimate t.
  $|t - 36.5| \le 1.5$ or equivalently $|36.5 - t| \le 1.5$

  b. Solve the inequality and interpret the answer.
  [35, 38]; If the refrigerator is set to 36.5°F, the actual temperature would be between 35°F and 38°F, inclusive.

80. The variation between the measured value t and 60 min is less than 0.2 min.
  a. $|t - 60| < 0.2$ or equivalently $|60 - t| < 0.2$ **b.** (59.8, 60.2)

82. The value of y differs from 10 by more than 2 units.
  a. $|y - 10| > 2$ or equivalently $|10 - y| > 2$ **b.** $(-\infty, 8) \cup (12, \infty)$

84. A box of cereal is labeled to contain 16 oz. A consumer group takes a sample of 50 boxes and measures the contents of each box. The individual content of each box differs slightly from 16 oz, but by no more than 0.5 oz.

  a. If x represents the exact weight of the contents of a box of cereal, write an absolute value inequality that represents an interval in which to estimate x.
  $|x - 16| \le 0.5$ or equivalently $|16 - x| \le 0.5$

  b. Solve the inequality and interpret the answer. [15.5, 16.5]; The boxes of cereal vary in weight between 15.5 oz and 16.5 oz, inclusive.

85. The results of a political poll indicate that the leading candidate will receive 51% of the votes with a margin of error of no more than 3%. Let x represent the true percentage of votes received by this candidate.

   a. Write an absolute value inequality that represents an interval in which to estimate x.
   $|x - 0.51| \leq 0.03$ or equivalently $|0.51 - x| \leq 0.03$
   b. Solve the inequality and interpret the answer.
   $[0.48, 0.54]$; The candidate is expected to receive between 48%
   of the vote and 54% of the vote, inclusive.

86. A police officer uses a radar detector to determine that a motorist is traveling 34 mph in a 25-mph school zone. The driver goes to court and argues that the radar detector is not accurate. The manufacturer claims that the radar detector is calibrated to be in error by no more than 3 mph.

   a. If x represents the motorist's actual speed, write an inequality that represents an interval in which to estimate x. $|x - 34| \leq 3$ or equivalently $|34 - x| \leq 3$

   b. Solve the inequality and interpret the answer. Should the motorist receive a ticket?

Mixed Exercises

For Exercises 87–92, determine the set of values for x for which the radical expression would produce a real number. For example, the expression $\sqrt{x - 1}$ is a real number if $x - 1 \geq 0$ or equivalently, $x \geq 1$.

87. a. $\sqrt{x - 2}$ $\{x \mid x \geq 2\}$ **b.** $\sqrt{2 - x}$ $\{x \mid x \leq 2\}$ **88. a.** $\sqrt{x - 6}$ $\{x \mid x \geq 6\}$ **b.** $\sqrt{6 - x}$ $\{x \mid x \leq 6\}$

89. a. $\sqrt{x + 4}$ $\{x \mid x \geq -4\}$ **b.** $\sqrt[3]{x + 4}$ $\mathbb{R}$ **90. a.** $\sqrt{x + 7}$ $\{x \mid x \geq -7\}$ **b.** $\sqrt[3]{x + 7}$ $\mathbb{R}$

91. a. $\sqrt{2x - 9}$ $\{x \mid x \geq \frac{9}{2}\}$ **b.** $\sqrt[4]{2x - 9}$ $\{x \mid x \geq \frac{9}{2}\}$ **92. a.** $\sqrt{3x - 7}$ $\{x \mid x \geq \frac{7}{3}\}$ **b.** $\sqrt[4]{3x - 7}$ $\{x \mid x \geq \frac{7}{3}\}$

For Exercises 93–96, answer true or false given that $a > 0$, $b < 0$, $c > 0$, and $d < 0$.

93. $cd > a$ False   **94.** $ab < c$ True   **95.** If $a > c$, then $ad < cd$.   True   **96.** If $a < c$, then $ab < bc$.   False

Write About It

97. How is the process to solve a linear inequality different from the process to solve a linear equation?

98. Explain why $8 < x < 2$ has no solution.

99. Explain the difference between the solution sets for the following inequalities:

$$|x - 3| \leq 0 \quad \text{and} \quad |x - 3| > 0$$

100. Explain why $x^2 = 4$ is equivalent to the equation $|x| = 2$.

Expanding Your Skills

For Exercises 101–106, solve the inequality and write the solution set in interval notation.

101. $|x| + x < 11$ (*Hint*: Use the definition of $|x|$ to consider two cases.)
   Case 1: $x + x < 11$ if $x \geq 0$.
   Case 2: $-x + x < 11$ if $x < 0$. $\left(-\infty, \frac{11}{2}\right)$

102. $|x| - x > 10$ $(-\infty, -5)$

103. $1 < |x| < 9$ $(-9, -1) \cup (1, 9)$

104. $2 < |y| < 11$ $(-11, -2) \cup (2, 11)$

105. $5 \leq |2x + 1| \leq 7$ $[-4, -3] \cup [2, 3]$

106. $7 \leq |3x - 5| \leq 13$ $\left[-\frac{8}{3}, -\frac{2}{3}\right] \cup [4, 6]$

107. Solve the inequality for p: $|p - \hat{p}| < z\sqrt{\dfrac{\hat{p}\hat{q}}{n}}$. (Do not rationalize the denominator.) $\hat{p} - z\sqrt{\dfrac{\hat{p}\hat{q}}{n}} < p < \hat{p} + z\sqrt{\dfrac{\hat{p}\hat{q}}{n}}$

108. Solve the inequality for μ: $|\mu - \bar{x}| < \dfrac{z\sigma}{\sqrt{n}}$. (Do not rationalize the denominator.) $\bar{x} - \dfrac{z\sigma}{\sqrt{n}} < \mu < \bar{x} + \dfrac{z\sigma}{\sqrt{n}}$

PROBLEM RECOGNITION EXERCISES

Recognizing and Solving Equations and Inequalities

For Exercises 1–20,

    a. Identify the type of equation or inequality (some may fit more than one category).

    b. Solve the equation or inequality. Write the solution sets to the inequalities in interval notation if possible.

- Linear equation or inequality
- Quadratic equation
- Rational equation
- Absolute value equation or inequality

- Radical equation
- Equation in quadratic form
- Polynomial equation (degree > 2)
- Compound inequality

1. $(x^2 - 5)^2 - 5(x^2 - 5) + 4 = 0$ **a.** Equation in quadratic form and a polynomial equation   **b.** $\{\pm 3, \pm \sqrt{6}\}$

2. $2 \le |3t - 1| - 6$ **a.** Absolute value inequality   **b.** $\left(-\infty, -\frac{7}{3}\right] \cup [3, \infty)$

3. $\sqrt[3]{2y - 5} - 4 = -1$ **a.** Radical equation   **b.** $\{16\}$

4. $-9|3z - 7| + 1 = 4$ **a.** Absolute value equation   **b.** $\{\ \}$

5. $\dfrac{2}{w - 3} + \dfrac{5}{w + 1} = 1$ **a.** Rational equation   **b.** $\left\{\dfrac{9 \pm \sqrt{41}}{2}\right\}$

6. $48x^3 + 80x^2 - 3x - 5 = 0$ **a.** Polynomial equation   **b.** $\left\{-\dfrac{5}{3}, \pm\dfrac{1}{4}\right\}$

7. $-2(m + 2) < -m + 5$ and $6 \ge m + 3$ **a.** Compound inequality   **b.** $(-9, 3]$

8. $6 \le -2c + 8$ or $\dfrac{1}{3}c - 2 < 2$ **a.** Compound inequality   **b.** $(-\infty, 12)$

9. $(2p + 1)(p + 5) = 2p + 40$ **a.** Quadratic equation   **b.** $\left\{\dfrac{5}{2}, -7\right\}$

10. $2x(x - 4) + 7 = 2x^2 - 3[x + 5 - (2 + x)]$ **a.** Linear equation   **b.** $\{2\}$

11. $\dfrac{a - 4}{2} - \dfrac{3a + 1}{4} \le -\dfrac{a}{8}$ **a.** Linear inequality   **b.** $[-18, \infty)$

12. $3x^2 + 11 = 4$ **a.** Quadratic equation   **b.** $\left\{\pm\dfrac{\sqrt{21}}{3}i\right\}$

13. $-1 \le \dfrac{6 - x}{-5} \le 7$ **a.** Compound inequality   **b.** $[1, 41]$

14. $5 = \sqrt{5 + 2n} + \sqrt{2 + n}$ **a.** Radical equation   **b.** $\{2\}$; The value 142 does not check.

15. $|4x - 5| = |3x - 2|$ **a.** Absolute value equation   **b.** $\{1, 3\}$

16. $\dfrac{1}{d} - \dfrac{1}{2d - 1} + \dfrac{2d}{2d - 1} = 0$ **a.** Rational equation   **b.** $\{-1\}$; The value $\frac{1}{2}$ does not check.

17. $-|x + 4| + 8 > 3$ **a.** Absolute value inequality   **b.** $(-9, 1)$

18. $y - 4\sqrt{y} - 12 = 0$ **a.** Radical equation and an equation in quadratic form   **b.** $\{36\}$

19. $c^{2/3} = 16$ **a.** Radical equation   **b.** $\{\pm 64\}$

20. $2|z - 14| + 8 > 4$ **a.** Absolute value inequality   **b.** $(-\infty, \infty)$

ALGEBRA FOR CALCULUS

For Exercises 1–2, write an inequality using an absolute value that represents the given condition.

1. The distance between y and L is less than ε (epsilon). $|y - L| < \varepsilon$ or $|L - y| < \varepsilon$

2. The distance between x and c is less than δ (delta). $|x - c| < \delta$ or $|c - x| < \delta$

3. Simplify $\dfrac{x + 8}{|x + 8|}$

    a. for $x > -8$.   1   **b.** for $x < -8$. -1

4. Simplify $\dfrac{14 - x}{|x - 14|}$

    a. for $x > 14$. -1 **b.** for $x < 14$.   1

For Exercises 5–10,

    a. Simplify the expression.

    b. Substitute 0 for h in the simplified expression.

5. $\dfrac{2(x + h)^2 + 3(x + h) - (2x^2 + 3x)}{h}$ **a.** $4x + 2h + 3$ **b.** $4x + 3$

6. $\dfrac{3(x + h)^2 - 4(x + h) - (3x^2 - 4x)}{h}$ **a.** $6x + 3h - 4$ **b.** $6x - 4$

7. $\dfrac{\dfrac{1}{(x + h) - 2} - \dfrac{1}{x - 2}}{h}$ **a.** $\dfrac{-1}{(x - 2)(x + h - 2)}$ **b.** $\dfrac{-1}{(x - 2)^2}$

8. $\dfrac{\dfrac{1}{2(x + h) + 5} - \dfrac{1}{2x + 5}}{h}$ **a.** $\dfrac{-2}{(2x + 5)(2x + 2h + 5)}$ **b.** $\dfrac{-2}{(2x + 5)^2}$

9. $\dfrac{(x + h)^3 - x^3}{h}$ **a.** $3x^2 + 3xh + h^2$ **b.** $3x^2$

10. $\dfrac{(x + h)^4 - x^4}{h}$ **a.** $4x^3 + 6x^2h + 4xh^2 + h^3$ **b.** $4x^3$

For Exercises 11–12,

    a. Rationalize the numerator of the expression and simplify.

    b. Substitute 0 for h in the simplified expression.

11. $\dfrac{\sqrt{x + h} + 1 - (\sqrt{x} + 1)}{h}$ a. $\dfrac{1}{\sqrt{x} + \sqrt{x + h}}$ b. $\dfrac{1}{2\sqrt{x}}$

12. $\dfrac{\sqrt{2(x + h)} - \sqrt{2x}}{h}$ a. $\dfrac{2}{\sqrt{2x + 2h} + \sqrt{2x}}$ b. $\dfrac{1}{\sqrt{2x}}$

For Exercises 13–22, factor completely and write the answer with no negative exponents. Do not rationalize the denominator.

13. $\dfrac{3}{2}x^{1/2} + \dfrac{5}{2}x^{3/2}$ $\dfrac{x^{1/2}(5x + 3)}{2}$

14. $\dfrac{7}{6}x^{1/6} - \dfrac{1}{6}x^{-5/6}$ $\dfrac{7x - 1}{6x^{5/6}}$

15. $4(3x + 1)^3(3)(x^2 + 2)^3 + (3x + 1)^4(3)(x^2 + 2)^2(2x)$ $6(3x + 1)^3(x^2 + 2)^2(5x^2 + x + 4)$

16. $3(-2x + 3)^2(-2)(4x^2 - 5)^2 + (-2x + 3)^3(2)(4x^2 - 5)(8x)$ $-2(-2x + 3)^2(4x^2 - 5)(28x^2 - 24x - 15)$

17. $\dfrac{6(t - 1)^5(2t + 5)^6 - 6(2t + 5)^5(2)(t - 1)^6}{[(2t + 5)^6]^2}$ $\dfrac{42(t - 1)^5}{(2t + 5)^7}$

18. $\dfrac{6x^5(x^2 + 4)^3 - 3(x^2 + 4)^2(2x)x^6}{[(x^2 + 4)^3]^2}$ $\dfrac{24x^5}{(x^2 + 4)^4}$

19. $(x^2 + 4)^{1/2} + x \cdot \dfrac{1}{2}(x^2 + 4)^{-1/2}(2x)$ $\dfrac{2(x^2 + 2)}{(x^2 + 4)^{1/2}}$

20. $(2 - x^2)^{1/2} + x \cdot \dfrac{1}{2}(2 - x^2)^{-1/2}(-2x)$ $-\dfrac{2(x - 1)(x + 1)}{(2 - x^2)^{1/2}}$

21. $(x^2 + 1)^{-1/2} + x\left(-\dfrac{1}{2}\right)(x^2 + 1)^{-3/2}(2x)$ $\dfrac{1}{(x^2 + 1)^{3/2}}$

22. $6(3x - 1)^{1/3} + 6x\left(-\dfrac{1}{3}\right)(3x - 1)^{-2/3}(3)$ $\dfrac{6(2x - 1)}{(3x - 1)^{2/3}}$

For Exercises 23–28, write the answer as a single term and simplify. It is not necessary to rationalize the denominator.

23. $\dfrac{2\sqrt{x + 4} - \dfrac{x}{\sqrt{x + 4}}}{\left(\sqrt{x + 4}\right)^2}$ $\dfrac{x + 8}{(x + 4)\sqrt{x + 4}}$

24. $\dfrac{2x\sqrt{16 - x^2} + \dfrac{x^3}{\sqrt{16 - x^2}}}{\left(\sqrt{16 - x^2}\right)^2}$ $-\dfrac{x(x^2 - 32)}{(16 - x^2)\sqrt{16 - x^2}}$

25. $(x - 4)^{1/3} + \dfrac{x}{3(x - 4)^{2/3}}$ $\dfrac{4(x - 3)}{3(x - 4)^{2/3}}$

26. $(x + 5)^{1/4} + \dfrac{x}{4(x + 5)^{3/4}}$ $\dfrac{5(x + 4)}{4(x + 5)^{3/4}}$

27. $\dfrac{1}{2}\left(\dfrac{2x}{x + 1}\right)^{-1/2}\left[\dfrac{(x + 1)(2) - 2x(1)}{(x + 1)^2}\right]$ $\dfrac{1}{(2x)^{1/2}(x + 1)^{3/2}}$

28. $\dfrac{1}{3}\left(\dfrac{3x}{x^2 + 1}\right)^{-2/3}\left[\dfrac{(x^2 + 1) \cdot 3 - 3x(2x)}{(x^2 + 1)^2}\right]$ $-\dfrac{(x - 1)(x + 1)}{(3x)^{2/3}(x^2 + 1)^{4/3}}$

EQUATIONS AND INEQUALITIES FOR CALCULUS

In Calculus you will see the symbol y'. For Exercises 1–4, treat y' as a variable and solve the equation for y'.

1. $\dfrac{2x}{25} + \dfrac{2y}{9}y' = 0$ $y' = -\dfrac{9x}{25y}$

2. $2xy^3 + 3x^2y^2y' - y' = 1$ $y' = \dfrac{1 - 2xy^3}{3x^2y^2 - 1}$

3. $3y^2y' + 6xy + 3x^2y' = 2y^2 + 4xyy'$ $y' = \dfrac{2y(y - 3x)}{3x^2 - 4xy + 3y^2}$

4. $3(x + y)^2 + 3(x + y)^2\,y' - 3y^2y' = 3x^2$ $y' = -\dfrac{y(2x + y)}{x(2y + x)}$

For Exercises 5–7, simplify the expression. Do not rationalize the denominator.

5. $2x\sqrt{2x - 3} + x^2\left(\dfrac{1}{2}\right)\dfrac{1}{\sqrt{2x - 3}}(2)$ $\dfrac{x(5x - 6)}{\sqrt{2x - 3}}$

6. $\dfrac{2x(2x - 7)^{1/2} - x^2\left(\dfrac{1}{2}\right)(2x - 7)^{-1/2}(2)}{\left[(2x - 7)^{1/2}\right]^2}$ $\dfrac{x(3x - 14)}{(2x - 7)^{3/2}}$

7. $\dfrac{(1)(x^2 - 9)^{1/2} - x\left(\dfrac{1}{2}\right)(x^2 - 9)^{-1/2}(2x)}{\left[(x^2 - 9)^{1/2}\right]^2}$ $-\dfrac{9}{(x^2 - 9)^{3/2}}$

For Exercises 8–10,

   a. Simplify the expression. Do not rationalize the denominator.

   b. Find the values of x for which the expression equals zero.

   c. Find the values of x for which the denominator is zero.

8. $\dfrac{4x(4x-5)-2x^2(4)}{(4x-5)^2}$ **a.** $\dfrac{4x(2x-5)}{(4x-5)^2}$ **b.** $x=0,\ x=\dfrac{5}{2}$ **c.** $x=\dfrac{5}{4}$

9. $\dfrac{-6x(6x+1)-(-3x^2)(6)}{(6x+1)^2}$ **a.** $-\dfrac{6x(3x+1)}{(6x+1)^2}$

   b. $x=0,\ x=-\dfrac{1}{3}$ **c.** $x=-\dfrac{1}{6}$

10. $\sqrt{4-x^2}-x\left(\dfrac{1}{2}\right)\dfrac{1}{\sqrt{4-x^2}}(2x)$ **a.** $\dfrac{2(2-x^2)}{\sqrt{4-x^2}}$

   b. $x=\pm\sqrt{2}$ **c.** $x=2$ and $x=-2$

Some applications of calculus use a mathematical structure called a power series. To find the interval of convergence of a power series, it is often necessary to solve an absolute value inequality. For Exercises 11–12, solve the absolute value inequality to find the interval of convergence.

11. $\left|\dfrac{x+1}{2}\right|<1$ $(-3, 1)$

12. $\left|-\dfrac{x}{2}\right|<1$ $(-2, 2)$

13. A 6-ft man walks away from a lamppost. At the instant the man is 14 ft away from the lamppost, the shadow is 10 ft. Find the height of the lamppost.   14.4 ft

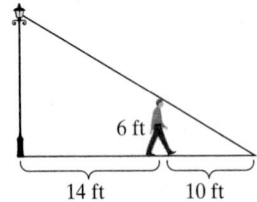

14 ft   10 ft

14. A water trough has a cross section in the shape of an equilateral triangle with sides of length 1 m. The length is 3 m. Determine the volume of water when the water level is $\frac{1}{2}$ m. $\dfrac{\sqrt{3}}{4}$ m^3

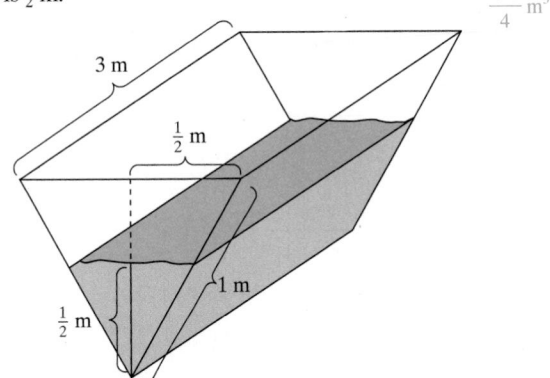

3 m

$\frac{1}{2}$ m

1 m

$\frac{1}{2}$ m

15. A contractor builds a swimming pool with cross section in the shape of a trapezoid. The deep end is 8 ft deep and the shallow end is 3 ft deep. The length of the pool is 50 ft and the width is 20 ft. As the pool is being filled, find the volume of water when the depth is 4 ft.   1600 ft^3

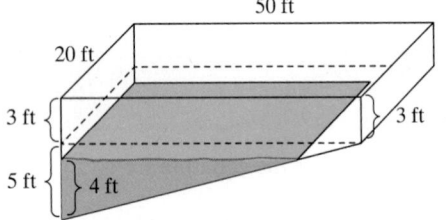

50 ft

20 ft

3 ft   3 ft

5 ft   4 ft

CHAPTER R   KEY CONCEPTS

SECTION R.1 Sets and the Real Number Line	Reference
Natural Numbers: $\mathbb{N}=\{1, 2, 3, \ldots\}$ **Whole Numbers:** $\mathbb{W}=\{0, 1, 2, 3, \ldots\}$ **Integers:** $\mathbb{Z}=\{\ldots, -3, -2, -1, 0, 1, 2, 3, \ldots\}$ **Rational Numbers:** $\mathbb{Q}=\left\{\frac{p}{q}\,\middle|\,p, q\in\mathbb{Z}\text{ and }q\neq 0\right\}$ **Irrational Numbers:** $\mathbb{H}$ is the set of real numbers that cannot be expressed as a ratio of integers.	p. 2
$A\cup B$ is the **union** of A and B. This is the set of elements that belong to set A or set B or to both sets A and B. $A\cap B$ is the **intersection** of A and B. This is the set of elements common to both A and B. $A\cup B$ $A\cap B$	p. 4

| For a real number x, the **absolute value of x** is $\|x\| = \begin{cases} x & \text{if } x \geq 0 \\ -x & \text{if } x < 0 \end{cases}$ | p. 6 |

The distance between two points a and b on a number line is given by $\|a - b\|$ or $\|b - a\|$. — p. 7

		p. 9
• Commutative property of addition	$a + b = b + a$	
• Commutative property of multiplication	$a \cdot b = b \cdot a$	
• Associative property of addition	$(a + b) + c = a + (b + c)$	
• Associative property of multiplication	$(a \cdot b) \cdot c = a \cdot (b \cdot c)$	
• Identity property of addition	$a + 0 = a$ and $0 + a = a$	
• Identity property of multiplication	$a \cdot 1 = a$ and $1 \cdot a = a$	
• Inverse property of addition	$a + (-a) = 0$ and $(-a) + a = 0$	
• Inverse property of multiplication	$a \cdot \frac{1}{a} = 1$ and $\frac{1}{a} \cdot a = 1$ where $a \neq 0$	
• Distributive property of multiplication over addition	$a(b + c) = ab + ac$	

SECTION R.2 Exponents and Radicals — Reference

Properties of exponents and key definitions: — pp. 16–17

$$b^m \cdot b^n = b^{m+n} \qquad (b^m)^n = b^{m \cdot n} \qquad \left(\frac{a}{b}\right)^m = \frac{a^m}{b^m} \qquad b^{-n} = \frac{1}{b^n}$$

$$\frac{b^m}{b^n} = b^{m-n} \qquad (ab)^m = a^m b^m \qquad b^0 = 1$$

A number expressed in the form $a \times 10^n$, where $1 \leq \|a\| < 10$ and n is an integer is said to be in **scientific notation.** — p. 19

b is an nth-root of a if $b^n = a$.

$\sqrt[n]{a}$ represents the principal nth-root of a. — p. 20

If $n > 1$ is an integer and $\sqrt[n]{a}$ is a real number, then — p. 21
- $a^{1/n} = \sqrt[n]{a}$
- $a^{m/n} = \left(\sqrt[n]{a}\right)^m$ and $a^{m/n} = \sqrt[n]{a^m}$

Let $n > 1$ be an integer and a be a real number. — p. 22
- If n is *even* then $\sqrt[n]{a^n} = \|a\|$.   • If n is *odd* then $\sqrt[n]{a^n} = a$.

Product property of radicals: $\sqrt[n]{a} \cdot \sqrt[n]{b} = \sqrt[n]{ab}$

Quotient property of radicals: $\dfrac{\sqrt[n]{a}}{\sqrt[n]{b}} = \sqrt[n]{\dfrac{a}{b}}$

Property of nested radicals: $\sqrt[m]{\sqrt[n]{a}} = \sqrt[m \cdot n]{a}$

SECTION R.3 Polynomials and Factoring — Reference

Special Case Products: — p. 31
$(a + b)(a - b) = a^2 - b^2$
$(a + b)^2 = a^2 + 2ab + b^2$
$(a - b)^2 = a^2 - 2ab + b^2$

General factoring strategy:	p. 37
1. Factor out the GCF.	
2. Identify the number of terms in the polynomial.	
4 terms: Factor by grouping either 2 terms with 2 terms or 3 terms with 1 term.	
3 terms: If the trinomial is a perfect square trinomial, factor as the square of a binomial.	
$a^2 + 2ab + b^2 = (a + b)^2$	
$a^2 - 2ab + b^2 = (a - b)^2$	
Otherwise, factor by the trial-and-error method.	
2 terms: Determine whether the binomial fits one of the following patterns.	
$a^2 - b^2 = (a + b)(a - b)$	
$a^3 + b^3 = (a + b)(a^2 - ab + b^2)$	
$a^3 - b^3 = (a - b)(a^2 + ab + b^2)$	

SECTION R.4 Rational Expressions and More Operations on Radicals	Reference
A **rational expression** is a ratio of two polynomials. Values of the variable that make the denominator equal to zero are called restricted values of the variable.	p. 43
To simplify a rational expression, use the property of equivalent algebraic fractions. $$\frac{ac}{bc} = \frac{a}{b} \quad \text{for } b \neq 0, c \neq 0$$	p. 44
To multiply or divide rational expressions, $$\frac{a}{b} \cdot \frac{c}{d} = \frac{ac}{bd} \quad \text{for } b \neq 0, d \neq 0 \quad \text{and} \quad \frac{a}{b} \div \frac{c}{d} = \frac{a}{b} \cdot \frac{d}{c} = \frac{ad}{bc} \quad \text{for } b \neq 0, c \neq 0, d \neq 0$$	p. 45
To add or subtract rational expressions, write each fraction as an equivalent fraction with a common denominator. Then apply the following properties. $$\frac{a}{b} + \frac{c}{b} = \frac{a + c}{b} \quad \text{and} \quad \frac{a}{b} - \frac{c}{b} = \frac{a - c}{b} \quad \text{for } b \neq 0$$	p. 46
Removing a radical from the denominator of a fraction is called **rationalizing the denominator.**	p. 50

SECTION R.5 Equations with Real Solutions	Reference
A **linear equation in one variable** is an equation that can be written in the form $ax + b = 0$, where a and b are real numbers and $a \neq 0$.	p. 54
A **conditional equation** is an equation that is true for some values of the variable but false for others. A **contradiction** is false for all values of the variable. An **identity** is true for all values of the variable for which the expressions in the equation are defined.	p. 56
Let a, b, and c represent real numbers. A **quadratic equation** in the variable x is an equation of the form $ax^2 + bx + c = 0$, where $a \neq 0$.	p. 57
Zero product property: If $mn = 0$, then $m = 0$ or $n = 0$.	p. 57
Square root property: If $x^2 = k$, then $x = \pm\sqrt{k}$.	p. 57
A quadratic equation can be solved by completing the square and applying the square root property.	p. 58
The solutions to $ax^2 + bx + c = 0$ $(a \neq 0)$ are given by the quadratic formula. $$x = \frac{-b \pm \sqrt{b^2 - 4ac}}{2a}$$	p. 60

Rational equations: Solve a rational equation by multiplying both sides of the equation by the LCD of all fractions in the equation.	p. 61
Absolute value equations: To solve an absolute value equation, isolate the absolute value. Then use one of the following properties for a positive real number k. 1. $\|u\| = k$ is equivalent to $u = k$ or $u = -k$. 2. $\|u\| = 0$ is equivalent to $u = 0$. 3. $\|u\| = -k$ has no solution. 4. $\|u\| = \|w\|$ is equivalent to $u = w$ or $u = -w$.	p. 63
Solving radical equations: 1. Isolate the radical. If an equation has more than one radical, choose one of the radicals to isolate. 2. Raise each side of the equation to a power equal to the index of the radical. 3. Solve the resulting equation. If the equation still has a radical, repeat steps 1 and 2. 4. Check the potential solutions in the original equation and write the solution set.	p. 64

SECTION R.6 Complex Numbers and More Quadratic Equations	**Reference**
$i = \sqrt{-1}$ and $i^2 = -1$. For a real number $b > 0$, $\sqrt{-b} = i\sqrt{b}$.	p. 71
To add or subtract complex numbers, combine the real parts, and combine the imaginary parts.	p. 73
Multiply complex numbers by using the distributive property.	p. 74
The product of complex conjugates: $(a + bi)(a - bi) = a^2 + b^2$	p. 75
Divide complex numbers by multiplying the numerator and denominator by the conjugate of the denominator.	p. 76
The discriminant to the equation $ax^2 + bx + c = 0$ $(a \neq 0)$ is given by $b^2 - 4ac$. The discriminant indicates the number of and type of solutions to the equation. • If $b^2 - 4ac < 0$, the equation has 2 nonreal solutions. • If $b^2 - 4ac = 0$, the equation has 1 real solution. • If $b^2 - 4ac > 0$, the equation has 2 real solutions.	p. 78
Substitution can be used to solve equations that are in quadratic form.	p. 79

SECTION R.7 Applications with Linear and Rational Equations	**Reference**
Equations in algebra can be used to organize information from a physical situation.	p. 82
Quadratic equations are used to model applications with the Pythagorean theorem, volume, area, and objects moving vertically under the influence of gravity.	p. 87
The vertical position s of an object moving vertically under the influence of gravity is approximated by $s = -\frac{1}{2}gt^2 + v_0t + s_0$, where • g is the acceleration due to gravity (at sea level on Earth: $g = 32$ ft/sec^2 or 9.8 m/sec^2). • t is the time after the start of the experiment. • v_0 is the initial velocity. • s_0 is the initial position (height). • s is the position of the object at time t.	p. 88

SECTION R.8 Linear, Compound, and Absolute Value Inequalities	Reference
An inequality that can be written in one of the following forms is a **linear inequality in one variable.** $ax + b < 0$, $ax + b \leq 0$, $ax + b > 0$, or $ax + b \geq 0$	p. 95
Solving compound inequalities: • If two inequalities are joined by the word "and," take the *intersection* of the individual solution sets. • The inequality $a < x < b$ is equivalent to $a < x$ and $x < b$. • If two inequalities are joined by the word "or," take the *union* of the individual solution sets.	p. 97
Solving absolute value inequalities: For a positive real number k, 1. $|u| < k$ is equivalent to $-k < u < k$. 2. $|u| > k$ is equivalent to $u < -k$ or $u > k$.	p. 99

Expanded Chapter Summary available at www.mhhe.com/millerprecalculus.

CHAPTER R   Review Exercises

SECTION R.1

1. Complete the table.

	Graph	Interval Notation	Set-Builder Notation
a.	−3 to 7	$[-3, 7)$	$\{x \mid -3 \leq x < 7\}$
b.	2.1	$(2.1, \infty)$	$\{x \mid x > 2.1\}$
c.	4	$(-\infty, 4]$	$\{x \mid 4 \geq x\}$

2. Given $A = \{10, 11, 12, 13\}$, $B = \{10, 12, 14, 16\}$, and $C = \{7, 8, 9, 10, 11\}$, find

   a. $A \cup B$ **b.** $A \cap B$ **c.** $A \cup C$

   d. $A \cap C$ **e.** $B \cup C$ **f.** $B \cap C$

3. Given $X = \{x \mid x < 7\}$, $Y = \{x \mid x \geq -2\}$, and $Z = \{x \mid x < -3\}$, write the union or intersection in set-builder notation.

   a. $X \cup Y$ **b.** $X \cap Y$ **c.** $X \cup Z$

   d. $X \cap Z$ **e.** $Y \cup Z$ **f.** $Y \cap Z$

4. Simplify without absolute value bars.

   a. $|w - 4|$ for $w < 4$ $4 - w$

   b. $|w - 4|$ for $w \geq 4$ $w - 4$

5. a. Write an absolute value expression to represent the distance between 2 and $\sqrt{5}$ on the number line.
   $|2 - \sqrt{5}|$ or $|\sqrt{5} - 2|$
   b. Simplify the expression from part (a) without absolute value bars. $\sqrt{5} - 2$

For Exercises 6–8, simplify the expression.

   6. a. 16^2 256 **b.** $(-16)^2$ 256 **c.** -16^2 -256

   d. $\sqrt{16}$ 4 **e.** $-\sqrt{16}$ -4 **f.** $\sqrt{-16}$ Not a real number

7. $\dfrac{32 - |-11 + 3|}{36 \div 2 \div 3 - 2}$ 6

8. $-5 - 3\left[8 - 2\sqrt{4^2 + (2 - 5)^2}\right]$ 1

9. Jesse makes $150 more per week than Ethan.

   a. Write a model for Jesse's salary J in terms of Ethan's salary E. $J = E + 150$

   b. Write a model for Ethan's salary E in terms of Jesse's salary J. $E = J - 150$

For Exercises 10–11, simplify the expression.

   10. $15.2c^2d - 11.1cd + 8.7c^2d - 5.4cd$ $23.9c^2d - 16.5cd$

   11. $8 - \left[4x^2 - \dfrac{1}{2}(6 - 4x^2) + 3\right] + 13x^2$ $7x^2 + 8$

SECTION R.2

For Exercises 12–15, simplify completely. Write the answers with positive exponents only.

   12. a. 9^0 1 **b.** -9^0 -1 **c.** $9x^0$ 9 **d.** $(9x)^0$ 1

   13. a. $\dfrac{1}{m^{-5}}$ m^5 **b.** m^{-5} $\dfrac{1}{m^5}$ **c.** $8m^{-9}n^2$ $\dfrac{8n^2}{m^9}$ **d.** $8m^9n^{-2}$ $\dfrac{8m^9}{n^2}$

   14. $(-12a^{-3}b^4)^2$ $\dfrac{144b^8}{a^6}$

   15. $\left(\dfrac{1}{2u^5v^{-2}}\right)^{-3}\left(\dfrac{4}{u^{-3}v^2}\right)^{-1}$ $\dfrac{2u^{12}}{v^4}$

   16. Write the numbers in scientific notation.

     a. 4920 **b.** 0.00492 **c.** 4.92
     4.92×10^3 4.92×10^{-3} 4.92×10^0

   17. Write the numbers in standard decimal notation.

     a. 9.8×10^{-1} **b.** 9.8×10^0 **c.** 9.8×10^1
     0.98 9.8 98

   18. Simplify. $\dfrac{(8.6 \times 10^{-3})(4.1 \times 10^8)}{2.0 \times 10^{-6}}$ 1.763×10^{12}

Additional answers can be found in the Instructor Answer Appendix.

For Exercises 19–21, simplify the expression. Write exponential expressions with positive exponents only.

19. a. $-\sqrt[4]{256}$ $\quad$ _−4_ $\qquad$ **b.** $\sqrt[4]{-256}$ $\quad$ _Not a real number_

20. $\sqrt[3]{-\dfrac{8}{125}}$ $\quad$ _$-\dfrac{2}{5}$_

21. a. $10{,}000^{3/4}$ $\quad$ _1000_ $\qquad$ **b.** $10{,}000^{-3/4}$ $\quad$ _$\dfrac{1}{1000}$_

$\quad$ **c.** $-10{,}000^{3/4}$ $\quad$ _−1000_ $\qquad$ **d.** $-10{,}000^{-3/4}$ $\quad$ _$-\dfrac{1}{1000}$_

$\quad$ **e.** $(-10{,}000)^{3/4}$ $\quad$ _Undefined_ **f.** $(-10{,}000)^{-3/4}$ $\quad$ _Undefined_

For Exercises 22–26, simplify the radical expressions. Assume that all variables represent positive real numbers.

22. $\sqrt[3]{54xy^{12}z^{14}}$ $\quad$ _$3y^4z^4\sqrt[3]{2xz^2}$_

23. $\sqrt[4]{32b^3c^{15}d^8}$ $\quad$ _$2c^3d^2\sqrt[4]{2b^3c^3}$_

24. $\dfrac{\sqrt[3]{3xy^5}}{\sqrt[3]{81x^7y^2}}$ $\quad$ _$\dfrac{y}{3x^2}$_

25. $\sqrt{10}\cdot\sqrt{35}$ $\quad$ _$5\sqrt{14}$_

26. $\left(-5\sqrt[5]{a^3}\right)\left(6\sqrt[5]{a^4}\right)$ $\quad$ _$-30a\sqrt[5]{a^2}$_

For Exercises 27–29, add or subtract as indicated. Assume that all variables represent positive real numbers.

27. $\dfrac{1}{5}\sqrt{125}+\dfrac{3}{2}\sqrt{20}-\dfrac{1}{4}\sqrt{80}$ $\quad$ _$3\sqrt{5}$_

28. $-2c\sqrt[3]{54c^2d^3}+5cd\sqrt[3]{2c^2}-10d\sqrt[3]{250c^5}$ $\quad$ _$-51cd\sqrt[3]{2c^2}$_

29. $3\sqrt{5t}+8t\sqrt{5t}$ $\quad$ _$(3+8t)\sqrt{5t}$_

SECTION R.3

For Exercises 30–38, perform the indicated operations.

30. $(4x^3-3x^2+7)+(-3x^3-6x^2+2x)$ $\quad$ _x^3-9x^2+2x+7_

31. $(-7.2a^2b^3+4.1ab^2-3.9b)-(0.8a^2b^3-3.2ab^2-b)$ $\quad$ _$-8a^2b^3+7.3ab^2-2.9b$_

32. $(5w^3+6y^2)(2w^3-y^2)$ $\quad$ _$10w^6+7w^3y^2-6y^4$_

33. $(4p-6)\left(\dfrac{1}{2}p^2-p+4\right)$ $\quad$ _$2p^3-7p^2+22p-24$_

34. $(9t-4)(9t+4)$ $\quad$ _$81t^2-16$_

35. $\left(\dfrac{1}{3}m-\dfrac{1}{4}n^3\right)\left(\dfrac{1}{3}m+\dfrac{1}{4}n^3\right)$ $\quad$ _$\dfrac{1}{9}m^2-\dfrac{1}{16}n^6$_

36. $(5k-3)^2$ $\quad$ _$25k^2-30k+9$_

37. $(6c^2+5d^3)^2$ $\quad$ _$36c^4+60c^2d^3+25d^6$_

38. $[(2v-1)+w][(2v-1)-w]$ $\quad$ _$4v^2-4v+1-w^2$_

For Exercises 39–42, multiply the radicals and simplify. Assume that all variable expressions represent positive real numbers.

39. $\left(6\sqrt{5}-2\sqrt{3}\right)\left(2\sqrt{5}+5\sqrt{3}\right)$ $\quad$ _$30+26\sqrt{15}$_

40. $\left(7\sqrt{2}-2\sqrt{11}\right)\left(7\sqrt{2}+2\sqrt{11}\right)$ $\quad$ _54_

41. $\left(2c^2\sqrt{d}-5d^2\sqrt{c}\right)^2$ $\quad$ _$4c^4d-20c^2d^2\sqrt{cd}+25cd^4$_

42. $\left(\sqrt{x+2}+4\right)^2$ $\quad$ _$x+18+8\sqrt{x+2}$_

For Exercises 43–59, factor completely and write the answer with no negative exponents.

43. $80m^4n^8-48m^5n^3-16m^2n$ $\quad$ _$16m^2n(5m^2n^7-3m^3n^2-1)$_

44. $11p^2(2p+1)-22p(2p+1)$ $\quad$ _$11p(2p+1)(p-2)$_

45. $15ac-14b-10a+21bc$ $\quad$ _$(5a+7b)(3c-2)$_

46. $-t+12t^2-6$ $\quad$ _$(4t-3)(3t+2)$_

47. $8x^3-40x^2y+50xy^2$ $\quad$ _$2x(2x-5y)^2$_

48. $256a^4-625$ $\quad$ _$(16a^2+25)(4a+5)(4a-5)$_

49. $3k^4-81k$ $\quad$ _$3k(k-3)(k^2+3k+9)$_

50. $(c+2)^3+d^3$ $\quad$ _$(c+2+d)(c^2+4c+4-cd-2d+d^2)$_

51. $25n^2-m^2-12m-36$ $\quad$ _$(5n+m+6)(5n-m-6)$_

52. $(x^2-7)^2+8(x^2-7)+15$ $\quad$ _$(x^2-2)(x+2)(x-2)$_

53. $(2p-5)^2-(4p+1)^2$ $\quad$ _$-4(3p-2)(p+3)$_

54. m^6+9m^3+8 $\quad$ _$(m+1)(m^2-m+1)(m+2)(m^2-2m+4)$_

55. $x^4+6x^2y+9y^2-x^2-3y$ $\quad$ _$(x^2+3y)(x^2+3y-1)$_

56. $7w^{-8}+5w^{-7}+w^{-6}$ $\quad$ _$\dfrac{7+5w+w^2}{w^8}$_

57. $12x^{7/2}-4x^{5/2}$ $\quad$ _$4x^{5/2}(3x-1)$_

58. $x(2x+5)^{-3/4}+(2x+5)^{1/4}$ $\quad$ _$\dfrac{3x+5}{(2x+5)^{3/4}}$_

59. $2x(1-x^2)^{1/2}+x^2\cdot\dfrac{1}{2}(1-x^2)^{-1/2}(-2x)$ $\quad$ _$\dfrac{x(3x^2-2)}{(1-x^2)^{1/2}}$_

SECTION R.4

For Exercises 60–61, simplify and state the restricted values on the variable.

60. $-\dfrac{24a^2c^5d^2}{16c^6d^7}$ $\quad$ _$-\dfrac{3a^2}{2cd^5}$; $c\ne 0$, $d\ne 0$_

61. $\dfrac{m^2-16}{m^2-m-12}$ $\quad$ _$\dfrac{m+4}{m+3}$; $m\ne 4$, $m\ne -3$_

For Exercises 62–70, perform the indicated operations.

62. $\dfrac{2x-5y}{x^3y}\cdot\dfrac{x^5y^7}{6x-15y}$ $\quad$ _$\dfrac{x^2y^6}{3}$_

63. $\dfrac{4ac}{a^2+4ac}\cdot\dfrac{a^2+8ac+16c^2}{8a+32c}$ $\quad$ _$\dfrac{c}{2}$_

64. $\dfrac{5x^2+25x+125}{16x-x^3}\div\dfrac{x^3-125}{x^2-9x+20}$ $\quad$ _$-\dfrac{5}{x(x+4)}$_

65. $\dfrac{7}{20x}+\dfrac{2}{15x^4}$ $\quad$ _$\dfrac{21x^3+8}{60x^4}$_

66. $\dfrac{6}{9x^2-y^2}-\dfrac{1}{3x^2-xy}$ $\quad$ _$\dfrac{1}{x(3x+y)}$_

67. $\dfrac{4}{x^2}+\dfrac{3}{x+3}-\dfrac{2}{x}$ $\quad$ _$\dfrac{x^2-2x+12}{x^2(x+3)}$_

68. $\dfrac{x-1}{x-6}+\dfrac{5}{6-x}$ $\quad$ _1_

69. $\dfrac{\dfrac{1}{16x} + \dfrac{1}{8}}{\dfrac{1}{4} + \dfrac{1}{8x}}$ $\dfrac{1}{2}$

70. $\dfrac{5m^{-1} - n^{-1}}{25m^{-2} - n^{-2}}$ $\dfrac{mn}{5n + m}$

For Exercises 71–74, simplify the expression. Assume that all variable expressions represent positive real numbers.

71. $\dfrac{5}{\sqrt{k}}$ $\dfrac{5\sqrt{k}}{k}$

72. $\dfrac{5}{\sqrt[4]{k}}$ $\dfrac{5\sqrt[4]{k^3}}{k}$

73. $\dfrac{15}{\sqrt{10} - \sqrt{7}}$ $5(\sqrt{10} + \sqrt{7})$

74. $\dfrac{x - 4}{\sqrt{x} + 2}$ $\sqrt{x} - 2$

SECTION R.5

For Exercises 75–90, solve the equation.

75. $-8(t - 4) + 7 = 4[t - 3(1 - t)] + 6$ $\left\{\dfrac{15}{8}\right\}$

76. $\dfrac{m + 2}{3} - \dfrac{m - 4}{4} = \dfrac{m + 1}{6} - 1$ $\{30\}$

77. $x - 5 + 2(x - 4) = 3(x + 1) - 5$ $\{\,\}$

78. $0.2x + 1.6 = x - 0.8(x - 2)$ $\mathbb{R}$

79. $\dfrac{x + 3}{5x} + 2 = \dfrac{x - 4}{x}$ $\left\{-\dfrac{23}{6}\right\}$

80. $3y^2 - 4y = 8 - 6y$ $\left\{\dfrac{4}{3}, -2\right\}$ **81.** $(2v + 3)^2 - 1 = 6$ $\left\{\dfrac{-3 \pm \sqrt{7}}{2}\right\}$

82. $2d(d - 3) = 1 + 4d$

83. $4x^3 - 6x^2 - 20x + 30 = 0$ $\left\{\pm\sqrt{5}, \dfrac{3}{2}\right\}$

84. $\dfrac{n}{3n + 2} + 1 = \dfrac{4}{n - 2}$ $\left\{\dfrac{9 \pm \sqrt{129}}{4}\right\}$

85. $\sqrt{k + 7} - \sqrt{3 - k} = 2$ $\{2\}$; The value -6 does not check.

86. $\sqrt{51 - 14x} + 4 = x - 2$

87. $(x - 11)^{2/3} = 9$ $\{ \}$; The values -5 and 3 do not check. $\{-16, 38\}$

88. $-2|3y - 10| + 4 = -6$ $\left\{\dfrac{5}{3}, 5\right\}$

89. $|6 - w| + 7 = 2$ $\{\,\}$

90. $10w^{2/3} = \dfrac{1}{10}$ $\left\{\pm\dfrac{1}{1000}\right\}$

For Exercises 91–96, solve for the indicated variable.

91. $4x - 3y = 6$ for y $y = \dfrac{4}{3}x - 2$

92. $t_a = \dfrac{t_1 + t_2}{2}$ for t_2 $t_2 = 2t_a - t_1$

93. $4x + 6y = ax + c$ for x $x = \dfrac{c - 6y}{4 - a}$ or $x = \dfrac{6y - c}{a - 4}$

94. $s = a_0t^2 + v_0t + s_0$ for t $t = \dfrac{-v_0 \pm \sqrt{v_0^2 - 4a_0(s_0 - s)}}{2a_0}$

95. $m = \dfrac{1}{2}\sqrt{2a^2 + 2b^2 - 2c^2}$ for $a > 0$ $a = \sqrt{2m^2 + c^2 - b^2}$

96. $\dfrac{1}{a} = \dfrac{1}{b} + \dfrac{1}{c}$ for b $b = \dfrac{ac}{c - a}$ or $b = -\dfrac{ac}{a - c}$

SECTION R.6

For Exercises 97–98, simplify the expressions.

97. $\sqrt{-16} \cdot \sqrt{-4}$ -8

98. Simplify the powers of i.

a. i^{35} $-i$ **b.** i^{56} 1 **c.** i^{62} -1

d. i^{17} i **e.** i^{-5} $-i$

For Exercises 99–103, perform the indicated operations.

99. $\left(\dfrac{2}{3} + \dfrac{3}{5}i\right) - \left(\dfrac{1}{6} + \dfrac{2}{5}i\right)$ **100.** $(4 - 7i)(5 + i)$ $27 - 31i$

101. $(4 - 6i)^2$ $-20 - 48i$ **102.** $(8 - 3i)(8 + 3i)$ 73

103. $\dfrac{4 + 3i}{3 - i}$ $\dfrac{9}{10} + \dfrac{13}{10}i$

For Exercises 104–108, solve the equation.

104. $10t^2 + 1210 = 0$ $\{\pm 11i\}$

105. $x(x - 4) = -9$ $\{2 \pm i\sqrt{5}\}$

106. $6d^{2/3} - 7d^{1/3} - 3 = 0$ $\left\{-\dfrac{1}{27}, \dfrac{27}{8}\right\}$

107. $(2u^2 - 1)^2 - 10(2u^2 - 1) + 9 = 0$ $\{\pm\sqrt{5}, \pm 1\}$

108. $2\left(\dfrac{4}{w} + 1\right)^2 - 10\left(\dfrac{4}{w} + 1\right) = 0$ $\{1, -4\}$

For Exercises 109–111, (a) evaluate the discriminant and (b) determine the number and type of solutions to each equation.

109. $4x^2 - 20x + 25 = 0$ a. 0 b. 1 real solution

110. $-2y^2 = 5y - 1$ a. 33 b. 2 real solutions

111. $5t(t + 1) = 4t - 11$ a. -219 b. 2 nonreal solutions

SECTION R.7

112. Shawna invested a total of $12,000 in two mutual funds: an international fund and a real estate fund. After 1 yr, the international fund earned the equivalent of 8.2% simple interest and the real estate fund returned 1.5%. If the total earnings at the end of the year was $749.50, determine the amount invested in each fund. Shawna invested $8500 in the international fund and $3500 in the real estate fund.

113. Cassandra bought a 10-yr Treasury note that paid the equivalent of 3.5% simple interest. She invested $4000 more in a 15-yr bond earning 4.1% than she did in the Treasury note. If the total amount of interest from these investments is $10,180, determine the amount of principal for each investment. Assume that each investment was held to maturity. Cassandra invested $8000 in the Treasury note and $12,000 in the bond.

114. A chemist mixed 100 cc of a 60% acid solution with a 20% acid solution to produce a 25% acid solution. How much of the 20% solution did he use? 700 cc

115. Suppose that 250 ft³ of dry concrete mixture is 50% sand by volume. How much pure sand must be added to form a new mixture that is 70% sand by volume? $166\frac{2}{3}$ ft³

116. When Kevin commuted to work one morning, his average speed was 45 mph. He averaged only 30 mph for the return trip because of an accident on the highway.

If the total time for the round trip was 50 min ($\frac{5}{6}$ hr), determine the distance between his place of work and his home.   15 mi

117. Two boats leave a marina at the same time. One travels south and the other travels north. The southbound boat travels 6 mph faster than the northbound boat. After 3 hr, the distance between the boats is 66 mi. Determine the speed of each boat.   The northbound boat travels 8 mph and the southbound boat travels 14 mph.

118. A tablet computer has a 7-in. diagonal screen. The length is 2.7 in. more than the width. Find the length and width of the screen. Round to 1 decimal place.   The length is 6.1 in. and the width is 3.4 in.

119. A fireworks mortar is shot straight upward with an initial velocity of 200 ft/sec from a platform 2 ft off the ground.

    a. Use the formula $s = -\frac{1}{2}gt^2 + v_0t + s_0$ to write a model for the height of the mortar s (in ft) at a time t seconds after launch. Assume that $g = 32$ ft/sec^2. $s = -16t^2 + 200t + 2$

    b. How long will it take the mortar (on the way up) to clear a tree line that is 80 ft high? Round to the nearest tenth of a second.   0.4 sec

120. Petra and Dawn are typesetters for a publishing company. Petra can typeset 50 pages in 4 days and Dawn can typeset 50 pages in 3.5 days. How long would it take them to typeset a 150-page manuscript if they work together?   5.6 days

121. One pump can drain a pond in 22 hr. Working with a second pump, the two pumps together can drain the pond in 10 hr. How fast can the second pump drain the pond by itself? $\frac{55}{3}$ hr = 18.3 hr

SECTION R.8

For Exercises 122–127, solve the inequality. Graph the solution set and write the solution set in set-builder notation and interval notation.

122. $-4 \le -\frac{2}{3}p + 14$

123. $\frac{2+y}{3} - \frac{y-1}{4} < \frac{y}{6}$

124. a. $t + 2 \le 8$ or $\frac{1}{3}t < -4$

    b. $t + 2 \le 8$ and $\frac{1}{3}t < -4$

125. a. $-2(x - 1) + 4 < x + 3$ or $5(x + 2) - 3 \le 4x + 1$

    b. $-2(x - 1) + 4 < x + 3$ and $5(x + 2) - 3 \le 4x + 1$

126. $-11 \le -4x - 1 \le 7$ 127. $0 < \dfrac{-3x + 9}{-4} < 6$

128. Write an inequality to represent the following statement. A pilot is instructed to keep a plane at an altitude a of over 29,000 ft, but not to exceed 31,000 ft. $29,000 < a \le 31,000$

129. The months of June, July, August, and September are the wettest months in Miami, Florida, averaging 7.83 in./month. If Miami gets 8.54 in. in June, 5.79 in. in July, and 8.63 in. in August, how much rain is needed in September to exceed the monthly average for these 4 months?   More than 8.36 in. is needed.

For Exercises 130–136, solve the equation or inequality. Write the solution set to the inequalities in interval notation if possible.

130. a. $|w + 2| + 1 = 6$ $\{-7, 3\}$
    b. $|w + 2| + 1 < 6$ $(-7, 3)$
    c. $|w + 2| + 1 \ge 6$ $(-\infty, -7] \cup [3, \infty)$

131. a. $3 = |7x + 1| + 4$ $\{\ \}$
    b. $3 < |7x + 1| + 4$ $(-\infty, \infty)$
    c. $3 \ge |7x + 1| + 4$ $\{\ \}$

132. a. $|y + 5| - 3 = -3$ $\{-5\}$
    b. $|y + 5| - 3 < -3$ $\{\ \}$
    c. $|y + 5| - 3 \le -3$ $\{-5\}$
    d. $|y + 5| - 3 > -3$ $(-\infty, -5) \cup (-5, \infty)$
    e. $|y + 5| - 3 \ge -3$ $(-\infty, \infty)$

133. a. $|x - 1| = |3x + 5|$ $\{-3, -1\}$
    b. $|x - 1| = |x + 5|$ $\{-2\}$
    c. $|x - 1| = |1 - x|$ $(-\infty, \infty)$

134. $4|x + 2| - 10 \ge -6$ $(-\infty, -3] \cup [-1, \infty)$

135. $|0.5x - 8| < 0.01$ $(15.98, 16.02)$

136. $-9 \le 4 - |2k - 1|$ $[-6, 7]$

For Exercises 137–138, (a) write an absolute value inequality that represents the given statement and (b) solve the inequality.

137. The distance between x and 3 on a number line is no more than 0.5.   a. $|x - 3| \le 0.5$ or $|3 - x| \le 0.5$ b. $[2.5, 3.5]$

138. The distance between t and -2 on a number line exceeds 0.01.   a. $|t + 2| > 0.01$ or $|-2 - t| > 0.01$ b. $(-\infty, -2.01) \cup (-1.99, \infty)$

CHAPTER R   Test

1. Write the interval notation representing $\{x \,|\, 2.7 < x\}$. $(2.7, \infty)$

2. Write the set-builder notation representing the interval $[-3, 5)$. $\{x \,|\, -3 \le x < 5\}$

3. Given $A = \{x \,|\, x < 2\}$, $B = \{x \,|\, x \ge 0\}$, and $C = (x \,|\, x < -1\}$, write the union or intersection in set notation.

    a. $A \cup B$ $\mathbb{R}$
    b. $A \cap B$ $\{x \,|\, 0 \le x < 2\}$
    c. $A \cup C$ $\{x \,|\, x < 2\}$
    d. $A \cap C$ $\{x \,|\, x < -1\}$
    e. $B \cup C$ $\{x \,|\, x < -1 \text{ or } x \ge 0\}$
    f. $B \cap C$ $\{\ \}$

4. Simplify without absolute value bars. $|5 - t|$ for $t > 5$. $t - 5$

5. a. Find the distance between $\sqrt{2}$ and 2 on the number line. $|\sqrt{2} - 2|$ or $|2 - \sqrt{2}|$

    b. Simplify without absolute value bars. $2 - \sqrt{2}$

6. a. Determine the restrictions on y. $y \ne 11, y \ne -2$ $\dfrac{2y - 22}{y^2 - 9y - 22}$

    b. Simplify the expression. $\dfrac{2}{y + 2}$

For Exercises 7–23, perform the indicated operations and simplify the expression. Assume that all radical expressions represent real numbers.

7. $-\{6 + 4[9x - 2(3x - 5)] + 3\}$ $-12x - 49$

8. $\dfrac{7^2 - 4[8 - (-2)] - 4^2}{-2|-4 + 6|}$ $\dfrac{7}{4}$

9. $(2a^3b^{-4})^2(4a^{-2}b^7)^{-1}$ $\dfrac{a^8}{b^{15}}$

10. $\left(\dfrac{1}{2}\right)^0 + \left(\dfrac{1}{3}\right)^{-3} + \left(\dfrac{1}{4}\right)^{-1}$ 32

11. $\sqrt[3]{80k^{15}m^2n^7}$ $2k^5n^2\sqrt[3]{10m^2n}$

12. $\dfrac{3}{4}\sqrt[4]{8p^2q^3} \cdot \sqrt[4]{2p^3q}$ $\dfrac{3}{2}pq\sqrt[4]{p}$

13. $\sqrt{125ab^3} - 3b\sqrt{20ab}$ $-b\sqrt{5ab}$

14. $(12n^2 - 4)\left(\dfrac{1}{2}n^2 - 3n + 5\right)$ $6n^4 - 36n^3 + 58n^2 + 12n - 20$

15. $[(3a + b) - c][(3a + b) + c]$ $9a^2 + 6ab + b^2 - c^2$

16. $\left(\dfrac{1}{4}\sqrt{z} - p^2\right)\left(\dfrac{1}{4}\sqrt{z} + p^2\right)$ $\dfrac{1}{16}z - p^4$

17. $(5\sqrt{x} + \sqrt{2})(3\sqrt{x} - 2\sqrt{2})$ $15x - 7\sqrt{2x} - 4$

18. $(\sqrt{x} - 6z)^2$ $x - 12z\sqrt{x} + 36z^2$

19. $\dfrac{3x^2}{x^3 + 14x^2 + 49x} \div \dfrac{8x - 4}{4x^2 + 26x - 14}$ $\dfrac{3x}{2(x + 7)}$

20. $\dfrac{x^2}{x - 5} + \dfrac{10x - 25}{5 - x}$ $x - 5$

21. $\dfrac{-12}{y^3 + 4y^2} + \dfrac{1}{y} - \dfrac{3}{y^2 + 4y}$ $\dfrac{y - 3}{y^2}$

22. $\dfrac{\dfrac{x}{4} - \dfrac{9}{4x}}{\dfrac{1}{4} - \dfrac{3}{4x}}$ $x + 3$

23. $\dfrac{6}{\sqrt{13} + \sqrt{10}}$ $2(\sqrt{13} - \sqrt{10})$

For Exercises 24–31, factor completely. Write the answer with positive exponents only.

24. $30x^3 + 2x^2 - 4x$ $2x(3x - 1)(5x + 2)$

25. $xy + 5ay + 10ac + 2xc$ $(x + 5a)(y + 2c)$

26. $x^5 + 2x^4 - 81x - 162$ $(x^2 + 9)(x - 3)(x + 3)(x + 2)$

27. $c^2 - 4a^2 - 44a - 121$ $(c - 2a - 11)(c + 2a + 11)$

28. $27u^3 - v^6$ $(3u - v^2)(9u^2 + 3uv^2 + v^4)$

29. $4w^{-6} + 2w^{-5} + 7w^{-4}$ $\dfrac{7w^2 + 2w + 4}{w^6}$

30. $y(2y - 1)^{-3/4} + (2y - 1)^{1/4}$ $\dfrac{3y - 1}{(2y - 1)^{3/4}}$

31. $2x(9 - x^2)^{1/2} + x^2 \cdot \dfrac{1}{2}(9 - x^2)^{-1/2}(-2x)$ $\dfrac{3x(6 - x^2)}{(9 - x^2)^{1/2}}$

32. Perform the indicated operations.
$$\dfrac{(8.4 \times 10^{11})(6.0 \times 10^{-3})}{(4.2 \times 10^{-5})}$$ 1.2×10^{14}

33. Write the expression in terms of i and simplify.
$\sqrt{-25} \cdot \sqrt{-4}$ -10

34. Simplify the powers of i.

   a. i^{89} i b. i^{46} -1 c. i^{35} $-i$ d. i^{120} 1 e. i^{-11} i

For Exercises 35–56, perform the indicated operations. Write the answers in the form $a + bi$.

35. $(3 - 5i)^2$ $-16 - 30i$ 36. $\dfrac{4 + 3i}{2 - 5i}$ $-\dfrac{7}{29} + \dfrac{26}{29}i$

For Exercises 37–39, (a) evaluate the discriminant and (b) determine the number and type of solutions to the equation.

37. $2x^2 - 4x + 7 = 0$ a. -40 b. 2 nonreal solutions

38. $x^2 + 25 = 10x$ a. 0 b. 1 real solution

39. $3x(x + 4) = 2x - 2$ a. 76 b. 2 real solutions

For Exercises 40–50, solve the equation.

40. $3y + 2[5(y - 4) - 2] = 5y + 6(7 + y) - 3$ $\left\{\dfrac{83}{2}\right\}$

41. $(3x - 4)^2 - 2 = 11$ $\left\{\dfrac{4 \pm \sqrt{13}}{3}\right\}$

42. $y^2 + 10y = 4$ $\{-5 \pm \sqrt{29}\}$

43. $6t(2t + 1) = 5 - 5t$ $\left\{\dfrac{1}{3}, -\dfrac{5}{4}\right\}$

44. $\dfrac{3x^2}{4} - x = -\dfrac{1}{2}$ $\left\{\dfrac{2 \pm i\sqrt{2}}{3}\right\}$

45. $(2y - 3)^{1/3} - (4y + 5)^{1/3} = 0$ $\{-4\}$

46. $\sqrt{2d} = 1 - \sqrt{d + 7}$ $\{\ \}$; The values 2 and 18 do not check.

47. $\dfrac{c}{c + 6} - 4 = \dfrac{72}{c^2 - 36}$ $\{4\}$; The value -6 does not check.

48. $\left(5 - \dfrac{2}{k}\right)^2 - 6\left(5 - \dfrac{2}{k}\right) - 27 = 0$ $\left\{\dfrac{1}{4}, -\dfrac{1}{2}\right\}$

49. $-2 = |x - 3| - 6$ $\{-1, 7\}$

50. $|2v + 5| = |2v - 1|$ $\{-1\}$

For Exercises 51–53, solve for the indicated variable.

51. $aP - 4 = Pt + 2$ for P $P = \dfrac{6}{a - t}$ or $P = -\dfrac{6}{t - a}$

52. $-16t^2 + v_0t + 2 = 0$ for t

53. $a^2 + b^2 + c^2 = 49$ for $c > 0$ $c = \sqrt{49 - a^2 - b^2}$

For Exercises 54–60, solve the equations and inequalities. Write the answers to the inequalities in interval notation if possible.

54. $-2 \le \dfrac{4 - x}{3} \le 6$ $[-14, 10]$

55. $-\dfrac{4}{3}y < -24$ or $y + 7 \le 2y - 3$ $[10, \infty)$

56. $3(x - 5) + 1 \le 4(x + 2) + 6$ and $0.3x - 1.6 > 0.2$ $(6, \infty)$

57. $2 < -1 + |4w - 3|$ $(-\infty, 0) \cup (\frac{3}{2}, \infty)$ 58. $-|8 - v| \ge -6$ $[2, 14]$

59. a. $|7x + 4| + 11 = 2$ $\{\ \}$ b. $|7x + 4| + 11 < 2$ $\{\ \}$

   c. $|7x + 4| + 11 > 2$ $\mathbb{R}$

60. a. $|x - 13| + 4 = 4$ $\{13\}$ **b.** $|x - 13| + 4 < 4$ $\{\ \}$

c. $|x - 13| + 4 \leq 4$ $\{13\}$ **d.** $|x - 13| + 4 > 4$
$(-\infty, 13) \cup (13, \infty)$

e. $|x - 13| + 4 \geq 4$ $(-\infty, \infty)$

61. How much of an 80% antifreeze solution should be mixed with 2 gal of a 50% antifreeze solution to make a 60% antifreeze solution?   1 gal of 80% antifreeze should be used.

62. Two passengers leave the airport at Denver, Colorado. One takes a 2.3-hr flight to Seattle, Washington, and the other takes a 3.3-hr flight to New York City. The plane flying to New York flies 60 mph faster than the plane flying to Seattle. If the total distance traveled by both planes is 2662 mi, determine the average speed of each plane.
The plane flying to Seattle flies 440 mph, and the plane flying to New York flies 500 mph.

63. Kelly has an aboveground swimming pool. Using water from one hose requires 3 hr to fill the pool. If a second hose is turned on, the pool can be filled in 1.2 hr. How long would it take the second hose to fill the pool if it worked alone?   The second hose can fill the pool in 2 hr.

64. A garden area is configured in the shape of a rectangle with two right triangles of equal size attached at the ends. The length of the rectangle is 18 ft. The height of the right triangles is 7 ft longer than the base. If the total area of the garden is 276 ft², determine the base and height of the triangular portions.
The base of the triangular portions is 5 ft and the height is 12 ft.

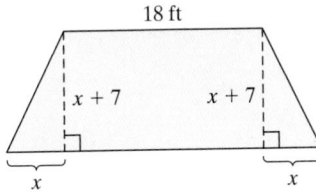

65. A varsity soccer player kicks a soccer ball approximately straight upward with an initial velocity of 60 ft/sec. The ball leaves the player's foot at a height of 2 ft.

a. Use the formlua
$s = -\frac{1}{2}gt^2 + v_0 t + s_0$
to write a model representing the height of the ball s (in ft), t seconds after being kicked. Assume that the acceleration due to gravity is $g = 32$ ft/sec². $s = -16t^2 + 60t + 2$

b. Determine the times at which the ball is 52 ft in the air.   The ball will be at a height of 52 ft at times 1.25 sec and 2.5 sec after being kicked.

66. A golfer plays 5 rounds of golf with the following scores: 92, 88, 85, 90, and 89. What score would he need on his sixth round to have an average below 88?
The golfer would need to score less than 84.

Functions and Relations

Each year the IRS (Internal Revenue Service) publishes tax rates that tell us how much federal income tax we need to pay based on our taxable income. For example, for a recent year, a single person with taxable income of more than \$36,250 but not more than \$87,850 pays \$4991.25 plus 25% of the amount over \$36,250 in federal income tax. However, finding taxable income is not always trivial. There are numerous variables that come into play. The IRS takes into account exemptions, deductions, and tax credits among other things.

In Chapter 1, we will look at mathematical relationships involving two or more variables, including the relationship between taxable income and federal income tax. To fully appreciate the connection among several variables, we will investigate their relationships algebraically, numerically, and graphically.

Schedule X—If your filling status is Single

| If your taxable income is | | | |
over—	but not over—	The tax is	of the amount over—
\$0	\$8925	\$0 + 10%	\$0
\$8925	\$36,250	\$892.50 + 15%	\$8925
\$36,250	\$87,850	\$4991.25 + 25%	\$36,250
\$87,850	\$183,250	\$17,891.25 + 28%	\$87,850
\$183,250	\$398,350	\$44,603.25 + 33%	\$183,250
\$398,350	\$400,000	\$115,586.25 + 35%	\$398,350
\$400,000	——	\$116,163.75 + 39.6%	\$400,000

Source: Internal Revenue Service, www.irs.gov

SECTION 1.1   The Rectangular Coordinate System and Graphing Utilities

Websites, newspapers, sporting events, and the workplace all utilize graphs and tables to present data. Therefore, it is important to learn how to create and interpret meaningful graphs. Understanding how points are located relative to a fixed origin is important for many graphing applications. For example, computer game developers use a rectangular coordinate system to define the locations of objects moving around the screen.

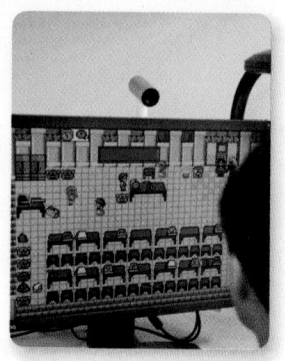

1. Plot Points on a Rectangular Coordinate System

Mathematician René Descartes (pronounced "day cart") (1597–1650) was the first to identify points in a plane by a pair of coordinates. He did this by intersecting two perpendicular number lines with the point of intersection called the **origin.** These lines form a **rectangular coordinate system** (also known in his honor as the **Cartesian coordinate system**) or simply a **coordinate plane.** The horizontal line is called the **x-axis** and the vertical line is called the **y-axis.** The *x*- and *y*-axes divide the plane into four **quadrants.** The quadrants are labeled counterclockwise as I, II, III, and IV (Figure 1-1).

Every point in the plane can be uniquely identified by using an ordered pair (x, y) to specify its coordinates with respect to the origin. In an ordered pair, the first coordinate is called the **x-coordinate,** and the second is called the **y-coordinate.** The origin is identified as $(0, 0)$. In Figure 1-2, six points have been graphed. The point $(-3, 5)$, for example, is placed 3 units in the negative *x* direction (to the left) and 5 units in the positive *y* direction (upward).

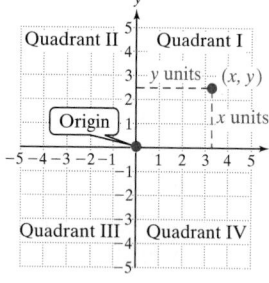

Figure 1-1

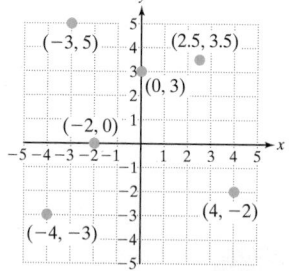

Figure 1-2

2. Use the Distance and Midpoint Formulas

Recall that the distance between two points A and B on a number line can be represented by $|A - B|$ or $|B - A|$. Now we want to find the distance between two points in a coordinate plane. For example, consider the points $(1, 5)$ and $(4, 9)$. The distance d between the points is labeled in Figure 1-3. The dashed horizontal and vertical line segments form a right triangle with hypotenuse d.

The horizontal distance between the points is $|4 - 1| = 3$.
The vertical distance between the points is $|9 - 5| = 4$.

Figure 1-3

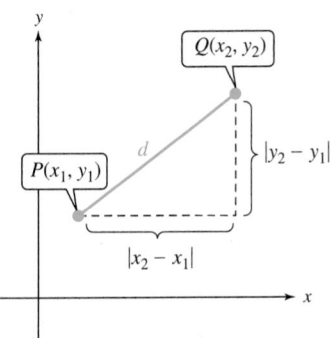

Figure 1-4

Applying the Pythagorean theorem, we have

$$d^2 = (3)^2 + (4)^2$$
$$d = \sqrt{(3)^2 + (4)^2} = \sqrt{25} = 5$$

Since d is a distance, reject the negative square root.

The distance between the points is 5 units.

We can make this process generic by labeling the points $P(x_1, y_1)$ and $Q(x_2, y_2)$. See Figure 1-4.

- The horizontal leg of the right triangle is $|x_2 - x_1|$ or equivalently $|x_1 - x_2|$.
- The vertical leg of the right triangle is $|y_2 - y_1|$ or equivalently $|y_1 - y_2|$.

Applying the Pythagorean theorem, we have

$$d^2 = (x_2 - x_1)^2 + (y_2 - y_1)^2$$
$$d = \sqrt{(x_2 - x_1)^2 + (y_2 - y_1)^2}$$

We can drop the absolute value bars because $|a|^2 = (a)^2$ for all real numbers a. Likewise $|x_2 - x_1|^2 = (x_2 - x_1)^2$ and $|y_2 - y_1|^2 = (y_2 - y_1)^2$.

TIP Since

$$(x_2 - x_1)^2 = (x_1 - x_2)^2 \text{ and }$$
$$(y_2 - y_1)^2 = (y_1 - y_2)^2,$$

the distance formula can also be expressed as

$$d = \sqrt{(x_1 - x_2)^2 + (y_1 - y_2)^2}.$$

Distance Formula

The distance between points (x_1, y_1) and (x_2, y_2) is given by

$$d = \sqrt{(x_2 - x_1)^2 + (y_2 - y_1)^2}$$

Classroom Example: p. 127
Exercise 14

EXAMPLE 1   Finding the Distance Between Two Points

Find the distance between the points $(-5, 1)$ and $(7, -3)$. Give the exact distance and an approximation to 2 decimal places.

Solution:

$(-5, 1) \quad \text{and} \quad (7, -3)$

$(x_1, y_1) \quad \text{and} \quad (x_2, y_2)$

Label the points. Note that the choice for (x_1, y_1) and (x_2, y_2) will not affect the outcome.

$$d = \sqrt{[7 - (-5)]^2 + (-3 - 1)^2}$$

Apply the distance formula.
$d = \sqrt{(x_2 - x_1)^2 + (y_2 - y_1)^2}$

$$= \sqrt{(12)^2 + (-4)^2}$$

Simplify the radical.

$$= \sqrt{160}$$
$$= 4\sqrt{10} \approx 12.65$$

The exact distance is $4\sqrt{10}$ units. This is approximately 12.65 units.

Avoiding Mistakes

A statement of the form "if p, then q" is called a **conditional statement.** Its **converse** is the statement "if q, then p." The converse of a statement is not necessarily true. However, in the case of the Pythagorean theorem, the converse is a true statement.

Skill Practice 1 Find the distance between the points $(-1, 4)$ and $(3, -6)$. Give the exact distance and an approximation to 2 decimal places.

The Pythagorean theorem tells us that if a right triangle has legs of lengths a and b and hypotenuse of length c, then $a^2 + b^2 = c^2$. The following related statement is also true: If $a^2 + b^2 = c^2$, then a triangle with sides of lengths a, b, and c is a right triangle. We use this important concept in Example 2.

Answer

1. $2\sqrt{29}$ units ≈ 10.77 units

EXAMPLE 2 Determining if Three Points Form the Vertices of a Right Triangle

Determine if the points $M(-2, -3)$, $P(4, 1)$, and $Q(-1, 7)$ form the vertices of a right triangle.

Solution:

Determine the distance between each pair of points.

$d(M, P) = \sqrt{[4 - (-2)]^2 + [1 - (-3)]^2} = \sqrt{52}$

$d(P, Q) = \sqrt{(-1 - 4)^2 + (7 - 1)^2} = \sqrt{61}$

$d(M, Q) = \sqrt{[-1 - (-2)]^2 + [7 - (-3)]^2} = \sqrt{101}$

The line segment $\overline{MQ}$ is the longest and would potentially be the hypotenuse, c. Label the shorter sides as a and b.

Check the condition that $a^2 + b^2 = c^2$.

$(\sqrt{52})^2 + (\sqrt{61})^2 \overset{?}{=} (\sqrt{101})^2$

$52 + 61 \neq 101$

The points M, P, and Q do not form the vertices of a right triangle.

> **Skill Practice 2** Determine if the points $X(-6, -4)$, $Y(2, -2)$, and $Z(0, 5)$ form the vertices of a right triangle.

> **TIP** We denote the distance between points P and Q as
>
> $d(P, Q)$ or PQ.
>
> The second notation is the length of the line segment with endpoints P and Q.

Now suppose that we want to find the midpoint of the line segment between the distinct points (x_1, y_1) and (x_2, y_2). The **midpoint** of a line segment is the point equidistant (the same distance) from the endpoints (Figure 1-5).

The x-coordinate of the midpoint is the average of the x-coordinates from the endpoints. Likewise, the y-coordinate of the midpoint is the average of the y-coordinates from the endpoints.

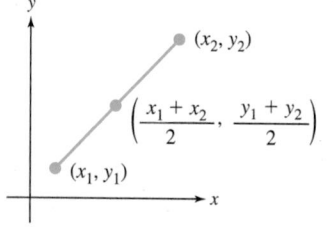

Figure 1-5

Midpoint Formula

The midpoint of the line segment with endpoints (x_1, y_1) and (x_2, y_2) is

$$M = \left(\frac{x_1 + x_2}{2}, \frac{y_1 + y_2}{2} \right)$$

average of x-coordinates   average of y-coordinates

Avoiding Mistakes

The midpoint of a line segment is an ordered pair (with two coordinates), not a single number.

EXAMPLE 3 Finding the Midpoint of a Line Segment

Find the midpoint of the line segment with endpoints $(4.2, -4)$ and $(-2.8, 3)$.

Solution:

$(4.2, -4)$ and $(-2.8, 3)$

(x_1, y_1) and (x_2, y_2) .... Label the points.

$M = \left(\dfrac{4.2 + (-2.8)}{2}, \dfrac{-4 + 3}{2} \right)$ .... Apply the midpoint formula.

$= \left(0.7, -\dfrac{1}{2} \right)$ or $(0.7, -0.5)$ .... Simplify.

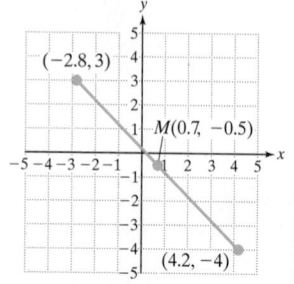

Answer

2. No

> **Skill Practice 3**   Find the midpoint of the line segment with endpoints $(-1.5, -9)$ and $(-8.7, 4)$.

3. Graph Equations by Plotting Points

The relationship between two variables can often be expressed as a graph or expressed algebraically as an equation. For example, suppose that two variables, x and y, are related such that y is 2 more than x. An equation to represent this relationship is $y = x + 2$. A **solution to an equation** in the variables x and y is an ordered pair (x, y) that when substituted into the equation makes the equation a true statement.

For example, the following ordered pairs are solutions to the equation $y = x + 2$.

Solution	$y = x + 2$
$(0, 2)$	$2 = 0 + 2$ ✓
$(-4, -2)$	$-2 = -4 + 2$ ✓
$(2, 4)$	$4 = 2 + 2$ ✓

The set of all solutions to an equation is called the **solution set of the equation.** The graph of all solutions to an equation is called the **graph of the equation.** The graph of $y = x + 2$ is shown in Figure 1-6.

One of the goals of this text is to identify families of equations and the characteristics of their graphs. As we proceed through the text, we will develop tools to graph equations efficiently. For now, we present the point-plotting method to graph the solution set of an equation. In Example 4, we start by selecting several values of x and using the equation to calculate the corresponding values of y. Then we plot the points to form a general outline of the curve.

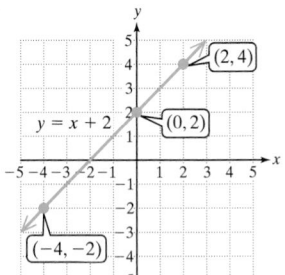

Figure 1-6

Classroom Example: p. 128
Exercise 36

> **EXAMPLE 4**   **Graphing an Equation by Plotting Points**
>
> Graph the equation by plotting points. $y - |x| = -1$
>
> **Solution:**
>
> $y - |x| = -1$ Solve for y in terms of x.
>
> $\qquad y = |x| - 1$ Arbitrarily select negative and positive values for x such as -3, -2, -1, 0, 1, 2, and 3. Then use the equation to calculate the corresponding y values.
>
> | x | y | $y = |x| - 1$ | Ordered pair |
> |-----|-----|---------------|--------------|
> | -3 | 2 | $y = |-3| - 1 = 2$ | $(-3, 2)$ |
> | -2 | 1 | $y = |-2| - 1 = 1$ | $(-2, 1)$ |
> | -1 | 0 | $y = |-1| - 1 = 0$ | $(-1, 0)$ |
> | 0 | -1 | $y = |0| - 1 = -1$ | $(0, -1)$ |
> | 1 | 0 | $y = |1| - 1 = 0$ | $(1, 0)$ |
> | 2 | 1 | $y = |2| - 1 = 1$ | $(2, 1)$ |
> | 3 | 2 | $y = |3| - 1 = 2$ | $(3, 2)$ |
>
>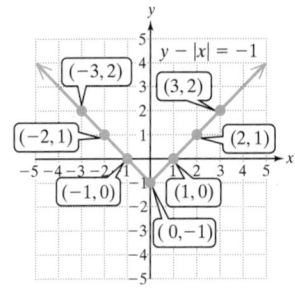
>
> **Skill Practice 4**   Graph the equation by plotting points. $x^2 + y = 4$

Answers

3. $\left(-5.1, -\dfrac{5}{2}\right)$ or $(-5.1, -2.5)$

4.

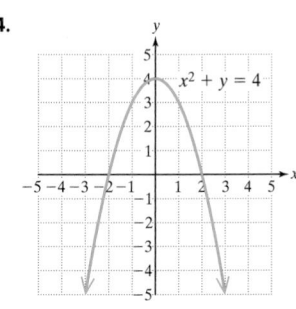

The graph of an equation in the variables x and y represents a relationship between a real number x and a corresponding real number y. Therefore, the values of x must be chosen so that when substituted into the equation, they produce a real number for y. Sometimes the values of x must be restricted to produce real numbers for y. This is demonstrated in Example 5.

Classroom Example: p. 128
Exercise 40

EXAMPLE 5   Graphing an Equation by Plotting Points

Graph the equation by plotting points. $y^2 - 1 = x$

Solution:

$$y^2 - 1 = x \qquad \text{Solve for } y \text{ in terms of } x.$$
$$y^2 = x + 1$$
$$y = \pm\sqrt{x + 1} \qquad \text{Apply the square root property.}$$

Choose $x \geq -1$ so that the radicand is nonnegative.

> **TIP** In Example 5, we choose several convenient values of x such as -1, 0, 3, and 8 so that the radicand will be a perfect square.

x	y	$y = \pm\sqrt{x+1}$	Ordered pairs
-1	0	$y = \pm\sqrt{(-1)+1} = 0$	$(-1, 0)$
0	±1	$y = \pm\sqrt{(0)+1} = \pm1$	$(0, 1), (0, -1)$
1	$\pm\sqrt{2}$	$y = \pm\sqrt{(1)+1} = \pm\sqrt{2}$ ≈ 1.4	$(1, \sqrt{2}), (1, -\sqrt{2})$
3	±2	$y = \pm\sqrt{(3)+1} = \pm2$	$(3, 2), (3, -2)$
8	±3	$y = \pm\sqrt{(8)+1} = \pm3$	$(8, 3), (8, -3)$

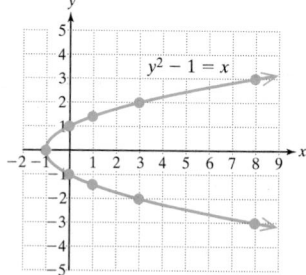

Skill Practice 5   Graph the equation by plotting points. $x + y^2 = 2$

4. Identify x- and y-Intercepts

When analyzing graphs, we want to examine their most important features. Two key features are the x- and y-intercepts of a graph. These are the points where a graph intersects the x- and y-axes.

Any point on the x-axis has a y-coordinate of zero. Therefore, an ***x*-intercept** is a point $(a, 0)$ where a graph intersects the x-axis (Figure 1-7). Any point on the y-axis has an x-coordinate of zero. Therefore, a ***y*-intercept** is a point $(0, b)$ where a graph intersects the y-axis (Figure 1-7).

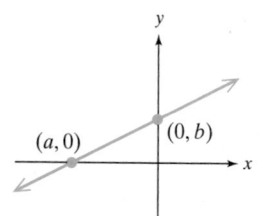

Figure 1-7

> **TIP** In some applications, we may refer to an x-intercept as the x-*coordinate* of a point of intersection that a graph makes with the x-axis. For example, if an x-intercept is $(-4, 0)$, then the x-intercept may be stated simply as -4 (the y-coordinate is understood to be zero). Similarly, we may refer to a y-intercept as the y-*coordinate* of a point of intersection that a graph makes with the y-axis. For example, if a y-intercept is $(0, 2)$, then it may be stated simply as 2.

Answer

5.

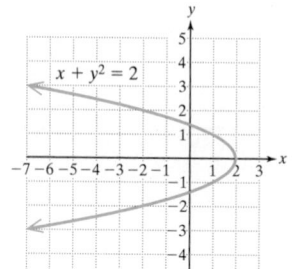

To find the x- and y-intercepts from an equation in x and y, follow these steps.

Determining x- and y-Intercepts from an Equation

Given an equation in x and y,

- Find the x-intercept(s) by substituting 0 for y in the equation and solving for x.
- Find the y-intercept(s) by substituting 0 for x in the equation and solving for y.

EXAMPLE 6   Finding *x*- and *y*-Intercepts

Given the equation $y = |x| - 1$,

a. Find the *x*-intercept(s).   **b.** Find the *y*-intercept(s).

Solution:

a. $y = |x| - 1$
$0 = |x| - 1$ To find the *x*-intercept(s), substitute 0 for *y* and solve for *x*.
$|x| = 1$ Isolate the absolute value.
$x = 1$ or $x = -1$ Recall that for $k > 0$, $|x| = k$ is equivalent to $x = k$ or $x = -k$.

The *x*-intercepts are $(1, 0)$ and $(-1, 0)$.

b. $y = |x| - 1$
$= |0| - 1$ To find the *y*-intercept(s), substitute 0 for *x* and solve for *y*.
$= -1$

The *y*-intercept is $(0, -1)$.

The intercepts $(1, 0)$, $(-1, 0)$, and $(0, -1)$ are consistent with the graph of the equation $y = |x| - 1$ found in Example 4 (Figure 1-8).

Skill Practice 6   Given the equation $y = x^2 - 4$,

a. Find the *x*-intercept(s).   **b.** Find the *y*-intercept(s).

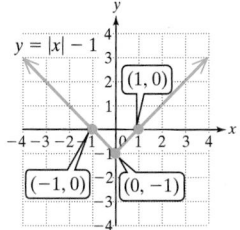

Figure 1-8

Classroom Example: p. 129
Exercise 54

TIP   Sometimes when solving for an *x*- or *y*-intercept, we encounter an equation with an imaginary solution. In such a case, the graph has no *x*- or *y*-intercept.

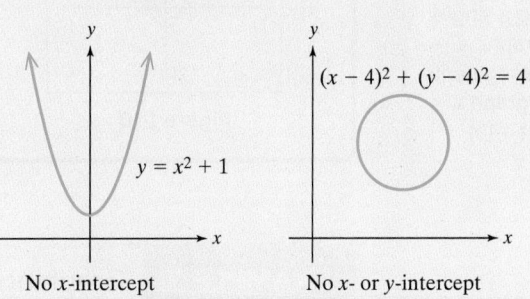

5. Graph Equations Using a Graphing Utility

Graphing by the point-plotting method should only be considered a beginning strategy for creating the graphs of equations in two variables. We will quickly enhance this method with other techniques that are less cumbersome and use more analysis and strategy.

One weakness of the point-plotting method is that it may be slow to execute by pencil and paper. Also, the selected points must fairly represent the shape of the graph. Otherwise the sketch will be inaccurate. Graphing utilities can help with both of these weaknesses. They can graph many points quickly, and the more points that are plotted, the greater the likelihood that we see the key features of the graph. Graphing utilities include graphing calculators, spreadsheets, specialty graphing programs, and apps on phones.

Figures 1-9 and 1-10 show a table and a graph for $y = x^2 - 3$.

Answers
6. a. $(2, 0)$ and $(-2, 0)$
   b. $(0, -4)$

TECHNOLOGY CONNECTIONS

Using the Table Feature and Graphing an Equation

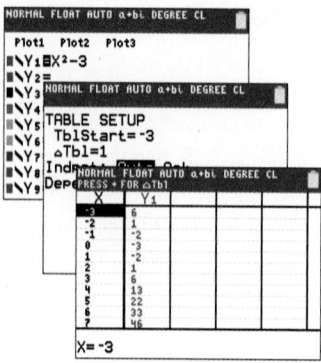

Figure 1-9

In Figure 1-9, we first enter the equation into the graphing editor. Notice that the calculator expects the equation represented with the y variable isolated.

To set up a table, enter the starting value for x, in this case, -3. Then set the increment by which to increase x, in this case 1. The x-increment is entered as ΔTbl (read "delta table"). Using the "Auto" setting means that the table of values for X and Y_1 will be automatically generated.

The table shows eleven x-y pairs but more can be accessed by using the up and down arrow keys on the keypad.

The graph in Figure 1-10 is shown between x and y values from -10 to 10. The tick marks on the axes are 1 unit apart. The viewing window with these parameters is denoted $[-10, 10, 1]$ by $[-10, 10, 1]$.

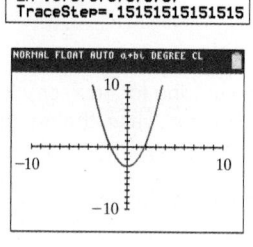

Figure 1-10

| minimum x value | maximum x value | minimum y value | maximum y value |

$[-10,\ 10,\ 1]$ by $[-10,\ 10,\ 1]$.

| distance between tick marks x-axis | distance between tick marks y-axis |

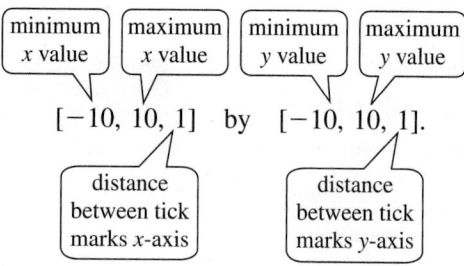

TIP The Greek letter Δ ("delta") written before a variable represents an increment of change in that variable. In this context, it represents the change from one value of x to the next.

TIP The calculator plots a large number of points and then connects the points. So instead of graphing a single smooth curve, it graphs a series of short line segments. This may give the graph a jagged look (Figure 1-10).

Classroom Examples: p. 131
Exercises 92, 94

EXAMPLE 7   Graphing Equations Using a Graphing Utility

Use a graphing utility to graph $y = |x| - 15$ and $y = -x^2 + 12$ on the viewing window defined by $[-20, 20, 2]$ by $[-15, 15, 3]$.

Solution:

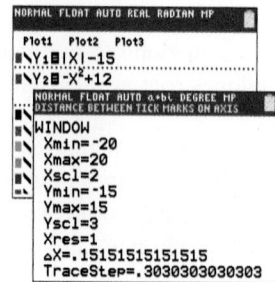

Enter the equations using the Y= editor.

Use the WINDOW editor to change the viewing window parameters. The variables Xmin, Xmax, and Xscl relate to $[-20, 20, 2]$. The variables Ymin, Ymax, and Yscl relate to $[-15, 15, 3]$.

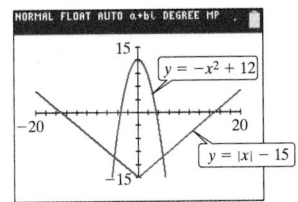

Select the **GRAPH** feature. Notice that the graphs of both equations appear. This provides us with a tool for visually examining two different models at the same time.

Answer

7.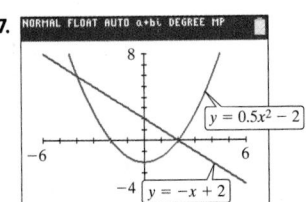

Skill Practice 7   Use a graphing utility to graph $y = -x + 2$ and $y = 0.5x^2 - 2$ on the viewing window $[-6, 6, 1]$ by $[-4, 8, 1]$.

SECTION 1.1   Practice Exercises

Prerequisite Review

R.1. Simplify the radical. $\sqrt{48}$ $4\sqrt{3}$

R.2. Given a right triangle with a leg of length 7 km and hypotenuse of length 25 km, find the length of the unknown leg.   24 km

R.3. Solve for y. $ax + by = c$ $y = \dfrac{c - ax}{b}$ or $y = \dfrac{c}{b} - \dfrac{ax}{b}$

R.4. Evaluate $x^2 + 4x + 5$ for $x = -5$ 10

Concept Connections

1. In a rectangular coordinate system, the point where the x- and y-axes meet is called the ____origin____.

2. The x- and y-axes divide the coordinate plane into four regions called ____quadrants____.

3. The distance between two distinct points (x_1, y_1) and (x_2, y_2) is given by the formula _____. $d = \sqrt{(x_2 - x_1)^2 + (y_2 - y_1)^2}$

4. The midpoint of the line segment with endpoints (x_1, y_1) and (x_2, y_2) is given by the formula _____. $M = \left(\dfrac{x_1 + x_2}{2}, \dfrac{y_1 + y_2}{2} \right)$

5. A ____solution____ to an equation in the variables x and y is an ordered pair (x, y) that makes the equation a true statement.

6. An x-intercept of a graph has a y-coordinate of ____0____.

7. A y-intercept of a graph has an x-coordinate of ____0____.

8. Given an equation in the variables x and y, find the y-intercept by substituting ____0____ for x and solving for ____y____.

Objective 1: Plot Points on a Rectangular Coordinate System

For Exercises 9–10, plot the points on a rectangular coordinate system.

9. $A(-3, -4)$ $B\left(\dfrac{5}{3}, \dfrac{7}{4}\right)$ $C(-1.2, 3.8)$ $D(\pi, -5)$ $E(0, 4.5)$ $F(\sqrt{5}, 0)$

10. $A(-2, -5)$ $B\left(\dfrac{9}{2}, \dfrac{7}{3}\right)$ $C(-3.6, 2.1)$ $D(5, -\pi)$ $E(3.4, 0)$ $F(0, \sqrt{3})$

Objective 2: Use the Distance and Midpoint Formulas

For Exercises 11–18,

a. Find the exact distance between the points. (**See Example 1**)

b. Find the midpoint of the line segment whose endpoints are the given points. (**See Example 3**)

11. $(-2, 7)$ and $(-4, 11)$
a. $2\sqrt{5}$ b. $(-3, 9)$

12. $(-1, -3)$ and $(3, -7)$ a. $4\sqrt{2}$ b. $(1, -5)$

13. $(-7, -4)$ and $(2, 5)$ a. $9\sqrt{2}$ b. $\left(-\dfrac{5}{2}, \dfrac{1}{2}\right)$

14. $(3, 6)$ and $(-4, -1)$ a. $7\sqrt{2}$ b. $\left(-\dfrac{1}{2}, \dfrac{5}{2}\right)$

15. $(2.2, -2.4)$ and $(5.2, -6.4)$
a. 5   b. $(3.7, -4.4)$

16. $(37.1, -24.7)$ and $(31.1, -32.7)$
a. 10   b. $(34.1, -28.7)$

17. $(\sqrt{5}, -\sqrt{2})$ and $(4\sqrt{5}, -7\sqrt{2})$
a. $\sqrt{117}$ b. $\left(\dfrac{5\sqrt{5}}{2}, -4\sqrt{2}\right)$

18. $(\sqrt{7}, -3\sqrt{5})$ and $(2\sqrt{7}, \sqrt{5})$
a. $\sqrt{87}$ b. $\left(\dfrac{3\sqrt{7}}{2}, -\sqrt{5}\right)$

For Exercises 19–22, determine if the given points form the vertices of a right triangle. (See Example 2)

19. $(1, 3)$, $(3, 1)$, and $(0, -2)$ Yes

20. $(1, 2)$, $(3, 0)$, and $(-3, -2)$ Yes

21. $(-2, 4)$, $(5, 0)$, and $(-5, 1)$ No

22. $(-6, 2)$, $(3, 1)$, and $(1, -2)$ No

Objective 3: Graph Equations by Plotting Points

For Exercises 23–24, determine if the given points are solutions to the equation.

23. $x^2 + y = 1$

   a. $(-2, -3)$ Yes   **b.** $(4, -17)$ No   **c.** $\left(\dfrac{1}{2}, \dfrac{3}{4}\right)$ Yes

24. $|x - 3| - y = 4$

   a. $(1, -2)$ Yes   **b.** $(-2, -3)$ No   **c.** $\left(\dfrac{1}{10}, -\dfrac{11}{10}\right)$ Yes

For Exercises 25–30, identify the set of values x for which y will be a real number.

25. $y = \dfrac{2}{x - 3}$ $\{x \mid x \neq 3\}$

26. $y = \dfrac{2}{x + 7}$ $\{x \mid x \neq -7\}$

27. $y = \sqrt{x - 10}$ $\{x \mid x \geq 10\}$

28. $y = \sqrt{x + 11}$ $\{x \mid x \geq -11\}$

29. $y = \sqrt{1.5 - x}$ $\{x \mid x \leq 1.5\}$

30. $y = \sqrt{2.2 - x}$ $\{x \mid x \leq 2.2\}$

For Exercises 31–44, graph the equations by plotting points. (See Examples 4–5)

31. $y = x$

32. $y = x^2$

33. $y = \sqrt{x}$

34. $y = |x|$

35. $y = x^3$

36. $y = \dfrac{1}{x}$

37. $y - |x| = 2$

38. $|x| + y = 3$

39. $y^2 - x - 2 = 0$

40. $y^2 - x + 1 = 0$

41. $x = |y| + 1$

42. $x = |y| - 3$

43. $y = |x + 1|$

44. $y = |x - 2|$

Objective 4: Identify x- and y-Intercepts

For Exercises 45–50, estimate the x- and y-intercepts from the graph.

45.
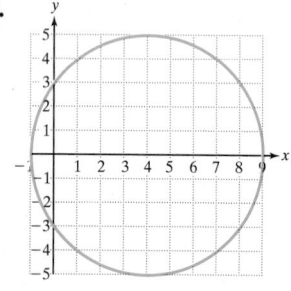

46.
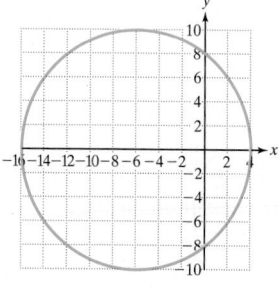

47.
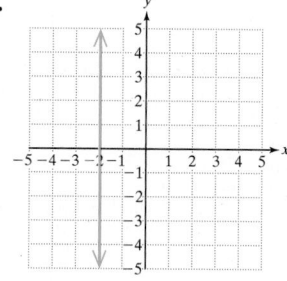

48.
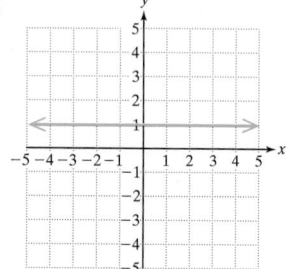

49.
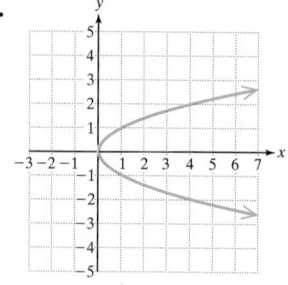

50.
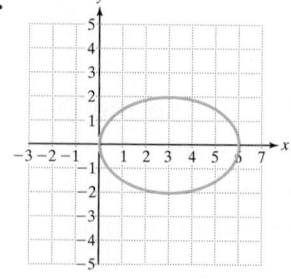

For Exercises 51–62, find the *x*- and *y*-intercepts. (See Example 6)

51. $-2x + 4y = 12$
  x-intercept: $(-6, 0)$; *y*-intercept: $(0, 3)$

52. $-3x - 5y = 60$
  x-intercept: $(-20, 0)$; *y*-intercept: $(0, -12)$

53. $x^2 + y = 9$
  x-intercepts: $(-3, 0)$, $(3, 0)$; *y*-intercept: $(0, 9)$

54. $x^2 = -y + 16$
  x-intercepts: $(-4, 0)$, $(4, 0)$; *y*-intercept: $(0, 16)$

55. $y = |x - 5| - 2$
  x-intercepts: $(3, 0)$, $(7, 0)$; *y*-intercept: $(0, 3)$

56. $y = |x + 4| - 3$
  x-intercepts: $(-7, 0)$, $(-1, 0)$; *y*-intercept: $(0, 1)$

57. $x = y^2 - 1$
  x-intercept: $(-1, 0)$; *y*-intercepts: $(0, -1)$, $(0, 1)$

58. $x = y^2 - 4$
  x-intercept: $(-4, 0)$; *y*-intercepts: $(0, -2)$, $(0, 2)$

59. $|x| = |y|$
  x-intercept: $(0, 0)$; *y*-intercept: $(0, 0)$

60. $x = |5y|$
  x-intercept: $(0, 0)$; *y*-intercept: $(0, 0)$

61. $\dfrac{(x - 3)^2}{4} + \dfrac{(y - 4)^2}{9} = 1$
  x-intercept: None; *y*-intercept: None

62. $\dfrac{(x + 6)^2}{16} + \dfrac{(y + 3)^2}{4} = 1$
  x-intercept: None; *y*-intercept: None

Mixed Exercises

63. A map of a wilderness area is drawn with the origin placed at the parking area. Two fire observation platforms are located at points *A* and *B*. If a fire is located at point *C*, determine the distance to the fire from each observation platform.
$d(A, C) = 2\sqrt{26}$ mi and $d(B, C) = 2\sqrt{17}$ mi

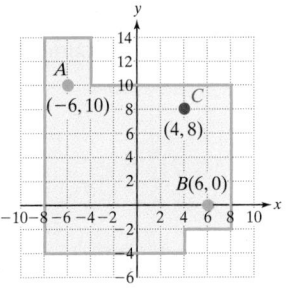

64. A map of a state park is drawn so that the origin is placed at the visitor center. The distance between grid lines is 1 mi. Suppose that two hikers are located at points *A* and *B*.

a. Determine the distance between the hikers.
$3\sqrt{5}$ mi ≈ 6.7 mi

b. If the hikers want to meet for lunch, determine the location of the midpoint between the hikers. $\left(-\dfrac{1}{2}, 0\right)$

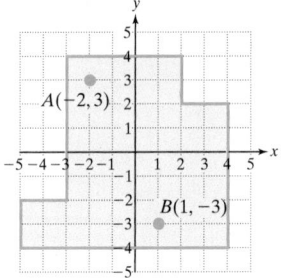

The position of an object in a video game is represented by an ordered pair. The coordinates of the ordered pair give the number of pixels horizontally and vertically from the origin. Use this scenario for Exercises 65–66.

65. a. Suppose that player A is located at (36, 315) and player B is located at (410, 53). How far apart are the players? Round to the nearest pixel.   457 pixels

b. If the two players move directly toward each other at the same speed, where will they meet?   (223, 184)

c. If player A moves three times faster than player B, where will they meet? Round to the nearest pixel.
(317, 119)

66. Suppose that a player is located at point *A*(460, 420) and must move in a direct line to point *B*(80, 210) and then in a direct line to point *C*(120, 60) to pick up prizes before a 5-sec timer runs out. If the player moves at 120 pixels per second, will the player have enough time to pick up both prizes? Explain.   Yes. The total distance from *A* to *B* to *C* is approximately 589 pixels. At 120 pixels per second, the time required is only about 4.9 sec.

67. Verify that the points $A(0, 0)$, $B(x, 0)$, and $C\left(\dfrac{1}{2}x, \dfrac{\sqrt{3}}{2}x\right)$ make up the vertices of an equilateral triangle.
From the distance formula, $d(A, B) = |x|$, $d(A, C) = |x|$, and $d(B, C) = |x|$.

68. Verify that the points $A(0, 0)$, $B(x, 0)$, and $C(0, x)$ make up the vertices of an isosceles right triangle (an isosceles triangle has two sides of equal length).
To show that the triangle is isosceles: $d(A, B) = |x|$ and $d(A, C) = |x|$. The triangle is a right triangle because $d(B, C) = \sqrt{2}|x|$ and $x^2 + x^2 = (\sqrt{2}|x|)^2$.

For Exercises 69–70, assume that the units shown in the grid are in feet.

a. Determine the exact length and width of the rectangle shown.

b. Determine the perimeter and area.

69.

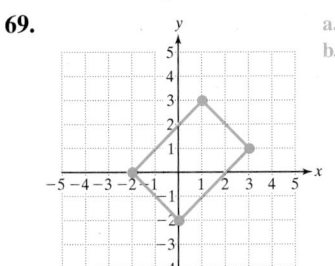

a. Length: $3\sqrt{2}$ ft; Width: $2\sqrt{2}$ ft
b. Perimeter: $10\sqrt{2}$ ft; Area: 12 ft^2

70.

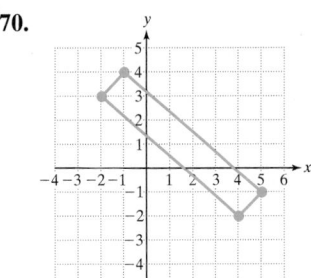

a. Length: $\sqrt{61}$ ft; Width: $\sqrt{2}$ ft
b. Perimeter: $2(\sqrt{61} + \sqrt{2})$ ft; Area: $\sqrt{122}$ ft^2

For Exercises 71–72, the endpoints of a diameter of a circle are shown. Find the center and radius of the circle.

71.   Center: $(1, 2)$; Radius: $\sqrt{10}$

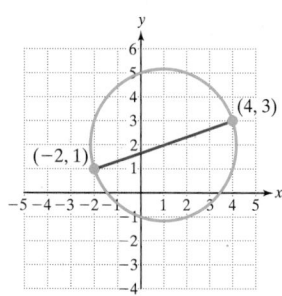

72.   Center: $\left(-\dfrac{3}{2}, 1\right)$; Radius: $\dfrac{\sqrt{65}}{2}$

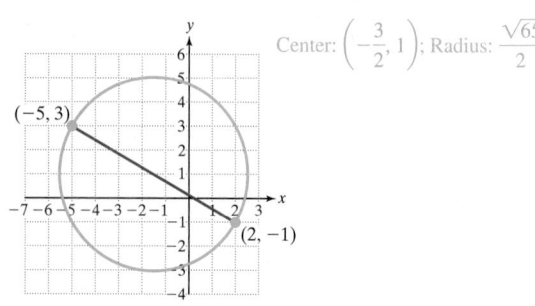

For Exercises 73–74, an isosceles triangle is shown. Find the area of the triangle. Assume that the units shown in the grid are in meters.

73.   Area: $25\ \text{m}^2$

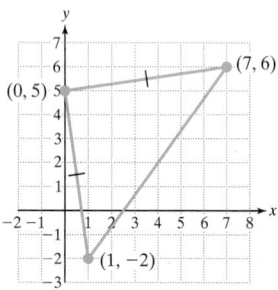

74.   Area: $\dfrac{45}{2}\ \text{m}^2$ or $22.5\ \text{m}^2$

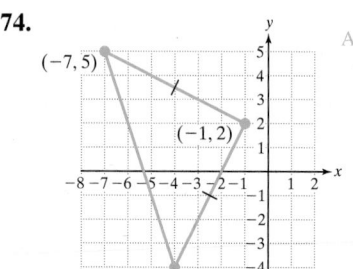

For Exercises 75–78, determine if points A, B, and C are collinear. Three points are collinear if they all fall on the same line. There are several ways that we can determine if three points, A, B, and C are collinear. One method is to determine if the sum of the lengths of the line segments $\overline{AB}$ and $\overline{BC}$ equals the length of $\overline{AC}$.

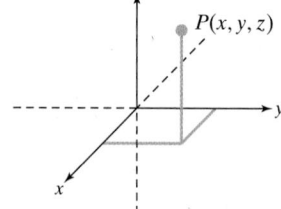

75. $(2, 2)$, $(4, 3)$, and $(8, 5)$ Collinear

76. $(2, 1.5)$, $(4, 2)$, and $(8, 3)$ Collinear

77. $(-2, 8)$, $(1, 2)$, and $(4, -3)$ Not collinear

78. $(-1, 5)$, $(0, 3)$, and $(5, -13)$ Not collinear

Write About It

79. Suppose that d represents the distance between two points (x_1, y_1) and (x_2, y_2). Explain how the distance formula is developed from the Pythagorean theorem.

80. Explain how you might remember the midpoint formula to find the midpoint of the line segment between (x_1, y_1) and (x_2, y_2).

81. Explain how to find the x- and y-intercepts from an equation in the variables x and y.

82. Given an equation in the variables x and y, what does the graph of the equation represent?

Expanding Your Skills

A point in three-dimensional space can be represented in a three-dimensional coordinate system. In such a case, a z-axis is taken perpendicular to both the x- and y-axes. A point P is assigned an ordered triple $P(x, y, z)$ relative to a fixed origin where the three axes meet. For Exercises 83–86, determine the distance between the two given points in space. Use the distance formula $d = \sqrt{(x_2 - x_1)^2 + (y_2 - y_1)^2 + (z_2 - z_1)^2}$.

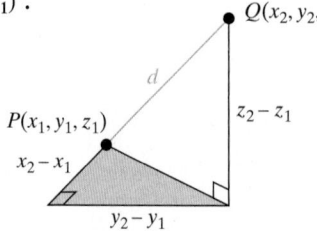

83. $(5, -3, 2)$ and $(4, 6, -1)$ $\sqrt{91}$

84. $(6, -4, -1)$ and $(2, 3, 1)$ $\sqrt{69}$

85. $(3, 7, -2)$ and $(0, -5, 1)$ $9\sqrt{2}$

86. $(9, -5, -3)$ and $(2, 0, 1)$ $3\sqrt{10}$

Objective 5: Graph Equations Using a Graphing Utility (Technology Connections)

87. What is meant by a viewing window on a graphing device?   The viewing window is part of the Cartesian plane shown in the display
screen of a calculator. The boundaries of the window are often denoted by [Xmin, Xmax, Xscl] by [Ymin, Ymax, Yscl].
88. Which of the viewing windows would show both the x- and y-intercepts of the graph of $780x - 42y = 5460$?   Window d

   a. $[-20, 20, 2]$ by $[-40, 40, 10]$

   b. $[-10, 10, 1]$ by $[-10, 10, 1]$

   c. $[-10, 10, 1]$ by $[-10, 150, 10]$

   d. $[-10, 10, 1]$ by $[-150, 10, 10]$

For Exercises 89–92, graph the equation with a graphing utility on the given viewing window. (See Example 7)

89. $y = 2x - 5$ on $[-10, 10, 1]$ by $[-10, 10, 1]$ **90.** $y = -4x + 1$ on $[-10, 10, 1]$ by $[-10, 10, 1]$

91. $y = 1400x^2 - 1200x$ on $[-5, 5, 1]$ **92.** $y = -800x^2 + 600x$ on $[-5, 5, 1]$
    by $[-1000, 2000, 500]$ by $[-1000, 500, 200]$

For Exercises 93–94, graph the equations on the standard viewing window. (See Example 7)

93. **a.** $y = x^3$ **94.** **a.** $y = \sqrt{x + 4}$

    b. $y = |x| - 9$ **b.** $y = |x - 2|$

Additional answers can be found in the Instructor Answer Appendix.

SECTION 1.2   Circles

OBJECTIVES

1. **Write an Equation of a Circle in Standard Form**
2. **Write the General Form of an Equation of a Circle**

1. Write an Equation of a Circle in Standard Form

In addition to graphing equations by plotting points, we will learn to recognize specific categories of equations and the characteristics of their graphs. We begin by presenting the definition of a circle.

> ### Definition of a Circle
>
> A **circle** is the set of all points in a plane that are equidistant from a fixed point called the **center.** The fixed distance from any point on the circle to the center is called the **radius.**

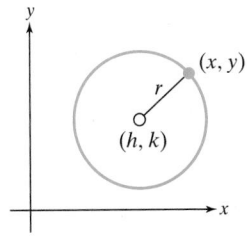

Figure 1-11

The radius of a circle is often denoted by r, where $r > 0$. It is also important to note that the center is not actually part of the graph of a circle. It will be drawn in the text as an open dot for reference only.

    Suppose that a circle is centered at the point (h, k) and has radius r (Figure 1-11). The distance formula can be used to derive an equation of the circle. Let (x, y) be an arbitrary point on the circle. Then by definition the distance between (h, k) and (x, y) must be r.

Apply the distance formula. $\sqrt{(x_2 - x_1)^2 + (y_2 - y_1)^2} = d$

$$\sqrt{(x - h)^2 + (y - k)^2} = r \qquad \text{Distance between } (h, k) \text{ and } (x, y)$$

$$(x - h)^2 + (y - k)^2 = r^2 \qquad \text{Squaring both sides of the equation results in the standard form of an equation of a circle.}$$

Standard Form of an Equation of a Circle

Given a circle centered at (h, k) with radius r, the **standard form** of an equation of the circle (also called **center-radius form**) is given by

$$(x - h)^2 + (y - k)^2 = r^2 \quad \text{where } r > 0.$$

Examples	Standard form	Center	Radius
$(x - 4)^2 + (y + 3)^2 = 25$	$(x - 4)^2 + [y - (-3)]^2 = (5)^2$	$(4, -3)$	5
$x^2 + \left(y - \frac{1}{2}\right)^2 = 12$	$(x - 0)^2 + \left(y - \frac{1}{2}\right)^2 = \left(\sqrt{12}\right)^2$	$\left(0, \frac{1}{2}\right)$	$2\sqrt{3}$
$x^2 + y^2 = 7$	$(x - 0)^2 + (y - 0)^2 = \left(\sqrt{7}\right)^2$	$(0, 0)$	$\sqrt{7}$

In Example 1, we write an equation of a circle in standard form.

Classroom Example: p. 135
Exercise 22

Point of Interest

Among his many contributions to mathematics, René Descartes discovered analytic geometry, which uses algebraic equations to describe geometric shapes. For example, a circle can be described by the algebraic equation $(x - h)^2 + (y - k)^2 = r^2$.

EXAMPLE 1 Writing an Equation of a Circle in Standard Form

a. Write the standard form of an equation of the circle with center $(-4, 6)$ and radius 2.

b. Graph the circle.

Solution:

a. $(h, k) = (-4, 6)$ and $r = 2$ 　　Label the center (h, k) and the radius r.

$[x - (-4)]^2 + (y - 6)^2 = (2)^2$ 　　Standard form: $(x - h)^2 + (y - k)^2 = r^2$

$(x + 4)^2 + (y - 6)^2 = 4$ 　　Simplify.

b.

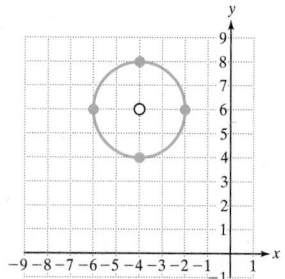

To graph the circle, first locate the center and draw a small open dot. Then plot points r units to the left, right, above, and below the center.

Draw the circle through the points.

Skill Practice 1

a. Write an equation of the circle with center $(3, -1)$ and radius 4.

b. Graph the circle.

Classroom Example: p. 135
Exercise 26

Answers

1. a. $(x - 3)^2 + (y + 1)^2 = 16$

b.

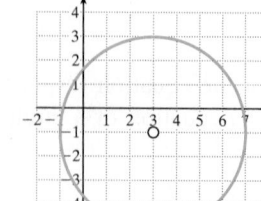

EXAMPLE 2 Writing an Equation of a Circle in Standard Form

Write the standard form of an equation of the circle with endpoints of a diameter $(-1, 0)$ and $(3, 4)$.

Solution:

A sketch of this scenario is given in Figure 1-12. Notice that the midpoint of the diameter is the center of the circle.

$(-1, 0)$ and $(3, 4)$

(x_1, y_1) and (x_2, y_2) 　　　　　　Label the points.

The center is $\left(\dfrac{-1 + 3}{2}, \dfrac{0 + 4}{2}\right) = (1, 2)$.

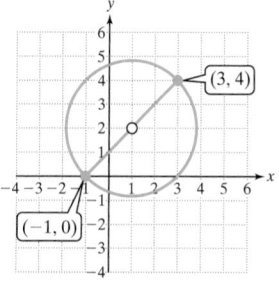

Figure 1-12

The radius of the circle is the distance between either endpoint of the diameter and the center. Using the endpoint $(-1, 0)$ as (x_1, y_1) and the center $(1, 2)$ as (x_2, y_2), apply the distance formula.

$$d = \sqrt{(x_2 - x_1)^2 + (y_2 - y_1)^2}$$
$$r = \sqrt{[1 - (-1)]^2 + (2 - 0)^2} = \sqrt{(2)^2 + (2)^2} = \sqrt{8}$$

An equation of the circle is: $(x - h)^2 + (y - k)^2 = r^2$.

$$(x - 1)^2 + (y - 2)^2 = (\sqrt{8})^2$$
$$(x - 1)^2 + (y - 2)^2 = 8 \quad \text{(Standard form)}$$

> **Skill Practice 2**   Write the standard form of an equation of the circle with endpoints of a diameter $(-3, 3)$ and $(-1, -1)$.

2. Write the General Form of an Equation of a Circle

In Example 2 we have the equation $(x - 1)^2 + (y - 2)^2 = 8$. If we expand the binomials and combine like terms, we can write the equation in *general form*.

$$(x - 1)^2 + (y - 2)^2 = 8 \qquad \text{Standard form (center-radius form)}$$
$$x^2 - 2x + 1 + y^2 - 4y + 4 = 8 \qquad \text{Expand the binomials}$$
$$x^2 + y^2 - 2x - 4y - 3 = 0 \qquad \text{General form}$$

> **General Form of an Equation of a Circle**
>
> An equation of a circle written in the form $x^2 + y^2 + Ax + By + C = 0$ is called the **general form** of an equation of a circle.

By completing the square we can write an equation of a circle given in general form as an equation in standard form. The purpose of writing an equation of a circle in standard form is to identify the radius and center. This is demonstrated in Example 3.

Classroom Example: p. 136
Exercise 42

> **EXAMPLE 3**   Writing an Equation of a Circle in Standard Form
>
> Write the equation of the circle in standard form. Then identify the center and radius.
>
> $$x^2 + y^2 + 10x - 6y + 25 = 0$$
>
> **Solution:**
>
> $$x^2 + y^2 + 10x - 6y + 25 = 0$$
> $$(x^2 + 10x \quad) + (y^2 - 6y \quad) = -25 \qquad \begin{array}{l}\text{Group the } x \text{ terms. Group the } y \text{ terms.}\\ \text{Move the constant term to the right.}\end{array}$$
> $$(x^2 + 10x + 25) + (y^2 - 6y + 9) \qquad \begin{array}{l}\text{Complete the squares.}\\ \textit{Note: } [\tfrac{1}{2}(10)]^2 = 25, [\tfrac{1}{2}(-6)]^2 = 9\end{array}$$
> $$= -25 + 25 + 9$$
> $$(x + 5)^2 + (y - 3)^2 = 9 \qquad \text{Factor.}$$
>
> The center is $(-5, 3)$, and the radius is $\sqrt{9} = 3$. See Figure 1-13.

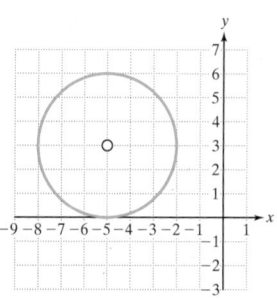

Figure 1-13

> **Skill Practice 3**   Write the equation of the circle in standard form. Then identify the center and radius. $x^2 + y^2 - 8x + 2y - 8 = 0$

Answers

2. $(x + 2)^2 + (y - 1)^2 = 5$

3. $(x - 4)^2 + (y + 1)^2 = 25$;
   Center: $(4, -1)$; Radius: 5

Not all equations of the form $x^2 + y^2 + Ax + By + C = 0$ represent the graph of a circle. Completing the square results in an equation of the form $(x - h)^2 + (y - k)^2 = c$, where c is a constant. In the case where $c > 0$, the graph of the equation is a circle with radius $r = \sqrt{c}$. However, if $c = 0$, or if $c < 0$, the graph will be a single point or nonexistent. These are called **degenerate cases.**

- If $c > 0$, then the graph will be a circle with radius $r = \sqrt{c}$.
- If $c = 0$, then the graph will be a single point, (h, k). The solution set is $\{(h, k)\}$.
- If $c < 0$, then the solution set is the empty set $\{\ \}$.

Classroom Example: p. 136
Exercise 52

EXAMPLE 4   Determining if an Equation Represents the Graph of a Circle

Write the equation in the form $(x - h)^2 + (y - k)^2 = r^2$, and identify the solution set.

$$x^2 + y^2 - 14y + 49 = 0$$

Solution:

$$x^2 + y^2 - 14y + 49 = 0$$
$$x^2 + (y^2 - 14y \quad) = -49$$
Group the y terms and complete the square. Note that the x^2 term is already a perfect square: $(x - 0)^2$.

$$x^2 + (y^2 - 14y + 49) = -49 + 49$$
Complete the square: $\left[\frac{1}{2}(-14)\right]^2 = 49$.

$$x^2 + (y - 7)^2 = 0$$
Factor.

Since $r^2 = 0$, the solution set is $\{(0, 7)\}$. The sum of two squares will equal zero only if each individual term is zero. Therefore, $x = 0$ and $y = 7$.

Skill Practice 4   Write the equation in the form $(x - h)^2 + (y - k)^2 = r^2$, and identify the solution set. $x^2 + y^2 + 2x + 5 = 0$

TECHNOLOGY CONNECTIONS

Setting a Square Viewing Window and Graphing a Circle

A graphing calculator expects an equation with the y variable isolated. Therefore, to graph an equation of a circle such as $(x + 5)^2 + (y - 3)^2 = 9$, from Example 3, we first solve for y.

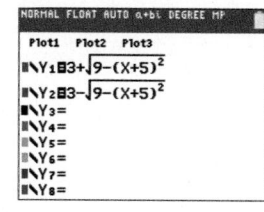

$$(x + 5)^2 + (y - 3)^2 = 9$$
$$(y - 3)^2 = 9 - (x + 5)^2$$
$$y - 3 = \pm\sqrt{9 - (x + 5)^2}$$
$$y = 3 \pm \sqrt{9 - (x + 5)^2}$$

Notice that the graph looks more oval-shaped than circular. This is because the calculator has a rectangular screen. If the scaling is the same on the x- and y-axes, the graph will appear elongated horizontally. To eliminate this distortion, use a ZSquare option, located in the Zoom menu.

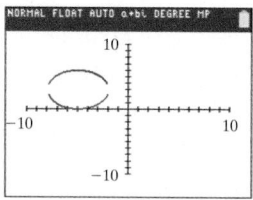

Also notice that the calculator display may not show the upper and lower semicircles connecting. The viewing window between $x = -16.1$ and $x = 16.1$ is divided by the number of pixels displayed horizontally to get the values of x used to graph the equation. These may not include x values at the leftmost and rightmost points on the circle. That is, the calculator may graph points *close* to $(-8, 3)$ and $(-2, 3)$ but not exactly at $(-8, 3)$ and $(-2, 3)$. Therefore, the upper and lower semicircles may not "hook up."

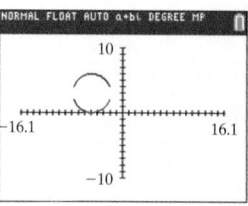

Answer

4. $(x + 1)^2 + y^2 = -4$; The solution set is $\{\ \}$.

SECTION 1.2   Practice Exercises

Prerequisite Review

For Exercises R.1–R.2, find the value of n so that the expression is a perfect square trinomial. Then factor the trinomial.

R.1. $c^2 - 8c + n$ $n = 16; (c - 4)^2$

R.2. $x^2 + \dfrac{2}{7}x + n$ $n = \dfrac{1}{49}; \left(x + \dfrac{1}{7}\right)^2$

R.3. Find the distance between $(2, 3)$ and $(-3, -2)$. Express your answer in simplified radical form. $5\sqrt{2}$

R.4. Multiply by using the special case products. Simplify. $(x - 2)^2$ $x^2 - 4x + 4$

Concept Connections

1. A _____circle_____ is the set of all points in a plane equidistant from a fixed point called the _____center_____.

2. The distance from the center of a circle to any point on the circle is called the _____radius_____ and is often denoted by r.

3. The standard form of an equation of a circle with center (h, k) and radius r is given by $\underline{(x - h)^2 + (y - k)^2 = r^2}$.

4. An equation of a circle written in the form $x^2 + y^2 + Ax + By + C = 0$ is called the _____general_____ form of an equation of a circle.

Objective 1: Write an Equation of a Circle in Standard Form

5. Is the point $(2, 7)$ on the circle defined by $(x - 2)^2 + (y - 7)^2 = 4$?   No

6. Is the point $(3, 5)$ on the circle defined by $(x - 3)^2 + (y - 5)^2 = 36$?   No

7. Is the point $(-4, 7)$ on the circle defined by $(x + 1)^2 + (y - 3)^2 = 25$?   Yes

8. Is the point $(2, -7)$ on the circle defined by $(x + 6)^2 + (y + 1)^2 = 100$?   Yes

For Exercises 9–16, determine the center and radius of the circle.

9. $(x - 4)^2 + (y + 2)^2 = 81$
   Center: $(4, -2)$; Radius: 9

10. $(x + 3)^2 + (y - 1)^2 = 16$
    Center: $(-3, 1)$; Radius: 4

11. $x^2 + (y - 2.5)^2 = 6.25$
    Center: $(0, 2.5)$; Radius: 2.5

12. $(x - 1.5)^2 + y^2 = 2.25$
    Center: $(1.5, 0)$; Radius: 1.5

13. $x^2 + y^2 = 20$ Center: $(0, 0)$; Radius: $2\sqrt{5}$

14. $x^2 + y^2 = 28$
    Center: $(0, 0)$; Radius: $2\sqrt{7}$

15. $\left(x - \dfrac{3}{2}\right)^2 + \left(y + \dfrac{3}{4}\right)^2 = \dfrac{81}{49}$
    Center: $\left(\dfrac{3}{2}, -\dfrac{3}{4}\right)$; Radius: $\dfrac{9}{7}$

16. $\left(x + \dfrac{1}{7}\right)^2 + \left(y - \dfrac{3}{5}\right)^2 = \dfrac{25}{9}$ Center: $\left(-\dfrac{1}{7}, \dfrac{3}{5}\right)$; Radius: $\dfrac{5}{3}$

For Exercises 17–32, information about a circle is given.

a. Write an equation of the circle in standard form.

b. Graph the circle. (See Examples 1–2)

17. Center: $(-2, 5)$; Radius: 1

18. Center: $(-3, 2)$; Radius: 4

19. Center: $(-4, 1)$; Radius: 3

20. Center: $(6, -2)$; Radius: 6

21. Center: $(-4, -3)$; Radius: $\sqrt{11}$

22. Center: $(-5, -2)$; Radius: $\sqrt{21}$

23. Center: $(0, 0)$; Radius: 2.6

24. Center: $(0, 0)$; Radius: 4.2

25. The endpoints of a diameter are $(-2, 4)$ and $(6, -2)$.

26. The endpoints of a diameter are $(7, 3)$ and $(5, -1)$.

27. The center is $(-2, -1)$ and another point on the circle is $(6, 5)$.

28. The center is $(3, 1)$ and another point on the circle is $(6, 5)$.

29. The center is $(4, 6)$ and the circle is tangent to the y-axis. (Informally, a line is tangent to a circle if it touches the circle in exactly one point.)

30. The center is $(-2, -4)$ and the circle is tangent to the x-axis.

31. The center is in Quadrant IV, the radius is 5, and the circle is tangent to both the x- and y-axes.

32. The center is in Quadrant II, the radius is 3, and the circle is tangent to both the x- and y-axes.

33. Write an equation that represents the set of points that are 5 units from $(8, -11)$. $(x - 8)^2 + (y + 11)^2 = 25$

34. Write an equation that represents the set of points that are 9 units from $(-4, 16)$. $(x + 4)^2 + (y - 16)^2 = 81$

35. Write an equation of the circle that is tangent to both axes with radius $\sqrt{7}$ and center in Quadrant I.
    $(x - \sqrt{7})^2 + (y - \sqrt{7})^2 = 7$

36. Write an equation of the circle that is tangent to both axes with radius $\sqrt{11}$ and center in Quadrant III.
    $(x + \sqrt{11})^2 + (y + \sqrt{11})^2 = 11$

Objective 2: Write the General Form of an Equation of a Circle

37. Determine the solution set for the equation
$(x + 1)^2 + (y - 5)^2 = 0.$ $\{(-1, 5)\}$

38. Determine the solution set for the equation
$(x - 3)^2 + (y + 12)^2 = 0.$ $\{(3, -12)\}$

39. Determine the solution set for the equation
$(x - 17)^2 + (y + 1)^2 = -9.$ $\{\ \}$

40. Determine the solution set for the equation
$(x + 15)^2 + (y - 3)^2 = -25.$ $\{\ \}$

For Exercises 41–54, write the equation in the form $(x - h)^2 + (y - k)^2 = c$. Then if the equation represents a circle, identify the center and radius. If the equation represents a degenerate case, give the solution set. (See Examples 3–4)

41. $x^2 + y^2 + 6x - 2y + 6 = 0$
$(x + 3)^2 + (y - 1)^2 = 4$; Center: $(-3, 1)$; Radius: 2

42. $x^2 + y^2 + 12x - 14y + 84 = 0$
$(x + 6)^2 + (y - 7)^2 = 1$; Center: $(-6, 7)$; Radius: 1

43. $x^2 + y^2 - 22x + 6y + 129 = 0$
$(x - 11)^2 + (y + 3)^2 = 1$; Center: $(11, -3)$; Radius: 1

44. $x^2 + y^2 - 10x + 4y - 20 = 0$
$(x - 5)^2 + (y + 2)^2 = 49$; Center: $(5, -2)$; Radius: 7

45. $x^2 + y^2 - 20y - 4 = 0$
$x^2 + (y - 10)^2 = 104$; Center: $(0, 10)$; Radius: $2\sqrt{26}$

46. $x^2 + y^2 + 22x - 4 = 0$
$(x + 11)^2 + y^2 = 125$; Center: $(-11, 0)$; Radius: $5\sqrt{5}$

47. $10x^2 + 10y^2 - 80x + 200y + 920 = 0$
(*Hint*: Divide by 10 to make the x^2 and y^2
term coefficients equal to 1.)
$(x - 4)^2 + (y + 10)^2 = 24$; Center: $(4, -10)$; Radius: $2\sqrt{6}$

48. $2x^2 + 2y^2 - 32x + 12y + 90 = 0$
$(x - 8)^2 + (y + 3)^2 = 28$; Center: $(8, -3)$; Radius: $2\sqrt{7}$

49. $x^2 + y^2 - 4x - 18y + 89 = 0$
$(x - 2)^2 + (y - 9)^2 = -4$; Degenerate case: $\{\ \}$

50. $x^2 + y^2 - 10x - 22y + 155 = 0$
$(x - 5)^2 + (y - 11)^2 = -9$; Degenerate case: $\{\ \}$

51. $4x^2 + 4y^2 - 20y + 25 = 0$ $x^2 + \left(y - \dfrac{5}{2}\right)^2 = 0$; Degenerate case (single point): $\left\{\left(0, \dfrac{5}{2}\right)\right\}$

52. $4x^2 + 4y^2 - 12x + 9 = 0$ $\left(x - \dfrac{3}{2}\right)^2 + y^2 = 0$; Degenerate case (single point): $\left\{\left(\dfrac{3}{2}, 0\right)\right\}$

53. $x^2 + y^2 - x - \dfrac{3}{2}y - \dfrac{3}{4} = 0$
$\left(x - \dfrac{1}{2}\right)^2 + \left(y - \dfrac{3}{4}\right)^2 = \dfrac{25}{16}$; Center: $\left(\dfrac{1}{2}, \dfrac{3}{4}\right)$; Radius: $\dfrac{5}{4}$

54. $x^2 + y^2 - \dfrac{2}{3}x - \dfrac{5}{3}y - \dfrac{5}{9} = 0$
$\left(x - \dfrac{1}{3}\right)^2 + \left(y - \dfrac{5}{6}\right)^2 = \dfrac{49}{36}$; Center: $\left(\dfrac{1}{3}, \dfrac{5}{6}\right)$; Radius: $\dfrac{7}{6}$

Mixed Exercises

55. A cell tower is a site where antennas, transmitters, and receivers are placed to create a cellular network. Suppose that a cell tower is located at a point $A(4, 6)$ on a map and its range is 1.5 mi. Write an equation that represents the boundary of the area that can receive a signal from the tower. Assume that all distances are in miles. $(x - 4)^2 + (y - 6)^2 = 2.25$

56. A radar transmitter on a ship has a range of 20 nautical miles. If the ship is located at a point $(-32, 40)$ on a map, write an equation for the boundary of the area within the range of the ship's radar. Assume that all distances on the map are represented in nautical miles.
$(x + 32)^2 + (y - 40)^2 = 400$

57. Suppose that three geological study areas are set up on a map at points $A(-4, 12)$, $B(11, 3)$, and $C(0, 1)$, where all units are in miles. Based on the speed of compression waves, scientists estimate the distances from the study areas to the epicenter of an earthquake to be 13 mi, 5 mi, and 10 mi, respectively. Graph three circles whose centers are located at the study areas and whose radii are the given distances to the earthquake. Then estimate the location of the earthquake.

58. Three fire observation towers are located at points $A(-6, -14)$, $B(14, 10)$, and $C(-3, 13)$ on a map where all units are in kilometers. A fire is located at distances of 17 km, 15 km, and 13 km, respectively, from the observation towers. Graph three circles whose centers are located at the observation towers and whose radii are the given distances to the fire. Then estimate the location of the fire.

Write About It

59. State the definition of a circle.
A circle is the set of all points in a plane that are equidistant from a fixed point called the center.

60. What are the advantages of writing an equation of a circle in standard form?   The center and radius can easily be identified from an equation of a circle written in standard form.

Expanding Your Skills

61. Find all values of y such that the distance between $(4, y)$ and $(-2, 6)$ is 10 units. $y = -2$ and $y = 14$

62. Find all values of x such that the distance between $(x, -1)$ and $(4, 2)$ is 5 units. $x = 0$ and $x = 8$

63. Find all points on the line $y = x$ that are 6 units from $(2, 4)$. $(3 + \sqrt{17}, 3 + \sqrt{17})$ and $(3 - \sqrt{17}, 3 - \sqrt{17})$

64. Find all points on the line $y = -x$ that are 4 units from $(-4, 6)$. $(-5 + \sqrt{7}, 5 - \sqrt{7})$ and $(-5 - \sqrt{7}, 5 + \sqrt{7})$

The general form of an equation of a circle is $(x - h)^2 + (y - k)^2 = r^2$. If we solve the equation for x we get equations of the form $x = h \pm \sqrt{r^2 - (y - k)^2}$. The equation $x = h + \sqrt{r^2 - (y - k)^2}$ represents the graph of the corresponding right-side semicircle, and the equation $x = h - \sqrt{r^2 - (y - k)^2}$ represents the graph of the left-side semicircle. Likewise, if we solve for y, we have $y = k \pm \sqrt{r^2 - (x - h)^2}$. These equations represent the top and bottom semicircles. For Exercises 65–68, graph the equations.

65. a. $y = \sqrt{16 - x^2}$
 b. $y = -\sqrt{16 - x^2}$
 c. $x = \sqrt{16 - y^2}$
 d. $x = -\sqrt{16 - y^2}$

66. a. $y = \sqrt{9 - x^2}$
 b. $y = -\sqrt{9 - x^2}$
 c. $x = \sqrt{9 - y^2}$
 d. $x = -\sqrt{9 - y^2}$

67. a. $x = -1 - \sqrt{9 - (y - 2)^2}$
 b. $x = -1 + \sqrt{9 - (y - 2)^2}$
 c. $y = 2 - \sqrt{9 - (x + 1)^2}$
 d. $y = 2 + \sqrt{9 - (x + 1)^2}$

68. a. $x = 3 - \sqrt{4 - (y + 2)^2}$
 b. $x = 3 + \sqrt{4 - (y + 2)^2}$
 c. $y = -2 - \sqrt{4 - (x - 3)^2}$
 d. $y = -2 + \sqrt{4 - (x - 3)^2}$

69. Find the shortest distance from the origin to a point on the circle defined by $x^2 + y^2 - 6x - 12y + 41 = 0$.
$\sqrt{49 - 12\sqrt{5}}$ or $3\sqrt{5} - 2$

70. Find the shortest distance from the origin to a point on the circle defined by $x^2 + y^2 + 4x - 12y + 31 = 0$.
$\sqrt{49 - 12\sqrt{10}}$ or $2\sqrt{10} - 3$

Technology Connections

For Exercises 71–74, use a graphing calculator to graph the circles on an appropriate square viewing window.

71. $x^2 + y^2 = 36$

72. $x^2 + y^2 = 49$

73. $(x - 18)^2 + (y + 20)^2 = 80$

74. $(x + 0.04)^2 + (y - 0.02)^2 = 0.01$

Additional answers can be found in the Instructor Answer Appendix.

SECTION 1.3 | Functions and Relations

OBJECTIVES

1. Determine Whether a Relation Is a Function
2. Apply Function Notation
3. Determine x- and y-Intercepts of a Function Defined by $y = f(x)$
4. Determine Domain and Range of a Function
5. Interpret a Function Graphically

1. Determine Whether a Relation Is a Function

In the physical world, many quantities that are subject to change are related to other variables. For example:

- The cost of mailing a package is related to the weight of a package.
- The minimum braking distance of a car depends on the speed of the car.
- The perimeter of a rectangle is a function of its length and width.
- The test score that a student earns is related to the number of hours of study.

In mathematics we can express the relationship between two values as a set of ordered pairs.

> ### Definition of a Relation
>
> A set of ordered pairs (x, y) is called a **relation** in x and y.
>
> - The set of x values in the ordered pairs is called the **domain** of the relation.
> - The set of y values in the ordered pairs is called the **range** of the relation.

Classroom Example: p. 146
Exercise 10

EXAMPLE 1   Writing a Relation from Observed Data Points

Table 1-1 shows the score y that a student earned on an algebra test based on the number of hours x spent studying one week prior to the test.

a. Write the set of ordered pairs that defines the relation given in Table 1-1.

b. Write the domain.

c. Write the range.

Hours of Study, x	Test Score, y
8	92
3	58
11	98
5	72
8	86

Table 1-1

Avoiding Mistakes

Do not list the elements in a set more than once. The value 8 is listed in the domain one time only.

Solution:

a. Relation: {(8, 92), (3, 58), (11, 98), (5, 72), (8, 86)}

b. Domain: {8, 3, 11, 5}

c. Range: {92, 58, 98, 72, 86}

Skill Practice 1   For the table shown,

a. Write the set of ordered pairs that defines the relation.

b. Write the domain.

c. Write the range.

x	3	−2	5	1
y	−4	0	3	0

The data in Table 1-1 show two different test scores for 8 hr of study. That is, for $x = 8$, there are two different y values. In many applications, we prefer to work with relations that assign one and only one y value for a given value of x. Such a relation is called a function.

Definition of a Function

Given a relation in x and y, we say that **y is a function of x** if for each value of x in the domain, there is exactly one value of y in the range.

Classroom Example: p. 146
Exercise 12

EXAMPLE 2   Determining if a Relation Is a Function

Determine if the relation defines y as a function of x.

a. {(3, 1), (2, 5), (−4, 2), (−1, 0), (3, −4)}

b. {(−1, 4), (2, 3), (3, 4), (−4, 5)}

Solution:

a.

same x values

{(3, 1), (2, 5), (−4, 2), (−1, 0), (3, −4)}

different y values

When $x = 3$, there are two different y values: $y = 1$ and $y = -4$.

This relation is *not* a function.

TIP   A function may not
have the same x value paired with different y values. However, it is acceptable for a function to have two or more x values paired with the same y value, as shown in Example 2(b).

b. {(−1, 4), (2, 3), (3, 4), (−4, 5)}

No two ordered pairs have the same x value but different y values.

This relation *is* a function.

Skill Practice 2   Determine if the relation defines y as a function of x.

a. {(8, 4), (3, −1), (5, 4)}   **b.** {(−3, 2), (9, 5), (1, 0), (−3, 1)}

Answers

1. a. {(3, −4), (−2, 0), (5, 3), (1, 0)}
   b. Domain: {3, −2, 5, 1}
   c. Range: {−4, 0, 3}
2. a. Yes   **b.** No

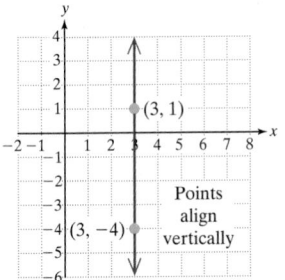

Figure 1-14

Classroom Examples: p. 146
Exercises 18, 22

A relation that is not a function has at least one domain element x paired with more than one range element y. For example, the ordered pairs $(3, 1)$ and $(3, -4)$ do not make up a function. On a graph, these two points are aligned vertically. A vertical line drawn through one point also intersects the other point (Figure 1-14). This observation leads to the vertical line test.

Using the Vertical Line Test

Consider a relation defined by a set of points (x, y) graphed on a rectangular coordinate system. The graph defines y as a function of x if no vertical line intersects the graph in more than one point.

EXAMPLE 3   Applying the Vertical Line Test

The graphs of three relations are shown in blue. In each case, determine if the relation defines y as a function of x.

Solution:

a.

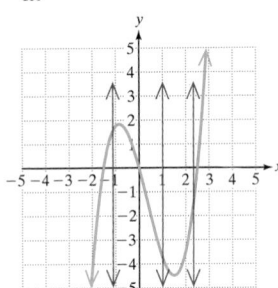

This is a function.

No vertical line intersects the graph in more than one point.

b.

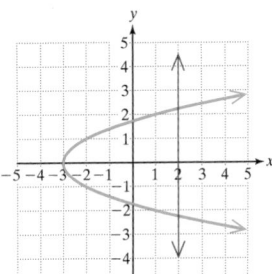

This is not a function.

There is at least one vertical line that intersects the graph in more than one point.

c.

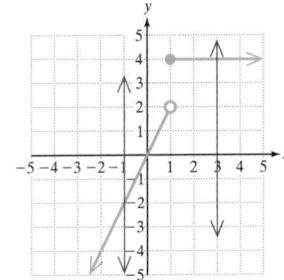

This is a function.

No vertical line intersects the graph in more than one point.

TIP In Example 3(c) there is only one y value assigned to $x = 1$. This is because the point $(1, 2)$ is *not* included in the graph of the function as denoted by the open dot.

Skill Practice 3   Determine if the given relation defines y as a function of x.

a.

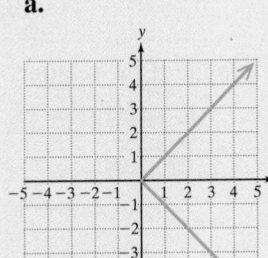

b.

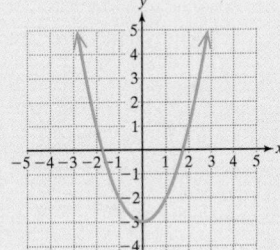

c.

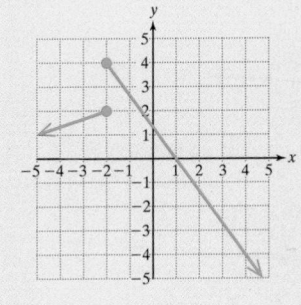

A relation can also be defined by a figure showing a "mapping" between x and y, or by an equation in x and y.

Answers
3. a. No   **b.** Yes   **c.** No

Classroom Examples: p. 147
Exercises 26, 28, 32

EXAMPLE 4 Determining if a Relation Is a Function

Determine if the relation defines y as a function of x.

a.

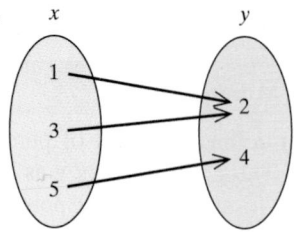

b. $y^2 = x$ c. $(x - 2)^2 + (y + 1)^2 = 9$

Solution:

a. This mapping defines the set of ordered pairs: $\{(1, 2), (3, 2), (5, 4)\}$. This relation *is* a function.

No two ordered pairs have the same x value but different y values.

b. $y^2 = x$

$y = \pm\sqrt{x}$

Solve the equation for y. For any $x > 0$, there are two corresponding y values.

x	y	Ordered pairs
0	0	$(0, 0)$
1	$1, -1$	$(1, 1), (1, -1)$
4	$2, -2$	$(4, 2), (4, -2)$
9	$3, -3$	$(9, 3), (9, -3)$

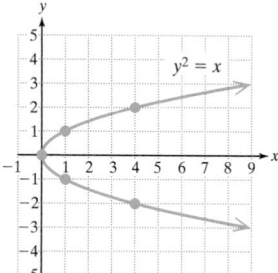

This relation is *not* a function.

c. $(x - 2)^2 + (y + 1)^2 = 9$

This equation represents the graph of a circle with center $(2, -1)$ and radius 3.

This relation is *not* a function because it fails the vertical line test.

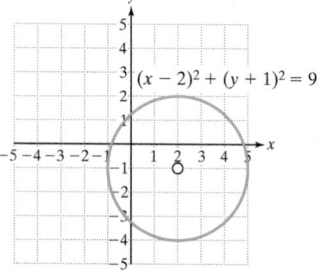

Skill Practice 4 Determine if the relation defines y as a function of x.

a.

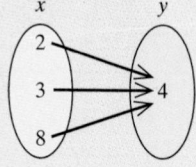

b. $|y + 1| = x$ c. $x^2 + y^2 = 25$

2. Apply Function Notation

A function may be defined by an equation with two variables. For example, the equation $y = x - 2$ defines y as a function of x. This is because for any real number x, the value of y is the unique number that is 2 less than x.

When a function is defined by an equation, we often use function notation. For example, the equation $y = x - 2$ may be written in function notation as

$$f(x) = x - 2 \text{ read as "} f \text{ of } x \text{ equals } x - 2."$$

Answers

4. a. Yes   b. No   c. No

With function notation,

- f is the name of the function,
- x is an input variable from the domain,
- $f(x)$ is the function value (or y value) corresponding to x.

A function may be evaluated at different values of x by using substitution.

$$f(x) = x - 2$$
$$f(4) = (4) - 2 = 2 \qquad f(4) = 2 \text{ can be interpreted as } (4, 2).$$
$$f(1) = (1) - 2 = -1 \qquad f(1) = -1 \text{ can be interpreted as } (1, -1).$$

Classroom Example: p. 147
Exercise 38

EXAMPLE 5   Evaluating a Function

Evaluate the function defined by $g(x) = 2x + 1$ for the given values of x.

   a. $g(-2)$ **b.** $g(-1)$ **c.** $g(0)$ **d.** $g(1)$ **e.** $g(2)$

Solution:

a. $g(-2) = 2(-2) + 1$ Substitute -2 for x.
       $= -3$ $g(-2) = -3$

b. $g(-1) = 2(-1) + 1$ Substitute -1 for x.
       $= -1$ $g(-1) = -1$

c. $g(0) = 2(0) + 1$ Substitute 0 for x.
       $= 1$ $g(0) = 1$

d. $g(1) = 2(1) + 1$ Substitute 1 for x.
       $= 3$ $g(1) = 3$

e. $g(2) = 2(2) + 1$ Substitute 2 for x.
       $= 5$ $g(2) = 5$

The function values represent the ordered pairs $(-2, -3)$, $(-1, -1)$, $(0, 1)$, $(1, 3)$, and $(2, 5)$. The line through the points represents all ordered pairs defined by this function. This is the graph of the function.

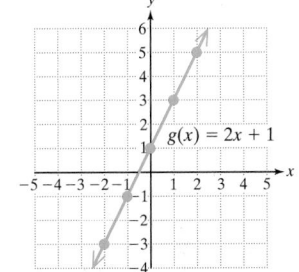

TIP The name of a function can be represented by any letter or symbol. However, lowercase letters such as f, g, h, and so on are often used.

Skill Practice 5   Evaluate the function defined by $h(x) = 4x - 3$ for the given values of x.

   a. $h(-3)$ **b.** $h(-1)$ **c.** $h(0)$ **d.** $h(1)$ **e.** $h(3)$

Classroom Examples: p. 147
Exercises 48, 58

EXAMPLE 6   Evaluating a Function

Evaluate the function defined by $f(x) = 3x^2 + 2x$ for the given values of x.
   a. $f(a)$ **b.** $f(x + h)$

Solution:

a. $f(a) = 3a^2 + 2a$ Substitute a for x.

b. $f(x + h) = 3(x + h)^2 + 2(x + h)$ Substitute $x + h$ for x.
               $= 3(x^2 + 2xh + h^2) + 2x + 2h$ Simplify.
                                        Recall: $(a + b)^2 = a^2 + 2ab + b^2$
               $= 3x^2 + 6xh + 3h^2 + 2x + 2h$

Answers

5. a. $h(-3) = -15$
   b. $h(-1) = -7$ **c.** $h(0) = -3$
   d. $h(1) = 1$ **e.** $h(3) = 9$
6. a. $f(t) = -t^2 + 4t$
   b. $f(a + h)$
      $= -a^2 - 2ah - h^2 + 4a + 4h$

Skill Practice 6   Evaluate the function defined by $f(x) = -x^2 + 4x$ for the given values of x.

   a. $f(t)$ **b.** $f(a + h)$

3. Determine *x*- and *y*-Intercepts of a Function Defined by $y = f(x)$

Recall that to find an *x*-intercept(s) of the graph of an equation, we substitute 0 for *y* in the equation and solve for *x*. Using function notation, $y = f(x)$, this is equivalent to finding the real solutions of the equation $f(x) = 0$. To find the *y*-intercept, substitute 0 for *x* and solve the equation for *y*. Using function notation, this is equivalent to finding $f(0)$.

Finding Intercepts Using Function Notation

Given a function defined by $y = f(x)$,

- The *x*-intercepts are the real solutions to the equation $f(x) = 0$.
- The *y*-intercept is given by $f(0)$.

Classroom Example: p. 148
Exercise 78

EXAMPLE 7   **Finding the *x*- and *y*-Intercepts of a Function**

Find the *x*- and *y*-intercepts of the function defined by $f(x) = x^2 - 4$.

Solution:

To find the *x*-intercept(s), solve the equation $f(x) = 0$.

$$f(x) = x^2 - 4$$
$$0 = x^2 - 4$$
$$x^2 = 4$$
$$x = \pm 2 \qquad \text{The \textit{x}-intercepts are } (2, 0) \text{ and } (-2, 0).$$

To find the *y*-intercept, evaluate $f(0)$.

$$f(0) = (0)^2 - 4$$
$$= -4 \qquad \text{The \textit{y}-intercept is } (0, -4).$$

Skill Practice 7   Find the *x*- and *y*-intercepts of the function defined by $f(x) = |x| - 5$.

---

4. Determine Domain and Range of a Function

Given a relation defining *y* as a function of *x*, the **domain** is the set of *x* values in the function, and the **range** is the set of *y* values in the function. In Example 8, we find the domain and range from the graph of a function.

Classroom Examples: pp. 148–149
Exercises 88, 90, 94

EXAMPLE 8   **Determining Domain and Range**

Determine the domain and range for the functions shown.

a. 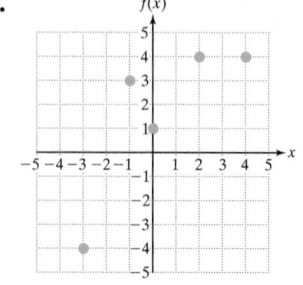   **b.** 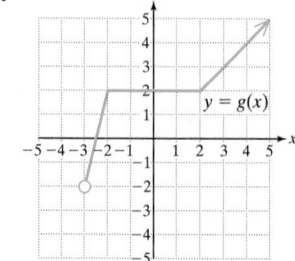   **c.**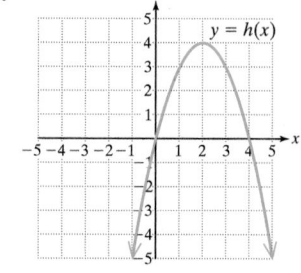

Answer

7. *x*-intercepts: (5, 0) and (−5, 0);
   y-intercept: (0, −5)

Solution:

a. The graph defines the set of ordered pairs:
$\{(-3, -4), (-1, 3), (0, 1), (2, 4), (4, 4)\}$

Domain: $\{-3, -1, 0, 2, 4\}$ The domain is the set of x values.

Range: $\{-4, 1, 3, 4\}$ The range is the set of y values.

b.

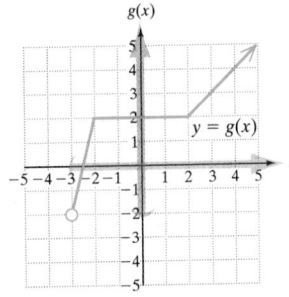

The domain is shown on the x-axis in green tint.

Domain: $\{x \mid x > -3\}$ or in interval notation: $(-3, \infty)$.

The range is shown on the y-axis in red tint.

Range: $\{y \mid y > -2\}$ or in interval notation: $(-2, \infty)$.

c.

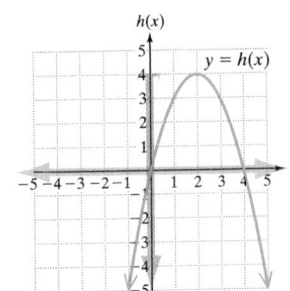

The graph extends infinitely far downward and infinitely far to the left and right. Therefore, the domain is the set of all real numbers, x.

The domain is shown on the x-axis in green tint.

Domain: $\mathbb{R}$ or in interval notation: $(-\infty, \infty)$.

The range is shown on the y-axis in red tint.

Range: $\{y \mid y \le 4\}$ or in interval notation: $(-\infty, 4]$.

Skill Practice 8   Determine the domain and range for the functions shown.

a.

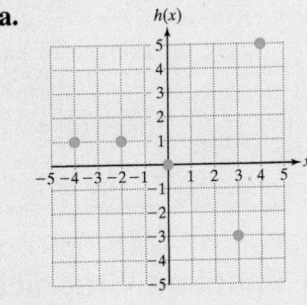

b.

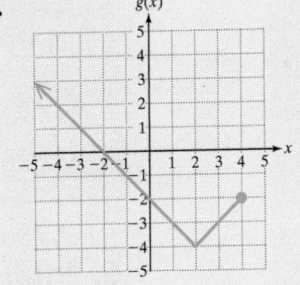

In some cases, a function may have restrictions on the domain. For example, consider the function defined by

$$f(x) = x^2 + 2 \quad \text{for} \quad x \ge 0$$

The restriction on x (that is, $x \ge 0$) is explicitly stated along with the definition of the function. If no such restriction is stated, then by default, the domain is all real numbers that when substituted into the function produce real numbers in the range.

Guidelines to Find Domain of a Function

To determine the implied domain of a function defined by $y = f(x)$,

- Exclude values of x that make the denominator of a fraction zero.
- Exclude values of x that make the radicand negative within an even-indexed root.

Answers

8. a. Domain: $\{-4, -2, 0, 3, 4\}$;
Range: $\{-3, 0, 1, 5\}$

b. Domain: $\{x \mid x \le 4\}$ or $(-\infty, 4]$;
Range: $\{y \mid y \ge -4\}$ or $[-4, \infty)$

Classroom Examples: p. 149
Exercises 98, 100, 106

EXAMPLE 9   Determining the Domain of a Function

Write the domain of each function in interval notation.

a. $f(x) = \dfrac{x + 3}{2x - 5}$ **b.** $g(x) = \dfrac{x}{x^2 + 4}$

c. $h(t) = \sqrt{2 - t}$ **d.** $m(a) = |4 + a|$

Solution:

Instructor Note:
Although Example 9 shows students that they can find the domain of a function analytically, also consider showing the students the graphs of these functions in class or have them graph the functions themselves on a graphing utility.

a. $f(x) = \dfrac{x + 3}{2x - 5}$ The domain is all real numbers except those that make the denominator zero.

The variable x has the restriction that $2x - 5 \neq 0$. Therefore, $x \neq \frac{5}{2}$.

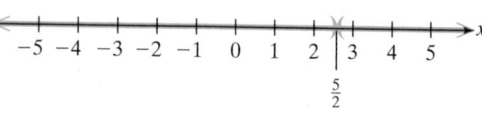

Domain: $\left(-\infty, \dfrac{5}{2}\right) \cup \left(\dfrac{5}{2}, \infty\right)$

b. $g(x) = \dfrac{x}{x^2 + 4}$ [Denominator always positive (never zero)]   The expression $x^2 \geq 0$ for all real numbers x. Therefore, $x^2 + 4 > 0$ for all real numbers x.

Domain: $(-\infty, \infty)$

c. $h(t) = \sqrt{2 - t}$ The domain is restricted to the real numbers that make the radicand greater than or equal to zero.

$2 - t \geq 0$

$-t \geq -2$ Divide by -1 and reverse the inequality sign.

$t \leq 2$

Domain: $(-\infty, 2]$

d. $m(a) = |4 + a|$ There are no fractions or radicals that would restrict the domain.

Domain: $(-\infty, \infty)$ The expression $|4 + a|$ is a real number for all real numbers a.

Skill Practice 9   Write the domain of each function in interval notation.

a. $f(x) = \dfrac{x - 2}{3x + 1}$ **b.** $g(x) = \dfrac{x^2}{5}$

c. $k(x) = \sqrt{x + 3}$ **d.** $p(x) = 2x^2 + 3x$

5. Interpret a Function Graphically

In Example 10, we will review the key concepts studied in this section by identifying characteristics of a function based on its graph.

Classroom Example: p. 150
Exercise 112

EXAMPLE 10   Identifying Characteristics of a Function

Use the function f pictured to answer the questions.

a. Determine $f(2)$.

b. Determine $f(-5)$.

c. Find all x for which $f(x) = 0$.

d. Find all x for which $f(x) = 3$.

e. Determine the x-intercept(s).

f. Determine the y-intercept.

g. Determine the domain of f.

h. Determine the range of f.

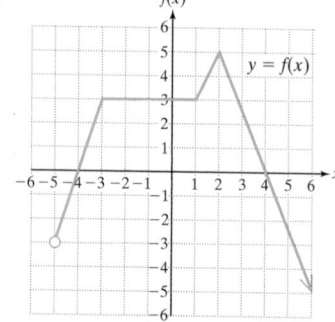

Answers

9. a. $\left(-\infty, -\dfrac{1}{3}\right) \cup \left(-\dfrac{1}{3}, \infty\right)$

b. $(-\infty, \infty)$

c. $[-3, \infty)$

d. $(-\infty, \infty)$

Solution:

a. $f(2) = 5$

$f(2) = 5$ because the function contains the point $(2, 5)$.

b. $f(-5)$ is not defined.

The point $(-5, -3)$ is not included in the function as indicated by the open dot.

c. $f(x) = 0$ for $x = -4$ and $x = 4$.

The points $(-4, 0)$ and $(4, 0)$ represent the points where $f(x) = 0$.

d. $f(x) = 3$ for all x on the interval $[-3, 1]$ and for $x = \frac{14}{5}$.

e. The x-intercepts are $(-4, 0)$ and $(4, 0)$.

f. The y-intercept is $(0, 3)$.

g. The domain is $(-5, \infty)$.

h. The range is $(-\infty, 5]$.

Skill Practice 10   Use the function f pictured to find:

a. $f(-2)$.

b. $f(4)$.

c. All x for which $f(x) = 3$.

d. All x for which $f(x) = 1$.

e. The x-intercept(s).

f. The y-intercept.

g. The domain of f.

h. The range of f.

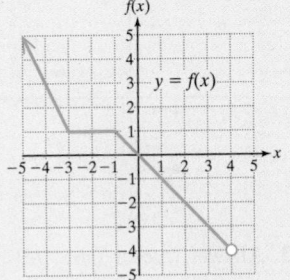

Answers

10. a. $f(-2) = 1$
  b. $f(4)$ is not defined.
  c. $x = -4$
  d. All x on the interval $[-3, -1]$
  e. $(0, 0)$
  f. $(0, 0)$
  g. $(-\infty, 4)$
  h. $(-4, \infty)$

SECTION 1.3   Practice Exercises

Prerequisite Review

R.1. Solve the equation using the square root property. $8x^2 - 40 = 0$ $\{-\sqrt{5}, \sqrt{5}\}$

R.2. Solve. $4x^2 - 7x - 15 = 0$ $\left\{-\frac{5}{4}, 3\right\}$

R.3. Solve. Write the solution set in interval notation. $-3y - 9 \le 15$ $[-8, \infty)$

R.4. Solve. $|2n + 5| = 2$ $\left\{-\frac{7}{2}, -\frac{3}{2}\right\}$

R.5. Given $2x - 5y = 20$,

  a. Find the x-intercept. $(10, 0)$ **b.** Find the y-intercept. $(0, -4)$

Concept Connections

1. A set of ordered pairs (x, y) is called a ___relation___ in x and y. The set of x values in the relation is called the ___domain___ of the relation. The set of ___y___ values is called the range of the relation.

2. Given a function defined by $y = f(x)$, the statement $f(2) = 4$ is equivalent to what ordered pair? $(2, 4)$

3. Given a function defined by $y = f(x)$, to find the ___y___-intercept, evaluate $f(0)$.

4. Given a function defined by $y = f(x)$, to find the x-intercept(s), substitute 0 for ___$f(x)$___ and solve for x.

5. Given $f(x) = \dfrac{x + 1}{x + 5}$, the domain is restricted so that $x \ne$ ___-5___.

6. Given $g(x) = \sqrt{x - 5}$, the domain is restricted so that $x \ge$ ___5___.

7. Consider a relation that defines the height y of a tree for a given time t after it is planted. Does this relation define y as a function of t? Explain.   Yes. For a given time after the tree is planted, there cannot be two or more different heights. That is, the height is unique at any given time.

8. Consider a relation that defines a time y during the course of a year when the temperature T in Fort Collins, Colorado, is 70°. Does this relation define y as a function of T? Explain.   No. There are numerous times during the year when the temperature in Fort Collins, Colorado is 70°. Therefore, given an input value of 70°, there is more than one output value (time).

Objective 1: Determine Whether a Relation Is a Function

For Exercises 9–12,

a. Write a set of ordered pairs (x, y) that defines the relation.

b. Write the domain of the relation.

c. Write the range of the relation.

d. Determine if the relation defines y as a function of x. (**See Examples 1–2**)

9.

Actor x	Number of Oscar Nominations y
Tom Hanks	5
Jack Nicholson	12
Sean Penn	5
Dustin Hoffman	7

a. {(Tom Hanks, 5), (Jack Nicholson, 12), (Sean Penn, 5), (Dustin Hoffman, 7)}
b. {Tom Hanks, Jack Nicholson, Sean Penn, Dustin Hoffman}
c. {5, 12, 7}
d. Yes

10.

City x	Elevation at Airport (ft) y
Albany	285
Denver	5883
Miami	11
San Francisco	11

a. {(Albany, 285), (Denver, 5883), (Miami, 11), (San Francisco, 11)}
b. {Albany, Denver, Miami, San Francisco}
c. {285, 5883, 11}
d. Yes

11.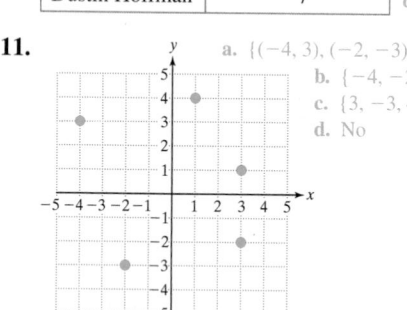

a. {(−4, 3), (−2, −3), (1, 4), (3, −2), (3, 1)}
b. {−4, −2, 1, 3}
c. {3, −3, 4, −2, 1}
d. No

12.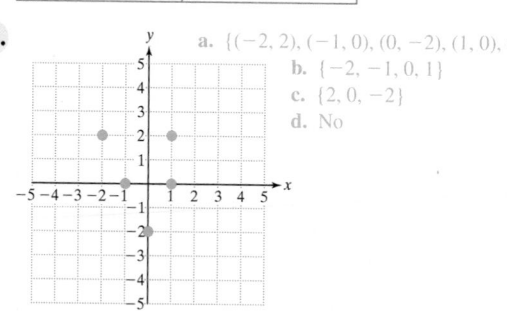

a. {(−2, 2), (−1, 0), (0, −2), (1, 0), (1, 2)}
b. {−2, −1, 0, 1}
c. {2, 0, −2}
d. No

13. Answer true or false. All relations are functions.   False

14. Answer true or false. All functions are relations.   True

For Exercises 15–32, determine if the relation defines y as a function of x. (**See Examples 3–4**)

15.   Yes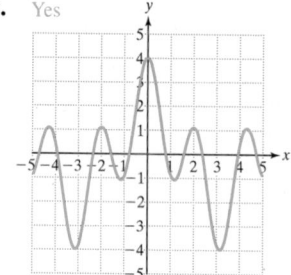

16.   Yes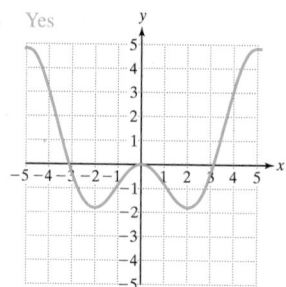

17.   No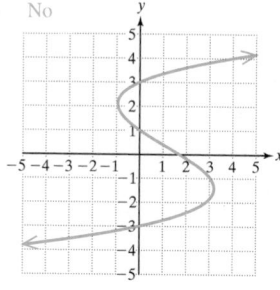

18.   Yes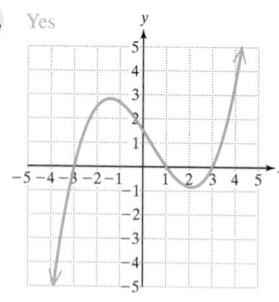

19.   Yes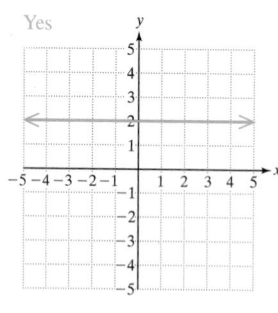

20.   No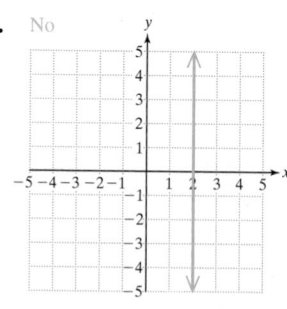

21.   Yes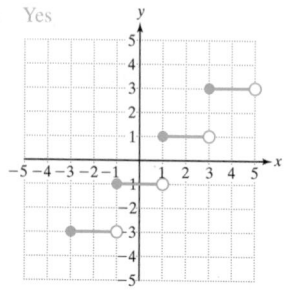

22.   No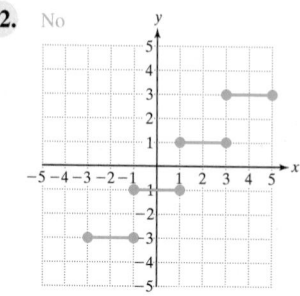

23.   No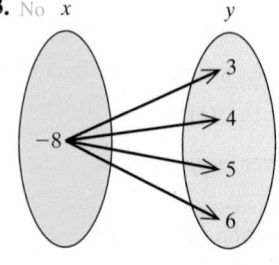

24. No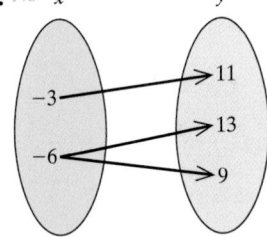

25. Yes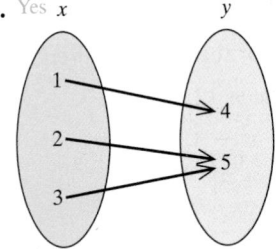

26. Yes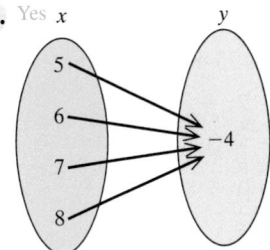

27. $(x + 1)^2 + (y + 5)^2 = 25$ No

28. $(x + 3)^2 + (y + 4)^2 = 1$ No

29. $y = x + 3$ Yes

30. $y = x - 4$ Yes

31. a. $y = x^2$ Yes

b. $x = y^2$ No

32. a. $y = |x|$ Yes

b. $x = |y|$ No

Objective 2: Apply Function Notation

33. The statement $f(4) = 1$ corresponds to what ordered pair? $(4, 1)$

34. The statement $g(7) = -5$ corresponds to what ordered pair? $(7, -5)$

For Exercises 35–56, evaluate the function for the given value of x. (See Examples 5–6)

$$f(x) = x^2 + 3x \qquad g(x) = \frac{1}{x} \qquad h(x) = 5 \qquad k(x) = \sqrt{x + 1}$$

35. a. $f(-2)$ -2 **b.** $f(-1)$ -2 **c.** $f(0)$ 0 **d.** $f(1)$ 4 **e.** $f(2)$ 10

36. a. $g(-2)$ $-\frac{1}{2}$ **b.** $g(-1)$ -1 **c.** $g\left(-\frac{1}{2}\right)$ -2 **d.** $g\left(\frac{1}{2}\right)$ 2 **e.** $g(2)$ $\frac{1}{2}$

37. a. $h(-2)$ 5 **b.** $h(-1)$ 5 **c.** $h(0)$ 5 **d.** $h(1)$ 5 **e.** $h(2)$ 5

38. a. $k(-2)$ Undefined   **b.** $k(-1)$ 0 **c.** $k(0)$ 1 **d.** $k(1)$ $\sqrt{2}$ **e.** $k(3)$ 2

39. $g(3)$ $\frac{1}{3}$

40. $h(-7)$ 5

41. $g\left(\frac{1}{3}\right)$ 3

42. $h(7)$ 5

43. $k(-5)$ Undefined

44. $f(5)$ 40

45. $k(8)$ 3

46. $f(-5)$ 10

47. $g(t)$ $\frac{1}{t}$

48. $f(a)$ $a^2 + 3a$

49. $k(x + h)$ $\sqrt{x + h + 1}$

50. $h(x + h)$ 5

51. $f(a + 4)$ $a^2 + 11a + 28$

52. $f(t - 3)$ $t^2 - 3t$

53. $g(0)$ Undefined

54. $k(-10)$ Undefined

55. $f(x + h)$ $x^2 + 2xh + h^2 + 3x + 3h$

56. $g(x + h)$ $\frac{1}{x + h}$

For Exercises 57–62, find and simplify $f(x + h)$. (See Example 6)

57. $f(x) = -4x^2 - 5x + 2$
$-4x^2 - 8xh - 4h^2 - 5x - 5h + 2$

58. $f(x) = -2x^2 + 6x - 3$
$-2x^2 - 4xh - 2h^2 + 6x + 6h - 3$

59. $f(x) = 7 - 3x^2$ $-3x^2 - 6xh - 3h^2 + 7$

60. $f(x) = 11 - 5x^2$
$-5x^2 - 10xh - 5h^2 + 11$

61. $f(x) = x^3 + 2x - 5$
$x^3 + 3x^2h + 3xh^2 + h^3 + 2x + 2h - 5$

62. $f(x) = x^3 - 4x + 2$
$x^3 + 3x^2h + 3xh^2 + h^3 - 4x - 4h + 2$

For Exercises 63–70, refer to the function $f = \{(2, 3), (9, 7), (3, 4), (-1, 6)\}$.

63. Determine $f(9)$. 7

64. Determine $f(-1)$. 6

65. Determine $f(3)$. 4

66. Determine $f(2)$. 3

67. For what value of x is $f(x) = 6$? -1

68. For what value of x is $f(x) = 7$? 9

69. For what value of x is $f(x) = 3$? 2

70. For what value of x is $f(x) = 4$? 3

71. Joe rides his bicycle an average of 18 mph. The distance Joe rides $d(t)$ (in mi) is given by $d(t) = 18t$, where t is the time in hours that he rides.

  a. Evaluate $d(2)$ and interpret the meaning.
  $d(2) = 36$; Joe rides 36 mi in 2 hr.

  b. Determine the distance Joe travels in 40 min.   12 mi

72. Frank needs to drive 250 mi from Daytona Beach to Miami. After having driven x miles, the distance remaining $r(x)$ (in mi) is given by $r(x) = 250 - x$.

  a. Evaluate $r(50)$ and interpret the meaning.
  $r(50) = 200$; After having driven 50 mi, Frank still has 200 mi remaining.

  b. Determine the distance remaining after 122 mi.
  128 mi

73. At a restaurant, if a party has eight or more people, the gratuity is automatically added to the bill. If x is the cost of the meal, then the total bill $C(x)$ with an 18% gratuity and a 6% sales tax is given by: $C(x) = x + 0.06x + 0.18x$. Evaluate $C(225)$ and interpret the meaning in the context of this problem.
$C(225) = 279$; If the cost of the food is $225, then the total bill including tax and tip is $279.

74. A bookstore marks up the price of a book by 40% of the cost from the publisher. Therefore, the bookstore's price to the student, $P(x)$ (in $) after a 7.5% sales tax, is given by $P(x) = 1.075(x + 0.40x)$, where x is the cost of the book from the publisher. Evaluate $P(60)$ and interpret the meaning in the context of this problem.
$P(60) = \$90.30$; If the cost of the book from the publisher is $60, the cost to the student after the bookstore markup and sales tax is $90.30.

Objective 3: Determine *x*- and *y*-Intercepts of a Function Defined by $y = f(x)$

For Exercises 75–84, determine the *x*- and *y*-intercepts for the given function. (See Example 7)

75. $f(x) = 2x - 4$
x-intercept: (2, 0); *y*-intercept: (0, −4)

76. $g(x) = 3x - 12$
x-intercept: (4, 0); *y*-intercept: (0, −12)

77. $h(x) = |x| - 8$
x-intercepts: (8, 0), (−8, 0); *y*-intercept: (0, −8)

78. $k(x) = -|x| + 2$
x-intercepts: (2, 0), (−2, 0); *y*-intercept: (0, 2)

79. $p(x) = -x^2 + 12$
x-intercepts: $(2\sqrt{3}, 0)$, $(-2\sqrt{3}, 0)$; *y*-intercept: (0, 12)

80. $q(x) = x^2 - 8$
x-intercepts: $(2\sqrt{2}, 0)$, $(-2\sqrt{2}, 0)$; *y*-intercept: (0, −8)

81. $r(x) = |x - 8|$
x-intercept: (8, 0); *y*-intercept: (0, 8)

82. $s(x) = |x + 3|$
x-intercept: (−3, 0); *y*-intercept: (0, 3)

83. $f(x) = \sqrt{x} - 2$
x-intercept: (4, 0); *y*-intercept: (0, −2)

84. $g(x) = -\sqrt{x} + 3$
x-intercept: (9, 0); *y*-intercept: (0, 3)

85. A student decides to finance a used car over a 5-yr (60-month) period. After making a down payment of $2000, the remaining cost of the car including tax and interest is $14,820. The amount owed $y = A(t)$ (in $) is given by $A(t) = 14{,}820 - 247t$, where t is the number of months after purchase and $0 \le t \le 60$. Determine the *t*-intercept and *y*-intercept and interpret their meanings in context.

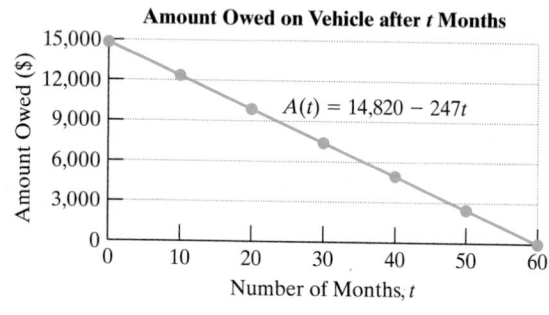

Amount Owed on Vehicle after *t* Months

$A(t) = 14{,}820 - 247t$

The *y*-intercept is (0, 14,820) and means that the amount owed after the initial down payment is $14,820. The *t*-intercept is (60, 0) and means that after 60 months, the amount owed is $0.

86. The amount spent on video games per person in the United States has been increasing since 2006. (*Source*: www.census.gov) The function defined by $f(x) = 9.4x + 35.7$ represents the amount spent $f(x)$ (in $) x years since 2006. Determine the *y*-intercept and interpret its meaning in context.

Amount Spent on Video Games per Person by Year

$f(x) = 9.4x + 35.7$

Year ($x = 0$ represents 2006)

(0, 35.7); The *y*-intercept means that for $x = 0$ (the year 2006), the average amount spent on video games per person in the United States was $35.70.

Objective 4: Determine Domain and Range of a Function

For Exercises 87–96, determine the domain and range of the function. (See Example 8)

87.

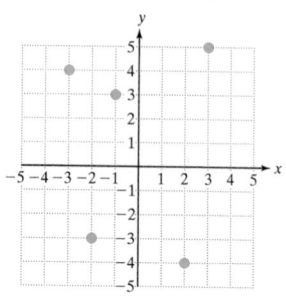

Domain: $\{-3, -2, -1, 2, 3\}$;
Range: $\{-4, -3, 3, 4, 5\}$

88.

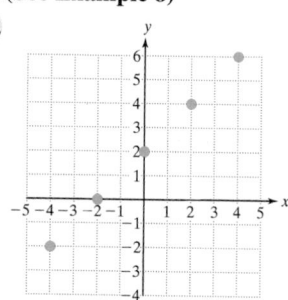

Domain: $\{-4, -2, 0, 2, 4\}$;
Range: $\{-2, 0, 2, 4, 6\}$

89.

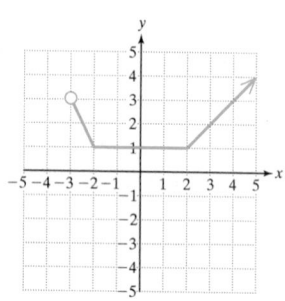

Domain: $(-3, \infty)$;
Range: $[1, \infty)$

90.

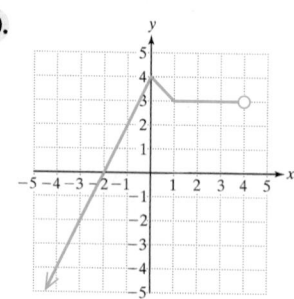

Domain: $(-\infty, 4)$;
Range: $(-\infty, 4]$

91. 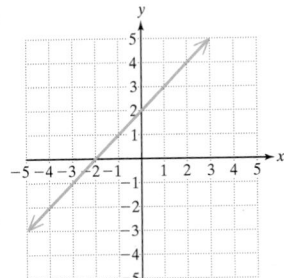 Domain: $(-\infty, \infty)$;
Range: $(-\infty, \infty)$

92. 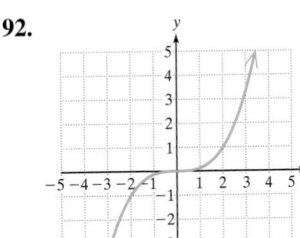 Domain: $(-\infty, \infty)$;
Range: $(-\infty, \infty)$

93. 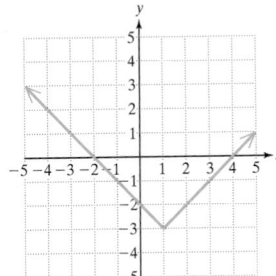 Domain: $(-\infty, \infty)$;
Range: $[-3, \infty)$

94. 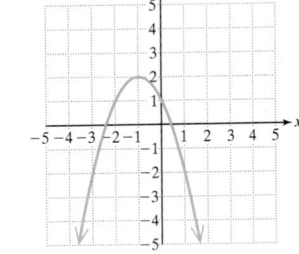 Domain: $(-\infty, \infty)$;
Range: $(-\infty, 2]$

95. 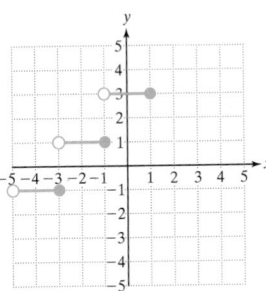 Domain: $(-5, 1]$;
Range: $\{-1, 1, 3\}$

96. 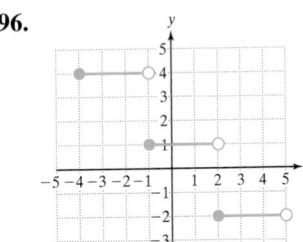 Domain: $[-4, 5)$;
Range: $\{-2, 1, 4\}$

For Exercises 97–110, write the domain in interval notation. (See Example 9)

97. a. $f(x) = \dfrac{x - 3}{x - 4}$ $(-\infty, 4) \cup (4, \infty)$ **b.** $g(x) = \dfrac{x - 3}{x^2 - 4}$ $(-\infty, -2) \cup (-2, 2) \cup (2, \infty)$ **c.** $h(x) = \dfrac{x - 3}{x^2 + 4}$ $(-\infty, \infty)$

98. a. $k(x) = \dfrac{x + 6}{x - 2}$ $(-\infty, 2) \cup (2, \infty)$ **b.** $j(x) = \dfrac{x + 6}{x^2 + 2}$ $(-\infty, \infty)$ **c.** $p(x) = \dfrac{x + 6}{x^2 - 2}$
$(-\infty, -\sqrt{2}) \cup (-\sqrt{2}, \sqrt{2}) \cup (\sqrt{2}, \infty)$

99. a. $a(x) = \sqrt{x + 9}$ $[-9, \infty)$ **b.** $b(x) = \sqrt{9 - x}$ $(-\infty, 9]$ **c.** $c(x) = \dfrac{1}{\sqrt{x + 9}}$ $(-9, \infty)$

100. a. $y(t) = \sqrt{16 - t}$ $(-\infty, 16]$ **b.** $w(t) = \sqrt{t - 16}$ $[16, \infty)$ **c.** $z(t) = \dfrac{1}{\sqrt{16 - t}}$ $(-\infty, 16)$

101. a. $f(t) = \sqrt[3]{t - 5}$ $(-\infty, \infty)$ **b.** $g(t) = \sqrt[3]{5 - t}$ $(-\infty, \infty)$ **c.** $h(t) = \dfrac{1}{\sqrt[3]{t - 5}}$ $(-\infty, 5) \cup (5, \infty)$

102. a. $k(x) = \sqrt[5]{3 + x}$ $(-\infty, \infty)$ **b.** $m(x) = \sqrt[5]{x - 3}$ $(-\infty, \infty)$ **c.** $n(x) = \dfrac{1}{\sqrt[5]{x - 3}}$ $(-\infty, 3) \cup (3, \infty)$

103. a. $f(x) = x^2 - 3x - 28$ $(-\infty, \infty)$ **b.** $g(x) = \dfrac{x + 2}{x^2 - 3x - 28}$ **c.** $h(x) = \dfrac{x^2 - 3x - 28}{x + 2}$
$(-\infty, -4) \cup (-4, 7) \cup (7, \infty)$ $(-\infty, -2) \cup (-2, \infty)$

104. a. $r(x) = x^2 - 4x - 12$ $(-\infty, \infty)$ **b.** $s(x) = \dfrac{x^2 - 4x - 12}{x + 1}$ **c.** $t(x) = \dfrac{x + 1}{x^2 - 4x - 12}$
$(-\infty, -1) \cup (-1, \infty)$ $(-\infty, -2) \cup (-2, 6) \cup (6, \infty)$

105. a. $w(x) = |x + 1| + 4$ $(-\infty, \infty)$ **b.** $y(x) = \dfrac{x}{|x + 1| + 4}$ $(-\infty, \infty)$ **c.** $z(x) = \dfrac{x}{|x + 1| - 4}$
$(-\infty, -5) \cup (-5, 3) \cup (3, \infty)$

106. a. $f(a) = 8 - |a - 2|$ $(-\infty, \infty)$ **b.** $g(a) = \dfrac{5}{8 - |a - 2|}$ **c.** $h(a) = \dfrac{5}{8 + |a - 2|}$ $(-\infty, \infty)$
$(-\infty, -6) \cup (-6, 10) \cup (10, \infty)$

107. a. $f(x) = \sqrt{x + 15}$ $[-15, \infty)$ **b.** $g(x) = \sqrt{x + 15} - 2$ $[-15, \infty)$ **c.** $k(x) = \dfrac{5}{\sqrt{x + 15} - 2}$

$[-15, -11) \cup (-11, \infty)$

108. a. $f(c) = \sqrt{c + 20}$ $[-20, \infty)$ **b.** $g(c) = \sqrt{c + 20} - 1$ $[-20, \infty)$ **c.** $h(c) = \dfrac{-4}{\sqrt{c + 20} - 1}$

$[-20, -19) \cup (-19, \infty)$

109. a. $p(x) = 2x + 1$ $(-\infty, \infty)$ **b.** $q(x) = 2x + 1; x \geq 0$ $[0, \infty)$ **c.** $r(x) = 2x + 1; 0 \leq x < 7$ $[0, 7)$

110. a. $m(x) = 3x - 7$ $(-\infty, \infty)$ **b.** $n(x) = 3x - 7; x < 0$ $(-\infty, 0)$ **c.** $n(x) = 3x - 7; -2 < x < 2$

$(-2, 2)$

Objective 5: Interpret a Function Graphically

For Exercises 111–114, use the graph of $y = f(x)$ to answer the following. (See Example 10)

a. Determine $f(-2)$.   **b.** Determine $f(3)$.   **c.** Find all x for which $f(x) = -1$.

d. Find all x for which $f(x) = -4$.   **e.** Determine the x-intercept(s).   **f.** Determine the y-intercept.

g. Determine the domain of f.   **h.** Determine the range of f.

111.

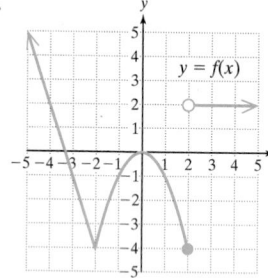

a. -4
b. 2
c. $x = -3, x = -1, x = 1$
d. $x = -2, x = 2$
e. $(0, 0)$ and $\left(-\dfrac{10}{3}, 0\right)$
f. $(0, 0)$
g. $(-\infty, \infty)$
h. $[-4, \infty)$

112.

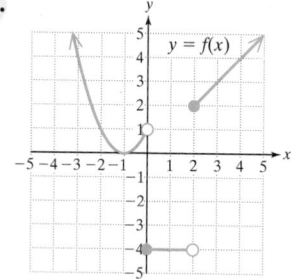

a. 1
b. 3
c. None
d. All x on the interval $[0, 2)$
e. $(-1, 0)$
f. $(0, -4)$
g. $(-\infty, \infty)$
h. $\{-4\} \cup [0, \infty)$

113.

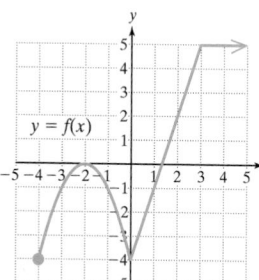

a. 0
b. 5
c. $x = -3, x = -1, x = 1$
d. $x = -4, x = 0$
e. $(-2, 0)$ and $\left(\dfrac{4}{3}, 0\right)$
f. $(0, -4)$
g. $[-4, \infty)$
h. $[-4, 5]$

114.

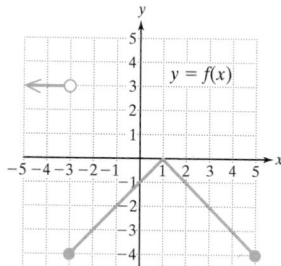

a. -3
b. -2
c. $x = 0, x = 2$
d. $x = -3, x = 5$
e. $(1, 0)$
f. $(0, -1)$
g. $(-\infty, 5]$
h. $[-4, 0] \cup \{3\}$

Mixed Exercises

For Exercises 115–122, write a function that represents the given statement.

115. Suppose that a phone card has 400 min. Write a relationship that represents the number of minutes remaining $r(x)$ as a function of the number of minutes already used x. $r(x) = 400 - x$

116. Suppose that a roll of wire has 200 ft. Write a relationship that represents the amount of wire remaining $w(x)$ as a function of the number of feet of wire x already used. $w(x) = 200 - x$

117. Given an equilateral triangle with sides of length x, write a relationship that represents the perimeter $P(x)$ as a function of x. $P(x) = 3x$

118. In an isosceles triangle, two angles are equal in measure. If the third angle is x degrees, write a relationship that represents the measure of one of the equal angles $A(x)$ as a function of x. $A(x) = \dfrac{180 - x}{2}$

119. Two adjacent angles form a right angle. If the measure of one angle is x degrees, write a relationship representing the measure of the other angle $C(x)$ as a function of x. $C(x) = 90 - x$

120. Two adjacent angles form a straight angle (180°). If the measure of one angle is x degrees, write a relationship representing the measure of the other angle $S(x)$ as a function of x. $S(x) = 180 - x$

121. Write a relationship for a function whose $f(x)$ values are 2 less than three times the square of x. $f(x) = 3x^2 - 2$

122. Write a relationship for a function whose $f(x)$ values are 3 more than the principal square root of x. $f(x) = \sqrt{x} + 3$

Write About It

123. If two points align vertically then the points do not define y as a function of x. Explain why.

124. Given a function defined by $y = f(x)$, explain how to determine the x- and y-intercepts.   To find the x-intercept(s), find the real solutions to the equation $f(x) = 0$. To find the y-intercept, evaluate $f(0)$.

Expanding Your Skills

125. Given a square with sides of length s, diagonal of length d, perimeter P, and area A,

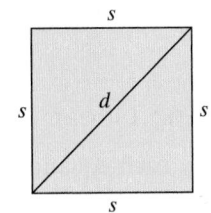

   a. Write P as a function of s.
   $P(s) = 4s$
   b. Write A as a function of s.
   $A(s) = s^2$
   c. Write A as a function of P.
   d. Write P as a function of A. $P(A) = 4\sqrt{A}$
   e. Write d as a function of s. $d(s) = \sqrt{2}s$
   f. Write s as a function of d. $s(d) = \dfrac{d}{\sqrt{2}}$ or $s(d) = \dfrac{d\sqrt{2}}{2}$
   g. Write P as a function of d. $P(d) = 2\sqrt{2}d$
   h. Write A as a function of d. $A(d) = \dfrac{d^2}{2}$

126. Given a circle with radius r, diameter d, circumference C, and area A,

   a. Write C as a function of r.
   $C(r) = 2\pi r$
   b. Write A as a function of r.
   $A(r) = \pi r^2$
   c. Write r as a function of d. $r(d) = \dfrac{d}{2}$
   d. Write d as a function of r. $d(r) = 2r$
   e. Write C as a function of d. $C(d) = \pi d$
   f. Write A as a function of d. $A(d) = \dfrac{\pi d^2}{4}$
   g. Write A as a function of C. $A(C) = \dfrac{C^2}{4\pi}$
   h. Write C as a function of A. $C(A) = 2\sqrt{\pi A}$

Additional answers can be found in the Instructor Answer Appendix.

SECTION 1.4   # Linear Equations in Two Variables and Linear Functions

OBJECTIVES

1. **Graph Linear Equations in Two Variables**
2. **Determine the Slope of a Line**
3. **Apply the Slope-Intercept Form of a Line**
4. **Compute Average Rate of Change**
5. **Solve Equations and Inequalities Graphically**

1. Graph Linear Equations in Two Variables

The median incomes for individuals for all levels of education have shown an increasing trend since 1990. However, the median income for individuals with a bachelor's degree is consistently greater than for individuals whose highest level of education is a high school degree or equivalent (Figure 1-15). (*Source*: U.S. Census Bureau, www.census.gov)

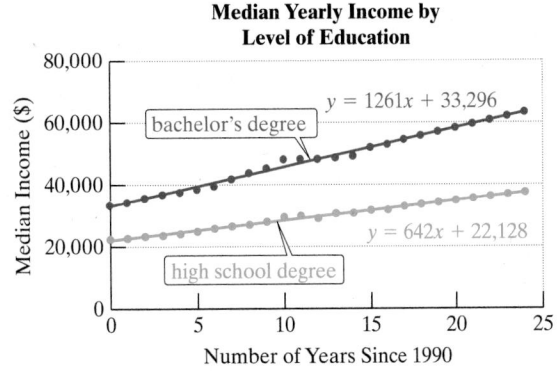

Figure 1-15

The graph in Figure 1-15 is called a scatter plot. A **scatter plot** is a visual representation of a set of points. In this case, the x values represent the number of years since 1990, and the y values represent the median income in dollars. The line that models each set of data is called a **regression line** and is found by using techniques taught in a first course in statistics. The equations that represent the two lines are called linear equations in two variables.

TIP For an equation in standard form, the value of A, B, and C are usually taken to be integers where A, B, and C share no common factors.

Linear Equation in Two Variables

Let A, B, and C represent real numbers such that A and B are not both zero. A **linear equation** in the variables x and y is an equation that can be written in the form:

$Ax + By = C$ This is called the **standard form** of an equation of a line.

Note: A linear equation $Ax + By = C$ has variables x and y each of first degree.

In Example 1, we demonstrate that the graph of a linear equation $Ax + By = C$ is a line. The line may be slanted, horizontal, or vertical depending on the coefficients A, B, and C.

Classroom Examples: p. 161
Exercises 10, 14, 16

EXAMPLE 1   **Graphing Linear Equations**

Graph the line represented by each equation.

a. $2x + 3y = 6$ **b.** $x = -3$ **c.** $2y = 4$

Solution:

a. Solve the equation for y. Then substitute arbitrary values of x into the equation and solve for the corresponding values of y.

$$2x + 3y = 6 \qquad \text{Solve the equation for } y.$$
$$3y = -2x + 6$$
$$y = -\frac{2}{3}x + 2$$

In the table we have selected convenient values of x that are multiples of 3.

x	y
-3	4
0	2
3	0
6	-2

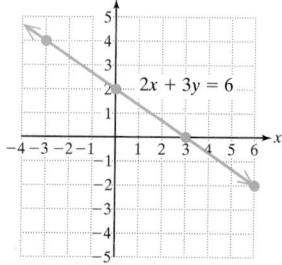

> **Avoiding Mistake**
>
> The graph of a linear equation is a line. Therefore, a minimum of two points is needed to graph the line. A third point can be used to verify that the line is graphed correctly. The points must all line up.

b. $x = -3$

The solutions to this equation must have an x-coordinate of -3. The y variable can be *any* real number.

x	y
-3	-2
-3	0
-3	2
-3	4

x must be -3. y can be any real number.

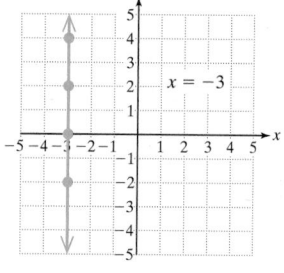

> **TIP** The graph of a vertical line will have no *y*-intercept unless the line is the *y*-axis itself.

c. $2y = 4$ Solve for y.
$$y = 2$$

The solutions to this equation must have a y-coordinate of 2. The x variable can be *any* real number.

x	y
-2	2
0	2
2	2
4	2

x can be any real number. y must be 2.

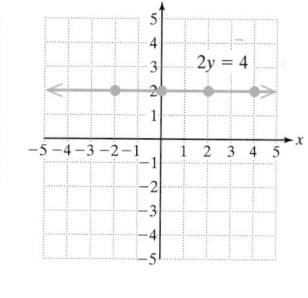

> **TIP** The graph of a horizontal line will have no *x*-intercept unless the line is the *x*-axis itself.

Skill Practice 1   Graph the line represented by each equation.

a. $4x + 2y = 2$ **b.** $y = 1$ **c.** $-3x = 12$

Answer

1. a.–c.

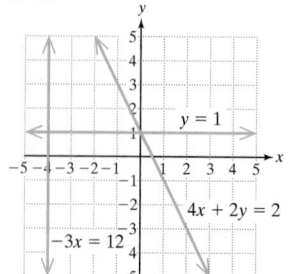

2. Determine the Slope of a Line

One of the important characteristics of a nonvertical line is that for every 1 unit of change in the horizontal variable, the vertical change is a constant m called the **slope** of the line. For example, consider the line representing the median income for individuals with a bachelor's degree, x years since the year 1990. The line in Figure 1-16 has a slope of $1261. This means that median income for individuals with a bachelor's degree increased on average by $1261 per year during this time period.

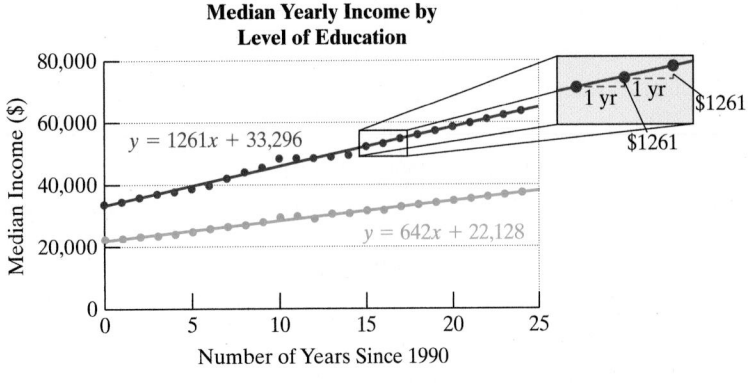

Figure 1-16

Consider any two distinct points (x_1, y_1) and (x_2, y_2) on a line (Figure 1-17). The slope m of the line through the points is the ratio between the change in the y values $(y_2 - y_1)$ and the change in the x values $(x_2 - x_1)$. In many applications in the sciences, the change in a variable is denoted by the Greek letter Δ (delta). Therefore, $(y_2 - y_1)$ can be represented by Δy and $(x_2 - x_1)$ can be represented by Δx.

Figure 1-17

Slope Formula

The **slope** of a line passing through the distinct points (x_1, y_1) and (x_2, y_2) is

change in y (rise)

$$m = \frac{\Delta y}{\Delta x} = \frac{y_2 - y_1}{x_2 - x_1} \text{ provided that } x_2 - x_1 \neq 0$$

change in x (run)

Classroom Examples: p. 162
Exercises 30, 34

EXAMPLE 2   Finding the Slope of a Line Through Two Points

Find the slope of the line passing through the given points.

a. $(-3, -2)$ and $(2, 5)$ **b.** $\left(-\frac{5}{2}, 0\right)$ and $(1, -7)$

Solution:

a. $(-3, -2)$ and $(2, 5)$

(x_1, y_1) and (x_2, y_2) Label the points.

$$m = \frac{y_2 - y_1}{x_2 - x_1} = \frac{5 - (-2)}{2 - (-3)} = \frac{7}{5}$$

A line with a positive slope *"rises"* upward from left to right.

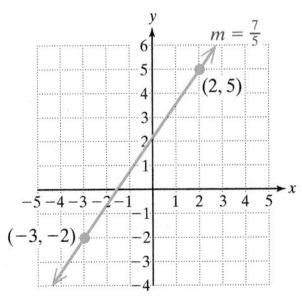

b. $\left(-\dfrac{5}{2}, 0\right)$ and $(1, -7)$

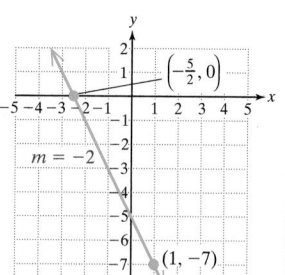

(x_1, y_1) and (x_2, y_2) Label the points.

$$m = \frac{y_2 - y_1}{x_2 - x_1} = \frac{-7 - 0}{1 - \left(-\frac{5}{2}\right)} = \frac{-7}{\frac{7}{2}} = -7 \cdot \frac{2}{7} = -2$$

A line with a negative slope *"falls"* downward from left to right.

Skill Practice 2   Find the slope of the line passing through the given points.

a. $(-4, 1)$ and $(2, -2)$ **b.** $\left(\dfrac{3}{4}, 2\right)$ and $(-3, 17)$

Classroom Examples: pp. 162–163
Exercises 42, 44

EXAMPLE 3   Finding the Slope of Horizontal and Vertical Lines

Find the slope of each line.

Solution:

a.

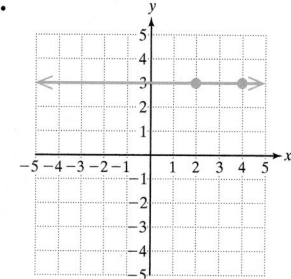

By inspection, we see that between any two points on the graph, the vertical change is zero, so the slope is zero.

   To compute this numerically, select any two points on the line such as $(2, 3)$ and $(4, 3)$.

$$m = \frac{y_2 - y_1}{x_2 - x_1} = \frac{3 - 3}{4 - 2} = \frac{0}{2} = 0$$

b.

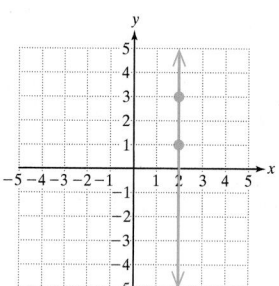

To find the slope, select any two points on the line such as $(2, 1)$ and $(2, 3)$.

$$m = \frac{y_2 - y_1}{x_2 - x_1} = \frac{3 - 1}{2 - 2} = \frac{2}{0} \quad \text{(undefined)}$$

By inspection, we see that between any two points on the line, the change in x is zero. This makes the slope undefined because the ratio representing the slope has a divisor of zero.

Skill Practice 3   Fill in the blank.

a. The slope of a vertical line is _____.
b. The slope of a horizontal line is _____.

From Example 1, we see that a linear equation may represent the graph of a slanted line, a horizontal line, or a vertical line. From Examples 2 and 3, we see that a line may have a positive slope, a negative slope, a zero slope, or an undefined slope.

Answers

2. a. $-\dfrac{1}{2}$ **b.** -4

3. a. Undefined   **b.** 0

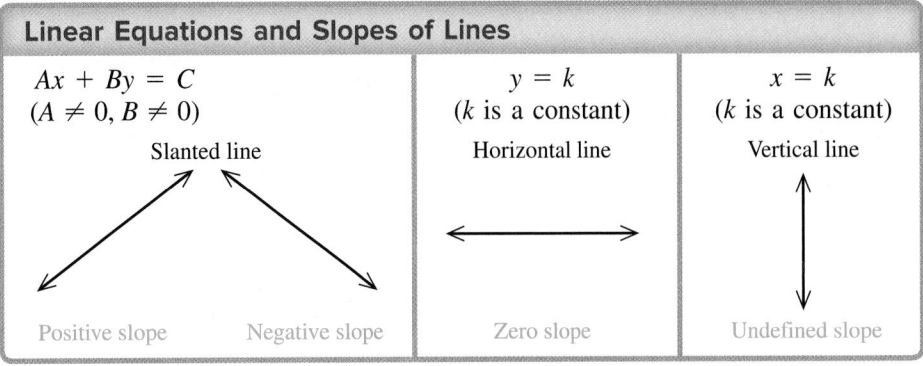

Linear Equations and Slopes of Lines

$Ax + By = C$ ($A \neq 0, B \neq 0$) Slanted line	$y = k$ (k is a constant) Horizontal line	$x = k$ (k is a constant) Vertical line
Positive slope Negative slope	Zero slope	Undefined slope

3. Apply the Slope-Intercept Form of a Line

The slope formula can be used to develop the slope-intercept form of a line. Suppose that a line has a slope m and y-intercept $(0, b)$. Let (x, y) be any other point on the line. From the slope formula, we have:

$$\frac{y - b}{x - 0} = m \qquad \text{Slope formula}$$

$$y - b = mx \qquad \text{Multiply by } x.$$

$$y = mx + b \qquad \text{This is slope-intercept form. Slope intercept-form has the } y \text{ variable isolated.}$$

Classroom Example: p. 163
Exercise 60

Avoiding Mistakes

An equation of a vertical line takes the form $x = k$, where k is a constant. Because there is no y variable and because the slope is undefined, an equation of a vertical line cannot be written in slope-intercept form.

Slope-Intercept Form of a Line

Given a line with slope m and y-intercept $(0, b)$, the **slope-intercept form** of the line is given by $y = mx + b$.

The slope-intercept form of a line is particularly useful because we can identify the slope and y-intercept by inspection. For example:

$$y = \frac{2}{3}x - 5 \qquad\qquad m = \frac{2}{3} \qquad y\text{-intercept: } (0, -5)$$

$$y = x + 4 \qquad\qquad m = 1 \qquad y\text{-intercept: } (0, 4)$$

$$y = 2x \quad (\text{or } y = 2x + 0) \qquad m = 2 \qquad y\text{-intercept: } (0, 0)$$

$$y = 6 \quad (\text{or } y = 0x + 6) \qquad m = 0 \qquad y\text{-intercept: } (0, 6)$$

If the slope and y-intercept of a line are known, we can graph the line. This is demonstrated in Example 4.

EXAMPLE 4   **Using the Slope and y-Intercept to Graph a Line**

Given $3x + 4y = 4$,

a. Write the equation in slope-intercept form.

b. Determine the slope and y-intercept.

c. Graph the line by using the slope and y-intercept.

Solution:

a. $3x + 4y = 4$

$$4y = -3x + 4 \qquad \text{To write an equation in slope-intercept form, isolate the } y \text{ variable.}$$

$$y = -\frac{3}{4}x + 1 \qquad \text{Slope-intercept form}$$

b. $m = -\dfrac{3}{4}$ and the y-intercept is $(0, 1)$.   The slope is the coefficient on x.
The constant term gives the y-intercept.

c.

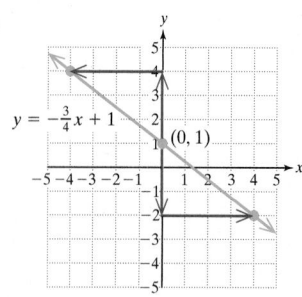

$y = -\frac{3}{4}x + 1$

(0, 1)

To graph the line, first plot the y-intercept $(0, 1)$.

Then begin at the y-intercept, and use the slope to find a second point on the line. In this case, the slope can be interpreted as the following two ratios:

$m = \dfrac{-3}{4}$ ◄—— Move down 3 units.
◄—— Move to the right 4 units.

$m = \dfrac{3}{-4}$ ◄—— Move up 3 units.
◄—— Move to the left 4 units.

> **Skill Practice 4**   Given $2x + 4y = 8$,
>
> **a.** Write the equation in slope-intercept form.
> **b.** Determine the slope and y-intercept.
> **c.** Graph the line by using the slope and y-intercept.

Notice that the slope-intercept form of a line $y = mx + b$ has the y variable isolated and defines y in terms of x. Therefore, an equation written in slope-intercept form defines y as a function of x. In Example 4, $y = -\frac{3}{4}x + 1$ can be written using function notation as $f(x) = -\frac{3}{4}x + 1$.

> **Definition of Linear and Constant Functions**
>
> Let m and b represent real numbers where $m \neq 0$. Then,
>
> - A function defined by $f(x) = mx + b$ is a **linear function.** The graph of a linear function is a slanted line.
> - A function defined by $f(x) = b$ is a **constant function.** The graph of a constant function is a horizontal line.

The slope-intercept form of a line can be used as a tool to define a linear function given a point on the line and the slope.

EXAMPLE 5   **Writing an Equation of a Line Given a Point and the Slope**

Write an equation of the line that passes through the point $(2, -3)$ and has slope -4. Then write the linear equation using function notation, where $y = f(x)$.

Solution:

Given $m = -4$ and $(2, -3)$.   We need to find an equation of the form $y = mx + b$.

$$y = mx + b$$
$$y = -4x + b$$ The value of m is given as -4.
$$-3 = -4(2) + b$$ Substitute $x = 2$ and $y = -3$ from the given point $(2, -3)$.
$$-3 = -8 + b$$ Solve for b.
$$b = 5$$
$$y = mx + b$$
$$y = -4x + 5$$ Substitute $m = -4$ and $b = 5$ into the equation $y = mx + b$.
$$f(x) = -4x + 5$$ Write the relation using function notation.

From the graph, we see that the graph of $f(x) = -4x + 5$ does indeed pass through the point $(2, -3)$ and has slope -4.

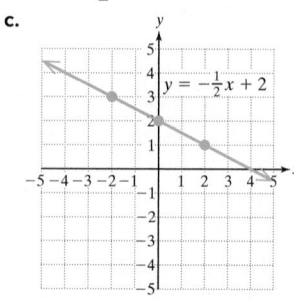

> **Skill Practice 5**   Write an equation of the line that passes through the point $(-1, -4)$ and has slope 3. Then write the equation using function notation.

4. Compute Average Rate of Change

The graphs of many functions are not linear. However, we often use linear approximations to analyze nonlinear functions on small intervals. For example, the graph in Figure 1-18 shows the blood alcohol concentration (BAC) for an individual over a period of 9 hr.

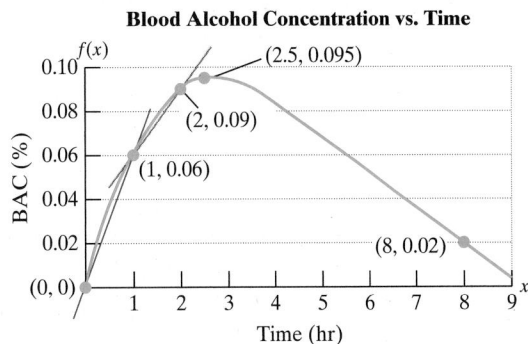

Figure 1-18

A line drawn through two points on a curve is called a **secant line.** In Figure 1-18, the average rate of change in BAC between two points on the graph is the slope of the secant line through the points. Notice that the slope of the secant line between $x = 0$ and $x = 1$ (shown in red) is greater than the slope of the secant line between $x = 1$ and $x = 2$ (shown in green). This means that the average increase in BAC is greater over the first hour than over the second hour.

Average Rate of Change of a Function

Suppose that the points (x_1, y_1) and (x_2, y_2) are points on the graph of a function f. Using function notation, these are the points $(x_1, f(x_1))$ and $(x_2, f(x_2))$.

If f is defined on the interval $[x_1, x_2]$, then the **average rate of change** of f on the interval $[x_1, x_2]$ is the slope of the secant line containing $(x_1, f(x_1))$ and $(x_2, f(x_2))$.

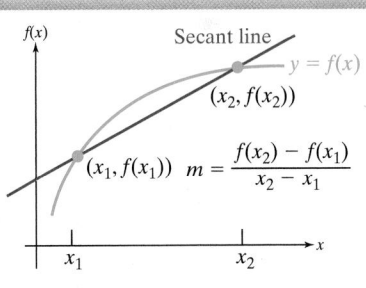

Average rate of change: $m = \dfrac{\Delta y}{\Delta x} = \dfrac{y_2 - y_1}{x_2 - x_1}$ or $m = \dfrac{f(x_2) - f(x_1)}{x_2 - x_1}$

Classroom Example: p. 164
Exercise 82

EXAMPLE 6   Computing Average Rate of Change

Determine the average rate of change of blood alcohol level

a. from $x_1 = 0$ to $x_2 = 1$.

b. from $x_1 = 1$ to $x_2 = 2$.

c. Interpret the results from parts (a) and (b).

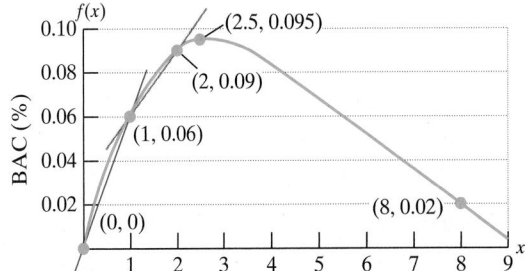

Answer

5. $y = 3x - 1$; $f(x) = 3x - 1$

Solution:

a. Average rate of change $= \dfrac{f(x_2) - f(x_1)}{x_2 - x_1} = \dfrac{f(1) - f(0)}{1 - 0} = \dfrac{0.06 - 0}{1}$

$\qquad\qquad\qquad\qquad\qquad\qquad = 0.06$

b. Average rate of change $= \dfrac{f(x_2) - f(x_1)}{x_2 - x_1} = \dfrac{f(2) - f(1)}{2 - 1} = \dfrac{0.09 - 0.06}{1}$

$\qquad\qquad\qquad\qquad\qquad\qquad = 0.03$

c. The blood alcohol concentration rose by an average of 0.06% per hour during the first hour.

The blood alcohol concentration rose by an average of 0.03% per hour during the second hour.

Skill Practice 6   Refer to the graph in Example 6.

a. Determine the average rate of change of blood alcohol level from $x_1 = 2.5$ to $x_2 = 8$. Round to 3 decimal places.

b. Interpret the results from part (a).

Classroom Example: p. 164
Exercise 90

EXAMPLE 7   **Computing Average Rate of Change**

Given the function defined by $f(x) = x^2 - 1$, determine the average rate of change from $x_1 = -2$ to $x_2 = 0$.

Solution:

Average rate of change $= \dfrac{f(x_2) - f(x_1)}{x_2 - x_1}$

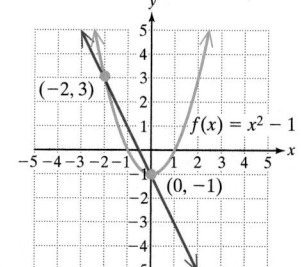

$= \dfrac{f(0) - f(-2)}{0 - (-2)} = \dfrac{-1 - 3}{2} = -2$

The average rate of change is -2.

Skill Practice 7   Given the function defined by $f(x) = x^3 + 2$, determine the average rate of change from $x_1 = -3$ to $x_2 = 0$.

---

5. Solve Equations and Inequalities Graphically

In many settings, the use of technology can provide a numerical and visual interpretation of an algebraic problem. For example, consider the equation $-x - 1 = x + 5$.

$$-x - 1 = x + 5$$
$$-6 = 2x$$
$$-3 = x \qquad \text{The solution set is } \{-3\}.$$

Now suppose that we create two functions from the left and right sides of the equation. We have $Y_1 = -x - 1$ and $Y_2 = x + 5$. Figure 1-19 shows that the graphs of the two lines intersect at $(-3, 2)$. The x-coordinate of the point of intersection is the solution to the equation $-x - 1 = x + 5$. That is, $Y_1 = Y_2$ when $x = -3$.

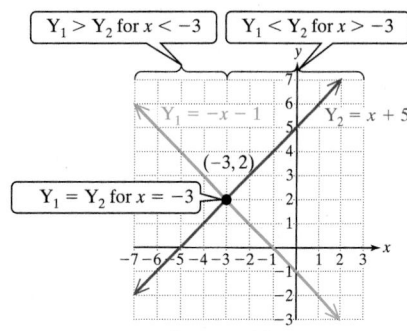

Figure 1-19

The graphs $Y_1 = -x - 1$ and $Y_2 = x + 5$ can also be used to find the solution sets to the related inequalities.

$-x - 1 < x + 5$ The solution set is the set of x values for which $Y_1 < Y_2$. This is the interval where the blue line is below the red line. The solution set is $(-3, \infty)$.

$-x - 1 > x + 5$ The solution set is the set of x values for which $Y_1 > Y_2$. This is the interval where the blue line is *above* the red line. The solution set is $(-\infty, -3)$.

Classroom Example: p. 165
Exercise 92

EXAMPLE 8   **Solving Equations and Inequalities Graphically**

Solve the equations and inequalities graphically.

   a. $2x - 3 = x - 1$ **b.** $2x - 3 < x - 1$ **c.** $2x - 3 > x - 1$

Solution:

a. The left side of the equation is graphed as $Y_1 = 2x - 3$. The right side of the equation is graphed as $Y_2 = x - 1$. The point of intersection is $(2, 1)$. Therefore, $Y_1 = Y_2$ for $x = 2$.
The solution set is $\{2\}$.

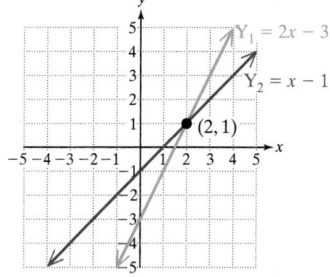

b. $Y_1 < Y_2$ to the *left* of $x = 2$. (That is, the blue line is below the red line for $x < 2$.)
In interval notation the solution set is $(-\infty, 2)$.

c. $Y_1 > Y_2$ to the *right* of $x = 2$. (That is, the blue line is above the red line for $x > 2$.)
In interval notation the solution set is $(2, \infty)$.

TIP The solution set to the inequality $2x - 3 \leq x - 1$ includes equality, so the right endpoint would be included: $(-\infty, 2]$.

The solution set to the inequality $2x - 3 \geq x - 1$ includes equality, so the left endpoint would be included: $[2, \infty)$.

Skill Practice 8   Use the graph to solve the equations and inequalities.

   a. $x + 1 = 2x - 2$
   b. $x + 1 \leq 2x - 2$
   c. $x + 1 \geq 2x - 2$

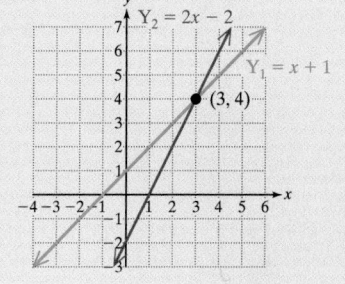

Answers
8. a. $\{3\}$ **b.** $[3, \infty)$ **c.** $(-\infty, 3]$

TECHNOLOGY CONNECTIONS

Verifying Solutions to an Equation

We can verify the solutions to the equations and inequalities from Example 8 on a graphing calculator.

The solutions can be verified numerically by using the Table feature on the calculator. First enter $Y_1 = 2x - 3$ and $Y_2 = x - 1$.

Then display the table values for Y_1 and Y_2 for $x = 2$ and for x values less than and greater than 2.

Display the graphs of Y_1 and Y_2 and use the Intersect feature to determine the point of intersection.

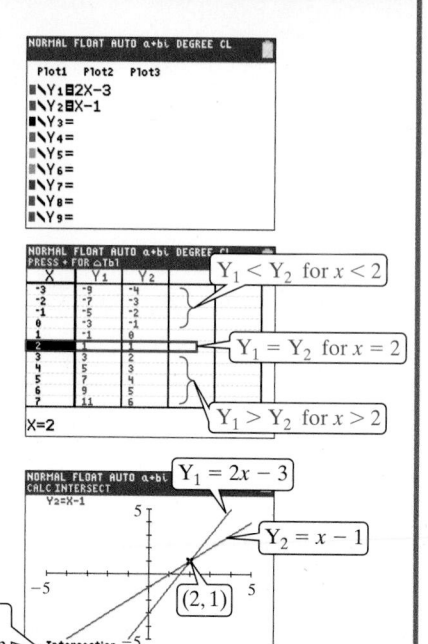

x-coordinate of the point of intersection

In Example 9 we solve the equation $6x - 2(x + 2) - 5 = 0$. Notice that one side is zero. We can check the solution graphically by determining where the related function $Y_1 = 6x - 2(x + 2) - 5$ intersects the x-axis.

Classroom Example: p. 166
Exercise 110

EXAMPLE 9   Solving Equations and Inequalities Graphically

a. Solve the equation $6x - 2(x + 2) - 5 = 0$ and verify the solution graphically on a graphing utility.

b. Use the graph to find the solution set to the inequality $6x - 2(x + 2) - 5 \leq 0$.

c. Use the graph to find the solution set to the inequality $6x - 2(x + 2) - 5 \geq 0$.

Solution:

a.
$$6x - 2(x + 2) - 5 = 0$$
$$6x - 2x - 4 - 5 = 0$$
$$4x - 9 = 0$$
$$x = \frac{9}{4}$$

The solution set is $\left\{ \dfrac{9}{4} \right\}$.

To verify the solution graphically enter the left side of the equation as $Y_1 = 6x - 2(x + 2) - 5$.

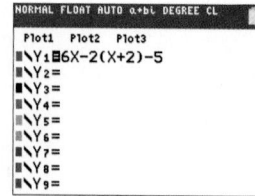

Using the Zero feature, we have $Y_1 = 0$ for $x = 2.25$. This is consistent with the solution $x = \frac{9}{4}$.

b. To solve $6x - 2(x + 2) - 5 \leq 0$ determine the values of x for which $Y_1 \leq 0$ (where the function is on or below the x-axis).

The solution set is $\left(-\infty, \dfrac{9}{4} \right]$.

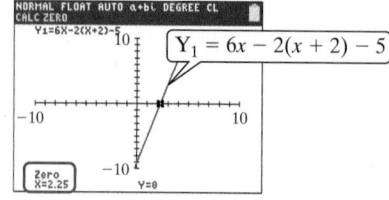

Answers

9. a. $\left\{\dfrac{5}{2}\right\}$

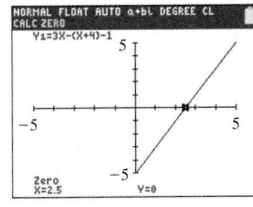

b. $\left(-\infty, \dfrac{5}{2}\right]$

c. $\left[\dfrac{5}{2}, \infty\right)$

c. To solve $6x - 2(x + 2) - 5 \geq 0$ determine the values of x for which $Y_1 \geq 0$ (where the function is on or above the x-axis).

The solution set is $\left[\dfrac{9}{4}, \infty\right)$.

Skill Practice 9

a. Solve the equation $3x - (x + 4) - 1 = 0$ and verify the solution graphically on a graphing utility.

b. Use the graph to find the solution set to the inequality $3x - (x + 4) - 1 \leq 0$.

c. Use the graph to find the solution set to the inequality $3x - (x + 4) - 1 \geq 0$.

SECTION 1.4   Practice Exercises

Prerequisite Review

R.1. Determine the x- and y-intercepts for $h(x) = 6x - 42$. x-intercept $(7, 0)$; y-intercept $(0, -42)$

R.2. Solve $-7x - 8y = 1$ for y. $y = -\dfrac{7}{8}x - \dfrac{1}{8}$

For Exercises R.3–R.4, solve the inequality. Write the solution set in interval notation.

R.3. $-4t + 5 < 13$ $(-2, \infty)$

R.4. $6p - 2 \geq 5p + 8$ $[10, \infty)$

R.5. Given the function defined by $g(x) = -x^2 + 3x + 2$, find $g(-1)$. -2

Concept Connections

1. A _____linear_____ equation in the variables x and y can be written in the form $Ax + By = C$, where A and B are not both zero.

2. An equation of the form $x = k$ where k is a constant represents the graph of a _____vertical_____ line.

3. An equation of the form $y = k$ where k is a constant represents the graph of a _____horizontal_____ line.

4. Write the formula for the slope of a line between the two distinct points (x_1, y_1) and (x_2, y_2). $m = \dfrac{y_2 - y_1}{x_2 - x_1}$

5. The slope of a horizontal line is _____zero_____ and the slope of a vertical line is _____undefined_____.

6. A function f is a linear function if $f(x) = $ _____$mx + b$_____, where m represents the slope and $(0, b)$ represents the y-intercept.

7. If f is defined on the interval $[x_1, x_2]$, then the average rate of change of f on the interval $[x_1, x_2]$ is given by the formula _____. $m = \dfrac{f(x_2) - f(x_1)}{x_2 - x_1}$

8. The graph of a constant function defined by $f(x) = b$ is a (horizontal/vertical) line.   horizontal

Objective 1: Graph Linear Equations in Two Variables

For Exercises 9–20, graph the equation and identify the x- and y-intercepts. (See Example 1)

9. $-3x + 4y = 12$

10. $-2x + y = 4$

11. $2y = -5x + 2$

12. $3y = -4x + 6$

13. $x = -6$

14. $y = 4$

15. $5y + 1 = 11$

16. $3x - 2 = 4$

17. $0.02x + 0.05y = 0.1$

18. $0.03x + 0.07y = 0.21$

19. $2x = 3y$

20. $2x = -5y$

Objective 2: Determine the Slope of a Line

21. Find the average slope of the hill.

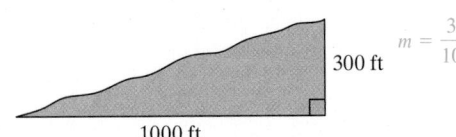

$$m = \frac{3}{10}$$

300 ft

1000 ft

22. Find the absolute value of the slope of the storm drainage pipe. $m = \frac{1}{16}$

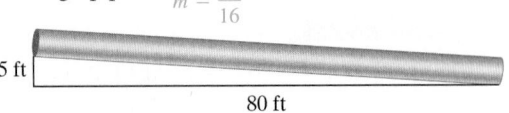

5 ft

80 ft

23. The road sign shown in the figure indicates the percent grade of a hill. This gives the slope of the road as the change in elevation per 100 horizontal feet. Given a 2.5% grade, write this as a slope in fractional form.

$$m = \frac{1}{40}$$

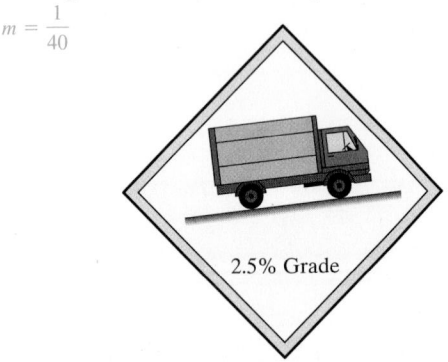

2.5% Grade

24. The pitch of a roof is defined as $\dfrac{\text{rafter rise}}{\text{rafter run}}$ and the fraction is typically written with a denominator of 12. Determine the pitch of the roof from point A to point C.

$\dfrac{7}{12}$

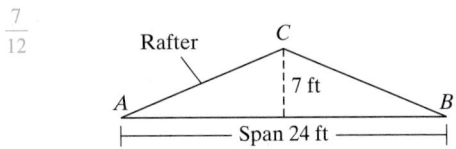

Rafter

C

7 ft

A B

Span 24 ft

For Exercises 25–36, determine the slope of the line passing through the given points. (See Example 2)

25. $(4, -7)$ and $(2, -1)$ $m = -3$

26. $(-3, -8)$ and $(4, 6)$ $m = 2$

27. $(17, 9)$ and $(42, -6)$ $m = -\dfrac{3}{5}$

28. $(-9, 4)$ and $(-1, -6)$ $m = -\dfrac{5}{4}$

29. $(30, -52)$ and $(-22, -39)$ $m = -\dfrac{1}{4}$

30. $(-100, -16)$ and $(84, 30)$ $m = \dfrac{1}{4}$

31. $(2.6, 4.1)$ and $(9.5, -3.7)$ $m = -\dfrac{26}{23}$

32. $(8.5, 6.2)$ and $(-5.1, 7.9)$ $m = -\dfrac{1}{8}$

33. $\left(\dfrac{3}{4}, 6\right)$ and $\left(\dfrac{5}{2}, 1\right)$ $m = -\dfrac{20}{7}$

34. $\left(-3, \dfrac{2}{5}\right)$ and $\left(4, \dfrac{3}{10}\right)$ $m = -\dfrac{1}{70}$

35. $(3\sqrt{6}, 2\sqrt{5})$ and $(\sqrt{6}, \sqrt{5})$ $m = \dfrac{\sqrt{30}}{12}$

36. $(2\sqrt{11}, -3\sqrt{3})$ and $(\sqrt{11}, -5\sqrt{3})$ $m = \dfrac{2\sqrt{33}}{11}$

For Exercises 37–42, determine the slope of the line. (See Examples 2–3)

37. $m = 3$

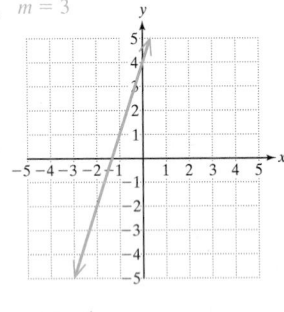

38. $m = \dfrac{1}{2}$

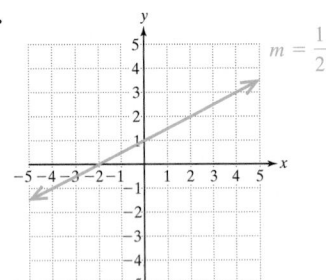

39. $m = -\dfrac{1}{3}$

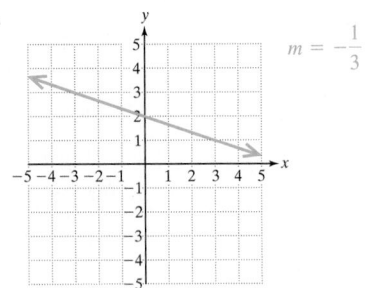

40. $m = -4$

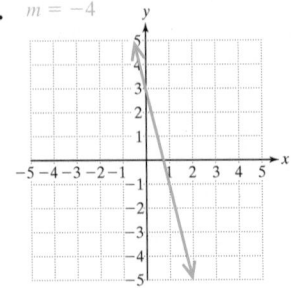

41. $m = 0$

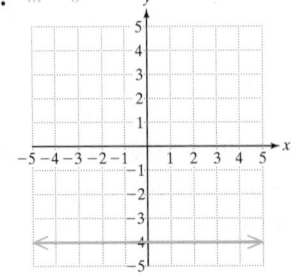

42.

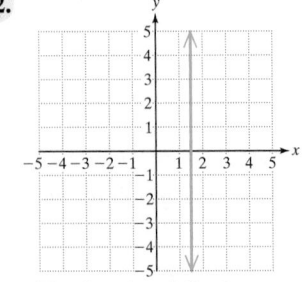

The slope is undefined.

43. What is the slope of a line perpendicular to the *x*-axis?
Undefined

44. What is the slope of a line parallel to the *x*-axis?   0

45. What is the slope of a line defined by $y = -7$?   0

46. What is the slope of a line defined by $x = 2$?   Undefined

47. If the slope of a line is $\frac{4}{5}$, how much vertical change will be present for a horizontal change of 52 ft?   41.6 ft

48. If the slope of a line is $\frac{5}{8}$, how much horizontal change will be present for a vertical change of 216 m?   345.6 m

49. Suppose that $y = P(t)$ represents the population of a city at time *t*. What does $\frac{\Delta P}{\Delta t}$ represent?
Change in population over change in time

50. Suppose that $y = d(t)$ represents the distance that an object travels in time *t*. What does $\frac{\Delta d}{\Delta t}$ represent?
Change in distance over change in time which is speed

Objective 3: Apply the Slope-Intercept Form of a Line

For Exercises 51–62,

a. Write the equation in slope-intercept form if possible, and determine the slope and *y*-intercept.

b. Graph the equation using the slope and *y*-intercept. (**See Example 4**)

51. $2x - 4y = 8$

52. $3x - y = 6$

53. $3x = 2y - 4$

54. $5x = 3y - 6$

55. $3x = 4y$

56. $-2x = 3y$

57. $2y - 6 = 8$

58. $3y + 9 = 6$

59. $0.02x + 0.06y = 0.06$

60. $0.03x + 0.04y = 0.12$

61. $\frac{x}{4} + \frac{y}{7} = 1$

62. $\frac{x}{3} + \frac{y}{4} = 1$

For Exercises 63–64, determine if the function is linear, constant, or neither.

63. a. $f(x) = -\frac{3}{4}x$ Linear   **b.** $g(x) = -\frac{3}{4}x - 3$ Linear   **c.** $h(x) = -\frac{3}{4x}$ Neither   **d.** $k(x) = -\frac{3}{4}$ Constant

64. a. $m(x) = 5x + 1$ Linear   **b.** $n(x) = \frac{5}{x} + 1$ Neither   **c.** $p(x) = 5$ Constant   **d.** $q(x) = 5x$ Linear

For Exercises 65–74,

a. Use slope-intercept form to write an equation of the line that passes through the given point and has the given slope.

b. Write the equation using function notation where $y = f(x)$. (**See Example 5**)

65. $(0, 9)$; $m = \frac{1}{2}$ a. $y = \frac{1}{2}x + 9$
b. $f(x) = \frac{1}{2}x + 9$

66. $(0, -4)$; $m = \frac{1}{3}$ a. $y = \frac{1}{3}x - 4$
b. $f(x) = \frac{1}{3}x - 4$

67. $(1, -6)$; $m = -3$
a. $y = -3x - 3$
b. $f(x) = -3x - 3$

68. $(2, -8)$; $m = -5$
a. $y = -5x + 2$
b. $f(x) = -5x + 2$

69. $(-5, -3)$; $m = \frac{2}{3}$
a. $y = \frac{2}{3}x + \frac{1}{3}$ b. $f(x) = \frac{2}{3}x + \frac{1}{3}$

70. $(-4, -2)$; $m = \frac{3}{2}$
a. $y = \frac{3}{2}x + 4$ b. $f(x) = \frac{3}{2}x + 4$

71. $(2, 5)$; $m = 0$
a. $y = 5$
b. $f(x) = 5$

72. $(-1, -3)$; $m = 0$
a. $y = -3$
b. $f(x) = -3$

73. $(3.6, 5.1)$; $m = 1.2$
a. $y = 1.2x + 0.78$ b. $f(x) = 1.2x + 0.78$

74. $(1.2, 2.8)$; $m = 2.4$
a. $y = 2.4x - 0.08$ b. $f(x) = 2.4x - 0.08$

For Exercises 75–78,

a. Use slope-intercept form to write an equation of the line that passes through the two given points.

b. Then write the equation using function notation where $y = f(x)$.

75. $(4, 2)$ and $(0, -6)$

76. $(-8, 1)$ and $(0, -3)$

77. $(7, -3)$ and $(4, 1)$

78. $(2, -4)$ and $(-1, 3)$

Objective 4: Compute Average Rate of Change

For Exercises 79–80, find the slope of the secant line pictured in red. (See Example 6)

79.

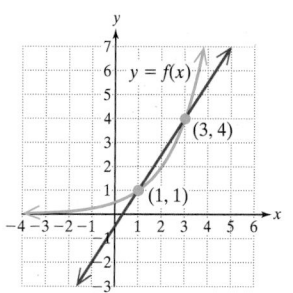

$m = \frac{3}{2}$

80.

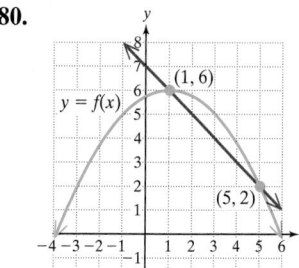

$m = -1$

81. The function given by $y = f(x)$ shows the value of $5000 invested at 5% interest compounded continuously, x years after the money was originally invested.

a. Find the average amount earned per year between the 5th year and 10th year. $364.80/yr

b. Find the average amount earned per year between the 20th year and the 25th year. $772.20/yr

c. Based on the answers from parts (a) and (b), does it appear that the rate at which annual income increases is increasing or decreasing with time?   Increasing

Value of $5000 with Continuous Compounding at 5%

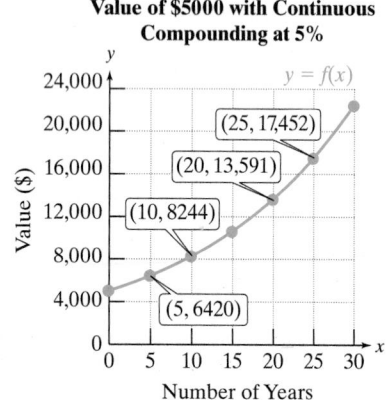

Number of Years

82. The function given by $y = f(x)$ shows the average monthly temperature (in °F) for Cedar Key. The value of x is the month number and $x = 1$ represents January.

a. Find the average rate of change in temperature between months 3 and 5 (March and May).   5.5°F/month

b. Find the average rate of change in temperature between months 9 and 11 (September and November).   −7.5°F/month

c. Comparing the results in parts (a) and (b), what does a positive rate of change mean in the context of this problem? What does a negative rate of change mean?   A positive rate of change means that average monthly temperature is increasing. A negative rate of change means that average monthly temperature is decreasing.

Average Monthly Temperature for Cedar Key, Florida

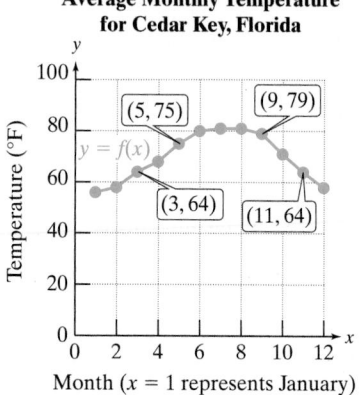

Month ($x = 1$ represents January)

83. The number $N(t)$ of new cases of a flu outbreak for a given city is given by $N(t) = 5000 \cdot 2^{-0.04t^2}$, where t is the number of months since the outbreak began.

a. Find the average rate of change in the number of new flu cases between months 0 and 2, and interpret the result. Round to the nearest whole unit.

b. Find the average rate of change in the number of new flu cases between months 4 and 6, and between months 10 and 12.

c. Use a graphing utility to graph the function. Use the graph and the average rates of change found in parts (a) and (b) to discuss the pattern of the number of new flu cases.

84. The speed $v(L)$ (in m/sec) of an ocean wave in deep water is approximated by $v(L) = 1.2\sqrt{L}$, where L (in meters) is the wavelength of the wave. (The wavelength is the distance between two consecutive wave crests.)

a. Find the average rate of change in speed between waves that are between 1 m and 4 m in length.

b. Find the average rate of change in speed between waves that are between 4 m and 9 m in length.

c. Use a graphing utility to graph the function. Using the graph and the results from parts (a) and (b), what does the difference in the rates of change mean?

For Exercises 85–90, determine the average rate of change of the function on the given interval. (See Example 7)

85. $f(x) = x^2 - 3$
a. on $[0, 1]$ 1
b. on $[1, 3]$ 4
c. on $[-2, 0]$ −2

86. $g(x) = 2x^2 + 2$
a. on $[0, 1]$ 2
b. on $[1, 3]$ 8
c. on $[-2, 0]$ −4

87. $h(x) = x^3$
a. on $[-1, 0]$ 1
b. on $[0, 1]$ 1
c. on $[1, 2]$ 7

88. $k(x) = x^3 - 2$
a. on $[-1, 0]$ 1
b. on $[0, 1]$ 1
c. on $[1, 2]$ 7

89. $m(x) = \sqrt{x}$
a. $[0, 1]$ 1
b. $[1, 4]$ $\frac{1}{3}$
c. $[4, 9]$ $\frac{1}{5}$

90. $n(x) = \sqrt{x - 1}$
a. $[1, 2]$ 1
b. $[2, 5]$ $\frac{1}{3}$
c. $[5, 10]$ $\frac{1}{5}$

Objective 5: Solve Equations and Inequalities Graphically

For Exercises 91–98, use the graph to solve the equation and inequalities. Write the solutions to the inequalities in interval notation. (See Examples 8–9)

91. a. $2x + 4 = -x + 1$ $\{-1\}$

    b. $2x + 4 < -x + 1$ $(-\infty, -1)$

    c. $2x + 4 \geq -x + 1$ $[-1, \infty)$

92. a. $4x - 2 = -3x + 5$ $\{1\}$

    b. $4x - 2 < -3x + 5$ $(-\infty, 1)$

    c. $4x - 2 \geq -3x + 5$ $[1, \infty)$

93. a. $-3x + 1 = -x - 3$ $\{2\}$

    b. $-3x + 1 > -x - 3$ $(-\infty, 2)$

    c. $-3x + 1 \leq -x - 3$ $[2, \infty)$

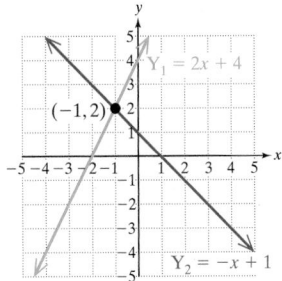
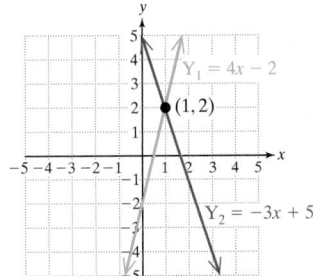
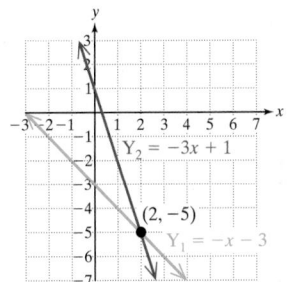

94. a. $-x - 2 = 2x - 5$ $\{1\}$

    b. $-x - 2 \leq 2x - 5$ $[1, \infty)$

    c. $-x - 2 > 2x - 5$ $(-\infty, 1)$

95. a. $-3(x + 2) + 1 = -x + 5$ $\{-5\}$

    b. $-3(x + 2) + 1 \leq -x + 5$ $[-5, \infty)$

    c. $-3(x + 2) + 1 \geq -x + 5$ $(-\infty, -5]$

96. a. $-4(x - 5) + 3x = -3x + 1$ $\{9.5\}$

    b. $-4(x - 5) + 3x \leq -3x + 1$ $(-\infty, -9.5]$

    c. $-4(x - 5) + 3x \geq -3x + 1$ $[-9.5, \infty)$

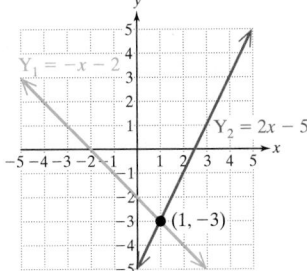
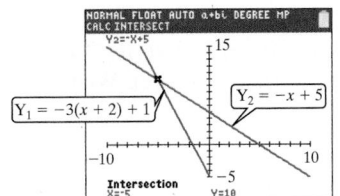
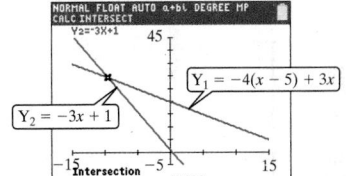

97. a. $4 - 2(x + 1) + 12 + x = 0$ $\{14\}$

    b. $4 - 2(x + 1) + 12 + x < 0$ $(14, \infty)$

    c. $4 - 2(x + 1) + 12 + x > 0$ $(-\infty, 14)$

98. a. $8 - 4(1 - x) - 7 - 2x = 0$ $\{1.5\}$

    b. $8 - 4(1 - x) - 7 - 2x < 0$ $(-\infty, 1.5)$

    c. $8 - 4(1 - x) - 7 - 2x > 0$ $(1.5, \infty)$

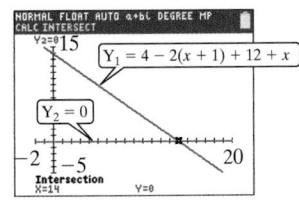
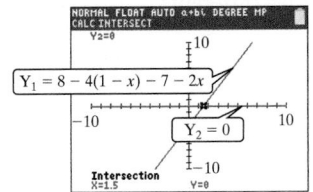

Write About It

99. Explain how you can determine from a linear equation $Ax + By = C$ (A and B not both zero) whether the line is slanted, horizontal, or vertical.

101. What is the benefit of writing an equation of a line in slope-intercept form?   The slope and y-intercept are easily determined by inspection of the equation.

100. Explain how you can determine from a linear equation $Ax + By = C$ (A and B not both zero) whether the line passes through the origin.   If C is zero, then the line passes through the origin.

102. Explain how the average rate of change of a function f on the interval $[x_1, x_2]$ is related to slope.   The average rate of change of f on the interval $[x_1, x_2]$ is the slope of the secant line passing through the points $(x_1, f(x_1))$ and $(x_2, f(x_2))$.

Expanding Your Skills

103. Determine the area in the second quadrant enclosed by the equation $y = 2x + 4$ and the x- and y-axes.
4 units^2

104. Determine the area enclosed by the equations.   27 units^2

$$y = x + 6$$
$$y = -2x + 6$$
$$y = 0$$

105. Determine the area enclosed by the equations.   10 units^2

$$y = -\frac{1}{2}x - 2$$
$$y = \frac{1}{3}x - 2$$
$$y = 0$$

106. Determine the area enclosed by the equations. 2π units^2

$$y = \sqrt{4 - (x - 2)^2}$$
$$y = 0$$

107. Consider the standard form of a linear equation $Ax + By = C$ in the case where $B \neq 0$.

    a. Write the equation in slope-intercept form. $y = -\frac{A}{B}x + \frac{C}{B}$

$m = -\frac{A}{B}$ **b.** Identify the slope in terms of the coefficients A and B.

    c. Identify the y-intercept in terms of the coefficients B and C. $\left(0, \frac{C}{B}\right)$

108. Use the results from Exercise 107 to determine the slope and y-intercept for the graphs of the lines.

    a. $5x - 9y = 6$ $m = \frac{5}{9}$; y-intercept: $\left(0, -\frac{2}{3}\right)$

    b. $0.052x - 0.013y = 0.39$
      $m = 4$; y-intercept: $(0, -30)$

Technology Connections

For Exercises 109–112, solve the equation in part (a) and verify the solution on a graphing calculator. Then use the graph to find the solution set to the inequalities in parts (b) and (c). Write the solution sets to the inequalities in interval notation. (See Example 9)

109. a. $3.1 - 2.2(t + 1) = 6.3 + 1.4t$

    b. $3.1 - 2.2(t + 1) > 6.3 + 1.4t$

    c. $3.1 - 2.2(t + 1) < 6.3 + 1.4t$

111. a. $|2x - 3.8| - 4.6 = 7.2$

    b. $|2x - 3.8| - 4.6 \geq 7.2$

    c. $|2x - 3.8| - 4.6 \leq 7.2$

110. a. $-11.2 - 4.6(c - 3) + 1.8c = 0.4(c + 2)$

    b. $-11.2 - 4.6(c - 3) + 1.8c > 0.4(c + 2)$

    c. $-11.2 - 4.6(c - 3) + 1.8c < 0.4(c + 2)$

112. a. $|x - 1.7| + 4.95 = 11.15$

    b. $|x - 1.7| + 4.95 \geq 11.15$

    c. $|x - 1.7| + 4.95 \leq 11.15$

For Exercises 113–114, graph the lines in (a)–(c) on the standard viewing window. Compare the graphs. Are they exactly the same? If not, how are they different?

113. a. $y = 3x + 1$

    b. $y = 2.99x + 1$

    c. $y = 3.01x + 1$

114. a. $y = x + 3$

    b. $y = x + 2.99$

    c. $y = x + 3.01$

SECTION 1.5 **Applications of Linear Equations and Modeling**

OBJECTIVES

1. Apply the Point-Slope Formula
2. Determine the Slopes of Parallel and Perpendicular Lines
3. Create Linear Functions to Model Data
4. Create Models Using Linear Regression

1. Apply the Point-Slope Formula

The slope formula can be used to develop the point-slope form of an equation of a line. Suppose that a line has a slope m and passes through a known point (x_1, y_1). Let (x, y) be any other point on the line. From the slope formula, we have

$$\frac{y - y_1}{x - x_1} = m \qquad \text{Slope formula}$$

$$\left(\frac{y - y_1}{x - x_1}\right)(x - x_1) = m(x - x_1) \qquad \text{Clear fractions.}$$

$$y - y_1 = m(x - x_1) \qquad \text{This is called the point slope formula for a line.}$$

The point-slope formula is useful to build an equation of a line given a point on the line and the slope of the line.

Point-Slope Formula

The **point-slope formula** for a line is given by $y - y_1 = m(x - x_1)$, where m is the slope of the line and (x_1, y_1) is a point on the line.

Classroom Example: p. 176
Exercise 6

EXAMPLE 1   Writing an Equation of a Line Given a Point on the Line and the Slope

Use the point-slope formula to find an equation of the line passing through the point $(2, -3)$ and having slope -4. Write the answer in slope-intercept form.

Solution:

Label $(2, -3)$ as (x_1, y_1) and $m = -4$.

$y - y_1 = m(x - x_1)$	Apply the point-slope formula.
$y - (-3) = -4(x - 2)$	Substitute $x_1 = 2$, $y_1 = -3$, and $m = -4$.
$y + 3 = -4x + 8$	Simplify.
$y = -4x + 5$ (slope-intercept form)	

TIP The slope-intercept form of a line can also be used to write an equation of a line if a point on the line and the slope are known. See Example 5 in Section 1.4.

Skill Practice 1   Use the point-slope formula to find an equation of the line passing through the point $(-5, 2)$ and having slope -3. Write the answer in slope-intercept form.

Classroom Example: p. 176
Exercise 12

EXAMPLE 2   Writing an Equation of a Line Given Two Points

Use the point-slope formula to write an equation of the line passing through the points $(4, -6)$ and $(-1, 2)$. Write the answer in slope-intercept form.

Answer

1. $y = -3x - 13$

Solution:

To apply the point-slope formula, we first need to know the slope of the line.

$(4, -6)$ and $(-1, 2)$	Label the points. Either point can be
(x_1, y_1) and (x_2, y_2)	labeled (x_1, y_1).

$$m = \frac{y_2 - y_1}{x_2 - x_1} = \frac{2 - (-6)}{-1 - 4} = \frac{8}{-5} = -\frac{8}{5}$$ Apply the slope formula.

$$y - y_1 = m(x - x_1)$$ Apply the point-slope formula.

$$y - (-6) = -\frac{8}{5}(x - 4)$$ Substitute $y_1 = -6$, $x_1 = 4$, and $m = -\frac{8}{5}$.

$$y + 6 = -\frac{8}{5}x + \frac{32}{5}$$

$$y = -\frac{8}{5}x + \frac{32}{5} - 6$$

$$y = -\frac{8}{5}x + \frac{32}{5} - \frac{30}{5}$$

$$y = -\frac{8}{5}x + \frac{2}{5} \text{ (slope-intercept form)}$$

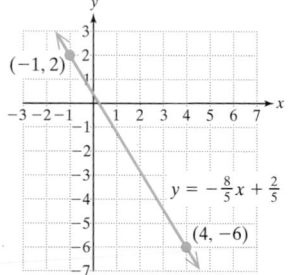

To check, we see that the graph of the line passes through $(4, -6)$ and $(-1, 2)$ as expected.

> **TIP** In Example 2, the slope-intercept form of a line can also be used to find an equation of the line. Substitute $-\frac{8}{5}$ for m and $(4, -6)$ for (x, y).
>
> $$y = mx + b$$
> $$-6 = -\frac{8}{5}(4) + b$$
> $$-6 = -\frac{32}{5} + b$$
> $$-6 + \frac{32}{5} = b$$
> $$\frac{2}{5} = b$$
>
> Therefore, $y = mx + b$ is $y = -\frac{8}{5}x + \frac{2}{5}$.

Skill Practice 2   Write an equation of the line passing through the points $(2, -5)$ and $(7, -3)$.

---

2. Determine the Slopes of Parallel and Perpendicular Lines

Lines in the same plane that do not intersect are **parallel lines.** Nonvertical parallel lines have the same slope and different y-intercepts (Figure 1-20).

Lines that intersect at a right angle are **perpendicular lines.** If two nonvertical lines are perpendicular, then the slope of one line is the opposite of the reciprocal of the slope of the other line (Figure 1-21).

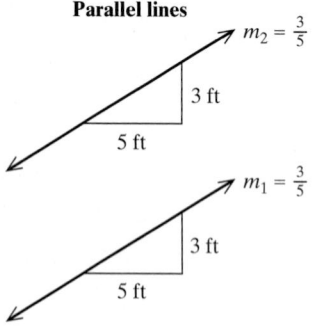

Figure 1-20

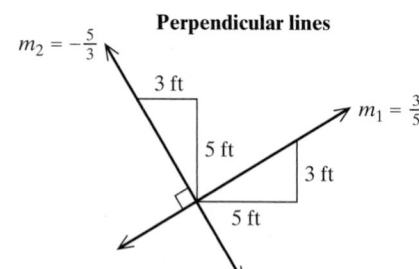

Figure 1-21

Slopes of Parallel and Perpendicular Lines

- If m_1 and m_2 represent the slopes of two nonvertical parallel lines, then $m_1 = m_2$.
- If m_1 and m_2 represent the slopes of two nonvertical perpendicular lines, then $m_1 = -\dfrac{1}{m_2}$ or equivalently $m_1 m_2 = -1$.

Answer

2. $y = \dfrac{2}{5}x - \dfrac{29}{5}$

In Examples 3 and 4, we use the point-slope formula to find an equation of a line through a specified point and parallel or perpendicular to another line.

EXAMPLE 3   Writing an Equation of a Line Parallel to Another Line

Write an equation of the line passing through the point $(-4, 1)$ and parallel to the line defined by $x + 4y = 3$. Write the answer in slope-intercept form and in standard form.

Solution:

$x + 4y = 3$ The slope of the given line can be found from its slope-intercept form. Solve for y.

$\quad 4y = -x + 3$

$\quad\quad y = -\dfrac{1}{4}x + \dfrac{3}{4}$ The slope of both lines is $-\frac{1}{4}$.

$y - y_1 = m(x - x_1)$ Apply the point-slope formula with $x_1 = -4$, $y_1 = 1$, and $m = -\frac{1}{4}$.

$y - 1 = -\dfrac{1}{4}[x - (-4)]$

$y - 1 = -\dfrac{1}{4}(x + 4)$

$y - 1 = -\dfrac{1}{4}x - 1$

$\quad\quad y = -\dfrac{1}{4}x$ (slope-intercept form)

$4(y) = 4\left(-\dfrac{1}{4}x\right)$

$\quad 4y = -x$

$x + 4y = 0$ (standard form)

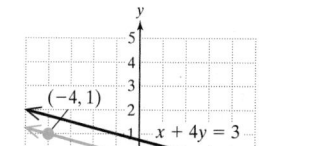

From the graph, we see that the line $y = -\frac{1}{4}x$ passes through the point $(-4, 1)$ and is parallel to the graph of $x + 4y = 3$.

Clearing fractions, and collecting the x and y terms on one side of the equation gives us standard form.

Skill Practice 3   Write an equation of the line passing through the point $(-3, 2)$ and parallel to the line defined by $x + 3y = 6$. Write the answer in slope-intercept form and in standard form.

EXAMPLE 4   Writing an Equation of a Line Perpendicular to Another Line

Write an equation of the line passing through the point $(2, -3)$ and perpendicular to the line defined by $y = \frac{1}{2}x - 4$. Write the answer in slope-intercept form and in standard form.

Answer

3. $y = -\dfrac{1}{3}x + 1$; $x + 3y = 3$

The solution to Example 4 can be checked by graphing both lines and verifying that they are perpendicular and that the line $y = -2x + 1$ passes through the point $(2, -3)$.

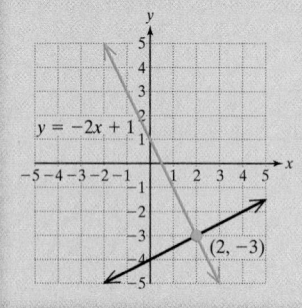

Solution:

From the slope-intercept form, $y = \frac{1}{2}x - 4$, the slope of given line is $\frac{1}{2}$.

$y - y_1 = m(x - x_1)$	The slope of a line perpendicular to the given line is -2.
$y - (-3) = -2(x - 2)$	Apply the point-slope formula with $x_1 = 2$, $y_1 = -3$, and $m = -2$.
$y + 3 = -2x + 4$	Simplify.
$y = -2x + 1$ (slope-intercept form)	Write the equation in slope-intercept form by solving for y.
$2x + y = 1$ (standard form)	Write the equation in standard form by collecting the x and y terms on one side of the equation.

Skill Practice 4   Write an equation of the line passing through the point $(-8, -4)$ and perpendicular to the line defined by $y = \frac{1}{6}x + 3$.

3. Create Linear Functions to Model Data

In many day-to-day applications, two variables are related linearly. By finding an equation of the line, we produce a model that relates the two variables. This is demonstrated in Example 5.

Classroom Example: p. 177
Exercise 54

EXAMPLE 5   **Using a Linear Function in an Application**

A family plan for a cell phone has a monthly base price of $99 plus $12.99 for each additional family member added beyond the primary account holder.

a. Write a linear function to model the monthly cost $C(x)$ (in $) of a family plan for x additional family members added.

b. Evaluate $C(4)$ and interpret the meaning in the context of this problem.

Solution:

a. $C(x) = mx + b$	The base price $99 is the fixed cost with zero additional family members added. So the constant b is 99.
$C(x) = 12.99x + 99$	The rate of increase, $12.99 per additional family member, is the slope.
b. $C(4) = 12.99(4) + 99$	Substitute 4 for x.
$= 150.96$	

The total monthly cost of the plan with 4 additional family members beyond the primary account holder is $150.96.

Skill Practice 5   A speeding ticket is $100 plus $5 for every 1 mph over the speed limit.

a. Write a linear function to model the cost $S(x)$ (in $) of a speeding ticket for a person caught driving x mph over the speed limit.

b. Evaluate $S(15)$ and interpret the meaning in the context of this problem.

Answers

4. $y = -6x - 52$; $6x + y = -52$

5. a. $S(x) = 5x + 100$

b. $S(15) = 175$ means that a ticket costs $175 for a person caught speeding 15 mph over the speed limit.

Linear functions can sometimes be used to model the cost, revenue, and profit of producing and selling x items.

Linear Cost, Revenue, and Profit Functions

A **linear cost function** models the cost $C(x)$ to produce x items.

$C(x) = mx + b$

m is the variable cost per item.
b is the fixed cost.

The fixed cost does not change relative to the number of items produced. For example, the cost to rent an office is a fixed cost. The variable cost per item is the rate at which cost increases for each additional unit produced. Variable costs include labor, material, and shipping.

A **linear revenue function** models revenue $R(x)$ for selling x items.

$R(x) = px$

The product px represents the price per item p times the number of items sold x.

A **linear profit function** models the profit for producing and selling x items.

$P(x) = R(x) - C(x)$

Subtract the cost to produce x items from the revenue brought in from selling x items.

Classroom Example: p. 178
Exercise 60

EXAMPLE 6   Writing Linear Cost, Revenue, and Profit Functions

At a summer art show a vendor sells lemonade for \$2.00 per cup. The cost to rent the booth is \$120. Furthermore, the vendor knows that the lemons, sugar, and cups collectively cost \$0.50 for each cup of lemonade produced.

a. Write a linear cost function to produce x cups of lemonade.
b. Write a linear revenue function for selling x cups of lemonade.
c. Write a linear profit function for producing and selling x cups of lemonade.
d. How much profit will the vendor make if 50 cups of lemonade are produced and sold?
e. How much profit will be made for producing and selling 128 cups?
f. Determine the break-even point.

Solution:

a. $C(x) = 0.50x + 120$

The fixed cost is \$120 because it does not change relative to the number of cups of lemonade produced. The variable cost is \$0.50 per lemonade.

b. $R(x) = 2.00x$

The price per cup of lemonade is \$2.00. Therefore, the product $2.00x$ gives the amount of revenue for x cups of lemonade sold.

c. $P(x) = R(x) - C(x)$
$P(x) = 2.00x - (0.50x + 120)$
$P(x) = 1.50x - 120$

Profit is defined as the difference of revenue and cost.

d. $P(50) = 1.50(50) - 120$
$= -45$

Substitute 50 for x.
The vendor will lose \$45.

e. $P(128) = 1.50(128) - 120$ Substitute 128 for x.

$\qquad\qquad = 72$ The vendor will make $72.

f. For what value of x will $R(x) = C(x)$ or $P(x) = 0$?

The break-even point is defined as the point where revenue equals cost. Alternatively, this can be stated as the point where profit equals zero: $P(x) = 0$.

$$P(x) = 0$$
$$1.50x - 120 = 0 \qquad \text{Solve for } x.$$
$$1.50x = 120$$
$$x = 80$$

If the vendor produces and sells 80 cups of lemonade, the cost and revenue will be equal, resulting in a profit of $0. This is the break-even point.

Skill Practice 6 Repeat Example 6 in the case where the vendor can cut the cost to $0.40 per cup of lemonade, and sell lemonades for $1.50 per cup.

Figure 1-22 shows the graphs of the revenue and cost functions from Example 6. Notice that R and C intersect at $(80, 160)$. This means that if 80 cups of lemonade are produced and sold, the revenue and cost are both $160. That is, $R(x) = C(x)$ and the company breaks even. The graph of the profit function P is consistent with this result. The value of $P(x)$ is 0 for 80 lemonades produced and sold (Figure 1-23).

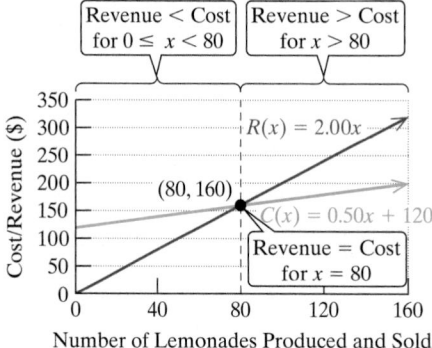

Figure 1-22   **Figure 1-23**

From Figures 1-22 and 1-23, we can draw the following conclusions.

- The company experiences a loss if fewer than 80 cups of lemonade are produced and sold. That is, $R(x) < C(x)$, or equivalently $P(x) < 0$.
- The company experiences a profit if more than 80 cups of lemonade are produced and sold. That is, $R(x) > C(x)$, or equivalently $P(x) > 0$.
- The company breaks even if exactly 80 cups of lemonade are produced and sold. That is, $R(x) = C(x)$, or equivalently $P(x) = 0$.

Answers

6. a. $C(x) = 0.40x + 120$
  b. $R(x) = 1.50x$
  c. $P(x) = 1.10x - 120$
  d. $-\$65$
  e. $20.80
  f. Approximately 109 cups

EXAMPLE 7   Writing a Linear Model to Relate Two Variables

The data shown in the graph represent the age and systolic blood pressure for a sample of 12 randomly selected healthy adults.

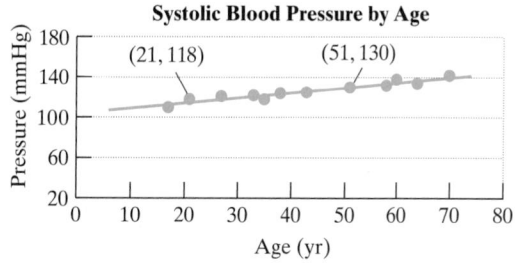

Systolic Blood Pressure by Age

a. Suppose that x represents the age of an adult (in yr), and y represents the systolic blood pressure (in mmHg). Use the points (21, 118) and (51, 130) to write a linear model relating y as a function of x.

b. Interpret the meaning of the slope in the context of this problem.

c. Use the model to estimate the systolic blood pressure for a 55-year-old. Round to the nearest whole unit.

TIP The equation $y = -0.4x + 109.6$ can also be expressed in function notation. For example, we can rename y as $S(x)$.

$$S(x) = -0.4x + 109.6$$

The value $S(x)$ represents the estimated systolic blood pressure for an adult of age x years.

Solution:

a. (21, 118) and (51, 130)

   (x_1, y_1) and (x_2, y_2) Label the points.

   $m = \dfrac{130 - 118}{51 - 21} = 0.4$ Determine the slope of the line.

   $y - y_1 = m(x - x_1)$ Apply the point-slope formula.
   $y - 118 = 0.4(x - 21)$
   $y = 0.4x + 109.6$ The equation $y = 0.4x + 109.6$ relates an individual's age to an estimated systolic blood pressure for that age.

b. The slope is 0.4. This means that the average increase in systolic blood pressure for adults is 0.4 mmHg per year of age.

c. $y = 0.4x + 109.6$
   $y = 0.4(55) + 109.6$ Substitute 55 for x.
   $y = 131.6$

Based on the sample of data, the estimated systolic blood pressure for a 55-year-old is 132 mmHg.

Instructor Note:
Point out that two different points on the line may produce a slightly different equation. However, the model will still approximate the trend of the data.

Skill Practice 7   Suppose that y represents the average consumer spending on television services per year (in dollars), and that x represents the number of years since 2004.

a. Use the data points (2, 308) and (6, 408) to write a linear equation relating y to x.

b. Interpret the meaning of the slope in the context of this problem.

c. Interpret the meaning of the y-intercept in the context of this problem.

d. Use the model from part (a) to estimate the average consumer spending on television services for the year 2007.

Answers

7. a. $y = 25x + 258$
   b. The slope is 25 and means that consumer spending on television services rose $25 per year during this time period.
   c. (0, 258); The average consumer spending on television services for the year 2004 was $258.
   d. $333

4. Create Models Using Linear Regression

In Example 7, we used two given data points to determine a linear model for systolic blood pressure versus age. There are two drawbacks to this method. First, the equation is not necessarily unique. If we use two different data points, we may get a different equation. Second, it is generally preferable to write a model that is based on *all* the data points, rather than just two points. One such model is called the least-squares regression line.

The procedure to find the least-squares regression line is discussed in detail in a statistics course. Here we will give the basic premise and use a graphing utility to perform the calculations. Consider a set of data points (x_1, y_1), (x_2, y_2), (x_3, y_3), ... , (x_n, y_n). The **least-squares regression line,** $\hat{y} = mx + b$, is the unique line that minimizes the sum of the squared vertical deviations from the observed data points to the line (Figure 1-24).

On a calculator or spreadsheet, the equation $\hat{y} = mx + b$ may be denoted as $y = ax + b$ or as $y = b_0 + b_1 x$. In any event, the coefficient of x is the slope of the line, and the constant gives us the y-intercept. Although the exact keystrokes on different calculators and graphing utilities may vary, we will use the following guidelines to find the least-squares regression line.

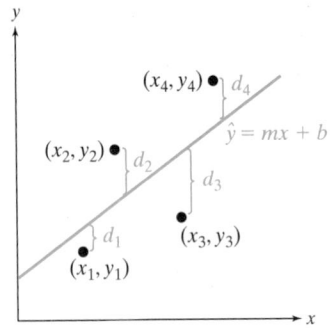

Figure 1-24

Creating a Linear Regression Model

1. Graph the data in a scatter plot.
2. Inspect the data visually to determine if the data suggest a linear trend.
3. Invoke the linear regression feature on a calculator, graphing utility, or spreadsheet.
4. Check the result by graphing the line with the data points to verify that the line passes through or near the data points.

Classroom Example: p. 180
Exercise 72

EXAMPLE 8   Finding a Least-Squares Regression Line

The data given in the table represent the age and systolic blood pressure for a sample of 12 randomly selected healthy adults.

Age (yr)	17	21	27	33	35	38	43	51	58	60	64	70
Systolic blood pressure (mmHg)	110	118	121	122	118	124	125	130	132	138	134	142

a. Make a scatter plot of the data using age as the independent variable x and systolic pressure as the dependent variable y.

b. Based on the graph, does a linear model seem appropriate?

c. Determine the equation of the least-squares regression line.

d. Use the least-squares regression line to approximate the systolic blood pressure for a healthy 55-year-old. Round to the nearest whole unit.

Solution:

a. On a graphing calculator hit the STAT button and select EDIT to enter the x and y data into two lists (shown here as L1 and L2).

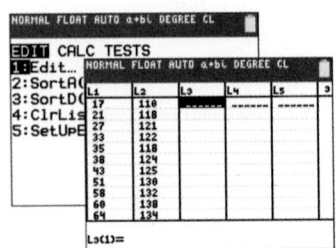

Select the STAT PLOT option and turn Plot1 to On. For the type of graph, select the scatter plot image.

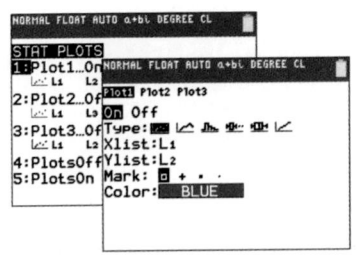

Be sure that the window is set to accommodate x values between 17 and 70, and y values between 110 and 142, inclusive. Then hit the GRAPH key. The window settings shown here are [0, 80, 10] by [0, 200, 20].

b. From the graph, the data appear to follow a linear trend.

c. Under the STAT menu, select CALC and then the LinReg(ax + b) option.

The command LinReg(ax + b) prompts the user to enter the list names (L_1 and L_2) containing the x and y data values. Then highlight Calculate and hit ENTER.

> **TIP** The linear equation found in Example 7 was based on two data points. The least-squares regression line is based on all available data points. The estimate from each model for systolic blood pressure for a 55-year-old rounds to 132 mmHg.

In the regression model $y = ax + b$, the values for the coefficients a and b are placed on the home screen.

Rounding the values of a and b gives us $y = 0.511x + 104$.

Enter the equation $Y_1 = 0.511x + 104$ into the equation editor and hit the GRAPH key. The graph of the regression line passes near or through the observed data points.

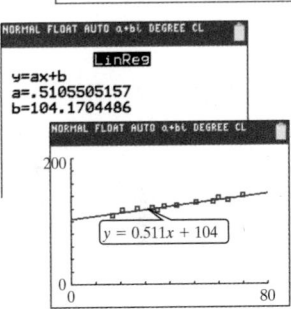

d. $y = 0.511x + 104$

$y = 0.511(55) + 104$ To approximate the systolic blood pressure for a 55-year-old, substitute 55 for x.

$= 132.105$

The systolic blood pressure for a healthy 55-year-old would be approximately 132 mmHg.

> **TIP** In Example 8(d), the value of the function at $x = 55$ can also be found by selecting the CALC menu and selecting the VALUE function. Enter 55 for x and press the ENTER key.

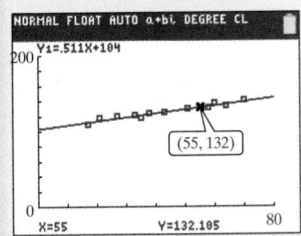

Skill Practice 8   The data given represent the class averages for individual students based on the number of absences from class.

Number of Absences (x)	3	7	1	11	2	14	2	5
Average in Class (y)	88	67	96	62	90	56	97	82

a. Find the equation of the least-squares regression line.

b. Use the model from part (a) to approximate the average for a student who misses 6 classes.

Answers

8. a. $y = -3.27x + 98.1$
 b. The student's average would be approximately 78.5.

SECTION 1.5   Practice Exercises

Prerequisite Review

R.1. Use slope-intercept form to write an equation of the line that passes through $(3, -7)$ with slope -5. $y = -5x + 8$

R.2. Write the equation in slope-intercept form and determine the slope and y-intercept $3x - 5y = -15$. $y = \frac{3}{5}x + 3$; Slope $= \frac{3}{5}$, y-intercept is $(0, 3)$

R.3. Determine the slope of the line containing the points $(-4, -2)$ and $(-4, -7)$.   Undefined

R.4. Determine the slope of the line containing the points $(3, -2)$ and $(5, -2)$.   0

R.5. The cost C (in dollars) to rent an apartment is \$850 per month, plus a \$450 nonrefundable security deposit, plus a \$250 deposit for each dog or cat. Write a formula for the total cost to rent an apartment for m months with n cats/dogs. $C = 450 + 850m + 250n$

Concept Connections

1. Given a point (x_1, y_1) on a line with slope m, the point-slope formula is given by __$y - y_1 = m(x - x_1)$__.

2. If two nonvertical lines have the same slope but different y-intercepts, then the lines are (parallel/perpendicular).   parallel

3. If m_1 and m_2 represent the slopes of two nonvertical perpendicular lines, then $m_1m_2 = $ ____-1____.

4. Suppose that $y = C(x)$ represents the cost to produce x items, and that $y = R(x)$ represents the revenue for selling x items. The profit $P(x)$ of producing and selling x items is defined by $P(x) = $ ____$R(x) - C(x)$____.

Objective 1: Apply the Point-Slope Formula

For Exercises 5–20, use the point-slope formula to write an equation of the line having the given conditions. Write the answer in slope-intercept form (if possible). (See Examples 1–2)

5. Passes through $(-3, 5)$ and $m = -2$. $y = -2x - 1$

6. Passes through $(4, -6)$ and $m = 3$. $y = 3x - 18$

7. Passes through $(-1, 0)$ and $m = \frac{2}{3}$. $y = \frac{2}{3}x + \frac{2}{3}$

8. Passes through $(-4, 0)$ and $m = \frac{3}{5}$. $y = \frac{3}{5}x + \frac{12}{5}$

9. Passes through $(3.4, 2.6)$ and $m = 1.2$. $y = 1.2x - 1.48$

10. Passes through $(2.2, 4.1)$ and $m = 2.4$. $y = 2.4x - 1.18$

11. Passes through $(6, 2)$ and $(-3, 1)$. $y = \frac{1}{9}x + \frac{4}{3}$

12. Passes through $(-4, 8)$ and $(-7, -3)$. $y = \frac{11}{3}x + \frac{68}{3}$

13. Passes through $(0, 8)$ and $(5, 0)$. $y = -\frac{8}{5}x + 8$

14. Passes through $(0, -6)$ and $(11, 0)$. $y = \frac{6}{11}x - 6$

15. Passes through $(2.3, 5.1)$ and $(1.9, 3.7)$. $y = 3.5x - 2.95$

16. Passes through $(1.6, 4.8)$ and $(0.8, 6)$. $y = -1.5x + 7.2$

17. Passes through $(3, -4)$ and $m = 0$. $y = -4$

18. Passes through $(-5, 1)$ and $m = 0$. $y = 1$

19. Passes through $\left(\frac{2}{3}, \frac{1}{5}\right)$ and the slope is undefined. $x = \frac{2}{3}$

20. Passes through $\left(-\frac{4}{7}, \frac{3}{10}\right)$ and the slope is undefined. $x = -\frac{4}{7}$

21. Given a line defined by $x = 4$, what is the slope of the line?   Undefined

22. Given a line defined by $y = -2$, what is the slope of the line?   0

Objective 2: Determine the Slopes of Parallel and Perpendicular Lines

For Exercises 23–28, the slope of a line is given.

a. Determine the slope of a line parallel to the given line, if possible.

b. Determine the slope of a line perpendicular to the given line, if possible.

23. $m = \frac{3}{11}$ a. $m = \frac{3}{11}$ b. $m = -\frac{11}{3}$

24. $m = \frac{6}{7}$ a. $m = \frac{6}{7}$ b. $m = -\frac{7}{6}$

25. $m = -6$ a. $m = -6$ b. $m = \frac{1}{6}$

26. $m = -10$ a. $m = -10$ b. $m = \frac{1}{10}$

27. $m = 1$ a. $m = 1$ b. $m = -1$

28. m is undefined   a. m is undefined   b. $m = 0$

For Exercises 29–36, determine if the lines defined by the given equations are parallel, perpendicular, or neither.

29. $y = 2x - 3$
    $y = -\frac{1}{2}x + 7$ Perpendicular

30. $y = \frac{4}{3}x - 1$
    $y = -\frac{3}{4}x + 5$ Perpendicular

31. $8x - 5y = 3$
    $2x = \frac{5}{4}y + 1$ Parallel

32. $2x + 3y = 7$
    $4x = -6y + 2$ Parallel

33. $2x = 6$
    $5 = y$ Perpendicular

34. $3y = 5$
    $x = 1$ Perpendicular

35. $6x = 7y$
    $\frac{7}{2}x - 3y = 0$ Neither

36. $5y = 2x$
    $\frac{5}{2}x - y = 0$ Neither

For Exercises 37–44, write an equation of the line satisfying the given conditions. Write the answer in slope-intercept form (if possible) and in standard form ($Ax + By = C$) with no fractional coefficients. (See Examples 3–4)

37. Passes through $(2, 5)$ and is parallel to the line defined by $2x + y = 6$. $y = -2x + 9;\ 2x + y = 9$

38. Passes through $(3, -1)$ and is parallel to the line defined by $-3x + y = 4$. $y = 3x - 10;\ 3x - y = 10$

39. Passes through $(6, -4)$ and is perpendicular to the line defined by $x - 5y = 1$. $y = -5x + 26;\ 5x + y = 26$

40. Passes through $(5, 4)$ and is perpendicular to the line defined by $x - 2y = 7$. $y = -2x + 14;\ 2x + y = 14$

41. Passes through $(6, 8)$ and is parallel to the line defined by $3x = 7y + 5$. $y = \frac{3}{7}x + \frac{38}{7};\ 3x - 7y = -38$

42. Passes through $(7, -6)$ and is parallel to the line defined by $2x = 5y - 4$. $y = \frac{2}{5}x - \frac{44}{5};\ 2x - 5y = 44$

43. Passes through $(2.2, 6.4)$ and is perpendicular to the line defined by $2x = 4 - y$. $y = 0.5x + 5.3;\ 5x - 10y = -53$

44. Passes through $(3.6, 1.2)$ and is perpendicular to the line defined by $4x = 9 - y$. $y = 0.25x + 0.3;\ 5x - 20y = -6$

For Exercises 45–50, write an equation of the line that satisfies the given conditions.

45. Passes through $(8, 6)$ and is parallel to the x-axis. $y = 6$

46. Passes through $(-11, 13)$ and is parallel to the y-axis. $x = -11$

47. Passes through $\left(\frac{5}{11}, -\frac{3}{4}\right)$ and is perpendicular to the y-axis. $y = -\frac{3}{4}$

48. Passes through $\left(-\frac{7}{9}, \frac{7}{3}\right)$ and is perpendicular to the x-axis. $x = -\frac{7}{9}$

49. Passes through $(-61.5, 47.6)$ and is parallel to the line defined by $x = -12$. $x = -61.5$

50. Passes through $(-0.004, 0.009)$ and is parallel to the line defined by $y = 6$. $y = 0.009$

Objective 3: Create Linear Functions to Model Data

51. A sales person makes a base salary of \$400 per week plus 12% commission on sales. **(See Example 5)**

   a. Write a linear function to model the sales person's weekly salary $S(x)$ for x dollars in sales. $S(x) = 0.12x + 400$ for $x \geq 0$

   b. Evaluate $S(8000)$ and interpret the meaning in the context of this problem. $S(8000) = 1360$ means that the sales person will make \$1360 if \$8000 in merchandise is sold for the week.

52. At a parking garage in a large city, the charge for parking consists of a flat fee of \$2.00 plus \$1.50/hr.

   a. Write a linear function to model the cost for parking $P(t)$ for t hours. $P(t) = 1.5t + 2$ for $t > 0$

   b. Evaluate $P(1.6)$ and interpret the meaning in the context of this problem. $P(1.6) = 4.4$ means that it costs \$4.40 to park for 1.6 hr (1 hr 36 min).

53. Millage rate is the amount per \$1000 that is often used to calculate property tax. For example, a home with a \$60,000 taxable value in a municipality with a 19 mil tax rate would require $(0.019)(\$60,000) = \$1140$ in property taxes. In one county, homeowners pay a flat tax of \$172 plus a rate of 19 mil on the taxable value of a home.

   a. Write a linear function that represents the total property tax $T(x)$ for a home with a taxable value of x dollars. $T(x) = 0.019x + 172$ for $x > 0$

   b. Evaluate $T(80,000)$ and interpret the meaning in the context of this problem. $T(80,000) = 1692$ means that the property tax is \$1692 for a home with a taxable value of \$80,000.

54. The average water level in a retention pond is 6.8 ft. During a time of drought, the water level decreases at a rate of 3 in./day.

   a. Write a linear function W that represents the water level $W(t)$ (in ft) t days after a drought begins. $W(t) = -0.25t + 6.8$ for $0 \leq t \leq 27.2$

   b. Evaluate $W(20)$ and interpret the meaning in the context of this problem. $W(20) = 1.8$ means that after 20 days of drought, the water level will be 1.8 ft.

For Exercises 55–56, the fixed and variable costs to produce an item are given along with the price at which an item is sold. (See Example 6)

   a. Write a linear cost function that represents the cost $C(x)$ to produce x items.

   b. Write a linear revenue function that represents the revenue $R(x)$ for selling x items.

   c. Write a linear profit function that represents the profit $P(x)$ for producing and selling x items.

   d. Determine the break-even point.

55. Fixed cost: \$2275
   Variable cost per item: \$34.50
   Price at which the item is sold: \$80.00

   a. $C(x) = 34.5x + 2275$
   b. $R(x) = 80x$
   c. $P(x) = 45.5x - 2275$
   d. 50 items

56. Fixed cost: \$5625
   Variable cost per item: \$0.40
   Price at which the item is sold: \$1.30

   a. $C(x) = 0.4x + 5625$
   b. $R(x) = 1.3x$
   c. $P(x) = 0.9x - 5625$
   d. 6250 items

57. The profit function P is shown for producing and selling x items. Determine the values of x for which

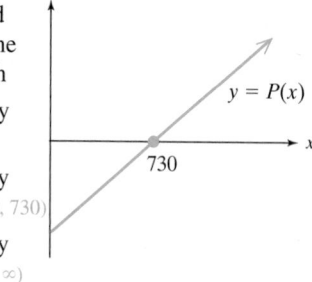

a. $P(x) = 0$ (the company breaks even) {730}

b. $P(x) < 0$ (the company experiences a loss) [0, 730)

c. $P(x) > 0$ (the company makes a profit) (730, ∞)

58. The cost and revenue functions C and R are shown for producing and selling x items. Determine the values of x for which

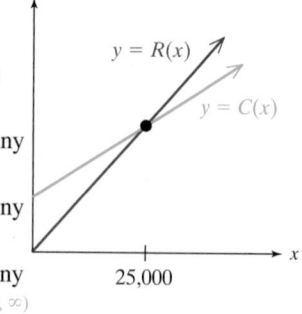

a. $R(x) = C(x)$ (the company breaks even) {25,000}

b. $R(x) < C(x)$ (the company experiences a loss) [0, 25,000)

c. $R(x) > C(x)$ (the company makes a profit) (25,000, ∞)

59. A small business makes cookies and sells them at the farmer's market. The fixed monthly cost for use of a Health Department–approved kitchen and rental space at the farmer's market is $790. The cost of labor, taxes, and ingredients for the cookies amounts to $0.24 per cookie, and the cookies sell for $6.00 per dozen. **(See Example 6)**

a. Write a linear cost function representing the cost $C(x)$ to produce x dozen cookies per month. $C(x) = 2.88x + 790$

b. Write a linear revenue function representing the revenue $R(x)$ for selling x dozen cookies. $R(x) = 6x$

c. Write a linear profit function representing the profit for producing and selling x dozen cookies in a month. $P(x) = 3.12x - 790$

d. Determine the number of cookies (in dozens) that must be produced and sold for a monthly profit. The business will make a profit if it produces and sells 254 dozen or more cookies.

e. If 150 dozen cookies are sold in a given month, how much money will the business make or lose? The business will lose $322.

60. A lawn service company charges $60 for each lawn maintenance call. The fixed monthly cost of $680 includes telephone service and depreciation of equipment. The variable costs include labor, gasoline, and taxes and amount to $36 per lawn.

a. Write a linear cost function representing the monthly cost $C(x)$ for x maintenance calls. $C(x) = 36x + 680$

b. Write a linear revenue function representing the monthly revenue $R(x)$ for x maintenance calls. $R(x) = 60x$

c. Write a linear profit function representing the monthly profit $P(x)$ for x maintenance calls. $P(x) = 24x - 680$

d. Determine the number of lawn maintenance calls needed per month for the company to make money. The business will make a profit if 29 or more lawn maintenance calls are made per month.

e. If 42 maintenance calls are made for a given month, how much money will the lawn service make or lose? The business will make $328.

61. The data in the graph show the wind speed y (in mph) for Hurricane Katrina versus the barometric pressure x (in millibars, mb). (*Source*: NOAA: www.noaa.gov) **(See Example 7)**

a. Use the points (950, 110) and (1000, 50) to write a linear model for these data. $y = -1.2x + 1250$

b. Interpret the meaning of the slope in the context of this problem. $m = -1.2$ mph/mb means that for an increase of 1 mb in pressure, the wind speed decreases by 1.2 mph.

c. Use the model from part (a) to estimate the wind speed for a hurricane with a pressure of 900 mb. 170 mph

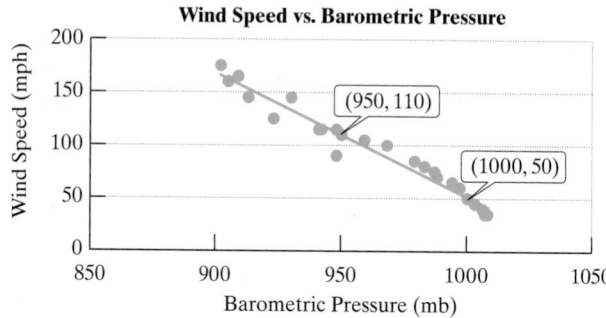

d. The lowest barometric pressure ever recorded for an Atlantic hurricane was 882 mb for Hurricane Wilma in 2005. Would it be reasonable to use the model from part (a) to estimate the wind speed for a hurricane with a pressure of 800 mb? No. There is no guarantee that the linear trend continues outside the interval of the observed data points.

62. Caroline adopted a puppy named Dodger from an animal shelter in Chicago. She recorded Dodger's weight during the first two months. The data in the graph show Dodger's weight y (in lb), x days after adoption.

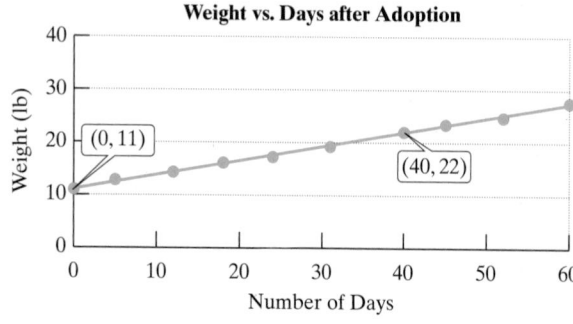

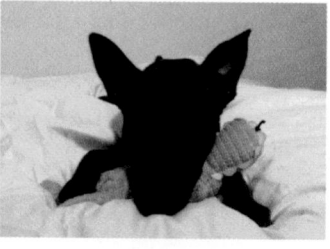

a. Use the points (0, 11) and (40, 22) to write a linear model for these data. $y = 0.275x + 11$

b. Interpret the meaning of the slope in context. $m = 0.275$ means that Dodger's weight increased by 0.275 lb per day during this period.

c. Interpret the meaning of the *y*-intercept in context.   The *y*-intercept is (0, 11) and indicates that Dodger's weight the day
he was adopted was 11 lb.

d. If this linear trend continues during Dodger's growth period, how long will it take Dodger to reach 90% of his expected
full-grown weight of 70 lb? Round to the nearest day.   189 days

e. Is the model from part (a) reasonable long term?   No. Dodger will eventually stop growing (or at least Caroline hopes so).

63. A pediatrician records the age *x* (in yr) and average height *y*
(in inches) for girls between the ages of 2 and 10.

a. Use the points (2, 35) and (6, 46) to write a linear model for
these data.   *y* = 2.75*x* + 29.5

b. Interpret the meaning of the slope in context.   *m* = 2.75 means
that the average height of girls increased by 2.75 in. per year during this time period.

c. Use the model to forecast the average height of 11-yr-old
girls.   59.75 in.

d. If the height of a girl at age 11 is 90% of her full-grown adult
height, use the result of part (c) to estimate the average height
of adult women. Round to the nearest tenth of an inch.
66.4 in.

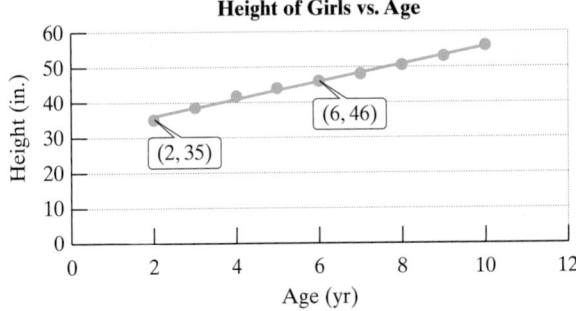

Height of Girls vs. Age

64. The graph shows the number of students enrolled in public
colleges for selected years (*Source*: U.S. National Center for
Education Statistics, www.nces.ed.gov). The *x* variable
represents the number of years since 1990 and the *y* variable
represents the number of students (in millions).

a. Use the points (4, 11.2) and (14, 13.0) to write a linear model
for these data.   *y* = 0.18*x* + 10.48

b. Interpret the meaning of the slope in the context of this
problem.   *m* = 0.18 means that enrollment in public colleges increased
at an average rate of 0.18 million per yr (180,000 per yr).

c. Interpret the meaning of the *y*-intercept in the context of this
problem.   The *y*-intercept is (0, 10.48) and means that in the year 1990, there were approximately 10,480,000 students enrolled in public colleges.

d. In the event that the linear trend continues beyond the last observed data point, use the model in part (a) to predict the
number of students enrolled in public colleges for the year 2020.   Approximately 15.88 million

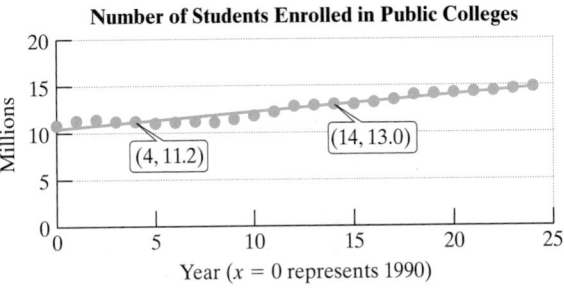

Number of Students Enrolled in Public Colleges

65. The table gives the number of calories and the amount of cholesterol for
selected fast food hamburgers.

a. Graph the data in a scatter plot using the number of calories as the
independent variable *x* and the amount of cholesterol as the dependent
variable *y*.

b. Use the data points (480, 60) and (720, 90) to write a linear function that
defines the amount of cholesterol *c*(*x*) as a linear function of the number of
calories *x*.   *c*(*x*) = 0.125*x*

c. Interpret the meaning of the slope in the context of this problem.   *m* = 0.125
means that the amount of cholesterol increases at an average rate of 0.125 mg per calorie of hamburger.

d. Use the model from part (b) to predict the amount of cholesterol for a
hamburger with 650 calories.   81.25 mg

Hamburger Calories	Cholesterol (mg)
220	35
420	50
460	50
480	60
560	70
590	105
610	65
680	80
720	90

66. The table gives the average gestation period for selected animals and
their corresponding average longevity.

a. Graph the data in a scatter plot using the number of days for gestation
as the independent variable *x* and the longevity as the dependent
variable *y*.

b. Use the data points (44, 8.5) and (620, 35) to write a linear function that
defines longevity *L*(*x*) as a linear function of the length of the gestation
period *x*. Round the slope to 3 decimal places and the *y*-intercept to 2
decimal places.   *L*(*x*) = 0.046*x* + 6.48

c. Interpret the meaning of the slope in the context of this
problem.   *m* = 0.046 means that longevity increases at an average rate of 0.046 yr
per 1 day increase in the gestation period.

d. Use the model from part (b) to predict the longevity for an animal with
an 80-day gestation period. Round to the nearest year.
Approximately 10 yr

Animal	Gestation Period (days)	Longevity (yr)
Rabbit	33	7.0
Squirrel	44	8.5
Fox	57	9.0
Cat	60	11.0
Dog	62	11.0
Lion	109	10.0
Pig	115	10.0
Goat	148	12.0
Horse	337	23.0
Elephant	620	35.0

Objective 4: Create Models Using Linear Regression

For Exercises 67–70, use the scatter plot to determine if a linear regression model appears to be appropriate.

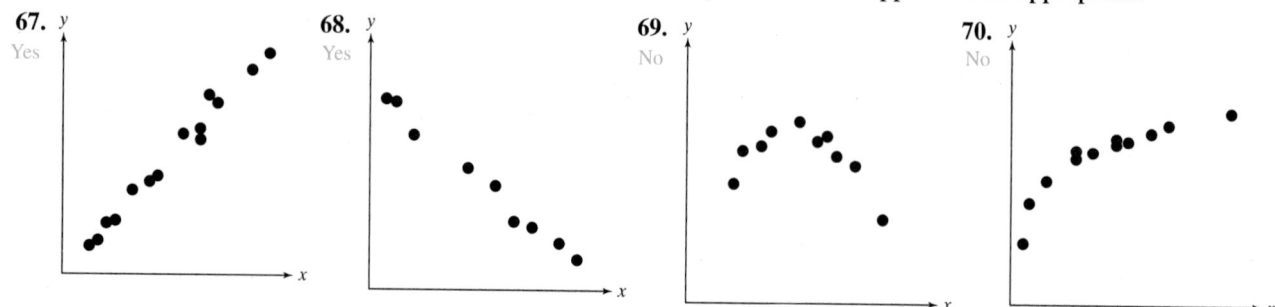

67. *y*   Yes   **68.** *y*   Yes   **69.** *y*   No   **70.** *y*   No

71. The graph in Exercise 61 shows the wind speed *y* (in mph) of a hurricane versus the barometric pressure *x* (in mb). The table gives a partial list of data from the graph. (**See Example 8**)

    a. Use the data in the table to find the least-squares regression line. Round the slope to 2 decimal places and the *y*-intercept to the nearest whole unit.
        $y = -1.22x + 1273$

    b. Use a graphing utility to graph the regression line and the observed data.

    c. Use the model in part (a) to approximate the wind speed of a hurricane with a barometric pressure of 900 mb.   175 mph

    d. By how much do the results of part (c) differ from the result of Exercise 61(c)?
        5 mph

Barometric Pressure (mb) (*x*)	Wind Speed (mph) (*y*)
1007	35
1003	45
1000	50
994	65
983	80
968	100
950	110
930	145
905	160

72. The graph in Exercise 62 shows the weight of Dodger, a puppy recently adopted from an animal shelter. The data in the table give Dodger's weight *y* (in lb), *x* days after adoption.

    a. Use the data in the table to find the least-squares regression line. Round the slope to 2 decimal places and the *y*-intercept to 1 decimal place.
        $y = 0.27x + 11.1$

    b. Use a graphing utility to graph the regression line and the observed data.

    c. Use the model in part (a) to approximate the time required for Dodger to reach 90% of his full-grown weight of 70 lb. Round to the nearest day.   192 days

    d. By how much do the results of part (c) differ from the result of Exercise 62(d)?   3 days

Number of Days (*x*)	Weight (lb) (*y*)
0	11.0
5	12.8
12	14.3
18	16.1
24	17.2
31	19.2
40	22.0
45	23.4
52	24.7
60	27.5

73. The graph in Exercise 63 shows the average height of girls based on their age. The data in the table give the average height *y* (in inches) for girls of age *x* (in yr).

    a. Use the data in the table to find the least-squares regression line. Round the slope to 2 decimal places and the *y*-intercept to 1 decimal place.
        $y = 2.48x + 31.0$

    b. Use a graphing utility to graph the regression line and the observed data.

    c. Use the model in part (a) to approximate the average height of 11-yr-old girls.
        58.28 in.

    d. If the height of a girl at age 11 is 90% of her full-grown adult height, use the result of part (c) to estimate the average height of adult women. Round to the nearest tenth of an inch.   64.8 in.

    e. By how much do the results of part (d) differ from the result of Exercise 63(d)?   1.6 in.

Age (yr) (*x*)	Height (in.) (*y*)
2	35.00
3	38.50
4	41.75
5	44.00
6	46.00
7	48.00
8	50.50
9	53.00
10	56.00

74. The graph in Exercise 64 shows the number of students y enrolled in public colleges for selected years x, where x is the number of years since 1990. The table gives a partial list of data from the graph.

Number of Years Since 1990 (x)	Enrollment (millions) (y)
0	10.8
4	11.2
8	11.1
12	12.8
16	13.2
20	14.2
24	14.8

   a. Use the data in the table to find the least-squares regression line. Round the slope to 2 decimal places and the y-intercept to 1 decimal place. $y = 0.18x + 10.4$

   b. Use a graphing utility to graph the regression line and the observed data.

   c. Assuming that the linear trend continues, use the model from part (a) to predict the number of students enrolled in public colleges for the year 2020.   15.8 million

   d. By how much do the results of part (c) differ from the result of Exercise 64(d)?   0.08 million (or 80,000)

75. The data in Exercise 65 give the amount of cholesterol y for a hamburger with x calories.

   a. Use these data to find the least-squares regression line. Round the slope to 3 decimal places and the y-intercept to 2 decimal places. $y = 0.118x + 4.97$

   b. Use a graphing utility to graph the regression line and the observed data.

   c. Use the regression line to predict the amount of cholesterol in a hamburger with 650 calories. Round to the nearest milligram.   Approximately 82 mg

76. The data in Exercise 66 give the average gestation period x (in days) for selected animals and their corresponding average longevity y (in yr).

   a. Use these data to find the least-squares regression line. Round the slope to 3 decimal places and the y-intercept to 2 decimal places. $y = 0.046x + 6.28$

   b. Use a graphing utility to graph the regression line and the observed data.

   c. Use the regression line to predict the longevity for an animal with an 80-day gestation period. Round to the nearest year.   Approximately 10 yr

Mixed Exercises

77. Suppose that a line passes through the points $(4, -6)$ and $(2, -1)$. Where will it pass through the x-axis? $\left(\frac{8}{5}, 0\right)$

78. Suppose that a line passes through the point $(2, -5)$ and $(-4, 7)$. Where will it pass through the x-axis? $\left(-\frac{1}{2}, 0\right)$

79. Write a rule for a linear function $y = f(x)$, given that $f(0) = 4$ and $f(3) = 11$. $f(x) = \frac{7}{3}x + 4$

80. Write a rule for a linear function $y = g(x)$, given that $g(0) = 7$ and $g(-2) = 4$. $g(x) = \frac{3}{2}x + 7$

81. Write a rule for a linear function $y = h(x)$, given that $h(1) = 6$ and $h(-3) = 2$. $h(x) = x + 5$

82. Write a rule for a linear function $y = k(x)$, given that $k(-2) = 10$ and $k(5) = -18$. $k(x) = -4x + 2$

Write About It

83. Explain how you can use slope to determine if two nonvertical lines are parallel or perpendicular.

84. State one application of using the point-slope formula.
The point-slope formula is used to construct an equation of a line if a point on the line and the slope of the line are known.

85. Explain how cost and revenue are related to profit.
Profit is equal to revenue minus cost.

86. Explain how to determine the break-even point.
The break-even point is found by determining where revenue equals cost or where profit equals zero.

Expanding Your Skills

87. Find an equation of line L. $y = -x + 4$

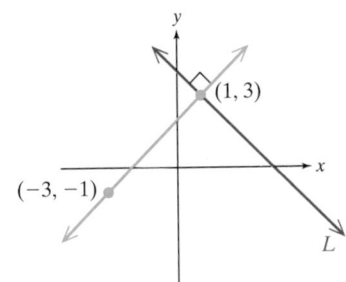

88. In geometry, it is known that the tangent line to a circle at a given point A on the circle is perpendicular to the radius drawn to point A. Suppose that line L is tangent to the given circle at the point $(4, 3)$. Write an equation representing line L.

$y = -\frac{4}{3}x + \frac{25}{3}$

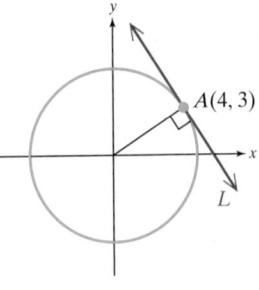

89. In calculus, we can show that the slope of the line drawn tangent to the curve $y = x^3 + 1$ at the point $(c, c^3 + 1)$ is given by $3c^2$. Find an equation of the line tangent to $y = x^3 + 1$ at the point $(-2, -7)$.

$y = 12x + 17$

90. In calculus, we can show that the slope of the line drawn tangent to the curve $y = \frac{1}{x}$ at the point $\left(c, \frac{1}{c}\right)$ is given by $-\frac{1}{c^2}$. Find an equation of the line tangent to $y = \frac{1}{x}$ at the point $\left(2, \frac{1}{2}\right)$.

$y = -\frac{1}{4}x + 1$

For Exercises 91–92, use the fact that a median of a triangle is a line segment drawn from a vertex of the triangle to the midpoint of the opposite side of the triangle.

91. Find an equation of the median of a triangle drawn from vertex $A(5, -2)$ to the side formed by $B(-2, 9)$ and $C(4, 7)$.

$y = -\frac{5}{2}x + \frac{21}{2}$ for $1 \le x \le 5$

92. Find an equation of the median of a triangle drawn from vertex $A(6, -5)$ to the side formed by $B(-4, 1)$ and $C(12, 3)$.

$y = -\frac{7}{2}x + 16$ for $4 \le x \le 6$

PROBLEM RECOGNITION EXERCISES

Comparing Graphs of Equations

In Section 1.6, we will learn additional techniques to graph functions by recognizing characteristics of the functions. In many cases, we can also graph families of functions by relating them to one of several basic graphs. To prepare for the discussion in Section 1.6, use a graphing utility or plot points to graph the basic functions in Exercises 1–8.

1. $y = 1$

2. $y = x$

3. $y = x^2$

4. $y = x^3$

5. $y = \sqrt{x}$

6. $y = \sqrt[3]{x}$

7. $y = |x|$

8. $y = \dfrac{1}{x}$

For Exercises 9–18, graph the functions by plotting points or by using a graphing utility. Explain how the graphs are related.

9. a. $f(x) = x^2$
   b. $g(x) = x^2 + 2$
   c. $h(x) = x^2 - 4$

10. a. $f(x) = |x|$
   b. $g(x) = |x| + 2$
   c. $h(x) = |x| - 4$

11. a. $f(x) = \sqrt{x}$
   b. $g(x) = \sqrt{x - 2}$
   c. $h(x) = \sqrt{x + 4}$

12. a. $f(x) = x^2$
   b. $g(x) = (x - 2)^2$
   c. $h(x) = (x + 3)^2$

13. a. $f(x) = |x|$
   b. $g(x) = -|x|$

14. a. $f(x) = \sqrt{x}$
   b. $g(x) = -\sqrt{x}$

15. a. $f(x) = x^2$
   b. $g(x) = \frac{1}{2}x^2$
   c. $h(x) = 2x^2$

16. a. $f(x) = |x|$
   b. $g(x) = \frac{1}{3}|x|$
   c. $h(x) = 3|x|$

17. a. $f(x) = \sqrt{x}$
   b. $g(x) = \sqrt{-x}$

18. a. $f(x) = \sqrt[3]{x}$
   b. $g(x) = \sqrt[3]{-x}$

SECTION 1.6   Transformations of Graphs

OBJECTIVES

1. Recognize Basic Functions
2. Apply Vertical and Horizontal Translations (Shifts)
3. Apply Vertical and Horizontal Shrinking and Stretching
4. Apply Reflections Across the *x*- and *y*-Axes
5. Summarize Transformations of Graphs

1. Recognize Basic Functions

A function defined by $f(x) = mx + b$ is a linear function and its graph is a line in a rectangular coordinate system. In addition to linear functions, we will learn to identify other categories of functions and the shapes of their graphs (Table 1-2).

Table 1-2   Basic Functions and Their Graphs

1. Linear functions   Constant functions
$f(x) = mx + b$ $f(x) = b$

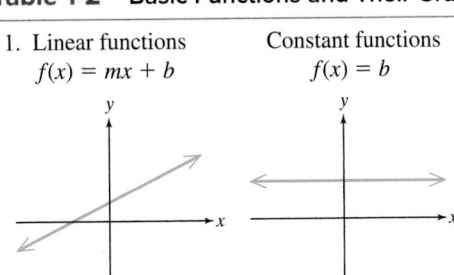

2. Identity function: $f(x) = x$

x	$f(x)$
-2	-2
-1	-1
0	0
1	1
2	2

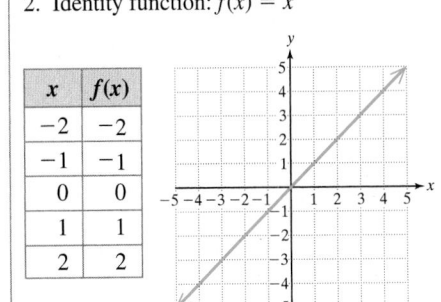

3. Quadratic function: $f(x) = x^2$
(graph is a parabola)

x	$f(x)$
-2	4
-1	1
0	0
1	1
2	4

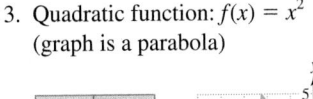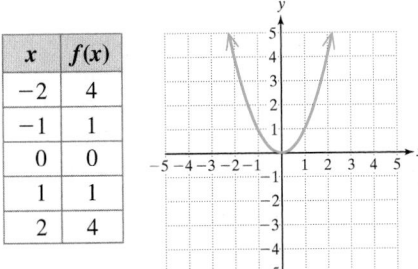

4. Cube function: $f(x) = x^3$

x	$f(x)$
-2	-8
-1	-1
0	0
1	1
2	8

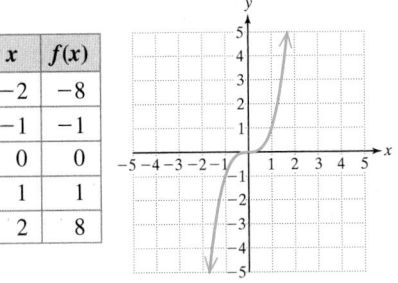

5. Square root function: $f(x) = \sqrt{x}$

x	$f(x)$
0	0
1	1
4	2
9	3
16	4

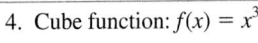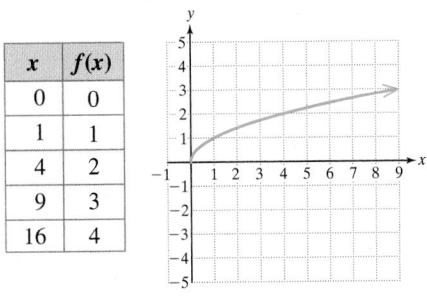

6. Cube root function: $f(x) = \sqrt[3]{x}$

x	$f(x)$
-8	-2
-1	-1
0	0
1	1
8	2

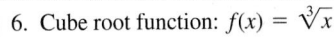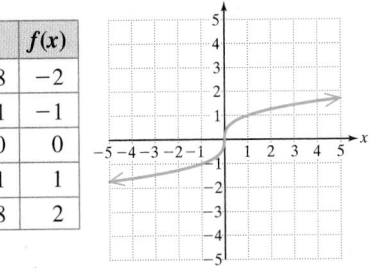

7. Absolute value function: $f(x) = |x|$

x	$f(x)$
-2	2
-1	1
0	0
1	1
2	2

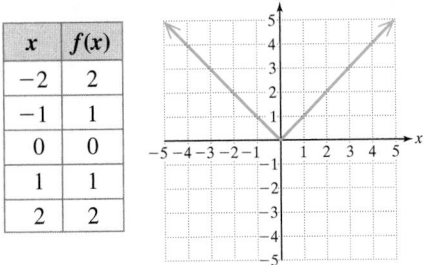

8. Reciprocal function: $f(x) = \dfrac{1}{x}$

x	$f(x)$
-2	$-\frac{1}{2}$
-1	-1
$-\frac{1}{2}$	-2
$\frac{1}{2}$	2
1	1
2	$\frac{1}{2}$

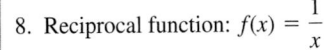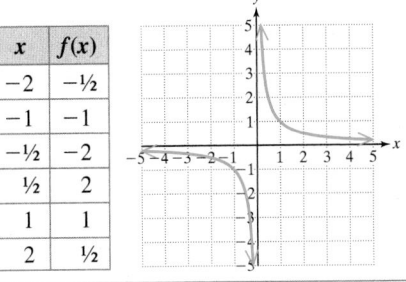

TIP The functions given in Table 1-2 were introduced in Section 1.1, Exercises 31–36, and in the Problem Recognition Exercises on page 182.

Notice that the graph of $f(x) = \frac{1}{x}$ gets close to (but never touches) the y-axis as x gets close to zero. Likewise, as x approaches ∞ and $-\infty$, the graph approaches the x-axis without touching the x-axis. The x- and y-axes are called **asymptotes** of f and will be studied in detail in Section 2.5.

2. Apply Vertical and Horizontal Translations (Shifts)

We will call the eight basic functions pictured in Table 1-2 "parent" functions. Other functions that share the characteristics of a parent function are grouped as a "family" of functions. For example, consider the functions defined by $g(x) = x^2 + 2$ and $h(x) = x^2 - 4$, pictured in Figure 1-25.

x	$f(x) = x^2$	$g(x) = x^2 + 2$	$h(x) = x^2 - 4$
-3	9	11	5
-2	4	6	0
-1	1	3	-3
0	0	2	-4
1	1	3	-3
2	4	6	0
3	9	11	5

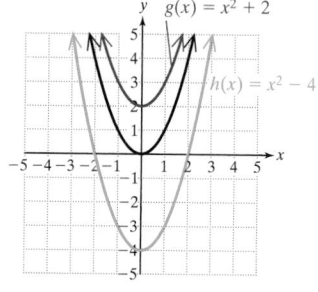

Figure 1-25

The graphs of g and h both resemble the graph of $f(x) = x^2$, but are shifted vertically upward or downward. The table of points reveals that for corresponding x values, the values of $g(x)$ are 2 more than the values of $f(x)$. Thus, the graph is shifted *upward* 2 units. Likewise, the values of $h(x)$ are 4 less than the values of $f(x)$ and the graph is shifted *downward* 4 units. Such shifts are called **translations.** These observations are consistent with the following rules.

> **TIP** For each ordered pair (x, y) on the graph of $y = f(x)$, the corresponding point
> - $(x, y + k)$ is on the graph of $y = f(x) + k$.
> - $(x, y - k)$ is on the graph of $y = f(x) - k$.

Vertical Translations of Graphs

Consider a function defined by $y = f(x)$. Let k represent a positive real number.

- The graph of $y = f(x) + k$ is the graph of $y = f(x)$ shifted k units *upward*.
- The graph of $y = f(x) - k$ is the graph of $y = f(x)$ shifted k units *downward*.

Classroom Examples: p. 193
Exercises 16, 18

EXAMPLE 1 Translating a Graph Vertically

Use translations to graph the given functions.

  a. $g(x) = |x| - 3$ **b.** $h(x) = x^3 + 2$

Solution:

a.

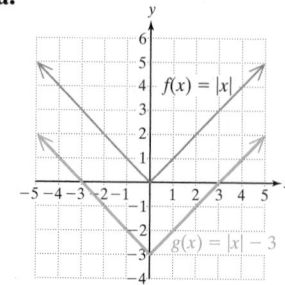

The parent function for $g(x) = |x| - 3$ is $f(x) = |x|$.

The graph of g (shown in blue) is the graph of f shifted *downward* 3 units. For example the point $(0, 0)$ on the graph of $f(x) = |x|$ corresponds to $(0, -3)$ on the graph of $g(x) = |x| - 3$.

b.

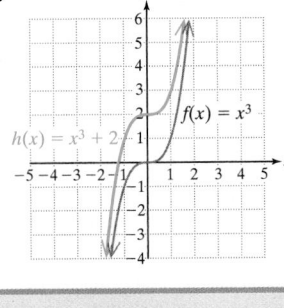

The parent function for $h(x) = x^3 + 2$ is $f(x) = x^3$.

The graph of h (shown in blue) is the graph of f shifted *upward* 2 units. For example:

The point $(0, 0)$ on the graph of $f(x) = x^3$ corresponds to $(0, 2)$ on the graph of $h(x) = x^3 + 2$.

The point $(1, 1)$ on the graph of $f(x) = x^3$ corresponds to $(1, 3)$ on the graph of $h(x) = x^3 + 2$.

Skill Practice 1   Use translations to graph the given functions.

a. $g(x) = \sqrt{x} - 2$ **b.** $h(x) = \sqrt{x} + 3$

The graph of a function will be shifted to the right or left if a constant is added to or subtracted from the input variable x. In Example 2, we consider $g(x) = (x + 3)^2$.

Classroom Example: p. 193
Exercise 20

Instructor Note:
As an alternative to identifying the x-intercept of the graph of g, have the students construct a table of points for $f(x) = x^2$ and $g(x) = (x + 3)^2$ for several values of x.

EXAMPLE 2   **Translating a Graph Horizontally**

Graph the function defined by $g(x) = (x + 3)^2$.

Solution:

Because 3 is added to the x variable, we might expect the graph of $g(x) = (x + 3)^2$ to be the same as the graph of $f(x) = x^2$, but shifted in the x direction (horizontally). To determine whether the shift is to the left or right, we can locate the x-intercept of the graph of $g(x) = (x + 3)^2$. Substituting 0 for $g(x)$, we have:

$$0 = (x + 3)^2$$
$$x = -3 \qquad \text{The } x\text{-intercept is } (-3, 0).$$

Therefore, the new x-intercept (and also the vertex of the parabola) is $(-3, 0)$. This means that the graph is shifted to the left.

Skill Practice 2   Graph the function defined by $g(x) = |x + 2|$.

Using similar logic as in Example 2, we can show that the graph of $h(x) = (x - 3)^2$ is the graph of $f(x) = x^2$ translated to the *right* 3 units. These observations are consistent with the following rules.

Answers

1. a.–b.

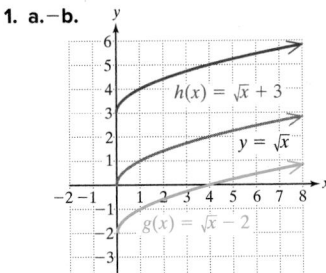

2.

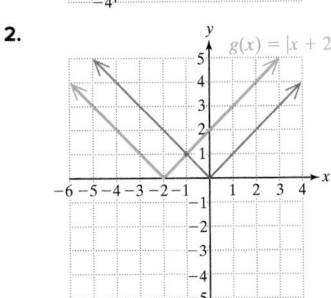

Horizontal Translations of Graphs

Consider a function defined by $y = f(x)$. Let h represent a positive real number.

- The graph of $y = f(x - h)$ is the graph of $y = f(x)$ shifted h units to the *right*.
- The graph of $y = f(x + h)$ is the graph of $y = f(x)$ shifted h units to the *left*.

TIP Consider a positive real number h. To graph $y = f(x - h)$ or $y = f(x + h)$, shift the graph of $y = f(x)$ horizontally in the opposite direction of the sign within parentheses. The graph of $y = f(x - h)$ is a shift in the positive x direction. The graph of $y = f(x + h)$ is a shift in the negative x direction.

Classroom Example: p. 193
Exercise 24

EXAMPLE 3 Translating a Function Horizontally and Vertically

Use translations to graph the function defined by $p(x) = \sqrt{x - 3} - 2$.

Solution:

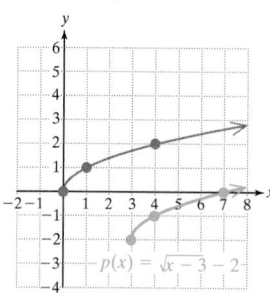

The parent function for $p(x) = \sqrt{x - 3} - 2$ is $f(x) = \sqrt{x}$.

The graph of p (shown in blue) is the graph of f shifted to the right 3 units and downward 2 units. We can plot several strategic points as an outline for the new curve.

- The point $(0, 0)$ on the graph of f corresponds to $(0 + 3, 0 - 2) = (3, -2)$ on the graph of p.
- The point $(1, 1)$ on the graph of f corresponds to $(1 + 3, 1 - 2) = (4, -1)$ on the graph of p.
- The point $(4, 2)$ on the graph of f corresponds to $(4 + 3, 2 - 2) = (7, 0)$ on the graph of p.

Skill Practice 3 Use translations to graph the function defined by $q(x) = \sqrt{x + 2} - 5$.

3. Apply Vertical and Horizontal Shrinking and Stretching

Horizontal and vertical translations of functions are called **rigid transformations** because the shape of the graph is not affected. We now look at **nonrigid transformations**. These operations cause a distortion of the graph (either an elongation or contraction in the horizontal or vertical direction). We begin by investigating the functions defined by $y = f(x)$ and $y = a \cdot f(x)$, where a is a positive real number.

Classroom Examples: p. 194
Exercises 28, 30

EXAMPLE 4 Graphing a Function with a Vertical Stretch or Shrink

Graph the functions.

  a. $f(x) = |x|$ **b.** $g(x) = 2|x|$ **c.** $h(x) = \dfrac{1}{2}|x|$

Solution:

Answers

3.

graph showing $y = \sqrt{x}$ and $q(x) = \sqrt{x + 2} - 5$

4.

graph showing $y = x^2$, $g(x) = 3x^2$, and $h(x) = \frac{1}{3}x^2$

x	$f(x) = \|x\|$	$g(x) = 2\|x\|$	$h(x) = \frac{1}{2}\|x\|$
-3	3	6	$\frac{3}{2}$
-2	2	4	1
-1	1	2	$\frac{1}{2}$
0	0	0	0
1	1	2	$\frac{1}{2}$
2	2	4	1
3	3	6	$\frac{3}{2}$

double

multiply by $\frac{1}{2}$

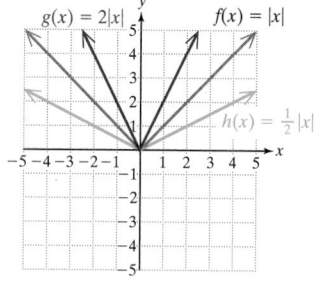

For a given value of x, the value of $g(x)$ is twice the value of $f(x)$. Therefore, the graph of g is elongated or stretched vertically by a factor of 2.

For a given value of x, the value of $h(x)$ is one-half that of $f(x)$. Therefore, the graph of h is shrunk vertically.

Skill Practice 4 Graph the functions.

  a. $f(x) = x^2$ **b.** $g(x) = 3x^2$ **c.** $h(x) = \dfrac{1}{3}x^2$

Vertical Shrinking and Stretching of Graphs

Consider a function defined by $y = f(x)$. Let a represent a positive real number.

- If $a > 1$, then the graph of $y = af(x)$ is the graph of $y = f(x)$ stretched vertically by a factor of a.
- If $0 < a < 1$, then the graph of $y = af(x)$ is the graph of $y = f(x)$ shrunk vertically by a factor of a.

Note: For any point (x, y) on the graph of $y = f(x)$, the point (x, ay) is on the graph of $y = af(x)$.

A function may also be stretched or shrunk horizontally.

Horizontal Shrinking and Stretching of Graphs

Consider a function defined by $y = f(x)$. Let a represent a positive real number.

- If $a > 1$, then the graph of $y = f(ax)$ is the graph of $y = f(x)$ shrunk horizontally by a factor of $\frac{1}{a}$.
- If $0 < a < 1$, then the graph of $y = f(ax)$ is the graph of $y = f(x)$ stretched horizontally by a factor of $\frac{1}{a}$.

Note: For any point (x, y) on the graph of $y = f(x)$, the point $\left(\frac{x}{a}, y\right)$ is on the graph of $y = f(ax)$.

A point (x, y) on the graph of $y = f(x)$ corresponds to the point $\left(\frac{x}{a}, y\right)$ on the graph of $y = f(ax)$. Since the x-coordinate is multiplied by the *reciprocal* of a, values of a greater than 1 actually compress (shrink) the graph horizontally toward the y-axis. Values of a between 0 and 1 *stretch* the graph horizontally away from the y-axis. This is demonstrated in Example 5.

Classroom Examples: p. 194
Exercises 36, 40

EXAMPLE 5   **Graphing a Function with a Horizontal Shrink or Stretch**

The graph of $y = f(x)$ is shown. Graph

a. $y = f(2x)$

b. $y = f\left(\frac{1}{2}x\right)$

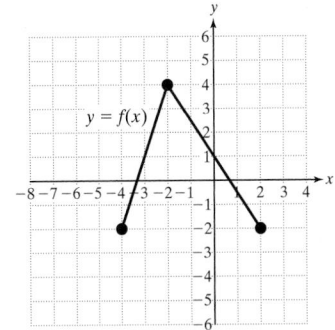

Solution:

a. $f(2x)$ is in the form $f(ax)$ with $a = 2 > 1$. The graph of $y = f(2x)$ is the graph of $y = f(x)$ horizontally compressed.

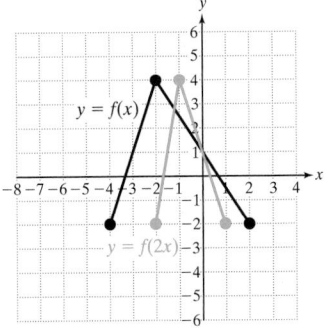

The graph of f has the following "strategic" points that define the shape of the function: $(-4, -2)$, $(-2, 4)$, and $(2, -2)$.

To graph $y = f(2x)$, divide each x value by 2.

$(-4, -2)$ becomes $\left(\frac{-4}{2}, -2\right) = (-2, -2)$.

$(-2, 4)$ becomes $\left(\frac{-2}{2}, 4\right) = (-1, 4)$.

$(2, -2)$ becomes $\left(\frac{2}{2}, -2\right) = (1, -2)$.

The graph of $y = f(2x)$ is shown in blue.

TIP Dividing the x values by $\frac{1}{2}$ is the same as multiplying the x values by 2.

b. $f\left(\frac{1}{2}x\right)$ is in the form $f(ax)$ with $a = \frac{1}{2}$. The graph of $y = f\left(\frac{1}{2}x\right)$ is the graph of $y = f(x)$ stretched horizontally.

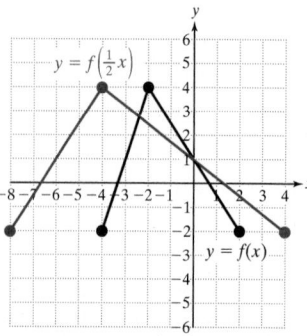

To graph $y = f\left(\frac{1}{2}x\right)$, divide each x value on the graph of $y = f(x)$ by $\frac{1}{2}$. For example:

$(-4, -2)$ becomes $\left(\dfrac{-4}{\frac{1}{2}}, -2\right) = (-8, -2)$.

$(-2, 4)$ becomes $\left(\dfrac{-2}{\frac{1}{2}}, 4\right) = (-4, 4)$.

$(2, -2)$ becomes $\left(\dfrac{2}{\frac{1}{2}}, -2\right) = (4, -2)$.

The graph of $y = f\left(\frac{1}{2}x\right)$ is shown in red.

Skill Practice 5   The graph of $y = f(x)$ is shown.

Graph.   **a.** $y = f(2x)$ **b.** $y = f\left(\dfrac{1}{2}x\right)$

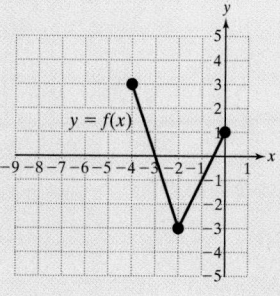

4. Apply Reflections Across the x- and y-Axes

The graphs of $f(x) = x^2$ (in black) and $g(x) = -x^2$ (in blue) are shown in Figure 1-26. Notice that a point (x, y) on the graph of f corresponds to the point $(x, -y)$ on the graph of g. Therefore, the graph of g is the graph of f reflected across the x-axis.

The graphs of $f(x) = \sqrt{x}$ (in black) and $g(x) = \sqrt{-x}$ (in blue) are shown in Figure 1-27. Notice that a point (x, y) on the graph of f corresponds to the point $(-x, y)$ on g. Therefore, the graph of g is the graph of f reflected across the y-axis.

Answers

5. a.

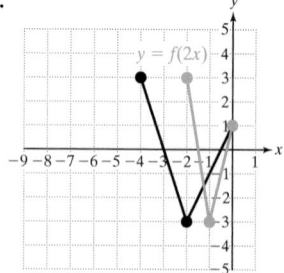

b.

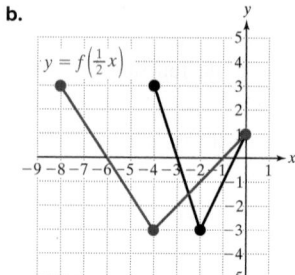

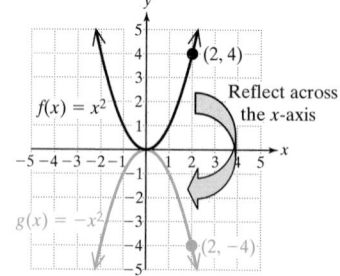

Figure 1-26

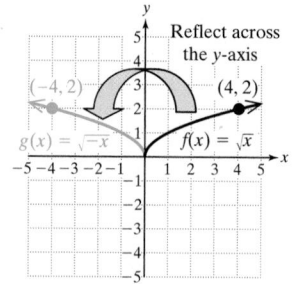

Figure 1-27

> **Reflections Across the *x*- and *y*-Axes**
>
> Consider a function defined by $y = f(x)$.
>
> - The graph of $y = -f(x)$ is the graph of $y = f(x)$ reflected across the *x*-axis.
> - The graph of $y = f(-x)$ is the graph of $y = f(x)$ reflected across the *y*-axis.

Classroom Examples: p. 194
Exercises 48, 50

EXAMPLE 6   **Reflecting the Graph of a Function Across the *x*- and *y*-Axes**

The graph of $y = f(x)$ is given.

  a. Graph $y = -f(x)$.
  b. Graph $y = f(-x)$.

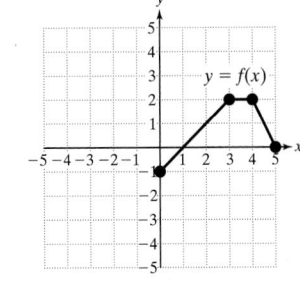

TIP For a given point (x, y), notice that $(-x, y)$ is on the opposite side of and equidistant to the *y*-axis. Likewise, $(x, -y)$ is on the opposite side of and equidistant from the *x*-axis as (x, y).

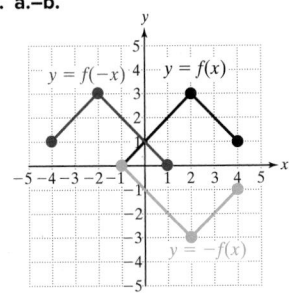

Solution:

  a. Reflect $y = f(x)$ across the *x*-axis.

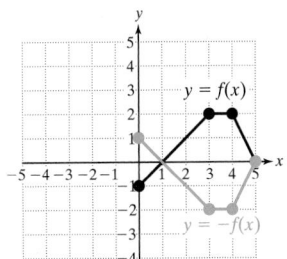

  b. Reflect $y = f(x)$ across the *y*-axis.

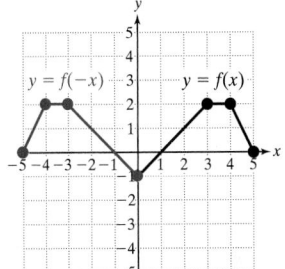

Skill Practice 6   The graph of $y = f(x)$ is given.

  a. Graph $y = -f(x)$.
  b. Graph $y = f(-x)$.

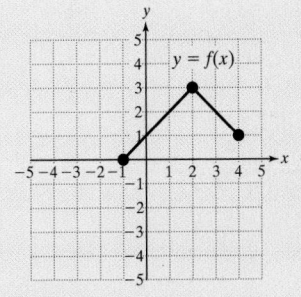

5. Summarize Transformations of Graphs

The operations of reflecting a graph of a function about an axis and shifting, stretching, and shrinking a graph are called **transformations**. Transformations give us tools to graph families of functions that are built from basic "parent" functions.

Answers

6. a.–b.

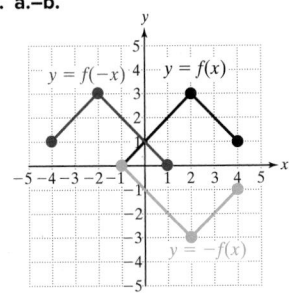

Transformations of Functions

Consider a function defined by $y = f(x)$. If h, k, and a represent positive real numbers, then the graphs of the following functions are related to $y = f(x)$ as follows.

Transformation	Effect on the Graph of f	Changes to Points on f
Vertical translation (shift) $y = f(x) + k$ $y = f(x) - k$	Shift upward k units Shift downward k units	Replace (x, y) by $(x, y + k)$. Replace (x, y) by $(x, y - k)$.
Horizontal translation (shift) $y = f(x - h)$ $y = f(x + h)$	Shift to the right h units Shift to the left h units	Replace (x, y) by $(x + h, y)$. Replace (x, y) by $(x - h, y)$.
Vertical stretch/shrink $y = a[f(x)]$	Vertical stretch (if $a > 1$) Vertical shrink (if $0 < a < 1$) Graph is stretched/shrunk vertically by a factor of a.	Replace (x, y) by (x, ay).
Horizontal stretch/shrink $y = f(a \cdot x)$	Horizontal shrink (if $a > 1$) Horizontal stretch (if $0 < a < 1$) Graph is shrunk/stretched horizontally by a factor of $\frac{1}{a}$.	Replace (x, y) by $\left(\frac{x}{a}, y\right)$.
Reflection $y = -f(x)$ $y = f(-x)$	Reflection across the x-axis Reflection across the y-axis	Replace (x, y) by $(x, -y)$. Replace (x, y) by $(-x, y)$.

When graphing a function requiring multiple transformations on the parent function, it is important to follow the correct sequence of steps.

Steps for Graphing Multiple Transformations of Functions

To graph a function requiring multiple transformations, use the following order.

1. Horizontal translation (shift)
2. Horizontal and vertical stretch and shrink
3. Reflections across the x- and y-axes
4. Vertical translation (shift)

EXAMPLE 7   Using Transformations to Graph a Function

Use transformations to graph the function defined by $n(x) = -\frac{1}{2}(x-2)^2 + 3$.

Solution:

The graph of $n(x) = -\frac{1}{2}(x-2)^2 + 3$ is the same as the graph of $f(x) = x^2$, with four transformations in the following order.

1. Shift the graph to the right 2 units.
2. Apply a vertical shrink (multiply the y values by $\frac{1}{2}$).
3. Reflect the graph over the x-axis.
4. Shift the graph upward 3 units.

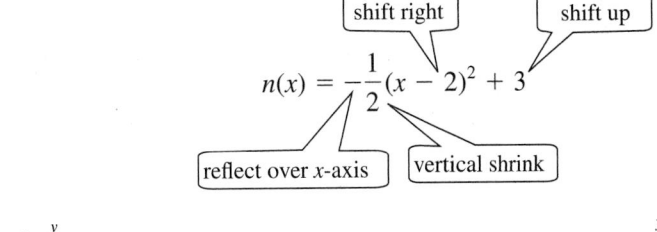

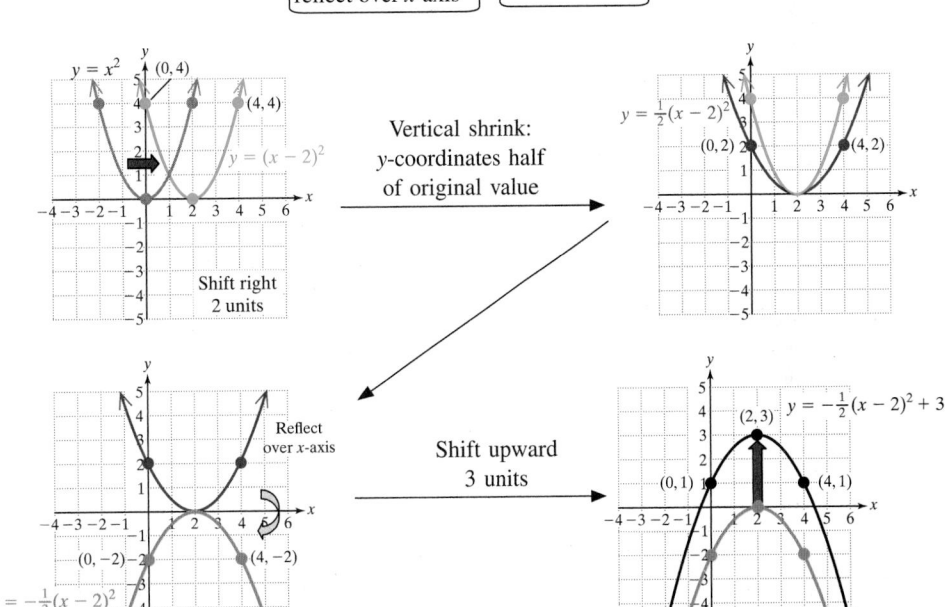

Avoiding Mistakes

As a means to check the graph of $y = n(x)$, substitute the x-coordinates of the strategic points $(0, 1)$, $(2, 3)$, and $(4, 1)$ into the function.

$n(0) = -\frac{1}{2}(0-2)^2 + 3 = 1$ ✓
$n(2) = -\frac{1}{2}(2-2)^2 + 3 = 3$ ✓
$n(4) = -\frac{1}{2}(4-2)^2 + 3 = 1$ ✓

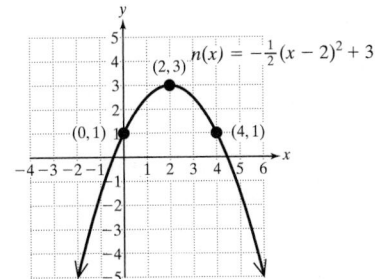

Classroom Example: p. 195
Exercise 72

Skill Practice 7   Use transformations to graph the function defined by $m(x) = -3|x - 2| - 4$.

EXAMPLE 8   Using Transformations to Graph a Function

Use transformations to graph the function defined by $v(x) = -\sqrt{-x + 2}$.

Solution:

The graph of $v(x) = -\sqrt{-x + 2}$ is the same as the graph of $f(x) = \sqrt{x}$, with three transformations in the following order.

1. Shift the graph to the left 2 units.
2. Reflect the graph across the y-axis.
3. Reflect the graph across the x-axis.
   (Note that the reflections in steps 2 and 3 can be applied in either order.)

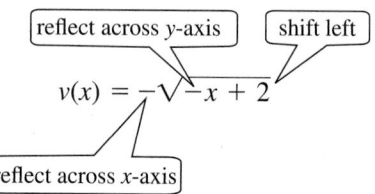

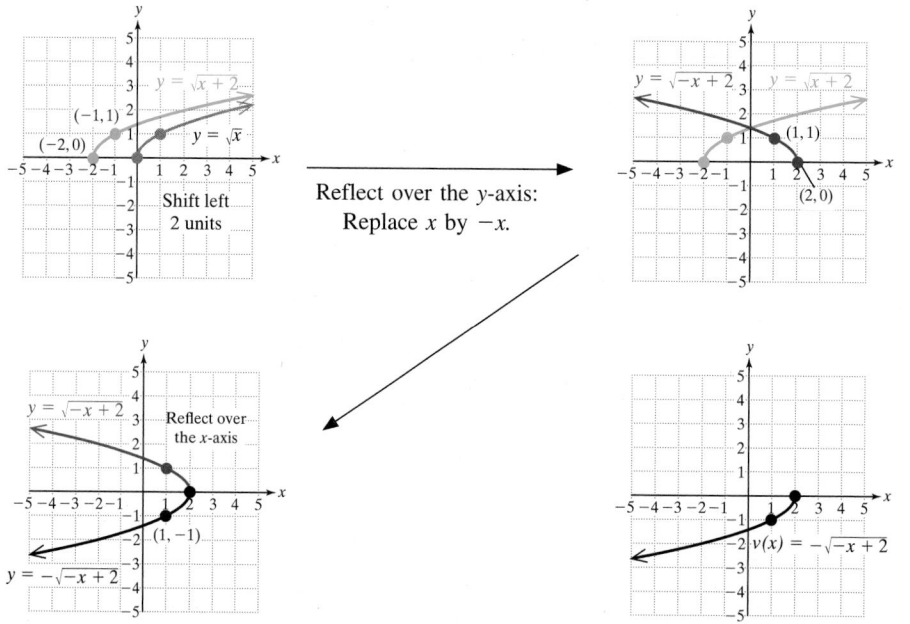

Skill Practice 8   Use transformations to graph the function defined by $r(x) = \sqrt[3]{-x} + 1$.

Answers

7.

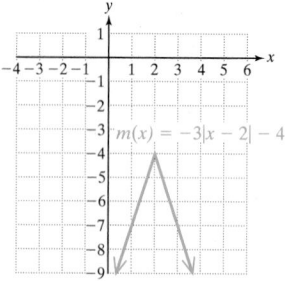

8.

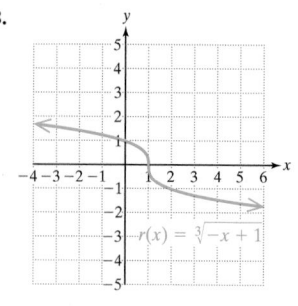

Avoiding Mistakes

Transformations involving a horizontal shrink, stretch, or reflection often introduce confusion when coupled with a horizontal shift. To further illustrate the rationale for the order of steps taken in Example 8, begin with the parent function $y = \sqrt{x}$. Performing a horizontal shift first means that we replace x by $x + 2$. This gives us $y = \sqrt{x + 2}$. Then to perform the reflection across the y-axis, we replace x by $-x$ to get $y = \sqrt{-x + 2}$. Performing these two transformations in the reverse order, would *not* result in the function we want. We would first have $y = \sqrt{-x}$, and then replacing x by $x + 2$ would give $y = \sqrt{-(x + 2)} = \sqrt{-x - 2}$ rather than $y = \sqrt{-x + 2}$.

SECTION 1.6   Practice Exercises

Prerequisite Review

For Exercises R.1–R.3, graph each equation.

R.1. $y = -3x - 1$

R.2. $y = \dfrac{3}{5}x + 2$

R.3. $y = 1$

Concept Connections

1. Let c represent a positive real number. The graph of $y = f(x + c)$ is the graph of $y = f(x)$ shifted (up/down/left/right) c units.   left

2. Let c represent a positive real number. The graph of $y = f(x - c)$ is the graph of $y = f(x)$ shifted (up/down/left/right) c units.   right

3. Let c represent a positive real number. The graph of $y = f(x) - c$ is the graph of $y = f(x)$ shifted (up/down/left/right) c units.   down

4. The graph of $y = 3f(x)$ is the graph of $y = f(x)$ with a (choose one: vertical stretch, vertical shrink, horizontal stretch, horizontal shrink).   vertical stretch

5. The graph of $y = f(3x)$ is the graph of $y = f(x)$ with a (choose one: vertical stretch, vertical shrink, horizontal stretch, horizontal shrink).   horizontal shrink

6. The graph of $y = f\left(\frac{1}{3}x\right)$ is the graph of $y = f(x)$ with a (choose one: vertical stretch, vertical shrink, horizontal stretch, horizontal shrink).   horizontal stretch

7. The graph of $y = \frac{1}{3}f(x)$ is the graph of $y = f(x)$ with a (choose one: vertical stretch, vertical shrink, horizontal stretch, horizontal shrink).   vertical shrink

8. The graph of $y = -f(x)$ is the graph of $y = f(x)$ reflected across the _____ x _____ -axis.

Objective 1: Recognize Basic Functions

For Exercises 9–14, from memory match the equation with its graph.

9. $f(x) = \sqrt{x}$ e

10. $f(x) = \sqrt[3]{x}$ f

11. $f(x) = x^3$ b

12. $f(x) = x^2$ c

13. $f(x) = |x|$ a

14. $f(x) = \dfrac{1}{x}$ d

a.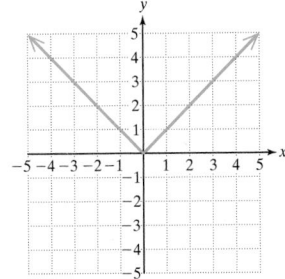

b.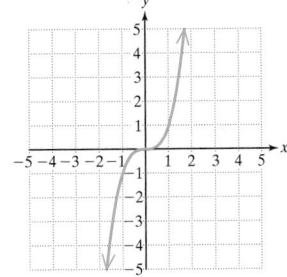

c.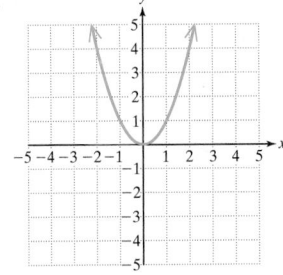

d.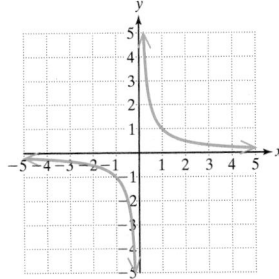

e.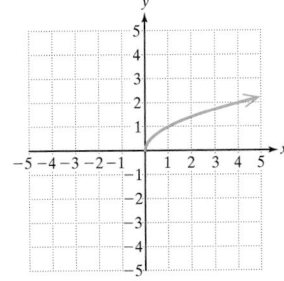

f.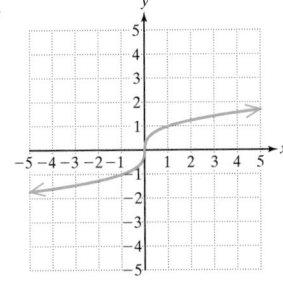

Objective 2: Apply Vertical and Horizontal Translations (Shifts)

For Exercises 15–26, use translations to graph the given functions. (See Examples 1–3)

15. $f(x) = |x| + 1$

16. $g(x) = \sqrt{x} + 2$

17. $k(x) = x^3 - 2$

18. $h(x) = \dfrac{1}{x} - 2$

19. $g(x) = \sqrt{x + 5}$

20. $m(x) = |x + 1|$

21. $r(x) = (x - 4)^2$

22. $t(x) = \sqrt[3]{x - 2}$

23. $a(x) = \sqrt{x + 1} - 3$

24. $b(x) = |x - 2| + 4$

25. $c(x) = \dfrac{1}{x - 3} + 1$

26. $d(x) = \dfrac{1}{x + 4} - 1$

Additional answers can be found in the Instructor Answer Appendix.

Objective 3: Apply Vertical and Horizontal Shrinking and Stretching

For Exercises 27–32, use transformations to graph the functions. (See Example 4)

27. $m(x) = 4\sqrt[3]{x}$

28. $n(x) = 3|x|$

29. $r(x) = \frac{1}{2}x^2$

30. $t(x) = \frac{1}{3}|x|$

31. $p(x) = |2x|$

32. $q(x) = \sqrt{2x}$

For Exercises 33–40, use the graphs of $y = f(x)$ and $y = g(x)$ to graph the given function. (See Example 5)

33. $y = \frac{1}{3}f(x)$

34. $y = \frac{1}{2}g(x)$

35. $y = 3f(x)$

36. $y = 2g(x)$

37. $y = f(3x)$

38. $y = g(2x)$

39. $y = f\left(\frac{1}{3}x\right)$

40. $y = g\left(\frac{1}{2}x\right)$

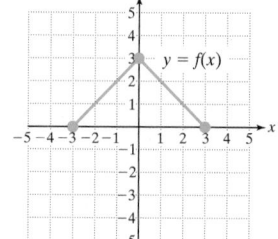 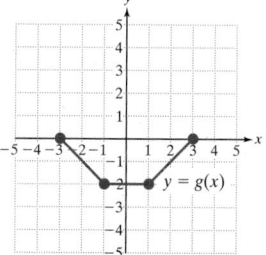

Objective 4: Apply Reflections Across the *x*- and *y*-Axes

For Exercises 41–46, graph the function by applying an appropriate reflection.

41. $f(x) = -\frac{1}{x}$

42. $g(x) = -\sqrt{x}$

43. $h(x) = -x^3$

44. $k(x) = -|x|$

45. $p(x) = (-x)^3$

46. $q(x) = \sqrt[3]{-x}$

For Exercises 47–50, use the graphs of $y = f(x)$ and $y = g(x)$ to graph the given function. (See Example 6)

47. $y = f(-x)$

48. $y = g(-x)$

49. $y = -f(x)$

50. $y = -g(x)$

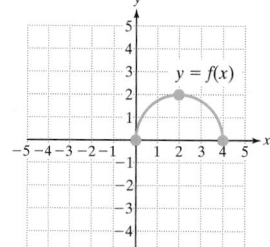 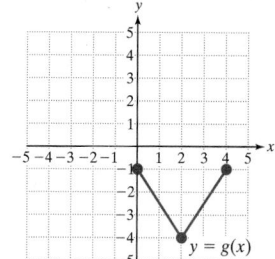

For Exercises 51–54, use the graphs of $y = f(x)$ and $y = g(x)$ to graph the given function. (See Example 6)

51. $y = f(-x)$

52. $y = g(-x)$

53. $y = -f(x)$

54. $y = -g(x)$

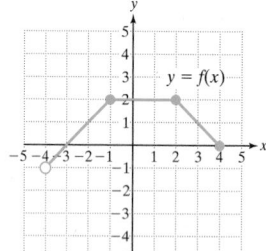 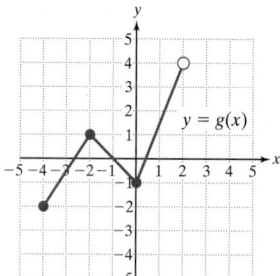

Objective 5: Summarize Transformations of Graphs

For Exercises 55–62, a function *g* is given. Identify the parent function from Table 1-2 on page 183. Then use the steps for graphing multiple transformations of functions on page 190 to list, in order, the transformations applied to the parent function to obtain the graph of *g*.

55. $g(x) = \frac{3}{1+x} - 2$

56. $g(x) = \frac{5}{x-4} + 1$

57. $g(x) = \frac{1}{3}(x - 2.1)^2 + 7.9$

58. $g(x) = \frac{1}{2}\sqrt{x + 4.3} - 8.4$

59. $g(x) = 2\sqrt{-2x + 5}$

60. $g(x) = 3\left|-\frac{1}{2}x - 4\right|$

61. $g(x) = -\sqrt{\frac{1}{3}x - 6}$

62. $g(x) = -|2x| + 8$

For Exercises 63–78, use transformations to graph the functions. (See Examples 7–8)

63. $v(x) = -(x + 2)^2 + 1$

64. $u(x) = -(x - 1)^2 - 2$

65. $f(x) = 2\sqrt{x + 3} - 1$

66. $g(x) = 2\sqrt{x - 1} + 3$

67. $p(x) = \frac{1}{2}|x - 1| - 2$

68. $q(x) = \frac{1}{3}|x + 2| - 1$

69. $r(x) = -\sqrt{-x} + 1$

70. $s(x) = -\sqrt{-x} - 2$

71. $f(x) = \sqrt{-x + 3}$

72. $g(x) = \sqrt{-x - 4}$

73. $n(x) = -\left|\frac{1}{2}x - 3\right|$

74. $m(x) = -\left|\frac{1}{3}x + 1\right|$

75. $f(x) = -\frac{1}{2}(x - 3)^2 + 8$

76. $g(x) = -\frac{1}{3}(x + 2)^2 + 3$

77. $p(x) = -4|x + 2| - 1$

78. $q(x) = -2|x - 1| + 4$

Mixed Exercises

For Exercises 79–86, the graph of $y = f(x)$ is given. Graph the indicated function.

79. Graph $y = -f(x - 1) + 2$.

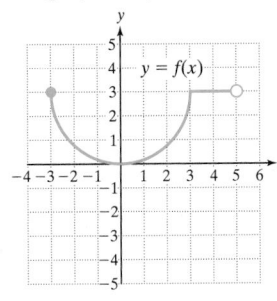

80. Graph $y = -f(x + 1) - 2$.

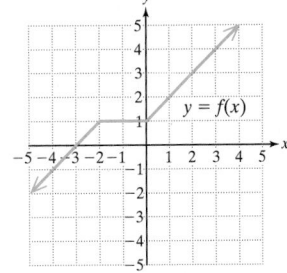

81. Graph $y = 2f(x - 2) - 3$.

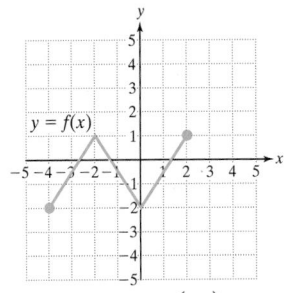

82. Graph $y = 2f(x + 2) - 4$.

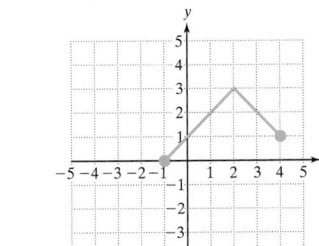

83. Graph $y = -3f(2x)$.

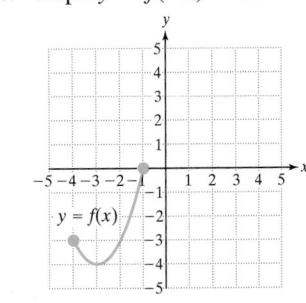

84. Graph $y = -\frac{1}{2}f\left(\frac{1}{2}x\right)$.

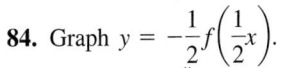
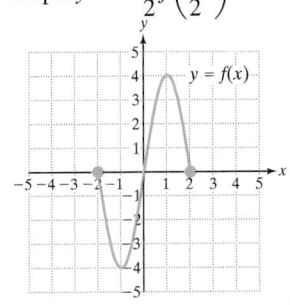

85. Graph $y = f(-x) - 2$.

86. Graph $y = f(-x) + 3$.

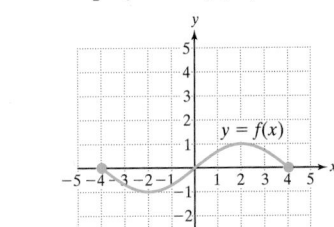

For Exercises 87–90, write a function based on the given parent function and transformations in the given order.

87. Parent function: $y = x^3$

1. Shift 4.5 units to the left. $y = (-x + 4.5)^3 + 2.1$
2. Reflect across the y-axis.
3. Shift upward 2.1 units.

88. Parent function $y = \sqrt[3]{x}$

1. Shift 1 unit to the left. $y = -\sqrt[3]{\frac{1}{4}x + 1}$
2. Stretch horizontally by a factor of 4.
3. Reflect across the x-axis.

89. Parent function: $y = \frac{1}{x}$

1. Stretch vertically by a factor of 2. $y = -\frac{2}{x} - 3$
2. Reflect across the x-axis.
3. Shift downward 3 units.

90. Parent function: $y = |x|$

1. Shift 3.7 units to the right. $y = |-3x - 3.7|$
2. Shrink horizontally by a factor of $\frac{1}{3}$.
3. Reflect across the y-axis.

Write About It

91. Explain why the graph of $g(x) = |2x|$ can be interpreted as a horizontal shrink of the graph of $f(x) = |x|$ or as a vertical stretch of the graph of $f(x) = |x|$.

92. Explain why the graph of $h(x) = \sqrt{\frac{1}{2}x}$ can be interpreted as a horizontal stretch of the graph of $f(x) = \sqrt{x}$ or as a vertical shrink of the graph of $f(x) = \sqrt{x}$.

93. Explain the difference between the graphs of $f(x) = |x - 2| - 3$ and $g(x) = |x - 3| - 2$.

The graph of *f* is the same as the graph of $y = |x|$ with a horizontal shift to the right 2 units and a vertical shift downward 3 units. By contrast, the graph of *g* is the graph of $y = |x|$ with a horizontal shift to the right 3 units and a vertical shift downward 2 units.

94. Explain why $g(x) = \dfrac{1}{-x + 1}$ can be graphed by shifting the graph of $f(x) = \dfrac{1}{x}$ one unit to the left and reflecting across the *y*-axis, or by shifting the graph of *f* one unit to the right and reflecting across the *x*-axis.

Expanding Your Skills

For Exercises 95–100, use transformations on the basic functions presented in Table 1-2 to write a rule $y = f(x)$ that would produce the given graph.

95.
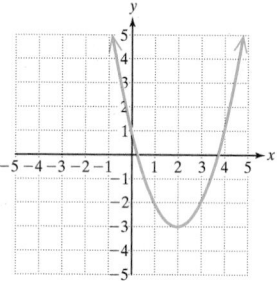

96.
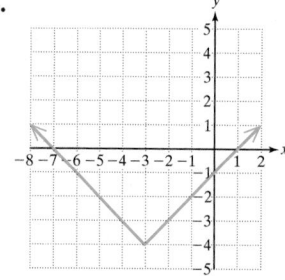

97.
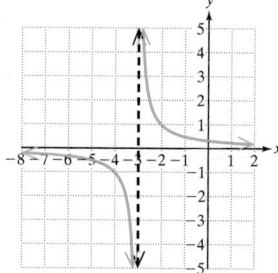

98.
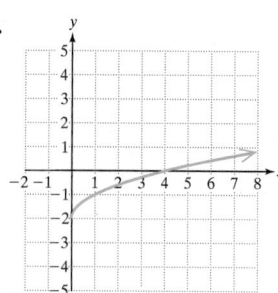

99.
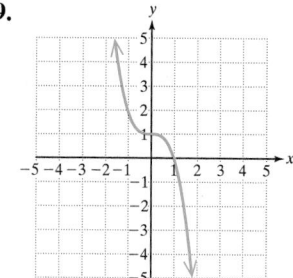

100.
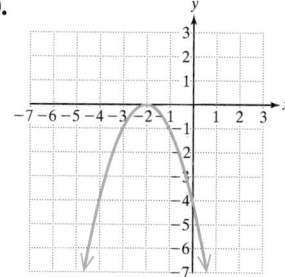

101. The graph shows the number of views *y* (in thousands) for a new online video, *t* days after it was posted. Use transformations on one of the parent functions from Table 1-2 on page 183 to model these data.

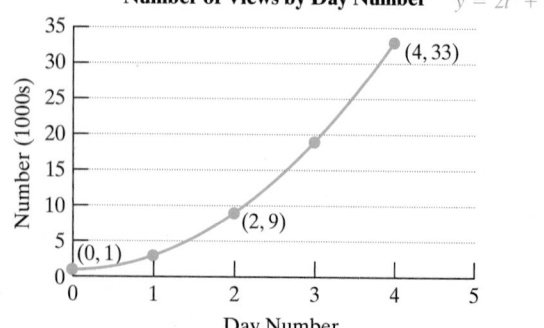

Number of Views by Day Number $y = 2t^2 + 1$

102. The graph shows the cumulative number *y* of flu cases among passengers on a 25-day cruise, *t* days after the cruise began. Use transformations on one of the parent functions from Table 1-2 on page 183 to model these data.

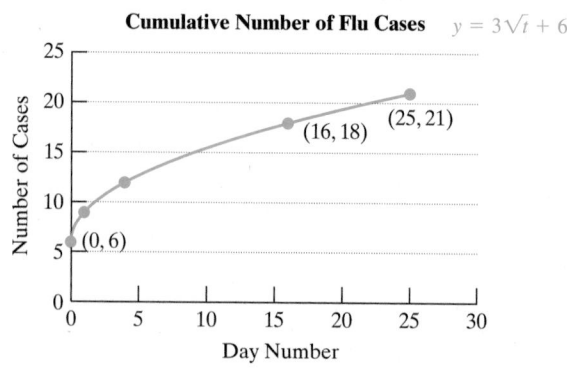

Cumulative Number of Flu Cases $y = 3\sqrt{t} + 6$

Technology Connections

103. a. Graph the functions on the viewing window [−5, 5, 1] by [−2, 8, 1].

$$y = x^2$$
$$y = x^4$$
$$y = x^6$$

c. Describe the general shape of the graph of $y = x^n$ where n is an even integer greater than 1.

b. Graph the functions on the viewing window [−4, 4, 1] by [−10, 10, 1].

$$y = x^3$$
$$y = x^5$$
$$y = x^7$$

d. Describe the general shape of the graph of $y = x^n$ where n is an odd integer greater than 1.

Additional answers can be found in the Instructor Answer Appendix.

SECTION 1.7 Analyzing Graphs of Functions and Piecewise-Defined Functions

OBJECTIVES

1. Test for Symmetry
2. Identify Even and Odd Functions
3. Graph Piecewise-Defined Functions
4. Investigate Increasing, Decreasing, and Constant Behavior of a Function
5. Determine Relative Minima and Maxima of a Function

1. Test for Symmetry

The photos in Figures 1-28 through 1-30 each show a type of symmetry.

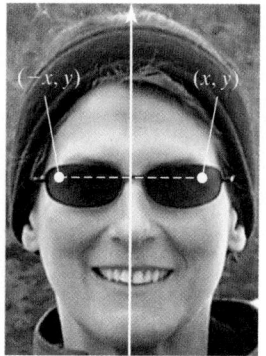
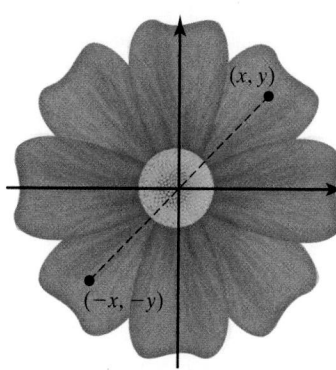

| **Figure 1-28** | **Figure 1-29** | **Figure 1-30** |

The photo of the kingfisher (Figure 1-28) shows an image of the bird reflected in the water. Suppose that we superimpose the x-axis at the waterline. Every point (x, y) on the bird has a mirror image $(x, -y)$ below the x-axis. Therefore, this image is symmetric with respect to the x-axis.

A human face is symmetric with respect to a vertical line through the center (Figure 1-29). If we place the y-axis along this line, a point (x, y) on one side has a mirror image at $(-x, y)$. This image is symmetric with respect to the y-axis.

The flower shown in Figure 1-30 is symmetric with respect to the point at its center. Suppose that we place the origin at the center of the flower. Notice that a point (x, y) on the image has a corresponding point $(-x, -y)$ on the image. This image is symmetric with respect to the origin.

Given an equation in the variables x and y, use the following rules to determine if the graph is symmetric with respect to the x-axis, the y-axis, or the origin.

Tests for Symmetry

Consider an equation in the variables x and y.

- The graph of the equation is symmetric with respect to the y-axis if substituting $-x$ for x in the equation results in an equivalent equation.
- The graph of the equation is symmetric with respect to the x-axis if substituting $-y$ for y in the equation results in an equivalent equation.
- The graph of the equation is symmetric with respect to the origin if substituting $-x$ for x and $-y$ for y in the equation results in an equivalent equation.

Classroom Examples: p. 209
Exercises 8, 10

EXAMPLE 1 Testing for Symmetry

Determine whether the graph is symmetric with respect to the y-axis, x-axis, or origin.

a. $y = |x|$ **b.** $x = y^2 - 4$

Solution:

a.

$y = |x|$ ← Same equation:
$y = |-x|$ Graph is symmetric
$y = |x|$ with respect to the y-axis.

Test for symmetry with respect to the y-axis.
Replace x by $-x$. Note that $|-x| = |x|$.
The resulting equation *is* equivalent to the original equation.

$y = |x|$ ←
$-y = |x|$ not the same
$y = -|x|$

Test for symmetry with respect to the x-axis.
Replace y by $-y$. The resulting equation is *not* equivalent to the original equation.

$y = |x|$ ←
$-y = |-x|$ not the same
$-y = |x|$
$y = -|x|$

Test for symmetry with respect to the origin.
Replace x by $-x$ and y by $-y$.
The resulting equation is *not* equivalent to the original equation.

The graph is symmetric with respect to the y-axis only.

b.

$x = y^2 - 4$ ←
$-x = y^2 - 4$ not the same
$x = -y^2 + 4$

Test for symmetry with respect to the y-axis.
Replace x by $-x$. The resulting equation is *not* equivalent to the original equation.

$x = y^2 - 4$ ← Same equation:
$x = (-y)^2 - 4$ Graph is symmetric
$x = y^2 - 4$ with respect to the x-axis.

Test for symmetry with respect to the x-axis.
Replace y by $-y$. The resulting equation *is* equivalent to the original equation.

$x = y^2 - 4$ ←
$-x = (-y)^2 - 4$
$-x = y^2 - 4$ not the same
$x = -y^2 + 4$

Test for symmetry with respect to the origin.
Replace x by $-x$ and y by $-y$.
The resulting equation is *not* equivalent to the original equation.

The graph is symmetric with respect to the x-axis only (Figure 1-31).

TIP The graph of $y = |x|$ is one of the basic graphs presented in Section 1.6. From our familiarity with the graph we can visualize the symmetry with respect to the y-axis.

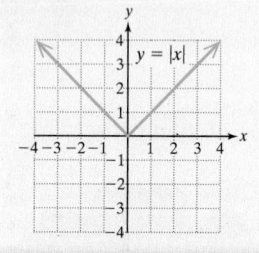

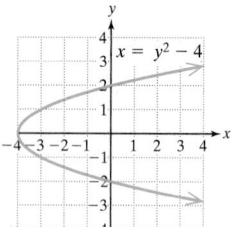

Figure 1-31

> **Skill Practice 1** Determine whether the graph is symmetric with respect to the y-axis, x-axis, or origin.
> **a.** $y = x^2$ **b.** $|y| = x + 1$

Classroom Example: p. 209
Exercise 16

EXAMPLE 2 Testing for Symmetry

Determine whether the graph is symmetric with respect to the y-axis, x-axis, or origin.

$$x^2 + y^2 = 9$$

Solution:

The graph of $x^2 + y^2 = 9$ is a circle with center at the origin and radius 3. By inspection, we can see that the graph is symmetric with respect to both axes and the origin.

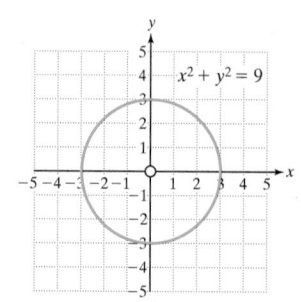

Answers
1. a. y-axis
   b. x-axis

Test for y-axis symmetry.
Replace x by $-x$.

$$x^2 + y^2 = 9$$
$$(-x)^2 + y^2 = 9$$ same
$$x^2 + y^2 = 9$$

Test for x-axis symmetry.
Replace y by $-y$.

$$x^2 + y^2 = 9$$
$$x^2 + (-y)^2 = 9$$ same
$$x^2 + y^2 = 9$$

Test for origin symmetry.
Replace x by $-x$ and y by $-y$.

$$x^2 + y^2 = 9$$
$$(-x)^2 + (-y)^2 = 9$$ same
$$x^2 + y^2 = 9$$

The graph is symmetric with respect to the y-axis, the x-axis, and the origin.

> **Skill Practice 2**   Determine whether the graph is symmetric with respect to the y-axis, x-axis, or origin.
>
> $$\frac{x^2}{4} + \frac{y^2}{9} = 1$$

2. Identify Even and Odd Functions

A function may be symmetric with respect to the y-axis or to the origin. A function that is symmetric with respect to the y-axis is called an *even* function. A function that is symmetric with respect to the origin is called an *odd* function.

Avoiding Mistakes

The only functions that are symmetric with respect to the x-axis are functions whose points lie solely on the x-axis.

> **Even and Odd Functions**
>
> - A function f is an **even function** if $f(-x) = f(x)$ for all x in the domain of f. The graph of an even function is symmetric with respect to the y-axis.
> - A function f is an **odd function** if $f(-x) = -f(x)$ for all x in the domain of f. The graph of an odd function is symmetric with respect to the origin.

Classroom Examples: p. 210
Exercises 22, 24, 26

> **EXAMPLE 3**   Identifying Even and Odd Functions

By inspection determine if the function is even, odd, or neither.

Solution:

a.

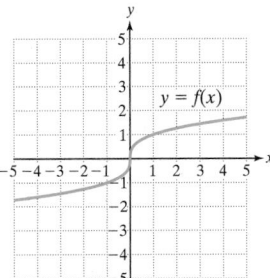

The function is symmetric with respect to the origin. Therefore, the function is an *odd* function.

b.

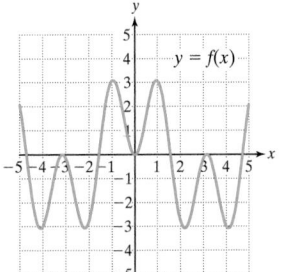

The function is symmetric with respect to the y-axis. Therefore, the function is an *even* function.

c.

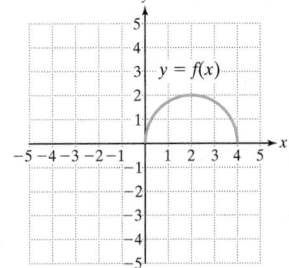

The function is not symmetric with respect to either the y-axis or the origin. Therefore, the function is *neither* even nor odd.

Answer

2. y-axis, x-axis, and origin

Skill Practice 3   Determine if the function is even, odd, or neither.

a.   **b.**   **c.**

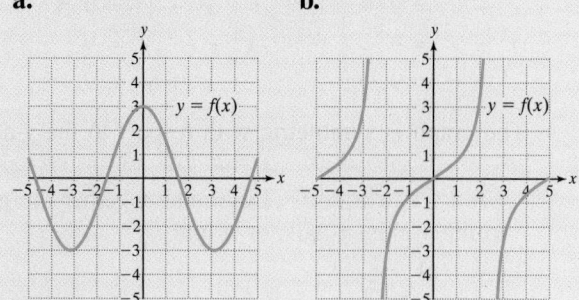

Classroom Examples: p. 210
Exercises 34, 36, 44

EXAMPLE 4   **Identifying Even and Odd Functions**

Determine if the function is even, odd, or neither.

    a. $f(x) = -2x^4 + 5|x|$ **b.** $g(x) = 4x^3 - x$ **c.** $h(x) = 2x^2 + x$

Solution:

> **TIP** In Example 4(a), we suspect that f is an even function because each term is of the form x^{even} or $|x|$. In each case, replacing x by $-x$ results in an equivalent term.

a. $f(x) = -2x^4 + 5|x|$ Determine whether the function is even.

$f(-x) = -2(-x)^4 + 5|-x|$ same   Replace x by $-x$ to determine if $f(-x) = f(x)$.

       $= -2x^4 + 5|x|$

Since $f(-x) = f(x)$, the function f is an even function.

There is no need to test whether f is an odd function because a function cannot be both even and odd unless all points are on the x-axis.

> **TIP** In Example 4(b), we suspect that g is an odd function because each term is of the form x^{odd}. In each case, replacing x by $-x$ results in the *opposite* of the original term.

b. $g(x) = 4x^3 - x$ Each term has x raised to an odd power. Therefore, replacing x by $-x$ will result in the *opposite* of the original term. Therefore, test whether g is an odd function. That is, test whether $g(-x) = -g(x)$.

Evaluate: $g(-x)$ Evaluate: $-g(x)$

$g(-x) = 4(-x)^3 - (-x)$ $-g(x) = -(4x^3 - x)$

       $= -4x^3 + x$ same $= -4x^3 + x$

Since $g(-x) = -g(x)$, the function g is an odd function.

> **TIP** In Example 4(c), $h(x)$ has a mixture of terms of the form x^{odd} and x^{even}. Therefore, we might suspect that the function is neither even nor odd.

c. $h(x) = 2x^2 + x$ Determine whether the function is even.

$h(-x) = 2(-x)^2 + (-x)$ not the same   Replace x by $-x$ to determine if $h(-x) = h(x)$.

       $= 2x^2 - x$

Since $h(-x) \neq h(x)$, the function is not even.

Next, test whether h is an odd function. Test whether $h(-x) = -h(x)$.

Evaluate: $h(-x)$ Evaluate: $-h(x)$

$h(-x) = 2(-x)^2 + (-x)$ $-h(x) = -(2x^2 + x)$

       $= 2x^2 - x$ not the same $= -2x^2 - x$

Since $h(-x) \neq -h(x)$, the function is not an odd function. Therefore, h is neither even nor odd.

Answers

3. a. Even function
   b. Odd function
   c. Neither even nor odd
4. a. Odd function
   b. Even function
   c. Neither even nor odd

Skill Practice 4   Determine if the function is even, odd, or neither.

    a. $m(x) = -x^5 + x^3$ **b.** $n(x) = x^2 - |x| + 1$ **c.** $p(x) = 2|x| + x$

3. Graph Piecewise-Defined Functions

Suppose that a car is stopped for a red light. When the light turns green, the car undergoes a constant acceleration for 20 sec until it reaches a speed of 45 mph. It travels 45 mph for 1 min (60 sec), and then decelerates for 30 sec to stop at another red light. The graph of the car's speed y (in mph) versus the time x (in sec) after leaving the first red light is shown in Figure 1-32.

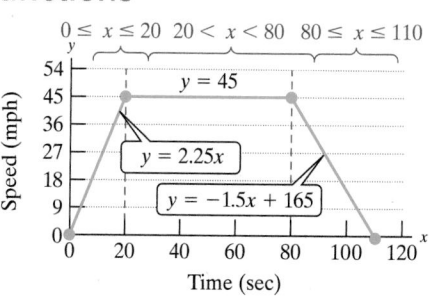

Figure 1-32

Notice that the graph can be segmented into three pieces. The first 20 sec is represented by a linear function with a positive slope, $y = 2.25x$. The next 60 sec is represented by the constant function $y = 45$. And the last 30 sec is represented by a linear function with a negative slope, $y = -1.5x + 165$.

To write a rule defining this function we use a **piecewise-defined function** in which we define each "piece" on a restricted domain.

$$f(x) = \begin{cases} 2.25x & \text{for } 0 \le x \le 20 \\ 45 & \text{for } 20 < x < 80 \\ -1.5x + 165 & \text{for } 80 \le x \le 110 \end{cases}$$

Classroom Example: p. 210
Exercise 48

EXAMPLE 5   Interpreting a Piecewise-Defined Function

Evaluate the function for the given values of x.

$$f(x) = \begin{cases} -x - 1 & \text{for } x < -1 \\ -3 & \text{for } -1 \le x < 2 \\ \sqrt{x - 2} & \text{for } x \ge 2 \end{cases}$$

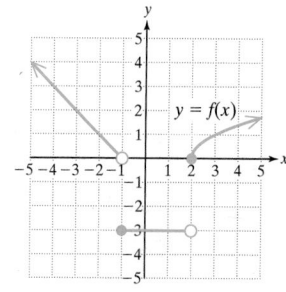

a. $f(-3)$ **b.** $f(-1)$
c. $f(2)$ **d.** $f(6)$

Solution:

a. $f(x) = -x - 1$ $x = -3$ is on the interval $x < -1$. Use the first rule
   $f(-3) = -(-3) - 1$ in the function: $f(x) = -x - 1$.
   $= 2$

b. $f(x) = -3$ $x = -1$ is on the interval $-1 \le x < 2$. Use the second rule in
   $f(-1) = -3$ the function: $f(x) = -3$.

c. $f(x) = \sqrt{x - 2}$ $x = 2$ is on the interval $x \ge 2$. Use the third rule in the
   $f(2) = \sqrt{2 - 2}$ function: $f(x) = \sqrt{x - 2}$.
   $= 0$

d. $f(x) = \sqrt{x - 2}$ $x = 6$ is on the interval $x \ge 2$. Use the third rule in the
   $f(6) = \sqrt{6 - 2}$ function: $f(x) = \sqrt{x - 2}$.
   $= 2$

Skill Practice 5   Evaluate the function for the given values of x.

$$f(x) = \begin{cases} x + 7 & \text{for } x < -2 \\ x^2 & \text{for } -2 \le x < 1 \\ 3 & \text{for } x \ge 1 \end{cases}$$

a. $f(-3)$ **b.** $f(-2)$ **c.** $f(1)$ **d.** $f(4)$

Answers
5. a. 4   **b.** 4   **c.** 3   **d.** 3

TECHNOLOGY CONNECTIONS

Graphing a Piecewise-Defined Function

A graphing calculator can be used to graph a piecewise-defined function. The format to enter the function is as follows.

Y_1 = (first piece)/(first condition)

Y_2 = (second piece)/(second condition)

⋮

Each condition in parentheses is an inequality and the calculator assigns it a value of 1 or 0 depending on whether the inequality is true or false. If an inequality is true, the function is divided by 1 on that interval and is "turned on." If an inequality is false, then the function is divided by 0. Since division by zero is undefined, the calculator does not graph the function on that interval, and the function is effectively "turned off."

Enter the function from Example 5 as shown. Note that the inequality symbols can be found in the TEST menu.

$$f(x) = \begin{cases} -x - 1 & \text{for } x < -1 \\ -3 & \text{for } -1 \le x < 2 \\ \sqrt{x - 2} & \text{for } x \ge 2 \end{cases}$$

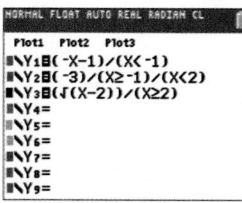 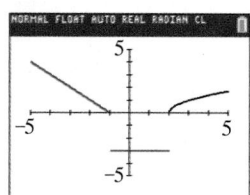

Notice that the individual "pieces" of the graph do not "hook-up." For this reason, it is also a good practice to put the calculator in DOT mode in the MODE menu.

In Examples 6 and 7, we graph piecewise-defined functions.

Classroom Example: p. 211
Exercise 58

EXAMPLE 6   Graphing a Piecewise-Defined Function

Graph the function defined by $f(x) = \begin{cases} -3x & \text{for } x < 1 \\ -3 & \text{for } x \ge 1 \end{cases}$.

Solution:

- The first rule $f(x) = -3x$ defines a line with slope -3 and y-intercept $(0, 0)$. This line should be graphed only to the left of $x = 1$. The point $(1, -3)$ is graphed as an open dot, because the point is not part of the rule $f(x) = -3x$. See the blue portion of the graph in Figure 1-33.

- The second rule $f(x) = -3$ is a horizontal line for $x \ge 1$. The point $(1, -3)$ is a closed dot to show that it is part of the rule $f(x) = -3$. The closed dot from the red segment of the graph "overrides" the open dot from the blue segment. Taken together, the closed dot "plugs" the hole in the graph.

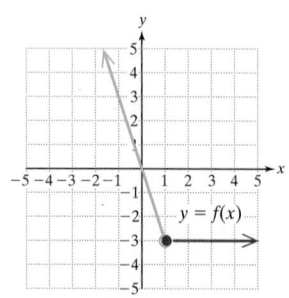

Figure 1-33

Skill Practice 6   Graph the function.

$$f(x) = \begin{cases} 2 & \text{for } x \le -1 \\ -2x & \text{for } x > -1 \end{cases}$$

TIP The function in Example 6 has no "gaps," and therefore we say that the function is **continuous.** Informally, this means that we can draw the function without lifting our pencil from the page. The formal definition of a continuous function will be studied in calculus.

Classroom Example: p. 212
Exercise 70

EXAMPLE 7   Graphing a Piecewise-Defined Function

Graph the function. $f(x) = \begin{cases} x + 3 & \text{for } x < -1 \\ x^2 & \text{for } -1 \le x < 2 \end{cases}$

Solution:

The first rule $f(x) = x + 3$ defines a line with slope 1 and y-intercept $(0, 3)$. This line should be graphed only for $x < -1$ (that is to the left of $x = -1$). The point $(-1, 2)$ is graphed as an open dot, because the point is not part of the function. See the red portion of the graph in Figure 1-34.

The second rule $f(x) = x^2$ is one of the basic functions learned in Section 1.6. It is a parabola with vertex at the origin. We sketch this function only for x values on the interval $-1 \le x < 2$. The point $(-1, 1)$ is a closed dot to show that it is part of the function. The point $(2, 4)$ is displayed as an open dot to indicate that it is not part of the function.

TIP The function in Example 7 has a gap at $x = -1$, and therefore, we say that f is **discontinuous** at -1.

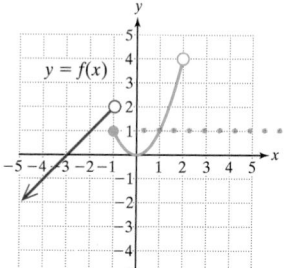

Avoiding Mistakes

Note that the function cannot have a closed dot at both $(-1, 1)$ and $(-1, 2)$ because it would not pass the vertical line test.

Figure 1-34

Skill Practice 7   Graph the function.

$$f(x) = \begin{cases} |x| & \text{for } -4 \le x < 2 \\ -x + 2 & \text{for } x \ge 2 \end{cases}$$

We now look at a special category of piecewise-defined functions called **step functions.** The graph of a step function is a series of discontinuous "steps." One important step function is called the **greatest integer function** or **floor function.** It is defined by

$$f(x) = [\![x]\!] \text{ where } [\![x]\!] \text{ is the greatest integer less than or equal to } x.$$

The operation $[\![x]\!]$ may also be denoted as **int(x)** or by **floor(x).** These alternative notations are often used in computer programming.

In Example 8, we graph the greatest integer function.

Answers

6.

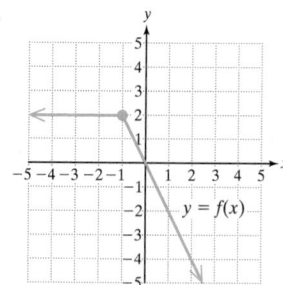

7.

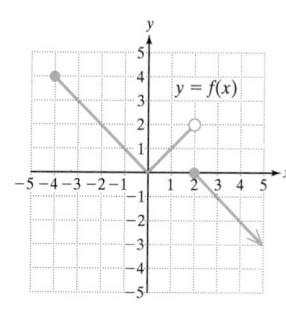

Classroom Example: p. 212
Exercise 82

EXAMPLE 8 **Graphing the Greatest Integer Function**

Graph the function defined by $f(x) = [\![x]\!]$.

Solution:

x	$f(x) = [\![x]\!]$	
-1.7	-2	Evaluate f for several values of x. Greatest integer less than or equal to -1.7 is -2.
-1	-1	Greatest integer less than or equal to -1 is -1.
-0.6	-1	Greatest integer less than or equal to -0.6 is -1.
0	0	Greatest integer less than or equal to 0 is 0.
0.4	0	Greatest integer less than or equal to 0.4 is 0.
1	1	Greatest integer less than or equal to 1 is 1.
1.8	1	Greatest integer less than or equal to 1.8 is 1.
2	2	Greatest integer less than or equal to 2 is 2.
2.5	2	Greatest integer less than or equal to 2.5 is 2.

TIP On many graphing calculators, the greatest integer function is denoted by int( ) and is found under the MATH menu followed by NUM.

From the table, we see a pattern and from the pattern, we form the graph.

If $-3 \le x < -2$, then $[\![x]\!] = -3$
If $-2 \le x < -1$, then $[\![x]\!] = -2$
If $-1 \le x < 0$, then $[\![x]\!] = -1$
If $0 \le x < 1$, then $[\![x]\!] = 0$
If $1 \le x < 2$, then $[\![x]\!] = 1$
If $2 \le x < 3$, then $[\![x]\!] = 2$
$\cdots$

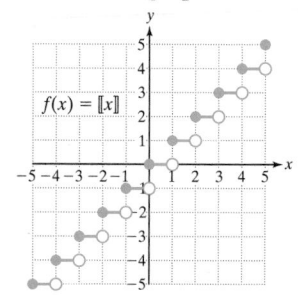

$f(x) = [\![x]\!]$

Skill Practice 8 Evaluate $f(x) = [\![x]\!]$ for the given values of x.
a. $f(1.7)$ **b.** $f(5.5)$ **c.** $f(-4)$ **d.** $f(-4.2)$

In Example 9, we use a piecewise-defined function to model an application.

Classroom Example: p. 212
Exercise 88

EXAMPLE 9 **Using a Piecewise-Defined Function in an Application**

A salesperson makes a monthly salary of $3000 along with a 5% commission on sales over $20,000 for the month. Write a piecewise-defined function to represent the salesperson's monthly income $I(x)$ (in $) for x dollars in sales.

Solution:

Let x represent the amount in sales.
Then $x - 20,000$ represents the amount in sales over $20,000.

There are two scenarios for the salesperson's income.
Scenario 1: The salesperson sells $20,000 or less. In this case, the monthly income is a constant $3000. This is represented by

$$y = 3000 \quad \text{for } 0 \le x \le 20,000$$

Scenario 2: The salesperson sells over $20,000. In this case, the monthly income is $3000 plus 5% of sales over $20,000. This is represented by

$$y = 3000 + 0.05(x - 20,000) \quad \text{for } x > 20,000$$

Answers
8. **a.** 1   **b.** 5
   c. -4 **d.** -5

Therefore, a piecewise-defined function for monthly income is

$$I(x) = \begin{cases} 3000 & \text{for } 0 \le x \le 20,000 \\ 3000 + 0.05(x - 20,000) & \text{for } x > 20,000 \end{cases}$$

Alternatively, we can simplify to get

$$I(x) = \begin{cases} 3000 & \text{for } 0 \le x \le 20,000 \\ 0.05x + 2000 & \text{for } x > 20,000 \end{cases}$$

A graph of $y = I(x)$ is shown in Figure 1-35. Notice that for $x = \$20,000$, both equations within the piecewise-defined function yield a monthly salary of $3000. Therefore, the two line segments in the graph meet at $(20,000, 3000)$.

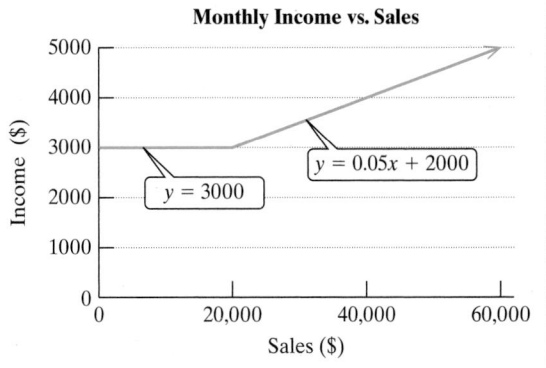

Monthly Income vs. Sales

$y = 3000$

$y = 0.05x + 2000$

Figure 1-35

Skill Practice 9   A retail store buys T-shirts from the manufacturer. The cost is $7.99 per shirt for 1 to 100 shirts, inclusive. Then the price is decreased to $6.99 per shirt thereafter. Write a piecewise-defined function that expresses the cost $C(x)$ (in $) to buy x shirts.

---

4. Investigate Increasing, Decreasing, and Constant Behavior of a Function

The graph in Figure 1-36 approximates the altitude of an airplane, $f(t)$, at a time t minutes after takeoff.

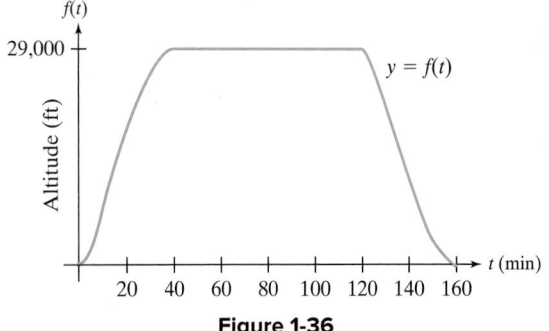

Figure 1-36

Notice that the plane's altitude increases up to the first 40 min of the flight. So we say that the function f is increasing on the interval $(0, 40)$. The plane flies at a constant altitude for the next 1 hr 20 min, so we say that f is constant on the interval $(40, 120)$. Finally, the plane's altitude decreases for the last 40 min, so we say that f is decreasing on the interval $(120, 160)$.

Informally, a function is increasing on an interval in its domain if its graph rises from left to right. A function is decreasing on an interval in its domain if the graph "falls" from left to right. A function is constant on an interval in its domain if its graph is horizontal over the interval. These ideas are stated formally using mathematical notation.

Answer

9. $C(x) = \begin{cases} 7.99x & \text{for } 1 \le x \le 100 \\ 799 + 6.99(x - 100) & \text{for } x > 100 \end{cases}$

Intervals of Increasing, Decreasing, and Constant Behavior

Suppose that I is an interval contained within the domain of a function f.

- f is increasing on I if $f(x_1) < f(x_2)$ for all $x_1 < x_2$ on I.
- f is decreasing on I if $f(x_1) > f(x_2)$ for all $x_1 < x_2$ on I.
- f is constant on I if $f(x_1) = f(x_2)$ for all x_1 and x_2 on I.

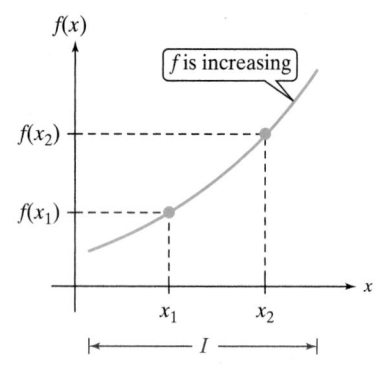

For all $x_1 < x_2$ on $I, f(x_1) < f(x_2)$

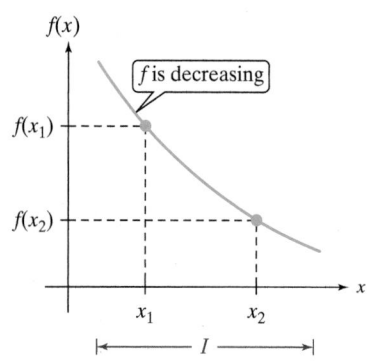

For all $x_1 < x_2$ on $I, f(x_1) > f(x_2)$

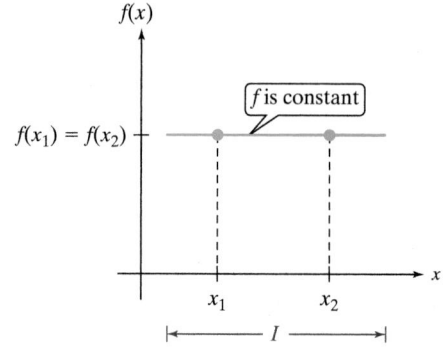

For all x_1 and x_2 on $I, f(x_1) = f(x_2)$

Classroom Example: p. 212
Exercise 90

EXAMPLE 10   **Determining the Intervals Over Which a Function Is Increasing, Decreasing, and Constant**

Use interval notation to write the interval(s) over which f is

   a. Increasing   **b.** Decreasing

   c. Constant

Solution:

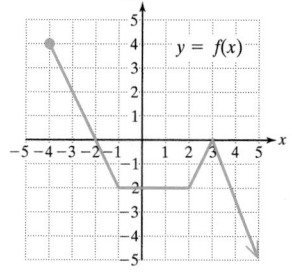

  a. f is increasing on the interval $(2, 3)$.
     (Highlighted in red tint.)

  b. f is decreasing on the interval
     $(-4, -1) \cup (3, \infty)$.
     (Highlighted in orange tint.)

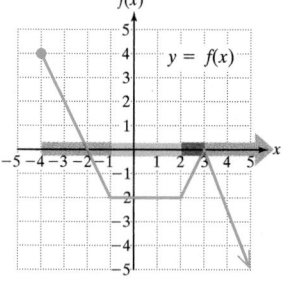

  c. f is constant on the interval $(-1, 2)$.
     (Highlighted in green tint.)

Skill Practice 10   Use interval notation to write the interval(s) over which f is

   a. Increasing

   b. Decreasing

   c. Constant

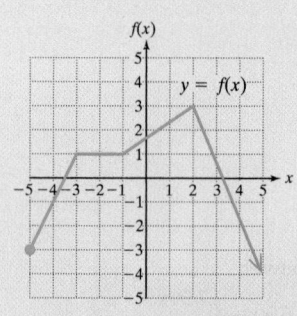

Answers

10. a. $(-5, -3) \cup (-1, 2)$
   b. $(2, \infty)$ **c.** $(-3, -1)$

5. Determine Relative Minima and Maxima of a Function

The intervals over which a function changes from increasing to decreasing behavior or vice versa tell us where to look for relative maximum values and relative minimum values of a function. Consider the function pictured in Figure 1-37.

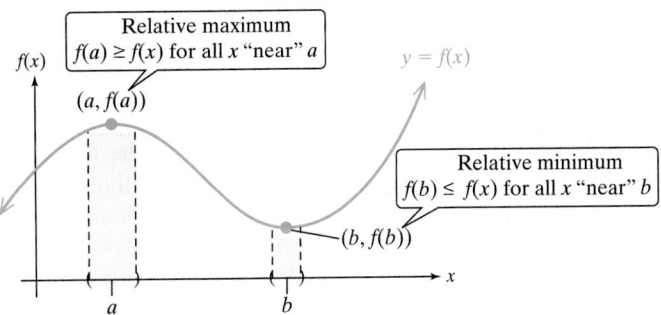

Figure 1-37

- The function has a relative maximum of $f(a)$. Informally, this means that $f(a)$ is the greatest function value relative to other points on the function nearby.
- The function has a relative minimum of $f(b)$. Informally, this means that $f(b)$ is the smallest function value relative to other points on the function nearby.

This is stated formally in the following definition.

TIP The plural of maximum and minimum are **maxima** and **minima**.

Note that relative maxima and minima are also called *local* maxima and minima.

Relative Minimum and Relative Maximum Values

- $f(a)$ is a **relative maximum** of f if there exists an open interval containing a such that $f(a) \geq f(x)$ for all x in the interval.
- $f(b)$ is a **relative minimum** of f if there exists an open interval containing b such that $f(b) \leq f(x)$ for all x in the interval.

Note: An open interval is an interval in which the endpoints are not included.

If an ordered pair $(a, f(a))$ corresponds to a relative minimum or relative maximum, we interpret the coordinates of the ordered pair as follows.

- The x-coordinate is the *location* of the relative maximum or minimum within the domain of the function.
- The y-coordinate is the *value* of the relative maximum or minimum. This tells us how "high" or "low" the graph is at that point.

Classroom Example: p. 213
Exercise 100

EXAMPLE 11   Finding Relative Maxima and Minima

For the graph of $y = g(x)$ shown,

a. Determine the location and value of any relative maxima.

b. Determine the location and value of any relative minima.

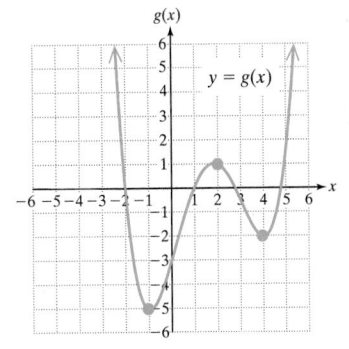

Solution:

a. The point (2, 1) is the highest point in a small interval surrounding $x = 2$. Therefore, at $x = 2$, the function has a relative maximum of 1.

b. The point $(-1, -5)$ is the lowest point in a small interval surrounding $x = -1$. Therefore, at $x = -1$, the function has a relative minimum of -5.

The point $(4, -2)$ is the lowest point in a small interval surrounding $x = 4$. Therefore, at $x = 4$, the function has a relative minimum of -2.

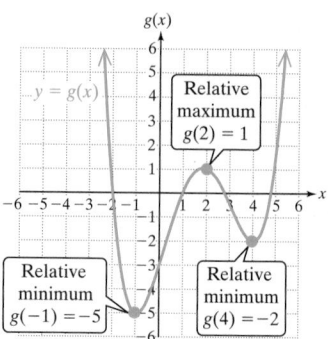

Skill Practice 11   For the graph shown,

a. Determine the location and value of any relative maxima.

b. Determine the location and value of any relative minima.

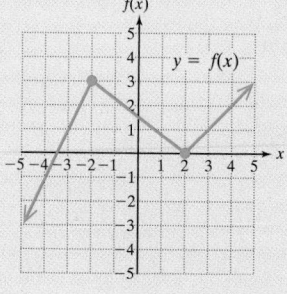

TECHNOLOGY CONNECTIONS

Determining Relative Maxima and Minima

Relative maxima and relative minima are often difficult to find analytically and require techniques from calculus. However, a graphing utility can be used to approximate the location and value of relative maxima and minima. To do so, we use the Minimum and Maximum features.

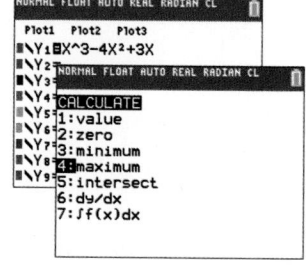

For example, enter the function defined by $Y_1 = x^3 - 4x^2 + 3x$. Then access the Maximum feature from the CALC menu.

The calculator asks for a left bound. This is a point slightly to the left of the relative maximum. Then hit ENTER.

The calculator asks for a right bound. This is a point slightly to the right of the relative maximum. Hit ENTER.

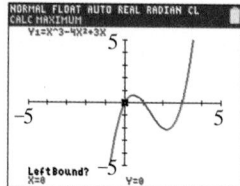

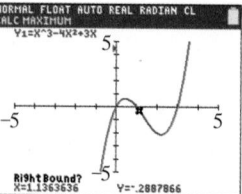

Answers

11. a. At $x = -2$, the function has a relative maximum of 3.

   b. At $x = 2$, the function has a relative minimum of 0.

The calculator asks for a guess. This is a point close to the relative maximum. Hit ENTER and the approximate coordinates of the relative maximum point are shown (0.45, 0.63).

To find the relative minimum, repeat these steps using the Minimum feature. The coordinates of the relative minimum point are approximately (2.22, −2.11).

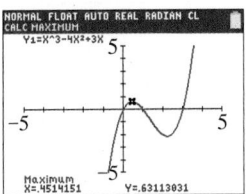

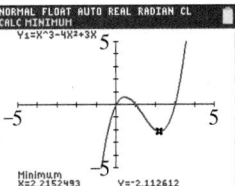

SECTION 1.7   Practice Exercises

Prerequisite Review

R.1. Given the function defined by $f(x) = 7x - 2$, find $f(-a)$. $f(-a) = -7a - 2$

For Exercises R.2–R.4, graph the set and express the set in interval notation.

R.2. $\{x \mid x < 8\}$

R.3. $\{x \mid -2.4 \le x < 5.8\}$

R.4. $\left\{ x \mid x \ge -\dfrac{9}{2} \right\}$

Concept Connections

1. A graph of an equation is symmetric with respect to the _____ _y_ _____ -axis if replacing x by $-x$ results in an equivalent equation.

2. A graph of an equation is symmetric with respect to the _____ _x_ _____ -axis if replacing y by $-y$ results in an equivalent equation.

3. A graph of an equation is symmetric with respect to the _____ origin _____ if replacing x by $-x$ and y by $-y$ results in an equivalent equation.

4. An even function is symmetric with respect to the _____ y-axis _____ .

5. An odd function is symmetric with respect to the _____ origin _____ .

6. The expression _____ $\lfloor x \rfloor$ or int(x) or floor(x) _____ represents the greatest integer, less than or equal to x.

Objective 1: Test for Symmetry

For Exercises 7–18, determine whether the graph of the equation is symmetric with respect to the x-axis, y-axis, origin, or none of these. (See Examples 1–2)

7. $y = x^2 + 3$ y-axis

8. $y = -|x| - 4$ y-axis

9. $x = -|y| - 4$ x-axis

10. $x = y^2 + 3$ x-axis

11. $x^2 + y^2 = 3$ x-axis, y-axis, and origin

12. $|x| + |y| = 4$ x-axis, y-axis, and origin

13. $y = |x| + 2x + 7$ None of these

14. $y = x^2 + 6x + 1$ None of these

15. $x^2 = 5 + y^2$ x-axis, y-axis, and origin

16. $y^4 = 2 + x^2$ x-axis, y-axis, and origin

17. $y = \dfrac{1}{2}x - 3$ None of these

18. $y = \dfrac{2}{5}x + 1$ None of these

Objective 2: Identify Even and Odd Functions

19. What type of symmetry does an even function have?   y-axis symmetry

20. What type of symmetry does an odd function have?   Origin symmetry

For Exercises 21–26, use the graph to determine if the function is even, odd, or neither. (See Example 3)

21.   Odd

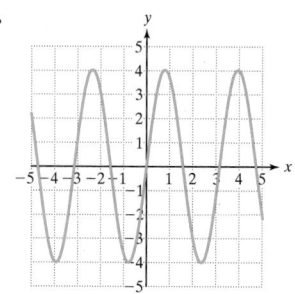

22.   Odd

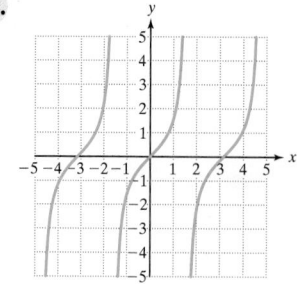

23.   Even

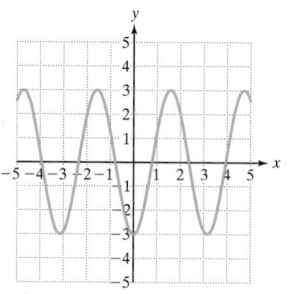

24.   Even

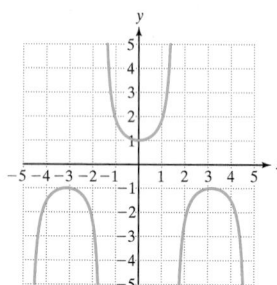

25.   Neither even nor odd

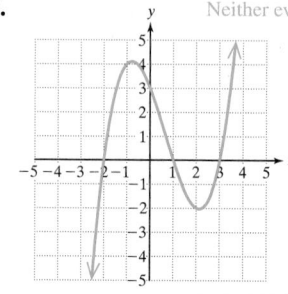

26.   Neither even nor odd

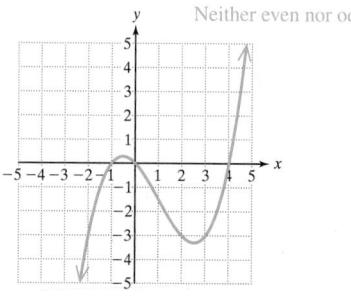

27. a. Given $f(x) = 4x^2 - 3|x|$, find $f(-x)$. $f(-x) = 4x^2 - 3|x|$
   b. Is $f(-x) = f(x)$?   Yes
   c. Is this function even, odd, or neither?   Even

28. a. Given $g(x) = -x^8 + |3x|$, find $g(-x)$.
   b. Is $g(-x) = g(x)$?   Yes $g(-x) = -x^8 + |3x|$
   c. Is this function even, odd, or neither?   Even

29. a. Given $h(x) = 4x^3 - 2x$, find $h(-x)$. $h(-x) = -4x^3 + 2x$
   b. Find $-h(x)$. $-h(x) = -4x^3 + 2x$
   c. Is $h(-x) = -h(x)$?   Yes
   d. Is this function even, odd, or neither?   Odd

30. a. Given $k(x) = -8x^5 - 6x^3$, find $k(-x)$.
   b. Find $-k(x)$. $-k(x) = 8x^5 + 6x^3$ $k(-x) = 8x^5 + 6x^3$
   c. Is $k(-x) = -k(x)$?   Yes
   d. Is this function even, odd, or neither?   Odd

31. a. Given $m(x) = 4x^2 + 2x - 3$, find $m(-x)$.
   b. Find $-m(x)$. $-m(x) = -4x^2 - 2x + 3$ $m(-x) = 4x^2 - 2x - 3$
   c. Is $m(-x) = m(x)$?   No
   d. Is $m(-x) = -m(x)$?   No
   e. Is this function even, odd, or neither?   Neither

32. a. Given $n(x) = 7|x| + 3x - 1$, find $n(-x)$.
   b. Find $-n(x)$. $-n(x) = -7|x| - 3x + 1$ $n(-x) = 7|x| - 3x - 1$
   c. Is $n(-x) = n(x)$?   No
   d. Is $n(-x) = -n(x)$?   No
   e. Is this function even, odd, or neither?   Neither

For Exercises 33–46, determine if the function is even, odd, or neither. (See Example 4)

33. $f(x) = 3x^6 + 2x^2 + |x|$ Even   **34.** $p(x) = -|x| + 12x^{10} + 5$ Even   **35.** $k(x) = 13x^3 + 12x$ Odd

36. $m(x) = -4x^5 + 2x^3 + x$ Odd   **37.** $n(x) = \sqrt{16 - (x - 3)^2}$ Neither   **38.** $r(x) = \sqrt{81 - (x + 2)^2}$ Neither

39. $q(x) = \sqrt{16 + x^2}$ Even   **40.** $z(x) = \sqrt{49 + x^2}$ Even   **41.** $h(x) = 5x$ Odd

42. $g(x) = -x$ Odd   **43.** $f(x) = \dfrac{x^2}{3(x - 4)^2}$ Neither   **44.** $g(x) = \dfrac{x^3}{2(x - 1)^3}$ Neither

45. $v(x) = \dfrac{-x^5}{|x| + 2}$ Odd   **46.** $w(x) = \dfrac{-\sqrt[3]{x}}{x^2 + 1}$ Odd

Objective 3: Graph Piecewise-Defined Functions

For Exercises 47–50, evaluate the function for the given values of x. (See Example 5)

47. $f(x) = \begin{cases} -3x + 7 & \text{for } x < -1 \\ x^2 + 3 & \text{for } -1 \le x < 4 \\ 5 & \text{for } x \ge 4 \end{cases}$

   a. $f(3)$ 12   **b.** $f(-2)$ 13   **c.** $f(-1)$ 4
   d. $f(4)$ 5   **e.** $f(5)$ 5

48. $g(x) = \begin{cases} -2|x| - 3 & \text{for } x \le -2 \\ 5x + 6 & \text{for } -2 < x < 3 \\ 4 & \text{for } x \ge 3 \end{cases}$

   a. $g(-3)$ -9 **b.** $g(3)$ 4   **c.** $g(-2)$ -7
   d. $g(0)$ 6   **e.** $g(4)$ 4

49. $h(x) = \begin{cases} 2 & \text{for } -3 \le x < -2 \\ 1 & \text{for } -2 \le x < -1 \\ 0 & \text{for } -1 \le x < 0 \\ -1 & \text{for } 0 \le x < 1 \end{cases}$

a. $h(-1.7)$ 1   **b.** $h(-2.5)$ 2   **c.** $h(0.05)$ -1

d. $h(-2)$ 1   **e.** $h(0)$ -1

50. $t(x) = \begin{cases} x & \text{for } 0 < x \le 1 \\ 2x & \text{for } 1 < x \le 2 \\ 3x & \text{for } 2 < x \le 3 \\ 4x & \text{for } 3 < x \le 4 \end{cases}$

a. $t(1.99)$ 3.98   **b.** $t(0.4)$ 0.4   **c.** $t(3)$ 9

d. $t(1)$ 1   **e.** $t(3.001)$ 12.004

51. A sled accelerates down a hill and then slows down after it reaches a flat portion of ground. The speed of the sled $s(t)$ (in ft/sec) at a time t (in sec) after movement begins can be approximated by:

$$s(t) = \begin{cases} 1.5t & \text{for } 0 \le t \le 20 \\ \dfrac{30}{t-19} & \text{for } 20 < t \le 40 \end{cases}$$

Determine the speed of the sled after 10 sec, 20 sec, 30 sec, and 40 sec. Round to 1 decimal place if necessary.   15 ft/sec; 30 ft/sec; 2.7 ft/sec; and 1.4 ft/sec

52. A car starts from rest and accelerates to a speed of 60 mph in 12 sec. It travels 60 mph for 1 min and then decelerates for 20 sec until it comes to rest. The speed of the car $s(t)$ (in mph) at a time t (in sec) after the car begins motion can be modeled by: 15 mph, 60 mph, 60 mph, and 21.6 mph

$$s(t) = \begin{cases} \dfrac{5}{12}t^2 & \text{for } 0 \le t \le 12 \\ 60 & \text{for } 12 < t \le 72 \\ \dfrac{3}{20}(92 - t)^2 & \text{for } 72 < t \le 92 \end{cases}$$

Determine the speed of the car 6 sec, 12 sec, 45 sec, and 80 sec after the car begins motion.

For Exercises 53–56, match the function with its graph.

53. $f(x) = x + 1$ for $x < 2$ c

54. $f(x) = x + 1$ for $-1 < x \le 2$ a

55. $f(x) = x + 1$ for $-1 \le x < 2$ d

56. $f(x) = x + 1$ for $x \ge 2$ b

a. 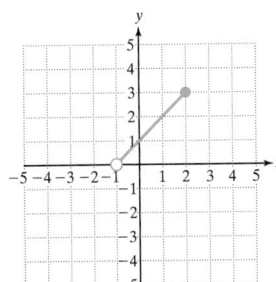   **b.** 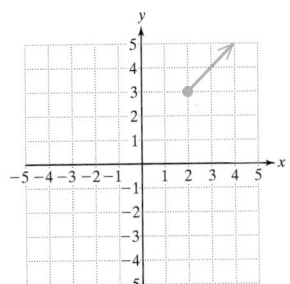   **c.** 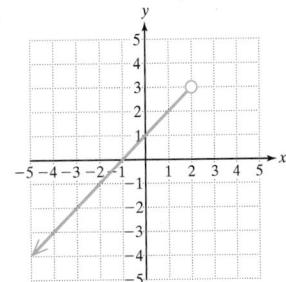   **d.**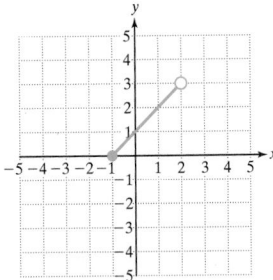

57. **a.** Graph $p(x) = x + 2$ for $x \le 0$. (See Examples 6–7)

 b. Graph $q(x) = -x^2$ for $x > 0$.

 c. Graph $r(x) = \begin{cases} x + 2 & \text{for } x \le 0 \\ -x^2 & \text{for } x > 0 \end{cases}$

58. **a.** Graph $f(x) = |x|$ for $x < 0$.

 b. Graph $g(x) = \sqrt{x}$ for $x \ge 0$.

 c. Graph $h(x) = \begin{cases} |x| & \text{for } x < 0 \\ \sqrt{x} & \text{for } x \ge 0 \end{cases}$

59. **a.** Graph $m(x) = \dfrac{1}{2}x - 2$ for $x \le -2$.

 b. Graph $n(x) = -x + 1$ for $x > -2$.

 c. Graph $t(x) = \begin{cases} \dfrac{1}{2}x - 2 & \text{for } x \le -2 \\ -x + 1 & \text{for } x > -2 \end{cases}$

60. **a.** Graph $a(x) = x$ for $x < 1$.

 b. Graph $b(x) = \sqrt{x - 1}$ for $x \ge 1$.

 c. Graph $c(x) = \begin{cases} x & \text{for } x < 1 \\ \sqrt{x - 1} & \text{for } x \ge 1 \end{cases}$

For Exercises 61–70, graph the function. (See Examples 6–7)

61. $f(x) = \begin{cases} |x| & \text{for } x < 2 \\ -x + 4 & \text{for } x \ge 2 \end{cases}$

62. $h(x) = \begin{cases} -2x & \text{for } x < 0 \\ \sqrt{x} & \text{for } x \ge 0 \end{cases}$

63. $g(x) = \begin{cases} x + 2 & \text{for } x < -1 \\ -x + 2 & \text{for } x \ge -1 \end{cases}$

64. $k(x) = \begin{cases} 3x & \text{for } x < 1 \\ -3x & \text{for } x \ge 1 \end{cases}$

65. $r(x) = \begin{cases} x^2 - 4 & \text{for } x \le 2 \\ 2x - 4 & \text{for } x > 2 \end{cases}$

66. $s(x) = \begin{cases} -x - 1 & \text{for } x \le -1 \\ \sqrt{x + 1} & \text{for } x > -1 \end{cases}$

67. $t(x) = \begin{cases} -3 & \text{for } -4 \le x < -2 \\ -1 & \text{for } -2 \le x < 0 \\ 1 & \text{for } 0 \le x < 2 \end{cases}$

68. $z(x) = \begin{cases} -1 & \text{for } -3 < x \le -1 \\ 1 & \text{for } -1 < x \le 1 \\ 3 & \text{for } 1 < x \le 3 \end{cases}$

69. $m(x) = \begin{cases} 3 & \text{for } -4 < x < -1 \\ -x & \text{for } -1 \le x < 3 \\ \sqrt{x-3} & \text{for } x \ge 3 \end{cases}$

70. $n(x) = \begin{cases} -4 & \text{for } -3 < x < -1 \\ x & \text{for } -1 \le x < 2 \\ -x^2 + 4 & \text{for } x \ge 2 \end{cases}$

71. a. Graph $f(x) = \begin{cases} -x & \text{for } x < 0 \\ x & \text{for } x \ge 0 \end{cases}$

b. To what basic function from Section 1.6 is the graph of f equivalent?

For Exercises 72–80, evaluate the step function defined by $f(x) = [\![x]\!]$ for the given values of x. **(See Example 8)**

72. $f(-3.7)$ -4 **73.** $f(-4.2)$ -5 **74.** $f(-0.5)$ -1 **75.** $f(-0.09)$ -1 **76.** $f(0.5)$ 0

77. $f(0.09)$ 0 **78.** $f(6)$ 6 **79.** $f(-9)$ -9 **80.** $f(-5)$ -5

For Exercises 81–84, graph the function. **(See Example 8)**

81. $f(x) = [\![x + 3]\!]$ **82.** $g(x) = [\![x - 3]\!]$ **83.** $k(x) = \text{int}\left(\dfrac{1}{2}x\right)$ **84.** $h(x) = \text{int}(2x)$

85. For a recent year, the rate for first class postage was as follows. **(See Example 9)**

Weight not Over	Price
1 oz	$0.44
2 oz	$0.61
3 oz	$0.78
3.5 oz	$0.95

Write a piecewise-defined function to model the cost $C(x)$ to mail a letter first class if the letter is x ounces.

86. The water level in a retention pond started at 5 ft (60 in.) and decreased at a rate of 2 in./day during a 14-day drought. A tropical depression moved through at the beginning of the 15th day and produced rain at an average rate of 2.5 in./day for 5 days. Write a piecewise-defined function to model the water level $L(x)$ (in inches) as a function of the number of days x since the beginning of the drought.

$L(x) = \begin{cases} 60 - 2x & \text{for } 0 \le x \le 14 \\ 32 + 2.5(x - 14) & \text{for } 14 < x \le 19 \end{cases}$

87. A salesperson makes a base salary of $2000 per month. Once he reaches $40,000 in total sales, he earns an additional 5% commission on the amount in sales over $40,000. Write a piecewise-defined function to model the salesperson's total monthly salary $S(x)$ (in $) as a function of the amount in sales x.

88. A cell phone plan charges $49.95 per month plus $14.02 in taxes, plus $0.40 per minute for calls beyond the 600-min monthly limit. Write a piecewise-defined function to model the monthly cost $C(x)$ (in $) as a function of the number of minutes used x for the month.

$C(x) = \begin{cases} 63.97 & \text{for } 0 \le x \le 600 \\ 63.97 + 0.40(x - 600) & \text{for } x > 600 \end{cases}$

Objective 4: Investigate Increasing, Decreasing, and Constant Behavior of a Function

For Exercises 89–96, use interval notation to write the intervals over which f is **(a)** increasing, **(b)** decreasing, and **(c)** constant. **(See Example 10)**

89.

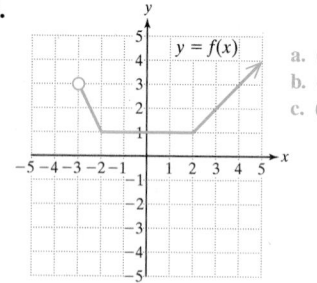

a. $(2, \infty)$
b. $(-3, -2)$
c. $(-2, 2)$

90.

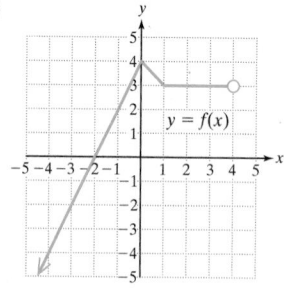

a. $(-\infty, 0)$
b. $(0, 1)$
c. $(1, 4)$

91.

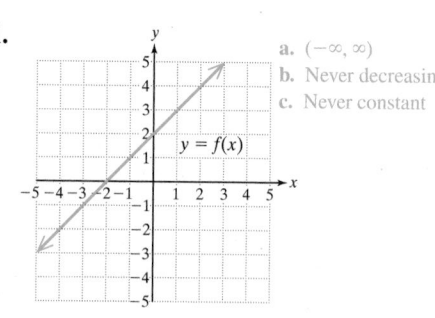

a. $(-\infty, \infty)$
b. Never decreasing
c. Never constant

92.

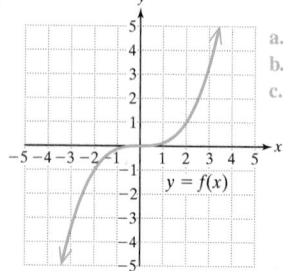

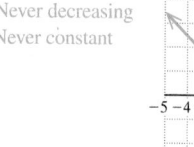

a. $(-\infty, \infty)$
b. Never decreasing
c. Never constant

93.

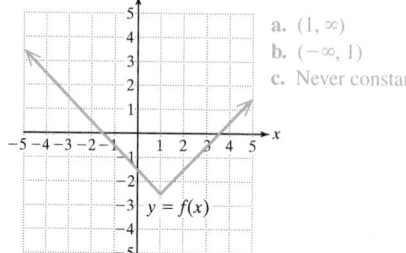

a. $(1, \infty)$
b. $(-\infty, 1)$
c. Never constant

94.

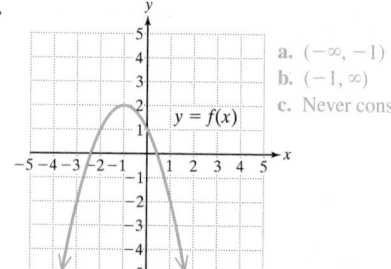

a. $(-\infty, -1)$
b. $(-1, \infty)$
c. Never constant

95.

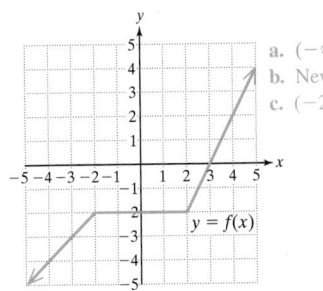

a. $(-\infty, -2) \cup (2, \infty)$
b. Never decreasing
c. $(-2, 2)$

96.

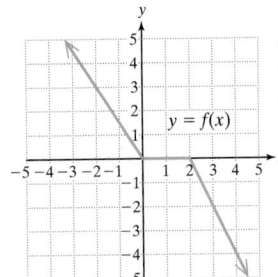

a. Never increasing
b. $(-\infty, 0) \cup (2, \infty)$
c. $(0, 2)$

Objective 5: Determine Relative Minima and Maxima of a Function

For Exercises 97–102, identify the location and value of any relative maxima or minima of the function. (See Example 11)

97.

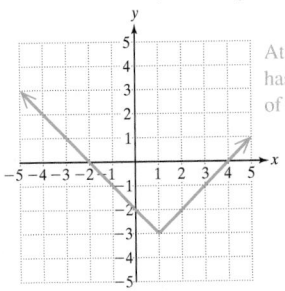

98.

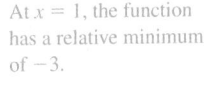

At $x = 1$, the function has a relative minimum of -3.

99.

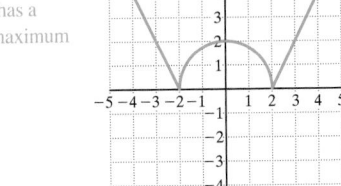

At $x = -1$, the function has a relative maximum of 2.

100.

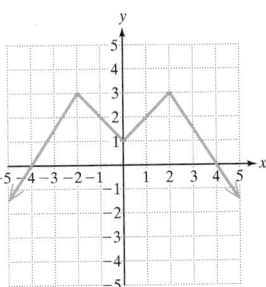

101.

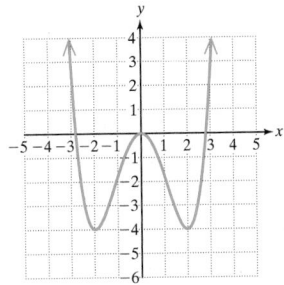

102.

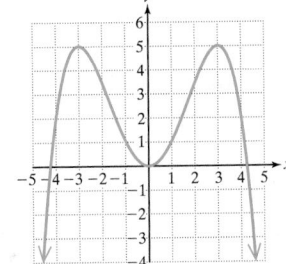

103. The graph shows the depth d (in ft) of a retention pond, t days after recording began.

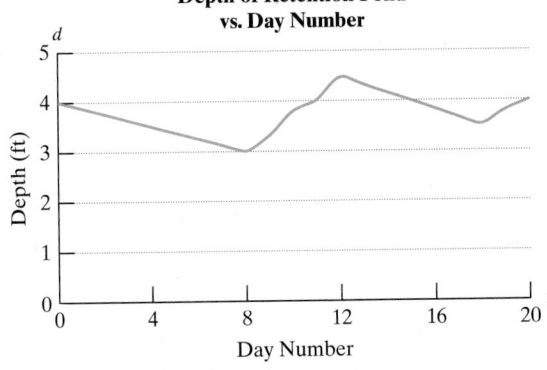

Depth of Retention Pond vs. Day Number

Day Number

a. Over what interval(s) does the depth increase?
   (8, 12) and (18, 20)
b. Over what interval(s) does the depth decrease?
   (0, 8) and (12, 18)
c. Estimate the times and values of any relative maxima or minima on the interval (0, 20).

d. If rain is the only water that enters the pond, explain what the intervals of increasing and decreasing behavior mean in the context of this problem.

104. The graph shows the height h (in meters) of a roller coaster t seconds after the ride starts.

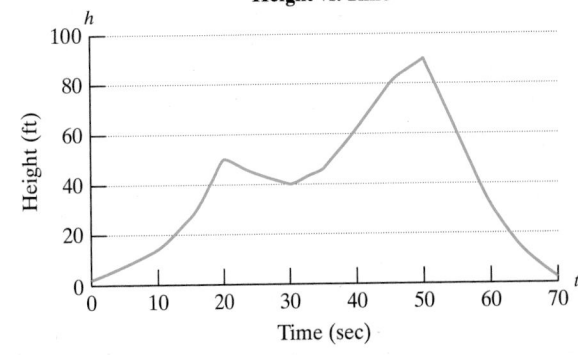

Height vs. Time

Time (sec)

a. Over what interval(s) does the height increase?
   (0, 20) and (30, 50)
b. Over what interval(s) does the height decrease?
   (20, 30) and (50, 70)
c. Estimate the times and values of any relative maxima or minima on the interval (0, 70).   The function has relative maxima of 50 m and 90 m at approximately 20 sec and 50 sec into the ride. The function has a relative minimum of 40 m at a time 30 sec into the ride.

Mixed Exercises

For Exercises 105–110, produce a rule for the function whose graph is shown. (*Hint*: Consider using the basic functions learned in Section 1.6 and transformations of their graphs.)

105.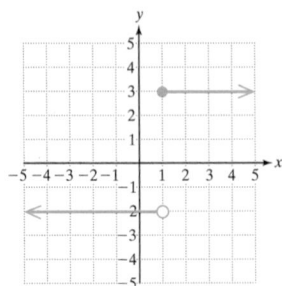

106.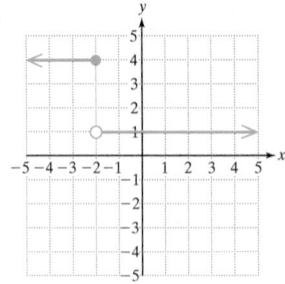

107.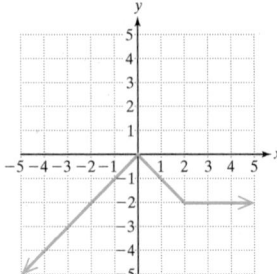

108.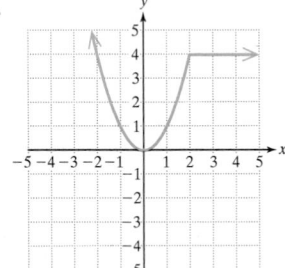

109.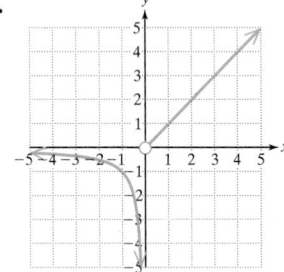

110.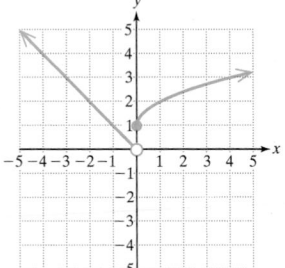

For Exercises 111–112,

    a. Graph the function.

    b. Write the domain in interval notation.

    c. Write the range in interval notation.

    d. Evaluate $f(-1)$, $f(1)$, and $f(2)$.

    e. Find the value(s) of x for which $f(x) = 6$.

    f. Find the value(s) of x for which $f(x) = -3$.

    g. Use interval notation to write the intervals over which f is increasing, decreasing, or constant.

111. $f(x) = \begin{cases} -x^2 + 1 & \text{for } x \leq 1 \\ 2x & \text{for } x > 1 \end{cases}$

112. $f(x) = \begin{cases} |x| & \text{for } x < 2 \\ -x & \text{for } x \geq 2 \end{cases}$

In computer programming, the greatest integer function is sometimes called the "floor" function. Programmers also make use of the "ceiling" function, which returns the smallest integer not less than x. For example: ceil(3.1) = 4. For Exercises 113–114, evaluate the floor and ceiling functions for the given value of x.

floor(x) is the greatest integer less than or equal to x.
ceil(x) is the smallest integer not less than x.

113. a. floor(2.8) 2   **b.** floor(−3.1) −4 **c.** floor(4) 4   **114. a.** floor(5.5) 5 **b.** floor(−0.1) −1 **c.** floor(−2) −2

     d. ceil(2.8) 3   **e.** ceil(−3.1) −3 **f.** ceil(4) 4   **d.** ceil(5.5) 6   **e.** ceil(−0.1) 0 **f.** ceil(−2) −2

Write About It

115. From an equation in x and y, explain how to determine whether the graph of the equation is symmetric with respect to the x-axis, y-axis, or origin.

116. From the graph of a function, how can you determine if the function is even or odd?

If the graph is symmetric with respect to the y-axis, then the function is even. If the graph is symmetric to the origin, then the function is odd.

117. Explain why the relation defined by

$$y = \begin{cases} 2x & \text{for } x \leq 1 \\ 3 & \text{for } x \geq 1 \end{cases}$$

At $x = 1$, there are two different y values. The relation contains the ordered pairs (1, 2) and (1, 3).

is not a function.

118. Explain why the function is discontinuous at $x = 1$.

$$f(x) = \begin{cases} 3x & \text{for } x < 1 \\ 3 & \text{for } x > 1 \end{cases}$$

The two rules defining $f(x)$ are stated for $x < 1$ and for $x > 1$. The function is not defined at $x = 1$. Therefore, there is a "hole" in the function at the point (1, 3).

119. Provide an informal explanation of a relative maximum. A relative maximum of a function is the greatest function value relative to other points on the function nearby.

120. Explain what it means for a function to be increasing on an interval. A function is increasing on an interval I if for all $x_1 < x_2$ on I, $f(x_1) < f(x_2)$. In other words, the function "rises" from left to right over the interval.

Expanding Your Skills

121. Suppose that the average rate of change of a continuous function between any two points to the left of $x = a$ is negative, and the average rate of change of the function between any two points to the right of $x = a$ is positive. Does the function have a relative minimum or maximum at a? Relative minimum

122. Suppose that the average rate of change of a continuous function between any two points to the left of $x = a$ is positive, and the average rate of change of the function between any two points to the right of $x = a$ is negative. Does the function have a relative minimum or maximum at a? Relative maximum

A graph is *concave up* on a given interval if it "bends" upward. A graph is *concave down* on a given interval if it "bends" downward. For Exercises 123–126, determine whether the curve is (a) concave up or concave down and (b) increasing or decreasing.

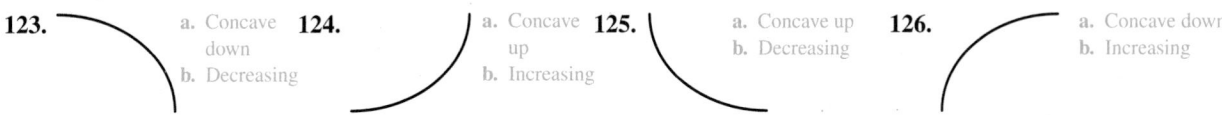

123.   a. Concave down   b. Decreasing

124.   a. Concave up   b. Increasing

125.   a. Concave up   b. Decreasing

126.   a. Concave down   b. Increasing

127. For a recent year, the federal income tax owed by a taxpayer (single—no dependents) was based on the individual's taxable income. (*Source*: Internal Revenue Service, www.irs.gov)

If your taxable income is			
over—	but not over—	The tax is	of the amount over—
$0	$8925	$0 + 10%	$0
$8925	$36,250	$892.50 + 15%	$8925
$36,250	$87,850	$4991.25 + 25%	$36,250

Write a piecewise-defined function that expresses an individual's federal income tax $f(x)$ (in $) as a function of the individual's taxable income x (in $).

Technology Connections

For Exercises 128–131, use a graphing utility to graph the piecewise-defined function.

128. $f(x) = \begin{cases} 2.5x + 2 & \text{for } x \le 1 \\ x^2 - x - 1 & \text{for } x > 1 \end{cases}$

129. $g(x) = \begin{cases} -3.1x - 4 & \text{for } x < -2 \\ -x^3 + 4x - 1 & \text{for } x \ge -2 \end{cases}$

130. $k(x) = \begin{cases} -2.7x - 4.1 & \text{for } x \le -1 \\ -x^3 + 2x + 5 & \text{for } -1 < x < 2 \\ 1 & \text{for } x \ge 2 \end{cases}$

131. $z(x) = \begin{cases} 2.5x + 8 & \text{for } x < -2 \\ -2x^2 + x + 4 & \text{for } -2 \le x < 2 \\ -2 & \text{for } x \ge 2 \end{cases}$

For Exercises 132–135, use a graphing utility to

a. Find the locations and values of the relative maxima and relative minima of the function on the standard viewing window. Round to 3 decimal places.

b. Use interval notation to write the intervals over which f is increasing or decreasing.

132. $f(x) = -0.6x^2 + 2x + 3$

133. $f(x) = 0.4x^2 - 3x - 2.2$

134. $f(x) = 0.5x^3 + 2.1x^2 - 3x - 7$

135. $f(x) = -0.4x^3 - 1.1x^2 + 2x + 3$

SECTION 1.8   Algebra of Functions and Function Composition

OBJECTIVES

1. Perform Operations on Functions
2. Evaluate a Difference Quotient
3. Compose and Decompose Functions

1. Perform Operations on Functions

In Section 1.5, we learned that a profit function can be constructed from the difference of a revenue function and a cost function according to the following rule.

$$P(x) = R(x) - C(x)$$

As this example illustrates, the difference of two functions makes up a new function. New functions can also be formed from the sum, product, and quotient of two functions.

Sum, Difference, Product, and Quotient of Functions

Given functions f and g, the functions $f + g, f - g, f \cdot g$, and $\frac{f}{g}$ are defined by

$$(f + g)(x) = f(x) + g(x)$$
$$(f - g)(x) = f(x) - g(x)$$
$$(f \cdot g)(x) = f(x) \cdot g(x)$$
$$\left(\frac{f}{g}\right)(x) = \frac{f(x)}{g(x)} \text{ provided that } g(x) \neq 0$$

The domains of the functions $f + g, f - g, f \cdot g$, and $\frac{f}{g}$ are all real numbers in the intersection of the domains of the individual functions f and g. For $\frac{f}{g}$, we further restrict the domain to exclude values of x for which $g(x) = 0$.

Classroom Example: p. 225
Exercise 8

EXAMPLE 1   Adding Two Functions

Given $f(x) = \sqrt{25 - x^2}$ and $g(x) = 5$, find $(f + g)(x)$.

Solution:

By definition $(f + g)(x) = f(x) + g(x)$.

$$= \sqrt{25 - x^2} + 5$$

Skill Practice 1   Given $m(x) = -|x|$ and $n(x) = 4$, find $(m + n)(x)$.

In Example 1, the graph of function f is a semicircle and the graph of function g is a horizontal line (Figure 1-38). Therefore, the graph of $y = (f + g)(x)$ is the graph of f with a vertical shift (shown in blue). Notice that each y value on $f + g$ is the sum of the y values from the individual functions f and g.

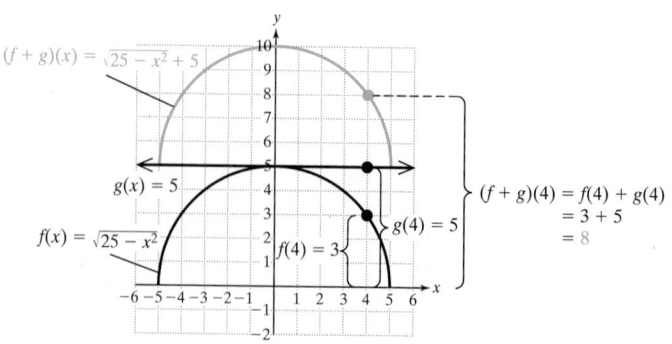

Answer

1. $(m + n)(x) = -|x| + 4$

Figure 1-38

In Example 2, we evaluate the difference, product, and quotient of functions for given values of x.

Classroom Examples: p. 225
Exercises 10, 12, 18

> **EXAMPLE 2**   Evaluating Functions for Given Values of x
>
> Given $m(x) = 4x$, $n(x) = |x - 3|$, and $p(x) = \dfrac{1}{x + 1}$, determine the function values if possible.
>
> **a.** $(m - n)(-2)$ **b.** $(m \cdot p)(1)$ **c.** $\left(\dfrac{p}{n}\right)(3)$
>
> **Solution:**
>
> **a.** $(m - n)(-2) = m(-2) - n(-2)$ **b.** $(m \cdot p)(1) = m(1) \cdot p(1)$
> $$= 4(-2) - |-2 - 3|$$
> $$= -8 - 5$$ $$= 4(1) \cdot \dfrac{1}{1 + 1}$$
> $$= -13$$ $$= 2$$
>
> **c.** $\left(\dfrac{p}{n}\right)(3) = \dfrac{p(3)}{n(3)} = \dfrac{\dfrac{1}{3 + 1}}{|3 - 3|}$ The domain of $\dfrac{p}{n}$ excludes any values of x that make $n(x) = 0$. In this case, $x = 3$ is excluded from the domain.
>
> $$= \dfrac{\frac{1}{4}}{0} \text{ (undefined)}$$

> **Skill Practice 2**   Use the functions defined in Example 2 to find
>
> **a.** $(n - m)(-6)$ **b.** $(n \cdot p)(0)$ **c.** $\left(\dfrac{p}{m}\right)(0)$

When combining two or more functions to create a new function, always be sure to determine the domain of the new function. Notice that in Example 2(c), the function $\frac{p}{n}$ is not defined for $x = -1$ or for $x = 3$.

$$\left(\dfrac{p}{n}\right)(x) = \dfrac{p(x)}{n(x)} = \dfrac{\dfrac{1}{x + 1}}{|x - 3|}$$ ⟵ Denominator is zero for $x = -1$.
⟵ Denominator of the complex fraction is zero for $x = 3$.

Classroom Examples: p. 225
Exercises 20, 22

> **EXAMPLE 3**   Combining Functions and Determining Domain
>
> Given $g(x) = 2x$, $h(x) = x^2 - 4x$, and $k(x) = \sqrt{x - 1}$,
>
> **a.** Find $(g - h)(x)$ and write the domain of $g - h$ in interval notation.
> **b.** Find $(g \cdot k)(x)$ and write the domain of $g \cdot k$ in interval notation.
>
> **c.** Find $\left(\dfrac{k}{h}\right)(x)$ and write the domain of $\dfrac{k}{h}$ in interval notation.
>
> **Solution:**
>
> **a.** $(g - h)(x) = g(x) - h(x)$
> $$= 2x - (x^2 - 4x)$$ The domain of g is $(-\infty, \infty)$.
> $$= -x^2 + 6x$$ The domain of h is $(-\infty, \infty)$.
> Therefore, the intersection of their
> The domain is $(-\infty, \infty)$.   domains is $(-\infty, \infty)$.

Answers

2. a. 33   **b.** 3   **c.** Undefined

b. $(g \cdot k)(x) = g(x) \cdot k(x)$

$\qquad\qquad = 2x\sqrt{x-1}$

The domain is $[1, \infty)$.

> The domain of g is $(-\infty, \infty)$.
> The domain of k is $[1, \infty)$.
> Therefore, the intersection of their domains is $[1, \infty)$.

c. $\left(\dfrac{k}{h}\right)(x) = \dfrac{k(x)}{h(x)} = \dfrac{\sqrt{x-1}}{x^2 - 4x}$

$\qquad\qquad = \dfrac{\sqrt{x-1}}{x(x-4)}$

The domain is $[1, 4) \cup (4, \infty)$.

> The domain of k is $[1, \infty)$.
> The domain of h is $(-\infty, \infty)$.
> The intersection of their domains is $[1, \infty)$.
>
> However, we must also exclude values of x that make the denominator zero. In this case, exclude $x = 0$ and $x = 4$. The value $x = 0$ is already excluded because it is not on the interval $[1, \infty)$. Excluding $x = 4$, the domain of $\frac{k}{h}$ is $[1, 4) \cup (4, \infty)$.

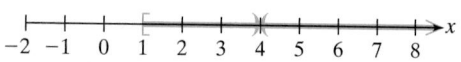

Skill Practice 3 Given $m(x) = x + 3$, $n(x) = x^2 - 9$, and $p(x) = \sqrt{x + 1}$,

a. Find $(n - m)(x)$ and write the domain of $n - m$ in interval notation.

b. Find $(m \cdot p)(x)$ and write the domain of $m \cdot p$ in interval notation.

c. Find $\left(\dfrac{p}{n}\right)(x)$ and write the domain of $\dfrac{p}{n}$ in interval notation.

---

2. Evaluate a Difference Quotient

In Section 1.4, we learned that if f is defined on an interval $[x_1, x_2]$, then the average rate of change of f between $(x_1, f(x_1))$ and $(x_2, f(x_2))$ is given by

$$m = \frac{f(x_2) - f(x_1)}{x_2 - x_1}. \quad \text{(Figure 1-39)}$$

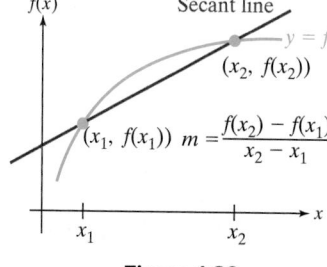

Figure 1-39

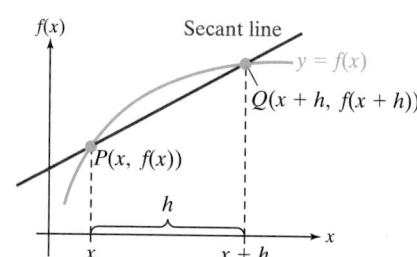

Figure 1-40

> **TIP** h is taken to be a positive real number, implying that $h \neq 0$.

Now we look at a related idea. Let P be an arbitrary point $(x, f(x))$ on the function f. Let h be a positive real number and let Q be the point $(x + h, f(x + h))$. See Figure 1-40. The average rate of change between P and Q is the slope of the secant line and is given by:

$$m = \frac{f(x + h) - f(x)}{(x + h) - x}$$

$$= \frac{f(x + h) - f(x)}{h} \quad \text{(Difference quotient)}$$

Answers

3. a. $(n - m)(x) = x^2 - x - 12$;
   Domain: $(-\infty, \infty)$

b. $(m \cdot p)(x) = (x + 3)\sqrt{x + 1}$;
   Domain: $[-1, \infty)$

c. $\left(\dfrac{p}{n}\right)(x) = \dfrac{\sqrt{x + 1}}{x^2 - 9}$;
   Domain: $[-1, 3) \cup (3, \infty)$

The expression on the right is called the **difference quotient** and is very important for the foundation of calculus. In Examples 4 and 5, we practice evaluating the difference quotient for two functions.

Classroom Example: p. 225
Exercise 34

EXAMPLE 4   **Finding a Difference Quotient**

Given $f(x) = 3x - 5$,

  a. Find $f(x + h)$.

  b. Find the difference quotient, $\dfrac{f(x + h) - f(x)}{h}$.

Solution:

  a. $f(x + h) = 3(x + h) - 5$ Substitute $(x + h)$ for x.

               $= 3x + 3h - 5$

  b. $\dfrac{f(x + h) - f(x)}{h} = \dfrac{\overbrace{(3x + 3h - 5)}^{f(x+h)} - \overbrace{(3x - 5)}^{f(x)}}{h}$

                    $= \dfrac{3x + 3h - 5 - 3x + 5}{h}$ Clear parentheses.

                    $= \dfrac{3h}{h}$ Combine like terms.

                    $= 3$ Simplify the fraction.

Skill Practice 4   Given $f(x) = 4x - 2$,

  a. Find $f(x + h)$.

  b. Find the difference quotient, $\dfrac{f(x + h) - f(x)}{h}$.

Classroom Example: p. 226
Exercise 40

EXAMPLE 5   **Finding a Difference Quotient**

Given $f(x) = -2x^2 + 4x - 1$,

  a. Find $f(x + h)$.

  b. Find the difference quotient, $\dfrac{f(x + h) - f(x)}{h}$.

Solution:

  a. $f(x + h) = -2(x + h)^2 + 4(x + h) - 1$ Substitute $(x + h)$ for x.

               $= -2(x^2 + 2xh + h^2) + 4x + 4h - 1$

               $= -2x^2 - 4xh - 2h^2 + 4x + 4h - 1$

  b. $\dfrac{f(x + h) - f(x)}{h} = \dfrac{\overbrace{(-2x^2 - 4xh - 2h^2 + 4x + 4h - 1)}^{f(x+h)} - \overbrace{(-2x^2 + 4x - 1)}^{f(x)}}{h}$

          $= \dfrac{-2x^2 - 4xh - 2h^2 + 4x + 4h - 1 + 2x^2 - 4x + 1}{h}$ Clear parentheses.

          $= \dfrac{-4xh - 2h^2 + 4h}{h}$ Combine like terms.

          $= \dfrac{\overset{1}{\cancel{h}}(-4x - 2h + 4)}{\cancel{h}}$ Factor numerator and denominator, and simplify the fraction.

          $= -4x - 2h + 4$

Answers

4. a. $4x + 4h - 2$ **b.** 4

> **Skill Practice 5**   Given $f(x) = -x^2 - 5x + 2$,
>
> **a.** Find $f(x + h)$.
>
> **b.** Find the difference quotient, $\dfrac{f(x + h) - f(x)}{h}$.

3. Compose and Decompose Functions

The next operation on functions we present is called the composition of functions. Informally, this involves a substitution process in which the output from one function becomes the input to another function.

> **Composition of Functions**
>
> The **composition of f and g,** denoted $f \circ g$ is defined by $(f \circ g)(x) = f(g(x))$. The domain of $f \circ g$ is the set of real numbers x in the domain of g such that $g(x)$ is in the domain of f.

To visualize the composition of functions $(f \circ g)(x) = f(g(x))$, consider Figure 1-41.

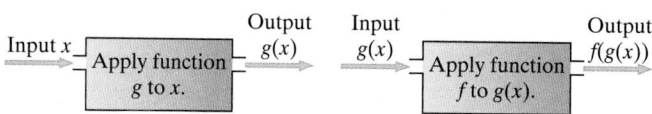

Figure 1-41

Classroom Examples: p. 226
Exercises 50, 56, 58

> **EXAMPLE 6   Composing Functions**
>
> Given $f(x) = x^2 + 2x$ and $g(x) = x - 4$, find
>
> **a.** $f(g(6))$ **b.** $g(f(-3))$ **c.** $(f \circ g)(0)$ **d.** $(g \circ f)(5)$
>
> **Solution:**
>
> **a.** $f(\overset{\frown}{g(6)}) = f(2)$ Evaluate $g(6)$ first. $g(6) = 6 - 4 = 2$.
> $\quad\quad\quad\quad = 8$ $f(2) = (2)^2 + 2(2) = 8$
>
> **b.** $g(\overset{\frown}{f(-3)}) = g(3)$ Evaluate $f(-3)$ first. $f(-3) = (-3)^2 + 2(-3) = 3$.
> $\quad\quad\quad\quad\quad = -1$ $g(3) = 3 - 4 = -1$
>
> **c.** $(f \circ g)(0) = f(g(0))$ Evaluate $g(0)$ first. $g(0) = 0 - 4 = -4$.
> $\quad\quad\quad\quad\quad = f(-4)$ $f(-4) = (-4)^2 + 2(-4) = 8$
> $\quad\quad\quad\quad\quad = 8$
>
> **d.** $(g \circ f)(5) = g(f(5))$ Evaluate $f(5)$ first. $f(5) = (5)^2 + 2(5) = 35$.
> $\quad\quad\quad\quad\quad = g(35)$ $g(35) = 35 - 4 = 31$
> $\quad\quad\quad\quad\quad = 31$

> **Skill Practice 6**   Refer to functions f and g given in Example 6. Find
>
> **a.** $f(g(-4))$ **b.** $g(f(-5))$ **c.** $(f \circ g)(9)$ **d.** $(g \circ f)(10)$

Answers

5. a. $-x^2 - 2xh - h^2 - 5x - 5h + 2$
  b. $-2x - h - 5$
6. a. 48   **b.** 11   **c.** 35
  d. 116

In Example 7, we practice composing functions and identifying the domain of the composite function. This example also illustrates that function composition is not commutative. That is, $(f \circ g)(x) \neq (g \circ f)(x)$ for all functions f and g.

Classroom Example: p. 226
Exercise 70

EXAMPLE 7   Composing Functions and Determining Domain

Given $f(x) = 2x - 6$ and $g(x) = \frac{1}{x + 4}$, write a rule for each function and write the domain in interval notation.

a. $(f \circ g)(x)$ **b.** $(g \circ f)(x)$

Solution:

a. $(f \circ g)(x) = f(g(x)) = 2(g(x)) - 6$

$$= 2\left(\frac{1}{x + 4}\right) - 6$$

Function g has the restriction that $x \neq -4$.

$$= \frac{2}{x + 4} - 6 \quad \text{provided } x \neq -4$$

The domain of f is all real numbers. Therefore, no further restrictions need to be imposed.

The domain is $(-\infty, -4) \cup (-4, \infty)$.

b. $(g \circ f)(x) = g(f(x)) = \dfrac{1}{f(x) + 4}$

The domain of f has no restrictions.

$$= \frac{1}{(2x - 6) + 4} \qquad f(x) \neq -4$$

However, function g must not have an input value of -4. Therefore, we have the restriction $f(x) \neq -4$. Thus,

$$= \frac{1}{2x - 2} \quad \text{provided } x \neq 1$$

$$2x - 6 = -4$$
$$2x = 2$$

The domain is $(-\infty, 1) \cup (1, \infty)$.

$x = 1$ must be excluded.

Skill Practice 7   Given $f(x) = 3x + 4$ and $g(x) = \frac{1}{x - 1}$, write a rule for each function and write the domain in interval notation.

a. $(f \circ g)(x)$ **b.** $(g \circ f)(x)$

Classroom Example: p. 226
Exercise 72

EXAMPLE 8   Composing Functions and Determining Domain

Given $m(x) = \frac{1}{x - 5}$ and $p(x) = \sqrt{x - 2}$, find $(m \circ p)(x)$ and write the domain in interval notation.

Solution:

$$(m \circ p)(x) = m(p(x)) = \frac{1}{p(x) - 5}$$

First note that function p has the restriction that $x \geq 2$.

$$p(x) \neq 5$$

The input value for function m must not be 5. Therefore, $p(x) \neq 5$. Thus,

$$= \frac{1}{\sqrt{x - 2} - 5}$$

$$\sqrt{x - 2} = 5$$
$$(\sqrt{x - 2})^2 = (5)^2$$

$$(m \circ p)(x) = \frac{1}{\sqrt{x - 2} - 5}$$

$$x - 2 = 25$$
$$x = 27 \quad \text{must be excluded}$$

The domain is $[2, 27) \cup (27, \infty)$.

Skill Practice 8   Given $f(x) = \sqrt{x - 1}$ and $g(x) = \frac{1}{x - 3}$, find $(g \circ f)(x)$ and write the domain of $g \circ f$ in interval notation.

Answers

7. a. $(f \circ g)(x) = \dfrac{3}{x - 1} + 4$;

  Domain: $(-\infty, 1) \cup (1, \infty)$

b. $(g \circ f)(x) = \dfrac{1}{3x + 3}$;

  Domain: $(-\infty, -1) \cup (-1, \infty)$

8. $(g \circ f)(x) = \dfrac{1}{\sqrt{x - 1} - 3}$;

  Domain: $[1, 10) \cup (10, \infty)$

Classroom Example: p. 226
Exercise 80

EXAMPLE 9 Composing Functions and Determining Domain

Given $k(x) = \dfrac{x}{x-2}$ and $m(x) = \dfrac{6}{x^2-1}$, find $(k \circ m)(x)$ and write the domain in interval notation.

Solution:

$(k \circ m)(x) = k(m(x))$ — Evaluate k at $m(x)$.

$= \dfrac{\left(\dfrac{6}{x^2-1}\right)}{\left(\dfrac{6}{x^2-1}\right) - 2}$ — Substitute $\dfrac{6}{x^2-1}$ for x in $k(x)$.

- m has the restriction that $x^2 - 1 \neq 0$. Therefore, $x \neq \pm 1$.

$= \dfrac{(x^2-1)}{(x^2-1)} \cdot \dfrac{\left(\dfrac{6}{x^2-1}\right)}{\left(\dfrac{6}{x^2-1} - 2\right)}$ — Simplify the complex fraction by multiplying numerator and denominator by the LCD $x^2 - 1$.

$= \dfrac{6}{6 - 2(x^2-1)}$ — Apply the distributive property.

$= \dfrac{6}{-2x^2 + 8}$ — Simplify the denominator.

$= -\dfrac{6}{2x^2 - 8}$ — Factor out -1 from the denominator.

$= -\dfrac{3}{x^2 - 4}$ — Simplify the rational expression.

- Note the added restriction that $x^2 - 4 \neq 0$ which means that $x \neq \pm 2$.

The domain is $(-\infty, -2) \cup (-2, -1) \cup (-1, 1) \cup (1, 2) \cup (2, \infty)$.

Skill Practice 9 Given $r(x) = \dfrac{x}{x+1}$ and $t(x) = \dfrac{5}{x^2-9}$, find $(r \circ t)(x)$ and write the domain in interval notation.

Classroom Example: p. 227
Exercise 88

EXAMPLE 10 Applying Function Composition

At a popular website, the cost to download individual songs is $1.49 per song. In addition, a first-time visitor to the website has a one-time coupon for $1.00 off.

a. Write a function to represent the cost $C(x)$ (in $) for a first-time visitor to purchase x songs.

b. The sales tax for online purchases depends on the location of the business and customer. If the sales tax rate on a purchase is 6%, write a function to represent the total cost $T(a)$ for a first-time visitor who buys a dollars in songs.

c. Find $(T \circ C)(x)$ and interpret the meaning in context.

d. Evaluate $(T \circ C)(10)$ and interpret the meaning in context.

Answer

9. $(r \circ t)(x) = \dfrac{5}{x^2-4}$; Domain: $(-\infty, -3) \cup (-3, -2) \cup (-2, 2) \cup (2, 3) \cup (3, \infty)$

Solution:

a. $C(x) = 1.49x - 1.00; \ x \geq 1$ — The cost function is a linear function with $1.49 as the variable rate per song.

b. $T(a) = a + 0.06a$ — The total cost is the sum of the cost of the songs plus the sales tax.
$= 1.06a$

c. $(T \circ C)(x) = T(C(x))$
$= 1.06(C(x))$
$= 1.06(1.49x - 1.00)$ Substitute $1.49x - 1.00$ for $C(x)$.
$= 1.5794x - 1.06$

$(T \circ C)(x) = 1.5794x - 1.06$ represents the total cost to buy x songs for a first-time visitor to the website.

d. $(T \circ C)(x) = 1.5794x - 1.06$
$(T \circ C)(10) = 1.5794(10) - 1.06$
$= 14.734$

The total cost for a first-time visitor to buy 10 songs is $14.73.

Skill Practice 10   An artist shops online for tubes of watercolor paint. The cost is $16 for each 14-mL tube.

a. Write a function representing the cost $C(x)$ (in $) for x tubes of paint.

b. There is a 5.5% sales tax on the cost of merchandise and a fixed cost of $4.99 for shipping. Write a function representing the total cost $T(a)$ for a dollars spent in merchandise.

c. Find $(T \circ C)(x)$ and interpret the meaning in context.

d. Evaluate $(T \circ C)(18)$ and interpret the meaning in context.

The composition of two functions creates a new function in which the output from one function becomes the input to the other. We can also reverse this process. That is, we can decompose a composite function into two or more simpler functions.

For example, consider the function h defined by $h(x) = (x - 3)^2$. To write h as a composition of two functions, we have $h(x) = (f \circ g)(x) = f(g(x))$. The function g is the "inside" function and f is the "outside" function. So one natural choice for g and f would be:

$g(x) = x - 3$ Function g subtracts 3 from the input value.
$f(x) = x^2$ Function f squares the result.
$h(x) = f(g(x)) = (g(x))^2 = (x - 3)^2$

EXAMPLE 11   Decomposing Two Functions

Given $h(x) = |2x^2 - 5|$, find two functions f and g such that $h(x) = (f \circ g)(x)$.

Solution:

We need to find two functions f and g such that $h(x) = (f \circ g)(x) = f(g(x))$. The function h first evaluates the expression $2x^2 - 5$, and then takes the absolute value. Therefore, it would be natural to take the absolute value of $g(x) = 2x^2 - 5$.

We have: $g(x) = 2x^2 - 5$ and $f(x) = |x|$

Check: $h(x) = (f \circ g)(x) = f(g(x)) = |g(x)|$
$= |2x^2 - 5|$

Skill Practice 11   Given $m(x) = \sqrt[3]{5x + 1}$, find two functions f and g such that $m(x) = (f \circ g)(x)$.

In Example 12, we have the graphs of two functions, and we apply function addition, subtraction, multiplication, and composition for selected values of x.

Classroom Example: p. 227
Exercise 102

EXAMPLE 12   Estimating Function Values from a Graph

The graphs of f and g are shown. Evaluate the functions at the given values of x if possible.

a. $(f + g)(1)$
b. $(fg)(0)$
c. $(g - f)(-3)$
d. $(f \circ g)(3)$
e. $(g \circ f)(4)$
f. $f(g(1))$

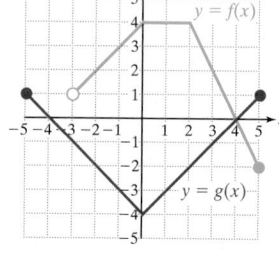

Solution:

a. $(f + g)(1) = f(1) + g(1)$
$= 4 + (-3)$
$= 1$

b. $(fg)(0) = f(0) \cdot g(0)$
$= (4)(-4)$
$= -16$

c. $(g - f)(-3) = g(-3) - f(-3)$ $f(-3)$ is undefined.
$(g - f)(-3)$ is undefined.

d. $(f \circ g)(3) = f(g(3))$
$= f(-1)$
$= 3$

e. $(g \circ f)(4) = g(f(4))$
$= g(0)$
$= -4$

f. $f(g(1)) = f(-3)$ is undefined.

The open dot at $(-3, 1)$ indicates that -3 is not in the domain of f. The value $g(1) = -3$, but $f(-3)$ is undefined. Therefore, $f(g(1))$ is undefined.

Skill Practice 12   Refer to the functions f and g pictured in Example 12. Evaluate the functions at the given values of x if possible.

a. $(f - g)(-2)$
b. $\left(\dfrac{f}{g}\right)(3)$
c. $(gf)(5)$
d. $(g \circ f)(5)$
e. $(f \circ g)(5)$
f. $f(g(0))$

Answers
11. $g(x) = 5x + 1$ and $f(x) = \sqrt[3]{x}$
12. a. 4   **b.** -2 **c.** -2
    d. -2 **e.** 4   **f.** Undefined

SECTION 1.8   Practice Exercises

Prerequisite Review

For Exercises R.1–R.4, write the domain in interval notation.

R.1. $f(x) = \dfrac{x - 1}{x + 2}$ $(-\infty, -2) \cup (-2, \infty)$

R.2. $r(x) = \sqrt{x + 3}$ $[-3, \infty)$

R.3. $h(x) = \dfrac{4}{\sqrt{3 - x}}$ $(-\infty, 3)$

R.4. $p(x) = 2x^2 - 3x + 1$ $(-\infty, \infty)$

R.5. Given $k(x) = x^2 - 2x + 3$, find $k(x + 3)$. $k(x + 3) = x^2 + 4x + 6$

Concept Connections

1. The function $f + g$ is defined by $(f + g)(x) = \underline{\quad f(x) \quad} + \underline{\quad g(x) \quad}$.

2. The function $\dfrac{f}{g}$ is defined by $\left(\dfrac{f}{g}\right)(x) = \underline{\quad \dfrac{f(x)}{g(x)} \quad}$ provided that $\underline{\quad g(x) \quad} \neq 0$.

3. Let h represent a positive real number. Given a function defined by $y = f(x)$, the difference quotient is given by $\underline{\quad \dfrac{f(x + h) - f(x)}{h} \quad}$.

4. The composition of f and g, denoted by $f \circ g$, is defined by $(f \circ g)(x) = \underline{\quad f(g(x)) \quad}$.

Objective 1: Perform Operations on Functions

For Exercises 5–8, find $(f + g)(x)$ and identify the graph of $f + g$. (See Example 1)

5. $f(x) = |x|$ and $g(x) = 3$ $(f + g)(x) = |x| + 3$; Graph d

6. $f(x) = |x|$ and $g(x) = -4$ $(f + g)(x) = |x| - 4$; Graph b

7. $f(x) = x^2$ and $g(x) = -4$ $(f + g)(x) = x^2 - 4$; Graph a

8. $f(x) = x^2$ and $g(x) = 3$ $(f + g)(x) = x^2 + 3$; Graph c

a. 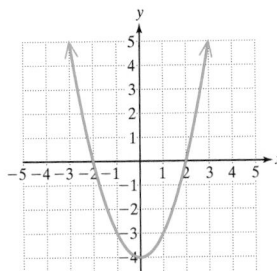   **b.** 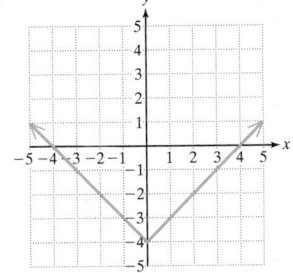   **c.** 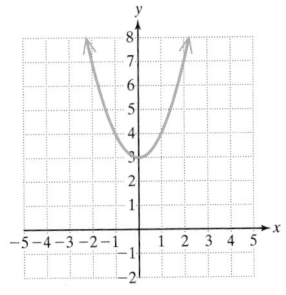   **d.**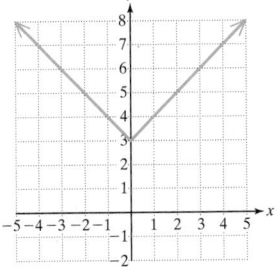

For Exercises 9–18, evaluate the functions for the given values of x. (See Example 2)

$$f(x) = -2x \qquad g(x) = |x + 4| \qquad h(x) = \dfrac{1}{x - 3}$$

9. $(f - g)(3)$ -13

10. $(g - h)(2)$ 7

11. $(f \cdot g)(-1)$ 6

12. $(h \cdot g)(4)$ 8

13. $(g + h)(0)$ $\dfrac{11}{3}$

14. $(f + h)(5)$ $-\dfrac{19}{2}$

15. $\left(\dfrac{f}{g}\right)(8)$ $-\dfrac{4}{3}$

16. $\left(\dfrac{h}{f}\right)(7)$ $-\dfrac{1}{56}$

17. $\left(\dfrac{g}{f}\right)(0)$ Undefined

18. $\left(\dfrac{h}{g}\right)(-4)$ Undefined

For Exercises 19–26, refer to the functions r, p, and q. Find the indicated function and write the domain in interval notation. (See Example 3)

$$r(x) = -3x \qquad p(x) = x^2 + 3x \qquad q(x) = \sqrt{1 - x}$$

19. $(r - p)(x)$

20. $(p - r)(x)$

21. $(p \cdot q)(x)$

22. $(r \cdot q)(x)$

23. $\left(\dfrac{q}{p}\right)(x)$

24. $\left(\dfrac{q}{r}\right)(x)$

25. $\left(\dfrac{p}{q}\right)(x)$

26. $\left(\dfrac{r}{q}\right)(x)$

For Exercises 27–32, refer to functions s, t, and v. Find the indicated function and write the domain in interval notation. (See Example 3)

$$s(x) = \dfrac{x - 2}{x^2 - 9} \qquad t(x) = \dfrac{x - 3}{2 - x} \qquad v(x) = \sqrt{x + 3}$$

27. $(s \cdot t)(x)$

28. $\left(\dfrac{s}{t}\right)(x)$

29. $(s + t)(x)$

30. $(s - t)(x)$

31. $(s \cdot v)(x)$

32. $\left(\dfrac{v}{s}\right)(x)$

Objective 2: Evaluate a Difference Quotient

For Exercises 33–36, a function is given. (See Examples 4–5)

a. Find $f(x + h)$.

b. Find $\dfrac{f(x + h) - f(x)}{h}$.

33. $f(x) = 5x + 9$

34. $f(x) = 8x + 4$

35. $f(x) = x^2 + 4x$

36. $f(x) = x^2 - 3x$

For Exercises 37–44, find the difference quotient and simplify. (See Examples 4–5)

37. $f(x) = -2x + 5$ -2 **38.** $f(x) = -3x + 8$ -3 **39.** $f(x) = -5x^2 - 4x + 2$ **40.** $f(x) = -4x^2 - 2x + 6$
                                                                 $-10x - 5h - 4$ $-8x - 4h - 2$

41. $f(x) = x^3 + 5$ **42.** $f(x) = x^3 - 2$ **43.** $f(x) = \dfrac{1}{x}$ $-\dfrac{1}{x(x+h)}$
    $3x^2 + 3xh + h^2$ $3x^2 + 3xh + h^2$ **44.** $f(x) = \dfrac{1}{x + 2}$
                                                                     $-\dfrac{1}{(x+2)(x+h+2)}$

45. Given $f(x) = 4\sqrt{x}$,

    a. Find the difference quotient (do not simplify). $\dfrac{4\sqrt{x+h} - 4\sqrt{x}}{h}$

    b. Evaluate the difference quotient for $x = 1$, and the following values of h: $h = 1$, $h = 0.1$, $h = 0.01$, and $h = 0.001$. Round to 4 decimal places.
      1.6569; 1.9524; 1.9950; 1.9995

    c. What value does the difference quotient seem to be approaching as h gets close to 0? 2

46. Given $f(x) = \dfrac{12}{x}$,

    a. Find the difference quotient (do not simplify). $\dfrac{\frac{12}{x+h} - \frac{12}{x}}{h}$

    b. Evaluate the difference quotient for $x = 2$, and the following values of h: $h = 0.1$, $h = 0.01$, $h = 0.001$, and $h = 0.0001$. Round to 4 decimal places.
      -2.8571; -2.9851; -2.9985; -2.9999

    c. What value does the difference quotient seem to be approaching as h gets close to 0? -3

Objective 3: Compose and Decompose Functions

For Exercises 47–62, refer to functions f, g, and h. Evaluate the functions for the given values of x. (See Example 6)

$$f(x) = x^3 - 4x \qquad g(x) = \sqrt{2x} \qquad h(x) = 2x + 3$$

47. $f(g(8))$ 48 **48.** $h(g(2))$ 7 **49.** $h(f(1))$ -3 **50.** $g(f(3))$ $\sqrt{30}$

51. $(f \circ g)(18)$ 192 **52.** $(f \circ h)(-1)$ -3 **53.** $(g \circ f)(5)$ $\sqrt{210}$ **54.** $(h \circ f)(-2)$ 3

55. $(h \circ f)(-3)$ -27 **56.** $(h \circ g)(72)$ 27 **57.** $(g \circ f)(1)$ Undefined (not a real number)   **58.** $(g \circ f)(-4)$ Undefined (not a real number)

59. $(f \circ f)(3)$ 3315 **60.** $(h \circ h)(-4)$ -7 **61.** $(f \circ h \circ g)(2)$ 315 **62.** $(f \circ h \circ g)(8)$ 1287

63. Given $f(x) = 2x + 4$ and $g(x) = x^2$,

    a. Find $(f \circ g)(x)$.   **b.** Find $(g \circ f)(x)$.
    $(f \circ g)(x) = 2x^2 + 4$ $(g \circ f)(x) = 4x^2 + 16x + 16$
    c. Is the operation of function composition commutative?   No

64. Given $k(x) = -3x + 1$ and $m(x) = \dfrac{1}{x}$,

    a. Find $(k \circ m)(x)$. $\dfrac{3}{x} + 1$ **b.** Find $(m \circ k)(x)$. $(m \circ k)(x) = \dfrac{1}{-3x + 1}$
                     $(k \circ m)(x) = -\dfrac{3}{x} + 1$
    c. Is $(k \circ m)(x) = (m \circ k)(x)$?   No

For Exercises 65–76, refer to functions m, n, p, q, and r. Find the indicated function and write the domain in interval notation. (See Examples 7–9)

$$m(x) = \sqrt{x + 8} \qquad n(x) = x - 5 \qquad p(x) = x^2 - 9x \qquad q(x) = \dfrac{1}{x - 10} \qquad r(x) = |2x + 3|$$

65. $(n \circ p)(x)$ **66.** $(p \circ n)(x)$ **67.** $(m \circ n)(x)$ **68.** $(n \circ m)(x)$

69. $(q \circ n)(x)$ **70.** $(q \circ p)(x)$ **71.** $(q \circ r)(x)$ **72.** $(q \circ m)(x)$

73. $(n \circ r)(x)$ **74.** $(r \circ n)(x)$ **75.** $(q \circ q)(x)$ **76.** $(p \circ p)(x)$

For Exercises 77–80, find $(f \circ g)(x)$ and write the domain in interval notation. (See Example 9)

77. $f(x) = \dfrac{3}{x^2 - 16}$, $g(x) = \sqrt{2 - x}$ **78.** $f(x) = \dfrac{4}{x^2 - 9}$, $g(x) = \sqrt{3 - x}$

79. $f(x) = \dfrac{x}{x - 1}$, $g(x) = \dfrac{9}{x^2 - 16}$ **80.** $f(x) = \dfrac{x}{x + 4}$, $g(x) = \dfrac{3}{x^2 - 1}$

81. Given $f(x) = \dfrac{1}{x - 2}$, find $(f \circ f)(x)$ and write the domain in interval notation.

82. Given $g(x) = \sqrt{x - 3}$, find $(g \circ g)(x)$ and write the domain in interval notation. $(g \circ g)(x) = \sqrt{\sqrt{x - 3} - 3}$; $[12, \infty)$

For Exercises 83–86, find the indicated functions.

$$f(x) = 2x + 1 \qquad g(x) = x^2 \qquad h(x) = \sqrt[3]{x}$$

83. $(f \circ g \circ h)(x)$ $(f \circ g \circ h)(x) = 2(\sqrt[3]{x})^2 + 1$ **84.** $(g \circ f \circ h)(x)$ $(g \circ f \circ h)(x) = (2\sqrt[3]{x} + 1)^2$

85. $(h \circ g \circ f)(x)$ $(h \circ g \circ f)(x) = \sqrt[3]{(2x + 1)^2}$ **86.** $(g \circ h \circ f)(x)$ $(g \circ h \circ f)(x) = (\sqrt[3]{2x + 1})^2$

87. A law office orders business stationery. The cost is $21.95 per box. **(See Example 10)**

    a. Write a function that represents the cost $C(x)$ (in $) for x boxes of stationery. $C(x) = 21.95x$

    b. There is a 6% sales tax on the cost of merchandise and $10.99 for shipping. Write a function that represents the total cost $T(a)$ for a dollars spent in merchandise and shipping. $T(a) = 1.06a + 10.99$

    c. Find $(T \circ C)(x)$. $(T \circ C)(x) = 23.267x + 10.99$

    d. Find $(T \circ C)(4)$ and interpret its meaning in the context of this problem. $(T \circ C)(4) = 104.058$; The total cost to purchase 4 boxes of stationery is $104.06.

89. A bicycle wheel turns at a rate of 80 revolutions per minute (rpm).

    a. Write a function that represents the number of revolutions $r(t)$ in t minutes. $r(t) = 80t$

    b. For each revolution of the wheels, the bicycle travels 7.2 ft. Write a function that represents the distance traveled $d(r)$ (in ft) for r revolutions of the wheel. $d(r) = 7.2r$

    c. Find $(d \circ r)(t)$ and interpret the meaning in the context of this problem. $(d \circ r)(t) = 576t$ represents the distance traveled (in ft) in t minutes.

    d. Evaluate $(d \circ r)(30)$ and interpret the meaning in the context of this problem. $(d \circ r)(30) = 17,280$ means that the bicycle will travel 17,280 ft (approximately 3.27 mi) in 30 min.

88. The cost to buy tickets online for a dance show is $60 per ticket.

    a. Write a function that represents the cost $C(x)$ (in $) for x tickets to the show. $C(x) = 60x$

    b. There is a sales tax of 5.5% and a processing fee of $8.00 for a group of tickets. Write a function that represents the total cost $T(a)$ for a dollars spent on tickets. $T(a) = 1.055a + 8$

    c. Find $(T \circ C)(x)$. $(T \circ C)(x) = 63.3x + 8$

    d. Find $(T \circ C)(6)$ and interpret its meaning in the context of this problem. $(T \circ C)(6) = 387.8$; The total cost to purchase 6 tickets is $387.80.

90. While on vacation in France, Sadie bought a box of almond croissants. Each croissant cost €2.4 (euros).

    a. Write a function that represents the cost $C(x)$ (in euros) for x croissants. $C(x) = 2.4x$

    b. At the time of the purchase, the exchange rate was $1 = €0.80. Write a function that represents the amount $D(C)$ (in $) for C euros spent. $D(C) = \dfrac{C}{0.8}$

    c. Find $(D \circ C)(x)$ and interpret the meaning in the context of this problem. $(D \circ C)(x) = 3x$ represents the cost in dollars to buy x croissants.

    d. Evaluate $(D \circ C)(12)$ and interpret the meaning in the context of this problem. $(D \circ C)(12) = 36$ means that 12 croissants cost $36.

For Exercises 91–98, find two functions f and g such that $h(x) = (f \circ g)(x)$. (See Example 11)

91. $h(x) = (x + 7)^2$
$f(x) = x^2$ and $g(x) = x + 7$

92. $h(x) = (x - 8)^2$
$f(x) = x^2$ and $g(x) = x - 8$

93. $h(x) = \sqrt[3]{2x + 1}$
$f(x) = \sqrt[3]{x}$ and $g(x) = 2x + 1$

94. $h(x) = \sqrt[4]{9x - 5}$
$f(x) = \sqrt[4]{x}$ and $g(x) = 9x - 5$

95. $h(x) = |2x^2 - 3|$
$f(x) = |x|$ and $g(x) = 2x^2 - 3$

96. $h(x) = |4 - x^2|$
$f(x) = |x|$ and $g(x) = 4 - x^2$

97. $h(x) = \dfrac{5}{x + 4}$
$f(x) = \dfrac{5}{x}$ and $g(x) = x + 4$

98. $h(x) = \dfrac{11}{x - 3}$
$f(x) = \dfrac{11}{x}$ and $g(x) = x - 3$

Mixed Exercises

For Exercises 99–102, the graphs of two functions are shown. Evaluate the function at the given values of x, if possible. (See Example 12)

99. **a.** $(f + g)(0)$ 1

    b. $(g - f)(2)$ −1

    c. $(g \cdot f)(-1)$ −6

    d. $\left(\dfrac{g}{f}\right)(1)$ $-\dfrac{1}{2}$

    e. $(f \circ g)(4)$ 1

    f. $(g \circ f)(0)$ 1

    g. $g(f(4))$ −2

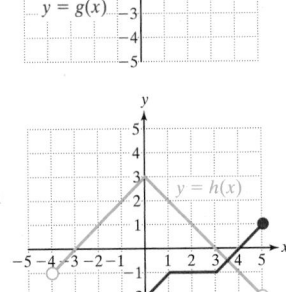

100. **a.** $(f + g)(0)$ −2

    b. $(g - f)(1)$ −3

    c. $(g \cdot f)(2)$ −2

    d. $\left(\dfrac{f}{g}\right)(-3)$ 2

    e. $(f \circ g)(3)$ 1

    f. $(g \circ f)(0)$ −2

    g. $g(f(-4))$ −1

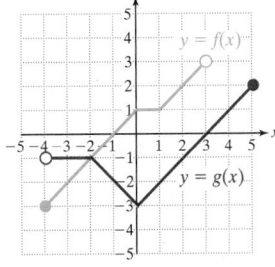

101. **a.** $(h + k)(-1)$ −1

    b. $(h \cdot k)(4)$ 0

    c. $\left(\dfrac{k}{h}\right)(-3)$ Undefined

    d. $(k - h)(1)$ −3

    e. $(k \circ h)(4)$ −3

    f. $(h \circ k)(-2)$ 0

    g. $h(k(3))$ 2

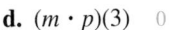

102. **a.** $(m + p)(1)$ 6

    b. $(p - m)(-4)$ 6

    c. $\left(\dfrac{m}{p}\right)(3)$ Undefined

    d. $(m \cdot p)(3)$ 0

    e. $(m \circ p)(0)$ 3

    f. $(p \circ m)(0)$ Undefined

    g. $p(m(-4))$ 4

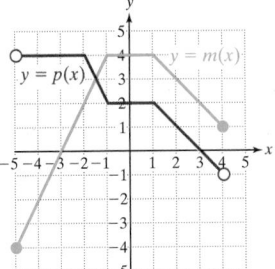

For Exercises 103–110, refer to the functions f and g and evaluate the functions for the given values of x.

$$f = \{(2, 4), (6, -1), (4, -2), (0, 3), (-1, 6)\} \quad \text{and} \quad g = \{(4, 3), (0, 6), (5, 7), (6, 0)\}$$

103. $(f + g)(4)$ 1

104. $(g \cdot f)(0)$ 18

105. $(g \circ f)(2)$ 3

106. $(f \circ g)(0)$ −1

107. $(g \circ g)(6)$ 6

108. $(f \circ f)(-1)$ −1

109. $(f \circ g)(5)$ Undefined

110. $(g \circ f)(0)$ Undefined

111. The diameter d of a sphere is twice the radius r. The volume of the sphere as a function of its radius is given by $V(r) = \frac{4}{3}\pi r^3$.

   a. Write the diameter d of the sphere as a function of the radius r. $d(r) = 2r$

   b. Write the radius r as a function of the diameter d. $r(d) = \frac{d}{2}$

   c. Find $(V \circ r)(d)$ and interpret its meaning.

     $(V \circ r)(d) = \frac{1}{6}\pi d^3$ is the volume of the sphere as a function of its diameter.

112. Consider a right circular cone with given height h. The volume of the cone as a function of its radius r is given by $V(r) = \frac{1}{3}\pi r^2 h$. Consider a right circular cone with fixed height $h = 6$ in.

   a. Write the diameter d of the cone as a function of the radius r. $d(r) = 2r$

   b. Write the radius r as a function of the diameter d. $r(d) = \frac{d}{2}$

   c. Find $(V \circ r)(d)$ and interpret its meaning. Assume that $h = 6$ in. $(V \circ r)(d) = \frac{1}{2}\pi d^2$ is the volume of the cone as a function of its diameter.

113. An investment earns 4.5% interest paid at the end of 1 yr. If x is the amount of money initially invested, then $A(x) = 1.045x$ represents the amount of money in the account 1 yr later. Find $(A \circ A)(x)$ and interpret the result.

114. The population in a certain town has been decreasing at a rate of 2% per year. If x is the population at a certain fixed time, then $P(x) = 0.98x$ represents the population 1 yr later. Find $(P \circ P)(x)$ and interpret the result.

115. Suppose that a function H gives the high temperature $H(x)$ (in °F) for day x. Suppose that a function L gives the low temperature $L(x)$ (in °F) for day x. What does $\left(\frac{H+L}{2}\right)(x)$ represent? $\left(\frac{H+L}{2}\right)(x)$ represents the average of the high and low temperatures for day x.

116. For the given figure,

   a. What does $A_1(x) = \pi(x + 5)^2$ represent?

   b. What does $A_2(x) = \pi x^2$ represent?

   c. Find $(A_1 - A_2)(x)$ and interpret its meaning.

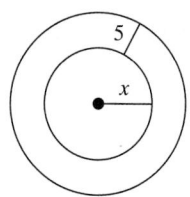

117. For the given figure,

   a. Write an expression $S_1(x)$ that represents the area of the rectangle. $S_1(x) = x^2 + 4x$

   b. Write an expression $S_2(x)$ that represents the area of the semicircle. $S_2(x) = \frac{1}{8}\pi x^2$

   c. Find $(S_1 - S_2)(x)$ and interpret its meaning.

     $(S_1 - S_2)(x) = x^2 + 4x - \frac{1}{8}\pi x^2$ and represents the area of the region outside the semicircle, but inside the rectangle.

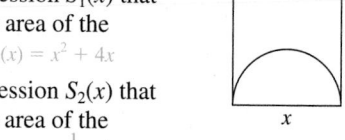

Write About It

118. Given functions f and g, explain how to determine the domain of $\left(\frac{f}{g}\right)(x)$.

119. Given functions f and g, explain how to determine the domain of $(f \circ g)(x)$.

120. Explain what the difference quotient represents.

Expanding Your Skills

121. Given $f(x) = \sqrt{x + 3}$,

   a. Find the difference quotient. $\frac{\sqrt{x + h + 3} - \sqrt{x + 3}}{h}$

   b. Rationalize the numerator of the expression in part (a) and simplify. $\frac{1}{\sqrt{x + h + 3} + \sqrt{x + 3}}$

   c. Evaluate the expression in part (b) for $h = 0$. $\frac{1}{2\sqrt{x + 3}}$

122. Given $f(x) = \sqrt{x - 4}$,

   a. Find the difference quotient. $\frac{\sqrt{x + h - 4} - \sqrt{x - 4}}{h}$

   b. Rationalize the numerator of the expression in part (a) and simplify. $\frac{1}{\sqrt{x + h - 4} + \sqrt{x - 4}}$

   c. Evaluate the expression in part (b) for $h = 0$. $\frac{1}{2\sqrt{x - 4}}$

123. A car traveling 60 mph (88 ft/sec) undergoes a constant deceleration until it comes to rest approximately 9.09 sec later. The distance $d(t)$ (in ft) that the car travels t seconds after the brakes are applied is given by $d(t) = -4.84t^2 + 88t$, where $0 \le t \le 9.09$. **(See Example 5)**

   a. Find the difference quotient $\frac{d(t + h) - d(t)}{h}$. $-9.68t - 4.84h + 88$

   Use the difference quotient to determine the average rate of speed on the following intervals for t.

   b. $[0, 2]$ (*Hint*: $t = 0$ and $h = 2$) 78.32 ft/sec

   c. $[2, 4]$ (*Hint*: $t = 2$ and $h = 2$) 58.96 ft/sec

   d. $[4, 6]$ (*Hint*: $t = 4$ and $h = 2$) 39.6 ft/sec

   e. $[6, 8]$ (*Hint*: $t = 6$ and $h = 2$) 20.24 ft/sec

124. A car accelerates from 0 to 60 mph (88 ft/sec) in 8.8 sec. The distance $d(t)$ (in ft) that the car travels t seconds after motion begins is given by $d(t) = 5t^2$, where $0 \le t \le 8.8$.

   a. Find the difference quotient $\frac{d(t + h) - d(t)}{h}$. $10t + 5h$

   Use the difference quotient to determine the average rate of speed on the following intervals for t.

   b. $[0, 2]$ 10 ft/sec

   c. $[2, 4]$ 30 ft/sec

   d. $[4, 6]$ 50 ft/sec

   e. $[6, 8]$ 70 ft/sec

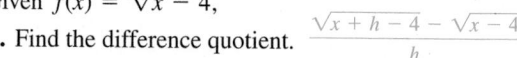

125. If a is b plus eight, and c is the square of a, write c as a function of b. $c(b) = (b + 8)^2$

126. If q is r minus seven, and s is the square root of q, write s as a function of r. $s(r) = \sqrt{r - 7}$

127. If x is twice y, and z is four less than x, write z as a function of y. $z(y) = 2y - 4$

128. If m is one-third of n, and p is two less than m, write p as a function of n. $p(n) = \dfrac{1}{3}n - 2$

129. Given $f(x) = \sqrt[3]{4x^2 + 1}$, define functions m, n, h, and k such that $f(x) = (m \circ n \circ h \circ k)(x)$.
$m(x) = \sqrt[3]{x}, n(x) = x + 1, h(x) = 4x, k(x) = x^2$

130. Given $f(x) = |-2x^3 - 4|$, define functions m, n, h, and k such that $f(x) = (m \circ n \circ h \circ k)(x)$.
$m(x) = |x|, n(x) = x - 4, h(x) = -2x, k(x) = x^3$

CHAPTER 1   KEY CONCEPTS

SECTION 1.1 The Rectangular Coordinate System and Graphing Utilities	Reference
The **distance** between two points (x_1, y_1) and (x_2, y_2) in a rectangular coordinate system is given by $$d = \sqrt{(x_2 - x_1)^2 + (y_2 - y_1)^2}.$$	p. 121
The **midpoint** between the points is given by $M = \left(\dfrac{x_1 + x_2}{2}, \dfrac{y_1 + y_2}{2}\right)$.	p. 122
• To find an x-intercept $(a, 0)$ of the graph of an equation, substitute 0 for y and solve for x. • To find a y-intercept $(0, b)$ of the graph of an equation, substitute 0 for x and solve for y.	p. 124

SECTION 1.2 Circles	Reference
The **standard form** of an equation of a circle with radius r and center (h, k) is $(x - h)^2 + (y - k)^2 = r^2$.	p. 132
An equation of a circle written in the form $x^2 + y^2 + Ax + By + C = 0$ is called the **general form** of an equation of a circle.	p. 133

SECTION 1.3 Functions and Relations	Reference
A set of ordered pairs (x, y) is called a **relation** in x and y. The set of x values is the **domain** of the relation, and the set of y values is the **range** of the relation.	p. 137
Given a relation in x and y, we say that **y is a function of x** if for each value of x in the domain, there is exactly one value of y in the range.	p. 138
The vertical line test tells us that the graph of a relation defines y as a function of x if no vertical line intersects the graph in more than one point.	p. 139

Given a function defined by $y = f(x)$,
  • The x-intercepts are the real solutions to $f(x) = 0$.
  • The y-intercept is given by $f(0)$.
	p. 142

Given $y = f(x)$, the domain of f is the set of real numbers x that when substituted into the function produce a real number. This excludes
  • Values of x that make the denominator zero.
  • Values of x that make a radicand negative within an even-indexed root.
	p. 143

SECTION 1.4 Linear Equations in Two Variables and Linear Functions	Reference
Let A, B, and C represent real numbers where A and B are not both zero. A **linear equation** in the variables x and y is an equation that can be written as $Ax + By = C$.	p. 151
The slope of a line passing through the distinct points (x_1, y_1) and (x_2, y_2) is given by $m = \dfrac{\Delta y}{\Delta x} = \dfrac{y_2 - y_1}{x_2 - x_1}$.	p. 153
Given a line with slope m and y-intercept $(0, b)$, the **slope-intercept form** of the line is given by $y = mx + b$.	p. 155

If f is defined on the interval $[x_1, x_2]$, then the **average rate of change** of f on the interval $[x_1, x_2]$ is the slope of the secant line containing $(x_1, f(x_1))$ and $(x_2, f(x_2))$ and is given by $$m = \frac{f(x_2) - f(x_1)}{x_2 - x_1}$$	p. 157
The x-coordinates of the points of intersection between the graphs of $y = f(x)$ and $y = g(x)$ are the solutions to the equation $f(x) = g(x)$.	p. 159

SECTION 1.5 Applications of Linear Equations and Modeling	Reference
The **point-slope formula** for a line is given by $y - y_1 = m(x - x_1)$ where m is the slope of the line and (x_1, y_1) is a point on the line.	p. 167
• If m_1 and m_2 represent the slopes of two nonvertical parallel lines, then $m_1 = m_2$. • If m_1 and m_2 represent the slopes of two nonvertical perpendicular lines, then $m_1 = -\dfrac{1}{m_2}$ or equivalently $m_1 m_2 = -1$.	p. 168
In many-day-to-day applications, two variables are related linearly.	p. 173
• A linear model can be made from two data points that represent the general trend of the data. • Alternatively, the least-squares regression line is a model that utilizes *all* observed data points.	p. 173 p. 174

SECTION 1.6 Transformations of Graphs	Reference
Consider a function defined by $y = f(x)$. Let h, k, and a represent positive real numbers. The graphs of the following functions are related to $y = f(x)$ as follows.	pp. 184–185
• Vertical translation (shift) $y = f(x) + k$ Shift upward $y = f(x) - k$ Shift downward • Horizontal translation (shift) $y = f(x - h)$ Shift to the right $y = f(x + h)$ Shift to the left	

• Vertical stretch/shrink
$y = af(x)$ Vertical stretch (if $a > 1$)
                Vertical shrink (if $0 < a < 1$)
• Horizontal stretch/shrink
$y = f(ax)$ Horizontal shrink (if $a > 1$)
                Horizontal stretch (if $0 < a < 1$)
	p. 187

• Reflection $y = -f(x)$ Reflection across the x-axis $y = f(-x)$ Reflection across the y-axis	p. 189
To graph a function requiring multiple transformations, use the following order. 1. Horizontal translation (shift) 2. Horizontal and vertical stretch and shrink 3. Reflections across the x- and y-axes 4. Vertical translation (shift)	p. 190

SECTION 1.7 Analyzing Graphs of Functions and Piecewise-Defined Functions	Reference
Consider the graph of an equation in x and y. • The graph of the equation is symmetric to the y-axis if substituting $-x$ for x results in an equivalent equation. • The graph of the equation is symmetric to the x-axis if substituting $-y$ for y results in an equivalent equation. • The graph of the equation is symmetric to the origin if substituting $-x$ for x and $-y$ for y results in an equivalent equation.	p. 197
• f is an even function if $f(-x) = f(x)$ for all x in the domain of f. • f is an odd function if $f(-x) = -f(x)$ for all x in the domain of f.	p. 199
To graph a piecewise-defined function, graph each individual function on its domain.	p. 201

The **greatest integer function,** denoted by $f(x) = [\![x]\!]$ or $f(x) = \text{int}(x)$ or $f(x) = \text{floor}(x)$ defines $f(x)$ as the greatest integer less than or equal to x.	p. 204
Suppose that I is an interval contained within the domain of a function f. • f is increasing on I if $f(x_1) < f(x_2)$ for all $x_1 < x_2$ on I. • f is decreasing on I if $f(x_1) > f(x_2)$ for all $x_1 < x_2$ on I. • f is constant on I if $f(x_1) = f(x_2)$ for all x_1 and x_2 on I.	p. 206
• $f(a)$ is a **relative maximum** of f if there exists an open interval containing a such that $f(a) \geq f(x)$ for all x in the interval. • $f(b)$ is a **relative minimum** of f if there exists an open interval containing b such that $f(b) \leq f(x)$ for all x in the interval.	p. 207

SECTION 1.8 Algebra of Functions and Function Composition	Reference
Given functions f and g, the functions $f + g, f - g, f \cdot g$, and $\frac{f}{g}$ are defined by $$(f + g)(x) = f(x) + g(x)$$ $$(f - g)(x) = f(x) - g(x)$$ $$(f \cdot g)(x) = f(x) \cdot g(x)$$ $$\left(\frac{f}{g}\right)(x) = \frac{f(x)}{g(x)} \text{ provided that } g(x) \neq 0$$	p. 216
The **difference quotient** represents the average rate of change of a function f between two points $(x, f(x))$ and $(x + h, f(x + h))$. $$\frac{f(x + h) - f(x)}{h} \quad \text{Difference quotient}$$	p. 218
The **composition of f and g,** denoted $f \circ g$ is defined by $(f \circ g)(x) = f(g(x))$. The domain of $f \circ g$ is the set of real numbers x in the domain of g such that $g(x)$ is in the domain of f.	p. 220

Expanded Chapter Summary available at www.mhhe.com/millerprecalculus.

CHAPTER 1 Review Exercises

SECTION 1.1

For Exercises 1–2,

a. Find the exact distance between the points.

b. Find the midpoint of the line segment whose endpoints are the given points.

1. $(-1, 8)$ and $(4, -2)$ a. $5\sqrt{5}$ b. $\left(\frac{3}{2}, 3\right)$

2. $\left(\sqrt{3}, -\sqrt{6}\right)$ and $\left(3\sqrt{3}, 4\sqrt{6}\right)$ a. $9\sqrt{2}$ b. $\left(2\sqrt{3}, \frac{3\sqrt{6}}{2}\right)$

3. Determine if the given ordered pair is a solution to the equation $4|x - 1| + y = 18$.

    a. $(-3, 2)$ Yes   **b.** $(5, -2)$ No

For Exercises 4–6, determine the x- and y-intercepts of the graph of the equation.

4. $-3y + 4x = 6$ x-intercept: $\left(\frac{3}{2}, 0\right)$; y-intercept: $(0, -2)$

5. $x = |y + 7| - 3$ x-intercept: $(4, 0)$; y-intercepts: $(0, -4), (0, -10)$

6. $\dfrac{(x + 4)^2}{9} + \dfrac{y^2}{4} = 1$ x-intercepts: $(-1, 0), (-7, 0)$; y-intercept: None

7. Graph the equation by plotting points. $y = x^2 - 2x$

8. Find the length of the diagonal shown.

   $4\sqrt{5}$ units

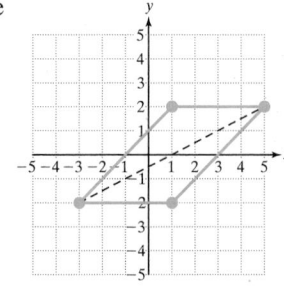

SECTION 1.2

For Exercises 9–10, determine the center and radius of the circle.

9. $(x - 4)^2 + (y + 3)^2 = 4$ Center: $(4, -3)$; Radius: 2

10. $x^2 + \left(y - \dfrac{3}{2}\right)^2 = 17$ Center: $\left(0, \dfrac{3}{2}\right)$; Radius: $\sqrt{17}$

For Exercises 11–14, information about a circle is given.

a. Write an equation of the circle in standard form.

b. Graph the circle.

11. Center: $(-3, 1)$; Radius: $\sqrt{11}$

12. Center: $(0, 0)$; Radius: 3.2

13. Endpoints of a diameter $(7, 5)$ and $(1, -3)$

14. The center is in quadrant IV, the radius is 4, and the circle is tangent to both the x- and y-axes.

For Exercises 15–16, (a) write the equation of the circle in standard form and (b) identify the center and radius.

15. $x^2 + y^2 + 10x - 2y + 17 = 0$
   a. $(x + 5)^2 + (y - 1)^2 = 9$ b. Center: $(-5, 1)$; Radius: 3
16. $x^2 + y^2 - 8y + 3 = 0$
   a. $x^2 + (y - 4)^2 = 13$ b. Center: $(0, 4)$; Radius: $\sqrt{13}$

For Exercises 17–18, determine the solution set to the equation.

17. $(x + 3)^2 + (y - 5)^2 = 0$ $\{(-3, 5)\}$

18. $x^2 + y^2 + 6x - 4y + 15 = 0$ $\{\ \}$

SECTION 1.3

19. The table lists four Olympic athletes and the number of Olympic medals won by the athlete.

Athlete (x)	Number of Medals (y)
Dara Torres (swimming)	12
Carl Lewis (track and field)	10
Bonnie Blair (speed skating)	6
Michael Phelps (swimming)	16

a. Write a set of ordered pairs (x, y) that defines the relation. {(Dara Torres, 12), (Carl Lewis, 10), (Bonnie Blair, 6), (Michael Phelps, 16)}

b. Write the domain of the relation.
{Dara Torres, Carl Lewis, Bonnie Blair, Michael Phelps}

c. Write the range of the relation.   {12, 10, 6, 16}

d. Determine if the relation defines y as a function of x.   Yes

For Exercises 20–23, determine if the relation defines y as a function of x.

20.   Yes   **21.**   No

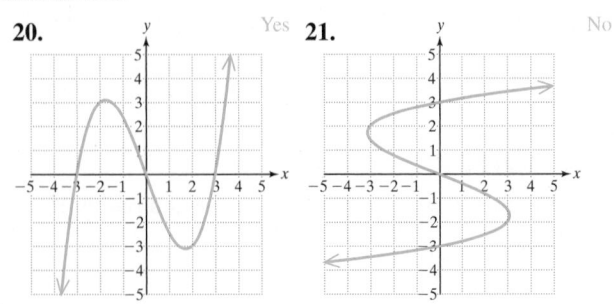

22. $x^2 + (y - 3)^2 = 4$ No   **23.** $x^2 + y - 3 = 4$ Yes

24. Evaluate $f(x) = -2x^2 + 4x$ for the values of x given.

   a. $f(0)$ 0   **b.** $f(-1)$ -6 **c.** $f(3)$ -6

   d. $f(t)$ $-2t^2 + 4t$ **e.** $f(x + 4)$ $-2x^2 - 12x - 16$

25. Given $f = \{(3, -1), (1, 5), (-2, 4), (0, 4)\}$,

   a. Determine $f(1)$.   5

   b. Determine $f(0)$.   4

   c. For what value(s) of x is $f(x) = -1$? $x = 3$

26. A department store marks up the price of a power drill by 32% of the price from the manufacturer. The price $P(x)$ (in \$) to a customer after a 6.5% sales tax is given by $P(x) = 1.065(x + 0.32x)$, where x is the cost of the drill from the manufacturer. Evaluate $P(189)$ and interpret the meaning in the context of this problem.

For Exercises 27–28, determine the x- and y-intercepts for the given function.

27. $p(x) = |x - 3| - 1$ x-intercepts: $(4, 0)$, $(2, 0)$; y-intercept: $(0, 2)$

28. $q(x) = -\sqrt{x} + 2$ x-intercept: $(4, 0)$; y-intercept: $(0, 2)$

For Exercises 29–30, determine the domain and range of the function.

29.   **30.**

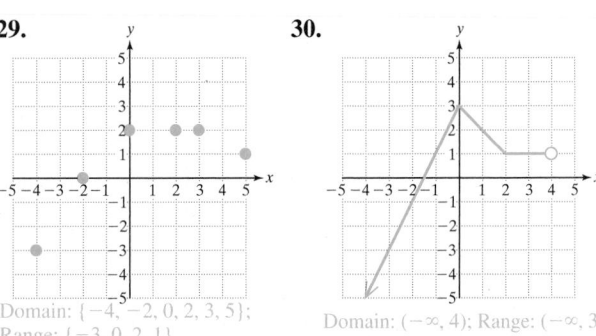

Domain: $\{-4, -2, 0, 2, 3, 5\}$;
Range: $\{-3, 0, 2, 1\}$
                    Domain: $(-\infty, 4)$; Range: $(-\infty, 3]$

For Exercises 31–34, write the domain in interval notation.

31. $f(x) = \dfrac{x - 2}{x - 5}$ **32.** $g(x) = \dfrac{6}{|x| - 3}$
   $(-\infty, 5) \cup (5, \infty)$

33. $m(x) = 2x^2 - 4x + 1$ **34.** $n(x) = \dfrac{10}{\sqrt{2 - x}}$ $(-\infty, 2)$
   $(-\infty, \infty)$

35. Use the graph of $y = f(x)$ to

   a. Determine $f(-2)$. -2

   b. Determine $f(3)$. -1

   c. Find all x for which $f(x) = -1$. $x = -1, x = 3$

   d. Find all x for which $f(x) = -4$. $x = -4$

   e. Determine the x-intercept(s). $(0, 0), (2, 0)$

   f. Determine the y-intercept. $(0, 0)$

   g. Determine the domain of f. $(-\infty, \infty)$

   h. Determine the range of f. $(-\infty, 1]$

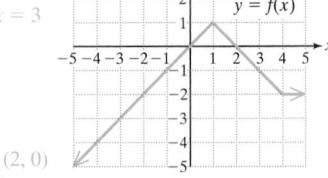

36. Write a relationship for a function whose $f(x)$ value is 4 less than two times the square of x. $f(x) = 2x^2 - 4$

SECTION 1.4

For Exercises 37–40, graph the equation and determine the x- and y-intercepts.

37. $-2x + 4y = 8$ **38.** $-4x = 5y$

39. $y = 2$ **40.** $3x = 5$

For Exercises 41–43, determine the slope of the line passing through the given points.

41. $(4, -2)$ and $(-12, -4)$ $m = \dfrac{1}{8}$

42. $\left(-3, \dfrac{2}{3}\right)$ and $\left(1, -\dfrac{4}{3}\right)$ $m = -\dfrac{1}{2}$

43. $(a, f(a))$ and $(b, f(b))$ $m = \dfrac{f(b) - f(a)}{b - a}$

44. What is the slope of a line parallel to the x-axis?   0

45. What is the slope of a line with equation $x = -2$?
Undefined

46. What is the slope of a line perpendicular to a line with equation $y = 1$?   Undefined

47. Suppose that $y = C(t)$ represents the average cost of a gallon of milk in the United States t years since 1980. What does $\dfrac{\Delta C}{\Delta t}$ represent? $\dfrac{\Delta C}{\Delta t}$ represents the change in cost per change in time.

48. Determine if the function is linear, constant, or neither.

a. $f(x) = -\dfrac{3}{2}x$ **b.** $g(x) = -\dfrac{3}{2x}$ **c.** $h(x) = -\dfrac{3}{2}$
   Linear   Neither   Constant

For Exercises 49–50, use slope-intercept form to write an equation of the line that passes through the given point and has the given slope. Then write the equation using function notation where $y = f(x)$.

49. $(1, -5)$ and $m = -\dfrac{2}{3}$ $y = -\dfrac{2}{3}x - \dfrac{13}{3}; f(x) = -\dfrac{2}{3}x - \dfrac{13}{3}$

50. $\left(2, \dfrac{1}{4}\right)$ and $m = 0$ $y = \dfrac{1}{4}; f(x) = \dfrac{1}{4}$

51. Find the slope of the secant line pictured in red. $m = \dfrac{3}{7}$

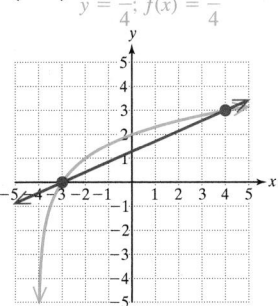

52. The function given by $y = f(x)$ shows the value of $8000 invested at 6% interest compounded continuously, x years after the money was originally invested.

Value of $8000 with Continuous Compounding at 6%

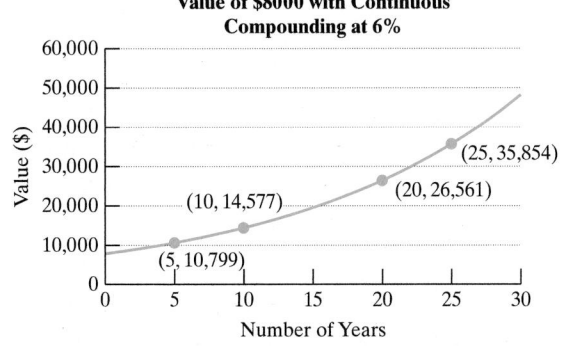

a. Find the average amount earned per year between the 5th year and the 10th year. $755.60 per year

b. Find the average amount earned per year between the 20th year and the 25th year. $1858.60 per year

c. Based on the answers from parts (a) and (b), does it appear that the rate at which annual income increases is increasing or decreasing with time?   Increasing

53. Given $f(x) = -x^3 + 4$, determine the average rate of change of the function on the given intervals.

a. $[0, 2]$ -4 **b.** $[2, 4]$ -28

54. Use the graph to solve the equation and inequalities. Write the solutions to the inequalities in interval notation.

a. $2x + 1 = -x + 4$ $\{1\}$

b. $2x + 1 < -x + 4$ $(-\infty, 1)$

c. $2x + 1 \geq -x + 4$ $[1, \infty)$

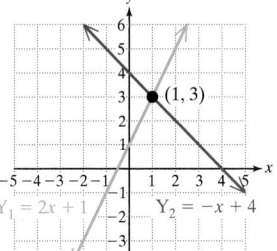

SECTION 1.5

55. If the slope of a line is $\dfrac{2}{3}$,

a. Determine the slope of a line parallel to the given line. $\dfrac{2}{3}$

b. Determine the slope of a line perpendicular to the given line. $-\dfrac{3}{2}$

56. Given a line L_1 defined by $L_1: 2x - 4y = 3$, determine if the equations given in parts (a)–(c) represent a line parallel to L_1, perpendicular to L_1, or neither parallel nor perpendicular to L_1.

a. $12x + 6y = 6$ Perpendicular   **b.** $3y = 1.5x - 5$ Parallel

c. $4x + 8y = 8$ Neither

For Exercises 57–63, write an equation of the line having the given conditions. Write the answer in slope-intercept form if possible.

57. Passes through $(-2, -7)$ and $m = 3$. $y = 3x - 1$

58. Passes through $(0, 5)$ and $m = -\dfrac{2}{5}$. $y = -\dfrac{2}{5}x + 5$

59. Passes through $(1.1, 5.3)$ and $(-0.9, 7.1)$. $y = -0.9x + 6.29$

60. Passes through $(5, -7)$ and the slope is undefined. $x = 5$

61. Passes through $(2, -6)$ and is parallel to the line defined by $2x - y = 4$. $y = 2x - 10$

62. Passes through $(-2, 3)$ and is perpendicular to the line defined by $5y = 2x$. $y = -\dfrac{5}{2}x - 2$

63. The line is perpendicular to the y-axis and the y-intercept is $(0, 7)$. $y = 7$

64. A car has a 15-gal tank for gasoline and gets 30 mpg on a highway while driving 60 mph. Suppose that the driver starts a trip with a full tank of gas and travels 450 mi on the highway at an average speed of 60 mph.

a. Write a linear model representing the amount of gas $G(t)$ left in the tank t hours into a trip. $G(t) = 15 - 2t$

b. Evaluate $G(4.5)$ and interpret the meaning in the context of this problem. $G(4.5) = 6$ means that after 4.5 hr of driving, the tank has 6 gal left.

65. A dance studio has fixed monthly costs of $1500 that include rent, utilities, insurance, and advertising. The studio charges $60 for each private lesson, but has a variable cost for each lesson of $35 to pay the instructor.

a. Write a linear cost function representing the cost to the studio $C(x)$ to hold x private lessons for a given month. $C(x) = 1500 + 35x$

b. Write a linear revenue function representing the revenue $R(x)$ for holding x private lessons for the month. $R(x) = 60x$

c. Write a linear profit function representing the profit $P(x)$ for holding x private lessons for the month. $P(x) = 25x - 1500$

d. Determine the number of private lessons that must be held for the studio to make a profit. The studio needs more than 60 private lessons per month to make a profit.

e. If 82 private lessons are held during a given month, how much money will the studio make or lose? $550

66. The height y (in meters) of a volcano in the southeast Pacific Ocean is recorded in the table for selected years since 1960.

Number of Years Since 1960, x	Height (m) y
0	166
10	290
20	408
30	526
40	650
50	760
54	813

a. Graph the data in a scatter plot.

b. Use the points (0, 166) and (40, 650) to write a linear function that defines the height y of the volcano, x years since 1960. $y = 12.1x + 166$

c. Interpret the meaning of the slope in the context of this problem. The slope is 12.1 and means that the volcano's height increases by 12.1 m per yr.

d. Use the model in part (b) to predict the height of the volcano in the year 2030 assuming that the linear trend continues. 1013 m

67. Refer to the data given in Exercise 66.

a. Use a graphing utility to find the least-squares regression line. Round the slope to 1 decimal place and the y-intercept to the nearest whole unit. $y = 11.9x + 169$

b. Use a graphing utility to graph the regression line and the observed data.

c. In the event that the linear trend continues, use the model from part (a) to predict the height of the volcano in the year 2030. 1002 m

SECTION 1.6

68. Write a function based on the given parent function and transformations in the given order. $y = (-x + 5)^2 - 2$

Parent function: $y = x^2$

1. Shift 5 units to the left.

2. Reflect across the y-axis.

3. Shift downward 2 units.

For Exercises 69–78, use translations to graph the given functions.

69. $f(x) = |x| - 2$

70. $g(x) = \sqrt{x} + 1$

71. $h(x) = (x - 2)^2$

72. $k(x) = \sqrt[3]{x} + 1$

73. $r(x) = \sqrt{x - 3} + 1$

74. $s(x) = (x + 2)^2 - 3$

75. $t(x) = -2|x|$

76. $v(x) = -\dfrac{1}{2}|x|$

77. $m(x) = \sqrt{-x} + 5$

78. $n(x) = \sqrt{-x} - 1$

For Exercises 79–84, use the graph of $y = f(x)$ to graph the given function.

79. $y = f(2x)$

80. $y = f(\frac{1}{2}x)$

81. $y = -f(x + 1) - 3$

82. $y = -f(x - 4) - 1$

83. $y = 2f(x - 3) + 1$

84. $y = \frac{1}{2}f(x + 2) - 3$

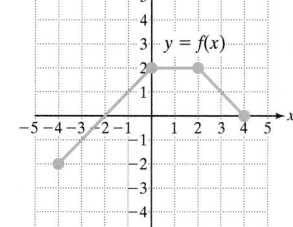

SECTION 1.7

For Exercises 85–88, determine if the graph of the equation is symmetric to the y-axis, x-axis, origin, or none of these.

85. $y = x^4 - 3$ y-axis

86. $x = |y| + y^2$ x-axis

87. $y = \dfrac{1}{3}x - 1$ None of these

88. $x^2 = y^2 + 1$ y-axis, x-axis, and origin

For Exercises 89–94, determine if the function is even, odd, or neither.

89. $f(x) = -4x^3 + x$ Odd

90. $g(x) = \sqrt[3]{x}$ Odd

91. $p(x) = \sqrt{4 - x^2}$ Even

92. $q(x) = -|x|$ Even

93. $k(x) = (x - 3)^2$ Neither

94. $m(x) = |x + 2|$ Neither

95. Evaluate the function for the given values of x.

$$f(x) = \begin{cases} -4x + 2 & \text{for } x < -1 \\ x^2 & \text{for } -1 \le x \le 2 \\ 5 & \text{for } x > 2 \end{cases}$$

a. $f(-4)$ 18   **b.** $f(-1)$ 1   **c.** $f(3)$ 5   **d.** $f(2)$ 4

For Exercises 96–98, graph the function.

96. $f(x) = \begin{cases} -4x - 3 & \text{for } x < 0 \\ x^2 & \text{for } x \ge 0 \end{cases}$

97. $g(x) = \begin{cases} |x| & \text{for } x \le 2 \\ 2 & \text{for } x > 2 \end{cases}$

98. $h(x) = \begin{cases} -3 & \text{for } x < -2 \\ 1 & \text{for } -2 \le x < 0 \\ \sqrt{x} & \text{for } x \ge 0 \end{cases}$

99. Evaluate $f(x) = [\![x - 1]\!]$ for the given values of x.

    a. $f(-1.5)$ -3 **b.** $f(-2)$ -3 **c.** $f(0.1)$ -1 **d.** $f(6.3)$ 5

For Exercises 100–101, use interval notation to write the interval(s) over which f is

    a. Increasing.   **b.** Decreasing.   **c.** Constant.

100.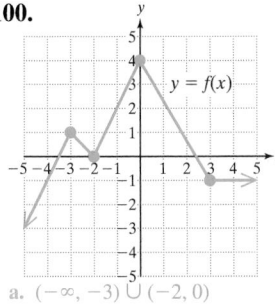

a. $(-\infty, -3) \cup (-2, 0)$
b. $(-3, -2) \cup (0, 3)$ c. $(3, \infty)$

101.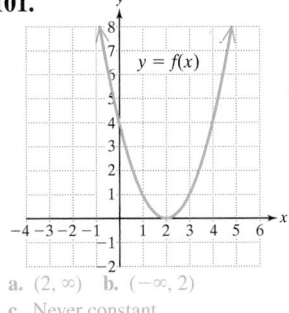

a. $(2, \infty)$ b. $(-\infty, 2)$
c. Never constant

For Exercises 102–103, identify the location and value of any relative maxima or minima of the function.

102.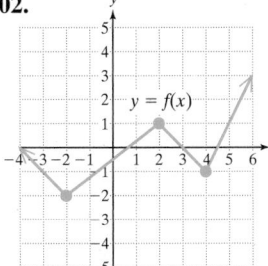

103.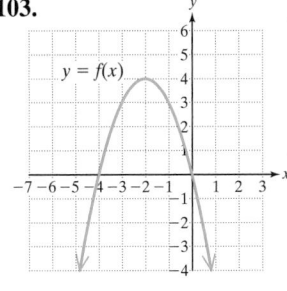

104. Write a rule for the graph of the function. Answers may vary.

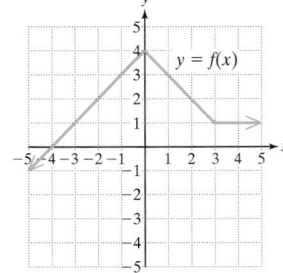

$f(x) = \begin{cases} -|x| + 4 & \text{for } x \le 3 \\ 1 & \text{for } x > 3 \end{cases}$

SECTION 1.8

For Exercises 105–109, evaluate the function for the given values of x.

$$f(x) = -3x \qquad g(x) = |x - 2| \qquad h(x) = \frac{1}{x+1}$$

105. $(f - h)(2)$ $-\dfrac{19}{3}$ **106.** $(g \cdot h)(3)$ $\dfrac{1}{4}$ **107.** $\left(\dfrac{g}{h}\right)(-5)$ -28

108. $(f \circ g)(5)$ -9 **109.** $(g \circ f)(5)$ 17

110. Use the graphs of f and g to find the function values for the given values of x.

    a. $(f + g)(2)$ 1

    b. $(g \cdot f)(-4)$ -2

    c. $\left(\dfrac{g}{f}\right)(-3)$ Undefined

    d. $f[g(-4)]$ 3

    e. $(g \circ f)(-4)$ 1

    f. $(g \circ f)(5)$ Undefined

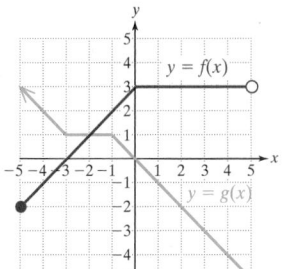

For Exercises 111–116, refer to the functions m, n, p, and q. Find the function and write the domain in interval notation.

$$m(x) = -4x \qquad\qquad n(x) = x^2 - 4x$$
$$p(x) = \sqrt{x - 2} \qquad\quad q(x) = \frac{1}{x - 5}$$

111. $(n - m)(x)$ **112.** $\left(\dfrac{p}{n}\right)(x)$ **113.** $\left(\dfrac{n}{p}\right)(x)$

114. $(m \cdot p)(x)$ **115.** $(q \circ n)(x)$ **116.** $(q \circ p)(x)$

For Exercises 117–118, find the difference quotient,
$$\frac{f(x + h) - f(x)}{h}.$$

117. $f(x) = -6x - 5$ -6 **118.** $f(x) = 3x^2 - 4x + 9$
                                            $6x + 3h - 4$

For Exercises 119–120, find two functions, f and g such that $h(x) = (f \circ g)(x)$.

119. $h(x) = (x - 4)^2$ **120.** $h(x) = \dfrac{12}{x + 5}$ $f(x) = \dfrac{12}{x}$
  $f(x) = x^2$ and $g(x) = x - 4$ and $g(x) = x + 5$

121. A car traveling 60 mph on the highway gets 28 mpg.

    a. Write a function that represents the distance $d(t)$ (in miles) that the car travels in t hours. $d(t) = 60t$

    b. Write a function that represents the number of gallons of gasoline $n(d)$ used for d miles traveled. $n(d) = \dfrac{d}{28}$

    c. Find $(n \circ d)(t)$ and interpret the meaning in the context of this problem.

    d. Evaluate $(n \circ d)(7)$ and interpret the meaning in the context of this problem. $(n \circ d)(7) = 15$ means that 15 gal of gasoline is used in 7 hr.

CHAPTER 1   Test

1. The endpoints of a diameter of a circle are $(-2, 3)$ and $(8, -5)$.

   a. Determine the center of the circle. $(3, -1)$

   b. Determine the radius of the circle. $\sqrt{41}$

   c. Write an equation of the circle in standard form.
   $(x - 3)^2 + (y + 1)^2 = 41$

2. Given $x = |y| - 4$,

   a. Determine the x- and y-intercepts of the graph of the equation. x-intercept: $(-4, 0)$; y-intercepts: $(0, 4)$, $(0, -4)$

   b. Does the equation define y as a function of x?   No

3. Given $x^2 + y^2 + 14x - 10y + 70 = 0$,

   a. Write the equation of the circle in standard form.
   $(x + 7)^2 + (y - 5)^2 = 4$

   b. Identify the center and radius.   Center: $(-7, 5)$; Radius: 2

For Exercises 4–5, determine if the relation defines y as a function of x.

4. x y Yes   5.   No

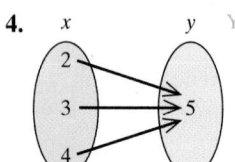

6. Given $f(x) = -2x^2 + 7x - 3$, find

   a. $f(-1)$. -12

   b. $f(x + h)$. $-2x^2 - 4xh - 2h^2 + 7x + 7h - 3$

   c. The difference quotient: $\dfrac{f(x + h) - f(x)}{h}$.
   $-4x - 2h + 7$

   d. The x-intercepts of the graph of f. $(\frac{1}{2}, 0)$ and $(3, 0)$

   e. The y-intercept of the graph of f. $(0, -3)$

   f. The average rate of change of f on the interval $[1, 3]$.
   -1

7. Use the graph of $y = f(x)$ to estimate

   a. $f(0)$. -2

   b. $f(-4)$. 0

   c. The values of x for which $f(x) = 2$.
   $x = -2$ and $x = 2$

   d. The interval(s) over which f is increasing.
   $(-\infty, -2) \cup (0, 2)$

   e. The interval(s) over which f is decreasing.
   $(-2, 0) \cup (2, \infty)$

   f. Determine the location and value of any relative minima.   At $x = 0$, the function has a relative minimum of -2.

   g. Determine the location and value of any relative maxima.   At $x = -2$, the function has a relative maximum of 2.
   At $x = 2$, the function has a relative maximum of 2.

   h. The domain. $(-\infty, \infty)$

   i. The range. $(-\infty, 2]$

   j. Whether f is even, odd, or neither.   Even

For Exercises 8–9, write the domain in interval notation.

8. $f(w) = \dfrac{2w}{3w + 7}$ $\left(-\infty, -\dfrac{7}{3}\right) \cup \left(-\dfrac{7}{3}, \infty\right)$

9. $f(c) = \sqrt{4 - c}$ $(-\infty, 4]$

10. Given $3x = -4y + 8$,

   a. Identify the slope. $m = -\dfrac{3}{4}$

   b. Identify the y-intercept. $(0, 2)$

   c. Graph the line.

   d. What is the slope of a line perpendicular to this line? $\dfrac{4}{3}$

   e. What is the slope of a line parallel to this line? $-\dfrac{3}{4}$

11. Write an equation of the line passing through the point $(-2, 6)$ and perpendicular to the line defined by $x + 3y = 4$. $y = 3x + 12$

12. Use the graph to solve the equation and inequalities. Write the solutions to the inequalities in interval notation.

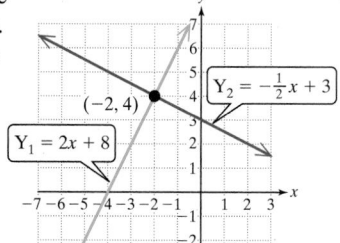

   a. $2x + 8 = -\dfrac{1}{2}x + 3$
   $\{-2\}$

   b. $2x + 8 < -\dfrac{1}{2}x + 3$
   $(-\infty, -2)$

   c. $2x + 8 \geq -\dfrac{1}{2}x + 3$
   $[-2, \infty)$

For Exercises 13–16, graph the equation.

13. $x^2 + \left(y + \dfrac{5}{2}\right)^2 = 9$

14. $f(x) = 2|x + 3|$

15. $g(x) = -\sqrt{x + 4} + 3$

16. $h(x) = \begin{cases} -x + 3 & \text{for } x < 1 \\ \sqrt{x - 1} & \text{for } x \geq 1 \end{cases}$

17. Determine if the graph of the equation is symmetric to the y-axis, x-axis, origin, or none of these.
$$x^2 + |y| = 8$$
Symmetric to the y-axis, x-axis, and origin.

For Exercises 18–19, determine if the function is even, odd, or neither.

18. $f(x) = x^3 - x$ Odd

19. $g(x) = x^4 + x^3 + x$ Neither

20. Evaluate the greatest integer function for the following values of x.

   a. 4.27   4   b. -4.27 -5

For Exercises 21–26, refer to the functions *f*, *g*, and *h* defined here.

$$f(x) = x - 4 \qquad g(x) = \frac{1}{x - 3} \qquad h(x) = \sqrt{x - 5}$$

21. Evaluate $(f - h)(6)$.   1

22. Evaluate $(g \cdot h)(5)$.   0

23. Evaluate $(h \circ f)(1)$.   Undefined (not a real number)

24. Find $(f \cdot g)(x)$ and state the domain in interval notation. $(f \cdot g)(x) = \dfrac{x - 4}{x - 3}$; Domain: $(-\infty, 3) \cup (3, \infty)$

25. Find $\left(\dfrac{g}{f}\right)(x)$ and state the domain in interval notation.

26. Find $(g \circ h)(x)$ and state the domain in interval notation.

27. Write two functions *f* and *g* such that $h(x) = (f \circ g)(x)$.
$$h(x) = \sqrt[3]{x - 7} \quad f(x) = \sqrt[3]{x} \text{ and } g(x) = x - 7$$

28. For *f* and *g* pictured, estimate the following.

    a. $(f + g)(3)$ −1

    b. $(f \cdot g)(0)$ −1

    c. $g(f(3))$ −2

    d. $(f \circ g)(2)$ Undefined

    e. The interval(s) over which *f* is increasing. $(-3, 0)$

    f. The interval(s) over which *g* is decreasing. $(-\infty, 2)$

29. The number of people *y* that attend a weekly bingo game at an adult recreation center is given in the table for selected weeks, *x*.

Week Number, *x*	Number of attendees, *y*
1	8
3	21
6	30
9	40
12	46
15	56
18	68

    a. Graph the data in a scatter plot.

    b. Use the points $(1, 8)$ and $(9, 40)$ to write a linear function that defines the number of attendees as a function of week number. $y = 4x + 4$

    c. Interpret the meaning of the slope in the context of this problem. The slope is 4 and means that the number of attendees has increased by approximately 4 people per week.

    d. Use the model in part (b) to predict the number of attendees in week 24 assuming that the linear trend continues.   100 people

30. Refer to the data given in Exercise 29.

    a. Use a graphing utility to find the least-squares regression line. Round the slope and *y*-intercept to 1 decimal place. $y = 3.3x + 8.5$

    b. Use a graphing utility to graph the regression line and the observed data.

    c. In the event that the linear trend continues, use the model from part (a) to predict the number of attendees in week 24.   88 people

CHAPTER 1   Cumulative Review Exercises

1. Use the graph of $y = f(x)$ to

    a. Evaluate $f(2)$.   −1

    b. Find all *x* such that $f(x) = 0$. $x = 0$ and $x = 3$

    c. Determine the domain of *f*. $(-4, \infty)$

    d. Determine the range of *f*. $[-2, \infty)$

    e. Determine the interval(s) over which *f* is increasing. $(1, \infty)$

    f. Determine the interval(s) over which *f* is decreasing. $(-1, 1)$

    g. Determine the intervals(s) over which *f* is constant. $(-4, -1)$

    h. Evaluate $(f \circ f)(-1)$.   −1

2. Given the equation of the circle $x^2 + y^2 + 12x - 4y + 31 = 0$,

    a. Write the equation in standard form. $(x + 6)^2 + (y - 2)^2 = 9$

    b. Identify the center and radius.   Center: $(-6, 2)$; Radius: 3

For Exercises 3–7, refer to the functions *f*, *g*, and *h* defined here.

$$f(x) = -x^2 + 3x \qquad g(x) = \frac{1}{x} \qquad h(x) = \sqrt{x + 2}$$

3. Find $(g \circ f)(x)$ and write the domain in interval notation.

4. Find $(g \cdot h)(x)$ and write the domain in interval notation.

5. Find the difference quotient. $\dfrac{f(x + h) - f(x)}{h}$
$$-2x - h + 3$$

6. Find the average rate of change of f over the interval $[0, 3]$. $\quad 0$

7. Determine the x- and y-intercepts of f.
x-intercepts: $(0, 0)$ and $(3, 0)$; y-intercept: $(0, 0)$

For Exercises 8–9, graph the function.

8. $f(x) = -\sqrt{x + 3}$

9. $g(x) = \begin{cases} -4 & \text{for } x < -2 \\ 1 & \text{for } -2 \le x < 0 \\ x^2 + 1 & \text{for } x \ge 0 \end{cases}$

10. Write an equation of the line passing through the points $(8, -3)$ and $(-2, 1)$. Write the final answer in slope-intercept form. $\quad y = -\dfrac{2}{5}x + \dfrac{1}{5}$

11. Write an absolute value expression that represents the distance between the points x and 7 on the number line.
$|x - 7|$ or $|7 - x|$

12. Factor. $\quad 2x^3 - 128 \quad 2(x - 4)(x^2 + 4x + 16)$

For Exercises 13–17, solve the equation or inequality. Write the solutions to the inequalities in interval notation.

13. $-3t(t - 1) = 2t + 6 \quad \left\{ \dfrac{1}{6} \pm \dfrac{\sqrt{71}}{6}i \right\}$

14. $7 = |4x - 2| + 5 \quad \{0, 1\}$ **15.** $x^{2/5} - 3x^{1/5} + 2 = 0 \quad \{1, 32\}$

16. $|3a + 1| - 2 \le 9 \quad \left[-4, \dfrac{10}{3} \right]$ **17.** $3 \le -2x + 1 < 7 \quad (-3, -1]$

For Exercises 18–20, perform the indicated operations and simplify.

18. $\dfrac{6}{\sqrt{15} + \sqrt{11}} \quad \dfrac{3\sqrt{15} - 3\sqrt{11}}{2}$

19. $3c\sqrt{8c^2d^3} + c^2\sqrt{50d^3} - 2d\sqrt{2c^4d} \quad 9c^2d\sqrt{2d}$

20. $\dfrac{2u^{-1} - w^{-1}}{4u^{-2} - w^{-2}} \quad \dfrac{uw}{2w + u}$

2

Polynomial and Rational Functions

Chapter Outline

Meteorology and the study of weather have a strong basis in mathematics. The factors impacting weather are not constant and change over time. For example, during the summer months, hot ocean temperatures in the Atlantic Ocean often produce breeding grounds for hurricanes off the coast of Africa or in the Caribbean. To predict the path of a hurricane, meteorologists collect data from satellites, weather stations around the world, and weather buoys in the ocean. Piecing together the data requires a variety of techniques of mathematical modeling using powerful computers. In the end, scientists combine a series of simple curves to approximate weather patterns that closely fit complicated models.

In this chapter, we study polynomial and rational functions. Both types of functions represent simple curves that can be used for modeling in a wide range of applications, including predictions for the path of a hurricane.

OBJECTIVES

1. Graph a Quadratic Function Written in Vertex Form
2. Write $f(x) = ax^2 + bx + c$ $(a \neq 0)$ in Vertex Form
3. Find the Vertex of a Parabola by Using the Vertex Formula
4. Solve Applications Involving Quadratic Functions
5. Create Quadratic Models Using Regression

1. Graph a Quadratic Function Written in Vertex Form

In Chapter 1, we defined a function of the form $f(x) = mx + b$ $(m \neq 0)$ as a linear function. The function defined by $f(x) = ax^2 + bx + c$ $(a \neq 0)$ is called a *quadratic function*. Notice that a quadratic function has a leading term of second degree. We are already familiar with the graph of $f(x) = x^2$ (Figure 2-1). The graph is a parabola opening upward with vertex at the origin. Also note that the graph is symmetric with respect to the vertical line through the vertex called the **axis of symmetry.**

We can write $f(x) = ax^2 + bx + c$ $(a \neq 0)$ in the form $f(x) = a(x - h)^2 + k$ by completing the square. Furthermore, from Section 1.6 we know that the graph of $f(x) = a(x - h)^2 + k$ is related to the graph of $y = x^2$ by a vertical shrink or stretch determined by a, a horizontal shift determined by h, and a vertical shift determined by k. Therefore, the graph of a quadratic function is a parabola with vertex at (h, k).

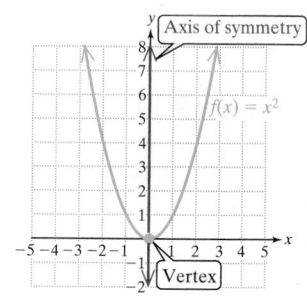

Figure 2-1

TIP A quadratic function is often used as a model for projectile motion. This is motion followed by an object influenced by an initial force and by the force of gravity.

Quadratic Function

A function defined by $f(x) = ax^2 + bx + c$ $(a \neq 0)$ is called a **quadratic function.** By completing the square, $f(x)$ can be expressed in **vertex form** as $f(x) = a(x - h)^2 + k$.

- The graph of f is a parabola with vertex (h, k).
- If $a > 0$, the parabola opens upward, and the vertex is the minimum point. The minimum *value* of f is k.
- If $a < 0$, the parabola opens downward, and the vertex is the maximum point. The maximum *value* of f is k.
- The axis of symmetry is $x = h$. This is the vertical line that passes through the vertex.

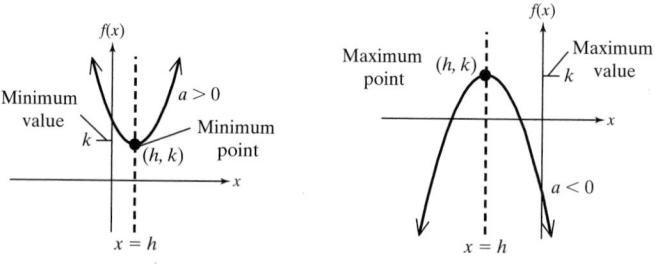

In Example 1, we analyze and graph a quadratic function by identifying the vertex, axis of symmetry, and x- and y-intercepts. From the graph, the minimum or maximum value of the function is readily apparent.

Classroom Example: p. 249
Exercise 8

| EXAMPLE 1 | Analyzing and Graphing a Quadratic Function |

Given $f(x) = -2(x - 1)^2 + 8$,

a. Determine whether the graph of the parabola opens upward or downward.
b. Identify the vertex.
c. Determine the x-intercept(s).
d. Determine the y-intercept.
e. Sketch the function.
f. Determine the axis of symmetry.
g. Determine the maximum or minimum value of f.
h. Write the domain and range in interval notation.

Solution:

Instructor Note:
As a check, consider mentioning to students that the x-coordinate of the vertex is halfway between the x-intercepts. The corresponding y-coordinate of the vertex can be evaluated from $f(x)$.

a. $f(x) = -2(x - 1)^2 + 8$
The parabola opens downward.

The function is written as $f(x) = a(x - h)^2 + k$, where $a = -2$, $h = 1$, and $k = 8$. Since $a < 0$, the parabola opens downward.

b. The vertex is $(1, 8)$.

The vertex is (h, k), which is $(1, 8)$.

c.
$$f(x) = -2(x - 1)^2 + 8$$
$$0 = -2(x - 1)^2 + 8$$
$$-8 = -2(x - 1)^2$$
$$4 = (x - 1)^2$$
$$\pm\sqrt{4} = x - 1$$
$$1 \pm 2 = x$$
$$x = 3 \quad \text{or} \quad x = -1$$
The x-intercepts are $(3, 0)$ and $(-1, 0)$.

To find the x-intercept(s), find all real solutions to the equation $f(x) = 0$.

d. $f(0) = -2(0 - 1)^2 + 8$
$= 6$
The y-intercept is $(0, 6)$.

To find the y-intercept, evaluate $f(0)$.

e. The graph of f is shown in Figure 2-2.

f. The axis of symmetry is the vertical line through the vertex: $x = 1$.

g. The maximum value is 8.

h. The domain is $(-\infty, \infty)$.
The range is $(-\infty, 8]$.

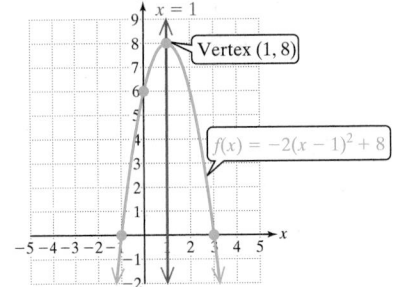

Figure 2-2

Answers

1. a. Upward   **b.** $(-2, -1)$
c. $(-3, 0)$ and $(-1, 0)$ **d.** $(0, 3)$
e.

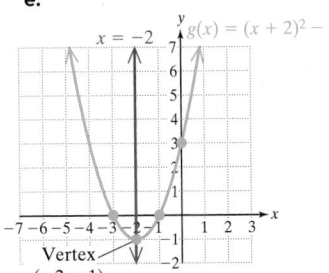

f. $x = -2$
g. The minimum value is -1.
h. The domain is $(-\infty, \infty)$.
The range is $[-1, \infty)$.

Skill Practice 1   Repeat Example 1 with $g(x) = (x + 2)^2 - 1$.

2. Write $f(x) = ax^2 + bx + c$ $(a \neq 0)$ in Vertex Form

In Section 1.2, we learned how to complete the square to write an equation of a circle $x^2 + y^2 + Ax + By + C = 0$ in standard form $(x - h)^2 + (y - k)^2 = r^2$. We use the same process to write a quadratic function $f(x) = ax^2 + bx + c$ $(a \neq 0)$ in vertex form $f(x) = a(x - h)^2 + k$. However, we will work on the right side of the equation only. This is demonstrated in Example 2.

Classroom Example: p. 249
Exercise 18

EXAMPLE 2   **Writing a Quadratic Function in Vertex Form**

Given $f(x) = 3x^2 + 12x + 5$,

a. Write the function in vertex form: $f(x) = a(x - h)^2 + k$.
b. Identify the vertex.
c. Identify the x-intercept(s).
d. Identify the y-intercept.
e. Sketch the function.
f. Determine the axis of symmetry.
g. Determine the minimum or maximum value of f.
h. Write the domain and range in interval notation.

Solution:

a. $f(x) = 3x^2 + 12x + 5$

$= 3(x^2 + 4x \qquad) + 5$

Factor out the leading coefficient of the x^2 term from the two terms containing x. The leading term within parentheses now has a coefficient of 1.

$= 3(x^2 + 4x + 4 - 4) + 5$

Complete the square within parentheses. Add and subtract $\left[\frac{1}{2}(4)\right]^2 = 4$ within parentheses.

$= 3(x^2 + 4x + 4) + 3(-4) + 5$

Remove -4 from within parentheses, along with a factor of 3.

$= 3(x + 2)^2 - 7$ (vertex form)

b. The vertex is $(-2, -7)$.

c. $f(x) = 3x^2 + 12x + 5$

$0 = 3x^2 + 12x + 5$

To find the x-intercept(s), find the real solutions to the equation $f(x) = 0$.

$x = \dfrac{-12 \pm \sqrt{(12)^2 - 4(3)(5)}}{2(3)}$

The right side is not factorable. Apply the quadratic formula.

$= \dfrac{-12 \pm \sqrt{84}}{6}$

$= \dfrac{-12 \pm 2\sqrt{21}}{6}$

The x-intercepts are $\left(\dfrac{-6 + \sqrt{21}}{3}, 0\right)$ and $\left(\dfrac{-6 - \sqrt{21}}{3}, 0\right)$ or approximately

$= \dfrac{-6 \pm \sqrt{21}}{3}$ $\begin{array}{l} x \approx -0.47 \\ x \approx -3.53 \end{array}$

$(-0.47, 0)$ and $(-3.53, 0)$.

d. $f(0) = 3(0)^2 + 12(0) + 5$

$= 5$

To find the y-intercept, evaluate $f(0)$. The y-intercept is $(0, 5)$.

e. The graph of f is shown in Figure 2-3.

f. The axis of symmetry is $x = -2$.

g. The minimum value is -7.

h. The domain is $(-\infty, \infty)$.

The range is $[-7, \infty)$.

Answers
2. a. $f(x) = 3(x - 1)^2 - 2$ **b.** $(1, -2)$
c. $\left(\dfrac{3 \pm \sqrt{6}}{3}, 0\right)$
d. $(0, 1)$
e.

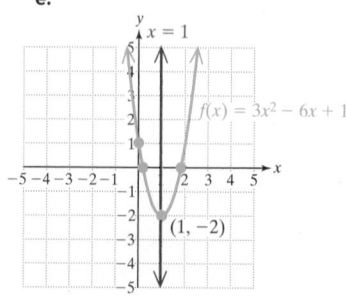

f. $x = 1$
g. The minimum value is -2.
h. The domain is $(-\infty, \infty)$.
   The range is $[-2, \infty)$.

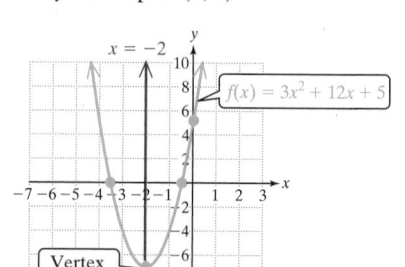

Figure 2-3

Skill Practice 2   Repeat Example 2 with $f(x) = 3x^2 - 6x + 1$.

3. Find the Vertex of a Parabola by Using the Vertex Formula

Completing the square and writing a quadratic function in the form $f(x) = a(x - h)^2 + k$ is one method to find the vertex of a parabola. Another method is to use the vertex formula. The vertex formula can be derived by completing the square on $f(x) = ax^2 + bx + c$.

$f(x) = ax^2 + bx + c \ (a \neq 0)$

$= a\left(x^2 + \dfrac{b}{a}x + \dfrac{b^2}{4a^2} - \dfrac{b^2}{4a^2}\right) + c$

Factor out a from the x terms, and complete the square within parentheses. $\left[\dfrac{1}{2}\left(\dfrac{b}{a}\right)\right]^2 = \dfrac{b^2}{4a^2}$

$= a\left(x^2 + \dfrac{b}{a}x + \dfrac{b^2}{4a^2}\right) + a\left(-\dfrac{b^2}{4a^2}\right) + c$

Remove the term $-\dfrac{b^2}{4a^2}$ from within parentheses along with a factor of a.

$= a\left(x + \dfrac{b}{2a}\right)^2 - \dfrac{b^2}{4a} + c$

Factor the trinomial.

$= a\left(x + \dfrac{b}{2a}\right)^2 + \dfrac{4ac - b^2}{4a}$

Obtain a common denominator and add the terms outside parentheses.

$= a\left[x - \left(\dfrac{-b}{2a}\right)\right]^2 + \dfrac{4ac - b^2}{4a}$

$f(x)$ is now written in vertex form.

$f(x) = a(x - h)^2 \quad + \quad k$

$h = \dfrac{-b}{2a}$ and $k = \dfrac{4ac - b^2}{4a}$

The vertex is $\left(\dfrac{-b}{2a}, \dfrac{4ac - b^2}{4a}\right)$.

The y-coordinate of the vertex is given by $\dfrac{4ac - b^2}{4a}$ and is often hard to remember. Therefore, it is usually easier to evaluate the x-coordinate first from $\dfrac{-b}{2a}$, and then evaluate $f\left(\dfrac{-b}{2a}\right)$.

Vertex Formula to Find the Vertex of a Parabola

For $f(x) = ax^2 + bx + c \ (a \neq 0)$, the vertex is given by $\left(\dfrac{-b}{2a}, f\left(\dfrac{-b}{2a}\right)\right)$.

Classroom Example: p. 250
Exercise 36

EXAMPLE 3 **Using the Vertex Formula**

Given $f(x) = -x^2 + 4x - 5$,

a. State whether the graph of the parabola opens upward or downward.

b. Determine the vertex of the parabola by using the vertex formula.

c. Determine the x-intercept(s).

d. Determine the y-intercept.

e. Sketch the graph.

f. Determine the axis of symmetry.

g. Determine the minimum or maximum value of f.

h. Write the domain and range in interval notation.

Solution:

a. $f(x) = -x^2 + 4x - 5$

The parabola opens downward.

> The function is written as $f(x) = ax^2 + bx + c$ where $a = -1$. Since $a < 0$, the parabola opens downward.

b. x-coordinate: $\dfrac{-b}{2a} = \dfrac{-(4)}{2(-1)} = 2$

y-coordinate: $f(2) = -(2)^2 + 4(2) - 5$
$$= -1$$

The vertex is $(2, -1)$.

c. Since the vertex of the parabola is below the x-axis and the parabola opens downward, the parabola cannot cross or touch the x-axis.

Therefore, there are no x-intercepts.

> Solving the equation $f(x) = 0$ to find the x-intercepts results in imaginary solutions:
>
> $0 = -x^2 + 4x - 5$
>
> $x = \dfrac{-(4) \pm \sqrt{(4)^2 - 4(-1)(-5)}}{2(-1)}$
>
> $x = 2 \pm i$

> **TIP** For more accuracy in the graph, plot one or two points near the vertex. Then use the symmetry of the curve to find additional points on the graph.
>
> For example, the points $(1, -2)$ and $(0, -5)$ are on the left branch of the parabola. The corresponding points to the right of the axis of symmetry are $(3, -2)$ and $(4, -5)$.

d. To find the y-intercept, evaluate $f(0)$.

$f(0) = -(0)^2 + 4(0) - 5 = -5$

The y-intercept is $(0, -5)$.

e. The graph of f is shown in Figure 2-4.

f. The axis of symmetry is $x = 2$.

g. The maximum value of f is -1.

h. The domain is $(-\infty, \infty)$.

The range is $(-\infty, -1]$.

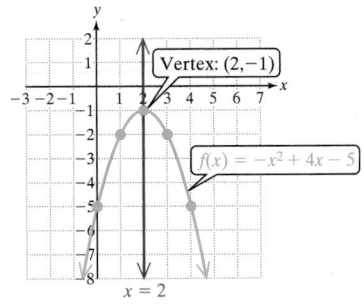

Figure 2-4

> **Skill Practice 3** Repeat Example 3 with $f(x) = -x^2 - 4x - 7$.

The x-intercepts of a quadratic function defined by $f(x) = ax^2 + bx + c$ are the real solutions to the equation $f(x) = 0$. The discriminant $b^2 - 4ac$ enables us to determine the number of real solutions to the equation and thus, the number of x-intercepts of the graph of the function.

Answers

3. a. Downward
b. $(-2, -3)$
c. No x-intercepts
d. $(0, -7)$
e.

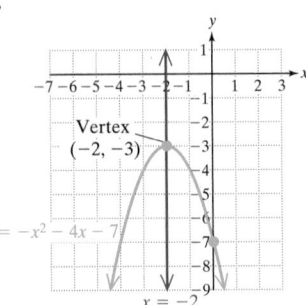

f. $x = -2$
g. The maximum value is -3.
h. The domain is $(-\infty, \infty)$.
The range is $(-\infty, -3]$.

> **Using the Discriminant to Determine the Number of x-Intercepts**
>
> Given a quadratic function defined by $f(x) = ax^2 + bx + c$ $(a \neq 0)$,
>
> - If $b^2 - 4ac = 0$, the graph of $y = f(x)$ has one x-intercept.
> - If $b^2 - 4ac > 0$, the graph of $y = f(x)$ has two x-intercepts.
> - If $b^2 - 4ac < 0$, the graph of $y = f(x)$ has no x-intercept.

From Example 2, the discriminant of $3x^2 + 12x + 5 = 0$ is $(12)^2 - 4(3)(5) = 84 > 0$. Therefore, the graph of $f(x) = 3x^2 + 12x + 5$ has two x-intercepts (Figure 2-3).

From Example 3, the discriminant of $-x^2 + 4x - 5 = 0$ is $(4)^2 - 4(-1)(-5) = -4 < 0$. Therefore, the graph of $f(x) = -x^2 + 4x - 5$ has no x-intercept (Figure 2-4).

4. Solve Applications Involving Quadratic Functions

Quadratic functions can be used in a variety of applications in which a variable is optimized. That is, the vertex of a parabola gives the maximum or minimum value of the dependent variable. We show three such applications in Examples 4–6.

Classroom Example: p. 250
Exercise 44

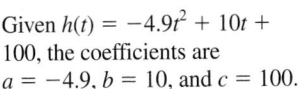

EXAMPLE 4   Using a Quadratic Function for Projectile Motion

A stone is thrown from a 100-m cliff at an initial speed of 20 m/sec at an angle of 30° from the horizontal. The height of the stone can be modeled by $h(t) = -4.9t^2 + 10t + 100$, where $h(t)$ is the height in meters and t is the time in seconds after the stone is released.

a. Determine the time at which the stone will be at its maximum height. Round to 2 decimal places.

b. Determine the maximum height. Round to the nearest meter.

c. Determine the time at which the stone will hit the ground.

Point of Interest

The movie *Apollo 13* starring Tom Hanks was filmed in part in a "Vomit Comet," an aircraft that uses a parabolic flight trajectory to produce weightlessness. As the plane climbs toward the top of the parabolic path, occupants experience a force of nearly 2 Gs (twice their body weight). Once the plane goes over the vertex of the parabola, flyers free fall inside the plane. Such motion often produces motion sickness, thus earning the aircraft its name.

Solution:

a. The time at which the stone will be at its maximum height is the t-coordinate of the vertex.

$$t = \frac{-b}{2a} = \frac{-10}{2(-4.9)} \approx 1.02$$

The stone will be at its maximum height approximately 1.02 sec after release.

Given $h(t) = -4.9t^2 + 10t + 100$, the coefficients are $a = -4.9$, $b = 10$, and $c = 100$.

The vertex is given by
$$\left(\frac{-b}{2a}, h\left(\frac{-b}{2a}\right)\right).$$

b. The maximum height is the value of $h(t)$ at the vertex.
$$h(1.02) = -4.9(1.02)^2 + 10(1.02) + 100$$
$$\approx 105 \text{ The maximum height is 105 m.}$$

c. The stone will hit the ground when $h(t) = 0$.

$$h(t) = -4.9t^2 + 10t + 100$$
$$0 = -4.9t^2 + 10t + 100$$
$$t = \frac{-10 \pm \sqrt{(10)^2 - 4(-4.9)(100)}}{2(-4.9)}$$

$t \approx 5.65$ or $t \approx -3.61$ Reject the negative solution. The stone will hit the ground in approximately 5.65 sec.

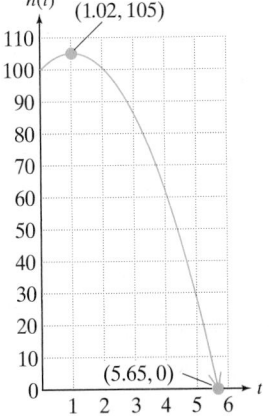

Skill Practice 4   A quarterback throws a football with an initial velocity of 72 ft/sec at an angle of 25°. The height of the ball can be modeled by $h(t) = -16t^2 + 30.4t + 5$, where $h(t)$ is the height (in ft) and t is the time in seconds after release.

a. Determine the time at which the ball will be at its maximum height.

b. Determine the maximum height of the ball.

c. Determine the amount of time required for the ball to reach the receiver's hands if the receiver catches the ball at a point 3 ft off the ground.

Answers
4. a. 0.95 sec   **b.** 19.44 ft
   c. Approximately 1.96 sec

TECHNOLOGY CONNECTIONS

Compute Solutions to a Quadratic Equation

The syntax to compute the expressions from Example 4(c) is shown for a calculator in Classic mode and in Mathprint mode. In Classic mode, parentheses are required around the numerator and denominator of the fraction and around the radicand within the square root. In Mathprint mode, select the (ALPHA) key followed by F1 to access the fraction template.

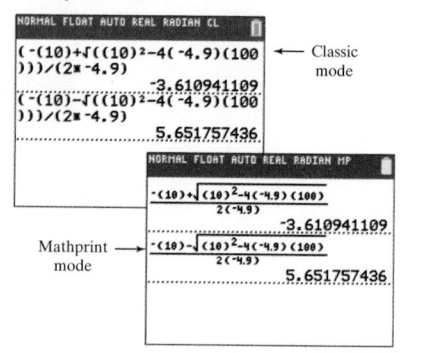

In Example 5, we present a type of application called an optimization problem. The goal is to maximize or minimize the value of the dependent variable by finding an optimal value of the independent variable.

Classroom Example: p. 251
Exercise 54

EXAMPLE 5   Applying a Quadratic Function to Geometry

A parking area is to be constructed adjacent to a road. The developer has purchased 340 ft of fencing. Determine dimensions for the parking lot that would maximize the area. Then find the maximum area.

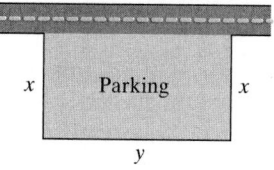

Solution:

Let x represent the width of the parking area.

Let y represent the length.

Let A represent the area.

Read the problem carefully, draw a representative diagram, and label the unknowns.

We need to find the values of x and y that maximize the area A of the rectangular region. The area is given by $A = (\text{length})(\text{width}) = yx = xy$.

To write the area as a function of one variable only, we need an equation that relates x and y. We know that the parking area is limited by a fixed amount of fencing. That is, the sum of the lengths of the three sides to be fenced can be at most 340 ft.

Instructor Note:
Help students understand that the variable y represents the length of the parking lot rather than the y-coordinate of function A.

$2x + y = 340$ ⎤
                 ⎥ Solve
                 ⎥ for y.
$y = 340 - 2x$ ⎦

The equation $2x + y = 340$ is called a **constraint equation.** This equation gives an implied restriction on x and y due to the limited amount of fencing.

Solve the constraint equation, $2x + y = 340$ for either x or y. In this case, we have solved for y.

$A = xy$

$A(x) = x(340 - 2x)$

Substitute $340 - 2x$ for y in the equation $A = xy$.

$\quad = -2x^2 + 340x$

Function A is a quadratic function with a negative leading coefficient. The graph of the parabola opens downward, so the vertex is the maximum point on the function.

x-coordinate of vertex:

$$x = \frac{-b}{2a} = \frac{-340}{2(-2)} = 85$$

$$y = 340 - 2(85) = 170$$

The x-coordinate of the vertex $\frac{-b}{2a}$ is the value of x that will maximize the area.

The second dimension of the parking lot can be determined from the constraint equation.

The values of x and y that would maximize the area are $x = 85$ ft and $y = 170$ ft.

$$A(85) = -2(85)^2 + 340(85) = 14{,}450$$

The value of the function at $x = 85$ gives the maximum area.

The maximum area is $14{,}450 \text{ ft}^2$.

Skill Practice 5   A farmer has 200 ft of fencing and wants to build three adjacent rectangular corrals. Determine the dimensions that should be used to maximize the area, and find the area of each individual corral.

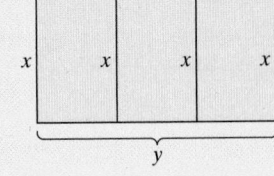

5. Create Quadratic Models Using Regression

In Section 1.5, we introduced linear regression. A regression line is a linear model based on all observed data points. In a similar fashion, we can create a quadratic function using regression. For example, suppose that a scientist growing bacteria measures the population of bacteria as a function of time. A scatter plot reveals that the data follow a curve that is approximately parabolic (Figure 2-5). In Example 6, we use a graphing calculator to find a quadratic function that models the population of the bacteria as a function of time.

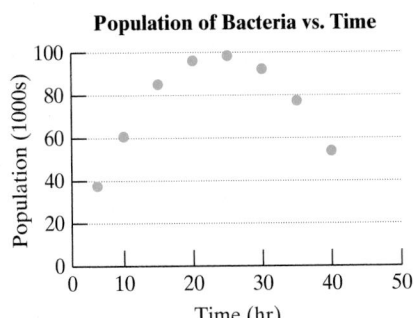

Figure 2-5

Classroom Example: p. 251
Exercise 58

EXAMPLE 6	Creating a Quadratic Function Using Regression

The data in the table represent the population of bacteria $P(t)$ (in 1000s) versus the number of hours t since the culture was started.

a. Use regression to find a quadratic function to model the data. Round the coefficients to 3 decimal places.

b. Use the model to determine the time at which the population is the greatest. Round to the nearest hour.

c. What is the maximum population? Round to the nearest hundred.

Time (hr) t	Population (1000s) $P(t)$
5	37.7
10	60.9
15	85.3
20	96.3
25	98.6
30	92.4
35	77.5
40	54.1

Answer

5. The dimensions should be $x = 25$ ft and $y = 50$ ft. The area of each individual corral is
$$\frac{1250}{3} = 416.\overline{6} \text{ ft}^2.$$

Solution:

a. From the graph in Figure 2-5, it appears that the data follow a parabolic curve. Therefore, a quadratic model would be reasonable.

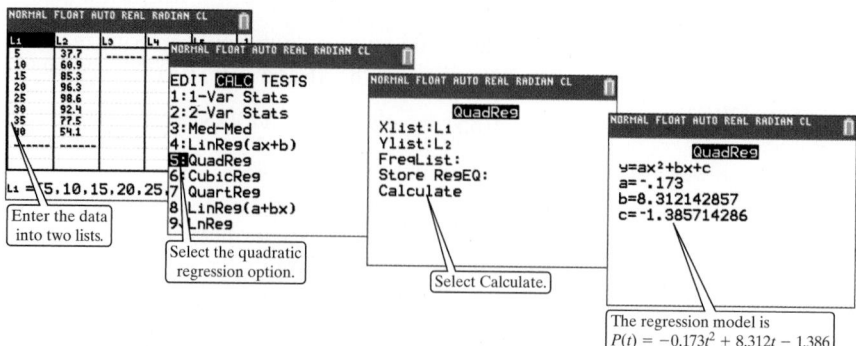

The regression model is
$P(t) = -0.173t^2 + 8.312t - 1.386$

b. From the graph, the time when the population is greatest is the t-coordinate of the vertex.

$$t = \frac{-b}{2a} = \frac{-(8.312)}{2(-0.173)} \approx 24$$

The population is greatest 24 hr after the culture is started.

c. The maximum population of the bacteria is the $P(t)$ value at the vertex.

$$P(24) = -0.173(24)^2 + 8.312(24) - 1.386$$

≈ 98.5 The maximum number of bacteria is approximately 98,500.

Skill Practice 6   The funding $f(t)$ (in $ millions) for a drug rehabilitation center is given in the table for selected years t.

t	0	3	6	9	12	15
$f(t)$	3.5	2.2	2.1	3	4.9	8

a. Use regression to find a quadratic function to model the data.
b. During what year is the funding the least? Round to the nearest year.
c. What is the minimum yearly amount of funding received? Round to the nearest million.

Answers
6. a. $f(t) = 0.060t^2 - 0.593t + 3.486$
   b. Year 5   c. $2 million

SECTION 2.1   Practice Exercises

Prerequisite Review

R.1. Solve the equation. $x^2 + 3x - 18 = 0$ {−6, 3}

R.2. a. Find the values of x for which $f(x) = 0$. $x = -\dfrac{5}{3}$ or $x = 4$

   b. Find $f(0)$. $f(0) = -20$

$$f(x) = 3x^2 - 7x - 20$$

R.3. Solve the equation by completing the square and applying the square root property.

$$x^2 + 8x + 12 = 0$$ {−2, −6}

R.4. Find $g\left(-\dfrac{1}{2}\right)$ for $g(x) = -x^2 + 2x - 4$. $-\dfrac{21}{4}$

R.5. Write the domain and range in interval notation.

Domain $(-4, 4]$; Range $[-1, 4]$

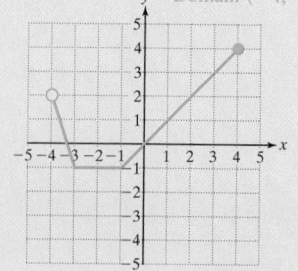

Concept Connections

1. A function defined by $f(x) = ax^2 + bx + c$ $(a \neq 0)$ is called a _____quadratic_____ function.

2. The vertical line drawn through the vertex of a quadratic function is called the _____axis_____ of symmetry.

3. Given $f(x) = a(x - h)^2 + k$ $(a \neq 0)$, the vertex of the parabola is the point _____(h, k)_____.

4. Given $f(x) = a(x - h)^2 + k$, if $a < 0$, the parabola opens (upward/downward) and the (minimum/maximum) value is _____downward; maximum; k_____.

5. Given $f(x) = a(x - h)^2 + k$, if $a > 0$, the parabola opens (upward/downward) and the (minimum/maximum) value is _____upward; minimum; k_____.

6. The graph of $f(x) = a(x - h)^2 + k$, $a \neq 0$, is a parabola and the axis of symmetry is the line given by _____$x = h$_____.

Objective 1: Graph a Quadratic Function Written in Vertex Form

For Exercises 7–14,

a. Determine whether the graph of the parabola opens upward or downward.

b. Identify the vertex.

c. Determine the x-intercept(s).

d. Determine the y-intercept.

e. Sketch the function.

f. Determine the axis of symmetry.

g. Determine the minimum or maximum value of the function.

h. Write the domain and range in interval notation.
   (See Example 1)

7. $f(x) = -(x - 4)^2 + 1$

8. $g(x) = -(x + 2)^2 + 4$

9. $h(x) = 2(x + 1)^2 - 8$

10. $k(x) = 2(x - 3)^2 - 2$

11. $m(x) = 3(x - 1)^2$

12. $n(x) = \dfrac{1}{2}(x + 2)^2$

13. $p(x) = -\dfrac{1}{5}(x + 4)^2 + 1$

14. $q(x) = -\dfrac{1}{3}(x - 1)^2 + 1$

Objective 2: Write $f(x) = ax^2 + bx + c$ $(a \neq 0)$ in Vertex Form

For Exercises 15–24,

a. Write the function in vertex form.

b. Identify the vertex.

c. Determine the x-intercept(s).

d. Determine the y-intercept.

e. Sketch the function.

f. Determine the axis of symmetry.

g. Determine the minimum or maximum value of the function.

h. Write the domain and range in interval notation.
   (See Example 2)

15. $f(x) = x^2 + 6x + 5$

16. $g(x) = x^2 + 8x + 7$

17. $p(x) = 3x^2 - 12x - 7$

18. $q(x) = 2x^2 - 4x - 3$

19. $c(x) = -2x^2 - 10x + 4$

20. $d(x) = -3x^2 - 9x + 8$

21. $h(x) = -2x^2 + 7x$

22. $k(x) = 3x^2 - 8x$

23. $p(x) = x^2 + 9x + 17$

24. $q(x) = x^2 + 11x + 26$

Objective 3: Find the Vertex of a Parabola by Using the Vertex Formula

For Exercises 25–32, find the vertex of the parabola by applying the vertex formula.

25. $f(x) = 3x^2 - 42x - 91$ (7, −238)

26. $g(x) = 4x^2 - 64x + 107$ (8, −149)

27. $k(a) = -\dfrac{1}{3}a^2 + 6a + 1$ (9, 28)

28. $j(t) = -\dfrac{1}{4}t^2 + 10t - 5$ (20, 95)

29. $f(c) = 4c^2 - 5$ (0, −5)

30. $h(a) = 2a^2 + 14$ (0, 14)

31. $P(x) = 1.2x^2 + 1.8x - 3.6$ (−0.75, −4.275)
   (Write the coordinates of the vertex as decimals.)

32. $Q(x) = 7.5x^2 - 2.25x + 4.75$ (0.15, 4.58125)
   (Write the coordinates of the vertex as decimals.)

For Exercises 33–42,

a. State whether the graph of the parabola opens upward or downward.

b. Identify the vertex.

c. Determine the x-intercept(s).

d. Determine the y-intercept.

e. Sketch the graph.

f. Determine the axis of symmetry.

g. Determine the minimum or maximum value of the function.

h. Write the domain and range in interval notation. (See Example 3)

33. $g(x) = -x^2 + 2x - 4$

34. $h(x) = -x^2 - 6x - 10$

35. $f(x) = 5x^2 - 15x + 3$

36. $k(x) = 2x^2 - 10x - 5$

37. $f(x) = 2x^2 + 3$

38. $g(x) = -x^2 - 1$

39. $f(x) = -2x^2 - 20x - 50$

40. $m(x) = 2x^2 - 8x + 8$

41. $n(x) = x^2 - x + 3$

42. $r(x) = x^2 - 5x + 7$

Objective 4: Solve Applications Involving Quadratic Functions

43. The monthly profit for a small company that makes long-sleeve T-shirts depends on the price per shirt. If the price is too high, sales will drop. If the price is too low, the revenue brought in may not cover the cost to produce the shirts. After months of data collection, the sales team determines that the monthly profit is approximated by $f(p) = -50p^2 + 1700p - 12,000$, where p is the price per shirt and $f(p)$ is the monthly profit based on that price. (See Example 4)

    a. Find the price that generates the maximum profit. $17

    b. Find the maximum profit. $2450

    c. Find the price(s) that would enable the company to break even. $10 and $24

44. The monthly profit for a company that makes decorative picture frames depends on the price per frame. The company determines that the profit is approximated by $f(p) = -80p^2 + 3440p - 36,000$, where p is the price per frame and $f(p)$ is the monthly profit based on that price.

    a. Find the price that generates the maximum profit. $21.50

    b. Find the maximum profit. $980

    c. Find the price(s) that would enable the company to break even. $18 and $25

45. A long jumper leaves the ground at an angle of 20° above the horizontal, at a speed of 11 m/sec. The height of the jumper can be modeled by $h(x) = -0.046x^2 + 0.364x$, where h is the jumper's height in meters and x is the horizontal distance from the point of launch.

    a. At what horizontal distance from the point of launch does the maximum height occur? Round to 2 decimal places. 3.96 m

    b. What is the maximum height of the long jumper? Round to 2 decimal places. 0.72 m

    c. What is the length of the jump? Round to 1 decimal place. 7.9 m

46. A firefighter holds a hose 3 m off the ground and directs a stream of water toward a burning building. The water leaves the hose at an initial speed of 16 m/sec at an angle of 30°. The height of the water can be approximated by $h(x) = -0.026x^2 + 0.577x + 3$, where $h(x)$ is the height of the water in meters at a point x meters horizontally from the firefighter to the building.

    a. Determine the horizontal distance from the firefighter at which the maximum height of the water occurs. Round to 1 decimal place. 11.1 m

    b. What is the maximum height of the water? Round to 1 decimal place. 6.2 m

    c. The flow of water hits the house on the downward branch of the parabola at a height of 6 m. How far is the firefighter from the house? Round to the nearest meter. 14 m

47. The population $P(t)$ of a culture of the bacterium *Pseudomonas aeruginosa* is given by $P(t) = -1718t^2 + 82,000t + 10,000$, where t is the time in hours since the culture was started.

    a. Determine the time at which the population is at a maximum. Round to the nearest hour. 24 hr

    b. Determine the maximum population. Round to the nearest thousand. 988,000

48. The gas mileage $m(x)$ (in mpg) for a certain vehicle can be approximated by $m(x) = -0.028x^2 + 2.688x - 35.012$, where x is the speed of the vehicle in mph.

    a. Determine the speed at which the car gets its maximum gas mileage. 48 mph

    b. Determine the maximum gas mileage. 29.5 mpg

49. The sum of two positive numbers is 24. What two numbers will maximize the product? (See Example 5) The numbers are 12 and 12.

50. The sum of two positive numbers is 1. What two numbers will maximize the product? The numbers are 0.5 and 0.5.

51. The difference of two numbers is 10. What two numbers will minimize the product?
The numbers are 5 and −5.

53. Suppose that a family wants to fence in an area of their yard for a vegetable garden to keep out deer. One side is already fenced from the neighbor's property. (**See Example 5**)

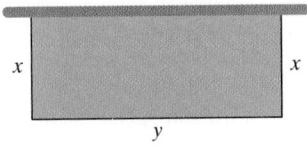

  a. If the family has enough money to buy 160 ft of fencing, what dimensions would produce the maximum area for the garden? 40 ft by 80 ft

  b. What is the maximum area? 3200 ft²

52. The difference of two numbers is 30. What two numbers will minimize the product?
The numbers are 15 and −15.

54. Two chicken coops are to be built adjacent to one another using 120 ft of fencing.

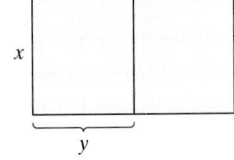

  a. What dimensions should be used to maximize the area of an individual coop?
  20 ft by 15 ft

  b. What is the maximum area of an individual coop? 300 ft²

55. A trough at the end of a gutter spout is meant to direct water away from a house. The homeowner makes the trough from a rectangular piece of aluminum that is 20 in. long and 12 in. wide. He makes a fold along the two long sides a distance of x inches from the edge.

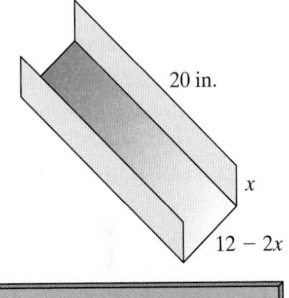

  a. Write a function to represent the volume in terms of x. $V(x) = -40x^2 + 240x$

  b. What value of x will maximize the volume of water that can be carried by the gutter?
  x = 3; The sheet of aluminum should be folded 3 in. from each end.

  c. What is the maximum volume? 360 in.³

56. A rectangular frame of uniform depth for a shadow box is to be made from a 36-in. piece of wood.

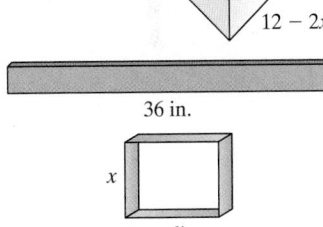

  a. Write a function to represent the display area in terms of x. $A(x) = -x^2 + 18x$

  b. What dimensions should be used to maximize the display area?
  9 in. by 9 in.; That is, the shadow box should be square.

  c. What is the maximum area? 81 in.²

Objective 5: Create Quadratic Models Using Regression

57. *Tetanus bacillus* bacteria are cultured to produce tetanus toxin used in an inactive form for the tetanus vaccine. The amount of toxin produced per batch increases with time and then decreases as the culture becomes unstable. The variable t is the time in hours after the culture has started, and y(t) is the yield of toxin in grams. (**See Example 6**)

t	8	16	24	32	40	48
y(t)	0.60	1.12	1.60	1.78	1.90	2.00

t	56	64	72	80	88	96
y(t)	1.94	1.80	1.48	1.30	0.66	0.10

  a. Use regression to find a quadratic function to model the data. $y = -0.000838t^2 + 0.0812t + 0.040$

  b. At what time is the yield the greatest? Round to the nearest hour. 48 hr

  c. What is the maximum yield? Round to the nearest gram.
  2 g

58. Gas mileage is tested for a car under different driving conditions. At lower speeds, the car is driven in stop-and-go traffic. At higher speeds, the car must overcome more wind resistance. The variable x given in the table represents the speed (in mph) for a compact car, and m(x) represents the gas mileage (in mpg).

x	25	30	35	40	45
m(x)	22.7	25.1	27.9	30.8	31.9

x	50	55	60	65
m(x)	30.9	28.4	24.2	21.9

  a. Use regression to find a quadratic function to model the data. $y = -0.0235x^2 + 2.10x - 15.9$

  b. At what speed is the gas mileage the greatest? Round to the nearest mile per hour. 45 mph

  c. What is the maximum gas mileage? Round to the nearest mile per gallon. 31 mpg

59. Fluid runs through a drainage pipe with a 10-cm radius and a length of 30 m (3000 cm). The velocity of the fluid gradually decreases from the center of the pipe toward the edges as a result of friction with the walls of the pipe. For the data shown, $v(x)$ is the velocity of the fluid (in cm/sec) and x represents the distance (in cm) from the center of the pipe toward the edge.

x	0	1	2	3	4
$v(x)$	195.6	195.2	194.2	193.0	191.5

x	5	6	7	8	9
$v(x)$	189.8	188.0	185.5	183.0	180.0

  a. The pipe is 30 m long (3000 cm). Determine how long it will take fluid to run the length of the pipe through the center of the pipe. Round to 1 decimal place. 15.3 sec

  b. Determine how long it will take fluid at a point 9 cm from the center of the pipe to run the length of the pipe. Round to 1 decimal place. 16.7 sec

  c. Use regression to find a quadratic function to model the data. $v(x) = -0.1436x^2 - 0.4413x + 195.7$

  d. Use the model from part (c) to predict the velocity of the fluid at a distance 5.5 cm from the center of the pipe. Round to 1 decimal place. 188.9 cm/sec

60. The braking distance required for a car to stop depends on numerous variables such as the speed of the car, the weight of the car, reaction time of the driver, and the coefficient of friction between the tires and the road. For a certain vehicle on one stretch of highway, the braking distances $d(s)$ (in ft) are given for several different speeds s (in mph).

s	30	35	40	45	50
$d(s)$	109	134	162	191	223

s	55	60	65	70	75
$d(s)$	256	291	328	368	409

  a. Use regression to find a quadratic function to model the data. $d(s) = 0.039s^2 + 2.53s - 2.51$

  b. Use the model from part (a) to predict the stopping distance for the car if it is traveling 62 mph before the brakes are applied. Round to the nearest foot. 304 ft

  c. Suppose that the car is traveling 53 mph before the brakes are applied. If a deer is standing in the road at a distance of 245 ft from the point where the brakes are applied, will the car hit the deer? No. The stopping distance required is 242 ft (rounding up), so the car would miss the deer by approximately 3 ft.

Mixed Exercises

For Exercises 61–64, given a quadratic function defined by $f(x) = ax^2 + bx + c$ ($a \neq 0$), answer true or false. If an answer is false, explain why.

61. The graph of f can have two y-intercepts.
  False. If a relation has two y-intercepts, then it would fail the vertical line test and the relation would not define y as a function of x.

62. The graph of f can have two x-intercepts. True

63. If $a < 0$, then the vertex of the parabola is the maximum point on the graph of f. True

64. The axis of symmetry of the graph of f is the line defined by $y = c$.
  False. The axis of symmetry is a vertical line, whereas the graph of $y = c$ is a horizontal line. The axis of symmetry should be $x = -\dfrac{b}{2a}$.

For Exercises 65–70, determine the number of x-intercepts of the graph of $f(x) = ax^2 + bx + c$ ($a \neq 0$), based on the discriminant of the related equation $f(x) = 0$. (*Hint*: Recall that the discriminant is $b^2 - 4ac$.)

65. $f(x) = 4x^2 + 12x + 9$ Discriminant is 0; one x-intercept **66.** $f(x) = 25x^2 - 20x + 4$ Discriminant is 0; one x-intercept

67. $f(x) = -x^2 - 5x + 8$ Discriminant is 57; two x-intercepts **68.** $f(x) = -3x^2 + 4x + 9$ Discriminant is 124; two x-intercepts

69. $f(x) = -3x^2 + 6x - 11$
  Discriminant is −96; no x-intercepts
70. $f(x) = -2x^2 + 5x - 10$ Discriminant is −55; no x-intercepts

For Exercises 71–78, given a quadratic function defined by $f(x) = a(x - h)^2 + k$ ($a \neq 0$), match the graph with the function based on the conditions given.

71. $a > 0, h < 0, k > 0$ Graph g

72. $a > 0, h < 0, k < 0$ Graph d

73. $a < 0, h < 0, k < 0$ Graph h

74. $a < 0, h < 0, k > 0$ Graph b

75. $a > 0$, axis of symmetry $x = 2, k < 0$ Graph f

76. $a < 0$, axis of symmetry $x = 2, k > 0$ Graph e

77. $a < 0, h = 2$, maximum value equals −2 Graph c

78. $a > 0, h = 2$, minimum value equals 2 Graph a

a. $f(x)$

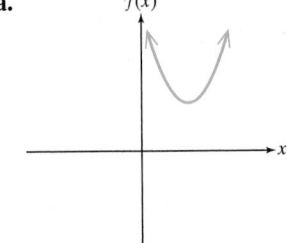

b. $f(x)$

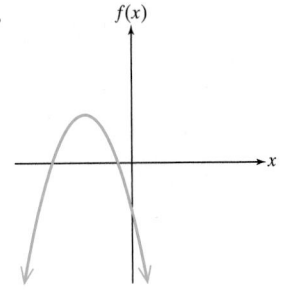

c. $f(x)$

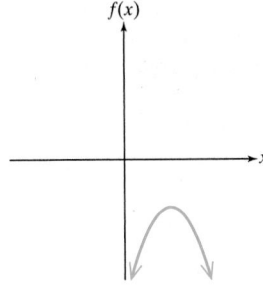

d. $f(x)$

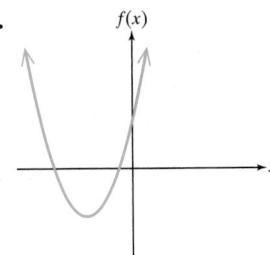

e. $f(x)$

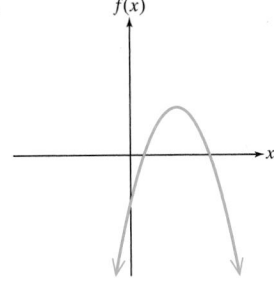

f. $f(x)$

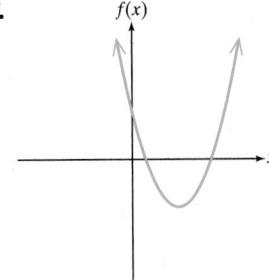

g. $f(x)$

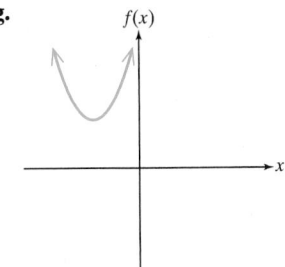

h. $f(x)$

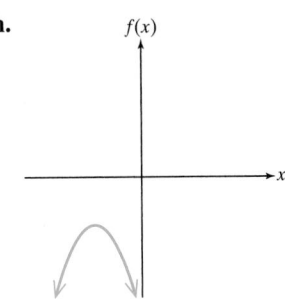

Write About It

79. Explain why a parabola opening upward has a minimum value but no maximum value. Use the graph of $f(x) = x^2$ to explain.

80. Explain why a quadratic function whose graph opens downward with vertex $(4, -3)$ has no x-intercept.

81. Explain why a quadratic function given by $f(x) = ax^2 + bx + c$ cannot have two y-intercepts.

82. Explain how to use the discriminant to determine the number of x-intercepts for the graph of $f(x) = ax^2 + bx + c$.

83. If a quadratic function given by $y = f(x)$ has x-intercepts of $(2, 0)$ and $(6, 0)$, explain why the vertex must be $(4, f(4))$.

84. Given an equation of a parabola in the form $y = a(x - h)^2 + k$, explain how to determine by inspection if the parabola has no x-intercepts.

Expanding Your Skills

For Exercises 85–88, define a quadratic function $y = f(x)$ that satisfies the given conditions.

85. Vertex $(2, -3)$ and passes through $(0, 5)$
$f(x) = 2(x - 2)^2 - 3$

86. Vertex $(-3, 1)$ and passes through $(0, -17)$
$f(x) = -2(x + 3)^2 + 1$

87. Axis of symmetry $x = 4$, maximum value 6, passes through $(1, 3)$ $f(x) = -\dfrac{1}{3}(x - 4)^2 + 6$

88. Axis of symmetry $x = -2$, minimum value 5, passes through $(2, 13)$ $f(x) = \dfrac{1}{2}(x + 2)^2 + 5$

For Exercises 89–92, find the value of b or c that gives the function the given minimum or maximum value.

89. $f(x) = 2x^2 + 12x + c$; minimum value -9 $c = 9$

90. $f(x) = 3x^2 + 12x + c$; minimum value -4 $c = 8$

91. $f(x) = -x^2 + bx + 4$; maximum value 8 $b = 4$ or $b = -4$

92. $f(x) = -x^2 + bx - 2$; maximum value 7
$b = 6$ or $b = -6$

SECTION 2.2   Introduction to Polynomial Functions

1. Determine the End Behavior of a Polynomial Function

A solar oven is to be made from an open box with reflective sides. Each box is made from a 30-in. by 24-in. rectangular sheet of aluminum with squares of length x (in inches) removed from each corner. Then the flaps are folded up to form an open box.

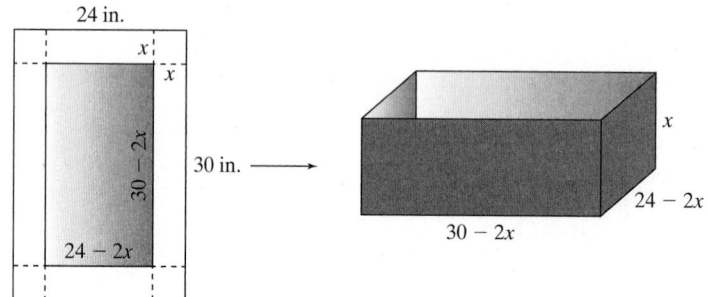

The volume $V(x)$ (in cubic inches) of the box is given by

$$V(x) = 4x^3 - 108x^2 + 720x, \text{ where } 0 < x < 12.$$

From the graph of $y = V(x)$ (Figure 2-6), the maximum volume appears to occur when squares of approximately 4 inches in length are cut from the corners of the sheet of aluminum. See Exercise 99.

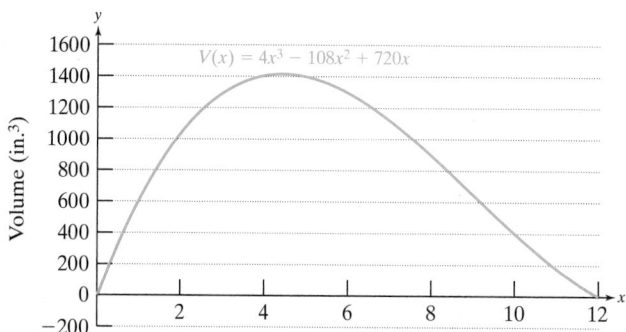

Figure 2-6

The function defined by $V(x) = 4x^3 - 108x^2 + 720x$ is an example of a polynomial function of degree 3.

Definition of a Polynomial Function

Let n be a whole number and $a_n, a_{n-1}, a_{n-2}, \dots, a_1, a_0$ be real numbers, where $a_n \neq 0$. Then a function defined by

$$f(x) = a_n x^n + a_{n-1}x^{n-1} + a_{n-2}x^{n-2} + \cdots + a_1 x + a_0$$

is called a **polynomial function of degree n.**

The coefficients of each term of a polynomial function are real numbers, and the exponents on x must be whole numbers.

Polynomial Function	**Not a Polynomial Function**
$f(x) = 4x^5 - 3x^4 + 2x^2$	$f(x) = 4\sqrt{x} - \dfrac{3}{x} + (3 + 2i)x^2$

$\sqrt{x} = x^{1/2}$	$3/x = 3x^{-1}$	$(3 + 2i)$
Exponent not a whole number	Exponent not a whole number	Coefficient not a real number

TIP A third-degree polynomial function is referred to as a *cubic* polynomial function.

A fourth-degree polynomial function is referred to as a *quartic* polynomial function.

We have already studied several special cases of polynomial functions. For example:

$f(x) = 2$	constant function	(polynomial function, degree 0)
$g(x) = 3x + 1$	linear function	(polynomial function, degree 1)
$h(x) = 4x^2 + 7x - 1$	quadratic function	(polynomial function, degree 2)

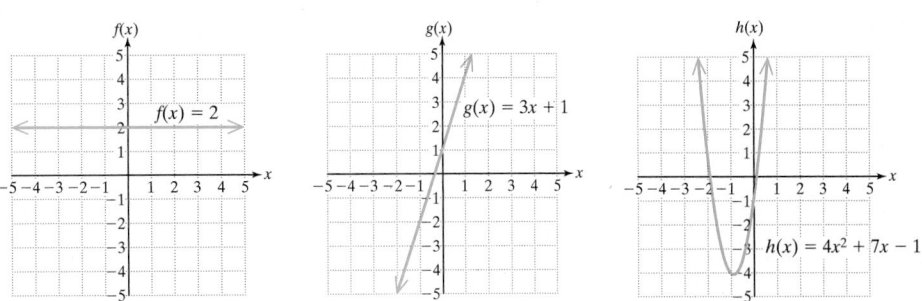

The domain of a polynomial function is all real numbers. Furthermore, the graph of a polynomial function is both continuous and smooth. Informally, a continuous function can be drawn without lifting the pencil from the paper. A smooth function has no sharp corners or points. For example, the first curve shown here could be a polynomial function, but the last three are not polynomial functions.

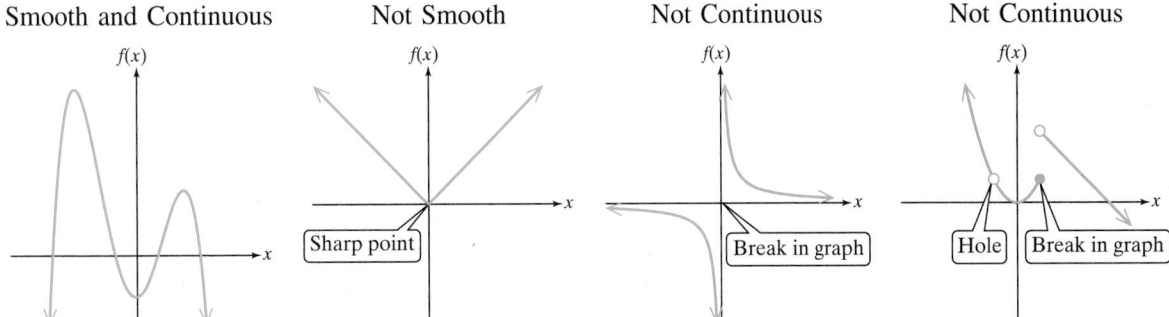

To begin our analysis of polynomial functions, we first consider the graphs of functions of the form $f(x) = ax^n$, where a is a real number and n is a positive integer. These fall into a category of functions called **power functions.** The graphs of three power functions with even degrees and positive coefficients are shown in Figure 2-7. The graphs of three power functions with odd degrees and positive coefficients are shown in Figure 2-8.

TIP For a positive integer n, the graph of the power function $y = x^n$ becomes "flatter" near the *x*-intercept for higher powers of n.

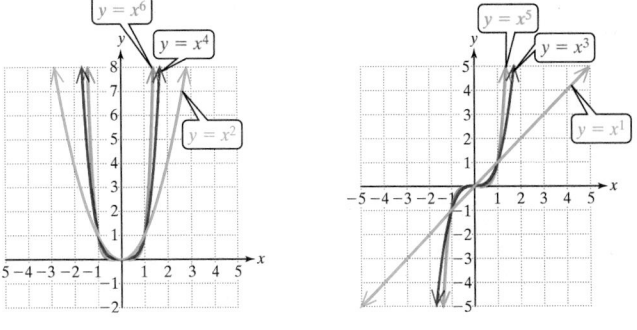

Figure 2-7   Figure 2-8

From Figure 2-7, notice that for even powers of n, the behavior of $y = x^n$ is similar to the graph of $y = x^2$ with variations on the "steepness" of the curve. Figure 2-8 shows that for odd powers, the behavior of $y = x^n$ with $n \geq 3$ is similar to the graph of $y = x^3$. For any power function $y = ax^n$, the coefficient a will impose a vertical shrink or stretch on the graph of $y = x^n$ by a factor of $|a|$. If $a < 0$, then the graph is reflected across the x-axis.

Power functions are helpful to analyze the "end behavior" of a polynomial function with multiple terms. The end behavior is the general direction that the fuction follows as x approaches ∞ or $-\infty$. To describe end behavior, we have the following notation.

Notation for Infinite Behavior of $y = f(x)$	
$x \to \infty$	is read as "x approaches infinity." This means that x becomes infinitely large in the positive direction.
$x \to -\infty$	is read as "x approaches negative infinity." This means that x becomes infinitely "large" in the negative direction.
$f(x) \to \infty$	is read as "$f(x)$ approaches infinity." This means that the y value becomes infinitely large in the positive direction.
$f(x) \to -\infty$	is read as "$f(x)$ approaches negative infinity." This means that the y value becomes infinitely "large" in the negative direction.

Consider the function defined by

$$f(x) = a_n x^n + a_{n-1} x^{n-1} + a_{n-2} x^{n-2} + \cdots + a_1 x + a_0$$

> The leading term has the greatest exponent on x.

The leading term has the greatest exponent on x. Therefore, as $|x|$ gets large (that is, as $x \to \infty$ or as $x \to -\infty$), the leading term will be relatively larger in absolute value than all other terms. In fact, x^n will eventually be greater in absolute value than the *sum* of all other terms. Therefore, the end behavior of the function is dictated only by the leading term, and the graph of the function far to the left and far to the right will follow the general behavior of the power function $y = ax^n$.

The Leading Term Test

Consider a polynomial function given by

$$f(x) = a_n x^n + a_{n-1} x^{n-1} + a_{n-2} x^{n-2} + \cdots + a_1 x + a_0.$$

As $x \to \infty$ or as $x \to -\infty$, f eventually becomes forever increasing or forever decreasing and will follow the general behavior of $y = a_n x^n$.

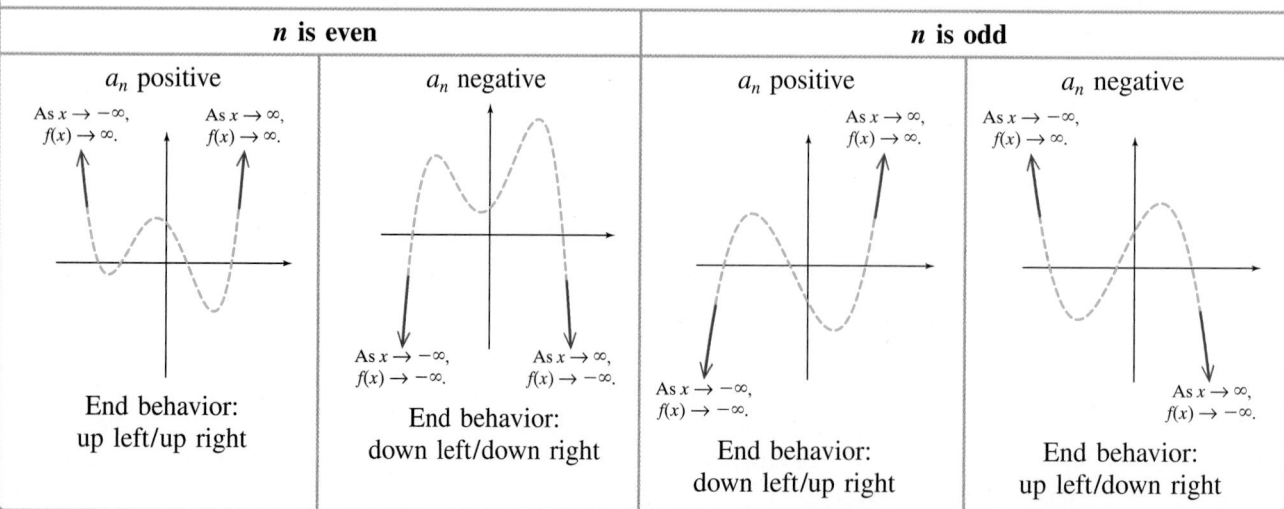

n is even		n is odd	
a_n positive	a_n negative	a_n positive	a_n negative
As $x \to -\infty$, $f(x) \to \infty$. As $x \to \infty$, $f(x) \to \infty$.	As $x \to -\infty$, $f(x) \to -\infty$. As $x \to \infty$, $f(x) \to -\infty$.	As $x \to \infty$, $f(x) \to \infty$. As $x \to -\infty$, $f(x) \to -\infty$.	As $x \to -\infty$, $f(x) \to \infty$. As $x \to \infty$, $f(x) \to -\infty$.
End behavior: up left/up right	End behavior: down left/down right	End behavior: down left/up right	End behavior: up left/down right

Classroom Examples: p. 265
Exercises 14, 20

EXAMPLE 1   **Determining End Behavior**

Use the leading term to determine the end behavior of the graph of the function.

a. $f(x) = -4x^5 + 6x^4 + 2x$ **b.** $g(x) = \frac{1}{4}x(2x - 3)^3(x + 4)^2$

TIP The graph of $y = f(x)$ from Example 1(a) will exhibit the same behavior as the graph of the power function $y = -4x^5$ for values of x far to the right and far to the left. This is similar to the graph of $y = x^5$ reflected across the x-axis.

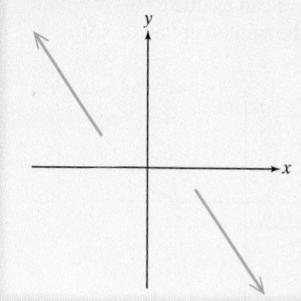

Solution:

a. $f(x) = -4x^5 + 6x^4 + 2x$

negative   odd

The leading coefficient is negative and the degree is odd. By the leading term test, the end behavior is up to the left and down to the right.

As $x \to -\infty$, $f(x) \to \infty$.
As $x \to \infty$, $f(x) \to -\infty$.

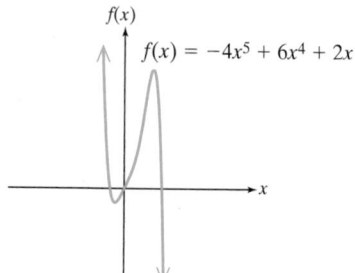

b. $g(x) = \frac{1}{4}x(2x - 3)^3(x + 4)^2$

positive   even

$g(x) = \frac{1}{4}x(2x - 3)^3(x + 4)^2 = 2x^6 + \cdots$

To determine the leading term, multiply the leading terms from each factor. That is,

$\frac{1}{4}x(2x)^3(x)^2 = 2x^6$.

The leading coefficient is positive and the degree is even. By the leading term test, the end behavior is up to the left and up to the right.

As $x \to -\infty$, $f(x) \to \infty$.
As $x \to \infty$, $f(x) \to \infty$.

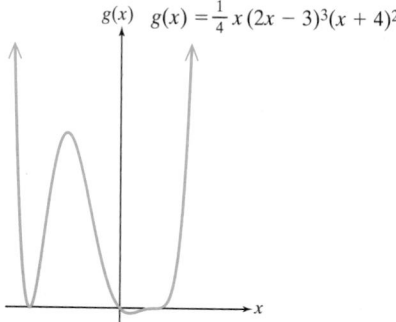

Skill Practice 1   Use the leading term to determine the end behavior of the graph of the function.

a. $f(x) = -0.3x^4 - 5x^2 - 3x + 4$ **b.** $g(x) = \frac{6}{7}(x - 9)^4(x + 4)^2(3x - 5)$

2. Identify Zeros and Multiplicities of Zeros

Consider a polynomial function defined by $y = f(x)$. The values of x in the domain of f for which $f(x) = 0$ are called the **zeros** of the function. These are the real solutions (or **roots**) of the equation $f(x) = 0$ and correspond to the x-intercepts of the graph of $y = f(x)$.

Answers

1. a. Down to the left, down to the right.
   As $x \to -\infty$, $f(x) \to -\infty$.
   As $x \to \infty$, $f(x) \to -\infty$.

b. Down to the left, up to the right.
   As $x \to -\infty$, $f(x) \to -\infty$.
   As $x \to \infty$, $f(x) \to \infty$.

Classroom Example: p. 266
Exercise 24

EXAMPLE 2   Determining the Zeros of a Polynomial Function

Find the zeros of the function defined by $f(x) = x^3 + x^2 - 9x - 9$.

Solution:

$$f(x) = x^3 + x^2 - 9x - 9$$

To find the zeros of f, set $f(x) = 0$ and solve for x.

$$0 = x^3 + x^2 - 9x - 9$$

$$0 = x^2(x + 1) - 9(x + 1)$$

Factor by grouping.

$$0 = (x + 1)(x^2 - 9)$$

$$0 = (x + 1)(x - 3)(x + 3)$$

Factor the difference of squares.

$$x = -1, x = 3, x = -3$$

Set each factor equal to zero and solve for x.

The zeros of f are -1, 3, and -3.

The graph of f is shown in Figure 2-9. The zeros of the function are real numbers and correspond to the x-intercepts of the graph. By inspection, we can evaluate $f(0) = -9$, indicating that the y-intercept is $(0, -9)$.

Check:

A table of points can be used to check that $f(-1)$, $f(3)$, and $f(-3)$ all equal 0.

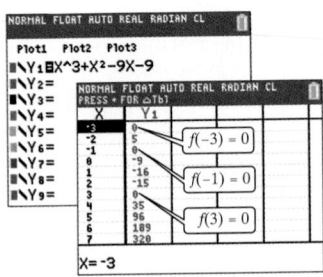

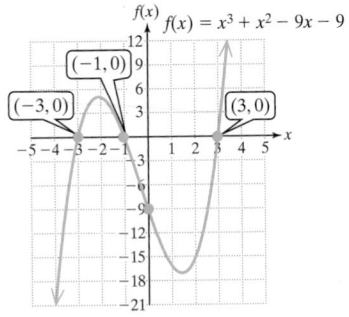

Figure 2-9

Skill Practice 2   Find the zeros of the function defined by
$$f(x) = 4x^3 - 4x^2 - 25x + 25.$$

Classroom Example: p. 266
Exercise 26

EXAMPLE 3   Determining the Zeros of a Polynomial Function

Find the zeros of the function defined by $f(x) = -x^3 + 8x^2 - 16x$.

Solution:

$$f(x) = -x^3 + 8x^2 - 16x$$

To find the zeros of f, set $f(x) = 0$ and solve for x.

$$0 = -x(x^2 - 8x + 16)$$

Factor out the GCF.

$$0 = -x(x - 4)^2$$

Factor the perfect square trinomial.

$$x = 0, x = 4$$

Set each factor equal to zero and solve for x.

The zeros of f are 0 and 4.

Instructor Note:
We restrict the graphs of polynomial functions to R × R. Therefore, imaginary zeros will not be discussed from a graphical perspective. These are left for a general discussion on polynomials and polynomial equations in Sections 2.3 and 2.4.

The graph of f is shown in Figure 2-10. The zeros of the function are real numbers and correspond to the x-intercepts $(0, 0)$ and $(4, 0)$.

The leading term of $f(x)$ is $-x^3$. The coefficient is negative and the exponent is odd. The graph shows the end behavior up to the left and down to the right as expected.

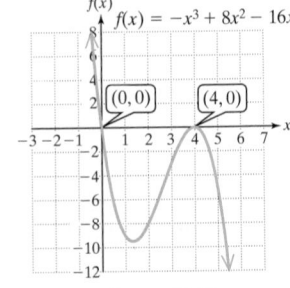

Figure 2-10

Answers

2. $1, \dfrac{5}{2}, -\dfrac{5}{2}$

3. $0, -5$

Skill Practice 3   Find the zeros of the function defined by
$$f(x) = x^3 + 10x^2 + 25x.$$

From Example 3, $f(x) = -x^3 + 8x^2 - 16x$ can be written as a product of linear factors:

$$f(x) = -x(x - 4)^2$$

Notice that the factor $(x - 4)$ appears to the second power. Therefore, we say that the corresponding zero, 4, has a multiplicity of 2. In general, we say that if a polynomial function has a factor $(x - c)$ that appears exactly k times, then c is a **zero of multiplicity** k. For example, consider:

$$g(x) = x^2(x - 2)^3(x + 4)^7$$

0 is a zero of multiplicity 2.

2 is a zero of multiplicity 3.

-4 is a zero of multiplicity 7.

The graph of a polynomial function behaves in the following manner based on the multiplicity of the zeros.

Touch Points and Cross Points

Let f be a polynomial function and let c be a real zero of f. Then the point $(c, 0)$ is an x-intercept of the graph of f. Furthermore,

- If c is a zero of odd multiplicity, then the graph *crosses* the x-axis at c. The point $(c, 0)$ is called a **cross point.**
- If c is a zero of even multiplicity, then the graph *touches* the x-axis at c and turns back around (does not cross the x-axis). The point $(c, 0)$ is called a **touch point.**

To illustrate the behavior of a polynomial function at its real zeros, consider the graph of $f(x) = -x(x - 4)^2$ from Example 3 (Figure 2-11).

- 0 has a multiplicity of 1 (odd multiplicity). The graph *crosses* the x-axis at $(0, 0)$.
- 4 has a multiplicity of 2 (even multiplicity). The graph *touches* the x-axis at $(4, 0)$ and turns back around.

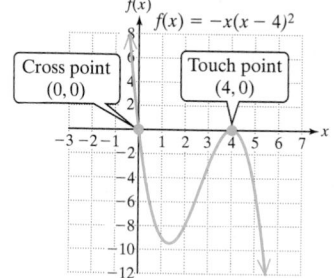

Figure 2-11

Classroom Example: p. 266
Exercise 30

EXAMPLE 4 **Determining Zeros and Multiplicities**

Determine the zeros and their multiplicities for the given functions.

a. $m(x) = \dfrac{1}{10}(x - 4)^2(2x + 5)^3$ **b.** $n(x) = x^4 - 2x^2$

Solution:

a. $m(x) = \dfrac{1}{10}(x - 4)^2(2x + 5)^3$ [even] [odd]

The function is factored into linear factors. The zeros are 4 and $-\frac{5}{2}$.

Instructor Note:
Mention to students that a multiplicity of 1 causes the behavior around the zero to be linear. A multiplicity of 2 causes quadratic behavior, a multiplicity of 3 causes cubic behavior, and so on.

The function has a zero of 4 with multiplicity 2 (even). The graph has a touch point at $(4, 0)$.

   The function has a zero of $-\frac{5}{2}$ with multiplicity 3 (odd). The graph has a cross point at $\left(-\frac{5}{2}, 0\right)$.

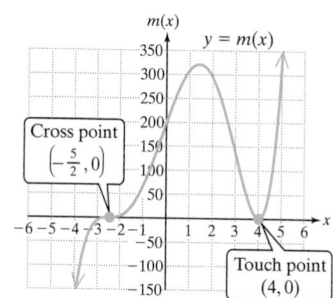

b. $n(x) = x^4 - 2x^2$
$= x^2(x^2 - 2)$
$= x^2(x - \sqrt{2})(x + \sqrt{2})$

The function has a zero of 0 with multiplicity 2 (even). The graph has a touch point at $(0, 0)$.

The function has a zero of $\sqrt{2}$ with multiplicity 1 (odd). The graph has a cross point at $(\sqrt{2}, 0) \approx (1.41, 0)$.

The function has a zero of $-\sqrt{2}$ with multiplicity 1 (odd). The graph has a cross point at $(-\sqrt{2}, 0) \approx (-1.41, 0)$.

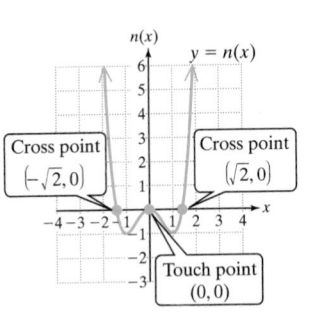

Skill Practice 4   Determine the zeros and their multiplicities for the given functions.

a. $p(x) = -\dfrac{3}{5}(x + 3)^4(5x - 1)^5$ **b.** $q(x) = 2x^6 - 14x^4$

Classroom Example: p. 266
Exercise 42

3. Apply the Intermediate Value Theorem

In Examples 2–4, the zeros of the functions were easily identified by first factoring the polynomial. However, in most cases, the real zeros of a polynomial are difficult or impossible to determine algebraically. For example, the function given by $f(x) = x^4 + 6x^3 - 26x + 15$ has zeros of $-1 \pm \sqrt{6}$ and $-2 \pm \sqrt{7}$. At this point, we do not have the tools to find the zeros of this function analytically. However, we can use the intermediate value theorem to help us search for zeros of a polynomial function and approximate their values.

> **Intermediate Value Theorem**
>
> Let f be a polynomial function. For $a < b$, if $f(a)$ and $f(b)$ have opposite signs, then f has at least one zero on the interval $[a, b]$.

EXAMPLE 5   **Applying the Intermediate Value Theorem**

Show that $f(x) = x^4 + 6x^3 - 26x + 15$ has a zero on the interval $[1, 2]$.

Solution:

$f(x) = x^4 + 6x^3 - 26x + 15$
$f(1) = (1)^4 + 6(1)^3 - 26(1) + 15 = -4$
$f(2) = (2)^4 + 6(2)^3 - 26(2) + 15 = 27$

Since $f(1)$ and $f(2)$ have opposite signs, then by the intermediate value theorem, we know that the function must have at least one zero on the interval $[1, 2]$.

The actual value of the zero on the interval $[1, 2]$ is $-1 + \sqrt{6} \approx 1.45$.

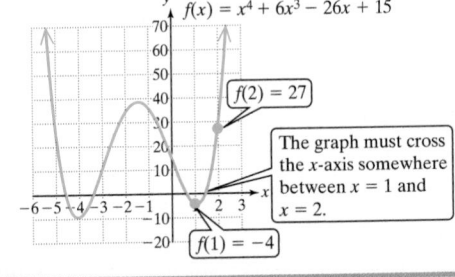

Skill Practice 5   Show that $f(x) = x^4 + 6x^3 - 26x + 15$ has a zero on the interval $[-4, -3]$.

Sidebar (left column)

Avoiding Mistakes

For a polynomial function f, if $f(a)$ and $f(b)$ have opposite signs, then f must have at least one zero on the interval $[a, b]$. This includes the possibility that f may have more than one zero on $[a, b]$.

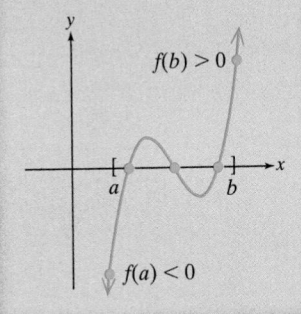

TIP It is important to note that if the signs of $f(a)$ and $f(b)$ are the same, then the intermediate value theorem is inconclusive.

Answers

4. a. -3 (multiplicity 4) and $\dfrac{1}{5}$ (multiplicity 5)

   b. 0 (multiplicity 4), $\sqrt{7}$ (multiplicity 1), and $-\sqrt{7}$ (multiplicity 1)

5. $f(-4) = -9$ and $f(-3) = 12$. Since $f(-4)$ and $f(-3)$ have opposite signs, then the intermediate value theorem guarantees the existence of at least one zero on the interval $[-4, -3]$.

The intermediate value theorem can be used repeatedly in a technique called the bisection method to approximate the value of a zero. See the online group activity "Investigating the Bisection Method for Finding Zeros."

Point of Interest

The modern definition of a computer is a programmable device designed to carry out a sequence of arithmetic or logical operations. However, the word "computer" originally referred to a person who did such calculations using paper and pencil. "Human computers" were notably used in the eighteenth century to predict the path of Halley's comet and to produce astronomical tables critical to surveying and navigation. Later, during World Wars I and II, human computers developed ballistic firing tables that would describe the trajectory of a shell.

Computing tables of values was very time consuming, and the "computers" would often interpolate to find intermediate values within a table. Interpolation is a method by which intermediate values between two numbers are estimated. Often the interpolated values were based on a polynomial function.

4. Sketch a Polynomial Function

TIP Even with advanced techniques from calculus or the use of a graphing utility, it is often difficult or impossible to find the exact location of the turning points of a polynomial function.

The graph of a polynomial function may also have "turning points." These correspond to relative maxima and minima. For example, consider $f(x) = x(x + 2)(x - 2)^2$. See Figure 2-12.

Multiplying the leading terms within the factors, we have a leading term of $(x)(x)(x)^2 = x^4$. Therefore, the end behavior of the graph is up to the left and up to the right.

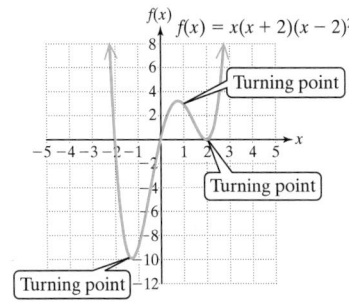

$$f(x) = x(x + 2)(x - 2)^2$$

Figure 2-12

Avoiding Mistakes

A polynomial of degree n may have fewer than $n - 1$ turning points. For example, $f(x) = x^3$ is a degree 3 polynomial function (indicating that it could have a maximum of two turning points), yet the graph has no turning points.

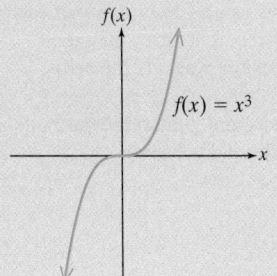

Starting from the far left, the graph of f decreases to the x-intercept of -2. Since -2 is a zero with an odd multiplicity, the graph must cross the x-axis at -2. For the same reason, the graph must cross the x-axis again at the origin. Therefore, somewhere between $x = -2$ and $x = 0$, the graph must "turn around." This point is called a "turning point."

The turning points of a polynomial function are the points where the function changes from increasing to decreasing or vice versa.

Number of Turning Points of a Polynomial Function

Let f represent a polynomial function of degree n. Then the graph of f has at most $n - 1$ turning points.

At this point we are ready to outline a strategy for sketching a polynomial function.

Graphing a Polynomial Function

To graph a polynomial function defined by $y = f(x)$,

1. Use the leading term to determine the end behavior of the graph.
2. Determine the y-intercept by evaluating $f(0)$.
3. Determine the real zeros of f and their multiplicities (these are the x-intercepts of the graph of f).
4. Plot the x- and y-intercepts and sketch the end behavior.
5. Draw a sketch starting from the left-end behavior. Connect the x- and y-intercepts in the order that they appear from left to right using these rules:
   - The curve will cross the x-axis at an x-intercept if the corresponding zero has an odd multiplicity.
   - The curve will touch but not cross the x-axis at an x-intercept if the corresponding zero has an even multiplicity.
6. If a test for symmetry is easy to apply, use symmetry to plot additional points. Recall that
   - f is an even function (symmetric to the y-axis) if $f(-x) = f(x)$.
   - f is an odd function (symmetric to the origin) if $f(-x) = -f(x)$.
7. Plot more points if a greater level of accuracy is desired. In particular, to estimate the location of turning points, find several points between two consecutive x-intercepts.

In Examples 6 and 7, we demonstrate the process of graphing a polynomial function.

Classroom Example: p. 267
Exercise 60

EXAMPLE 6   Graphing a Polynomial Function

Graph $f(x) = x^3 - 9x$.

Solution:

$f(x) = x^3 - 9x$

1. The leading term is x^3. The end behavior is down to the left and up to the right.

   The exponent on the leading term is odd and the leading coefficient is positive.

2. $f(0) = (0)^3 - 9(0) = 0$

   The y-intercept is $(0, 0)$.

   Determine the y-intercept by evaluating $f(0)$.

3. $0 = x^3 - 9x$

   $0 = x(x^2 - 9)$

   $0 = x(x - 3)(x + 3)$

   The zeros of the function are 0, 3, and -3, and each has a multiplicity of 1.

   Find the real zeros of f by solving for the real solutions to the equation $f(x) = 0$.

   The zeros are real numbers and correspond to x-intercepts on the graph. Since the multiplicity of each zero is an odd number, the graph will cross the x-axis at the zeros.

4.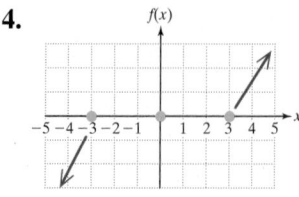

   Plot the x- and y-intercepts and sketch the end behavior.

5. Moving from left to right, the curve increases from the far left and then crosses the x-axis at -3. The graph must have a turning point between $x = -3$ and $x = 0$ so that the curve can pass through the next x-intercept of $(0, 0)$. ⟶ The graph crosses the x-axis at $x = 0$. The graph must then have another turning point between $x = 0$ and $x = 3$ so that the curve can pass through the next x-intercept of $(3, 0)$. Finally, the graph crosses the x-axis at $x = 3$ and continues to increase to the far right.

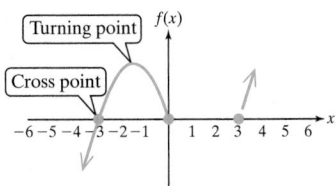

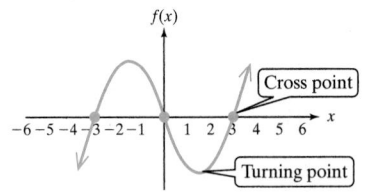

6. $f(x) = x^3 - 9x$

$f(-x) = (-x)^3 - 9(-x) \qquad -f(x) = -(x^3 - 9x)$
$\qquad = -x^3 + 9x \xleftrightarrow{\quad f(-x) = -f(x) \quad} = -x^3 + 9x$
$\qquad\qquad\qquad\qquad\text{(same)}$

Testing for symmetry, we see that $f(-x) = -f(x)$. Therefore, f is an odd function and is symmetric with respect to the origin.

> **TIP** Techniques of calculus can be used to find the exact coordinates of the turning points of the polynomial function in Example 6.

7. If more accuracy is desired, plot additional points. In this case, since f is symmetric to the origin, if a point (x, y) is on the graph, then so is $(-x, -y)$. The graph of f is shown in Figure 2-13.

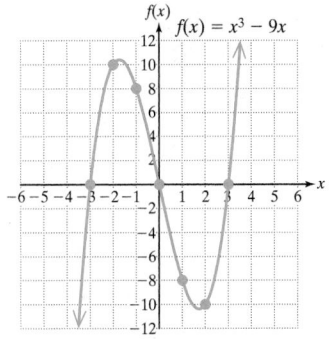

x	$f(x)$
1	-8
2	-10
4	28

Use symmetry. ⟶

x	$f(x)$
-1	8
-2	10
-4	-28

Figure 2-13

Skill Practice 6 Graph $g(x) = -x^3 + 4x$.

Classroom Example: p. 267
Exercise 66

EXAMPLE 7   Graphing a Polynomial Function

Graph $g(x) = -0.1(x - 1)(x + 2)(x - 4)^2$.

Solution:

$g(x) = -0.1(x - 1)(x + 2)(x - 4)^2$

1. Multiplying the leading terms within the factors, we have a leading term of $-0.1(x)(x)(x)^2 = -0.1x^4$. The end behavior is down to the left and down to the right.

The exponent on the leading term is even and the leading coefficient is negative.

2. $g(0) = -0.1(0 - 1)(0 + 2)(0 - 4)^2 = 3.2$
The y-intercept is $(0, 3.2)$.

Determine the y-intercept by evaluating $g(0)$.

3. $0 = -0.1(x - 1)(x + 2)(x - 4)^2$
The zeros of the function are 1, -2, and 4.
The multiplicity of 1 is 1.
The multiplicity of -2 is 1.
The multiplicity of 4 is 2.

Find the real zeros of g by solving for the real solutions of the equation $g(x) = 0$.

The zeros are real numbers and correspond to x-intercepts on the graph: $(1, 0)$, $(-2, 0)$, and $(4, 0)$.

Answer

6.

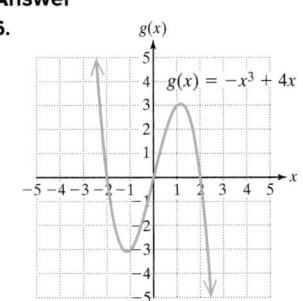

4.

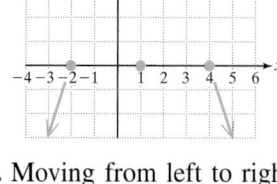

Plot the x- and y-intercepts and sketch the end behavior.

5. Moving from left to right, the curve increases from the far left. It then crosses the x-axis at $x = -2$ and turns back around to pass through the next x-intercept at $x = 1$.

   The curve has another turning point between $x = 1$ and $x = 4$ so that it can touch the x-axis at 4. From there it turns back downward and continues to decrease to the far right.

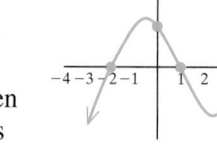

6. From our preliminary sketch in step 5, we see that the function is not symmetric with respect to either the y-axis or origin.

7. If more accuracy is desired, plot additional points. The graph is shown in Figure 2-14.

x	$g(x)$
-3	-19.6
-1	5
2	-1.6
3	-1
5	-2.8

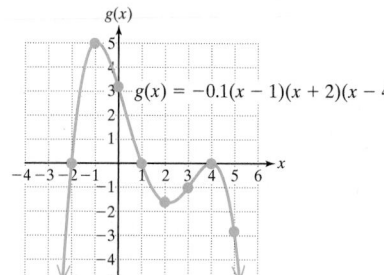

Figure 2-14

Skill Practice 7 Graph $h(x) = 0.5x(x - 1)(x + 3)^2$.

TECHNOLOGY CONNECTIONS

Using a Graphing Utility to Graph a Polynomial Function

It is important to have a strong knowledge of algebra to use a graphing utility effectively. For example, consider the graph of $f(x) = 0.005(x - 2)(x + 3)(x - 5)(x + 15)$ on the standard viewing window.

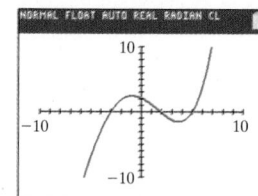

From the leading term, $0.005x^4$, we know that the end behavior should be up to the left and up to the right. Furthermore, the function has four real zeros (2, -3, 5, and -15), and should have four corresponding x-intercepts. Therefore, on the standard viewing window, the calculator does not show the key features of the graph.

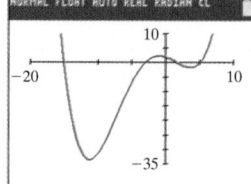

By graphing f on the window $[-20, 10, 2]$ by $[-35, 10, 5]$, we see the end behavior displayed correctly, all four x-intercepts, and the turning points (there should be at most 3).

Answer

7.

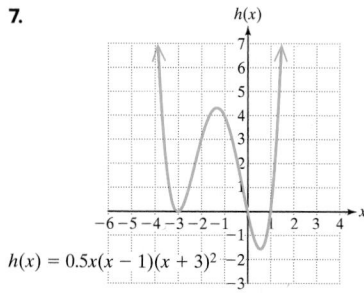

$h(x) = 0.5x(x - 1)(x + 3)^2$

SECTION 2.2   Practice Exercises

Prerequisite Review

For Exercises R.1–R.2, solve the equation.

R.1. $3x^3 + 21x^2 - 54x = 0$ $\{-9, 0, 2\}$ **R.2.** $5x^3 + 6x^2 - 20x - 24 = 0$ $\left\{-2, -\frac{6}{5}, 2\right\}$

For Exercises R.3–R.5, use transformations to graph the given function.

R.3. $m(x) = x^3 - 5$ **R.4.** $f(x) = (x + 2)^2 - 4$ **R.5.** $g(x) = (3x - 6)^2$

Concept Connections

1. A function defined by $f(x) = a_n x^n + a_{n-1} x^{n-1} + a_{n-2} x^{n-2} + \cdots + a_1 x + a_0$ where $a_n, a_{n-1}, a_{n-2}, \ldots, a_1, a_0$ are real numbers and $a_n \neq 0$ is called a _____polynomial_____ function.

2. The function given by $f(x) = -3x^5 + \sqrt{2}x + \frac{1}{2}x$ (is/is not) a polynomial function.   is

3. The function given by $f(x) = -3x^5 + 2\sqrt{x} + \frac{2}{x}$ (is/is not) a polynomial function.   is not

4. A quadratic function is a polynomial function of degree ____2____.

5. A linear function is a polynomial function of degree ____1____.

6. The values of x in the domain of a polynomial function f for which $f(x) = 0$ are called the ____zeros____ of the function.

7. What is the maximum number of turning points of the graph of $f(x) = -3x^6 - 4x^5 - 5x^4 + 2x^2 + 6$?   5

8. If the graph of a polynomial function has 3 turning points, what is the minimum degree of the function?   4

9. If c is a real zero of a polynomial function and the multiplicity is 3, does the graph of the function cross the x-axis or touch the x-axis (without crossing) at $(c, 0)$?   cross

10. If c is a real zero of a polynomial function and the multiplicity is 6, does the graph of the function cross the x-axis or touch the x-axis (without crossing) at $(c, 0)$?   touch (without crossing)

11. Suppose that f is a polynomial function and that $a < b$. If $f(a)$ and $f(b)$ have opposite signs, then what conclusion can be drawn from the intermediate value theorem? f has at least one zero on the interval $[a, b]$.

12. What is the leading term of $f(x) = -\frac{1}{3}(x - 3)^4(3x + 5)^2$? $-3x^6$

Objective 1: Determine the End Behavior of a Polynomial Function

For Exercises 13–20, determine the end behavior of the graph of the function. (See Example 1)

13. $f(x) = -3x^4 - 5x^2 + 2x - 6$ 14. $g(x) = -\frac{1}{2}x^6 + 8x^4 - x^3 + 9$ 15. $h(x) = 12x^5 + 8x^4 - 4x^3 - 8x + 1$

16. $k(x) = 11x^7 - 4x^2 + 9x + 3$ 17. $m(x) = -4(x - 2)(2x + 1)^2(x + 6)^4$ 18. $n(x) = -2(x + 4)(3x - 1)^3(x + 5)$

19. $p(x) = -2x^2(3 - x)(2x - 3)^3$ 20. $q(x) = -5x^4(2 - x)^3(2x + 5)$

Objective 2: Identify Zeros and Multiplicities of Zeros

21. Given the function defined by $g(x) = -3(x - 1)^3(x + 5)^4$, the value 1 is a zero with multiplicity ____3____, and the value -5 is a zero with multiplicity ____4____.

22. Given the function defined by $h(x) = \frac{1}{2}x^5(x + 0.6)^3$, the value 0 is a zero with multiplicity ____5____, and the value -0.6 is a zero with multiplicity ____3____.

For Exercises 23–38, find the zeros of the function and state the multiplicities. (See Examples 2–4)

23. $f(x) = x^3 + 2x^2 - 25x - 50$ **24.** $g(x) = x^3 + 5x^2 - x - 5$ **25.** $h(x) = -6x^3 - 9x^2 + 60x$

26. $k(x) = -6x^3 + 26x^2 - 28x$ **27.** $m(x) = x^5 - 10x^4 + 25x^3$ **28.** $n(x) = x^6 + 4x^5 + 4x^4$

29. $p(x) = -3x(x + 2)^3(x + 4)$ **30.** $q(x) = -2x^4(x + 1)^3(x - 2)^2$

31. $t(x) = 5x(3x - 5)(2x + 9)(x - \sqrt{3})(x + \sqrt{3})$ **32.** $z(x) = 4x(5x - 1)(3x + 8)(x - \sqrt{5})(x + \sqrt{5})$

33. $c(x) = [x - (3 - \sqrt{5})][x - (3 + \sqrt{5})]$ **34.** $d(x) = [x - (2 - \sqrt{11})][x - (2 + \sqrt{11})]$

35. $f(x) = 4x^4 - 37x^2 + 9$ **36.** $k(x) = 4x^4 - 65x^2 + 16$

37. $n(x) = x^6 - 7x^4$ **38.** $m(x) = x^5 - 5x^3$

Objective 3: Apply the Intermediate Value Theorem

For Exercises 39–40, determine whether the intermediate value theorem guarantees that the function has a zero on the given interval. (See Example 5)

39. $f(x) = 2x^3 - 7x^2 - 14x + 30$ **40.** $g(x) = 2x^3 - 13x^2 + 18x + 5$

    a. $[1, 2]$ Yes   **b.** $[2, 3]$ No   **a.** $[1, 2]$ No   **b.** $[2, 3]$ Yes

    c. $[3, 4]$ No   **d.** $[4, 5]$ Yes   **c.** $[3, 4]$ No   **d.** $[4, 5]$ Yes

For Exercises 41–42, a table of values is given for $Y_1 = f(x)$. Determine whether the intermediate value theorem guarantees that the function has a zero on the given interval.

41. $Y_1 = 21x^4 + 46x^3 - 238x^2 - 506x + 77$ **42.** $Y_1 = 10x^4 + 21x^3 - 119x^2 - 147x + 343$

    a. $[-4, -3]$ Yes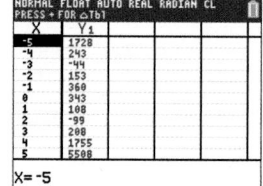

    b. $[-3, -2]$ Yes

    c. $[-2, -1]$ No

    d. $[-1, 0]$ No

    a. $[-4, -3]$ Yes

    b. $[-3, -2]$ Yes

    c. $[-2, -1]$ No

    d. $[-1, 0]$ No

43. Given $f(x) = 4x^3 - 8x^2 - 25x + 50$,   **44.** Given $f(x) = 9x^3 - 18x^2 - 100x + 200$,

    a. Determine if f has a zero on the interval $[-3, -2]$. Yes

    b. Find a zero of f on the interval $[-3, -2]$. $-\dfrac{5}{2}$

    a. Determine if f has a zero on the interval $[-4, -3]$. Yes

    b. Find a zero of f on the interval $[-4, -3]$. $-\dfrac{10}{3}$

Objective 4: Sketch a Polynomial Function

For Exercises 45–52, determine if the graph can represent a polynomial function. If so, assume that the end behavior and all turning points are represented in the graph.

a. Determine the minimum degree of the polynomial.

b. Determine whether the leading coefficient is positive or negative based on the end behavior and whether the degree of the polynomial is odd or even.

c. Approximate the real zeros of the function, and determine if their multiplicities are even or odd.

45.
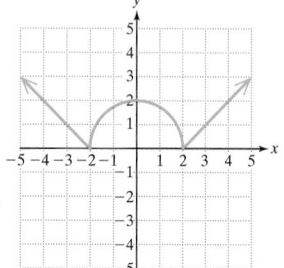

46.
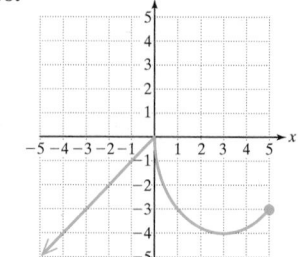

47.
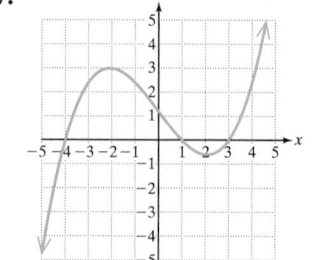

48.
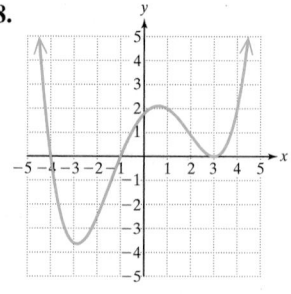

49. 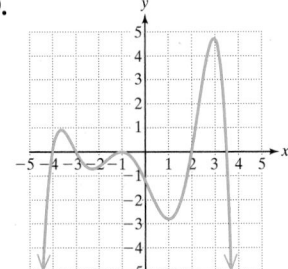   **50.** 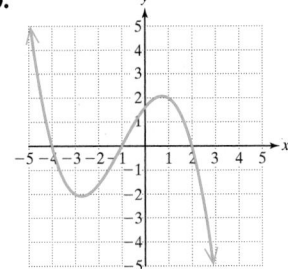   **51.** 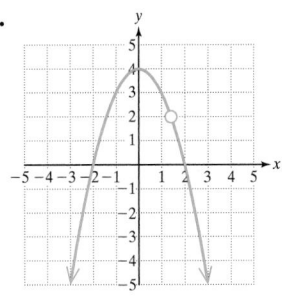   **52.**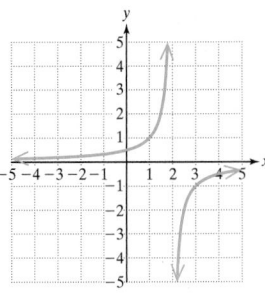

For Exercises 53–58,

a. Identify the power function of the form $y = x^n$ that is the parent function to the given graph.

b. In order, outline the transformations that would be required on the graph of $y = x^n$ to make the graph of the given function. See Section 1.6, page 190.

c. Match the function with the graph of i–vi.

53. $g(x) = -\dfrac{1}{3}x^6 - 2$

54. $f(x) = -\dfrac{1}{2}(x - 3)^4$

55. $k(x) = -(x + 2)^3 + 3$

56. $p(x) = 2(x + 4)^3 - 3$

57. $m(x) = (-x - 3)^5 + 1$

58. $n(x) = (-x + 3)^4 - 1$

i. 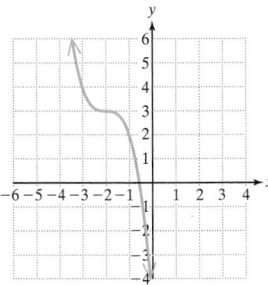   **ii.** 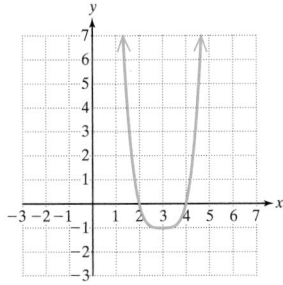   **iii.**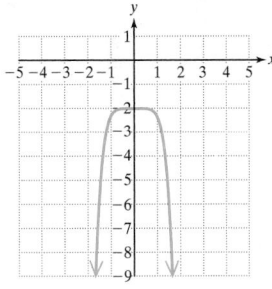

iv. 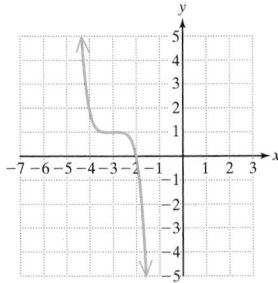   **v.** 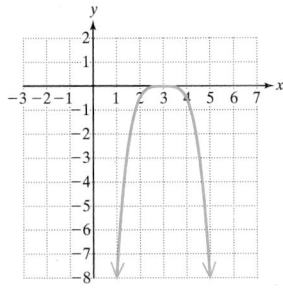   **vi.**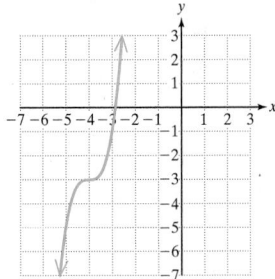

For Exercises 59–76, sketch the function. (See Examples 6–7)

59. $f(x) = x^3 - 5x^2$

60. $g(x) = x^5 - 2x^4$

61. $f(x) = \dfrac{1}{2}(x - 2)(x + 1)(x + 3)$

62. $h(x) = \dfrac{1}{4}(x - 1)(x - 4)(x + 2)$

63. $k(x) = x^4 + 2x^3 - 8x^2$

64. $h(x) = x^4 - x^3 - 6x^2$

65. $k(x) = 0.2(x + 2)^2(x - 4)^3$

66. $m(x) = 0.1(x - 3)^2(x + 1)^3$

67. $p(x) = 9x^5 + 9x^4 - 25x^3 - 25x^2$

68. $q(x) = 9x^5 + 18x^4 - 4x^3 - 8x^2$

69. $t(x) = -x^4 + 11x^2 - 28$

70. $v(x) = -x^4 + 15x^2 - 44$

71. $g(x) = -x^4 + 5x^2 - 4$

72. $h(x) = -x^4 + 10x^2 - 9$

73. $c(x) = 0.1x(x - 2)^4(x + 2)^3$

74. $d(x) = 0.05x(x - 2)^4(x + 3)^2$

75. $m(x) = -\dfrac{1}{10}(x + 3)(x - 3)(x + 1)^3$

76. $f(x) = -\dfrac{1}{10}(x - 1)(x + 3)(x - 4)^2$

Mixed Exercises

For Exercises 77–88, determine if the statement is true or false. If a statement is false, explain why.

77. The function defined by $f(x) = (x + 1)^5(x - 5)^2$ crosses the x-axis at 5. False. The value 5 is a zero with even multiplicity. Therefore, the graph touches but does not cross the x-axis at 5.

78. The function defined by $g(x) = -3(x + 4)(2x - 3)^4$ touches but does not cross the x-axis at $\left(\frac{3}{2}, 0\right)$. True

79. A third-degree polynomial has three turning points.

80. A third-degree polynomial has two turning points.

81. There is more than one polynomial function with zeros of 1, 2, and 6. True

82. There is exactly one polynomial with integer coefficients with zeros of 2, 4, and 6.

83. The graph of a polynomial function with leading term of even degree is up to the far left and up to the far right.

84. If c is a real zero of an even polynomial function, then $-c$ is also a zero of the function. True

85. The graph of $f(x) = x^3 - 27$ has three x-intercepts. False. The only real solution to the equation $x^3 - 27 = 0$ is $x = 3$. Therefore, the graph of $f(x) = x^3 - 27$ has only one x-intercept.

86. The graph of $f(x) = 3x^2(x - 4)^4$ has no points in Quadrants III or IV. True

87. The graph of $p(x) = -5x^4(x + 1)^2$ has no points in Quadrants I or II. True

88. A fourth-degree polynomial has exactly two relative minima and two relative maxima. False. A fourth-degree polynomial may have at most 3 turning points and consequently 3 relative maxima or minima.

89. A rocket will carry a communications satellite into low Earth orbit. Suppose that the thrust during the first 200 sec of flight is provided by solid rocket boosters at different points during liftoff.

The graph shows the acceleration in G-forces (that is, acceleration in 9.8-m/sec² increments) versus time after launch.

a. Approximate the interval(s) over which the acceleration is increasing. $(0, 12) \cup (68, 184)$

b. Approximate the interval(s) over which the acceleration is decreasing. $(12, 68) \cup (184, 200)$

c. How many turning points does the graph show? 3

d. Based on the number of turning points, what is the minimum degree of a polynomial function that could be used to model acceleration versus time? Would the leading coefficient be positive or negative? Degree 4; leading coefficient negative

e. Approximate the time when the acceleration was the greatest. 184 sec after launch

f. Approximate the value of the maximum acceleration. 2.85 G-forces

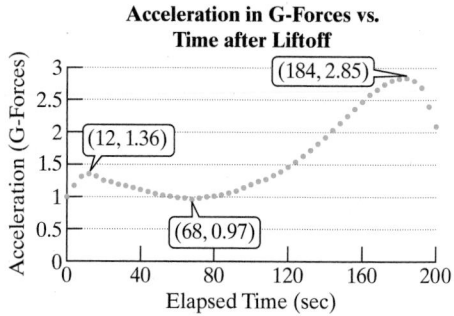

Acceleration in G-Forces vs. Time after Liftoff

90. Data from a 20-yr study show the number of new AIDS cases diagnosed among 20- to 24-yr-olds in the United States x years after the study began.

a. Approximate the interval(s) over which the number of new AIDS cases among 20- to 24-yr-olds increased. $(0, 8) \cup (14, 20)$

b. Approximate the interval(s) over which the number of new AIDS cases among 20- to 24-yr-olds decreased. $(8, 14)$

c. How many turning points does the graph show? 2

d. Based on the number of turning points, what is the minimum degree of a polynomial function that could be used to model the data? Would the leading coefficient be positive or negative? Degree 3; Positive leading coefficient

e. How many years after the study began was the number of new AIDS cases among 20- to 24-yr-olds the greatest? 8

f. What was the maximum number of new cases diagnosed in a single year? 2648 new cases

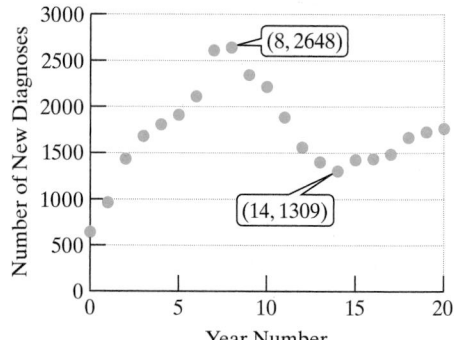

AIDS Diagnoses, 20- to 24-Yr-Olds United States, 1985–2005

Write About It

91. Given a polynomial function defined by $y = f(x)$, explain how to find the x-intercepts.

92. Given a polynomial function, explain how to determine whether an x-intercept is a touch point or a cross point.

93. Write an informal explanation of what it means for a function to be continuous.

94. Write an informal explanation of the intermediate value theorem.

Expanding Your Skills

The intermediate value theorem given on page 260 is actually a special case of a broader statement of the theorem. Consider the following:

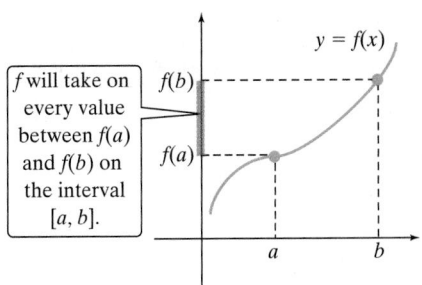

> Let f be a polynomial function. For $a < b$, if $f(a) \neq f(b)$,
> then f takes on every value between $f(a)$ and $f(b)$ on the interval $[a, b]$.

Use this broader statement of the intermediate value theorem for Exercises 95–96.

95. Given $f(x) = x^2 - 3x + 2$,

   a. Evaluate $f(3)$ and $f(4)$.

   b. Use the intermediate value theorem to show that there exists at least one value of x for which $f(x) = 4$ on the interval $[3, 4]$.

   c. Find the value(s) of x for which $f(x) = 4$ on the interval $[3, 4]$.

96. Given $f(x) = -x^2 - 4x + 3$,

   a. Evaluate $f(-4)$ and $f(-3)$.

   b. Use the intermediate value theorem to show that there exists at least one value of x for which $f(x) = 5$ on the interval $[-4, -3]$.

   c. Find the value(s) of x for which $f(x) = 5$ on the interval $[-4, -3]$.

Technology Connections

97. For a certain individual, the volume (in liters) of air in the lungs during a 4.5-sec respiratory cycle is shown in the table for 0.5-sec intervals. Graph the points and then find a third-degree polynomial function to model the volume $V(t)$ for t between 0 sec and 4.5 sec. (*Hint*: Use a CubicReg option or polynomial degree 3 option on a graphing utility.)

Time (sec)	Volume (L)
0.0	0.00
0.5	0.11
1.0	0.29
1.5	0.47
2.0	0.63
2.5	0.76
3.0	0.81
3.5	0.75
4.0	0.56
4.5	0.20

98. The torque (in ft-lb) produced by a certain automobile engine turning at x thousand revolutions per minute is shown in the table. Graph the points and then find a third-degree polynomial function to model the torque $T(x)$ for $1 \leq x \leq 5$.

Engine speed (1000 rpm)	Torque (ft-lb)
1.0	165
1.5	180
2.0	188
2.5	190
3.0	186
3.5	176
4.0	161
4.5	142
5.0	120

99. A solar oven is to be made from an open box with reflective sides. Each box is made from a 30-in. by 24-in. rectangular sheet of aluminum with squares of length x (in inches) removed from each corner. Then the flaps are folded up to form an open box.

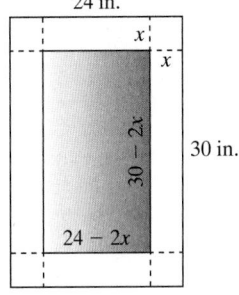

   a. Show that the volume of the box is given by

$$V(x) = 4x^3 - 108x^2 + 720x \quad \text{for } 0 < x < 12.$$

   b. Graph the function from part (a) and use a "Maximum" feature on a graphing utility to approximate the length of the sides of the squares that should be removed to maximize the volume. Round to the nearest tenth of an inch.

   c. Approximate the maximum volume. Round to the nearest cubic inch. 1418 in.^{3}

For Exercises 100–101, two viewing windows are given for the graph of $y = f(x)$. Choose the window that best shows the key features of the graph.

100. $f(x) = 2(x - 0.5)(x - 0.1)(x + 0.2)$

   a. $[-10, 10, 1]$ by $[-10, 10, 1]$

   b. $[-1, 1, 0.1]$ by $[-0.05, 0.05, 0.01]$

101. $g(x) = 0.08(x - 16)(x + 2)(x - 3)$

   a. $[-10, 10, 1]$ by $[-10, 10, 1]$

   b. $[-5, 20, 5]$ by $[-50, 30, 10]$

For Exercises 102–103, graph the function defined by $y = f(x)$ on an appropriate viewing window.

102. $k(x) = \dfrac{1}{100}(x - 20)(x + 1)(x + 8)(x - 6)$

103. $p(x) = (x - 0.4)(x + 0.5)(x + 0.1)(x - 0.8)$

TIP Take a minute to review long division of whole numbers: $2273 \div 5$

$$
\begin{array}{r}
454 \longleftarrow \text{Quotient} \\
5\overline{)2273} \\
-20 \\
\hline
27 \\
-25 \\
\hline
23 \\
-20 \\
\hline
3 \longleftarrow \text{Remainder}
\end{array}
$$

Answer: $454 + \frac{3}{5}$ or $454\frac{3}{5}$

SECTION 2.3   Division of Polynomials and the Remainder and Factor Theorems

1. Divide Polynomials Using Long Division

In this section, we use the notation $f(x)$, $g(x)$, and so on to represent polynomials in x. We also present two types of polynomial division: long division and synthetic division. Polynomial division can be used to factor a polynomial, solve a polynomial equation, and find the zeros of a polynomial.

When dividing polynomials, if the divisor has two or more terms we can use a long division process similar to the division of real numbers. This is demonstrated in Examples 1–3.

EXAMPLE 1   Dividing Polynomials Using Long Division

Use long division to divide. $(6x^3 - 5x^2 - 3) \div (3x + 2)$

Solution:

First note that the dividend can be written as $6x^3 - 5x^2 + 0x - 3$. The term $0x$ is used as a place holder for the missing power of x. The place holder is helpful to keep the powers of x lined up. We also set up long division with both the dividend and divisor written in descending order.

$$3x + 2\overline{)6x^3 - 5x^2 + 0x - 3}$$

Divide the leading term in the dividend by the leading term in the divisor.

$$\frac{6x^3}{3x} = 2x^2$$ This is the first term in the quotient.

$$
\begin{array}{r}
2x^2 \\
3x + 2\overline{)6x^3 - 5x^2 + 0x - 3} \\
-(6x^3 + 4x^2)
\end{array}
$$

Multiply the divisor by $2x^2$:
$2x^2(3x + 2) = 6x^3 + 4x^2$, and subtract the result.

Subtract.

$$
\begin{array}{r}
2x^2 \\
3x + 2\overline{)6x^3 - 5x^2 + 0x - 3} \\
-(6x^3 + 4x^2) \\
\hline
-9x^2 + 0x
\end{array}
$$

Bring down the next term from the dividend and repeat the process.

$$
\begin{array}{r}
2x^2 - 3x \\
3x + 2\overline{)6x^3 - 5x^2 + 0x - 3} \\
-(6x^3 + 4x^2) \\
\hline
-9x^2 + 0x \\
-(-9x^2 - 6x)
\end{array}
$$

Divide $-9x^2$ by the first term in the divisor.

$$\frac{-9x^2}{3x} = -3x$$

Subtract.

Multiply the divisor by $-3x$:
$-3x(3x + 2) = -9x^2 - 6x$,
and subtract the result.

$$
\begin{array}{r}
2x^2 - 3x + 2 \\
3x + 2\overline{)6x^3 - 5x^2 + 0x - 3} \\
-(6x^3 + 4x^2) \\
\hline
-9x^2 + 0x \\
-(-9x^2 - 6x) \\
\hline
6x - 3 \\
-(6x + 4) \\
\hline
-7
\end{array}
$$

Bring down the next term from the dividend and repeat the process.

Divide $6x$ by the first term in the divisor.
$\frac{6x}{3x} = 2$. This is the next term in the quotient.

Multiply the divisor by 2: $2(3x + 2) = 6x + 4$, and subtract the result.

The remainder is -7.

Long division is complete when the remainder is either zero or has degree less than the degree of the divisor.

The quotient is $2x^2 - 3x + 2$.

The remainder is -7.

The divisor is $3x + 2$.

The dividend is $6x^3 - 5x^2 - 3$.

The result of a long division problem is usually written as the quotient plus the remainder divided by the divisor.

$$\underbrace{\frac{\boxed{\text{Dividend}}\rightarrow 6x^3 - 5x^2 - 3}{\boxed{\text{Divisor}}\rightarrow 3x + 2}} = \overset{\boxed{\text{Quotient}}}{2x^2 - 3x + 2} + \frac{-7}{3x + 2}\overset{\leftarrow\boxed{\text{Remainder}}}{\underset{\leftarrow\boxed{\text{Divisor}}}{}}$$

Skill Practice 1   Use long division to divide $(4x^3 - 23x + 3) \div (2x - 5)$.

By clearing fractions, the result of Example 1 can be checked by multiplication.

$$\text{Dividend} = (\text{Divisor})(\text{Quotient}) \quad + \text{Remainder}$$
$$6x^3 - 5x^2 - 3 \overset{?}{=} (3x + 2)(2x^2 - 3x + 2) + (-7)$$
$$\overset{?}{=} 6x^3 - 5x^2 + 4 + (-7)$$
$$\overset{?}{=} 6x^3 - 5x^2 - 3 \checkmark$$

This result illustrates the division algorithm.

Division Algorithm

Suppose that $f(x)$ and $d(x)$ are polynomials where $d(x) \neq 0$ and the degree of $d(x)$ is less than or equal to the degree of $f(x)$. Then there exists unique polynomials $q(x)$ and $r(x)$ such that

$$f(x) = d(x) \cdot q(x) + r(x)$$

where either the degree of $r(x)$ is less than $d(x)$, or $r(x)$ is the zero polynomial.

Note: The polynomial $f(x)$ is the **dividend,** $d(x)$ is the **divisor,** $q(x)$ is the **quotient,** and $r(x)$ is the **remainder.**

Classroom Example: p. 279
Exercise 16

EXAMPLE 2   **Dividing Polynomials Using Long Division**

Use long division to divide $(-5 + x + 4x^2 + 2x^3 + 3x^4) \div (x^2 + 2)$.

Solution:

Write the dividend and divisor in descending order and insert place holders for missing powers of x. $(3x^4 + 2x^3 + 4x^2 + x - 5) \div (x^2 + 0x + 2)$

$$
\begin{array}{r}
3x^2 + 2x - 2 \\
x^2 + 0x + 2{\overline{\smash{\big)}\,3x^4 + 2x^3 + 4x^2 + x - 5}} \\
\underline{-(3x^4 + 0x^3 + 6x^2)} \\
2x^3 - 2x^2 + x \\
\underline{-(2x^3 + 0x^2 + 4x)} \\
-2x^2 - 3x - 5 \\
\underline{-(-2x^2 + 0x - 4)} \\
-3x - 1
\end{array}
$$

To begin, divide the leading term in the dividend by the leading term in the divisor.

$$\frac{3x^4}{x^2} = 3x^2$$

Multiply the divisor by $3x^2$ and subtract the result.

Bring down the next term from the dividend and repeat the process.

The process is complete when the remainder is either 0 or has degree less than the degree of the divisor.

The result is $3x^2 + 2x - 2 + \dfrac{-3x - 1}{x^2 + 2}$.

Answer

1. $2x^2 + 5x + 1 + \dfrac{8}{2x - 5}$

Check by using the division algorithm.

$$3x^4 + 2x^3 + 4x^2 + x - 5 \overset{?}{=} (x^2 + 2)(3x^2 + 2x - 2) + (-3x - 1)$$
$$\overset{?}{=} 3x^4 + 2x^3 - 2x^2 + 6x^2 + 4x - 4 + (-3x - 1)$$
$$\overset{?}{=} 3x^4 + 2x^3 + 4x^2 + x - 5 ✓$$

Skill Practice 2   Use long division to divide.
$$(1 - 7x + 5x^2 - 3x^3 + 2x^4) \div (x^2 + 3)$$

In Example 3, we discuss the implications of obtaining a remainder of zero when performing division of polynomials.

Classroom Example: p. 279
Exercise 20

EXAMPLE 3   **Dividing Polynomials Using Long Division**

Use long division to divide. $\dfrac{2x^2 + 3x - 14}{x - 2}$

Solution:

$$\begin{array}{r} 2x + 7 \\ x - 2 \overline{)2x^2 + 3x - 14} \\ -(2x^2 - 4x) \\ \hline 7x - 14 \\ -(7x - 14) \\ \hline 0 \end{array}$$

$$\frac{2x^2 + 3x - 14}{x - 2} = 2x + 7$$

To begin, divide the leading term in the dividend by the leading term in the divisor.

$$\frac{2x^2}{x} = 2x$$

Multiply the divisor by $2x$ and subtract the result.

Bring down the next term from the dividend and repeat the process.

The process is complete when the remainder is either 0 or has degree less than the degree of the divisor.

The remainder is zero. This implies that the divisor divides evenly into the dividend. Therefore, both the divisor and quotient are factors of the dividend. This is easily verified by the division algorithm.

| Dividend | Divisor | Quotient | Remainder |

$$2x^2 + 3x - 14 \overset{?}{=} (x - 2)(2x + 7) + 0$$
$$\overset{?}{=} (x - 2)(2x + 7)$$

Factored form of $2x^2 + 3x - 14$

Skill Practice 3   Use long division to divide.
$$(3x^2 - 14x + 15) \div (x - 3)$$

2. Divide Polynomials Using Synthetic Division

When dividing polynomials where the divisor is a binomial of the form $(x - c)$ and c is a constant, we can use synthetic division. Synthetic division enables us to find the quotient and remainder more quickly than long division. It uses an algorithm that manipulates the coefficients of the dividend, divisor, and quotient without the accompanying variable factors.

The division of polynomials from Example 3 is shown at the top left of page 273. The equivalent synthetic division is shown on the right. Notice that the same coefficients are used in both cases.

Answers

2. $2x^2 - 3x - 1 + \dfrac{2x + 4}{x^2 + 3}$

3. $3x - 5$

$$\begin{array}{r} 2x + 7 \\ x - 2\overline{)2x^2 + 3x - 14} \\ -(2x^2 - 4x) \\ \hline 7x - 14 \\ -(7x - 14) \\ \hline 0 \end{array}$$

Coefficients of dividend

$$\begin{array}{r|rrr} 2| & 2 & 3 & -14 \\ & & 4 & 14 \\ \hline & 2 & 7 & \boxed{0} \longleftarrow \text{Remainder} \end{array}$$

Coefficients of quotient

In Example 4, we demonstrate the process to divide polynomials by synthetic division.

Classroom Example: p. 280
Exercise 30

EXAMPLE 4   **Dividing Polynomials Using Synthetic Division**

Use synthetic division to divide. $(-10x^2 + 2x^3 - 5) \div (x - 4)$

Solution:

As with long division, the terms of the dividend and divisor must be written in descending order with place holders for missing powers of x.

$$(2x^3 - 10x^2 + 0x - 5) \div (x - 4)$$

To use synthetic division, the divisor must be in the form $x - c$. In this case, $c = 4$.

Avoiding Mistakes

It is important to check that the divisor is in the form $(x - c)$ before applying synthetic division. The variable x in the divisor must be of first degree, and its coefficient must be 1.

Step 1: Write the value of c in a box.

Step 2: Write the coefficients of the dividend to the right of the box.

$$\begin{array}{r|rrrr} 4| & 2 & -10 & 0 & -5 \\ \hline & 2 & & & \end{array}$$

Step 3: Skip a line and draw a horizontal line below the list of coefficients.

Step 4: Bring down the leading coefficient from the dividend and write it below the line.

Step 5: Multiply the value of c by the number below the line ($4 \times 2 = 8$). Write the result in the next column above the line.

$$\begin{array}{r|rrrr} 4| & 2 & -10 & 0 & -5 \\ & & 8 & & \\ \hline & 2 & -2 & & \end{array}$$

Step 6: Add the numbers in the column above the line ($-10 + 8 = -2$), and write the result below the line.

Repeat steps 5 and 6 until all columns have been completed.

$$\begin{array}{r|rrrr} 4| & 2 & -10 & 0 & -5 \\ & & 8 & -8 & -32 \\ \hline & 2 & -2 & -8 & \boxed{-37} \\ & x^2 & x & \text{constant} \end{array}$$

A box is often drawn around the remainder.

The rightmost number below the line is the remainder. The other numbers below the line are the coefficients of the quotient in order by the degree of the term.

Since the divisor is linear (first degree), the degree of the quotient is 1 less than the degree of the dividend. In this case, the dividend is of degree 3. Therefore, the quotient will be of degree 2.

The quotient is $2x^2 - 2x - 8$ and the remainder is -37. Therefore,

$$\frac{2x^3 - 10x^2 - 5}{x - 4} = 2x^2 - 2x - 8 + \frac{-37}{x - 4}$$

Skill Practice 4   Use synthetic division to divide.

$(4x^3 - 28x - 7) \div (x - 3)$

Answer

4. $4x^2 + 12x + 8 + \dfrac{17}{x - 3}$

Classroom Example: p. 280
Exercise 32

TIP Given a divisor of the form $(x - c)$, we can determine the value of c by setting the divisor equal to zero and solving for x. In Example 5, we have $x + 2 = 0$, which implies that $x = -2$. The value of c is -2.

EXAMPLE 5 Dividing Polynomials Using Synthetic Division

Use synthetic division to divide. $\quad (-2x + 4x^3 + 18 + x^4) \div (x + 2)$

Solution:

Write the dividend and divisor in descending order and insert place holders for missing powers of x. $\quad (x^4 + 4x^3 + 0x^2 - 2x + 18) \div (x + 2)$

To use synthetic division, the divisor must be of the form $x - c$. In this case, we have $x + 2 = x - (-2)$. Therefore, $c = -2$.

The dividend is a fourth-degree polynomial and the divisor is a first-degree polynomial. Therefore, the quotient is a third-degree polynomial. The coefficients of the quotient are found below the line: 1, 2, -4, 6. The quotient is $x^3 + 2x^2 - 4x + 6$, and the remainder is 6.

$$\frac{x^4 + 4x^3 - 2x + 18}{x + 2} = x^3 + 2x^2 - 4x + 6 + \frac{6}{x + 2}$$

Skill Practice 5 Use synthetic division to divide.
$(-3x + 7x^3 + 5 + 2x^4) \div (x + 1)$

3. Apply the Remainder and Factor Theorems

Consider the special case of the division algorithm where $f(x)$ is the dividend and $(x - c)$ is the divisor.

$$f(x) = (x - c) \cdot q(x) + r$$

Now evaluate $f(c)$:

$$f(c) = (c - c) \cdot q(c) + r$$
$$f(c) = 0 \cdot q(c) + r$$
$$f(c) = r$$

The remainder r is constant because its degree must be one less than the degree of $x - c$.

This result is stated formally as the remainder theorem.

Remainder Theorem

If a polynomial $f(x)$ is divided by $x - c$, then the remainder is $f(c)$.

Note: The remainder theorem tells us that the value of $f(c)$ is the same as the remainder we get from dividing $f(x)$ by $x - c$.

Answer

5. $2x^3 + 5x^2 - 5x + 2 + \dfrac{3}{x + 1}$

Classroom Example: p. 280
Exercise 46

EXAMPLE 6 Using the Remainder Theorem to Evaluate a Polynomial

Given $f(x) = x^4 + 6x^3 - 12x^2 - 30x + 35$, use the remainder theorem to evaluate

a. $f(2)$ **b.** $f(-7)$

Solution:

a. If $f(x)$ is divided by $x - 2$, then the remainder is $f(2)$.

$$
\begin{array}{r|rrrrr}
2 & 1 & 6 & -12 & -30 & 35 \\
  &   & 2 & 16 & 8   & -44 \\
\hline
  & 1 & 8 & 4   & -22 & \underline{-9}
\end{array}
$$

By the remainder theorem, $f(2) = -9$.

b. If $f(x)$ is divided by $x - (-7)$ or equivalently $x + 7$, then the remainder is $f(-7)$.

$$
\begin{array}{r|rrrrr}
-7 & 1 & 6 & -12 & -30 & 35 \\
   &   & -7 & 7   & 35 & -35 \\
\hline
   & 1 & -1 & -5 & 5   & \underline{0}
\end{array}
$$

By the remainder theorem, $f(-7) = 0$.

The results can be checked by direct substitution.

$$f(2) = (2)^4 + 6(2)^3 - 12(2)^2 - 30(2) + 35 = -9 ✓$$
$$f(-7) = (-7)^4 + 6(-7)^3 - 12(-7)^2 - 30(-7) + 35 = 0 ✓$$

Skill Practice 6 Given $f(x) = x^4 + x^3 - 6x^2 - 5x - 15$, use the remainder theorem to evaluate

a. $f(5)$ **b.** $f(-3)$

TIP From Example 6, the values $f(2) = -9$ and $f(-7) = 0$, imply that $(2, -9)$ and $(-7, 0)$ are on the graph of $y = f(x)$.

TIP Polynomials with complex coefficients include polynomials with real coefficients and with imaginary coefficients. The following are complex polynomials.

$$f(x) = (2 + 3i)x^2 + 4i$$
$$g(x) = \sqrt{2}x^2 + 3x + 4i$$
$$h(x) = 2x^2 + 3x + 4$$

The division algorithm and remainder theorem can be extended over the set of complex numbers. The definition of a polynomial was given in Section R.3.

$$f(x) = a_n x^n + a_{n-1} x^{n-1} + a_{n-2} x^{n-2} + \cdots + a_1 x + a_0$$

where $a_n \neq 0$ and the coefficients $a_n, a_{n-1}, a_{n-2}, \ldots, a_0$ are real numbers. We now extend our discussion to **complex polynomials.** These are polynomials with complex coefficients.

We will also evaluate polynomials over the set of complex numbers rather than restricting x to the set of real numbers. A complex number $a + bi$ is a zero of a polynomial $f(x)$ if $f(a + bi) = 0$. For example, given $f(x) = x - (5 + 2i)$, we see that the imaginary number $5 + 2i$ is a zero of $f(x)$.

Classroom Examples: p. 280
Exercises 50, 54

EXAMPLE 7 Using the Remainder Theorem to Identify Zeros of a Polynomial

Use the remainder theorem to determine if the given number c is a zero of the polynomial.

a. $f(x) = 2x^3 - 4x^2 - 13x - 9$; $c = 4$
b. $f(x) = x^3 + x^2 - 3x - 3$; $c = \sqrt{3}$
c. $f(x) = x^3 + x + 10$; $c = 1 + 2i$

Answers
6. a. 560 **b.** 0

Solution:

In each case, divide $f(x)$ by $x - c$ to determine the remainder. If the remainder is 0, then the value c is a zero of the polynomial.

a. Divide $f(x) = 2x^3 - 4x^2 - 13x - 9$ by $x - 4$.

$$
\begin{array}{r|rrrr}
4 & 2 & -4 & -13 & -9 \\
  &   & 8 & 16 & 12 \\
\hline
  & 2 & 4 & 3 & \underline{|3}
\end{array}
$$

By the remainder theorem, $f(4) = 3$.
Since $f(4) \neq 0$, 4 is not a zero of $f(x)$.

b. Divide $f(x) = x^3 + x^2 - 3x - 3$ by $x - \sqrt{3}$.

$$
\begin{array}{r|rrrr}
\sqrt{3} & 1 & 1 & -3 & -3 \\
  &   & \sqrt{3} & 3 + \sqrt{3} & 3 \\
\hline
  & 1 & 1 + \sqrt{3} & \sqrt{3} & \underline{|0}
\end{array}
$$

By the remainder theorem, $f(\sqrt{3}) = 0$.
Therefore, $\sqrt{3}$ is a zero of $f(x)$.

c. Divide $f(x) = x^3 + x + 10$ by $x - (1 + 2i)$.

$$
\begin{array}{r|rrrr}
1 + 2i & 1 & 0 & 1 & 10 \\
  &   & 1 + 2i & -3 + 4i & -10 \\
\hline
  & 1 & 1 + 2i & -2 + 4i & \underline{|0}
\end{array}
$$

Note that $(1 + 2i)(1 + 2i)$
$= 1 + 2i + 2i + 4i^2$
$= 1 + 4i + 4(-1)$
Recall that $i^2 = -1$.
$= -3 + 4i$

Note that $(1 + 2i)(-2 + 4i)$
$= -2 + 4i - 4i + 8i^2$
$= -2 - 8$
$= -10$

By the remainder theorem, $f(1 + 2i) = 0$.
Therefore, $1 + 2i$ is a zero of $f(x)$.

Instructor Note:
Students may need a refresher regarding the multiplication of radical expressions and complex numbers.

Skill Practice 7   Use the remainder theorem to determine if the given number, c, is a zero of the function.

a. $f(x) = 2x^4 - 3x^2 + 5x - 11$; $c = 2$
b. $f(x) = 2x^3 + 5x^2 - 14x - 35$; $c = \sqrt{7}$
c. $f(x) = x^3 - 7x^2 + 16x - 10$; $c = 3 + i$

Suppose that we again apply the division algorithm to a dividend of $f(x)$ and a divisor of $x - c$, where c is a complex number.

$$f(x) = (x - c) \cdot q(x) + r$$
$$\qquad = (x - c) \cdot q(x) + f(c)$$

By the remainder theorem, $r = f(c)$.

If $f(c) = 0$, then $f(x) = (x - c) \cdot q(x)$

This tells us that if $f(c)$ is a zero of $f(x)$, then $(x - c)$ is a factor of $f(x)$.

Now suppose that $x - c$ is a factor of $f(x)$. Then for some polynomial $q(x)$,

$$f(x) = (x - c) \cdot q(x)$$
$$f(c) = (c - c) \cdot q(x)$$
$$\quad = 0$$

This tells us that if $(x - c)$ is a factor of $f(x)$, then c is a zero of $f(x)$.

These results can be summarized in the factor theorem.

Factor Theorem

Let $f(x)$ be a polynomial.

1. If $f(c) = 0$, then $(x - c)$ is a factor of $f(x)$.
2. If $(x - c)$ is a factor of $f(x)$, then $f(c) = 0$.

Answers
7. a. No   **b.** Yes   **c.** Yes

Classroom Example: p. 280
Exercise 56

EXAMPLE 8   Identifying Factors of a Polynomial

Use the factor theorem to determine if the given polynomials are factors of $f(x) = x^4 - x^3 - 11x^2 + 11x + 12$.

a. $x - 3$ **b.** $x + 2$

Solution:

Instructor Note:
Consider a discussion about the
contrapositive of an if-then
statement. In Example 8(b), the
contrapositive of statement 2 of the
factor theorem is used.

a. If $f(3) = 0$, then $x - 3$ is a factor of $f(x)$. Using synthetic division we have:

$$
\begin{array}{r|rrrrr}
3 & 1 & -1 & -11 & 11 & 12 \\
  &   & 3 & 6 & -15 & -12 \\
\hline
  & 1 & 2 & -5 & -4 & \underline{|0} \\
\end{array}
$$

By the factor theorem, since $f(3) = 0$, $x - 3$ is a factor of $f(x)$. $\boxed{f(3) = 0}$

b. If $f(-2) = 0$, then $x + 2$ is a factor of $f(x)$. Using synthetic division we have:

$$
\begin{array}{r|rrrrr}
-2 & 1 & -1 & -11 & 11 & 12 \\
   &   & -2 & 6 & 10 & -42 \\
\hline
   & 1 & -3 & -5 & 21 & \underline{|-30} \\
\end{array}
$$

By the factor theorem, since $f(-2) \neq 0$, $x + 2$ is not a factor of $f(x)$. $\boxed{f(-2) = -30}$

Skill Practice 8   Use the factor theorem to determine if the given polynomials are factors of $f(x) = 2x^4 - 13x^3 + 10x^2 - 25x + 6$.

a. $x - 6$ **b.** $x + 3$

In Example 9, we illustrate the relationship between the zeros of a polynomial and the solutions (roots) of a polynomial equation.

Classroom Example: p. 281
Exercise 66

EXAMPLE 9   Factoring a Polynomial Given a Known Zero

a. Factor $f(x) = 3x^3 + 25x^2 + 42x - 40$, given that -5 is a zero of $f(x)$.
b. Solve the equation. $3x^3 + 25x^2 + 42x - 40 = 0$

Solution:

a. The value -5 is a zero of $f(x)$, which means that $f(-5) = 0$. By the factor theorem, $x - (-5)$ or equivalently $x + 5$ is a factor of $f(x)$. Using synthetic division, we have

$$
\begin{array}{r|rrrr}
-5 & 3 & 25 & 42 & -40 \\
   &   & -15 & -50 & 40 \\
\hline
   & 3 & 10 & -8 & \underline{|0} \\
\end{array}
$$

divisor   quotient   remainder

This means that $3x^3 + 25x^2 + 42x - 40 = (x + 5)(3x^2 + 10x - 8) + 0$
Therefore, $f(x) = (x + 5)(3x - 2)(x + 4)$.

factors as $(3x - 2)(x + 4)$

b. $3x^3 + 25x^2 + 42x - 40 = 0$ To solve the equation, set one side equal to zero.

$(x + 5)(3x - 2)(x + 4) = 0$ Factor the left side.

$x = -5, \ x = \frac{2}{3}, \ x = -4$ Set each factor equal to zero and solve for x.

The solution set is $\left\{ -5, \frac{2}{3}, -4 \right\}$.

Answers
8. a. Yes   **b.** No
9. a. $f(x) = (x + 4)(x + 2)(2x - 5)$
b. $\left\{ -4, -2, \frac{5}{2} \right\}$

Skill Practice 9

a. Factor $f(x) = 2x^3 + 7x^2 - 14x - 40$, given that -4 is a zero of f.
b. Solve the equation. $2x^3 + 7x^2 - 14x - 40 = 0$

Classroom Example: p. 281
Exercise 76

EXAMPLE 10   Using the Factor Theorem to Build a Polynomial

Write a polynomial $f(x)$ of degree 3 that has the zeros $\frac{1}{2}$, $\sqrt{6}$, and $-\sqrt{6}$.

Solution:

By the factor theorem, if $\frac{1}{2}$, $\sqrt{6}$, and $-\sqrt{6}$ are zeros of a polynomial $f(x)$, then $\left(x - \frac{1}{2}\right)$, $\left(x - \sqrt{6}\right)$, and $\left(x + \sqrt{6}\right)$ are factors of $f(x)$. Therefore, $f(x) = \left(x - \frac{1}{2}\right)\left(x - \sqrt{6}\right)\left(x + \sqrt{6}\right)$ is a third-degree polynomial with the given zeros.

$$f(x) = \left(x - \frac{1}{2}\right)(x^2 - 6) \qquad \text{Multiply conjugates.}$$

$$f(x) = x^3 - \frac{1}{2}x^2 - 6x + 3$$

Skill Practice 10   Write a polynomial $f(x)$ of degree 3 that has the zeros $\frac{1}{3}$, $\sqrt{3}$, and $-\sqrt{3}$.

In Example 10, the polynomial $f(x)$ is not unique. If we multiply $f(x)$ by any nonzero constant a, the polynomial will still have the desired factors and zeros.

$$g(x) = a\left(x - \frac{1}{2}\right)\left(x - \sqrt{6}\right)\left(x + \sqrt{6}\right) \qquad \text{The zeros are still } \frac{1}{2}, \sqrt{6}, \text{ and } -\sqrt{6}.$$

If a is any nonzero multiple of 2, then the polynomial will have integer coefficients. For example:

$$g(x) = 2\left(x - \frac{1}{2}\right)\left(x - \sqrt{6}\right)\left(x + \sqrt{6}\right)$$

$$= 2\left(x^3 - \frac{1}{2}x^2 - 6x + 3\right)$$

$$= 2x^3 - x^2 - 12x + 6$$

The zeros of $f(x)$ and $g(x)$ are real numbers and correspond to the x-intercepts of the graphs of the related functions. The graphs of $y = f(x)$ and $y = g(x)$ are shown in Figure 2-15. Notice that the graphs have the same x-intercepts and differ only by a vertical stretch.

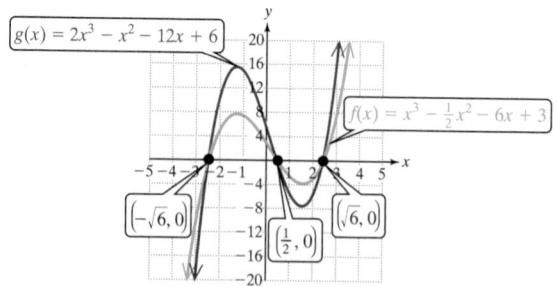

Figure 2-15

Answer

10. $f(x) = x^3 - \dfrac{1}{3}x^2 - 3x + 1$

SECTION 2.3   Practice Exercises

Prerequisite Review

R.1. Evaluate $(3 + 4i)^2$ and write the answer in standard form, $a + bi$. $-7 + 24i$

R.2. Verify by substitution that the given values of x are solutions to $x^2 - 12x + 39 = 0$.

   a. $6 - i\sqrt{3}$ **b.** $6 + i\sqrt{3}$

R.3. Solve by using the quadratic formula.

   $z^2 + 8z + 19 = 0$ $\{-4 \pm i\sqrt{3}\}$

Concept Connections

1. Given the division algorithm, identify the polynomials representing the dividend, divisor, quotient, and remainder.

   $f(x) = d(x) \cdot q(x) + r(x)$ Dividend: $f(x)$; Divisor: $d(x)$; Quotient: $q(x)$; Remainder: $r(x)$

2. Given $\dfrac{2x^3 - 5x^2 - 6x + 1}{x - 3} = 2x^2 + x - 3 + \dfrac{-8}{x - 3}$, use the division algorithm to check the result.

   $(x - 3)(2x^2 + x - 3) + (-8) = 2x^3 - 5x^2 - 6x + 1$ ✓

3. The remainder theorem indicates that if a polynomial $f(x)$ is divided by $x - c$, then the remainder is _____ $f(c)$.

4. Given a polynomial $f(x)$, the factor theorem indicates that if $f(c) = 0$, then $x - c$ is a _____ factor _____ of $f(x)$. Furthermore, if $x - c$ is a factor of $f(x)$, then $f(c) = $ _____ 0 _____ .

5. Answer true or false. If $\sqrt{5}$ is a zero of a polynomial, then $(x - \sqrt{5})$ is a factor of the polynomial.   True

6. Answer true or false. If $(x + 3)$ is a factor of a polynomial, then 3 is a zero of the polynomial.   False

Objective 1: Divide Polynomials Using Long Division

For Exercises 7–8, (See Example 1)

a. Use long division to divide.

b. Identify the dividend, divisor, quotient, and remainder.

c. Check the result from part (a) with the division algorithm.

7. $(6x^2 + 9x + 5) \div (2x - 5)$

8. $(12x^2 + 10x + 3) \div (3x + 4)$

For Exercises 9–22, use long division to divide. (See Examples 1–3)

9. $(3x^3 - 11x^2 - 10) \div (x - 4)$ $3x^2 + x + 4 + \dfrac{6}{x - 4}$

10. $(2x^3 - 7x^2 - 65) \div (x - 5)$ $2x^2 + 3x + 15 + \dfrac{10}{x - 5}$

11. $(8 + 30x - 27x^2 - 12x^3 + 4x^4) \div (x + 2)$ $4x^3 - 20x^2 + 13x + 4$

12. $(-48 - 28x + 20x^2 + 17x^3 + 3x^4) \div (x + 3)$ $3x^3 + 8x^2 - 4x - 16$

13. $(-20x^2 + 6x^4 - 16) \div (2x + 4)$ $3x^3 - 6x^2 + 2x - 4$

14. $(-60x^2 + 8x^4 - 108) \div (2x - 6)$ $4x^3 + 12x^2 + 6x + 18$

15. $(x^5 + 4x^4 + 18x^2 - 20x - 10) \div (x^2 + 5)$ $x^3 + 4x^2 - 5x - 2 + \dfrac{5x}{x^2 + 5}$

16. $(x^5 - 2x^4 + x^3 - 8x + 18) \div (x^2 - 3)$ $x^3 - 2x^2 + 4x - 6 + \dfrac{4x}{x^2 - 3}$

17. $\dfrac{6x^4 + 3x^3 - 7x^2 + 6x - 5}{2x^2 + x - 3}$ $3x^2 + 1 + \dfrac{5x - 2}{2x^2 + x - 3}$

18. $\dfrac{12x^4 - 4x^3 + 13x^2 + 2x + 1}{3x^2 - x + 4}$ $4x^2 - 1 + \dfrac{x + 5}{3x^2 - x + 4}$

19. $\dfrac{x^3 - 27}{x - 3}$ $x^2 + 3x + 9$

20. $\dfrac{x^3 + 64}{x + 4}$ $x^2 - 4x + 16$

21. $(5x^3 - 2x^2 + 3) \div (2x - 1)$ $\dfrac{5}{2}x^2 + \dfrac{1}{4}x + \dfrac{1}{8} + \dfrac{\frac{25}{8}}{2x - 1}$

22. $(2x^3 + x^2 + 1) \div (3x + 1)$ $\dfrac{2}{3}x^2 + \dfrac{1}{9}x - \dfrac{1}{27} + \dfrac{\frac{28}{27}}{3x + 1}$

Objective 2: Divide Polynomials Using Synthetic Division

For Exercises 23–26, consider the division of two polynomials: $f(x) \div (x - c)$. The result of the synthetic division process is shown here. Write the polynomials representing the

   a. Dividend.   **b.** Divisor.   **c.** Quotient.   **d.** Remainder.

23. $\underline{3|}$ 2 -5 -5 -4 29

           6   3 -6 -30

     $\overline{2 \quad 1 \quad -2 \quad -10 \quad |-1}$

   a. $2x^4 - 5x^3 - 5x^2 - 4x + 29$
   b. $x - 3$
   c. $2x^3 + x^2 - 2x - 10$
   d. -1

24. $\underline{2|}$ 1 -5 2 -1 20

           2   -6 -8 -18

     $\overline{1 \quad -3 \quad -4 \quad -9 \quad |2}$

   a. $x^4 - 5x^3 + 2x^2 - x + 20$
   b. $x - 2$
   c. $x^3 - 3x^2 - 4x - 9$
   d. 2

25. $-4\,|$ $\quad 1 \quad -2 \quad -25 \quad -4$ a. $x^3 - 2x^2 - 25x - 4$
$\qquad\qquad\quad -4 \quad\;\; 24 \quad\;\; 4$ b. $x + 4$
$\qquad\qquad \overline{\;\;1 \quad -6 \quad\;\; -1 \quad\;\; |\,0\,}$ c. $x^2 - 6x - 1$
$\qquad\qquad\qquad\qquad\qquad\qquad\qquad$ d. 0

26. $-5\,|$ $\quad 3 \quad\;\; 13 \quad -14 \quad -20$ a. $3x^3 + 13x^2 - 14x - 20$
$\qquad\qquad\quad\; -15 \quad\;\; 10 \quad\;\; 20$ b. $x + 5$
$\qquad\qquad \overline{\;\;3 \quad -2 \quad\;\; -4 \quad\;\; |\,0\,}$ c. $3x^2 - 2x - 4$
$\qquad\qquad\qquad\qquad\qquad\qquad\qquad$ d. 0

For Exercises 27–38, use synthetic division to divide the polynomials. (See Examples 4–5)

27. $(4x^2 + 15x + 1) \div (x + 6)$ $4x - 9 + \dfrac{55}{x + 6}$

28. $(6x^2 + 25x - 19) \div (x + 5)$ $6x - 5 + \dfrac{6}{x + 5}$

29. $(5x^2 - 17x - 12) \div (x - 4)$ $5x + 3$

30. $(2x^2 + x - 21) \div (x - 3)$ $2x + 7$

31. $(4 - 8x - 3x^2 - 5x^4) \div (x + 2)$ $-5x^3 + 10x^2 - 23x + 38 + \dfrac{-72}{x + 2}$

32. $(-5 + 2x + 5x^3 - 2x^4) \div (x + 1)$ $-2x^3 + 7x^2 - 7x + 9 + \dfrac{-14}{x + 1}$

33. $\dfrac{4x^5 - 25x^4 - 58x^3 + 232x^2 + 198x - 63}{x - 3}$ $4x^4 - 13x^3 - 97x^2 - 59x + 21$

34. $\dfrac{2x^5 + 13x^4 - 3x^3 - 58x^2 - 20x + 24}{x - 2}$ $2x^4 + 17x^3 + 31x^2 + 4x - 12$

35. $\dfrac{x^5 + 32}{x + 2}$ $x^4 - 2x^3 + 4x^2 - 8x + 16$

36. $\dfrac{x^4 - 81}{x + 3}$ $x^3 - 3x^2 + 9x - 27$

37. $(2x^4 - 7x^3 - 56x^2 + 37x + 84) \div \left(x - \dfrac{3}{2}\right)$ $2x^3 - 4x^2 - 62x - 56$

38. $(-5x^4 - 18x^3 + 63x^2 + 128x - 60) \div \left(x - \dfrac{2}{5}\right)$ $-5x^3 - 20x^2 + 55x + 150$

Objective 3: Apply the Remainder and Factor Theorems

39. The value $f(-6) = 39$ for a polynomial $f(x)$. What can be concluded about the remainder or quotient of $\dfrac{f(x)}{x + 6}$? The remainder is 39.

40. Given a polynomial $f(x)$, the quotient $\dfrac{f(x)}{x - 2}$ has a remainder of 12. What is the value of $f(2)$?   12

41. Given $f(x) = 2x^4 - 5x^3 + x^2 - 7$,
   a. Evaluate $f(4)$.   201
   b. Determine the remainder when $f(x)$ is divided by $(x - 4)$.   201

42. Given $g(x) = -3x^5 + 2x^4 + 6x^2 - x + 4$,
   a. Evaluate $g(2)$. -38
   b. Determine the remainder when $g(x)$ is divided by $(x - 2)$. -38

For Exercises 43–46, use the remainder theorem to evaluate the polynomial for the given values of x. (See Example 6)

43. $f(x) = 2x^4 + x^3 - 49x^2 + 79x + 15$
   a. $f(-1)$ -112 **b.** $f(3)$ 0 **c.** $f(4)$ 123 **d.** $f\left(\dfrac{5}{2}\right)$ 0

44. $g(x) = 3x^4 - 22x^3 + 51x^2 - 42x + 8$
   a. $g(-1)$ 126 **b.** $g(2)$ 0 **c.** $g(1)$ -2 **d.** $g\left(\dfrac{4}{3}\right)$ 0

45. $h(x) = 5x^3 - 4x^2 - 15x + 12$
   a. $h(1)$ -2 **b.** $h\left(\dfrac{4}{5}\right)$ 0 **c.** $h(\sqrt{3})$ 0 **d.** $h(-1)$ 18

46. $k(x) = 2x^3 - x^2 - 14x + 7$
   a. $k(2)$ -9 **b.** $k\left(\dfrac{1}{2}\right)$ 0 **c.** $k(\sqrt{7})$ 0 **d.** $k(-2)$ 15

For Exercises 47–54, use the remainder theorem to determine if the given number c is a zero of the polynomial. (See Example 7)

47. $f(x) = x^4 + 3x^3 - 7x^2 + 13x - 10$
   a. $c = 2$ No   **b.** $c = -5$ Yes

48. $g(x) = 2x^4 + 13x^3 - 10x^2 - 19x + 14$
   a. $c = -2$ No   **b.** $c = -7$ Yes

49. $p(x) = 2x^3 + 3x^2 - 22x - 33$
   a. $c = -2$ No   **b.** $c = -\sqrt{11}$ Yes

50. $q(x) = 3x^3 + x^2 - 30x - 10$
   a. $c = -3$ No   **b.** $c = -\sqrt{10}$ Yes

51. $m(x) = x^3 - 2x^2 + 25x - 50$
   a. $c = 5i$ Yes   **b.** $c = -5i$ Yes

52. $n(x) = x^3 + 4x^2 + 9x + 36$
   a. $c = 3i$ Yes   **b.** $c = -3i$ Yes

53. $g(x) = x^3 - 11x^2 + 25x + 37$
   a. $c = 6 + i$ Yes   **b.** $c = 6 - i$ Yes

54. $f(x) = 2x^3 - 5x^2 + 54x - 26$
   a. $c = 1 + 5i$ Yes   **b.** $c = 1 - 5i$ Yes

For Exercises 55–60, use the factor theorem to determine if the given binomial is a factor of $f(x)$. (See Example 8)

55. $f(x) = x^4 + 11x^3 + 41x^2 + 61x + 30$
   a. $x + 5$ Yes   **b.** $x - 2$ No

56. $g(x) = x^4 - 10x^3 + 35x^2 - 50x + 24$
   a. $x - 4$ Yes   **b.** $x + 1$ No

57. $f(x) = x^3 + 64$
   a. $x - 4$ No   **b.** $x + 4$ Yes

58. $f(x) = x^4 - 81$
   a. $x - 3$ Yes   **b.** $x + 3$ Yes

59. $f(x) = 2x^3 + x^2 - 16x - 8$
   a. $x - 1$ No   **b.** $x - 2\sqrt{2}$ Yes

60. $f(x) = 3x^3 - x^2 - 54x + 18$
   a. $x - 2$ No   **c.** $x - 3\sqrt{2}$ Yes

61. Given $g(x) = x^4 - 14x^2 + 45$,

    a. Evaluate $g(\sqrt{5})$.   0

    b. Evaluate $g(-\sqrt{5})$.   0

    c. Solve $g(x) = 0$. $\{-3, -\sqrt{5}, \sqrt{5}, 3\}$

63. a. Use synthetic division and the factor theorem to determine if $[x - (2 + 5i)]$ is a factor of $f(x) = x^2 - 4x + 29$.   Yes

    b. Use synthetic division and the factor theorem to determine if $[x - (2 - 5i)]$ is a factor of $f(x) = x^2 - 4x + 29$.   Yes

    c. Use the quadratic formula to solve the equation. $x^2 - 4x + 29 = 0$ $\{2 \pm 5i\}$

    d. Find the zeros of the polynomial $f(x) = x^2 - 4x + 29$. $2 \pm 5i$

65. a. Factor $f(x) = 2x^3 + x^2 - 37x - 36$, given that -1 is a zero. (**See Example 9**) $f(x) = (x + 1)(2x - 9)(x + 4)$

    b. Solve. $2x^3 + x^2 - 37x - 36 = 0$ $\left\{-1, \frac{9}{2}, -4\right\}$

67. a. Factor $f(x) = 20x^3 + 39x^2 - 3x - 2$, given that $\frac{1}{4}$ is a zero. $f(x) = 4\left(x - \frac{1}{4}\right)(5x + 1)(x + 2)$
or $f(x) = (4x - 1)(5x + 1)(x + 2)$

    b. Solve. $20x^3 + 39x^2 - 3x - 2 = 0$ $\left\{\frac{1}{4}, -\frac{1}{5}, -2\right\}$

69. a. Factor $f(x) = 9x^3 - 33x^2 + 19x - 3$, given that 3 is a zero. $f(x) = (x - 3)(3x - 1)^2$

    b. Solve. $9x^3 - 33x^2 + 19x - 3 = 0$ $\left\{3, \frac{1}{3}\right\}$

62. Given $h(x) = x^4 - 15x^2 + 44$,

    a. Evaluate $h(\sqrt{11})$.   0

    b. Evaluate $h(-\sqrt{11})$.   0

    c. Solve $h(x) = 0$. $\{-\sqrt{11}, -2, 2, \sqrt{11}\}$

64. a. Use synthetic division and the factor theorem to determine if $[x - (3 + 4i)]$ is a factor of $f(x) = x^2 - 6x + 25$.   Yes

    b. Use synthetic division and the factor theorem to determine if $[x - (3 - 4i)]$ is a factor of $f(x) = x^2 - 6x + 25$.   Yes

    c. Use the quadratic formula to solve the equation. $x^2 - 6x + 25 = 0$ $\{3 \pm 4i\}$

    d. Find the zeros of the polynomial $f(x) = x^2 - 6x + 25$. $3 \pm 4i$

66. a. Factor $f(x) = 3x^3 + 16x^2 - 5x - 50$, given that -2 is a zero. $f(x) = (x + 2)(3x - 5)(x + 5)$

    b. Solve. $3x^3 + 16x^2 - 5x - 50 = 0$ $\left\{-2, \frac{5}{3}, -5\right\}$

68. a. Factor $f(x) = 8x^3 - 18x^2 - 11x + 15$, given that $\frac{3}{4}$ is a zero. $f(x) = 4\left(x - \frac{3}{4}\right)(2x - 5)(x + 1)$
or $f(x) = (4x - 3)(2x - 5)(x + 1)$

    b. Solve. $8x^3 - 18x^2 - 11x + 15 = 0$ $\left\{\frac{3}{4}, \frac{5}{2}, -1\right\}$

70. a. Factor $f(x) = 4x^3 - 20x^2 + 33x - 18$, given that 2 is a zero. $f(x) = (x - 2)(2x - 3)^2$

    b. Solve. $4x^3 - 20x^2 + 33x - 18 = 0$ $\left\{2, \frac{3}{2}\right\}$

For Exercises 71–82, write a polynomial $f(x)$ that meets the given conditions. Answers may vary. (See Example 10)

71. Degree 3 polynomial with zeros 2, 3, and -4.
$f(x) = x^3 - x^2 - 14x + 24$

73. Degree 4 polynomial with zeros 1, $\frac{3}{2}$ (each with multiplicity 1), and 0 (with multiplicity 2).

75. Degree 2 polynomial with zeros $2\sqrt{11}$ and $-2\sqrt{11}$.
$f(x) = x^2 - 44$

77. Degree 3 polynomial with zeros -2, $3i$, and $-3i$.
$f(x) = x^3 + 2x^2 + 9x + 18$

79. Degree 3 polynomial with integer coefficients and zeros of $-\frac{2}{3}, \frac{1}{2}$, and 4. $f(x) = 6x^3 - 23x^2 - 6x + 8$

81. Degree 2 polynomial with zeros of $7 + 8i$ and $7 - 8i$.
$f(x) = x^2 - 14x + 113$

72. Degree 3 polynomial with zeros 1, -6, and 3.
$f(x) = x^3 + 2x^2 - 21x + 18$

74. Degree 5 polynomial with zeros 2, $\frac{5}{2}$ (each with multiplicity 1), and 0 (with multiplicity 3).

76. Degree 2 polynomial with zeros $5\sqrt{2}$ and $-5\sqrt{2}$.
$f(x) = x^2 - 50$

78. Degree 3 polynomial with zeros 4, $2i$, and $-2i$.
$f(x) = x^3 - 4x^2 + 4x - 16$

80. Degree 3 polynomial with integer coefficients and zeros of $-\frac{2}{5}, \frac{3}{2}$, and 6.
$f(x) = 10x^3 - 71x^2 + 60x + 36$

82. Degree 2 polynomial with zeros of $5 + 6i$ and $5 - 6i$.
$f(x) = x^2 - 10x + 61$

Mixed Exercises

83. Given $p(x) = 2x^{452} - 4x^{92}$, is it easier to evaluate $p(1)$ by using synthetic division or by direct substitution? Find the value of $p(1)$.   Direct substitution; -2

84. Given $q(x) = 5x^{721} - 2x^{450}$, is it easier to evaluate $q(-1)$ by using synthetic division or by direct substitution? Find the value of $q(-1)$.   Direct substitution; -7

85. a. Is $(x - 1)$ a factor of $x^{100} - 1$?   Yes

    c. Is $(x - 1)$ a factor of $x^{99} - 1$?   Yes

    e. If n is a positive even integer, is $(x - 1)$ a factor of $x^n - 1$?   Yes

    b. Is $(x + 1)$ a factor of $x^{100} - 1$?   Yes

    d. Is $(x + 1)$ a factor of $x^{99} - 1$?   No

    f. If n is a positive odd integer, is $(x + 1)$ a factor of $x^n - 1$?   No

86. If a fifth-degree polynomial is divided by a second-degree polynomial, the quotient is a _____third_____ -degree polynomial.

87. Determine if the statement is true or false: Zero is a zero of the polynomial $3x^5 - 7x^4 - 2x^3 - 14$.   False

88. Determine if the statement is true or false: Zero is a zero of the polynomial $-2x^4 + 5x^3 + 6x$.   True

89. Find m so that $x + 4$ is a factor of $4x^3 + 13x^2 - 5x + m$. $m = 28$

90. Find m so that $x + 5$ is a factor of $-3x^4 - 10x^3 + 20x^2 - 22x + m$. $m = 15$

91. Find m so that $x + 2$ is a factor of $4x^3 + 5x^2 + mx + 2$. $m = -5$

92. Find m so that $x - 3$ is a factor of $2x^3 - 7x^2 + mx + 6$. $m = 1$

93. For what value of r is the statement an identity?
$$\frac{x^2 - x - 12}{x - 4} = x + 3 + \frac{r}{x - 4} \text{ provided that } x \neq 4$$
$r = 0$

94. For what value of r is the statement an identity?
$$\frac{x^2 - 5x - 8}{x - 2} = x - 3 + \frac{r}{x - 2} \text{ provided that } x \neq 2$$
$r = -14$

95. A metal block is formed from a rectangular solid with a rectangular piece cut out.

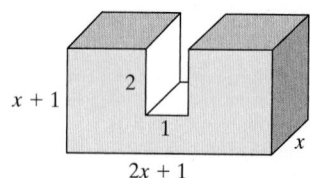

a. Write a polynomial $V(x)$ that represents the volume of the block. All distances in the figure are in centimeters. $V(x) = 2x^3 + 3x^2 - x$

b. Use synthetic division to evaluate the volume if x is 6 cm.   534 cm³

96. A wedge is cut from a rectangular solid.

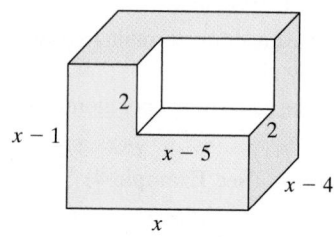

a. Write a polynomial $V(x)$ that represents the volume of the remaining part of the solid. All distances in the figure are in feet. $V(x) = x^3 - 5x^2 + 20$

b. Use synthetic division to evaluate the volume if x is 10 ft.   520 ft³

Write About It

97. Under what circumstances can synthetic division be used to divide polynomials?
The divisor must be of the form $(x - c)$, where c is a constant.

98. How can the division algorithm be used to check the result of polynomial division?   Multiply the quotient times the divisor and add the remainder. The result should be equal to the dividend.

99. Given a polynomial $f(x)$ and a constant c, state two methods by which the value $f(c)$ can be computed.   Compute $f(c)$ either by direct substitution or by using the remainder theorem. The remainder theorem states that $f(c)$ is equal to the remainder obtained after dividing $f(x)$ by $(x - c)$.

100. Write an informal explanation of the factor theorem.   Given a polynomial $f(x)$, if c is a zero of $f(x)$, then $(x - c)$ is a factor of $f(x)$. The converse is also true. If $(x - c)$ is a factor of $f(x)$, then c is a zero of $f(x)$.

Expanding Your Skills

101. a. Factor $f(x) = x^3 - 5x^2 + x - 5$ into factors of the form $(x - c)$, given that 5 is a zero. $f(x) = (x - 5)(x - i)(x + i)$

b. Solve. $x^3 - 5x^2 + x - 5 = 0$ $\{5, i, -i\}$

102. a. Factor $f(x) = x^3 - 3x^2 + 100x - 300$ into factors of the form $(x - c)$, given that 3 is a zero. $f(x) = (x - 3)(x - 10i)(x + 10i)$

b. Solve. $x^3 - 3x^2 + 100x - 300 = 0$ $\{3, 10i, -10i\}$

103. a. Factor $f(x) = x^4 + 2x^3 - 2x^2 - 6x - 3$ into factors of the form $(x - c)$, given that -1 is a zero. $f(x) = (x + 1)^2(x - \sqrt{3})(x + \sqrt{3})$

b. Solve. $x^4 + 2x^3 - 2x^2 - 6x - 3 = 0$ $\{-1, \sqrt{3}, -\sqrt{3}\}$

104. a. Factor $f(x) = x^4 + 4x^3 - x^2 - 20x - 20$ into factors of the form $(x - c)$, given that -2 is a zero. $f(x) = (x + 2)^2(x - \sqrt{5})(x + \sqrt{5})$

b. Solve. $x^4 + 4x^3 - x^2 - 20x - 20 = 0$ $\{-2, \sqrt{5}, -\sqrt{5}\}$

Technology Connections

For Exercises 105–106,

a. Use the graph to determine a solution to the given equation.

b. Verify your answer from part (a) using the remainder theorem.

c. Find the remaining solutions to the equation.

105. $5x^3 + 7x^2 - 58x - 24 = 0$ a. 3   b. 3 is a solution.   c. $\left\{3, -4, -\dfrac{2}{5}\right\}$

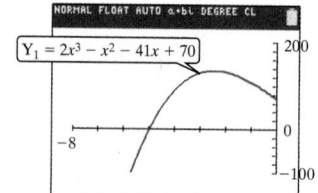

106. $2x^3 - x^2 - 41x + 70 = 0$ a. -5 b. -5 is a solution.   c. $\left\{-5, \dfrac{7}{2}, 2\right\}$

| SECTION 2.4 | Zeros of Polynomials |

OBJECTIVES

1. Apply the Rational Zero Theorem
2. Apply the Fundamental Theorem of Algebra
3. Apply Descartes' Rule of Signs
4. Find Upper and Lower Bounds

1. Apply the Rational Zero Theorem

The **zeros of a polynomial** $f(x)$ are the solutions (roots) to the corresponding polynomial equation $f(x) = 0$. For a polynomial function defined by $y = f(x)$, the real zeros of $f(x)$ are the x-intercepts of the graph of the function. Applications of polynomials and polynomial functions arise throughout mathematics. For this reason, it is important to learn techniques to find or approximate the zeros of a polynomial.

The zeros of a polynomial may be real numbers or imaginary numbers. The real zeros can be further categorized as rational or irrational numbers. For example, consider

$$f(x) = 2x^6 - 3x^5 - 7x^4 + 102x^3 - 88x^2 - 279x + 273$$

In factored form this is:

$$f(x) = (x - 1)(2x + 7)(x - \sqrt{3})(x + \sqrt{3})[x - (2 + 3i)][x - (2 - 3i)]$$

The zeros are:

$$\underbrace{1, -\frac{7}{2}}_{\text{rational zeros}}, \underbrace{\sqrt{3}, -\sqrt{3}}_{\text{irrational zeros}}, \underbrace{2 + 3i, 2 - 3i}_{\text{nonreal zeros}}$$

rational zeros | irrational zeros

real zeros | nonreal zeros

In this section, we develop tools to search for the zeros of polynomials. First we will consider polynomials with integer coefficients. For example, the factored form of $f(x) = 6x^2 + 13x - 5$ is $f(x) = (2x + 5)(3x - 1)$, which leads to zeros of $-\frac{5}{2}$ and $\frac{1}{3}$. The polynomial $f(x) = 6x^2 + 13x - 5$ is in the form $f(x) = ax^2 + bx + c$, where $a = 6$, $b = 13$, and $c = -5$. Notice that the numerator of each zero is a factor of c, and the denominator of each zero is a factor of a. This observation is consistent with the following theorem to search for zeros that are rational numbers.

> **TIP** Recall that a rational number is a number that can be expressed as a ratio of two integers.

Rational Zero Theorem

If $f(x) = a_n x^n + a_{n-1} x^{n-1} + a_{n-2} x^{n-2} + \cdots + a_1 x + a_0$ has integer coefficients and $a_n \neq 0$, and if $\frac{p}{q}$ (written in lowest terms) is a rational zero of f, then

- p is a factor of the constant term a_0.
- q is a factor of the leading coefficient a_n.

The rational zero theorem does not guarantee the existence of rational zeros. Rather, it indicates that *if* a rational zero exists for a polynomial, then it must be of the form

$$\frac{p}{q} = \frac{\text{Factors of } a_0 \text{ (constant term)}}{\text{Factors of } a_n \text{ (leading coefficient)}}$$

The rational zero theorem is important because it limits our search to find rational zeros (if they exist) to a finite number of choices.

Classroom Example: p. 295
Exercise 10

EXAMPLE 1 **Listing All Possible Rational Zeros**

List all possible rational zeros. $f(x) = -2x^5 + 3x^2 - 2x + 10$

Solution:

First note that the polynomial has integer coefficients.

$$f(x) = -2x^5 + 3x^2 - 2x + 10$$

The constant term is 10. $\pm 1, \pm 2, \pm 5, \pm 10$ ——Factors of 10

The leading coefficient is -2. $\pm 1, \pm 2$ ——Factors of -2

$$\frac{\text{Factors of } 10}{\text{Factors of } -2} = \frac{\pm 1, \pm 2, \pm 5, \pm 10}{\pm 1, \pm 2} = \pm\frac{1}{1}, \pm\frac{2}{1}, \pm\frac{5}{1}, \pm\frac{10}{1}, \pm\frac{1}{2}, \pm\frac{2}{2}, \pm\frac{5}{2}, \pm\frac{10}{2}$$

The values $\pm\frac{2}{2}$ and $\pm\frac{10}{2}$ are redundant. They equal ± 1 and ± 5, respectively. The possible rational zeros are $\pm 1, \pm 2, \pm 5, \pm 10, \pm\frac{1}{2}$, and $\pm\frac{5}{2}$.

Skill Practice 1 List all possible rational zeros.
$f(x) = -4x^4 + 5x^3 - 7x^2 + 8$

Classroom Example: p. 295
Exercise 20

EXAMPLE 2 **Finding the Zeros of a Polynomial**

Find the zeros. $f(x) = x^3 - 4x^2 + 3x + 2$

Solution:

We begin by first looking for rational zeros. We can apply the rational zero theorem because the polynomial has integer coefficients.

$$f(x) = 1x^3 - 4x^2 + 3x + 2$$

Possible rational zeros: $\dfrac{\text{Factors of } 2}{\text{Factors of } 1} = \dfrac{\pm 1, \pm 2}{\pm 1} = \pm 1, \pm 2$

Next, use synthetic division and the remainder theorem to determine if any of the numbers in the list is a zero of f.

> **TIP** Since $f(1) = 2$ and $f(-1) = -6$, we know from the intermediate value theorem that $f(x)$ must have a zero between -1 and 1.

Test $x = 1$:

$$\begin{array}{r|rrrr} 1 & 1 & -4 & 3 & 2 \\ & & 1 & -3 & 0 \\ \hline & 1 & -3 & 0 & \underline{\hspace{0.3em}}2 \end{array}$$

The remainder is not zero. Therefore, 1 is not a zero of $f(x)$.

Test $x = -1$:

$$\begin{array}{r|rrrr} -1 & 1 & -4 & 3 & 2 \\ & & -1 & 5 & -8 \\ \hline & 1 & -5 & 8 & \underline{\hspace{0.3em}}-6 \end{array}$$

The remainder is not zero. Therefore, -1 is not a zero of $f(x)$.

Test $x = 2$:

$$\begin{array}{r|rrrr} 2 & 1 & -4 & 3 & 2 \\ & & 2 & -4 & -2 \\ \hline & 1 & -2 & -1 & \underline{\hspace{0.3em}}0 \end{array}$$

The remainder *is* zero. Therefore, 2 is a zero of $f(x)$.

By the factor theorem, since $f(2) = 0$, then $(x - 2)$ is a factor of $f(x)$. The quotient $(x^2 - 2x - 1)$ is also a factor of $f(x)$. We have

$$f(x) = x^3 - 4x^2 + 3x + 2 = (x - 2)(x^2 - 2x - 1)$$

We now have a third-degree polynomial written as the product of a first-degree polynomial and a quadratic polynomial. The quotient $x^2 - 2x - 1$ is called a **reduced polynomial** (or a **depressed polynomial**) of $f(x)$. It has degree 1 less than the degree of $f(x)$, and the remaining zeros of $f(x)$ are the zeros of the reduced polynomial.

Answer

1. $\pm 1, \pm 2, \pm 4, \pm 8, \pm\dfrac{1}{2}$, and $\pm\dfrac{1}{4}$

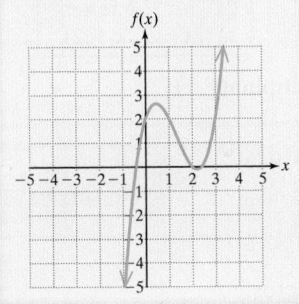

TIP The graph of $f(x) = x^3 - 4x^2 + 3x + 2$ is shown here. The zeros are $1 + \sqrt{2} \approx 2.41$, $1 - \sqrt{2} \approx -0.41$, and 2.

At this point, we no longer need to test for more rational zeros. The reason is that any remaining zeros (whether they be rational, irrational, or imaginary) are the solutions to the quadratic equation $x^2 - 2x - 1 = 0$. There is no guess work because the equation can be solved by using the quadratic formula.

$$x^2 - 2x - 1 = 0$$

$$x = \frac{-(-2) \pm \sqrt{(-2)^2 - 4(1)(-1)}}{2(1)} = \frac{2 \pm \sqrt{8}}{2} = \frac{2 \pm 2\sqrt{2}}{2}$$

$$= 1 \pm \sqrt{2}$$

The zeros of $f(x)$ are 2, $1 + \sqrt{2}$, and $1 - \sqrt{2}$.

Skill Practice 2   Find the zeros. $f(x) = x^3 - x^2 - 4x - 2$

The polynomial in Example 2 has one rational zero. If we had continued testing for more rational zeros, we would not have found others because the remaining two zeros are irrational numbers. In Example 3, we illustrate the case where a function has multiple rational zeros.

Classroom Example: p. 295
Exercise 24

EXAMPLE 3   **Finding the Zeros of a Polynomial**

Find the zeros and their multiplicities. $f(x) = 2x^4 + 5x^3 - 2x^2 - 11x - 6$

Solution:

Begin by searching for rational zeros.

$$f(x) = 2x^4 + 5x^3 - 2x^2 - 11x - 6$$

Possible rational zeros:

TIP It does not matter in which order you test the potential rational zeros.

$$\frac{\text{Factors of } -6}{\text{Factors of } 2} = \frac{\pm 1, \pm 2, \pm 3, \pm 6}{\pm 1, \pm 2} = \pm 1, \pm 2, \pm 3, \pm 6, \pm \frac{1}{2}, \pm \frac{3}{2}$$

We can work methodically through the list of possible rational zeros to determine which if any are actual zeros of $f(x)$. After trying several possibilities, we find that -1 is a zero of $f(x)$.

$$\begin{array}{r|rrrrr} -1 & 2 & 5 & -2 & -11 & -6 \\ & & -2 & -3 & 5 & 6 \\ \hline & 2 & 3 & -5 & -6 & \underline{|0} \end{array}$$

The quotient is $2x^3 + 3x^2 - 5x - 6$.

Since $f(-1) = 0$, then $x + 1$ is a factor of $f(x)$.

$$f(x) = 2x^4 + 5x^3 - 2x^2 - 11x - 6 = (x + 1)\underbrace{(2x^3 + 3x^2 - 5x - 6)}$$

Now find the zeros of the quotient.

The zeros of $f(x)$ are -1 along with the roots of the equation $2x^3 + 3x^2 - 5x - 6 = 0$. Therefore, we need to find the zeros of $g(x) = 2x^3 + 3x^2 - 5x - 6$.

TIP The zeros of the polynomial in Example 3 are rational numbers (and therefore real numbers). They correspond to x-intercepts on the graph of $y = f(x)$. Also notice that the graph has a touch point at $(-1, 0)$ because -1 has an even multiplicity.

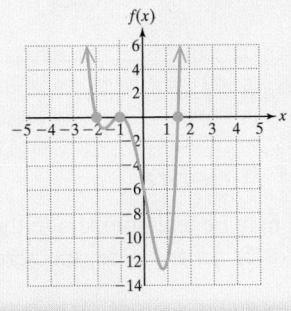

The possible rational zeros are ± 1, ± 2, ± 3, ± 6, $\pm \frac{1}{2}$, $\pm \frac{3}{2}$.

We will test -1 again because it may have multiplicity greater than 1.

$$\begin{array}{r|rrrr} -1 & 2 & 3 & -5 & -6 \\ & & -2 & -1 & 6 \\ \hline & 2 & 1 & -6 & \underline{|0} \end{array}$$

The quotient is $2x^2 + x - 6$.

Answer

2. $-1, 1 - \sqrt{3}$, and $1 + \sqrt{3}$

The value -1 is a repeated zero. We have:

$$f(x) = 2x^4 + 5x^3 - 2x^2 - 11x - 6 = (x + 1)(x + 1)(2x^2 + x - 6)$$
$$= (x + 1)^2(2x - 3)(x + 2)$$

> $2x^2 + x - 6$ factors as $(2x - 3)(x + 2)$.

The zeros of $f(x)$ are

-1 (multiplicity 2), $\frac{3}{2}$ (multiplicity 1), and -2 (multiplicity 1).

Skill Practice 3 Find the zeros and their multiplicities.

$f(x) = 2x^4 + 3x^3 - 15x^2 - 32x - 12$

In Example 4, we have a polynomial with no rational zeros.

Classroom Example: p. 295
Exercise 28

EXAMPLE 4 Finding the Zeros of a Polynomial

Find the zeros. $f(x) = x^4 - 2x^2 - 3$

Solution:

$f(x) = 1x^4 - 2x^2 - 3$ The possible rational zeros are $\dfrac{\pm 1, \pm 3}{\pm 1} = \pm 1, \pm 3$.

If we apply the rational zero theorem, we see that $f(x)$ has no *rational* zeros.

$$\begin{array}{c|rrrrr} 1 & 1 & 0 & -2 & 0 & -3 \\ & & 1 & 1 & -1 & -1 \\ \hline & 1 & 1 & -1 & -1 & \underline{-4} \end{array} \qquad \begin{array}{c|rrrrr} -1 & 1 & 0 & -2 & 0 & -3 \\ & & -1 & 1 & 1 & -1 \\ \hline & 1 & -1 & -1 & 1 & \underline{-4} \end{array}$$

$$\begin{array}{c|rrrrr} 3 & 1 & 0 & -2 & 0 & -3 \\ & & 3 & 9 & 21 & 63 \\ \hline & 1 & 3 & 7 & 21 & \underline{60} \end{array} \qquad \begin{array}{c|rrrrr} -3 & 1 & 0 & -2 & 0 & -3 \\ & & -3 & 9 & -21 & 63 \\ \hline & 1 & -3 & 7 & -21 & \underline{60} \end{array}$$

However, finding the zeros of $f(x)$ is equivalent to finding the roots of the equation $x^4 - 2x^2 - 3 = 0$.

$$x^4 - 2x^2 - 3 = 0$$
$$(x^2 - 3)(x^2 + 1) = 0 \qquad \text{Factor the trinomial.}$$
$$x^2 - 3 = 0 \quad \text{or} \quad x^2 + 1 = 0 \qquad \text{Set each factor equal to zero.}$$
$$x = \pm\sqrt{3} \quad \text{or} \quad x = \pm i \qquad \text{Apply the square root property.}$$

> **TIP** From the factor theorem,
>
> $f(x) = (x - \sqrt{3})(x + \sqrt{3})(x - i)(x + i)$.

The zeros are $\sqrt{3}$, $-\sqrt{3}$, i, and $-i$.

TIP The graph of the function defined by $f(x) = x^4 - 2x^2 - 3$ shows the real zeros of $f(x)$ as x-intercepts.

Instructor Note:
We restrict the graphs of polynomial functions to R × R. The imaginary numbers i and $-i$ are not in the domain of a function defined over the real numbers.

Skill Practice 4 Find the zeros. $f(x) = x^4 - x^2 - 20$

---

2. Apply the Fundamental Theorem of Algebra

From Examples 2–4, we see that the zeros of a polynomial may be real numbers (either rational or irrational) or nonreal numbers such as $2 + 3i$ or $5i$. In any case, the zeros are all complex numbers.

To find the zeros of a polynomial, it is important to know how many zeros to expect. This is answered by the following three theorems. The first is called the fundamental theorem of algebra because it is so basic to the foundation of algebra.

Answers

3. -2 (multiplicity 2);

$-\dfrac{1}{2}$ (multiplicity 1); and

3 (multiplicity 1)

4. $\sqrt{5}, -\sqrt{5}, 2i$, and $-2i$

Fundamental Theorem of Algebra

If $f(x)$ is a polynomial of degree $n \geq 1$ with complex coefficients, then $f(x)$ has at least one complex zero.

The fundamental theorem of algebra, first proved by German mathematician Carl Friedrich Gauss (1777–1855), guarantees that every polynomial of degree $n \geq 1$ has at least one zero.

Now suppose that $f(x)$ is a polynomial of degree $n \geq 1$ with complex coefficients. The fundamental theorem of algebra guarantees the existence of at least one complex zero, call this c_1. By the factor theorem, we have

$$f(x) = (x - c_1) \cdot q_1(x) \qquad \text{where } q_1(x) \text{ is a polynomial of degree } n - 1.$$

If $q_1(x)$ is of degree 1 or more, then the fundamental theorem of algebra guarantees that $q_1(x)$ must have at least one complex zero, call this c_2. Then,

$$f(x) = (x - c_1) \cdot (x - c_2) \cdot q_2(x) \quad \text{where } q_2(x) \text{ is a polynomial of degree } n - 2.$$

We can continue with this reasoning until the quotient polynomial, $q_n(x)$, is a constant equal to the leading coefficient of $f(x)$.

TIP The set of complex numbers includes the set of real numbers. Therefore, theorems relating to polynomials with complex coefficients also apply to polynomials with real coefficients.

Linear Factorization Theorem

If $f(x) = a_n x^n + a_{n-1} x^{n-1} + a_{n-2} x^{n-2} + \cdots + a_1 x + a_0$, where $n \geq 1$ and $a_n \neq 0$, then

$f(x) = a_n(x - c_1)(x - c_2) \ldots (x - c_n)$, where $c_1, c_2, \ldots, c_n$ are complex numbers.

Note: The complex numbers $c_1, c_2, \ldots, c_n$ are not necessarily unique.

The linear factorization theorem tells us that a polynomial of degree $n \geq 1$ with complex coefficients has exactly n linear factors of the form $(x - c)$, where some of the factors may be repeated. The value of c in each factor is a zero of the function, so the function must also have n zeros provided that the zeros are counted according to their multiplicities.

TIP Refer to Examples 2–4. In each case, the number of zeros (including multiplicities) is the same as the degree of the polynomial.

Number of Zeros of a Polynomial

If $f(x)$ is a polynomial of degree $n \geq 1$ with complex coefficients, then $f(x)$ has exactly n complex zeros provided that each zero is counted by its multiplicity.

Now consider the polynomial from Example 4.

$$f(x) = x^4 - 2x^2 - 3 \quad \text{Zeros: } \sqrt{3}, -\sqrt{3}, i, -i$$

Notice that the polynomial has real coefficients. Furthermore, the zeros i and $-i$ appear as a pair. This is not a coincidence. For a polynomial with real coefficients, if $a + bi$ is a zero, then $a - bi$ is a zero.

Conjugate Zeros Theorem

If $f(x)$ is a polynomial with real coefficients and if $a + bi$ ($b \neq 0$) is a zero of $f(x)$, then its conjugate $a - bi$ is also a zero of $f(x)$.

Classroom Example: p. 295
Exercise 34

EXAMPLE 5 Finding Zeros and Factoring a Polynomial

Given $f(x) = x^4 - 6x^3 + 28x^2 - 18x + 75$, and that $3 - 4i$ is a zero of $f(x)$,

a. Find the remaining zeros.

b. Factor $f(x)$ as a product of linear factors.

c. Solve the equation. $x^4 - 6x^3 + 28x^2 - 18x + 75 = 0$

Solution:

$f(x)$ is a fourth-degree polynomial, so we expect to find four zeros (including multiplicities). Further note that because $f(x)$ has real coefficients and because $3 - 4i$ is a zero, then the conjugate $3 + 4i$ must also be a zero. This leaves only two remaining zeros to find.

$$
\begin{array}{c|ccccc}
3 - 4i & 1 & -6 & 28 & -18 & 75 \\
& & 3 - 4i & -25 & 9 - 12i & -75 \\
\hline
& 1 & -3 - 4i & 3 & -9 - 12i & \underline{|0}
\end{array}
$$

Divide by $[x - (3 - 4i)]$.

coefficients of the quotient

One strategy is to use synthetic division twice using the two known zeros.

Note: $(3 - 4i)(-3 - 4i)$
   $= -9 - 12i + 12i + 16i^2$
   $= -25$

Note: $(3 - 4i)(-9 - 12i)$
   $= -27 - 36i + 36i + 48i^2$
   $= -75$

Since $3 + 4i$ is a zero of $f(x)$ it must also be a zero of the quotient.

$$
\begin{array}{c|ccccc}
3 + 4i & 1 & -3 - 4i & 3 & -9 - 12i \\
& & 3 + 4i & 0 & 9 + 12i \\
\hline
& 1 & 0 & 3 & \underline{|0}
\end{array}
$$

Divide by $[x - (3 + 4i)]$.

Divide the quotient by $[x - (3 + 4i)]$.

The resulting quotient is quadratic: $x^2 + 3$.

Now we have $f(x) = [x - (3 - 4i)][x - (3 + 4i)](x^2 + 3)$.
The remaining two zeros are found by solving $x^2 + 3 = 0$.

$$x^2 + 3 = 0$$
$$x^2 = -3$$
$$x = \pm i\sqrt{3}$$

a. The zeros of $f(x)$ are: $3 - 4i$, $3 + 4i$, $i\sqrt{3}$, and $-i\sqrt{3}$.

b. $f(x)$ factors as four linear factors.

$$f(x) = [x - (3 - 4i)][x - (3 + 4i)](x - i\sqrt{3})(x + i\sqrt{3})$$

c. The solution set for $x^4 - 6x^3 + 28x^2 - 18x + 75 = 0$ is $\{3 \pm 4i, \pm i\sqrt{3}\}$.

Skill Practice 5 Given $f(x) = x^4 - 2x^3 + 28x^2 - 4x + 52$, and that $1 + 5i$ is a zero of $f(x)$,

a. Find the zeros.

b. Factor $f(x)$ as a product of linear factors.

c. Solve the equation. $x^4 - 2x^3 + 28x^2 - 4x + 52 = 0$

Answers

5. a. Zeros:
$1 + 5i, 1 - 5i, i\sqrt{2}, -i\sqrt{2}$

b. $f(x) = [x - (1 + 5i)][x - (1 - 5i)]$
$(x - i\sqrt{2})(x + i\sqrt{2})$

c. $\{1 \pm 5i, \pm i\sqrt{2}\}$

Classroom Examples: p. 296
Exercises 44, 46

Instructor Note:
Point out that the factors $(ax - b)$ and $\left(x - \frac{b}{a}\right)$ have the same zero, but the factor $\left(x - \frac{b}{a}\right)$ fits the form outlined in the linear factorization theorem.

EXAMPLE 6   Building a Polynomial with Specified Conditions

a. Find a third-degree polynomial $f(x)$ with integer coefficients and with zeros of $\frac{2}{3}$ and $4 + 2i$.

b. Find a polynomial $g(x)$ of lowest degree with zeros of -2 (multiplicity 1) and 4 (multiplicity 3), and satisfying the condition that $g(0) = 256$.

Solution:

a. $f(x)$ is to be a polynomial with integer coefficients (and therefore real coefficients). If $4 + 2i$ is a zero, then $4 - 2i$ must also be a zero. By the linear factorization theorem we have:

$$f(x) = a\left(x - \frac{2}{3}\right)[x - (4 + 2i)][x - (4 - 2i)] \quad a \text{ is a nonzero number.}$$

$$= a\left(x - \frac{2}{3}\right)[x^2 - x(4 - 2i) - x(4 + 2i) + (4 + 2i)(4 - 2i)]$$

$$= a\left(x - \frac{2}{3}\right)(x^2 - 4x + 2xi - 4x - 2xi + 16 - 4i^2)$$

$$= 3\left(x - \frac{2}{3}\right)(x^2 - 8x + 20)$$

$$= (3x - 2)(x^2 - 8x + 20)$$

$$= 3x^3 - 24x^2 + 60x - 2x^2 + 16x - 40$$

$$= 3x^3 - 26x^2 + 76x - 40$$

To give $f(x)$ integer coefficients, choose a to be any multiple of 3. We have chosen $a = 3$.

Instructor Note:
Comment that the coefficient "a" has no effect on the zeros of the polynomial. It serves to stretch or shrink the graph vertically, or to reflect the graph across the x-axis.

b. $g(x) = a(x + 2)^1(x - 4)^3$

$$= a(x + 2)(x^3 - 12x^2 + 48x - 64)$$

$$= a(x^4 - 10x^3 + 24x^2 + 32x - 128)$$

-2 is a zero of multiplicity 1.
4 is a zero of multiplicity 3.
Note:
$(x - 4)^3 = (x - 4)^2(x - 4)$
$\qquad = (x^2 - 8x + 16)(x - 4)$
$\qquad = x^3 - 12x^2 + 48x - 64$

We also have the condition that $g(0) = 256$.

$$g(0) = a[(0)^4 - 10(0)^3 + 24(0)^2 + 32(0) - 128] = 256$$

$$-128a = 256$$

$$a = -2$$

Therefore, $g(x) = -2(x^4 - 10x^3 + 24x^2 + 32x - 128)$

$$g(x) = -2x^4 + 20x^3 - 48x^2 - 64x + 256$$

Skill Practice 6

a. Find a third-degree polynomial $f(x)$ with integer coefficients and with zeros of $2 + i$ and $\frac{4}{3}$.

b. Find a polynomial $g(x)$ of lowest degree with zeros of -3 (multiplicity 2) and 5 (multiplicity 2), and satisfying the condition that $g(0) = 450$.

3. Apply Descartes' Rule of Signs

Finding the zeros of a polynomial analytically can be a difficult (or impossible) task. For example, consider

$$f(x) = x^5 - 18x^4 + 128x^3 - 450x^2 + 783x - 540$$

Answers
6. a. $f(x) = 3x^3 - 16x^2 + 31x - 20$
   b. $g(x) = 2x^4 - 8x^3 - 52x^2 + 120x + 450$

Applying the rational zero theorem gives us the following possible rational zeros.

$$\pm 1, \pm 2, \pm 3, \pm 4, \pm 5, \pm 6, \pm 9, \pm 10, \pm 12, \pm 15, \pm 18, \pm 20,$$
$$\pm 27, \pm 30, \pm 36, \pm 45, \pm 54, \pm 60, \pm 90, \pm 108, \pm 135, \pm 180, \pm 270, \pm 540$$

However, we can use the upper and lower bound theorem to show that the real zeros of $f(x)$ are between -1 and 18. Furthermore, we can use a tool called Descartes' rule of signs to show that none of the zeros is negative. This eliminates all possible rational zeros except for 1, 2, 3, 4, 5, 6, 9, 10, 12, and 15.

To study Descartes' rule of signs, we need to establish what is meant by "sign changes" between consecutive terms in a polynomial. For example, the following polynomial is written in descending order and has three changes in sign between consecutive coefficients.

$$2x^6 - 3x^4 - x^3 + 5x^2 - 6x - 4 \qquad \text{(3 sign changes)}$$

| positive to negative | negative to positive | positive to negative |

TIP For a polynomial with real coefficients, the reason we reduce the number of possible real zeros in increments of 2 is because the alternative, nonreal zeros, occur in conjugate pairs.

Descartes' Rule of Signs

Let $f(x)$ be a polynomial with real coefficients and a nonzero constant term. Then,

1. The number of *positive* real zeros is either
   - the same as the number of sign changes in $f(x)$ or
   - less than the number of sign changes in $f(x)$ by a positive even integer.
2. The number of *negative* real zeros is either
   - the same as the number of sign changes in $f(-x)$ or
   - less than the number of sign changes in $f(-x)$ by a positive even integer.

Descartes' rule of signs is demonstrated in Examples 7 and 8.

Classroom Example: p. 296
Exercise 52

EXAMPLE 7   Applying Descartes' Rule of Signs

Determine the number of possible positive and negative real zeros.
$$f(x) = x^5 - 6x^4 + 12x^3 - 12x^2 + 11x - 6$$

Solution:

$f(x)$ has real coefficients and the constant term is nonzero.

To determine the number of possible positive real zeros, determine the number of sign changes in $f(x)$.

$$f(x) = x^5 - 6x^4 + 12x^3 - 12x^2 + 11x - 6 \qquad \text{(5 sign changes)}$$

The number of possible positive real zeros is either 5, 3, or 1.

To determine the number of possible negative real zeros, determine the number of sign changes in $f(-x)$.

$$f(-x) = (-x)^5 - 6(-x)^4 + 12(-x)^3 - 12(-x)^2 + 11(-x) - 6$$
$$= -x^5 - 6x^4 - 12x^3 - 12x^2 - 11x - 6 \qquad \text{(0 sign changes)}$$

There are no sign changes in $f(-x)$. Therefore, $f(x)$ has no negative real zeros.

TIP Nonreal zeros are numbers that contain the imaginary number i such as $-3i$, $3i$, $4 + 7i$, $4 - 7i$ and so on.

Number of possible positive real zeros	5	3	1
Number of possible negative real zeros	0	0	0
Number of nonreal zeros	0	2	4
Total (including multiplicities)	5	5	5

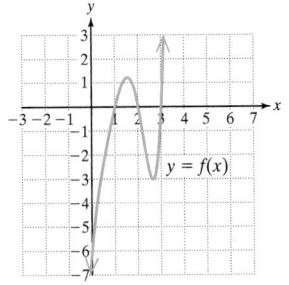

Figure 2-16

Classroom Example: p. 296
Exercise 56

The graph of $f(x) = x^5 - 6x^4 + 12x^3 - 12x^2 + 11x - 6$ is shown in Figure 2-16. Notice that there are three positive x-intercepts and therefore, three positive real zeros. There are no negative x-intercepts as expected. The remaining zeros of the polynomial $x^5 - 6x^4 + 12x^3 - 12x^2 + 11x - 6$ are not real numbers.

Skill Practice 7 Determine the number of possible positive and negative real zeros. $f(x) = 4x^5 + 6x^3 + 2x^2 + 6$

EXAMPLE 8   Applying Descartes' Rule of Signs

Determine the number of possible positive and negative real zeros.

$$g(x) = 2x^6 - 5x^4 - 3x^3 + 7x^2 + 2x + 5$$

Solution:

$g(x)$ has real coefficients and the constant term is nonzero.

$$g(x) = 2x^6 - 5x^4 - 3x^3 + 7x^2 + 2x + 5 \qquad \text{2 sign changes in } g(x)$$

The number of possible positive real zeros is either 2 or 0.

$$g(-x) = 2(-x)^6 - 5(-x)^4 - 3(-x)^3 + 7(-x)^2 + 2(-x) + 5$$
$$= 2x^6 - 5x^4 + 3x^3 + 7x^2 - 2x + 5 \qquad \text{4 sign changes in } g(-x)$$

The number of possible negative real zeros is either 4, 2, or 0.

Number of possible positive real zeros	2	2	2	0	0	0
Number of possible negative real zeros	4	2	0	4	2	0
Number of nonreal zeros	0	2	4	2	4	6
Total (including multiplicities)	6	6	6	6	6	6

Skill Practice 8 Determine the number of possible positive and negative real zeros. $g(x) = 8x^6 - 5x^7 + 3x^5 - x^2 - 3x + 1$

Descartes' rule of signs stipulates that the constant term of the polynomial $f(x)$ is nonzero. If the constant term is 0, we can factor out the lowest power of x and apply Descartes' rule of signs to the resulting factor.

$$f(x) = x^7 - 8x^6 + 15x^5$$
$$f(x) = x^5(x^2 - 8x + 15)$$

Descartes' rule of signs can be applied to $x^2 - 8x + 15$ to show that there may be 2 or 0 remaining positive real zeros.

The value 0 is a zero of $f(x)$ of multiplicity 5.

4. Find Upper and Lower Bounds

The next theorem helps us limit our search for the real zeros of a polynomial. First we define two key terms.

- A real number b is called an **upper bound** of the real zeros of a polynomial if all real zeros are less than or equal to b.
- A real number a is called a **lower bound** of the real zeros of a polynomial if all real zeros are greater than or equal to a.

Answers

7. Positive: 0; Negative: 1
8. Positive: 4, 2, or 0;
     Negative: 2 or 0

Upper and Lower Bound Theorem for the Real Zeros of a Polynomial

Let $f(x)$ be a polynomial of degree $n \geq 1$ with real coefficients and a positive leading coefficient. Further suppose that $f(x)$ is divided by $(x - c)$.

1. If $c > 0$ and if both the remainder and the coefficients of the quotient are nonnegative, then c is an upper bound for the real zeros of $f(x)$.
2. If $c < 0$ and the coefficients of the quotient and the remainder alternate in sign (with 0 being considered either positive or negative as needed), then c is a lower bound for the real zeros of $f(x)$.

The rules for finding upper and lower bounds are stated for polynomial functions having a positive leading coefficient. However, $f(x) = 0$ and $-f(x) = 0$ are equivalent equations. Therefore, if $f(x)$ has a negative leading coefficient, we can factor out -1 from $f(x)$ and apply the rule for upper and lower bounds accordingly.

EXAMPLE 9 Applying the Upper and Lower Bound Theorem

Given $f(x) = 2x^5 + x^4 + 9x^2 - 32x + 20,$

a. Determine if the upper bound theorem identifies 2 as an upper bound for the real zeros of $f(x)$.

b. Determine if the lower bound theorem identifies -2 as a lower bound for the real zeros of $f(x)$.

Solution:

a. Divide $f(x)$ by $(x - 2)$.

First note that the leading coefficient of the polynomial is positive.

```
 2|  2   1    0    9   -32   20
         4   10   20    58   52
    ─────────────────────────────
     2   5   10   29    26  |72
```

This row nonnegative

The remainder and all coefficients of the quotient are nonnegative.

2 is an upper bound for the real zeros of $f(x)$.

b. Divide $f(x)$ by $(x + 2)$.

```
-2|  2   1    0     9   -32   20
        -4   6   -12     6    52
    ─────────────────────────────
     2  -3   6    -3   -26  |72
     +   -   +            +
```

No sign change

The signs of the quotient do not alternate. Therefore, we cannot conclude that -2 is a lower bound for the real zeros of $f(x)$.

Skill Practice 9 Given $f(x) = x^4 - 2x^3 - 13x^2 - 4x - 30,$

a. Determine if the upper bound theorem identifies 4 as an upper bound for the real zeros of $f(x)$.

b. Determine if the lower bound theorem identifies -4 as a lower bound for the real zeros of $f(x)$.

Answers

9. a. No   **b.** Yes

Classroom Example: p. 296
Exercise 72

TIP From Example 9, although we cannot conclude that -2 is a lower bound for the real zeros of $f(x)$, we can try other negative real numbers. For example, -3 is a lower bound for the real zeros of $f(x)$.

$$
\begin{array}{r|rrrrrr}
-3 & 2 & 1 & 0 & 9 & -32 & 20 \\
   &   & -6 & 15 & -45 & 108 & -228 \\
\hline
   & 2 & -5 & 15 & -36 & 76 & \underline{-208} \\
\end{array}
$$

Therefore, -3 is a lower bound.

signs all alternate

In Example 10, we will use the tools presented in Sections 2.2–2.4 to find all zeros of a polynomial.

EXAMPLE 10   **Finding the Zeros of a Polynomial**

Find the zeros and their multiplicities. $f(x) = 2x^5 + x^4 + 9x^2 - 32x + 20$

Solution:

$f(x)$ is a fifth-degree polynomial and must have five zeros (including multiplicities). We begin by finding the rational zeros (if any exist). By the rational zero theorem, the possible rational zeros are

$$
\pm 1, \pm 2, \pm 4, \pm 5, \pm 10, \pm 20, \pm\frac{1}{2}, \pm\frac{5}{2}
$$

However, we also know from Example 9 that 2 is not a zero of $f(x)$, but is an upper bound for the real zeros. From the Tip following Example 9, we know that -3 is not a zero of $f(x)$, but is a lower bound for the real zeros. Therefore, we can restrict the list of possible rational zeros to those on the interval $(-3, 2)$.

$$
-\frac{5}{2}, -2, \pm\frac{1}{2}, \pm 1
$$

After testing several possible rational zeros, we find that 1 is a zero.

$$
\begin{array}{r|rrrrrr}
1 & 2 & 1 & 0 & 9 & -32 & 20 \\
  &   & 2 & 3 & 3 & 12 & -20 \\
\hline
  & 2 & 3 & 3 & 12 & -20 & \underline{0} \\
\end{array}
$$

We have $f(x) = (x - 1)(2x^4 + 3x^3 + 3x^2 + 12x - 20)$. Now look for the zeros of the reduced polynomial. We will try 1 again because it may be a repeated zero.

$$
\begin{array}{r|rrrrr}
1 & 2 & 3 & 3 & 12 & -20 \\
  &   & 2 & 5 & 8 & 20 \\
\hline
  & 2 & 5 & 8 & 20 & \underline{0} \\
\end{array}
$$

The value 1 is a repeated zero.

We have $f(x) = (x - 1)^2(2x^3 + 5x^2 + 8x + 20)$.

Because the polynomial $2x^3 + 5x^2 + 8x + 20$ has no sign changes, Descartes' rule of signs indicates that there are no other positive real zeros. Now the list of possible rational zeros is restricted to $-\frac{5}{2}$, -2, and -1. We find that $-\frac{5}{2}$ is a zero of $f(x)$.

$$
\begin{array}{r|rrrr}
-\frac{5}{2} & 2 & 5 & 8 & 20 \\
             &   & -5 & 0 & -20 \\
\hline
             & 2 & 0 & 8 & \underline{0} \\
\end{array}
$$

Thus, $f(x) = (x - 1)^2(x + \frac{5}{2})(2x^2 + 8)$.

$$2x^2 + 8 = 0$$

$$x^2 = -4$$

$$x = \pm 2i$$

The remaining two zeros are found by solving the equation $2x^2 + 8 = 0$.

The zeros are: 1 (multiplicity of 2), $-\frac{5}{2}$, $2i$, and $-2i$ (each with multiplicity of 1).

Skill Practice 10   Find the zeros and their multiplicities.

$$f(x) = x^5 + 6x^3 - 2x^2 - 27x - 18$$

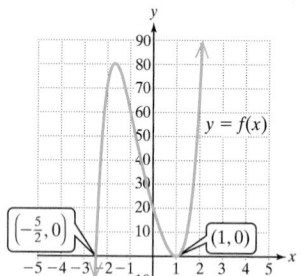

Figure 2-17

The graph of $f(x) = 2x^5 + x^4 + 9x^2 - 32x + 20$ from Example 10 is shown in Figure 2-17. The end behavior is down to the left and up to the right as expected.

The real zeros of $f(x)$ correspond to the x-intercepts $\left(-\frac{5}{2}, 0\right)$ and $(1, 0)$. The point $(1, 0)$ is a touch point because 1 is a zero with an even multiplicity. The graph crosses the x-axis at $-\frac{5}{2}$ because $-\frac{5}{2}$ is a zero with an odd multiplicity.

TECHNOLOGY CONNECTIONS

Applications of Graphing Utilities to Polynomials

A graphing utility can help us analyze a polynomial. For example, given $f(x) = 2x^3 - 11x^2 - 5x + 50$, the possible rational zeros are: ± 1, ± 2, ± 5, ± 10, ± 25, ± 50, $\pm \frac{1}{2}$, $\pm \frac{5}{2}$, and $\pm \frac{25}{2}$. By graphing the function f on a graphing utility, it appears that the function may cross the x-axis at -2, 2.5, and 5. So we might consider testing these values first.

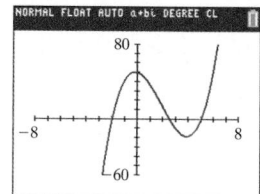

Our knowledge of algebra can also help us use a graphing device effectively. Consider $f(x) = 2x^5 + x^4 + 9x^2 - 32x + 20$ from Examples 9 and 10. We know that -3 is a lower bound for the real zeros of $f(x)$ and that 2 is an upper bound. Therefore, we can set a viewing window showing x between -3 and 2 and be guaranteed to see all the real zeros of $f(x)$.

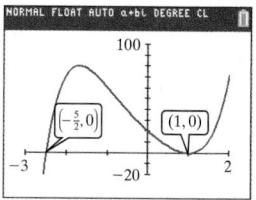

Finally, when analytical methods fail, we can use the Zero feature on a graphing utility to approximate the real zeros of a polynomial. Given $f(x) = x^3 - 7.14x^2 + 25.6x - 40.8$, the calculator approximates a zero of 3.1258745. The real zeros of a function can also be approximated by repeated use of the intermediate value theorem. (See the online group activity "Investigating the Bisection Method for Finding Zeros.")

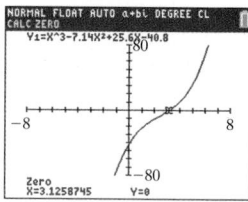

Answer

10. -1 (multiplicity 2), 2, $3i$, $-3i$ (each with multiplicity 1)

SECTION 2.4   Practice Exercises

Prerequisite Review

For Exercises R.1–R.3, perform the indicated operations. Write the answers in standard form $a + bi$.

R.1. $(9i)(-9i)$ $81 + 0i$ **R.2.** $\left(6 + \sqrt{13}i\right)\left(6 - \sqrt{13}i\right)$ $49 + 0i$ **R.3.** $(5 - 6i)^2 + (5 + 6i)^2$ $-22 + 0i$

Concept Connections

1. The _____*zeros*_____ of a polynomial $f(x)$ are the solutions (or roots) of the equation $f(x) = 0$.

2. If $f(x)$ is a polynomial of degree $n \geq 1$ with complex coefficients, then $f(x)$ has exactly _____*n*_____ complex zeros, provided that each zero is counted by its multiplicity.

3. The conjugate zeros theorem states that if $f(x)$ is a polynomial with real coefficients, and if $a + bi$ is a zero of $f(x)$, then _____*$a - bi$*_____ is also a zero of $f(x)$.

4. A real number b is called an _____*upper*_____ bound of the real zeros of a polynomial $f(x)$ if all real zeros of $f(x)$ are less than or equal to b.

5. A real number a is called a lower bound of the real zeros of a polynomial $f(x)$ if all real zeros of $f(x)$ are _____*greater than*_____ or equal to a.

6. Explain why the number 7 cannot be a rational zero of the polynomial $f(x) = 2x^3 + 5x^2 - x + 6$.
   7 is not the ratio of any of the factors of 6 over the factors of 2.

Objective 1: Apply the Rational Zero Theorem

For Exercises 7–12, list the possible rational zeros. (See Example 1)

7. $f(x) = x^5 - 2x^3 + 7x^2 + 4$ $\pm 1, \pm 2, \pm 4$ 8. $g(x) = x^3 - 5x^2 + 2x - 9$ $\pm 1, \pm 3, \pm 9$ 9. $h(x) = 4x^4 + 9x^3 + 2x - 6$

10. $k(x) = 25x^7 + 22x^4 - 3x^2 + 10$ 11. $m(x) = -12x^6 + 4x^3 - 3x^2 + 8$ 12. $n(x) = -16x^4 - 7x^3 + 2x + 6$

13. Which of the following is *not* a possible zero of $f(x) = 2x^3 - 5x^2 + 12$?
    $$1, 7, \frac{5}{3}, \frac{3}{2}$$ *7 and $\frac{5}{3}$*

14. Which of the following is *not* a possible zero of $f(x) = 4x^5 - 2x^3 + 10$?
    $$3, 5, \frac{5}{2}, \frac{3}{2}$$ *3 and $\frac{3}{2}$*

For Exercises 15–16, find all the rational zeros.

15. $p(x) = 2x^4 - x^3 - 5x^2 + 2x + 2$ $-\frac{1}{2}, 1$

16. $q(x) = x^4 + x^3 - 7x^2 - 5x + 10$ $1, -2$

For Exercises 17–28, find all the zeros. (See Examples 2–4)

17. $c(x) = 2x^4 - 7x^3 - 17x^2 + 58x - 24$ 18. $d(x) = 3x^4 - 2x^3 - 21x^2 - 4x + 12$ 19. $f(x) = x^3 - 7x^2 + 6x + 20$ $5, \frac{1 \pm \sqrt{5}}{?}$

20. $g(x) = x^3 - 7x^2 + 14x - 6$ 21. $h(x) = 5x^3 - x^2 - 35x + 7$ $\frac{1}{5}, \pm\sqrt{7}$ 22. $k(x) = 7x^3 - x^2 - 21x + 3$ $\frac{1}{7}, \pm\sqrt{3}$

23. $m(x) = 3x^4 - x^3 - 36x^2 + 60x - 16$ $3, 2 \pm \sqrt{2}$ 24. $n(x) = 2x^4 + 9x^3 - 5x^2 - 57x - 45$ 25. $q(x) = x^3 - 4x^2 - 2x + 20$ $-2, 3 \pm i$

26. $p(x) = x^3 - 8x^2 + 29x - 52$ $4, 2 \pm 3i$ 27. $t(x) = x^4 - x^2 - 90$ $\pm\sqrt{10}, \pm 3i$ 28. $v(x) = x^4 - 12x^2 - 13$ $\pm\sqrt{13}, \pm i$

Objective 2: Apply the Fundamental Theorem of Algebra

29. Given a polynomial $f(x)$ of degree $n \geq 1$, the fundamental theorem of algebra guarantees at least _____*one*_____ complex zero.

30. The number of zeros of $f(x) = 4x^3 - 5x^2 + 6x - 3$ is _____*3*_____, provided that each zero is counted according to its multiplicity.

31. If $f(x)$ is a polynomial with real coefficients and zeros of 5 (multiplicity 2), -1 (multiplicity 1), $2i$, and $3 + 4i$, what is the minimum degree of $f(x)$?   *7*

32. If $g(x)$ is a polynomial with real coefficients and zeros of -4 (multiplicity 3), 6 (multiplicity 2), $1 + i$, and $2 - 7i$, what is the minimum degree of $g(x)$?   *9*

For Exercises 33–38, a polynomial $f(x)$ and one or more of its zeros is given.

a. Find all the zeros.

b. Factor $f(x)$ as a product of linear factors.

c. Solve the equation $f(x) = 0$. (See Example 5)

33. $f(x) = x^4 - 4x^3 + 22x^2 + 28x - 203$; $2 - 5i$ is a zero

34. $f(x) = x^4 - 6x^3 + 5x^2 + 30x - 50$; $3 - i$ is a zero

35. $f(x) = 3x^3 - 28x^2 + 83x - 68$; $4 + i$ is a zero

36. $f(x) = 5x^3 - 54x^2 + 170x - 104$; $5 + i$ is a zero

37. $f(x) = 4x^5 + 37x^4 + 117x^3 + 87x^2 - 193x - 52$;
    $-3 + 2i$ and $-\frac{1}{4}$ are zeros

38. $f(x) = 2x^5 - 5x^4 - 4x^3 - 22x^2 + 50x + 75$;
    $-1 - 2i$ and $\frac{5}{2}$ are zeros

For Exercises 39–48, write a polynomial $f(x)$ that satisfies the given conditions. (See Example 6)

39. Degree 3 polynomial with integer coefficients with zeros $6i$ and $\frac{4}{5}$ $f(x) = 5x^3 - 4x^2 + 180x - 144$

40. Degree 3 polynomial with integer coefficients with zeros $-4i$ and $\frac{3}{2}$ $f(x) = 2x^3 - 3x^2 + 32x - 48$

41. Polynomial of lowest degree with zeros of -4 (multiplicity 1), 2 (multiplicity 3) and with $f(0) = 160$
$f(x) = -5x^4 + 10x^3 + 60x^2 - 200x + 160$

42. Polynomial of lowest degree with zeros of 5 (multiplicity 2) and -3 (multiplicity 2) and with $f(0) = -450$ $f(x) = -2x^4 + 8x^3 + 52x^2 - 120x - 450$

43. Polynomial of lowest degree with zeros of $-\frac{4}{3}$ (multiplicity 2) and $\frac{1}{2}$ (multiplicity 1) and with $f(0) = -16$ $f(x) = 18x^3 + 39x^2 + 8x - 16$

44. Polynomial of lowest degree with zeros of $-\frac{5}{6}$ (multiplicity 2) and $\frac{1}{3}$ (multiplicity 1) and with $f(0) = -25$ $f(x) = 108x^3 + 144x^2 + 15x - 25$

45. Polynomial of lowest degree with real coefficients and with zeros $7 - 4i$ and 0 (multiplicity 4)
$f(x) = x^6 - 14x^5 + 65x^4$

46. Polynomial of lowest degree with real coefficients and with zeros $5 - 10i$ and 0 (multiplicity 3)
$f(x) = x^5 - 10x^4 + 125x^3$

47. Polynomial of lowest degree with real coefficients and zeros of $5i$ and $6 - i$. $f(x) = x^4 - 12x^3 + 62x^2 - 300x + 925$

48. Polynomial of lowest degree with real coefficients and zeros of $-3i$ and $5 + 2i$. $f(x) = x^4 - 10x^3 + 38x^2 - 90x + 261$

Objective 3: Apply Descartes' Rule of Signs

For Exercises 49–56, determine the number of possible positive and negative real zeros for the given function. (See Examples 7–8)

49. $f(x) = x^6 - 2x^4 + 4x^3 - 2x^2 - 5x - 6$
Positive: 3 or 1; Negative: 3 or 1

50. $g(x) = 3x^7 + 4x^4 - 6x^3 + 5x^2 - 6x + 1$
Positive: 4, 2, or 0; Negative: 1

51. $k(x) = -8x^7 + 5x^6 - 3x^4 + 2x^3 - 11x^2 + 4x - 3$
Positive: 6, 4, 2, or 0; Negative: 1

52. $h(x) = -4x^9 + 6x^8 - 5x^5 - 2x^4 + 3x^2 - x + 8$
Positive: 5, 3, or 1; Negative: 2 or 0

53. $p(x) = 0.11x^4 + 0.04x^3 + 0.31x^2 + 0.27x + 1.1$
Positive: 0; Negative: 4, 2, or 0

54. $q(x) = -0.6x^4 + 0.8x^3 - 0.6x^2 + 0.1x - 0.4$
Positive: 4, 2, or 0; Negative: 0

55. $v(x) = \frac{1}{8}x^6 + \frac{1}{6}x^4 + \frac{1}{3}x^2 + \frac{1}{10}$
Positive: 0; Negative: 0

56. $t(x) = \frac{1}{1000}x^6 + \frac{1}{100}x^4 + \frac{1}{10}x^2 + 1$
Positive: 0; Negative: 0

For Exercises 57–58, use Descartes' rule of signs to determine the total number of real zeros and the number of positive and negative real zeros. (_Hint_: First factor out x to its lowest power.)

57. $f(x) = x^8 + 5x^6 + 6x^4 - x^3$ 4 real zeros; $f(x)$ has 1 positive real zero, no negative real zeros, and the number 0 is a zero of multiplicity 3.

58. $f(x) = -5x^8 - 3x^6 - 4x^2 + x$ 2 real zeros; $f(x)$ has 1 positive real zero, no negative real zeros, and the number 0 is a zero of multiplicity 1.

Objective 4: Find Upper and Lower Bounds

For Exercises 59–64, (See Example 9)

a. Determine if the upper bound theorem identifies the given number as an upper bound for the real zeros of $f(x)$.

b. Determine if the lower bound theorem identifies the given number as a lower bound for the real zeros of $f(x)$.

59. $f(x) = x^5 + 6x^4 + 5x^2 + x - 3$
  a. 2   Yes   **b.** -5 No

60. $f(x) = x^4 + 8x^3 - 4x^2 + 7x - 3$
  a. 3   Yes   **b.** -4 No

61. $f(x) = 8x^3 - 42x^2 + 33x + 28$
  a. 6   Yes   **b.** -1 Yes

62. $f(x) = 6x^3 - x^2 - 57x + 70$
  a. 4   Yes   **b.** -4 Yes

63. $f(x) = 2x^5 + 11x^4 - 63x^2 - 50x + 40$
  a. 3   Yes   **b.** -6 Yes

64. $f(x) = 3x^5 - 16x^4 + 5x^3 + 90x^2 - 138x + 36$
  a. 6   Yes   **b.** -3 Yes

For Exercises 65–68, determine if the statement is true or false. If a statement is false, explain why.

65. If 5 is an upper bound for the real zeros of $f(x)$, then 6 is also an upper bound.   True

66. If 5 is an upper bound for the real zeros of $f(x)$, then 4 is also an upper bound.   False. Only numbers greater than 5 are also guaranteed to be upper bounds.

67. If -3 is a lower bound for the real zeros of $f(x)$, then -2 is also a lower bound.   False. Only numbers less than -3 are also guaranteed to be lower bounds.

68. If -3 is a lower bound for the real zeros of $f(x)$, then -4 is also a lower bound.   True

For Exercises 69–84, find the zeros and their multiplicities. Consider using Descartes' rule of signs and the upper and lower bound theorem to limit your search for rational zeros. (See Example 10)

69. $f(x) = 8x^3 - 42x^2 + 33x + 28$
(_Hint_: See Exercise 61.) $\frac{7}{4}, -\frac{1}{2}$, and 4 (each with multiplicity 1)

70. $f(x) = 6x^3 - x^2 - 57x + 70$
(_Hint_: See Exercise 62.) $-\frac{7}{2}, \frac{5}{3}$, and 2 (each with multiplicity 1)

71. $f(x) = 2x^5 + 11x^4 - 63x^2 - 50x + 40$
(_Hint_: See Exercise 63.) $\pm\sqrt{5}, \frac{1}{2}, -2$, and -4 (each with multiplicity 1)

72. $f(x) = 3x^5 - 16x^4 + 5x^3 + 90x^2 - 138x + 36$
(_Hint_: See Exercise 64.) $\pm\sqrt{6}, \frac{1}{3}, 2$, and 3 (each with multiplicity 1)

73. $f(x) = 4x^4 + 20x^3 + 13x^2 - 30x + 9$

74. $f(x) = 9x^4 + 30x^3 + 13x^2 - 20x + 4$

75. $f(x) = 2x^4 - 11x^3 + 27x^2 - 41x + 15$

76. $g(x) = 3x^4 - 20x^3 + 51x^2 - 56x + 20$

77. $h(x) = 4x^4 - 28x^3 + 73x^2 - 90x + 50$

78. $k(x) = 9x^4 - 42x^3 + 70x^2 - 34x + 5$

79. $f(x) = x^6 + 2x^5 + 11x^4 + 20x^3 + 10x^2$
0 (multiplicity 2), -1 (multiplicity 2), and $\pm i\sqrt{10}$ (each multiplicity 1)

80. $f(x) = x^6 + 6x^5 + 12x^4 + 18x^3 + 27x^2$
0 (multiplicity 2), -3 (multiplicity 2), and $\pm i\sqrt{3}$ (each multiplicity 1)

81. $f(x) = x^5 - 10x^4 + 34x^3$
0 (multiplicity 3) and $5 \pm 3i$ (each multiplicity 1)

82. $f(x) = x^6 - 12x^5 + 40x^4$
0 (multiplicity 4) and $6 \pm 2i$ (each multiplicity 1)

83. $f(x) = -x^3 + 3x^2 - 9x - 13$
-1 and $2 \pm 3i$ (each multiplicity 1)

84. $f(x) = -x^3 + 5x^2 - 11x + 15$
3 and $1 \pm 2i$ (each multiplicity 1)

Mixed Exercises

For Exercises 85–90, determine if the statement is true or false. If a statement is false, explain why.

85. A polynomial with real coefficients of degree 4 must have at least one real zero.

86. Given $f(x) = 2ix^4 - (3 + 6i)x^3 + 5x^2 + 7$, if $a + bi$ is a zero of $f(x)$, then $a - bi$ must also be a zero.

87. The graph of a 10th-degree polynomial must cross the x-axis exactly once.

88. Suppose that $f(x)$ is a polynomial, and that a and b are real numbers where $a < b$. If $f(a) < 0$ and $f(b) < 0$, then $f(x)$ has no real zeros on the interval $[a, b]$.

89. If c is a zero of a polynomial $f(x)$, with degree $n \geq 2$ then all other zeros of $f(x)$ are zeros of $\dfrac{f(x)}{x - c}$.   True

90. If b is an upper bound for the real zeros of a polynomial, then $-b$ is a lower bound for the real zeros of the polynomial.

91. Given that $x - c$ divides evenly into a polynomial $f(x)$, which statements are true?   All statements are true.

a. $x - c$ is a factor of $f(x)$.

b. c is a zero of $f(x)$.

c. The remainder of $f(x) \div (x - c)$ is 0.

d. c is a solution (root) of the equation $f(x) = 0$.

92. a. Use the quadratic formula to solve $x^2 - 7x + 5 = 0$. $\left\{ \dfrac{7 \pm \sqrt{29}}{2} \right\}$

b. Write $x^2 - 7x + 5$ as a product of linear factors.
$\left(x - \dfrac{7 + \sqrt{29}}{2} \right)\left(x - \dfrac{7 - \sqrt{29}}{2} \right)$

93. a. Use the intermediate value theorem to show that $f(x) = 2x^2 - 7x + 4$ has a real zero on the interval $[2, 3]$.

b. Find the zeros.

94. Show that $x - a$ is a factor of $x^n - a^n$ for any positive integer n and constant a.
Let $f(x) = x^n - a^n$. Then $f(a) = a^n - a^n = 0$. By the factor theorem, since $f(a) = 0$, then $x - a$ is a factor of $x^n - a^n$.

Write About It

95. Explain why a polynomial with real coefficients of degree 3 must have at least one real zero.

96. Why is it not necessary to apply the rational zero theorem, Descartes' rule of signs, or the upper and lower bound theorem to find the zeros of a second-degree polynomial?
The zeros of a second-degree polynomial can be found by applying the quadratic formula.

97. Explain why $f(x) = 5x^6 + 7x^4 + x^2 + 9$ has no real zeros. $f(x)$ has no variation in sign, nor does $f(-x)$. By Descartes' rule of signs, there are no positive or negative real zeros. Furthermore, 0 itself is not a zero of $f(x)$ because x is not a factor of $f(x)$. Therefore, there are no real zeros of $f(x)$.

98. Explain why the fundamental theorem of algebra does not apply to $f(x) = \sqrt{x} + 3$. That is, no complex number c exists such that $f(c) = 0$.   The expression $\sqrt{x} + 3$ is not a polynomial. The term $\sqrt{x} = x^{1/2}$ has an exponent that is not a positive integer.

Expanding Your Skills

99. Let n be a positive even integer. Determine the greatest number of possible nonreal zeros of $f(x) = x^n - 1$.
$n - 2$ possible nonreal zeros

100. Let n be a positive odd integer. Determine the greatest number of possible nonreal zeros of $f(x) = x^n - 1$.
$n - 1$ possible nonreal zeros

101. The front face of a tent is triangular and the height of the triangle is two-thirds of the base. The length of the tent is 3 ft more than the base of the triangular face. If the tent holds a volume of 108 ft³, determine its dimensions.

102. An underground storage tank for gasoline is in the shape of a right circular cylinder with hemispheres on each end. If the total volume of the tank is $\dfrac{104\pi}{3}$ ft³, find the radius of the tank.   The radius is 2 ft.

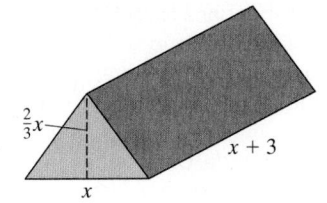

The triangular front has a base of 6 ft and a height of 4 ft. The length is 9 ft.

$\frac{2}{3}x$ $x + 3$ x

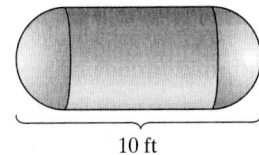

10 ft

103. A food company originally sells cereal in boxes with dimensions 10 in. by 7 in. by 2.5 in. To make more profit, the company decreases each dimension of the box by x inches but keeps the price the same. If the new volume is 81 in.³ by how much was each dimension decreased? *Each dimension was decreased by 1 in.*

104. A truck rental company rents a 12-ft by 8-ft by 6-ft truck for $69.95 per day plus mileage. A customer prefers to rent a less expensive smaller truck whose dimensions are x ft smaller on each side. If the volume of the smaller truck is 240 ft³, determine the dimensions of the smaller truck. *The smaller truck is 10 ft by 6 ft by 4 ft.*

105. A rectangle is bounded by the x-axis and a parabola defined by $y = 4 - x^2$. What are the dimensions of the rectangle if the area is 6 cm²? Assume that all units of length are in centimeters.

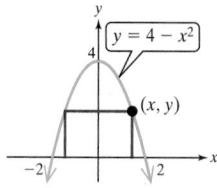

The dimensions are either 2 cm by 3 cm or $-1 + \sqrt{13}$ cm by $\dfrac{1 + \sqrt{13}}{2}$ cm.

106. A rectangle is bounded by the parabola defined by $y = x^2$, the x-axis, and the line $x = 5$ as shown in the figure. If the area of the rectangle is 12 in.² determine the dimensions of the rectangle.

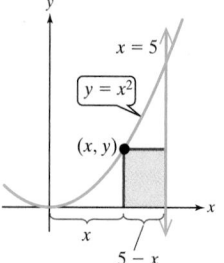

The rectangle is either 3 in. by 4 in. or $\dfrac{7 - \sqrt{33}}{2}$ in. by $\dfrac{21 + 3\sqrt{33}}{2}$ in.

The linear factorization theorem tells us that a polynomial of degree $n \geq 1$ factors into n linear factors over the complex numbers. If we do not factor over the set of complex numbers, then a polynomial with real coefficients can be factored into linear factors and irreducible quadratic factors. An *irreducible quadratic factor* is a quadratic polynomial that does not factor further over the set of real numbers.

For example, consider the polynomial $f(x) = x^4 - 5x^3 + 5x^2 + 25x - 26$.

Factoring over the real numbers, we have two linear factors and one irreducible quadratic factor:

$$\underbrace{}_{\substack{\text{2 linear}\\\text{factors}}} \quad \underbrace{}_{\substack{\text{irreducible}\\\text{quadratic factor}}}$$

$$x^4 - 5x^3 + 5x^2 + 25x - 26 = \overbrace{(x + 2)(x - 1)}\overbrace{(x^2 - 6x + 13)}$$

Factoring over the complex numbers, we have four linear factors as guaranteed by the linear factor theorem.

$$x^4 - 5x^3 + 5x^2 + 25x - 26 = (x + 2)(x - 1)[x - (3 + 2i)][x - (3 - 2i)]$$

For Exercises 107–110,

a. Factor the polynomial over the set of real numbers.

b. Factor the polynomial over the set of complex numbers.

107. $f(x) = x^4 + 2x^3 + x^2 + 8x - 12$ a. $(x + 3)(x - 1)(x^2 + 4)$
b. $(x + 3)(x - 1)(x + 2i)(x - 2i)$

108. $f(x) = x^4 - 6x^3 + 9x^2 - 6x + 8$ a. $(x - 4)(x - 2)(x^2 + 1)$
b. $(x - 4)(x - 2)(x + i)(x - i)$

109. $f(x) = x^4 + 2x^2 - 35$ a. $(x - \sqrt{5})(x + \sqrt{5})(x^2 + 7)$
b. $(x - \sqrt{5})(x + \sqrt{5})(x + \sqrt{7}i)(x - \sqrt{7}i)$

110. $f(x) = x^4 + 8x^2 - 33$ a. $(x - \sqrt{3})(x + \sqrt{3})(x^2 + 11)$
b. $(x - \sqrt{3})(x + \sqrt{3})(x + \sqrt{11}i)(x - \sqrt{11}i)$

111. Find all fourth roots of 1, by solving the equation $x^4 = 1$. (*Hint:* Find the zeros of the polynomial $f(x) = x^4 - 1$.)
The fourth roots of 1 are 1, -1, i, and $-i$.

112. Find all sixth roots of 1, by solving the equation $x^6 = 1$. [*Hint:* Find the zeros of the polynomial $f(x) = x^6 - 1$. Begin by factoring $x^6 - 1$ as $(x^3 - 1)(x^3 + 1)$.] *The sixth roots of 1 are 1, -1, $\dfrac{-1 \pm i\sqrt{3}}{2}$, and $\dfrac{1 \pm i\sqrt{3}}{2}$.*

113. Use the rational zero theorem to show that $\sqrt{5}$ is an irrational number. (*Hint:* Show that $f(x) = x^2 - 5$ has no rational zeros.)

114. a. Given a linear equation $ax + b = 0$ $(a \neq 0)$, the solution is given by $x = $ _____ $\dfrac{-b}{a}$.

b. Given a quadratic equation $ax^2 + bx + c = 0$ $(a \neq 0)$, the solutions are given by $x = $ _____ $\dfrac{-b \pm \sqrt{b^2 - 4ac}}{2a}$.

From Exercise 114, we see that linear and quadratic equations have generic formulas that can be used to find the solution sets. But what about a cubic polynomial equation? Mathematicians struggled for centuries to find such a formula. Finally, Italian mathematician Niccolo Tartaglia (1500–1557) developed a method to solve a cubic equation of the form

$$x^3 + mx = n$$

The result was later published in *Ars Magna*, by Gerolamo Cardano (1501–1576).

Foe Exercises 115–116, use the formula

$$x = \sqrt[3]{\sqrt{\left(\frac{n}{2}\right)^2 + \left(\frac{m}{3}\right)^3} + \frac{n}{2}} - \sqrt[3]{\sqrt{\left(\frac{n}{2}\right)^2 + \left(\frac{m}{3}\right)^3} - \frac{n}{2}}$$

to find a solution to the equation $x^3 + mx = n$.

115. $x^3 - 3x = -2$ -2

116. $x^3 + 9x = 26$ 2

Point of Interest

Early in the sixteenth century, Italian mathematicians Niccolo Tartaglia and Gerolamo Cardano solved a general cubic equation in terms of the constants appearing in the equation. Cardano's pupil, Ludovico Ferrari, then solved a general equation of fourth degree. Despite decades of work, no general solution to a fifth-degree equation was found. Finally, Norwegian mathematician Niels Abel and French mathematician Evariste Galois proved that no such solution exists.

SECTION 2.5   Rational Functions

1. Apply Notation Describing Infinite Behavior of a Function

In this chapter, we have studied polynomials and polynomial functions. Now we look at functions that are defined as the ratio of two polynomials. These are called rational functions.

Definition of a Rational Function

Let $p(x)$ and $q(x)$ be polynomials where $q(x) \neq 0$. A function f defined by

$$f(x) = \frac{p(x)}{q(x)} \text{ is called a } \textbf{rational function.}$$

Note: The domain of a rational function is all real numbers excluding the real zeros of $q(x)$.

Function	Factored Form	Domain
$f(x) = \dfrac{1}{x}$	$f(x) = \dfrac{1}{x}$	$\{x \mid x \neq 0\}$ $(-\infty, 0) \cup (0, \infty)$
$g(x) = \dfrac{5x^2}{2x^2 + 5x - 12}$	$g(x) = \dfrac{5x^2}{(2x - 3)(x + 4)}$	$\{x \mid x \neq \frac{3}{2}, x \neq -4\}$ $(-\infty, -4) \cup \left(-4, \frac{3}{2}\right) \cup \left(\frac{3}{2}, \infty\right)$
$k(x) = \dfrac{x + 3}{x^2 + 4}$	$k(x) = \dfrac{x + 3}{x^2 + 4}$	$\mathbb{R}$ $(-\infty, \infty)$

In this section, we will analyze rational functions. To do so, we want to determine the behavior of the function as x approaches ∞ or $-\infty$ and as x approaches values for which the function is undefined. We will use the following notation (Table 2-1).

Table 2-1

Notation	Meaning
$x \to c^+$	x approaches c from the right (but will not equal c).
$x \to c^-$	x approaches c from the left (but will not equal c).
$x \to \infty$	x approaches infinity (x increases without bound).
$x \to -\infty$	x approaches negative infinity (x decreases without bound).

For example, consider the reciprocal function $f(x) = \frac{1}{x}$ first introduced in Section 1.6 (Figure 2-18).

x	$f(x) = \frac{1}{x}$
-1	-1
-10	-0.1
-100	-0.01
-1000	-0.001

x	$f(x) = \frac{1}{x}$
-1	-1
-0.1	-10
-0.01	-100
-0.001	-1000

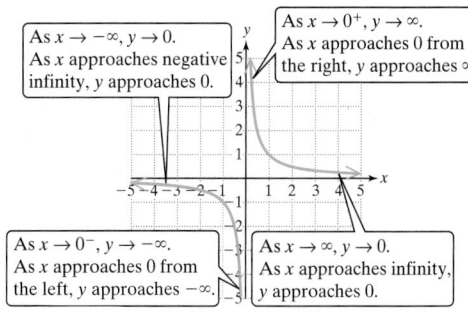

x	$f(x) = \frac{1}{x}$
1	1
0.1	10
0.01	100
0.001	1000

x	$f(x) = \frac{1}{x}$
1	1
10	0.1
100	0.01
1000	0.001

As $x \to -\infty$, $y \to 0$.
As x approaches negative infinity, y approaches 0.

As $x \to 0^+$, $y \to \infty$.
As x approaches 0 from the right, y approaches ∞.

As $x \to 0^-$, $y \to -\infty$.
As x approaches 0 from the left, y approaches $-\infty$.

As $x \to \infty$, $y \to 0$.
As x approaches infinity, y approaches 0.

Figure 2-18

In Example 1, we study the graph of another basic rational function, $f(x) = \dfrac{1}{x^2}$. From the definition of the function, we make the following observations.

- The domain of $f(x) = \dfrac{1}{x^2}$ is all real numbers excluding zero.

- f is an even function and the graph is symmetric to the y-axis.
- The values of $f(x)$ are positive over the domain of f.

Classroom Example: p. 316
Exercise 14

EXAMPLE 1   **Investigating the Behavior of a Rational Function**

The graph of $f(x) = \dfrac{1}{x^2}$ is given.

Complete the statements.

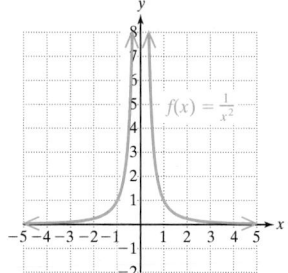

a. As $x \to -\infty, f(x) \to$ _____.
b. As $x \to 0^-,\ f(x) \to$ _____.
c. As $x \to 0^+,\ f(x) \to$ _____.
d. As $x \to \infty,\ f(x) \to$ _____.

Solution:

a. As $x \to -\infty, f(x) \to 0$.

b. As $x \to 0^-,\ f(x) \to \infty$.

c. As $x \to 0^+,\ f(x) \to \infty$.

d. As $x \to \infty,\ f(x) \to 0$.

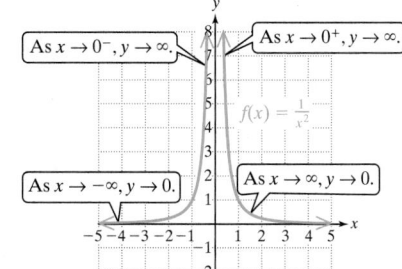

Skill Practice 1   The graph of $f(x) = \dfrac{1}{x - 2}$ is given.

Complete the statements.

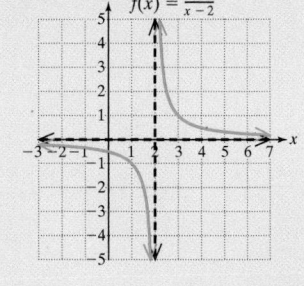

a. As $x \to -\infty, f(x) \to$ _____.
b. As $x \to 2^-,\ f(x) \to$ _____.
c. As $x \to 2^+,\ f(x) \to$ _____.
d. As $x \to \infty,\ f(x) \to$ _____.

---

2. Identify Vertical Asymptotes

The graphs of $f(x) = \dfrac{1}{x}$ and $f(x) = \dfrac{1}{x^2}$ both approach the y-axis, but do not touch the y-axis. The y-axis is called a vertical asymptote of the graphs of the functions.

Answers
1. a. 0   **b.** $-\infty$ **c.** ∞ **d.** 0

Definition of a Vertical Asymptote

The line $x = c$ is a **vertical asymptote** of the graph of a function f if $f(x)$ approaches infinity or negative infinity as x approaches c from either side.

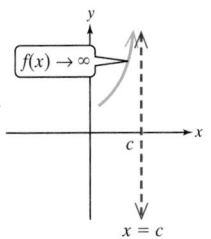
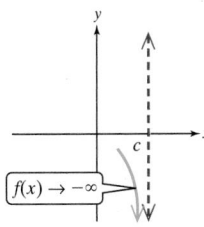
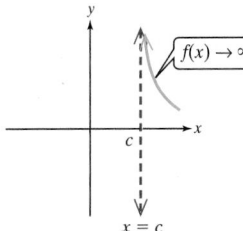
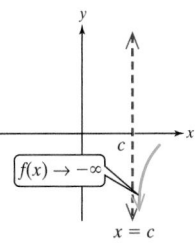

A function may have no vertical asymptotes, one vertical asymptote, or many vertical asymptotes. To locate the vertical asymptotes of a function, determine the real numbers x where the denominator is zero, but the numerator is nonzero.

Identifying Vertical Asymptotes of a Rational Function

TIP The case where $p(x)$ and $q(x)$ share a common factor is addressed in Exercises 111–114 on page 321.

Consider a rational function f defined by $f(x) = \dfrac{p(x)}{q(x)}$, where $p(x)$ and $q(x)$ have no common factors other than 1. If c is a real zero of $q(x)$, then $x = c$ is a vertical asymptote of the graph of f.

Classroom Examples: p. 317
Exercises 18, 20, 22

EXAMPLE 2   Identifying Vertical Asymptotes

Identify the vertical asymptotes.

a. $f(x) = \dfrac{2}{x - 3}$ **b.** $g(x) = \dfrac{x - 4}{3x^2 + 5x - 2}$ **c.** $k(x) = \dfrac{4x^2}{x^2 + 4}$

Solution:

a. $f(x) = \dfrac{2}{x - 3}$

The expression $\frac{2}{x-3}$ is written in lowest terms. The denominator is zero for $x = 3$.

f has a vertical asymptote of $x = 3$.

Avoiding Mistakes

A vertical asymptote is a line and should be identified by an equation of the form $x = c$, where c is a constant.

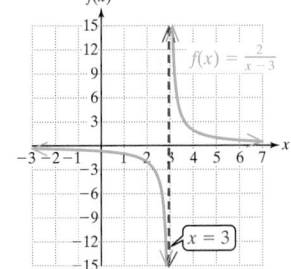

b. $g(x) = \dfrac{x - 4}{3x^2 + 5x - 2}$

Factor the numerator and denominator.

$= \dfrac{x - 4}{(3x - 1)(x + 2)}$

The numerator and denominator share no common factors other than 1. The zeros of the denominator are $\frac{1}{3}$ and -2.

The vertical asymptotes are

$x = \dfrac{1}{3}$ and $x = -2$.

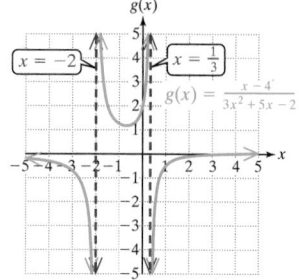

c. $k(x) = \dfrac{4x^2}{x^2 + 4}$

The numerator and denominator are already factored over the real numbers and the rational expression is in lowest terms.

The denominator has no real zeros, so the graph of k has no vertical asymptotes.

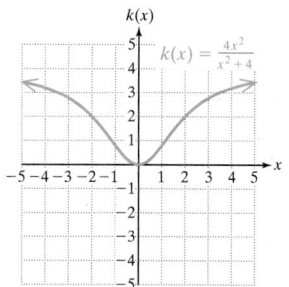

Skill Practice 2   Identify the vertical asymptotes.

a. $f(x) = \dfrac{3}{x + 1}$ **b.** $h(x) = \dfrac{x + 7}{2x^2 - x - 10}$ **c.** $m(x) = \dfrac{5x}{x^4 + 1}$

Instructor Note:
Emphasize to students that both the numerator and denominator should be factored to identify possible vertical asymptotes or "holes."

Avoiding Mistakes

The procedure to find the vertical asymptotes of a rational function is given under the condition that the numerator and denominator share no common factors. This important observation can be illustrated by the graph of

$f(x) = \dfrac{2x^2 + 5x + 3}{x + 1}$ (Figure 2-19). The numerator and

denominator share a common factor of $(x + 1)$.

$$f(x) = \dfrac{2x^2 + 5x + 3}{x + 1} = \dfrac{(2x + 3)(x + 1)}{x + 1}$$

The value $x = -1$ is not in the domain of f, but the graph of f has a "hole" at $x = -1$ rather than a vertical asymptote.

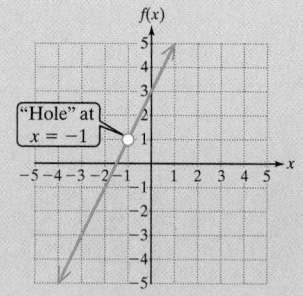

Figure 2-19

Answers

2. a. $x = -1$

b. $x = \dfrac{5}{2}$ and $x = -2$

c. No vertical asymptotes

3. Identify Horizontal Asymptotes

Refer to the graph of $f(x) = \frac{1}{x}$ (Figure 2-20). Toward the far left and far right of the graph, $f(x)$ approaches the line $y = 0$ (the x-axis). The x-axis is called a horizontal asymptote of the graph of f.

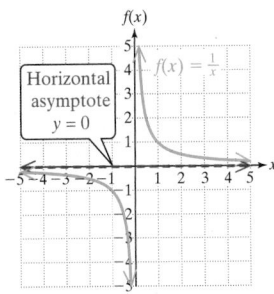

Figure 2-20

TIP While the graph of a function may not cross a vertical asymptote, it may cross a horizontal asymptote.

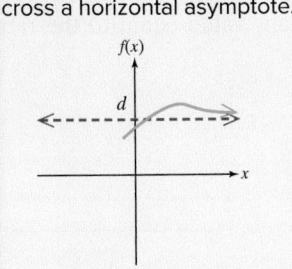

Definition of a Horizontal Asymptote

The line $y = d$ is a **horizontal asymptote** of the graph of a function f if $f(x)$ approaches d as x approaches infinity or negative infinity.

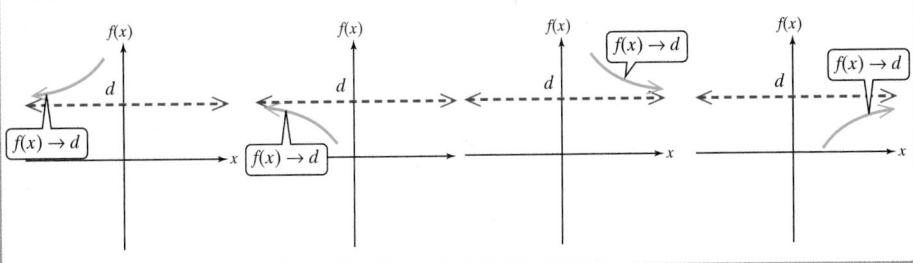

Recall that the leading term determines the far left and far right behavior of the graph of a polynomial function. Since a rational function is the ratio of two polynomials, it seems reasonable that the leading terms of the numerator and denominator determine the end behavior of a rational function.

TIP A rational function may have many vertical asymptotes, but at most one horizontal asymptote.

Identifying Horizontal Asymptotes of a Rational Function

Let f be a rational function defined by

$$f(x) = \frac{a_n x^n + a_{n-1} x^{n-1} + a_{n-2} x^{n-2} + \cdots + a_1 x + a_0}{b_m x^m + b_{m-1} x^{m-1} + b_{m-2} x^{m-2} + \cdots + b_1 x + b_0}$$

The definition of $f(x)$ indicates that n is the degree of the numerator and m is the degree of the denominator.

1. If $n > m$, then f has no horizontal asymptote.
2. If $n < m$, then the line $y = 0$ (the x-axis) is the horizontal asymptote of f.
3. If $n = m$, then the line $y = \dfrac{a_n}{b_m}$ is the horizontal asymptote of f.

1. If the degree of the numerator is greater than the degree of the denominator ($n > m$), then the numerator will "dominate" the quotient. For example:

$$f(x) = \frac{x^4 + \cdots}{x^2 + \cdots}$$ will behave like $y = x^2$ as $|x|$ becomes large. Therefore, f has no horizontal asymptote.

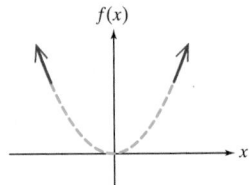

2. If the degree of the numerator is less than the degree of the denominator ($n < m$), then the denominator will "dominate" the quotient. For example:

$$f(x) = \frac{x^2 + \cdots}{x^4 + \cdots}$$ will behave like $y = \frac{1}{x^2}$. The ratio $\frac{1}{x^2}$ tends toward 0 as $|x|$ becomes large. Therefore, f has a horizontal asymptote of $y = 0$.

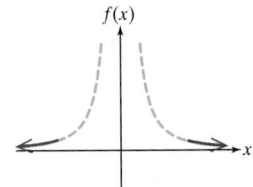

3. If the degree of the numerator is equal to the degree of the denominator ($n = m$), then the magnitude of the numerator and denominator somewhat "offset" each other. As a result, the function tends toward a constant value equal to the ratio of the leading coefficients. For example:

$$f(x) = \frac{4x^2 + \cdots}{3x^2 + \cdots}$$ will behave like $y = \frac{4}{3}$ as $|x|$ becomes large. Therefore, f has a horizontal asymptote of $y = \frac{4}{3}$.

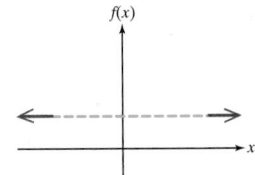

Classroom Examples: p. 317
Exercises 32, 34

EXAMPLE 3   Identifying Horizontal Asymptotes

Find the horizontal asymptotes (if any) for the given functions.

a. $f(x) = \dfrac{8x^2 + 1}{x^4 + 1}$ **b.** $g(x) = \dfrac{2x^3 - 6x}{x^2 + 4}$ **c.** $h(x) = \dfrac{8x^2 + 9x - 5}{2x^2 + 1}$

Solution:

a. $f(x) = \dfrac{8x^2 + 1}{x^4 + 1}$

The degree of the numerator is 2 ($n = 2$).
The degree of the denominator is 4 ($m = 4$).

Since $n < m$, then the line $y = 0$ (the x-axis) is a horizontal asymptote of f.

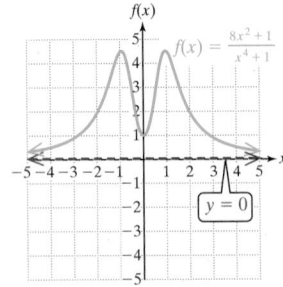

b. $g(x) = \dfrac{2x^3 - 6x}{x^2 + 4}$

The degree of the numerator is 3 ($n = 3$).
The degree of the denominator is 2 ($m = 2$).

Since $n > m$, then the function has no horizontal asymptotes.

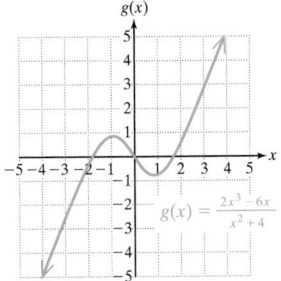

c. $h(x) = \dfrac{8x^2 + 9x - 5}{2x^2 + 1}$

The degree of the numerator is 2 ($n = 2$).
The degree of the denominator is 2 ($m = 2$).

Since $n = m$, then the line $y = \frac{8}{2}$ or equivalently $y = 4$ is a horizontal asymptote of the graph of f.

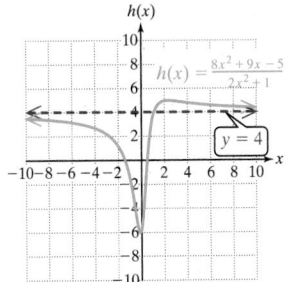

Skill Practice 3 Find the horizontal asymptotes (if any) for the given functions.

a. $f(x) = \dfrac{7x^2 + 2x}{4x^2 - 3}$ **b.** $m(x) = \dfrac{4x^3 + 2}{2x - 1}$ **c.** $n(x) = \dfrac{5}{4x^2 + 9}$

The graph of a rational function may not cross a vertical asymptote. However, as demonstrated in Example 3(c), the graph may cross a horizontal asymptote. For the purpose of graphing a rational function, it is helpful to determine where a graph crosses a horizontal asymptote.

Suppose that the line $y = d$ is a horizontal asymptote of a rational function $y = f(x)$. The solutions to the equation $f(x) = d$ are the values of x where the graph of f crosses its horizontal asymptote. If the equation has no real solution, then the graph does not cross its horizontal asymptote.

Classroom Example: p. 317
Exercises 30

EXAMPLE 4 Determining Where a Graph Crosses a Horizontal Asymptote

Given $h(x) = \dfrac{8x^2 + 9x - 5}{2x^2 + 1}$, determine the point where the graph of h crosses its horizontal asymptote.

Answers

3. a. $y = \dfrac{7}{4}$
  b. No horizontal asymptotes
  c. $y = 0$

Solution:

$$h(x) = \frac{8x^2 + 9x - 5}{2x^2 + 1}$$ From Example 3(c), the horizontal asymptote is $y = 4$.

$$\frac{8x^2 + 9x - 5}{2x^2 + 1} = 4$$ Set $h(x)$ equal to 4.

$$\frac{8x^2 + 9x - 5}{2x^2 + 1} \cdot (2x^2 + 1) = 4 \cdot (2x^2 + 1)$$ Clear fractions by multiplying by the LCD.

$$8x^2 + 9x - 5 = 8x^2 + 4$$
$$9x - 5 = 4$$
$$x = 1$$

The function crosses its horizontal asymptote at $(1, 4)$.

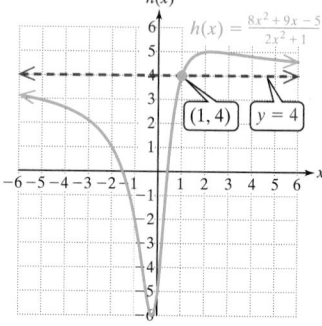

Answer

4. Horizontal asymptote $y = 3$;
   Crosses at (3, 3)

Skill Practice 4   Given $g(x) = \dfrac{3x^2 + 4x - 3}{x^2 + 3}$, determine the horizontal asymptote and the point where the graph crosses the horizontal asymptote.

4. Identify Slant Asymptotes

Consider the function defined by $f(x) = \dfrac{x^2 + 1}{x - 1}$. The graph has a vertical asymptote of $x = 1$, but no horizontal asymptote (the degree of the numerator is greater than the degree of the denominator). However, as x approaches infinity and negative infinity, the graph approaches the graph of $y = x + 1$ (shown in red in Figure 2-21). This line is called a **slant asymptote** because it is neither horizontal nor vertical.

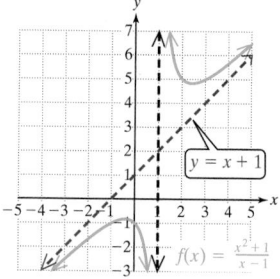

Figure 2-21

Identifying a Slant Asymptote of a Rational Function

- A rational function will have a slant asymptote if the degree of the numerator is exactly one greater than the degree of the denominator.
- To find an equation of a slant asymptote, divide the numerator of the function by the denominator. The quotient will be linear and the slant asymptote will be of the form $y =$ quotient.

TIP Because the divisor is of the form $x - c$, we can also use synthetic division to determine the slant asymptote.

$$\begin{array}{r} 1\rfloor \;\; 1 \quad 0 \quad 1 \\ \underline{\quad\;\; 1 \quad 1} \\ 1 \quad 1 \;\; \lfloor 2 \end{array}$$

The quotient is $x + 1$. The slant asymptote is $y = x + 1$.

For $f(x) = \dfrac{x^2 + 1}{x - 1}$, divide $(x^2 + 1)$ by $(x - 1)$ using long division or synthetic division.

$$\begin{array}{r} x + 1 \\ x - 1 \overline{) x^2 + 0x + 1} \\ \underline{-(x^2 - x)} \\ x + 1 \\ \underline{-(x - 1)} \\ 2 \end{array}$$

The slant asymptote is $y = x + 1$.

Using the division algorithm, $f(x) = \dfrac{x^2 + 1}{x - 1} = x + 1 + \dfrac{2}{x - 1}$. The expression $\dfrac{2}{x - 1}$ will approach 0 as $|x|$ approaches infinity. Therefore, $f(x)$ will approach the line $y = x + 1$ as $|x|$ approaches infinity.

Classroom Example: p. 318
Exercise 42

EXAMPLE 5   **Identifying the Asymptotes of a Rational Function**

Determine the asymptotes. $f(x) = \dfrac{2x^2 - 5x - 3}{x - 2}$

Solution:

$f(x) = \dfrac{2x^2 - 5x - 3}{x - 2}$

f has a vertical asymptote of $x = 2$.

The expression $\dfrac{2x^2 - 5x - 3}{x - 2} = \dfrac{(2x + 1)(x - 3)}{x - 2}$ is in lowest terms, and the denominator is zero at $x = 2$.

f has no horizontal asymptote.

The degree of the numerator is exactly one greater than the degree of the denominator. Therefore, f has no horizontal asymptote, but does have a slant asymptote.

To find the slant asymptote divide $(2x^2 - 5x - 3)$ by $(x - 2)$.

$$
\begin{array}{r}
2x - 1 \\
x - 2 \overline{\smash{)}\, 2x^2 - 5x - 3} \\
-(2x^2 - 4x) \\
\hline
-x - 3 \\
-(-x + 2) \\
\hline
-5
\end{array}
$$

The quotient is $2x - 1$.
The slant asymptote is given by $y = 2x - 1$.

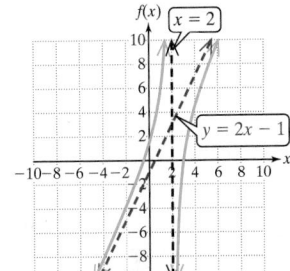

Skill Practice 5   Determine the asymptotes. $g(x) = \dfrac{2x^2 - 9}{x + 1}$

5. Graph Rational Functions

We now turn our attention to graphing rational functions. The transformations used in Section 1.6 can be applied to the basic rational functions $y = \dfrac{1}{x}$ and $y = \dfrac{1}{x^2}$.

Classroom Example: p. 318
Exercise 54

EXAMPLE 6   **Using Transformations to Graph a Rational Function**

Use transformations to graph $f(x) = \dfrac{1}{(x + 2)^2} + 3$.

Answer

5. Vertical asymptote: $x = -1$;
No horizontal asymptote; Slant asymptote: $y = 2x - 2$

Solution:

$$f(x) = \frac{1}{(x+2)^2} + 3$$

The graph of f is the graph of $y = \dfrac{1}{x^2}$ with a shift to the left 2 units and a shift upward 3 units.

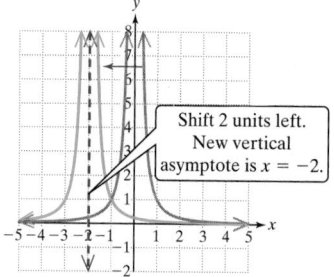

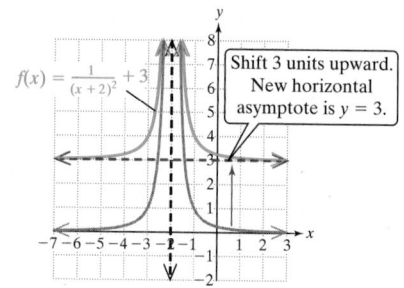

> **Skill Practice 6** Use transformations to graph $g(x) = \dfrac{1}{x-3} - 2$.

To graph a rational function that is not a simple transformation of $y = \dfrac{1}{x}$ or $y = \dfrac{1}{x^2}$, more steps must be employed. Our strategy is to find all asymptotes and key points (intercepts and points where the function crosses a horizontal asymptote). Then determine the behavior of the function on the intervals defined by these key points and the vertical asymptotes.

Graphing a Rational Function

Consider a rational function f defined by $f(x) = \dfrac{p(x)}{q(x)}$, where $p(x)$ and $q(x)$ are polynomials with no common factors.

1. Determine the y-intercept by evaluating $f(0)$.
2. Determine the x-intercept(s) by finding the real solutions of $f(x) = 0$. The value $f(x)$ equals zero when the numerator $p(x) = 0$.
3. Identify any vertical asymptotes and graph them as dashed lines.
4. Determine whether the function has a horizontal asymptote or a slant asymptote (or neither), and graph the asymptote as a dashed line.
5. Determine where the function crosses the horizontal or slant asymptote (if applicable).
6. If a test for symmetry is easy to apply, use symmetry to plot additional points. Recall:
   - f is an even function (symmetric to the y-axis) if $f(-x) = f(x)$.
   - f is an odd function (symmetric to the origin) if $f(-x) = -f(x)$.
7. Plot at least one point on the intervals defined by the x-intercepts, vertical asymptotes, and points where the function crosses a horizontal or slant asymptote.
8. Sketch the function based on the information found in steps 1–7.

Answer

6.

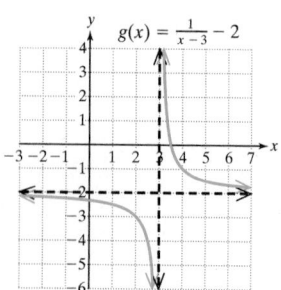

Classroom Example: p. 319
Exercise 70

EXAMPLE 7   Graphing a Rational Function

Graph $f(x) = \dfrac{x + 3}{x - 2}$.

Solution:

1. **Determine the y-intercept.**

   $$f(0) = \frac{(0) + 3}{(0) - 2} = -\frac{3}{2}$$

   The y-intercept is $\left(0, -\frac{3}{2}\right)$.

2. **Determine the x-intercept(s).**

   $\dfrac{x + 3}{x - 2} = 0$ when $x + 3 = 0$ or $x = -3$.

   The x-intercept is $(-3, 0)$.

3. **Identify the vertical asymptotes.**

   The polynomial $x - 2$ has a zero at $x = 2$, and the numerator $x + 3$ is nonzero for $x = 2$.

   The vertical asymptotes occur at the values of x for which the denominator is zero and the numerator is nonzero.

   The graph has one vertical asymptote, $x = 2$.

4. **Determine whether f has a horizontal or slant asymptote.**

   The degree of the numerator is equal to the degree of the denominator. Therefore, the graph has a horizontal asymptote given by the ratio of leading coefficients of the numerator and denominator.

   $y = \dfrac{1}{1} = 1$ is the horizontal asymptote.

5. **Determine where f crosses its horizontal asymptote (if at all).**

   Solve the equation $f(x) = 1$.

   $$\frac{x + 3}{x - 2} = 1 \Rightarrow x + 3 = x - 2 \Rightarrow 3 = -2 \text{ (contradiction)}$$

   The graph of f does not cross its horizontal asymptote.

6. **Test for symmetry.**

   $f(-x) = \dfrac{-x + 3}{-x - 2}$ does not equal $f(x)$ or $-f(x)$. The function is neither even nor odd and is not symmetric with respect to the y-axis or origin.

7. **Determine the behavior of f on each interval.**

   Determine the sign of the function on the intervals (shown in red) defined by the x-intercept at $x = -3$ and the vertical asymptote $x = 2$.

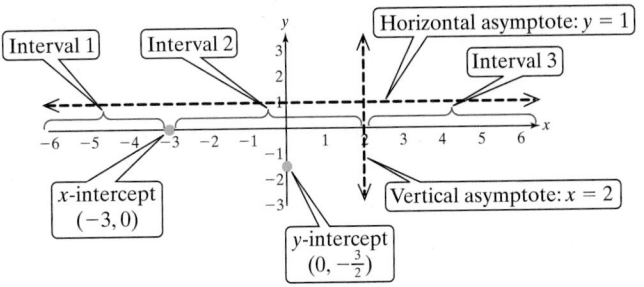

Interval	Test Point	Comments
$(-\infty, -3)$	$\left(-4, \frac{1}{6}\right)$	• Since $f(x)$ is positive on this interval, $f(x)$ must approach the horizontal asymptote $y = 1$ from below as $x \to -\infty$.
$(-3, 2)$	$\left(0, -\frac{3}{2}\right)$	• Since $f(x)$ is negative on this interval, the graph crosses the x-axis at the intercept $(-3, 0)$ and continues downward (through the y-intercept). As x approaches the vertical asymptote $x = 2$ from the left, $f(x) \to -\infty$.
$(2, \infty)$	$(3, 6)$	• Since $f(x)$ is positive on this interval, as x approaches the vertical asymptote $x = 2$ from the right, $f(x) \to \infty$. • Since $f(x)$ is positive on this interval, $f(x)$ must approach the horizontal asymptote from above as $x \to \infty$.

8. Sketch the function (Figure 2-22).

Plot the x- and y-intercepts $(-3, 0)$ and $\left(0, -\frac{3}{2}\right)$, and the additional points $\left(-4, \frac{1}{6}\right)$ and $(3, 6)$.

Graph the horizontal asymptote ($y = 1$) and vertical asymptote $x = 2$ as dashed lines.

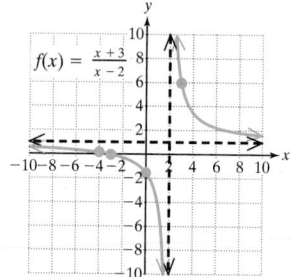

Figure 2-22

> **TIP** Using long division, we can rewrite $f(x) = \frac{x+3}{x-2}$ as $f(x) = 1 + \frac{5}{x-2}$. Then we can graph f by shifting the graph of $y = \frac{1}{x}$ two units to the right, stretching by a factor of 5, and shifting the graph upward 1 unit.

Skill Practice 7   Graph $f(x) = \dfrac{x-1}{x+4}$.

Classroom Example: p. 319
Exercise 74

EXAMPLE 8   Graphing a Rational Function

Graph $f(x) = \dfrac{4x}{x^2 - 4}$.

Solution:

1. Determine the y-intercept.

$$f(0) = \frac{4(0)}{(0)^2 - 4} = 0 \qquad \text{The } y\text{-intercept is } (0, 0).$$

2. Determine the x-intercept(s).

$$\frac{4x}{x^2 - 4} = 0 \quad \text{for } x = 0. \qquad \text{The } x\text{-intercept is } (0, 0).$$

3. Identify the vertical asymptotes.

The zeros of $x^2 - 4$ are 2 and -2. Vertical asymptotes: $x = 2$ and $x = -2$.

4. Determine whether f has a horizontal or slant asymptote.

The degree of the numerator is less than the degree of the denominator. The horizontal asymptote is $y = 0$.

Answer

7.

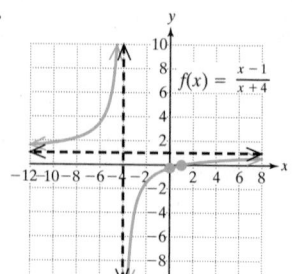

5. Determine where f crosses its horizontal asymptote.

Set $f(x) = 0$. We have $\dfrac{4x}{x^2 - 4} = 0$ for $x = 0$.

Therefore, f crosses its horizontal asymptote at $(0, 0)$.

6. Test for symmetry.

f is an odd function because $f(-x) = \dfrac{4(-x)}{(-x)^2 - 4} = -\dfrac{4x}{x^2 - 4} = -f(x)$.

7. Determine the behavior of f on each interval.

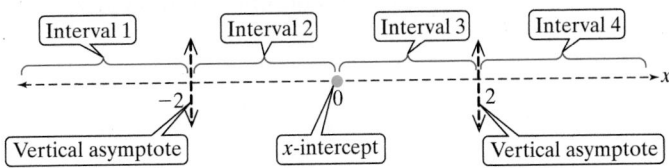

Interval	Test Point	Comments
$(-\infty, -2)$	$\left(-3, -\dfrac{12}{5}\right)$	• Since $f(x)$ is negative on this interval, $f(x)$ must approach the horizontal asymptote $y = 0$ from below as $x \to -\infty$. • Since $f(x)$ is negative on this interval, as x approaches the vertical asymptote $x = -2$ from the left, $f(x) \to -\infty$.
$(-2, 0)$	$\left(-1, \dfrac{4}{3}\right)$	• Since $f(x)$ is positive on this interval, as x approaches the vertical asymptote $x = -2$ from the right, $f(x) \to \infty$.
$(0, 2)$	$\left(1, -\dfrac{4}{3}\right)$	• Since $f(x)$ is negative on this interval, as x approaches the vertical asymptote $x = 2$ from the left, $f(x) \to -\infty$.
$(2, \infty)$	$\left(3, \dfrac{12}{5}\right)$	• Since $f(x)$ is positive on this interval, $f(x)$ must approach the horizontal asymptote from above as $x \to \infty$. • Since $f(x)$ is positive on this interval, as x approaches the vertical asymptote $x = 2$ from the right, $f(x) \to \infty$.

TIP The graph of f is symmetric to the origin because $f(-x) = -f(x)$. Therefore, if $\left(-3, -\dfrac{12}{5}\right)$ and $\left(-1, \dfrac{4}{3}\right)$ are points on the graph of f, then $\left(3, \dfrac{12}{5}\right)$ and $\left(1, -\dfrac{4}{3}\right)$ are also points on the graph.

8. Sketch the function.

Plot the x- and y-intercept $(0, 0)$.

Graph the asymptotes as dashed lines.

Plot the points.

$$\left(-3, -\frac{12}{5}\right), \left(-1, \frac{4}{3}\right), \left(1, -\frac{4}{3}\right), \text{ and } \left(3, \frac{12}{5}\right)$$

Sketch the curve.

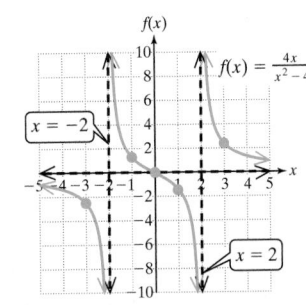

Skill Practice 8   Graph $g(x) = \dfrac{-5x}{x^2 - 9}$.

Answer

8.

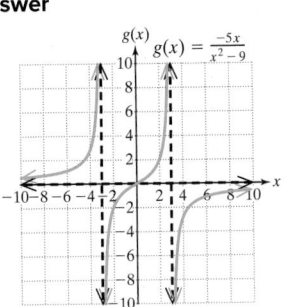

Classroom Example: p. 319
Exercise 80

EXAMPLE 9   Graphing a Rational Function

Graph $g(x) = \dfrac{2x^2 - 3x - 5}{x^2 + 1}$.

Solution:

1. Determine the y-intercept.

$$g(0) = \frac{2(0)^2 - 3(0) - 5}{(0)^2 + 1} = -5 \qquad \text{The y-intercept is } (0, -5).$$

2. Determine the x-intercept(s).

$$\frac{2x^2 - 3x - 5}{x^2 + 1} = 0$$
$$2x^2 - 3x - 5 = 0$$
$$(2x - 5)(x + 1) = 0$$
$$x = \frac{5}{2}, \; x = -1 \qquad \text{The x-intercepts are } \left(\tfrac{5}{2}, 0\right) \text{ and } (-1, 0).$$

3. Identify the vertical asymptotes.

$x^2 + 1$ is nonzero for all real numbers. The graph of g has no vertical asymptotes.

4. Determine whether g has a horizontal or slant asymptote.

The degree of the numerator is equal to the degree of the denominator. The horizontal asymptote is $y = \frac{2}{1}$ or simply $y = 2$.

5. Determine where g crosses its horizontal asymptote.

Find the real solutions to the equation $\dfrac{2x^2 - 3x - 5}{x^2 + 1} = 2$.

$$2x^2 - 3x - 5 = 2(x^2 + 1)$$
$$2x^2 - 3x - 5 = 2x^2 + 2$$
$$-3x - 5 = 2$$
$$x = -\frac{7}{3} \qquad g \text{ crosses its horizontal asymptote at } \left(-\frac{7}{3}, 2\right).$$

6. Test for symmetry.

g is neither even nor odd because $g(-x) \neq g(x)$ and $g(-x) \neq -g(x)$.

7. Determine the behavior of g on each interval.

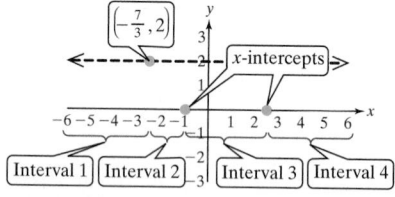

Interval	Test Point	Comments
$\left(-\infty, -\dfrac{7}{3}\right)$	$\left(-3, \dfrac{11}{5}\right)$	• Since $g(-3) = \frac{11}{5}$ is above the horizontal asymptote $y = 2$, $g(x)$ must approach the horizontal asymptote from above as $x \to -\infty$.
$\left(-\dfrac{7}{3}, -1\right)$	$\left(-2, \dfrac{9}{5}\right)$	• Plot the point $\left(-2, \frac{9}{5}\right)$ between the horizontal asymptote and the x-intercept of $(-1, 0)$.
$\left(-1, \dfrac{5}{2}\right)$	$(0, -5)$	• The point $(0, -5)$ is the y-intercept.
$\left(\dfrac{5}{2}, \infty\right)$	$\left(3, \dfrac{2}{5}\right)$	• Since $g(3) = \frac{2}{5}$ is below the horizontal asymptote $y = 2$, $g(x)$ must approach the horizontal asymptote from below as $x \to \infty$.

8. Sketch the function.

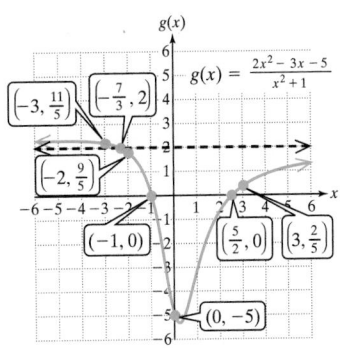

$g(x) = \dfrac{2x^2 - 3x - 5}{x^2 + 1}$

$\left(-3, \dfrac{11}{5}\right)$ $\left(-\dfrac{7}{3}, 2\right)$ $\left(-2, \dfrac{9}{5}\right)$ $(-1, 0)$ $\left(\dfrac{5}{2}, 0\right)$ $\left(3, \dfrac{2}{5}\right)$ $(0, -5)$

Skill Practice 9   Graph $g(x) = \dfrac{4x^2 + 7x - 2}{x^2 + 4}$.

Classroom Example: p. 319
Exercise 84

EXAMPLE 10   **Graphing a Rational Function**

Graph. $h(x) = \dfrac{2x^2 + 9x + 4}{x + 3}$

Solution:

1. Determine the y-intercept.

$$h(0) = \frac{2(0)^2 + 9(0) + 4}{(0) + 3} = \frac{4}{3}$$

The y-intercept is $\left(0, \frac{4}{3}\right)$.

2. Determine the x-intercept(s).

$$\frac{2x^2 + 9x + 4}{x + 3} = 0$$

$$2x^2 + 9x + 4 = 0$$

$$(2x + 1)(x + 4) = 0$$

$$x = -\frac{1}{2}, \ x = -4$$

The x-intercepts are $\left(-\frac{1}{2}, 0\right)$ and $(-4, 0)$.

3. Identify the vertical asymptotes.

$\dfrac{2x^2 + 9x + 4}{x + 3}$ is in lowest terms, and $x + 3$ is zero for $x = -3$.

4. Determine whether h has a horizontal or slant asymptote.

The degree of the numerator is one greater than the degree of the denominator. To find the slant asymptote, divide $(2x^2 + 9x + 4)$ by $(x + 3)$.

$$\begin{array}{r|rrr} -3 & 2 & 9 & 4 \\ & & -6 & -9 \\ \hline & 2 & 3 & \underline{-5} \end{array}$$

The quotient is $2x + 3$.
The slant asymptote is $y = 2x + 3$.

5. Determine where h will cross the slant asymptote.

Set $h(x) = 2x + 3$. We have $\dfrac{2x^2 + 9x + 4}{x + 3} = 2x + 3$.

$$2x^2 + 9x + 4 = (2x + 3)(x + 3)$$
$$2x^2 + 9x + 4 = 2x^2 + 9x + 9$$
$$4 = 9 \quad \text{(No solution)}$$

> The equation has no solution. Therefore, the graph does not cross the slant asymptote.

Answer

9.

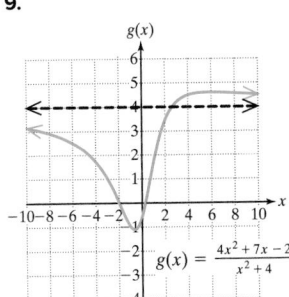

$g(x) = \dfrac{4x^2 + 7x - 2}{x^2 + 4}$

6. Test for symmetry.

h is neither even nor odd because $h(-x) \neq h(x)$ and $h(-x) \neq -h(x)$.

7. Plot test points. Pick values of x on the intervals defined by the x-intercepts and vertical asymptote.

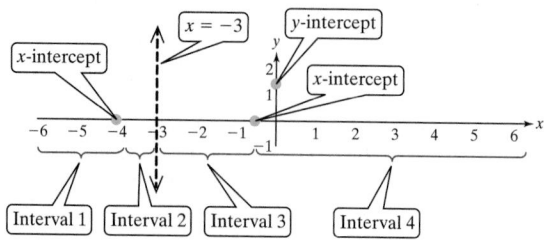

Select test points from each interval.

The graph of $y = h(x)$ passes through $(-5, -4.5)$, $(-3.5, 6)$, $(-2, -6)$, and $(2, 6)$.

8. Sketch the graph.

Plot the x-intercepts: $(-4, 0)$ and $\left(-\frac{1}{2}, 0\right)$.

Plot the y-intercept: $\left(0, \frac{4}{3}\right)$.

Graph the asymptotes as dashed lines.
Plot the points:
$(-5, -4.5)$, $(-3.5, 6)$, $(-2, -6)$, and $(2, 6)$.

Sketch the graph.

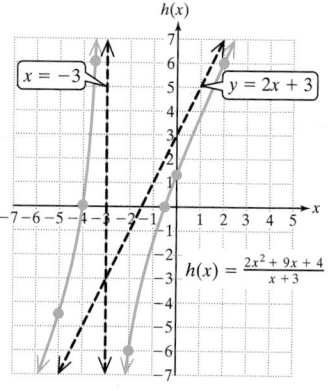

Skill Practice 10   Graph $k(x) = \dfrac{2x^2 - 7x + 3}{x - 2}$.

6. Use Rational Functions in Applications

In Section 1.5, we presented a linear model for the cost for a business to manufacture x items. The model is $C(x) = mx + b$, where m is the variable cost to produce an individual item, and b is the fixed cost.

The average cost $\overline{C}(x)$ per item manufactured is the sum of all costs (variable and fixed) divided by the total number of items produced x. This is given by

$$\overline{C}(x) = \frac{C(x)}{x}$$

The average cost per item will decrease as more items are produced because the fixed cost will be distributed over a greater number of items. This is demonstrated in Example 11.

TIP Recall that variable costs include items such as materials and labor. Fixed costs include overhead costs such as rent and utilities.

Answer

10.

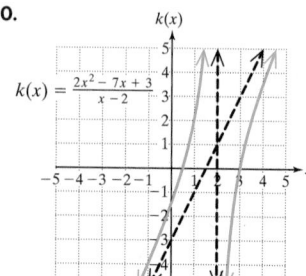

Classroom Example: p. 319
Exercise 92

EXAMPLE 11 Investigating Average Cost

A cleaning service cleans homes. For each house call, the cost to the company is approximately $40 for cleaning supplies, gasoline, and labor. The business also has fixed monthly costs of $300 from phone service, advertising, and depreciation on the vehicles.

a. Write a cost function to represent the cost $C(x)$ (in dollars) for x house calls per month.

b. Write the average cost function that represents the average cost $\overline{C}(x)$ (in $) for x house calls per month.

c. Evaluate $\overline{C}(5)$, $\overline{C}(20)$, $\overline{C}(30)$, and $\overline{C}(100)$.

d. The cleaning service can realistically make a maximum of 160 calls per month. However, if the number of calls were unlimited, what value would the average cost approach? What does this mean in the context of the problem?

Solution:

a. $C(x) = 40x + 300$ The variable cost is $40 per call ($m = 40$), and the fixed cost is $300 ($b = 300$). $C(x) = mx + b$.

b. $\overline{C}(x) = \dfrac{40x + 300}{x}$ The average cost per item is the total cost divided by the total number of items produced.

c. $\overline{C}(5) = 100$ The average cost per house call is $100 if 5 calls are made.

$\overline{C}(20) = 55$ The average cost per house call is $55 if 20 calls are made.

$\overline{C}(30) = 50$ The average cost per house call is $50 if 30 calls are made.

$\overline{C}(100) = 43$ The average cost per house call is $43 if 100 calls are made.

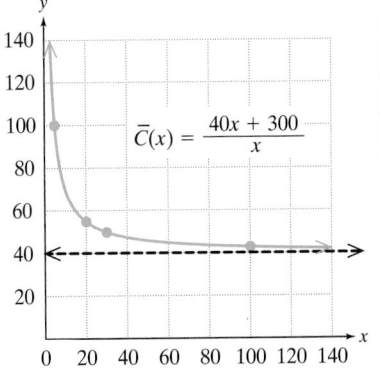

d. As x approaches infinity, $\overline{C}(x)$ will approach its horizontal asymptote $y = 40$. This is the cost per house call in the absence of other fixed costs.

Answers

11. a. $C(x) = 50x + 200$

b. $\overline{C}(x) = \dfrac{50x + 200}{x}$

c. $\overline{C}(5) = 90$; $\overline{C}(20) = 60$; $\overline{C}(30) = 56.67$; $\overline{C}(100) = 52$

d. The average cost $\overline{C}(x)$ will approach $50 per house call.

Skill Practice 11 Repeat Example 11 under the assumption that the company cuts its fixed costs to $200 per month and pays its employees more, leading to a variable cost per house call of $50.

SECTION 2.5   Practice Exercises

Prerequisite Review

For Exercises R.1–R.3, simplify the expression and state the restrictions on the variable.

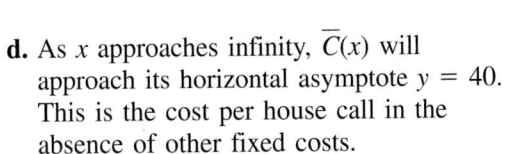

R.1. $\dfrac{q^2 - 36}{q^2 - 4q - 12}$ $\dfrac{q+6}{q+2}; q \neq 6, q \neq -2$ **R.2.** $\dfrac{9 - u^2}{2u^2 - 6u}$ $-\dfrac{u+3}{2u}; u \neq 0, u \neq 3$ **R.3.** $\dfrac{3p - 3m}{xm - xp + 3m - 3p}$ $-\dfrac{3}{x+3}; m \neq p, x \neq -3$

Concept Connections

1. The domain of a rational function defined by $f(x) = \dfrac{p(x)}{q(x)}$ is all real numbers excluding the zeros of _____ $q(x)$ _____.

2. The notation $x \to \infty$ is read as __x approaches infinity__.

3. The notation $x \to 5^-$ is read as __x approaches 5 from the left__.

4. The line $x = c$ is a _____ vertical _____ asymptote of the graph of a function f if $f(x)$ approaches infinity or negative infinity as x approaches _____ c _____ from either the left or right.

5. To locate the vertical asymptotes of a function, determine the real numbers x where the denominator is zero, but the numerator is _____ nonzero _____.

6. Consider a rational function in which the degree of the numerator is n and the degree of the denominator is m. If n _____ $<$ _____ m, then the x-axis is the horizontal asymptote. If n _____ $>$ _____ m, then the function has no horizontal asymptote.

Objective 1: Apply Notation Describing Infinite Behavior of a Function

For Exercises 7–12, write the domain of the function in interval notation.

7. $f(x) = \dfrac{x^2 - 25}{x - 5}$ $(-\infty, 5) \cup (5, \infty)$

8. $g(x) = \dfrac{x^2 - 9}{x - 3}$ $(-\infty, 3) \cup (3, \infty)$

9. $r(x) = \dfrac{2x - 3}{4x^2 + 3x - 1}$ $(-\infty, -1) \cup \left(-1, \frac{1}{4}\right) \cup \left(\frac{1}{4}, \infty\right)$

10. $p(x) = \dfrac{3x - 5}{2x^2 + 5x - 7}$ $\left(-\infty, -\frac{7}{2}\right) \cup \left(-\frac{7}{2}, 1\right) \cup (1, \infty)$

11. $h(x) = \dfrac{18x}{x^2 + 100}$ $(-\infty, \infty)$

12. $k(x) = \dfrac{14}{x^2 + 49}$ $(-\infty, \infty)$

For Exercises 13–16, refer to the graph of the function and complete the statement. (See Example 1)

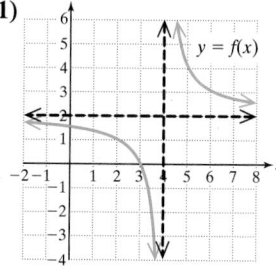

13. a. As $x \to -\infty$, $f(x) \to$ ____ 2 ____.
  c. As $x \to 4^+$, $f(x) \to$ ____ ∞ ____.
  e. The graph is increasing over the interval(s) Never increasing.
  g. The domain is $(-\infty, 4) \cup (4, \infty)$
  i. The vertical asymptote is the line ____ $x = 4$ ____.

  b. As $x \to 4^-$, $f(x) \to$ ____ $-\infty$ ____.
  d. As $x \to \infty$, $f(x) \to$ ____ 2 ____.
  f. The graph is decreasing over the interval(s) $(-\infty, 4) \cup (4, \infty)$
  h. The range is $(-\infty, 2) \cup (2, \infty)$
  j. The horizontal asymptote is the line ____ $y = 2$ ____.

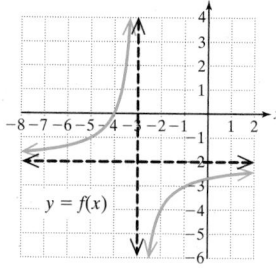

14. a. As $x \to -\infty$, $f(x) \to$ ____ -2 ____.
  c. As $x \to -3^+$, $f(x) \to$ ____ $-\infty$ ____.
  e. The graph is increasing over the interval(s) $(-\infty, -3) \cup (-3, \infty)$
  g. The domain is $(-\infty, -3) \cup (-3, \infty)$
  i. The vertical asymptote is the line ____ $x = -3$ ____.

  b. As $x \to -3^-$, $f(x) \to$ ____ ∞ ____.
  d. As $x \to \infty$, $f(x) \to$ ____ -2 ____.
  f. The graph is decreasing over the interval(s) Never decreasing.
  h. The range is $(-\infty, -2) \cup (-2, \infty)$.
  j. The horizontal asymptote is the line ____ $y = -2$ ____.

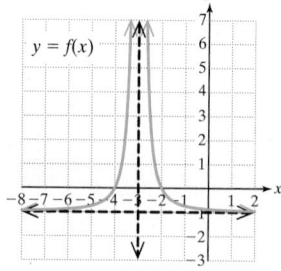

15. a. As $x \to -\infty$, $f(x) \to$ ____ -1 ____.
  c. As $x \to -3^+$, $f(x) \to$ ____ ∞ ____.
  e. The graph is increasing over the interval(s) ____ $(-\infty, -3)$ ____.
  g. The domain is $(-\infty, -3) \cup (-3, \infty)$
  i. The vertical asymptote is the line ____ $x = -3$ ____.

  b. As $x \to -3^-$, $f(x) \to$ ____ ∞ ____.
  d. As $x \to \infty$, $f(x) \to$ ____ -1 ____.
  f. The graph is decreasing over the interval(s) ____ $(-3, \infty)$ ____.
  h. The range is ____ $(-1, \infty)$ ____.
  j. The horizontal asymptote is the line ____ $y = -1$ ____.

16. a. As $x \to -\infty$, $f(x) \to$ ___4___.

b. As $x \to 1^-$, $f(x) \to$ ___$-\infty$___.

c. As $x \to 1^+$, $f(x) \to$ ___$-\infty$___.

d. As $x \to \infty$, $f(x) \to$ ___4___.

e. The graph is increasing over the interval(s) ___$(1, \infty)$___.

f. The graph is decreasing over the interval(s) ___$(-\infty, 1)$___.

g. The domain is ___$(-\infty, 1) \cup (1, \infty)$___

h. The range is ___$(-\infty, 4)$___.

i. The vertical asymptote is the line ___$x = 1$___.

j. The horizontal asymptote is the line ___$y = 4$___.

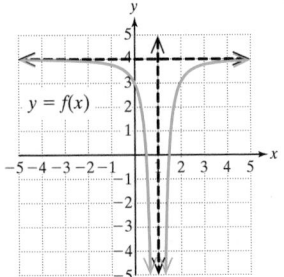

Objective 2: Identify Vertical Asymptotes

For Exercises 17–24, determine the vertical asymptotes of the graph of the function. **(See Example 2)**

17. $f(x) = \dfrac{8}{x - 4}$ $x = 4$

18. $g(x) = \dfrac{2}{x + 7}$ $x = -7$

19. $h(x) = \dfrac{x - 3}{2x^2 - 9x - 5}$
$x = 5$ and $x = -\dfrac{1}{2}$

20. $k(x) = \dfrac{x + 2}{3x^2 + 8x - 3}$
$x = -3$ and $x = \dfrac{1}{3}$

21. $m(x) = \dfrac{x}{x^2 + 5}$ None

22. $n(x) = \dfrac{6}{x^4 + 1}$ None

23. $f(t) = \dfrac{t^2 + 2}{2t^2 + 4t - 3}$
$t = \dfrac{-2 + \sqrt{10}}{2}$ and $t = \dfrac{-2 - \sqrt{10}}{2}$

24. $k(a) = \dfrac{5 + a^4}{3a^2 + 4a - 1}$
$a = \dfrac{-2 + \sqrt{7}}{3}$ and $a = \dfrac{-2 - \sqrt{7}}{3}$

Objective 3: Identify Horizontal Asymptotes

25. The graph of $f(x) = \dfrac{-x^2 + 8}{2x^2 - 3}$ will behave like which function for large values of $|x|$?   a

a. $y = -\dfrac{1}{2}$

b. $y = -\dfrac{x}{2}$

c. $y = -\dfrac{8}{3}$

d. $y = -\dfrac{1}{2}x - \dfrac{8}{3}$

26. The graph of $f(x) = \dfrac{2x^3 + 7}{5x^3}$ will behave like which function for large values of $|x|$?   c

a. $y = \dfrac{2}{5x}$

b. $y = \dfrac{2x}{5}$

c. $y = \dfrac{2}{5}$

d. $y = \dfrac{2}{5}x^3$

27. The graph of $f(x) = \dfrac{-3x^4 - 2x + 5}{x^5 + x^2 - 2}$ will behave like which function for large values of $|x|$?   d

a. $y = -3$

b. $y = -3x$

c. $y = -\dfrac{5}{2}$

d. $y = 0$

28. The graph of $f(x) = \dfrac{x^2 + 7x - 3}{6x^4 + 2}$ will behave like which function for large values of $|x|$?   b

a. $y = \dfrac{x^2}{6}$

b. $y = 0$

c. $y = -\dfrac{3}{2}$

d. $y = \dfrac{1}{6}x - \dfrac{3}{2}$

For Exercises 29–36,

a. Identify the horizontal asymptotes (if any). **(See Example 3)**

b. If the graph of the function has a horizontal asymptote, determine the point (if any) where the graph crosses the horizontal asymptote. **(See Example 4)**

29. $p(x) = \dfrac{5}{x^2 + 2x + 1}$
a. $y = 0$
b. Graph does not cross $y = 0$.

30. $q(x) = \dfrac{8}{x^2 + 4x + 4}$
a. $y = 0$
b. Graph does not cross $y = 0$.

31. $h(x) = \dfrac{3x^2 + 8x - 5}{x^2 + 3}$
a. $y = 3$
b. $\left(\dfrac{7}{4}, 3\right)$

32. $r(x) = \dfrac{-4x^2 + 5x - 1}{x^2 + 2}$
a. $y = -4$
b. $\left(-\dfrac{7}{5}, -4\right)$

33. $m(x) = \dfrac{x^4 + 2x + 1}{5x + 2}$
a. No horizontal asymptote   b. Not applicable

34. $n(x) = \dfrac{x^3 - x^2 + 1}{2x - 3}$
a. No horizontal asymptote
b. Not applicable

35. $t(x) = \dfrac{2x + 4}{x^2 + 7x - 4}$
a. $y = 0$
b. $(-2, 0)$

36. $s(x) = \dfrac{x + 3}{2x^2 - 3x - 5}$
a. $y = 0$ b. $(-3, 0)$

37. Consider the expression $\dfrac{x^2 + 3x + 1}{2x^2 + 5}$.

   a. Divide the numerator and denominator by the greatest power of x that appears in the denominator. That is, divide numerator and denominator by x^2.

   b. As $|x| \to \infty$ what value will $\dfrac{3}{x}$, $\dfrac{1}{x^2}$, and $\dfrac{5}{x^2}$ approach?

     (*Hint*: Substitute large values of x such as 100, 1000, 10,000, and so on to help you understand the behavior of each expression.)   0

   c. Use the results from parts (a) and (b) to identify the horizontal asymptote for the graph of

$$f(x) = \dfrac{x^2 + 3x + 1}{2x^2 + 5}. \quad y = \dfrac{1}{2}$$

38. Consider the expression $\dfrac{3x^3 - 2x^2 + 7x}{5x^3 + 1}$.

   a. Divide the numerator and denominator by the greatest power of x that appears in the denominator.

   b. As $|x| \to \infty$ what value will $-\dfrac{2}{x}$, $\dfrac{7}{x^2}$, and $\dfrac{1}{x^3}$ approach?   0

   c. Use the results from parts (a) and (b) to identify the horizontal asymptote for the graph of

$$f(x) = \dfrac{3x^3 - 2x^2 + 7x}{5x^3 + 1}. \quad y = \dfrac{3}{5}$$

Objective 4: Identify Slant Asymptotes

For Exercises 39–48, identify the asymptotes. (See Example 5)

39. $f(x) = \dfrac{2x^2 + 3}{x}$

40. $g(x) = \dfrac{3x^2 + 2}{x}$

41. $h(x) = \dfrac{-3x^2 + 4x - 5}{x + 6}$

42. $k(x) = \dfrac{-2x^2 - 3x + 7}{x + 3}$

43. $p(x) = \dfrac{x^3 + 5x^2 - 4x + 1}{x^2 - 5}$ Vertical asymptotes: $x = \sqrt{5}$ and $x = -\sqrt{5}$; Slant asymptote: $y = x + 5$

44. $q(x) = \dfrac{x^3 + 3x^2 - 2x - 4}{x^2 - 7}$ Vertical asymptotes: $x = \sqrt{7}$ and $x = -\sqrt{7}$; Slant asymptote: $y = x + 3$

45. $r(x) = \dfrac{2x + 1}{x^3 + x^2 - 4x - 4}$ Vertical asymptotes: $x = 2, x = -2$, and $x = -1$; Horizontal asymptote: $y = 0$

46. $t(x) = \dfrac{3x - 4}{x^3 + 2x^2 - 9x - 18}$ Vertical asymptotes: $x = 3, x = -3$, and $x = -2$; Horizontal asymptote: $y = 0$

47. $f(x) = \dfrac{4x^3 - 2x^2 + 7x - 3}{2x^2 + 4x + 3}$ Slant asymptote: $y = 2x - 5$

48. $a(x) = \dfrac{9x^3 - 5x + 4}{3x^2 + 2x + 1}$ Slant asymptote: $y = 3x - 2$

Objective 5: Graph Rational Functions

For Exercises 49–56, graph the functions by using transformations of the graphs of $y = \dfrac{1}{x}$ and $y = \dfrac{1}{x^2}$. (See Example 6)

49. $f(x) = \dfrac{1}{x - 3}$

50. $g(x) = \dfrac{1}{x + 4}$

51. $h(x) = \dfrac{1}{x^2} + 2$

52. $k(x) = \dfrac{1}{x^2} - 3$

53. $m(x) = \dfrac{1}{(x + 4)^2} - 3$

54. $n(x) = \dfrac{1}{(x - 1)^2} + 2$

55. $p(x) = -\dfrac{1}{x}$

56. $q(x) = -\dfrac{1}{x^2}$

For Exercises 57–62, for the graph of $y = f(x)$,

   a. Identify the x-intercepts.

   c. Identify the horizontal asymptote or slant asymptote if applicable.

   b. Identify any vertical asymptotes.

   d. Identify the y-intercept.

57. $f(x) = \dfrac{(x + 3)(2x - 7)}{(x + 2)(4x + 1)}$

58. $f(x) = \dfrac{(3x - 4)(x - 6)}{(2x - 3)(x + 5)}$

59. $f(x) = \dfrac{4x - 9}{x^2 - 9}$

60. $f(x) = \dfrac{5x - 8}{x^2 - 4}$

61. $f(x) = \dfrac{(5x - 1)(x + 3)}{x + 2}$

62. $f(x) = \dfrac{(4x + 3)(x + 2)}{x + 3}$

For Exercises 63–66, sketch a rational function subject to the given conditions. Answers may vary.

63. Horizontal asymptote: $y = 2$

   Vertical asymptote: $x = 3$

   y-intercept: $\left(0, \frac{8}{3}\right)$

   x-intercept: $(4, 0)$

64. Horizontal asymptote: $y = 0$

   Vertical asymptote: $x = -1$

   y-intercept: $(0, 1)$

   No x-intercepts

   Range: $(0, \infty)$

65. Horizontal asymptote: $y = 0$

Vertical asymptotes: $x = -2$ and $x = 2$

y-intercept: $(0, 1)$

No x-intercepts

Symmetric to the y-axis

Passes through the point $\left(3, -\frac{4}{5}\right)$

66. Horizontal asymptote: $y = 3$

Vertical asymptotes: $x = -1$ and $x = 1$

y-intercept: $(0, 0)$

x-intercept: $(0, 0)$

Symmetric to the y-axis

Passes through the point $(2, 4)$

For Exercises 67–90, graph the function. (See Examples 7–10)

67. $n(x) = \dfrac{-3}{2x + 7}$

68. $m(x) = \dfrac{-4}{2x - 5}$

69. $f(x) = \dfrac{x - 4}{x - 2}$

70. $g(x) = \dfrac{x - 3}{x - 1}$

71. $h(x) = \dfrac{2x - 4}{x + 3}$

72. $k(x) = \dfrac{3x - 9}{x + 2}$

73. $p(x) = \dfrac{6}{x^2 - 9}$

74. $q(x) = \dfrac{4}{x^2 - 16}$

75. $r(x) = \dfrac{5x}{x^2 - x - 6}$

76. $t(x) = \dfrac{4x}{x^2 - 2x - 3}$

77. $k(x) = \dfrac{5x - 3}{2x - 7}$

78. $h(x) = \dfrac{4x + 3}{3x - 5}$

79. $g(x) = \dfrac{3x^2 - 5x - 2}{x^2 + 1}$

80. $c(x) = \dfrac{2x^2 - 5x - 3}{x^2 + 1}$

81. $n(x) = \dfrac{x^2 + 2x + 1}{x}$

82. $m(x) = \dfrac{x^2 - 4x + 4}{x}$

83. $f(x) = \dfrac{x^2 + 7x + 10}{x + 3}$

84. $d(x) = \dfrac{x^2 - x - 12}{x - 2}$

85. $w(x) = \dfrac{-4x^2}{x^2 + 4}$

86. $u(x) = \dfrac{-3x^2}{x^2 + 1}$

87. $f(x) = \dfrac{x^3 + x^2 - 4x - 4}{x^2 + 3x}$

88. $g(x) = \dfrac{x^3 + 3x^2 - x - 3}{x^2 - 2x}$

89. $v(x) = \dfrac{2x^4}{x^2 + 9}$

90. $g(x) = \dfrac{4x^4}{x^2 + 8}$

Objective 6: Use Rational Functions in Applications

91. A sports trainer has monthly costs of $69.95 for phone service and $39.99 for his website and advertising. In addition he pays a $20 fee to the gym for each session in which he trains a client. **(See Example 11)**

a. Write a cost function to represent the cost $C(x)$ for x training sessions for a given month.

b. Write a function representing the average cost $\overline{C}(x)$ for x sessions.

c. Evaluate $\overline{C}(5)$, $\overline{C}(30)$, and $\overline{C}(120)$.

d. The trainer can realistically have 120 sessions per month. However, if the number of sessions were unlimited, what value would the average cost approach? What does this mean in the context of the problem?

92. An on-demand printing company has monthly overhead costs of $1200 in rent, $420 in electricity, $100 for phone service, and $200 for advertising and marketing. The printing cost is $40 per thousand pages for paper and ink.

a. Write a cost function to represent the cost $C(x)$ for printing x thousand pages for a given month.

b. Write a function representing the average cost $\overline{C}(x)$ for printing x thousand pages for a given month.

c. Evaluate $\overline{C}(20)$, $\overline{C}(50)$, $\overline{C}(100)$, and $\overline{C}(200)$.

d. Interpret the meaning of $\overline{C}(200)$.

e. For a given month, if the printing company could print an unlimited number of pages, what value would the average cost per thousand pages approach? What does this mean in the context of the problem?

93. A parallel circuit is one with several paths through which electricity can travel. The total resistance in a parallel circuit is always less than the resistance in any single branch. The total resistance R for the circuit shown can be computed from the formula $\dfrac{1}{R} = \dfrac{1}{R_1} + \dfrac{1}{R_2}$, where R_1 and R_2 are the resistances in the individual branches. Suppose that a resistor with a fixed resistance of 6 Ω (ohms) is placed in parallel with a variable resistor of resistance x.

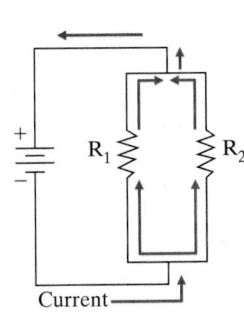

a. Write R as a function of x.

b. Complete the table.

x	6	12	18	30
$R(x)$				

c. What value does $R(x)$ approach as $x \to \infty$? Discuss the significance of this result.

94. The total resistance R of three resistors in parallel is given by

$$R = \frac{R_1 R_2 R_3}{R_1 R_2 + R_1 R_3 + R_2 R_3}$$

Suppose that an 8-Ω and a 12-Ω resistor are placed in parallel with a variable resistor of resistance x.

a. Write R as a function of x. $R(x) = \dfrac{96x}{20x + 96}$

b. What value does $R(x)$ approach as $x \to \infty$? Write the value in decimal form. $4.8 \ \Omega$

95. The concentration $C(t)$ (in milligrams per liter, mg/L) of a drug in the blood stream t hours after the drug is administered is modeled by

$$C(t) = \frac{10t}{2t^2 + 1}$$

a. Use a graphing utility to graph the function.

b. What are the domain restrictions on the function? $t \geq 0$

c. Use the graph to approximate the maximum concentration. Round to the nearest mg/L.   4 mg/L

d. What is the limiting concentration?   0 mg/L

96. A certain diet pill is designed to delay the administration of the active ingredient for several hours. The concentration $C(t)$ (in mg/L) of the active ingredient in the blood stream t hours after taking the pill is modeled by

$$C(t) = \frac{3t}{2t^2 - 20t + 51}$$

a. Use a graphing utility to graph the function.

b. What are the domain restrictions on the function? $t \geq 0$

c. Use the graph to approximate the maximum concentration. Round to the nearest mg/L.   15 mg/L

d. What is the limiting concentration?   0 mg/L

97. A power company burns coal to generate electricity. The cost $C(x)$ (in \$1000) to remove $x\%$ of the air pollutants is given by

$$C(x) = \frac{600x}{100 - x}$$

a. Compute the cost to remove 25% of the air pollutants. (*Hint*: $x = 25$.)   \$200,000

b. Determine the cost to remove 50%, 75%, and 90% of the air pollutants.   \$600,000; \$1,800,000; \$5,400,000

c. If the power company budgets \$1.4 million for pollution control, what percentage of the air pollutants can be removed?   70%

98. The cost $C(x)$ (in \$1000) for a city to remove $x\%$ of the waste from a polluted river is given by

$$C(x) = \frac{80x}{100 - x}$$

a. Determine the cost to remove 20%, 40%, and 90% of the waste. Round to the nearest thousand dollars.   \$20,000; \$53,000; \$720,000

b. If the city has \$320,000 budgeted for river cleanup, what percentage of the waste can be removed?   80%

The Doppler effect is a change in the observed frequency of a wave (such as a sound wave or light wave) when the source of the wave and observer are in motion relative to each other. The Doppler effect explains why an observer hears a change in pitch of an ambulance siren as the ambulance passes by the observer. The frequency $F(v)$ of a sound relative to an observer is given by $F(v) = f_a\left(\dfrac{s_0}{s_0 - v}\right)$, where f_a is the actual frequency of the sound at the source, s_0 is the speed of sound in air (772.4 mph), and v is the speed at which the source of sound is moving toward the observer. Use this relationship for Exercises 99–100.

99. Suppose that an ambulance moves toward an observer.

a. Write F as a function of v if the actual frequency of sound emitted by the ambulance is 560 Hz.

b. Use a graphing utility to graph the function from part (a) on the window [0, 1000, 100] by [0, 5000, 1000].

c. As the speed of the ambulance increases, what is the effect of the frequency of sound?   The frequency increases, making the pitch of the siren higher to the observer.

100. Suppose the frequency of sound emitted by a police car siren is 600 Hz.

a. Write F as a function of v if the police car is moving toward an observer.

b. Suppose that the frequency of the siren as heard by an observer is 664 Hz. Determine the velocity of the police car. Round to the nearest tenth of a mph.   74.4 mph

c. Although a police car cannot travel close to the speed of sound, interpret the meaning of the vertical asymptote.   If the speed of the police car were to approach the speed of sound in air, the frequency would increase infinitely and a sonic boom would occur.

Mixed Exercises

101. a. Write an equation for a rational function f whose graph is the same as the graph of $y = \dfrac{1}{x^2}$ shifted up 3 units and to the left 1 unit. $f(x) = \dfrac{1}{(x + 1)^2} + 3$

b. Write the domain and range of the function in interval notation.   Domain: $(-\infty, -1) \cup (-1, \infty)$; Range: $(3, \infty)$

102. a. Write an equation for a rational function f whose graph is the same as the graph of $y = \dfrac{1}{x}$ shifted to the right 4 units and down 3 units. $f(x) = \dfrac{1}{x - 4} - 3$

b. Write the domain and range of the function in interval notation.   Domain: $(-\infty, 4) \cup (4, \infty)$; Range: $(-\infty, -3) \cup (-3, \infty)$

For Exercises 103–104, given $y = f(x)$,

a. Divide the numerator by the denominator to write $f(x)$ in the form $f(x) = \text{quotient} + \dfrac{\text{remainder}}{\text{divisor}}$.

b. Use transformations of $y = \dfrac{1}{x}$ to graph the function.

103. $f(x) = \dfrac{2x + 7}{x + 3}$

104. $f(x) = \dfrac{5x + 11}{x + 2}$

Write About It

105. Explain why $x = -2$ is not a vertical asymptote of the graph of $f(x) = \dfrac{x^2 + 7x + 10}{x + 2}$. The numerator and denominator share a common factor of $x + 2$. The value –2 is not in the domain of f. The graph will have a "hole" at $x = -2$ rather than a vertical asymptote.

106. Write an informal definition of a horizontal asymptote of a rational function. A horizontal asymptote is a horizontal line that the function approaches as $|x|$ becomes very large—that is, as x approaches infinity or as x approaches negative infinity.

Expanding Your Skills

For Exercises 107–110, write an equation of a function that meets the given conditions. Answers may vary.

107. x-intercepts: $(-3, 0)$ and $(-1, 0)$

vertical asymptote: $x = 2$

horizontal asymptote: $y = 1$ $f(x) = \dfrac{x^2 + 4x + 3}{x^2 - 4x + 4}$

y-intercept: $\left(0, \frac{3}{4}\right)$

108. x-intercepts: $(4, 0)$ and $(2, 0)$

vertical asymptote: $x = 1$

horizontal asymptote: $y = 1$ $f(x) = \dfrac{x^2 - 6x + 8}{x^2 - 2x + 1}$

y-intercept: $(0, 8)$

109. x-intercept: $\left(\frac{3}{2}, 0\right)$

vertical asymptotes: $x = -2$ and $x = 5$

horizontal asymptote: $y = 0$ $f(x) = \dfrac{20x - 30}{x^2 - 3x - 10}$

y-intercept: $(0, 3)$

110. x-intercept: $\left(\frac{4}{3}, 0\right)$

vertical asymptotes: $x = -3$ and $x = -4$

horizontal asymptote: $y = 0$ $f(x) = \dfrac{9x - 12}{x^2 + 7x + 12}$

y-intercept: $(0, -1)$

Graphs with "Holes"

The rational functions studied in this section all have the characteristic that the numerator and denominator do not share a common variable factor. We now investigate rational functions for which this is not the case. For Exercises 111–114,

a. Write the domain of f in interval notation.

b. Simplify the rational expression defining the function.

c. Identify any vertical asymptotes.

d. Identify any other values of x (other than those corresponding to vertical asymptotes) for which the function is discontinuous.

e. Identify the graph of the function.

111. $f(x) = \dfrac{x^2 + x - 6}{x - 2}$

112. $f(x) = \dfrac{-x^2 + 2x + 3}{x + 1}$

113. $f(x) = \dfrac{2x + 10}{x^2 + 9x + 20}$

114. $f(x) = \dfrac{2x - 2}{x^2 + 2x - 3}$

i.

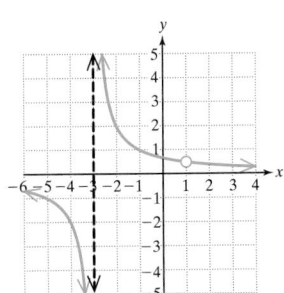

ii.

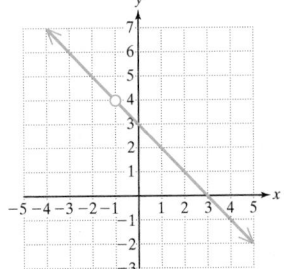

iii.

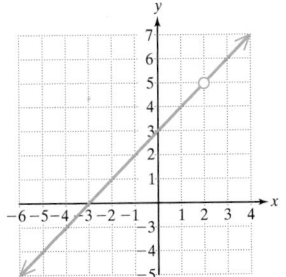

iv.

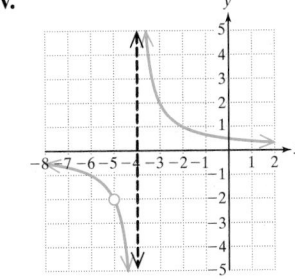

PROBLEM RECOGNITION EXERCISES

Polynomial and Rational Functions

For Exercises 1–8, refer to $p(x) = x^3 + 3x^2 - 6x - 8$ and $q(x) = x^3 - 2x^2 - 5x + 6$.

1. Find the zeros of $p(x)$. $2, -1,$ and -4

2. Find the zeros of $q(x)$. $-2, 1,$ and 3

3. Find the x-intercept(s) of the graph of $y = q(x)$.
$(-2, 0), (1, 0),$ and $(3, 0)$

4. Find the x-intercept(s) of the graph of $y = p(x)$.
$(2, 0), (-1, 0),$ and $(-4, 0)$

5. Find the x-intercepts of the graph of
$$f(x) = \frac{p(x)}{q(x)} = \frac{x^3 + 3x^2 - 6x - 8}{x^3 - 2x^2 - 5x + 6}.$$ $(2, 0), (-1, 0),$ and $(-4, 0)$

6. Find the vertical asymptotes of the graph of
$$f(x) = \frac{p(x)}{q(x)} = \frac{x^3 + 3x^2 - 6x - 8}{x^3 - 2x^2 - 5x + 6}.$$
$x = -2, x = 1,$ and $x = 3$

7. Find the horizontal asymptote or slant asymptote of the graph of $f(x) = \dfrac{p(x)}{q(x)} = \dfrac{x^3 + 3x^2 - 6x - 8}{x^3 - 2x^2 - 5x + 6}.$
Horizontal asymptote: $y = 1$

8. Determine where the graph of $f(x) = \dfrac{x^3 + 3x^2 - 6x - 8}{x^3 - 2x^2 - 5x + 6}$ crosses its horizontal or slant asymptote.

For Exercises 9–16, refer to $c(x) = x^3 - 4x^2 - 2x + 8$ and $d(x) = x^3 + 3x^2 - 4$.

9. Find the zeros of $c(x)$. $4, \sqrt{2},$ and $-\sqrt{2}$

10. Find the zeros of $d(x)$. $1, -2$ (multiplicity 2)

11. Find the x-intercept(s) of the graph of $y = d(x)$.
$(1, 0)$ and $(-2, 0)$

12. Find the x-intercept(s) of the graph of $y = c(x)$.
$(4, 0), (\sqrt{2}, 0),$ and $(-\sqrt{2}, 0)$

13. Find the x-intercepts of the graph of
$$g(x) = \frac{c(x)}{d(x)} = \frac{x^3 - 4x^2 - 2x + 8}{x^3 + 3x^2 - 4}.$$
$(4, 0), (\sqrt{2}, 0),$ and $(-\sqrt{2}, 0)$

14. Find the vertical asymptotes of the graph of
$$g(x) = \frac{c(x)}{d(x)} = \frac{x^3 - 4x^2 - 2x + 8}{x^3 + 3x^2 - 4}.$$ $x = 1$ and $x = -2$

15. Find the horizontal asymptote or slant asymptote of the graph of $g(x) = \dfrac{c(x)}{d(x)} = \dfrac{x^3 - 4x^2 - 2x + 8}{x^3 + 3x^2 - 4}.$
Horizontal asymptote: $y = 1$

16. Determine where the graph of $g(x) = \dfrac{x^3 - 4x^2 - 2x + 8}{x^3 + 3x^2 - 4}$ crosses its horizontal or slant asymptote.

For Exercises 17–18, use the results from Exercises 5–8 and 13–16 to match the function with its graph.

17. $f(x) = \dfrac{x^3 + 3x^2 - 6x - 8}{x^3 - 2x^2 - 5x + 6}$ Graph b

18. $g(x) = \dfrac{x^3 - 4x^2 - 2x + 8}{x^3 + 3x^2 - 4}$ Graph a

a.

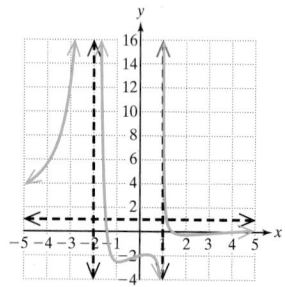

b.

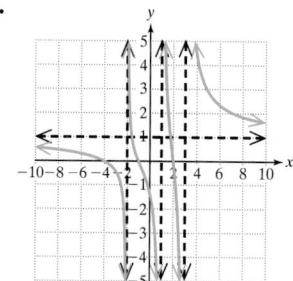

19. Divide $(2x^3 - 4x^2 - 10x + 12) \div (x^2 - 11)$ by using an appropriate method.

  a. Identify the quotient $q(x)$. $q(x) = 2x - 4$ **b.** Idenitfy the remainder $r(x)$. $r(x) = 12x - 32$

20. Identify the slant asymptote of $f(x) = \dfrac{2x^3 - 4x^2 - 10x + 12}{x^2 - 11}.$ $y = 2x - 4$

21. Identify the point where the graph of $f(x) = \dfrac{2x^3 - 4x^2 - 10x + 12}{x^2 - 11}$ crosses its slant asymptote. $\left(\dfrac{8}{3}, \dfrac{4}{3}\right)$

22. Refer to Exercise 19. Solve the equation $r(x) = 0$. How does the solution to the equation $r(x) = 0$ relate to the point where the graph of f crosses its slant asymptote?
$\left\{\dfrac{8}{3}\right\}$; The solution to $r(x) = 0$ gives the x-coordinate of the point where the graph of f crosses its slant asymptote.

OBJECTIVES

1. Solve Polynomial Inequalities
2. Solve Rational Inequalities
3. Solve Applications Involving Polynomial and Rational Inequalities

SECTION 2.6   Polynomial and Rational Inequalities

1. Solve Polynomial Inequalities

An engineer for a food manufacturer must design an aluminum container for a hot drink mix. The container is to be a right circular cylinder 5.5 in. in height. The surface area represents the amount of aluminum used and is given by

$$S(r) = 2\pi r^2 + 11\pi r \qquad \text{where } r \text{ is the radius of the can.}$$

The engineer wants to limit the surface area so that at most 90 in.^{2} of aluminum is used. To determine the restrictions on the radius, the engineer must solve the inequality $2\pi r^2 + 11\pi r \leq 90$ (see Exercise 123). This inequality is a quadratic inequality in the variable r. It is also categorized as a polynomial inequality of degree 2.

> **TIP** If $f(x)$ is a quadratic polynomial, then the inequalities $f(x) < 0$, $f(x) > 0$, $f(x) \leq 0$, and $f(x) \geq 0$ are called quadratic inequalities.

Definition of a Polynomial Inequality

Let $f(x)$ be a polynomial. Then an inequality of the form

$f(x) < 0, f(x) > 0, f(x) \leq 0$, or $f(x) \geq 0$ is called a **polynomial inequality.**

Note: A polynomial inequality is nonlinear if $f(x)$ is a polynomial of degree greater than 1.

Consider the polynomial inequalities $f(x) < 0$ and $f(x) > 0$. We need to determine the intervals over which $f(x)$ is negative or positive. For example, consider the graph of $f(x) = x^2 - 6x + 5$ (Figure 2-23).

The graph shows the solution sets for the following equation and inequalities.

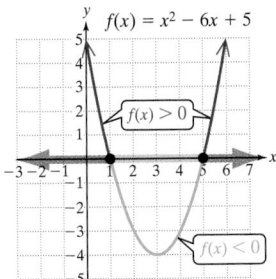

Figure 2-23

- $f(x) = 0$ for $\{1, 5\}$.
- $f(x) < 0$ on the interval $(1, 5)$. (shown in blue)
- $f(x) > 0$ on the interval $(-\infty, 1) \cup (5, \infty)$. (shown in red)

Notice that the x-intercepts define the endpoints (or "boundary" points) for the solution sets of the inequalities. We can solve a polynomial inequality if we can identify the *sign* of the polynomial for each interval defined by the boundary points. This is the basis on which we solve any nonlinear inequality.

Procedure to Solve a Nonlinear Inequality

1. Express the inequality as $f(x) < 0, f(x) > 0, f(x) \leq 0$, or $f(x) \geq 0$. That is, rearrange the terms of the inequality so that one side is set to zero.
2. Find the real solutions of the related equation $f(x) = 0$ and any values of x that make $f(x)$ undefined. These are the "boundary" points for the solution set to the inequality.
3. Determine the sign of $f(x)$ on the intervals defined by the boundary points.
   - If $f(x)$ is positive, then the values of x on the interval are solutions to $f(x) > 0$.
   - If $f(x)$ is negative, then the values of x on the interval are solutions to $f(x) < 0$.
4. Determine whether the boundary points are included in the solution set.
5. Write the solution set in interval notation or set-builder notation.

Classroom Example: p. 333
Exercise 24

> **EXAMPLE 1**   Solving a Quadratic Inequality
>
> Solve the inequality. $3x(x - 1) > 10 - 2x$
>
> **Solution:**
>
> $$3x(x - 1) > 10 - 2x$$ **Step 1:** Write the inequality in the form $f(x) > 0$.
>
> $$3x^2 - 3x > 10 - 2x$$
>
> $$\overset{f(x)}{\overbrace{3x^2 - x - 10}} > 0$$
>
> $$3x^2 - x - 10 = 0$$ **Step 2:** Find the real solutions to the related equation $f(x) = 0$.
>
> $$(3x + 5)(x - 2) = 0$$
>
> $$x = -\frac{5}{3} \quad \text{and} \quad x = 2$$ The boundary points are $-\frac{5}{3}$ and 2.
>
> **Step 3:** Divide the x-axis into intervals defined by the boundary points.
>
>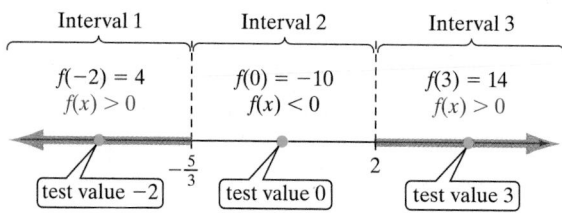
>
> Determine the sign of $f(x) = 3x^2 - x - 10$ on each interval. One method is to evaluate $f(x)$ for a test value x on each interval.
>
> **Step 4:** The solution set does not include the boundary points because the inequality is strict.
>
> The solution set is $\left(-\infty, -\frac{5}{3}\right) \cup (2, \infty)$ or equivalently in set-builder notation $\left\{x \mid x < -\frac{5}{3} \text{ or } x > 2\right\}$.
>
> **Step 5:** Write the solution set.

TIP To evaluate the polynomial $f(x) = 3x^2 - x - 10$ at the test points, we can perform direct substitution such as:

$$f(3) = 3(3)^2 - (3) - 10$$
$$= 14$$

Or use synthetic division and the remainder theorem.

$$\begin{array}{r} 3\underline{| \;\; 3 \quad -1 \quad -10} \\ 9 \quad\;\; 24 \\ \hline 3 \quad\;\; 8 \quad\;\; \underline{|14} \end{array}$$

> **Skill Practice 1**   Solve the inequality. $2x(x - 1) < 21 - x$

From Example 1, the key step is to determine the sign of $f(x)$ on the intervals $\left(-\infty, -\frac{5}{3}\right)$, $\left(-\frac{5}{3}, 2\right)$, and $(2, \infty)$. We can avoid the arithmetic from evaluating $f(x)$ at the test points by creating a sign chart. The inequality $3x^2 - x - 10 > 0$ is equivalent to $(3x + 5)(x - 2) > 0$. We have

TIP The sign of the product in the bottom row of the sign chart is determined by the signs of the individual factors from the rows above.

Sign of $(3x + 5)$:	$-$	$+$	$+$
Sign of $(x - 2)$:	$-$	$-$	$+$
Sign of $(3x + 5)(x - 2)$:	$+$	$-$	$+$

$$-\frac{5}{3} \qquad 2$$

The sign chart organizes the signs of each factor on the given intervals. Then the sign of the product of factors is given in the bottom row. We see that $f(x) = (3x + 5)(x - 2) > 0$ for $\left(-\infty, -\frac{5}{3}\right) \cup (2, \infty)$.

The result of Example 1 can also be viewed graphically. From Section 2.1, the graph of $f(x) = 3x^2 - x - 10$ is a parabola opening upward (Figure 2-24).

From the factored form $f(x) = (3x + 5)(x - 2)$, the x-intercepts $\left(-\frac{5}{3}, 0\right)$ and $(2, 0)$ mark the points of transition between the intervals where $f(x)$ potentially changes sign.

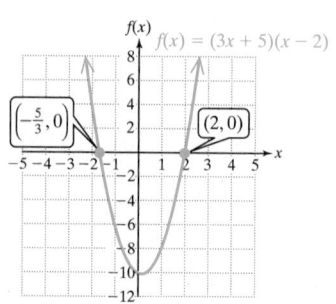

Figure 2-24

Answer

1. Interval notation: $\left(-3, \frac{7}{2}\right)$;
   Set-builder notation:
   $\left\{x \mid -3 < x < \frac{7}{2}\right\}$

EXAMPLE 2   Solving a Polynomial Inequality

Solve the inequality. $x^4 - 12x \geq 8x^2 - x^3$

Solution:

$x^4 - 12x \geq 8x^2 - x^3$ **Step 1:** Write the inequality in the form $f(x) \geq 0$.

$$\overbrace{x^4 + x^3 - 8x^2 - 12x}^{f(x)} \geq 0$$

$x^4 + x^3 - 8x^2 - 12x = 0$ **Step 2:** Find the real solutions to the related equation $f(x) = 0$.

$x(x^3 + x^2 - 8x - 12) = 0$ Factor the left side of the equation.

The possible rational zeros of $x^3 + x^2 - 8x - 12$ are $\pm 1,\ \pm 2,\ \pm 3,\ \pm 4,\ \pm 6,\ \pm 12$.

$$\begin{array}{r|rrrr} 3 & 1 & 1 & -8 & -12 \\ & & 3 & 12 & 12 \\ \hline & 1 & 4 & 4 & \underline{0} \end{array}$$

After testing several potential rational zeros, we find that 3 is a zero of $f(x)$.

$x(x - 3)(x^2 + 4x + 4) = 0$ Now factor the quadratic polynomial.

$x(x - 3)(x + 2)^2 = 0$

$x = 0,\ x = 3,\ x = -2$ The boundary points are 0, 3, and -2.

Step 3: The inequality $x^4 + x^3 - 8x^2 - 12x \geq 0$ is equivalent to $x(x - 3)(x + 2)^2 \geq 0$. Divide the x-axis into intervals defined by the boundary points and determine the sign of $f(x)$ on each interval.

Evaluate: $f(x) = x(x-3)(x+2)^2$	Interval 1 $f(-3) = 18$ $f(x) > 0$	Interval 2 $f(-1) = 4$ $f(x) > 0$	Interval 3 $f(1) = -18$ $f(x) < 0$	Interval 4 $f(4) = 144$ $f(x) > 0$
Sign of x	$-$	$-$	$+$	$+$
Sign of $(x - 3)$	$-$	$-$	$-$	$+$
Sign of $(x + 2)^2$	$+$	$+$	$+$	$+$
Sign of $x(x-3)(x+2)^2$	$+$	$+$	$-$	$+$

                                                   -2 0 3

The solution set is $(-\infty, 0] \cup [3, \infty)$.
In set-builder notation this is $\{x \mid x \leq 0 \text{ or } x \geq 3\}$.

Step 4: The solution set includes the boundary points because the inequality sign includes equality. Therefore, the union of intervals 1 and 2 becomes $(-\infty, 0]$.

Step 5: Write the solution set.

Skill Practice 2   Solve the inequality. $x^4 - 18x \geq 3x^2 - 4x^3$

Answer

2. Interval notation: $(-\infty, 0] \cup [2, \infty)$;
Set-builder notation: $\{x \mid x \leq 0$ or $x \geq 2\}$

The result of Example 2 can also be interpreted graphically. From Section 2.2, the graph of $f(x) = x^4 + x^3 - 8x^2 - 12x$ is up to the far left and up to the far right.

In factored form $f(x) = x(x - 3)(x + 2)^2$. The x-intercepts are $(0, 0)$, $(3, 0)$, and $(-2, 0)$. Furthermore, the factors x and $(x - 3)$ have odd exponents. This means that the corresponding zeros have odd multiplicities, and that the graph will cross the x-axis at $(0, 0)$ and $(3, 0)$ and change sign. The factor $(x + 2)$ has an even exponent meaning that the corresponding zero has an even multiplicity. The graph will touch the x-axis at $(-2, 0)$ but will *not* change sign. From the sketch in Figure 2-25, we see that $f(x) \geq 0$ on the intervals $(-\infty, 0]$ and $[3, \infty)$.

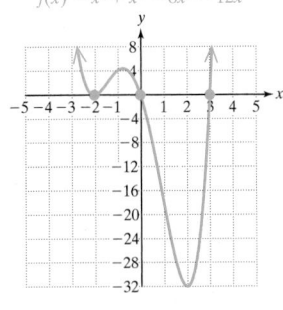

$f(x) = x^4 + x^3 - 8x^2 - 12x$

Figure 2-25

In some situations, the sign of a polynomial may be easily determined by inspection. In such a case, we can abbreviate the procedure to solve a polynomial inequality. This is demonstrated in Example 3.

Classroom Example: p. 332
Exercise 20

EXAMPLE 3 Solving Polynomial Inequalities

Solve the inequalities.

a. $4x^2 - 12x + 9 < 0$ **b.** $4x^2 - 12x + 9 \le 0$

c. $4x^2 - 12x + 9 > 0$ **d.** $4x^2 - 12x + 9 \ge 0$

Solution:

a. $4x^2 - 12x + 9 < 0$ Factor $4x^2 - 12x + 9$ as $(2x - 3)^2$.

$(2x - 3)^2 < 0$ The square of any real number is nonnegative. Therefore, this

The solution set is { }.   inequality has no solution.

b. $4x^2 - 12x + 9 \le 0$ The inequality in part (b) is the same as the inequality in part (a) except that equality is included.

$(2x - 3)^2 \le 0$

The solution set is $\left\{\frac{3}{2}\right\}$.   The expression $(2x - 3)^2 = 0$ for $x = \frac{3}{2}$.

c. $4x^2 - 12x + 9 > 0$ The expression $(2x - 3)^2 > 0$ for

$(2x - 3)^2 > 0$ all real numbers except where

The solution set is $\left(-\infty, \frac{3}{2}\right) \cup \left(\frac{3}{2}, \infty\right)$. $(2x - 3)^2 = 0$. Therefore, the solution set is all real numbers

In set-builder notation: $\left\{x \mid x < \frac{3}{2} \text{ or } x > \frac{3}{2}\right\}$.   except $\frac{3}{2}$.

d. $4x^2 - 12x + 9 \ge 0$ The square of any real number is

$(2x - 3)^2 \ge 0$ greater than or equal to zero.

The solution set is $(-\infty, \infty)$.   Therefore, the solution set is all real numbers.

Skill Practice 3   Solve the inequalities.

a. $25x^2 - 10x + 1 < 0$ **b.** $25x^2 - 10x + 1 \le 0$

c. $25x^2 - 10x + 1 > 0$ **d.** $25x^2 - 10x + 1 \ge 0$

2. Solve Rational Inequalities

We now turn our attention to solving rational inequalities.

Definition of a Rational Inequality

Let $f(x)$ be a rational expression. Then an inequality of the form $f(x) < 0$, $f(x) > 0$, $f(x) \le 0$, or $f(x) \ge 0$ is called a **rational inequality.**

We solve polynomial and rational inequalities in the same way with one exception. With a rational inequality such as $f(x) < 0$, the list of boundary points must include the real solutions to the related equation $f(x) = 0$ along with the values of x that make $f(x)$ undefined.

Answers

3. a. { }   **b.** $\left\{\dfrac{1}{5}\right\}$

c. Interval notation:
$\left(-\infty, \frac{1}{5}\right) \cup \left(\frac{1}{5}, \infty\right)$; Set-builder
notation: $\left\{x \mid x < \frac{1}{5} \text{ or } x > \frac{1}{5}\right\}$

d. $(-\infty, \infty)$

The graph of $f(x) = \dfrac{2x - 5}{x + 2}$ is shown in Figure 2-26. Notice that the function changes sign to the left and right of the vertical asymptote $x = -2$ and to the left and right of the x-intercept $\left(\frac{5}{2}, 0\right)$.

From the graph of $f(x) = \dfrac{2x - 5}{x + 2}$ we can determine the solution sets for the following inequalities.

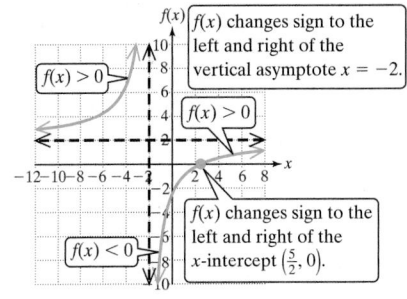

Figure 2-26

$$\frac{2x - 5}{x + 2} < 0 \qquad \text{on the interval } \left(-2, \tfrac{5}{2}\right)$$

$$\frac{2x - 5}{x + 2} > 0 \qquad \text{on the intervals } (-\infty, -2) \cup \left(\tfrac{5}{2}, \infty\right)$$

$$\frac{2x - 5}{x + 2} \leq 0 \qquad \text{on the interval } \left(-2, \tfrac{5}{2}\right]$$

$$\frac{2x - 5}{x + 2} \geq 0 \qquad \text{on the intervals } (-\infty, -2) \cup \left[\tfrac{5}{2}, \infty\right)$$

Classroom Example: p. 334
Exercise 74

Instructor Note:
Caution students about multiplying both sides of the inequality in Example 4 by $(x - 2)$. We do not know the sign of $(x - 2)$ and would not know if the direction of the inequality sign would need to be reversed.

EXAMPLE 4 **Solving a Rational Inequality**

Solve the inequality. $\dfrac{4x - 5}{x - 2} \leq 3$

Solution:

$$\frac{4x - 5}{x - 2} \leq 3$$

$$\overbrace{\frac{4x - 5}{x - 2} - 3}^{f(x)} \leq 0 \qquad$$ **Step 1:** First write the inequality in the form $f(x) \leq 0$. That is, set one side to 0.

$$\frac{4x - 5}{x - 2} - 3 \cdot \frac{x - 2}{x - 2} \leq 0 \qquad$$ Write each term with a common denominator.

$$\frac{4x - 5 - 3(x - 2)}{x - 2} \leq 0 \qquad$$ Simplify.

$$\frac{4x - 5 - 3x + 6}{x - 2} \leq 0$$

$$\frac{x + 1}{x - 2} \leq 0 \qquad$$ The expression $\dfrac{x + 1}{x - 2}$ is undefined for $x = 2$. Therefore, the value $x = 2$ is *not* part of the solution set.
However, 2 *is* a boundary point for the solution set.

$$\frac{x + 1}{x - 2} = 0 \qquad$$ **Step 2:** Solve for the real solutions to the equation $f(x) = 0$. The solution is -1, and this is another boundary point.

The boundary points are $x = 2$ and $x = -1$.

$$x = -1 \qquad$$ **Step 3:** Divide the x-axis into intervals defined by the boundary points and determine the sign of $f(x)$ on each interval.

	Interval 1	Interval 2	Interval 3
Evaluate: $f(x) = \frac{x+1}{x-2}$	$f(-2) = \frac{1}{4}$ $f(x) > 0$	$f(1) = -2$ $f(x) < 0$	$f(3) = 4$ $f(x) > 0$
Sign of $(x+1)$	$-$	$+$	$+$
Sign of $(x-2)$	$-$	$-$	$+$
Sign of $\frac{(x+1)}{(x-2)}$	$+$	$-$	$+$

$$\begin{array}{ccc} & -1 & 2 \end{array}$$

Instructor Note:
Emphasize to students never to include a value of *x* for which the inequality is undefined.

Step 4: Substituting the boundary point $x = -1$ into the inequality $\frac{x+1}{x-2} \leq 0$ makes a true statement, $0 \leq 0$. Thus, $x = -1$ is part of the solution set. The boundary point 2 is excluded because $\frac{x+1}{x-2}$ is undefined for $x = 2$.

Step 5: Write the solution set.
The solution set is $[-1, 2)$.
In set-builder notation this is $\{x \mid -1 \leq x < 2\}$.

> **Skill Practice 4** Solve the inequality. $\dfrac{5-x}{x-1} \geq -2$

Classroom Example: p. 334
Exercise 78

EXAMPLE 5 Solving a Rational Inequality

Solve the inequality. $\dfrac{1}{x-2} \geq \dfrac{1}{x+3}$

Solution:

$\dfrac{1}{x-2} \geq \dfrac{1}{x+3}$ **Step 1:** First write the inequality in the form $f(x) \geq 0$.

$\dfrac{1}{x-2} - \dfrac{1}{x+3} \geq 0$ Set one side to zero.

$\dfrac{1}{x-2} \cdot \dfrac{x+3}{x+3} - \dfrac{1}{x+3} \cdot \dfrac{x-2}{x-2} \geq 0$ Write each term with a common denominator.

$\dfrac{x+3}{(x-2)(x+3)} - \dfrac{x-2}{(x-2)(x+3)} \geq 0$ Simplify and write the numerator and denominator in factored form.

$\dfrac{x+3-x+2}{(x-2)(x+3)} \geq 0$ The expression $\frac{5}{(x-2)(x+3)}$ is undefined for $x = 2$ and $x = -3$. The values 2 and -3 are not included in the solution set, but they are boundary points for the solution set.

$\dfrac{5}{(x-2)(x+3)} \geq 0$

$\dfrac{5}{(x-2)(x+3)} = 0$ **Step 2:** Solve the related equation to determine any additional boundary points.

$\dfrac{5}{(x-2)(x+3)} \cdot (x-2)(x+3) = 0 \cdot (x-2)(x+3)$

$5 = 0$ Clearing fractions results in the contradiction $5 = 0$. There are no
(contradiction)   solutions to the related equation.

Answer
4. $(-\infty, -3] \cup (1, \infty)$

Step 3: Divide the x-axis into intervals defined by the boundary points and determine the sign of $f(x) = \frac{5}{(x-2)(x+3)}$ on each interval.

	Interval 1	Interval 2	Interval 3
Evaluate: $f(x) = \frac{5}{(x-2)(x+3)}$	$f(-4) = \frac{5}{6}$ $f(x) > 0$	$f(0) = -\frac{5}{6}$ $f(x) < 0$	$f(3) = \frac{5}{6}$ $f(x) > 0$
Sign of 5	+	+	+
Sign of $(x-2)$	−	−	+
Sign of $(x+3)$	−	+	+
Sign of $f(x) = \frac{5}{(x-2)(x+3)}$	+	−	+

$$\overset{\longleftarrow}{} \underset{-3}{} \underset{2}{} \overset{\longrightarrow}{}$$

Now interpret the results to solve $\dfrac{5}{(x-2)(x+3)} \geq 0$.

Step 4: Neither boundary point is included because $\frac{5}{(x-2)(x+3)}$ is undefined for $x = 2$ and $x = -3$.

Step 5: Write the solution set. The expression $\frac{5}{(x-2)(x+3)}$ is greater than zero for $x < -3$ and $x > 2$.

The solution set is $(-\infty, -3) \cup (2, \infty)$.

In set-builder notation, this is $\{x \mid x < -3 \text{ or } x > 2\}$.

Skill Practice 5   Solve the inequality. $\dfrac{1}{x+4} \geq \dfrac{1}{x+2}$

In Example 6 we encounter an inequality in which the signs of the numerator and denominator of the rational expression can be determined by inspection.

Classroom Example: p. 333
Exercise 62

EXAMPLE 6   Solving Rational Inequalities

Solve the inequalities.

a. $\dfrac{x^2}{x^2+4} \geq 0$ **b.** $\dfrac{x^2}{x^2+4} > 0$

c. $\dfrac{x^2}{x^2+4} \leq 0$ **d.** $\dfrac{x^2}{x^2+4} < 0$

Solution:

The solution to the related equation $\dfrac{x^2}{x^2+4} = 0$ is $x = 0$.

The denominator is nonzero for all real numbers.

Therefore, the only boundary point is $x = 0$.

Sign of x^2	+	+
Sign of $x^2 + 4$	+	+
Sign of $\dfrac{x^2}{x^2+4}$	+	+

$$\underset{0}{}$$

Answer

5. $(-4, -2)$

Therefore, $\dfrac{x^2}{x^2 + 4} = 0$ at $x = 0$ and is positive for all other real numbers.

a. $\dfrac{x^2}{x^2 + 4} \geq 0$ Solution set: $(-\infty, \infty)$

b. $\dfrac{x^2}{x^2 + 4} > 0$ Solution set: $(-\infty, 0) \cup (0, \infty)$

c. $\dfrac{x^2}{x^2 + 4} \leq 0$ Solution set: $\{0\}$

d. $\dfrac{x^2}{x^2 + 4} < 0$ Solution set: $\{\ \}$

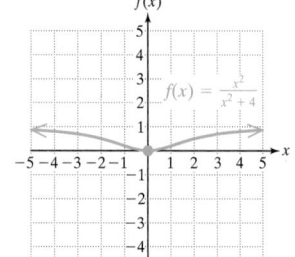

Skill Practice 6   Solve the inequalities.

a. $\dfrac{x^2}{x^4 + 1} \geq 0$ **b.** $\dfrac{x^2}{x^4 + 1} > 0$ **c.** $\dfrac{x^2}{x^4 + 1} \leq 0$ **d.** $\dfrac{x^2}{x^4 + 1} < 0$

3. Solve Applications Involving Polynomial and Rational Inequalities

In Section R.7, we studied the vertical position $s(t)$ of an object moving upward or downward under the influence of gravity. We use this model to solve the application in Example 7.

$$s(t) = -\frac{1}{2}gt^2 + v_0 t + s_0$$

where

g is the acceleration due to gravity (32 ft/sec^2 or 9.8 m/sec^2).

t is the time of travel.

v_0 is the initial speed.

s_0 is the initial vertical position.

Classroom Example: p. 334
Exercise 86

EXAMPLE 7   **Solving an Application of a Polynomial Inequality**

A toy rocket is shot straight upward from a launch pad 1 ft above ground level with an initial speed of 64 ft/sec.

a. Write a model to express the vertical position $s(t)$ (in ft) of the rocket t seconds after launch.

b. Determine the times at which the rocket is above a height of 50 ft.

TIP Choose $g = 32$ ft/sec^2 because the height is given in feet and speed is given in feet per second.

Solution:

a. $s(t) = -\dfrac{1}{2}gt^2 + v_0 t + s_0$ In this example,

$\qquad$ $s_0 = 1$ ft

$s(t) = -\dfrac{1}{2}(32)t^2 + (64)t + (1)$ $\qquad v_0 = 64$ ft/sec

$\qquad g = 32$ ft/sec^2

$s(t) = -16t^2 + 64t + 1$

Answers

6. a. $(-\infty, \infty)$

$\quad$ **b.** $(-\infty, 0) \cup (0, \infty)$

$\quad$ **c.** $\{0\}$

$\quad$ **d.** $\{\ \}$

b. $-16t^2 + 64t + 1 > 50$

$$\overbrace{-16t^2 + 64t - 49}^{f(t)} > 0$$

$-16t^2 + 64t - 49 > 0$ Write the inequality in the form $f(t) > 0$.

$-16t^2 + 64t - 49 = 0$ Use the quadratic formula to solve the related equation $f(t) = 0$.

$$t = \frac{-64 \pm \sqrt{(64)^2 - 4(-16)(-49)}}{2(-16)}$$ Evaluate $f(t) = -16t^2 + 64t - 49$ for test points in each interval.

$$= \frac{-64 \pm \sqrt{960}}{-32}$$

$$= \frac{-64 \pm 8\sqrt{15}}{-32}$$

$$= \frac{8 \pm \sqrt{15}}{4} \quad\begin{array}{l}\approx 2.97 \\ \approx 1.03\end{array}$$

$$\begin{array}{ccc} f(1) = -1 & f(2) = 15 & f(3) = -1 \\ f(t) < 0 & f(t) > 0 & f(t) < 0 \end{array}$$

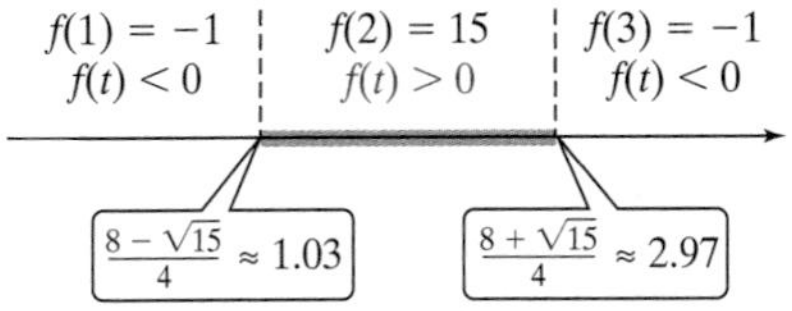

$\boxed{\frac{8-\sqrt{15}}{4} \approx 1.03}$ $\boxed{\frac{8+\sqrt{15}}{4} \approx 2.97}$

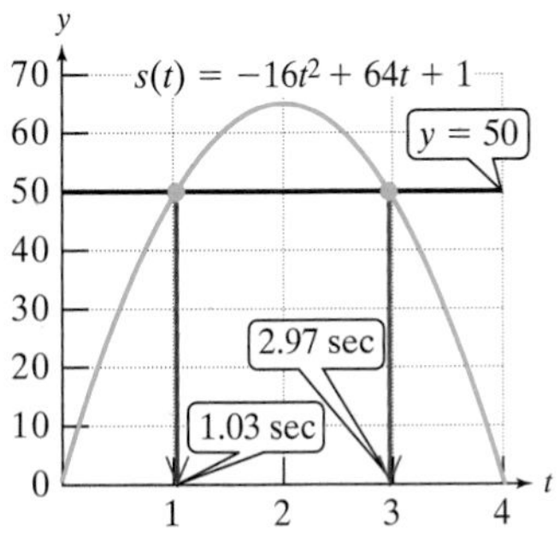

The solution set is $\left(\dfrac{8 - \sqrt{15}}{4}, \dfrac{8 + \sqrt{15}}{4}\right)$ or approximately $(1.03, 2.97)$.

The rocket will be above 50 ft high between 1.03 sec and 2.97 sec after launch.

The graph of $s(t) = -16t^2 + 64t + 1$ is a parabola opening downward (Figure 2-27). We see that $s(t) > 50$ for t between 1.03 and 2.97 as expected.

Figure 2-27

$s(t) = -16t^2 + 64t + 1$
$y = 50$
2.97 sec
1.03 sec

Answers

7. a. $s(t) = -16t^2 + 80t + 5$

b. $\left(\dfrac{10 - \sqrt{55}}{4}, \dfrac{10 + \sqrt{55}}{4}\right)$

Skill Practice 7   Repeat Example 7 under the assumption that the rocket is launched with an initial speed of 80 ft/sec from a height of 5 ft.

SECTION 2.6   Practice Exercises

Prerequisite Review

For Exercises R.1–R.3, solve the inequality and write the solution in interval notation when possible.

R.1. $1 - 3(z - 5) \geq 3 + 3(2z + 7)$ $\left(-\infty, -\dfrac{8}{9}\right]$ **R.2.** $-4(3x + 2) < 1 - (x - 4) - 9x$ $\left(-\dfrac{13}{2}, \infty\right)$ **R.3.** $2w - 3 \geq 9$ or $w < -1$ $(-\infty, -1) \cup [6, \infty)$

For Exercises R.4–R.5, add the rational expressions.

R.4. $\dfrac{2}{w + 2} + \dfrac{1}{3w + 2}$ $\dfrac{7w + 6}{(w + 2)(3w + 2)}$ **R.5.** $y + 5 + \dfrac{1}{y - 5}$ $\dfrac{y^2 - 24}{y - 5}$

Concept Connections

1. Let $f(x)$ be a polynomial. An inequality of the form $f(x) < 0, f(x) > 0, f(x) \geq 0$, or $f(x) \leq 0$ is called a ___polynomial___ inequality. If the polynomial is of degree ___2___, then the inequality is also called a quadratic inequality.

2. Let $f(x)$ be a rational expression. An inequality of the form $f(x) < 0, f(x) > 0, f(x) \geq 0$, or $f(x) \leq 0$ is called a ___rational___ inequality.

3. The solution set for the inequality $(x + 10)^2 \geq -4$ is ___$(-\infty, \infty)$___, whereas the solution set for the inequality $(x + 10)^2 \leq -4$ is ___{ }___.

4. The solutions to an inequality $f(x) < 0$ are the values of x on the intervals where $f(x)$ is (positive/negative). negative

Objective 1: Solve Polynomial Inequalities

For Exercises 5–14, the graph of $y = f(x)$ is given. Solve the inequalities.

   a. $f(x) < 0$ **b.** $f(x) \le 0$ **c.** $f(x) > 0$ **d.** $f(x) \ge 0$

5.

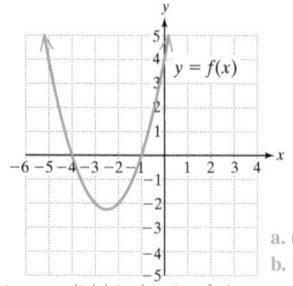

   a. $(-4, -1)$
   b. $[-4, -1]$
   c. $(-\infty, -4) \cup (-1, \infty)$ d. $(-\infty, -4] \cup [-1, \infty)$

6.

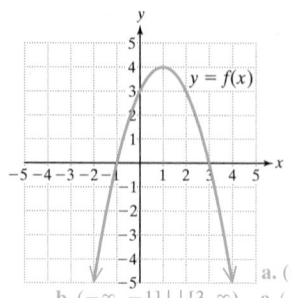

   a. $(-\infty, -1) \cup (3, \infty)$
   b. $(-\infty, -1] \cup [3, \infty)$ c. $(-1, 3)$ d. $[-1, 3]$

7.

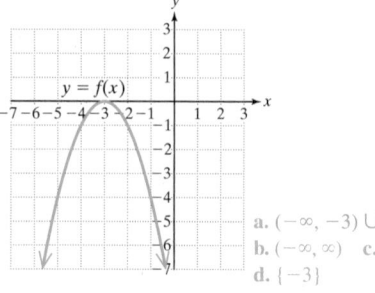

   a. $(-\infty, -3) \cup (-3, \infty)$
   b. $(-\infty, \infty)$ c. $\{\ \}$
   d. $\{-3\}$

8.

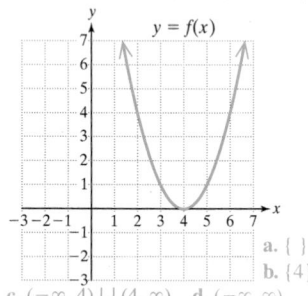

   a. $\{\ \}$
   b. $\{4\}$
   c. $(-\infty, 4) \cup (4, \infty)$ d. $(-\infty, \infty)$

9.

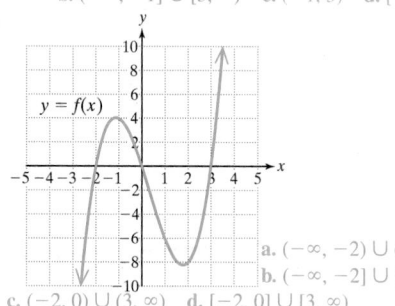

   a. $(-\infty, -2) \cup (0, 3)$
   b. $(-\infty, -2] \cup [0, 3]$
   c. $(-2, 0) \cup (3, \infty)$ d. $[-2, 0] \cup [3, \infty)$

10.

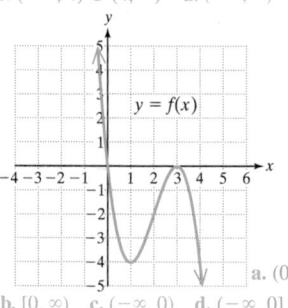

   a. $(-\infty, -4) \cup (0, 1)$
   b. $(-\infty, -4] \cup [0, 1]$
   c. $(-4, 0) \cup (1, \infty)$
   d. $[-4, 0] \cup [1, \infty)$

11.

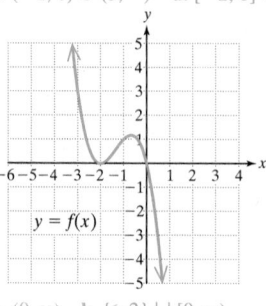

   a. $(0, 3) \cup (3, \infty)$
   b. $[0, \infty)$ c. $(-\infty, 0)$ d. $(-\infty, 0] \cup \{3\}$

12.

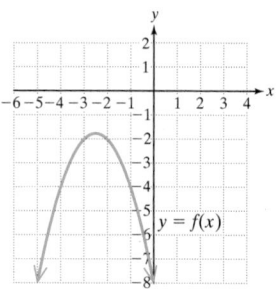

   a. $(0, \infty)$ b. $\{-2\} \cup [0, \infty)$
   c. $(-\infty, -2) \cup (-2, 0)$ d. $(-\infty, 0]$

13.

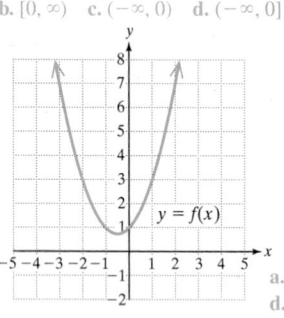

   a. $(-\infty, \infty)$ b. $(-\infty, \infty)$ c. $\{\ \}$ d. $\{\ \}$

14.

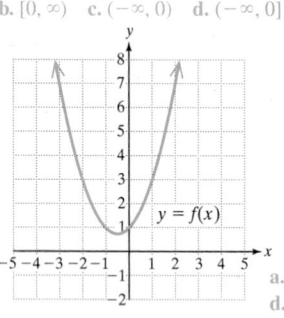

   a. $\{\ \}$ b. $\{\ \}$ c. $(-\infty, \infty)$
   d. $(-\infty, \infty)$

For Exercises 15–20, solve the equations and inequalities. (See Example 3)

15. a. $(5x - 3)(x - 5) = 0$
   b. $(5x - 3)(x - 5) < 0$
   c. $(5x - 3)(x - 5) \le 0$
   d. $(5x - 3)(x - 5) > 0$
   e. $(5x - 3)(x - 5) \ge 0$

16. a. $(3x + 7)(x - 2) = 0$
   b. $(3x + 7)(x - 2) < 0$
   c. $(3x + 7)(x - 2) \le 0$
   d. $(3x + 7)(x - 2) > 0$
   e. $(3x + 7)(x - 2) \ge 0$

17. a. $-x^2 + x + 12 = 0$ $\{-3, 4\}$
   b. $-x^2 + x + 12 < 0$ $(-\infty, -3) \cup (4, \infty)$
   c. $-x^2 + x + 12 \le 0$ $(-\infty, -3] \cup [4, \infty)$
   d. $-x^2 + x + 12 > 0$ $(-3, 4)$
   e. $-x^2 + x + 12 \ge 0$ $[-3, 4]$

18. a. $-x^2 - 10x - 9 = 0$ $\{-9, -1\}$
   b. $-x^2 - 10x - 9 < 0$
   c. $-x^2 - 10x - 9 \le 0$
   d. $-x^2 - 10x - 9 > 0$ $(-9, -1)$
   e. $-x^2 - 10x - 9 \ge 0$ $[-9, -1]$

19. a. $a^2 + 12a + 36 = 0$ $\{-6\}$
   b. $a^2 + 12a + 36 < 0$ $\{\ \}$
   c. $a^2 + 12a + 36 \le 0$ $\{-6\}$
   d. $a^2 + 12a + 36 > 0$
   e. $a^2 + 12a + 36 \ge 0$ $(-\infty, \infty)$

20. a. $t^2 - 14t + 49 = 0$ $\{7\}$
   b. $t^2 - 14t + 49 < 0$ $\{\ \}$
   c. $t^2 - 14t + 49 \le 0$ $\{7\}$
   d. $t^2 - 14t + 49 > 0$
   e. $t^2 - 14t + 49 \ge 0$ $(-\infty, \infty)$

For Exercises 21–54, solve the inequalities. (See Examples 1–2)

21. $4w^2 - 9 \geq 0$ $\left(-\infty, -\frac{3}{2}\right] \cup \left[\frac{3}{2}, \infty\right)$ **22.** $16z^2 - 25 < 0$ $\left(-\frac{5}{4}, \frac{5}{4}\right)$ **23.** $3w^2 + w < 2(w + 2)$ $\left(-1, \frac{4}{3}\right)$

24. $5y^2 + 7y < 3(y + 4)$ $\left(-2, \frac{6}{5}\right)$ **25.** $a^2 \geq 3a$ $(-\infty, 0] \cup [3, \infty)$ **26.** $d^2 \geq 6d$ $(-\infty, 0] \cup [6, \infty)$

27. $10 - 6x > 5x^2$ **28.** $6 - 4x > 3x^2$ **29.** $m^2 < 49$ $(-7, 7)$

30. $y^2 \geq 9$ $(-\infty, -3] \cup [3, \infty)$ **31.** $16p^2 \geq 2$ $\left(-\infty, -\frac{\sqrt{2}}{4}\right] \cup \left[\frac{\sqrt{2}}{4}, \infty\right)$ **32.** $54q^2 \leq 50$ $\left[-\frac{5\sqrt{3}}{9}, \frac{5\sqrt{3}}{9}\right]$

33. $(x + 4)(x - 1)(x - 3) \geq 0$ **34.** $(x + 2)(x + 5)(x - 4) \geq 0$ **35.** $-5c(c + 2)^2(4 - c) > 0$
$[-4, 1] \cup [3, \infty)$ $[-5, -2] \cup [4, \infty)$ $(-\infty, -2) \cup (-2, 0) \cup (4, \infty)$

36. $-6u(u + 1)^2(3 - u) > 0$ **37.** $t^4 - 10t^2 + 9 \leq 0$ $[-3, -1] \cup [1, 3]$ **38.** $w^4 - 20w^2 + 64 \leq 0$
$(-\infty, -1) \cup (-1, 0) \cup (3, \infty)$ $[-4, -2] \cup [2, 4]$

39. $2x^3 + 5x^2 < 8x + 20$ **40.** $3x^3 - 3x < 4x^2 - 4$ **41.** $-2x^4 + 10x^3 - 6x^2 - 18x \geq 0$
$[-1, 0] \cup \{3\}$

42. $-4x^4 + 4x^3 + 64x^2 + 80x \geq 0$ **43.** $-5u^6 + 28u^5 - 15u^4 \leq 0$ **44.** $-3w^6 + 8w^5 - 4w^4 \leq 0$ $\left(-\infty, \frac{2}{3}\right] \cup [2, \infty)$
$\{-2\} \cup [0, 5]$

45. $6x(2x - 5)^4(3x + 1)^5(x - 4) < 0$ **46.** $5x(3x - 2)^2(4x + 1)^3(x - 3)^4 < 0$ **47.** $(5x - 3)^2 > -2$ $(-\infty, \infty)$

48. $(4x + 1)^2 > -6$ $(-\infty, \infty)$ **49.** $-4 \geq (x - 7)^2$ $\{\ \}$ **50.** $-1 \geq (x + 2)^2$ $\{\ \}$

51. $16y^2 > 24y - 9$ **52.** $4w^2 > 20w - 25$ $\left(-\infty, \frac{5}{2}\right) \cup \left(\frac{5}{2}, \infty\right)$ **53.** $(x + 3)(x + 1) \leq -1$ $\{-2\}$

54. $(x + 2)(x + 4) \leq -1$ $\{-3\}$

Objective 2: Solve Rational Inequalities

For Exercises 55–58, the graph of $y = f(x)$ is given. Solve the inequalities.

  a. $f(x) < 0$ **b.** $f(x) \leq 0$ **c.** $f(x) > 0$ **d.** $f(x) \geq 0$

55. 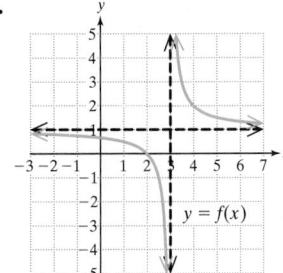   a. $(2, 3)$
   b. $[2, 3)$
   c. $(-\infty, 2) \cup (3, \infty)$
   d. $(-\infty, 2] \cup (3, \infty)$

56.   a. $(-\infty, -2) \cup (2, \infty)$
   b. $(-\infty, -2] \cup (2, \infty)$
   c. $(-2, 2)$
   d. $[-2, 2)$

57.   a. $(-\infty, 2) \cup (2, \infty)$
   b. $(-\infty, 2) \cup (2, \infty)$
   c. $\{\ \}$
   d. $\{\ \}$

58.   a. $\{\ \}$
   b. $\{\ \}$
   c. $(-\infty, -3) \cup (-3, \infty)$
   d. $(-\infty, -3) \cup (-3, \infty)$

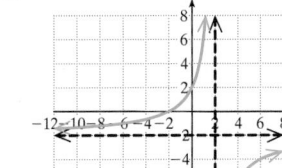
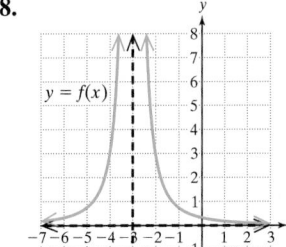

For Exercises 59–62, solve the inequalities. (See Example 6)

59. a. $\dfrac{x + 2}{x - 3} \leq 0$ $[-2, 3)$ **60. a.** $\dfrac{x + 4}{x - 1} \leq 0$ $[-4, 1)$ **61. a.** $\dfrac{x^4}{x^2 + 9} \leq 0$ $\{0\}$ **62. a.** $\dfrac{-x^2}{x^4 + 16} \leq 0$ $(-\infty, \infty)$

b. $\dfrac{x + 2}{x - 3} < 0$ $(-2, 3)$ **b.** $\dfrac{x + 4}{x - 1} < 0$ $(-4, 1)$ **b.** $\dfrac{x^4}{x^2 + 9} < 0$ $\{\ \}$ **b.** $\dfrac{-x^2}{x^4 + 16} < 0$ $(-\infty, 0) \cup (0, \infty)$

c. $\dfrac{x + 2}{x - 3} \geq 0$ $(-\infty, -2] \cup (3, \infty)$ **c.** $\dfrac{x + 4}{x - 1} \geq 0$ $(-\infty, -4] \cup (1, \infty)$ **c.** $\dfrac{x^4}{x^2 + 9} \geq 0$ $(-\infty, \infty)$ **c.** $\dfrac{-x^2}{x^4 + 16} \geq 0$ $\{0\}$

d. $\dfrac{x + 2}{x - 3} > 0$ $(-\infty, -2) \cup (3, \infty)$ **d.** $\dfrac{x + 4}{x - 1} > 0$ $(-\infty, -4) \cup (1, \infty)$ **d.** $\dfrac{x^4}{x^2 + 9} > 0$ $(-\infty, 0) \cup (0, \infty)$ **d.** $\dfrac{-x^2}{x^4 + 16} > 0$ $\{\ \}$

For Exercises 63–84, solve the inequalities. (See Examples 4–5)

63. $\dfrac{5 - x}{x + 1} \geq 0$ $(-1, 5]$ **64.** $\dfrac{2 - x}{x + 6} \geq 0$ $(-6, 2]$ **65.** $\dfrac{4 - 2x}{x^2} \leq 0$ $[2, \infty)$

66. $\dfrac{9 - 3x}{x^2} \le 0$ $[3, \infty)$

67. $\dfrac{w^2 - w - 2}{w + 3} \ge 0$ $(-3, -1] \cup [2, \infty)$

68. $\dfrac{p^2 - 2p - 8}{p - 1} \ge 0$ $[-2, 1) \cup [4, \infty)$

69. $\dfrac{5}{2t - 7} > 1$ $\left(\dfrac{7}{2}, 6\right)$

70. $\dfrac{4}{3c - 8} > 1$ $\left(\dfrac{8}{3}, 4\right)$

71. $\dfrac{2x}{x - 2} \le 2$ $(-\infty, 2)$

72. $\dfrac{3x}{3x - 7} \le 1$ $\left(-\infty, \dfrac{7}{3}\right)$

73. $\dfrac{4 - x}{x + 5} \ge 2$ $(-5, -2]$

74. $\dfrac{3 - x}{x + 2} \ge 4$ $(-2, -1]$

75. $\dfrac{a - 2}{a^2 + 4} \le 0$ $(-\infty, 2]$

76. $\dfrac{d - 3}{d^2 + 1} \le 0$ $(-\infty, 3]$

77. $\dfrac{10}{x + 2} \ge \dfrac{2}{x + 2}$ $(-2, \infty)$

78. $\dfrac{4}{x - 3} \ge \dfrac{1}{x - 3}$ $(3, \infty)$

79. $\dfrac{4}{y + 3} > -\dfrac{2}{y}$ $(-3, -1) \cup (0, \infty)$

80. $\dfrac{2}{z - 1} > -\dfrac{4}{z}$ $\left(0, \dfrac{2}{3}\right) \cup (1, \infty)$

81. $\dfrac{3}{4 - x} \le \dfrac{6}{1 - x}$ $(-\infty, 1) \cup (4, 7]$

82. $\dfrac{5}{2 - x} \le \dfrac{3}{3 - x}$ $(2, 3) \cup \left[\dfrac{9}{2}, \infty\right)$

83. $\dfrac{(2 - x)(2x + 1)^2}{(x - 4)^4} \le 0$ $[2, 4) \cup (4, \infty)$

84. $\dfrac{(3 - x)(4x - 1)^4}{(x + 2)^2} \le 0$ $[3, \infty) \cup \left\{\dfrac{1}{4}\right\}$

Objective 3: Solve Applications Involving Polynomial and Rational Inequalities

85. A professional fireworks team shoots an 8-in. mortar straight upwards from ground level with an initial speed of 216 ft/sec. (**See Example 7**)

  a. Write a function modeling the vertical position $s(t)$ (in ft) of the shell at a time t seconds after launch. $s(t) = -16t^2 + 216t$

  b. The mortar is designed to explode when the shell is at its maximum height. How long after launch will the shell explode? (*Hint*: Consider the vertex formula from Section 2.1.)   The shell will explode 6.75 sec after launch.

  c. The spectators can see the shell rising once it clears a 200-ft tree line. For what period of time after launch is the shell visible before it explodes?   The spectators can see the shell between 1 sec and 6.75 sec after launch.

86. Suppose that a basketball player jumps straight up for a rebound.

  a. If his initial speed leaving the ground is 16 ft/sec, write a function modeling his vertical position $s(t)$ (in ft) at a time t seconds after leaving the ground. $s(t) = -16t^2 + 16t$

  b. Find the times after leaving the ground when the player will be at a height of more than 3 ft in the air.   The player will be more than 3 ft in the air between 0.25 sec and 0.75 sec after leaving the ground.

87. For a certain stretch of road, the distance d (in ft) required to stop a car that is traveling at speed v (in mph) before the brakes are applied can be approximated by $d(v) = 0.06v^2 + 2v$. Find the speeds for which the car can be stopped within 250 ft.   The car will stop within 250 ft if the car is traveling less than 50 mph.

88. The population $P(t)$ of a bacteria culture is given by $P(t) = -1500t^2 + 60{,}000t + 10{,}000$, where t is the time in hours after the culture is started. Determine the time(s) at which the population will be greater than 460,000 organisms.   The population will be greater than 460,000 organisms between 10 hr and 30 hr after the culture is started.

89. Suppose that an object that is originally at room temperature of 32°C is placed in a freezer. The temperature $T(x)$ (in °C) of the object can be approximated by the model $T(x) = \dfrac{320}{x^2 + 3x + 10}$, where x is the time in hours after the object is placed in the freezer.

  a. What is the horizontal asymptote of the graph of this function and what does it represent in the context of this problem?   The horizontal asymptote is $y = 0$ and means that the temperature will approach 0°C as time increases without bound.

  b. A chemist needs a compound cooled to less than 5°C. Determine the amount of time required for the compound to cool so that its temperature is less than 5°C.   More than 6 hr is required for the temperature to fall below 5°C.

90. The average round trip speed S (in mph) of a vehicle traveling a distance of d miles in each direction is given by

$$S = \dfrac{2d}{\dfrac{d}{r_1} + \dfrac{d}{r_2}}$$

where r_1 and r_2 are the rates of speed for the initial trip and the return trip, respectively.

  a. Suppose that a motorist travels 200 mi from her home to an athletic event and averages 50 mph for the trip to the event. Determine the speeds necessary if the motorist wants the average speed for the round trip to be at least 60 mph?   The motorist must travel at least 75 mph on the return trip to average at least 60 mph for the round trip.

  b. Would the motorist be traveling within the speed limit of 70 mph?   No.

91. A rectangular quilt is to be made so that the length is 1.2 times the width. The quilt must be between 72 ft² and 96 ft² to cover the bed. Determine the restrictions on the width so that the dimensions of the quilt will meet the required area. Give exact values and the approximated values to the nearest tenth of a foot.   The width should be between $2\sqrt{15}$ ft and $4\sqrt{5}$ ft. This is between approximately 7.7 ft and 8.9 ft.

92. A landscaping team plans to build a rectangular garden that is between 480 yd² and 720 yd² in area. For aesthetic reasons, they also want the length to be 1.5 times the width. Determine the restrictions on the width so that the dimensions of the garden will meet the required area. Give exact values and the approximated values to the nearest tenth of a yard.   The width should be between $8\sqrt{5}$ yd and $4\sqrt{30}$ yd. This is between approximately 17.9 yd and 21.9 yd.

Mixed Exercises

For Exercises 93–102, write the domain of the function in interval notation.

93. $f(x) = \sqrt{9 - x^2}$ $[-3, 3]$

94. $g(t) = \sqrt{1 - t^2}$ $[-1, 1]$

95. $h(a) = \sqrt{a^2 - 5}$ $(-\infty, -\sqrt{5}] \cup [\sqrt{5}, \infty)$

96. $f(u) = \sqrt{u^2 - 7}$ $(-\infty, -\sqrt{7}] \cup [\sqrt{7}, \infty)$

97. $p(x) = \sqrt{2x^2 + 9x - 18}$ $(-\infty, -6] \cup \left[\dfrac{3}{2}, \infty\right)$

98. $q(x) = \sqrt{4x^2 + 7x - 2}$ $(-\infty, -2] \cup \left[\dfrac{1}{4}, \infty\right)$

99. $r(x) = \dfrac{1}{\sqrt{2x^2 + 9x - 18}}$ $(-\infty, -6) \cup \left(\dfrac{3}{2}, \infty\right)$

100. $s(x) = \dfrac{1}{\sqrt{4x^2 + 7x - 2}}$ $(-\infty, -2) \cup \left(\dfrac{1}{4}, \infty\right)$

101. $h(x) = \sqrt{\dfrac{3x}{x + 2}}$ $(-\infty, -2) \cup [0, \infty)$

102. $k(x) = \sqrt{\dfrac{2x}{x + 1}}$ $(-\infty, -1) \cup [0, \infty)$

103. Let a, b, and c represent positive real numbers, where $a < b < c$, and let $f(x) = (x - a)^2(b - x)(x - c)^3$.

a. Complete the sign chart.

Sign of $(x - a)^2$:			
Sign of $(b - x)$:			
Sign of $(x - c)^3$:			
Sign of $(x - a)^2(b - x)(x - c)^3$:			
	a	b	c

b. Solve $f(x) > 0$.   **c.** Solve $f(x) < 0$.

104. Let a, b, and c represent positive real numbers, where $a < b < c$, and let $g(x) = \dfrac{(a - x)(x - b)^2}{(c - x)^5}$.

a. Complete the sign chart.

Sign of $(a - x)$:			
Sign of $(x - b)^2$:			
Sign of $(c - x)^5$:			
Sign of $\dfrac{(a - x)(x - b)^2}{(c - x)^5}$:			
	a	b	c

b. Solve $g(x) > 0$.   **c.** Solve $g(x) < 0$.

Write About It

105. Explain how the solution set to the inequality $f(x) < 0$ is related to the graph of $y = f(x)$.

106. Explain how the solution set to the inequality $f(x) \geq 0$ is related to the graph of $y = f(x)$.

107. Explain why $\dfrac{x^2 + 2}{x^2 + 1} < 0$ has no solution.

108. Given $\dfrac{x - 3}{x - 1} \leq 0$, explain why the solution set includes 3, but does not include 1.

Expanding Your Skills

The procedure to solve a polynomial or rational inequality may be applied to all inequalities of the form $f(x) > 0$, $f(x) < 0$, $f(x) \geq 0$, and $f(x) \leq 0$. That is, find the real solutions to the related equation and determine restricted values of x. Then determine the sign of $f(x)$ on each interval defined by the boundary points. Use this process to solve the inequalities in Exercises 109–120.

109. $\sqrt{2x - 6} - 2 < 0$ $[3, 5)$

110. $\sqrt{3x - 5} - 4 < 0$ $\left[\dfrac{5}{3}, 7\right)$

111. $\sqrt{4 - x} - 6 \geq 0$ $(-\infty, -32]$

112. $\sqrt{5 - x} - 7 \geq 0$ $(-\infty, -44]$

113. $\dfrac{1}{\sqrt{x - 2} - 4} \leq 0$ $[2, 18)$

114. $\dfrac{1}{\sqrt{x - 3} - 5} \leq 0$ $[3, 28)$

115. $-3 < x^2 - 6x + 5 \leq 5$ $[0, 2) \cup (4, 6]$

116. $8 \leq x^2 + 4x + 3 < 15$ $(-6, -5] \cup [1, 2)$

117. $|x^2 - 4| < 5$ $(-3, 3)$

118. $|x^2 + 1| < 17$ $(-4, 4)$

119. $|x^2 - 18| > 2$ $(-\infty, -2\sqrt{5}) \cup (-4, 4) \cup (2\sqrt{5}, \infty)$

120. $|x^2 - 6| > 3$ $(-\infty, -3) \cup (-\sqrt{3}, \sqrt{3}) \cup (3, \infty)$

Technology Connections

121. Given the inequality,
$0.552x^3 + 4.13x^2 - 1.84x - 3.5 < 6.7$,

a. Write the inequality in the form $f(x) < 0$.

b. Graph $y = f(x)$ on a suitable viewing window.

c. Use the **Zero** feature to approximate the real zeros of $f(x)$. Round to 1 decimal place.

d. Use the graph to approximate the solution set for the inequality $f(x) < 0$.

122. Given the inequality,
$0.24x^4 + 1.8x^3 + 3.3x^2 + 2.84x - 1.8 > 4.5$,

a. Write the inequality in the form $f(x) > 0$.

b. Graph $y = f(x)$ on a suitable viewing window.

c. Use the **Zero** feature to approximate the real zeros of $f(x)$. Round to 1 decimal place.

d. Use the graph to approximate the solution set for the inequality $f(x) > 0$.

123. An engineer for a food manufacturer designs an aluminum container for a hot drink mix. The container is to be a right circular cylinder 5.5 in. in height. The surface area represents the amount of aluminum used and is given by

$$S(r) = 2\pi r^2 + 11\pi r,$$ where r is the radius of the can.

  a. Graph the function $y = S(r)$ and the line $y = 90$ on the viewing window [0, 3, 1] by [0, 150, 10].

  b. Use the **Intersect** feature to approximate point of intersection of $y = S(r)$ and $y = 90$. Round to 1 decimal place if necessary.

  c. Determine the restrictions on r so that the amount of aluminum used is at most 90 in.^{2}. Round to 1 decimal place.

124. The concentration $C(t)$ (in ng/mL) of a drug in the bloodstream t hours after ingestion is modeled by

$$C(t) = \frac{500t}{t^3 + 100}$$

  a. Graph the function $y = C(t)$ and the line $y = 4$ on the window [0, 32, 4] by [0, 15, 3].

  b. Use the **Intersect** feature to approximate the point(s) of intersection of $y = C(t)$ and $y = 4$. Round to 1 decimal place if necessary.

  c. To avoid toxicity, a physician may give a second dose of the medicine once the concentration falls below 4 ng/mL for increasing values of t. Determine the times at which it is safe to give a second dose. Round to 1 decimal place.

PROBLEM RECOGNITION EXERCISES

Solving Equations and Inequalities

At this point in the text, we have studied several categories of equations and related inequalities. These include

- linear equations and inequalities
- quadratic equations and inequalities
- polynomial equations and inequalities
- rational equations and inequalities

- radical equations and inequalities
- absolute value equations and inequalities
- compound inequalities

For Exercises 1–30, solve the equations and inequalities. Write the solution sets to the inequalities in interval notation if possible.

1. $-\dfrac{1}{2} \le -\dfrac{1}{4}x - 5 < 2$ $(-28, -18]$

2. $2x^2 - 6x = 5$ $\left\{\dfrac{3 \pm \sqrt{19}}{2}\right\}$

3. $50x^3 - 25x^2 - 2x + 1 = 0$ $\left\{\pm\dfrac{1}{5}, \dfrac{1}{2}\right\}$

4. $\dfrac{-5x(x-3)^2}{2+x} \le 0$ $(-\infty, -2) \cup [0, \infty)$

5. $\sqrt[4]{m+4} - 5 = -2$ $\{77\}$

6. $-5 < y$ and $-3y + 4 \ge 7$ $(-5, -1]$

7. $|5t - 4| + 2 = 7$ $\left\{-\dfrac{1}{5}, \dfrac{9}{5}\right\}$

8. $3 - 4\{x - 5[x + 2(3 - 2x)]\} = -2[4 - (x - 1)]$ $\left\{\dfrac{133}{66}\right\}$

9. $10x(2x - 14) = -29x^2 - 100$ $\left\{\dfrac{10}{7}\right\}$

10. $\dfrac{5}{y-4} = \dfrac{3y}{y+2} - \dfrac{2y^2 - 14y}{y^2 - 2y - 8}$ $\{5\}$; The value -2 does not check.

11. $x(x - 14) \le -40$ $[4, 10]$

12. $\dfrac{1}{x^2 - 14x + 40} \le 0$ $(4, 10)$

13. $|x - 0.15| = |x + 0.05|$ $\{0.05\}$

14. $\sqrt{t - 1} - 5 \le 1$ $[1, 37]$

15. $n^{1/2} + 7 = 10$ $\{9\}$

16. $-4x(3 - x)(x + 2)^2(x - 5)^3 \ge 0$ $\{-2\} \cup [0, 3] \cup [5, \infty)$

17. $-2x - 5(x + 3) = -4(x + 2) - 3x$ $\{\}$

18. $\sqrt{7x + 29} - 3 = x$ $\{5\}$; The value -4 does not check.

19. $(x^2 - 9)^2 - 5(x^2 - 9) - 14 = 0$ $\{\pm 4, \pm\sqrt{7}\}$

20. $2 + 7x^{-1} - 15x^{-2} = 0$ $\left\{\dfrac{3}{2}, -5\right\}$

21. $|8x - 3| + 10 \le 7$ $\{\}$

22. $2(x - 1)^{3/4} = 16$ $\{17\}$

23. $x^3 - 3x^2 < 6x - 8$ $(-\infty, -2) \cup (1, 4)$

24. $\dfrac{3 - x}{x + 5} \ge 1$ $(-5, -1]$

25. $15 - 3(x - 1) = -2x - (x - 18)$ $(-\infty, \infty)$

26. $25x^2 + 70x > -49$ $\left(-\infty, -\dfrac{7}{5}\right) \cup \left(-\dfrac{7}{5}, \infty\right)$

27. $2 < |3 - x| - 9$ $(-\infty, -8) \cup (14, \infty)$

28. $-4(x - 3) < 8$ or $-7 > x - 3$ $(-\infty, -4) \cup (1, \infty)$

29. $\dfrac{1}{3}x + \dfrac{2}{5} > \dfrac{5}{6}x - 1$ $\left(-\infty, \dfrac{14}{5}\right)$

30. $2|2x + 1| - 2 \le 8$ $[-3, 2]$

SECTION 2.7 Variation

1. Write Models Involving Direct, Inverse, and Joint Variation

The familiar relationship $d = rt$ tells us that distance traveled equals the rate of speed times the time of travel. For a car traveling 60 mph, we have $d = 60t$. From Table 2-2, notice that as the time of travel increases, the distance increases proportionally. We say that d is directly proportional to t, or that d varies directly as t. This is shown graphically in Figure 2-28.

Table 2-2

t (hr)	d (mi)
1	60
2	120
3	180
4	240
5	300
6	360

$d = 60t$

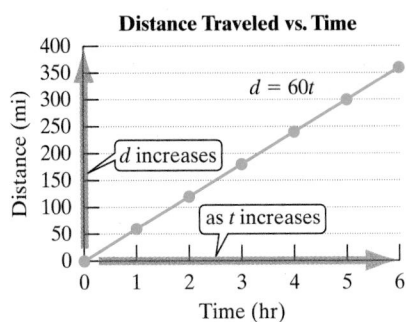

Figure 2-28

Now suppose that a motorist travels a fixed distance of 240 mi. We have

$$d = rt$$

$$240 = rt \longrightarrow t = \frac{240}{r}$$

The time of travel t varies *inversely* as the rate of speed. As the rate r increases, the time of travel will decrease proportionally. Likewise, for slower rates, the time of travel is greater. See Table 2-3 and Figure 2-29.

Table 2-3

r (mph)	t (hr)
10	24
20	12
30	8
40	6
50	4.8
60	4

$t = \dfrac{240}{r}$

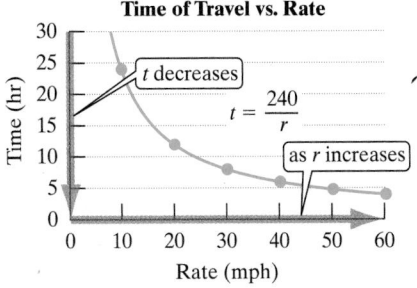

Figure 2-29

Direct and Inverse Variation

Let k be a nonzero constant real number. The statements on the left are equivalent to the equation on the right.

1. y varies **directly** as x.
   y is **directly** proportional to x. $\Big\} y = kx$

2. y varies **inversely** as x.
   y is **inversely** proportional to x. $\Big\} y = \dfrac{k}{x}$

Note: The value of k is called the **constant of variation.**

The first step in using a variation model is to write an English statement as an equivalent mathematical equation.

Classroom Examples: p. 342
Exercises 12, 14, 18

EXAMPLE 1 **Writing a Variation Model**

Write a variation model using k as the constant of variation.

Solution:

a. The amount of medicine A prescribed by a physician varies directly as the weight of the patient w.

$A = kw$

Since the variables are directly related, set up the *product* of k and w.

b. The frequency f in a vibrating string is inversely proportional to the length L of the string.

$f = \dfrac{k}{L}$

Since the variables are inversely related, set up the *quotient* of k and L.

c. The variable y varies directly as the square of x and inversely as the square root of z.

$y = \dfrac{kx^2}{\sqrt{z}}$

Since the square of the variable x is directly related to y, set up the *product* of k and x^2. And since the square root of z is inversely related to y, set up the *quotient* of k and $\sqrt{z}$.

TIP Notice that in each variation model, the constant of variation, k, is always in the numerator.

Skill Practice 1 Write a variation model using k as the constant of variation.

a. The distance d that a spring stretches varies directly as the force F applied to the spring.

b. The force F required to keep a car from skidding on a curved road varies inversely as the radius r of the curve.

c. The variable a varies directly as b and inversely as the cube root of c.

Sometimes a variable varies directly as the product of two or more other variables. In such a case we have joint variation.

Joint Variation

Let k be a nonzero constant real number. The statements on the left are equivalent to the equation on the right.

y varies **jointly** as w and x.
y is **jointly** proportional to w and x. $\Big\}$ $y = kwx$

Classroom Example: p. 342
Exercise 20

EXAMPLE 2 **Writing a Joint Variation Model**

Write a variation model using k as the constant of variation.

Solution:

a. y varies jointly as t and the cube root of u.

$y = kt\sqrt[3]{u}$

The variable t and the quantity $\sqrt[3]{u}$ are jointly related to y. Set up the product of k, t, and $\sqrt[3]{u}$.

b. The gravitational force of attraction between two planets varies jointly as the product of their masses and inversely as the square of the distance between them.

$F = \dfrac{km_1m_2}{d^2}$

Let m_1 and m_2 represent the masses of the planets, let d represent the distance between the planets, and let F represent the gravitational force between the planets.

Answers

1. a. $d = kF$ **b.** $F = \dfrac{k}{r}$

c. $a = \dfrac{kb}{\sqrt[3]{c}}$

Skill Practice 2   Write a variation model using k as the constant of variation.

> **a.** The kinetic energy E of an object varies jointly as the object's mass m and the square of its velocity v.
>
> **b.** z varies jointly as x and y and inversely as the square root of w.

2. Solve Applications Involving Variation

Consider the variation models $y = kx$ and $y = \dfrac{k}{x}$. In either case, if values for x and y are known, we can solve for k. Once k is known, we can write a variation model and use it to find y if x is known, or to find x if y is known. This concept is the basis for solving many applications involving variation.

> **Procedure to Solve an Application Involving Variation**
>
> **Step 1**   Write a general variation model that relates the variables given in the problem. Let k represent the constant of variation.
> **Step 2**   Solve for k by substituting known values of the variables into the model from step 1.
> **Step 3**   Substitute the value of k into the original variation model from step 1.
> **Step 4**   Use the variation model from step 3 to solve the application.

Classroom Example: p. 343
Exercise 30

EXAMPLE 3   **Solving an Application Involving Direct Variation**

The amount of an allergy medicine that a physician prescribes for a child varies directly as the weight of the child. Clinical research suggests that 13.5 mg of the drug should be given for a 30-lb child.

> **a.** How much should be prescribed for a 50-lb child?
> **b.** How much should be prescribed for a 60-lb child?
> **c.** A nurse wants to double check the dosage on a doctor's order of 18 mg. For a child of what weight is this dosage appropriate?

Solution:

Let A represent the amount of medicine.   Label the variables.
Let w represent the weight of the child.

$A = kw$ **Step 1:** Write a general variation model.

$13.5 = k(30)$ **Step 2:** Substitute known values of A and w into the variation model.

$\dfrac{13.5}{30} = k$ Solve for k by dividing both sides by 30.

$k = 0.45$

$A = 0.45w$ **Step 3:** Substitute the value of k into the original variation model.

a. $A = 0.45(50)$ **Step 4:** Solve the application by substituting 50 for w.

$A = 22.5$ A 50-lb child would require 22.5 mg of the drug.

Answers

2. a. $E = kmv^2$ **b.** $z = \dfrac{kxy}{\sqrt{w}}$

TIP Notice from Example 3 that more medicine is given in proportion to a patient's weight.

A 60-lb patient weighs twice as much as a 30-lb patient and the amount of medicine given is also twice as much. This is consistent with direct variation.

b. $A = 0.45(60)$ — Substitute 60 for w.

$A = 27$ — A 60-lb child would require 27 mg of the drug.

c. $A = 0.45w$

$18 = 0.45w$ — Substitute 18 mg for the amount of medicine.

$40 = w$ — Solve for the weight w. A 40-lb child would receive 18 mg.

Skill Practice 3 The amount of the medicine ampicillin that a physician prescribes for a child varies directly as the weight of the child. A physician prescribes 420 mg for a 35-lb child.

a. How much should be prescribed for a 30-lb child?

b. How much should be prescribed for a 40-lb child?

Classroom Example: p. 343
Exercise 40

EXAMPLE 4 Solving an Application Involving Inverse Variation

The loudness of sound measured in decibels (dB) varies inversely as the square of the distance between the listener and the source of the sound. If the loudness of sound is 17.92 dB at a distance of 10 ft from a stereo speaker, what is the decibel level 20 ft from the speaker?

Solution:

Let L represent the loudness of sound in decibels and d represent the distance in feet. The inverse relationship between decibel level and the square of the distance is modeled by:

$$L = \frac{k}{d^2}$$

$$17.92 = \frac{k}{(10)^2}$$ — Substitute $L = 17.92$ dB and $d = 10$ ft.

$$17.92 = \frac{k}{100}$$

$$(17.92)100 = \frac{k}{100} \cdot 100$$ — Solve for k (clear fractions).

$$k = 1792$$

$$L = \frac{1792}{d^2}$$ — Substitute $k = 1792$ into the original model $L = \frac{k}{d^2}$.

With the value of k known, we can find L for any value of d.

$$L = \frac{1792}{(20)^2}$$ — Find the loudness when $d = 20$ ft.

$$= 4.48 \text{ dB}$$

Notice that the loudness of sound is 17.92 dB at a distance 10 ft from the speaker. When the distance from the speaker is increased to 20 ft, the decibel level decreases to 4.48 dB. This is consistent with an inverse relationship. For $k > 0$, as one variable is increased, the other is decreased. It also seems reasonable that the farther one moves away from the source of a sound, the softer the sound becomes.

Skill Practice 4 The yield on a bond varies inversely as the price. The yield on a particular bond is 4% when the price is $100. Find the yield when the price is $80.

Answers

3. a. 360 mg **b.** 480 mg

4. 5%

Classroom Example: p. 344
Exercise 44

EXAMPLE 5   Solving an Application Involving Joint Variation

In the early morning hours of August 29, 2005, Hurricane Katrina plowed into the Gulf Coast of the United States, bringing unprecedented destruction to southern Louisiana, Mississippi, and Alabama. The winds of a hurricane are strong enough to send a piece of plywood through a tree.

The kinetic energy of an object varies jointly as the weight of the object at sea level and as the square of its velocity. During a hurricane, a 0.5-lb stone traveling 60 mph has 81 joules (J) of kinetic energy. Suppose the wind speed doubles to 120 mph. Find the kinetic energy.

Solution:

Let E represent the kinetic energy, let w represent the weight, and let v represent the velocity of the stone. The variation model is

$$E = kwv^2$$
$$81 = k(0.5)(60)^2 \quad \text{Substitute } E = 81 \text{ J, } w = 0.5 \text{ lb, and } v = 60 \text{ mph.}$$
$$81 = k(0.5)(3600) \quad \text{Simplify the exponent.}$$
$$81 = k(1800)$$
$$\frac{81}{1800} = \frac{k(1800)}{1800} \quad \text{Divide by 1800.}$$
$$0.045 = k \quad \text{Solve for } k.$$

With the value of k known, the model $E = kwv^2$ can be written as $E = 0.045wv^2$. We now find the kinetic energy of a 0.5-lb stone traveling 120 mph.

$$E = 0.045(0.5)(120)^2$$
$$= 324$$

The kinetic energy of a 0.5-lb stone traveling 120 mph is 324 J.

Skill Practice 5   The amount of simple interest earned in an account varies jointly as the interest rate and time of the investment. An account earns $200 in 2 yr at 4% interest. How much interest would be earned in 3 yr at a rate of 5%?

Answer

5. $375

SECTION 2.7   Practice Exercises

Prerequisite Review

For Exercises R.1–R.3, solve for the indicated variable.

R.1. $A = Ptr$ for r $\quad r = \dfrac{A}{Pt}$

R.2. $V = \dfrac{1}{3}\pi r^2 h$ for h $\quad h = \dfrac{3V}{\pi r^2}$

R.3. $K = \dfrac{IR}{E}$ for E $\quad E = \dfrac{IR}{K}$

R.4. Solve for x. $\dfrac{16}{12} = \dfrac{24}{x}$ {18}

Concept Connections

1. If k is a nonzero constant real number, then the statement $y = kx$ implies that y varies _____directly_____ as x.

2. If k is a nonzero constant real number, then the statement $y = \frac{k}{x}$ implies that y varies _____inversely_____ as x.

3. The value of k in the variation models $y = kx$ and $y = \frac{k}{x}$ is called the _____constant_____ of _____variation_____.

4. If y varies directly as two or more other variables such as x and w, then $y = kxw$, and we say that y varies _____jointly_____ as x and w.

5. a. Given $y = 2x$, evaluate y for the given values of x: $x = 1$, $x = 2$, $x = 3$, $x = 4$, and $x = 5$. 2; 4; 6; 8; 10

   b. How does y change when x is doubled? y is also doubled.

   c. How does y change when x is tripled? y is also tripled.

   d. Complete the statement. Given $y = 2x$, when x increases, y (increases/decreases) proportionally. increases

   e. Complete the statement. Given $y = 2x$, when x decreases, y (increases/decreases) proportionally. decreases

6. a. Given $y = \frac{24}{x}$, evaluate y for the given values of x: $x = 1$, $x = 2$, $x = 3$, $x = 4$, and $x = 6$. 24; 12; 8; 6; 4

   b. How does y change when x is doubled? y is one-half its original value.

   c. How does y change when x is tripled? y is one-third its original value.

   d. Complete the statement. Given $y = \frac{24}{x}$, when x increases, y (increases/decreases) proportionally. decreases

   e. Complete the statement. Given $y = \frac{24}{x}$, when x decreases, y (increases/decreases) proportionally. increases

7. The time required to drive from Atlanta, Georgia, to Nashville, Tennessee, varies _____inversely_____ as the average speed at which a vehicle travels.

8. The amount of a person's paycheck varies _____directly_____ as the number of hours worked.

9. The volume of a right circular cone varies _____jointly_____ as the square of the radius of the cylinder and as the height of the cylinder.

10. A student's grade on a test varies _____directly_____ as the number of hours the student spends studying for the test.

Objective 1: Write Models Involving Direct, Inverse, and Joint Variation

For Exercises 11–20, write a variation model using k as the constant of variation. (See Examples 1–2)

11. The circumference C of a circle varies directly as its radius r. $C = kr$

12. Simple interest I on a loan or investment varies directly as the amount A of the loan. $I = kA$

13. The average cost per minute $\overline{C}$ for a flat rate cell phone plan is inversely proportional to the number of minutes used n. $\overline{C} = \dfrac{k}{n}$

14. The time of travel t is inversely proportional to the rate of travel r. $t = \dfrac{k}{r}$

15. The volume V of a right circular cylinder varies jointly as the height h of the cylinder and as the square of the radius r of the cylinder. $V = khr^2$

16. The volume V of a rectangular solid varies jointly as the length l and width w of the solid. $V = klw$

17. The variable E is directly proportional to s and inversely proportional to the square root of n. $E = \dfrac{ks}{\sqrt{n}}$

18. The variable n is directly proportional to the square of σ and inversely proportional to the square of E. $n = \dfrac{k\sigma^2}{E^2}$

19. The variable c varies jointly as m and n and inversely as the cube of t. $c = \dfrac{kmn}{t^3}$

20. The variable d varies jointly as u and v and inversely as the cube root of T. $d = \dfrac{kuv}{\sqrt[3]{T}}$

Objective 2: Solve Applications Involving Variation

For Exercises 21–26, find the constant of variation k.

21. y varies directly as x. When x is 8, y is 20. $k = \dfrac{5}{2}$

22. m varies directly as x. When x is 10, m is 42. $k = \dfrac{21}{5}$

23. p is inversely proportional to q. When q is 18, p is 54. $k = 972$

24. T is inversely proportional to x. When x is 50, T is 200. $k = 10,000$

25. y varies jointly as w and v. When w is 40 and v is 0.2, y is 40. $k = 5$

26. N varies jointly as t and p. When t is 2 and p is 2.5, N is 15. $k = 3$

27. The value of y equals 4 when $x = 10$. Find y when $x = 5$ if

   a. y varies directly as x. 2

   b. y varies inversely as x. 8

28. The value of y equals 24 when x is $\frac{1}{2}$. Find y when $x = 3$ if

   a. y varies directly as x. 144

   b. y varies inversely as x. 4

For Exercises 29–48, use a variation model to solve for the unknown value.

29. The amount of a pain reliever that a physician prescribes for a child varies directly as the weight of the child. A physician prescribes 180 mg of the medicine for a 40-lb child. (**See Example 3**)

   a. How much medicine would be prescribed for a 50-lb child?   225 mg

   b. How much would be prescribed for a 60-lb child?   270 mg

   c. How much would be prescribed for a 70-lb child?   315 mg

   d. If 135 mg of medicine is prescribed, what is the weight of the child?   30 lb

30. The number of people that a ham can serve varies directly as the weight of the ham. An 8-lb ham feeds 20 people.

   a. How many people will a 10-lb ham serve?   25 people

   b. How many people will a 15-lb ham serve?   37 people (rounded down to have more food per person)

   c. How many people will an 18-lb ham serve?   45 people

   d. If a ham feeds 30 people, what is the weight of the ham?   12 lb

31. A rental car company charges a fixed amount to rent a car per day. Therefore, the cost per mile to rent a car for a given day is inversely proportional to the number of miles driven. If 100 mi is driven, the average daily cost is $0.80 per mile.

   a. Find the cost per mile if 200 mi is driven. $0.40 per mile

   b. Find the cost per mile if 300 mi is driven. $0.27 per mile

   c. Find the cost per mile if 400 mi is driven. $0.20 per mile

   d. If the cost per mile is $0.16, how many miles were driven?   500 mi

32. A chef self-publishes a cookbook and finds that the number of books she can sell per month varies inversely as the price of the book. The chef can sell 1500 books per month when the price is set at $8 per book.

   a. How many books would she expect to sell per month if the price were $12?   1000 books

   b. How many books would she expect to sell per month if the price were $15?   800 books

   c. How many books would she expect to sell per month if the price were $6?   2000 books

   d. If the chef sells 1200 books, what price was set? $10

33. The distance that a bicycle travels in 1 min varies directly as the number of revolutions per minute (rpm) that the wheels are turning. A bicycle with a 14-in. radius travels approximately 440 ft in 1 min if the wheels turn at 60 rpm. How far will the bicycle travel in 1 min if the wheels turn at 87 rpm?   638 ft

34. The amount of pollution entering the atmosphere varies directly as the number of people living in an area. If 100,000 people create 71,000 tons of pollutants, how many tons enter the atmosphere in a city with 750,000 people?   532,500 tons

35. The stopping distance of a car is directly proportional to the square of the speed of the car.

   a. If a car traveling 50 mph has a stopping distance of 170 ft, find the stopping distance of a car that is traveling 70 mph.   333.2 ft

   b. If it takes 244.8 ft for a car to stop, how fast was it traveling before the brakes were applied?   60 mph

36. The area of a picture projected on a wall varies directly as the square of the distance from the projector to the wall.

   a. If a 15-ft distance produces a 36 ft^2 picture, what is the area of the picture when the projection unit is moved to a distance of 25 ft from the wall?   100 ft^2

   b. If the projected image is 144 ft^2, how far is the projector from the wall?   30 ft

37. The time required to complete a job varies inversely as the number of people working on the job. It takes 8 people 12 days to do a job. (**See Example 4**)

   a. How many days will it take if 15 people work on the job?   6.4 days

   b. If the contractor wants to complete the job in 8 days, how many people should work on the job?   12 people

38. The yield on a bond varies inversely as the price. The yield on a particular bond is 5% when the price is $120.

   a. Find the yield when the price is $100.   6%

   b. What price is necessary for a yield of 7.5%? $80

39. The current in a wire varies directly as the voltage and inversely as the resistance. If the current is 9 amperes (A) when the voltage is 90 volts (V) and the resistance is 10 ohms (Ω), find the current when the voltage is 160 V and the resistance is 5 Ω.   32 A

40. The resistance of a wire varies directly as its length and inversely as the square of its diameter. A 50-ft wire with a 0.2-in. diameter has a resistance of 0.0125 Ω. Find the resistance of a 40-ft wire with a diameter of 0.1 in.   0.04 Ω

41. The amount of simple interest owed on a loan varies jointly as the amount of principal borrowed and the amount of time the money is borrowed. If $4000 in principal results in $480 in interest in 2 yr, determine how much interest will be owed on $6000 in 4 yr. $1440

42. The amount of simple interest earned in an account varies jointly as the amount of principal invested and the amount of time the money is invested. If $5000 in principal earns $750 in 6 yr, determine how much interest will be earned on $8000 in 4 yr. $800

43. The body mass index (BMI) of an individual varies directly as the weight of the individual and inversely as the square of the height of the individual. The body mass index for a 150-lb person who is 70 in. tall is 21.52. Determine the BMI for an individual who is 68 in. tall and 180 lb. Round to 2 decimal places. (**See Example 5**)   27.37

44. The strength of a wooden beam varies jointly as the width of the beam and the square of the thickness of the beam, and inversely as the length of the beam. A beam that is 48 in. long, 6 in. wide, and 2 in. thick can support a load of 417 lb. Find the maximum load that can be safely supported by a board that is 12 in. wide, 72 in. long, and 4 in. thick.   2224 lb

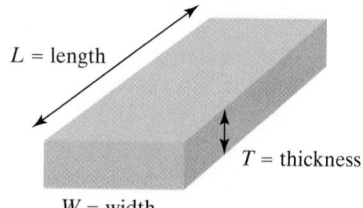

L = length

T = thickness

W = width

45. The speed of a racing canoe in still water varies directly as the square root of the length of the canoe. A 16-ft canoe can travel 6.2 mph in still water. Find the speed of a 25-ft canoe.   7.75 mph

46. The period of a pendulum is the length of time required to complete one swing back and forth. The period varies directly as the square root of the length of the pendulum. If it takes 1.8 sec for a 0.81-m pendulum to complete one period, what is the period of a 1-m pendulum?   2 sec

47. The cost to carpet a rectangular room varies jointly as the length of the room and the width of the room. A 10-yd by 15-yd room costs \$3870 to carpet. What is the cost to carpet a room that is 18 yd by 24 yd?   \$11,145.60

48. The cost to tile a rectangular kitchen varies jointly as the length of the kitchen and the width of the kitchen. A 10-ft by 12-ft kitchen costs \$1104 to tile. How much will it cost to tile a kitchen that is 20 ft by 14 ft?   \$2576

Mixed Exercises

For Exercises 49–52, use the given data to find a variation model relating y to x.

49. 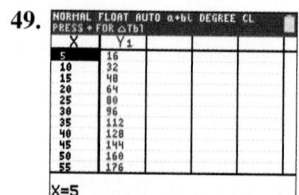 $y = 3.2x$

50.
NORMAL FLOAT AUTO a+bi DEGREE CL	
PRESS + FOR △Tbl	
X	Y1
5	24
15	72
25	120
35	168
45	216
55	264
65	312
75	360
85	408
95	456
105	504
X=5	

$y = 4.8x$

51.
	A	B	C	D	E
1	x	2	4	12	48
2	y	6	3	1	0.25

$y = \dfrac{12}{x}$

52.
	A	B	C	D	E
1	x	4	8	32	100
2	y	2	1	0.25	0.08

$y = \dfrac{8}{x}$

53. Which formula(s) can represent a variation model?

a. $y = kxyz$ **b.** $y = kx + yz$ a and c

c. $y = \dfrac{kx}{yz}$ **d.** $y = kx - yz$

54. Which formula(s) can represent a variation model?

a. $y = k\sqrt{x} - z^2$ **b.** $y = \dfrac{k\sqrt{x}}{z^2}$ b and c

c. $y = k\sqrt{x}z^2$ **d.** $y = k + \sqrt{x}z^2$

Write About It

For Exercises 55–56, write a statement in words that describes the variation model given. Use k as the constant of variation.

55. $P = \dfrac{kv^2}{t}$ The variable P varies directly as the square of v and inversely as t.

56. $E = \dfrac{kc^2}{\sqrt{b}}$ The variable E varies directly as the square of c and inversely as the square root of b.

Expanding Your Skills

57. The light from a lightbulb radiates outward in all directions.

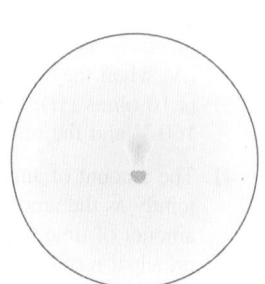

a. Consider the interior of an imaginary sphere on which the light shines. The surface area of the sphere is directly proportional to the square of the radius. If the surface area of a sphere with a 10-m radius is 400π m^2, determine the exact surface area of a sphere with a 20-m radius. 1600π m^2

b. Explain how the surface area changed when the radius of the sphere increased from 10 m to 20 m.   The surface area is 4 times as great. Doubling the radius results in $(2)^2$ times the surface area of the sphere.

c. Based on your answer from part (b) how would you expect the intensity of light to change from a point 10 m from the lightbulb to a point 20 m from the lightbulb?   The intensity at 20 m should be $\frac{1}{4}$ the intensity at 10 m. This is because the energy from the light is distributed across an area 4 times as great.

d. The intensity of light from a light source varies inversely as the square of the distance from the source. If the intensity of a lightbulb is 200 lumen/m^2 (lux) at a distance of 10 m, determine the intensity at 20 m.   50 lux

58. Kepler's third law states that the square of the time T required for a planet to complete one orbit around the Sun is directly proportional to the cube of the average distance d of the planet to the Sun. For the Earth assume that $d = 9.3 \times 10^7$ mi and $T = 365$ days.

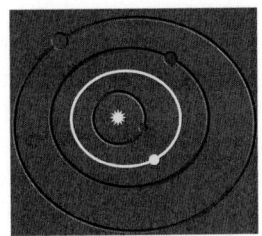

  a. Find the period of Mars, given that the distance between Mars and the Sun is 1.5 times the distance from the Earth to the Sun. Round to the nearest day.   671 Earth days

  b. Find the average distance of Venus to the Sun, given that Venus revolves around the Sun in 223 days. Round to the nearest million miles.   67 million miles

59. The intensity of radiation varies inversely as the square of the distance from the source to the receiver. If the distance is increased to 10 times its original value, what is the effect on the intensity to the receiver?
The intensity is $\frac{1}{100}$ as great.

60. Suppose that y varies inversely as the cube of x. If the value of x is decreased to $\frac{1}{4}$ of its original value, what is the effect on y? y will be 64 times as great.

61. Suppose that y varies directly as x^2 and inversely as w^4. If both x and w are doubled, what is the effect on y?
y will be $\frac{1}{4}$ its original value.

62. Suppose that y varies directly as x^5 and inversely as w^2. If both x and w are doubled, what is the effect on y?
y will be 8 times its original value.

63. Suppose that y varies jointly as x and w^3. If x is replaced by $\frac{1}{3}x$ and w is replaced by $3w$, what is the effect on y? y will be 9 times its original value.

64. Suppose that y varies jointly as x^4 and w. If x is replaced by $\frac{1}{4}x$ and w is replaced by $4w$, what is the effect on y? y will be $\frac{1}{64}$ its original value.

CHAPTER 2   KEY CONCEPTS

SECTION 2.1 Quadratic Functions and Applications	**Reference**
Quadratic function: The function defined by $f(x) = ax^2 + bx + c \ (a \neq 0)$ is called a **quadratic function.** A quadratic function can be written in **vertex form:** $f(x) = a(x - h)^2 + k$ by completing the square.	p. 240
• The graph of a quadratic function $f(x) = a(x - h)^2 + k$ is a parabola. • The vertex is (h, k). • If $a > 0$, the parabola opens upward, and the minimum value of the function is k. • If $a < 0$, the parabola opens downward, and the maximum value of the function is k. • The axis of symmetry is the line $x = h$. • The x-intercepts are determined by the real solutions to the equation $f(x) = 0$. • The y-intercept is determined by $f(0)$.	p. 240
Given $f(x) = ax^2 + bx + c \ (a \neq 0)$, the vertex of the parabola is $\left(\dfrac{-b}{2a}, f\left(\dfrac{-b}{2a} \right) \right)$.	p. 243

SECTION 2.2 Introduction to Polynomial Functions	**Reference**
Let n be a whole number and $a_n, a_{n-1}, a_{n-2}, \ldots, a_1, a_0$ be real numbers, where $a_n \neq 0$. Then a function defined by $$f(x) = a_n x^n + a_{n-1} x^{n-1} + a_{n-2} x^{n-2} + \cdots + a_1 x + a_0$$ is called a **polynomial function of degree n.**	p. 254
The far left and far right behavior of the graph of a polynomial function is determined by the leading term of the polynomial, $a_n x^n$. n is even and $a_n > 0$ ⟋⟍ ⟍⟋ n is even and $a_n < 0$ ⟍⟋⟍⟋ n is odd and $a_n > 0$ ⟋⟋ n is odd and $a_n < 0$ ⟍⟍	p. 256
The **zeros** of a polynomial function defined by $y = f(x)$ are the values of x in the domain of f for which $f(x) = 0$. These are the real solutions (or **roots**) of the equation $f(x) = 0$.	p. 257

If a polynomial function f has a factor $(x - c)$ that appears exactly k times, then c is a **zero of multiplicity k.** • If c is a real zero of odd multiplicity, then the graph of $y = f(x)$ *crosses* the x-axis at c. • If c is a real zero of even multiplicity, then the graph of $y = f(x)$ *touches* the x-axis (but does not cross) at c.	p. 259
Intermediate value theorem: Let f be a polynomial function. For $a < b$, if $f(a)$ and $f(b)$ have opposite signs, then f has at least one zero on the interval $[a, b]$.	p. 260
The graph of a polynomial function of degree n will have at most $n - 1$ turning points.	p. 261

SECTION 2.3 Division of Polynomials and the Remainder and Factor Theorems	Reference
Long division can be used to divide two polynomials.	p. 270
Synthetic division can be used to divide polynomials if the divisor is of the form $x - c$.	p. 273
Remainder theorem: If a polynomial $f(x)$ is divided by $x - c$, then the remainder is $f(c)$.	p. 274
Factor theorem: Let $f(x)$ be a polynomial. 1. If $f(c) = 0$, then $(x - c)$ is a factor of $f(x)$. 2. If $(x - c)$ is a factor of $f(x)$, then $f(c) = 0$.	p. 276

SECTION 2.4 Zeros of Polynomials	Reference
Rational zero theorem: If $f(x) = a_n x^n + a_{n-1} x^{n-1} + \cdots + a_1 x + a_0$ has integer coefficients and $a_n \neq 0$, and if $\frac{p}{q}$ (written in lowest terms) is a rational zero of f, then • p is a factor of the constant term, a_0. • q is a factor of the leading coefficient a_n.	p. 283
Fundamental theorem of algebra: If $f(x)$ is a polynomial of degree $n \geq 1$ with complex coefficients, then $f(x)$ has at least one complex zero.	p. 287
Linear factorization theorem: If $f(x) = a_n x^n + a_{n-1} x^{n-1} + \cdots + a_1 x + a_0$, where $n \geq 1$ and $a_n \neq 0$, then $f(x) = a_n(x - c_1)(x - c_2) \ldots (x - c_n)$, where $c_1, c_2, \ldots, c_n$ are complex numbers.	p. 287
If $f(x)$ is a polynomial of degree $n \geq 1$ with complex coefficients, then $f(x)$ has exactly n complex zeros provided that each zero is counted by its multiplicity.	p. 287
Conjugate zeros theorem: If $f(x)$ is a polynomial with real coefficients and if $a + bi$ is a zero of $f(x)$, then its conjugate $a - bi$ is also a zero of $f(x)$.	p. 287
Descartes' rule of signs: Let $f(x)$ be a polynomial with real coefficients and a nonzero constant term. Then, 1. The number of *positive* real zeros is either the same as the number of sign changes in $f(x)$ or less than the number of sign changes in $f(x)$ by a positive even integer. 2. The number of *negative* real zeros is either the same as the number of sign changes in $f(-x)$ or less than the number of sign changes in $f(-x)$ by a positive even integer.	p. 290
Upper and lower bounds: Let $f(x)$ be a polynomial of degree $n \geq 1$ with real coefficients and a positive leading coefficient. Further suppose that $f(x)$ is divided by $(x - c)$. 1. If $c > 0$ and if both the remainder and the coefficients of the quotient are nonnegative, then c is an upper bound for the real zeros of $f(x)$. 2. If $c < 0$ and the coefficients of the quotient and the remainder alternate in sign (with 0 being considered either positive or negative as needed), then c is a lower bound for the real zeros of $f(x)$.	p. 292

SECTION 2.5 Rational Functions	Reference
Let $p(x)$ and $q(x)$ be polynomials where $q(x) \neq 0$. A function f defined by $f(x) = \dfrac{p(x)}{q(x)}$ is called a **rational function.**	p. 299
The line $x = c$ is a **vertical asymptote** of the graph of $y = f(x)$ if $f(x)$ approaches infinity or negative infinity as x approaches c from either side. To locate the vertical asymptotes of a function, determine the real numbers x where the denominator is zero, but the numerator is nonzero.	p. 301
The line $y = d$ is a **horizontal asymptote** of the graph of $y = f(x)$ if $f(x)$ approaches d as x approaches infinity or negative infinity.	p. 303
Let f be a rational function defined by $$f(x) = \frac{a_n x^n + a_{n-1} x^{n-1} + a_{n-2} x^{n-2} + \cdots + a_1 x + a_0}{b_m x^m + b_{m-1} x^{m-1} + b_{m-2} x^{m-2} + \cdots + b_1 x + b_0}$$ 1. If $n > m$, then f has no horizontal asymptote. 2. If $n < m$, then the line $y = 0$ (the x-axis) is the horizontal asymptote of f. 3. If $n = m$, then the line $y = \dfrac{a_n}{b_m}$ is the horizontal asymptote of f.	p. 303
A rational function will have a slant asymptote if the degree of the numerator is exactly one greater than the degree of the denominator. To find an equation of a slant asymptote of a rational function, divide the numerator by the denominator. The quotient will be linear and the slant asymptote will be of the form $y = $ quotient.	p. 306

SECTION 2.6 Polynomial and Rational Inequalities	Reference
Let $f(x)$ be a polynomial. Then an inequality of the form $f(x) < 0, \quad f(x) > 0, \quad f(x) \leq 0, \quad$ or $\quad f(x) \geq 0$ is called a **polynomial inequality.**	p. 323
Let $f(x)$ be a rational expression. Then an inequality of the form $f(x) < 0, \quad f(x) > 0, \quad f(x) \leq 0, \quad$ or $\quad f(x) \geq 0$ is called a **rational inequality.**	p. 326

Solving nonlinear inequalities:
 	
 1. Express the inequality as $f(x) < 0, f(x) > 0, f(x) \leq 0,$ or $f(x) \geq 0$.
 2. Find the real solutions of the related equation $f(x) = 0$ and the values of x where $f(x)$ is undefined. These are the "boundary" points for the solution set to the inequality.
 3. Determine the sign of $f(x)$ on the intervals defined by the boundary points.
   • If $f(x)$ is positive, then the values of x on the interval are solutions to $f(x) > 0$.
   • If $f(x)$ is negative, then the values of x on the interval are solutions to $f(x) < 0$.
 4. Determine whether the boundary points are included in the solution set.
 5. Write the solution set.
	p. 323

SECTION 2.7 Variation	Reference
Let k be a nonzero constant real number. Then the statements on the left are equivalent to the equations on the right.	p. 337
1. y varies **directly** as x. y is **directly** proportional to x. $\qquad \Big\} \; y = kx$	
2. y varies **inversely** as x. y is **inversely** proportional to x. $\qquad \Big\} \; y = \dfrac{k}{x}$	
3. y varies **jointly** as w and x. y is **jointly** proportional to w and x. $\Big\} \; y = kwx$	p. 338
The value of k is called the **constant of variation.**	

CHAPTER 2   Review Exercises

SECTION 2.1

1. Given $f(x) = -(x + 5)^2 + 2$, identify the vertex of the graph of the parabola. $(-5, 2)$

For Exercises 2–3,

a. Write the equation in vertex form: $f(x) = a(x - h)^2 + k$.

b. Determine whether the parabola opens upward or downward.

c. Identify the vertex.

d. Identify the x-intercepts.

e. Identify the y-intercept.

f. Sketch the function.

g. Determine the axis of symmetry.

h. Determine the minimum or maximum value of the function.

i. State the domain and range.

  2. $f(x) = x^2 - 8x + 15$

  3. $f(x) = -2x^2 + 4x + 6$

4. a. Use the vertex formula to determine the vertex of $f(x) = 2x^2 + 12x + 19$. $(-3, 1)$

  b. Based on the location of the vertex and the orientation of the parabola, how many x-intercepts will the graph of $f(x) = 2x^2 + 12x + 19$ have?
  The graph will have no x-intercepts.

5. Suppose that a farmer encloses a corral for cattle adjacent to a river. No fencing is used by the river.

  a. If he has 180 yd of fencing, what dimensions should he use to maximize the area?
  45 yd by 90 yd

  b. What is the maximum area?
  4050 yd^2

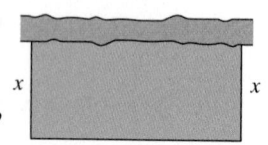

6. Suppose that p is the probability that a randomly selected person is left-handed. The value $(1 - p)$ is the probability that the person is not left-handed. In a sample of 100 people, the function $V(p) = 100p(1 - p)$ represents the variance of the number of left-handed people in a group of 100.

  a. What value of p maximizes the variance? $p = \frac{1}{2}$

  b. What is the maximum variance?   25

7. The annual expenditure for cell phones and cellular service varies in part by the age of an individual. The average annual expenditure $E(a)$ (in \$) for individuals of age a (in yr) is given in the table.
(*Source*: U.S. Bureau of Labor Statistics, www.bls.gov)

a	20	30	40	50	60	70
$E(a)$	502	658	649	627	476	213

  a. Use regression to find a quadratic function to model the data. $E(a) = -0.476a^2 + 37.0a - 44.6$

  b. At what age is the yearly expenditure for cell phones and cellular service the greatest? Round to the nearest year.
  39 yr

  c. What is the maximum yearly expenditure? Round to the nearest dollar.   \$674

SECTION 2.2

For Exercises 8–11,

a. Determine the end behavior of the graph of the function.

b. Find all the zeros of the function and state their multiplicities.

c. Determine the x-intercepts.

d. Determine the y-intercept.

e. Is the function even, odd, or neither?

f. Graph the function.

  8. $f(x) = -4x^3 + 16x^2 + 25x - 100$

  9. $f(x) = x^4 - 10x^2 + 9$

  10. $f(x) = x^4 + 3x^3 - 3x^2 - 11x - 6$

  11. $f(x) = x^5 - 8x^4 + 13x^3$

12. Determine whether the intermediate value theorem guarantees that the function has a zero on the given interval.

$$f(x) = 2x^3 - 5x^2 - 6x + 2$$

  a. $[-2, -1]$ **b.** $[-1, 0]$ **c.** $[0, 1]$ **d.** $[1, 2]$
  Yes   No   Yes   No

For Exercises 13–16, determine if the statement is true or false. If a statement is false, explain why.

13. A fourth-degree polynomial has exactly three turning points.   False. It may have three or fewer turning points.

14. A fourth-degree polynomial has at most three turning points.   True

15. There is exactly one polynomial with zeros of 2, 3, and 4.

16. If c is a real zero of an odd polynomial function, then $-c$ is also a zero.   True

SECTION 2.3

For Exercises 17–18,

a. Divide the polynomials.

b. Identify the dividend, divisor, quotient, and remainder.

  17. $(-2x^4 + x^3 + 4x - 1) \div (x^2 + x - 3)$

  18. $\dfrac{3x^4 - 2x^3 - 15x^2 + 22x - 8}{3x - 2}$

For Exercises 19–20, use synthetic division to divide the polynomials.

  19. $(2x^5 + x^2 - 5x + 1) \div (x + 2)$ $2x^4 - 4x^3 + 8x^2 - 15x + 25 + \dfrac{-49}{x + 2}$

  20. $\dfrac{x^4 + 3x^3 - x^2 + 7x + 2}{x - 3}$ $x^3 + 6x^2 + 17x + 58 + \dfrac{176}{x - 3}$

For Exercises 21–22, use the remainder theorem to evaluate the polynomial for the given values of x.

  21. $f(x) = 3x^4 + 2x^2 - 4x + 1; f(-2)$ 65

  22. $f(x) = x^4 + 2x^3 - 4x^2 - 10x - 5; f(\sqrt{5})$ 0

For Exercises 23–24, use the remainder theorem to determine if the given number c is a zero of the polynomial.

23. $f(x) = 3x^4 + 13x^3 + 2x^2 + 52x - 40$

    a. $c = 2$ No   **b.** $c = \frac{2}{3}$ Yes

24. $f(x) = x^4 + 6x^3 + 9x^2 + 24x + 20$

    a. $c = -5$ Yes   **b.** $c = 2i$ Yes

For Exercises 25–26, use the factor theorem to determine if the given binomial is a factor of the polynomial.

25. $f(x) = x^3 + 4x^2 + 9x + 36$

    a. $x + 4$ Yes   **b.** $x - 3i$ Yes

26. $f(x) = x^2 - 4x - 46$

    a. $x + 2$ No   **b.** $x - \left(2 - 5\sqrt{2}\right)$ Yes

27. Factor $f(x) = 15x^3 - 67x^2 + 26x + 8$, given that $\frac{2}{3}$ is a zero of $f(x)$. $f(x) = (3x - 2)(5x + 1)(x - 4)$

28. Write a third-degree polynomial $f(x)$ with zeros -1, $3\sqrt{2}$, and $-3\sqrt{2}$. $f(x) = x^3 + x^2 - 18x - 18$

29. Write a third-degree polynomial $f(x)$ with integer coefficients and zeros of $\frac{1}{4}$, $-\frac{1}{2}$, and 3.
  $f(x) = 8x^3 - 22x^2 - 7x + 3$

SECTION 2.4

30. Given $f(x) = 2x^5 - 7x^4 + 9x^3 - 18x^2 + 4x + 40$,

    a. How many zeros does $f(x)$ have (including multiplicities)?   5

    b. List the possible rational zeros of $f(x)$.
       $\pm 1, \pm 2, \pm 4, \pm 5, \pm 8, \pm 10, \pm 20, \pm 40, \pm \frac{1}{2}, \pm \frac{5}{2}$

    c. Find all rational zeros of $f(x)$. $\frac{5}{2}, 2, -1$

    d. Find all the zeros of $f(x)$. $\frac{5}{2}, 2, -1, 2i, -2i$

31. Given $f(x) = x^4 + 4x^3 + 2x^2 - 8x - 8$,

    a. How many zeros does $f(x)$ have (including multiplicities)?   4

    b. List the possible rational zeros of $f(x)$. $\pm 1, \pm 2, \pm 4, \pm 8$

    c. Find all rational zeros of $f(x)$. -2 (multiplicity 2)

    d. Find all the zeros of $f(x)$. -2 (multiplicity 2), $\pm \sqrt{2}$

32. If $f(x)$ is a polynomial with real coefficients and zeros of 4 (multiplicity 3), -2 (multiplicity 1), and $2 + 7i$ (multiplicity 1), what is the minimum degree of $f(x)$?   6

33. Given $f(x) = x^4 - 22x^3 + 119x^2 + 66x - 366$ and that $11 - i$ is a zero of $f(x)$,

    a. Find all the zeros of $f(x)$. $11 \pm i, \pm \sqrt{3}$

    b. Factor $f(x)$ as a product of linear factors.
       $[x - (11 - i)][x - (11 + i)](x - \sqrt{3})(x + \sqrt{3})$

    c. Solve the equation $f(x) = 0$. $\{11 \pm i, \pm \sqrt{3}\}$

34. Write a polynomial $f(x)$ of lowest degree with real coefficients and with zeros $2 - 3i$ (multiplicity 1) and 0 (multiplicity 2). $f(x) = x^4 - 4x^3 + 13x^2$

35. Write a third-degree polynomial $f(x)$ with integer coefficients and with zeros of $-2i$ and $\frac{5}{3}$.
  $f(x) = 3x^3 - 5x^2 + 12x - 20$

For Exercises 36–37, determine the number of possible positive and negative real zeros for the given function.

36. $g(x) = -3x^7 + 4x^6 - 2x^2 + 5x - 4$
    Positive: 4, 2, or 0; Negative: 1

37. $n(x) = x^6 + \frac{1}{3}x^4 + \frac{2}{7}x^3 + 4x^2 + 3$
    Positive: 0; Negative: 2 or 0

For Exercises 38–39,

    a. Determine if the upper bound theorem identifies the given number as an upper bound for the real zeros of $f(x)$.

    b. Determine if the lower bound theorem identifies the given number as a lower bound for the real zeros of $f(x)$.

38. $f(x) = x^4 - 3x^3 + 2x - 3$

    a. 2   No   **b.** -2 Yes

39. $f(x) = x^3 - 4x^2 + 2x + 1$

    a. 5   Yes   **b.** -2 Yes

SECTION 2.5

40. Refer to the graph of $y = f(x)$ and complete the statements.

    a. As $x \to -\infty$,
       $f(x) \to$ _____ -3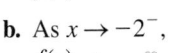

    b. As $x \to -2^-$,
       $f(x) \to$ _____ ∞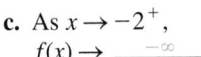

    c. As $x \to -2^+$,
       $f(x) \to$ _____ $-\infty$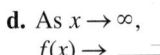

    d. As $x \to \infty$,
       $f(x) \to$ _____ -3

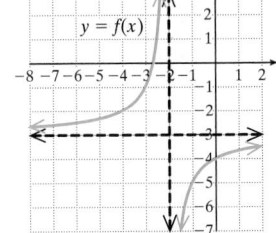

    e. The graph is increasing over the interval(s) _____.
       $(-\infty, -2) \cup (-2, \infty)$

    f. The graph is decreasing over the interval(s) _____.
       Never decreasing

    g. The domain is _____. $(-\infty, -2) \cup (-2, \infty)$

    h. The range is _____. $(-\infty, -3) \cup (-3, \infty)$

    i. The vertical asymptote is the line _____. $x = -2$

    j. The horizontal asymptote is the line _____. $y = -3$

For Exercises 41–42, determine the vertical asymptotes of the graph of the function.

41. $f(x) = \dfrac{x + 4}{2x^2 + x - 15}$
    $x = \frac{5}{2}, x = -3$

42. $g(x) = \dfrac{5}{x^2 + 3}$
    No vertical asymptotes

For Exercises 43–45,

    a. Determine the horizontal asymptotes (if any).

    b. If the graph of the function has a horizontal asymptote, determine the point where the graph crosses the horizontal asymptote.

43. $r(x) = \dfrac{3}{x^2 + 2x + 1}$
    a. $y = 0$ **b.** Graph does not cross $y = 0$.

44. $q(x) = \dfrac{-2x^2 - 3x + 4}{x^2 + 1}$
    a. $y = -2$ **b.** $(2, -2)$

45. $k(x) = \dfrac{x^3 + 4}{x + 1}$
    a. No horizontal asymptote   **b.** Not applicable

For Exercises 46–47, identify all asymptotes (vertical, horizontal, and slant).

46. $m(x) = \dfrac{2x^3 - x^2 - 6x + 7}{x^2 - 3}$ Vertical asymptotes: $x = \sqrt{3}$, $x = -\sqrt{3}$;

Slant asymptote: $y = 2x - 1$

47. $n(x) = \dfrac{-4x^2 + 5}{3x^2 - 14x - 5}$ Vertical asymptotes: $x = -\dfrac{1}{3}$, $x = 5$;

Horizontal asymptote: $y = -\dfrac{4}{3}$

For Exercises 48–51, graph the function.

48. $f(x) = \dfrac{1}{x - 4} + 2$

49. $k(x) = \dfrac{x^2}{x^2 - x - 12}$

50. $m(x) = \dfrac{x^2 + 6x + 9}{x}$

51. $q(x) = \dfrac{12}{x^2 + 6}$

52. After taking a certain class, the percentage of material retained $P(t)$ decreases with the number of months t after taking the class. $P(t)$ can be approximated by

$$P(t) = \dfrac{t + 90}{0.16t + 1}$$

  a. Determine the percentage retained after 1 month, 4 months, and 6 months. Round to the nearest percent. $P(1) = 78\%$ $P(4) = 57\%$ $P(6) = 49\%$

  b. As t becomes infinitely large, what percentage of material will be retained? $P(t)$ will approach 6.25%.

SECTION 2.6

53. The graph of $y = f(x)$ is given. Solve the inequalities.

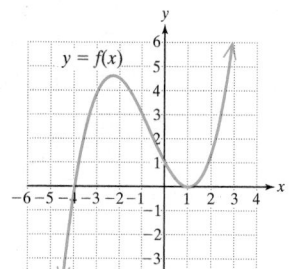

  a. $f(x) < 0$ $(-\infty, -4)$

  b. $f(x) \leq 0$ $(-\infty, -4] \cup \{1\}$

  c. $f(x) > 0$ $(-4, 1) \cup (1, \infty)$

  d. $f(x) \geq 0$ $[-4, \infty)$

54. The graph of $y = f(x)$ is given. Solve the inequalities.

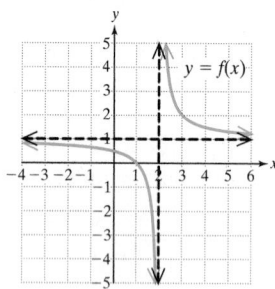

  a. $f(x) < 0$ $(1, 2)$

  b. $f(x) \leq 0$ $[1, 2)$

  c. $f(x) > 0$ $(-\infty, 1) \cup (2, \infty)$

  d. $f(x) \geq 0$ $(-\infty, 1] \cup (2, \infty)$

55. Solve the equation and inequalities.

  a. $x^2 + 7x + 10 = 0$ $\{-5, -2\}$

  b. $x^2 + 7x + 10 < 0$ $(-5, -2)$

  c. $x^2 + 7x + 10 \leq 0$ $[-5, -2]$

  d. $x^2 + 7x + 10 > 0$ $(-\infty, -5) \cup (-2, \infty)$

  e. $x^2 + 7x + 10 \geq 0$ $(-\infty, -5] \cup [-2, \infty)$

56. Solve the inequalities.

  a. $\dfrac{x + 1}{x - 5} \leq 0$ $[-1, 5)$

  b. $\dfrac{x + 1}{x - 5} < 0$ $(-1, 5)$

  c. $\dfrac{x + 1}{x - 5} \geq 0$ $(-\infty, -1] \cup (5, \infty)$

  d. $\dfrac{x + 1}{x - 5} > 0$ $(-\infty, -1) \cup (5, \infty)$

For Exercises 57–66, solve the inequalities.

57. $t(t - 3) \geq 18$ $(-\infty, -3] \cup [6, \infty)$

58. $w^3 + w^2 - 9w - 9 > 0$ $(-3, -1) \cup (3, \infty)$

59. $x^2 - 2x + 4 \leq 3$ $\{1\}$

60. $-6x^4(3x - 4)^2(x + 2)^3 \leq 0$ $[-2, \infty)$

61. $z^3 - 3z^2 > 10z - 24$ $(-3, 2) \cup (4, \infty)$

62. $(4x - 5)^4 > 0$ $\left(-\infty, \dfrac{5}{4}\right) \cup \left(\dfrac{5}{4}, \infty\right)$

63. $\dfrac{6 - 2x}{x^2} \geq 0$ $(-\infty, 0) \cup (0, 3]$

64. $\dfrac{8}{3x - 4} \leq 1$ $\left(-\infty, \dfrac{4}{3}\right) \cup [4, \infty)$

65. $\dfrac{3}{x - 2} < -\dfrac{2}{x}$ $(-\infty, 0) \cup \left(\dfrac{4}{5}, 2\right)$

66. $\dfrac{(1 - x)(3x + 5)^2}{(x - 3)^4} < 0$ $(1, 3) \cup (3, \infty)$

67. A sports trainer has monthly costs of $80 for phone service and $40 for his website and advertising. In addition he pays a $15 fee to the gym for each session in which he works with a client.

  a. Write a function representing the average cost $\overline{C}(x)$ (in $) for x training sessions. $\overline{C}(x) = \dfrac{120 + 15x}{x}$

  b. Find the number of sessions the trainer needs if he wants the average cost to drop below $16 per session. The trainer must have more than 120 sessions with his clients for his average cost to drop below $16 per session.

68. A child throws a ball straight upwards to his friend who is sitting in a tree 18 ft above ground level.

  a. If the ball leaves the child's hand at a height of 2 ft with an initial speed of 40 ft/sec, write a function representing the vertical position of the ball $s(t)$ (in ft) in terms of the time t after the ball leaves the child's hand. (*Hint:* Use the model $s(t) = -\frac{1}{2}gt^2 + v_0 t + s_0$ with $g = 32$ ft/sec². See page 330.) $s(t) = -16t^2 + 40t + 2$

  b. Determine the time interval for which the ball will be more than 18 ft high. $0.5 < t < 2$ sec

SECTION 2.7

For Exercises 69–71, write a variation model using k as the constant of variation.

69. The mass m of an animal varies directly as the weight w of the animal's heart. $m = kw$

70. The value of x varies inversely to the square of p. $x = \dfrac{k}{p^2}$

71. The variable y is jointly proportional to x and the square root of z, and inversely proportional to the cube of t. $y = \dfrac{kx\sqrt{z}}{t^3}$

For Exercises 72–73, determine the constant of variation k.

72. The variable Q varies jointly as p and the square root of t. The value of Q is 132 when p is 11 and t is 9. $k = 4$

73. The variable d is directly proportional to c and inversely proportional to the square of x. The value of d is 1.8 when c is 3 and x is 2. $k = 2.4$

74. The weight of a ball varies directly as the cube of its radius. A weighted exercise ball of radius 3 in. weighs 3.24 lb. How much would a ball weigh if its radius were 5 in.? 15 lb

75. In karate, the force F required to break a board varies inversely as the length L of the board. If it takes 6.25 lb of force to break a board 1.6 ft long, determine how much force is required to break a 2-ft board. 5 lb

76. The power in an electric circuit varies jointly as the current and the square of the resistance. If the power is 144 watts (W) when the current is 4 A and the resistance is 6 Ω, find the power when the current is 3 A and the resistance is 10 Ω.   300 W

77. Coulomb's law states that the force F of attraction between two oppositely charged particles varies jointly as the magnitude of their electrical charges q_1 and q_2 and inversely as the square of the distance d between the particles. Find the effect on F of doubling q_1 and q_2 and halving the distance between them.   The force will be 16 times as great.

CHAPTER 2   Test

1. Given $f(x) = 2x^2 - 12x + 16$,

  a. Write the equation in vertex form: $f(x) = a(x - h)^2 + k$.
    $f(x) = 2(x - 3)^2 - 2$

  b. Determine whether the parabola opens upward or downward.   Upward

  c. Identify the vertex. $(3, -2)$

  d. Identify the x-intercepts. $(2, 0), (4, 0)$

  e. Identify the y-intercept. $(0, 16)$

  f. Sketch the function.

  g. Determine the axis of symmetry. $x = 3$

  h. Determine the minimum or maximum value of the function.   Minimum value: -2

  i. State the domain and range.   Domain: $(-\infty, \infty)$; Range: $[-2, \infty)$

2. Given $f(x) = 2x^4 - 5x^3 - 17x^2 + 41x - 21$,

  a. Determine the end behavior of the graph of the function.   Up to the left and up to the right; As $x \to -\infty$, $f(x) \to \infty$, and as $x \to \infty$, $f(x) \to \infty$.

  b. List all possible rational zeros. $\pm 1, \pm 3, \pm 7, \pm 21, \pm\frac{1}{2}, \pm\frac{3}{2}, \pm\frac{7}{2}, \pm\frac{21}{2}$

  c. Find all the zeros of the function and state their multiplicities. $\frac{7}{2}$ (multiplicity 2), -3 (each multiplicity 1), and 1

  d. Determine the x-intercepts. $\left(\frac{7}{2}, 0\right)$, $(-3, 0)$, $(1, 0)$

  e. Determine the y-intercept. $(0, -21)$

  f. Is the function even, odd, or neither?   Neither even nor odd

  g. Graph the function.

3. Given $f(x) = -0.25x^3(x - 2)^2(x + 1)^4$,

  a. Identify the leading term. $-0.25x^9$

  b. Determine the end behavior of the graph of the function.   Up to the left and down to the right; As $x \to -\infty$, $f(x) \to \infty$, and as $x \to \infty$, $f(x) \to -\infty$.

  c. Find all the zeros of the function and state their multiplicities.   0 (multiplicity 3), 2 (multiplicity 2), -1 (multiplicity 4)

4. Given $f(x) = x^4 + 5x^2 - 36$,

  a. How many zeros does $f(x)$ have (including multiplicities)?   4

  b. Find the zeros of $f(x)$. $2, -2, 3i, -3i$

  c. Identify the x-intercepts of the graph of f. $(2, 0)$ and $(-2, 0)$

  d. Is the function even, odd, or neither?   Even

5. Determine whether the intermediate value theorem guarantees that the function has a zero on the given interval.

$$f(x) = x^3 - 5x^2 + 2x + 5$$

  a. $[-2, -1]$ No   **b.** $[-1, 0]$ Yes   **c.** $[0, 1]$ No   **d.** $[1, 2]$ Yes

6. a. Divide the polynomials. $\dfrac{2x^4 - 4x^3 + x - 5}{x^2 - 3x + 1}$ $2x^2 + 2x + 4 + \dfrac{11x - 9}{x^2 - 3x + 1}$

  b. Identify the dividend, divisor, quotient, and remainder.   Dividend: $2x^4 - 4x^3 + x - 5$; Divisor: $x^2 - 3x + 1$; Quotient: $2x^2 + 2x + 4$; Remainder: $11x - 9$

7. Given $f(x) = 5x^4 + 47x^3 + 80x^2 - 51x - 9$,

  a. Is $\dfrac{3}{5}$ a zero of $f(x)$?   Yes

  b. Is -1 a zero of $f(x)$?   No

  c. Is $(x + 1)$ a factor of $f(x)$?   No

  d. Is $(x + 3)$ a factor of $f(x)$?   Yes

  e. Use the remainder theorem to evaluate $f(-2)$.   117

8. Given $f(x) = x^4 - 8x^3 + 21x^2 - 32x + 68$ and that $2i$ is a zero of $f(x)$,

  a. Find all zeros of $f(x)$. $\pm 2i, 4 \pm i$

  b. Factor $f(x)$ as a product of linear factors. $(x - 2i)(x + 2i)[x - (4 + i)][x - (4 - i)]$

  c. Solve the equation $f(x) = 0$. $\{\pm 2i, 4 \pm i\}$

9. Given $f(x) = 3x^4 + 7x^3 - 12x^2 - 14x + 12$,

  a. How many zeros does $f(x)$ have (including multiplicities)?   4

  b. List the possible rational zeros. $\pm 1, \pm 2, \pm 3, \pm 4, \pm 6, \pm 12, \pm\frac{1}{3}, \pm\frac{2}{3}, \pm\frac{4}{3}$

  c. Determine if the upper bound theorem identifies 2 as an upper bound for the real zeros of $f(x)$.   Yes

  d. Determine if the lower bound theorem identifies -4 as a lower bound for the real zeros of $f(x)$.   Yes

  e. Revise the list of possible rational zeros based on the answers to parts (c) and (d).

  f. Find the rational zeros. $\dfrac{2}{3}$ and -3

  g. Find all the zeros. $\dfrac{2}{3}, -3, \sqrt{2}, -\sqrt{2}$

  h. Graph the function.

10. Write a third-degree polynomial $f(x)$ with integer coefficients and zeros of $\frac{1}{5}$, $-\frac{2}{3}$, and 4.

$f(x) = 15x^3 - 53x^2 - 30x + 8$

11. Determine the number of possible positive and negative real zeros for $f(x) = -6x^7 - 4x^5 + 2x^4 - 3x^2 + 1$.

Positive: 3 or 1; Negative: 2 or 0

For Exercises 12–14, determine the asymptotes (vertical, horizontal, and slant).

12. $r(x) = \dfrac{2x^2 - 3x + 5}{x - 7}$

13. $p(x) = \dfrac{-3x + 1}{4x^2 - 1}$

14. $n(x) = \dfrac{5x^2 - 2x + 1}{3x^2 + 4}$ Horizontal asymptote: $y = \dfrac{5}{3}$

For Exercises 15–17, graph the function.

15. $m(x) = -\dfrac{1}{x^2} + 3$

16. $h(x) = \dfrac{-4}{x^2 - 4}$

17. $k(x) = \dfrac{x^2 + 2x + 1}{x}$

For Exercises 18–24, solve the inequality.

18. $c^2 < c + 20$ $(-4, 5)$

19. $y^3 > 13y - 12$ $(-4, 1) \cup (3, \infty)$

20. $-2x(x - 4)^2(x + 1)^3 \le 0$ $(-\infty, -1] \cup [0, \infty)$

21. $9x^2 + 42x + 49 > 0$ $\left(-\infty, -\dfrac{7}{3}\right) \cup \left(-\dfrac{7}{3}, \infty\right)$

22. $\dfrac{x + 3}{2 - x} \le 0$ $(-\infty, -3] \cup (2, \infty)$

23. $\dfrac{-4}{x^2 - 9} \ge 0$ $(-3, 3)$

24. $\dfrac{4}{x - 1} < -\dfrac{3}{x}$ $(-\infty, 0) \cup \left(\dfrac{3}{7}, 1\right)$

25. Write a variation model using k as the constant of variation: Energy E varies directly as the square of the velocity v of the wind. $E = kv^2$

26. Solve for the constant of variation k: The variable w varies jointly as y and the square root of x, and inversely as z. The value of w is 7.2 when x is 4, y is 6, and z is 7. $k = 4.2$

27. The surface area of a cube varies directly as the square of the length of an edge. The surface area is 24 ft^2 when the length of an edge is 2 ft. Find the surface area of a cube with an edge that is 7 ft. 294 ft^2

28. The weight of a body varies inversely as the square of its distance from the center of the Earth. The radius of the Earth is approximately 4000 mi. How much would a 180-lb man weigh 20 mi above the surface of the Earth? Round to the nearest pound. 178 lb

29. The pressure of wind on a wall varies jointly as the area of the wall and the square of the velocity of the wind. If the velocity of the wind is tripled, what is the effect on the pressure on the wall? The pressure is 9 times as great.

30. The population $P(t)$ of rabbits in a wildlife area t years after being introduced to the area is given by

$$P(t) = \dfrac{2000t}{t + 1}$$

a. Determine the number of rabbits after 1 yr, 5 yr, and 10 yr. Round to the nearest whole unit.
1000 rabbits after 1 yr, 1667 rabbits after 5 yr, and 1818 after 10 yr.

b. What will the rabbit population approach as t approaches infinity? The rabbit population will approach 2000 as t increases.

31. An agricultural school wants to determine the number of corn plants per acre that will produce the maximum yield. The model $y(n) = -0.103n^2 + 8.32n + 15.1$ represents the yield $y(n)$ (in bushels per acre) based on n thousand plants per acre.

a. Evaluate $y(20)$, $y(30)$, and $y(60)$ and interpret their meaning in the context of this problem.

b. Determine the number of plants per acre that will maximize yield. Round to the nearest hundred plants. 40,400

c. What is the maximum yield? Round to the nearest bushel per acre. 183 bushels

32. Suppose that a rocket is shot straight upward from ground level with an initial speed of 98 m/sec.

a. Write a model that represents the height of the rocket $s(t)$ (in meters) t seconds after launch. (*Hint*: Use the model $s(t) = -\dfrac{1}{2}gt^2 + v_0t + s_0$ with $g = 9.8$ m/sec^2. See page 330.) $s(t) = -4.9t^2 + 98t$

b. When will the rocket reach its maximum height? 10 sec after launch

c. What is the maximum height? 490 m

d. Determine the time interval for which the rocket will be more than 200 m high. Round to the nearest tenth of a second. $2.3 < t < 17.7$ sec

33. The number of yearly visits to physicians' offices varies in part by the age of the patient. For the data shown in the table, a represents the age of patients (in yr) and $n(a)$ represents the corresponding number of visits to physicians' offices per year. (*Source*: Centers for Disease Control, www.cdc.gov)

a	8	20	35	55	65	85
$n(a)$	2.7	2.0	2.5	3.7	6.7	7.6

a. Use regression to find a quadratic function to model the data. $n(a) = 0.0011a^2 - 0.027a + 2.46$

b. At what age is the number of yearly visits to physicians' offices the least? Round to the nearest year of age. 12 yr

c. What is the minimum number of yearly visits? Round to 1 decimal place. 2.3 visits per year

CHAPTER 2   Cumulative Review Exercises

1. Given $r(x) = \dfrac{2x^2 - 3}{x^2 - 16}$,

   a. Find the vertical asymptotes. $x = 4, x = -4$

   b. Find the horizontal asymptote or slant asymptote.
     Horizontal asymptote: $y = 2$

2. Find a polynomial $f(x)$ of lowest degree with real coefficients and with zeros of $3 + 2i$ and 2.
   $f(x) = x^3 - 8x^2 + 25x - 26$

3. Given $f(x) = 2x^3 - x^2 - 8x - 5$,

   a. Determine the end behavior of the graph of f.

   b. Find the zeros and their multiplicities. $\dfrac{5}{2}$ (multiplicity 1) and -1 (multiplicity 2)

   c. Find the x-intercepts. $\left(\dfrac{5}{2}, 0\right)$ and $(-1, 0)$

   d. Find the y-intercept. $(0, -5)$

   e. Graph the function.

4. Divide and write the answer in the form $a + bi$.

$$\frac{3 + 2i}{4 - i} \qquad \frac{10}{17} + \frac{11}{17}i$$

5. Determine the center and radius of the circle given by

$$x^2 + y^2 + 8x - 14y + 56 = 0$$
   Center: $(-4, 7)$; Radius: 3

6. Write an equation of the line passing through the points $(4, -8)$ and $(2, -3)$. Write the answer in slope-intercept form. $y = -\dfrac{5}{2}x + 2$

7. Determine the x- and y-intercepts of the graph of $x = y^2 - 9$. x-intercept: $(-9, 0)$; y-intercepts: $(0, 3), (0, -3)$

8. Graph $f(x) = \begin{cases} -x - 1 & \text{for } x < 1 \\ \sqrt{x - 1} & \text{for } x \ge 1 \end{cases}$

9. Solve the equation for m. $v_0 t = \sqrt{m - t}$ $m = (v_0 t)^2 + t$

10. Given $f(x) = 2x^2 - 6x + 1$,

   a. Find the x-intercepts. $\left(\dfrac{3 + \sqrt{7}}{2}, 0\right), \left(\dfrac{3 - \sqrt{7}}{2}, 0\right)$

   b. Find the y-intercept. $(0, 1)$

   c. Find the vertex of the parabola. $\left(\dfrac{3}{2}, -\dfrac{7}{2}\right)$

11. Factor. $125x^6 - y^9$ $(5x^2 - y^3)(25x^4 + 5x^2 y^3 + y^6)$

For Exercises 12–14, simplify the expression.

12. $\left(\dfrac{4x^3 y^{-5}}{z^{-2}}\right)^{-3}\left(\dfrac{4y^{-6}}{x^{-12}}\right)^{1/2}$ $\dfrac{y^{12}}{32x^3 z^6}$

13. $\sqrt[3]{250z^5 xy^{21}}$ $5zy^7\sqrt[3]{2z^2 x}$

14. $\dfrac{\dfrac{1}{3x} - \dfrac{1}{x^2}}{\dfrac{1}{3} - \dfrac{3}{x^2}}$ $\dfrac{1}{x + 3}$

For Exercises 15–20, solve the equations and inequalities. Write the solutions to the inequalities in interval notation if possible.

15. $-5 \le -\dfrac{1}{4}x + 3 < \dfrac{1}{2}$ $(10, 32]$

16. $|x - 3| + 4 \le 10$ $[-3, 9]$

17. $|2x + 1| = |x - 4|$ $\{-5, 1\}$

18. $c^2 - 5c + 9 < c(c + 3)$ $\left(\dfrac{9}{8}, \infty\right)$

19. $\dfrac{49x^2 + 14x + 1}{x} > 0$ $(0, \infty)$

20. $\sqrt{4x - 3} - \sqrt{x + 12} = 0$ $\{5\}$

3

Exponential and Logarithmic Functions

Chapter Outline

Visible light from the Sun is vitally important for the health of an ocean, lake, or any body of water. In particular, light penetrating through a body of water provides the energy to fuel vast amounts of microscopic plants called phytoplankton that are an essential source of food and oxygen for an aquatic ecosystem. Phytoplankton converts energy from the Sun to usable energy for plant growth. Thus, the amount of light directly affects plant productivity at the base of the food chain and ultimately animal life farther up the food chain.

With increasing depth, the percentage of visible light from the surface of a body of water drops exponentially. This means that light intensity drops quickly at first and then drops more slowly with increasing depth. (More specifically, light intensity decreases at a rate proportional to the intensity at a particular depth.) To study this phenomenon, scientists use exponential functions and their inverses, logarithmic functions. These two important categories of functions have many applications including the study of the decay of radioactive substances, short-term population growth, and the growth of investments subject to compound interest.

SECTION 3.1   Inverse Functions

1. Identify One-to-One Functions

Throughout our study of algebra, we have made use of the fact that the operations of addition and subtraction are inverse operations. For example, adding 5 to a number and then subtracting 5 from the result gives us the original number. Likewise, multiplication and division are inverse operations. We now look at the concept of an inverse function.

Changing currency is an important consideration when traveling abroad. For example, traveling between the United States and several countries in Europe would involve changing American dollars to Euros and then changing back for the return trip. Fortunately, we can use a function to change from one currency to the other, and then use the function's *inverse* to change back again.

Suppose that \$1 (American dollar) can be exchanged for 0.8 € (Euro). Then,

$f(x) = 0.8x$ gives the number of Euros $f(x)$ that can be bought from x dollars.

$g(x) = \dfrac{x}{0.8}$ gives the number of dollars $g(x)$ that can be bought from x Euros.

Tables 3-1 and 3-2 show the values of $f(x)$ and $g(x)$ for several values of x.

Table 3-1

x (Dollars)	$f(x) = 0.8x$ (Euros)
100	80
150	120
200	160
250	200

Table 3-2

x (Euros)	$g(x) = \dfrac{x}{0.8}$ (Dollars)
80	100
120	150
160	200
200	250

In this example, functions f and g are inverses of each other, and we observe several interesting characteristics about inverse functions.

- By listing the ordered pairs from Tables 3-1 and 3-2, notice that the x and y values are reversed.

f: $\{(100, 80), (150, 120), (200, 160), (250, 200)\}$

g: $\{(80, 100), (120, 150), (160, 200), (200, 250)\}$

- For a function and its inverse, the values of x and y are interchanged. This tells us that the domain of a function is the same as the range of its inverse and vice versa.
- From the graphs of f and g (Figure 3-1), we see that the corresponding points on f and g are symmetric with respect to the line $y = x$.
- When we compose functions f and g in both directions, the result is the input value x. In a sense, what function f does to x, function g "undoes" and vice versa.

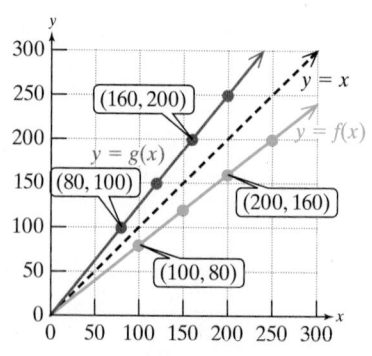

Figure 3-1

$$(f \circ g)(x) = f[g(x)] = 0.8\left(\frac{x}{0.8}\right) = x$$

$$(g \circ f)(x) = g[f(x)] = \frac{(0.8x)}{0.8} = x$$

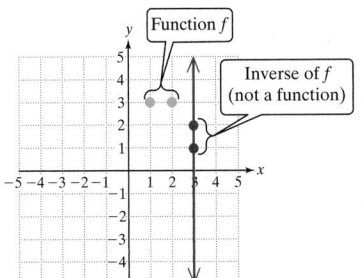

Figure 3-2

The inverse of any relation is found by interchanging the values of x and y in the relation. However, the inverse of a function may itself not be a function. For example, consider $f = \{(1, 3), (2, 3)\}$ shown in blue in Figure 3-2. The inverse is the set of ordered pairs $\{(3, 1), (3, 2)\}$ shown in red. Notice that the relation defining the inverse of f is not a function because it fails the vertical line test.

The function $f = \{(1, 3), (2, 3)\}$ has two points that are aligned horizontally. When the x and y values are reversed to form the inverse, the resulting points will be aligned vertically and will fail the vertical line test. Thus, a function will have an inverse function only if no points in the original function are aligned horizontally. That is, no two distinct points on the function may have the same y value. In such a case, the function is said to be one-to-one.

Definition of a One-to-One Function

A function f is a **one-to-one function,** if for a and b in the domain of f,

if $a \neq b$, then $f(a) \neq f(b)$, or equivalently, if $f(a) = f(b)$, then $a = b$.

The definition of a one-to-one function tells us that each y value in the range is associated with only one x value in the domain. This implies that the graph of a one-to-one function will have no two points aligned horizontally.

Horizontal Line Test for a One-to-One Function

A function defined by $y = f(x)$ is a one-to-one function if no horizontal line intersects the graph in more than one point.

Classroom Examples: p. 364
Exercises 8, 10

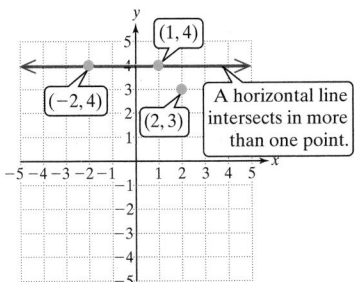

Figure 3-3

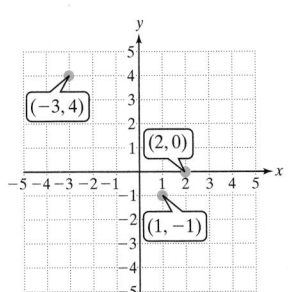

Figure 3-4

EXAMPLE 1   **Determining Whether a Function is One-to-One**

Determine whether the function is one-to-one.

a. $f = \{(1, 4), (2, 3), (-2, 4)\}$ **b.** $g = \{(-3, 4), (1, -1), (2, 0)\}$

Solution:

same y value

a. $f = \{(1, 4), (2, 3), (-2, 4)\}$

different x value

f is not a one-to-one function.

The ordered pairs $(1, 4)$ and $(-2, 4)$ have the same y value but different x values. That is, $f(1) = f(-2)$, but $1 \neq -2$.

A horizontal line passes through the function in more than one point (Figure 3-3).

All points have different y-values.

b. $g = \{(-3, 4), (1, -1), (2, 0)\}$

g is a one-to-one function.

Each unique ordered pair has a different y value, so the function is one-to-one.

No horizontal line passes through the function in more than one point (Figure 3-4).

Skill Practice 1   Determine whether the function is one-to-one.

a. $h = \{(4, -5), (6, 1), (2, 4), (0, -3)\}$ **b.** $k = \{(1, 0), (3, 0), (4, -5)\}$

Answers
1. a. Yes   **b.** No

> **EXAMPLE 2**   Using the Horizontal Line Test

Use the horizontal line test to determine if the graph in blue defines y as a one-to-one function of x.

Solution:

a.

b.

c.

The graph does not define y as a one-to-one function of x because a horizontal line intersects the graph in more than one point.

The graph does define y as a one-to-one function of x because no horizontal line intersects the graph in more than one point.

The relation does not define y as a function of x, because it fails the vertical line test. If the relation is not a function, it is not a one-to-one function.

Skill Practice 2   Use the horizontal line test to determine if the graph defines y as a one-to-one function of x.

a.

b.

c.

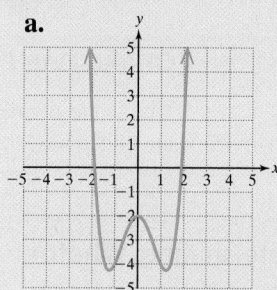

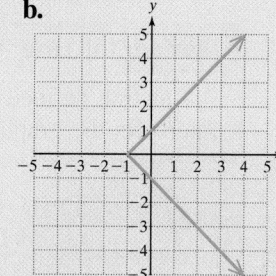

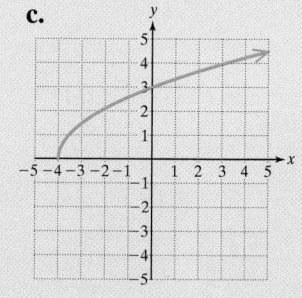

In Example 3, we use algebraic methods to determine whether a function is one-to-one.

> **EXAMPLE 3**   Determining Whether a Function Is One-to-One

Use the definition of a one-to-one function to determine whether the function is one-to-one.

a. $f(x) = 2x - 3$ **b.** $f(x) = x^2 + 1$

Solution:

a. We must show that if $f(a) = f(b)$, then $a = b$.

Assume that $f(a) = f(b)$. That is,

$$2a - 3 = 2b - 3$$
$$2a - 3 + 3 = 2b - 3 + 3$$
$$\frac{2a}{2} = \frac{2b}{2}$$
$$a = b$$

The logic of this algebraic proof begins with the assumption that $f(a) = f(b)$, that is, that two y values are equal. For a one-to-one function, this can happen only if the x values (in this case a and b) are the same.

Otherwise, if $a \neq b$, we would have the same y value with two different x values and f would not be one-to-one.

Since $f(a) = f(b)$ implies that $a = b$, then f is one-to-one.

Answers

2. a. No   **b.** No   **c.** Yes

b. $f(x) = x^2 + 1$

Assume that $f(a) = f(b)$.
$$a^2 + 1 = b^2 + 1$$
$$a^2 = b^2$$
$$a = \pm b$$

For nonzero values of b, $f(a) = f(b)$ does not necessarily imply that $a = b$. Therefore, f is not one-to-one.

From the graph of $f(x) = x^2 + 1$, we see that f is not one-to-one (Figure 3-5). We can also show this algebraically by finding two ordered pairs with the same y value but different x values. From the graph, we have arbitrarily selected $(-2, 5)$ and $(2, 5)$.

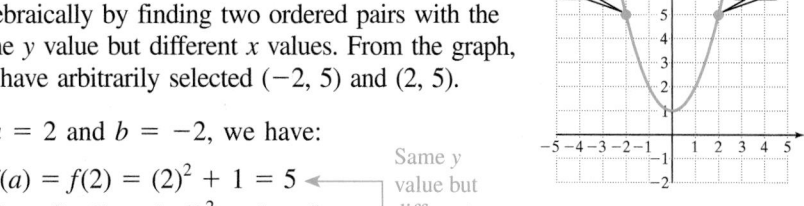

Figure 3-5

If $a = 2$ and $b = -2$, we have:

$$f(a) = f(2) = (2)^2 + 1 = 5 \leftarrow$$ Same y value but
$$f(b) = f(-2) = (-2)^2 + 1 = 5 \leftarrow$$ different x values

We have that $f(a) = f(b)$, but $a \neq b$.

Therefore, f fails to be a one-to-one function.

Skill Practice 3   Determine whether the function is one-to-one.
a. $f(x) = -4x + 1$ **b.** $f(x) = |x| - 3$

---

2. Determine Whether Two Functions Are Inverses

We now have enough background to define an inverse function.

Definition of an Inverse Function

Let f be a one-to-one function. Then g is the **inverse of f** if the following conditions are both true.

1. $(f \circ g)(x) = x$ for all x in the domain of g.
2. $(g \circ f)(x) = x$ for all x in the domain of f.

We should also note that if g is the inverse of f, then f is the inverse of g. Furthermore, given a function f, we often denote its inverse as f^{-1}. So given a function f and its inverse f^{-1}, the definition implies that

$$(f \circ f^{-1})(x) = x \text{ and } (f^{-1} \circ f)(x) = x$$

EXAMPLE 4   Determining Whether Two Functions Are Inverses

Determine whether the functions are inverses.

a. $f(x) = 100 + 12x$ and $g(x) = \dfrac{x - 100}{12}$

b. $h(x) = \sqrt[3]{x - 1}$ and $k(x) = -1 + x^3$

Answers

3. a. Yes   **b.** No

Solution:

a. $(f \circ g)(x) = f(g(x))$

$$= f\left(\frac{x - 100}{12}\right)$$

$$= 100 + 12\left(\frac{x - 100}{12}\right)$$

$$= 100 + (x - 100)$$

$$= x \checkmark$$

$(g \circ f)(x) = g(f(x))$

$$= g(100 + 12x)$$

$$= \frac{(100 + 12x) - 100}{12}$$

$$= \frac{12x}{12}$$

$$= x \checkmark$$

Since $(f \circ g)(x) = (g \circ f)(x) = x$, f and g are inverses.

b. $(h \circ k)(x) = h(k(x))$

$$= \sqrt[3]{(-1 + x^3)} - 1$$

$$= \sqrt[3]{x^3 - 2} \neq x$$

If either $(h \circ k)(x) \neq x$ or $(k \circ h)(x) \neq x$, then h and k are *not* inverses.

Since $(h \circ k)(x) \neq x$, h and k are not inverses.

Skill Practice 4 Determine whether the functions are inverses.

a. $f(x) = \dfrac{x + 6}{2}$ and $g(x) = 2(x - 6)$ **b.** $m(x) = \dfrac{5}{x - 2}$ and $n(x) = \dfrac{2x + 5}{x}$

---

3. Find the Inverse of a Function

For a one-to-one function defined by $y = f(x)$, the inverse is a function $y = f^{-1}(x)$ that performs the inverse operations in the reverse order. The function given by $f(x) = 100 + 12x$ multiplies x by 12 first, and then adds 100 to the result. Therefore, the inverse function must *subtract* 100 from x first and then *divide* by 12.

$$f^{-1}(x) = \frac{x - 100}{12}$$

To facilitate the process of finding an equation of the inverse of a one-to-one function, we offer the following steps.

Procedure to Find an Equation of an Inverse of a Function

For a one-to-one function defined by $y = f(x)$, the equation of the inverse can be found as follows.

Step 1 Replace $f(x)$ by y.
Step 2 Interchange x and y.
Step 3 Solve for y.
Step 4 Replace y by $f^{-1}(x)$.

Classroom Example: p. 365
Exercise 46

EXAMPLE 5   **Finding an Equation of an Inverse Function**

Write an equation for the inverse function for $f(x) = 3x - 1$.

Answers
4. a. No   **b.** Yes

Solution:

Function f is a linear function, and its graph is a nonvertical line. Therefore, f is a one-to-one function.

$$f(x) = 3x - 1$$
$$y = 3x - 1 \qquad \textbf{Step 1:} \text{ Replace } f(x) \text{ by } y.$$
$$x = 3y - 1 \qquad \textbf{Step 2:} \text{ Interchange } x \text{ and } y.$$
$$x + 1 = 3y \qquad \textbf{Step 3:} \text{ Solve for } y. \text{ Add 1 to both sides and divide by 3.}$$
$$\frac{x + 1}{3} = y$$
$$f^{-1}(x) = \frac{x + 1}{3} \qquad \textbf{Step 4:} \text{ Replace } y \text{ by } f^{-1}(x).$$

To check the result, verify that $(f \circ f^{-1})(x) = x$ and $(f^{-1} \circ f)(x) = x$.

$$(f \circ f^{-1})(x) = 3\left(\frac{x + 1}{3}\right) - 1 = x \checkmark \quad \text{and} \quad (f^{-1} \circ f)(x) = \frac{(3x - 1) + 1}{3} = x \checkmark$$

Skill Practice 5   Write an equation for the inverse function for $f(x) = 4x + 3$.

TIP We can sometimes find an equation of an inverse function by mentally reversing the operations given in the original function. In Example 5, the function f multiplies x by 3 and then subtracts 1. Therefore, f^{-1} must add 1 to x and then divide by 3.

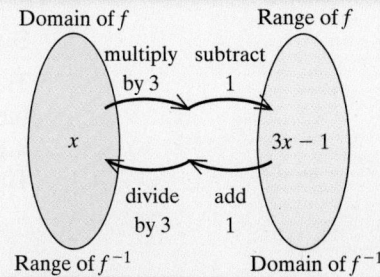

The key step in determining the equation of the inverse of a function is to interchange x and y. By so doing, a point (a, b) on f corresponds to a point (b, a) on f^{-1}. This is why the graphs of f and f^{-1} are symmetric with respect to the line $y = x$. From Example 5, notice that the point $(2, 5)$ on the graph of f corresponds to the point $(5, 2)$ on the graph of f^{-1} (Figure 3-6).

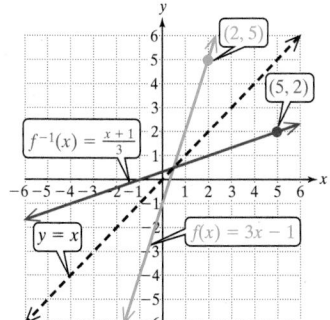

Figure 3-6

Classroom Example: p. 365
Exercise 50

EXAMPLE 6   Finding an Equation of an Inverse Function

Write an equation for the inverse function for the one-to-one function defined by $f(x) = \dfrac{3 - x}{x + 3}$.

TIP In Example 6, we can show that f is a one-to-one function by graphing the function (see Section 2.5). Or we can show that $f(a) = f(b)$ implies that $a = b$ by solving the equation $\dfrac{3 - a}{a + 3} = \dfrac{3 - b}{b + 3}$ for a or b to show that $a = b$.

Solution:

$$f(x) = \frac{3 - x}{x + 3}$$
$$y = \frac{3 - x}{x + 3} \qquad \textbf{Step 1:} \text{ Replace } f(x) \text{ by } y.$$
$$x = \frac{3 - y}{y + 3} \qquad \textbf{Step 2:} \text{ Interchange } x \text{ and } y.$$

Answer

5. $f^{-1}(x) = \dfrac{x - 3}{4}$

$$x(y + 3) = 3 - y$$

Step 3: Solve for y.

Clear fractions (multiply both sides by $y + 3$).

$$xy + 3x = 3 - y$$

Apply the distributive property.

$$xy + y = 3 - 3x$$

Collect the y terms on one side.

$$y(x + 1) = 3 - 3x$$

Factor out y as the greatest common factor.

$$y = \frac{3 - 3x}{x + 1}$$

Divide both sides by $x + 1$.

$$f^{-1}(x) = \frac{3 - 3x}{x + 1}$$

Step 4: Replace y by $f^{-1}(x)$.

Instructor Note:
In Example 6, f has vertical asymptote $x = -3$ and horizontal asymptote $y = -1$. The domain is $(-\infty, -3) \cup (-3, \infty)$ and the range is $(-\infty, -1) \cup (-1, \infty)$. Consider asking the students to find the domain and range of f^{-1}.

Skill Practice 6 Write an equation for the inverse function for the one-to-one function defined by $f(x) = \dfrac{x - 2}{x + 2}$.

For a function that is not one-to-one, sometimes we restrict its domain to create a new function that is one-to-one. This is demonstrated in Example 7.

Classroom Example: p. 365
Exercise 54

EXAMPLE 7 Finding an Equation of an Inverse Function

Given $m(x) = x^2 + 4$ for $x \geq 0$, write an equation of the inverse.

Solution:

The graph of $y = x^2 + 4$ is a parabola with vertex $(0, 4)$. See Figure 3-7. The function is not one-to-one. However, with the restriction on the domain that $x \geq 0$, the graph consists of only the right branch of the parabola (Figure 3-8). This *is* a one-to-one function.

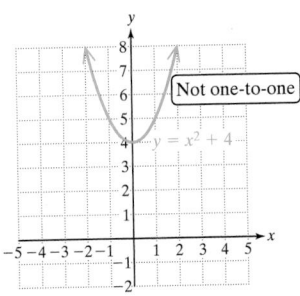

Figure 3-7

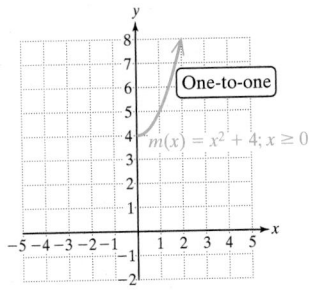

Figure 3-8

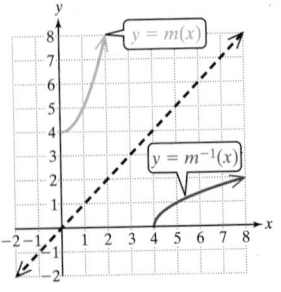

Figure 3-9

To find the inverse, we have

$$m(x) = x^2 + 4; \quad x \geq 0$$

$$y = x^2 + 4; \quad x \geq 0$$

Step 1: Replace $m(x)$ by y.

$$x = y^2 + 4 \quad y \geq 0$$

Step 2: Interchange x and y. Notice that the restriction $x \geq 0$ becomes $y \geq 0$.

$$x - 4 = y^2$$

Step 3: Solve for y by subtracting 4 from both sides.

$$y = \pm\sqrt{x - 4}$$

Apply the square root property.

$$y = +\sqrt{x - 4}$$

Choose the positive square root of $(x - 4)$ because of the restriction $y \geq 0$.

$$m^{-1}(x) = \sqrt{x - 4}$$

Step 4: Replace y by $m^{-1}(x)$.

The graphs of m and m^{-1} are symmetric with respect to the line $y = x$ as expected (Figure 3-9).

Answer

6. $f^{-1}(x) = -\dfrac{2x + 2}{x - 1}$

> **Skill Practice 7** Given $n(x) = x^2 + 1$ for $x \leq 0$, write an equation of the inverse.

Classroom Example: p. 366
Exercise 56

EXAMPLE 8 **Finding an Equation of an Inverse Function**

Given $f(x) = \sqrt{x - 1}$, find an equation of the inverse.

Solution:

TIP When finding the inverse of a function, the key step of interchanging x and y has the effect of interchanging the domain and range between the function and its inverse.

The function f is a one-to-one function and the graph is the same as the graph of $y = \sqrt{x}$ with a shift 1 unit to the right. The domain of f is $\{x \mid x \geq 1\}$ and the range is $\{y \mid y \geq 0\}$. When defining the inverse, we will have the conditions that $x \geq 0$ and $y \geq 1$.

$$f(x) = \sqrt{x - 1} \quad \text{Note that } x \geq 1 \text{ and } y \geq 0.$$
$$y = \sqrt{x - 1}$$
$$x = \sqrt{y - 1} \quad \text{Interchange } x \text{ and } y.$$
$$\text{Note that } y \geq 1 \text{ and } x \geq 0.$$
$$x^2 = y - 1 \quad \text{Square both sides.}$$
$$y = x^2 + 1$$

$$f^{-1}(x) = x^2 + 1, \quad x \geq 0 \qquad \text{The restriction } x \geq 0 \text{ on } f^{-1} \text{ is necessary because } f \text{ has the restriction that } y \geq 0.$$
Furthermore, $y = x^2 + 1$ is not a one-to-one function without a restricted domain.

Answers
7. $n^{-1}(x) = -\sqrt{x - 1}$
8. $g^{-1}(x) = x^2 - 2; x \geq 0$

> **Skill Practice 8** Given $g(x) = \sqrt{x + 2}$, find an equation of the inverse.

SECTION 3.1 Practice Exercises

Prerequisite Review

For Exercises R.1–R.3, find the domain. Write the answer in interval notation.

R.1. $f(x) = \dfrac{x + 4}{x + 1}$ $(-\infty, -1) \cup (-1, \infty)$ **R.2.** $k(p) = \dfrac{p + 1}{p^2 + 4}$ $(-\infty, \infty)$ **R.3.** $m(x) = \sqrt{2 - 8x}$ $\left(-\infty, \dfrac{1}{4}\right]$

For Exercises R.4–R.6, refer to functions n and p to evaluate the function.

$n(x) = x + 1 \quad p(x) = x^2 - 3x$

R.4. $(n \circ p)(x)$ $(n \circ p)(x) = x^2 - 3x + 1$ **R.5.** $(p \circ n)(x)$ $(p \circ n)(x) = x^2 - x - 2$ **R.6.** $(p \circ p)(x)$ $(p \circ p)(x) = x^4 - 6x^3 + 6x^2 + 9x$

Concept Connections

1. Given the function $f = \{(1, 2), (2, 3), (3, 4)\}$ write the set of ordered pairs representing f^{-1}. $\{(2, 1), (3, 2), (4, 3)\}$

2. The graph of a function and its inverse are symmetric with respect to the line ____$y = x$____.

3. If no horizontal line intersects the graph of a function f in more than one point, then f is a ___one___ - ___to___ - ___one___ function.

4. Given a one-to-one function f, if $f(a) = f(b)$, then a ___=___ b.

5. Let f be a one-to-one function and let g be the inverse of f. Then $(f \circ g)(x) =$ ___x___ and $(g \circ f)(x) =$ ___x___.

6. If (a, b) is a point on the graph of a one-to-one function f, then the corresponding ordered pair ___(b, a)___ is a point on the graph of f^{-1}.

Objective 1: Identify One-to-One Functions

For Exercises 7–12, a relation in x and y is given. Determine if the relation defines y as a one-to-one function of x. (See Example 1)

7. $\{(6, -5), (4, 2), (3, 1), (8, 4)\}$ Yes

8. $\{(-14, 1), (-2, 3), (7, 4), (-9, -2)\}$ Yes

9.

x	y
0.6	1.8
1	−1.1
0.5	1.8
2.4	0.7

No

10.

x	y
12.5	3.21
5.75	−4.5
2.34	7.25
−12.7	3.21

No

11.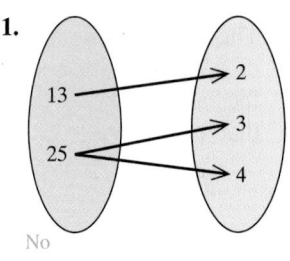

No

12.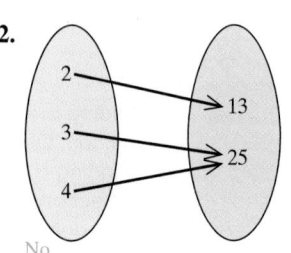

No

For Exercises 13–22, determine if the relation defines y as a one-to-one function of x. (See Example 2)

13.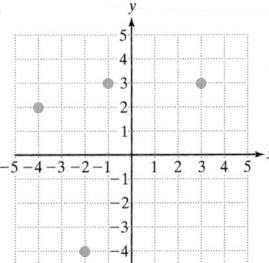

No

14.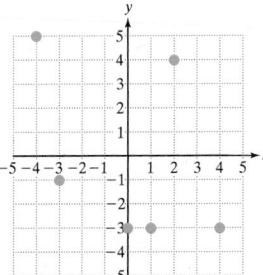

No

15.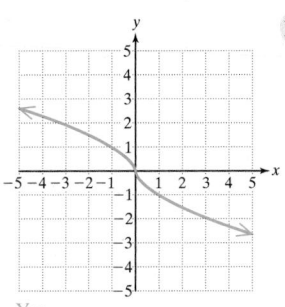

Yes

16.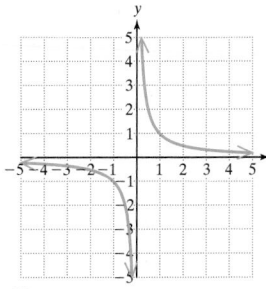

Yes

17.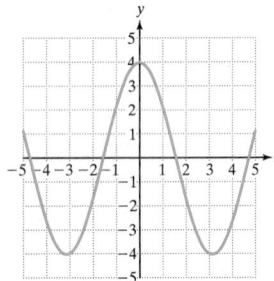

No

18.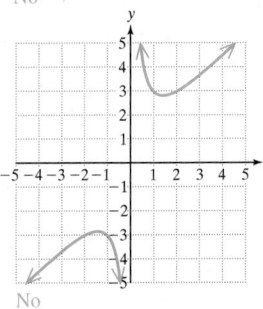

No

19.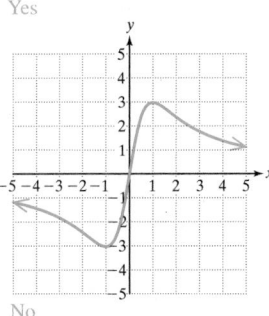

No

20.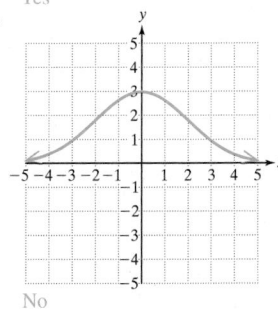

No

21.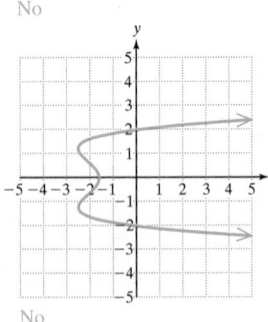

No

22.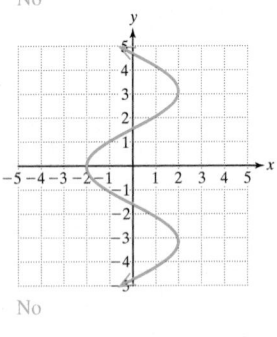

No

For Exercises 23–30, use the definition of a one-to-one function to determine if the function is one-to-one. (See Example 3)

23. $f(x) = 4x - 7$

24. $h(x) = -3x + 2$

25. $g(x) = x^3 + 8$

26. $k(x) = x^3 - 27$

27. $m(x) = x^2 - 4$
No; For example the points $(1, -3)$ and $(-1, -3)$ have the same y values but different x values. That is, $m(a) = m(b) = -3$, but $a \neq b$.

28. $n(x) = x^2 + 1$
No; For example, the points $(3, 10)$ and $(-3, 10)$ have the same y values but different x values. That is, $n(a) = n(b) = 10$, but $a \neq b$.

29. $p(x) = |x + 1|$
No; For example, the points $(2, 3)$ and $(-4, 3)$ have the same y values but different x values. That is, $p(a) = p(b) = 3$, but $a \neq b$.

30. $q(x) = |x - 3|$
No; For example, the points $(0, 3)$ and $(6, 3)$ have the same y value but different x values. That is, $q(a) = q(b) = 3$, but $a \neq b$.

Objective 2: Determine Whether Two Functions Are Inverses

For Exercises 31–36, determine whether the two functions are inverses. (See Example 4)

31. $f(x) = 5x + 4$ and $g(x) = \dfrac{x - 4}{5}$ Yes

32. $h(x) = 7x - 3$ and $k(x) = \dfrac{x + 3}{7}$ Yes

33. $m(x) = \dfrac{-2 + x}{6}$ and $n(x) = 6x - 2$ No

34. $p(x) = \dfrac{-3 + x}{4}$ and $q(x) = 4x - 3$ No

35. $t(x) = \dfrac{4}{x - 1}$ and $v(x) = \dfrac{x + 4}{x}$ Yes

36. $w(x) = \dfrac{6}{x + 2}$ and $z(x) = \dfrac{6 - 2x}{x}$ Yes

37. There were 2000 applicants for enrollment to the freshman class at a small college in the year 2010. The number of applications has risen linearly by roughly 150 per year. The number of applications $f(x)$ is given by $f(x) = 2000 + 150x$, where x is the number of years since 2010.

    a. Determine if the function $g(x) = \dfrac{x - 2000}{150}$ is the inverse of f.   Yes

    b. Interpret the meaning of function g in the context of this problem.   The value $g(x)$ represents the number of years since the year 2010 based on the number of applicants to the freshman class, x.

38. The monthly sales for January for a whole foods market was $60,000 and has increased linearly by $2500 per month. The amount in sales $f(x)$ (in $) is given by $f(x) = 60,000 + 2500x$, where x is the number of months since January.

    a. Determine if the function $g(x) = \dfrac{x - 60,000}{2500}$ is the inverse of f.   Yes

    b. Interpret the meaning of function g in the context of this problem.   The value $g(x)$ represents the number of months since January based on the monthly amount in sales, x.

Objective 3: Find the Inverse of a Function

39. **a.** Show that $f(x) = 2x - 3$ defines a one-to-one function.
    b. Write an equation for $f^{-1}(x)$.
    c. Graph $y = f(x)$ and $y = f^{-1}(x)$ on the same coordinate system.

40. **a.** Show that $f(x) = 4x + 4$ defines a one-to-one function.
    b. Write an equation for $f^{-1}(x)$.
    c. Graph $y = f(x)$ and $y = f^{-1}(x)$ on the same coordinate system.

For Exercises 41–52, a one-to-one function is given. Write an equation for the inverse function. (See Examples 5–6)

41. $f(x) = \dfrac{4 - x}{9}$ $f^{-1}(x) = 4 - 9x$

42. $g(x) = \dfrac{8 - x}{3}$ $g^{-1}(x) = 8 - 3x$

43. $h(x) = \sqrt[3]{x - 5}$ $h^{-1}(x) = x^3 + 5$

44. $k(x) = \sqrt[3]{x + 8}$ $k^{-1}(x) = x^3 - 8$

45. $m(x) = 4x^3 + 2$ $m^{-1}(x) = \sqrt[3]{\dfrac{x - 2}{4}}$

46. $n(x) = 2x^3 - 5$ $n^{-1}(x) = \sqrt[3]{\dfrac{x + 5}{2}}$

47. $c(x) = \dfrac{5}{x + 2}$ $c^{-1}(x) = \dfrac{5 - 2x}{x}$

48. $s(x) = \dfrac{2}{x - 3}$ $s^{-1}(x) = \dfrac{3x + 2}{x}$

49. $t(x) = \dfrac{x - 4}{x + 2}$ $t^{-1}(x) = -\dfrac{2x + 4}{x - 1}$

50. $v(x) = \dfrac{x - 5}{x + 1}$ $v^{-1}(x) = -\dfrac{x + 5}{x - 1}$

51. $f(x) = \dfrac{(x - a)^3}{b} - c$ $f^{-1}(x) = \sqrt[3]{b(x + c)} + a$

52. $g(x) = b(x + a)^3 + c$ $g^{-1}(x) = \sqrt[3]{\dfrac{x - c}{b}} - a$

53. **a.** Graph $f(x) = x^2 - 3; x \le 0$. (See Example 7)
    b. From the graph of f, is f a one-to-one function?   Yes
    c. Write the domain of f in interval notation. $(-\infty, 0]$
    d. Write the range of f in interval notation. $[-3, \infty)$
    e. Write an equation for $f^{-1}(x)$. $f^{-1}(x) = -\sqrt{x + 3}$
    f. Graph $y = f(x)$ and $y = f^{-1}(x)$ on the same coordinate system.
    g. Write the domain of f^{-1} in interval notation. $[-3, \infty)$
    h. Write the range of f^{-1} in interval notation. $(-\infty, 0]$

54. **a.** Graph $f(x) = x^2 + 1; x \le 0$.
    b. From the graph of f, is f a one-to-one function?   Yes
    c. Write the domain of f in interval notation. $(-\infty, 0]$
    d. Write the range of f in interval notation. $[1, \infty)$
    e. Write an equation for $f^{-1}(x)$. $f^{-1}(x) = -\sqrt{x - 1}$
    f. Graph $y = f(x)$ and $y = f^{-1}(x)$ on the same coordinate system.
    g. Write the domain of f^{-1} in interval notation. $[1, \infty)$
    h. Write the range of f^{-1} in interval notation. $(-\infty, 0]$

55. a. Graph $f(x) = \sqrt{x + 1}$. (**See Example 8**)

  b. From the graph of f, is f a one-to-one function?   Yes

  c. Write the domain of f in interval notation. $[-1, \infty)$

  d. Write the range of f in interval notation. $[0, \infty)$

  e. Write an equation for $f^{-1}(x)$. $f^{-1}(x) = x^2 - 1; x \geq 0$

  f. Explain why the restriction $x \geq 0$ is placed on f^{-1}.
  The range of f is $[0, \infty)$. Therefore, the domain of f^{-1} must be $[0, \infty)$.

  g. Graph $y = f(x)$ and $y = f^{-1}(x)$ on the same coordinate system.

  h. Write the domain of f^{-1} in interval notation. $[0, \infty)$

  i. Write the range of f^{-1} in interval notation. $[-1, \infty)$

56. a. Graph $f(x) = \sqrt{x - 2}$.

  b. From the graph of f, is f a one-to-one function?   Yes

  c. Write the domain of f in interval notation. $[2, \infty)$

  d. Write the range of f in interval notation. $[0, \infty)$

  e. Write an equation for $f^{-1}(x)$. $f^{-1}(x) = x^2 + 2; x \geq 0$

  f. Explain why the restriction $x \geq 0$ is placed on f^{-1}.
  The range of f is $[0, \infty)$. Therefore, the domain of f^{-1} must be $[0, \infty)$.

  g. Graph $y = f(x)$ and $y = f^{-1}(x)$ on the same coordinate system.

  h. Write the domain of f^{-1} in interval notation. $[0, \infty)$

  i. Write the range of f^{-1} in interval notation. $[2, \infty)$

57. Given that the domain of a one-to-one function f is $[0, \infty)$ and the range of f is $[0, 4)$, state the domain and range of f^{-1}.   Domain: $[0, 4)$; Range: $[0, \infty)$

58. Given that the domain of a one-to-one function f is $[-3, 5)$ and the range of f is $(-2, \infty)$, state the domain and range of f^{-1}.   Domain: $(-2, \infty)$; Range: $[-3, 5)$

59. Given $f(x) = |x| + 3; x \leq 0$, write an equation for f^{-1}. (*Hint*: Sketch $f(x)$ and note the domain and range.) $f^{-1}(x) = 3 - x; x \geq 3$

60. Given $f(x) = |x| - 3; x \geq 0$, write an equation for f^{-1}. (*Hint*: Sketch $f(x)$ and note the domain and range.) $f^{-1}(x) = x + 3; x \geq -3$

For Exercises 61–66, fill in the blanks and determine an equation for $f^{-1}(x)$ mentally.

61. If function f adds 6 to x, then f^{-1} ___subtracts___ 6 from x. Function f is defined by $f(x) = x + 6$, and function f^{-1} is defined by $f^{-1}(x) = $ ___$x - 6$___.

62. If function f multiplies x by 2, then f^{-1} ___divides___ x by 2. Function f is defined by $f(x) = 2x$, and function f^{-1} is defined by $f^{-1}(x) = $ ___$\frac{x}{2}$___.

63. Suppose that function f multiplies x by 7 and subtracts 4. Write an equation for $f^{-1}(x)$. $f^{-1}(x) = \frac{x + 4}{7}$

64. Suppose that function f divides x by 3 and adds 11. Write an equation for $f^{-1}(x)$. $f^{-1}(x) = 3(x - 11)$

65. Suppose that function f cubes x and adds 20. Write an equation for $f^{-1}(x)$. $f^{-1}(x) = \sqrt[3]{x - 20}$

66. Suppose that function f takes the cube root of x and subtracts 10. Write an equation for $f^{-1}(x)$. $f^{-1}(x) = (x + 10)^3$

For Exercises 67–70, find the inverse mentally.

67. $f(x) = 8x + 1$ $f^{-1}(x) = \frac{x - 1}{8}$

68. $p(x) = 2x - 10$ $p^{-1}(x) = \frac{x + 10}{2}$

69. $q(x) = \sqrt[5]{x - 4} + 1$ $q^{-1}(x) = (x - 1)^5 + 4$

70. $m(x) = \sqrt[3]{4x} + 3$ $m^{-1}(x) = \frac{(x - 3)^3}{4}$

Mixed Exercises

For Exercises 71–74, the graph of a function is given. Graph the inverse function.

71.
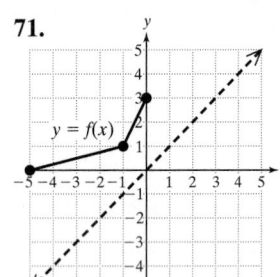

72.
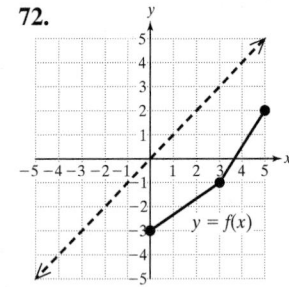

73.
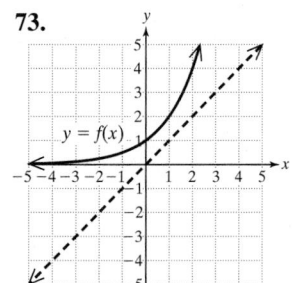

74.
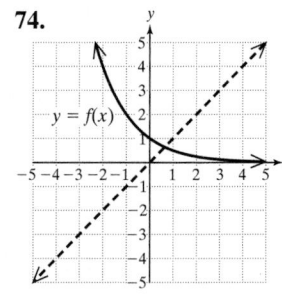

For Exercises 75–76, the table defines $Y_1 = f(x)$ as a one-to-one function of x. Find the values of f^{-1} for the selected values of x.

75. a. $f^{-1}(32)$ 12

  b. $f^{-1}(-2.5)$ 0.5

  c. $f^{-1}(26)$ 10

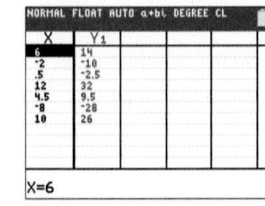

76. a. $f^{-1}(5)$ 20

  b. $f^{-1}(9.45)$ 2.2

  c. $f^{-1}(8)$ 8

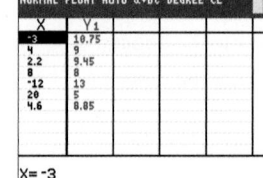

For Exercises 77–80, determine if the statement is true or false. If a statement is false, explain why.

77. All linear functions with a nonzero slope have an inverse function.   True

78. The domain of any one-to-one function is the same as the domain of its inverse function.   False. The domain of a one-to-one function is the same as the range of its inverse.

79. The range of a one-to-one function is the same as the range of its inverse function. False. The range of a one-to-one function is the same as the domain of its inverse.

81. Based on data from Hurricane Katrina, the function defined by $w(x) = -1.17x + 1220$ gives the wind speed $w(x)$ (in mph) based on the barometric pressure x (in millibars, mb).

  a. Approximate the wind speed for a hurricane with a barometric pressure of 1000 mb. 50 mph

  b. Write a function representing the inverse of w and interpret its meaning in context.

  c. Approximate the barometric pressure for a hurricane with wind speed 100 mph. Round to the nearest mb. 957 mb

83. Suppose that during normal respiration, the volume of air inhaled per breath (called "tidal volume") by a mammal of any size is 6.33 mL per kilogram of body mass.

  a. Write a function representing the tidal volume $T(x)$ (in mL) of a mammal of mass x (in kg). $T(x) = 6.33x$

  b. Write an equation for $T^{-1}(x)$.

  c. What does the inverse function represent in the context of this problem?

  d. Find $T^{-1}(170)$ and interpret its meaning in context. Round to the nearest whole unit.

85. The millage rate is the amount of property tax per $1000 of the taxable value of a home. For a certain county the millage rate is 24 mil ($24 in tax per $1000 of taxable value of the home). A city within the county also imposes a flat fee of $108 per home.

  a. Write a function representing the total amount of property tax $T(x)$ (in $) for a home with a taxable value of x thousand dollars. $T(x) = 24x + 108$

  b. Write an equation for $T^{-1}(x)$. $T^{-1}(x) = \dfrac{x - 108}{24}$

  c. What does the inverse function represent in the context of this problem?

  d. Evaluate $T^{-1}(2988)$ and interpret its meaning in context.

  c. $T^{-1}(x)$ represents the taxable value of a home (in $1000) based on x dollars of property tax paid on the home.

  d. $T^{-1}(2988) = 120$ means that if a homeowner is charged $2998 in property taxes, then the taxable value of the home is $120,000.

Write About It

87. Explain the relationship between the domain and range of a one-to-one function f and its inverse f^{-1}.
The domain and range of a function and its inverse are reversed.

89. Explain why if a horizontal line intersects the graph of a function in more than one point, then the function is not one-to-one.
If a horizontal line intersects the graph of a function in more than one point, then the function has at least two ordered pairs with the same y-coordinate but different x-coordinates. This conflicts with the definition of a one-to-one function.

80. No quadratic function defined by $f(x) = ax^2 + bx + c$ ($a \neq 0$) is one-to-one. True

82. The function defined by $F(x) = \dfrac{9}{5}x + 32$ gives the temperature $F(x)$ (in degrees Fahrenheit) based on the temperature x (in Celsius).

  a. Determine the temperature in Fahrenheit if the temperature in Celsius is 25°C. 77°F

  b. Write a function representing the inverse of F and interpret its meaning in context.

  c. Determine the temperature in Celsius if the temperature in Fahrenheit is 5°F. $-15°C$

84. At a cruising altitude of 35,000 ft, a certain airplane travels 555 mph.

  a. Write a function representing the distance $d(x)$ (in mi) for x hours at cruising altitude. $d(x) = 555x$

  b. Write an equation for $d^{-1}(x)$. $d^{-1}(x) = \dfrac{x}{555}$

  c. What does the inverse function represent in the context of this problem?

  d. Evaluate $d^{-1}(2553)$ and interpret its meaning in context.

  c. $d^{-1}(x)$ represents the amount of time (in hr) at which the aircraft has been at its cruising altitude based on the distance traveled at that altitude.

  d. $d^{-1}(2553) = 4.6$ means that if 2553 mi is traveled at cruising altitude, then the plane has been at its cruising altitude for 4.6 hr.

86. Beginning on January 1, park rangers in Everglades National Park began recording the water level for one particularly dry area of the park. The water level was initially 2.5 ft and decreased by approximately 0.015 ft/day.

  a. Write a function representing the water level $L(x)$ (in ft), x days after January 1. $L(x) = 2.5 - 0.015x$

  b. Write an equation for $L^{-1}(x)$. $L^{-1}(x) = \dfrac{2.5 - x}{0.015}$

  c. What does the inverse function represent in the context of this problem? $L^{-1}(x)$ represents the number of days since January 1 based on the water level, x.

  d. Evaluate $L^{-1}(1.9)$ and interpret its meaning in context.
$L^{-1}(1.9) = 40$ means that given a water level of 1.9 ft, the time elapsed since January 1 is 40 days.

88. Write an informal definition of a one-to-one function.
A one-to-one function $y = f(x)$ is a function in which no two ordered pairs have the same y-coordinate but different x-coordinates.

90. Explain why the domain of $f(x) = x^2 + k$ must be restricted to find an inverse function.
The graph of $f(x) = x^2 + k$ is a parabola opening upward or downward. The function is not one-to-one. However, if the domain is restricted to x values strictly to one side or the other of the axis of symmetry, then the resulting function is one-to-one and the inverse function can be found.

Expanding Your Skills

91. Consider a function defined as follows. Given x, the value $f(x)$ is the exponent above the base of 2 that produces x. For example, $f(16) = 4$ because $2^4 = 16$. Evaluate

a. $f(8)$ $f(8) = 3$

b. $f(32)$ $f(32) = 5$

c. $f(2)$ $f(2) = 1$

d. $f\left(\frac{1}{8}\right)$ $f\left(\frac{1}{8}\right) = -3$

92. Consider a function defined as follows. Given x, the value $f(x)$ is the exponent above the base of 3 that produces x. For example, $f(9) = 2$ because $3^2 = 9$. Evaluate

a. $f(27)$ $f(27) = 3$

b. $f(81)$ $f(81) = 4$

c. $f(3)$ $f(3) = 1$

d. $f\left(\frac{1}{9}\right)$ $f\left(\frac{1}{9}\right) = -2$

93. Show that every increasing function is one-to-one.

Let f be an increasing function. Then for every value a and b in the domain of f such that $a < b$ we have $f(a) < f(b)$. Now if $u \neq v$, then either $u < v$ or $v < u$. Then either $f(u) < f(v)$ or $f(v) < f(u)$. In either case, $f(u) \neq f(v)$, and f is one-to-one.

94. A function is said to be periodic if there exists some nonzero real number p, called the period, such that $f(x + p) = f(x)$ for all real numbers x in the domain of f. Explain why no periodic function is one-to-one.

Let f be a periodic function and let $p \neq 0$ be the period. Then $f(x + p) = f(x)$ for all x in the domain of f. Since $p \neq 0$, then $x + p \neq x$, so f is not one-to-one.

SECTION 3.2   Exponential Functions

OBJECTIVES

1. Graph Exponential Functions
2. Evaluate the Exponential Function Base e
3. Use Exponential Functions to Compute Compound Interest
4. Use Exponential Functions in Applications

1. Graph Exponential Functions

The concept of a function was first introduced in Section 1.3. Since then we have learned to recognize several categories of functions. In this section and the next, we will define two new types of functions called exponential functions and logarithmic functions.

To introduce exponential functions, consider two salary plans for a new job. Plan A pays $1 million for 1 month's work. Plan B starts with 2¢ on the first day, and every day thereafter the salary is doubled. At first glance, the million-dollar plan appears to be more favorable. However, Table 3-3 shows otherwise. The daily payments for 30 days are listed for Plan B.

Table 3-3

Day	Payment	Day	Payment	Day	Payment
1	2¢	11	$20.48	21	$20,971.52
2	4¢	12	$40.96	22	$41,943.04
3	8¢	13	$81.92	23	$83,886.08
4	16¢	14	$163.84	24	$167,772.16
5	32¢	15	$327.68	25	$335,554.32
6	64¢	16	$655.36	26	$671,088.64
7	$1.28	17	$1310.72	27	$1,342,177.28
8	$2.56	18	$2621.44	28	$2,684,354.56
9	$5.12	19	$5242.88	29	$5,368,709.12
10	$10.24	20	$10,485.76	30	$10,737,418.24

TIP Consider the pattern involved for the payment for day $x = 1, 2, 3, 4, 5, \ldots$

2^1 ¢ = 2¢
2^2 ¢ = 4¢
2^3 ¢ = 8¢
2^4 ¢ = 16¢
2^5 ¢ = 32¢

The salary for the 30th day for Plan B is over $10 million. Taking the sum of the payments, we see that the total salary for the 30-day period is $21,474,836.46.

The daily salary $S(x)$ (in ¢) for Plan B can be represented by the function $S(x) = 2^x$, where x is the number of days on the job. An interesting characteristic of this function is that for every positive 1-unit change in x, the function value doubles. The function $S(x) = 2^x$ is called an exponential function.

Definition of an Exponential Function

Let b be a constant real number such that $b > 0$ and $b \neq 1$. Then for any real number x, a function of the form $f(x) = b^x$ is called an **exponential function of base b.**

An exponential function is recognized as a function with a constant base (positive and not equal to 1) with a variable exponent, x.

Exponential Functions	**Not Exponential Functions**	
$f(x) = 3^x$	$m(x) = x^2$	base is not constant
$g(x) = \left(\dfrac{1}{3}\right)^x$	$n(x) = \left(-\dfrac{1}{3}\right)^x$	base is negative
$h(x) = \left(\sqrt{2}\right)^x$	$p(x) = 1^x$	base is 1

Avoiding Mistakes

- The base of an exponential function must not be negative to avoid situations where the function values are not real numbers. For example, $f(x) = (-4)^x$ is not defined for $x = \frac{1}{2}$ because $\sqrt{-4}$ is not a real number.
- The base of an exponential function must not equal 1 because $f(x) = 1^x = 1$ for all real numbers x. This is a constant function, not an exponential function.

At this point in the text, we have evaluated exponential expressions with integer exponents and with rational exponents. For example,

$$4^2 = 16 \qquad 4^{1/2} = \sqrt{4} = 2$$

$$4^{-1} = \frac{1}{4} \qquad 4^{10/23} = \sqrt[23]{4^{10}} \approx 1.827112184$$

However, how do we evaluate an exponential expression with an *irrational* exponent such as 4^π? In such a case, the exponent is a nonterminating, nonrepeating decimal. We define an exponential expression raised to an irrational exponent as a sequence of approximations using rational exponents. For example:

$$4^{3.14} \approx 77.7084726$$
$$4^{3.141} \approx 77.81627412$$
$$4^{3.1415} \approx 77.87023095$$
$$\cdots$$
$$4^\pi \approx 77.88023365$$

With this definition of a base raised to an irrational exponent, we can define an exponential function over the entire set of real numbers. In Example 1, we graph two exponential functions by plotting points.

Classroom Examples: p. 377
Exercises 16, 18

EXAMPLE 1   Graphing Exponential Functions

Graph the functions.

a. $f(x) = 2^x$ **b.** $g(x) = \left(\dfrac{1}{2}\right)^x$

Solution:

Table 3-4 shows several function values $f(x)$ and $g(x)$ for both positive and negative values of x.

Table 3-4

x	$f(x) = 2^x$	$g(x) = \left(\frac{1}{2}\right)^x$
-3	$\frac{1}{8}$	8
-2	$\frac{1}{4}$	4
-1	$\frac{1}{2}$	2
0	1	1
1	2	$\frac{1}{2}$
2	4	$\frac{1}{4}$
3	8	$\frac{1}{8}$

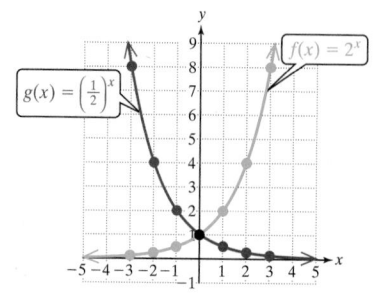

Figure 3-10

> **TIP** The values of $f(x)$ become closer and closer to 0 as $x \to -\infty$. This means that the x-axis is a horizontal asymptote.
> Likewise, the values of $g(x)$ become closer to 0 as $x \to \infty$. The x-axis is a horizontal asymptote.

Notice that $g(x) = \left(\frac{1}{2}\right)^x$ is equivalent to $g(x) = 2^{-x}$. Therefore, the graph of $g(x) = \left(\frac{1}{2}\right)^x = 2^{-x}$ is the same as the graph of $f(x) = 2^x$ with a reflection across the y-axis (Figure 3-10).

Skill Practice 1 Graph the functions.

a. $f(x) = 5^x$ **b.** $g(x) = \left(\frac{1}{5}\right)^x$

The graphs in Figure 3-10 illustrate several important features of exponential functions.

Graphs of $f(x) = b^x$

The graph of an exponential function defined by $f(x) = b^x$ ($b > 0$ and $b \neq 1$) has the following properties.

1. If $b > 1$, f is an *increasing* exponential function, sometimes called an **exponential growth function.**
   If $0 < b < 1$, f is a *decreasing* exponential function, sometimes called an **exponential decay function.**

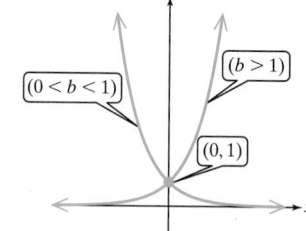

2. The domain is the set of all real numbers, $(-\infty, \infty)$.
3. The range is $(0, \infty)$.
4. The line $y = 0$ (x-axis) is a horizontal asymptote.
5. The function passes through the point $(0, 1)$ because $f(0) = b^0 = 1$.

Answer

1.

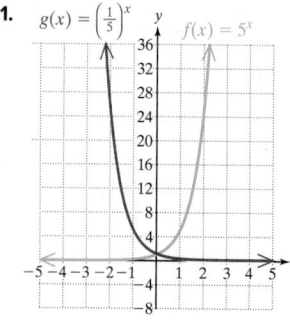

These properties indicate that the graph of an exponential function is an increasing function if the base is greater than 1. Furthermore, the base affects the rate of increase. Consider the graphs of $f(x) = 2^x$ and $k(x) = 5^x$ (Figure 3-11). For every positive 1-unit change in x, $f(x) = 2^x$ is 2 times as great and $k(x) = 5^x$ is 5 times as great (Table 3-5).

Table 3-5

x	$f(x) = 2^x$	$k(x) = 5^x$
-3	$\frac{1}{8}$	$\frac{1}{125}$
-2	$\frac{1}{4}$	$\frac{1}{25}$
-1	$\frac{1}{2}$	$\frac{1}{5}$
0	1	1
1	2	5
2	4	25
3	8	125

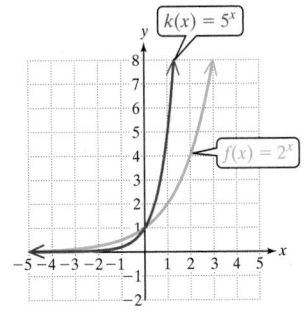

Figure 3-11

In Example 2, we use the transformations of functions learned in Section 1.6 to graph an exponential function.

> If $h > 0$, shift to the right.
> If $h < 0$, shift to the left.

$$f(x) = ab^{x-h} + k$$

> If $a < 0$, reflect across the x-axis.
> Shrink vertically if $0 < |a| < 1$.
> Stretch vertically if $|a| > 1$.

> If $k > 0$, shift upward.
> If $k < 0$, shift downward.

EXAMPLE 2   **Graphing an Exponential Function**

Graph. $f(x) = 3^{x-2} + 4$

Solution:

The graph of f is the graph of the parent function $y = 3^x$ shifted 2 units to the right and 4 units upward.

The parent function $y = 3^x$ is an increasing exponential function. We can plot a few points on the graph of $y = 3^x$ and use these points and the horizontal asymptote to form the outline of the transformed graph.

x	$y = 3^x$
-2	$\frac{1}{9}$
-1	$\frac{1}{3}$
0	1
1	3
2	9

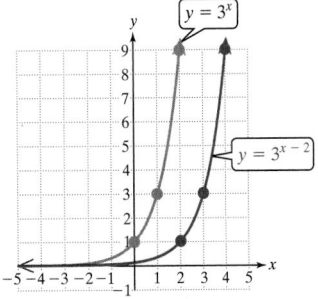

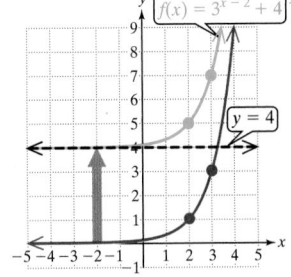

Shift 2 units to the right. For example, the point $(0, 1)$ on $y = 3^x$ corresponds to $(2, 1)$ on $y = 3^{x-2}$.

Shift the graph of $y = 3^{x-2}$ up 4 units. Notice that with the vertical shift, the new horizontal asymptote is $y = 4$.

Answer

2.

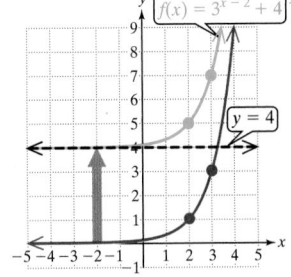

Skill Practice 2 Graph. $g(x) = 2^{x+2} - 1$

2. Evaluate the Exponential Function Base e

We now introduce an important exponential function whose base is an irrational number called e. Consider the expression $\left(1 + \dfrac{1}{x}\right)^x$. The value of the expression for increasingly large values of x approaches a constant (Table 3-6).

As $x \to \infty$, the expression $\left(1 + \dfrac{1}{x}\right)^x$ approaches a constant value that we call e. From Table 3-6, this value is approximately 2.718281828.

$$e \approx 2.718281828$$

The value of e is an irrational number (a non-terminating, nonrepeating decimal) and like the number π, it is a universal constant. The function defined by $f(x) = e^x$ is called the exponential function base e or the **natural exponential function.**

Table 3-6

x	$\left(1 + \dfrac{1}{x}\right)^x$
100	2.70481382942
1000	2.71692393224
10,000	2.71814592683
100,000	2.71826823717
1,000,000	2.71828046932
1,000,000,000	2.71828182710

Classroom Example: p. 378
Exercise 40

EXAMPLE 3 Graphing $f(x) = e^x$

Graph the function. $f(x) = e^x$

Solution:

Because the base e is greater than 1 ($e \approx 2.718281828$), the graph is an increasing exponential function. We can use a calculator to evaluate $f(x) = e^x$ at several values of x. On many calculators, the exponential function, base e, is invoked by selecting **2ND** **LN** or by accessing e^x on the keyboard.

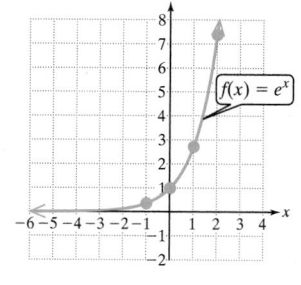

x	$f(x) = e^x$
-3	0.050
-2	0.135
-1	0.368
0	1.000
1	2.718
2	7.389
3	20.086

Figure 3-12

The graph of $f(x) = e^x$ is shown in Figure 3-12.

Skill Practice 3 Explain how the graph of $f(x) = -e^{x-1}$ is related to the graph of $y = e^x$.

TIP In Section 3.3, we will see that the exponential function base e is the inverse of the natural logarithmic function, $y = \ln x$. This is why the exponential function base e is accessed with the **2ND** **LN** keys.

---

3. Use Exponential Functions to Compute Compound Interest

Recall that simple interest is interest computed on the principal amount invested (or borrowed). Compound interest is interest computed on both the original principal and the interest already accrued.

Answer

3. The graph of $f(x) = -e^{x-1}$ is the graph of $y = e^x$ with a shift to the right 1 unit and a reflection across the x-axis.

Suppose that interest is compounded annually (one time per year) on an investment of P dollars at an annual interest rate r for t years. Then the amount A (in \$) in the account after 1, 2, and 3 yr is computed as follows.

After 1 yr: $\begin{pmatrix} \text{Total} \\ \text{amount} \end{pmatrix} = \begin{pmatrix} \text{Initial} \\ \text{principal} \end{pmatrix} + (\text{Interest})$

$A = P + Pr$ The interest is given by $I = Prt$, where $t = 1$ yr. So $I = Pr$.

$\qquad = P(1 + r)$ Factor out P.

After 2 yr: $\begin{pmatrix} \text{Total} \\ \text{amount} \end{pmatrix} = \begin{pmatrix} \text{Year 1} \\ \text{balance} \end{pmatrix} + \begin{pmatrix} \text{Interest on} \\ \text{Year 1 balance} \end{pmatrix}$

$A = P(1 + r) + [P(1 + r)]r$

$\qquad = P(1 + r)(1 + r) \qquad$ Factor out $P(1 + r)$.

$\qquad = P(1 + r)^2$

After 3 yr: $\begin{pmatrix} \text{Total} \\ \text{amount} \end{pmatrix} = \begin{pmatrix} \text{Year 2} \\ \text{balance} \end{pmatrix} + \begin{pmatrix} \text{Interest on} \\ \text{Year 2 balance} \end{pmatrix}$

$A = P(1 + r)^2 + [P(1 + r)^2]r$

$\qquad = P(1 + r)^2(1 + r) \qquad$ Factor out $P(1 + r)^2$.

$\qquad = P(1 + r)^3$

...

After t years: $A = P(1 + r)^t$ Amount in an account with interest compounded annually.

Compound interest is often computed more frequently during the course of 1 yr. Let n represent the number of compounding periods per year. For example:

$n = 1$ for interest compounded annually
$n = 4$ for interest compounded quarterly
$n = 12$ for interest compounded monthly
$n = 365$ for interest compounded daily

Each compounding period represents a fraction of a year and the interest rate is scaled accordingly for each compounding period as $\frac{1}{n} \cdot r$ or $\frac{r}{n}$. The number of compounding periods over the course of the investment is nt. Therefore, to determine the amount in an account where interest is compounded n times per year we have

replace t by nt

$A = P(1 + r)^t \qquad\qquad A = P\left(1 + \dfrac{r}{n}\right)^{nt}$ Amount in an account with interest compounded n times per year.

replace r by $\frac{r}{n}$

Now suppose it were possible to compute interest continuously, that is, for $n \to \infty$. If we use the substitution $x = \frac{n}{r}$ (which implies that $n = xr$) the formula for compound interest becomes

$$A = P\left(1 + \frac{r}{n}\right)^{nt} \xrightarrow{\text{Substitute } x = \frac{n}{r}} P\left(1 + \frac{1}{x}\right)^{xrt} = P\left[\left(1 + \frac{1}{x}\right)^x\right]^{rt}$$

For a fixed interest rate r, as n approaches infinity, x also approaches infinity. Since the expression $\left(1 + \frac{1}{x}\right)^x$ approaches e as $x \to \infty$, we have

$$A = Pe^{rt}$$ Amount in an account with interest compounded continuously.

Summary of Formulas Relating to Simple and Compound Interest

Suppose that P dollars in principal is invested (or borrowed) at an annual interest rate r for t years. Then

- $I = Prt$ — Amount of simple interest I (in \$).
- $A = P\left(1 + \dfrac{r}{n}\right)^{nt}$ — The future value A (in \$) of the account after t years with n compounding periods per year.
- $A = Pe^{rt}$ — The future value A (in \$) of the account after t years under continuous compounding.

In Example 4, we compare the value of an investment after 10 yr under several different compounding options.

Classroom Example: p. 378
Exercise 46

EXAMPLE 4 Computing the Balance on an Account

Suppose that \$5000 is invested and pays 6.5% per year under the following compounding options.

a. Compounded annually   **b.** Compounded quarterly
c. Compounded monthly   **d.** Compounded daily
e. Compounded continuously

Determine the total amount in the account after 10 yr with each option.

Solution:

Using $A = P\left(1 + \dfrac{r}{n}\right)^{nt}$ and $A = Pe^{rt}$, we have

Compounding Option	n Value	Formula	Result
Annually	$n = 1$	$A = 5000\left(1 + \dfrac{0.065}{1}\right)^{(1 \cdot 10)}$	\$9385.69
Quarterly	$n = 4$	$A = 5000\left(1 + \dfrac{0.065}{4}\right)^{(4 \cdot 10)}$	\$9527.79
Monthly	$n = 12$	$A = 5000\left(1 + \dfrac{0.065}{12}\right)^{(12 \cdot 10)}$	\$9560.92
Daily	$n = 365$	$A = 5000\left(1 + \dfrac{0.065}{365}\right)^{(365 \cdot 10)}$	\$9577.15
Continuously	Not applicable	$A = 5000e^{(0.065 \cdot 10)}$	\$9577.70

Notice that there is a \$192.01 difference in the account balance between annual compounding and continuous compounding. The table also supports our finding that

$$A = P\left(1 + \frac{r}{n}\right)^{nt} \quad \text{converges to} \quad A = Pe^{rt} \quad \text{as } n \to \infty.$$

Skill Practice 4 Suppose that \$8000 is invested and pays 4.5% per year under the following compounding options.

a. Compounded annually   **b.** Compounded quarterly
c. Compounded monthly   **d.** Compounded daily
e. Compounded continuously

Determine the total amount in the account after 5 yr with each option.

Answers
4. a. \$9969.46   **b.** \$10,006.00
c. \$10,014.37   **d.** \$10,018.44
e. \$10,018.58

4. Use Exponential Functions in Applications

Increasing and decreasing exponential functions can be used in a variety of real-world applications. For example:

- Population growth can often be modeled by an exponential function.
- The growth of an investment under compound interest increases exponentially.
- The mass of a radioactive substance decreases exponentially with time.
- The temperature of a cup of coffee decreases exponentially as it approaches room temperature.

A substance that undergoes radioactive decay is said to be radioactive. The **half-life** of a radioactive substance is the amount of time it takes for one-half of the original amount of the substance to change into something else. That is, after each half-life, the amount of the original substance decreases by one-half.

Classroom Example: p. 379
Exercise 54

EXAMPLE 5 Using an Exponential Function in an Application

The half-life of radium 226 is 1620 yr. In a sample originally having 1 g of radium 226, the amount $A(t)$ (in grams) of radium 226 present after t years is given by $A(t) = \left(\frac{1}{2}\right)^{t/1620}$ where t is the time in years after the start of the experiment. How much radium will be present after

a. 1620 yr?   **b.** 3240 yr?   **c.** 4860 yr?

Solution:

a. $A(t) = \left(\frac{1}{2}\right)^{t/1620}$

$A(1620) = \left(\frac{1}{2}\right)^{1620/1620}$

$= \left(\frac{1}{2}\right)^{1}$

$= 0.5$

b. $A(t) = \left(\frac{1}{2}\right)^{t/1620}$

$A(3240) = \left(\frac{1}{2}\right)^{3240/1620}$

$= \left(\frac{1}{2}\right)^{2}$

$= 0.25$

c. $A(t) = \left(\frac{1}{2}\right)^{t/1620}$

$A(4860) = \left(\frac{1}{2}\right)^{4860/1620}$

$= \left(\frac{1}{2}\right)^{3}$

$= 0.125$

The half-life of radium is 1620 yr. Therefore, we can interpret these results as follows.

After 1620 yr (1 half-life), 0.5 g remains $\left(\frac{1}{2}\right.$ of the original amount remains$\left.\right)$.

After 3240 yr (2 half-lives), 0.25 g remains $\left(\frac{1}{4}\right.$ of the original amount remains$\left.\right)$.

After 4860 yr (3 half-lives), 0.125 g remains $\left(\frac{1}{8}\right.$ of the original amount remains$\left.\right)$.

Skill Practice 5 Cesium-137 is a radioactive metal with a short half-life of 30 yr. In a sample originally having 2 g of cesium-137, the amount $A(t)$ (in grams) of cesium-137 present after t years is given by $A(t) = 2\left(\frac{1}{2}\right)^{t/30}$. How much cesium-137 will be present after

a. 30 yr?   **b.** 60 yr?   **c.** 90 yr?

Answers
5. a. 1 g   **b.** 0.5 g   **c.** 0.25 g

TECHNOLOGY CONNECTIONS

Graphing an Exponential Function

A graphing utility can be used to graph and analyze exponential functions. The table shows several values of $A(x) = \left(\frac{1}{2}\right)^{x/1620}$ for selected values of x.

The graph of $A(x) = \left(\frac{1}{2}\right)^{x/1620}$ is shown on the viewing window [0, 10,000, 1000] by [0, 1.2, 0.2]. Notice that the graph is a decreasing exponential function because the base is between 0 and 1.

SECTION 3.2   Practice Exercises

Prerequisite Review

R.1. Determine the domain and range.   Domain $(-\infty, 3)$,
Range $(-\infty, 0]$

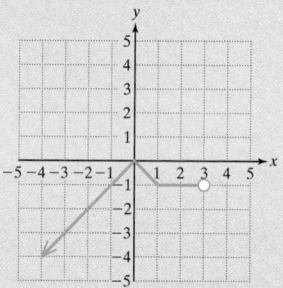

R.3. Use transformations to graph $q(x) = -(x-3)^2 + 4$.

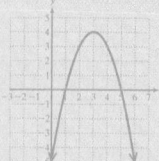

R.2. Refer to the graph of the function and complete the statement.

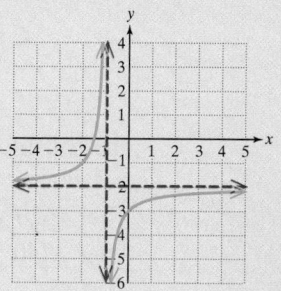

a. As $x \to -\infty, f(x) \to$ _____ -2

b. As $x \to -1^+, f(x) \to$ _____ $-\infty$

c. As $x \to \infty, f(x) \to$ _____ -2

d. As $x \to -1^-, f(x) \to$ _____ ∞

e. The graph is decreasing over the interval(s)
_____ Never decreasing _____

f. The graph is increasing over the interval(s)
_____ $(-\infty, -1) \cup (-1, \infty)$ _____

Concept Connections

1. The function defined by $y = x^3$ (is/is not) an exponential function, whereas the function defined by $y = 3^x$ (is/is not) an exponential function.   is not; is

2. The graph of $f(x) = \left(\dfrac{5}{3}\right)^x$ is (increasing/decreasing) over its domain.   increasing

3. The graph of $f(x) = \left(\dfrac{3}{5}\right)^x$ is (increasing/decreasing) over its domain.   decreasing

4. The domain of an exponential function $f(x) = b^x$ is _____. $(-\infty, \infty)$

5. The range of an exponential function $f(x) = b^x$ is _____. $(0, \infty)$

6. All exponential functions $f(x) = b^x$ pass through the point _____. $(0, 1)$

7. The horizontal asymptote of an exponential function $f(x) = b^x$ is the line _____. $y = 0$

8. As $x \to \infty$, the value of $\left(1 + \dfrac{1}{x}\right)^x$ approaches _____. e

Objective 1: Graph Exponential Functions

For Exercises 9–12, evaluate the functions at the given values of x. Round to 4 decimal places if necessary.

9. $f(x) = 5^x$
  a. $f(-1)$ 0.2
  b. $f(4.8)$ 2264.9364
  c. $f(\sqrt{2})$ 9.7385
  d. $f(\pi)$ 156.9925

10. $g(x) = 7^x$
  a. $g(-2)$ 0.0204
  b. $g(5.9)$ 96,845.2749
  c. $g(\sqrt{11})$ 635.1453
  d. $g(e)$ 198.2507

11. $h(x) = \left(\dfrac{1}{4}\right)^x$
  a. $h(-3)$ 64
  b. $h(1.4)$ 0.1436
  c. $h(\sqrt{3})$ 0.0906
  d. $h(0.5e)$ 0.1520

12. $k(x) = \left(\dfrac{1}{6}\right)^x$
  a. $k(-3)$ 216
  b. $k(1.4)$ 0.0814
  c. $k(\sqrt{0.5})$ 0.2817
  d. $h(0.5\pi)$ 0.0599

13. Which functions are exponential functions?   a, d
  a. $f(x) = 4.2^x$ **b.** $g(x) = x^{4.2}$ **c.** $h(x) = 4.2x$
  d. $k(x) = \left(\sqrt{4.2}\right)^x$ **e.** $m(x) = (-4.2)^x$

14. Which functions are exponential functions?   b, d
  a. $v(x) = (-\pi)^x$ **b.** $t(x) = \pi^x$ **c.** $w(x) = \pi x$
  d. $n(x) = \left(\sqrt{\pi}\right)^x$ **e.** $p(x) = x^\pi$

For Exercises 15–22, graph the functions and write the domain and range in interval notation. (See Example 1)

15. $f(x) = 3^x$

16. $g(x) = 4^x$

17. $h(x) = \left(\dfrac{1}{3}\right)^x$

18. $k(x) = \left(\dfrac{1}{4}\right)^x$

19. $m(x) = \left(\dfrac{3}{2}\right)^x$

20. $n(x) = \left(\dfrac{5}{4}\right)^x$

21. $b(x) = \left(\dfrac{2}{3}\right)^x$

22. $c(x) = \left(\dfrac{4}{5}\right)^x$

For Exercises 23–32,
  a. Use transformations of the graphs of $y = 3^x$ (see Exercise 15) and $y = 4^x$ (see Exercise 16) to graph the given function. (**See Example 2**)
  b. Write the domain and range in interval notation.
  c. Write an equation of the asymptote.

23. $f(x) = 3^x + 2$

24. $g(x) = 4^x - 3$

25. $m(x) = 3^{x+2}$

26. $n(x) = 4^{x-3}$

27. $p(x) = 3^{x-4} - 1$

28. $q(x) = 4^{x+1} + 2$

29. $k(x) = -3^x$

30. $h(x) = -4^x$

31. $t(x) = 3^{-x}$

32. $v(x) = 4^{-x}$

For Exercises 33–36,
  a. Use transformations of the graphs of $y = \left(\frac{1}{3}\right)^x$ (see Exercise 17) and $y = \left(\frac{1}{4}\right)^x$ (see Exercise 18) to graph the given function. (**See Example 2**)
  b. Write the domain and range in interval notation.
  c. Write an equation of the asymptote.

33. $f(x) = \left(\dfrac{1}{3}\right)^{x+1} - 3$

34. $g(x) = \left(\dfrac{1}{4}\right)^{x-2} + 1$

35. $k(x) = -\left(\dfrac{1}{3}\right)^x + 2$

36. $h(x) = -\left(\dfrac{1}{4}\right)^x - 2$

Objective 2: Evaluate the Exponential Function Base e

For Exercises 37–38, evaluate the functions for the given values of x. Round to 4 decimal places.

37. $f(x) = e^x$

   a. $f(4)$ 54.5982   **b.** $f(-3.2)$ 0.0408

   c. $f\left(\sqrt{13}\right)$ 36.8020   **d.** $f(\pi)$ 23.1407

38. $f(x) = e^x$

   a. $f(-3)$ 0.0498   **b.** $f(6.8)$ 897.8473

   c. $f\left(\sqrt{7}\right)$ 14.0940   **d.** $f(e)$ 15.1543

For Exercises 39–44,

   a. Use transformations of the graph of $y = e^x$ to graph the given function. (**See Example 3**)

   b. Write the domain and range in interval notation.

   c. Write an equation of the asymptote.

39. $f(x) = e^{x-4}$ **40.** $g(x) = e^{x-2}$ **41.** $h(x) = e^x + 2$

42. $k(x) = e^x - 1$ **43.** $m(x) = -e^x - 3$ **44.** $n(x) = -e^x + 4$

Objective 3: Use Exponential Functions to Compute Compound Interest

For Exercises 45–46, complete the table to determine the effect of the number of compounding periods when computing interest. (**See Example 4**)

45. Suppose that $10,000 is invested at 4% interest for 5 yr under the following compounding options. Complete the table.

	Compounding Option	n Value	Result
a.	Annually	1	$12,166.53
b.	Quarterly	4	$12,201.90
c.	Monthly	12	$12,209.97
d.	Daily	365	$12,213.89
e.	Continuously	n/a	$12,214.03

46. Suppose that $8000 is invested at 3.5% interest for 20 yr under the following compounding options. Complete the table.

	Compounding Option	n Value	Result
a.	Annually	1	$15,918.31
b.	Quarterly	4	$16,061.05
c.	Monthly	12	$16,093.62
d.	Daily	365	$16,109.48
e.	Continuously	n/a	$16,110.02

For Exercises 47–48, suppose that P dollars in principal is invested for t years at the given interest rates with continuous compounding. Determine the amount that the investment is worth at the end of the given time period.

47. $P = \$20,000$, $t = 10$ yr

   a. 3% interest $26,997.18

   b. 4% interest $29,836.49

   c. 5.5% interest $34,665.06

48. $P = \$6000$, $t = 12$ yr

   a. 1% interest $6764.98

   b. 2% interest $7627.49

   c. 4.5% interest $10,296.04

49. Bethany needs to borrow $10,000. She can borrow the money at 5.5% simple interest for 4 yr or she can borrow at 5% with interest compounded continuously for 4 yr.

   a. How much total interest would Bethany pay at 5.5% simple interest? $2200

   b. How much total interest would Bethany pay at 5% interest compounded continuously? $2214.03

   c. Which option results in less total interest?
      5.5% simple interest results in less interest.

51. Jerome wants to invest $25,000 as part of his retirement plan. He can invest the money at 5.2% simple interest for 30 yr, or he can invest at 3.8% interest compounded continuously for 30 yr. Which option results in more total interest?
3.8% compounded continuously for 30 yr results in more interest.

50. Al needs to borrow $15,000 to buy a car. He can borrow the money at 6.7% simple interest for 5 yr or he can borrow at 6.4% interest compounded continuously for 5 yr.

   a. How much total interest would Al pay at 6.7% simple interest? $5025

   b. How much total interest would Al pay at 6.4% interest compounded continuously? $5656.92

   c. Which option results in less total interest?
      6.7% simple interest results in less interest.

52. Heather wants to invest $35,000 of her retirement. She can invest at 4.8% simple interest for 20 yr, or she can choose an option with 3.6% interest compounded continuously for 20 yr. Which option results in more total interest?
3.6% compounded continuously for 20 yr results in more interest.

Objective 4: Use Exponential Functions in Applications

53. Strontium-90 (^{90}Sr) is a by-product of nuclear fission with a half-life of approximately 28.9 yr. After the Chernobyl nuclear reactor accident in 1986, large areas surrounding the site were contaminated with ^{90}Sr. If 10 μg (micrograms) of ^{90}Sr is present in a sample, the function $A(t) = 10\left(\dfrac{1}{2}\right)^{t/28.9}$ gives the amount $A(t)$ (in μg) present after t years. Evaluate the function for the given values of t and interpret the meaning in context. Round to 3 decimal places if necessary. (**See Example 5**)

   a. $A(28.9)$ **b.** $A(57.8)$ **c.** $A(100)$

54. In 2006, the murder of Alexander Litvinenko, a Russian dissident, was thought to be by poisoning from the rare and highly radioactive element polonium-210 (^{210}Po). The half-life of ^{210}Po is 138.4 yr. If 0.1 mg of ^{210}Po is present in a sample then $A(t) = 0.1\left(\dfrac{1}{2}\right)^{t/138.4}$ gives the amount $A(t)$ (in mg) present after t years. Evaluate the function for the given values of t and interpret the meaning in context. Round to 3 decimal places if necessary.

   a. $A(138.4)$ **b.** $A(276.8)$ **c.** $A(500)$

55. According to the CIA's *World Fact Book*, in 2010, the population of the United States was approximately 310 million with a 0.97% annual growth rate. (*Source*: www.cia.gov) At this rate, the population $P(t)$ (in millions) can be approximated by $P(t) = 310(1.0097)^t$, where t is the time in years since 2010.

   a. Is the graph of P an increasing or decreasing exponential function?   Increasing

   b. Evaluate $P(0)$ and interpret its meaning in the context of this problem.

   c. Evaluate $P(10)$ and interpret its meaning in the context of this problem. Round the population value to the nearest million.

   d. Evaluate $P(20)$ and $P(30)$. $P(20) = 376$; $P(30) = 414$

   e. Evaluate $P(200)$ and use this result to determine if it is reasonable to expect this model to continue indefinitely.

56. The population of Canada in 2010 was approximately 34 million with an annual growth rate of 0.804%. At this rate, the population $P(t)$ (in millions) can be approximated by $P(t) = 34(1.00804)^t$, where t is the time in years since 2010. (*Source*: www.cia.gov)

   a. Is the graph of P an increasing or decreasing exponential function?   Increasing

   b. Evaluate $P(0)$ and interpret its meaning in the context of this problem.

   c. Evaluate $P(5)$ and interpret its meaning in the context of this problem. Round the population value to the nearest million.

   d. Evaluate $P(15)$ and $P(25)$.
   $P(15) = 38$;
   $P(25) = 42$

57. The atmospheric pressure on an object decreases as altitude increases. If a is the height (in km) above sea level, then the pressure $P(a)$ (in mmHg) is approximated by $P(a) = 760e^{-0.13a}$.

   a. Find the atmospheric pressure at sea level.   760 mmHg

   b. Determine the atmospheric pressure at 8.848 km (the altitude of Mt. Everest). Round to the nearest whole unit.   241 mmHg

58. The function defined by $A(t) = 100e^{0.0318t}$ approximates the equivalent amount of money needed t years after the year 2010 to equal $100 of buying power in the year 2010. The value 0.0318 is related to the average rate of inflation.

   a. Evaluate $A(15)$ and interpret its meaning in the context of this problem.

   b. Verify that by the year 2032, more than $200 will be needed to have the same buying power as $100 in 2010.
   $A(22) = 201.29$

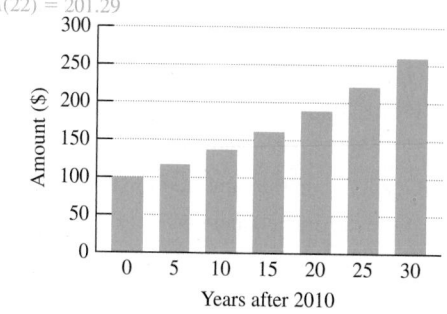

Newton's law of cooling indicates that the temperature of a warm object, such as a cake coming out of the oven, will decrease exponentially with time and will approach the temperature of the surrounding air. The temperature $T(t)$ is modeled by $T(t) = T_a + (T_0 - T_a)e^{-kt}$. In this model, T_a represents the temperature of the surrounding air, T_0 represents the initial temperature of the object, and t is the time after the object starts cooling. The value of k is a constant of proportion relating the temperature of the object to its rate of temperature change. Use this model for Exercises 59–60.

59. A cake comes out of the oven at 350°F and is placed on a cooling rack in a 78°F kitchen. After checking the temperature several minutes later, the value of k is measured as 0.046.

a. Write a function that models the temperature $T(t)$ (in °F) of the cake t minutes after being removed from the oven. $T(t) = 78 + 272e^{-0.046t}$

b. What is the temperature of the cake 10 min after coming out of the oven? Round to the nearest degree. 250°F

c. It is recommended that the cake should not be frosted until it has cooled to under 100°F. If Jessica waits 1 hr to frost the cake, will the cake be cool enough to frost? Yes; after 60 min, the cake will be approximately 95.2°F.

61. A farmer depreciates a $120,000 tractor. He estimates that the resale value $V(t)$ (in $1000) of the tractor t years after purchase is 80% of its value from the previous year. Therefore, the resale value can be approximated by $V(t) = 120(0.8)^t$.

a. Find the resale value 5 yr after purchase. Round to the nearest $1000. $39,000

b. The farmer estimates that the cost to run the tractor is $18/hr in labor, $36/hr in fuel, and $22/hr in overhead costs (for maintenance and repair). Estimate the farmer's cost to run the tractor for the first year if he runs the tractor for a total of 800 hr. Include hourly costs and depreciation. It costs the farmer $84,800 to run the tractor for 800 hr during the first year.

60. Water in a water heater is originally 122°F. The water heater is shut off and the water cools to the temperature of the surrounding air, which is 60°F. The water cools slowly because of the insulation inside the heater, and the value of k is measured as 0.00351.

a. Write a function that models the temperature $T(t)$ (in °F) of the water t hours after the water heater is shut off. $T(t) = 60 + 62e^{-0.00351t}$

b. What is the temperature of the water 12 hr after the heater is shut off? Round to the nearest degree. 119°F

c. Dominic does not like to shower with water less than 115°F. If Dominic waits 24 hr, will the water still be warm enough for a shower? Yes; the temperature after 24 hr will be approximately 117°F.

62. A veterinarian depreciates a $10,000 X-ray machine. He estimates that the resale value $V(t)$ (in $) after t years is 90% of its value from the previous year. Therefore, the resale value can be approximated by $V(t) = 10,000(0.9)^t$.

a. Find the resale value after 4 yr. $6561

b. If the veterinarian wants to sell his practice 8 yr after the X-ray machine was purchased, how much is the machine worth? Round to the nearest $100. $4300

Mixed Exercises

For Exercises 63–64, solve the equations in parts (a)–(c) by inspection. Then estimate the solutions to parts (d) and (e) between two consecutive integers.

63. a. $2^x = 4$ $\{2\}$

b. $2^x = 8$ $\{3\}$

c. $2^x = 16$ $\{4\}$

d. $2^x = 7$ x is between 2 and 3.

e. $2^x = 10$ x is between 3 and 4.

65. a. Graph $f(x) = 2^x$. (**See Example 1**)

b. Is f a one-to-one function? Yes

c. Write the domain and range of f in interval notation. Domain: $(-\infty, \infty)$; Range: $(0, \infty)$

d. Graph f^{-1} on the same coordinate system as f.

e. Write the domain and range of f^{-1} in interval notation. Domain: $(0, \infty)$; Range: $(-\infty, \infty)$

f. From the graph evaluate $f^{-1}(1)$, $f^{-1}(2)$, and $f^{-1}(4)$. $f^{-1}(1) = 0$; $f^{-1}(2) = 1$; $f^{-1}(4) = 2$

67. Refer to the graphs of $f(x) = 2^x$ and the inverse function, $y = f^{-1}(x)$ from Exercise 65. Fill in the blanks.

a. As $x \to \infty$, $f(x) \to$ _____ ∞ .

b. As $x \to -\infty$, $f(x) \to$ _____ 0 .

c. As $x \to \infty$, $f^{-1}(x) \to$ _____ ∞ .

d. As $x \to 0^+$, $f^{-1}(x) \to$ _____ $-\infty$.

64. a. $3^x = 3$ $\{1\}$

b. $3^x = 9$ $\{2\}$

c. $3^x = 27$ $\{3\}$

d. $3^x = 7$ x is between 1 and 2.

e. $3^x = 10$ x is between 2 and 3.

66. a. Graph $g(x) = 3^x$ (see Exercise 15).

b. Is g a one-to-one function? Yes

c. Write the domain and range of g in interval notation. Domain: $(-\infty, \infty)$; Range: $(0, \infty)$

d. Graph g^{-1} on the same coordinate system as g.

e. Write the domain and range of g^{-1} in interval notation. Domain: $(0, \infty)$; Range: $(-\infty, \infty)$

f. From the graph evaluate $g^{-1}(1)$, $g^{-1}(3)$, and $g^{-1}(\frac{1}{3})$. $g^{-1}(1) = 0$; $g^{-1}(3) = 1$; $g^{-1}(\frac{1}{3}) = -1$

68. Refer to the graphs of $g(x) = 3^x$ and the inverse function, $y = g^{-1}(x)$ from Exercise 66. Fill in the blanks.

a. As $x \to \infty$, $g(x) \to$ _____ ∞ .

b. As $x \to -\infty$, $g(x) \to$ _____ 0 .

c. As $x \to \infty$, $g^{-1}(x) \to$ _____ ∞ .

d. As $x \to 0^+$, $g^{-1}(x) \to$ _____ $-\infty$.

Write About It

69. Explain why the equation $2^x = -2$ has no solution.

70. Explain why the $f(x) = x^2$ is not an exponential function.
The base is variable and the exponent is constant rather than the
other way around.

Expanding Your Skills

For Exercises 71–72, find the real solutions to the equation.

71. $3x^2e^{-x} - 6xe^{-x} = 0$ $\{0, 2\}$

72. $x^2e^x - e^x = 0$ $\{-1, 1\}$

73. Use the properties of exponents to simplify.

    a. e^xe^h **b.** $(e^x)^2$ **c.** $\dfrac{e^x}{e^h}$ e^{x-h}
     e^{x+h} e^{2x}

    d. $e^x \cdot e^{-x}$ **e.** e^{-2x} $\dfrac{1}{e^{2x}}$
     1

74. Factor.

    a. $e^{x+h} - e^x$ $e^x(e^h - 1)$

    b. $e^{4x} - e^{2x}$ $e^{2x}(e^x - 1)(e^x + 1)$

75. Multiply. $(e^x + e^{-x})^2$ $e^{2x} + 2 + e^{-2x}$ or $\dfrac{e^{4x} + 2e^{2x} + 1}{e^{2x}}$

76. Multiply. $(e^x - e^{-x})^2$ $e^{2x} - 2 + e^{-2x}$ or $\dfrac{e^{4x} - 2e^{2x} + 1}{e^{2x}}$

77. Show that $\left(\dfrac{e^x + e^{-x}}{2}\right)^2 - \left(\dfrac{e^x - e^{-x}}{2}\right)^2 = 1$.

78. Show that $2\left(\dfrac{e^x - e^{-x}}{2}\right)\left(\dfrac{e^x + e^{-x}}{2}\right) = \dfrac{e^{2x} - e^{-2x}}{2}$.

For Exercises 79–80, find the difference quotient $\dfrac{f(x + h) - f(x)}{h}$. Write the answers in factored form.

79. $f(x) = e^x$ $\dfrac{e^x(e^h - 1)}{h}$

80. $f(x) = 2^x$ $\dfrac{2^x(2^h - 1)}{h}$

Technology Connections

81. Graph the following functions on the window $[-3, 3, 1]$ by $[-1, 8, 1]$ and comment on the behavior of the graphs near $x = 0$.

    $Y_1 = e^x$

    $Y_2 = 1 + x + \dfrac{x^2}{2}$

    $Y_3 = 1 + x + \dfrac{x^2}{2} + \dfrac{x^3}{6}$

Additional answers can be found in the Instructor Answer Appendix.

SECTION 3.3   ## Logarithmic Functions

OBJECTIVES

1. Convert Between Logarithmic and Exponential Forms
2. Evaluate Logarithmic Expressions
3. Apply Basic Properties of Logarithms
4. Graph Logarithmic Functions
5. Use Logarithmic Functions in Applications

1. Convert Between Logarithmic and Exponential Forms

Consider the following equations in which the variable is located in the exponent of an expression. In some cases, the solution can be found by inspection.

Equation	Solution
$5^x = 5$	$x = 1$
$5^x = 20$	$x = ?$
$5^x = 25$	$x = 2$
$5^x = 60$	$x = ?$
$5^x = 125$	$x = 3$

The equation $5^x = 20$ cannot be solved by inspection. However, we suspect that x is between 1 and 2 because $5^1 = 5$ and $5^2 = 25$. To solve for x explicitly, we must isolate x by performing the inverse operation of 5^x. Fortunately, all exponential functions $y = b^x$ ($b > 0$, $b \neq 1$) are one-to-one and therefore have inverse functions. The inverse of an exponential function, base b, is the *logarithmic* function base b which we define here.

Definition of a Logarithmic Function

If x and b are positive real numbers such that $b \neq 1$, then $y = \log_b x$ is called the **logarithmic function base b,** where

$$y = \log_b x \text{ is equivalent to } b^y = x$$

Notes:

- Given $y = \log_b x$, the value y is the exponent to which b must be raised to obtain x.
- The value of y is called the **logarithm,** b is called the **base,** and x is called the **argument.**
- The equations $y = \log_b x$ and $b^y = x$ both define the same relationship between x and y. The expression $y = \log_b x$ is called the **logarithmic form,** and $b^y = x$ is called the **exponential form.**

The logarithmic function base b is defined as the inverse of the exponential function base b.

exponential function	$f(x) = b^x$	First replace $f(x)$ by y.
	$y = b^x$	Next, interchange x and y.
inverse of exponential function	$x = b^y$	This equation provides an implicit relationship between x and y. To solve for y explicitly (that is, to isolate y), we must use logarithmic notation.
logarithmic function	$y = \log_b x$	

To be able to solve equations involving logarithms, it is often advantageous to write a logarithmic expression in its exponential form.

Classroom Examples: p. 392
Exercises 10, 12, 14

EXAMPLE 1 Writing Logarithmic Form and Exponential Form

Write each equation in exponential form.

a. $\log_2 16 = 4$ **b.** $\log_{10}\left(\dfrac{1}{100}\right) = -2$ **c.** $\log_7 1 = 0$

Solution:

Logarithmic form $y = \log_b x$ **Exponential form $b^y = x$**

The logarithm is the exponent to which the base is raised to obtain x.

a. $\log_2 16 = 4$ $\Leftrightarrow$ $2^4 = 16$

b. $\log_{10}\left(\dfrac{1}{100}\right) = -2$ $\Leftrightarrow$ $10^{-2} = \dfrac{1}{100}$

c. $\log_7 1 = 0$ $\Leftrightarrow$ $7^0 = 1$

Skill Practice 1 Write each equation in exponential form.

a. $\log_3 9 = 2$ **b.** $\log_{10}\left(\dfrac{1}{1000}\right) = -3$ **c.** $\log_6 1 = 0$

In Example 2 we reverse this process and write an exponential equation in its logarithmic form.

Answers

1. a. $3^2 = 9$ **b.** $10^{-3} = \dfrac{1}{1000}$

c. $6^0 = 1$

Classroom Examples: p. 392
Exercises 18, 20, 24

EXAMPLE 2   **Writing Exponential Form and Logarithmic Form**

Write each equation in logarithmic form.

a. $3^4 = 81$ **b.** $10^6 = 1,000,000$ **c.** $\left(\dfrac{1}{5}\right)^{-1} = 5$

Solution:

Exponential form $b^y = x$

logarithm

a. $3^4 = 81$

base   argument

Logarithmic form $\log_b x = y$

logarithm (power)

$\Leftrightarrow$ $\log_3 81 = 4$

base   argument

b. $10^6 = 1,000,000$ $\Leftrightarrow$ $\log_{10} 1,000,000 = 6$

c. $\left(\dfrac{1}{5}\right)^{-1} = 5$ $\Leftrightarrow$ $\log_{1/5} 5 = -1$

Skill Practice 2   Write each equation in logarithmic form.

a. $2^5 = 32$ **b.** $10^4 = 10,000$ **c.** $\left(\dfrac{1}{8}\right)^{-2} = 64$

2. Evaluate Logarithmic Expressions

To evaluate a logarithmic expression, we can write the expression in exponential form. Then we make use of the equivalence property of exponential expressions. This states that if two exponential expressions of the same base are equal, then their exponents must be equal.

> **Equivalence Property of Exponential Expressions**
>
> If b, x, and y are real numbers, with $b > 0$ and $b \neq 1$, then
>
> $$b^x = b^y \text{ implies that } x = y.$$

In Example 3, we evaluate several logarithmic expressions.

Classroom Examples: p. 392
Exercises 26, 30, 32

EXAMPLE 3   **Evaluating a Logarithmic Expression**

Evaluate each expression.

a. $\log_4 16$ **b.** $\log_2 8$ **c.** $\log_{1/2} 8$

Solution:

Let y represent the value of the logarithm.

a. $\log_4 16$ is the exponent to which 4 must be raised to equal 16. That is, $4^{\square} = 16$.

$\log_4 16 = y$

$4^y = 16$ or equivalently $4^y = 4^2$ Write the equivalent exponential form.

$y = 2$

Therefore, $\log_4 16 = 2$.   Check: $4^2 = 16$ ✓

Answers

2. a. $\log_2 32 = 5$
   b. $\log_{10} 10,000 = 4$
   c. $\log_{1/8} 64 = -2$

b. $\log_2 8$ is the exponent to which 2 must be raised to obtain 8. That is, $2^{\square} = 8$.

$$\log_2 8 = y$$
$$2^y = 8 \text{ or equivalently } 2^y = 2^3 \qquad \text{Write the equivalent exponential form.}$$
$$y = 3$$

Therefore, $\log_2 8 = 3$. Check: $2^3 = 8$ ✓

c. $\log_{1/2} 8 = y$
$$\left(\tfrac{1}{2}\right)^y = 8 \text{ or equivalently } \left(\tfrac{1}{2}\right)^y = \left(\tfrac{1}{2}\right)^{-3} \qquad \text{Write the equivalent exponential form.}$$
$$y = -3$$

Therefore, $\log_{1/2} 8 = -3$. Check: $\left(\tfrac{1}{2}\right)^{-3} = 8$ ✓

Skill Practice 3   Evaluate each expression.

a. $\log_5 125$ **b.** $\log_3 81$ **c.** $\log_4\left(\dfrac{1}{64}\right)$

The statement $y = \log_b x$ represents a family of logarithmic functions where the base is any positive real number except 1. Two specific logarithmic functions that come up often in applications are the logarithmic functions base 10 and base e.

Definition of Common and Natural Logarithmic Functions

- The logarithmic function base 10 is called the **common logarithmic function**. The common logarithmic function is denoted by $y = \log x$. Notice that the base 10 is not explicitly written; that is, $y = \log_{10} x$ is written simply as $y = \log x$.
- The logarithmic function base e is called the **natural logarithmic function**. The natural logarithmic function is denoted by $y = \ln x$; that is, $y = \log_e x$ is written as $y = \ln x$.

EXAMPLE 4   Evaluating Common and Natural Logarithms

Evaluate.

a. $\log 100{,}000$ **b.** $\log 0.001$ **c.** $\ln e^4$ **d.** $\ln\left(\dfrac{1}{e}\right)$

Solution:

Let y represent the value of the logarithm.

a. $\log 100{,}000 = y$
$$10^y = 100{,}000 \text{ or equivalently } 10^y = 10^5 \qquad \text{Write the exponential form.}$$
$$y = 5$$

Thus, $\log 100{,}000 = 5$ because $10^5 = 100{,}000$.

b. $\log 0.001 = y$
$$10^y = 0.001 \text{ or equivalently } 10^y = 10^{-3} \qquad \text{Write the exponential form.}$$
$$y = -3$$

Thus, $\log 0.001 = -3$ because $10^{-3} = 0.001$.

Classroom Examples: p. 392–393
Exercises 52, 54

TIP In Example 4, to evaluate:

a. log 100,000 we ask
$10^{\square} = 100,000$.

b. log 0.001 we ask
$10^{\square} = 0.001$.

c. $\ln e^4$ we ask $e^{\square} = e^4$.

d. $\ln\left(\dfrac{1}{e}\right)$ we ask $e^{\square} = e^{-1}$.

c. $\ln e^4 = y$

$e^y = e^4$ Write the equivalent exponential form.

$y = 4$

Therefore, $\ln e^4 = 4$.

d. $\ln\left(\dfrac{1}{e}\right) = y$

$e^y = \left(\dfrac{1}{e}\right)$ or equivalently $e^y = e^{-1}$ Write the equivalent exponential form.

$y = -1$

Therefore, $\ln\left(\dfrac{1}{e}\right) = -1$.

Skill Practice 4   Evaluate.

   a. log 10,000,000   **b.** log 0.1   **c.** $\ln e^5$ **d.** $\ln e$

Most scientific calculators have a key for the common logarithmic function **LOG** and a key for the natural logarithmic function **LN**. We demonstrate their use in Example 5.

EXAMPLE 5   **Approximating Common and Natural Logarithms**

Approximate the logarithms.

   a. log 5809   **b.** $\log(4.6 \times 10^7)$ **c.** log 0.003

   d. ln 472   **e.** ln 0.05   **f.** $\ln\sqrt{87}$

Solution:

For parts (a)−(c), use the **LOG** key.   For parts (d)−(f), use the **LN** key.

TIP There are an infinite number of logarithmic functions. The calculator has keys for the base 10 and base e logarithmic functions only.

To approximate a logarithm of different base, we need to use the change-of-base formula presented in Section 3.4.

```
NORMAL FLOAT AUTO a+bi DEGREE CL
log(5809)
              3.764101376
log(4.6ε7)
              7.662757832
log(.003)
             -2.522878745
```

```
NORMAL FLOAT AUTO a+bi DEGREE CL
ln(472)
              6.156978986
ln(0.05)
             -2.995732274
ln(√(87))
              2.232954059
```

When using a calculator, there is always potential for user-input error. Therefore, it is good practice to estimate values when possible to confirm the reasonableness of an answer from a calculator. For example,

For part (a), $10^3 < 5809 < 10^4$. Therefore, $3 < \log 5809 < 4$.

For part (b), $10^7 < 4.6 \times 10^7 < 10^8$. Therefore, $7 < \log(4.6 \times 10^7) < 8$.

For part (c), $10^{-3} < 0.003 < 10^{-2}$. Therefore, $-3 < \log 0.003 < -2$.

Skill Practice 5   Approximate the logarithms. Round to 4 decimal places.

   a. log 229   **b.** $\log(3.76 \times 10^{12})$ **c.** log 0.0216

   d. ln 87   **e.** ln 0.0032   **f.** $\ln \pi$

Answers

4. a. 7   **b.** −1   **c.** 5   **d.** 1

5. a. 2.3598   **b.** 12.5752

   c. −1.6655   **d.** 4.4659

   e. −5.7446   **f.** 1.1447

3. Apply Basic Properties of Logarithms

From the definition of a logarithmic function, we have the following basic properties.

> **Basic Properties of Logarithms**
>
> | Property | Example |
> |---|---|
> | 1. $\log_b 1 = 0$ because $b^0 = 1$ | $\log_5 1 = 0$ because $5^0 = 1$ |
> | 2. $\log_b b = 1$ because $b^1 = b$ | $\log_3 3 = 1$ because $3^1 = 3$ |
> | 3. $\log_b b^x = x$ because $b^x = b^x$ | $\log_2 2^x = x$ because $2^x = 2^x$ |
> | 4. $b^{\log_b x} = x$ because $\log_b x = \log_b x$ | $7^{\log_7 x} = x$ because $\log_7 x = \log_7 x$ |

Properties 3 and 4 follow from the fact that a logarithmic function is the inverse of an exponential function of the same base. Given $f(x) = b^x$ and $f^{-1}(x) = \log_b x$,

$$(f \circ f^{-1})(x) = b^{(\log_b x)} = x \qquad \text{(Property 4)}$$

$$(f^{-1} \circ f)(x) = \log_b(b^x) = x \qquad \text{(Property 3)}$$

Classroom Examples: p. 393
Exercises 56, 58, 60, 62

> **EXAMPLE 6   Applying the Properties of Logarithms**
>
> Simplify.
>
> **a.** $\log_3 3^{10}$ **b.** $\ln e^2$ **c.** $\log_{11} 11$ **d.** $\log 10$
>
> **e.** $\log_{\sqrt{7}} 1$ **f.** $\ln 1$ **g.** $5^{\log_5(c^2 + 4)}$ **h.** $10^{\log(a^2 + b^2)}$
>
> **Solution:**
>
> **a.** $\log_3 3^{10} = 10$ Property 3   **b.** $\ln e^2 = \log_e e^2 = 2$ Property 3
>
> **c.** $\log_{11} 11 = 1$ Property 2   **d.** $\log 10 = \log_{10} 10 = 1$ Property 2
>
> **e.** $\log_{\sqrt{7}} 1 = 0$ Property 1   **f.** $\ln 1 = 0$ Property 1
>
> **g.** $5^{\log_5(c^2+4)} = c^2 + 4$ Property 4   **h.** $10^{\log(a^2+b^2)} = a^2 + b^2$ Property 4
>
> ---
>
> **Skill Practice 6**   Simplify.
>
> **a.** $\log_{13} 13$ **b.** $\ln e$ **c.** $a^{\log_a 3}$ **d.** $e^{\ln 6}$
>
> **e.** $\log_\pi 1$ **f.** $\log 1$ **g.** $\log_9 9^{\sqrt{2}}$ **h.** $\log 10^e$

4. Graph Logarithmic Functions

Since a logarithmic function $y = \log_b x$ is the inverse of the corresponding exponential function $y = b^x$, their graphs must be symmetric with respect to the line $y = x$. See Figures 3-13 and 3-14.

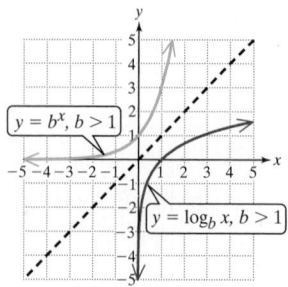

Figure 3-13

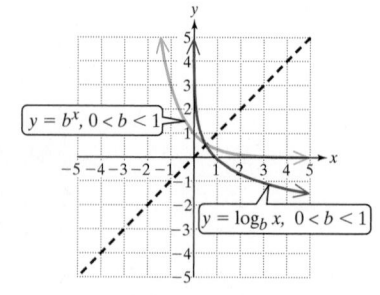

Figure 3-14

Answers

6. a. 1   **b.** 1   **c.** 3   **d.** 6

e. 0   **f.** 0   **g.** $\sqrt{2}$ **h.** e

From Figures 3-13 and 3-14, the range of $y = b^x$ is the set of positive real numbers. As expected, the domain of its inverse function $y = \log_b x$ is the set of positive real numbers.

Classroom Example: p. 393
Exercise 66

EXAMPLE 7   Graphing Logarithmic Functions

Graph the functions.

a. $y = \log_2 x$ **b.** $y = \log_{1/4} x$

Solution:

To find points on a logarithmic function, we can interchange the x- and y-coordinates of the ordered pairs on the corresponding exponential function.

a. To graph $y = \log_2 x$, interchange the x- and y-coordinates of the ordered pairs from its inverse function $y = 2^x$. The graph of $y = \log_2 x$ is shown in Figure 3-15.

Exponential Function   **Logarithmic Function**

x	$y = 2^x$
-3	$\frac{1}{8}$
-2	$\frac{1}{4}$
-1	$\frac{1}{2}$
0	1
1	2
2	4
3	8

x	$y = \log_2 x$
$\frac{1}{8}$	-3
$\frac{1}{4}$	-2
$\frac{1}{2}$	-1
1	0
2	1
4	2
8	3

Switch x and y.

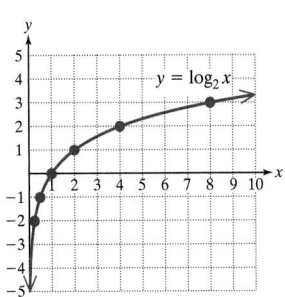

Figure 3-15

b. To graph $y = \log_{1/4} x$, interchange the x- and y-coordinates of the ordered pairs from its inverse function $y = \left(\frac{1}{4}\right)^x$. See Figure 3-16.

Exponential Function   **Logarithmic Function**

x	$y = \left(\dfrac{1}{4}\right)^x$
-3	64
-2	16
-1	4
0	1
1	$\frac{1}{4}$
2	$\frac{1}{16}$
3	$\frac{1}{64}$

x	$y = \log_{1/4} x$
64	-3
16	-2
4	-1
1	0
$\frac{1}{4}$	1
$\frac{1}{16}$	2
$\frac{1}{64}$	3

Switch x and y.

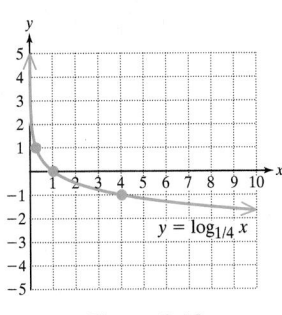

Figure 3-16

Answers

7. a.

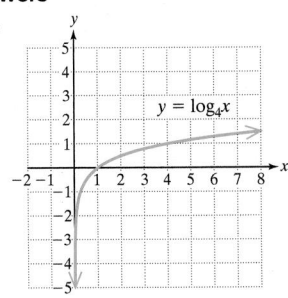

b.

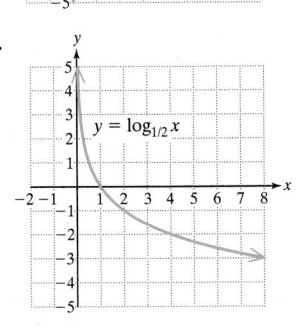

Skill Practice 7   Graph the functions.

a. $y = \log_4 x$ **b.** $y = \log_{1/2} x$

Based on the graphs in Example 7 and our knowledge of exponential functions, we offer the following summary of the characteristics of logarithmic and exponential functions.

Graphs of Exponential and Logarithmic Functions

Exponential Functions	Logarithmic Functions
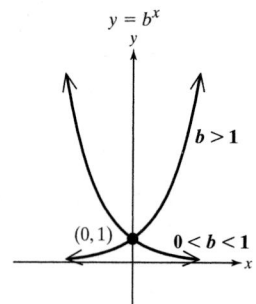	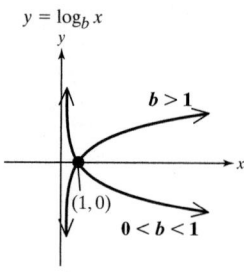
Domain: $(-\infty, \infty)$	Domain: $(0, \infty)$
Range: $(0, \infty)$	Range: $(-\infty, \infty)$
Horizontal asymptote: $y = 0$	Vertical asymptote: $x = 0$
Passes through $(0, 1)$	Passes through $(1, 0)$
If $b > 1$, the function is increasing.	If $b > 1$, the function is increasing.
If $0 < b < 1$, the function is decreasing.	If $0 < b < 1$, the function is decreasing.

The roles of x and y are reversed between a function and its inverse. Therefore, it is not surprising that the domain and range are reversed between exponential and logarithmic functions. Furthermore, an exponential function passes through $(0, 1)$, whereas a logarithmic function passes through $(1, 0)$. An exponential function has a horizontal asymptote of $y = 0$, whereas a logarithmic function has a vertical asymptote of $x = 0$.

In Example 8 we use the transformations of functions learned in Section 1.6 to graph a logarithmic function.

> If $h > 0$, shift to the right.
> If $h < 0$, shift to the left.

> If $k > 0$, shift upward.
> If $k < 0$, shift downward.

$$f(x) = a \log_b(x - h) + k$$

> If $a < 0$, reflect across the x-axis.
> Shrink vertically if $0 < |a| < 1$.
> Stretch vertically if $|a| > 1$.

Classroom Example: p. 393
Exercise 76

EXAMPLE 8　Using Transformations to Graph Logarithmic Functions

Graph the function. Identify the vertical asymptote and write the domain in interval notation.

$$f(x) = \log_2(x + 3) - 2$$

Solution:

The graph of the "parent" function $y = \log_2 x$ was presented in Example 7. The graph of $f(x) = \log_2(x + 3) - 2$ is the graph of $y = \log_2 x$ shifted to the left 3 units and down 2 units.

We can plot a few points on the graph of $y = \log_2 x$ and use these points and the vertical asymptote to form an outline of the transformed graph.

x	$y = \log_2 x$
$\frac{1}{4}$	-2
$\frac{1}{2}$	-1
1	0
2	1
4	2

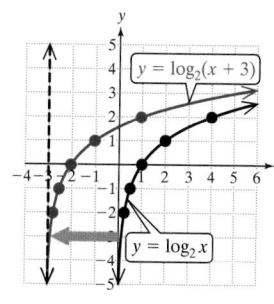

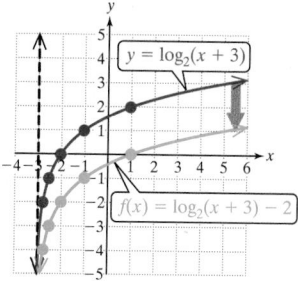

The graph of $f(x) = \log_2(x + 3) - 2$ is shown in blue.
The vertical asymptote is $x = -3$. The domain is $(-3, \infty)$.

Skill Practice 8   Graph the function. Identify the vertical asymptote and write the domain in interval notation. $g(x) = \log_3(x - 4) + 1$

The domain of $f(x) = \log_b x$ is restricted to $x > 0$. In Example 8, this graph was shifted to the left 3 units, restricting the domain of $f(x) = \log_2(x + 3) - 2$ to $x > -3$. The domain of a logarithmic function is the set of real numbers that make the argument positive.

Classroom Examples: p. 393
Exercises 80, 82, 90

EXAMPLE 9   **Identifying the Domain of a Logarithmic Function**

Write the domain in interval notation.

a. $f(x) = \log_2(2x + 4)$ **b.** $g(x) = \ln(5 - x)$ **c.** $h(x) = \log(x^2 - 9)$

Solution:

a. $f(x) = \log_2(2x + 4)$

$\quad 2x + 4 > 0$ Set the argument greater than zero.

$\quad\quad 2x > -4$ Solve for x.

$\quad\quad\quad x > -2$

The domain is $(-2, \infty)$.
The graph of f is shown in Figure 3-17.
The vertical asymptote is $x = -2$.

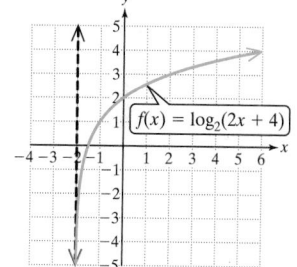

Figure 3-17

b. $g(x) = \ln(5 - x)$

$\quad 5 - x > 0$ Set the argument greater than zero.

$\quad\quad -x > -5$ Subtract 5 and divide by -1

$\quad\quad\quad x < 5$ (reverse the inequality sign).

The domain is $(-\infty, 5)$.
The graph of g is shown in Figure 3-18.
The vertical asymptote is $x = 5$.

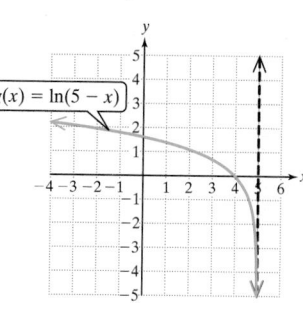

Figure 3-18

Answer

8. Vertical asymptote: $x = 4$
Domain: $(4, \infty)$

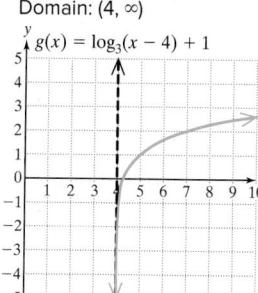

c. $h(x) = \log(x^2 - 9)$

$x^2 - 9 > 0$

Set the argument greater than zero. The result is a polynomial inequality (Section 2.6).

$(x - 3)(x + 3) = 0$

Solve the related equation by setting one side equal to zero and factoring the other side.

Sign of $(x - 3)$:	$-$	$-$	$+$
Sign of $(x + 3)$:	$-$	$+$	$+$
Sign of $(x - 3)(x + 3)$:	$+$	$-$	$+$

$\qquad\qquad\quad -3 \qquad 3$

The boundary points for the solution set are the solutions to the equation: $x = 3$ and $x = -3$.

The domain is $(-\infty, -3) \cup (3, \infty)$.

The graph of h is shown in Figure 3-19.

The vertical asymptotes are $x = -3$ and $x = 3$.

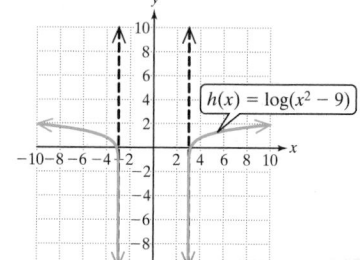

$h(x) = \log(x^2 - 9)$

Figure 3-19

Skill Practice 9   Write the domain in interval notation.

a. $\log_4(1 - 3x)$ **b.** $\log(2 + x)$ **c.** $m(x) = \ln(64 - x^2)$

---

5. Use Logarithmic Functions in Applications

When physical quantities vary over a large range, it is often convenient to take a logarithm of the quantity to have a more manageable set of numbers. For example, suppose a set of data values consists of 10, 100, 1000, and 10,000. The corresponding common logarithms are 1, 2, 3, and 4. The latter list of numbers is easier to manipulate and to visualize on a graph. For this reason, logarithmic scales are used in applications such as

- measuring pH (representing hydrogen ion concentration from 10^{-14} to 1 moles per liter).
- measuring wave energy from an earthquake (often ranging from 10^6 J to 10^{17} J).
- measuring loudness of sound on the decibel scale (representing sound intensity from 10^{-12} to 10^2 Watts per square meter).

Point of Interest

In 1935, American geologist Charles Richter developed the local magnitude (M_L) scale, or Richter scale, for measuring the intensity of moderate-sized earthquakes ($3 < M_L < 7$) in southern California. Today, seismologists no longer follow Richter's methodology because it does not give reliable results for earthquakes of higher magnitude. The magnitudes of modern earthquakes are based on a variety of data types from numerous seismic stations. However, both the Richter scale and modern magnitude scales use a base 10 logarithmic scale to compare amplitudes of waves on a seismogram. This means that an increase of 1 unit in magnitude represents a 10-fold increase in the amplitude of the waves on a seismogram.

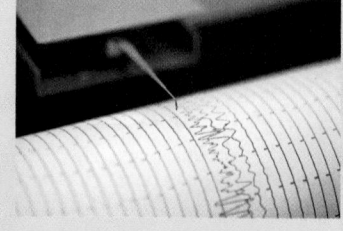

Answers

9. a. $\left(-\infty, \dfrac{1}{3}\right)$ **b.** $(-2, \infty)$

c. $(-8, 8)$

Classroom Example: p. 393
Exercise 94

EXAMPLE 10   Using a Logarithmic Function in an Application

The intensity I of an earthquake is measured by a seismograph—a device that measures amplitudes of shock waves. I_0 is a minimum reference intensity of a "zero-level" earthquake against which the intensities of other earthquakes may be compared. The magnitude M of an earthquake of intensity I is given by

$$M = \log\left(\frac{I}{I_0}\right).$$

a. Determine the magnitude of the earthquake that devastated Haiti on January 12, 2010, if the intensity was approximately $10^{7.0}$ times I_0.

b. Determine the magnitude of the earthquake that occurred near Washington, D.C., on August 23, 2011, if the intensity was approximately $10^{5.8}$ times I_0.

c. How many times more intense was the earthquake that hit Haiti than the earthquake that hit Washington, D.C.? Round to the nearest whole unit.

Solution:

a. $M = \log\left(\dfrac{I}{I_0}\right)$ **b.** $M = \log\left(\dfrac{I}{I_0}\right)$

$M = \log\left(\dfrac{10^{7.0} \cdot I_0}{I_0}\right)$ $M = \log\left(\dfrac{10^{5.8} \cdot I_0}{I_0}\right)$

$ = \log 10^{7.0}$ $ = \log 10^{5.8}$

$ = 7.0$ $ = 5.8$

c. Using the intensities given in parts (a) and (b) we have

$$\frac{10^{7.0} I_0}{10^{5.8} I_0} = 10^{7.0-5.8} = 10^{1.2} \approx 16$$

The earthquake in Haiti was approximately 16 times more intense.

Skill Practice 10

a. Determine the magnitude of an earthquake that is $10^{5.2}$ times I_0.

b. Determine the magnitude of an earthquake that is $10^{4.2}$ times I_0.

c. How many times more intense is a 5.2-magnitude earthquake than a 4.2-magnitude earthquake?

Answers
10. a. 5.2   **b.** 4.2
  c. 10 times more intense

SECTION 3.3   Practice Exercises

Prerequisite Review

For Exercises R.1–R.4, simplify each expression.

R.1. $9^{-3/2}$ $\dfrac{1}{27}$ **R.2.** $64^{2/3}$ 16 **R.3.** $(-8)^{2/3}$ 4 **R.4.** $\dfrac{1}{25^{-1/2}}$ 5

For Exercises R.5–R.6, solve the polynomial inequality. Write the answer in interval notation.

R.5. $y^2 + 20y \geq -75$ $(-\infty, -15] \cup [-5, \infty)$ **R.6.** $x^2 - 8x + 16 > 0$ $(-\infty, 4) \cup (4, \infty)$

Concept Connections

1. Given positive real numbers x and b such that $b \neq 1$, $y = \log_b x$ is the _____ logarithmic _____ function base b and is equivalent to $b^y = x$.

2. Given $y = \log_b x$, the value y is called the _____ logarithm _____, b is called the _____ base _____, and x is called the _____ argument _____.

3. The logarithmic function base 10 is called the _____ common _____ logarithmic function, and the logarithmic function base e is called the _____ natural _____ logarithmic function.

4. Given $y = \log x$, the base is understood to be _____ 10 _____. Given $y = \ln x$, the base is understood to be _____ e _____.

5. $\log_b 1 = $ _____ 0; 0 _____ because $b^\square = 1$.

6. $\log_b b = $ _____ 1; 1 _____ because $b^\square = b$.

7. $f(x) = \log_b x$ and $g(x) = b^x$ are inverse functions. Therefore, $\log_b b^x = $ _____ x _____ and $b^{\log_b x} = $ _____ x _____.

8. The graph of $y = \log_b x$ passes through the point $(1, 0)$ and the line _____ $x = 0$; vertical _____ is a (horizontal/vertical) asymptote.

For the exercises in this set, assume that all variable expressions represent positive real numbers.

Objective 1: Convert Between Logarithmic and Exponential Forms

For Exercises 9–16, write the equation in exponential form. (See Example 1)

9. $\log_8 64 = 2$ $8^2 = 64$

10. $\log_9 81 = 2$ $9^2 = 81$

11. $\log\left(\dfrac{1}{10,000}\right) = -4$ $10^{-4} = \dfrac{1}{10,000}$

12. $\log\left(\dfrac{1}{1,000,000}\right) = -6$ $10^{-6} = \dfrac{1}{1,000,000}$

13. $\ln 1 = 0$ $e^0 = 1$

14. $\log_8 1 = 0$ $8^0 = 1$

15. $\log_a b = c$ $a^c = b$

16. $\log_x M = N$ $x^N = M$

For Exercises 17–24, write the equation in logarithmic form. (See Example 2)

17. $5^3 = 125$ $\log_5 125 = 3$

18. $2^5 = 32$ $\log_2 32 = 5$

19. $\left(\dfrac{1}{5}\right)^{-3} = 125$ $\log_{1/5} 125 = -3$

20. $\left(\dfrac{1}{2}\right)^{-5} = 32$ $\log_{1/2} 32 = -5$

21. $10^9 = 1,000,000,000$ $\log 1,000,000,000 = 9$

22. $e^1 = e$ $\ln e = 1$

23. $a^7 = b$ $\log_a b = 7$

24. $M^3 = N$ $\log_M N = 3$

Objective 2: Evaluate Logarithmic Expressions

For Exercises 25–50, simplify the expression without using a calculator. (See Examples 3–4)

25. $\log_3 9$ 2

26. $\log_2 16$ 4

27. $\log_5 5$ 1

28. $\log_6 6$ 1

29. $\log 100,000,000$ 8

30. $\log 10,000,000$ 7

31. $\log_2\left(\dfrac{1}{16}\right)$ -4

32. $\log_3\left(\dfrac{1}{9}\right)$ -2

33. $\log\left(\dfrac{1}{10}\right)$ -1

34. $\log\left(\dfrac{1}{10,000}\right)$ -4

35. $\ln e^6$ 6

36. $\ln e^{10}$ 10

37. $\ln\left(\dfrac{1}{e^3}\right)$ -3

38. $\ln\left(\dfrac{1}{e^8}\right)$ -8

39. $\log_{1/7} 49$ -2

40. $\log_{1/4} 16$ -2

41. $\log_{1/2}\left(\dfrac{1}{32}\right)$ 5

42. $\log_{1/6}\left(\dfrac{1}{36}\right)$ 2

43. $\log 0.00001$ -5

44. $\log 0.0001$ -4

45. $\log_{3/2}\dfrac{4}{9}$ -2

46. $\log_{3/2}\dfrac{9}{4}$ 2

47. $\log_3 \sqrt[5]{3}$ $\dfrac{1}{5}$

48. $\log_2 \sqrt[3]{2}$ $\dfrac{1}{3}$

49. $\log_5 \sqrt{\dfrac{1}{5}}$ $-\dfrac{1}{2}$

50. $\log \sqrt{\dfrac{1}{1000}}$ $-\dfrac{3}{2}$

For Exercises 51–52, estimate the value of each logarithm between two consecutive integers. Then use a calculator to approximate the value to 4 decimal places. For example, $\log 8970$ is between 3 and 4 because $10^3 < 8970 < 10^4$. (See Example 5)

51. a. $\log 46,832$ Between 4 and 5; 4.6705

b. $\log 1,247,310$ Between 6 and 7; 6.0960

c. $\log 0.24$ Between -1 and 0; -0.6198

d. $\log 0.0000032$ Between -6 and -5; -5.4949

e. $\log(5.6 \times 10^5)$ Between 5 and 6; 5.7482

f. $\log(5.1 \times 10^{-3})$ Between -3 and -2; -2.2924

52. a. $\log 293,416$ Between 5 and 6; 5.4675

b. $\log 897$ Between 2 and 3; 2.9528

c. $\log 0.038$ Between -2 and -1; -1.4202

d. $\log 0.00061$ Between -4 and -3; -3.2147

e. $\log(9.1 \times 10^8)$ Between 8 and 9; 8.9590

f. $\log(8.2 \times 10^{-2})$ Between -2 and -1; -1.0862

For Exercises 53–54, approximate $f(x) = \ln x$ for the given values of x. Round to 4 decimal places. (See Example 5)

53. a. $f(94)$ 4.5433

    b. $f(0.182)$ -1.7037

    c. $f(\sqrt{155})$ 2.5217

    d. $f(4\pi)$ 2.5310

    e. $f(3.9 \times 10^9)$ 22.0842

    f. $f(7.1 \times 10^{-4})$ -7.2502

54. a. $f(1860)$ 7.5283

    b. $f(0.0694)$ -2.6679

    c. $f(\sqrt{87})$ 2.2330

    d. $f(2\pi)$ 1.8379

    e. $f(1.3 \times 10^{12})$ 27.8934

    f. $f(8.5 \times 10^{-17})$ -37.0039

Objective 3: Apply Basic Properties of Logarithms

For Exercises 55–64, simplify the expression without using a calculator. (See Example 6)

55. $\log_4 4^{11}$ 11

56. $\log_6 6^7$ 7

57. $\log_c c$ 1

58. $\log_d d$ 1

59. $5^{\log_5(x+y)}$ $x + y$

60. $4^{\log_4(a-c)}$ $a - c$

61. $\ln e^{a+b}$ $a + b$

62. $\ln e^{x^2+1}$ $x^2 + 1$

63. $\log_{\sqrt{5}} 1$ 0

64. $\log_\pi 1$ 0

Objective 4: Graph Logarithmic Functions

For Exercises 65–70, graph the function. (See Example 7)

65. $y = \log_3 x$

66. $y = \log_5 x$

67. $y = \log_{1/3} x$

68. $y = \log_{1/5} x$

69. $y = \ln x$

70. $y = \log x$

For Exercises 71–78, (See Example 8)

a. Use transformations of the graphs of $y = \log_2 x$ (see Example 7) and $y = \log_3 x$ (see Exercise 65) to graph the given functions.

b. Write the domain and range in interval notation.

c. Write an equation of the asymptote.

71. $y = \log_3(x + 2)$

72. $y = \log_2(x + 3)$

73. $y = 2 + \log_3 x$

74. $y = 3 + \log_2 x$

75. $y = \log_3(x - 1) - 3$

76. $y = \log_2(x - 2) - 1$

77. $y = -\log_3 x$

78. $y = -\log_2 x$

For Exercises 79–92, write the domain in interval notation. (See Example 9)

79. $f(x) = \log(8 - x)$ $(-\infty, 8)$

80. $g(x) = \log(3 - x)$
$(-\infty, 3)$

81. $h(x) = \log_2(6x + 7)$ $\left(-\frac{7}{6}, \infty\right)$

82. $k(x) = \log_3(5x + 6)$ $\left(-\frac{6}{5}, \infty\right)$

83. $m(x) = \ln(x^2 + 14)$ $(-\infty, \infty)$

84. $n(x) = \ln(x^2 + 11)$
$(-\infty, \infty)$

85. $f(x) = \log_4(x^2 - 16)$
$(-\infty, -4) \cup (4, \infty)$

86. $g(x) = \log_7(x^2 - 49)$
$(-\infty, -7) \cup (7, \infty)$

87. $m(x) = 3 + \ln \dfrac{1}{\sqrt{11 - x}}$
$(-\infty, 11)$

88. $n(x) = 4 - \log \dfrac{1}{\sqrt{x + 5}}$
$(-5, \infty)$

89. $p(x) = \log(x^2 - x - 12)$
$(-\infty, -3) \cup (4, \infty)$

90. $q(x) = \log(x^2 + 10x + 9)$
$(-\infty, -9) \cup (-1, \infty)$

91. $r(x) = \log_3(4 - x)^2$
$(-\infty, 4) \cup (4, \infty)$

92. $s(x) = \log_5(3 - x)^2$
$(-\infty, 3) \cup (3, \infty)$

Objective 5: Use Logarithmic Functions in Applications

93. In 1989, the Loma Prieta earthquake damaged the city of San Francisco with an intensity of approximately $10^{6.9}I_0$. Film footage of the 1989 earthquake was captured on a number of video cameras including a broadcast of Game 3 of the World Series played at Candlestick Park. (**See Example 10**)

    a. Determine the magnitude of the Loma Prieta earthquake.
      6.9

    b. Smaller earthquakes occur daily in the San Francisco area and most are not detectable without a seismograph. Determine the magnitude of an earthquake with an intensity of $10^{3.2} I_0$. 3.2

    c. How many times more intense was the Loma Prieta earthquake than an earthquake with a magnitude of 3.2? Round to the nearest whole unit.
    Approximately 5012 times more intense

94. The intensities of earthquakes are measured with seismographs all over the world at different distances from the epicenter. Suppose that the intensity of a medium earthquake is originally reported as $10^{5.4}$ times I_0. Later this value is revised as $10^{5.8}$ times I_0.

    a. Determine the magnitude of the earthquake using the original estimate for intensity. 5.4

    b. Determine the magnitude using the revised estimate for intensity. 5.8

    c. How many times more intense was the earthquake than originally thought? Round to 1 decimal place.
    Approximately 2.5 times more intense.

Sounds are produced when vibrating objects create pressure waves in some medium such as air. When these variations in pressure reach the human eardrum, it causes the eardrum to vibrate in a similar manner and the ear detects sound. The intensity of sound is measured as power per unit area. The threshold for hearing (minimum sound detectable by a young, healthy ear) is defined to be $I_0 = 10^{-12}$ W/m^2 (watts per square meter). The sound level L, or "loudness" of sound, is measured in decibels (dB) as $L = 10 \log\left(\dfrac{I}{I_0}\right)$, where I is the intensity of the given sound. Use this formula for Exercises 95–96.

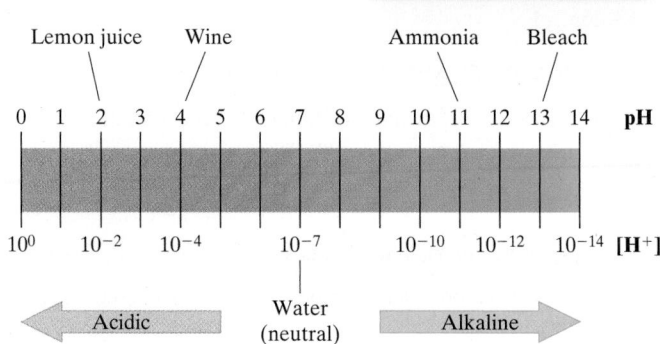

95. **a.** Find the sound level of a jet plane taking off if its intensity is 10^{15} times the intensity of I_0.   150 dB

   b. Find the sound level of the noise from city traffic if its intensity is 10^9 times I_0.   90 dB

   c. How many times more intense is the sound of a jet plane taking off than noise from city traffic?   1,000,000 times more intense

96. **a.** Find the sound level of a motorcycle if its intensity is 10^{10} times I_0.   100 dB

   b. Find the sound level of a vacuum cleaner if its intensity is 10^7 times I_0.   70 dB

   c. How many times more intense is the sound of a motorcycle than a vacuum cleaner?   1000 times more intense

Scientists use the pH scale to represent the level of acidity or alkalinity of a liquid. This is based on the molar concentration of hydrogen ions, $[\text{H}^+]$. Since the values of $[\text{H}^+]$ vary over a large range, 1×10^0 mole per liter to 1×10^{-14} mole per liter (mol/L), a logarithmic scale is used to compute pH. The formula

$$\text{pH} = -\log[\text{H}^+]$$

represents the pH of a liquid as a function of its concentration of hydrogen ions, $[\text{H}^+]$.

The pH scale ranges from 0 to 14. Pure water is taken as neutral having a pH of 7. A pH less than 7 is acidic. A pH greater than 7 is alkaline (or basic). For Exercises 97–98, use the formula for pH. Round pH values to 1 decimal place.

97. Vinegar and lemon juice are both acids. Their $[\text{H}^+]$ values are 5.0×10^{-3} mol/L and 1×10^{-2} mol/L, respectively.

   a. Find the pH for vinegar.   2.3

   b. Find the pH for lemon juice.   2

   c. Which substance is more acidic?   Lemon juice is more acidic.

98. Bleach and milk of magnesia are both bases. Their $[\text{H}^+]$ values are 2.0×10^{-13} mol/L and 4.1×10^{-10} mol/L, respectively.

   a. Find the pH for bleach.   12.7

   b. Find the pH for milk of magnesia.   9.4

   c. Which substance is more basic?   Bleach is more basic.

Mixed Exercises

For Exercises 99–102,

a. Write the equation in exponential form.

b. Solve the equation from part (a).

c. Verify that the solution checks in the original equation.

a. $3^4 = x + 1$ b. $\{80\}$

99. $\log_3(x + 1) = 4$

c. $\log_3(80 + 1) = \log_3 81 = 4 \checkmark$

a. $2^4 = x - 5$ b. $\{21\}$

100. $\log_2(x - 5) = 4$

c. $\log_2(21 - 5) = \log_2 16 = 4 \checkmark$

a. $4^3 = 7x - 6$ b. $\{10\}$

101. $\log_4(7x - 6) = 3$

c. $\log_4(7 \cdot 10 - 6) = \log_4 64 = 3 \checkmark$

a. $5^2 = 9x - 11$ b. $\{4\}$

102. $\log_5(9x - 11) = 2$

c. $\log_5(9 \cdot 4 - 11) = \log_5 25 = 2 \checkmark$

For Exercises 103–106, evaluate the expressions.

103. $\log_3(\log_4 64)$ 1

104. $\log_2\left[\log_{1/2}\left(\dfrac{1}{4}\right)\right]$ 1

105. $\log_{16}(\log_{81} 3)$ $-\dfrac{1}{2}$

106. $\log_4(\log_{16} 4)$ $-\dfrac{1}{2}$

107. a. Evaluate $\log_2 2 + \log_2 4$ 3

   b. Evaluate $\log_2(2 \cdot 4)$ 3

   c. How do the values of the expressions in parts (a) and (b) compare?   They are the same.

108. a. Evaluate $\log_3 3 + \log_3 27$ 4

   b. Evaluate $\log_3(3 \cdot 27)$ 4

   c. How do the values of the expressions in parts (a) and (b) compare?   They are the same.

109. a. Evaluate $\log_4 64 - \log_4 4$ 2

   b. Evaluate $\log_4\left(\dfrac{64}{4}\right)$ 2

   c. How do the values of the expressions in parts (a) and (b) compare?   They are the same.

110. a. Evaluate $\log 100{,}000 - \log 100$ 3

   b. Evaluate $\log\left(\dfrac{100{,}000}{100}\right)$ 3

   c. How do the values of the expressions in parts (a) and (b) compare?   They are the same.

111. a. Evaluate $\log_2 2^5$ 5

   b. Evaluate $5 \cdot \log_2 2$ 5

   c. How do the values of the expressions in parts (a) and (b) compare?   They are the same.

112. a. Evaluate $\log_7 7^6$ 6

   b. Evaluate $6 \cdot \log_7 7$ 6

   c. How do the values of the expressions in parts (a) and (b) compare?   They are the same.

113. The time t (in years) required for an investment to double with interest compounded continuously depends on the interest rate r according to the function $t(r) = \dfrac{\ln 2}{r}$.

   a. If an interest rate of 3.5% is secured, determine the length of time needed for an initial investment to double. Round to 1 decimal place.   19.8 yr

   b. Evaluate $t(0.04)$, $t(0.06)$, and $t(0.08)$. $t(0.04) = 17.3$; $t(0.06) = 11.6$; $t(0.08) = 8.7$

114. The number n of monthly payments of P dollars each required to pay off a loan of A dollars in its entirety at interest rate r is given by

$$n = -\frac{\log\left(1 - \dfrac{Ar}{12P}\right)}{\log\left(1 + \dfrac{r}{12}\right)}$$

   a. A college student wants to buy a car and realizes that he can only afford payments of $200 per month. If he borrows $3000 and pays it off at 6% interest, how many months will it take him to retire the loan? Round to the nearest month.   16 months

   b. Determine the number of monthly payments of $611.09 that would be required to pay off a home loan of $128,000 at 4% interest.   360 months (30 yr)

For Exercises 115–116, use a calculator to approximate the given logarithms to 4 decimal places.

115. a. Avogadro's number is 6.022×10^{23}. Approximate $\log(6.022 \times 10^{23})$.   23.7797

   b. Planck's constant is 6.626×10^{-34} J $\cdot$ sec. Approximate $\log(6.626 \times 10^{-34})$. -33.1787

   c. Compare the value of the common logarithm to the power of 10 used in scientific notation.   Given a number $a \times 10^n$, $\log(a \times 10^n)$ is between n and $n + 1$, inclusive.

116. a. The speed of light is 2.9979×10^8 m/sec. Approximate $\log(2.9979 \times 10^8)$.   8.4768

   b. An elementary charge is 1.602×10^{-19} C. Approximate $\log(1.602 \times 10^{-19})$. -18.7953

   c. Compare the value of the common logarithm to the power of 10 used in scientific notation. Given a number $a \times 10^n$, $\log(a \times 10^n)$ is between n and $n + 1$, inclusive.

Expanding Your Skills

For Exercises 117–122, write the domain in interval notation.

117. $t(x) = \log_4\left(\dfrac{x - 1}{x - 3}\right)$

   $(-\infty, 1) \cup (3, \infty)$

118. $r(x) = \log_5\left(\dfrac{x + 2}{x - 4}\right)$ $(-\infty, -2) \cup (4, \infty)$

119. $s(x) = \ln\left(\sqrt{x + 5} - 1\right)$ $(-4, \infty)$

120. $v(x) = \ln\left(\sqrt{x - 8} - 1\right)$ $(9, \infty)$

121. $c(x) = \log\left(\dfrac{1}{\sqrt{x - 6}}\right)$ $(6, \infty)$

122. $d(x) = \log\left(\dfrac{1}{\sqrt{x + 8}}\right)$ $(-8, \infty)$

Technology Connections

123. a. Graph $f(x) = \ln x$ and

$$g(x) = (x - 1) - \frac{(x - 1)^2}{2} + \frac{(x - 1)^3}{3} - \frac{(x - 1)^4}{4}$$

   on the viewing window $[-2, 4, 1]$ by $[-5, 2, 1]$. How do the graphs compare on the interval $(0, 2)$?

   b. Use function g to approximate $\ln 1.5$. Round to 4 decimal places. $\ln 1.5 \approx 0.4010$

124. Compare the graphs of $Y_1 = \dfrac{e^x - e^{-x}}{2}$,

   $Y_2 = \ln(x + \sqrt{x^2 + 1})$, and $Y_3 = x$ on the viewing window $[-16.1, 16.1, 1]$ by $[-10, 10, 1]$. Based on the graphs, how do you suspect that the functions are related?

125. Compare the graphs of the functions.

   $Y_1 = \ln(2x)$ and $Y_2 = \ln 2 + \ln x$

   The graphs are the same.

126. Compare the graphs of the functions.

   $Y_1 = \ln\left(\dfrac{x}{2}\right)$ and $Y_2 = \ln x - \ln 2$

   The graphs are the same.

PROBLEM RECOGNITION EXERCISES

Analyzing Functions

For Exercises 1–14,

  a. Write the domain.   **b.** Write the range.   **c.** Find the *x*-intercept(s).   **d.** Find the *y*-intercept.
  e. Determine the asymptotes if applicable.   **f.** Determine the intervals over which the function is increasing.
  g. Determine the intervals over which the function is decreasing.   **h.** Match the function with its graph.

1. $f(x) = 3$

2. $g(x) = 2x - 3$

3. $d(x) = (x - 3)^2 - 4$

4. $h(x) = \sqrt[3]{x - 2}$

5. $k(x) = \dfrac{2}{x - 1}$

6. $z(x) = \dfrac{3x}{x + 2}$

7. $p(x) = \left(\dfrac{4}{3}\right)^x$

8. $q(x) = -x^2 - 6x - 9$

9. $m(x) = |x - 4| - 1$

10. $n(x) = -|x| + 3$

11. $r(x) = \sqrt{3 - x}$

12. $s(x) = \sqrt{x - 3}$

13. $t(x) = e^x + 2$

14. $v(x) = \ln(x + 2)$

A.

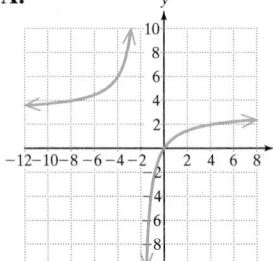

B.

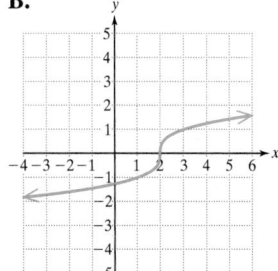

C.

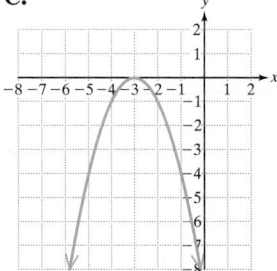

D.

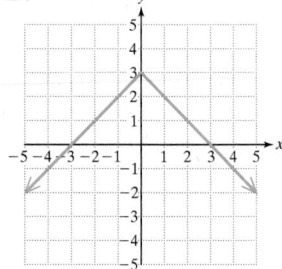

E.

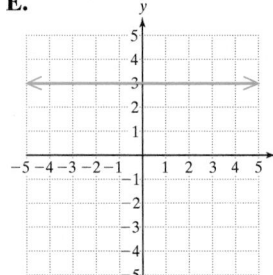

F.

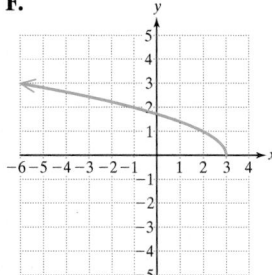

G.

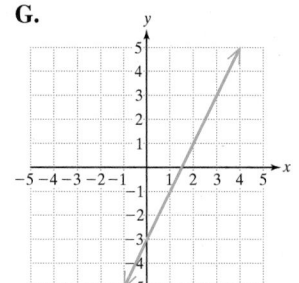

H.

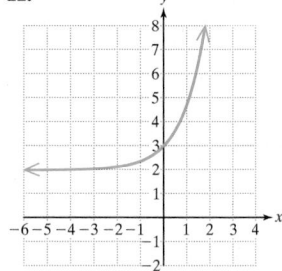

I.

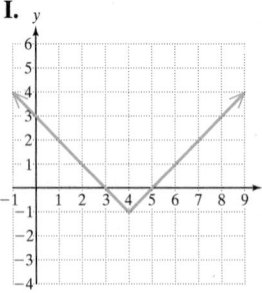

J.

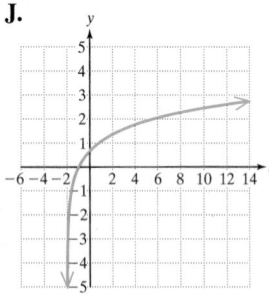

K.

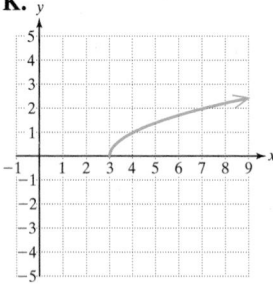

L.

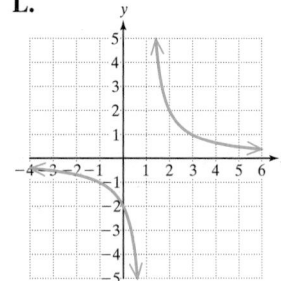

M.

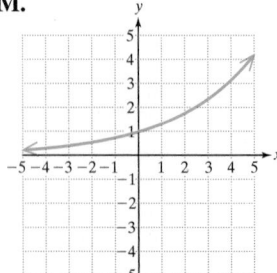

N.

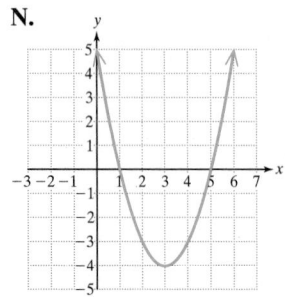

SECTION 3.4 **Properties of Logarithms**

1. Apply the Product, Quotient, and Power Properties of Logarithms

By definition, $y = \log_b x$ is equivalent to $b^y = x$. Because a logarithm is an exponent, the properties of exponents can be applied to logarithms. The first is called the product property of logarithms.

Product Property of Logarithms

Let b, x, and y be positive real numbers where $b \neq 1$. Then

$$\log_b(xy) = \log_b x + \log_b y.$$

The logarithm of a product equals the sum of the logarithms of the factors.

TIP When two factors of the same base are multiplied, the base is unchanged and we add the exponents. This is the underlying principle for the product property of logarithms.

Proof:

Let $M = \log_b x$, which implies $b^M = x$.
Let $N = \log_b y$, which implies $b^N = y$.
Then $xy = b^M b^N = b^{M+N}$.

Writing the expression $xy = b^{M+N}$ in logarithmic form, we have,

$$\log_b(xy) = M + N$$
$$\log_b(xy) = \log_b x + \log_b y \checkmark$$

To demonstrate the product property of logarithms, simplify the following expressions by using the order of operations.

$$\log_3(3 \cdot 9) \overset{?}{=} \log_3 3 + \log_3 9$$
$$\log_3 27 \overset{?}{=} 1 + 2$$
$$3 \overset{?}{=} 3 \checkmark \text{ True}$$

Classroom Examples: p. 404
Exercises 8, 10

EXAMPLE 1   **Applying the Product Property of Logarithms**

Write the logarithm as a sum and simplify if possible. Assume that x and y represent positive real numbers.

   a. $\log_2(8x)$ **b.** $\ln(5xy)$

Solution:

   a. $\log_2(8x) = \log_2 8 + \log_2 x$ Product property of logarithms
   $= 3 + \log_2 x$ Simplify. $\log_2 8 = \log_2 2^3 = 3$

   b. $\ln(5xy) = \ln 5 + \ln x + \ln y$

Skill Practice 1   Write the logarithm as a sum and simplify if possible. Assume that a, c, and d represent positive real numbers.

   a. $\log_4(16a)$ **b.** $\log(12cd)$

Answers
1. a. $2 + \log_4 a$
   b. $\log 12 + \log c + \log d$

The quotient rule of exponents tells us that $\dfrac{b^M}{b^N} = b^{M-N}$ for $b \neq 0$. This property can be applied to logarithms.

> ### Quotient Property of Logarithms
>
> Let b, x, and y be positive real numbers where $b \neq 1$. Then
>
> $$\log_b\left(\frac{x}{y}\right) = \log_b x - \log_b y.$$
>
> The logarithm of a quotient equals the difference of the logarithm of the numerator and the logarithm of the denominator.

The proof of the quotient property for logarithms is similar to the proof of the product property (see Exercise 107). To demonstrate the quotient property for logarithms, simplify the following expressions by using the order of operations.

$$\log\left(\frac{1,000,000}{100}\right) \overset{?}{=} \log 1,000,000 - \log 100$$

$$\log 10,000 \overset{?}{=} 6 - 2$$

$$4 \overset{?}{=} 4 \checkmark \text{ True}$$

Classroom Examples: p. 404
Exercises 14, 18

Instructor Note:
Consider steering students clear of the following common errors by using numerical examples.

$\log_b(M + N)$
    $\neq \log_b M + \log_b N$
$\log_b(M - N)$
    $\neq \log_b M - \log_b N$
$\log_b(MN)$
    $\neq (\log_b M) \cdot (\log_b N)$
$\log_b\left(\dfrac{M}{N}\right) \neq \dfrac{\log_b M}{\log_b N}$

> **EXAMPLE 2   Applying the Quotient Property of Logarithms**
>
> Write the logarithm as the difference of logarithms and simplify if possible. Assume that the variables represent positive real numbers.
>
> **a.** $\log_3\left(\dfrac{c}{d}\right)$ **b.** $\log\left(\dfrac{x}{1000}\right)$
>
> **Solution:**
>
> **a.** $\log_3\left(\dfrac{c}{d}\right) = \log_3 c - \log_3 d$ Quotient property of logarithms.
>
> **b.** $\log\left(\dfrac{x}{1000}\right) = \log x - \log 1000$ Quotient property of logarithms.
>
> $\qquad\qquad\quad = \log x - 3$ Simplify. $\log 1000 = \log 10^3 = 3$

> **Skill Practice 2**   Write the logarithm as the difference of logarithms and simplify if possible. Assume that t represents a positive real number.
>
> **a.** $\log_6\left(\dfrac{8}{t}\right)$ **b.** $\ln\left(\dfrac{e}{12}\right)$

The last property we present here is the power property of logarithms. The power property of exponents tells us that $(b^M)^N = b^{MN}$. The same principle can be applied to logarithms.

> ### Power Property of Logarithms
>
> Let b and x be positive real numbers where $b \neq 1$. Let p be any real number. Then
>
> $$\log_b x^p = p \log_b x.$$

The power property of logarithms is proved in Exercise 108.

Answers
2. a. $\log_6 8 - \log_6 t$ **b.** $1 - \ln 12$

EXAMPLE 3   Applying the Power Property of Logarithms

Apply the power property of logarithms.   **a.** $\ln \sqrt[5]{x^2}$ **b.** $\log x^2$

Solution:

a. $\ln \sqrt[5]{x^2} = \ln x^{2/5}$ Write $\sqrt[5]{x^2}$ using rational exponents.

$\qquad\qquad = \dfrac{2}{5} \ln x$ provided that $x > 0$ Apply the power rule.

b. $\log x^2 = 2 \log x$ provided that $x > 0$ Apply the power rule.

In both parts (a) and (b), the condition that $x > 0$ is mandatory. The properties of logarithms hold true only for values of the variable for which the logarithms are defined. That is, the arguments must be positive.

From the graphs of $y = \log x^2$ and $y = 2 \log x$, we see that the domains are different. Therefore, the statement $\log x^2 = 2 \log x$ is true only for $x > 0$.

$y = \log x^2$ Domain: $(-\infty, 0) \cup (0, \infty)$ $y = 2 \log x$ Domain: $(0, \infty)$

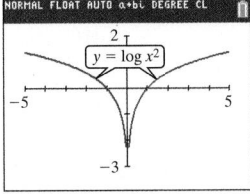
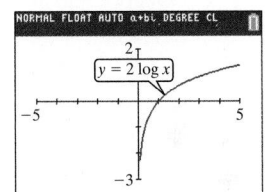

Skill Practice 3   Apply the power property of logarithms.

a. $\log_5 \sqrt[5]{x^4}$ **b.** $\ln x^4$

At this point, we have learned seven properties of logarithms. The properties hold true for values of the variable for which the logarithms are defined. **Therefore, in the examples and exercises, we will assume that the variable expressions within the logarithms represent positive real numbers.**

Properties of Logarithms

Let b, x, and y be positive real numbers where $b \neq 1$, and let p be a real number. Then the following properties of logarithms are true.

1. $\log_b 1 = 0$	**5.** $\log_b(xy) = \log_b x + \log_b y$	**Product property**
2. $\log_b b = 1$	**6.** $\log_b\left(\dfrac{x}{y}\right) = \log_b x - \log_b y$	**Quotient property**
3. $\log_b b^p = p$	**7.** $\log_b x^p = p \log_b x$	**Power property**
4. $b^{\log_b x} = x$		

---

2. Write a Logarithmic Expression in Expanded Form

Properties 5, 6, and 7 can be used in either direction. For example,

$$\log\left(\frac{ab}{c}\right) = \log a + \log b - \log c \qquad \text{or} \qquad \log a + \log b - \log c = \log\left(\frac{ab}{c}\right).$$

In some applications of algebra and calculus, the "condensed" form of the logarithm is preferred. In other applications, the "expanded" form is preferred. In Examples 4–6, we practice manipulating logarithmic expressions in both forms.

Classroom Examples: p. 404
Exercises 32, 38

EXAMPLE 4 **Writing a Logarithmic Expression in Expanded Form**

Write the expression as the sum or difference of logarithms.

a. $\log_2\left(\dfrac{z^3}{xy^5}\right)$ **b.** $\log\sqrt[3]{\dfrac{(x+y)^2}{10}}$

Solution:

a. $\log_2\left(\dfrac{z^3}{xy^5}\right) = \log_2 z^3 - \log_2(xy^5)$ Apply the quotient property.

$= \log_2 z^3 - (\log_2 x + \log_2 y^5)$ Apply the product property.

$= \log_2 z^3 - \log_2 x - \log_2 y^5$ Apply the distributive property.

$= 3\log_2 z - \log_2 x - 5\log_2 y$ Apply the power property.

b. $\log\sqrt[3]{\dfrac{(x+y)^2}{10}} = \log\left[\dfrac{(x+y)^2}{10}\right]^{1/3}$ Write the radical expression with rational exponents.

$= \dfrac{1}{3}\log\left[\dfrac{(x+y)^2}{10}\right]$ Apply the power property.

$= \dfrac{1}{3}[\log(x+y)^2 - \log 10]$ Apply the quotient property.

$= \dfrac{1}{3}[2\log(x+y) - 1]$ Apply the power property and simplify: $\log 10 = 1$.

$= \dfrac{2}{3}\log(x+y) - \dfrac{1}{3}$ Apply the distributive property.

> **Avoiding Mistakes**
>
> In Example 4(b) do not try to simplify $\log(x+y)$. The argument contains a sum, not a product.
>
> $\log(\underbrace{x+y}_{\text{sum}})$ cannot be simplified.
>
> Compare to the logarithm of a product which can be simplified.
>
> $\log(\underbrace{xy}_{\text{product}}) = \log x + \log y$

> **Skill Practice 4** Write the expression as the sum or difference of logarithms.
>
> **a.** $\ln\left(\dfrac{a^4 b}{c^9}\right)$ **b.** $\log_5\sqrt[3]{\dfrac{25}{(a^2+b)^2}}$

3. Write a Logarithmic Expression as a Single Logarithm

In Examples 5 and 6, we demonstrate how to write a sum or difference of logarithms as a single logarithm. We apply Properties 5, 6, and 7 of logarithms in reverse.

Classroom Example: p. 405
Exercise 52

EXAMPLE 5 **Writing the Sum or Difference of Logarithms as a Single Logarithm**

Write the expression as a single logarithm and simplify the result if possible.

$$\log_2 560 - \log_2 7 - \log_2 5$$

Solution:

$\log_2 560 - \log_2 7 - \log_2 5$

$= \log_2 560 - (\log_2 7 + \log_2 5)$ Factor out -1 from the last two terms.

$= \log_2 560 - \log_2(7 \cdot 5)$ Apply the product property.

$= \log_2\left(\dfrac{560}{7 \cdot 5}\right)$ Apply the quotient property.

$= \log_2 16$ Simplify within the argument.

$= 4$ Simplify. $\log_2 16 = \log_2 2^4 = 4$

Answers

4. a. $4\ln a + \ln b - 9\ln c$

b. $\dfrac{2}{3} - \dfrac{2}{3}\log_5(a^2 + b)$

Skill Practice 5 Write the expression as a single logarithm and simplify the result if possible. $\log_3 54 + \log_3 10 - \log_3 20$

EXAMPLE 6 Writing the Sum or Difference of Logarithms as a Single Logarithm

Write the expression as a single logarithm and simplify the result if possible.

a. $3 \log a - \dfrac{1}{2}\log b - \dfrac{1}{2}\log c$ **b.** $\dfrac{1}{2}\ln x + \ln(x^2 - 1) - \ln(x + 1)$

Solution:

a. $3 \log a - \dfrac{1}{2}\log b - \dfrac{1}{2}\log c$

$\quad = 3 \log a - \dfrac{1}{2}(\log b + \log c)$ Factor out $-\frac{1}{2}$ from the last two terms.

$\quad = 3 \log a - \dfrac{1}{2}\log(bc)$ Apply the product property.

$\quad = \log a^3 - \log(bc)^{1/2}$

$\quad = \log a^3 - \log\sqrt{bc}$ Apply the power property.

$\quad = \log\left(\dfrac{a^3}{\sqrt{bc}}\right)$ Apply the quotient property.

Avoiding Mistakes

In all examples and exercises in which we manipulate logarithmic expressions, it is important to note that the equivalences are true only for the values of the variables that make the expressions defined. In Example 6(b) we have the restriction that $x > 1$.

b. $\dfrac{1}{2}\ln x + \ln(x^2 - 1) - \ln(x + 1)$

$\quad = \ln x^{1/2} + \ln(x^2 - 1) - \ln(x + 1)$ Apply the power property.

$\quad = \ln[x^{1/2}(x^2 - 1)] - \ln(x + 1)$ Apply the product property.

$\quad = \ln\left[\dfrac{\sqrt{x}(x^2 - 1)}{x + 1}\right]$ Apply the quotient property.

$\quad = \ln\left[\dfrac{\sqrt{x}(x + 1)(x - 1)}{x + 1}\right]$ Factor the numerator of the argument.

$\quad = \ln\left[\sqrt{x}(x - 1)\right]$ Simplify the argument.

Skill Practice 6 Write the expression as a single logarithm and simplify the result if possible.

a. $3 \log x - \dfrac{1}{3}\log y - \dfrac{2}{3}\log z$ **b.** $\dfrac{1}{3}\ln t + \ln(t^2 - 9) - \ln(t - 3)$

EXAMPLE 7 Applying Properties of Logarithms

Given that $\log_b 2 \approx 0.356$ and $\log_b 3 \approx 0.565$, approximate the value of $\log_b 36$.

Solution:

$\log_b 36 = \log_b(2 \cdot 3)^2$ Write the argument as a product of the factors 2 and 3.

$\quad = 2 \log_b(2 \cdot 3)$ Apply the power property of logarithms.

$\quad = 2(\log_b 2 + \log_b 3)$ Apply the product property of logarithms.

$\quad \approx 2(0.356 + 0.565)$ Simplify.

$\quad \approx 1.842$

Answers

5. $\log_3 27 = 3$

6. a. $\log\left(\dfrac{x^3}{\sqrt[3]{yz^2}}\right)$

$\quad$ **b.** $\ln\left[\sqrt[3]{t}(t + 3)\right]$

> **Skill Practice 7** Given that $\log_b 2 \approx 0.356$ and $\log_b 3 \approx 0.565$, approximate the value of $\log_b 24$.

4. Apply the Change-of-Base Formula

A calculator can be used to approximate the value of a logarithm base 10 or base e by using the **LOG** key or the **LN** key, respectively. However, to use a calculator to evaluate a logarithmic expression with a different base, we must use the change-of-base formula.

> ### Change-of-Base Formula
>
> Let a and b be positive real numbers such that $a \neq 1$ and $b \neq 1$. Then for any positive real number x,
>
> $$\log_b x = \frac{\log_a x}{\log_a b}$$
>
> *Note*: The change-of-base formula converts a logarithm of one base to a ratio of logarithms of a different base. For the purpose of using a calculator, we often apply the change-of-base formula with base 10 or base e.
>
> $$\log_b x = \frac{\log x}{\log b}$$
> Original base is b. Ratio of base 10 logarithms
>
> $$\log_b x = \frac{\ln x}{\ln b}$$
> Original base is b. Ratio of base e logarithms

To derive the change-of-base formula, assume that a and b are positive real numbers with $a \neq 1$ and $b \neq 1$. Begin by letting $y = \log_b x$. If $y = \log_b x$, then

$$b^y = x \qquad \text{Write the original logarithm in exponential form.}$$
$$\log_a b^y = \log_a x \qquad \text{Take the logarithm base } a \text{ on both sides.}$$
$$y \cdot \log_a b = \log_a x \qquad \text{Apply the power property of logarithms.}$$
$$y = \frac{\log_a x}{\log_a b} \qquad \text{Solve for } y.$$
$$\log_b x = \frac{\log_a x}{\log_a b} \qquad \begin{array}{l}\text{Replace } y \text{ by } \log_b x.\\ \text{This is the change-of-base formula.}\end{array}$$

Classroom Example: p. 405
Exercise 82

EXAMPLE 8 Applying the Change-of-Base Formula

a. Estimate $\log_4 153$ between two consecutive integers.
b. Use the change-of-base formula to approximate $\log_4 153$ by using base 10. Round to 4 decimal places.
c. Use the change-of-base formula to approximate $\log_4 153$ by using base e.
d. Check the result by using the related exponential form.

Solution:

a. $64 < 153 < 256$
  $4^3 < 153 < 4^4$
  $3 < \log_4 153 < 4$ $\qquad \log_4 153$ is between 3 and 4.

TIP Although the numerators and denominators in parts (b) and (c) are different, their ratios are the same.

b. $\log_4 153 = \dfrac{\log 153}{\log 4} \approx \dfrac{2.184691431}{0.6020599913} \approx 3.6287$

c. $\log_4 153 = \dfrac{\ln 153}{\ln 4} \approx \dfrac{5.030437921}{1.386294361} \approx 3.6287$

d. Check: $4^{3.6287} \approx 153$ ✓

Skill Practice 8

a. Estimate $\log_6 23$ between two consecutive integers.

b. Use the change-of-base formula to evaluate $\log_6 23$ by using base 10. Round to 4 decimal places.

c. Use the change-of-base formula to evaluate $\log_6 23$ by using base e. Round to 4 decimal places.

d. Check the result by using the related exponential form.

TECHNOLOGY CONNECTIONS

Using the Change-of-Base Formula to Graph a Logarithmic Function

The change-of-base formula can be used to graph logarithmic functions using a graphing utility. For example, to graph $Y_1 = \log_2 x$, enter the function as

$$Y = \log(x)/\log(2) \quad \text{or} \quad Y = \ln(x)/\ln(2)$$

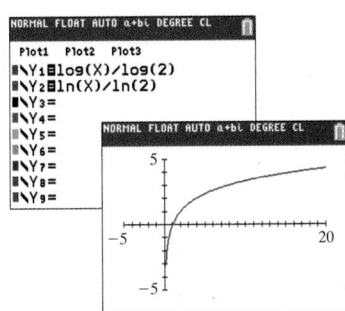

Point of Interest

The slide rule, first built in England in the early 17th century, is a mechanical computing device that uses logarithmic scales to perform operations involving multiplication, division, roots, logarithms, exponentials, and trigonometry. Amazingly, slide rules were used into the space age by engineers in the 1960's to help send astronauts to the moon.
It was only with the invention of the pocket calculator that slide rules were replaced by modern computing devices.

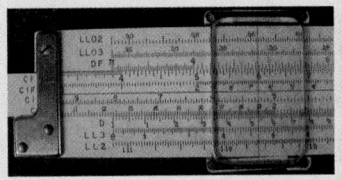

Answers

8. a. Between 1 and 2
   b. 1.7500
   c. 1.7500
   d. $6^{1.7500} \approx 23$

SECTION 3.4   Practice Exercises

Prerequisite Review

For Exercises R.1–R.4, use the properties of exponents to simplify the expression.

R.1. $x^{-3} \cdot x^5 \cdot x^7$ x^9

R.2. $\dfrac{y^{-2}\,y^{10}}{y^3}$ y^5

R.3. $\left(4w^{-3}\,z^4\right)^2$ $\dfrac{16z^8}{w^6}$

R.4. $\left(\dfrac{7k^4}{n}\right)^{-3}$ $\dfrac{n^3}{343k^{12}}$

Concept Connections

1. The product property of logarithms states that $\log_b(xy) = $ _____$\log_b x + \log_b y$_____ for positive real numbers b, x, and y, where $b \neq 1$.

2. The _____quotient_____ property of logarithms states that $\log_b\left(\dfrac{x}{y}\right) = $ _____$\log_b x - \log_b y$_____ for positive real numbers b, x, and y, where $b \neq 1$.

3. The power property of logarithms states that for any real number p, $\log_b x^p = $ _____$p \log_b x$_____ for positive real numbers b, x, and y, where $b \neq 1$.

4. The change-of-base formula states that $\log_b x$ can be written as a ratio of logarithms with base a as
$$\log_b x = \frac{\boxed{}}{\boxed{}} \quad \frac{\log_a x}{\log_a b}$$

5. The change-of-base formula is often used to convert a logarithm to a ratio of logarithms with base _____10_____ or base _____e_____ so that a calculator can be used to approximate the logarithm.

6. To use a graphing utility to graph the function defined by $y = \log_5 x$, use the change-of-base formula to write the function as $y = $ _____$\dfrac{\log x}{\log 5}$_____ or $y = $ _____$\dfrac{\ln x}{\ln 5}$_____.

For the exercises in this set, assume that all variable expressions represent positive real numbers.

Objective 1: Apply the Product, Quotient, and Power Properties of Logarithms

For Exercises 7–12, use the product property of logarithms to write the logarithm as a sum of logarithms. Then simplify if possible. (See Example 1)

7. $\log_5(125z)$ $3 + \log_5 z$

8. $\log_7(49k)$ $2 + \log_7 k$

9. $\log(8cd)$ $\log 8 + \log c + \log d$

10. $\log(24vw)$ $\log 24 + \log v + \log w$

11. $\log_2[(x + y) \cdot z]$ $\log_2(x + y) + \log_2 z$

12. $\log_3[(a + b) \cdot c]$ $\log_3(a + b) + \log_3 c$

For Exercises 13–18, use the quotient property of logarithms to write the logarithm as a difference of logarithms. Then simplify if possible. (See Example 2)

13. $\log_{12}\left(\dfrac{p}{q}\right)$ $\log_{12} p - \log_{12} q$

14. $\log_9\left(\dfrac{m}{n}\right)$ $\log_9 m - \log_9 n$

15. $\ln\left(\dfrac{e}{5}\right)$ $1 - \ln 5$

16. $\ln\left(\dfrac{x}{e}\right)$ $\ln x - 1$

17. $\log\left(\dfrac{m^2 + n}{100}\right)$ $\log(m^2 + n) - 2$

18. $\log\left(\dfrac{1000}{c^2 + 1}\right)$ $3 - \log(c^2 + 1)$

For Exercises 19–24, apply the power property of logarithms. (See Example 3)

19. $\log(2x - 3)^4$ $4 \log(2x - 3)$

20. $\log(8t - 3)^2$ $2 \log(8t - 3)$

21. $\log_6 \sqrt[7]{x^3}$ $\dfrac{3}{7}\log_6 x$

22. $\log_8 \sqrt[4]{x^3}$ $\dfrac{3}{4}\log_8 x$

23. $\ln 2^{kt}$ $kt \ln 2$

24. $\ln(0.5)^{rt}$ $rt \ln 0.5$

Objective 2: Write a Logarithmic Expression in Expanded Form

For Exercises 25–44, write the logarithm as a sum or difference of logarithms. Simplify each term as much as possible. (See Example 4)

25. $\log_4(7yz)$ $\log_4 7 + \log_4 y + \log_4 z$

26. $\log_2(5ab)$ $\log_2 5 + \log_2 a + \log_2 b$

27. $\log_7\left(\dfrac{1}{7}mn^2\right)$ $-1 + \log_7 m + 2\log_7 n$

28. $\log_4\left(\dfrac{1}{16}t^3 v\right)$ $-2 + 3\log_4 t + \log_4 v$

29. $\log_2\left(\dfrac{x^{10}}{yz}\right)$ $10\log_2 x - \log_2 y - \log_2 z$

30. $\log_5\left(\dfrac{p^5}{mn}\right)$ $5\log_5 p - \log_5 m - \log_5 n$

31. $\log_6\left(\dfrac{p^5}{qt^3}\right)$ $5\log_6 p - \log_6 q - 3\log_6 t$

32. $\log_8\left(\dfrac{a^4}{b^9 c}\right)$ $4\log_8 a - 9\log_8 b - \log_8 c$

33. $\log\left(\dfrac{10}{\sqrt{a^2 + b^2}}\right)$ $1 - \dfrac{1}{2}\log(a^2 + b^2)$

34. $\log\left(\dfrac{\sqrt{d^2 + 1}}{10,000}\right)$ $-4 + \dfrac{1}{2}\log(d^2 + 1)$

35. $\ln\left(\dfrac{\sqrt[3]{xy}}{wz^2}\right)$ $\dfrac{1}{3}\ln x + \dfrac{1}{3}\ln y - \ln w - 2\ln z$

36. $\ln\left(\dfrac{\sqrt[4]{pq}}{t^3 m}\right)$ $\dfrac{1}{4}\ln p + \dfrac{1}{4}\ln q - 3\ln t - \ln m$

37. $\ln\sqrt[4]{\dfrac{a^2 + 4}{e^3}}$ $\dfrac{1}{4}\ln(a^2 + 4) - \dfrac{3}{4}$

38. $\ln\sqrt[5]{\dfrac{e^2}{c^2 + 5}}$ $\dfrac{2}{5} - \dfrac{1}{5}\ln(c^2 + 5)$

39. $\log\left[\dfrac{2x(x^2 + 3)^8}{\sqrt{4 - 3x}}\right]$

40. $\log\left[\dfrac{5y(4x + 1)^7}{\sqrt[3]{2 - 7x}}\right]$

41. $\log_5 \sqrt[3]{x\sqrt{5}}$ $\dfrac{1}{3}\log_5 x + \dfrac{1}{6}$

42. $\log_2 \sqrt[4]{y\sqrt{2}}$ $\dfrac{1}{4}\log_2 y + \dfrac{1}{8}$

43. $\log_2\left[\dfrac{4a^2\sqrt{3 - b}}{c(b + 4)^2}\right]$

44. $\log_3\left[\dfrac{27x^3\sqrt{y^2 - 1}}{y(x - 1)^2}\right]$

Objective 3: Write a Logarithmic Expression as a Single Logarithm

For Exercises 45–68, write the logarithmic expression as a single logarithm with coefficient 1, and simplify as much as possible. (See Examples 5–6)

45. $\ln y + \ln 4$ $\ln 4y$

46. $\log 5 + \log p$ $\log 5p$

47. $\log_{15} 3 + \log_{15} 5$ 1

48. $\log_{12} 8 + \log_{12} 18$ 2

49. $\log_7 98 - \log_7 2$ 2

50. $\log_6 144 - \log_6 4$ 2

51. $\log 150 - \log 3 - \log 5$ 1

52. $\log_3 693 - \log_3 33 - \log_3 7$ 1

53. $2 \log_2 x + \log_2 t$ $\log_2(x^2 t)$

54. $5 \log_4 y + \log_4 w$ $\log_4(y^5 w)$

55. $4 \log_8 m - 3 \log_8 n - 2 \log_8 p$

56. $8 \log_3 x - 2 \log_3 z - 7 \log_3 y$

57. $3[\ln x - \ln(x + 3) - \ln (x - 3)]$

58. $2[\log(p - 4) - \log(p - 1) - \log(p + 4)]$ $\log\left(\dfrac{p - 4}{p^2 + 3p - 4}\right)^2$

59. $\dfrac{1}{2}\ln(x + 1) - \dfrac{1}{2}\ln(x - 1)$ $\ln\sqrt{\dfrac{x + 1}{x - 1}}$

60. $\dfrac{1}{3}\ln(x^2 + 1) - \dfrac{1}{3}\ln(x + 1)$ $\ln\sqrt[3]{\dfrac{x^2 + 1}{x + 1}}$

61. $6 \log x - \dfrac{1}{3}\log y - \dfrac{2}{3}\log z$ $\log\left(\dfrac{x^6}{\sqrt[3]{yz^2}}\right)$

62. $15 \log c - \dfrac{1}{4}\log d - \dfrac{3}{4}\log k$ $\log\left(\dfrac{c^{15}}{\sqrt[4]{dk^3}}\right)$

63. $\dfrac{1}{3}\log_4 p + \log_4(q^2 - 16) - \log_4(q - 4)$ $\log_4[\sqrt[3]{p}(q + 4)]$

64. $\dfrac{1}{4}\log_2 w + \log_2(w^2 - 100) - \log_2(w + 10)$ $\log_2[\sqrt[4]{w}(w - 10)]$

65. $\dfrac{1}{2}[6 \ln(x + 2) + \ln x - \ln x^2]$ $\ln\left[\dfrac{(x + 2)^3}{\sqrt{x}}\right]$

66. $\dfrac{1}{3}[12 \ln(x - 5) + \ln x - \ln x^3]$ $\ln\left[\dfrac{(x - 5)^4}{\sqrt[3]{x^2}}\right]$

67. $\log(8y^2 - 7y) + \log y^{-1}$ $\log(8y - 7)$

68. $\log(9t^3 - 5t) + \log t^{-1}$ $\log(9t^2 - 5)$

For Exercises 69–78, use $\log_b 2 \approx 0.356$, $\log_b 3 \approx 0.565$, and $\log_b 5 \approx 0.827$ to approximate the value of the given logarithms. (See Example 7)

69. $\log_b 15$ 1.392

70. $\log_b 10$ 1.183

71. $\log_b 81$ 2.26

72. $\log_b 125$ 2.481

73. $\log_b 50$ 2.01

74. $\log_b 12$ 1.277

75. $\log_b\left(\dfrac{15}{2}\right)$ 1.036

76. $\log_b\left(\dfrac{6}{5}\right)$ 0.094

77. $\log_b 100$ 2.366

78. $\log_b 225$ 2.784

Objective 4: Apply the Change-of-Base Formula

For Exercises 79–84, (See Example 8)

a. Estimate the value of the logarithm between two consecutive integers. For example, $\log_2 7$ is between 2 and 3 because $2^2 < 7 < 2^3$.

b. Use the change-of-base formula and a calculator to approximate the logarithm to 4 decimal places.

c. Check the result by using the related exponential form.

79. $\log_2 15$
   a. Between 3 and 4   b. 3.9069   c. $2^{3.9069} \approx 15$

80. $\log_3 15$
   a. Between 2 and 3   b. 2.4650   c. $3^{2.4650} \approx 15$

81. $\log_5 3$
   a. Between 0 and 1   b. 0.6826   c. $5^{0.6826} \approx 3$

82. $\log_8 5$
   a. Between 0 and 1   b. 0.7740   c. $8^{0.7740} \approx 5$

83. $\log_2 0.3$
   a. Between -2 and -1 b. -1.7370
   c. $2^{-1.7370} \approx 0.3$

84. $\log_2 0.2$
   a. Between -3 and -2 b. -2.3219
   c. $2^{-2.3219} \approx 0.2$

For Exercises 85–88, use the change-of-base formula and a calculator to approximate the given logarithms. Round to 4 decimal places. Then check the answer by using the related exponential form. (See Example 8)

85. $\log_2(4.68 \times 10^7)$
   25.4800; $2^{25.4800} \approx 4.68 \times 10^7$

86. $\log_2(2.54 \times 10^{10})$
   34.5641; $2^{34.5641} \approx 2.54 \times 10^{10}$

87. $\log_4(5.68 \times 10^{-6})$
   -8.7128; $4^{-8.7128} \approx 5.68 \times 10^{-6}$

88. $\log_4(9.84 \times 10^{-5})$
   -6.6555; $4^{-6.6555} \approx 9.84 \times 10^{-5}$

Mixed Exercises

For Exercises 89–98, determine if the statement is true or false. For each false statement, provide a counterexample. For example, $\log(x + y) \neq \log x + \log y$ because $\log(2 + 8) \neq \log 2 + \log 8$ (the left side is 1 and the right side is approximately 1.204).

89. $\log e = \dfrac{1}{\ln 10}$ True

90. $\ln 10 = \dfrac{1}{\log e}$ True

91. $\log_5\left(\dfrac{1}{x}\right) = \dfrac{1}{\log_5 x}$

92. $\log_6\left(\dfrac{1}{t}\right) = \dfrac{1}{\log_6 t}$

93. $\log_4\left(\dfrac{1}{p}\right) = -\log_4 p$ True

94. $\log_8\left(\dfrac{1}{w}\right) = -\log_8 w$ True

95. $\log(xy) = (\log x)(\log y)$

96. $\log\left(\dfrac{x}{y}\right) = \dfrac{\log x}{\log y}$

97. $\log_2(7y) + \log_2 1 = \log_2(7y)$ True

98. $\log_4(3d) + \log_4 1 = \log_4(3d)$ True

Write About It

99. Explain why the product property of logarithms does not apply to the following statement.

$\log_5(-5) + \log_5(-25)$ The given statement $\log_5(-5) +$
$= \log_5[(-5)(-25)]$ $\log_5(-25)$ is not defined because the
$= \log_5 125 = 3$ arguments to the logarithmic expressions are not positive real numbers.

100. Explain how to use the change-of-base formula and explain why it is important.

The change-of-base formula enables us to write a logarithm of one base in terms of a ratio of logarithms of a different base. This is important among other reasons because it enables us to approximate the value of a logarithm of a base other than 10 or e.

Expanding Your Skills

101. a. Write the difference quotient for $f(x) = \ln x$.

    b. Show that the difference quotient from part (a) can be written as $\ln\left(\dfrac{x+h}{x}\right)^{1/h}$.

102. Show that
$$-\ln\left(x - \sqrt{x^2 - 1}\right) = \ln\left(x + \sqrt{x^2 - 1}\right)$$

103. Show that
$$\log\left(\frac{-b + \sqrt{b^2 - 4ac}}{2a}\right) + \log\left(\frac{-b - \sqrt{b^2 - 4ac}}{2a}\right)$$
$$= \log c - \log a$$

104. Show that
$$\ln\left(\frac{c + \sqrt{c^2 - x^2}}{c - \sqrt{c^2 - x^2}}\right) = 2\ln\left(c + \sqrt{c^2 - x^2}\right) - 2\ln x$$

105. Use the change-of-base formula to write $(\log_2 5)(\log_5 9)$ as a single logarithm. $\log_2 9$

106. Use the change-of-base formula to write $(\log_3 11)(\log_{11} 4)$ as a single logarithm. $\log_3 4$

107. Prove the quotient property of logarithms:
$$\log_b\left(\frac{x}{y}\right) = \log_b x - \log_b y.$$

(*Hint*: Modify the proof of the product property given on page 397.)

108. Prove the power property of logarithms:
$$\log_b x^p = p \log_b x.$$

Let $y = \log_b x$, which implies that $b^y = x$. Raising both sides to the p power yields $(b^y)^p = (x)^p$ or $b^{py} = x^p$. Converting back to logarithmic form, we have $\log_b x^p = py$, but since $y = \log_b x$, we have $\log_b x^p = p \log_b x$ as desired.

Technology Connections

For Exercises 109–112, graph the function.

109. $f(x) = \log_5(x + 4)$ **110.** $g(x) = \log_7(x - 3)$

111. $k(x) = -3 + \log_{1/2} x$ **112.** $h(x) = 4 + \log_{1/3} x$

113. a. Graph $Y_1 = \log|x|$ and $Y_2 = \dfrac{1}{2}\log x^2$. How are the graphs related?

    b. Show algebraically that $\dfrac{1}{2}\log x^2 = \log|x|$.

114. Graph $Y_1 = \ln(0.1x)$, $Y_2 = \ln(0.5x)$, $Y_3 = \ln x$, and $Y_4 = \ln(2x)$. How are the graphs related? Support your answer algebraically.

Additional answers can be found in the Instructor Answer Appendix.

SECTION 3.5

Exponential and Logarithmic Equations and Applications

OBJECTIVES

1. Solve Exponential Equations
2. Solve Logarithmic Equations
3. Use Exponential and Logarithmic Equations in Applications

1. Solve Exponential Equations

A couple invests $8000 in a bond fund. The expected yield is 4.5% and the earnings are reinvested monthly. The growth of the investment is modeled by

$$A = 8000\left(1 + \frac{0.045}{12}\right)^{12t}$$

where A is the amount in the account after t years.

If the couple wants to know how long it will take for the investment to double, they would solve the equation:

$$16{,}000 = 8000\left(1 + \frac{0.045}{12}\right)^{12t} \quad \text{(See Example 11.)}$$

This equation is called an **exponential equation** because the equation contains a variable in the exponent. To solve an exponential equation first note that all exponential functions are one-to-one. Therefore, $b^x = b^y$ implies that $x = y$. This is called the equivalence property of exponential expressions.

TIP The equivalence property tells us that if two exponential expressions with the same base are equal, then their exponents must be equal.

Classroom Examples: p. 416
Exercises 10, 12

Equivalence Property of Exponential Expressions

If b, x, and y are real numbers with $b > 0$ and $b \neq 1$, then

$$b^x = b^y \quad \text{implies that} \quad x = y.$$

EXAMPLE 1   Solving Exponential Equations Using the Equivalence Property

Solve.   **a.** $3^{2x-6} = 81$ **b.** $25^{4-t} = \left(\dfrac{1}{5}\right)^{3t+1}$

Solution:

a.
$$3^{2x-6} = 81$$
$$3^{2x-6} = 3^4 \qquad \text{Write 81 as an exponential expression with a base of 3.}$$
$$2x - 6 = 4 \qquad \text{Equate the exponents.}$$
$$x = 5$$

Check: $3^{2x-6} = 81$
$$3^{2(5)-6} \stackrel{?}{=} 81$$
$$3^4 \stackrel{?}{=} 81 \checkmark$$

The solution set is $\{5\}$.

Avoiding Mistakes

When writing the expression $(5^2)^{4-t}$ as $5^{2(4-t)}$, it is important to use parentheses around the quantity $(4 - t)$. The exponent of 2 must be multiplied by the entire quantity $(4 - t)$. Likewise, parentheses are used around $(3t + 1)$ in the expression $5^{-1(3t + 1)}$.

b.
$$25^{4-t} = \left(\frac{1}{5}\right)^{3t+1}$$
$$(5^2)^{4-t} = (5^{-1})^{3t+1} \qquad \text{Express both 25 and } \tfrac{1}{5} \text{ as integer powers of 5.}$$
$$5^{2(4-t)} = 5^{-1(3t+1)} \qquad \text{Apply the power property of exponents: } (b^m)^n = b^{m \cdot n}.$$
$$5^{8-2t} = 5^{-3t-1} \qquad \text{Apply the distributive property within the exponents.}$$
$$8 - 2t = -3t - 1 \qquad \text{Equate the exponents.}$$
$$t = -9 \qquad \text{The solution checks in the original equation.}$$

The solution set is $\{-9\}$.

Skill Practice 1   Solve.   **a.** $4^{2x-3} = 64$ **b.** $27^{2w+5} = \left(\dfrac{1}{3}\right)^{2-5w}$

In Example 1, we were able to write the left and right sides of the equation with a common base. However, most exponential equations cannot be written in this form by inspection. For example:

$$7^x = 60 \qquad \text{60 is not a recognizable power of 7.}$$
$$7^x = 7^?$$

To solve such an equation, we can take a logarithm of the same base on each side of the equation, and then apply the power property of logarithms. This is demonstrated in Examples 2–4.

Answers
1. a. $\{3\}$ **b.** $\{-17\}$

> **Steps to Solve Exponential Equations by Using Logarithms**
> 1. Isolate the exponential expression on one side of the equation.
> 2. Take a logarithm of the same base on both sides of the equation.
> 3. Use the power property of logarithms to "bring down" the exponent.
> 4. Solve the resulting equation.

Classroom Example: p. 417
Exercise 18

EXAMPLE 2   Solving an Exponential Equation Using Logarithms

Solve. $7^x = 60$

Solution:

$7^x = 60$	The exponential expression 7^x is isolated.
$\log 7^x = \log 60$	Take a logarithm of the same base on both sides of the equation. In this case, we have chosen base 10.
$x \log 7 = \log 60$	Apply the power property of logarithms.
$x = \dfrac{\log 60}{\log 7} \approx 2.1041$	Divide both sides by $\log 7$.

This equation is now linear.

It is important to note that the exact solution to this equation is $\dfrac{\log 60}{\log 7}$ or equivalently by the change-of-base formula, $\log_7 60$. The value 2.1041 is merely an approximation.

The solution set is $\left\{ \dfrac{\log 60}{\log 7} \right\}$ or $\{\log_7 60\}$.

Avoiding Mistakes

While 2.1041 is only an approximation, it is useful to check the result.

$$7^{2.1041} \approx 60$$

Skill Practice 2   Solve. $5^x = 83$

To solve the equation from Example 2, we can take a logarithm of any base. For example:

$$7^x = 60 \qquad\qquad 7^x = 60$$
$$\log_7 7^x = \log_7 60 \qquad \ln 7^x = \ln 60$$
$$x = \log_7 60 \text{ (solution)} \qquad x \ln 7 = \ln 60$$
$$x = \dfrac{\ln 60}{\ln 7} \text{ (solution)}$$

Take the logarithm base 7 on both sides.

Take the natural logarithm on both sides.

The values $\log_7 60$, $\dfrac{\log 60}{\log 7}$, and $\dfrac{\ln 60}{\ln 7}$ are all equivalent. However, common logarithms and natural logarithms are often used to express the solution to an exponential equation so that the solution can be approximated on a calculator.

Answer

2. $\left\{ \dfrac{\log 83}{\log 5} \right\}$ or $\{\log_5 83\}$

Classroom Examples: p. 417
Exercises 22, 24

EXAMPLE 3 Solving Exponential Equations Using Logarithms

Solve.　**a.** $10^{5+2x} + 820 = 49{,}600$　　**b.** $2000 = 18{,}000e^{-0.4t}$

Solution:

a. $10^{5+2x} + 820 = 49{,}600$　　Isolate the exponential expression on the left by subtracting 820 on both sides.
$10^{5+2x} = 48{,}780$

$\log 10^{5+2x} = \log 48{,}780$　　Since the exponential expression on the left has a base of 10, take the log base 10 on both sides.

$5 + 2x = \log 48{,}780$　　On the left, $\log 10^{5+2x} = 5 + 2x$.
$2x = \log 48{,}780 - 5$　　Solve the linear equation by subtracting 5 and dividing by 2.

$x = \dfrac{\log 48{,}780 - 5}{2} \approx -0.1559$　　The solution checks in the original equation.

The solution set is $\left\{ \dfrac{\log 48{,}780 - 5}{2} \right\}$.

b.　$2000 = 18{,}000e^{-0.4t}$　　Isolate the exponential expression on the right by dividing both sides by 18,000.

$\dfrac{1}{9} = e^{-0.4t}$　　Since the exponential expression on the left has a base of e, take the log base e on both sides.

$\ln\left(\dfrac{1}{9}\right) = \ln e^{-0.4t}$　　On the right, $\ln e^{-0.4t} = -0.4t$.

$\ln\left(\dfrac{1}{9}\right) = -0.4t$　　Solve the linear equation by dividing by -0.4.
(linear equation)

$\dfrac{\ln\left(\dfrac{1}{9}\right)}{-0.4} = t$　　The exact solution to the equation can be written in a variety of forms by applying the properties of logarithms:

$$\dfrac{\ln\left(\dfrac{1}{9}\right)}{-0.4} = \dfrac{\ln 1 - \ln 9}{-0.4} = \dfrac{0 - \ln 9}{-0.4} = \dfrac{\ln 9}{0.4} \approx 5.4931$$

Alternatively, $\dfrac{\ln 9}{0.4} = \dfrac{\ln 9}{\frac{2}{5}} = \dfrac{5\ln 9}{2} \approx 5.4931$

The solution set is $\left\{\dfrac{\ln 9}{0.4}\right\}$ or $\left\{\dfrac{5\ln 9}{2}\right\}$.

Skill Practice 3 Solve.
a. $400 + 10^{4x-1} = 63{,}000$　　**b.** $100 = 700e^{-0.2k}$

Answers
3. a. $\left\{ \dfrac{\log 62{,}600 + 1}{4} \right\}$
b. $\left\{ \dfrac{\ln 7}{0.2} \right\}$　or　$\{5\ln 7\}$

In Example 4, we have an equation with two exponential expressions involving different bases.

Classroom Example: p. 417

Exercise 30

EXAMPLE 4 **Solving an Exponential Equation**

Solve. $4^{2x-7} = 5^{3x+1}$

Solution:

$$4^{2x-7} = 5^{3x+1}$$

$$\ln 4^{2x-7} = \ln 5^{3x+1}$$ Take a logarithm of the same base on both sides.

$$(2x - 7)\ln 4 = (3x + 1)\ln 5$$ Apply the power property of logarithms.

$$2x \ln 4 - 7 \ln 4 = 3x \ln 5 + \ln 5$$ Apply the distributive property.

$$2x \ln 4 - 3x \ln 5 = \ln 5 + 7 \ln 4$$ Collect x terms on one side of the equation.

$$x(2 \ln 4 - 3 \ln 5) = \ln 5 + 7 \ln 4$$ Factor out x on the left.

$$x = \frac{\ln 5 + 7 \ln 4}{2 \ln 4 - 3 \ln 5} \approx -5.5034$$ Divide by $(2 \ln 4 - 3 \ln 5)$.

The solution set is $\left\{ \dfrac{\ln 5 + 7 \ln 4}{2 \ln 4 - 3 \ln 5} \right\}$.   The solution checks in the original equation.

Skill Practice 4 Solve. $3^{5x-6} = 2^{4x+1}$

In Example 5, we look at an exponential equation in quadratic form.

Classroom Example: p. 417

Exercise 32

EXAMPLE 5 **Solving an Exponential Equation in Quadratic Form**

Solve. $e^{2x} + 5e^x - 36 = 0$

Solution:

$$e^{2x} + 5e^x - 36 = 0$$

$$(e^x)^2 + 5(e^x) - 36 = 0$$ Note that $e^{2x} = (e^x)^2$.

$$u^2 + 5u - 36 = 0$$ The equation is in quadratic form. Let $u = e^x$.

$$(u - 4)(u + 9) = 0$$ Factor.

$$u = 4 \quad \text{or} \quad u = -9$$

$$e^x = 4 \quad \text{or} \quad e^x = -9$$ Back substitute. The second equation $e^x = -9$ has no solution.

$$\ln e^x = \ln 4$$

$$x = \ln 4 \approx 1.3863$$ No solution to this equation because $\ln(-9)$ is undefined.

The solution set is $\{\ln 4\}$.   The solution checks in the original equation.

Avoiding Mistakes

Recall that the range of $f(x) = e^x$ is the set of positive real numbers. Therefore, $e^x \neq -9$.

Skill Practice 5 Solve. $e^{2x} - 5e^x - 14 = 0$

Answers

4. $\left\{ \dfrac{\ln 2 + 6 \ln 3}{5 \ln 3 - 4 \ln 2} \right\}$

5. $\{\ln 7\}$

2. Solve Logarithmic Equations

An equation containing a variable within a logarithmic expression is called a **logarithmic equation.** For example:

$$\log_2(3x - 4) = \log_2(x + 2) \quad \text{and} \quad \ln(x + 4) = 7 \quad \text{are logarithmic equations.}$$

Given an equation in which two logarithms of the same base are equated, we can apply the equivalence property of logarithms. Since all logarithmic functions are one-to-one, $\log_b x = \log_b y$ implies that $x = y$.

> **TIP** The equivalence property tells us that if two logarithmic expressions with the same base are equal, then their arguments must be equal.

Equivalence Property of Logarithmic Expressions

If b, x, and y are positive real numbers with $b \neq 1$, then

$$\log_b x = \log_b y \quad \text{implies that } x = y.$$

Classroom Example: p. 417
Exercise 38

EXAMPLE 6 Solving a Logarithmic Equation Using the Equivalence Property

Solve. $\quad \log_2(3x - 4) = \log_2(x + 2)$

Solution:

$\log_2(3x - 4) = \log_2(x + 2)$	Two logarithms of the same base are equated.
$3x - 4 = x + 2$	Equate the arguments.
$2x = 6$	Solve for x.
$x = 3$	Because the domain of a logarithmic function is restricted, it is mandatory that we check all potential solutions to a logarithmic equation.

$$\underline{\text{Check:}} \ \log_2(3x - 4) = \log_2(x + 2)$$
$$\log_2[3(3) - 4] \overset{?}{=} \log_2[(3) + 2]$$

The solution set is $\{3\}$.
$$\log_2 5 \overset{?}{=} \log_2 5 \ \checkmark$$

Skill Practice 6 Solve. $\quad \log_2(7x - 4) = \log_2(2x + 1)$

TECHNOLOGY CONNECTIONS

Using a Calculator to View the Potential Solutions to a Logarithmic Equation

The solution to the equation in Example 6 is the x-coordinate of the point of intersection of $Y_1 = \log_2(3x - 4)$ and $Y_2 = \log_2(x + 2)$. The domain of $Y_1 = \log_2(3x - 4)$ is $\{x \mid x > \frac{4}{3}\}$ and the domain of $Y_2 = \log_2(x + 2)$ is $\{x \mid x > -2\}$. The solution to the equation $Y_1 = Y_2$ may not lie outside the domain of either function. This is why it is mandatory to check all potential solutions to a logarithmic equation.

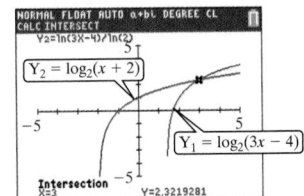

Answer

6. $\{1\}$

In Example 7, we encounter a logarithmic equation in which one or more solutions does not check.

Classroom Example: p. 417
Exercise 56

EXAMPLE 7   Solving a Logarithmic Equation

Solve. $\ln(x - 4) = \ln(x + 6) - \ln x$

Solution:

$$\ln(x - 4) = \ln(x + 6) - \ln x$$

$$\ln(x - 4) = \ln\left(\frac{x + 6}{x}\right)$$ Combine the two logarithmic terms on the right.

$$x - 4 = \frac{x + 6}{x}$$ Apply the equivalence property of logarithms.

$$x^2 - 4x = x + 6$$ Clear fractions by multiplying both sides by x.

$$x^2 - 5x - 6 = 0$$ The resulting equation is quadratic.

$$(x - 6)(x + 1) = 0$$

$$x = 6 \quad \text{or} \quad x = -1$$ The potential solutions are 6 and -1.

Check:

$\ln(x - 4) = \ln(x + 6) - \ln x$	$\ln(x - 4) = \ln(x + 6) - \ln x$
$\ln(6 - 4) \stackrel{?}{=} \ln(6 + 6) - \ln 6$	$\ln(-1 - 4) \stackrel{?}{=} \ln(-1 + 6) - \ln(-1)$
$\ln 2 \stackrel{?}{=} \ln 12 - \ln 6$	$\ln(-5) \stackrel{?}{=} \ln 5 - \ln(-1)$
$\ln 2 \stackrel{?}{=} \ln\left(\frac{12}{6}\right)$ ✓	undefined undefined

The only solution that checks is 6.
The solution set is $\{6\}$.

Skill Practice 7   Solve. $\ln x + \ln(x - 8) = \ln(x - 20)$

Many logarithmic equations, such as $4 \log_3 (2t - 7) = 8$ and $\log_2 x = 3 - \log_2 (x - 2)$, involve logarithmic terms and constant terms. In such a case, we can apply the properties of logarithms to write the equation in the form $\log_b x = k$, where k is a constant. At this point, we can solve for x by writing the equation in its equivalent exponential form $x = b^k$.

Solving Logarithmic Equations by Using Exponential Form

Step 1   Given a logarithmic equation, isolate the logarithms on one side of the equation.

Step 2   Use the properties of logarithms to write the equation in the form $\log_b x = k$, where k is a constant.

Step 3   Write the equation in exponential form.

Step 4   Solve the equation from step 3.

Step 5   Check the potential solution(s) in the original equation.

Answer

7. $\{\ \}$; The values 4 and 5 do not check.

Classroom Example: p. 417
Exercise 42

EXAMPLE 8   Solving a Logarithmic Equation

Solve. $4 \log_3(2t - 7) = 8$

Solution:

$4 \log_3(2t - 7) = 8$

$\log_3(2t - 7) = 2$ Isolate the logarithm by dividing both sides by 4.

  The equation is in the form $\log_b x = k$, where $x = 2t - 7$.

$2t - 7 = 3^2$ Write the equation in exponential form.

$2t - 7 = 9$ Check: $4 \log_3(2t - 7) = 8$

$t = 8$ $4 \log_3[2(8) - 7] \overset{?}{=} 8$

        $4 \log_3 9 \overset{?}{=} 8$

        $4 \cdot 2 \overset{?}{=} 8 \checkmark$

The solution set is $\{8\}$.

Skill Practice 8   Solve. $8 \log_4(w + 6) = 24$

Classroom Example: p. 417
Exercise 46

EXAMPLE 9   Solving a Logarithmic Equation

Solve. $\log(w + 47) = 2.6$

Solution:

$\log(w + 47) = 2.6$ The equation is in the form $\log_b x = k$ where $x = w + 47$ and $b = 10$.

$w + 47 = 10^{2.6}$ Write the equation in exponential form.

$w = 10^{2.6} - 47 \approx 351.1072$ Solve the resulting linear equation.

      Check: $\log(w + 47) = 2.6$

      $\log[(10^{2.6} - 47) + 47] \overset{?}{=} 2.6$

The solution set is $\{10^{2.6} - 47\}$. $\log 10^{2.6} \overset{?}{=} 2.6 \checkmark$

Skill Practice 9   Solve. $\log(t - 18) = 1.4$

Example 10 contains multiple logarithmic terms and a constant term. We apply the strategy of collecting the logarithmic terms on one side and the constant term on the other side. Then after combining the logarithmic terms, we write the equation in exponential form.

Answers
8. $\{58\}$
9. $\{10^{1.4} + 18\}$

Classroom Example: p. 417
Exercise 54

EXAMPLE 10 Solving a Logarithmic Equation

Solve. $\log_2 x = 3 - \log_2(x - 2)$

Solution:

$\log_2 x = 3 - \log_2(x - 2)$

$\log_2 x + \log_2(x - 2) = 3$ Isolate the logarithms on one side of the equation.

$\log_2[x(x - 2)] = 3$ Use the product property of logarithms to write a single logarithm.

$x(x - 2) = 2^3$ Write the equation in exponential form.

$x^2 - 2x = 8$

$x^2 - 2x - 8 = 0$ Set one side equal to zero.

$(x - 4)(x + 2) = 0$

$x = 4 \quad x = -2$ Check:

$\log_2 x = 3 - \log_2(x - 2)$	$\log_2 x = 3 - \log_2(x - 2)$
$\log_2 4 \overset{?}{=} 3 - \log_2(4 - 2)$	$\log_2(-2) \overset{?}{=} 3 - \log_2(-2 - 2)$
$\log_2 4 \overset{?}{=} 3 - \log_2 2$	$\log_2(-2) \overset{?}{=} 3 - \log_2(-4)$
$2 \overset{?}{=} 3 - 1 ✓$	undefined undefined

The only solution that checks is $x = 4$.
The solution set is $\{4\}$.

Skill Practice 10 Solve. $2 - \log_7 x = \log_7(x - 48)$

3. Use Exponential and Logarithmic Equations in Applications

In Examples 11 and 12, we solve applications involving exponential and logarithmic equations.

Classroom Example: p. 417
Exercise 66

EXAMPLE 11 Using an Exponential Equation in a Finance Application

A couple invests $8000 in a bond fund. The expected yield is 4.5% and the earnings are reinvested monthly.

a. Use $A = P\left(1 + \dfrac{r}{n}\right)^{nt}$ to write a model representing the amount A (in $) in the account after t years. The value r is the interest rate and n is the number of times interest is compounded per year.

b. Determine how long it will take the initial investment to double. Round to 1 decimal place.

TIP Recall that monthly compounding indicates that interest is computed $n = 12$ times per year.

Solution:

a. $A = P\left(1 + \dfrac{r}{n}\right)^{nt}$

$A = 8000\left(1 + \dfrac{0.045}{12}\right)^{12t}$ Substitute $P = 8000$, $r = 0.045$, and $n = 12$.

Answer

10. $\{49\}$; The value -1 does not check.

Instructor Note:
Remind students that expressions
such as $\ln\left(1 + \dfrac{0.045}{12}\right)$ may look
daunting, but are simply constants.

b. $16{,}000 = 8000\left(1 + \dfrac{0.045}{12}\right)^{12t}$ The couple wants to double their money from $8000 to $16,000. Substitute $A = 16{,}000$ and solve for t.

$2 = \left(1 + \dfrac{0.045}{12}\right)^{12t}$ Isolate the exponential expression by dividing both sides by 8000.

$\ln 2 = \ln\left(1 + \dfrac{0.045}{12}\right)^{12t}$ Take a logarithm of the same base on both sides. We have chosen to use the natural logarithm.

$\ln 2 = 12t \ln\left(1 + \dfrac{0.045}{12}\right)$ Apply the power property of logarithms. The equation is now linear in the variable t.

$\dfrac{\ln 2}{12 \ln\left(1 + \dfrac{0.045}{12}\right)} = t$ Divide both sides by $12 \ln\left(1 + \frac{0.045}{12}\right)$.

$t \approx 15.4$

It will take approximately 15.4 yr for the investment to double.

Skill Practice 11   Determine how long it will take $8000 compounded monthly at 6% to double. Round to 1 decimal place.

In Example 12, we use a logarithmic equation in an application.

Classroom Example: p. 419
Exercise 80

EXAMPLE 12   Using a Logarithmic Equation in a Medical Application

Suppose that the sound at a rock concert measures 124 dB (decibels).

a. Use the formula $L = 10 \log\left(\frac{I}{I_0}\right)$ to find the intensity of sound I (in W/m^2). The variable L represents the loudness of sound (in dB) and $I_0 = 10^{-12}$ W/m^2.

b. If the threshold at which sounds become painful is 1 W/m^2, will the music at this concert be physically painful? (Ignore the quality of the music.)

Solution:

a. $L = 10 \log\left(\dfrac{I}{I_0}\right)$

$124 = 10 \log\left(\dfrac{I}{10^{-12}}\right)$ Substitute 124 for L and 10^{-12} for I_0.

$12.4 = \log\left(\dfrac{I}{10^{-12}}\right)$ Divide both sides by 10. The logarithm is now isolated.

$10^{12.4} = \dfrac{I}{10^{-12}}$ Write the equation in exponential form.

$10^{12.4} \cdot 10^{-12} = I$ Multiply both sides by 10^{-12}.

$I = 10^{0.4} \approx 2.5$ W/m^2 Simplify.

b. The intensity of sound at the rock concert is approximately 2.5 W/m^2. This is above the threshold for pain.

Skill Practice 12

a. Find the intensity of sound from a leaf blower if the decibel level is 115 dB.

b. Is the intensity of sound from a leaf blower above the threshold for pain?

Answers
11. 11.6 yr
12. a. $10^{-0.5}$ W/m$^2 \approx 0.3$ W/m^2
   b. No

TECHNOLOGY CONNECTIONS

Using a Calculator to Approximate the Solutions to Exponential and Logarithmic Equations

There are many situations in which analytical methods fail to give a solution to a logarithmic or exponential equation. To find solutions graphically,

Enter the left side of the equation as Y_1.

Enter the right side of the equation as Y_2.

Then determine the point(s) of intersection of the graphs.

Example: $4 \ln x - 3x = -8$
$Y_1 = 4 \ln x - 3x$
$Y_2 = -8$ Solutions: $x \approx 0.1516$ and $x \approx 4.7419$

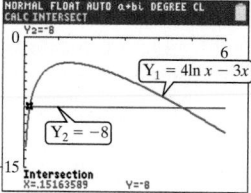 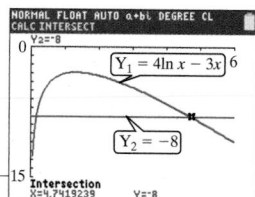

SECTION 3.5   Practice Exercises

Prerequisite Review

For Exercises R.1–R.6, solve the equation.

R.1. $7 = 5 - 2(4q - 1)$ $\{0\}$

R.2. $15n(n + 3) = 14n - 10$ $\left\{ -\dfrac{5}{3}, -\dfrac{2}{5} \right\}$

R.3. $y^2 - 5y - 9 = 0$ $\left\{ \dfrac{5 \pm \sqrt{61}}{2} \right\}$

R.4. $\dfrac{20}{n^2 - 2n} + 5 = \dfrac{10}{n - 2}$ $\{\ \}$; The value 2 does not check.

R.5. $\sqrt{a + 18} + 2 = a$ $\{7\}$; The value -2 does not check.

R.6. $36d^{-2} - 5d^{-1} - 1 = 0$ $\{-9, 4\}$

Concept Connections

1. An equation such as $4^x = 9$ is called an __exponential__ equation because the equation contains a variable in the exponent.

2. The equivalence property of exponential expressions states that if $b^x = b^y$, then ____x____ = ____y____.

3. The equivalence property of logarithmic expressions states that if $\log_b x = \log_b y$, then ____x____ = ____y____.

4. An equation containing a variable within a logarithmic expression is called a __logarithmic__ equation.

Objective 1: Solve Exponential Equations

For Exercises 5–16, solve the equation. (See Example 1)

5. $3^x = 81$ $\{4\}$

6. $2^x = 32$ $\{5\}$

7. $\sqrt[3]{5} = 5^t$ $\left\{ \dfrac{1}{3} \right\}$

8. $\sqrt{3} = 3^w$ $\left\{ \dfrac{1}{2} \right\}$

9. $2^{-3y+1} = 16$ $\{-1\}$

10. $5^{2z+2} = 625$ $\{1\}$

11. $11^{3c+1} = \left(\dfrac{1}{11} \right)^{c-5}$ $\{1\}$

12. $7^{2x-3} = \left(\dfrac{1}{49} \right)^{x+1}$ $\left\{ \dfrac{1}{4} \right\}$

13. $8^{2x-5} = 32^{x-6}$ $\{-15\}$

14. $27^{x-4} = 9^{2x+1}$ $\{-14\}$

15. $100^{3t-5} = 1000^{3-t}$ $\left\{ \dfrac{19}{9} \right\}$

16. $100{,}000^{2w+1} = 10{,}000^{4-w}$ $\left\{ \dfrac{11}{14} \right\}$

For Exercises 17–34, solve the equation. Write the solution set with the exact values given in terms of common or natural logarithms. Also give approximate solutions to 4 decimal places. (See Examples 2–5)

17. $6^t = 87$ $\left\{\dfrac{\ln 87}{\ln 6}\right\}; t \approx 2.4925$

18. $2^z = 70$ $\left\{\dfrac{\ln 70}{\ln 2}\right\}; z \approx 6.1293$

19. $1024 = 19^x + 4$ $\left\{\dfrac{\ln 1020}{\ln 19}\right\}; x \approx 2.3528$

20. $801 = 23^y + 6$ $\left\{\dfrac{\ln 795}{\ln 23}\right\}; y \approx 2.1299$

21. $10^{3+4x} - 8100 = 120{,}000$

22. $10^{5+8x} + 4200 = 84{,}000$

23. $21{,}000 = 63{,}000e^{-0.2t}$

24. $80 = 320e^{-0.5t}$

25. $4e^{2n-5} + 3 = 11$ $\left\{\dfrac{5 + \ln 2}{2}\right\}; n \approx 2.8466$

26. $5e^{4m-3} - 7 = 13$

27. $3^{6x+5} = 5^{2x}$

28. $7^{4x-1} = 3^{5x}$

29. $2^{1-6x} = 7^{3x+4}$

30. $11^{1-8x} = 9^{2x+3}$

31. $e^{2x} - 9e^x - 22 = 0$ $\{\ln 11\}; x \approx 2.3979$

32. $e^{2x} - 6e^x - 16 = 0$
$\{\ln 8\}; x \approx 2.0794$

33. $e^{2x} = -9e^x$ $\{\ \}$

34. $e^{2x} = -7e^x$ $\{\ \}$

Objective 2: Solve Logarithmic Equations

For Exercises 35–36, determine if the given value of x is a solution to the logarithmic equation.

35. $\log_2(x - 31) = 5 - \log_2 x$

   a. $x = 16$ No

   b. $x = 32$ Yes

   c. $x = -1$ No

36. $\log_4 x = 3 - \log_4(x - 63)$

   a. $x = 64$ Yes

   b. $x = -1$ No

   c. $x = 32$ No

For Exercises 37–60, solve the equation. Write the solution set with the exact solutions. Also give approximate solutions to 4 decimal places if necessary. (See Examples 6–10)

37. $\log_4(3w + 11) = \log_4(3 - w)$
$\{-2\}$

38. $\log_7(12 - t) = \log_7(t + 6)$
$\{3\}$

39. $\log(x^2 + 7x) = \log 18$
$\{-9, 2\}$

40. $\log(p^2 + 6p) = \log 7$
$\{-7, 1\}$

41. $6 \log_5(4p - 3) - 2 = 16$
$\{32\}$

42. $5 \log_6(7w + 1) + 3 = 13$
$\{5\}$

43. $2 \log_8(3y - 5) + 20 = 24$
$\{23\}$

44. $5 \log_3(7 - 5z) + 2 = 17$
$\{-4\}$

45. $\log(p + 17) = 4.1$
$\{10^{4.1} - 17\}; p \approx 12{,}572.2541$

46. $\log(q - 6) = 3.5$
$\{10^{3.5} + 6\}; q \approx 3168.2777$

47. $2 \ln(4 - 3t) + 1 = 7$

48. $4 \ln(6 - 5t) + 2 = 22$

49. $\log_2 w - 3 = -\log_2(w + 2)$
$\{2\}$; The value -4 does not check.

50. $\log_3 y + \log_3(y + 6) = 3$
$\{3\}$; The value -9 does not check.

51. $\log_6(7x - 2) = 1 + \log_6(x + 5)$
$\{32\}$

52. $\log_4(5x - 13) = 1 + \log_4(x - 2)$
$\{5\}$

53. $\log_5 z = 3 - \log_5(z - 20)$
$\{25\}$; The value -5 does not check.

54. $\log_2 x = 4 - \log_2(x - 6)$
$\{8\}$; The value -2 does not check.

55. $\ln x + \ln(x - 4) = \ln(3x - 10)$
$\{5\}$; The value 2 does not check.

56. $\ln x + \ln(x - 3) = \ln(5x - 7)$
$\{7\}$; The value 1 does not check.

57. $\log x + \log(x - 7) = \log(x - 15)$
$\{\ \}$; The values 5 and 3 do not check.

58. $\log x + \log(x - 10) = \log(x - 18)$
$\{\ \}$; The values 2 and 9 do not check.

59. $\log_8(6 - m) + \log_8(-m - 1) = 1$
$\{-2\}$; The value 7 does not check.

60. $\log_3(n - 5) + \log_3(n + 3) = 2$
$\{6\}$; The value -4 does not check.

Objective 3: Use Exponential and Logarithmic Equations in Applications

For Exercises 61–70, use the model $A = Pe^{rt}$ or $A = P\left(1 + \dfrac{r}{n}\right)^{nt}$, where A is the future value of P dollars invested at interest rate r compounded continuously or n times per year for t years. (See Example 11)

61. If \$10,000 is invested in an account earning 5.5% interest compounded continuously, determine how long it will take the money to triple. Round to the nearest year.   20 yr

62. If a couple has \$80,000 in a retirement account, how long will it take the money to grow to \$1,000,000 if it grows by 6% compounded continuously? Round to the nearest year.   42 yr

63. A \$2500 bond grows to \$3729.56 in 10 yr under continuous compounding. Find the interest rate. Round to the nearest whole percent.   4%

64. \$5000 grows to \$5438.10 in 2 yr under continuous compounding. Find the interest rate. Round to the nearest tenth of a percent.   4.2%

65. An \$8000 investment grows to \$9289.50 at 3% interest compounded quarterly. For how long was the money invested? Round to the nearest year.   5 yr

66. \$20,000 is invested at 3.5% interest compounded monthly. How long will it take for the investment to double? Round to the nearest tenth of a year.   19.8 yr

67. A \$25,000 inheritance is invested for 15 yr compounded quarterly and grows to \$52,680. Find the interest rate. Round to the nearest percent.   5%

68. A \$10,000 investment grows to \$11,273 in 4 yr compounded monthly. Find the interest rate. Round to the nearest percent.   3%

69. If \$4000 is put aside in a money market account with interest compounded continuously at 2.2%, find the time required for the account to *earn* \$1000. Round to the nearest month.   10 yr, 2 months

70. Victor puts aside \$10,000 in an account with interest compounded continuously at 2.7%. How long will it take for him to *earn* \$2000? Round to the nearest month.   6 yr, 9 months

71. Physicians often treat thyroid cancer with a radioactive form of iodine called iodine-131 (^{131}I). The radiological half-life of ^{131}I is approximately 8 days, but the biological half-life for most individuals is 4.2 days. The biological half-life is shorter because in addition to ^{131}I being lost to decay, the iodine is also excreted from the body in urine, sweat, and saliva.

For a patient treated with 100 mCi (millicuries) of ^{131}I, the radioactivity level R (in mCi) after t days is given by $R = 100(2)^{-t/4.2}$.

a. State law mandates that the patient stay in an isolated hospital room for 2 days after treatment with ^{131}I. Determine the radioactivity level at the end of 2 days. Round to the nearest whole unit.   72 mCi

b. After the patient is released from the hospital, the patient is directed to avoid direct human contact until the radioactivity level drops below 30 mCi. For how many days *after* leaving the hospital will the patient need to stay in isolation? Round to the nearest tenth of a day.   5.3 days

72. Caffeine occurs naturally in a variety of food products such as coffee, tea, and chocolate. The kidneys filter the blood and remove caffeine and other drugs through urine. The biological half-life of caffeine is approximately 6 hr. If one cup of coffee has 80 mg of caffeine, then the amount of caffeine C (in mg) remaining after t hours is given by $C = 80(2)^{-t/6}$.

a. How long will it take for the amount of caffeine to drop below 60 mg? Round to 1 decimal place.   2.5 hr

b. Laura has trouble sleeping if she has more than 30 mg of caffeine in her bloodstream. How many hours after drinking a cup of coffee would Laura have to wait so that the coffee would not disrupt her sleep? Round to 1 decimal place.   8.5 hr

Sunlight is absorbed in water, and as a result the light intensity in oceans, lakes, and ponds decreases exponentially with depth. The percentage of visible light, P (in decimal form), at a depth of x meters is given by $P = e^{-kx}$, where k is a constant related to the clarity and other physical properties of the water. The graph shows models for the open ocean, Lake Tahoe, and Lake Erie for data taken under similar conditions. Use these models for Exercises 73–76.

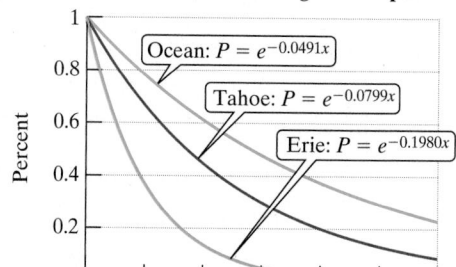

Percent of Surface Light vs. Depth

Ocean: $P = e^{-0.0491x}$

Tahoe: $P = e^{-0.0799x}$

Erie: $P = e^{-0.1980x}$

73. Determine the depth at which the light intensity is half the value from the surface for each body of water given. Round to the nearest tenth of a meter.
   Ocean: 14.1 m; Tahoe: 8.7 m; Erie: 3.5 m

75. The *euphotic* depth is the depth at which light intensity falls to 1% of the value at the surface. This depth is of interest to scientists because no appreciable photosynthesis takes place. Find the euphotic depth for the open ocean. Round to the nearest tenth of a meter.   Ocean: 93.8 m

77. Forge welding is a process in which two pieces of steel are joined together by heating the pieces of steel and hammering them together. A welder takes a piece of steel from a forge at 1600°F and places it on an anvil where the outdoor temperature is 50°F. The temperature of the steel T (in °F) can be modeled by $T = 50 + 1550e^{-0.05t}$, where t is the time in minutes after the steel is removed from the forge. How long will it take for the steel to reach a temperature of 100°F so that it can be handled without heat protection? Round to the nearest minute.
   69 min (1 hr 9 min)

74. Determine the depth at which the light intensity is 20% of the value from the surface for each body of water given. Round to the nearest tenth of a meter.
   Ocean: 32.8 m; Tahoe: 20.1 m; Erie: 8.1 m

76. Refer to Exercise 75, and find the euphotic depth for Lake Tahoe and for Lake Erie. Round to the nearest tenth of a meter.   Tahoe: 57.6 m; Erie: 23.3 m

78. A pie comes out of the oven at 325°F and is placed to cool in a 70°F kitchen. The temperature of the pie T (in °F) after t minutes is given by $T = 70 + 255e^{-0.017t}$. The pie is cool enough to cut when the temperature reaches 110°F. How long will this take? Round to the nearest minute.   109 min (1 hr 49 min)

For Exercises 79–80, the formula $L = 10 \log\left(\frac{I}{I_0}\right)$ gives the loudness of sound L (in dB) based on the intensity of sound I (in W/m²). The value $I_0 = 10^{-12}$ W/m² is the minimal threshold for hearing for midfrequency sounds. Hearing impairment is often measured according to the minimal sound level (in dB) detected by an individual for sounds at various frequencies. For one frequency, the table depicts the level of hearing impairment.

Category	Loudness (dB)
Mild	$26 \leq L \leq 40$
Moderate	$41 \leq L \leq 55$
Moderately severe	$56 \leq L \leq 70$
Severe	$71 \leq L \leq 90$
Profound	$L > 90$

79. **a.** If the minimum intensity heard by an individual is 3.4×10^{-8} W/m², determine if the individual has a hearing impairment. 3.4×10^{-8} W/m² corresponds to 45.3 dB which indicates a moderate hearing impairment.
   b. If the minimum loudness of sound detected by an individual is 30 dB, determine the corresponding intensity of sound. (**See Example 12**) 10^{-9} W/m²

80. Determine the range that represents the intensity of sound that can be heard by an individual with severe hearing impairment. $1.26 \times 10^{-5} \leq I \leq 1 \times 10^{-3}$ W/m²

For Exercises 81–82, use the formula pH $= -\log[\text{H}^+]$. The variable pH represents the level of acidity or alkalinity of a liquid on the pH scale, and H^+ is the concentration of hydronium ions in the solution. Determine the value of H^+ (in mol/L) for the following liquids, given their pH values.

81. **a.** Seawater pH $= 8.5$ 3.16×10^{-9} mol/L
   b. Acid rain pH $= 2.3$ 5.01×10^{-3} mol/L

82. **a.** Milk pH $= 6.2$ 6.31×10^{-7} mol/L
   b. Sodium bicarbonate pH $= 8.4$ 3.98×10^{-9} mol/L

83. A new teaching method to teach vocabulary to sixth-graders involves having students work in groups on an assignment to learn new words. After the lesson was completed, the students were tested at 1-month intervals. The average score for the class $S(t)$ can be modeled by
$$S(t) = 94 - 18 \ln(t + 1)$$
where t is the time in months after completing the assignment. If the average score is 65, how many months had passed since the students completed the assignment? Round to the nearest month. 4 months

84. A company spends x hundred dollars on an advertising campaign. The amount of money in sales $S(x)$ (in \$1000) for the 4-month period after the advertising campaign can be modeled by
$$S(x) = 5 + 7 \ln(x + 1)$$
If the sales total \$19,100, how much was spent on advertising? Round to the nearest dollar. \$650

85. Radiated seismic energy from an earthquake is estimated by $\log E = 4.4 + 1.5M$, where E is the energy in Joules (J) and M is surface wave magnitude.

   a. How many times more energy does an 8.2-magnitude earthquake have than a 5.5-magnitude earthquake? Round to the nearest thousand. 11,000

   b. How many times more energy does a 7-magnitude earthquake have than a 6-magnitude earthquake? Round to the nearest whole number. 32

86. On August 31, 1854, an epidemic of cholera was discovered in London, England, resulting from a contaminated community water pump. By the end of September, more than 600 citizens who drank water from the pump had died. The cumulative number of deaths $D(t)$ at a time t days after August 31 is given by $D(t) = 91 + 160 \ln(t + 1)$.

   a. Determine the cumulative number of deaths by September 15. Round to the nearest whole unit. 535

   b. Approximately how many days after August 31 did the cumulative number of deaths reach 600? 23 days

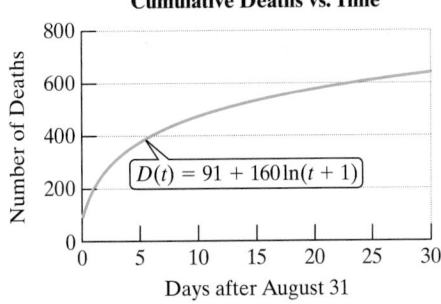

Cumulative Deaths vs. Time

$D(t) = 91 + 160\ln(t + 1)$

Mixed Exercises

For Exercises 87–94, find an equation for the inverse function.

87. $f(x) = 2^x - 7$ $f^{-1}(x) = \log_2(x + 7)$

88. $f(x) = 5^x + 6$ $f^{-1}(x) = \log_5(x - 6)$

89. $f(x) = \ln(x + 5)$ $f^{-1}(x) = e^x - 5$

90. $f(x) = \ln(x - 7)$ $f^{-1}(x) = e^x + 7$

91. $f(x) = 10^{x-3} + 1$ $f^{-1}(x) = \log(x - 1) + 3$

92. $f(x) = 10^{x+2} - 4$ $f^{-1}(x) = \log(x + 4) - 2$

93. $f(x) = \log(x + 7) - 9$ $f^{-1}(x) = 10^{x+9} - 7$

94. $f(x) = \log(x - 11) + 8$ $f^{-1}(x) = 10^{x-8} + 11$

For Exercises 95–112, solve the equation. Write the solution set with exact solutions. Also give approximate solutions to 4 decimal places if necessary.

95. $5^{|x|} - 3 = 122$ $\{-3, 3\}$

96. $11^{|x|} + 9 = 130$ $\{-2, 2\}$

97. $\log x - 2 \log 3 = 2$ $\{900\}$

98. $\log y - 3 \log 5 = 3$ $\{125,000\}$

99. $6^{x^2-2} = 36$ $\{2, -2\}$

100. $8^{y^2-7} = 64$ $\{3, -3\}$

101. $\log_9 |x + 4| = \log_9 6$ $\{-10, 2\}$

102. $\log_8 |3 - x| = \log_8 5$ $\{-2, 8\}$

103. $x^2 e^x = 9e^x$ $\{3, -3\}$

104. $x^2 6^x = 6^x$ $\{1, -1\}$

105. $\log_3(\log_3 x) = 0$ $\{3\}$

106. $\log_5(\log_5 x) = 1$ $\{3125\}$

107. $3|\ln x| - 12 = 0$

108. $7|\ln x| - 14 = 0$

109. $\log_3 x - \log_3(2x + 6) = \dfrac{1}{2}\log_3 4$

$\{\ \}$; The value -4 does not check.

110. $\log_5 x - \log_5(x + 1) = \dfrac{1}{3}\log_5 8$

$\{\ \}$; The value -2 does not check.

111. $2e^x(e^x - 3) = 3e^x - 4$

$\left\{\ln\dfrac{1}{2}, \ln 4\right\}$; $x \approx -0.6931$, $x \approx 1.3863$

112. $3e^x(e^x - 6) = 4e^x - 7$

$\left\{\ln\dfrac{1}{3}, \ln 7\right\}$; $x \approx -1.0986$, $x \approx 1.9459$

Write About It

113. Explain the process to solve the equation $4^x = 11$.

114. Explain the process to solve the equation $\log_b 5 + \log_b(x - 3) = 4$.

Expanding Your Skills

For Exercises 115–126, solve the equation.

115. $\dfrac{10^x - 13 \cdot 10^{-x}}{3} = 4$ $\{\log 13\}$; $x \approx 1.1139$

116. $\dfrac{e^x - 9e^{-x}}{2} = 4$ $\{\ln 9\}$; $x \approx 2.1972$

117. $(\ln x)^2 - \ln x^5 = -4$ $\{e, e^4\}$; $x \approx 2.7183$, $x \approx 54.5982$

118. $(\ln x)^2 + \ln x^3 = -2$

119. $(\log x)^2 = \log x^2$ $\{100, 1\}$

120. $(\log x)^2 = \log x^3$ $\{1000, 1\}$

121. $\log w + 4\sqrt{\log w} - 12 = 0$ $\{10,000\}$

122. $\ln x + 3\sqrt{\ln x} - 10 = 0$ $\{e^4\}$; $x \approx 54.5982$

123. $e^{2x} - 8e^x + 6 = 0$ $\{\ln(4 \pm \sqrt{10})\}$; $x \approx 1.9688$, $x \approx -0.1771$

124. $e^{2x} - 6e^x + 4 = 0$ $\{\ln(3 \pm \sqrt{5})\}$; $x \approx 1.6556$, $x \approx -0.2693$

125. $\log_5 \sqrt{6c + 5} + \log_5 \sqrt{c} = 1$ $\left\{\dfrac{5}{3}\right\}$; The value $-\dfrac{5}{2}$ does not check.

126. $\log_3 \sqrt{x - 8} + \log_3 \sqrt{x} = 1$ $\{9\}$; The value -1 does not check.

Technology Connections

For Exercises 127–130, an equation is given in the form $Y_1(x) = Y_2(x)$. Graph Y_1 and Y_2 on a graphing utility on the window $[10, 10, 1]$ by $[10, 10, 1]$. Then approximate the point(s) of intersection to approximate the solution(s) to the equation. Round to 4 decimal places.

127. $4x - e^x + 6 = 0$ $\{-1.4408, 2.8584\}$

128. $x^3 - e^{2x} + 4 = 0$ $\{-1.5818, 0.7417\}$

129. $x^2 + 5 \log x = 6$ $\{2.0960\}$

130. $x^2 - 0.05 \ln x = 4$ $\{2.0087\}$

Additional answers can be found in the Instructor Answer Appendix.

SECTION 3.6 Modeling with Exponential and Logarithmic Functions

OBJECTIVES

1. Solve Literal Equations for a Specified Variable
2. Create Models for Exponential Growth and Decay
3. Apply Logistic Growth Models
4. Create Exponential and Logarithmic Models Using Regression

1. Solve Literal Equations for a Specified Variable

A short-term model to predict the U.S. population P is $P = 310e^{0.00965t}$, where t is the number of years since 2010. If we solve this equation for t, we have

$$t = \frac{\ln\left(\dfrac{P}{310}\right)}{0.00965} \quad \text{or equivalently} \quad t = \frac{\ln P - \ln 310}{0.00965}.$$

This is a model that predicts the time required for the U.S. population to reach a value P. Manipulating an equation for a specified variable was first introduced in Section R.5. In Example 1, we revisit this skill using exponential and logarithmic equations.

Classroom Examples: p. 430
Exercises 6, 10

> **EXAMPLE 1**   Solving an Equation for a Specified Variable
>
> **a.** Given $P = 100e^{kx} - 100$, solve for x. (Used in geology)
> **b.** Given $L = 8.8 + 5.1 \log D$, solve for D. (Used in astronomy)
>
> **Solution:**
>
> **a.** $P = 100e^{kx} - 100$
>
> $$P + 100 = 100e^{kx} \qquad \text{Add 100 to both sides to isolate the } x \text{ term.}$$
>
> $$\frac{P + 100}{100} = e^{kx} \qquad \text{Divide by 100.}$$
>
> $$\ln\left(\frac{P + 100}{100}\right) = \ln e^{kx} \qquad \text{Take the natural logarithm of both sides.}$$
>
> $$\ln\left(\frac{P + 100}{100}\right) = kx \qquad \text{Simplify: } \ln e^{kx} = kx$$
>
> $$x = \frac{\ln\left(\dfrac{P + 100}{100}\right)}{k} \qquad \text{Divide by } k.$$
>
> $$x = \frac{\ln\left(\dfrac{P + 100}{100}\right)}{k} \quad \text{or equivalently } x = \frac{\ln(P + 100) - \ln 100}{k}$$
>
> **b.** $L = 8.8 + 5.1 \log D$
>
> $$\frac{L - 8.8}{5.1} = \log D \qquad \text{Subtract 8.8 from both sides and divide by 5.1.}$$
>
> $$D = 10^{(L-8.8)/5.1} \qquad \text{Write the equation in exponential form.}$$

> **Skill Practice 1**
>
> **a.** Given $T = 78 + 272e^{-kt}$, solve for k.
> **b.** Given $S = 90 - 20 \ln(t + 1)$, solve for t.

2. Create Models for Exponential Growth and Decay

In Section 3.2, we defined an exponential function as $y = b^x$, where $b > 0$ and $b \neq 1$. Throughout the chapter, we have used transformations of basic exponential functions to solve a variety of applications. The following variation of the general exponential form is used to solve applications involving exponential growth and decay.

Answers

1. a. $k = -\dfrac{\ln\left(\dfrac{T - 78}{272}\right)}{t}$ or

$\quad k = \dfrac{\ln 272 - \ln(T - 78)}{t}$

b. $e^{(90-S)/20} - 1$

Exponential Growth and Decay Models

Let y be a variable changing exponentially with respect to t, and let y_0 represent the initial value of y when $t = 0$. Then for a constant k:

If $k > 0$, then $y = y_0 e^{kt}$ is a model for exponential growth.	If $k < 0$, then $y = y_0 e^{kt}$ is a model for exponential decay.
Example:	**Example:**
$y = 2000e^{0.06t}$ represents the value of a \$2000 investment after t years with interest compounded continuously.	$y = 100e^{-0.165t}$ represents the radioactivity level t hours after a patient is treated for thyroid cancer with 100 mCi of radioactive iodine.
(*Note:* $k = 0.06 > 0$)	(*Note:* $k = -0.165 < 0$)

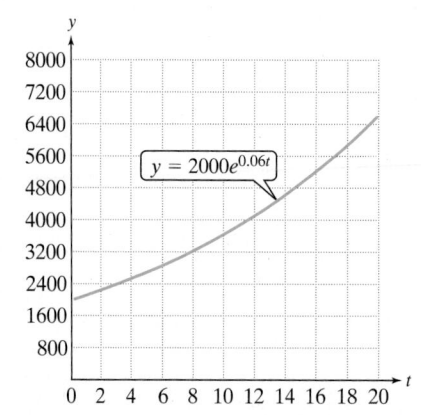
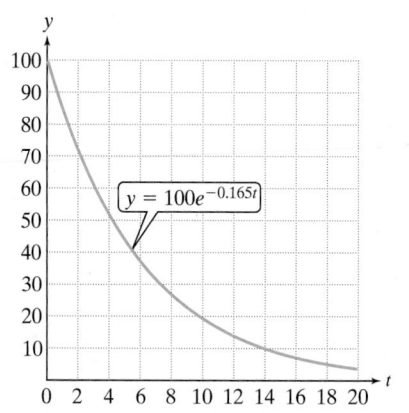

The model $y = y_0 e^{kt}$ is often presented with different letters or symbols in place of y, y_0, k, and t to convey their meaning in the context of the application. For example, to compute the value of an investment under continuous compounding, we have

$$A = Pe^{rt}$$

P (for principal) is used in place of y_0.
r (for the annual interest rate) is used in place of k.
A (for the future value of the investment) is used in place of y.

We can also use function notation when expressing a model for exponential growth or decay. For example, consider the model for population growth.

$$P(t) = P_0 e^{kt}$$

P_0 (for initial population) is used in place of y_0.
$P(t)$ represents the population as a function of time and is used in place of y.

Classroom Example: p. 430
Exercise 16

EXAMPLE 2 Creating a Model for Growth of an Investment

Suppose that \$15,000 is invested and at the end of 3 yr, the value of the account is \$19,356.92. Use the model $A = Pe^{rt}$ to determine the average rate of return r under continuous compounding.

Solution:

$$A = Pe^{rt}$$ Begin with an appropriate model.

$$A = 15{,}000e^{rt}$$ P represents the initial value of the account (initial principal). Substitute 15,000 for P.

$$19{,}356.92 = 15{,}000e^{r(3)}$$ We have a known data point where $A = 19{,}356.92$ when $t = 3$. Substituting these values into the formula enables us to solve for r.

$$\frac{19{,}356.92}{15{,}000} = e^{3r}$$ Divide both sides by 15,000.

$$\ln\left(\frac{19{,}356.92}{15{,}000}\right) = \ln(e^{3r})$$ Take the natural logarithm of both sides.

$$\ln\left(\frac{19{,}356.92}{15{,}000}\right) = 3r$$ Simplify: $\ln e^{3r} = 3r$

$$r = \frac{\ln\left(\frac{19{,}356.92}{15{,}000}\right)}{3}$$ Divide by 3 to isolate r.

$$r \approx 0.085$$

The average rate of return is approximately 8.5%.

Skill Practice 2 Suppose that $10,000 is invested and at the end of 5 yr, the value of the account is $13,771.28. Use the model $A = Pe^{rt}$ to determine the average rate of return r under continuous compounding.

In Example 3, we build a model to predict short-term population growth.

EXAMPLE 3 **Creating a Model for Population Growth**

Classroom Example: p. 431
Exercise 20

On January 1, 2000, the population of California was approximately 34 million. On January 1, 2010, the population was 37.3 million.

a. Write a function of the form $P(t) = P_0e^{kt}$ to represent the population of California $P(t)$ (in millions), t years after January 1, 2000. Round k to 5 decimal places.

b. Use the function in part (a) to predict the population on January 1, 2018. Round to 1 decimal place.

c. Use the function from part (a) to determine the year during which the population of California will be twice the value from the year 2000.

Solution:

TIP The value of k in the model $P(t) = P_0e^{kt}$ is called a parameter and is related to the growth rate of the population being studied. The value of k will be different for different populations.

a. $$P(t) = P_0e^{kt}$$ Begin with an appropriate model.

$$P(t) = 34e^{kt}$$ The initial population is $P_0 = 34$ million.

$$37.3 = 34e^{k(10)}$$ We have a known data point $P(10) = 37.3$. Substituting these values into the function enables us to solve for k.

$$\frac{37.3}{34} = e^{k(10)}$$ Divide both sides by 34.

$$\ln\left(\frac{37.3}{34}\right) = 10k$$ Take the natural logarithm of both sides.

$$k = \frac{\ln\left(\frac{37.3}{34}\right)}{10} \approx 0.00926$$ Divide by 10 to isolate k.

$$P(t) = 34e^{0.00926t}$$ This model gives the population as a function of time.

Answer

2. 6.4%

b. $P(t) = 34e^{0.00926t}$

$P(18) = 34e^{0.00926(18)}$ Substitute 18 for t.

       $= 40.2$

The population in California on January 1, 2018, will be approximately 40.2 million if this trend continues.

c. $P(t) = 34e^{0.00926t}$

$68 = 34e^{0.00926t}$ Substitute 68 for $P(t)$.

$\dfrac{68}{34} = e^{0.00926t}$ Divide both sides by 34.

$\ln 2 = 0.00926t$ Take the natural logarithm of both sides.

$t = \dfrac{\ln 2}{0.00926} \approx 74.85$ Divide by 0.00926 to isolate t.

The population of California will reach 68 million toward the end of the year 2074 if this trend continues.

Skill Practice 3   On January 1, 2000, the population of Texas was approximately 21 million. On January 1, 2010, the population was 25.2 million.

   a. Write a function of the form $P(t) = P_0 e^{kt}$ to represent the population $P(t)$ of Texas t years after January 1, 2000. Round k to 5 decimal places.

   b. Use the function in part (a) to predict the population on January 1, 2020. Round to 1 decimal place.

   c. Use the function in part (a) to determine the year during which the population of Texas will reach 40 million if this trend continues.

An exponential model can be presented with a base other than base e. For example, suppose that a culture of bacteria begins with 5000 organisms and the population doubles every 4 hr. Then the population $P(t)$ can be modeled by

    $P(t) = 5000(2)^{t/4}$, where t is the time in hours after the culture was started.

Notice that this function is defined using base 2. It is important to realize that any exponential function of one base can be rewritten in terms of an exponential function of another base. In particular we are interested in expressing the function with base e.

Writing an Exponential Expression Using Base e

Let t and b be real numbers, where $b > 0$ and $b \neq 1$. Then,

    b^t is equivalent to $e^{(\ln b)t}$.

To show that $e^{(\ln b)t} = b^t$, use the power property of exponents; that is,

$$e^{(\ln b)t} = (e^{\ln b})^t = b^t$$

Classroom Example: p. 431
Exercise 22

Answers

3. a. $P(t) = 21e^{0.01823t}$
   b. 30.2 million
   c. 2035

EXAMPLE 4   **Writing an Exponential Function with Base e**

a. The population $P(t)$ of a culture of bacteria is given by $P(t) = 5000(2)^{t/4}$, where t is the time in hours after the culture was started. Write the rule for this function using base e.

b. Find the population after 12 hr using both forms of the function from part (a).

Solution:

a. $P(t) = 5000(2)^{t/4}$

Note that $2^{t/4} = (2^t)^{1/4}$

$\qquad\qquad = \left[e^{(\ln 2)t}\right]^{1/4}$ Apply the property that $e^{(\ln b)t} = b^t$.

$\qquad\qquad = e^{[(\ln 2)/4]t}$ Apply the power rule of exponents.

Therefore, $P(t) = 5000(2)^{t/4}$

$\qquad\qquad\quad = 5000e^{[(\ln 2)/4]t}$

$\qquad\qquad\quad \approx 5000e^{0.17329t}$

b. $P(t) = 5000(2)^{t/4}$ $P(t) \approx 5000e^{0.17329t}$

$\quad P(12) = 5000(2)^{(12)/4}$ $P(12) \approx 5000e^{0.17329(12)}$

$\qquad\qquad = 40{,}000$ $\approx 40{,}000$

Skill Practice 4

a. Given $P(t) = 10{,}000(2)^{-0.4t}$, write the rule for this function using base e.

b. Find the function value for $t = 10$ for both forms of the function from part (a).

In Example 5, we apply an exponential decay function to determine the age of a bone through radiocarbon dating. Animals ingest carbon through respiration and through the food they eat. Most of the carbon is carbon-12 (^{12}C), an abundant and stable form of carbon. However, a small percentage of carbon is the radioactive isotope, carbon-14 (^{14}C). The ratio of carbon-12 to carbon-14 is constant for all living things. When an organism dies, it no longer takes in carbon from the environment. Therefore, as the carbon-14 decays, the ratio of carbon-12 to carbon-14 changes. Scientists know that the half-life of ^{14}C is 5730 years and from this, they can build a model to represent the amount of ^{14}C remaining t years after death. This is illustrated in Example 5.

Classroom Example: p. 431
Exercise 24

EXAMPLE 5   Creating a Model for Exponential Decay

a. Carbon-14 has a half-life of 5730 yr. Write a model of the form $Q(t) = Q_0 e^{-kt}$ to represent the amount $Q(t)$ of carbon-14 remaining after t years if no additional carbon is ingested.

b. An archeologist uncovers human remains at an ancient Roman burial site and finds that 76.6% of the carbon-14 still remains in the bone. How old is the bone? Round to the nearest hundred years.

> **TIP** Given the half-life of a radioactive substance, we can also write an exponential model using base $\frac{1}{2}$. The format is
>
> $$Q(t) = Q_0\left(\frac{1}{2}\right)^{t/h}$$
>
> where h is the half-life of the substance.
>
> In Example 5, we have
>
> $$Q(t) = Q_0\left(\frac{1}{2}\right)^{t/5730}$$

Solution:

a. $Q(t) = Q_0 e^{-kt}$ Begin with a general exponential decay model.

$\quad 0.5Q_0 = Q_0 e^{-k(5730)}$ Substitute the known data value. One-half of the original quantity Q_0 is present after 5730 yr.

$\qquad 0.5 = e^{-k(5730)}$ Divide by Q_0 on both sides.

$\quad \ln 0.5 = -5730k$ Take the natural logarithm of both sides.

$\qquad\quad k = \dfrac{\ln 0.5}{-5730}$ Divide by -5730.

$\qquad\qquad \approx 0.000121$

$\quad Q(t) = Q_0 e^{-0.000121t}$

Answers

4. a. $P(t) = 10{,}000e^{-0.27726t}$

    b. 625

b. $0.766Q_0 = Q_0e^{-0.000121t}$ 　　　The quantity $Q(t)$ of carbon-14 in the bone is 76.6% of Q_0.

$0.766 = e^{-0.000121t}$ 　　　Divide by Q_0 on both sides.

$\ln 0.766 = -0.000121t$ 　　　Take the natural logarithm of both sides.

$t = \dfrac{\ln 0.766}{-0.000121} \approx 2200$ 　　　Divide by -0.000121 to isolate t.

The bone is approximately 2200 years old.

Skill Practice 5   Use the function $Q(t) = Q_0e^{-0.000121t}$ to determine the age of a piece of wood that has 42% of its carbon-14 remaining. Round to the nearest 10 yr.

3. Apply Logistic Growth Models

In Examples 3 and 4, we used a model of the form $P(t) = P_0e^{kt}$ to predict population as an exponential function of time. However, unbounded population growth is not possible due to limited resources. A growth model that addresses this problem is called logistic growth. In particular, a logistic growth model imposes a limiting value on the dependent variable.

Logistic Growth Model

A logistic growth model is a function written in the form

$$y = \frac{c}{1 + ae^{-bt}}$$

where a, b, and c are positive constants.

The general logistic growth equation can be written with a complex fraction.

$$y = \frac{c}{1 + \dfrac{a}{e^{bt}}}$$ This term approaches 0 as t approaches ∞.

In this form, we can see that for large values of t, the term $\dfrac{a}{e^{bt}}$ approaches 0, and the function value y approaches $\frac{c}{1}$.

The line $y = c$ is a horizontal asymptote of the graph, and c represents the limiting value of the function (Figure 3-20).

Notice that the graph of a logistic curve is increasing over its entire domain. However, the *rate* of increase begins to decrease as the function levels off and approaches the horizontal asymptote $y = c$.

In Example 3 we created a function to approximate the population of California assuming unlimited growth. In Example 6, we use a logistic growth model.

TIP The rate of increase of a logistic curve changes from increasing to decreasing to the left and right of a point called the *point of inflection*.

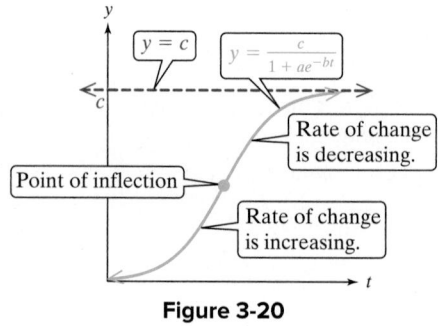

Figure 3-20

Classroom Example: p. 433
Exercise 32

EXAMPLE 6 Using Logistic Growth to Model Population

The population of California $P(t)$ (in millions) can be approximated by the logistic growth function

$$P(t) = \frac{95.2}{1 + 1.8e^{-0.018t}}, \text{ where } t \text{ is the number of years since the year 2000.}$$

a. Determine the population in the year 2000.

b. Use this function to determine the time required for the population of California to double from its value in 2000. Compare this with the result from Example 3(c).

c. What is the limiting value of the population of California under this model?

Solution:

a. $P(t) = \dfrac{95.2}{1 + 1.8e^{-0.018t}}$

$$P(0) = \frac{95.2}{1 + 1.8e^{-0.018(0)}} = \frac{95.2}{1 + 1.8(1)} = 34 \qquad \text{Substitute 0 for } t. \text{ Recall that } e^0 = 1.$$

The population was approximately 34 million in the year 2000.

b. $68 = \dfrac{95.2}{1 + 1.8e^{-0.018t}}$ Substitute 68 for $P(t)$.

$68(1 + 1.8e^{-0.018t}) = 95.2$ Multiply both sides by $(1 + 1.8e^{-0.018t})$.

$1 + 1.8e^{-0.018t} = 1.4$ Divide by 68 on both sides.

$1.8e^{-0.018t} = 1.4 - 1$ Subtract 1 from both sides.

$e^{-0.018t} = \dfrac{0.4}{1.8}$ Divide by 1.8 on both sides.

$-0.018t = \ln\left(\dfrac{0.4}{1.8}\right)$ Take the natural logarithm of both sides.

$t = \dfrac{\ln\left(\dfrac{0.4}{1.8}\right)}{-0.018} \approx 83.6$ Divide by -0.018 on both sides.

The population will double in approximately 83.6 yr. This is 9 yr later than the predicted value from Example 3(c).

The graphs of $P(t) = \dfrac{95.2}{1 + 1.8e^{-0.018t}}$ and $P(t) = 34e^{0.00926t}$ are shown in

Figure 3-21. Notice that the two models agree relatively closely for short-term population growth (out to about 2060). However, in the long term, the unbounded exponential model breaks down. The logistic growth model approaches a limiting population, which is reasonable due to the limited resources to sustain a large human population.

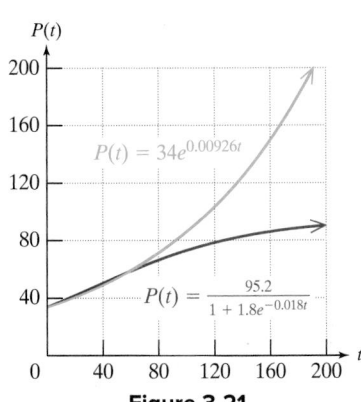

$P(t)$

$P(t) = 34e^{0.00926t}$

$P(t) = \dfrac{95.2}{1 + 1.8e^{-0.018t}}$

Figure 3-21

c. $P(t) = \dfrac{95.2}{1 + 1.8e^{-0.018t}} = \dfrac{95.2}{1 + \dfrac{1.18}{e^{0.018t}}}$ As $t \to \infty$, the term $\dfrac{1.18}{e^{0.018t}} \to 0$.

As t becomes large, the denominator of $\dfrac{1.18}{e^{0.018t}}$ also becomes large. This causes the quotient to approach zero. Therefore, as t approaches infinity, $P(t)$ approaches 95.2. Under this model, the limiting value for the population of California is 95.2 million.

> **Skill Practice 6** The score on a test of dexterity is given by
>
> $$P(t) = \frac{100}{1 + 19e^{-0.354x}},$$ where x is the number of times the test is taken.
>
> **a.** Determine the initial score.
> **b.** Use the function to determine the minimum number of times required for the score to exceed 90.
> **c.** What is the limiting value of the scores?

4. Create Exponential and Logarithmic Models Using Regression

In Examples 7 and 8, we use a graphing utility and regression techniques to find an exponential model or logarithmic model based on observed data.

Classroom Example: p. 435

Exercise 48

EXAMPLE 7 Creating an Exponential Model from Observed Data

The amount of sunlight y [in langleys (Ly)—a unit used to measure solar energy in calories/cm^2] is measured for six different depths x (in meters) in Lake Lyndon B. Johnson in Texas.

x (m)	1	3	5	7	9	11
y (Ly)	300	161	89	50	27	15

a. Graph the data.
b. From visual inspection of the graph, which model would best represent the data? Choose from $y = mx + b$ (linear), $y = ab^x$ (exponential), or $y = a + b \ln x$ (logarithmic).
c. Use a graphing utility to find a regression equation that fits the data.

Solution:

a. Enter the data in two lists.

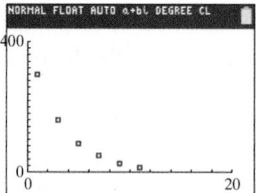

b. Note that for large depths, the amount of sunlight approaches 0. Therefore, the curve is asymptotic to the x-axis. This is consistent with a decreasing exponential model. The exponential model $y = ab^x$ appears to fit.
c. Under the STAT menu, choose CALC, ExpReg, and then Calculate.

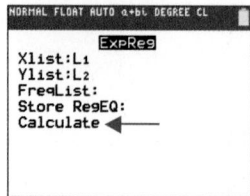
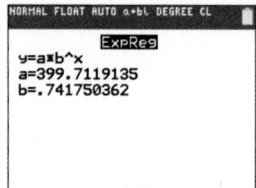
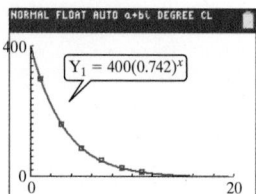

Answers

6. a. 5 **b.** 15 **c.** 100

The equation $y = ab^x$ is $y = 400(0.742)^x$.

Skill Practice 7 For the given data,

x	1	3	5	7	9	11
y	2.9	5.6	11.1	22.4	43.0	85.0

a. Graph the data points.

b. Use a graphing utility to find a model of the form $y = ab^x$ to fit the data.

Classroom Example: p. 435
Exercise 50

Instructor Note:
Ask students if other functions may provide a reasonable representation of the data. In this case, perhaps a model of the form $y = a\sqrt{x}$ will apply.

EXAMPLE 8 Creating a Logarithmic Model from Observed Data

The diameter x (in mm) of a sugar maple tree, along with the corresponding age y (in yr) of the tree is given for six different trees.

x (mm)	1	50	100	200	300	400
y (yr)	4	60	72	82	89	94

a. Graph the data.

b. From visual inspection of the graph, which model would best represent the data? Choose from $y = mx + b$ (linear), $y = ab^x$ (exponential), or $y = a + b \ln x$ (logarithmic).

c. Use a graphing utility to find a regression equation that fits the data.

Solution:

a. Enter the data into two lists.

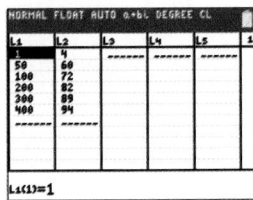 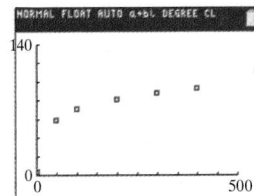

b. By inspection of the graph, the logarithmic model $y = a + b \ln x$ appears to fit.

c. Under the STAT menu, choose CALC, and then LnReg.

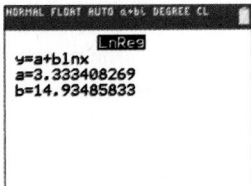 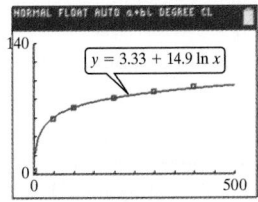

Answers

7. a–b.

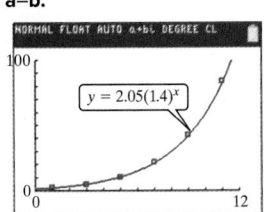

8. a–b.

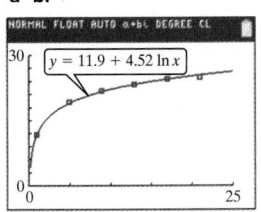

Skill Practice 8 For the given data,

x	1	5	9	13	17	21
y	11.9	19.3	21.9	23.5	24.7	25.7

a. Graph the data points.

b. Use a graphing utility to find a model of the form $y = a + b \ln x$ to fit the data.

SECTION 3.6   Practice Exercises

Prerequisite Review

For Exercises R.1–R.3, solve for the indicated variable.

R.1. $P = \dfrac{1}{3}Lm$ for L $\quad L = \dfrac{3P}{m}$ **R.2.** $Q = \dfrac{3}{4}m^3$ for m, $m > 0$ $\quad m = \sqrt[3]{\dfrac{4Q}{3}}$ **R.3.** $9 + \sqrt{a^2 - b^2} = c$ for b $\quad b = \pm\sqrt{a^2 - (c-9)^2}$

R.4. Given $f(x) = 5x^2 + 2x$, evaluate

    a. $f(-2)$ 16   **b.** $f(-1)$ 3   **c.** $f(0)$ 0   **d.** $f(1)$ 7

Concept Connections

1. If $k > 0$, the equation $y = y_0 e^{kt}$ is a model for exponential (growth/decay), whereas if $k < 0$, the equation is a model for exponential (growth/decay).
growth; decay

2. A function defined by $y = ab^x$ can be written in terms of an exponential function base e as _____. $y = ae^{(\ln b)x}$

3. A function defined by $y = \dfrac{c}{1 + ae^{-bt}}$ is called a _____ growth model and imposes a limiting value on y. logistic

4. Given a logistic growth function $y = \dfrac{c}{1 + ae^{-bt}}$, the limiting value of y is _____. c

Objective 1: Solve Literal Equations for a Specified Variable

For Exercises 5–14, solve for the indicated variable. (See Example 1)

5. $Q = Q_0 e^{-kt}$ for k (used in chemistry)

6. $N = N_0 e^{-0.025t}$ for t (used in chemistry)

7. $M = 8.8 + 5.1 \log D$ for D (used in astronomy)
$D = 10^{(M-8.8)/5.1}$

8. $\log E - 12.2 = 1.44M$ for E (used in geology)
$E = 10^{1.44M+12.2}$

9. $pH = -\log[H^+]$ for $[H^+]$ (used in chemistry) $\quad [H^+] = 10^{-pH}$

10. $L = 10 \log\left(\dfrac{I}{I_0}\right)$ for I (used in medicine) $\quad I = I_0 \cdot 10^{(L/10)}$

11. $A = P(1 + r)^t$ for t (used in finance)

12. $A = Pe^{rt}$ for r (used in finance)

13. $\ln\left(\dfrac{k}{A}\right) = \dfrac{-E}{RT}$ for k (used in chemistry) $\quad k = Ae^{-E/(RT)}$

14. $-\dfrac{1}{k}\ln\left(\dfrac{P}{14.7}\right) = A$ for P (used in meteorology)
$P = 14.7e^{-kA}$

Objective 2: Create Models for Exponential Growth and Decay

15. Suppose that \$12,000 is invested in a bond fund and the account grows to \$14,309.26 in 4 yr. (See Example 2)

    a. Use the model $A = Pe^{rt}$ to determine the average rate of return under continuous compounding. Round to the nearest tenth of a percent.   4.4%

    b. How long will it take the investment to reach \$20,000 if the rate of return continues? Round to the nearest tenth of a year.   11.6 yr

16. Suppose that \$50,000 from a retirement account is invested in a large cap stock fund. After 20 yr, the value is \$194,809.67.

    a. Use the model $A = Pe^{rt}$ to determine the average rate of return under continuous compounding. Round to the nearest tenth of a percent.   6.8%

    b. How long will it take the investment to reach one-quarter million dollars? Round to the nearest tenth of a year.   23.7 yr

17. Suppose that P dollars in principal is invested in an account earning 3.2% interest compounded continuously. At the end of 3 yr, the amount in the account has earned \$806.07 in interest.

    a. Find the original principal. Round to the nearest dollar. (*Hint:* Use the model $A = Pe^{rt}$ and substitute $P + 806.07$ for A.)   \$8000

    b. Using the original principal from part (a) and the model $A = Pe^{rt}$, determine the time required for the investment to reach \$10,000. Round to the nearest year.   7 yr

18. Suppose that P dollars in principal is invested in an account earning 2.1% interest compounded continuously. At the end of 2 yr, the amount in the account has earned \$193.03 in interest.

    a. Find the original principal. Round to the nearest dollar. (*Hint:* Use the model $A = Pe^{rt}$ and substitute $P + 193.03$ for A.)   \$4500

    b. Using the original principal from part (a) and the model $A = Pe^{rt}$, determine the time required for the investment to reach \$6000. Round to the nearest tenth of a year.   13.7 yr

19. The populations of two countries are given for January 1, 2000, and for January 1, 2010.

  a. Write a function of the form $P(t) = P_0 e^{kt}$ to model each population $P(t)$ (in millions) t years after January 1, 2000. (**See Example 3**)

Country	Population in 2000 (millions)	Population in 2010 (millions)	$P(t) = P_0 e^{kt}$
Australia	19.0	22.6	$P(t) = 19e^{0.01735t}$
Taiwan	22.9	23.7	$P(t) = 22.9e^{0.00343t}$

  b. Use the models from part (a) to approximate the population on January 1, 2020, for each country. Round to the nearest hundred thousand.
    Australia: 26.9 million; Taiwan: 24.5 million

  c. Australia had fewer people than Taiwan in the year 2000, yet from the result of part (b), Australia would have more people in the year 2020? Why?
    The population growth rate for Australia is greater.

  d. Use the models from part (a) to predict the year during which each population would reach 30 million if this trend continues.   Australia: 2026; Taiwan: 2078

21. A function of the form $P(t) = ab^t$ represents the population (in millions) of the given country t years after January 1, 2000. (**See Example 4**)

  a. Write an equivalent function using base e; that is, write a function of the form $P(t) = P_0 e^{kt}$. Also, determine the population of each country for the year 2000.

Country	$P(t) = ab^t$	$P(t) = P_0 e^{kt}$	Population in 2000
Costa Rica	$P(t) = 4.3(1.0135)^t$	$P(t) = 4.3e^{0.01341t}$	4.3 million
Norway	$P(t) = 4.6(1.0062)^t$	$P(t) = 4.6e^{0.00618t}$	4.6 million

  b. The population of the two given countries is very close for the year 2000, but their growth rates are different. Use the model to approximate the year during which the population of each country reached 5 million.
    Costa Rica: 2011; Norway: 2013

  c. Costa Rica had fewer people in the year 2000 than Norway. Why would Costa Rica reach a population of 5 million sooner than Norway?
    The population growth rate for Costa Rica is greater.

20. The populations of two countries are given for January 1, 2000, and for January 1, 2010.

  a. Write a function of the form $P(t) = P_0 e^{kt}$ to model each population $P(t)$ (in millions) t years after January 1, 2000.

Country	Population in 2000 (millions)	Population in 2010 (millions)	$P(t) = P_0 e^{kt}$
Switzerland	7.3	7.8	$P(t) = 7.3e^{0.00662t}$
Israel	6.7	7.7	$P(t) = 6.7e^{0.01391t}$

  b. Use the models from part (a) to approximate the population on January 1, 2020, for each country. Round to the nearest hundred thousand.
    Switzerland: 8.3 million; Israel: 8.8 million

  c. Israel had fewer people than Switzerland in the year 2000, yet from the result of part (b), Israel would have more people in the year 2020? Why?
    The population growth rate is greater for Israel.

  d. Use the models from part (a) to predict the year during which each population would reach 10 million if this trend continues.   Switzerland: 2047; Israel: 2028

22. A function of the form $P(t) = ab^t$ represents the population (in millions) of the given country t years after January 1, 2000.

  a. Write an equivalent function using base e; that is, write a function of the form $P(t) = P_0 e^{kt}$. Also, determine the population of each country for the year 2000.

Country	$P(t) = ab^t$	$P(t) = P_0 e^{kt}$	Population in 2000
Haiti	$P(t) = 8.5(1.0158)^t$	$P(t) = 8.5e^{0.01568t}$	8.5 million
Sweden	$P(t) = 9.0(1.0048)^t$	$P(t) = 9.0e^{0.00479t}$	9.0 million

  b. The population of the two given countries is very close for the year 2000, but their growth rates are different. Use the model to approximate the year during which the population of each country would reach 10.5 million.
    Haiti: 2013; Sweden: 2032

  c. Haiti had fewer people in the year 2000 than Sweden. Why would Haiti reach a population of 10.5 million sooner?   The population growth rate for Haiti is greater.

For Exercises 23–24, refer to the model $Q(t) = Q_0 e^{-0.000121t}$ used in Example 5 for radiocarbon dating.

23. A sample from a mummified bull was taken from a pyramid in Dashur, Egypt. The sample shows that 78% of the carbon-14 still remains. How old is the sample? Round to the nearest year. (**See Example 5**)   2053 yr

24. At the "Marmes Man" archeological site in southeastern Washington State, scientists uncovered the oldest human remains yet to be found in Washington State. A sample from a human bone taken from the site showed that 29.4% of the carbon-14 still remained. How old is the sample? Round to the nearest year.   10,117 yr

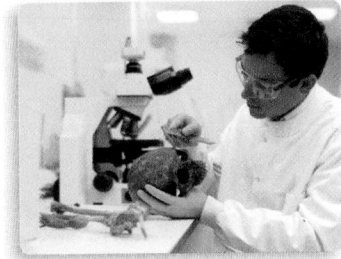

25. The isotope of plutonium ^{238}Pu is used to make thermoelectric power sources for spacecraft. Suppose that a space probe was launched in 2012 with 2.0 kg of ^{238}Pu.

a. If the half-life of ^{238}Pu is 87.7 yr, write a function of the form $Q(t) = Q_0 e^{-kt}$ to model the quantity $Q(t)$ of ^{238}Pu left after t years. $Q(t) = 2e^{-0.0079t}$

b. If 1.6 kg of ^{238}Pu is required to power the spacecraft's data transmitter, for how long after launch would scientists be able to receive data? Round to the nearest year. 28 yr

26. Technetium-99 (^{99m}Tc) is a radionuclide used widely in nuclear medicine. ^{99m}Tc is combined with another substance that is readily absorbed by a targeted body organ. Then, special cameras sensitive to the gamma rays emitted by the technetium are used to record pictures of the organ. Suppose that a technician prepares a sample of ^{99m}Tc-pyrophosphate to image the heart of a patient suspected of having had a mild heart attack.

a. At noon, the patient is given 10 mCi (millicuries) of ^{99m}Tc. If the half-life of ^{99m}Tc is 6 hr, write a function of the form $Q(t) = Q_0 e^{-kt}$ to model the radioactivity level $Q(t)$ after t hours. $Q(t) = 10e^{-0.1155t}$

b. At what time will the level of radioactivity reach 3 mCi? Round to the nearest tenth of an hour. 10.4 hr

27. Fluorodeoxyglucose is a derivative of glucose that contains the radionuclide fluorine-18 (^{18}F). A patient is given a sample of this material containing 300 MBq of ^{18}F (a megabecquerel is a unit of radioactivity). The patient then undergoes a PET scan (positron emission tomography) to detect areas of metabolic activity indicative of cancer. After 174 min, one-third of the original dose remains in the body.

a. Write a function of the form $Q(t) = Q_0 e^{-kt}$ to model the radioactivity level $Q(t)$ of fluorine-18 at a time t minutes after the initial dose. $Q(t) = 300e^{-0.0063t}$

b. What is the half-life of ^{18}F? Round to the nearest minute. 110 min

28. Painful bone metastases are common in advanced prostate cancer. Physicians often order treatment with strontium-89 (^{89}Sr), a radionuclide with a strong affinity for bone tissue. A patient is given a sample containing 4 mCi of ^{89}Sr.

a. If 20% of the ^{89}Sr remains in the body after 90 days, write a function of the form $Q(t) = Q_0 e^{-kt}$ to model the amount $Q(t)$ of radioactivity in the body t days after the initial dose. $Q(t) = Q_0 e^{-0.01788t}$

b. What is the biological half-life of ^{89}Sr under this treatment? Round to the nearest tenth of a day. 38.8 days

29. Two million *E. coli* bacteria are present in a laboratory culture. An antibacterial agent is introduced and the population of bacteria $P(t)$ decreases by half every 6 hr. The population can be represented by $P(t) = 2{,}000{,}000\left(\frac{1}{2}\right)^{t/6}$.

a. Convert this to an exponential function using base e. $P(t) = 2{,}000{,}000e^{-0.1155t}$

b. Verify that the original function and the result from part (a) yield the same result for $P(0)$, $P(6)$, $P(12)$, and $P(60)$. (*Note*: There may be round-off error.) $P(0) = 2{,}000{,}000;$ $P(6) = 1{,}000{,}000;$ $P(12) = 500{,}000;$ $P(60) = 1953$

30. The half-life of radium-226 is 1620 yr. Given a sample of 1 g of radium-226, the quantity left $Q(t)$ (in g) after t years is given by $Q(t) = \left(\frac{1}{2}\right)^{t/1620}$.

a. Convert this to an exponential function using base e. $Q(t) = e^{-0.000428t}$

b. Verify that the original function and the result from part (a) yield the same result for $Q(0)$, $Q(1620)$, and $Q(3240)$. (*Note*: There may be round-off error.) $Q(0) = 1;$ $Q(1620) = 0.5;$ $Q(3240) = 0.25$

Objective 3: Apply Logistic Growth Models

31. The population of the United States $P(t)$ (in millions) since January 1, 1900, can be approximated by

$$P(t) = \frac{725}{1 + 8.295e^{-0.0165t}}$$

where t is the number of years since January 1, 1900. (**See Example 6**)

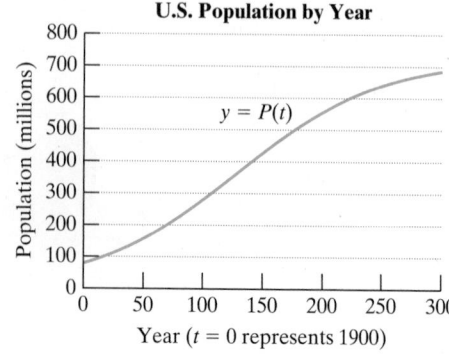

U.S. Population by Year

Year ($t = 0$ represents 1900)

a. Evaluate $P(0)$ and interpret its meaning in the context of this problem. $P(0) = 78$ means that on January 1, 1900, the U.S. population was approximately 78 million.

b. Use the function to approximate the U.S. population on January 1, 2020. Round to the nearest million. 338 million

c. Use the function to approximate the U.S. population on January 1, 2050. 427 million

d. From the model, during which year would the U.S. population reach 500 million? 2076

e. What value will the term $\dfrac{8.295}{e^{0.0165t}}$ approach as $t \to \infty$? 0

f. Determine the limiting value of $P(t)$. 725 million

32. The population of Canada $P(t)$ (in millions) since January 1, 1900, can be approximated by

$$P(t) = \frac{55.1}{1 + 9.6e^{-0.02515t}}$$

where t is the number of years since January 1, 1900.

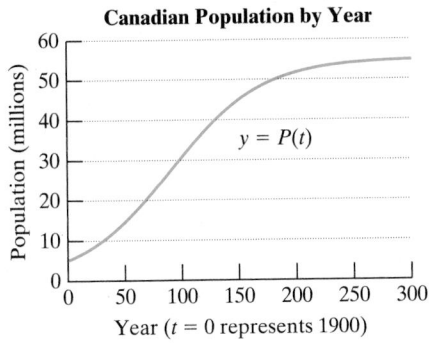

Canadian Population by Year

Population (millions)

$y = P(t)$

Year ($t = 0$ represents 1900)

a. Evaluate $P(0)$ and interpret its meaning in the context of this problem. $P(0) = 5.2$ means that on January 1, 1900, the Canadian population was approximately 5.2 million.

b. Use the function to approximate the Canadian population on January 1, 2015. Round to the nearest tenth of a million.   36.0 million

c. Use the function to approximate the Canadian population on January 1, 2040.   42.9 million

d. From the model, during which year would the Canadian population reach 45 million?   2049

e. What value will the term $\dfrac{9.6}{e^{0.02515t}}$ approach as $t \to \infty$?   0

f. Determine the limiting value of $P(t)$.   55.1 million

33. The number of computers $N(t)$ (in millions) infected by a computer virus can be approximated by

$$N(t) = \frac{2.4}{1 + 15e^{-0.72t}}$$

where t is the time in months after the virus was first detected.

a. Determine the number of computers initially infected when the virus was first detected.   150,000

b. How many computers were infected after 6 months? Round to the nearest hundred thousand.   2,000,000

c. Determine the amount of time required after initial detection for the virus to affect 1 million computers. Round to the nearest tenth of a month.   3.3 months

d. What is the limiting value of the number of computers infected according to this model?   2,400,000

34. After a new product is launched the cumulative sales $S(t)$ (in $1000) t weeks after launch is given by

$$S(t) = \frac{72}{1 + 9e^{-0.36t}}$$

a. Determine the cumulative amount in sales 3 weeks after launch. Round to the nearest thousand. $18,000

b. Determine the amount of time required for the cumulative sales to reach $70,000.   16 weeks

c. What is the limiting value in sales? $72,000

Objective 4: Create Exponential and Logarithmic Models Using Regression

For Exercises 35–38, a graph of data is given. From visual inspection, which model would best fit the data? Choose from

$y = mx + b$ (linear) $y = ab^x$ (exponential)

$y = a + b \ln x$ (logarithmic) $y = \dfrac{c}{1 + ae^{-bx}}$ (logistic)

35.   exponential

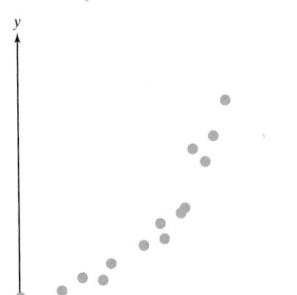

36.   linear

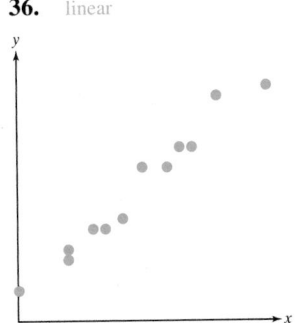

37.   logarithmic

38.   logistic

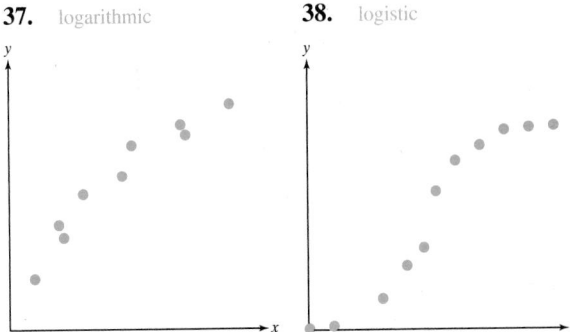

For Exercises 39–46, a table of data is given.

a. Graph the points and from visual inspection, select the model that would best fit the data. Choose from

$$y = mx + b \text{ (linear)} \qquad y = ab^x \text{ (exponential)}$$

$$y = a + b \ln x \text{ (logarithmic)} \qquad y = \frac{c}{1 + ae^{-bx}} \text{ (logistic)}$$

b. Use a graphing utility to find a function that fits the data. (*Hint:* For a logistic model, go to STAT, CALC, Logistic.)

39.

x	y
0	2.3
4	3.6
8	5.7
12	9.1
16	14
20	22

40.

x	y
0	52
1	67
2	87
3	114
4	147
5	195

41.

x	y
3	2.7
7	12.2
13	25.7
15	30
17	34
21	44.4

42.

x	y
0	640
20	530
40	430
50	360
80	210
100	90

43.

x	y
10	43.3
20	50
30	53
40	56.8
50	58.8
60	60.8

44.

x	y
5	29
10	40
15	45.6
20	50
25	53.3
30	56

45.

x	y
2	0.326
4	2.57
6	10.8
8	16.8
10	17.9
5	6
7	14.8

46.

x	y
0	0.05
2	0.45
4	2.94
5	5.8
6	8.8
7	10.6
8	11.5
10	11.9

47. During a recent outbreak of Ebola in western Africa, the cumulative number of cases *y* was reported *t* months after April 1. (**See Example 7**)

Month Number (*t*)	Cumulative Cases (*y*)
0	18
1	105
2	230
3	438
4	752
5	1437
6	2502

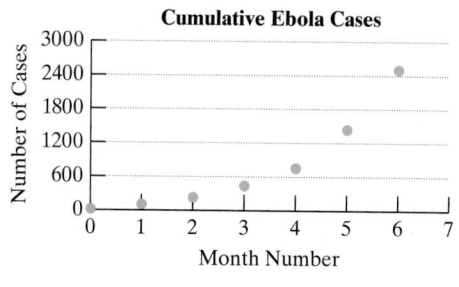

Cumulative Ebola Cases

a. Use a graphing utility to find a model of the form $y = ab^t$. Round *a* to 1 decimal place and *b* to 3 decimal places. $y = 34.9(2.134)^t$

b. Write the function from part (a) as an exponential function of the form $y = ae^{bt}$. $y = 34.9e^{0.758t}$

c. Use either model to predict the number of Ebola cases 8 months after April 1 if this trend continues. Round to the nearest thousand. 15,000 cases

d. Would it seem reasonable for this trend to continue indefinitely? No, eventually the number of cases would exceed the human population.

e. Use a graphing utility to find a logistic model $y = \dfrac{c}{1 + ae^{-bt}}$. Round *a* and *c* to the nearest whole number and *b* to 2 decimal places. $y = \dfrac{11,731}{1 + 205e^{-0.67t}}$

f. Use the logistic model from part (e) to predict the number of Ebola cases 8 months after April 1. Round to the nearest thousand. 6000 cases

48. The monthly costs for a small company to do business has been increasing over time due in part to inflation. The table gives the monthly cost y (in $) for the month of January for selected years. The variable t represents the number of years since 2016.

Year (t = 0 is 2016)	Monthly Costs ($) y
0	12,000
1	12,400
2	12,800
3	13,300

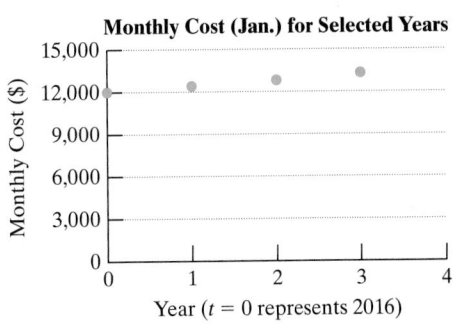

Monthly Cost (Jan.) for Selected Years

a. Use a graphing utility to find a model of the form $y = ab^t$. Round a to the nearest whole unit and b to 3 decimal places. $y = 11,988(1.035)^t$

b. Write the function from part (a) as an exponential function with base e. $y = 11,988e^{0.0344t}$

c. Use either model to predict the monthly cost for January in the year 2023 if this trend continues. Round to the nearest hundred dollars. $15,300

49. The age of a tree t (in yr) and its corresponding height $H(t)$ are given in the table. (**See Example 8**)

Age of Tree (yr) t	Height (ft) H(t)
1	5
2	8.3
3	11.6
4	14.6
5	15.4
6	16.5
7	17.5
8	18.3
9	19
10	19.4
11	19.7
12	20

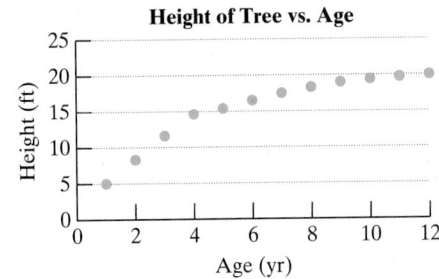

Height of Tree vs. Age

a. Write a model of the form $H(t) = a + b \ln t$. Round a and b to 2 decimal places.
$H(t) = 4.86 + 6.35 \ln t$

b. Use the model to predict the age of a tree if it is 25 ft high. Round to the nearest year.   24 yr

c. Is it reasonable to assume that this logarithmic trend will continue indefinitely? Why or why not?   No, the tree will eventually die.

50. The sales of a book tend to increase over the short-term as word-of-mouth makes the book "catch on." The number of books sold $N(t)$ for a new novel t weeks after release at a certain book store is given in the table for the first 6 weeks.

Weeks t	Number Sold N(t)
1	20
2	27
3	31
4	35
5	38
6	39

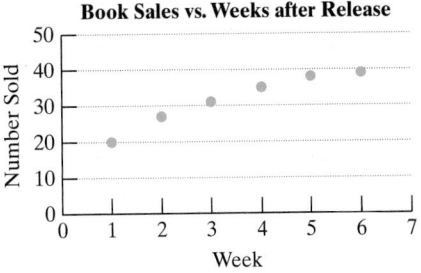

Book Sales vs. Weeks after Release

a. Find a model of the form $N(t) = a + b \ln t$. Round a and b to 1 decimal place. $N(t) = 19.7 + 10.9 \ln t$

b. Use the model to predict the sales in week 7. Round to the nearest whole unit.   41 books

c. Is it reasonable to assume that this logarithmic trend will continue? Why or why not?
No. The trend cannot continue indefinitely. At some point, book sales will begin to decrease as most reading enthusiasts have read the book.

Mixed Exercises

51. A van is purchased new for $29,200.

a. Write a linear function of the form $y = mt + b$ to represent the value y of the vehicle t years after purchase. Assume that the vehicle is depreciated by $2920 per year. $y = -2920t + 29,200$

b. Suppose that the vehicle is depreciated so that it holds only 80% of its value from the previous year. Write an exponential function of the form $y = V_0 b^t$, where V_0 is the initial value and t is the number of years after purchase. $y = 29,200(0.8)^t$

c. To the nearest dollar, determine the value of the vehicle after 5 yr and after 10 yr using the linear model. $14,600 and $0

d. To the nearest dollar, determine the value of the vehicle after 5 yr and after 10 yr using the exponential model. $9568 and $3135

52. A delivery truck is purchased new for $54,000.

a. Write a linear function of the form $y = mt + b$ to represent the value y of the vehicle t years after purchase. Assume that the vehicle is depreciated by $6750 per year. $y = -6750t + 54,000$

b. Suppose that the vehicle is depreciated so that it holds 70% of its value from the previous year. Write an exponential function of the form $y = V_0 b^t$, where V_0 is the initial value and t is the number of years after purchase. $y = 54,000(0.7)^t$

c. To the nearest dollar, determine the value of the vehicle after 4 yr and after 8 yr using the linear model. $27,000 and $0

d. To the nearest dollar, determine the value of the vehicle after 4 yr and after 8 yr using the exponential model. $12,965 and $3113

Write About It

53. Why is it important to graph a set of data before trying to find an equation or function to model the data.

55. Explain the difference between an exponential growth model and a logistic growth model.

54. How does the average rate of change differ for a linear function versus an increasing exponential function?

56. Explain how to convert an exponential expression b^t to an exponential expression base e.

Expanding Your Skills

57. The monthly payment P (in $) to pay off a loan of amount A (in $) at an interest rate r in t years is given by

$$P = \frac{\dfrac{Ar}{12}}{1 - \left(1 + \dfrac{r}{12}\right)^{-12t}}.$$

a. Solve for t (note that there are numerous equivalent algebraic forms for the result).

b. Interpret the meaning of the resulting relationship.

58. Suppose that a population follows a logistic growth pattern, with a limiting population N. If the initial population is denoted by P_0, and t is the amount of time elapsed, then the population P can be represented by

$$P = \frac{P_0 N}{P_0 + (N - P_0)e^{-kt}}.$$

where k is a constant related to the growth rate.

a. Solve for t (note that there are numerous equivalent algebraic forms for the result).

b. Interpret the meaning of the resulting relationship.

Additional answers can be found in the Instructor Answer Appendix.

CHAPTER 3   KEY CONCEPTS

SECTION 3.1 Inverse Functions	Reference
A function f is **one-to-one** if for a and b in the domain of f, if $a \neq b$, then $f(a) \neq f(b)$, or equivalently, if $f(a) = f(b)$, then $a = b$.	p. 357
Horizontal line test: A function defined by $y = f(x)$ is one-to-one if no horizontal line intersects the graph in more than one point.	p. 357
Function g is the **inverse of** f if $(f \circ g)(x) = x$ for all x in the domain of g and $(g \circ f)(x) = x$ for all x in the domain of f.	p. 359
Procedure to find $f^{-1}(x)$: 1. Replace $f(x)$ by y. 2. Interchange x and y. 3. Solve for y. 4. Replace y by $f^{-1}(x)$.	p. 360

SECTION 3.2 Exponential Functions	Reference
Let b be a real number with $b > 0$ and $b \neq 1$. Then for any real number x, a function of the form $f(x) = b^x$ is an **exponential function of base b.**	p. 369
For the graph of an exponential function $f(x) = b^x$, • If $b > 1$, f is an increasing function. • If $0 < b < 1$, f is a decreasing function. • The domain is $(-\infty, \infty)$. • The range is $(0, \infty)$. • The line $y = 0$ is a horizontal asymptote. • The function passes through $(0, 1)$.	p. 370
The irrational number e is the limiting value of the expression $\left(1 + \frac{1}{x}\right)^x$ as x approaches ∞. $e \approx 2.71828$	p. 372
If P dollars in principal is invested or borrowed at an annual interest rate r for t years, then $I = Prt$ Simple interest $A = P\left(1 + \dfrac{r}{n}\right)^{nt}$ Future value A with interest compounded n times per year. $A = Pe^{rt}$ Future value A with interest compounded continuously.	p. 374

SECTION 3.3 Logarithmic Functions	Reference
If x and b are positive real numbers such that $b \neq 1$, then $y = \log_b x$ is called the **logarithmic function** base b, where $$y = \log_b x \text{ is equivalent to } b^y = x.$$ logarithmic form exponential form	p. 382
The functions $f(x) = \log_b x$ and $g(x) = b^x$ are inverses.	p. 382
Basic properties of logarithms: 1. $\log_b 1 = 0$ 2. $\log_b b = 1$ 3. $\log_b b^x = x$ 4. $b^{\log_b x} = x$	p. 386
$y = \log_{10} x$ is written as $y = \log x$ and is called the **common logarithmic function.** $y = \log_e x$ is written as $y = \ln x$ and is called the **natural logarithmic function.**	p. 384
Given $f(x) = \log_b x$, • If $b > 1$, f is an increasing function. • If $0 < b < 1$, f is a decreasing function. • The domain is $(0, \infty)$. • The range is $(-\infty, \infty)$. • The line $x = 0$ is a vertical asymptote. • The function passes through $(1, 0)$.	p. 388
The domain of $f(x) = \log_b x$ is $\{x \mid x > 0\}$.	p. 389

	Reference
SECTION 3.4 Properties of Logarithms	
Let b, x, and y be positive real numbers with $b \neq 1$. Then, $\log_b(xy) = \log_b x + \log_b y$ (Product property) $\log_b\left(\dfrac{x}{y}\right) = \log_b x - \log_b y$ (Quotient property) $\log_b x^p = p \log_b x$ (Power property)	p. 399
Change-of-base formula: For positive real numbers a and b, where $a \neq 1$ and $b \neq 1$, $\log_b x = \dfrac{\log_a x}{\log_a b}$.	p. 402
SECTION 3.5 Exponential and Logarithmic Equations and Applications	Reference
Equivalence property of exponential expressions: If b, x, and y are real numbers with $b > 0$ and $b \neq 1$, then $b^x = b^y$ implies that $x = y$.	p. 407
Equivalence property of logarithmic expressions: If b, x, and y are positive real numbers and $b \neq 1$, then $\log_b x = \log_b y$ implies that $x = y$.	p. 411
Steps to solve exponential equations by using logarithms: 1. Isolate the exponential expression on one side of the equation. 2. Take a logarithm of the same base on both sides of the equation. 3. Use the power property of logarithms to "bring down" the exponent. 4. Solve the resulting equation.	p. 408
Guidelines to solve a logarithmic equation: 1. Isolate the logarithms on one side of the equation. 2. Use the properties of logarithms to write the equation in the form $\log_b x = k$, where k is a constant. 3. Write the equation in exponential form. 4. Solve the equation from step 3. 5. Check the potential solution(s) in the original equation.	p. 412
SECTION 3.6 Modeling with Exponential and Logarithmic Functions	Reference
The function defined by $y = y_0 e^{kt}$ represents exponential growth if $k > 0$ and exponential decay if $k < 0$.	p. 422
An exponential expression can be rewritten as an expression of a different base. In particular, to convert to base e, we have b^t is equivalent to $e^{(\ln b)\, t}$.	p. 424
A **logistic growth function** is a function of the form $$y = \frac{c}{1 + ae^{-bt}}.$$ A logistic growth function imposes a limiting value on the dependent variable.	p. 426

Expanded Chapter Summary available at www.mhhe.com/millerprecalculus.

CHAPTER 3 Review Exercises

SECTION 3.1

For Exercises 1–2, determine if the relation defines y as a one-to-one function of x.

1. No

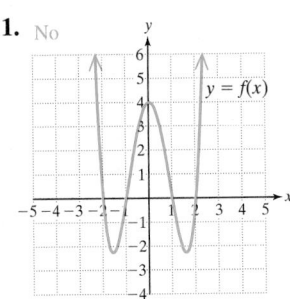

$y = f(x)$

2. Yes

x	y
5	7
−3	1
−4	−2
6	0

For Exercises 3–4, use the definition of a one-to-one function to determine if the function is one-to-one. Recall that f is one-to-one if a ≠ b implies that f(a) ≠ f(b), or equivalently, if f(a) = f(b), then a = b.

3. $f(x) = x^3 - 1$ **4.** $f(x) = x^2 - 1$

For Exercises 5–6, determine if the functions are inverses.

5. $f(x) = 4x - 3$ and $g(x) = \dfrac{x + 3}{4}$

Yes, because $(f \circ g)(x) = (g \circ f)(x) = x$

6. $m(x) = \sqrt[3]{x + 1}$ and $n(x) = (x - 1)^3$

No, because $(m \circ n)(x) \neq x$. Likewise, $(n \circ m)(x) \neq x$.

For Exercises 7–8, a one-to-one function is given. Write an equation for the inverse function.

7. $f(x) = 2x^3 - 5$ **8.** $f(x) = \dfrac{2}{x + 7}$ $f^{-1}(x) = \dfrac{2}{x} - 7$

$f^{-1}(x) = \sqrt[3]{\dfrac{x + 5}{2}}$

9. a. Graph $f(x) = x^2 - 9, x \leq 0$.

  b. Is f a one-to-one function?   Yes

  c. Write the domain of f in interval notation. $(-\infty, 0]$

  d. Write the range of f in interval notation. $[-9, \infty)$

  e. Find an equation for f^{-1}. $f^{-1}(x) = -\sqrt{x + 9}$

  f. Graph $y = f(x)$ and $y = f^{-1}(x)$ on the same coordinate system.

  g. Write the domain of f^{-1} in interval notation. $[-9, \infty)$

  h. Write the range of f^{-1} in interval notation. $(-\infty, 0]$

10. a. Graph $g(x) = \sqrt{x + 4}$.

  b. Is g a one-to-one function?   Yes

  c. Write the domain of g in interval notation. $[-4, \infty)$

  d. Write the range of g in interval notation. $[0, \infty)$

  e. Find an equation for g^{-1}. $g^{-1}(x) = x^2 - 4; x \geq 0$

  f. Graph $y = g(x)$ and $y = g^{-1}(x)$ on the same coordinate system.

  g. Write the domain of g^{-1} in interval notation. $[0, \infty)$

  h. Write the range of g^{-1} in interval notation. $[-4, \infty)$

11. The function $f(x) = 5280x$ provides the conversion from x miles to $f(x)$ feet.

  a. Write an equation for f^{-1}. $f^{-1}(x) = \dfrac{x}{5280}$

  b. What does the inverse function represent in the context of this problem?
  f^{-1} represents the conversion from x feet to $f^{-1}(x)$ miles.

  c. Determine the number of miles represented by 22,176 ft.
  4.2 mi

SECTION 3.2

12. Which of the functions is an exponential function?
  b and c only

  a. $f(x) = x^4$ **b.** $h(x) = 4^{-x}$ **c.** $g(x) = \left(\dfrac{4}{3}\right)^x$

  d. $k(x) = \dfrac{4x}{3}$ **e.** $n(x) = \dfrac{4}{3x}$ **f.** $r(x) = \left(-\dfrac{4}{3}\right)^x$

For Exercises 13–16,

  a. Graph the function.

  b. Write the domain in interval notation.

  c. Write the range in interval notation.

  d. Write an equation of the asymptote.

13. $f(x) = \left(\dfrac{5}{2}\right)^x$ **14.** $g(x) = \left(\dfrac{5}{2}\right)^{-x}$

15. $k(x) = -3^x + 1$ **16.** $h(x) = 2^{x-3} - 4$

17. Is the graph of $y = e^x$ an increasing or decreasing exponential function?   Increasing

For Exercises 18–19, use the formulas on page 374.

18. Suppose that $24,000 is invested at the given interest rates and compounding options. Determine the amount that the investment is worth at the end of t years.

  a. 5% interest compounded monthly for 10 yr $39,528.23

  b. 4.5% interest compounded continuously for 30 yr $92,578.21

19. Jorge needs to borrow $12,000 to buy a car. He can borrow the money at 7.2% simple interest for 4 yr or he can borrow at 6.5% interest compounded continuously for 4 yr.

  a. How much total interest would Jorge pay at 7.2% simple interest? $3456

  b. How much total interest would Jorge pay at 6.5% interest compounded continuously? $3563.16

  c. Which option results in less total interest?
  7.2% simple interest results in less interest.

20. A patient is treated with 128 mCi (millicuries) of iodine-131 (^{131}I). The radioactivity level $R(t)$ (in mCi) after t days is given by $R(t) = 128(2)^{-t/4.2}$. (In this model, the value 4.2 is related to the biological half-life of radioactive iodine in the body.)

  a. Determine the radioactivity level of ^{131}I in the body after 6 days. Round to the nearest whole unit.   48 mCi

b. Evaluate $R(4.2)$ and interpret its meaning in the context of this problem.

c. After how many half-lives will the radioactivity level be 16 mCi?   3 half-lives

SECTION 3.3

For Exercises 21–22, write the equation in exponential form.

21. $\log_b(x^2 + y^2) = 4$ **22.** $\ln x = (c + d)$ $e^{c+d} = x$
$b^4 = x^2 + y^2$

For Exercises 23–24, write the equation in logarithmic form.

23. $10^6 = 1{,}000{,}000$ **24.** $8^{-1/3} = \dfrac{1}{2}$ $\log_8\left(\dfrac{1}{2}\right) = -\dfrac{1}{3}$
$\log 1{,}000{,}000 = 6$

For Exercises 25–32, simplify the logarithmic expression without using a calculator.

25. $\log_3 81$ 4   **26.** $\log 100{,}000$ 5

27. $\log_2\left(\dfrac{1}{64}\right)$ -6 **28.** $\log_{1/4}(16)$ -2

29. $\log_{11} 1$ 0   **30.** $\log_5 5$ 1

31. $4^{\log_4 7}$ 7   **32.** $\ln e^{11}$ 11

For Exercises 33–37, write the domain of the function in interval notation.

33. $f(x) = \log(x - 4)$ $(4, \infty)$ **34.** $g(x) = \ln(3 - 2x)$ $\left(-\infty, \dfrac{3}{2}\right)$

35. $h(x) = \log_2(x^2 + 4)$ **36.** $k(x) = \log_2(x^2 - 4)$
$(-\infty, \infty)$ $(-\infty, -2) \cup (2, \infty)$

37. $m(x) = \log_2(x - 4)^2$
$(-\infty, 4) \cup (4, \infty)$

For Exercises 38–39,

a. Graph the function.

b. Write the domain in interval notation.

c. Write the range in interval notation.

d. Write an equation of the asymptote.

38. $f(x) = \log_2(x - 3)$ **39.** $g(x) = 2 + \ln x$

For Exercise 40–41, use the formula $\text{pH} = -\log[H^+]$ to compute the pH of a liquid as a function of its concentration of hydronium ions, $[H^+]$ in mol/L. If the pH is less than 7, then the substance is acidic. If the pH is greater than 7, then the substance is alkaline (or basic).

a. Find the pH. Round to 1 decimal place.

b. Determine whether the substance is acidic or alkaline.

40. Baking soda: $[H^+] = 5.0 \times 10^{-9}$ mol/L   pH ≈ 8.3; alkaline

41. Tomatoes: $[H^+] = 3.16 \times 10^{-5}$ mol/L   pH ≈ 4.5; acidic

SECTION 3.4

For Exercises 42–48, fill in the blanks to state the basic properties of logarithms. Assume that x, y, and b are positive real numbers with $b \neq 1$.

42. $\log_b 1 =$ ____   0

43. $\log_b b =$ ____   1

44. $\log_b b^p =$ ____ p

45. $b^{\log_b x} =$ ____ x

46. $\log_b(xy) = \dfrac{\log_b x + \log_b y}{}$ **47.** $\log_b\left(\dfrac{x}{y}\right) = \dfrac{\log_b x - \log_b y}{}$

48. $\log_b x^p = \dfrac{p \cdot \log_b x}{}$

For Exercises 49–52, write the logarithm as a sum or difference of logarithms. Simplify each term as much as possible.

49. $\log\left(\dfrac{100}{\sqrt{c^2 + 10}}\right)$ **50.** $\log_2\left(\dfrac{1}{8}a^2 b\right)$

51. $\ln\left(\dfrac{\sqrt[3]{ab^2}}{cd^5}\right)$ **52.** $\log\left(\dfrac{x^2(2x + 1)^5}{\sqrt{1 - x}}\right)$

For Exercises 53–55, write the logarithmic expression as a single logarithm with coefficient 1, and simplify as much as possible.

53. $4 \log_5 y - 3 \log_5 x + \dfrac{1}{2}\log_5 z$ $\log_5\left(\dfrac{y^4\sqrt{z}}{x^3}\right)$

54. $\log 250 + \log 2 - \log 5$ 2

55. $\dfrac{1}{4}\ln(x^2 - 9) - \dfrac{1}{4}\ln(x - 3)$ $\ln\sqrt[4]{x + 3}$

For Exercises 56–58, use $\log_b 2 \approx 0.289$, $\log_b 3 \approx 0.458$, and $\log_b 5 \approx 0.671$ to approximate the value of the given logarithms.

56. $\log_b 8$ 0.867   **57.** $\log_b 45$ 1.587   **58.** $\log_b\left(\dfrac{1}{9}\right)$ -0.916

For Exercises 59–60, use the change-of-base formula and a calculator to approximate the given logarithms. Round to 4 decimal places. Then check the answer by using the related exponential form.

59. $\log_7 596$
3.2839; $7^{3.2839} \approx 596$

60. $\log_4 0.982$
-0.0131; $4^{-0.0131} \approx 0.982$

SECTION 3.5

For Exercises 61–80, solve the equation. Write the solution set with exact values and give approximate solutions to 4 decimal places.

61. $4^{2y-7} = 64$ {5}   **62.** $1000^{2x+1} = \left(\dfrac{1}{100}\right)^{x-4}$ $\left\{\dfrac{5}{8}\right\}$

63. $7^x = 51$ **64.** $516 = 11^w - 21$

65. $3^{2x+1} = 4^{3x}$ **66.** $2^{c+3} = 7^{2c+5}$

67. $400e^{-2t} = 2.989$ **68.** $2 \cdot 10^{1.2t} = 58$

69. $e^{2x} - 3e^x - 40 = 0$ **70.** $e^{2x} = -10e^x$ { }
{ln 8}; $x \approx 2.0794$

71. $\log_5(4p + 7) = \log_5(2 - p)$ {−1}

72. $\log_2(m^2 + 10m) = \log_2 11$ {−11, 1}

73. $2\log_6(4 - 8y) + 6 = 10$ {−4}

74. $5 = -4\log_3(2 - 5x) + 1$ $\left\{\dfrac{1}{3}\right\}$

75. $3\ln(n - 8) = 6.3$ $\{e^{2.1} + 8\}$; $n \approx 16.1662$

76. $-4 + \log_2 x = -\log_2(x + 6)$ {2}; The value −8 does not check.

77. $\log_6(3x + 2) = \log_6(x + 4) + 1$ { }; The value $-\dfrac{22}{3}$ does not check.

78. $\ln x + \ln(x + 2) = \ln(x + 6)$ {2}; The value −3 does not check.

79. $\log_5(\log_2 x) = 1$ {32}

80. $(\log x)^2 - \log x^2 = 35$ $\left\{10{,}000{,}000, \dfrac{1}{100{,}000}\right\}$

For Exercises 81–82, find the inverse of the function.

81. $f(x) = 4^x$ $f^{-1}(x) = \log_4 x$ **82.** $g(x) = \log(x-5) - 1$
$g^{-1}(x) = 10^{x+1} + 5$

83. The percentage of visible light P (in decimal form) at a depth of x meters for Long Island Sound can be approximated by $P = e^{-0.5x}$.

   a. Determine the depth at which the light intensity is half the value from the surface. Round to the nearest hundredth of a meter. Based on your answer, would you say that Long Island Sound is murky or clear water?
   1.39 m; murky

   b. Determine the euphotic depth for Long Island Sound. That is, find the depth at which the light intensity falls below 1%. Round to the nearest tenth of a meter.   9.2 m

SECTION 3.6

For Exercises 84–85, solve for the indicated variable.

84. $\log B - 1.7 = 2.3M$ for B $B = 10^{2.3M + 1.7}$

85. $T = T_f + T_0 e^{-kt}$ for t $t = -\dfrac{1}{k}\ln\left(\dfrac{T - T_f}{T_0}\right)$ or $\dfrac{1}{k}[\ln T_0 - \ln(T - T_f)]$

86. Suppose that $18,000 is invested in a bond fund and the account grows to $23,344.74 in 5 yr.

   a. Use the model $A = Pe^{rt}$ to determine the average rate of return under continuous compounding. Round to the nearest tenth of a percent.   5.2%

   b. How long will it take the investment to reach $30,000 if the rate of return continues? Round to the nearest tenth of a year.   9.8 yr

87. The population of Germany in 2011 was approximately 85.5 million. The model $P = 85.5e^{-0.00208t}$ represents a short-term model for the population, t years after 2011.

   a. Based on this model, is the population of Germany increasing or decreasing?   Decreasing

   b. Determine the number of years after 2011 at which the population of Germany would decrease to 80 million if this trend continues. Round to the nearest year.   32 yr

88. The population of Chile was approximately 16.9 million in the year 2011, with an annual growth rate of 0.836%. The population $P(t)$ (in millions) can be modeled by

$P(t) = 16.9(1.00836)^t$, where t is the number of years since 2011.

   a. Write a function of the form $P(t) = P_0 e^{kt}$ to model the population. $P(t) = 16.9e^{0.00833t}$

   b. Determine the amount of time required for the population to grow to 20 million if this trend continues. Round to the nearest year.   20 yr

89. A sample from human remains found near Stonehenge in England shows that 71.2% of the carbon-14 still remains. Use the model $Q(t) = Q_0 e^{-0.000121t}$ to determine the age of the sample. In this model, $Q(t)$ represents the amount of carbon-14 remaining t years after death, and Q_0 represents the initial amount of carbon-14 at the time of death. Round to the nearest 100 yr.   2800 yr

90. A lake is stocked with bass by the U.S. Park Service. The population of bass is given by $P(t) = \dfrac{3000}{1 + 2e^{-0.37t}}$, where t is the time in years after the lake was stocked.

   a. Evaluate $P(0)$ and interpret its meaning in the context of this problem.
   $P(0) = 1000$; The lake was initially stocked with 1000 bass.

   b. Use the function to predict the bass population 2 yr after being stocked. Round to the nearest whole unit.   1535 bass

   c. Use the function to predict the bass population 4 yr after being stocked.   2061 bass

   d. Determine the number of years required for the bass population to reach 2800. Round to the nearest year.   9 yr

   e. What value will the term $\dfrac{2}{e^{0.37t}}$ approach as $t \to \infty$.   0

   f. Determine the limiting value of $P(t)$.   3000 bass

91. For the given data,

   a. Use a graphing utility to find an exponential function $Y_1 = ab^x$ that fits the data. $Y_1 = 2.38(1.5)^x$

   b. Graph the data and the function from part (a) on the same coordinate system.

x	y
0	2.4
1	3.5
2	5.5
3	8.1
4	12.0
5	18.4

CHAPTER 3   Test

1. Given $f(x) = 4x^3 - 1$,

   a. Write an equation for $f^{-1}(x)$. $f^{-1}(x) = \sqrt[3]{\dfrac{x+1}{4}}$

   b. Verify that $(f \circ f^{-1})(x) = (f^{-1} \circ f)(x) = x$.

2. The graph of f is given.

   a. Is f a one-to-one function?   Yes

   b. If f is a one-to-one function, graph f^{-1} on the same coordinate system as f.

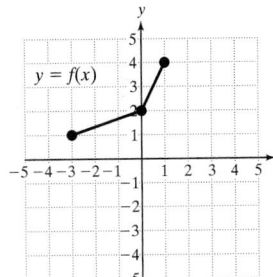

3. Given $f(x) = \dfrac{x+3}{x-4}$, write an equation for the inverse function. $f^{-1}(x) = \dfrac{4x+3}{x-1}$

For Exercises 4–7,

   a. Write the domain and range of f in interval notation.

   b. Write an equation of the inverse function.

   c. Write the domain and range of f^{-1} in interval notation.

4. $f(x) = -x^2 + 1, x \le 0$ **5.** $f(x) = \log x$

6. $f(x) = 3^x + 1$ **7.** $f(x) = \sqrt{x+5}$

For Exercises 8–11,

a. Graph the function.

b. Write the domain in interval notation.

c. Write the range in interval notation.

d. Write an equation of the asymptote.

8. $f(x) = \left(\dfrac{1}{3}\right)^x + 2$ **9.** $g(x) = 2^{x-4}$

10. $h(x) = -\ln x$ **11.** $k(x) = \log_2(x + 1) - 3$

12. Write the statement in exponential form. $\ln(x + y) = a$
$e^a = x + y$

For Exercises 13–18, evaluate the logarithmic expression without using a calculator.

13. $\log_9 \dfrac{1}{81}$ -2 **14.** $\log_6 216$ 3 **15.** $\ln e^8$ 8

16. $\log 10^{-4}$ -4 **17.** $10^{\log(a^2 + b^2)}$ $a^2 + b^2$ **18.** $\log_{1/2} 1$ 0

For Exercises 19–20, write the domain of the function in interval notation.

19. $f(x) = \log(7 - 2x)$ $\left(-\infty, \dfrac{7}{2}\right)$ **20.** $g(x) = \log_4(x^2 - 25)$
$(-\infty, -5) \cup (5, \infty)$

For Exercises 21–22, write the logarithm as a sum or difference of logarithms. Simplify each term as much as possible.

21. $\ln\left(\dfrac{x^5 y^2}{w \sqrt[3]{z}}\right)$ **22.** $\log\left(\dfrac{\sqrt{a^2 + b^2}}{10^4}\right)$
$5 \ln x + 2 \ln y - \ln w - \dfrac{1}{3}\ln z$ $\dfrac{1}{2}\log(a^2 + b^2) - 4$

For Exercises 23–24, write the logarithmic expression as a single logarithm with coefficient 1, and simplify as much as possible.

23. $6 \log_2 a - 4 \log_2 b + \dfrac{2}{3}\log_2 c$ $\log_2\left(\dfrac{a^6 \sqrt[3]{c^2}}{b^4}\right)$

24. $\dfrac{1}{2}\ln(x^2 - x - 12) - \dfrac{1}{2}\ln(x - 4)$ $\ln \sqrt{x + 3}$

For Exercises 25–26, use $\log_b 2 \approx 0.289$, $\log_b 3 \approx 0.458$, and $\log_b 5 \approx 0.671$ to approximate the value of the given logarithms.

25. $\log_b 72$ 1.783 **26.** $\log_b\left(\dfrac{1}{125}\right)$ -2.013

For Exercises 27–36, solve the equation. Write the solution set with exact values and give approximate solutions to 4 decimal places.

27. $2^{5y+1} = 4^{y-3}$ $\left\{-\dfrac{7}{3}\right\}$ **28.** $5^{x+3} + 3 = 56$

29. $2^{c+7} = 3^{2c+3}$ **30.** $7e^{4x} - 2 = 12$

31. $e^{2x} + 7e^x - 8 = 0$ $\{0\}$

32. $\log_5(3 - x) = \log_5(x + 1)$ $\{1\}$

33. $5 \ln(x + 2) + 1 = 16$ $\{e^3 - 2\}; x \approx 18.0855$

34. $\log x + \log(x - 1) = \log 12$
$\{4\}$; The value -3 does not check.
35. $-3 + \log_4 x = -\log_4(x + 30)$
$\{2\}$; The value -32 does not check.
36. $\log 3 + \log(x + 3) = \log(4x + 5)$ $\{4\}$

For Exercises 37–38, solve for the indicated variable.

37. $S = 92 - k \ln(t + 1)$ for t $t = e^{(92 - S)/k} - 1$

38. $A = P\left(1 + \dfrac{r}{n}\right)^{nt}$ for t $t = \dfrac{\ln\left(\frac{A}{P}\right)}{n \ln\left(1 + \frac{r}{n}\right)}$ or $\dfrac{\ln A - \ln P}{n \ln\left(1 - \frac{r}{n}\right)}$

39. Suppose that $10,000$ is invested and the account grows to $13,566.25$ in 5 yr.

  a. Use the model $A = Pe^{rt}$ to determine the average rate of return under continuous compounding. Round to the nearest tenth of a percent.   6.1%

  b. Using the interest rate from part (a), how long will it take the investment to reach $50,000$? Round to the nearest tenth of a year.   26.4 yr

40. The number of bacteria in a culture begins with approximately 10,000 organisms at the start of an experiment. If the bacteria doubles every 5 hr, the model $P(t) = 10,000(2)^{t/5}$ represents the population $P(t)$ after t hours.

  a. Write a function of the form $P(t) = P_0 e^{kt}$ to model the population. $P(t) = 10,000e^{0.1386t}$

  b. Determine the amount of time required for the population to grow to 5 million. Round to the nearest hour.   Approximately 45 hr

41. The population $P(t)$ of a herd of deer on an island can be modeled by $P(t) = \dfrac{1200}{1 + 2e^{-0.12t}}$, where t represents the number of years since the park service has been tracking the herd.

  a. Evaluate $P(0)$ and interpret its meaning in the context of this problem.   400 deer were present when the park service began tracking the herd.

  b. Use the function to predict the deer population after 4 yr. Round to the nearest whole unit.   536 deer

  c. Use the function to predict the deer population after 8 yr.   680 deer

  d. Determine the number of years required for the deer population to reach 900. Round to the nearest year.   15 yr

  e. What value will the term $\dfrac{2}{e^{0.12t}}$ approach as $t \to \infty$.   0

  f. Determine the limiting value of $P(t)$.   1200 deer

42. The number N of visitors to a new website is given in the table t weeks after the website was launched.

t	0	1	2	3	4
N	24	50	121	270	640

  a. Use a graphing utility to find an equation of the form $N = ab^t$ to model the data. Round a to 1 decimal place and b to 3 decimal places. $N = 23.1(2.283)^t$

  b. Use a graphing utility to graph the data and the model from part (a).

  c. Use the model to predict the number of visitors to the website 10 weeks after launch. Round to the nearest thousand.   89,000

CHAPTER 3 Cumulative Review Exercises

For Exercises 1–2, simplify the expression.

1. $\dfrac{3x^{-1} - 6x^{-2}}{2x^{-2} - x^{-1}}$ -3

2. $\dfrac{5}{\sqrt[3]{2x^2}}$ $\dfrac{5\sqrt[3]{4x}}{2x}$

3. Factor. $a^3 - b^3 - a + b$ $(a - b)(a^2 + ab + b^2 - 1)$

4. Perform the operations and write the answer in scientific notation. $\dfrac{(3.0 \times 10^7)(8.2 \times 10^{-3})}{1.23 \times 10^{-5}}$ 2.0×10^{10}

For Exercises 5–13, solve the equations and inequalities. Write the solution sets to the inequalities in interval notation.

5. $5 \le 3 + |2x - 7|$ $\left(-\infty, \dfrac{5}{2}\right] \cup \left[\dfrac{9}{2}, \infty\right)$

6. $3x(x - 1) = x + 6$ $\left\{\dfrac{2 \pm \sqrt{22}}{3}\right\}$

7. $\sqrt{t + 3} + 4 = t + 1$ $\{6\}$; The value 1 does not check.

8. $9^{2m - 3} = 27^{m + 1}$ $\{9\}$

9. $-x^3 - 5x^2 + 4x + 20 < 0$ $(-5, -2) \cup (2, \infty)$

10. $|5x - 1| = |3 - 4x|$ $\left\{\dfrac{4}{9}, -2\right\}$

11. $(x^2 - 9)^2 - 2(x^2 - 9) - 35 = 0$ $\{\pm 4, \pm 2\}$

12. $\log_2(3x - 1) = \log_2(x + 1) + 3$ $\{\ \}$; The value $-\dfrac{9}{5}$ does not check.

13. $\dfrac{x - 4}{x + 2} \le 0$ $(-2, 4]$

14. Find all the zeros of $f(x) = x^4 + 10x^3 + 10x^2 + 10x + 9$ $-1, -9, \pm i$

15. Given $f(x) = x^2 - 16x + 55$,

a. Does the graph of the parabola open upward or downward?   Upward

b. Find the vertex of the parabola. $(8, -9)$

c. Identify the maximum or minimum point.   Minimum point: $(8, -9)$

d. Identify the maximum or minimum value of the function.   Minimum value: -9

e. Identify the x-intercept(s). $(5, 0)$ and $(11, 0)$

f. Identify the y-intercept. $(0, 55)$

g. Write an equation for the axis of symmetry. $x = 8$

h. Write the domain in interval notation. $(-\infty, \infty)$

i. Write the range in interval notation. $[-9, \infty)$

16. Graph. $f(x) = -1.5x^2(x - 2)^3(x + 1)$

17. Given $f(x) = \dfrac{3x + 6}{x - 2}$,

a. Write an equation of the vertical asymptote(s). $x = 2$

b. Write an equation of the horizontal or slant asymptote.   Horizontal asymptote: $y = 3$

c. Graph the function.

18. Given $f(x) = 2^{x + 2} - 3$,

a. Write an equation of the asymptote. $y = -3$

b. Write the domain in interval notation. $(-\infty, \infty)$

c. Write the range in interval notation. $(-3, \infty)$

19. Write the expression as a single logarithm and simplify. $\log 40 + \log 50 - \log 2$ 3

20. Given the one-to-one function defined by $f(x) = \sqrt[3]{x - 4} + 1$, write an equation for $f^{-1}(x)$. $f^{-1}(x) = (x - 1)^3 + 4$

Trigonometric Functions

Chapter Outline

With the luxury of the modern world, we have become accustomed to having GPS navigation systems in our cars, boats, aircraft, and even smart phones and cameras. But how do these systems work? And how did people in earlier times navigate across continents and oceans or approximate large distances? The answer is by using a branch of mathematics called trigonometry.

Trigonometry, from the Greek roots "trigon" (three-sided) and "metron" (measure), is the study of the relationships among the sides and angles of a triangle. The building blocks of trigonometry were developed in antiquity, going back as far as the early Egyptians and Babylonians who studied the ratios of lengths of sides of similar triangles. Today, trigonometry is used in a wide variety of fields, including navigation. The Global Positioning System (GPS) is a network of 24 satellites and their ground stations. Using trigonometry and the known location of GPS satellites in their precise orbits enables mathematicians to locate points on Earth accurate to within a few meters.

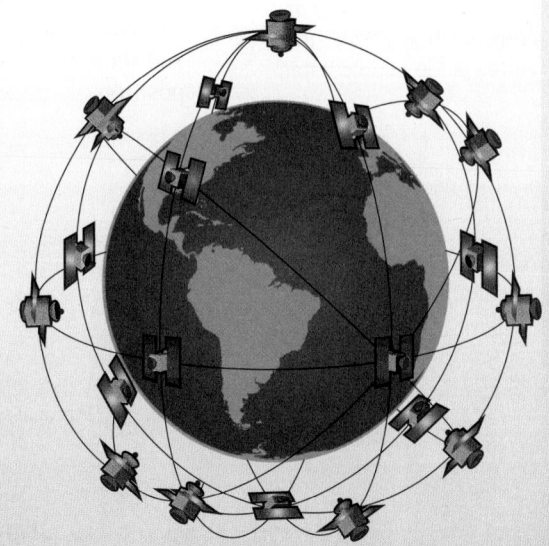

SECTION 4.1 | Angles and Their Measure

1. Find Degree Measure

The Tour de France is the most famous bicycle race in the world spanning 3500 km (2175 mi) in 21 days through France and two mountain ranges. The speed of a racer depends on a number of variables including the gear ratio and the cadence. The gear ratio determines the number of times the rear wheel turns with each pedal stroke, and the cadence is the number of revolutions of the pedals per minute. In this section, we study angles and their measure, and the relationship between angular and linear speeds to study such applications as the speed of a bicycle.

A **ray** is a part of a line that consists of an endpoint and all points on the line to one side of the endpoint. In Figure 4-1, ray $\overrightarrow{PQ}$ is named by using the endpoint P and another point Q on the ray. Notice that the rays $\overrightarrow{PQ}$ and $\overrightarrow{QP}$ are different because the initial points are different and the rays extend in opposite directions.

> **TIP** A ray has only one endpoint, which is always written first when naming the ray.

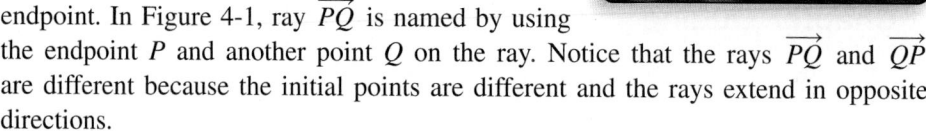

Figure 4-1

An **angle** is formed by rotating a ray about its endpoint. The starting position of the ray is called the **initial side** of the angle, and the final position of the ray is called the **terminal side.** The common endpoint is called the **vertex** of the angle and the vertex is often named by a capital letter such as A (Figure 4-2).

Angle A in Figure 4-2 can be denoted by $\angle A$ (read as "angle A") or by $\angle BAC$, where B is a point on the initial side of the angle, A represents the vertex, and C is a point on the terminal side. Alternatively, Greek letters such as θ (theta), α (alpha), β (beta), and γ (gamma) are often used to denote angles.

An angle is in **standard position** if its vertex is at the origin in the xy-plane, and its initial side is the positive x-axis. In Figure 4-3, angle α is drawn in standard position.

> **TIP** When denoting an angle such as $\angle BAC$, the vertex is the middle letter.

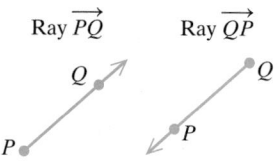

Figure 4-2

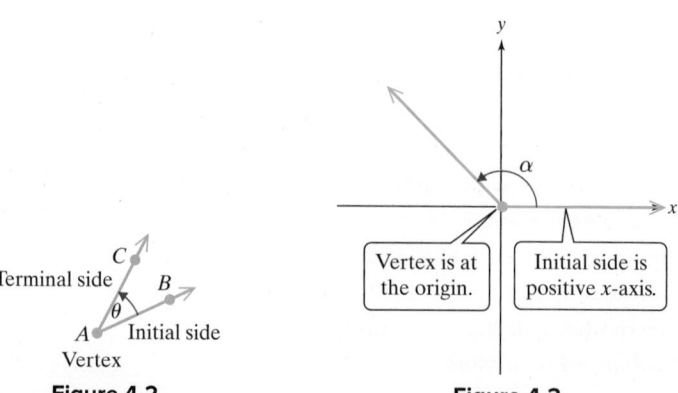

Figure 4-3

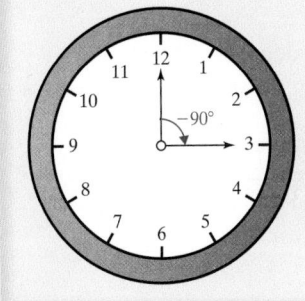

TIP The hands of a clock each resemble a ray. When the minute hand moves from 12:00 to 12:15, the rotation is −90°.

The **measure** of an angle quantifies the direction and amount of rotation from the initial side to the terminal side. The measure of an angle is *positive* if the rotation is counterclockwise, and the measure is *negative* if the rotation is clockwise. One unit with which to measure an angle is the **degree.** One full rotation of a ray about its endpoint is 360 degrees, denoted 360°. Therefore, 1° is $\frac{1}{360}$ of a full rotation. Figure 4-4 shows a variety of angles and their measures.

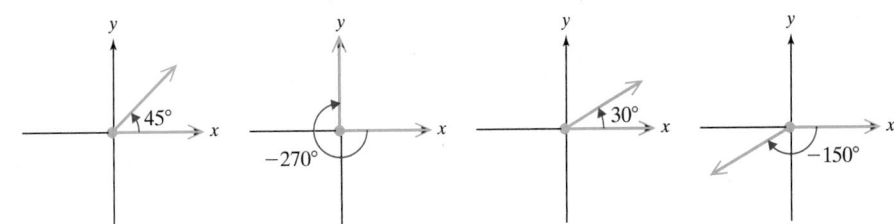

Figure 4-4

If the measure of an angle θ is 30°, we denote the measure of θ as $m(\theta) = 30°$ or simply $\theta = 30°$. We may also refer to θ as a 30° angle, rather than using the more formal, but cumbersome language "an angle whose measure is 30°." We also introduce several key terms associated with the measure of an angle (Figure 4-5).

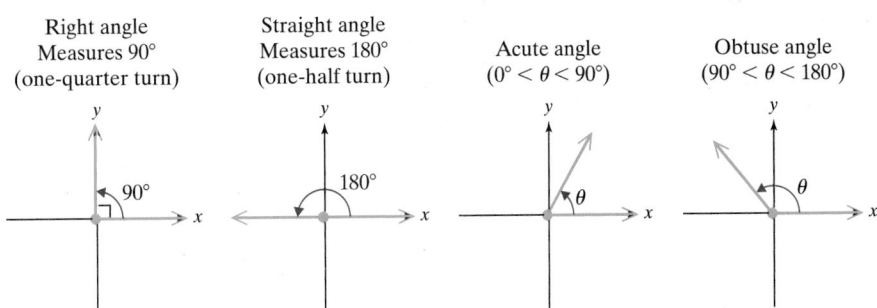

Figure 4-5

If the sum of the measures of two angles is 90°, we say that the angles are **complementary** (for example, the *complement* of a 20° angle is a 70° angle and vice versa). If the sum of the measures of two angles is 180°, we say that the angles are **supplementary** (for example, the supplement of a 20° angle is a 160° angle and vice versa).

A degree can be divided into 60 equal parts called **minutes** (min or ′), and each minute is divided into 60 equal parts called **seconds** (sec or ″).

- 1 min $= \left(\dfrac{1}{60}\right)°$ or $1' = \left(\dfrac{1}{60}\right)°$

- 1 sec $= \left(\dfrac{1}{60}\right)' = \left(\dfrac{1}{3600}\right)°$ or $1'' = \left(\dfrac{1}{60}\right)' = \left(\dfrac{1}{3600}\right)°$

For example, 74 degrees, 42 minutes, 15 seconds is denoted 74°42′15″.

Instructor Note:
Point out to students that the relationships among angular measures of degrees-minutes-seconds is similar to hours-minutes-seconds.

Classroom Example: p. 457
Exercise 30

EXAMPLE 1   Converting from Degrees, Minutes, Seconds (DMS) to Degree Decimal Form

Convert $74°42'15''$ to decimal degrees. Round to 4 decimal places.

Solution:

Convert the minute and second portions of the angle to degrees. Choose the conversion factors so that the original units (minutes and seconds) "cancel," leaving the measurement in degrees.

$$74°42'15'' = 74° + (42 \text{ min}) \cdot \left(\frac{1°}{60 \text{ min}}\right) + (15 \text{ sec}) \cdot \left(\frac{1°}{3600 \text{ sec}}\right)$$

$$= 74° + 0.7° + 0.0041\overline{6}°$$

$$\approx 74.7042°$$

Skill Practice 1   Convert $131°12'33''$ to decimal degrees. Round to 4 decimal places.

Classroom Example: p. 457
Exercise 34

EXAMPLE 2   Converting from Decimal Degrees to Degrees, Minutes, Seconds (DMS) Form

Convert $159.26°$ to degree, minute, second form.

Solution:

$159.26° = 159° + 0.26°$	Write the decimal as a whole number part plus a fractional part. The fractional part of $1°$ needs to be converted from degrees to minutes and seconds.
$= 159° + 0.26° \cdot \left(\frac{60'}{1°}\right)$	Use the conversion factor $60' = 1°$ to convert to minutes.
$= 159° + 15.6'$	Now convert the fractional part of 1 minute to seconds.
$= 159° + 15' + 0.6'$	Write $15.6'$ as a whole number part plus a fractional part.
$= 159° + 15' + 0.6' \cdot \left(\frac{60''}{1'}\right)$	Use the conversion factor $60'' = 1'$ to convert to seconds.
$= 159° + 15' + 36''$	
$= 159°15'36''$	

Skill Practice 2   Convert $26.48°$ to degree, minute, second form.

2. Find Radian Measure

Degree measure is used extensively in many applications of engineering, surveying, and navigation. Another type of angular measure that is better suited for applications in trigonometry and calculus is **radian measure.** To begin, we define a **central angle** as an angle with the vertex at the center of a circle.

Definition of One Radian

A central angle that intercepts an arc on the circle with length equal to the radius of the circle has a measure of **1 radian** (Figure 4-6).

Note: One radian may be denoted as 1 radian, 1 rad, or simply 1. That is, radian measure carries no units.

Answers
1. $131.2092°$
2. $26°28'48''$

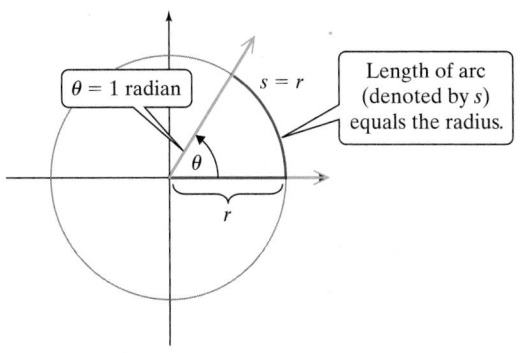

Figure 4-6

> **TIP** When two lines or rays cross a circle, the part of the circle between the intersection points is called the *intercepted arc* and is often denoted by *s*.

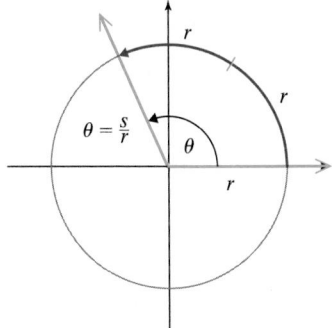

Figure 4-7

> **TIP** Radian measure carries no units because it is measured as a ratio of two lengths with the same units (the units associated with *s/r* "cancel"). Thus, 2π rad is simply written as 2π. It is universally understood that the measure is in radians. Sometimes the notation "rad" is included for emphasis, but is not necessary.

Any central angle can be measured in radians by dividing the length *s* of the intercepted arc by the radius *r*. For example, in Figure 4-7, the length of the red arc is $2r$ (twice the radius). Therefore, the measure of angle θ is given by

$$\theta = \frac{s}{r} = \frac{2r}{r} = 2 \quad (2 \text{ radians})$$

Definition of Radian Measure of an Angle

The radian measure of a central angle θ subtended by an arc of length *s* on a circle of radius *r* is given by $\theta = \dfrac{s}{r}$.

You may have an intuitive feel for angles measured in degrees (for example, 90° is one-quarter of a full rotation). However, radian measure is unfamiliar. From Figure 4-6, notice that an angle of 1 radian (1 rad) is approximately 57.3°. If we divide the circumference of a circle into arcs of length *r* (Figure 4-8), we see that there are just over 6 rad in one full rotation. In fact, we can show that there are exactly 2π rad in one full rotation.

Recall that π is defined as the ratio of the circumference of a circle to its diameter *d*. Therefore, the circumference *C* is given by $C = \pi d$ or equivalently $C = 2\pi r$, where *r* is the radius of the circle. The circumference is the arc length of a full circle and dividing this by the radius gives the number of radians in one revolution.

One revolution
= 2π radians
≈ 6.28 rad

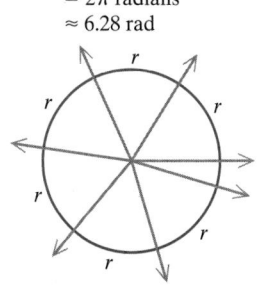

Figure 4-8

Number of radians in one revolution

Arc length of one revolution

$$\theta = \frac{s}{r} = \frac{\text{circumference}}{\text{radius}} = \frac{2\pi r}{r} = 2\pi \approx 6.28$$

The angular measure of one full rotation is 2π (2π rad). Therefore, we have the following relationships.

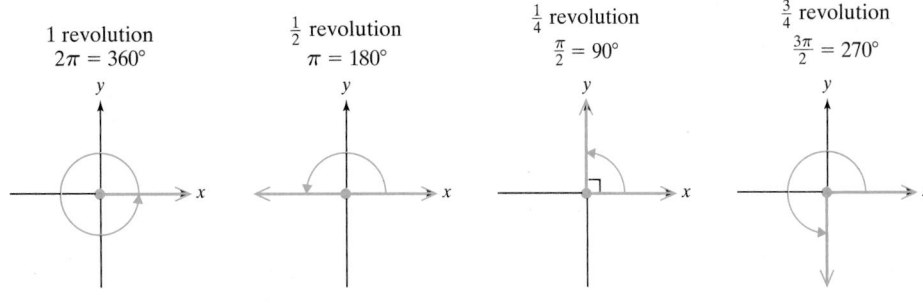

1 revolution
$2\pi = 360°$

$\frac{1}{2}$ revolution
$\pi = 180°$

$\frac{1}{4}$ revolution
$\frac{\pi}{2} = 90°$

$\frac{3}{4}$ revolution
$\frac{3\pi}{2} = 270°$

TIP Some angles are used frequently in the study of trigonometry. Their degree measures and equivalent radian measures are worth memorizing.

$30° = \frac{\pi}{6}$ $\quad 45° = \frac{\pi}{4}$

$60° = \frac{\pi}{3}$ $\quad 90° = \frac{\pi}{2}$

The statement $\pi = 180°$ gives us a conversion factor to convert between degree measure and angular measure.

Converting Between Degree and Radian Measure

- To convert from degrees to radians, multiply the degree measure by $\dfrac{\pi}{180°}$.
- To convert from radians to degrees, multiply the radian measure by $\dfrac{180°}{\pi}$.

Classroom Example: p. 457
Exercise 40

EXAMPLE 3 Converting from Degrees to Radians

Convert from degrees to radians.

   a. $210°$ **b.** $-135°$

Solution:

a. $210° \cdot \left(\dfrac{\pi \text{ rad}}{180°}\right) = \dfrac{\overset{7}{\cancel{210}}\pi}{\underset{6}{\cancel{180}}} \text{ rad} = \dfrac{7\pi}{6} \text{ rad} = \dfrac{7\pi}{6}$ The units of degrees "cancel" in the numerator and denominator in the first step, leaving units of radians.

b. $-135° \cdot \left(\dfrac{\pi}{180°}\right) = -\dfrac{\overset{3}{\cancel{135}}\pi}{\underset{4}{\cancel{180}}} = -\dfrac{3\pi}{4}$ The units of "rad" are implied in the numerator of the conversion factor $\frac{\pi}{180°}$.

Skill Practice 3 Convert from degrees to radians.

   a. $300°$ **b.** $-70°$

Classroom Example: p. 457
Exercise 48

EXAMPLE 4 Converting from Radians to Degrees

Convert from radians to degrees.

   a. $\dfrac{\pi}{12}$ **b.** $-\dfrac{4\pi}{3}$

Solution:

TIP To help determine which conversion factor to use, remember that you want the original units to "cancel," leaving the new unit of measurement in the numerator.

a. $\dfrac{\pi}{12} \text{ rad} \cdot \left(\dfrac{180°}{\pi \text{ rad}}\right) = \left(\dfrac{\overset{15}{\cancel{180}}\pi}{\underset{1}{\cancel{12}}\pi}\right)° = 15°$ The units of rad "cancel" in the numerator and denominator in the first step, leaving units of degrees.

b. $-\dfrac{4\pi}{3} \cdot \left(\dfrac{180°}{\pi}\right) = \left(-\dfrac{\overset{240}{\cancel{720}}\pi}{\underset{1}{\cancel{3}}\pi}\right)° = -240°$ The units of "rad" are implied in the denominator of the conversion factor $\frac{180°}{\pi}$.

Skill Practice 4 Convert from radians to degrees.

   a. $\dfrac{\pi}{18}$ **b.** $-\dfrac{7\pi}{4}$

---

3. Determine Coterminal Angles

Two angles in standard position with the same initial side and same terminal side are called **coterminal angles.** Figure 4-9 illustrates three angles in standard position that are coterminal to 30°. Notice that each angle is 30° plus or minus some number of full revolutions clockwise or counterclockwise.

Answers

3. a. $\dfrac{5\pi}{3}$ **b.** $-\dfrac{7\pi}{18}$

4. a. $10°$ **b.** $-315°$

Classroom Example: p. 457
Exercise 58

TIP The number of revolutions contained in 960° can be found by dividing 960° by 360°.

$$\begin{array}{r} 2 \\ 360\overline{)960} \\ 720 \\ \overline{240} \end{array}$$

2 revolutions

remainder

$\theta = 960°$ is two full revolutions plus 240°.

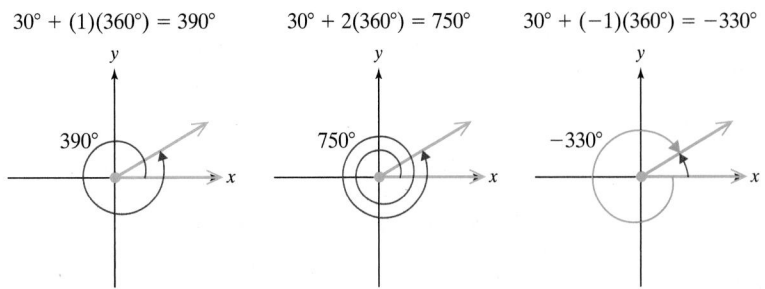

$30° + (1)(360°) = 390°$ $30° + 2(360°) = 750°$ $30° + (-1)(360°) = -330°$

Figure 4-9   Coterminal Angles

EXAMPLE 5   **Finding Coterminal Angles**

Find an angle coterminal to θ between 0° and 360°.

    a. $\theta = 960°$ **b.** $\theta = -225°$

Solution:

a. $\theta = 960°$ is more than 360°. Therefore, we will subtract some multiple of 360° to get an angle coterminal to θ between 0° and 360°.

- One revolution measures 360°.
- Two revolutions measure 720°.
- Three revolutions measure 1080°.

$360°(2)$ is subtracted from 960° because θ is between 2 and 3 revolutions. Therefore,

$$960° - (360°)(2) = 960° - 720° = 240°$$

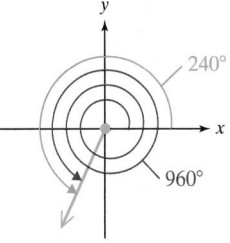

b. An angle of $-225°$ is less than one revolution. Therefore, we will add $360°(1)$ to obtain a positive angle between 0° and 360° and coterminal to $-225°$.

$$-225° + 360° = 135°$$

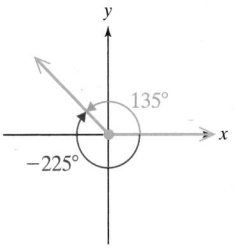

Skill Practice 5   Find an angle coterminal to θ between 0° and 360°.

    a. $\theta = 1230°$ **b.** $\theta = -315°$

Two angles in standard position are coterminal if their measures differ by a multiple of 360° or 2π rad. The angles shown in Figure 4-10 are coterminal to an angle of $\frac{\pi}{6}$ rad (or 30°).

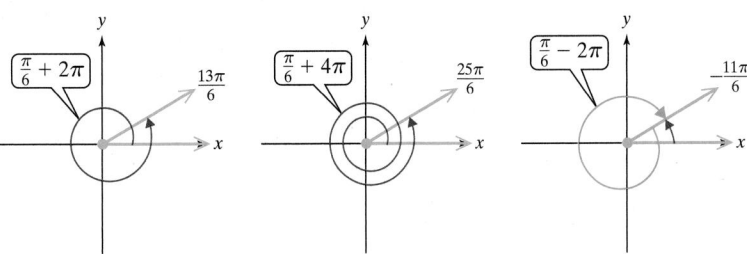

Figure 4-10   Coterminal Angles

Answers
5. a. 150°   **b.** 45°

EXAMPLE 6 **Finding Coterminal Angles**

Find an angle coterminal to θ on the interval $[0, 2\pi)$.

a. $\theta = -\dfrac{5\pi}{6}$ **b.** $\theta = \dfrac{13\pi}{2}$

Solution:

a. One revolution is 2π rad or equivalently $\dfrac{12\pi}{6}$.

Therefore, adding any multiple of $\dfrac{12\pi}{6}$ to $-\dfrac{5\pi}{6}$

results in an angle coterminal to $-\dfrac{5\pi}{6}$.

$$-\dfrac{5\pi}{6} + \dfrac{12\pi}{6} = \dfrac{7\pi}{6}$$

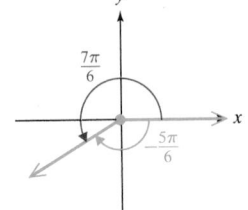

b. One revolution is 2π rad, or equivalently $\dfrac{4\pi}{2}$.

We can divide $\dfrac{13\pi}{2} \div \dfrac{4\pi}{2} = \dfrac{13\pi}{2} \cdot \dfrac{2}{4\pi} = \dfrac{13}{4} = 3\dfrac{1}{4}$.

Therefore, $\dfrac{13\pi}{2}$ is three full revolutions plus $\dfrac{1}{4}$ of

a revolution. Subtracting three revolutions results in an

angle coterminal to $\dfrac{13\pi}{2}$.

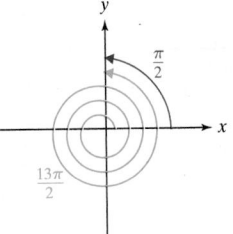

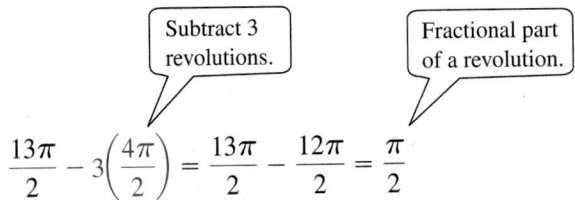

| Subtract 3 revolutions. | | Fractional part of a revolution. |

$$\dfrac{13\pi}{2} - 3\left(\dfrac{4\pi}{2}\right) = \dfrac{13\pi}{2} - \dfrac{12\pi}{2} = \dfrac{\pi}{2}$$

Skill Practice 6 Find an angle coterminal to θ on the interval $[0, 2\pi)$.

a. $\theta = -\dfrac{\pi}{8}$ **b.** $\theta = \dfrac{19\pi}{4}$

4. Compute Arc Length of a Sector of a Circle

By definition, the measure of a central angle θ in radians equals the length s of the intercepted arc divided by the radius r, that is, $\theta = \dfrac{s}{r}$. Solving this relationship for s gives $s = r\theta$, which enables us to compute arc length if the measure of the central angle and radius are known.

Arc Length

Given a circle of radius r, the length s of an arc intercepted by a central angle θ (in radians) is given by

$$s = r\theta$$

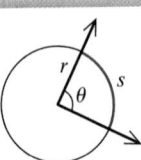

Answers

6. a. $\dfrac{15\pi}{8}$ **b.** $\dfrac{3\pi}{4}$

Classroom Example: p. 458
Exercise 74

EXAMPLE 7 Determining Arc Length

Find the length of the arc made by an angle of $105°$ on a circle of radius 15 cm. Give the exact arc length and round to the nearest tenth of a centimeter.

Solution:

$$\theta = 105° \cdot \frac{\pi}{180°} = \frac{7\pi}{12} \qquad \text{Convert } \theta \text{ to radians.}$$

$$s = r\theta = (15 \text{ cm}) \cdot \frac{7\pi}{12} = 8.75\pi \text{ cm} \qquad \text{Apply the arc length formula.}$$

The arc is $8.75\pi \approx 27.5$ cm.

Skill Practice 7 Find the length of the arc made by an angle of $220°$ on a circle of radius 9 in.

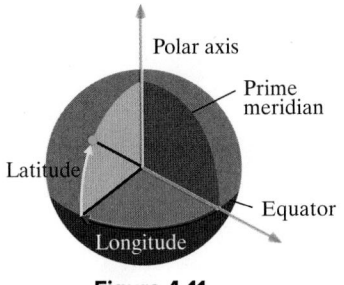

Figure 4-11

Instructor Note:

To help remember the difference between the terms "latitude" and "longitude," tell students that the Earth is not a perfect sphere, but rather is "fatter" at the equator. Therefore, "longitude" measures the "long" way around the Earth.

The Earth is approximately spherical and the most common way to locate points on the surface is by using *latitude* and *longitude*. These coordinates are measured in degrees and represent central angles measured from the center of the Earth.

The **equator** is an imaginary circle around the Earth equidistant between the north and south poles. **Latitude** is the angular measure of a central angle measuring north (N) or south (S) from the equator. There are $90°$ of latitude measured north from the equator and $90°$ of latitude measured south from the equator. The equator has a latitude of $0°$, the north pole has a latitude of $90°$N, and the south pole has a latitude of $90°$S (Figure 4-11).

Lines of **longitude,** called meridians, are circles that pass through both poles and run perpendicular to the equator. By international agreement, $0°$ longitude is taken to be the meridian line through Greenwich, England. This is called the **prime meridian.** Thus, longitude is the angular measure of a central angle east (E) or west (W) of the prime meridian. The Earth is divided into $360°$ of longitude. There are $180°$ east (E) of the prime meridian and $180°$ west (W) of the prime meridian.

For example, New York City is located at $40.7°$N, $74.0°$W. This means that New York City is located $40.7°$ north of the equator and $74.0°$ west of the prime meridian.

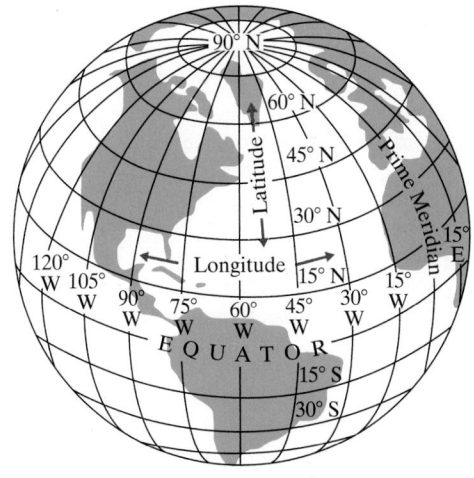

Classroom Example: p. 458
Exercise 80

EXAMPLE 8 Determining the Distance Between Cities

Seattle, Washington, is located at $47.6°$N, $122.4°$W, and San Francisco, California, is located at $37.8°$N, $122.3°$W. Since the longitudes are nearly the same, the cities are roughly due north-south of each other. Using the difference in latitude, approximate the distance between the cities assuming that the radius of the earth is 3960 mi. Round to the nearest mile.

Answer

7. $11\pi \approx 34.6$ in.

Solution:

$$47.6° - 37.8° = 9.8°$$

$$\theta = 9.8° \cdot \frac{\pi}{180°} \approx 0.17104$$

$$s = r\theta$$
$$s = (3960)(0.17104) \approx 677 \text{ mi}$$

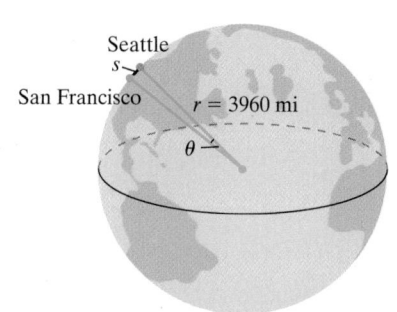

Skill Practice 8 Lincoln, Nebraska, is located at 40.8°N, 96.7°W and Dallas, Texas, is located at 32.8°N, 96.7°W. Since the longitudes are the same, the cities are north-south of each other. Using the difference in latitudes, approximate the distance between the cities assuming that the radius of the Earth is 3960 mi. Round to the nearest mile.

5. Compute Linear and Angular Speed

Consider a ceiling fan rotating at a constant rate. For a small increment of time, suppose that point A on the tip of a blade travels a distance s_1. In the same amount of time, point B will travel a shorter distance s_2 (Figure 4-12). Therefore, point A on the tip has a greater linear speed v than point B. However, each point has the same *angular* speed ω (omega) because they move through the same angle for a given unit of time.

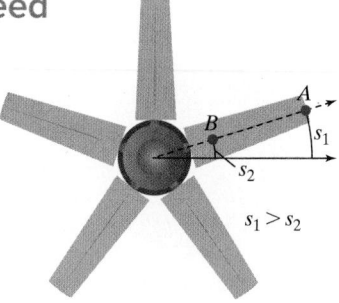

Figure 4-12

Angular and Linear Speed

If a point on a circle of radius r moves through an angle of θ radians in time t, the angular and linear speeds of the point are

$$\text{angular speed:} \quad \omega = \frac{\theta}{t}$$

$$\text{linear speed:} \quad v = \frac{s}{t} \quad \text{or} \quad v = \frac{r\theta}{t} \quad \text{or} \quad v = r\omega$$

EXAMPLE 9 Finding Linear and Angular Speed

A ceiling fan rotates at 90 rpm (revolutions per minute). For a point at the tip of a 2-ft blade,

a. Find the angular speed.

b. Find the linear speed. Round to the nearest whole unit.

Solution:

a. For each revolution of the blade, the point moves through an angle of 2π radians.

$$\omega = \frac{\theta}{t} = \frac{90 \text{ rev}}{\text{min}} \cdot \frac{2\pi \text{ rad}}{\text{rev}} = 180\pi \text{ rad/min} = 180\pi/\text{min}$$

b. $v = r\omega$

$$v = (2 \text{ ft})\left(\frac{180\pi}{\text{min}}\right) = 360\pi \text{ ft/min} \approx 1131 \text{ ft/min}$$

Skill Practice 9   A bicycle wheel rotates at 2 revolutions per second.

  a. Find the angular speed.
  b. How fast does the bicycle travel (in ft/sec) if the wheel is 2.2 ft in diameter? Round to the nearest tenth.

6. Compute the Area of a Sector of a Circle

A sector of a circle is a "pie-shaped" wedge of a circle bounded by the sides of a central angle and the intercepted arc (Figure 4-13).

   The area of a sector of a circle is proportional to the measure of the central angle. The expression $\dfrac{\theta}{2\pi}$ is the fractional amount of a full rotation represented by angle θ (in radians). So the area of a sector formed by θ is

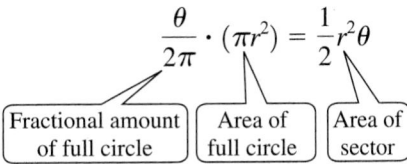

$$\frac{\theta}{2\pi} \cdot (\pi r^2) = \frac{1}{2}r^2\theta$$

Fractional amount of full circle · Area of full circle = Area of sector

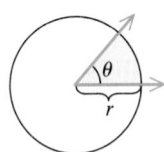

Figure 4-13

Area of a Sector

The area A of a sector of a circle of radius r with central angle θ (in radians) is given by

$$A = \frac{1}{2}r^2\theta$$

Classroom Example: p. 460
Exercise 100

EXAMPLE 10   Determining the Area of a Sector

A crop sprinkler rotates through an angle of $150°$ and sprays water a distance of 90 ft. Find the amount of area watered. Round to the nearest whole unit.

Solution:

To use the formula $A = \dfrac{1}{2}r^2\theta$, we need to convert θ to radians.

$$150° \cdot \frac{\pi}{180°} = \frac{5\pi}{6}$$

$$A = \frac{1}{2}r^2\theta = \frac{1}{2}(90 \text{ ft})^2\left(\frac{5\pi}{6}\right) = 3375\pi \text{ ft}^2 \approx 10{,}603 \text{ ft}^2$$

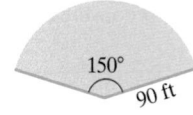

Skill Practice 10   A sprinkler rotates through an angle of $120°$ and sprays water a distance of 30 ft. Find the amount of area watered. Round to the nearest whole unit.

Answers
9. a. 4π/sec
   b. ≈ 13.8 ft/sec (≈ 9.4 mph)
10. 300π ft$^2 \approx 942$ ft^2

SECTION 4.1   Practice Exercises

Prerequisite Review

R.1. Solve for the specified variable.

$p = hz$ for z $z = \dfrac{p}{h}$

R.2. If a plane travels 280 mph for 3.5 hr, find the distance traveled.   980 mi

For Exercises R.3–R.5, convert the unit of time.

R.3. 210 min = _____3.5_____ hr

R.4. 120 sec = _____2_____ min

R.5. 7200 sec = _____2_____ hr

R.6. Find the circumference of a circle with a radius of $2\dfrac{1}{2}$ m.

    a. Give the exact answer in terms of π. 5π m

    b. Approximate the answer by using 3.14 for π. Round to 1 decimal place. ≈ 15.7 m

R.7. Determine the area of a circle with a diameter of 40 ft. Use 3.14 for π. Round to the nearest whole unit. ≈ 1256 ft^2

Concept Connections

1. The measure of an angle is (positive/negative) _____negative_____ if its rotation from the initial side to the terminal side is clockwise. If the rotation is counterclockwise, then the measure is _____positive_____.

2. An angle with its vertex at the origin of an xy-coordinate plane and with initial side on the positive x-axis is in _____standard_____ position.

3. Two common units used to measure angles are _____degrees_____ and _____radians_____.

4. One degree is what fractional amount of a full rotation? $\dfrac{1}{360}$

5. An angle that measures 360° has a measure of _____2π_____ radians.

6. Two angles are called _____complementary_____ if the sum of their measures is 90°. Two angles are called _____supplementary_____ if the sum of their measures is 180°.

7. An angle with measure _____90_____° or _____$\dfrac{\pi}{2}$_____ radians is a right angle. A straight angle has a measure of _____180_____° or _____π_____ radians.

8. A(n) _____acute_____ angle has a measure between 0° and 90°, whereas a(n) _____obtuse_____ angle has a measure between 90° and 180°.

9. One degree is equally divided into 60 parts called _____minutes_____.

10. One minute is equally divided into 60 parts called _____seconds_____.

11. 1° = _____60_____′ = _____3600_____″

12. An angle with its vertex at the center of a circle is called a(n) _____central_____ angle.

13. A central angle of a circle that intercepts an arc equal in length to the radius of the circle has a measure of _____1 radian_____.

14. Which angle has a greater measure, 2° or 2 radians?   2 radians

15. To convert from radians to degrees, multiply by _____$\dfrac{180°}{\pi}$_____. To convert from degrees to radians, multiply by _____$\dfrac{\pi}{180°}$_____.

16. Two angles are _____coterminal_____ if they have the same initial side and same terminal side.

17. The measure of all angles coterminal to $\dfrac{7\pi}{4}$ differ from $\dfrac{7\pi}{4}$ by a multiple of _____2π_____.

18. The measure of all angles coterminal to 112° differ from 112° by a multiple of _____360°_____.

19. The length s of an arc made by an angle θ on a circle of radius r is given by the formula _____$s = r\theta$_____, where θ is measured in _____radians_____.

20. To locate points on the surface of the Earth that are north or south of the equator, we measure the _____latitude_____ of the point. To locate points that are east or west of the prime meridian, we measure the _____longitude_____ of the point.

21. The relationship $v = \dfrac{\text{arc length}}{\text{time}}$ represents the _____linear_____ speed of a point traveling in a circular path.

22. The symbol ω is typically used to denote _____angular_____ speed and represents the number of radians per unit time that an object rotates.

23. A wedge of a circle, similar in shape to a slice of pie, is called a(n) _____sector_____ of the circle.

24. The area A of a sector of a circle of radius r with central angle θ is given by the formula _____$A = \frac{1}{2}r^2\theta$_____, where θ is measured in _____radians_____.

Objective 1: Find Degree Measure

For Exercises 25–26, sketch the angles in standard position.

25. a. $60°$ **b.** $225°$ **c.** $-210°$ **d.** $-86°$

26. a. $30°$ **b.** $120°$ **c.** $-135°$ **d.** $-73°$

For Exercises 27–30, convert the given angle to decimal degrees. Round to 4 decimal places. (See Example 1)

27. $17°34'$ 17.5667°   **28.** $215°47'$ 215.7833°   **29.** $54°36'55''$ 54.6153°   **30.** $23°42'48''$ 23.7133°

For Exercises 31–34, convert the given angle to DMS (degree-minute-second) form. Round to the nearest second if necessary. (See Example 2)

31. $46.418°$ 46°25'5''   **32.** $82.074°$ 82°4'26''   **33.** $-84.64°$ −84°38'24''   **34.** $-61.46°$ −61°27'36''

Objective 2: Find Radian Measure

35. Use the conversion factor $\pi = 180°$ along with the symmetry of the circle to complete the degree or radian measure of the missing angles.

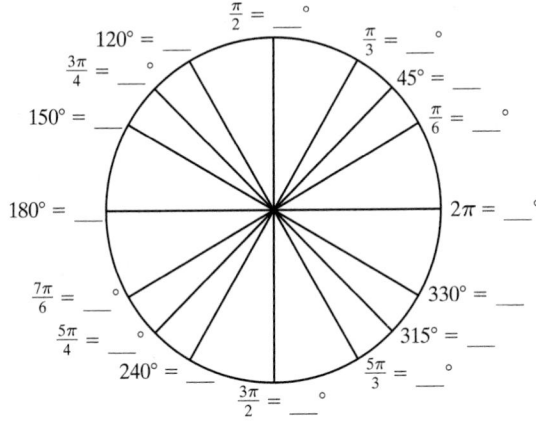

36. Insert the appropriate symbol $<$, $>$, or $=$ in the blank.

a. $\dfrac{5\pi}{6}$ ☐ $120°$ $>$

b. $-\dfrac{4\pi}{3}$ ☐ $-270°$ $>$

For Exercises 37–40, convert from degrees to radians. Give the answers in exact form in terms of π. (See Example 3)

37. $75°$ $\dfrac{5\pi}{12}$ **38.** $240°$ $\dfrac{4\pi}{3}$ **39.** $-210°$ $-\dfrac{7\pi}{6}$ **40.** $-195°$ $-\dfrac{13\pi}{12}$

For Exercises 41–44, convert from degrees to radians. Round to 4 decimal places.

41. $-64.6°$ −1.1275   **42.** $-312.4°$ −5.4524   **43.** $12°6'36''$ 0.2114   **44.** $108°42'9''$ 1.8972

For Exercises 45–56, convert from radians to decimal degrees. Round to 1 decimal place if necessary. (See Example 4)

45. $\dfrac{\pi}{4}$ 45°   **46.** $\dfrac{11\pi}{6}$ 330°   **47.** $-\dfrac{5\pi}{3}$ −300°   **48.** $-\dfrac{7\pi}{6}$ −210°

49. $\dfrac{5\pi}{18}$ 50°   **50.** $\dfrac{7\pi}{9}$ 140°   **51.** $-\dfrac{2\pi}{5}$ −72°   **52.** $-\dfrac{3\pi}{8}$ −67.5°

53. 2.7 154.7°   **54.** 5.3 303.7°   **55.** $\dfrac{9\pi}{2}$ 810°   **56.** 7π 1260°

Objective 3: Determine Coterminal Angles

For Exercises 57–64, find a positive angle and a negative angle that is coterminal to the given angle. (See Examples 5 and 6)

57. $57°$ For example: 417°, −303°   **58.** $313°$ For example: 673°, −47°   **59.** $-105°$ For example: 255°, −465°   **60.** $-12°$ For example: 348°, −372°

61. $\dfrac{5\pi}{6}$ For example: $\dfrac{17\pi}{6}$, $-\dfrac{7\pi}{6}$ **62.** $\dfrac{3\pi}{4}$ For example: $\dfrac{11\pi}{4}$, $-\dfrac{5\pi}{4}$ **63.** $-\dfrac{3\pi}{2}$ For example: $\dfrac{\pi}{2}$, $-\dfrac{7\pi}{2}$ **64.** $-\pi$ For example: π, -3π

Additional answers can be found in the Instructor Answer Appendix.

For Exercises 65–70, find an angle between 0° and 360° or between 0 and 2π that is coterminal to the given angle. (See Examples 5 and 6)

65. $1521°$ 81°

66. $-603°$ 117°

67. $\dfrac{17\pi}{4}$ $\dfrac{\pi}{4}$

68. $\dfrac{11\pi}{3}$ $\dfrac{5\pi}{3}$

69. $-\dfrac{7\pi}{18}$ $\dfrac{29\pi}{18}$

70. $-\dfrac{5\pi}{9}$ $\dfrac{13\pi}{9}$

Objective 4: Compute Arc Length of a Sector of a Circle

For Exercises 71–74, find the exact length of the arc intercepted by a central angle θ on a circle of radius r. Then round to the nearest tenth of a unit. (See Example 7)

71. $\theta = \dfrac{\pi}{3}$, $r = 12$ cm 4π cm ≈ 12.6 cm

72. $\theta = \dfrac{5\pi}{6}$, $r = 4$ m $\dfrac{10\pi}{3}$ m ≈ 10.5 m

73. $\theta = 135°$, $r = 10$ in. $\dfrac{15\pi}{2}$ in. ≈ 23.6 in.

74. $\theta = 315°$, $r = 2$ yd $\dfrac{7\pi}{2}$ yd ≈ 11.0 yd

75. A 6-ft pendulum swings through an angle of 40°36'. What is the length of the arc that the tip of the pendulum travels? Round to the nearest hundredth of a foot.   4.25 ft

76. A gear with a 1.2-cm radius moves through an angle of 220°15'. What distance does a point on the edge of the gear move? Round to the nearest tenth of a centimeter.   4.6 cm

77. Which location would best fit the coordinates 12°S, 77°W?   b. Lima, Peru

   a. Paris, France   **b.** Lima, Peru

   c. Miami, Florida   **d.** Moscow, Russia

78. a. What is the geographical relationship between two points that have the same latitude?
   The points are due east-west of each other.

   b. What is the geographical relationship between two points that have the same longitude?
   The points are due north-south of each other.

For Exercises 79–82, assume that the Earth is approximately spherical with radius 3960 mi. Approximate the distances to the nearest mile. (See Example 8)

79. Barrow, Alaska (71.3°N, 156.8°W), and Kailua, Hawaii (19.7°N, 156.1°W), have approximately the same longitude, which means that they are roughly due north-south of each other. Use the difference in latitude to approximate the distance between the cities.   ≈ 3566 mi

80. Rochester, New York (43.2°N, 77.6°W), and Richmond, Virginia (37.5°N, 77.5°W), have approximately the same longitude, which means that they are roughly due north-south of each other. Use the difference in latitude to approximate the distance between the cities.   ≈ 394 mi

81. Raleigh, North Carolina (35.8°N, 78.6°W), is located north of the equator, and Quito, Ecuador (0.3°S, 78.6°W), is located south of the equator. The longitudes are the same indicating that the cities are due north-south of each other. Use the difference in latitude to approximate the distance between the cities.   ≈ 2495 mi

82. Trenton, New Jersey (40.2°N, 74.8°W), is located north of the equator, and Ayacucho, Peru (13.2°S, 74.2°W), is located south of the equator. The longitudes are nearly the same indicating that the cities are roughly due north-south of each other. Use the difference in latitude to approximate the distance between the cities.   ≈ 3691 mi

83. A pulley is 16 cm in diameter.

   a. Find the distance the load will rise if the pulley is rotated 1350°. Find the exact distance in terms of π and then round to the nearest centimeter. 60π cm ≈ 188 cm

   b. Through how many degrees should the pulley rotate to lift the load 100 cm? Round to the nearest degree.   716°

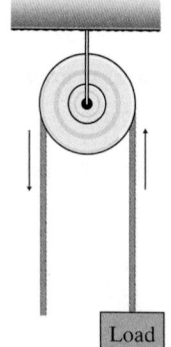

Load

84. A pulley is 1.2 ft. in diameter.

   a. Find the distance the load will rise if the pulley is rotated 630°. Find the exact distance in terms of π and then round to the nearest tenth of a foot. 2.1π ft ≈ 6.6 ft

   b. Through how many degrees should the pulley rotate to lift the load 24 ft? Round to the nearest degree.   2292°

85. A hoist is used to lift a palette of bricks. The drum on the hoist is 15 in. in diameter. How many degrees should the drum be rotated to lift the palette a distance of 6 ft? Round to the nearest degree.   550°

15 in.

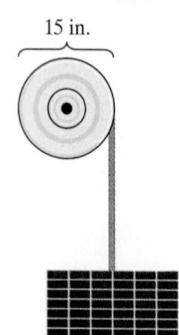

86. A winch on a sailboat is 8 in. in diameter and is used to pull in the "sheets" (ropes used to control the corners of a sail). To the nearest degree, how far should the winch be turned to pull in 2 ft of rope?   344°

Before the widespread introduction of electronic devices to measure distances, surveyors used a **subtense bar** to measure a distance x that is not directly measurable. A subtense bar is a bar of known length h with marks or "targets" at either end. The surveyor measures the angle θ formed by the location of the surveyor's scope and the top and bottom of the bar (this is the angle *subtended* by the bar). Since the angle and height of the bar are known, right triangle trigonometry can be used to find the horizontal distance. Alternatively, if the distance from the surveyor to the bar is large, then the distance can be approximated by the radius r of the arc s intercepted by the bar. Use this information for Exercises 87–88.

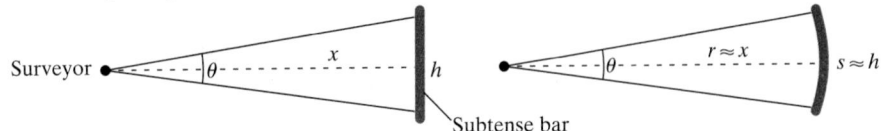

87. A surveyor uses a subtense bar to find the distance across a river. If the angle of sight between the bottom and top marks on a 2-m bar is $57'18''$, approximate the distance across the river between the surveyor and the bar. Round to the nearest meter.   120 m

88. A surveyor uses a subtense bar to find the distance across a canyon. If the angle of sight between the bottom and top marks on a 2-m bar is $24'33''$, approximate the distance across the river between the surveyor and the bar. Round to the nearest meter.   280 m

Objective 5: Compute Linear and Angular Speed

89. A circular paddle wheel of radius 3 ft is lowered into a flowing river. The current causes the wheel to rotate at a speed of 12 rpm. To 1 decimal place,

  a. What is the angular speed? (**See Example 9**)
    24π/min ≈ 75.4/min
  b. Find the speed of the current in ft/min.
    72π ft/min ≈ 226.2 ft/min
  c. Find the speed of the current in mph. $\dfrac{9\pi}{11}$ mph ≈ 2.6 mph

90. An energy-efficient hard drive has a 2.5-in. diameter and spins at 4200 rpm.

  a. What is the angular speed?
    8400π/min
  b. How fast in in./min does a point on the edge of the hard drive spin? Give the exact speed and the speed rounded to the nearest in./min.
    $10{,}500\pi$ in./min $\approx 32{,}987$ in./min

91. A $7\frac{1}{4}$-in.-diameter circular saw has 24 teeth and spins at 5800 rpm.

  a. What is the angular speed? $11{,}600\pi$/min

  b. What is the linear speed of one of the "teeth" on the outer edge of the blade? Round to the nearest inch per minute.   132,104 in./min

92. On a weed-cutting device, a thick nylon line rotates on a spindle at 3000 rpm.

  a. Determine the angular speed. 6000π/min

  b. Determine the linear speed (to the nearest inch per minute) of a point on the tip of the line if the line is 5 in.   94,248 in./min

93. A truck has 2.5-ft tires (in diameter).

  a. What distance will the truck travel with one rotation of the wheels? Give the exact distance and an approximation to the nearest tenth of a foot.
    2.5π ft ≈ 7.9 ft
  b. How far will the truck travel with 10,000 rotations of the wheels? Give the exact distance and an approximation to the nearest foot. $25{,}000\pi$ ft $\approx 78{,}540$ ft

  c. If the wheels turn at 672 rpm, what is the angular speed? 1344π/min

  d. If the wheels turn at 672 rpm, what is the linear speed in feet per minute? Give the exact distance and an approximation to the nearest whole unit.
    1680π ft/min ≈ 5278 ft/min
  e. If the wheels turn at 672 rpm, what is the linear speed in miles per hour? Round to the nearest mile per hour. (*Hint*: 1 mi = 5280 ft and 1 hr = 60 min.)   60 mph

94. A bicycle has 25-in. wheels (in diameter).

  a. What distance will the bicycle travel with one rotation of the wheels? Give the exact distance and an approximation to the nearest tenth of an inch.
    25π in. ≈ 78.5 in.
  b. How far will the bicycle travel with 200 rotations of the wheels? Give the exact distance and approximations to the nearest inch and nearest foot.
    5000π in. $\approx 15{,}708$ in. ≈ 1309 ft
  c. If the wheels turn at 80 rpm, what is the angular speed? 160π/min

  d. If the wheels turn at 80 rpm, what is the linear speed in inches per minute? Give the exact speed and an approximation to the nearest inch per minute.
    2000π in./min ≈ 6283 in./min
  e. If the wheels turn at 80 rpm, what is the linear speed in miles per hour? Round to the nearest mile per hour. (*Hint*: 1 ft = 12 in., 1 mi = 5280 ft, and 1 hr = 60 min.)
    6 mph

Objective 6: Compute the Area of a Sector of a Circle

For Exercises 95–98, find the exact area of the sector. Then round the result to the nearest tenth of a unit. (See Example 10)

95.

$\dfrac{5\pi}{3}$ $r = 6$ m

30π m$^2 \approx 94.2$ m^2

96.

$\dfrac{\pi}{6}$
$r = 1.2$ ft

$\dfrac{3\pi}{25}$ ft$^2 \approx 0.4$ ft^2

97.

$120°$ $r = 3$ cm

3π cm$^2 \approx 9.4$ cm^2

98.

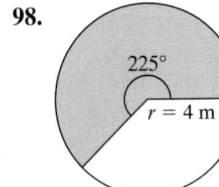

$225°$ $r = 4$ m

10π m$^2 \approx 31.4$ m^2

99. A slice of a circular pizza 12 in. in diameter is cut into a wedge with a 45° angle. Find the area and round to the nearest tenth of a square inch. (**See Example 10**)
14.1 in.²

100. A circular cheesecake 9 in. in diameter is cut into a slice with a 20° angle. Find the area and round to the nearest tenth of a square inch.   3.5 in.²

101. The back wiper blade on an SUV extends 3 in. from the pivot point to a distance of 17 in. from the pivot point. If the blade rotates through an angle of 175°, how much area does it cover? Round to the nearest square inch.   428 in.²

102. A robotic arm rotates through an angle of 160°. It sprays paint between a distance of 0.5 ft and 3 ft from the pivot point. Determine the amount of area that the arm makes. Round to the nearest square foot.   12 ft²

Mixed Exercises

For Exercises 103–108, find the (a) complement and (b) supplement of the given angle.

103. 16.21°   a. 73.79°   b. 163.79°   **104.** 49.87°   a. 40.13°   b. 130.13°   **105.** 18°13′37″   a. 71°46′23″   b. 161°46′23″

106. 22°9′54″   a. 67°50′6″   b. 157°50′6″   **107.** 9°42′7″   a. 80°17′53″   b. 170°17′53″   **108.** 82°15′3″   a. 7°44′57″   b. 97°44′57″

109. The second hand of a clock moves from 12:10 to 12:30.

   a. How many degrees does it move during this time?
   120°
   b. How many radians does it move during this time? $\frac{2\pi}{3}$

   c. If the second hand is 10 in. in length, determine the exact distance that the tip of the second hand travels during this time. $\frac{20\pi}{3}$ in.

   d. Determine the exact angular speed of the second hand in radians per second. $\frac{\pi}{30}$/sec

   e. What is the exact linear speed (in inches per second) of the tip of the second hand? $\frac{\pi}{3}$ in./sec

   f. What is the amount of area that the second hand sweeps out during this time? Give the exact area in terms of π and then approximate to the nearest square inch. $\frac{100\pi}{3}$ in.² ≈ 105 in.²

110. The minute hand of a clock moves from 12:10 to 12:15.

   a. How many degrees does it move during this time?   30°
   b. How many radians does it move during this time? $\frac{\pi}{6}$

   c. If the minute hand is 9 in. in length, determine the exact distance that the tip of the minute hand travels during this time. $\frac{3\pi}{2}$ in.

   d. Determine the exact angular speed of the minute hand in radians per minute. $\frac{\pi}{30}$/min

   e. What is the exact linear speed (in inches per minute) of the tip of the minute hand? $\frac{3\pi}{10}$ in./min

   f. What is the amount of area that the minute hand sweeps out during this time? Give the exact area in terms of π and then approximate to the nearest square inch. $\frac{27\pi}{4}$ in.² ≈ 21 in.²

111. The Earth's orbit around the Sun is elliptical (oval shaped). However, the elongation is small, and for our discussion here, we take the orbit to be circular with a radius of approximately 93,000,000 mi.

   a. Find the linear speed (in mph) of the Earth through its orbit around the Sun. Round to the nearest hundred miles per hour.   66,700 mph

   b. How far does the Earth travel in its orbit in one day? Round to the nearest thousand miles.   1,601,000 mi

112. The Earth completes one full rotation around its axis (poles) each day.

   a. Determine the angular speed (in radians per hour) of the Earth during its rotation around its axis. $\frac{\pi}{12}$/hr

   b. The Earth is nearly spherical with a radius of approximately 3960 mi. Find the linear speed of a point on the surface of the Earth rounded to the nearest mile per hour.   1037 mph

113. Two gears are calibrated so that the smaller gear drives the larger gear. For each rotation of the smaller gear, how many degrees will the larger gear rotate?   144°

$r = 2$ cm $r = 5$ cm

114. Two gears are calibrated so that the larger gear drives the smaller gear. The larger gear has a 6-in. radius, and the smaller gear has a 1.5-in. radius. For each rotation of the larger gear, by how many degrees will the smaller gear rotate?   1440°

115. A spinning-disc confocal microscope contains a rotating disk with multiple small holes arranged in a series of nested Archimedean spirals. An intense beam of light is projected through the holes, enabling biomedical researchers to obtain detailed video images of live cells. The spinning disk has a diameter of 55 mm and rotates at a rate of 1800 rpm. At the edge of the disk,

   a. Find the angular speed.   3600π/min

   b. Find the linear speed. Round to the nearest whole unit.   311,018 mm/min

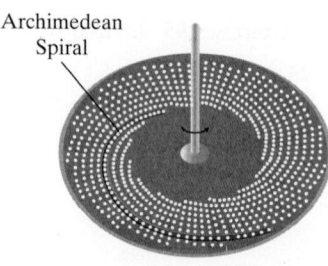

Archimedean Spiral

For Exercises 116–119, approximate the area of the shaded region to 1 decimal place. In the figure, *s* represents arc length, and *r* represents the radius of the circle.

116. $s = 28$ in.   320.9 in.²   **117.** $s = 9$ cm   21.5 cm²   **118.** $s = 18$ m   40.5 m²   **119.** $s = 4.5$ in.   3.4 in.²

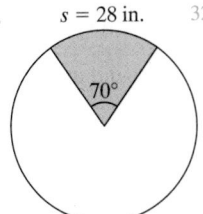 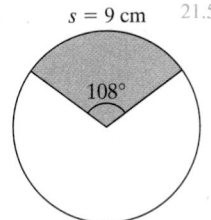 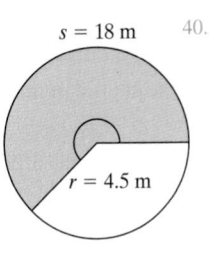 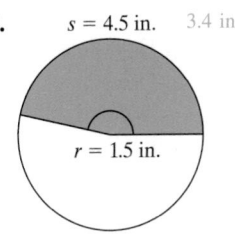

Write About It

120. Explain what is meant by 1 radian. Explain what is meant by 1°.   One radian is the angular measure of a central angle that intercepts an arc equal in length to the radius of the circle. One degree is the angular measure equal to $\frac{1}{360}$ of a full rotation.

121. For an angle drawn in standard position, explain how to determine in which quadrant the terminal side lies.

122. As the fan rotates (see figure), which point *A* or *B* has a greater angular speed? Which point has a greater linear speed? Why?

123. If an angle of a sector is held constant, but the radius is doubled, how will the arc length of the sector and area of the sector be affected?   The arc length will also double because it varies directly as the radius $(s = r\theta)$. However, the area will increase by four times because the area varies directly as the *square* of the radius $(A = \frac{1}{2}r^2\theta)$.

124. If an angle of a sector is doubled, but the radius is held constant, how will the arc length of the sector and the area of the sector be affected?
Both the arc length $(s = r\theta)$ and the area $(A = \frac{1}{2}r^2\theta)$ will double because they vary directly as the angle θ.

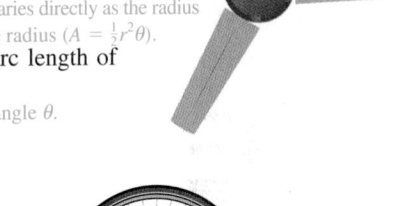

Expanding Your Skills

125. When a person pedals a bicycle, the front sprocket moves a chain that drives the back wheel and propels the bicycle forward. For each rotation of the front sprocket, the chain moves a distance equal to the circumference of the front sprocket. The back sprocket is smaller, so it will simultaneously move through a greater rotation. Furthermore, since the back sprocket is rigidly connected to the back wheel, each rotation of the back sprocket generates a rotation of the wheel.

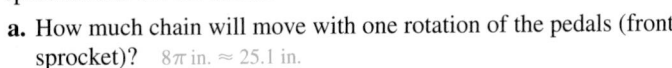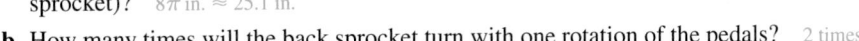

Suppose that the front sprocket of a bicycle has a 4-in. radius and the back sprocket has a 2-in. radius.

  a. How much chain will move with one rotation of the pedals (front sprocket)? 8π in. ≈ 25.1 in.

  b. How many times will the back sprocket turn with one rotation of the pedals?   2 times

  c. How many times will the wheels turn with one rotation of the pedals?   2 times

  d. If the wheels are 27 in. in diameter, how far will the bicycle travel with one rotation of the pedals? 54π in. ≈ 169.6 in. or equivalently 4.5π ft ≈ 14.1 ft

  e. If the bicyclist pedals 80 rpm, what is the linear speed (in ft/min) of the bicycle? 360π ft/min ≈ 1131 ft/min

  f. If the bicyclist pedals 80 rpm, what is the linear speed (in mph) of the bicycle? (*Hint*: 1 mi = 5280 ft, and 1 hr = 60 min) $\frac{45\pi}{11}$ mph ≈ 12.9 mph

126. In the third century B.C., the Greek astronomer Eratosthenes approximated the Earth's circumference. On the summer solstice at noon in Alexandria, Egypt, Eratosthenes measured the angle α of the Sun relative to a line perpendicular to the ground. At the same time in Syene (now Aswan), located on the Tropic of Cancer, the Sun was directly overhead.

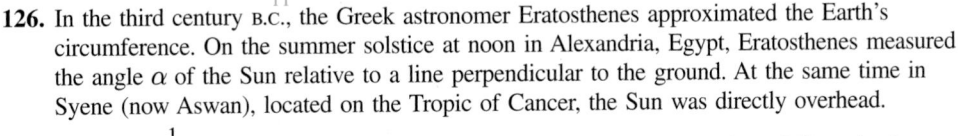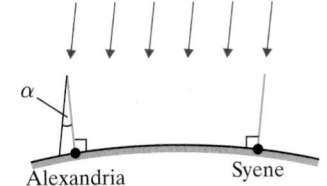

  a. If $\alpha = \dfrac{1}{50}$ of a circle, find the measure of α in degrees. (In Eratosthenes' time, the degree measure had not yet been defined.)   7.2°

  b. If the distance between Alexandria and Syene is 5000 stadia, find the circumference of the Earth measured in stadia.   250,000 stadia

  c. If 10 stadia $\approx$ 1 mi, find Eratosthenes' approximation of the circumference of the Earth in miles (the modern-day approximation at the equator is 24,900 mi).   25,000 mi

127. The space shuttle program involved 135 manned space flights in 30 yr. In addition to supplying and transporting astronauts to the International Space Station, space shuttle missions serviced the Hubble Space Telescope and deployed satellites. For a particular mission, a space shuttle orbited the Earth in 1.5 hr at an altitude of 200 mi.

   a. Determine the angular speed (in radians per hour) of the shuttle. $\frac{4\pi}{3}$/hr

   b. Determine the linear speed of the shuttle in miles per hour. Assume that the Earth's radius is 3960 mi. Round to the nearest hundred miles per hour.   17,400 mph

128. What is the first time (to the nearest second) after 12:00 midnight for which the minute hand and hour hand of a clock make a 120° angle?   12:21:49

Technology Connections

For Exercises 129–130, use the functions in the ANGLE menu on your calculator to

a. Convert from decimal degrees to DMS (degree-minute-second) form and

b. Convert from DMS form to decimal degrees.

On some calculators, the ″ symbol is accessed by hitting (ALPHA) followed by (+).

129. **a.** −216.479°

   b. 42°13′5.9″

130. **a.** −14.908°

   b. 71°19′4.7″

For Exercises 131–132, use a calculator to convert from degrees to radians.

131. **a.** 147°26′9″

   b. −228.459°

132. **a.** 36°4′47″

   b. −25.716°

For Exercises 133–134, use a calculator to convert from radians to degrees.

133. **a.** $\frac{4\pi}{9}$

   b. −5.718

134. **a.** $\frac{11\pi}{18}$

   b. −1.356

Additional answers can be found in the Instructor Answer Appendix.

SECTION 4.2   Trigonometric Functions Defined on the Unit Circle

OBJECTIVES

1. Evaluate Trigonometric Functions Using the Unit Circle
2. Identify the Domains of the Trigonometric Functions
3. Use Fundamental Trigonometric Identities
4. Apply the Periodic and Even and Odd Function Properties of Trigonometric Functions
5. Approximate Trigonometric Functions on a Calculator

1. Evaluate Trigonometric Functions Using the Unit Circle

The functions we have studied thus far have been used to model phenomena such as exponential growth and decay, projectile motion, and profit and cost, to name a few. However, none of the functions in our repertoire represent cyclical behavior such as the orbits of the moon and planets, the variation in air pressure that produces sound, the back-and-forth oscillations of a stretched spring, and so on. To model these behaviors, we introduce six new functions called *trigonometric functions*.

Historically, the study of trigonometry arose from the need to study relationships among the angles and sides of a triangle. An alternative approach is to define trigonometric functions as *circular functions*. To begin, we define the **unit circle** as the circle of radius 1 unit, centered at the origin. The unit circle consists of all points (x, y) that satisfy the equation $x^2 + y^2 = 1$. For example, $\left(-\frac{1}{2}, \frac{\sqrt{3}}{2}\right)$ is a point on the unit circle because $\left(-\frac{1}{2}\right)^2 + \left(\frac{\sqrt{3}}{2}\right)^2 = 1$ (Figure 4-14).

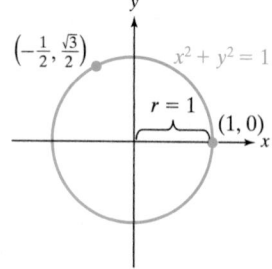

Figure 4-14

TIP The circumference of
the unit circle is $2\pi r = 2\pi(1)$
$= 2\pi$. Thus, the value $t = \pi$
represents one-half of a
revolution and falls on the
point $(-1, 0)$. Likewise the
value $t = \frac{\pi}{2}$ represents
one-quarter of a revolution
and falls on the point $(0, 1)$.

Now suppose we wrap the real number line around
the unit circle by placing the point $t = 0$ from the number
line on the point $(1, 0)$ on the circle. Positive real numbers
from the number line $(t > 0)$ wrap onto the unit circle in
the counterclockwise direction. Negative numbers from
the number line $(t < 0)$ wrap onto the unit circle in the
clockwise direction (Figure 4-15). For a value of t greater
than 2π (or less than -2π) more than one revolution
around the unit circle is required.

The purpose of wrapping the real number line around
the unit circle is to associate each real number t with a
unique point (x, y) on the unit circle.

For a real number t corresponding to a point $P(x, y)$
on the unit circle, we can use the coordinates of P to
define six trigonometric functions of t (Table 4-1). Notice
that rather than using letters such as f, g, h, and so on, trigonometric functions are
given the word names sine, cosine, tangent, cosecant, secant, and cotangent. These
functions are abbreviated as "sin," "cos," "tan," "csc," "sec," and "cot," respectively.
The value t is the input value or **argument** of each function.

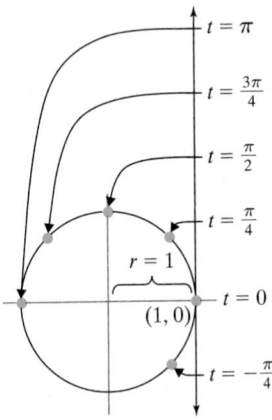

Figure 4-15

Table 4-1

Unit Circle Definitions of the Trigonometric Functions

Let $P(x, y)$ be the point associated with a real number t measured along the circumference
of the unit circle from the point $(1, 0)$.

Function Name	Definition
sine	$\sin t = y$
cosine	$\cos t = x$
tangent	$\tan t = \dfrac{y}{x}\,(x \neq 0)$
cosecant	$\csc t = \dfrac{1}{y}\,(y \neq 0)$
secant	$\sec t = \dfrac{1}{x}\,(x \neq 0)$
cotangent	$\cot t = \dfrac{x}{y}\,(y \neq 0)$

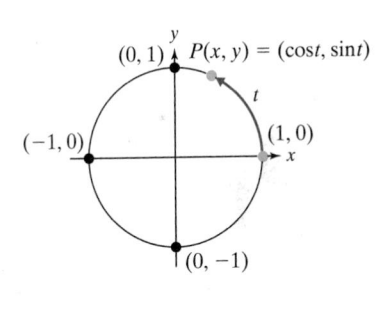

- If P is on the y-axis [either $(0, 1)$ or $(0, -1)$], then $x = 0$ and the tangent and secant
  functions are undefined.
- If P is on the x-axis [either $(1, 0)$ or $(-1, 0)$], then $y = 0$ and the cotangent and cosecant
  functions are undefined.

Classroom Example: p. 477
Exercise 16

EXAMPLE 1   **Evaluating Trigonometric Functions**

Suppose that the real number t corresponds to the point $P\left(-\dfrac{2}{3}, -\dfrac{\sqrt{5}}{3}\right)$ on the
unit circle. Evaluate the six trigonometric functions of t.

TIP In the figure in Example 1, t was arbitrarily taken to be positive as noted by the red arc wrapping counterclockwise. However, we can just as easily use a negative value of t, terminating at the same point in Quadrant III.

Solution:

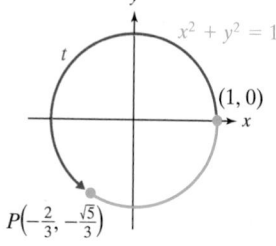

$$\sin t = y = -\frac{\sqrt{5}}{3} \qquad \cos t = x = -\frac{2}{3}$$

$$\tan t = \frac{y}{x} = \frac{-\dfrac{\sqrt{5}}{3}}{-\dfrac{2}{3}} = -\frac{\sqrt{5}}{3} \cdot \left(-\frac{3}{2}\right) = \frac{\sqrt{5}}{2}$$

$$\csc t = \frac{1}{y} = \frac{1}{-\dfrac{\sqrt{5}}{3}} = -\frac{3}{\sqrt{5}} = -\frac{3}{\sqrt{5}} \cdot \frac{\sqrt{5}}{\sqrt{5}} = -\frac{3\sqrt{5}}{5}$$

$$\sec t = \frac{1}{x} = \frac{1}{-\dfrac{2}{3}} = -\frac{3}{2}$$

$$\cot t = \frac{x}{y} = \frac{-\dfrac{2}{3}}{-\dfrac{\sqrt{5}}{3}} = -\frac{2}{3} \cdot \left(-\frac{3}{\sqrt{5}}\right) = \frac{2}{\sqrt{5}} = \frac{2}{\sqrt{5}} \cdot \frac{\sqrt{5}}{\sqrt{5}} = \frac{2\sqrt{5}}{5}$$

Skill Practice 1 Suppose that the real number t corresponds to the point $P\left(\dfrac{\sqrt{5}}{5}, \dfrac{2\sqrt{5}}{5}\right)$ on the unit circle. Evaluate the six trigonometric functions of t.

Let $P(x, y)$ be the point on the unit circle associated with a real number $t \geq 0$. Let $\theta \geq 0$ be the central angle in standard position measured in radians with terminal side through P and arc length t (Figure 4-16). Because the radius of the unit circle is 1, the arc length formula $s = r\theta$ becomes $t = 1 \cdot \theta$ or simply $t = \theta$.

A similar argument can be made for $t < 0$ and $\theta < 0$. The arc length formula is $s = r(-\theta)$, or equivalently $-t = r(-\theta)$, which also implies that $t = \theta$ (Figure 4-17).

Avoiding Mistakes

If θ is negative, then the opposite of θ is used in the arc length formula to ensure that the length of the arc is positive.

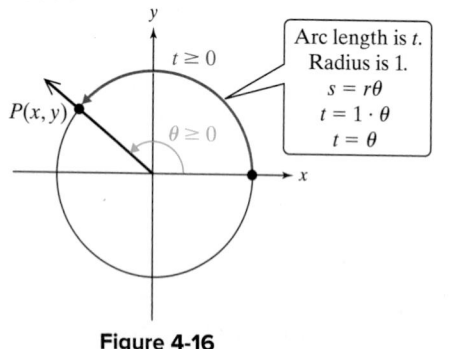

Figure 4-16

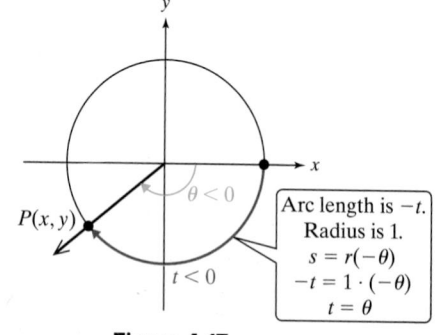

Figure 4-17

From this discussion, we have the following important result. The real number t taken along the circumference of the unit circle gives the radian measure of the corresponding central angle. That is, $\theta = t$ radians. Furthermore, this one-to-one correspondence between the real number t and the radian measure of the central angle θ means that the trigonometric functions can also be defined as functions of θ.

Answer

1. $\sin t = \dfrac{2\sqrt{5}}{5}, \cos t = \dfrac{\sqrt{5}}{5},$

$\tan t = 2, \csc t = \dfrac{\sqrt{5}}{2},$

$\sec t = \sqrt{5}, \cot t = \dfrac{1}{2}$

Trigonometric Functions of Real Numbers and Angles

If $\theta = t$ radians, then

$\sin t = \sin \theta$	$\cos t = \cos \theta$	$\tan t = \tan \theta$
$\csc t = \csc \theta$	$\sec t = \sec \theta$	$\cot t = \cot \theta$

We now want to determine the values of the trigonometric functions for several "special" values of t corresponding to central angles θ that are integer multiples of $\frac{\pi}{6}$ and $\frac{\pi}{4}$. Consider the point $P(x, y)$ on the unit circle corresponding to $t = \frac{\pi}{4}$ (Figure 4-18). Point P lies on the line $y = x$, and the coordinates of P can be written as (x, x). Substituting the coordinates of (x, x) into the equation $x^2 + y^2 = 1$, we have

$$x^2 + x^2 = 1$$
$$2x^2 = 1$$
$$x^2 = \frac{1}{2}$$

> Choose x positive for a first quadrant point.

$$x = \sqrt{\frac{1}{2}}$$
$$x = \frac{1}{\sqrt{2}}$$

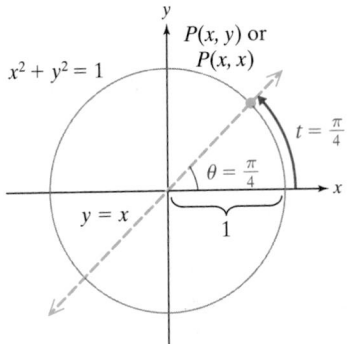

Figure 4-18

Rationalizing the denominator, we have

$$x = \frac{1}{\sqrt{2}} \cdot \frac{\sqrt{2}}{\sqrt{2}} = \frac{\sqrt{2}}{2}.$$

Since $y = x$, point P has coordinates $\left(\frac{1}{\sqrt{2}}, \frac{1}{\sqrt{2}}\right)$ or $\left(\frac{\sqrt{2}}{2}, \frac{\sqrt{2}}{2}\right)$.

From the symmetry of the circle, the points corresponding to $t = \frac{3\pi}{4}, t = \frac{5\pi}{4}$, and $t = \frac{7\pi}{4}$ have coordinates $\left(-\frac{\sqrt{2}}{2}, \frac{\sqrt{2}}{2}\right), \left(-\frac{\sqrt{2}}{2}, -\frac{\sqrt{2}}{2}\right),$ and $\left(\frac{\sqrt{2}}{2}, -\frac{\sqrt{2}}{2}\right)$, respectively (Figure 4-19).

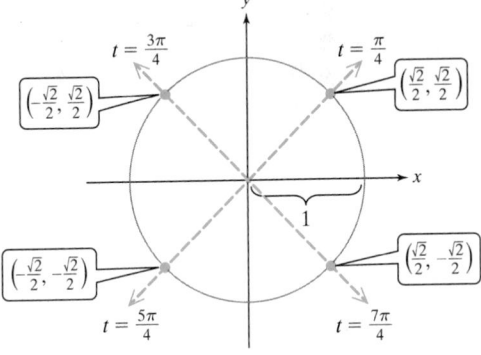

Figure 4-19

Now consider the point $Q(x, y)$ on the unit circle corresponding to $t = \frac{\pi}{6}$ (Figure 4-20). Dropping a line segment from Q perpendicular to the x-axis at point A, we construct a right triangle (ΔOAQ). The acute angles in ΔOAQ are 30° and 60°, and the hypotenuse of the triangle is 1 unit. Placing two such triangles adjacent to one another on opposite sides of the x-axis we have an equilateral triangle (ΔOBQ) with sides of 1 unit and angles of 60° (Figure 4-21).

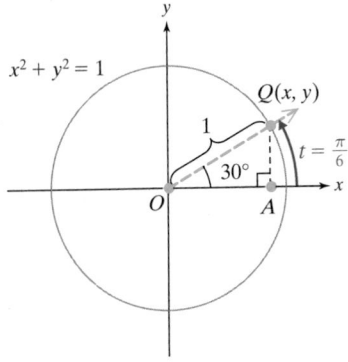

Figure 4-20

From $\triangle OBQ$ we have $2y = 1$. Thus, $y = \dfrac{1}{2}$. Since $Q(x, y)$ is a point on the unit circle, we can substitute $y = \dfrac{1}{2}$ into the equation $x^2 + y^2 = 1$ and solve for x.

$$x^2 + y^2 = 1$$
$$x^2 + \left(\frac{1}{2}\right)^2 = 1$$
$$x^2 + \frac{1}{4} = 1$$
$$x^2 = \frac{3}{4}$$
$$x = \sqrt{\frac{3}{4}} = \frac{\sqrt{3}}{2}$$

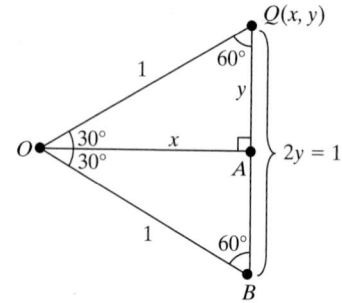

Figure 4-21

> **TIP** The value $t = \dfrac{\pi}{3}$ corresponds to the point $\left(\dfrac{1}{2}, \dfrac{\sqrt{3}}{2}\right)$ on the unit circle. $\triangle OAR$ is congruent to the triangle shown in Figure 4-20, but is oriented with the 60° angle in standard position rather than the 30° angle in standard position. As a result, notice that the x- and y-coordinates of points Q and R are reversed.
>
>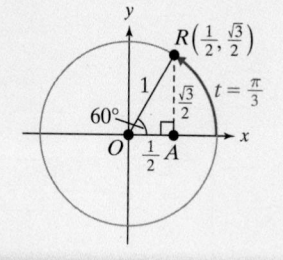

Thus, Q has coordinates $\left(\dfrac{\sqrt{3}}{2}, \dfrac{1}{2}\right)$. Using symmetry, we can also find coordinates of the points on the unit circle corresponding to $t = \dfrac{5\pi}{6}$, $t = \dfrac{7\pi}{6}$, and $t = \dfrac{11\pi}{6}$ (Figure 4-22).

Using similar reasoning, we can show that the point $R(x, y)$ corresponding to $t = \dfrac{\pi}{3}$ has coordinates $\left(\dfrac{1}{2}, \dfrac{\sqrt{3}}{2}\right)$.

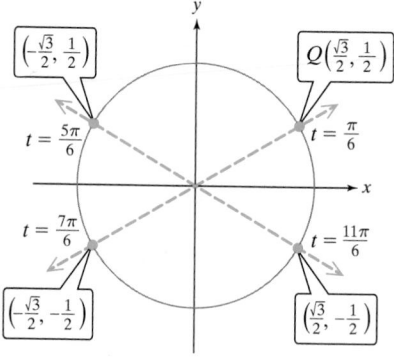

Figure 4-22

We summarize our findings in Figure 4-23 for selected values of t and the corresponding central angle θ.

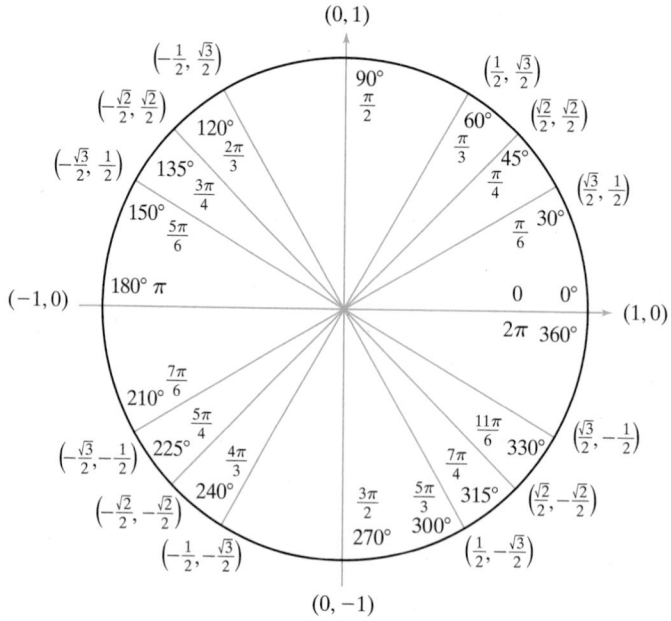

Figure 4-23

Figure 4-23 looks daunting, but using the symmetry of the circle and recognizing several patterns can make the coordinates of these special points easy to determine. First we recommend memorizing the coordinates of the three special points in Quadrant I.

- Note that for $t = \frac{\pi}{4}$, the x- and y-coordinates are both $\frac{\sqrt{2}}{2}$, or equivalently $\frac{1}{\sqrt{2}}$.
- For $t = \frac{\pi}{6}$ and $t = \frac{\pi}{3}$, the x- and y-coordinates are reversed.
- The value $\frac{\sqrt{3}}{2} \approx 0.866$ is greater than the value $\frac{1}{2} = 0.5$. For $t = \frac{\pi}{6}$, the x-coordinate is greater than the y-coordinate. Therefore, x must be $\frac{\sqrt{3}}{2}$. For $t = \frac{\pi}{3}$, the x-coordinate is less than the y-coordinate. Therefore, x must be $\frac{1}{2}$.
- For the special points in Quadrants II, III, and IV, use the values $\frac{1}{2}$, $\frac{\sqrt{3}}{2}$, and $\frac{\sqrt{2}}{2}$ with the appropriate signs attached.

Classroom Example: p. 477
Exercise 22

EXAMPLE 2   Evaluating Trigonometric Functions

Evaluate the six trigonometric functions of the real number t.

a. $t = \dfrac{5\pi}{3}$ **b.** $t = -\dfrac{5\pi}{4}$

Solution:

a. $t = \dfrac{5\pi}{3}$ corresponds to the point $(x, y) = \left(\dfrac{1}{2}, -\dfrac{\sqrt{3}}{2} \right)$ on the unit circle.

$$\sin \frac{5\pi}{3} = y = -\frac{\sqrt{3}}{2}$$

$$\csc \frac{5\pi}{3} = \frac{1}{y} = \frac{1}{-\dfrac{\sqrt{3}}{2}} = -\frac{2}{\sqrt{3}} = -\frac{2}{\sqrt{3}} \cdot \frac{\sqrt{3}}{\sqrt{3}} = -\frac{2\sqrt{3}}{3}$$

$$\cos \frac{5\pi}{3} = x = \frac{1}{2}$$

$$\sec \frac{5\pi}{3} = \frac{1}{x} = \frac{1}{\dfrac{1}{2}} = 2$$

$$\tan \frac{5\pi}{3} = \frac{y}{x} = \frac{-\dfrac{\sqrt{3}}{2}}{\dfrac{1}{2}} = -\frac{\sqrt{3}}{2} \cdot \frac{2}{1} = -\sqrt{3}$$

$$\cot \frac{5\pi}{3} = \frac{x}{y} = \frac{\dfrac{1}{2}}{-\dfrac{\sqrt{3}}{2}} = \frac{1}{2} \cdot \left(-\frac{2}{\sqrt{3}} \right) = -\frac{1}{\sqrt{3}} \cdot \frac{\sqrt{3}}{\sqrt{3}} = -\frac{\sqrt{3}}{3}$$

$\theta = \frac{5\pi}{3} = 300°$

$P\left(\frac{1}{2}, -\frac{\sqrt{3}}{2}\right)$

$t = \frac{5\pi}{3}$

b. Since the circumference of the unit circle is 2π, the values $t = -\dfrac{5\pi}{4}$ and $t_1 = -\dfrac{5\pi}{4} + 2\pi = \dfrac{3\pi}{4}$ have the same location in Quadrant II on the unit circle. Both correspond to the point $(x, y) = \left(-\dfrac{\sqrt{2}}{2}, \dfrac{\sqrt{2}}{2}\right)$, or equivalently $\left(-\dfrac{1}{\sqrt{2}}, \dfrac{1}{\sqrt{2}}\right)$. Thus,

$$\sin\left(-\frac{5\pi}{4}\right) = y = \frac{\sqrt{2}}{2} \qquad\qquad \csc\left(-\frac{5\pi}{4}\right) = \frac{1}{y} = \frac{1}{\frac{1}{\sqrt{2}}} = \sqrt{2}$$

$$\cos\left(-\frac{5\pi}{4}\right) = x = -\frac{\sqrt{2}}{2} \qquad\qquad \sec\left(-\frac{5\pi}{4}\right) = \frac{1}{x} = \frac{1}{-\frac{1}{\sqrt{2}}} = -\sqrt{2}$$

$$\tan\left(-\frac{5\pi}{4}\right) = \frac{y}{x} = \frac{\frac{\sqrt{2}}{2}}{-\frac{\sqrt{2}}{2}} = \frac{\sqrt{2}}{2} \cdot \left(-\frac{2}{\sqrt{2}}\right) = -1$$

$$\cot\left(-\frac{5\pi}{4}\right) = \frac{x}{y} = \frac{-\frac{\sqrt{2}}{2}}{\frac{\sqrt{2}}{2}} = -\frac{\sqrt{2}}{2} \cdot \frac{2}{\sqrt{2}} = -1$$

Skill Practice 2 Evaluate the six trigonometric functions of the real number t.

a. $t = \dfrac{5\pi}{6}$ **b.** $t = -\dfrac{3\pi}{4}$

2. Identify the Domains of the Trigonometric Functions

From Table 4-1, $\sin t = y$ and $\cos t = x$ have no restrictions on their domain. However, $\tan t = \dfrac{y}{x}$ and $\sec t = \dfrac{1}{x}$ are undefined for all values of t corresponding to the points $(0, 1)$ and $(0, -1)$ on the unit circle. These are $t = \dfrac{\pi}{2}$, $t = \dfrac{3\pi}{2}$, and all other *odd* multiples of $\dfrac{\pi}{2}$ (Figure 4-24).

- $\tan t$ and $\sec t$ are undefined for $\ldots -\dfrac{3\pi}{2}, -\dfrac{\pi}{2}, \dfrac{\pi}{2}, \dfrac{3\pi}{2}, \dfrac{5\pi}{2}, \ldots, \dfrac{(2n+1)\pi}{2}$, for all integers n.

Likewise, $\cot t = \dfrac{x}{y}$ and $\csc t = \dfrac{1}{y}$ are undefined for all values of t corresponding to the points $(1, 0)$ and $(-1, 0)$ on the unit circle. These are $t = 0$, $t = \pi$, $t = 2\pi$, and all other multiples of π (Figure 4-24).

- $\cot t$ and $\csc t$ are undefined for $\ldots -3\pi, -2\pi, -\pi, 0, \pi, 2\pi, 3\pi, \ldots, n\pi$ for all integers n.

Answers

2. a. $\sin\dfrac{5\pi}{6} = \dfrac{1}{2}$, $\cos\dfrac{5\pi}{6} = -\dfrac{\sqrt{3}}{2}$,
$\tan\dfrac{5\pi}{6} = -\dfrac{\sqrt{3}}{3}$, $\csc\dfrac{5\pi}{6} = 2$,
$\sec\dfrac{5\pi}{6} = -\dfrac{2\sqrt{3}}{3}$,
$\cot\dfrac{5\pi}{6} = -\sqrt{3}$

b. $\sin\left(-\dfrac{3\pi}{4}\right) = -\dfrac{\sqrt{2}}{2}$,
$\cos\left(-\dfrac{3\pi}{4}\right) = -\dfrac{\sqrt{2}}{2}$,
$\tan\left(-\dfrac{3\pi}{4}\right) = 1$,
$\csc\left(-\dfrac{3\pi}{4}\right) = -\sqrt{2}$,
$\sec\left(-\dfrac{3\pi}{4}\right) = -\sqrt{2}$,
$\cot\left(-\dfrac{3\pi}{4}\right) = 1$

For integer values of n, the formula $\dfrac{(2n + 1)\pi}{2}$ generates odd multiples of $\dfrac{\pi}{2}$. For example, if $n = 6$,

$$\frac{[2(6) + 1]\pi}{2} = \frac{13\pi}{2}.$$

The formula $n\pi$ generates multiples of π. For example, if $n = 6$,

$$n\pi = (6)\pi = 6\pi.$$

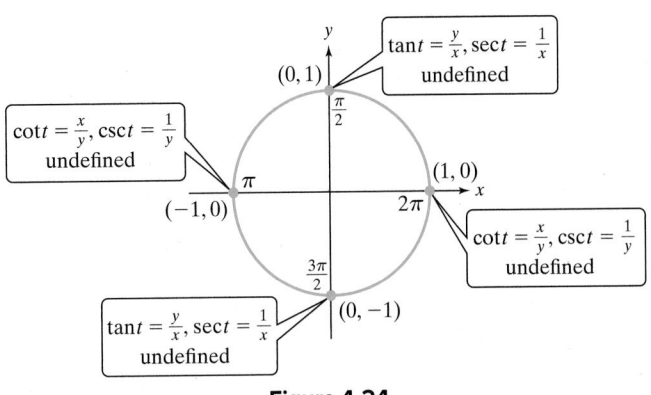

Figure 4-24

Domains of the Trigonometric Functions

Function	Domain*	Notes
$f(t) = \sin t$	All real numbers	No restrictions
$f(t) = \cos t$	All real numbers	No restrictions
$f(t) = \tan t$	$\left\{t \middle| t \neq \dfrac{(2n + 1)\pi}{2} \text{ for all integers } n\right\}$	Exclude odd multiples of $\dfrac{\pi}{2}$.
$f(t) = \cot t$	$\{t \mid t \neq n\pi \text{ for all integers } n\}$	Exclude multiples of π.
$f(t) = \sec t$	$\left\{t \middle| t \neq \dfrac{(2n + 1)\pi}{2} \text{ for all integers } n\right\}$	Exclude odd multiples of $\dfrac{\pi}{2}$.
$f(t) = \csc t$	$\{t \mid t \neq n\pi \text{ for all integers } n\}$	Exclude multiples of π.

*The range of each trigonometric function will be discussed in Sections 4.5 and 4.6 when we cover the graphs of the functions.

EXAMPLE 3 **Evaluate the Trigonometric Functions for Given Values of t**

Evaluate the six trigonometric functions of the real number t.

a. $t = \pi$ **b.** $t = \dfrac{5\pi}{2}$

Solution:

a. The value $t = \pi$ corresponds to $(-1, 0)$ on the unit circle.

$$\sin \pi = y = 0 \qquad \cos \pi = x = -1 \qquad \tan \pi = \frac{y}{x} = \frac{0}{-1} = 0$$

$$\csc \pi = \frac{1}{y} \text{ is undefined because } \frac{1}{0} \text{ is undefined.}$$

$$\sec \pi = \frac{1}{x} = \frac{1}{-1} = -1$$

$$\cot \pi = \frac{x}{y} \text{ is undefined because } \frac{-1}{0} \text{ is undefined.}$$

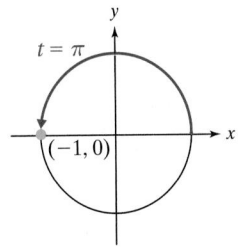

b. The circumference of the unit circle is 2π. Therefore, since $t = \dfrac{5\pi}{2} = 2\pi + \dfrac{\pi}{2}$, the value $t = \dfrac{5\pi}{2}$ corresponds to the point (0, 1) on the unit circle.

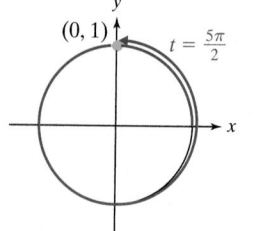

$$\sin\frac{5\pi}{2} = y = 1 \qquad \cos\frac{5\pi}{2} = x = 0$$

$$\tan\frac{5\pi}{2} = \frac{y}{x} \text{ is undefined because } \frac{1}{0} \text{ is undefined.}$$

$$\csc\frac{5\pi}{2} = \frac{1}{y} = \frac{1}{1} = 1$$

$$\sec\frac{5\pi}{2} = \frac{1}{x} \text{ is undefined because } \frac{1}{0} \text{ is undefined.}$$

$$\cot\frac{5\pi}{2} = \frac{x}{y} = \frac{0}{1} = 0$$

Skill Practice 3   Evaluate the six trigonometric functions of the real number t.

a. $t = -2\pi$ **b.** $\dfrac{3\pi}{2}$

3. Use Fundamental Trigonometric Identities

You may have already noticed several relationships among the six trigonometric functions that follow directly from their definitions. For example, for a real number t corresponding to a point (x, y) on the unit circle, $\sin t = y$ and $\csc t = \dfrac{1}{y}$. Thus, the sine and cosecant functions are reciprocals for all t in their common domain. Table 4-2 summarizes the reciprocal relationships among the trigonometric functions along with two important quotient properties. These relationships should be committed to memory.

Table 4-2

Reciprocal and Quotient Identities	
$\csc t = \dfrac{1}{\sin t}$ or $\sin t = \dfrac{1}{\csc t}$	$\sin t$ and $\csc t$ are reciprocals.
$\sec t = \dfrac{1}{\cos t}$ or $\cos t = \dfrac{1}{\sec t}$	$\cos t$ and $\sec t$ are reciprocals.
$\cot t = \dfrac{1}{\tan t}$ or $\tan t = \dfrac{1}{\cot t}$	$\tan t$ and $\cot t$ are reciprocals.
$\tan t = \dfrac{\sin t}{\cos t}$	$\tan t$ is the ratio of $\sin t$ and $\cos t$.
$\cot t = \dfrac{\cos t}{\sin t}$	$\cot t$ is the ratio of $\cos t$ and $\sin t$.

Answers

3. a. $\sin(-2\pi) = 0$, $\cos(-2\pi) = 1$, $\tan(-2\pi) = 0$, $\csc(-2\pi)$ is undefined, $\sec(-2\pi) = 1$, $\cot(-2\pi)$ is undefined

b. $\sin\dfrac{3\pi}{2} = -1$, $\cos\dfrac{3\pi}{2} = 0$, $\tan\dfrac{3\pi}{2}$ is undefined, $\csc\dfrac{3\pi}{2} = -1$, $\sec\dfrac{3\pi}{2}$ is undefined, $\cot\dfrac{3\pi}{2} = 0$

Classroom Example: p. 478
Exercise 44

EXAMPLE 4 Using the Reciprocal and Quotient Identities

Given that $\sin t = \dfrac{5}{8}$ and $\cos t = \dfrac{\sqrt{39}}{8}$, use the reciprocal and quotient identities to find the values of the other trigonometric functions of t.

Solution:

Given the values of $\sin t$ and $\cos t$, we can use the quotient identities to find $\tan t$ and $\cot t$.

$$\tan t = \frac{\sin t}{\cos t} = \frac{\dfrac{5}{8}}{\dfrac{\sqrt{39}}{8}} = \frac{5}{8} \cdot \frac{8}{\sqrt{39}} = \frac{5}{\sqrt{39}} = \frac{5}{\sqrt{39}} \cdot \frac{\sqrt{39}}{\sqrt{39}} = \frac{5\sqrt{39}}{39}$$

$$\cot t = \frac{\cos t}{\sin t} = \frac{\dfrac{\sqrt{39}}{8}}{\dfrac{5}{8}} = \frac{\sqrt{39}}{8} \cdot \frac{8}{5} = \frac{\sqrt{39}}{5}$$

The remaining two functions can be found by using the reciprocal identities.

$$\csc t = \frac{1}{\sin t} = \frac{1}{\dfrac{5}{8}} = \frac{8}{5}$$

$$\sec t = \frac{1}{\cos t} = \frac{1}{\dfrac{\sqrt{39}}{8}} = \frac{8}{\sqrt{39}} = \frac{8}{\sqrt{39}} \cdot \frac{\sqrt{39}}{\sqrt{39}} = \frac{8\sqrt{39}}{39}$$

Skill Practice 4 Given $\sin t = \dfrac{3}{7}$ and $\cos t = \dfrac{2\sqrt{10}}{7}$ use the reciprocal and quotient identities to find the values of the other trigonometric functions of t.

Consider a real number t corresponding to the point $P(x, y)$ on the unit circle. Since $\cos t = x$ and $\sin t = y$, point P can be labeled $P(\cos t, \sin t)$. See Figure 4-25. Furthermore, since P is on the unit circle, it satisfies the equation $x^2 + y^2 = 1$. Thus,

$$(\cos t)^2 + (\sin t)^2 = 1 \text{ or } (\sin t)^2 + (\cos t)^2 = 1$$

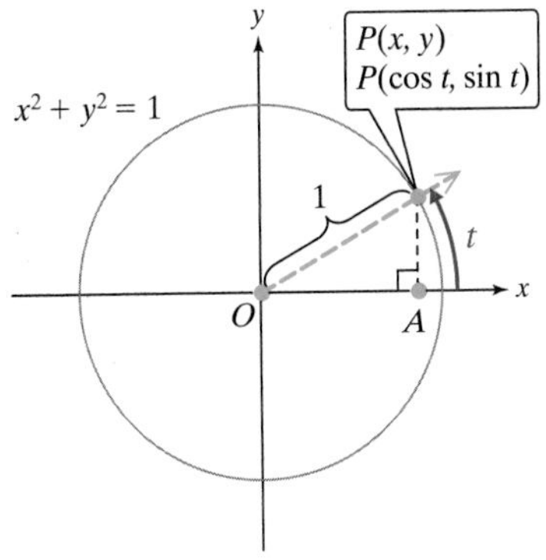

It is universally preferred to write the exponent associated with a trigonometric function between the name of the function and its argument. For example, $(\sin t)^2$ is most often written as $\sin^2 t$. Thus, the relationship $(\sin t)^2 + (\cos t)^2 = 1$ is written as $\sin^2 t + \cos^2 t = 1$. This relationship is called a "Pythagorean identity" because $\sin^2 t + \cos^2 t = 1$ is a statement of the Pythagorean theorem for $\triangle OAP$.

Figure 4-25

Table 4-3 summarizes three Pythagorean identities. The second and third relationships can be derived by dividing both sides of the equation $\sin^2 t + \cos^2 t = 1$ by $\cos^2 t$ and $\sin^2 t$, respectively (see Exercises 47 and 48).

Answer

4. $\tan t = \dfrac{3\sqrt{10}}{20}$, $\cot t = \dfrac{2\sqrt{10}}{3}$,

$\csc t = \dfrac{7}{3}$, $\sec t = \dfrac{7\sqrt{10}}{20}$

Table 4-3

Pythagorean Identities		
$\sin^2 t + \cos^2 t = 1$	$\tan^2 t + 1 = \sec^2 t$	$1 + \cot^2 t = \csc^2 t$

Classroom Example: p. 478
Exercise 50

EXAMPLE 5   **Using the Pythagorean Identities**

Given that $\tan t = \dfrac{12}{5}$ for $\pi < t < \dfrac{3\pi}{2}$, use an appropriate identity to find the value of $\sec t$.

Solution:

The real number t corresponds to a point $P(x, y)$ on the unit circle in Quadrant III. Since the value of $\tan t$ is known, we can use the relationship $\tan^2 t + 1 = \sec^2 t$ to find the value of $\sec t$.

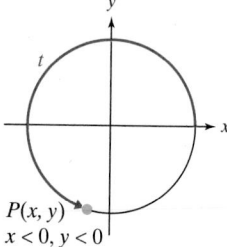

$$\tan^2 t + 1 = \sec^2 t$$

$$\sec t = \pm\sqrt{\tan^2 t + 1}$$

$$\sec t = -\sqrt{\tan^2 t + 1}$$

Since t is between π and $\frac{3\pi}{2}$, it corresponds to a point $P(x, y)$ on the unit circle in Quadrant III. Since P is in Quadrant III, the values of both x and y are negative. Hence $\sec t = \frac{1}{x}$ is negative.

$$\sec t = -\sqrt{\left(\dfrac{12}{5}\right)^2 + 1} = -\sqrt{\dfrac{144}{25} + \dfrac{25}{25}} = -\sqrt{\dfrac{169}{25}} = -\dfrac{13}{5}$$

Skill Practice 5   Given that $\csc t = \dfrac{5}{4}$ for $\dfrac{\pi}{2} < t < \pi$, use an appropriate identity to find the value of $\cot t$.

Sometimes it is beneficial to express one trigonometric function in terms of another. This is demonstrated in Example 6.

Classroom Example: p. 478
Exercise 56

EXAMPLE 6   **Expressing a Trigonometric Function in Terms of Another Trigonometric Function**

For a given real number t, express $\sin t$ in terms of $\cos t$.

Solution:

Let $P(x, y) = (\cos t, \sin t)$ be the point on the unit circle determined by t.

$$x^2 + y^2 = 1 \qquad \text{Equation of unit circle}$$

$$\cos^2 t + \sin^2 t = 1 \qquad \text{Substitute } x = \cos t$$

$$\sin^2 t = 1 - \cos^2 t \qquad \text{and } y = \sin t.$$

$$\sin t = \pm\sqrt{1 - \cos^2 t} \qquad \text{The sign is determined by the quadrant in which } P \text{ lies.}$$

Skill Practice 6   For a given real number t, express $\tan t$ in terms of $\sec t$.

Answers

5. $\cot t = -\dfrac{3}{4}$

6. $\tan t = \pm\sqrt{\sec^2 t - 1}$

4. Apply the Periodic and Even and Odd Function Properties of Trigonometric Functions

Many cyclical patterns occur in nature such as the changes of the seasons, the rise and fall of the tides, and the phases of the moon. In many cases, we can predict these behaviors by determining the period of one complete cycle. For example, an observer on the Earth would note that the moon changes from a new moon to a full moon and back again to a new moon in approximately 29.5 days. If $m(t)$ represents the percentage of the moon seen on day t, then

$$m(t + 29.5) = m(t) \text{ for all } t \text{ in the domain of } m.$$

We say that function m is *periodic* because it repeats at regular intervals. In this example, the period is 29.5 days because this is the shortest time required to complete one full cycle.

Definition of a Periodic Function

A function f is **periodic** if $f(t + p) = f(t)$ for some constant p.
The smallest positive value p for which f is periodic is called the **period** of f.

The values of the six trigonometric functions of t are determined by the corresponding point $P(x, y)$ on the unit circle. Since the circumference of the unit circle is 2π, adding (or subtracting) 2π to t results in the same terminal point (x, y). Consequently, the values of the trigonometric functions are the same for t and $t + 2n\pi$.

In Exercises 121 and 122, we show that the sine and cosine functions are periodic with period 2π. Likewise, their reciprocal functions, cosecant and secant, are periodic with period 2π. However, the period of the tangent and cotangent functions is π, which we show in Exercises 119 and 120.

Periodic Properties of Trigonometric Functions

Function	Period	Property
Sine	2π	$\sin(t + 2\pi) = \sin t$
Cosine	2π	$\cos(t + 2\pi) = \cos t$
Cosecant	2π	$\csc(t + 2\pi) = \csc t$
Secant	2π	$\sec(t + 2\pi) = \sec t$
Tangent	π	$\tan(t + \pi) = \tan t$
Cotangent	π	$\cot(t + \pi) = \cot t$

Given a periodic function f, if the period is p, then $f(t + p) = f(t)$. It is also true that $f(t + np) = f(t)$ for any integer n. That is, adding any integer multiple of the period to a domain element of a periodic function results in the same function value. This is demonstrated in Example 7.

EXAMPLE 7 Applying the Periodic Properties of the Trigonometric Functions

Given $\sin\dfrac{\pi}{12} = \dfrac{\sqrt{6} - \sqrt{2}}{4}$, determine the value of $\sin\dfrac{49\pi}{12}$.

Solution:

The period of the sine function is 2π or equivalently $\dfrac{24\pi}{12}$. Therefore, adding or subtracting any multiple of $\dfrac{24\pi}{12}$ to the argument $\dfrac{\pi}{12}$ results in the same value of the sine function.

$$\sin\frac{49\pi}{12} = \sin\left(\frac{\pi}{12} + \frac{48\pi}{12}\right) = \sin\left[\frac{\pi}{12} + 2\left(\frac{24\pi}{12}\right)\right] = \sin\left(\frac{\pi}{12} + 2(2\pi)\right)$$

$$= \sin\frac{\pi}{12} = \frac{\sqrt{6} - \sqrt{2}}{4}$$

Skill Practice 7 Given $\sin\dfrac{\pi}{8} = \dfrac{\sqrt{2 - \sqrt{2}}}{2}$, determine the value of $\sin\left(-\dfrac{15\pi}{8}\right)$.

Recall that a function f is even if $f(-x) = f(x)$ and odd if $f(-x) = -f(x)$. Suppose that t is a real number associated with a point $P(x, y)$ on the unit circle. Then $-t$ is associated with the point $Q(x, -y)$. See Figure 4-26.

Notice that $\cos(-t) = \cos t = x$, so the cosine function is an even function. The sine function, however, is an odd function because $\sin t = y$, but $\sin(-t) = -y$. Thus, $\sin(-t) = -\sin t$. The tangent function is also an odd function because $\tan(-t) = \dfrac{-y}{x}$ and $\tan t = \dfrac{y}{x}$. Therefore, $\tan(-t) = -\tan t$.

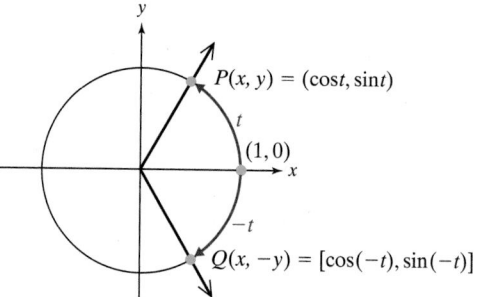

Figure 4-26

The functions secant, cosecant, and cotangent carry the same even and odd properties as their reciprocals: cosine, sine, and tangent, respectively.

Even and Odd Properties of Trigonometric Functions		
Function	**Evaluate at t and $-t$**	**Property**
Sine	$\sin t = y$ and $\sin(-t) = -y$	$\sin(-t) = -\sin t$ (odd function)
Cosine	$\cos t = x$ and $\cos(-t) = x$	$\cos(-t) = \cos t$ (even function)
Cosecant	$\csc t = \dfrac{1}{y}$ and $\csc(-t) = \dfrac{1}{-y}$	$\csc(-t) = -\csc t$ (odd function)
Secant	$\sec t = \dfrac{1}{x}$ and $\sec(-t) = \dfrac{1}{x}$	$\sec(-t) = \sec t$ (even function)
Tangent	$\tan t = \dfrac{y}{x}$ and $\tan(-t) = \dfrac{-y}{x}$	$\tan(-t) = -\tan t$ (odd function)
Cotangent	$\cot t = \dfrac{x}{y}$ and $\cot(-t) = \dfrac{x}{-y}$	$\cot(-t) = -\cot t$ (odd function)

Answer

7. $\dfrac{\sqrt{2 - \sqrt{2}}}{2}$

Classroom Examples: p. 479
Exercises 70, 72

Avoiding Mistakes

A factor from the argument of a function cannot be factored out in front of the function. For example,

$$f(2x) \neq 2f(x)$$

In step 2 of Example 8(b), the value -1 that appears in front of the first term was not factored out, but rather is a result of the odd function property of the tangent function.

| EXAMPLE 8 | Simplifying Expressions Using the Even-Odd Properties of Trigonometric Functions |

Use the properties of the trigonometric functions to simplify.

a. $4\cos t + \cos(-t)$ **b.** $\tan(-3t) - \tan(-3t + \pi)$

Solution:

a. $4\cos t + \cos(-t)$

$\qquad = 4\cos t + \cos t$ The cosine function is even. $\cos(-t) = \cos t$

$\qquad = 5\cos t$

b. $\tan(-3t) - \tan(3t + \pi)$ The tangent function is odd. $\tan(-3t) = -\tan 3t$

$\qquad = -\tan 3t - \tan 3t$ The period of the tangent function is π. Therefore,

$\qquad = -2\tan 3t$ $\tan(3t + \pi) = \tan 3t$.

Skill Practice 8   Use the properties of the trigonometric functions to simplify.

a. $\sec t - 2\sec(-t)$ **b.** $\sin(2t + 2\pi) - \sin(-2t)$

5. Approximate Trigonometric Functions on a Calculator

From Figure 4-23, we display the coordinates of the points on the unit circle that correspond to t values in multiples of $\dfrac{\pi}{6}$ and $\dfrac{\pi}{4}$. From this information we can find the values of the six trigonometric functions for these real numbers. However, it is often the case that the exact value of a trigonometric function for a real number t cannot be found analytically. In such a case, we can use a calculator to approximate the function values.

Classroom Example: p. 479
Exercise 76

| EXAMPLE 9 | Approximating Trigonometric Functions Using a Calculator |

Use a calculator to approximate the function values. Round to 4 decimal places.

a. $\cos\dfrac{2\pi}{7}$ **b.** $\csc 0.92$

Solution:

The real number t corresponds to a point $P(x, y)$ on the unit circle. The value of t is the same as the radian measure of the central angle θ formed by the positive x-axis and the ray from the origin to point P. Therefore, we must first be sure that the calculator is in radian mode.

Graphing Calculator

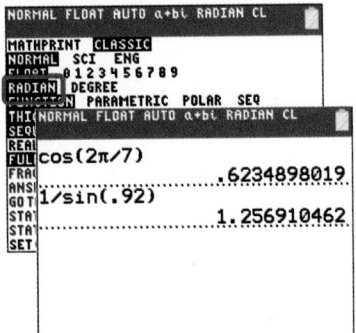

Scientific Calculator (radian mode)   Rounded

a. $2 \times \pi \div 7 =$ **COS**   0.6235

b. 0.92 **SIN** **x⁻¹**   1.2569

This takes the reciprocal of the sine function (cosecant).

Answers

8. a. $-\sec t$ **b.** $2\sin 2t$

Skill Practice 9   Use a calculator to approximate the function values. Round to 4 decimal places.

a. $\tan 1.4$

b. $\cot \dfrac{\pi}{8}$

Point of Interest (Cyclical Data)

In September 1859, a gigantic *coronal mass ejection* erupted from the Sun, sending a huge amount of electromagnetic radiation toward the Earth. This geomagnetic storm was the largest on record to have struck the Earth, causing telegraph lines to short out and compasses to go haywire. The aurora borealis or "northern lights" following the event could be seen as far south as Cuba.

Solar storms are natural phenomena that occur as a result of the rise and fall of the Sun's magnetic activity. Although the cause of solar storms is not completely understood, they seem to be cyclical. Scientists have established a correlation between solar storms and the number of sun spots that cycle over a period of approximately 11 yr. The effects of a massive solar storm today could bring potentially devastating disruption to power grids, satellite communication, and air travel. As a result, solar activity and "space weather" are studied intensely by NASA and NOAA.

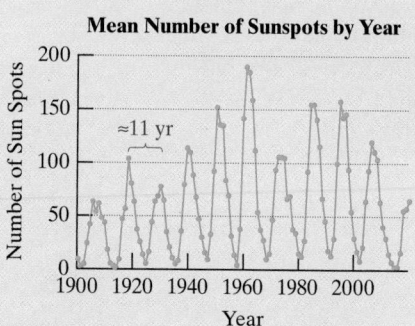

Mean Number of Sunspots by Year

(*Source:* SILSO data/image, Royal Observatory of Belgium, Brussels)

Answers
9. a. 5.7979   **b.** 2.4142

SECTION 4.2   Practice Exercises

Prerequisite Review

For Exercises R.1–R.3, determine if the function is even, odd, or neither.

R.1. $k(x) = 11x^3 + 12x$ Odd

R.2. $r(x) = \sqrt{25 - (x + 2)^2}$ Neither

R.3. $q(x) = \sqrt{23 + x^2}$ Even

For Exercises R.4–R.5, write the domain of the function in set-builder notation.

R.4. $p(v) = \dfrac{v + 3}{v + 4}$ $\{v \mid v \text{ is a real number and } v \ne -4\}$

R.5. $k(q) = \dfrac{q^2}{2q^2 + 3q - 27}$ $\left\{q \mid q \text{ is a real number and } q \ne -\dfrac{9}{2}, q \ne 3\right\}$

Concept Connections

1. The graph of $x^2 + y^2 = 1$ is known as the _____unit_____ circle. It has a radius of length ____1____ and center at the _____origin_____.

2. The circumference of the unit circle is _____2π_____. Thus, the value $t = \frac{\pi}{2}$ represents ____$\frac{1}{4}$____ of a revolution.

3. When $\tan t = 0$, the value of $\cot t$ is _____undefined_____.

4. The value of $\sec t = \dfrac{1}{\cos t}$. Therefore, when $\cos t = 0$, the value of $\sec t$ is _____undefined_____.

5. On the interval $[0, 2\pi)$, $\sin t = 0$ for $t =$ _____0_____ and $t =$ _____π_____. Since $\csc t = \dfrac{1}{\sin t}$, the value of $\csc t$ is _____undefined_____ for these values of t.

6. The domain of the trigonometric functions _____sine_____ and _____cosine_____ is all real numbers.

7. Which trigonometric functions have domain $\left\{t \mid t \ne \dfrac{(2n + 1)\pi}{2} \text{ for all integers } n\right\}$?   tangent and secant

8. A function f is _____periodic_____ if $f(t + p) = f(t)$ for some constant p.

9. The period of the tangent and cotangent functions is _____π_____. The period of the sine, cosine, cosecant, and secant functions is _____2π_____.

10. The cosine function is an _____even_____ function because $\cos(-t) = \cos t$. The sine function is an _____odd_____ function because $\sin(-t) = -\sin t$.

Objective 1: Evaluate Trigonometric Functions Using the Unit Circle

For Exercises 11–14, determine if the point lies on the unit circle.

11. $\left(\dfrac{\sqrt{10}}{5}, \dfrac{\sqrt{15}}{5}\right)$ Yes   **12.** $\left(\dfrac{\sqrt{61}}{8}, -\dfrac{\sqrt{2}}{8}\right)$ No   **13.** $\left(-\dfrac{\sqrt{7}}{10}, -\dfrac{2\sqrt{23}}{10}\right)$ No   **14.** $\left(-\dfrac{2\sqrt{34}}{17}, \dfrac{3\sqrt{17}}{17}\right)$ Yes

For Exercises 15–18, the real number t corresponds to the point P on the unit circle. Evaluate the six trigonometric functions of t. (See Example 1)

15.
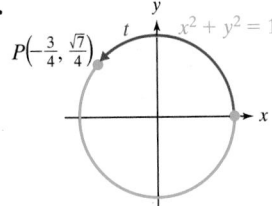
$P\left(-\dfrac{3}{4}, \dfrac{\sqrt{7}}{4}\right)$

16.
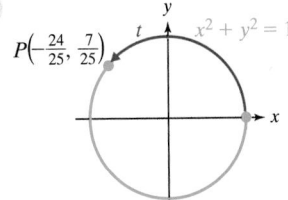
$P\left(-\dfrac{24}{25}, \dfrac{7}{25}\right)$

17.
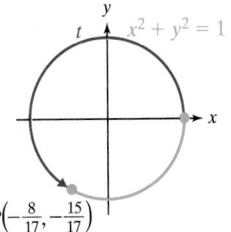
$P\left(-\dfrac{8}{17}, -\dfrac{15}{17}\right)$

18.
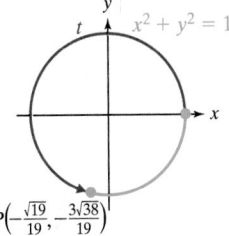
$P\left(-\dfrac{\sqrt{19}}{19}, -\dfrac{3\sqrt{38}}{19}\right)$

19. Fill in the ordered pairs on the unit circle corresponding to each real number t.

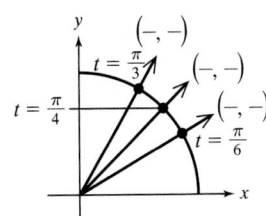
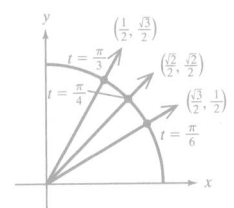

For Exercises 20–23, identify the coordinates of point P. Then evaluate the six trigonometric functions of t. (See Example 2)

20.
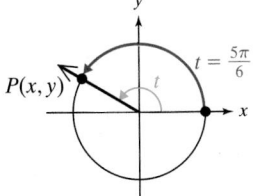

21.

22.
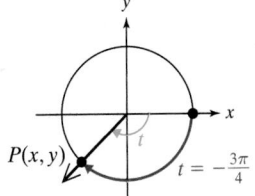

23.
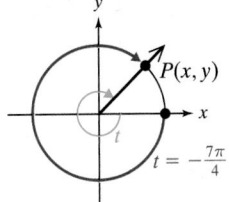

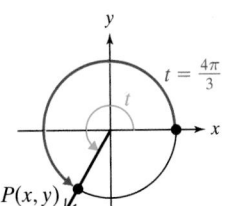

For Exercises 24–26, identify the ordered pairs on the unit circle corresponding to each real number t.

24. a. $t = \dfrac{2\pi}{3}$ $\left(-\dfrac{1}{2}, \dfrac{\sqrt{3}}{2}\right)$ **b.** $t = -\dfrac{5\pi}{4}$ $\left(-\dfrac{\sqrt{2}}{2}, \dfrac{\sqrt{2}}{2}\right)$ **c.** $t = \dfrac{5\pi}{6}$ $\left(-\dfrac{\sqrt{3}}{2}, \dfrac{1}{2}\right)$

25. a. $t = \dfrac{7\pi}{6}$ $\left(-\dfrac{\sqrt{3}}{2}, -\dfrac{1}{2}\right)$ **b.** $t = -\dfrac{3\pi}{4}$ $\left(-\dfrac{\sqrt{2}}{2}, -\dfrac{\sqrt{2}}{2}\right)$ **c.** $t = \dfrac{4\pi}{3}$ $\left(-\dfrac{1}{2}, -\dfrac{\sqrt{3}}{2}\right)$

26. a. $t = \dfrac{5\pi}{3}$ $\left(\dfrac{1}{2}, -\dfrac{\sqrt{3}}{2}\right)$ **b.** $t = \dfrac{7\pi}{4}$ $\left(\dfrac{\sqrt{2}}{2}, -\dfrac{\sqrt{2}}{2}\right)$ **c.** $t = -\dfrac{\pi}{6}$ $\left(\dfrac{\sqrt{3}}{2}, -\dfrac{1}{2}\right)$

For Exercises 27–32, evaluate the trigonometric function at the given real number.

27. $f(t) = \cos t; t = \dfrac{2\pi}{3}$ $f\left(\dfrac{2\pi}{3}\right) = -\dfrac{1}{2}$ **28.** $g(t) = \sin t; t = \dfrac{5\pi}{4}$ $g\left(\dfrac{5\pi}{4}\right) = -\dfrac{\sqrt{2}}{2}$ **29.** $h(t) = \cot t; t = \dfrac{11\pi}{6}$ $h\left(\dfrac{11\pi}{6}\right) = -\sqrt{3}$

30. $s(t) = \sec t; t = \dfrac{5\pi}{3}$ $s\left(\dfrac{5\pi}{3}\right) = 2$ **31.** $z(t) = \csc t; t = \dfrac{7\pi}{4}$ $z\left(\dfrac{7\pi}{4}\right) = -\sqrt{2}$ **32.** $r(t) = \tan t; t = \dfrac{5\pi}{6}$ $r\left(\dfrac{5\pi}{6}\right) = -\dfrac{\sqrt{3}}{3}$

Objective 2: Identify the Domains of the Trigonometric Functions

For Exercises 33–38, select the domain of the trigonometric function.

a. All real numbers   **b.** $\left\{ t \mid t \neq \dfrac{(2n + 1)\pi}{2} \text{ for all integers } n \right\}$ **c.** $\{t \mid t \neq n\pi \text{ for all integers } n\}$

33. $f(t) = \sin t$ a   **34.** $f(t) = \tan t$ b   **35.** $f(t) = \cot t$ c

36. $f(t) = \cos t$ a   **37.** $f(t) = \sec t$ b   **38.** $f(t) = \csc t$ c

For Exercises 39–42, evaluate the function if possible. (See Example 3)

39. a. $\sin 0$ 0   **b.** $\cot \pi$ Undefined   **c.** $\tan 3\pi$ 0

    d. $\sec \pi$ -1 **e.** $\csc 0$ Undefined   **f.** $\cos \pi$ -1

40. a. $\cos\left(-\dfrac{\pi}{2}\right)$ 0   **b.** $\csc \dfrac{\pi}{2}$ 1   **c.** $\cot \dfrac{5\pi}{2}$ 0

    d. $\tan \dfrac{\pi}{2}$ Undefined   **e.** $\sec \dfrac{3\pi}{2}$ Undefined   **f.** $\sin \dfrac{\pi}{2}$ 1

41. a. $\sin \dfrac{3\pi}{2}$ -1 **b.** $\cos \dfrac{7\pi}{2}$ 0   **c.** $\tan \dfrac{3\pi}{2}$ Undefined

    d. $\csc\left(-\dfrac{\pi}{2}\right)$ -1 **e.** $\sec 1.5\pi$ Undefined   **f.** $\cot \dfrac{\pi}{2}$ 0

42. a. $\cot 0$ Undefined   **b.** $\cos 2\pi$ 1   **c.** $\csc \pi$ Undefined

    d. $\tan 0$ 0   **e.** $\sin(-3\pi)$ 0   **f.** $\sec 0$ 1

Objective 3: Use Fundamental Trigonometric Identities

For Exercises 43–46, given the values for $\sin t$ and $\cos t$, use the reciprocal and quotient identities to find the values of the other trigonometric functions of t. (See Example 4)

43. $\sin t = \dfrac{\sqrt{5}}{3}$ and $\cos t = \dfrac{2}{3}$

44. $\sin t = \dfrac{3}{4}$ and $\cos t = \dfrac{\sqrt{7}}{4}$

45. $\sin t = -\dfrac{\sqrt{39}}{8}$ and $\cos t = -\dfrac{5}{8}$

46. $\sin t = \dfrac{28}{53}$ and $\cos t = -\dfrac{45}{53}$

For Exercises 47–48, derive the given identity from the Pythagorean identity, $\sin^2 t + \cos^2 t = 1$.

47. $\tan^2 t + 1 = \sec^2 t$

48. $1 + \cot^2 t = \csc^2 t$

For Exercises 49–54, use an appropriate Pythagorean identity to find the indicated value. (See Example 5)

49. Given $\cos t = -\dfrac{7}{25}$ for $\dfrac{\pi}{2} < t < \pi$, find the value of $\sin t$. $\dfrac{24}{25}$

50. Given $\sin t = -\dfrac{8}{17}$ for $\pi < t < \dfrac{3\pi}{2}$, find the value of $\cos t$. $-\dfrac{15}{17}$

51. Given $\cot t = \dfrac{45}{28}$ for $\pi < t < \dfrac{3\pi}{2}$, find the value of $\csc t$. $-\dfrac{53}{28}$

52. Given $\csc t = -\dfrac{41}{40}$ for $\dfrac{3\pi}{2} < t < 2\pi$, find the value of $\cot t$. $-\dfrac{9}{40}$

53. Given $\tan t = -\dfrac{11}{60}$ for $\dfrac{3\pi}{2} < t < 2\pi$, find the value of $\sec t$. $\dfrac{61}{60}$

54. Given $\sec t = -\dfrac{37}{35}$ for $\dfrac{\pi}{2} < t < \pi$, find the value of $\tan t$. $-\dfrac{12}{35}$

55. Write $\sin t$ in terms of $\cos t$ for

    a. t in Quadrant I. (See Example 6) $\sin t = \sqrt{1 - \cos^2 t}$

    b. t in Quadrant III. $\sin t = -\sqrt{1 - \cos^2 t}$

56. Write $\tan t$ in terms of $\sec t$ for

    a. t in Quadrant II. $\tan t = -\sqrt{\sec^2 t - 1}$

    b. t in Quadrant IV. $\tan t = -\sqrt{\sec^2 t - 1}$

57. Write $\cot t$ in terms of $\csc t$ for

    a. t in Quadrant I. $\cot t = \sqrt{\csc^2 t - 1}$

    b. t in Quadrant III. $\cot t = \sqrt{\csc^2 t - 1}$

58. Write $\cos t$ in terms of $\sin t$ for

    a. t in Quadrant II. $\cos t = -\sqrt{1 - \sin^2 t}$

    b. t in Quadrant IV. $\cos t = \sqrt{1 - \sin^2 t}$

Objective 4: Apply the Periodic and Even and Odd Function Properties of Trigonometric Functions

59. Given that $\cos\dfrac{29\pi}{12} = \dfrac{\sqrt{6} - \sqrt{2}}{4}$, determine the value of $\cos\dfrac{5\pi}{12}$. **(See Example 7)** $\dfrac{\sqrt{6} - \sqrt{2}}{4}$

60. Given that $\sec\dfrac{11\pi}{12} = \sqrt{2} - \sqrt{6}$, determine the value of $\sec\left(-\dfrac{13\pi}{12}\right)$. $\sqrt{2} - \sqrt{6}$

61. Given $\tan\left(-\dfrac{\pi}{8}\right) = 1 - \sqrt{2}$, determine the value of $\cot\dfrac{7\pi}{8}$. $-1 - \sqrt{2}$

62. Given that $\sin\dfrac{\pi}{10} = \dfrac{\sqrt{5} - 1}{4}$, determine the value of $\csc\dfrac{21\pi}{10}$. $1 + \sqrt{5}$

For exercises 63–66, use the periodic properties of the trigonometric functions to simplify each expression to a single function of t.

63. $\sin(t + 2\pi) \cdot \cot(t + \pi)$ $\cos t$

64. $\sin(t + 2\pi) \cdot \sec(t + 2\pi)$ $\tan t$

65. $\csc(t + 2\pi) \cdot \cos(t + 2\pi)$ $\cot t$

66. $\tan(t + \pi) \cdot \csc(t + 2\pi)$ $\sec t$

For Exercises 67–74, use the even-odd and periodic properties of the trigonometric functions to simplify. **(See Example 8)**

67. $\csc t - 4\csc(-t)$ $5\csc t$

68. $\tan(-t) - 3\tan t$ $-4\tan t$

69. $-\cot(-t + \pi) - \cot t$ 0

70. $\sec(t + 2\pi) - \sec(-t)$ 0

71. $-2\sin(3t + 2\pi) - 3\sin(-3t)$ $\sin 3t$

72. $\cot(-3t) - 3\cot(3t + \pi)$ $-4\cot 3t$

73. $\cos(-2t) - \cos 2t$ 0

74. $\sec(-2t) + 3\sec 2t$ $4\sec 2t$

Objective 5: Approximate Trigonometric Functions on a Calculator

Use a calculator to approximate the function values. Round to 4 decimal places. **(See Example 9)**

75. a. $\sin(-0.15)$ -0.1494

    b. $\cos\dfrac{2\pi}{5}$ 0.3090

76. a. $\cos\left(-\dfrac{7\pi}{11}\right)$ -0.4154

    b. $\sin 0.96$ 0.8192

77. a. $\cot\dfrac{12\pi}{7}$ -0.7975

    b. $\sec 5.43$ 1.5207

78. a. $\csc 7.58$ 1.0387

    b. $\tan\dfrac{3\pi}{8}$ 2.4142

Mixed Exercises

For Exercises 79–80, evaluate $\sin t$, $\cos t$, and $\tan t$ for the real number t.

79. a. $t = \dfrac{2\pi}{3}$

    b. $t = -\dfrac{4\pi}{3}$

80. a. $t = -\dfrac{11\pi}{6}$

    b. $t = \dfrac{\pi}{6}$

For Exercises 81–86, identify the values of t on the interval $[0, 2\pi]$ that make the function undefined (if any).

81. $y = \sin t$ None

82. $y = \cot t$ $0, \pi, 2\pi$

83. $y = \tan t$ $\dfrac{\pi}{2}, \dfrac{3\pi}{2}$

84. $y = \cos t$ None

85. $y = \csc t$ $0, \pi, 2\pi$

86. $y = \sec t$ $\dfrac{\pi}{2}, \dfrac{3\pi}{2}$

For Exercises 87–92, select all properties that apply to the trigonometric function.

a. The function is even.   **b.** The function is odd.

c. The period is 2π.   **d.** The period is π.

e. The domain is all real numbers.

f. The domain is all real numbers excluding odd multiples of $\dfrac{\pi}{2}$.

g. The domain is all real numbers excluding multiples of π.

87. $f(t) = \sin(t)$ b, c, e

88. $f(t) = \tan(t)$ b, d, f

89. $f(t) = \sec(t)$ a, c, f

90. $f(t) = \cot(t)$ b, d, g

91. $f(t) = \csc(t)$ b, c, g

92. $f(t) = \cos(t)$ a, c, e

For Exercises 93–98, simplify using properties of trigonometric functions.

93. $\sin^2(t + 2\pi) + \cos^2 t + \tan^2(t + \pi)$　$\sec^2 t$

94. $\cot(-t) \cdot \sin(t + 2\pi) + \cos^2 t \cdot \sec(-t + 2\pi)$　0

95. $\sec^2\left(-\dfrac{\pi}{3}\right) + \tan^2\left(-\dfrac{\pi}{3}\right)$　7

96. $\sin^2\left(\dfrac{\pi}{6}\right) - \cos^2\left(-\dfrac{\pi}{6}\right)$　$-\dfrac{1}{2}$

97. $\sec^2\left(-\dfrac{\pi}{4}\right) - \csc^2\left(\dfrac{7\pi}{6}\right)$　-2

98. $\tan^2\left(\dfrac{2\pi}{3}\right) + \csc^2\left(-\dfrac{5\pi}{4}\right)$　5

99. In a small coastal town, the monthly revenue received from tourists rises and falls throughout the year. The tourist revenue peaks during the months when the town holds seafood cooking competitions. The function

$$f(t) = 5.6\cos\left(\dfrac{\pi}{3}t\right) + 11.2$$

represents the monthly revenue $f(t)$, in tens of thousands of dollars, for a month t, where $t = 0$ represents April.

a. Complete the table and give the period of the function. Round to 1 decimal place.　Period = 6 months

t	0	1	2	3	4	5
$f(t)$	16.8	14	8.4	5.6	8.4	14

t	6	7	8	9	10	11
$f(t)$	16.8	14	8.4	5.6	8.4	14

b. What are the months of peak revenue and what is the revenue for those months?　Peak revenue is in month 0 (April) and month 6 (October). The revenue is $168,000.

100. The fluctuating brightness of a distant star is given by the function

$$f(d) = 3.8 + 0.25\sin\left(\dfrac{2\pi}{3}d\right)$$

where d is the number of days and $f(d)$ is the apparent brightness.

Complete the table and give the period of the function. Round to 2 decimal places.　Period = 3 days

d	0	1	2	3	4	5
$f(d)$	3.80	4.02	3.58	3.80	4.02	3.58

For Exercises 101–104, identify each function as even, odd, or neither.

101. $f(t) = t^2\sin t$　Odd

102. $g(t) = t\cos t$　Odd

103. $z(t) = t^3\tan t$　Even

104. $h(t) = t^3 + \sec t$　Neither

Write About It

105. Describe the changes in $\sin t$ and $\cos t$ as t increases from 0 to $\dfrac{\pi}{2}$.

106. Describe the changes in $\cos t$ and $\sec t$ as t increases from 0 to $\dfrac{\pi}{2}$.

107. Explain why $-1 \le \cos t \le 1$ and $-1 \le \sin t \le 1$ for all real numbers t.

108. Do the trigonometric functions satisfy the definition of a function learned in Chapter 1? Explain your answer.

Expanding Your Skills

For Exercises 109–114, use the figure to estimate the value of (a) sin *t* and (b) cos *t* for the given value of *t*.

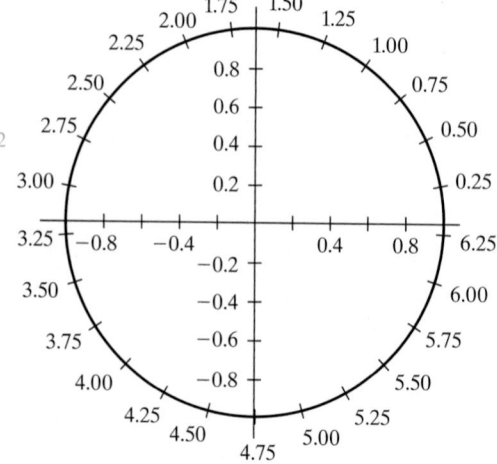

109. $t = 0.5$
　a. 0.48　b. 0.88

110. $t = 1.25$
　a. 0.95　b. 0.32

111. $t = 3.75$
　a. -0.57　b. -0.82

112. $t = 2.75$
　a. 0.38　b. -0.92

113. $t = 5$
　a. -0.96　b. 0.28

114. $t = 3$
　a. 0.14　b. -0.99

For Exercises 115–120, use the figure from Exercises 109–114 to approximate the solutions to the equation over the interval $[0, 2\pi)$.

115. $\sin t = 0.2$　$t \approx 0.2$ and $t \approx 2.9$

116. $\sin t = -0.4$　$t \approx 3.6$ and $t \approx 5.9$

117. $\cos t = -0.4$　$t \approx 2.0$ and $t \approx 4.3$

118. $\cos t = 0.6$　$t \approx 0.9$ and $t \approx 5.4$

119. Show that $\tan(t + \pi) = \tan t$.

120. Show that $\cot(t + \pi) = \cot t$.

121. Prove that the period of $f(t) = \sin t$ is 2π.

122. Prove that the period of $f(t) = \cos t$ is 2π.

Additional answers can be found in the Instructor Answer Appendix.

Technology Connections

123. a. Complete the table to show that $\sin t \approx t$ for values of t close to zero. Round to 5 decimal places.

t	$\sin t$	$\dfrac{\sin t}{t}$
0.2	0.19867	0.99335
0.1	0.09983	0.99833
0.01	0.01000	0.99998
0.001	0.00100	1.00000

b. What value does the ratio $\dfrac{\sin t}{t}$ seem to approach as $t \to 0$?   1

124. Consider the expression $\dfrac{1 - \cos t}{t}$.

a. What is the value of the numerator for $t = 0$?   0

b. What is the value of the denominator for $t = 0$?   0

c. The expression $\dfrac{1 - \cos t}{t}$ is undefined at $t = 0$. Complete the table to investigate the value of the expression *close* to $t = 0$. Round to 5 decimal places.

t	$\dfrac{1 - \cos t}{t}$
0.2	0.09967
0.1	0.04996
0.01	0.00500
0.001	0.00050

SECTION 4.3 — Right Triangle Trigonometry

OBJECTIVES

1. Evaluate Trigonometric Functions of Acute Angles
2. Use Fundamental Trigonometric Identities
3. Use Trigonometric Functions in Applications

1. Evaluate Trigonometric Functions of Acute Angles

The science of land surveying encompasses the measurement and mapping of land using mathematics and specialized equipment and technology. Surveyors provide data relevant to the shape and contour of the Earth's surface for engineering, mapmaking, and construction projects.

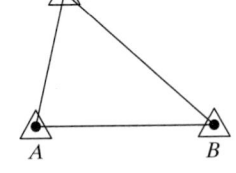

One technique used by surveyors to map a landscape is called *triangulation*. The surveyor first measures the distance between two fixed points, A and B. Then the surveyor measures the bearing from A and the bearing from B to a third point P. From this information, the surveyor can compute the measures of the angles of the triangle formed by the points, and can use trigonometry to find the lengths of the unknown sides.

To use trigonometry in such applications, we present alternative definitions of the trigonometric functions using right triangles. Consider a right triangle with an acute angle θ. The longest side in the triangle is the hypotenuse ("hyp") and is opposite the right angle (Figure 4-27). The two legs of the triangle will be distinguished by their relative positions to θ. The leg that lies on one ray of angle θ is called the adjacent leg ("adj") and the leg that lies across the triangle from θ is called the opposite leg ("opp").

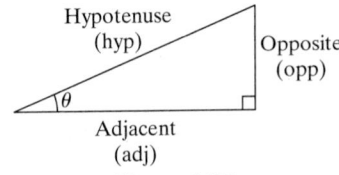

Figure 4-27

We now define the six trigonometric functions as functions whose input (or **argument**) is an acute angle θ. The output value for each function will be one of the six possible ratios of the lengths of the sides of the triangle (Table 4-4).

The definitions of the six trigonometric functions of acute angles are given in Table 4-4 along with examples based on Figure 4-28.

TIP The mnemonic device "SOH-CAH-TOA" may help you remember the ratios for $\sin\theta$, $\cos\theta$, and $\tan\theta$, respectively.

Sine: Opp over Hyp
Cosine: Adj over Hyp
Tangent: Opp over Adj

Table 4-4

Definition of Trigonometric Functions of Acute Angles		
Function Name	**Definition**	**Example**
sine	$\sin\theta = \dfrac{\text{opp}}{\text{hyp}}$	$\sin\theta = \dfrac{3\text{ ft}}{5\text{ ft}} = \dfrac{3}{5}$
cosine	$\cos\theta = \dfrac{\text{adj}}{\text{hyp}}$	$\cos\theta = \dfrac{4\text{ ft}}{5\text{ ft}} = \dfrac{4}{5}$
tangent	$\tan\theta = \dfrac{\text{opp}}{\text{adj}}$	$\tan\theta = \dfrac{3\text{ ft}}{4\text{ ft}} = \dfrac{3}{4}$
cosecant	$\csc\theta = \dfrac{\text{hyp}}{\text{opp}}$	$\csc\theta = \dfrac{5\text{ ft}}{3\text{ ft}} = \dfrac{5}{3}$
secant	$\sec\theta = \dfrac{\text{hyp}}{\text{adj}}$	$\sec\theta = \dfrac{5\text{ ft}}{4\text{ ft}} = \dfrac{5}{4}$
cotangent	$\cot\theta = \dfrac{\text{adj}}{\text{opp}}$	$\cot\theta = \dfrac{4\text{ ft}}{3\text{ ft}} = \dfrac{4}{3}$

Hypotenuse hyp = 5 ft Opposite opp = 3 ft θ Adjacent adj = 4 ft

Figure 4-28

The notation $\sin\theta$ is read as "sine of theta" (likewise, $\cos\theta$ is read as "cosine of theta" and so on). Also notice that the output value of a trigonometric function is unitless because the common units of length "cancel" within each ratio.

The definitions of the trigonometric functions given in Table 4-4 are consistent with the definitions based on the unit circle from Section 4.2. For example, consider a real number $0 < t < \dfrac{\pi}{2}$ that corresponds to the point $P(x, y)$ on the unit circle (Figure 4-29). Suppose that we drop a perpendicular line segment from P to the x-axis at point A. Then $\triangle OAP$ is a right triangle. The lengths of the legs are x and y, and the hypotenuse is 1 unit. Furthermore, the radian measure of $\angle AOP$ is equal to t. If we let θ represent the degree measure of $\angle AOP$, we have

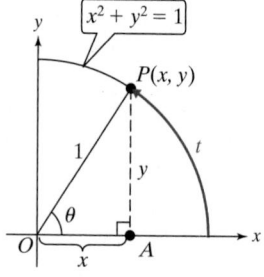

Figure 4-29

Unit circle definition

$\sin t = y$

$\cos t = x$

$\tan t = \dfrac{y}{x}$

Right triangle definition

$\sin\theta = \dfrac{\text{opp}}{\text{hyp}} = \dfrac{y}{1} = y$

$\cos\theta = \dfrac{\text{adj}}{\text{hyp}} = \dfrac{x}{1} = x$

$\tan\theta = \dfrac{\text{opp}}{\text{adj}} = \dfrac{y}{x}$

It is also very important to note that the values of the trigonometric functions depend only on the measure of the angle, not the size of the triangle. For any given acute angle θ, a series of right triangles can be formed by drawing a line segment perpendicular to the initial side of θ with endpoints on the initial and terminal sides of θ (Figure 4-30). Triangles $\triangle ABC$ and $\triangle ADE$ are similar triangles with common angle θ. Therefore, the corresponding sides are proportional, and the value of each trigonometric function of θ will be the same regardless of which triangle is used. For example, $\sin \theta = \dfrac{x}{b} = \dfrac{y}{d}$.

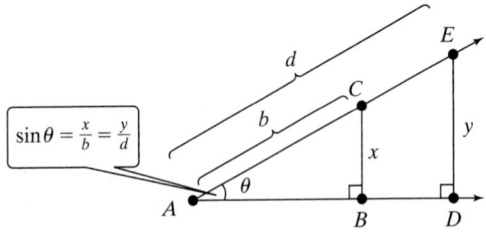

Figure 4-30

In Example 1, we find the values of the six trigonometric functions of an acute angle within a right triangle where the lengths of two legs are given.

Classroom Example: p. 493
Exercise 20

EXAMPLE 1   Evaluating Trigonometric Functions by First Applying the Pythagorean Theorem

Suppose that a right triangle has legs of length 4 cm and 7 cm. Evaluate the six trigonometric functions for the smaller acute angle.

Solution:

A right triangle $\triangle ABC$ meeting the conditions of this example is drawn in Figure 4-31. Since $\angle B$ is opposite the shorter leg, $\angle B$ is the smaller acute angle.

To find the values of all six trigonometric functions, we also need to know the length of the hypotenuse. The hypotenuse is always opposite the right angle.

$c^2 = 4^2 + 7^2$ Apply the Pythagorean theorem.

$c^2 = 16 + 49$ Simplify terms with exponents.

$c^2 = 65$ Since c represents the length of a side of

$c = \sqrt{65}$ a triangle, take the positive square root of 65.

Figure 4-31

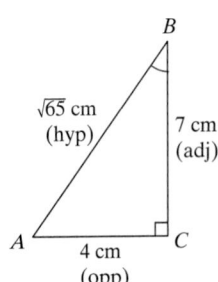

Avoiding Mistakes

In Example 1, the opposite (opp) and adjacent (adj) legs are labeled in the triangle relative to angle B.

Relative to angle B, the 4-cm side is the opposite side. The 7-cm side is the adjacent leg, and the hypotenuse is $c = \sqrt{65}$.

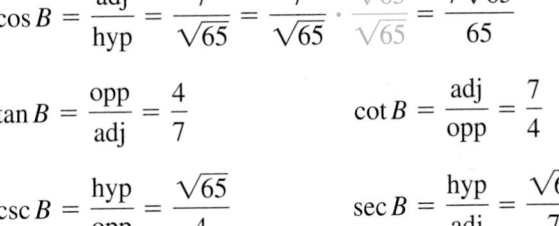

Instructor Note:

Ask your students to label opp, adj, and hyp for angle A. Then ask them to find the values of the six trigonometric functions of angle A.

$\sin B = \dfrac{\text{opp}}{\text{hyp}} = \dfrac{4}{\sqrt{65}} = \dfrac{4}{\sqrt{65}} \cdot \dfrac{\sqrt{65}}{\sqrt{65}} = \dfrac{4\sqrt{65}}{65}$

$\cos B = \dfrac{\text{adj}}{\text{hyp}} = \dfrac{7}{\sqrt{65}} = \dfrac{7}{\sqrt{65}} \cdot \dfrac{\sqrt{65}}{\sqrt{65}} = \dfrac{7\sqrt{65}}{65}$

$\tan B = \dfrac{\text{opp}}{\text{adj}} = \dfrac{4}{7}$ $\cot B = \dfrac{\text{adj}}{\text{opp}} = \dfrac{7}{4}$

$\csc B = \dfrac{\text{hyp}}{\text{opp}} = \dfrac{\sqrt{65}}{4}$ $\sec B = \dfrac{\text{hyp}}{\text{adj}} = \dfrac{\sqrt{65}}{7}$

> **Skill Practice 1** Suppose that a right triangle $\triangle ABC$ has legs of length 6 cm and 3 cm. Evaluate the six trigonometric functions for angle B, where angle B is the larger acute angle.

In Example 1, we were given the lengths of the sides of a right triangle and asked to find the values of the six trigonometric functions. This process can be reversed. Given the value of one trigonometric function of an acute angle, we can find the lengths of the sides of a representative triangle, and thus the values of the other trigonometric functions.

Classroom Example: p. 493
Exercise 26

EXAMPLE 2 Using a Known Value of a Trigonometric Function to Determine Other Function Values

Suppose that $\cos\theta = \dfrac{\sqrt{5}}{3}$ for the acute angle θ. Evaluate $\tan\theta$.

Solution:

Since $\cos\theta$ is the ratio of the length of the leg adjacent to θ and the length of the hypotenuse, it is convenient to construct a triangle with adjacent leg $\sqrt{5}$ units and hypotenuse 3 units. Using the Pythagorean theorem, we can find the length of the opposite leg.

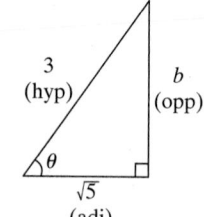

$$\left(\sqrt{5}\right)^2 + b^2 = (3)^2 \qquad \text{Apply the Pythagorean theorem.}$$

$$5 + b^2 = 9 \qquad \text{Simplify terms with exponents.}$$

$$b^2 = 4$$

$$b = 2 \qquad \text{Take the positive square root of 4.}$$

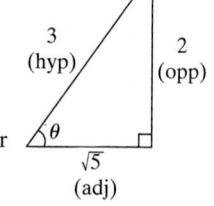

$$\tan\theta = \frac{\text{opp}}{\text{adj}} = \frac{2}{\sqrt{5}} \cdot \frac{\sqrt{5}}{\sqrt{5}} = \frac{2\sqrt{5}}{5} \qquad \begin{array}{l}\text{Once the proper ratio is identified for}\\ \tan\theta, \text{ rationalize the denominator.}\end{array}$$

> **Skill Practice 2** Suppose that $\sin\theta = \dfrac{3}{4}$ for the acute angle θ. Evaluate $\sec\theta$.

In Example 3, we find the sine, cosine, and tangent of 45° by using a right triangle approach.

EXAMPLE 3 Determine the Trigonometric Function Values of a 45° Angle

Evaluate $\sin 45°$, $\cos 45°$, and $\tan 45°$.

Answers

1. $\sin B = \dfrac{2\sqrt{5}}{5}$, $\cos B = \dfrac{\sqrt{5}}{5}$,

   $\tan B = 2$, $\csc B = \dfrac{\sqrt{5}}{2}$,

   $\sec B = \sqrt{5}$, $\cot B = \dfrac{1}{2}$

2. $\sec\theta = \dfrac{4\sqrt{7}}{7}$

Solution:

A right triangle with an acute angle of 45° must have a second acute angle of 45° because the sum of the angles must equal 180°. This is called an isoceles right triangle or a 45°-45°-90° triangle. Since all such triangles are similar, we can choose one of the sides to be of arbitrary length. We have chosen the hypotenuse to be 1 unit. The sides opposite the 45° angles are equal in length and have been labeled a.

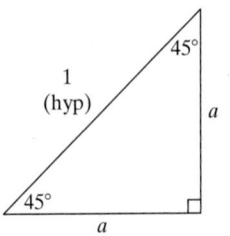

$$a^2 + a^2 = 1^2 \qquad \text{Apply the Pythagorean theorem.}$$
$$2a^2 = 1$$
$$a^2 = \frac{1}{2}$$
$$a = \sqrt{\frac{1}{2}} = \frac{1}{\sqrt{2}} = \frac{1}{\sqrt{2}} \cdot \frac{\sqrt{2}}{\sqrt{2}} = \frac{\sqrt{2}}{2} \qquad \begin{array}{l}\text{Take the positive square root of } \frac{1}{2}.\\ \text{Rationalize the denominator.}\end{array}$$

TIP The values of sin 45°, cos 45°, and tan 45° found in Example 3 are consistent with the values of $\sin\frac{\pi}{4}$, $\cos\frac{\pi}{4}$, and $\tan\frac{\pi}{4}$ found by using the unit circle.

$$\sin 45° = \frac{\text{opp}}{\text{hyp}} = \frac{\frac{\sqrt{2}}{2}}{1} = \frac{\sqrt{2}}{2}$$

$$\cos 45° = \frac{\text{adj}}{\text{hyp}} = \frac{\frac{\sqrt{2}}{2}}{1} = \frac{\sqrt{2}}{2}$$

$$\tan 45° = \frac{\text{opp}}{\text{adj}} = \frac{\frac{\sqrt{2}}{2}}{\frac{\sqrt{2}}{2}} = 1$$

Skill Practice 3   Evaluate csc 45°, sec 45°, and cot 45°.

A right triangle with a 30° angle will also have a 60° angle so that the sum of the angular measures is 180°. Such a triangle is called a 30°-60°-90° triangle. Interestingly, as we will show in Example 4, the length of the shorter leg of a 30°-60°-90° triangle is always one-half the length of the hypotenuse.

Classroom Example: p. 493
Exercise 32

EXAMPLE 4   Determine the Trigonometric Function Values of a 60° Angle

Evaluate sin 60°, cos 60°, and tan 60°.

Solution:

Suppose that we start with an equilateral triangle with sides of length L. All angles in the triangle are equal to 60°. Any altitude a of this triangle bisects a 60° angle as well as the opposite side. Thus, we can create two congruent right triangles with legs of length a and $\frac{1}{2}L$ and hypotenuse of length L.

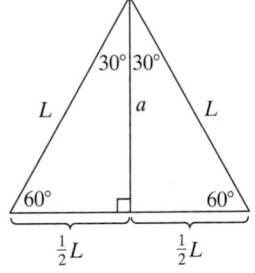

Answer
3. csc 45° = $\sqrt{2}$, sec 45° = $\sqrt{2}$, and cot 45° = 1

Use the Pythagorean theorem to find the altitude a in terms of L.

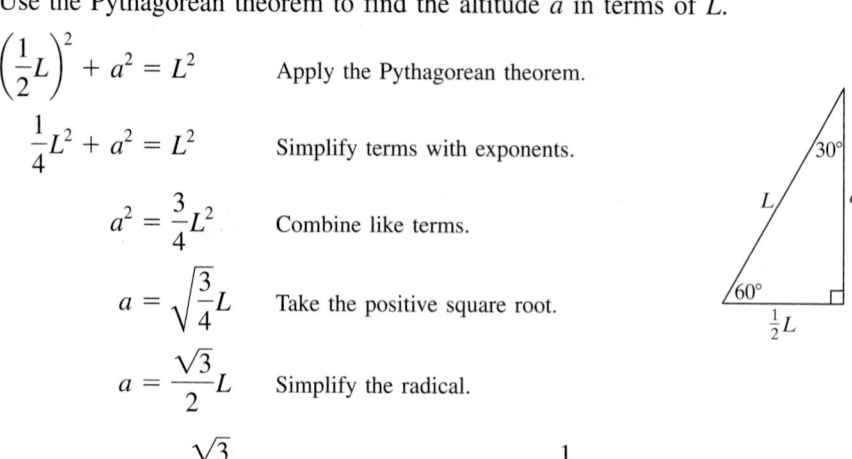

$$\left(\frac{1}{2}L\right)^2 + a^2 = L^2 \qquad \text{Apply the Pythagorean theorem.}$$

$$\frac{1}{4}L^2 + a^2 = L^2 \qquad \text{Simplify terms with exponents.}$$

$$a^2 = \frac{3}{4}L^2 \qquad \text{Combine like terms.}$$

$$a = \sqrt{\frac{3}{4}}L \qquad \text{Take the positive square root.}$$

$$a = \frac{\sqrt{3}}{2}L \qquad \text{Simplify the radical.}$$

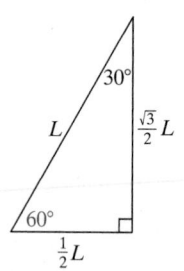

$$\sin 60° = \frac{\text{opp}}{\text{hyp}} = \frac{\frac{\sqrt{3}}{2}L}{L} = \frac{\sqrt{3}}{2} \qquad \cos 60° = \frac{\text{adj}}{\text{hyp}} = \frac{\frac{1}{2}L}{L} = \frac{1}{2}$$

$$\tan 60° = \frac{\text{opp}}{\text{adj}} = \frac{\frac{\sqrt{3}}{2}L}{\frac{1}{2}L} = \left(\frac{\sqrt{3}}{2}\right) \cdot \left(\frac{2}{1}\right) = \sqrt{3}$$

Skill Practice 4   Evaluate $\sin 30°$, $\cos 30°$, and $\tan 30°$.

TIP As a memory device, note the 1-2-3 pattern in the numerator for the sine function for the special angles.

$$\sin 30° = \frac{\sqrt{1}}{2}$$

$$\sin 45° = \frac{\sqrt{2}}{2}$$

$$\sin 60° = \frac{\sqrt{3}}{2}$$

For cosine, the order is reversed, 3-2-1.

$$\cos 30° = \frac{\sqrt{3}}{2}$$

$$\cos 45° = \frac{\sqrt{2}}{2}$$

$$\cos 60° = \frac{\sqrt{1}}{2}$$

From Examples 3 and 4, we can summarize the trigonometric function values of the "special" angles 30°, 45°, and 60°, or equivalently $\frac{\pi}{6}$, $\frac{\pi}{4}$, and $\frac{\pi}{3}$ radians (Table 4-5).

Table 4-5

Trigonometric Function Values of Special Angles						
θ	$\sin\theta$	$\cos\theta$	$\tan\theta$	$\csc\theta$	$\sec\theta$	$\cot\theta$
$30° = \frac{\pi}{6}$	$\frac{1}{2}$	$\frac{\sqrt{3}}{2}$	$\frac{\sqrt{3}}{3}$	2	$\frac{2\sqrt{3}}{3}$	$\sqrt{3}$
$45° = \frac{\pi}{4}$	$\frac{\sqrt{2}}{2}$	$\frac{\sqrt{2}}{2}$	1	$\sqrt{2}$	$\sqrt{2}$	1
$60° = \frac{\pi}{3}$	$\frac{\sqrt{3}}{2}$	$\frac{1}{2}$	$\sqrt{3}$	$\frac{2\sqrt{3}}{3}$	2	$\frac{\sqrt{3}}{3}$

To help you remember the values of the trigonometric functions for 30°, 45°, and 60°, you can refer to the first quadrant of the unit circle. The values $t = \frac{\pi}{6}$, $t = \frac{\pi}{4}$, and $t = \frac{\pi}{3}$ correspond to the central angles $\theta = 30°$, $\theta = 45°$, and $\theta = 60°$, respectively (Figure 4-32).

Answer

4. $\sin 30° = \frac{1}{2}$, $\cos 30° = \frac{\sqrt{3}}{2}$,

and $\tan 30° = \frac{\sqrt{3}}{3}$

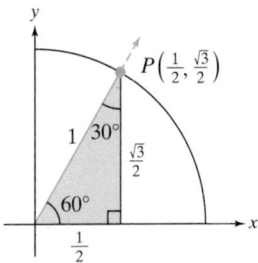 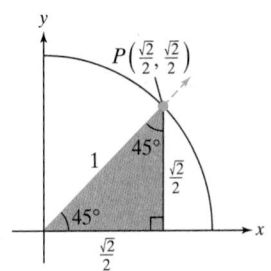 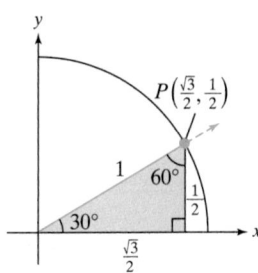

Figure 4-32

EXAMPLE 5   Simplifying Expressions Involving Trigonometric Functions of Special Angles

Simplify the expressions.

a. $\tan 60° - \tan 30°$

b. $2 \sin \dfrac{\pi}{6} \cos \dfrac{\pi}{6}$

Solution:

a. $\tan 60° - \tan 30° = \sqrt{3} - \dfrac{\sqrt{3}}{3}$ Evaluate $\tan 60°$ and $\tan 30°$.

$ = \dfrac{3}{3} \cdot \dfrac{\sqrt{3}}{1} - \dfrac{\sqrt{3}}{3}$ Write the first term with a common denominator of 3.

$ = \dfrac{2\sqrt{3}}{3}$ Combine like terms.

b. $2 \sin \dfrac{\pi}{6} \cos \dfrac{\pi}{6} = 2 \left(\dfrac{1}{2} \right) \left(\dfrac{\sqrt{3}}{2} \right)$ Evaluate $\sin \dfrac{\pi}{6}$ and $\cos \dfrac{\pi}{6}$.

$\phantom{2 \sin \dfrac{\pi}{6} \cos \dfrac{\pi}{6}} = \dfrac{\sqrt{3}}{2}$ Multiply.

Skill Practice 5   Simplify the expressions.

a. $\cot 60° - \cot 30°$

b. $\sin \dfrac{\pi}{3} \cos \dfrac{\pi}{6} + \cos \dfrac{\pi}{3} \sin \dfrac{\pi}{6}$

2. Use Fundamental Trigonometric Identities

In Section 4.2, we observed several relationships among the trigonometric functions that follow directly from the definitions of the functions. The reciprocal and quotient identities also follow from the right triangle defintions of trigonometric functions of acute angles (Table 4-6). For example, since $\sin \theta = \dfrac{\text{opp}}{\text{hyp}}$ and $\csc \theta = \dfrac{\text{hyp}}{\text{opp}}$ for an acute angle θ, we know that $\sin \theta$ and $\csc \theta$ are reciprocals.

Answers

5. a. $-\dfrac{2\sqrt{3}}{3}$ **b.** 1

Table 4-6

Reciprocal and Quotient Identities	
$\csc\theta = \dfrac{1}{\sin\theta}$ or $\sin\theta = \dfrac{1}{\csc\theta}$	$\sin\theta$ and $\csc\theta$ are reciprocals.
$\sec\theta = \dfrac{1}{\cos\theta}$ or $\cos\theta = \dfrac{1}{\sec\theta}$	$\cos\theta$ and $\sec\theta$ are reciprocals.
$\cot\theta = \dfrac{1}{\tan\theta}$ or $\tan\theta = \dfrac{1}{\cot\theta}$	$\tan\theta$ and $\cot\theta$ are reciprocals.
$\tan\theta = \dfrac{\sin\theta}{\cos\theta}$	$\tan\theta$ is the ratio of $\sin\theta$ and $\cos\theta$.
$\cot\theta = \dfrac{\cos\theta}{\sin\theta}$	$\cot\theta$ is the ratio of $\cos\theta$ and $\sin\theta$.

We can also show that the Pythagorean identities follow from the right triangle definitions of the trigonometric functions of an acute angle θ. Consider a right triangle with legs of length a and b and hypotenuse c (Figure 4-33). From the Pythagorean theorem, we know that $a^2 + b^2 = c^2$. Dividing both sides by the positive real number c^2 we have

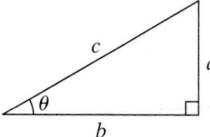

Figure 4-33

$$\frac{a^2}{c^2} + \frac{b^2}{c^2} = \frac{c^2}{c^2} \qquad \text{Divide by } c^2.$$

$$\left(\frac{a}{c}\right)^2 + \left(\frac{b}{c}\right)^2 = \left(\frac{c}{c}\right)^2 \qquad \text{Rewrite each term as the power of a quotient.}$$

$$(\sin\theta)^2 + (\cos\theta)^2 = 1 \qquad \text{From Figure 4-33, } \sin\theta = \frac{a}{c} \text{ and } \cos\theta = \frac{b}{c}.$$

$$\sin^2\theta + \cos^2\theta = 1$$

Table 4-7 summarizes this and two other Pythagorean identities. The second and third relationships can be derived by dividing both sides of the equation $a^2 + b^2 = c^2$ by b^2 and a^2, respectively.

Table 4-7

Pythagorean Identities		
$\sin^2\theta + \cos^2\theta = 1$	$\tan^2\theta + 1 = \sec^2\theta$	$1 + \cot^2\theta = \csc^2\theta$

Visualing the ratios of the lengths of the sides of a right triangle can also help us understand the cofunction identities.

The two acute angles in a right triangle are complementary because the sum of their measures is 90°. Symbolically, the complement of angle θ is $(90° - \theta)$. In Figure 4-34, notice that side a is opposite angle θ but is the adjacent leg to angle $(90° - \theta)$. Likewise, side b is adjacent to angle θ but opposite angle $(90° - \theta)$. From these relationships, we have

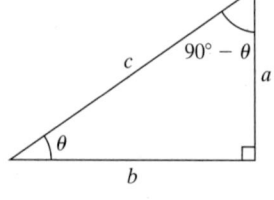

Figure 4-34

$$\sin\theta = \cos(90° - \theta) = \frac{a}{c}$$

$$\tan\theta = \cot(90° - \theta) = \frac{a}{b}$$

$$\sec\theta = \csc(90° - \theta) = \frac{c}{b}$$

Notice that the sine of angle θ equals the cosine of the complement $(90° - \theta)$. For this reason, the sine and cosine functions are called *cofunctions*. More generally, for an acute angle θ, two trigonometric functions f and g are **cofunctions** if

$$f(\theta) = g(90° - \theta) \text{ and } g(\theta) = f(90° - \theta)$$

Cofunction Identities

Cofunctions of complementary angles are equal.

$\sin\theta = \cos(90° - \theta)$	$\cos\theta = \sin(90° - \theta)$	Sine and cosine are cofunctions.
$\sin\theta = \cos\left(\dfrac{\pi}{2} - \theta\right)$	$\cos\theta = \sin\left(\dfrac{\pi}{2} - \theta\right)$	
$\tan\theta = \cot(90° - \theta)$	$\cot\theta = \tan(90° - \theta)$	Tangent and cotangent are cofunctions.
$\tan\theta = \cot\left(\dfrac{\pi}{2} - \theta\right)$	$\cot\theta = \tan\left(\dfrac{\pi}{2} - \theta\right)$	
$\sec\theta = \csc(90° - \theta)$	$\csc\theta = \sec(90° - \theta)$	Secant and cosecant are cofunctions.
$\sec\theta = \csc\left(\dfrac{\pi}{2} - \theta\right)$	$\csc\theta = \sec\left(\dfrac{\pi}{2} - \theta\right)$	

TIP *Cosine* means "complement's sine."

Cotangent means "complement's tangent."

Cosecant means "complement's secant."

Classroom Examples: p. 493
Exercises 46, 48

EXAMPLE 6 Using the Cofunction Identities

For each function value, find a cofunction with the same value.

  a. $\cot 15° = 2 + \sqrt{3}$ **b.** $\cos\dfrac{\pi}{6} = \dfrac{\sqrt{3}}{2}$

Solution:

  a. The complement of $15°$ is $(90° - 15°) = 75°$.
  Cotangent and tangent are cofunctions.
  Therefore, $\cot 15° = \tan 75° = 2 + \sqrt{3}$.

  b. The complement of $\dfrac{\pi}{6}$ is $\left(\dfrac{\pi}{2} - \dfrac{\pi}{6}\right) = \left(\dfrac{3\pi}{6} - \dfrac{\pi}{6}\right) = \dfrac{2\pi}{6} = \dfrac{\pi}{3}$.
  Cosine and sine are cofunctions.
  Therefore, $\cos\dfrac{\pi}{6} = \sin\dfrac{\pi}{3} = \dfrac{\sqrt{3}}{2}$.

Skill Practice 6 For each function value, find a cofunction with the same value.

  a. $\tan 22.5° = \sqrt{2} - 1$ **b.** $\sec\dfrac{\pi}{3} = 2$

3. Use Trigonometric Functions in Applications

In many applications of right triangle trigonometry, angles are measured relative to an imaginary horizontal line of reference. An angle of elevation is an angle measured upward from a horizontal line of reference. An angle of depression is an angle measured downward from a horizontal line of reference (Figure 4-35).

Answers

6. a. $\cot 67.5° = \sqrt{2} - 1$

  b. $\csc\dfrac{\pi}{6} = 2$

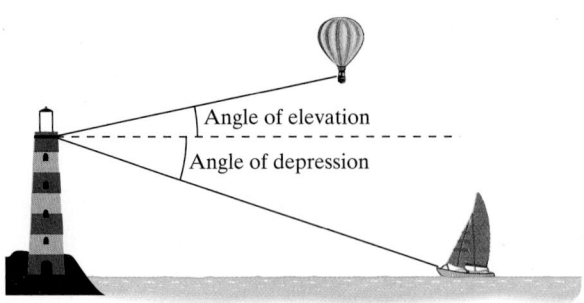

Figure 4-35

Classroom Example: p. 494
Exercise 56

Point of Interest

A variety of instruments are used to measure angles. Historically, quadrants and sextants were used for early celestial navigation. Theodolites (or transits) are used for surveying, and clinometers are used in construction to measure slope.

Nineteenth-century sextant, courtesy of Geoff Miller

TIP To approximate $\tan 30.2°$ on a calculator, be sure that the calculator is in degree mode.

| **EXAMPLE 7** | **Using Trigonometry to Find the Height of a Tree** |

Palm trees are easy to transplant relative to similarly sized broad-leaved trees. At one tree farm, palm trees are harvested once they reach a height of 20 ft. Suppose a farm worker determines that the distance along the ground from his position to the base of a palm tree is 22 ft. He then uses an instrument called a clinometer held at an eye level of 6 ft to measure the angle of elevation to the top of the tree as 30.2°. Is the tree tall enough to harvest?

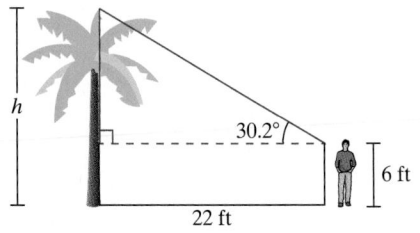

Solution:

To find the height of the tree, we will use the right triangle to find the vertical leg y and then add 6 ft.

Relative to the 30.2° angle, we know the adjacent side is 22 ft and we want to know the length of the opposite side y. The two trigonometric functions that are defined by the opposite and adjacent legs are tangent and cotangent. To solve for y, we can use the relationship

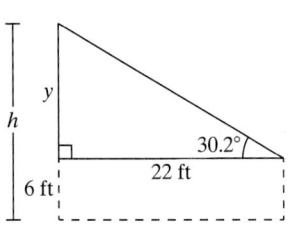

$\tan 30.2° = \dfrac{y}{22}$ or $\cot 30.2° = \dfrac{22}{y}$. Using the tangent function we have

$\tan 30.2° = \dfrac{y}{22}$	We use the tangent function because it is easily approximated on a calculator.
$y = 22 \tan 30.2°$	Multiply both sides by 22.
$y \approx 12.8 \text{ ft}$	Use a calculator to approximate $22 \tan 30.2°$.
$h \approx 12.8 \text{ ft} + 6 \text{ ft}$	To find the height of the tree, add 6 ft.
$\approx 18.8 \text{ ft}$	The tree is approximately 18.8 ft and is not ready to harvest.

Skill Practice 7 Suppose that a farm worker determines that the distance along the ground from her position to the base of a palm tree is 16 ft. She measures the angle of elevation from an eye level of 5 ft to the top of the tree as 45.9°. Is the tree tall enough to harvest (at least 20 ft tall)?

Answer

7. Yes; the tree is approximately 21.5 ft.

EXAMPLE 8   Using Trigonometry in Aeronautical Science

A pilot flying an airplane at an
altitude of 1 mi sights a point at
the end of a runway. The angle
of depression is 3°. What is the
distance d from the plane to the
point on the runway? Round to
the nearest tenth of a mile.

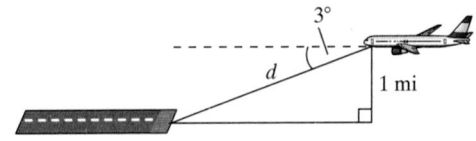

Solution:

The complement of the 3° angle
of depression is the acute angle
87° within the right triangle.

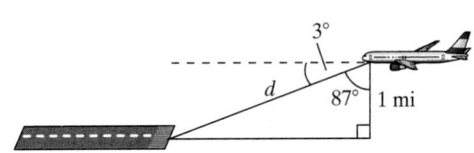

Relative to the 87° angle, we
know that the adjacent side is 1 mi
and we want to find the length of
the hypotenuse. The two trigonometric functions that are defined by the adjacent
leg and hypotenuse are the cosine and secant functions.

Using the relationship $\cos 87° = \dfrac{1}{d}$, we have $d = \dfrac{1}{\cos 87°} \approx 19.1$.

Therefore, the plane is approximately 19.1 mi from the end of the runway.

Skill Practice 8   If a 15-ft ladder is leaning against a wall at an angle of
62° with the ground, how high up the wall will the ladder reach? Round to
the nearest tenth of a foot.

Answer
8. 13.2 ft

SECTION 4.3   Practice Exercises

Prerequisite Review

R.1. Simplify the radical. $\sqrt{45}$ $3\sqrt{5}$

R.2. Rationalize the denominator. $\dfrac{14}{\sqrt{7}}$ $2\sqrt{7}$

R.3. Rationalize the denominator. $\dfrac{9}{\sqrt{18}}$ $\dfrac{3\sqrt{2}}{2}$

R.4. Use the Pythagorean theorem to find the length of
the missing side c.   25

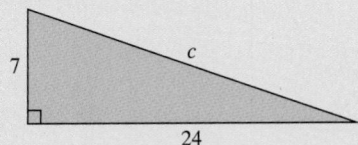

R.5. In a right triangle, one leg measures 24 ft and the
hypotenuse measures 26 ft. Find the length of the
missing side.   10 ft

R.6. Find the measure of angle a.   43°

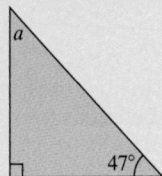

Concept Connections

1. In a right triangle with an acute angle θ, the longest side in the triangle is the _____hypotenuse_____ and is opposite the _____right_____ angle.

2. The leg of a right triangle that lies on one ray of angle θ is called the _____adjacent_____ leg, and the leg that lies across the triangle from θ is called the _____opposite_____ leg.

3. The mnemonic device "SOH-CAH-TOA" stands for the ratios
$\sin\theta = \dfrac{\square}{\square}$, $\cos\theta = \dfrac{\square}{\square}$, and $\tan\theta = \dfrac{\square}{\square}$. $\dfrac{\text{opp}}{\text{hyp}}, \dfrac{\text{adj}}{\text{hyp}}, \dfrac{\text{opp}}{\text{adj}}$

4. Complete the reciprocal and quotient identities.
$\csc\theta = \dfrac{1}{\square}$, $\sec\theta = \dfrac{1}{\square}$, $\cot\theta = \dfrac{1}{\square}$ or $\dfrac{\square}{\square}$. $\sin\theta, \cos\theta, \tan\theta$ or $\dfrac{\cos\theta}{\sin\theta}$

5. Given the lengths of two sides of a right triangle, we can find the length of the third side by using the _____Pythagorean_____ theorem.

6. An _____isosceles_____ right triangle is a right triangle in which the two legs are of equal length. The two acute angles in this triangle each measure _____45_____ °.

7. The length of the shorter leg of a 30°–60°–90° triangle is always _____one-half_____ the length of the hypotenuse.

8. For the six trigonometric functions $\sin\theta$, $\cos\theta$, $\tan\theta$, $\csc\theta$, $\sec\theta$, and $\cot\theta$, identify the three reciprocal pairs. $\sin\theta$ and $\csc\theta$, $\cos\theta$ and $\sec\theta$, and $\tan\theta$ and $\cot\theta$

9. $\tan\theta$ is the _____ratio_____ of $\sin\theta$ and $\cos\theta$.

10. Complete the Pythagorean identities.
$\sin^2\theta + \underline{\quad} = 1$, $\underline{\quad} + 1 = \sec^2\theta$, $1 + \cot^2\theta = \underline{\quad}$. $\cos^2\theta, \tan^2\theta, \csc^2\theta$

11. The two acute angles in a right triangle are complementary because the sum of their measures is _____90_____ °.

12. The sine of angle θ equals the cosine of _____$90° - \theta$ or $\frac{\pi}{2} - \theta$_____. For this reason, the sine and cosine functions are called _____cofunctions_____.

Objective 1: Evaluate Trigonometric Functions of Acute Angles

For Exercises 13–14, find the exact values of the six trigonometric functions for angle θ.

13. a.
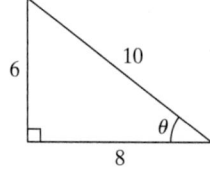
$\sin\theta = \dfrac{3}{5}$, $\cos\theta = \dfrac{4}{5}$, $\tan\theta = \dfrac{3}{4}$,
$\csc\theta = \dfrac{5}{3}$, $\sec\theta = \dfrac{5}{4}$, $\cot\theta = \dfrac{4}{3}$

b.
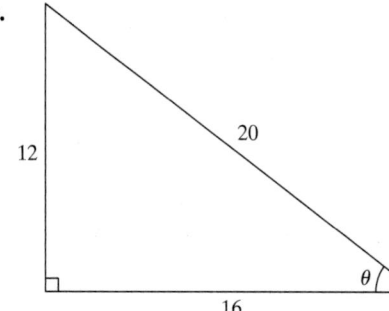
$\sin\theta = \dfrac{3}{5}$, $\cos\theta = \dfrac{4}{5}$, $\tan\theta = \dfrac{3}{4}$, .
$\csc\theta = \dfrac{5}{3}$, $\sec\theta = \dfrac{5}{4}$, $\cot\theta = \dfrac{4}{3}$

14. a.
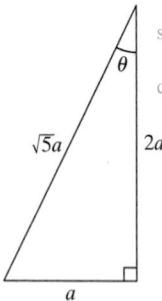
$\sin\theta = \dfrac{\sqrt{5}}{5}$, $\cos\theta = \dfrac{2\sqrt{5}}{5}$, $\tan\theta = \dfrac{1}{2}$,
$\csc\theta = \sqrt{5}$, $\sec\theta = \dfrac{\sqrt{5}}{2}$, $\cot\theta = 2$

b.
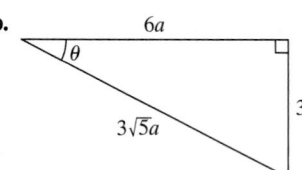
$\sin\theta = \dfrac{\sqrt{5}}{5}$, $\cos\theta = \dfrac{2\sqrt{5}}{5}$, $\tan\theta = \dfrac{1}{2}$,
$\csc\theta = \sqrt{5}$, $\sec\theta = \dfrac{\sqrt{5}}{2}$, $\cot\theta = 2$

For Exercises 15–18, first use the Pythagorean theorem to find the length of the missing side. Then find the exact values of the six trigonometric functions for angle θ. (See Example 1)

15.
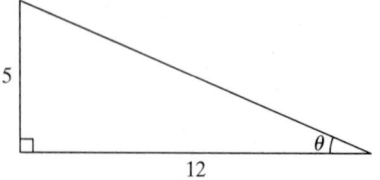

16.
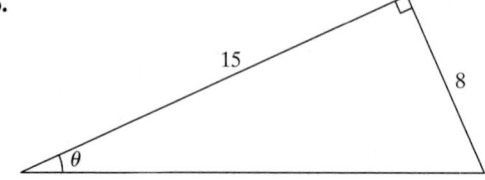

17.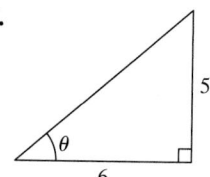

18.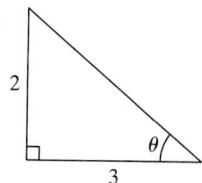

For Exercises 19–22, first use the Pythagorean theorem to find the length of the missing side of the right triangle. Then find the exact values of the six trigonometric functions for the angle θ opposite the shortest side. (See Example 1)

19. Leg = 2 cm, leg = 6 cm

20. Leg = 3 cm, leg = 15 cm

21. Leg = $2\sqrt{7}$ m, hypotenuse = $2\sqrt{11}$ m

22. Leg = $5\sqrt{3}$ in., hypotenuse = $2\sqrt{21}$ in.

In Exercises 23–24, given the value of one trigonometric function of an acute angle θ, find the values of the remaining five trigonometric functions of θ. (See Example 2)

23. $\tan\theta = \dfrac{4}{7}$

24. $\cos\theta = \dfrac{\sqrt{10}}{10}$

For Exercises 25–30, assume that θ is an acute angle. (See Example 2)

25. If $\cos\theta = \dfrac{\sqrt{21}}{7}$, find $\csc\theta$. $\dfrac{\sqrt{7}}{2}$

26. If $\sin\theta = \dfrac{\sqrt{17}}{17}$, find $\cot\theta$.   4

27. If $\sec\theta = \dfrac{3}{2}$, find $\sin\theta$. $\dfrac{\sqrt{5}}{3}$

28. If $\csc\theta = 3$, find $\cos\theta$. $\dfrac{2\sqrt{2}}{3}$

29. If $\tan\theta = \dfrac{\sqrt{15}}{9}$, find $\cos\theta$. $\dfrac{3\sqrt{6}}{8}$

30. If $\cot\theta = \dfrac{\sqrt{3}}{2}$, find $\cos\theta$. $\dfrac{\sqrt{21}}{7}$

For Exercise 31, use the isosceles right triangle and the 30°–60°–90° triangle to complete the table. (See Examples 3–4)

31.

θ	$\sin\theta$	$\cos\theta$	$\tan\theta$	$\csc\theta$	$\sec\theta$	$\cot\theta$
$30° = \dfrac{\pi}{6}$	$\dfrac{1}{2}$	$\dfrac{\sqrt{3}}{2}$	$\dfrac{\sqrt{3}}{3}$	2	$\dfrac{2\sqrt{3}}{3}$	$\sqrt{3}$
$45° = \dfrac{\pi}{4}$	$\dfrac{\sqrt{2}}{2}$	$\dfrac{\sqrt{2}}{2}$	1	$\sqrt{2}$	$\sqrt{2}$	1
$60° = \dfrac{\pi}{3}$	$\dfrac{\sqrt{3}}{2}$	$\dfrac{1}{2}$	$\sqrt{3}$	$\dfrac{2\sqrt{3}}{3}$	2	$\dfrac{\sqrt{3}}{3}$

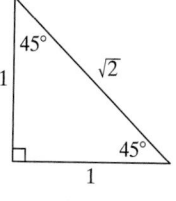

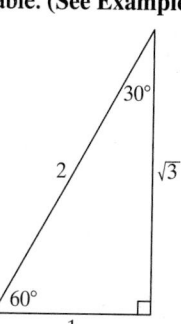

32. a. Evaluate $\sin 60°$. $\dfrac{\sqrt{3}}{2}$

  b. Evaluate $\sin 30° + \sin 30°$.   1

  c. Are the values in parts (a) and (b) the same?   No

For Exercises 33–38, find the exact value of each expression without the use of a calculator. (See Example 5)

33. $\sin\dfrac{\pi}{4} \cdot \cot\dfrac{\pi}{3}$ $\dfrac{\sqrt{6}}{6}$

34. $\tan\dfrac{\pi}{6} \cdot \csc\dfrac{\pi}{4}$ $\dfrac{\sqrt{6}}{3}$

35. $3\cos\dfrac{\pi}{3} + 4\sin\dfrac{\pi}{6}$ $\dfrac{7}{2}$

36. $2\cos\dfrac{\pi}{6} - 5\sin\dfrac{\pi}{3}$ $-\dfrac{3\sqrt{3}}{2}$

37. $\csc^2 60° - \sin^2 45°$ $\dfrac{5}{6}$

38. $\cos^2 45° + \tan^2 60°$ $\dfrac{7}{2}$

Objective 2: Use Fundamental Trigonometric Identities

For Exercises 39–44, determine whether the statement is true or false for an acute angle θ by using the fundamental identities. If the statement is false, provide a counterexample by using a special angle: $\dfrac{\pi}{3}, \dfrac{\pi}{4},$ or $\dfrac{\pi}{6}$.

39. $\sin\theta \cdot \tan\theta = 1$

40. $\cos^2\theta \cdot \tan^2\theta = \sin^2\theta$ True

41. $\sin^2\theta + \tan^2\theta + \cos^2\theta = \sec^2\theta$ True

42. $\csc\theta \cdot \cot\theta = \sec\theta$

43. $\dfrac{1}{\tan\theta} \cdot \cot\theta + 1 = \csc^2\theta$ True

44. $\sin\theta \cdot \cos\theta \cdot \tan\theta + 1 = \cos^2\theta$

For Exercises 45–50, given the function value, find a cofunction of another angle with the same value. (See Example 6)

45. $\tan 75° = 2 + \sqrt{3}$ $\cot 15°$

46. $\sec\dfrac{\pi}{12} = \sqrt{6} - \sqrt{3}$ $\csc\dfrac{5\pi}{12}$

47. $\csc\dfrac{\pi}{3} = \dfrac{2\sqrt{3}}{3}$ $\sec\dfrac{\pi}{6}$

48. $\sin 15° = \dfrac{\sqrt{6} - \sqrt{2}}{4}$ $\cos 75°$

49. $\cos\dfrac{\pi}{4} = \dfrac{\sqrt{2}}{2}$ $\sin\dfrac{\pi}{4}$

50. $\cot\dfrac{\pi}{6} = \sqrt{3}$ $\tan\dfrac{\pi}{3}$

For Exercises 51–54, use a calculator to approximate the function values to 4 decimal places. Be sure that your calculator is in the correct mode.

51. a. $\cos 48.2°$ 0.6665   **b.** $\sin 2°55'42''$ 0.0511   **c.** $\tan \dfrac{3\pi}{8}$ 2.4142

52. a. $\sin 12.6°$ 0.2181   **b.** $\tan 19°36'18''$ 0.3562   **c.** $\cos\left(\dfrac{5\pi}{22}\right)$ 0.7557

53. a. $\csc 39.84°$ 1.5609   **b.** $\sec \dfrac{\pi}{18}$ 1.0154   **c.** $\cot 0.8$ 0.9712

54. a. $\cot 18.46°$ 2.9956   **b.** $\csc \dfrac{2\pi}{9}$ 1.5557   **c.** $\sec 1.25$ 3.1714

Objective 3: Use Trigonometric Functions in Applications

55. An observer at the top of a 462-ft cliff measures the angle of depression from the top of a cliff to a point on the ground to be 5°. What is the distance from the base of the cliff to the point on the ground? Round to the nearest foot. **(See Example 7)** 5281 ft

56. A lamppost casts a shadow of 18 ft when the angle of elevation of the Sun is 33.7°. How high is the lamppost? Round to the nearest foot. 12 ft

57. A 30-ft boat ramp makes a 7° angle with the water. What is the height of the ramp above the water at the ramp's highest point? Round to the nearest tenth of a foot. **(See Example 8)** 3.7 ft

58. A backyard slide is designed for a child's playground. If the top of the 10-ft slide makes an angle of 58° from the vertical, how far out from the base of the steps will the slide extend? Round to the nearest tenth of a foot. 8.5 ft

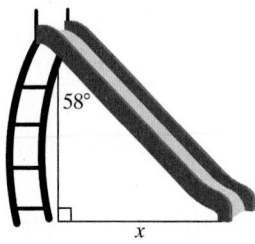

59. The Lookout Mountain Incline Railway, located in Chattanooga, Tennessee, is 4972 ft long and runs up the side of the mountain at an average incline of 17°. What is the gain in altitude? Round to the nearest foot. 1454 ft

60. A 12-ft ladder leaning against a house makes a 64° angle with the ground. Will the ladder reach a window sill that is 10.5 ft up from the base of the house?
Yes. The top of the ladder reaches the house approximately 10.8 ft above ground level.

61. According to National Football League (NFL) rules, all crossbars on goalposts must be 10 ft from the ground. However, teams are allowed some freedom on how high the vertical posts on each end may extend, as long as they measure at least 30 ft. A measurement on an NFL field taken 100 yd from the goalposts yields an angle of 7.8° from the ground to the top of the posts. If the crossbar is 10 ft from the ground, do the goalposts satisfy the NFL rules?

62. A zip line is to be built between two towers labeled A and B across a wetland area. To approximate the distance of the zip line, a surveyor marks a third point C, a distance of 175 ft from one end of the zip line and perpendicular to the zip line. The measure of $\angle ACB$ is 74.5°. How long is the zip line? Round to the nearest foot. 631 ft

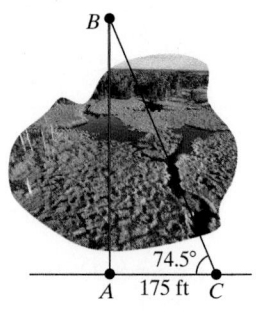

63. To determine the width of a river from point A to point B, a surveyor walks downriver 50 ft along a line perpendicular to $\overline{AB}$ to a new position at point C. The surveyor determines that the measure of $\angle ACB$ is 60°. Find the exact width of the river from point A to point B. $50\sqrt{3}$ ft

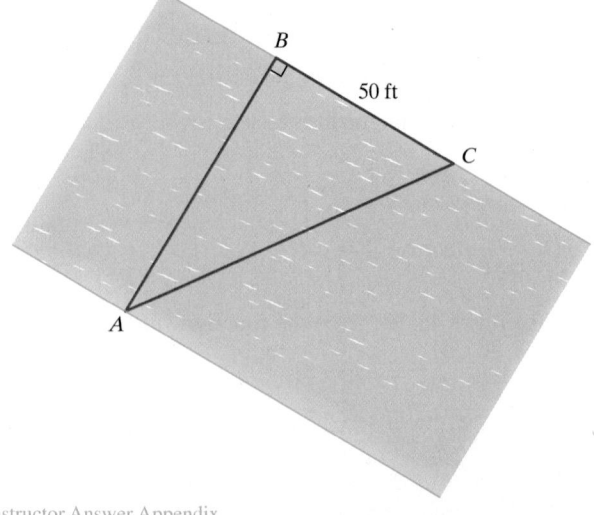

Mixed Exercises

For Exercises 64–68, use the fundamental trigonometric identities as needed.

64. Given that $\sin x° \approx 0.3746$, approximate the given function values. Round to 4 decimal places.

    a. $\cos(90 - x)°$ 0.3746
    b. $\cos x°$ 0.9272
    c. $\tan x°$ 0.4040

    d. $\sin(90 - x)°$ 0.9272
    e. $\cot(90 - x)°$ 0.4040
    f. $\csc x°$ 2.6695

65. Given that $\cos x \approx 0.6691$, approximate the given function values. Round to 4 decimal places.

    a. $\sin x$ 0.7432
    b. $\sin\left(\dfrac{\pi}{2} - x\right)$ 0.6691
    c. $\tan x$ 1.1107

    d. $\cos\left(\dfrac{\pi}{2} - x\right)$ 0.7432
    e. $\sec x$ 1.4945
    f. $\cot\left(\dfrac{\pi}{2} - x\right)$ 1.1107

66. Given that $\cos\dfrac{\pi}{12} = \dfrac{\sqrt{2} + \sqrt{6}}{4}$, give the exact function values.

    a. $\sin\dfrac{5\pi}{12}$ $\dfrac{\sqrt{2} + \sqrt{6}}{4}$
    b. $\sin\dfrac{\pi}{12}$ $\dfrac{\sqrt{2} - \sqrt{3}}{2}$
    c. $\sec\dfrac{\pi}{12}$ $\sqrt{6} - \sqrt{2}$

67. Given that $\tan 36° = \sqrt{5 - 2\sqrt{5}}$, give the exact function values.

    a. $\sec 36°$ $\sqrt{6 - 2\sqrt{5}}$
    b. $\csc 54°$ $\sqrt{6 - 2\sqrt{5}}$
    c. $\cot 54°$ $\sqrt{5 - 2\sqrt{5}}$

68. Simplify each expression to a single trigonometric function.

    a. $\sec\theta \cdot \tan\left(\dfrac{\pi}{2} - \theta\right)$ $\csc\theta$
    b. $\cot^2\theta \cdot \csc(90° - \theta) \cdot \sin\theta$ $\cot\theta$

69. A scenic overlook along the Pacific Coast Highway in Big Sur, California, is 280 ft above sea level. A 6-ft-tall hiker standing at the overlook sees a sailboat and estimates the angle of depression to be 30°. Approximately how far off the coast is the sailboat? Round to the nearest foot.   495 ft

70. An airplane traveling 400 mph at a cruising altitude of 6.6 mi begins its descent. If the angle of descent is 2° from the horizontal, determine the new altitude after 15 min. Round to the nearest tenth of a mile.   3.1 mi

71. A scientist standing at the top of a mountain 2 mi above sea level measures the angle of depression to the ocean horizon to be 1.82°. Use this information to approximate the radius of the Earth to the nearest mile. (*Hint*: The line of sight $\overleftrightarrow{AB}$ is tangent to the Earth and forms a right angle with the radius at the point of tangency.)   3963 mi

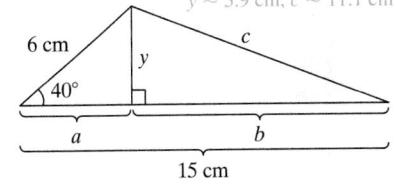

72. Find the lengths a, b, y, and c. Round to the nearest tenth of a centimeter. $a \approx 4.6$ cm, $b \approx 10.4$ cm, $y \approx 3.9$ cm, $c \approx 11.1$ cm

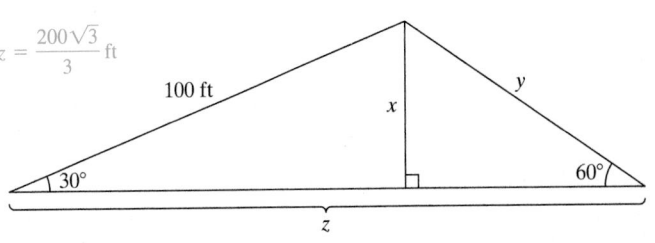

73. Find the exact lengths x, y, and z. $x = 50$ ft, $y = \dfrac{100\sqrt{3}}{3}$ ft, $z = \dfrac{200\sqrt{3}}{3}$ ft

Write About It

74. Use the figure to explain why $\tan\theta = \cot(90° - \theta)$.

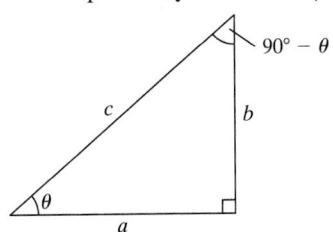

75. Explain the difference between an angle of elevation and an angle of depression.   An angle of elevation is an angle measured upward from a horizontal line of reference. An angle of depression is an angle measured downward from a horizontal line of reference.

Expanding Your Skills

76. An athlete is in a boat at point A, $\frac{1}{4}$ mi from the nearest point D on a straight shoreline. She can row at a speed of 3 mph and run at a speed of 6 mph. Her planned workout is to row to point D and then run to point C farther down the shoreline. However, the current pushes her at an angle of 24° from her original path so that she comes ashore at point B, 2 mi from her final destination at point C. How many minutes will her trip take? Round to the nearest minute.   *25 minutes*

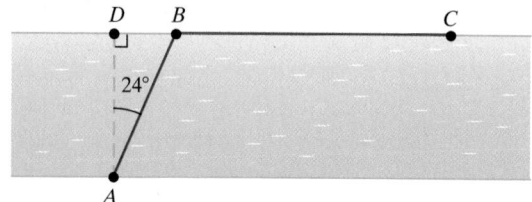

77. Given that the distance from A to B is 60 ft, find the distance from B to C. Round to the nearest foot.   *72 ft*

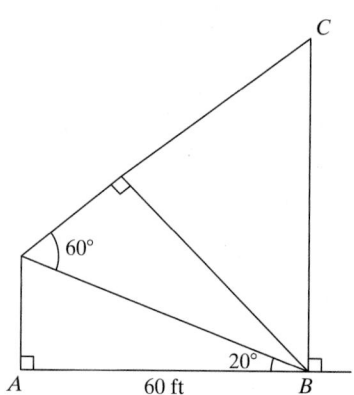

78. In the figure, $\overline{CD} = 15$, $\overline{DE} = 8$, $\tan\alpha = \dfrac{4}{3}$, and $\sin\beta = \dfrac{3}{5}$. Find the lengths of

a. $\overline{AC}$ *$AC = 20$*   **b.** $\overline{AD}$ *$AD = 25$*   **c.** $\overline{DB}$ *$DB = 10$*

d. $\overline{BE}$ *$BE = 6$*   **e.** $\overline{AB}$ *$AB = 5\sqrt{29}$*

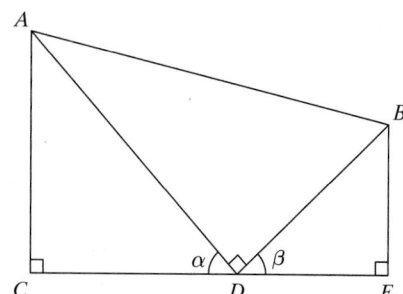

79. For the given triangle, show that $a = h\cot A - h\cot B$.

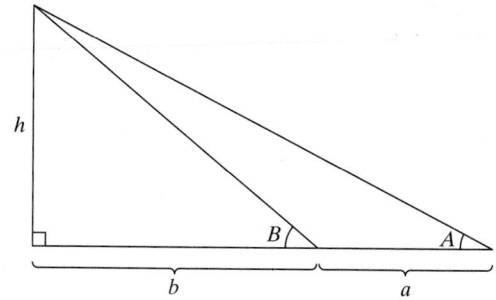

80. Show that the area A of the triangle is given by

$$A = \frac{1}{2}bc\sin A.$$

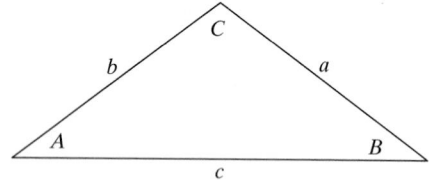

81. Use a cofunction relationship to show that the product $(\tan 1°)(\tan 2°)(\tan 3°) \cdot \cdots \cdot (\tan 87°)(\tan 88°)(\tan 89°)$ is equal to 1.

$(\tan 1°)(\tan 2°)(\tan 3°) \cdot \cdots \cdot (\tan 88°)(\tan 89°)$
$= (\tan 1°)(\tan 89°)(\tan 2°)(\tan 88°)(\tan 3°)(\tan 87°) \cdots (\tan 45°)$
$= [(\tan 1°)(\cot 1°)][(\tan 2°)(\cot 2°)][(\tan 3°)(\cot 3°)] \cdots (\tan 45°)$
$= [1][1][1] \cdots (\tan 45°)$
$= 1$

Technology Connections

For Exercises 82–85, use a calculator to approximate the values of the left- and right-hand sides of each statement for $A = 30°$ and $B = 45°$. Based on the approximations from your calculator, determine if the statement appears to be true or false.

82. a. $\sin(A + B) = \sin A + \sin B$ *False*

b. $\sin(A + B) = \sin A\cos B + \cos A\sin B$ *True*

83. a. $\tan(A - B) = \tan A - \tan B$ *False*

b. $\tan(A - B) = \dfrac{\tan A - \tan B}{1 + \tan A\tan B}$ *True*

84. a. $\cos\left(\dfrac{A}{2}\right) = \sqrt{\dfrac{1 + \cos A}{2}}$ *True*

b. $\cos\left(\dfrac{A}{2}\right) = \dfrac{1}{2}\cos A$ *False*

85. a. $\tan\left(\dfrac{B}{2}\right) = \dfrac{1 - \cos B}{\sin B}$ *True*

b. $\tan\left(\dfrac{B}{2}\right) = \dfrac{\sin B}{1 + \cos B}$ *True*

| SECTION 4.4 | **Trigonometric Functions of Any Angle** |

OBJECTIVES

1. Evaluate Trigonometric Functions of Any Angle
2. Determine Reference Angles
3. Evaluate Trigonometric Functions Using Reference Angles

1. Evaluate Trigonometric Functions of Any Angle

In many applications we encounter angles that are not acute. For example, a robotic arm may have a range of motion of 360°, or an object may rotate in a repeated circular pattern through all possible angles. For these situations, we need to extend the definition of trigonometric functions to any angle.

In Figure 4-36, the real number t corresponds to a point $Q(a, b)$ on the unit circle. Angle θ is a second quadrant angle passing through point $P(x, y)$ with terminal side passing through Q. Suppose that we drop perpendicular line segments from points P and Q to the x-axis to form two similar right triangles. For $\triangle ORP$, the hypotenuse r is given by $r = \sqrt{x^2 + y^2}$ and the lengths of the two legs are $|x|$ and y. For $\triangle OSQ$, the hypotenuse is 1 unit and the lengths of the legs are $|a|$ and b.

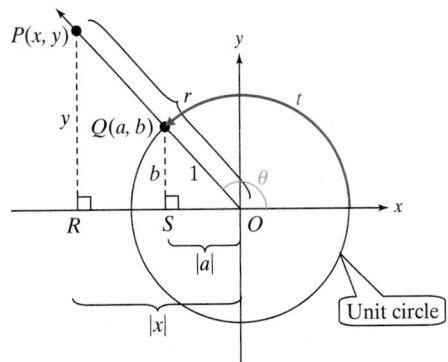

Figure 4-36

Since $\triangle ORP$ and $\triangle OSQ$ are similar triangles, the ratios of their corresponding sides are proportional, which implies that $\dfrac{b}{1} = \dfrac{y}{r}$. Furthermore, since $\sin t = b$, we have $\sin t = \dfrac{y}{r}$. We also know that there is a one-to-one correspondence between the real number t and the radian measure of the central angle θ in standard position with terminal side through P. From this correspondence, we define $\sin \theta = \sin t = \dfrac{y}{r}$.

Similar logic can be used to define the other five trigonometric functions for a point P in any quadrant or on the x- or y-axes. The results are summarized in Table 4-8.

Table 4-8

Trigonometric Functions of Any Angle

Let θ be an angle in standard position with point $P(x, y)$ on the terminal side, and let $r = \sqrt{x^2 + y^2} \neq 0$ represent the distance from $P(x, y)$ to $(0, 0)$. Then,

$\sin \theta = \dfrac{y}{r}$	$\csc \theta = \dfrac{r}{y}\,(y \neq 0)$	
$\cos \theta = \dfrac{x}{r}$	$\sec \theta = \dfrac{r}{x}\,(x \neq 0)$	
$\tan \theta = \dfrac{y}{x}\,(x \neq 0)$	$\cot \theta = \dfrac{x}{y}\,(y \neq 0)$	

Point P is on the terminal side of angle θ and not at the origin, which means that $r = \sqrt{x^2 + y^2} \neq 0$. Therefore, the signs of the values of the trigonometric functions depend solely on the signs of x and y. For an acute angle, the values of the trigonometric functions are positive because x and y are both positive in the first quadrant. For angles in the other quadrants, we determine the signs of the trigonometric functions by analyzing the signs of x and y (Figure 4-37). Note that the reciprocal functions cosecant, secant, and cotangent will have the same signs as sine, cosine, and tangent, respectively for a given value of θ.

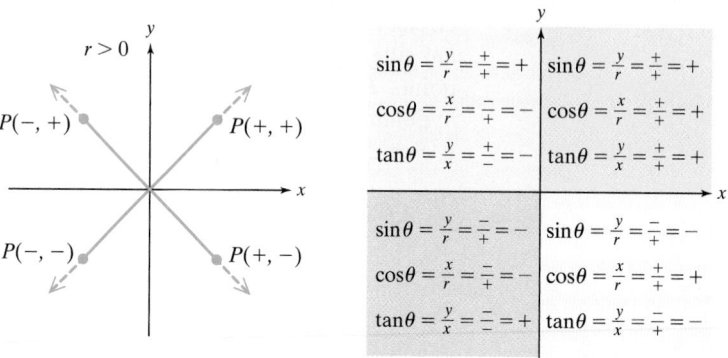

Figure 4-37

Classroom Example: p. 505
Exercise 18

> ### EXAMPLE 1   Evaluating Trigonometric Functions
>
> Let $P(-2, -5)$ be a point on the terminal side of angle θ drawn in standard position. Find the values of the six trigonometric functions of θ.
>
> **Solution:**
>
> We first draw θ in standard position and label point P.
> The distance between $P(-2, -5)$ and $(0, 0)$ is
>
> $$r = \sqrt{x^2 + y^2} = \sqrt{(-2)^2 + (-5)^2} = \sqrt{4 + 25} = \sqrt{29}.$$
>
> We have $x = -2$, $y = -5$, and $r = \sqrt{29}$.
>
> $$\sin\theta = \frac{y}{r} = \frac{-5}{\sqrt{29}} = \frac{-5}{\sqrt{29}} \cdot \frac{\sqrt{29}}{\sqrt{29}} = -\frac{5\sqrt{29}}{29} \qquad \csc\theta = \frac{r}{y} = \frac{\sqrt{29}}{-5} = -\frac{\sqrt{29}}{5}$$
>
> $$\cos\theta = \frac{x}{r} = \frac{-2}{\sqrt{29}} = \frac{-2}{\sqrt{29}} \cdot \frac{\sqrt{29}}{\sqrt{29}} = -\frac{2\sqrt{29}}{29} \qquad \sec\theta = \frac{r}{x} = \frac{\sqrt{29}}{-2} = -\frac{\sqrt{29}}{2}$$
>
> $$\tan\theta = \frac{y}{x} = \frac{-5}{-2} = \frac{5}{2} \qquad\qquad\qquad\qquad \cot\theta = \frac{x}{y} = \frac{-2}{-5} = \frac{2}{5}$$
>
> **Skill Practice 1** Let $P(-5, -7)$ be a point on the terminal side of angle θ drawn in standard position. Find the values of the six trigonometric functions of θ.

Answer

1. $\sin\theta = -\dfrac{7\sqrt{74}}{74}$, $\cos\theta = -\dfrac{5\sqrt{74}}{74}$,

$\tan\theta = \dfrac{7}{5}$, $\csc\theta = -\dfrac{\sqrt{74}}{7}$,

$\sec\theta = -\dfrac{\sqrt{74}}{5}$, $\cot\theta = \dfrac{5}{7}$

2. Determine Reference Angles

For nonacute angles θ, we often find values of the six trigonometric functions by using the related *reference angle*.

> **Definition of a Reference Angle**
>
> Let θ be an angle in standard position. The **reference angle** for θ is the acute angle θ' formed by the terminal side of θ and the horizontal axis.

Figure 4-38 shows the reference angles θ' for angles on the interval $[0, 2\pi)$ drawn in standard position for each of the four quadrants.

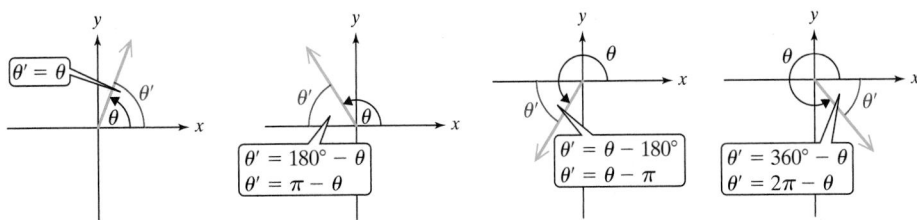

Figure 4-38

Classroom Examples: p. 505
Exercises 24, 26

> **EXAMPLE 2**   **Finding Reference Angles**
>
> Find the reference angle θ'.
>
> **a.** $\theta = 315°$ **b.** $\theta = -195°$ **c.** $\theta = 3.5$ **d.** $\theta = \dfrac{25\pi}{4}$
>
> **Solution:**
>
> **a.** $\theta = 315°$ is a fourth quadrant angle. The acute angle it makes with the x-axis is $\theta' = 360° - 315° = 45°$.
>
>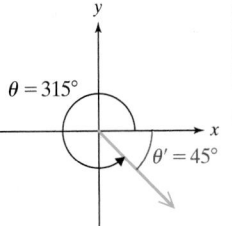
>
> **b.** $\theta = -195°$ is a second quadrant angle coterminal to $165°$. The actue angle it makes with the x-axis is $\theta' = 180° - 165° = 15°$.
>
>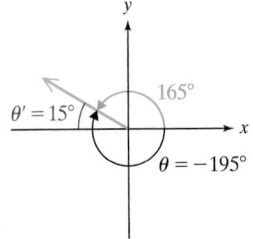

c. $\theta = 3.5$ is measured in radians. Since $\pi \approx 3.14$ and $\dfrac{3\pi}{2} \approx 4.71$, we know that $\pi < 3.5 < \dfrac{3\pi}{2}$ implying that 3.5 is in the third quadrant. The reference angle is $\theta' = 3.5 - \pi \approx 0.3584$.

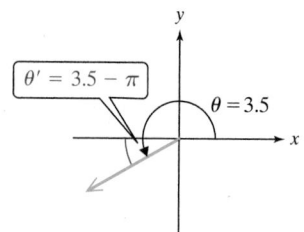

d. $\theta = \dfrac{25\pi}{4}$ is a first quadrant angle coterminal to $\dfrac{\pi}{4}$. The angle $\dfrac{\pi}{4}$ is an acute angle and is its own reference angle. Therefore, $\theta' = \dfrac{\pi}{4}$.

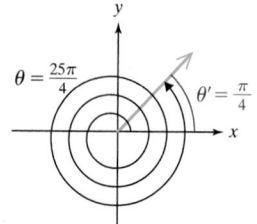

Skill Practice 2   Find the reference angle θ'.

    a. $\theta = 150°$ **b.** $\theta = -157.5°$ **c.** $\theta = 5$ **d.** $\theta = \dfrac{13\pi}{3}$

---

3. Evaluate Trigonometric Functions Using Reference Angles

In Figure 4-39, an angle θ is drawn in standard position along with its accompanying reference angle θ'. A right triangle can be formed by dropping a perpendicular line segment from a point $P(x, y)$ on the terminal side of θ to the x-axis. The length of the vertical leg of the triangle is $|y|$, the length of the horizontal leg is $|x|$, and the hypotenuse is r.

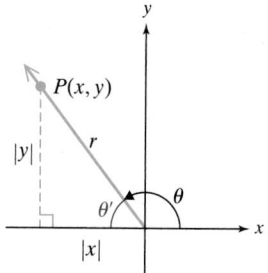

    Now suppose we compare the values of the cosine function of θ and θ'.

$$\cos\theta = \frac{x}{r} \quad \text{and} \quad \cos\theta' = \frac{\text{adj}}{\text{hyp}} = \frac{|x|}{r}$$

Figure 4-39

Notice that the ratio for $\cos\theta$ has x in the numerator and the ratio for $\cos\theta'$ has $|x|$. It follows that $\cos\theta$ and $\cos\theta'$ are equal except possibly for their signs. A similar argument holds for the other five trigonometric functions. This leads to the following procedure to evaluate a trigonometric function based on reference angles.

Evaluating Trigonometric Functions Using Reference Angles

To find the value of a trigonometric function of a given angle θ,

    1. Determine the function value of the reference angle θ'.
    2. Affix the appropriate sign based on the quadrant in which θ lies.

Answers

2. a. $\theta' = 30°$
   b. $\theta' = 22.5°$
   c. $\theta' = 2\pi - 5$
   d. $\theta' = \dfrac{\pi}{3}$

Before we present examples of this process, take a minute to review the values of the trigonometric functions of 30°, 45°, and 60° (Table 4-9).

Table 4-9

θ	$\sin\theta$	$\cos\theta$	$\tan\theta$	$\csc\theta$	$\sec\theta$	$\cot\theta$
$30° = \dfrac{\pi}{6}$	$\dfrac{1}{2}$	$\dfrac{\sqrt{3}}{2}$	$\dfrac{\sqrt{3}}{3}$	2	$\dfrac{2\sqrt{3}}{3}$	$\sqrt{3}$
$45° = \dfrac{\pi}{4}$	$\dfrac{\sqrt{2}}{2}$	$\dfrac{\sqrt{2}}{2}$	1	$\sqrt{2}$	$\sqrt{2}$	1
$60° = \dfrac{\pi}{3}$	$\dfrac{\sqrt{3}}{2}$	$\dfrac{1}{2}$	$\sqrt{3}$	$\dfrac{2\sqrt{3}}{3}$	2	$\dfrac{\sqrt{3}}{3}$

Classroom Examples: p. 505
Exercises 34, 36, 38

EXAMPLE 3   Using Reference Angles to Evaluate Functions

Evaluate the functions.   **a.** $\sin 240°$ **b.** $\tan(-225°)$ **c.** $\sec\dfrac{11\pi}{6}$

Solution:

a. $\theta = 240°$ is in Quadrant III. The reference

angle is $240° - 180° = 60°$

Since $\sin\theta$ is negative in Quadrant III,

$$\sin 240° = -\sin 60° = -\left(\frac{\sqrt{3}}{2}\right) = -\frac{\sqrt{3}}{2}$$

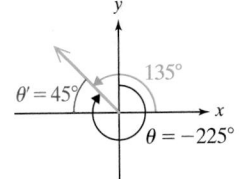

b. $\theta = -225°$ is an angle in Quadrant II, coterminal to $135°$. The reference angle is $180° - 135° = 45°$. Since $\tan\theta$ is negative in Quadrant II,

$$\tan(-225°) = -\tan 45° = -(1) = -1$$

c. $\theta = \dfrac{11\pi}{6}$ is in Quadrant IV. The reference angle is

$$2\pi - \frac{11\pi}{6} = \frac{12\pi}{6} - \frac{11\pi}{6} = \frac{\pi}{6}$$

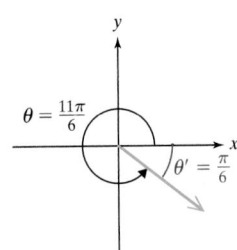

Since $\cos\theta$ and its reciprocal $\sec\theta$ are positive in Quadrant IV,

$$\sec\frac{11\pi}{6} = \sec\frac{\pi}{6} = \frac{2\sqrt{3}}{3}$$

Skill Practice 3   Evaluate the functions.

a. $\cos\dfrac{5\pi}{6}$ **b.** $\cot(-120°)$ **c.** $\csc\dfrac{7\pi}{4}$

Answers

3. a. $-\dfrac{\sqrt{3}}{2}$ **b.** $\dfrac{\sqrt{3}}{3}$ **c.** $-\sqrt{2}$

Classroom Examples: p. 505
Exercises 44, 54

EXAMPLE 4   **Evaluate Trigonometric Functions**

Evaluate the functions.   **a.** $\sec\dfrac{9\pi}{2}$ **b.** $\sin(-510°)$

Solution:

a. $\dfrac{9\pi}{2}$ is coterminal to $\dfrac{\pi}{2}$. The terminal side of $\dfrac{\pi}{2}$ is on the positive y-axis, where we have selected the arbitrary point $(0, 1)$.

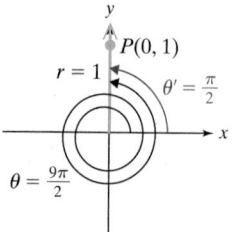

$$\sec\frac{9\pi}{2} = \sec\frac{\pi}{2} = \frac{r}{x} \to \frac{1}{0} \text{ is undefined.}$$

b. $-510°$ is coterminal to $210°$, which is a third-quadrant angle. The reference angle is $210° - 180° = 30°$.

Since $\sin\theta$ is negative in Quadrant III,

$$\sin(-510°) = -\sin 30° = -\left(\frac{1}{2}\right) = -\frac{1}{2}$$

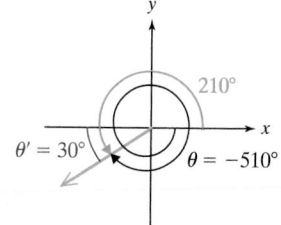

Skill Practice 4   Evaluate the functions.

  a. $\csc(11\pi)$ **b.** $\cos(-600°)$

In Example 5, we use a reference angle to determine the values of trigonometric functions based on given information about θ.

Classroom Example: p. 506
Exercise 64

EXAMPLE 5   **Evaluating Trigonometric Functions**

Given $\sin\theta = -\dfrac{4}{7}$ and $\cos\theta > 0$, find $\cos\theta$ and $\tan\theta$.

Solution:

First note that $\sin\theta < 0$ and $\cos\theta > 0$ in Quadrant IV. We can label the reference angle θ' and then draw a representative triangle with opposite leg of length 4 units and hypotenuse of length 7 units.

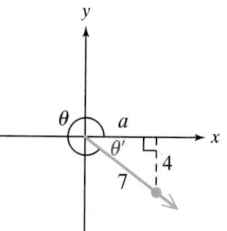

Using the Pythagorean theorem, we can determine the length of the adjacent leg.

$$a^2 + (4)^2 = (7)^2$$
$$a^2 + 16 = 49$$
$$a^2 = 33$$
$$a = \sqrt{33}$$

Choose the positive square root for the length of a side of a triangle.

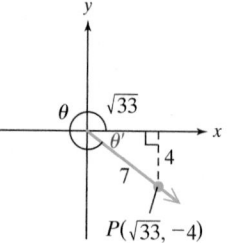

Answers

4. a. Undefined   **b.** $-\dfrac{1}{2}$

Classroom Example: p. 506
Exercise 70

> **TIP** With the reference angle and representative triangle drawn in Quadrant IV, we can find point $P(\sqrt{33}, -4)$ on the terminal side of θ.
>
> $$\cos\theta = \frac{x}{r} = \frac{\sqrt{33}}{7}$$
>
> $$\tan\theta = \frac{y}{x} = \frac{-4}{\sqrt{33}}$$

$$\cos\theta = \cos\theta' = \frac{\sqrt{33}}{7}$$ $\cos\theta$ is positive in Quadrant IV.

$$\tan\theta = -\tan\theta' = -\frac{4}{\sqrt{33}} = -\frac{4}{\sqrt{33}} \cdot \frac{\sqrt{33}}{\sqrt{33}} = -\frac{4\sqrt{33}}{33}$$ $\tan\theta$ is negative in Quadrant IV.

Skill Practice 5   Given $\cos\theta = -\dfrac{3}{8}$ and $\sin\theta < 0$, find $\sin\theta$ and $\tan\theta$.

The identities involving trigonometric functions of acute angles presented in Section 4.3 are also true for trigonometric functions of non-acute angles provided that the functions are well defined at θ.

Pythagorean Identities	Reciprocal Identities	Quotient Identities
$\sin^2\theta + \cos^2\theta = 1$	$\csc\theta = \dfrac{1}{\sin\theta}$	$\tan\theta = \dfrac{\sin\theta}{\cos\theta}$
$\tan^2\theta + 1 = \sec^2\theta$	$\sec\theta = \dfrac{1}{\cos\theta}$	$\cot\theta = \dfrac{\cos\theta}{\sin\theta}$
$1 + \cot^2\theta = \csc^2\theta$	$\cot\theta = \dfrac{1}{\tan\theta}$	

In Example 6, we are given information about an angle θ and will find the values of the trigonometric functions by applying the fundamental identities and by using reference angles.

EXAMPLE 6   Using Fundamental Identities

Given $\cos\theta = -\dfrac{3}{5}$ for θ in Quadrant II, find $\sin\theta$ and $\tan\theta$.

Solution:

Using Identities

$$\sin^2\theta + \cos^2\theta = 1$$

$$\sin^2\theta + \left(-\frac{3}{5}\right)^2 = 1$$

$$\sin^2\theta + \frac{9}{25} = 1$$

$$\sin^2\theta = 1 - \frac{9}{25}$$

$$\sin^2\theta = \frac{25}{25} - \frac{9}{25}$$

$$\sin^2\theta = \frac{16}{25}$$

$$\sin\theta = \pm\frac{4}{5}$$

Alternative Approach

Label the reference angle θ' and then draw a representative triangle with adjacent leg of length 3 units and hypotenuse of length 5 units.

Using the Pythagorean theorem, we can determine the length of the opposite leg.

$$(3)^2 + b^2 = (5)^2$$

$$9 + b^2 = 25$$

$$b^2 = 16$$

$$b = 4$$

> **TIP** With the reference angle and representative triangle drawn in Quadrant II, we can find point $P(-3, 4)$ on the terminal side of θ. Therefore, $x = -3$, $y = 4$, and $r = 5$.
>
> $$\sin\theta = \frac{y}{r} = \frac{4}{5}$$
>
> $$\tan\theta = \frac{y}{x} = \frac{4}{-3}$$

Answer

5. $\sin\theta = -\dfrac{\sqrt{55}}{8}$ and $\tan\theta = \dfrac{\sqrt{55}}{3}$

In Quadrant II, $\sin \theta > 0$.

Therefore, choose $\sin \theta = \dfrac{4}{5}$.

$$\tan \theta = \frac{\sin \theta}{\cos \theta} = \frac{\dfrac{4}{5}}{-\dfrac{3}{5}} = -\frac{4}{3}$$

In Quadrant II, $\sin \theta > 0$.

Therefore, $\sin \theta = \sin \theta' = \dfrac{4}{5}$.

In Quadrant II, $\tan \theta < 0$.

Therefore, $\tan \theta = -\tan \theta' = -\dfrac{4}{3}$.

Answer

6. $\cos \theta = \dfrac{12}{13}$, $\tan \theta = -\dfrac{5}{12}$

Skill Practice 6   Given $\sin \theta = -\dfrac{5}{13}$ for θ in Quadrant IV, find $\cos \theta$ and $\tan \theta$.

SECTION 4.4   Practice Exercises

Prerequisite Review

R.1. Find the length of the missing side of the right triangle. $c = 29$ m

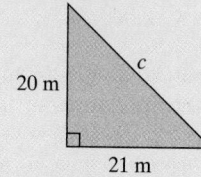

20 m

c

21 m

R.2. Find the length of the third side using the Pythagorean theorem. Round the answer to the nearest tenth of a foot.

6.4 ft

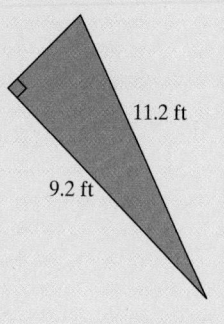

11.2 ft

9.2 ft

R.3. Use the distance formula to find the distance from $(2, 5)$ to $(-3, 10)$. $5\sqrt{2}$

Concept Connections

1. The distance from the origin to a point $P(x, y)$ is given by __$\sqrt{x^2 + y^2}$__.

2. Angles with terminal sides on the coordinate axes are referred to as ___quadrantal___ angles.

3. If θ is an angle in standard position, the ___reference___ angle for θ is the acute angle θ' formed by the terminal side of θ and the horizontal axis.

4. The values $\tan \theta$ and $\sec \theta$ are undefined for odd multiples of _____. $\dfrac{\pi}{2}$ or 90°

5. The values $\cot \theta$ and $\csc \theta$ are undefined for multiples of ____π or 180°____.

6. For what values of θ is $\sin \theta$ greater than 1? There are no values for which $\sin \theta$ is greater than 1.

Objective 1: Evaluate Trigonometric Functions of Any Angle

7. Let $P(x, y)$ be a point on the terminal side of an angle θ drawn in standard position and let r be the distance from P to the origin. Fill in the boxes to form the ratios defining the six trigonometric functions.

a. $\sin \theta = \dfrac{\square}{\square}$ $\dfrac{y}{r}$ **b.** $\cos \theta = \dfrac{\square}{\square}$ $\dfrac{x}{r}$ **c.** $\tan \theta = \dfrac{\square}{\square}$ $\dfrac{y}{x}$

d. $\csc \theta = \dfrac{\square}{\square}$ $\dfrac{r}{y}$ **e.** $\sec \theta = \dfrac{\square}{\square}$ $\dfrac{r}{x}$ **f.** $\cot \theta = \dfrac{\square}{\square}$ $\dfrac{x}{y}$

8. Fill in the cells in the table with the appropriate sign for each trigonometric function for θ in Quadrants I, II, III, and IV. The signs for the sine function are done for you.

	Quadrant			
	I	**II**	**III**	**IV**
$\sin \theta$	+	+	−	−
$\cos \theta$	+	−	−	+
$\tan \theta$	+	−	+	−
$\csc \theta$	+	+	−	−
$\sec \theta$	+	−	−	+
$\cot \theta$	+	−	+	−

For Exercises 9–14, given the stated conditions, identify the quadrant in which θ lies.

9. $\sin \theta < 0$ and $\tan \theta > 0$
   Quadrant III

10. $\csc \theta > 0$ and $\cot \theta < 0$
    Quadrant II

11. $\sec \theta < 0$ and $\tan \theta < 0$
    Quadrant II

12. $\cot \theta < 0$ and $\sin \theta < 0$
    Quadrant IV

13. $\cos \theta > 0$ and $\cot \theta < 0$
    Quadrant IV

14. $\cot \theta < 0$ and $\sec \theta < 0$
    Quadrant II

For Exercises 15–20, a point is given on the terminal side of an angle θ drawn in standard position. Find the values of the six trigonometric functions of θ. (See Example 1)

15. $(5, -12)$

16. $(-8, 15)$

17. $(-3, -5)$

18. $(2, -3)$

19. $\left(-\dfrac{3}{2}, 2\right)$

20. $\left(5, -\dfrac{8}{3}\right)$

21. Complete the table for the given angles.

θ	$\sin \theta$	$\cos \theta$	$\tan \theta$	$\csc \theta$	$\sec \theta$	$\cot \theta$
$0° = 0$	0	1	0	undefined	1	undefined
$90° = \dfrac{\pi}{2}$	1	0	undefined	1	undefined	0
$180° = \pi$	0	-1	0	undefined	-1	undefined
$270° = \dfrac{3\pi}{2}$	-1	0	undefined	-1	undefined	0
$360° = 2\pi$	0	1	0	undefined	1	undefined

22. Evaluate the expression for $A = 90°$, $B = 180°$, and $C = 270°$.

$$\frac{\sin A + 2\cos B}{\sec B - 3\csc C} \qquad -\frac{1}{2}$$

Objective 2: Determine Reference Angles

For Exercises 23–30, find the reference angle for the given angle. (See Example 2)

23. a. $135°$ $45°$ b. $-330°$ $30°$ c. $660°$ $60°$ d. $-690°$ $30°$

24. a. $-120°$ $60°$ b. $225°$ $45°$ c. $-1035°$ $45°$ d. $510°$ $30°$

25. a. $\dfrac{2\pi}{3}$ $\dfrac{\pi}{3}$ b. $-\dfrac{5\pi}{6}$ $\dfrac{\pi}{6}$ c. $\dfrac{13\pi}{4}$ $\dfrac{\pi}{4}$ d. $-\dfrac{10\pi}{3}$ $\dfrac{\pi}{3}$

26. a. $-\dfrac{5\pi}{4}$ $\dfrac{\pi}{4}$ b. $\dfrac{11\pi}{6}$ $\dfrac{\pi}{6}$ c. $\dfrac{17\pi}{3}$ $\dfrac{\pi}{3}$ d. $\dfrac{19\pi}{6}$ $\dfrac{\pi}{6}$

27. a. $\dfrac{20\pi}{17}$ $\dfrac{3\pi}{17}$ b. $-\dfrac{99\pi}{20}$ $\dfrac{\pi}{20}$ c. $110°$ $70°$ d. $-422°$ $62°$

28. a. $\dfrac{25\pi}{11}$ $\dfrac{3\pi}{11}$ b. $-\dfrac{27\pi}{14}$ $\dfrac{\pi}{14}$ c. $-512°$ $28°$ d. $1280°$ $20°$

29. a. 1.8 $\pi - 1.8$ b. 1.8π 0.2π c. 5.1 $2\pi - 5.1$ d. $5.1°$ $5.1°$

30. a. 0.6 0.6 b. 0.6π 0.4π c. 100 $32\pi - 100$ d. $100°$ $80°$

Objective 3: Evaluate Trigonometric Functions Using Reference Angles

For Exercises 31–54, use reference angles to find the exact value. (See Examples 3 and 4)

31. $\sin 120°$ $\dfrac{\sqrt{3}}{2}$

32. $\cos 225°$ $-\dfrac{\sqrt{2}}{2}$

33. $\cos \dfrac{4\pi}{3}$ $-\dfrac{1}{2}$

34. $\sin \dfrac{5\pi}{6}$ $\dfrac{1}{2}$

35. $\sec(-330°)$ $\dfrac{2\sqrt{3}}{3}$

36. $\csc(-225°)$ $\sqrt{2}$

37. $\sec \dfrac{13\pi}{3}$ 2

38. $\csc \dfrac{5\pi}{3}$ $-\dfrac{2\sqrt{3}}{3}$

39. $\cot 240°$ $\dfrac{\sqrt{3}}{3}$

40. $\tan(-150°)$ $\dfrac{\sqrt{3}}{3}$

41. $\tan \dfrac{5\pi}{4}$ 1

42. $\cot\left(-\dfrac{3\pi}{4}\right)$ 1

43. $\cos(-630°)$ 0

44. $\sin 630°$ -1

45. $\sin \dfrac{17\pi}{6}$ $\dfrac{1}{2}$

46. $\cos\left(-\dfrac{11\pi}{4}\right)$ $-\dfrac{\sqrt{2}}{2}$

47. $\sec 1170°$
    Undefined

48. $\csc 750°$ 2

49. $\csc(-5\pi)$
    Undefined

50. $\sec 5\pi$ -1

51. $\tan(-2400°)$
    $-\sqrt{3}$

52. $\cot 900°$
    Undefined

53. $\cot \dfrac{42\pi}{8}$ 1

54. $\tan \dfrac{18\pi}{4}$
    Undefined

For Exercises 55–58, find two angles between 0° and 360° for the given condition.

55. $\sin\theta = \dfrac{1}{2}$ 30° and 150°   **56.** $\cos\theta = -\dfrac{\sqrt{2}}{2}$ 135° and 225°   **57.** $\cot\theta = -\sqrt{3}$ 150° and 330°   **58.** $\tan\theta = \sqrt{3}$ 60° and 240°

For Exercises 59–62, find two angles between 0 and 2π for the given condition.

59. $\sec\theta = \sqrt{2}$ $\dfrac{\pi}{4}$ and $\dfrac{7\pi}{4}$ **60.** $\csc\theta = -\dfrac{2\sqrt{3}}{3}$ $\dfrac{4\pi}{3}$ and $\dfrac{5\pi}{3}$ **61.** $\tan\theta = -\dfrac{\sqrt{3}}{3}$ $\dfrac{5\pi}{6}$ and $\dfrac{11\pi}{6}$ **62.** $\cot\theta = 1$ $\dfrac{\pi}{4}$ and $\dfrac{5\pi}{4}$

In Exercises 63–68, find the values of the trigonometric functions from the given information. (See Example 5)

63. Given $\tan\theta = -\dfrac{20}{21}$ and $\cos\theta < 0$, find $\sin\theta$ and $\cos\theta$.
$\sin\theta = \dfrac{20}{29}, \cos\theta = -\dfrac{21}{29}$

64. Given $\cot\theta = \dfrac{11}{60}$ and $\sin\theta < 0$, find $\cos\theta$ and $\sin\theta$.
$\cos\theta = -\dfrac{11}{61}, \sin\theta = -\dfrac{60}{61}$

65. Given $\sin\theta = -\dfrac{3}{10}$ and $\tan\theta > 0$, find $\cos\theta$ and $\cot\theta$.
$\cos\theta = -\dfrac{\sqrt{91}}{10}, \cot\theta = \dfrac{\sqrt{91}}{3}$

66. Given $\cos\theta = -\dfrac{5}{8}$ and $\csc\theta > 0$, find $\sin\theta$ and $\tan\theta$.
$\sin\theta = \dfrac{\sqrt{39}}{8}, \tan\theta = -\dfrac{\sqrt{39}}{5}$

67. Given $\sec\theta = \dfrac{\sqrt{58}}{7}$ and $\cot\theta < 0$, find $\csc\theta$ and $\cos\theta$.
$\csc\theta = -\dfrac{\sqrt{58}}{3}, \cos\theta = \dfrac{7\sqrt{58}}{58}$

68. Given $\csc\theta = -\dfrac{\sqrt{26}}{5}$ and $\cos\theta < 0$, find $\sin\theta$ and $\cot\theta$.
$\sin\theta = -\dfrac{5\sqrt{26}}{26}, \cot\theta = \dfrac{1}{5}$

For Exercises 69–74, use fundamental trigonometric identities to find the values of the functions. (See Example 6)

69. Given $\cos\theta = \dfrac{20}{29}$ for θ in Quadrant IV, find $\sin\theta$ and $\tan\theta$.
$\sin\theta = -\dfrac{21}{29}, \tan\theta = -\dfrac{21}{20}$

70. Given $\sin\theta = -\dfrac{8}{17}$ for θ in Quadrant III, find $\cos\theta$ and $\cot\theta$.
$\cos\theta = -\dfrac{15}{17}, \cot\theta = \dfrac{15}{8}$

71. Given $\tan\theta = -4$ for θ in Quadrant II, find $\sec\theta$ and $\cot\theta$.
$\sec\theta = -\sqrt{17}, \cot\theta = -\dfrac{1}{4}$

72. Given $\sec\theta = 5$ for θ in Quadrant IV, find $\csc\theta$ and $\cos\theta$.
$\csc\theta = -\dfrac{5\sqrt{6}}{12}, \cos\theta = \dfrac{1}{5}$

73. Given $\cot\theta = \dfrac{5}{4}$ for θ in Quadrant III, find $\csc\theta$ and $\sin\theta$.
$\csc\theta = -\dfrac{\sqrt{41}}{4}, \sin\theta = -\dfrac{4\sqrt{41}}{41}$

74. Given $\csc\theta = \dfrac{7}{3}$ for θ in Quadrant II, find $\cot\theta$ and $\cos\theta$.
$\cot\theta = -\dfrac{2\sqrt{10}}{3}, \cos\theta = -\dfrac{2\sqrt{10}}{7}$

Mixed Exercises

For Exercises 75–76, find the sign of the expression for θ in each quadrant.

75.

		I	II	III	IV
a.	$\sin\theta\cos\theta$	+	−	+	−
b.	$\dfrac{\tan\theta}{\cos\theta}$	+	+	−	−

76.

		I	II	III	IV
a.	$\dfrac{\cot\theta}{\sin\theta}$	+	−	−	+
b.	$\tan\theta\sec\theta$	+	+	−	−

For Exercises 77–84, suppose that θ is an acute angle. Identify each statement as true or false. If the statement is false, rewrite the statement to give the correct answer for the right side.

77. $\cos(180° - \theta) = -\cos\theta$ True

78. $\tan(180° - \theta) = \tan\theta$ False; $\tan(180° - \theta) = -\tan\theta$

79. $\tan(180° + \theta) = \tan\theta$ True

80. $\sin(180° + \theta) = -\sin\theta$ True

81. $\csc(\pi - \theta) = -\csc\theta$ False; $\csc(\pi - \theta) = \csc\theta$

82. $\sec(\pi + \theta) = \sec\theta$ False; $\sec(\pi + \theta) = -\sec\theta$

83. $\cos(\pi + \theta) = -\cos\theta$ True

84. $\sin(\pi + \theta) = -\sin\theta$ True

For Exercises 85–90, find the value of each expression.

85. $\sin 30° \cdot \cos 150° \cdot \sec 60° \cdot \csc 120°$ -1

86. $\cos 45° \cdot \sin 240° \cdot \tan 135° \cdot \cot 60°$ $\dfrac{\sqrt{2}}{4}$

87. $\cos^2\dfrac{5\pi}{4} - \sin^2\dfrac{2\pi}{3}$ $-\dfrac{1}{4}$

88. $\sin^2\dfrac{11\pi}{6} + \cos^2\dfrac{4\pi}{3}$ $\dfrac{1}{2}$

89. $\dfrac{2\tan\dfrac{11\pi}{6}}{1 - \tan^2\dfrac{11\pi}{6}}$ $-\sqrt{3}$

90. $\dfrac{\cot^2\dfrac{4\pi}{3} - 1}{2\cot\dfrac{4\pi}{3}}$ $\dfrac{\sqrt{3}}{3}$

For Exercises 91–94, verify the statement for the given values.

91. $\sin(A - B) = \sin A \cos B - \cos A \sin B$; $A = 240°$, $B = 120°$

92. $\cos(B - A) = \cos B \cos A + \sin B \sin A$; $A = 330°$, $B = 120°$

93. $\tan(A + B) = \dfrac{\tan A + \tan B}{1 - \tan A \tan B}$; $A = 210°$, $B = 120°$

94. $\cot(A + B) = \dfrac{\cot A \cot B - 1}{\cot A + \cot B}$; $A = 300°$, $B = 150°$

For Exercises 95–96, give the exact values if possible. Otherwise, use a calculator and approximate the result to 4 decimal places.

95. a. $\sin 30°$ $\dfrac{1}{2}$

b. $\sin(30\pi)$ 0

c. $\sin 30$ -0.9880

96. a. $\cos(0.25\pi)$ $\dfrac{\sqrt{2}}{2}$

b. $\dfrac{\cos \pi}{0.25}$ -4

c. $\cos(25°)$ 0.9063

Write About It

97. Explain why neither $\sin \theta$ nor $\cos \theta$ can be greater than 1. Refer to the figure for your explanation.

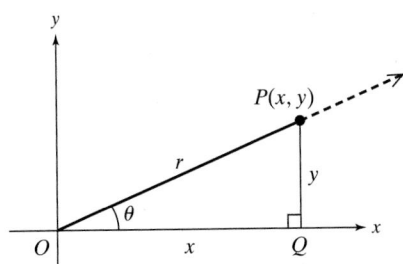

98. Explain why $\tan \theta$ is undefined at $\theta = \dfrac{\pi}{2}$ but $\cot \theta$ is defined at $\theta = \dfrac{\pi}{2}$.

When $\theta = \dfrac{\pi}{2}$, $\cos \theta = 0$ and $\sin \theta = 1$. In the case of $\tan \theta$, we would be dividing by zero, which is undefined. In the case of $\cot \theta$, we have zero in the numerator and 1 in the denominator, so the quotient is defined.

Expanding Your Skills

99. a. Graph the point (3, 4) on a rectangular coordinate system and draw a line segment connecting the point to the origin. Find the slope of the line segment.

b. Draw another line segment from the point (3, 4) to meet the x-axis at a right angle, thus forming a right triangle with the x-axis as one side. Find the tangent of the acute angle that has the x-axis as its initial side.

c. Compare the results in part (a) and part (b).

100. The circle shown is centered at the origin with a radius of 1. The segment $\overline{BD}$ is tangent to the circle at D. Match the length of each segment with the appropriate trigonometric function.

a. $\overline{AC}$ iii   **i.** $\tan \theta$

b. $\overline{BD}$ i   **ii.** $\cos \theta$

c. $\overline{OB}$ iv   **iii.** $\sin \theta$

d. $\overline{OC}$ ii   **iv.** $\sec \theta$

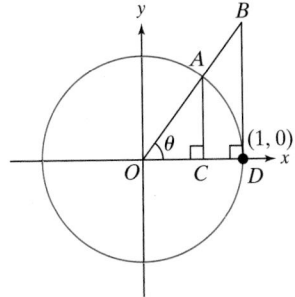

101. Circle A, with radius a, and circle B, with radius b, are tangent to each other and to $\overline{PQ}$ (see figure). $\overline{PR}$ passes through the center of each circle. Let x be the distance from point P to a point S where $\overline{PR}$ intersects circle A on the left. Let θ denote $\angle RPQ$.

a. Show that $\sin \theta = \dfrac{a}{x + a}$ and $\sin \theta = \dfrac{b}{x + 2a + b}$.

b. Use the results from part (a) to show that $\sin \theta = \dfrac{b - a}{b + a}$.

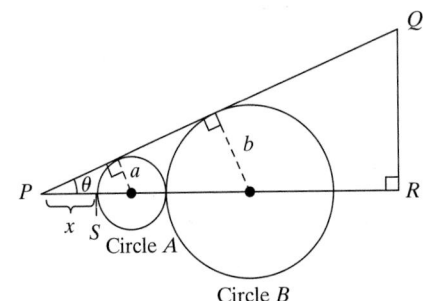

OBJECTIVES

1. Graph $y = \sin x$ and $y = \cos x$
2. Graph $y = A\sin x$ and $y = A\cos x$
3. Graph $y = A\sin Bx$ and $y = A\cos Bx$
4. Graph $y = A\sin(Bx - C) + D$ and $y = A\cos(Bx - C) + D$
5. Model Sinusoidal Behavior

1. Graph $y = \sin x$ and $y = \cos x$

Throughout this text, we have been progressively introducing categories of functions and analyzing their properties and graphs. In each case, we begin with a fundamental "parent" function such as $y = x^2$ and then develop interesting variations on the graph by applying transformations. For example, $y = a(x - h)^2 + k$ is a family of quadratic functions whose properties help us model physical phenomena such as projectile motion.

In a similar manner we will graph the basic functions $y = \sin x$ and $y = \cos x$ and then analyze their variations. To begin, note that we will use variable x (rather than θ or t) to represent the independent variable.

To develop the graph of $y = \sin x$ we return to the notion of a number line wrapped onto the unit circle. The points on the number line correspond to the real numbers x in the domain of the sine function. The value of $\sin x$ is the y-coordinate of the corresponding point on the unit circle (Figure 4-40).

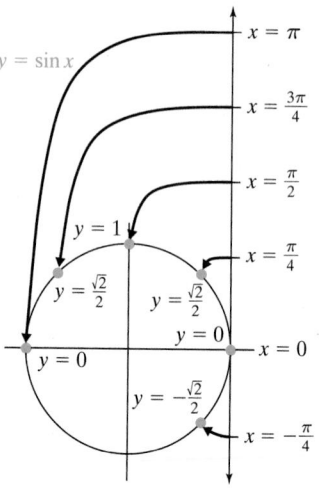

Figure 4-40

Table 4-10 gives several values of x and the corresponding values of $\sin x$. Plotting these as ordered pairs gives the graph of $y = \sin x$ in Figure 4-41.

Table 4-10

x	0	$\dfrac{\pi}{6}$	$\dfrac{\pi}{4}$	$\dfrac{\pi}{3}$	$\dfrac{\pi}{2}$	$\dfrac{2\pi}{3}$	$\dfrac{3\pi}{4}$	$\dfrac{5\pi}{6}$	π	$\dfrac{7\pi}{6}$	$\dfrac{5\pi}{4}$	$\dfrac{4\pi}{3}$	$\dfrac{3\pi}{2}$	$\dfrac{5\pi}{3}$	$\dfrac{7\pi}{4}$	$\dfrac{11\pi}{6}$	2π
$\sin x$	0	$\dfrac{1}{2}$	$\dfrac{\sqrt{2}}{2}$	$\dfrac{\sqrt{3}}{2}$	1	$\dfrac{\sqrt{3}}{2}$	$\dfrac{\sqrt{2}}{2}$	$\dfrac{1}{2}$	0	$-\dfrac{1}{2}$	$-\dfrac{\sqrt{2}}{2}$	$-\dfrac{\sqrt{3}}{2}$	-1	$-\dfrac{\sqrt{3}}{2}$	$-\dfrac{\sqrt{2}}{2}$	$-\dfrac{1}{2}$	0

TIP The decimal approximations of the coordinates involving irrational numbers can be used to sketch the graph. For example,
$$\left(\frac{\pi}{4}, \frac{\sqrt{2}}{2}\right) \approx (0.79, 0.71).$$

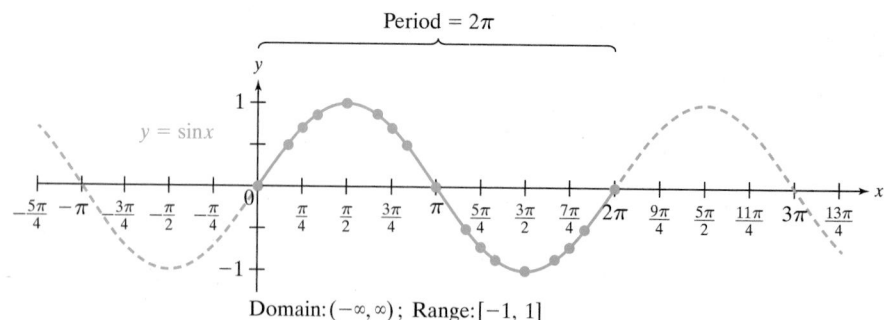

Domain: $(-\infty, \infty)$; Range: $[-1, 1]$

Figure 4-41

Recall that the sine function is periodic with period 2π. This means that the pattern shown between 0 and 2π repeats infinitely far to the left and right as denoted by the dashed portion of the curve (Figure 4-41). In Figure 4-42, one complete cycle of the graph of $y = \sin x$ is shown over the interval $[0, 2\pi]$. Notice the relationship between the y-coordinates on the unit circle and the graph of $y = \sin x$.

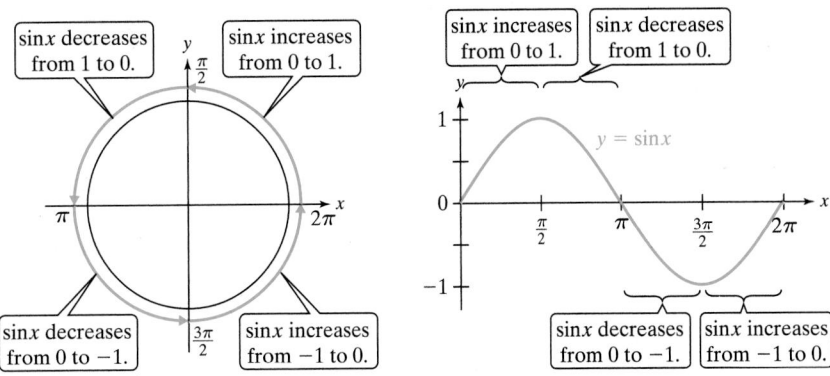

Figure 4-42

In a similar fashion, we can graph $y = \cos x$. Table 4-11 gives several values of x with the corresponding values of $\cos x$. Plotting these as ordered pairs gives the graph of $y = \cos x$ in Figure 4-43.

Table 4-11

x	0	$\frac{\pi}{6}$	$\frac{\pi}{4}$	$\frac{\pi}{3}$	$\frac{\pi}{2}$	$\frac{2\pi}{3}$	$\frac{3\pi}{4}$	$\frac{5\pi}{6}$	π	$\frac{7\pi}{6}$	$\frac{5\pi}{4}$	$\frac{4\pi}{3}$	$\frac{3\pi}{2}$	$\frac{5\pi}{3}$	$\frac{7\pi}{4}$	$\frac{11\pi}{6}$	2π
$\cos x$	1	$\frac{\sqrt{3}}{2}$	$\frac{\sqrt{2}}{2}$	$\frac{1}{2}$	0	$-\frac{1}{2}$	$-\frac{\sqrt{2}}{2}$	$-\frac{\sqrt{3}}{2}$	-1	$-\frac{\sqrt{3}}{2}$	$-\frac{\sqrt{2}}{2}$	$-\frac{1}{2}$	0	$\frac{1}{2}$	$\frac{\sqrt{2}}{2}$	$\frac{\sqrt{3}}{2}$	1

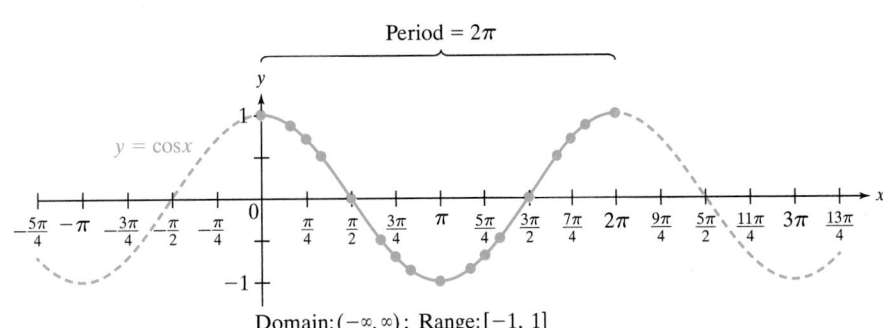

Domain: $(-\infty, \infty)$; Range: $[-1, 1]$

Figure 4-43

Avoiding Mistakes

When we define $\sin t = y$ and $\cos t = x$, the values of x and y refer to the coordinates of a point on the unit circle $x^2 + y^2 = 1$.

When we define $y = \sin x$ and $y = \cos x$, the variable x has taken the role of t (a real number on a number line wrapped onto the unit circle) and y is the corresponding sine or cosine value.

In Figure 4-44, the graphs of $y = \sin x$ and $y = \cos x$ are shown on the interval $[-2\pi, 2\pi]$ for comparison. From the graphs, we make the following observations.

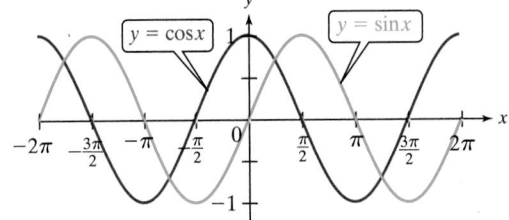

Figure 4-44

> **Characteristics of the Graphs of $y = \sin x$ and $y = \cos x$**
>
> - The domain is $(-\infty, \infty)$.
> - The range is $[-1, 1]$.
> - The period is 2π.
> - The graph of $y = \sin x$ is symmetric with respect to the origin. $y = \sin x$ is an odd function.
> - The graph of $y = \cos x$ is symmetric with respect to the y-axis. $y = \cos x$ is an even function.
> - The graphs of $y = \sin x$ and $y = \cos x$ differ by a horizontal shift of $\dfrac{\pi}{2}$.

2. Graph $y = A\sin x$ and $y = A\cos x$

Figures 4-45 and 4-46 show one complete cycle of the graphs of $y = \sin x$ and $y = \cos x$. When we graph variations of these functions we want to graph the key points. These are the relative minima, relative maxima, and x-intercepts. Notice that the graphs of both the sine and cosine function alternate between a relative minimum or maximum and an x-intercept for each quarter of a period.

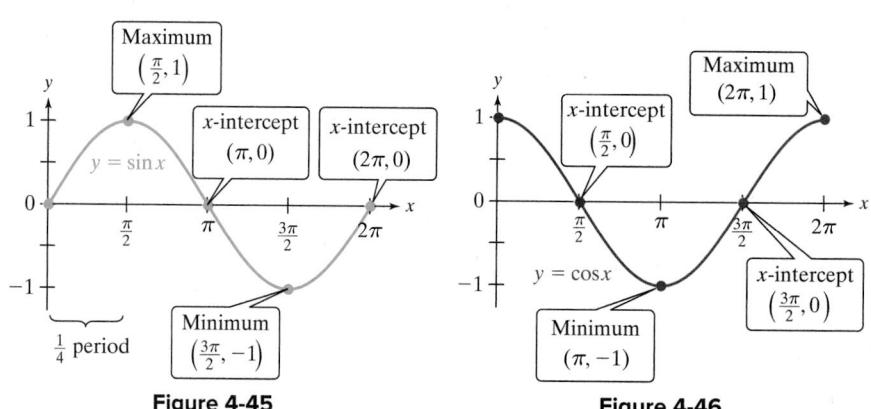

Figure 4-45   Figure 4-46

We begin analyzing variations of the sine and cosine graphs by graphing functions of the form $y = A\sin x$ and $y = A\cos x$.

Classroom Examples: p. 521
Exercises 20, 22

EXAMPLE 1   Graphing $y = A\sin x$ and $y = A\cos x$

Graph the function and identify the key points on one full period.

a. $y = 3\sin x$ **b.** $y = -\dfrac{1}{2}\cos x$

Instructor Note:
Before beginning this section, consider assigning a few problems where students review the transformations of the six basic graphs presented in Section 1.6.

Solution:

Recall from Section 1.6 that the graph of $y = A \cdot f(x)$ is the graph of $y = f(x)$ with
- A vertical stretch if $|A| > 1$ or
- A vertical shrink if $0 < |A| < 1$.
- A reflection across the x-axis if $A < 0$.

a. Given $y = 3\sin x$, the value of A is 3, so the graph of $y = 3\sin x$ is the graph of $y = \sin x$ with a vertical stretch by a factor of 3.

The dashed curve $y = \sin x$ is shown for comparison.

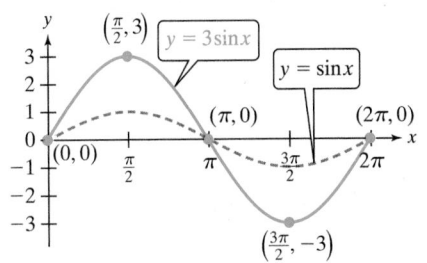

b. Given $y = -\frac{1}{2}\cos x$, the value of A is $-\frac{1}{2}$, so the graph of $y = -\frac{1}{2}\cos x$ is the graph of $y = \cos x$ with a vertical shrink and a reflection across the x-axis.

The dashed curve, $y = \cos x$, is shown for comparison.

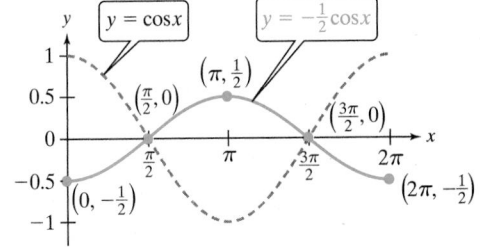

Skill Practice 1 Graph the function and identify the key points on one full period.

a. $y = 2\cos x$ **b.** $y = -\frac{1}{3}\sin x$

Notice that the graph of $y = 3\sin x$ from Example 1(a) deviates 3 units above and below the x-axis. For the graphs of $y = |A|\sin x$ and $y = |A|\cos x$, the value of the vertical scaling factor $|A|$ is called the **amplitude** of the function. The amplitude represents the amount of deviation from the central position of the sine wave. The amplitude is half the distance between the maximum value of the function and the minimum value of the function. In Example 1(a), the amplitude is $|3| = 3$, and in Example 1(b), the amplitude is $\left|-\dfrac{1}{2}\right| = \dfrac{1}{2}$.

3. Graph $y = A\sin Bx$ and $y = A\cos Bx$

Recall from Section 1.6 that the graph of $y = f(Bx)$ is the graph of $y = f(x)$ with a horizontal shrink or stretch. This means that for $y = \sin x$ and $y = \cos x$, the factor B will affect the period of the graph. For example, Figures 4-47 and 4-48 show the graphs of $y = \sin 2x$ and $y = \sin\dfrac{1}{2}x$, respectively, as compared to the parent function $y = \sin x$.

Answers

1. a.

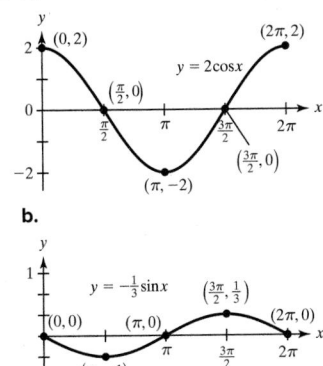

b.

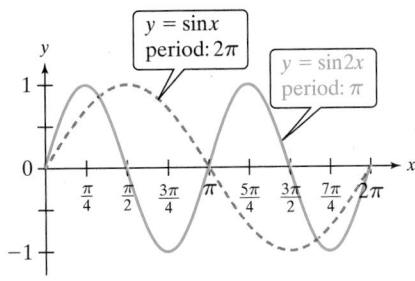

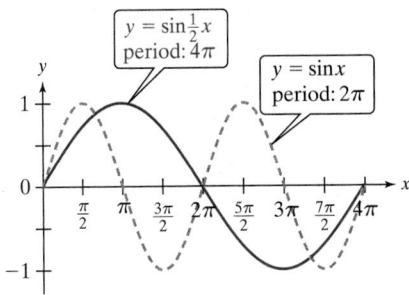

Figure 4-47   Figure 4-48

The period of the functions $y = \sin x$ and $y = \cos x$ is 2π. That is, the graphs of $y = \sin x$ and $y = \cos x$ show one complete cycle on the interval $0 \le x \le 2\pi$. To determine a comparable interval for one complete cycle of the graphs of $y = \sin Bx$ and $y = \cos Bx$ for $B > 0$, we can solve the following inequality.

$$0 \le Bx \le 2\pi$$

"Starting" value for the cycle $\dfrac{0}{B} \le \dfrac{Bx}{B} \le \dfrac{2\pi}{B}$ "Ending" value for the cycle

$$0 \le x \le \dfrac{2\pi}{B}$$

This tells us that for $B > 0$, the period of $y = \sin Bx$ and $y = \cos Bx$ is $\dfrac{2\pi}{B}$.

Amplitude and Period of the Sine and Cosine Functions

For $y = A\sin Bx$ and $y = A\cos Bx$ and $B > 0$, the amplitude and period are

$$\text{Amplitude} = |A| \quad \text{and} \quad \text{Period} = \dfrac{2\pi}{B}$$

To analyze the period of a sine or cosine function with a negative coefficient on x, we would first rewrite $y = A\sin(-Bx)$ and $y = A\cos(-Bx)$ using the odd and even properties. That is, for $B > 0$,

Rewrite $y = \sin(-Bx)$ as $y = -\sin Bx$ because $y = \sin x$ is an odd function.
Rewrite $y = \cos(-Bx)$ as $y = \cos Bx$ because $y = \cos x$ is an even function.

Classroom Example: p. 521

Exercise 28

EXAMPLE 2   **Graphing** $y = A\sin Bx$

Given $f(x) = 4\sin 3x$,
    a. Identify the amplitude and period.
    b. Graph the function and identify the key points on one full period.

Solution:

    a. $f(x) = 4\sin 3x$ is in the form $f(x) = A\sin Bx$ with $A = 4$ and $B = 3$.
       The amplitude is $|A| = |4| = 4$.

       The period is $\dfrac{2\pi}{B} = \dfrac{2\pi}{3}$.

    b. The period $\dfrac{2\pi}{3}$ is shorter than 2π, which tells us that the graph is compressed horizontally. For $y = \sin x$, one complete cycle can be graphed for x on the interval $0 \le x \le 2\pi$. For $y = 4\sin 3x$, one complete cycle can be graphed on the interval defined by $0 \le 3x \le 2\pi$.

$$0 \le 3x \le 2\pi$$

$$\dfrac{0}{3} \le \dfrac{3x}{3} \le \dfrac{2\pi}{3}$$

$$0 \le x \le \dfrac{2\pi}{3}$$

Dividing the period into fourths, we have increments of $\frac{1}{4}\left(\frac{2\pi}{3}\right) = \frac{\pi}{6}$.

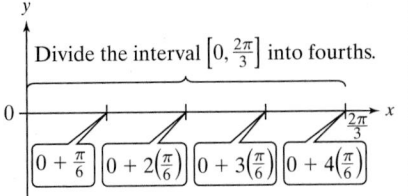

Since the amplitude is 4, the maximum and minimum points have y-coordinates of 4 and -4.

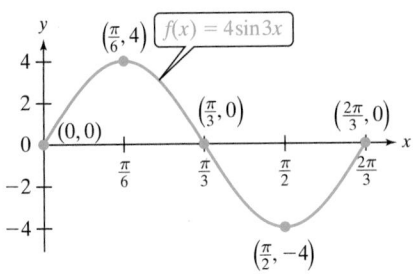

Skill Practice 2   Given $f(x) = 2\sin 4x$,

a. Identify the amplitude and period.

b. Graph the function and identify the key points on one full period.

TECHNOLOGY CONNECTIONS

Graphing a Trigonometric Function

To graph a trigonometric function, first be sure that the calculator is in radian mode. The graph of $f(x) = 4\sin 3x$ is shown along with the graph of $g(x) = \sin x$ for comparison. Notice that the graph of f deviates more from the x-axis than g because the amplitude is greater. The period of f is smaller and as a result, shows the sine wave "compressed" horizontally.

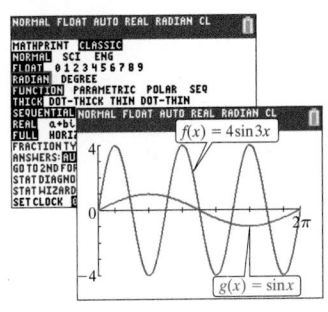

4. Graph $y = A\sin(Bx - C) + D$ and $y = A\cos(Bx - C) + D$

Recall that the graph of $y = f(x - h)$ is the graph of $y = f(x)$ with a horizontal shift of $|h|$ units. If $h > 0$, the shift is to the right, and if $h < 0$, the shift is to the left. The graphs of $y = A\sin(Bx - C)$ and $y = A\cos(Bx - C)$ may have both a change in period and a horizontal shift. In Example 3, we illustrate how to graph such a function.

Answers

2. a. Amplitude: 2, Period: $\dfrac{\pi}{2}$

b.

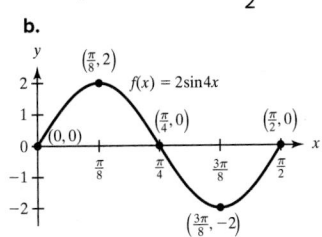

Classroom Example: p. 522
Exercise 40

EXAMPLE 3   Graphing $y = A\cos(Bx - C)$

Given $y = \cos\left(2x + \dfrac{\pi}{2}\right)$,

a. Identify the amplitude and period.

b. Graph the function and identify the key points on one full period.

Solution:

The function is of the form $y = A\cos(Bx - C)$, where $A = 1$, $B = 2$, and $C = -\dfrac{\pi}{2}$.

a. $y = \cos\left(2x + \dfrac{\pi}{2}\right)$ can be written as $y = 1 \cdot \cos\left(2x + \dfrac{\pi}{2}\right)$.

The amplitude is $|A| = |1| = 1$.

The period is $\dfrac{2\pi}{B} = \dfrac{2\pi}{2} = \pi$.

b. To find an interval over which this function completes one cycle, solve the inequality.

$$0 \le 2x + \frac{\pi}{2} \le 2\pi$$

$$-\frac{\pi}{2} \le 2x \le 2\pi - \frac{\pi}{2} \qquad \text{Subtract } \frac{\pi}{2}.$$

$$-\frac{\pi}{2} \le 2x \le \frac{4\pi}{2} - \frac{\pi}{2} \qquad \begin{array}{l}\text{Write the right side with a common} \\ \text{denominator. Then divide by 2.}\end{array}$$

$$-\frac{\pi}{2} \le 2x \le \frac{3\pi}{2}$$

$$-\frac{\pi}{4} \le x \le \frac{3\pi}{4}$$

"Starting" value for the cycle   "Ending" value for the cycle

Dividing the period into fourths, we have increments of $\dfrac{1}{4}(\pi) = \dfrac{\pi}{4}$.

Divide the interval $\left[-\frac{\pi}{4}, \frac{3\pi}{4}\right]$ into fourths.

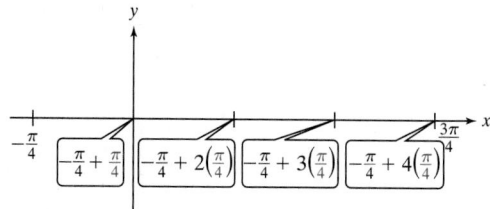

TIP Note that the distance between the "ending" point and "starting" point of the cycle shown is

$$\frac{3\pi}{4} - \left(-\frac{\pi}{4}\right) = \frac{4\pi}{4}$$
$$= \pi$$

which is the period of the function.

Period is π.

Shift is $\frac{\pi}{4}$ to the left.

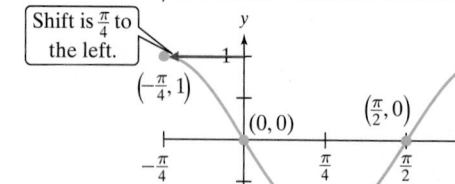
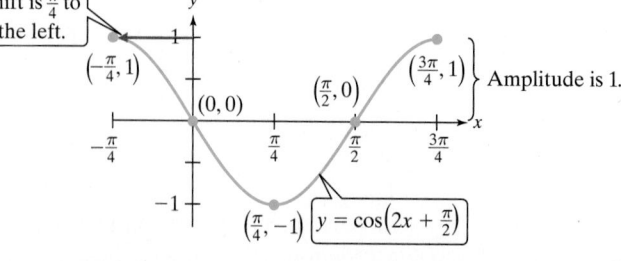

Amplitude is 1.

$y = \cos\left(2x + \dfrac{\pi}{2}\right)$

Answers

3. a. Amplitude: 1,
Period: $\dfrac{2\pi}{3}$

b.

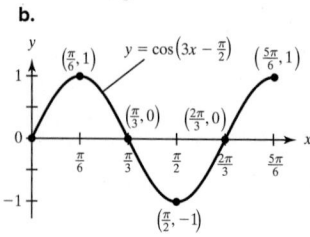

Skill Practice 3   Given $y = \cos\left(3x - \dfrac{\pi}{2}\right)$,

a. Identify the amplitude and period.

b. Graph the function and identify the key points on one full period.

In Example 3, the graph of $y = \cos\left(2x + \dfrac{\pi}{2}\right)$ is the graph of $y = \cos x$ compressed horizontally by a factor of 2 and shifted to the left $\dfrac{\pi}{4}$ units. The shift to the left is called the *phase shift*. In general, given $y = A\sin(Bx - C)$ and $y = A\cos(Bx - C)$ for $B > 0$, the horizontal transformations are controlled by the variables B and C. The value of B controls the horizontal shrink or stretch, and thus the period of the function. For $B > 0$, the phase shift can be determined by solving the following inequality.

$$0 \le Bx - C \le 2\pi$$

$$C \le Bx \le 2\pi + C \qquad \text{Add } C \text{ to each part.}$$

$$\frac{C}{B} \le \frac{Bx}{B} \le \frac{2\pi + C}{B} \qquad \text{Divide by } B \; (B > 0).$$

> The phase shift is the left endpoint.

$$\frac{C}{B} \le x \le \frac{2\pi + C}{B} \qquad \text{The graph is shifted horizontally } \frac{C}{B} \text{ units.}$$

$$\text{The phase shift is } \frac{C}{B}.$$

The phase shift is a horizontal shift of a trigonometric function. To find the vertical shift, recall that the graph of $y = f(x) + D$ is the graph of $y = f(x)$ shifted $|D|$ units upward for $D > 0$ and $|D|$ units downward if $D < 0$. We are now ready to summarize the properties of the graphs of $y = A\sin(Bx - C) + D$ and $y = A\cos(Bx - C) + D$.

Properties of the General Sine and Cosine Functions

Consider the graphs of $y = A\sin(Bx - C) + D$ and $y = A\cos(Bx - C) + D$ with $B > 0$.

1. The amplitude is $|A|$.
2. The period is $\dfrac{2\pi}{B}$.
3. The phase shift is $\dfrac{C}{B}$.
4. The vertical shift is D.
5. One full cycle is given on the interval $0 \le Bx - C \le 2\pi$.
6. The domain is $-\infty < x < \infty$.
7. The range is $-|A| + D \le y \le |A| + D$.

Classroom Example: p. 523
Exercise 56

EXAMPLE 4   Graphing $y = A\cos(Bx - C) + D$

Given $y = 2\cos(4x - 3\pi) + 5$,

a. Identify the amplitude, period, phase shift, and vertical shift.

b. Graph the function and identify the key points on one full period.

Solution:

a. $y = 2\cos(4x - 3\pi) + 5$ has the form $y = A\cos(Bx - C) + D$, where $A = 2$, $B = 4$, $C = 3\pi$, and $D = 5$.

The amplitude is $|A| = |2| = 2$.

The period is $\dfrac{2\pi}{B} = \dfrac{2\pi}{4} = \dfrac{\pi}{2}$.

The phase shift is $\dfrac{C}{B} = \dfrac{3\pi}{4}$.

The vertical shift is D. Since $D = 5 > 0$, the shift is upward 5 units.

b. To find an interval over which this function completes one cycle, solve the inequality.

$$0 \le 4x - 3\pi \le 2\pi$$

$$3\pi \le 4x \le 2\pi + 3\pi \qquad \text{Add } 3\pi.$$

$$\frac{3\pi}{4} \le \frac{4x}{4} \le \frac{5\pi}{4} \qquad \text{Divide by 4.}$$

"Starting" value for the cycle → $\dfrac{3\pi}{4} \le x \le \dfrac{5\pi}{4}$ ← "Ending" value for the cycle

Dividing the period into fourths, we have increments of $\frac{1}{4}\left(\frac{\pi}{2}\right) = \frac{\pi}{8}$.

Divide the interval $\left[\frac{3\pi}{4}, \frac{5\pi}{4}\right]$ into fourths.

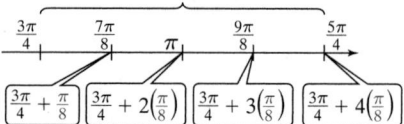

First sketch the function on the interval $\left[\frac{3\pi}{4}, \frac{5\pi}{4}\right]$ without the vertical shift (solid gray curve).

The dashed curve is a continuation of this pattern.

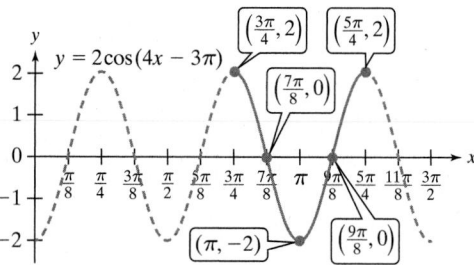

Now apply the vertical shift upwards 5 units.

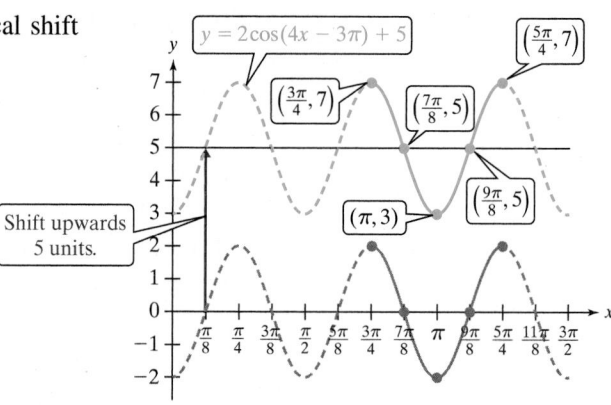

Skill Practice 4   Given $y = 2\cos(3x - \pi) + 3$,

a. Identify the amplitude, period, phase shift, and vertical shift.

b. Graph the function and identify the key points on one full period.

In Example 4, notice that the x-intercepts of the graph before the vertical shift become the points where the graph intersects the line $y = 5$ after the vertical shift. The curve oscillates above and below the line $y = 5$ rather than the x-axis. In a sense, $y = 5$ is the "central" value or "equilibrium" value of the function. This line represents the midpoint of the range, and the amplitude tells us by how much the function deviates from this line.

Answers

4. a. Amplitude: 2,

Period: $\dfrac{2\pi}{3}$,

Phase shift: $\dfrac{\pi}{3}$,

Vertical shift: 3

b.

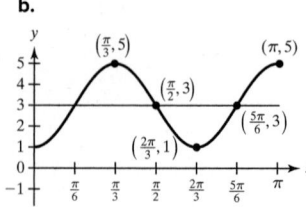

Classroom Example: p. 523
Exercise 68

EXAMPLE 5 Graphing $y = A\sin(Bx - C) + D$

Given $y = 3\sin\left(-\dfrac{\pi}{4}x - \dfrac{\pi}{2}\right) - 4$,

a. Identify the amplitude, period, phase shift, and vertical shift.

b. Graph the function and identify the key points on one full period.

Solution:

a. First note that the coefficient on x in the argument is not positive.
We want to write the function in the form $y = A\sin(Bx - C) + D$,
where $B > 0$.

$$y = 3\sin\left[-1\left(\dfrac{\pi}{4}x + \dfrac{\pi}{2}\right)\right] - 4 \qquad \text{Factor out } -1 \text{ from the argument.}$$

$$y = -3\sin\left(\dfrac{\pi}{4}x + \dfrac{\pi}{2}\right) - 4 \qquad \begin{array}{l}\text{The sine function is an odd function.}\\ \sin(-x) = -\sin(x)\end{array}$$

$$y = -3\sin\left[\dfrac{\pi}{4}x - \left(-\dfrac{\pi}{2}\right)\right] + (-4) \qquad \begin{array}{l}\text{Write the equation in the form}\\ y = A\sin(Bx - C) + D, \text{ where } B > 0.\end{array}$$

$$A = -3,\ B = \dfrac{\pi}{4},\ C = -\dfrac{\pi}{2},\text{ and } D = -4$$

The amplitude is $|A| = |-3| = 3$.

The period is $\dfrac{2\pi}{B} = \dfrac{2\pi}{\dfrac{\pi}{4}} = 2\pi \cdot \dfrac{4}{\pi} = 8$.

The phase shift is $\dfrac{C}{B} = \dfrac{-\dfrac{\pi}{2}}{\dfrac{\pi}{4}} = -\dfrac{\pi}{2} \cdot \dfrac{4}{\pi} = -2$.

The vertical shift is D. Since $D = -4 < 0$, the shift is downward 4 units.

b. To find an interval over which this function completes one cycle,
solve the inequality.

$$0 \le \dfrac{\pi}{4}x - \left(-\dfrac{\pi}{2}\right) \le 2\pi$$

$$-\dfrac{\pi}{2} \le \dfrac{\pi}{4}x \le \dfrac{3\pi}{2} \qquad \text{Add } -\dfrac{\pi}{2}.$$

$$\dfrac{4}{\pi}\cdot\left(-\dfrac{\pi}{2}\right) \le \dfrac{4}{\pi}\cdot\left(\dfrac{\pi}{4}x\right) \le \dfrac{4}{\pi}\cdot\left(\dfrac{3\pi}{2}\right) \qquad \text{Multiply by } \dfrac{4}{\pi}.$$

$$-2 \le x \le 6$$

"Starting" value for the cycle ⟶   ⟵ "Ending" value for the cycle

Dividing the period into fourths, we have increments of $\dfrac{1}{4}(8) = 2$.

Divide the interval $[-2, 6]$ into fourths.

$-2 \quad 0 \quad 2 \quad 4 \quad 6$

$\boxed{-2 + (2)}\ \boxed{-2 + 2(2)}\ \boxed{-2 + 3(2)}\ \boxed{-2 + 4(2)}$

Avoiding Mistakes

A factor from the argument of a function cannot be factored out in front of the function. For example,

$$f(2x) \neq 2f(x)$$

In step 2 of Example 5(a), the value -1 that appears in front of the function was not factored out, but rather is a result of the odd function property of the sine function.

$$\sin(-x) = -\sin(x)$$

TIP The negative sign on the phase shift indicates that the shift is to the left.

First sketch the function without the vertical shift.

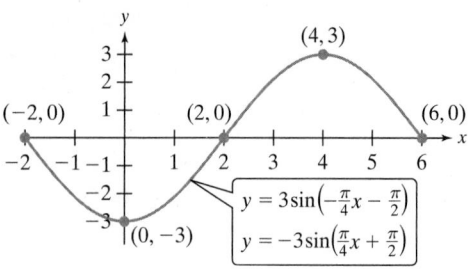

$$y = 3\sin\left(-\tfrac{\pi}{4}x - \tfrac{\pi}{2}\right)$$
$$y = -3\sin\left(\tfrac{\pi}{4}x + \tfrac{\pi}{2}\right)$$

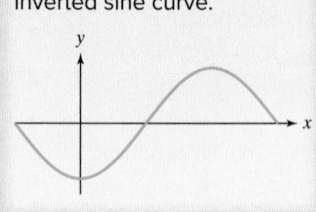

Now apply the vertical shift downward 4 units.

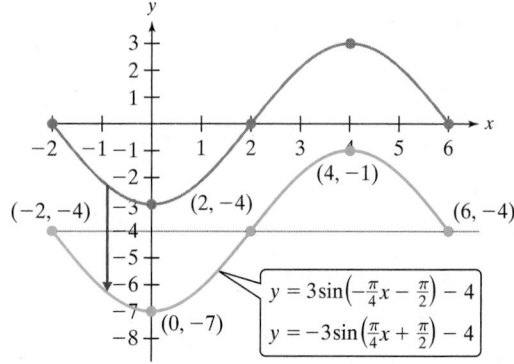

$$y = 3\sin\left(-\tfrac{\pi}{4}x - \tfrac{\pi}{2}\right) - 4$$
$$y = -3\sin\left(\tfrac{\pi}{4}x + \tfrac{\pi}{2}\right) - 4$$

Skill Practice 5 Given $y = -2\sin\left(-\dfrac{\pi}{6}x - \dfrac{\pi}{2}\right) + 1,$

a. Identify the amplitude, period, phase shift, and vertical shift.

b. Graph the function and identify the key points on one full period.

5. Model Sinusoidal Behavior

To this point, we have taken an equation of a sine or cosine function and sketched its graph. Now we reverse the process. In Example 6, we take observed data that follow a "wavelike" pattern similar to a sine or cosine graph and build a model. When the graph of a data set is shaped like a sine or cosine graph, we say the graph is **sinusoidal.**

EXAMPLE 6   Modeling the Level of the Tide

The water level relative to the top of a boat dock varies with the tides. One particular day, low tide occurs at midnight and the water level is 7 ft below the dock. The first high tide of the day occurs at approximately 6:00 A.M., and the water level is 3 ft below the dock. The next low tide occurs at noon and the water level is again 7 ft below the dock.

Assuming that this pattern continues indefinitely and behaves like a cosine wave, write a function of the form $w(t) = A\cos(Bt - C) + D$. The value $w(t)$ is the water level (in ft) relative to the top of the dock, t hours after midnight.

Answers

5. a. Amplitude: 2,
Period: 12,
Phase shift: -3,
Vertical shift: 1

b.

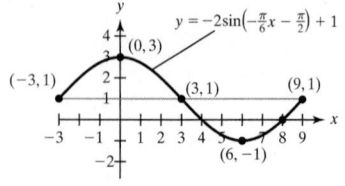

$$y = -2\sin\left(-\tfrac{\pi}{6}x - \tfrac{\pi}{2}\right) + 1$$

Solution:

Plotting the water level at midnight, 6:00 A.M., and noon helps us visualize the curve. By inspection, the curve behaves like a cosine reflected across the t-axis and shifted down 5 units.

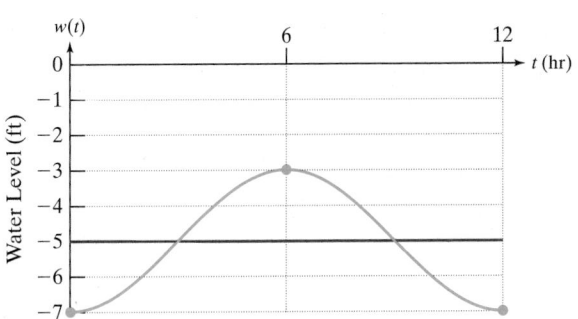

$$|A| = \frac{1}{2}[-3 - (-7)] = \frac{1}{2}(4) = 2$$

The amplitude of the curve is half the distance between the highest value and lowest value.

Vertical shift: $D = \dfrac{-7 + (-3)}{2} = -5$

The midpoint of the range gives us the vertical shift.

$P = \dfrac{2\pi}{B}$, which implies that $12 = \dfrac{2\pi}{B}$.

Therefore, $B = \dfrac{2\pi}{12} = \dfrac{\pi}{6}$.

One complete cycle takes place between $t = 0$ and $t = 12$ (from low tide to low tide). Therefore, the period P is $12 - 0 = 12$.

Since a minimum value of the curve occurs at $t = 0$, there is no phase shift for this cosine function, implying that $C = 0$.

Finally, the amplitude is 2, however, we take A to be *negative* because the graph was reflected across the t-axis before being shifted downward.

$$w(t) = -2\cos\left(\frac{\pi}{6}t\right) - 5$$

Substitute $A = -2$, $B = \dfrac{\pi}{6}$, $C = 0$, and $D = -5$, into $w(t) = A\cos(Bt - C) + D$.

Skill Practice 6   A mechanical metronome uses an inverted pendulum that makes a regular, rhythmic click as it swings to the left and right. With each swing, the pendulum moves 3 in. to the left and right of the center position. The pendulum is initially pulled to the right 3 in. and then released. It returns to its starting position in 0.8 sec. Assuming that this pattern continues indefinitely and behaves like a cosine wave, write a function of the form $x(t) = A\cos(Bt - C) + D$. The value $x(t)$ is the horizontal position (in inches) relative to the center line of the pendulum.

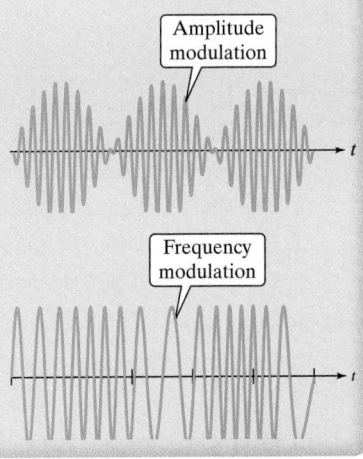

Point of Interest

AM (amplitude modulation) and FM (frequency modulation) radio are ways of broadcasting radio signals by sending electromagnetic waves through space from a transmitter to a receiver. Each method transmits information in the form of electromagnetic waves. AM works by modulating (or varying) the amplitude of the signal while the frequency remains constant. FM works by varying the frequency and keeping the amplitude constant. In 1873, James Maxwell showed mathematically that electromagnetic waves could propagate through free space. Today the far-reaching applications of wireless radio technology include use in televisions, computers, cell phones, and even deep-space radio communications.

Answer

6. $x(t) = 3\cos\left(\dfrac{5\pi}{2}t\right)$

SECTION 4.5 Practice Exercises

Prerequisite Review

For Exercises R.1–R.5, use translations to graph the function.

R.1. $h(x) = |x| - 1$ **R.2.** $n(x) = |x + 1|$ **R.3.** $b(x) = 2|x|$ **R.4.** $h(x) = \dfrac{1}{3}|x|$ **R.5.** $p(x) = -|x|$

Concept Connections

1. The value of $\sin x$ (increases/decreases) _____increases_____ on $\left(0, \dfrac{\pi}{2}\right)$ and (increases/decreases) _____decreases_____ on $\left(\dfrac{\pi}{2}, \pi\right)$.

2. The value of $\cos x$ (increases/decreases) _____decreases_____ on $\left(0, \dfrac{\pi}{2}\right)$ and (increases/decreases) _____decreases_____ on $\left(\dfrac{\pi}{2}, \pi\right)$.

3. The graph of $y = \sin x$ and $y = \cos x$ differ by a horizontal shift of _____$\dfrac{\pi}{2}$_____ units.

4. Given $y = A\sin(Bx - C) + D$ or $y = A\cos(Bx - C) + D$, for $B > 0$ the amplitude is _____$|A|$_____, the period is _____$\dfrac{2\pi}{B}$_____, the phase shift is _____$\dfrac{C}{B}$_____, and the vertical shift is _____D_____.

5. The sine function is an (even/odd) _____odd_____ function because $\sin(-x) = $ _____$-\sin(x)$_____. The cosine function is an (even/odd) _____even_____ function because $\cos(-x) = $ _____$\cos(x)$_____.

6. Given $B > 0$, how would the equation $y = A\sin(-Bx - C) + D$ be rewritten to obtain a positive coefficient on x?
$y = -A\sin(Bx + C) + D$

7. Given $B > 0$, how would the equation $y = A\cos(-Bx - C) + D$ be rewritten to obtain a positive coefficient on x?
$y = A\cos(Bx + C) + D$

8. Given $y = \sin(Bx)$ and $y = \cos(Bx)$, for $B > 1$, is the period less than or greater than 2π? If $0 < B < 1$, is the period less than or greater than 2π? less than 2π; greater than 2π

Objective 1: Graph $y = \sin x$ and $y = \cos x$

9. From memory, sketch $y = \sin x$ on the interval $[0, 2\pi]$. 10. From memory, sketch $y = \cos x$ on the interval $[0, 2\pi]$.

11. For $y = \cos x$,
   a. The domain is _____$(-\infty, \infty)$_____.
   b. The range is _____$[-1, 1]$_____.
   c. The amplitude is _____1_____.
   d. The period is _____2π_____.
   e. The cosine function is symmetric to the _____y_____-axis.
   f. On the interval $[0, 2\pi]$, the x-intercepts are _____$\left(\dfrac{\pi}{2}, 0\right)$, $\left(\dfrac{3\pi}{2}, 0\right)$_____.
   g. On the interval $[0, 2\pi]$, the maximum points are _____$(0, 1)$_____ and _____$(2\pi, 1)$_____, and the minimum point is _____$(\pi, -1)$_____.

12. For $y = \sin x$,
   a. The domain is _____$(-\infty, \infty)$_____.
   b. The range is _____$[-1, 1]$_____.
   c. The amplitude is _____1_____.
   d. The period is _____2π_____.
   e. The sine function is symmetric to the _____origin_____.
   f. On the interval $[0, 2\pi]$, the x-intercepts are _____$(0, 0)$, $(\pi, 0)$, $(2\pi, 0)$_____.
   g. On the interval $[0, 2\pi]$, the maximum point is _____$\left(\dfrac{\pi}{2}, 1\right)$_____ and the minimum point is _____$\left(\dfrac{3\pi}{2}, -1\right)$_____.

13. a. Over what interval(s) taken between 0 and 2π is the graph of $y = \sin x$ increasing? $\left(0, \dfrac{\pi}{2}\right) \cup \left(\dfrac{3\pi}{2}, 2\pi\right)$
   b. Over what interval(s) taken between 0 and 2π is the graph of $y = \sin x$ decreasing? $\left(\dfrac{\pi}{2}, \dfrac{3\pi}{2}\right)$

14. a. Over what interval(s) taken between 0 and 2π is the graph of $y = \cos x$ increasing? $(\pi, 2\pi)$
   b. Over what interval(s) taken between 0 and 2π is the graph of $y = \cos x$ decreasing? $(0, \pi)$

Objective 2: Graph $y = A\sin x$ and $y = A\cos x$

For Exercises 15–16, identify the amplitude of the function.

15. a. $y = 7\sin x$ 7

    b. $y = \dfrac{1}{7}\sin x$ $\dfrac{1}{7}$

    c. $y = -7\sin x$ 7

16. a. $y = 2\cos x$ 2

    b. $y = \dfrac{1}{2}\cos x$ $\dfrac{1}{2}$

    c. $y = -2\cos x$ 2

17. By how many units does the graph of $y = \dfrac{1}{4}\cos x$ deviate from the x-axis? $\dfrac{1}{4}$ unit

18. By how many units does the graph of $y = -5\sin x$ deviate from the x-axis?   5 units

For Exercises 19–24, graph the function and identify the key points on one full period. (See Example 1)

19. $y = 5\cos x$

20. $y = 4\sin x$

21. $y = \dfrac{1}{2}\sin x$

22. $y = \dfrac{1}{4}\cos x$

23. $y = -2\cos x$

24. $y = -3\sin x$

Objective 3: Graph $y = A\sin Bx$ and $y = A\cos Bx$

For Exercises 25–26, identify the period.

25. a. $\sin 2x$ π

    b. $\sin 2\pi x$ 1

    c. $\sin\left(-\dfrac{2}{3}x\right)$ 3π

26. a. $\cos\dfrac{1}{3}x$ 6π

    b. $\cos(-3\pi x)$ $\dfrac{2}{3}$

    c. $\cos\dfrac{1}{3}\pi x$ 6

For Exercises 27–32

a. Identify the amplitude and period.

b. Graph the function and identify the key points on one full period. (**See Example 2**)

27. $y = 2\cos 3x$

28. $y = 6\sin 4x$

29. $y = 4\sin\dfrac{\pi}{3}x$

30. $y = 5\cos\dfrac{\pi}{6}x$

31. $y = \sin\left(-\dfrac{1}{3}x\right)$

32. $y = \cos\left(-\dfrac{1}{2}x\right)$

33. Write a function of the form $f(x) = A\cos Bx$ for the given graph. $f(x) = 2\cos\dfrac{1}{2}x$

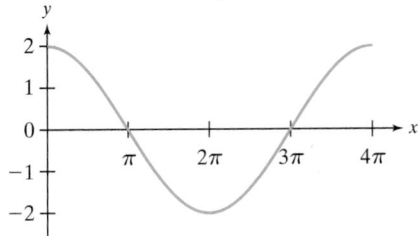

34. Write a function of the form $g(x) = A\sin Bx$ for the given graph. $g(x) = -\dfrac{3}{2}\sin\dfrac{1}{4}x$

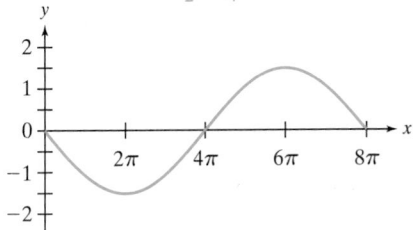

35. The graph shows the percentage in decimal form of the moon illuminated for the first 40 days of a recent year. (*Source*: Astronomical Applications Department, U.S. Naval Observatory: http://aa.usno.navy.mil)

    a. On approximately which days during this time period did a full moon occur? (A full moon corresponds to 100% or 1.0.)   Day 5 and Day 35

    b. On which day was there a new moon (no illumination)?   Day 20

    c. From the graph, approximate the period of a synodic month. A synodic month is the period of one lunar cycle (full moon to full moon).   Approximately 30 days

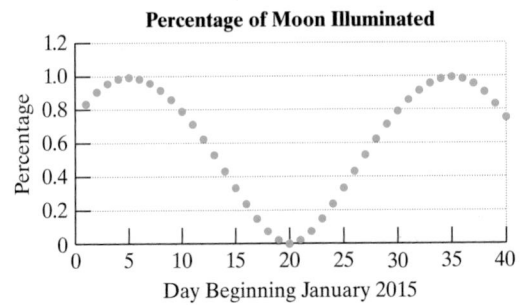

Percentage of Moon Illuminated

36. A respiratory cycle is defined as the beginning of one breath to the beginning of the next breath. The rate of air intake r (in L/sec) during a respiratory cycle for a physically fit male can be approximated by $r(t) = 0.9 \sin \dfrac{\pi}{3.5} t$, where t is the number of seconds into the cycle. A positive value for r represents inhalation and a negative value represents exhalation.

  a. How long is the respiratory cycle?   7 sec
  b. What is the maximum rate of air intake?   0.9 L/sec
  c. Graph one cycle of the function. On what interval does inhalation occur? On what interval does exhalation occur?

Objective 4: Graph $y = A\sin(Bx - C) + D$ and $y = A\cos(Bx - C) + D$

For Exercises 37–38, identify the phase shift and indicate whether the shift is to the left or to the right.

37. a. $\cos\left(x - \dfrac{\pi}{3}\right)$ $\dfrac{\pi}{3}$, right   **b.** $\cos\left(2x - \dfrac{\pi}{3}\right)$ $\dfrac{\pi}{6}$, right   **c.** $\cos\left(3\pi x + \dfrac{\pi}{3}\right)$ $-\dfrac{1}{9}$, left

38. a. $\sin\left(x + \dfrac{\pi}{8}\right)$ $-\dfrac{\pi}{8}$, left   **b.** $\sin\left(2\pi x - \dfrac{\pi}{8}\right)$ $\dfrac{1}{16}$, right   **c.** $\sin\left(4x - \dfrac{\pi}{8}\right)$ $\dfrac{\pi}{32}$, right

For Exercises 39–44,
a. Identify the amplitude, period, and phase shift.
b. Graph the function and identify the key points on one full period. **(See Example 3)**

39. $y = 2\cos(x + \pi)$

40. $y = 4\sin\left(x + \dfrac{\pi}{2}\right)$

41. $y = \sin\left(2x - \dfrac{\pi}{3}\right)$

42. $y = \cos\left(3x - \dfrac{\pi}{4}\right)$

43. $y = -6\cos\left(\dfrac{1}{2}x + \dfrac{\pi}{4}\right)$

44. $y = -5\sin\left(\dfrac{1}{3}x + \dfrac{\pi}{6}\right)$

45. Write a function of the form $f(x) = A\cos(Bx - C)$ for the given graph. $y = 2\cos\left(x - \dfrac{\pi}{3}\right)$

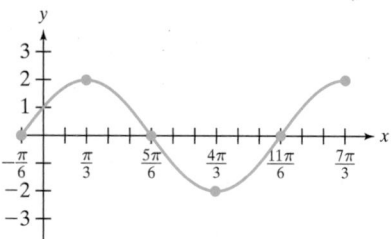

46. Write a function of the form $f(x) = A\sin(Bx - C)$ for the given graph. $y = \dfrac{1}{2}\sin\left(x - \dfrac{\pi}{4}\right)$

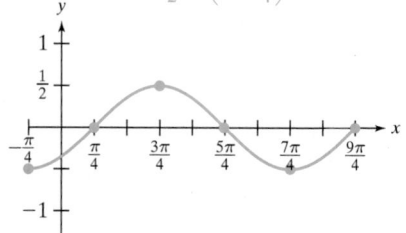

47. Given $y = -2\sin\left(2x - \dfrac{\pi}{6}\right) - 7$,

  a. Is the period less than or greater than 2π?   Less than
  b. Is the phase shift to the left or right?   Right
  c. Is the vertical shift upward or downward?   Downward

48. Given $y = \cos\left(\dfrac{1}{2}x + \pi\right) + 4$,

  a. Is the period less than or greater than 2π?   Greater than
  b. Is the phase shift to the left or right?   Left
  c. Is the vertical shift upward or downward?   Upward

For Exercises 49–52, rewrite the equation so that the coefficient on x is positive.

49. $y = \cos\left(-2x + \dfrac{\pi}{6}\right) - 4$ $y = \cos\left(2x - \dfrac{\pi}{6}\right) - 4$

50. $y = 4\cos(-3x - \pi) + 5$ $y = 4\cos(3x + \pi) + 5$

51. $y = \sin\left(-2x + \dfrac{\pi}{6}\right) - 4$ $y = -\sin\left(2x - \dfrac{\pi}{6}\right) - 4$

52. $y = 4\sin(-3x - \pi) + 5$ $y = -4\sin(3x + \pi) + 5$

53. Given $y = 2\sin\left(-\dfrac{\pi}{6}x + \dfrac{\pi}{2}\right) - 3$, is the phase shift to the right or left?   Right

54. Given $y = -4\cos\left(-\dfrac{\pi}{6}x - \dfrac{\pi}{2}\right) + 1$, is the phase shift to the right or left?   Left

For Exercises 55–68,

a. Identify the amplitude, period, phase shift, and vertical shift.

b. Graph the function and identify the key points on one full period. (**See Examples 4–5**)

55. $h(x) = 3\sin(4x - \pi) + 5$

56. $g(x) = 2\sin(3x - \pi) - 4$

57. $f(x) = 4\cos\left(3x - \dfrac{\pi}{2}\right) - 1$

58. $k(x) = 5\cos\left(2x - \dfrac{\pi}{2}\right) + 1$

59. $y = \dfrac{1}{2}\sin\left(-\dfrac{1}{3}x\right)$

60. $y = \dfrac{2}{3}\sin\left(-\dfrac{1}{2}x\right)$

61. $v(x) = 1.6\cos(-\pi x)$

62. $m(x) = 2.4\cos(-4\pi x)$

63. $y = 2\sin(-2x - \pi) + 5$

64. $y = 3\sin(-4x - \pi) - 7$

65. $p(x) = -\cos\left(-\dfrac{\pi}{2}x - \pi\right) + 2$

66. $q(x) = -\cos\left(-\dfrac{\pi}{3}x - \pi\right) - 2$

67. $y = \sin\left(-\dfrac{\pi}{4}x - \dfrac{\pi}{2}\right) - 3$

68. $y = \sin\left(-\dfrac{\pi}{3}x - \dfrac{\pi}{2}\right) + 4$

69. The temperature T (in °F) for Kansas City, Missouri, over a several day period in April can be approximated by $T(t) = -5.9\cos(0.262t - 1.245) + 48.2$, where t is the number of hours since midnight on day 1.

   a. What is the period of the function? Round to the nearest hour.   24 hr

   b. What is the significance of the term 48.2 in this model?   48.2°F represents the average temperature for the day.

   c. What is the significance of the factor 5.9 in this model?   The factor 5.9 is the amplitude of the curve and represents the amount that the temperature deviates above or below the daily average.

   d. What was the minimum temperature for the day? When did it occur?   42.3° at approximately 4:45 A.M.

   e. What was the maximum temperature for the day? When did it occur?   54.1° at approximately 4:45 P.M.

70. The duration of daylight and darkness varies during the year due to the angle of the Sun in the sky. The model $d(t) = 2.65\sin(0.51t - 1.32) + 12$ approximates the amount of daylight $d(t)$ (in hours) for Sacramento, California, as a function of the time t (in months) after January 1 for a recent year; that is, $t = 0$ is January 1, $t = 1$ is February 1, and so on. The model $y = n(t)$ represents the amount of darkness as a function of t.

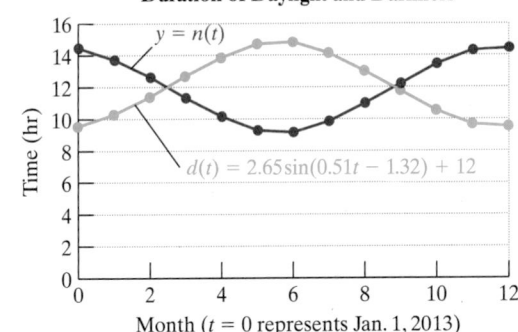

Duration of Daylight and Darkness

   a. Describe the relationship between the graphs of the functions and the line $y = 12$.

   b. Use the result of part (a) and a transformation of $y = d(t)$ to write an equation representing n as a function of t.

   c. What do the points of intersection of the two graphs represent?

   d. What do the relative minima and relative maxima of the graphs represent?

   e. What does $T(t) = d(t) + n(t)$ represent?

Objective 5: Model Sinusoidal Behavior

71. The probability of precipitation in Modesto, California, varies from a peak of 0.34 (34%) in January to a low of 0.04 (4%) in July. Assume that the percentage of precipitation varies monthly and behaves like a cosine curve.

   a. Write a function of the form $P(t) = A\cos(Bt - C) + D$ to model the precipitation probability. The value $P(t)$ is the probability of precipitation (as a decimal), for month t, with January as $t = 1$.

   b. Graph the function from part (a) on the interval $[0, 13]$ and plot the points $(1, 0.34)$, $(7, 0.04)$, and $(13, 0.34)$ to check the accuracy of your model.

72. The monthly high temperature for Atlantic City, New Jersey, peaks at an average high of 86° in July and goes down to an average high of 64° in January. Assume that this pattern for monthly high temperatures continues indefinitely and behaves like a cosine wave.

   a. Write a function of the form $H(t) = A\cos(Bt - C) + D$ to model the average high temperature. The value $H(t)$ is the average high temperature for month t, with January as $t = 0$.

   b. Graph the function from part (a) on the interval $[0, 13]$ and plot the points $(0, 64)$, $(6, 86)$, and $(12, 64)$ to check the accuracy of your model.

73. An adult human at rest inhales and exhales approximately 500 mL of air (called the tidal volume) in approximately 5 sec. However, at the end of each exhalation, the lungs still contain a volume of air, called the functional residual capacity (FRC), which is approximately 2000 mL. **(See Example 6)**

a. What volume of air is in the lungs after inhalation?
  2500 mL

b. What volume of air is in the lungs after exhalation?
  2000 mL

c. What is the period of a complete respiratory cycle?
  5 sec

d. Write a function $V(t) = A \cos Bt + D$ to represent the volume of air in the lungs t seconds after the end of an inhalation. $V(t) = 250 \cos \dfrac{2\pi}{5}t + 2250$

e. What is the average amount of air in the lungs during one breathing cycle? 2250 mL

f. During hyperventilation, breathing is more rapid with deep inhalations and exhalations. What parts of the equation from part (d) change? The period would shorten and the amplitude would increase.

75. The data in the table represent the monthly power bills (in dollars) for a homeowner in southern California.

a. Enter the data in a graphing utility and use the sinusoidal regression tool (SinReg) to find a model of the form $A(t) = a\sin(bt + c) + d$, where $A(t)$ represents the amount of the bill for month t ($t = 1$ represents January, $t = 2$ represents February, and so on).
  $A(t) = 35.4\sin(0.753t + 2.18) + 92.2$

b. Graph the data and the resulting function.

Month, t	1	2	3	4	5	6
Amount ($)	104.73	66.13	48.99	56.04	85.51	98.57

Month, t	7	8	9	10	11	12
Amount ($)	125.08	124.48	113.93	81.06	63.30	71.85

74. The times for high and low tides are given in the table for a recent day in Jacksonville Beach, Florida. The times are rounded to the nearest hour and the tide levels are measured relative to mean sea level (MSL).

a. Write a model $h(t) = A\cos(Bt - C)$ to represent the tide level $h(t)$ (in feet) in terms of the amount of time t elapsed since midnight. $h(t) = 3.4\cos\dfrac{\pi t}{6}$

b. Use the model from part (a) to estimate the tide level at 3:00 P.M. 0 ft

	Time (hr after midnight)	Height Relative to MSL (ft)
High tide	0	3.4
Low tide	6	−3.4
High tide	12	3.4
Low tide	18	−3.4
High tide	24	3.4

76. The data in the table represent the duration of daylight $d(t)$ (in hours) for Houston, Texas, for the first day of the month, t months after January 1 for a recent year. (*Source*: Astronomical Applications Department, U.S. Naval Observatory: http://aa.usno.navy.mil)

a. Enter the data in a graphing utility and use the sinusoidal regression tool (SinReg) to find a model of the form $d(t) = a\sin(bt + c) + d$.
  $d(t) = 1.9\sin(0.5t - 1.3) + 12.2$

b. Graph the data and the resulting function.

Month, t	0	1	2	3	4	5
Time (hr)	10.28	10.80	11.57	12.48	13.65	13.95

Month, t	6	7	8	9	10	11
Time (hr)	14.02	13.55	12.73	11.87	11.00	10.38

Mixed Exercises

For Exercises 77–78, write the range of the function in interval notation.

77. a. $y = 8\cos(2x - \pi) + 4$ [−4, 12]

b. $y = -3\cos\left(x + \dfrac{\pi}{3}\right) - 5$ [−8, −2]

78. a. $y = -6\sin\left(3x - \dfrac{\pi}{2}\right) - 2$ [−8, 4]

b. $y = 2\sin(3x + 2\pi) + 12$ [10, 14]

79. Given $f(x) = \cos x$ and $h(x) = 3x + 2$, find $(h \circ f)(x)$ and for $(h \circ f)(x)$, $(h \circ f)(x) = 3\cos x + 2$

a. Find the amplitude. 3

b. Find the period. 2π

c. Write the domain in interval notation. $(-\infty, \infty)$

d. Write the range in interval notation. [−1, 5]

80. Given $g(x) = \sin x$ and $k(x) = 6x$, find $(g \circ k)(x)$ and for $(g \circ k)(x)$, $(g \circ k)(x) = \sin 6x$

a. Find the amplitude. 1

b. Find the period. $\dfrac{\pi}{3}$

c. Write the domain in interval notation. $(-\infty, \infty)$

d. Write the range in interval notation. [−1, 1]

81. Write a function of the form $y = A\sin(Bx - C) + D$ that has period $\dfrac{\pi}{3}$, amplitude 4, phase shift $\dfrac{\pi}{2}$, and vertical shift 5.

82. Write a function of the form $y = A\cos(Bx - C) + D$ that has period $\dfrac{\pi}{4}$, amplitude 2, phase shift $-\dfrac{\pi}{3}$, and vertical shift 7.

83. Write a function of the form $y = A\cos(Bx - C) + D$ that has period 16, phase shift −4, and range $3 \le y \le 7$.

$y = 2\cos\left(\dfrac{\pi}{8}x + \dfrac{\pi}{2}\right) + 5$ or $y = -2\cos\left(\dfrac{\pi}{8}x + \dfrac{\pi}{2}\right) + 5$

84. Write a function of the form $y = A\sin(Bx - C) + D$ that has period 8, phase shift −2, and range $-14 \le y \le -6$.

$y = 4\sin\left(\dfrac{\pi}{4}x + \dfrac{\pi}{2}\right) - 10$ or $y = -4\sin\left(\dfrac{\pi}{4}x + \dfrac{\pi}{2}\right) - 10$

Additional answers can be found in the Instructor Answer Appendix.

For Exercises 85–86,

a. Write an equation of the form $y = A\cos(Bx - C) + D$ with $A > 0$ to model the graph.

b. Write an equation of the form $y = A\sin(Bx - C) + D$ with $A > 0$ to model the graph.

85.

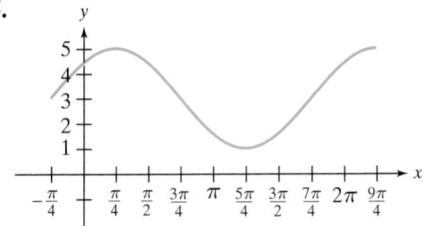

86.

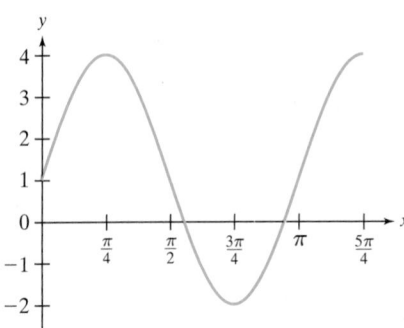

For Exercises 87–90, explain how to graph the given function by performing transformations on the "parent" graphs $y = \sin x$ **and** $y = \cos x$.

87. a. $y = \sin 2x$

    b. $y = 2\sin x$

88. a. $y = \dfrac{1}{3}\cos x$

    b. $y = \cos\dfrac{1}{3}x$

89. a. $y = \sin(x + 2)$

    b. $y = \sin x + 2$

90. a. $y = \cos x - 4$

    b. $y = \cos(x - 4)$

Write About It

91. Is $f(x) = \sin x$ one-to-one? Explain why or why not.

92. Is $f(x) = \cos x$ one-to-one? Explain why or why not.

93. If f and g are both periodic functions with period P, is $(f + g)(x)$ also periodic? Explain why or why not.

Expanding Your Skills

94. Explain why a function that is increasing on its entire domain cannot be periodic.

For Exercises 95–96, find the average rate of change on the given interval. Give the exact value and an approximation to 4 decimal places. Verify that your results are reasonable by comparing the results to the slopes of the lines given in the graph.

95. $f(x) = \sin x$

  a. $\left[0, \dfrac{\pi}{6}\right]$ **b.** $\left[\dfrac{\pi}{6}, \dfrac{\pi}{3}\right]$ **c.** $\left[\dfrac{\pi}{3}, \dfrac{\pi}{2}\right]$

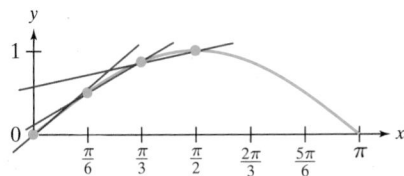

96. $f(x) = \cos x$

  a. $\left[0, \dfrac{\pi}{6}\right]$ **b.** $\left[\dfrac{\pi}{6}, \dfrac{\pi}{3}\right]$ **c.** $\left[\dfrac{\pi}{3}, \dfrac{\pi}{2}\right]$

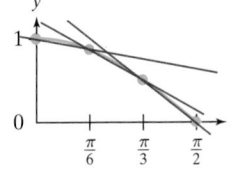

For Exercises 97–100, use your knowledge of the graphs of the sine function and linear functions to determine the number of solutions to the equation.

97. $\sin x = x - 2$
  One solution

98. $\cos x = -x$
  One solution

99. $\sin 2x = -2$
  No solution

100. $2\sin 2x = -2$
  Infinitely many solutions

For Exercises 101–102, graph the piecewise-defined function.

101. $g(x) = \begin{cases} \sin x & \text{for} \quad 0 \le x \le \pi \\ -\sin x & \text{for} \quad \pi < x \le 2\pi \end{cases}$

102. $f(x) = \begin{cases} \cos x & \text{for} \quad 0 \le x \le \dfrac{\pi}{4} \\ \sin x & \text{for} \quad \dfrac{\pi}{4} < x \le \dfrac{\pi}{2} \end{cases}$

Technology Connections

103. Functions a and m approximate the duration of daylight, respectively, for Albany, New York, and Miami, Florida, for a given year for day t. The value $t = 1$ represents January 1, $t = 2$ represents February 1, and so on.

$$a(t) = 12 + 3.1 \sin\left[\frac{2\pi}{365}(t - 80)\right] \qquad m(t) = 12 + 1.6 \sin\left[\frac{2\pi}{365}(t - 80)\right]$$

a. Graph the two functions with a graphing utility and comment on the difference between the two graphs.

b. Both functions have a constant term of 12. What does this represent graphically and in the context of this problem?

c. What do the factors 3.1 and 1.6 represent in the two functions?

d. What is the period of each function?

e. What does the horizontal shift of 80 units represent in the context of this problem.

f. Use the Intersect feature to approximate the points of intersection.

g. Interpret the meaning of the points of intersection.

For Exercises 104–105, we demonstrate that trigonometric functions can be approximated by polynomial functions over a given interval in the domain.

Graph functions f, g, h, and k on the viewing window $-4 \le x \le 4$, $-4 \le y \le 4$. Then use a Table feature on a graphing utility to evaluate each function for the given values of x. How do functions g, h, and k compare to function f for x values farther from 0? [*Hint*: For a given natural number n, the value $n!$, read as "n factorial," is defined as $n! = n(n - 1)(n - 2) \cdots 1$. For example, $3! = 3 \cdot 2 \cdot 1 = 6$.]

104.

Function	$x = 0.1$	$x = 0.5$	$x = 1.0$	$x = 1.5$
$f(x) = \cos x$				
$g(x) = 1 - \dfrac{x^2}{2!}$				
$h(x) = 1 - \dfrac{x^2}{2!} + \dfrac{x^4}{4!}$				
$k(x) = 1 - \dfrac{x^2}{2!} + \dfrac{x^4}{4!} - \dfrac{x^6}{6!}$				

105.

Function	$x = 0.1$	$x = 0.5$	$x = 1.0$	$x = 1.5$
$f(x) = \sin x$				
$g(x) = x - \dfrac{x^3}{3!}$				
$h(x) = x - \dfrac{x^3}{3!} + \dfrac{x^5}{5!}$				
$k(x) = x - \dfrac{x^3}{3!} + \dfrac{x^5}{5!} - \dfrac{x^7}{7!}$				

For Exercises 106–107, use a graph to solve the equation on the given interval.

106. $\cos\left(2x - \dfrac{\pi}{3}\right) = 0.5$ on $[0, \pi]$

Viewing window: $\left[0, \pi, \dfrac{\pi}{3}\right]$ by $\left[-1, 1, \dfrac{1}{2}\right]$

107. $\sin\left(2x + \dfrac{\pi}{4}\right) = 1$ on $[0, 2\pi]$

Viewing window: $\left[0, 2\pi, \dfrac{\pi}{8}\right]$ by $[-2, 2, 1]$

For Exercises 108–109, use a graph to solve the equation on the given interval. Round the answer to 2 decimal places.

108. $\sin\left(x - \dfrac{\pi}{4}\right) = -e^x$ on $[-\pi, \pi]$

Viewing window: $\left[-\pi, \pi, \dfrac{\pi}{2}\right]$ by $[-2, 2, 1]$

109. $6\cos\left(x + \dfrac{\pi}{6}\right) = \ln x$ on $[0, 2\pi]$

Viewing window: $\left[0, 2\pi, \dfrac{\pi}{2}\right]$ by $[-7, 7, 1]$

110. Graph the functions on the window provided.

a. $y = 2 \qquad y = -2 \qquad y = 2\sin x$

Viewing window: $\left[-2\pi, 2\pi, \dfrac{\pi}{2}\right]$ by $[-3, 3, 1]$

b. $y = 0.5x \qquad y = -0.5x \qquad y = 0.5x \sin x$

Viewing window: $[-5\pi, 5\pi, \pi]$ by $[-8, 8, 1]$

c. $y = \cos x \qquad y = -\cos x \qquad y = \cos x \cdot \sin 12x$

Viewing window: $[0, 2.5\pi, 0.25\pi]$ by $[-1, 1, 0.5]$

d. Explain the relationship among the three functions in parts (a), (b), and (c). The first two functions are the upper and lower bounds of the third function.

Graphs of Other Trigonometric Functions

1. Graph the Secant and Cosecant Functions

In this section, we turn our attention to the graphs of the secant, cosecant, tangent, and cotangent functions. Each of these functions has vertical asymptotes and, for this reason, we often use them to model patterns involving unbounded behavior.

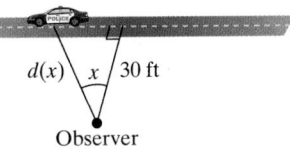

For example, the distance $d(x)$ between an observer 30 ft from a straight highway and a police car traveling down the highway is given by $d(x) = 30\sec x$. The independent variable x is the angle formed from the perpendicular line from the road to the observer and from the observer to the police car (Figure 4-49). Notice that for x values close to $90°$ $\left(\frac{\pi}{2}\text{ radians}\right)$, the distance $d(x)$ approaches infinity. This means that the graph of $d(x) = 30\sec x$ will have a vertical asymptote at $x = \frac{\pi}{2}$.

Figure 4-49

To graph the cosecant and secant functions, we first review their reciprocal relationships.

$$y = \csc x = \frac{1}{\sin x} \quad \text{and} \quad y = \sec x = \frac{1}{\cos x}$$

- If $\sin x = 0$, then $\dfrac{1}{\sin x}$ is undefined, and the graph of $y = \csc x$ has a vertical asymptote (the numerator is nonzero and the denominator is zero). This occurs when x is a multiple of π: $x = n\pi$ for all integers n.

- If $\cos x = 0$, then $\dfrac{1}{\cos x}$ is undefined, and the graph of $y = \sec x$ has a vertical asymptote (the numerator is nonzero and the denominator is zero). This occurs when x is an odd multiple of $\dfrac{\pi}{2}$: $x = \dfrac{(2n + 1)\pi}{2}$ for all integers n.

To sketch the graph of $y = \csc x$, we construct a table with several values of x and the corresponding values of $\sin x$ and $\csc x$ (Table 4-12). The cosecant function is an odd function, so the points plotted in Quadrants I and II each have a mirror image in Quadrants III and IV. Furthermore, both the sine and cosecant functions are periodic, and their graphs show a repeated pattern in intervals of 2π (Figure 4-50).

Table 4-12

x	0	$\dfrac{\pi}{6}$	$\dfrac{\pi}{4}$	$\dfrac{\pi}{3}$	$\dfrac{\pi}{2}$	$\dfrac{2\pi}{3}$	$\dfrac{3\pi}{4}$	$\dfrac{5\pi}{6}$	π
$\sin x$	0	$\dfrac{1}{2}$	$\dfrac{\sqrt{2}}{2} \approx 0.71$	$\dfrac{\sqrt{3}}{2} \approx 0.87$	1	$\dfrac{\sqrt{3}}{2} \approx 0.87$	$\dfrac{\sqrt{2}}{2} \approx 0.71$	$\dfrac{1}{2}$	0
$\csc x$	Undefined	2	$\sqrt{2} \approx 1.4$	$\dfrac{2\sqrt{3}}{3} \approx 1.2$	1	$\dfrac{2\sqrt{3}}{3} \approx 1.2$	$\sqrt{2} \approx 1.4$	2	Undefined

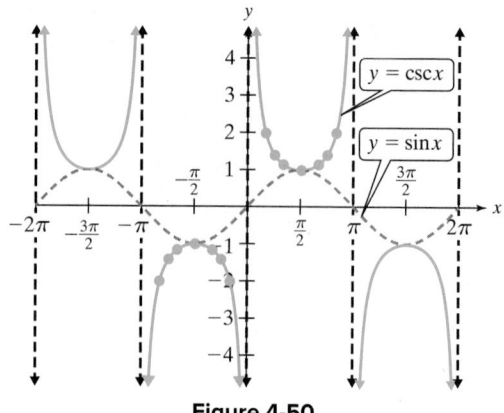

Figure 4-50

Instructor Note:
Presenting the graphs of the secant
and cosecant functions immediately
after graphing the sine and cosine
functions provides reinforcement to
students' homework from the
previous class.

Notice that the relative maxima and minima from the sine function correspond to the relative minima and maxima for the cosecant function. The x-intercepts of the sine graph give the location of the vertical asymptotes for the graph of $y = \csc x$.

The graphs of the secant function and cosine function are similarly related. The secant function is an even function, and therefore its graph is symmetric with respect to the y-axis. The asymptotes of the graph of $y = \sec x$ occur at odd multiples of $\frac{\pi}{2}$ where $\cos x = 0$ (Figure 4-51).

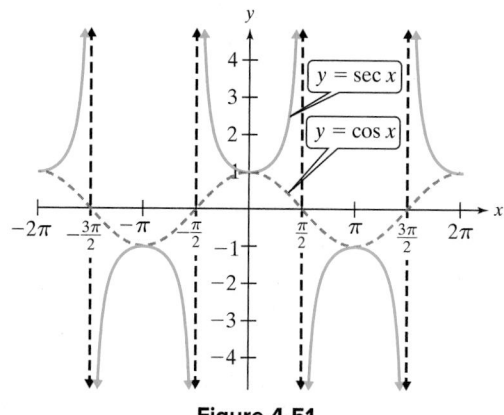

Figure 4-51

Table 4-13 summarizes key properties of the graphs of $y = \csc x$ and $y = \sec x$.

Table 4-13

Graphs of the Cosecant and Secant Functions		
	$y = \csc x$	$y = \sec x$
Domain	$\{x \,\vert\, x \neq n\pi \text{ for all integers } n\}$	$\{x \,\vert\, x \neq \frac{(2n+1)\pi}{2} \text{ for all integers } n\}$
Range	$\{y \,\vert\, y \leq -1 \text{ or } y \geq 1\}$	$\{y \,\vert\, y \leq -1 \text{ or } y \geq 1\}$
Amplitude	None ($y = \csc x$ increases and decreases without bound)	None ($y = \sec x$ increases and decreases without bound)
Period	2π	2π
Vertical Asymptotes	$x = n\pi$ (multiples of π)	$x = \frac{(2n+1)\pi}{2}$ (odd mulitples of $\frac{\pi}{2}$)
Symmetry	Origin (The cosecant function is an odd function.)	y-axis (The secant function is an even function.)

To graph variations of $y = \csc x$ or $y = \sec x$, we use the graph of the related reciprocal function for reference.

Classroom Example: p. 536
Exercise 22

EXAMPLE 1 Graphing a Variation of $y = \sec x$

Graph $y = 3\sec 2x$.

Solution:

We first graph the reciprocal function $y = 3\cos 2x$ (shown in gray) using the techniques from Section 4.5.

The amplitude is $|3| = 3$.

The period is $\dfrac{2\pi}{2} = \pi$.

There is no phase shift or vertical shift.

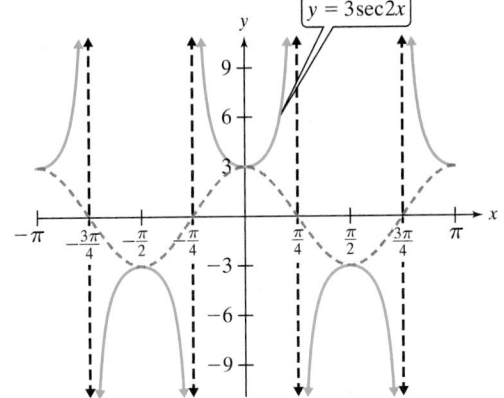

The graph of $y = 3\sec 2x$ (shown in blue) has vertical asymptotes at the odd multiples of $\frac{\pi}{4}$ where the graph of $y = 3\cos 2x$ has x-intercepts.

Skill Practice 1 Graph $y = 2\sec 4x$.

Instructor Note:
Emphasize to students that the reciprocal relationships among the six trigonometric functions are crucial when determining the equations of the vertical asymptotes.

In Example 1 we built the graph of $y = 3\sec 2x$ by using reference points (x-intercepts and relative extrema) from the graph of $y = 3\cos 2x$. To graph a cosecant or secant function involving a vertical shift, we recommend first graphing the function *without* the vertical shift. This is to make use of the x-intercepts of the reciprocal function to find the vertical asymptotes of the cosecant or secant function. This is demonstrated in Example 2.

Classroom Example: p. 536
Exercise 32

EXAMPLE 2 Graphing a Cosecant Function with a Vertical Shift

Graph $y = \csc\left(x - \dfrac{\pi}{4}\right) + 3$.

Solution:

First graph the function without the vertical shift, $y = \csc\left(x - \dfrac{\pi}{4}\right)$.

For the related reciprocal function $y = \sin\left(x - \dfrac{\pi}{4}\right)$, we have:

The amplitude is 1.
The period is 2π.

The phase shift is $\dfrac{\pi}{4}$.

The x-intercepts occur at

$x = \ldots, -\dfrac{3\pi}{4}, \dfrac{\pi}{4}, \dfrac{5\pi}{4}, \dfrac{9\pi}{4}, \ldots.$

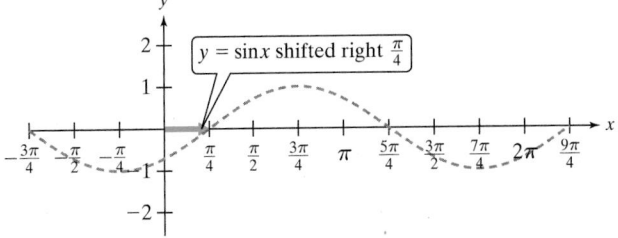

Answer
1. See page SA 34

The vertical asymptotes of $y = \csc\left(x - \dfrac{\pi}{4}\right)$ pass through the x-intercepts of the related sine function.

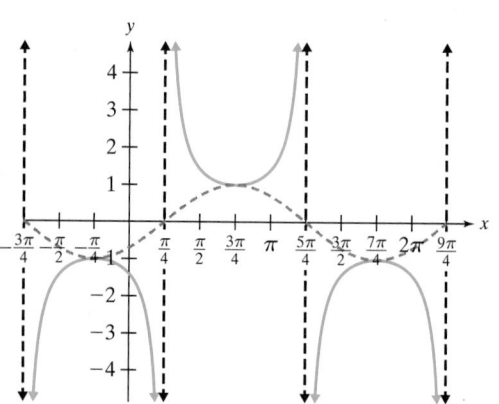

Next, apply the vertical shift 3 units upward to graph $y = \csc\left(x - \dfrac{\pi}{4}\right) + 3$.

Notice that the range of $y = \csc\left(x - \dfrac{\pi}{4}\right) + 3$ is $(-\infty, 2] \cup [4, \infty)$.

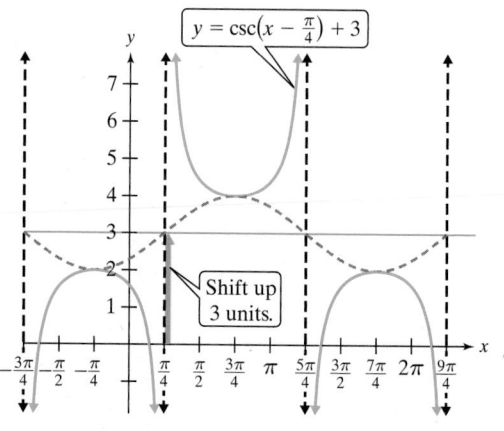

Skill Practice 2 Graph $y = \csc(x + \pi) - 2$.

2. Graph the Tangent and Cotangent Functions

We now investigate the graph $y = \tan x$ and begin with several observations. The graph of $y = \tan x = \dfrac{\sin x}{\cos x}$ will have

- An x-intercept where the numerator equals zero. The value $\sin x = 0$ when x is a multiple of π; that is, $x = n\pi$ for all integers n (Figure 4-52).
- A vertical asymptote where the denominator equals zero. The value $\cos x = 0$ when x is an odd multiple of $\dfrac{\pi}{2}$; that is, at $x = \dfrac{(2n + 1)\pi}{2}$ for all integers n (Figure 4-52).

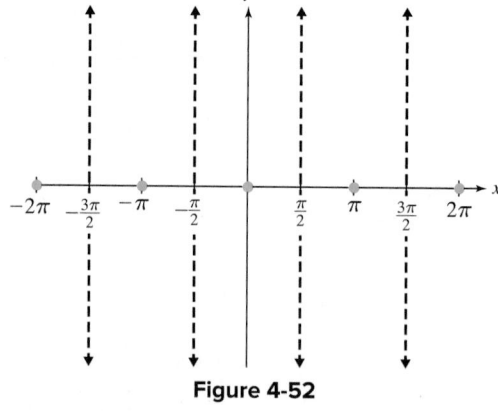

Figure 4-52

The tangent function is an odd function and is symmetric with respect to the origin. Therefore, each Quadrant I point given in Table 4-14 has a mirror image in Quadrant III (Figure 4-53). The period of the tangent function is π, so one complete period occurs between two consecutive vertical asymptotes. We can then sketch more cycles to the left and right as desired.

Answer

2.

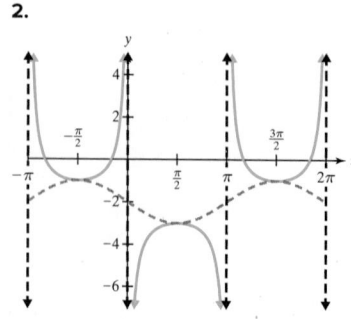

Table 4-14

x	0	$\dfrac{\pi}{6}$	$\dfrac{\pi}{4}$	$\dfrac{\pi}{3}$	$\dfrac{\pi}{2}$
$\tan x$	0	$\dfrac{\sqrt{3}}{3} \approx 0.58$	1	$\sqrt{3} \approx 1.73$	Undefined

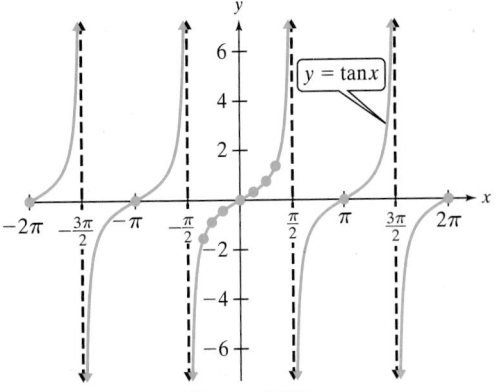

Figure 4-53

We need to use particular care graphing the function near the asymptotes; for example, we can find a value of $\tan x$ arbitrarily large if x is taken sufficiently close to $\frac{\pi}{2}$ from the left. Therefore, $\tan x$ approaches ∞ as x approaches $\frac{\pi}{2}$ from the left (Table 4-15). Likewise, $\tan x$ approaches $-\infty$ as x approaches $\frac{\pi}{2}$ from the right.

As $x \to \left(\frac{\pi}{2}\right)^-$, $\tan x \to \infty$

As $x \to \left(\frac{\pi}{2}\right)^+$, $\tan x \to -\infty$

Table 4-15

x	1.52	1.54	1.56	$\dfrac{\pi}{2}$	1.58	1.60	1.62
$\tan x$	≈ 19.7	≈ 32.5	≈ 92.6	Undefined	≈ -108.6	≈ -34.2	≈ -20.3

To graph the cotangent function, we make the following observations. The graph of $y = \cot x = \dfrac{\cos x}{\sin x}$ will have

- An x-intercept where the numerator equals zero. The value $\cos x = 0$ when x is an odd multiple of $\dfrac{\pi}{2}$; that is, at $x = \dfrac{(2n + 1)\pi}{2}$ for all integers n.
- A vertical asymptote where the denominator equals zero. The value $\sin x = 0$ when x is a multiple of π; that is, at $x = n\pi$ for all integers n.

The cotangent function is an odd function and is symmetric with respect to the origin. Therefore, each Quadrant I point given in Table 4-16 has a mirror image in Quadrant III (Figure 4-54 on page 532). The period of the cotangent function is π, so one complete period occurs between two consecutive vertical asymptotes. We can then sketch more cycles to the left and right as desired.

Table 4-16

x	0	$\dfrac{\pi}{6}$	$\dfrac{\pi}{4}$	$\dfrac{\pi}{3}$	$\dfrac{\pi}{2}$
$\cot x$	Undefined	$\sqrt{3} \approx 1.73$	1	$\dfrac{\sqrt{3}}{3} \approx 0.58$	0

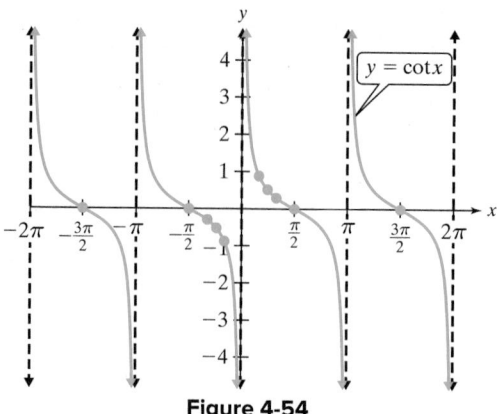

Figure 4-54

Take a few minutes to review the key characteristics of the tangent and cotangent functions and their graphs.

Graphs of the Tangent and Cotangent Functions		
	$y = \tan x$	$y = \cot x$
Domain	$\left\{x \mid x \neq \frac{(2n+1)\pi}{2} \text{ for all integers } n\right\}$	$\{x \mid x \neq n\pi \text{ for all integers } n\}$
Range	All real numbers	All real numbers
Amplitude	None ($y = \tan x$ is unbounded.)	None ($y = \cot x$ is unbounded.)
Period	π	π
Vertical Asymptotes	$x = \frac{(2n+1)\pi}{2}$ $\left(\text{odd multiples of } \frac{\pi}{2}\right)$	$x = n\pi$ (multiples of π)
Symmetry	Origin (The tangent function is an odd function.)	Origin (The cotanget function is an odd function.)

Now consider variations on the tangent and cotangent functions $y = A\tan(Bx - C) + D$ and $y = A\cot(Bx - C) + D$ with $B > 0$. (If $B < 0$, we can rewrite the equations using the odd function property of the tangent and cotangent functions.)

- $|A|$ is the vertical scaling factor.
- If $A < 0$, the graph is reflected across the x-axis.
- The period is $\dfrac{\pi}{B}$.
- The phase shift is $\dfrac{C}{B}$.
- The vertical shift is D.

To graph variations of the tangent or cotangent functions, we graph the functions first without the vertical shift using the guidelines in Table 4-17. The vertical shift is applied once the other transformations are accounted for.

Table 4-17

Guidelines to Graph $y = A\tan(Bx - C)$ and $y = A\cot(Bx - C)$

1. Find an interval between two consecutive vertical asymptotes and graph the asymptotes as dashed lines. For the graph of

   • $y = A\tan(Bx - C)$ and $B > 0$, solve the inequality $-\dfrac{\pi}{2} < Bx - C < \dfrac{\pi}{2}$.

   • $y = A\cot(Bx - C)$ and $B > 0$, solve the inequality $0 < Bx - C < \pi$.

2. Plot an x-intercept halfway between the asymptotes from step 1.

3. Sketch the general shape of the "parent" function between the asymptotes (if $A < 0$, the parent function is reflected across the x-axis). If more accuracy is desired, consider evaluating the function for x values midway between the x-intercept and the asymptotes found in step 1.

   • For $y = \tan x$, the y-coordinates will be $-A$ and A.

   • For $y = \cot x$, the y-coordinates will be A and $-A$.

4. Sketch additional cycles to the right or left as desired.

Classroom Example: p. 537
Exercise 42

EXAMPLE 3   Graphing a Variation of $y = \tan x$

Graph $f(x) = 3\tan\dfrac{\pi}{4}x$.

Solution:

One complete period of the "parent" function $y = \tan x$ occurs between

consecutive asymptotes on the interval $\left(-\dfrac{\pi}{2}, \dfrac{\pi}{2}\right)$. To find the related interval

for $f(x) = 3\tan\dfrac{\pi}{4}x$, solve the following inequality.

> **TIP** In Example 3, we know by inspection that the factor of 3 will stretch the graph of $y = \tan x$ vertically, and the factor of $\frac{\pi}{4}$ will affect the period.

$$-\dfrac{\pi}{2} < \dfrac{\pi}{4}x < \dfrac{\pi}{2} \qquad \text{Multiply by } \tfrac{4}{\pi}.$$

$$\dfrac{4}{\pi} \cdot \left(-\dfrac{\pi}{2}\right) < \dfrac{4}{\pi} \cdot \left(\dfrac{\pi}{4}x\right) < \dfrac{4}{\pi} \cdot \left(\dfrac{\pi}{2}\right)$$

One complete cycle of $f(x) = 3\tan\frac{\pi}{4}x$ occurs between the asymptotes $x = -2$ and $x = 2$.

$$-2 < x < 2$$

vertical asymptote $x = -2$ vertical asymptote $x = 2$

The midpoint of the interval $(-2, 2)$ is $\dfrac{-2 + 2}{2} = 0.$

The function has an x-intercept at $(0, 0)$.

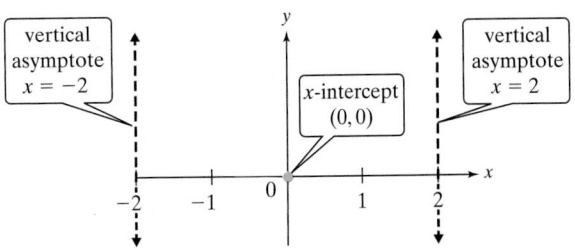

$$f(-1) = 3\tan\left[\frac{\pi}{4}(-1)\right] = 3\tan\left(-\frac{\pi}{4}\right) = 3(-1) = -3$$

$$f(1) = 3\tan\left[\frac{\pi}{4}(1)\right] = 3\tan\left(\frac{\pi}{4}\right) = 3(1) = 3$$

$x = -1$ and $x = 1$ are midway between the x-intercept and the vertical asymptotes.

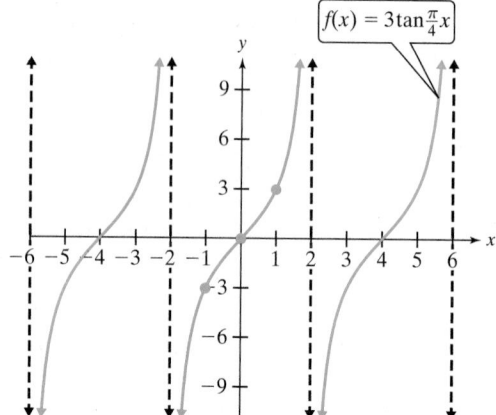

$f(x) = 3\tan\frac{\pi}{4}x$

Skill Practice 3   Graph $y = 4\tan\dfrac{\pi}{2}x$.

If the graph of a tangent or cotangent function involves a vertical shift, we recommend graphing the function without the vertical shift first.

Classroom Example: p. 537
Exercise 58

EXAMPLE 4   **Graphing a Variation of $y = \cot x$**

Graph $y = \cot\left(x + \dfrac{\pi}{4}\right) + 2$.

Solution:

We first graph $g(x) = \cot\left(x + \dfrac{\pi}{4}\right)$.

One complete period of the "parent" function $y = \cot x$ occurs between consecutive asymptotes on the interval $(0, \pi)$. To find the related interval for $g(x) = \cot\left(x + \dfrac{\pi}{4}\right)$, solve the following inequality.

$$0 < x + \frac{\pi}{4} < \pi$$

$$-\frac{\pi}{4} < x < \pi - \frac{\pi}{4}$$ Subtract $\frac{\pi}{4}$.

vertical asymptote $x = -\frac{\pi}{4}$ $-\dfrac{\pi}{4} < x < \dfrac{3\pi}{4}$ vertical asymptote $x = \frac{3\pi}{4}$

One complete cycle of $g(x) = \cot\left(x + \frac{\pi}{4}\right)$ occurs between the asymptotes $x = -\frac{\pi}{4}$ and $x = \frac{3\pi}{4}$.

Answer

3.

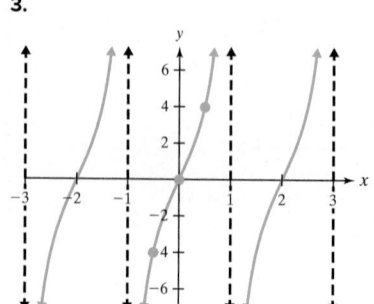

The midpoint of the interval $\left(-\dfrac{\pi}{4}, \dfrac{3\pi}{4}\right)$ is $\dfrac{1}{2}\left(-\dfrac{\pi}{4} + \dfrac{3\pi}{4}\right) = \dfrac{1}{2}\left(\dfrac{2\pi}{4}\right) = \dfrac{\pi}{4}$.

$g(x) = \cot\left(x + \dfrac{\pi}{4}\right)$ has an x-intercept at $\left(\dfrac{\pi}{4}, 0\right)$.

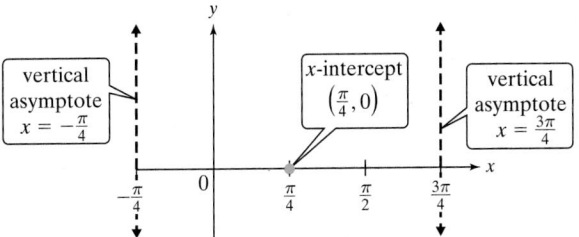

$g(0) = \cot\left[0 + \dfrac{\pi}{4}\right] = \cot\left(\dfrac{\pi}{4}\right) = 1$

$g\left(\dfrac{\pi}{2}\right) = \cot\left[\dfrac{\pi}{2} + \dfrac{\pi}{4}\right] = \cot\left(\dfrac{3\pi}{4}\right) = -1$

If more accuracy in the graph is desired, plot a few more points. The values $x = 0$ and $x = \dfrac{\pi}{2}$ are midway between the x-intercept and the vertical asymptotes.

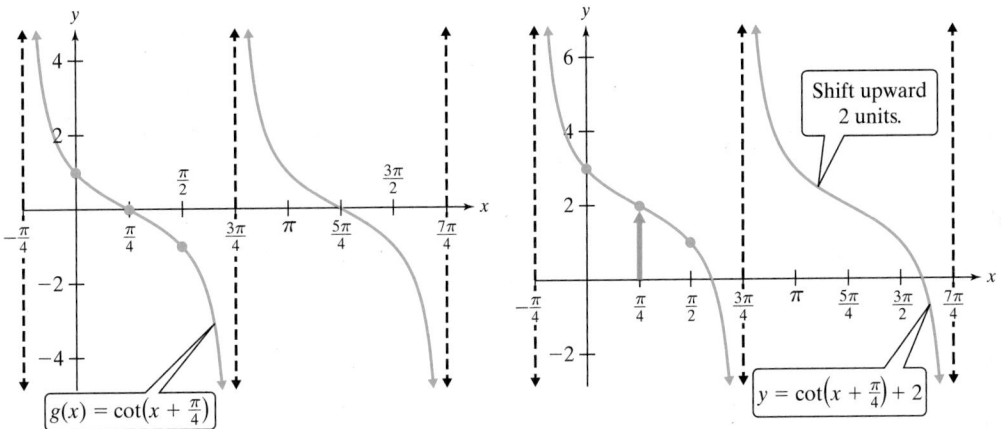

Skill Practice 4 Graph $y = \cot\left(x + \dfrac{\pi}{2}\right) - 3$.

Answer

4. See page SA 34

SECTION 4.6   Practice Exercises

Prerequisite Review

R.1. Write the domain of the function in interval notation.

$f(x) = \dfrac{x^2 - 16}{x - 4}$ $(-\infty, 4) \cup (4, \infty)$

R.3. Determine the vertical asymptotes of the graph of the function.

$f(x) = \dfrac{6}{x - 5}$ $x = 5$

R.2. Refer to the graph of the function to complete the statement.

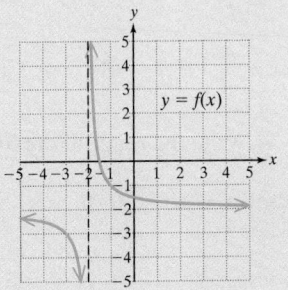

a. As $x \to -2^-, f(x) \to$ _____. $-\infty$

b. As $x \to -2^+, f(x) \to$ _____. ∞

Concept Connections

1. At values of x for which $\sin x = 0$, the graph of $y = \csc x$ will have a ___vertical___ ___asymptote___. This occurs for $x =$ ___$n\pi$___ for all integers n.

2. At values of x for which $\cos x = 0$, the graph of $y = \sec x$ will have a ___vertical___ ___asymptote___. This occurs for x at odd multiples of ___$\dfrac{\pi}{2}$___.

3. The relative maxima on the graph of $y = \sin x$ correspond to the ___relative___ ___minima___ on the graph of $y = \csc x$.

4. The graph of $y = \csc x$ is symmetric with respect to the ___origin___. The graph of $y = \sec x$ is symmetric with respect to the ___y___-axis.

5. If a function is an odd function, then each point (x, y) in Quadrant I will have a corresponding point (___$-x$___, ___$-y$___) in Quadrant ___III___.

6. The range of $y = \tan x$ and $y = \cot x$ is ___all real numbers or $(-\infty, \infty)$ or $\mathbb{R}$___.

7. The graphs of both $y = \tan x$ and $y = \cot x$ are symmetric with respect to the ___origin___.

8. For the functions $y = A\tan(Bx - C)$ and $y = A\cot(Bx - C)$ with $B > 0$, the vertical scaling factor is ___$|A|$___, the period is ___$\dfrac{\pi}{B}$___, and the phase shift is ___$\dfrac{C}{B}$___.

Objective 1: Graph the Secant and Cosecant Functions

9. Sketch the graph of $y = \csc x$ from memory. Use the graph of $y = \sin x$ for reference.

10. Sketch the graph of $y = \sec x$ from memory. Use the graph of $y = \cos x$ for reference.

For Exercises 11–16, identify the statements among a–h that follow directly from the given condition about x.

a. $\csc x$ is undefined.

b. $\sec x$ is undefined.

c. The graph of $y = \sec x$ has a relative maximum at x.

d. The graph of $y = \csc x$ has a relative minimum at x.

e. The graph of $y = \sec x$ has a vertical asymptote.

f. The graph of $y = \csc x$ has a vertical asymptote.

g. The graph of $y = \csc x$ has a relative maximum at x.

h. The graph of $y = \sec x$ has a relative minimum at x.

11. $\sin x = 0$　a, f

12. $\cos x = 0$　b, e

13. The graph of $y = \cos x$ has a relative maximum at x.　h

14. The graph of $y = \sin x$ has a relative minimum at x.　g

15. The graph of $y = \cos x$ has a relative minimum at x.　c

16. The graph of $y = \sin x$ has a relative maximum at x.　d

For Exercises 17–32, graph one period of the function. (See Examples 1–2)

17. $y = 2\csc x$

18. $y = \dfrac{1}{4}\sec x$

19. $y = -5\sec x$

20. $y = -\dfrac{1}{3}\csc x$

21. $y = 3\csc\dfrac{x}{3}$

22. $y = -4\sec\dfrac{x}{2}$

23. $y = \sec 2\pi x$

24. $y = \csc 3\pi x$

25. $y = \csc\left(x - \dfrac{\pi}{4}\right)$

26. $y = \sec\left(x + \dfrac{\pi}{3}\right)$

27. $y = -2\sec(2\pi x + \pi)$

28. $y = -\csc\left(\dfrac{\pi}{3}x + \dfrac{\pi}{2}\right)$

29. $y = 2\csc\left(2x + \dfrac{\pi}{4}\right) + 1$

30. $y = 3\sec\left(x - \dfrac{\pi}{3}\right) - 3$

31. $y = \sec\left(-x - \dfrac{\pi}{4}\right) - 2$

32. $y = -\csc(-x + \pi) + 4$

For Exercises 33–34, write the range of the function in interval notation.

33. a. $y = -4\csc(2x - \pi) + 7$　b. $y = 2\csc\left(\dfrac{\pi}{3}x\right) - 10$　$(-\infty, -12] \cup [-8, \infty)$

　　$(-\infty, 3] \cup [11, \infty)$

34. a. $y = -3\sec(5\pi x) - 1$　b. $y = 5\sec\left(3x - \dfrac{\pi}{2}\right) - 2$　$(-\infty, -7] \cup [3, \infty)$

　　$(-\infty, -4] \cup [2, \infty)$

35. Write a function of the form $y = A \sec Bx$ for the given graph. $y = 2\sec 3x$

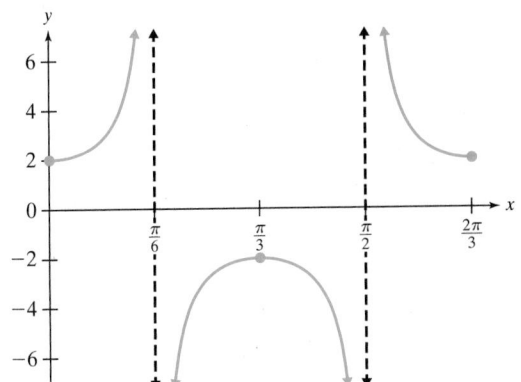

36. Write a function of the form $y = \csc(Bx - C)$ for the given graph. $y = \csc\left(2x - \dfrac{\pi}{4}\right)$

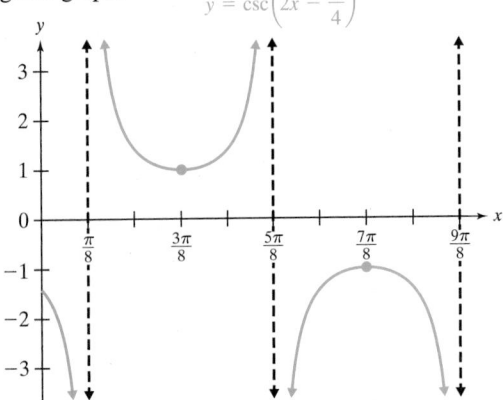

37. A plane flying at an altitude of 5 mi travels on a path directly over a radar tower.

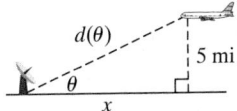

a. Express the distance $d(\theta)$ (in miles) between the plane and the tower as a function of the angle θ in standard position from the tower to the plane. $d(\theta) = 5\csc\theta$

b. If $d(\theta) = 5$, what is the measure of the angle and where is the plane located relative to the tower?
$\theta = 90°$ and the plane is directly overhead.

c. Can the value of θ be π? Explain your answer in terms of the function d.   No. π is not in the domain of d.

38. The distance $d(x)$ (in feet) between an observer 30 ft from a straight highway and a police car traveling down the highway is given by $d(x) = 30\sec x$, where x is the angle (in degrees) between the observer and the police car.

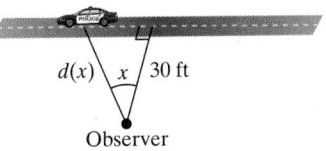

Observer

a. Use a calculator to evaluate $d(x)$ for the given values of x. Round to the nearest foot.

x	45	60	70	80	89
$d(x)$	42	60	88	173	1719

b. Try experimenting with values of x closer to $90°$. What happens as $x \to 90°$?
The distance approaches infinity.

Objective 2: Graph the Tangent and Cotangent Functions

39. a. Graph $y = \tan x$ on the interval $[-\pi, \pi]$.

   b. How many periods of the tangent function are shown on the interval $[-\pi, \pi]$?   2

40. a. Graph $y = \cot x$ on the interval $[-\pi, \pi]$.

   b. How many periods of the cotangent function are shown on the interval $[-\pi, \pi]$?   2

For Exercises 41–42, graph one complete period of the function. Identify the x-intercept and evaluate the function for x values midway between the x-intercept and the asymptotes. (See Example 3)

41. a. $y = \dfrac{1}{2}\tan \pi x$ **42. a.** $y = 2\cot \pi x$

   b. $y = -3\tan \pi x$ **b.** $y = -\dfrac{1}{4}\cot \pi x$

For Exercises 43–58, graph the function. (See Examples 3–4)

43. $y = \tan 2x$ **44.** $y = \cot 3x$ **45.** $y = \cot(-2x)$ **46.** $y = \tan(-x)$

47. $y = \tan\left(\dfrac{1}{2}x\right)$ **48.** $y = \cot\left(\dfrac{1}{3}x\right)$ **49.** $y = 4\cot 2x$ **50.** $y = -3\tan 4x$

51. $y = -2\tan\dfrac{\pi}{3}x$ **52.** $y = 5\cot\dfrac{\pi}{4}x$ **53.** $y = \cot\left(2x + \dfrac{\pi}{3}\right)$ **54.** $y = \tan\left(3x - \dfrac{\pi}{4}\right)$

55. $y = -\tan 3x + 4$ **56.** $y = -\cot 2x - 3$ **57.** $y = 3\tan(\pi x + \pi) - 2$ **58.** $y = 2\cot\left(2\pi x - \dfrac{\pi}{2}\right) + 1$

59. Write a function of the form $y = \tan(Bx - C)$ for the given graph.

$y = \tan\left(x - \dfrac{\pi}{4}\right)$

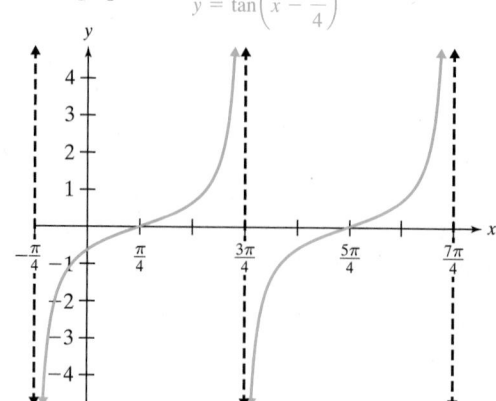

60. Write a function of the form $y = \cot(Bx)$ for the given graph. $y = \cot(4\pi x)$

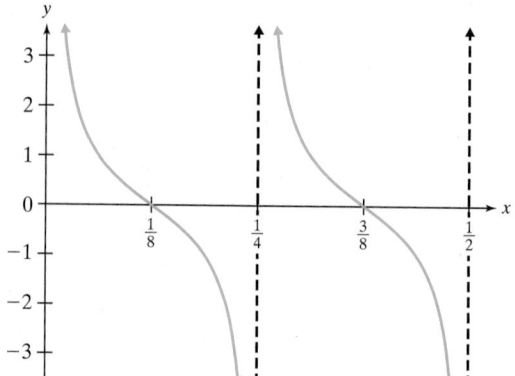

Mixed Exercises

For Exercises 61–64, given $y = f(x)$ and $y = g(x)$,

a. Find $(f \circ g)(x)$ and graph the resulting function.

b. Find $(g \circ f)(x)$ and graph the resulting function.

61. $f(x) = \tan x$ and $g(x) = \dfrac{x}{4}$

62. $f(x) = -2x$ and $g(x) = \cot x$

63. $f(x) = -3x$ and $g(x) = \csc x$

64. $f(x) = \dfrac{x}{3}$ and $g(x) = \sec x$

For Exercises 65–68, complete the statements for the function provided.

65. $f(x) = \tan x$

    a. As $x \to -\dfrac{\pi}{2}^{-}$, $f(x) \to$ _____ ∞

    b. As $x \to -\dfrac{\pi}{2}^{+}$, $f(x) \to$ _____ $-\infty$

66. $f(x) = \cot x$

    a. As $x \to 0^{-}$, $f(x) \to$ _____ $-\infty$

    b. As $x \to 0^{+}$, $f(x) \to$ _____ ∞

67. $f(x) = \csc x$

    a. As $x \to 0^{-}$, $f(x) \to$ _____ $-\infty$

    b. As $x \to 0^{+}$, $f(x) \to$ _____ ∞

68. $f(x) = \sec x$

    a. As $x \to \dfrac{\pi}{2}^{-}$, $f(x) \to$ _____ ∞

    b. As $x \to \dfrac{\pi}{2}^{+}$, $f(x) \to$ _____ $-\infty$

Write About It

69. Explain how to find two consecutive vertical asymptotes of $y = A\tan(Bx - C)$ for $B > 0$.

70. Explain how to find two consecutive vertical asymptotes of $y = A\cot(Bx - C)$ for $B > 0$.

71. Explain how to graph $y = A\sec(Bx - C) + D$.

72. Explain how to graph $y = A\csc(Bx - C) + D$.

Expanding Your Skills

For Exercises 73–76, solve each equation for x on the interval $[0, 2\pi)$.

73. $\tan x = 1$ $\left\{\dfrac{\pi}{4}, \dfrac{5\pi}{4}\right\}$

74. $\sec x = -1$ $\{\pi\}$

75. $\csc x = 1$ $\left\{\dfrac{\pi}{2}\right\}$

76. $\cot x = -1$ $\left\{\dfrac{3\pi}{4}, \dfrac{7\pi}{4}\right\}$

77. Show that the maximum length L (in feet) of a beam that can fit around the corner shown in the figure is $L = 5\sec\theta + 4\csc\theta$.

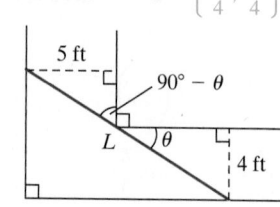

Technology Connections

78. Graph the functions $y = \tan x$ and $y = x + \dfrac{x^3}{3} + \dfrac{2x^5}{15}$

on the interval $\left[-\dfrac{\pi}{2}, \dfrac{\pi}{2}\right]$. How do the functions compare for values of x taken close to 0?

79. Graph the functions $y = \sec x$ and $y = 1 + \dfrac{x^2}{2} + \dfrac{5x^4}{24}$

on the interval $[-\pi, \pi]$. How do the functions compare for values of x taken close to 0?

80. Given $f(x) = x^2$, $g(x) = \tan x$, and $h(x) = \sec x$,

   a. Find $(f \circ h)(x)$. $(f \circ h)(x) = \sec^2 x$

   b. Graph $g(x)$ and $(f \circ h)(x)$ together using the ZTRIG window. The relationship between the two graphs will be studied in calculus. For a given value of x in the domain of $g(x) = \tan x$, $y = \sec^2 x$ gives the slope of a line tangent to g at x.

81. Given $r(x) = -x^2$, $s(x) = \cot x$, and $t(x) = \csc x$,

   a. Find $(r \circ t)(x)$ $(r \circ t)(x) = -\csc^2 x$

   b. Graph $s(x)$ and $(r \circ t)(x)$ together on the ZTRIG window. The relationship between the two graphs will be studied in calculus. For a given value of x in the domain of $s(x) = \cot x$, $y = -\csc^2 x$ gives the slope of a line tangent to s at x.

PROBLEM RECOGNITION EXERCISES

Comparing Graphical Characteristics of Trigonometric Functions

For Exercises 1–16, identify which functions shown here (f, g, h, and so on) have the given characteristics.

$$f(x) = \sin\left(\frac{\pi}{2}x\right) + 3$$

$$g(x) = -3\cos\left(\frac{1}{2}x - \frac{\pi}{4}\right)$$

$$h(x) = 3\sin\left(-\frac{1}{2}x - \frac{\pi}{4}\right)$$

$$k(x) = -3\sec(2x + \pi)$$

$$m(x) = 2\csc\left(2x - \frac{\pi}{2}\right) - 3$$

$$n(x) = 3\tan\left(x - \frac{\pi}{2}\right)$$

$$p(x) = -2\cot\left(\frac{1}{2}x + \pi\right)$$

$$t(x) = -3 + 2\cos x$$

1. Has an amplitude of 3 g, h

2. Has an amplitude of 2 t

3. Has no amplitude k, m, n, p

4. Has a period of 2π p, t

5. Has a period of 4π g, h

6. Has a period of π k, m, n

7. Has a vertical shift upward from the parent graph f

8. Has a vertical shift downward from the parent graph m, t

9. Has no asymptotes f, g, h, t

10. Has no x-intercepts f, k, t

11. Has no y-intercept n, p

12. Has a range of all real numbers n, p

13. Has domain of all real numbers f, g, h, t

14. Has a phase shift of $-\dfrac{\pi}{2}$ h, k

15. Has a phase shift of $\dfrac{\pi}{2}$ g, n

16. Has no phase shift f, t

SECTION 4.7   Inverse Trigonometric Functions

Instructor Note:
Remind students of the difference between an inverse and an inverse *function*. The graph of $x = \sin y$ does not define y as a function of x.

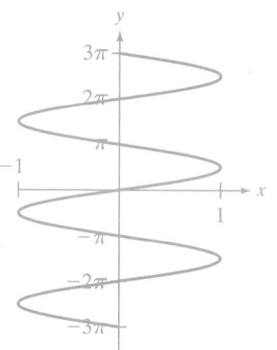

This may help them understand the need for a restricted domain.

1. Evaluate the Inverse Sine Function

Suppose that a yardstick casts a 4-ft shadow when the Sun is at an angle of elevation θ (Figure 4-55). It seems reasonable that we should be able to determine the angle of elevation from the relationship $\tan\theta = \dfrac{3}{4}$.

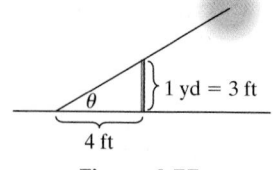

Figure 4-55

Until now, we have always been given an angle and then asked to find the sine, cosine, or tangent of the angle. However, finding the angle of elevation of the Sun from $\tan\theta = \dfrac{3}{4}$ requires that we reverse this process. Given the value of the tangent, we must find an angle that produced it. Therefore, we need to use the inverse of the tangent function.

We begin our study of the inverse trigonometric functions with the inverse of the sine function. First recall that a function must be one-to-one to have an inverse function. From the graph of $y = \sin x$ (Figure 4-56), we see that any horizontal line taken between $-1 \leq y \leq 1$ intersects the graph infinitely many times. Therefore, $y = \sin x$ is not a one-to-one function.

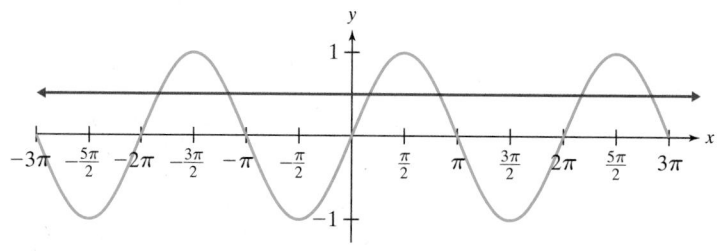

Figure 4-56

However, suppose we restrict the domain of $y = \sin x$ to the interval $-\dfrac{\pi}{2} \leq x \leq \dfrac{\pi}{2}$ (shown in blue in Figure 4-57). The graph of the restricted sine function is one-to-one and contains the entire range of y values $-1 \leq y \leq 1$. The inverse of this restricted sine function is called the inverse sine function and is denoted by $\sin^{-1}$ or arcsin (shown in red in Figure 4-58). Recall that a point (a, b) on the graph of a function f corresponds to the point (b, a) on its inverse. Thus, points on the graph of $y = \sin x$ such as $\left(-\dfrac{\pi}{2}, -1\right)$ and $\left(\dfrac{\pi}{2}, 1\right)$ have their coordinates reversed on the graph of $y = \sin^{-1} x$. Also notice that the graphs of $y = \sin x$ and $y = \sin^{-1} x$ are symmetric with respect to the line $y = x$ as expected.

TIP The horizontal line test indicates that if no horizontal line intersects the graph of a function more than once, the function is one-to-one and has an inverse function. The restricted sine function is one-to-one (shown in blue in Figure 4-57).

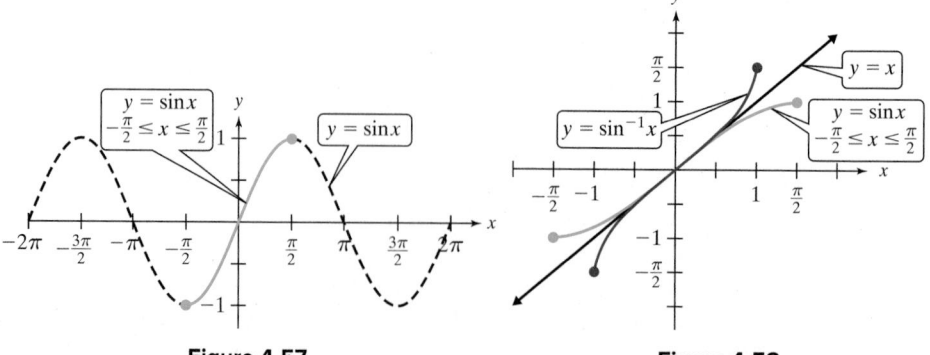

Figure 4-57

Figure 4-58

TIP The double arrow symbol ⇔ means that the statements to the left and right of the arrow are logically equivalent. That is, one statement follows from the other and vice versa.

The Inverse Sine Function

The **inverse sine function** (or arcsine) denoted by $\sin^{-1}$ or arcsin is the inverse of the restricted sine function $y = \sin x$ for $-\dfrac{\pi}{2} \leq x \leq \dfrac{\pi}{2}$. Therefore,

$$y = \sin^{-1} x \quad \Leftrightarrow \quad \sin y = x$$

$$y = \arcsin x \quad \Leftrightarrow \quad \sin y = x$$

where $-1 \leq x \leq 1$ and $-\dfrac{\pi}{2} \leq y \leq \dfrac{\pi}{2}$.

Avoiding Mistakes

The notation $\sin^{-1} x$ represents the inverse of the sine function, not the reciprocal. That is,

$$\sin^{-1} x \neq \frac{1}{\sin x}.$$

- $y = \sin^{-1} x$ is read as "y equals the inverse sine of x" and $y = \arcsin x$ is read as "y equals the arcsine of x."

- To evaluate $y = \sin^{-1} x$ or $y = \arcsin x$ means to find an angle y between $-\dfrac{\pi}{2}$ and $\dfrac{\pi}{2}$, inclusive, whose sine value is x.

Classroom Examples: p. 551
Exercises 8, 10, 12

EXAMPLE 1   Evaluating the Inverse Sine Function

Find the exact values or state that the expression is undefined.

a. $\sin^{-1}\left(-\dfrac{\sqrt{3}}{2}\right)$ **b.** $\arcsin\dfrac{1}{2}$ **c.** $\sin^{-1} 2$

Solution:

Avoiding Mistakes

There are infinitely many values x for which $\sin x = -\dfrac{\sqrt{3}}{2}$, such as $x = \dfrac{4\pi}{3}$, $\dfrac{5\pi}{3}$, $-\dfrac{\pi}{3}$, and $-\dfrac{2\pi}{3}$ to name a few. However, the inverse sine function requires that x be between $-\dfrac{\pi}{2}$ and $\dfrac{\pi}{2}$.

a. Let $y = \sin^{-1}\left(-\dfrac{\sqrt{3}}{2}\right)$.

Then $\sin y = -\dfrac{\sqrt{3}}{2}$ for $-\dfrac{\pi}{2} \leq y \leq \dfrac{\pi}{2}$.

$y = -\dfrac{\pi}{3}$. Therefore $\sin^{-1}\left(-\dfrac{\sqrt{3}}{2}\right) = -\dfrac{\pi}{3}$.

Find an angle y on the interval $\left[-\dfrac{\pi}{2}, \dfrac{\pi}{2}\right]$ such that $\sin y = -\dfrac{\sqrt{3}}{2}$.

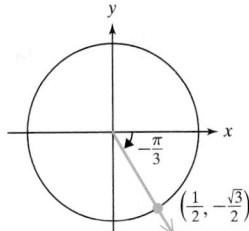

Instructor Note:
Perhaps the most common student mistake when evaluating an inverse trigonometric function is failing to recognize the range of the function. Emphasize this point again and again.

b. Let $y = \arcsin\dfrac{1}{2}$.

Then $\sin y = \dfrac{1}{2}$ for $-\dfrac{\pi}{2} \leq y \leq \dfrac{\pi}{2}$.

$y = \dfrac{\pi}{6}$. Therefore $\arcsin\dfrac{1}{2} = \dfrac{\pi}{6}$.

Find an angle y on the interval $\left[-\dfrac{\pi}{2}, \dfrac{\pi}{2}\right]$ such that $\sin y = \dfrac{1}{2}$.

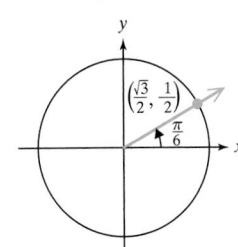

c. Let $y = \sin^{-1} 2$.

To evaluate $y = \sin^{-1} 2$ would mean that we find an angle y such that $\sin y = 2$ for $-\dfrac{\pi}{2} \leq y \leq \dfrac{\pi}{2}$. However, recall that $-1 \leq \sin y \leq 1$ for any angle y. Therefore $\sin^{-1} 2$ is undefined.

> **Skill Practice 1**   Find the exact values or state that the expression is undefined.
>
> **a.** $\sin^{-1}\dfrac{\sqrt{2}}{2}$ **b.** $\arcsin(-1)$ **c.** $\sin^{-1}(-3)$

2. Evaluate the Inverse Cosine and Tangent Functions

The inverse cosine function and the inverse tangent function are defined in a similar way. First, the domain of $y = \cos x$ and $y = \tan x$ must each be restricted to create a one-to-one function on an interval containing all values in the range.

The restricted cosine function is defined on $0 \le x \le \pi$ (Figure 4-59). The graph of the inverse cosine or "arccosine" (denoted by $\cos^{-1}$ or arccos) is shown in red in Figure 4-60.

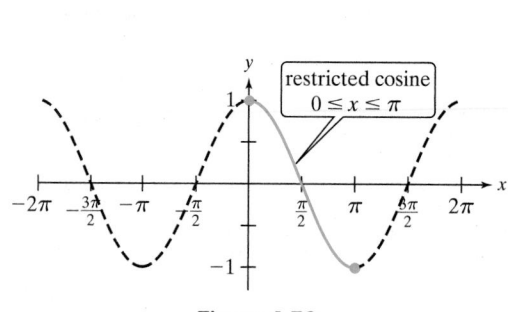

Figure 4-59   **Figure 4-60**

The restricted tangent function is defined on $-\dfrac{\pi}{2} < x < \dfrac{\pi}{2}$ (Figure 4-61). The graph of the inverse tangent or "arctangent" (denoted by $\tan^{-1}$ or arctan) is shown in red in Figure 4-62.

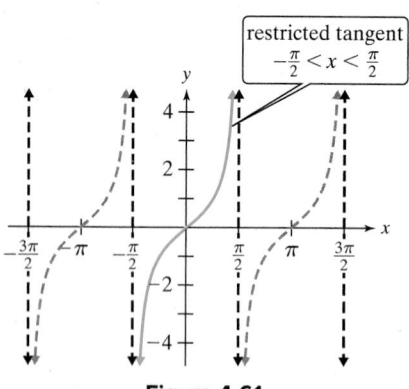
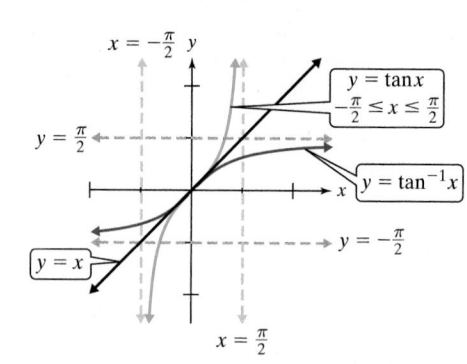

Figure 4-61   **Figure 4-62**

The restricted tangent function has vertical asymptotes at $x = -\dfrac{\pi}{2}$ and $x = \dfrac{\pi}{2}$. The inverse tangent function has horizontal asymptotes at $y = -\dfrac{\pi}{2}$ and $y = \dfrac{\pi}{2}$. We are now ready to summarize the definitions of the inverse functions for sine, cosine, and tangent.

Answers

1. a. $\dfrac{\pi}{4}$ **b.** $-\dfrac{\pi}{2}$ **c.** Undefined

Inverse Trigonometric Functions

Restricted Function	Inverse Function	Graph
$y = \sin x$ $-\dfrac{\pi}{2} \le x \le \dfrac{\pi}{2}$ $-1 \le y \le 1$	The **inverse sine function** (or arcsine), denoted by $\sin^{-1}$ or arcsin, is defined by $y = \sin^{-1} x \iff \sin y = x$ $y = \arcsin x \iff \sin y = x$ $-1 \le x \le 1$ and $-\dfrac{\pi}{2} \le y \le \dfrac{\pi}{2}$	$y = \sin^{-1}x$
$y = \cos x$ $0 \le x \le \pi$ $-1 \le y \le 1$	The **inverse cosine function** (or arccosine), denoted by $\cos^{-1}$ or arccos, is defined by $y = \cos^{-1} x \iff \cos y = x$ $y = \arccos x \iff \cos y = x$ $-1 \le x \le 1$ and $0 \le y \le \pi$	$y = \cos^{-1}x$
$y = \tan x$ $-\dfrac{\pi}{2} < x < \dfrac{\pi}{2}$ $y \in \mathbb{R}$ Vertical asymptotes: $x = -\dfrac{\pi}{2},\ x = \dfrac{\pi}{2}$	The **inverse tangent function** (or arctangent), denoted by $\tan^{-1}$ or arctan, is defined by $y = \tan^{-1} x \iff \tan y = x$ $y = \arctan x \iff \tan y = x$ $x \in \mathbb{R}$ and $-\dfrac{\pi}{2} < y < \dfrac{\pi}{2}$ Horizontal asymptotes: $y = -\dfrac{\pi}{2},\ y = \dfrac{\pi}{2}$	$y = \tan^{-1}x$

Classroom Examples: p. 552
Exercises 18, 22, 24

EXAMPLE 2   Evaluating Inverse Trigonometric Functions

Find the exact values.

a. $\cos^{-1}\left(-\dfrac{1}{2}\right)$ **b.** $\tan^{-1}\sqrt{3}$ **c.** $\arctan(-1)$

Solution:

a. Let $y = \cos^{-1}\left(-\dfrac{1}{2}\right)$.

Then $\cos y = -\dfrac{1}{2}$ for $0 \le y \le \pi$.

$y = \dfrac{2\pi}{3}$. Therefore, $\cos^{-1}\left(-\dfrac{1}{2}\right) = \dfrac{2\pi}{3}$.

Find an angle y on the interval $[0, \pi]$ such that $\cos y = -\dfrac{1}{2}$.

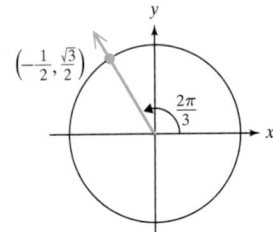

$\left(-\dfrac{1}{2}, \dfrac{\sqrt{3}}{2}\right)$

Avoiding Mistakes

Perhaps the most common error when evaluating the inverse trigonometric functions is to fail to recognize the restrictions on the range. For instance, in Example 2(a), the result of the inverse cosine function must be an angle between 0 and π.

b. Let $y = \tan^{-1}\sqrt{3}$.

Then $\tan y = \sqrt{3}$ for $-\dfrac{\pi}{2} < y < \dfrac{\pi}{2}$.

$y = \dfrac{\pi}{3}$. Therefore, $\tan^{-1}\sqrt{3} = \dfrac{\pi}{3}$.

Find an angle y on the interval $\left(-\frac{\pi}{2}, \frac{\pi}{2}\right)$ such that $\tan y = \sqrt{3}$.

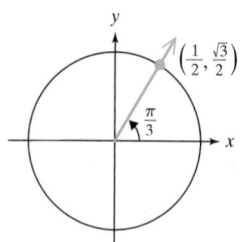

c. Let $y = \arctan(-1)$.

Then $\tan y = -1$ for $-\dfrac{\pi}{2} < y < \dfrac{\pi}{2}$.

$y = -\dfrac{\pi}{4}$. Therefore, $\arctan(-1) = -\dfrac{\pi}{4}$

Find an angle y on the interval $\left(-\frac{\pi}{2}, \frac{\pi}{2}\right)$ such that $\tan y = -1$.

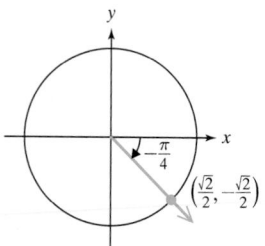

Skill Practice 2 Find the exact values.

a. $\cos^{-1}(-1)$ **b.** $\tan^{-1}\left(-\dfrac{\sqrt{3}}{3}\right)$ **c.** $\arctan 0$

---

3. Approximate Inverse Trigonometric Functions on a Calculator

On a calculator we press the ⓷ⓝⓓ key followed by the **SIN** key, **COS** key, or **TAN** key to invoke $\sin^{-1}$, $\cos^{-1}$, or $\tan^{-1}$. By definition, the values of the inverse trigonometric functions are in radians. However, we often use the inverse functions in applications where the degree measure of an angle is desired. In Example 3, we approximate the values of several inverse trigonometric functions in both radians and degrees.

Classroom Example: p. 552
Exercise 34

Answers

2. **a.** π **b.** $-\dfrac{\pi}{6}$ **c.** 0

3. In radians:

```
NORMAL FLOAT AUTO REAL RADIAN CL
tan⁴(-7.92)
              -1.445198325
cos⁴(2/7)
               1.281044625
sin⁴(-0.81)
               -.9441521152
```

In degrees:

```
NORMAL FLOAT AUTO REAL DEGREE CL
tan⁴(-7.92)
              -82.80376457
cos⁴(2/7)
               73.3984504
sin⁴(-0.81)
               -54.09593142
```

EXAMPLE 3 **Approximating Values of Inverse Functions**

Use a calculator to approximate the function values in both radians and degrees.

a. $\tan^{-1}5.69$ **b.** $\cos^{-1}\left(-\dfrac{3}{8}\right)$ **c.** $\arcsin(-0.6)$

Solution:

Calculator in radian mode   Calculator in degree mode

Skill Practice 3 Use a calculator to approximate the function values in both radians and degrees.

a. $\tan^{-1}(-7.92)$ **b.** $\arccos\dfrac{2}{7}$ **c.** $\sin^{-1}(-0.81)$

We sometimes encounter applications in which we use the inverse trigonometric functions to find the value of an angle where the desired angle is not within the range of the inverse function. In such cases, we need to adjust the output value from the calculator to obtain an angle in the desired quadrant. This is demonstrated in Example 4.

Classroom Examples: p. 552
Exercises 38, 40, 44

EXAMPLE 4   Approximating Angles Based on Characteristics About the Angle

Use a calculator to approximate the degree measure (to 1 decimal place) or radian measure (to 4 decimal places) of the angle θ subject to the given conditions.

a. $\cos \theta = -\dfrac{8}{11}$ and $180° \le \theta \le 270°$ **b.** $\tan \theta = -\dfrac{9}{7}$ and $\dfrac{\pi}{2} < \theta < \pi$

Solution:

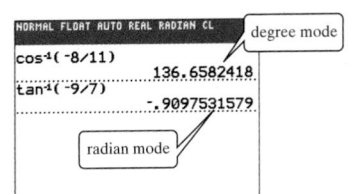

a. $\cos^{-1}\left(-\dfrac{8}{11}\right) \approx 136.7°$

The calculator returns a second quadrant angle. The reference angle is

$$180° - \cos^{-1}\left(-\dfrac{8}{11}\right) \approx 43.3°.$$

The corresponding third quadrant angle is
$\theta \approx 180° + 43.3°$
$ \approx 223.3°.$

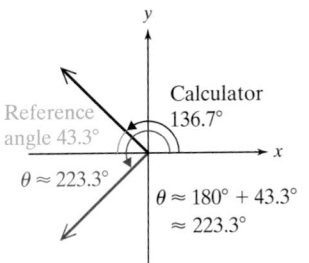

Avoiding Mistakes

The results of Example 4 can be checked on a calculator. Evaluate $\cos(223.3°)$ in degree mode and evaluate $\tan(2.2318)$ in radian mode. The results are approximately equal to $-8/11$ and $-9/7$, respectively.

b. $\tan^{-1}\left(-\dfrac{9}{7}\right) \approx -0.9098$ (radians)

The calculator returns a negative fourth quadrant angle. The reference angle is 0.9098. The corresponding angle in Quadrant II is $\theta \approx \pi - 0.9098 \approx 2.2318$.

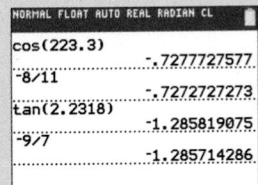

Skill Practice 4   Use a calculator to approximate the degree measure (to 1 decimal place) or radian measure (to 4 decimal places) of the angle θ subject to the given conditions.

a. $\sin \theta = -\dfrac{3}{7}$ and $180° \le \theta \le 270°$ **b.** $\tan \theta = -\dfrac{8}{3}$ and $\dfrac{\pi}{2} < \theta < \pi$

4. Compose Trigonometric Functions and Inverse Trigonometric Functions

Recall from Section 3.1 that inverse functions are defined such that

$$(f \circ f^{-1})(x) = f\left[f^{-1}(x)\right] = x \text{ and } (f^{-1} \circ f)(x) = f^{-1}\left[f(x)\right] = x$$

Answers
4. a. 205.4°   **b.** 1.9296

When composing a trigonometric function with its inverse and vice versa, particular care must be taken regarding their domains.

Composing Trigonometric Functions and Their Inverses

$\sin(\sin^{-1}x) = x$ for $-1 \le x \le 1$ $\qquad \tan(\tan^{-1}x) = x$ for $x \in \mathbb{R}$

$\sin^{-1}(\sin x) = x$ for $-\dfrac{\pi}{2} \le x \le \dfrac{\pi}{2}$ $\qquad \tan^{-1}(\tan x) = x$ for $-\dfrac{\pi}{2} < x < \dfrac{\pi}{2}$

$\cos(\cos^{-1}x) = x$ for $-1 \le x \le 1$

$\cos^{-1}(\cos x) = x$ for $0 \le x \le \pi$

Classroom Examples: p. 552
Exercises 48, 54

EXAMPLE 5 Composing Inverse Trigonometric Functions

Find the exact values.

a. $\sin(\sin^{-1}1)$ $\qquad$ **b.** $\sin^{-1}\left(\sin\dfrac{7\pi}{6}\right)$

Solution:

a. $\sin(\sin^{-1}1) = 1$ $\qquad$ The value $x = 1$ lies in the domain of the inverse sine function. Therefore, we can apply the inverse property $\sin(\sin^{-1}x) = x$.

b. $\sin^{-1}\left(\sin\dfrac{7\pi}{6}\right)$

Since $\dfrac{7\pi}{6}$ is not on the interval $[-\frac{\pi}{2}, \frac{\pi}{2}]$, it is not in the domain of the restricted sine function. Therefore, we cannot conclude that $\sin^{-1}(\sin x) = x$.

$= \sin^{-1}\left[\sin\left(-\dfrac{\pi}{6}\right)\right]$ $\qquad$ Rewrite $\sin\frac{7\pi}{6}$ as an equivalent expression with an angle on the interval $[-\frac{\pi}{2}, \frac{\pi}{2}]$.

$= -\dfrac{\pi}{6}$ $\qquad$ Apply the property $\sin^{-1}(\sin x) = x$.

> **TIP** As an alternative method, we can simplify the expressions from Example 5 by applying the order of operations.
>
> $\sin[\sin^{-1}(1)] = \sin\dfrac{\pi}{2} = 1$
>
> and
>
> $\sin^{-1}\left(\sin\dfrac{7\pi}{6}\right) = \sin^{-1}\left(-\dfrac{1}{2}\right)$
>
> $= -\dfrac{\pi}{6}$

Skill Practice 5 Find the exact values.

a. $\cos[\cos^{-1}(-1)]$ $\qquad$ **b.** $\cos^{-1}\left(\cos\dfrac{4\pi}{3}\right)$

Classroom Example: p. 553
Exercise 68

EXAMPLE 6 Composing Trigonometric Functions and Inverse Trigonometric Functions

Find the exact value of $\cos\left[\tan^{-1}\left(-\dfrac{8}{15}\right)\right]$.

Solution:

Let $\theta = \tan^{-1}\left(-\dfrac{8}{15}\right)$. Since $-\dfrac{8}{15} < 0$, angle θ is on the interval $\left(-\dfrac{\pi}{2}, 0\right)$. Let $P(15, -8)$ be a point on the terminal side of θ. Then, $r = \sqrt{(15)^2 + (-8)^2} = \sqrt{289} = 17$.

$\cos\left[\tan^{-1}\left(-\dfrac{8}{15}\right)\right] = \cos\theta = \dfrac{15}{17}$

Answers

5. a. -1 $\qquad$ **b.** $\dfrac{2\pi}{3}$

Skill Practice 6 Find the exact value of $\sin\left[\tan^{-1}\left(\frac{12}{5}\right)\right]$.

Avoiding Mistakes

You can confirm your answer from Example 6 on a calculator by applying the order of operations.

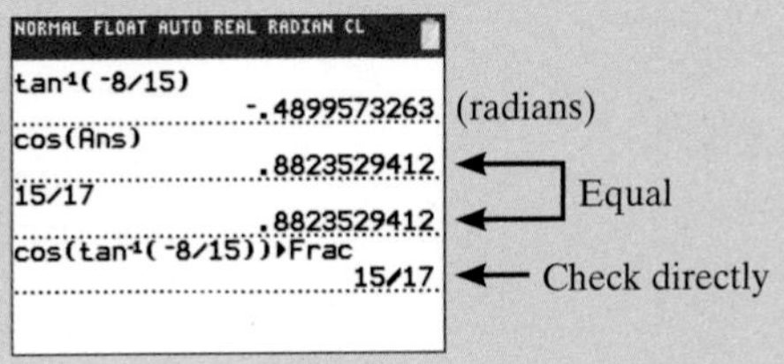
NORMAL FLOAT AUTO REAL RADIAN CL
tan⁻¹(-8/15)
-.4899573263
(radians)
cos(Ans)
.8823529412
15/17
.8823529412
Equal
cos(tan⁻¹(-8/15))▸Frac
15/17
Check directly

Classroom Example: p. 553 Exercise 62

EXAMPLE 7 Composing Trigonometric Functions and Inverse Trigonometric Functions

Find the exact value of $\sin\left[\cos^{-1}\left(-\frac{3}{7}\right)\right]$.

Solution:

Let $\theta = \cos^{-1}\left(-\frac{3}{7}\right)$. Since $-\frac{3}{7} < 0$, angle θ is on the interval $\left(\frac{\pi}{2}, \pi\right)$. Let $P(-3, y)$ be a point on the terminal side of θ. From Figure 4-63:

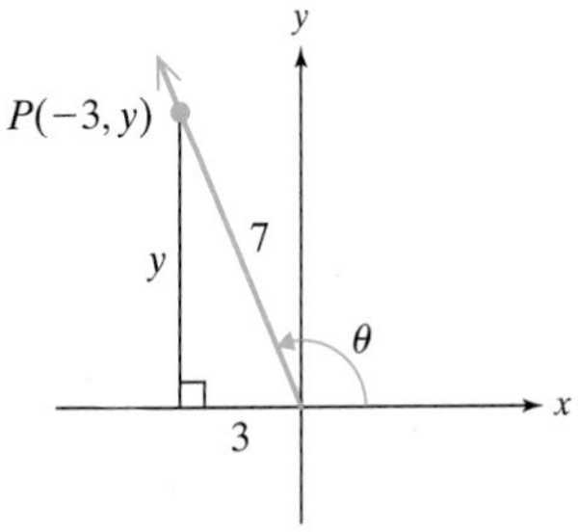
y
P(−3, y)
7
y
θ
3
x

Figure 4-63

$$(-3)^2 + y^2 = 7^2$$
$$9 + y^2 = 49$$
$$y^2 = 40$$
$$y = \sqrt{40} = 2\sqrt{10}$$

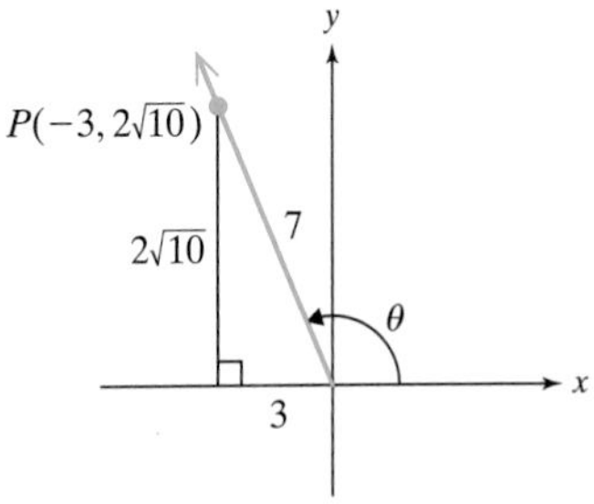
y
P(−3, 2√10)
7
2√10
θ
3
x

Therefore, $\sin\left[\cos^{-1}\left(-\frac{3}{7}\right)\right] = \sin\theta = \frac{2\sqrt{10}}{7}$.

Skill Practice 7 Find the exact value of $\cos\left[\sin^{-1}\left(-\frac{2}{11}\right)\right]$.

In Example 8, we use a right triangle to visualize the composition of a trigonometric function and an inverse function, and then write an equivalent algebraic expression. This skill is used often in calculus.

Classroom Example: p. 553 Exercise 72

Answers

6. $\frac{12}{13}$

7. $\frac{\sqrt{117}}{11}$

EXAMPLE 8 Writing a Trigonometric Expression as an Algebraic Expression

Write the expression $\tan\left(\sin^{-1}\frac{\sqrt{x^2 - 16}}{x}\right)$ as an algebraic expression for $x > 4$.

Solution:

Let $\theta = \sin^{-1}\dfrac{\sqrt{x^2 - 16}}{x}$. Since $x > 4$, $\dfrac{\sqrt{x^2 - 16}}{x} > 0$,

and we know that θ is an acute angle. We can set up a representative right triangle using the relationship

$$\sin\theta = \frac{\text{opp}}{\text{hyp}} = \frac{\sqrt{x^2 - 16}}{x}$$

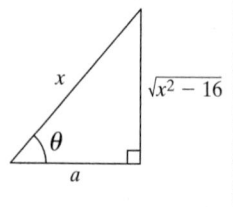

To find an expression for the adjacent side a, apply the Pythagorean theorem.

$$a^2 + \left(\sqrt{x^2 - 16}\right)^2 = x^2$$
$$a^2 + x^2 - 16 = x^2$$
$$a^2 = 16$$
$$a = 4$$

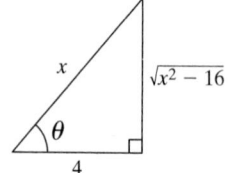

Therefore, $\tan\left(\sin^{-1}\dfrac{\sqrt{x^2 - 16}}{x}\right) = \tan\theta = \dfrac{\sqrt{x^2 - 16}}{4}$.

Skill Practice 8 Write the expression $\tan\left(\sin^{-1}\dfrac{x}{\sqrt{x^2 + 9}}\right)$ as an algebraic expression for $x > 0$.

---

5. Apply Inverse Trigonometric Functions

We now revisit the scenario that we used to introduce this section. Example 9 shows how inverse trigonometric functions can help us find the measure of an unknown angle in an application.

Classroom Example: p. 553
Exercise 80

EXAMPLE 9 Applying an Inverse Trigonometric Function

A yardstick casts a 4-ft shadow when the Sun is at an angle of elevation θ. Determine the angle of elevation of the Sun to the nearest tenth of a degree.

Solution:

The tip of the shadow and the points at the top and bottom of the yardstick form a right triangle with θ representing the angle of elevation of the Sun.

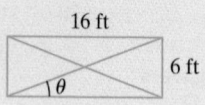

$$\tan\theta = \frac{3}{4}$$

$\theta = \tan^{-1}\dfrac{3}{4}$ Use the inverse tangent to find the acute angle θ whose tangent is $\frac{3}{4}$.

$\theta \approx 36.9°$ Be sure that your calculator is in degree mode.

Skill Practice 9 For the construction of a house, a 16-ft by 6-ft wooden frame is made. Find the angle that the diagonal beam makes with the base of the frame. Round to the nearest tenth of a degree.

Answers

8. $\dfrac{x}{3}$ **9.** 20.6°

6. Evaluate the Inverse Secant, Cosecant, and Cotangent Functions

We need to establish intervals over which the secant, cosecant, and cotangent functions are one-to-one before we can define their corresponding inverse functions. These intervals are chosen to accommodate the entire range of function values and to make the restricted functions one-to-one. The graphs of the restricted secant, cosecant, and cotangent functions and their inverses are shown in Table 4-18.

Table 4-18

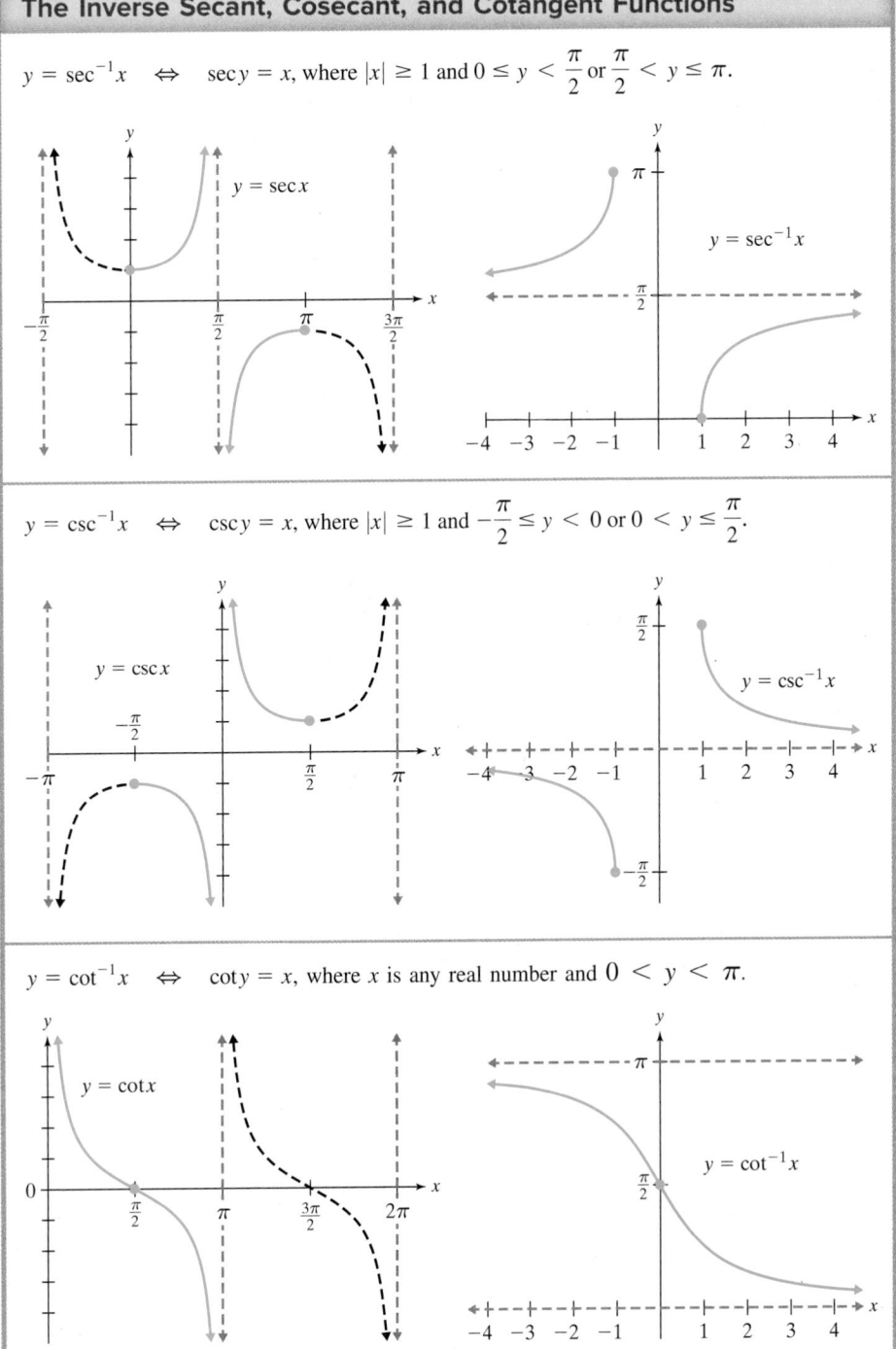

| **TIP** | It is important to note that there are infinitely many intervals over which the secant, cosecant, and cotangent functions can be restricted to define the inverse functions. These restrictions are not universally agreed upon. |

The Inverse Secant, Cosecant, and Cotangent Functions

$y = \sec^{-1}x \iff \sec y = x$, where $|x| \geq 1$ and $0 \leq y < \dfrac{\pi}{2}$ or $\dfrac{\pi}{2} < y \leq \pi$.

$y = \csc^{-1}x \iff \csc y = x$, where $|x| \geq 1$ and $-\dfrac{\pi}{2} \leq y < 0$ or $0 < y \leq \dfrac{\pi}{2}$.

$y = \cot^{-1}x \iff \cot y = x$, where x is any real number and $0 < y < \pi$.

A calculator does not have keys for inverse secant, cosecant, or cotangent. Therefore, to evaluate an inverse secant, cosecant, or cotangent on a calculator, we first rewrite the expression using inverse sine or cosine.

Classroom Examples: p. 554
Exercises 96, 100

EXAMPLE 10   Approximating the Value of an Inverse Secant, Cosecant, or Cotangent

Approximate each expression in radians, rounded to 4 decimal places.

a. $\csc^{-1} 3$ **b.** $\cot^{-1}\left(-\dfrac{3}{4}\right)$

Solution:

a. Let $\theta = \csc^{-1} 3$. Then $\csc\theta = 3$ for $-\dfrac{\pi}{2} \le \theta \le \dfrac{\pi}{2}, \theta \ne 0$.

Since $y = \sin^{-1} x$ has the same range as $y = \csc^{-1} x$ for $x \ne 0$, we have

$\csc\theta = 3$ implies that $\dfrac{1}{\sin\theta} = 3$ and that $\sin\theta = \dfrac{1}{3}$.

Therefore, $\theta = \sin^{-1}\dfrac{1}{3} \approx 0.3398$.

b. Let $\theta = \cot^{-1}\left(-\dfrac{3}{4}\right)$. Then $\cot\theta = -\dfrac{3}{4}$ for $0 < \theta < \pi$. Furthermore, since

the argument $-\dfrac{3}{4}$ is negative, $\theta = \cot^{-1}\left(-\dfrac{3}{4}\right)$ must be a second quadrant angle.

The related expression using the inverse tangent function will *not* return a second quadrant angle but rather an angle on the

interval $-\dfrac{\pi}{2} < \theta < 0$. Therefore, we will rewrite

the expression $\cot\theta = -\dfrac{3}{4}$ using the cosine function.

From Figure 4-64, $\cos\theta = -\dfrac{3}{5}$. Therefore,

$\theta = \cos^{-1}\left(-\dfrac{3}{5}\right) \approx 2.2143.$

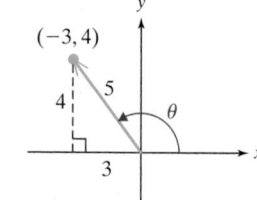

Figure 4-64

Skill Practice 10   Approximate each expression in radians, rounded to 4 decimal places.

a. $\sec^{-1}(-4)$ **b.** $\cot^{-1}\left(-\dfrac{5}{12}\right)$

Answers
10. a. 1.8235   **b.** 1.9656

SECTION 4.7   Practice Exercises

Prerequisite Review

R.1. Determine if the relation defines y as a one-to-one function of x.   Yes

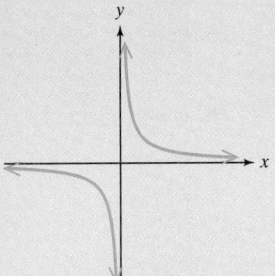

R.2. Determine if the relation defines y as a one-to-one function of x.   No

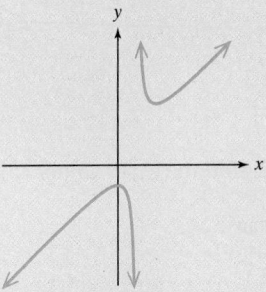

R.3. A one-to-one function is given. Write an equation for the inverse function. $g^{-1}(x) = 3 - 4x$

$$g(x) = \frac{3 - x}{4}$$

R.4. The graph of a function is given. Graph the inverse function.

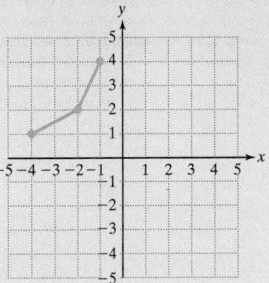

Concept Connections

1. A function must be _____one-to-one_____ on its entire domain to have an inverse function.

2. If $\left(\dfrac{2\pi}{3}, -\dfrac{1}{2}\right)$ is on the graph of $y = \cos x$, what is the related point on $y = \cos^{-1} x$? $\left(-\dfrac{1}{2}, \dfrac{2\pi}{3}\right)$

3. The graph of $y = \tan^{-1} x$ has two ___horizontal___ (horizontal/vertical) asymptotes represented by the equations ___$y = -\dfrac{\pi}{2}$___ and ___$y = \dfrac{\pi}{2}$___.

4. The domain of $y = \arctan x$ is _____$\mathbb{R}$_____. The output is a real number (or angle in radians) between ___$-\dfrac{\pi}{2}$___ and ___$\dfrac{\pi}{2}$___.

5. In interval notation, the domain of $y = \cos^{-1} x$ is ___$[-1, 1]$___. The output is a real number (or angle in radians) between ___0___ and ___π___, inclusive.

6. In interval notation, the domain of $y = \sin^{-1} x$ is ___$[-1, 1]$___. The output is a real number (or angle in radians) between ___$-\dfrac{\pi}{2}$___ and ___$\dfrac{\pi}{2}$___, inclusive.

Objective 1: Evaluate the Inverse Sine Function

For Exercises 7–12, find the exact value or state that the expression is undefined. (See Example 1)

7. $\arcsin \dfrac{\sqrt{2}}{2}$ $\dfrac{\pi}{4}$

8. $\sin^{-1}\left(-\dfrac{1}{2}\right)$ $-\dfrac{\pi}{6}$

9. $\sin^{-1} \pi$ Undefined

10. $\sin^{-1}\dfrac{3}{2}$ Undefined

11. $\arcsin\left(-\dfrac{\sqrt{2}}{2}\right)$ $-\dfrac{\pi}{4}$

12. $\arcsin \dfrac{\sqrt{3}}{2}$ $\dfrac{\pi}{3}$

For Exercises 13–16, find the exact value.

13. $\sin^{-1}\dfrac{\sqrt{3}}{2} + \sin^{-1}\dfrac{1}{2}$ $\dfrac{\pi}{2}$ **14.** $\sin^{-1}\dfrac{\sqrt{2}}{2} - \sin^{-1}(-1)$ $\dfrac{3\pi}{4}$ **15.** $2\sin^{-1}\left(-\dfrac{\sqrt{2}}{2}\right) - \dfrac{\pi}{3}$ $-\dfrac{5\pi}{6}$ **16.** $\dfrac{\pi}{2} + 3\sin^{-1}\left(-\dfrac{\sqrt{3}}{2}\right)$ $-\dfrac{\pi}{2}$

Objective 2: Evaluate the Inverse Cosine and Tangent Functions

For Exercises 17–28, find the exact value or state that the expression is undefined. (See Example 2)

17. $\arccos\left(-\dfrac{\sqrt{2}}{2}\right)$ $\dfrac{3\pi}{4}$

18. $\tan^{-1}\left(-\sqrt{3}\right)$ $-\dfrac{\pi}{3}$

19. $\cos^{-1}0$ $\dfrac{\pi}{2}$

20. $\tan^{-1}0$ 0

21. $\cos^{-1}(-2)$ Undefined

22. $\arccos\dfrac{4}{3}$ Undefined

23. $\arctan\dfrac{\sqrt{3}}{3}$ $\dfrac{\pi}{6}$

24. $\cos^{-1}\dfrac{\sqrt{2}}{2}$ $\dfrac{\pi}{4}$

25. $\tan^{-1}1$ $\dfrac{\pi}{4}$

26. $\arctan\left(-\dfrac{\sqrt{3}}{3}\right)$ $-\dfrac{\pi}{6}$

27. $\arccos\dfrac{\sqrt{3}}{2}$ $\dfrac{\pi}{6}$

28. $\cos^{-1}\left(-\dfrac{\sqrt{3}}{2}\right)$ $\dfrac{5\pi}{6}$

For Exercises 29–32, find the exact value.

29. $\tan^{-1}(-1) + \tan^{-1}\sqrt{3}$ $\dfrac{\pi}{12}$

30. $\cos^{-1}\dfrac{\sqrt{2}}{2} + \cos^{-1}\dfrac{1}{2}$ $\dfrac{7\pi}{12}$

31. $2\cos^{-1}\left(-\dfrac{\sqrt{3}}{2}\right) - \tan^{-1}\dfrac{\sqrt{3}}{3}$ $\dfrac{3\pi}{2}$

32. $3\tan^{-1}1 + \cos^{-1}\left(-\dfrac{\sqrt{2}}{2}\right)$ $\dfrac{3\pi}{2}$

Objective 3: Approximate Inverse Trigonometric Functions on a Calculator

For Exercises 33–36, use a calculator to approximate the function values in both radians and degrees. (See Example 3)

33. a. $\cos^{-1}\dfrac{3}{8}$

  b. $\tan^{-1}25$

  c. $\arcsin 0.05$

34. a. $\sin^{-1}0.93$

  b. $\arccos 0.17$

  c. $\arctan\dfrac{7}{4}$

35. a. $\tan^{-1}(-28)$

  b. $\arccos\dfrac{\sqrt{3}}{5}$

  c. $\sin^{-1}(-0.14)$

36. a. $\cos^{-1}(-0.75)$

  b. $\tan^{-1}\dfrac{8}{3}$

  c. $\arcsin\dfrac{\pi}{7}$

For Exercises 37–46, use a calculator to approximate the degree measure (to 1 decimal place) or radian measure (to 4 decimal places) of the angle θ subject to the given conditions. (See Example 4)

37. $\cos\theta = -\dfrac{5}{6}$ and $180° < \theta < 270°$ $213.6°$

38. $\sin\theta = -\dfrac{4}{5}$ and $180° < \theta < 270°$ $233.1°$

39. $\tan\theta = -\dfrac{12}{5}$ and $90° < \theta < 180°$ $112.6°$

40. $\cos\theta = -\dfrac{2}{13}$ and $180° < \theta < 270°$ $261.2°$

41. $\sin\theta = \dfrac{12}{19}$ and $90° < \theta < 180°$ $140.8°$

42. $\tan\theta = \dfrac{7}{15}$ and $180° < \theta < 270°$ $205.0°$

43. $\cos\theta = -\dfrac{5}{8}$ and $\pi < \theta < \dfrac{3\pi}{2}$ 4.0373

44. $\tan\theta = -\dfrac{9}{5}$ and $\dfrac{\pi}{2} < \theta < \pi$ 2.0779

45. $\sin\theta = -\dfrac{17}{20}$ and $\pi < \theta < \dfrac{3\pi}{2}$ 4.1576

46. $\cos\theta = \dfrac{1}{17}$ and $\dfrac{3\pi}{2} < \theta < 2\pi$ 4.7712

Objective 4: Compose Trigonometric Functions and Inverse Trigonometric Functions

For Exercises 47–58, find the exact values. (See Example 5)

47. $\sin\left(\sin^{-1}\dfrac{\sqrt{2}}{2}\right)$ $\dfrac{\sqrt{2}}{2}$

48. $\arcsin\left(\sin\dfrac{5\pi}{3}\right)$ $-\dfrac{\pi}{3}$

49. $\sin^{-1}\left(\sin\dfrac{5\pi}{4}\right)$ $-\dfrac{\pi}{4}$

50. $\sin\left[\sin^{-1}\left(-\dfrac{1}{2}\right)\right]$ $-\dfrac{1}{2}$

51. $\cos\left(\cos^{-1}\dfrac{2}{3}\right)$ $\dfrac{2}{3}$

52. $\arccos\left(\cos\dfrac{11\pi}{6}\right)$ $\dfrac{\pi}{6}$

53. $\cos^{-1}\left(\cos\dfrac{4\pi}{3}\right)$ $\dfrac{2\pi}{3}$

54. $\cos\left[\cos^{-1}\left(-\dfrac{1}{2}\right)\right]$ $-\dfrac{1}{2}$

55. $\tan^{-1}\left(\tan\dfrac{2\pi}{3}\right)$ $-\dfrac{\pi}{3}$

56. $\tan\left(\tan^{-1}2\right)$ 2

57. $\tan\left[\arctan(-\pi)\right]$ $-\pi$

58. $\tan^{-1}\left[\tan\left(-\dfrac{\pi}{6}\right)\right]$ $-\dfrac{\pi}{6}$

For Exercises 59–70, find the exact values. (See Examples 6–7)

59. $\cos\left(\tan^{-1}\dfrac{\sqrt{3}}{3}\right)$ $\dfrac{\sqrt{3}}{2}$

60. $\sin\left(\cos^{-1}\dfrac{1}{2}\right)$ $\dfrac{\sqrt{3}}{2}$

61. $\tan\left[\sin^{-1}\left(-\dfrac{2}{3}\right)\right]$ $-\dfrac{2\sqrt{5}}{5}$

62. $\sin\left[\cos^{-1}\left(-\dfrac{2}{3}\right)\right]$ $\dfrac{\sqrt{5}}{3}$

63. $\sin\left(\cos^{-1}\dfrac{3}{4}\right)$ $\dfrac{\sqrt{7}}{4}$

64. $\sin\left(\tan^{-1}\dfrac{4}{3}\right)$ $\dfrac{4}{5}$

65. $\sin\left[\tan^{-1}(-1)\right]$ $-\dfrac{\sqrt{2}}{2}$

66. $\cos\left[\tan^{-1}(-1)\right]$ $\dfrac{\sqrt{2}}{2}$

67. $\cos\left[\sin^{-1}\left(-\dfrac{2}{7}\right)\right]$ $\dfrac{3\sqrt{5}}{7}$

68. $\cos\left[\tan^{-1}\left(-\dfrac{5}{12}\right)\right]$ $\dfrac{12}{13}$

69. $\tan\left[\cos^{-1}\left(-\dfrac{5}{6}\right)\right]$ $-\dfrac{\sqrt{11}}{5}$

70. $\tan\left[\sin^{-1}\left(-\dfrac{\sqrt{5}}{3}\right)\right]$ $-\dfrac{\sqrt{5}}{2}$

For Exercises 71–76, write the given expression as an algebraic expression. It is not necessary to rationalize the denominator. (See Example 8)

71. $\cos\left(\sin^{-1}\dfrac{x}{\sqrt{25+x^2}}\right)$ for $x > 0$. $\dfrac{5}{\sqrt{25+x^2}}$

72. $\cot\left(\cos^{-1}\dfrac{\sqrt{x^2-1}}{x}\right)$ for $x > 1$. $\sqrt{x^2-1}$

73. $\sin(\tan^{-1}x)$ for $x > 0$. $\dfrac{x}{\sqrt{x^2+1}}$

74. $\tan(\sin^{-1}x)$ for $|x| < 1$. $\dfrac{x}{\sqrt{1-x^2}}$

75. $\tan\left(\cos^{-1}\dfrac{3}{x}\right)$ for $x > 3$. $\dfrac{\sqrt{x^2-9}}{3}$

76. $\sin\left(\cos^{-1}\dfrac{\sqrt{x^2-25}}{x}\right)$ for $x > 5$. $\dfrac{5}{x}$

Objective 5: Apply Inverse Trigonometric Functions

77. To meet the requirements of the Americans with Disabilities Act (ADA) a wheelchair ramp must have a slope of 1:12 or less. That is, for every 1 in. of "rise," there must be at least 12 in. of "run." **(See Example 9)**

   a. If a wheelchair ramp is constructed with the maximum slope, what angle does the ramp make with the ground? Round to the nearest tenth of a degree.   4.8°

   b. If the ramp is 22 ft long, how much elevation does the ramp provide? Round to the nearest tenth of a foot.   1.8 ft

78. A student measures the length of the shadow of the Washington Monument to be 620 ft. If the Washington Monument is 555 ft tall, approximate the angle of elevation of the Sun to the nearest tenth of a degree.   41.8°

79. A balloon advertising an open house is stabilized by two cables of lengths 20 ft and 40 ft tethered to the ground. If the perpendicular distance from the balloon to the ground is $10\sqrt{3}$ ft, what is the degree measure of the angle each cable makes with the ground? Round to the nearest tenth of a degree if necessary.

The short cable makes an angle of 60° with the ground, and the long cable makes an angle of 25.7° with the ground.

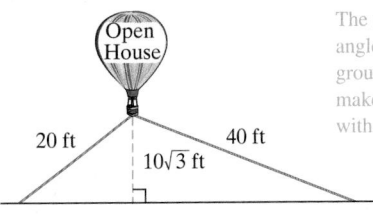

80. A group of campers hikes down a steep path. One member of the group has an altimeter on his watch to measure altitude. If the path is 1250 yd and the amount of altitude lost is 480 yd, what is the angle of incline? Round to the nearest tenth of a degree.   22.6°

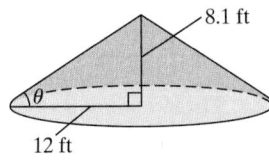

81. Navajo Tube Hill, a snow tubing hill in Utah, is 550 ft long and has a 75-ft vertical drop. Find the angle of incline of the hill. Round to the nearest tenth of a degree.   7.8°

82. A ski run on Giant Steps Mountain in Utah is 1475 m long. The difference in altitude from the beginning to the end of the run is 350 m. Find the angle of the ski run. Round to the nearest tenth of a degree.   13.7°

83. When granular material such as sand or gravel is poured onto a horizontal surface it forms a right circular cone. The angle that the surface of the cone makes with the horizontal is called the **angle of repose**. The angle of repose depends on a number of variables such as the shape of the particles and the amount of friction between them. "Stickier" particles have a greater angle of repose, and "slippery" particles have a smaller angle of repose. Find the angle of repose for the pile of dry sand. Round to the nearest degree.   34°

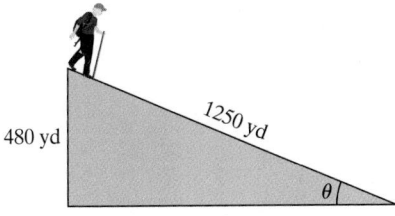

84. For the construction of a bookcase, a 5-ft by 3-ft wooden frame is made with a cross-brace in the back for stability. Find the angle that the diagonal brace makes with the base of the frame. Round to the nearest tenth of a degree.   59.0°

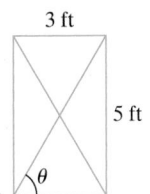

Objective 6: Evaluate the Inverse Secant, Cosecant, and Cotangent Functions

85. Show that $\sec^{-1}x = \cos^{-1}\dfrac{1}{x}$ for $x \geq 1$.

86. Show that $\csc^{-1}x = \sin^{-1}\dfrac{1}{x}$ for $x \geq 1$.

87. Show that $\sec^{-1}x + \csc^{-1}x = \dfrac{\pi}{2}$ for $x \geq 1$.

88. Complete the table, giving the domain and range in interval notation for each inverse function.

Inverse Function	Domain	Range
$y = \sin^{-1}x$		
$y = \csc^{-1}x$		
$y = \cos^{-1}x$		
$y = \sec^{-1}x$		
$y = \tan^{-1}x$		
$y = \cot^{-1}x$		

For Exercises 89–94, find the exact values.

89. $\sec^{-1}\dfrac{2\sqrt{3}}{3}$ $\dfrac{\pi}{6}$

90. $\sec^{-1}(-\sqrt{2})$ $\dfrac{3\pi}{4}$

91. $\csc^{-1}(-1)$ $-\dfrac{\pi}{2}$

92. $\csc^{-1}(2)$ $\dfrac{\pi}{6}$

93. $\cot^{-1}\sqrt{3}$ $\dfrac{\pi}{6}$

94. $\cot^{-1}(1)$ $\dfrac{\pi}{4}$

For Exercises 95–100, use a calculator to approximate each expression in radians, rounded to 4 decimal places. (See Example 10)

95. $\sec^{-1}\dfrac{7}{4}$ 0.9626

96. $\csc^{-1}\dfrac{6}{5}$ 0.9851

97. $\csc^{-1}(-8)$ -0.1253

98. $\sec^{-1}(-6)$ 1.7382

99. $\cot^{-1}\left(-\dfrac{8}{15}\right)$ 2.0608

100. $\cot^{-1}\left(-\dfrac{24}{7}\right)$ 2.8578

Mixed Exercises

For Exercises 101–104, find the exact value if possible. Otherwise find an approximation to 4 decimal places or state that the expression is undefined.

101. a. $\sin\dfrac{\pi}{4}$ $\dfrac{\sqrt{2}}{2}$

b. $\sin^{-1}\dfrac{\pi}{4}$ 0.9033

c. $\sin^{-1}\dfrac{\sqrt{2}}{2}$ $\dfrac{\pi}{4}$

102. a. $\cos\dfrac{2\pi}{3}$ $-\dfrac{1}{2}$

b. $\cos^{-1}\dfrac{2\pi}{3}$ Undefined

c. $\cos^{-1}\left(-\dfrac{1}{2}\right)$ $\dfrac{2\pi}{3}$

103. a. $\cos^{-1}\left(-\dfrac{\pi}{6}\right)$ 2.1219

b. $\cos\left(-\dfrac{\pi}{6}\right)$ $\dfrac{\sqrt{3}}{2}$

c. $\cos^{-1}\left(-\dfrac{\sqrt{3}}{2}\right)$ $\dfrac{5\pi}{6}$

104. a. $\sin^{-1}\dfrac{7\pi}{6}$ Undefined

b. $\sin\dfrac{7\pi}{6}$ $-\dfrac{1}{2}$

c. $\sin^{-1}\left(-\dfrac{1}{2}\right)$ $-\dfrac{\pi}{6}$

For Exercises 105–108, find the inverse function and its domain and range.

105. $f(x) = 3\sin x + 2$ for $-\dfrac{\pi}{2} \leq x \leq \dfrac{\pi}{2}$

106. $g(x) = 6\cos x - 4$ for $0 \leq x \leq \pi$

107. $h(x) = \dfrac{\pi}{4} + \tan x$ for $-\dfrac{\pi}{2} < x < \dfrac{\pi}{2}$

108. $k(x) = \pi + \sin x$ for $-\dfrac{\pi}{2} \leq x \leq \dfrac{\pi}{2}$

109. A video camera located at ground level follows the liftoff of an Atlas V Rocket from the Kennedy Space Center. Suppose that the camera is 1000 m from the launch pad.

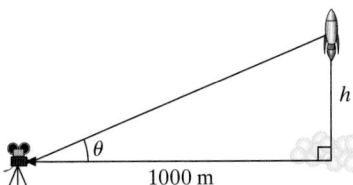

a. Write the angle of elevation θ from the camera to the rocket as a function of the rocket's height, h. $\theta = \tan^{-1}\dfrac{h}{1000}$

b. Without the use of a calculator, will the angle of elevation be less than 45° or greater than 45° when the rocket is 400 m high? θ is less than 45° because the ratio $\dfrac{400}{1000}$ is less than 1.

c. Use a calculator to find θ to the nearest tenth of a degree when the rocket's height is 400 m, 1500 m, and 3000 m.   21.8°, 56.3°, 71.6°

110. The effective focal length f of a camera is the distance required for the lens to converge light to a single focal point. The angle of view α of a camera describes the angular range (either horizontally, vertically, or diagonally) that is imaged by a camera.

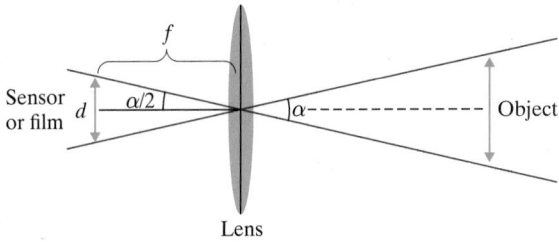

a. Show that $\alpha = 2\arctan\dfrac{d}{2f}$ where d is the dimension of the image sensor or film.

b. A typical 35-mm camera has image dimensions of 24 mm (vertically) by 36 mm (horizontally). If the focal length is 50 mm, find the vertical and horizontal viewing angles. Round to the nearest tenth of a degree.   Vertical viewing angle: 27.0°; Horizontal viewing angle: 39.6°

For Exercises 111–114, use the relationship given in the right triangle and the inverse sine, cosine, and tangent functions to write θ as a function of x in three different ways. It is not necessary to rationalize the denominator.

111.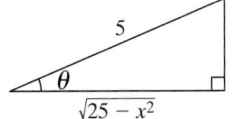

112.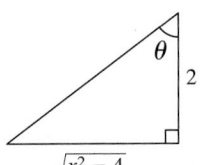

113.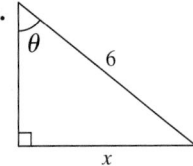

114.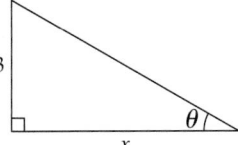

For Exercises 115–120, find the exact solution to each equation.

115. $-2\sin^{-1}x - \pi = 0$ $\{-1\}$

116. $3\cos^{-1}x - \pi = 0$ $\left\{\dfrac{1}{2}\right\}$

117. $6\cos^{-1}x - 3\pi = 0$ $\{0\}$

118. $4\sin^{-1}x + \pi = 0$ $\left\{-\dfrac{\sqrt{2}}{2}\right\}$

119. $4\tan^{-1}2x = \pi$ $\left\{\dfrac{1}{2}\right\}$

120. $6\tan^{-1}2x = 2\pi$ $\left\{\dfrac{\sqrt{3}}{2}\right\}$

Write About It

121. Explain the difference between the reciprocal of a function and the inverse of a function.   Given $y = f(x)$, the reciprocal $y = [f(x)]^{-1}$ is the quotient of 1 and $f(x)$, that is $y = \dfrac{1}{f(x)}$. The inverse of the function, denoted by $y = f^{-1}(x)$, is the function that reverses the operations performed by the original function.

123. In terms of angles, explain what is meant when we find $\sin^{-1}\left(-\dfrac{1}{2}\right)$.   We want to find an angle such that the sine of the angle is $-\frac{1}{2}$. However, the angle must be between $-\frac{\pi}{2}$ and $\frac{\pi}{2}$.

122. Explain the flaw in the logic: $\cos\left(-\dfrac{\pi}{4}\right) = \dfrac{\sqrt{2}}{2}$. Therefore, $\cos^{-1}\dfrac{\sqrt{2}}{2} = -\dfrac{\pi}{4}$.   By the definition of the inverse cosine function, the range (output) is a number between 0 and π, inclusive.

124. Why must the domains of the sine, cosine, and tangent functions be restricted in order to define their inverse functions?   The domains must be restricted because the sine, cosine, and tangent functions are not one-to-one on their entire domains.

Expanding Your Skills

125. In calculus, we can show that the area below the graph of $f(x) = \dfrac{1}{1+x^2}$, above the x-axis, and between the lines $x = a$ and $x = b$ for $a < b$, is given by

$$\tan^{-1}b - \tan^{-1}a$$

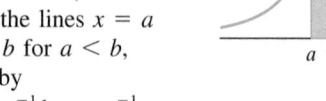

a. Find the area under the curve between $x = 0$ and $x = 1$. $\frac{\pi}{4}$ square units

b. Evaluate $f(0)$ and $f(1)$. $f(0) = 1, f(1) = \frac{1}{2}$

c. Find the area of the trapezoid defined by the points $(0, 0)$, $(1, 0)$, $[0, f(0)]$, and $[1, f(1)]$ to confirm that your answer from part (a) is reasonable.

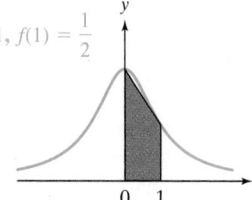

126. In calculus, we can show that the area below the graph of $f(x) = \dfrac{1}{\sqrt{1-x^2}}$, above the x-axis, and between the lines $x = a$ and $x = b$ for $a < b$, is given by

$$\sin^{-1}b - \sin^{-1}a$$

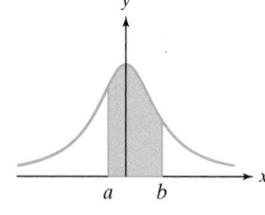

a. Find the area under the curve between $x = 0$ and $x = 0.5$. $\frac{\pi}{6}$ square units

b. Evaluate $f(0)$ and $f(0.5)$. $f(0) = 1, f(0.5) = \frac{2\sqrt{3}}{3}$

c. Find the area of the trapezoid defined by the points $(0, 0)$, $(1, 0)$, $[0, f(0)]$, and $[0.5, f(0.5)]$ to confirm that your answer from part (a) is reasonable.

127. The vertical viewing angle θ to a movie screen is the angle formed from the bottom of the screen to a viewer's eye to the top of the screen. Suppose that the viewer is sitting x horizontal feet from an IMAX screen 53 ft high and that the bottom of the screen is 10 vertical feet above the viewer's eye level. Let α be the angle of elevation to the bottom of the screen.

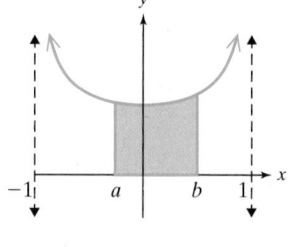

a. Write an expression for $\tan\alpha$. $\tan\alpha = \dfrac{10}{x}$

b. Write an expression for $\tan(\alpha + \theta)$. $\tan(\theta + \alpha) = \dfrac{63}{x}$

c. Using the relationships found in parts (a) and (b), show that $\theta = \tan^{-1}\dfrac{63}{x} - \tan^{-1}\dfrac{10}{x}$.

Technology Connections

128. Refer to the movie screen and observer in Exercise 127. The vertical viewing angle is given by

$$\theta = \tan^{-1}\dfrac{63}{x} - \tan^{-1}\dfrac{10}{x}.$$

a. Find the vertical viewing angle (in radians) for an observer sitting 15, 25, and 35 ft away. Round to 2 decimal places. 0.75, 0.81, 0.79

b. Graph $y = \tan^{-1}\dfrac{63}{x} - \tan^{-1}\dfrac{10}{x}$ on a graphing utility on the window $[0, 100, 10]$ by $[0, 1, 0.1]$.

c. Use the Maximum feature on the graphing utility to estimate the distance that an observer should sit from the screen to produce the maximum viewing angle. Round to the nearest tenth of a foot.

129. a. Graph the functions $y = \tan^{-1}x$ and

$$y = x - \dfrac{x^3}{3} + \dfrac{x^5}{5}$$ on the window

$$\left[-\dfrac{\pi}{2}, \dfrac{\pi}{2}, \dfrac{\pi}{4}\right]$$ by $[-2, 2, 1]$.

b. Graph the functions $y = \tan^{-1}x$ and $y = \dfrac{x}{1 + \dfrac{x^2}{3 - x^2}}$

on the window $\left[-\dfrac{\pi}{2}, \dfrac{\pi}{2}, \dfrac{\pi}{4}\right]$ by $[-2, 2, 1]$.

c. How do the functions in parts (a) and (b) compare for values of x taken close to 0? The function values are close for x taken near 0.

CHAPTER 4 KEY CONCEPTS

SECTION 4.1 Angles and Their Measure	Reference
The measure of an angle may be given in **degrees,** where $1°$ is $\frac{1}{360}$ of a full rotation. A measure of $1°$ can be further divided into 60 equal parts called **minutes.** Each minute can be divided into 60 equal parts called **seconds.**	p. 447
The measure of an angle may be given in **radians,** where 1 radian is the measure of the central angle that intercepts an arc equal in length to the radius of the circle.	p. 448

$\theta = \dfrac{s}{r}$ gives the measure of a central angle θ (in radians), where s is the length of the arc intercepted by θ and r is the radius of the circle.	p. 449
The relationship $180° = \pi$ radians provides the conversion factors to convert from radians to degrees and vice versa.	p. 450
Two angles with the same initial and terminal sides are called **coterminal angles.** Coterminal angles differ in measure by an integer multiple of $360°$ or 2π radians.	p. 450
Given a circle of radius r, the length s of an arc intercepted by a central angle θ (in radians) is given by $s = r\theta$.	p. 452
Points on the Earth are identified by their latitude and longitude. **Latitude** is the angular measure of a central angle measuring north or south from the equator. Lines of **longitude,** also called meridians, are circles passing through both poles and running perpendicular to the equator.	p. 453
Suppose that a point on a circle of radius r moves through an angle of θ radians in time t. Then the angular and linear speeds of the point are given by angular speed: $\quad \omega = \dfrac{\theta}{t}$ linear speed: $\quad v = \dfrac{s}{t} \quad$ or $\quad v = \dfrac{r\theta}{t} \quad$ or $\quad v = r\omega$	p. 454
The area A of a sector of a circle of radius r with central angle θ (in radians) is given by $A = \dfrac{1}{2}r^2\theta$.	p. 455

SECTION 4.2 Trigonometric Functions Defined on the Unit Circle	Reference
The **unit circle** is a circle of radius r with center at the origin of a rectangular coordinate system: $x^2 + y^2 = 1$.	p. 462
The six trigonometric functions can be defined as functions of real numbers by using the unit circle. Let $P(x, y)$ be the point associated with a real number t measured along the circumference of the unit circle from the point $(1, 0)$. $\sin t = y \qquad\qquad \cos t = x \qquad\qquad \tan t = \dfrac{y}{x} \ (x \neq 0)$ $\csc t = \dfrac{1}{y} \ (y \neq 0) \qquad \sec t = \dfrac{1}{x} \ (x \neq 0) \qquad \cot t = \dfrac{x}{y} \ (y \neq 0)$	p. 463
The real number t taken along the circumference of the unit circle gives the number of radians of the corresponding central angle θ; that is, $\theta = t$ radians.	p. 465
• The domain of both the sine function and cosine function is all real numbers. • The domain of both the tangent function and secant function excludes real numbers t that are odd multiples of $\frac{\pi}{2}$. • The domain of both the cotangent function and cosecant function excludes real numbers t that are multiples of π.	p. 469
A function f is **periodic** if $f(t + p) = f(t)$ for some constant p. The smallest positive value p for which f is periodic is called the **period** of f. • The period of the sine, cosine, secant, and cosecant functions is 2π. • The period of the tangent and cotangent functions is π.	p. 473
• The cosine and secant functions are even functions. $f(-x) = f(x)$. • The sine, cosecant, tangent, and cotangent functions are odd functions. $f(-x) = -f(x)$.	p. 474

SECTION 4.3 Right Triangle Trigonometry	Reference
Each trigonometric function of an acute angle θ is one of the six possible ratios of the sides of a right triangle containing θ. $\sin\theta = \dfrac{\text{opp}}{\text{hyp}}, \cos\theta = \dfrac{\text{adj}}{\text{hyp}}, \tan\theta = \dfrac{\text{opp}}{\text{adj}}, \csc\theta = \dfrac{\text{hyp}}{\text{opp}}, \sec\theta = \dfrac{\text{hyp}}{\text{adj}}, \cot\theta = \dfrac{\text{adj}}{\text{opp}}$	p. 482
Reciprocal Identities: $\quad \csc\theta = \dfrac{1}{\sin\theta}, \sec\theta = \dfrac{1}{\cos\theta}, \cot\theta = \dfrac{1}{\tan\theta}$ **Quotient Identities:** $\quad \tan\theta = \dfrac{\sin\theta}{\cos\theta}, \cot\theta = \dfrac{\cos\theta}{\sin\theta}$	p. 488
Pythagorean Identities: $\quad \sin^2\theta + \cos^2\theta = 1, \tan^2\theta + 1 = \sec^2\theta, 1 + \cot^2\theta = \csc^2\theta$	p. 488
The **cofunction identities** indicate that cofunctions of complementary angles are equal.	p. 489
An **angle of elevation** or **depression** from an observer to an object is used in many applications of trigonometric functions.	p. 490

SECTION 4.4 Trigonometric Functions of Any Angle	Reference
$\sin\theta = \dfrac{y}{r} \qquad \csc\theta = \dfrac{r}{y} \ (y \neq 0)$ $\cos\theta = \dfrac{x}{r} \qquad \sec\theta = \dfrac{r}{x} \ (x \neq 0)$ $\tan\theta = \dfrac{y}{x} \ (x \neq 0) \qquad \cot\theta = \dfrac{x}{y} \ (y \neq 0)$ Let $P(x, y)$ be a point on the terminal side of angle θ drawn in standard position, and r be the distance from P to the origin.	p. 497
Let θ be an nonquadrantal angle in standard position. The **reference angle** for θ is the acute angle θ' formed by the terminal side of θ and the horizontal axis.	p. 499
The value of each trigonometric function of θ is the same as the corresponding function of the reference angle θ' except possibly for the sign.	p. 500

SECTION 4.5 Graphs of Sine and Cosine Functions		Reference
The graph of $y = \sin x$ (for one period): 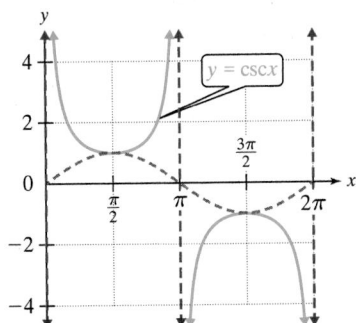	The graph of $y = \cos x$ (for one period): 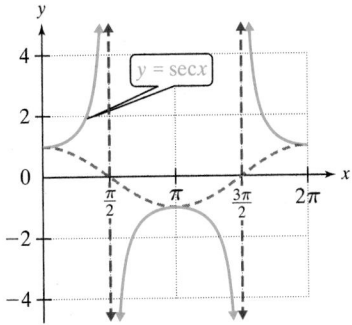	p. 510

Period: 2π　　　　Amplitude: 1

Domain: $(-\infty, \infty)$　　Range: $[-1, 1]$

Symmetric to the origin

Period: 2π　　　　Amplitude: 1

Domain: $(-\infty, \infty)$　　Range: $[-1, 1]$

Symmetric to the y-axis

	Reference
Characteristics of $y = A\sin(Bx - C) + D$ and $y = A\cos(Bx - C) + D$ with $B > 0$:	p. 515

1. The amplitude is $|A|$.　　　　2. The period is $\dfrac{2\pi}{B}$.

3. The phase shift is $\dfrac{C}{B}$.　　　　4. The vertical shift is D.

5. One full cycle is given on the interval $0 \leq Bx - C \leq 2\pi$.

6. The domain is the set of real numbers.

7. The range is $-|A| + D \leq y \leq |A| + D$.

SECTION 4.6 Graphs of Other Trigonometric Functions		Reference
The graph of $y = \csc x = \dfrac{1}{\sin x}$ has vertical asymptotes where $\sin x = 0$. 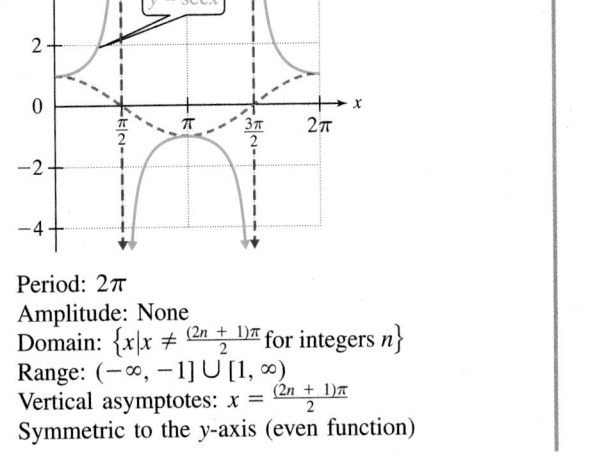	The graph of $y = \sec x = \dfrac{1}{\cos x}$ has vertical asymptotes where $\cos x = 0$.	p. 528

Period: 2π

Amplitude: None

Domain: $\{x \mid x \neq n\pi \text{ for integers } n\}$

Range: $(-\infty, -1] \cup [1, \infty)$

Vertical asymptotes: $x = n\pi$

Symmetric to the origin (odd function)

Period: 2π

Amplitude: None

Domain: $\left\{x \mid x \neq \dfrac{(2n + 1)\pi}{2} \text{ for integers } n\right\}$

Range: $(-\infty, -1] \cup [1, \infty)$

Vertical asymptotes: $x = \dfrac{(2n + 1)\pi}{2}$

Symmetric to the y-axis (even function)

The graph of $y = \tan x = \dfrac{\sin x}{\cos x}$ has vertical asymptotes where $\cos x = 0$.	The graph of $y = \cot x = \dfrac{\cos x}{\sin x}$ has vertical asymptotes where $\sin x = 0$.	p. 532

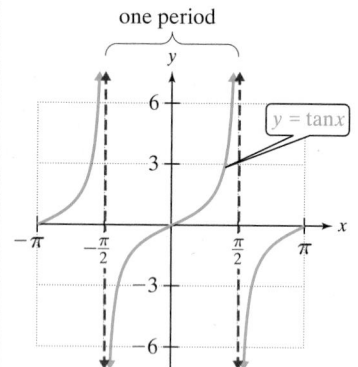

one period

$y = \tan x$

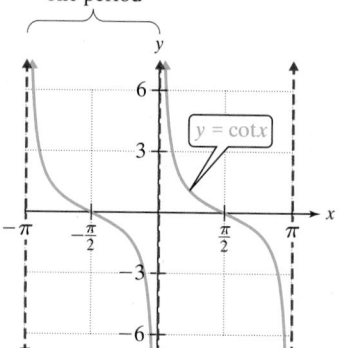

one period

$y = \cot x$

Period: π
Amplitude: None
Domain: $\left\{x \mid x \neq \dfrac{(2n + 1)\pi}{2} \text{ for integers } n\right\}$
Range: $(-\infty, \infty)$
Vertical asymptotes: $x = \dfrac{(2n + 1)\pi}{2}$
Symmetric to the origin (odd function)

Period: π
Amplitude: None
Domain: $\{x \mid x \neq n\pi \text{ for integers } n\}$
Range: $(-\infty, \infty)$
Vertical asymptotes: $x = n\pi$
Symmetric to the origin (odd function)

To graph variations of $y = \csc x$ or $y = \sec x$, use the graph of the related reciprocal function for reference.	p. 529
To graph variations of $y = \tan x$ and $y = \cot x$, 1. First graph two consecutive asymptotes. 2. Plot an x-intercept halfway between the asymptotes. 3. Sketch the general shape of the "parent" function between the asymptotes. 4. Apply a vertical shift if applicable. 5. Sketch additional cycles to the right or left as desired.	p. 533

SECTION 4.7 Inverse Trigonometric Functions | Reference

The **inverse sine function** (or arcsine), denoted by $\sin^{-1}$ or arcsin, is defined by $y = \sin^{-1} x \iff \sin y = x$ for $-1 \leq x \leq 1$ and $-\dfrac{\pi}{2} \leq y \leq \dfrac{\pi}{2}$. To compose the sine function and its inverse, $\qquad \sin(\sin^{-1} x) = x$ for $-1 \leq x \leq 1 \qquad$ and $\sin^{-1}(\sin x) = x$ for $-\dfrac{\pi}{2} \leq x \leq \dfrac{\pi}{2}$.	p. 541
The **inverse cosine function** (or arccosine), denoted by $\cos^{-1}$ or arccos, is defined by $y = \cos^{-1} x \iff \cos y = x$ for $-1 \leq x \leq 1$ and $0 \leq y \leq \pi$. To compose the cosine function and its inverse, $\cos(\cos^{-1} x) = x$ for $-1 \leq x \leq 1$ and $\cos^{-1}(\cos x) = x$ for $0 \leq x \leq \pi$.	p. 543
The **inverse tangent function** (or arctangent), denoted by $\tan^{-1}$ or arctan, is defined by $y = \tan^{-1} x \iff \tan y = x$ for $x \in \mathbb{R}$ and $-\dfrac{\pi}{2} < y < \dfrac{\pi}{2}$. The graph of the inverse tangent function has horizontal asymptotes $y = -\dfrac{\pi}{2}$ and $y = \dfrac{\pi}{2}$. To compose the tangent function and its inverse, $\qquad \tan(\tan^{-1} x) = x$ for $x \in \mathbb{R}$ and $\tan^{-1}(\tan x) = x$ for $-\dfrac{\pi}{2} < x < \dfrac{\pi}{2}$.	p. 543

Expanded Chapter Summary available at www.mhhe.com/millerprecalculus.

CHAPTER 4 Review Exercises

SECTION 4.1

1. Convert $96°28'$ to decimal degrees. Round to 2 decimal places. $96.47°$

2. Convert $225.24°$ to DMS (degree-minute-second) form. Round to the nearest second if necessary. $225°14'\ 24''$

3. Convert $124°$ to radians. Give the answer in exact form in terms of π. $\dfrac{31\pi}{45}$

4. Convert $-73.8°$ to radians. Round to 2 decimal places. -1.29

5. Convert $\dfrac{5\pi}{8}$ radians to decimal degrees. Round to 1 decimal place if necessary. $112.5°$

For Exercises 6–8, find a positive angle coterminal with the given angle.

6. $48°$ For example: $408°$

7. $-110°$ For example: $250°$

8. $\dfrac{3\pi}{5}$ For example: $\dfrac{13\pi}{5}$

For Exercises 9–11, find a negative angle coterminal with the given angle.

9. $\dfrac{7\pi}{4}$ For example: $-\dfrac{\pi}{4}$

10. $-\dfrac{5\pi}{6}$ For example: $-\dfrac{17\pi}{6}$

11. $126°$ For example: $-234°$

12. Find an angle between $0°$ and $360°$ that is coterminal to $745°$. $25°$

13. Find an angle between 0 and 2π that is coterminal to $-\dfrac{19\pi}{6}$. $\dfrac{5\pi}{6}$

14. A unicycle with a wheel diameter of 24 in. moves through an angle of $140°$. What distance does a point on the edge of the wheel move? Round the answer to the nearest tenth of an inch. 29.3 in.

15. A pulley is 1.5 ft in diameter. Find the distance the load will move if the pulley is rotated $750°$. Find the exact distance in terms of π and then round the answer to the nearest tenth of a foot. $\dfrac{25\pi}{8}$ ft ≈ 9.8 ft

16. Seattle, Washington ($47.61°$N, $122.33°$W), and San Francisco, California ($37.78°$N, $122.42°$W), have approximately the same longitude, which means that they are roughly due north-south of each other. Use the difference in latitude to approximate the distance between the cities assuming that the radius of the Earth is 3960 mi. Round the answer to the nearest mile. 679 mi

17. A spinning disk has radius of 10 in. and rotates at 2800 rpm. For a point at the edge of the disk,

    a. Find the exact value of the angular speed. 5600π/min

    b. Find the linear speed. Round the answer to the nearest inch per minute. $175{,}929$ in./min

18. A bicycle has wheels 26 inches in diameter. If the wheels turn at 90 rpm, what is the linear speed in inches per minute? Give the exact speed and an approximation to the nearest inch per minute. 2340π in./min ≈ 7351 in./min

19. A sprinkler rotates through an angle of $75°$ spraying water outward for a distance of 6 ft. Find the exact area watered, then round the result to the nearest tenth of a square foot. $\dfrac{15}{2}\pi$ ft$^2 \approx 23.6$ ft^2

20. A round pie 10 in. in diameter is cut into a slice with a $30°$ angle. Find the exact area of the slice, then round the result to the nearest tenth of a square inch. $\dfrac{25}{12}\pi$ in.$^2 \approx 6.5$ in.^{2}

SECTION 4.2

21. The real number t corresponds to the point $P\left(-\dfrac{3\sqrt{5}}{7},\dfrac{2}{7}\right)$ on the unit circle. Evaluate each expression.

    a. $\sin t$ b. $\cos t$ c. $\tan t$

    d. $\csc t$ e. $\sec t$ f. $\cot t$

22. Identify the ordered pair on the unit circle corresponding to each real number t.

    a. $t = \dfrac{2\pi}{3}$ b. $t = \dfrac{5\pi}{4}$ c. $t = \dfrac{11\pi}{6}$

23. Identify the domain for each pair of functions.

    a. $f(t) = \sin t$, $g(t) = \cos t$ $(-\infty, \infty)$ or All real numbers

    b. $f(t) = \tan t$, $f(t) = \sec t$ $\left\{t\,\middle|\,t \neq \dfrac{(2n+1)\pi}{2} \text{ for all integers } n\right\}$

    c. $f(t) = \cot t$, $f(t) = \csc t$ $\{t\,|\,t \neq n\pi \text{ for all integers } n\}$

For Exercises 24–29, use the unit circle and the period of the function to evaluate the function or state that the function is undefined at the given value.

24. $\cos\dfrac{15\pi}{2}$ 0

25. $\cot 540°$ Undefined

26. $\csc(-240°)$ $\dfrac{2\sqrt{3}}{3}$

27. $\sin\left(-\dfrac{14\pi}{3}\right)$ $-\dfrac{\sqrt{3}}{2}$

28. $\tan\left(-\dfrac{23\pi}{6}\right)$ $\dfrac{\sqrt{3}}{3}$

29. $\sec(480°)$ -2

For Exercises 30–31, use the even-odd and periodic properties of the trigonometric functions to simplify.

30. $\tan(\pi - t) \cdot \cos(2\pi - t)$ $-\sin t$

31. $\sin(-t - \pi) + \sin(t + 2\pi)$ $2\sin t$

32. Write $\cos t$ in terms of $\sin t$ for t in Quadrant IV. $\cos t = \sqrt{1 - \sin^2 t}$

33. Write $\cot t$ in terms of $\csc t$ for t in Quadrant II. $\cot t = -\sqrt{\csc^2 t - 1}$

SECTION 4.3

34. Suppose that a right triangle $\triangle ABC$ has legs of length 5 cm and $\sqrt{2}$ cm. Evaluate the six trigonometric functions for angle θ, where angle θ is the larger acute angle.

35. Determine the values of the six trigonometric functions of θ for the given right triangle.

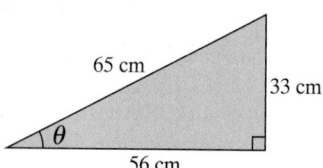

For Exercises 36–37, construct a right triangle to find the indicated values. Assume θ is an acute angle.

36. If $\cos\theta = \dfrac{5}{7}$, find $\csc\theta$ and $\tan\theta$.

37. If $\tan\theta = 3$, find $\sin\theta$ and $\sec\theta$.

For Exercises 38–39, evaluate the expression without the use of a calculator.

38. $\sin\dfrac{\pi}{3} + \tan\dfrac{\pi}{6}$ $\dfrac{5\sqrt{3}}{6}$

39. $\cos 45° \cdot \csc 60°$ $\dfrac{\sqrt{6}}{3}$

40. Given $\sin\theta = \dfrac{99}{101}$ and $\cos\theta = \dfrac{20}{101}$, use the reciprocal and quotient identities to find the values of the other trigonometric functions of θ.

For Exercises 41–43, use an appropriate Pythagorean identity to find the indicated value for an acute angle θ.

41. Given $\cos\theta = \dfrac{40}{41}$, find the value of $\sin\theta$. $\sin\theta = \dfrac{9}{41}$

42. Given $\sec\theta = \dfrac{29}{20}$, find the value of $\tan\theta$. $\tan\theta = \dfrac{21}{20}$

43. Given $\cot\theta = \dfrac{13}{84}$, find the value of $\csc\theta$. $\csc\theta = \dfrac{85}{84}$

For Exercises 44–45, given the function value, find a cofunction of another angle with the same function value.

44. $\cos 15° = \dfrac{\sqrt{6}+\sqrt{2}}{4}$ $\sin 75°$

45. $\tan\dfrac{\pi}{12} = 2 - \sqrt{3}$ $\cot\dfrac{5\pi}{12}$

46. Use a calculator to approximate the function values. Round to 4 decimal places.

   a. $\tan 23.8°$ 0.4411 **b.** $\cos\dfrac{5}{8}$ 0.8110 **c.** $\sin\dfrac{\pi}{8}$ 0.3827

47. An observer at the top of a 48-ft building measures the angle of depression from the top of the building to a point on the ground to be 27°. What is the distance from the base of the building to the point on the ground? Round to the nearest foot.   94 ft

48. During the first quarter moon, the Earth, Sun, and Moon form a right triangle. The distance between the Sun and the Earth is approximately 92,900,000 mi and the measure of $\angle SEM$ is approximately 89.85°. Determine the distance between the Earth and the Moon. Round to the nearest thousand miles.   243,000 mi

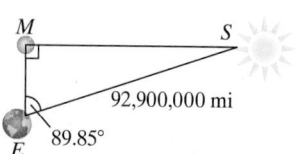

SECTION 4.4

49. Let $P(3, -5)$ be a point on the terminal side of angle θ drawn in standard position. Find the values of the six trigonometric functions of θ.

For Exercises 50–55, find the reference angle for the given angle.

50. $\dfrac{5\pi}{6}$ $\dfrac{\pi}{6}$ **51.** $260°$ $80°$ **52.** $-200°$ $20°$

53. 5 $2\pi - 5$ **54.** $\dfrac{7\pi}{4}$ $\dfrac{\pi}{4}$ **55.** $750°$ $30°$

For Exercises 56–61, use reference angles to find the exact value.

56. $\cos\dfrac{11\pi}{6}$ $\dfrac{\sqrt{3}}{2}$ **57.** $\sin\left(-\dfrac{5\pi}{3}\right)$ $\dfrac{\sqrt{3}}{2}$ **58.** $\tan\dfrac{17\pi}{6}$ $-\dfrac{\sqrt{3}}{3}$

59. $\cot\left(-\dfrac{3\pi}{4}\right)$ 1 **60.** $\csc(-120°)$ $-\dfrac{2\sqrt{3}}{3}$ **61.** $\sec 240°$ -2

62. Identify which expressions are undefined.   a, b, c

   a. $\sec 270°$ **b.** $\tan\dfrac{\pi}{2}$ **c.** $\cot 180°$ **d.** $\csc\left(-\dfrac{\pi}{2}\right)$

63. Given $\tan\theta = -\dfrac{2}{3}$ and $\sin\theta > 0$, find $\sec\theta$. $-\dfrac{\sqrt{13}}{3}$

64. Given $\cos\theta = -\dfrac{3}{7}$ and $\cot\theta > 0$, find $\sin\theta$. $-\dfrac{2\sqrt{10}}{7}$

65. Given $\sin\theta = -\dfrac{60}{61}$ and θ in Quadrant III, find $\cot\theta$. $\dfrac{11}{60}$

SECTION 4.5

For Exercises 66–69, determine the amplitude and period.

66. $y = 4\sin 2x$ Amplitude = 4, Period = π **67.** $y = -2\cos\dfrac{x}{2}$ Amplitude = 2, Period = 4π

68. $y = \dfrac{1}{3}\cos\pi x$ Amplitude = $\dfrac{1}{3}$, Period = 2   **69.** $y = \sin 3\pi x$ Amplitude = 1, Period = $\dfrac{2}{3}$

For Exercises 70–72, graph one period of the function.

70. $y = \cos 3\pi x$

71. $y = 3\sin 2x$

72. $y = -2\cos\dfrac{x}{4}$

73. Determine the amplitude, period, and phase shift for each function.

   a. $y = -\dfrac{1}{3}\cos(2x + \pi)$ **b.** $y = 5\sin(2\pi x - \pi)$

For Exercises 74–75, graph one period of the function.

74. $y = \dfrac{1}{3}\cos\left(x - \dfrac{\pi}{4}\right)$ **75.** $y = -2\sin\left(\pi x + \dfrac{\pi}{3}\right)$

For Exercises 76–77, determine the amplitude, period, phase shift, and vertical shift for each function.

76. $y = 4\sin(3x + \pi) - 2$ **77.** $y = -\dfrac{1}{4}\cos\left(\dfrac{\pi}{3}x - \dfrac{\pi}{2}\right) + 5$

For Exercises 78–79, graph one period of the function.

78. $y = \cos\left(x - \dfrac{\pi}{3}\right) + 2$ **79.** $y = 2\sin\left(2x + \dfrac{\pi}{4}\right) - 1$

80. The depth of water along a coastal inlet varies with the tides. On a summer day, the water depth at the end of a pier is 23 ft at 6:30 A.M., 21 ft at 12:30 P.M., and 23 ft at 6:30 P.M. Assuming that this pattern continues indefinitely and behaves like a cosine wave, write a function of the form $h(t) = A\cos(Bt - C) + D$. The value $h(t)$ is the water depth (in feet), t hours after 6:30 A.M. $h(t) = \cos\dfrac{\pi}{6}t + 22$

81. The data in the table represent the percentage of the moon illuminated for selected days in January for a recent year. The value 0.0 = 0% represents a new moon and 1.0 = 100% represents a full moon. (*Source*: Astronomical Applications Department, U.S. Naval Observatory: http://aa.usno.navy.mil)

  a. Enter the data in a graphing utility and use the sinusoidal regression tool (SinReg) to find a model of the form $y = a\sin(bx + c) + d$. Round a, b, c, and d to 1 decimal place. $y = 0.5\sin(0.2x - 1.6) + 0.5$

  b. Graph the data and the resulting function.

Day	1	3	5	7	9	11	13	15
Percent	0.00	0.05	0.20	0.40	0.61	0.79	0.92	0.99

Day	17	19	21	23	25	27	29	31
Percent	0.99	0.92	0.79	0.61	0.40	0.19	0.05	0.00

SECTION 4.6

For Exercises 82–85, give the period of the function and the equations of two asymptotes.

82. $y = \csc 3x$ **83.** $y = \tan 3x$

84. $y = \cot(\pi x + 2\pi)$ **85.** $y = -2\sec(x + \pi)$

For Exercises 86–89, graph one period of the function.

86. $y = 5\sec 3x$ **87.** $y = -2\csc\dfrac{x}{2}$

88. $y = 2\sec(\pi x + \pi)$ **89.** $y = 3\csc 4x + 2$

For Exercises 90–91, graph two periods of the function.

90. $y = \tan 4x$ **91.** $y = -2\tan\left(x - \dfrac{\pi}{3}\right)$

92. $y = \cot 2\pi x$ **93.** $y = \cot 2x + 3$

SECTION 4.7

For Exercises 94–109, evaluate the expression.

94. $\cos^{-1}\left(-\dfrac{\sqrt{2}}{2}\right)$ $\dfrac{3\pi}{4}$ **95.** $\sin^{-1}\left(-\dfrac{\sqrt{2}}{2}\right)$ $-\dfrac{\pi}{4}$

96. $\arctan\left(\dfrac{\sqrt{3}}{3}\right)$ $\dfrac{\pi}{6}$ **97.** $\arccos\left(\dfrac{\sqrt{3}}{2}\right)$ $\dfrac{\pi}{6}$

98. $\arcsin\left(\dfrac{\sqrt{3}}{2}\right)$ $\dfrac{\pi}{3}$ **99.** $\cos^{-1}\left(\cos\dfrac{7\pi}{6}\right)$ $\dfrac{5\pi}{6}$

100. $\tan\left[\tan^{-1}(7)\right]$ 7   **101.** $\cos\left(\cos^{-1}0.35\right)$ 0.35

102. $\arcsin\left(\sin\dfrac{7\pi}{6}\right)$ $-\dfrac{\pi}{6}$ **103.** $\sin\left[\sin^{-1}\left(-\dfrac{2}{3}\right)\right]$ $-\dfrac{2}{3}$

104. $\cos\left[\sin^{-1}\left(-\dfrac{\sqrt{3}}{2}\right)\right]$ $\dfrac{1}{2}$ **105.** $\sin\left[\arccos\left(-\dfrac{\sqrt{3}}{2}\right)\right]$ $\dfrac{1}{2}$

106. $\sin\left(\arctan\dfrac{3}{2}\right)$ $\dfrac{3\sqrt{13}}{13}$ **107.** $\cos\left(\sin^{-1}\dfrac{2}{5}\right)$ $\dfrac{\sqrt{21}}{5}$

108. $\tan\left[\arccos\left(-\dfrac{3}{4}\right)\right]$ $-\dfrac{\sqrt{7}}{3}$ **109.** $\sin\left[\cos^{-1}\left(-\dfrac{1}{4}\right)\right]$ $\dfrac{\sqrt{15}}{4}$

110. Use a calculator to approximate the function values in both radians and degrees.

  a. $\sin^{-1}(-0.35)$ **b.** $\cos^{-1}\dfrac{\sqrt{2}}{5}$ **c.** $\arctan 10$

For Exercises 111–114, use a calculator to approximate the degree measure (to 1 decimal place) or radian measure (to 4 decimal places) of the angle θ subject to the given conditions.

111. $\tan\theta = \dfrac{4}{19}$ and $180° \le \theta \le 270°$ 191.9°

112. $\cos\theta = \dfrac{2}{15}$ and $270° \le \theta \le 360°$ 277.7°

113. $\sin\theta = -\dfrac{5}{11}$ and $\pi < \theta < \dfrac{3\pi}{2}$ 3.6135

114. $\tan\theta = 4$ and $\pi < \theta < \dfrac{3\pi}{2}$ 4.4674

115. Write the given expression as an algebraic expression. It is not necessary to rationalize the denominator.

  a. $\cos(\sin^{-1}x)$ for $|x| < 1$. $\sqrt{1 - x^2}$

  b. $\sin(\arctan x)$ $\dfrac{x}{\sqrt{x^2 + 1}}$

  c. $\tan\left(\arccos\dfrac{x}{\sqrt{x^2 + 9}}\right)$ for $x \neq 0$. $\dfrac{3}{x}$

116. Find the lengths of sides a, b, and c to the nearest tenth of a meter and the measure of angle α to the nearest tenth of a degree. $a \approx 9.0$ m, $b \approx 12.1$ m, $c \approx 8.3$ m, $\alpha \approx 25.1°$

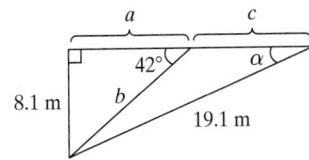

117. The length of the perpendicular line segment $\overline{BP}$ from a rotating beacon to a straight shoreline is 100 ft. The beam of light emitted from the beacon strikes the shoreline at a point Q, a distance of x feet from point P. Let θ represent $\angle QBP$.

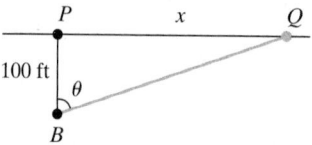

a. Write θ as a function of x. $\theta = \tan^{-1}\dfrac{x}{100}$

b. Find θ for $x = 100$, 200, and 300 ft. Round to the nearest degree.   45°, 63°, 72°

c. What happens to θ as $x \to \infty$? $\theta \to 90°$

118. Find the angle of repose θ for the pile of coarse gravel. Round to the nearest degree.   40°

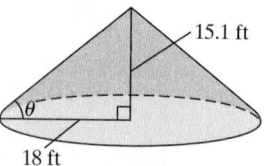

15.1 ft

18 ft

CHAPTER 4   Test

1. Convert 15.36° to DMS (degree-minute-second) form. Round to the nearest second if necessary.   15° 21′ 36″

2. Convert 130.3° to radians. Round to 2 decimal places.   2.27

3. Find the exact length of the arc intercepted by a central angle of 27° on a circle of radius 5 ft. $\dfrac{3\pi}{4}$ ft

4. A skateboard designed for rough surfaces has wheels with diameter 60 mm. If the wheels turn at 2200 rpm, what is the linear speed in mm per minute?   132,000π mm/min

5. A pulley is 20 in. in diameter. Through how many degrees should the pulley rotate to lift a load 3 ft? Round to the nearest degree.   206°

6. A circle has a radius of 9 yd. Find the area of a sector with a central angle of 120°.   27π yd²

7. For an acute angle θ, if $\sin \theta = \dfrac{5}{6}$, evaluate $\cos \theta$ and $\tan \theta$.

8. Evaluate $\tan \dfrac{\pi}{6} - \cot \dfrac{\pi}{6}$ without the use of a calculator. $-\dfrac{2\sqrt{3}}{3}$

9. Given $\sin \theta = \dfrac{5}{13}$, use a Pythagorean identity to find $\cos \theta$. $\cos \theta = \dfrac{12}{13}$

10. Given $\sec 75° = \sqrt{2} + \sqrt{6}$, find a cofunction of another angle with the same function value.   csc 15°

11. Use a calculator to approximate $\sin \dfrac{2\pi}{11}$ to 4 decimal places.   0.5406

12. A flag pole casts a shadow of 20 ft when the angle of elevation of the Sun is 40°. How tall is the flag pole? Round to the nearest foot.   17 ft

13. A newly planted tree is anchored by a covered wire running from the top of the tree to a post in the ground 5 ft from the base of the tree. If the angle between the wire and the top of the tree is 20°, what is the length of the wire? Round to the nearest foot.   15 ft

For Exercises 14–21, evaluate the function or state that the function is undefined at the given value.

14. $\sin\left(-\dfrac{3\pi}{4}\right)$ $-\dfrac{\sqrt{2}}{2}$

15. $\tan 930°$ $\dfrac{\sqrt{3}}{3}$

16. $\sec \dfrac{11\pi}{6}$ $\dfrac{2\sqrt{3}}{3}$

17. $\csc(-150°)$ -2

18. $\cot 20\pi$ Undefined

19. $\cos 690°$ $\dfrac{\sqrt{3}}{2}$

20. $\tan \dfrac{7\pi}{3}$ $\sqrt{3}$

21. $\sec(-630°)$ Undefined

22. Given $\sin \theta = \dfrac{5}{8}$ and $\cos \theta < 0$, find $\tan \theta$. $-\dfrac{5\sqrt{39}}{39}$

23. Given $\sec \theta = \dfrac{4}{3}$ and $\sin \theta < 0$, find $\csc \theta$. $-\dfrac{4\sqrt{7}}{7}$

24. Given $\tan \theta = -\dfrac{4}{3}$ and θ is in Quadrant IV, find $\sec \theta$. $\dfrac{5}{3}$

For Exercises 25–28, select the trigonometric function, $f(t) = \sin t$, $g(t) = \cos t$, $h(t) = \tan t$ or $r(t) = \cot t$, with the given properties.

25. The function is odd, with period 2π, and domain of all real numbers. $f(t) = \sin t$

26. The function is odd, with period π, and domain of all real numbers excluding odd multiples of $\dfrac{\pi}{2}$. $h(t) = \tan t$

27. The function is even, with period 2π, and domain of all real numbers. $g(t) = \cos t$

28. The function is odd, with period π, and domain of all real numbers excluding multiples of π. $r(t) = \cot t$

29. Use the even-odd and periodic properties of the trigonometric functions to simplify
$\cos(-\theta - 2\pi) - \sin(\theta - 2\pi) \cdot \cot(\theta + \pi)$ 0

30. Suppose that a rectangle is bounded by the *x*-axis and the graph of $y = \sin x$ on the interval $[0, \pi]$.

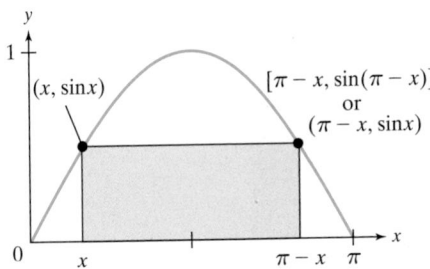

$(x, \sin x)$

$[\pi - x, \sin(\pi - x)]$
or
$(\pi - x, \sin x)$

a. Write a function that represents the area $A(x)$ of the rectangle for $0 < x < \dfrac{\pi}{2}$. $A(x) = (\pi - 2x)\sin x$

b. Determine the area of the rectangle for $x = \dfrac{\pi}{6}$ and $x = \dfrac{\pi}{4}$. $A\left(\dfrac{\pi}{6}\right) = \dfrac{\pi}{3}$ square units and $A\left(\dfrac{\pi}{4}\right) = \dfrac{\sqrt{2}\,\pi}{4}$ square units

For Exercises 31–32, determine the amplitude, period, phase shift, and vertical shift for each function.

31. $y = \dfrac{3}{4}\sin\left(2\pi x - \dfrac{\pi}{6}\right)$ **32.** $y = -5\cos(5x + \pi) + 7$

For Exercises 33–36, graph one period of the function.

33. $y = -2\sin x$ **34.** $y = 3\cos 2x$

35. $y = \sin\left(2x - \dfrac{\pi}{4}\right)$ **36.** $y = 3 - 2\cos(2\pi x - \pi)$

37. Write a function of the form $f(x) = A\cos(Bx)$ for the given graph. $f(x) = -3\cos 3x$

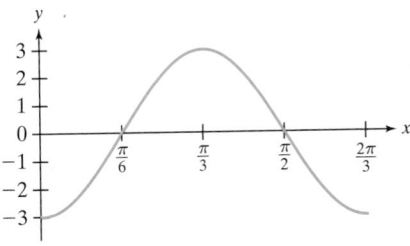

38. Identify each statement as true or false. If a statement is false, explain why.

a. The relative maxima of the graph of $y = \sin x$ correspond to the relative minima of the graph of $y = \csc x$.   True

b. The period of $y = \tan 2x$ is π.   False; The period is $\dfrac{\pi}{2}$.

c. The amplitude of $y = 2\cot x$ is 2.   False; The cotangent function has no amplitude because it is unbounded.

d. The period of $y = \sec 2x$ is π.   True

e. The vertical asymptotes of the graph of $y = \cot x$ occur where the graph of $y = \cos x$ has *x*-intercepts.   False; The asymptotes occur where the graph of $y = \sin x$ has *x*-intercepts.

For Exercises 39–40, graph one period of the function.

39. $y = -4\sec 2x$ **40.** $y = \csc\left(2x - \dfrac{\pi}{4}\right)$

For Exercises 41–42, graph two periods of the function.

41. $y = \tan 3x$ **42.** $y = -4\cot\left(x + \dfrac{\pi}{4}\right)$

43. Simplify each expression.

a. $\sin^{-1}\left(-\dfrac{\sqrt{3}}{2}\right) + \cos^{-1}\dfrac{\sqrt{2}}{2}$ $-\dfrac{\pi}{12}$

b. $\cos\left[\arctan\left(-\sqrt{3}\right)\right]$ $\dfrac{1}{2}$

44. Use a calculator to approximate the degree measure (to 1 decimal place) of the angle θ subject to the given conditions.   158.0°

$$\sin\theta = \dfrac{3}{8} \text{ and } 90° \le \theta \le 180°$$

45. Find the exact value of $\sin^{-1}\left(\sin\dfrac{11\pi}{6}\right)$. $-\dfrac{\pi}{6}$

46. Find the exact value of $\tan\left(\cos^{-1}\dfrac{7}{8}\right)$. $\dfrac{\sqrt{15}}{7}$

47. Find the exact value of $\cos\left[\arcsin\left(-\dfrac{5}{6}\right)\right]$. $\dfrac{\sqrt{11}}{6}$

48. Write the expression $\tan\left(\cos^{-1}\dfrac{x}{\sqrt{x^2 + 25}}\right)$ for $x \ne 0$ as an algebraic expression. $\dfrac{5}{x}$

49. A radar station tracks a plane flying at a constant altitude of 6 mi on a path directly over the station. Let θ be the angle of elevation from the radar station to the plane.

a. Write θ as a function of the plane's ground distance $x > 0$ from the station.

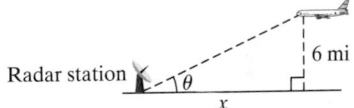

Radar station   6 mi

θ x

b. Without the use of a calculator, will the angle of elevation be less than 45° or greater than 45° when the plane's ground distance is 3.2 mi away?

c. Use a calculator to find θ to the nearest degree for $x = 3.2, 1.6,$ and 0.5 mi.   62°, 75°, 85°

CHAPTER 4   Cumulative Review Exercises

For Exercises 1–9, solve the equations and inequalities. Write the solution set to inequalities in interval notation.

1. $|3x - 5| + 3 \geq 6$ $\left(-\infty, \dfrac{2}{3}\right] \cup \left[\dfrac{8}{3}, \infty\right)$

2. $(x - 2)^2(x + 5) < 0$ $(-\infty, -5)$

3. $\dfrac{x + 2}{x - 3} \geq 2$ $(3, 8]$

4. $3x^4 - x^2 - 10 = 0$ $\left\{\pm\sqrt{2}, \pm\dfrac{\sqrt{15}}{3}i\right\}$

5. $\sqrt{x + 7} = 4 - \sqrt{x - 1}$ $\{2\}$

6. $2x^4 + 5x^3 - 29x^2 - 17x + 15 = 0$ $\left\{-5, -1, \dfrac{1}{2}, 3\right\}$

7. $e^{3x} = 2$ $\left\{\dfrac{\ln 2}{3}\right\}$

8. $\log(3x + 2) - \log x = 3$ $\left\{\dfrac{2}{997}\right\}$

9. $\ln(x - 1) + \ln x = \ln 2$ $\{2\}$

10. Given $x^2 + y^2 - 4x + 6y + 9 = 0$

   a. Write the equation of the circle in standard form.
   $(x - 2)^2 + (y + 3)^2 = 4$
   b. Identify the center and radius.
   Center: $(2, -3)$; Radius: 2
   c. Write the domain and range in interval notation.
   Domain: $[0, 4]$; Range: $[-5, -1]$

11. Given $y = -x^2 + 2x - 4$

   a. Identify the vertex. $(1, -3)$

   b. Write the domain and range in interval notation.
   Domain: $(-\infty, \infty)$; Range: $(-\infty, -3]$

12. Given $f(x) = x^4 - 2x^3 - 4x^2 + 8x$,

   a. Find the x-intercepts of the graph of f.
   $(-2, 0), (0, 0), (2, 0)$
   b. Determine the end behavior of the graph of f.
   Up to the left and up to the right

13. a. Graph $y = \dfrac{3x^2 + x - 5}{3x - 2}$.

   b. Identify the asymptotes.   Vertical asymptote: $x = \dfrac{2}{3}$;
   Slant asymptote: $y = x + 1$

14. Given $y = -2\sin(4x - \pi) - 6$, identify the

   a. Domain and range in interval notation.
   Domain: $(-\infty, \infty)$; Range: $[-8, -4]$
   b. Amplitude.   2

   c. Period. $\dfrac{\pi}{2}$

   d. Phase shift. $\dfrac{\pi}{4}$

   e. Vertical shift. -6

15. Given $f(x) = \tan x$ and $g(x) = \dfrac{x}{2}$, evaluate

   a. $(f \circ g)\left(\dfrac{\pi}{2}\right)$ 1

   b. $(g \circ f)(\pi)$ 0

16. Evaluate $\cos(-450°)$.   0

17. Evaluate $\tan\left[\arcsin\left(-\dfrac{3}{8}\right)\right]$. $-\dfrac{3\sqrt{55}}{55}$

18. Write the logarithm as the sum or difference of logarithms. Simplify as much as possible.

   $\log\left(\dfrac{x^2 y}{100z}\right)$ $2\log x + \log y - \log z - 2$

19. Divide. Write the answer in standard form, $a + bi$.
   $\dfrac{2 - 8i}{3 - 5i}$ $\dfrac{23}{17} - \dfrac{7}{17}i$

20. Suppose that y varies inversely as x and directly as z. If y is 12 when x is 8 and z is 3, find the constant of variation k. $k = 32$

Analytic Trigonometry

Chapter Outline

Most applications of trigonometry require the use of technology for the computation of the six trigonometric functions of a given angle. However, ancient and medieval mathematicians and astronomers had no such luxury. Instead, they built a table of sines: a table giving the sine of every integer angle between 1° and 90°. The cofunction and reciprocal identities can then be used to calculate the values of the other five trigonometric functions from the value of the sine.

For certain angles, such as 30°, 45°, and 60°, the sine of the angle is easy to compute geometrically (see Section 4.3). The trigonometric identities presented in this chapter then give us the means to compute the sine of many other angles.* For example, if θ is the sum of two angles u and v with known sine values, we can use the identity $\sin(u + v) = \sin u \cos v + \cos u \sin v$ to find $\sin \theta$ (Section 5.2). For example,

$$\sin 75° = \sin(45° + 30°) = \sin 45° \cos 30° + \cos 45° \sin 30°.$$

In addition to evaluating trigonometric functions of certain angles, the trigonometric identities are very useful in simplifying complex trigonometric expressions into more manageable forms and for solving trigonometric equations.

Table of Sines

θ	$\sin \theta$	θ	$\sin \theta$
1°	0.0174	31°	0.5150
2°	0.0349	32°	0.5299
3°	0.0523	33°	0.5446
4°	0.0698	34°	0.5592
5°	0.0872	35°	0.5736
6°	0.1045	36°	0.5878
...	...	...	...

---

*Not all integer angles between 1° and 90° can be found using the fundamental identities and the sine values of the special angles. In fact, only angles that are a multiple of 3° can be found in this way. For ancient astronomers and mathematicians, finding the value of $\sin 1°$ became a problem of critical importance and was addressed throughout the ages by using numerical methods of approximation.

OBJECTIVES

1. Simplify Trigonometric Expressions
2. Verify Trigonometric Identities
3. Write an Algebraic Expression as a Trigonometric Expression

1. Simplify Trigonometric Expressions

In Chapter 4 we presented the six trigonometric functions by using two different approaches. We used the lengths of the sides of right triangles, and we used coordinates taken from points on the unit circle. The resulting definitions yield consistent results and give rise to thousands of relationships among the functions. In this section, we use algebra skills to simplify trigonometric expressions and verify identities.

You have already been introduced to the fundamental trigonometric identities given in Table 5-1.

Table 5-1

Fundamental Identities			
Reciprocal Identities	$\csc x = \dfrac{1}{\sin x}$	$\sec x = \dfrac{1}{\cos x}$	$\cot x = \dfrac{1}{\tan x}$
	$\sin x = \dfrac{1}{\csc x}$	$\cos x = \dfrac{1}{\sec x}$	$\tan x = \dfrac{1}{\cot x}$
Quotient Identities	$\tan x = \dfrac{\sin x}{\cos x}$	$\cot x = \dfrac{\cos x}{\sin x}$	
Pythagorean Identities	$\sin^2 x + \cos^2 x = 1$	$\tan^2 x + 1 = \sec^2 x$	$1 + \cot^2 x = \csc^2 x$
Even and Odd Identities	$\sin(-x) = -\sin x$	$\cos(-x) = \cos x$	
	$\csc(-x) = -\csc x$	$\sec(-x) = \sec x$	
	$\tan(-x) = -\tan x$	$\cot(-x) = -\cot x$	

TIP It is often helpful to recognize alternative forms of the Pythagorean identities. For example, $\sin^2 x = 1 - \cos^2 x$ and $\cos^2 x = 1 - \sin^2 x$.

Each statement in Table 5-1 is true for every value of x for which the expressions within the statement are defined. For example,

$$\sin x = \frac{1}{\csc x} \text{ is true provided that } x \neq n\pi \text{ for integers } n.$$

The identities in Table 5-1 are tools to help us simplify trigonometric expressions and to verify other identities.

Classroom Example: p. 575
Exercise 26

EXAMPLE 1 Simplifying an Expression Using the Quotient and Reciprocal Identities

Simplify the expression. Write the final form with no fractions. $\sec^2 x \cot x \cos x$

Solution:

$\sec^2 x \cot x \cos x$

$= \dfrac{1}{\cos^2 x} \cdot \dfrac{\cos x}{\sin x} \cdot \dfrac{\cos x}{1}$ Write the trigonometric functions in terms of sine and cosine.

$= \dfrac{1}{\cancel{\cos^2 x}} \cdot \dfrac{\cancel{\cos x}}{\sin x} \cdot \dfrac{\cancel{\cos x}}{1}$ Simplify common factors from the numerator and denominator.

$= \dfrac{1}{\sin x}$ Simplify.

$= \csc x$ Apply the reciprocal identity relating the sine and cosecant functions.

Skill Practice 1 Simplify. Write the final form with no fractions.
$\tan x \cos^2 x \sec x$

Answer

1. $\sin x$

In Example 2, we simplify an expression by writing each term with a common denominator and adding the terms.

Classroom Example: p. 576
Exercise 30

| **EXAMPLE 2** | **Simplifying by Adding Fractional Expressions** |

Simplify the expression. Write the final form with no fractions. $\dfrac{\cos\theta}{1+\sin\theta}+\tan\theta$

Solution:

$\dfrac{\cos\theta}{1+\sin\theta}+\tan\theta$

$=\dfrac{\cos\theta}{1+\sin\theta}+\dfrac{\sin\theta}{\cos\theta}$

Write the trigonometric functions in terms of sine and cosine.

$=\dfrac{\cos\theta}{\cos\theta}\cdot\dfrac{\cos\theta}{(1+\sin\theta)}+\dfrac{\sin\theta}{\cos\theta}\cdot\dfrac{(1+\sin\theta)}{(1+\sin\theta)}$

The least common denominator is $\cos\theta(1+\sin\theta)$.

$=\dfrac{\cos^2\theta+\sin\theta+\sin^2\theta}{\cos\theta(1+\sin\theta)}$

Add the fractions.

$=\dfrac{\sin^2\theta+\cos^2\theta+\sin\theta}{\cos\theta(1+\sin\theta)}$

Group the squared terms because we recognize the Pythagorean identity, $\sin^2\theta+\cos^2\theta=1$.

$=\dfrac{1+\sin\theta}{\cos\theta(1+\sin\theta)}$

Substitute 1 for $\sin^2\theta+\cos^2\theta$.

$=\dfrac{1+\sin\theta}{\cos\theta(1+\sin\theta)}$

Simplify common factors from the numerator and denominator.

$=\dfrac{1}{\cos\theta}$

$=\sec\theta$

Apply the reciprocal identity relating the cosine and secant functions.

> **TIP** Recall that two fractions with different denominators b and d can be added as $\dfrac{a}{b}+\dfrac{c}{d}=\dfrac{ad+bc}{bd}$ provided that $b\neq0$ and $d\neq0$.

> **Skill Practice 2**   Simplify. Write the final form with no fractions.
> $\dfrac{\cos\theta}{\sin\theta}+\dfrac{\sin\theta}{1+\cos\theta}$

In Example 3, we make use of factoring techniques to simplify a trigonometric expression.

Classroom Example: p. 576
Exercise 36

| **EXAMPLE 3** | **Simplifying a Trigonometric Expression by Factoring** |

Simplify. Write the final form with no fractions. $\dfrac{\tan^2 t-1}{\tan t\sin t+\sin t}$

Answer

2. $\csc\theta$

Solution:

$$\frac{\tan^2 t - 1}{\tan t \sin t + \sin t}$$

$$= \frac{(\tan t - 1)(\tan t + 1)}{\sin t(\tan t + 1)}$$

The numerator factors as a difference of squares. The terms in the denominator share a common factor of sin t.

$$= \frac{(\tan t - 1)\cancel{(\tan t + 1)}}{\sin t\cancel{(\tan t + 1)}}$$

Simplify common factors from the numerator and denominator.

$$= \frac{\tan t - 1}{\sin t}$$

Simplify.

$$= \frac{\dfrac{\sin t}{\cos t} - 1}{\sin t}$$

Write $\tan t$ as $\dfrac{\sin t}{\cos t}$.

$$= \left(\frac{\sin t}{\cos t} - 1\right) \cdot \frac{1}{\sin t}$$

Multiply the numerator of the complex fraction by the reciprocal of the denominator.

$$= \frac{1}{\cos t} - \frac{1}{\sin t}$$

Apply the distributive property.

$$= \sec t - \csc t$$

Apply the reciprocal identities.

Skill Practice 3 Simplify. Write the final form with no fractions.

$$\frac{1 - \sec^2 t}{\tan t - \tan t \sec t}$$

---

2. Verify Trigonometric Identities

In Example 1, we showed that:

$$\sec^2 x \cot x \cos x = \csc x \qquad (1)$$

Example 1 also illustrates that trigonometric expressions can be written in many different equivalent forms. Equation (1) is called an **identity** because it is true for all values of x for which the expressions on the left and right are defined.

One method to verify that an equation is an identity is to manipulate one side of the equation (usually the more complicated side) until it takes the form of the other side of the equation. This process is the same as simplifying an expression, except that we know the final form.

While there is no explicit procedure to verify a trigonometric identity, we recommend some helpful guidelines.

> **TIP** Recall that a **conditional equation** is an equation that is true only for *some* values of the variable. For example, $\tan x = 1$ is true only for $x = \frac{\pi}{4} + n\pi$.

Guidelines for Verifying a Trigonometric Identity

1. Work with one side of the equation (usually the more complicated side) and keep the other side in mind as your final goal.
2. Look for opportunities to apply the fundamental identities.
   - If the expression is a product or quotient of factors, consider the reciprocal and quotient identities.
   - If squared terms are present, look to see if the terms can be grouped in one of the forms of a Pythagorean identity.
   - If an expression involves a negative argument, consider using the even or odd function identities.
3. Apply basic algebraic techniques such as factoring, multiplying terms, combining like terms, and writing fractions with a common denominator.
4. Consider writing expressions explicitly in terms of sine and cosine.

Answer

3. $\cot t + \csc t$

Classroom Example: p. 576
Exercise 40

EXAMPLE 4 Verifying an Identity Containing Negative Arguments

Verify that the equation is an identity. $\dfrac{\cos(-x)\tan(-x)}{\sin x} = -1$

Solution:

$\dfrac{\cos(-x)\tan(-x)}{\sin x} = -1$ The left-hand side (LHS) is the more complicated side.

$= \dfrac{\cos x(-\tan x)}{\sin x}$ To simplify, we require the same arguments for each factor. The cosine function is an even function. Thus, $\cos(-x) = \cos x$. The tangent function is an odd function. Thus, $\tan(-x) = -\tan x$.

$= \dfrac{\cos x\left(-\dfrac{\sin x}{\cos x}\right)}{\sin x}$ Write tangent as the ratio of sine and cosine.

$= \dfrac{-\sin x}{\sin x}$

$= -1$ (RHS)   This equals the right-hand side (RHS) of the original equation. Therefore, we have verified the identity.

Skill Practice 4 Verify that the equation is an identity. $\dfrac{\sin(-x)\cot(-x)}{\cos x} = 1$

When we verify an identity, we are trying to show that the expression on one side equals the expression on the other. Therefore, we cannot use the equality of the statement because we have not yet proved it. For this reason, we work with the left or right side of the equation, but not both. In Example 4, we worked only with the left side of the equation.

Classroom Example: p. 576
Exercise 46

EXAMPLE 5 Verifying an Identity by Combining Fractions

Verify that the equation is an identity. $\dfrac{1}{1-\cos x} - \dfrac{1}{1+\cos x} = 2\cot x\csc x$

Solution:

$\dfrac{1}{1-\cos x} - \dfrac{1}{1+\cos x}$ The LHS has two terms and the RHS has one term. Therefore, a good strategy is to combine the two terms on the left by adding the fractions.

$= \dfrac{(1+\cos x)}{(1+\cos x)} \cdot \dfrac{1}{(1-\cos x)} - \dfrac{1}{(1+\cos x)} \cdot \dfrac{(1-\cos x)}{(1-\cos x)}$ The LCD is $(1+\cos x)(1-\cos x)$.

$= \dfrac{(1+\cos x) - (1-\cos x)}{(1+\cos x)(1-\cos x)}$ Subtract the fractions.

$= \dfrac{1+\cos x - 1 + \cos x}{1-\cos^2 x}$ Apply the distributive property in the numerator. Multiply conjugates in the denominator.

$= \dfrac{2\cos x}{\sin^2 x}$ In the denominator, from the Pythagorean identity $\sin^2 x + \cos^2 x = 1$, we know that $\sin^2 x = 1 - \cos^2 x$.

$= 2 \cdot \dfrac{\cos x}{\sin x} \cdot \dfrac{1}{\sin x}$ The RHS of the original equation has factors of $\cot x$ and $\csc x$. Therefore, we group factors in the numerator and denominator conveniently to reach the desired result.

$= 2\cot x\csc x$ (RHS)

Answer

4. $\dfrac{\sin(-x)\cot(-x)}{\cos x} = \dfrac{-\sin x(-\cot x)}{\cos x}$

$= \dfrac{\sin x\left(\dfrac{\cos x}{\sin x}\right)}{\cos x} = \dfrac{\cos x}{\cos x} = 1$

Skill Practice 5   Verify that the equation is an identity.

$$\frac{1}{1 - \sin x} - \frac{1}{1 + \sin x} = 2\tan x \sec x$$

Keep in mind that the Pythagorean identities have several alternative forms.

$$\sin^2 x + \cos^2 x = 1 \qquad \tan^2 x + 1 = \sec^2 x \qquad 1 + \cot^2 x = \csc^2 x$$
$$1 - \cos^2 x = \sin^2 x \qquad \sec^2 x - 1 = \tan^2 x \qquad \csc^2 x - 1 = \cot^2 x$$
$$1 - \sin^2 x = \cos^2 x \qquad \sec^2 x - \tan^2 x = 1 \qquad \csc^2 x - \cot^2 x = 1$$

These alternative forms each involve a difference of squares that factors as a product of conjugates. This was illustrated in steps 2–5 of Example 5.

$$(1 + \cos x)(1 - \cos x) = 1 - \cos^2 x = \sin^2 x$$

Notice that upon simplification, the expression ultimately results in a single term. This pattern is sometimes helpful when we want to simplify a fraction with a denominator containing two terms as a fraction with a single-term denominator.

Classroom Example: p. 576
Exercise 52

EXAMPLE 6   Verifying an Identity by Using Conjugates

Verify that the equation is an identity. $\dfrac{1 - \sin t}{1 + \sin t} = (\sec t - \tan t)^2$

Solution:

$$\frac{1 - \sin t}{1 + \sin t} = (\sec t - \tan t)^2$$

Since the RHS has no fractions, one strategy might be to multiply the numerator and denominator of the LHS by the conjugate of the denominator. The purpose is to create the difference of squares $1 - \sin^2 t$ which then simplifies to the single term $\cos^2 t$.

$$\frac{1 - \sin t}{1 + \sin t} = \frac{(1 - \sin t)}{(1 + \sin t)} \cdot \frac{(1 - \sin t)}{(1 - \sin t)}$$

$$= \frac{(1 - \sin t)^2}{1 - \sin^2 t}$$

The product of conjugates in the denominator results in a difference of squares.

$$= \frac{(1 - \sin t)^2}{\cos^2 t}$$

Substitute $\cos^2 t$ for $1 - \sin^2 t$.

$$= \left(\frac{1 - \sin t}{\cos t}\right)^2$$

Write the quotient of squares as the square of a quotient.

$$= \left(\frac{1}{\cos t} - \frac{\sin t}{\cos t}\right)^2$$

Use the property $\dfrac{a - b}{c} = \dfrac{a}{c} - \dfrac{b}{c}$.

$$= (\sec t - \tan t)^2 \quad \text{(RHS)}$$

Apply reciprocal and quotient identities.

Skill Practice 6   Verify that the equation is an identity. $1 - \dfrac{\sin^2 t}{1 + \cos t} = \cos t$

It is important to understand that there are often many paths to simplifying a trigonometric identity. A different set of algebraic steps can be used, or we can manipulate the other side of the expression. For example, the expression in Example 6 can also be simplified by manipulating the RHS.

Answers

5. See page SA-40.

6. See page SA-40.

$$(\sec t - \tan t)^2 = \left(\frac{1}{\cos t} - \frac{\sin t}{\cos t}\right)^2 = \left(\frac{1 - \sin t}{\cos t}\right)^2 = \frac{(1 - \sin t)^2}{\cos^2 t}$$

Write the expression in terms of $\sin t$ and $\cos t$. Combine terms on the left by subtracting fractions.

$$= \frac{(1 - \sin t)^2}{1 - \sin^2 t} = \frac{(1 - \sin t)^2}{(1 - \sin t)(1 + \sin t)} = \frac{1 - \sin t}{1 + \sin t} = \text{LHS}$$

Factor the denominator and simplify common factors in the numerator and denominator.

Classroom Example: p. 576
Exercise 54

TIP Recall that
1. $\log_b 1 = 0$
2. $\log_b b = 1$
3. $\log_b b^p = p$
4. $b^{\log_b x} = x$
5. $\log_b(xy) = \log_b x + \log_b y$
6. $\log_b\left(\dfrac{x}{y}\right) = \log_b x - \log_b y$
7. $\log_b x^p = p\log_b x$

EXAMPLE 7 Verifying an Identity Involving Logarithms

Verify that the equation is an identity. $\ln|\sin x| + \ln|\sec x| = \ln|\tan x|$

Solution:

$\ln|\sin x| + \ln|\sec x|$ The LHS is the more complicated side.

$= \ln|\sin x| + \ln\left|\dfrac{1}{\cos x}\right|$ Write $\sec x$ as $\dfrac{1}{\cos x}$.

$= \ln\left(|\sin x| \cdot \left|\dfrac{1}{\cos x}\right|\right)$ Apply the property $\ln x + \ln y = \ln(xy)$.

$= \ln\left|\dfrac{\sin x}{\cos x}\right|$ Apply the property of absolute value $|a| \cdot |b| = |ab|$.

$= \ln|\tan x|$ (RHS)   Write $\dfrac{\sin x}{\cos x}$ as $\tan x$.

Skill Practice 7 Verify that the equation is an identity.
$\ln|\cot x| + \ln|\sin x| = \ln|\cos x|$

When verifying an identity, we generally work on one side of the equation only and manipulate the expression until it takes the form of the other side. It is logically equivalent, however, to verify an identity by transforming each side of the equation *independently* until we reach a common equivalent expression. This is demonstrated in Example 8.

Classroom Example: p. 576
Exercise 58

EXAMPLE 8 Verifying an Identity by Working Each Side Independently

Verify that the equation is an identity by manipulating each side of the equation independently. $\dfrac{\sin x}{1 + \sin x} = \dfrac{\csc x - 1}{\cot^2 x}$

Solution:

LHS: $\dfrac{\sin x}{1 + \sin x}$

$= \dfrac{(\sin x)}{(1 + \sin x)} \cdot \dfrac{(1 - \sin x)}{(1 - \sin x)}$

$= \dfrac{\sin x(1 - \sin x)}{1 - \sin^2 x}$

$= \dfrac{\sin x - \sin^2 x}{\cos^2 x}$

RHS: $\dfrac{\csc x - 1}{\cot^2 x}$

$= \dfrac{\dfrac{1}{\sin x} - 1}{\dfrac{\cos^2 x}{\sin^2 x}} = \dfrac{\dfrac{\sin^2 x}{1} \cdot \left(\dfrac{1}{\sin x} - 1\right)}{\dfrac{\sin^2 x}{1} \cdot \left(\dfrac{\cos^2 x}{\sin^2 x}\right)}$

$= \dfrac{\sin x - \sin^2 x}{\cos^2 x}$

TIP On the left side of the identity, we multiplied numerator and denominator by the conjugate of the denominator. Recall that the product of conjugates results in a difference of squares. The motivation is to use the Pythagorean relationship
$1 - \sin^2 x = \cos^2 x$.

Answer
7. See page SA-40.

Since the left-hand side equals the right-hand side, we have verified that the equation is an identity.

Skill Practice 8 Verify the equation is an identity by manipulating each side of the equation independently. $\dfrac{1 - \cos x}{\cos x} = \dfrac{\tan x \sin x}{1 + \cos x}$

The process used in Example 8 was to manipulate each side of the equation independently into a third, equivalent expression. In a rigorous setting, this method is frowned upon. In Example 8, we could have manipulated only one side (say the LHS) by first reaching the intermediate step $\dfrac{\sin x - \sin^2 x}{\cos^2 x}$. The remaining steps to verify the identity would be to reverse the steps used to manipulate the RHS.

3. Write an Algebraic Expression as a Trigonometric Expression

In some applications of calculus, it is advantageous to write an algebraic expression of the form $\sqrt{u^2 + a^2}$, $\sqrt{a^2 - u^2}$, or $\sqrt{u^2 - a^2}$ as a trigonometric expression by way of an appropriate substitution. Notice that each algebraic expression looks similar to an expression that might be used to find the length of a side of a right triangle using the Pythagorean theorem. Therefore, it seems reasonable that these algebraic expressions may be linked to trigonometric expressions.

Classroom Example: p. 576
Exercise 62

EXAMPLE 9 Applying Trigonometric Substitution

Write $\sqrt{x^2 + 9}$ as a function of θ, where $0 < \theta < \dfrac{\pi}{2}$, by making the substitution $x = 3 \tan \theta$.

Solution:

$$\sqrt{x^2 + 9} = \sqrt{(3 \tan \theta)^2 + 9} \qquad \text{Substitute } 3\tan\theta \text{ for } x.$$
$$= \sqrt{9 \tan^2 \theta + 9} \qquad \text{Square the expression } 3\tan\theta.$$
$$= \sqrt{9(1 + \tan^2 \theta)} \qquad \text{Factor.}$$
$$= \sqrt{9 \sec^2 \theta} \qquad \text{Apply the Pythagorean identity } 1 + \tan^2 \theta = \sec^2 \theta.$$
$$= 3 \sec \theta \qquad \text{Take the positive square root, since } 0 < \theta < \dfrac{\pi}{2}.$$

Skill Practice 9 Write $\sqrt{x^2 - 121}$ as a function of θ, where $0 < \theta < \dfrac{\pi}{2}$, by making the substitution $x = 11 \sec \theta$.

Answers

8. LHS:
$$\frac{1 - \cos x}{\cos x} = \frac{(1 - \cos x)}{\cos x} \cdot \frac{(1 + \cos x)}{(1 + \cos x)}$$
$$= \frac{1 - \cos^2 x}{\cos x(1 + \cos x)} = \frac{\sin^2 x}{\cos x(1 + \cos x)}$$
RHS:
$$\frac{\tan x \sin x}{1 + \cos x} = \frac{\dfrac{\sin x}{\cos x} \cdot \sin x}{1 + \cos x} = \frac{\sin^2 x}{\cos x(1 + \cos x)}$$

9. $11 \tan \theta$

From Example 9, we can illustrate the link between the algebraic form $\sqrt{x^2 + 9}$ and the trigonometric form $3 \sec \theta$ by using a right triangle. The expression $\sqrt{x^2 + 9}$ represents the hypotenuse of a right triangle with legs of lengths x and 3 (Figure 5-1).

From the triangle, $\sec \theta = \dfrac{\sqrt{x^2 + 9}}{3}$, so $\sqrt{x^2 + 9} = 3 \sec \theta$.

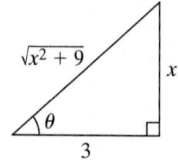

Figure 5-1

SECTION 5.1   Practice Exercises

Prerequisite Review

R.1. Factor. $5r^2 + 33rs + 18s^2$ $(5r + 3s)(r + 6s)$

R.2. Find the least common denominator.

$$\frac{3}{(c-3)(c+3)}; \frac{-6}{(c-3)(c-1)} \quad (c-3)(c+3)(c-1)$$

R.3. Subtract. $\dfrac{3a}{a+9} - \dfrac{2}{a-5}$ $\dfrac{3a^2 - 17a - 18}{(a+9)(a-5)}$

R.4. Add. $\dfrac{y-4}{y-5} + \dfrac{5y^2 - 38y + 55}{y^2 - 25}$ $\dfrac{6y-7}{y+5}$

R.5. Simplify. $\dfrac{\frac{8x^2}{19y^2}}{\frac{3x}{xy^2}}$ $\dfrac{8x^2}{57}$

R.6. Simplify. $\dfrac{\frac{8}{n} + \frac{3}{n^2}}{\frac{6}{n^2} - \frac{7}{n}}$ $\dfrac{8n+3}{6-7n}$ or $-\dfrac{8n+3}{7n-6}$

R.7. Assume that y represents a positive real number. What is the conjugate of $\sqrt{y} - 24$? $\sqrt{y} + 24$

R.8. Rationalize the denominator. $\dfrac{5}{6 - \sqrt{3}}$ $\dfrac{30 + 5\sqrt{3}}{33}$

Concept Connections

1. The value $\tan\theta$ is the quotient of ____$\sin\theta$____ and ____$\cos\theta$____, and $\cot\theta$ is the quotient of ____$\cos\theta$____ and ____$\sin\theta$____.

2. The value $\csc\theta$ is the reciprocal of ____$\sin\theta$____, $\cot\theta$ is the reciprocal of ____$\tan\theta$____, and $\sec\theta$ is the reciprocal of ____$\cos\theta$____.

3. Given the Pythagorean identity $\sin^2 x + \cos^2 x = 1$, write two alternative forms that involve the difference of squares. $\sin^2 x = 1 - \cos^2 x$, $\cos^2 x = 1 - \sin^2 x$, $-\sin^2 x = \cos^2 x - 1$, $-\cos^2 x = \sin^2 x - 1$

4. Given the Pythagorean identity $\tan^2 x + 1 = \sec^2 x$, write two alternative forms that involve the difference of squares. $\tan^2 x = \sec^2 x - 1$, $\sec^2 x - \tan^2 x = 1$, $-\tan^2 x = 1 - \sec^2 x$, $\tan^2 x - \sec^2 x = -1$

5. $\sin(-x) =$ ____$-\sin x$____, $\cos(-x) =$ ____$\cos x$____, and $\tan(-x) =$ ____$-\tan x$____.

6. An equation that is true for all values of the variable except where the individual expressions are not defined is called an ____identity____.

Objective 1: Simplify Trigonometric Expressions

For Exercises 7–10, find the least common denominator. Then rewrite each expression with the new denominator.

7. a. $\dfrac{1}{a}; \dfrac{1}{b}$

b. $\dfrac{1}{\cos x}; \dfrac{1}{\sin x}$

8. a. $\dfrac{a}{b}; a$

b. $\dfrac{\sin x}{\cos x}; \sin x$

9. a. $\dfrac{1}{1+a}; \dfrac{a}{b}$

b. $\dfrac{1}{1+\sin x}; \dfrac{\sin x}{\cos x}$

10. a. $\dfrac{1}{a}; \dfrac{a}{1-a}$

b. $\dfrac{1}{\cos x}; \dfrac{\cos x}{1-\cos x}$

For Exercises 11–12, multiply.

11. a. $(a+b)^2$ $a^2 + 2ab + b^2$ **b.** $(\cos x + \cot x)^2$ $\cos^2 x + 2\cos x \cot x + \cot^2 x$

12. a. $(a-b)^2$ $a^2 - 2ab + b^2$ **b.** $(\sin x - \tan x)^2$ $\sin^2 x - 2\sin x \tan x + \tan^2 x$

For Exercises 13–20, factor each expression.

13. a. $a^2 - b^2$ **b.** $\sin^2 x - \cos^2 x$

14. a. $a^2 + b^2$ **b.** $\tan^2 x + \sec^2 x$

15. a. $a^3 + b^3$ **b.** $\sin^3 x + \sec^3 x$

16. a. $a^3 - b^3$ **b.** $\cos^3 x - \cot^3 x$

17. a. $a^2 + 2ab + b^2$ **b.** $\cos^2 x + 2\cos x \csc x + \csc^2 x$

18. a. $a^2 - 2ab + b^2$ **b.** $\sin^2 x - 2\sin x \tan x + \tan^2 x$

19. a. $3a^2 - 4ab - 4b^2$ **b.** $3\sin^2 x - 4\sin x \cos x - 4\cos^2 x$

20. a. $4a^2 - 12ab + 5b^2$ **b.** $4\cos^2 x - 12\cos x \sin x + 5\sin^2 x$

For Exercises 21–36, simplify the expression. Write the final form with no fractions. (See Examples 1–3)

21. $\sin x \sec x \cot x$ 1

22. $\cos x \tan x \csc x$ 1

23. $\dfrac{\sin x}{\sec x \tan x}$ $\cos^2 x$

24. $\dfrac{\sin x}{\sec x \cot x}$ $\sin^2 x$

25. $\tan^2 x \cot x \csc x$ $\sec x$

26. $\csc^2 x \tan^2 x \sec x$ $\sec^3 x$

27. $\dfrac{\tan^2 x}{\sin x \sec^2 x}$ $\sin x$

28. $\dfrac{\tan x}{\sin x \sec^2 x}$ $\cos x$

29. $\dfrac{\cos^2 x + 1}{\sin x} + \sin x$ $2\csc x$

30. $\dfrac{\sin^2 x + 1}{\cos^2 x} + 1$ $2\sec^2 x$

31. $\dfrac{\csc\theta - \sin\theta}{\csc\theta}$ $\cos^2\theta$

32. $\dfrac{\cos\theta - \sec\theta}{\sec\theta}$ $-\sin^2\theta$

33. $\dfrac{\sin^2 x + 2\sin x + 1}{\sin^2 x + \sin x}$ $1 + \csc x$

34. $\dfrac{\sin x \sec^2 x - \sin x}{\sin x \sec x + \sin x}$ $\sec x - 1$

35. $\dfrac{\cos^2 x - 2\cos x + 1}{\sin x \cos x - \sin x}$ $\cot x - \csc x$

36. $\dfrac{\tan^2 x + 2\tan x + 1}{\sin x \tan x + \sin x}$ $\sec x + \csc x$

Objective 2: Verify Trigonometric Identities

For Exercises 37–40, verify that the equation is an identity. (See Example 4)

37. $\sin(-x) + \csc x = \cot x \cos x$

38. $\sin(-x) + \cot(-x)\cos(-x) = -\csc x$

39. $\cot x[\cot(-x) + \tan(-x)] = -\csc^2 x$

40. $\dfrac{\csc^2(-x)\tan x}{\cot(-x)} = -\sec^2 x$

For Exercises 41–46, verify that the equation is an identity. (See Example 5)

41. $\dfrac{\sec x}{\tan x} - \dfrac{\tan x}{\sec x} = \cos x \cot x$

42. $\dfrac{\cot x}{\csc x} - \dfrac{\csc x}{\cot x} = -\sin x \tan x$

43. $\dfrac{1}{1 + \sin t} + \dfrac{1}{1 - \sin t} = 2\sec^2 t$

44. $\dfrac{1}{\sec t - 1} - \dfrac{1}{\sec t + 1} = 2\cot^2 t$

45. $\dfrac{1}{1 + \sec x} - \dfrac{1}{1 - \sec x} = 2\cot x \csc x$

46. $\dfrac{1}{1 - \csc x} - \dfrac{1}{1 + \csc x} = -2\tan x \sec x$

For Exercises 47–52, verify that the equation is an identity. (See Example 6)

47. $\dfrac{\sin\theta}{\csc\theta - \cot\theta} = 1 + \cos\theta$

48. $\dfrac{\cos\theta}{\tan\theta + \sec\theta} = 1 - \sin\theta$

49. $\dfrac{1}{\cos x + \sin x \cos x} = \dfrac{1 - \sin x}{\cos^3 x}$

50. $\dfrac{1}{\sin x - \sin^2 x} = \dfrac{1 + \sin x}{\sin x \cos^2 x}$

51. $\dfrac{\sec x + 1}{\sec x - 1} = (\csc x + \cot x)^2$

52. $\dfrac{\sin x + \cos x}{\sin x - \cos x} = \dfrac{1 + 2\sin x \cos x}{1 - 2\cos^2 x}$

For Exercises 53–56, verify that the equation is an identity. (See Example 7)

53. $\ln|\cos t| - \ln|\cot t| = \ln|\sin t|$

54. $\ln|\cot t| - \ln|\tan t| = 2\ln|\cot t|$

55. $\ln|\sec\theta + \tan\theta| = -\ln|\sec\theta - \tan\theta|$

56. $\ln|\csc\theta + \cot\theta| = -\ln|\csc\theta - \cot\theta|$

For Exercises 57–60, simplify each side of the equation independently to reach a common equivalent expression. (See Example 8)

57. $\dfrac{\cos x}{1 - \sin x} = \sec x + \tan x$

58. $\dfrac{1 + \tan^2 x}{\tan x} = \dfrac{\cos x}{\sin x - \sin^3 x}$

59. $\cot^2 t - \cos^2 t = \csc^2 t \cos^2 t - \cos^2 t$

60. $\tan^2 t + \sin^2 t = \sec^2 t - \cos^2 t$

Objective 3: Write an Algebraic Expression as a Trigonometric Expression

For Exercises 61–68, write the given algebraic expression as a function of θ, where $0 < \theta < \dfrac{\pi}{2}$, by making the given substitution. (See Example 9)

61. $\sqrt{x^2 + 25}$; Substitute $x = 5\tan\theta$. $\sqrt{x^2 + 25} = 5\sec\theta$

62. $\sqrt{49 - x^2}$; Substitute $x = 7\sin\theta$. $\sqrt{49 - x^2} = 7\cos\theta$

63. $\sqrt{16 - (x - 1)^2}$; Substitute $x = 4\sin\theta + 1$. $\sqrt{16 - (x - 1)^2} = 4\cos\theta$

64. $\sqrt{(x + 2)^2 - 36}$; Substitute $x = 6\sec\theta - 2$. $\sqrt{(x + 2)^2 - 36} = 6\tan\theta$

65. $\dfrac{\sqrt{9 - x^2}}{x}$; Substitute $x = 3\sin\theta$. $\dfrac{\sqrt{9 - x^2}}{x} = \cot\theta$

66. $\dfrac{\sqrt{x^2 + 16}}{x}$; Substitute $x = 4\tan\theta$. $\dfrac{\sqrt{x^2 + 16}}{x} = \csc\theta$

67. $\dfrac{1}{(49x^2 - 1)^{3/2}}$; Substitute $x = \dfrac{1}{7}\sec\theta$. $\dfrac{1}{(49x^2 - 1)^{3/2}} = \cot^3\theta$

68. $\dfrac{1}{(1 - 100x^2)^{3/2}}$; Substitute $x = \dfrac{1}{10}\sin\theta$. $\dfrac{1}{(1 - 100x^2)^{3/2}} = \sec^3\theta$

Mixed Exercises

For Exercises 69–72, factor the expression and simplify factors using fundamental identities if possible.

69. $2\sin^2 x + 5\sin x - 3$ $\quad (2\sin x - 1)(\sin x + 3)$

70. $2\cos^2 x - \cos x - 1$ $\quad (\cos x - 1)(2\cos x + 1)$

71. $\sin^4 x - \cos^4 x$ $\quad (\sin x - \cos x)(\sin x + \cos x)$

72. $1 - \cot^4 x$ $\quad \csc^2 x(1 + \cot x)(1 - \cot x)$

For Exercises 73–104, verify that the equation is an identity.

73. $(\cos^2 \theta - 1)(\csc^2 \theta - 1) = -\cos^2 \theta$

74. $(1 - \sin^2 \theta)(1 + \tan^2 \theta) = 1$

75. $\sec x \cdot \dfrac{\tan x}{\sin x} = \sec^2 x$

76. $\sec x \cdot \dfrac{\cot x}{\sin x} = \csc^2 x$

77. $\dfrac{\sec x \sin x}{\csc x \cos x} = \tan^2 x$

78. $\dfrac{\sin x \cot x}{\cos x \tan x} = \cot x$

79. $\dfrac{2 + \sin^2 t - 3\sin^4 t}{\cos^2 t} = 2 + 3\sin^2 t$

80. $\dfrac{2\tan^4 x + 7\tan^2 x + 5}{\sec^2 x} = 2\tan^2 x + 5$

81. $\dfrac{\tan x \sin x}{\tan x - \sin x} = \dfrac{\sin x}{1 - \cos x}$

82. $\dfrac{\csc x \cos x}{\csc x - \sin x} = \sec x$

83. $(\sec \theta + \tan \theta)^2 = \dfrac{1 + \sin \theta}{1 - \sin \theta}$

84. $(\csc t + \cot t)^2 = \dfrac{1 + \cos t}{1 - \cos t}$

85. $(1 + \sin x)[1 + \sin(-x)] = \cos^2 x$

86. $[1 + \csc(-x)](1 + \csc x) = -\cot^2 x$

87. $\dfrac{\sin x}{\cos x + 1} = \dfrac{1 - \cos x}{\sin x}$

88. $\dfrac{\tan x}{\sec x - 1} = \dfrac{\sec x + 1}{\tan x}$

89. $\dfrac{1 + \sin \theta}{\tan \theta} = \dfrac{\csc \theta + 1}{\sec \theta}$

90. $\dfrac{\sec \theta + 1}{\csc \theta} = \dfrac{1 + \cos \theta}{\cot \theta}$

91. $\dfrac{\sin^3 t - \cos^3 t}{\cos t \sin t - \cos^2 t} = \sec t + \sin t$

92. $\dfrac{\cos^3 t + \sin^3 t}{\cos t \sin t + \sin^2 t} = \csc t - \cos t$

93. $\dfrac{\tan^4 \theta - 4}{\sec^2 \theta + 1} = \sec^2 \theta - 3$

94. $\dfrac{\sin^4 \theta - 9}{\cos^2 \theta + 2} = \cos^2 \theta - 4$

95. $\dfrac{\cot^3 z - 1}{\cot z - 1} - \cot z = \csc^2 z$

96. $\dfrac{\tan^3 z + 1}{\tan z + 1} + \tan z = \sec^2 z$

97. $\cos x \tan x - \sec x \cot x = -\cot x \cos x$

98. $\sin x \cot x + \cos x \tan^2 x = \sec x$

99. $\log|\cos x| - \log|\sin x| + \log|\sec x| = \log|\csc x|$

100. $\ln|\cot x| + \ln|\sec x| + \ln|\sin x| = 0$

101. $e^{\ln(\sin^2 x + \cos^2 x)} = 1$

102. $\log 100^{\sin x} = 2\sin x$

103. $\dfrac{1}{\sin x + \cos x} + \dfrac{1}{\sin x - \cos x} = \dfrac{2\sin x}{1 - 2\cos^2 x}$

104. $\dfrac{1}{\sin x + \cos x} - \dfrac{\sin x}{(\sin x + \cos x)^2} = \dfrac{\cos x}{1 + 2\sin x \cos x}$

105. Given $f(x) = \sqrt{1 - x^2}$ and $g(x) = \cos x$, show that $(f \circ g)(x) = |\sin x|$.

106. Given $f(x) = \dfrac{1}{\sqrt{x^2 + 1}}$ and $g(x) = \tan x$, show that $(f \circ g)(x) = |\cos x|$.

Write About It

107. Determine if the following logic is correct and explain why or why not.

$$\sin x \sec x = 1 \text{ because } \sin\frac{\pi}{4}\sec\frac{\pi}{4} = \frac{\sqrt{2}}{2} \cdot \sqrt{2} = \frac{2}{2} = 1$$

108. Determine if the following procedure to verify an identity is correct and explain why or why not.

$$\frac{\csc\theta - 1}{\cot\theta} = \frac{\cot\theta}{\csc\theta + 1}$$

Multiply both sides by the LCD: $\cot\theta(\csc\theta + 1)$

$$\cot\theta(\csc\theta + 1)\left(\frac{\csc\theta - 1}{\cot\theta}\right) = \left(\frac{\cot\theta}{\csc\theta + 1}\right)\cot\theta(\csc\theta + 1)$$

$$(\csc\theta + 1)(\csc\theta - 1) = \cot^2\theta$$
$$\csc^2\theta - 1 = \cot^2\theta \; \checkmark$$

109. Explain how to verify a trigonometric identity.

Expanding Your Skills

For Exercises 110–116, verify that the equation is an identity.

110. $\sqrt{\dfrac{1 + \cos x}{1 - \cos x}} = \csc x + \cot x$, for $0 < x < \dfrac{\pi}{2}$.

111. $\sqrt{\dfrac{\csc x - \cot x}{\csc x + \cot x}} = \dfrac{1 - \cos x}{\sin x}$, for $0 < x < \dfrac{\pi}{2}$.

112. $(\sin x + \cos y)^2 + (\sin x - \cos y)(\sin x + \cos y) = 2\sin x(\sin x + \cos y)$

113. $(\tan x + \tan y)(\tan x - \tan y) + 2\tan y(\tan y - \tan x) = (\tan x - \tan y)^2$

114. $(A\cot\theta - B)(A\cot\theta + B) - (B\cot\theta - A)(B\cot\theta + A) = (A^2 - B^2)\csc^2\theta$

115. $\dfrac{\sec^2 x\tan x - 25\tan x}{\sec^3 x + 5\sec^2 x - \sec x - 5} = \csc x - 5\cot x$

116. $\sin^6 x - \cos^6 x = (\sin^2 x - \cos^2 x)(1 - \sin^2 x\cos^2 x)$

Technology Connections

117. An equation of the form $f(x) = g(x)$ is given. Use a graphing utility to graph $y = f(x)$ and $y = g(x)$ on the same screen on the interval $[0, 2\pi]$ (be sure to use radian mode). Make a conjecture as to whether the equation is an identity, a conditional equation, or a contradiction.

   a. $\sin 2x = 2\sin x$

   b. $\cos 2x = 2\cos^2 x - 1$

   c. $\dfrac{\sec x\sin x}{\tan x} = 2$

118. Graph the function in the standard viewing window to determine the constant value it might be equivalent to. Confirm your conjecture algebraically.

$$y = (2\cos x - \sin x)^2 + (\cos x + 2\sin x)^2$$

Additional answers can be found in the Instructor Answer Appendix.

SECTION 5.2   Sum and Difference Formulas

OBJECTIVES

1. Apply the Sum and Difference Formulas for Sine and Cosine
2. Apply the Sum and Difference Formulas for Tangent
3. Use Sum and Difference Formulas to Verify Identities
4. Write a Sum of $A\sin x$ and $B\cos x$ as a Single Term

1. Apply the Sum and Difference Formulas for Sine and Cosine

In Exercise 32 from Section 4.3, we evaluated the expressions $\sin 60°$ and $\sin 30° + \sin 30°$ and found that they are not equal. That is,

$$\sin 60° = \frac{\sqrt{3}}{2}, \text{ whereas } \sin 30° + \sin 30° = \frac{1}{2} + \frac{1}{2} = 1.$$

So the question arises, can we express $\sin(u \pm v)$ and $\cos(u \pm v)$ in terms of trigonometric functions of u and v alone?

We begin by investigating $\cos(u - v)$. Let points P and Q be the points on the unit circle associated with the real numbers u and v, measured from the point $S(1, 0)$. For simplicity, we show the case where u and v are positive and $u - v > 0$ (Figure 5-2). Let R be the point on the unit circle associated with the real number $u - v$ (Figure 5-3).

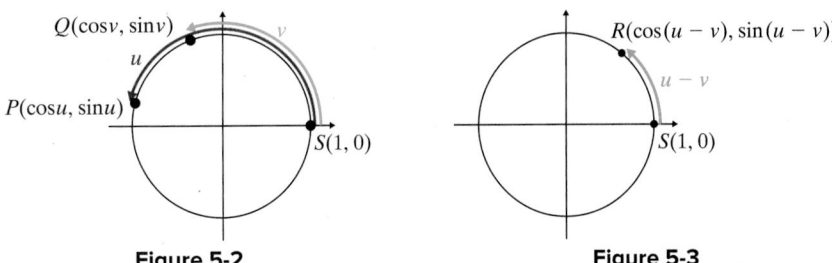

| Figure 5-2 | Figure 5-3 |

TIP A chord of a circle is a line segment with endpoints on the circle.

The lengths of the arcs between P and Q and between R and S are the same. Consequently, the lengths of the chords $\overline{PQ}$ and $\overline{RS}$ are also equal. By the distance formula we have:

$$\sqrt{(\cos u - \cos v)^2 + (\sin u - \sin v)^2} = \sqrt{[\cos(u - v) - 1]^2 + [\sin(u - v) - 0]^2}$$

Squaring both sides and expanding the squared expressions yields

LHS: $\cos^2 u - 2\cos u \cos v + \cos^2 v + \sin^2 u - 2\sin u \sin v + \sin^2 v$

$= \cos^2 u + \sin^2 u - 2\cos u \cos v + \cos^2 v + \sin^2 v - 2\sin u \sin v$ Group the squared terms.

$= 1 - 2\cos u \cos v + 1 - 2\sin u \sin v$ Substitute $\sin^2 u + \cos^2 u = 1$, $\sin^2 v + \cos^2 v = 1$.

$= 2 - 2\cos u \cos v - 2\sin u \sin v$

RHS: $\cos^2(u - v) - 2\cos(u - v) + 1 + \sin^2(u - v)$

$= \cos^2(u - v) + \sin^2(u - v) - 2\cos(u - v) + 1$ Group the squared terms.

$= 1 - 2\cos(u - v) + 1$ Substitute $\cos^2(u - v) + \sin^2(u - v) = 1$.

$= 2 - 2\cos(u - v)$

Thus,

$$2 - 2\cos u \cos v - 2\sin u \sin v = 2 - 2\cos(u - v)$$

$$-2\cos u \cos v - 2\sin u \sin v = -2\cos(u - v)$$ Subtract 2.

$$\cos u \cos v + \sin u \sin v = \cos(u - v)$$ Divide by -2.

Similar identities for $\cos(u + v)$, $\sin(u \pm v)$, and $\tan(u \pm v)$ are given in Table 5-2. The derivations for these identities are addressed in Exercises 73–76.

Instructor Note:
Emphasize that trigonometric identities may be applied only to values for which each constituent trigonometric function is defined. For example, the sum formula for the tangent function cannot be used to show that $\tan\frac{\pi}{2} = \tan\left(\frac{\pi}{6} + \frac{\pi}{3}\right)$ because $\tan\frac{\pi}{2}$ is not defined.

Table 5-2

Sum and Difference Formulas		
Sine Formulas	$\sin(u + v) = \sin u \cos v + \cos u \sin v$	
	$\sin(u - v) = \sin u \cos v - \cos u \sin v$	
Cosine Formulas	$\cos(u + v) = \cos u \cos v - \sin u \sin v$	
	$\cos(u - v) = \cos u \cos v + \sin u \sin v$	
Tangent Formulas	$\tan(u + v) = \dfrac{\tan u + \tan v}{1 - \tan u \tan v}$	$\tan(u - v) = \dfrac{\tan u - \tan v}{1 + \tan u \tan v}$

If an angle θ is the sum or difference of two angles for which the trigonometric function values are known, we can use the identities from Table 5-2 to find the exact values of $\sin\theta$, $\cos\theta$, and $\tan\theta$. For example, consider the angles $15°$, $105°$, and $\dfrac{7\pi}{12}$ as a sum or difference of other "special" angles.

$$15° = 45° - 30° \qquad 105° = 135° - 30° \qquad \frac{7\pi}{12} = \frac{3\pi}{12} + \frac{4\pi}{12} = \frac{\pi}{4} + \frac{\pi}{3}$$

Classroom Examples: p. 587
Exercises 8, 18

EXAMPLE 1   Applying the Addition and Subtraction Formulas

Find the exact values.   **a.** $\cos 15°$ **b.** $\sin\dfrac{11\pi}{12}$

Solution:

a. $\cos 15° = \cos(45° - 30°)$

Write $15°$ as the sum or difference of angles for which the sine and cosine functions are known (that is, integer multiples of $30°$ or $45°$).

$= \cos 45°\cos 30° + \sin 45°\sin 30°$

Apply $\cos(u - v) = \cos u \cos v + \sin u \sin v$ with $u = 45°$ and $v = 30°$.

$= \dfrac{\sqrt{2}}{2}\cdot\dfrac{\sqrt{3}}{2} + \dfrac{\sqrt{2}}{2}\cdot\dfrac{1}{2}$

Substitute the known function values.

$= \dfrac{\sqrt{6} + \sqrt{2}}{4}$

Multiply radicals and write the result over the common denominator.

> **TIP** The combination $45° - 30°$ is not the only way to make a sum or difference of $15°$ from our "special" angles. For example, the same result is obtained from
> $\cos(-45° + 60°)$ and
> $\cos(135° - 120°)$.

b. First note that $\dfrac{11\pi}{12} = \dfrac{3\pi}{12} + \dfrac{8\pi}{12} = \dfrac{\pi}{4} + \dfrac{2\pi}{3}$.   Look for a combination of angles that are integer multiples of $\frac{\pi}{6}$ or $\frac{\pi}{4}$.

$\sin\dfrac{11\pi}{12} = \sin\left(\dfrac{\pi}{4} + \dfrac{2\pi}{3}\right)$

$= \sin\dfrac{\pi}{4}\cos\dfrac{2\pi}{3} + \cos\dfrac{\pi}{4}\sin\dfrac{2\pi}{3}$

Apply $\sin(u + v) = \sin u \cos v + \cos u \sin v$ with $u = \frac{\pi}{4}$ and $v = \frac{2\pi}{3}$.

$= \dfrac{\sqrt{2}}{2}\cdot\left(-\dfrac{1}{2}\right) + \left(\dfrac{\sqrt{2}}{2}\right)\cdot\left(\dfrac{\sqrt{3}}{2}\right)$

Substitute the known function values.

$= -\dfrac{\sqrt{2}}{4} + \dfrac{\sqrt{6}}{4}$

$= \dfrac{\sqrt{6} - \sqrt{2}}{4}$

Add the fractions.

Skill Practice 1   Find the exact values.

a. $\sin 195°$ **b.** $\cos\dfrac{5\pi}{12}$

Answers

1. a. $\dfrac{\sqrt{2} - \sqrt{6}}{4}$ **b.** $\dfrac{\sqrt{6} - \sqrt{2}}{4}$

Avoiding Mistakes

You can support your results from Example 1 by using a calculator. Confirm that the decimal approximations agree to the level of accuracy provided by the calculator.

Degree mode $\left\{\vphantom{\begin{array}{c}a\\b\end{array}}\right.$

Radian mode $\left\{\vphantom{\begin{array}{c}a\\b\end{array}}\right.$

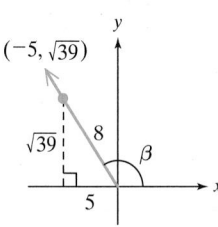

```
NORMAL FLOAT AUTO REAL RADIAN CL
cos(15)
                    .9659258263
(√(6)+√(2))/4
                    .9659258263
sin(11π/12)
                    .2588190451
(√(6)−√(2))/4
                    .2588190451
```

Classroom Example: p. 587
Exercise 20

EXAMPLE 2   Applying the Sum Formula for Sine

Find the exact value of the expression. $\sin 25^\circ \cos 35^\circ + \cos 25^\circ \sin 35^\circ$

Solution:

$\sin 25^\circ \cos 35^\circ + \cos 25^\circ \sin 35^\circ$ Recognize that this expression is in the form $\sin u \cos v + \cos u \sin v$ for $u = 25^\circ$ and $v = 35^\circ$.

$= \sin(25^\circ + 35^\circ)$ Apply the formula $\sin(u + v) = \sin u \cos v + \cos u \sin v$.

$= \sin 60^\circ$

$= \dfrac{\sqrt{3}}{2}$

Skill Practice 2   Find the exact value of the expression.
$\cos 99^\circ \cos 36^\circ - \sin 99^\circ \sin 36^\circ$

Classroom Example: p. 587
Exercise 28

EXAMPLE 3   Applying the Difference Formula for Cosine

Find the exact value of $\cos(\alpha - \beta)$ given that $\sin\alpha = -\dfrac{4}{5}$ and $\cos\beta = -\dfrac{5}{8}$ for α in Quadrant III and β in Quadrant II.

Solution:

$\cos(\alpha - \beta) = \cos\alpha\cos\beta + \sin\alpha\sin\beta$

To find the value of $\cos(\alpha - \beta)$, we need to know the values of the factors on the right side of the identity. We can set up a representative triangle for each angle.

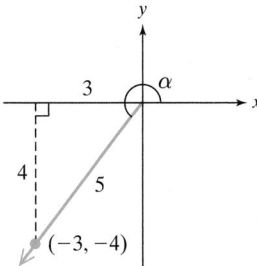

$\cos(\alpha - \beta) = \cos\alpha\cos\beta + \sin\alpha\sin\beta$ Apply the difference formula for cosine.

$= -\dfrac{3}{5} \cdot \left(-\dfrac{5}{8}\right) + \left(-\dfrac{4}{5}\right)\left(\dfrac{\sqrt{39}}{8}\right)$ Substitute values for the sine and cosine of α and β.

$= \dfrac{15 - 4\sqrt{39}}{40}$ Simplify.

Answer

2. $-\dfrac{\sqrt{2}}{2}$

Skill Practice 3 Find the exact value of $\cos(\alpha + \beta)$ given that $\sin\alpha = \dfrac{5}{13}$ and $\cos\beta = \dfrac{5}{6}$ for α in Quadrant II and β in Quadrant IV.

Classroom Example: p. 587
Exercise 38

EXAMPLE 4 **Evaluating the Sine of the Sum of Two Angles Given as Inverse Functions**

Find the exact value of $\sin\left[\tan^{-1}\left(-\dfrac{9}{40}\right) + \cos^{-1}\dfrac{8}{17}\right]$.

Solution:

Let $u = \tan^{-1}\left(-\dfrac{9}{40}\right)$ and $v = \cos^{-1}\dfrac{8}{17}$.

It follows that $\tan u = -\dfrac{9}{40}$ and $\cos v = \dfrac{8}{17}$.

Draw angles u and v in standard position and set up representative triangles with the hypotenuse on the terminal sides of u and v (Figures 5-4 and 5-5).

> **TIP** The values of $\sin u$, $\cos u$, $\sin v$, and $\cos v$ can also be found by using trigonometric identities.
>
> Because $\tan u = -\frac{9}{40}$ for u in Quadrant IV, $\sec^2 u = 1 + \tan^2 u$ and
>
> $$\sec u = \sqrt{1 + \left(-\tfrac{9}{40}\right)^2}$$
> $$= \tfrac{41}{40}$$
>
> Thus, $\cos u = \frac{40}{41}$
> Using $\tan u = \frac{\sin u}{\cos u}$, we have $\sin u = \tan u \cos u$ and $\sin u = -\frac{9}{40} \cdot \frac{40}{41}$
> $$= -\tfrac{9}{41}$$
>
> We also know that $\cos v = \frac{8}{17}$ for v in Quadrant I. Using the Pythagorean identity $\sin^2 v + \cos^2 v = 1$, we find that $\sin v = \frac{15}{17}$.

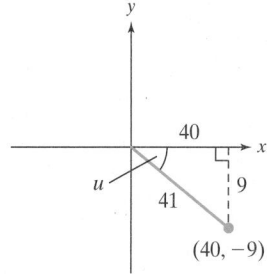

Figure 5-4

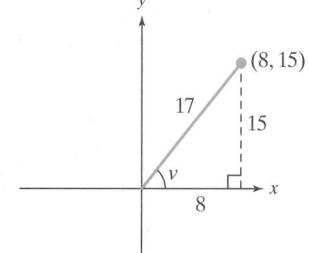

Figure 5-5

$$\sin\left[\tan^{-1}\left(-\dfrac{9}{40}\right) + \cos^{-1}\dfrac{8}{17}\right]$$

$$= \sin(u + v) = \sin u \cos v + \cos u \sin v$$

$$= -\dfrac{9}{41} \cdot \dfrac{8}{17} + \dfrac{40}{41} \cdot \dfrac{15}{17}$$

$$= \dfrac{528}{697}$$

Apply the formula for the sine of a sum. The result checks on a calculator.

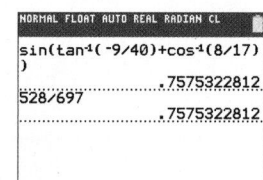

Skill Practice 4 Find the exact value of $\cos\left[\sin^{-1}\left(-\dfrac{12}{37}\right) + \tan^{-1}\left(\dfrac{5}{12}\right)\right]$.

2. Apply the Sum and Difference Formulas for Tangent

We can use the identities for $\sin(u \pm v)$ and $\cos(u \pm v)$ to derive similar identities for $\tan(u \pm v)$.

Answers

3. $\dfrac{-60 + 5\sqrt{11}}{78}$

4. $\dfrac{480}{481}$

$$\tan(u + v) = \frac{\sin(u + v)}{\cos(u + v)} = \frac{\sin u \cos v + \cos u \sin v}{\cos u \cos v - \sin u \sin v}$$

To write this identity in terms of tan u and tan v, multiply numerator and denominator by

$$\frac{1}{\cos u \cos v}.$$

$$= \frac{\dfrac{1}{\cos u \cos v} \cdot (\sin u \cos v + \cos u \sin v)}{\dfrac{1}{\cos u \cos v} \cdot (\cos u \cos v - \sin u \sin v)}$$

$$= \frac{\dfrac{\sin u \overset{1}{\cancel{\cos v}}}{\cos u \cancel{\cos v}} + \dfrac{\overset{1}{\cancel{\cos u}} \sin v}{\cancel{\cos u} \cos v}}{\dfrac{\cancel{\cos u \cos v}}{\underset{1}{\cancel{\cos u \cos v}}} - \dfrac{\sin u \sin v}{\cos u \cos v}}$$

Simplify common factors within each term of the complex fraction.

$$= \frac{\tan u + \tan v}{1 - \tan u \tan v}$$

Thus, $\tan(u + v) = \dfrac{\tan u + \tan v}{1 - \tan u \tan v}$.

Applying this result to $\tan(u - v) = \tan[u + (-v)]$ and applying the odd-function identity $\tan(-x) = -\tan x$, we can derive the identity $\tan(u - v) = \dfrac{\tan u - \tan v}{1 + \tan u \tan v}$.

Classroom Examples: p. 587
Exercises 10, 24

EXAMPLE 5   Applying the Subtraction Formula for Tangent

Find the exact value of $\tan 255°$.

Solution:

$$\tan 255° = \tan(300° - 45°)$$

Write 255° as the sum or difference of integer multiples of 30° or 45°.

$$= \frac{\tan 300° - \tan 45°}{1 + \tan 300° \tan 45°}$$

Apply the formula $\tan(u - v) = \dfrac{\tan u - \tan v}{1 + \tan u \tan v}$ with $u = 300°$ and $v = 45°$.

$$= \frac{-\sqrt{3} - 1}{1 + (-\sqrt{3})(1)}$$

Substitute known values for $\tan 300°$ and $\tan 45°$.

$$= -\frac{(\sqrt{3} + 1)}{(1 - \sqrt{3})} \cdot \frac{(1 + \sqrt{3})}{(1 + \sqrt{3})}$$

Rationalize the denominator by multiplying numerator and denominator by the conjugate of the denominator.

$$= -\frac{4 + 2\sqrt{3}}{1 - 3}$$

Multiply.

$$= -\frac{2(2 + \sqrt{3})}{-2}$$

Factor the numerator and simplify the denominator.

$$= 2 + \sqrt{3}$$

Simplify common factors.

Skill Practice 5   Find the exact value of $\tan 165°$.

3. Use Sum and Difference Formulas to Verify Identities

The sum and difference formulas for sine, cosine, and tangent can be used to verify the cofunction identities and periodic identities of the six trigonometric functions.

Answer

5. $\sqrt{3} - 2$

Cofunction Identities*		Periodic Identities*
$\sin\theta = \cos\left(\dfrac{\pi}{2} - \theta\right)$ $\qquad \cos\theta = \sin\left(\dfrac{\pi}{2} - \theta\right)$		$\sin(\theta + 2\pi) = \sin\theta$ $\cos(\theta + 2\pi) = \cos\theta$
$\tan\theta = \cot\left(\dfrac{\pi}{2} - \theta\right)$ $\qquad \cot\theta = \tan\left(\dfrac{\pi}{2} - \theta\right)$		$\tan(\theta + \pi) = \tan\theta$ $\cot(\theta + \pi) = \cot\theta$
$\sec\theta = \csc\left(\dfrac{\pi}{2} - \theta\right)$ $\qquad \csc\theta = \sec\left(\dfrac{\pi}{2} - \theta\right)$		$\csc(\theta + 2\pi) = \csc\theta$ $\sec(\theta + 2\pi) = \sec\theta$

*All statements can be made using 90° for $\dfrac{\pi}{2}$, 180° for π, and 360° for 2π.

Classroom Example: p. 588
Exercise 44

EXAMPLE 6   Verifying a Cofunction Identity

Verify the cofunction identity $\sin\left(\dfrac{\pi}{2} - \theta\right) = \cos\theta$.

Solution:

$\sin\left(\dfrac{\pi}{2} - \theta\right) = \sin\dfrac{\pi}{2}\cos\theta - \cos\dfrac{\pi}{2}\sin\theta$ $\quad$ Apply $\sin(u - v) = \sin u\cos v - \cos u\sin v$ with

$\qquad\qquad\qquad = (1)\cos\theta - (0)\sin\theta$ $\qquad\qquad\qquad u = \dfrac{\pi}{2}$ and $v = \theta$.

$\qquad\qquad\qquad = \cos\theta$

Skill Practice 6   Verify the identity. $\cos\left(\dfrac{3\pi}{2} - \theta\right) = -\sin\theta$

Classroom Example: p. 588
Exercise 48

EXAMPLE 7   Verifying an Identity

Verify the identity. $\cos(x + y)\cos(x - y) = \cos^2 x - \sin^2 y$

Solution:

$\cos(x + y)\cos(x - y)$

$\qquad = (\cos x\cos y - \sin x\sin y)(\cos x\cos y + \sin x\sin y)$ $\quad$ Apply the sum and difference formulas for cosine.

$\qquad = \cos^2 x\cos^2 y + \cos x\cos y\sin x\sin y - \sin x\sin y\cos x\cos y - \sin^2 x\sin^2 y$

$\qquad = \cos^2 x\cos^2 y - \sin^2 x\sin^2 y$ $\qquad\qquad\qquad\qquad \cos^2 y = 1 - \sin^2 y$ and

$\qquad = \cos^2 x(1 - \sin^2 y) - (1 - \cos^2 x)\sin^2 y$ $\qquad\quad \sin^2 x = 1 - \cos^2 x$.

$\qquad = \cos^2 x - \cos^2 x\sin^2 y - \sin^2 y + \cos^2 x\sin^2 y$ $\qquad$ Apply the distributive

$\qquad = \cos^2 x - \sin^2 y$ $\qquad\qquad\qquad\qquad\qquad\qquad$ property and simplify.

Skill Practice 7   Verify the identity. $\sin(x + y) - \sin(x - y) = 2\cos x\sin y$

Answers

6. $\cos\left(\dfrac{3\pi}{2} - \theta\right)$

$\quad = \cos\dfrac{3\pi}{2}\cos\theta + \sin\dfrac{3\pi}{2}\sin\theta$

$\quad = (0)\cos\theta + (-1)\sin\theta = -\sin\theta$

7. $\sin(x + y) - \sin(x - y)$

$\quad = \sin x\cos y + \cos x\sin y$

$\qquad - (\sin x\cos y - \cos x\sin y)$

$\quad = \sin x\cos y + \cos x\sin y$

$\qquad - \sin x\cos y + \cos x\sin y$

$\quad = 2\cos x\sin y$

5. Write $\dfrac{13\pi}{12}$ as a sum or difference of angles that are multiples of $\dfrac{\pi}{6}$ or $\dfrac{\pi}{4}$.   For example: $\dfrac{3\pi}{4} + \dfrac{\pi}{3}$

6. Write $285°$ as a sum or difference of angles that are multiples of $30°$ or $45°$.   For example: $315° - 30°$

Objectives 1–2: Apply the Sum and Difference Formulas for Sine, Cosine, and Tangent.

For Exercises 7–18, use an addition or subtraction formula to find the exact value. (See Examples 1 and 5)

7. $\cos 165°$ $-\dfrac{\sqrt{6} + \sqrt{2}}{4}$

8. $\sin \dfrac{5\pi}{12}$ $\dfrac{\sqrt{6} + \sqrt{2}}{4}$

9. $\sin\left(-\dfrac{\pi}{12}\right)$ $\dfrac{-\sqrt{6} + \sqrt{2}}{4}$

10. $\tan \dfrac{19\pi}{12}$ $-2 - \sqrt{3}$

11. $\cos\left(-\dfrac{19\pi}{12}\right)$ $\dfrac{\sqrt{6} - \sqrt{2}}{4}$

12. $\sin(-15°)$ $\dfrac{-\sqrt{6} + \sqrt{2}}{4}$

13. $\tan \dfrac{7\pi}{12}$ $-2 - \sqrt{3}$

14. $\cos \dfrac{7\pi}{12}$ $\dfrac{-\sqrt{6} + \sqrt{2}}{4}$

15. $\sin 105°$ $\dfrac{\sqrt{6} + \sqrt{2}}{4}$

16. $\tan(-105°)$ $2 + \sqrt{3}$

17. $\tan 15°$ $2 - \sqrt{3}$

18. $\cos(-195°)$ $\dfrac{\sqrt{6} + \sqrt{2}}{4}$

For Exercises 19–26, use an addition or subtraction formula to find the exact value. (See Examples 2 and 5)

19. $\sin 140° \cos 20° - \cos 140° \sin 20°$ $\dfrac{\sqrt{3}}{2}$

20. $\cos \dfrac{35\pi}{18} \cos \dfrac{5\pi}{18} + \sin \dfrac{35\pi}{18} \sin \dfrac{5\pi}{18}$ $\dfrac{1}{2}$

21. $\dfrac{\tan \dfrac{5\pi}{4} + \tan \dfrac{\pi}{12}}{1 - \tan \dfrac{5\pi}{4} \tan \dfrac{\pi}{12}}$ $\sqrt{3}$

22. $\sin \dfrac{35\pi}{36} \cos \dfrac{13\pi}{18} - \cos \dfrac{35\pi}{36} \sin \dfrac{13\pi}{18}$ $\dfrac{\sqrt{2}}{2}$

23. $\cos \dfrac{2\pi}{3} \cos \dfrac{7\pi}{6} - \sin \dfrac{2\pi}{3} \sin \dfrac{7\pi}{6}$ $\dfrac{\sqrt{3}}{2}$

24. $\dfrac{\tan 15° - \tan 45°}{1 + \tan 15° \tan 45°}$ $\dfrac{\sqrt{3}}{3}$

25. $\cos 200° \cos 25° - \sin 200° \sin 25°$ $-\dfrac{\sqrt{2}}{2}$

26. $\sin \dfrac{10\pi}{9} \cos \dfrac{7\pi}{18} + \cos \dfrac{10\pi}{9} \sin \dfrac{7\pi}{18}$ -1

For Exercises 27–32, find the exact value for the expression under the given conditions. (See Examples 3 and 5)

27. $\cos(\alpha + \beta)$; $\sin\alpha = -\dfrac{3}{5}$ for α in Quadrant III and $\cos\beta = -\dfrac{3}{4}$ for β in Quadrant II. $\dfrac{12 + 3\sqrt{7}}{20}$

28. $\sin(\alpha - \beta)$; $\sin\alpha = \dfrac{3}{8}$ for α in Quadrant II and $\cos\beta = \dfrac{12}{13}$ for β in Quadrant IV. $\dfrac{36 - 5\sqrt{55}}{104}$

29. $\tan(\alpha - \beta)$; $\sin\alpha = \dfrac{8}{17}$ for α in Quadrant II and $\cos\beta = -\dfrac{9}{41}$ for β in Quadrant III. $\dfrac{672}{185}$

30. $\cos(\alpha - \beta)$; $\sin\alpha = \dfrac{2}{3}$ for α in Quadrant II and $\cos\beta = -\dfrac{1}{4}$ for β in Quadrant III. $\dfrac{\sqrt{5} - 2\sqrt{15}}{12}$

31. $\sin(\alpha + \beta)$; $\cos\alpha = \dfrac{3}{7}$ for α in Quadrant IV and $\sin\beta = \dfrac{7}{25}$ for β in Quadrant II. $\dfrac{48\sqrt{10} + 21}{175}$

32. $\tan(\alpha + \beta)$; $\cos\alpha = -\dfrac{13}{85}$ for α in Quadrant III and $\sin\beta = -\dfrac{63}{65}$ for β in Quadrant IV. $\dfrac{21}{220}$

For Exercises 33–40, find the exact value. (See Example 4)

33. $\sin\left(\arcsin \dfrac{1}{2} - \arccos \dfrac{\sqrt{2}}{2}\right)$ $\dfrac{-\sqrt{6} + \sqrt{2}}{4}$

34. $\cos\left(\tan^{-1}\sqrt{3} - \sin^{-1}\dfrac{\sqrt{3}}{2}\right)$ 1

35. $\tan\left[\sin^{-1}\left(-\dfrac{\sqrt{2}}{2}\right) + \cos^{-1}\left(-\dfrac{\sqrt{3}}{2}\right)\right]$ $-2 - \sqrt{3}$

36. $\sin\left[\arccos\left(-\dfrac{1}{2}\right) + \arcsin\left(-\dfrac{\sqrt{2}}{2}\right)\right]$ $\dfrac{\sqrt{6} + \sqrt{2}}{4}$

37. $\cos\left(\tan^{-1}\dfrac{8}{15} - \cos^{-1}\dfrac{3}{5}\right)$ $\dfrac{77}{85}$

38. $\sin\left(\cos^{-1}\dfrac{12}{13} + \tan^{-1}\dfrac{4}{3}\right)$ $\dfrac{63}{65}$

39. $\cos\left(\sin^{-1}\dfrac{4}{5} + \tan^{-1}2\right)$ $-\dfrac{\sqrt{5}}{5}$

40. $\tan\left(\arccos \dfrac{1}{5} - \arcsin \dfrac{3}{5}\right)$ $\dfrac{6 - \sqrt{6}}{4}$

Objective 3: Use Sum and Difference Formulas to Verify Identities

For Exercises 41–62, verify the identity. (See Examples 6–7)

41. $\cos\left(\dfrac{\pi}{2} + x\right) = -\sin x$

42. $\sin\left(\dfrac{\pi}{2} + x\right) = \cos x$

43. $\tan(\pi + x) = \tan x$

44. $\cos(x - \pi) = -\cos x$

45. $\sin\left(\dfrac{3\pi}{2} + x\right) = -\cos x$

46. $\tan(\pi - x) = -\tan x$

47. $\sin(x + y) + \sin(x - y) = 2\sin x \cos y$

48. $\cos(x - y) - \cos(x + y) = 2\sin x \sin y$

49. $\dfrac{\cos(x + y)}{\sin x \cos y} = \cot x - \tan y$

50. $\dfrac{\sin(x + y)}{\cos x \sin y} = \tan x \cot y + 1$

51. $\sin\left(x + \dfrac{\pi}{4}\right) + \sin\left(x - \dfrac{\pi}{4}\right) = \sqrt{2}\sin x$

52. $\cos\left(x + \dfrac{\pi}{4}\right) - \cos\left(x - \dfrac{\pi}{4}\right) = -\sqrt{2}\sin x$

53. $\tan\left(x - \dfrac{\pi}{4}\right) = \dfrac{\sin x - \cos x}{\sin x + \cos x}$

54. $\tan\left(x + \dfrac{\pi}{4}\right) = \dfrac{\cos x + \sin x}{\cos x - \sin x}$

55. $\dfrac{\cos(\alpha - \beta)}{\cos(\alpha + \beta)} = \dfrac{\cot\alpha\cot\beta + 1}{\cot\alpha\cot\beta - 1}$

56. $\dfrac{\sin(\alpha - \beta)}{\sin(\alpha + \beta)} = \dfrac{\tan\alpha - \tan\beta}{\tan\alpha + \tan\beta}$

57. $\csc(x + y) = \dfrac{\sec x \sec y}{\tan x + \tan y}$

58. $\sec(x - y) = \dfrac{\csc x \sec y}{\cot x + \tan y}$

59. $\cos(A + B)\cos(A - B) = \cos^2 A + \cos^2 B - 1$

60. $\sin(A + B)\sin(A - B) = \sin^2 A - \sin^2 B$

61. $\tan(x + y) - \tan(x - y) = \dfrac{2\tan y \sec^2 x}{1 - \tan^2 x \tan^2 y}$

62. $\tan(x + y) + \tan(x - y) = \dfrac{2\tan x \sec^2 y}{1 - \tan^2 x \tan^2 y}$

Objective 4: Write a Sum of *A*sin*x* and *B*cos*x* as a Single Term

For Exercises 63–66,

   a. Write the given expression in the form $k\sin(x + \alpha)$ for $0 \le \alpha < 2\pi$. Round α to 3 decimal places. **(See Example 8)**

   b. Verify the result from part (a) by applying the sum formula for sine.

63. $8\sin x - 6\cos x$

64. $-15\sin x + 8\cos x$

65. $-\sin x - 3\cos x$

66. $2\sin x + 9\cos x$

Often graphing a function of the form $y = A\sin x + B\cos x$ is easier by using its reduction formula $y = k\sin(x + \alpha)$. For Exercises 67–70,

   a. Use the reduction formula to write the given function as a sine function.

   b. Graph the function.

67. $y = \sqrt{3}\sin x + \cos x$

68. $y = \sqrt{2}\sin x - \sqrt{2}\cos x$

69. $y = -\dfrac{1}{2}\sin x + \dfrac{\sqrt{3}}{2}\cos x$

70. $y = -\dfrac{\sqrt{3}}{2}\sin x - \dfrac{1}{2}\cos x$

Mixed Exercises

71. a. Is it true that $\sin(2 \cdot 30°) = 2\sin 30°$?   No

    b. Expand $\sin(2 \cdot 30°)$ as $\sin(30° + 30°)$ using the sum formula for sine. Then simplify the result.

      $\sin(30° + 30°) = \sin 30°\cos 30° + \cos 30°\sin 30° = \dfrac{\sqrt{3}}{2}$

72. a. Is it true that $\cos(2 \cdot 45°) = 2\cos 45°$?   No

    b. Expand $\cos(2 \cdot 45°)$ as $\cos(45° + 45°)$ using the sum formula for cosine. Then simplify the result.

      $\cos(45° + 45°) = \cos 45°\cos 45° - \sin 45°\sin 45° = 0$

73. Derive $\cos(u + v) = \cos u \cos v - \sin u \sin v$ by using the identity for $\cos(u - v)$ and the odd and even function identities for sine and cosine.

74. Derive $\tan(u - v) = \dfrac{\tan u - \tan v}{1 + \tan u \tan v}$ by using the identity for $\tan(u + v)$ and the odd function identity for tangent.

75. Derive $\sin(u + v) = \sin u \cos v + \cos u \sin v$. {*Hint*: Write $\sin(u + v) = \cos\left[\dfrac{\pi}{2} - (u + v)\right]$ and apply the cofunction identities.}

76. Derive $\sin(u - v) = \sin u \cos v - \cos u \sin v$.

77. For $f(x) = \sin x$, show that
$$\frac{f(x + h) - f(x)}{h} = -\sin x\left(\frac{1 - \cos h}{h}\right) + \cos x\left(\frac{\sin h}{h}\right)$$

78. For $f(x) = \cos x$, show that
$$\frac{f(x + h) - f(x)}{h} = -\cos x\left(\frac{1 - \cos h}{h}\right) - \sin x\left(\frac{\sin h}{h}\right)$$

79. A human ear detects sound as a result of fluctuations in air pressure against the eardrum, which are in turn transmitted into nerve impulses that the brain translates as sound. Suppose that the pressure $P(t)$ of a sound wave on a person's eardrum from a source 20 ft away is given by $P(t) = 0.02\cos(1000t - 10\pi)$. In this model $P(t)$ is in pounds per square foot and t is the time in seconds after the sound begins. Simplify the right side by applying the difference formula for cosine. $P(t) = 0.02\cos 1000t$

80. The raised part of a sundial, called a *gnomon*, casts a shadow of length l when the angle of elevation of the Sun is θ. The length of the shadow is given by
$$l = \frac{h\sin(90° - \theta)}{\sin\theta}, \text{ where } h \text{ is the height of the}$$
gnomon. Simplify the right side of this equation.
$l = h\cot\theta$

For Exercises 81–82, consider a 1-kg object oscillating at the end of a horizontal spring. The horizontal position $x(t)$ of the object is given by

$$x(t) = \frac{v_0}{\omega}\sin(\omega t) + x_0\cos(\omega t)$$

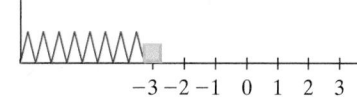

$$-3\ -2\ -1\quad 0\quad 1\quad 2\quad 3$$

where v_0 is the initial velocity, x_0 is the initial position, and ω is the number of back-and-forth cycles that the object makes per unit time t.

81. At time $t = 0$ sec, the object is moved 3 ft to the left of the equilibrium position and then given a velocity of 4 ft/sec to the right ($v_0 = 4$ ft/sec).

   a. If the object completes 1 cycle in 1 sec ($\omega = 1$), write a model of the form $x(t) = \dfrac{v_0}{\omega}\sin(\omega t) + x_0\cos(\omega t)$ to represent the horizontal motion of the spring. $x(t) = 4\sin t - 3\cos t$

   b. Use the reduction formula to write the function in part (a) in the form $x(t) = k\sin(t + \alpha)$. Round α to 2 decimal places. $x(t) = 5\sin(t + 5.64)$

   c. What is the maximum displacement of the object from its equilibrium position?   5 ft

82. At time $t = 0$ sec, the object is moved 2 ft to the right of the equilibrium position and then given a velocity of 3 ft/sec to the left ($v_0 = -3$ ft/sec).

   a. If the block completes 1 cycle in 1 sec ($\omega = 1$), write a model of the form $x(t) = \dfrac{v_0}{\omega}\sin(\omega t) + x_0\cos(\omega t)$ to represent the horizontal motion of the spring. $x(t) = -3\sin t + 2\cos t$

   b. Use the reduction formula to write the function in part (a) in the form $x(t) = k\sin(t + \alpha)$. Round α to 2 decimal places. $x(t) = \sqrt{13}\sin(t + 2.55)$

   c. What is the maximum displacement of the object from its equilibrium position? Round to 2 decimal places. $\sqrt{13} \approx 3.61$ ft

Write About It

83. Explain why we cannot use $\tan\left(\dfrac{\pi}{2} - \theta\right) = \dfrac{\tan\dfrac{\pi}{2} - \tan\theta}{1 - \tan\dfrac{\pi}{2}\cdot\tan\theta}$ to prove the cofunction identity $\cot\theta = \tan\left(\dfrac{\pi}{2} - \theta\right)$.   The value $\tan\dfrac{\pi}{2}$ is undefined.

84. Describe the pattern for the expansions of $\cos(u + v)$ and $\cos(u - v)$.   Answers will vary.

85. Describe the pattern for the expansions of $\sin(u + v)$ and $\sin(u - v)$. Answers will vary.

86. The sum and difference formulas for cosine can be written in a single statement as $\cos(u \pm v) = \cos u \cos v \mp \sin u \sin v$. Explain why the symbol $\mp$ is used on the right. The sign between the terms in the expansion is opposite the sign between the angles u and v.

Expanding Your Skills

For Exercises 87–90, write the trigonometric expression as an algebraic expression in x and y. Assume that x and y are Quadrant I angles.

87. $\cos(\sin^{-1}x + \tan^{-1}y)$ $\dfrac{\sqrt{1-x^2} - xy}{\sqrt{1+y^2}}$

88. $\sin(\cos^{-1}x + \tan^{-1}y)$ $\dfrac{\sqrt{1-x^2} + xy}{\sqrt{1+y^2}}$

89. $\sin(\sin^{-1}x + \sin^{-1}y)$ $x\sqrt{1-y^2} + y\sqrt{1-x^2}$

90. $\cos(\sin^{-1}x - \cos^{-1}y)$ $y\sqrt{1-x^2} + x\sqrt{1-y^2}$

For Exercises 91–96, verify the identity.

91. $\cos(a + b + c) = \cos a \cos b \cos c - \sin a \sin b \cos c - \sin a \cos b \sin c - \cos a \sin b \sin c$

92. $\sin(a + b + c) = \sin a \cos b \cos c + \cos a \sin b \cos c + \cos a \cos b \sin c - \sin a \sin b \sin c$

93. $\cos(x + y)\cos y + \sin(x + y)\sin y = \cos x$

94. $\cos(x - y)\sin y + \sin(x - y)\cos y = \sin x$

95. $\sec(x + y) = \dfrac{\cos(x - y)}{\cos^2 x - \sin^2 y}$

96. $\sec(x - y) = \dfrac{\cos(x + y)}{\cos^2 x - \sin^2 y}$

97. Suppose that $\triangle ABC$ is a right triangle.

Show that

$$\sin(A - B) = \frac{a^2 - b^2}{c^2} \quad \text{and} \quad \cos(A - B) = \frac{2ab}{c^2}$$

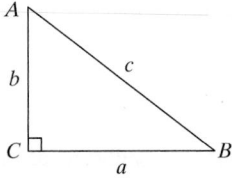

98. Suppose that $\triangle ABC$ contains no right angle. Show that $\tan A + \tan B + \tan C = (\tan A)(\tan B)(\tan C)$.

(*Hint*: $A + B + C = 180°$ which implies that $A + B = 180° - C$. Then take the tangent of both sides of the equation.)

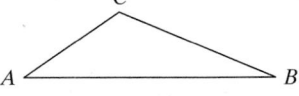

99. Let L be a line defined by $y = mx + b$ with a positive slope, and let θ be the acute angle formed by L and the horizontal. Let $P(x_1, mx_1 + b)$ and $Q(x_2, mx_2 + b)$ be arbitrary points on L. Show that $m = \tan\theta$.

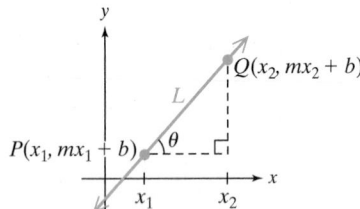

100. Let L_1 and L_2 be nonparallel lines with positive slopes m_1 and m_2, respectively, where $m_2 > m_1$.

a. Show that the acute angle α formed by L_1 and L_2 must satisfy $\tan\alpha = \dfrac{m_2 - m_1}{1 + m_2 m_1}$.

b. Find the measure of the acute angle formed by the lines $y = \frac{3}{4}x + 2$ and $y = 2x - 3$. Round to the nearest tenth of a degree.

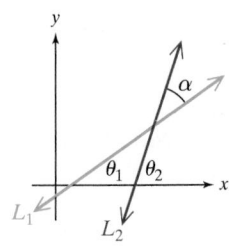

Technology Connections

101. **a.** Graph $y = \cos\left(x + \dfrac{\pi}{2}\right) - \cos\left(x - \dfrac{\pi}{2}\right)$ on the interval $[0, 2\pi]$. What simpler function $y = f(x)$ does this graph appear to represent.

b. Simplify $\cos\left(x + \dfrac{\pi}{2}\right) - \cos\left(x - \dfrac{\pi}{2}\right)$ to confirm your response to part (a).

102. **a.** Graph $y = \sin 3x \cos 2x - \cos 3x \sin 2x$ on the interval $[0, 2\pi]$. What simpler function $y = f(x)$ does this graph appear to represent.

b. Simplify $\sin 3x \cos 2x - \cos 3x \sin 2x$ to confirm your response to part (a).

SECTION 5.3 **Double-Angle, Power-Reducing, and Half-Angle Formulas**

OBJECTIVES

1. Apply the Double-Angle Formulas
2. Apply the Power-Reducing Formulas
3. Apply the Half-Angle Formulas

1. Apply the Double-Angle Formulas

A projectile launched from ground level at an angle of elevation θ has a maximum horizontal range given by $x_{\max} = \dfrac{v_0^2 \sin 2\theta}{g}$ (see Exercises 85–86). This expression involves the double angle 2θ. We can write a trigonometric function of a double angle 2θ in terms of $\sin \theta$, $\cos \theta$, and $\tan \theta$ by applying the sum and difference formulas.

- $\sin 2\theta = \sin(\theta + \theta) = \sin \theta \cos \theta + \cos \theta \sin \theta \qquad \sin(u + v) = \sin u \cos v + \cos u \sin v$

  $\qquad\quad = 2 \sin \theta \cos \theta$

- $\cos 2\theta = \cos(\theta + \theta) = \cos \theta \cos \theta - \sin \theta \sin \theta \qquad \cos(u + v) = \cos u \cos v - \sin u \sin v$

  $\qquad\quad = \cos^2 \theta - \sin^2 \theta$

- $\tan 2\theta = \tan(\theta + \theta) = \dfrac{\tan \theta + \tan \theta}{1 - \tan \theta \tan \theta} \qquad\qquad \tan(u + v) = \dfrac{\tan u + \tan v}{1 - \tan u \tan v}$

  $\qquad\quad = \dfrac{2 \tan \theta}{1 - \tan^2 \theta}$

Notice that the double angle formula for cosine contains the expressions $\cos^2 \theta$ and $\sin^2 \theta$. From the Pythagorean identity $\sin^2 \theta + \cos^2 \theta = 1$, we can express the double angle formula for $\cos 2\theta$ in alternative forms (Table 5-3).

Substitute $1 - \sin^2 \theta$ for $\cos^2 \theta$.

$$\cos^2 \theta - \sin^2 \theta$$
$$= (1 - \sin^2 \theta) - \sin^2 \theta$$
$$= 1 - 2 \sin^2 \theta$$

Substitute $1 - \cos^2 \theta$ for $\sin^2 \theta$.

$$\cos^2 \theta - \sin^2 \theta$$
$$= \cos^2 \theta - (1 - \cos^2 \theta)$$
$$= \cos^2 \theta - 1 + \cos^2 \theta$$
$$= 2 \cos^2 \theta - 1$$

Table 5-3

Double-Angle Formulas

$$\sin 2\theta = 2 \sin \theta \cos \theta \qquad\qquad \cos 2\theta = \cos^2 \theta - \sin^2 \theta$$
$$\tan 2\theta = \dfrac{2 \tan \theta}{1 - \tan^2 \theta} \qquad\qquad\quad = 1 - 2 \sin^2 \theta$$
$$= 2 \cos^2 \theta - 1$$

Classroom Example: p. 598
Exercise 10

EXAMPLE 1	**Using the Double-Angle Formulas**

Given that $\sin \theta = \dfrac{2}{3}$ for θ in Quadrant II, find the exact function values.

a. $\sin 2\theta$ **b.** $\cos 2\theta$ **c.** $\tan 2\theta$

Solution:

To find $\sin 2\theta = 2\sin\theta\cos\theta$, we need to know the value of $\cos\theta$. Using the Pythagorean identity $\sin^2\theta + \cos^2\theta = 1$, we have $\cos\theta = \pm\sqrt{1 - \sin^2\theta}$.

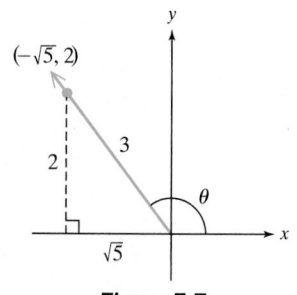

Figure 5-7

In Quadrant II,
$\cos\theta < 0$.

$$\cos\theta = -\sqrt{1 - \sin^2\theta} = -\sqrt{1 - \left(\frac{2}{3}\right)^2} = -\sqrt{1 - \frac{4}{9}}$$

The cosine function is negative in Quadrant II (Figure 5-7).

$$= -\sqrt{\frac{5}{9}} = -\frac{\sqrt{5}}{3}$$

a. $\sin 2\theta = 2\sin\theta\cos\theta = 2\left(\frac{2}{3}\right)\left(-\frac{\sqrt{5}}{3}\right) = -\frac{4\sqrt{5}}{9}$

b. $\cos 2\theta = \cos^2\theta - \sin^2\theta = \left(-\frac{\sqrt{5}}{3}\right)^2 - \left(\frac{2}{3}\right)^2 = \frac{5}{9} - \frac{4}{9} = \frac{1}{9}$ Use any of the three formulas for $\cos 2\theta$.

c. With the values of $\sin 2\theta$ and $\cos 2\theta$ known, the easiest way to find $\tan 2\theta$ is to divide.

$$\tan 2\theta = \frac{\sin 2\theta}{\cos 2\theta} = \frac{-\dfrac{4\sqrt{5}}{9}}{\dfrac{1}{9}} = -\frac{4\sqrt{5}}{9} \cdot \frac{9}{1} = -4\sqrt{5}$$

Alternatively, from Figure 5-7, $\tan\theta = -\dfrac{2}{\sqrt{5}} = -\dfrac{2\sqrt{5}}{5}$. Therefore,

$$\tan 2\theta = \frac{2\tan\theta}{1 - \tan^2\theta} = \frac{2\left(-\dfrac{2\sqrt{5}}{5}\right)}{1 - \left(-\dfrac{2\sqrt{5}}{5}\right)^2} = \frac{-\dfrac{4\sqrt{5}}{5}}{1 - \dfrac{4}{5}} = \frac{-\dfrac{4\sqrt{5}}{5}}{\dfrac{1}{5}}$$

$$= -\frac{4\sqrt{5}}{5} \cdot \frac{5}{1} = -4\sqrt{5}$$

Skill Practice 1   Given that $\sin\theta = \dfrac{4}{5}$ for θ in Quadrant II, find the exact function values.

    a. $\sin 2\theta$ **b.** $\cos 2\theta$ **c.** $\tan 2\theta$

The double-angle formulas can be used with angles other than 2θ. For example:

$$\sin 4\theta = 2\sin 2\theta\cos 2\theta \qquad\qquad \cos 6x = \cos^2 3x - \sin^2 3x$$

$$\sin(\alpha - 1) = 2\sin\left(\frac{\alpha - 1}{2}\right)\cos\left(\frac{\alpha - 1}{2}\right) \qquad\qquad \tan x = \frac{2\tan\dfrac{x}{2}}{1 - \tan^2\dfrac{x}{2}}$$

In Example 2, we apply the sum formula for sine as well as the double-angle formulas for sine and cosine to prove an identity.

Answers

1. a. $-\dfrac{24}{25}$ **b.** $-\dfrac{7}{25}$ **c.** $\dfrac{24}{7}$

Classroom Example: p. 598
Exercise 26

EXAMPLE 2   **Applying the Sum and Double-Angle Formulas**

Verify the identity. $\sin 3x = 3\sin x - 4\sin^3 x$

Solution:

$\sin 3x = \sin(2x + x)$ Apply $\sin(u + v) = \sin u \cos v + \cos u \sin v$ with $u = 2x$ and $v = x$.

$= \sin 2x \cos x + \cos 2x \sin x$

$= (2\sin x \cos x)\cos x + (\cos^2 x - \sin^2 x)\sin x$ Simplify $\sin 2x = 2\sin x \cos x$ and $\cos 2x = \cos^2 x - \sin^2 x$.

$= 2\sin x \cos^2 x + \sin x \cos^2 x - \sin^3 x$

$= 3\sin x \cos^2 x - \sin^3 x$ Combine like terms.

$= 3\sin x(1 - \sin^2 x) - \sin^3 x$ Substitute $1 - \sin^2 x$ for $\cos^2 x$.

$= 3\sin x - 3\sin^3 x - \sin^3 x$ Apply the distributive property.

$= 3\sin x - 4\sin^3 x$

Skill Practice 2   Verify the identity. $\cos 3x = 4\cos^3 x - 3\cos x$

2. Apply the Power-Reducing Formulas

The double-angle formulas for cosine can be used to express even powers of sine and cosine in terms of the first power of cosine only. For example, solving $\cos 2\theta = 1 - 2\sin^2 \theta$ for $\sin^2 \theta$ yields

> RHS expressed in terms of first power of cosine

$$\cos 2\theta = 1 - 2\sin^2 \theta \quad \Rightarrow \quad \sin^2 \theta = \frac{1 - \cos 2\theta}{2} \qquad (1)$$

Equation (1) is called a **power-reducing** formula. The derivations of the power-reducing formulas in Table 5-4 are addressed in Exercises 57–58.

Table 5-4

Power-Reducing Formulas		
$\sin^2 \theta = \dfrac{1 - \cos 2\theta}{2}$	$\cos^2 \theta = \dfrac{1 + \cos 2\theta}{2}$	$\tan^2 \theta = \dfrac{1 - \cos 2\theta}{1 + \cos 2\theta}$

TIP The power-reducing formulas are important in integral calculus.

Answer

2. $\cos 3x = \cos(2x + x)$

$= \cos 2x \cos x - \sin 2x \sin x$

$= (1 - 2\sin^2 x)\cos x$
$\quad - (2\sin x \cos x)\sin x$

$= \cos x - 2\sin^2 x \cos x - 2\sin^2 x \cos x$

$= \cos x - 4\sin^2 x \cos x$

$= \cos x - 4(1 - \cos^2 x)\cos x$

$= \cos x - 4\cos x + 4\cos^3 x$

$= 4\cos^3 x - 3\cos x$

Classroom Example: p. 598

Exercise 38

EXAMPLE 3 Applying the Power-Reducing Formulas

Write $\sin^4 x + \cos^2 x$ in terms of first powers of cosine.

Solution:

$\sin^4 x + \cos^2 x = (\sin^2 x)^2 + \cos^2 x$ — Write each term as a power of $\sin^2 x$ and $\cos^2 x$.

$= \left(\dfrac{1 - \cos 2x}{2}\right)^2 + \dfrac{1 + \cos 2x}{2}$ — Apply the power-reducing formulas.

$= \dfrac{1 - 2\cos 2x + \cos^2 2x}{4} + \dfrac{1 + \cos 2x}{2}$ — Square the first term.

$= \dfrac{1 - 2\cos 2x + \cos^2 2x}{4} + \dfrac{1 + \cos 2x}{2} \cdot \dfrac{2}{2}$ — Write each term with a common denominator.

$= \dfrac{1 - 2\cos 2x + \cos^2 2x}{4} + \dfrac{2 + 2\cos 2x}{4}$

$= \dfrac{1}{4}[1 - 2\cos 2x + \cos^2 2x + 2 + 2\cos 2x]$ — Factor out $\frac{1}{4}$.

$= \dfrac{1}{4}[\cos^2 2x + 3]$ — Combine like terms.

$= \dfrac{1}{4}\left[\left(\dfrac{1 + \cos 4x}{2}\right) + 3\right]$ — Apply the power-reducing formula for cosine $\cos^2 2x = \dfrac{1 + \cos 4x}{2}$. Notice that the angle on the RHS is doubled.

$= \dfrac{1}{4}\left[\dfrac{1}{2} + \dfrac{\cos 4x}{2} + \dfrac{6}{2}\right]$ — Write each term with a common denominator.

$= \dfrac{1}{4}\left(\dfrac{\cos 4x}{2} + \dfrac{7}{2}\right)$ — Combine like terms.

$= \dfrac{1}{8}\cos 4x + \dfrac{7}{8}$ — Apply the distributive property.

Skill Practice 3 Write $\cos^4 x$ in terms of first powers of cosine.

3. Apply the Half-Angle Formulas

Another set of formulas called the half-angle formulas can be derived from the power-reducing formulas. Substituting $\dfrac{\alpha}{2}$ for θ in the power-reducing formulas in Table 5-4 results in the half-angle formulas given in Table 5-5.

Point of Interest

In ancient times, trigonometric relationships were based on the chord function, denoted crd. The chord of a central angle $\angle POQ$ is the length of the line segment with endpoints P and Q where the angle intercepts the circle. The chord function is related to the modern sine function by

$$\text{crd}\,\alpha = d\sin\frac{\alpha}{2}$$

where d is the diameter of the circle.

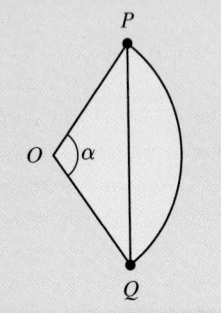

Table 5-5

Half-Angle Formulas	
$\sin\dfrac{\alpha}{2} = \pm\sqrt{\dfrac{1 - \cos\alpha}{2}}$	$\cos\dfrac{\alpha}{2} = \pm\sqrt{\dfrac{1 + \cos\alpha}{2}}$
$\tan\dfrac{\alpha}{2} = \pm\sqrt{\dfrac{1 - \cos\alpha}{1 + \cos\alpha}} = \dfrac{1 - \cos\alpha}{\sin\alpha} = \dfrac{\sin\alpha}{1 + \cos\alpha}$	
The sign $+$ or $-$ is determined by the quadrant in which the angle $\dfrac{\alpha}{2}$ lies.	

Answer

3. $\dfrac{1}{2}\cos 2x + \dfrac{1}{8}\cos 4x + \dfrac{3}{8}$

To derive the formula for $\sin\dfrac{\alpha}{2}$ start with the power-reducing formula for $\sin^2\theta$.

$$\sin^2\theta = \frac{1 - \cos 2\theta}{2}$$

$$\sin^2\frac{\alpha}{2} = \frac{1 - \cos\left[2\left(\dfrac{\alpha}{2}\right)\right]}{2} \qquad \text{Substitute } \frac{\alpha}{2} \text{ for } \theta.$$

$$\sin\frac{\alpha}{2} = \pm\sqrt{\frac{1 - \cos\alpha}{2}} \qquad \text{Simplify the argument and take the square root of both sides.}$$

The sign is chosen based on the quadrant of $\dfrac{\alpha}{2}$ in standard position.

Classroom Example: p. 598
Exercise 42

EXAMPLE 4 Using the Half-Angle Formula for Cosine

Use the half-angle formula to find the exact value of $\cos 157.5°$.

Solution:

$$\cos 157.5° = \cos\frac{315°}{2} \qquad \frac{\alpha}{2} = 157.5° \text{ which means that } \alpha = 2(157.5°) = 315°.$$

In Quadrant II, $\cos\theta < 0$.   Apply the formula $\cos\dfrac{\alpha}{2} = \pm\sqrt{\dfrac{1 + \cos\alpha}{2}}$ for $\alpha = 315°$.

$$= -\sqrt{\frac{1 + \cos 315°}{2}} \qquad \cos\frac{\alpha}{2} < 0 \text{ because } 157.5° \text{ is a second quadrant angle.}$$

$$= -\sqrt{\frac{1 + \dfrac{\sqrt{2}}{2}}{2}} \qquad \cos 315° = \frac{\sqrt{2}}{2}$$

$$= -\sqrt{\frac{2 \cdot \left(1 + \dfrac{\sqrt{2}}{2}\right)}{2 \cdot 2}} \qquad \text{Multiply numerator and denominator by 2 to simplify the complex fraction.}$$

$$= -\sqrt{\frac{2 + \sqrt{2}}{4}}$$

$$= -\frac{\sqrt{2 + \sqrt{2}}}{2} \qquad \text{Simplify the radical in the denominator.}$$

Skill Practice 4 Use the half-angle formula to find the exact value of $\sin 165°$.

Instructor Note:

Ask students to find the exact value of $\sin 15°$ using

(a) the sum or difference formulas and
(b) the half-angle formula.

Then ask them to show algebraically that the resulting expressions are equal. Alternatively, they can use a calculator to show that the two expressions are equal to each other and to $\sin 15°$.

Avoiding Mistakes

The results from Example 4 can be supported on a calculator.

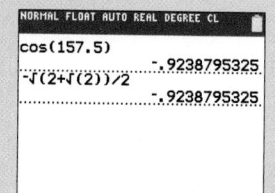

In Table 5-5 notice that we present three forms for the half-angle tangent formula. The formula $\tan\dfrac{\alpha}{2} = \pm\sqrt{\dfrac{1 - \cos\alpha}{1 + \cos\alpha}}$ follows directly from $\tan\dfrac{\alpha}{2} = \dfrac{\sin\dfrac{\alpha}{2}}{\cos\dfrac{\alpha}{2}}$. However, in practice we prefer to apply the more convenient alternative forms without the radical sign or the $\pm$ sign. In Example 5, we derive $\tan\dfrac{\alpha}{2} = \dfrac{1 - \cos\alpha}{\sin\alpha}$.

Answer

4. $\dfrac{\sqrt{2 - \sqrt{3}}}{2}$

Classroom Example: p. 598
Exercise 46

EXAMPLE 5 Verifying an Identity for the Tangent of a Half-Angle

Show that $\tan\dfrac{\alpha}{2} = \pm\sqrt{\dfrac{1-\cos\alpha}{1+\cos\alpha}} = \dfrac{1-\cos\alpha}{\sin\alpha}$.

Solution:

$\pm\sqrt{\dfrac{1-\cos\alpha}{1+\cos\alpha}} = \pm\sqrt{\dfrac{(1-\cos\alpha)}{(1+\cos\alpha)} \cdot \dfrac{(1-\cos\alpha)}{(1-\cos\alpha)}}$ Rationalize the denominator by multiplying numerator and denominator by the conjugate of the denominator.

$= \pm\sqrt{\dfrac{(1-\cos\alpha)^2}{1-\cos^2\alpha}}$ Multiply.

$= \pm\sqrt{\dfrac{(1-\cos\alpha)^2}{\sin^2\alpha}}$ From the Pythagorean identity $\sin^2\alpha + \cos^2\alpha = 1$, substitute $\sin^2\alpha$ for $1-\cos^2\alpha$.

$= \pm\dfrac{|1-\cos\alpha|}{|\sin\alpha|}$ Take the principal square root.
- Since $-1 \le \cos\alpha \le 1$, the value $1-\cos\alpha \ge 0$ for all α.

$= \dfrac{1-\cos\alpha}{\sin\alpha}$
- With the numerator positive or zero, the denominator controls the sign of the quotient. The value of $\sin\alpha$ has the same sign as $\tan\frac{\alpha}{2}$ for α in any quadrant (see Exercise 45). Therefore, the absolute value bars and $\pm$ are not needed.

Thus, $\tan\dfrac{\alpha}{2} = \dfrac{1-\cos\alpha}{\sin\alpha}$.

Skill Practice 5   Show that $\pm\sqrt{\dfrac{1-\cos\alpha}{1+\cos\alpha}} = \dfrac{\sin\alpha}{1+\cos\alpha}$.

Classroom Example: p. 599
Exercise 48

EXAMPLE 6 Using the Half-Angle Formulas

If $\sin\alpha = -\dfrac{4}{5}$ and $\pi < \alpha < \dfrac{3\pi}{2}$, find the exact values of each expression.

   a. $\sin\dfrac{\alpha}{2}$ **b.** $\cos\dfrac{\alpha}{2}$ **c.** $\tan\dfrac{\alpha}{2}$

Solution:

To apply the half-angle formulas, we need to know the value of $\cos\alpha$. The angle α is in the third quadrant with $\sin\alpha = -\dfrac{4}{5}$. From Figure 5-8, we have $\cos\alpha = -\dfrac{3}{5}$. Next, since α is on the interval $\left(\pi, \dfrac{3\pi}{2}\right)$, angle $\dfrac{\alpha}{2}$ is on the interval $\left(\dfrac{\pi}{2}, \dfrac{3\pi}{4}\right)$. Therefore $\dfrac{\alpha}{2}$ is in the second quadrant.

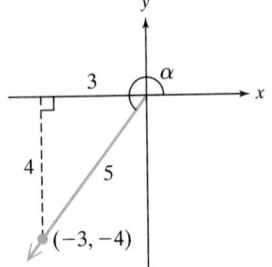

Figure 5-8

Answer

5. Rationalize the numerator.

$\pm\sqrt{\dfrac{1-\cos\alpha}{1+\cos\alpha}} = \pm\sqrt{\dfrac{(1-\cos\alpha)}{(1+\cos\alpha)} \cdot \dfrac{(1+\cos\alpha)}{(1+\cos\alpha)}}$

$= \pm\sqrt{\dfrac{1-\cos^2\alpha}{(1+\cos\alpha)^2}} = \pm\sqrt{\dfrac{\sin^2\alpha}{(1+\cos\alpha)^2}} = \dfrac{\sin\alpha}{1+\cos\alpha}$

In Quadrant II, $\sin\frac{\alpha}{2} > 0$.

a. $\sin\dfrac{\alpha}{2} = +\sqrt{\dfrac{1 - \cos\alpha}{2}} = \sqrt{\dfrac{1 - \left(-\dfrac{3}{5}\right)}{2}} = \sqrt{\dfrac{\dfrac{8}{5}}{2}} = \sqrt{\dfrac{4}{5}} = \dfrac{2}{\sqrt{5}} = \dfrac{2\sqrt{5}}{5}$

In Quadrant II, $\cos\frac{\alpha}{2} < 0$.

b. $\cos\dfrac{\alpha}{2} = -\sqrt{\dfrac{1 + \cos\alpha}{2}} = -\sqrt{\dfrac{1 + \left(-\dfrac{3}{5}\right)}{2}} = -\sqrt{\dfrac{\dfrac{2}{5}}{2}} = -\sqrt{\dfrac{2}{5}\cdot\dfrac{1}{2}}$

$= -\sqrt{\dfrac{1}{5}} = -\dfrac{1}{\sqrt{5}} = -\dfrac{\sqrt{5}}{5}$

c. Use any of the three half-angle formulas for tangent.

$\tan\dfrac{\alpha}{2} = \dfrac{1 - \cos\alpha}{\sin\alpha} = \dfrac{1 - \left(-\dfrac{3}{5}\right)}{-\dfrac{4}{5}} = \dfrac{\dfrac{8}{5}}{-\dfrac{4}{5}} = \dfrac{8}{5}\cdot\left(-\dfrac{5}{4}\right) = -2$

TIP Alternatively in Example 6(c),

$\tan\dfrac{\alpha}{2} = \dfrac{\sin\dfrac{\alpha}{2}}{\cos\dfrac{\alpha}{2}} = \dfrac{\dfrac{2\sqrt{5}}{5}}{-\dfrac{\sqrt{5}}{5}}$

$= \dfrac{2\sqrt{5}}{5}\cdot\left(-\dfrac{5}{\sqrt{5}}\right) = -2$

Skill Practice 6   If $\sin\alpha = -\dfrac{5}{13}$ and $\dfrac{3\pi}{2} < \alpha < 2\pi$, find the exact values of each expression.

a. $\sin\dfrac{\alpha}{2}$ **b.** $\cos\dfrac{\alpha}{2}$ **c.** $\tan\dfrac{\alpha}{2}$

Answers

6. **a.** $\dfrac{\sqrt{26}}{26}$ **b.** $-\dfrac{5\sqrt{26}}{26}$ **c.** $-\dfrac{1}{5}$

SECTION 5.3   Practice Exercises

Prerequisite Review

R.1. Simplify the radical expression. $\sqrt{(a+3)^2}$ $|a+3|$

R.2. Simplify the radical expression. $\sqrt[4]{u^4 v^8}$ $|u|v^2$

R.3. Rationalize the denominator. Assume the variable represents a positive number.

$$\sqrt{\dfrac{4}{w}} \qquad \dfrac{2\sqrt{w}}{w}$$

R.4. Rationalize the denominator. $\dfrac{-14}{\sqrt{2}-3}$ $2\sqrt{2}+6$

Concept Connections

1. $\cos 2\theta = \underline{\quad \cos^2\theta - \sin^2\theta \quad}$ or $\underline{\quad 2\cos^2\theta - 1 \quad}$ or $\underline{\quad 1 - 2\sin^2\theta \quad}$

2. $\sin 2\theta = \underline{\quad 2\sin\theta\cos\theta \quad}$

3. From the relationship $\cos 2\theta = 2\cos^2\theta - 1$, it follows that $\cos^2\theta = \underline{\quad \dfrac{1 + \cos 2\theta}{2} \quad}$.

4. From the relationship $\cos 2\theta = 1 - 2\sin^2\theta$, it follows that $\sin^2\theta = \underline{\quad \dfrac{1 - \cos 2\theta}{2} \quad}$.

5. From the relationship $\sin^2\theta = \dfrac{1 - \cos 2\theta}{2}$, it follows that $\sin\dfrac{u}{2} = \underline{\quad \pm\sqrt{\dfrac{1 - \cos u}{2}} \quad}$.

6. From the relationship $\cos^2\theta = \dfrac{1 + \cos 2\theta}{2}$, it follows that $\cos\dfrac{u}{2} = \underline{\quad \pm\sqrt{\dfrac{1 + \cos u}{2}} \quad}$.

Objective 1: Apply the Double-Angle Formulas

For Exercises 7–14, use the given information to find the exact function values. (See Example 1)

  a. $\sin 2\theta$ **b.** $\cos 2\theta$ **c.** $\tan 2\theta$

7. $\sin\theta = -\dfrac{12}{13}$, θ in Quadrant IV

8. $\sin\theta = \dfrac{4}{7}$, $\tan\theta > 0$

9. $\cos\theta = -\dfrac{\sqrt5}{5}$, $\sin\theta > 0$

10. $\cos\theta = -\dfrac{5}{8}$, θ in Quadrant III

11. $\tan\theta = -\dfrac{5}{2}$, θ in Quadrant II

12. $\cot\theta = \dfrac{\sqrt2}{6}$, $\cos\theta < 0$

13. $\sec\theta = \dfrac{37}{12}$, θ in Quadrant IV

14. $\csc\theta = \dfrac{25}{24}$, $\cos\theta > 0$

For Exercises, 15–20, find the exact value of the expression.

15. $\dfrac{2\tan 67.5°}{1 - \tan^2 67.5°}$ -1

16. $\dfrac{2\tan 15°}{1 - \tan^2 15°}$ $\dfrac{\sqrt3}{3}$

17. $2\sin\dfrac{\pi}{12}\cos\dfrac{\pi}{12}$ $\dfrac{1}{2}$

18. $2\sin(-22.5°)\cos(-22.5°)$ $-\dfrac{\sqrt2}{2}$

19. $\cos^2\dfrac{\pi}{8} - \sin^2\dfrac{\pi}{8}$ $\dfrac{\sqrt2}{2}$

20. $2\cos^2(-165°) - 1$ $\dfrac{\sqrt3}{2}$

For Exercises, 21–34, verify the identity. (See Example 2)

21. $\dfrac{\sin 2\theta}{1 - \cos 2\theta} = \cot\theta$

22. $\dfrac{\sin 2\theta}{\sin 4\theta} = \dfrac{1}{2}\sec 2\theta$

23. $\cos^4 x - \sin^4 x = \cos 2x$

24. $(\sin x - \cos x)^2 = 1 - \sin 2x$

25. $\dfrac{\cos 3x}{\sin 2x} = \dfrac{1}{2}\csc x - 2\sin x$

26. $\dfrac{\sin 3x}{\sin 2x} = 2\cos x - \dfrac{1}{2}\sec x$

27. $\sin 2A - \tan A = \tan A\cos 2A$

28. $\tan 2A - \tan A = \tan A\sec 2A$

29. $\dfrac{2\tan x}{2 - \sec^2 x} = \tan 2x$

30. $\cot x - \tan x = 2\cot 2x$

31. $\dfrac{\cos 2x}{\sin x} - \dfrac{\sin 2x}{\cos x} = \csc x - 4\sin x$

32. $\dfrac{\sin 2x}{\sin x} - \dfrac{\cos 2x}{\cos x} = \sec x$

33. $\sin 4\theta = 4\cos^3\theta\sin\theta - 4\sin^3\theta\cos\theta$

34. $\cos 4\theta = \sin^4\theta - 6\sin^2\theta\cos^2\theta + \cos^4\theta$

Objective 2: Apply the Power-Reducing Formulas

For Exercises 35–38, write the expression in terms of first powers of cosine. (See Example 3)

35. $\sin^4 x$ $\dfrac{1}{8}\cos 4x - \dfrac{1}{2}\cos 2x + \dfrac{3}{8}$ **36.** $\tan^4 x$ $\dfrac{\cos 4x - 4\cos 2x + 3}{\cos 4x + 4\cos 2x + 3}$ **37.** $\cos^2 x\sin^2 x$ $\dfrac{1}{8} - \dfrac{1}{8}\cos 4x$ **38.** $\cos^4 x + \sin^2 x$ $\dfrac{1}{8}\cos 4x + \dfrac{7}{8}$

Objective 3: Apply the Half-Angle Formulas

For Exercises 39–44, use the half-angle formula to find the exact value. (See Example 4)

39. $\sin 165°$ $\dfrac{\sqrt{2 - \sqrt3}}{2}$

40. $\tan\dfrac{7\pi}{8}$ $1 - \sqrt2$

41. $\cos\dfrac{7\pi}{12}$ $-\dfrac{\sqrt{2 - \sqrt3}}{2}$

42. $\cos 112.5°$ $-\dfrac{\sqrt{2 - \sqrt2}}{2}$

43. $\tan 75°$ $2 + \sqrt3$

44. $\sin\dfrac{\pi}{8}$ $\dfrac{\sqrt{2 - \sqrt2}}{2}$

45. Fill in the table for α in the given quadrant. What do you notice about the signs of $\sin\alpha$ and $\tan\dfrac{\alpha}{2}$? (See Example 5)

Quadrant for α	I	II	III	IV
Quadrant for $\dfrac{\alpha}{2}$				
Sign of $\sin\alpha$				
Sign of $\tan\dfrac{\alpha}{2}$				

46. What are the advantages to using the formula

$$\tan\dfrac{\alpha}{2} = \dfrac{1 - \cos\alpha}{\sin\alpha} \text{ or } \tan\dfrac{\alpha}{2} = \dfrac{\sin\alpha}{1 + \cos\alpha} \text{ as opposed to}$$

$$\tan\dfrac{\alpha}{2} = \pm\sqrt{\dfrac{1 - \cos\alpha}{1 + \cos\alpha}}?$$

Using the first two formulas eliminates the need to determine the sign of the result. Furthermore, using the last formula may involve simplifying a "messy" radical.

For Exercises 47–50, use the given information to find the exact function values. (See Example 6)

a. $\sin\dfrac{\alpha}{2}$ b. $\cos\dfrac{\alpha}{2}$ c. $\tan\dfrac{\alpha}{2}$

47. $\sin\alpha = \dfrac{11}{61},\ 0 < \alpha < \dfrac{\pi}{2}$ **48.** $\cos\alpha = \dfrac{12}{13},\ \dfrac{3\pi}{2} < \alpha < 2\pi$ **49.** $\cos\alpha = -\dfrac{12}{37},\ \pi < \alpha < \dfrac{3\pi}{2}$

50. $\sin\alpha = \dfrac{33}{65},\ \dfrac{\pi}{2} < \alpha < \pi$

For Exercises 51–56, verify the identity.

51. $\sin^2\dfrac{\theta}{2} = \dfrac{\csc\theta - \cot\theta}{2\csc\theta}$ **52.** $\cos^2\dfrac{\theta}{2} = \dfrac{\sec\theta + 1}{2\sec\theta}$ **53.** $\tan\dfrac{x}{2} + \cot\dfrac{x}{2} = 2\csc x$

54. $\tan\dfrac{x}{2} = \csc x - \cot x$ **55.** $2\sin^2\dfrac{x}{2} + 4\cos^2\dfrac{x}{2} - 3 = \cos x$ **56.** $\cos^2\dfrac{x}{2} - 3\sin^2\dfrac{x}{2} + 1 = 2\cos x$

Mixed Exercises

57. Show that $\cos^2\theta = \dfrac{1 + \cos 2\theta}{2}$.

58. Show that $\tan^2\theta = \dfrac{1 - \cos 2\theta}{1 + \cos 2\theta}$.

For Exercises 59–64, given restrictions on x, determine whether Expression 1 is less than, greater than, or equal to Expression 2 or whether the relationship cannot be determined.

	x	Expression 1	Expression 2	
59.	$\dfrac{5\pi}{4} < x < \dfrac{3\pi}{2}$	$\tan x$	$\tan\dfrac{x}{2}$	Greater than
60.	$\dfrac{\pi}{4} < x < \dfrac{\pi}{2}$	$\cos x$	$\cos 2x$	Greater than
61.	$\dfrac{\pi}{2} < x < \dfrac{3\pi}{4}$	$\cos x$	$\cos\dfrac{x}{2}$	Less than
62.	$\dfrac{\pi}{2} < x < \dfrac{3\pi}{4}$	$\tan x$	$\sin\dfrac{x}{2}$	Less than
63.	$\dfrac{5\pi}{4} < x < \dfrac{3\pi}{2}$	$\tan x$	$\tan 2x$	Greater than
64.	$0 < x < \dfrac{\pi}{4}$	$\sin x$	$\sin 2x$	Less than

For Exercises 65–70, use the information about α and β given in the figures to find the exact function values. Write the answers as decimals or as square roots of decimals.

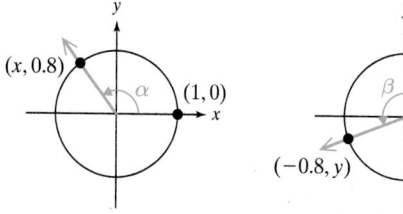

65. a. $\sin 2\alpha$ -0.96 **b.** $\cos 2\alpha$ -0.28

66. a. $\sin 2\beta$ 0.96 **b.** $\cos 2\beta$ 0.28

67. a. $\sin\dfrac{\alpha}{2}$ $\sqrt{0.8}$ **b.** $\cos\dfrac{\alpha}{2}$ $\sqrt{0.2}$

68. a. $\sin\dfrac{\beta}{2}$ $\sqrt{0.9}$ **b.** $\cos\dfrac{\beta}{2}$ $-\sqrt{0.1}$

69. a. $\sin(\alpha + \beta)$ -0.28 **b.** $\cos(\alpha + \beta)$ 0.96

70. a. $\sin(\alpha - \beta)$ -1 **b.** $\cos(\alpha - \beta)$ 0

For Exercises 71–74,

a. Rewrite the function as a single trigonometric function raised to the first power.

b. Graph the result over one period.

71. $y = 4\cos^2 x - 4\sin^2 x$ **72.** $y = -2\sin 2x\cos 2x$ **73.** $y = -6\sin^2 x$ **74.** $y = 4\cos^2 x$
  a. $y = 4\cos 2x$ a. $y = -\sin 4x$ a. $y = -3 + 3\cos 2x$ a. $y = 2 + 2\cos 2x$

For Exercises 75–80, find the exact value.

75. $\cos\left(2\sin^{-1}\dfrac{4}{5}\right)$ $-\dfrac{7}{25}$ **76.** $\cos\left(\dfrac{1}{2}\arcsin\dfrac{3}{8}\right)$ $\dfrac{\sqrt{8+\sqrt{55}}}{4}$ **77.** $\sin\left(\dfrac{1}{2}\cos^{-1}\dfrac{2}{5}\right)$ $\dfrac{\sqrt{30}}{10}$

78. $\sin\left(2\tan^{-1}\dfrac{12}{5}\right)$ $\dfrac{120}{169}$ **79.** $\tan\left[2\arcsin\left(-\dfrac{1}{3}\right)\right]$ $-\dfrac{4\sqrt{2}}{7}$ **80.** $\tan\left[\dfrac{1}{2}\cos^{-1}\left(-\dfrac{15}{17}\right)\right]$ 4

81. Find an algebraic expression representing $\cos(2\sin^{-1}x)$ for $0 < x \le 1$. $1 - 2x^2$

82. Find an algebraic expression representing $\sin(2\cos^{-1}x)$ for $0 < x \le 1$. $2x\sqrt{1 - x^2}$

83. A feeding trough for cattle is made from a metal sheet 3 ft wide. The sides of the trough are made by folding up a flap 1 ft wide from each end of the strip. Each flap makes an angle θ with the horizontal.

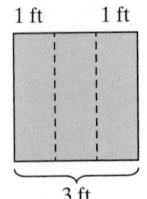
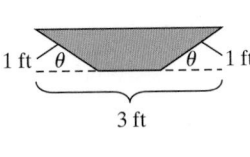

a. Write an expression representing the area of a cross section of the trough in terms of θ.

b. Show that the area can be written as $A = \dfrac{1}{2}\sin 2\theta + \sin\theta$.

84. Consider the triangular area of the roof truss.

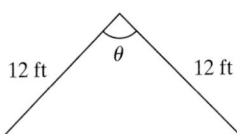

a. Write the area as a function of $\sin\dfrac{\theta}{2}$ and $\cos\dfrac{\theta}{2}$.
$A = 144\sin\dfrac{\theta}{2}\cos\dfrac{\theta}{2}$

b. Write the area as a function of $\sin\theta$.
$A = 72\sin\theta$

85. Consider an object launched from an initial height h_0 with an initial velocity v_0 at an angle θ from the horizontal. The path of the object is given by

$$y = -\frac{g}{2v_0^2\cos^2\theta}x^2 + (\tan\theta)x + h_0$$

where x (in ft) is the horizontal distance from the launching point, y (in ft) is the height above ground level, and g is the acceleration due to gravity ($g = 32$ ft/sec^2 or 9.8 m/sec^2). Show that the horizontal distance traveled by a soccer ball kicked from ground level with velocity v_0 at angle θ is $x = \dfrac{v_0^2\sin 2\theta}{g}$.

86. Refer to Exercise 85. Suppose that you kick a soccer ball from ground level with an initial velocity of 80 ft/sec.

a. Can you make the ball travel 200 horizontal feet? Yes. Kick the ball at a 45° angle.

b. Determine the smallest positive angle at which the ball can be kicked to make it travel 100 ft downfield to a teammate. 15°

87. Write $\cos 3x$ as a third-degree polynomial in $\cos x$. [*Hint*: Write $\cos 3x$ as $\cos(2x + x)$.] $4\cos^3 x - 3\cos x$

88. Write $\cos 4x$ as a fourth-degree polynomial in $\cos x$. $8\cos^4 x - 8\cos^2 x + 1$

89. Write $\cos 5x$ in terms of powers of $\cos x$.
$16\cos^5 x - 20\cos^3 x + 5\cos x$

90. Write $\sin 5x$ in terms of powers of $\sin x$.
$16\sin^5 x - 20\sin^3 x + 5\sin x$

Write About It

91. When using the half-angle formulas, explain how to determine whether the $+$ or $-$ sign is used.

92. Explain the process to derive the double-angle formulas for $\sin 2\theta$ and $\cos 2\theta$.

93. Explain the process to derive the power-reducing formulas for $\sin^2\theta$ and $\cos^2\theta$.

94. Explain the process to derive the half-angle formulas for $\sin\dfrac{\alpha}{2}$ and $\cos\dfrac{\alpha}{2}$.

Expanding Your Skills

For Exercises 95–100, verify the identities for a right $\triangle ABC$

95. $\sin 2A = \sin 2B$

96. $\cos 2A = -\cos 2B$

97. $\tan 2A = \dfrac{2ab}{b^2 - a^2}$

98. $\cos 2A = \dfrac{b^2 - a^2}{c^2}$

99. $\cos\dfrac{A}{2} = \sqrt{\dfrac{c + b}{2c}}$

100. $\tan\dfrac{A}{2} = \dfrac{c - b}{a}$

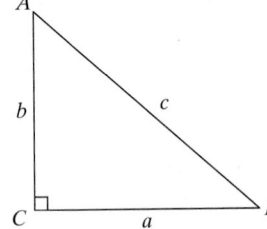

SECTION 5.4 **Product-to-Sum and Sum-to-Product Formulas**

OBJECTIVES

1. **Apply the Product-to-Sum Formulas**
2. **Apply the Sum-to-Product Formulas**

1. Apply the Product-to-Sum Formulas

We now introduce a set of identities that enables us to write a product of factors of sine and cosine as a sum. These are called the product-to-sum formulas (Table 5-6).

Table 5-6

Product-to-Sum Formulas	
$\sin u \sin v = \dfrac{1}{2}[\cos(u - v) - \cos(u + v)]$	(1)
$\cos u \cos v = \dfrac{1}{2}[\cos(u - v) + \cos(u + v)]$	(2)
$\sin u \cos v = \dfrac{1}{2}[\sin(u + v) + \sin(u - v)]$	(3)
$\cos u \sin v = \dfrac{1}{2}[\sin(u + v) - \sin(u - v)]$	(4)

TIP In applications of integral calculus, often working with a sum of terms involving sine or cosine is easier than working with a product.

The derivations of the product-to-sum formulas are straightforward and follow from the sum and difference formulas for sine and cosine. To verify any of the formulas (1)–(4), simplify the right-hand side. For example, to verify (1), we have:

$$\text{RHS: } \frac{1}{2}[\cos(u - v) - \cos(u + v)]$$

$$= \frac{1}{2}[\cos u \cos v + \sin u \sin v - (\cos u \cos v - \sin u \sin v)]$$

$$= \frac{1}{2}(\cos u \cos v + \sin u \sin v - \cos u \cos v + \sin u \sin v)$$

$$= \frac{1}{2}(2 \sin u \sin v)$$

$$= \sin u \sin v$$

The remaining product-to-sum formulas are verified in a similar way (see Exercises 37–38). Although the product-to-sum formulas are difficult to memorize they are easy to derive. Each is $\frac{1}{2}$ the sum or difference of $\cos(u - v)$ and $\cos(u + v)$ or $\frac{1}{2}$ the sum or difference of $\sin(u + v)$ and $\sin(u - v)$.

Classroom Examples: p. 605
Exercises 4, 6

EXAMPLE 1   **Writing a Product of Sine and Cosine as a Sum**

Write the product as a sum or difference.
  a. $\sin 7x \cos 3x$ 　　　　　　　　　　**b.** $\sin(-x)\sin 4x$

Solution:

  a. $\sin 7x \cos 3x$

$$= \frac{1}{2}[\sin(7x + 3x) + \sin(7x - 3x)] \qquad \text{Apply product-to-sum formula (3) with } u = 7x \text{ and } v = 3x.$$

$$= \frac{1}{2}[\sin 10x + \sin 4x] \qquad \text{Combine terms within the arguments.}$$

$$= \frac{1}{2}\sin 10x + \frac{1}{2}\sin 4x \qquad \text{Apply the distributive property.}$$

Classroom Example: p. 605
Exercise 12

TIP By convention, when we apply the product-to-sum formulas, we simplify as much as possible. In Example 1(b), we simplified $\cos(-5x)$ as $\cos 5x$ by using the even function property of the cosine function.

b. $\sin(-x)\sin 4x$

$= \dfrac{1}{2}\{\cos[(-x) - 4x] - \cos[(-x) + 4x]\}$ Apply product-to-sum formula (1) with $u = -x$ and $v = 4x$.

$= \dfrac{1}{2}[\cos(-5x) - \cos 3x]$ Combine terms within the arguments.

$= \dfrac{1}{2}(\cos 5x - \cos 3x)$ Apply the even function property of cosine, $\cos(-x) = \cos x$.

$= \dfrac{1}{2}\cos 5x - \dfrac{1}{2}\cos 3x$ Apply the distributive property.

Skill Practice 1 Write the product as a sum or difference.

a. $\cos 6x \sin 2x$ **b.** $\cos 4x \cos 5x$

EXAMPLE 2   **Using a Product-to-Sum Formula to Find an Exact Value**

Use a product-to-sum formula to find the exact value of $\cos 75° \sin 15°$.

Solution:

$\cos 75° \sin 15°$

$= \dfrac{1}{2}[\sin(75° + 15°) - \sin(75° - 15°)]$ Apply product-to-sum formula (4) with $u = 75°$ and $v = 15°$.

$= \dfrac{1}{2}(\sin 90° - \sin 60°)$ Combine terms within the arguments.

$= \dfrac{1}{2}\left(1 - \dfrac{\sqrt{3}}{2}\right)$ Evaluate $\sin 90° = 1$ and $\sin 60° = \frac{\sqrt{3}}{2}$.

$= \dfrac{1}{2}\left(\dfrac{2 - \sqrt{3}}{2}\right)$ Write the terms inside parentheses with a common denominator and subtract.

$= \dfrac{2 - \sqrt{3}}{4}$ Multiply.

TIP The result of Example 2 can be supported on a calculator.

```
NORMAL FLOAT AUTO REAL DEGREE CL
cos(75)sin(15)
              .0669872981
(2-√(3))/4
              .0669872981
```
The decimal approximations are the same.

Skill Practice 2 Use a product-to-sum formula to find the exact value of $\sin 165° \cos 75°$.

2. Apply the Sum-to-Product Formulas

The product-to-sum formulas can also be used in reverse to write a sum or difference of sine and cosine terms as a product. We call these the sum-to-product formulas (Table 5-7).

TIP The sum-to-product formulas are often helpful when we solve trigonometric equations (Section 5.5).

Table 5-7

Sum-to-Product Formulas	
$\cos x - \cos y = -2\sin\dfrac{x+y}{2}\sin\dfrac{x-y}{2}$	(5)
$\cos x + \cos y = 2\cos\dfrac{x+y}{2}\cos\dfrac{x-y}{2}$	(6)
$\sin x + \sin y = 2\sin\dfrac{x+y}{2}\cos\dfrac{x-y}{2}$	(7)
$\sin x - \sin y = 2\cos\dfrac{x+y}{2}\sin\dfrac{x-y}{2}$	(8)

Answers

1. a. $\dfrac{1}{2}\sin 8x - \dfrac{1}{2}\sin 4x$

b. $\dfrac{1}{2}\cos x + \dfrac{1}{2}\cos 9x$

2. $\dfrac{2 - \sqrt{3}}{4}$

The sum-to-product formulas follow directly from the product-to-sum formulas. For example, to derive formula (5), start with the product-to-sum formula (1) and let $u = \dfrac{x + y}{2}$ and $v = \dfrac{x - y}{2}$.

$$\sin u \sin v = \frac{1}{2}[\cos(u - v) - \cos(u + v)] \tag{1}$$

$$\sin\frac{x + y}{2}\sin\frac{x - y}{2} = \frac{1}{2}\left[\cos\left(\frac{x + y}{2} - \frac{x - y}{2}\right) - \cos\left(\frac{x + y}{2} + \frac{x - y}{2}\right)\right]$$

$$= \frac{1}{2}(\cos y - \cos x)$$

$$= -\frac{1}{2}(\cos x - \cos y) \qquad \text{Factor out } -1.$$

Therefore, $\cos x - \cos y = -2\sin\dfrac{x + y}{2}\sin\dfrac{x - y}{2}.$ (5)

Classroom Examples: p. 605
Exercises 20, 22

EXAMPLE 3   Writing a Sum or Difference as a Product

Write each expression as a product.
  a. $\sin 7\theta + \sin 4\theta$
  b. $\cos\alpha - \cos 3\alpha$

Solution:

a. $\sin 7\theta + \sin 4\theta$

$= 2\sin\dfrac{7\theta + 4\theta}{2}\cos\dfrac{7\theta - 4\theta}{2}$ Apply sum-to-product formula (7) with $x = 7\theta$ and $y = 4\theta$.

$= 2\sin\dfrac{11\theta}{2}\cos\dfrac{3\theta}{2}$ Simplify the arguments.

b. $\cos\alpha - \cos 3\alpha$

$= -2\sin\dfrac{\alpha + 3\alpha}{2}\sin\dfrac{\alpha - 3\alpha}{2}$ Apply sum-to-product formula (5) with $x = \alpha$ and $y = 3\alpha$.

$= -2\sin 2\alpha\sin(-\alpha)$ Simplify the arguments.

$= -2\sin 2\alpha(-\sin\alpha)$ Apply the odd function property $\sin(-\alpha) = -\sin\alpha$.

$= 2\sin 2\alpha\sin\alpha$

> **TIP** By convention, when we apply the sum-to-product formulas, we simplify as much as possible. In Example 3(b), we simplified $\sin(-\alpha)$ as $-\sin\alpha$ by using the odd function property of the sine function.

Skill Practice 3   Write each expression as a product.
  a. $\cos 5x + \cos 2x$
  b. $\sin\beta - \sin 5\beta$

Classroom Example: p. 605
Exercise 26

EXAMPLE 4   Using a Sum-to-Product Formula to Find an Exact Value

Use a sum-to-product formula to find the exact value of $\cos 255° + \cos 195°$.

Solution:

$\cos 255° + \cos 195°$

$= 2\cos\dfrac{255° + 195°}{2}\cos\dfrac{255° - 195°}{2}$ Apply sum-to-product formula (6) with $x = 255°$ and $y = 195°$.

$= 2\cos 225°\cos 30°$ Simplify the arguments.

$= 2\left(-\dfrac{\sqrt{2}}{2}\right)\left(\dfrac{\sqrt{3}}{2}\right)$ Evaluate $\cos 225° = -\dfrac{\sqrt{2}}{2}$ and $\cos 30° = \dfrac{\sqrt{3}}{2}$.

$= -\dfrac{\sqrt{6}}{2}$ Simplify.

Answers

3. a. $2\cos\dfrac{7x}{2}\cos\dfrac{3x}{2}$
  b. $-2\cos 3\beta\sin 2\beta$

Skill Practice 4   Use a sum-to-product formula to find the exact value of $\sin 75° - \sin 15°$.

In Example 5, we use the sum-to-product formulas to verify an identity.

Classroom Example: p. 605
Exercise 28

EXAMPLE 5   **Using Sum-to-Product Formulas to Verify an Identity**

Verify the identity. $\dfrac{\sin 5x + \sin x}{\cos 5x + \cos x} = \tan 3x$

Solution:

$$\frac{\sin 5x + \sin x}{\cos 5x + \cos x} = \frac{2\sin\dfrac{5x + x}{2}\cos\dfrac{5x - x}{2}}{2\cos\dfrac{5x + x}{2}\cos\dfrac{5x - x}{2}}$$

In the numerator apply formula (7),
$$\sin x + \sin y = 2\sin\frac{x + y}{2}\cos\frac{x - y}{2}.$$
In the denominator, apply formula (6),
$$\cos x + \cos y = 2\cos\frac{x + y}{2}\cos\frac{x - y}{2}.$$

$$= \frac{2\sin 3x\cos 2x}{2\cos 3x\cos 2x}$$

$$= \frac{\sin 3x}{\cos 3x}$$

$$= \tan 3x$$

Skill Practice 5   Verify the identity. $\dfrac{\cos 4x - \cos 2x}{\sin 2x - \sin 4x} = \tan 3x$

Answers

4. $\dfrac{\sqrt{2}}{2}$

5. $\dfrac{\cos 4x - \cos 2x}{\sin 2x - \sin 4x}$

$$= \frac{-2\sin\dfrac{4x + 2x}{2}\sin\dfrac{4x - 2x}{2}}{2\cos\dfrac{4x + 2x}{2}\sin\dfrac{2x - 4x}{2}}$$

$$= \frac{-2\sin 3x\sin x}{-2\cos 3x\sin x} = \tan 3x$$

Point of Interest

Before the introduction of logarithms, the product-to-sum formulas (historically known as the "prosthapheretic formulae") were used in an algorithm to approximate the products of many-digit numbers. Each factor was scaled to a number between −1 and 1 by dividing the factor by a power of 10. Then using a table of cosines, two angles u and v were found whose cosines were closest to the scaled factors. Taking the sum and difference of u and v and applying the formula $\cos u\cos v = \frac{1}{2}[\cos(u - v) + \cos(u + v)]$ resulted in the product of the two scaled factors. At this point, the product was multiplied by an appropriate power of 10 to reverse the original scaling factor in the first step.

This process is very similar to the process using logarithms to approximate the product of two numbers and is the basis of multiplication using a slide rule: scale down each factor, add the logarithms of the scaled down factors, take the inverse logarithm, and reverse the scaling factor.

SECTION 5.4   **Practice Exercises**

Concept Connections

1. The formula $\sin u\sin v = \dfrac{1}{2}[\cos(u - v) - \cos(u + v)]$ converts a product of sines to a

_____difference_____ of _____cosines_____.

2. The formula $\cos x - \cos y = -2\sin\dfrac{x + y}{2}\sin\dfrac{x - y}{2}$ converts a difference of cosines to a

_____product_____ of _____sines_____.

Objective 1: Apply the Product-to-Sum Formulas

For Exercises 3–10, write the product as a sum or difference. (See Example 1)

3. $\sin 5x \sin 8x$ $\quad \frac{1}{2}\cos 3x - \frac{1}{2}\cos 13x$

4. $\sin(-2\theta)\sin 4\theta$ $\quad \frac{1}{2}\cos 6\theta - \frac{1}{2}\cos 2\theta$

5. $\cos(-3t)\cos 7t$ $\quad \frac{1}{2}\cos 10t + \frac{1}{2}\cos 4t$

6. $\cos\frac{x}{2}\cos\frac{3x}{2}$ $\quad \frac{1}{2}\cos x + \frac{1}{2}\cos 2x$

7. $\sin\frac{x}{4}\cos\frac{5x}{4}$ $\quad \frac{1}{2}\sin\frac{3x}{2} - \frac{1}{2}\sin x$

8. $\sin 4\alpha \cos 5\alpha$ $\quad \frac{1}{2}\sin 9\alpha - \frac{1}{2}\sin\alpha$

9. $\cos(-x)\sin 2x$ $\quad \frac{1}{2}\sin x + \frac{1}{2}\sin 3x$

10. $\cos 5x \sin 2x$ $\quad \frac{1}{2}\sin 7x - \frac{1}{2}\sin 3x$

In Exercises 11–14, use a product-to-sum formula to find the exact value. (See Example 2)

11. $\sin\frac{17\pi}{12}\sin\frac{\pi}{12}$ $\quad -\frac{1}{4}$

12. $\sin 82.5°\cos 37.5°$ $\quad \frac{\sqrt{3}+\sqrt{2}}{4}$

13. $\cos 97.5°\cos 37.5°$ $\quad \frac{1-\sqrt{2}}{4}$

14. $\cos\frac{25\pi}{24}\sin\frac{17\pi}{24}$ $\quad -\frac{\sqrt{3}+\sqrt{2}}{4}$

For Exercises 15–18, verify the identities.

15. $\sin(x+y)\sin(x-y) = \sin^2 x - \sin^2 y$

16. $\cos(x+y)\cos(x-y) = \cos^2 x - \sin^2 y$

17. $2\sin\left(\frac{\pi}{4}+\frac{x-y}{2}\right)\cos\left(\frac{\pi}{4}-\frac{x+y}{2}\right) = \cos y + \sin x$

18. $2\sin\left(\frac{\pi}{4}+\frac{x-y}{2}\right)\cos\left(\frac{\pi}{4}+\frac{x+y}{2}\right) = \cos x - \sin y$

Objective 2: Apply the Sum-to-Product Formulas

For Exercises 19–22, write each expression as a product. (See Example 3)

19. $\sin 7x - \sin 9x$ $\quad -2\cos 8x \sin x$

20. $\sin 3\theta + \sin 10\theta$ $\quad 2\sin\frac{13\theta}{2}\cos\frac{7\theta}{2}$

21. $\cos 5\beta + \cos 6\beta$ $\quad 2\cos\frac{11\beta}{2}\cos\frac{\beta}{2}$

22. $\cos 20x - \cos 14x$ $\quad -2\sin 17x \sin 3x$

For Exercises 23–26, use a sum-to-product formula to find the exact value. (See Example 4)

23. $\sin\frac{11\pi}{12} + \sin\frac{5\pi}{12}$ $\quad \frac{\sqrt{6}}{2}$

24. $\sin 165° - \sin 105°$ $\quad -\frac{\sqrt{2}}{2}$

25. $\cos 285° - \cos 15°$ $\quad -\frac{\sqrt{2}}{2}$

26. $\cos\frac{23\pi}{12} + \cos\frac{5\pi}{12}$ $\quad \frac{\sqrt{6}}{2}$

For Exercises 27–34, verify the identity. (See Example 5)

27. $\dfrac{\cos 3x - \cos x}{\cos 3x + \cos x} = -\tan 2x \tan x$

28. $\dfrac{\cos x - \cos 3x}{\sin x - \sin 3x} = -\tan 2x$

29. $\sin 2x + \sin 4x + \sin 6x = 4\cos x \cos 2x \sin 3x$

30. $\cos 2x + \cos 4x + \cos 6x = 4\cos x \cos 2x \cos 3x - 1$

31. $\sin 5x - \sin 3x + 2\sin x \cos 2x = 2\sin 2x \cos 3x$

32. $\sin t + \sin 3t + \sin 5t + \sin 7t = 4\cos t \cos 2t \sin 4t$

33. $\dfrac{\cos 3x - \cos 2x + \cos x}{\sin 3x - \sin 2x + \sin x} = \cot 2x$

34. $\dfrac{\sin 4t + \sin 3t + \sin 2t}{\cos 4t + \cos 3t + \cos 2t} = \tan 3t$

Mixed Exercises

35. Change $\sin(x+h) - \sin x$ to a product. $\quad 2\cos\left(\frac{2x+h}{2}\right)\sin\frac{h}{2}$

36. Change $\cos(x+h) - \cos x$ to a product. $\quad -2\sin\frac{2x+h}{2}\sin\frac{h}{2}$

37. Derive formula (2) on page 601. $\cos u \cos v = \frac{1}{2}[\cos(u-v) + \cos(u+v)]$

38. Derive formula (3) on page 601. $\sin u \cos v = \frac{1}{2}[\sin(u+v) + \sin(u-v)]$

39. Derive formula (6) on page 602. $\cos x + \cos y = 2\cos\frac{x+y}{2}\cos\frac{x-y}{2}$

40. Derive formula (7) on page 602. $\sin x + \sin y = 2\sin\frac{x+y}{2}\cos\frac{x-y}{2}$

Touch-tone telephones use DTMF (dual-tone multi-frequency) signaling over analog telephone lines. The DTMF keypad is a four-row, three-column grid in which each row represents a low frequency and each column represents a high frequency. Pressing a single key sends a signal formed by the sum of two sine waves. The tone is then decoded by the switching center to determine the key pressed by the user.

The signal produced by pressing any key is given by

$$y = \sin(2\pi lt) + \sin(2\pi ht)$$

where l and h represent the low and high frequencies for that key. Frequencies are measured in cycles per second or hertz (Hz).

For Exercises 41–42, use the information given in the figure.

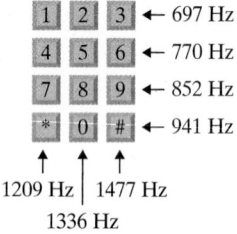

1209 Hz | 1477 Hz
1336 Hz

41. Suppose a user makes a long distance call and presses the number 1 on the keypad.

a. Write a function representing this sound as the sum of two sine waves. $y = \sin 1394\pi t + \sin 2418\pi t$

b. Write a function representing this sound as the product of sines and cosines. $y = 2\sin 1906\pi t \cos 512\pi t$

c. Use a graphing utility to graph the signal on the interval $0 \le t \le 0.01$.

42. Suppose that the number 9 is pressed on a phone keypad.

a. Write a function representing this sound as the sum of two sine waves. $y = \sin 1704\pi t + \sin 2954\pi t$

b. Write a function representing this sound as the product of sines and cosines. $y = 2\sin 2329\pi t \cos 625\pi t$

c. Use a graphing utility to graph the signal on the interval $0 \le t \le 0.01$.

Write About It

43. Explain how the formulas (2) and (6) on pages 601 and 602 are related.

44. In Section 5.2, we derived the reduction formula $A\sin x + B\cos x = k\sin(x + \alpha)$ to write a linear combination of $\sin x$ and $\cos x$ as a single term. Explain how this is different from the sum-to-product formulas.

Expanding Your Skills

For Exercises 45–46, verify the identity.

45. $\cos A \sin(B - C) + \cos B \sin(C - A) + \cos C \sin(A - B) = 0$

46. $\cos A \sin(B - C) - \cos B \sin(C - A) - \cos C \sin(A - B) = 2\cos A \sin(B - C)$

For Exercises 47–48, let A, B, and C represent the angles in $\triangle ABC$. Use the fact that $C = \pi - (A + B)$ to prove the identities.

47. $\sin A + \sin B + \sin C = 4\cos\dfrac{A}{2}\cos\dfrac{B}{2}\cos\dfrac{C}{2}$

48. $\cos A + \cos B + \cos C = 1 + 4\sin\dfrac{A}{2}\sin\dfrac{B}{2}\sin\dfrac{C}{2}$

Technology Connections

In this section, we used the product-to-sum formulas to write sums and differences of trigonometric functions. Another type of expression involving the sum or difference of sine and cosine terms is a Fourier series, named after Jean-Baptiste Joseph Fourier (1768–1830). A Fourier series is an expression of the form $A_0 + (A_1\cos nx + B_1\sin nx) + (A_2\cos nx + B_2\sin nx) + \cdots$ where A_i and B_i are constants and n is an integer. Fourier proved that any continuous function can be represented by a Fourier series. In Exercises 49–50, we use Fourier polynomials (a finite number of terms from a Fourier series) to model a "saw-tooth" wave and a "square" wave.

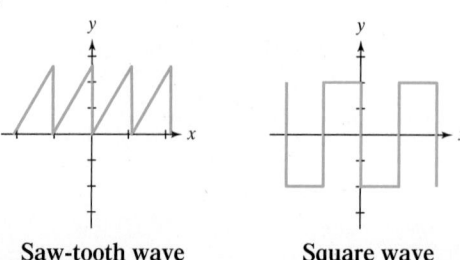

Saw-tooth wave   Square wave

49. A "saw-tooth" wave is a periodic wave that rises linearly upward and then drops sharply. In music, such waves are generated in digital synthesizers to give high-quality sound without distortion. Given,

$$f(x) = \frac{1}{2} - \frac{1}{\pi}\left(\frac{\sin \pi x}{1} + \frac{\sin 2\pi x}{2} + \frac{\sin 3\pi x}{3} + \cdots\right)$$

a. Graph the first three terms of the function on the window $[-4, 4, 1]$ by $[-1, 1.5, 0.5]$.

b. Graph the first five terms of the function on the window $[-4, 4, 1]$ by $[-1, 1.5, 0.5]$.

50. A "square" wave is a periodic wave that alternates between two fixed values with equal time spent at each value and with negligible transition time between them. Square waves have practical uses in electronics and music, and because of their rectangular pattern, they are used in timing devices to synchronize circuits. Given,

$$f(x) = \frac{4}{\pi}\left(\frac{1}{1}\sin \pi x + \frac{1}{3}\sin 3\pi x + \frac{1}{5}\sin 5\pi x + \cdots\right)$$

a. Graph the first three terms of the Fourier series on the window $[-4, 4, 1]$ by $[-2, 2, 1]$.

b. Graph the first five terms of the Fourier series on the window $[-4, 4, 1]$ by $[-2, 2, 1]$.

51. The sound wave generated by the note middle C on a piano is modeled by $y_C = k\sin 524\pi t$, and the sound wave for A above middle C is modeled by $y_A = k\sin 880\pi t$. In each case, y is the pressure of the sound wave, and the amplitude k is related to the loudness of sound. When two keys on a piano are struck simultaneously, a sound is made called a musical chord. The chord C-A is represented by $y = k(\sin 524\pi t + \sin 880\pi t)$.

a. Use the sum-to-product formula to represent this function as a product of factors for $k = 1$. $y = 2\sin 702\pi t \cos 178\pi t$

b. Graph the function $y = \sin 524\pi t + \sin 880\pi t$ and your result from part (a) on the window $[0, 0.01, 0.001]$ by $[-2, 2, 1]$ and verify that the graphs are the same.

Additional answers can be found in the Instructor Answer Appendix.

SECTION 5.5   Trigonometric Equations

1. Solve Trigonometric Equations in Linear Form

In this chapter we have worked extensively with trigonometric identities. These are equations that are true for all values of the variable for which the expressions within the equation are defined. Now we look at conditional trigonometric equations. These are equations that are true for some values of the variable but not for others.

For example, from the unit circle we know that the equation $\cos x = \frac{1}{2}$ has solutions $x = \frac{\pi}{3}$ and $x = \frac{5\pi}{3}$ on the interval $[0, 2\pi)$ (Figure 5-9). However, because the cosine function is periodic, any integer multiple of 2π added to $\frac{\pi}{3}$ or $\frac{5\pi}{3}$ also has a cosine value of $\frac{1}{2}$. Therefore, there are infinitely many solutions to the equation. For any integer n, the solutions are:

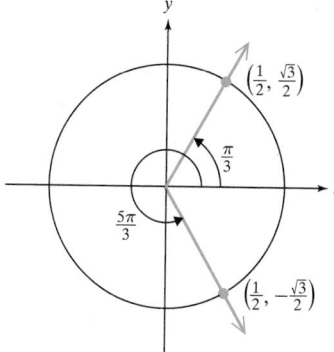

Figure 5-9

$$x = \frac{\pi}{3} + 2n\pi \qquad \left(\frac{\pi}{3} \text{ plus any integer multiple of } 2\pi\right)$$

$$x = \frac{5\pi}{3} + 2n\pi \qquad \left(\frac{5\pi}{3} \text{ plus any integer multiple of } 2\pi\right)$$

The graphs of $y = \cos x$ and $y = \frac{1}{2}$ illustrate several such solutions (Figure 5-10).

> **TIP** Solutions to trigonometric equations such as $\cos x = \frac{1}{2}$ are in radians.

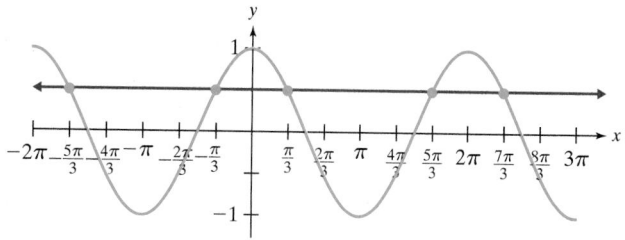

Figure 5-10

We can write the solution set to the equation $\cos x = \frac{1}{2}$ over a specific interval such as $[0, 2\pi)$, or we can write the solution set for the general solution.

The solution set over the interval $[0, 2\pi)$ is $\left\{ \dfrac{\pi}{3}, \dfrac{5\pi}{3} \right\}$.

The general solution set is $\left\{ x \middle| x = \dfrac{\pi}{3} + 2n\pi, x = \dfrac{5\pi}{3} + 2n\pi \right\}$.

In Example 1, we solve a trigonometric equation that is linear in the variable $\tan x$.

Classroom Example: p. 618
Exercise 14

EXAMPLE 1   Solving a Trigonometric Equation in Linear Form

Solve $2\tan x = \sqrt{3} - \tan x$

a. Over $[0, 2\pi)$.   **b.** Over the set of real numbers.

Solution:

a. This equation is linear in the variable $\tan x$. Therefore, isolate $\tan x$.

$2\tan x = \sqrt{3} - \tan x$

$3\tan x = \sqrt{3}$ Add $\tan x$ to both sides.

$\tan x = \dfrac{\sqrt{3}}{3}$ Divide by 3.

$x = \dfrac{\pi}{6}, x = \dfrac{7\pi}{6}$ On the interval $[0, 2\pi)$, $\tan x$ equals $\frac{\sqrt{3}}{3}$ for $x = \frac{\pi}{6}$ and $x = \frac{7\pi}{6}$.

The solution set is $\left\{ \dfrac{\pi}{6}, \dfrac{7\pi}{6} \right\}$.   Write the solution set.

Instructor Note:
Tell students that although the solutions to trigonometric equations are given in radians, the corresponding values can easily be converted to degrees. This may be particularly helpful when solving applications. The solution set in Example 1 is $\{x | x = 30° + 180°n\}$.

b. The period of the tangent function is π. Therefore, adding $n\pi$ to either solution for any integer n also gives a solution.

$$x = \frac{\pi}{6} + n\pi \text{ and } x = \frac{7\pi}{6} + n\pi$$

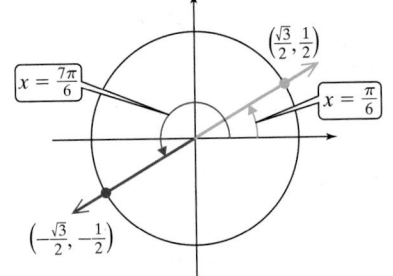

However, notice that for $n = 1$, $\dfrac{\pi}{6} + (1)\pi = \dfrac{7\pi}{6}$. This means that the formula $x = \dfrac{\pi}{6} + n\pi$ is sufficient to generate all solutions to the equation.

Thus, the general solution is $\left\{ x \middle| x = \dfrac{\pi}{6} + n\pi \right\}$.

Skill Practice 1   Solve $\sin x = \sqrt{2} - \sin x$
   a. Over $[0, 2\pi)$.   **b.** Over the set of real numbers.

Answers

1. a. $\left\{ \dfrac{\pi}{4}, \dfrac{3\pi}{4} \right\}$

b. $\left\{ x \middle| x = \dfrac{\pi}{4} + 2n\pi, x = \dfrac{3\pi}{4} + 2n\pi \right\}$

2. Solve Trigonometric Equations Involving Multiple or Compound Angles

On the interval $[0, 2\pi)$, there are two solutions to the equation $\cos x = \frac{1}{2}$ (Figure 5-11).

Now consider the equation $\cos 4x = \frac{1}{2}$. The factor of 4 compresses the graph of $y = \cos 4x$ horizontally, and the period is $\frac{\pi}{2}$. Therefore, the interval $[0, 2\pi)$ contains four complete cycles of $y = \cos 4x$ and eight solutions to the equation $\cos 4x = \frac{1}{2}$ (Figure 5-12).

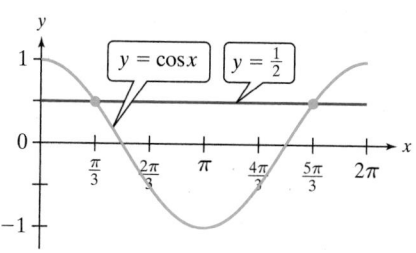

Figure 5-11

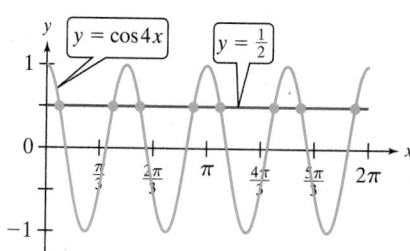

Figure 5-12

This discussion illustrates that the period plays an important role in determining the number of solutions to a trigonometric equation containing a multiple angle.

Classroom Example: p. 618
Exercise 24

EXAMPLE 2   **Solving an Equation Containing a Multiple Angle**

Given $2\sin 2x - \sqrt{3} = 0$,

   a. Write the solution set for the general solution.
   b. Write the solution set on the interval $[0, 2\pi)$.

Solution:

a. To solve $2\sin 2x - \sqrt{3} = 0$, begin by isolating $\sin 2x$.

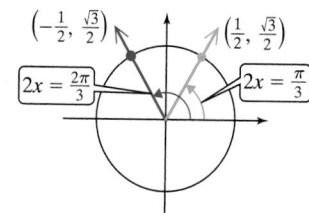

$$\sin 2x = \frac{\sqrt{3}}{2}$$

$$2x = \frac{\pi}{3} + 2n\pi, \qquad\qquad 2x = \frac{2\pi}{3} + 2n\pi$$

From the unit circle,
$\sin 2x = \frac{\sqrt{3}}{2}$ for $2x = \frac{\pi}{3}$
and $2x = \frac{2\pi}{3}$.

$$\frac{1}{2}\cdot 2x = \frac{1}{2}\cdot\left(\frac{\pi}{3} + 2n\pi\right), \quad \frac{1}{2}\cdot 2x = \frac{1}{2}\cdot\left(\frac{2\pi}{3} + 2n\pi\right) \quad \text{Solve for } x.$$

$$x = \frac{\pi}{6} + n\pi, \qquad\qquad x = \frac{\pi}{3} + n\pi$$

General solution: $\left\{ x \,\middle|\, x = \dfrac{\pi}{6} + n\pi,\ x = \dfrac{\pi}{3} + n\pi \right\}$

TIP As an alternative approach, we can use a substitution to help solve an equation with a compound argument. In Example 2, letting $u = 2x$, gives the simpler equation $2\sin u - \sqrt{3} = 0$.
The solutions are
$u = \dfrac{\pi}{3} + 2n\pi$, and
$u = \dfrac{2\pi}{3} + 2n\pi$.
Then using back substitution we can solve for x.
$2x = \dfrac{\pi}{3} + 2n\pi$ and
$2x = \dfrac{2\pi}{3} + 2n\pi$

b. To find solutions on the interval $[0, 2\pi)$, substitute $n = \ldots, -1, 0, 1, 2, \ldots$

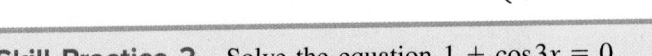

n	$x = \dfrac{\pi}{6} + n\pi$	$x = \dfrac{\pi}{3} + n\pi$
-1	$x = \dfrac{\pi}{6} + (-1)\pi = -\dfrac{5\pi}{6}$ Not on $[0, 2\pi)$	$x = \dfrac{\pi}{3} + (-1)\pi = -\dfrac{2\pi}{3}$ Not on $[0, 2\pi)$
0	$x = \dfrac{\pi}{6} + (0)\pi = \dfrac{\pi}{6}$	$x = \dfrac{\pi}{3} + (0)\pi = \dfrac{\pi}{3}$
1	$x = \dfrac{\pi}{6} + (1)\pi = \dfrac{7\pi}{6}$	$x = \dfrac{\pi}{3} + (1)\pi = \dfrac{4\pi}{3}$
2	$x = \dfrac{\pi}{6} + (2)\pi = \dfrac{13\pi}{6}$ Not on $[0, 2\pi)$	$x = \dfrac{\pi}{3} + (2)\pi = \dfrac{7\pi}{3}$ Not on $[0, 2\pi)$

The solution set on the interval $[0, 2\pi)$ is $\left\{\dfrac{\pi}{6}, \dfrac{\pi}{3}, \dfrac{7\pi}{6}, \dfrac{4\pi}{3}\right\}$ (Figure 5-13).

Figure 5-13

Skill Practice 2 Solve the equation $1 + \cos 3x = 0$.
a. Write the solution set for the general solution.
b. Write the solution set on the interval $[0, 2\pi)$.

The solutions to the equation $2\sin 2x - \sqrt{3} = 0$ are the x-coordinates of the points of intersection of the graphs of $y = 2\sin 2x - \sqrt{3}$ and $y = 0$. See Figure 5-14.

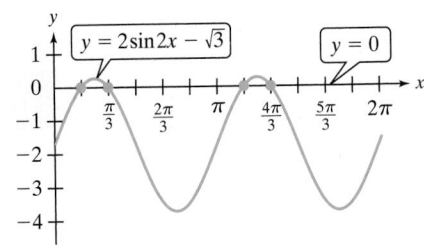

Figure 5-14

Classroom Example: p. 618
Exercise 28

TIP When solving a trigonometric equation with a compound argument, sometimes it is helpful to use a graphing utility to illustrate the number of solutions on a given interval. In Example 3, we can graph $Y = -1 + \sin\frac{x}{2}$ on the interval $[0, 2\pi)$. From the graph it appears that only one solution exists on the interval.

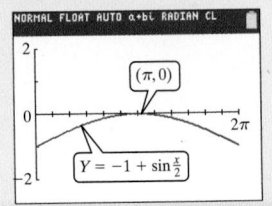

EXAMPLE 3 **Solving an Equation Involving a Half-Angle**

Given $-1 + \sin\dfrac{x}{2} = 0$,

a. Write the solution set for the general solution.
b. Write the solution set on the interval $[0, 2\pi)$.

Solution:

a. $-1 + \sin\dfrac{x}{2} = 0$

$\sin\dfrac{x}{2} = 1$ \qquad Isolate $\sin\frac{x}{2}$.

$\dfrac{x}{2} = \dfrac{\pi}{2} + 2n\pi$ \qquad From the unit circle, $\sin\frac{x}{2} = 1$ for $\frac{x}{2} = \ldots, -\frac{3\pi}{2}, \frac{\pi}{2}, \frac{5\pi}{2}, \ldots$.

$2 \cdot \left(\dfrac{x}{2}\right) = 2 \cdot \left(\dfrac{\pi}{2} + 2n\pi\right)$ \qquad Solve for x.

$x = \pi + 4n\pi$

The general solution is $\{x \,|\, x = \pi + 4n\pi\}$.

Answers

2. a. $\left\{x \,\middle|\, x = \dfrac{\pi}{3} + \dfrac{2n\pi}{3}\right\}$ **b.** $\left\{\dfrac{\pi}{3}, \pi, \dfrac{5\pi}{3}\right\}$

b. Substituting integer values of n gives $x = \ldots -3\pi,\ \pi,\ 5\pi,\ 9\pi, \ldots$.

The solution set on the interval $[0, 2\pi)$ is $\{\pi\}$.

Skill Practice 3 Solve the equation $1 + 2\cos\dfrac{x}{2} = 0$.

a. Write the solution set for the general solution.

b. Write the solution set on the interval $[0, 2\pi)$.

In Example 4, we solve a trigonometric equation where the argument to the trigonometric function has more than one term.

Classroom Example: p. 618
Exercise 32

EXAMPLE 4   **Solving an Equation Involving a Compound Angle**

Given $\sin\left(x - \dfrac{\pi}{3}\right) = -\dfrac{\sqrt{2}}{2}$,

a. Write the solution set for the general solution.

b. Write the solution set on the interval $[0, 2\pi)$.

Solution:

a. $\sin\left(x - \dfrac{\pi}{3}\right) = -\dfrac{\sqrt{2}}{2}$

$$\sin u = -\dfrac{\sqrt{2}}{2} \qquad \text{Let } u = x - \tfrac{\pi}{3}.$$

$$u = \dfrac{5\pi}{4} + 2n\pi, \qquad u = \dfrac{7\pi}{4} + 2n\pi \qquad \text{From the unit circle, recognize that } \sin u = -\tfrac{\sqrt{2}}{2} \text{ for } u = \tfrac{5\pi}{4} \text{ and } u = \tfrac{7\pi}{4}.$$

$$x - \dfrac{\pi}{3} = \dfrac{5\pi}{4} + 2n\pi, \qquad x - \dfrac{\pi}{3} = \dfrac{7\pi}{4} + 2n\pi \qquad \text{Back substitute.}$$

$$x = \dfrac{5\pi}{4} + \dfrac{\pi}{3} + 2n\pi, \qquad x = \dfrac{7\pi}{4} + \dfrac{\pi}{3} + 2n\pi \qquad \text{Add } \tfrac{\pi}{3} \text{ to both sides.}$$

$$x = \dfrac{15\pi}{12} + \dfrac{4\pi}{12} + 2n\pi, \qquad x = \dfrac{21\pi}{12} + \dfrac{4\pi}{12} + 2n\pi \qquad \text{Write terms with a common denominator.}$$

$$x = \dfrac{19\pi}{12} + 2n\pi, \qquad x = \dfrac{25\pi}{12} + 2n\pi$$

The general solution is $\left\{ x \,\middle|\, x = \dfrac{19\pi}{12} + 2n\pi,\ x = \dfrac{25\pi}{12} + 2n\pi \right\}$.

b. To find solutions on the interval $[0, 2\pi)$, substitute $n = \ldots, -1, 0, 1, \ldots$

Answers

3. a. $\left\{ x \,\middle|\, x = \dfrac{4\pi}{3} + 4n\pi, \right.$

$\left. x = \dfrac{8\pi}{3} + 4n\pi \right\}$

b. $\left\{ \dfrac{4\pi}{3} \right\}$

n	$x = \dfrac{19\pi}{12} + 2n\pi$		$x = \dfrac{25\pi}{12} + 2n\pi$	
-1	$x = \dfrac{19\pi}{12} + 2(-1)\pi = -\dfrac{5\pi}{12}$	Not on $[0, 2\pi)$	$x = \dfrac{25\pi}{12} + 2(-1)\pi = \dfrac{\pi}{12}$	
0	$x = \dfrac{19\pi}{12} + 2(0)\pi = \dfrac{19\pi}{12}$		$x = \dfrac{25\pi}{12} + 2(0)\pi = \dfrac{25\pi}{12}$	Not on $[0, 2\pi)$
1	$x = \dfrac{19\pi}{12} + 2(1)\pi = \dfrac{43\pi}{12}$	Not on $[0, 2\pi)$	$x = \dfrac{25\pi}{12} + 2(1)\pi = \dfrac{49\pi}{12}$	Not on $[0, 2\pi)$

The solution set on the interval $[0, 2\pi)$ is $\left\{\dfrac{\pi}{12}, \dfrac{19\pi}{12}\right\}$.

The graphs of $y = \sin\left(x - \dfrac{\pi}{3}\right)$ and $y = -\dfrac{\sqrt{2}}{2}$ are shown in Figure 5-15.

The x-coordinates of the points of intersection of the graphs make up the solution set to the equation $\sin\left(x - \dfrac{\pi}{3}\right) = -\dfrac{\sqrt{2}}{2}$.

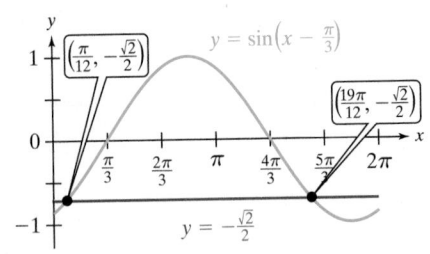

Figure 5-15

Skill Practice 4   Solve the equation $\cot\left(x - \dfrac{\pi}{4}\right) = -1$.

  a. Write the solution set for the general solution.
  b. Write the solution set on the interval $[0, 2\pi)$.

---

3. Solve Higher-Degree Trigonometric Equations

In Examples 5 and 6, we solve trigonometric equations that are quadratic in $\sin x$, $\cos x$, or $\tan x$. As with standard algebraic quadratic equations, we set one side equal to zero and factor the other side. If the nonzero side is not factorable, we can use the square root property or apply the quadratic formula.

Classroom Example: p. 618
Exercise 38

EXAMPLE 5   **Solving a Quadratic Trigonometric Equation**

Solve the equation on the interval $[0, 2\pi)$. $2\sin^2 x + 7\sin x - 4 = 0$

Solution:

$2\sin^2 x + 7\sin x - 4 = 0$	This equation is quadratic in form. $2u^2 + 7u - 4 = 0$, where $u = \sin x$.
$(2\sin x - 1)(\sin x + 4) = 0$	Factor as $(2u - 1)(u + 4) = 0$.
$2\sin x - 1 = 0$ or $\sin x + 4 = 0$	Set each factor equal to zero and solve the resulting equation.
$\sin x = \dfrac{1}{2}$ or $\sin x = -4$	The only solutions are those for which $\sin x = \frac{1}{2}$.
$x = \dfrac{\pi}{6}, x = \dfrac{5\pi}{6}$ No solution here because $-1 \le \sin x \le 1$.	From the unit circle, recognize that $\sin x = \frac{1}{2}$ for $x = \frac{\pi}{6}$ and $x = \frac{5\pi}{6}$ on the interval $[0, 2\pi)$.

The solution set is $\left\{\dfrac{\pi}{6}, \dfrac{5\pi}{6}\right\}$.

TIP In Example 5, if we were asked to find the general solution, it would consist of $x = \frac{\pi}{6} + 2n\pi$ and $x = \frac{5\pi}{6} + 2n\pi$.

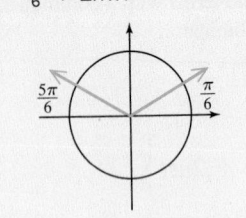

Skill Practice 5   Solve the equation on the interval $[0, 2\pi)$.
$2\cos^2 x - \cos x - 3 = 0$

In Example 6, we apply the square root property to solve a trigonometric equation of the form $u^2 = k$.

Answers
4. a. $\{x \mid x = n\pi\}$ **b.** $\{0, \pi\}$

5. $\{\pi\}$

Classroom Example: p. 618
Exercise 36

EXAMPLE 6   Solving a Quadratic Trigonometric Equation

Solve the equation on the interval $[0, 2\pi)$. $\cot^2 x - 3 = 0$

Solution:

$$\cot^2 x - 3 = 0$$

$$\cot^2 x = 3$$
Isolate $\cot^2 x$ to write the equation in the form $u^2 = k$.

$$\cot x = \pm\sqrt{3}$$
Apply the square root property.

$\cot x = \sqrt{3}$ $\qquad\qquad$ $\cot x = -\sqrt{3}$ $\qquad$ From the unit circle, recognize that $\cot x = \sqrt{3}$ for $x = \frac{\pi}{6}$ and $x = \frac{7\pi}{6}$ and that $\cot x = -\sqrt{3}$ for $x = \frac{5\pi}{6}$ and $x = \frac{11\pi}{6}$.

$$x = \frac{\pi}{6}, x = \frac{7\pi}{6} \qquad\qquad x = \frac{5\pi}{6}, x = \frac{11\pi}{6}$$

The solution set is $\left\{ \dfrac{\pi}{6}, \dfrac{5\pi}{6}, \dfrac{7\pi}{6}, \dfrac{11\pi}{6} \right\}$.

TIP In Example 6, if we were asked to find the general solution, notice that each pair of solutions differs by π radians (180°). The general solution consists of $x = \frac{\pi}{6} + n\pi$ and $x = \frac{5\pi}{6} + n\pi$.

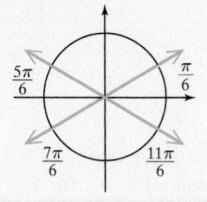

Skill Practice 6   Solve the equation on the interval $[0, 2\pi)$.
$$4\cos^2 x - 3 = 0$$

For trigonometric equations containing more than one trigonometric function, sometimes we can set one side equal to zero and factor the other side to separate factors.

Classroom Example: p. 618
Exercise 42

EXAMPLE 7   Solving a Trigonometric Equation by Factoring

Solve the equation on the interval $[0, 2\pi)$. $\tan x \sin^2 x = \tan x$

Solution:

$$\tan x \sin^2 x = \tan x$$

$$\tan x \sin^2 x - \tan x = 0$$
Set one side equal to zero.

$$\tan x (\sin^2 x - 1) = 0$$
Factor the left side.

$\tan x = 0$ $\quad$ or $\quad$ $\sin^2 x - 1 = 0$ $\qquad$ Set each factor equal to zero.

$x = 0, x = \pi$ $\qquad\qquad\qquad$ $\sin^2 x = 1$ $\qquad$ The equation $\sin^2 x - 1 = 0$ is quadratic.

$\qquad\qquad\qquad\qquad\qquad$ $\sin x = \pm 1$ $\qquad$ Isolate $\sin^2 x$ and apply the square root property.

$$x = \frac{\pi}{2}, x = \frac{3\pi}{2}$$

TIP In Example 7, notice that the solutions differ by π. Therefore, we can generate the general solution with one statement.
$$x = n\pi$$

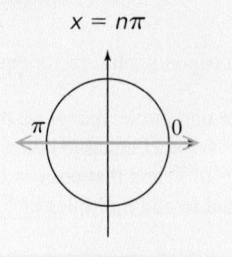

At this point, particular care must be taken before stating the solution set. Recall that the tangent function is undefined at odd multiples of $\dfrac{\pi}{2}$. Therefore, the values $\dfrac{\pi}{2}$ and $\dfrac{3\pi}{2}$ do not check in the original equation and must be excluded from the solution set.

The solution set is $\{0, \pi\}$.

Skill Practice 7   Solve the equation on the interval $[0, 2\pi)$.
$$\cot x \cos^2 x = \cot x$$

Answers

6. $\left\{ \dfrac{\pi}{6}, \dfrac{5\pi}{6}, \dfrac{7\pi}{6}, \dfrac{11\pi}{6} \right\}$

7. $\left\{ \dfrac{\pi}{2}, \dfrac{3\pi}{2} \right\}$

4. Use Identities to Solve Trigonometric Equations

The identities studied earlier in this chapter can be of great service when solving a trigonometric equation as shown in Examples 8–10.

Classroom Example: p. 619
Exercise 44

TIP In Example 8, if we were asked to find the general solution, notice that the pairs of solutions differ by π radians (180°) and can be combined into one statement. The general solution consists of $x = n\pi$ and $x = \frac{\pi}{4} + n\pi$.

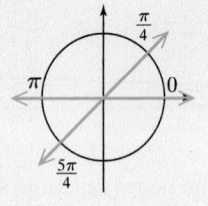

EXAMPLE 8 Using Identities to Solve an Equation

Solve the equation on the interval $[0, 2\pi)$. $\sec^2 x - \tan x = 1$

Solution:

$\sec^2 x - \tan x = 1$	Use the Pythagorean identity $\tan^2 x + 1 = \sec^2 x$ and substitute $\tan^2 x + 1$ for $\sec^2 x$. The left side is now quadratic in the variable $\tan x$.
$(\tan^2 x + 1) - \tan x = 1$	
$\tan^2 x - \tan x = 0$	Set one side of the equation equal to zero.
$\tan x(\tan x - 1) = 0$	Factor.

$\tan x = 0$ or $\tan x - 1 = 0$ Set each factor equal to zero.

$x = 0, x = \pi$ $\tan x = 1$

$x = \dfrac{\pi}{4}, x = \dfrac{5\pi}{4}$ From the unit circle, recognize that $\tan x = 0$ for $x = 0$ and $x = \pi$, and that $\tan x = 1$ for $x = \frac{\pi}{4}$ and $x = \frac{5\pi}{4}$.

The solution set is $\left\{ 0, \dfrac{\pi}{4}, \pi, \dfrac{5\pi}{4} \right\}$.

> **Skill Practice 8** Solve the equation on the interval $[0, 2\pi)$.
> $\csc^2 x - 1 = \cot x$

Classroom Example: p. 619
Exercise 50

TIP In Example 9, notice that the odd multiples of $\frac{\pi}{4}$ differ by $\frac{\pi}{2}$. The odd multiples of $\frac{\pi}{2}$ differ by π. Therefore, the general solution consists of $x = \frac{\pi}{4} + \frac{n\pi}{2}$ and $x = \frac{\pi}{2} + n\pi$.

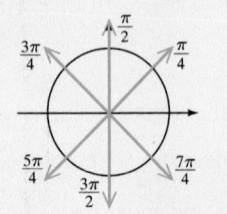

EXAMPLE 9 Using Identities to Solve an Equation

Solve the equation on the interval $[0, 2\pi)$. $\cos 3x + \cos x = 0$

Solution:

$\cos 3x + \cos x = 0$	Apply the sum-to-product formula
$2\cos\dfrac{3x + x}{2}\cos\dfrac{3x - x}{2} = 0$	$\cos x + \cos y = 2\cos\dfrac{x + y}{2}\cos\dfrac{x - y}{2}$.
$2\cos 2x\cos x = 0$	

$\cos 2x = 0$ or $\cos x = 0$ Set each trigonometric factor equal to zero.

$2x = \dfrac{\pi}{2}, \dfrac{3\pi}{2}, \dfrac{5\pi}{2}, \dfrac{7\pi}{2}, \ldots$ or $x = \dfrac{\pi}{2}, \dfrac{3\pi}{2}, \ldots$ From the unit circle, recognize that $\cos 2x = 0$ for $2x$ equal to odd multiples of $\frac{\pi}{2}$, and that $\cos x = 0$ for x equal to odd multiples of $\frac{\pi}{2}$.

$x = \dfrac{\pi}{4}, \dfrac{3\pi}{4}, \dfrac{5\pi}{4}, \dfrac{7\pi}{4}, \ldots$ or $x = \dfrac{\pi}{2}, \dfrac{3\pi}{2}, \ldots$

The solution set on the interval $[0, 2\pi)$ is $\left\{ \dfrac{\pi}{4}, \dfrac{\pi}{2}, \dfrac{3\pi}{4}, \dfrac{5\pi}{4}, \dfrac{3\pi}{2}, \dfrac{7\pi}{4} \right\}$.

> **Skill Practice 9** Solve the equation on the interval $[0, 2\pi)$.
> $\sin 3x + \sin x = 0$

Answers

8. $\left\{ \dfrac{\pi}{4}, \dfrac{\pi}{2}, \dfrac{5\pi}{4}, \dfrac{3\pi}{2} \right\}$

9. $\left\{ 0, \dfrac{\pi}{2}, \pi, \dfrac{3\pi}{2} \right\}$

In Example 10, we encounter a situation where squaring both sides of an equation results in a quadratic equation that is easily factored and solved. However, we must remember to check all potential solutions to eliminate any extraneous solutions.

Classroom Example: p. 619
Exercise 54

EXAMPLE 10   Squaring Both Sides of a Trigonometric Equation

Solve the equation on the interval $[0, 2\pi)$. $\sin x + 1 = \cos x$

Solution:

$$\sin x + 1 = \cos x$$

$$(\sin x + 1)^2 = (\cos x)^2 \qquad$$ Squaring both sides of the equation will result in $\sin^2 x$ and $\cos^2 x$ terms.

$$\sin^2 x + 2\sin x + 1 = \cos^2 x$$

$$\sin^2 x + 2\sin x + 1 = 1 - \sin^2 x \qquad$$ Replacing $\cos^2 x$ by $1 - \sin^2 x$ results in a quadratic equation in the variable $\sin x$.

$$2\sin^2 x + 2\sin x = 0$$

$$2\sin x(\sin x + 1) = 0 \qquad$$ Set one side equal to zero and factor the other side.

$\sin x = 0 \qquad$ or $\qquad \sin x + 1 = 0 \qquad$ Set each factor equal to zero.

$x = 0, x = \pi \qquad\qquad\qquad \sin x = -1 \qquad$ From the unit circle, $\sin x = 0$ for $x = 0, x = \pi$.

$$x = \frac{3\pi}{2} \qquad \sin x = -1 \text{ for } x = \frac{3\pi}{2}.$$

After squaring both sides of an equation, we must check all potential solutions to eliminate any extraneous solutions.

Check: $x = 0$ | Check: $x = \pi$ | Check: $x = \dfrac{3\pi}{2}$

$\sin(0) + 1 \overset{?}{=} \cos(0)$ | $\sin(\pi) + 1 \overset{?}{=} \cos(\pi)$ | $\sin\left(\dfrac{3\pi}{2}\right) + 1 \overset{?}{=} \cos\left(\dfrac{3\pi}{2}\right)$

$0 + 1 \overset{?}{=} 1 \checkmark$ (True) | $0 + 1 \overset{?}{=} -1$ (False) | $-1 + 1 \overset{?}{=} 0 \checkmark$ (True)

The solution set is $\left\{0, \dfrac{3\pi}{2}\right\}$.

The graphs of $y = \sin x + 1$ and $y = \cos x$ are shown in Figure 5-16. The solutions to the equation $\sin x + 1 = \cos x$ are the x-coordinates of the points of intersection.

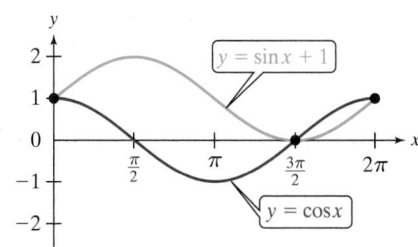

Figure 5-16

Skill Practice 10   Solve the equation on the interval $[0, 2\pi)$.
$\tan x + 1 = \sec x$

In Example 10, we could have taken an alternative approach to solving $\sin x + 1 = \cos x$ by first writing the equation as $\sin x - \cos x = -1$. The left side of the equation can then be written as a single term of the form $k\sin(x + \alpha)$ by using the technique offered on page 585. This approach is examined in Exercise 113, on page 621.

Answer
10. $\{0\}$

5. Use Inverse Functions to Express Solutions to Trigonometric Equations

In Example 11, we encounter equations in which the solutions are not "special" angles (multiples of $\frac{\pi}{6}$ or $\frac{\pi}{4}$). Therefore, we will use inverse functions to express the exact values of the solutions.

Classroom Examples: p. 619
Exercises 58, 60

EXAMPLE 11   Using Inverse Functions to Solve Trigonometric Equations

Solve the equation on the interval $[0, 2\pi)$. Give the exact solutions in radians and give approximations in degrees rounded to 1 decimal place.

a. $\tan x = 3$ **b.** $\sin x = -0.8$

Solution:

a. No multiples of the "special" angles $\left(\dfrac{\pi}{6}, \dfrac{\pi}{4}, \dfrac{\pi}{3}, \dfrac{\pi}{2}, \cdots\right)$ have a tangent value equal to 3. Therefore, our strategy is to find one solution to the equation by using the inverse tangent function. Then use the periodicity of the tangent function to find other solutions.

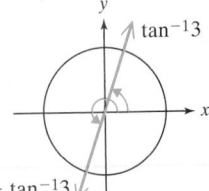

Given $\tan x = 3$, two solutions are

$x = \tan^{-1}3 \approx 1.249$ ($\approx 71.6°$)   The value $\tan^{-1}3$ is a first quadrant angle.
$x = \pi + \tan^{-1}3 \approx 4.391$ ($\approx 251.6°$)   The corresponding angle in Quadrant III is $\pi + \tan^{-1}3$.

The solution set is $\{\tan^{-1}3, \pi + \tan^{-1}3\}$ or approximately $\{71.6°, 251.6°\}$.

b. Given $\sin x = -0.8$, one solution is
$x = \sin^{-1}(-0.8) \approx -0.927 \ (\approx -53.1°)$.

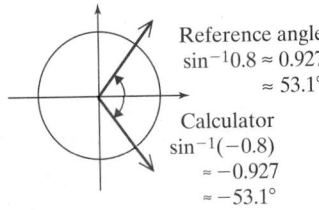

Notice that $x = \sin^{-1}(-0.8) \approx -0.927$ is a negative fourth quadrant angle. The reference angle is $|\sin^{-1}(-0.8)| = \sin^{-1}0.8 \approx 0.927$. Therefore, the solutions on the interval $[0, 2\pi)$ are

$\pi + \sin^{-1}0.8 \approx 4.069 \ (\approx 233.1°)$
$2\pi - \sin^{-1}0.8 \approx 5.356 \ (\approx 306.9°)$

The solution set is $\{\pi + \sin^{-1}0.8, 2\pi - \sin^{-1}0.8\}$ or approximately $\{233.1°, 306.9°\}$.

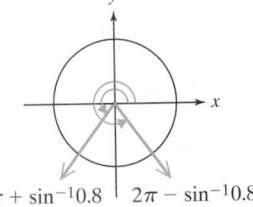

> **Skill Practice 11**   Solve the equation on the interval $[0, 2\pi)$. Give the exact solutions in radians and give approximations in degrees rounded to 1 decimal place.
>
> **a.** $\cos x = 0.6$ **b.** $\tan x = -5$

6. Approximate the Solutions to Trigonometric Equations on a Calculator

Answers
11. a. $\{\cos^{-1}0.6, 2\pi - \cos^{-1}0.6\}$ or approximately $\{53.1°, 306.9°\}$
b. $\{\pi - \tan^{-1}5, 2\pi - \tan^{-1}5\}$ or approximately $\{101.3°, 281.3°\}$

Sometimes finding the solutions to a trigonometric equation is not possible using analytical methods. In particular, this happens when there is a combination of trigonometric expressions and variable algebraic expressions within the equation.

In Example 12, we will solve the equation $x^2 - \sin 2x = 2$ by graphing $Y_1 = x^2 - \sin 2x$ and $Y_2 = 2$. The x-coordinates of the points of intersection represent solutions to the equation.

Classroom Example: p. 619
Exercise 64

> **EXAMPLE 12** Using a Graphing Utility to Solve an Equation
>
> Use a graphing utility to solve the equation $x^2 - \sin 2x = 2$. Round the solutions to 4 decimal places.
>
> **Solution:**
>
> Graph $Y_1 = x^2 - \sin 2x$ and $Y_2 = 2$.
>
>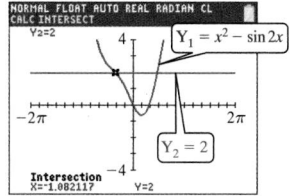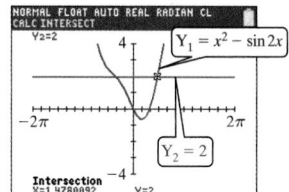
>
> The x-coordinates of the points of intersection are approximately -1.0821, and 1.4780. Since $-1 \leq \sin 2x \leq 1$, the end behavior of the graph of $Y_1 = x^2 - \sin 2x$ is dictated by the term x^2. Furthermore, the graph of $y = x^2$ does not cross the line $y = 2$ more than two times, so we know that these are the only solutions.
>
> The solution set is $\{-1.0821, 1.4780\}$.
>
> **Skill Practice 12** Use a graphing utility to solve the equation $\cos 3x = 0.6x$. Round the solutions to 4 decimal places.

TIP As an alternative method to solve the equation in Example 12, set one side of the equation equal to zero.

$x^2 - \sin 2x - 2 = 0$

Then set Y_1 equal to the nonzero expression on the left.

$Y_1 = x^2 - \sin 2x - 2$

Use the ZERO function on a graphing utility to find the zeros of the function Y_1. These are the solutions to the equation.

Answer
12. $\{-1.2792, -0.6591, 0.4355\}$

SECTION 5.5 Practice Exercises

Prerequisite Review

R.1. Factor by grouping. $15tx + 10t + 12x + 8$ $(5t + 4)(3x + 2)$ **R.2.** Factor the trinomial. $39x + 10x^2 + 35$ $(2x + 5)(5x + 7)$

For Exercises R.3–R.6, solve the equation.

R.3. $4v^2 - 3v - 7 = 0$ $\left\{-1, \dfrac{7}{4}\right\}$ **R.4.** $p^3 = 9p$ $\{-3, 0, 3\}$

R.5. $36x^3 + 9x^2 - 16x - 4 = 0$ $\left\{-\dfrac{1}{4}, -\dfrac{2}{3}, \dfrac{2}{3}\right\}$ **R.6.** $4y^4 - 29y^2 + 7 = 0$ $\left\{-\dfrac{1}{2}, \dfrac{1}{2}, -\sqrt{7}, \sqrt{7}\right\}$

Concept Connections

1. How many solutions to the equation $\cos x = \dfrac{\sqrt{3}}{2}$ exist?

How many solutions exist on the interval $[0, 2\pi)$.
Infinitely many; 2

2. The solutions to the equation $\tan^2 x = 1$ on the interval $[0, 2\pi)$ are _____. The formula
$x =$ _____ can be used to generate *all* solutions to the equation over the set of real numbers.

3. Given $\sin 2x = \dfrac{\sqrt{2}}{2}$, the general solution consists of

$\dfrac{\pi}{8} + n\pi$ and $\dfrac{3\pi}{8} + n\pi$. What are the solutions on the interval $[0, 2\pi)$? $\dfrac{\pi}{8}, \dfrac{3\pi}{8}, \dfrac{9\pi}{8}, \dfrac{11\pi}{8}$

4. Given the equation $\sin x = 0.2$, one solution is $x = \sin^{-1}(0.2)$. What is the other solution on the interval $[0, 2\pi)$? $\pi - \sin^{-1}(0.2)$

Objective 1: Solve Trigonometric Equations in Linear Form

For Exercises 5–8, determine if the given value is a solution to the equation.

5. $2\sin x + \dfrac{\sqrt{3}}{2} = \sin x$ **a.** $\dfrac{4\pi}{3}$ Yes   **b.** $-\dfrac{\pi}{3}$ Yes   **6.** $4(\cot x - \sqrt{3}) = \cot x - \sqrt{3}$ **a.** $\dfrac{5\pi}{6}$ No **b.** $\dfrac{7\pi}{6}$ Yes

7. $10\cos x - \sqrt{2} = 8\cos x$ **a.** $\dfrac{7\pi}{4}$ Yes   **b.** $\dfrac{3\pi}{4}$ No   **8.** $3\tan x - 2\sqrt{3} = 2\tan x - \sqrt{3}$ **a.** $\dfrac{\pi}{3}$ Yes **b.** $\dfrac{4\pi}{3}$ Yes

For Exercises 9–12, solve the equation over the interval $[0, 2\pi)$.

9. a. $\sin\theta = \dfrac{\sqrt{2}}{2}$ $\left\{\dfrac{\pi}{4}, \dfrac{3\pi}{4}\right\}$ **b.** $\sin\theta = -\dfrac{\sqrt{2}}{2}$ $\left\{\dfrac{5\pi}{4}, \dfrac{7\pi}{4}\right\}$ **10. a.** $\cos\alpha = -\dfrac{\sqrt{3}}{2}$ $\left\{\dfrac{5\pi}{6}, \dfrac{7\pi}{6}\right\}$ **b.** $\cos\alpha = \dfrac{\sqrt{3}}{2}$ $\left\{\dfrac{\pi}{6}, \dfrac{11\pi}{6}\right\}$

11. a. $\cot\beta = -\dfrac{\sqrt{3}}{3}$ $\left\{\dfrac{2\pi}{3}, \dfrac{5\pi}{3}\right\}$ **b.** $\cot\beta = \dfrac{\sqrt{3}}{3}$ $\left\{\dfrac{\pi}{3}, \dfrac{4\pi}{3}\right\}$ **12. a.** $\tan x = 1$ $\left\{\dfrac{\pi}{4}, \dfrac{5\pi}{4}\right\}$ **b.** $\tan x = -1$ $\left\{\dfrac{3\pi}{4}, \dfrac{7\pi}{4}\right\}$

For Exercises 13–18, solve the equation

a. over the interval $[0, 2\pi)$. **(See Example 1)**   **b.** over the set of real numbers.

13. $2\sin x + 5 = 6$ **14.** $3\sqrt{2} + 6\cos x = 0$ **15.** $5\sec x + 10 = 3\sec x + 14$

16. $3(2\csc x - \sqrt{3}) = \sqrt{3}$ **17.** $3\tan x + 1 = -2(2 + \tan x)$ **18.** $5\cot x + 2\sqrt{3} = -2\cot x - 5\sqrt{3}$

Objective 2: Solve Trigonometric Equations Involving Multiple or Compound Angles

For Exercises 19–20, identify the number of solutions to the equation on the interval $[0, 2\pi)$.

19. a. $\cos x = \dfrac{1}{2}$ 2   **b.** $\cos 4x = \dfrac{1}{2}$ 8 **c.** $\cos\dfrac{x}{2} = \dfrac{1}{2}$ 1   **20. a.** $\sin x = -1$ 1 **b.** $\sin 2x = -1$ 2 **c.** $\sin\dfrac{x}{2} = -1$ 0

21. a. Write the solution set to the equation $\cos u = 0$ on the interval $[0, 2\pi)$. $\left\{\dfrac{\pi}{2}, \dfrac{3\pi}{2}\right\}$
   b. Write the solution set to the equation $\cos 2x = 0$ on the interval $[0, 2\pi)$. $\left\{\dfrac{\pi}{4}, \dfrac{3\pi}{4}, \dfrac{5\pi}{4}, \dfrac{7\pi}{4}\right\}$

22. a. Write the solution set to the equation $\sin u = 1$ on the interval $[0, 2\pi)$. $\left\{\dfrac{\pi}{2}\right\}$
   b. Write the solution set to the equation $\sin\left(x - \dfrac{\pi}{2}\right) = 1$ on the interval $[0, 2\pi)$. $\{\pi\}$

For Exercises 23–32, solve the equation.

a. Write the solution set for the general solution.

b. Write the solution set on the interval $[0, 2\pi)$. **(See Examples 2–4)**

23. $2\cos 2x = \sqrt{3}$ **24.** $5 + \sin 3x = 4$ **25.** $\tan(-3x) = \sqrt{3}$

26. $\sec(-2x) = -\sqrt{2}$ **27.** $\sqrt{2}\sin\dfrac{x}{2} - 1 = 0$ **28.** $\sqrt{5}\cot\dfrac{x}{2} = \sqrt{5}$

29. $\csc\left(x - \dfrac{\pi}{3}\right) = 2$ **30.** $\cos\left(x + \dfrac{\pi}{4}\right) = \dfrac{1}{2}$ **31.** $\tan\left(x - \dfrac{\pi}{2}\right) + 1 = 0$

32. $\cot\left(x - \dfrac{\pi}{2}\right) + \sqrt{3} = 0$

Objective 3: Solve Higher-Degree Trigonometric Equations

For Exercises 33–42, solve the equation on the interval $[0, 2\pi)$. (See Examples 5–7)

33. $(\tan\theta + 1)(\sec\theta - 2) = 0$ **34.** $(\cot x - 1)(2\sin x + 1) = 0$ **35.** $\sec^2\theta - 2 = 0$ $\left\{\dfrac{\pi}{4}, \dfrac{3\pi}{4}, \dfrac{5\pi}{4}, \dfrac{7\pi}{4}\right\}$

36. $2\sin^2 x - 1 = 0$ $\left\{\dfrac{\pi}{4}, \dfrac{3\pi}{4}, \dfrac{5\pi}{4}, \dfrac{7\pi}{4}\right\}$ **37.** $\cos^2 x + 2\cos x - 3 = 0$ $\{0\}$ **38.** $2\csc^2 x - 5\csc x + 2 = 0$

39. $2\sin^2\alpha + \sin\alpha - 1 = 0$ **40.** $2\cos^2 x + 5\cos x + 2 = 0$ $\left\{\dfrac{2\pi}{3}, \dfrac{4\pi}{3}\right\}$ **41.** $\cos x\tan^2 x = 3\cos x$

42. $4\sin^2 x\cot x - 3\cot x = 0$ $\left\{\dfrac{\pi}{3}, \dfrac{\pi}{2}, \dfrac{2\pi}{3}, \dfrac{4\pi}{3}, \dfrac{3\pi}{2}, \dfrac{5\pi}{3}\right\}$

Objective 4: Use Identities to Solve Trigonometric Equations

For Exercises 43–56, solve the equation on the interval $[0, 2\pi)$. (See Examples 8–10)

43. $6\cos^2 x - 7\sin x - 1 = 0$

44. $2\sin^2 x - \cos x + 1 = 0$ $\{0\}$

45. $2\cos x - 5 = 3\sec x$ $\left\{\dfrac{2\pi}{3}, \dfrac{4\pi}{3}\right\}$

46. $4\csc x + \sin x + 5 = 0$ $\left\{\dfrac{3\pi}{2}\right\}$

47. $2\cot x = \csc x$ $\left\{\dfrac{\pi}{3}, \dfrac{5\pi}{3}\right\}$

48. $2\tan x + \sec x = 0$ $\left\{\dfrac{7\pi}{6}, \dfrac{11\pi}{6}\right\}$

49. $\cos 4x - \cos 2x = 0$

50. $\sin\theta - \sin 3\theta = 0$

51. $\sin 3x + \sin x = \cos x$

52. $\cos 3\theta - \cos\theta = \sin 2\theta$

53. $1 - \sin x = \cos x$ $\left\{0, \dfrac{\pi}{2}\right\}$

54. $\sec x + 1 = \tan x$ $\{\pi\}$

55. $\tan 2x + 1 = \sec 2x$ $\{0, \pi\}$

56. $\sin 2x = 1 - \cos 2x$ $\left\{0, \dfrac{\pi}{4}, \pi, \dfrac{5\pi}{4}\right\}$

Objective 5: Use Inverse Functions to Express Solutions to Trigonometric Equations

For Exercises 57–62, solve the equations on the interval $[0, 2\pi)$. Give the exact solutions in radians and give approximations in degrees rounded to 1 decimal place. (See Example 11)

57. $\cos x = \dfrac{3}{11}$

58. $\tan x = -3$

59. $4\tan x = -7$

60. $6\sin x = 1$

61. $(\sin x - 1)(4\sin x - 3) = 0$

62. $(3\cos x + 2)(\cos x - 1) = 0$

Objective 6: Approximate the Solutions to Trigonometric Equations on a Calculator

For Exercises 63–68, use a graphing utility to solve the equation. Round the solutions to 4 decimal places. (See Example 12)

63. $\sin x = 1 - x^3$ $\{0.7057\}$

64. $\cos 2x = x^2 - 2$ $\{-1.1530, 1.1530\}$

65. $\cos x = \ln x^2$ $\{-1.1991, 1.1991\}$

66. $\sin x = (\ln x)^2$ $\{0.5003, 2.3370\}$

67. $3^{\sin x} = 0.5x + 1$ $\{0, -1.3076, 2.3546\}$

68. $3^{\cos x} = -0.5x + 1$ $\{-1.1458\}$

Mixed Exercises

For Exercises 69–88, solve the equation on the interval $[0, 2\pi)$.

69. $\sin 2x = 2\sin x$ $\{0, \pi\}$

70. $\cos 2x = \cos x$ $\left\{0, \dfrac{2\pi}{3}, \dfrac{4\pi}{3}\right\}$

71. $\cos 6x - 5\sin 3x + 2 = 0$

72. $\cos 4x - 3\cos 2x - 1 = 0$

73. $\cos 2x - 3\sin x + 1 = 0$ $\left\{\dfrac{\pi}{6}, \dfrac{5\pi}{6}\right\}$

74. $4\sin^2 x = 1 - 4\cos 2x$ $\left\{\dfrac{\pi}{3}, \dfrac{2\pi}{3}, \dfrac{4\pi}{3}, \dfrac{5\pi}{3}\right\}$

75. $\dfrac{\sin 3x - \sin x}{\cos 3x + \cos x} = 1$ $\left\{\dfrac{\pi}{4}, \dfrac{5\pi}{4}\right\}$

76. $\dfrac{\cos 3x + \cos x}{2\cos 2x} = -1$ $\{\pi\}$

77. $\sin 5x + \sin x + \cos 4x - \cos 2x = 0$

78. $\sin 4x - \sin 2x - \cos 3x = 0$

79. $\sin 3\theta = \sin 2\theta$ $\left\{0, \dfrac{\pi}{5}, \dfrac{3\pi}{5}, \pi, \dfrac{7\pi}{5}, \dfrac{9\pi}{5}\right\}$

80. $\tan 2\theta \tan\theta = 1$

81. $15\cos^2 x - 7\cos x - 2 = 0$

82. $20\sin^2 x - 7\sin x - 3 = 0$

83. $16\cos^2 x - 8\cos x - 1 = 0$

84. $16\sin^2 x - 12\sin x + 1 = 0$

85. $\sin^2 x - 4\sin x + 1 = 0$

86. $\cos^2 x - 4\cos x - 1 = 0$
$\{\cos^{-1}(2 - \sqrt{5}), 2\pi - \cos^{-1}(2 - \sqrt{5})\}$

87. $\sin(\cos x) = 0$ $\left\{\dfrac{\pi}{2}, \dfrac{3\pi}{2}\right\}$

88. $\cos(\sin x) = 0$ $\{\ \}$

89. The height $h(t)$ (in feet) of the seat of a child's swing above ground level is given by

$$h(t) = -1.1\cos\left(\dfrac{2\pi}{3}t\right) + 3.1$$

where t is the time in seconds after the swing is set in motion.

a. Find the maximum and minimum height of the swing.   Maximum height: 4.2 ft; Minimum height: 2 ft

b. When is the first time after $t = 0$ that the swing is at a height of 3 ft? Round to 1 decimal place. $t \approx 0.7$ sec

c. When is the second time after $t = 0$ that the swing is at a height of 3 ft? Round to 1 decimal place. $t \approx 2.3$ sec

90. A vertical spring is attached to the ceiling. The height h of a block attached to the spring relative to ground level is given by

$$h(t) = -0.8\cos\left(\dfrac{10\pi}{3}t\right) + 5.8$$

where t is the time in seconds and $h(t)$ is in feet.

a. Find the maximum and minimum height of the block.   Maximum height: 6.6 ft; Minimum height: 5 ft

b. When is the first time after $t = 0$ that the block is at a height of 6 ft? Round to 3 decimal places. $t \approx 0.174$ sec

c. When is the second time after $t = 0$ that the block is at a height of 6 ft? Round to 3 decimal places. $t \approx 0.426$ sec

91. The monthly sales of winter coats follow a periodic cycle with sales peaking late in the year and before the winter holidays. The monthly sales total S (in $1000) is given by

$$S(t) = 350\cos\left(\frac{\pi}{6}t + \frac{\pi}{6}\right) + 500$$

where t is the month number ($t = 1$ corresponds to January).

a. Write the range of this function in interval notation.
[150, 850]

b. During which months ($0 \le t \le 12$) are sales at their maximum?   Month 11 (November)

c. During which months ($0 \le t \le 12$) are sales at their minimum?   Month 5 (May)

d. In which months, between 1 and 12 inclusive, are the sales $675,000?
Month 1 and month 9 (January and September)

92. With each heartbeat, blood pressure increases as the heart contracts, then decreases as the heart rests between beats. The maximum blood pressure is called the systolic pressure and the minimum blood pressure is called the diastolic pressure. When a doctor records an individual's blood pressure such as "120 over 80," it is understood as "systolic over diastolic." Suppose that the blood pressure for a certain individual is approximated by $p(t) = 90 + 20\sin(140\pi t)$ where p is the blood pressure in mmHg (millimeters of mercury) and t is the time in minutes after recording begins.

a. Find the period of the function and interpret the results.

b. Find the maximum and minimum values and interpret this as a blood pressure reading.

c. Find the times at which the blood pressure is at its maximum.

The refractive index n of a substance is a dimensionless measure of how much light bends (refracts) when passing from one medium to another.

$$n = \frac{c}{v}$$

where c is the speed of light in a vacuum (a constant) and v is the speed of light in the medium. For example, the refractive index of diamond is 2.42 which means that light travels 2.42 times as fast in a vacuum as it does in a diamond.

Snell's law is an equation that relates the indices of refraction of two different mediums to the angle of incidence θ_1 and angle of refraction θ_2. Angles θ_1 and θ_2 are measured from a line perpendicular to the boundary between the two mediums. Use this relationship and the table of refraction indices to complete Exercises 93–94.

Index of Refraction	
Medium	Index
Air	1.00029
Water	1.33
Acetone	1.36
Human cornea	1.38
Glycerin	1.47
Sodium chloride	1.54
Sapphire	1.77
Cubic zirconia	2.15
Diamond	2.42

$$n_1 \sin\theta_1 = n_2 \sin\theta_2$$

93. Assume that a beam of light travels from air into water.

a. If the incidence angle θ_1 is 40°, what is the angle of refraction θ_2? Round to the nearest tenth of a degree.   28.9°

b. Alina attempts to spear a lobster from her boat. If she aims her spear directly at the lobster, will she hit it? Explain your answer.   No. As a result of the refraction of light the actual lobster is not where it appears from Alina's position above the water. (She should aim below the image.)

94. If a beam of light traveling through air enters a sapphire at an angle of incidence of 23°, what is the angle of refraction? Round to the nearest tenth of a degree.   12.8°

Write About It

95. Explain why $\cos^2 x - \cos x - 12 = 0$ has no solution.

96. Explain why $x = \sin^{-1}(-0.4)$ is not a solution to the equation $\sin x = -0.4$ on the interval $[0, 2\pi)$.
The value of $\sin^{-1}(-0.4) \approx -0.4115$ which is not on the interval $[0, 2\pi)$.

97. What is the difference between the general solution to a trigonometric equation and the solution over the interval $[0, 2\pi)$?

98. Explain two different methods to solve the equation $\sin 3x + x = 1$ by using a graphing utility.

Expanding Your Skills

For Exercises 99–112, solve the equation on the interval $[0, 2\pi)$.

99. $\sqrt{3}\tan^2 x - 2\tan x - \sqrt{3} = 0$ $\left\{\dfrac{\pi}{3}, \dfrac{5\pi}{6}, \dfrac{4\pi}{3}, \dfrac{11\pi}{6}\right\}$

100. $\sqrt{2}\sin^2 x + \sin x - \sqrt{2} = 0$ $\left\{\dfrac{\pi}{4}, \dfrac{3\pi}{4}\right\}$

101. $4\sin^2 x\cos x - 2\cos x + 2\sin^2 x - 1 = 0$

102. $\tan^2 x\sec x - 2\tan^2 x - 3\sec x + 6 = 0$ $\left\{\dfrac{\pi}{3}, \dfrac{2\pi}{3}, \dfrac{4\pi}{3}, \dfrac{5\pi}{3}\right\}$

103. $4\sin^4 x - 7\sin^2 x + 3 = 0$ $\left\{\dfrac{\pi}{3}, \dfrac{\pi}{2}, \dfrac{2\pi}{3}, \dfrac{4\pi}{3}, \dfrac{3\pi}{2}, \dfrac{5\pi}{3}\right\}$

104. $8\cos^4 x - 6\cos^2 x + 1 = 0$

105. $2\sin^3 x + 7\sin^2 x + 2\sin x - 3 = 0$

106. $6\sec^3 x - 7\sec^2 x - 11\sec x + 2 = 0$ $\left\{\dfrac{\pi}{3}, \pi, \dfrac{5\pi}{3}\right\}$

107. $\sqrt{\sin x + 1} + \sqrt{2\sin x + 3} = 1$ $\left\{\dfrac{3\pi}{2}\right\}$

108. $\sqrt{5\cos x + 4} - 2 = \sqrt{\cos x}$ $\left\{0, \dfrac{\pi}{2}, \dfrac{3\pi}{2}\right\}$

Additional answers can be found in the Instructor Answer Appendix.

109. $4\sin^{-1}(x - 2) + \pi = 2\pi$ $\left\{\dfrac{4 + \sqrt{2}}{2}\right\}$

110. $\dfrac{3}{2}\cos^{-1}(x + 1) = \dfrac{\pi}{2}$ $\left\{-\dfrac{1}{2}\right\}$

111. $2\sin\left(2x - \dfrac{\pi}{2}\right) - \sqrt{2} = 0$ $\left\{\dfrac{3\pi}{8}, \dfrac{5\pi}{8}, \dfrac{11\pi}{8}, \dfrac{13\pi}{8}\right\}$

112. $2\cos(2x - \pi) - \sqrt{3} = 0$ $\left\{\dfrac{5\pi}{12}, \dfrac{7\pi}{12}, \dfrac{17\pi}{12}, \dfrac{19\pi}{12}\right\}$

113. Consider the equation $\sin x + 1 = \cos x$ from Example 10.

  a. Write the equation in the form $\sin x - \cos x = -1$. Then write the equation with the left side of the equation in the form $k\sin(x + \alpha)$. (*Hint*: See page 585.)

  b. Solve the equation over the interval $[0, 2\pi)$ using the form from part (a).

For Exercises 114–116, use the technique from Exercise 113 to solve the equation over the interval $[0, 2\pi)$.

114. $\sqrt{3}\sin x + \cos x = 1$ $\left\{0, \dfrac{2\pi}{3}\right\}$

115. $\sqrt{2}\sin x - \sqrt{2}\cos x = -1$ $\left\{\dfrac{\pi}{12}, \dfrac{17\pi}{12}\right\}$

116. $-\dfrac{1}{2}\sin x + \dfrac{\sqrt{3}}{2}\cos x = 1$ $\left\{\dfrac{11\pi}{6}\right\}$

117. In a six-cylinder engine, the height of the pistons above the crankshaft can be modeled by sine or cosine functions. There are three piston pairs, and the firing sequence is 1, 5, 3, 2, 6, 4. Therefore the height of the piston pairs differs by a phase shift. Suppose that the height of piston 1 is given by $h_1 = 3.5\sin(100t) + 7$ and the height of piston 2 is given by $h_2 = 3.5\sin\left(100t + \dfrac{4000\pi}{3}\right) + 7$. The values of h_1 and h_2 are measured in inches and t is measured in seconds.

Determine the times for which pistons 1 and 2 are at the same height. (*Hint*: Set $h_1 = h_2$. Then use the sum formula for the sine function to expand the right side of the equation.) $t = \dfrac{\pi}{120} + \dfrac{n\pi}{100}$

118. From Exercise 83 in Section 5.3, the cross-sectional area of a feeding trough in the figure is given by $A = \dfrac{1}{2}\sin 2\theta + \sin\theta$.

1 ft θ ⎯⎯⎯ θ 1 ft

3 ft

In calculus, you will learn that to find the value of θ to maximize the area, we need to solve the equation $\cos 2\theta + \cos\theta = 0$ for $0 < \theta < \dfrac{\pi}{2}$.

  a. Find the value of θ to maximize the area. $\theta = \dfrac{\pi}{3}$

  b. Find the maximum area. $\dfrac{3\sqrt{3}}{4}$ ft^2

  c. If the trough is 20 ft long, what is the maximum volume that it can hold? $15\sqrt{3}$ ft^3

119. Consider an isosceles triangle with two sides of length x and angle θ included between them. We have already shown that the area of the triangle is given by $A = \dfrac{1}{2}x^2\sin\theta$. If an isosceles triangle has two 6-in. sides, what angle is required to make the area 10 in.^{2}? Give the exact value and approximate to the nearest one-hundredth of a degree.

$\sin^{-1}\dfrac{5}{9} \approx 33.75°$ or $180° - \sin^{-1}\dfrac{5}{9} \approx 146.25°$

For Exercises 120–121, consider a projectile launched from ground level at an angle of elevation θ with an initial velocity v_0. The maximum horizontal range is given by $x_{max} = \dfrac{v_0^2\sin 2\theta}{g}$, where g is the acceleration due to gravity ($g = 32$ ft/sec^2 or $g = 9.8$ m/sec^2).

120. If a soccer ball is kicked from ground level with an initial velocity of 28 m/sec, what is the smallest positive angle at which the player should kick the ball to reach a teammate 48 m down the field? Assume that the ball reaches the teammate at ground level on the fly. Round to the nearest tenth of a degree. 18.4°

121. A quarterback throws a football with an initial velocity of 62 ft/sec to a receiver 40 yd (120 ft) down the field. At what angle could the ball be released so that it hits the receiver's hands at the same height that it left the quarterback's hand? Round to the nearest tenth of a degree. 43.7° or 46.3°

Technology Connections

122. Suppose that a rectangle is bounded by the x-axis and the graph of $y = \cos x$.

   a. Write a function that represents the area $A(x)$ of the rectangle for $0 < x < \dfrac{\pi}{2}$.

   b. Complete the table.

x	$A(x)$
0.2	
0.4	
0.6	
0.8	
1.0	
1.2	
1.4	

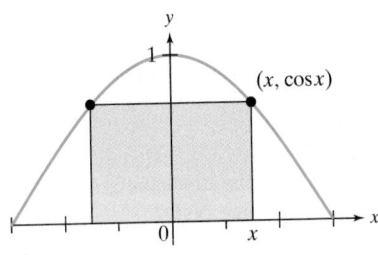

$(x, \cos x)$

   c. Graph the function from part (a) on the viewing window: $\left[0, \dfrac{\pi}{2}, \dfrac{\pi}{6}\right]$ by $[-3, 3, 1]$ and approximate the values of x for which the area is 1 square unit. Round to 2 decimal places.

   d. In calculus, we can show that the maximum value of the area of the rectangle will occur at values of x for which $2\cos x - 2x\sin x = 0$. Confirm this result by graphing $y = 2\cos x - 2x\sin x$ and the function from part (a) on the same viewing window. What do you notice?

PROBLEM RECOGNITION EXERCISES

Trigonometric Identities and Trigonometric Equations

For Exercises 1–6, prove the identity in part (a), and solve the equation in part (b) on the interval $[0, 2\pi)$.

1. a. Prove that $\sin^4 x - \cos^4 x = -\cos 2x$.
   b. Solve $\sin^4 x - \cos^4 x = 1$.

2. a. Prove that $\tan x \sin x + \cos x = \sec x$.
   b. Solve $\tan x \sin x + \cos x = 2$.

3. a. Prove that $\cot x - \sec x \csc x \cos 2x = \tan x$.
   b. Solve $\cot x - \sec x \csc x \cos 2x = 1$.

4. a. Prove that $\dfrac{\sec x + \tan x}{\cos x + \cot x} = \sec x \tan x$.
   b. Solve $\dfrac{\sec x + \tan x}{\cos x + \cot x} = 0$.

5. a. Prove that $\dfrac{2\cot x}{1 + \cot^2 x} = \sin 2x$.
   b. Solve $\dfrac{2\cot x}{1 + \cot^2 x} = 0$.

6. a. Prove that $\cos 2x - \cos 4x = 2\sin 3x \sin x$.
   b. Solve $\cos 2x - \cos 4x = 0$.

7. a. Compute $\sin 75°$ by using $\sin(u + v) = \sin u \cos v + \cos u \sin v$.
   b. Compute $\sin 75°$ by using the half-angle formula $\sin\dfrac{\alpha}{2} = \pm\sqrt{\dfrac{1 - \cos\alpha}{2}}$.
   c. Show that the results for parts (a) and (b) are equivalent.

8. a. Compute $\cos\dfrac{\pi}{12}$ by applying a sum or difference formula for cosine.
   b. Use the result of part (a) to find $\sin\dfrac{\pi}{24}$ and $\cos\dfrac{\pi}{24}$.

CHAPTER 5   KEY CONCEPTS

SECTION 5.1 Fundamental Trigonometric Identities	Reference

The fundamental trigonometric identities are used to simplify trigonometric expressions and to prove other trigonometric relationships.   *p. 568*

Reciprocal and Quotient Identities:

$$\csc x = \frac{1}{\sin x} \qquad \sec x = \frac{1}{\cos x} \qquad \cot x = \frac{1}{\tan x} \qquad \tan x = \frac{\sin x}{\cos x}$$

$$\sin x = \frac{1}{\csc x} \qquad \cos x = \frac{1}{\sec x} \qquad \tan x = \frac{1}{\cot x} \qquad \cot x = \frac{\cos x}{\sin x}$$

The Pythagorean Identities can be used in several forms.   *p. 568*

$$\sin^2 x + \cos^2 x = 1 \qquad \tan^2 x + 1 = \sec^2 x \qquad 1 + \cot^2 x = \csc^2 x$$
$$1 - \cos^2 x = \sin^2 x \qquad \sec^2 x - 1 = \tan^2 x \qquad \csc^2 x - 1 = \cot^2 x$$
$$1 - \sin^2 x = \cos^2 x \qquad \sec^2 x - \tan^2 x = 1 \qquad \csc^2 x - \cot^2 x = 1$$

Even and Odd Identities: *p. 568*

$$\sin(-x) = -\sin x \qquad \csc(-x) = -\csc x \qquad \tan(-x) = -\tan x$$
$$\cos(-x) = \cos x \qquad \sec(-x) = \sec x \qquad \cot(-x) = -\cot x$$

To verify a trigonometric identity, work on one side of the equation (usually the more complicated side) until it matches the other side of the equation. Although there is no explicit procedure that works for every identity, helpful guidelines are offered on page 570.   *p. 570*

SECTION 5.2 Sum and Difference Formulas	Reference

The sine, cosine, or tangent of a sum or difference of angles can be expressed in terms of the sine, cosine, or tangent of the individual angles.   *p. 579*

Sum and Difference Formulas:

$$\sin(u + v) = \sin u \cos v + \cos u \sin v \qquad \cos(u + v) = \cos u \cos v - \sin u \sin v$$
$$\sin(u - v) = \sin u \cos v - \cos u \sin v \qquad \cos(u - v) = \cos u \cos v + \sin u \sin v$$
$$\tan(u + v) = \frac{\tan u + \tan v}{1 - \tan u \tan v} \qquad \tan(u - v) = \frac{\tan u - \tan v}{1 + \tan u \tan v}$$

The cofunction and periodic properties were first introduced in Chapter 4, and can be verified using the sum and difference formulas.   *p. 584*

Cofunction Identities:

$$\sin\theta = \cos\left(\frac{\pi}{2} - \theta\right) \qquad \cos\theta = \sin\left(\frac{\pi}{2} - \theta\right) \qquad \tan\theta = \cot\left(\frac{\pi}{2} - \theta\right)$$

$$\sec\theta = \csc\left(\frac{\pi}{2} - \theta\right) \qquad \csc\theta = \sec\left(\frac{\pi}{2} - \theta\right) \qquad \cot\theta = \tan\left(\frac{\pi}{2} - \theta\right)$$

Periodic Identities: *p. 584*

$$\sin(\theta + 2\pi) = \sin\theta \qquad \cos(\theta + 2\pi) = \cos\theta \qquad \tan(\theta + \pi) = \tan\theta$$
$$\csc(\theta + 2\pi) = \csc\theta \qquad \sec(\theta + 2\pi) = \sec\theta \qquad \cot(\theta + \pi) = \cot\theta$$

A linear combination of $\sin x$ and $\cos x$ can be written as a sine function involving a phase shift. For the real numbers A, B, and x,   *p. 585*

$$A\sin x + B\cos x = k\sin(x + \alpha) \text{ where } k = \sqrt{A^2 + B^2} \text{ and } \alpha \text{ satisfies}$$

$$\cos\alpha = \frac{A}{\sqrt{A^2 + B^2}} \text{ and } \sin\alpha = \frac{B}{\sqrt{A^2 + B^2}}.$$

	Reference
SECTION 5.3 Double-Angle, Power-Reducing, and Half-Angle Formulas	

The sine, cosine, or tangent of a double angle follows directly from the sum and difference formulas.

Double-Angle Formulas: — *p. 591*

$$\sin 2\theta = 2\sin\theta\cos\theta \qquad\qquad \cos 2\theta = \cos^2\theta - \sin^2\theta$$
$$\tan 2\theta = \frac{2\tan\theta}{1 - \tan^2\theta} \qquad\qquad\qquad\quad = 1 - 2\sin^2\theta$$
$$\qquad\qquad\qquad\qquad\qquad\qquad\quad = 2\cos^2\theta - 1$$

The double-angle cosine formulas are used to derive the power-reducing formulas. These convert the square of a sine or cosine term to a first power of a cosine term. — *p. 593*

Power-Reducing Formulas:

$$\sin^2\theta = \frac{1 - \cos 2\theta}{2} \qquad \cos^2\theta = \frac{1 + \cos 2\theta}{2} \qquad \tan^2\theta = \frac{1 - \cos 2\theta}{1 + \cos 2\theta}$$

Substituting $\dfrac{\alpha}{2}$ for θ in the power-reducing formulas results in the half-angle formulas. — *p. 594*

Half-Angle Formulas:

$$\sin\frac{\alpha}{2} = \pm\sqrt{\frac{1 - \cos\alpha}{2}} \qquad \cos\frac{\alpha}{2} = \pm\sqrt{\frac{1 + \cos\alpha}{2}}$$

The + or − sign is determined by the quadrant of angle $\dfrac{\alpha}{2}$.

$$\tan\frac{\alpha}{2} = \pm\sqrt{\frac{1 - \cos\alpha}{1 + \cos\alpha}} = \frac{1 - \cos\alpha}{\sin\alpha} = \frac{\sin\alpha}{1 + \cos\alpha}$$

	Reference
SECTION 5.4 Product-to-Sum and Sum-to-Product Formulas	

Product-to-Sum Formulas: — *pp. 601–602*

$$\sin u \sin v = \frac{1}{2}[\cos(u - v) - \cos(u + v)]$$
$$\cos u \cos v = \frac{1}{2}[\cos(u - v) + \cos(u + v)]$$
$$\sin u \cos v = \frac{1}{2}[\sin(u + v) + \sin(u - v)]$$
$$\cos u \sin v = \frac{1}{2}[\sin(u + v) - \sin(u - v)]$$

Sum-to-Product Formulas:

$$\cos x - \cos y = -2\sin\frac{x + y}{2}\sin\frac{x - y}{2}$$
$$\cos x + \cos y = 2\cos\frac{x + y}{2}\cos\frac{x - y}{2}$$
$$\sin x + \sin y = 2\sin\frac{x + y}{2}\cos\frac{x - y}{2}$$
$$\sin x - \sin y = 2\cos\frac{x + y}{2}\sin\frac{x - y}{2}$$

	Reference
SECTION 5.5 Trigonometric Equations	

To solve a conditional trigonometric equation, we manipulate both sides of the equation into a simpler form to find the values of the variable that make the equation true. For example, this equation is linear in the variable $\sin x$. — *p. 608*

$$2\sin x + 1 = 0$$
$$\sin x = -\frac{1}{2} \text{ is an equivalent equation.}$$

The solution set on the interval $[0, 2\pi)$ is $\left\{\dfrac{7\pi}{6}, \dfrac{11\pi}{6}\right\}$.

The general solution is $\left\{x \,\middle|\, x = \dfrac{7\pi}{6} + 2n\pi, x = \dfrac{11\pi}{6} + 2n\pi \text{ for integers } n\right\}$.

This equation is quadratic in the variable $\cos x$. — *p. 612*

$$2\cos^2 x + 5\cos x - 3 = 0$$
$$(2\cos x - 1)(\cos x + 3) = 0$$
$$\cos x = \frac{1}{2} \text{ or } \cos x = -3$$

Setting each factor equal to zero results in two linear equations in the variable $\cos x$. The second of the two equations, $\cos x = -3$ has no solution because -3 is not in the range of the cosine function.

The solution set over the interval $[0, 2\pi)$ is $\left\{\dfrac{\pi}{3}, \dfrac{5\pi}{3}\right\}$.

An equation involving a multiple angle may have numerous solutions on the interval $[0, 2\pi)$. $\tan 3x = 1$ The solution set over $[0, 2\pi)$ is $\left\{\dfrac{\pi}{12}, \dfrac{5\pi}{12}, \dfrac{3\pi}{4}, \dfrac{13\pi}{12}, \dfrac{17\pi}{12}, \dfrac{7\pi}{4}\right\}$.	p. 609
Trigonometric identities are also used to manipulate an equation into a more manageable form. $\begin{aligned} \sin^2 x &= 2\cos x + 2 \\ 1 - \cos^2 x &= 2\cos x + 2 \qquad \text{Use the Pythagorean identity } \sin^2 x = 1 - \cos^2 x. \\ \cos^2 x + 2\cos x + 1 &= 0 \qquad \text{Combine like terms.} \\ (\cos x + 1)^2 &= 0 \qquad \text{Factor.} \\ \cos x + 1 &= 0 \qquad \text{Set each factor equal to zero.} \\ \cos x &= -1 \quad \text{The solution set over } [0, 2\pi) \text{ is } \{\pi\}. \end{aligned}$	p. 614
Other algebraic techniques such as factoring, squaring both sides of the equation (being sure to check the potential solutions), applying the square root property, and applying the quadratic formula may also be used.	p. 615
Often the solutions to a trigonometric equation are not multiples of the special angles $0, \dfrac{\pi}{6}, \dfrac{\pi}{4}, \dfrac{\pi}{3},$ or $\dfrac{\pi}{2}$. In such a case, use the inverse trigonometric functions to find the exact value. $\tan x = 5 \qquad$ The solution set over $[0, 2\pi)$ is $\{\tan^{-1} 5, \pi + \tan^{-1} 5\}$.	p. 615

Expanded Chapter Summary available at www.mhhe.com/millerprecalculus.

CHAPTER 5   Review Exercises

SECTION 5.1

For Exercises 1–2, factor the expression completely.

1. $\sin^4 x - \sin^2 x \cos^2 x$
$\sin^2 x(\sin x - \cos x)(\sin x + \cos x)$

2. $12\tan^2 x + 11\tan x - 15$
$(3\tan x + 5)(4\tan x - 3)$

For Exercises 3–4, find the LCD of the expressions.

3. $\dfrac{\sin x}{\tan x - 1}; \dfrac{\cos x}{\tan x + 1}$
$(\tan x - 1)(\tan x + 1)$ or $\tan^2 x - 1$

4. $\dfrac{\sin x}{\cos x}; \dfrac{1}{\cos^2 x}; \dfrac{\cos x}{\sin x}$
$\cos^2 x \sin x$

For Exercises 5–6, simplify the expression.

5. $\sin x(\cot x - \csc x) + 1$
$\cos x$

6. $\sec^2 x(1 - \sin^2 x)$ 1

For Exercises 7–14, verify the identity.

7. $\dfrac{\sin x \tan^2 x - \sin x}{\cos x \tan x + \cos x} = \tan x(\tan x - 1)$

8. $\dfrac{3\sin^2 x - \sin x - 2}{\cos x \sin x - \cos x} = 3\tan x + 2\sec x$

9. $\dfrac{1}{\csc x - 1} + \dfrac{1}{\csc x + 1} = 2\tan x \sec x$

10. $\dfrac{1 - \sin x}{\cos x} = \dfrac{\cos x}{1 + \sin x}$

11. $\ln|\csc x| + \ln|\tan x| = \ln|\sec x|$

12. $\ln|\tan x| - \ln|\sin x| + \ln|\cos x| = 0$

13. $\cos x + \sin(-x)\tan(-x) = \sec x$

14. $[\csc(-x) + 1](\csc x + 1) = -\cot^2 x$

15. Write $\sqrt{16 - x^2}$ as a function of θ by making the substitution $x = 4\cos\theta$ for $0 < \theta < \dfrac{\pi}{2}$. $\sqrt{16 - x^2} = 4\sin\theta$

16. Write $\sqrt{(x + 1)^2 + 49}$ as a function of θ by making the substitution $x = 7\tan\theta - 1$ for $0 < \theta < \dfrac{\pi}{2}$.
$\sqrt{(x + 1)^2 + 49} = 7\sec\theta$

SECTION 5.2

In Exercises 17–26, use an addition or subtraction formula to find the exact value.

17. $\sin\dfrac{7\pi}{12}$ $\dfrac{\sqrt{2} + \sqrt{6}}{4}$

18. $\cos 105°$ $\dfrac{\sqrt{2} - \sqrt{6}}{4}$

19. $\tan(-15°)$ $\sqrt{3} - 2$

20. $\sin(-195°)$ $\dfrac{\sqrt{6} - \sqrt{2}}{4}$

21. $\cos\dfrac{13\pi}{12}$ $-\dfrac{\sqrt{2} + \sqrt{6}}{4}$

22. $\tan\dfrac{17\pi}{12}$ $2 + \sqrt{3}$

23. $\cos 40°\cos 80° - \sin 40°\sin 80°$ $-\dfrac{1}{2}$

24. $\dfrac{\tan\dfrac{29\pi}{18} + \tan\dfrac{5\pi}{36}}{1 - \tan\dfrac{29\pi}{18}\tan\dfrac{5\pi}{36}}$ -1

25. $\tan\left[\sin^{-1}\left(\dfrac{28}{53}\right) + \cos^{-1}\left(-\dfrac{5}{13}\right)\right]$ $-\dfrac{400}{561}$

26. $\sin\left[\arccos\left(\dfrac{3}{4}\right) + \arcsin\left(\dfrac{1}{2}\right)\right]$ $\dfrac{\sqrt{21} + 3}{8}$

27. Find the exact value for $\sin(\alpha - \beta)$ given $\sin\alpha = \dfrac{8}{17}$ for α in Quadrant II and $\cos\beta = \dfrac{6}{7}$ for β in Quadrant IV.

28. Find the exact value for $\cos(\alpha - \beta)$ given $\sin\alpha = \dfrac{21}{29}$ for α in Quadrant I and $\cos\beta = -\dfrac{24}{25}$ for β in Quadrant III. $-\dfrac{627}{725}$

For Exercises 29–32, verify the identity.

29. $\dfrac{\cos(x - y)}{\cos x \sin y} = \cot y + \tan x$

Additional answers can be found in the Instructor Answer Appendix.

30. $\dfrac{\sin(x-y)}{\cos x \cos y} = \tan x - \tan y$

31. $\dfrac{\tan\left(x - \dfrac{3\pi}{4}\right)}{\tan\left(x + \dfrac{\pi}{4}\right)} = 1$

32. $\sin\left(x + \dfrac{\pi}{4}\right) - \cos\left(x + \dfrac{\pi}{4}\right) = \sqrt{2}\sin x$

33. Write $\sqrt{3}\sin x - \cos x$ in the form $k\sin(x + \alpha)$ for $0 \le \alpha < 2\pi$. $2\sin\left(x + \dfrac{11\pi}{6}\right)$

34. Write $3\cos x - 4\sin x$ in the form $k\sin(x + \alpha)$ for $0 \le \alpha < 2\pi$. Round α to 3 decimal places.
$5\sin(x + 2.498)$

SECTION 5.3

For Exercises 35–38, verify the identity.

35. $\dfrac{2\tan x}{1 + \tan^2 x} = \sin 2x$ **36.** $\dfrac{\cos 2x}{\sin x} = \csc x - 2\sin x$

37. $\sin^2\dfrac{x}{2} - \cos^2\dfrac{x}{2} = -\cos x$ **38.** $\tan x \tan\dfrac{x}{2} = \sec x - 1$

39. Write $16\cos^4 x$ in terms of first powers of cosine.
$8\cos 2x + 2\cos 4x + 6$

40. Write $\sin^4 x + 2\cos^2 x$ in terms of first powers of cosine.
$\dfrac{1}{2}\cos 2x + \dfrac{1}{8}\cos 4x + \dfrac{11}{8}$

For Exercises 41–44, use the given information to find the exact value of each expression.

a. $\sin 2\theta$ **b.** $\cos 2\theta$ **c.** $\sin\dfrac{\theta}{2}$ **d.** $\cos\dfrac{\theta}{2}$

41. $\sin\theta = \dfrac{5}{13}, \dfrac{\pi}{2} < \theta < \pi$

42. $\cos\theta = -\dfrac{4}{5}, \pi < \theta < \dfrac{3\pi}{2}$

43. $\tan\theta = -2, \sin\theta > 0$

44. $\tan\theta = -3, \cos\theta > 0$

For Exercises 45–46, use the given information to find the exact value of each expression.

a. $\tan 2\theta$ **b.** $\tan\dfrac{\theta}{2}$

45. $\cos\theta = \dfrac{3}{5}, \dfrac{3\pi}{2} < \theta < 2\pi$ **a.** $\dfrac{24}{7}$ **b.** $-\dfrac{1}{2}$

46. $\sin\theta = -\dfrac{3}{8}, \cos\theta < 0$ **a.** $\dfrac{3\sqrt{55}}{23}$ **b.** $-\dfrac{8 + \sqrt{55}}{3}$

SECTION 5.4

For Exercises 47–50, write the product as a sum or difference.

47. $\sin 6x \sin 4x$ **48.** $\cos(-2x)\sin 6x$

49. $\sin(-3x)\cos 6x$ **50.** $\cos 10x \cos 5x$

For Exercises 51–54, write each expression as a product.

51. $\cos 2x + \cos 8x$ **52.** $\sin 10x + \sin 2x$
$2\cos 5x\cos 3x$ $2\sin 6x\cos 4x$

53. $\sin 6x - \sin 2x$ **54.** $\cos 5x - \cos 2x$
$2\cos 4x\sin 2x$ $-2\sin\dfrac{7x}{2}\sin\dfrac{3x}{2}$

For Exercises 55–56, use a product-to-sum formula to find the exact value.

55. $\sin 37.5°\cos 7.5°$ $\dfrac{1 + \sqrt{2}}{4}$ **56.** $\cos\dfrac{5\pi}{4}\cos\dfrac{5\pi}{12}$ $\dfrac{1 - \sqrt{3}}{4}$

For Exercises 57–58, use a sum-to-product formula to find the exact value.

57. $\sin\dfrac{31\pi}{12} - \sin\dfrac{13\pi}{12}$ $\dfrac{\sqrt{6}}{2}$ **58.** $\cos 345° + \cos 105°$ $\dfrac{\sqrt{2}}{2}$

For Exercises 59–60, use the sum-to-product formulas to verify the identity.

59. $\cos\left(\dfrac{\pi}{4} + t\right) - \cos\left(\dfrac{\pi}{4} - t\right) = -\sqrt{2}\sin t$

60. $\sin\left(\theta - \dfrac{\pi}{4}\right) + \sin\left(\theta + \dfrac{\pi}{4}\right) = \sqrt{2}\sin\theta$

For Exercises 61–62, verify the identity.

61. $\dfrac{\sin 4x - \sin 2x}{\cos 4x - \cos 2x} = -\cot 3x$

62. $\sin 3x + \sin 5x + \sin 8x = 4\sin 4x\cos\dfrac{5x}{2}\cos\dfrac{3x}{2}$

SECTION 5.5

For Exercises 63–70,

a. Write the solution set for the general solution.
b. Write the solution set on the interval $[0, 2\pi)$.

63. $4\sin x + 2\sqrt{3} = 0$ **64.** $4\sqrt{2} = 2\cos x + 3\sqrt{2}$

65. $3\sec^2 x = 4$ **66.** $3\cot^2 x - 1 = 0$

67. $7\sin 2x + 10 = 3$ **68.** $\cos 3x = -\dfrac{1}{2}$

69. $\tan\dfrac{x}{2} = 1$ **70.** $\sin\dfrac{x}{2} = \dfrac{\sqrt{3}}{2}$

For Exercises 71–88, solve the equations on the interval $[0, 2\pi)$.

71. $\cos\left(x + \dfrac{\pi}{4}\right) = -1$ **72.** $\tan\left(x - \dfrac{\pi}{2}\right) = \dfrac{\sqrt{3}}{3}$

73. $\sin\theta = \dfrac{1}{5}$ **74.** $\cos\theta = \dfrac{3}{4}$

75. $8\tan x + 3 = 0$ **76.** $4\cos x = -7$ $\{\ \}$

77. $6\cos^2 x + 7\cos x - 5 = 0$

78. $6\csc^2 x + 11\csc x - 2 = 0$

79. $17\cos^2 x + 4\cos x - 1 = 0$

80. $10\sin^2 x - 3\sin x - 4 = 0$

81. $2\cos^2 x - 5\sin x + 1 = 0$

82. $\tan^2 x - 4\sec x + 5 = 0$

83. $\sin 2x + \cos x = 0$ **84.** $\sin x = \cos 2x$

85. $\cos\theta - \cos 3\theta = 0$ **86.** $\sin 2x + \sin 4x = 0$

87. $\cos x + 1 = \sin x$ **88.** $\cot x = \csc x + 1$

89. Use a graphing utility to solve the equation $\cos 3x = x^4$. Round the solutions to 4 decimal places. $\{-0.5024, 0.5024\}$

90. Use a graphing utility to solve the equation $\sin^2 x - x = 3$. Round the solution to 4 decimal places. $\{-2.9713\}$

91. A beam of light traveling through air enters a diamond at an angle of incidence of 35°. Use Snell's law $n_1 \sin \theta_1 = n_2 \sin \theta_2$ to find the angle of refraction if the index of refraction of air is 1.00029 and the index of refraction of the diamond is 2.42. Round to the nearest tenth of a degree. (*Hint*: See Section 5.5, Exercises 93–94 on page 620.)   13.7°

92. A researcher uses Snell's law $n_1 \sin \theta_1 = n_2 \sin \theta_2$ to determine if a large clear stone is a diamond, with a refractive index of 2.42, or a quartz crystal, with a refractive index of 1.54. If the angle of incidence of a laser beam is 75° and the angle of refraction is 23.5°, is the stone more likely to be diamond or quartz? Take the refraction index of air to be 1.00029.   The refractive index is 2.42; thus the stone is more likely to be a diamond.

93. The monthly sales of a math textbook follow a periodic cycle with sales peaking in January and September. The monthly sales total S (in \$100) is given by

$$S(t) = 30 \sin\left(\frac{\pi}{4}t + \frac{\pi}{4}\right) + 75$$

where t is the month number ($t = 1$ corresponds to January).

a. Write the range of this function in interval notation.   [45, 105]

b. In which months, between 1 and 12 inclusive, are the sales \$7500?   Months 3, 7, and 11 (March, July, and November)

CHAPTER 5   Test

For Exercises 1–2, simplify the expression.

1. $\dfrac{1}{\cot^2 \theta + 1} + \dfrac{1}{\tan^2 \theta + 1}$ 1

2. $\cos\left(x + \dfrac{\pi}{6}\right) + \cos\left(x - \dfrac{\pi}{6}\right)$ $\sqrt{3}\cos x$

For Exercises 3–8, verify the identity.

3. $(\sin \theta - \cos \theta)(\sec \theta + \csc \theta) = \tan \theta - \cot \theta$

4. $\dfrac{\cot x - \tan x}{\cot x + \tan x} = \cos 2x$

5. $\dfrac{\sin 7x - \sin 5x}{\cos 7x + \cos 5x} = \tan x$

6. $\dfrac{\sin^3 x - \cos^3 x}{\sin x - \cos x} = \dfrac{1}{2}\sin 2x + 1$

7. $4\cos^2 \dfrac{x}{2} + 2\sin^2 \dfrac{x}{2} - 3 = \cos x$

8. $\dfrac{1 - \cos x}{1 + \cos x} = (\csc x - \cot x)^2$

9. Write $\sqrt{144 - x^2}$ as a function of θ by making the substitution $x = 12 \sin \theta$ for $0 < \theta < \dfrac{\pi}{2}$. $\sqrt{144 - x^2} = 12\cos\theta$

10. Write $8\cos x - 15 \sin x$ in the form $k\sin(x + \alpha)$ for $0 \le \alpha < 2\pi$. Round α to 3 decimal places. $17\sin(x + 2.652)$

11. Write $\cos^2 x + 3\sin^4 x$ in terms of first powers of cosine.

For Exercises 12–17, find the exact value.

12. $\sin 250° \cos 10° - \cos 250° \sin 10°$ $-\dfrac{\sqrt{3}}{2}$

13. $\tan \dfrac{\pi}{12}$ $2 - \sqrt{3}$

14. $\cos\left(\arctan 3 + \arcsin \dfrac{4}{5}\right)$ $\dfrac{9\sqrt{10}}{50}$

15. $\cos 105°$ $\dfrac{-\sqrt{6} + \sqrt{2}}{4}$ **16.** $\sin \dfrac{17\pi}{24} \sin \dfrac{\pi}{24}$ $\dfrac{\sqrt{2} - 1}{4}$

17. $\sin 555° + \sin 105°$ $\dfrac{\sqrt{2}}{2}$

18. Find the exact value for $\tan(\alpha + \beta)$ given $\sin \alpha = \dfrac{24}{25}$ for α in Quadrant II and $\cos \beta = \dfrac{3\sqrt{73}}{73}$ for β in Quadrant IV. $\dfrac{128}{171}$

19. Given $\sin \theta = \dfrac{2\sqrt{5}}{5}$ and θ in Quadrant II find the exact function values.

   a. $\sin 2\theta$ $-\dfrac{4}{5}$ **b.** $\cos 2\theta$ $-\dfrac{3}{5}$ **c.** $\tan 2\theta$ $\dfrac{4}{3}$

20. Given $\tan \theta = \dfrac{15}{8}$ and $\pi < \theta < \dfrac{3\pi}{2}$ find the exact function values.

   a. $\sin \dfrac{\theta}{2}$ $\dfrac{5\sqrt{34}}{34}$ **b.** $\cos \dfrac{\theta}{2}$ $\dfrac{-3\sqrt{34}}{34}$ **c.** $\tan \dfrac{\theta}{2}$ $-\dfrac{5}{3}$

For Exercises 21–22,

a. Write the solution set for the general solution.

b. Write the solution set on the interval $[0, 2\pi)$.

21. $\sqrt{3}\tan x + 5 = 6$ **22.** $2\cos 2x = \sqrt{2}$

For Exercises 23–30, solve the equation on the interval $[0, 2\pi)$.

23. $2\cos^3 x - \cos x = 0$

24. $6\csc^2 x - 17\csc x + 10 = 0$ $\left\{\dfrac{\pi}{6}, \dfrac{5\pi}{6}\right\}$

25. $6\sin^2 x - 5\cos x - 2 = 0$ $\left\{\dfrac{\pi}{3}, \dfrac{5\pi}{3}\right\}$

26. $2\sin^2 2x - \sin 2x - 1 = 0$

27. $18\cos^2 x + 3\cos x - 4 = 0$

28. $\cot\left(x - \dfrac{\pi}{4}\right) = \sqrt{3}$ $\left\{\dfrac{5\pi}{12}, \dfrac{17\pi}{12}\right\}$

29. $8\sin x + 5 = 0$

30. $\sin x + \sin 3x - \sin 2x = 0$

31. Use a graphing utility to solve $\sin 2x = 3x^3$. Round the solutions to 4 decimal places.

32. For a projectile launched from ground level at an angle of elevation θ with an initial velocity v_0, the maximum horizontal range is given by $x_{max} = \dfrac{v_0^2 \sin 2\theta}{g}$, where g is the acceleration due to gravity ($g = 32$ ft/sec^2 or $g = 9.8$ m/sec^2). If a toy rocket is launched from the ground with an initial velocity of 50 ft/sec and lands 73 ft from the launch point, find the angle of elevation of the rocket at launch. Round to the nearest tenth of a degree. 34.6° or 55.4°

CHAPTER 5 Cumulative Review Exercises

For Exercises 1–8, solve the equations and inequalities. Write the solution set to inequalities in interval notation. Solve trigonometric equations on the interval $[0, 2\pi)$.

1. $5 + 3|2x - 1| = 6$

2. $4x^4 + 3x^2 - 1 = 0$

3. $4\sin^4 x + 3\sin^2 x - 1 = 0$

4. $4x^2 + 3x - 1 \geq 0$

5. $4x^2 + 3x - 1 < 0$

6. $\sqrt{x + 2} - 2x = 1$

7. $16^{x^2 - 1} = \left(\dfrac{1}{8}\right)^{x+1}$ $\left\{-1, \dfrac{1}{4}\right\}$

8. $\ln 2x + \ln(2x + 1) - \ln(1 - x) = 0$

9. Given $P(x) = 4x^4 + 3x^3 + 15x^2 + 12x - 4$

  a. Use synthetic division to show that $x = 2i$ is a zero of the polynomial.

  b. Use the result from part (a) to factor the polynomial completely. $P(x) = (4x - 1)(x + 1)(x - 2i)(x + 2i)$

10. Given $L(x) = -\dfrac{2}{3}x + 5$

  a. Find the equation of a line perpendicular to L that passes through the point $(-1, 3)$ $y = \dfrac{3}{2}x + \dfrac{9}{2}$

  b. Find the equation of a line parallel to L that has an undefined slope.   There is no line parallel to L with an undefined slope.

11. Determine if the function is even, odd, or neither.

  a. $f(x) = 6x^4 + 7|x|$ Even

  b. $g(x) = 6x^4 + 7x$ Neither

  c. $h(x) = 6x^3 + 7x$ Odd

  d. $r(x) = \cos x + \sec x$ Even

  e. $t(x) = x^2 \sin x$ Odd

For Exercises 12–15, write the domain and range of each function in interval notation.

12. $f(x) = (x - 2)^2(x + 2)^2$

13. $g(x) = -2\sin 4x$

14. $h(x) = 3 + \sqrt{x - 2}$

15. $r(x) = \sin^{-1} x$

16. Given $y = \dfrac{2x^2}{2x^2 - x - 3}$

  a. Identify the asymptotes.   Vertical asymptotes: $x = -1, x = \dfrac{3}{2}$; Horizontal asymptote: $y = 1$

  b. Determine if the graph crosses the horizontal asymptote.   Yes. The graph crosses at $(-3, 1)$.

17. Given $f(x) = \log_3 x$,

  a. Write the domain and range in interval notation.   Domain: $(0, \infty)$; Range: $(-\infty, \infty)$

  b. Write an equation of the asymptote. $x = 0$

  c. Find the intercepts. $(1, 0)$

  d. Determine the intervals over which f is increasing, decreasing, and constant.   Increasing on $(0, \infty)$

For Exercises 18–19, verify the identity.

18. $\cot^2 \theta - \cos^2 \theta = \cot^2 \theta \cos^2 \theta$

19. $\sin 2x \cos x - \cos 2x \sin x = \sin x$

20. Given the right triangle, find a and b. $a = 10, b = 5\sqrt{3}$

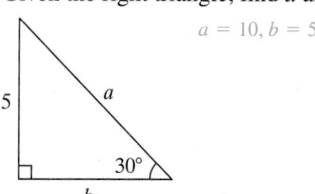

6

Applications of Trigonometric Functions

Chapter Outline

The study of trigonometry helps us understand the cycles in many naturally occurring phenomena. One such example is the sinusoidal change in the angle of the Sun at different times during the year.

The Earth's axis is tilted 23.5° relative to the plane of its orbit around the Sun. This combined with the Earth's revolution around the Sun causes the seasons. At the summer solstice, the Earth is at a position in its orbit where the tilt of the Earth results in the greatest amount of sunlight in the northern hemisphere. The Sun appears higher in the sky on this day than on any other day of the year. At the winter solstice, the reverse is true. The Earth is positioned so that the least amount of sunlight hits the northern hemisphere, and the path of the Sun in the sky is at its lowest. This is the first day of winter and the shortest day of the year.

For a point on the Earth at a given latitude α, the angle of elevation of the Sun at noon in the northern hemisphere varies sinusoidally between $90° - \alpha - 23.5°$ (winter low) and $90° - \alpha + 23.5°$ (summer high). For this reason, a sine or cosine model can be used to track the angle of elevation of the Sun in the sky and the length of daylight for any point on the Earth (Exercises 59–60 in Section 6.4).

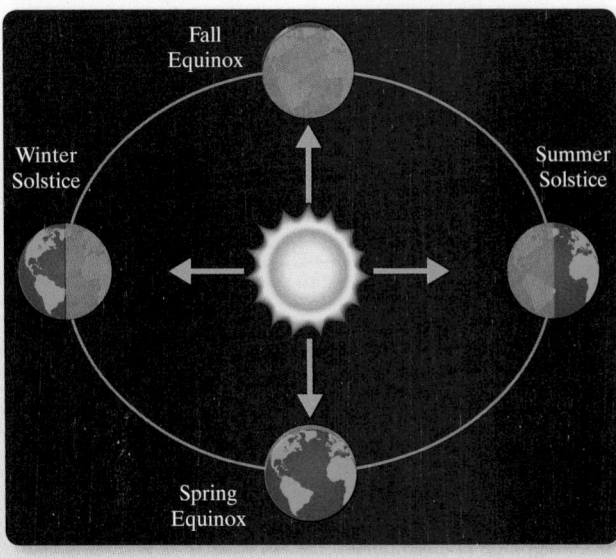

Applications of Right Triangles

OBJECTIVES

1. Solve a Right Triangle
2. Solve Applications of Right Triangles
3. Compute the Bearing of an Object

1. Solve a Right Triangle

In Section 4.3, we introduced the trigonometric functions as functions of acute angles of a right triangle. We saw immediate applications in finding the measures of the angles and sides of a right triangle when two sides or a side and an angle are known. This is called **solving a right triangle.** In this section, we revisit this concept. Then in Sections 6.2 and 6.3, we expand our study to solve *oblique* triangles (triangles that do not contain a right angle).

As a labeling convention, we will use capital letters to label the vertices of a triangle. The same letters in lower case will be used to denote the lengths of the sides opposite the corresponding vertices. Thus, $\triangle ABC$ has sides of length a, b, and c, opposite angles A, B, and C, respectively (Figure 6-1).

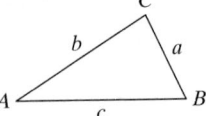

Figure 6-1

Classroom Examples: p. 636
Exercises 14, 16

EXAMPLE 1   **Solving a Right Triangle**

Solve the right triangle given in Figure 6-2.
Round the lengths of the sides to 1 decimal place.

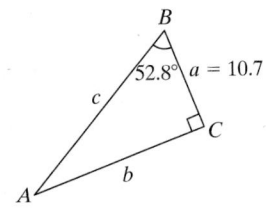

Figure 6-2

Solution:

We begin by noting that $B = 52.8°$, $C = 90°$, and $A + B + C = 180°$.

Therefore, $A + 52.8° + 90° = 180°$
$$A = 37.2°$$

To find b, we need to establish a relationship between a known angle and the known side a.

$$\tan 52.8° = \frac{b}{10.7} \Rightarrow b = 10.7\tan 52.8° \approx 14.1$$

To find c, we establish a relationship between a known angle and side a.

$$\cos 52.8° = \frac{10.7}{c} \Rightarrow c = \frac{10.7}{\cos 52.8°} \approx 17.7$$

Since $\triangle ABC$ is a right triangle, its sides must satisfy the Pythagorean theorem. Therefore to double check, we have

$$a^2 + b^2 = c^2$$
$$(10.7)^2 + (14.1)^2 \overset{?}{=} (17.7)^2$$
$$313.3 \approx 313.29 ✓$$

In summary, $A = 37.2°$, $B = 52.8°$, $C = 90°$, $a = 10.7$, $b \approx 14.1$, and $c \approx 17.7$.

Avoiding Mistakes

In Example 1, after finding the length of side b, you can use it to find the length of c by applying the relationship $\sin 52.8° = \dfrac{14.1}{c}$. However, the length of side b was rounded. Using a rounded value in further calculations potentially compounds the error. Therefore, we recommend using relationships involving the given information (in this case, angle B and side a) rather than the computed information.

Skill Practice 1   Solve the right triangle.
Round the lengths of the sides to 1 decimal place.

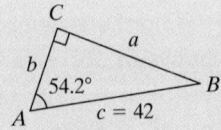

Answer

1. $A = 54.2°$, $B = 35.8°$, $C = 90°$,
$a \approx 34.1$, $b \approx 24.6$, $c = 42$

2. Solve Applications of Right Triangles

Historically, trigonometry was developed to measure distances that were difficult or impossible to measure directly. We examine such scenarios in Examples 2 and 3.

Classroom Example: p. 636
Exercise 20

Instructor Note:
Encourage students to draw sketches with reasonably accurate angles if possible.

EXAMPLE 2   Computing Distances in an Application

A beachgoer flies a kite on a windy day. He holds the spool of string at a height of 4 ft above the ground and notes that the string makes an angle of 42° from the horizontal. If 120 ft of string is out and the sag in the string is negligible, find the height of the kite to the nearest foot.

Solution:

A sketch of this scenario is given in Figure 6-3. ΔABC is a right triangle and the height of the kite is 4 ft + a. To solve for a, notice that a is the side opposite the given angle, and that $c = 120$ ft is the length of the hypotenuse. Therefore,

$$\sin 42° = \frac{a}{120} \Rightarrow a = 120 \sin 42° \approx 80$$

The height of the kite is 4 ft + 80 ft = 84 ft.

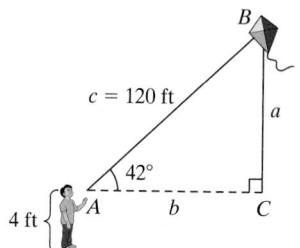

Figure 6-3

Skill Practice 2   A 40-ft ladder on a fire truck leans against a burning building such that the top of the ladder makes an angle of 20° with the building. If the bottom of the ladder is anchored to the top of the fire truck 6 ft off the ground, how far above ground does the ladder reach? Round to the nearest tenth of a foot.

Classroom Example: p. 636
Exercise 26

EXAMPLE 3   Computing Distances in an Application

A mathematics enthusiast visiting London wants to know the diameter of the clock face on Big Ben. She stands 200 ft down the road from the tower. Then from an eye level of 5 ft, she measures the angle of elevation to the bottom of the clock face at 42° and the angle of elevation to the top of the clock face at 45.4°. Find the diameter of the clock face to the nearest foot.

Solution:

Figure 6-4 reveals that the diameter of the clock face is given by $h_2 - h_1$. From the lower right triangle,

$$\tan 42° = \frac{h_1}{200} \Rightarrow h_1 = (200)\tan 42° \approx 180$$

From the larger right triangle,

$$\tan 45.4° = \frac{h_2}{200} \Rightarrow h_2 = (200)\tan 45.4° \approx 203$$

Therefore, the diameter of the clock face is $h_2 - h_1 \approx 203$ ft $- 180$ ft ≈ 23 ft.

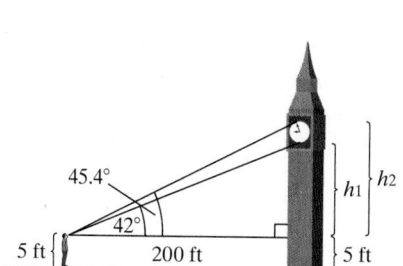

Figure 6-4

Answer
2. 43.6 ft

Skill Practice 3   From an eye level of 6 ft, an observer standing 30 ft away from a building sights the bottom of a window at an angle of elevation of 6.1°. He sights the top of the window at an angle of elevation of 20.1°. Find the vertical dimension of the window. Round to the nearest tenth of a foot.

In Example 4, we use the inverse trigonometric functions to find the measure of an unknown angle in an application.

Classroom Example: p. 637
Exercise 30

EXAMPLE 4   Computing an Angle in a Construction Application

Find the pitch angle of the roof to the nearest tenth of a degree.

Solution:

The pitch angle is the angle that the roof makes with the horizontal. Using either right triangle formed by the 6-ft altitude and half the width of the house we have

$$\tan \theta = \frac{6}{8}$$

$$\theta = \tan^{-1} \frac{6}{8} \approx 36.9°$$

The pitch angle of the roof is approximately 36.9°.

Skill Practice 4   A straight hiking trail is 2500 ft long with a gain in altitude of 310 ft. To the nearest degree, what is the average angle of ascent?

Point of Interest

In Examples 2–4, we used trigonometry to determine distances and angles that are difficult to measure directly. One distance impossible for us to measure directly is the distance between the Earth and a star. Astronomers use an indirect method called the method of parallax. They measure the angle represented by the apparent displacement of a star caused by the change in the position of an observer on Earth. This is done at a 6-month interval when the Earth is at opposite points in its orbit around the Sun. Then using the right triangle formed by the star, Earth, and Sun, astronomers can calculate the distance from the Earth to the star (see Exercises 31–32).

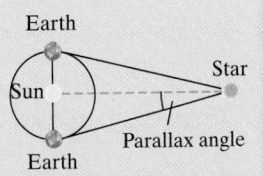

You can simulate parallax by holding your index finger at arm's length in front of your face. View your finger with one eye open and the other eye closed. Then alternate. Your finger will seem to move relative to objects in the background. This demonstrates parallax.

3. Compute the Bearing of an Object

In navigation and surveying applications, the term *bearing* is used to convey the angular relationship from one point to another. The **bearing** from point A to point B is the degree measure of the acute angle formed by ray $\overrightarrow{AB}$ and the north-south line. Figure 6-5 provides four examples of the bearing from A to B.

Answers
3. 7.8 ft
4. 7°

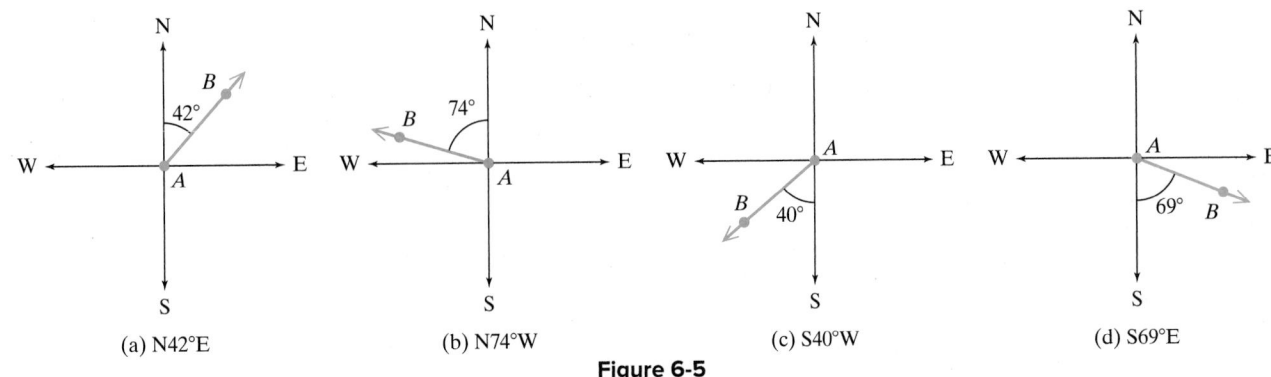

(a) N42°E   (b) N74°W   (c) S40°W   (d) S69°E

Figure 6-5

Notice that a bearing is denoted by the letter N or S (north or south), followed by the measure of the angle in degrees, followed by the letter E or W (east or west). For example, a ship bearing S40°W is heading in a southwesterly direction. Specifically, the ship is traveling 40° to the west of due south.

Classroom Example: p. 638
Exercise 38

EXAMPLE 5   Finding and Interpreting Bearing

Refer to Figure 6-6.

a. Find the bearing from O to C.

b. Find the bearing from O to B.

c. Find the bearing from A to O.

Solution:

a. Ray $\overrightarrow{OC}$ is directed 40° to the east of due south.

Therefore, the bearing from O to C is S40°E.

b. Ray $\overrightarrow{OB}$ is directed $118° - 90° = 28°$ to the west of due north.

Therefore, the bearing from O to B is N28°W.

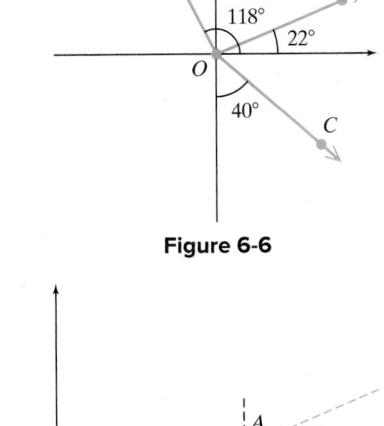

Figure 6-6

> **TIP** Remember that bearing is always measured from the north-south line. Therefore, the letter N or S is written first.

c. To find the bearing from A to O, we need to find the acute angle between ray $\overrightarrow{AO}$ and the north-south line.

It is helpful to sketch a coordinate system with origin at the initial point of the ray (Figure 6-7).

Ray $\overrightarrow{AO}$ is directed $90° - 22° = 68°$ to the west of due south.

Therefore, the bearing from A to O is S68°W.

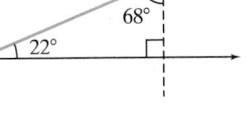

Figure 6-7

Skill Practice 5   Refer to Figure 6-6.

a. Find the bearing from C to O.

b. Find the bearing from B to O.

Answers

5. a. N40° W   **b.** S28° E

TIP One nautical mile (1 nmi) is defined as the distance along any meridian through 1' of latitude. A circle has 360° and 60' per degree. Thus, 1 nmi is the circumference of the Earth divided by (360)(60).

$$1 \text{ nmi} = \frac{24{,}860 \text{ mi}}{(360)(60)}$$
$$\approx 1.15 \text{ mi}$$

1 nmi ≈ 1.15 mi

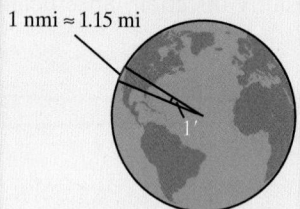

The unit of nautical miles is often used in navigation for ships and planes. The related unit of speed is 1 nautical mile per hour or 1 knot.

EXAMPLE 6 Interpreting Bearing

A plane leaves an airport heading S32°W at 320 knots (nautical miles per hour). Determine the number of nautical miles (nmi) south and the number of nautical miles west that the plane travels in 1.2 hr. Round to the nearest nautical mile.

Solution:

First sketch the path of the plane from the airport A to a point P, 1.2 hr after take-off (Figure 6-8). After 1.2 hr, the number of nautical miles traveled from A to P is

$$AP = (320 \text{ knots})(1.2 \text{ hr}) = 384 \text{ nmi}$$

The distance that the plane flies south is the vertical leg of $\triangle APB$, and the distance that the plane flies west is the horizontal leg of $\triangle APB$.

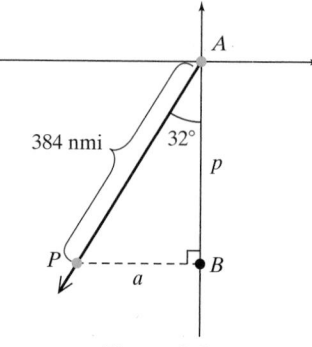

Figure 6-8

$$\text{Distance south: } \cos 32° = \frac{p}{384} \Rightarrow p = 384 \cos 32° \approx 326 \text{ nmi}$$

$$\text{Distance west: } \sin 32° = \frac{a}{384} \Rightarrow a = 384 \sin 32° \approx 203 \text{ nmi}$$

Skill Practice 6 A boat leaves a dock heading N27°W at 6 knots (nautical miles per hour). Determine the number of nautical miles (nmi) north and the number of nautical miles west that the boat travels in 0.5 hr. Round to the nearest tenth of a nautical mile.

EXAMPLE 7 Computing Bearing

A boat leaves a marina heading N52°E at 6 mph. After 1 hr, the boat makes a 90° counterclockwise turn to a new bearing of N38°W. If the boat travels this new course for 2 hr,

a. Find the boat's distance from the marina to the nearest tenth of a mile.

b. To a tenth of a degree, find the bearing required for the boat to return to the marina.

Solution:

a. The distance that the boat travels during the first hour is (6 mph)(1 hr) = 6 mi.

The distance that the boat travels for the last 2 hr is (6 mph)(2 hr) = 12 mi.

Since the boat makes a 90° turn, the path of the boat forms two legs of a right triangle. The length of the hypotenuse is the distance a from the boat to the marina (Figure 6-9). Using the Pythagorean theorem,

$$a^2 = (6)^2 + (12)^2$$
$$a = \sqrt{(6)^2 + (12)^2} = \sqrt{180} = 6\sqrt{5} \approx 13.4$$

The boat is approximately 13.4 mi from the marina.

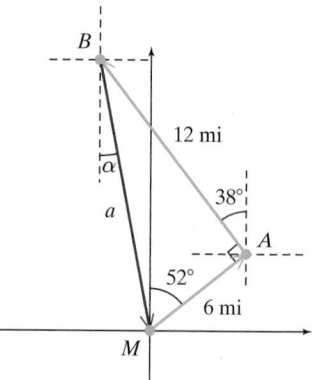

Figure 6-9

b. To find the bearing from the boat to the marina we need to find the acute angle formed by ray $\overrightarrow{BM}$ and the north-south line. This is labeled α. To find α, our strategy is to find the measure of angle M in $\triangle BMA$ and subtract $52°$. That is, $\alpha = M - 52°$.

We have $\tan M = \dfrac{12}{6} \Rightarrow M = \tan^{-1} 2 \approx 63.4°$.

Therefore, $\alpha = 63.4° - 52° = 11.4°$.

The boat should follow a bearing of S11.4°E to return to the marina.

Skill Practice 7   A plane leaves an airport heading S50°E at 450 mph. After 2 hr, the plane makes a 90° clockwise turn to a new bearing of S40°W. If the plane travels this new course for $\frac{1}{2}$ hr,

  a. Find the plane's distance from the airport to the nearest mile.

  b. To the nearest degree, find the bearing from the airport to the plane.

Answers

7. a. 928 mi   **b.** S36°E

Prerequisite Review

R.1. In a right triangle, one leg measures 15 ft and the hypotenuse measures 25 ft. Find the length of the missing side.   20 ft

R.2. Find the exact values of the six trigonometric functions for the angle θ.

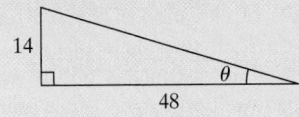

$\sin\theta = \dfrac{7}{25}$, $\cos\theta = \dfrac{24}{25}$, $\tan\theta = \dfrac{7}{24}$, $\csc\theta = \dfrac{25}{7}$, $\sec\theta = \dfrac{25}{24}$, $\cot\theta = \dfrac{24}{7}$

Concept Connections

For Exercises 1–2, refer to triangle ABC.

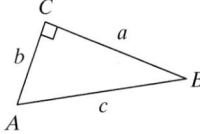

1. If the lengths of sides a and b are known, the inverse _____tangent_____ or _____cotangent_____ function can be used directly to find the value of angle A.

2. The measure of angle B and the length of side c can be used along with which trigonometric functions to find the length of side a?   cosine; secant

3. If the bearing from A to C is N12°E, what is the bearing from C to A?   S12°W

4. The acute angle used to express the bearing between two points is measured relative to the (north-south or east-west) line.
   north-south

Objective 1: Solve a Right Triangle

For Exercises 5–12, given right triangle ABC, determine if the expression is true or false.

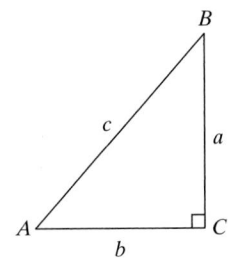

5. $\cot B = \dfrac{a}{b}$ True

6. $\sin^{-1}\dfrac{a}{b} = A$ False

7. $\csc B = \dfrac{c}{a}$ False

8. $\sec A = \dfrac{c}{b}$ True

9. $\cos^{-1}\dfrac{a}{c} = B$ True

10. $A + B = 90°$ True

11. $c^2 - b^2 = a^2$ True

12. $\tan A = \dfrac{b}{a}$ False

Note: *Due to round-off error, the answers to exercises involving calculations to determine unknown angles and sides may differ slightly from those given.*

For Exercises 13–18, solve the right triangle for the unknown sides and angles. Round values to 1 decimal place if necessary. (See Example 1)

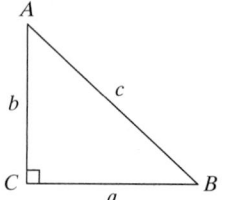

13. $B = 24°, c = 18$
   $A = 66°, b \approx 7.3, a \approx 16.4$

14. $A = 28°, c = 15$
   $B = 62°, b \approx 13.2, a \approx 7.0$

15. $a = 10, b = 7$
   $A \approx 55.0°, B \approx 35.0°, c \approx 12.2$

16. $a = 2, b = 8.6$
   $A \approx 13.1°, B \approx 76.9°, c \approx 8.8$

17. $A = 38.2°, b = 17.8$
   $B = 51.8°, a \approx 14, c \approx 22.7$

18. $A = 25.6°, a = 34$
   $B = 64.4°, b \approx 71.0, c \approx 78.7$

Objective 2: Solve Applications of Right Triangles

19. A boat is anchored off Elliot Key, Florida. From the bow of the boat, 36 ft of anchor line is out with 4 ft of line above the water. The angle that the line makes with the water line is 22°. **(See Example 2)**

   a. How deep is the water? Round to the nearest foot.   12 ft

   b. What is the horizontal distance between the anchor and the bow of the boat? Round to the nearest foot.   33 ft

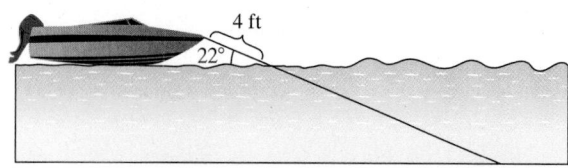

20. A swimming pool has a depth of 4 ft at the shallow end and 8 ft at the deep end. The bottom of the pool slopes downward at an angle of 5.7°. How long is the pool? Round to the nearest foot.   40 ft

21. An airplane climbs at an angle of 11° at an average speed of 420 mph. How long will it take for the plane to reach its cruising altitude of 6.5 mi? Round to the nearest minute.   5 min

22. An airplane begins its descent with an average speed of 240 mph at an angle of depression of 4°. How much altitude will the plane lose in 5 min. Round to the nearest tenth of a mile.   1.4 mi

23. A railroad bridge over New Scotland Road in Slingerlands, New York, has a low clearance for trucks. An engineer standing 20 ft away measures a 15.4° angle of elevation from her eye level of 5.5 ft to the bottom of the bridge. If the road is flat between the engineer and the bridge, how high over the roadway is the bottom of the bridge? Round to the nearest inch.
11 ft 0 in.

24. At 11:30 A.M. the angle of elevation of the sun for Washington, D.C., is 55.7°. If the height of the Washington monument is approximately 555 ft, what is the length of the shadow it will cast at that time? Round to the nearest foot.   379 ft

25. The suggested size for a United States flag is often determined by the height of the flagpole from which it will fly. Federal recommendations on flag size for flagpole heights are provided in the table. (www.publications.usa.gov)

Flagpole height (ft)	Flag size (ft)
25	5 by 8
40	6 by 10
60	10 by 15
125	20 by 30

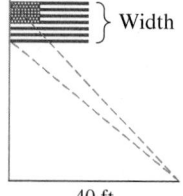

Width

40 ft

Measurements taken 40 ft from the base of a flagpole show the angle of elevation to the top of the flagpole to be 56.3° and the angle of elevation to the bottom of the flag to be 51.3°. **(See Example 3)**

   a. Determine the height of the flagpole. Round to the nearest foot.   60 ft

   b. Determine the vertical width of the flag. Round to the nearest foot.   10 ft

   c. What other information is needed to determine if the display conforms to the federal guidelines?
   The horizontal length of the flag is needed to determine if the display conforms to federal guidelines.

26. Roberto has plans to build a tree house in a large oak tree in his backyard. He takes measurements from the ground and uses trigonometry to estimate how tall the tree house can be. Twenty feet from the base of the tree, Roberto measures a 26.6° angle of elevation to the proposed bottom of the tree house and a 42° angle of elevation to the proposed top of the tree house. If Roberto is 6 ft tall, will there be enough room for him to stand inside the tree house?   Yes; the estimated height of the tree house from floor to roof is 8 ft.

27. A police officer hiding between two bushes 50 ft from a straight highway sights two points *A*, and *B*. The angle from the police car to *A* is 62°, and the angle to point *B* is 72°.

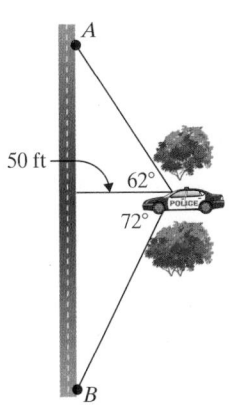

   a. Find the distance between *A* and *B*. Round to the nearest foot. 248 ft

   b. Suppose that a motorist takes 2.7 sec to pass from *A* to *B*. Using the rounded distance from part (a), find the motorist's speed in ft/sec. Round to 1 decimal place. 91.9 ft/sec

   c. Determine the motorist's speed in mph. Round to the nearest mph. 63 mph

29. The Duquesne Incline is a century-old railroad scaling Mt. Washington in Pittsburgh's south side neighborhood. The incline is 244 m long and rises 120 m in elevation. What is the angle of the incline to the nearest tenth of a degree? **(See Example 4)** 29.5°

28. A large weather balloon is tethered by two ropes. One rope measures 23 ft and attaches to the balloon at an angle of 32° from the ground. The second rope attaches to the base of the balloon at an angle of 15° with the ground.

   a. How far from the ground is the balloon floating? Round to the nearest tenth of a foot. 12.2 ft

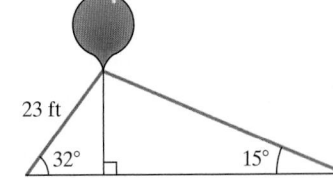

   b. Find the length of the second rope. Round to the nearest tenth of a foot. 47.1 ft

   c. If both ropes suddenly detach and the balloon rises straight up at a rate of 3 ft/sec, how long will it take the balloon to reach a height of 50 ft from the ground? Round to the nearest tenth of a second. 12.6 sec

30. The Jackson Hole Aerial Tram takes passengers from an elevation of 6311 ft to an elevation of 10,450 ft up the slope of Rendezvous Mountain in Wyoming. If the tram runs on a cable measuring 12,463 ft in length, what is the angle of incline of the tram to the nearest tenth of a degree? 19.4°

To approximate the distance from the Earth to stars relatively close by, astronomers often use the method of parallax. Parallax is the apparent displacement of an object caused by a change in the observer's point of view. As the Earth orbits the Sun, a nearby star will appear to move against the more distant background stars. Astronomers measure a star's position at times exactly 6 months apart when the Earth is at opposite points in its orbit around the Sun. The Sun, Earth, and star form the vertices of a right triangle with $\angle PSE = 90°$. The length of $\overline{SE}$ is the distance between the Earth and Sun, approximately 92,900,000 mi. The parallax angle (or simply parallax) is denoted by *p*. Use this information for Exercises 31–32.

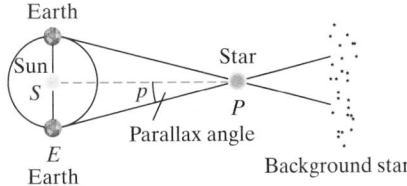

31. **a.** Find the distance between the Earth and Proxima Centauri (the closest star to the Earth beyond the Sun) if the parallax angle is 0.772″ (arcseconds). Round to the nearest hundred billion miles. 2.48×10^{13} mi

   b. Write the distance in part (a) in light-years. Round to 1 decimal place. (*Hint*: 1 light-year is the distance that light travels in 1 yr and is approximately 5.878×10^{12} mi.) 4.2 light-years

33. A sewer line must have a minimum slope of 0.25 in. per horizontal foot but not more than 3 in. per horizontal foot. A slope less than 0.25 in. per foot will cause drain clogs, and a slope of more than 3 in. per foot will allow water to drain without the solids.

   a. To the nearest tenth of a degree, find the angle of depression for the minimum slope of a sewer line. 1.2°

   b. Find the angle of depression for the maximum slope of a sewer line. Round to the nearest tenth of a degree. 14.0°

32. **a.** Find the distance between the Earth and Barnard's Star if the parallax angle is 0.547 arcseconds. Round to the nearest hundred billion miles. 3.50×10^{13} mi

   b. Write the distance in part (a) in light-years. Round to 1 decimal place. (*Hint*: 1 light-year is the distance that light travels in 1 yr and is approximately 5.878×10^{12} mi.) 6.0 light-years

34. In order for gravel roads to have proper drainage, the highest point on the road (the crown) should slope downward on either side to the shoulders of the road. EPA guidelines for maintaining gravel roads with a low volume of traffic suggest that a 20-ft wide road should have a centerline crown that is 5 to 7 in. high. (*Source*: http://water.epa.gov)

   a. To the nearest tenth of a degree, find the angle of depression for a 5-in. centerline crown. 2.4°

   b. To the nearest tenth of a degree, find the angle of depression for a 7-in. centerline crown. 3.3°

35. The weather satellite NOAA-15 orbits the Earth every 101 min at an altitude of 807 km as shown in the figure. The cameras on the satellite have a maximum range defined by the set of lines tangent to the Earth from the satellite. Using the radius of the Earth as 6357 km, approximate the length of the arc from A to B to the nearest kilometer. This is the maximum distance along the surface of the Earth "seen" by the satellite's cameras. 6093 km

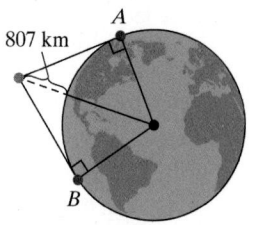

36. A communications satellite is in geosynchronous orbit. This means that its orbit coincides with the rotation period of the Earth so that it remains above a fixed point on the surface of the Earth at all times. The altitude of the satellite is 35,786 km. The sender and receiver of a signal must both be within the line of sight of the satellite. What is the maximum distance along the surface of the Earth for which the sender and receiver can communicate? Take the radius of the Earth to be 6357 km and round to the nearest kilometer. 18,046 km

Objective 3: Compute the Bearing of an Object

37. **a.** Find the bearing from O to A.
   (**See Example 5**) N12°E

   b. Find the bearing from O to B. S75°W

   c. Find the bearing from O to C. S51°E

   d. Find the bearing from C to O. N51°W

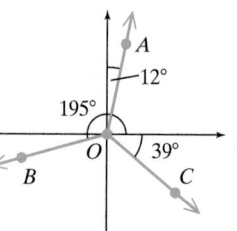

38. **a.** Find the bearing from O to A. N25°W

   b. Find the bearing from O to B. S42°W

   c. Find the bearing from O to C. S27°E

   d. Find the bearing from A to O. S25°E

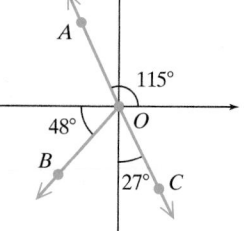

For Exercises 39–42, make a sketch to illustrate the bearing.

39. N45°W

40. S10°W

41. S67°E

42. N75°E

For Exercises 43–46, find the bearing of the ship. Round to the nearest tenth of a degree if necessary.

43.   N68°E

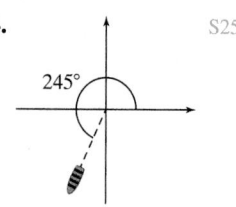

44.   S25°W

45.   N54.8°W

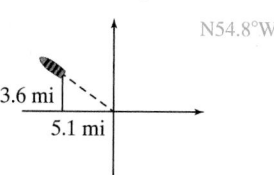

46.   S61.4°E

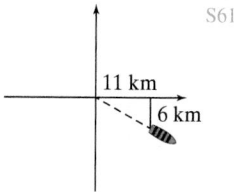

For Exercises 47–50, find the bearing to the nearest tenth of a degree.

47. A runner jogs 5 mi east then 2 mi south. What is the bearing from his starting point? S68.2°E

48. A cyclist rides west for 7 mi and then north 10 mi. What is the bearing from her starting point? N35.0°W

49. A boat leaves Matheson Hammock Marina at a constant speed of 3.5 mph. The boat travels south for 36 min and then east for 24 min to a favorite fishing spot. After a day of fishing, find the bearing that the captain should use to travel back to the marina. N33.7°W

50. A plane leaves Atlanta's Hartsfield Airport and flies north for 1 hr and west for 2 hr at an average speed of 400 mph. Find the bearing that the plane should take for the return trip. S63.4°E

For Exercises 51–54, round answers to the nearest unit.

51. A fishing boat leaves a dock at a bearing of N34°E. After running for 1.4 hr at 25 mph, how far north and east has it traveled? (**See Example 6**) 29 mi north and 20 mi east

52. A helicopter takes off from the roof of a building and travels at 100 mph on a bearing of S53°E. The flight takes 3.2 hr. How far south and how far east has the helicopter traveled? 193 mi south and 256 mi east

53. An airplane flying 310 knots has a bearing of S44°W. After flying for 2.4 hr, how many nautical miles (nmi) south and west has it traveled? 535 nmi south and 517 nmi west

54. A ship traveling 6.4 knots has a bearing of N11°W. After 3 hr, how many nautical miles (nmi) north and west has it traveled? 19 nmi north and 4 nmi west

55. A ship leaves port at 8 knots heading N27°W. After 2 hr it makes a 90° clockwise turn to a new bearing of N63°E and travels for 1.4 hr. (**See Example 7**)

   a. Find the ship's distance from port to the nearest tenth of a nautical mile. 19.5 nmi

   b. Find the bearing required for the ship to return to port. Round to the nearest degree. S8°W

56. A plane leaves an airport heading S82°E at 320 mph. After 1.2 hr, it makes a 90° counterclockwise turn to a new bearing of N8°E and travels for 45 min.

   a. Find the plane's distance from the airport to the nearest mile. 453 mi

   b. Find the bearing required for the plane to return to the airport. Round to the nearest degree. S66°W

57. A boat leaves a dock at a bearing of N42°W and travels 24 mph for 1 hr. The boat then makes a 90° left turn and travels 8 mi to its destination.

   a. Find the exact distance from the dock to the boat's destination. $8\sqrt{10}$ mi

   b. Find the bearing required to locate the boat from the dock. Round to the nearest degree.   N60.4°W

58. A runner leaves the starting point of a race and heads N10°W at 6.3 mph. After running for 20 min she makes a 90° right turn and runs 3 more miles at a speed of 6.8 mph.

   a. Find the total time of her run to the nearest minute.   46 min

   b. Find the bearing required to locate the runner from the starting point. Round to the nearest degree.   N45.0°E

Mixed Exercises

For Exercises 59–60, find the lengths of the sides to the nearest tenth of a unit and the measures of the angles to the nearest tenth of a degree if necessary.

59. $A =$ ___ $A = 55°$ $C =$ ___ $C = 70°$

  $AB =$ ___ $AB \approx 42.8$ $BC =$ ___ $BC \approx 22.2$

  $AD =$ ___ $AD \approx 74.6$ $DB =$ ___ $DB \approx 61.1$

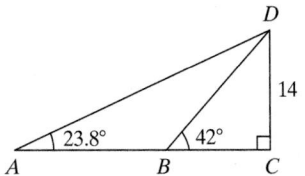

60. $\angle ADC =$ $\angle ADC = 66.2°$ $\angle ADB =$ $\angle ADB = 18.2°$

  $\angle BDC =$ $\angle BDC = 48°$ $BC =$ $BC \approx 15.5$

  $AC =$ $AC \approx 31.7$ $AD =$ $AD \approx 34.7$

  $BD =$ $BD \approx 20.9$ $AB =$ $AB \approx 16.2$

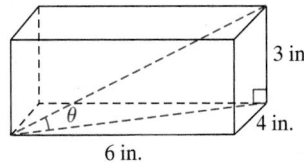

61. Find the measures of the sides and angles. Round angle measures to the nearest tenth of a degree. Give exact answers for the lengths of sides.

  $\angle ADC =$ ___ $\angle ADC \approx 66.7°$

  $\angle CBA =$ ___ $\angle CBA \approx 150.9°$

  $\angle DCB =$ ___ $\angle DCB \approx 52.4°$

  $BD =$ ___ $BD = 2\sqrt{106}$; $BC =$ ___ $BC = 6\sqrt{7}$

62. A rectangular prism measures 6 in. by 4 in. by 3 in. Find the angle θ formed by the diagonal of the prisim and the diagonal of the base, as shown. Round to 1 decimal place.   22.6°

63. Two children sit on either end of an 18-ft teeter totter. The center of the teeter totter is 2 ft off the ground, and the position of each child is 8 ft from the center.

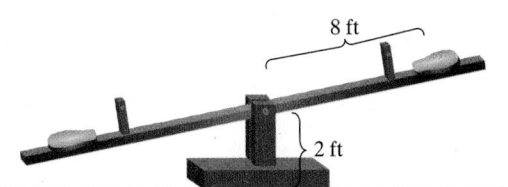

   a. When one end of the teeter totter is on the ground, find the angle that the teeter totter makes with the ground. Round to the nearest tenth of a degree.   12.8°

   b. When one end of the teeter totter is on the ground, how high is the child's seat at the other end? Round to 1 decimal place.   3.8 ft

64. A light fixture mounted 24 ft above the ground illuminates a cone of light with an angle of 42° at the top.

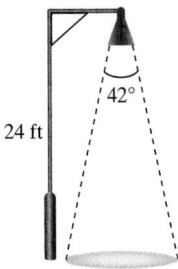

   a. What is the radius of the circle of light on the ground? Round to 1 decimal place.   9.2 ft

   b. How many square feet will be lit on the ground? Round to the nearest whole unit.   266 ft²

For Exercises 65–66, refer to the table giving the angle of the Sun at noon at the first day of each season based on the latitude of the observer in the northern hemisphere. The term 23.5° represents the tilt of the Earth to the ecliptic plane (plane of the Earth's orbit around the Sun).

Date	Sun's angle at noon
First day of spring: March equinox	90° – latitude
First day of fall: September equinox	90° – latitude
First day of summer: Summer solstice	90° – latitude + 23.5°
First day of winter: Winter solstice	90° – latitude – 23.5°

65. In 1987, a group of protesters lead by Jacqueline Kennedy Onassis defeated a plan to build two high rises that would have blocked sunlight to a significant swath of New York City's Central Park. Eventually the design was scaled down and the buildings' height reduced from approximately 1020 ft to what is now the Time Warner Center at 750 ft.

   a. The latitude of New York City is 40.7°N. Find the angle of elevation of the Sun at noon on the winter solstice (shortest day of the year).   25.8°

   b. At noon on the winter solstice, how long is the shadow cast by a 1020-ft building?   2110 ft

   c. At noon on the winter solstice, how long is the shadow cast by a 750-ft building?   1551 ft

66. The Earth is tilted 23.5° on its axis. Therefore, at noon on the summer solstice, the Earth is tilted 23.5° toward the Sun and all points on the Earth with latitude 23.5°N will see the Sun directly overhead at noon. Points on the Earth with latitude 23.5°N are on the Tropic of Cancer. Key West, Florida, has latitude 24.6°N, just above the Tropic of Cancer.

   a. Find the angle of elevation of the Sun at noon in Key West on the first day of summer and on the first day of winter.   Summer: 88.9°; Winter: 41.9°

   b. Find the length of a shadow cast by a 105-ft building at noon on the first day of summer and on the first day of winter. Round to the nearest tenth of a foot.   Summer: 2.0 ft; Winter: 117.0 ft

   c. Find the length of a shadow cast by a 40-ft tree at noon on the first day of fall or spring.   18.3 ft

67. OSHA safety regulations require that the base of a ladder be placed 1 ft from the wall for every 4 ft of ladder length. To the nearest tenth of a degree, find the angle that the ladder forms with the ground and the angle that it forms with the wall.   Angle with ground: 75.5°; Angle with wall: 14.5°

68. An A-frame storage shed is constructed such that the slanted roof line is always 3 times as long as the distance from the bottom corner of the roof to the centerline of the shed. To the nearest degree, what is the angle at the top of the roof?   39°

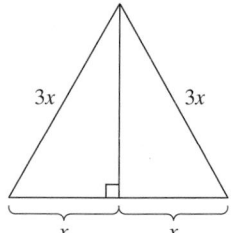

69. An airplane needs to fly to an airfield located 189 mi east and 568 mi north of its current location. To the nearest tenth of a degree, determine the bearing at which the plane should fly.   N18.4°E

70. Software used to program video games often uses an origin at the top left of the display canvas. The positive *x*-axis is to the right and the positive *y*-axis is downward. Suppose that a player moves on a direct path from the origin to a point *P* with pixel location (135, 200). Then the player moves directly to point *Q* at a pixel location of (420, 150).

   a. Find the player's bearing from the origin to point *P*. Round to one-hundredth of a degree.   S34.02°E

   b. Find the player's bearing from point *P* to point *Q*. Round to one-hundredth of a degree.   N80.05°E

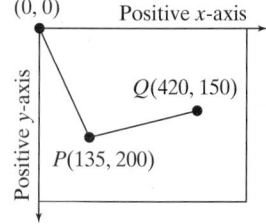

71. Two ships leave port at the same time. One travels N72°E at 20 knots and the other travels S18°E at a rate of 30 knots.

   a. After 2 hr, how far apart are the two ships? Give an exact answer in nautical miles. $20\sqrt{13}$ nmi

   b. Find the bearing from the ship farther to the north to the ship farther to the south. Round to the nearest tenth of a degree.   S15.7°W

72. Donna lives 17 mi due west of her friend Julie. They both like to bicycle and decide to meet for lunch at a restaurant 13 mi from Donna's house. If the bearing from Donna's house to the restaurant is N20.2°E, how far does Julie have to ride? Round to the nearest tenth of a mile.   17.5 mi

Write About It

73. Explain how to represent the bearing from point *P* to point *Q*.

74. What is meant by solving a triangle?

Expanding Your Skills

75. A contractor building eight ocean-front condominiums wants to maximize the view of the ocean for each unit. The side of the building facing the ocean is not built in a straight line parallel to the ocean, but instead is built in a zigzag pattern as shown in the figure. Each condo has a window of length *x* facing the ocean at an angle of 70° from a line perpendicular to the ocean.

  a. Find the length *x* of each window. Round to the nearest foot.   34 ft

  b. The windows facing the ocean are 8 ft high and *x* feet wide. By using the zigzag pattern how much more ocean-front viewing area does each window provide than if the windows were parallel to the ocean? Round to the nearest square foot.   16 ft²

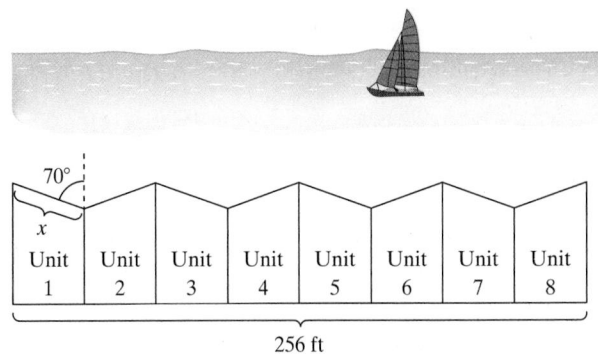

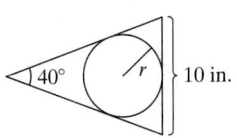

76. A circle is inscribed within an isosceles triangle such that the circle is tangent to all three sides as shown in the figure. Using the fact that the radius *r* of the circle is perpendicular to any line tangent to the circle, find the value of *r* to the nearest tenth of an inch.   3.5 in.

77. A tool requires a V-shaped block to hold two rollers of radius 3 cm and 2 cm. Determine the angle of the "V" cut to the nearest tenth of a degree.   23.1°

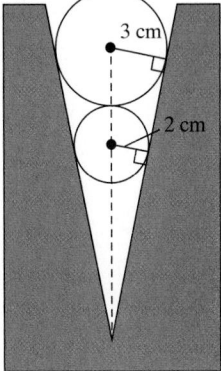

Technology Connections

78. A 5-ft (60-in.) belt is to be looped around a rotating drum of radius 8 in. and a small spindle located *d* inches from the rim of the drum. Find the distance *d* between the spindle and the rim of the drum to the nearest tenth of an inch. (*Hint:* Set up an equation and use a graphing utility to approximate the solution.)   7.3 in.

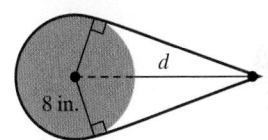

79. Pipe for a water line must be installed from a main water line at point *A* to a building on Hontoon Island State Park at point *B* as shown in the figure. The cost to install water pipe over land is $10 per foot and the cost to install pipe under water is $20 per foot.

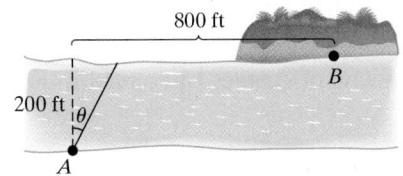

  a. Write an expression in terms of θ to represent the total cost *c* (in dollars) to lay pipe from point *A* to point *B*. $4000\sec\theta + 10(800 - 200\tan\theta)$

  b. Use the TABLE function on a calculator to find the cost for $\theta = 20°, 25°, 30°, 35°,$ and $40°$. Round to 1 decimal place.

  c. Which angle from part (b) yields the least cost?   30°

  d. Using calculus, we can show that the angle needed to minimize the total cost is a solution to the equation $4000\sec\theta\tan\theta - 2000\sec^2\theta = 0$. Solve the equation for θ, where $0° < \theta < 90°$.   30.0°

80. A woman participating in a triathlon can run 11 ft/sec and swim 3 ft/sec. She is at point *A*, 900 ft from a straight shoreline and must swim to shore and run to point *B*, 3000 ft down the beach.

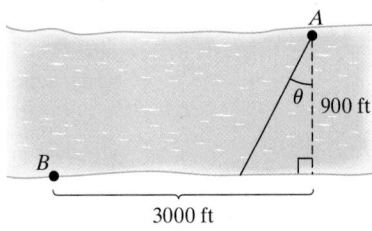

  a. Write an expression representing the total time *t* (in seconds) for her to get from point *A* to point *B* as a function of θ. $t = 300\sec\theta + \dfrac{3000 - 900\tan\theta}{11}$

  b. Use the TABLE function on a calculator to find the time *t* for $\theta = 0°, 5°, 10°, 15°, 20°,$ and $25°$. Round to 1 decimal place.

  c. Which angle from part (b) gives the shortest total time?   15°

  d. Using calculus, we can show that the angle needed to minimize the total time is a solution to the equation $300\sec\theta\tan\theta - \dfrac{900}{11}\sec^2\theta = 0$. Solve the equation for θ, where $0° < \theta < 90°$. Round to the nearest tenth of a degree.   15.8°

SECTION 6.2 The Law of Sines

1. Solve a Triangle Using the Law of Sines (SAA or ASA)

A mechanical engineer designing a new tool will use computer software for digital prototyping. It is imperative that the precise measure of each line segment, angle, and curve is entered and represented correctly before manufacturing takes place. The software uses mathematical relationships among the sides and angles in 2- and 3-dimensional figures as a means to check the engineer's measurements.

As a simple example, suppose that an engineer wants to draw a triangle with sides of lengths 3 cm, 5 cm, and 15 cm. No such triangle is possible because the length of the longest side exceeds the sum of the lengths of the other two sides. The law of sines and law of cosines presented in this chapter enable us to determine conditions under which a triangle cannot be constructed.

A triangle that does not contain a right angle is called an **oblique triangle.** An oblique triangle will have either three acute angles or one obtuse angle and two acute angles (Figure 6-10).

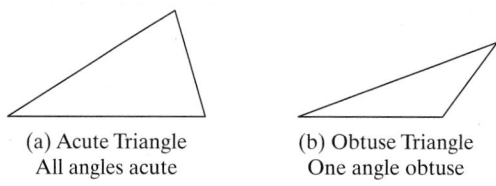

(a) Acute Triangle
All angles acute

(b) Obtuse Triangle
One angle obtuse

Figure 6-10

For an oblique triangle, given the length of a side and any other two parts of the triangle we can solve the triangle. To organize this discussion, we outline four distinct cases involving known angles (A) and known sides (S).

1. Two angles and any side are known (ASA or SAA).
2. Two sides and an angle opposite one of them are known (SSA).
3. Three sides are known (SSS).
4. Two sides and the angle between them are known (SAS).

The first two cases can be solved by using the law of sines (presented in this section), whereas cases 3 and 4 require the law of cosines (presented in Section 6.3).

> **TIP** Notation such as ASA is a shortcut used to convey what parts of an oblique triangle are known. For example, ASA stands for "Angle-Side-Angle."

Law of Sines

If $\triangle ABC$ is a triangle with sides of lengths a, b, and c opposite angles A, B, and C, respectively, then

$$\frac{a}{\sin A} = \frac{b}{\sin B} = \frac{c}{\sin C} \quad \text{or} \quad \frac{\sin A}{a} = \frac{\sin B}{b} = \frac{\sin C}{c}$$

In words: The ratio of the length of a side of a triangle to the sine of the angle opposite that side is the same for all three side-angle combinations.

To prove the law of sines, refer to Figure 6-11. The altitude h drawn from vertex C to side c is a common leg of two right triangles. From the right triangles, we have

$$\sin A = \frac{h}{b} \Rightarrow h = b \sin A$$

$$\sin B = \frac{h}{a} \Rightarrow h = a \sin B$$

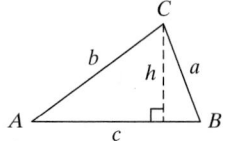

Figure 6-11

Equating the expressions for h gives

$$b \sin A = a \sin B \Rightarrow \frac{b}{\sin B} = \frac{a}{\sin A}$$

Drawing an altitude from vertex A to side a and applying similar reasoning can be used to show that $\dfrac{b}{\sin B} = \dfrac{c}{\sin C}$, thus completing the law of sines.

To apply the law of sines, we need to have one complete ratio and one other piece of information about the triangle. That is, we need to know the measure of one side and the angle opposite it, along with one other part of the triangle.

Classroom Example: p. 651
Exercise 14

TIP Notice that the units of measurement were left off the calculations in Example 1. Since $\sin \theta$ is unitless, the values of a and b will have the same units of measurement as c. For simplicity, when applying the law of sines or cosines, the units are left off in the equations, and applied later once a solution is found.

EXAMPLE 1 Solving an Oblique Triangle Given SAA

Solve the triangle with $A = 31°$, $C = 76°$, and $c = 19$. Round the lengths of the sides to 1 decimal place.

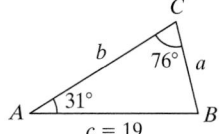

Solution:

In this situation, side c and angle C provide a complete ratio and angle A is a third piece of information. To find the length of side a, we have

$$\frac{c}{\sin C} = \frac{a}{\sin A} \Rightarrow \frac{19}{\sin 76°} = \frac{a}{\sin 31°} \Rightarrow a = \frac{19 \sin 31°}{\sin 76°} \approx 10.1$$

The measure of angle B is given by $B = 180° - (31° + 76°) = 73°$.

To find the length of side b, we can use $\dfrac{c}{\sin C} = \dfrac{b}{\sin B}$ or $\dfrac{a}{\sin A} = \dfrac{b}{\sin B}$.

However, we choose the equation on the left because it does not include any values that were rounded in previous calculations.

$$\frac{c}{\sin C} = \frac{b}{\sin B} \Rightarrow \frac{19}{\sin 76°} = \frac{b}{\sin 73°} \Rightarrow b = \frac{19 \sin 73°}{\sin 76°} \approx 18.7$$

In summary, $A = 31°$, $B = 73°$, $C = 76°$, $a \approx 10.1$, $b \approx 18.7$, and $c = 19$.

Skill Practice 1 Solve $\triangle ABC$ with $B = 70°$, $C = 48°$, and $b = 15$ cm. Round the lengths of the sides to 1 decimal place.

In Example 2, we solve an oblique triangle with two given angles and the side between them known (ASA).

Answer

1. $A = 62°$, $B = 70°$, $C = 48°$, $a \approx 14.1$, $b = 15$, $c \approx 11.9$

Classroom Example: p. 651

Exercise 16

EXAMPLE 2   Solving an Oblique Triangle Given ASA

Solve the triangle with $A = 132°$, $B = 28°$, and $c = 6.8$. Round the lengths of the sides to 1 decimal place.

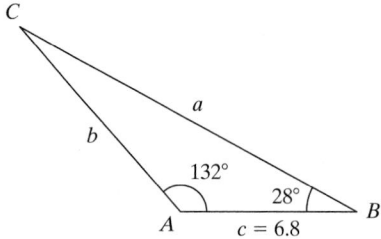

Solution:

First note that $C = 180° - (132° + 28°) = 20°$.

With the measure of angle C known, angle C and side c provide a complete ratio. To find the length of side a, we have

$$\frac{c}{\sin C} = \frac{a}{\sin A} \Rightarrow \frac{6.8}{\sin 20°} = \frac{a}{\sin 132°} \Rightarrow a = \frac{6.8 \sin 132°}{\sin 20°} \approx 14.8$$

To find the length of side b, we have

$$\frac{c}{\sin C} = \frac{b}{\sin B} \Rightarrow \frac{6.8}{\sin 20°} = \frac{b}{\sin 28°} \Rightarrow b = \frac{6.8 \sin 28°}{\sin 20°} \approx 9.3$$

In summary, $A = 132°$, $B = 28°$, $C = 20°$, $a \approx 14.8$, $b \approx 9.3$, and $c = 6.8$.

Skill Practice 2   Solve $\triangle ABC$ with $A = 33°$, $B = 112°$, and $c = 16$. Round the lengths of the sides to 1 decimal place.

2. Solve a Triangle Using the Law of Sines (SSA) Ambiguous Case

In Examples 1 and 2, the measures of two angles and a side of a triangle were given and we found a unique solution for the measures of the remaining sides and angle. Now we address the case where the lengths of two sides and the measure of an angle opposite one of the known sides are given. A great deal of care must be given to this scenario because the given information may define a unique triangle, two different triangles, or no triangle.

Suppose that the measure of angle A and the lengths of sides a and b of a triangle are given (Figure 6-12). The altitude h of the triangle is given by

$$\sin A = \frac{h}{b} \Rightarrow h = b \sin A$$

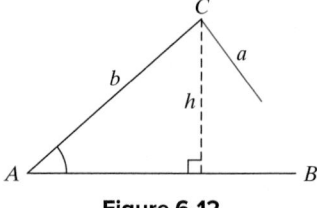

Figure 6-12

Depending on the length of side a (shown in red) relative to the length of the altitude h, we have four different scenarios.

Law of Sines Ambiguous Case

Two triangles: One acute
and one obtuse triangle

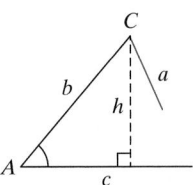

No triangle

(a) $a < h$
Side a is too short
to meet side c.

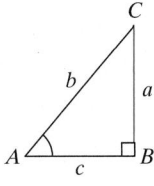

One right triangle

(b) $a = h$
Side a exactly
equals the
altitude.

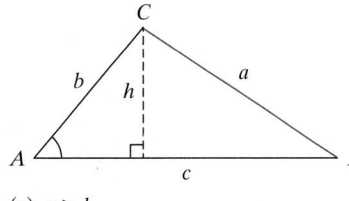

One oblique triangle

(c) $a > b$
One triangle is possible.

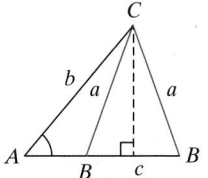

(d) $h < a < b$
Side a intersects
side c at an obtuse
angle or acute angle.

Classroom Example: p. 651
Exercise 22

EXAMPLE 3   Solving a Triangle Given SSA (One Solution)

Solve the triangle with $A = 57°$, $a = 62$, and $b = 50$. Round the measures of
the unknown angles and side to 1 decimal place.

Solution:

We begin with a sketch. For convenience we draw side
c as a horizontal line segment and angle A in standard
position relative to side c (Figure 6-13). Since $a > b$,
side a is long enough to meet side c. Furthermore, side
a must extend to the right as shown in Figure 6-13.

Figure 6-13

From the law of sines,

$$\frac{a}{\sin A} = \frac{b}{\sin B} \Rightarrow \frac{62}{\sin 57°} = \frac{50}{\sin B} \Rightarrow \sin B = \frac{50 \sin 57°}{62} \approx 0.6763$$

Using a calculator we determine that $\sin^{-1}\left(\dfrac{50 \sin 57°}{62}\right) \approx 42.6°$. However, there
are two angles between $0°$ and $180°$ for which $\sin B \approx 0.6763$.

$$B \approx 42.6° \text{ or } B \approx 180° - 42.6° \approx 137.4°$$

However, angle B cannot equal $137.4°$ because taken together with $A = 57°$,
the sum exceeds $180°$ ($137.4° + 57° = 194.4°$). This tells us that $B \approx 42.6°$,
and consequently, there is one unique triangle that can be formed from the
given information.

With $A = 57°$ and $B \approx 42.6°$, we have $C = 180° - (57° + 42.6°) \approx 80.4°$.

$$\frac{a}{\sin A} = \frac{c}{\sin C} \Rightarrow \frac{62}{\sin 57°} \approx \frac{c}{\sin 80.4°}$$

$$c \approx \frac{62 \sin 80.4°}{\sin 57°}$$

$$c \approx 72.9$$

> **TIP** In Example 3, since
> $a > b$, side a must extend to
> the right of the altitude. This
> means that angle B must be
> acute. Therefore, angle B
> is the first quadrant angle
> equal to $\sin^{-1}(50\sin 57°/62)$.

In summary, $A = 57°$, $B \approx 42.6°$, $C \approx 80.4°$, $a = 62$, $b = 50$, and $c \approx 72.9$.

> **Skill Practice 3** Solve $\triangle ABC$ with $A = 49°$, $a = 32$, and $b = 28$. Round the measures of the unknown angles and side to 1 decimal place.

Classroom Example: p. 651
Exercise 24

EXAMPLE 4 Solving a Triangle Given SSA (No Solution)

Solve the triangle with $C = 39°$, $c = 6$, and $a = 15$. Round the measures of the unknown angles and side to 1 decimal place.

Solution:

We begin with a sketch. For convenience we draw side b as a horizontal line segment and angle C in standard position relative to side b (Figure 6-14). Intuitively, we might question whether the length of side c is long enough to intersect side b.

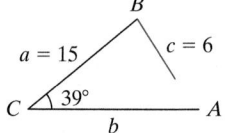

Figure 6-14

From the law of sines,

$$\frac{c}{\sin C} = \frac{a}{\sin A} \Rightarrow \frac{6}{\sin 39°} = \frac{15}{\sin A} \Rightarrow \sin A = \frac{15\sin 39°}{6} \approx 1.5733$$

This is not possible because $-1 \leq \sin A \leq 1$ for any angle A. No triangle can be constructed from the given information.

<u>Alternative Approach:</u>

The altitude h drawn from vertex B to side b can be computed by

$$\sin 39° = \frac{h}{15} \Rightarrow h = 15\sin 39°$$

$$h \approx 9.4$$

The length of side c would need to be at least 9.4 units long to reach side b. Therefore, no triangle is possible.

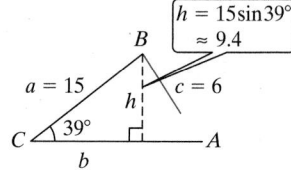

> **Skill Practice 4** Solve $\triangle ABC$ with $B = 31°$, $b = 3$, $a = 6.2$. Round the measures of the unknown angles and side to 1 decimal place.

Classroom Example: p. 651
Exercise 26

EXAMPLE 5 Solving a Triangle Given SSA (Two Solutions)

Solve the triangle with $A = 29°$, $a = 7$, and $b = 10$. Round the measures of the unknown angles and side to 1 decimal place.

Answers

3. $A = 49°$, $B \approx 41.3°$, $C \approx 89.7°$,
$a = 32$, $b = 28$, $c \approx 42.4$

4. No triangle

Instructor Note:
Spend some time identifying the number of possible triangles from the given information. For example:
1. $A = 40°$, $a = 7$, $b = 12$ (no triangle)
2. $B = 67°$, $b = 10$, $c = 7$ (one triangle)
3. $C = 50°$, $a = 18$, $c = 16$ (two triangles)

Solution:

From a preliminary sketch, we do not know if side a is long enough to reach side c, or if side a makes up the third side of one or two triangles.

To determine whether the given information defines zero, one, or two triangles, we can apply the law of sines to determine the possible values for angle B. Or we can compare the length of side a to the altitude h.

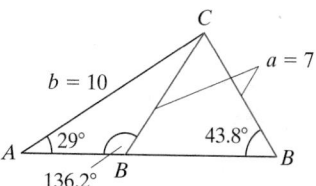

<u>Using the law of sines:</u>

$$\frac{a}{\sin A} = \frac{b}{\sin B} \Rightarrow \frac{7}{\sin 29°} = \frac{10}{\sin B}$$

$$\sin B = \frac{10 \sin 29°}{7} \approx 0.6926$$

Using a calculator we determine that

$$\sin^{-1}\left(\frac{10 \sin 29°}{7}\right) \approx 43.8°. \text{ However,}$$

there are two angles between $0°$ and $180°$ for which $\sin^{-1} 0.6926$.

$$B \approx 43.8° \text{ or } B \approx 180° - 43.8° \approx 136.2°$$

<u>Finding the altitude h:</u>

$$\sin A = \frac{h}{10}$$

$$h = 10 \sin 29° \approx 4.8$$

Since $h < a < b$ (that is, $4.8 < 7 < 10$), there are two possible triangles.

Instructor Note:
Remind students that finding the first unknown angle in the ambiguous case results in two possible angles that are supplementary.

Adding either value for angle B to the given angle results in a sum less than $180°$.

$$A + B = 29° + 43.8° = 72.8° < 180°$$
$$A + B = 29° + 136.2° = 165.2° < 180°$$

With two viable values for angle B, we can construct two triangles.

Triangle 1: $\angle B$ acute   Triangle 2: $\angle B$ obtuse

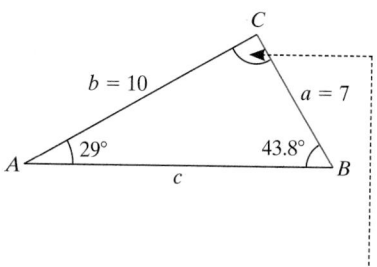   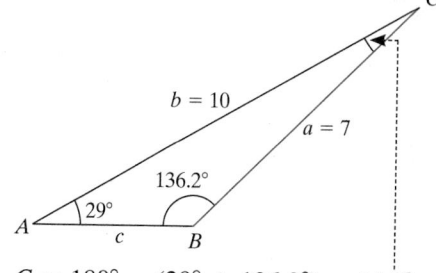

Avoiding Mistakes

After solving a triangle, confirm that the longest side is opposite the largest angle and that the shortest side is opposite the smallest angle. To further confirm your results, you can double check that the three ratios $a/\sin A$, $b/\sin B$, and $c/\sin C$ are the same. For example, for Triangle 1 in Example 5, $7/\sin 29° \approx 14.4$, $10/\sin 43.8° \approx 14.4$, and $13.8/\sin 107.2° \approx 14.4$.

$C \approx 180° - (29° + 43.8°) \approx 107.2°$ $C \approx 180° - (29° + 136.2°) \approx 14.8°$

$$\frac{a}{\sin A} = \frac{c}{\sin C} \qquad\qquad \frac{a}{\sin A} = \frac{c}{\sin C}$$

$$\frac{7}{\sin 29°} = \frac{c}{\sin 107.2°} \qquad\qquad \frac{7}{\sin 29°} = \frac{c}{\sin 14.8°}$$

$$c = \frac{7 \sin 107.2°}{\sin 29°} \approx 13.8 \qquad c = \frac{7 \sin 14.8°}{\sin 29°} \approx 3.7$$

In summary we have
Triangle 1: $A \approx 29°$, $B \approx 43.8°$, $C \approx 107.2°$, $a = 7$, $b = 10$, $c \approx 13.8$
Triangle 2: $A \approx 29°$, $B \approx 136.2°$, $C \approx 14.8°$, $a = 7$, $b = 10$, $c \approx 3.7$

Answer
5. Triangle 1: $A = 40°$, $B \approx 46.9°$, $C \approx 93.1°$, $a = 22$, $b = 25$, $c \approx 34.2$;
Triangle 2: $A = 40°$, $B \approx 133.1°$, $C \approx 6.9°$, $a = 22$, $b = 25$, $c \approx 4.1$

Skill Practice 5   Solve $\triangle ABC$ with $A = 40°$, $a = 22$, and $b = 25$. Round the measures of the unknown angles and the third side to 1 decimal place.

3. Compute the Area of a Triangle Given SAS

The approach we used to prove the law of sines can be used to find a formula for the area of a triangle given the lengths of two sides and the angle between them.

In Figure 6-15, we have the relationship

$$\sin A = \frac{h}{b} \Rightarrow h = b \sin A$$

The area of $\triangle ABC$ is $\frac{1}{2}(\text{base})(\text{height}) = \frac{1}{2}cb\sin A$.

The area is one-half the product of the lengths of the two sides times the sine of the angle between them.

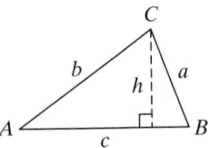

Figure 6-15

Area of a Triangle Given Two Sides and the Angle Between Them

For a triangle $\triangle ABC$, with sides of lengths a, b, and c across from angles A, B, and C, respectively, the area of the triangle is given by

$$\text{Area} = \frac{1}{2}bc\sin A \quad \text{or} \quad \text{Area} = \frac{1}{2}ab\sin C \quad \text{or} \quad \text{Area} = \frac{1}{2}ac\sin B$$

The proofs of the second and third relationships are found in a similar way by drawing the altitude from vertex B and vertex A, respectively.

Classroom Example: p. 652
Exercise 30

EXAMPLE 6 **Finding the Area of an Oblique Triangle (SAS)**

A triangle has sides of 5 cm and 8 cm, and the angle between them is 121°. Find the area of the triangle. Round to 1 decimal place.

Solution:

Let A denote the vertex of the given angle, and let sides b and c represent the lengths of the two given sides (Figure 6-16).

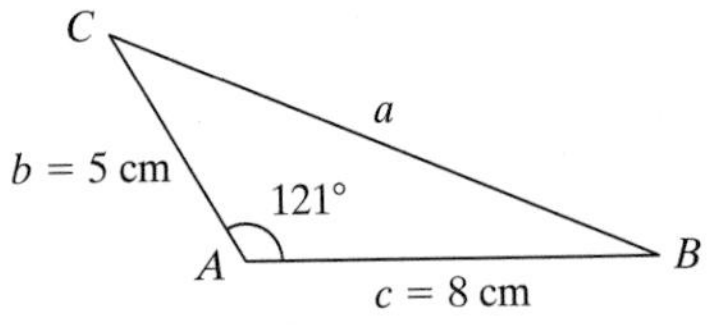

$$\text{Area} = \frac{1}{2}bc\sin A = \frac{1}{2}(5 \text{ cm})(8 \text{ cm})\sin 121°$$
$$\approx 17.1 \text{ cm}^2$$

The area is approximately 17.1 cm².

Figure 6-16

Skill Practice 6 A triangle has sides of 12 m and 6 m, and the angle between them is 76°. Find the area of the triangle. Round to 1 decimal place.

4. Apply the Law of Sines

The law of sines can be used in applications in which the value of an angle in a triangle and the length of the side opposite it are known, along with one other part of the triangle.

Answer

6. 34.9 m²

| **EXAMPLE 7**   Apply the Law of Sines to Find Height |

To estimate the height of a hill, the
angle of elevation is measured at two
points 480 m apart on a direct line to
the hill. The angle of elevation from
the closer point is 42.1°. If the angle
of elevation from the farther point is
32°, find the height of the hill. Round
to the nearest meter.

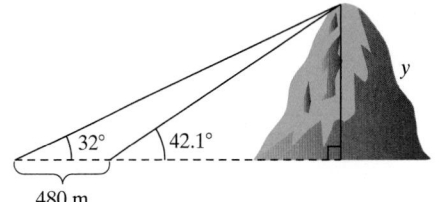

Solution:

A sketch is provided with the relevant angles and
sides labeled (Figure 6-17). $\angle CBA$ is supplementary
to $\angle DBC$. Therefore,

$$\angle CBA = 180° - 42.1° = 137.9°$$

From $\triangle ABC$, $C = 180° - (32° + 137.9°) = 10.1°$.

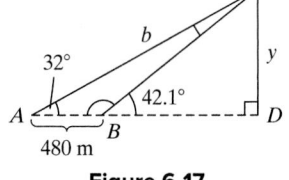

Figure 6-17

Using $\triangle ABC$ and the law of sines,

$$\frac{480}{\sin 10.1°} = \frac{b}{\sin 137.9°}$$

$$b = \frac{480 \sin 137.9°}{\sin 10.1°} \approx 1835 \text{ m}$$

Finally, we can use the right triangle ADC to find the
value of y (Figure 6-18).

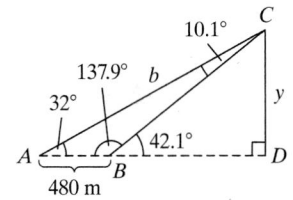

Figure 6-18

$$\frac{y}{b} = \sin 32° = \frac{y}{1835} \Rightarrow y = 1835 \sin 32° \approx 972$$

The height of the hill is approximately 972 m.

Skill Practice 7   To measure the maximum height of the water in a
fountain, a trigonometry student measures the angle of elevation of the
maximum point to be 25°. At a point 30 ft farther away along the same
line of sight, the angle of elevation is 18°. If the eye level of the student
is 5.5 ft, find the maximum height of the water. Round to the nearest tenth
of a foot.

| **EXAMPLE 8**   Apply the Law of Sines to Find Angles |

A 90-ft cell tower is located at the top of a hill.
An observer 200 ft down the hill measures the
angle between the bottom and top of the tower
to be 19.6°. Determine the angle of inclination of
the hill to the nearest tenth of a degree.

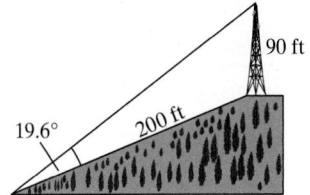

Answer

7. 37.6 ft

Solution:

To begin, we can solve $\triangle ABC$ for angles B and C (Figure 6-19).

$$\frac{\sin B}{200} = \frac{\sin 19.6°}{90} \Rightarrow \sin B = \frac{200\sin 19.6°}{90}$$

Since the observer is looking uphill, angle C must be greater than 90°. Therefore, angle B must be acute.

$$B = \sin^{-1}\left(\frac{200\sin 19.6°}{90}\right) \approx 48.2° \text{ and}$$

$$C \approx 180° - (19.6° + 48.2°) \approx 112.2°$$

Let α represent the angle of inclination of the hill. From Figure 6-20,

$$\beta \approx 180° - 112.2° \approx 67.8° \text{ and}$$

$$\alpha = 90° - \beta \approx 90° - 67.8° \approx 22.2°$$

The angle of the hill is approximately 22.2°.

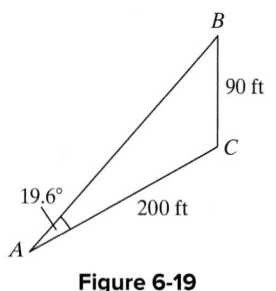

Figure 6-19

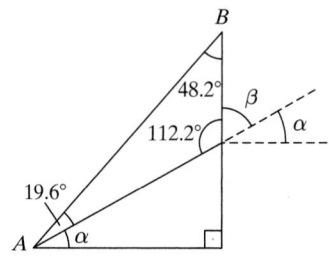

Figure 6-20

Skill Practice 8   An 80-ft flagpole is located at the top of a hill. An observer 220 ft down the hill measures the angle formed by the bottom of the pole to the top of the pole to be 17.7°. Determine the angle of inclination of the hill to the nearest tenth of a degree.

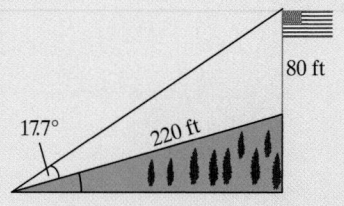

Point of Interest

In addition to the six trigonometric functions, other related functions were used by early mathematicians and astronomers to address specific applications. For example, the *haversine* function is given by haversin $\theta = \sin^2\frac{\theta}{2}$. The haversine function is used in navigation to compute great-circle distances between two points on a sphere based on the latitude and longitude of each point. Furthermore, the *law of haversines* is used to relate the sides and angles of a triangle whose vertices are on a spherical surface rather than a flat surface.

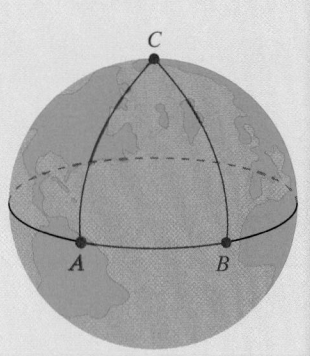

Answer

8. 15.6°

SECTION 6.2   Practice Exercises

Prerequisite Review

R.1. Use a calculator to approximate the value of $\sin^{-1} 0.76$ to the nearest tenth of a degree.   49.5°

R.2. Use a calculator to approximate the degree measure of θ for $\sin\theta = \frac{3}{4}$ and $90° \leq \theta \leq 180°$.   131.4°

Concept Connections

1. To apply the law of sines, an angle must be given and a side (opposite/adjacent) the given angle must be given, along with one other known side or angle.   opposite

2. In which cases can the law of sines be used to solve a triangle? Choose from ASA, SAA, SSA, SSS, SAS.   ASA, SAA, SSA

3. Suppose that ΔABC is a triangle with sides of length a, b, and c opposite angles A, B, and C, respectively. Write the law of sines. $\dfrac{a}{\sin A} = \dfrac{b}{\sin B} = \dfrac{c}{\sin C}$ or $\dfrac{\sin A}{a} = \dfrac{\sin B}{b} = \dfrac{\sin C}{c}$

4. Suppose that the lengths of two sides a and b in a triangle are known along with the measure of the angle C between them. Then the area is given by Area = _____. $\dfrac{1}{2}ab\sin C$

5. Suppose that we know the measure of angle A and the lengths of sides a and b of a triangle. If $a > b$, how many possible triangles can be formed?   one

6. Suppose that we know the measure of angle A and the lengths of sides a and b of a triangle. If $b\sin A < a < b$, how many possible triangles can be formed?   two

Instructor Note:
Exercises with no solution: 23, 24, 58, 62, 65
Exercises with two solutions: 25, 26, 27, 28, 51, 52, 55, 59

Objective 1: Solve a Triangle Using the Law of Sines (SAA or ASA)

Note: Due to round-off error, the answers to exercises involving calculations to determine unknown angles and sides may differ slightly from those given.

For Exercises 7–12, solve the triangle. For the sides, give an expression for the exact value of the length and an approximation to 1 decimal place. (See Examples 1–2)

7.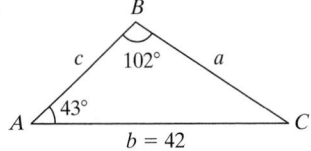

8.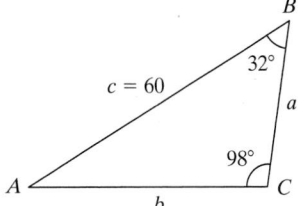

9.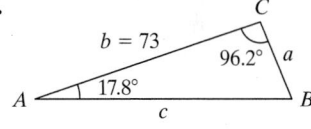

10.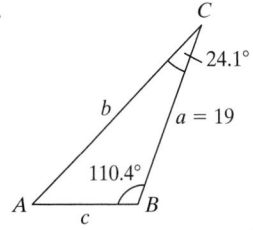

11.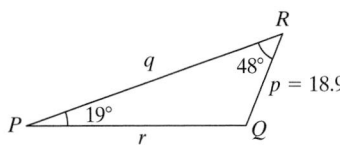

12.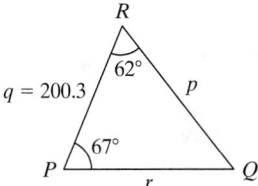

For Exercises 13–20, solve ΔABC subject to the given conditions. Round the lengths of sides and measures of the angles to 1 decimal place if necessary. (See Examples 1–2)

13. $A = 127°$, $B = 34°$, $a = 42$ $C = 19°, b \approx 29.4, c \approx 17.1$

14. $A = 27°$, $B = 71°$, $b = 186$
    $C = 82°, a \approx 89.3, c \approx 194.8$

15. $A = 122.1°$, $B = 24.3°$, $c = 102$
    $C = 33.6°, a \approx 156.1, b \approx 75.8$

16. $A = 10.6°$, $C = 98.8°$, $b = 59$
    $B = 70.6°, a \approx 11.5, c \approx 61.8$

17. $A = 34.7°$, $C = 68.1°$, $c = 43.9$ $B = 77.2°, a \approx 26.9, b \approx 46.1$

18. $A = 112.7°$, $C = 50.2°$, $a = 12.3$
    $B = 17.1°, b \approx 3.9, c \approx 10.2$

19. $B = 28.1°$, $C = 45°$, $a = \dfrac{2}{3}$ $A = 106.9°, b \approx 0.3, c \approx 0.5$

20. $A = 53.8°$, $B = 60°$, $c = \dfrac{17}{9}$ $C = 66.2°, a \approx 1.7, b \approx 1.8$

Objective 2: Solve a Triangle Using the Law of Sines (SSA) Ambiguous Case

For Exercises 21–28, information is given about ΔABC. Determine if the information gives one triangle, two triangles, or no triangle. Solve the resulting triangle(s). Round the lengths of sides and measures of the angles to 1 decimal place if necessary. (See Examples 3–5)

21. $b = 33$, $c = 25$, $B = 38°$
    One triangle; $A \approx 114.2°, C \approx 27.8°, a \approx 48.9$

22. $b = 5$, $c = 12$, $C = 73°$
    One triangle; $A \approx 83.5°, B \approx 23.5°, a \approx 12.5$

23. $a = 185$, $c = 132$, $C = 63°$ No triangle

24. $b = 6$, $c = 12$, $B = 38°$ No triangle

25. $a = 13$, $b = 18$, $A = 45°$ Two triangles; $B \approx 78.3°, C \approx 56.7°,$
    $c \approx 15.4; B \approx 101.7°, C \approx 33.3°, c \approx 10.1$

26. $a = 3$, $b = 1$, $B = 17°$ Two triangles; $A \approx 61.3°,$
    $C \approx 101.7°, c \approx 3.3; A \approx 118.7°, C \approx 44.3°, c \approx 2.4$

27. $a = 132.5$, $b = 108.2$, $B = 13.1°$
    Two triangles; $A \approx 16.1°, C \approx 150.8°, c \approx 232.9$
    $A \approx 163.9°, C \approx 3.0°, c \approx 25.1$

28. $a = 48.8$, $c = 39.9$, $C = 22°$
    Two triangles; $A \approx 27.3°, B \approx 130.7°, b \approx 80.7$
    $A \approx 152.7°, B \approx 5.3°, b \approx 9.8$

Additional answers can be found in the Instructor Answer Appendix.

Objective 3: Compute the Area of a Triangle Given SAS

For Exercises 29–34, find the area of a triangle with the given measurements. Round to 1 decimal place. (See Example 6)

29. $A = 107°$, $b = 17$ ft, $c = 3$ ft   24.4 ft²

30. $A = 143°$, $b = 20$ m, $c = 16$ m   96.3 m²

31. $B = 98.8°$, $a = 2.1$ in., $c = 5.3$ in.   5.5 in.²

32. $B = 2.6°$, $a = 3.5$ cm, $c = 10.8$ cm   0.9 cm²

33. $C = 74.6°$, $a = \dfrac{1}{3}$ mm, $b = \dfrac{2}{3}$ mm   0.1 mm²

34. $C = 125.7°$, $a = \dfrac{17}{8}$ mi, $b = \dfrac{23}{9}$ mi   2.2 mi²

35. The area in the Atlantic Ocean known as the Bermuda Triangle is defined by an imaginary triangle connecting Miami, Florida; San Juan, Puerto Rico; and the island of Bermuda. Measuring on a map, the distance from both Miami to San Juan and from Miami to Bermuda is approximately 1033 mi. Assuming that the angle from Bermuda to Miami to San Juan is approximately 65°, what is the area of the Bermuda Triangle? Round to the nearest square mile.   483,556 mi²

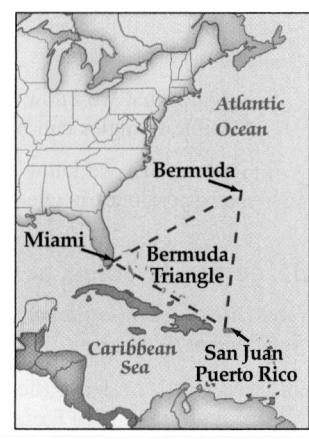

36. The triangular face of a gabled roof measures 33.8 ft on each sloping side with an angle of 134.8° at the top of the roof. What is the area of the face? Round to the nearest square foot.   405 ft²

Objective 4: Apply the Law of Sines

37. Two fire lookout towers at points A and B are 2 mi apart. Find the distances from points A and B to a fire at point C if $\angle ABC$ is 78°48′ and $\angle BAC$ is 84°36′. Round to the nearest tenth of a mile. $AC ≈ 6.9$ mi, $BC ≈ 7.0$ mi

38. A helicopter is on a path directly overhead line AB when it is simultaneously observed from locations A and B separated by 900 ft. The angle of elevation from A is 42°30′ and the angle of elevation from B is 30°12′.

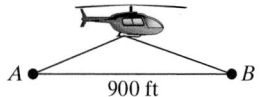

a. What is the distance from each location to the helicopter? Round to the nearest foot.

b. How high is the helicopter from the ground at the moment of observation? Round to the nearest foot.   320 ft

39. A surveyor wants to measure the distance across a lake from a point A due north to a point B. He runs a 1500-ft line from point A to point C in the direction N52°E. He then runs a second line from C to B in the direction N41°24′W.

a. To the nearest foot, how long is the second line?   1787 ft

b. To the nearest foot, what is the distance between A and B?   2264 ft

40. A catamaran sails due north parallel to a straight shoreline at a constant speed of 8 mph. The captain records a bearing of N30°E to a lighthouse on shore. Then, 1.5 hr later, the captain records the bearing to the lighthouse as S42°E. To the nearest tenth of a mile, what is the distance from the catamaran to the lighthouse at the time the second bearing is recorded?   6.3 mi

41. From a point along a straight road, the angle of elevation to the top of a hill is 33°. From 300 ft farther down the road, the angle of elevation to the top of the hill is 24°. How high is the hill? Round to the nearest foot. **(See Example 7)**   425 ft

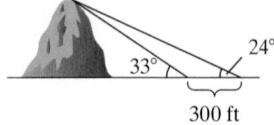

42. An observer on the ground sites a plane at an angle of elevation of 41.2°. At the same time, a second observer 3000 m farther away along the same line of site measures the angle of elevation as 35°. How high is the plane? Round to the nearest 100 m.   10,500 m

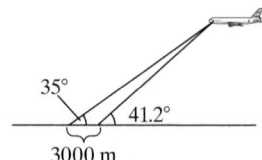

43. A wire is fastened to a point *T* on a tree and to point *A* located 8.8 ft from the base of the tree along level ground (see figure). The angle that the wire makes with level ground is 44°, and the tree leans 12° from vertical away from point *A*. How high off the ground is the point where the wire is fastened to the tree? Round to the nearest tenth of a foot.   10.7 ft

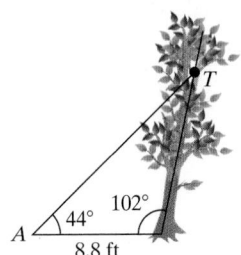

44. The Leaning Tower of Suurhusen is a medieval steeple in Suurhasen, Germany. The tower leans at an angle of 5.1939° from the vertical. (In comparison, the Leaning Tower of Pisa leans at an angle of 3.97°.) The angle of elevation to the top of the tower is 44.2° when measured 100 ft from the base of the tower. Find the distance *h* from the base of the tower to the top of the tower. Round to the nearest tenth of a foot.   89.7 ft

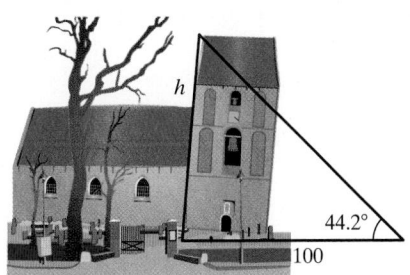

45. A 25-ft flag pole is oriented vertically at the top of a hill. An observer standing 100 ft down hill measures the angle formed between the top and bottom of the pole as 12.2°. To the nearest degree, determine the angle of inclination of the hill. (**See Example 8**)   20

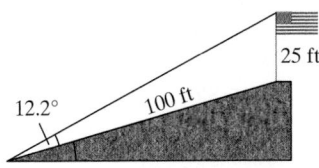

46. A mountain cabin is built on the side of a hill with a porch extending over a portion of the downhill slope. An observer 20 ft downhill from one of the 8-ft vertical support beams measures the angle formed from the top of the beam to the bottom as 10°. Find the angle of inclination of the hill to the nearest tenth of a degree.   54.3°

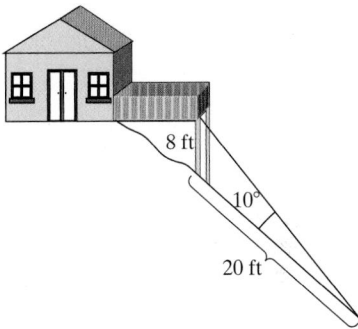

47. Two residential buildings are to be constructed with a grassy recreational area between them. The taller building is 700 ft high. From the roof of the shorter building, the angle of elevation to the top of the taller building is 78° and the angle of depression to the base of the taller building is 62°.

   a. How tall is the shorter building? Round to the nearest foot.   200 ft

   b. What is the distance between the buildings? Round to the nearest foot.   106 ft

48. A hiker wants to estimate the height of a mountain before attempting a climb to the top. Her first measurement shows an angle of elevation to the top of the mountain as 63.4°. Her second measurement, taken 950 ft closer to the base of the mountain, yields an angle of elevation of 75.3°. From these measurements, estimate the height of the mountain. Round to the nearest hundred feet.   4000 ft

49. A surveyor fixes a 420-ft baseline between points *A* and *B,* where *B* is due east of *A*. The bearing from *A* to a third point *P* is N52°E and the bearing from *B* to *P* is N27°W. Find the perimeter and area of the triangular plot of land formed by *A*, *B*, and *P*. Round the perimeter to the nearest foot and the area to the nearest hundred square feet.   Perimeter: 1065 ft; Area: 49,300 ft²

50. A surveyor fixes a 200-ft baseline between points *A* and *B,* where *B* is due east of *A*. The bearing from *A* to a third point *P* is S37°W and the bearing from *B* to *P* is S62°W. Find the perimeter and area of the triangular plot of land formed by *A*, *B*, and *P*. Round the perimeter to the nearest foot and the area to the nearest 100 square feet.   Perimeter: 800 ft; Area: 17,700 ft²

51. The connector rod from the piston to the crankshaft in a certain 2.0-L engine is 6.4 in. The radius of the crank circle is 2.8 in. If the angle made by the connector rod with the horizontal at the wrist pin P is 20°, how far is the wrist pin from the center C of the crankshaft? Round to the nearest tenth of an inch.   7.8 in. or 4.3 in.

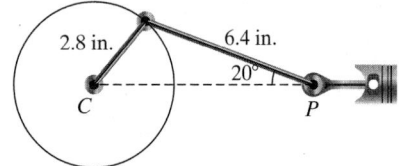

52. Two planets follow a circular orbit around a central star in the same plane. The distance between the star at point S and one planet at point A is 135 million miles. The distance between the star and the other planet at point B is 100 million miles. If an observer on the first planet at point A sights the second planet such that $\angle SAB = 42°$, find the distance between the planets. Round to the nearest million miles.   57 million miles or 143 million miles

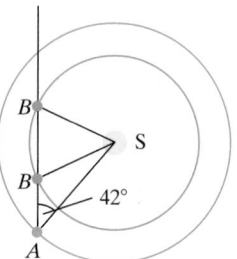

Mixed Exercises

For Exercises 53–66, solve $\triangle ABC$. Round the lengths of sides and measures of the angles to 1 decimal place if necessary.

53. $A = 117°$, $B = 32°$, $b = 8.2$ $C = 31°, a \approx 13.8, c \approx 8.0$

54. $a = 42$, $c = 60$, $C = 17°$
$A \approx 11.8°, B \approx 151.2°, b \approx 98.9$

55. $b = 6$, $c = 8$, $B = 24°$ $A \approx 123.2°, C \approx 32.8°, a \approx 12.3;$
$A \approx 8.8°, C \approx 147.2°, a \approx 2.3$

56. $A = 103.1°$, $B = 13°$, $c = 3.8$
$C = 63.9°, a \approx 4.1, b \approx 1.0$

57. $B = 13.3°$, $C = 68.4°$, $c = 12$ $A = 98.3°, a \approx 12.8, b \approx 3.0$

58. $a = 11$, $c = 3$, $C = 142°$ No triangle

59. $a = 118$, $c = 112$, $C = 24.6°$ $A \approx 26.0°, B \approx 129.4°, b \approx 207.9;$
$A \approx 154.0°, B \approx 1.4°, b \approx 6.6$

60. $B = 47.8°$, $C = 99.3°$, $a = 183$
$A = 32.9°, b \approx 249.6, c \approx 332.5$

61. $a = 16$, $b = 15.1$, $A = 113°$ $B \approx 60.3°, C \approx 6.7°, c \approx 2.0$

62. $b = 15.1$, $c = 18.6$, $B = 113°$ No triangle

63. $A = 4.6°$, $C = 1.2°$, $a = 23$ $B = 174.2°, b \approx 29.0, c \approx 6.0$

64. $B = 12°$, $C = 136°$, $b = 800$
$A = 32°, a \approx 2039.0, c \approx 2672.9$

65. $a = 325$, $c = 221$, $C = 78.8°$ No triangle

66. $A = 153°$, $C = 2°$, $b = 2$ $B = 25°, a \approx 2.1, c \approx 0.2$

67. After a hurricane, a homeowner examines trees on his property for damage. He thinks a 20-ft palm tree is leaning slightly from its original vertical position. From a point 23 ft away, he measures the angle of elevation to the top of the palm tree as 43°. Is the palm tree leaning? If so, by how many degrees from the vertical? Round to the nearest tenth of a degree.   Yes, it is leaning by 4.7° from the vertical.

68. A company manufactures pennants in the shape of an isosceles triangle. The long sides of each triangle are 18 in., and the angle between the long sides is 40°. The weatherproof fabric from which the pennants are made costs $6.95/yd^2. How much will it cost the company to make 10,000 pennants? Round to the nearest dollar. $5584

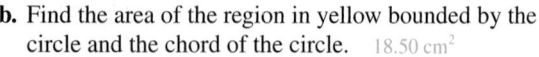

69. Consider a sector of a circle of radius 6 cm with central angle 112°.

a. Find the area of the triangular region shaded in green. Round to the nearest hundredth of a square centimeter.   16.69 cm^2

b. Find the area of the region in yellow bounded by the circle and the chord of the circle.   18.50 cm^2

70. Refer to the figure.

a. Write expressions for the exact values of a, b, and c.

b. Approximate the values of a, b, and c to the nearest tenth of a centimeter.

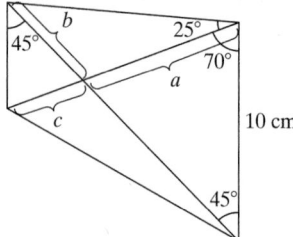

71. Refer to the figure.

   a. Write expressions for the exact values of a, b, and c.

   b. Approximate the values of a, b, and c to the nearest tenth of an inch.

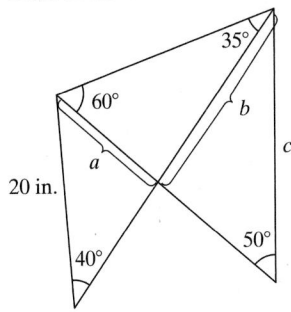

72. Given $\triangle ABC$ with sides of lengths a, b, and c opposite angles A, B, and C, respectively, use the law of sines to show that

$$\frac{a+b}{b} = \frac{\sin A + \sin B}{\sin B}.$$

$$\frac{a}{\sin A} = \frac{b}{\sin B}$$
$$a\sin B = b\sin A$$
$$a\sin B + b\sin B = b\sin A + b\sin B$$
$$(a+b)\sin B = (\sin A + \sin B)b$$
$$\frac{a+b}{b} = \frac{\sin A + \sin B}{\sin B}$$

Write About It

73. Explain why the law of sines can be used to solve a triangle if the measure of an angle is given along with the length of its opposite side and the measure of one other side or angle.   The measure of an angle and the length of the side opposite it form a complete ratio. Then using the third piece of information (the measure of another angle or side), a proportion can be set up containing only one unknown.

74. Given SSA (the *ambiguous case*), outline the four possible scenarios regarding the number of triangles that are possible.

75. List various methods to confirm your answers after solving a triangle using the law of sines.

76. A triangle $\triangle ABC$ has sides a, b, and c across from angles A, B, and C, respectively. Explain the relationship given by Area $= \frac{1}{2}bc\sin A$.   The area of a triangle is one-half the product of the lengths of two known sides and the sine of the angle between them.

Expanding Your Skills

77. From the figure, show that $\sin(45° - \alpha) = \frac{2}{3}\sin 85°$.

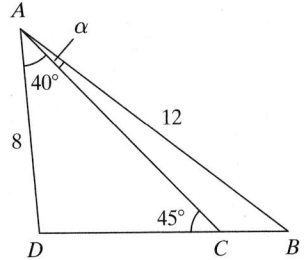

78. Use the law of sines in the form $\dfrac{\sin A}{\sin B} = \dfrac{a}{b}$ (1)

to prove the law of tangents: $\dfrac{a-b}{a+b} = \dfrac{\tan\left(\frac{A-B}{2}\right)}{\tan\left(\frac{A+B}{2}\right)}$.

[*Hint*: Subtract 1 from both sides of equation (1) and simplify to get a second equation. Then add 1 to both sides of (1) and simplify to get a third equation. Then divide the results.]

79. Use the relationships $\dfrac{a}{c} = \dfrac{\sin A}{\sin C}$ and $\dfrac{b}{c} = \dfrac{\sin B}{\sin C}$ from the law of sines to show that $\dfrac{a+b}{c} = \dfrac{\cos\frac{A-B}{2}}{\sin\frac{C}{2}}$. This is called a Mollweide equation, named after German mathematician and astronomer Karl Mollweide (1774–1825). Notice that the equation uses all six parts of a triangle, and for this reason, is sometimes used to check the solution to a triangle.

80. Given $\triangle ABC$ circumscribed by a circle, show that the diameter d of the circle is $d = \dfrac{b}{\sin B}$. (*Hint*: Draw two radii from the center of the circle to points A and C. Use a theorem from geometry relating the measure of an inscribed angle of a circle to the corresponding central angle.) Note that it follows from the law of sines that $d = \dfrac{a}{\sin A} = \dfrac{b}{\sin B} = \dfrac{c}{\sin C}$.

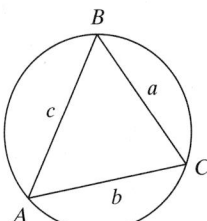

SECTION 6.3   The Law of Cosines

1. Solve a Triangle Using the Law of Cosines (SAS)

Given the measure of an angle in a triangle, the side opposite the angle, and one other part of the triangle, we can use the law of sines to solve the triangle. The law of sines applies in such a case because the known angle and opposite side form a complete ratio such as $\dfrac{a}{\sin A}$. We now look at the cases in which a complete ratio is not given.

- Two sides and the angle between them are known (SAS).
- Three sides are known (SSS).

In these cases, we use the law of cosines.

Law of Cosines

If $\triangle ABC$ has sides of lengths a, b, and c opposite vertices A, B, and C, respectively, then

Law of Cosines	Alternative Form
(1) $a^2 = b^2 + c^2 - 2bc\cos A$	$\cos A = \dfrac{b^2 + c^2 - a^2}{2bc}$
(2) $b^2 = a^2 + c^2 - 2ac\cos B$	$\cos B = \dfrac{a^2 + c^2 - b^2}{2ac}$
(3) $c^2 = a^2 + b^2 - 2ab\cos C$	$\cos C = \dfrac{a^2 + b^2 - c^2}{2ab}$

The formulas for the law of cosines looks daunting, but they all share the same pattern.

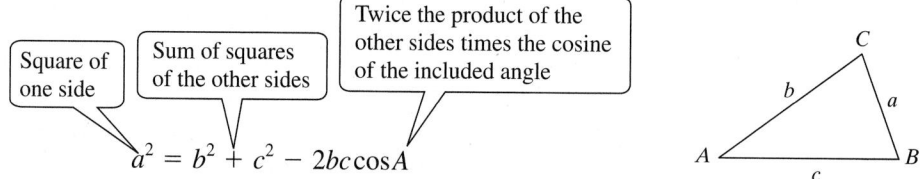

The alternative forms of equations (1)–(3) have the cosine factor isolated for the purpose of computing the measure of an unknown angle in a triangle given the lengths of the sides.

We will prove the first of the three formulas given by the law of cosines. Formulas (2) and (3) are proved likewise.

First place $\triangle ABC$ in a rectangular coordinate system with vertex A at the origin, side c on the positive x-axis, and vertex B at $(c, 0)$. Let (x, y) represent point C (Figure 6-21).

From the right triangle on the left,

$$\cos A = \frac{x}{b} \Rightarrow x = b\cos A$$

$$\sin A = \frac{y}{b} \Rightarrow y = b\sin A$$

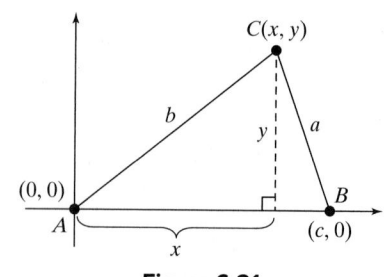

Figure 6-21

Therefore, the ordered pair representing point C is $(b\cos A, b\sin A)$. The distance from C to B is the length of side a.

$a = \sqrt{(x - c)^2 + (y - 0)^2}$	Apply the distance formula.
$a^2 = (x - c)^2 + y^2$	Square both sides.
$a^2 = (b\cos A - c)^2 + (b\sin A)^2$	Substitute $x = b\cos A$ and $y = b\sin A$.
$a^2 = b^2\cos^2 A - 2bc\cos A + c^2 + b^2\sin^2 A$	Square each term.
$a^2 = b^2\cos^2 A + b^2\sin^2 A - 2bc\cos A + c^2$	Group terms with b^2.
$a^2 = b^2(\cos^2 A + \sin^2 A) - 2bc\cos A + c^2$	Factor out b^2.
$a^2 = b^2(1) - 2bc\cos A + c^2$	Replace $\sin^2 A + \cos^2 A$ with 1.
$a^2 = b^2 + c^2 - 2bc\cos A$	Law of cosines

In Example 1, we solve an oblique triangle given two sides and the angle between them (SAS). We use the following guidelines.

Guidelines for Solving an Oblique Triangle Given SAS

1. Find the length of the side opposite the known angle by using the law of cosines.
2. Use the law of sines to find the measure of the angle opposite the shorter of the two given sides. This angle will always be acute. Alternatively, use the law of cosines (alternative form) to find either remaining angle.
3. Find the measure of the third angle by subtracting the sum of the measures of the other two angles from 180°.

The rationale for step 2 is as follows.

- The angles opposite the shorter two sides in a triangle are always acute angles. Therefore, the measure of the smaller of the two given angles must be acute. Applying the law of sines to find this angle will not involve the ambiguous case.
- The law of cosines can be used to find the measure of *either* remaining angle because the inverse cosine will return an acute angle or obtuse angle depending on whether the argument is positive or negative.

Classroom Example: p. 662
Exercise 12

EXAMPLE 1   Solving an Oblique Triangle Given SAS

Solve the triangle with $A = 52°$, $b = 22$, and $c = 35$. Round the measures of the unknown side and angles to 1 decimal place.

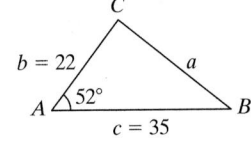

Solution:

We begin by finding the length of the side opposite the known angle.

$a^2 = b^2 + c^2 - 2bc\cos A$	Apply the law of cosines.
$a^2 = (22)^2 + (35)^2 - 2(22)(35)\cos 52°$	Substitute $b = 22$, $c = 35$, $A = 52°$.
$a = \sqrt{(22)^2 + (35)^2 - 2(22)(35)\cos 52°}$	Take the principal square root.
≈ 27.6	

Of the two remaining unknown angles, angle B is opposite the shortest side. Using the law of sines,

$$\frac{a}{\sin A} = \frac{b}{\sin B} \Rightarrow \frac{27.6}{\sin 52°} = \frac{22}{\sin B}$$

$$\sin B \approx \frac{22 \sin 52°}{27.6}$$

$$B = \sin^{-1}\left(\frac{22 \sin 52°}{27.6}\right) \approx 38.9°$$

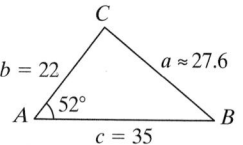

With the measures of angle A and B known,

$$C = 180° - (A + B) \approx 180° - (52° + 38.9°) \approx 89.1°.$$

In summary $A = 52°$, $B \approx 38.9°$, $C \approx 89.1°$, $a \approx 27.6$, $b = 22$, and $c = 35$.

Skill Practice 1 Solve $\triangle ABC$ with $B = 61°$, $c = 19$, and $a = 28$. Round the measures of the unknown side and angles to 1 decimal place.

TIP In Example 1, we used a calculator to approximate the length of side a. Notice that no rounding was done until all calculations were completed. Then we used the value of side a to find the measure of angle B. The level of rounding used for side a can affect the result of subsequent calculations. For example,

$$B = \sin^{-1}\left(\frac{22 \sin 52°}{27.6}\right) \approx 38.91188771 \text{ and } B = \sin^{-1}\left(\frac{22 \sin 52°}{27.58407744}\right) \approx 38.93859082°$$

The result of your calculations may differ slightly from those given in the text. This may be due to round-off error or the order in which you solve for the sides and angles in a triangle.

2. Solve a Triangle Using the Law of Cosines (SSS)

In Example 2, we apply the law of cosines to find the measures of the angles of a triangle where the lengths of the three sides are known. We recommend the following guidelines.

Guidelines for Solving an Oblique Triangle Given SSS

1. Use the alternative formulas for the law of cosines to find the largest angle of the triangle (angle opposite the longest side).
2. Find either of the remaining two angles using either the law of sines or law of cosines. Since the largest angle was found in step 1, the two remaining angles are guaranteed to be acute.

   • The law of sines may be easier for you to remember and apply, but the drawback is that you will have to use a rounded value from a prior calculation.

   • The benefit of using the law of cosines is that it involves using the lengths of the three given sides, rather than a rounded value from a prior calculation.

3. Find the measure of the third angle by subtracting the sum of the measures of the other two angles from 180°.

Avoiding Mistakes
You can also use the law of cosines to find the measure of the third angle as a check.

Answer
1. $A = 77.5°$, $B = 61°$, $C = 41.5°$, $a = 28$, $b \approx 25.1$, $c = 19$

Classroom Example: p. 662
Exercise 14

EXAMPLE 2 Solving an Oblique Triangle Given SSS

Solve the triangle with $a = 24.6$, $b = 16.2$, and $c = 33.5$. Round the measures of the angles to the nearest tenth of a degree.

Solution:

We begin by finding angle C because it is opposite the longest side.

$$\cos C = \frac{a^2 + b^2 - c^2}{2ab}$$

Apply the alternative form of the law of cosines.

$$\cos C = \frac{(24.6)^2 + (16.2)^2 - (33.5)^2}{2(24.6)(16.2)}$$

Substitute $a = 24.6$, $b = 16.2$, $c = 33.5$.

$$C = \cos^{-1}\left[\frac{(24.6)^2 + (16.2)^2 - (33.5)^2}{2(24.6)(16.2)}\right]$$

Take the inverse cosine.

$$\approx 108.6°$$

Now solve for either of the other two angles. We will solve for angle B.

TIP In Example 2, using the law of sines to find the second angle requires that we use the rounded value 108.6° from the previous calculation. Using the law of cosines does not require the use of a rounded value.

Using the law of cosines:	Using the law of sines:
$\cos B = \dfrac{a^2 + c^2 - b^2}{2ac}$	$\dfrac{c}{\sin C} = \dfrac{b}{\sin B}$
$\cos B = \dfrac{(24.6)^2 + (33.5)^2 - (16.2)^2}{2(24.6)(33.5)}$	$\dfrac{33.5}{\sin 108.6°} = \dfrac{16.2}{\sin B}$
$B = \cos^{-1}\left[\dfrac{(24.6)^2 + (33.5)^2 - (16.2)^2}{2(24.6)(33.5)}\right]$	$\sin B = \dfrac{16.2\sin 108.6°}{33.5}$
$\approx 27.3°$	$B = \sin^{-1}\left(\dfrac{16.2\sin 108.6°}{33.5}\right)$
	$\approx 27.3°$

With the measures of angle B and C known,

$$A = 180° - (B + C) \approx 180° - (27.3° + 108.6°) \approx 44.1°$$

In summary $A \approx 44.1°$, $B \approx 27.3°$, $C \approx 108.6°$, $a = 24.6$, $b = 16.2$, and $c = 33.5$.

Skill Practice 2 Solve $\triangle ABC$ with $a = 62$, $b = 48$, and $c = 41$. Round the measures of the angles to the nearest tenth of a degree.

3. Apply the Law of Cosines

Classroom Example: p. 662
Exercise 24

EXAMPLE 3 Applying the Law of Cosines

A Coast Guard ship on a routine training run leaves its base at 16 knots (nautical miles per hour) on a course of N82°E. Three hours later, a distress call comes in from a cargo ship located 180 nmi (nautical miles) from base at a bearing of N48°E.

a. How far is the Coast Guard ship from the cargo ship? Round to the nearest tenth of a nautical mile.

b. What bearing should the Coast Guard ship follow to reach the cargo ship? Round to the nearest tenth of a degree.

Answer

2. $A \approx 87.9°$, $B \approx 50.7°$, $C \approx 41.4°$, $a = 62$, $b = 48$, $c = 41$

Solution:

a. Let A represent the location of the base, let C represent the location of the Coast Guard ship after 3 hr, and let B be the location of the cargo ship (Figure 6-22). The distance c between the base and the cargo ship is given as 180 nmi. The distance b between the base and the Coast Guard ship is

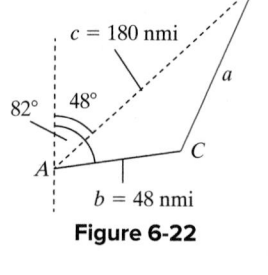

Figure 6-22

$$b = (16 \text{ knots})(3 \text{ hr}) = 48 \text{ nmi}$$

From Figure 6-22, $\angle BAC = 82° - 48° = 34°$.

In $\triangle ABC$ we know the lengths of sides b and c and the angle between them, so we can apply the law of cosines SAS case (Figure 6-23). To find the length of side a, we have

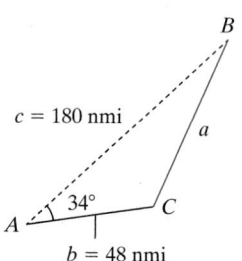

Figure 6-23

$$a^2 = b^2 + c^2 - 2bc\cos A$$

$$a^2 = (48)^2 + (180)^2 - 2(48)(180)\cos 34°$$

$$a = \sqrt{(48)^2 + (180)^2 - 2(48)(180)\cos 34°}$$

$$\approx 142.8 \text{ nmi}$$

The Coast Guard ship is approximately 142.8 nmi from the cargo ship.

b. To find the bearing α from the Coast Guard ship to the cargo ship, we will first find the measure of angle C (Figure 6-24).

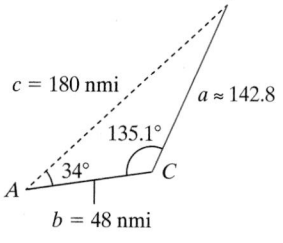

Figure 6-24

$$\cos C = \frac{a^2 + b^2 - c^2}{2ab} = \frac{(142.8)^2 + (48)^2 - (180)^2}{2(142.8)(48)}$$

$$C = \cos^{-1}\left[\frac{(142.8)^2 + (48)^2 - (180)^2}{2(142.8)(48)}\right]$$

$$\approx 135.1°$$

From Figure 6-25,

$$\alpha = C - 98°$$

$$\alpha \approx 135.1° - 98° \approx 37.1°$$

The bearing from the Coast Guard ship to the cargo ship is N37.1°E.

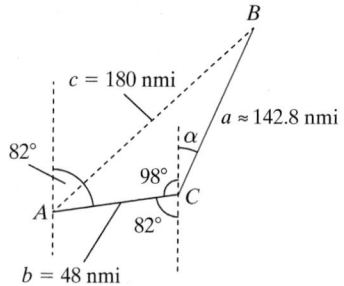

Figure 6-25

Skill Practice 3   Repeat Example 3 with the initial bearing of the Coast Guard ship N17°W at a speed of 18 mph for 3 hr.

Answers
3. a. 164.6 nmi   **b.** N65.3°E

4. Use Heron's Formula to Compute the Area of a Triangle

Suppose that the lengths of three sides of a triangle are known (SSS). We can use the law of cosines to find the measures of the three angles. Then from Section 6.2, we can use any of the expressions $\frac{1}{2}ab\sin C$, $\frac{1}{2}bc\sin A$, or $\frac{1}{2}ac\sin B$ to find the area of $\triangle ABC$. The question then arises, is there a more direct route to finding the area? The answer is yes. We can use Heron's formula, named in honor of its discoverer, Greek mathematician Heron of Alexandria (c. A.D. 62), to find the area of a triangle given the lengths of its sides.

> **Heron's Formula for the Area of a Triangle**
>
> Given a triangle with sides of lengths a, b, and c, the area is given by
>
> $$\text{Area} = \sqrt{s(s-a)(s-b)(s-c)}$$
>
> where s is the semi-perimeter: $s = \frac{1}{2}(a+b+c)$.

Not surprisingly, the proof of Heron's formula uses the law of cosines as outlined in Exercise 60.

Classroom Example: p. 664
Exercise 36

> **EXAMPLE 4** | **Finding the Area of a Triangle with Heron's Formula**
>
> Find the area of a triangle with sides of length $a = 12.2$ cm, $b = 16$ cm, and $c = 9.8$ cm. Round to the nearest tenth of a square centimeter.
>
> **Solution:**
>
> The semi-perimeter is given by:
>
> $$s = \frac{1}{2}(a+b+c) = \frac{1}{2}(12.2+16+9.8) = 19$$
>
> The area is given by:
>
> $$\text{Area} = \sqrt{s(s-a)(s-b)(s-c)}$$
> $$= \sqrt{19(19-12.2)(19-16)(19-9.8)}$$
> $$\approx 59.7$$
>
> The area is approximately 59.7 cm^2.

> **Skill Practice 4**   Find the area of a triangle with sides of length $a = 15.4$ m, $b = 22.6$ m, and $c = 26$ m. Round to the nearest square meter.

Answer

4. 173 m^2

SECTION 6.3 Practice Exercises

Prerequisite Review

R.1. Find the exact distance between the points $(-7, 2)$ and $(-6, -5)$. $5\sqrt{2}$

R.2. Determine if the points $(-3, 7)$, $(-1, 5)$ and $(-7, 3)$ form the vertices of a right triangle. Yes

R.3. Evaluate $\cos^{-1}\left(\dfrac{5}{8}\right)$ in degrees to 2 decimal places. 51.32°

Concept Connections

1. In which situation(s) can the law of cosines be used to solve for the remaining angles and sides of an oblique triangle? Choose from AAS, SAS, ASS, SSS. SAS, SSS

2. Given $\triangle ABC$ with sides a, b, and c opposite vertices A, B, and C, write an expression for a^2 in terms of the lengths of the other sides and opposite angle. $a^2 = b^2 + c^2 - 2bc\cos A$

3. Given $\triangle ABC$ with sides a, b, and c opposite vertices A, B, and C, write an expression for $\cos B$ in terms of the lengths of the sides of the triangle. $\cos B = \dfrac{a^2 + c^2 - b^2}{2ac}$

4. Given a triangle of sides of lengths a, b, and c, Heron's formula gives the area of the triangle as _____, where s represents half the perimeter of the triangle. That is $s = $ _____. $\text{Area} = \sqrt{s(s-a)(s-b)(s-c)}$; $\dfrac{1}{2}(a+b+c)$

Objectives 1 and 2: Solve Triangles Using the Law of Cosines

For Exercises 5–22, solve $\triangle ABC$ subject to the given conditions if possible. Round the lengths of sides and measures of the angles (in degrees) to 1 decimal place if necessary. (See Examples 1–2)

5.
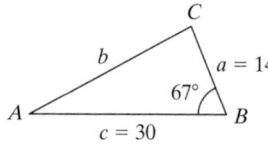
$b \approx 27.7$, $A \approx 27.7°$, $C \approx 85.3°$

6.
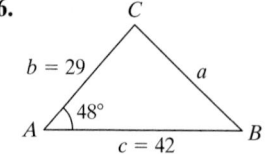
$a \approx 31.2$, $B \approx 43.6°$, $C \approx 88.4°$

7.
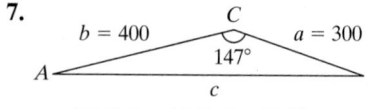
$c \approx 671.8$, $A \approx 14.1°$, $B \approx 18.9°$

8.
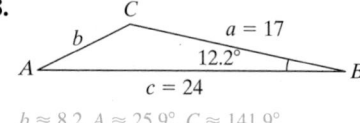
$b \approx 8.2$, $A \approx 25.9°$, $C \approx 141.9°$

9.
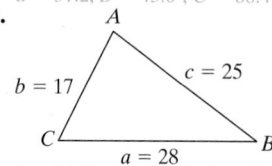
$A \approx 81.2°$, $B \approx 36.9°$, $C \approx 61.9°$

10.
$A \approx 41.8°$, $B \approx 22.8°$, $C \approx 115.4°$

11. $a = 28.3$, $c = 17.4$, $B = 11.3°$
$A \approx 151.8°$, $C \approx 16.9°$, $b \approx 11.7$

12. $b = 89.2$, $c = 23.1$, $A = 108°$
$B \approx 59.2°$, $C \approx 12.8°$, $a \approx 98.8$

13. $a = 15$, $b = 12$, $c = 15$
$A \approx 66.4°$, $B \approx 47.2°$, $C \approx 66.4°$

14. $a = 25$, $b = 30$, $c = 35$
$A \approx 44.4°$, $B \approx 57.1°$, $C \approx 78.5°$

15. $a = 27$, $c = 26$, $B = 67.8°$
$A \approx 57.7°$, $C \approx 54.5°$, $b \approx 29.6$

16. $a = 40.5$, $b = 38.1$, $C = 73.2°$
$A \approx 55.8°$, $B \approx 51.0°$, $c \approx 46.9$

17. $a = 2.3$, $b = 10.8$, $c = 9.7$
$A \approx 11.3°$, $B \approx 112.8°$, $C \approx 55.9°$

18. $a = 500$, $b = 200$, $c = 400$
$A \approx 108.2°$, $B \approx 22.3°$, $C \approx 49.5°$

19. $b = 146.8$, $c = 122.7$, $A = 110.4°$
$a \approx 221.7$, $B \approx 38.4°$, $C \approx 31.2°$

20. $b = 802.5$, $c = 436.1$, $A = 103.7°$
$a \approx 1000$, $B \approx 51.2°$, $C \approx 25.1°$

21. $a = 4.4$, $b = 6.2$, $c = 11.1$
No such triangle is possible.

22. $a = 18$, $b = 32$, $c = 10$
No such triangle is possible.

Objective 3: Apply the Law of Cosines

23. A boat leaves port and follows a course of N77°E at 9 knots for 3 hr and 20 min. Then, the boat changes to a new course of S28°E at 12 knots for 5 hr. **(See Example 3)**

a. How far is the boat from port? 73.7 nmi

b. Suppose that the boat becomes disabled. How long will it take a rescue boat to arrive if the rescue boat leaves from port and travels 18 knots? Round to the nearest minute. 4 hr 6 min

c. What bearing should the rescue boat follow? S51.2°E

24. A fishing boat leaves a marina and follows a course of S62°W at 6 knots for 20 min. Then the boat changes to a new course of S30°W at 4 knots for 1.5 hr.

a. How far is the boat from the marina? 7.8 nmi

b. What course should the boat follow for its return trip to the marina? N37.8°E

25. Two planes leave the same airport. The first plane leaves at 1:00 P.M. and averages 480 mph at a bearing of S62°E. The second plane leaves at 1:15 P.M. and averages 410 mph at a bearing of N12°W.

   a. How far apart are the planes at 2:45 P.M.?   1322 mi

   b. What is the bearing from the first plane to the second plane at that time? Round to the nearest degree.
   N41°W

27. A 40-ft boom on a crane is attached to the crane platform at point *D*. A cable is attached to the end of the boom at point *B* and to a 12-ft A-frame at point *A* anchored by the counterweight. If ∠*BDA* is 85.2°, find the length of the cable to the nearest tenth of a foot.   40.8 ft

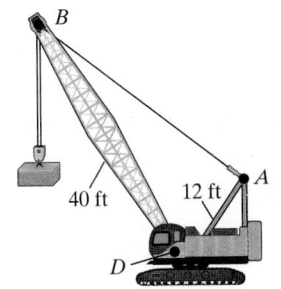

26. Two boats leave a marina at the same time. The first boat travels 6 knots at a bearing of N39°E, and the second boat travels 4 knots at a bearing of S87°W.

   a. How far apart are the boats at the end of 2 hr?
   18.3 nmi

   b. What is the bearing from the first boat to the second boat at that time?   S57.9°W

28. In a public park, a triangular region of grass is bounded by walkways. Sprinkler heads are placed at the vertices of the triangle as shown in the figure. Each sprinkler head can be set to limit its angle of rotation. For each sprinkler head, how large should the angle of rotation be set so as to keep the walkways dry and to conserve water? Round to the nearest degree.
   ∠*A* ≈ 78°, ∠*B* ≈ 41°, ∠*C* ≈ 61°

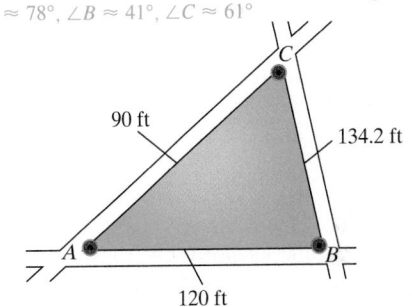

29. A regulation fast-pitch softball diamond for high school competition is a square, 60 ft on a side. The pitcher's mound is colinear with home plate and second base. Furthermore, the distance from the back of home plate to the center of the pitcher's mound is 43 ft. To the nearest tenth of a foot, find the distance between

   a. The pitcher's mound and first base.   42.4 ft

   b. The pitcher's mound and second base.   41.9 ft

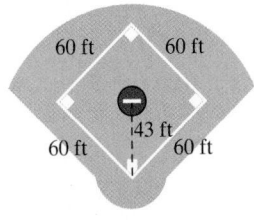

30. The Flatiron Building in New York City is said to have gotten its name because its cross-section resembles the shape of the household appliance of the same name. The perimeter of the building is bounded by Broadway, 5th Avenue, and East 22nd and 23rd Streets. Using side lengths of approximately 87 ft, 190 ft, and 173 ft, determine the angle at the "point" of the Flatiron Building at the corner of 5th Avenue and East 23rd Street. Round to the nearest tenth of a degree.   27.2°

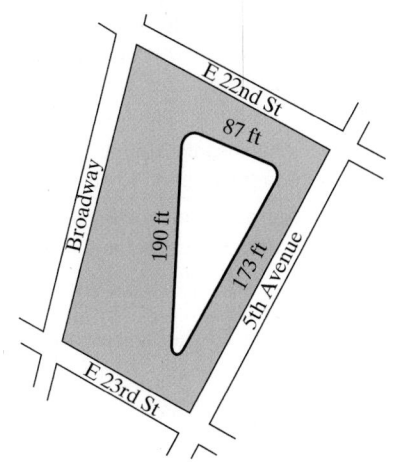

31. The distance between Dallas, Texas, and Atlanta, Georgia, on a map is 11.75 in. The distance from Atlanta to Chicago, Illinois, is 9.5 in., and the distance from Chicago to Dallas is 13.25 in. If the bearing from Dallas to Atlanta is N85°E, find the bearing from Chicago to Dallas. Round to the nearest tenth of a degree. S40.8°W

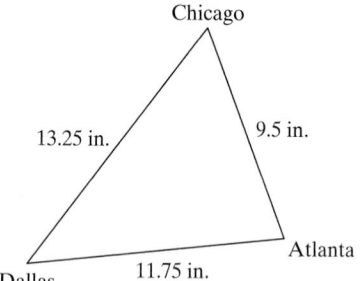

33. A 150-ft tower is anchored on a hill by two guy wires. The angle of elevation of the hill is 8°. Each guy wire extends from the top of the tower to a ground anchor 60 ft from the base of the tower and in line with the tower. Find the length of each guy wire. Round to the nearest foot.

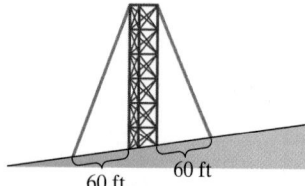

The downhill wire length is 169 ft and the uphill wire length is 154 ft.

32. A bicyclist is at point A on a paved road and must ride to point C on another paved road. The two roads meet at an angle of 38° at point B. The distance from A to B is 18 mi, and the distance from B to C is 12 mi (see the figure). If the bicyclist can ride 22 mph on the paved roads and 6.8 mph off-road, would it be faster for the bicyclist to ride from A to C on the paved roads or to ride a direct line from A to C off-road? Explain.

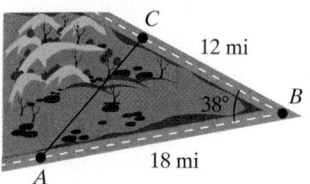

The time required to ride off-road is approximately 1 hr 40 min, whereas the time required to ride on the paved road is 1 hr 22 min. It takes less time to ride on paved roads.

34. A 75-ft tower is located on the side of a hill that is inclined 26° to the horizontal. A cable is attached to the top of the tower and anchored uphill a distance of 35 ft from the base of the tower. Find the length of the cable. Round to the nearest foot. 67 ft

Objective 4: Use Heron's Formula to Compute the Area of a Triangle

In Exercises 35–38, use Heron's formula to find the area of the triangle with sides of the given lengths. Round to the nearest tenth of a square unit. (See Example 4)

35. $a = 13$ in., $b = 7$ in., $c = 8$ in. 24.2 in.²

36. $a = 10$ mi, $b = 17$ mi, $c = 9$ mi 36.0 mi²

37. $a = 18.6$ cm, $b = 12.3$ cm, $c = 25.9$ cm 105.8 cm²

38. $a = 7.9$ yd, $b = 12.1$ yd, $c = 19.3$ yd 24.7 yd²

39. A triangular sail for a Bolger Tortise sailboat (named after boat designer Phil Bolger) has sides of length 8′1″, 10′1″, and 9′10″. What is the sail area? 36.8 ft²

40. Find the area of the grassy region from Exercise 28. Round to the nearest square foot. 5282 ft²

41. A farmer wants to fertilize a triangular field with sides of length 300 yd, 220 yd, and 180 yd. Fertilizer costs $180 per acre (1 acre = 4840 yd²). Furthermore, the time required to fertilizer 1 acre is approximately 2.8 hr with combined labor and equipment costs of $23.00 per hour.

a. To the nearest acre, how big is the field? 4 acres

b. Using the number of acres from part (a), estimate the cost to fertilize the field. Round to the nearest dollar. $978

42. An artist lays down a white triangular background on a building on which she will paint a Penrose triangle. The triangular area measures 20 ft on each side and will be covered with three layers of primer paint.

a. To the nearest square foot, what is the area of the triangular region to be primed? 173 ft²

b. If primer paint covers 200 ft²/gal, how many gallons of primer must be purchased? 3 gal

Mixed Exercises

For Exercises 43–46, the vertices of a triangle are defined by the given points. To the nearest tenth, determine

a. the perimeter of the triangle.

b. the area of the triangle.

c. the measure of the angles in the triangle.

43. $A(2, 2)$, $B(4, 7)$, $C(8, 1)$ a. 18.7   b. 16.0
   c. $A \approx 77.7°$, $B \approx 55.5°$, $C \approx 46.8°$

45. $A(-3, -2)$, $B(-1, 5)$, $C(6, 1)$ a. 24.8   b. 28.5
   c. $A \approx 55.6°$, $B \approx 76.2°$, $C \approx 48.2°$

44. $A(1, 5)$, $B(5, 8)$, $C(10, 3)$ a. 21.3   b. 17.5
   c. $A \approx 49.4°$, $B \approx 98.1°$, $C \approx 32.5°$

46. $A(-2, -1)$, $B(10, 2)$, $C(5, -4)$ a. 27.8   b. 28.5
   c. $A \approx 37.2°$, $B \approx 36.2°$, $C \approx 106.6°$

47. Three mutually tangential circles with radii located at A, B, and C, have radii of 2 ft, 3 ft, and 4 ft, respectively.

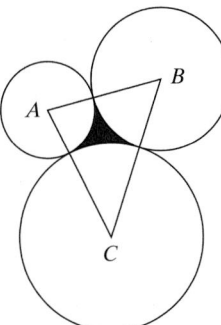

a. Find the area of $\triangle ABC$. Round to the nearest tenth of a square foot.   14.7 ft²

b. Find the measures of angles A, B, and C. Round to the nearest tenth of a degree. $A \approx 78.5°$, $B \approx 57.1°$, $C \approx 44.4°$

c. Find the area of the shaded region. Round to the nearest tenth of a square foot.   1.3 ft²

49. A parallelogram has adjacent sides of 12.2 cm and 9.6 cm and the included angle is 62.5°. To the nearest tenth of a centimeter,

a. Find the length of the shorter diagonal.   11.5 cm

b. Find the length of the longer diagonal.   18.7 cm

48. The rectangular box shown in the figure is 8 in. long, 5 in. wide, and 3 in. high. Line segment $\overline{AC}$ connects opposite vertices on the left side of the box. Line segment $\overline{CB}$ connects opposite vertices at the back of the box, and line segment $\overline{AB}$ connects opposite vertices on the bottom of the box.

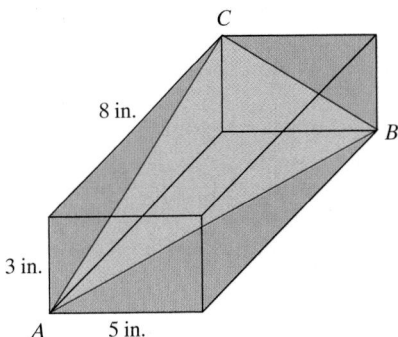

a. Find the measure of $\angle ABC$. Round to the nearest degree. $\angle ABC \approx 63°$

b. Find the area of the triangle.   24.5 in.²

50. A regular octagon (8-sided figure with sides of equal length) is inscribed in a circle of radius 10 in. Find the perimeter of the octagon to the nearest inch.   61 in.

Write About It

51. Why is the law of sines not an option to solve a triangle given SAS or SSS?
Neither case (SAS or SSS) provides the measure of an angle and a side opposite the angle to form a complete ratio such as
$\dfrac{a}{\sin A}$. Therefore, the law of sines cannot be used.

53. Given the lengths of the three sides of a triangle (SSS), why is the law of cosines used to find the measure of the largest angle first?

55. Suppose that $\triangle ABC$ is a right triangle with right angle at vertex C. What is the result when the law of cosines is used to find the length of side c?

52. To solve a triangle given SAS, the law of cosines is used first to find the length of the side opposite the known angle. Then, if you choose to use the law of sines to solve for one of the two remaining angles, the guidelines suggest solving for the angle opposite the *shorter* of the two given sides. Why is this the case?

54. When applying the law of cosines (or law of sines) how can you minimize round-off error?

56. If the measures of the three angles in a triangle are known, can you solve for the lengths of the sides?

Expanding Your Skills

57. A piston in an engine is attached to a connector rod of length L at the wrist pin at point P. As the piston travels back and forth, the connector cap at point Q travels counterclockwise along a circular path of radius r. If $L > 2r$, and α represents $\angle PCQ$,

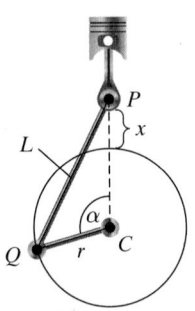

a. Use the law of cosines to show that the distance x from the wrist pin to the crank circle is given by
$x = r\cos\alpha + \sqrt{r^2\cos^2\alpha - r^2 + L^2} - r$.

b. Find the distance between the wrist pin and crank circle for a truck engine with connector rod of length 5.9 in. and crank radius of 1.7 in. when $\alpha = 105°$. Round to the nearest tenth of an inch.   3.5 in

c. Find the maximum and minimum distances that the wrist pin is to the crank circle, using the values from part (b).

58. Use the law of cosines to show that
$$\frac{\cos A}{a} + \frac{\cos B}{b} + \frac{\cos C}{c} = \frac{a^2 + b^2 + c^2}{2abc}.$$

59. Use the law of cosines to show that
$a = b\cos C + c\cos B$

60. Complete the proof of Heron's formula.

a. Begin with the formula $\text{Area} = \frac{1}{2}bc\sin A$ and square both sides.

b. In the equation from part (a), replace $\sin^2 A$ by $1 - \cos^2 A$.

c. In the equation from part (b), factor $1 - \cos^2 A$ as a difference of squares.

d. Take the square root of both sides and write the area as

$$\text{Area} = \sqrt{\left[\frac{1}{2}bc(1 + \cos A)\right]\left[\frac{1}{2}bc(1 - \cos A)\right]} \qquad (1)$$

e. Use the law of cosines to show that

$$\frac{1}{2}bc(1 + \cos A) = \frac{a + b + c}{2} \cdot \frac{-a + b + c}{2} \qquad (2)$$

Note: Using similar logic, we can also show that

$$\frac{1}{2}bc(1 - \cos A) = \frac{a - b + c}{2} \cdot \frac{a + b - c}{2} \qquad (3)$$

f. Use the substitution $s = \frac{1}{2}(a + b + c)$ to rewrite equation (2) as $\frac{1}{2}bc(1 + \cos A) = s(s - a)$ (4)

Note: Using similar logic, we can also show that $\frac{1}{2}bc(1 - \cos A) = (s - b)(s - c)$ (5)

g. Substitute equations (4) and (5) into equation (1).

PROBLEM RECOGNITION EXERCISES

Solving Triangles Using a Variety of Tools

For Exercises 1–12, solve $\triangle ABC$ subject to the given conditions, if possible. Round the lengths of the sides and the measures of the angles to 2 decimal places, if necessary.

1. $a = 20, b = 29, B = 90°$
$c = 21, A \approx 43.60°, C \approx 46.40°$

2. $b = 16.2, c = 12, A = 113°$
$a \approx 23.63, B \approx 39.13°, C \approx 27.87°$

3. $a = 16.2, b = 18.5, c = 28.6$
$A \approx 31.96°, B \approx 37.19°, C \approx 110.84°$

4. $a = 19, b = 26, A = 72°$
No triangle is possible.

5. $c = 20, b = 17, A = 90°$
$a \approx 26.25, B \approx 40.36°, C \approx 49.64°$

6. $a = 14.1, A = 28°, B = 110°$
$b \approx 28.22, c \approx 20.10, C = 42°$

7. $a = 7, b = 8.8, A = 40°$

8. $b = 18, c = 26, C = 56°$
$a \approx 31.36, A \approx 88.97°, B \approx 35.03°$

9. $b = 108, B = 62°, C = 54°$
$a \approx 109.94, c \approx 98.96, A = 64°$

10. $a = 12, c = 48, B = 62°$
$b \approx 43.67, A \approx 14.04°, C \approx 103.96°$

11. $a = 106, b = 58, c = 172$
No triangle is possible.

12. $a = 46, b = 61, B = 136°$
$c \approx 18.87, A \approx 31.59°, C \approx 12.41°$

13. Refer to the figure. Find the indicated values to the nearest hundredth of a centimeter or hundredth of a degree.

  a. Find the lengths of $BC, AC, DB, DC, ED, EC,$ and EA.

  b. Find the measure of $\angle AEC$ and $\angle CAE$.

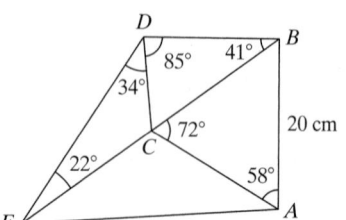

14. Use the right triangles $\triangle APB$ and $\triangle APC$ to show that for $\triangle ABC$, the length of side a is given by $a = b\cos C + c\cos B$.

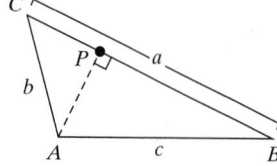

15. A hiking trail is roughly in the shape of a circle with radius 2000 ft. Suppose that a rectangular coordinate system is set up with the origin at the center of the circle. Donna and Eddie start walking from point $A(2000, 0)$ at 220 ft/min. Eddie goes off-trail and walks directly toward the center of the circle and Donna walks along the trail counterclockwise.

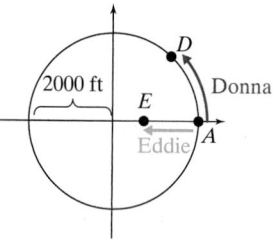

  a. Find the coordinates of each person after 5 min. Round the coordinates to the nearest foot.

  b. How far apart are Eddie and Donna after 5 min? Round to the nearest foot.   1319 ft

  c. If Eddie can no longer walk because he sprains his ankle after foolishly leaving the trail, at what bearing would Donna walk to find him? Round to the nearest tenth of a degree.   S37.6°W

1. Model Simple Harmonic Motion

The wavelike appearance of the graphs of the sine and cosine functions suggests that they would be good models to describe the motion of an oscillating or rotating object. For example, suppose that a block is attached to a spring hanging from the ceiling. If the block is pushed upward to compress the spring or pulled downward to stretch the spring, the spring applies a force on the block in the opposite direction. This is called a *restorative force* because it acts to return the block toward the equilibrium position (natural resting state). The greater the displacement d of the block from its equilibrium position, the more force F the spring applies to the block. In the absence of other forces called damping forces such as internal friction in the spring, the block will travel down and up indefinitely in a motion we call **simple harmonic motion.**

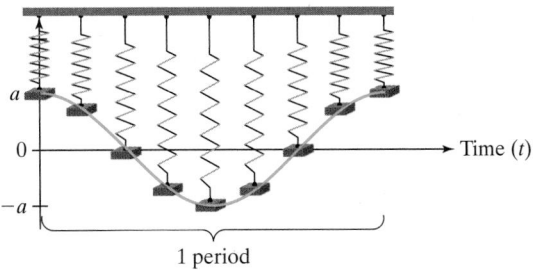

Figure 6-26   Spring-Mass System in Simple Harmonic Motion

TIP In physics, you will often see the period of an object in simple harmonic motion represented by the letter T. This is because the period is a value of time.

In Figure 6-26, a block at the end of the spring is displaced a units upwards from equilibrium position and then released. In the absence of damping, the model $d = a\cos\omega t$ describes the vertical displacement d as a function of the time t after release. The constant ω (omega) is generally taken to be positive and is related to the **period** P (time required to complete one up-and-down cycle). From our discussion in Section 4.5, $P = \dfrac{2\pi}{\omega}$. The **frequency** f with which the block oscillates up and down is the number of complete cycles per unit time. Therefore, the frequency is the reciprocal of the period.

Instructor Note:
Understanding the difference between the two models for simple harmonic motion is sometimes a problem for students. Show them that

$d = a\sin(\omega \cdot 0) = 0$

$d = a\cos(\omega \cdot 0) = a$

Use the sine function if the object "starts" at equilibrium. Use the cosine function if the object "starts" at maximum or minimum displacement.

Simple Harmonic Motion

If the displacement d of an object at time t is given by

$$d = a\sin\omega t \quad \text{or} \quad d = a\cos\omega t$$

then the object is said to have **simple harmonic motion.**

- The amplitude of motion is $|a|$.
- The period is given by $P = \dfrac{2\pi}{\omega}$ (time required for one complete cycle).
- The frequency f is the reciprocal of the period, $f = \dfrac{1}{P}$ or equivalently, $f = \dfrac{\omega}{2\pi}$ and is the number of cycles per unit time.
- The relationship $\omega = \dfrac{2\pi}{P}$ implies that $\omega = 2\pi f$. Therefore, simple harmonic motion can also be modeled by $d = a\cos 2\pi f t$ and $d = a\sin 2\pi f t$.

The difference between the models $d = a \cos \omega t$ and $d = a \sin \omega t$ is the position of the object at $t = 0$. Initially, if $d = 0$, then the motion "starts" with a displacement of 0 which is modeled most conveniently by the sine function. If $d = a$ or $d = -a$ initially, then the motion "starts" at a high or low point in the curve and is best modeled by the cosine function. If d is something other than 0, a, or $-a$ initially, then either function can be used with a phase shift.

Classroom Example: p. 673
Exercise 24

EXAMPLE 1   **Writing a Model for Simple Harmonic Motion**

Suppose that the block at the end of the spring in Figure 6-26 is originally compressed 10 in. upward. After release, the block makes one complete down-and-up cycle in 0.5 sec.

   a. What is the period of motion?
   b. What is the frequency of motion?
   c. What is the amplitude?
   d. Write a function of the form $d = a \cos \omega t$ to model the displacement d (in inches) of the block as a function of the time t (in seconds) after release.
   e. Find the position of the block 6.25 sec after release.

Solution:

   a. The period P is given as 0.5 sec.
       This means that one up-and-down cycle is completed in $\frac{1}{2}$ sec.

   b. To find the frequency, we have $f = \dfrac{1}{P} = \dfrac{1}{0.5} = 2$.

Instructor Note:
Remind students that Hz is the abbreviation for Hertz.
1 Hz = 1 cycle/sec

       A frequency of 2 means that 2 cycles are completed in 1 sec. The units of measurement assigned to frequency is "cycles per second" or Hertz (abbreviated Hz). Therefore, $f = 2$ Hz.

   c. The amplitude is the maximum deviation from the equilibrium position. $|a| = |10| = 10$. The amplitude is 10 in.

   d. The starting position is positive 10 in. above equilibrium. Therefore, we will use $d = a \cos \omega t$ with $a = 10$.
       Since $P = \dfrac{2\pi}{\omega}$, we have $\omega = \dfrac{2\pi}{P} = \dfrac{2\pi}{0.5} = 4\pi$.
       The model is $d = 10 \cos 4\pi t$.

   e. When $t = 6.25$ sec, we have: $d = 10 \cos 4\pi (6.25)$
$$= 10 \cos 25\pi = 10(-1) = -10$$
       Therefore, 6.25 sec after release, the block is 10 in. *below* the equilibrium position.

Skill Practice 1   Suppose the block from Example 1 is initially pulled 6 in. *below* the equilibrium position. Repeat Example 1 if it takes the block 0.25 sec to return to its starting position.

In Example 2, we demonstrate that the position of an object rotating at a uniform speed can also be modeled by simple harmonic motion.

Answers
1.   **a.** $P = 0.25$ sec
    b. $f = 4$ Hz   **c.** $a = 6$ in.
    d. $d = -6 \cos 8\pi t$ **e.** -6 in.

Classroom Example: p. 674
Exercise 28

EXAMPLE 2   Writing and Interpreting an Application of Simple Harmonic Motion

A Ferris wheel is 150 ft in diameter with its lowest point 3 ft off the ground. Once all the passengers have been loaded, the wheel makes one full rotation counterclockwise in 2.5 min. Suppose that two children are seated at the lowest point on the wheel and are the last passengers to be loaded. Choose a coordinate system with the origin at ground level below the center of the wheel.

a. Write a model representing the children's horizontal position x (in feet) at a time t min after the ride starts.

b. Write a model representing the children's height y (in feet) above ground level, t min after the ride starts.

c. Describe the children's position 2 min into the ride.

Solution:

a. Figure 6-27 shows the Ferris wheel and a coordinate system with origin 3 units below the lowest point on the wheel. The children's seat on the Ferris wheel is initially at $(0, 3)$. Notice that the x-coordinate of the children's position is initially zero and then turns positive as the wheel begins its counterclockwise motion. Therefore, we will use the sine function $x = a \sin 2\pi f t$ with $a > 0$ to model the children's position to the right or left of the center of the wheel.

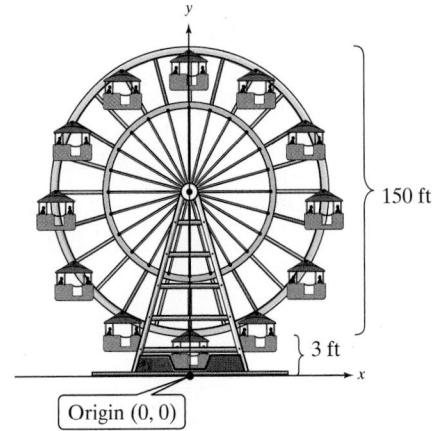

150 ft

3 ft

Origin $(0, 0)$

Figure 6-27

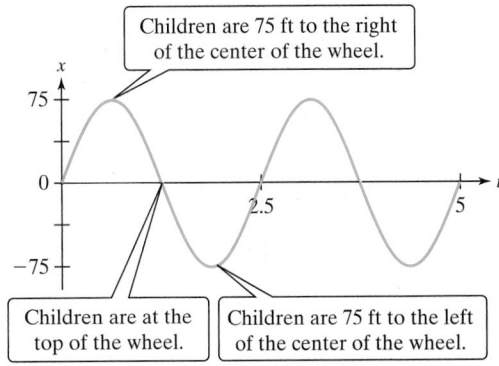

Children are 75 ft to the right of the center of the wheel.

Children are at the top of the wheel.

Children are 75 ft to the left of the center of the wheel.

- The maximum deviation from $x = 0$ is equal to the radius of the wheel. Therefore, $a = 75$.

- The period is 2.5 min, and the frequency is $f = \dfrac{1}{2.5} = 0.4$ rpm.

Therefore, the horizontal position of the children is $x = 75 \sin(2\pi \cdot 0.4t)$ or simply $x = 75 \sin 0.8\pi t$.

TIP Recall from Section 4.5 that the vertical shift is the midpoint of the range of a sine or cosine function.

The amplitude is half the difference between the maximum and minimum values.

b. The children's height starts at the lowest point $y = 3$ at $t = 0$ and increases to the highest point $y = 153$ at the half-way point during the revolution. Therefore, we will use the cosine function $y = a\cos 2\pi ft$ with a vertical shift of $(3 + 153)/2 = 78$.

The amplitude is $(153 - 3)/2 = 75$, but the value of a is taken to be -75 because the starting point is at the lowest point on the wheel. Therefore, the height is $y = -75\cos(2\pi \cdot 0.4t) + 78$ or simply $y = 78 - 75\cos 0.8\pi t$.

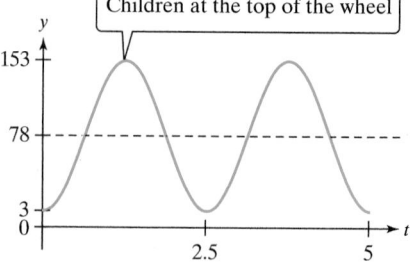

Children at the top of the wheel

c. At $t = 2$ min,
$$x = 75\sin[0.8\pi(2)] \approx -71.3$$
$$y = 78 - 75\cos[0.8\pi(2)] \approx 54.8$$

Two minutes into the ride, the children have completed $\frac{2}{2.5} = 0.8$ revolution. They are on the way down from the high point to the low point at a position 71.3 ft to the left of center and 54.8 ft above ground. This is $(-71.3, 54.8)$.

Skill Practice 2   Repeat Example 2 with a Ferris wheel 120 ft in diameter that completes one revolution in 1.25 min.

2. Interpret Damped Harmonic Motion

The spring-mass system in Example 1 would realistically be subject to internal friction within the spring and a small degree of air resistance. As a result, if the block were pushed up or pulled down from its natural resting position and then released, the distance it moves with each recovery will decrease until the block ultimately comes to rest. In such a case, the amplitude decreases with time and we say that the block undergoes **damped harmonic motion.** Air resistance and the force of friction are called **damping forces** because they dissipate energy in the form of heat causing a decrease of movement.

TIP The negative factor in the model for damped harmonic motion is a result of the damping force acting in the opposite direction of motion.

Damped Harmonic Motion

The displacement d of an object oscillating under damped harmonic motion at time t is given by

$$d = ae^{-ct}\cos \omega t \quad \text{or} \quad d = ae^{-ct}\sin \omega t \quad \text{for } c > 0$$

- a is the displacement of the object at $t = 0$.
- c is a constant called the **damping factor** and is related to the damping force opposing motion.*
- The value of ω is related to the period P and and frequency f by $P = 2\pi/\omega$ and $\omega = 2\pi f$.**

*If the damping factor $c = 0$, then there is no damping and the model simplifies to that of simple harmonic motion: $d = a\cos \omega t$ or $d = a\sin \omega t$.

**Damped harmonic motion is not actually periodic because the amplitude decreases over time. However, we use the word "period" in this context as the time required for one complete back-and-forth cycle.

Answers

2. a. $60\sin 1.6\pi t$
   b. $63 - 60\cos 1.6\pi t$
   c. Two minutes into the ride, the children have completed 1.6 revolutions. Their position is $(-35.3, 111.5)$ on the way down from the maximum height for the second time.

Classroom Example: p. 676
Exercise 44

EXAMPLE 3   Comparing Models Involving Damped Harmonic Motion

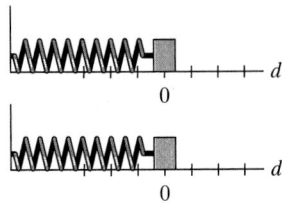

Suppose that two blocks of equal mass are attached to two horizontal springs and pulled 8 cm to the right of the natural rest positions of the springs. Due to internal friction in the springs and friction on the surface of movement, the blocks move in damped harmonic motion. The damping factor for the first spring is 0.6 and the damping factor for the second spring is 0.2.

a. If each spring-mass system oscillates at 1 Hz, write a function to model the displacement for each system t seconds after release. Assume that a displacement to the right of equilibrium is positive.

b. Graph the functions from part (a) on a graphing utility.

c. Describe the differences in the graphs.

Solution:

a. The frequency f is given as 1 Hz. Therefore, $\omega = 2\pi f = 2\pi(1) = 2\pi$. Since the initial displacement is 8 cm from the equilibrium position, we choose the cosine model $d = ae^{-ct}\cos\omega t$ with $a = 8$.

Spring 1: $d_1 = 8e^{-0.6t}\cos 2\pi t$
Spring 2: $d_2 = 8e^{-0.2t}\cos 2\pi t$

b. Spring 1   Spring 2

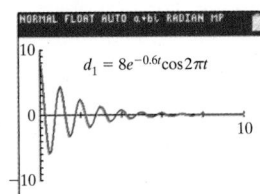   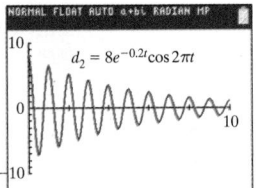

c. The graphs show that the "amplitude" decreases more rapidly for a function with a greater damping factor (graph of Spring 1).

Skill Practice 3   Two spring-mass systems have an initial displacement of 6 cm and then oscillate at 0.25 Hz. The damping factors are 0.4 and 0.1, for spring 1 and spring 2, respectively. Write a function to model the displacement for each system t seconds after release.

> **TIP** If the restorative force of the spring (the force that causes recoil) is greater than the damping forces, the system is said to be *lightly damped*. In such a case the motion of the object oscillates, but with progressively smaller amplitudes (as in Example 3). If the damping forces are greater than the restorative force of the spring, then motion is said to be *over damped* and follows a path of exponential decay.

From Example 3, the graphs of $d_1 = 8e^{-0.6t}\cos 2\pi t$ and $d_2 = 8e^{-0.2t}\cos 2\pi t$ both show a decrease in the "amplitude" for increasing time. In fact, plotting points at the relative maxima reveals a pattern of exponential decay. In Exercise 42, we show that the displacement of an object in damped harmonic motion is bounded above and below by $d = ae^{-ct}$ and $d = -ae^{-ct}$ (Figures 6-28 and 6-29).

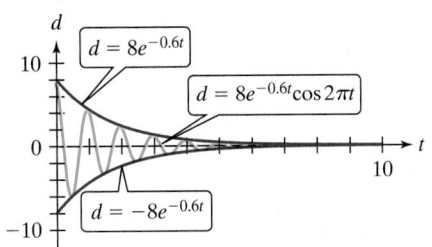

Figure 6-28

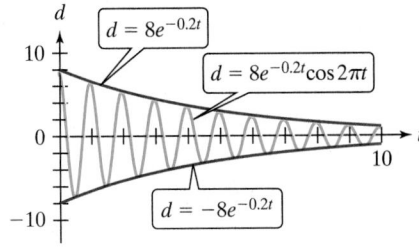

Figure 6-29

Answer
3. Spring 1: $d_1 = 6e^{-0.4t}\cos 0.5\pi t$;
Spring 2: $d_2 = 6e^{-0.1t}\cos 0.5\pi t$

Point of Interest

Harmonic motion is in evidence in the world all around us. For example, a shock absorber mounted in the suspension system of a vehicle consists of a piston in a reservoir of hydraulic fluid. When a vehicle encounters vibration or a bump in the road, the piston moves through the fluid, which introduces a damping force. This turns the kinetic energy of suspension movement into heat energy that is dissipated through the hydraulic fluid. Ultimately, the shock absorber keeps the vehicle from bouncing up and down on its springs.

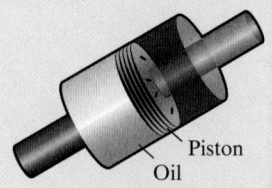
Piston
Oil

When a guitar string is plucked, the string vibrates and produces sound. But the sound fades away as internal friction in the string causes a decrease in the amplitude of vibration.

Scientists study earthquake waves as a means to build stronger structures in areas prone to earthquakes. *P* waves and *S* waves are the fundamental waves that result after an earthquake. *P* waves (primary waves) alternatively push (compress) and pull (dilate) the rock through which it travels. *S* waves (secondary waves) shear rock at right angles to the direction of travel. Areas affected by earthquakes feel the *P* waves first because they travel faster than *S* waves. However, *S* waves tend to cause more damage because they have greater amplitude, and because the undulation of the earth caused by *S* waves is particularly stressful to man-made structures.

S waves: Shear waves

P waves: Push and pull

Extension   Compression

SECTION 6.4   Practice Exercises

Prerequisite Review

R.1. Identify the amplitude and period.
$$y = 2\cos 4x$$ Amplitude = 2, Period = $\dfrac{\pi}{2}$

R.2. Identify the amplitude and period.
$$y = 3\sin \dfrac{\pi}{3}x$$ Amplitude = 3, Period = 6

R.3. Solve the equation. Write the solution set with the exact value given in terms of common or natural logarithms. Also give the approximate solution to 4 decimal places.
$$2600 = 20{,}800e^{-0.2r}$$ {5 ln 8}, {10.3972}

Concept Connections

1. The ____period____ of an object in simple harmonic motion is the amount of time required for one complete cycle.

2. For an object in simple harmonic motion, the number of cycles per unit time is called the ____frequency____.

3. Given an object in damped harmonic motion, the amplitude (increases/decreases) with time.   decreases

4. The term "ultrasound" refers to sound waves of a frequency greater than those detectable by the human ear ($> 20{,}000$ Hz). Suppose that a medical imaging device produces ultrasound with a period of 5 μs (microseconds: note that 5 μs = 0.000005 sec). What is the frequency?   200,000 Hz

5. The frequency of middle C on a piano is 264 Hz. What is the time required for one complete cycle? $\frac{1}{264}$ sec

6. If the displacement d of an object moving under simple harmonic motion is maximized at time $t = 0$, which model would be most convenient? $d = a\sin\omega t$ or $d = a\cos\omega t$ $d = a\cos\omega t$

Objective 1: Model Simple Harmonic Motion

For Exercises 7–14, suppose that an object moves in simple harmonic motion with displacement d (in centimeters) at time t (in seconds). Determine the

a. Amplitude

b. Period

c. Frequency

d. Phase shift

e. Least positive value of t for which $d = 0$

7. $d = 6\cos 80\pi t$

8. $d = -2\cos 120\pi t$

9. $d = -0.25\sin\dfrac{\pi}{4}t$

10. $d = 0.125\sin\dfrac{\pi}{2}t$

11. $d = -\cos\left(t - \dfrac{\pi}{6}\right)$

12. $d = -\cos\left(t + \dfrac{\pi}{4}\right)$

13. $d = \dfrac{1}{4}\sin\left(\dfrac{1}{2}t + 1\right)$

14. $d = \dfrac{1}{3}\sin\left(\dfrac{1}{4}t + \dfrac{1}{2}\right)$

For Exercises 15–22, suppose that an object is attached to a horizontal spring subject to the given conditions. Find a model for the displacement d as a function of the time t. (See Example 1)

	Initial Displacement (d at $t = 0$)	Amplitude	Period or Frequency	
15.	2.5 in.	2.5 in.	1.2 sec	$d = 2.5\cos\dfrac{5\pi}{3}t$
16.	4.1 cm	4.1 cm	2 sec	$d = 4.1\cos\pi t$
17.	-6 ft	6 ft	2 Hz	$d = -6\cos 4\pi t$
18.	-12 m	12 m	4 Hz	$d = -12\cos 8\pi t$
19.	0 cm Initially moving to the right	8.5 cm	0.2 sec	$d = 8.5\sin 10\pi t$
20.	0 cm Initially moving to the right	15 cm	0.125 sec	$d = 15\sin 16\pi t$
21.	0 ft Initially moving to the left	0.25 ft	0.1 Hz	$d = -0.25\sin\dfrac{\pi}{5}t$
22.	0 m Initially moving to the left	1 m	0.01 Hz	$d = -\sin\dfrac{\pi}{50}t$

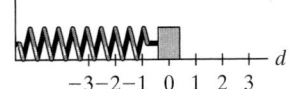

23. A block hangs on a spring attached to the ceiling and is pulled down 5 in. below its equilibrium position. After release, the block makes one complete up-and-down cycle in 1 sec and follows simple harmonic motion. (**See Example 1**)

a. What is the period of motion?   1 sec

b. What is the frequency?   1 Hz

c. What is the amplitude?   5 in.

d. Write a function to model the displacement d (in inches) as a function of the time t (in seconds) after release. Assume that a displacement above the equilibrium position is positive. $d = -5\cos 2\pi t$

e. Find the displacement of the block and direction of movement at $t = 0.125$ sec. $d = -\dfrac{5\sqrt{2}}{2} \approx -3.5$ in.

The block is approximately 3.5 in. below its natural resting state and moving upwards.

24. The bob on a simple pendulum is pulled to the left 4 in. from its equilibrium position. After release, the pendulum makes one complete back-and-forth cycle in 2 sec and follows simple harmonic motion.

a. What is the period of motion?   2 sec

b. What is the frequency? $\dfrac{1}{2}$ Hz

c. What is the amplitude?   4 in.

d. Write a function to model the displacement d (in inches) of the bob as a function of the time t (in seconds) after release. Assume that a displacement to the right of the equilibrium position is positive. $d = -4\cos\pi t$

e. Find the position and direction of movement of the bob at $t = 1.25$ sec. $d = 2\sqrt{2} \approx 2.8$ in.

The bob is 2.8 in. to the right of the equilibrium position and moving to the left.

25. A 7-in. connecting rod in an engine connects the wrist pin of a piston at point A to the end of a second rod at point B on the crankshaft. As the piston moves forward and back, the crankshaft is turned counterclockwise around point C. The stroke length of the engine is the distance that the piston travels (one way) within its cylinder. If the stroke length is 4 in., and the engine turns at 1800 rpm, write a model representing the distance d (in inches) from A to the center of the crank shaft C as a function of the time t (in minutes). Assume that at $t = 0$, the piston is at its farthest point from the crankshaft. $d = 2\cos 3600\pi t + 7$

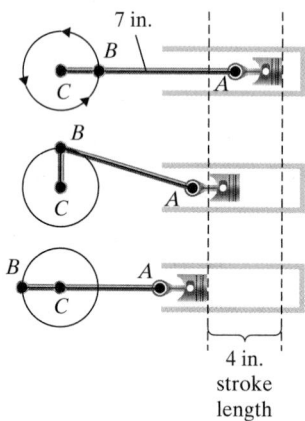

7 in.

4 in.
stroke
length

27. A Ferris wheel at a county Fair is 180 ft in diameter with its lowest point 2.5 ft off the ground. Once all the passengers have been loaded, the wheel makes one full rotation counterclockwise in 1.2 min. Suppose that two children are seated at the lowest point on the wheel and are the last passengers to be loaded when the wheel starts. **(See Example 2)**

a. Write a model representing the children's horizontal position x (in feet) relative to the center of the Ferris wheel, t minutes after the ride starts. $x = 90\sin\dfrac{5\pi}{3}t$

b. Write a model representing the children's height y (in feet) above ground level, t minutes after the ride starts. $y = -90\cos\dfrac{5\pi}{3}t + 92.5$

c. Give the coordinates of the children's position 1 min into the ride and describe the location.

29. Refer to the piston and crankshaft from Exercise 25. The stroke length of the engine is 4 in. and the engine turns at 1800 rpm. The connecting rod cap at point B rotates around the crank circle centered at point C. Assume that at $t = 0$, the piston is at its farthest point from the center of the crank circle.

a. How many revolutions per second is the engine turning? 30 rev/sec

b. Write a model representing the horizontal position x (in inches) of the connecting rod cap (point B) relative to the center of the crank circle (point C) at time t (in seconds). $x = 2\cos 60\pi t$

c. Write a model representing the vertical position y (in inches) of the connecting rod cap relative to the center of the crank circle at time t (in seconds). $y = 2\sin 60\pi t$

26. The blood pressure p for a certain individual follows a pattern of simple harmonic motion during the pumping cycle between heartbeats. The minimum pressure is 80 mmHg (millimeters of mercury) and the maximum pressure is 120 mmHg. The individual's pulse is 60 beats per minute or equivalently 1 beat per second. Write a model representing the blood pressure p at a time t seconds into the cycle. Assume that at $t = 0$, the blood pressure is 100 mmHg and is initially increasing. $p = 100 + 20\sin 2\pi t$

28. The 30-ft diameter waterwheel at Deleon Springs State Park is an old sugar mill that turns at a rate of 2 rev/min, counterclockwise. Suppose that a coordinate system is chosen with the origin at the bottom of the wheel (at water level). At $t = 0$, suppose that a point on the wheel is located at (0, 0).

a. Write a function that represents the x position (in feet) of the point as a function of the time t in minutes. $x = 15\sin 4\pi t$

b. Write a function that represents the y position (in feet) of the point as a function of the time t in minutes. $y = 15 - 15\cos 4\pi t$

c. Give the coordinates of the point at $t = 40$ sec and describe the location.

30. A laboratory centrifuge is a piece of equipment that spins blood samples at a high speed to separate substances within the blood of greater or lesser density. Suppose that a test tube is placed at a point 10 cm to the right of the center of rotation and that the centrifuge spins at 3600 rpm counterclockwise.

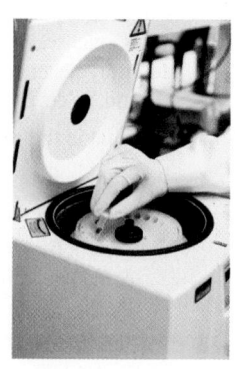

a. How many revolutions does the centrifuge make per second? 60 rev/sec

b. Write a model representing the horizontal position x (in centimeters) of the test tube relative to the center of the centrifuge at time t (in seconds). $x = 10\cos 120\pi t$

c. Write a model representing the vertical position y (in centimeters) of the test tube relative to the center of the centrifuge at time t (in seconds). $y = 10\sin 120\pi t$

31. An alternating current generator generates current with a frequency of 60 Hz. Suppose that initially, the current is at its maximum of 5 amperes. If the current varies in simple harmonic motion over time, write a model for the current I (in amperes) as a function of the time t (in seconds). $I = 5\cos 120\pi t$

32. An alternating current generator in the United Kingdom generates current with a frequency of 50 Hz. Suppose that initially, the current is at its maximum of 12 amperes. If the current varies in simple harmonic motion over time, write a model for the current I (in amperes) as a function of the time t (in seconds). $I = 12\cos 100\pi t$

The angular displacement θ (in radians) of a pendulum swinging through a small angle with limited air resistance and internal friction can be approximated by the simple harmonic motion models $\theta = a\sin\omega t$ or $\theta = a\cos\omega t$. The period T (time required for one back-and-forth swing) is given by $T = 2\pi\sqrt{\dfrac{L}{g}}$, where L is the length of the pendulum and g is the acceleration due to gravity. The maximum angular displacement is $|a|$ and positive angles are taken to the right of the center position. Use this information for Exercises 33–34.

33. A 4-ft pendulum is initially at its right-most position of 12°.

 a. Determine the period for one back-and-forth swing. Use $g = 32$ ft/sec². $\dfrac{\sqrt{2}\pi}{2}$ sec

 b. Write a model for the angular displacement θ of the pendulum after t seconds. (*Hint*: Be sure to convert the initial position to radians.) $\theta = \dfrac{\pi}{15}\cos 2\sqrt{2}t$

34. A 1.4-m pendulum is initially at its left-most position of $-9°$.

 a. Determine the period for one back-and-forth swing. Use $g = 9.8$ m/sec². $\dfrac{2\sqrt{7}\pi}{7}$ sec

 b. Write a model for the angular displacement θ of the pendulum after t seconds. $\theta = -\dfrac{\pi}{20}\cos\sqrt{7}t$

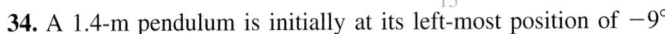

35. The brightness of a "young" star sometimes increases and decreases as a result of regional areas of "hot" and "cold" on the star's surface as well as variations in the density of the star's planet-forming debris, which can obstruct light. Suppose that for a particular star, the average magnitude (measure of brightness) is 4.3 with a variation of ±0.31 (on the magnitude scale, brighter objects have a smaller magnitude than dimmer objects). Furthermore, the magnitude of a star is initially observed to be 4.61, and the time between minimum brightness and maximum brightness is 6.4 days. Write a simple harmonic motion model to describe the magnitude M of the star for day t. $M = 4.3 + 0.31\cos\dfrac{5\pi}{32}t$

36. The magnitude of a star named Delta Cephei varies from an apparent magnitude of 3.6 to an apparent magnitude of 4.3 with a period of 5.4 days. At $t = 0$ days, the star is at its brightest with a magnitude of 3.6 (on the magnitude scale, brighter objects have a smaller magnitude than dimmer objects). Write a simple harmonic motion model to describe the magnitude M of the star for day t. $m = 3.95 - 0.35\cos\dfrac{10\pi}{27}$

For Exercises 37–38, the ordered pair (t, d) gives the displacement d (in centimeters) of an object undergoing simple harmonic motion at time t (in seconds).

37. Suppose that the object has a minimum at $(24, 32)$ and next consecutive maximum at $(48, 54)$.

 a. What is the period?   48 sec

 b. What is the frequency? $\dfrac{1}{48}$ Hz

 c. What is the amplitude?   11 cm

 d. Write a model representing the displacement as a function of time. $d = 43 + 11\cos\dfrac{\pi}{24}t$

38. Suppose that the object has a maximum at $(4, 16)$ and then returns to its next equilibrium position at $(6, 11)$.

 a. What is the period?   8 sec

 b. What is the frequency? $\dfrac{1}{8}$ Hz

 c. What is the amplitude?   5 cm

 d. Write a model representing the displacement as a function of time. $d = 11 - 5\cos\dfrac{\pi}{4}t$

39. Write a function $y = f(t)$ for simple harmonic motion whose graph has a minimum at $\left(\dfrac{\pi}{4}, -6\right)$ and next consecutive equilibrium at $\left(\dfrac{\pi}{2}, 2\right)$. $y = 2 - 8\sin 2t$

40. Write a function $y = f(t)$ for simple harmonic motion whose graph has a maximum at $\left(\dfrac{\pi}{3}, 8\right)$ and next consecutive minimum at $(\pi, -2)$. $y = 3 + 5\sin\dfrac{3}{2}t$

Objective 2: Interpret Damped Harmonic Motion

41. Use a graphing utility to graph the functions of the form $d = ae^{-ct}\cos\omega t$ and comment on the role of the damping factor $c > 0$ in the equation. (**See Example 3**)

 a. $d = 6e^{-0.1t}\cos\pi t$

 b. $d = 6e^{-0.3t}\cos\pi t$

 c. $d = 6e^{-0.5t}\cos\pi t$

42. Show that the graph of $d = ae^{-ct}\cos\omega t$ is bounded by $y = ae^{-ct}$ and $y = -ae^{-ct}$ for $a > 0$ and $c > 0$.

For Exercises 43–46, use a graphing utility to graph the function and bounding curves for $t \geq 0$.

43. $d = -10e^{-0.125t} \sin 1.25t$

44. $d = 2e^{-0.55t} \sin 16t$

45. $d = 8e^{-1.2t} \cos 4\pi t$

46. $d = -4e^{-0.008t} \cos \dfrac{\pi}{8} t$

47. Suppose that a guitar string is plucked such that the center of the string is initially displaced 10 mm and then vibrates under damped harmonic motion. The note produced has a frequency of 110 Hz. The note is no longer audible to a normal human ear once the displacement at the middle of the string is less than 0.1 mm. What is the damping factor if the sound is no longer audible after 2.5 sec? Round to 2 decimal places. $c \approx 1.84$

48. A tuning fork is struck, and the tips of the prongs (tines) are initially displaced 0.4 mm from their natural position at rest. The prongs oscillate at 440 Hz under damped harmonic motion. The sound is no longer audible once the displacement is less than 0.1 mm. What is the damping factor if the sound is no longer audible after 4.0 sec? Round to 2 decimal places. $c \approx 0.35$

Write About It

49. Explain the criteria under which you would choose the model $d = a \sin \omega t$ versus $d = a \cos \omega t$ to represent the displacement d of an object moving in simple harmonic motion.

50. Explain the difference between an object moving in simple harmonic motion versus damped harmonic motion.

Expanding Your Skills

51. Consider the function defined by $y = x + 2 \sin x$.

   a. As $x \to \infty$, which term in the function will contribute most to the y value? x

   b. Before graphing the function, speculate about its general behavior.

   c. Graph $y = x$ and $y = x + 2 \sin x$ on a graphing utility.

52. Consider the function defined by $y = e^x + \cos x$.

   a. As $x \to \infty$, which term in the function will contribute most to the y value? e^x

   b. Before graphing the function, speculate about its general behavior.

   c. Graph $y = e^x$ and $y = e^x + \cos x$ on a graphing utility.

For Exercises 53–54, use the model $d = ae^{-ct} \cos \omega t$ or $d = ab^{kt} \cos \omega t$ to represent damped harmonic motion.

53. Suppose that a mass at the end of a spring is initially pulled in the positive direction 8 in. from its equilibrium position. Upon release, the displacement of the mass follows a pattern of damped harmonic motion with each cycle lasting 1 sec. If the maximum displacement for each cycle decreases by 25%, find a function that models the displacement of the mass t sec after being released. $d = 8(0.75)^t \cos 2\pi t$ or $d = 8e^{\ln(0.75)t} \cos 2\pi t$

54. A pendulum is pulled $\dfrac{\pi}{18}$ radians to one side and then released. The angular displacement θ follows a pattern of damped harmonic motion with each cycle lasting 2 sec. If the maximum displacement for each cycle decreases by 20%, find a function that models the angular displacement t sec after being released. $d = \dfrac{\pi}{18}(0.8)^{t/2} \cos \pi t$ or $d = \dfrac{\pi}{18}e^{\ln(0.8)t/2} \cos \pi t$

In a course in differential equations or physics, a more specific model for a mass moving in damped harmonic motion is found as $d = ae^{-bt/(2m)} \cos\left(t\sqrt{\omega^2 - \left(\dfrac{b}{2m}\right)^2}\right)$, where b is a damping constant unique to the physical system, m is the mass of the object, and $|a|$ is the displacement at time $t = 0$. The value $\dfrac{2\pi}{\omega}$ is the related period under simple harmonic motion. Use this model for Exercises 55–56.

55. a. Write a model for damped harmonic motion for a mass of 5 kg oscillating on a spring with initial displacement 0.8 m, period of 6 sec, and damping constant $b = 0.3 \frac{\text{N·s}}{\text{m}}$ (Newton-second/meter).

   b. Graph the model from part (a) on the window $[0, 10, 1]$ by $[-1, 1, 0.1]$.

56. a. Write a model for damped harmonic motion for the bob of a pendulum of mass of 2 kg originally pulled to the right of its equilibrium position 0.6 m. Upon release, the bob makes one complete swing back and forth in 2 sec with a damping constant $b = 0.2 \frac{\text{N·s}}{\text{m}}$.

   b. Graph the model from part (a) on the window $[0, 10, 1]$ by $[-0.8, 0.8, 0.1]$.

57. Given the model $d = 20e^{-0.6t} \cos 2\pi t$, for an object in damped harmonic motion,

   a. Determine the initial displacement, d (in centimeters).
    20 cm
   b. Graph the model on a graphing utility.

   c. Determine the time t (in seconds) between two consecutive relative maxima.   1 sec

   d. By what percentage is the displacement of the object decreased with each successive oscillation between consecutive maxima? Round to the nearest tenth of a percent.   45.1%

58. Given the model $d = -8e^{-0.4t} \cos \pi t$, for an object in damped harmonic motion,

   a. Determine the initial displacement, d (in centimeters).
    −8 cm
   b. Graph the model on a graphing utility.

   c. Determine the time t (in seconds) between two consecutive relative maxima.   2 sec

   d. By what percentage is the displacement of the object decreased with each successive oscillation between consecutive maxima? Round to the nearest tenth of a percent.   55.1%

Given a point on the Earth of latitude α, the angle of elevation of the Sun is at its minimum of $90° - \alpha - 23.5°$ at noon at the winter solstice. The angle of elevation of the Sun is at its maximum of $90° - \alpha + 23.5°$ at noon on the summer solstice. The angle of elevation follows simple harmonic motion as it varies from its minimum to its maximum over the course of 365 days. Use this information for Exercises 59–60.

59. Dallas, Texas, has latitude 32.8°N.

   a. Find the angle of elevation of the Sun at noon on the winter solstice and summer solstice.   Winter: 33.7°; Summer: 80.7°

   b. Suppose that the winter solstice is 10 days before the end of a given year. Write a function representing the angle A (in degrees) of the Sun as a function of the day number t for the following year. (*Hint:* Use a phase shift of -10 days.)

   c. Graph the function from part (b) on a graphing utility.

   d. Determine the angle of elevation of the Sun at noon on February 1 for Dallas. Round to the nearest tenth of a degree.   39.6°

   e. How many days into the year will the spring and fall equinoxes occur? Round to the nearest day. (*Hint:* The angle of elevation of the Sun at noon on an equinox will be halfway between its minimum and maximum value. This is the day of the year when the duration of daylight equals the duration of darkness.)   81 days for the spring equinox and 264 days for the fall equinox

60. Denver, Colorado, has latitude 39.7°N.

   a. Find the angle of elevation of the Sun at noon on the winter solstice and summer solstice.   Winter: 26.8°; Summer: 73.8°

   b. Suppose that the winter solstice is 10 days before the end of a given year. Write a function representing the angle A (in degrees) of the Sun as a function of the day number t for the following year. (*Hint:* Use a phase shift of -10 days.)

   c. Graph the function from part (b) on a graphing utility.

   d. Determine the angle of elevation of the Sun at noon on January 15 for Denver. Round to the nearest tenth of a degree.   28.9°

CHAPTER 6   KEY CONCEPTS

SECTION 6.1 Applications of Right Triangles	Reference
Finding the measures of the angles and sides of a right triangle when two sides or a side and an angle are known is called **solving a right triangle**.	p. 630
The **bearing** from point A to point B is the degree measure of the acute angle formed by ray $\overrightarrow{AB}$ and the north-south line. In the figure, the bearing from A to B is S39°W and the bearing from B to A is N39°E.	p. 632

SECTION 6.2 The Law of Sines	Reference
A triangle that does not contain a right angle is called an **oblique triangle.** To solve an oblique triangle, • Use the law of sines if two angles and a side are known (ASA or SAA) or if two sides and angle opposite one of them are known (SSA). • Use the law of cosines if three sides are known (SSS) or if two sides and the angle between them are known (SAS).	p. 642

Law of Sines	p. 642
If $\triangle ABC$ is a triangle with sides of lengths a, b, and c opposite angles A, B, and C, respectively, then $$\frac{a}{\sin A} = \frac{b}{\sin B} = \frac{c}{\sin C} \text{ or equivalently } \frac{\sin A}{a} = \frac{\sin B}{b} = \frac{\sin C}{c}$$	

Law of Sines—Ambiguous Case	p. 645
When the lengths of two sides and the measure of an angle opposite one of the known sides are given, then four cases are possible regarding the number of triangles that can be constructed. No triangle One right triangle One oblique triangle Two oblique triangles (a) $a < h$ (b) $a = h$ (c) $a > b$ (d) $h < a < b$	

Given two sides and the angle between them, the area of triangle $\triangle ABC$ is given by $$\text{Area} = \frac{1}{2}bc\sin A \quad \text{or} \quad \text{Area} = \frac{1}{2}ab\sin C \quad \text{or} \quad \text{Area} = \frac{1}{2}ac\sin B$$	p. 648

SECTION 6.3 The Law of Cosines	Reference
Use the law of cosines to solve an oblique triangle when • Two sides and the angle between them are known (SAS). • Three sides are known (SSS).	p. 656

Law of Cosines	p. 656
If $\triangle ABC$ has sides of lengths a, b, and c opposite vertices A, B, and C, respectively, then	

	Law of Cosines	Alternative Form
(1)	$a^2 = b^2 + c^2 - 2bc\cos A$	$\cos A = \dfrac{b^2 + c^2 - a^2}{2bc}$
(2)	$b^2 = a^2 + c^2 - 2ac\cos B$	$\cos B = \dfrac{a^2 + c^2 - b^2}{2ac}$
(3)	$c^2 = a^2 + b^2 - 2ab\cos C$	$\cos C = \dfrac{a^2 + b^2 - c^2}{2ab}$

For a triangle with sides of lengths a, b, and c, the area is given by $$\text{Area} = \sqrt{s(s - a)(s - b)(s - c)}$$ where s is the semi-perimeter of the triangle: $s = \dfrac{1}{2}(a + b + c)$.	p. 661

SECTION 6.4 Harmonic Motion	Reference
An object moving indefinitely in a sinusoidal manner with constant amplitude is said to be moving in **simple harmonic motion.** The **period** is the time required to complete one cycle and the **frequency** is the number of cycles per unit time.	p. 667

The displacement d of an object undergoing simple harmonic motion is given by $$d = a\sin\omega t \text{ or } d = a\cos\omega t \text{ or equivalently } d = a\sin 2\pi ft \text{ or } d = a\cos 2\pi ft$$ where t is the time after recording begins. • The amplitude of motion is $|a|$. • The period is given by $P = \dfrac{2\pi}{\omega}$. The frequency f is the reciprocal of the period, $f = \dfrac{1}{P}$.	p. 667
An object undergoes **damped harmonic motion** if the amplitude of oscillation decreases with time. The displacement of an object under damped harmonic motion at time t is given by $$d = ae^{-ct}\cos\omega t \quad \text{or} \quad d = ae^{-ct}\sin\omega t \quad \text{for } c > 0$$ • a is the displacement of the object at $t = 0$. • c is a constant called the **damping factor** and is related to the damping force opposing motion. • The value of ω is related to the period P and frequency f by $P = 2\pi/\omega$ and $\omega = 2\pi f$.	p. 670

Expanded Chapter Summary available at www.mhhe.com/millerprecalculus.

CHAPTER 6   Review Exercises

SECTION 6.1

For Exercises 1–4, solve the right triangle for the unknown sides and angles. Round values to 1 decimal place if necessary.

1. $B = 37°$, $a = 17$
   $A = 53°,\ b \approx 12.8,\ c \approx 21.3$
2. $A = 7.8°$, $c = 29$
   $B = 82.2°,\ a \approx 3.9,\ b \approx 28.7$
3. $b = 12.6$, $c = 40.8$
   $A \approx 72.0°,\ B \approx 18.0°,\ a \approx 38.8$
4. $a = 427$, $b = 120$
   $A \approx 74.3°,\ B \approx 15.7°,\ c \approx 443.5$

5. A 20-ft-tall light post casts a 25-ft shadow along level ground. What is the angle of elevation of the sun at that time? Round to the nearest degree.   39°

6. A passenger lift runs 680 ft up the side of a mountain from the base to an exit station. If the change in elevation of the lift from the bottom to the top of the mountain is 228 ft, what is the angle of incline to the nearest degree?   20°

7. A 75-ft guy wire is attached to the top of a tower. If the guy wire makes an angle of 38.2° with the ground, how tall is the tower? Round to the nearest tenth of a foot.   46.4 ft

8. A 125-ft kite string is anchored to the ground. If the string makes an angle of 49° with the ground, what is the distance from the anchor site to a point directly beneath the kite? Round to the nearest foot.   82 ft

For Exercises 9–10, find the bearing.

9.   N35°W   **10.**   S15°W

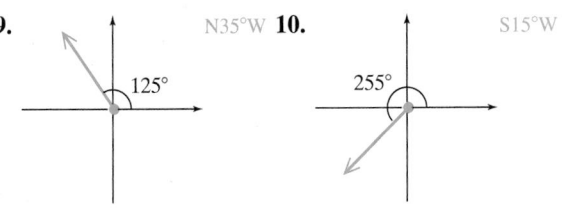
125°   255°

11. A plane flies due west for 2 hr then due south for 1 hr at an average speed of 375 mph. To the nearest degree, what bearing should the plane take for the return trip?   N63°E

12. A boat travels 3 mi east then 7 mi south. To the nearest tenth of a degree, what is the bearing from its starting point?   S23.2°E

13. A ship leaves a dock on a bearing of S48.2°W. After traveling for 1.6 hr at 28 mph, how far south and west has the ship traveled? Round to the nearest tenth of a mile.   29.9 mi south; 33.4 mi west

14. A plane flying 275 mph on a bearing of N73.5°W travels for 4 hr. How far north and west has the plane traveled? Round to the nearest mile.   312 mi north; 1055 west

SECTION 6.2

For Exercises 15–20, solve $\triangle ABC$ subject to the given conditions. Round the lengths of sides and measures of the angles (in degrees) to 1 decimal place if necessary.

15. $B = 63°$, $a = 26$, $b = 42$ $A \approx 33.5°,\ C \approx 83.5°,\ c \approx 46.8$

16. $C = 140°$, $b = 100$, $c = 200$ $A \approx 21.3°,\ B \approx 18.7°,\ a \approx 112.8$

17. $A = 26°$, $a = 8$, $b = 10$
   $B \approx 33.2°,\ C \approx 120.8°,\ c \approx 15.7$ or $B \approx 146.8°,\ C \approx 7.2°,\ c \approx 2.3$

18. $B = 12°$, $C = 122°$, $b = 17$ $A = 46°,\ a \approx 58.8,\ c \approx 69.3$

19. $A = 78°$, $a = 4$, $b = 18$ No triangle

20. $A = 113°$, $B = 28°$, $a = 20$ $C = 39°,\ b \approx 10.2,\ c \approx 13.7$

For Exercises 21–22, find the area of a triangle with the given measurements. Round to 1 decimal place.

21. $A = 129°$, $b = 14.8$ mi, $c = 24$ mi   138.0 mi²

22. $C = 16.3°$, $a = 2.1$ cm, $b = 1.4$ cm   0.4 cm²

23. Points on a straight shoreline, *A* and *B*, are separated by 700 yd. Observers at each point record the bearing to a ship at sea. The observer at point *A* locates the ship at N50°E, while the observer at point *B* notes the ship on a bearing of N38°W. If point *A* is due west of point *B*, how far is the ship from each observer? Round to the nearest yard.
From *A* to the ship ≈ 552 yd, from *B* to the ship ≈ 450 yd

24. Two landing points, *A* and *B*, lie on the straight bank of a river and are separated by 45 ft. Find the distance from each landing point to a boat pulled ashore on the opposite bank at a point *C* if ∠*ABC* is 68° and ∠*BAC* is 73°. Round to the nearest foot.   *AC* ≈ 66 ft, *BC* ≈ 68 ft

25. A ranger station at a point *R* in a wilderness area is located 5.6 mi from a campground at point *C* (see figure). A camper hikes 4 mi in a linear path away from the campground to point *H*, and then shoots a flare straight up as a distress signal. The signal is seen from the ranger station such that ∠*HRC* = 35°. To the nearest tenth of a mile, how far is the camper from the ranger station?
2.2 mi or 7.0 mi

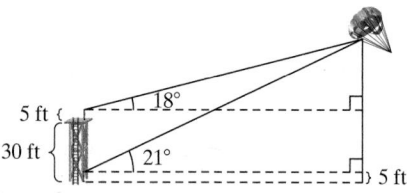

26. Standing on top of a 30-ft lifeguard tower, an observer measures the angle of elevation of a parasail to be 18°. The same observer standing on the ground next to the tower measures the angle of elevation as 21°. If the eye level of the observer is 5 ft, determine the height of the parasail to the nearest foot.   200 ft

27. Points *A*, *B*, and *P* are collinear points along a hillside. A blimp located at point *Q* is directly overhead point *P*. Points *A* and *B* are 200 yd apart, and the angle of elevation (relative to the horizontal) from *B* to the blimp is 48°. The angle of elevation from point *A* farther down the hill to the blimp is 44°.

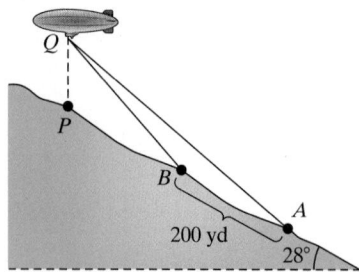

a. To the nearest yard, approximate the distance between point *A* and the blimp and the distance between point *B* and the blimp.   *AQ* = 981 yd, *BQ* = 790 yd

b. Find the exact height of the blimp relative to ground level (distance between *P* and *Q*). $PQ = \dfrac{200\sin 16°\sin 20°}{\sin 4°\sin 118°}$

c. Approximate the height from part (b).   Height ≈ 306 yd

28. A surveyor fixes a baseline of 64 ft between two points *A* and *B* on a plot of land.

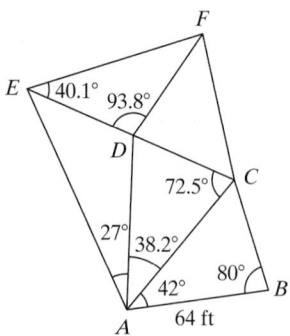

a. Write an expression for the exact length of *AD*.

b. Approximate *AD* to the nearest tenth of a foot.
AD ≈ 75.8 ft

SECTION 6.3

For Exercises 29–32, solve △ABC subject to the given conditions if possible. Round the lengths of the sides and measures of the angles (in degrees) to 1 decimal place if necessary.

29. *a* = 42, *b* = 66, *c* = 31   *A* ≈ 29.7°, *B* ≈ 128.8°, *C* ≈ 21.5°

30. *a* = 3.1, *b* = 2.3, *c* = 5   *A* ≈ 26.0°, *B* ≈ 18.9°, *C* ≈ 135.1°

31. *a* = 17.3, *b* = 24.2, *C* = 63°   *c* ≈ 22.5, *A* ≈ 43.3°, *B* ≈ 73.7°

32. *b* = 231, *c* = 478, *A* = 12.1°
a ≈ 256.7, *B* ≈ 10.9°, *C* ≈ 157.0°

For Exercises 33–34, find the measures of the angles within the triangle formed by the given points. Round to the nearest tenth of a degree.

33. *A*(1, 0), *B*(5, 2), *C*(2, 5)   **34.** *A*(0, 3), *B*(5, 1), *C*(2, 4)
∠*A* ≈ 52.1°, ∠*B* ≈ 71.6°, ∠*C* ≈ 56.3°

In Exercises 35–36, find the area of the triangle with sides of the given lengths. Round to the nearest tenth of a square unit.

35. *a* = 26.8 ft, *b* = 12.2 ft, *c* = 20 ft   114.4 ft²

36. *a* = 47 mm, *b* = 32 mm, *c* = 60 mm   746.4 mm²

37. Two straight jogging paths begin at a kiosk in a park, and the angle between the paths is 73°. Two runners leave the kiosk at the same time each taking one of the paths. One person runs at a rate of 5 mph, the other at a rate of 6.2 mph. After 30 min, how far apart are the runners? Round to the nearest tenth of a mile.   3.4 mi

38. Given points *A*(6, 2) and *B*(1, 4) on the coordinate plane, find the measure of angle ∠*AOB*, where point *O* is at the origin. Round to the nearest tenth of a degree.   57.5°

39. A boat leaves a marina and travels due south for 1 hr. The boat then changes course to a bearing of S47°E and travels for another 2 hr.

a. If the boat keeps a constant speed of 15 mph, how far from the marina is the boat after 3 hr? Round to the nearest tenth of a mile.   41.7 mi

b. Find the bearing from the boat back to the marina. Round to the nearest tenth of a degree.   N31.7°W

40. Two planes take off from an airport at the same time. The first plane averages 480 mph and flies on a bearing of N20°E. The second plane averages 360 mph and heads out at a bearing of N35°W.

    a. After 1.5 hr, how far apart are the two planes? Round to the nearest tenth of a mile.   603.3 mi

    b. Find the bearing from the first plane to the second plane. Round to the nearest tenth of a degree.   S67.2°W

41. Two sides of a triangular plot of land measure 510 ft and 237 ft and the angle between them is 78.1°.

    a. If the land is $5500 per acre, what is the cost for this plot? Round to the nearest dollar and use the fact that 1 acre is 4840 yd^2. $7467

    b. If materials for a chain-link fence cost $4.20 per linear foot, what is the cost to completely enclose the property? Round to the nearest dollar. $5305

42. A gardener digs out a space for a triangular flower bed by her front entryway. The triangular plot measures 10 ft by 12 ft by 14 ft and is 6 in. deep. How many cubic feet of potting mix will she need to fill her plot? Round to the nearest tenth of a cubic foot.   29.4 ft^3

SECTION 6.4

For Exercises 43–46, suppose that an object moves in simple harmonic motion with displacement d (in centimeters) at time t (in seconds). Determine the

    a. Amplitude   **b.** Period   **c.** Frequency   **d.** Phase shift

43. $d = -8\cos\dfrac{\pi}{6}t$ **44.** $d = 5\sin 30\pi t$

45. $d = 2\sin\left(t - \dfrac{\pi}{3}\right)$ **46.** $d = \dfrac{1}{2}\cos\left(\dfrac{1}{3}t + \dfrac{1}{6}\right)$

For Exercises 47–48, a block is attached to a horizontal spring.

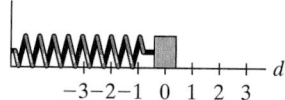

47. At time $t = 0$, the block is initially pulled to the right 6 cm and then released. If the block completes one back-and-forth cycle in 2 sec, write a model for the displacement d as a function of the time t (in seconds) after release. $d = 6\cos\pi t$

48. At time $t = 0$, if the block is moving to the left with an initial displacement of 0 ft, a maximum displacement of $\dfrac{1}{3}$ ft and a frequency of 3 Hz, find a model for the displacement d as a function of time t. $d = -\dfrac{1}{3}\sin 6\pi t$

49. The angular displacement θ (in radians) for a simple pendulum is given by $\theta = 0.175\sin\pi t$.

    a. Determine the period of the pendulum.   2 sec

    b. How many swings are completed in 1 sec? $\dfrac{1}{2}$ swing

    c. To the nearest degree, what is the maximum displacement of the pendulum?   10°

    d. The length L of a pendulum is related to the period T by the equation $T = 2\pi\sqrt{\dfrac{L}{g}}$, where g is the acceleration due to gravity. Find the length of the pendulum to the nearest meter ($g = 9.8$ m/sec^2).   1 m

50. A Ferris wheel is 200 ft in diameter with its lowest point 3 ft off the ground. Once all the passengers have been loaded, the wheel makes one full rotation counterclockwise in 3 min. Suppose that a couple is seated at the lowest point on the wheel and are the last passengers to be loaded.

    a. Write a model representing the couple's horizontal position x (in feet) relative to the center of the Ferris wheel, t minutes after the ride starts. $x = 100\sin\dfrac{2\pi}{3}t$

    b. Write a model representing the couple's height y (in feet) above ground level, t minutes after the ride starts.

    c. Give the coordinates of the couple's position 2 min into the ride and describe the position.

For Exercises 51–52, use a graphing utility to graph the function and bounding curves for $t \geq 0$.

51. $d = 4e^{-0.03t}\cos 1.4t$ **52.** $d = -6e^{-1.3t}\sin 5\pi t$

CHAPTER 6   Test

For Exercises 1–8, solve $\triangle ABC$ subject to the given conditions. Round the lengths of sides and measures of the angles to 1 decimal place if necessary.

1. $A = 62°$, $C = 90°$, $a = 17$ $B = 28°$, $b \approx 9.0$, $c \approx 19.3$

2. $a = 17.6$, $b = 20$, $c = 12.8$ $A \approx 60.2°$, $B \approx 80.6°$, $C \approx 39.2°$

3. $B = 97.3°$, $C = 12.7°$, $b = 77.1$ $A = 70.0°$, $a \approx 73.0$, $c \approx 17.1$

4. $B = 14.8°$, $C = 90°$, $a = 213$ $A = 75.2°$, $b \approx 56.3$, $c \approx 220.3$

5. $C = 81°$, $a = 59$, $b = 44$ $A \approx 59.2°$, $B \approx 39.8°$, $c \approx 67.9$

6. $A = 120°$, $B = 30°$, $a = 51$ $C = 30°$, $b \approx 29.4$, $c \approx 29.4$

7. $B = 13°$, $a = 7$, $b = 2$ $A \approx 51.9°$, $C \approx 115.1°$, $c \approx 8.1$ or $A \approx 128.1°$, $C \approx 38.9°$, $c \approx 5.6$

8. $B = 13°$, $a = 7$, $b = 1$ No triangle

For Exercises 9–12, find the area of $\triangle ABC$ with the given measurements. Round to 1 decimal place.

9. $A = 102°$, $b = 230$ yd, $c = 188$ yd   21,147.6 yd²

10. $C = 90°$, $c = 56$ cm, $a = 12$ cm   328.2 cm²

11. $a = 28$ m, $b = 46$ m, $c = 23$ m   251.8 m²

12. $B = 43°$, $a = 1.8$ ft, $c = 6.8$ ft   4.2 ft²

13. A ship travels west at 25 knots. At 7:30 P.M., the captain sees a floating becon due south of the ship. By 9:15 P.M., the captain sights the becon at a bearing of S32°E. What was the distance from the ship to the becon at the time of the first sighting? Round to the nearest nautical mile.
70 nmi

14. Over the course of a day, as the angle of inclination of the Sun decreases from 64° to 49°, the length of the shadow of a building increases by 60 ft. How tall is the building? Round to the nearest foot.   157 ft

15. To determine the height of a tower, a surveyor notes that the angle of elevation to the top of the tower is 48° when measured at a point on the ground 45 ft from the base of the tower. To the nearest tenth of a foot, how tall is the tower?   50.0 ft

16. In addition to the light-year to quantify large distances in outer space, astronomers often use two other units of measurement.

   • the **astronomical unit** (AU), which is the average distance between the Earth and Sun (1 AU ≈ 93,000,000 mi).

   • the **parsec** (pc), which is the distance from the Earth to an object having a parallax angle of 1″ (1 second of arc).

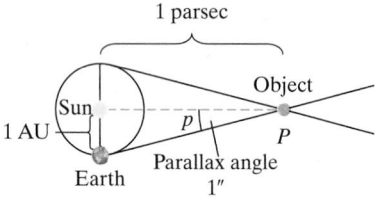

1 parsec

a. To the nearest whole unit, how many astronomical units are in 1 parsec?   206,265 AU

b. How many light-years are in 1 parsec? Round to 2 decimal places. (*Hint*: 1 light-year ≈ 5.878×10^{12} mi.)
3.26 light-years

17. A triangular shade sail has sides of length 12′3″ and 15′6″, and the angle between them is 73°18′.

a. If the cost of the sail material is \$8.75/ft², how much would this sail cost? Round to the nearest dollar.   \$796

b. If the cost of wrapping the edges of the sail with a decorative border in a contrasting color is \$1.25/yd, what is the added cost to the sail? Round to the nearest dollar.   \$19

18. The distance between Los Angeles, California, and Salt Lake City, Utah, on a map is 9.5 in. The distance from Salt Lake City to Albuquerque, New Mexico, is 8.0 in., and the distance from Albuquerque to Los Angeles is 11.0 in. If the bearing from Salt Lake City to Los Angeles is S34°W, find the bearing from Albuquerque to Los Angeles. Round to the nearest tenth of a degree.   S79.2°W

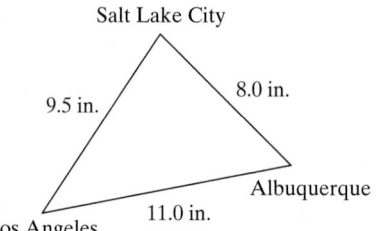

19. The connector rod from the piston to the crankshaft in a certain engine is 8.5 in. The radius of the crank circle is 3 in. If the angle made by the connector rod with the vertical at the wrist pin P is 12°, how far is the wrist pin from the center C of the crankshaft? Round to the nearest tenth of an inch.
5.9 in. or 10.7 in.

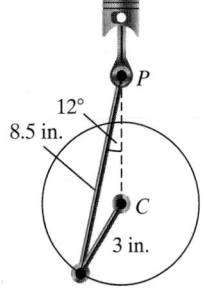

20. Refer to the figure.

a. Write expressions for the exact values of a and b.

b. Approximate a and b to the nearest tenth of a foot.
$a \approx 6.9$ ft; $b \approx 3.2$ ft

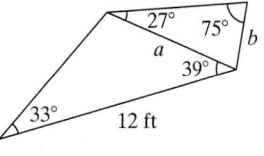

21. Explain what the relationships $\dfrac{a}{\sin A} = \dfrac{b}{\sin B} = \dfrac{c}{\sin C}$ given in the law of sines represent.

22. The displacement of an object attached to a spring that moves in simple harmonic motion is given by $d = -1.8\cos 12\pi t$, where the displacement d is in centimeters at time t in seconds. Determine the amplitude and frequency for the motion of the object.
Amplitude = 1.8 cm; Frequency = 6 Hz

23. The displacement of an object attached to a spring that moves in simple harmonic motion is given by

$$d = 4\sin\left(2t - \frac{\pi}{4}\right),$$ where the displacement d is in centimeters at time t is in seconds. Determine the period and phase shift for the motion of the object.

24. A block hangs on a spring attached to the ceiling and is pulled down 8 in. below its equilibrium position. After release, the block makes one complete up-and-down cycle in 1.2 sec and follows simple harmonic motion.

  a. Write a function to model the displacement d (in inches) as a function of the time t (in seconds) after release. Assume that a displacement above the equilibrium position is positive. $d = -8\cos\dfrac{5\pi}{3}t$

  b. Find the displacement of the block and direction of movement at $t = 0.3$ sec. $d = 0$; The block is at its equilibrium position and moving upward.

25. The angular displacement θ (in radians) for a simple pendulum is given by $\theta = 0.14\sin\dfrac{\pi}{2}t$.

  a. Determine the period of the pendulum.   4 sec

  b. How many swings are completed in 1 sec? $\dfrac{1}{4}$ swing

  c. To the nearest degree, what is the maximum displacement of the pendulum?   8°

  d. The length L of a pendulum is related to the period T by the equation $T = 2\pi\sqrt{\dfrac{L}{g}}$, where g is the acceleration due to gravity. Find the length of the pendulum to the nearest foot ($g = 32$ ft/sec²).   13 ft

26. A 20-ft-diameter waterwheel operates at a grain mill and turns at a rate of 1.5 rev/min, counterclockwise. Suppose that a coordinate system is chosen with the origin at the bottom of the wheel (at water level). At $t = 0$, suppose that a point on the wheel is located at $(0, 0)$.

  a. Write a function that represents the x position (in feet) of the point as a function of the time t in minutes. $x = 10\sin 3\pi t$

  b. Write a function that represents the y position (in feet) of the point as a function of the time t in minutes. $y = 10 - 10\cos 3\pi t$

  c. Give the coordinates of the point at $t = 10$ sec and describe the location. $(10, 10)$; The point is 10 ft to the right of and 10 ft above the bottom of the wheel.

27. Use a graphing utility to graph the function $d = 4e^{-0.25t}\sin 1.3t$ and bounding curves for $t \geq 0$.

CHAPTER 6   Cumulative Review Exercises

1. Write the equation of the line in slope-intercept form that passes through the point $\left(-3, \dfrac{1}{2}\right)$ with slope $m = \dfrac{2}{3}$.

2. Write the equation of a circle in standard form with a center of $(-8, 12)$ and that passes through the point $(-15, 12)$. $(x + 8)^2 + (y - 12)^2 = 49$

3. Write the function defined by $f(x) = -2x^2 - 12x$ in vertex form and identify the vertex of the parabola. $f(x) = -2(x + 3)^2 + 18$; $(-3, 18)$

4. Write a polynomial $f(x)$ of degree 5, with integer coefficients, and with zeros -2 (multiplicity 2), 6, and $2i$. $f(x) = x^5 - 2x^4 - 16x^3 - 32x^2 - 80x - 96$

5. Write equations for the asymptotes for the graph of $f(x) = \dfrac{2x^2 + 5x - 3}{x - 8}$.   Vertical asymptote: $x = 8$; Slant asymptote: $y = 2x + 21$

6. Write equations for the asymptotes for the graph of $f(x) = \sec 2x$ on the interval $\left[-\dfrac{\pi}{2}, \dfrac{\pi}{2}\right]$.

7. Given $f(x) = \dfrac{x + 1}{x - 3}$, write an equation for the inverse function. $f^{-1}(x) = \dfrac{3x + 1}{x - 1}$

8. Write an equation for a sine function with an amplitude of 6, a period of π, and a phase shift of $\dfrac{\pi}{6}$.

9. Write an equation for a tangent function with a period of $\dfrac{1}{3}$ and a phase shift of $-\dfrac{4}{3}$. $y = \tan(3\pi x + 4\pi)$

10. A motorist drives on a toll road to and from work each day and pays \$4.25 in tolls one-way.

  a. Write a model for the cost C (in \$) for tolls in terms of the number of working days, x. $C(x) = 8.5x$

  b. The department of transportation offers a prepaid toll pass for \$150 a month. How many working days are required for the motorist to save money by buying the pass?   The motorist will save money beginning on the 18th working day.

11. a. Write a variation model using k as the constant of variation if Z is directly proportional to the cube of A and inversely proportional to the square root of B.

  b. If $Z = 100$ when $A = 2$ and $B = 25$, what is the constant of proportionality k? $k = 62.5$

12. a. Write an equation representing the fact that the product of two consecutive odd integers is 323. $x(x + 2) = 323$

  b. Solve the equation from part (a) to find the two integers.   The integers are 17 and 19 or -17 and -19.

13. A sales person makes a base salary of \$2500 per month. Once she reaches \$30,000 in total sales, she earns an additional 5% commission on the amount in sales over \$30,000. Write a piecewise-defined function to model the sales person's total monthly salary $S(x)$ (in \$) as a function of the amount in sales x.

14. The cost to buy tickets to a concert is $50 per ticket.

    a. Write a function that represents the cost $C(x)$ (in $) for x tickets to the concert. $C(x) = 50x$

    b. There is a sales tax of 6% and a processing fee of $5 for a group of tickets. Write a function that represents the total cost $T(a)$ for a dollars spent on tickets. $T(a) = 1.06a + 5$

    c. Find $(T \circ C)(x)$. $(T \circ C)(x) = 53x + 5$

    d. Find $(T \circ C)(10)$ and interpret its meaning in the context of this problem. $(T \circ C)(10) = 535$; The total cost to purchase 10 tickets is $535.

15. Suppose that the population of a country in the year 2000 was 19.0 million and grew to 22.6 million in 2010. Write a model of the form $P(t) = P_0 e^{kt}$, where $P(t)$ is the population in millions, t years after the year 2000. Round the growth rate to 5 decimal places. $P(t) = 19.0e^{0.01735t}$

16. The bob on a simple pendulum is pulled to the left 6 in. from its equilibrium position. After release, the pendulum makes one complete back-and-forth cycle in 3 sec and follows simple harmonic motion. Write a function to model the displacement d (in inches) of the bob as a function of the time t (in seconds) after release. Assume that a displacement to the right of the equilibrium position is positive. $d = -6\cos\dfrac{2\pi}{3}t$

17. Simplify to a single trigonometric function.

$$\frac{1}{1 - \cos x} + \frac{1}{1 + \cos x} \qquad 2\csc^2 x$$

18. Find the exact value of
$\cos 200° \cos 50° + \sin 200° \sin 50°$. $-\dfrac{\sqrt{3}}{2}$

For Exercises 19–20, solve $\triangle ABC$ subject to the given conditions. Round the lengths of sides and measures of the angles to 1 decimal place if necessary.

19. $a = 12.8, b = 20, c = 17.1$ $A \approx 39.4°, B \approx 82.6°, C \approx 58.0°$

20. $a = 12.8, b = 20, B = 98.7°$ $c \approx 13.6, A \approx 39.2°, C \approx 42.1°$

7

Trigonometry Applied to Polar Coordinate Systems and Vectors

Chapter Outline

Suppose that a mathematically inclined friend tells you that a bag with $1000 in cash is yours if you can find the bag. Your friend also tells you that the bag is located 5 mi from your current location. As a clever student of mathematics, you would know that the bag is located somewhere on a circle of radius 5 mi centered at your current location. Unfortunately, walking the circumference of the circle to find the bag might be a long walk. Now suppose that your friend tells you that the bag is located 5 mi from your current location in a specific direction, such as northeast. Then after walking 5 mi in the designated direction, you will be $1000 richer.

The distance 5 mi by itself is called a *scalar* quantity because it tells us how much of something (in this case, distance). However, the quantity 5 mi northeast is a *vector* quantity because it tells us how far *and* in what direction. The study of vectors is critically important in mathematics, physics, navigation, and many branches of engineering.

In this chapter we will study vectors, along with two other topics that may seem unrelated, but are actually very similar mathematically. The first is the study of a polar coordinate system in which points in a plane are located according to their distance from the origin and direction from a line of reference. The second is the study of the complex plane, where we can graph complex numbers and visualize their properties geometrically.

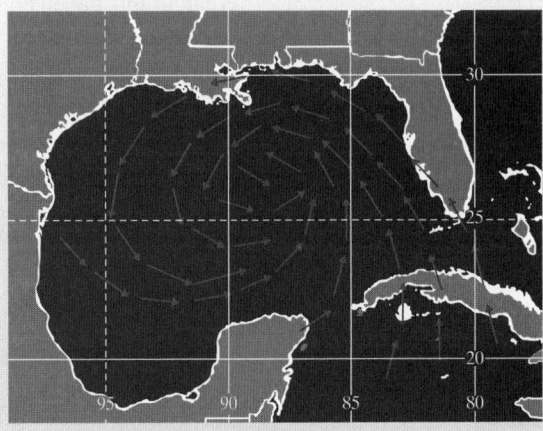

Meteorologists use vectors to describe the speed and direction of wind in a weather system.

SECTION 7.1 **Polar Coordinates**

OBJECTIVES

1. **Plot Points Using Polar Coordinates**
2. **Convert Ordered Pairs Between Polar and Rectangular Coordinates**
3. **Convert Between Equations in Polar and Rectangular Coordinates**

> **TIP** The polar axis in a polar coordinate system is usually aligned with the positive x-axis in a rectangular coordinate system.

1. Plot Points Using Polar Coordinates

Suppose that a ship leaves port and sends out a distress call after traveling 10 mi at a bearing of N30°E. Although the rescue team does not have the x- and y-coordinates of the ship relative to the port (the origin), the exact location of the ship is known. The ship's location is defined by its distance from port, 10 mi, and its direction, 30° east of north.

Defining a point in a plane by using distance and direction is the basis of a coordinate system called a polar coordinate system.

Polar Coordinate System

A **polar coordinate system** consists of a fixed point O called the **pole** (or origin), and a ray called the **polar axis**, with endpoint at the pole (Figure 7-1). Each point P in the plane is defined by an ordered pair of the form (r, θ), where r is the directed distance from the pole to P.

- If $r > 0$, point P is located r units from the pole in the direction of θ.
- If $r < 0$, point P is located $|r|$ units from the pole in the direction of $\theta + \pi$ (the direction *opposite* θ).
- If $r = 0$, point P is located at the pole.

θ is a directed angle from the polar axis to line $\overleftrightarrow{OP}$.

- $\theta > 0$ is measured counterclockwise from the polar axis.
- $\theta < 0$ is measured clockwise from the polar axis.

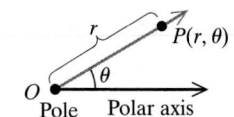

Figure 7-1

Classroom Examples: p. 694
Exercises 16, 18

EXAMPLE 1 Plotting Points in a Polar Coordinate System

Plot the points whose polar coordinates are given.

a. $\left(3, \dfrac{5\pi}{6}\right)$ **b.** $\left(5, -\dfrac{\pi}{4}\right)$ **c.** $(3.5, 3\pi)$ **d.** $\left(-4, \dfrac{\pi}{3}\right)$

Solution:

Polar graph paper consists of concentric circles centered at the pole whose radii represent incremental distances from the pole.

> **TIP** Polar coordinates are sometimes given with θ in degree measure. For example, the point $\left(3, \dfrac{5\pi}{6}\right)$ can also be represented by $(3, 150°)$.

a. $\left(3, \dfrac{5\pi}{6}\right)$

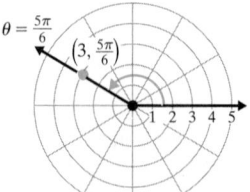

To locate $\left(3, \dfrac{5\pi}{6}\right)$, draw the terminal side of angle $\theta = \dfrac{5\pi}{6}$. Then measure 3 units from the pole in that direction.

b. $\left(5, -\dfrac{\pi}{4}\right)$

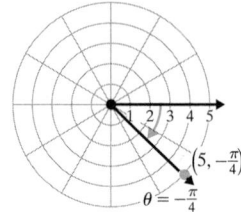

To locate $\left(5, -\dfrac{\pi}{4}\right)$, draw the terminal side of angle $\theta = -\dfrac{\pi}{4}$. Then measure 5 units from the pole in that direction.

c. $(3.5, 3\pi)$

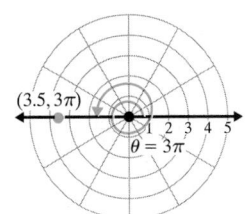

To locate $(3.5, 3\pi)$, draw the terminal side of angle $\theta = 3\pi$. Then measure 3.5 units from the pole in that direction.

d. $\left(-4, \dfrac{\pi}{3}\right)$

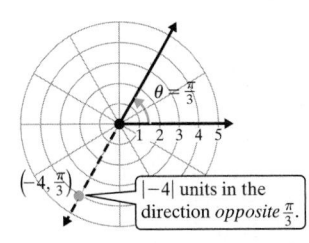

To locate $\left(-4, \frac{\pi}{3}\right)$, draw the terminal side of angle $\theta = \frac{\pi}{3}$. Then measure $|-4| = 4$ units in the direction opposite $\theta = \frac{\pi}{3}$. That is, measure 4 units from the pole in the direction of $\frac{\pi}{3} + \pi$.

Skill Practice 1 Plot the points whose polar coordinates are given.

a. $\left(3, \dfrac{5\pi}{4}\right)$ **b.** $\left(4, -\dfrac{7\pi}{6}\right)$ **c.** $(2.5, 4\pi)$ **d.** $\left(-3, \dfrac{4\pi}{3}\right)$

The location of a point is uniquely described by an ordered pair (x, y) in a rectangular coordinate system. However, with polar coordinates, an infinite number of ordered pairs can represent a point in the plane. For example, the ordered pairs (r, θ) and $(r, \theta + 2n\pi)$ for any integer n represent the same point in polar coordinates. The coordinate $\theta + 2n\pi$ represents infinitely many angles coterminal to θ. One such example is shown in Figure 7-2. Likewise, (r, θ) and $(-r, \theta + (2n + 1)\pi)$ represent the same point. The coordinate $\theta \pm (2n + 1)\pi$ represents θ plus or minus an odd multiple of π. This is an angle with terminal side in the opposite direction of θ (Figure 7-3).

Instructor Note:
As a quick classroom exercise to check student understanding, ask students to find three ordered pairs that represent a given point such as $\left(2, \frac{\pi}{4}\right)$ in polar coordinates.

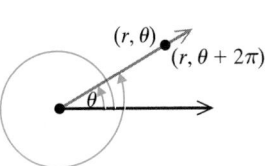

Figure 7-2   **Figure 7-3**

Classroom Example: p. 694
Exercise 20

EXAMPLE 2 **Finding Multiple Representations of a Point in Polar Coordinates**

Given $\left(4, \dfrac{\pi}{6}\right)$ in polar coordinates, find two other polar coordinate representations

a. with $r > 0$.   **b.** with $r < 0$.

Solution:

a. Using a positive value for r, we have $r = 4$. Since r is unchanged, any angle coterminal to $\theta = \dfrac{\pi}{6}$ will

complete the ordered pair: $\left(4, \dfrac{\pi}{6} + 2n\pi\right)$.

For example: $\left(4, \dfrac{\pi}{6} + 2\pi\right) = \left(4, \dfrac{13\pi}{6}\right)$

$\left(4, \dfrac{\pi}{6} - 2\pi\right) = \left(4, -\dfrac{11\pi}{6}\right)$

Answers

1. a. – d.

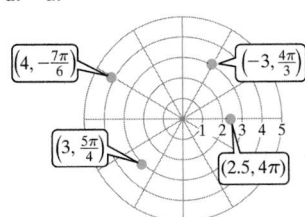

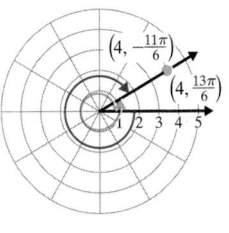

b. Using a negative value for r, we have $r = -4$. Any angle coterminal to $\theta = \dfrac{\pi}{6} + \pi$ will complete the ordered pair. For example:

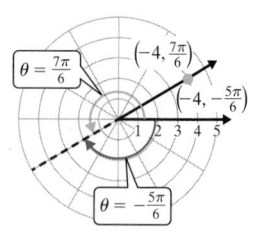

$$\left(-4, \frac{\pi}{6} + \pi\right) = \left(-4, \frac{7\pi}{6}\right)$$

$$\left(-4, \frac{\pi}{6} - \pi\right) = \left(-4, -\frac{5\pi}{6}\right)$$

Skill Practice 2 Given $\left(7, \dfrac{3\pi}{2}\right)$ in polar coordinates, find two other polar coordinate representations

a. with $r > 0$.   **b.** with $r < 0$.

2. Convert Ordered Pairs Between Polar and Rectangular Coordinates

A point in a plane can be described by an ordered pair in rectangular coordinates or by an ordered pair in polar coordinates. It seems reasonable that we should be able to convert between the two coordinate systems.

To convert the coordinates of a point P in polar coordinates (r, θ) to rectangular coordinates (x, y), first consider the case with $r > 0$. From Figure 7-4,

$$x = r\cos\theta \quad \text{and} \quad y = r\sin\theta \qquad (1)$$

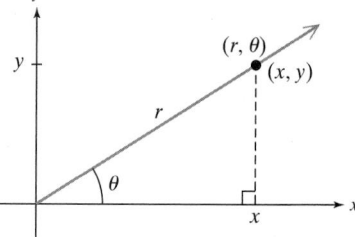

If $r = 0$, then P is at the pole with rectangular coordinates $(0, 0)$ and equation (1) holds true.

If $r < 0$, then point P can be represented by $(-r, \theta + \pi)$, where $-r > 0$. Since $\cos(\theta + \pi) = -\cos\theta = \dfrac{x}{-r}$, we have $x = r\cos\theta$.

Likewise, since $\sin(\theta + \pi) = -\sin\theta = \dfrac{y}{-r}$, we have $y = r\sin\theta$. Thus, equation (1) holds true for any real number r.

Figure 7-4

Convert Between Rectangular and Polar Coordinates

1. To convert the polar coordinates (r, θ) to rectangular coordinates (x, y), use $x = r\cos\theta$ and $y = r\sin\theta$.
2. To convert the rectangular coordinates (x, y) to polar coordinates (r, θ), use $r^2 = x^2 + y^2$ and $\tan\theta = \dfrac{y}{x}$ for $x \neq 0$.

Classroom Examples: p. 694
Exercises 26, 28

EXAMPLE 3   **Converting from Polar to Rectangular Coordinates**

Convert the ordered pairs in polar coordinates to rectangular coordinates.

a. $\left(4, \dfrac{3\pi}{2}\right)$

b. $\left(-2, -\dfrac{5\pi}{6}\right)$

Answers

2. a. $\left(7, \dfrac{7\pi}{2}\right), \left(7, -\dfrac{\pi}{2}\right)$

b. $\left(-7, \dfrac{\pi}{2}\right), \left(-7, -\dfrac{3\pi}{2}\right)$

Solution:

a. Given $\left(4, \dfrac{3\pi}{2}\right)$,

$$x = r\cos\theta = 4\cos\dfrac{3\pi}{2} = 4(0) = 0$$

$$y = r\sin\theta = 4\sin\dfrac{3\pi}{2} = 4(-1) = -4.$$

The rectangular coordinates are $(0, -4)$.

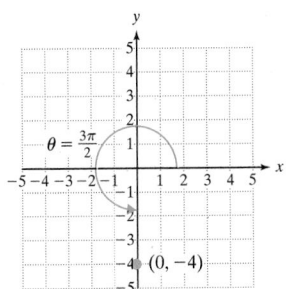

b. First note that $\left(-2, -\dfrac{5\pi}{6}\right)$ defines a point in Quadrant I. Therefore, we expect the x- and y-coordinates to be positive.

$$x = r\cos\theta = -2\cos\left(-\dfrac{5\pi}{6}\right) = -2\left(-\dfrac{\sqrt{3}}{2}\right) = \sqrt{3}$$

$$y = r\sin\theta = -2\sin\left(-\dfrac{5\pi}{6}\right) = -2\left(-\dfrac{1}{2}\right) = 1$$

The rectangular coordinates are $\left(\sqrt{3}, 1\right)$.

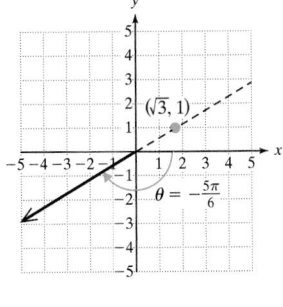

> **Skill Practice 3**   Convert the ordered pairs in polar coordinates to rectangular coordinates.
>
> **a.** $(5, \pi)$ **b.** $\left(-10, -\dfrac{2\pi}{3}\right)$

Changing from polar coordinates to rectangular coordinates as in Example 3 is generally easier than the reverse process because the rectangular coordinates of a point are unique. In Example 4, we will convert from rectangular coordinates to polar coordinates, where we see that the value of r (whether taken positive or negative) depends on the choice of θ and vice versa.

Classroom Examples: p. 695
Exercises 34, 40

> **EXAMPLE 4**   Converting from Rectangular to Polar Coordinates
>
> Convert $(-3, 3)$ from rectangular coordinates to polar coordinates. Give two representations, one with $r > 0$ and one with $r < 0$.

Solution:

The point $(-3, 3)$ is in Quadrant II with $x = -3$ and $y = 3$.

$$r^2 = x^2 + y^2 = (-3)^2 + (3)^2 = 18,$$
so $r = \pm\sqrt{18} = \pm 3\sqrt{2}$

$$\tan\theta = \dfrac{y}{x} = \dfrac{3}{-3} = -1$$

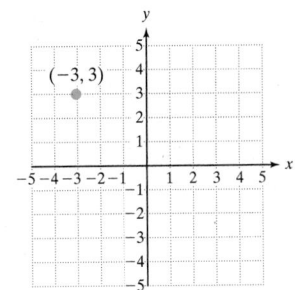

- Taking $r > 0$, we require $\tan\theta = -1$ for θ in Quadrant II. One possibility is $r = 3\sqrt{2}$ and $\theta = \dfrac{3\pi}{4}$.

- Taking $r < 0$, we require $\tan\theta = -1$ for θ in Quadrant IV. One possibility is $r = -3\sqrt{2}$ and $\theta = -\dfrac{\pi}{4}$.

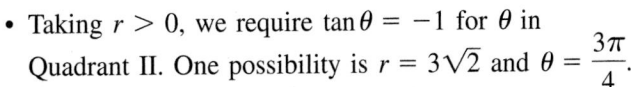

Answers

3. a. $(-5, 0)$ **b.** $(5, 5\sqrt{3})$

In polar coordinates, we have $\left(3\sqrt{2}, \dfrac{3\pi}{4}\right)$ and $\left(-3\sqrt{2}, -\dfrac{\pi}{4}\right)$.

> **Skill Practice 4** Convert $\left(-2, -2\sqrt{3}\right)$ from rectangular coordinates to polar coordinates. Give two representations, one with $r > 0$ and one with $r < 0$.

Classroom Examples: p. 695
Exercises 38, 44

EXAMPLE 5 Converting from Rectangular to Polar Coordinates

Convert $(-5, -2)$ from rectangular coordinates to polar coordinates. Give two representations, one with $r > 0$ and one with $r < 0$.

Solution:

The point $(-5, -2)$ is in Quadrant III with $x = -5$ and $y = -2$.
$r^2 = x^2 + y^2 = (-5)^2 + (-2)^2 = 29$, so $r = \pm\sqrt{29}$.

$\tan\theta = \dfrac{y}{x} = \dfrac{-2}{-5} = \dfrac{2}{5}$. Therefore, $\theta = n\pi + \tan^{-1}\dfrac{2}{5}$.

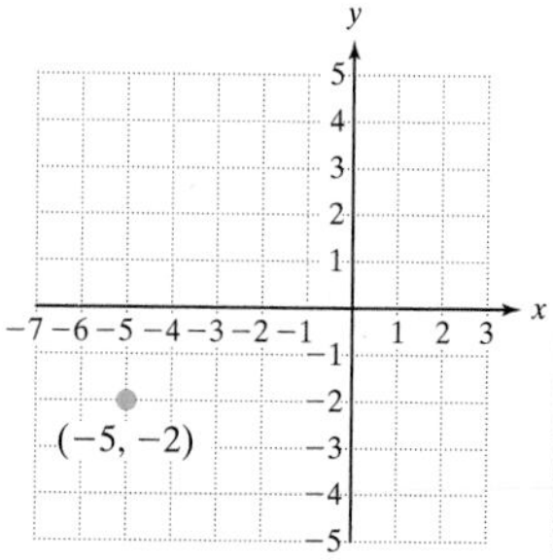

- Taking $r > 0$, we require $\tan\theta = \dfrac{2}{5}$ for θ in Quadrant III.

  One possibility is $r = \sqrt{29}$ and $\theta = \pi + \tan^{-1}\dfrac{2}{5}$.

- Taking $r < 0$, we require $\tan\theta = \dfrac{2}{5}$ for θ in Quadrant I. One possibility is $r = -\sqrt{29}$ and $\theta = \tan^{-1}\dfrac{2}{5}$.

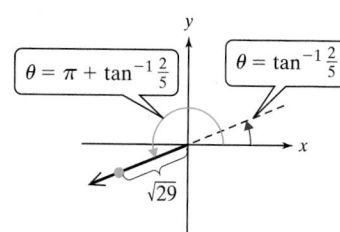

In polar coordinates we have $\left(\sqrt{29},\ \pi + \tan^{-1}\dfrac{2}{5}\right)$ and $\left(-\sqrt{29},\ \tan^{-1}\dfrac{2}{5}\right)$.

> **Skill Practice 5** Convert $(2, -3)$ from rectangular coordinates to polar coordinates. Give two representations, one with $r > 0$ and one with $r < 0$.

3. Convert Between Equations in Polar and Rectangular Coordinates

TIP Notice that the curves in Figures 7-5 and 7-6 fail the vertical line test and do not define y as a function of x in a rectangular coordinate system.

Answers

4. $\left(4, \dfrac{4\pi}{3}\right), \left(-4, \dfrac{\pi}{3}\right)$

5. $\left(\sqrt{13}, \tan^{-1}\left(-\dfrac{3}{2}\right)\right),$
$\left(-\sqrt{13}, \pi + \tan^{-1}\left(-\dfrac{3}{2}\right)\right)$

Why do we introduce a polar coordinate system when a rectangular coordinate system (at least to this point) seems to suffice? The answer is that many curves are more easily represented by a polar equation than by an equation in rectangular coordinates. A **polar equation** is an equation described in a polar coordinate system with variables r and θ rather than x and y.

The relationships $x = r\cos\theta$ and $y = r\sin\theta$ between polar and rectangular coordinates are based on trigonometric functions of the unit circle. Therefore, it seems reasonable that circles, ellipses, and other curves with loops such as those shown in Figures 7-5 and 7-6 may be easier to represent in polar form.

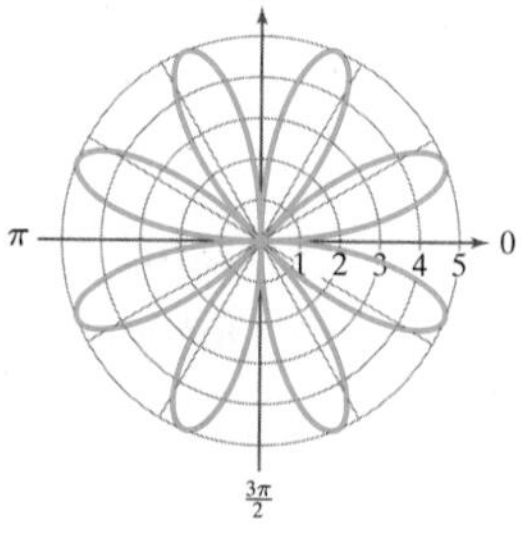

$r = 5\sin 4\theta$

Figure 7-5

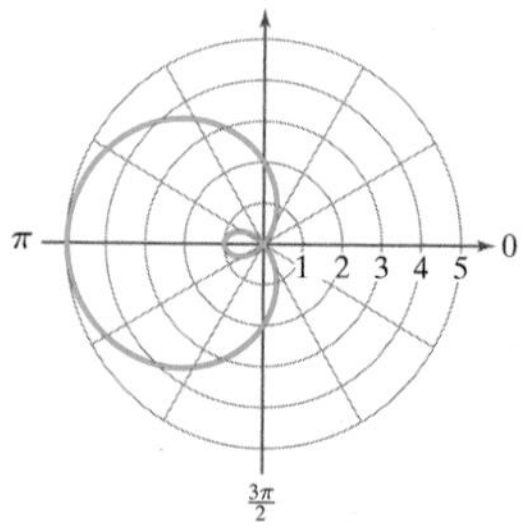

$r = 2 - 3\cos\theta$

Figure 7-6

We convert equations from rectangular coordinates to polar coordinates by substituting $x = r\cos\theta$ and $y = r\sin\theta$. This is demonstrated in Example 6.

Classroom Examples: p. 695
Exercises 46, 54

EXAMPLE 6   Converting an Equation from Rectangular to Polar Coordinates

Write an equivalent equation using polar coordinates.

a. $2x + y = 4$ **b.** $x^2 + (y - 2)^2 = 4$

Solution:

a. $2x + y = 4$ The equation represents the line $y = -2x + 4$ in rectangular coordinates.

$2r\cos\theta + r\sin\theta = 4$ Substitute $x = r\cos\theta$ and $y = r\sin\theta$.

$r(2\cos\theta + \sin\theta) = 4$ Factor out r.

$r = \dfrac{4}{2\cos\theta + \sin\theta}$ Isolate r to write r as a function of θ.

The polar equation is $r = \dfrac{4}{2\cos\theta + \sin\theta}$.

b. $x^2 + (y - 2)^2 = 4$ This equation in rectangular coordinates represents a circle with center $(0, 2)$ and radius 2.

$(r\cos\theta)^2 + (r\sin\theta - 2)^2 = 4$ Substitute $x = r\cos\theta$ and $y = r\sin\theta$.

$r^2\cos^2\theta + r^2\sin^2\theta - 4r\sin\theta + 4 = 4$ Square the binomial $(a - b)^2 = a^2 - 2ab + b^2$.

$r^2(\cos^2\theta + \sin^2\theta) - 4r\sin\theta = 0$ Factor out r^2 and subtract 4 from both sides.

$r^2 - 4r\sin\theta = 0$ Substitute 1 for $\cos^2\theta + \sin^2\theta$.

$r(r - 4\sin\theta) = 0$ Factor out r.

$r = 0$ or $r = 4\sin\theta$ Set each factor equal to zero.

The graph of $r = 0$ is simply the pole. This point is included in the graph of $r = 4\sin\theta$ because $r = 0$ when $\theta = n\pi$. Therefore, the equation $r = 4\sin\theta$ by itself is sufficient to represent the equation $x^2 + (y - 2)^2 = 4$.

The polar equation is $r = 4\sin\theta$.

> **TIP** In Example 6, notice that the linear equation $2x + y = 4$ is simpler in rectangular coordinates, but an equation of a circle such as $r = 4\sin\theta$ is simpler in polar coordinates.

Skill Practice 6   Write an equivalent equation using polar coordinates.
    a. $y = -3$ **b.** $y = x^2$

TECHNOLOGY CONNECTIONS

Graphing a Polar Equation

Most graphing utilities have the capability to graph polar equations. To begin, hit the MODE button and verify that the calculator is in radian mode and that the POLAR setting is highlighted (Figure 7-7). Then, in the graphing editor, enter the equation with r in terms of θ (Figure 7-8). In the WINDOW menu,

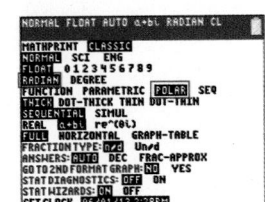

Figure 7-7

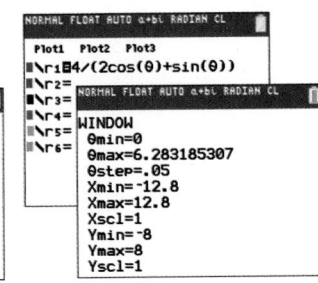

Figure 7-8

Answers
6. a. $r = -3\csc\theta$
   b. $r = \tan\theta\sec\theta$

give a range for θ, x, and y. For now, choose θmin $= 0$ and θmax to be 2π (Figure 7-8). More will be discussed about the range for θ in Section 7.2. The value of θstep gives the increment by which θ is increased to compute successive values of r. Small values of θstep produce more points and render a smoother curve than a graph rendered with a large value of θstep. A large value of θstep

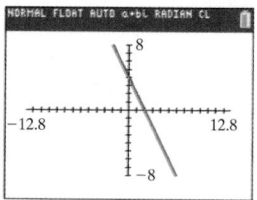

Figure 7-9

yields fewer points and a more jagged curve but takes less time to graph.

The graph of $r = \dfrac{4}{2\cos\theta + \sin\theta}$ from Example 6(a) is shown in Figure 7-9.

Notice that the graph of the polar equation is a line with slope -2 and y-intercept 4 as expected.

To convert an equation from polar coordinates to rectangular coordinates, the goal is to write an equation exclusively with x and y rather than r and θ. This is sometimes cumbersome and to help with the task, we can use one or more of the following strategies.

1. Apply one or more of the relationships

$$x = r\cos\theta, \quad y = r\sin\theta, \quad x^2 + y^2 = r^2, \quad \text{or} \quad \tan\theta = \frac{y}{x} \text{ for } (x \neq 0).$$

2. Multiply both sides by r.
3. Square both sides of the equation.

Classroom Examples: p. 695
Exercises 58, 60, 62, 64

EXAMPLE 7   **Converting an Equation from Polar to Rectangular Coordinates**

Write an equivalent equation using rectangular coordinates and identify the type of curve that the equation represents.

a. $r = 2$ **b.** $\theta = \dfrac{\pi}{3}$ **c.** $r = 4\csc\theta$ **d.** $r = -8\cos\theta$

Solution:

a. $r = 2$ Square both sides to get r^2 on the left, which in turn can be written as $x^2 + y^2$.
           $r^2 = 4$
   $x^2 + y^2 = 4$ This is an equation of a circle centered at the origin with radius 2.

b. $\theta = \dfrac{\pi}{3}$

   $\tan\theta = \tan\dfrac{\pi}{3}$ Take the tangent of both sides so that $\tan\theta$ can be written in rectangular coordinates as $\frac{y}{x}$.

   $\dfrac{y}{x} = \sqrt{3}$ Substitute $\frac{y}{x}$ for $\tan\theta$.

   $y = \sqrt{3}x$ Write the equation in slope-intercept form. The equation is a line through the origin with a slope of $\sqrt{3}$.

c. $r = 4\csc\theta$

   $r = \dfrac{4}{\sin\theta}$ Write the expression on the right in terms of $\sin\theta$ for the possibility of applying the substitution $y = r\sin\theta$.

   $r\sin\theta = 4$ Multiply both sides by $\sin\theta$.

   $y = 4$ Replace $r\sin\theta$ by y. The equation is a horizontal line.

d.
$$r = -8\cos\theta$$
$$r^2 = -8r\cos\theta$$
$$x^2 + y^2 = -8x$$

Multiply both sides by r. By so doing, we have r^2 on the left, which can be replaced by $x^2 + y^2$. On the right, $r\cos\theta$ can be replaced by x.

$$x^2 + 8x + y^2 = 0$$

Collect variable terms on the left.

$$x^2 + 8x + 16 + y^2 = 16$$

Complete the square on the left. Add $\left[\frac{1}{2}(8)\right]^2$ to both sides.

$$(x + 4)^2 + y^2 = 16$$

The equation is a circle centered at $(-4, 0)$ with radius 4.

Skill Practice 7   Write an equivalent equation using rectangular coordinates and identify the type of curve the equation represents.

a. $r = -3$ **b.** $\theta = -\dfrac{\pi}{4}$ **c.** $r = 6\sec\theta$ **d.** $r = 6\sin\theta$

- The result of Example 7(a) suggests that a polar equation of the form $r = a$ for $a \neq 0$ is a circle centered at the pole with radius $|a|$ (see Exercise 72). The equation $r = a$ represents all points $|a|$ units from the pole regardless of the value of θ (regardless of the direction taken).

- The result of Example 7(b) suggests that a polar equation of the form $\theta = k$ is a line through the origin making an angle k with the polar axis (see Exercise 71). Of particular interest, are the graphs of $\theta = 0$ and $\theta = \dfrac{\pi}{2}$. If the polar axis is positioned to coincide with the positive x-axis in a rectangular coordinate system, these equations represent the x- and y-axes, respectively (Figures 7-10 and 7-11).

Answers

7. a. $x^2 + y^2 = 9$; Circle centered at the origin, radius 3
  b. $y = -x$; Line through the origin with a slope of -1
  c. $x = 6$; Vertical line
  d. $x^2 + (y - 3)^2 = 9$; Circle centered at $(0, 3)$ with radius 3

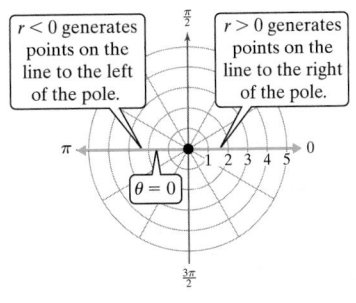

Figure 7-10

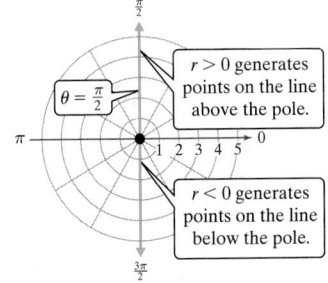

Figure 7-11

SECTION 7.1   Practice Exercises

Prerequisite Review

R.1. Given a circle with center $(-7, 3)$ and radius 3, write an equation of the circle in standard form. $(x + 7)^2 + (y - 3)^2 = 9$

R.2. Write the equation in slope-intercept form and determine the slope. $y = -\dfrac{4}{3}x + 4; \; m = -\dfrac{4}{3}$

$$4x + 3y = 12$$

R.3. Find the exact distance between the points $(-9, 4)$ and $(-4, -1)$. $5\sqrt{2}$

Concept Connections

1. The origin of a polar coordinate system is called the _____pole_____.

2. Given a point P represented by the ordered pair (r, θ) in polar coordinates, the distance from P to $(0, 0)$ is _____$|r|$_____.

3. To convert an ordered pair (r, θ) in polar coordinates to rectangular coordinates, use the relationships $x = $ _____$r\cos\theta$_____ and $y = $ _____$r\sin\theta$_____.

4. To convert an ordered pair (x, y) from rectangular coordinates to polar coordinates, use the relationships $r^2 = $ _____$x^2 + y^2$_____ and $\tan\theta = $ _____$\frac{y}{x}$ for $x \neq 0$_____.

5. Use an equation in polar coordinates to describe the set of points 4 units from the pole. $r = 4$

6. Write three ordered pairs in polar coordinates that represent the pole. For example: $(0, 0)$, $\left(0, \dfrac{\pi}{6}\right)$, and $\left(0, \dfrac{\pi}{4}\right)$

Objective 1: Plot Points Using Polar Coordinates

For Exercises 7–14, match each point in polar coordinates to the points A, B, C, or D.

$$A\left(3, \frac{\pi}{6}\right) \qquad B\left(3, \frac{3\pi}{4}\right) \qquad C\left(3, -\frac{\pi}{6}\right) \qquad D\left(-3, \frac{\pi}{4}\right)$$

7. $\left(3, -\dfrac{5\pi}{4}\right)$ B

8. $\left(3, \dfrac{11\pi}{6}\right)$ C

9. $\left(3, -\dfrac{11\pi}{6}\right)$ A

10. $\left(3, \dfrac{5\pi}{4}\right)$ D

11. $\left(3, \dfrac{13\pi}{4}\right)$ D

12. $\left(-3, \dfrac{7\pi}{4}\right)$ B

13. $\left(-3, \dfrac{5\pi}{6}\right)$ C

14. $\left(-3, \dfrac{7\pi}{6}\right)$ A

For Exercises 15–18, plot the points whose polar coordinates are given. (See Example 1)

15. a. $\left(5, \dfrac{\pi}{2}\right)$ **b.** $\left(2.5, \dfrac{8\pi}{3}\right)$ **c.** $\left(-2, \dfrac{\pi}{6}\right)$ **d.** $\left(-4, \dfrac{5\pi}{4}\right)$

16. a. $\left(3, \dfrac{\pi}{3}\right)$ **b.** $\left(4.5, \dfrac{11\pi}{4}\right)$ **c.** $\left(-5, \dfrac{5\pi}{6}\right)$ **d.** $\left(-3, \dfrac{\pi}{4}\right)$

17. a. $\left(4, -\dfrac{5\pi}{6}\right)$ **b.** $\left(3, -\dfrac{5\pi}{4}\right)$ **c.** $\left(-2, -\dfrac{\pi}{3}\right)$ **d.** $\left(-4.5, -\dfrac{2\pi}{3}\right)$

18. a. $\left(2, -\dfrac{2\pi}{3}\right)$ **b.** $(3, -\pi)$ **c.** $\left(-4, -\dfrac{\pi}{4}\right)$ **d.** $\left(-5, -\dfrac{5\pi}{6}\right)$

For Exercises 19–22, given a point in polar coordinates, find two other polar coordinate representations. (See Example 2)

   a. with $r > 0$.   **b.** with $r < 0$.

19. $\left(7, \dfrac{5\pi}{6}\right)$

20. $\left(3.6, \dfrac{\pi}{2}\right)$

21. $\left(-8, \dfrac{7\pi}{4}\right)$

22. $\left(-2, \dfrac{4\pi}{3}\right)$

Objective 2: Convert Ordered Pairs Between Polar and Rectangular Coordinates

For Exercises 23–32, convert the ordered pair in polar coordinates to rectangular coordinates. (See Example 3)

23. $\left(8, \dfrac{5\pi}{6}\right)$ $(-4\sqrt{3}, 4)$

24. $\left(6, \dfrac{2\pi}{3}\right)$ $(-3, 3\sqrt{3})$

25. $\left(-1.2, \dfrac{5\pi}{4}\right)$ $(0.6\sqrt{2}, 0.6\sqrt{2})$

26. $\left(-\dfrac{3}{4}, \dfrac{7\pi}{4}\right)$ $\left(-\dfrac{3\sqrt{2}}{8}, \dfrac{3\sqrt{2}}{8}\right)$

27. $\left(17, -\dfrac{\pi}{3}\right)$ $\left(\dfrac{17}{2}, -\dfrac{17\sqrt{3}}{2}\right)$

28. $\left(5, -\dfrac{7\pi}{6}\right)$ $\left(-\dfrac{5\sqrt{3}}{2}, \dfrac{5}{2}\right)$

29. $\left(-4, -\dfrac{11\pi}{6}\right)$ $(-2\sqrt{3}, -2)$

30. $\left(-8, -\dfrac{4\pi}{3}\right)$ $(4, -4\sqrt{3})$

31. $\left(7, \dfrac{\pi}{2}\right)$ $(0, 7)$

32. $(16, \pi)$ $(-16, 0)$

For Exercises 33–38, convert the ordered pair in rectangular coordinates to polar coordinates with $r > 0$ and $0 \le \theta < 2\pi$. (See Examples 4–5)

33. $(-3, 3\sqrt{3})$ $\left(6, \dfrac{2\pi}{3}\right)$

34. $(-7\sqrt{3}, 7)$ $\left(14, \dfrac{5\pi}{6}\right)$

35. $(8\sqrt{2}, -8\sqrt{2})$ $\left(16, \dfrac{7\pi}{4}\right)$

36. $(5, -5\sqrt{3})$ $\left(10, \dfrac{5\pi}{3}\right)$

37. $(2, -6)$ $(2\sqrt{10}, \tan^{-1}(-3) + 2\pi)$

38. $(4, -10)$ $\left(2\sqrt{29}, \tan^{-1}\left(-\dfrac{5}{2}\right) + 2\pi\right)$

For Exercises 39–44, convert the ordered pair in rectangular coordinates to polar coordinates with $r < 0$ and $0 \le \theta < 2\pi$. (See Examples 4–5)

39. $\left(\dfrac{5\sqrt{3}}{2}, \dfrac{5}{2}\right)$ $\left(-5, \dfrac{7\pi}{6}\right)$

40. $\left(\dfrac{3\sqrt{2}}{2}, -\dfrac{3\sqrt{2}}{2}\right)$ $\left(-3, \dfrac{3\pi}{4}\right)$

41. $(-1, \sqrt{3})$ $\left(-2, \dfrac{5\pi}{3}\right)$

42. $(10\sqrt{2}, 10\sqrt{2})$ $\left(-20, \dfrac{5\pi}{4}\right)$

43. $(-3, -8)$ $\left(-\sqrt{73}, \tan^{-1}\dfrac{8}{3}\right)$

44. $(-5, -1)$ $\left(-\sqrt{26}, \tan^{-1}\dfrac{1}{5}\right)$

Objective 3: Convert Between Equations in Polar and Rectangular Coordinates

For Exercises 45–56, write an equivalent equation using polar coordinates. (See Example 6)

45. $x - y = 10$ $r = \dfrac{10}{\cos\theta - \sin\theta}$

46. $3x + 2y = 6$ $r = \dfrac{6}{3\cos\theta + 2\sin\theta}$

47. $x = -6$ $r = -6\sec\theta$

48. $y = 2$ $r = 2\csc\theta$

49. $x^2 + y^2 = 16$ $r = 4$

50. $2x^2 + 2y^2 = 50$ $r = 5$

51. $x^2 = 3y$ $r = 3\tan\theta\sec\theta$

52. $x = 2y^2$ $r = \dfrac{1}{2}\cot\theta\csc\theta$

53. $(x + 2)^2 + y^2 = 4$ $r = -4\cos\theta$

54. $x^2 + (y - 1)^2 = 1$ $r = 2\sin\theta$

55. $(x - 5)^5 + (y + 6)^2 = 61$ $r = 10\cos\theta - 12\sin\theta$

56. $(x + 3)^2 + (y - 4)^2 = 25$ $r = 8\sin\theta - 6\cos\theta$

For Exercises 57–70, convert the polar equation to rectangular form and identify the type of curve represented. (See Example 7)

57. $r = 3$

58. $r = 7$

59. $r = 2\csc\theta$

60. $r = \sec\theta$

61. $r = 4\cos\theta$

62. $r = -2\sin\theta$

63. $\theta = \dfrac{3\pi}{4}$

64. $\theta = \dfrac{\pi}{6}$

65. $r = a\sin\theta \ (a \ne 0)$

66. $r = a\cos\theta \ (a \ne 0)$

67. $r = \dfrac{a}{\sin\theta} \ (a \ne 0)$

68. $r = \dfrac{a}{\cos\theta} \ (a \ne 0)$

69. $r = 2h\cos\theta + 2k\sin\theta$

70. $r = h\cos\theta + k\sin\theta$

71. Suppose that the polar axis is positioned to coincide with the positive x-axis in a rectangular coordinate system. Describe the graphs of the polar equations as they relate to a rectangular coordinate system.

   a. $\theta = 0$ $\theta = 0$ represents the x-axis.   **b.** $\theta = \dfrac{\pi}{4}$ $\theta = \dfrac{\pi}{4}$ represents the line $y = x$.   **c.** $\theta = \dfrac{\pi}{2}$ $\theta = \dfrac{\pi}{2}$ represents the y-axis.

72. Show that a polar equation of the form $r = a \ (a \ne 0)$ represents a circle centered at the pole with radius $|a|$.
   $r = a \Rightarrow r^2 = a^2 \Rightarrow x^2 + y^2 = a^2$ is a circle centered at $(0, 0)$ with radius $|a|$.

Mixed Exercises

For Exercises 73–76, determine whether the two ordered pairs in polar coordinates represent the same point in the plane. If not, explain the change needed to make the two ordered pairs represent the same point. Assume that n is any integer.

73. (r, θ) and $(-r, -\theta)$

74. $(-r, \theta)$ and $(r, \theta + 2n\pi)$

75. $(-r, \theta)$ and $(r, \pi - \theta)$

76. (r, θ) and $(r, \theta + 2n\pi)$ Yes

For Exercises 77–78, determine whether the statement is true or false for two points $P(r_1, \theta_1)$ and $Q(r_2, \theta_2)$ represented in polar coordinates. If the statement is false, give a counterexample.

77. If $|r_1| = |r_2|$, then P and Q are the same distance from the pole.   True

78. If $|\theta_1| = |\theta_2|$, then P and Q lie on the same line through the pole.   False; $P\left(1, \dfrac{\pi}{4}\right)$ and $Q\left(1, -\dfrac{\pi}{4}\right)$

For Exercises 79–82, a point in polar coordinates is given. Let n be an integer and find *all* other polar representations of the given point.

79. $\left(3, \dfrac{\pi}{4}\right)$

80. $\left(-2, -\dfrac{4\pi}{3}\right)$

81. $\left(-4, \dfrac{5\pi}{6}\right)$

82. $\left(6, -\dfrac{7\pi}{4}\right)$

$\left(3, \dfrac{\pi}{4} + 2n\pi\right)$ and $\left(2, \dfrac{5\pi}{3} + 2n\pi\right)$ and $\left(4, \dfrac{11\pi}{6} + 2n\pi\right)$ and $\left(6, \dfrac{\pi}{4} + 2n\pi\right)$ and

$\left(-3, \dfrac{5\pi}{4} + 2n\pi\right)$ $\left(-2, \dfrac{2\pi}{3} + 2n\pi\right)$ $\left(-4, \dfrac{5\pi}{6} + 2n\pi\right)$ $\left(-6, \dfrac{5\pi}{4} + 2n\pi\right)$

Additional answers can be found in the Instructor Answer Appendix.

For Exercises 83–88, describe the graph of the set of points (r, θ) represented by the polar inequality. Assume that the polar axis is oriented to coincide with the positive x-axis in a rectangular coordinate system.

83. $r < 8$ **84.** $r \geq 5$ **85.** $2 \leq r \leq 4$

86. $1 < r < 2$ **87.** $0 < \theta < \dfrac{\pi}{4}$ **88.** $\dfrac{\pi}{2} < \theta < \pi$

89. Suppose that seating in a theater is in an area defined in polar coordinates where the pole is located at the front and center of the stage labeled as point A. The seating area is defined by $-\dfrac{\pi}{4} \leq \theta \leq \dfrac{\pi}{4}$ and $30 \leq r \leq 100$, and the values of r are in feet.

  a. Sketch the seating area.

  b. Determine the amount of area for seating. Write the exact answer in terms of π and give an approximation to the nearest square foot. $2275\pi \text{ ft}^2 \approx 7147 \text{ ft}^2$

90. A rotating sprinkler spreads water over an area defined by $\dfrac{\pi}{6} \leq \theta \leq \dfrac{5\pi}{6}$ and $4 \leq r \leq 24$ relative to the sprinkler head at the pole at point A and an imaginary line defined by $\theta = 0$. The values of r are in feet.

  a. Sketch the area that is watered.

  b. Determine the amount of area that is watered. Write the exact answer in terms of π and give an approximation to the nearest square foot. $\dfrac{560\pi}{3} \text{ ft}^2 \approx 586 \text{ ft}^2$

91. A player in a video game must knock out a target located 84 pixels above and 156 pixels to the left of his position. Choose a polar coordinate system with the player at the pole and the polar axis extending to the player's right. Find the polar coordinates of the target (this determines the distance and angle at which the player should fire his gun). Find r to the nearest pixel and θ in degree measure to the nearest tenth of a degree. $(177, 151.7°)$

92. A gunner on a naval ship sights a target located 2.1 mi north and 0.8 mi east of the ship's position. Choose a polar coordinate system with the gunner at the pole and the polar axis extending to the east. Find the polar coordinates of the target. Find r to the nearest hundredth of a mile and θ in degree measure to the nearest hundredth of a degree. $(2.25, 69.15°)$

Write About It

93. Explain the significance of $r < 0$ in an ordered pair (r, θ) in polar coordinates.

94. Explain why a point in the plane can be represented by infinitely many ordered pairs in polar coordinates.

Expanding Your Skills

95. a. Given points $P(r_1, \theta_1)$ and $Q(r_2, \theta_2)$ represented in polar coordinates, write the ordered pairs in rectangular coordinates.

  b. Use the rectangular coordinates from part (a) and the distance formula to show that the distance d between P and Q is given by $d = \sqrt{r_1^2 + r_2^2 - 2r_1 r_2 \cos(\theta_2 - \theta_1)}$.

96. Use the results of Exercise 95.

  a. If $\theta_1 = \theta_2$, what is the distance between P and Q? Explain the significance of the answer.

  b. If $\theta_1 - \theta_2 = \dfrac{\pi}{2}$, what is the distance between P and Q? Explain the significance of the answer.

97. Given points $P(r_1, \theta_1)$ and $Q(r_2, \theta_2)$ represented in polar coordinates with $\theta_1 \neq \theta_2$ and $r_1 > 0$ and $r_2 > 0$, use the law of cosines to show that the distance d between P and Q is given by $d = \sqrt{r_1^2 + r_2^2 - 2r_1 r_2 \cos(\theta_2 - \theta_1)}$.

98. Use the results of Exercise 97.

Find the distance between the points $\left(2, \dfrac{\pi}{6}\right)$ and $\left(4, \dfrac{\pi}{3}\right)$. $\sqrt{20 - 8\sqrt{3}}$ or $2\sqrt{5 - 2\sqrt{3}} \approx 2.5$

Technology Connections

For Exercises 99–102, use a graphing utility to graph the polar equation for the given intervals of θ, x, and y to confirm your answer to the given exercise. Use the window setting for $0 \leq \theta \leq \pi$, $-4.83 \leq x \leq 4.83$, and $-3 \leq y \leq 3$.

99. $r = 2\csc\theta$; Exercise 59   **100.** $r = \sec\theta$; Exercise 60

101. $r = 4\cos\theta$; Exercise 61   **102.** $r = -2\sin\theta$; Exercise 62

SECTION 7.2 ## Graphs of Polar Equations

1. Graph Polar Equations by Plotting Points

Galaxies are clusters of millions of stars, planets, and space debris held together by gravitational attraction that when seen from a very large distance often appear in organized patterns. Perhaps the most beautiful pattern is the classic spiral galaxy. Interestingly, the graph of a spiral can be represented by a simple equation represented in a polar coordinate system (see Example 6).

A **solution** to a polar equation is an ordered pair (r, θ) that satisfies the equation. The graph of a polar equation is the graph of all solutions to the equation. In some cases, we can convert a polar equation to rectangular coordinates and then graph the equation from its rectangular form. For example, in Section 7.1, we determined that the equation $r = 2$ is equivalent to the rectangular equation $x^2 + y^2 = 4$. Thus, the graph of $r = 2$ is a circle with center at the pole and radius 2 units.

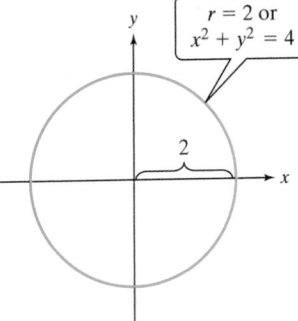

Converting a polar equation to an equation in rectangular coordinates does not always produce a rectangular equation that we recognize. As a result, it is sometimes necessary to make a table of solutions and plot the points to graph a polar equation.

Classroom Example: p. 707
Exercise 12

| EXAMPLE 1 | Graph a Polar Equation by Plotting Points |

Graph $r = 6\sin\theta$.

Solution:

The equation $r = 6\sin\theta$ is written with r as a function of θ. Therefore, we organize a table with arbitrary values of θ and solve for r (Table 7-1).

Table 7-1

θ	0	$\frac{\pi}{6}$	$\frac{\pi}{4}$	$\frac{\pi}{3}$	$\frac{\pi}{2}$	$\frac{2\pi}{3}$	$\frac{3\pi}{4}$	$\frac{5\pi}{6}$	π
r	0	3	$3\sqrt{2}$	$3\sqrt{3}$	6	$3\sqrt{3}$	$3\sqrt{2}$	3	0
$r \approx$	0	3	4.2	5.2	6	5.2	4.2	3	0

θ	$\frac{7\pi}{6}$	$\frac{5\pi}{4}$	$\frac{4\pi}{3}$	$\frac{3\pi}{2}$	$\frac{5\pi}{3}$	$\frac{7\pi}{4}$	$\frac{11\pi}{6}$	2π
r	-3	$-3\sqrt{2}$	$-3\sqrt{3}$	-6	$-3\sqrt{3}$	$-3\sqrt{2}$	-3	0
$r \approx$	-3	-4.2	-5.2	-6	-5.2	-4.2	-3	0

The graph of $r = 6\sin\theta$ is shown in Figure 7-12. The graph is a circle with radius 3, centered at $\left(3, \frac{\pi}{2}\right)$ [or equivalently $(0, 3)$ in rectangular coordinates].

Notice that the graph of the circle is traced completely by graphing solutions with $0 \le \theta \le \pi$. The points in Table 7-1 with $\pi < \theta \le 2\pi$ are redundant. For example, $\left(-3, \frac{7\pi}{6}\right)$ is the same point in the plane as $\left(3, \frac{\pi}{6}\right)$.

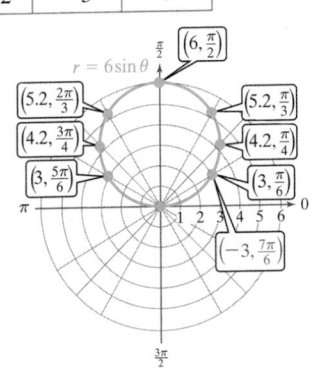

Figure 7-12

> **Skill Practice 1**   Graph $r = -4\cos\theta$.

The graph from Example 1 is consistent with the result obtained by converting the equation to rectangular form.

$$r = 6\sin\theta$$

$$r^2 = 6r\sin\theta \qquad \text{Multiply both sides by } r.$$

$$x^2 + y^2 = 6y \qquad \text{Replace } r^2 \text{ by } x^2 + y^2 \text{ and } r\sin\theta \text{ by } y.$$

$$x^2 + y^2 - 6y + 9 = 9 \qquad \text{Complete the square. Add } [\tfrac{1}{2}(-6)]^2 = 9 \text{ to both sides.}$$

$$x^2 + (y - 3)^2 = 9 \qquad \text{The equation is a circle centered at } (0, 3) \text{ with radius 3.}$$

We also notice in Figure 7-12 that the graph of $r = 6\sin\theta$ is symmetric with respect to the line $\theta = \dfrac{\pi}{2}$ (y-axis). Knowing that the graph of $r = 6\sin\theta$ is traced completely for $0 \le \theta \le \pi$ and that the graph is symmetric to the line $\theta = \dfrac{\pi}{2}$ means that we can significantly reduce the number of points needed to construct the graph. We can graph $r = 6\sin\theta$ by plotting points for θ on the interval $\left[0, \dfrac{\pi}{2}\right]$ and using symmetry.

2. Test Polar Equations for Symmetry

Figures 7-13, 7-14, and 7-15 show a point P and a corresponding point Q that is symmetric to P with respect to the polar axis, the line $\theta = \dfrac{\pi}{2}$, and the pole, respectively.

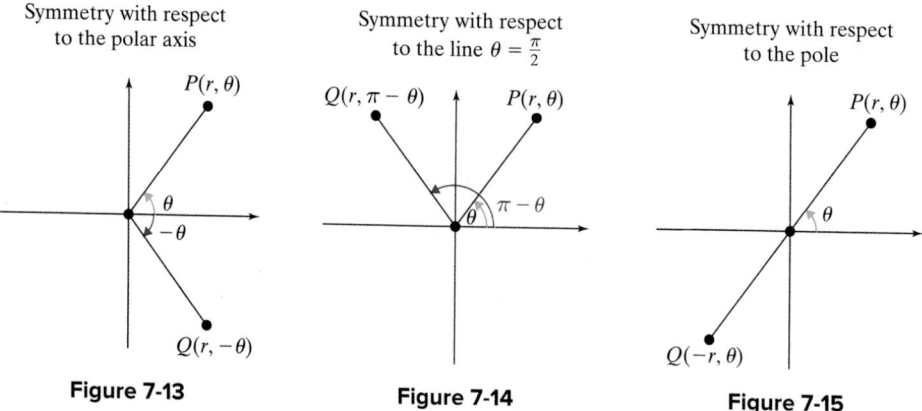

Symmetry with respect to the polar axis

$P(r, \theta)$
θ
$-\theta$
$Q(r, -\theta)$

Figure 7-13

Symmetry with respect to the line $\theta = \frac{\pi}{2}$

$Q(r, \pi - \theta)$ $P(r, \theta)$
$\pi - \theta$
θ

Figure 7-14

Symmetry with respect to the pole

$P(r, \theta)$
θ
$Q(-r, \theta)$

Figure 7-15

Answer

1.

$r = -4\cos\theta$

Table 7-2

> ### Tests for Symmetry in Polar Coordinates
>
> For an equation in polar coordinates, if the indicated substitution produces an equivalent equation, then the graph has the indicated type of symmetry.
>
> **(1)** Symmetry with respect to the polar axis: Replace (r, θ) by $(r, -\theta)$
>
> **(2)** Symmetry with respect to the line $\theta = \dfrac{\pi}{2}$: Replace (r, θ) by $(r, \pi - \theta)$
>
> **(3)** Symmetry with respect to the pole: Replace (r, θ) by $(-r, \theta)$

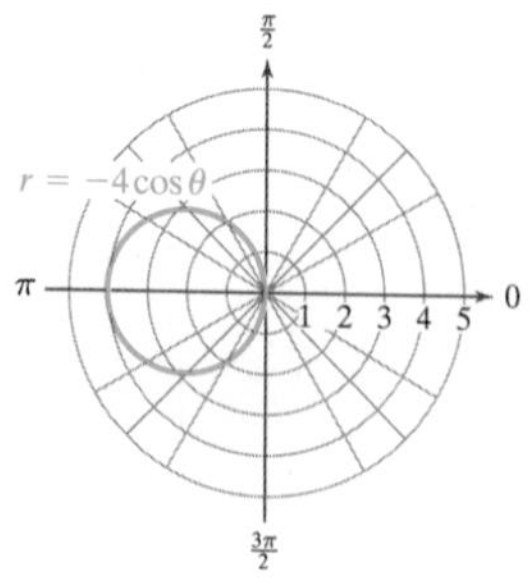

It is very important to note that the three tests for symmetry given in Table 7-2 are inconclusive if the given substitution fails to produce an equivalent equation. For example, a graph may be symmetric with respect to the polar axis, but a different representation of the point $(r, -\theta)$ satisfies the equation determining the graph. A rigorous test for symmetry would require that we test *all* representations of the point $(r, -\theta)$ in the original equation. That is, we would need to substitute $(r, -\theta + 2n\pi)$ and $(-r, (\pi - \theta) + 2n\pi)$ to determine if we reach an equivalent equation for some value of n.

For example, consider the graph of $r = 4\sin\dfrac{\theta}{2}$. From Figure 7-16, the graph is symmetric with respect to the polar axis. However, applying (1) from Table 7-2 results in a different equation, giving us no information about symmetry.

$$r = 4\sin\frac{\theta}{2}$$

$$r = 4\sin\frac{-\theta}{2}$$

$$r = -4\sin\frac{\theta}{2} \quad \text{Different}$$

TIP Apply the odd function property of the sine function.

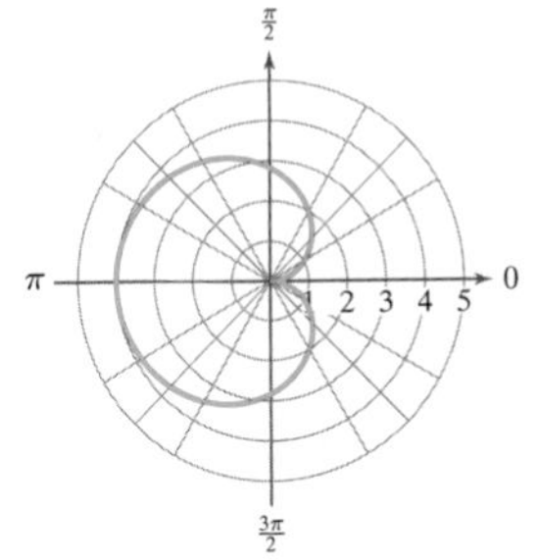

$$\sin(-\theta) = -\sin\theta$$

Figure 7-16

Although the graph is actually symmetric with respect to the polar axis, the test is inconclusive.

However, substituting a *different* representation of the point $(r, -\theta)$ such as $(r, 2\pi - \theta)$ may result in an equivalent equation and indicate that symmetry exists.

$$r = 4\sin\frac{\theta}{2} \qquad\qquad \text{Replace } (r, \theta) \text{ by } (r, 2\pi - \theta) \text{ and apply the difference formula for } \sin(a-b) = \sin a \cos b - \cos a \sin b$$

$$r = 4\sin\frac{2\pi - \theta}{2} = 4\sin\left(\pi - \frac{\theta}{2}\right)$$

$$= 4\left(\sin\pi\cos\frac{\theta}{2} - \cos\pi\sin\frac{\theta}{2}\right) = 4\left[0 \cdot \cos\frac{\theta}{2} - (-1) \cdot \sin\frac{\theta}{2}\right] = 4\sin\frac{\theta}{2}$$

Since substituting $(r, 2\pi - \theta)$ into the original equation results in an equivalent equation, we confirm that the graph is symmetric with respect to the polar axis.

EXAMPLE 2   **Using Symmetry to Graph a Polar Equation**

Classroom Example: p. 707
Exercise 24

Graph $r = 2 - 2\cos\theta$.

Solution:

First test for symmetry.

Test for symmetry with respect to the polar axis:

$$r = 2 - 2\cos\theta$$
$$r = 2 - 2\cos(-\theta) \qquad\qquad \text{Replace } (r, \theta) \text{ by } (r, -\theta).$$
$$= 2 - 2\cos\theta \quad \text{Same} \qquad \text{The cosine function is even: } \cos\theta = \cos(-\theta).$$

The graph is symmetric with respect to the polar axis.

Test for symmetry with respect to the line $\theta = \frac{\pi}{2}$:

$$r = 2 - 2\cos\theta$$
$$r = 2 - 2\cos(\pi - \theta) \qquad\qquad \text{Replace } (r, \theta) \text{ by } (r, \pi - \theta).$$
$$= 2 - 2(-\cos\theta) \qquad\qquad \cos(\pi - \theta) = -\cos\theta$$
$$= 2 + 2\cos\theta \quad \text{Different}$$

The test is inconclusive.

Test for symmetry with respect to the pole:

$$r = 2 - 2\cos\theta$$

$$-r = 2 - 2\cos\theta$$

$$r = -2 + 2\cos\theta \quad \text{Different}$$

Replace (r, θ) by $(-r, \theta)$.

The test is inconclusive.

Since the graph of $r = 2 - 2\cos\theta$ is symmetric with respect to the polar axis, we can plot several points for $0 \le \theta \le \pi$ (Figure 7-17) and then use symmetry to plot additional points for $\pi < \theta < 2\pi$ (Figure 7-18).

θ	0	$\frac{\pi}{6}$	$\frac{\pi}{4}$	$\frac{\pi}{3}$	$\frac{\pi}{2}$	$\frac{2\pi}{3}$	$\frac{3\pi}{4}$	$\frac{5\pi}{6}$	π
r	0	$2 - \sqrt{3}$	$2 - \sqrt{2}$	1	2	3	$2 + \sqrt{2}$	$2 + \sqrt{3}$	4
$r \approx$	0	0.27	0.59	1	2	3	3.41	3.73	4

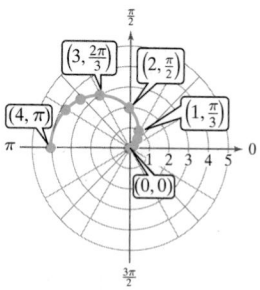

Figure 7-17 Figure 7-18

The graph of $r = 2 - 2\cos\theta$ is called a limaçon (pronounced "lee-ma-sohn") and is further categorized as a cardioid because it resembles the shape of a heart.

Skill Practice 2 Graph $r = 3 + 3\sin\theta$.

Another tool in our analysis of polar equations is identifying the maximum value of $|r|$ if it exists. In Example 2, the value of $|r|$ in the equation $r = 2 - 2\cos\theta$ is maximized when $\cos\theta = -1$. Therefore, for $\theta = \pi$,

$$|r| = |2 - 2\cos(\pi)| = |2 - 2(-1)| = 4 \text{ (maximum value of } |r|).$$

The maximum value of $|r|$ tells us the distance from the pole of the point or points on a graph that are farthest from the pole. This can be helpful when setting up a viewing window to graph a polar equation by hand or by using a graphing utility.

Answer

2.

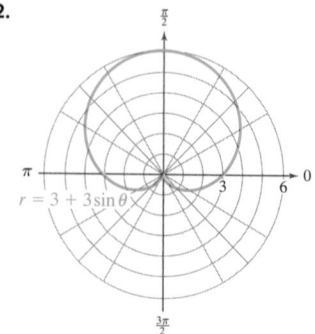

TECHNOLOGY CONNECTIONS

Graphing a Polar Equation

To graph a polar equation on a graphing utility, the calculator must be in RADIAN mode with the POLAR option highlighted (Figure 7-19). It is also important to set appropriate window parameters. To graph $r = 2 - 2\cos\theta$ from Example 2, we know that the maximum value of $|r|$ is 4. We need a window at least as large as $-4 \le x \le 4$ and $-4 \le y \le 4$. However, we choose $-6.4 \le x \le 6.4$ and $-4 \le y \le 4$ to accommodate an aspect ratio of width (x) to height (y) of 1.6 to 1 (Figure 7-20). This ensures a viewing window without distortion (Figure 7-21). Check the user's manual to determine the correct aspect ratio for your device.

Figure 7-19

The value of θmin and θmax are the endpoints of an interval over which the graph is traced once, in this case $[0, 2\pi)$. The value of θstep is the incremental change in θ between consecutive points. Smaller values of θstep result in a smoother curve and a graph traced more slowly than graphs traced with larger values of θstep.

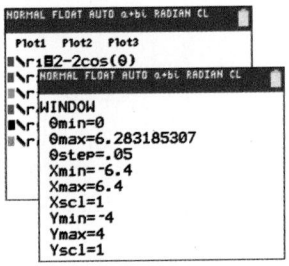

Figure 7-20

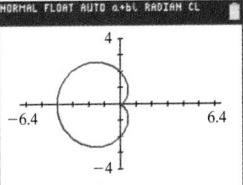

Figure 7-21

Classroom Example: p. 707
Exercise 26

EXAMPLE 3   Using Symmetry to Graph a Polar Equation

Graph $r = 2 + 3\sin\theta$.

Solution:

Test for symmetry with respect to the polar axis:

$$r = 2 + 3\sin\theta$$
$$r = 2 + 3\sin(-\theta) \qquad \text{Replace } (r, \theta) \text{ by } (r, -\theta).$$
$$= 2 + 3(-\sin\theta)$$
$$= 2 - 3\sin\theta \qquad \text{Different} \qquad \text{The sine function is odd: } \sin(-\theta) = -\sin\theta.$$

The test is inconclusive.

Test for symmetry with respect to the line $\theta = \dfrac{\pi}{2}$:

$r = 2 + 3\sin\theta$

$r = 2 + 3\sin(\pi - \theta)$ Replace (r, θ) by $(r, \pi - \theta)$.

$ = 2 + 3\sin\theta$ Same $\sin(\pi - \theta) = \sin\theta$

The graph is symmetric with respect to the line $\theta = \dfrac{\pi}{2}$.

Test for symmetry with respect to the pole:

$r = 2 + 3\sin\theta$

$-r = 2 + 3\sin\theta$ Replace (r, θ) by $(-r, \theta)$.

$r = -2 - 3\sin\theta$ Different

The test is inconclusive.

Since the graph of $r = 2 + 3\sin\theta$ is symmetric with respect to the line $\theta = \dfrac{\pi}{2}$, we can plot points for $0 \le \theta \le \dfrac{\pi}{2}$ and $\dfrac{3\pi}{2} \le \theta \le 2\pi$ (Figure 7-22) and then use symmetry to plot additional points for $\dfrac{\pi}{2} < \theta < \dfrac{3\pi}{2}$ (Figure 7-23).

θ	0	$\frac{\pi}{6}$	$\frac{\pi}{4}$	$\frac{\pi}{3}$	$\frac{\pi}{2}$	$\frac{3\pi}{2}$	$\frac{5\pi}{3}$	$\frac{7\pi}{4}$	$\frac{11\pi}{6}$	2π
r	2	$\frac{7}{2}$	$2 + \frac{3\sqrt{2}}{2}$	$2 + \frac{3\sqrt{3}}{2}$	5	-1	$2 - \frac{3\sqrt{3}}{2}$	$2 - \frac{3\sqrt{2}}{2}$	$\frac{1}{2}$	2
$r \approx$	2	3.5	4.1	4.6	5	-1	-0.6	-0.1	0.5	2

Shown in Figure 7-22 in blue   Shown in Figure 7-22 in red

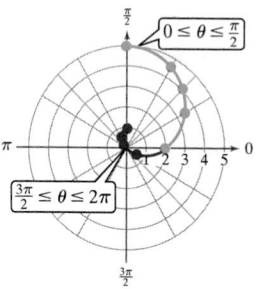

Figure 7-22

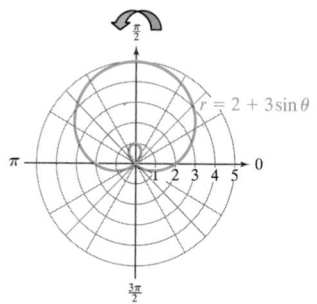

Figure 7-23

The graph of $r = 2 + 3\sin\theta$ (Figure 7-23) is another type of limaçon and is further categorized as a limaçon with a loop.

Answer

3.

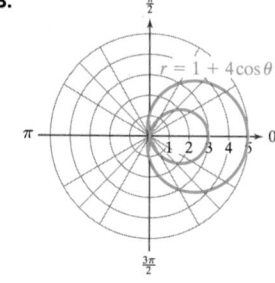

Skill Practice 3   Graph $r = 1 + 4\cos\theta$.

In Example 3, the graph passes through the pole twice for θ on the interval $[0, 2\pi)$. The values of θ for which $r = 0$ sometimes tell us where a loop "begins" or "ends" and is helpful to graph a polar equation. We demonstrate this in Example 4.

EXAMPLE 4 **Using Symmetry to Graph a Polar Equation**

Graph $r = 4\cos 2\theta$.

Solution:

The graph is at the pole when $r = 0$.

$$r = 4\cos 2\theta = 0 \text{ for } \theta = \frac{\pi}{4}, \frac{3\pi}{4}, \frac{5\pi}{4}, \frac{7\pi}{4}, \dots$$

Test for symmetry.

Polar axis:

$r = 4\cos 2\theta$ ⟵
$r = 4\cos 2(-\theta)$
$r = 4\cos 2\theta$ ⟶ Same

The graph is symmetric
with respect to the
polar axis.

Line $\theta = \frac{\pi}{2}$:

$r = 4\cos 2\theta$ ⟵
$r = 4\cos[2(\pi - \theta)]$
$r = 4\cos(2\pi - 2\theta)$
$r = 4\cos 2\theta$ ⟶ Same

The graph is symmetric
with respect to the line
$\theta = \frac{\pi}{2}$.

Pole:

$r = 4\cos 2\theta$ ⟵
$-r = 4\cos 2\theta$
$r = -4\cos 2\theta$ ⟶ Different

The test for symmetry
about the pole is
inconclusive.

Since the graph of $r = 4\cos 2\theta$ is symmetric with
respect to both the polar axis and the line $\theta = \frac{\pi}{2}$, we
can plot points for $0 \le \theta \le \frac{\pi}{2}$ (Figure 7-24) and then
use symmetry to plot additional points.

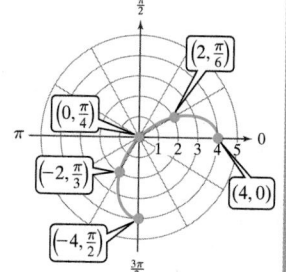

θ	0	$\frac{\pi}{6}$	$\frac{\pi}{4}$	$\frac{\pi}{3}$	$\frac{\pi}{2}$
r	4	2	0	-2	-4

Figure 7-24

Applying symmetry with respect to the polar axis we
can sketch a mirror image of the part of the curve
above the polar axis. Applying symmetry with respect to the line $\theta = \frac{\pi}{2}$ we can
sketch a mirror image of the curve to the left of $\theta = \frac{\pi}{2}$ (Figure 7-25).

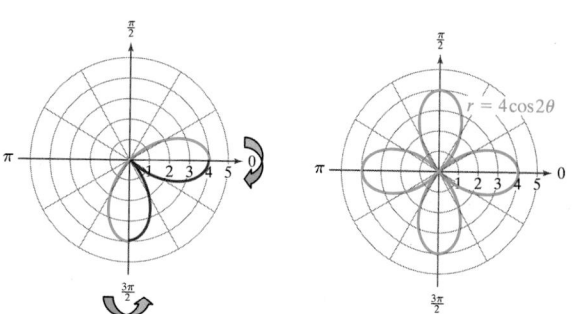

Figure 7-25   **Figure 7-26**

Applying symmetry with respect to the polar axis and line $\theta = \frac{\pi}{2}$ results in a
graph of a "rose" with four petals (Figure 7-26).

Skill Practice 4 Graph $r = 3\sin 2\theta$.

3. Categorize Polar Equations and Their Graphs

Throughout the text, we have categorized a number of different types of equations in a rectangular coordinate system along with their graphs. For example, the graph of a linear equation in two variables $Ax + By = C$ is a line, the graph of a quadratic equation $y = ax^2 + bx + c$ is a parabola, and so on. We now present a library of polar equations and their associated graphs (Table 7-3).

Table 7-3

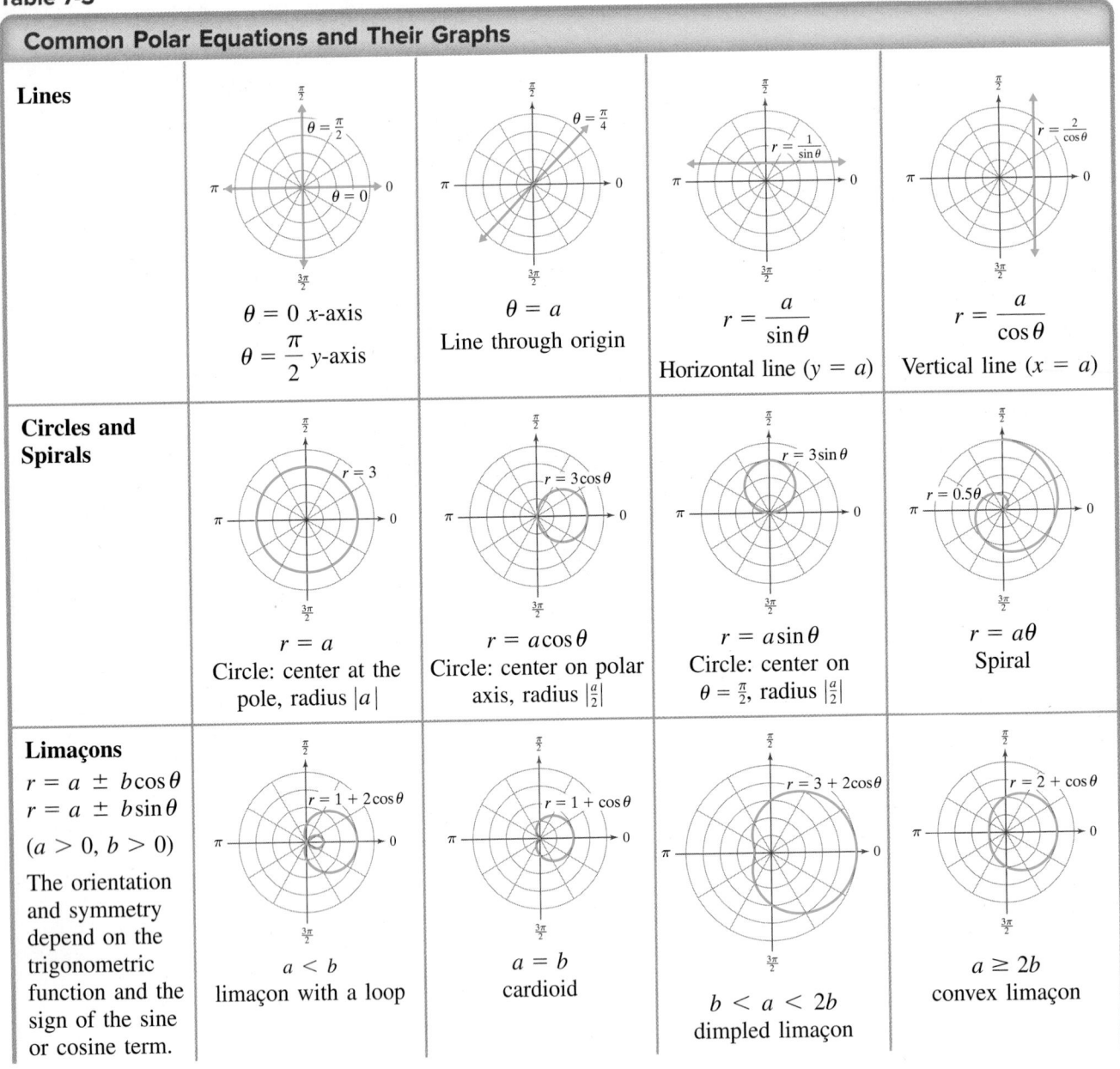

Common Polar Equations and Their Graphs			
Lines	$\theta = 0$ x-axis $\theta = \dfrac{\pi}{2}$ y-axis	$\theta = a$ Line through origin	$r = \dfrac{a}{\sin\theta}$ Horizontal line ($y = a$)	$r = \dfrac{a}{\cos\theta}$ Vertical line ($x = a$)
Circles and Spirals	$r = a$ Circle: center at the pole, radius $|a|$	$r = a\cos\theta$ Circle: center on polar axis, radius $\left|\dfrac{a}{2}\right|$	$r = a\sin\theta$ Circle: center on $\theta = \frac{\pi}{2}$, radius $\left|\dfrac{a}{2}\right|$	$r = a\theta$ Spiral
Limaçons $r = a \pm b\cos\theta$ $r = a \pm b\sin\theta$ $(a > 0, b > 0)$ The orientation and symmetry depend on the trigonometric function and the sign of the sine or cosine term.	$a < b$ limaçon with a loop	$a = b$ cardioid	$b < a < 2b$ dimpled limaçon	$a \geq 2b$ convex limaçon

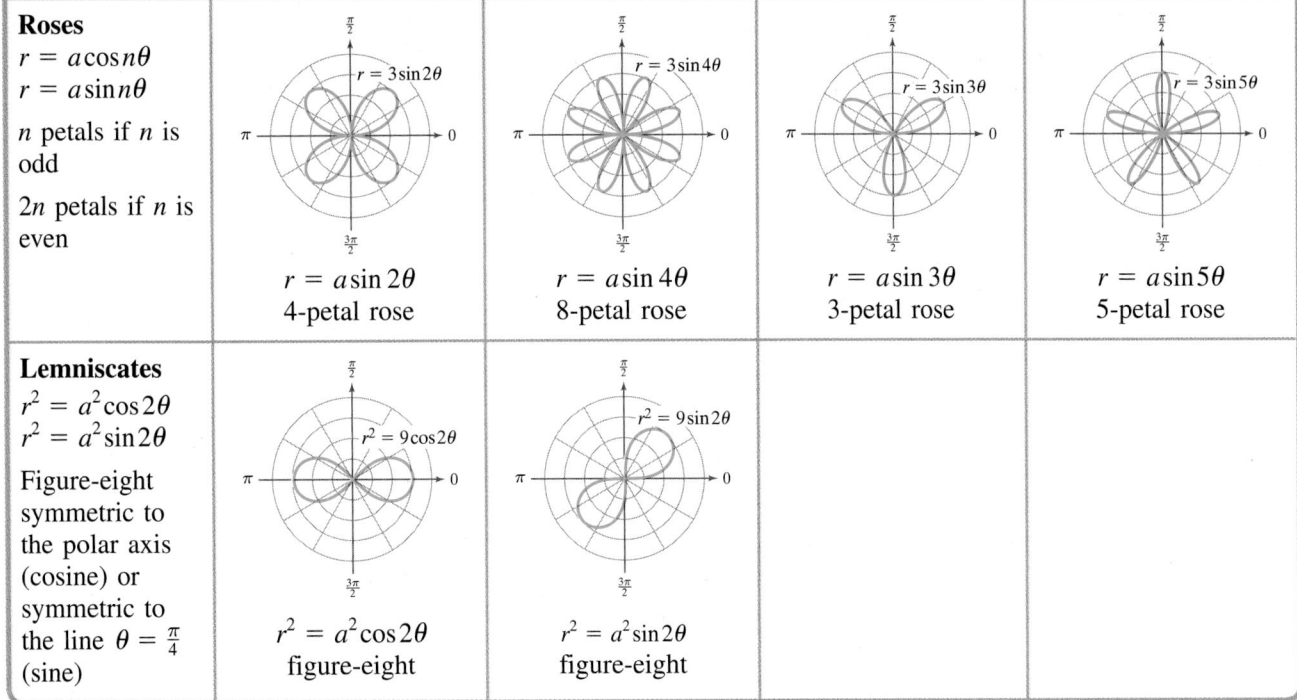

| **Roses**
$r = a\cos n\theta$
$r = a\sin n\theta$

n petals if n is odd

$2n$ petals if n is even | $r = a\sin 2\theta$
4-petal rose | $r = a\sin 4\theta$
8-petal rose | $r = a\sin 3\theta$
3-petal rose | $r = a\sin 5\theta$
5-petal rose |
| **Lemniscates**
$r^2 = a^2\cos 2\theta$
$r^2 = a^2\sin 2\theta$

Figure-eight symmetric to the polar axis (cosine) or symmetric to the line $\theta = \frac{\pi}{4}$ (sine) | $r^2 = a^2\cos 2\theta$
figure-eight | $r^2 = a^2\sin 2\theta$
figure-eight | | |

Classroom Example: p. 708
Exercise 44

EXAMPLE 5   Identifying and Graphing a Polar Equation

Graph $r^2 = 16\sin 2\theta$.

Solution:

- The graph of $r^2 = 16\sin 2\theta$ is a lemniscate of the form $r^2 = a^2\sin 2\theta$. It is a figure-eight curve symmetric with respect to the line $\theta = \dfrac{\pi}{4}$ ($y = x$ in rectangular coordinates) and to the pole.

- The graph is at the pole when $r = 0$. We have $\pm\sqrt{16\sin 2\theta} = 0$ for $\theta = \dfrac{\pi}{2}, \pi, \dfrac{3\pi}{2}, \ldots$.

- The maximum value of $|r|$ is 4 and occurs when $\sin 2\theta = 1$ or $\theta = \dfrac{\pi}{4}, \dfrac{5\pi}{4}, \ldots$.

The value of $16\sin 2\theta$ is negative for $\dfrac{\pi}{2} < \theta < \pi$, and consequently, the equation has no solution on this interval (r^2 cannot be negative). Therefore, we can plot points for $0 \le \theta \le \dfrac{\pi}{2}$ and apply symmetry to graph the curve (Figure 7-27).

TIP The graph of a lemniscate looks like a propeller on an airplane. The two loops meet at the pole for values of θ for which $r = 0$.

θ	0	$\frac{\pi}{6}$	$\frac{\pi}{4}$	$\frac{\pi}{3}$	$\frac{\pi}{2}$
r	0	$\pm 2\sqrt{2\sqrt{3}}$	± 4	$\pm 2\sqrt{2\sqrt{3}}$	0
$\approx r$	0	± 3.7	± 4	± 3.7	0

Since a lemniscate of the form $r^2 = a^2\sin 2\theta$ is symmetric with respect to the line $\theta = \dfrac{\pi}{4}$, we could have limited the table to $0 \le \theta \le \dfrac{\pi}{4}$ and then used symmetry to complete the curve.

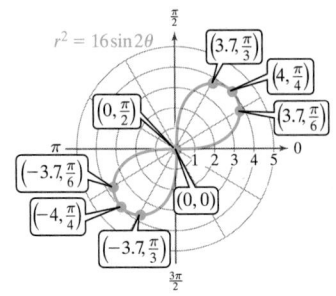

Figure 7-27

> **Skill Practice 5** Graph $r^2 = 25\cos 2\theta$.

Classroom Example: p. 708
Exercise 46

EXAMPLE 6   Identifying and Graphing a Polar Equation

Graph $r = -0.4\theta$ for $\theta \geq 0$.

Solution:

- The graph of $r = -0.4\theta$ is a spiral and is not symmetric with respect to the polar axis, line $\theta = \dfrac{\pi}{2}$, or the pole.
- The value of $|r|$ increases for increasing values of $|\theta|$ and has no upper bound.
- The graph is at the pole only for $\theta = 0$.

To see several revolutions of the spiral, we take values of θ and r on the interval $0 \leq \theta \leq 4\pi$. The graph is shown in Figure 7-28.

θ	0	$\dfrac{\pi}{2}$	π	$\dfrac{3\pi}{2}$	2π	$\dfrac{5\pi}{2}$	3π	$\dfrac{7\pi}{2}$	4π
$r \approx$	0	-0.63	-1.26	-1.88	-2.51	-3.14	-3.77	-4.40	-5.03

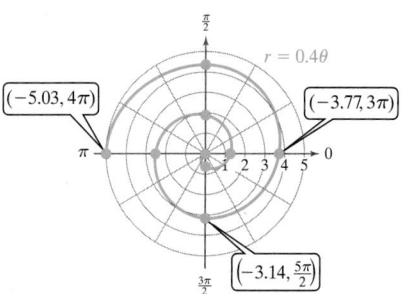

Figure 7-28

> **Skill Practice 6** Graph $r = 0.8\theta$ for $\theta \geq 0$.

Answers

5.

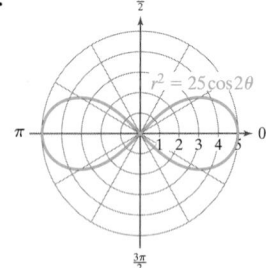

6.

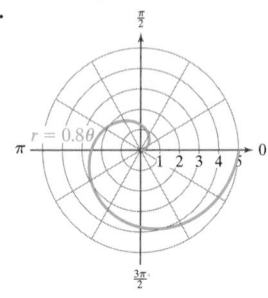

Point of Interest

The logarithmic spiral, represented by $r = ae^{b\theta}$, was referred to by the Swiss mathematician Jacob Bernoulli (1654–1705) as *spira mirabilis* (marvelous spiral). Bernoulli was so fascinated with the properties of the logarithmic spiral that he asked to have it inscribed on his tombstone. Ironically, the spiral found on Bernoulli's tomb was carved incorrectly. It is *not* a logarithmic spiral, but rather an Archimedean spiral given by $r = a\theta$.

Logarithmic spiral   Archimedean spiral

SECTION 7.2   Practice Exercises

Prerequisite Review

For Exercises R.1–R.4, determine whether the graph of the equation is symmetric with respect to the x-axis, y-axis, origin, or none of these.

R.1. $y = -|x| - 6$ y-axis

R.2. $x = y^2 + 3$ x-axis

R.3. $x^2 + y^2 = 7$ x-axis, y-axis, and origin

R.4. $y = x^2 + 2x + 5$ None of these

R.5. Solve $2\sin 2x = 1$ on the interval $(0, \pi]$. $\left\{\dfrac{\pi}{12}, \dfrac{5\pi}{12}\right\}$

R.6. Solve $\cos 3x = 1$ on the interval $[0, 2\pi)$. $\left\{0, \dfrac{2\pi}{3}, \dfrac{4\pi}{3}\right\}$

Concept Connections

1. If replacing (r, θ) by $(r, -\theta)$ in a polar equation results in an equivalent equation, then the graph of the equation is symmetric with respect to the ____polar axis____.

2. If replacing (r, θ) by $(r, \pi - \theta)$ in a polar equation results in an equivalent equation, then the graph of the equation is symmetric with respect to the _____. line $\theta = \dfrac{\pi}{2}$

3. If replacing (r, θ) by $(-r, \theta)$ in a polar equation results in an equivalent equation, then the graph of the equation is symmetric with respect to the ____pole____.

4. The graph of a polar equation of the form $r = a\cos\theta$ is symmetric with respect to the ____polar axis____, whereas the graph of a polar equation of the form $r = a\sin\theta$ is symmetric with respect to the ____line $\theta = \dfrac{\pi}{2}$____.

5. Determining the ordered pairs for which $r = 0$ in a polar equation gives what information?
It gives us the values of θ for which the graph passes through the pole.

6. What is the maximum value of r for the equation $r = a\sin\theta + 2$? What is the minimum value of r? $a + 2$; $2 - a$

Objective 1: Graph Polar Equations by Plotting Points

For Exercises 7–14, graph the polar equation. (See Example 1)

7. $r = 3$

8. $r = 4$

9. $\theta = \dfrac{3\pi}{4}$

10. $\theta = \dfrac{\pi}{4}$

11. $r = 4\cos\theta$

12. $r = -4\sin\theta$

13. $r\sin\theta = -2$

14. $r\cos\theta = 3$

Objective 2: Test Polar Equations for Symmetry

For Exercises 15–22, determine whether the tests for symmetry in Table 7-2 detect symmetry with respect to

a. The polar axis. Replace (r, θ) by $(r, -\theta)$.

b. The line $\theta = \dfrac{\pi}{2}$. Replace (r, θ) by $(r, \pi - \theta)$.

c. The pole. Replace (r, θ) by $(-r, \theta)$.

Otherwise, indicate that the test is inconclusive.

15. $r^2 = 16\cos 2\theta$
a. Yes   b. Yes   c. Yes

16. $r = 5\cos 4\theta$
a. Yes   b. Yes   c. Inconclusive

17. $r = 3\cos 3\theta$ a. Yes
b. Inconclusive   c. Inconclusive

18. $r = 6 - 4\cos\theta$

19. $r = \dfrac{3}{4 + \sin\theta}$

20. $r = 3 + 2\sin\theta$

21. $r = 2\theta$

22. $r^2 = 9\sin 2\theta$

For Exercises 23–30, use symmetry to graph the polar curve and identify the type of curve. (See Examples 2–4)

23. $r = -2 - 2\sin\theta$

24. $r = 3 + 3\cos\theta$

25. $r = 2 - 4\cos\theta$

26. $r = 1 + 7\sin\theta$

27. $r = 3\cos 5\theta$

28. $r = 8\sin 4\theta$

29. $r = 3 - 2\sin\theta$

30. $r = -3 + 2\cos\theta$

Objective 3: Categorize Polar Equations and Their Graphs

For Exercises 31–40,

a. Identify the type of polar curve from among those listed in Table 7-3.

b. Write a polar equation that represents the given graph.

31.

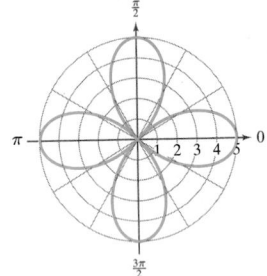

32.

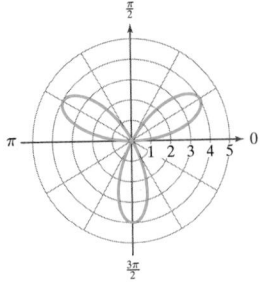

33.

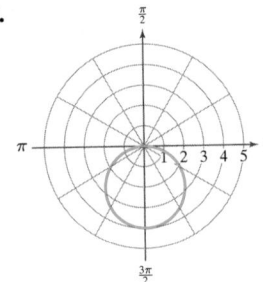

34.

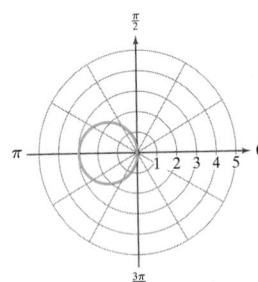

35.

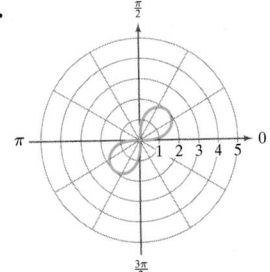

36.

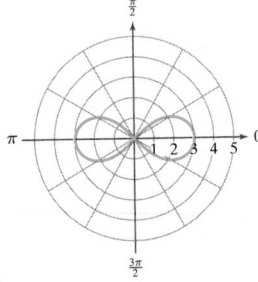

37.

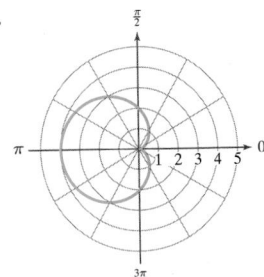

38.

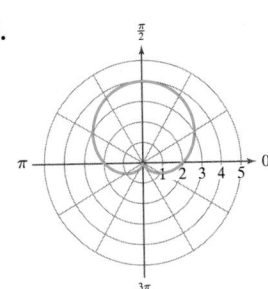

39.

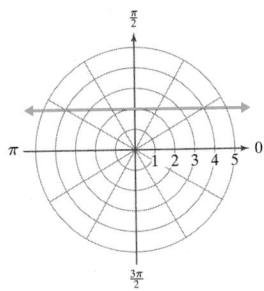

40.

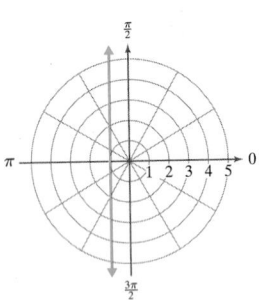

41. Identify the values of θ on the interval $[0, 2\pi)$ for which there are no points on the graph of $r^2 = 100\cos\theta$. $\dfrac{\pi}{2} < \theta < \dfrac{3\pi}{2}$

42. Identify values of θ on the interval $[0, 2\pi)$ for which there are no points on the graph of $r^2 = 49\sin\theta$. $\pi < \theta < 2\pi$

For Exercises 43–46, graph the polar curve. (See Examples 5–6)

43. $r^2 = 81\sin 2\theta$

44. $r^2 = 36\cos 2\theta$

45. $r = 0.5\theta$ for $\theta \geq 0$

46. $r = -0.6\theta$ for $\theta \geq 0$

For Exercises 47–52, for the purpose of determining a suitable viewing window for the graph of the equation,

a. Find the maximum value of $|r|$.

b. Experiment with a graphing utility to find an interval for θ over which the graph is traced only once.

47. $r = 6\cos\theta$ a. 6   b. $[0, \pi)$

48. $r = -4\sin\theta$ a. 4   b. $[0, \pi)$

49. $r = 3 - \sin\theta$ a. 4   b. $[0, 2\pi)$

50. $r = 5 + 2\cos\theta$ a. 7   b. $[0, 2\pi)$

51. $r = 3\cos\dfrac{\theta}{2}$ a. 3   b. $[0, 4\pi)$

52. $r = 7\sin\dfrac{\theta}{4}$ a. 7   b. $[0, 8\pi)$

Mixed Exercises

For Exercises 53–56, use a graphing utility or construct a table of values to match each polar equation with a graph.

a.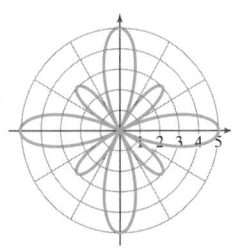

b.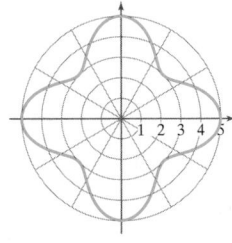

c.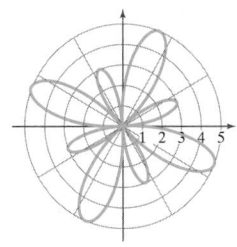

d.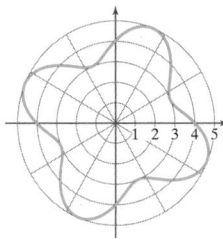

53. $r = 4 - \sin 4\theta$ d

54. $r = 1 - 4\sin 4\theta$ c

55. $r = 1 + 4\cos 4\theta$ a

56. $r = 4 + \cos 4\theta$ b

57. A quilter wants to make a design in the shape of a daisy where the "petals" are 3.5 in. long. She plans to use a computer to print a graph of the outline of the figure. Determine an equation she could use. $r = 3.5\cos 4\theta$

3.5 in.

58. The outline of a propeller on an airplane is in the shape of a lemniscate. Write an equation that could be used for the outline.

$r^2 = 3.24\sin 2\theta$

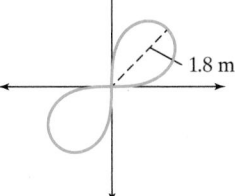

1.8 m

59. a. Given $r = 8\sin 4\theta$, determine the set of values of θ for which $r = 0$ on the interval $[0, 2\pi)$.

   b. Use a graphing utility to graph $r = 8\sin 4\theta$ on the given intervals.

   i. $0 \le \theta \le \dfrac{\pi}{2}$ **ii.** $\dfrac{\pi}{2} \le \theta \le \pi$

   iii. $\pi \le \theta \le \dfrac{3\pi}{2}$ **iv.** $\dfrac{3\pi}{2} \le \theta \le 2\pi$

60. a. Given $r = 6\sin 3\theta$, determine the set of values of θ for which $r = 0$ on the interval $[0, \pi)$.

   b. Use a graphing utility to graph $r = 6\sin 3\theta$ on the given intervals.

   i. $0 \le \theta \le \dfrac{\pi}{3}$ **ii.** $\dfrac{\pi}{3} \le \theta \le \dfrac{2\pi}{3}$

   iii. $\dfrac{2\pi}{3} \le \theta \le \pi$

61. If $r = f(\sin \theta)$, that is, if r is expressed as a function of $\sin \theta$, what type of symmetry does the graph have? Explain.

62. If $r = f(\cos \theta)$, that is, if r is expressed as a function of $\cos \theta$, what type of symmetry does the graph have? Explain.

For Exercises 63–66, use the results of Exercises 61–62 to determine by inspection whether the graph of the given equation is symmetric with respect to the polar axis or to the line $\theta = \dfrac{\pi}{2}$.

63. $r = 2\sin \theta + 3$ Symmetric with respect to the line $\theta = \dfrac{\pi}{2}$

64. $r = \sin^2\theta + \sin \theta - 1$ Symmetric with respect to the line $\theta = \dfrac{\pi}{2}$

65. $r = 2\cos^2\theta - 3\cos \theta + 1$
   Symmetric with respect to the polar axis

66. $r = 1 - 5\cos \theta$
   Symmetric with respect to the polar axis

67. Given $r = 1 + \sec \theta$ (*conchoid*),

   a. For what value(s) of θ on the interval $[0, 2\pi)$ is r undefined?

   b. Use a graphing utility to graph $r = 1 + \sec \theta$ on the interval $[0, 2\pi)$.

   c. Discuss the behavior of the graph for values of θ near those found in part (a).

68. Given $r = \dfrac{2}{\theta}$ (*hyperbolic spiral*),

   a. For what value of θ is r undefined? $\theta = 0$

   b. Use a graphing utility to graph $r = \dfrac{2}{\theta}$ on the interval $[-4\pi, 4\pi]$.

   c. Discuss the behavior of the graph for values of θ near 0.
   The graph appears to have a horizontal asymptote of $y = 2$ (in rectangular coordinates).

69. Use a graphing utility to graph the equations on the same viewing window for $0 \leq \theta \leq 2\pi$.

  a. $r = 6\sin 2\theta$

  b. $r = 6\sin\left[2\left(\theta + \dfrac{\pi}{4}\right)\right]$

  c. How do the graphs of $r = 6\sin 2\theta$ and $r = 6\sin\left[2\left(\theta + \dfrac{\pi}{4}\right)\right]$ differ?

71. A logarithmic spiral is represented by $r = ae^{b\theta}$.

  a. Graph $r = 0.5e^{0.1\theta}$ and $r = 0.5e^{-0.1\theta}$ over the interval $0 \leq \theta \leq 4\pi$.

  b. What appears to be the difference between the graphs for $b = 0.1$ and $b = -0.1$?

70. Use a graphing utility to graph the equations on the same viewing window for $0 \leq \theta \leq 2\pi$.

  a. $r = 2 + 4\cos\theta$

  b. $r = 2 + 4\cos\left(\theta - \dfrac{\pi}{4}\right)$

  c. $r = 2 + 4\cos\left(\theta - \dfrac{\pi}{2}\right)$

  d. Based on the results of parts (a)–(c), make a hypothesis about the effect of α on the graph of $r = f(\theta - \alpha)$.

72. An Archimedean spiral is represented by $r = a\theta$.

  a. Graph $r = 0.5\theta$ and $r = -0.5\theta$ over the interval $0 \leq \theta \leq 8\pi$ and use a ZOOM square viewing window.

  b. Archimedean spirals have the property that a ray through the origin will intersect successive turns of the spiral at a constant distance of $2\pi|a|$. What is the distance between each point of the spiral $r = 0.5\theta$ along the line $\theta = \dfrac{\pi}{2}$? π

Write About It

73. Graph the limaçons given by the equation $r = a + b\cos\theta$ for $a = 1, 2, 3, 4, 5,$ and 6, and $b = 3$. Comment on the effect of $\dfrac{a}{b}$ on the graph.

74. Graph the limaçons given by the equation $r = a + b\cos\theta$ for $a = 1, 2,$ and 3 and $b = 4$. Comment on the effect of $\dfrac{a}{b}$ on the graph.

Expanding Your Skills

75. Find the values of θ on the interval $[0, 2\pi)$ for which the graphs of $r = 1 + \sin\theta$ and $r = 1 - \sin\theta$ intersect. $\theta = 0$ and $\theta = \pi$

77. Given $r = 6\cos\dfrac{\theta}{2}$,

  a. Replace (r, θ) by $(r, \pi - \theta)$ in the equation. Does this show that the graph of the equation is symmetric with respect to the line $\theta = \dfrac{\pi}{2}$?   No

  b. The ordered pair $(-r, 2\pi - \theta)$ is another representation of the point $(r, \pi - \theta)$. Replace (r, θ) by $(-r, 2\pi - \theta)$ in the equation. Does this show that the graph of the equation is symmetric with respect to the line $\theta = \dfrac{\pi}{2}$?   Yes

  c. What other type of symmetry (if any) does the graph of the equation have?   The graph is also symmetric with respect to the polar axis and to the pole.

  d. Use a graphing utility to graph the equation on a graphing utility for $0 \leq \theta \leq 4\pi$.

76. Find the values of θ on the interval $[0, 2\pi)$ for which the graphs of $r = \cos\theta$ and $r = 1 - \cos\theta$ intersect. $\theta = \dfrac{\pi}{3}$ and $\theta = \dfrac{5\pi}{3}$

78. Given $r = 4\sin\dfrac{\theta}{4}$,

  a. Replace (r, θ) by $(r, -\theta)$ in the equation. Does this show that the graph of the equation is symmetric with respect to the polar axis?   No

  b. The ordered pair $(r, 4\pi - \theta)$ is another representation of the point $(r, -\theta)$. Replace (r, θ) by $(r, 4\pi - \theta)$ in the equation. Does this show that the graph of the equation is symmetric with respect to the polar axis?   Yes

  c. What other type of symmetry (if any) does the graph of the equation have?

  d. Use a graphing utility to graph the equation on a graphing utility for $0 \leq \theta \leq 8\pi$.

79. The graph of $r = 4\cos 2\theta$ is shown.

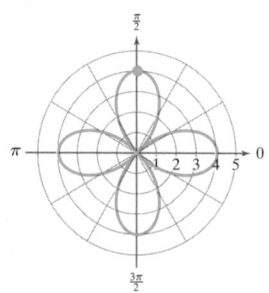

a. Is $\left(4, \dfrac{\pi}{2}\right)$ a solution to the equation?   No

b. The point $\left(4, \dfrac{\pi}{2}\right)$ appears to lie on the curve.

How do you explain this?

Any point in a plane has an infinite number of representations in polar coordinates. While the curve physically passes through the point defined by $\left(4, \dfrac{\pi}{2}\right)$, it is a different representation of this point that satisfies the equation. For example, $\left(-4, \dfrac{3\pi}{2}\right)$ is a solution to $r = 4\cos 2\theta$ and represents the same physical location as $\left(4, \dfrac{\pi}{2}\right)$.

80. Given the equations of two curves in rectangular coordinates, we can find the points of intersection by solving the equations simultaneously. In polar coordinates, solving systems of equations simultaneously may reveal some intersection points without finding others. For example, the graphs of $r = 1 - \sin\theta$ (shown in blue) and $r^2 = 16\sin\theta$ (shown in red) on the interval $0 \le \theta < 2\pi$ intersect in four points.

$$r = 1 - \sin\theta$$
$$r^2 = 16\sin\theta$$

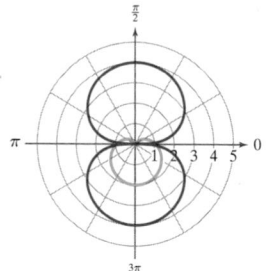

a. Solve the system simultaneously for r. (*Hint*: From the second equation, $\sin\theta = \dfrac{r^2}{16}$. Substitute $\dfrac{r^2}{16}$ for $\sin\theta$ in the first equation and solve for r.)
$r = -8 \pm 4\sqrt{5}$

b. From the graph, are both values of r reasonable?
No. For $r = -8 - 4\sqrt{5}$, $|r| \approx 16.9 > 4$.

c. Find the points of intersection corresponding to the reasonable value of r found in part (a). Round r and θ to 3 decimal places.   (0.944, 0.056) and (0.944, 3.086)

d. Do the results in part (c) give all points of intersection between the two curves?   No. The results from part (c) only give points above the polar axis. The two points of intersection below the polar axis were not found.

PROBLEM RECOGNITION EXERCISES

Comparing Equations in Polar and Rectangular Form

For Exercises 1–8,

a. Match the polar or rectangular equation on the left with the letter of the rectangular or polar equation on the right.
b. Describe the graph.
c. Determine if the graph is symmetric with respect to the origin (pole), x-axis (polar axis), or y-axis $\left(\text{line } \theta = \dfrac{\pi}{2}\right)$.

1. $x^2 + y^2 = 144$

2. $x = -3$

3. $r = \dfrac{4}{2\cos\theta + \sin\theta}$

4. $r = -6\cos\theta$

5. $y = -\sqrt{3}x$

6. $x^2 + (y + 5)^2 = 25$

7. $r = 5\csc\theta$

8. $r = 2\sin\theta + 2\cos\theta$

A. $2x + y = 4$

B. $r = -10\sin\theta$

C. $r = 12$

D. $\theta = \dfrac{2\pi}{3}$

E. $y = 5$

F. $r = -3\sec\theta$

G. $(x - 1)^2 + (y - 1)^2 = 2$

H. $(x + 3)^2 + y^2 = 9$

1. a. Equation C   **b.** Circle centered at the origin (pole), radius 12
   c. Origin, x-axis, and y-axis
2. a. Equation F   **b.** Vertical line through the x-axis at -3
   c. x-axis
3. a. Equation A   **b.** Line with slope -2 and y-intercept $(0, 4)$
   c. None of these
4. a. Equation H   **b.** Circle centered at $(-3, 0)$ with radius 3
   c. x-axis
5. a. Equation D   **b.** Line through the origin with slope $-\sqrt{3}$
   c. Origin
6. a. Equation B   **b.** Circle centered at $(0, -5)$ with radius 5
   c. y-axis
7. a. Equation E   **b.** Horizontal line through the y-axis at 5   **c.** y-axis
8. a. Equation G   **b.** Circle centered at $(1, 1)$, with radius $\sqrt{2}$
   c. None of these

For Exercises 9–10, use a graphing utility to graph the polar equation and describe the graph in a rectangular coordinate system.

9. $r = \dfrac{3}{1 - \sin\theta}$

10. $r = \dfrac{2}{1 + \cos\theta}$

For Exercises 11–13, points A and B make up a graph that has an indicated type of symmetry for all polar representations of point B. Fill in an ordered pair for point B that would satisfy the condition of symmetry. Note that for ordered pairs in polar form, there is more than one correct answer.

11. a. Symmetry: x-axis $(x, -y)$

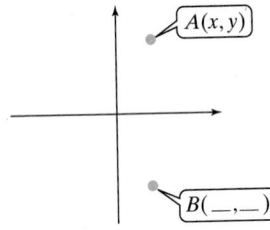

b. Symmetry: polar axis $(r, -\theta)$

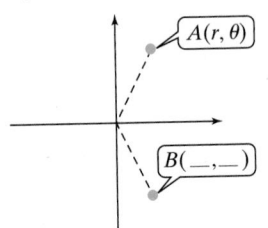

12. a. Symmetry: y-axis $(-x, y)$

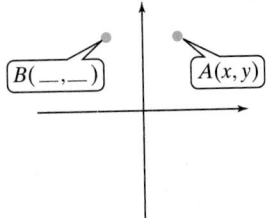

b. Symmetry: Line $\theta = \dfrac{\pi}{2}$ $(r, \pi - \theta)$

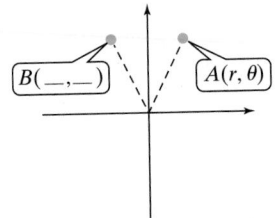

13. a. Symmetry: Origin $(-x, -y)$

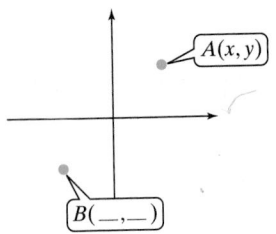

b. Symmetry: Pole $(-r, \theta)$

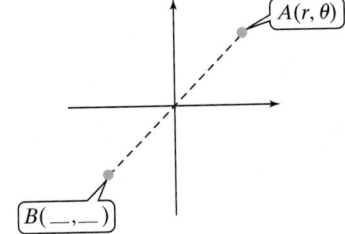

SECTION 7.3   Complex Numbers in Polar Form

Classroom Example: p. 722
Exercise 8

1. Plot Complex Numbers in the Complex Plane

A real number can be represented as a point on the real number line. Since a complex number $a + bi$ has two parts, the real part a and the imaginary part b, two axes are required to represent $a + bi$ graphically. The horizontal axis (called the **real axis**) represents the real part of a complex number, and the vertical axis (called the **imaginary axis**) represents the imaginary part. Together, the real and imaginary axes form the **complex plane.** A complex number $z = a + bi$ is written as an ordered pair (a, b) and can then be graphed as such in the complex plane (Figure 7-29).

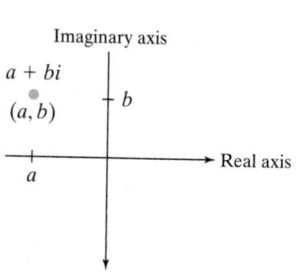

Figure 7-29   Complex Plane

EXAMPLE 1   **Graphing Complex Numbers**

Graph the complex numbers in the complex plane.

  a. $3 + 4i$ **b.** $4 - 2i$ **c.** -5 **d.** $-2i$

Solution:

  a. $3 + 4i$ is represented by the point $(3, 4)$.

  b. $4 - 2i$ is represented by the point $(4, -2)$.

  c. $-5 = -5 + 0i$ is represented by the point $(-5, 0)$ on the real axis.

  d. $-2i = 0 - 2i$ is represented by the point $(0, -2)$ on the imaginary axis.

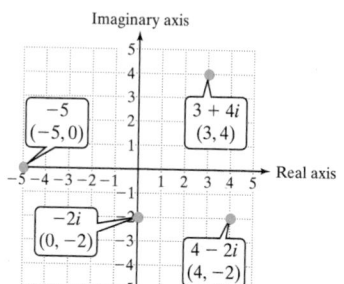

Skill Practice 1   Graph the complex numbers in the complex plane.

  a. $-1 + 3i$ **b.** $-4 - 5i$ **c.** 3 **d.** $4i$

> **TIP** Complex numbers are often denoted by the variable *z* such as $z = 3 + 4i$.

> **TIP** Notice that for a complex number of the form $a + 0i$, the absolute value is $\sqrt{a^2 + 0^2} = |a|$. This is consistent with the absolute value of a real number *a*.

Answers

1. **a. – d.**

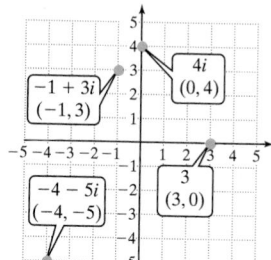

Geometrically, the absolute value of a real number is its distance from zero on the real number line. Given a complex number $z = a + bi$, the absolute value of z, denoted $|z|$, is the distance between (a, b) and $(0, 0)$ in the complex plane (Figure 7-30). The absolute value of a complex number is more commonly called the *modulus* of the number.

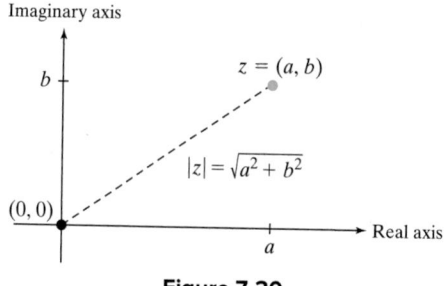

Figure 7-30

> ### Absolute Value of a Complex Number
>
> If $z = a + bi$ is a complex number, then the absolute value of z (also called the **modulus** or **magnitude** of z) is given by
>
> $$|z| = \sqrt{a^2 + b^2}.$$

Classroom Example: p. 722
Exercise 12

EXAMPLE 2   Finding the Modulus (Absolute Value) of a Complex Number

Find the modulus.

a. $z = 3 - 4i$ **b.** $z = -8 + 2i$

Solution:

a. $|z| = |3 - 4i| = \sqrt{(3)^2 + (-4)^2} = \sqrt{9 + 16} = \sqrt{25} = 5$

b. $|z| = |-8 + 2i| = \sqrt{(-8)^2 + (2)^2} = \sqrt{64 + 4} = \sqrt{68} = 2\sqrt{17}$

> **Skill Practice 2**   Find the modulus.
>
> **a.** $z = -5 - 12i$ **b.** $z = 4 + 6i$

2. Write the Polar Form of a Complex Number

The complex number $z = a + bi$ is said to be in rectangular (or Cartesian) form because a and b are the coordinates of the point (a, b) in the rectangular coordinate system formed by the complex plane. For a and b not both zero, let r represent the length of the line segment between (a, b) and $(0, 0)$. Thus, $r = |z| = \sqrt{a^2 + b^2} \geq 0$ (Figure 7-31). Taking angle θ to be in standard position with terminal side through (a, b), we have $a = r \cos \theta$ and $b = r \sin \theta$.

Therefore, $z = a + bi$
$= r \cos \theta + i \cdot r \sin \theta$
$= r(\cos \theta + i \sin \theta)$

This is called the polar form (or trigonometric form) of a complex number.

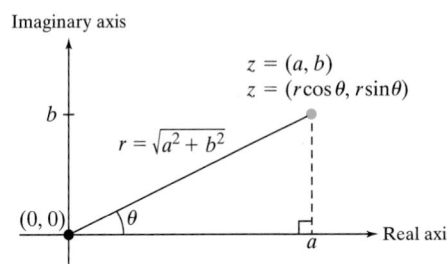

Figure 7-31

> ### Polar Form of a Complex Number
>
> The **polar form** (or **trigonometric form**) of a complex number $z = a + bi$ is
>
> $$z = r(\cos \theta + i \sin \theta)$$
>
> where $a = r \cos \theta$, $b = r \sin \theta$, $r = \sqrt{a^2 + b^2}$, and $\tan \theta = \dfrac{b}{a}$ $(a \neq 0)$. The value of r is the modulus of z, and we say that θ is an argument of z. There are infinitely many choices for θ, but normally, θ is taken on the interval $0 \leq \theta < 2\pi$ or $0° \leq \theta < 360°$.

Answers

2. a. 13   **b.** $2\sqrt{13}$

EXAMPLE 3 **Convert a Complex Number from Rectangular Form to Polar Form**

Write the complex number in polar form.
  a. $z = -2 + 2i$
  b. $z = 8 - 15i$

Solution:

a. The point $z = -2 + 2i$ is in Quadrant II.
$$r = \sqrt{a^2 + b^2} = \sqrt{(-2)^2 + (2)^2} = \sqrt{8} = 2\sqrt{2}$$
$$\tan\theta = \frac{b}{a} = \frac{2}{-2} = -1$$

Taking θ in Quadrant II on the interval
$0 \le \theta < 2\pi$, we have $\theta = \dfrac{3\pi}{4}$.

Modulus   Argument

Polar form: $z = 2\sqrt{2}\left(\cos\dfrac{3\pi}{4} + i\sin\dfrac{3\pi}{4}\right)$

b. The point $z = 8 - 15i$ is in Quadrant IV.
$$r = \sqrt{a^2 + b^2} = \sqrt{(8)^2 + (-15)^2} = \sqrt{289} = 17$$
$$\tan\theta = \frac{-15}{8} \Rightarrow \theta = \tan^{-1}\left(-\frac{15}{8}\right)$$

Taking θ in Quadrant IV on the interval
$0 \le \theta < 2\pi$, we have $\theta = 2\pi + \tan^{-1}\left(-\dfrac{15}{8}\right)$.

Polar form: $z = 17\left[\cos\left(2\pi + \tan^{-1}\left(-\dfrac{15}{8}\right)\right) + i\sin\left(2\pi + \tan^{-1}\left(-\dfrac{15}{8}\right)\right)\right]$
$$z \approx 17(\cos 5.2 + i\sin 5.2)$$

Skill Practice 3 Write the complex number in polar form.
  a. $z = 1 - i$
  b. $z = 3 + 4i$

Point of Interest

The complex number $z = r(\cos\theta + i\sin\theta)$ is sometimes shortened to $z = r\operatorname{cis}\theta$, where "cis" stands for "cosine i sine." For example,

$$z = 5\left(\cos\frac{\pi}{4} + i\sin\frac{\pi}{4}\right) \text{ would be written as } z = 5\operatorname{cis}\frac{\pi}{4}.$$

Another form in which a complex number can be written makes use of Euler's formula. Swiss mathematician Leonhard Euler (pronounced "Oy-ler") (1707–1783), discovered a relationship between the exponential function of a complex number and the trigonometric functions. For a real number θ, Euler's formula states that $e^{i\theta} = \cos\theta + i\sin\theta$. Thus, the complex number

$$z = 5\left(\cos\frac{\pi}{4} + i\sin\frac{\pi}{4}\right) \text{ can be written as } z = 5e^{\frac{\pi}{4}i}.$$

Answers

3. a. $z = \sqrt{2}\left(\cos\dfrac{7\pi}{4} + i\sin\dfrac{7\pi}{4}\right)$

  b.
$z = 5\left[\cos\left(\tan^{-1}\dfrac{4}{3}\right) + i\sin\left(\tan^{-1}\dfrac{4}{3}\right)\right]$

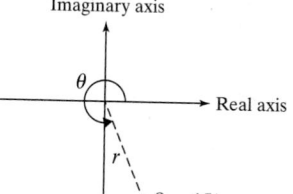

In Example 4, we perform the simpler task of writing a complex number in rectangular form.

Classroom Example: p. 723

Exercise 24

EXAMPLE 4   Convert a Complex Number from Polar Form to Rectangular Form

Write $z = 4(\cos 120° + i\sin 120°)$ in rectangular form, $a + bi$.

Solution:

$$z = 4(\cos 120° + i\sin 120°) = 4\left(-\frac{1}{2} + i \cdot \frac{\sqrt{3}}{2}\right) = -2 + 2\sqrt{3}i$$

Skill Practice 4   Write $z = 6(\cos 150° + i\sin 150°)$ in rectangular form.

3. Compute Products and Quotients of Complex Numbers in Polar Form

In Chapter R, we learned how to add, subtract, multiply, and divide complex numbers in rectangular form. We now present the procedure to multiply and divide complex numbers in polar form.

Products and Quotients of Complex Numbers in Polar Form

If $z_1 = r_1(\cos\theta_1 + i\sin\theta_1)$ and $z_2 = r_2(\cos\theta_2 + i\sin\theta_2)$, then

$$z_1 z_2 = r_1 r_2[\cos(\theta_1 + \theta_2) + i\sin(\theta_1 + \theta_2)] \tag{1}$$

$$\frac{z_1}{z_2} = \frac{r_1}{r_2}[\cos(\theta_1 - \theta_2) + i\sin(\theta_1 - \theta_2)] \quad (z_2 \neq 0) \tag{2}$$

TIP The word "moduli" is the plural of the word "modulus."

In words, to multiply two complex numbers in polar form, multiply their moduli and add their arguments. To divide two complex numbers in polar form, divide their moduli and subtract their arguments.

To prove (1), we have

Recall that $i^2 = -1$.

$$\begin{aligned}
z_1 z_2 &= r_1(\cos\theta_1 + i\sin\theta_1) \cdot r_2(\cos\theta_2 + i\sin\theta_2) \\
&= r_1 r_2(\cos\theta_1\cos\theta_2 + i\cos\theta_1\sin\theta_2 + i\sin\theta_1\cos\theta_2 + i^2\sin\theta_1\sin\theta_2) \\
&= r_1 r_2[(\cos\theta_1\cos\theta_2 - \sin\theta_1\sin\theta_2) + i(\sin\theta_1\cos\theta_2 + \cos\theta_1\sin\theta_2)] \\
&= r_1 r_2[\cos(\theta_1 + \theta_2) + i\sin(\theta_1 + \theta_2)]
\end{aligned}$$

Apply the sum formulas for sine and cosine.

The proof of (2) is similar and left as an exercise (Exercise 85).

Classroom Example: p. 723

Exercise 34

EXAMPLE 5   Multiplying Complex Numbers in Polar Form

Given $z_1 = 4\left(\cos\frac{4\pi}{3} + i\sin\frac{4\pi}{3}\right)$ and $z_2 = 2\left(\cos\frac{\pi}{2} + i\sin\frac{\pi}{2}\right)$, find $z_1 z_2$ and write the product in polar form.

Answer

4. $-3\sqrt{3} + 3i$

TIP In Example 5, the product is left in polar form. To find the corresponding rectangular form,

$$8\left(\cos\frac{11\pi}{6} + i\sin\frac{11\pi}{6}\right)$$

$$= 8\left(\frac{\sqrt{3}}{2} - \frac{1}{2}i\right)$$

$$= 4\sqrt{3} - 4i$$

Solution:

$$z_1z_2 = 4\left(\cos\frac{4\pi}{3} + i\sin\frac{4\pi}{3}\right)\cdot 2\left(\cos\frac{\pi}{2} + i\sin\frac{\pi}{2}\right)$$

$$= 4\cdot 2\left[\cos\left(\frac{4\pi}{3} + \frac{\pi}{2}\right) + i\sin\left(\frac{4\pi}{3} + \frac{\pi}{2}\right)\right]$$

Multiply the moduli and add the arguments. Note that $\frac{4\pi}{3} + \frac{\pi}{2} = \frac{8\pi}{6} + \frac{3\pi}{6} = \frac{11\pi}{6}$

$$= 8\left(\cos\frac{11\pi}{6} + i\sin\frac{11\pi}{6}\right)\quad\text{(polar form)}$$

> **Skill Practice 5**   Given $z_1 = 3\left(\cos\dfrac{\pi}{12} + i\sin\dfrac{\pi}{12}\right)$ and
>
> $z_2 = 2\left(\cos\dfrac{\pi}{6} + i\sin\dfrac{\pi}{6}\right)$, find z_1z_2 and write the product in polar form.

In Exercise 83, we verify the result of Example 5 by converting the complex numbers to rectangular form and then multiplying.

Classroom Example: p. 723
Exercise 36

EXAMPLE 6   Dividing Complex Numbers in Polar Form

Given $z_1 = 4(\cos 15° + i\sin 15°)$ and $z_2 = 12(\cos 285° + i\sin 285°)$, find $\dfrac{z_1}{z_2}$ and write the quotient in polar form.

Solution:

$$\frac{z_1}{z_2} = \frac{4(\cos 15° + i\sin 15°)}{12(\cos 285° + i\sin 285°)}$$

$$= \frac{1}{3}[\cos(15° - 285°) + i\sin(15° - 285°)]$$

Divide the moduli and subtract the arguments.

$$= \frac{1}{3}[\cos(-270°) + i\sin(-270°)]$$

$$= \frac{1}{3}[\cos(90°) + i\sin(90°)]$$

$90°$ is coterminal to $-270°$ and on the interval $0° \le \theta < 360°$.

> **Skill Practice 6**   Given $z_1 = 18(\cos 75° + i\sin 75°)$ and
>
> $z_2 = 2(\cos 255° + i\sin 255°)$, find $\dfrac{z_1}{z_2}$ and write the quotient in polar form.

4. Compute Powers of Complex Numbers (De Moivre's Theorem)

Given $z = a + bi$, the value of z^n, where n is a positive integer, can be computed by using repeated multiplication of the base z in its polar form.

Let $z = r(\cos\theta + i\sin\theta)$. Then,

$$z^2 = r(\cos\theta + i\sin\theta)\cdot r(\cos\theta + i\sin\theta) = r^2(\cos 2\theta + i\sin 2\theta)$$

$$z^3 = r^2(\cos 2\theta + i\sin 2\theta)\cdot r(\cos\theta + i\sin\theta) = r^3(\cos 3\theta + i\sin 3\theta)$$

$$z^4 = r^3(\cos 3\theta + i\sin 3\theta)\cdot r(\cos\theta + i\sin\theta) = r^4(\cos 4\theta + i\sin 4\theta)$$

$$\vdots$$

$$z^n = r^{n-1}[\cos(n-1)\theta + i\sin(n-1)\theta]\cdot r(\cos\theta + i\sin\theta) = r^n(\cos n\theta + i\sin n\theta).$$

Answers

5. $6\left(\cos\dfrac{\pi}{4} + i\sin\dfrac{\pi}{4}\right)$

6. $9[\cos(180°) - i\sin(180°)]$

This pattern involving integer powers of a complex number is known as De Moivre's theorem named after the French mathematician Abraham de Moivre (pronounced "duh mwah-vruh") (1667–1754).

> ### De Moivre's Theorem
>
> If $z = r(\cos\theta + i\sin\theta)$ and n is a positive integer, then
>
> $$z^n = [r(\cos\theta + i\sin\theta)]^n = r^n(\cos n\theta + i\sin n\theta).$$
>
> In words, to raise z to a positive integer power n, raise the modulus to the nth power and multiply the argument by n.

Classroom Example: p. 723
Exercise 44

EXAMPLE 7 Raising a Complex Number to a Power

Compute $[5(\cos 75° + i\sin 75°)]^4$ and write the result in rectangular form $a + bi$.

Solution:

$$[5(\cos 75° + i\sin 75°)]^4 = 5^4[\cos(4 \cdot 75°) + i\sin(4 \cdot 75°)]$$

$$= 625(\cos 300° + i\sin 300°) \quad \text{Polar form}$$

$$= 625\left(\frac{1}{2} - i\frac{\sqrt{3}}{2}\right) \quad \begin{array}{l}\text{Evaluate } \cos 300° \text{ and} \\ \sin 300° \text{ to convert to} \\ \text{rectangular form.}\end{array}$$

$$= \frac{625}{2} - \frac{625\sqrt{3}}{2}i \quad \text{Rectangular form}$$

> **TIP** Recall that $a + (-bi)$ is usually written as $a - bi$. Likewise, a real number or pure imaginary number is usually simplified. For example, $a + 0i$ is usually written as a, and $0 + bi$ is usually written as bi.

Skill Practice 7 Compute $[4(\cos 70° + i\sin 70°)]^3$ and write the result in rectangular form $a + bi$.

Classroom Example: p. 723
Exercise 52

EXAMPLE 8 Raising a Complex Number to a Power

Compute $(-3 + 3i)^6$ and write the result in rectangular form $a + bi$.

Solution:

Computing the product in rectangular form would be very tedious to carry out:

$$(-3 + 3i)^6 = (-3 + 3i)(-3 + 3i)(-3 + 3i)(-3 + 3i)(-3 + 3i)(-3 + 3i)$$

Instead, we will write $-3 + 3i$ in polar form and then apply De Moivre's theorem.

$z = -3 + 3i = (-3, 3)$ is in Quadrant II.

$r = \sqrt{(-3)^2 + (3)^2} = \sqrt{18} = 3\sqrt{2}$

$\tan\theta = \dfrac{3}{-3} = -1 \Rightarrow \theta = \dfrac{3\pi}{4}$ for θ in Quadrant II.

Thus, $(-3 + 3i)^6 = \left[(3\sqrt{2})\left(\cos\dfrac{3\pi}{4} + i\sin\dfrac{3\pi}{4}\right)\right]^6$

$$= (3\sqrt{2})^6\left[\cos\left(6 \cdot \dfrac{3\pi}{4}\right) + i\sin\left(6 \cdot \dfrac{3\pi}{4}\right)\right]$$

$$= 3^6 \cdot 2^3\left(\cos\dfrac{9\pi}{2} + i\sin\dfrac{9\pi}{2}\right)$$

$$= 5832(0 + i)$$

$$= 5832i$$

Answer

7. $-32\sqrt{3} - 32i$

TECHNOLOGY CONNECTIONS

The calculation from Example 8 performed on a graphing utility results in a small round-off error. The real part is displayed as 5.832×10^{-10} when in fact it should be zero.

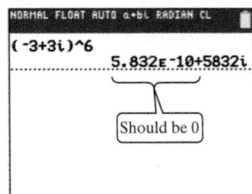

Skill Practice 8   Compute $(-2 - 2i)^4$ and write the result in rectangular form $a + bi$.

5. Find the nth Roots of a Complex Number

For a positive integer $n \geq 2$, an nth root of a complex number z is a complex number w such that $w^n = z$. In Example 7, we found that

$$[5(\cos 75° + i\sin 75°)]^4 = 625(\cos 300° + i\sin 300°)$$

This means that $5(\cos 75° + i\sin 75°)$ is a fourth root of $625(\cos 300° + i\sin 300°)$. Likewise, in Example 8, we found that $(-3 + 3i)^6 = 5832i$. This means that $-3 + 3i$ is a sixth root of $5832i$. In general, there are n distinct nth roots of a complex number.

---

nth Roots of a Complex Number

If $z = r(\cos\theta + i\sin\theta)$, then for an integer $n \geq 2$, z has n distinct nth roots of the form

$$(3) \quad w_k = \sqrt[n]{r}\left[\cos\left(\frac{\theta + 2\pi k}{n}\right) + i\sin\left(\frac{\theta + 2\pi k}{n}\right)\right] \text{ for } k = 0, 1, 2, \ldots, n - 1$$

or

$$w_k = \sqrt[n]{r}\left[\cos\left(\frac{\theta + 360°k}{n}\right) + i\sin\left(\frac{\theta + 360°k}{n}\right)\right] \text{ for } k = 0, 1, 2, \ldots, n - 1$$

Equation (3) generates n nth roots of a complex number z, and the argument of each root corresponds to a value of k between 0 and $n - 1$, inclusive. Values of $k \geq n$ do not need to be considered because they generate arguments that are coterminal to those generated by $k = 0, 1, 2, \ldots, n - 1$ and would therefore be redundant.

While we will not take up a formal proof of equation (3), we can use De Moivre's theorem to show that the complex numbers listed in equation (3) are indeed roots of $z = r(\cos\theta + i\sin\theta)$. We require that $w^n = z$.

Point of Interest

In addition to his discovery of the roots of a complex number, French mathematician Abraham de Moivre (1667–1754) was noted for his study of probability and statistics. He made numerous publications involving the probabilities associated with games of chance and had investigated mortality statistics that laid the foundation of the theory of annuities. As de Moivre aged, a popular anecdote suggests that he slept 15 min longer each night, and that he correctly predicted the date of his death by linear extrapolation to the day that he would sleep for 24 hr.

Answer

8. -64

$$(w_k)^n = \left(\sqrt[n]{r}\left[\cos\left(\frac{\theta + 2\pi k}{n}\right) + i\sin\left(\frac{\theta + 2\pi k}{n}\right) \right] \right)^n \qquad \text{Definition of an } n\text{th root of a complex number.}$$

$$= (\sqrt[n]{r})^n\left[\cos\left(n \cdot \frac{\theta + 2\pi k}{n}\right) + i\sin\left(n \cdot \frac{\theta + 2\pi k}{n}\right) \right] \qquad \text{Apply De Moivre's theorem.}$$

$$= r\left[\cos(\theta + 2\pi k) + i\sin(\theta + 2\pi k)\right] \qquad \text{Simplify.}$$

$$= r(\cos\theta + i\sin\theta) \qquad \theta \text{ and } \theta + 2\pi k \text{ are coterminal for all integers } k.$$

$$= z$$

Classroom
Example: p. 724
Exercise 56

EXAMPLE 9   Finding the nth Roots of a Complex Number

Find the sixth roots of 64 over the set of complex numbers. Write the results in rectangular form $a + bi$.

Solution:

First write $z = 64 + 0i$ in polar form.

$r = \sqrt{(64)^2 + (0)^2} = 64$ and $\theta = 0$

Therefore, $z = 64(\cos 0 + i\sin 0)$.

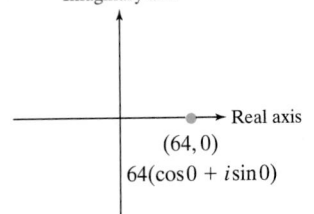

The sixth roots of $z = 64(\cos 0 + i\sin 0)$ are

$$w_k = \sqrt[6]{64}\left[\cos\left(\frac{0 + 2\pi k}{6}\right) + i\sin\left(\frac{0 + 2\pi k}{6}\right) \right] \text{ for } k = 0, 1, 2, 3, 4, \text{ and } 5$$

$$= 2\left[\cos\left(\frac{\pi}{3}k\right) + i\sin\left(\frac{\pi}{3}k\right) \right] \text{ for } k = 0, 1, 2, 3, 4, \text{ and } 5$$

$$w_0 = 2\left[\cos\left(\frac{\pi}{3} \cdot 0\right) + i\sin\left(\frac{\pi}{3} \cdot 0\right) \right] = 2(\cos 0 + i\sin 0) = 2(1 + 0i) = 2$$

$$w_1 = 2\left[\cos\left(\frac{\pi}{3} \cdot 1\right) + i\sin\left(\frac{\pi}{3} \cdot 1\right) \right] = 2\left(\cos\frac{\pi}{3} + i\sin\frac{\pi}{3} \right) = 2\left(\frac{1}{2} + \frac{\sqrt{3}}{2}i\right) = 1 + \sqrt{3}i$$

$$w_2 = 2\left[\cos\left(\frac{\pi}{3} \cdot 2\right) + i\sin\left(\frac{\pi}{3} \cdot 2\right) \right] = 2\left(\cos\frac{2\pi}{3} + i\sin\frac{2\pi}{3} \right) = 2\left(-\frac{1}{2} + \frac{\sqrt{3}}{2}i\right) = -1 + \sqrt{3}i$$

$$w_3 = 2\left[\cos\left(\frac{\pi}{3} \cdot 3\right) + i\sin\left(\frac{\pi}{3} \cdot 3\right) \right] = 2(\cos\pi + i\sin\pi) = 2(-1 + 0i) = -2$$

$$w_4 = 2\left[\cos\left(\frac{\pi}{3} \cdot 4\right) + i\sin\left(\frac{\pi}{3} \cdot 4\right) \right] = 2\left(\cos\frac{4\pi}{3} + i\sin\frac{4\pi}{3} \right) = 2\left(-\frac{1}{2} - \frac{\sqrt{3}}{2}i\right) = -1 - \sqrt{3}i$$

$$w_5 = 2\left[\cos\left(\frac{\pi}{3} \cdot 5\right) + i\sin\left(\frac{\pi}{3} \cdot 5\right) \right] = 2\left(\cos\frac{5\pi}{3} + i\sin\frac{5\pi}{3} \right) = 2\left(\frac{1}{2} - \frac{\sqrt{3}}{2}i\right) = 1 - \sqrt{3}i$$

Notice that the six roots all have the same magnitude 2. This means that they all lie on a circle centered at the origin of the complex plane with radius 2. Furthermore, the argument of each root differs by a constant angular measure of $\dfrac{\pi}{3}$ (Figure 7-32). Geometrically, this means that the six 6th roots are all equally spaced around the circle. This is true in general when we find the nth roots of a complex number.

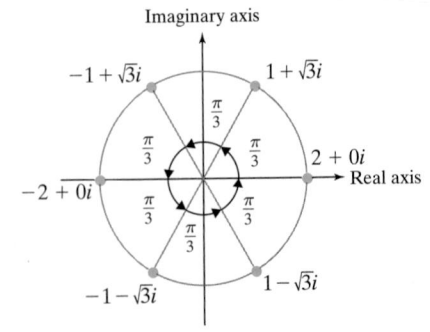

Figure 7-32

Skill Practice 9   Find the cube roots of -64 over the set of complex numbers. Write the results in rectangular form $a + bi$.

From the fundamental theorem of algebra presented in Section 2.4, we know that a polynomial equation of degree n has n complex solutions. Therefore, the six 6th roots of 64 can also be found by solving the equation $x^6 = 64$ (Exercise 93).

Classroom Example: p. 724
Exercise 70

EXAMPLE 10   Finding the *n*th Roots of a Complex Number

Find the fourth roots of $-\sqrt{3} - i$ over the set of complex numbers. Write the results in polar form using degree measure for the argument.

Solution:

First write $z = -\sqrt{3} - i$ in polar form.

$r = \sqrt{(-\sqrt{3})^2 + (-1)^2} = \sqrt{3 + 1} = 2$

$\tan\theta = \dfrac{-1}{-\sqrt{3}} = \dfrac{\sqrt{3}}{3}$

Since $z = -\sqrt{3} - i$ is in Quadrant III, $\theta = 210°$.
Thus, $z = 2(\cos 210° + i\sin 210°)$

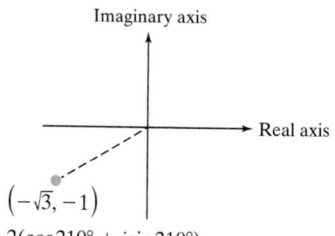

$2(\cos 210° + i\sin 210°)$

The fourth roots of $z = 2(\cos 210° + i\sin 210°)$ are

$$w_k = \sqrt[4]{2}\left[\cos\left(\frac{210° + 360°k}{4}\right) + i\sin\left(\frac{210° + 360°k}{4}\right)\right] \text{ for } k = 0, 1, 2, \text{ and } 3$$

$$= \sqrt[4]{2}[\cos(52.5° + 90°k) + i\sin(52.5° + 90°k)] \text{ for } k = 0, 1, 2, \text{ and } 3.$$

$w_0 = \sqrt[4]{2}[\cos(52.5° + 90° \cdot 0) + i\sin(52.5° + 90° \cdot 0)] = \sqrt[4]{2}(\cos 52.5° + i\sin 52.5°)$

$w_1 = \sqrt[4]{2}[\cos(52.5° + 90° \cdot 1) + i\sin(52.5° + 90° \cdot 1)] = \sqrt[4]{2}(\cos 142.5° + i\sin 142.5°)$

$w_2 = \sqrt[4]{2}[\cos(52.5° + 90° \cdot 2) + i\sin(52.5° + 90° \cdot 2)] = \sqrt[4]{2}(\cos 232.5° + i\sin 232.5°)$

$w_3 = \sqrt[4]{2}[\cos(52.5° + 90° \cdot 3) + i\sin(52.5° + 90° \cdot 3)] = \sqrt[4]{2}(\cos 322.5° + i\sin 322.5°)$

The fourth roots are equally spaced around a circle of radius $\sqrt[4]{2}$ centered at the origin in the complex plane. The arguments of the polar forms of the roots differ by

$$\frac{360°}{n} = \frac{360°}{4} = 90°.$$

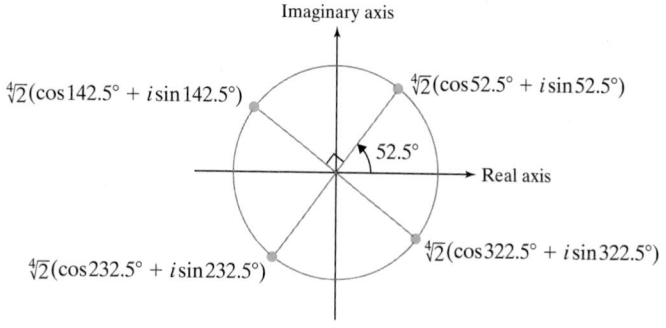

Skill Practice 10   Find the fourth roots of $3 + 3\sqrt{3}i$ over the set of complex numbers. Write the results in polar form using degree measure for the argument.

Answer

10. $\sqrt[4]{6}(\cos 15° + i\sin 15°),$
$\sqrt[4]{6}(\cos 105° + i\sin 105°),$
$\sqrt[4]{6}(\cos 195° + i\sin 195°),$
$\sqrt[4]{6}(\cos 285° + i\sin 285°)$

SECTION 7.3 Practice Exercises

Prerequisite Review

For Exercises R.1–R.5, perform the indicated operations and write the result in standard form $a + bi$.

R.1. $\left(\frac{1}{3} - \frac{1}{2}i\right) - \left(\frac{1}{6} - \frac{7}{12}i\right)$ $\quad \frac{1}{6} + \frac{1}{12}i$ **R.2.** $3i(8 - 5i)$ $\quad 15 + 24i$ **R.3.** $(2 + 3i)(-3 - 9i)$ $\quad 21 - 27i$

R.4. $(10 - 3i)^2$ $\quad 91 - 60i$ **R.5.** $\dfrac{-5 + 8i}{3 - 9i}$ $\quad \dfrac{29}{30} - \dfrac{7}{30}i$

Concept Connections

1. The complex plane has two axes. The horizontal axis is called the _____real_____ axis, and the vertical axis is called the _____imaginary_____ axis.

2. Given $z = a + bi$, the value $|z| = $ _____$\sqrt{a^2 + b^2}$_____.

3. The rectangular form of a complex number z is $a + bi$ and the polar form (or trigonometric form) is given by $z = $ _____$r(\cos\theta + i\sin\theta)$_____, where $r = $ _____$\sqrt{a^2 + b^2}$_____ and $\tan\theta = $ _____$\frac{b}{a}$_____ for _____a_____ $\neq 0$.

4. Given $z = r(\cos\theta + i\sin\theta)$, the value r is called the ___modulus (or magnitude)___ of z and θ is called a(n) _____argument_____ of z.

5. If $z_1 = r_1(\cos\theta_1 + i\sin\theta_1)$ and $z_2 = r_2(\cos\theta_2 + i\sin\theta_2)$, then $z_1 \cdot z_2 = $ ___$r_1 r_2[\cos(\theta_1 + \theta_2) + i\sin(\theta_1 + \theta_2)]$___ and $\dfrac{z_1}{z_2} = $ ___$\frac{r_1}{r_2}[\cos(\theta_1 - \theta_2) + i\sin(\theta_1 - \theta_2)]$___ for $z_2 \neq 0$.

6. If $z = r(\cos\theta + i\sin\theta)$, then $z^n = $ ___$r^n(\cos n\theta + i\sin n\theta)$___ for a positive integer n. An nth root of z is a complex number w such that _____w^n_____ $= z$.

Objective 1: Plot Complex Numbers in the Complex Plane

For Exercises 7–8, graph the complex numbers in the complex plane. (See Example 1)

7. a. $z = -3 + 4i$ **b.** $z = 2 - 5i$ **c.** $z = -2i$ **d.** $z = 3$

8. a. $z = -2 - i$ **b.** $z = 5 + i$ **c.** $z = -4$ **d.** $z = 3i$

For Exercises 9–14, find the modulus of each complex number. (See Example 2)

9. a. $20 - 21i$ $\quad 29$ **b.** $-11 + 60i$ $\quad 61$

10. a. $5 + 12i$ $\quad 13$ **b.** $7 - 24i$ $\quad 25$

11. a. $4 - 6i$ $\quad 2\sqrt{13}$ **b.** $-5 + 5i$ $\quad 5\sqrt{2}$

12. a. $-14 + 6i$ $\quad 2\sqrt{58}$ **b.** $3 + 9i$ $\quad 3\sqrt{10}$

13. a. $-\dfrac{1}{3} - \dfrac{2}{3}i$ $\quad \dfrac{\sqrt{5}}{3}$ **b.** $\dfrac{3}{8} + \dfrac{7}{8}i$ $\quad \dfrac{\sqrt{58}}{8}$

14. a. $\dfrac{\sqrt{2}}{2} - 2i$ $\quad \dfrac{3\sqrt{2}}{2}$ **b.** $-1 + \dfrac{\sqrt{3}}{3}i$ $\quad \dfrac{2\sqrt{3}}{3}$

Objective 2: Write the Polar Form of a Complex Number

For Exercises 15–22, write the complex number in polar form with $0 \leq \theta < 2\pi$. (See Example 3)

15. a. $3 + 3i$ **b.** $-3 - 3i$

16. a. $2 - 2\sqrt{3}i$ **b.** $-2 + 2\sqrt{3}i$

17. a. $-4\sqrt{3} + 4i$ **b.** $10\sqrt{3} - 10i$

18. a. $3\sqrt{2} - 3\sqrt{2}i$ **b.** $-\sqrt{2} + \sqrt{2}i$

19. a. $24 + 7i$ **b.** $-20 + 21i$

20. a. $35 - 12i$ **b.** $-28 - 45i$

21. a. 17 **b.** $-4i$

22. a. $10i$ **b.** -12

For Exercises 23–30, convert the complex number from polar form to rectangular form $a + bi$. (See Example 4)

23. $18\left(\cos\dfrac{5\pi}{3} + i\sin\dfrac{5\pi}{3}\right)$ $\quad 9 - 9\sqrt{3}\,i$

24. $35(\cos 330° + i\sin 330°)$ $\quad \dfrac{35\sqrt{3}}{2} - \dfrac{35}{2}i$

25. $15(\cos 315° + i\sin 315°)$ $\quad \dfrac{15\sqrt{2}}{2} - \dfrac{15\sqrt{2}}{2}i$

26. $24\left(\cos\dfrac{\pi}{6} + i\sin\dfrac{\pi}{6}\right)$ $\quad 12\sqrt{3} + 12i$

27. $18.6\left(\cos\dfrac{4\pi}{3} + i\sin\dfrac{4\pi}{3}\right)$ $\quad -9.3 - 9.3\sqrt{3}\,i$

28. $12.3\left(\cos\dfrac{\pi}{4} + i\sin\dfrac{\pi}{4}\right)$ $\quad 6.15\sqrt{2} + 6.15\sqrt{2}\,i$

29. $43\pi(\cos 90° + i\sin 90°)$ $\quad 43\pi i$

30. $\sqrt{5}(\cos 180° + i\sin 180°)$ $\quad -\sqrt{5}$

Objective 3: Compute Products and Quotients of Complex Numbers in Polar Form

For Exercises 31–42, given complex numbers z_1 and z_2,

a. Find $z_1 z_2$ and write the product in polar form.

b. Find $\dfrac{z_1}{z_2}$ and write the quotient in polar form. (See Examples 5–6)

31. $z_1 = 20(\cos 31° + i\sin 31°)$, $z_2 = 40(\cos 14° + i\sin 14°)$

32. $z_1 = 27(\cos 67° + i\sin 67°)$, $z_2 = 9(\cos 53° + i\sin 53°)$

33. $z_1 = 3\left(\cos\dfrac{7\pi}{4} + i\sin\dfrac{7\pi}{4}\right)$, $z_2 = 6\left(\cos\dfrac{7\pi}{12} + i\sin\dfrac{7\pi}{12}\right)$

34. $z_1 = 10\left(\cos\dfrac{11\pi}{12} + i\sin\dfrac{11\pi}{12}\right)$, $z_2 = 2\left(\cos\dfrac{5\pi}{4} + i\sin\dfrac{5\pi}{4}\right)$

35. $z_1 = \cos 40° + i\sin 40°$, $z_2 = \cos 120° + i\sin 120°$

36. $z_1 = \cos 17° + i\sin 17°$, $z_2 = \cos 73° + i\sin 73°$

37. $z_1 = \dfrac{3}{4}\left(\cos\dfrac{\pi}{12} + i\sin\dfrac{\pi}{12}\right)$, $z_2 = \dfrac{1}{12}\left(\cos\dfrac{5\pi}{12} + i\sin\dfrac{5\pi}{12}\right)$

38. $z_1 = \dfrac{5}{6}\left(\cos\dfrac{\pi}{4} + i\sin\dfrac{\pi}{4}\right)$, $z_2 = 30\left(\cos\dfrac{3\pi}{4} + i\sin\dfrac{3\pi}{4}\right)$

39. $z_1 = 2 - 2i$ and $z_2 = 3 + 3i$

40. $z_1 = -5 - 5i$ and $z_2 = 10 + 10i$

41. $z_1 = -12\sqrt{3} - 12i$, $z_2 = -6 + 6\sqrt{3}\,i$

42. $z_1 = 24 + 24\sqrt{3}\,i$, $z_2 = 4\sqrt{3} - 4i$

Objective 4: Compute Powers of Complex Numbers (De Moivre's Theorem)

For Exercises 43–52, use De Moivre's theorem to find the indicated power. Write the result in rectangular form $a + bi$. (See Examples 7–8)

43. $\left[6\left(\cos\dfrac{\pi}{36} + i\sin\dfrac{\pi}{36}\right)\right]^6$ $\quad 23{,}328\sqrt{3} + 23{,}328i$

44. $\left[4\left(\cos\dfrac{4\pi}{15} + i\sin\dfrac{4\pi}{15}\right)\right]^5$ $\quad -512 - 512\sqrt{3}\,i$

45. $\left[2\left(\cos\dfrac{2\pi}{3} + i\sin\dfrac{2\pi}{3}\right)\right]^3$ $\quad 8$

46. $\left[3\left(\cos\dfrac{3\pi}{8} + i\sin\dfrac{3\pi}{8}\right)\right]^4$ $\quad -81i$

47. $\left[\sqrt{3}(\cos 35° + i\sin 35°)\right]^9$ $\quad \dfrac{81\sqrt{6}}{2} - \dfrac{81\sqrt{6}}{2}i$

48. $\left[\sqrt{5}(\cos 60° + i\sin 60°)\right]^5$ $\quad \dfrac{25\sqrt{5}}{2} - \dfrac{25\sqrt{15}}{2}i$

49. $\left[2(\cos 26.25° + i\sin 26.25°)\right]^8$ $\quad -128\sqrt{3} - 128i$

50. $\left[4(\cos 33.75° + i\sin 33.75°)\right]^4$ $\quad -128\sqrt{2} + 128\sqrt{2}\,i$

51. $\left(2\sqrt{3} - 2i\right)^5$ $\quad -512\sqrt{3} - 512i$

52. $\left(4 + 4\sqrt{3}\,i\right)^3$ $\quad -512$

Objective 5: Find the nth Roots of a Complex Number

53. When expressing the fourth roots w_k of a complex number generated by $w_k = \sqrt[n]{r}\left[\cos\left(\dfrac{\theta + 360°k}{n}\right) + i\sin\left(\dfrac{\theta + 360°k}{n}\right)\right]$ for $k = 0, 1, 2, \ldots, n - 1$, by how many degrees will consecutive roots differ? In general, given an integer $n \geq 2$, by how many degrees will consecutive nth roots differ? $\quad 90°; \dfrac{360°}{n}$

54. When expressing the fifth roots w_k of a complex number generated by $w_k = \sqrt[n]{r}\left[\cos\left(\dfrac{\theta + 2\pi k}{n}\right) + i\sin\left(\dfrac{\theta + 2\pi k}{n}\right)\right]$ for $k = 0, 1, 2, \ldots, n - 1$, by how many radians will consecutive roots differ? $\quad \dfrac{2\pi}{5}$

For Exercises 55–60, find the indicated complex roots by first writing the number in polar form. Write the results in rectangular form $a + bi$. (See Example 9)

55. The sixth roots of 729

56. The fourth roots of 625

57. The cube roots of $27i$

58. The cube roots of $-125i$

59. The square roots of $-8 + 8\sqrt{3}i$

60. The square roots of $50 - 50\sqrt{3}i$

For Exercises 61–64, find the indicated complex roots. Write the results in polar form.

61. The square roots of $16\left(\cos\dfrac{3\pi}{4} + i\sin\dfrac{3\pi}{4}\right)$

62. The square roots of $49\left(\cos\dfrac{11\pi}{6} + i\sin\dfrac{11\pi}{6}\right)$

63. The fifth roots of $16\sqrt{3} + 16i$

64. The sixth roots of $\dfrac{\sqrt{2}}{2} - \dfrac{\sqrt{2}}{2}i$

For Exercises 65–70, find the indicated complex roots. Write the results in polar form using degree measure for the argument. (See Example 10)

65. The cube roots of $27(\cos 114° + i\sin 114°)$

66. The fourth roots of $256(\cos 228° + i\sin 228°)$

67. The square roots of $\dfrac{81\sqrt{2}}{2} - \dfrac{81\sqrt{2}}{2}i$

68. The square roots of $-\dfrac{3\sqrt{3}}{2} + \dfrac{3}{2}i$

69. The sixth roots of $-2 - 2\sqrt{3}i$

70. The fifth roots of $5\sqrt{3} - 5i$.

71. The square roots, cube roots, and fourth roots of 1 are shown in the figure from left to right, respectively. Use the pattern to describe how to find the nth roots of 1.

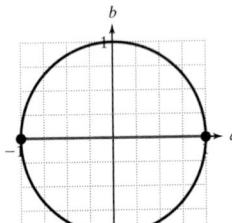 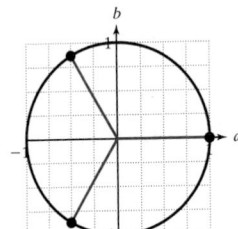 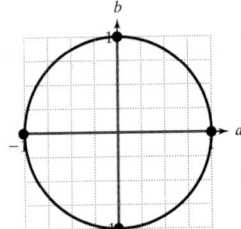

72. Use the results of Exercise 71 to find the fifth roots of 1.

73. Use De Moivre's theorem to show that $2\left(\cos\dfrac{2\pi}{9} + i\sin\dfrac{2\pi}{9}\right)$ is a cube root of $8\left(\cos\dfrac{2\pi}{3} + i\sin\dfrac{2\pi}{3}\right)$.

74. Use De Moivre's theorem to show that $1 - i$ is a cube root of $2\sqrt{2}\left(\cos\dfrac{5\pi}{4} + i\sin\dfrac{5\pi}{4}\right)$.

Mixed Exercises

For Exercises 75–78, shade the area in the complex plane.

75. $\{z = a + bi \mid a < 3, b \geq 2\}$

76. $\{z = a + bi \mid a \geq -1, b < 3\}$

77. $\{z = a + bi \mid |z| < 3\}$

78. $\{z = a + bi \mid |z| \geq 2\}$

For Exercises 79–82, convert the complex number from polar form to rectangular form, $a + bi$. (Hint: to evaluate the sine or cosine of the given angle, use the sum or difference formulas for sine and cosine.)

79. $4\left(\cos\dfrac{13\pi}{12} + i\sin\dfrac{13\pi}{12}\right)$

80. $36\left(\cos\dfrac{7\pi}{12} + i\sin\dfrac{7\pi}{12}\right)$

81. $\cos 15° + i\sin 15°$

82. $\cos 285° + i\sin 285°$

83. Verify the results of Example 5 by converting z_1 and z_2 to rectangular form and then computing the product.

84. Given $z_1 = 2 + 2i$ and $z_2 = 3i$,

   a. Find $\dfrac{z_1}{z_2}$ by dividing the numbers in rectangular form and then converting the quotient to polar form.

   b. Find $\dfrac{z_1}{z_2}$ by dividing the numbers in polar form.

85. Let $z_1 = r_1(\cos\theta_1 + i\sin\theta_1)$ and $z_2 = r_2(\cos\theta_2 + i\sin\theta_2)$. Prove that $\dfrac{z_1}{z_2} = \dfrac{r_1}{r_2}[\cos(\theta_1 - \theta_2) + i\sin(\theta_1 - \theta_2)]\,(z_2 \neq 0)$.

86. Show that $z_1 = r(\cos\theta + i\sin\theta)$ and $z_2 = r[\cos(-\theta) + i\sin(-\theta)]$ are conjugates.

87. Let $z = r(\cos\theta + i\sin\theta)$.

   a. Prove that $z^{-1} = r^{-1}[\cos(-\theta) + i\sin(-\theta)]$

   b. Prove that $z^{-2} = r^{-2}[\cos(-2\theta) + i\sin(-2\theta)]$

   Repeating this argument, we can show that De Moivre's theorem

   $z^n = [r(\cos\theta + i\sin\theta)]^n = r^n(\cos n\theta + i\sin n\theta)$ also holds for negative integers.

88. Use De Moivre's theorem to find the indicated power. Write the result in rectangular form $a + bi$.

   a. $(-4\sqrt{2} - 4\sqrt{2}i)^{-2}$ $-\dfrac{1}{64}i$ **b.** $(-1 + \sqrt{3}i)^{-4}$ $-\dfrac{1}{32} - \dfrac{\sqrt{3}}{32}i$ **c.** $\left(\dfrac{\sqrt{3}}{2} - \dfrac{1}{2}i\right)^{-3}$ i

For Exercises 89–92, simplify and write the solution in rectangular form, $a + bi$. (*Hint*: Convert the complex numbers to polar form before simplifying.)

89. $(1 + i)^3(-1 - i)^4$ $8 - 8i$ **90.** $(1 + i)^4(-3i)^5$ $972i$ **91.** $\dfrac{(-3 + \sqrt{3}i)^5}{(1 - \sqrt{3}i)^2}$ **92.** $\dfrac{(5\sqrt{2} - 5\sqrt{2}i)^4}{(-\sqrt{3} - i)^3}$ $-1250i$

                                                                              $-108 + 36\sqrt{3}i$

93. Solve $x^6 = 64$ by factoring and applying the zero product rule to show that the results are the same as the six 6th roots of 64 as found in Example 9 on page 720.

94. Solve $x^3 = -64$ by factoring and applying the zero product rule to show that the results are the same as the three cube roots of -64 as found in Skill Practice Exercise 9 on page 720.

For Exercises 95–98, find all complex solutions to the equation. Write the solutions in polar form.

95. $z^4 + 8\sqrt{2} = -8\sqrt{2}i$ **96.** $z^5 = 16\sqrt{2}(1 + i)$ **97.** $2z^3 + \sqrt{3} = i$ **98.** $z^3 = 8i$

Write About It

99. How is the absolute value of a complex number $z = a + bi$ similar to the absolute value of a real number x?

100. Is the polar representation of a complex number $z = a + bi$ unique? Explain.

Expanding Your Skills

101. Euler's formula states that $e^{i\theta} = \cos\theta + i\sin\theta$.

   a. Evaluate $e^{(\pi/4)i}$.

   b. Use Euler's formula to show that $e^{\pi i} + 1 = 0$. This equation relates five fundamental numbers used in mathematics $(0, 1, e, \pi, \text{and } i)$.

102. In calculus, we can show that

$$e^x = 1 + \frac{x}{1} + \frac{x^2}{2 \cdot 1} + \frac{x^3}{3 \cdot 2 \cdot 1} + \frac{x^4}{4 \cdot 3 \cdot 2 \cdot 1} + \cdots,$$

$$\cos x = 1 - \frac{x^2}{2 \cdot 1} + \frac{x^4}{4 \cdot 3 \cdot 2 \cdot 1} - \frac{x^6}{6 \cdot 5 \cdot 4 \cdot 3 \cdot 2 \cdot 1} + \cdots, \text{ and}$$

$$\sin x = x - \frac{x^3}{3 \cdot 2 \cdot 1} + \frac{x^5}{5 \cdot 4 \cdot 3 \cdot 2 \cdot 1} - \frac{x^7}{7 \cdot 6 \cdot 5 \cdot 4 \cdot 3 \cdot 2 \cdot 1} + \cdots$$

   Write the first eight terms of the expansion of e^{ix} to illustrate that $e^{ix} = \cos x + i\sin x$.

103. Show that $3e^{\pi i/12}$ is a solution to $z^8 - 81z^4 + 6561 = 0$.

104. Show that $e^{7\pi i/12}$ is a solution to $z^8 - z^4 + 1 = 0$.

Vectors

1. Interpret Vectors Geometrically

In physics, the mathematical quantities that are used to describe the motion of an object can be divided into two quantities, called scalars and vectors. **Scalars** are quantities that are described by a single value called the **magnitude** (the size or measure of the quantity). **Vectors** are quantities that are described by both magnitude *and* direction. For example, suppose that a bus travels 30 mi. The value 30 mi is a scalar quantity, because it tells us how much distance the bus covered. However, if we say that the bus travels 30 mi east, we know both its magnitude, 30 mi, and direction, east. So this information makes up a vector quantity. This is called a displacement vector because the bus is displaced 30 mi to the east of its starting point.

The **speed** of an object is a scalar quantity that tells us how fast an object travels. By contrast, the **velocity** of an object is a vector quantity, because it tells us how fast the object travels *and* in which direction. For example, suppose that a car travels at a constant speed around a turn at the Daytona International Speedway. Although the speed is constant, the velocity of the car is continuously changing because its direction of travel around the curve changes.

A variety of scalar and vector quantities are used in physical science courses. Some common scalar quantities are distance, speed, area, mass, volume, and temperature. Common vector quantities are displacement, velocity, acceleration, and force.

Vector quantities are so named because they can be represented by a vector in a plane or in space. A **vector** in a plane is a line segment with a specified direction. For example, in Figure 7-33, we show an arrow with endpoints P and Q pointing from P to Q. We denote this vector as $\overrightarrow{PQ}$. We call P the **initial point** of the vector and Q the **terminal point** of the vector. The **magnitude** of $\overrightarrow{PQ}$, denoted $\|\overrightarrow{PQ}\|$, is the length of the line segment. Using the context of the **displacement** of an object, the vector $\overrightarrow{PQ}$ refers to the object's overall change in position in moving from point P to point Q.

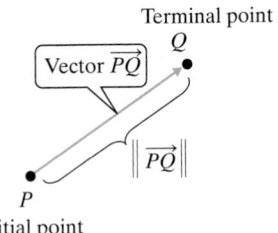

Figure 7-33

Two vectors are equal if they have the same magnitude and direction. The vectors in Figure 7-34 are all equal to $\overrightarrow{AB}$. Notice that the vectors in Figure 7-34 each represent the same displacement left and upward from the initial point to the terminal point.

To denote vectors, we often use boldface lower case letters such as **u**, **v**, and **w**. If $\mathbf{v} = \overrightarrow{AB}$, then **v** represents the set of *all* vectors equal to $\overrightarrow{AB}$.

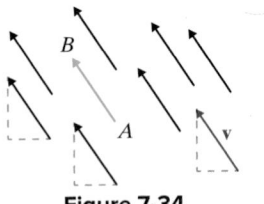

Figure 7-34

> **TIP** Because it is difficult to depict boldface when using a pencil and paper, we sometimes use an arrow over a lowercase letter to denote a vector such as $\vec{v}$.

Classroom Example: p. 739
Exercise 12

EXAMPLE 1 Determining if Two Vectors Are Equal

Vector **v** extends from $(0, 0)$ to $(-4, 2)$. Vector **w** extends from $(1, 3)$ to $(-3, 5)$.

a. Find the magnitude of **v** and the magnitude of **w**.

b. Are **v** and **w** equal?

Solution:

a. Find the magnitude (length) of each vector by applying the distance formula.

$$\|\mathbf{v}\| = \sqrt{(-4-0)^2 + (2-0)^2} = \sqrt{16+4}$$
$$= \sqrt{20} = 2\sqrt{5}$$

$$\|\mathbf{w}\| = \sqrt{(-3-1)^2 + (5-3)^2} = \sqrt{16+4}$$
$$= \sqrt{20} = 2\sqrt{5}$$

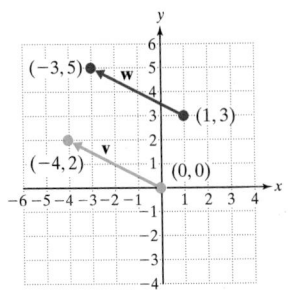

Instructor Note:
Show that a vector is independent of its initial point. As an example, show that $\langle -2, 3 \rangle$ drawn in standard position is the same vector drawn from (3, 1) to (1, 4).

b. The magnitudes of **v** and **w** are equal, but we must also determine if they have the same direction. Both vectors are directed from the lower right to the upper left. Furthermore, the slopes of the line segments representing **v** and **w** are equal. Therefore, **v** = **w**.

Slope of line through **v**: $m = \dfrac{2-0}{-4-0} = -\dfrac{1}{2}$

Slope of line through **w**: $m = \dfrac{5-3}{-3-1} = -\dfrac{1}{2}$

Skill Practice 1   Vector **v** extends from (0, 0) to (3, 4). Vector **w** extends from (−2, 3) to (−5, −1).

  a. Find the magnitude of **v** and magnitude of **w**.

  b. Are **v** and **w** equal?

We can envision the sum or difference of two vectors by regarding their displacements from their initial points to their terminal points. If vector **v** represents the displacement from P to Q and vector **w** represents the vector from Q to R, then the vector **v** + **w** represents the displacement from P to R (Figure 7-35).

Intuitively, the statement $\mathbf{v} + \mathbf{w} = \overrightarrow{PR}$ means that a path directly from P to R represents the same displacement from the starting point as the path from P to Q to R.

Notice that the vector **v** + **w** is formed by placing **v** and **w** "tip to tail" (the initial point of **w** is placed at the terminal point of **v**).

The sum **v** + **w** can also be visualized by placing **v** and **w** at a common initial point ("tail" to "tail"). The sum **v** + **w** is the diagonal of the parallelogram from the initial point of **v** and **w** to the opposite vertex (Figure 7-36).

If a vector **v** is multiplied by a real number k, the magnitude of $k\mathbf{v}$ is $|k|\|\mathbf{v}\|$ (the absolute value of k times the magnitude of **v**). The direction of $k\mathbf{v}$ is

- in the same direction as **v** if $k > 0$ (Figure 7-37).
- in the opposite direction as **v** if $k < 0$.

The value k is a scalar, and for this reason, the vector $k\mathbf{v}$ is called a **scalar multiple of v**. Notice that the value of k either stretches or shrinks the length (magnitude) of **v** by a factor of $|k|$. If $k = 0$, then $k\mathbf{v}$ is the zero vector, denoted **0**. The **zero vector** has a magnitude of 0 and is assigned no direction.

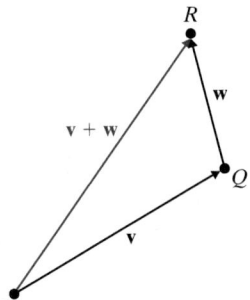

Figure 7-35

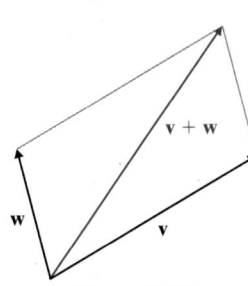

Figure 7-36

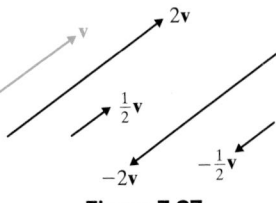

Figure 7-37

Answers

1. a. $\|\mathbf{v}\| = 5$, $\|\mathbf{w}\| = 5$
  b. No (their directions are opposite)

The vector $-\mathbf{v}$ is $-1\mathbf{v}$ and represents the opposite of $\mathbf{v}$ (same magnitude, but opposite direction). We use the opposite of a vector to define vector subtraction. The difference of vectors $\mathbf{v}$ and $\mathbf{w}$ is given by $\mathbf{v} - \mathbf{w} = \mathbf{v} + (-\mathbf{w})$. Geometrically, $\mathbf{v} - \mathbf{w}$ is found by adding the opposite of $\mathbf{w}$ to $\mathbf{v}$ (Figure 7-38). With $\mathbf{v}$ and $\mathbf{w}$ placed at a common initial point, the difference $\mathbf{v} - \mathbf{w}$ is visualized as the diagonal of the parallelogram from the "tip" of $\mathbf{w}$ to the "tip" of $\mathbf{v}$ (Figure 7-39).

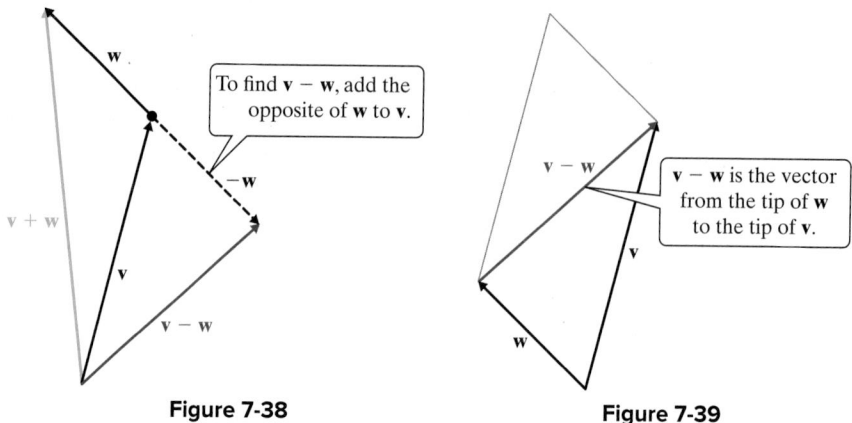

Figure 7-38   **Figure 7-39**

2. Represent Vectors in Component Form

Working with vectors geometrically can be a cumbersome process. Fortunately, we have a means to represent vectors algebraically. Suppose that for a vector $\mathbf{v} = \overrightarrow{PQ}$, the horizontal displacement moving from P to Q is a units, and the vertical displacement is b units (Figure 7-40). We can represent $\mathbf{v}$ by a pair of numbers of the form $\langle a, b \rangle$. The notation $\langle a, b \rangle$ is called the **component form** of a vector.

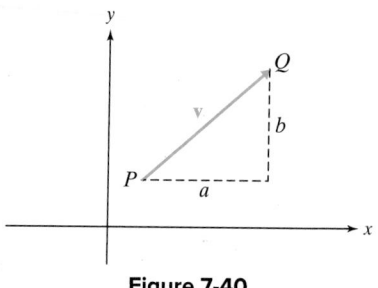

Figure 7-40

Component Form of a Vector

Let $P(x_1, y_1)$ and $Q(x_2, y_2)$ be points in the plane, and let $\mathbf{v} = \overrightarrow{PQ}$. The component form of $\mathbf{v}$ is given by

$$\mathbf{v} = \langle x_2 - x_1, y_2 - y_1 \rangle$$

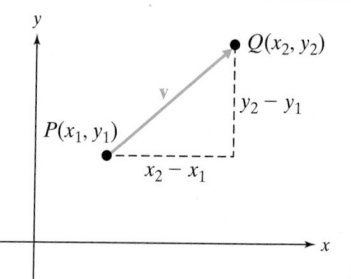

Sketching a vector with its initial point at the origin of a rectangular coordinate system is often a convenient location to represent a vector. In such a case, the vector is said to be drawn in **standard position**. For a vector $\mathbf{v} = \langle a, b \rangle$ drawn in standard position, the terminal point of $\mathbf{v}$ is (a, b) (Figure 7-41).

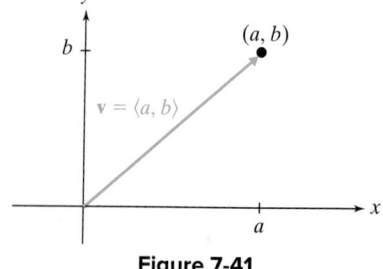

Figure 7-41

Classroom Example: p. 739
Exercise 22

> **EXAMPLE 2**   Finding the Component Form of a Vector
>
> Suppose that vector **v** has initial point $(-2, 1)$ and terminal point $(2, -5)$.
>
>   **a.** Find the component form of **v**.
>   **b.** Sketch **v** in standard position.
>   **c.** Sketch **v** with its initial point at $(2, 3)$.
>
> **Solution:**
>
>   **a.** Label $(-2, 1)$ as (x_1, y_1) and $(2, -5)$ as (x_2, y_2).
>
>   Then $\mathbf{v} = \langle x_2 - x_1, y_2 - y_1 \rangle = \langle 2 - (-2), -5 - 1 \rangle$
> $= \langle 4, -6 \rangle$
>
>   **b.** In standard position, the initial point of **v** is $(0, 0)$ and the terminal point is $(4, -6)$, shown in red.
>
>   **c.** If the initial point is $(2, 3)$, then the terminal point is $(2 + 4, 3 + (-6))$, which is $(6, -3)$, shown in blue.

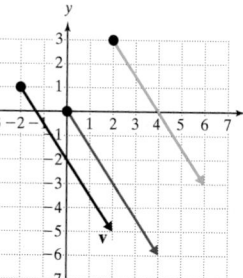

> **Skill Practice 2**   Suppose that vector **v** has initial point $(-4, 5)$ and terminal point $(-1, 12)$.
>
>   **a.** Find the component form of **v**.
>   **b.** If **v** is placed with initial point at $(-2, -4)$, what is the terminal point of **v**?

3. Perform Operations on Vectors in Component Form

With an understanding of the component form of a vector, we have the tools to perform algebraic operations on vectors and to develop important related properties. First, we can state the equality of vectors in terms of their components. Two vectors are equal if and only if their corresponding components are equal.

Equality of Vectors

If $\mathbf{v} = \langle a_1, b_1 \rangle$ and $\mathbf{w} = \langle a_2, b_2 \rangle$, then $\mathbf{v} = \mathbf{w}$ if and only if $a_1 = a_2$ and $b_1 = b_2$.

To add two vectors geometrically, we add their displacements. In component form, this is equivalent to adding their horizontal components and adding their vertical components. Likewise to subtract vectors, we subtract their corresponding components.

Operations on Vectors

Let $\mathbf{v} = \langle a_1, b_1 \rangle$, $\mathbf{w} = \langle a_2, b_2 \rangle$, and c be a real number.

Vector addition: $\mathbf{v} + \mathbf{w} = \langle a_1 + a_2, b_1 + b_2 \rangle$

Vector subtraction: $\mathbf{v} - \mathbf{w} = \langle a_1 - a_2, b_1 - b_2 \rangle$

Multiplication of a vector by a scalar: $c\mathbf{v} = \langle ca_1, cb_1 \rangle$

Answers
2. a. $\langle 3, 7 \rangle$ **b.** $(1, 3)$

Classroom Examples: p. 740
Exercises 32, 36

EXAMPLE 3   Applying Operations on Vectors Algebraically

Given $\mathbf{v} = \langle 1, 4 \rangle$ and $\mathbf{w} = \langle 3, -2 \rangle$, find

  a. $\mathbf{v} + \mathbf{w}$ **b.** $\mathbf{v} - \mathbf{w}$

Solution:

a. $\mathbf{v} + \mathbf{w}$
$$= \langle 1, 4 \rangle + \langle 3, -2 \rangle$$
$$= \langle 1 + 3, 4 + (-2) \rangle$$
$$= \langle 4, 2 \rangle$$

b. $\mathbf{v} - \mathbf{w}$
$$= \langle 1, 4 \rangle - \langle 3, -2 \rangle$$
$$= \langle 1 - 3, 4 - (-2) \rangle$$
$$= \langle -2, 6 \rangle$$

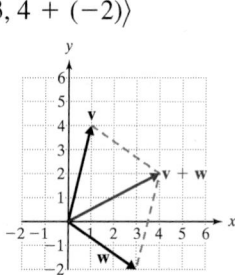

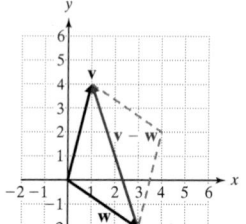

Skill Practice 3   Given $\mathbf{r} = \langle -3, 10 \rangle$ and $\mathbf{s} = \langle 12, -7 \rangle$, find

  a. $\mathbf{r} + \mathbf{s}$ **b.** $\mathbf{s} - \mathbf{r}$

Classroom Example: p. 740
Exercise 40

EXAMPLE 4   Applying Operations on Vectors Algebraically

Given $\mathbf{r} = \langle -3, 8 \rangle$ and $\mathbf{s} = \langle 2, -5 \rangle$, find

  a. $5\mathbf{r}$ **b.** $2\mathbf{r} - 3\mathbf{s}$

Solution:

  a. $5\mathbf{r} = 5\langle -3, 8 \rangle = \langle 5(-3), 5(8) \rangle = \langle -15, 40 \rangle$

  b. $2\mathbf{r} - 3\mathbf{s} = 2\langle -3, 8 \rangle - 3\langle 2, -5 \rangle$
$$= \langle 2(-3), 2(8) \rangle - \langle 3(2), 3(-5) \rangle$$
$$= \langle -6, 16 \rangle - \langle 6, -15 \rangle = \langle -6 - 6, 16 - (-15) \rangle = \langle -12, 31 \rangle$$

Skill Practice 4   Given $\mathbf{a} = \langle 11, -7 \rangle$ and $\mathbf{b} = \langle -4, -6 \rangle$, find

  a. $6\mathbf{a}$ **b.** $-3\mathbf{a} + 10\mathbf{b}$

Several properties of vectors follow directly from the definition of vector addition, vector subtraction, and multiplication of a vector by a scalar.

Properties of Vectors	
If $\mathbf{u}$, $\mathbf{v}$, and $\mathbf{w}$ are vectors, and c and d are scalars, then	
1. $\mathbf{v} + \mathbf{w} = \mathbf{w} + \mathbf{v}$	**3.** $\mathbf{v} + \mathbf{0} = \mathbf{v}$
2. $(\mathbf{v} + \mathbf{w}) + \mathbf{u} = \mathbf{v} + (\mathbf{w} + \mathbf{u})$	**4.** $\mathbf{v} + (-\mathbf{v}) = \mathbf{0}$
5. $c(\mathbf{v} + \mathbf{w}) = c\mathbf{v} + c\mathbf{w}$	**8.** $1(\mathbf{v}) = \mathbf{v}$
6. $(c + d)\mathbf{v} = c\mathbf{v} + d\mathbf{v}$	**9.** $0(\mathbf{v}) = \mathbf{0}$
7. $(cd)\mathbf{v} = c(d\mathbf{v})$	**10.** $c(\mathbf{0}) = \mathbf{0}$
11. $\|c\mathbf{v}\| = |c|\|\mathbf{v}\|$	

Answers

3. a. $\langle 9, 3 \rangle$ **b.** $\langle 15, -17 \rangle$

4. a. $\langle 66, -42 \rangle$ **b.** $\langle -73, -39 \rangle$

To show that property 1 is true, let $\mathbf{v} = \langle a_1, b_1 \rangle$ and $\mathbf{w} = \langle a_2, b_2 \rangle$, Then

$$\mathbf{v} + \mathbf{w} = \langle a_1 + a_2, b_1 + b_2 \rangle$$

$$= \langle a_2 + a_1, b_2 + b_1 \rangle \qquad \text{Commutative property of real numbers}$$

$$= \mathbf{w} + \mathbf{v}$$

Several other properties of vectors are left for verification in the exercises.

4. Represent Vectors in Terms of **i** and **j**

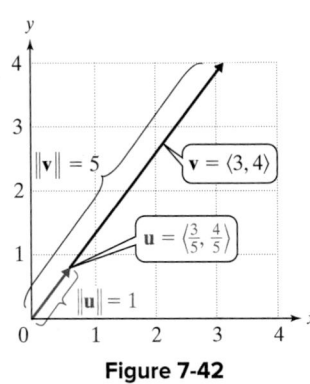

Figure 7-42

TIP Notice that the length of **u** is $\frac{1}{5}$ the length of **v** and in the same direction as **v**. Therefore, $\mathbf{v} = 5\mathbf{u}$.

A vector that has a magnitude of 1 unit is called a **unit vector.** Finding a unit vector in the same direction as a given vector is an important skill in many applications. For example, consider $\mathbf{v} = \langle 3, 4 \rangle$ pictured in standard position in Figure 7-42. The magnitude of **v** is $\|\mathbf{v}\| = \sqrt{3^2 + 4^2} = \sqrt{25} = 5$. We want to find a vector **u** (pictured in red in Figure 7-42) such that **u** is 1 unit in length and in the same direction as **v**. To do this, we apply a scaling factor of $\dfrac{1}{\|\mathbf{v}\|}$ to vector **v**.

That is, $\mathbf{u} = \dfrac{1}{5}\langle 3, 4 \rangle = \left\langle \dfrac{3}{5}, \dfrac{4}{5} \right\rangle$. The vectors **u** and **v** have the same direction because **u** is a positive scalar multiple of **v**. Furthermore, **u** is a unit vector because its magnitude is 1.

$$\|\mathbf{u}\| = \sqrt{\left(\frac{3}{5}\right)^2 + \left(\frac{4}{5}\right)^2} = \sqrt{\frac{9}{25} + \frac{16}{25}} = \sqrt{\frac{25}{25}} = 1$$

We generalize this process as follows.

Find a Unit Vector in the Direction of a Given Vector

If $\mathbf{v} = \langle a, b \rangle$, then a unit vector $\mathbf{u}_v$ in the direction of **v** is given by

$$\mathbf{u}_v = \frac{1}{\|\mathbf{v}\|}\mathbf{v} = \frac{1}{\|\mathbf{v}\|}\langle a, b \rangle = \left\langle \frac{a}{\|\mathbf{v}\|}, \frac{b}{\|\mathbf{v}\|} \right\rangle$$

Classroom Example: p. 740
Exercise 60

EXAMPLE 5   Finding a Unit Vector

Find a unit vector in the direction of $\mathbf{w} = \langle -3, 2 \rangle$.

Solution:

$$\|\mathbf{w}\| = \sqrt{(-3)^2 + (2)^2} = \sqrt{9 + 4} = \sqrt{13}$$

$$\mathbf{u}_w = \frac{1}{\|\mathbf{w}\|}\mathbf{w} = \frac{1}{\sqrt{13}}\langle -3, 2 \rangle = \left\langle -\frac{3}{\sqrt{13}}, \frac{2}{\sqrt{13}} \right\rangle \text{ or } \left\langle -\frac{3\sqrt{13}}{13}, \frac{2\sqrt{13}}{13} \right\rangle$$

Skill Practice 5   Find a unit vector in the direction of $\mathbf{v} = \langle 5, -1 \rangle$.

Answer

5. $\left\langle \dfrac{5}{\sqrt{26}}, -\dfrac{1}{\sqrt{26}} \right\rangle$ or $\left\langle \dfrac{5\sqrt{26}}{26}, -\dfrac{\sqrt{26}}{26} \right\rangle$

Avoiding Mistakes

We can verify the result from Example 5 by showing that $\|\mathbf{u_w}\| = 1$.

$$\|\mathbf{u_w}\| = \sqrt{\left(-\frac{3}{\sqrt{13}}\right)^2 + \left(\frac{2}{\sqrt{13}}\right)^2} = \sqrt{\frac{9}{13} + \frac{4}{13}} = \sqrt{\frac{13}{13}} = 1 \checkmark$$

Two important unit vectors are $\mathbf{i} = \langle 1, 0 \rangle$ and $\mathbf{j} = \langle 0, 1 \rangle$ because they are unit vectors in the direction of the positive x- and y-axes, respectively (Figure 7-43). The importance of vectors $\mathbf{i}$ and $\mathbf{j}$ is that any other vector can be expressed as a linear combination of $\mathbf{i}$ and $\mathbf{j}$. Given $\mathbf{v} = \langle a, b \rangle$ we have

$$\mathbf{v} = \langle a, b \rangle$$
$$= a\langle 1, 0 \rangle + b\langle 0, 1 \rangle$$
$$= a\mathbf{i} + b\mathbf{j}$$

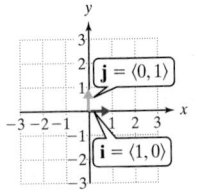

Figure 7-43

Instructor Note:
Point out the difference between the vector $\mathbf{i} = \langle 1, 0 \rangle$ and the imaginary number $i = \sqrt{-1}$.

Represent Vectors in Terms of i and j

The representation of $\mathbf{v} = \langle a, b \rangle$ in terms of $\mathbf{i}$ and $\mathbf{j}$ is $\mathbf{v} = a\mathbf{i} + b\mathbf{j}$.

The values a and b are called the **scalar horizontal and vertical components** of $\mathbf{v}$, respectively.

Classroom Example: p. 740
Exercise 66

EXAMPLE 6   Writing Vectors in Terms of i and j

a. Write $\langle -3, 5 \rangle$ in terms of $\mathbf{i}$ and $\mathbf{j}$.
b. Given $\mathbf{w} = 6.2\mathbf{i} - 3.4\mathbf{j}$ and $\mathbf{v} = -1.7\mathbf{i} + 2.2\mathbf{j}$, write $3\mathbf{w} - 5\mathbf{v}$ in terms of $\mathbf{i}$ and $\mathbf{j}$.

Solution:

a. $\boxed{\langle -3, 5 \rangle = -3\mathbf{i} + 5\mathbf{j}}$

b. $3\mathbf{w} - 5\mathbf{v} = 3(6.2\mathbf{i} - 3.4\mathbf{j}) - 5(-1.7\mathbf{i} + 2.2\mathbf{j})$
$$= 18.6\mathbf{i} - 10.2\mathbf{j} + 8.5\mathbf{i} - 11\mathbf{j}$$
$$= (18.6 + 8.5)\mathbf{i} + (-10.2 - 11)\mathbf{j}$$
$$= 27.1\mathbf{i} - 21.2\mathbf{j}$$

Skill Practice 6

a. Write $\langle 12, -1 \rangle$ in terms of $\mathbf{i}$ and $\mathbf{j}$.
b. Given $\mathbf{a} = -9.4\mathbf{i} + 7.1\mathbf{j}$ and $\mathbf{b} = -0.6\mathbf{i} + 3.2\mathbf{j}$, write $-2\mathbf{a} - 4\mathbf{b}$ in terms of $\mathbf{i}$ and $\mathbf{j}$.

5. Relate Component Form and Direction and Magnitude of a Vector

In many applications of vectors we are given the magnitude and direction of a vector rather than its component form. Given a vector in standard position, its **direction angle** θ is measured counterclockwise from the positive x-axis over the interval

Answers
6. a. $12\mathbf{i} - \mathbf{j}$ **b.** $21.2\mathbf{i} - 27\mathbf{j}$

$0 \le \theta < 2\pi$ or $0° \le \theta < 360°$. The relationships among the magnitude, direction, and components of a vector are summarized as follows.

Magnitude, Direction, and Components of a Vector

Let $\mathbf{v} = \langle a, b \rangle$ be a vector in standard position, and let $0° \le \theta < 360°$ be the direction of $\mathbf{v}$ measured counterclockwise from the positive x-axis.

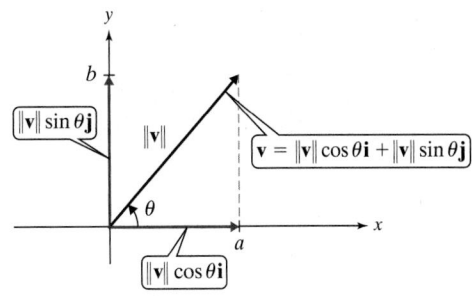

$\boxed{\|\mathbf{v}\| \sin\theta\, \mathbf{j}}$ $\|\mathbf{v}\|$ $\boxed{\mathbf{v} = \|\mathbf{v}\| \cos\theta\, \mathbf{i} + \|\mathbf{v}\| \sin\theta\, \mathbf{j}}$

$\boxed{\|\mathbf{v}\| \cos\theta\, \mathbf{i}}$

- $\|\mathbf{v}\| = \sqrt{a^2 + b^2}$ and $\tan\theta = \dfrac{b}{a}$ $(a \ne 0)$ (magnitude and direction of $\mathbf{v}$)
- $a = \|\mathbf{v}\|\cos\theta$ and $b = \|\mathbf{v}\|\sin\theta$
- $\mathbf{v} = \langle a, b \rangle = \langle \|\mathbf{v}\|\cos\theta, \|\mathbf{v}\|\sin\theta \rangle$ or $\mathbf{v} = a\mathbf{i} + b\mathbf{j} = \|\mathbf{v}\|\cos\theta\, \mathbf{i} + \|\mathbf{v}\|\sin\theta\, \mathbf{j}$

TIP One Newton is the force required to accelerate a mass of 1 kg at 1 m/sec². Therefore, $1\,\text{N} = 1\,\frac{\text{kg}\cdot\text{m}}{\text{sec}^2}$.

One dyne is the force required to accelerate a mass of 1 g at 1 cm/sec². Therefore, $1\,\text{dyn} = 1\,\frac{\text{g}\cdot\text{cm}}{\text{sec}^2}$.

In Examples 7–12 and in the exercises, we present vectors in an applied setting with several of the following common vector quantities.

- A vector representing the displacement of an object is called a position or displacement vector.
- A vector representing the speed and direction of an object is called a velocity vector.
- A vector representing the force acting on an object in a given direction is called a force vector. Common units of force are pounds (lb), Newtons (N), and dynes (dyn). Force vectors are often denoted by $\mathbf{F}$, by $\mathbf{W}$ (if the force represents a weight), or by $\mathbf{T}$ (if the force represents the tension in a rope or wire).

Classroom Example: p. 740
Exercise 80

EXAMPLE 7   Finding the Component Form of a Force Vector

A force of 100 lb is applied to a hook on the ceiling at an angle of 45° with the horizontal. Write the force vector $\mathbf{F}$ in terms of $\mathbf{i}$ and $\mathbf{j}$.

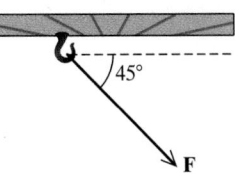

Solution:

The magnitude of the force is $\|\mathbf{F}\| = 100$ lb. Measuring the direction angle counterclockwise from the positive x-axis, we have $\theta = 360° - 45° = 315°$. Then,

$$\mathbf{F} = \|\mathbf{F}\|\cos\theta\, \mathbf{i} + \|\mathbf{F}\|\sin\theta\, \mathbf{j}$$
$$= 100\cos 315°\, \mathbf{i} + 100\sin 315°\, \mathbf{j}$$
$$= 100 \cdot \frac{\sqrt{2}}{2}\mathbf{i} + 100 \cdot \left(-\frac{\sqrt{2}}{2}\right)\mathbf{j}$$
$$= 50\sqrt{2}\mathbf{i} - 50\sqrt{2}\mathbf{j}$$

Force $\mathbf{F}$ is applied in a direction to the right and downward. Thus, the horizontal component is positive and the vertical component is negative.

TIP The direction angle θ is usually taken between 0° and 360°, inclusive. However, any angle coterminal to θ results in an equivalent vector. In Example 7,

$F = 100\cos(-45°)\mathbf{i} + 100\sin(-45°)\mathbf{j}$

also results in the vector

$50\sqrt{2}\mathbf{i} - 50\sqrt{2}\mathbf{j}$

Skill Practice 7   A man exerts a force of 80 lb on an object by pulling on a rope at an angle of 30° from the horizontal. Write the force vector **F** in terms of **i** and **j**.

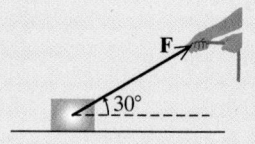

In Example 8, we are given a velocity vector in terms of **i** and **j** and asked to find the direction and magnitude of the vector.

EXAMPLE 8   Finding the Magnitude and Direction of a Vector

A ball is thrown and at the time of release, the velocity of the ball is given by **v** = 22.6**i** + 8.2**j**, where each component is given in m/sec. Find the initial velocity v_0 (magnitude of the velocity vector at the time of release) and the angle above the horizontal (in degrees) at which the ball was thrown. Round to 1 decimal place.

Solution:

The initial velocity is the magnitude of the velocity vector.

$$v_0 = \|\mathbf{v}\| = \sqrt{(22.6)^2 + (8.2)^2} \approx 24.0 \text{ m/sec}$$

To find the direction angle, we have $\tan\theta = \dfrac{8.2}{22.6}$.

For θ in Quadrant I, $\theta = \tan^{-1}\dfrac{8.2}{22.6} \approx 19.9°$.

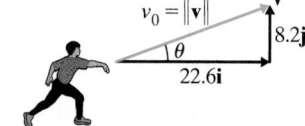

The ball was thrown with an initial velocity of 24.0 m/sec at an angle of 19.9° above the horizontal.

Skill Practice 8   A ball is thrown from a height of 1 m. At the time of release, the velocity of the ball is given by **v** = 27.2**i** + 12.7**j**, where each component is given in m/sec. Find the initial velocity v_0 and the angle θ above the horizontal (in degrees) at which the ball was thrown. Round to 1 decimal place.

6. Use Vectors in Applications

In Example 9, a ship travels in open ocean with velocity **v** relative to the water. The ship then encounters a current **c** that alters its course. The resulting velocity **r** of the ship is given by **r** = **v** + **c**. As demonstrated in Example 9, finding the component form of vectors **v** and **c** enables us to find the resultant velocity vector.

EXAMPLE 9   Finding a Resultant Vector in an Application

A ship in open ocean travels 18 mph at a bearing of N64°W and encounters a current of 4 mph acting in the direction of N45°E. Rounding to 1 decimal place,

   a. Express the velocity of the ship **v** relative to the water in terms of **i** and **j**.

   b. Express the velocity of the current **c** in terms of **i** and **j**.

   c. Find the true speed and bearing of the ship.

Answers

7. $40\sqrt{3}\mathbf{i} + 40\mathbf{j}$

8. $v_0 \approx 30.0$ m/sec, $\theta \approx 25.0°$

Solution:

A diagram of the vectors is shown in Figure 7-44.

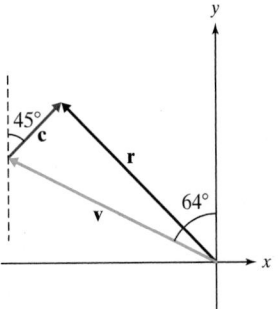

a. The direction angle of the ship is $90° + 64° = 154°$, and the magnitude (speed) is 18 mph. Therefore,

$$\mathbf{v} = \|\mathbf{v}\|\cos\theta\mathbf{i} + \|\mathbf{v}\|\sin\theta\mathbf{j}$$
$$= 18\cos 154°\mathbf{i} + 18\sin 154°\mathbf{j}$$
$$\approx -16.2\mathbf{i} + 7.9\mathbf{j}$$

b. The direction angle of the current is $90° - 45° = 45°$, and the magnitude is 4 mph. Therefore,

Figure 7-44

$$\mathbf{c} = \|\mathbf{c}\|\cos\theta\mathbf{i} + \|\mathbf{c}\|\sin\theta\mathbf{j}$$
$$= 4\cos 45°\mathbf{i} + 4\sin 45°\mathbf{j}$$
$$= 4 \cdot \frac{\sqrt{2}}{2}\mathbf{i} + 4 \cdot \frac{\sqrt{2}}{2}\mathbf{j} = 2\sqrt{2}\mathbf{i} + 2\sqrt{2}\mathbf{j} \approx 2.8\mathbf{i} + 2.8\mathbf{j}$$

c. The true velocity of the ship is given by the sum of vectors $\mathbf{v}$ and $\mathbf{c}$. We call this the resultant vector $\mathbf{r}$.

$$\mathbf{r} = \mathbf{v} + \mathbf{c} \approx (-16.2\mathbf{i} + 7.9\mathbf{j}) + (2.8\mathbf{i} + 2.8\mathbf{j}) \approx -13.4\mathbf{i} + 10.7\mathbf{j}$$

The true speed of the ship is the magnitude of $\mathbf{r}$.

$$\|\mathbf{r}\| \approx \sqrt{(-13.4)^2 + (10.7)^2} \approx 17.1 \text{ mph}$$

The true direction of the ship is the direction angle of $\mathbf{r}$.

$$\tan\theta \approx \left(\frac{10.7}{-13.4}\right)$$

For θ in Quadrant II, we have $\theta \approx \tan^{-1}\left(-\frac{10.7}{13.4}\right) + 180° \approx 141.4°$.

The ship travels approximately 17.1 mph at a bearing of N51.4°W.

Skill Practice 9   A plane travels with an air speed of 400 knots at S20°E and encounters a wind of 50 knots acting in the direction of S30°W. Rounding to the nearest whole unit,

   a. Express the velocity of the plane $\mathbf{p}$ relative to the air in terms of $\mathbf{i}$ and $\mathbf{j}$.

   b. Express the velocity of the wind $\mathbf{w}$ in terms of $\mathbf{i}$ and $\mathbf{j}$.

   c. Find the true speed (ground speed) and bearing of the plane.

In Example 10, we use vectors to find an unknown direction angle.

Classroom Example: p. 741
Exercise 88

EXAMPLE 10   Calculate a Direction Angle

Joe enters a triathlon and must swim to a buoy offshore and then back again. A current of 0.6 mph flows due west.

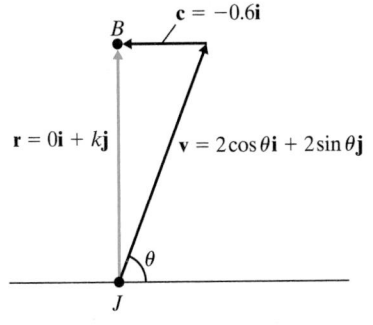

   a. If Joe starts at the closest point J on the shoreline to the buoy B, in what direction θ should he swim in order to arrive at the buoy? Assume that Joe's speed swimming in still water is 2 mph. Round to the nearest tenth of a degree.

   b. What is Joe's true speed? Round to 1 decimal place.

Answers
9. a. $\mathbf{p} \approx 137\mathbf{i} - 376\mathbf{j}$
  b. $\mathbf{w} \approx -25\mathbf{i} - 43\mathbf{j}$
  c. True speed: 434 knots;
    Bearing: S15°E

Solution:

From the diagram, the sum of Joe's velocity **v** in still water and the velocity **c** of the current must equal the resulting velocity **r**. The magnitude of **r** is Joe's actual speed k, and the direction of **r** is due north. Thus, vector **r** is of the form $\mathbf{r} = 0\mathbf{i} + k\mathbf{j}$.

a. $\mathbf{v} + \mathbf{c} = \mathbf{r}$

$$(2\cos\theta\,\mathbf{i} + 2\sin\theta\,\mathbf{j}) + (-0.6\mathbf{i}) = 0\mathbf{i} + k\mathbf{j}$$

$$(2\cos\theta - 0.6)\mathbf{i} = 0\mathbf{i} \qquad \text{Equating the horizontal components results in an equation with one unknown and enables us to solve for } \theta.$$

$$2\cos\theta - 0.6 = 0$$

$$\cos\theta = \frac{0.6}{2} \Rightarrow \theta = \cos^{-1}\frac{0.6}{2}$$

In Quadrant I, we have $\theta \approx 72.5°$.

b. $2\sin\theta\,\mathbf{j} = k\mathbf{j}$ Joe's true speed is the vertical component k of the resultant
$2\sin 72.5° = k$ vector. To solve for k, equate the vertical components.
$k \approx 1.9$ mph

Skill Practice 10   Repeat Example 10 with Joe's speed swimming in still water as 2.5 mph, and a current of 0.5 mph due west.

In many applications of physics, we encounter systems with numerous forces acting on an object. If forces $\mathbf{F}_1$, $\mathbf{F}_2$, $\mathbf{F}_3$, ..., $\mathbf{F}_n$ act on an object, the sum of the forces is called the resultant force **R**.

$$\mathbf{R} = \mathbf{F}_1 + \mathbf{F}_2 + \mathbf{F}_3 + \cdots + \mathbf{F}_n$$

In Example 11, we find the resultant force of three forces acting on an object.

Classroom Example: p. 741
Exercise 94

EXAMPLE 11   **Finding the Resultant Force**

Three forces, $\mathbf{F}_1$, $\mathbf{F}_2$, and $\mathbf{F}_3$, act on an object in the directions shown in Figure 7-45. If the magnitude of $\mathbf{F}_1$ is 60 lb, the magnitude of $\mathbf{F}_2$ is 100 lb, and the magnitude of $\mathbf{F}_3$ is 85 lb, write the resultant force **R** in terms of **i** and **j**. Round the components to the nearest tenth of a pound.

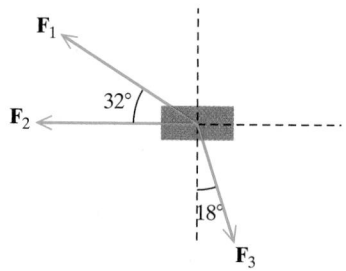

Figure 7-45

Solution:

From Figure 7-45, the direction angle of $\mathbf{F}_1$ is $180° - 32° = 148°$.
The direction angle of $\mathbf{F}_2$ is $180°$.
The direction angle of $\mathbf{F}_3$ is $270° + 18° = 288°$.

> **TIP** Force $\mathbf{F}_2$ acts in the negative x direction in a coordinate plane. Therefore, there is no vertical component of force, and by inspection, we have that $\mathbf{F}_2 = -100\mathbf{i}$.

$$\mathbf{F}_1 = \|\mathbf{F}_1\|\cos\theta\,\mathbf{i} + \|\mathbf{F}_1\|\sin\theta\,\mathbf{j} = 60\cos 148°\mathbf{i} + 60\sin 148°\mathbf{j}$$
$$\approx -50.9\mathbf{i} + 31.8\mathbf{j}$$

Write $\mathbf{F}_1$, $\mathbf{F}_2$, and $\mathbf{F}_3$ in terms of **i** and **j**. Then add the results to get the resultant force.

$$\mathbf{F}_2 = \|\mathbf{F}_2\|\cos\theta\,\mathbf{i} + \|\mathbf{F}_2\|\sin\theta\,\mathbf{j} = 100\cos 180°\mathbf{i} + 100\sin 180°\mathbf{j}$$
$$= -100\mathbf{i} + 0\mathbf{j}$$
$$= -100\mathbf{i}$$

$$\mathbf{F}_3 = \|\mathbf{F}_3\|\cos\theta\,\mathbf{i} + \|\mathbf{F}_3\|\sin\theta\,\mathbf{j} = 85\cos 288°\mathbf{i} + 85\sin 288°\mathbf{j}$$
$$\approx 26.3\mathbf{i} - 80.8\mathbf{j}$$

$$\mathbf{R} = \mathbf{F}_1 + \mathbf{F}_2 + \mathbf{F}_3 \approx (-50.9\mathbf{i} + 31.8\mathbf{j}) + (-100\mathbf{i}) + (26.3\mathbf{i} - 80.8\mathbf{j})$$
$$\approx -124.6\mathbf{i} - 49\mathbf{j}$$

Answers
10. a. 78.5°   **b.** 2.4 mph

Classroom Example: p. 743
Exercise 114

Skill Practice 11   If the magnitude of $\mathbf{F}_1$ is 32 lb, the magnitude of $\mathbf{F}_2$ is 24 lb, and the magnitude of $\mathbf{F}_3$ is 15 lb, write the resultant force $\mathbf{R}$ in terms of $\mathbf{i}$ and $\mathbf{j}$.

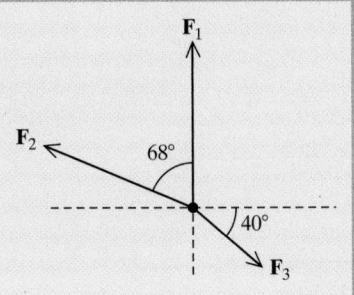

Suppose that forces $\mathbf{F}_1, \mathbf{F}_2, \mathbf{F}_3, \dots, \mathbf{F}_n$ act on an object. The object is said to be in **static equilibrium** if the resultant force $\mathbf{R}$ is the zero vector. That is,

$$\mathbf{R} = \mathbf{F}_1 + \mathbf{F}_2 + \mathbf{F}_3 + \cdots + \mathbf{F}_n = \mathbf{0}$$

For example, the sum of the three forces in Example 11 is $-124.6\mathbf{i} - 49\mathbf{j}$. If a fourth force $\mathbf{F}_4 = 124.6\mathbf{i} + 49\mathbf{j}$ acted on the object, then the sum of the forces would be $0\mathbf{i} + 0\mathbf{j}$ and the object would be in static equilibrium. Intuitively, this means that the forces counteract one another.

EXAMPLE 12   **Solving for Force Vectors in a System in Static Equilibrium**

Two cranes support a 3000-lb crate in static equilibrium, as shown in the figure (not drawn to scale). Find the magnitudes of the tension vectors $\mathbf{T}_1$ and $\mathbf{T}_2$. Round to the nearest pound.

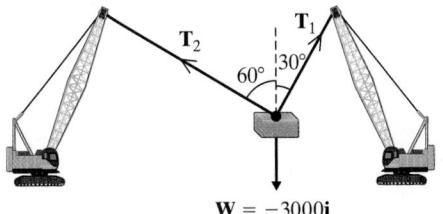

Solution:

The weight $\mathbf{W}$ of the crate is directed straight downward and can be expressed as the vector $\mathbf{W} = -3000\mathbf{j}$. We need to solve for $\|\mathbf{T}_1\|$ and $\|\mathbf{T}_2\|$. Since the crate is in static equilibrium, we know that the sum of the force vectors is $\mathbf{0}$. That is, $\mathbf{T}_1 + \mathbf{T}_2 + \mathbf{W} = \mathbf{0}$.

$(\|\mathbf{T}_1\|\cos 60°\mathbf{i} + \|\mathbf{T}_1\|\sin 60°\mathbf{j}) + (\|\mathbf{T}_2\|\cos 150°\mathbf{i} + \|\mathbf{T}_2\|\sin 150°\mathbf{j}) + (-3000\mathbf{j}) = 0\mathbf{i} + 0\mathbf{j}$

For two vectors to be equal, their corresponding components must be equal.

$(\|\mathbf{T}_1\|\cos 60° + \|\mathbf{T}_2\|\cos 150°)\mathbf{i} = 0\mathbf{i}$ (1)   Equate the horizontal components.

$(\|\mathbf{T}_1\|\sin 60° + \|\mathbf{T}_2\|\sin 150° - 3000)\mathbf{j} = 0\mathbf{j}$ (2)   Equate the vertical components.

Simplifying equations (1) and (2), we have

$$\frac{1}{2}\|\mathbf{T}_1\| - \frac{\sqrt{3}}{2}\|\mathbf{T}_2\| = 0 \qquad (3)$$

$$\frac{\sqrt{3}}{2}\|\mathbf{T}_1\| + \frac{1}{2}\|\mathbf{T}_2\| - 3000 = 0 \qquad (4)$$

Answer

11. $\mathbf{R} = -10.8\mathbf{i} + 31.3\mathbf{j}$

From (3) we have $\|\mathbf{T}_1\| = \sqrt{3}\|\mathbf{T}_2\|$. Substituting $\sqrt{3}\|\mathbf{T}_2\|$ for $\|\mathbf{T}_1\|$ in equation (4) we have

$$\frac{\sqrt{3}}{2}\left(\sqrt{3}\|\mathbf{T}_2\|\right) + \frac{1}{2}\|\mathbf{T}_2\| - 3000 = 0$$

$$\frac{3}{2}\|\mathbf{T}_2\| + \frac{1}{2}\|\mathbf{T}_2\| - 3000 = 0$$

$$2\|\mathbf{T}_2\| = 3000$$

$$\|\mathbf{T}_2\| = 1500 \text{ lb}$$

Using the relationship $\|\mathbf{T}_1\| = \sqrt{3}\|\mathbf{T}_2\|$, we have $\|\mathbf{T}_1\| = \sqrt{3}(1500) \approx 2598$ lb. The tensions in the cables are 2598 lb and 1500 lb.

> **Skill Practice 12** Repeat Example 12, with the two cranes supporting a 1-ton (2000-lb) beam in static equilibrium. The direction angles for $\mathbf{T}_1$ and $\mathbf{T}_2$ are 45° and 120°, respectively. Round to the nearest pound.

Point of Interest

The study of *statics* deals with bodies in equilibrium, whereas the study of *dynamics* focuses on bodies under acceleration. Early principles of statics were known in ancient times. For example, the writings of Archimedes show that the ancient Greeks were familiar with the concept of a lever. The study of dynamics came much later because accurate measurements of time were required to investigate bodies in motion. Major contributions were made by Galileo Galilei (1564–1642) and later by Isaac Newton (1642–1727) who formulated three fundamental laws of motion.

Answer
12. $\|\mathbf{T}_1\| \approx 1035$ lb and
$\|\mathbf{T}_2\| \approx 1464$ lb

SECTION 7.4   Practice Exercises

Prerequisite Review

For Exercises R.1–R.2, use the distance formula to find the distance between the two points.

R.1. (2, 4) and (−2, 2) $2\sqrt{5}$

R.2. (3, 7) and (−5, 7)   8

For Exercises R.3–R.4, make a sketch to illustrate the bearing.

R.3. N53°W

R.4. S38°E

Concept Connections

1. A _____scalar_____ is a quantity that has magnitude only, whereas a vector has both magnitude and _____direction_____.

2. Given a vector $\overrightarrow{AB}$, the initial point is _____A_____ and the terminal point is _____B_____.
   The _____magnitude_____ of $\overrightarrow{AB}$ is the length of the vector and is denoted by _____$\|\overrightarrow{AB}\|$_____.

3. Suppose that Dale Earnhardt Jr. drives around the 2.5-mi oval track at the Daytona International Speedway.

   a. What is the distance traveled? Is this a scalar or a vector?   2.5 mi; scalar

   b. What is his displacement? Is this a scalar or a vector?   **0** (the zero vector); vector

4. For a real number k, the vector $k\mathbf{v}$ is in the same direction as $\mathbf{v}$ if k is _____positive_____ and in the opposite direction as $\mathbf{v}$ if k is _____negative_____.

5. Two vectors $\mathbf{r} = \langle a_1, b_1 \rangle$ and $\mathbf{s} = \langle a_2, b_2 \rangle$ are equal if $\underline{a_1 = a_2 \text{ and } b_1 = b_2}$.

6. Given a nonzero vector $\mathbf{v} = \langle a, b \rangle$, a unit vector $\mathbf{u}$ in the same direction as $\mathbf{v}$ is given by $\mathbf{u} = \underline{\frac{1}{\|\mathbf{v}\|}\mathbf{v} \text{ or } \frac{1}{\|\mathbf{v}\|}\langle a, b\rangle \text{ or } \left\langle \frac{a}{\|\mathbf{v}\|}, \frac{b}{\|\mathbf{v}\|}\right\rangle}$.

7. In component form, the vector $\mathbf{i} = \underline{\langle 1, 0 \rangle}$ and $\mathbf{j} = \underline{\langle 0, 1 \rangle}$.

8. If a nonzero vector $\mathbf{v} = \langle a, b \rangle$ has magnitude $\|\mathbf{v}\|$ and direction θ, then $a =$ _____$\|\mathbf{v}\|\cos\theta$_____ and $b =$ _____$\|\mathbf{v}\|\sin\theta$_____.

9. If a nonzero vector $\mathbf{v} = \langle a, b \rangle$, then the magnitude $\|\mathbf{v}\|$ is given by _____$\sqrt{a^2 + b^2}$_____, and the direction is given by $\tan\theta =$ _____$\dfrac{b}{a}$_____ for $0 \le \theta < 2\pi$ and $a \ne 0$.

10. Suppose that forces $\mathbf{F}_1, \mathbf{F}_2, \mathbf{F}_3, \dots, \mathbf{F}_n$ are acting on an object in static equilibrium. Then the sum of the forces equals _____the zero vector_____.

Objective 1: Interpret Vectors Geometrically

For Exercises 11–16, vector v has initial point P and terminal point Q. Vector w has initial point R and terminal point S. (See Example 1)

    a. Find the magnitude of **v**.   **b.** Find the magnitude of **w**.

    c. Determine whether $\mathbf{v} = \mathbf{w}$ and explain your reasoning.

11. $P(4, 7)$, $Q(5, 10)$ and $R(-2, 8)$, $S(-1, 11)$ **12.** $P(-2, -10)$, $Q(-5, -8)$ and $R(9, -3)$, $S(6, -1)$

13. $P(4, -1)$, $Q(7, -6)$ and $R(5, 7)$, $S(2, 12)$ **14.** $P(10, 18)$, $Q(27, -6)$ and $R(9, 12)$, $S(-8, -12)$

15. $P(-12, 10)$, $Q(-16, 7)$ and $R(9, -3)$, $S(1, -9)$ **16.** $P(-6, 5)$, $Q(10, 35)$ and $R(10, -36)$, $S(34, 9)$

For Exercises 17–20, refer to vectors v and w in the figure.

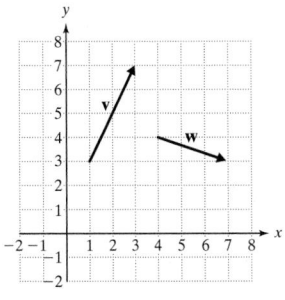

17. a. Sketch $\mathbf{v} + \mathbf{w}$ by placing the initial point of **w** at the terminal point of **v**.

    b. Sketch $\mathbf{v} + \mathbf{w}$ by drawing **v** and **w** with the same initial point.

18. a. Sketch $\mathbf{v} - \mathbf{w}$ by placing the initial point of –**w** at the terminal point of **v**.

    b. Sketch $\mathbf{v} - \mathbf{w}$ by drawing **v** and **w** with the same initial point.

19. Sketch –**v** and 2**w** in standard position.   **20.** Sketch 2**v** and –2**w** in standard position.

Objective 2: Represent Vectors in Component Form

For Exercises 21–22, given a vector v with initial point P and terminal point Q,

    a. Write the component form of **v**.

    b. Sketch **v** from P to Q and sketch **v** in standard position. **(See Example 2)**

21. $P(3, -2)$, $Q(5, -5)$ a. $\langle 2, -3 \rangle$ **22.** $P(-1, 4)$, $Q(-3, 8)$ a. $\langle -2, 4 \rangle$

For Exercises 23–26, write the component form of a vector v with initial point P and terminal point Q.

23. $P(-8, -3)$, $Q(-8, -6)$ $\langle 0, -3 \rangle$ **24.** $P(10, 10)$, $Q(12, 10)$ $\langle 2, 0 \rangle$

25. $P(8.4, -2.3)$, $Q(6.4, -5.8)$ $\langle -2, -3.5 \rangle$ **26.** $P(2.5, -1.9)$, $Q(10, 3.1)$ $\langle 7.5, 5 \rangle$

27. Given vector **v** with initial point $(16, -30)$ and terminal point $(-22, -10)$,

    a. Find the component form of **v**. $\langle -38, 20 \rangle$

    b. If **v** is placed with initial point at $(15, 8)$, what is the terminal point of **v**? $(-23, 28)$

28. Given vector **v** with initial point $(17, 80)$ and terminal point $(-72, 53)$,

    a. Find the component form of **v**. $\langle -89, -27 \rangle$

    b. If **v** is placed with initial point at $(-13, -12)$, what is the terminal point of **v**? $(-102, -39)$

29. Given $\mathbf{v} = \langle -17, 29 \rangle$ with initial point $(4, -10)$, find the terminal point of **v**. $(-13, 19)$

30. Given $\mathbf{v} = \langle 5, -8 \rangle$ with initial point $(-1, -7)$, find the terminal point of **v**. $(4, -15)$

Objective 3: Perform Operations on Vectors in Component Form

For Exercises 31–46, perform the indicated operations for the given vectors. (See Examples 3–4)

$\mathbf{v} = \langle 8, -10 \rangle$ $\mathbf{w} = \langle -3, 7 \rangle$ $\mathbf{s} = \langle -1, -2 \rangle$ $\mathbf{r} = \langle 2, 10 \rangle$

31. $\mathbf{v} + \mathbf{w}$ $\langle 5, -3 \rangle$ **32.** $\mathbf{s} + \mathbf{w}$ $\langle -4, 5 \rangle$ **33.** $\mathbf{v} + \mathbf{r}$ $\langle 10, 0 \rangle$ **34.** $\mathbf{r} + \mathbf{s}$ $\langle 1, 8 \rangle$

35. $\mathbf{w} - \mathbf{v}$ $\langle -11, 17 \rangle$ **36.** $\mathbf{w} - \mathbf{s}$ $\langle -2, 9 \rangle$ **37.** $\mathbf{r} - \mathbf{w}$ $\langle 5, 3 \rangle$ **38.** $\mathbf{v} - \mathbf{r}$ $\langle 6, -20 \rangle$

39. $5\mathbf{s}$ $\langle -5, -10 \rangle$ **40.** $-4\mathbf{w}$ $\langle 12, -28 \rangle$ **41.** $-\dfrac{1}{2}\mathbf{v}$ $\langle -4, 5 \rangle$ **42.** $\dfrac{2}{3}\mathbf{w}$ $\left\langle -2, \dfrac{14}{3} \right\rangle$

43. $\mathbf{v} + \mathbf{s} - \mathbf{r}$ $\langle 5, -22 \rangle$ **44.** $\mathbf{s} - \mathbf{r} + \mathbf{w}$ $\langle -6, -5 \rangle$ **45.** $2\mathbf{r} - (\mathbf{w} + \mathbf{s})$ $\langle 8, 15 \rangle$ **46.** $3(\mathbf{v} + 2\mathbf{w}) - \mathbf{s}$
                                                                             $\langle 7, 14 \rangle$

For Exercises 47–54, let c and d be scalars and let $\mathbf{v} = \langle a_1, b_1 \rangle$, $\mathbf{w} = \langle a_2, b_2 \rangle$, $\mathbf{u} = \langle a_3, b_3 \rangle$, and $\mathbf{0} = \langle 0, 0 \rangle$. Prove the given statement.

47. $(\mathbf{v} + \mathbf{w}) + \mathbf{u} = \mathbf{v} + (\mathbf{w} + \mathbf{u})$ **48.** $\mathbf{v} + \mathbf{0} = \mathbf{0} + \mathbf{v}$

49. $\mathbf{v} + (-\mathbf{v}) = \mathbf{0}$ **50.** $c(\mathbf{v} + \mathbf{w}) = c\mathbf{v} + c\mathbf{w}$

51. $(c + d)\mathbf{v} = c\mathbf{v} + d\mathbf{v}$ **52.** $(cd)\mathbf{v} = c(d\mathbf{v})$

53. $0(\mathbf{v}) = \mathbf{0}$ **54.** $\|c\mathbf{v}\| = |c|\|\mathbf{v}\|$

Objective 4: Represent Vectors in Terms of i and j

For Exercises 55–60, find a unit vector in the direction of $\mathbf{v}$. (See Example 5)

55. $\mathbf{v} = 20\mathbf{i} - 21\mathbf{j}$ $\mathbf{u}_v = \dfrac{20}{29}\mathbf{i} - \dfrac{21}{29}\mathbf{j}$ **56.** $\mathbf{w} = -11\mathbf{i} + 60\mathbf{j}$ $\mathbf{u}_w = -\dfrac{11}{61}\mathbf{i} + \dfrac{60}{61}\mathbf{j}$ **57.** $\mathbf{t} = -0.7\mathbf{i} - 2.4\mathbf{j}$ $\mathbf{u}_t = -0.28\mathbf{i} - 0.96\mathbf{j}$

58. $\mathbf{s} = 3.9\mathbf{i} + 5.2\mathbf{j}$ $\mathbf{u}_s = 0.6\mathbf{i} + 0.8\mathbf{j}$ **59.** $\mathbf{r} = -10\mathbf{i} + 14\mathbf{j}$ $\mathbf{u}_r = -\dfrac{5\sqrt{74}}{74}\mathbf{i} + \dfrac{7\sqrt{74}}{74}\mathbf{j}$ **60.** $\mathbf{v} = 6\mathbf{i} - 15\mathbf{j}$ $\mathbf{u}_v = \dfrac{2\sqrt{29}}{29}\mathbf{i} - \dfrac{5\sqrt{29}}{29}\mathbf{j}$

For Exercises 61–68, perform the indicated operations for the given vectors. (See Example 6)

$\mathbf{v} = -3.3\mathbf{i} + 11\mathbf{j}$ $\mathbf{w} = 4\mathbf{i} + \mathbf{j}$ $\mathbf{s} = 7.2\mathbf{i} - 5.4\mathbf{j}$ $\mathbf{r} = -12\mathbf{i} - 5\mathbf{j}$

61. $\mathbf{v} + \mathbf{s}$ $3.9\mathbf{i} + 5.6\mathbf{j}$ **62.** $\mathbf{w} + \mathbf{r}$ $-8\mathbf{i} - 4\mathbf{j}$ **63.** $\mathbf{s} - \mathbf{r}$ $19.2\mathbf{i} - 0.4\mathbf{j}$ **64.** $\mathbf{v} - \mathbf{w}$ $-7.3\mathbf{i} + 10\mathbf{j}$

65. $-2\mathbf{r} + 3\mathbf{w}$ $36\mathbf{i} + 13\mathbf{j}$ **66.** $6\mathbf{s} - 4\mathbf{v}$ $56.4\mathbf{i} - 76.4\mathbf{j}$ **67.** $\|\mathbf{r} + \mathbf{w}\|$ $4\sqrt{5}$ **68.** $\|\mathbf{r} - \mathbf{w}\|$ $2\sqrt{73}$

Objective 5: Relate Component Form and Direction and Magnitude of a Vector

For Exercises 69–74, write the vector $\mathbf{v}$ in the form $a\mathbf{i} + b\mathbf{j}$, where $\mathbf{v}$ has the given magnitude and direction angle.

69. $\|\mathbf{v}\| = 12$, $\theta = 30°$ $\mathbf{v} = 6\sqrt{3}\mathbf{i} + 6\mathbf{j}$ **70.** $\|\mathbf{v}\| = 18$, $\theta = 150°$ $\mathbf{v} = -9\sqrt{3}\mathbf{i} + 9\mathbf{j}$

71. $\|\mathbf{v}\| = \sqrt{5}$, $\theta = \dfrac{3\pi}{4}$ $\mathbf{v} = -\dfrac{\sqrt{10}}{2}\mathbf{i} + \dfrac{\sqrt{10}}{2}\mathbf{j}$ **72.** $\|\mathbf{v}\| = \sqrt{17}$, $\theta = \dfrac{4\pi}{3}$ $\mathbf{v} = -\dfrac{\sqrt{17}}{2}\mathbf{i} - \dfrac{\sqrt{51}}{2}\mathbf{j}$

73. $\|\mathbf{v}\| = \dfrac{3}{8}$, $\theta = 300°$ $\mathbf{v} = \dfrac{3}{16}\mathbf{i} - \dfrac{3\sqrt{3}}{16}\mathbf{j}$ **74.** $\|\mathbf{v}\| = \dfrac{6}{7}$, $\theta = 225°$ $\mathbf{v} = -\dfrac{3\sqrt{2}}{7}\mathbf{i} - \dfrac{3\sqrt{2}}{7}\mathbf{j}$

For Exercises 75–78, find the magnitude and direction angle ($0° \leq \theta < 360°$) for the given vector. Round to 1 decimal place.

75. $\mathbf{v} = 4\mathbf{i} + 6\mathbf{j}$ **76.** $\mathbf{w} = -5\mathbf{i} + 8\mathbf{j}$ **77.** $\mathbf{s} = -1.8\mathbf{i} - 3.5\mathbf{j}$ **78.** $\mathbf{r} = \mathbf{i} - 0.8\mathbf{j}$
    $\|\mathbf{v}\| \approx 7.2$, $\theta \approx 56.3°$ $\|\mathbf{w}\| \approx 9.4$, $\theta \approx 122.0°$ $\|\mathbf{s}\| \approx 3.9$, $\theta \approx 242.8°$ $\|\mathbf{r}\| \approx 1.3$, $\theta \approx 321.3°$

79. A kick boxer kicks a bag at an angle of 30° from the horizontal with a force of 250 N (Newtons). Write the force vector $\mathbf{F}$ in terms of $\mathbf{i}$ and $\mathbf{j}$. (See Example 7)
    $\mathbf{F} = 125\sqrt{3}\mathbf{i} + 125\mathbf{j}$

80. An athlete stretches his Achilles tendon by extending one leg behind him with his heel on the floor and pushing against a wall with his hands. If he pushes against the wall with a force of 120 N at an angle of 20° from the horizontal as shown, write the force vector $\mathbf{F}$ in terms of $\mathbf{i}$ and $\mathbf{j}$. Round the components to the nearest tenth of a Newton. $112.8\mathbf{i} + 41.0\mathbf{j}$

81. A ship travels 18 mph at a bearing of S30°W. Write the velocity vector **v** in terms of **i** and **j**. $v = -9i - 9\sqrt{3}j$

82. A baseball is thrown with a speed of 70 mph at an angle of 14° from the horizontal. Write the velocity vector **v** at the time of release in terms of **i** and **j**. Round the components to the nearest tenth of a mph. $v = 67.9i + 16.9j$

83. A punter kicks a football with an initial velocity given by **v** = 22.2**i** + 18.6**j** m/sec. (**See Example 8**)

  a. Find the magnitude of the velocity vector at the time the ball leaves the punter's foot. Round to the nearest m/sec.   29 m/sec

  b. Find the angle from the horizontal at which the ball was kicked. Round to the nearest degree.   40°

84. The velocity of a ship is given by the vector −6.4**i** + 7.7**j** mph.

  a. Find the speed of the ship. Round to the nearest mph.   10 mph

  b. Find the bearing of the ship. Round to the nearest degree.   N40°W

Objective 6: Use Vectors in Applications

85. A plane travels N30°W at 450 mph and encounters a wind blowing due west at 30 mph. (**See Example 9**)

  a. Express the velocity of the plane **v$_p$** relative to the air in terms of **i** and **j**. $v_p = -225i + 225\sqrt{3}j$

  b. Express the velocity of the wind **v$_w$** in terms of **i** and **j**. $v_w = -30i$

  c. Express the true velocity of the plane **v$_T$** in terms of **i** and **j** and find the true speed of the plane.
   $v_T = -255i + 225\sqrt{3}j$; speed = $30\sqrt{241}$ mph

86. A plane flies at a speed of 370 mph on a bearing of S45°E. Relative to the ground, the plane's speed is measured as 345 mph with a true bearing of S40°E. Rounding to the nearest whole unit,

  a. Express the velocity of the plane **v$_p$** relative to the air in terms of **i** and **j**. $v_p \approx 262i - 262j$

  b. Express the true velocity of the plane **v$_T$** in terms of **i** and **j**. $v_T \approx 222i - 264j$

  c. Express the velocity of the wind **v$_w$** in terms of **i** and **j** and find the speed of the wind. $v_w \approx -40i - 2j$; speed ≈ 40 mph

87. A swimmer swims 1.5 mph in still water and wants to swim to a point *B* due north from her starting point directly across a river. If the current is 0.5 mph due east, at what angle θ should the swimmer swim to reach point *B*? Round to the nearest tenth of a degree. (**See Example 10**)   109.5°

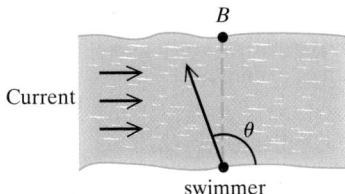

88. A boat that travels 6 mph in still water leaves a marina in Miami to travel to an island called Elliott Key due south of the marina. The prevailing current is directed northeast (N45°E) at 0.8 mph. At what bearing should the captain of the boat steer the ship?   S5.4°W

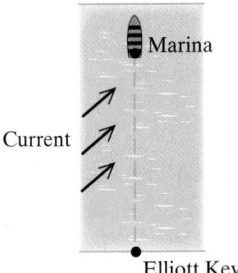

For Exercises 89–94, the given forces (in units of pounds) act on an object.

  a. Find the resultant force, **R**.

  b. What additional force **F** is needed for the object to be in static equilibrium? (**See Example 11**)

89. $F_1 = \langle 3, 11 \rangle$ and $F_2 = \langle -8, 6 \rangle$
  a. $R = \langle -5, 17 \rangle$
  b. $F = \langle 5, -17 \rangle$

90. $F_1 = \langle -5, 2 \rangle$ and $F_2 = \langle -7, -3 \rangle$
  a. $R = \langle -12, -1 \rangle$
  b. $F = \langle 12, 1 \rangle$

91. $F_1 = 5i - j$, $F_2 = 4i - 3j$, $F_3 = -9i + 2j$
  a. $R = -2j$
  b. $F = 2j$

92. $F_1 = -12i + 8j$, $F_2 = -9i - 15j$, $F_3 = 11i + 7j$
  a. $R = -10i$
  b. $F = 10i$

93.

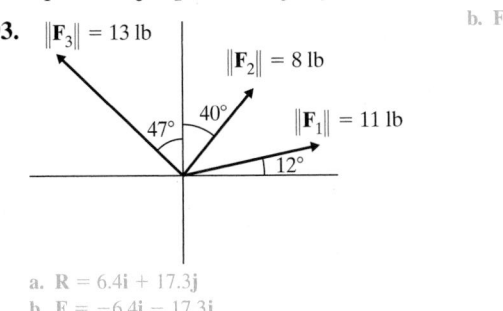

  a. $R = 6.4i + 17.3j$
  b. $F = -6.4i - 17.3j$

94.

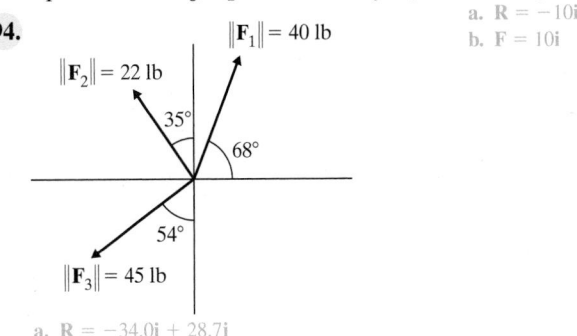

  a. $R = -34.0i + 28.7j$
  b. $F = 34.0i - 28.7j$

95. Two pickup trucks rescue a car stuck in the sand. The first truck pulls with a force of 5000 lb and the second truck pulls with a force of 4000 lb at an angle of 50° to the first truck. How much force can the combined pull of the two trucks generate? Round to the nearest pound.   8168 lb

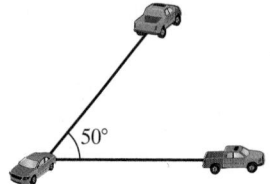

96. Two forces act on an object with an angle of 63° between them. If the magnitude of the first force is 48 N and the magnitude of the second force is 70 N, find the magnitude of the resultant force to the nearest Newton.   101 N

Mixed Exercises

For Exercises 97–102, refer to vectors t, u, v, and w in the figure and match each vector with an equivalent vector a–f.

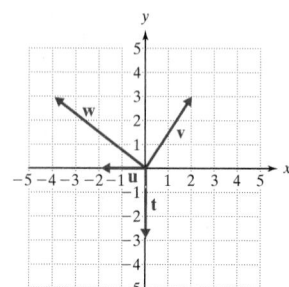

a. $-\mathbf{t}$

b. $3\mathbf{u}$

c. $\mathbf{w}$

d. $\mathbf{u} - 2\mathbf{t}$

e. $\mathbf{v}$

f. $-3\mathbf{u}$

97. $2\mathbf{u} - \mathbf{t}$ c (vector w)

98. $-\mathbf{u} - \mathbf{t}$ e (vector v)

99. $\mathbf{w} - \mathbf{v}$ b (vector 3u)

100. $\mathbf{v} + \mathbf{u}$ a (vector −t)

101. $\mathbf{v} + \mathbf{w}$ d (vector u − 2t)

102. $\mathbf{v} - \mathbf{w}$ f (vector −3u)

103. A kayaker paddles 3 mph in still water. If she heads due east but encounters a 0.8-mph current flowing south,

   a. Find the actual direction of travel. Write the direction as a bearing and round to the nearest degree.   S75°E

   b. How far will she travel in 2 hr? Round to the nearest tenth of a mile.   6.2 mi

104. A long-distance swimmer swims 2 mph in still water. If he heads north but encounters a 0.6-mph current flowing northeast,

   a. Find the actual direction of travel. Write the direction as a bearing and round to the nearest degree.   N10°E

   b. How far will he swim in 1.5 hr? Round to the nearest tenth of a mile.   3.7 mi

For Exercises 105–106, two vectors v and w act on a point in the plane with the indicated force.

   a. Use a parallelogram to sketch the resultant vector $\mathbf{v} + \mathbf{w}$.

   b. Use the law of cosines to find the magnitude of the resultant vector. (*Hint:* Adjacent angles in a parallelogram are supplementary.)

   c. Use the law of sines to find the angle that $\mathbf{v} + \mathbf{w}$ makes with $\mathbf{v}$.

Round answers to the nearest whole unit.

105.

$\|\mathbf{w}\| = 50$

$35°$

$\|\mathbf{v}\| = 38$

   b. $\|\mathbf{v} + \mathbf{w}\|^2 = 38^2 + 50^2 - 2 \cdot 38 \cdot 50 \cdot \cos 145°$; $\|\mathbf{v} + \mathbf{w}\| \approx 84$ **c.** 20°

106.

$\|\mathbf{w}\| = 10$

$28°$

$\|\mathbf{v}\| = 40$

   b. $\|\mathbf{v} + \mathbf{w}\|^2 = 40^2 + 10^2 - 2 \cdot 40 \cdot 10 \cdot \cos 152°$ $\|\mathbf{v} + \mathbf{w}\| \approx 49$ **c.** 5°

Write About It

107. How is the component form of a vector $\langle a, b \rangle$ related to the graph of the vector in standard position?

108. Explain how the sum $\mathbf{v} + \mathbf{w}$ is found geometrically.

109. Explain the geometric significance of the scalar multiple of a vector, such as $k\mathbf{v}$, where k is a real number.

110. Explain how vectors written in component form are added or subtracted.

Expanding Your Skills

For Exercises 111–112, assume that the system is in equilibrium with F_2 acting directly downward. Calculate the magnitude of the forces F_1 and F_2. Round to the nearest tenth of a pound.

111.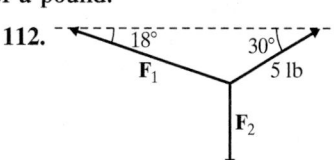
$\|F_1\| \approx 30.8$ lb, $\|F_2\| \approx 39.8$ lb

112.
$\|F_1\| \approx 4.6$ lb, $\|F_2\| \approx 3.9$ lb

113. A 50-lb box is supported by two ropes attached to the ceiling. Find the magnitudes of the tension vectors T_1 and T_2. Round to the nearest tenth of a pound. (**See Example 12**)

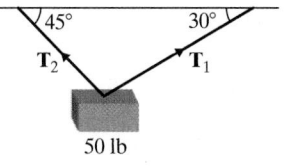

$\|T_1\| = 36.6$ lb, $\|T_2\| = 44.8$ lb

114. A tightrope walker and the balance bar together weigh 175 lb. Find the magnitudes of the tension vectors T_1 and T_2. Round to the nearest tenth of a pound. $\|T_1\| \approx 506.7$ lb, $\|T_2\| \approx 500.5$ lb

---

SECTION 7.5   Dot Product

OBJECTIVES

1. Compute Dot Product
2. Find the Angle Between Two Vectors
3. Decompose Vectors into Two Orthogonal Vectors
4. Use Vectors to Compute Work

1. Compute Dot Product

In this section, we present a special type of product of vectors called the dot product. The dot product of two vectors has a variety of applications including decomposing a vector into components parallel to and orthogonal (perpendicular) to another vector. We will also see in Examples 6–7 that the dot product is helpful in finding the amount of work done by a force in moving an object a specified distance.

> **Definition of Dot Product**
>
> If $\mathbf{v} = \langle a_1, b_1 \rangle$ and $\mathbf{w} = \langle a_2, b_2 \rangle$, the **dot product** $\mathbf{v} \cdot \mathbf{w}$ is defined as
>
> $$\mathbf{v} \cdot \mathbf{w} = a_1 a_2 + b_1 b_2$$

It is important to note that the dot product of two vectors is a scalar.

Classroom Examples: p. 752
Exercises 8, 12

EXAMPLE 1   Finding the Dot Product of Two Vectors

Given $\mathbf{v} = \langle 3, -7 \rangle$ and $\mathbf{w} = \langle -2, 5 \rangle$, find

  a. $\mathbf{v} \cdot \mathbf{w}$ **b.** $\mathbf{w} \cdot \mathbf{v}$ **c.** $\mathbf{v} \cdot \mathbf{v}$ **d.** $\|\mathbf{v}\|$

Instructor Note:
Point out the difference between the dot product and the product of a scalar with a vector. The dot product results in a scalar, whereas the product of a scalar and a vector is a vector quantity.

Solution:

  a. $\mathbf{v} \cdot \mathbf{w} = 3(-2) + (-7)(5) = -6 - 35 = -41$
  b. $\mathbf{w} \cdot \mathbf{v} = -2(3) + 5(-7) = -6 - 35 = -41$
  c. $\mathbf{v} \cdot \mathbf{v} = 3(3) + (-7)(-7) = 9 + 49 = 58$
  d. $\|\mathbf{v}\| = \sqrt{(3)^2 + (-7)^2} = \sqrt{9 + 49} = \sqrt{58}$

> **Skill Practice 1**   Given $\mathbf{v} = \langle -2, 6 \rangle$ and $\mathbf{w} = \langle -3, -1 \rangle$, find
>
>   **a.** $\mathbf{v} \cdot \mathbf{w}$ **b.** $\mathbf{w} \cdot \mathbf{v}$ **c.** $\mathbf{v} \cdot \mathbf{v}$ **d.** $\|\mathbf{v}\|$

Answers

1. a. 0   **b.** 0   **c.** 40   **d.** $2\sqrt{10}$

Notice from Examples 1(a) and 1(b) that $\mathbf{v} \cdot \mathbf{w} = \mathbf{w} \cdot \mathbf{v}$. We can show that this is true for any two vectors $\mathbf{v}$ and $\mathbf{w}$. That is, the dot product of two vectors is a commutative operation. Given $\mathbf{v} = \langle a_1, b_1 \rangle$ and $\mathbf{w} = \langle a_2, b_2 \rangle$:

$$\mathbf{v} \cdot \mathbf{w} = a_1 a_2 + b_1 b_2 \qquad \text{Dot product of } \mathbf{v} \text{ and } \mathbf{w}$$
$$= a_2 a_1 + b_2 b_1 \qquad \text{Commutative property of multiplication of real numbers}$$
$$= \mathbf{w} \cdot \mathbf{v} \qquad \text{Dot product of } \mathbf{w} \text{ and } \mathbf{v}$$

Examples 1(c) and 1(d) suggest that the dot product of a vector with itself is the square of the magnitude of the vector. That is, $\mathbf{v} \cdot \mathbf{v} = \|\mathbf{v}\|^2$. Given $\mathbf{v} = \langle a_1, b_1 \rangle$:

$$\mathbf{v} \cdot \mathbf{v} = a_1 a_1 + b_1 b_1 \qquad \text{Dot product of } \mathbf{v} \text{ and } \mathbf{v}$$
$$= a_1^2 + b_1^2$$
$$= \left(\sqrt{a_1^2 + b_1^2} \right)^2 \qquad \text{Recall that } \|\mathbf{v}\| = \sqrt{a_1^2 + b_1^2}.$$
$$= \|\mathbf{v}\|^2 \qquad \text{Square of the magnitude of } \mathbf{v}$$

Several important properties of the dot product are given in Table 7-4. The proofs of the remaining properties in Table 7-4 are addressed in Exercises 29–32.

Table 7-4

Properties of Dot Product
If $\mathbf{u}$, $\mathbf{v}$, and $\mathbf{w}$ are vectors, and c is a scalar, then

1. $\mathbf{v} \cdot \mathbf{w} = \mathbf{w} \cdot \mathbf{v}$	4. $\mathbf{v} \cdot \mathbf{v} = \|\mathbf{v}\|^2$
2. $(c\mathbf{v}) \cdot \mathbf{w} = c(\mathbf{v} \cdot \mathbf{w})$	5. $\mathbf{0} \cdot \mathbf{v} = 0$
3. $\mathbf{v} \cdot (\mathbf{w} + \mathbf{u}) = \mathbf{v} \cdot \mathbf{w} + \mathbf{v} \cdot \mathbf{u}$	6. $c(\mathbf{v} \cdot \mathbf{w}) = (c\mathbf{v}) \cdot \mathbf{w} = \mathbf{v} \cdot (c\mathbf{w})$

2. Find the Angle Between Two Vectors

Given two vectors $\mathbf{v}$ and $\mathbf{w}$ in standard position, the angle θ between $\mathbf{v}$ and $\mathbf{w}$ is the angle formed by the two line segments representing $\mathbf{v}$ and $\mathbf{w}$, where $0° \leq \theta \leq 180°$ (Figure 7-46). The vectors $\mathbf{v}$, $\mathbf{w}$, and $\mathbf{v} - \mathbf{w}$ form a triangle. Furthermore, by applying the law of cosines, θ can be expressed in terms of the dot product of $\mathbf{v}$ and $\mathbf{w}$ and the magnitudes of $\mathbf{v}$ and $\mathbf{w}$.

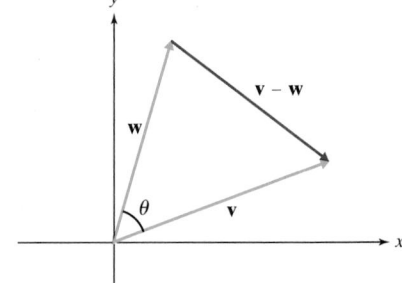

Figure 7-46

Angle Between Two Vectors
If θ is the angle between two nonzero vectors $\mathbf{v}$ and $\mathbf{w}$, then
$$\cos\theta = \frac{\mathbf{v} \cdot \mathbf{w}}{\|\mathbf{v}\|\|\mathbf{w}\|} \quad \text{and} \quad \theta = \cos^{-1}\left(\frac{\mathbf{v} \cdot \mathbf{w}}{\|\mathbf{v}\|\|\mathbf{w}\|} \right)$$

Proof:
Draw vectors **v** and **w** in standard position with angle θ between **v** and **w** (Figure 7-46). Applying the law of cosines on the triangle formed by **v**, **w**, and **v** − **w** we have,

$$\|\mathbf{v} - \mathbf{w}\|^2 = \|\mathbf{v}\|^2 + \|\mathbf{w}\|^2 - 2\|\mathbf{v}\|\|\mathbf{w}\|\cos\theta \quad \text{Apply the law of cosines.}$$

$$(\mathbf{v} - \mathbf{w}) \cdot (\mathbf{v} - \mathbf{w}) = \|\mathbf{v}\|^2 + \|\mathbf{w}\|^2 - 2\|\mathbf{v}\|\|\mathbf{w}\|\cos\theta \quad \text{Apply property 4.}$$

$$\mathbf{v} \cdot (\mathbf{v} - \mathbf{w}) - \mathbf{w} \cdot (\mathbf{v} - \mathbf{w}) = \|\mathbf{v}\|^2 + \|\mathbf{w}\|^2 - 2\|\mathbf{v}\|\|\mathbf{w}\|\cos\theta \quad \text{Apply property 3.}$$

$$\mathbf{v} \cdot \mathbf{v} - \mathbf{v} \cdot \mathbf{w} - \mathbf{w} \cdot \mathbf{v} + \mathbf{w} \cdot \mathbf{w} = \|\mathbf{v}\|^2 + \|\mathbf{w}\|^2 - 2\|\mathbf{v}\|\|\mathbf{w}\|\cos\theta \quad \text{Apply property 3.}$$

$$\|\mathbf{v}\|^2 - 2(\mathbf{v} \cdot \mathbf{w}) + \|\mathbf{w}\|^2 = \|\mathbf{v}\|^2 + \|\mathbf{w}\|^2 - 2\|\mathbf{v}\|\|\mathbf{w}\|\cos\theta \quad \text{Apply properties 1 and 4.}$$

$$\mathbf{v} \cdot \mathbf{w} = \|\mathbf{v}\|\|\mathbf{w}\|\cos\theta \quad \text{Simplify the equation.}$$

$$\cos\theta = \frac{\mathbf{v} \cdot \mathbf{w}}{\|\mathbf{v}\|\|\mathbf{w}\|} \quad \text{Solve for } \cos\theta.$$

Classroom Example: p. 753
Exercise 36

> **EXAMPLE 2**   **Finding the Angle Between Two Vectors**
>
> Find the angle between $\mathbf{v} = \langle -4, 3\rangle$ and $\mathbf{w} = \langle -6, -2\rangle$. Round to the nearest tenth of a degree.
>
> **Solution:**
>
> $$\mathbf{v} \cdot \mathbf{w} = -4(-6) + 3(-2) = 24 - 6 = 18$$
>
> $$\|\mathbf{v}\| = \sqrt{(-4)^2 + (3)^2} = \sqrt{16 + 9} = \sqrt{25} = 5$$
>
> $$\|\mathbf{w}\| = \sqrt{(-6)^2 + (-2)^2} = \sqrt{36 + 4}$$
> $$= \sqrt{40} = 2\sqrt{10}$$
>
>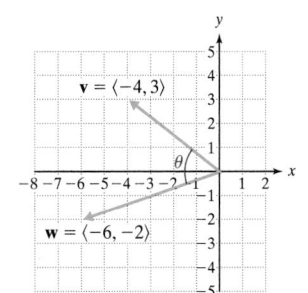
>
> $$\cos\theta = \frac{\mathbf{v} \cdot \mathbf{w}}{\|\mathbf{v}\|\|\mathbf{w}\|} = \frac{18}{5(2\sqrt{10})} = \frac{9}{5\sqrt{10}}$$
>
> $$\theta = \cos^{-1}\left(\frac{9}{5\sqrt{10}}\right) \approx 55.3°$$
>
> The angle between **v** and **w** is approximately 55.3°.
>
> > **Skill Practice 2**   Find the angle between $\mathbf{v} = \langle 5, -1\rangle$ and $\mathbf{w} = \langle -5, 12\rangle$. Round to the nearest tenth of a degree.

The relationship $\cos\theta = \dfrac{\mathbf{v} \cdot \mathbf{w}}{\|\mathbf{v}\|\|\mathbf{w}\|}$ can be written as $\mathbf{v} \cdot \mathbf{w} = \|\mathbf{v}\|\|\mathbf{w}\|\cos\theta$. This gives us an alternative method to compute the dot product of two vectors if their magnitudes and angle between them are known. Furthermore, since $\|\mathbf{v}\|$ and $\|\mathbf{w}\|$ are both positive, $\mathbf{v} \cdot \mathbf{w}$ and $\cos\theta$ must have the same sign. Therefore, the sign of $\mathbf{v} \cdot \mathbf{w}$ gives us information about the measure of θ.

- If $\mathbf{v} \cdot \mathbf{w} > 0$, then $\cos\theta > 0$, which implies that $0° \leq \theta < 90°$.
- If $\mathbf{v} \cdot \mathbf{w} < 0$, then $\cos\theta < 0$, which implies that $90° < \theta \leq 180°$.
- If $\mathbf{v} \cdot \mathbf{w} = 0$, then $\cos\theta = 0$, which implies that $\theta = 90°$.

If $\mathbf{v} \cdot \mathbf{w} = 0$, then $\theta = 90°$ and we say that **v** and **w** are **orthogonal**. Orthogonal vectors drawn "tail-to-tail" will form a right angle.

Answer
2. 123.9°

TIP The terms **orthogonal, perpendicular,** and **normal** all refer to objects where the measure of the angle between them is 90°. By convention, we say that two lines in a plane are perpendicular, two vectors in a plane are orthogonal, and a line or vector is normal to a plane.

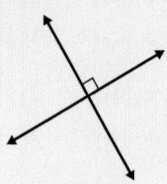

Perpendicular lines

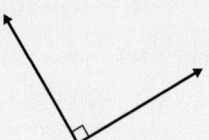

Orthogonal vectors

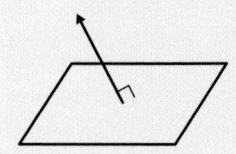

Vector normal to a plane

Orthogonal Vectors

Two vectors $\mathbf{v}$ and $\mathbf{w}$ are orthogonal if and only if $\mathbf{v} \cdot \mathbf{w} = 0$.

Classroom Examples: p. 753
Exercises 44, 48

EXAMPLE 3   Determining if Two Vectors Are Orthogonal

Determine if the given vectors are orthogonal.

a. $\mathbf{v} = 3\mathbf{i} - 6\mathbf{j}$ and $\mathbf{w} = 8\mathbf{i} + 4\mathbf{j}$ **b.** $\mathbf{s} = -2\mathbf{i} - 5\mathbf{j}$ and $\mathbf{r} = -4\mathbf{i} + 8\mathbf{j}$

Solution:

a. $\mathbf{v} \cdot \mathbf{w} = 3(8) + (-6)(4) = 24 - 24 = 0$
Since $\mathbf{v} \cdot \mathbf{w} = 0$, $\mathbf{v}$ and $\mathbf{w}$ are orthogonal.

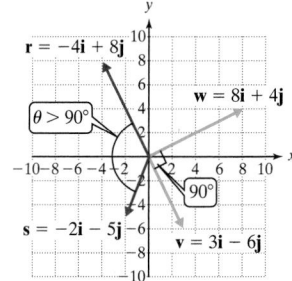

b. $\mathbf{s} \cdot \mathbf{r} = (-2)(-4) + (-5)(8) = 8 - 40 = -32$
Since $\mathbf{s} \cdot \mathbf{r} < 0$, we have $90° < \theta \leq 180°$ and $\mathbf{s}$ and $\mathbf{r}$ are *not* orthogonal.

Skill Practice 3   Determine if the given vectors are orthogonal.
a. $\mathbf{v} = 4\mathbf{i} + 3\mathbf{j}$ and $\mathbf{w} = 6\mathbf{i} - 10\mathbf{j}$ **b.** $\mathbf{s} = -2\mathbf{i} + \mathbf{j}$ and $\mathbf{t} = 11\mathbf{i} + 22\mathbf{j}$

Recall that two vectors $\mathbf{v}$ and $\mathbf{w}$ are parallel if $\mathbf{v} = c\mathbf{w}$ for some real number c. Likewise, if the angle θ between $\mathbf{v}$ and $\mathbf{w}$ is either 0° or 180°, then the vectors are parallel. If $\theta = 0°$, the vectors have the same direction, and if $\theta = 180°$ the vectors have opposite directions. Furthermore,

$$\text{If } \theta = 0°, \mathbf{v} \cdot \mathbf{w} = \|\mathbf{v}\|\|\mathbf{w}\|\cos 0° \qquad \text{If } \theta = 180°, \mathbf{v} \cdot \mathbf{w} = \|\mathbf{v}\|\|\mathbf{w}\|\cos 180°$$
$$= \|\mathbf{v}\|\|\mathbf{w}\|(1) \qquad\qquad\qquad = \|\mathbf{v}\|\|\mathbf{w}\|(-1)$$
$$= \|\mathbf{v}\|\|\mathbf{w}\| \qquad\qquad\qquad = -\|\mathbf{v}\|\|\mathbf{w}\|$$

This tells us that if $\mathbf{v}$ and $\mathbf{w}$ are parallel, then their dot product equals the product of their magnitudes or the opposite of the product of their magnitudes.

3. Decompose Vectors into Two Orthogonal Vectors

In many physical applications, it is important to find the contribution of a vector in a particular direction. For example, consider a car parked on a hill with angle of inclination θ. The force of gravity $\mathbf{F}$ pulls straight downward. However, $\mathbf{F}$ can be decomposed into two forces $\mathbf{F}_1$ and $\mathbf{F}_2$ acting parallel to the hill and orthogonal to the hill, respectively (Figure 7-47). $\mathbf{F}_1$ and $\mathbf{F}_2$ are called vector components of $\mathbf{F}$, and their sum is the original force vector ($\mathbf{F}_1 + \mathbf{F}_2 = \mathbf{F}$). The vector components of $\mathbf{F}$ are important because they determine the opposing forces needed to keep the car in place. The vector $-\mathbf{F}_1$ is the minimum amount of force needed by the brakes to keep the car from rolling down the hill. Force $\mathbf{F}_2$ tells us the amount of force that the tires must withstand against the ground.

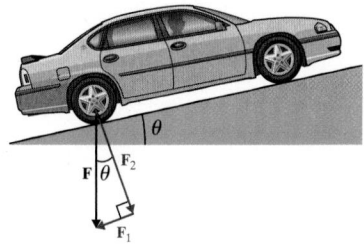

Figure 7-47

To decompose a vector into two orthogonal vectors, we use the idea of a vector projection. Consider two vectors $\mathbf{v}$ and $\mathbf{w}$ drawn with a common initial point A (Figures 7-48 and 7-49). Suppose we want to decompose $\mathbf{v}$ into vectors $\mathbf{v}_1$ and $\mathbf{v}_2$, parallel to $\mathbf{w}$ and orthogonal to $\mathbf{w}$, respectively. We call vector $\mathbf{v}_1$ the projection of $\mathbf{v}$ onto $\mathbf{w}$ and denote this by $\mathbf{v}_1 = \text{proj}_{\mathbf{w}}\mathbf{v}$.

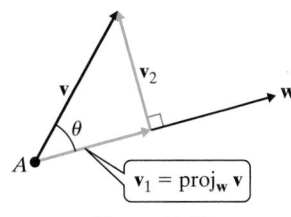

Figure 7-48

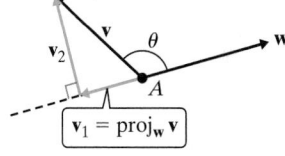

Figure 7-49

TIP To visualize the projection of **v** onto **w**, think of sunlight shining onto vector **w** at a right angle. The directed shadow cast by **v** onto **w** is the vector projection of **v** onto **w**.

$\mathbf{v}_1 = \text{proj}_{\mathbf{w}}\mathbf{v}$

We want to find the decomposition of $\mathbf{v}$ into two orthogonal vectors $\mathbf{v}_1$ and $\mathbf{v}_2$, where $\mathbf{v}_1 = \text{proj}_{\mathbf{w}}\mathbf{v}$, $\mathbf{v}_2$ is orthogonal to $\mathbf{w}$, and $\mathbf{v}_1 + \mathbf{v}_2 = \mathbf{v}$ (Figure 7-50). We first make two observations.

- $\mathbf{v}_1$ and $\mathbf{w}$ are parallel, so $\mathbf{v}_1 = c\mathbf{w}$ for some scalar c.
- $\mathbf{v}_2$ is orthogonal to $\mathbf{w}$, which means that $\mathbf{v}_2 \cdot \mathbf{w} = 0$.

Then since $\mathbf{v} = \mathbf{v}_1 + \mathbf{v}_2$,

$$
\begin{aligned}
\mathbf{v} \cdot \mathbf{w} &= (\mathbf{v}_1 + \mathbf{v}_2) \cdot \mathbf{w} \\
&= \mathbf{v}_1 \cdot \mathbf{w} + \mathbf{v}_2 \cdot \mathbf{w} \\
&= (c\mathbf{w}) \cdot \mathbf{w} + \mathbf{0} \\
&= c\|\mathbf{w}\|^2 .
\end{aligned}
$$

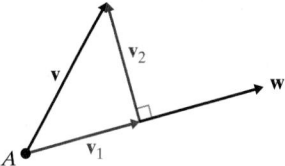

Figure 7-50

Solving the equation $\mathbf{v} \cdot \mathbf{w} = c\|\mathbf{w}\|^2$ for c gives $c = \dfrac{\mathbf{v} \cdot \mathbf{w}}{\|\mathbf{w}\|^2}$. Therefore,

$$
\mathbf{v}_1 = \text{proj}_{\mathbf{w}}\mathbf{v} = c\mathbf{w} = \frac{\mathbf{v} \cdot \mathbf{w}}{\|\mathbf{w}\|^2}\mathbf{w}.
$$

These results are summarized as follows.

Decomposition of a Vector into Orthogonal Vectors

If $\mathbf{v}$ and $\mathbf{w}$ are two nonzero vectors, then

1. The vector projection of $\mathbf{v}$ onto $\mathbf{w}$ is given by $\text{proj}_\mathbf{w}\mathbf{v} = \dfrac{\mathbf{v} \cdot \mathbf{w}}{\|\mathbf{w}\|^2}\mathbf{w}$.

2. To decompose $\mathbf{v}$ into vectors $\mathbf{v}_1$ and $\mathbf{v}_2$ parallel to $\mathbf{w}$ and orthogonal to $\mathbf{w}$, respectively, we have

$$\mathbf{v}_1 = \text{proj}_\mathbf{w}\mathbf{v} = \frac{\mathbf{v} \cdot \mathbf{w}}{\|\mathbf{w}\|^2}\mathbf{w} \quad \text{and} \quad \mathbf{v}_2 = \mathbf{v} - \mathbf{v}_1$$

Classroom Example: p. 753
Exercise 52

EXAMPLE 4   Decomposing a Vector into Orthogonal Vectors

Given $\mathbf{v} = \langle -5, 4 \rangle$ and $\mathbf{w} = \langle 2, 4 \rangle$,

a. Find $\text{proj}_\mathbf{w}\mathbf{v}$.

b. Find vectors $\mathbf{v}_1$ and $\mathbf{v}_2$ such that $\mathbf{v}_1$ is parallel to $\mathbf{w}$, $\mathbf{v}_2$ is orthogonal to $\mathbf{w}$, and $\mathbf{v}_1 + \mathbf{v}_2 = \mathbf{v}$.

Instructor Note:
Point out the difference in notation between $\mathbf{v} = a\mathbf{i} + b\mathbf{j}$ versus $\mathbf{v} = \mathbf{v}_1 + \mathbf{v}_2$.

Solution:

a. $\text{proj}_\mathbf{w}\mathbf{v} = \dfrac{\mathbf{v} \cdot \mathbf{w}}{\|\mathbf{w}\|^2}\mathbf{w} = \dfrac{(-5)(2) + (4)(4)}{(2)^2 + (4)^2}\langle 2, 4 \rangle$

$\qquad\qquad = \dfrac{6}{20}\langle 2, 4 \rangle$

$\qquad\qquad = \left\langle \dfrac{3}{5}, \dfrac{6}{5} \right\rangle$

$\mathbf{v}_1$ is the projection of $\mathbf{v}$ onto $\mathbf{w}$. Recall that for vector $\mathbf{w} = \langle b_1, b_2 \rangle$,

$\|\mathbf{w}\|^2 = \left(\sqrt{b_1{}^2 + b_2{}^2}\right)^2 = b_1{}^2 + b_2{}^2$

or equivalently

$\|\mathbf{w}\|^2 = \mathbf{w} \cdot \mathbf{w} = b_1{}^2 + b_2{}^2$.

b. $\mathbf{v}_1 = \text{proj}_\mathbf{w}\mathbf{v} = \left\langle \dfrac{3}{5}, \dfrac{6}{5} \right\rangle$

$\mathbf{v}_2 = \mathbf{v} - \mathbf{v}_1 = \langle -5, 4 \rangle - \left\langle \dfrac{3}{5}, \dfrac{6}{5} \right\rangle$

Use the relationship $\mathbf{v}_1 + \mathbf{v}_2 = \mathbf{v}$ to solve for $\mathbf{v}_2$.

$\qquad\qquad = \left\langle -\dfrac{25}{5}, \dfrac{20}{5} \right\rangle - \left\langle \dfrac{3}{5}, \dfrac{6}{5} \right\rangle$

$\qquad\qquad = \left\langle -\dfrac{28}{5}, \dfrac{14}{5} \right\rangle$

Skill Practice 4   Repeat Example 4 with $\mathbf{v} = \langle 7, -3 \rangle$ and $\mathbf{w} = \langle -2, -6 \rangle$.

We can check the solution to Example 4, by showing that vectors $\mathbf{v}_1$ and $\mathbf{v}_2$ meet all the necessary conditions.

1. $\mathbf{v}_1$ is parallel to $\mathbf{w}$ because $\mathbf{v}_1 = c\mathbf{w}$ for a constant c.

$$\mathbf{v}_1 = \left\langle \frac{3}{5}, \frac{6}{5} \right\rangle = \frac{3}{10}\langle 2, 4 \rangle = \frac{3}{10}\mathbf{w} \checkmark$$

Answers

4. a. $\text{proj}_\mathbf{w}\mathbf{v} = \left\langle -\dfrac{1}{5}, -\dfrac{3}{5} \right\rangle$

b. $\mathbf{v}_1 = \left\langle -\dfrac{1}{5}, -\dfrac{3}{5} \right\rangle$,

$\mathbf{v}_2 = \left\langle \dfrac{36}{5}, -\dfrac{12}{5} \right\rangle$

2. v_2 is orthogonal to **w** because $v_2 \cdot w = 0$.

$$v_2 \cdot w = \left\langle -\frac{28}{5}, \frac{14}{5} \right\rangle \cdot \langle 2, 4 \rangle$$

$$= \left(-\frac{28}{5} \right)(2) + \left(\frac{14}{5} \right)(4) = -\frac{56}{5} + \frac{56}{5} = 0 \checkmark \text{ Thus, } v \text{ and } w \text{ are orthogonal.}$$

3. $v_1 + v_2 \overset{?}{=} v$

$$\left\langle \frac{3}{5}, \frac{6}{5} \right\rangle + \left\langle -\frac{28}{5}, \frac{14}{5} \right\rangle \overset{?}{=} \langle -5, 4 \rangle$$

$$\left\langle -\frac{25}{5}, \frac{20}{5} \right\rangle \overset{?}{=} \langle -5, 4 \rangle \checkmark$$

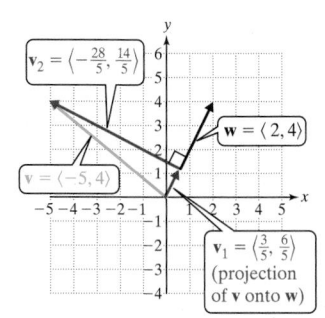

Classroom Example: p. 754
Exercise 56

> ### EXAMPLE 5   Finding a Component of Force in a Given Direction

A truck weighing 2000 lb is parked on a hill with a 10° incline.

a. Write the force vector **F** representing the weight of the truck through a single tire. Write **F** in terms of **i** and **j** and assume that the weight of the truck is evenly distributed among all four tires.

b. Find the component vector of **F** parallel to the hill. Round to 1 decimal place.

c. Find the magnitude of the force required by the brakes on each wheel to keep the truck from rolling down the hill.

Solution:

a. The weight of the truck is evenly distributed among all four tires and is a force acting straight downward. The magnitude of **F** is (2000 lb)/4. Thus, $F = -500j$ represents the force of gravity acting straight downward where a given tire meets the road.

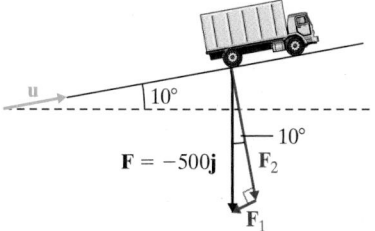

Figure 7-51

b. Let F_1 and F_2 be the components of **F** parallel and perpendicular to the hill, respectively. Let $u = \cos 10° i + \sin 10° j$ be the unit vector in the direction up the hill. The vector component of **F** parallel to the hill is the projection of **F** onto **u**, and is labeled as F_1 in Figure 7-51.

$$F_1 = \text{proj}_u F = \frac{F \cdot u}{\|u\|^2} u = \frac{(0i - 500j) \cdot (\cos 10° i + \sin 10° j)}{(1)^2}(\cos 10° i + \sin 10° j)$$

> Recall that a unit vector has a magnitude of 1.

$$= \frac{(0)(\cos 10°) + (-500)(\sin 10°)}{1}(\cos 10° i + \sin 10° j)$$

$$\approx -86.8(\cos 10° i + \sin 10° j)$$

$$\approx -85.5i - 15.1j$$

TIP As an alternative approach to part (c), use the right triangle formed by $\mathbf{F}$, $\mathbf{F_1}$, and $\mathbf{F_2}$ in Figure 7-51.

$$\|\mathbf{B}\| = \|\mathbf{F_1}\| = \|\mathbf{F}\|\sin 10°$$
$$= 500\sin 10° \approx 86.8 \text{ lb}$$

c. The brakes on each wheel represent the *opposing* force to $\mathbf{F_1}$. That is, the brakes provide a force $\mathbf{B} = -\mathbf{F_1} = -(-85.5\mathbf{i} - 15.1\mathbf{j}) = 85.5\mathbf{i} + 15.1\mathbf{j}$.

$$\|\mathbf{B}\| = \sqrt{(85.5)^2 + (15.1)^2} \approx 86.8 \text{ lb}$$

The brakes on each wheel must exert a force of approximately 86.8 lb to keep the truck from rolling down the hill.

Skill Practice 5   Repeat Example 5 with a 1400-lb SUV parked on a hill with an incline of 12°.

Point of Interest

A sailboat is propelled forward by the force of wind on the sail and by the force of water against the keel ("fin" that extends below the boat to prevent sideways "slippage"). The force of wind on the sail can be broken down into two vectors: one acting parallel to the sail and one acting perpendicular to the sail. The component acting parallel to the sail (called the drag component) "slips" by and does not contribute to the forward motion. The component force of the wind acting perpendicular to the sail (shown in red) contributes to the forward motion of the boat.

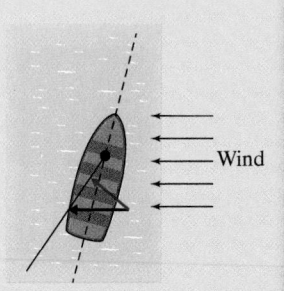

4. Use Vectors to Compute Work

Suppose a man lifts a 30-lb weight straight upward 4 ft. To quantify the amount of effort for this task, we can compute the amount of work done. In physics, the work W done by a constant force $\mathbf{F}$ in moving an object from point A to point B along the line of motion is given by

$$W = \left(\begin{array}{c}\text{Magnitude of force in}\\\text{the direction of motion}\end{array}\right)\left(\begin{array}{c}\text{Distance the}\\\text{object is moved}\end{array}\right)$$

Since a force of 30 lb is required to lift the weight, the work done raising it 4 ft is

$$W = (30 \text{ lb})(4 \text{ ft}) = 120 \text{ ft·lb}$$

Notice that work is a scalar quantity. In the U.S. customary units of measurement, the common unit for work is the foot-pound (ft·lb). In the metric system, the Newton (N) is a common unit of force, and the Newton·meter (N·m) is the associated unit of work.

It is important to note that if a force vector $\mathbf{F}$ is *not* directed along the line of motion, then only the component of force in the direction of motion is applied to the work done (Figure 7-52). Therefore, if $\mathbf{D}$ is the displacement vector of an object moved in a straight line from point A to point B by a force $\mathbf{F}$, the work done is given by

Magnitude of force in the direction of $\mathbf{D}$ Distance moved

$$W = \|\text{proj}_\mathbf{D}\mathbf{F}\|\|\mathbf{D}\| = \left(\frac{\mathbf{F} \cdot \mathbf{D}}{\|\mathbf{D}\|^2}\|\mathbf{D}\|\right)\|\mathbf{D}\|$$
$$= \mathbf{F} \cdot \mathbf{D}$$

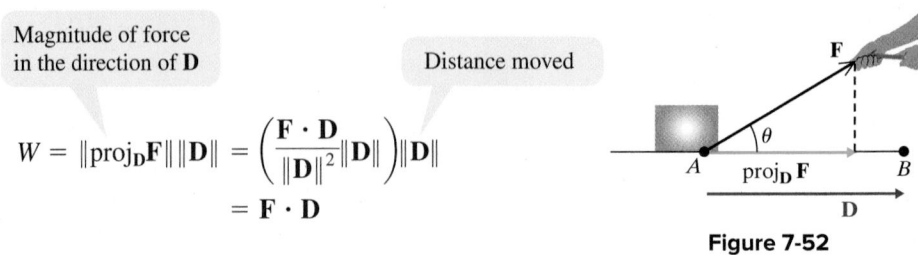

Figure 7-52

Definition of Work

If **D** is the displacement vector of an object in moving the object in a straight line from points A to B under a constant force **F**, then the work W done is computed by
1. $W = \mathbf{F} \cdot \mathbf{D}$.
2. $W = \|\text{proj}_{\mathbf{D}}\mathbf{F}\|\|\mathbf{D}\|$ or $W = \|\mathbf{F}\|\|\mathbf{D}\|\cos\theta$, where θ is the angle between **F** and **D**.

Classroom Example: p. 754
Exercise 72

EXAMPLE 6   Computing Work

Find the work done by a force $\mathbf{F} = 4\mathbf{i} + 9\mathbf{j}$ (in N) in moving an object in a straight line from point $A(1, 3)$ to point $B(4, 8)$. Assume that the units in the coordinate plane are in meters.

Solution:

The displacement vector for the object is $\mathbf{D} = \overrightarrow{AB} = (4 - 1)\mathbf{i} + (8 - 3)\mathbf{j} = 3\mathbf{i} + 5\mathbf{j}$.

The work done is the dot product of the force vector and displacement vector.
$W = \mathbf{F} \cdot \mathbf{D} = (4\mathbf{i} + 9\mathbf{j}) \cdot (3\mathbf{i} + 5\mathbf{j}) = (4)(3) + (9)(5) = 12 + 45 = 57$

The amount of work is 57 N·m.

> **Skill Practice 6**   Find the work done by a force $\mathbf{F} = 2\mathbf{i} + 7\mathbf{j}$ (in N) in moving an object in a straight line from point $A(-1, 5)$ to point $B(9, 11)$. Assume that the units in the coordinate plane are in meters.

Classroom Example: p. 755
Exercise 76

EXAMPLE 7   Computing Work

A man pushes a lawn mower 150 ft across a yard on a horizontal path. He exerts a constant force of 30 lb directed downward at an angle of 40° with the horizontal. Find the work done. Round to the nearest ft·lb.

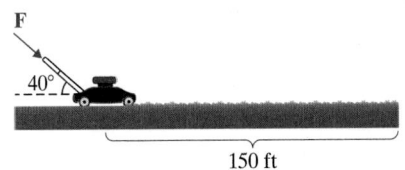

Instructor Note:
Remind students that the mathematical definition of work is different than the everyday usage. If a person holds a 100-lb weight for 5 min, the person might feel tired, but no mathematical work was done. This is because the weight was not moved.

Solution:

Let A and B designate the initial and final positions of the lawn mower. Let **D** represent the displacement vector from A to B. In Figure 7-53, we have drawn the force vector **F** with initial point at A. The angle θ between the vectors is 40°.

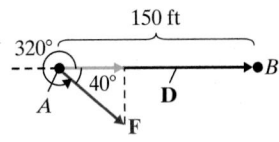

Figure 7-53

Using the projection of **F** onto **D** (shown in blue), we have

$$W = \|\text{proj}_{\mathbf{D}}\mathbf{F}\|\|\mathbf{D}\|$$
$$= \|\mathbf{F}\|\|\mathbf{D}\|\cos\theta$$
$$= (30 \text{ lb})(150 \text{ ft})(\cos 40°)$$
$$= 4500\cos 40°$$
$$\approx 3447 \text{ ft·lb}$$

> **TIP** As an alternative approach, in Example 7 we can write
> $\mathbf{F} = 30\cos 320°\mathbf{i} + 30\sin 320°\mathbf{j}$ and
> $\mathbf{D} = 150\mathbf{i} + 0\mathbf{j}$
>
> Then $W = \mathbf{F} \cdot \mathbf{D}$
> $= (30\cos 320°)(150) + (30\sin 320°)(0)$
> ≈ 3447 ft·lb

Answer
6. 62 N·m

> **Skill Practice 7**   A dog pulls a sled 200 ft across a yard on a horizontal path. She exerts a constant force of 25 lb directed upward 15° from the horizontal. Find the work done. Round to the nearest ft·lb.

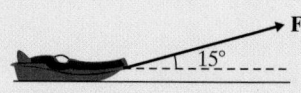

When the force and displacement vectors are in the same direction, the work done in moving the object is positive. If a force is applied in the opposite direction to a moving object (such as the force applied to brake a skidding car), the work done by the force is negative. In this section, we only address situations with positive work.

Answer

7. 4830 ft·lb

SECTION 7.5   Practice Exercises

Prerequisite Review

R.1. Simplify. $\sqrt{5^2 + 12^2}$ 13

For Exercises R.2–R.3, use a calculator to approximate the degree measure of θ to 1 decimal place.

R.2. $\cos \theta = \dfrac{5}{6}$ and $0° \le \theta \le 90°$ 33.6°

R.3. $\cos \theta = -\dfrac{8}{11}$ and $90° \le \theta \le 180°$ 136.7°

Concept Connections

1. The dot product of $\mathbf{v} = \langle a_1, b_1 \rangle$ and $\mathbf{w} = \langle a_2, b_2 \rangle$ equals _____$a_1 a_2 + b_1 b_2$_____, and the result is a (scalar/vector).   scalar

2. If θ is the angle between two nonzero vectors $\mathbf{v}$ and $\mathbf{w}$, then the cosine of θ can be found from the dot product of $\mathbf{v}$ and $\mathbf{w}$ as $\cos \theta =$ _____. $\dfrac{\mathbf{v} \cdot \mathbf{w}}{\|\mathbf{v}\|\,\|\mathbf{w}\|}$

3. If θ is the angle between two nonzero vectors $\mathbf{v}$ and $\mathbf{w}$, and $\mathbf{v} \cdot \mathbf{w} = 0$, then $\theta =$ _____90_____°.

4. Two vectors are said to be _____orthogonal_____ if and only if they meet at a right angle.

5. The projection of a nonzero vector $\mathbf{v}$ onto a nonzero vector $\mathbf{w}$ is given by $\text{proj}_{\mathbf{w}}\mathbf{v} =$ _____. $\dfrac{\mathbf{v} \cdot \mathbf{w}}{\|\mathbf{w}\|^2}\mathbf{w}$

6. Suppose that $\mathbf{D}$ is the displacement vector of an object when the object is moved in a straight line from points A to B under a constant force $\mathbf{F}$. Then the work done is given by $W =$ _____.

$\|\text{proj}_{\mathbf{D}}\mathbf{F}\|\,\|\mathbf{D}\|$, or $\mathbf{F} \cdot \mathbf{D}$, or $\|\mathbf{F}\|\,\|\mathbf{D}\|\cos\theta$, where θ is the angle between $\mathbf{F}$ and $\mathbf{D}$.

Objective 1: Compute Dot Product

For Exercises 7–14, for the given vectors, find

a. $\mathbf{v} \cdot \mathbf{w}$ **b.** $\mathbf{v} \cdot \mathbf{v}$ **c.** $\mathbf{w} \cdot \mathbf{w}$ (See Example 1)

7. $\mathbf{v} = \langle -3, 5 \rangle$, $\mathbf{w} = \langle 10, -8 \rangle$ a. −70   b. 34   c. 164

8. $\mathbf{v} = \langle 6, -7 \rangle$, $\mathbf{w} = \langle -1, -4 \rangle$ a. 22   b. 85   c. 17

9. $\mathbf{v} = \left\langle \dfrac{3}{8}, \dfrac{2}{3} \right\rangle$, $\mathbf{w} = \left\langle -\dfrac{1}{6}, -\dfrac{3}{4} \right\rangle$ a. $-\dfrac{9}{16}$ b. $\dfrac{337}{576}$ c. $\dfrac{85}{144}$

10. $\mathbf{v} = \left\langle -\dfrac{1}{3}, -\dfrac{4}{5} \right\rangle$, $\mathbf{w} = \left\langle \dfrac{1}{3}, -\dfrac{5}{8} \right\rangle$ a. $\dfrac{7}{18}$ b. $\dfrac{169}{225}$ c. $\dfrac{289}{576}$

11. $\mathbf{v} = 6\mathbf{i}$, $\mathbf{w} = -3\mathbf{j}$ a. 0   b. 36   c. 9

12. $\mathbf{v} = 2\mathbf{j}$, $\mathbf{w} = 14\mathbf{i}$ a. 0   b. 4   c. 196

13. $\mathbf{v} = 2.1\mathbf{i} - 6.8\mathbf{j}$, $\mathbf{w} = 0.4\mathbf{i} - 0.3\mathbf{j}$
a. 2.88   b. 50.65   c. 0.25

14. $\mathbf{v} = -0.5\mathbf{i} + 1.6\mathbf{j}$, $\mathbf{w} = -0.1\mathbf{i} + 2.3\mathbf{j}$
a. 3.73   b. 2.81   c. 5.3

For Exercises 15–18, use the dot product to find the magnitude of the vector.

15. $\mathbf{v} = 4\mathbf{i} - 8\mathbf{j}$ $4\sqrt{5}$

16. $\mathbf{w} = -18\mathbf{i} + 3\mathbf{j}$ $3\sqrt{37}$

17. $\mathbf{r} = \langle 5, -3 \rangle$ $\sqrt{34}$

18. $\mathbf{s} = \langle 7, -4 \rangle$ $\sqrt{65}$

For Exercises 19–28, use vectors $u = \langle -4, 1 \rangle$, $v = \langle 5, -2 \rangle$, and $w = \langle 0, 6 \rangle$ to perform the indicated operation. Then determine whether the result is a scalar or a vector.

19. $4v \cdot w$ -48; scalar

20. $3(u \cdot v)$ -66; scalar

21. $v \cdot v$ 29; scalar

22. $w \cdot w$ 36; scalar

23. $(v \cdot u)w$ $\langle 0, -132 \rangle$; vector

24. $(u \cdot w)v$ $\langle 30, -12 \rangle$; vector

25. $\|w\|u$ $\langle -24, 6 \rangle$; vector

26. $\|u\|w$ $\langle 0, 6\sqrt{17} \rangle$; vector

27. $u \cdot (v + w)$ -16; scalar

28. $v \cdot (w + u)$ -34; scalar

For Exercises 29–32, let $v = \langle a_1, b_1 \rangle$, $w = \langle a_2, b_2 \rangle$, $u = \langle a_3, b_3 \rangle$, and c be a real number. Prove the given statement.

29. $(cv) \cdot w = c(v \cdot w)$

30. $0 \cdot v = 0$ $0 \cdot v = \langle 0, 0 \rangle \cdot \langle a_1, b_1 \rangle = 0 \cdot a_1 + 0 \cdot b_1 = 0$

31. $c(v \cdot w) = cv \cdot w$

32. $v \cdot (w + u) = v \cdot w + v \cdot u$

Objective 2: Find the Angle Between Two Vectors

33. If θ is the angle between two nonzero vectors v and w, and $v \cdot w < 0$, what is the range of θ? Give the answer in degree measure. $90° < \theta \le 180°$

34. If θ is the angle between two nonzero vectors v and w, and $v \cdot w > 0$, what is the range of θ? Give the answer in degree measure. $0° \le \theta < 90°$

For Exercises 35–40, find the angle θ between v and w. If necessary, round to the nearest tenth of a degree. (See Example 2)

35. $v = -8i + 6j$, $w = 4i + 18j$ $\theta = 65.7°$

36. $v = 12i - 5j$, $w = -3i - 4j$ $\theta = 104.3°$

37. $v = \langle 1, 7 \rangle$, $w = \langle -12, -8 \rangle$ $\theta = 131.8°$

38. $v = \langle 8, 3 \rangle$, $w = \langle 4, -3 \rangle$ $\theta = 57.4°$

39. $v = \cos 45°i + \sin 45°j$ and $w = \cos 60°i + \sin 60°j$ $\theta = 15°$

40. $v = \cos 150°i + \sin 150°j$ and $w = \cos 225°i + \sin 225°j$ $\theta = 75°$

41. If $\|v\| = 10$ and $\|w\| = 6$ and the angle between v and w is $30°$, find $v \cdot w$. $30\sqrt{3}$

42. If $\|v\| = 2.6$ and $\|w\| = 8$ and the angle between v and w is $60°$, find $v \cdot w$. 10.4

For Exercises 43–50, determine if the given vectors are orthogonal, parallel, or neither. (See Example 3)

43. $r = \langle -12, 6 \rangle$, $s = \langle -8, -16 \rangle$ Orthogonal

44. $t = \langle -6, 18 \rangle$, $s = \langle 9, 3 \rangle$ Orthogonal

45. $w = \dfrac{1}{5}i - \dfrac{1}{4}j$, $v = \dfrac{15}{4}i + 3j$ Orthogonal

46. $b = \dfrac{5}{2}i + \dfrac{25}{6}j$, $a = -\dfrac{10}{9}i + \dfrac{2}{3}j$ Orthogonal

47. $p = \langle 2, 6 \rangle$, $q = \langle -1, -3 \rangle$ Parallel

48. $k = \left\langle -\dfrac{3}{2}, 2 \right\rangle$, $s = \langle 6, -8 \rangle$ Parallel

49. $v = 3i - j$, $w = i - 3j$ Neither

50. $z = 14i + 7j$, $t = -7i - 14j$ Neither

Objective 3: Decompose Vectors into Two Orthogonal Vectors

51. Given $v = \langle -3, 6 \rangle$ and $w = \langle 1, 3 \rangle$,

   a. Find $\text{proj}_w v$. (See Example 4)

   b. Find vectors v_1 and v_2 such that v_1 is parallel to w, v_2 is orthogonal to w, and $v_1 + v_2 = v$.

   c. Using the results from part (b) show that v_1 is parallel to w by finding a constant c such that $v_1 = cw$.

   d. Show that v_2 is orthogonal to w.

   e. Show that $v_1 + v_2 = v$.

52. Given $v = \langle 4, -3 \rangle$ and $w = \langle 2, -2 \rangle$,

   a. Find $\text{proj}_w v$.

   b. Find vectors v_1 and v_2 such that v_1 is parallel to w, v_2 is orthogonal to w, and $v_1 + v_2 = v$.

   c. Using the results from part (b) show that v_1 is parallel to w by finding a constant c such that $v_1 = cw$.

   d. Show that v_2 is orthogonal to w.

   e. Show that $v_1 + v_2 = v$.

53. Given $v = 4i + 9j$ and $w = 2i - j$,

   a. Find $\text{proj}_w v$.

   b. Find vectors v_1 and v_2 such that v_1 is parallel to w, v_2 is orthogonal to w, and $v_1 + v_2 = v$.

54. Given $v = -8i - 4j$ and $w = -6i + 4j$,

   a. Find $\text{proj}_w v$.

   b. Find vectors v_1 and v_2 such that v_1 is parallel to w, v_2 is orthogonal to w, and $v_1 + v_2 = v$.

55. A boat and trailer weighing a total of 450 lb are parked on a boat ramp with an 18° angle of inclination. Assume that the weight of the boat and trailer is evenly distributed between two wheels. (**See Example 5**)

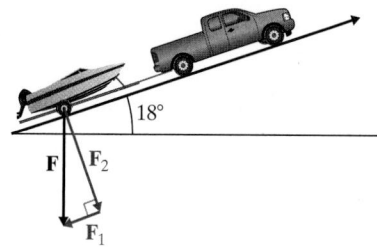

a. Write the force vector **F** in terms of **i** and **j** representing the weight of the boat and trailer for a single tire.

b. Find the component vector of **F** parallel to the ramp. Round values to 1 decimal place.

c. Find the magnitude of the force needed to keep the trailer from moving down the ramp. Round to the nearest pound.

57. Refer to Exercise 55.

a. Find the component vector of **F** perpendicular to the ramp. Round values to 1 decimal place.

b. Find the magnitude of the force that the ramp makes against each tire. Round to the nearest pound.

56. A forklift is used to offload freight from a delivery truck. Together the forklift and its contents weigh 800 lb, and the weight is evenly distributed among four wheels. The ramp is inclined 14° from the horizontal.

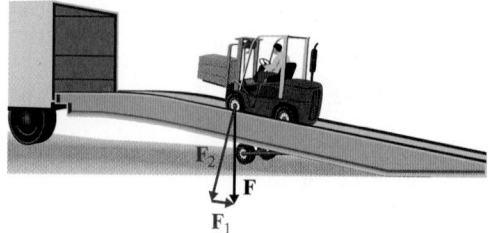

a. Write the force vector **F** in terms of **i** and **j** representing the weight against a single tire.

b. Find the component vector of **F** parallel to the ramp. Round values to 1 decimal place.

c. Find the magnitude of the force needed to keep the forklift from rolling down the ramp. Round to the nearest pound.

58. Refer to Exercise 56.

a. Find the component vector of **F** perpendicular to the ramp. Round values to 1 decimal place.

b. Find the magnitude of the force that the ramp makes against each tire. Round to the nearest pound.

Objective 4: Use Vectors to Compute Work

59. a. Find the amount of work done in lifting a 50-lb weight upward 5 ft.   250 ft·lb

b. If 200 N of force is applied in the direction of motion in moving an object 10 m, how much work is done?   2000 N·m

60. a. Find the amount of work done pulling an object horizontally 10 ft using a force of 40 lb in the direction of motion.   400 ft·lb

b. If 2250 N of force is applied in the direction of motion in moving an object 50 m, how much work is done?   112,500 N·m

For Exercises 61–64, find the work done by the given force F acting in the direction from point *A* to point *B*. Assume that the units in the coordinate plane are in meters.

61. $A(3, -1)$, $B(8, 11)$; $\|\mathbf{F}\| = 400$ N   5200 N·m

62. $A(-3, -2)$, $B(5, 4)$; $\|\mathbf{F}\| = 250$ N   2500 N·m

63. $A(12, 5)$, $B(-3, 10)$; $\|\mathbf{F}\| = 38$ N $190\sqrt{10}$ N·m

64. $A(17, -7)$, $B(27, 1)$; $\|\mathbf{F}\| = 18$ N $36\sqrt{41}$ N·m

For Exercises 65–68, find the work *W* done by a force F in moving an object in a straight line given by the displacement vector D. (See Example 6)

65. $\mathbf{F} = (40\mathbf{i} - 15\mathbf{j})$ lb; $\mathbf{D} = (30\mathbf{i} + 10\mathbf{j})$ ft $W = 1050$ ft·lb

66. $\mathbf{F} = (220\mathbf{i} + 350\mathbf{j})$ lb; $\mathbf{D} = (-12\mathbf{i} + 16\mathbf{j})$ ft $W = 2960$ ft·lb

67. $\mathbf{F} = (-26\mathbf{i} + 32\mathbf{j})$ N; $\mathbf{D} = (100\mathbf{i} + 120\mathbf{j})$ m $W = 1240$ N·m

68. $\mathbf{F} = (6\mathbf{i} + 3\mathbf{j})$ N; $\mathbf{D} = (14\mathbf{i} - 8\mathbf{j})$ m $W = 60$ N·m

For Exercises 69–72, find the work done by a force F (in lb) in moving an object in a straight line from point *A* to point *B*. Assume that the units in the coordinate plane are in feet. (See Example 6)

69. $A(0, 0)$, $B(2, 6)$; $\mathbf{F} = 3\mathbf{i} + 7\mathbf{j}$ 48 ft·lb

70. $A(0, 0)$, $B(-5, 4)$; $\mathbf{F} = -2\mathbf{i} + 10\mathbf{j}$ 50 ft·lb

71. $A(20, 35)$, $B(32, 57)$; $\mathbf{F} = 150\mathbf{i} + 200\mathbf{j}$ 6200 ft·lb

72. $A(-14, -20)$, $B(8, -35)$; $\mathbf{F} = 40\mathbf{i} + 26\mathbf{j}$ 490 ft·lb

73. Find the amount of work done pulling an object horizontally 50 ft if a force of 60 lb is directed 30° above the horizontal. $1500\sqrt{3}$ ft·lb

74. Find the amount of work done if an object is pushed horizontally 70 m by a force of 25 N directed 60° above the horizontal.   875 N·m

75. A tree removal service hooks a chain around the stump of a tree and attaches the other end of the chain to a pickup truck. The truck drags the stump horizontally 40 m to a wood chipper. The chain is directed upward 32° from the horizontal and the tension in the chain is 15,000 N. Find the amount of work done. Round to the nearest N·m. (**See Example 7**)   508,829 N·m

76. A child exerts a force of 28 N on the handle of a small wagon. If the handle of the wagon is directed upwards 18° from the horizontal, find the amount of work done in moving the wagon 20 m horizontally. Round to the nearest N·m.   533 N·m

77. A force of 16 lb is used to pull a block up a ramp. If the ramp is inclined 18° above the horizontal, and the force is directed 30° above the ramp, find the work done moving the block 15 ft along the ramp. $120\sqrt{3}$ ft·lb

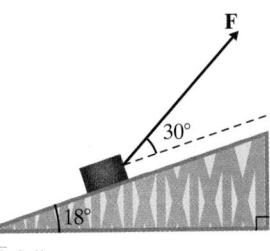

78. A force of 28 lb is used to pull a block up a ramp. If the ramp is inclined 10° above the horizontal, and the force is directed 45° above the ramp, find the work done moving the block 20 ft along the ramp. $280\sqrt{2}$ ft·lb

Mixed Exercises

79. Given $\mathbf{v} = 22\mathbf{i} - 55\mathbf{j}$ and $\mathbf{w} = -4\mathbf{i} + 10\mathbf{j}$,

   a. Find the angle between $\mathbf{v}$ and $\mathbf{w}$.   180°

   b. Are the vectors parallel? If yes, find a real number c such that $\mathbf{v} = c\mathbf{w}$.   Yes; $c = -5.5$

80. Given $\mathbf{v} = -21\mathbf{i} + 33.6\mathbf{j}$ and $\mathbf{w} = 5\mathbf{i} - 8\mathbf{j}$,

   a. Find the angle between $\mathbf{v}$ and $\mathbf{w}$.   180°

   b. Are the vectors parallel? If yes, find a real number c such that $\mathbf{v} = c\mathbf{w}$.   Yes; $c = -4.2$

81. Given $r = \langle -3, 4 \rangle$ and $s = \langle 9, -12 \rangle$,

   a. Find the product $\|\mathbf{r}\|\|\mathbf{s}\|$.   75

   b. Find $\mathbf{r} \cdot \mathbf{s}$.   −75

   c. Based on the results of parts (a) and (b), what do you know about the two vectors?   The vectors are parallel.

82. Given $c = \langle 5, -12 \rangle$ and $d = \langle 10, -24 \rangle$,

   a. Find the product $\|\mathbf{c}\|\|\mathbf{d}\|$.   338

   b. Find $\mathbf{c} \cdot \mathbf{d}$.   338

   c. Based on the results of parts (a) and (b), what do you know about the two vectors?   The vectors are parallel.

83. If $\text{proj}_{\mathbf{w}}\mathbf{v} = \mathbf{v}$, what do you know about $\mathbf{v}$ and $\mathbf{w}$?
   $\mathbf{v}$ and $\mathbf{w}$ are parallel.

84. If $\text{proj}_{\mathbf{w}}\mathbf{v} = \mathbf{0}$, what do you know about $\mathbf{v}$ and $\mathbf{w}$?
   $\mathbf{v}$ and $\mathbf{w}$ are orthogonal.

85. Given $\mathbf{v} = 4\mathbf{i} - 10\mathbf{j}$,

   a. Find two vectors parallel to $\mathbf{v}$, one in the same direction as $\mathbf{v}$ and one in the opposite direction as $\mathbf{v}$. Answers will vary.
   For example: $\mathbf{u} = 2\mathbf{i} - 5\mathbf{j}$ and $\mathbf{w} = -4\mathbf{i} + 10\mathbf{j}$

   b. Find two vectors orthogonal to $\mathbf{v}$. Answers will vary.
   For example: $\mathbf{u} = 5\mathbf{i} + 2\mathbf{j}$ and $\mathbf{w} = -10\mathbf{i} - 4\mathbf{j}$

86. Given $\mathbf{v} = -3\mathbf{i} - 9\mathbf{j}$,

   a. Find two vectors parallel to $\mathbf{v}$, one in the same direction as $\mathbf{v}$ and one in the opposite direction as $\mathbf{v}$. Answers will vary.
   For example: $\mathbf{u} = -\mathbf{i} - 3\mathbf{j}$ and $\mathbf{w} = 6\mathbf{i} + 18\mathbf{j}$

   b. Find two vectors orthogonal to $\mathbf{v}$. Answers will vary.
   For example: $\mathbf{u} = -3\mathbf{i} + \mathbf{j}$ and $\mathbf{w} = 9\mathbf{i} - 3\mathbf{j}$

The dot product of vectors can be used in business applications. For Exercises 87–88, find the dot product and interpret the results.

87. The components of $\mathbf{n} = \langle 500, 330 \rangle$ represent the number of T-shirts and hats, respectively, in the inventory of a surf shop. The components of $\mathbf{p} = \langle 15, 9 \rangle$ represent the price (in $) per T-shirt and hat, respectively. Find $\mathbf{n} \cdot \mathbf{p}$ and interpret the result.
$10,470; The dot product represents the total revenue if the entire inventory were sold.

88. The components of $\mathbf{n} = \langle 4, 2 \rangle$ represent the number of tacos and drinks, respectively, that a restaurant patron had for lunch. The components of $\mathbf{c} = \langle 120, 90 \rangle$ represent the number of calories per taco and number of calories per drink, respectively. Find $\mathbf{n} \cdot \mathbf{c}$ and interpret the result.   660; The dot product represents the total number of calories for the meal.

89. A force of 40 lb is used to pull a block up a ramp. If the ramp is inclined 10° above the horizontal, and the force is directed 25° from the horizontal, find the work done moving the block 25 ft along the ramp. Round to the nearest unit.   966 ft·lb

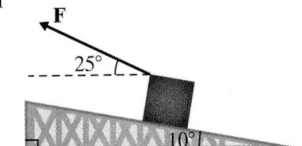

90. A force of 65 lb is used to pull a block up a ramp. If the ramp is inclined 18° above the horizontal, and the force is directed 38° from the horizontal, find the work done moving the block 12 ft along the ramp. Round to the nearest unit.   733 ft·lb

91. Suppose that the contact point between the femur (large bone in the thigh) and tibia (supporting bone from lower leg) meet at point A. The planar surface l between the two bones is directed $17°$ downward from the horizontal, and the individual's body weight is 200 lb directed downward. Rounding to the nearest tenth of a pound,

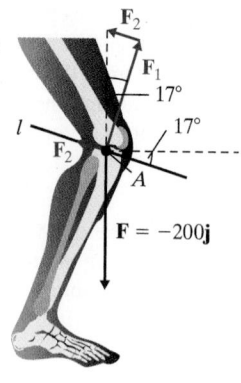

a. Find the magnitude of the force $\mathbf{F}_1$ that the tibia must exert normal to l to oppose the weight.   191.3 lb

b. Find the magnitude of force $\mathbf{F}_2$ within the knee joint that prevents the bones from slipping.   58.5 lb

92. The individual in Exercise 91 is directed to lose weight by his physician. After 6 months of exercise and dietary changes, his new weight is 175 lb. Recalculate parts (a) and (b) with the new body weight.
a. 167.4 lb   **b.** 51.2 lb

Write About It

93. Explain how the dot product can be used to determine whether the angle between two nonzero vectors is less than $90°$ or greater than $90°$.

94. Explain what is meant by decomposing a nonzero vector $\mathbf{v}$ into orthogonal vectors.   Vector $\mathbf{v}$ is broken down into a sum of vectors $\mathbf{v}_1 + \mathbf{v}_2$ where $\mathbf{v}_1$ and $\mathbf{v}_2$ are orthogonal.

Expanding Your Skills

95. Prove that $\|\mathbf{v} - \mathbf{w}\|^2 = \|\mathbf{v}\|^2 + \|\mathbf{w}\|^2 - 2\mathbf{v} \cdot \mathbf{w}$.

96. Prove that $\|\mathbf{v} + \mathbf{w}\|^2 - \|\mathbf{v} - \mathbf{w}\|^2 = 4(\mathbf{v} \cdot \mathbf{w})$.

97. Prove that $\|\mathbf{v} + \mathbf{w}\|^2 + \|\mathbf{v} - \mathbf{w}\|^2 = 2(\|\mathbf{v}\|^2 + \|\mathbf{w}\|^2)$.

98. If $\mathbf{v}$ and $\mathbf{w}$ are vectors with the same magnitude, show that $\mathbf{v} + \mathbf{w}$ and $\mathbf{v} - \mathbf{w}$ are orthogonal.

A vector $\mathbf{v}$ in a plane is a line segment with a specified direction, where the component form is given by two coordinates $\langle a, b \rangle$. Similarly, we may define a vector $\mathbf{w}$ in three-dimensional space as a line segment in space with a specified direction where the component form is given by three coordinates $\langle a, b, c \rangle$.

For example, a vector $\mathbf{w}$ from the origin to a point $P(2, 3, 3)$ is given in component form as $\mathbf{w} = \langle 2, 3, 3 \rangle$ or, in terms of the unit vectors $\mathbf{i} = \langle 1, 0, 0 \rangle$, $\mathbf{j} = \langle 0, 1, 0 \rangle$, and $\mathbf{k} = \langle 0, 0, 1 \rangle$, as $\mathbf{w} = 2\mathbf{i} + 3\mathbf{j} + 3\mathbf{k}$. Use this convention for Exercises 99–100.

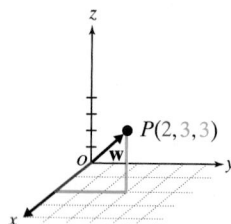

99. If $\mathbf{v} = \langle a_1, b_1, c_1 \rangle$ and $\mathbf{w} = \langle a_2, b_2, c_2 \rangle$, the **dot product** $\mathbf{v} \cdot \mathbf{w}$ is defined as $\mathbf{v} \cdot \mathbf{w} = a_1 a_2 + b_1 b_2 + c_1 c_2$. Evaluate $\mathbf{v} \cdot \mathbf{w}$ for
a. $\mathbf{v} = \langle -2, 1, 4 \rangle$, $\mathbf{w} = \langle 3, -1, 5 \rangle$ 13
b. $\mathbf{v} = -6\mathbf{j} + 3\mathbf{k}$, $\mathbf{w} = 10\mathbf{i} + 12\mathbf{k}$ 36

100. The magnitude of $\mathbf{v} = \langle a, b, c \rangle$ is given by $\|\mathbf{v}\| = \sqrt{a^2 + b^2 + c^2}$. Find the magnitude of each vector.
a. $\mathbf{v} = \langle -8, 2, 12 \rangle$ $2\sqrt{53}$
b. $\mathbf{w} = 3\mathbf{i} - 9\mathbf{j} + 12\mathbf{k}$ $3\sqrt{26}$

CHAPTER 7   KEY CONCEPTS

	Reference
SECTION 7.1 Polar Coordinates	
A **polar coordinate system** consists of a fixed point O called the **pole** (or origin), and a ray called the **polar axis,** with endpoint at the pole. Each point P in the plane is defined by an ordered pair of the form (r, θ), where r is the directed distance from the pole to P, and θ is a directed angle from the polar axis to line $\overleftrightarrow{OP}$.	p. 686
A point in a plane can be described by an ordered pair (x, y) in rectangular coordinates or by an ordered pair (r, θ) in polar coordinates. $x = r\cos\theta \qquad r^2 = x^2 + y^2$ $y = r\sin\theta \qquad \tan\theta = \dfrac{y}{x}$ for $x \neq 0$ 	p. 688
A **polar equation** is an equation described in a polar coordinate system with variables r and θ. Convert between rectangular coordinates and polar coordinates by using the relationships $x = r\cos\theta$, $y = r\sin\theta$, $r^2 = x^2 + y^2$, and $\tan\theta = \dfrac{y}{x}$ for $x \neq 0$.	p. 690

	Reference
SECTION 7.2 Graphs of Polar Equations	
A **solution** to a polar equation is an ordered pair (r, θ) that satisfies the equation. The graph of a polar equation is the graph of all solutions to the equation.	p. 697
The graphs of polar equations may be symmetric with respect to the polar axis, the line $\theta = \dfrac{\pi}{2}$, or the pole. For an equation in polar coordinates, if the given substitution produces an equivalent equation, then the graph has the indicated type of symmetry. (1) Symmetry with respect to the polar axis: Replace (r, θ) by $(r, -\theta)$. (2) Symmetry with respect to the line $\theta = \dfrac{\pi}{2}$: Replace (r, θ) by $(r, \pi - \theta)$. (3) Symmetry with respect to the pole: Replace (r, θ) by $(-r, \theta)$.	p. 698
Polar equations and their graphs may be members of specific categories of polar equations representing lines, circles, spirals, limaçons, roses, or lemniscates.	pp. 704–705

	Reference
SECTION 7.3 Complex Numbers in Polar Form	
In the **complex plane,** the horizontal axis (called the **real axis**) represents the real part of a complex number, and the vertical axis (called the **imaginary axis**) represents the imaginary part. The complex number $z = a + bi$ is written as an ordered pair (a, b) and can then be graphed as such in the complex plane.	p. 713
The absolute value or **modulus** of a complex number is its distance from the origin in the complex plane. If $z = a + bi$ is a complex number, then $|z| = \sqrt{a^2 + b^2}$.	p. 714
The **polar form** (or **trigonometric form**) of a complex number $z = a + bi$ is $z = r(\cos\theta + i\sin\theta)$, where $a = r\cos\theta$, $b = r\sin\theta$, $r = \sqrt{a^2 + b^2}$, and $\tan\theta = \dfrac{b}{a}$ $(a \neq 0)$. There are infinitely many choices for θ, but normally θ is taken on the interval $0 \leq \theta < 2\pi$.	p. 714

The product and quotient of complex numbers can be found using their polar forms. If $z_1 = r_1(\cos\theta_1 + i\sin\theta_1)$ and $z_2 = r_2(\cos\theta_2 + i\sin\theta_2)$, then $$z_1 z_2 = r_1 r_2[\cos(\theta_1 + \theta_2) + i\sin(\theta_1 + \theta_2)] \text{ and}$$ $$\frac{z_1}{z_2} = \frac{r_1}{r_2}[\cos(\theta_1 - \theta_2) + i\sin(\theta_1 - \theta_2)] \text{ for } (z_2 \neq 0)$$	p. 716
De Moivre's theorem enables us to raise a complex number to an integer power and to find its nth roots. Let $z = r(\cos\theta + i\sin\theta)$. If n is a positive integer, then $$z^n = [r(\cos\theta + i\sin\theta)]^n = r^n(\cos n\theta + i\sin n\theta)$$ For an integer $n \geq 2$, z has n distinct nth roots of the form $$w_k = \sqrt[n]{r}\left[\cos\left(\frac{\theta + 2\pi k}{n}\right) + i\sin\left(\frac{\theta + 2\pi k}{n}\right)\right] \text{ for } k = 0, 1, 2, \dots, n-1$$	pp. 718–719

SECTION 7.4 Vectors

	Reference
A **scalar** is a quantity that has magnitude only (such as speed). A **vector quantity** (such as velocity) has both magnitude *and* direction. A **vector** in the plane is a line segment with a specified direction.	p. 726
We denote a vector pointing from point P to point Q as $\overrightarrow{PQ}$. Point P is called the **initial point** of the vector and Q is called the **terminal point** of the vector. The **magnitude** of $\overrightarrow{PQ}$, denoted $\|\overrightarrow{PQ}\|$, is the length of the line segment. Two vectors are equal if they have the same magnitude and direction. A vector drawn in **standard position** has its initial point at the origin.	
Let $P(x_1, y_1)$ and $Q(x_2, y_2)$ be points in the plane, and let $\mathbf{v} = \overrightarrow{PQ}$. The component form of $\mathbf{v}$ is given by $$\mathbf{v} = \langle x_2 - x_1, y_2 - y_1 \rangle$$	p. 728
Two vectors may be added or subtracted geometrically, or by adding or subtracting the corresponding components. Multiplying a vector $\mathbf{v}$ by a scalar c results in a parallel vector $c\mathbf{v}$ and the length of $c\mathbf{v}$ is the length of $\mathbf{v}$ scaled by a factor of $|c|$.	p. 729
Several properties of vectors follow directly from the definition of vector addition, vector subtraction, and multiplication of a vector by a scalar.	p. 730
A unit vector has a length of 1 unit. If $\mathbf{v} = \langle a, b\rangle$, then a unit vector $\mathbf{u_v}$ in the direction of $\mathbf{v}$ is given by $$\mathbf{u_v} = \frac{1}{\|\mathbf{v}\|}\mathbf{v} = \frac{1}{\|\mathbf{v}\|}\langle a, b\rangle = \left\langle \frac{a}{\|\mathbf{v}\|}, \frac{b}{\|\mathbf{v}\|}\right\rangle.$$	p. 731
Two important unit vectors are $\mathbf{i} = \langle 1, 0\rangle$ and $\mathbf{j} = \langle 0, 1\rangle$ because they are unit vectors in the direction of the positive x- and y-axes, respectively. The representation of $\mathbf{v} = \langle a, b\rangle$ in terms of $\mathbf{i}$ and $\mathbf{j}$ is $\mathbf{v} = a\mathbf{i} + b\mathbf{j}$. The values a and b are called the **scalar horizontal and vertical components** of $\mathbf{v}$, respectively.	p. 732
It is often important to find the component form of a vector $\mathbf{v}$ given its magnitude $\|\mathbf{v}\|$ and direction θ: $\mathbf{v} = \langle \|\mathbf{v}\|\cos\theta, \|\mathbf{v}\|\sin\theta\rangle$. Likewise, given the component form $\mathbf{v} = \langle a, b\rangle$, the magnitude and direction of a vector are given by $$\|\mathbf{v}\| = \sqrt{a^2 + b^2} \text{ and } \tan\theta = \frac{b}{a} \ (a \neq 0).$$	p. 733
Vectors have many useful applications in physics, navigation, and engineering such as finding the true direction and speed of a boat or ship under the influence of a current or wind.	p. 734
Suppose that forces $\mathbf{F}_1, \mathbf{F}_2, \mathbf{F}_3, \dots, \mathbf{F}_n$ act on an object. The sum of these forces is called the **resultant force**. The object is said to be in **static equilibrium** if the resultant force $\mathbf{R}$ is the zero vector, that is, if $\mathbf{R} = \mathbf{F}_1 + \mathbf{F}_2 + \mathbf{F}_3 + \cdots + \mathbf{F}_n = \mathbf{0}$.	p. 736

SECTION 7.5 Dot Product	Reference
If $\mathbf{v} = \langle a_1, b_1 \rangle$ and $\mathbf{w} = \langle a_2, b_2 \rangle$, the **dot product** $\mathbf{v} \cdot \mathbf{w}$ is defined as the scalar quantity $\mathbf{v} \cdot \mathbf{w} = a_1 a_2 + b_1 b_2$.	p. 743

The following are important properties of the dot product.
 	
If $\mathbf{u}$, $\mathbf{v}$, and $\mathbf{w}$ are vectors, and c is a scalar, then
 	
    1. $\mathbf{v} \cdot \mathbf{w} = \mathbf{w} \cdot \mathbf{v}$ **4.** $\mathbf{v} \cdot \mathbf{v} = \|\mathbf{v}\|^2$
    2. $(c\mathbf{v}) \cdot \mathbf{w} = c(\mathbf{v} \cdot \mathbf{w})$ **5.** $\mathbf{0} \cdot \mathbf{v} = 0$
    3. $\mathbf{v} \cdot (\mathbf{w} + \mathbf{u}) = \mathbf{v} \cdot \mathbf{w} + \mathbf{v} \cdot \mathbf{u}$ **6.** $c(\mathbf{v} \cdot \mathbf{w}) = (c\mathbf{v}) \cdot \mathbf{w} = \mathbf{v} \cdot (c\mathbf{w})$
	p. 744

If θ is the angle between two nonzero vectors $\mathbf{v}$ and $\mathbf{w}$, then $$\cos \theta = \frac{\mathbf{v} \cdot \mathbf{w}}{\|\mathbf{v}\| \|\mathbf{w}\|} \text{ and } \theta = \cos^{-1}\left(\frac{\mathbf{v} \cdot \mathbf{w}}{\|\mathbf{v}\| \|\mathbf{w}\|}\right).$$	p. 744
Two vectors $\mathbf{v}$ and $\mathbf{w}$ are orthogonal if and only if $\mathbf{v} \cdot \mathbf{w} = 0$.	p. 746

In many applications it is important to decompose a vector into orthogonal vectors. If $\mathbf{v}$ and $\mathbf{w}$ are two nonzero vectors, then
 	
  1. The vector projection of $\mathbf{v}$ onto $\mathbf{w}$ is given by $\text{proj}_\mathbf{w}\mathbf{v} = \dfrac{\mathbf{v} \cdot \mathbf{w}}{\|\mathbf{w}\|^2}\mathbf{w}$.
 	
  2. To decompose $\mathbf{v}$ into vectors $\mathbf{v}_1$ and $\mathbf{v}_2$ parallel to $\mathbf{w}$ and orthogonal to $\mathbf{w}$, respectively, we have
      $\mathbf{v}_1 = \text{proj}_\mathbf{w}\mathbf{v} = \dfrac{\mathbf{v} \cdot \mathbf{w}}{\|\mathbf{w}\|^2}\mathbf{w}$ and $\mathbf{v}_2 = \mathbf{v} - \mathbf{v}_1$.
	p. 748

In physics, the work W done by a constant force $\mathbf{F}$ in moving an object from point A to point B along the line of motion is given by $$W = \begin{pmatrix} \text{Magnitude of force in} \\ \text{the direction of motion} \end{pmatrix} \begin{pmatrix} \text{Distance the} \\ \text{object is moved} \end{pmatrix}$$	p. 750

If a force vector $\mathbf{F}$ is *not* directed along the line of motion, then only the component of force in the direction of motion is applied to the work done. Therefore, if $\mathbf{D}$ is the displacement vector of an object moved in a straight line from point A to point B by a force $\mathbf{F}$, the work done is given by
 	
  1. $W = \|\text{proj}_\mathbf{D}\mathbf{F}\| \|\mathbf{D}\|$
  2. $W = \|\mathbf{F}\| \|\mathbf{D}\| \cos \theta$, where θ is the angle between $\mathbf{F}$ and $\mathbf{D}$
  3. $W = \mathbf{F} \cdot \mathbf{D}$
	p. 751

Expanded Chapter Summary available at www.mhhe.com/millerprecalculus.

CHAPTER 7 Review Exercises

SECTION 7.1

For Exercises 1–2, given a point in polar coordinates, find two other polar coordinate representations,

  a. with $r > 0$.   **b.** with $r < 0$.

  1. $\left(4, \dfrac{2\pi}{3}\right)$ **2.** $\left(-5, \dfrac{\pi}{6}\right)$

For Exercises 3–4, convert the ordered pair in polar coordinates to rectangular coordinates.

  3. $\left(6, \dfrac{\pi}{4}\right)$ $(3\sqrt{2}, 3\sqrt{2})$ **4.** $\left(-12, -\dfrac{\pi}{2}\right)$ $(0, 12)$

For Exercises 5–6, convert the ordered pair in rectangular coordinates to polar coordinates with $r > 0$ and $0 \leq \theta < 2\pi$.

  5. $(-9, 9\sqrt{3})$ $\left(18, \dfrac{2\pi}{3}\right)$ **6.** $\left(\dfrac{5\sqrt{3}}{2}, -\dfrac{5}{2}\right)$ $\left(5, \dfrac{11\pi}{6}\right)$

For Exercises 7–12, write an equivalent equation using polar coordinates.

  7. $x + 5y = 12$ **8.** $x = 12$ $r = 12 \sec \theta$

  9. $x^2 + (y - 3)^2 = 9$ **10.** $x^2 + y^2 = 100$ $r = 10$

      $r = 6\sin\theta$

  11. $y = -4$ $r = -4\csc\theta$ **12.** $y = x^2$ $r = \tan\theta \sec\theta$

For Exercises 13–16, convert the polar equation to rectangular form and identify the type of curve represented.

13. $r = 16$

14. $r = 16\cos\theta$

15. $\theta = -\dfrac{\pi}{3}$

16. $r = -\dfrac{10}{\sin\theta}$

SECTION 7.2

For Exercises 17–28, graph the polar equation and identify the type of curve.

17. $\theta = \dfrac{5\pi}{4}$

18. $r\cos\theta = -3$

19. $r = -1 - \cos\theta$

20. $r = 4 + 4\sin\theta$

21. $r = 3 + 2\sin\theta$

22. $r = 2 + 3\cos\theta$

23. $r = 2\sin 3\theta$

24. $r = 4\cos 7\theta$

25. $r^2 = 16\cos 2\theta$

26. $r = 0.3\theta$ for $\theta \geq 0$

For Exercises 27–30, determine whether the tests for symmetry in Table 7-2 detect symmetry with respect to
  a. The polar axis. Replace (r, θ) by $(r, -\theta)$.
  b. The line $\theta = \dfrac{\pi}{2}$. Replace (r, θ) by $(r, \pi - \theta)$.
  c. The pole. Replace (r, θ) by $(-r, \theta)$.

Otherwise, indicate that the test is inconclusive.

27. $r = -4\cos 2\theta$

28. $r = 12$

29. $r = -8\cos\theta$

30. $r = 8\sin\theta$

SECTION 7.3

For Exercises 31–32, find the modulus of each complex number and write the number in polar form with $0 \leq \theta < 2\pi$.

31. $6 - 6\sqrt{3}i$

32. $-7 - 7i$

For Exercises 33–34, convert the complex number from polar form to rectangular form, $a + bi$.

33. $17\sqrt{5}\left(\cos\dfrac{3\pi}{2} + i\sin\dfrac{3\pi}{2}\right)$ $-17\sqrt{5}i$

34. $18(\cos 330° + i\sin 330°)$ $9\sqrt{3} - 9i$

For Exercises 35–36, given complex numbers z_1 and z_2,
  a. Find $z_1 z_2$ and write the product in polar form.
  b. Find $\dfrac{z_1}{z_2}$ and write the quotient in polar form.

35. $z_1 = 8(\cos 60° + i\sin 60°)$, $z_2 = 4(\cos 140° + i\sin 140°)$

36. $z_1 = 10\left(\cos\dfrac{5\pi}{12} + i\sin\dfrac{5\pi}{12}\right)$, $z_2 = 15\left(\cos\dfrac{\pi}{4} + i\sin\dfrac{\pi}{4}\right)$

For Exercises 37–38, use De Moivre's theorem to find the indicated power. Write the result in rectangular form $a + bi$.

37. $\left[2\left(\cos\dfrac{\pi}{12} + i\sin\dfrac{\pi}{12}\right)\right]^4$ **38.** $(3 + 3i)^5$ $-972 - 972i$
    $8 + 8\sqrt{3}i$

For Exercises 39–40, find the indicated complex roots by first writing the number in polar form. Write the results in rectangular form $a + bi$.

39. The four roots of 2401.   **40.** The six roots of 4096.

For Exercises 41–42, find the indicated complex roots. Write the results in polar form using degree measure for the argument.

41. The square roots of $25(\cos 120° + i\sin 120°)$.

42. The cube roots of $32 - 32\sqrt{3}i$.

SECTION 7.4

For Exercises 43–44, write the component form of a vector **v** with initial point P and terminal point Q.

43. $P(-4, 6)$, $Q(2, -10)$ **44.** $P(-4, 6)$, $Q(-4, -10)$
     $\langle 6, -16\rangle$ $\langle 0, -16\rangle$

45.

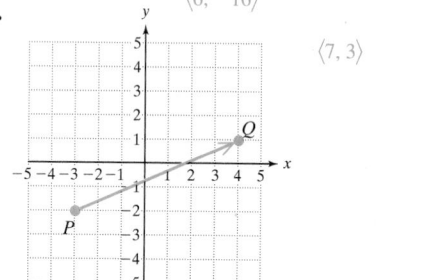

$\langle 7, 3\rangle$

46.

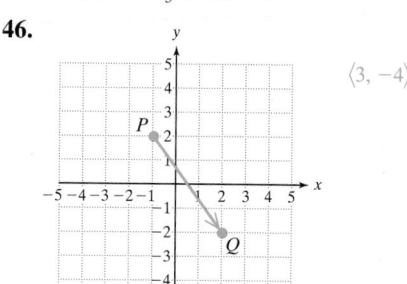

$\langle 3, -4\rangle$

For Exercises 47–48, refer to vectors **v** and **w** in the figure.

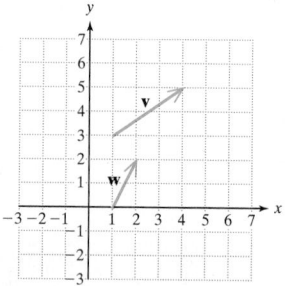

47. Sketch $\mathbf{v} + \mathbf{w}$ by placing the initial point of **w** at the terminal point of **v**.

48. Sketch $\mathbf{v} - \mathbf{w}$ by drawing **v** and **w** with the same initial point.

For Exercises 49–54, perform the indicated operations for the given vectors.

$\mathbf{v} = \langle 3, 5\rangle$ $\mathbf{w} = \langle -3, -2\rangle$ $\mathbf{s} = 10\mathbf{i} - 6\mathbf{j}$ $\mathbf{r} = 5\mathbf{i} + 3\mathbf{j}$

49. $2\mathbf{v} + \mathbf{w}$ $\langle 3, 8\rangle$ **50.** $-8\mathbf{s} + \mathbf{r}$ $-75\mathbf{i} + 51\mathbf{j}$

51. $\dfrac{1}{2}\mathbf{s} + 3\mathbf{r}$ $20\mathbf{i} + 6\mathbf{j}$ **52.** $\mathbf{w} - \dfrac{1}{3}\mathbf{v}$ $\left\langle -4, -\dfrac{11}{3}\right\rangle$

53. $\|\mathbf{v} + \mathbf{w}\|$ 3   **54.** $\|\mathbf{s} + \mathbf{r}\|$ $3\sqrt{26}$

For Exercises 55–56, find a unit vector in the direction of **v**.

55. $v = -8i + 15j$ **56.** $v = -9i - 40j$

For Exercises 57–58, write the vector **v** in the form $ai + bj$, where **v** has the given magnitude and direction angle.

57. $\|v\| = 20, \theta = 120°$ **58.** $\|v\| = 18, \theta = \dfrac{\pi}{4}$

$v = -10i + 10\sqrt{3}j$ $v = 9\sqrt{2}i + 9\sqrt{2}j$

For Exercises 59–60, find the magnitude and direction angle $(0° \le \theta < 360°)$ for the given vector. Round to 1 decimal place.

59. $v = -5i + 6j$ **60.** $v = 2i - 3j$

$\|v\| \approx 7.8, \theta \approx 129.8°$ $\|v\| \approx 3.6, \theta \approx 303.7°$

For Exercises 61–62, the given forces (in units of pounds) act on an object.

a. Find the resultant force, **R**.

b. What additional force **F** is needed for the object to be in static equilibrium?

61. $F_1 = \langle 7, -3 \rangle$ and $F_2 = \langle -10, -4 \rangle$

62. $F_1 = 8i - 2j, F_2 = i - j, F_3 = -3i + 10j$

63. A plane flies at a speed of 420 mph with a bearing of S45°E. Write the velocity vector in terms of **i** and **j**.

64. During a penalty kick, a soccer ball is kicked at an angle of 8° from the horizontal with a speed of 53 mph. Write the velocity vector **v** at the instant of the kick in terms of **i** and **j**. Round the components to the nearest tenth of a mph. $v = 52.5i + 7.4j$

65. A small plane travels 120 mph on a bearing of N40°E and encounters a wind of 20 mph acting in the direction of N45°W. Rounding to 1 decimal place,

a. Express the velocity of the plane **v** relative to the air in terms of **i** and **j**.

b. Express the velocity of the wind v_w in terms of **i** and **j**.

c. Find the true velocity v_T and true speed of the plane. Round the speed to the nearest mph.

SECTION 7.5

For Exercises 66–68, find the indicated value.

a. $v \cdot w$ **b.** $v \cdot v$

66. $v = \langle 12, 8 \rangle, w = \langle -3, -5 \rangle$ a. -76 b. 208

67. $v = \dfrac{1}{2}i + \dfrac{8}{3}j, w = -8i - \dfrac{3}{4}j$ a. -6 b. $\dfrac{265}{36}$

68. $v = -14i + j, w = 80i - 30j$ a. -1150 b. 197

For Exercises 69–72 use the dot product to determine if **v** and **w** are orthogonal. If the vectors are not orthogonal, find the angle between them. Round to the nearest tenth of a degree.

69. $v = \langle -0.2, -0.3 \rangle, w = \langle 0.6, -0.4 \rangle$ Orthogonal

70. $v = 20i + 4j, w = -i + 5j$ Orthogonal

71. $v = -3i + 10j, w = 6i - 4j$ Not orthogonal; $140.4°$

72. $v = \langle 2, 12 \rangle, w = \langle 18, 2 \rangle$ Not orthogonal; $74.2°$

73. Given $v = -i + 3j$ and $w = 2i - j$,

a. Find $proj_w v$. $proj_w v = -2i + j$

b. Find vectors v_1 and v_2 such that v_1 is parallel to **w**, v_2 is orthogonal to **w**, and $v_1 + v_2 = v$. $v_1 = -2i + j$, $v_2 = i + 2j$

74. Given $v = 10j$ and $w = 9i + 12j$,

a. Find $proj_w v$. $proj_w v = \dfrac{24}{5}i + \dfrac{32}{5}j$

b. Find vectors v_1 and v_2 such that v_1 is parallel to **w**, v_2 is orthogonal to **w**, and $v_1 + v_2 = v$.

75. A van weighing 2200 lb is parked on a street with an 8° incline.

a. Write the force vector **F** representing the weight against a single tire. Write **F** in terms of **i** and **j** and assume that the weight of the van is evenly distributed among all four tires. $F = -550j$

b. Find the component vector, F_1, of **F** parallel to the street. Round to 1 decimal place. $F_1 = -75.8i - 10.7j$

c. Find the magnitude of the force required by the brakes on each wheel to keep the truck from rolling down the street. Round to the nearest tenth of a pound. ≈ 76.5 lb

76. Refer to Exercise 75.

a. Find the component vector, F_2, of **F** perpendicular to the street. Round values to 1 decimal place.
$F_2 = 75.8i - 539.3j$

b. Find the magnitude of the force that the street makes against each tire. Round to the nearest pound. ≈ 545 lb

77. Find the work W done by a force $F = (-4i + 6j)$ lb in moving an object in a straight line given by the displacement vector $D = (-8i + 12j)$ ft. $W = 104$ ft·lb

78. Find the work W done by a force $F = (14i - 5j)$ N in moving an object in a straight line from point $A(14, -5)$ to point $B(26, -7)$. Assume that the units in the coordinate plane are in meters. $W = 178$ N·m

79. A gardener pulls on the handle of a cart with a force of 45 lb. If the handle of the cart is directed upwards 38° from the horizontal, find the amount of work done in moving the wagon 15 ft horizontally. Round to the nearest ft·lb. 532 ft·lb

80. Find the amount of work done by a man pushing an object horizontally 42 ft using a force of 520 lb in the direction of motion. $21,840$ ft·lb

CHAPTER 7   Test

1. Convert $\left(-5, \dfrac{5\pi}{4}\right)$ in polar coordinates to rectangular coordinates. $\left(\dfrac{5\sqrt{2}}{2}, \dfrac{5\sqrt{2}}{2}\right)$

2. Convert $(-6\sqrt{3}, -6)$ in rectangular coordinates to polar coordinates with $r > 0$ and $0 \le \theta < 2\pi$. $\left(12, \dfrac{7\pi}{6}\right)$

For Exercises 3–4, write an equivalent equation using polar coordinates.

3. $(x + 2)^2 + y^2 = 4$ $r = -4\cos\theta$

4. $y = x$ $\theta = \dfrac{\pi}{4}$

For Exercises 5–6, write an equivalent equation using rectangular coordinates.

5. $r = 8$ $x^2 + y^2 = 64$

6. $r = 5\sec\theta$ $x = 5$

For Exercises 7–10, graph the polar equation and identify the type of curve.

7. $r = 2\cos 3\theta$

8. $r = 2 - 6\cos\theta$

9. $r = 5 + 5\sin\theta$

10. $r = -5\sin 2\theta$

For Exercises 11–12, determine whether the tests for symmetry in Table 7-2 detect symmetry with respect to

a. The polar axis. Replace (r, θ) by $(r, -\theta)$.

b. The line $\theta = \dfrac{\pi}{2}$. Replace (r, θ) by $(r, \pi - \theta)$.

c. The pole. Replace (r, θ) by $(-r, \theta)$.

Otherwise, indicate that the test is inconclusive.

11. $r = 5\cos 3\theta$ a. Yes   b. Inconclusive   c. Inconclusive

12. $r = -7\sin\theta$ a. Inconclusive   b. Yes   c. Inconclusive

13. Find the modulus of $5 - 5i$ and write the number in polar form with $0 \le \theta < 2\pi$. $5\sqrt{2}; 5\sqrt{2}\left(\cos\dfrac{7\pi}{4} + i\sin\dfrac{7\pi}{4}\right)$

14. Convert $24\left(\cos\dfrac{4\pi}{3} + i\sin\dfrac{4\pi}{3}\right)$ from polar form to rectangular form, $a + bi$. $-12 - 12\sqrt{3}i$

15. Find $[12(\cos 143° + i\sin 143°)][8(\cos 220° + i\sin 220°)]$ and write the product in polar form. $96(\cos 3° + i\sin 3°)$

16. Find $\dfrac{18(\cos 100° + i\sin 100°)}{16(\cos 55° + i\sin 55°)}$ and write the quotient in polar form. $\dfrac{9}{8}(\cos 45° + i\sin 45°)$

17. Use De Moivre's theorem to find $[3(\cos 210° + i\sin 210°)]^4$ and write the result in rectangular form $a + bi$. $-\dfrac{81}{2} + \dfrac{81\sqrt{3}}{2}i$

18. Find the cube roots of $8\left(\cos\dfrac{5\pi}{6} + i\sin\dfrac{5\pi}{6}\right)$.

For Exercises 19–24, perform the indicated operations for $\mathbf{v} = \langle -8, 0 \rangle$ and $\mathbf{w} = \langle 5, -3 \rangle$.

19. $\mathbf{v} + 3\mathbf{w}$ $\langle 7, -9 \rangle$

20. $\|\mathbf{v} + \mathbf{w}\|$ $3\sqrt{2}$

21. $-\dfrac{6}{5}\mathbf{w}$ $\left\langle -6, \dfrac{18}{5} \right\rangle$

22. $8\mathbf{v} - 10\mathbf{w}$ $\langle -114, 30 \rangle$

23. $\mathbf{v} \cdot \mathbf{w}$ -40

24. $(\mathbf{v} \cdot \mathbf{v})\mathbf{w}$ $\langle 320, -192 \rangle$

25. Given $\mathbf{v} = \langle 3, -10 \rangle$,

a. Find $\|\mathbf{v}\|^2$.   109

b. Find $\mathbf{v} \cdot \mathbf{v}$.   109

c. How are $\|\mathbf{v}\|^2$ and $\mathbf{v} \cdot \mathbf{v}$ related?   They are the same.

26. Find a unit vector in the direction of $\mathbf{v} = -8\mathbf{i} + 6\mathbf{j}$.

27. Write the vector $\mathbf{v}$ in the form $a\mathbf{i} + b\mathbf{j}$, where $\|\mathbf{v}\| = 12$ and $\mathbf{v}$ has direction angle $\theta = 225°$. $\mathbf{v} = -6\sqrt{2}\mathbf{i} - 6\sqrt{2}\mathbf{j}$

28. Find the magnitude and direction angle $(0° \le \theta < 360°)$ for the vector $\mathbf{v} = -2\mathbf{i} + 7\mathbf{j}$. Round to 1 decimal place. $\|\mathbf{v}\| \approx 7.3, \theta \approx 105.9°$

29. Find the angle θ between $\mathbf{v} = -12\mathbf{i}$ and $\mathbf{w} = 4\mathbf{i} + 6\mathbf{j}$. Round to the nearest tenth of a degree. $\theta \approx 123.7°$

30. The forces $\mathbf{F}_1 = \langle -8, 1 \rangle$ and $\mathbf{F}_2 = \langle 5, 6 \rangle$ act on an object. Find the resultant force, $\mathbf{R}$. $\mathbf{R} = \langle -3, 7 \rangle$

31. The forces $\mathbf{F}_1 = 6\mathbf{i} - \mathbf{j}$, $\mathbf{F}_2 = -7\mathbf{i} - 2\mathbf{j}$, and $\mathbf{F}_3 = 9\mathbf{i} + 3\mathbf{j}$ act on an object. What additional force $\mathbf{F}$ is needed for the object to be in static equilibrium? $\mathbf{F} = -8\mathbf{i}$

32. Given $\mathbf{v} = \langle 7, -5 \rangle$ and $\mathbf{w} = \langle 10, 12 \rangle$, find $\text{proj}_{\mathbf{w}}\mathbf{v}$.

33. A plane heading S21°E with an an air speed of 375 mph encounters a wind blowing due east at 28 mph.

a. Express the velocity of the plane $\mathbf{v_p}$ relative to the air and the velocity of the wind $\mathbf{v_w}$ in terms of $\mathbf{i}$ and $\mathbf{j}$. Round components to 1 decimal place.

b. Find the velocity of the plane relative to the ground $\mathbf{v_g}$ and the speed of the plane relative to the ground. Round the speed to the nearest mph.

c. Find the bearing of the plane relative to the ground. Round to the nearest tenth of a degree.

34. Find the magnitude of the force needed to keep a 250-lb crate from sliding down a ramp inclined 28° from the horizontal. Round to the nearest pound.   117 lb

35. During a field goal kick, a football is kicked at an angle of 45° from the horizontal with a speed of 86 mph. Write the velocity vector $\mathbf{v}$ at the instant of the kick in terms of $\mathbf{i}$ and $\mathbf{j}$. $\mathbf{v} = 43\sqrt{2}\mathbf{i} + 43\sqrt{2}\mathbf{j}$

36. Find the work done by a force $\mathbf{F} = -7\mathbf{i} + 3\mathbf{j}$ (in N) in moving an object in a straight line from point $A(6, 1)$ to point $B(-5, 4)$. Assume that the units in the coordinate plane are meters.   86 N·m

37. Find the amount of work done if an object is pushed horizontally 120 m by a force of 15 N directed 40° above the horizontal. Round to the nearest N·m.   1379 N·m

CHAPTER 7   Cumulative Review Exercises

For Exercises 1–7, solve for *y*. Write solutions to inequalities in interval notation.

1. $A = \dfrac{x}{2}(x + y)$

2. $z = 10 - \sqrt{x^2 + y^2}$

3. $sy^2 + 2ty = r$

4. $A = A_0 e^{-ky}$

5. $R = 10\log\left(\dfrac{y}{y_0}\right)$

6. $\dfrac{y^2 - 3y - 4}{y + 2} \le 0$

7. $3 - |y - 10| \le 0$ $(-\infty, 7] \cup [13, \infty)$

8. Write an equation of a circle in rectangular coordinates and in standard form with center at $(-8, 5)$ and radius 8.
   $(x + 8)^2 + (y - 5)^2 = 64$

9. Write an equation of a circle in polar coordinates with center at $(-8, 0)$ and radius 8. $r = -16\cos\theta$

10. Write an equation of a line in slope-intercept form in rectangular coordinates that passes through the points $(-8, 5)$ and $(5, -8)$. $y = -x - 3$

11. Write an equation of a line in polar coordinates that passes through the points $(-8, 8)$ and $(8, -8)$. $\theta = \dfrac{3\pi}{4}$

For Exercises 12–14, solve each triangle. Round to the nearest tenth.

12.

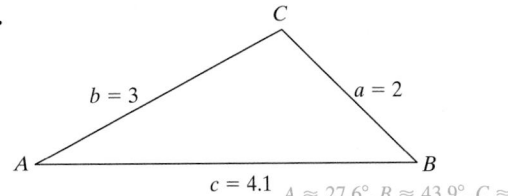

$c = 4.1$ $A \approx 27.6°, B \approx 43.9°, C \approx 108.5°$

13.

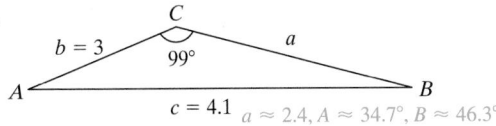

$c = 4.1$ $a \approx 2.4, A \approx 34.7°, B \approx 46.3°$

14.

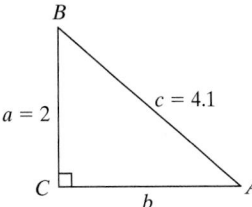

$b \approx 3.6, A \approx 29.2°, B \approx 60.8°$

15. Evaluate the difference quotient $\dfrac{f(x + h) - f(x)}{h}$ for $f(x) = \dfrac{3}{x}$. $-\dfrac{3}{x(x + h)}$

16. Give the domain and range of $f(x) = \dfrac{3}{x}$.

17. Find the period and phase shift of each function.

a. $y = \tan\left(2x + \dfrac{\pi}{3}\right)$

b. $y = \sin\left(2x + \dfrac{\pi}{3}\right)$

18. The height h (in feet) of a swing above ground level at time t (in sec) is modeled by

$$h(t) = -2\cos\left(\dfrac{2\pi}{3}t\right) + 3$$

a. Find the maximum and minimum height of the swing.
   Maximum height = 5 ft, Minimum height = 1 ft

b. When is the first time after $t = 0$ that the swing is at a height of 4 ft? $t = 1\,\text{sec}$

c. What is the period of the swing?   3 sec

d. What is the frequency of the swing? $\dfrac{1}{3}$ Hz

19. Given $\mathbf{v} = \langle 8, -3 \rangle$ and $\mathbf{w} = \langle 12, -4 \rangle$, find vector projection of $\mathbf{v}$ onto $\mathbf{w}$.

20. Find the amount of work done pulling an object horizontally 25 ft if a force of 120 lb is directed $38°$ above the horizontal. Round to the nearest ft · lb.   2364 ft·lb

8

Systems of Equations and Inequalities

Chapter Outline

The economy influences how people spend their money, but it also impacts how people save. When comparing options for investments such as stocks, bonds, and mutual funds, an individual can easily become overwhelmed by the possible scenarios. Collecting data on the rates of return on different investments is helpful in making an informed decision. In some scenarios, we turn to systems of linear equations to analyze such data.

In this chapter, we will solve systems involving both linear and nonlinear equations. In addition, we will use systems of linear inequalities in manufacturing applications. In such applications, the goal is to determine constraints on the production process (such as limits on the amount of material, labor, and equipment) and then maximize profit or minimize cost subject to these constraints.

Systems of Linear Equations in Two Variables and Applications

1. Identify Solutions to Systems of Linear Equations in Two Variables

The point of intersection between two curves is important in the analysis of supply and demand. For example, suppose that a small theater company wants to set an optimal price for tickets to a show. If the theater sells tickets for $1 each, there would be a high demand and many people would buy tickets. However, the revenue brought in would not cover the expense of the show. Therefore, the theater is not willing to offer (supply) tickets at this low price. If the theater sells tickets for $100 each, chances are that few people would buy tickets (demand would be low).

In an open market, the price y of an item is dependent on the number of items x supplied by the producer and demanded by the consumers. Competition between buyers and sellers steers the price toward an equilibrium price. This is the point where the supply curve and demand curve intersect (Figure 8-1).

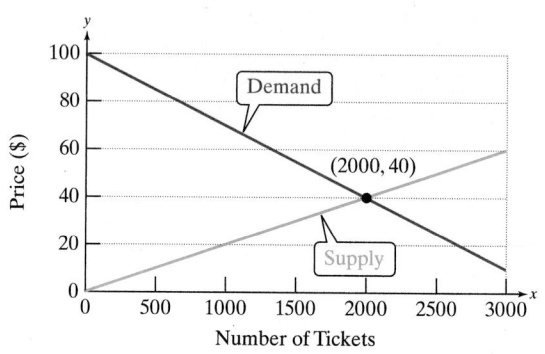

Figure 8-1

Suppose that the following linear equations represent the supply and demand curves for theater tickets. Taken together these equations form a *system* of linear equations.

$$\left.\begin{array}{l} \text{Supply: } y = 0.02x \\ \text{Demand: } y = -0.03x + 100 \end{array}\right\} \quad \begin{array}{l}\text{System of linear}\\\text{equations}\end{array}$$

Two or more linear equations make up a **system of linear equations**. A **solution** to a system of equations in two variables is an ordered pair that is a solution to each individual equation. Graphically, this is a point of intersection of the graphs of the equations. For the system given here, the solution is (2000, 40). This means that 2000 tickets are bought and sold at the equilibrium price of $40. The **solution set** to a system of equations is the set of all solutions to the system. In this case, the solution set is {(2000, 40)}.

EXAMPLE 1 Determining if an Ordered Pair Is a Solution to a System of Equations

Determine if the ordered pair is a solution to the system. $\begin{aligned} 6x + y &= -2 \\ 4x - 3y &= 17 \end{aligned}$

a. $\left(\dfrac{1}{2}, -5\right)$ **b.** $(0, -2)$

Solution:

a. <u>First equation</u>

$6x + y = -2$

$6\left(\frac{1}{2}\right) + (-5) \overset{?}{=} -2$

$3 + (-5) \overset{?}{=} -2 \checkmark$ true

<u>Second equation</u>

$4x - 3y = 17$

$4\left(\frac{1}{2}\right) - 3(-5) \overset{?}{=} 17$

$2 + 15 \overset{?}{=} 17 \checkmark$ true

Test the ordered pair $\left(\frac{1}{2}, -5\right)$ in each equation.

The ordered pair is a solution to both equations.

The ordered pair $\left(\frac{1}{2}, -5\right)$ is a solution to the system.

b. First equation   Second equation

$6x + y = -2$ $4x - 3y = 17$

$6(0) + (-2) \stackrel{?}{=} -2$ $4(0) - 3(-2) \stackrel{?}{=} 17$

$0 + (-2) \stackrel{?}{=} -2$ $0 + 6 \stackrel{?}{=} 17$

$-2 \stackrel{?}{=} -2 \checkmark$ true $6 \stackrel{?}{=} 17$ false

Test the ordered pair $(0, -2)$ in each equation.

The ordered pair is *not* a solution to the second equation.

The ordered pair $(0, -2)$ is *not* a solution to the system.

Skill Practice 1

Determine if the ordered pair is a solution to the system.

a. $(2, -4)$ **b.** $\left(\dfrac{1}{3}, -9\right)$

$3x - y = 10$

$x + \dfrac{1}{4}y = 1$

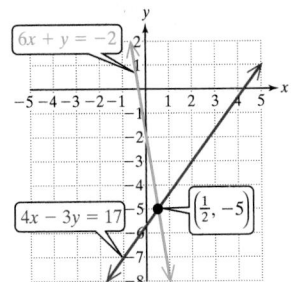

Figure 8-2

The lines from Example 1 are shown in Figure 8-2. From the graph, we can verify that $\left(\frac{1}{2}, -5\right)$ is the only solution to the system of equations.

There are three different possibilities regarding the number of solutions to a system of linear equations.

Solutions to Systems of Linear Equations in Two Variables

One Unique Solution	**No Solution**	**Infinitely Many Solutions**
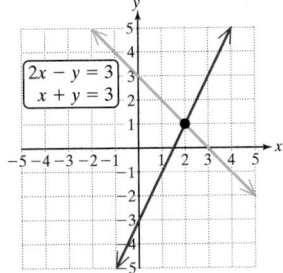	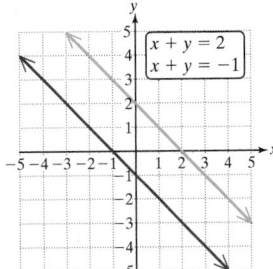	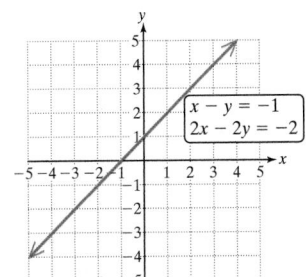
If a system of linear equations represents intersecting lines, then it has exactly one solution.	If a system of linear equations represents parallel lines, then the lines do not intersect, and the system has no solution. In such a case, we say that the system is **inconsistent.**	If a system of linear equations represents the same line, then all points on the common line satisfy each equation. Therefore, the system has infinitely many solutions. In such a case, we say that the equations are **dependent.**

---

2. Solve Systems of Linear Equations in Two Variables

Graphing a system of equations is one method to find the solution(s) to the system. However, sometimes it is difficult to determine the solution(s) using this method because of limitations in the accuracy of the graph. Instead we often use algebraic methods to solve a system of equations. The first method we present is called the **substitution method.**

Answers

1. a. Yes   **b.** No

Solving a System of Equations by Using the Substitution Method

Step 1 Isolate one of the variables from one equation.

Step 2 Substitute the expression found in step 1 into the *other* equation.

Step 3 Solve the resulting equation.

Step 4 Substitute the value found in step 3 back into the equation in step 1 to find the value of the remaining variable.

Step 5 Check the ordered pair in each equation and write the solution as an ordered pair in set notation.

Classroom Example: p. 775
Exercise 16

EXAMPLE 2   Solving a System of Equations by the Substitution Method

Solve the system by using the substitution method.

$$-5x - 4y = 2$$
$$4x + y = 5$$

Solution:

$$-5x - 4y = 2$$
$$4x + y = 5 \longrightarrow y = \underbrace{-4x + 5}$$

Step 1: Isolate one of the variables from one of the equations. A variable with coefficient 1 or −1 is easily isolated.

$$-5x - 4(-4x + 5) = 2$$

Step 2: Substitute the expression from step 1 into the other equation.

$$-5x + 16x - 20 = 2$$

Step 3: Solve for the remaining variable.

$$11x = 22$$
$$x = 2$$

TIP The lines from Example 2 are shown here. The graph shows the point of intersection at $(2, -3)$.

$$y = -4x + 5$$
$$y = -4(2) + 5$$
$$y = -3$$

Step 4: Substitute the known value of x into the equation where y is isolated. From step 1, this is $y = -4x + 5$.

Step 5: Check the ordered pair $(2, -3)$ in each original equation.

Check $(2, -3)$.

$$-5x - 4y = 2 \qquad\qquad 4x + y = 5$$
$$-5(2) - 4(-3) \stackrel{?}{=} 2 \qquad 4(2) + (-3) \stackrel{?}{=} 5$$
$$-10 + 12 \stackrel{?}{=} 2 \checkmark \text{ true} \qquad 8 - 3 \stackrel{?}{=} 5 \checkmark \text{ true}$$

The solution set is $\{(2, -3)\}$.

Skill Practice 2   Solve the system by using the substitution method.

$$3x + 4y = 5$$
$$x - 3y = 6$$

Now consider the following system of equations.

$$5x - 4y = 6$$
$$-3x + 7y = 1$$

None of the variable terms has a coefficient of 1 or −1. Therefore, if we isolate x or y from either equation, the resulting equation will have one or more terms with fractional coefficients. To avoid this scenario, we can use another method called the **addition method** (also called the elimination method).

Answer

2. $\{(3, -1)\}$

TIP The addition method is sometimes called the *elimination method*. However, since both the substitution method and addition method eliminate a variable, we use the name addition method to emphasize the technique used.

Solving a System of Equations by Using the Addition Method

Step 1 Write both equations in standard form: $Ax + By = C$.

Step 2 Clear fractions or decimals (optional).

Step 3 Multiply one or both equations by nonzero constants to create opposite coefficients for one of the variables.

Step 4 Add the equations from step 3 to eliminate one variable.

Step 5 Solve for the remaining variable.

Step 6 Substitute the known value found in step 5 into one of the original equations to solve for the other variable.

Step 7 Check the ordered pair in each equation and write the solution set.

Classroom Example: p. 776
Exercise 24

EXAMPLE 3   Solving a System of Equations by the Addition Method

Solve the system by using the addition method.

$$5x = 4y + 6$$
$$-3x + 7y = 1$$

Solution:

$$5x = 4y + 6 \xrightarrow{\text{Subtract } 4y.} 5x - 4y = 6$$
$$-3x + 7y = 1 \qquad\qquad\qquad -3x + 7y = 1$$

Step 1: Write each equation in standard form: $Ax + By = C$.

Step 2: There are no decimals or fractions.

TIP In Example 3, the variable x was eliminated.

Alternatively, we could have eliminated y by multiplying the first equation by 7 and the second equation by 4. This would create new equations with coefficients of -28 and 28 on the y terms.

$$5x - 4y = 6 \xrightarrow{\text{Multiply by 3.}} 15x - 12y = 18$$
$$-3x + 7y = 1 \xrightarrow[\text{Multiply by 5.}]{} \underline{-15x + 35y = 5}$$
$$23y = 23$$

Step 3: Multiply the first equation by 3. Multiply the second equation by 5.

Step 4: Add the equations to eliminate x.

$$y = 1 \qquad \textbf{Step 5: Solve for } y.$$

$$5x = 4y + 6$$
$$5x = 4(1) + 6$$
$$5x = 10$$
$$x = 2$$

Step 6: Substitute $y = 1$ into one of the original equations to solve for x.

Step 7: Check the ordered pair (2, 1) in each original equation.

$$5x = 4y + 6 \qquad\qquad -3x + 7y = 1$$
$$5(2) \stackrel{?}{=} 4(1) + 6 \qquad -3(2) + 7(1) \stackrel{?}{=} 1$$
$$10 \stackrel{?}{=} 4 + 6 \ \checkmark \text{ true} \qquad -6 + 7 \stackrel{?}{=} 1 \ \checkmark \text{ true}$$

The solution set is $\{(2, 1)\}$.

Skill Practice 3   Solve the system by using the addition method.

$$2x - 9y = 1$$
$$3x = 17 - 2y$$

Answer

3. $\{(5, 1)\}$

TECHNOLOGY CONNECTIONS

Solving a System of Linear Equations in Two Variables Using Intersect

The solution to a system of linear equations can be checked on a graphing calculator by first writing the equations in slope-intercept form. From Example 3 we have

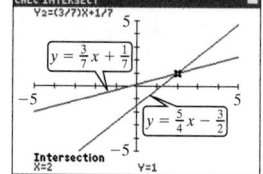

$$5x = 4y + 6 \longrightarrow y = \frac{5}{4}x - \frac{3}{2}$$

$$-3x + 7y = 1 \longrightarrow y = \frac{3}{7}x + \frac{1}{7}$$

Then graph the equations and use the Intersect feature to approximate the point of intersection. The Intersect feature gives the solution (2, 1).

It is important to write the individual equations in a system of equations in standard form so that the variables line up. Also consider clearing decimals or fractions within an equation to make integer coefficients.

Classroom Example: p. 776
Exercise 28

EXAMPLE 4   Solving a System of Equations by the Addition Method

Solve the system by using the addition method. $\dfrac{2}{5}x - y = \dfrac{19}{10}$

$$5(x + y) = -7y - 41$$

Solution:

$$\frac{2}{5}x - y = \frac{19}{10} \xrightarrow{\text{Multiply by 10.}} 4x - 10y = 19$$

$$5(x + y) = -7y - 41 \xrightarrow{\text{Simplify.}} 5x + 12y = -41$$

Clear the fractions in the first equation. Write the second equation in standard form.

$$4x - 10y = 19 \xrightarrow{\text{Multiply by } -5.} -20x + 50y = -95$$

$$5x + 12y = -41 \xrightarrow[\text{Multiply by 4.}]{} \underline{20x + 48y = -164}$$

$$98y = -259$$

The LCM of the x coefficients, 4 and 5, is 20. Create opposite coefficients on x of 20 and -20.

$$y = \frac{-259}{98}$$

Solve for y.

$$y = -\frac{37}{14}$$

Simplify to lowest terms.

Substituting $y = -\frac{37}{14}$ back into one of the original equations to solve for x would be cumbersome. Alternatively, we can solve for x by repeating the addition method. This time we will eliminate y by creating opposite coefficients on the y terms and then solving for x.

$$4x - 10y = 19 \xrightarrow{\text{Multiply by 6.}} 24x - 60y = 114$$
$$5x + 12y = -41 \xrightarrow{\text{Multiply by 5.}} \underline{25x + 60y = -205}$$
$$49x = -91$$

The LCM of 10 and 12 is 60. Create opposite coefficients of -60 and 60 on the y terms.

$$x = \frac{-91}{49}$$ Solve for y.

$$x = -\frac{13}{7}$$ Simplify.

The ordered pair $\left(-\frac{13}{7}, -\frac{37}{14}\right)$ checks in both original equations.

The solution set is $\left\{\left(-\frac{13}{7}, -\frac{37}{14}\right)\right\}$.

> **Skill Practice 4** Solve the system by using the addition method.
>
> $$2(x - 2y) = y + 14$$
> $$\frac{1}{2}x + \frac{7}{6}y = -\frac{13}{3}$$

The systems in Examples 1–4 each have one unique solution. That is, the lines represented by the two equations intersect in exactly one point. In Examples 5 and 6, we investigate systems with no solution or infinitely many solutions.

Classroom Example: p. 776
Exercise 30

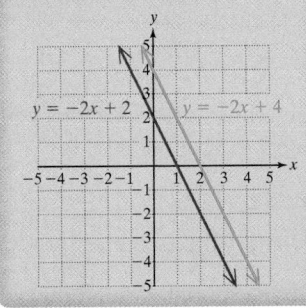
> **EXAMPLE 5** **Identifying a System of Equations with No Solution**
>
> Solve the system. $2x + y = 4$
> $6x + 3y = 6$
>
> **Solution:**
>
> $$2x + y = 4 \xrightarrow{\text{Multiply by } -3.} -6x - 3y = -12$$
> $$6x + 3y = 6 \longrightarrow \underline{6x + 3y = 6}$$
> $$0 = -6$$
>
> We can eliminate either the x terms or y terms by multiplying the first equation by -3.
>
> Both the x and y terms are eliminated, leading to the contradiction $0 = -6$.
>
> The system of equations reduces to a contradiction. This indicates that there is no solution and the system is inconsistent. The equations represent parallel lines and two parallel lines do not intersect.
>
> The solution set is { }.

> **Skill Practice 5** Solve the system. $3x - y = 2$
> $-9x + 3y = 4$

Classroom Example: p. 776
Exercise 32

| EXAMPLE 6 | Solving a System of Dependent Equations |

Solve the system.
$$y = 2x - 1$$
$$8x - 4y = 4$$

Solution:

$$y = 2x - 1$$
$$8x - 4y = 4 \qquad 8x - 4(2x - 1) = 4$$

With y already isolated in the first equation, apply the substitution method.

$$8x - 8x + 4 = 4 \qquad \text{Solve the resulting equation.}$$
$$4 = 4 \qquad \text{Identity.}$$

Notice that both the variable terms and the constant terms were eliminated. The system of equations is reduced to the identity $4 = 4$. Therefore, the two original equations are dependent and represent the same line. The solution set consists of an infinite number of ordered pairs (x, y) that fall on the common line of intersection, $y = 2x - 1$. Therefore, the solutions are ordered pairs of the form $(x, 2x - 1)$.

The solution set is $\{(x, 2x - 1) \mid x \text{ is any real number}\}$.

> **Skill Practice 6** Solve the system. $x = 5 - 3y$
> $$2x + 6y = 10$$

Avoiding Mistakes

To verify the solution to Example 6, write the two equations in slope-intercept form.

$$y = 2x - 1 \Rightarrow y = 2x - 1$$
$$8x - 4y = 4 \Rightarrow y = 2x - 1$$

The equations have the same slope-intercept form and define the same line.

TIP Sometimes the solution to a system of dependent equations is written with an arbitrary variable called a **parameter**. For example, letting t represent any real number, then the solution set to Example 6 can be written as $\{(t, 2t - 1) \mid t \text{ is a real number}\}$.

The solution set $\{(x, 2x - 1) \mid x \text{ is any real number}\}$ is called the **general solution** to the system in Example 6. By varying the value of x, we can produce any number of specific solutions to the system of equations. For example:

	$(x, 2x - 1)$	**Solution**
If $x = 1$	$(1, 2(1) - 1)$	$(1, 1)$
If $x = 2$	$(2, 2(2) - 1)$	$(2, 3)$
If $x = 3$	$(3, 2(3) - 1)$	$(3, 5)$
If $x = -1$	$(-1, 2(-1) - 1)$	$(-1, -3)$

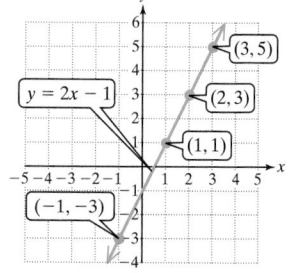

Figure 8-3

TIP The equations $y = 2x - 1$ and $8x - 4y = 4$ represent the same relationship between x and y. Therefore, we have only one unique equation, but two variables. As a result, y *depends* on the choice of x and vice versa.

Notice that the solutions fall on the common line of intersection as expected (Figure 8-3).

We should also note that the general solution to a system of dependent equations can be written in a variety of forms. For instance, the solutions to the system in Example 6 fall on the line $y = 2x - 1$. By solving the equation for x, we have $x = \dfrac{y + 1}{2}$. Thus, the solution set can also be written as $\left\{ \left(\dfrac{y + 1}{2}, y \right) \middle| y \text{ is any real number} \right\}$.

3. Use Systems of Linear Equations in Applications

When solving an application involving two unknowns, sometimes it is convenient to use a system of two independent equations as demonstrated in Examples 7–9.

Classroom Example: p. 776
Exercise 52

| EXAMPLE 7 | Solving an Application Involving Mixtures |

A hospital uses a 15% bleach solution to disinfect a quarantine area. How much 6% bleach solution must be mixed with an 18% bleach solution to make 50 L of a 15% bleach solution?

Answer

6. $\left\{ \left(x, \dfrac{-x + 5}{3} \right) \middle| x \text{ is any real number} \right\}$

or $\{(5 - 3y, y) \mid y \text{ is any real number}\}$

Solution:

Let x represent the amount of 6% bleach solution.

Let y represent the amount of 18% bleach solution.

The amount of pure bleach in each mixture is found by multiplying the amount of solution by the concentration rate. This information can be organized in a table.

	6% Solution	18% Solution	15% Solution
Amount of mixture	x	y	50
Amount of pure bleach	$0.06x$	$0.18y$	$0.15(50)$

There are two unknown quantities. We will set up a system of two independent equations relating x and y.

$$\left(\begin{array}{c}\text{Amount of}\\ \text{6\% mixture}\end{array}\right) + \left(\begin{array}{c}\text{Amount of}\\ \text{18\% mixture}\end{array}\right) = \left(\begin{array}{c}\text{Amount of}\\ \text{15\% mixture}\end{array}\right)$$

$$\left(\begin{array}{c}\text{Amount of}\\ \text{pure bleach in}\\ \text{6\% mixture}\end{array}\right) + \left(\begin{array}{c}\text{Amount of}\\ \text{pure bleach in}\\ \text{18\% mixture}\end{array}\right) = \left(\begin{array}{c}\text{Amount of}\\ \text{pure bleach in}\\ \text{15\% mixture}\end{array}\right)$$

$x + y = 50$
$0.06x + 0.18y = 0.15(50)$

$$
\begin{array}{lll}
x + y = 50 & x + y = 50 & x + y = 50 \\
0.06x + 0.18y = 7.5 \xrightarrow{\text{Multiply by 100.}} 6x + 18y = 750 \xrightarrow{\text{Divide by } -6.} & -x - 3y = -125 \\
& & \overline{ -2y = -75} \\
& & y = 37.5
\end{array}
$$

Substitute $y = 37.5$ back into the equation $x + y = 50$.

$$x + y = 50$$
$$x + 37.5 = 50$$
$$x = 12.5$$

Therefore, 12.5 L of the 6% bleach solution should be mixed with 37.5 L of the 18% bleach solution to make 50 L of 15% bleach solution.

> **Avoiding Mistakes**
>
> Check that the answer is reasonable. The total amount of the resulting solution is 12.5 L + 37.5 L, which is 50 L.
>
> Amount of pure bleach:
> 0.06(12.5 L) = 0.75 L
> 0.18(37.5 L) = 6.75 L
> 0.15(50 L) = 7.5 L ✓

> **Skill Practice 7** How many ounces of 20% and 35% acid solution should be mixed to produce 15 oz of 30% acid solution?

Classroom Example: p. 777
Exercise 64

EXAMPLE 8 Solving an Application Involving Uniform Motion

A riverboat traveling upstream against the current on the Mississippi River takes 3 hr to travel 24 mi. The return trip downstream with the current takes only 2 hr. Find the speed of the boat in still water and the speed of the current.

Solution:

Let b represent the speed of the boat in still water.

Let c represent the speed of the current.

The given information can be organized in a table.

Answer

7. Mix 5 oz of 20% acid solution with 10 oz of 35% acid solution to make 15 oz of 30% acid solution.

	Distance (mi)	Rate (mph)	Time (hr)
Upstream	24	$b - c$	3
Downstream	24	$b + c$	2

Use the relationship $d = rt$; that is, distance = (rate)(time).

$$\left(\begin{array}{c}\text{Distance} \\ \text{upstream}\end{array}\right) = \left(\begin{array}{c}\text{Rate} \\ \text{upstream}\end{array}\right)\left(\begin{array}{c}\text{Time} \\ \text{upstream}\end{array}\right) \longrightarrow 24 = (b - c) \cdot 3$$

$$\left(\begin{array}{c}\text{Distance} \\ \text{downstream}\end{array}\right) = \left(\begin{array}{c}\text{Rate} \\ \text{downstream}\end{array}\right)\left(\begin{array}{c}\text{Time} \\ \text{downstream}\end{array}\right) \longrightarrow 24 = (b + c) \cdot 2$$

$$24 = 3b - 3c \xrightarrow{\text{Divide by 3.}} 8 = b - c$$
$$24 = 2b + 2c \xrightarrow[\text{Divide by 2.}]{} 12 = b + c$$
$$\overline{} 20 = 2b$$
$$10 = b$$

Substitute $b = 10$ into the equation $12 = b + c$, which gives $c = 2$. The boat's speed in still water is 10 mph and the speed of the current is 2 mph.

Skill Practice 8 A boat takes 3 hr to go 24 mi upstream against the current. It can go downstream with the current a distance of 48 mi in the same amount of time. Determine the speed of the boat in still water and the speed of the current.

Classroom Example: p. 778
Exercise 68

EXAMPLE 9 Applying a System of Equations in Business

A lawn service company has fixed monthly costs of $500 and variable costs (labor, gasoline, and depreciation) of $40 per lawn. If the service charges $60 per lawn,

a. Write a cost function representing the cost $C(x)$ to the company to service x lawns per month.

b. Write a revenue function representing the revenue $R(x)$ to service x lawns per month.

c. Determine the number of lawns that must be serviced in a month for the company to break even.

TIP Notice that the result of Example 9(c) is found by solving the system of equations found in parts (a) and (b).

$$y = 500 + 40x$$
$$y = 60x$$

Solution:

Let x represent the number of lawns serviced for a given month.

a. $\left(\begin{array}{c}\text{Monthly} \\ \text{cost}\end{array}\right) = \left(\begin{array}{c}\text{Fixed} \\ \text{cost}\end{array}\right) + \left(\begin{array}{c}\text{Variable} \\ \text{cost}\end{array}\right) \longrightarrow C(x) = 500 + 40x$

b. $\left(\begin{array}{c}\text{Monthly} \\ \text{Revenue}\end{array}\right) = \left(\begin{array}{c}\text{Revenue} \\ \text{per lawn}\end{array}\right)\left(\begin{array}{c}\text{Number} \\ \text{of lawns}\end{array}\right) \longrightarrow R(x) = 60x$

c. To break even, the cost must equal revenue, $C(x) = R(x)$.

$$500 + 40x = 60x$$
$$500 = 20x$$
$$x = 25$$

To break even, the company must service 25 lawns.

Skill Practice 9 A storage company rents its units for $120 per month. The company has fixed monthly costs of $2100 and variable costs (air-conditioning and service) of $50 per unit.

a. Write a cost function representing the monthly cost $C(x)$ to the company for x units.

b. Write a revenue function representing the revenue $R(x)$ when x units per month are rented.

c. Determine the number of units that must be rented in a month for the company to break even.

Answers

8. The boat's speed in still water is 12 mph and the speed of the current is 4 mph.
9. a. $C(x) = 2100 + 50x$
   b. $R(x) = 120x$
   c. 30 units

| **SECTION 8.1** | **Practice Exercises** |

Concept Connections

1. Two or more linear equations taken together make up a ____system____ of linear equations.

2. A ____solution____ to a system of equations in two variables is an ordered pair that is a solution to each individual equation in the system.

3. Two algebraic methods to solve a system of linear equations in two variables are the ____substitution____ method and the ____addition____ method.

4. A system of linear equations in two variables may have no solution. In such a case, the equations represent ____parallel____ lines.

5. A system of equations that has no solution is called an ____inconsistent____ system.

6. A system of linear equations in two variables may have infinitely many solutions. In such a case, the equations are said to be ____dependent____.

Objective 1: Identify Solutions to Systems of Linear Equations in Two Variables

For Exercises 7–10, determine if the ordered pair is a solution to the system of equations. (See Example 1)

7. $3x - 5y = -7$
$x - 4y = -7$

 a. $(1, 2)$ Yes

 b. $\left(-\dfrac{2}{3}, 1\right)$ No

8. $-11x + 6y = -4$
$7x + 3y = 23$

 a. $\left(1, \dfrac{7}{6}\right)$ No

 b. $(2, 3)$ Yes

9. $y = \dfrac{3}{2}x - 5$
$6x - 4y = 20$

 a. $(2, -2)$ Yes

 b. $(-4, -11)$ Yes

10. $y = -\dfrac{1}{5}x + 2$
$2x + 10y = 10$

 a. $(5, 1)$ No

 b. $(-10, 4)$ No

For Exercises 11–14, a system of equations is given in which each equation is written in slope-intercept form. Determine the number of solutions. If the system does not have one unique solution, state whether the system is inconsistent or whether the equations are dependent.

11. $y = \dfrac{2}{5}x - 7$ One solution
$y = \dfrac{1}{4}x + 7$

12. $y = 6x - \dfrac{2}{3}$ No solution; The system is inconsistent.
$y = 6x + 4$

13. $y = 8x - \dfrac{1}{2}$ Infinitely many solutions; The equations are dependent.
$y = 8x - \dfrac{1}{2}$

14. $y = \dfrac{1}{2}x + 3$ One solution
$y = 2x + \dfrac{1}{3}$

Objective 2: Solve Systems of Linear Equations in Two Variables

For Exercises 15–20, solve the system of equations by using the substitution method. (See Example 2)

15. $x + 3y = 5$ $\{(-4, 3)\}$
$3x - 2y = -18$

16. $2x + y = 2$ $\{(-3, 8)\}$
$5x + 3y = 9$

17. $2x + 7y = 1$ $\{(-10, 3)\}$
$3y - 7 = 2$

18. $3x = 2y - 11$ $\{(-1, 4)\}$
$6 + 5x = 1$

19. $2(x + y) = 2 - y$ $\left\{\left(-\tfrac{1}{2}, 1\right)\right\}$
$4x - 1 = 2 - 5y$

20. $5(x + y) = 9 + 2y$ $\left\{\left(2, -\tfrac{1}{3}\right)\right\}$
$6y - 2 = 10 - 7x$

For Exercises 21–28, solve the system of equations by using the addition method. (See Examples 3–4)

21. $3x - 7y = 1$ $\{(-2, -1)\}$
$6x + 5y = -17$

22. $5x - 2y = -2$ $\{(2, 6)\}$
$3x + 4y = 30$

23. $11x = -5 - 4y$ $\{(1, -4)\}$
$2(x - 2y) = 22 + y$

24. $-3(x - y) = y - 14$ $\{(6, 2)\}$
$2x + 2 = 7y$

25. $0.6x + 0.1y = 0.4$ $\{(\frac{1}{2}, 1)\}$
$2x - 0.7y = 0.3$

26. $0.25x - 0.04y = 0.24$ $\{(1, \frac{1}{4})\}$
$0.15x - 0.12y = 0.12$

27. $2x + 11y = 4$ $\{(\frac{79}{45}, \frac{2}{45})\}$
$3x - 6y = 5$

28. $3x - 4y = 9$ $\{(\frac{89}{35}, -\frac{12}{35})\}$
$2x + 9y = 2$

For Exercises 29–34, solve the system by using any method. If a system does not have one unique solution, state whether the system is inconsistent or whether the equations are dependent. (See Examples 5–6)

29. $3x - 4y = 6$
$9x = 12y + 4$
$\{ \ \}$; The system is inconsistent.

30. $-4x - 8y = 2$
$2x = 8 - 4y$
$\{ \ \}$; The system is inconsistent.

31. $3x + y = 6$
$x + \dfrac{1}{3}y = 2$

32. $2x - y = 8$
$x - \dfrac{1}{2}y = 4$

33. $2x + 4 = 4 - 5y$ $\{(0, 0)\}$
$2 + 4(x + y) = 7y + 2$

34. $3(x - 3y) = 2y$ $\{(0, 0)\}$
$2x + 5 = 5 - 7y$

For Exercises 35–36,

a. Write the general solution.

b. Find three individual solutions. Answers will vary.

35. $-5x - y = 6$
$10x = -2(y + 6)$
 a. $\{(x, -5x - 6) | x \text{ is any real number}\}$
 or $\left\{ \left(-\dfrac{y + 6}{5}, y \right) \middle| y \text{ is any real number} \right\}$
 b. For example: $(0, -6)$, $(1, -11)$, $(-2, 4)$

36. $2y = 6 - 4x$
$8x = 12 - 4y$
 a. $\{(x, -2x + 3) | x \text{ is any real number}\}$
 or $\left\{ \left(\dfrac{3 - y}{2}, y \right) \middle| y \text{ is any real number} \right\}$
 b. For example: $(0, 3)$, $(1, 1)$, $(-3, 9)$

Mixed Exercises

For Exercises 37–50, solve the system using any method.

37. $3x - 10y = 1900$ $\{(300, -100)\}$
$5y + 800 = x$

38. $2x - 7y = 2400$ $\{(500, -200)\}$
$-4x + 1800 = y$

39. $5(2x + y) = y - x - 8$ $\left\{ \left(-\dfrac{4}{41}, -\dfrac{71}{41} \right) \right\}$
$x - \dfrac{3}{2}y = \dfrac{5}{2}$

40. $3(2x - y) = 2 - x$ $\left\{ \left(\dfrac{28}{47}, \dfrac{34}{47} \right) \right\}$
$x + \dfrac{5}{4}y = \dfrac{3}{2}$

41. $y = \dfrac{2}{3}x - 1$ $\{(6, 3)\}$
$y = \dfrac{1}{6}x + 2$

42. $y = -\dfrac{1}{4}x + 7$ $\{(8, 5)\}$
$y = -\dfrac{3}{2}x + 17$

43. $4(x - 2) = 6y + 3$ $\{ \ \}$
$\dfrac{1}{4}x - \dfrac{3}{8}y = -\dfrac{1}{2}$

44. $\dfrac{1}{14}x - \dfrac{1}{7}y = \dfrac{1}{2}$ $\{ \ \}$
$2(x - 2y) + 3 = 20$

45. $2x = \dfrac{y}{2} + 1$
$0.04x - 0.01y = 0.02$

46. $0.05x + 0.01y = 0.03$
$x + \dfrac{y}{5} = \dfrac{3}{5}$

47. $y = 2.4x - 1.54$ $\{(1.6, 2.3)\}$
$y = -3.5x + 7.9$

48. $y = -0.18x + 0.129$ $\{(0.05, 0.12)\}$
$y = -0.15x + 0.1275$

49. $\dfrac{x - 2}{8} + \dfrac{y + 1}{2} = -6$ $\{(18, -17)\}$
$\dfrac{x - 2}{2} - \dfrac{y + 1}{4} = 12$

50. $\dfrac{x + 1}{2} - \dfrac{y - 2}{10} = -1$ $\{(5, 42)\}$
$\dfrac{x + 1}{6} + \dfrac{y - 2}{2} = 21$

Objective 3: Use Systems of Linear Equations in Applications

51. One antifreeze solution is 36% alcohol and another is 20% alcohol. How much of each mixture should be added to make 40 L of a solution that is 30% alcohol? **(See Example 7)**   25 L of 36% solution and 15 L of 20% solution should be mixed.

52. A pharmacist wants to mix a 30% saline solution with a 10% saline solution to get 200 mL of a 12% saline solution. How much of each solution should she use? The pharmacist should mix 20 mL of the 30% solution and 180 mL of the 10% solution.

53. A radiator has 16 L of a 36% antifreeze solution. How much must be drained and replaced by pure antifreeze to bring the concentration level up to 50%?   3.5 L should be replaced.

54. Jonas performed an experiment for his science fair project. He learned that rinsing lettuce in vinegar kills more bacteria than rinsing with water or with a popular commercial product. As a follow-up to his project, he wants to determine the percentage of bacteria killed by rinsing with a diluted solution of vinegar.

   a. How much water and how much vinegar should be mixed to produce 10 cups of a mixture that is 40% vinegar?
   6 cups of water and 4 cups of vinegar should be mixed.
   b. How much pure vinegar and how much 40% vinegar solution should be mixed to produce 10 cups of a mixture that is 60% vinegar?   He should mix $6\frac{2}{3}$ cups of the 40% solution and $3\frac{1}{3}$ cups of pure vinegar.

55. Michelle borrows a total of $5000 in student loans from two lenders. One charges 4.6% simple interest and the other charges 6.2% simple interest. She is not required to pay off the principal or interest for 3 yr. However, at the end of 3 yr, she will owe a total of $762 for the interest from both loans. How much did she borrow from each lender?
She borrowed $3500 at 4.6% and $1500 at 6.2%.

56. Juan borrows $100,000 to pay for medical school. He borrows part of the money from the school whereby he will pay 4.5% simple interest. He borrows the rest of the money through a government loan that will charge him 6% interest. In both cases, he is not required to pay off the principal or interest during his 4 yr of medical school. However, at the end of 4 yr, he will owe a total of $19,200 for the interest from both loans. How much did he borrow from each source?   He borrowed $80,000 from the school and $20,000 through the government loan.

57. Stuart pays back two student loans over a 4-yr period. One loan charges the equivalent of 3% simple interest and the other charges the equivalent of 5.5% simple interest. If the total amount borrowed was $24,000 and the total amount of interest paid after 4 yr is $3280, find the amount borrowed from each loan.
He borrowed $20,000 at 3% and $4000 at 5.5%.

58. A total of $6000 is invested for 5 yr with a total return of $1080. Part of the money is invested in a fund that returns the equivalent of 2% simple interest. The rest of the money is invested at 4% simple interest. Determine the amount invested in each account.
$1200 was invested at 2% and $4800 was invested at 4%.

59. Monique and Tara each make an ice cream sundae. Monique gets 2 scoops of Cherry ice cream and 1 scoop of Mint Chocolate Chunk ice cream for a total of 43 g of fat. Tara has 1 scoop of Cherry and 2 scoops of Mint Chocolate Chunk for a total of 47 g of fat. How many grams of fat does 1 scoop of each type of ice cream have?
Cherry has 13 g of fat and Mint Chocolate Chunk has 17 g of fat.

60. Bryan and Jadyn had barbeque potato chips and soda at a football party. Bryan ate 3 oz of chips and drank 2 cups of soda for a total of 700 mg of sodium. Jadyn ate 1 oz of chips and drank 3 cups of soda for a total of 350 mg of sodium. How much sodium is in 1 oz of chips and how much is in 1 cup of soda?   1 oz of chips contains 200 mg of sodium and 1 cup of soda contains 50 mg of sodium.

61. The average weekly salary of two employees is $1350. One makes $300 more than the other. Find their salaries.   One makes $1200 and the other makes $1500.

62. The average of an electrician's hourly wage and a plumber's hourly wage is $33. One day a contractor hires the electrician for 8 hr of work and the plumber for 5 hr of work and pays a total of $438 in wages. Find the hourly wage for the electrician and for the plumber.   The electrician makes $36/hr and the plumber makes $30/hr.

63. A moving sidewalk in an airport moves people between gates. It takes Jason's 9-year-old daughter Josie 40 sec to travel 200 ft walking with the sidewalk. It takes her 30 sec to walk 90 ft against the moving sidewalk (in the opposite direction). Find the speed of the sidewalk and find Josie's speed walking on non-moving ground. **(See Example 8)**   The sidewalk moves at 1 ft/sec and Josie walks 4 ft/sec on nonmoving ground.

64. A fishing boat travels along the east coast of the United States and encounters the Gulf Stream current. It travels 44 mi north with the current in 2 hr. It travels 56 mi south against the current in 4 hr. Find the speed of the current and the speed of the boat in still water.

The current is 4 mph and the boat travels 18 mph in still water.

65. Two runners begin at the same point on a 390-m circular track and run at different speeds. If they run in opposite directions, they pass each other in 30 sec. If they run in the same direction, they meet each other in 130 sec. Find the speed of each runner.
The speeds are 8 m/sec and 5 m/sec.

66. Two particles begin at the same point and move at different speeds along a circular path of circumference 280 ft. Moving in opposite directions, they pass in 10 sec. Moving in the same direction, they pass in 70 sec. Find the speed of each particle.
The particles travel 16 ft/sec and 12 ft/sec.

67. A cleaning company charges $100 for each office it cleans. The fixed monthly cost of $480 for the company includes telephone service and the depreciation on cleaning equipment and a van. The variable cost is $52 per office and includes labor, gasoline, and cleaning supplies. **(See Example 9)**

   a. Write a linear cost function representing the cost $C(x)$ (in $) to the company to clean x offices per month. $C(x) = 52x + 480$

   b. Write a linear revenue function representing the revenue $R(x)$ (in $) for cleaning x offices per month. $R(x) = 100x$

   c. Determine the number of offices to be cleaned per month for the company to break even.   10 offices

   d. If 28 offices are cleaned, will the company make money or lose money?   The company will make money.

68. A vendor at a carnival sells cotton candy and caramel apples for $2.00 each. The vendor is charged $100 to set up his booth. Furthermore, the vendor's average cost for each product he produces is approximately $0.75.

    a. Write a linear cost function representing the cost $C(x)$ (in $) to the vendor to produce x products. $C(x) = 0.75x + 100$

    b. Write a linear revenue function representing the revenue $R(x)$ (in $) for selling x products. $R(x) = 2x$

    c. Determine the number of products to be produced and sold for the vendor to break even.   80 products

    d. If 60 products are sold, will the vendor make money or lose money?   The vendor will lose money.

For Exercises 69–70, refer to Figure 8-1 and the narrative at the beginning of this section.

69. Suppose that the price p (in $) of theater tickets is influenced by the number of tickets x offered by the theater and demanded by consumers.

    Supply: $p = 0.025x$

    Demand: $p = -0.04x + 104$

    a. Solve the system of equations defined by the supply and demand models.   {(1600, 40)}

    b. What is the equilibrium price? $40

    c. What is the equilibrium quantity?   1600 tickets

71. a. Sketch the lines defined by $y = 2x$ and $y = -\frac{1}{2}x + 5$.

    b. Find the area of the triangle bounded by the lines in part (a) and the x-axis.   20 square units

73. The **centroid** of a region is the geometric center. For the region shown, the centroid is the point of intersection of the diagonals of the parallelogram.

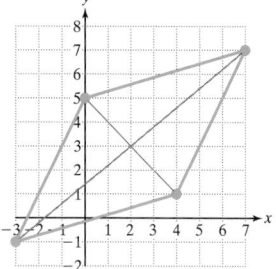

    a. Find an equation of the line through the points $(-3, -1)$ and $(7, 7)$. $y = \frac{4}{5}x + \frac{7}{5}$

    b. Find an equation of the line through the points $(0, 5)$ and $(4, 1)$. $y = -x + 5$

    c. Find the centroid of the region.   (2, 3)

75. Two angles are complementary. The measure of one angle is 6° less than twice the measure of the other angle. Find the measure of each angle.   The angles are 32° and 58°.

For Exercises 77–78, find the measure of angles x and y.

77.

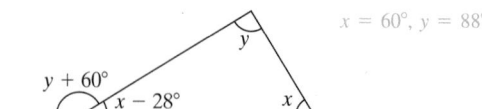

$x = 60°$, $y = 88°$

70. The price p (in $) of a cookbook is determined by the number of cookbooks x demanded by consumers and supplied by the publisher.

    Supply: $p = 0.002x$

    Demand: $p = -0.005x + 70$

    a. Solve the system of equations defined by the supply and demand models.   (10,000, 20)

    b. What is the equilibrium price? $20

    c. What is the equilibrium quantity?   10,000 cookbooks

72. a. Sketch the lines defined by $y = x + 2$ and $y = -\frac{1}{2}x + 2$.

    b. Find the area of the triangle bounded by the lines in part (a) and the x-axis.   6 square units

74. The centroid of the region shown is the point of intersection of the diagonals of the parallelogram.

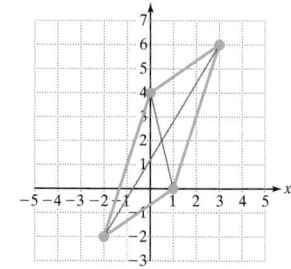

    a. Find an equation of the line through the points $(-2, -2)$ and $(3, 6)$. $y = \frac{8}{5}x + \frac{6}{5}$

    b. Find an equation of the line through the points $(1, 0)$ and $(0, 4)$. $y = -4x + 4$

    c. Find the centroid of the region. $\left(\frac{1}{2}, 2\right)$

76. Two angles are supplementary. The measure of one angle is 12° more than 5 times the measure of the other angle. Find the measure of each angle.   The angles are 28° and 152°.

78.

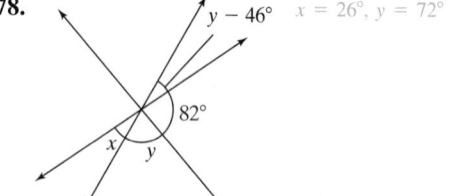

$x = 26°$, $y = 72°$

Mixed Exercises

79. Write a system of linear equations with solution set $\{(-3, 5)\}$.   For example: $x + y = 2$
$2x + y = -1$

81. Find C and D so that the solution set to the system is $\{(4, 1)\}$. $C = 2$ and $D = 3$

$$Cx + 5y = 13$$
$$-2x + Dy = -5$$

80. Write a system of linear equations with solution set $\{(4, -3)\}$.   For example: $x + y = 1$
$3x + 4y = 0$

82. Find A and B so that the solution set to the system is $\{(-5, 2)\}$. $A = 6$ and $B = 2$

$$3x + Ay = -3$$
$$Bx - y = -12$$

83. Given $f(x) = mx + b$, find m and b if $f(3) = -3$ and $f(-12) = -8$. $m = \frac{1}{3}$ and $b = -4$

84. Given $g(x) = mx + b$, find m and b if $g(2) = 1$ and $g(-4) = 10$. $m = -\frac{3}{2}$ and $b = 4$

For Exercises 85–86, use the substitution $u = \frac{1}{x}$ and $v = \frac{1}{y}$ to rewrite the equations in the system in terms of the variables u and v. Solve the system in terms of u and v. Then back substitute to determine the solution set to the original system in terms of x and y.

85.
$$\frac{1}{x} + \frac{2}{y} = 1 \quad \{(\tfrac{1}{3}, -1)\}$$
$$-\frac{1}{x} + \frac{4}{y} = -7$$

86.
$$-\frac{3}{x} + \frac{4}{y} = 11 \quad \{(-1, \tfrac{1}{2})\}$$
$$\frac{1}{x} - \frac{2}{y} = -5$$

87. During a race, Marta bicycled 12 mi and ran 4 mi in a total of 1 hr 20 min $\left(\frac{4}{3}\text{ hr}\right)$. In another race, she bicycled 21 mi and ran 3 mi in 1 hr 40 min $\left(\frac{5}{3}\text{ hr}\right)$. Determine the speed at which she bicycles and the speed at which she runs. Assume that her bicycling speed was the same in each race and that her running speed was the same in each race. Marta bicycles 18 mph and runs 6 mph.

88. Shelia swam 1 mi and ran 6 mi in a total of 1 hr 15 min $\left(\frac{5}{4}\text{ hr}\right)$. In another training session she swam 2 mi and ran 8 mi in a total of 2 hr. Determine the speed at which she swims and the speed at which she runs. Assume that her swimming speed was the same each day and that her running speed was the same each day. Sheila swims 2 mph and runs 8 mph.

89. A certain pickup truck gets 16 mpg in the city and 22 mpg on the highway. If a driver drives 254 mi on 14 gal of gas, determine the number of city miles and highway miles that the truck was driven. The truck was driven 144 mi in the city and 110 mi on the highway.

90. A sedan gets 12 mpg in the city and 18 mpg on the highway. If a driver drives a total of 420 mi on 26 gal of gas, how many miles in the city and how many miles on the highway did he drive? The driver drove 96 mi in the city and 324 mi on the highway.

Write About It

91. A system of linear equations in x and y can represent two intersecting lines, two parallel lines, or a single line. Describe the solution set to the system in each case.

92. When solving a system of linear equations in two variables using the substitution or addition method, explain how you can detect whether the equations are dependent.

93. When solving a system of linear equations in two variables using the substitution or addition method, explain how you can detect whether the system is inconsistent.

94. Consider a system of linear equations in two variables in which the solution set is $\{(x, x + 2)\,|\,x$ is any real number$\}$. Why do we say that the equations in the system are dependent?

Expanding Your Skills

95. A 50-lb weight is supported from two cables and the system is in equilibrium. The magnitudes of the forces on the cables are denoted by $|F_1|$ and $|F_2|$, respectively. An engineering student knows that the horizontal components of the two forces (shown in red) must be equal in magnitude. Furthermore, the sum of the magnitudes of the vertical components of the forces (shown in blue) must be equal to 50 lb to offset the downward force of the weight. Find the values of $|F_1|$ and $|F_2|$. Write the answers in exact form with no radical in the denominator. Also give approximations to 1 decimal place. $|F_1| = 50(\sqrt{3} - 1)$ lb ≈ 36.6 lb and $|F_2| = 25\sqrt{2}(3 - \sqrt{3})$ lb ≈ 44.8 lb

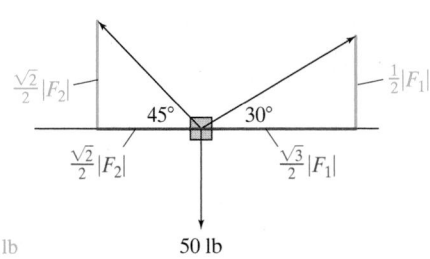

Technology Connections

For Exercises 96–99, use a graphing utility to approximate the solution to the system of equations. Round the x and y values to 3 decimal places.

96. $y = -3.729x + 6.958$
  $y = 2.615x - 8.713$

97. $y = -0.041x + 0.068$
  $y = 0.019x - 0.053$

98. $-0.25x + 0.04y = -0.42$
  $6.775x + 2.5y = -38.1$

99. $0.36x - 0.075y = -0.813$
  $0.066x + 0.008y = 0.194$

SECTION 8.2   # Systems of Linear Equations in Three Variables and Applications

OBJECTIVES

1. Identify Solutions to a System of Linear Equations in Three Variables
2. Solve Systems of Linear Equations in Three Variables
3. Use Systems of Linear Equations in Applications
4. Modeling with Linear Equations in Three Variables

1. Identify Solutions to a System of Linear Equations in Three Variables

In Section 8.1 we solved systems of linear equations in two variables. In this section, we expand the discussion to solving systems involving three variables. A **linear equation in three variables** is an equation that can be written in the form

$$Ax + By + Cz = D, \text{ where } A, B, \text{ and } C \text{ are not all zero.}$$

For example, $x + 2y + z = 4$ is a linear equation in three variables. A solution to a linear equation in three variables is an ordered triple (x, y, z) that satisfies the equation. For example, several solutions to $x + 2y + z = 4$ are given here.

Solution	Check: $x + 2y + z = 4$
$(1, 1, 1)$	$(1) + 2(1) + (1) \stackrel{?}{=} 4$ ✓ true
$(4, 0, 0)$	$(4) + 2(0) + (0) \stackrel{?}{=} 4$ ✓ true
$(0, 2, 0)$	$(0) + 2(2) + (0) \stackrel{?}{=} 4$ ✓ true
$(0, 0, 4)$	$(0) + 2(0) + (4) \stackrel{?}{=} 4$ ✓ true

There are infinitely many solutions to the equation $x + 2y + z = 4$. The set of all solutions to a linear equation in three variables can be represented graphically by a plane in space. Figure 8-4 shows a portion of the plane defined by $x + 2y + z = 4$.

In many applications, we are interested in determining the point or points of intersection of two or more planes. This is given by the solutions to a system of linear equations in three variables. For example:

$$2x + y - 3z = -2$$
$$x - 4y + z = 24$$
$$-3x - y + 4z = 0$$

A solution to a system of linear equations in three variables is an **ordered triple** (x, y, z) that satisfies each equation in the system. Geometrically, a solution is a point of intersection of the planes represented by the equations in the system (Figure 8-5).

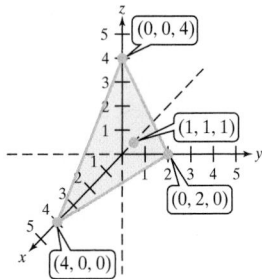

Figure 8-4

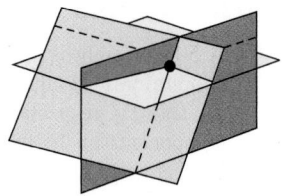

Figure 8-5

Point of Interest

Beautiful planar surfaces are seen in both art and architecture. For example, the great pyramid of Khufu outside Cairo, Egypt, has four triangular sides. The planes representing the sides form an angle of 52° with the ground.

Classroom Example: p. 788
Exercise 6

EXAMPLE 1   Determining if an Ordered Triple Is a Solution to a System of Equations

Determine if the ordered triple is a solution to the system.

$$2x + y - 3z = -2$$
$$x - 4y + z = 24$$
$$-3x - y + 4z = 0$$

a. $(3, -5, 1)$ **b.** $(2, -3, 1)$

Solution:

Test the ordered triple in each equation.

a. First equation

$2x + y - 3z = -2$

$2(3) + (-5) - 3(1) \overset{?}{=} -2$

$-2 \overset{?}{=} -2 \checkmark$ true

Second equation

$x - 4y + z = 24$

$(3) - 4(-5) + (1) \overset{?}{=} 24$

$24 \overset{?}{=} 24 \checkmark$ true

Third equation

$-3x - y + 4z = 0$

$-3(3) - (-5) + 4(1) \overset{?}{=} 0$

$0 \overset{?}{=} 0 \checkmark$ true

The ordered triple $(3, -5, 1)$ is a solution to the system of equations.

b. First equation

$2x + y - 3z = -2$

$2(2) + (-3) - 3(1) \overset{?}{=} -2$

$-2 \overset{?}{=} -2 \checkmark$ true

Second equation

$x - 4y + z = 24$

$(2) - 4(-3) + (1) \overset{?}{=} 24$

$15 \overset{?}{=} 24$ false

Third equation

$-3x - y + 4z = 0$

$-3(2) - (-3) + 4(1) \overset{?}{=} 0$

$1 \overset{?}{=} 0$ false

If an ordered triple fails to be a solution to any of the equations in the system, then it is not a solution to the system. The ordered triple $(2, -3, 1)$ is *not* a solution to the second or third equation. Therefore, $(2, -3, 1)$ is *not* a solution to the system of equations.

Skill Practice 1 Determine if the ordered triple is a solution to the system.

$$5x - y + 3z = -7$$
$$3x + 4y - z = 5$$
$$9x + 5y + 7z = 1$$

a. $(-2, -6, 3)$ **b.** $(-1, 2, 0)$

---

2. Solve Systems of Linear Equations in Three Variables

To solve a system of three linear equations in three variables, we first eliminate one variable. The system is then reduced to a two-variable system that can be solved by the techniques learned in Section 8.1.

Solving a System of Three Linear Equations in Three Variables

Step 1 Write each equation in standard form $Ax + By + Cz = D$.

Step 2 Choose a pair of equations and eliminate one of the variables by using the addition method.

Step 3 Choose a different pair of equations and eliminate the *same* variable.

Step 4 Once steps 2 and 3 are complete, the resulting system has two equations in two variables. Solve this system by using the substitution or addition method.

Step 5 Substitute the values of the variables found in step 4 into any of the three original equations that contain the third variable. Solve for the third variable.

Step 6 Check the ordered triple in each original equation. Then write the solution as an ordered triple in set notation.

Classroom Example: p. 788
Exercise 12

Answers

1. a. No   **b.** Yes

EXAMPLE 2 **Solving a System of Equations in Three Variables**

Solve the system.

$$3x - 2y + z = 2$$
$$5x + y - 2z = 1$$
$$4x - 3y + 3z = 7$$

TIP In Example 2, the y terms can also be eliminated easily because the coefficient on y in equation $\boxed{\text{B}}$ is 1. Therefore, y can be eliminated from equations $\boxed{\text{A}}$ and $\boxed{\text{B}}$ by multiplying equation $\boxed{\text{B}}$ by 2. Likewise y can be eliminated from equations $\boxed{\text{B}}$ and $\boxed{\text{C}}$ by multiplying equation $\boxed{\text{B}}$ by 3.

Solution:

$\boxed{\text{A}}$ $\quad 3x - 2y + z = 2$
$\boxed{\text{B}}$ $\quad 5x + y - 2z = 1$
$\boxed{\text{C}}$ $\quad 4x - 3y + 3z = 7$

Step 1: The equations are already in standard form.
- It is helpful to label the equations $\boxed{\text{A}}$, $\boxed{\text{B}}$, and $\boxed{\text{C}}$.
- The z variable can easily be eliminated from equations $\boxed{\text{A}}$ and $\boxed{\text{B}}$ and from equations $\boxed{\text{A}}$ and $\boxed{\text{C}}$. This is accomplished by creating opposite coefficients for the z terms and then adding the equations.

Step 2: Eliminate z from equations $\boxed{\text{A}}$ and $\boxed{\text{B}}$.

$\boxed{\text{A}}$ $\quad 3x - 2y + z = 2$ $\xrightarrow{\text{Multiply by 2.}}$ $\quad 6x - 4y + 2z = 4$
$\boxed{\text{B}}$ $\quad 5x + y - 2z = 1$ $\qquad\qquad\qquad\quad \dfrac{5x + y - 2z = 1}{11x - 3y \qquad\;\; = 5}$ $\boxed{\text{D}}$

Step 3: Eliminate z from equations $\boxed{\text{A}}$ and $\boxed{\text{C}}$.

$\boxed{\text{A}}$ $\quad 3x - 2y + z = 2$ $\xrightarrow{\text{Multiply by} -3.}$ $-9x + 6y - 3z = -6$
$\boxed{\text{C}}$ $\quad 4x - 3y + 3z = 7$ $\qquad\qquad\qquad \dfrac{4x - 3y + 3z = 7}{-5x + 3y \qquad\;\; = 1}$ $\boxed{\text{E}}$

Step 4: $\boxed{\text{D}}$ $\quad 11x - 3y = 5$ $\qquad\qquad$ $\boxed{\text{D}}$ $\quad 11(1) - 3y = 5$ $\qquad$ Solve the system
 $\boxed{\text{E}}$ $\quad \dfrac{-5x + 3y = 1}{6x \qquad\;\; = 6}$ $\qquad\qquad\qquad 11 - 3y = 5$ $\qquad$ of equations $\boxed{\text{D}}$
 $x = 1$ $\qquad\qquad\qquad\qquad\qquad -3y = -6$ $\qquad$ and $\boxed{\text{E}}$.
 $y = 2$

$\boxed{\text{A}}$ $\quad 3x - 2y + z = 2$
$\qquad\quad 3(1) - 2(2) + z = 2$
$\qquad\qquad\quad 3 - 4 + z = 2$
$\qquad\qquad\qquad -1 + z = 2$
$\qquad\qquad\qquad\qquad\; z = 3$

Step 5: Substitute the values of the known variables x and y back into one of the original equations. We have chosen equation $\boxed{\text{A}}$.

Step 6: Check the ordered triple $(1, 2, 3)$ in the three original equations.

$\boxed{\text{A}}$ $\; 3x - 2y + z = 2$ $\qquad$ $\boxed{\text{B}}$ $\; 5x + y - 2z = 1$ $\qquad$ $\boxed{\text{C}}$ $\; 4x - 3y + 3z = 7$
$\quad 3(1) - 2(2) + (3) \overset{?}{=} 2$ $\qquad 5(1) + (2) - 2(3) \overset{?}{=} 1$ $\qquad 4(1) - 3(2) + 3(3) \overset{?}{=} 7$
$\quad 2 \overset{?}{=} 2$ ✓ true $\qquad\qquad 1 \overset{?}{=} 1$ ✓ true $\qquad\qquad 7 \overset{?}{=} 7$ ✓ true

The solution set is $\{(1, 2, 3)\}$.

Skill Practice 2 Solve the system.
$$2x - y + 5z = -7$$
$$x + 4y - 2z = 1$$
$$3x + 2y + z = -7$$

In Example 3, we solve a system of linear equations in which one or more equations has a missing term.

Classroom Example: p. 788
Exercise 14

EXAMPLE 3   Solving a System of Equations in Three Variables

Solve the system. $\qquad 2x + y = -2$
$\qquad\qquad\qquad\qquad\quad 3y = 5z - 12$
$\qquad\qquad\qquad\quad 5(x + z) = 2z + 5$

Solution:

Step 1: Write the equations in standard form.

$\boxed{\text{A}}$ $\quad 2x + y = -2$ $\qquad\longrightarrow\qquad 2x + y \qquad\;\; = -2$ $\qquad$ Notice that the equations
$\boxed{\text{B}}$ $\quad 3y = 5z - 12$ $\qquad\longrightarrow\qquad\qquad 3y - 5z = -12$ $\qquad$ already have missing
$\boxed{\text{C}}$ $\quad 5(x + z) = 2z + 5$ $\longrightarrow\quad 5x \qquad\;\; + 3z = 5$ $\qquad$ variable terms.

Answer
2. $\{(-3, 1, 0)\}$

Steps 2 and 3: This system of equations has several missing terms. For example, equation $\boxed{C}$ is missing the variable y. If we eliminate y from equations $\boxed{A}$ and $\boxed{B}$, then we will have a second equation with variable y missing.

$$\boxed{A} \quad 2x + y \qquad = -2 \xrightarrow{\text{Multiply by } -3.} -6x - 3y \qquad = 6$$
$$\boxed{B} \qquad \quad 3y - 5z = -12 \qquad\qquad\qquad\qquad 3y - 5z = -12$$
$$\overline{\qquad\qquad -6x \qquad\quad - 5z = -6 \;\; \boxed{D}}$$

Step 4: Pair up equations $\boxed{C}$ and $\boxed{D}$. These equations form a system of linear equations in two variables. To solve the system with equations $\boxed{C}$ and $\boxed{D}$ we have chosen to eliminate the z variable.

$$\boxed{C} \qquad 5x + 3z = 5 \xrightarrow{\text{Multiply by 5.}} \quad 25x + 15z = 25$$
$$\boxed{D} \quad -6x - 5z = -6 \xrightarrow[\text{Multiply by 3.}]{} \quad -18x - 15z = -18$$
$$\overline{\qquad\qquad\qquad 7x \qquad\quad = 7}$$
$$x = 1$$

$$\boxed{C} \qquad 5x + 3z = 5$$
$$5(1) + 3z = 5$$
$$z = 0$$

$\boxed{B} \quad 3y = 5z - 12$ **Step 5:** Substitute the values of the known variables x
$\qquad\;\; 3y = 5(0) - 12$ and z back into one of the original equations containing
$\qquad\;\; 3y = -12$ y. We have chosen equation $\boxed{B}$.
$\qquad\quad\; y = -4$

Step 6: The ordered triple $(1, -4, 0)$ checks in each original equation.

The solution set is $\{(1, -4, 0)\}$.

Skill Practice 3   Solve the system.

$$a \qquad\quad + 3c = 4$$
$$b + 2c = -1$$
$$2a - 4b \qquad = 14$$

A system of linear equations in three variables may have no solution. This occurs if the equations represent planes that do not all intersect (Figure 8-6). In such a case, we say that the system is **inconsistent.**

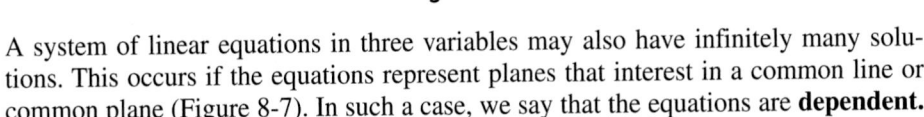

Figure 8-6

A system of linear equations in three variables may also have infinitely many solutions. This occurs if the equations represent planes that interest in a common line or common plane (Figure 8-7). In such a case, we say that the equations are **dependent.**

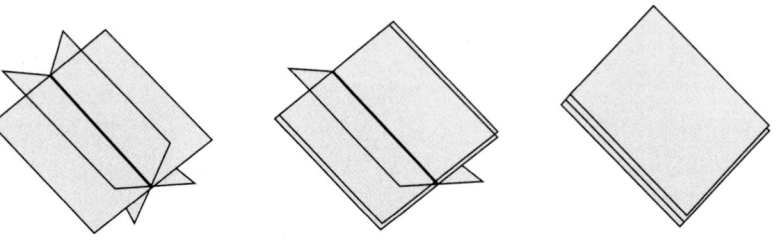

Figure 8-7

Classroom Example: p. 788
Exercise 16

EXAMPLE 4   Determining the Number of Solutions to a System

a. Determine the number of solutions to the system.
b. State whether the system is inconsistent or the equations are dependent.
c. Write the solution set.

$$-x + 6y - 3z = -8$$
$$x - 2y + 2z = 3$$
$$3x + 2y + 4z = -6$$

Solution:

Begin by eliminating a variable from two different pairs of equations. We will eliminate x from $\boxed{A}$ and $\boxed{B}$ and from $\boxed{A}$ and $\boxed{C}$.

$\boxed{A}$ $-x + 6y - 3z = -8$ ⟶ $-x + 6y - 3z = -8$
$\boxed{B}$ $x - 2y + 2z = 3$ ⟶ $\underline{x - 2y + 2z = 3}$ Add equations $\boxed{A}$ and $\boxed{B}$ to
$\boxed{C}$ $3x + 2y + 4z = -6$ $4y - z = -5$ $\boxed{D}$ eliminate x.

$\boxed{A}$ $-x + 6y - 3z = -8$ $\xrightarrow{\text{Multiply by 3.}}$ $-3x + 18y - 9z = -24$ Multiply
$\boxed{C}$ $3x + 2y + 4z = -6$ ⟶ $\underline{3x + 2y + 4z = -6}$ equation $\boxed{A}$ by 3 and add the
       $20y - 5z = -30$ $\boxed{E}$ result to $\boxed{C}$.

$\boxed{D}$ $4y - z = -5$ $\xrightarrow{\text{Multiply by }-5.}$ $-20y + 5z = 25$
$\boxed{E}$ $20y - 5z = -30$ $\underline{20y - 5z = -30}$ Solving the system of equations $\boxed{D}$ and $\boxed{E}$
       $0 = -5$ results in a contradiction.

The system of equations reduces to a contradiction.

a. There is no solution.

b. The system is inconsistent.

c. The solution set is { }.

Skill Practice 4   Repeat Example 4 with the given system.

$$x + y + 4z = -1$$
$$3x + y - 4z = 3$$
$$-4x - y + 8z = -2$$

In Example 5, we investigate the case in which a system of equations has infinitely many solutions.

Classroom Examples: pp. 788–789
Exercises 20, 34

EXAMPLE 5   Determining the Number of Solutions to a System

a. Determine the number of solutions to the system.
b. State whether the system is inconsistent, or the equations are dependent.
c. Write the solution set.

$$2x + y = -3$$
$$2y + 16z = -10$$
$$-7x - 3y + 4z = 8$$

Solution:

Eliminate variable z from equations $\boxed{B}$ and $\boxed{C}$.

$\boxed{A}$ $2x + y = -3$
$\boxed{B}$ $2y + 16z = -10$ ⟶ $2y + 16z = -10$
$\boxed{C}$ $-7x - 3y + 4z = 8$ $\xrightarrow{\text{Multiply by }-4.}$ $\underline{28x + 12y - 16z = -32}$
            $28x + 14y = -42$ $\boxed{D}$

Answers

4. a. There is no solution.
   b. The system is inconsistent.
   c. The solution set is { }.

Pair up equations $\boxed{A}$ and $\boxed{D}$ to solve for x and y.

$$\boxed{A} \quad 2x + \quad y = -3 \qquad\longrightarrow\qquad 2x + y = -3$$
$$\boxed{D} \quad 28x + 14y = -42 \underset{\text{Divide by } -14.}{\longrightarrow} \quad \underline{-2x - y = 3}$$
$$0 = 0$$

The system reduces to the identity $0 = 0$. This implies that

a. There are infinitely many solutions.

b. The equations are dependent.

c. To find the general solution, we need to express the dependency among the variables as an ordered triple. Note that from equation $\boxed{A}$, we can solve for x in terms of y, and from equation $\boxed{B}$, we can solve for z in terms of y. Thus,

$$\text{Equation } \boxed{A}: 2x = -y - 3 \Rightarrow x = \frac{-y-3}{2} \Rightarrow x = -\frac{y+3}{2}$$

$$\text{Equation } \boxed{B}: 16z = -2y - 10 \Rightarrow z = \frac{-2y-10}{16} \Rightarrow z = -\frac{y+5}{8}$$

Therefore, the solution set is $\left\{ \left(-\frac{y+3}{2}, y, -\frac{y+5}{8} \right) \,\middle|\, y \text{ is any real number} \right\}$.

Skill Practice 5   Repeat Example 5 with the given system.

$$5y + \quad z = 0$$
$$-x \quad\quad + 4z = 0$$
$$-x + 5y + 5z = 0$$

The general solution to the system in Example 5 can be written in a number of forms.

Solve for x and z in terms of y: $\left\{ \left(-\frac{y+3}{2}, y, -\frac{y+5}{8} \right) \,\middle|\, y \text{ is any real number} \right\}$

Solve for y and z in terms of x: $\left\{ \left(x, -2x - 3, \frac{x-1}{4} \right) \,\middle|\, x \text{ is any real number} \right\}$

Solve for x and y in terms of z: $\left\{ \left(4z + 1, -8z - 5, z \right) \,\middle|\, z \text{ is any real number} \right\}$

TIP Any form of the general solution can be checked by substitution in the original three equations.

Check: $\left\{ \left(-\frac{y+3}{2}, y, -\frac{y+5}{8} \right) \,\middle|\, y \text{ is any real number} \right\}$

$\boxed{A} \quad\quad 2x + y = -3$
$2\left(-\frac{y+3}{2} \right) + y \overset{?}{=} -3$
$-y - 3 + y \overset{?}{=} -3 \checkmark$

$\boxed{B} \quad\quad 2y + 16z = -10$
$2y + 16\left(-\frac{y+5}{8} \right) \overset{?}{=} -10$
$2y - 2y - 10 = -10 \checkmark$

$\boxed{C} \quad\quad -7x - 3y + 4z = 8$
$-7\left(-\frac{y+3}{2} \right) - 3y + 4\left(-\frac{y+5}{8} \right) \overset{?}{=} 8$
$\frac{7}{2}y + \frac{21}{2} - 3y - \frac{1}{2}y - \frac{5}{2} \overset{?}{=} 8$
$\frac{6}{2}y - 3y + \frac{16}{2} = 8 \checkmark$

Answers

5. a. There are infinitely many solutions.

b. The equations are dependent.

c. $\left\{ \left(4z, -\frac{1}{5}z, z \right) \,\middle|\, z \text{ is any real number} \right\}$ or $\left\{ \left(x, -\frac{x}{20}, \frac{x}{4} \right) \,\middle|\, x \text{ is any real number} \right\}$ or $\{(-20y, y, -5y) \,|\, y \text{ is any real number}\}$

Classroom
Exercise 5(

TIP
three po
are *not*
that the
pairs of

For (4, 2

n

For (4,

m

Answer

7. $y = -$

SECTION 8.2 Practice Exercises

Prerequisite Review

R.1. How much interest will Roxanne have to pay if she borrows $2000 for 2 yr at a simple interest rate of 3%? $120

R.2. Julie needs to have a toilet repaired in her house. The cost of the new plumbing fixtures is $110 and labor is $70/hr. Write a model that represents the cost of the repair C (in $) in terms of the number of hours of labor x. $C = 110 + 70x$

R.3. Solve the equation. $\dfrac{v-4}{5} + \dfrac{v}{8} = \dfrac{v-2}{5} - 7$ $\left\{ \dfrac{264}{5} \right\}$

Concept Connections

1. The graph of a linear equation in two variables is a line in a two-dimensional coordinate system. The graph of a linear equation in three variables is a ____plane____ in a three-dimensional coordinate system.

2. A solution to a system of linear equations in three variables is an ordered ____triple____ that satisfies each equation in the system. Graphically, this is a point of ____intersection____ of three planes.

Objective 1: Identify Solutions to a System of Linear Equations in Three Variables

For Exercises 3–4, find three ordered triples that are solutions to the linear equation in three variables.

3. $2x + 4y - 6z = 12$
For example: $(0, 0, -2)$, $(0, 3, 0)$, and $(6, 0, 0)$

4. $3x - 5y + z = 15$
For example: $(5, 0, 0)$, $(0, -3, 0)$, and $(0, 0, 15)$

For Exercises 5–8, determine if the ordered triple is a solution to the system of equations. (See Example 1)

5. $-x + 3y - 7z = 7$
$2x + 4y + z = 16$
$3x - 5y + 6z = -9$

a. $(2, 3, 0)$ Yes
b. $(-2, 4, 1)$ No

6. $2x - 3y + z = -12$
$x + y - 2z = 9$
$-3x + 2y - z = 7$

a. $(2, 5, -1)$ No
b. $(1, 4, -2)$ Yes

7. $x + y + z = 2$
$x + 2y - z = 2$
$3x + 5y - z = 6$

a. $(2, 0, 0)$ Yes
b. $(-1, 2, 1)$ Yes

8. $-x - y + z = 3$
$3x + 4y - z = 1$
$5x + 7y - z = -1$

a. $(1, 2, 6)$ No
b. $(3, -1, 5)$ No

Objective 2: Solve Systems of Linear Equations in Three Variables

For Exercises 9–32, solve the system. If a system has one unique solution, write the solution set. Otherwise, determine the number of solutions to the system, and determine whether the system is inconsistent, or the equations are dependent. (See Examples 2–5)

9. $x - 2y + z = -9$
$3x + 4y + 5z = 9$
$-2x + 3y - z = 12$
$\{(1, 4, -2)\}$

10. $2x - y + z = 6$
$-x + 5y - z = 10$
$3x + y - 3z = 12$
$\{(4, 3, 1)\}$

11. $4x = 3y - 2z - 5$
$2(x + y) = y + z - 6$
$6(x - y) + z = x - 5y - 8$
$\{(-2, 1, 3)\}$

12. $3x = 5y - z + 13$
$-(x - y) - z = x - 3$
$5(x + y) = 3y - 3z - 4$
$\{(-2, -3, 4)\}$

13. $2x + 5z = 2$
$3y - 7z = 9$
$-5x + 9y = 22$
$\{(1, 3, 0)\}$

14. $3x - 2y = -8$
$5y + 6z = 2$
$7x + 11z = -33$
$\{(0, 4, -3)\}$

15. $-4x - 3y = 0$
$3y + z = -1$
$4x - z = 12$
No solution; The system is inconsistent.

16. $4x - y + 2z = 1$
$3x + 5y - z = -2$
$-9x - 15y + 3z = 0$
No solution; The system is inconsistent.

17. $2x = 3y - 6z - 1$
$6y = 12z - 10x + 9$
$3z = 6y - 3x - 1$ $\left\{\left(\frac{1}{2}, \frac{1}{3}, -\frac{1}{6}\right)\right\}$

18. $5x = 2y - 3z - 3$
$4y = -1 - 10x - 5z$
$2z = 5x - 6y$ $\left\{\left(\frac{1}{5}, \frac{1}{2}, -1\right)\right\}$

19. $x + 2y + 4z = 3$
$y + 3z = 5$
$x - 2z = -7$
Infinitely many solutions; The equations are dependent.

20. $3x + 2y + 5z = 6$
$3y - z = 4$
$3x + 17y = 26$
Infinitely many solutions; The equations are dependent.

21. $0.2x = 0.1y - 0.6z$
$0.004x + 0.005y - 0.001z = 0$
$30x = 50z - 20y$
$\{(0, 0, 0)\}$

22. $0.3x = 0.5y - 1.2z$
$0.05x + 0.1y = 0.04z$
$100x = 300y - 700z$
$\{(0, 0, 0)\}$

23. $\frac{1}{12}x + \frac{1}{4}y + \frac{1}{3}z = \frac{7}{12}$ $\{(1, -2, 3)\}$
$-\frac{1}{10}x + \frac{1}{2}y - \frac{1}{5}z = -\frac{17}{10}$
$\frac{1}{2}x + \frac{1}{4}y + z = 3$

24. $x + \frac{7}{2}y + \frac{1}{2}z = 4$

$\frac{3}{4}x + y + \frac{1}{2}z = -1$

$\frac{1}{10}x - \frac{2}{5}y - \frac{3}{10}z = 1$ $\{(0, 2, -6)\}$

25. $3x + 2y + 5z = 12$

$3y + 8z = -8$

$10z = 20$ $\{(6, -8, 2)\}$

26. $-4x + 6y + z = -4$

$-2y - 4z = -24$

$6z = 48$ $\{(-3, -4, 8)\}$

27. $\frac{x+2}{3} + \frac{y-4}{2} + \frac{z+1}{6} = 8$

$-\frac{x+2}{3} + \frac{z+1}{2} = 8$

$\frac{y-4}{4} - \frac{z+1}{6} = -1$

$\{(1, 12, 17)\}$

28. $\frac{x-1}{7} + \frac{y-2}{3} + \frac{z+2}{4} = 13$

$\frac{y-2}{9} - \frac{z+2}{8} = 3$

$\frac{x-1}{7} + \frac{z+2}{2} = 3$

$\{(-6, 38, 6)\}$

29. $3(x + y) = 6 - 4z + y$

$4 = 6y + 5z$

$-3x + 4y + z = 0$

No solution; The system is inconsistent.

30. $4(x - y) = 8 - z - y$

$3 = 3x + 4z$

$-x + 3y + 3z = 1$

No solution; The system is inconsistent.

31. $-3x + 4y - z = -4$

$x + 2y + z = 4$

$-12x + 16y - 4z = -16$

Infinitely many solutions; The equations are dependent.

32. $x + 2y - 3z = 5$

$-2x - 5y + 4z = -6$

$3x + 6y - 9z = 15$

Infinitely many solutions; The equations are dependent.

For Exercises 33–36, solve the system from the indicated exercise and write the general solution. (See Example 5)

33. Exercise 19   **34.** Exercise 20   **35.** Exercise 31   **36.** Exercise 32

For Exercises 37–38, the general solution is given for a system of linear equations. Find three individual solutions to the system.

37. $-x + 4y + 2z = 4$ For example: (2, 1, 1),

$x - 3y - z = -2$ (0, 0, 2), (4, 2, 0)

$-x + y - z = -2$

Solution: $\{(-2z + 4, -z + 2, z) \mid z \text{ is any real number}\}$

38. $2x - 3y + z = 1$ For example: (1, 1, 2), (2, -2, -9),

$x + 4y - z = 3$ (0, 4, 13)

$5x - 2y + z = 5$

Solution: $\{(x, -3x + 4, -11x + 13) \mid x \text{ is any real number}\}$

Objective 3: Use Systems of Linear Equations in Applications

39. Devon invested $8000 in three different mutual funds. A fund containing large cap stocks made 6.2% return in 1 yr. A real estate fund lost 13.5% in 1 yr, and a bond fund made 4.4% in 1 yr. The amount invested in the large cap stock fund was twice the amount invested in the real estate fund. If Devon had a net return of $66 across all investments, how much did he invest in each fund? (See Example 6)

He invested $4000 in the large cap fund, $2000 in the real estate fund, and $2000 in the bond fund.

41. A basketball player scored 26 points in one game. In basketball, some baskets are worth 3 points, some are worth 2 points, and free-throws are worth 1 point. He scored four more 2-point baskets than he did 3-point baskets. The number of free-throws equaled the sum of the number of 2-point and 3-point shots made. How many free-throws, 2-point shots, and 3-point shots did he make?   He made eight free-throws, six 2-point shots, and two 3-point shots.

43. Plant fertilizers are categorized by the percentage of nitrogen (N), phosphorous (P), and potassium (K) they contain, by weight. For example, a fertilizer that has N-P-K numbers of 8-5-5 has 8% nitrogen, 5% phosphorous, and 5% potassium by weight. Suppose that a fertilizer has twice as much potassium by weight as phosphorous. The percentage of nitrogen equals the sum of the percentages of phosphorous and potassium. If nitrogen, phosphorous, and potassium make up 42% of the fertilizer, determine the proper N-P-K label on the fertilizer.   b

a. 14-7-14   **b.** 21-7-14   **c.** 14-7-21   **d.** 14-21-21

40. Pierre inherited $120,000 from his uncle and decided to invest the money. He put part of the money in a money market account that earns 2.2% simple interest. The remaining money was invested in a stock that returned 6% in the first year and a mutual fund that lost 2% in the first year. He invested $10,000 more in the stock than in the mutual fund, and his net gain for 1 yr was $2820. Determine the amount invested in each account.

$10,000 was put in the money market account. $60,000 was invested in the stock, and $50,000 was invested in the mutual fund.

42. A sawmill cuts boards for a lumber supplier. When saws A, B, and C all work for 6 hr, they cut 7200 linear board-ft of lumber. It would take saws A and B working together 9.6 hr to cut 7200 ft of lumber. Saws B and C can cut 7200 ft of lumber in 9 hr. Find the rate (in ft/hr) that each saw can cut lumber.

Saw A cuts 400 ft/hr, saw B cuts 350 ft/hr, and saw C cuts 450 ft/hr.

44. A theater charges $50 per ticket for seats in Section A, $30 per ticket for seats in Section B, and $20 per ticket for seats in Section C. For one play, 4000 tickets were sold for a total of $120,000 in revenue. If 1000 more tickets in Section B were sold than the other two sections combined, how many tickets in each section were sold?

500 tickets in Section A, 2500 tickets in Section B, and 1000 tickets in Section C were sold.

45. The perimeter of a triangle is 55 in. The shortest side is 7 in. less than the longest side. The middle side is 19 in. less than the combined lengths of the shortest and longest sides. Find the lengths of the three sides. The sides are 15 in., 18 in., and 22 in.

46. A package in the shape of a rectangular solid is to be mailed. The combination of the girth (perimeter of a cross section defined by w and h) and the length of the package is 48 in. The width is 2 in. greater than the height, and the length is 12 in. greater than the width. Find the dimensions of the package. The dimensions are $w = 8$ in., $h = 6$ in., and $l = 20$ in.

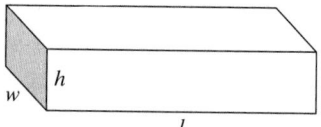

47. The measure of the largest angle in a triangle is 100° larger than the sum of the measures of the other two angles. The measure of the smallest angle is two-thirds the measure of the middle angle. Find the measure of each angle. The angles are 16°, 24°, and 140°.

48. The measure of the largest angle in a triangle is 18° more than the sum of the measures of the other two angles. The measure of the smallest angle is one-half the measure of the middle angle. Find the measure of each angle. The angles are 27°, 54°, and 99°.

Objective 4: Modeling with Linear Equations in Three Variables

49. a. Show that the points $(1, 0)$, $(3, 10)$, and $(-2, 15)$ are not collinear by finding the slope between $(1, 0)$ and $(3, 10)$, and the slope between $(3, 10)$ and $(-2, 15)$. **(See Example 7)** The slopes are 5 and -1.

 b. Find an equation of the form $y = ax^2 + bx + c$ that defines the parabola through the points. $y = 2x^2 - 3x + 1$

 c. Use a graphing utility to verify that the graph of the equation in part (b) passes through the given points.

50. a. Show that the points $(2, 9)$, $(-1, -6)$, and $(-4, -3)$ are not collinear by finding the slope between $(2, 9)$ and $(-1, -6)$, and the slope between $(2, 9)$ and $(-4, -3)$. The slopes are 5 and 2.

 b. Find an equation of the form $y = ax^2 + bx + c$ that defines the parabola through the points. $y = x^2 + 4x - 3$

 c. Use a graphing utility to verify that the graph of the equation in part (b) passes through the given points.

For Exercises 51–52, find an equation of the form $y = ax^2 + bx + c$ that defines the parabola through the three noncollinear points given.

51. $(0, 6)$, $(2, -6)$, $(-1, 9)$ $y = -x^2 - 4x + 6$

52. $(0, -4)$, $(2, -6)$, $(-3, -31)$ $y = -2x^2 + 3x - 4$

The motion of an object traveling along a straight path is given by $s(t) = \frac{1}{2}at^2 + v_0t + s_0$, where $s(t)$ is the position relative to the origin at time t. For Exercises 53–54, three observed data points are given. Find the values of a, v_0, and s_0.

53. $s(1) = 30$, $s(2) = 54$, $s(3) = 82$ $a = 4$, $v_0 = 18$, and $s_0 = 10$

54. $s(1) = -7$, $s(2) = 12$, $s(3) = 37$ $a = 6$, $v_0 = 10$, and $s_0 = -20$

Many statistics courses cover a topic called *multiple regression*. This provides a means to predict the value of a dependent variable y based on two or more independent variables $x_1, x_2, ..., x_n$. The model $y = ax_1 + bx_2 + c$ is a linear model that predicts y based on two independent variables x_1 and x_2. While statistical techniques may be used to find the values of a, b, and c based on a large number of data points, we can form a crude model given three data values (x_1, x_2, y). Use the information given in Exercises 55–56 to form a system of three equations and three variables to solve for a, b, and c.

55. The selling price of a home y (in \$1000) is given based on the living area x_1 (in 100 ft^2) and on the lot size x_2 (in acres).

Living Area (100 ft^2) x_1	Lot Size (acres) x_2	Selling Price (\$1000) y
28	0.5	225
25	0.8	207
18	0.4	154

 a. Use the data to create a model of the form $y = ax_1 + bx_2 + c$. $y = 7x_1 + 10x_2 + 24$

 b. Use the model from part (a) to predict the selling price of a home that is 2000 ft^2 on a 0.4-acre lot. \$168,000

56. The gas mileage y (in mpg) for city driving is given based on the weight of the vehicle x_1 (in lb) and on the number of cylinders.

Weight (lb) x_1	Cylinders x_2	Mileage (mpg) y
3500	6	20
3200	4	26
4100	8	18

 a. Use the data to create a model of the form $y = ax_1 + bx_2 + c$. $y = \frac{1}{75}x_1 - 5x_2 + \frac{10}{3}$

 b. Use the model from part (a) to predict the gas mileage of a vehicle that is 3800 lb and has 6 cylinders. 24 mpg

Write About It

57. Give a geometric description of the solution set to a linear equation in three variables. The set of all ordered pairs that are solutions to a linear equation in three variables forms a plane in space.

59. Explain the procedure presented in this section to solve a system of linear equations in three variables.

58. If a system of linear equations in three variables has no solution, then what can be said about the three planes represented by the equations in the system? The planes do not all intersect in a common point in space.

60. Explain how to check a solution to a system of linear equations in three variables. Substitute the ordered triple into each individual equation in the system and verify that it is a solution to each equation.

Additional answers can be found in the Instructor Answer Appendix.

Expanding Your Skills

For Exercises 61–62, find all solutions of the form (a, b, c, d).

61. $2a + b - c + d = 7$
$\qquad 3b + 2c - 2d = -11$
$\quad a \qquad + 3c + 3d = 14$
$\quad 4a + 2b - 5c \qquad = 6 \quad \{(2, -1, 0, 4)\}$

62. $3a - 4b + 2c + d = 8$
$\quad 2a + 3b \qquad + 2d = 7$
$\qquad 5b - 3c + 4d = -4$
$\quad -a + b - 2c \qquad = -7 \quad \{(2, 1, 3, 0)\}$

For Exercises 63–64, find all solutions of the form (u, v, w).

63. $\dfrac{u - 3}{4} + \dfrac{v + 1}{3} + \dfrac{w - 2}{8} = 1$

$\dfrac{u - 3}{2} + \dfrac{v + 1}{2} + \dfrac{w - 2}{4} = 0$

$\dfrac{u - 3}{4} - \dfrac{v + 1}{2} + \dfrac{w - 2}{2} = -6 \quad \{(-13, 11, 10)\}$

64. $\dfrac{u + 1}{6} + \dfrac{v - 1}{6} + \dfrac{w + 3}{4} = 11$

$\dfrac{u + 1}{3} - \dfrac{v - 1}{2} + \dfrac{w + 3}{4} = 7$

$\dfrac{u + 1}{2} - \dfrac{v - 1}{6} + \dfrac{w + 3}{2} = 20 \quad \{(23, 13, 17)\}$

Recall that an equation of a circle can be written in the form $(x - h)^2 + (y - k)^2 = r^2$, where (h, k) is the center and r is the radius. Expanding terms, the equation can also be written in the form $x^2 + y^2 + Ax + By + C = 0$. For Exercises 65–66,

a. Find an equation of the form $x^2 + y^2 + Ax + By + C = 0$ that represents the circle that passes through the given points.

b. Find the center and radius of the circle.

65. $(2, 2), (6, 0), (7, -3)$ a. $x^2 + y^2 - 4x + 6y - 12 = 0$
   b. Center: $(2, -3)$; Radius: 5

66. $(-1, 12), (5, 10), (9, 2)$ a. $x^2 + y^2 + 2x - 4y - 95 = 0$
   b. Center: $(-1, 2)$; Radius: 10

For Exercises 67–68, find the constants A and B so that the two polynomials are equal. (*Hint*: Create a system of linear equations by equating the constant terms and by equating the coefficients on the x terms and x^2 terms.)

67. $11x^2 + 26x - 5 = 2Ax^2 + 5Ax + 3A + Bx^2 - 2Bx - 8B + 2Cx^2 - 7Cx - 4C$ $A = 5, B = 3, C = -1$

68. $3x^2 + 37x - 82 = Ax^2 + Ax - 12A + 3Bx^2 - 10Bx + 3B + 3Cx^2 + 11Cx - 4C$ $A = 6, B = -2, C = 1$

SECTION 8.3   Partial Fraction Decomposition

OBJECTIVES

1. Set Up a Partial Fraction Decomposition
2. Decompose $\dfrac{f(x)}{g(x)}$, Where $g(x)$ Is a Product of Linear Factors
3. Decompose $\dfrac{f(x)}{g(x)}$, Where $g(x)$ Has Irreducible Quadratic Factors

1. Set Up a Partial Fraction Decomposition

In Section R.4 we learned how to add and subtract rational expressions. For example:

$$\frac{5}{x + 2} + \frac{3}{x - 5} = \frac{5(x - 5)}{(x + 2)(x - 5)} + \frac{3(x + 2)}{(x - 5)(x + 2)}$$

$$= \frac{5(x - 5) + 3(x + 2)}{(x + 2)(x - 5)}$$

$$= \frac{8x - 19}{(x + 2)(x - 5)}$$

The fraction $\dfrac{8x - 19}{(x + 2)(x - 5)}$ is the result of adding two simpler fractions, $\dfrac{5}{x + 2}$ and $\dfrac{3}{x - 5}$. The sum $\dfrac{5}{x + 2} + \dfrac{3}{x - 5}$ is called the **partial fraction decomposition** of $\dfrac{8x - 19}{(x + 2)(x - 5)}$. In some applications in higher mathematics, it is more convenient to work with the partial fraction decomposition than the more complicated single fraction. Therefore, in this section, we will learn the technique of partial fraction decomposition to write a rational expression as a sum of simpler fractions. That is, we will reverse the process of adding two or more fractions. There are two parts to this process.

I. First we set up the "form" or "structure" for the partial fraction decomposition into simpler fractions. For example, the denominator of $\dfrac{8x - 19}{(x + 2)(x - 5)}$ consists of the distinct linear factors $(x + 2)$ and $(x - 5)$. From the preceding discussion, the partial fraction decomposition must be of the form:

$$\frac{8x - 19}{(x + 2)(x - 5)} = \frac{A}{x + 2} + \frac{B}{x - 5}$$

> The expression on the right is the "form" or "structure" for the partial fraction decomposition of $\dfrac{8x - 19}{(x + 2)(x - 5)}$.

II. Next, we solve for the constants A and B. To do so, multiply both sides of the equation by the LCD, and set up a system of linear equations.

$$(x + 2)(x - 5) \cdot \left[\frac{8x - 19}{(x + 2)(x - 5)} \right] = (x + 2)(x - 5) \cdot \left[\frac{A}{x + 2} + \frac{B}{x - 5} \right]$$

Multiply by the LCD to clear fractions.

$8x - 19 = A(x - 5) + B(x + 2)$

$8x - 19 = Ax - 5A + Bx + 2B$ Apply the distributive property.

$8x - 19 = (A + B)x + (-5A + 2B)$ Simplify and combine like terms.

x coefficients are equal.

$8x - 19 = (A + B)x + (-5A + 2B)$ Two polynomials are equal if and only if the coefficients on like terms are equal.

Constants are equal.

$A + B = 8$ Equate the coefficients on x.

$-5A + 2B = -19$ Equate the constant terms.

Solve the system of linear equations. Then substitute the values of A and B into the partial fraction decomposition.

$\boxed{1}$ $A + B = 8$ Multiply by 5. $5A + 5B = 40$

$\boxed{2}$ $-5A + 2B = -19$ $\underline{-5A + 2B = -19}$

                                              $7B = 21$

$\boxed{1}$ $A + B = 8$ $B = 3$

        $A + 3 = 8$

          $A = 5$

$A = 5$ and $B = 3$

$$\frac{8x - 19}{(x + 2)(x - 5)} = \frac{A}{x + 2} + \frac{B}{x - 5} = \frac{5}{x + 2} + \frac{3}{x - 5}$$

We begin partial fraction decomposition by factoring the denominator into linear factors $(ax + b)$ and quadratic factors $(ax^2 + bx + c)$ that are irreducible over the integers. A quadratic factor that is irreducible over the integers cannot be factored as a product of binomials with integer coefficients. From the factorization of the denominator, we then determine the proper form for the partial fraction decomposition using the following guidelines.

Decomposition of $\dfrac{f(x)}{g(x)}$ into Partial Fractions

Consider a rational expression $\dfrac{f(x)}{g(x)}$, where $f(x)$ and $g(x)$ are polynomials with real coefficients, $g(x) \neq 0$, and the degree of $f(x)$ is less than the degree of $g(x)$.

PART I:

Step 1   Factor the denominator $g(x)$ completely into linear factors of the form $(ax + b)^m$ and quadratic factors of the form $(ax^2 + bx + c)^n$ that are not further factorible over the integers (irreducible over the integers).

Step 2   Set up the form for the decomposition. That is, write the original rational expression $\dfrac{f(x)}{g(x)}$ as a sum of simpler fractions using these guidelines. Note that $A_1, A_2, \ldots, A_m, B_1, B_2, \ldots, B_n$, and $C_1, C_2, \ldots, C_n$ are constants.

- **Linear factors of $g(x)$:** For each linear factor of $g(x)$, the partial fraction decomposition must include the sum:

$$\frac{A_1}{(ax + b)^1} + \frac{A_2}{(ax + b)^2} + \cdots + \frac{A_m}{(ax + b)^m}$$

- **Quadratic factors of $g(x)$:** For each quadratic factor of $g(x)$, the partial fraction decomposition must include the sum:

$$\frac{B_1x + C_1}{(ax^2 + bx + c)^1} + \frac{B_2x + C_2}{(ax^2 + bx + c)^2} + \cdots + \frac{B_nx + C_n}{(ax^2 + bx + c)^n}$$

PART II:

Step 3   With the form of the partial fraction decomposition set up, multiply both sides of the equation by the LCD to clear fractions.

Step 4   Using the equation from step 3, set up a system of linear equations by equating the constant terms and equating the coefficients of like powers of x.

Step 5   Solve the system of equations from step 4 and substitute the solutions to the system into the partial fraction decomposition.

In Examples 1 and 2, we focus on setting up the proper form for a partial fraction decomposition (Part I). In each example, note that the factors of the denominator will fall into one of the following categories:

Linear factors: $ax + b$

Repeated linear factors: $(ax + b)^m$ ($m \geq 2$, an integer)

Quadratic factors (irreducible over the integers): $ax^2 + bx + c$

Repeated quadratic factors (irreducible over the integers): $(ax^2 + bx + c)^n$ ($n \geq 2$, an integer)

EXAMPLE 1   **Setting Up the Form for a Partial Fraction Decomposition**

Set up the form for the partial fraction decomposition for the given rational expressions.

a. $\dfrac{4x - 15}{(2x + 3)(x - 2)}$ **b.** $\dfrac{4x^2 + 10x + 9}{x^3 + 6x^2 + 9x}$

Solution:

a. $\dfrac{4x - 15}{(2x + 3)^1(x - 2)^1} = \dfrac{A}{(2x + 3)^1} + \dfrac{B}{(x - 2)^1}$

The denominator has distinct linear factors of the form $(2x + 3)^1$ and $(x - 2)^1$. Since each linear factor is raised to the first power, only one fraction is needed for each factor.

b. $\dfrac{4x^2 + 10x + 9}{x^3 + 6x^2 + 9x} = \dfrac{4x^2 + 10x + 9}{x(x^2 + 6x + 9)}$

Factor the denominator completely.

The denominator has a linear factor of x^1 and a *repeated* linear factor $(x + 3)^2$.

$= \dfrac{4x^2 + 10x + 9}{x^1(x + 3)^2} = \dfrac{A}{x^1} + \dfrac{B}{(x + 3)^1} + \dfrac{C}{(x + 3)^2}$

For a repeated factor that occurs m times, one fraction must be given for each power less than or equal to m.

> Include one fraction with $(x + 3)$ raised to each positive integer up to and including 2.

Skill Practice 1 Set up the form for the partial fraction decomposition for the given rational expressions.

a. $\dfrac{-x + 18}{(3x + 1)(x + 4)}$ **b.** $\dfrac{-x^2 + 3x + 8}{x^3 + 4x^2 + 4x}$

In Example 2, we practice setting up the form for the partial decomposition of a rational expression that contains irreducible quadratic factors in the denominator.

Classroom Examples: p. 799
Exercises 14, 16

EXAMPLE 2   Setting Up the Form for a Partial Fraction Decomposition

Set up the form for the partial fraction decomposition for the given rational expressions.

a. $\dfrac{2x^2 - 3x + 4}{x^3 + 4x}$ **b.** $\dfrac{3x^2 + 8x + 14}{(x^2 + 2x + 5)^2}$

Solution:

a. $\dfrac{2x^2 - 3x + 4}{x^3 + 4x} = \dfrac{2x^2 - 3x + 4}{x(x^2 + 4)}$

Factor the denominator completely. The denominator has one linear factor x^1 and one irreducible quadratic factor $(x^2 + 4)^1$.

$= \dfrac{2x^2 - 3x + 4}{x^1(x^2 + 4)^1} = \dfrac{A}{x^1} + \dfrac{Bx + C}{(x^2 + 4)^1}$

Since each factor is raised to the first power, only one fraction is needed for each factor.

$= \dfrac{2x^2 - 3x + 4}{x(x^2 + 4)} = \dfrac{A}{x} + \dfrac{Bx + C}{(x^2 + 4)}$

> **TIP** For a first-degree (linear) denominator, the numerator is constant (degree 0). For a second-degree (quadratic) denominator, the numerator is linear (degree 1).

b. $\dfrac{3x^2 + 8x + 14}{(x^2 + 2x + 5)^2}$

The quadratic factor $x^2 + 2x + 5$ does not factor further over the integers.

$= \dfrac{Ax + B}{(x^2 + 2x + 5)^1} + \dfrac{Cx + D}{(x^2 + 2x + 5)^2}$

The factor $(x^2 + 2x + 5)$ appears to the *second* power in the denominator. Therefore, in the partial fraction composition, one fraction must have $(x^2 + 2x + 5)^1$ in the denominator, and one fraction must have $(x^2 + 2x + 5)^2$ in the denominator.

Answers

1. a. $\dfrac{A}{3x + 1} + \dfrac{B}{x + 4}$

b. $\dfrac{A}{x} + \dfrac{B}{x + 2} + \dfrac{C}{(x + 2)^2}$

Skill Practice 2   Set up the form for the partial fraction decomposition for the given rational expressions.

a. $\dfrac{7x^2 + 2x + 12}{x^3 + 3x}$ **b.** $\dfrac{-3x^2 - 5x - 19}{(x^2 + 3x + 6)^2}$

2. Decompose $\dfrac{f(x)}{g(x)}$, Where $g(x)$ Is a Product of Linear Factors

In Example 3, we find the partial fraction decomposition of a rational expression in which the denominator is a product of distinct linear factors.

Classroom Example: p. 799
Exercise 22

EXAMPLE 3   Decomposing $\dfrac{f(x)}{g(x)}$, Where $g(x)$ Has Distinct Linear Factors

Find the partial fraction decomposition. $\dfrac{4x - 15}{(2x + 3)(x - 2)}$

Solution:

$\dfrac{4x - 15}{(2x + 3)(x - 2)} = \dfrac{A}{2x + 3} + \dfrac{B}{x - 2}$ From Example 1(a), we have the form for the partial fraction decomposition.

$(2x + 3)(x - 2)\left[\dfrac{4x - 15}{(2x + 3)(x - 2)}\right] = (2x + 3)(x - 2)\left[\dfrac{A}{2x + 3} + \dfrac{B}{x - 2}\right]$ To solve for A and B, first multiply both sides by the LCD to clear fractions.

$4x - 15 = A(x - 2) + B(2x + 3)$ Apply the distributive property.

$4x - 15 = Ax - 2A + 2Bx + 3B$

$4x - 15 = (A + 2B)x + (-2A + 3B)$ Combine like terms.

[1] $4 = A + 2B$ Equate the x term coefficients.

[2] $-15 = -2A + 3B$ Equate the constant terms.

Solve the system of linear equations by using the substitution method or addition method.

$$\begin{array}{ll}
\boxed{1}\ \ \ \ 4 = A + 2B \xrightarrow{\text{Multiply by 2.}} & 2A + 4B = 8 \\
\boxed{2} -15 = -2A + 3B & \underline{-2A + 3B = -15} \\
 & \ \ \ \ \ \ \ \ \ 7B = -7 \\
 & \ \ \ \ \ \ \ \ \ \ B = -1
\end{array}$$

$$\begin{array}{l}
\boxed{1}\ \ 4 = A + 2B \\
\ \ \ \ 4 = A + 2(-1) \\
\ \ \ \ A = 6
\end{array}$$

$A = 6$ and $B = -1$.

$\dfrac{4x - 15}{(2x + 3)(x - 2)} = \dfrac{A}{2x + 3} + \dfrac{B}{x - 2}$ Substitute $A = 6$ and $B = -1$ into the partial fraction decomposition.

$\dfrac{4x - 15}{(2x + 3)(x - 2)} = \dfrac{6}{2x + 3} + \dfrac{-1}{x - 2}$ or equivalently $\dfrac{6}{2x + 3} - \dfrac{1}{x - 2}$

Answers

2. a. $\dfrac{A}{x} + \dfrac{Bx + C}{x^2 + 3}$

  b. $\dfrac{Ax + B}{x^2 + 3x + 6} + \dfrac{Cx + D}{(x^2 + 3x + 6)^2}$

3. $\dfrac{5}{3x + 1} + \dfrac{-2}{x + 4}$

Skill Practice 3   Find the partial fraction decomposition. $\dfrac{-x + 18}{(3x + 1)(x + 4)}$

TIP Always remember that the result of a partial fraction decomposition can be checked by adding the partial fractions and verifying that the sum equals the original rational expression.

To verify the result of Example 3, we can add the rational expressions.

$$\frac{6}{2x+3} + \frac{-1}{x-2} = \frac{6(x-2)}{(2x+3)(x-2)} + \frac{-1(2x+3)}{(x-2)(2x+3)}$$

$$= \frac{6(x-2) - 1(2x+3)}{(2x+3)(x-2)}$$

$$= \frac{4x-15}{(2x+3)(x-2)} \checkmark$$

In Example 4, we perform partial fraction decomposition with a rational expression that has repeated linear factors in the denominator.

Classroom Example: p. 799
Exercise 28

EXAMPLE 4 Decomposing $\dfrac{f(x)}{g(x)}$, Where $g(x)$ Has Repeated Linear Factors

Find the partial fraction decomposition. $\dfrac{4x^2 + 10x + 9}{x^3 + 6x^2 + 9x}$

Solution:

$$\frac{4x^2 + 10x + 9}{x(x+3)^2} = \frac{A}{x} + \frac{B}{(x+3)^1} + \frac{C}{(x+3)^2}$$

From Example 1(b), we have the form for the partial fraction decomposition.

$$x(x+3)^2 \left[\frac{4x^2 + 10x + 9}{x(x+3)^2} \right] = x(x+3)^2 \left[\frac{A}{x} + \frac{B}{(x+3)^1} + \frac{C}{(x+3)^2} \right]$$

$$4x^2 + 10x + 9 = A(x+3)^2 + Bx(x+3) + Cx$$
$$4x^2 + 10x + 9 = A(x^2 + 6x + 9) + Bx^2 + 3Bx + Cx$$
$$4x^2 + 10x + 9 = Ax^2 + 6Ax + 9A + Bx^2 + 3Bx + Cx$$
$$4x^2 + 10x + 9 = (A+B)x^2 + (6A + 3B + C)x + 9A$$

$$4 = A + B \qquad \text{Equate the } x^2 \text{ term coefficients.}$$
$$10 = 6A + 3B + C \qquad \text{Equate the } x \text{ term coefficients.}$$
$$9 = 9A \qquad \text{Equate the constant terms.}$$
$$A = 1, B = 3, \text{ and } C = -5 \qquad \text{Solve the system of linear equations.}$$

$$\frac{4x^2 + 10x + 9}{x(x+3)^2} = \frac{A}{x} + \frac{B}{(x+3)^1} + \frac{C}{(x+3)^2}$$

Substitute $A = 1$, $B = 3$, and $C = -5$ into the partial fraction decomposition.

$$\frac{4x^2 + 10x + 9}{x(x+3)^2} = \frac{1}{x} + \frac{3}{(x+3)^1} + \frac{-5}{(x+3)^2} = \frac{1}{x} + \frac{3}{x+3} - \frac{5}{(x+3)^2}$$

Skill Practice 4 Find the partial fraction decomposition. $\dfrac{-x^2 + 3x + 8}{x^3 + 4x^2 + 4x}$

3. Decompose $\dfrac{f(x)}{g(x)}$, Where $g(x)$ Has Irreducible Quadratic Factors

We now turn our attention to performing partial fraction decomposition where the denominator of a rational expression contains quadratic factors irreducible over the integers. In Example 5, we also address the situation in which the given rational expression is an **improper rational expression;** that is, the degree of the numerator

Answer

4. $\dfrac{2}{x} + \dfrac{-3}{x+2} + \dfrac{1}{(x+2)^2}$

is greater than or equal to the degree of the denominator. In such a case, we use long division to write the expression in the form:

$$\text{(polynomial)} + \text{(proper rational expression)}$$

where a **proper rational expression** is one in which the degree of the numerator is less than the degree of the denominator.

Classroom Example: p. 799
Exercise 32

EXAMPLE 5 Decomposing $\dfrac{f(x)}{g(x)}$, Where $g(x)$ Has an Irreducible Quadratic Factor

Find the partial fraction decomposition. $\dfrac{x^4 + 3x^3 + 6x^2 + 9x + 4}{x^3 + 4x}$

Solution:

First note that the degree of the numerator is not less than the degree of the denominator. Therefore, perform long division first.

$$\dfrac{x^4 + 3x^3 + 6x^2 + 9x + 4}{x^3 + 4x} \xrightarrow{\text{Long division}}$$

$$\begin{array}{r} x + 3 \\ x^3 + 4x \overline{)\,x^4 + 3x^3 + 6x^2 + 9x + 4} \\ -(x^4 \qquad\; + 4x^2) \\ \hline 3x^3 + 2x^2 + 9x \\ -(3x^3 \qquad + 12x) \\ \hline 2x^2 - 3x + 4 \end{array}$$

$$= x + 3 + \dfrac{2x^2 - 3x + 4}{x^3 + 4x} \xleftarrow{\text{Equivalent form}}$$

$$= x + 3 + \underbrace{\dfrac{2x^2 - 3x + 4}{x(x^2 + 4)}}_{\substack{\text{proper rational} \\ \text{expression}}} \qquad \text{Factor the denominator.}$$

with $x + 3$ labeled polynomial.

$$\dfrac{2x^2 - 3x + 4}{x(x^2 + 4)} = \dfrac{A}{x} + \dfrac{Bx + C}{(x^2 + 4)} \qquad$$ Perform partial fraction decomposition on the proper fraction. From Example 2(a), we have the form for the partial fraction decomposition.

$$x(x^2 + 4)\left[\dfrac{2x^2 - 3x + 4}{x(x^2 + 4)}\right] = x(x^2 + 4)\left[\dfrac{A}{x} + \dfrac{Bx + C}{(x^2 + 4)}\right] \qquad$$ To solve for A, B, and C, multiply both sides by the LCD to clear fractions.

$$2x^2 - 3x + 4 = A(x^2 + 4) + (Bx + C)x \qquad$$ Apply the distributive property.

$$2x^2 - 3x + 4 = Ax^2 + 4A + Bx^2 + Cx$$

$$2x^2 - 3x + 4 = (A + B)x^2 + Cx + 4A \qquad$$ Combine like terms.

$$\left.\begin{array}{l} 2 = A + B \\ -3 = C \\ 4 = 4A \end{array}\right\} \begin{array}{l} A = 1,\, B = 1, \\ \text{and } C = -3 \end{array}$$

Equate the x^2 term coefficients.
Equate the x term coefficients.
Equate the constant terms.
Solve the system of linear equations.

$$\dfrac{2x^2 - 3x + 4}{x(x^2 + 4)} = \dfrac{A}{x} + \dfrac{Bx + C}{(x^2 + 4)} \qquad$$ Substitute $A = 1$, $B = 1$, and $C = -3$.

$$\dfrac{2x^2 - 3x + 4}{x(x^2 + 4)} = \dfrac{1}{x} + \dfrac{1x + (-3)}{x^2 + 4} \quad \text{or} \quad \dfrac{1}{x} + \dfrac{x - 3}{x^2 + 4}$$

Therefore, $\dfrac{x^4 + 3x^3 + 6x^2 + 9x + 4}{x^3 + 4x} = x + 3 + \dfrac{1}{x} + \dfrac{x - 3}{x^2 + 4}$.

Answer

5. $x + 2 + \dfrac{4}{x} + \dfrac{3x + 2}{x^2 + 3}$

Skill Practice 5 Find the partial fraction decomposition.

$$\dfrac{x^4 + 2x^3 + 10x^2 + 8x + 12}{x^3 + 3x}$$

In Example 6, we demonstrate the case in which a rational expression contains a repeated quadratic factor.

Classroom Example: p. 799
Exercise 34

EXAMPLE 6 Decomposing $\dfrac{f(x)}{g(x)}$, Where $g(x)$ Has a Repeated Irreducible Quadratic Factor

Find the partial fraction decomposition. $\dfrac{3x^2 + 8x + 14}{(x^2 + 2x + 5)^2}$

Solution:

$\dfrac{3x^2 + 8x + 14}{(x^2 + 2x + 5)^2} = \dfrac{Ax + B}{(x^2 + 2x + 5)^1} + \dfrac{Cx + D}{(x^2 + 2x + 5)^2}$ — From Example 2(b), we have the form for the partial fraction decomposition.

To solve for A, B, C, and D, multiply both sides by the LCD to clear fractions.

$(x^2 + 2x + 5)^2 \left[\dfrac{3x^2 + 8x + 14}{(x^2 + 2x + 5)^2}\right] = (x^2 + 2x + 5)^2 \left[\dfrac{Ax + B}{x^2 + 2x + 5} + \dfrac{Cx + D}{(x^2 + 2x + 5)^2}\right]$

$3x^2 + 8x + 14 = (Ax + B)(x^2 + 2x + 5) + (Cx + D)$

$3x^2 + 8x + 14 = Ax^3 + 2Ax^2 + 5Ax + Bx^2 + 2Bx + 5B + Cx + D$

$3x^2 + 8x + 14 = Ax^3 + (2A + B)x^2 + (5A + 2B + C)x + 5B + D$ — Combine like terms.

$\begin{aligned} 0 &= A \\ 3 &= 2A + B \\ 8 &= 5A + 2B + C \\ 14 &= 5B + D \end{aligned}$

- Equate the x^3 term coefficients.
- Equate the x^2 term coefficients.
- Equate the x term coefficients.
- Equate the constant terms.

$A = 0$, $B = 3$, $C = 2$, and $D = -1$ — Solve the system of linear equations.

$\dfrac{3x^2 + 8x + 14}{(x^2 + 2x + 5)^2} = \dfrac{Ax + B}{x^2 + 2x + 5} + \dfrac{Cx + D}{(x^2 + 2x + 5)^2}$ — Substitute $A = 0$, $B = 3$, $C = 2$, and $D = -1$ into the partial fraction decomposition.

$\dfrac{3x^2 + 8x + 14}{(x^2 + 2x + 5)^2} = \dfrac{(0)x + (3)}{x^2 + 2x + 5} + \dfrac{(2)x + (-1)}{(x^2 + 2x + 5)^2}$ or $\dfrac{3}{x^2 + 2x + 5} + \dfrac{2x - 1}{(x^2 + 2x + 5)^2}$

Skill Practice 6 Find the partial fraction decomposition.

$\dfrac{-3x^2 - 5x - 19}{(x^2 + 3x + 6)^2}$

Answer

6. $\dfrac{-3}{x^2 + 3x + 6} + \dfrac{4x - 1}{(x^2 + 3x + 6)^2}$

SECTION 8.3 Practice Exercises

Prerequisite Review

For Exercises R.1–R.2, factor completely.

R.1. $12y^3 - 16y - 9y^2 + 12$
$(4y - 3)(3y^2 - 4)$

R.2. $64u^2 + 80u + 25$ $(8u + 5)^2$

R.3. Add. $\dfrac{4}{4a + 3} + \dfrac{4}{a - 3}$
$\dfrac{20a}{(4a + 3)(a - 3)}$

R.4. Use long division to divide.
$\dfrac{25x^4 - 5x^3 + 15x^2 + 4x - 2}{5x^2 - x + 4}$ $5x^2 - 1 + \dfrac{3x + 2}{5x^2 - x + 4}$

R.5. Solve. $\dfrac{16}{x} - \dfrac{16}{x - 2} = \dfrac{4}{x}$ $\{-6\}$

Concept Connections

1. The process of decomposing a rational expression into two or more simpler fractions is called partial <u>fraction decomposition</u>.

2. When setting up a partial fraction decomposition, if a fraction has a linear denominator, then the numerator should be (constant/linear). That is, should the numerator be set up as A or $Ax + B$?   constant; A

3. When setting up a partial fraction decomposition, if the denominator of a fraction is a quadratic polynomial irreducible over the integers, then the numerator should be (constant/linear). That is, should the numerator be set up as A or $Ax + B$?   linear; $Ax + B$

4. In what situation should long division be used before attempting to decompose a rational expression into partial fractions?   When the degree of the numerator is greater than or equal to the degree of the denominator, apply long division first.

Objective 1: Set Up a Partial Fraction Decomposition

For Exercises 5–20, set up the form for the partial fraction decomposition. Do not solve for A, B, C, and so on. (See Examples 1–2)

5. $\dfrac{-x - 37}{(x + 4)(2x - 3)}$ $\dfrac{A}{x + 4} + \dfrac{B}{2x - 3}$

6. $\dfrac{20x - 4}{(x - 5)(3x + 1)}$ $\dfrac{A}{x - 5} + \dfrac{B}{3x + 1}$

7. $\dfrac{8x - 10}{x^2 - 2x}$ $\dfrac{A}{x} + \dfrac{B}{x - 2}$

8. $\dfrac{y - 12}{y^2 + 3y}$ $\dfrac{A}{y} + \dfrac{B}{y + 3}$

9. $\dfrac{6w - 7}{w^2 + w - 6}$ $\dfrac{A}{w - 2} + \dfrac{B}{w + 3}$

10. $\dfrac{-10t - 11}{t^2 + 5t - 6}$ $\dfrac{A}{t + 6} + \dfrac{B}{t - 1}$

11. $\dfrac{x^2 + 26x + 100}{x^3 + 10x^2 + 25x}$ $\dfrac{A}{x} + \dfrac{B}{x + 5} + \dfrac{C}{(x + 5)^2}$

12. $\dfrac{-3x^2 + 2x + 8}{x^3 + 4x^2 + 4x}$ $\dfrac{A}{x} + \dfrac{B}{x + 2} + \dfrac{C}{(x + 2)^2}$

13. $\dfrac{13x^2 + 2x + 45}{2x^3 + 18x}$ $\dfrac{A}{2x} + \dfrac{Bx + C}{x^2 + 9}$

14. $\dfrac{17x^2 - 7x + 18}{7x^3 + 42x}$ $\dfrac{A}{7x} + \dfrac{Bx + C}{x^2 + 6}$

15. $\dfrac{2x^3 - x^2 + 13x - 5}{x^4 + 10x^2 + 25}$ $\dfrac{Ax + B}{x^2 + 5} + \dfrac{Cx + D}{(x^2 + 5)^2}$

16. $\dfrac{3x^3 - 4x^2 + 11x - 12}{x^4 + 6x^2 + 9}$ $\dfrac{Ax + B}{x^2 + 3} + \dfrac{Cx + D}{(x^2 + 3)^2}$

17. $\dfrac{5x^2 - 4x + 8}{(x - 4)(x^2 + x + 4)}$ $\dfrac{A}{x - 4} + \dfrac{Bx + C}{x^2 + x + 4}$

18. $\dfrac{x^2 + 15x - 6}{(x + 6)(x^2 + 2x + 6)}$ $\dfrac{A}{x + 6} + \dfrac{Bx + C}{x^2 + 2x + 6}$

19. $\dfrac{2x^5 + 3x^3 + 4x^2 + 5}{x(x + 2)^3(x^2 + 2x + 7)^2}$

$\dfrac{A}{x} + \dfrac{B}{x + 2} + \dfrac{C}{(x + 2)^2} + \dfrac{D}{(x + 2)^3} + \dfrac{Ex + F}{x^2 + 2x + 7} + \dfrac{Gx + H}{(x^2 + 2x + 7)^2}$

20. $\dfrac{6x^4 - 5x^3 + 2x^2 - 5}{(x - 3)(2x + 9)^2(x^2 + 1)^2}$

$\dfrac{A}{x - 3} + \dfrac{B}{2x + 9} + \dfrac{C}{(2x + 9)^2} + \dfrac{Dx + E}{x^2 + 1} + \dfrac{Fx + G}{(x^2 + 1)^2}$

Objectives 2 and 3: Decompose $\dfrac{f(x)}{g(x)}$ into Partial Fractions

For Exercises 21–42, find the partial fraction decomposition. (See Examples 3–6)

21. $\dfrac{-x - 37}{(x + 4)(2x - 3)}$ $\dfrac{3}{x + 4} + \dfrac{-7}{2x - 3}$

22. $\dfrac{20x - 4}{(x - 5)(3x + 1)}$ $\dfrac{6}{x - 5} + \dfrac{2}{3x + 1}$

23. $\dfrac{8x - 10}{x^2 - 2x}$ $\dfrac{5}{x} + \dfrac{3}{x - 2}$

24. $\dfrac{y - 12}{y^2 + 3y}$ $\dfrac{-4}{y} + \dfrac{5}{y + 3}$

25. $\dfrac{6w - 7}{w^2 + w - 6}$ $\dfrac{1}{w - 2} + \dfrac{5}{w + 3}$

26. $\dfrac{-10t - 11}{t^2 + 5t - 6}$ $\dfrac{-7}{t + 6} + \dfrac{-3}{t - 1}$

27. $\dfrac{x^2 + 26x + 100}{x^3 + 10x^2 + 25x}$ $\dfrac{4}{x} + \dfrac{-3}{x + 5} + \dfrac{1}{(x + 5)^2}$

28. $\dfrac{-3x^2 + 2x + 8}{x^3 + 4x^2 + 4x}$ $\dfrac{2}{x} + \dfrac{-5}{x + 2} + \dfrac{4}{(x + 2)^2}$

29. $\dfrac{13x^2 + 2x + 45}{2x^3 + 18x}$ $\dfrac{5}{2x} + \dfrac{4x + 1}{x^2 + 9}$

30. $\dfrac{17x^2 - 7x + 18}{7x^3 + 42x}$ $\dfrac{3}{7x} + \dfrac{2x - 1}{x^2 + 6}$

31. $\dfrac{x^4 - 3x^3 + 13x^2 - 28x + 28}{x^3 + 7x}$ $x - 3 + \dfrac{4}{x} + \dfrac{2x - 7}{x^2 + 7}$

32. $\dfrac{x^4 - 4x^3 + 11x^2 - 13x + 12}{x^3 + 2x}$ $x - 4 + \dfrac{6}{x} + \dfrac{3x - 5}{x^2 + 2}$

33. $\dfrac{2x^3 - x^2 + 13x - 5}{x^4 + 10x^2 + 25}$ $\dfrac{2x - 1}{x^2 + 5} + \dfrac{3x}{(x^2 + 5)^2}$

34. $\dfrac{3x^3 - 4x^2 + 11x - 12}{x^4 + 6x^2 + 9}$ $\dfrac{3x - 4}{x^2 + 3} + \dfrac{2x}{(x^2 + 3)^2}$

35. $\dfrac{5x^2 - 4x + 8}{(x - 4)(x^2 + x + 4)}$ $\dfrac{3}{x - 4} + \dfrac{2x + 1}{x^2 + x + 4}$

36. $\dfrac{x^2 + 15x - 6}{(x + 6)(x^2 + 2x + 6)}$ $\dfrac{-2}{x + 6} + \dfrac{3x + 1}{x^2 + 2x + 6}$

37. $\dfrac{4x^3 - 4x^2 + 11x - 7}{x^4 + 5x^2 + 6}$ $\dfrac{3x + 1}{x^2 + 2} + \dfrac{x - 5}{x^2 + 3}$

38. $\dfrac{3x^3 - 4x^2 + 6x - 7}{x^4 + 5x^2 + 4}$ $\dfrac{2x - 3}{x^2 + 4} + \dfrac{x - 1}{x^2 + 1}$

39. $\dfrac{2x^3 - 11x^2 - 4x + 24}{x^2 - 3x - 10}$ $2x - 5 + \dfrac{4}{x + 2} + \dfrac{-3}{x - 5}$

40. $\dfrac{3x^3 + 11x^2 + x + 10}{x^2 + 3x - 4}$ $3x + 2 + \dfrac{5}{x - 1} + \dfrac{2}{x + 4}$

41. $\dfrac{3x^3 + 2x^2 - x - 5}{x^2 + 2x + 1}$ $3x - 4 + \dfrac{4}{x + 1} + \dfrac{-5}{(x + 1)^2}$

42. $\dfrac{2x^3 - 17x^2 + 54x - 68}{x^2 - 6x + 9}$ $2x - 5 + \dfrac{6}{x - 3} + \dfrac{-5}{(x - 3)^2}$

43. a. Factor. $x^3 - x^2 - 21x + 45$
(*Hint*: Use the rational zero theorem.) $(x - 3)^2(x + 5)$

b. Find the partial fraction decomposition for
$$\frac{-3x^2 + 35x - 70}{x^3 - x^2 - 21x + 45}.$$ $\frac{2}{x - 3} + \frac{1}{(x - 3)^2} + \frac{-5}{x + 5}$

44. a. Factor. $x^3 + 2x^2 - 7x + 4$ $(x - 1)^2(x + 4)$

b. Find the partial fraction decomposition for
$$\frac{10x^2 + 17x - 17}{x^3 + 2x^2 - 7x + 4}.$$ $\frac{7}{x - 1} + \frac{2}{(x - 1)^2} + \frac{3}{x + 4}$

45. a. Factor. $x^3 + 6x^2 + 12x + 8$ $(x + 2)^3$

b. Find the partial fraction decomposition for
$$\frac{3x^2 + 8x + 5}{x^3 + 6x^2 + 12x + 8}.$$ $\frac{3}{x + 2} + \frac{-4}{(x + 2)^2} + \frac{1}{(x + 2)^3}$

46. a. Factor. $x^3 - 9x^2 + 27x - 27$ $(x - 3)^3$

b. Find the partial fraction decomposition for
$$\frac{2x^2 - 17x + 37}{x^3 - 9x^2 + 27x - 27}.$$ $\frac{2}{x - 3} + \frac{-5}{(x - 3)^2} + \frac{4}{(x - 3)^3}$

Write About It

47. Write an informal explanation of partial fraction decomposition. Partial fraction decomposition is a procedure in which a rational expression is written as a sum of two or more simpler rational expressions.

49. What is meant by a *proper* rational expression? A proper rational expression is a rational expression in which the degree of the numerator is less than the degree of the denominator.

48. Suppose that a proper rational expression has a single repeated linear factor $(ax + b)^3$ in the denominator. Explain how to set up the partial fraction decomposition.

50. Given an improper rational expression, what must be done first before the technique of partial fraction decomposition may be performed? Given an improper rational expression, use long division to divide the numerator by the denominator. Then perform partial fraction decomposition to the expression represented by the remainder/divisor.

Expanding Your Skills

51. a. Determine the partial fraction decomposition for $\frac{2}{n(n + 2)}$. $\frac{1}{n} - \frac{1}{n + 2}$

b. Use the partial fraction decomposition for $\frac{2}{n(n + 2)}$ to rewrite the infinite sum
$$\frac{2}{1(3)} + \frac{2}{2(4)} + \frac{2}{3(5)} + \frac{2}{4(6)} + \frac{2}{5(7)} \cdots$$

c. Determine the value of $\frac{1}{n + 2}$ as $n \to \infty$. 0

d. Find the value of the sum from part (b). $\frac{3}{2}$

52. a. Determine the partial fraction decomposition for $\frac{3}{n(n + 3)}$. $\frac{1}{n} - \frac{1}{n + 3}$

b. Use the partial fraction decomposition for $\frac{3}{n(n + 3)}$ to rewrite the infinite sum
$$\frac{3}{1(4)} + \frac{3}{2(5)} + \frac{3}{3(6)} + \frac{3}{4(7)} + \frac{3}{5(8)} \cdots$$

c. Determine the value of $\frac{1}{n + 3}$ as $n \to \infty$. 0

d. Find the value of the sum from part (b). $\frac{11}{6}$

For Exercises 53–54, find the partial fraction decomposition. Assume that a and b are nonzero constants.

53. $\frac{1}{x(a + bx)}$ $\frac{1}{ax} - \frac{b}{a(a + bx)}$

54. $\frac{1}{a^2 - x^2}$ $\frac{1}{2a(a - x)} + \frac{1}{2a(a + x)}$

For Exercises 55–56, find the partial fraction decomposition for the given expression. [*Hint*: Use the substitution $u = e^x$ and recall that $e^{2x} = (e^x)^2$.]

55. $\frac{5e^x + 7}{e^{2x} + 3e^x + 2}$ $\frac{2}{e^x + 1} + \frac{3}{e^x + 2}$

56. $\frac{-3e^x - 22}{e^{2x} + 3e^x - 4}$ $\frac{-5}{e^x - 1} + \frac{2}{e^x + 4}$

| SECTION 8.4 | **Systems of Nonlinear Equations in Two Variables** |

OBJECTIVES

1. Solve Nonlinear Systems of Equations by the Substitution Method
2. Solve Nonlinear Systems of Equations by the Addition Method
3. Use Nonlinear Systems of Equations to Solve Applications

1. Solve Nonlinear Systems of Equations by the Substitution Method

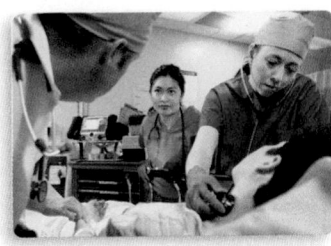

The attending physician in an emergency room treats an unconscious patient suspected of a drug overdose. The physician needs to know the concentration of the drug in the bloodstream at the time the drug was taken to determine the extent of damage to the kidneys. The patient's family does not know the original amount of the drug taken, but believes that he took the drug by injection 3 hr before arriving at the hospital. Blood work at the time of arrival ($t = 3$ hr after the patient had taken the drug) showed that the drug concentration in the bloodstream was 0.69 μg/dL. One hour later ($t = 4$ hr), the level had dropped to 0.655 μg/dL.

The physician can solve the following system of nonlinear equations to determine the concentration of the drug in the bloodstream at the time of injection. The value A_0 represents the initial concentration of the drug, and the value k is related to the rate at which the kidneys can remove the drug.

$$0.69 = A_0 e^{-3k}$$
$$0.655 = A_0 e^{-4k}$$ The solution to this problem is discussed in Exercise 59.

A **nonlinear system of equations** is a system in which one or more equations is nonlinear. For example:

$$\begin{aligned} -x + 7y &= 50 \\ x^2 + y^2 &= 100 \end{aligned}$$ Second equation nonlinear $$\left. \begin{aligned} 2x^2 + y^2 &= 17 \\ x^2 + 2y^2 &= 22 \end{aligned} \right\}$$ Both equations nonlinear

A **solution** to a nonlinear system of equations in two variables is an ordered pair with real-valued coordinates that satisfies each equation in the system. Graphically, these are the points of intersection of the graphs of the equations. A nonlinear system of equations may have no solution or one or more solutions. See Figure 8-9 through Figure 8-11.

TIP A nonlinear system of equations may also have infinitely many solutions. In the graph shown, the "wave" pattern extends infinitely far in both directions.

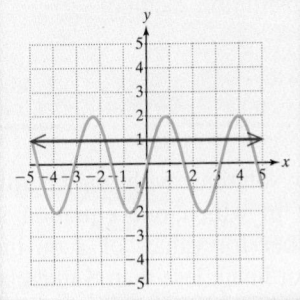

Two solutions

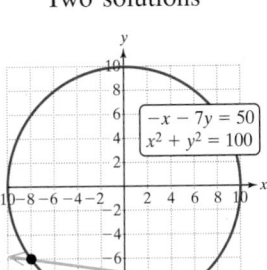

$-x - 7y = 50$
$x^2 + y^2 = 100$

Figure 8-9

Four solutions

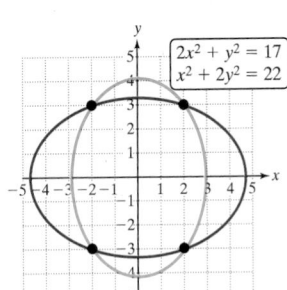

$2x^2 + y^2 = 17$
$x^2 + 2y^2 = 22$

Figure 8-10

No solution

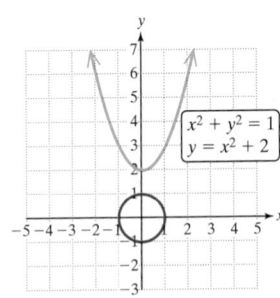

$x^2 + y^2 = 1$
$y = x^2 + 2$

Figure 8-11

We will solve nonlinear systems of equations by the substitution method and by the addition method. In Example 1, we begin with the substitution method.

Classroom Example: p. 806
Exercise 4

Figure 8-12

EXAMPLE 1   Solving a System of Nonlinear Equations by Using the Substitution Method

Solve the system by using the substitution method. $-x - 7y = 50$

$x^2 + y^2 = 100$

Solution:

$\boxed{A}$ $-x - 7y = 50$ Equation $\boxed{A}$ is a line and can be written in slope-intercept form as
$\boxed{B}$ $x^2 + y^2 = 100$ $y = -\frac{1}{7}x - \frac{50}{7}$.

Equation $\boxed{B}$ represents a circle centered at $(0, 0)$ with radius 10.

A sketch of the two equations suggests that the curves intersect at $(-8, -6)$ and $(6, -8)$. See Figure 8-12.

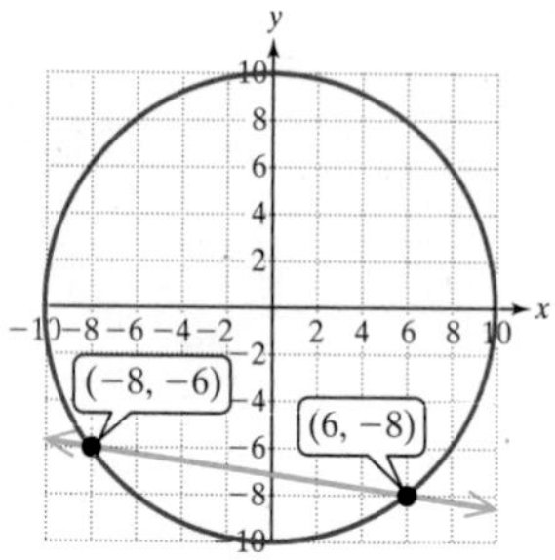

$\boxed{A}$ $-x - 7y = 50 \longrightarrow x = -7y - 50$ To solve the system algebraically by the
$\boxed{B}$ $x^2 + y^2 = 100$ substitution method, first solve for x or y from either equation.

$\boxed{B}$ $(-7y - 50)^2 + y^2 = 100$ Substitute $x = -7y - 50$ from equation $\boxed{A}$ into equation $\boxed{B}$.

$49y^2 + 700y + 2500 + y^2 = 100$

$50y^2 + 700y + 2400 = 0$ Solve the resulting equation for y.

$50(y^2 + 14y + 48) = 0$ Factor out the GCF of 50.

$50(y + 6)(y + 8) = 0$

$y = -6 \quad \text{or} \quad y = -8$

For each value of y, find the corresponding x value by substituting y into the equation in which x is isolated: $x = -7y - 50$

$y = -6: \quad x = -7(-6) - 50 = -8$ The solution is $(-8, -6)$.

$y = -8: \quad x = -7(-8) - 50 = 6$ The solution is $(6, -8)$.

Check: $(-8, -6)$ Check: $(6, -8)$

$\boxed{A}$ $-(-8) - 7(-6) = 50$ ✓ $\boxed{A}$ $-(6) - 7(-8) = 50$ ✓   The solutions
$\boxed{B}$ $(-8)^2 + (-6)^2 = 100$ ✓ $\boxed{B}$ $(6)^2 + (-8)^2 = 100$ ✓   both check in
each equation.

The solution set is $\{(-8, -6), (6, -8)\}$.

Skill Practice 1   Solve the system by using the substitution method.

$2x + y = 5$
$x^2 + y^2 = 50$

As we solve systems of equations, we will consider only solutions with real coordinates. In Example 2, we have a system of equations in which one equation is $y = 3\sqrt{x - 8}$. The expression $\sqrt{x - 8}$ is a real number for values of x on the interval $[8, \infty)$. Therefore, any ordered pair with an x-coordinate less than 8 must be rejected as a potential solution.

Classroom Example: p. 806
Exercise 8

EXAMPLE 2   Solving a System of Nonlinear Equations by Using the Substitution Method

Solve the system by using the substitution method. $(x - 5)^2 + y^2 = 25$

$y = 3\sqrt{x - 8}$

Answer

1. $\{(5, -5), (-1, 7)\}$

Solution:

A $(x - 5)^2 + y^2 = 25$
B $y = 3\sqrt{x - 8}$

Label the equations. The graphs of the equations are shown in Figure 8-13. The graph suggests that there is only one solution: (9, 3).

A $(x - 5)^2 + (3\sqrt{x - 8})^2 = 25$

Substitute $3\sqrt{x - 8}$ for y in equation A.

$x^2 - 10x + 25 + 9(x - 8) = 25$

Square each term.

$$x^2 - x - 72 = 0$$
$$(x - 9)(x + 8) = 0$$
$$x = 9 \quad \text{or} \quad x = -8$$

Reject $x = -8$ because $3\sqrt{x - 8}$ is not a real number for $x = -8$.

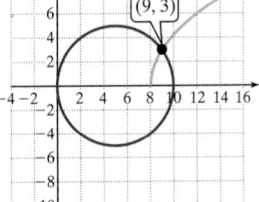

Figure 8-13

Given $x = 9$, solve for y:

B $y = 3\sqrt{x - 8}$
     $y = 3\sqrt{9 - 8} = 3$

The solution is (9, 3) and checks in each original equation.

The solution set is $\{(9, 3)\}$.

Skill Practice 2   Solve the system by using the substitution method.

$$x^2 + y^2 = 90$$
$$y = \sqrt{x}$$

---

2. Solve Nonlinear Systems of Equations by the Addition Method

The substitution method is used most often to solve a system of nonlinear equations. In some situations, however, the addition method is an efficient way to find a solution. Examples 3 and 4 demonstrate that we can eliminate a variable from both equations in a system provided the terms containing the corresponding variables are like terms.

EXAMPLE 3   **Solving a System of Nonlinear Equations by Using the Addition Method**

Solve the system by using the addition method. $2x^2 + y^2 = 17$
                                                        $x^2 + 2y^2 = 22$

Solution:

Using the addition method, the goal is to create opposite coefficients on either the x^2 terms or the y^2 terms. In this case, we have chosen to eliminate the x^2 terms.

A $2x^2 + y^2 = 17$ $2x^2 + y^2 = 17$
B $x^2 + 2y^2 = 22$ $\xrightarrow[\text{Multiply by } -2.}{}$ $\dfrac{-2x^2 - 4y^2 = -44}{-3y^2 = -27}$

$$y^2 = 9$$
$$y = \pm 3$$

$y = 3$:   B $x^2 + 2(3)^2 = 22$
                $x^2 = 4$
                $x = \pm 2$ The solutions are (2, 3), (−2, 3).

Substitute $y = \pm 3$ into either equation A or B to solve for the corresponding values of x.

$y = -3:$ $\boxed{B}$ $x^2 + 2(-3)^2 = 22$
$$x^2 = 4$$
$$x = \pm 2 \quad \text{The solutions are}$$
$$(2, -3), (-2, -3).$$

The solutions all check in the original equations.

The solution set is $\{(2, 3), (-2, 3), (2, -3), (-2, -3)\}$.

> **Skill Practice 3**   Solve the system by using the addition method.
>
> $x^2 + y^2 = 17$
> $x^2 - 2y^2 = -31$

TECHNOLOGY CONNECTIONS

Solving a System of Nonlinear Equations Using Intersect

The equations in Example 3 each represent a curve called an ellipse. We do not yet know how to graph an ellipse; however, we can graph the curves on a graphing calculator. First solve each equation for y. Enter the resulting functions in the calculator and use the **Intersect** feature to approximate the points of intersection (Figure 8-14).

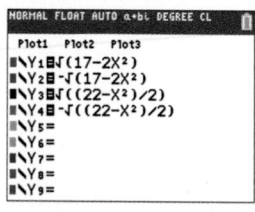

TIP The image produced by a graphing calculator in Figure 8-14 does not show the curves touching the x-axis, when indeed they do.

$$2x^2 + y^2 = 17 \xrightarrow{\text{Solve for } y.} y = \pm\sqrt{17 - 2x^2}$$

$$x^2 + 2y^2 = 22 \longrightarrow y = \pm\sqrt{\frac{22 - x^2}{2}}$$

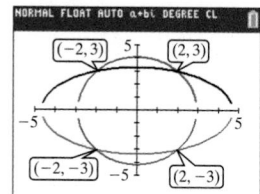

Figure 8-14

Example 4 illustrates that a nonlinear system of equations may have no solution.

Classroom Example: p. 807
Exercise 22

EXAMPLE 4   Solving an Inconsistent System by Using the Addition Method

Solve the system by using the addition method. $x^2 + 4y^2 = 4$
$x^2 - y^2 = 9$

Solution:

TIP Solutions to a system of equations are limited to ordered pairs with real-valued coordinates because we are interested in the points of intersection of the graphs of the equations.

$x^2 + 4y^2 = 4$ $x^2 + 4y^2 = 4$
$x^2 - y^2 = 9$ $\xrightarrow{\text{Multiply by } -1.}$ $\underline{-x^2 + y^2 = -9}$
                                          $5y^2 = -5$
                                          $y^2 = -1$
                                          $y = \pm i$

We use the addition method because like terms are aligned vertically.

The values for y are not real numbers. Therefore, there is no solution to the system of equations over the set of real numbers.

The solution set is $\{\ \}$.

> **Skill Practice 4**   Solve the system by using the addition method.
>
> $x^2 + y^2 = 16$
> $4x^2 + 9y^2 = 36$

Answers
3. $\{(1, 4), (-1, 4), (1, -4), (-1, -4)\}$
4. $\{\ \}$

TECHNOLOGY CONNECTIONS

Solving a Nonlinear System of Equations

The equations in Example 4 represent two curves that have not yet been studied: an ellipse and a hyperbola. However, we can graph the curves on a graphing calculator by solving for y and entering the functions into the calculator.

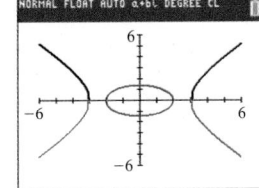

$$x^2 + 4y^2 = 4 \xrightarrow[\text{Solve for } y.]{} y = \pm\sqrt{\frac{4 - x^2}{4}}$$

$$x^2 - y^2 = 9 \longrightarrow y = \pm\sqrt{x^2 - 9}$$

From the graph, we see that the curves do not intersect.

3. Use Nonlinear Systems of Equations to Solve Applications

In Example 5, we set up a system of nonlinear equations to model an application involving two independent relationships between two variables.

Classroom Example: p. 807
Exercise 48

EXAMPLE 5   Solving an Application of a Nonlinear System

The perimeter of a television screen is 140 in. The area is 1200 in.^{2}.

a. Find the length and width of the screen.

b. Find the length of the diagonal.

Solution:

Let x represent the length of the screen.

Let y represent the width of the screen.

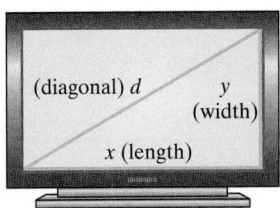

The statement of the problem gives two independent relationships between the length and width of the screen.

The perimeter of a television screen is 140 in. $\longrightarrow 2x + 2y = 140$
The area is 1200 in.^{2}. $\longrightarrow xy = 1200$

a. Solve the nonlinear system of equations for x and y.

$\boxed{A}$ $2x + 2y = 140$

$\boxed{B}$ $xy = 1200 \xrightarrow[\text{Solve for } y.]{} y = \dfrac{1200}{x}$ Using the substitution method, solve for x or y in either equation.

$\boxed{A}$ $2x + 2\left(\dfrac{1200}{x}\right) = 140$ Substitute $y = \frac{1200}{x}$ from equation $\boxed{B}$ into equation $\boxed{A}$.

$2x + \dfrac{2400}{x} = 140$

$x \cdot \left(2x + \dfrac{2400}{x}\right) = x \cdot (140)$ Multiply both sides by the LCD to clear fractions.

$2x^2 + 2400 = 140x$ The resulting equation is quadratic.

$2x^2 - 140x + 2400 = 0$

$2(x^2 - 70x + 1200) = 0$ Factor the left side.

$2(x - 40)(x - 30) = 0$

$x = 40 \quad \text{or} \quad x = 30$ There are two possible values for the length x.

Avoiding Mistakes

Check the solution to Example 5(a) by computing the perimeter and area.

Perimeter:
$2(40 \text{ in.}) + 2(30 \text{ in.})$
$= 140 \text{ in.}$ ✓

Area:
$(40 \text{ in.})(30 \text{ in.})$
$= 1200 \text{ in.}^2$ ✓

Substitute $x = 40$ and $x = 30$ into the equation $y = \dfrac{1200}{x}$.

$x = 40$: $\quad y = \dfrac{1200}{40} = 30 \quad$ If the length x is 40 in., then the width y is 30 in.

$x = 30$: $\quad y = \dfrac{1200}{30} = 40 \quad$ If the length x is 30 in., then the width y is 40 in.

Taking the length to be the longer side, we have that the length is 40 in. and the width is 30 in.

b. $x^2 + y^2 = d^2$ Use the Pythagorean theorem to determine the
$\quad (40)^2 + (30)^2 = d^2$ measure of the diagonal of the screen.
$\quad\quad 1600 + 900 = d^2$
$\quad\quad\quad\quad 2500 = d^2$
$\quad\quad\quad\quad\quad d = \pm 50$

Excluding the negative solution for d, the diagonal is 50 in.

Answer

5. The length of the rug is 12 ft and the width is 8 ft.

Skill Practice 5 The perimeter of a rectangular rug is 40 ft and the area is 96 ft^2. Find the dimensions of the rug.

SECTION 8.4   Practice Exercises

Prerequisite Review

For Exercises R.1–R.3, graph the equation.

R.1. $y = x^2 + 4x + 3$ **R.2.** $(x + 4)^2 + (y - 1)^2 = 4$ **R.3.** $y = \sqrt{x + 2}$

For Exercises R.4–R.5, solve the equation.

R.4. $\ln x = 17$ $\{e^{17}\}$ **R.5.** $3^x = 81$ $\{4\}$

Concept Connections

1. A _____nonlinear_____ system of equations in two variables is a system in which one or more equations in the system is nonlinear.

2. A solution to a nonlinear system of equations in two variables is an _____ordered_____ pair with real-valued coordinates that satisfies each equation in the system. Graphically, a solution is a point of _____intersection_____ of the graphs of the equations.

Objective 1: Solve Nonlinear Systems of Equations by the Substitution Method

For Exercises 3–14,

a. Graph the equations in the system.

b. Solve the system by using the substitution method. (**See Examples 1–2**)

3. $y = x^2 - 2$
$\quad 2x - y = 2$

4. $y = -x^2 + 3$
$\quad y - 2x = 0$

5. $x^2 + y^2 = 25$
$\quad x + y = 1$

6. $x^2 + y^2 = 25$
$\quad 3y = 4x$

7. $y = \sqrt{x}$
$\quad x^2 + y^2 = 20$

8. $x^2 + y^2 = 10$
$\quad y = \sqrt{x - 2}$

9. $(x + 2)^2 + y^2 = 9$
$\quad y = 2x - 4$

10. $x^2 + (y - 3)^2 = 4$
$\quad y = -x - 4$

11. $y = x^3$
$\quad y = x$

12. $y = \sqrt[3]{x}$
$\quad y = x$

13. $y = -(x - 2)^2 + 5$
$\quad y = 2x + 1$

14. $y = (x + 3)^2 - 1$
$\quad y = 2x + 5$

Objective 2: Solve Nonlinear Systems of Equations by the Addition Method

For Exercises 15–22, solve the system by using the addition method. (See Examples 3–4)

15. $2x^2 + 3y^2 = 11$
$x^2 + 4y^2 = 8$
$\{(2, 1), (2, -1), (-2, 1), (-2, -1)\}$

16. $3x^2 + y^2 = 21$
$4x^2 - 2y^2 = -2$
$\{(2, 3), (2, -3), (-2, 3), (-2, -3)\}$

17. $x^2 - xy = 20$
$-2x^2 + 3xy = -44$
$\{(4, -1), (-4, 1)\}$

18. $4xy + 3y^2 = -9$
$2xy + y^2 = -5$
$\{(-3, 1), (3, -1)\}$

19. $5x^2 - 2y^2 = 1$
$2x^2 - 3y^2 = -4$
$\{(1, \sqrt{2}), (1, -\sqrt{2}), (-1, \sqrt{2}), (-1, -\sqrt{2})\}$

20. $6x^2 + 5y^2 = 38$
$7x^2 - 3y^2 = 9$
$\{(\sqrt{3}, 2), (\sqrt{3}, -2), (-\sqrt{3}, 2), (-\sqrt{3}, -2)\}$

21. $x^2 = 1 - y^2$
$9x^2 - 4y^2 = 36$ $\{\ \}$

22. $4x^2 = 4 - y^2$
$16y^2 = 144 + 9x^2$ $\{\ \}$

Mixed Exercises

For Exercises 23–34, solve the system by using any method.

23. $x^2 - 4xy + 4y^2 = 1$
$x + y = 4$
$\{(3, 1), (\frac{7}{3}, \frac{5}{3})\}$

24. $x^2 - 6xy + 9y^2 = 0$
$x - y = 2$ $\{(3, 1)\}$

25. $y = x^2 + 4x + 5$
$y = 4x + 5$ $\{(0, 5)\}$

26. $y = x^2 - 6x + 9$
$y = -2x + 5$ $\{(2, 1)\}$

27. $y = x^2$
$y = \dfrac{1}{x}$ $\{(1, 1)\}$

28. $y = \dfrac{1}{x}$
$y = \sqrt{x}$ $\{(1, 1)\}$

29. $x^2 + (y - 4)^2 = 25$
$y = -x^2 + 9$
$\{(0, 9), (-3, 0), (3, 0)\}$

30. $(x - 10)^2 + y^2 = 100$
$x = y^2$
$\{(0, 0), (19, \sqrt{19}), (19, -\sqrt{19})\}$

31. $y = -x^2 + 6x - 7$
$y = x^2 - 10x + 23$
$\{(3, 2), (5, -2)\}$

32. $y = -x^2 + 6x - 9$
$y = x^2 - 2x - 3$
$\{(1, -4), (3, 0)\}$

33. $\dfrac{x^2}{4} + \dfrac{y^2}{16} = 1$
$x^2 + y = 4$
$\{(-2, 0), (0, 4), (2, 0)\}$

34. $\dfrac{x^2}{4} + y^2 = 1$
$x = -2y^2 + 2$
$\{(0, -1), (0, 1), (2, 0)\}$

For Exercises 35–36, use the substitutions $u = \dfrac{1}{x^2}$ and $v = \dfrac{1}{y^2}$ to solve the system of equations.

35. $\dfrac{4}{x^2} - \dfrac{3}{y^2} = -23$
$\dfrac{5}{x^2} + \dfrac{1}{y^2} = 14$
$\{(1, \frac{1}{3}), (-1, \frac{1}{3}), (1, -\frac{1}{3}), (-1, -\frac{1}{3})\}$

36. $-\dfrac{3}{x^2} + \dfrac{1}{y^2} = 13$
$\dfrac{5}{x^2} - \dfrac{1}{y^2} = -5$ $\{(\frac{1}{2}, \frac{1}{5}), (-\frac{1}{2}, \frac{1}{5}), (\frac{1}{2}, -\frac{1}{5}), (-\frac{1}{2}, -\frac{1}{5})\}$

Objective 3: Use Nonlinear Systems of Equations to Solve Applications

37. Find two numbers whose sum is 12 and whose product is 35. The numbers are 5 and 7.

38. Find two numbers whose sum is 9 and whose product is -36. The numbers are -3 and 12.

39. The sum of the squares of two positive numbers is 29 and the difference of the squares of the numbers is 21. Find the numbers. The numbers are 5 and 2.

40. The sum of the squares of two negative numbers is 145 and the difference of the squares of the numbers is 17. Find the numbers. The numbers are -9 and -8.

41. The difference of two positive numbers is 2 and the difference of their squares is 44. Find the numbers.
The numbers are 12 and 10.

42. The sum of two numbers is 4 and the difference of their squares is 64. Find the numbers.
The numbers are 10 and -6.

43. The ratio of two numbers is 3 to 4 and the sum of their squares is 225. Find the numbers.
The numbers are 9 and 12 or -9 and -12.

44. The ratio of two numbers is 5 to 12 and the sum of their squares is 676. Find the numbers.
The numbers are 10 and 24 or -10 and -24.

45. Find the dimensions of a rectangle whose perimeter is 36 m and whose area is 80 m^2.
The rectangle is 10 m by 8 m.

46. Find the dimensions of a rectangle whose perimeter is 56 cm and whose area is 192 cm^2.
The rectangle is 16 cm by 12 cm.

47. The floor of a rectangular bedroom requires 240 ft^2 of carpeting. Molding is placed around the base of the floor except at two 3-ft doorways. If 58 ft of molding is required around the base of the floor, determine the dimensions of the floor. **(See Example 5)**
The floor is 20 ft by 12 ft.

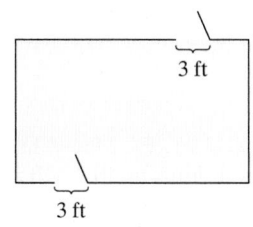
3 ft
3 ft

48. An electronic sign for a grocery store is in the shape of a rectangle. The perimeter of the sign is 72 ft and the area is 320 ft^2. Find the length and width of the sign. The sign is 20 ft by 16 ft.

49. A rental truck has a cargo capacity of 288 ft^3. A 10-ft pipe just fits resting diagonally on the floor of the truck. If the cargo space is 6 ft high, find the dimensions of the truck.
The truck is 6 ft by 6 ft by 8 ft.

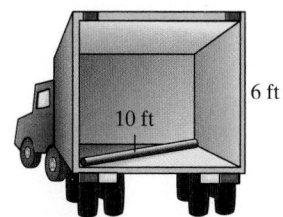
6 ft
10 ft

50. A rectangular window has a 15-yd diagonal and an area of 108 yd^2. Find the dimensions of the window.
The window is 9 yd by 12 yd.

51. An aquarium is 16 in. high with volume of 4608 in.^{3} (approximately 20 gal). If the amount of glass used for the bottom and four sides is 1440 in.^{2}, determine the dimensions of the aquarium.
The aquarium is 24 in. by 12 in. by 16 in.

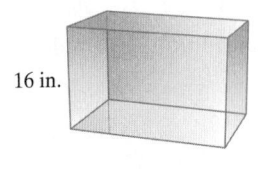
16 in.

52. A closed box is in the shape of a rectangular solid with height 3 m. Its surface area is 268 m^2. If the volume is 240 m^3, find the dimensions of the box.
The box is 8 m by 10 m by 3 m.

53. The hypotenuse of a right triangle is $\sqrt{65}$ ft. The sum of the lengths of the legs is 11 ft. Find the lengths of the legs. The legs are 4 ft and 7 ft.

54. The hypotenuse of a right triangle is $\sqrt{73}$ in. The sum of the lengths of the legs is 11 in. Find the lengths of the legs. The legs are 3 in. and 8 in.

55. A ball is kicked off the side of a hill at an angle of elevation of 30°. The hill slopes downward 30° from the horizontal. Consider a coordinate system in which the origin is the point on the edge of the hill from which the ball is kicked. The path of the ball and the line of declination of the hill can be approximated by

$$y = -\frac{x^2}{192} + \frac{\sqrt{3}}{3}x \qquad \text{Path of the ball}$$

$$y = -\frac{\sqrt{3}}{3}x \qquad \text{Line of declination of the hill}$$

Solve the system to determine where the ball will hit the ground.

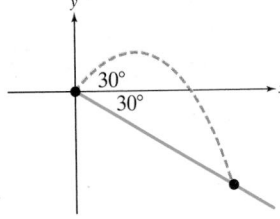

The ball will hit the ground at the point $\left(128\sqrt{3}, -128\right)$ or approximately $(221.7, -128)$.

56. A child kicks a rock off the side of a hill at an angle of elevation of 60°. The hill slopes downward 30° from the horizontal. Consider a coordinate system in which the origin is the point on the edge of the hill from which the rock is kicked. The path of the rock and the line of declination of the hill can be approximated by

$$y = -\frac{x^2}{36} + \sqrt{3}x \qquad \text{Path of the rock}$$

$$y = -\frac{\sqrt{3}}{3}x \qquad \text{Line of declination of the hill}$$

Solve the system to determine where the rock will hit the ground. The rock will hit the ground at a point $\left(48\sqrt{3}, -48\right)$ or approximately $(83.1, -48)$.

Write About It

57. What is the difference between a system of linear equations and a system of nonlinear equations?
A system of linear equations contains only linear equations, whereas a nonlinear system has one or more equations that are nonlinear.

58. Describe a situation in which the addition method is an efficient technique to solve a system of nonlinear equations. The addition method is an efficient technique to eliminate like terms from two or more equations.

Expanding Your Skills

59. The attending physician in an emergency room treats an unconscious patient suspected of a drug overdose. The physician does not know the initial concentration A_0 of the drug in the bloodstream at the time of injection. However, the physician knows that after 3 hr, the drug concentration in the blood is 0.69 μg/dL and after 4 hr, the concentration is 0.655 μg/dL. The model $A(t) = A_0e^{-kt}$ represents the drug concentration $A(t)$ (in μg/dL) in the bloodstream t hours after injection. The value of k is a constant related to the rate at which the drug is removed by the body.

a. Substitute 0.69 for $A(t)$ and 3 for t in the model and write the resulting equation. $0.69 = A_0e^{-3k}$

b. Substitute 0.655 for $A(t)$ and 4 for t in the model and write the resulting equation. $0.655 = A_0e^{-4k}$

c. Use the system of equations from parts (a) and (b) to solve for k. Round to 3 decimal places. $k \approx 0.052$

d. Use the system of equations from parts (a) and (b) to approximate the initial concentration A_0 (in μg/dL) at the time of injection. Round to 2 decimal places. $A_0 \approx 0.81$ μg/dL

e. Determine the concentration of the drug after 12 hr. Round to 2 decimal places. 0.43 μg/dL

60. A patient undergoing a heart scan is given a sample of fluorine-18 (^{18}F). After 4 hr, the radioactivity level in the patient is 44.1 MBq (megabecquerel). After 5 hr, the radioactivity level drops to 30.2 MBq. The radioactivity level $Q(t)$ can be approximated by $Q(t) = Q_0e^{-kt}$, where t is the time in hours after the initial dose Q_0 is administered.

a. Determine the value of k. Round to 4 decimal places. $k \approx 0.3786$

b. Determine the initial dose, Q_0. Round to the nearest whole unit. The original dose was 201 MBq.

c. Determine the radioactivity level after 12 hr. Round to 1 decimal place. After 12 hr, the radioactivity level is 2.1 MBq.

61. The population $P(t)$ of a culture of bacteria grows exponentially for the first 72 hr according to the model $P(t) = P_0 e^{kt}$. The variable t is the time in hours since the culture is started. The population of bacteria is 60,000 after 7 hr. The population grows to 80,000 after 12 hr.

a. Determine the constant k to 3 decimal places.
$k \approx 0.058$

b. Determine the original population P_0. Round to the nearest thousand.   The original population is 40,000.

c. Determine the time required for the population to reach 300,000. Round to the nearest hour.   The population will reach 300,000 approximately 35 hr after the culture is started.

62. An investment grows exponentially under continuous compounding. After 2 yr, the amount in the account is $7328.70. After 5 yr, the amount in the account is $8774.10. Use the model $A(t) = Pe^{rt}$ to

a. Find the interest rate r. Round to the nearest percent.
The interest rate is 6%.

b. Find the original principal P. Round to the nearest dollar.   The original principal is $6500.

c. Determine the amount of time required for the account to reach a value of $15,000. Round to the nearest year.   The investment will be worth $15,000 in approximately 14 yr.

For Exercises 63–64, determine the number of solutions to the system of equations.

63. $y = 2^{x+1}$
$-1 + \log_2 y = x$ Infinitely many solutions

64. $x^2 - y^2 = 0$
$|x| = |y|$ Infinitely many solutions

For Exercises 65–70, solve the system.

65. $\log x + 2 \log y = 5$
$2 \log x - \log y = 0$ $\{(10, 100)\}$

66. $\log_2 x + 3 \log_2 y = 6$
$\log_2 x - \log_2 y = 2$ $\{(8, 2)\}$

67. $2^x + 2^y = 6$
$4^x - 2^y = 14$ $\{(2, 1)\}$

68. $3^x - 9^y = 18$
$3^x + 3^y = 30$ $\{(3, 1)\}$

69. $(x - 1)^2 + (y + 1)^2 = 5$
$x^2 + (y + 4)^2 = 29$ $\{(2, 1), (\frac{7}{5}, \frac{6}{5})\}$

70. $(x + 3)^2 + (y - 2)^2 = 4$
$(x - 1)^2 + y^2 = 8$ $\{(-1, 2), (-\frac{9}{5}, \frac{2}{5})\}$

For Exercises 71–72, use substitution to solve the system for the set of ordered triples (x, y, λ) that satisfy the system.

71. $2 = 2\lambda x$
$6 = 2\lambda y$
$x^2 + y^2 = 10$ $\{(1, 3, 1), (-1, -3, -1)\}$

72. $8 = 4\lambda x$
$2 = 2\lambda y$
$2x^2 + y^2 = 9$ $\{(2, 1, 1), (-2, -1, -1)\}$

73. Two circles intersect as shown.

a. Find the points of intersection. $(0, -5)$ and $(4, 3)$
$x^2 + y^2 = 25$
$(x - 4)^2 + (y + 2)^2 = 25$

b. Find an equation of the chord common to both circles (shown in black). (*Hint*: A chord is a line segment on the interior of a circle with both endpoints on the circle.) $y = 2x - 5$ for $0 \le x \le 4$

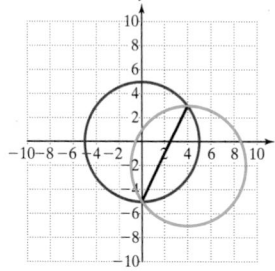

74. The minimum and maximum distances from a point P to a circle are found using the line determined by the given point and the center of the circle. Given the circle defined by $x^2 + y^2 = 9$ and the point $P(4, 5)$,

a. Find the point on the circle closest to the point $(4, 5)$. $\left(\dfrac{12\sqrt{41}}{41}, \dfrac{15\sqrt{41}}{41} \right)$

b. Find the point on the circle farthest from the point $(4, 5)$. $\left(-\dfrac{12\sqrt{41}}{41}, -\dfrac{15\sqrt{41}}{41} \right)$

Technology Connections

For Exercises 75–80, use a graphing utility to approximate the solution(s) to the system of equations. Round the coordinates to 3 decimal places.

75. $y = -0.6x + 7$
$y = e^x - 5$
$\{(2.359, 5.584)\}$

76. $y = -0.7x + 4$
$y = \ln x$
$\{(3.805, 1.336)\}$

77. $x^2 + y^2 = 40$
$y = -x^2 + 8.5$
$\{(1.538, 6.135),$
$(-1.538, 6.135),$
$(3.693, -5.135),$
$(-3.693, -5.135)\}$

78. $x^2 + y^2 = 32$
$y = 0.8x^2 - 9.2$
$\{(4.054, 3.946),$
$(-4.054, 3.946),$
$(2.237, -5.196),$
$(-2.237, -5.196)\}$

79. $y = x^2 - 8x + 20$
$y = 4 \log x$
$\{ \}$

80. $y = 0.2e^x$
$y = -0.6x^2 - 2x - 3$
$\{ \}$

Inequalities and Systems of Inequalities in Two Variables

OBJECTIVES

1. Solve Linear Inequalities in Two Variables
2. Solve Nonlinear Inequalities in Two Variables
3. Solve Systems of Inequalities in Two Variables

1. Solve Linear Inequalities in Two Variables

Adriana estimates that she has 12 hr of available study time before she takes tests in algebra and biology in back-to-back classes. Suppose that x represents the time she spends studying algebra and y represents the time she spends studying biology. Then the inequality $x + y \leq 12$, where $x \geq 0$ and $y \geq 0$ represents the distribution of time she can allocate studying for each subject.

An inequality of the form $Ax + By < C$, where A and B are not both zero, is called a **linear inequality in two variables.** (Note that the symbols $>$, $\leq$, and $\geq$ can be used in place of $<$ in the definition.) A **solution** to an inequality in two variables is an ordered pair that satisfies the inequality. The set of all such ordered pairs is called the **solution set** to the inequality. Graphically, the solution set is a region in the xy-plane.

To graph the solution set to a linear inequality in two variables, follow these guidelines.

TIP To graph the line in step 1, we can

- Use the slope-intercept form of the equation of the line.
- Graph the x- and y-intercepts.
- Create a table of points.

Graphing a Linear Inequality in Two Variables

Step 1 Graph the related equation. That is, replace the inequality sign with an $=$ sign and graph the line represented by the equation.

- If the inequality is strict (stated with the symbols $<$ or $>$), then draw the line as a dashed line to indicate that the line *is not* part of the solution set.
- If the inequality is stated with the symbols $\leq$ or $\geq$, then draw the line as a solid line to indicate that the line *is* part of the solution set.

Step 2 Choose a test point from either side of the line (not a point on the line itself) and substitute the ordered pair into the inequality.

- If a true statement results, then shade the region (half-plane) from which the test point was taken.
- If a false statement results, then shade the region (half-plane) on the opposite side of the line from which the test point was taken.

Classroom Example: p. 818
Exercise 12

EXAMPLE 1 **Graphing a Linear Inequality in Two Variables**

Graph the solution set. $3x - 2y < 6$

Solution:

$3x - 2y < 6 \xrightarrow[\text{equation}]{\text{related}} 3x - 2y = 6$ **Step 1:** Graph the related equation $3x - 2y = 6$ using any technique for graphing. In this case, we have chosen to find the x- and y-intercepts.

We can also graph the inequality from Example 1 by solving the inequality for y.

$$3x - 2y < 6$$
$$-2y < -3x + 6$$
$$y > \tfrac{3}{2}x - 3$$

Then shade the half-plane *above* the bounding line because this region contains points with y-coordinates greater than those on the bounding line.

x-intercept:

$$3x - 2(0) = 6$$
$$x = 2$$

y-intercept:

$$3(0) - 2y = 6$$
$$y = -3$$

The x- and y-intercepts are $(2, 0)$ and $(0, -3)$. Graph the line through the intercepts (Figure 8-15). Because the inequality is strict, draw the line as a dashed line.

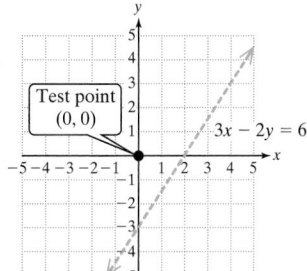

Figure 8-15

Test $(0, 0)$:

$$3x - 2y < 6$$
$$3(0) - 2(0) \overset{?}{<} 6$$
$$0 \overset{?}{<} 6 \checkmark \text{ true}$$

Step 2: Select a test point either above or below the line and test the ordered pair in the original inequality. In Figure 8-16, we have chosen $(0, 0)$ as a test point.

The test point $(0, 0)$ is a representative point above the line. Since $(0, 0)$ satisfies the original inequality, then it and all other points above the line are solutions.

The solution set is the set of ordered pairs in the region (half-plane) above the line (Figure 8-16).

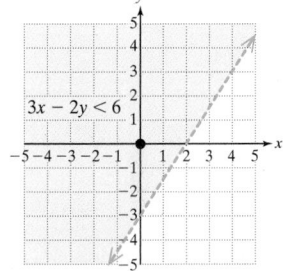

Figure 8-16

Skill Practice 1   Graph the solution set. $4x - y > 3$

TECHNOLOGY CONNECTIONS

Graphing a System of Inequalities in Two Variables

A graphing utility can be used to graph an inequality in two variables. In most cases, we solve for y first and enter the related equation in the graphing editor. Place the cursor to the left of Y_1 and press ENTER two times. This will set the graph style to shade the region above the line ◥. Notice that the calculator image does not differentiate between a solid and dashed bounding line (Figure 8-17).

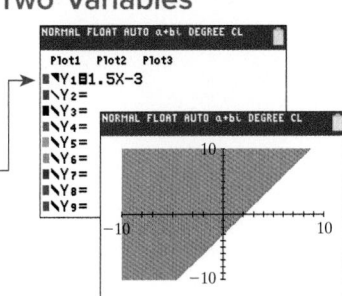

Figure 8-17

Note: With the cursor placed to the left of Y_1,

- Select the upper right triangle ◥ for inequalities of the form $Y_1 > f(x)$ and $Y_1 \geq f(x)$.
- Select the lower left triangle ◣ for inequalities of the form $Y_1 < f(x)$ and $Y_1 \leq f(x)$.

Answer

1.

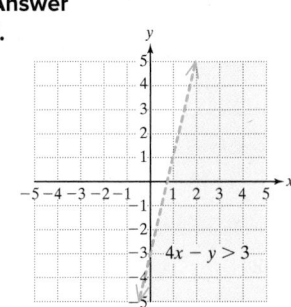

In Example 2, we graph the solutions to an inequality in which the bounding line passes through the origin.

Classroom Example: p. 818
Exercise 18

EXAMPLE 2 Graphing a Linear Inequality

Graph the solution set. $4y \leq 3x$

Solution:

$4y \leq 3x \xrightarrow[\text{related equation}]{} 4y = 3x$

$$y = \frac{3}{4}x + 0$$

Step 1: Graph the related equation $4y = 3x$. In this case, we have chosen to find the slope-intercept form of the equation and then graph the line using the slope and y-intercept.

Graph the line having y-intercept $(0, 0)$ and slope $\frac{3}{4}$ (Figure 8-18). Because the inequality symbol $\leq$ allows for equality, draw the line as a solid line.

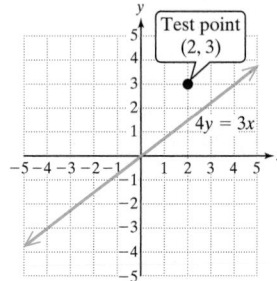

Figure 8-18

TIP As a check, we can select a test point *below* the line, such as (1, −1) and verify that it does indeed satisfy the original inequality.

Test (1, −1):

$4(-1) \overset{?}{\leq} 3(1)$

$-4 \overset{?}{\leq} 3$ ✓ true

Test (2, 3):

$4y \leq 3x$

$4(3) \overset{?}{\leq} 3(2)$

$12 \overset{?}{\leq} 6$ false

Step 2: Select a test point. We have chosen (2, 3).

The test point (2, 3) is a representative point above the line. Since (2, 3) does *not* satisfy the original inequality, then points on the other side of the line are solutions. Shade below the line.

The solution set is the set of ordered pairs on and below the line (Figure 8-19).

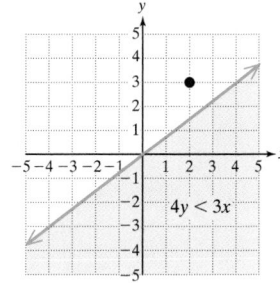

Figure 8-19

Skill Practice 2 Graph the solution set. $2y \geq 5x$

Recall that for a constant k, the equation $x = k$ represents a vertical line in the xy-plane. The inequalities $x < k$ and $x > k$ represent half-planes to the left or right of the vertical line $x = k$ (Figure 8-20). Likewise, $y = k$ represents a horizontal line in the xy-plane. The inequalities $y < k$ and $y > k$ represent half-planes below or above the line $y = k$ (Figure 8-21).

Answer

2.

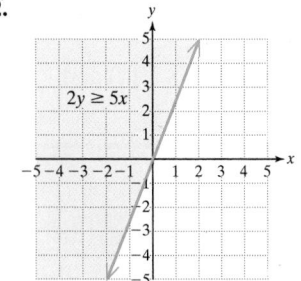

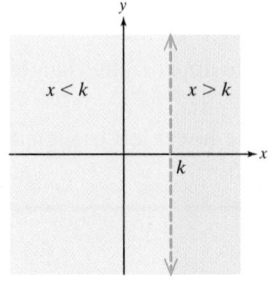

Figure 8-20

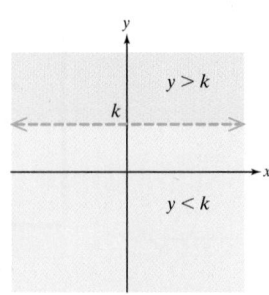

Figure 8-21

EXAMPLE 3   **Graphing Linear Inequalities with a Horizontal or Vertical Bounding Line**

Graph the solution set.

a. $x \leq -1$ **b.** $3y > 5$

Solution:

a. $x \leq -1$

- The related equation $x = -1$ is a vertical line.

- The inequality $x \leq -1$ represents all points to the *left* of or on the line $x = -1$.

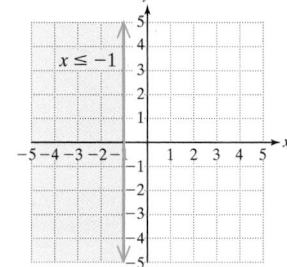

b. $3y > 5$

- The inequality is equivalent to $y > \frac{5}{3}$. The related equation $y = \frac{5}{3}$ is a horizontal line.

- The inequality $y > \frac{5}{3}$ represents all points strictly above the line $y = \frac{5}{3}$.

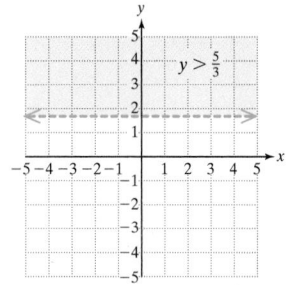

Skill Practice 3   Graph the solution set.

a. $2y < 6$ **b.** $x \geq \frac{9}{4}$

2. Solve Nonlinear Inequalities in Two Variables

The same approach used to graph a linear inequality in two variables is used to graph a *nonlinear* inequality in two variables. This is demonstrated in Example 4.

EXAMPLE 4   **Graphing a Nonlinear Inequality**

Graph the solution set. $(x - 2)^2 + y^2 > 9$

Solution:

$(x - 2)^2 + y^2 > 9 \xrightarrow[\text{equation}]{\text{related}} (x - 2)^2 + y^2 = 9$

Center: (2, 0)
Radius: 3

Step 1: Graph the related equation. The equation represents a circle centered at (2, 0) with radius 3. Because the inequality is strict, draw the circle as a dashed curve (Figure 8-22).

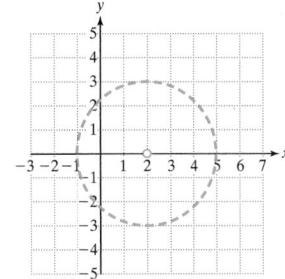

Figure 8-22

Answers

3. a.

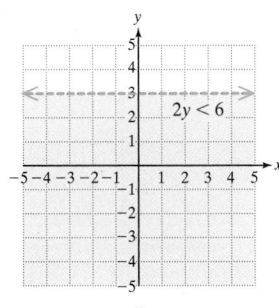

b.

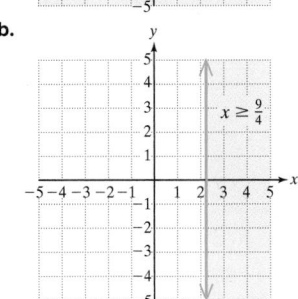

Test (2, 0):

$$(x - 2)^2 + y^2 > 9$$
$$(2 - 2)^2 + (0)^2 \overset{?}{>} 9$$
$$0 \overset{?}{>} 9 \text{ false}$$

The test point inside the circle does not satisfy the original inequality. Therefore, the solution set consists of the points strictly outside the circle (Figure 8-23).

Step 2: Select a test point. We have chosen the center (2, 0).

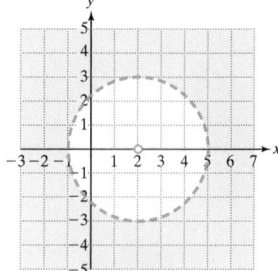

Figure 8-23

Skill Practice 4 Graph the solution set. $\quad x^2 + (y + 1)^2 < 16$

3. Solve Systems of Inequalities in Two Variables

Two or more inequalities in two variables make up a system of inequalities in two variables. The solution set is the set of ordered pairs that satisfy each inequality in the system. To graph the solution set to a system of inequalities, graph the solution sets to the individual inequalities first. The solution to the system of inequalities is the *intersection* of the graphs. This is demonstrated in Example 5.

Classroom Example: p. 819
Exercise 42

EXAMPLE 5 Graphing the Solution Set to a System of Linear Inequalities

Graph the solution set to the system of inequalities. $\quad y \le \frac{1}{2}x + 2$
$$3x - y < 3$$

Solution:

$$y \le \tfrac{1}{2}x + 2$$
$$3x - y < 3$$

First graph the solutions to the individual inequalities (Figures 8-24 and 8-25). Next, find the intersection (area of overlap) of the solution sets shown in purple in Figure 8-26.

Answer
4.

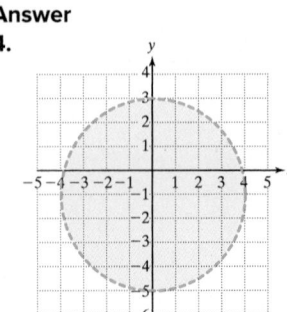

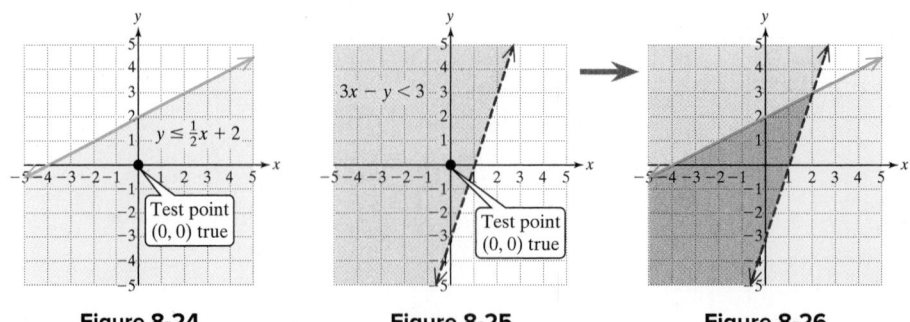

Figure 8-24   **Figure 8-25**   **Figure 8-26**

Notice that the point of intersection between the
two bounding lines is graphed as an open dot
(Figure 8-27). This indicates that it is *not* part of
the solution set. The reason is that it is not a
solution to the strict inequality $3x - y < 3$.

The point of intersection can be found by solving
the system of related equations. We have used the
substitution method.

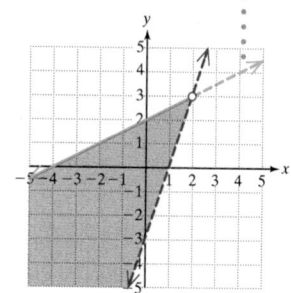

Figure 8-27

$$y = \tfrac{1}{2}x + 2$$
$$3x - y = 3 \qquad 3x - \left(\tfrac{1}{2}x + 2\right) = 3$$
$$3x - \tfrac{1}{2}x - 2 = 3$$
$$\tfrac{5}{2}x = 5$$
$$x = 2 \longrightarrow y = \tfrac{1}{2}(2) + 2$$
$$y = 3$$

The point of intersection is (2, 3) and is excluded from the solution set.

Skill Practice 5   Graph the solution set to the system of inequalities.
$$y < -\tfrac{1}{3}x + 1$$
$$-2x + y \le 1$$

TECHNOLOGY CONNECTIONS

Graphing a System of Inequalities in Two Variables

To graph the system of inequalities from Example 5
on a graphing calculator, solve each inequality
for y.

$$y \le \tfrac{1}{2}x + 2$$
$$3x - y < 3 \rightarrow -y < -3x + 3 \rightarrow y > 3x - 3$$

Then enter the related equations in the graphing
editor. Choose the appropriate graphing
style ◥ or ◣. For $Y_1 \le \tfrac{1}{2}x + 2$, choose ◣.
For $Y_2 > 3x - 3$, choose ◥.

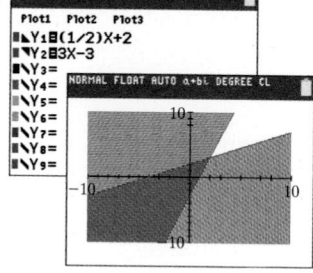

In Example 6, we graph a system of nonlinear inequalities in two variables. This
is a system of inequalities in which one or more of the individual inequalities is
nonlinear.

Answer

5.

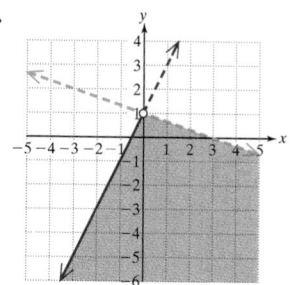

Classroom Example: p. 819
Exercise 52

EXAMPLE 6 **Solving a System of Nonlinear Inequalities**

Graph the solution set to the system of inequalities.

$$y \leq -x^2 + 4$$
$$x - y \geq -2$$
$$y > -5$$

Solution:

To graph the solution set to the given system, first graph each individual inequality.

$$y \leq -x^2 + 4 \qquad\qquad x - y \geq -2 \qquad\qquad y > -5$$

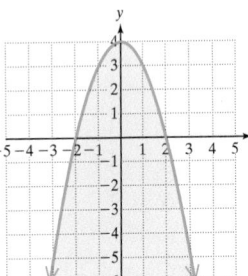 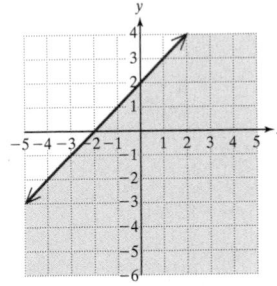 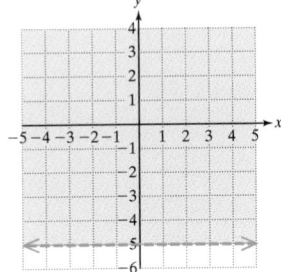

The solution set to the system is the intersection of the three shaded regions (Figure 8-28). The points of intersection are found by pairing up the related equations and solving the system of equations. The inequalities $y \leq -x^2 + 4$ and $x - y \geq -2$ include equality. Therefore, the intersection points between the parabola and slanted line are solutions to the system. These points are plotted as closed dots.

On the other hand, because the inequality $y > -5$ is strict, the intersection points between the parabola and horizontal line are *not* solutions to the system. These points are plotted as open dots.

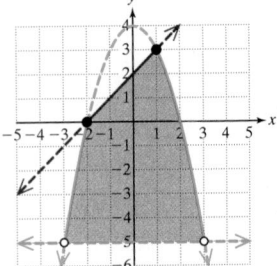

Figure 8-28

To find the points of intersection between the parabola $y = -x^2 + 4$ and the line $x - y = -2$, solve the system:

$$y = -x^2 + 4$$
$$x - y = -2 \qquad \text{The solutions are } (-2, 0) \text{ and } (1, 3).$$

To find the points of intersection between the parabola $y = -x^2 + 4$ and the line $y = -5$, solve the system:

$$y = -x^2 + 4$$
$$y = -5 \qquad \text{The solutions are } (-3, -5) \text{ and } (3, -5).$$

Skill Practice 6 Graph the solution set to the system of inequalities.

$$x^2 + y^2 \leq 25$$
$$-x + y < 1$$
$$y \geq -4$$

Answer

6.

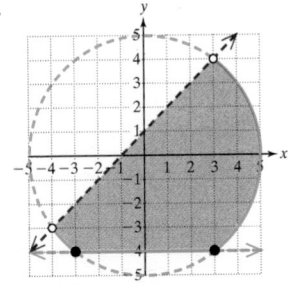

In Example 7, we refer to the problem addressed at the beginning of this section and set up a system of linear inequalities to model the allocation of time studying algebra and biology.

Classroom Example: p. 820
Exercise 66

EXAMPLE 7   Solving a System of Inequalities in an Application

Adriana has 12 available study hours for algebra and biology.

- Let x represent the number of hours she spends studying algebra.
- Let y represent the number of hours that she studies biology.

a. Set up an inequality that indicates that the number of hours spent studying algebra cannot be negative.

b. Set up an inequality that indicates that the number of hours spent studying biology cannot be negative.

c. Set up an inequality that indicates that the combined number of hours she spends studying for these two classes is at most 12 hr.

d. Graph the solution set to the system of inequalities from parts (a)–(c).

Solution:

a. $x \geq 0$ The number of hours spent studying algebra is 0 or more.

b. $y \geq 0$ The number of hours spent studying biology is 0 or more.

c. $x + y \leq 12$ The sum of time spent studying algebra and the time spent studying biology cannot exceed the maximum number of study hours available. Therefore, the sum is less than or equal to 12 hr.

TIP   The points of intersection of the bounding lines are closed dots because they are part of the solution set.

d. $x \geq 0$
$y \geq 0$
$x + y \leq 12$

The inequalities $x \geq 0$ and $y \geq 0$ together represent the set of points in the first quadrant, including the bounding axes.

To graph the inequality $x + y \leq 12$, we graph the related equation $x + y = 12$. A test point $(0, 0)$ taken below the line results in a true statement. Therefore, the inequality $x + y \leq 12$ represents the set of points on and below the line $x + y = 12$.

The solution to the system of inequalities is shown in Figure 8-29.

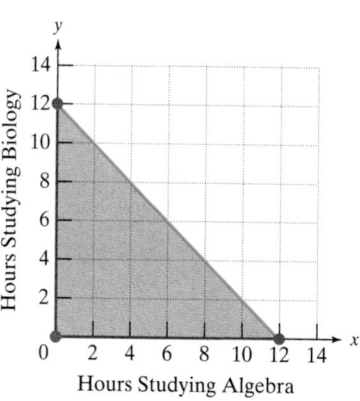

Figure 8-29

Skill Practice 7   A family plans to spend two 8-hr days at Disney World and will split time between the Magic Kingdom and Epcot Center. Let x represent the number of hours spent at the Magic Kingdom and let y represent the number of hours spent at Epcot Center.

a. Set up two inequalities that indicate that the number of hours spent at the Magic Kingdom and the number of hours spent at Epcot cannot be negative.

b. Set up an inequality that indicates that the combined number of hours spent at the two parks is at most 16 hr.

c. Graph the solution set to the system of inequalities.

Answers

7. a. $x \geq 0; y \geq 0$
b. $x + y \leq 16$
c.

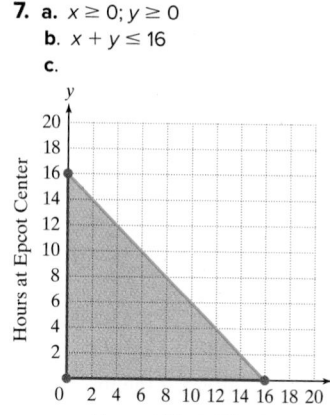

SECTION 8.5   Practice Exercises

Prerequisite Review

R.1. Use the graph to solve the equation and inequalities.
Write the solutions to the inequalities in interval notation.

   a. $x - 1 = 2x - 3$ $\{2\}$

   b. $x - 1 < 2x - 3$ $(2, \infty)$

   c. $x - 1 \geq 2x - 3$ $(-\infty, 2]$

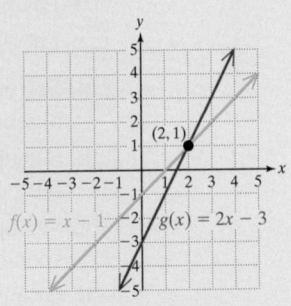

Concept Connections

1. An inequality that can be written in the form $Ax + By < C$ (where A and B are not both zero) is called a _____ linear _____ inequality in two variables.

2. For a constant real number k, the inequality $x < k$ represents the half-plane to the (left/right) of the (horizontal/vertical) line $x = k$.   left; vertical

3. For a constant real number k, the inequality $y > k$ represents the half-plane (above/below) the (horizontal/vertical) line $y = k$.   above; horizontal

4. Given the inequality $y \leq 2x + 1$, the bounding line $y = 2x + 1$ is drawn as a (dashed/solid) line.   solid

5. The solution set to the system of inequalities $x < 0$, $y > 0$ represents the points in quadrant (I, II, III, IV).   II

6. The equation $x^2 + y^2 = 4$ is a circle centered at _____ with radius _____. The solution set to the inequality $x^2 + y^2 < 4$ represents the set of points (inside/outside) the circle $x^2 + y^2 = 4$.   (0, 0); 2; inside

Objective 1: Solve Linear Inequalities in Two Variables

For Exercises 7–10, determine whether the ordered pair is a solution to the inequality.

7. $3x + 4y < 12$

   a. $(-1, 3)$ Yes

   b. $(5, 1)$ No

   c. $(4, 0)$ No

8. $2x + 3y > 6$

   a. $(-3, 3)$ No

   b. $(5, -1)$ Yes

   c. $(0, 2)$ No

9. $y \geq (x - 3)^2$

   a. $(-3, 30)$ No

   b. $(1, 4)$ Yes

   c. $(5, 5)$ Yes

10. $y \leq x^3 - 1$

   a. $(-1, -2)$ Yes

   b. $(2, 6)$ Yes

   c. $(-4, -50)$ No

11. a. Graph the solution set. $4x - 5y \leq 20$
(**See Example 1**)

   b. Explain how the graph would differ for the inequality $4x - 5y < 20$.

   c. Explain how the graph would differ for the inequality $4x - 5y > 20$.

12. a. Graph the solution set. $2x + 5y > 10$

   b. Explain how the graph would differ for the inequality $2x + 5y \geq 10$.

   c. Explain how the graph would differ for the inequality $2x + 5y < 10$.

For Exercises 13–24, graph the solution set. (**See Examples 1–3**)

13. $2x + 5y > 5$

16. $-400x < 100y + 8000$

19. $3 + 2(x + y) > y + 3$

22. $y \leq 5$

14. $-5x + 4y \leq 8$

17. $5x \leq 6y$

20. $-4 - 3(x - y) < 2y - 4$

23. $-\dfrac{1}{2}y + 4 \leq 5$

15. $-30x \geq 20y + 600$

18. $3x > 2y$

21. $x < 6$

24. $-\dfrac{1}{3}x + 2 < 4$

Objective 2: Solve Nonlinear Inequalities in Two Variables

25. a. Graph the solution set. $x^2 + y^2 < 4$ (**See Example 4**)

   b. Explain how the graph would differ for the inequality $x^2 + y^2 > 4$.   The region outside the circle would be shaded.

   c. Explain how the graph would differ for the inequality $x^2 + y^2 \geq 4$.   The shaded region would contain points on the circle (solid curve) and points outside the circle.

26. a. Graph the solution set. $y \geq x^2 - 1$

   b. Explain how the graph would differ for the inequality $y \leq x^2 - 1$.   The region on and below the parabola would be shaded.

   c. Explain how the graph would differ for the inequality $y > x^2 - 1$.   The region strictly above the parabola would be shaded (the parabola would be drawn as a dashed curve).

For Exercises 27–36, graph the solution set. (See Example 4)

27. $y < -x^2$
28. $x^2 + y^2 \geq 16$
29. $y \leq (x - 2)^2 + 1$
30. $y \geq -(x + 1)^2 - 2$
31. $|x| \leq 3$
32. $|y| \leq 2$
33. $2|y| > 2$
34. $|x| + 1 > 3$
35. $y \geq \sqrt{x}$
36. $y < \sqrt{x - 1}$

Objective 3: Solve Systems of Inequalities in Two Variables

37. a. Is the point $(2, 1)$ a solution to the inequality
$y < 2x + 3$? Yes
   b. Is the point $(2, 1)$ a solution to the inequality $x + y \leq 1$?
   No
   c. Is the point $(2, 1)$ a solution to the system of
   inequalities? No

$$y < 2x + 3$$
$$x + y \leq 1$$

38. a. Is the point $(3, 2)$ a solution to the inequality
$y < -x + 5$? No
   b. Is the point $(3, 2)$ a solution to the inequality
   $3x + y \geq 11$? Yes
   c. Is the point $(3, 2)$ a solution to the system of inequalities?
   No

$$y < -x + 5$$
$$3x + y \geq 11$$

For Exercises 39–40, determine whether the ordered pair is a solution to the system of inequalities.

39. $x + y < 4$
  $y \leq 2x + 1$

   a. $(0, 1)$ **b.** $(3, 1)$ **c.** $(2, 0)$ **d.** $(1, 4)$
   Yes   No   Yes   No

40. $y < -x^2 + 3$
  $x + 2y \leq 2$

   a. $(-2, -1)$ **b.** $(0, -2)$ **c.** $(0, 1)$ **d.** $(3, -6)$
   No   Yes   Yes   No

For Exercises 41–58, graph the solution set. If there is no solution, indicate that the solution set is the empty set. (See Examples 5–6)

41. $y < \dfrac{1}{2}x - 4$
  $y > -2x + 1$

42. $y \geq \dfrac{1}{3}x - 2$
  $y \leq x - 4$

43. $2x + 5y \leq 5$
  $-3x + 4y \geq 4$

44. $4x - 3y > 3$
  $x + 4y < -4$

45. $x^2 + y^2 \geq 9$
  $x^2 + y^2 \leq 16$

46. $x^2 + y^2 \geq 1$
  $x^2 + y^2 < 25$

47. $y \geq 3x + 3$
  $-3x + y < 1$
  The solution set is { }.

48. $y < 2x - 4$
  $-2x + y \geq 2$
  The solution set is { }.

49. $|x| < 3$
  $|y| < 3$

50. $|x| \geq 2$
  $|y| \geq 2$

51. $y \geq x^2 - 2$
  $y > x$
  $y \leq 4$

52. $y \leq -x^2 + 7$
  $y \leq -x + 5$
  $y > 1$

53. $x^2 + y^2 \leq 100$
  $y < \dfrac{4}{3}x$
  $x \leq 8$

54. $x^2 + y^2 < 100$
  $y \geq x$
  $y \geq 1$

55. $y < e^x$
  $y > 1$
  $x < 2$

56. $y \leq \dfrac{2}{x}$
  $y > 0$
  $y < x$

57. $(x + 2)^2 + (y - 3)^2 \leq 9$
  $x - y > 2$ The solution set is { }.

58. $(x - 4)^2 + (y + 1)^2 < 25$
  $2x - y < -4$ The solution set is { }.

Mixed Exercises

For Exercises 59–64, write an inequality to represent the statement.

59. x is at most 6. $x \leq 6$

60. y is no more than 7. $y \leq 7$

61. y is at least -2. $y \geq -2$

62. x is no less than $\frac{1}{2}$. $x \geq \frac{1}{2}$

63. The sum of x and y does not exceed 18. $x + y \leq 18$

64. The difference of x and y is not less than 4. $x - y \geq 4$

65. Let x represent the number of hours that Trenton spends studying algebra, and let y represent the number of hours he spends studying history. For parts (a)–(e), write an inequality to represent the given statement. (**See Example 7**)

   a. Trenton has a total of at most 9 hr to study for both algebra and history combined. $x + y \leq 9$

   b. Trenton will spend at least 3 hr studying algebra. $x \geq 3$

   c. Trenton will spend no more than 4 hr studying history. $y \leq 4$

   d. The number of hours spent studying algebra cannot be negative. $x \geq 0$

   e. The number of hours spent studying history cannot be negative. $y \geq 0$

   f. Graph the solution set to the system of inequalities from parts (a)–(e).

66. Let x represent the number of country songs that Sierra puts on a playlist on her portable media player. Let y represent the number of rock songs that she puts on the playlist. For parts (a)–(e), write an inequality to represent the given statement.

   a. Sierra will put at least 6 country songs on the playlist. $x \geq 6$

   b. Sierra will put no more than 10 rock songs on the playlist. $y \leq 10$

   c. Sierra wants to limit the length of the playlist to at most 20 songs. $x + y \leq 20$

   d. The number of country songs cannot be negative. $x \geq 0$

   e. The number of rock songs cannot be negative. $y \geq 0$

   f. Graph the solution set to the system of inequalities from parts (a)–(e).

67. A couple has \$60,000 to invest for retirement. They plan to put x dollars in stocks and y dollars in bonds. For parts (a)–(d), write an inequality to represent the given statement.

   a. The total amount invested is at most \$60,000.
   $x + y \leq 60,000$

   b. The couple considers stocks a riskier investment, so they want to invest at least twice as much in bonds as in stocks. $y \geq 2x$

   c. The amount invested in stocks cannot be negative.
   $x \geq 0$

   d. The amount invested in bonds cannot be negative.
   $y \geq 0$

   e. Graph the solution set to the system of inequalities from parts (a)–(d).

68. A college theater has a seating capacity of 2000. It reserves x tickets for students and y tickets for general admission. For parts (a)–(d) write an inequality to represent the given statement.

   a. The total number of seats available is at most 2000.
   $x + y \leq 2000$

   b. The college wants to reserve at least 3 times as many student tickets as general admission tickets.
   $x \geq 3y$

   c. The number of student tickets cannot be negative.
   $x \geq 0$

   d. The number of general admission tickets cannot be negative. $y \geq 0$

   e. Graph the solution set to the system of inequalities from parts (a)–(d).

69. Write a system of inequalities that represents the points in the first quadrant less than 3 units from the origin.
$x^2 + y^2 < 9,$ $x > 0,$ $y > 0$

70. Write a system of inequalities that represents the points in the second quadrant more than 4 units from the origin. $x^2 + y^2 > 16,$ $x < 0,$ $y > 0$

71. Write a system of inequalities that represents the points inside the triangle with vertices $(-3, -4)$, $(3, 2)$, and $(-5, 4)$.

72. Write a system of inequalities that represents the points inside the triangle with vertices $(-4, -4)$, $(1, 1)$, and $(5, -1)$.

73. A weak earthquake occurred roughly 9 km south and 12 km west of the center of Hawthorne, Nevada. The quake could be felt 16 km away. Suppose that the origin of a map is placed at the center of Hawthorne with the positive x-axis pointing east and the positive y-axis pointing north.

   a. Find an inequality that describes the points on the map for which the earthquake could be felt.
   $(x + 12)^2 + (y + 9)^2 \leq 256$

   b. Could the earthquake be felt at the center of Hawthorne?   Yes; The center of Hawthorne is 15 km from the earthquake.

74. A coordinate system is placed at the center of a town with the positive x-axis pointing east, and the positive y-axis pointing north. A cell tower is located 4 mi west and 5 mi north of the origin.

   a. If the tower has a 8-mi range, write an inequality that represents the points on the map serviced by this tower. $(x + 4)^2 + (y - 5)^2 \leq 64$

   b. Can a resident 5 mi east of the center of town get a signal from this tower?
   No; The resident is $\sqrt{106}$ mi ≈ 10.3 mi from the tower.

Write About It

75. Under what circumstances should a dashed line or curve be used when graphing the solution set to an inequality in two variables?

76. Explain how test points are used to determine the region of the plane that represents the solution to an inequality in two variables.

77. Explain how to find the solution set to a system of inequalities in two variables.

78. Describe the solution set to the system of inequalities.
$x \geq 0, y \geq 0, x \leq 1, y \leq 1$
The solution set is a unit square (a square with sides 1 unit in length) in the first quadrant with one vertex at the origin.

Expanding Your Skills

For Exercises 79–80, graph the solution set.

79. $|x| \geq |y|$

80. $|x| + |y| \leq 1$

Technology Connections

For Exercises 81–82, use a graphing utility to graph the solution set to the system of inequalities.

81. $y \geq 0.4e^x$
$y \leq 0.25x^3 - 4x$

82. $y < \dfrac{4}{x^2 + 1}$

   $y > \dfrac{-2}{x^2 + 0.5}$

Additional answers can be found in the Instructor Answer Appendix.

PROBLEM RECOGNITION EXERCISES

Equations and Inequalities in Two Variables

For Exercises 1–2, for parts (a) and (b), graph the equation. For part (c), solve the system of equations. For parts (d) and (e) graph the solution set to the system of inequalities. If there is no solution, indicate that the solution set is the empty set.

1. a. $y = -3x + 5$ **b.** $-2x + y = 0$ **c.** $y = -3x + 5$ **d.** $y > -3x + 5$ **e.** $y < -3x + 5$

                                                       $-2x + y = 0$ $-2x + y < 0$ $-2x + y > 0$

2. a. $y = 2x - 3$ **b.** $4x - 2y = -2$ **c.** $y = 2x - 3$ **d.** $y \geq 2x - 3$ **e.** $y \leq 2x - 3$

                                                       $4x - 2y = -2$ $4x - 2y \geq -2$ $4x - 2y \leq -2$

For Exercises 3–4, for part (a), graph the equations in the system and determine the solution set. For parts (b) and (c), graph the solution set to the inequality.

3. a. $y = x^2$ **b.** $y \leq x^2$ **c.** $y \geq x^2$

          $y = \frac{1}{2}x^2$ $y \geq \frac{1}{2}x^2$ $y \leq \frac{1}{2}x^2$

4. a. $x - y = 1$ **b.** $x - y \geq 1$ **c.** $x - y \leq 1$

          $y = (x - 3)^2$ $y \geq (x - 3)^2$ $y \leq (x - 3)^2$

Additional answers can be found in the Instructor Answer Appendix.

SECTION 8.6   Linear Programming

OBJECTIVES

1. Write an Objective Function
2. Solve a Linear Programming Application

TIP The notation $z = f(x, y)$ is read as "z is a function of x and y."

Classroom Example: p. 827
Exercise 6

1. Write an Objective Function

When a company manufactures a product, the goal is to obtain maximum profit at minimum cost. However, the production process is often limited by certain constraints such as the amount of labor available, the capacity of machinery, and the amount of money available for the company to invest in the process. In this section, we will study a process called **linear programming** that enables us to maximize or minimize a function under specified constraints.

The function to be optimized (maximized or minimized) in a linear programming application is called the **objective function**. The objective function often has two independent variables. For example, given $z = f(x, y)$, the variable z is dependent on the two independent variables x and y.

EXAMPLE 1   Writing an Objective Function

Suppose that a college wants to rent several buses to transport students to a championship college football game. A large bus costs $1200 to rent and a small bus costs $800 to rent.

> Let x represent the number of large buses.
>
> Let y represent the number of small buses.

Write an objective function that represents the total cost z (in $) to rent x large buses and y small buses.

Classroom Example: p. 828
Exercise 22

TIP An objective function represents a quantity that is to be maximized or minimized. In Example 1, the school will want to minimize the cost to transport the students.

Solution:

$$\begin{pmatrix}\text{Total}\\ \text{cost}\end{pmatrix} = \begin{pmatrix}\text{Cost to rent}\\ x \text{ large buses}\end{pmatrix} + \begin{pmatrix}\text{Cost to rent}\\ y \text{ small buses}\end{pmatrix}$$

Cost: z = $1200x$ + $800y$

The objective function is defined by $z = 1200x + 800y$.

Skill Practice 1   An office manager needs to staff the office. She hires full-time employees at \$36 per hour and part-time employees at \$24 per hour. Write an objective function that represents the total cost (in \$) to staff the office with x full-time employees and y part-time employees for 1 hr.

2. Solve a Linear Programming Application

In Example 2, we identify constraints imposed on the resources that affect the number of buses that the college can rent.

EXAMPLE 2   Writing a System of Constraints

Refer to the scenario from Example 1. We now set several constraints that affect the number of buses that can be rented.

Let x represent the number of large buses.
Let y represent the number of small buses.

For parts (a)–(d), write an inequality that represents the given statement.

a. The number of large buses cannot be negative.
b. The number of small buses cannot be negative.
c. Large buses can carry 60 people and small buses can carry 45 people. The college must transport at least 3600 students.
d. The number of available bus drivers is at most 75.

Solution:

a. $x \geq 0$ The number of large buses cannot be negative.
b. $y \geq 0$ The number of small buses cannot be negative.

c. $\begin{pmatrix}\text{Number of students}\\ \text{carried by large buses}\end{pmatrix} + \begin{pmatrix}\text{Number of students}\\ \text{carried by small buses}\end{pmatrix} \overset{\text{is at least}}{\geq} 3600$ Large buses hold 60 people and

   $60x$ + $45y$ ≥ 3600 small buses hold 45 people.

d. $\begin{pmatrix}\text{Number of}\\ \text{large buses}\end{pmatrix} + \begin{pmatrix}\text{Number of}\\ \text{small buses}\end{pmatrix} \overset{\text{is at most}}{\leq} 75$ Because each bus has only one driver, the total number of buses is the same as the

   x + y ≤ 75 total number of drivers.

Skill Practice 2   Refer to Skill Practice 1. Suppose that the office manager needs at least 20 employees, but not more than 24 full-time employees. Furthermore, to make the office run smoothly, the manager knows that the number of full-time employees must always be greater than or equal to the number of part-time employees. Write a system of inequalities that represents the constraints on the number of full-time employees x and the number of part-time employees y.

Answers
1. $z = 36x + 24y$
2. $x \geq 0, y \geq 0, x \leq 24,$
   $x + y \geq 20, x \geq y$

The constraints in Example 2 make up a system of linear inequalities.

The number of large buses is nonnegative.	$x \geq 0$
The number of small buses is nonnegative.	$y \geq 0$
The school must transport at least 3600 students.	$60x + 45y \geq 3600$
The number of drivers (and therefore buses) is at most 75.	$x + y \leq 75$

The region in the plane that represents the solution set to the system of constraints is called the **feasible region** (Figure 8-30). The points of intersection of the bounding lines in the feasible region are called the **vertices** of the feasible region.

The vertices in Figure 8-30 are (60, 0), (75, 0), and (15, 60).

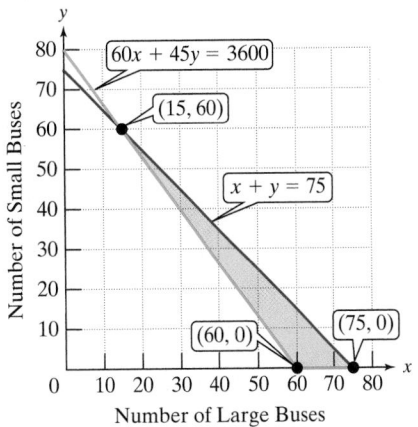

Figure 8-30

Points within the shaded region meet all the constraints in the problem. For example, the ordered pair (65, 5) represents 65 large buses and 5 small buses.

65 large buses and 5 small buses ⟶ 70 buses (≤ 75 drivers) ✓
65 large buses and 5 small buses ⟶ transports 4125 students (≥ 3600) ✓

However, points such as (65, 15) and (40, 20) are *outside* the feasible region and do not satisfy all constraints.

65 large buses and 15 small buses ⟶ 80 buses (exceeds the number of drivers)
40 large buses and 20 small buses ⟶ transports only 3300 students

The goal of a linear programming application is to find the maximum or minimum value of the objective function $z = f(x, y)$ when x and y are restricted to the ordered pairs in the feasible region. Fortunately, it has been proven mathematically that if a maximum or minimum value of a function exists, it occurs at one or more of the vertices of the feasible region. This is the basis for the following procedure to solve a linear programming application.

Solving an Application Involving Linear Programming

Step 1 Write an objective function $z = f(x, y)$.
Step 2 Write a system of inequalities defining the constraints on x and y.
Step 3 Graph the feasible region and identify the vertices.
Step 4 Evaluate the objective function at each vertex of the feasible region. Use the results to identify the values of x and y that optimize the objective function. Identify the optimal value of z.

In Example 3, we optimize the objective function found in Example 1 subject to the constraints defined in Example 2.

Classroom Example: p. 828
Exercise 22

EXAMPLE 3 Solving a Linear Programming Application

A college wants to rent buses to transport at least 3600 students to a championship college football game. Large buses hold 60 people and small buses hold 45 people. Furthermore, the number of available bus drivers is at most 75. Each large bus costs $1200 to rent and each small bus costs $800 to rent. Find the optimal number of large and small buses that will minimize cost.

Solution:

Let x represent the number of large buses.   Define the relevant variables.
Let y represent the number of small buses.

Cost: $z = 1200x + 800y$ **Step 1:** Write an objective function. Since cost is to be minimized, write a cost function. (See Example 1.)

Constraints: **Step 2:** Write a system of constraints on the relevant variables. (See Example 2.)

$x \geq 0$

$y \geq 0$

$60x + 45y \geq 3600$

$x + y \leq 75$

Step 3: Graph the feasible region and identify the vertices.

The vertices are found by identifying the points of intersection between the bounding lines.

Bounding lines:

$60x + 45y = 3600$

$x + \quad y = 75$ Point of intersection (15, 60)

$60x + 45y = 3600$

$\qquad y = 0$ Point of intersection (60, 0)

$x + y = 75$

$\qquad y = 0$ Point of intersection (75, 0)

Instructor Note:
Consider reviewing key phrases implying inequality, such as "at least," "at most," "no more than," and "no less than."

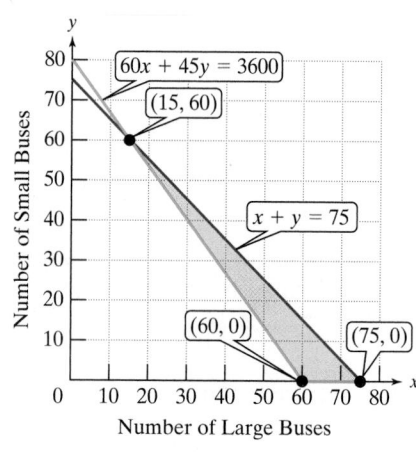

Number of Small Buses
Number of Large Buses

Cost function: $z = 1200x + 800y$

at (15, 60) $z = 1200(15) + 800(60) = \boxed{\$66{,}000}$ **Step 4:** Evaluate the objective function $z = 1200x + 800y$ at each vertex.
at (60, 0) $z = 1200(60) + 800(0) = \$72{,}000$
at (75, 0) $z = 1200(75) + 800(0) = \$90{,}000$

The cost would be minimized if the college rents 15 large buses and 60 small buses. The minimum cost is $66,000.

Skill Practice 3 Refer to Skill Practices 1 and 2. The office manager needs at least 20 employees, but no more than 24 full-time employees. Furthermore, to make the office run smoothly, the manager knows that the number of full-time employees must always be greater than or equal to the number of part-time employees. If she pays full-time employees $36 per hour and part-time employees $24 per hour, determine the number of full-time and part-time employees she should hire to minimize total labor cost per hour.

Answer

3. The cost would be minimized at $600/hr if she hires 10 full-time employees and 10 part-time employees.

In Example 4, we investigate a situation in which we maximize profit.

Classroom Example: p. 829
Exercise 28

Point of Interest

Linear programming was first introduced by Russian mathematician Leonid Kantorovich. The technique was used in World War II to minimize costs to the army. Later in 1975, Kantorovich won the Nobel Prize in economics for his contributions to the theory of optimum allocation of resources.

EXAMPLE 4   Solving a Linear Programming Application

A baker produces whole wheat bread and cheese bread to sell at the farmer's market. The whole wheat bread is denser and requires more baking time, whereas the cheese bread requires more labor. The baking times and average amount of labor per loaf are given in the table along with the profit for each loaf.

	Time to Bake	Labor	Profit
Wheat bread	1.5 hr	$\frac{1}{3}$ hr	$1.20
Cheese bread	1 hr	$\frac{1}{2}$ hr	$1.00

The oven space restricts the baker from baking more than 120 loaves. Furthermore, the amount of oven time for baking is no more than 165 hr and the amount of available labor is at most 55 hr. Determine the number of loaves of each type of bread that the baker should bake to maximize his profit. Assume that all loaves of bread produced are sold.

Solution:

Let x represent the number of loaves of wheat bread.

Let y represent the number of loaves of cheese bread.

$$\text{Profit} = \begin{pmatrix} \text{Profit from} \\ \text{wheat bread} \end{pmatrix} + \begin{pmatrix} \text{Profit from} \\ \text{cheese bread} \end{pmatrix}$$

Step 1: Write an objective function. In this example, we need to maximize profit.

$$z = 1.20x + 1.00y$$

Step 2: Write a system of constraints on the independent variables.

$x \geq 0$	Number of loaves of wheat bread cannot be negative.
$y \geq 0$	Number of loaves of cheese bread cannot be negative.
$x + y \leq 120$	Total number of loaves is no more than 120.
$1.5x + y \leq 165$	The total amount of baking time is no more than 165 hr.
$\frac{1}{3}x + \frac{1}{2}y \leq 55$	The total amount of available labor is at most 55 hr.

Step 3: Graph the feasible region and identify the vertices. Find the points of intersection between pairs of bounding lines:

$$x + y = 120$$
$$\frac{1}{3}x + \frac{1}{2}y = 55 \qquad \text{Point of intersection } (30, 90)$$

$$x + y = 120$$
$$1.5x + y = 165 \qquad \text{Point of intersection } (90, 30)$$

$$1.5x + y = 165$$
$$y = 0 \qquad \text{Point of intersection } (110, 0)$$

$$\frac{1}{3}x + \frac{1}{2}y = 55$$
$$x = 0 \qquad \text{Point of intersection } (0, 110)$$

$$x = 0$$
$$y = 0 \qquad \text{Point of intersection } (0, 0)$$

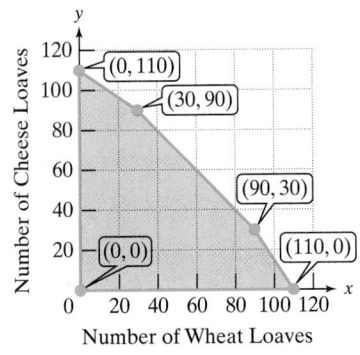

Number of Cheese Loaves (y-axis), Number of Wheat Loaves (x-axis)

Step 4: Evaluate the objective function at each vertex.

Profit: $z = 1.2x + y$

at $(0, 0)$ $z = 1.2(0) + (0) = \$0$

at $(0, 110)$ $z = 1.2(0) + (110) = \$110$

at $(30, 90)$ $z = 1.2(30) + (90) = \$126$

at $(90, 30)$ $z = 1.2(90) + (30) = \boxed{\$138}$

at $(110, 0)$ $z = 1.2(110) + (0) = \$132$

The profit will be maximized if the baker bakes 90 whole wheat loaves and 30 cheese loaves. The maximum profit is $138.

Skill Practice 4 A manufacturer produces two sizes of leather handbags. It takes longer to cut and dye the leather for the smaller bag, but it takes more time sewing the larger bag. The production constraints and profit for each type of bag are given in the table.

	Cutting and Dying	Sewing	Profit
Large bag	0.6 hr	2 hr	$30
Small bag	1 hr	1.5 hr	$25

The machinery limits the number of bags produced to at most 1000 per week. If the company has 900 hr per week available for cutting and dying and 1800 hr available per week for sewing, determine the number of each type of bag that should be produced weekly to maximize profit. Assume that all bags produced are also sold.

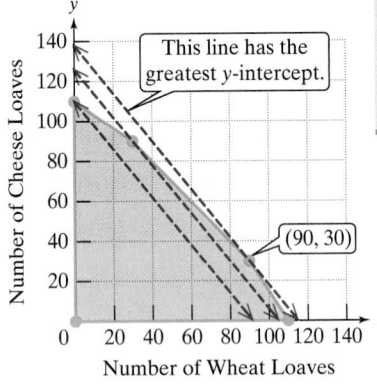

Figure 8-31

This line has the greatest y-intercept. (90, 30)

Number of Cheese Loaves (y-axis), Number of Wheat Loaves (x-axis)

To find the maximum or minimum value of an objective function, we evaluate the function at the vertices of the feasible region. It seems reasonable that the profit would be maximized at a point on the upper edge of the feasible region. These are the points in the feasible region where the combined values of x and y are the greatest.

The goal of Example 4 was to find the values of x and y that maximized the profit function $z = 1.2x + y$. To see why the profit z was maximized at a vertex of the feasible region, write the equation in slope-intercept form:

$$y = -1.2x + z$$

In this form, the objective function represents a family of parallel lines with slope -1.2 and y-intercept $(0, z)$. To maximize z, we want the line with the greatest y-intercept that still remains in contact with the feasible region. In Figure 8-31, we see that this occurs for the line passing through the point $(90, 30)$ as expected.

Answer

4. The maximum profit of $28,000 is realized when the company produces 600 large bags and 400 small bags.

SECTION 8.6   Practice Exercises

Prerequisite Review

R.1. A salesperson makes a base salary of $500 per week plus 11% commission on sales.

    a. Write a linear function to model the salesperson's weekly salary $S(x)$ for x dollars in sales. $S(x) = 0.11x + 500$, for $x \geq 0$

    b. Evaluate $S(8000)$ and interpret the meaning in the context of this problem. $S(8000) = 1380$; The salesperson will make $1380 if $8000 in merchandise is sold for the week.

R.2. A luxury car rental company charges $123 per day in addition to a flat fee of $75 for insurance.

    a. Write an equation that represents the cost y (in $) to rent the car for x days. $y = 123x + 75$

    b. What is the y-intercept and what does it mean in the context of this problem? $(0, 75)$; The base cost to rent the car is $75.

Concept Connections

1. The process that maximizes or minimizes a function subject to linear constraints is called ___linear___ programming.

2. The function to be optimized in a linear programming application is called the ___objective___ function.

3. The region in the plane that represents the solution set to a system of constraints is called the ___feasible___ region.

4. The points of intersection of a feasible region are called the ___vertices___ of the region.

Objective 1: Write an Objective Function

5. A diner makes a profit of $0.80 for a cup of coffee and $1.10 for a cup of tea. Write an objective function $z = f(x, y)$ that represents the total profit for selling x cups of coffee and y cups of tea. (**See Example 1**)
$z = 0.80x + 1.10y$

6. Rita burns 10 calories per minute running and 8 calories per minute lifting weights. Write an objective function $z = f(x, y)$ that represents the total number of calories burned by running for x minutes and lifting weights for y minutes. $z = 10x + 8y$

7. A courier company makes deliveries with two different trucks. Truck A costs $0.62/mi to operate and truck B costs $0.50/mi to operate. Write an objective function $z = f(x, y)$ that represents the total cost for driving truck A for x miles and driving truck B for y miles.
$z = 0.62x + 0.50y$

8. The cost for an animal shelter to spay a female cat is $82 and the cost to neuter a male cat is $55. Write an objective function $z = f(x, y)$ that represents the total cost for spaying x female cats and neutering y male cats. $z = 82x + 55y$

Objective 2: Solve a Linear Programming Application

For Exercises 9–12,

a. Determine the values of x and y that produce the maximum or minimum value of the objective function on the given feasible region.

b. Determine the maximum or minimum value of the objective function on the given feasible region.

9. Maximize: $z = 3x + 2y$ **a.** $x = 7, y = 5$
 b. Maximum value: 31

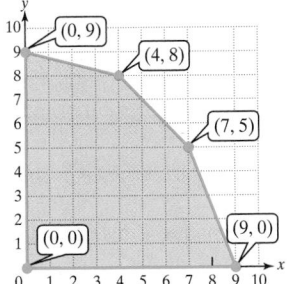

10. Maximize: $z = 1.8x + 2.2y$ **a.** $x = 18, y = 27$
 b. Maximum value: 91.8

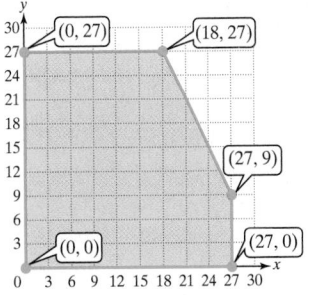

11. Minimize: $z = 1000x + 900y$ **a.** $x = 10, y = 30$
 b. Minimum value: 37,000

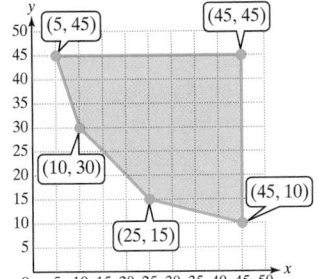

12. Minimize: $z = 6x + 9y$ **a.** $x = 14, y = 2$
 b. Minimum value: 102

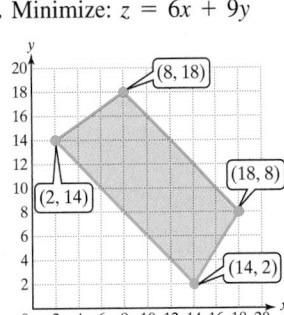

For Exercises 13–18,

a. For the given constraints, graph the feasible region and identify the vertices.

b. Determine the values of x and y that produce the maximum or minimum value of the objective function on the feasible region.

c. Determine the maximum or minimum value of the objective function on the feasible region.

13. $x \geq 0, y \geq 0$

$x + y \leq 60$

$y \leq 2x$

Maximize: $z = 250x + 150y$

14. $x \geq 0, y \geq 0$

$2x + y \leq 40$

$x + 2y \leq 50$

Maximize: $z = 9.2x + 8.1y$

15. $x \geq 0, y \geq 0$

$3x + y \geq 50$

$2x + y \geq 40$

Minimize: $z = 3x + 2y$

16. $x \geq 0, y \geq 0$

$4x + 3y \geq 60$

$2x + 3y \geq 36$

Minimize: $z = 4.5x + 6y$

17. $x \geq 0, y \geq 0$

$x \leq 36$

$y \leq 40$

$x + y \leq 48$

Maximize: $z = 150x + 90y$

18. $x \geq 0, y \geq 0$

$x \leq 10$

$y \leq 8$

$x + y \leq 12$

Maximize: $z = 50x + 70y$

For Exercises 19–20, use the given constraints to find the maximum value of the objective function and the ordered pair (x, y) that produces the maximum value.

19. $x \geq 0, y \geq 0$

$3x + 4y \leq 48$

$2x + y \leq 22$

$y \leq 9$

a. Maximize: $z = 100x + 120y$ 1520 at (8, 6)

b. Maximize: $z = 100x + 140y$ 1660 at (4, 9)

20. $x \geq 0, y \geq 0$

$x + y \leq 20$

$x + 2y \leq 36$

$x \leq 14$

a. Maximize: $z = 12x + 15y$ 288 at (4, 16)

b. Maximize: $z = 15x + 12y$ 282 at (14, 6)

21. A furniture manufacturer builds tables. The cost for materials and labor to build a kitchen table is $240 and the profit is $160. The cost to build a dining room table is $320 and the profit is $240. **(See Examples 2–3)**

Let x represent the number of kitchen tables produced per month. Let y represent the number of dining room tables produced per month.

a. Write an objective function representing the monthly profit for producing and selling x kitchen tables and y dining room tables.   Profit: $z = 160x + 240y$

b. The manufacturing process is subject to the following constraints. Write a system of inequalities representing the constraints.

$x \geq 0$
$y \geq 0$
$x \leq 120$
$y \leq 90$

- The number of each type of table cannot be negative. $240x + 320y \leq 48{,}000$
- Due to labor and equipment restrictions, the company can build at most 120 kitchen tables.
- The company can build at most 90 dining room tables.
- The company does not want to exceed a monthly cost of $48,000.

c. Graph the system of inequalities represented by the constraints.

d. Find the vertices of the feasible region.
(0, 0), (0, 90), (80, 90), (120, 60), (120, 0)

e. Test the objective function at each vertex.

f. How many kitchen tables and how many dining room tables should be produced to maximize profit? (Assume that all tables produced will be sold.)

g. What is the maximum profit?
The maximum profit is $34,400.

22. Guyton makes $24/hr tutoring chemistry and $20/hr tutoring math.

Let x represent the number of hours per week he spends tutoring chemistry. Let y represent the number of hours per week he spends tutoring math.

a. Write an objective function representing his weekly income for tutoring x hours of chemistry and y hours of math.   Income: $z = 24x + 20y$

b. The time that Guyton devotes to tutoring is limited by the following constraints. Write a system of inequalities representing the constraints.

- The number of hours spent tutoring each subject cannot be negative.
- Due to the academic demands of his own classes he tutors at most 18 hr per week.
- The tutoring center requires that he tutors math at least 4 hr per week.
- The demand for math tutors is greater than the demand for chemistry tutors. Therefore, the number of hours he spends tutoring math must be at least twice the number of hours he spends tutoring chemistry.

c. Graph the system of inequalities represented by the constraints.

d. Find the vertices of the feasible region.
(0, 18), (6, 12), (2, 4), (0, 4)

e. Test the objective function at each vertex.

f. How many hours tutoring math and how many hours tutoring chemistry should Guyton work to maximize his income?

g. What is the maximum income?
Guyton's maximum income is $384 per week.

h. Explain why Guyton's maximum income is found at a point on the line $x + y = 18$.

23. A plant nursery sells two sizes of oak trees to landscapers. Large trees cost the nursery $120 from the grower. Small trees cost the nursery $80. The profit for each large tree sold is $35 and the profit for each small tree sold is $30. The monthly demand is at most 400 oak trees. Furthermore, the nursery does not want to allocate more than $43,200 each month on inventory for oak trees.

a. Determine the number of large oak trees and the number of small oak trees that the nursery should have in its inventory each month to maximize profit. (Assume that all trees in inventory are sold.)
280 large trees and 120 small trees would maximize profit.
b. What is the maximum profit?
The maximum profit is $13,400.
c. If the profit on large trees were $50, and the profit on small trees remained the same, then how many of each should the nursery have to maximize profit?
In this case, the nursery should have 360 large trees and no small trees.

25. A paving company delivers gravel for a road construction project. The company has a large truck and a small truck. The large truck has a greater capacity, but costs more for fuel to operate. The load capacity and cost to operate each truck per load are given in the table.

	Load Capacity	Cost per Load
Small truck	18 yd^3	$120
Large truck	24 yd^3	$150

The company must deliver at least 288 yd^3 of gravel to stay on schedule. Furthermore, the large truck takes longer to load and cannot make as many trips as the small truck. As a result, the number of trips made by the large truck is at most $\frac{3}{4}$ times the number of trips made by the small truck.

a. Determine the number of trips that should be made by the large truck and the number of trips that should be made by the small truck to minimize cost. The company should make 8 trips with the small truck and 6 trips with the large truck.
b. What is the minimum cost to deliver gravel under these constraints? The minimum cost is $1860.

27. A manufacturer produces two models of a gas grill. Grill A requires 1 hr for assembly and 0.4 hr for packaging. Grill B requires 1.2 hr for assembly and 0.6 hr for packaging. The production information and profit for each grill are given in the table. (**See Example 4**)

	Assembly	Packaging	Profit
Grill A	1 hr	0.4 hr	$90
Grill B	1.2 hr	0.6 hr	$120

The manufacturer has 1200 hr of labor available for assembly and 540 hr of labor available for packaging.

a. Determine the number of grill A units and the number of grill B units that should be produced to maximize profit assuming that all grills will be sold. The manufacturer should produce 600 grill A units and 500 grill B units to maximize profit.
b. What is the maximum profit under these constraints?
The maximum profit is $114,000.
c. If the profit on grill A units is $110 and the profit on grill B units is unchanged, how many of each type of grill unit should the manufacturer produce to maximize profit?
In this case, the manufacturer should produce 1200 grill A units and 0 grill B units.

24. A sporting goods store sells two types of exercise bikes. The deluxe model costs the store $400 from the manufacturer and the standard model costs the store $300 from the manufacturer. The profit that the store makes on the deluxe model is $180 and the profit on the standard model is $120. The monthly demand for exercise bikes is at most 30. Furthermore, the store manager does not want to spend more than $9600 on inventory for exercise bikes.

a. Determine the number of deluxe models and the number of standard models that the store should have in its inventory each month to maximize profit. (Assume that all exercise bikes in inventory are sold.)
24 deluxe bikes and 0 standard model bikes would maximize profit.
b. What is the maximum profit?
The maximum profit is $4320.
c. If the profit on the deluxe bikes were $150 and the profit on the standard bikes remained the same, how many of each should the store have to maximize profit?
In this case, the store should have 6 standard bikes and 24 deluxe bikes.

26. A large department store needs at least 3600 labor hours covered per week. It employs full-time staff 40 hr/wk and part-time staff 25 hr/wk. The cost to employ a full-time staff member is more because the company pays benefits such as health care and life insurance.

	Hours per Week	Cost per Hour
Full time	40 hr	$25
Part time	25 hr	$18

The store manager also knows that to make the store run efficiently, the number of full-time employees must be at least 1.25 times the number of part-time employees.

a. Determine the number of full-time employees and the number of part-time employees that should be used to minimize the weekly labor cost. The store should have 60 full-time employees and 48 part-time employees.
b. What is the minimum weekly cost to staff the store under these constraints? The minimum labor cost for the week under these constraints is $81,600.

28. A manufacturer produces two models of patio furniture. Model A requires 2 hr for assembly and 1.2 hr for painting. Model B requires 3 hr for assembly and 1.5 hr for painting. The production information and profit for selling each model are given in the table.

	Assembly	Painting	Profit
Model A	2 hr	1.2 hr	$150
Model B	3 hr	1.5 hr	$200

The manufacturer has 1200 hr of labor available for assembly and 660 hr of labor available for painting.

a. Determine the number of model A units and the number of model B units that should be produced to maximize profit assuming that all furniture will be sold. The manufacturer should produce 300 model A units and 200 model B units.
b. What is the maximum profit under these constraints?
The maximum profit is $85,000.
c. If the profit on model A units is $180 and the profit on model B units remains the same, how many of each type should the manufacturer produce to maximize profit?
In this case, the manufacturer should produce 550 model A units and no model B units.

29. A farmer has 1200 acres of land and plans to plant corn and soybeans. The input cost (cost of seed, fertilizer, herbicide, and insecticide) for 1 acre for each crop is given in the table along with the cost of machinery and labor. The profit for 1 acre of each crop is given in the last column.

	Input Cost per Acre	Labor/Machinery Cost per Acre	Profit per Acre
Corn	$180	$80	$120
Soybeans	$120	$100	$100

Suppose the farmer has budgeted a maximum of $198,000 for input costs and a maximum of $110,000 for labor and machinery.

a. Determine the number of acres of each crop that the farmer should plant to maximize profit. (Assume that all crops will be sold.) *The farmer should plant 900 acres of corn and 300 acres of soybeans.*

b. What is the maximum profit? *The maximum profit is $138,000.*

c. If the profit per acre were reversed between the two crops (that is, $100 per acre for corn and $120 per acre for soybeans), how many acres of each crop should be planted to maximize profit? *In this case, 500 acres of corn and 700 acres of soybeans should be planted.*

Write About It

31. What is the purpose of linear programming?

33. How is the feasible region determined?

30. To protect soil from erosion, some farmers plant winter cover crops such as winter wheat and rye. In addition to conserving soil, cover crops often increase crop yields in the row crops that follow in spring and summer. Suppose that a farmer has 800 acres of land and plans to plant winter wheat and rye. The input cost for 1 acre for each crop is given in the table along with the cost for machinery and labor. The profit for 1 acre of each crop is given in the last column.

	Input Cost per Acre	Labor/Machinery Cost per Acre	Profit per Acre
Wheat	$90	$50	$42
Rye	$120	$40	$35

Suppose the farmer has budgeted a maximum of $90,000 for input costs and a maximum of $36,000 for labor and machinery.

a. Determine the number of acres of each crop that the farmer should plant to maximize profit. (Assume that all crops will be sold.) *The farmer should plant 400 acres of wheat and 400 acres of rye.*

b. What is the maximum profit? *The maximum profit is $30,800.*

c. If the profit per acre for wheat were $40 and the profit per acre for rye were $45, how many acres of each crop should be planted to maximize profit? *In this case, 200 acres of wheat and 600 acres of rye should be planted.*

32. What is an objective function?

34. If an optimal value exists for an objective function, it exists at one of the vertices of the feasible region. Explain how to find the vertices.

Additional answers can be found in the Instructor Answer Appendix.

CHAPTER 8   KEY CONCEPTS

SECTION 8.1 Systems of Linear Equations in Two Variables and Applications	Reference
Two or more linear equations taken together form a **system of linear equations.** A **solution** to a system of equations in two variables is an ordered pair that is a solution to each individual equation. Graphically, this is a point of intersection of the graphs of the equations.	p. 766
The substitution method and the addition method are often used to solve a system of linear equations in two variables.	pp. 768, 769
A system of linear equations in two variables will have no solution if the equations in the system represent parallel lines. In such a case, we say that the system is **inconsistent.**	p. 771
A system of linear equations in two variables will have infinitely many solutions if the equations represent the same line. In such a case, we say that the equations are **dependent.**	p. 772

SECTION 8.2 Systems of Linear Equations in Three Variables and Applications	Reference
A **linear equation in three variables** is an equation that can be written in the form $$Ax + By + Cz = D$$ where A, B, and C are not all zero.	p. 780
A solution to a system of linear equations in three variables is an **ordered triple** (x, y, z) that satisfies each equation in the system. Geometrically, a solution is a point of intersection of the planes represented by the equations in the system.	p. 780

SECTION 8.3 Partial Fraction Decomposition	Reference
Partial fraction decomposition is used to write a rational expression as a sum of simpler fractions.	p. 791
There are two basic parts to find the partial fraction decomposition of a rational expression.	p. 793
I. Factor the denominator of the expression into linear factors and quadratic factors that are not further factorable over the integers. Then set up the "form" or "structure" for the partial fraction decomposition into simpler fractions.	
II. Next, multiply both sides of the equation by the LCD. Then set up a system of linear equations to find the coefficients of the terms in the numerator of each fraction.	
Note: The numerator of the original rational expression must be of lesser degree than the denominator. If this is not the case, first use long division.	

SECTION 8.4 Systems of Nonlinear Equations in Two Variables	Reference
A **nonlinear system of equations** is a system in which one or more equations is nonlinear.	p. 801
The substitution method is often used to solve a nonlinear system of equations.	p. 802
In some cases, the addition method can be used provided that the terms containing the corresponding variables are like terms.	p. 803

SECTION 8.5 Inequalities and Systems of Inequalities in Two Variables	Reference
An inequality of the form $Ax + By < C$, where A and B are not both zero, is called a **linear inequality in two variables.** (The symbols $>$, $\leq$, and $\geq$ can be used in place of $<$ in the definition.)	p. 810
The basic steps to solve a linear inequality in two variables are as follows.	p. 810
1. Graph the related equation. The resulting line is drawn as a dashed line if the inequality is strict, and is otherwise drawn as a solid line.	
2. Select a test point from either side of the line. If the ordered pair makes the original inequality true, then shade the half-plane from which the point was taken. Otherwise, shade the other half-plane.	
A nonlinear inequality in two variables is solved using the same basic procedure.	p. 813
Two or more inequalities in two variables make up a system of inequalities in two variables. The solution set to the system is the region of overlap (intersection) of the solution sets of the individual inequalities.	p. 814

SECTION 8.6 Linear Programming	Reference
A process called **linear programming** enables us to maximize or minimize a function under specified constraints. The function to be maximized or minimized is called the **objective function.**	p. 821
The steps to solve a linear programming application are outlined here.	p. 823
Step 1 Write an objective function, $z = f(x, y)$.	
Step 2 Write a system of inequalities defining the constraints on x and y.	
Step 3 Graph the feasible region and identify the vertices.	
Step 4 Evaluate the objective function at each vertex of the feasible region. Use the results to identify the values of x and y that optimize the objective function and identify the optimal value of z.	

Expanded Chapter Summary available at www.mhhe.com/millerprecalculus.

CHAPTER 8 Review Exercises

SECTION 8.1

1. Determine if the ordered pair is a solution to the system.
$$2x - 3y = 0$$
$$-5x + 6y = -1$$

   a. $\left(1, \dfrac{2}{3}\right)$ Yes   **b.** $(6, 4)$ No

For Exercises 2–3, based on the slope-intercept form of the equations, determine the number of solutions to the system.

2. $y = -\dfrac{3}{5}x - 4$

   $y = -\dfrac{3}{5}x + 1$
   No solution

3. $y = 2x + 6$

   $y = \dfrac{1}{2}x - 6$
   One solution

For Exercises 4–8, solve the system by using any method. If the system does not have one unique solution, state whether the system is inconsistent, or whether the equations are dependent.

4. $4x - y = 7$
   $-2x + 5y = 19$ $\{(3, 5)\}$

5. $5(x - y) = 19 - 2y$
   $0.2x + 0.7y = -1.7$ $\{(2, -3)\}$

6. $9x - 2y = 4$
   $2x + 4y = 7$ $\left\{\left(\frac{3}{4}, \frac{11}{8}\right)\right\}$

7. $\frac{1}{10}x - \frac{1}{2}y = 1$
   $2x = 10y + 6$
   { }; The system is inconsistent.

8. $y = \frac{3}{4}x$
   $4(y - x) = -x$

9. Shenika wants to monitor her daily calcium intake. One day she had 3 cups of milk and 1 cup of cooked spinach for a total of 1140 mg of calcium. The next day, she had 2 cups of milk and $1\frac{1}{2}$ cups of cooked spinach for a total of 960 mg of calcium. How much calcium is in 1 cup of milk and how much is in 1 cup of cooked spinach?
Milk has 300 mg per cup and spinach has 240 mg per cup.

10. How many liters of a 40% acid mixture and how many liters of a 10% acid mixture should be mixed to obtain 20 L of a 22% acid mixture?
8 L of the 40% mixture and 12 L of the 10% mixture should be used.

11. A plane can travel 960 mi in 2 hr with a tail wind. The return trip against the wind takes 2 hr and 40 min. Find the speed of the plane in still air and the speed of the wind.
The speed of the plane in still air is 420 mph and the speed of the wind is 60 mph.

12. A fishing boat captain charges $250 for an excursion. His fixed monthly expenses are $1200 for insurance, rent for the dock, and minor office expenses. He also has variable costs of $100 per excursion to cover gasoline, bait, and other equipment.

   a. Write a linear cost function representing the cost $C(x)$ (in $) for the fishing boat captain to run x excursions per month. $C(x) = 100x + 1200$

   b. Write a linear revenue function representing the revenue $R(x)$ (in $) for x excursions per month.
   $R(x) = 250x$

   c. Determine the number of excursions per month for the captain to break even.   If eight excursions are run per month, then the captain will break even.

   d. If 18 excursions are run in a given month, how much money will the fishing boat captain earn or lose?
   The captain will earn $1500.

SECTION 8.2

For Exercises 13–16, solve the system. If a system has one unique solution, write the solution set. Otherwise, determine the number of solutions to the system, and determine whether the system is inconsistent, or the equations are dependent.

13. $3a - 4b + 2c = -17$
    $2a + 3b + c = 1$
    $4a + b - 3c = 7$
    $\{(-1, 2, -3)\}$

14. $6x = 24 - 5y$
    $14 = 7z - 3y$
    $4x - 3z = 10$
    $\{(4, 0, 2)\}$

15. $x + 2y + z = 5$
    $x + y - z = 1$
    $4x + 7y + 2z = 16$

16. $u + v + 2w = 1$
    $2v - 5w = 2$
    $3u + 5v + w = 1$

17. Solve the system and write the general solution.
    $$5x + 2y + z = 0$$
    $$-4x + 3y - z = 0$$
    $$6x + 7y + z = 0$$

18. An arena that hosts sporting events and concerts has three sections for three levels of seating. For a basketball game, seats in Section A cost $90, seats in Section B cost $65, and seats in Section C cost $40. The number of seats in Section C equals the number of seats in Sections A and B combined. The arena holds 12,000 seats and the game is sold out. If the total revenue from ticket sales is $655,000, determine the number of seats in each section.   There are 1000 seats in Section A, 5000 in Section B, and 6000 in Section C.

19. Emily receives an inheritance of $20,000 and decides to invest the money. She puts some money in her savings account that earns 1.5% simple interest per year. The remaining money is invested in a bond fund that returns 4.5% and a stock fund that returns 6.2%. She makes a total of $942 at the end of 1 yr. If she invested twice as much in the bond fund as the stock fund, determine the amount that she invested in each fund.
She put $2000 in savings, and invested $12,000 in the bond fund and $6000 in the stock fund.

For Exercises 20–21, use a system of linear equations in three variables to find an equation of the form $y = ax^2 + bx + c$ that defines the parabola through the points.

20. $(-1, -4), (1, 6), (3, 8)$
    $y = -x^2 + 5x + 2$

21. $(1, -2), (2, 1), (3, 10)$
    $y = 3x^2 - 6x + 1$

SECTION 8.3

For Exercises 22–27, set up the form for the partial fraction decomposition. Do not solve for A, B, C, and so on.

22. $\dfrac{5x + 22}{x^2 + 8x + 16}$

23. $\dfrac{-x - 11}{(x + 2)(x - 1)}$

24. $\dfrac{2x^2 + x - 10}{x^3 + 5x}$

25. $\dfrac{7x^2 + 19x + 15}{2x^3 + 3x^2}$

26. $\dfrac{4x^4 - 3x^2 + 2x + 5}{x(2x + 5)^3(x^2 + 2)^2}$

27. $\dfrac{2x^3 - x^2 + 8x - 16}{x^4 + 5x^2 + 4}$

For Exercises 28–32, perform the partial fraction decomposition.

28. $\dfrac{-x - 11}{(x + 2)(x - 1)}$

29. $\dfrac{5x + 22}{x^2 + 8x + 16}$

30. $\dfrac{2x^4 + 7x^3 + 13x^2 + 19x + 15}{2x^3 + 3x^2}$

31. $\dfrac{2x^2 + x - 10}{x^3 + 5x}$

32. $\dfrac{2x^3 - x^2 + 8x - 16}{x^4 + 5x^2 + 4}$

SECTION 8.4

For Exercises 33–34,

a. Graph the equations.

b. Solve the system.

33. $y - x^2 = 1$
$x - y = -3$

34. $y = \sqrt{x - 1}$
$x^2 + y^2 = 5$

For Exercises 35–37, solve the system.

35. $3x^2 - y^2 = -4$
$x^2 + 2y^2 = 36$
$\{(2, 4), (2, -4), (-2, 4), (-2, -4)\}$

36. $2x^2 - xy = 24$
$x^2 + 3xy = -9$
$\{(-3, 2), (3, -2)\}$

37. $y = \dfrac{8}{x}$
$y = \sqrt{x}$ $\{(4, 2)\}$

38. The sum of the squares of two negative numbers is 97 and the difference of their squares is 65. Find the numbers.   The numbers are -9 and -4.

39. The ratio of two numbers is 4 to 3. The sum of the squares of the numbers is 100. Find the numbers.
The numbers are 8 and 6 or -8 and -6.

40. The hypotenuse of a right triangle is $\sqrt{74}$ ft and the sum of the lengths of the legs is 12 ft. Find the lengths of the legs.   The legs are 5 ft and 7 ft.

41. A rectangular billboard has a perimeter of 72 ft and an area of 288 ft^2. Find the dimensions of the billboard.   The billboard is 12 ft by 24 ft.

SECTION 8.5

42. Graph the solution set to the inequality.

a. $3x + 4y \le 8$

b. $3x + 4y > 8$

43. Graph the solution set to the inequality.

a. $y < (x - 4)^2$

b. $y \ge (x - 4)^2$

For Exercises 44–48, graph the solution set.

44. $5(x + y) \ge 8x + 15$

45. $x \le 3.5$

46. $-\dfrac{3}{2}y + 1 < 4$

47. $x^2 + (y + 2)^2 < 4$

48. $|y| > 2$

49. Determine if the given ordered pair is a solution to the system of inequalities.

$$x + 2y < 4$$
$$3x - 4y \ge 6$$

a. $(0, 1)$ No

b. $(1, -4)$ Yes

For Exercises 50–53, graph the solution set. If there is no solution, indicate that the solution set is the empty set.

50. $y > \dfrac{1}{2}x + 1$
$3x + 2y < 4$

51. $x^2 + y^2 \le 9$
$(x - 1)^2 + y^2 \ge 4$

52. $y \ge x^2 - 3$
$y > 1$
$x + y \le 3$

53. $y > e^x$
$y < -x^2 - 1$ $\{\ \}$

54. Let x represent the number of hours that Gordon spends tutoring math, and let y represent the number of hours that he spends tutoring English. For parts (a)–(d), write an inequality to represent the given statement.

a. Gordon has at most 12 hr to tutor per week.
$x + y \le 12$

b. The amount of time that Gordon spends tutoring English is at least twice the amount of time he spends tutoring math. $y \ge 2x$

c. The number of hours spent tutoring math cannot be negative. $x \ge 0$

d. The number of hours spent tutoring English cannot be negative. $y \ge 0$

e. Graph the solution set to the system of inequalities from parts (a)–(d).

SECTION 8.6

55. At a home store, one sheet of $\frac{3}{8}$-in. sanded pine plywood costs \$24. One sheet of $\frac{1}{4}$-in. sanded pine plywood costs \$20. Write an objective function $z = f(x, y)$ that represents the total cost for x $\frac{3}{8}$-in. sheets and y $\frac{1}{4}$-in. sheets.
$z = 24x + 20y$

56. For the feasible region given in the figure and the objective function $z = 36x + 50y$,

a. Determine the values of x and y that produce the maximum value of the objective function.
$x = 16, y = 10$

b. Determine the maximum value of the objective function.

1076

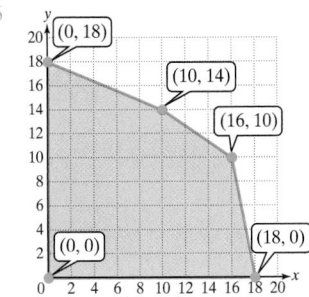

57. For the given constraints and the objective function, $z = 55x + 40y$,

    a. Graph the feasible region and identify the vertices.

$$x \geq 0, y \geq 0$$
$$2x + y \geq 18$$
$$5x + 4y \geq 60$$

    b. Determine the values of x and y that produce the minimum value of the objective function on the feasible region. $x = 4, y = 10$

    c. Determine the minimum value of the objective function on the feasible region.   620

58. A fitness instructor wants to mix two brands of protein powder to form a blend that limits the amount of fat and carbohydrate but maximizes the amount of fiber.

The nutritional information is given in the table for a single scoop of protein powder.

	Fat	Carbohydrates	Fiber
Brand A	3 g	3 g	10 g
Brand B	2 g	4 g	8 g

Suppose that the fitness instructor wants to make at most 180 scoops of the mixture. She also wants to limit the amount of fat to 480 g and she wants to limit the amount of carbohydrate to 696 g.

    a. Determine the number of scoops of each type of powder that will maximum the amount of fiber.
    She should use 120 scoops of brand A and 60 scoops of brand B.
    b. What is the maximum amount of fiber?
    The maximum amount of fiber is 1680 g.
    c. If the fiber content were reversed between the two brands (that is, 8 g for brand A and 10 g for brand B), then how much of each type of protein powder should be used to maximize the amount of fiber?
    In this case, 24 scoops of brand A and 156 scoops of brand B should be used.

CHAPTER 8   Test

For Exercises 1–3, determine if the ordered pair or ordered triple is a solution to the system.

1. $x - 5y = -3$
    $y = 2x - 12$
    a. $(7, 2)$ Yes
    b. $(-3, 0)$ No

2. $2x - 3y + z = -5$
    $5x + y - 3z = -18$
    $-x + 2y + 5z = 8$
    a. $(0, 1, -2)$ No
    b. $(-3, 0, 1)$ Yes

3. $2x - 4y < 9$
    $-3x + y \geq 4$
    a. $(-6, 1)$ Yes
    b. $(1, 4)$ No

For Exercises 4–14, solve the system. If the system does not have one unique solution, also state whether the system is inconsistent or whether the equations are dependent.

4. $x = 5 - 4y$
    $-3x + 7y = 4$
    $\{(1, 1)\}$

5. $0.2x = 0.35y - 2.5$
    $0.16x + 0.5y = 5.8$
    $\{(5, 10)\}$

6. $x - \dfrac{2}{5}y = \dfrac{3}{10}$
    $5x = 2y + \dfrac{3}{2}$

7. $7(x - y) = 3 - 5y$
    $4(3x - y) = -2x$
    $\{ \}$; The system is inconsistent.

8. $a + 6b + 3c = -14$
    $2a + b - 2c = -8$
    $-3a + 2b + c = -8$
    $\{(1, -4, 3)\}$

9. $x + 4z = 10$
    $3y - 2z = 9$
    $2x + 5y = 21$
    $\{(-2, 5, 3)\}$

10. $2x - y + z = -3$
    $x - 3y = 2$
    $x + 2y + z = -7$
    $\{ \}$; The system is inconsistent.

11. $(x - 4)^2 + y^2 = 25$
    $x - y = 3$
    $\{(0, -3), (7, 4)\}$

12. $5x^2 + y^2 = 14$
    $x^2 - 2y^2 = -17$
    $\{(1, 3), (1, -3), (-1, 3), (-1, -3)\}$

13. $2xy - y^2 = -24$
    $-3xy + 2y^2 = 38$
    $\{(5, -2), (-5, 2)\}$

14. $\dfrac{1}{x + 3} - \dfrac{2}{y - 1} = -7$
    $\dfrac{3}{x + 3} + \dfrac{1}{y - 1} = 7$ $\left\{\left(-2, \dfrac{5}{4}\right)\right\}$

15. Solve the system and write the general solution.
    $x - 2z = 6$
    $y + 3z = 2$
    $x + y + z = 8$

16. At a candy and nut shop, the manager wants to make a nut mixture that is 56% peanuts. How many pounds of peanuts must be added to an existing mixture of 45% peanuts to make 20 lb of a mixture that is 56% peanuts?

17. Two runners begin at the same point on a 400-m track. If they run in opposite directions they pass each other in 40 sec. If they run in the same direction, they will meet again in 200 sec. Find the speed of each runner.
One runner runs 6 m/sec and the other runs 4 m/sec.

18. Dylan invests $15,000 in three different stocks. One stock is very risky and after 1 yr loses 8%. The second stock returns 3.2%, and a third stock returns 5.8%. At the end of 1 yr, the total return is $274. If he invested $2000 more in the second stock than in the third stock, determine the amount he invested in each stock.

19. The difference of two positive numbers is 3 and the difference of their squares is 33. Find the numbers.
The numbers are 7 and 4.

20. A rectangular television screen has a perimeter of 154 in. and an area of 1452 in.^{2}. Find the dimensions of the screen.   The screen is 44 in. by 33 in.

21. Use a system of linear equations in three variables to find an equation of the form $y = ax^2 + bx + c$ that defines the parabola through the points $(1, -1)$, $(2, 1)$, and $(-1, 7)$.
$y = 2x^2 - 4x + 1$

For Exercises 22–23, set up the form for the partial fraction decomposition. Do not solve for A, B, C, and so on.

22. $\dfrac{-15x + 15}{3x^2 + x - 2}$

23. $\dfrac{5x^6 + 3x^5 - 4x^3 + x - 3}{x^3(x - 3)(x^2 + 5x + 1)^2}$

For Exercises 24–28, perform the partial fraction decomposition.

24. $\dfrac{-12x - 29}{2x^2 + 11x + 15}$

25. $\dfrac{6x + 8}{x^2 + 4x + 4}$

26. $\dfrac{x^4 - 6x^3 + 4x^2 + 20x - 32}{x^3 - 4x^2}$ $x - 2 - \dfrac{3}{x} + \dfrac{8}{x^2} - \dfrac{1}{x - 4}$

27. $\dfrac{x^2 - 2x - 21}{x^3 + 7x}$

28. $\dfrac{7x^3 + 4x^2 + 63x + 15}{x^4 + 11x^2 + 18}$

For Exercises 29–33, graph the solution set.

29. $2(x + y) > 6 - y$

30. $(x + 3)^2 + y^2 \geq 9$

31. $|x| < 4$

32. $x + y \leq 4$
  $2x - y > -2$

33. $y \leq -x^2 + 5$
  $y > 1$
  $x + y \leq 3$

34. A donut shop makes a profit of \$2.40 on a dozen donuts and \$0.55 per muffin. Write an objective function $z = f(x, y)$ that represents the total profit for selling x dozen donuts and y muffins. $z = 2.4x + 0.55y$

35. For the feasible region given and the objective function $z = 4x + 5y$,

a. Determine the values of x and y that produce the minimum value of the objective function on the feasible region.
  $x = 2, y = 8$
b. Determine the minimum value of the objective function on the feasible region.
  Minimum value: 48

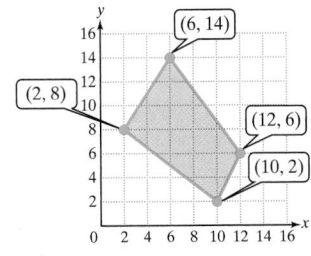

36. For the given constraints and objective function, $z = 600x + 850y$,

a. Graph the feasible region and identify the vertices.
$$x \geq 0, y \geq 0$$
$$x + y \leq 48$$
$$y \leq 3x$$

b. Determine the values of x and y that produce the maximum value of the objective function on the feasible region. $x = 12, y = 36$

c. Determine the maximum value of the objective function on the feasible region.
  Maximum value: 37,800

37. A weight lifter wants to mix two types of protein powder. One is a whey protein and one is a soy protein. The fat, carbohydrate, and protein content (in grams) for 1 scoop of each powder is given in the table.

	Fat	Carbohydrates	Protein
Whey	3 g	3 g	20 g
Soy	2 g	4 g	18 g

Suppose that the weight lifter wants to make at most 60 scoops of a protein powder mixture. Furthermore, he wants to limit the total fat content to at most 150 g and the total carbohydrate content to at most 216 g.

a. Determine the number of scoops of each type of powder that will maximize the total protein content under these constraints.   30 scoops of each type of protein powder should be mixed to maximize protein content.
b. What is the maximum total protein content?
  The maximum protein content is 1140 g.
c. If the protein content were reversed between the two brands (that is, 18 g for the whey protein and 20 g for the soy protein), then how much of each type of protein powder should be used to maximize the amount of protein?   In this case, 24 scoops of whey protein should be mixed with 36 scoops of soy protein.

CHAPTER 8   Cumulative Review Exercises

For Exercises 1–5, solve the equation.

1. $2x(x + 2) = 5x + 7$ $\left\{\dfrac{1 \pm \sqrt{57}}{4}\right\}$

2. $\sqrt{2t + 8} - t = 4$ $\{-2, -4\}$

3. $(x^2 - 4)^2 - 7(x^2 - 4) - 60 = 0$ $\{\pm 4, \pm i\}$

4. $\log_2(x - 4) = \log_2(x + 1) + 3$ { }; The value $-\dfrac{12}{7}$ does not check.

5. $50e^{2x+1} = 2000$ $\left\{\dfrac{-1 + \ln 40}{2}\right\}$

For Exercises 6–7, solve the inequality. Write the solution set in interval notation.

6. $\dfrac{x + 4}{x - 2} \leq 1$ $(-\infty, 2)$

7. $3|x + 2| - 1 > 8$ $(-\infty, -5) \cup (1, \infty)$

8. Find the partial fraction decomposition. $\dfrac{-5x + 17}{x^2 - 6x + 9}$
  $\dfrac{-5}{x - 3} + \dfrac{2}{(x - 3)^2}$

9. Given $f(x) = 2x^2 - 3x$ and $g(x) = 5x + 1$,

   a. Find $(f \circ g)(x)$. $(f \circ g)(x) = 50x^2 + 5x - 1$

   b. Find $(g \circ f)(x)$. $(g \circ f)(x) = 10x^2 - 15x + 1$

10. Given $f(x) = \sqrt[3]{x - 2}$, write an equation for $f^{-1}(x)$.
   $f^{-1}(x) = x^3 + 2$

11. Use a calculator to approximate the value of $\log_5 256$. Round to 4 decimal places.   3.4454

12. Given $f(x) = -\dfrac{1}{2}x^3 + 4x^2 + 2$, find the average rate of change on the interval $[1, 3]$. $\dfrac{19}{2}$

13. Write an equation of the line perpendicular to the line $x + 3y = 6$ and passing through the point $(2, -1)$.
   $y = 3x - 7$

14. Find all zeros of $f(x) = x^4 - 2x^3 + 10x^2 - 18x + 9$ and state the multiplicity of each zero.
   $3i$ (multiplicity 1); $-3i$ (multiplicity 1); 1 (multiplicity 2)

For Exercises 15–16,

a. Write the domain in interval notation.

b. Write the range in interval notation.

15. $f(x) = 2|x - 1| - 3$
   a. $(-\infty, \infty)$ **b.** $[-3, \infty)$

16. $f(x) = \ln(x - 3)$
   a. $(3, \infty)$ **b.** $(-\infty, \infty)$

For Exercises 17–19, solve the system.

17. $3x = 5y + 1$

    $y = \dfrac{3}{5}x + 4$ { }

18. $5a - 2b + 3c = 10$
    $-3a + b - 2c = -7$
    $a + 4b - 4c = -3$ $\{(1, 2, 3)\}$

19. $-2x^2 + 3y^2 = 10$
    $5x^2 + 2y^2 = 13$ $\{(1, 2), (1, -2), (-1, 2), (-1, -2)\}$

20. Given $f(x) = -x^2 + 5x + 1$, find the difference quotient.
   $-2x - h + 5$

21. Shen invested $8000. After 5 yr with interest compounded continuously, the account is worth $10,907.40.

   a. Write a model of the form $A(t) = Pe^{rt}$, where $A(t)$ represents the amount (in $) in the account if P dollars in principal is invested at interest rate r for t years.
   $A(t) = 8000e^{0.062t}$

   b. How long will it take for the investment to double? Round to the nearest tenth of a year.   11.2 yr

22. The variable y varies jointly as x and the square of z. If y is 36 when x is 10 and z is 3, find the value of y when $x = 12$ and z is 4. $y = 76.8$

9

Matrices and Determinants and Applications

Chapter Outline

For matters of security, e-mail messages, websites, and other electronic media often use encryption techniques to encode data into an obscure form. This prevents those who do not know the code from understanding what has been transmitted. A computer program can encrypt words by using a matrix (a two-dimensional array) to encode the message, and then use the inverse of that matrix to decode it.

In this chapter, we will use matrices in a variety of ways. Matrices can be used to model systems of linear equations. Furthermore, a variety of techniques involving the algebra and manipulation of matrices can be used to solve the systems. The first two techniques are called Gaussian elimination and Gauss-Jordan elimination. These methods reduce a system of equations to a simpler system of equations so that the solution can be found by inspection. Next, we use the inverse of a matrix to solve a related matrix equation. Finally, we present Cramer's rule. This method provides convenient formulas to solve a system of linear equations based on the coefficients of the system.

OBJECTIVES

1. Write an Augmented Matrix
2. Use Elementary Row Operations
3. Use Gaussian Elimination and Gauss-Jordan Elimination

1. Write an Augmented Matrix

The data in Table 9-1 represent the body mass index (BMI) for people of selected heights and weights. The National Institutes of Health suggests that a BMI should ideally be between 18.5 and 24.9. A table of data values such as Table 9-1 can be represented as a rectangular array of elements called a **matrix.**

Table 9-1   Body Mass Index

Weight / Height	140 lb	160 lb	180 lb	200 lb
5'7"	21.9	25.1	28.2	31.3
5'10"	20.1	23.0	25.8	28.7
6'1"	18.5	21.1	23.7	26.4

matrix $\longrightarrow$ $\begin{bmatrix} 21.9 & 25.1 & 28.2 & 31.3 \\ 20.1 & 23.0 & 25.8 & 28.7 \\ 18.5 & 21.1 & 23.7 & 26.4 \end{bmatrix}$

(*Source*: National Institutes of Health, www.nih.gov)

A matrix can be used to represent a system of linear equations written in standard form. To do so, we extract the coefficients of each term in the equation to form an **augmented matrix.** For example:

System of Equations	Augmented Matrix
$3x + 2y = 5$	
$x - y + 3z = 1$	$\begin{bmatrix} 3 & 2 & 0 & 5 \\ 1 & -1 & 3 & 1 \\ 2 & 1 & 1 & 4 \end{bmatrix}$
$2x + y + z = 4$	

The vertical bar within the augmented matrix separates the coefficients of the variable terms in the equations from the constant terms.

Classroom Examples: p. 846
Exercises 12, 20

EXAMPLE 1   Writing and Interpreting an Augmented Matrix

a. Write the augmented matrix for the system. $2x = 5y + 5$
$3(x + y) = 17 + y$

b. Write a system of linear equations represented by the augmented matrix. $\begin{bmatrix} 1 & 0 & 0 & 6 \\ 0 & 1 & 0 & -10 \\ 0 & 0 & 1 & 4 \end{bmatrix}$

Solution:

a. Write each equation in standard form.

$2x = 5y + 5 \longrightarrow 2x - 5y = 5$
$3(x + y) = 17 + y \longrightarrow 3x + 2y = 17$ augmented matrix $\begin{bmatrix} 2 & -5 & 5 \\ 3 & 2 & 17 \end{bmatrix}$

b. $\begin{bmatrix} 1 & 0 & 0 & 6 \\ 0 & 1 & 0 & -10 \\ 0 & 0 & 1 & 4 \end{bmatrix}$ $\begin{array}{l} \longrightarrow x + 0y + 0z = 6 \\ \longrightarrow 0x + y + 0z = -10 \\ \longrightarrow 0x + 0y + z = 4 \end{array}$ or simply $\begin{array}{l} x = 6 \\ y = -10 \\ z = 4 \end{array}$

Skill Practice 1

a. Write the augmented matrix. $7x = 9 + 2y$
$2(x - y) = 4$

b. Write a system of linear equations represented by the augmented matrix. $\begin{bmatrix} 1 & 0 & -8 \\ 0 & 1 & 3 \end{bmatrix}$

Answers

1. a. $\begin{bmatrix} 7 & -2 & 9 \\ 2 & -2 & 4 \end{bmatrix}$

b. $x = -8, y = 3$

2. Use Elementary Row Operations

We can use an augmented matrix to solve a system of linear equations in much the same way as we use the addition method. When using the addition method, notice that the following operations produce an equivalent system of equations.

- Interchange two equations.

$$\underline{\text{Example:}} \quad \begin{array}{l} -3x - 5y = -13 \\ x + 2y = 5 \end{array} \quad \xrightarrow[\text{equations.}]{\text{Interchange}} \quad \begin{array}{l} x + 2y = 5 \\ -3x - 5y = -13 \end{array}$$

- Multiply an equation by a nonzero constant.

$$\underline{\text{Example:}} \quad \begin{array}{l} x + 2y = 5 \\ -3x - 5y = -13 \end{array} \quad \xrightarrow{\text{Multiply by 3.}} \quad \begin{array}{l} 3x + 6y = 15 \\ -3x - 5y = -13 \end{array}$$

- Add a nonzero multiple of one equation to another equation.

$$\underline{\text{Example:}} \quad \begin{array}{l} x + 2y = 5 \\ -3x - 5y = -13 \end{array} \quad \xrightarrow{\text{Multiply by 3.}} \quad \left. \begin{array}{l} 3x + 6y = 15 \\ \underline{-3x - 5y = -13} \\ y = 2 \end{array} \right\} \begin{array}{l} \text{Add the} \\ \text{equations.} \end{array}$$

These same operations performed on a matrix are called **elementary row operations.**

Elementary Row Operations

1. Interchange two rows.

$$\underline{\text{Example:}} \begin{bmatrix} -3 & -5 & | & -13 \\ 1 & 2 & | & 5 \end{bmatrix} \xrightarrow[\substack{\text{Interchange rows} \\ \text{1 and 2.}}]{R_1 \Leftrightarrow R_2} \begin{bmatrix} 1 & 2 & | & 5 \\ -3 & -5 & | & -13 \end{bmatrix}$$

2. Multiply a row by a nonzero constant.

$$\underline{\text{Example:}} \begin{bmatrix} 1 & 2 & | & 5 \\ -3 & -5 & | & -13 \end{bmatrix} \xrightarrow[\substack{\text{Multiply} \\ \text{row 1 by 3.}}]{3R_1 \to R_1} \begin{bmatrix} 3 & 6 & | & 15 \\ -3 & -5 & | & -13 \end{bmatrix}$$

3. Add a nonzero multiple of one row to another row.

$$\underline{\text{Example:}} \begin{bmatrix} 1 & 2 & | & 5 \\ -3 & -5 & | & -13 \end{bmatrix} \xrightarrow[\substack{\text{Add 3 times} \\ \text{row 1 to row 2.}}]{3R_1 + R_2 \to R_2} \begin{bmatrix} 1 & 2 & | & 5 \\ 0 & 1 & | & 2 \end{bmatrix}$$

$$\begin{array}{lrr|r} 3R_1 & 3 & 6 & 15 \\ +R_2 & -3 & -5 & -13 \\ \hline \to R_2 & 0 & 1 & 2 \end{array}$$

> **Avoiding Mistakes**
>
> When we add a constant multiple of one row to another row, we do not change the row that was multiplied by the constant.

Two matrices are said to be **row equivalent** if one matrix can be transformed into the other matrix through a series of elementary row operations.

Classroom Examples: p. 846
Exercises 28, 30, 32

EXAMPLE 2   Performing Elementary Row Operations

Given $\begin{bmatrix} 4 & 3 & 0 & | & 5 \\ 1 & 4 & -1 & | & 9 \\ 2 & 0 & -3 & | & -8 \end{bmatrix}$, perform the following elementary row operations.

a. $R_1 \Leftrightarrow R_2$ **b.** $\dfrac{1}{4} R_1 \to R_1$ **c.** $-2R_2 + R_3 \to R_3$

Solution:

a. $\begin{bmatrix} 4 & 3 & 0 & | & 5 \\ 1 & 4 & -1 & | & 9 \\ 2 & 0 & -3 & | & -8 \end{bmatrix}$ $\xrightarrow[\substack{\text{Interchange rows} \\ \text{1 and 2.}}]{R_1 \Leftrightarrow R_2}$ $\begin{bmatrix} 1 & 4 & -1 & | & 9 \\ 4 & 3 & 0 & | & 5 \\ 2 & 0 & -3 & | & -8 \end{bmatrix}$

> **TIP** The notation $\frac{1}{4}R_1 \to R_1$ means to multiply row 1 by $\frac{1}{4}$ and then *replace* row 1 by the result.

b. $\begin{bmatrix} 4 & 3 & 0 & | & 5 \\ 1 & 4 & -1 & | & 9 \\ 2 & 0 & -3 & | & -8 \end{bmatrix}$ $\xrightarrow[\substack{\text{Multiply} \\ \text{row 1 by } \frac{1}{4}.}]{\frac{1}{4}R_1 \to R_1}$ $\begin{bmatrix} 1 & \frac{3}{4} & 0 & | & \frac{5}{4} \\ 1 & 4 & -1 & | & 9 \\ 2 & 0 & -3 & | & -8 \end{bmatrix}$

Instructor Note:
Emphasize to students that when a multiple of one row is added to another row, the row multiplied by the constant does not change.

c. $\begin{bmatrix} 4 & 3 & 0 & | & 5 \\ 1 & 4 & -1 & | & 9 \\ 2 & 0 & -3 & | & -8 \end{bmatrix}$

$\begin{array}{rrrrr} -2R_2 & -2 & -8 & 2 & | & -18 \\ + \quad R_3 & 2 & 0 & -3 & | & -8 \\ \hline \to R_3 & 0 & -8 & -1 & | & -26 \end{array}$ This now replaces row 3.

$\begin{bmatrix} 4 & 3 & 0 & | & 5 \\ 1 & 4 & -1 & | & 9 \\ 2 & 0 & -3 & | & -8 \end{bmatrix}$ $\xrightarrow[\substack{\text{Add } -2 \text{ times row 2 to row 3.} \\ \text{The result replaces row 3.}}]{-2R_2 + R_3 \to R_3}$ $\begin{bmatrix} 4 & 3 & 0 & | & 5 \\ 1 & 4 & -1 & | & 9 \\ 0 & -8 & -1 & | & -26 \end{bmatrix}$

> **Skill Practice 2** Use the matrix from Example 2 to perform the given row operations.
>
> **a.** $R_2 \Leftrightarrow R_3$ **b.** $-\frac{1}{2}R_3$ **c.** $-4R_2 + R_1 \to R_1$

3. Use Gaussian Elimination and Gauss-Jordan Elimination

When an elementary row operation is performed on an augmented matrix, a new row-equivalent augmented matrix is obtained that represents an equivalent system of equations. If we perform repeated row operations, we can form an augmented matrix that represents a system of equations that is easier to solve than the original system. In particular, it is easy to solve a system whose augmented matrix is in *row-echelon form* or *reduced row-echelon form*.

> **Row-Echelon Form and Reduced Row-Echelon Form**
>
> A matrix is in **row-echelon form** if it satisfies the following conditions.
>
> 1. Any rows consisting entirely of zeros are at the bottom of the matrix.
> 2. For all other rows, the first nonzero entry is 1. This is called the leading 1.
> 3. The leading 1 in each nonzero row is to the right of the leading 1 in the row immediately above.
>
> *Note*: A matrix is in **reduced row-echelon form** if it is in row-echelon form with the added condition that each row with a leading entry of 1 has zeros above the leading 1.

Answers

2. a. $\begin{bmatrix} 4 & 3 & 0 & | & 5 \\ 2 & 0 & -3 & | & -8 \\ 1 & 4 & -1 & | & 9 \end{bmatrix}$

b. $\begin{bmatrix} 4 & 3 & 0 & | & 5 \\ 1 & 4 & -1 & | & 9 \\ -1 & 0 & \frac{3}{2} & | & 4 \end{bmatrix}$

c. $\begin{bmatrix} 0 & -13 & 4 & | & -31 \\ 1 & 4 & -1 & | & 9 \\ 2 & 0 & -3 & | & -8 \end{bmatrix}$

The following matrices illustrate row-echelon form and reduced row-echelon form.

Row-Echelon Form

$\begin{bmatrix} 1 & 5 & | & 3 \\ 0 & 1 & | & 6 \end{bmatrix}$ $\begin{bmatrix} 1 & -4 & -\frac{1}{2} & | & 6 \\ 0 & 1 & 9 & | & -\frac{1}{3} \\ 0 & 0 & 1 & | & 2 \end{bmatrix}$ $\begin{bmatrix} 1 & 4.1 & 1.2 & | & 3.1 \\ 0 & 1 & 0.6 & | & 4.7 \\ 0 & 0 & 0 & | & 0 \end{bmatrix}$

Reduced Row-Echelon Form

$$\begin{bmatrix} 1 & 0 & | & -27 \\ 0 & 1 & | & 6 \end{bmatrix} \qquad \begin{bmatrix} 1 & 0 & 0 & | & 150 \\ 0 & 1 & 0 & | & -85 \\ 0 & 0 & 1 & | & 12 \end{bmatrix} \qquad \begin{bmatrix} 1 & 0 & 2 & | & 25 \\ 0 & 1 & -3 & | & -4 \\ 0 & 0 & 0 & | & 0 \end{bmatrix}$$

After writing an augmented matrix in row-echelon form, the corresponding system of linear equations can be solved using back substitution. This method is called **Gaussian elimination.**

> ### Solving a System of Linear Equations Using Gaussian Elimination
> 1. Write the augmented matrix for the system.
> 2. Use elementary row operations to write the augmented matrix in row-echelon form.
> 3. Use back substitution to solve the resulting system of equations.

When writing an augmented matrix in row-echelon form, the goal is to make the elements along the main diagonal 1 and the entries below the main diagonal 0. The **main diagonal** refers to the elements on the diagonal from the upper left to the lower right all to the left of the vertical bar.

The main diagonal stretches from the upper left to the lower right.

$$\begin{bmatrix} 1 & \square & \square & | & \square \\ 0 & 1 & \square & | & \square \\ 0 & 0 & 1 & | & \square \end{bmatrix}$$

To write an augmented matrix in row-echelon form, work one column at a time from left to right. This is demonstrated in Example 3.

Classroom Example: p. 847
Exercise 50

EXAMPLE 3 **Solving a System Using Gaussian Elimination**

Solve the system by using Gaussian elimination.

$$3x + 7y - 15z = -12$$
$$x + 2y - 4z = -3$$
$$-4x - 6y + 15z = 16$$

Solution:

Need a 1 here

$$\begin{array}{l} 3x + 7y - 15z = -12 \\ x + 2y - 4z = -3 \\ -4x - 6y + 15z = 16 \end{array} \qquad \begin{bmatrix} ③ & 7 & -15 & | & -12 \\ 1 & 2 & -4 & | & -3 \\ -4 & -6 & 15 & | & 16 \end{bmatrix}$$

Set up the augmented matrix.

Use elementary row operations to write the augmented matrix in row-echelon form.

$$R_1 \Leftrightarrow R_2 \longrightarrow \begin{bmatrix} 1 & 2 & -4 & | & -3 \\ 3 & 7 & -15 & | & -12 \\ -4 & -6 & 15 & | & 16 \end{bmatrix}$$

Begin working with column 1.

To obtain a leading 1 in the first row, interchange rows 1 and 2.

Now we want zeros below the leading 1 in the first row.

- Multiply row 1 by -3 and add the result to row 2.
- Multiply row 1 by 4 and add the result to row 3.

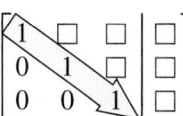

$$\begin{bmatrix} 1 & 2 & -4 & | & -3 \\ ③ & 7 & -15 & | & -12 \\ -4 & -6 & 15 & | & 16 \end{bmatrix}$$

Need 0's here

TIP To obtain a leading entry of 1 in the first row, we have the option of multiplying the first row by $\frac{1}{3}$. However, this would present fractions in the first row. Interchanging rows 1 and 2 is "cleaner."

$$\begin{array}{llllll} -3R_1 & -3 & -6 & 12 & | & 9 \\ +R_2 & 3 & 7 & -15 & | & -12 \\ \hline \to R_2 & 0 & 1 & -3 & | & -3 \end{array} \qquad \begin{array}{llllll} 4R_1 & 4 & 8 & -16 & | & -12 \\ +R_3 & -4 & -6 & 15 & | & 16 \\ \hline \to R_3 & 0 & 2 & -1 & | & 4 \end{array}$$

$$\begin{array}{l} -3R_1 + R_2 \to R_2 \\ 4R_1 + R_3 \to R_3 \end{array} \longrightarrow \begin{bmatrix} 1 & 2 & -4 & | & -3 \\ 0 & 1 & -3 & | & -3 \\ 0 & 2 & -1 & | & 4 \end{bmatrix}$$

Next, work with column 2.
We already have a leading 1 in row 2.
We need 0 below the leading 1 in row 2.

$$\begin{bmatrix} 1 & 2 & -4 & | & -3 \\ 0 & 1 & -3 & | & -3 \\ 0 & ② & -1 & | & 4 \end{bmatrix}$$

Need 0 here

- Multiply row 2 by -2 and add the result to row 3.

$$\begin{array}{rrrr|r} -2R_2 & 0 & -2 & 6 & | & 6 \\ +R_3 & 0 & 2 & -1 & | & 4 \\ \hline \rightarrow R_3 & 0 & 0 & 5 & | & 10 \end{array}$$

$-2R_2 + R_3 \rightarrow R_3 \longrightarrow \begin{bmatrix} 1 & 2 & -4 & | & -3 \\ 0 & 1 & -3 & | & -3 \\ 0 & 0 & 5 & | & 10 \end{bmatrix}$

$\frac{1}{5}R_3 \rightarrow R_3 \longrightarrow \begin{bmatrix} 1 & 2 & -4 & | & -3 \\ 0 & 1 & -3 & | & -3 \\ 0 & 0 & 1 & | & 2 \end{bmatrix}$

Now work with column 3. We need a leading 1 in column 3.
Multiply row 3 by $\frac{1}{5}$.

$\begin{bmatrix} 1 & 2 & -4 & | & -3 \\ 0 & 1 & -3 & | & -3 \\ 0 & 0 & 1 & | & 2 \end{bmatrix} \longrightarrow \begin{array}{l} x + 2y - 4z = -3 \\ y - 3z = -3 \\ z = 2 \end{array}$

We now have row-echelon form.
The corresponding system of equations has z isolated.

Use back substitution to find x and y.

$$\begin{array}{l} y - 3z = -3 \\ y - 3(2) = -3 \\ y = 3 \end{array}$$

Substitute $z = 2$ in the second equation.

$$\begin{array}{l} x + 2y - 4z = -3 \\ x + 2(3) - 4(2) = -3 \\ x + 6 - 8 = -3 \\ x = -1 \end{array}$$

Substitute $z = 2$ and $y = 3$ into the first equation.

The solution set is $\{(-1, 3, 2)\}$.

The solution $(-1, 3, 2)$ checks in each original equation.

Skill Practice 3 Solve the system by using Gaussian elimination.

$$\begin{array}{rrrr} 2x & + 7y & + & z & = 14 \\ x & + 3y & - & z & = 2 \\ x & + 7y & + & 12z & = 45 \end{array}$$

Example 3 illustrates that a system of equations represented by an augmented matrix in row-echelon form is easily solved by using back substitution. If we write an augmented matrix in *reduced* row-echelon form, we can solve the corresponding system of equations by inspection. This is called the Gauss-Jordan elimination method [named after Carl Friedrich Gauss (1777–1855) and German scientist Wilhelm Jordan (1842–1899)].

When writing an augmented matrix in *reduced* row-echelon form, the goal is to make the elements along the main diagonal 1 and the entries above and below the main diagonal 0. This is demonstrated in Examples 4 and 5.

$$\begin{bmatrix} 1 & 0 & | & \square \\ 0 & 1 & | & \square \end{bmatrix} \qquad \begin{bmatrix} 1 & 0 & 0 & | & \square \\ 0 & 1 & 0 & | & \square \\ 0 & 0 & 1 & | & \square \end{bmatrix}$$

Answer

3. $\{(2, 1, 3)\}$

Order of Row Operations to Obtain Reduced Row-Echelon Form

To write an augmented matrix with n rows in reduced row-echelon form, transform the entries in the matrix in the following order.

Column 1:   Obtain a leading element of 1 in row 1. Then obtain 0's below this element.

Column 2:   Obtain a leading element of 1 in row 2. Then obtain 0's above and below this element.

Column 3:   Obtain a leading element of 1 in row 3. Then obtain 0's above and below this element.

$\vdots$

Column n: Obtain a leading element of 1 in row n. Then obtain 0's above this element.

Classroom Example: p. 847
Exercise 46

EXAMPLE 4   Solving a System Using Gauss-Jordan Elimination

Solve the system by using Gauss-Jordan elimination.

$$x = 17 - 2y$$
$$3(x + 2y) = 47 - y$$

Solution:

$x = 17 - 2y \longrightarrow x + 2y = 17$

$3(x + 2y) = 47 - y \longrightarrow 3x + 7y = 47$

To use Gaussian elimination or Gauss-Jordan elimination, always begin by writing each equation in standard form.

$x + 2y = 17$

$3x + 7y = 47$

$\begin{bmatrix} 1 & 2 & | & 17 \\ ③ & 7 & | & 47 \end{bmatrix}$

Need 0 here

Set up the augmented matrix.
The leading element in row 1 is already 1.
Now we need a zero below the leading 1 in row 1.

Multiply row 1 by -3 and add the result to row 2.

$$\begin{array}{c|ccc} -3R_1 & -3 & -6 & | & -51 \\ +R_2 & 3 & 7 & | & 47 \\ \hline \rightarrow R_2 & 0 & 1 & | & -4 \end{array}$$

$-3R_1 + R_2 \rightarrow R_2 \longrightarrow \begin{bmatrix} 1 & 2 & | & 17 \\ 0 & 1 & | & -4 \end{bmatrix}$

Need 0 here

$\begin{bmatrix} 1 & ② & | & 17 \\ 0 & 1 & | & -4 \end{bmatrix}$

Now work with column 2. The leading element in row 2 is already 1.
Now we need a zero above the leading 1 in row 2.

Multiply row 2 by -2 and add the result to row 1.

$$\begin{array}{c|ccc} R_1 & 1 & 2 & | & 17 \\ + -2R_2 & 0 & -2 & | & 8 \\ \hline \rightarrow R_1 & 1 & 0 & | & 25 \end{array}$$

$-2R_2 + R_1 \rightarrow R_1 \longrightarrow \begin{bmatrix} 1 & 0 & | & 25 \\ 0 & 1 & | & -4 \end{bmatrix}$

$\begin{bmatrix} 1 & 0 & | & 25 \\ 0 & 1 & | & -4 \end{bmatrix} \longrightarrow \begin{array}{l} x = 25 \\ y = -4 \end{array}$

From the corresponding system of equations, we can determine the solution $(25, -4)$ by inspection.

The solution set is $\{(25, -4)\}$.

The solution checks in each original equation.

Classroom Example: p. 847
Exercise 56

Skill Practice 4   Solve the system by using Gauss-Jordan elimination.

$x - 2y = -1$

$4x - 7y = 1$

In Example 5, we use the Gauss-Jordan elimination method to solve a three-variable system of linear equations.

EXAMPLE 5   Solving a System Using Gauss-Jordan Elimination

Solve the system by using Gauss-Jordan elimination.

$$2x - 5y - 21z = 39$$
$$x - 3y - 10z = 22$$
$$x + 3y + 2z = -8$$

Solution:

$$\begin{aligned} 2x - 5y - 21z &= 39 \\ x - 3y - 10z &= 22 \\ x + 3y + 2z &= -8 \end{aligned} \qquad \left[\begin{array}{ccc|c} 2 & -5 & -21 & 39 \\ 1 & -3 & -10 & 22 \\ 1 & 3 & 2 & -8 \end{array}\right]$$

Set up the augmented matrix.

$$R_1 \Leftrightarrow R_2 \longrightarrow \left[\begin{array}{ccc|c} 1 & -3 & -10 & 22 \\ 2 & -5 & -21 & 39 \\ 1 & 3 & 2 & -8 \end{array}\right]$$

Use elementary row operations to write the augmented matrix in reduced row-echelon form.

To obtain a leading 1 in the first row, interchange rows 1 and 2.

$$\begin{aligned} -2R_1 + R_2 &\to R_2 \\ -R_1 + R_3 &\to R_3 \end{aligned} \longrightarrow \left[\begin{array}{ccc|c} 1 & -3 & -10 & 22 \\ 0 & 1 & -1 & -5 \\ 0 & 6 & 12 & -30 \end{array}\right]$$

Multiply row 1 by -2 and add the result to row 2. Multiply row 1 by -1 and add the result to row 3.

This results in zeros below the leading 1 in the first row.

$$\begin{aligned} 3R_2 + R_1 &\to R_1 \\ -6R_2 + R_3 &\to R_3 \end{aligned} \longrightarrow \left[\begin{array}{ccc|c} 1 & 0 & -13 & 7 \\ 0 & 1 & -1 & -5 \\ 0 & 0 & 18 & 0 \end{array}\right]$$

Row 2 already has a leading 1. Multiply row 2 by 3 and add the result to row 1. Multiply row 2 by -6 and add the result to row 3.

$$\tfrac{1}{18}R_3 \to R_3 \longrightarrow \left[\begin{array}{ccc|c} 1 & 0 & -13 & 7 \\ 0 & 1 & -1 & -5 \\ 0 & 0 & 1 & 0 \end{array}\right]$$

Multiply row 3 by $\frac{1}{18}$ to obtain a leading 1 in row 3.

$$\begin{aligned} 13R_3 + R_1 &\to R_1 \\ R_3 + R_2 &\to R_2 \end{aligned} \longrightarrow \left[\begin{array}{ccc|c} 1 & 0 & 0 & 7 \\ 0 & 1 & 0 & -5 \\ 0 & 0 & 1 & 0 \end{array}\right]$$

Multiply row 3 by 13 and add the result to row 1.

Multiply row 3 by 1 and add the result to row 2.

$$\left[\begin{array}{ccc|c} 1 & 0 & 0 & 7 \\ 0 & 1 & 0 & -5 \\ 0 & 0 & 1 & 0 \end{array}\right] \begin{array}{l} \to x = 7 \\ \to y = -5 \\ \to z = 0 \end{array}$$

The augmented matrix is in reduced row-echelon form.

From the corresponding system of equations, we can determine the solution $(7, -5, 0)$ by inspection.

The solution set is $\{(7, -5, 0)\}$.   The solution checks in each original equation.

Skill Practice 5   Solve the system by using Gauss-Jordan elimination.

$2x + 7y + 11z = 11$

$x + 2y + 8z = 14$

$x + 3y + 6z = 8$

Answers

4. $\{(9, 5)\}$

5. $\{(14, -4, 1)\}$

TECHNOLOGY CONNECTIONS

Finding the Row-Echelon Form and Reduced Row-Echelon Form of a Matrix

Many graphing utilities have the capability to enter and manipulate matrices. To enter the augmented matrix from Example 5, select the MATRIX menu and then the EDIT menu. Select a name for the matrix such as A. Then enter the dimensions of the matrix (in this case 3 × 4) followed by the individual elements.

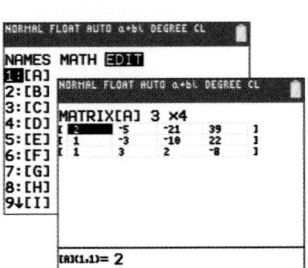

To access the commands for row-echelon form and reduced row-echelon form, select the **MATH** menu from within the MATRIX menu. Select either *ref* for row-echelon form or *rref* for reduced row-echelon form.

In the home screen, insert the matrix name within the parentheses for the *ref* and *rref* functions.

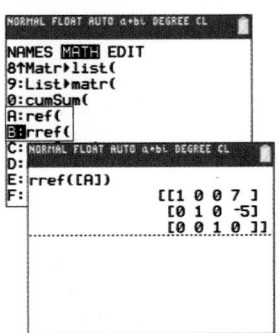

Alternatively, we can find the row-echelon form or reduced row-echelon form of a matrix if the calculator is in "MATHPRINT" mode. Select **MODE**, and using the arrow keys to navigate through the menu, highlight "MATHPRINT."

In the home screen, select the MATRIX menu, followed by MATH. Then select either *ref* or *rref*. Next, press the **ALPHA** key followed by F3. Enter the dimensions of the matrix (in this case, 3 × 4).

Then enter the elements of the matrix using the arrow keys to navigate through the matrix. Hit **ENTER** to complete the calculation.

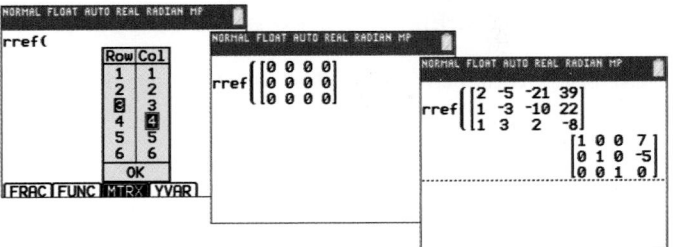

SECTION 9.1   Practice Exercises

Prerequisite Review

R.1. Solve the equation and check the solution.

$$3n + 25 = 73 \quad \{16\}$$

R.2. How much interest will Roxanne have to pay if she borrows \$1000 for 2 yr at a simple interest rate of 5%?   \$100

For Exercises R.3–R.4, set up the form for the partial fraction decomposition. Do not solve for *A, B, C,* **and so on.**

R.3. $\dfrac{-12t - 11}{4t^2 - 16t + 15}$ $\dfrac{A}{2t-5} + \dfrac{B}{2t-3}$

R.4. $\dfrac{-5x^2 - 10x - 13}{7x^3 + 7x}$ $\dfrac{A}{7x} + \dfrac{Bx + C}{x^2 + 1}$

Concept Connections

1. A rectangular array of elements is called a ___matrix___.

2. Identify the elements on the main diagonal.
$$\begin{bmatrix} -4 & -1 & 1 & | & 8 \\ 2 & 0 & 5 & | & 11 \\ 0 & 1 & -7 & | & -6 \end{bmatrix}$$ $-4; \ 0; \ -7$

3. Explain the meaning of the notation $R_2 \Leftrightarrow R_3$.
Interchange rows 2 and 3.

4. Explain the meaning of the notation $-R_2 \to R_2$.
Multiply row 2 by -1 and replace the original row 2 with the result.

5. Explain the meaning of the notation $3R_1 \to R_1$.
Multiply row 1 by 3 and replace the original row 1 with the result.

6. Explain the meaning of $R_1 \Leftrightarrow R_3$.
Interchange rows 1 and 3.

7. Explain the meaning of the notation $3R_1 + R_2 \to R_2$.
Add 3 times row 1 to row 2 and replace the original row 2 with the result.

8. Explain the meaning of the notation $4R_2 + R_3 \to R_3$.
Add 4 times row 2 to row 3 and replace the original row 3 with the result.

Objective 1: Write an Augmented Matrix

For Exercises 9–14, write the augmented matrix for the given system. (See Example 1)

9. $-3x + 2y - z = 4$
$8x \quad + 4z = 12$
$2y - 5z = 1$

10. $-4x - y + z = 8$
$2x \quad + 5z = 11$
$y - 7z = -6$

11. $4(x - 2y) = 6y + 2$
$3x = 5y + 7$
$\begin{bmatrix} 4 & -14 & | & 2 \\ 3 & -5 & | & 7 \end{bmatrix}$

12. $2(y - x) = 4 - 8x$
$5y = 6 - x$
$\begin{bmatrix} 6 & 2 & | & 4 \\ 1 & 5 & | & 6 \end{bmatrix}$

13. $x = 2$
$y = \frac{6}{7}$
$z = 12$
$\begin{bmatrix} 1 & 0 & 0 & | & 2 \\ 0 & 1 & 0 & | & \frac{6}{7} \\ 0 & 0 & 1 & | & 12 \end{bmatrix}$

14. $x = 4$
$y = 2$
$z = -\frac{1}{2}$
$\begin{bmatrix} 1 & 0 & 0 & | & 4 \\ 0 & 1 & 0 & | & 2 \\ 0 & 0 & 1 & | & -\frac{1}{2} \end{bmatrix}$

For Exercises 15–20, write a system of linear equations represented by the augmented matrix. (See Example 1)

15. $\begin{bmatrix} -4 & 6 & | & 11 \\ -3 & 9 & | & 1 \end{bmatrix}$
$-4x + 6y = 11$
$-3x + 9y = 1$

16. $\begin{bmatrix} -3 & 7 & | & -2 \\ -1 & 1 & | & 4 \end{bmatrix}$
$-3x + 7y = -2$
$-x + y = 4$

17. $\begin{bmatrix} 1 & 4 & 3 & | & 8 \\ 0 & 1 & 2 & | & 12 \\ 0 & 0 & 1 & | & 6 \end{bmatrix}$
$x + 4y + 3z = 8$
$y + 2z = 12$
$z = 6$

18. $\begin{bmatrix} 1 & 1 & -2 & | & -4 \\ 0 & 1 & 3 & | & 8 \\ 0 & 0 & 1 & | & 2 \end{bmatrix}$
$x + y - 2z = -4$
$y + 3z = 8$
$z = 2$

19. $\begin{bmatrix} 1 & 0 & 0 & | & 8 \\ 0 & 1 & 0 & | & -9 \\ 0 & 0 & 1 & | & \frac{3}{2} \end{bmatrix}$
$x = 8$
$y = -9$
$z = \frac{3}{2}$

20. $\begin{bmatrix} 1 & 0 & 0 & | & 2 \\ 0 & 1 & 0 & | & 6 \\ 0 & 0 & 1 & | & -\frac{1}{2} \end{bmatrix}$
$x = 2$
$y = 6$
$z = -\frac{1}{2}$

Objective 2: Use Elementary Row Operations

For Exercises 21–26, perform the elementary row operations on $\begin{bmatrix} 1 & 4 & | & 2 \\ -3 & 6 & | & 6 \end{bmatrix}$. **(See Example 2)**

21. $R_1 \Leftrightarrow R_2$ $\begin{bmatrix} -3 & 6 & | & 6 \\ 1 & 4 & | & 2 \end{bmatrix}$

22. $-\frac{1}{3}R_2 \to R_2$ $\begin{bmatrix} 1 & 4 & | & 2 \\ 1 & -2 & | & -2 \end{bmatrix}$

23. $3R_1 \to R_1$ $\begin{bmatrix} 3 & 12 & | & 6 \\ -3 & 6 & | & 6 \end{bmatrix}$

24. $-3R_1 \to R_1$ $\begin{bmatrix} -3 & -12 & | & -6 \\ -3 & 6 & | & 6 \end{bmatrix}$

25. $\frac{1}{3}R_2 + R_1 \to R_1$ $\begin{bmatrix} 0 & 6 & | & 4 \\ -3 & 6 & | & 6 \end{bmatrix}$

26. $3R_1 + R_2 \to R_2$ $\begin{bmatrix} 1 & 4 & | & 2 \\ 0 & 18 & | & 12 \end{bmatrix}$

For Exercises 27–32, perform the elementary row operations on $\begin{bmatrix} 1 & 5 & 6 & | & 2 \\ 2 & 1 & 5 & | & 1 \\ 4 & -2 & -3 & | & 10 \end{bmatrix}$. **(See Example 2)**

27. $R_2 \Leftrightarrow R_3$

28. $R_1 \Leftrightarrow R_2$

29. $\frac{1}{4}R_3 \to R_3$

30. $\frac{1}{2}R_2 \to R_2$

31. $-2R_1 + R_2 \to R_2$

32. $-4R_1 + R_3 \to R_3$

Objective 3: Use Gaussian Elimination and Gauss-Jordan Elimination

For Exercises 33–36, determine if the matrix is in row-echelon form. If not, explain why.

33. $\begin{bmatrix} 1 & 5 & | & 4 \\ 0 & 2 & | & 6 \end{bmatrix}$

No; The element on the main diagonal in the second row is not 1.

34. $\begin{bmatrix} 1 & 6 & 4 & | & 2 \\ 0 & 1 & 0 & | & -1 \\ 0 & 3 & 1 & | & 3 \end{bmatrix}$

No; The element below the leading 1 in the second row should be zero.

35. $\begin{bmatrix} 1 & 3 & 2 & | & 6 \\ 0 & 1 & 5 & | & 9 \\ 0 & 0 & 0 & | & 0 \end{bmatrix}$ Yes

36. $\begin{bmatrix} 1 & 4 & 2 & | & -6 \\ 0 & 1 & -3 & | & 2 \\ 0 & 0 & 1 & | & 0 \end{bmatrix}$ Yes

For Exercises 37–40, determine if the matrix is in reduced row-echelon form. If not, explain why.

37. $\begin{bmatrix} 1 & 0 & 0 & | & 3 \\ 1 & 0 & 0 & | & 4 \\ 1 & 0 & 0 & | & 5 \end{bmatrix}$
38. $\begin{bmatrix} 1 & 0 & 2 & | & 3 \\ 0 & 1 & 0 & | & 4 \\ 0 & 0 & 1 & | & 5 \end{bmatrix}$
39. $\begin{bmatrix} 1 & 0 & 0 & 0 & | & 1 \\ 0 & 1 & 0 & 0 & | & 2 \\ 0 & 0 & 1 & 0 & | & -7 \\ 0 & 0 & 0 & 1 & | & 4 \end{bmatrix}$ Yes
40. $\begin{bmatrix} 1 & 0 & 0 & -1 & | & 5 \\ 0 & 1 & 0 & 0 & | & 20 \\ 0 & 0 & 1 & 4 & | & -1 \\ 0 & 0 & 0 & 0 & | & 0 \end{bmatrix}$ Yes

For Exercises 41–60, solve the system by using Gaussian elimination or Gauss-Jordan elimination. (See Examples 3–5)

41. $2x + 3y = -13$ $\{(-2, -3)\}$
$x + 4y = -14$

42. $-3x + 11y = 58$ $\{(-1, 5)\}$
$x - 3y = -16$

43. $2x - 7y = -41$ $\{(4, 7)\}$
$3x - 9y = -51$

44. $-2x + 15y = 6$ $\{(-3, 0)\}$
$3x - 12y = -9$

45. $-3(x - 6y) = -167 - y$ $\{(5, -8)\}$
$14y = 2x - 122$

46. $2(x - y) = 4x + y - 40$ $\{(5, 10)\}$
$9y = 105 - 3x$

47. $3x + 7y + 22z = 83$
$x + 3y + 10z = 37$ $\{(1, 2, 3)\}$
$-2x - 5y - 18z = -66$

48. $3x + 5y + 9z = 3$
$x + 3y + 7z = 5$ $\{(-2, 0, 1)\}$
$-2x - 8y - 15z = -11$

49. $-2x + 4y + z = 7$
$4x - 13y + 10z = 17$ $\{(0, 1, 3)\}$
$3x - 9y + 6z = 9$

50. $2x - 8y + 54z = -4$
$x - 2y + 14z = -1$ $\{(-1, 7, 1)\}$
$x - 3y + 19z = -3$

51. $-3x + 4y - 15z = -44$
$x - y + 4z = 13$ $\{(7, -2, 1)\}$
$x - 3y + 14z = 27$

52. $-2x + 5y - 4z = -4$
$x - 2y + z = 3$ $\{(4, 0, -1)\}$
$x - 5y + 9z = -5$

53. $2x + 8z = 7y - 46$
$x = 3y - 3z - 18$ $\{(0, 2, -4)\}$
$6z = 5y - x - 34$

54. $2x = 7 - y - 3z$
$x + y = z + 5$ $\{(2, 3, 0)\}$
$16z = 2y - x - 4$

55. $11y + 65 = 3x + 13z$
$x + 3z = 3y + 15$ $\{(-6, -4, 3)\}$
$-2x + 4y - 7z = -25$

56. $2(x - 6z) = 3y + x + 17$
$2x - 19 = 3y + 18z$ $\{(2, -5, 0)\}$
$-3x + 7y + 36z = -41$

57. $w + 3x - 3z = -5$
$x - 2z = -6$ $\{(1, 2, 3, 4)\}$
$-2w - 4x + y + 2z = 1$
$x + y = 5$

58. $w - 2x + 5y = 20$
$-x + 2y = 9$ $\{(-2, -1, 4, 0)\}$
$x - y = -5$
$2w - x + 7y + z = 25$

59. $x_1 + x_2 + 5x_4 = -4$
$x_2 + 2x_4 = 3$
$-2x_2 + x_3 - 3x_4 = -5$ $\{(-10, 1, 0, 1)\}$
$3x_1 + 3x_2 + 17x_4 = -10$

60. $x_1 + x_3 - 5x_4 = 1$
$-2x_1 + x_2 - 2x_3 + 16x_4 = -3$
$x_1 + 2x_3 - 10x_4 = 5$ $\{(-3, -1, 4, 0)\}$
$x_1 - x_3 + 7x_4 = -7$

Mixed Exercises

For Exercises 61–64, set up a system of linear equations to represent the scenario. Solve the system by using Gaussian elimination or Gauss-Jordan elimination.

61. Andre borrowed $20,000 to buy a truck for his business. He borrowed from his parents who charge him 2% simple interest. He borrowed from a credit union that charges 4% simple interest, and he borrowed from a bank that charges 5% simple interest. He borrowed five times as much from his parents as from the bank, and the amount of interest he paid at the end of 1 yr was $620. How much did he borrow from each source?

62. Sylvia invested a total of $40,000. She invested part of the money in a certificate of deposit (CD) that earns 2% simple interest per year. She invested in a stock that returns the equivalent of 8% simple interest, and she invested in a bond fund that returns 5%. She invested twice as much in the stock as she did in the CD, and earned a total of $2300 at the end of 1 yr. How much principal did she put in each investment?

63. Danielle stayed in three different cities (Washington, D.C., Atlanta, Georgia, and Dallas, Texas) for a total of 14 nights. She spent twice as many nights in Dallas as she did in Washington. The total cost for 14 nights (excluding tax) was $2200. Determine the number of nights that she spent in each city.

City	Cost per Night
Washington	$200
Atlanta	$100
Dallas	$150

She spent 4 nights in Washington, 2 nights in Atlanta, and 8 nights in Dallas.

64. Three pumps (A, B, and C) work to drain water from a retention pond. Working together the pumps can pump 1500 gal/hr of water. Pump C works at a rate of 100 gal/hr faster than pump B. In 3 hr, pump C can pump as much water as pumps A and B working together in 2 hr. Find the rate at which each pump works.   Pump A works at 400 gal/hr, pump B works at 500 gal/hr, and pump C works at 600 gal/hr.

For Exercises 65–66, find the partial fraction decomposition for the given rational expression. Use the technique of Gaussian elimination to find A, B, and C.

65. $\dfrac{5x^2 - 6x - 13}{(x + 3)(x - 2)^2} = \dfrac{A}{x + 3} + \dfrac{B}{x - 2} + \dfrac{C}{(x - 2)^2}$

$\dfrac{2}{x + 3} + \dfrac{3}{x - 2} + \dfrac{-1}{(x - 2)^2}$

66. $\dfrac{2x^2 + 17x + 3}{(x + 5)(x + 1)^2} = \dfrac{A}{x + 5} + \dfrac{B}{x + 1} + \dfrac{C}{(x + 1)^2}$

$\dfrac{-2}{x + 5} + \dfrac{4}{x + 1} + \dfrac{-3}{(x + 1)^2}$

Write About It

67. Explain why interchanging two rows of an augmented matrix results in an augmented matrix that represents an equivalent system of equations.

68. Explain why multiplying a row of an augmented matrix by a nonzero constant results in an augmented matrix that represents an equivalent system of equations.

69. Explain the difference between a matrix in row-echelon form and reduced row-echelon form.

Reduced row-echelon form is the same format as row-echelon form with the added condition that all elements above the leading 1's must be 0's.

70. Consider the matrix $\begin{bmatrix} 5 & -9 & | & -57 \\ 1 & -2 & | & -12 \end{bmatrix}$. Identify two row operations that could be used to obtain a leading entry of 1 in the first row. Also indicate which operation would be less cumbersome as a first step toward writing the matrix in reduced row-echelon form.

Interchanging rows 1 and 2 would create a leading entry of 1 in the first row. Alternatively, multiplying row 1 by $\frac{1}{5}$ would also create a leading entry of 1. Interchanging rows 1 and 2 would be less cumbersome because it would not create fractional elements in row 1.

Technology Connections

For Exercises 71–72, use a calculator to approximate the reduced row-echelon form of the augmented matrix representing the given system. Give the solution set where x, y, and z are rounded to 2 decimal places.

71. $0.52x - 3.71y - 4.68z = 9.18$
$0.02x + 0.06y + 0.11z = 0.56$
$0.972x + 0.816y + 0.417z = 0.184$

72. $-3.61x + 8.17y - 5.62z = 30.2$
$8.04x - 3.16y + 9.18z = 28.4$
$-0.16x + 0.09y + 0.55z = 4.6$

73. A small grocer finds that the monthly sales y (in \$) can be approximated as a function of the amount spent advertising on the radio x_1 (in \$) and the amount spent advertising in the newspaper x_2 (in \$) according to $y = ax_1 + bx_2 + c$.

The table gives the amounts spent in advertising and the corresponding monthly sales for 3 months.

Radio Advertising, x_1	Newspaper Advertising, x_2	Monthly sales, y
\$2400	\$800	\$36,000
\$2000	\$500	\$30,000
\$3000	\$1000	\$44,000

a. Use the data to write a system of linear equations to solve for a, b, and c.

b. Use a graphing utility to find the reduced row-echelon form of the augmented matrix.

c. Write the model $y = ax_1 + bx_2 + c$. $y = 12x_1 + 4x_2 + 4000$

d. Predict the monthly sales if the grocer spends \$2500 advertising on the radio and \$500 advertising in the newspaper for a given month. \$36,000

74. The purchase price of a home y (in \$1000) can be approximated based on the annual income of the buyer x_1 (in \$1000) and on the square footage of the home x_2 (in 100 ft^2) according to $y = ax_1 + bx_2 + c$.

The table gives the incomes of three buyers, the square footages of the home purchased, and the corresponding purchase prices of the home.

Income (\$1000) x_1	Square Footage (100 ft^2) x_2	Price (\$1000) y
80	21	180
150	28	250
75	18	160

a. Use the data to write a system of linear equations to solve for a, b, and c.

b. Use a graphing utility to find the reduced row-echelon form of the augmented matrix.

c. Write the model $y = ax_1 + bx_2 + c$. $y = 0.4x_1 + 6x_2 + 22$

d. Predict the purchase price for a buyer who makes \$100,000 per year and wants a 2500 ft^2 home. \$212,000

For Exercises 75–76, the given function values satisfy a function defined by $f(x) = ax^2 + bx + c$.

a. Set up a system of equations to solve for a, b, and c.

b. Use a graphing utility to find the reduced row-echelon form of the augmented matrix.

c. Write a function of the form $f(x) = ax^2 + bx + c$ that fits the data.

75. $f(-3) = -7.28$
$f(-1) = 3.68$
$f(10) = 18.2$

a. $9a - 3b + c = -7.28$
$a - b + c = 3.68$
$100a + 10b + c = 18.2$
c. $f(x) = -0.32x^2 + 4.2x + 8.2$

76. $f(3) = 6.95$
$f(-2) = 20.2$
$f(12) = 39.8$

a. $9a + 3b + c = 6.95$
$4a - 2b + c = 20.2$
$144a + 12b + c = 39.8$
c. $f(x) = 0.45x^2 - 3.1x + 12.2$

SECTION 9.2   Inconsistent Systems and Dependent Equations

Classroom Example: p. 856
Exercise 24

1. Identify Inconsistent Systems

When we studied systems of linear equations in two and three variables in Chapter 8, we learned that a system may have no solution. Such a system is said to be **inconsistent,** and we recognize an inconsistent system if the system reduces to a contradiction.

EXAMPLE 1   Identifying an Inconsistent System

Solve the system.
$$\begin{aligned} x - 3y - 17z &= -59 \\ x - 2y - 12z &= -41 \\ -2y - 10z &= 20 \end{aligned}$$

Solution:

$$\begin{aligned} x - 3y - 17z &= -59 \\ x - 2y - 12z &= -41 \\ -2y - 10z &= 20 \end{aligned} \qquad \begin{bmatrix} 1 & -3 & -17 & | & -59 \\ 1 & -2 & -12 & | & -41 \\ 0 & -2 & -10 & | & 20 \end{bmatrix}$$

Write the augmented matrix.
The leading entry in row 1 is already 1.

$$-R_1 + R_2 \rightarrow R_2 \longrightarrow \begin{bmatrix} 1 & -3 & -17 & | & -59 \\ 0 & 1 & 5 & | & 18 \\ 0 & -2 & -10 & | & 20 \end{bmatrix}$$

Multiply row 1 by -1 and add the result to row 2.

$$\begin{aligned} 3R_2 + R_1 &\rightarrow R_1 \\ 2R_2 + R_3 &\rightarrow R_3 \end{aligned} \longrightarrow \begin{bmatrix} 1 & 0 & -2 & | & -5 \\ 0 & 1 & 5 & | & 18 \\ 0 & 0 & 0 & | & 56 \end{bmatrix}$$

Multiply row 2 by 3 and add the result to row 1.
Multiply row 2 by 2 and add the result to row 3.

The last row of the matrix cannot be written with a leading 1 along the main diagonal. The last row represents a contradiction, $0 = 56$.

$$\begin{bmatrix} 1 & 0 & -2 & | & -5 \\ 0 & 1 & 5 & | & 18 \\ 0 & 0 & 0 & | & 56 \end{bmatrix} \xrightarrow[\text{system}]{\text{equivalent}} \begin{cases} x - 2z = -5 \\ y + 5z = 18 \\ \qquad\quad 0 = 56 \end{cases}$$

The contradiction indicates that the system is inconsistent and that there is no solution. The solution set is $\{\ \}$.

Skill Practice 1   Solve the system.
$$\begin{aligned} 5x - 9y - 33z &= 3 \\ x - 2y - 7z &= 0 \\ -2x + y + 8z &= -12 \end{aligned}$$

Answer

1. $\{\ \}$

TECHNOLOGY CONNECTIONS

Recognizing an Inconsistent System on a Calculator

We can verify that the system of equations in Example 1 has no solution by using a calculator to find the reduced row-echelon form of the augmented matrix. Notice that the calculator also displays a contradiction in the third row, $0 = 1$.

In Example 1, once we reached the contradiction $0 = 56$, we stopped manipulating the augmented matrix. However, we have the option of simplifying the augmented matrix to reduced row-echelon form. To match the result given in the calculator, we would multiply row 3 by $\frac{1}{56}$. Then add -18 times row 3 to row 2, and add 5 times row 3 to row 1.

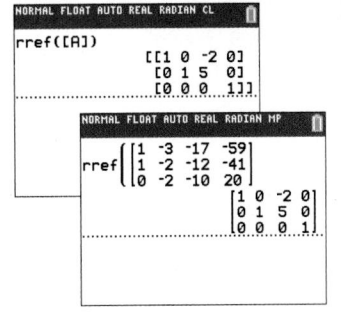

2. Solve Systems with Dependent Equations

Recall that a linear equation in two variables defines a line in a plane. A solution to a system of two linear equations in two variables is a point of intersection of the lines. If the two lines are parallel, the system is inconsistent and has no solution. If the equations in the system represent the same line, we say that the equations are **dependent** and the solution set is the set of all points on the line.

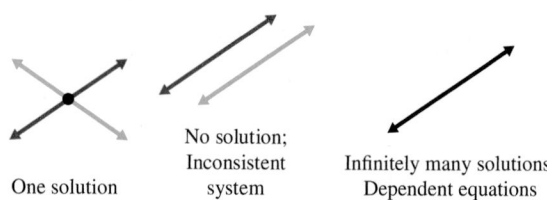

One solution

No solution;
Inconsistent
system

Infinitely many solutions;
Dependent equations

Classroom Example: p. 856
Exercise 22

EXAMPLE 2 Solving a System of Dependent Equations with Two Variables

Solve the system.
$$0.25x - 0.75y = 1$$
$$3y = x - 4$$

Solution:

$$0.25x - 0.75y = 1 \xrightarrow{\text{Multiply by 100.}} 25x - 75y = 100$$
$$3y = x - 4 \xrightarrow[\text{Standard form}]{} -x + 3y = -4$$

Write the equations in standard form. As an option, consider clearing decimals in the first equation.

$$\begin{pmatrix} 25 & -75 & | & 100 \\ -1 & 3 & | & -4 \end{pmatrix} \xrightarrow{\frac{1}{25}R_1 \to R_1} \begin{pmatrix} 1 & -3 & | & 4 \\ -1 & 3 & | & -4 \end{pmatrix}$$

To obtain a 1 in the first row, first column, multiply row 1 by $\frac{1}{25}$.

$$\xrightarrow{R_1 + R_2 \to R_2} \begin{pmatrix} 1 & -3 & | & 4 \\ 0 & 0 & | & 0 \end{pmatrix}$$

To obtain a 0 in the second row, first column, add row 1 to row 2.

The second row of the augmented matrix represents the equation $0 = 0$, which is true regardless of the values of x and y. This means that the original system reduces to the single equation $x - 3y = 4$. That is, the two original equations each represent the same line and all points on the line are solutions to the system. Solving the equation $x - 3y = 4$ for x or y illustrates the dependency between x and y.

Solving the equation $x - 3y = 4$ for x gives: $x = 3y + 4$ ← *x depends on the choice of y.*

Solving the equation $x - 3y = 4$ for y gives: $y = \dfrac{x - 4}{3}$ ← *y depends on the choice of x.*

The solution set can be written as

$$\{(3y + 4, y) \mid y \text{ is any real number}\} \text{ or } \left\{ \left(x, \frac{x - 4}{3} \right) \;\middle|\; x \text{ is any real number} \right\}.$$

Skill Practice 2 Solve the system.
$$0.3x - 0.1y = -2$$
$$y - 20 = 3x$$

TIP In Example 2, we wrote the solution set in two ways: one with an arbitrary value of y and one with an arbitrary value of x. However, when using reduced row-echelon form to find the solution set to a system of dependent equations, we generally let the *last* variable in the ordered pair (ordered triple, etc.) be arbitrary. In Example 2, this is $\{(3y + 4, y) \mid y \text{ is any real number}\}$.

Answer

2. $\left\{ \left(\dfrac{y - 20}{3}, y \right) \;\middle|\; y \text{ is any real number} \right\}$
or $\{(x, 3x + 20) \mid x \text{ is any real number}\}$

A linear equation in three variables represents a plane in space. A solution to a system of equations in three variables is a common point of intersection among all the planes in the system. From Example 1, we have a system with no solution. This means that the planes do not all intersect (Figure 9-1).

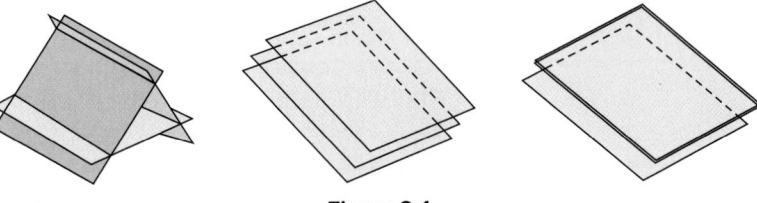

Figure 9-1

A system of linear equations may have infinitely many solutions. In such a case, the equations are dependent. For a system of three equations in three variables, this means that the planes intersect in a common line in space (Figure 9-2 and Figure 9-3), or the three planes all coincide (Figure 9-4).

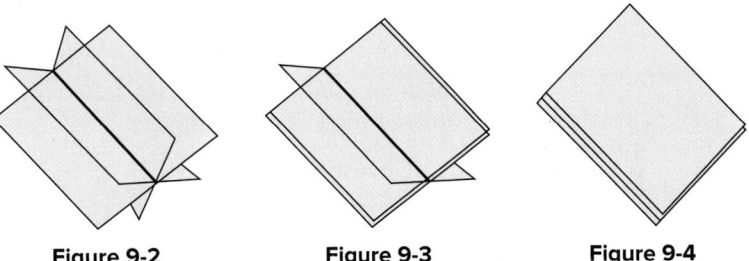

Figure 9-2   **Figure 9-3**   **Figure 9-4**

Classroom Example: p. 857
Exercise 26

EXAMPLE 3 **Solving a System of Dependent Equations with Three Variables**

Solve the system.
$$x - 4y + 7z = 14$$
$$-2x + 9y - 16z = -31$$
$$x - 7y + 13z = 23$$

Solution:

$$x - 4y + 7z = 14$$
$$-2x + 9y - 16z = -31$$
$$x - 7y + 13z = 23$$

$$\begin{bmatrix} 1 & -4 & 7 & | & 14 \\ -2 & 9 & -16 & | & -31 \\ 1 & -7 & 13 & | & 23 \end{bmatrix}$$ Set up the augmented matrix.

$$\xrightarrow[-R_1 + R_3 \to R_3]{2R_1 + R_2 \to R_2} \begin{bmatrix} 1 & -4 & 7 & | & 14 \\ 0 & 1 & -2 & | & -3 \\ 0 & -3 & 6 & | & 9 \end{bmatrix} \xrightarrow[3R_2 + R_3 \to R_3]{4R_2 + R_1 \to R_1} \begin{bmatrix} 1 & 0 & -1 & | & 2 \\ 0 & 1 & -2 & | & -3 \\ 0 & 0 & 0 & | & 0 \end{bmatrix}$$

The last row of the matrix cannot be written with a leading 1 along the main diagonal. The last row represents an identity, $0 = 0$.

$$\begin{bmatrix} 1 & 0 & -1 & | & 2 \\ 0 & 1 & -2 & | & -3 \\ 0 & 0 & 0 & | & 0 \end{bmatrix} \xrightarrow[\text{system}]{\text{equivalent}} \begin{cases} x - z = 2 \\ y - 2z = -3 \\ 0 = 0 \end{cases}$$

TIP A graphing utility can also be used to find the reduced row-echelon form of a system of dependent equations.

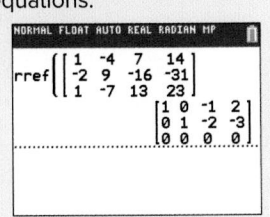

The system of three equations and three variables reduces to a system of two equations and three variables. The third equation, $0 = 0$, is true regardless of the values of x, y, and z. The top two equations each represent a plane in space. Furthermore, two nonparallel planes intersect in a line. All points on the line are solutions to the system, indicating that there are infinitely many solutions.

TIP In Example 3, we can also write the solution set by using a parameter. By choosing any real number t for z, we have:
$$\{(t + 2, 2t - 3, t)\}$$

Since the top two equations both contain the variable z, we can express x and y in terms of z.

$x - z = 2$ ——— Express x in terms of z. ——→ $x = z + 2$

$y - 2z = -3$ ———————————— ——→ $y = 2z - 3$
Express y in terms of z.

The solution set can be written as $\{(z + 2, 2z - 3, z) \mid z \text{ is any real number}\}$.

Skill Practice 3 Solve the system.
$$\begin{aligned} -4x - 11y + 3z &= -24 \\ x + 3y - z &= 7 \\ 3x + 11y - 5z &= 29 \end{aligned}$$

We can verify the solution to Example 3, by substituting $(z + 2, 2z - 3, z)$ back into each original equation.

$x - 4y + 7z = 14 \longrightarrow (z + 2) - 4(2z - 3) + 7z \stackrel{?}{=} 14 \longrightarrow 14 = 14 \checkmark$

$-2x + 9y - 16z = -31 \longrightarrow -2(z + 2) + 9(2z - 3) - 16z \stackrel{?}{=} -31 \longrightarrow -31 = -31 \checkmark$

$x - 7y + 13z = 23 \longrightarrow (z + 2) - 7(2z - 3) + 13z \stackrel{?}{=} 23 \longrightarrow 23 = 23 \checkmark$

In Example 3, the solution set $\{(z + 2, 2z - 3, z) \mid z \text{ is any real number}\}$ is the general solution representing an infinite number of ordered triples. To find individual solutions, substitute arbitrary real numbers for z, and construct the corresponding ordered triple. For example:

Choose z Arbitrarily	$(z + 2, 2z - 3, z)$	Solution
If $z = 1$ ——→	$(1 + 2, 2(1) - 3, 1)$	$(3, -1, 1)$
If $z = 2$ ——→	$(2 + 2, 2(2) - 3, 2)$	$(4, 1, 2)$
If $z = 3$ ——→	$(3 + 2, 2(3) - 3, 3)$	$(5, 3, 3)$
If $z = -1$ ——→	$(-1 + 2, 2(-1) - 3, -1)$	$(1, -5, -1)$

A system of linear equations that has the same number of equations as variables is called a **square system.** In Example 3, we were presented with three equations and three variables in the original system. However, after writing the augmented matrix in row-echelon form, we see that the system reduces to a system of two equations and three variables. A system of linear equations cannot have a unique solution unless there are at least the same number of equations as variables.

In Example 4, we investigate a nonsquare system that has fewer equations than variables.

Classroom Example: p. 857
Exercise 30

EXAMPLE 4 **Solving a System with Fewer Equations than Variables**

Solve the system.
$$\begin{aligned} 3x - 8y + 18z &= 15 \\ x - 3y + 4z &= 6 \end{aligned}$$

Solution:

The two given equations each represent a plane in space. By finding the row-echelon form of the augmented matrix, we can determine the geometrical relationship between the planes.

- If we encounter a contradiction, then the system has no solution and the planes are parallel.
- If we encounter an identity, such as $0 = 0$, then the two equations must represent the same plane.
- Otherwise, the planes must meet in a line.

Answer
3. $\{(-2z - 5, z + 4, z) \mid z \text{ is any real number}\}$

$$3x - 8y + 18z = 15 \longrightarrow \begin{bmatrix} 3 & -8 & 18 & | & 15 \\ 1 & -3 & 4 & | & 6 \end{bmatrix} \quad R_1 \Leftrightarrow R_2 \begin{bmatrix} 1 & -3 & 4 & | & 6 \\ 3 & -8 & 18 & | & 15 \end{bmatrix}$$

$$\xrightarrow{-3R_1 + R_2 \rightarrow R_2} \begin{bmatrix} 1 & -3 & 4 & | & 6 \\ 0 & 1 & 6 & | & -3 \end{bmatrix} \xrightarrow{3R_2 + R_1 \rightarrow R_1} \begin{bmatrix} 1 & 0 & 22 & | & -3 \\ 0 & 1 & 6 & | & -3 \end{bmatrix}$$

The augmented matrix is in reduced row-echelon form. $\begin{bmatrix} 1 & 0 & 22 & | & -3 \\ 0 & 1 & 6 & | & -3 \end{bmatrix} \longrightarrow \begin{cases} x + 22z = -3 \\ y + 6z = -3 \end{cases}$

The system did not reduce to a contradiction or an identity. Therefore, the two planes must intersect in a line and there must be infinitely many solutions.

To write the general solution, write x and y in terms of z.

TIP In parametric form, the solution to Example 4 is given by:
$\{(-22t - 3, -6t - 3, t)\}$

$$x + 22z = -3 \xrightarrow{\text{Express } x \text{ in terms of } z.} x = -22z - 3$$
$$y + 6z = -3 \xrightarrow{\text{Express } y \text{ in terms of } z.} y = -6z - 3$$

The solution set is $\{(-22z - 3, -6z - 3, z) \mid z \text{ is any real number}\}$.

> **Skill Practice 4** Solve the system. $\quad x - 3y - 17z = -17$
> $\qquad\qquad\qquad\qquad\qquad\qquad -2x + 7y + 38z = 40$

In Example 5, we investigate a situation in which the solution set to a system of three variables and three equations consists of all points on a common plane in space.

Classroom Example: p. 857
Exercise 32

> **EXAMPLE 5** Solving a System of Dependent Equations Representing the Same Plane
>
> Solve the system. $\quad x + 2y + 3z = 6$
> $\qquad\qquad\qquad -x - 2y - 3z = -6$
> $\qquad\qquad\qquad 2x + 4y + 6z = 12$

Solution:

$\boxed{A} \quad x + 2y + 3z = 6$
$\boxed{B} \quad -x - 2y - 3z = -6$
$\boxed{C} \quad 2x + 4y + 6z = 12$

Upon inspection, you might notice that each equation is a constant multiple of the others. That is, equation $\boxed{B}$ is -1 times equation $\boxed{A}$. Equation $\boxed{C}$ is 2 times equation $\boxed{A}$. This means that equations $\boxed{A}$, $\boxed{B}$, and $\boxed{C}$ represent the same plane in space.

TIP The solution to Example 5 tells us that $x = 6 - 2y - 3z$. That is, x depends on both y and z. If we write the solution set in parametric form, we would need *two* parameters.
$\{(6 - 2s - 3t, s, t)\}$

$$\begin{bmatrix} 1 & 2 & 3 & | & 6 \\ -1 & -2 & -3 & | & -6 \\ 2 & 4 & 6 & | & 12 \end{bmatrix} \xrightarrow[-2R_1 + R_3 \rightarrow R_3]{R_1 + R_2 \rightarrow R_2} \begin{bmatrix} 1 & 2 & 3 & | & 6 \\ 0 & 0 & 0 & | & 0 \\ 0 & 0 & 0 & | & 0 \end{bmatrix}$$

The reduced row-echelon form tells us that the system reduces to an equivalent system of one equation with three variables.

The system is equivalent to the single equation $x + 2y + 3z = 6$ and the solution set is the set of all ordered triples that satisfy this equation. Since we have one unique relationship among three variables, one variable is dependent on the other two variables. Letting two variables be arbitrary real numbers, the common equation defines the value of the third variable. For example, solving the equation $x + 2y + 3z = 6$ for x yields: $x = 6 - 2y - 3z$.

Answer
4. $\{(5z + 1, -4z + 6, z) \mid z$ is any real number}

The solution set is $\{(6 - 2y - 3z, y, z) \mid y \text{ and } z \text{ are any real numbers}\}$.

Alternatively, the solution set can be expressed as $\{(x, y, z) \mid x + 2y + 3z = 6\}$.

Skill Practice 5   Solve the system.

$$2x - 3y + 5z = 3$$
$$4x - 6y + 10z = 6$$
$$20x - 30y + 50z = 30$$

3. Solve Applications of Systems of Equations

A city planner can study the flow of traffic through a network of streets by measuring the flow rates (number of vehicles per unit time) at various points. For traffic to flow freely, we use the principle that the flow rate into an intersection is equal to the flow rate out of the intersection. This must be true for all intersections in the network.

Classroom Example: p. 857
Exercise 46

Point of Interest

The principle of equal flow rates into and out of a junction applies to many applications of networks, including communications systems, water systems, and even the circulatory system in the human body.

EXAMPLE 6   Solving an Application Involving Dependent Equations

Consider the network of four one-way streets shown in Figure 9-5. In the figure, x_1, x_2, x_3, and x_4 indicate flow rates (in vehicles per hour) along the stretches of roads AB, BC, CD, and DA, respectively. The other numbers in the figure indicate other flow rates moving into and out of intersections A, B, C, and D.

a. Set up a system of equations that represents traffic flowing freely.

b. Write the augmented matrix for the system of equations in reduced row-echelon form.

c. If the traffic between intersections D and A is 260 vehicles per hour, determine the flow rates x_1, x_2, and x_3.

d. If the traffic between intersections D and A is between 250 and 300 vehicles per hour, inclusive, determine the flow rates x_1, x_2, and x_3.

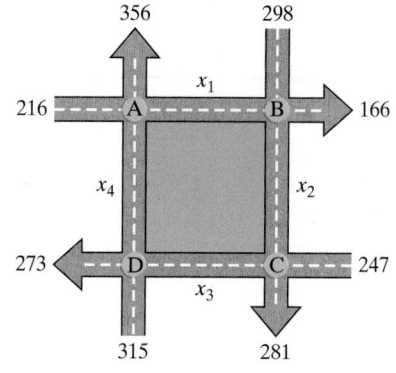

Figure 9-5

Solution:

a.

Intersection	Flow in	=	Flow out
A	$216 + x_4$	=	$356 + x_1$
B	$298 + x_1$	=	$166 + x_2$
C	$x_2 + 247$	=	$281 + x_3$
D	$x_3 + 315$	=	$273 + x_4$

The number of cars flowing into each intersection must equal the number of cars leaving the intersection for traffic to flow freely.

b.

$$\begin{aligned} -x_1 \qquad\qquad + x_4 &= 140 \\ x_1 - x_2 \qquad\qquad &= -132 \\ x_2 - x_3 \qquad &= 34 \\ x_3 - x_4 &= -42 \end{aligned} \Bigg\} \rightarrow \left[\begin{array}{cccc|c} -1 & 0 & 0 & 1 & 140 \\ 1 & -1 & 0 & 0 & -132 \\ 0 & 1 & -1 & 0 & 34 \\ 0 & 0 & 1 & -1 & -42 \end{array}\right]$$

Write the equations in standard form. Set up the augmented matrix.

Answer

5. $\left\{ \left(\dfrac{3 + 3y - 5z}{2}, y, z \right) \,\middle|\, y \text{ and } z \right.$

are any real numbers $\left. \right\}$ or

$\{(x, y, z) \mid 2x - 3y + 5z = 3\}$

$$\xrightarrow{-1R_1 \to R_1} \begin{bmatrix} 1 & 0 & 0 & -1 & | & -140 \\ 1 & -1 & 0 & 0 & | & -132 \\ 0 & 1 & -1 & 0 & | & 34 \\ 0 & 0 & 1 & -1 & | & -42 \end{bmatrix} \xrightarrow{-1R_1 + R_2 \to R_2} \begin{bmatrix} 1 & 0 & 0 & -1 & | & -140 \\ 0 & -1 & 0 & 1 & | & 8 \\ 0 & 1 & -1 & 0 & | & 34 \\ 0 & 0 & 1 & -1 & | & -42 \end{bmatrix}$$

$$\xrightarrow{-1R_2 \to R_2} \begin{bmatrix} 1 & 0 & 0 & -1 & | & -140 \\ 0 & 1 & 0 & -1 & | & -8 \\ 0 & 1 & -1 & 0 & | & 34 \\ 0 & 0 & 1 & -1 & | & -42 \end{bmatrix} \xrightarrow{-1R_2 + R_3 \to R_3} \begin{bmatrix} 1 & 0 & 0 & -1 & | & -140 \\ 0 & 1 & 0 & -1 & | & -8 \\ 0 & 0 & -1 & 1 & | & 42 \\ 0 & 0 & 1 & -1 & | & -42 \end{bmatrix}$$

$$\xrightarrow{-1R_3 \to R_3} \begin{bmatrix} 1 & 0 & 0 & -1 & | & -140 \\ 0 & 1 & 0 & -1 & | & -8 \\ 0 & 0 & 1 & -1 & | & -42 \\ 0 & 0 & 1 & -1 & | & -42 \end{bmatrix} \xrightarrow{-1R_3 + R_4 \to R_4} \begin{bmatrix} 1 & 0 & 0 & -1 & | & -140 \\ 0 & 1 & 0 & -1 & | & -8 \\ 0 & 0 & 1 & -1 & | & -42 \\ 0 & 0 & 0 & 0 & | & 0 \end{bmatrix}$$

c. From the reduced row-echelon form, we see that the flow rates x_1, x_2, and x_3 can all be expressed in terms of the flow rate x_4.

$$\begin{bmatrix} 1 & 0 & 0 & -1 & | & -140 \\ 0 & 1 & 0 & -1 & | & -8 \\ 0 & 0 & 1 & -1 & | & -42 \\ 0 & 0 & 0 & 0 & | & 0 \end{bmatrix}$$
$\longrightarrow x_1 - x_4 = -140 \longrightarrow x_1 = x_4 - 140$
$\longrightarrow x_2 - x_4 = -8 \longrightarrow x_2 = x_4 - 8$
$\longrightarrow x_3 - x_4 = -42 \longrightarrow x_3 = x_4 - 42$

If x_4 is 260 vehicles per hour, we have the following values for x_1, x_2, and x_3.

Flow rate between A and B: $x_1 = 260 - 140 \longrightarrow x_1 = 120$ vehicles per hour

Flow rate between B and C: $x_2 = 260 - 8 \longrightarrow x_2 = 252$ vehicles per hour

Flow rate between C and D: $x_3 = 260 - 42 \longrightarrow x_3 = 218$ vehicles per hour

d. If the traffic flow between intersections D and A is given by $250 \le x_4 \le 300$, we can solve the following inequalities to determine the flow rates x_1, x_2, and x_3.

$$250 \le x_1 + 140 \le 300 \longrightarrow 110 \le x_1 \le 160$$
$$250 \le x_2 + 8 \le 300 \longrightarrow 242 \le x_2 \le 292$$
$$250 \le x_3 + 42 \le 300 \longrightarrow 208 \le x_3 \le 258$$

Skill Practice 6   Refer to the figure. Assume that traffic flows freely with flow rates given in vehicles per hour.

a. If the traffic between intersections D and A is 400 vehicles per hour, determine the flow rates x_1, x_2, and x_3.

b. If the traffic between intersections D and A is between 380 and 420 vehicles per hour, determine the flow rates x_1, x_2, and x_3.

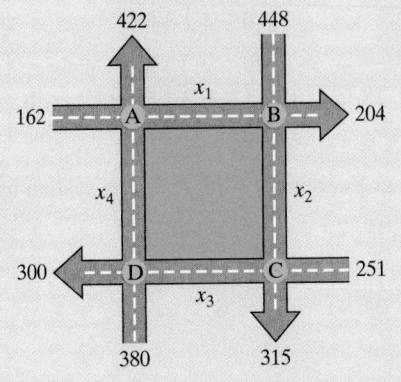

Answers

6. a. $x_1 = 140$, $x_2 = 384$, and $x_3 = 320$

 b. $120 \le x_1 \le 160$, $364 \le x_2 \le 404$, $300 \le x_3 \le 340$

SECTION 9.2   Practice Exercises

Prerequisite Review

For Exercises R.1–R.2, solve for the indicated variable.

R.1. $5x + y = -18$ for y $y = -5x - 18$

R.2. $3x - y = -8$ for x $x = \dfrac{y}{3} - \dfrac{8}{3}$

For Exercises R.3–R.4, determine whether the graph of the equation is symmetric with respect to the x-axis or y-axis.

R.3. $y = -|x| - 6$ y-axis

R.4. $x = y^2 + 7$ x-axis

Concept Connections

1. True or false? A system of linear equations in three variables may have no solution.   True

2. True or false? A system of linear equations in three variables may have exactly one solution.   True

3. True or false? A system of linear equations in three variables may have exactly two solutions.   False

4. True or false? A system of linear equations in three variables may have infinitely many solutions.   True

5. If a system of linear equations has no solution, then the system is said to be _____inconsistent_____.

6. If a system of linear equations has infinitely many solutions, then the equations are said to be _____dependent_____.

Objectives 1–2: Identify Inconsistent Systems and Solve Systems with Dependent Equations

For Exercises 7–14, an augmented matrix is given. Determine the number of solutions to the corresponding system of equations.

7. $\begin{bmatrix} 1 & 2 & | & 4 \\ 0 & 0 & | & 5 \end{bmatrix}$
No solution

8. $\begin{bmatrix} 1 & 0 & 2 & | & 5 \\ 0 & 1 & 4 & | & -2 \\ 0 & 0 & 0 & | & -1 \end{bmatrix}$
No solution

9. $\begin{bmatrix} 1 & 0 & 4 & | & 3 \\ 0 & 1 & -1 & | & 6 \\ 0 & 0 & 0 & | & 0 \end{bmatrix}$
Infinitely many solutions

10. $\begin{bmatrix} 1 & 3 & | & 5 \\ 0 & 0 & | & 0 \end{bmatrix}$
Infinitely many solutions

11. $\begin{bmatrix} 1 & 0 & 0 & | & -3 \\ 0 & 1 & 0 & | & 4 \\ 0 & 0 & 1 & | & 0 \end{bmatrix}$
One solution

12. $\begin{bmatrix} 1 & 0 & | & 3 \\ 0 & 1 & | & 0 \end{bmatrix}$
One solution

13. $\begin{bmatrix} 1 & 2 & 5 & | & -1 \\ 0 & 0 & 0 & | & 0 \\ 0 & 0 & 0 & | & 0 \end{bmatrix}$
Infinitely many solutions

14. $\begin{bmatrix} 1 & 0 & 6 & | & 7 \\ 0 & 0 & 0 & | & 0 \\ 0 & 0 & 0 & | & 0 \end{bmatrix}$
Infinitely many solutions

For Exercises 15–18, determine the solution set for the system represented by each augmented matrix.

15. a. $\begin{bmatrix} 1 & 2 & | & 5 \\ 0 & 1 & | & 0 \end{bmatrix}$ $\{(5, 0)\}$
b. $\begin{bmatrix} 1 & 2 & | & 5 \\ 0 & 0 & | & 0 \end{bmatrix}$ $\{(5 - 2y, y)\,|\,y \text{ is any real number}\}$
c. $\begin{bmatrix} 1 & 2 & | & 5 \\ 0 & 0 & | & 1 \end{bmatrix}$ $\{\ \}$

16. a. $\begin{bmatrix} 1 & 3 & | & -4 \\ 0 & 1 & | & 1 \end{bmatrix}$ $\{(-7, 1)\}$
b. $\begin{bmatrix} 1 & 3 & | & -4 \\ 0 & 0 & | & 1 \end{bmatrix}$ $\{\ \}$
c. $\begin{bmatrix} 1 & 3 & | & -4 \\ 0 & 0 & | & 0 \end{bmatrix}$ $\{(-3y - 4, y)\,|\,y \text{ is any real number}\}$

17. a. $\begin{bmatrix} 1 & 0 & 6 & | & 3 \\ 0 & 1 & 4 & | & 5 \\ 0 & 0 & 1 & | & 0 \end{bmatrix}$ $\{(3, 5, 0)\}$
b. $\begin{bmatrix} 1 & 0 & 6 & | & 3 \\ 0 & 1 & 4 & | & 5 \\ 0 & 0 & 0 & | & 1 \end{bmatrix}$ $\{\ \}$
c. $\begin{bmatrix} 1 & 0 & 6 & | & 3 \\ 0 & 1 & 4 & | & 5 \\ 0 & 0 & 0 & | & 0 \end{bmatrix}$ $\{(-6z + 3, -4z + 5, z)\,|\,z \text{ is any real number}\}$

18. a. $\begin{bmatrix} 1 & 0 & -2 & | & 3 \\ 0 & 1 & 3 & | & 5 \\ 0 & 0 & 1 & | & 1 \end{bmatrix}$ $\{(5, 2, 1)\}$
b. $\begin{bmatrix} 1 & 0 & -2 & | & 3 \\ 0 & 1 & 3 & | & 5 \\ 0 & 0 & 0 & | & 0 \end{bmatrix}$ $\{(2z + 3, -3z + 5, z)\,|\,z \text{ is any real number}\}$
c. $\begin{bmatrix} 1 & 0 & -2 & | & 3 \\ 0 & 1 & 3 & | & 5 \\ 0 & 0 & 0 & | & 1 \end{bmatrix}$ $\{\ \}$

For Exercises 19–38, solve the system by using Gaussian elimination or Gauss-Jordan elimination. (See Exercises 1–5)

19. $2x + 4y = 5$
$x + 2y = 4$ $\{\ \}$

20. $4x + 16y = 21$
$x + 4y = -1$ $\{\ \}$

21. $2x + 7y = 10$
$\frac{1}{5}x = 1 - \frac{7}{10}y$

22. $4x - 3y = 6$
$y = \frac{4}{3}x - 2$
$\left\{\left(\dfrac{3y + 6}{4}, y\right)\,\middle|\,y \text{ is any real number}\right\}$

23. $x - 3y + 14z = -9$
$-2x + 7y - 31z = 21$
$x - 5y + 20z = -14$ $\{\ \}$

24. $x - 3y + 17z = 1$
$x - y + 7z = 2$
$2x - 5y + 29z = 5$ $\{\ \}$

25. $5x + 7y - 11z = 45$
$3x + 5y - 9z = 23$
$x + y - z = 11$
$\{(-2z + 16, 3z - 5, z) \mid z \text{ is any real number}\}$

26. $x + 3y + 9z = 12$
$2x + 7y + 22z = 26$
$-5x - 17y - 53z = -64$
$\{(3z + 6, -4z + 2, z) \mid z \text{ is any real number}\}$

27. $2x = 5y - 16z + 40$
$2(x + y) = 4z$
$x - 2y + 7z = 18$
$\{(4, 0, 2)\}$

28. $x = 2y + 4z + 5$
$y - 4 = -3z$
$5(x - y) - 9z = 4x - 7$
$\{(13, 4, 0)\}$

29. $2x - 5y - 20z = -24$
$x - 3y - 11z = -15$
$\{(5z + 3, -2z + 6, z) \mid z \text{ is any real number}\}$

30. $2x - y - 5z = -3$
$x - 2y - 7z = -12$
$\{(z + 2, -3z + 7, z) \mid z \text{ is any real number}\}$

31. $2x + 3y + 4z = 12$
$-4x - 6y - 8z = -24$
$x + 1.5y + 2z = 6$

32. $-x + 2y + 7z = 14$
$10x - 20y - 70z = -140$
$-\frac{1}{7}x + \frac{2}{7}y + z = 2$

33. $2x - 3y + 9z = -2$
$x = 5y - 8z - 15$
$3(x - y) + 6z = 2x - 7$
$\{(-3z + 5, z + 4, z) \mid z \text{ is any real number}\}$

34. $3x + 11y - 3z = -13$
$x + y = z - 15$
$3(x + y) = z + 2x - 7$
$\{(z - 19, 4, z) \mid z \text{ is any real number}\}$

35. $-5x + 12y - 20z = -11$
$x + 4z = 3y + 1$
$\{(-4z + 7, 2, z) \mid z \text{ is any real number}\}$

36. $x = 4y + 20z - 1$
$-2x + 5y + 25z = -1$
$\{(3, -5z + 1, z) \mid z \text{ is any real number}\}$

37. $x_1 - 3x_2 + 9x_3 - 14x_4 = 32$
$x_2 - 3x_3 + 6x_4 = -10$
$x_2 - x_3 + 2x_4 = -4$
$x_1 - 2x_2 + 8x_3 - 12x_4 = 24$ $\{\ \}$

38. $x_1 - 3x_3 - 12x_4 = -15$
$x_2 + x_3 + 6x_4 = 8$
$x_2 - 2x_3 - 6x_4 = -7$
$-2x_1 + 4x_3 + 16x_4 = 22$ $\{\ \}$

For Exercises 39–40, the solution set to a system of dependent equations is given. Write the specific solutions corresponding to the given values of z.

39. $\{(2z + 1, z - 4, z) \mid z \text{ is any real number}\}$
   a. $z = 1$ **b.** $z = 4$ **c.** $z = -2$
   $(3, -3, 1)$ $(9, 0, 4)$ $(-3, -6, -2)$

40. $\{(z + 5, 3z - 2, z) \mid z \text{ is any real number}\}$
   a. $z = -3$ **b.** $z = 1$ **c.** $z = 0$
   $(2, -11, -3)$ $(6, 1, 1)$ $(5, -2, 0)$

For Exercises 41–44, the solution set to a system of dependent equations is given. Write three ordered triples that are solutions to the system. Answers may vary.

41. $\{(4z, 6 - z, z) \mid z \text{ is any real number}\}$
   For example: $(0, 6, 0), (4, 5, 1), (8, 4, 2)$

42. $\{(2z, z - 3, z) \mid z \text{ is any real number}\}$
   For example: $(0, -3, 0), (2, -2, 1), (4, -1, 2)$

43. $\left\{ \left(\dfrac{6 - 3y - 6z}{2}, y, z \right) \middle| y \text{ and } z \text{ are any real numbers} \right\}$
   For example: $(0, 0, 1), (0, 2, 0), (3, 0, 0)$

44. $\{(4y + 2z - 20, y, z) \mid y \text{ and } z \text{ are any real numbers}\}$
   For example: $(-20, 0, 0), (0, 5, 0), (0, 0, 10)$

Objective 3: Solve Applications of Systems of Equations

For Exercises 45–48, assume that traffic flows freely through the intersections A, B, C, and D. The values x_1, x_2, x_3, and x_4 and the other numbers in the figures represent flow rates in vehicles per hour. (See Example 6)

45.

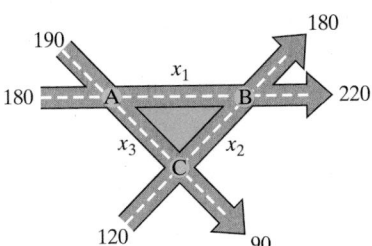

a. Write an equation representing equal flow into and out of intersection A. $180 + 190 = x_1 + x_3$

b. Write an equation representing equal flow into and out of intersection B. $x_1 + x_2 = 180 + 220$

c. Write an equation representing equal flow into and out of intersection C. $x_3 + 120 = x_2 + 90$

d. Write the system of equations from parts (a)–(c) in standard form.

e. Write the reduced row-echelon form of the augmented matrix representing the system of equations from part (d).

f. If the flow rate between intersections A and C is 120 vehicles per hour, determine the flow rates x_1 and x_2.

g. If the flow rate between intersections A and C is between 100 and 150 vehicles per hour, inclusive, determine the flow rates x_1 and x_2.

46.

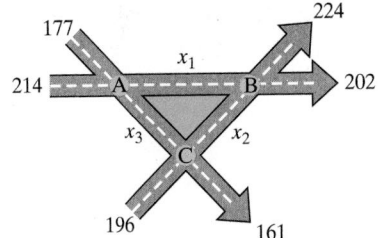

a. Write an equation representing equal flow into and out of intersection A. $214 + 177 = x_1 + x_3$

b. Write an equation representing equal flow into and out of intersection B. $x_1 + x_2 = 224 + 202$

c. Write an equation representing equal flow into and out of intersection C. $x_3 + 196 = x_2 + 161$

d. Write the system of equations from parts (a)–(c) in standard form.

e. Write the reduced row-echelon form of the augmented matrix representing the system of equations in part (d).

f. If the flow rate between intersections A and C is 156 vehicles per hour, determine the flow rates x_1 and x_2.

g. If the flow rate between intersections A and C is between 100 and 200 vehicles per hour, inclusive, determine the flow rates x_1 and x_2.

47.

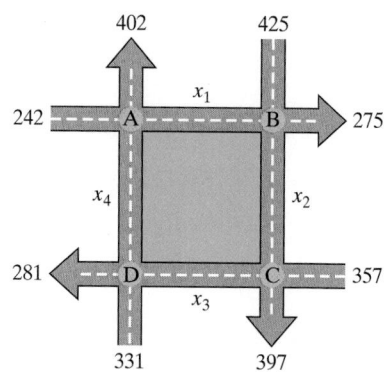

a. Assume that traffic flows at a rate of 220 vehicles per hour on the stretch of road between intersections D and A. Find the flow rates x_1, x_2, and x_3.

b. If traffic flows at a rate of between 200 and 250 vehicles per hour inclusive between intersections D and A, find the flow rates x_1, x_2, and x_3.

49. An accountant checks the reported earnings for a theater for three nightly performances against the number of tickets sold.

Night	Children Tickets	Student Tickets	General Admission	Total Revenue
1	80	400	480	$9280
2	50	350	400	$7800
3	75	525	600	$10,500

a. Let x, y, and z represent the cost for children tickets, student tickets, and general admission tickets, respectively. Set up a system of equations to solve for x, y, and z.

b. Set up the augmented matrix for the system and solve the system. (*Hint*: To make the augmented matrix simpler to work with, consider dividing each linear equation by an appropriate constant.)

c. Explain why the auditor knows that there was an error in the record keeping.

48.

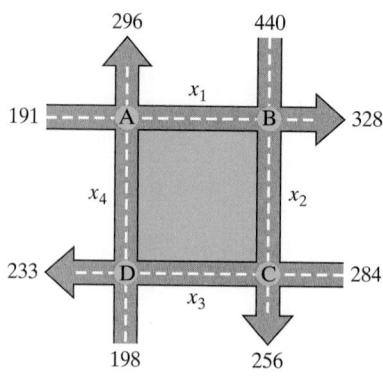

a. Assume that traffic flows at a rate of 180 vehicles per hour on the stretch of road between intersections D and A. Find the flow rates x_1, x_2, and x_3.

b. If traffic flows at a rate of between 150 and 200 vehicles per hour inclusive between intersections D and A, find the flow rates x_1, x_2, and x_3.

50. A concession stand at a city park sells hamburgers, hot dogs, and drinks. Three patrons buy the following food and drink combinations for the following prices.

Patron	Hamburgers	Hot Dogs	Drinks	Total Revenue
1	1	1	5	$11
2	0	1	2	$5
3	3	1	11	$22

a. Let x, y, and z represent the cost for a hamburger, a hot dog, and a drink, respectively. Set up a system of equations to solve for x, y, and z.

b. Set up the augmented matrix for the system and solve the system.

c. Explain why the concession stand manager knows that there was an error in the record keeping.

Mixed Exercises

51. The solution set to Exercise 25 is $\{(-2z + 16, 3z - 5, z) \mid z$ is any real number$\}$. Verify the solution by substituting the ordered triple into each individual equation.

52. The solution set to Exercise 26 is $\{(3z + 6, -4z + 2, z) \mid z$ is any real number$\}$. Verify the solution by substituting the ordered triple into each individual equation.

The systems in Exercises 53–56 are called **homogeneous systems** because each has $(0, 0, 0)$ as a solution. However, if a system is made up of dependent equations, it will have infinitely many more solutions. For each system, determine whether $(0, 0, 0)$ is the only solution or if the system has infinitely many solutions. If the system has infinitely many solutions, give the solution set.

53.
$$x + 2y - 8z = 0$$
$$-5x - 11y + 43z = 0$$
$$x + 5y - 12z = 0$$
$(0, 0, 0)$ is the only solution.

54.
$$x - 2y - 7z = 0$$
$$-3x + 8y + 31z = 0$$
$$-2x + 5y + 22z = 0$$
$(0, 0, 0)$ is the only solution.

55.
$$x - 5y + 13z = 0$$
$$-3x + 17y - 45z = 0$$
$$x - 4y + 10z = 0$$
Infinitely many solutions; $\{(2z, 3z, z) \mid z$ is any real number$\}$

56.
$$-2x + 15y - 83z = 0$$
$$x - 7y + 39z = 0$$
$$x - 5y + 29z = 0$$
Infinitely many solutions; $\{(-4z, 5z, z) \mid z$ is any real number$\}$

Write About It

57. Explain how you can determine from the reduced row-echelon form of a matrix whether the corresponding system of equations is inconsistent.

59. Consider the following system. By inspection describe the geometrical relationship among the planes represented by the three equations.

$$x + y + z = 1$$
$$2x + 2y + 2z = 2$$
$$3x + 3y + 3z = 3$$

58. What can you conclude about a system of equations if the corresponding reduced row-echelon form consists of a row entirely of zeros?

60. Explain why a system of two equations with three variables cannot have exactly one ordered triple as its solution. Each equation represents a plane in space. If the two equations represent different planes, then they must either be parallel planes (resulting in no solution) or they intersect in a line (resulting in infinitely many solutions). If the equations represent the same plane, then the solution set consists of infinitely many points on the common plane.

Technology Connections

For Exercises 61–66, use a graphing utility to find the reduced row-echelon form of the augmented matrix for the system in the given exercise. Use the result to verify the answer to the given exercise.

61. Exercise 23   **62.** Exercise 24   **63.** Exercise 25

64. Exercise 26   **65.** Exercise 31   **66.** Exercise 32

Additional answers can be found in the Instructor Answer Appendix.

| **SECTION 9.3** | **Operations on Matrices** |

Classroom Examples: p. 870
Exercises 14, 16

OBJECTIVES

1. **Determine the Order of a Matrix**
2. **Add and Subtract Matrices**
3. **Multiply a Matrix by a Scalar**
4. **Multiply Matrices**
5. **Apply Operations on Matrices**

1. Determine the Order of a Matrix

We have seen how the manipulation of an augmented matrix can be used to solve a system of linear equations. Matrices have many other useful mathematical applications, particularly when manipulating tables or databases. In particular, matrices play a useful role in digital photography, film, and computer animation.

The **order of a matrix** is determined by the number of rows and number of columns. A matrix with m rows and n columns is an $m \times n$ matrix (read as "m by n" matrix).

EXAMPLE 1   Determining the Order of a Matrix

Determine the order of each matrix.

a. $\begin{bmatrix} 3 & \pi & 1.7 \\ -1 & 6 & 10 \end{bmatrix}$ **b.** $\begin{bmatrix} -7 \\ 2 \\ 4 \\ \sqrt{2} \end{bmatrix}$ **c.** $\begin{bmatrix} 1 & 0 & 0 \\ 0 & 1 & 0 \\ 0 & 0 & 1 \end{bmatrix}$ **d.** $\begin{bmatrix} x & y & z \end{bmatrix}$

Solution:

a. The matrix has 2 rows and 3 columns. Therefore, it is a 2×3 matrix.

b. The matrix has 4 rows and 1 column. Therefore, it is a 4×1 matrix. A matrix with only 1 column is called a **column matrix.**

c. The matrix has 3 rows and 3 columns. Therefore, it is a 3×3 matrix. A matrix with the same number of rows and columns is called a **square matrix.**

d. The matrix has 1 row and 3 columns. Therefore, it is a 1×3 matrix. A matrix with only 1 row is called a **row matrix.**

Skill Practice 1 Determine the order of each matrix.

a. $\begin{bmatrix} -9 & 2 \\ 4.1 & -3 \\ \sqrt{3} & 4 \end{bmatrix}$ **b.** $\begin{bmatrix} -4 \\ 11 \end{bmatrix}$ **c.** $\begin{bmatrix} -0.1 & 0.4 \\ 0.5 & 0.2 \end{bmatrix}$ **d.** $[105 \quad 311]$

A matrix can be represented generically as follows:

$$[a_{ij}] = \begin{bmatrix} a_{11} & a_{12} & \cdots & a_{1n} \\ a_{21} & a_{22} & \cdots & a_{2n} \\ \vdots & & & \\ a_{m1} & a_{m2} & \cdots & a_{mn} \end{bmatrix}$$

TIP The notation $[a_{ij}]$ is a generic way to denote a matrix with elements of the form a_{ij}, where i and j represent generic row and column numbers.

Using the double subscript notation, a_{43} represents the element in the 4th row, 3rd column. The notation a_{ij} represents the element in the ith row, jth column.

Classroom Examples: p. 870
Exercises 18, 20

EXAMPLE 2 Using Matrix Notation

Consider $A = \begin{bmatrix} -4 & 5 & \sqrt{2} & -\pi \\ 10 & \frac{1}{2} & 0 & 6 \\ -\frac{7}{6} & 8 & 12 & -9 \end{bmatrix}$, where $A = [a_{ij}]$. Determine the value of each element.

a. a_{32} **b.** a_{23} **c.** a_{14}

Solution:

a. a_{32} represents the element in the 3rd row, 2nd column: 8

b. a_{23} represents the element in the 2nd row, 3rd column: 0

c. a_{14} represents the element in the 1st row, 4th column: $-\pi$

Skill Practice 2 Given matrix A from Example 2, determine the value of each element.

a. a_{13} **b.** a_{31} **c.** a_{34}

Answers

1. a. 3×2 **b.** 2×1
   c. 2×2 **d.** 1×2

2. a. $\sqrt{2}$ **b.** $-\dfrac{7}{6}$ **c.** -9

2. Add and Subtract Matrices

We now consider the conditions under which two matrices are equal.

> ### Equality of Matrices
>
> Two matrices are equal if and only if they have the same order and if their corresponding elements are equal.
>
> Example:
>
> - The statement $\begin{bmatrix} 3 & -4 \\ x & z \end{bmatrix} = \begin{bmatrix} 3 & y \\ 9 & 7 \end{bmatrix}$ implies that $x = 9$, $y = -4$, and $z = 7$.
>
> - Conversely, if $x = 9$, $y = -4$, and $z = 7$, then $\begin{bmatrix} 3 & -4 \\ x & z \end{bmatrix} = \begin{bmatrix} 3 & y \\ 9 & 7 \end{bmatrix}$.

Next, we define addition and subtraction of matrices.

> ### Addition and Subtraction of Matrices
>
> Let $A = [a_{ij}]$ and $B = [b_{ij}]$ be m by n matrices. Then,
>
> $$A + B = [a_{ij} + b_{ij}] \text{ for } i = 1, 2, \ldots, m \text{ and } j = 1, 2, \ldots, n$$
> $$A - B = [a_{ij} - b_{ij}] \text{ for } i = 1, 2, \ldots, m \text{ and } j = 1, 2, \ldots, n$$
>
> That is, to add or subtract two matrices, the matrices must have the same order, and the sum or difference is found by adding or subtracting the corresponding elements.

Classroom Examples: p. 871
Exercises 30, 32

EXAMPLE 3   Adding and Subtracting Matrices

Given $A = \begin{bmatrix} 3 & -7 & 0 \\ 1 & \frac{4}{3} & -6 \end{bmatrix}$, $B = \begin{bmatrix} 8 & -3 & 4 \\ 0 & 1 & -9 \end{bmatrix}$, and $C = \begin{bmatrix} 3 & -5 \\ 4 & 18 \end{bmatrix}$,

find the sum or difference if possible.

a. $A + B$ **b.** $B - A$ **c.** $A + C$

Solution:

a. $A + B = \begin{bmatrix} 3 & -7 & 0 \\ 1 & \frac{4}{3} & -6 \end{bmatrix} + \begin{bmatrix} 8 & -3 & 4 \\ 0 & 1 & -9 \end{bmatrix}$ Matrix A and matrix B both have the same order (2×3), so the sum is defined.

$= \begin{bmatrix} 3 + 8 & -7 + (-3) & 0 + 4 \\ 1 + 0 & \frac{4}{3} + 1 & -6 + (-9) \end{bmatrix}$ Add the elements in the corresponding positions.

$= \begin{bmatrix} 11 & -10 & 4 \\ 1 & \frac{7}{3} & -15 \end{bmatrix}$ Simplify.

b. $B - A = \begin{bmatrix} 8 & -3 & 4 \\ 0 & 1 & -9 \end{bmatrix} - \begin{bmatrix} 3 & -7 & 0 \\ 1 & \frac{4}{3} & -6 \end{bmatrix}$ Matrix A and matrix B both have the same order (2×3), so the difference is defined.

$= \begin{bmatrix} 8 - 3 & -3 - (-7) & 4 - 0 \\ 0 - 1 & 1 - \frac{4}{3} & -9 - (-6) \end{bmatrix}$ Subtract the elements in the corresponding positions.

$= \begin{bmatrix} 5 & 4 & 4 \\ -1 & -\frac{1}{3} & -3 \end{bmatrix}$ Simplify.

c. Matrix A is a 2×3 matrix, whereas matrix C is a 2×2 matrix. The matrices are of different orders and therefore cannot be added or subtracted.

Skill Practice 3 Given $A = \begin{bmatrix} 3 & 4 & -7 \\ 9 & 2 & 0 \end{bmatrix}$, $B = \begin{bmatrix} -8 & 0 \\ 1 & 5 \\ -4 & \frac{1}{2} \end{bmatrix}$, and $C = \begin{bmatrix} 4 & -5 \\ 12 & 3 \\ -2 & 2 \end{bmatrix}$, find the sum or difference if possible.

a. $A + B$ **b.** $B + C$ **c.** $C - B$

The opposite of a real number a is $-a$. The value $-a$ is also called the additive inverse of a. A matrix A also has an additive inverse, denoted by $-A$.

Additive Inverse of a Matrix

Given $A = [a_{ij}]$, the **additive inverse of A,** denoted by $-A$, is $-A = [-a_{ij}]$.

Verbal Interpretation	Example
The additive inverse of a matrix is found by taking the opposite of each element in the matrix.	Given $A = \begin{bmatrix} -4 & 5 & -1 \\ 0 & -3 & 2 \end{bmatrix}$, $-A = \begin{bmatrix} 4 & -5 & 1 \\ 0 & 3 & -2 \end{bmatrix}$.

The number 0 is the identity element under the addition of real numbers because $a + (-a) = 0$. A matrix in which all elements are zero is called a **zero matrix** and is denoted by 0. The sum of a matrix A and its additive inverse $-A$ is the zero matrix of the same order. For example,

$$\underset{A}{\begin{bmatrix} -4 & 5 & -1 \\ 0 & -3 & 2 \end{bmatrix}} + \underset{(-A)}{\begin{bmatrix} 4 & -5 & 1 \\ 0 & 3 & -2 \end{bmatrix}} = \underset{0}{\begin{bmatrix} 0 & 0 & 0 \\ 0 & 0 & 0 \end{bmatrix}} \quad \text{2} \times \text{3 zero matrix}$$

Properties of Matrix Addition

Let A, B, and C be matrices of order $m \times n$, and let 0 be the zero matrix of order $m \times n$. Then,

1. $A + B = B + A$ Commutative property of matrix addition
2. $A + (B + C) = (A + B) + C$ Associative property of matrix addition
3. $A + (-A) = 0$ Inverse property of matrix addition
4. $A + 0 = 0 + A = A$ Identity property of matrix addition

---

3. Multiply a Matrix by a Scalar

Next we look at the product of a real number k and a matrix. This is called **scalar multiplication.** The real number k is called a **scalar** to distinguish it from a matrix.

Answers

3. a. Not possible

b. $\begin{bmatrix} -4 & -5 \\ 13 & 8 \\ -6 & \frac{5}{2} \end{bmatrix}$

c. $\begin{bmatrix} 12 & -5 \\ 11 & -2 \\ 2 & \frac{3}{2} \end{bmatrix}$

Scalar Multiplication

Let $A = [a_{ij}]$ be an $m \times n$ matrix and let k be a real number. Then, $kA = [ka_{ij}]$.

Verbal Interpretation	Example
To multiply a matrix A by a scalar k, multiply each element in the matrix by k.	Given $A = \begin{bmatrix} -4 & 5 & -1 \\ 0 & -3 & 2 \end{bmatrix}$, $5A = \begin{bmatrix} -20 & 25 & -5 \\ 0 & -15 & 10 \end{bmatrix}$.

The following properties of scalar multiplication are similar to those of the multiplication of real numbers.

Properties of Scalar Multiplication

Let A and B be $m \times n$ matrices and let c and d be real numbers. Then,

1. $c(A + B) = cA + cB$ Distributive property of scalar multiplication
2. $(c + d)A = cA + dA$ Distributive property of scalar multiplication
3. $c(dA) = (cd)A$ Associative property of scalar multiplication

Classroom Example: p. 871
Exercise 42

EXAMPLE 4 Multiplying a Matrix by a Scalar

Given $A = \begin{bmatrix} 3 & -5 \\ 0 & 1 \\ -4 & 2 \end{bmatrix}$ and $B = \begin{bmatrix} -8 & 9 \\ 4 & -1 \\ 0 & 3 \end{bmatrix}$, find $2A + 4(A + B)$.

Solution:

$2A + 4(A + B) = 2A + 4A + 4B$ Apply the distributive property of scalar
$\qquad\qquad\qquad = 6A + 4B$ multiplication.

$6A + 4B = 6\begin{bmatrix} 3 & -5 \\ 0 & 1 \\ -4 & 2 \end{bmatrix} + 4\begin{bmatrix} -8 & 9 \\ 4 & -1 \\ 0 & 3 \end{bmatrix} = \begin{bmatrix} 18 & -30 \\ 0 & 6 \\ -24 & 12 \end{bmatrix} + \begin{bmatrix} -32 & 36 \\ 16 & -4 \\ 0 & 12 \end{bmatrix}$

$\qquad\qquad = \begin{bmatrix} -14 & 6 \\ 16 & 2 \\ -24 & 24 \end{bmatrix}$

Skill Practice 4   Given $A = [4 \quad -3 \quad 9]$ and $B = [-2 \quad 0 \quad 3]$, find $-5A - 2(A + B)$.

Answer
4. $[-24 \quad 21 \quad -69]$

TECHNOLOGY CONNECTIONS

Adding Matrices and Multiplying a Scalar by a Matrix

Many graphing utilities have the capability to perform operations on matrices. From Example 4, we can enter matrix A and matrix B into the calculator using the EDIT feature under the MATRIX menu. Then on the home screen enter $6[A] + 4[B]$ and press ⬤ ENTER.

Alternatively, in "MATHPRINT" mode, enter the matrices on the home screen using ⬤ALPHA F3.

```
NORMAL FLOAT AUTO REAL RADIAN MP
6[A]+4[B]
                    [-14  6 ]
                    [ 16  2 ]
                    [-24 24 ]
```
```
NORMAL FLOAT AUTO REAL RADIAN MP
   [ 3 -5]   [-8  9]
6  [ 0  1]+4[ 4 -1]
   [-4  2]   [ 0  3]
                    [-14  6 ]
                    [ 16  2 ]
                    [-24 24 ]
```

As seen, many of the same properties that are true for addition and multiplication of real numbers are true for the addition of matrices and for scalar multiplication. As a result, we can solve a matrix equation involving these operations in the same way as we solve a linear equation.

Classroom Example: p. 871
Exercise 48

EXAMPLE 5 Solving a Matrix Equation

Solve the matrix equation for X, given that $A = \begin{bmatrix} 2 & 5 \\ 1 & -4 \end{bmatrix}$ and $B = \begin{bmatrix} 6 & -1 \\ -3 & -8 \end{bmatrix}$.

$4X - A = B$

Solution:

> **TIP** We can substitute matrix X back into the equation $4X - A = B$ to verify that it is a solution to the equation.

$$4X - A = B$$
$$4X = A + B \qquad \text{Add matrix } A \text{ to both sides.}$$
$$X = \frac{1}{4}(A + B) \qquad \text{Perform scalar multiplication by } \tfrac{1}{4} \text{ on both sides.}$$
$$X = \frac{1}{4}\left(\begin{bmatrix} 2 & 5 \\ 1 & -4 \end{bmatrix} + \begin{bmatrix} 6 & -1 \\ -3 & -8 \end{bmatrix}\right) \qquad \text{Add the matrices within parentheses.}$$
$$X = \frac{1}{4}\begin{bmatrix} 8 & 4 \\ -2 & -12 \end{bmatrix} \qquad \text{Multiply the matrix by the scalar } \tfrac{1}{4}.$$
$$X = \begin{bmatrix} 2 & 1 \\ -\frac{1}{2} & -3 \end{bmatrix} \qquad \text{Simplify.}$$

Skill Practice 5 Given $A = \begin{bmatrix} 4 & -3 \\ 1 & 11 \end{bmatrix}$ and $B = \begin{bmatrix} 2 & -1 \\ 6 & 5 \end{bmatrix}$, solve $2X + A = B$ for X.

4. Multiply Matrices

Finding the product of two matrices is more complicated than finding the product of a scalar and a matrix. We will demonstrate the process to multiply two matrices and then offer a formal definition.

Answer

5. $X = \begin{bmatrix} -1 & 1 \\ \frac{5}{2} & -3 \end{bmatrix}$

Consider $A = \begin{bmatrix} 2 & -3 & 1 \\ -4 & 7 & 0 \end{bmatrix}$ and $B = \begin{bmatrix} -1 & -6 \\ 10 & 5 \\ 8 & -2 \end{bmatrix}$.

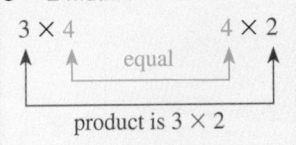

TIP For matrix multiplication, the number of rows of the first matrix does not need to be equal to the number of columns of the second matrix. For example, the product of a 3×4 and a 4×2 matrix results in a 3×2 matrix.

$$3 \times 4 \qquad\qquad 4 \times 2$$

equal

product is 3×2

- To multiply AB, we require that the number of columns in A be equal to the number of rows in B.

- The resulting matrix will have dimensions equal to the number of rows of A by the number of columns of B.

$$A \qquad\qquad\cdot\qquad\qquad B$$

$$\begin{bmatrix} 2 & -3 & 1 \\ -4 & 7 & 0 \end{bmatrix} \cdot \begin{bmatrix} -1 & -6 \\ 10 & 5 \\ 8 & -2 \end{bmatrix}$$

$$2 \times 3 \qquad\qquad 3 \times 2$$

must be equal

The product AB will be a 2×2 matrix.

The product will be a 2×2 matrix.

$$\begin{bmatrix} 2 & -3 & 1 \\ -4 & 7 & 0 \end{bmatrix} \cdot \begin{bmatrix} -1 & -6 \\ 10 & 5 \\ 8 & -2 \end{bmatrix} = \begin{bmatrix} \square & \square \\ \square & \square \end{bmatrix}$$

To find the entry in the first row, first column of the product, multiply the corresponding elements in the first row of A by the elements in the first column of B, and take the sum.

$$\begin{bmatrix} 2 & -3 & 1 \\ -4 & 7 & 0 \end{bmatrix} \cdot \begin{bmatrix} -1 & -6 \\ 10 & 5 \\ 8 & -2 \end{bmatrix} = \begin{bmatrix} 2(-1) + (-3)(10) + 1(8) & \square \\ \square & \square \end{bmatrix}$$

To find the entry in the first row, second column of the product, take the sum of the products of the corresponding elements in the first row of A and second column of B.

$$\begin{bmatrix} 2 & -3 & 1 \\ -4 & 7 & 0 \end{bmatrix} \cdot \begin{bmatrix} -1 & -6 \\ 10 & 5 \\ 8 & -2 \end{bmatrix} = \begin{bmatrix} 2(-1) + (-3)(10) + 1(8) & 2(-6) + (-3)(5) + 1(-2) \\ \square & \square \end{bmatrix}$$

To find the entry in the second row, first column of the product, take the sum of the products of the corresponding elements in the second row of A and first column of B.

$$\begin{bmatrix} 2 & -3 & 1 \\ -4 & 7 & 0 \end{bmatrix} \cdot \begin{bmatrix} -1 & -6 \\ 10 & 5 \\ 8 & -2 \end{bmatrix} = \begin{bmatrix} 2(-1) + (-3)(10) + 1(8) & 2(-6) + (-3)(5) + 1(-2) \\ -4(-1) + 7(10) + 0(8) & \square \end{bmatrix}$$

To find the entry in the second row, second column of the product, take the sum of the products of the corresponding elements in the second row of A and second column of B.

$$\begin{bmatrix} 2 & -3 & 1 \\ -4 & 7 & 0 \end{bmatrix} \cdot \begin{bmatrix} -1 & -6 \\ 10 & 5 \\ 8 & -2 \end{bmatrix} = \begin{bmatrix} 2(-1) + (-3)(10) + 1(8) & 2(-6) + (-3)(5) + 1(-2) \\ -4(-1) + 7(10) + 0(8) & -4(-6) + 7(5) + 0(-2) \end{bmatrix}$$

Simplifying, we have

$$AB = \begin{bmatrix} 2 & -3 & 1 \\ -4 & 7 & 0 \end{bmatrix} \cdot \begin{bmatrix} -1 & -6 \\ 10 & 5 \\ 8 & -2 \end{bmatrix} = \begin{bmatrix} -24 & -29 \\ 74 & 59 \end{bmatrix}$$

From this example, note that each element in the product AB is of the form $a_1b_1 + a_2b_2 + \cdots + a_nb_n$, where the elements $a_1, a_2, \ldots a_n$ are elements in a row of A and $b_1, b_2, \ldots b_n$ are elements in a column of B. The expression $a_1b_1 + a_2b_2 + \cdots + a_nb_n$ is called an **inner product.**

> ### Matrix Multiplication
>
> Let A be an $m \times p$ matrix and let B be a $p \times n$ matrix, then the product AB is an $m \times n$ matrix. For the matrix AB, the element in the ith row and jth column is the sum of the products of the corresponding elements in the ith row of A and the jth column of B (the inner product of the ith row and jth column).
>
> Formally, if
>
> $$A = \begin{bmatrix} a_{11} & a_{12} & \cdots & a_{1p} \\ a_{21} & a_{22} & \cdots & a_{2p} \\ \vdots & & & \\ a_{m1} & a_{m2} & \cdots & a_{mp} \end{bmatrix} \text{ and } B = \begin{bmatrix} b_{11} & b_{12} & \cdots & b_{1n} \\ b_{21} & b_{22} & \cdots & b_{2n} \\ \vdots & & & \\ b_{p1} & b_{p2} & \cdots & b_{pn} \end{bmatrix}$$
>
> then the elements in the matrix $C = AB$ are given by
>
> $$c_{ij} = a_{i1} \cdot b_{1j} + a_{i2} \cdot b_{2j} + \cdots + a_{ip} \cdot b_{pj}$$
>
> *Note*: If the number of columns in A does not equal the number of rows in B, then it is not possible to compute the product AB.

Classroom Examples: p. 871
Exercises 58, 60

EXAMPLE 6   Multiplying Matrices

Given $A = \begin{bmatrix} 2 & 5 & 6 \\ -3 & 0 & 1 \end{bmatrix}$, $B = \begin{bmatrix} 1 \\ 3 \\ -4 \end{bmatrix}$, and $C = \begin{bmatrix} -2 & 10 & 1 \end{bmatrix}$,

find the following products if possible.

a. AB **b.** BC **c.** AC

Solution:

a.

$$\begin{array}{ccc} A & \cdot & B \\ 2 \times 3 & & 3 \times 1 \end{array}$$

equal

The product is a 2×1 matrix.

Matrix A is a 2×3 matrix and matrix B is a 3×1 matrix. The number of columns in A is equal to the number of rows in B. The resulting matrix will be a 2×1 matrix.

$$= \begin{bmatrix} 2 & 5 & 6 \\ -3 & 0 & 1 \end{bmatrix} \cdot \begin{bmatrix} 1 \\ 3 \\ -4 \end{bmatrix} = \begin{bmatrix} 2(1) + 5(3) + 6(-4) \\ -3(1) + 0(3) + 1(-4) \end{bmatrix}$$

Multiply the elements in the first row of A by the elements in the first column of B.

Multiply the elements in the second row of A by the elements in the first column of B.

$$= \begin{bmatrix} -7 \\ -7 \end{bmatrix}$$

b.

$$\begin{array}{ccc} B & \cdot & C \\ 3 \times 1 & & 1 \times 3 \end{array}$$

equal

The product is a 3×3 matrix.

Matrix B is a 3×1 matrix and matrix C is a 1×3 matrix. The number of columns in B is equal to the number of rows in C. The resulting matrix will be a 3×3 matrix.

$$B \cdot C = \begin{bmatrix} 1 \\ 3 \\ -4 \end{bmatrix} \cdot \begin{bmatrix} -2 & 10 & 1 \end{bmatrix} = \begin{bmatrix} 1(-2) & 1(10) & 1(1) \\ 3(-2) & 3(10) & 3(1) \\ -4(-2) & -4(10) & -4(1) \end{bmatrix}$$

$$= \begin{bmatrix} -2 & 10 & 1 \\ -6 & 30 & 3 \\ 8 & -40 & -4 \end{bmatrix}$$

c. A $\cdot$ C

2×3 1×3

not equal

The number of columns of A does not equal the number of rows of C. Therefore, it is not possible to compute the product AC.

Skill Practice 6   Given $A = \begin{bmatrix} 2 & -5 \\ 1 & 0 \end{bmatrix}$, $B = \begin{bmatrix} 3 & 5 & 4 \\ 6 & 0 & -8 \end{bmatrix}$, and $C = \begin{bmatrix} 1 \\ 0 \\ 6 \end{bmatrix}$, find the products if possible.   **a.** AB **b.** BC **c.** AC

Classroom Example: p. 871
Exercise 54

EXAMPLE 7   **Multiplying Matrices**

Given $A = \begin{bmatrix} 3 & -4 \\ 1 & 5 \end{bmatrix}$ and $B = \begin{bmatrix} 1 & 0 \\ -3 & 6 \end{bmatrix}$, find the following products if possible.

a. AB **b.** BA

Solution:

a. A $\cdot$ B

2×2 2×2

equal

The product is a 2×2 matrix.

$$AB = \begin{bmatrix} 3 & -4 \\ 1 & 5 \end{bmatrix} \cdot \begin{bmatrix} 1 & 0 \\ -3 & 6 \end{bmatrix}$$

$$= \begin{bmatrix} 3(1) + (-4)(-3) & 3(0) + (-4)(6) \\ 1(1) + 5(-3) & 1(0) + 5(6) \end{bmatrix}$$

$$= \begin{bmatrix} 15 & -24 \\ -14 & 30 \end{bmatrix}$$

b. B $\cdot$ A

2×2 2×2

equal

The product is a 2×2 matrix.

$$BA = \begin{bmatrix} 1 & 0 \\ -3 & 6 \end{bmatrix} \cdot \begin{bmatrix} 3 & -4 \\ 1 & 5 \end{bmatrix}$$

$$= \begin{bmatrix} 1(3) + 0(1) & 1(-4) + 0(5) \\ -3(3) + 6(1) & -3(-4) + 6(5) \end{bmatrix}$$

$$= \begin{bmatrix} 3 & -4 \\ -3 & 42 \end{bmatrix}$$

Avoiding Mistakes

Example 7 shows that for two matrices A and B, the product AB does not necessarily equal BA. That is, matrix multiplication is *not* commutative.

Skill Practice 7   Given $A = \begin{bmatrix} 1 & -3 \\ 0 & 5 \end{bmatrix}$ and $B = \begin{bmatrix} 2 & 4 \\ 6 & 1 \end{bmatrix}$, find the following products if possible.   **a.** AB **b.** BA

TECHNOLOGY CONNECTIONS

Multiplying Matrices

To multiply two matrices on a graphing utility, first enter the matrices into the calculator using the EDIT feature under the MATRIX menu. Then on the home screen, enter the product. The products AB and BA from Example 7 are shown here.

Alternatively, in "MATHPRINT" mode, enter the matrices on the home screen using (ALPHA) F3. The products AB and BA are shown here.

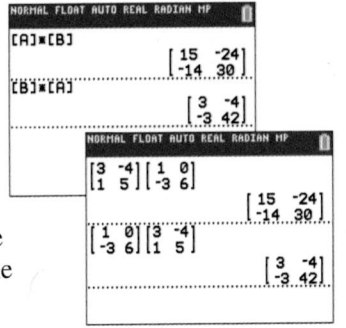

Answers

6. a. $AB = \begin{bmatrix} -24 & 10 & 48 \\ 3 & 5 & 4 \end{bmatrix}$

b. $BC = \begin{bmatrix} 27 \\ -42 \end{bmatrix}$

c. Not possible

7. a. $AB = \begin{bmatrix} -16 & 1 \\ 30 & 5 \end{bmatrix}$

b. $BA = \begin{bmatrix} 2 & 14 \\ 6 & -13 \end{bmatrix}$

5. Apply Operations on Matrices

In applications, a matrix gives mathematicians a systematic format in which to represent data. Multiplication of matrices is a tool to manipulate these data sets. For example, in business, we might multiply the number of items sold by the unit price per item to determine total revenue.

$$\underbrace{\begin{bmatrix} \text{Quantity} \\ \text{matrix} \end{bmatrix}}_{m \times p} \cdot \underbrace{\begin{bmatrix} \text{Unit price} \\ \text{matrix} \end{bmatrix}}_{p \times n} = \underbrace{\begin{bmatrix} \text{Total revenue} \\ \text{matrix} \end{bmatrix}}_{m \times n}$$

same

order of product

TIP The number of columns p of the quantity matrix must equal the number of rows p of the unit price matrix for the product to be defined.

Classroom Example: p. 872
Exercise 74

EXAMPLE 8   Multiplying Matrices in a Business Application

A company owns two coffee shops. The number of donuts, coffee cakes, hot drinks, and cold drinks sold for each shop is given in matrix Q. The price per item is given in matrix P. Find the product QP and interpret the result.

$$Q = \begin{bmatrix} 162 & 34 & 120 & 44 \\ 186 & 50 & 145 & 62 \end{bmatrix} \begin{matrix} \text{Shop 1} \\ \text{Shop 2} \end{matrix}$$

with columns: Donuts, Coffee cakes, Hot drinks, Cold drinks

$$P = \begin{bmatrix} \$0.40 \\ \$2.50 \\ \$3.50 \\ \$1.50 \end{bmatrix} \begin{matrix} \text{Donuts} \\ \text{Coffee cakes} \\ \text{Hot drinks} \\ \text{Cold drinks} \end{matrix}$$

Solution:

Matrix Q is a 2×4 matrix and matrix P is a 4×1 matrix. Matrix Q has 4 columns, which match the number of rows of P, so the product QP is defined. The product matrix will be a 2×1 matrix. The product QP is

162 donuts times \$0.40 per donut (Shop 1).

$$\begin{bmatrix} 162 & 34 & 120 & 44 \\ 186 & 50 & 145 & 62 \end{bmatrix} \cdot \begin{bmatrix} 0.40 \\ 2.50 \\ 3.50 \\ 1.50 \end{bmatrix} = \begin{bmatrix} 162(0.40) + 34(2.50) + 120(3.50) + 44(1.50) \\ 186(0.40) + 50(2.50) + 145(3.50) + 62(1.50) \end{bmatrix}$$

$$= \begin{bmatrix} \$635.80 \\ \$799.90 \end{bmatrix}$$

Row 1: Revenue for Shop 1
Row 2: Revenue for Shop 2

The product QP is a matrix representing the total revenue from these four items for each shop.

Skill Practice 8   A farmer sells organic zucchini, yellow squash, and corn in two different roadside stands. Matrix Q represents the number of pounds of each type of vegetable sold at each stand. Matrix P gives the price per pound of each item. Find the product QP and interpret the result.

$$Q = \begin{bmatrix} 42 & 40 & 84 \\ 30 & 36 & 90 \end{bmatrix} \begin{matrix} \text{Stand 1} \\ \text{Stand 2} \end{matrix}$$

with columns: Zucchini, Yellow squash, Corn

$$P = \begin{bmatrix} \$4.00 \\ \$3.40 \\ \$4.20 \end{bmatrix} \begin{matrix} \text{Zucchini} \\ \text{Yellow squash} \\ \text{Corn} \end{matrix}$$

Answer

8. $\begin{bmatrix} \$656.80 \\ \$620.40 \end{bmatrix}$; The product QP is a matrix representing the total revenue from these three items for each stand.

In Chapter 1 we learned how to apply transformations to the graphs of functions. In Example 9 we see how matrices can also be used to transform images in a rectangular coordinate system. This is particularly helpful to computer programmers who write software for computer gaming.

Classroom Example: p. 874
Exercise 80

EXAMPLE 9   Using Matrix Operations to Transform a Graph

Consider the triangle shown in Figure 9-6. The vertices of the triangle can be represented in a matrix where the first row represents the x-coordinates and the second row represents the corresponding y-coordinates.

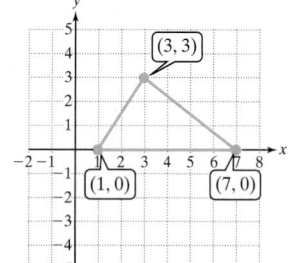

$$A = \begin{bmatrix} 1 & 3 & 7 \\ 0 & 3 & 0 \end{bmatrix} \begin{matrix} \longleftarrow x\text{-coordinates} \\ \longleftarrow y\text{-coordinates} \end{matrix}$$

a. Use addition of matrices to shift the triangle 3 units to the left and 2 units upward.

Figure 9-6

b. Find the product $\begin{bmatrix} -1 & 0 \\ 0 & 1 \end{bmatrix} \cdot A$ and determine the effect on the graph.

c. Find the product $\begin{bmatrix} 1 & 0 \\ 0 & -1 \end{bmatrix} \cdot A$ and determine the effect on the graph.

Solution:

a. To shift the graph to the left, subtract 3 from each x-coordinate.
To shift the graph upward, add 2 to each y-coordinate.

$$\begin{bmatrix} 1 & 3 & 7 \\ 0 & 3 & 0 \end{bmatrix} + \begin{bmatrix} -3 & -3 & -3 \\ 2 & 2 & 2 \end{bmatrix} \begin{matrix} \longleftarrow \text{This row subtracts 3 from each } x\text{-coordinate.} \\ \longleftarrow \text{This row adds 2 to each } y\text{-coordinate.} \end{matrix}$$

$$= \begin{bmatrix} -2 & 0 & 4 \\ 2 & 5 & 2 \end{bmatrix}$$

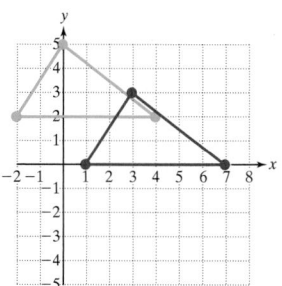

Figure 9-7

The vertices of the new triangle are $(-2, 2)$, $(0, 5)$, and $(4, 2)$. See Figure 9-7.

b. $\begin{bmatrix} -1 & 0 \\ 0 & 1 \end{bmatrix} \cdot A = \begin{bmatrix} -1 & 0 \\ 0 & 1 \end{bmatrix}\begin{bmatrix} 1 & 3 & 7 \\ 0 & 3 & 0 \end{bmatrix}$

$$= \begin{bmatrix} -1(1) + 0(0) & -1(3) + 0(3) & -1(7) + 0(0) \\ 0(1) + 1(0) & 0(3) + 1(3) & 0(7) + 1(0) \end{bmatrix}$$

$$= \begin{bmatrix} -1 & -3 & -7 \\ 0 & 3 & 0 \end{bmatrix} \quad \begin{matrix} \text{The corresponding points are } (-1, 0), \\ (-3, 3), \text{ and } (-7, 0). \end{matrix}$$

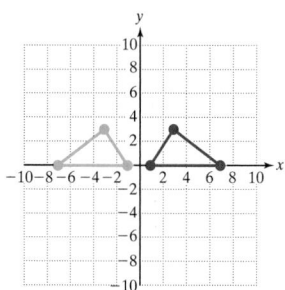

Figure 9-8

The triangle has been reflected across the y-axis. See Figure 9-8.

c. $\begin{bmatrix} 1 & 0 \\ 0 & -1 \end{bmatrix} \cdot A = \begin{bmatrix} 1 & 0 \\ 0 & -1 \end{bmatrix}\begin{bmatrix} 1 & 3 & 7 \\ 0 & 3 & 0 \end{bmatrix}$

$$= \begin{bmatrix} 1(1) + 0(0) & 1(3) + 0(3) & 1(7) + 0(0) \\ 0(1) + -1(0) & 0(3) + -1(3) & 0(7) + -1(0) \end{bmatrix}$$

$$= \begin{bmatrix} 1 & 3 & 7 \\ 0 & -3 & 0 \end{bmatrix} \quad \begin{matrix} \text{The corresponding points are} \\ (1, 0), (3, -3), \text{ and } (7, 0). \end{matrix}$$

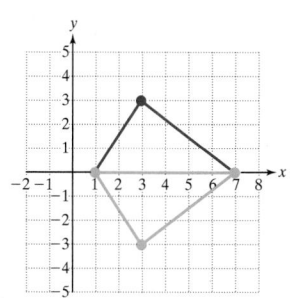

Figure 9-9

The triangle has been reflected across the x-axis. See Figure 9-9.

Answer

9. $\begin{bmatrix} -1 & -3 & -7 \\ 0 & -3 & 0 \end{bmatrix}$; The triangle is reflected across both the x- and y-axes.

Skill Practice 9   Use matrix A from Example 9. Find the product $\begin{bmatrix} -1 & 0 \\ 0 & -1 \end{bmatrix} \cdot A$ and determine the effect on the graph of the triangle in Figure 9-6.

SECTION 9.3   Practice Exercises

Prerequisite Review

R.1. Identify the additive inverse of 9. -9 **R.2.** Apply the commutative property of addition.
$$-9 + x \quad x + (-9)$$

R.3. Solve the equation. $5z - 2 = -22$ $\{-4\}$

Concept Connections

1. If the ____order____ of a matrix is $p \times q$, then p represents the number of ____rows____ and q represents the number of ____columns____.

2. A matrix with the same number of rows and columns is called a ____square____ matrix.

3. What are the requirements for two matrices to be equal?   The order of the matrices must be the same, and the corresponding elements must be equal.

4. An $m \times n$ matrix whose elements are all zero is called a ____zero____ matrix.

5. To multiply two matrices A and B, the number of ____columns____ of A must equal the number of ____rows____ of B.

6. If A is a 5×3 matrix and B is a 3×7 matrix, then the product AB will be a matrix of order ____5×7____. The product BA (is/is not) defined.   is not

7. True or false: Matrix multiplication is a commutative operation.   False

8. True or false: If a row matrix A and a column matrix B have the same number of elements, then the product AB is defined.   True

Objective 1: Determine the Order of a Matrix

9. What is a row matrix?
   A row matrix is a matrix with only one row.

10. What is a column matrix?
    A column matrix is a matrix with only one column.

For Exercises 11–16,

   a. Give the order of the matrix.

   b. Classify the matrix as a square matrix, row matrix, column matrix, or none of these. (**See Example 1**)

11. $\begin{bmatrix} 3 & 5 & -1 \\ \frac{1}{2} & \sqrt{3} & 1.7 \end{bmatrix}$ **a.** 2×3 **b.** None of these

12. $\begin{bmatrix} 1 & 5 & 6 & 2 \\ -1 & 3 & -\frac{2}{3} & \pi \\ 0 & 1 & -6.1 & 12 \end{bmatrix}$ **a.** 3×4 **b.** None of these

13. $\begin{bmatrix} 3 \\ 1 \\ 7 \end{bmatrix}$ **a.** 3×1 **b.** Column matrix

14. $\begin{bmatrix} 2.4 & 6.9 \end{bmatrix}$ **a.** 1×2 **b.** Row matrix

15. $\begin{bmatrix} 4 & 2 \\ 8 & 4 \end{bmatrix}$ **a.** 2×2 **b.** Square matrix

16. $\begin{bmatrix} -4 & 10 & 3 & 0 \\ 0 & 1 & 0 & 0 \\ 7 & 0 & 9 & -1 \\ 0 & 2 & 0 & 1 \end{bmatrix}$ **a.** 4×4 **b.** Square matrix

For Exercises 17–22, determine the value of the given element of matrix $A = [a_{ij}]$. (See Example 2)

$$A = \begin{bmatrix} 3 & -6 & \frac{1}{3} \\ 2 & 4 & 0 \\ \sqrt{5} & 11 & 8.6 \\ \frac{1}{2} & 4 & 2 \end{bmatrix}$$

17. a_{31} $\sqrt{5}$

18. a_{32} 11

19. a_{13} $\frac{1}{3}$

20. a_{23} 0

21. a_{43} 2

22. a_{42} 4

Objective 2: Add and Subtract Matrices

23. Given $A = \begin{bmatrix} 2 & x \\ z & -5 \end{bmatrix}$ and $B = \begin{bmatrix} y & 4 \\ 10 & -5 \end{bmatrix}$, for what values of x, y, and z will $A = B$? $x = 4, y = 2, z = 10$

24. Given $A = \begin{bmatrix} 4 & a \\ b & 7 \end{bmatrix}$ and $B = \begin{bmatrix} c & 12 \\ -1 & 6 \end{bmatrix}$, is it possible that $A = B$? Explain.   No; The elements in row 2, column 2 are not equal.

25. Given $B = \begin{bmatrix} -4 & 6 & 9 \\ \frac{3}{5} & 1 & 7 \end{bmatrix}$, find the additive inverse of B. $\begin{bmatrix} 4 & -6 & -9 \\ -\frac{3}{5} & -1 & -7 \end{bmatrix}$

26. Given $C = \begin{bmatrix} -1 & 6 \\ \sqrt{3} & 9 \end{bmatrix}$, find the additive inverse of C. $\begin{bmatrix} 1 & -6 \\ -\sqrt{3} & -9 \end{bmatrix}$

For Exercises 27–32, add or subtract the given matrices if possible. (See Example 3)

$$A = \begin{bmatrix} 6 & -1 \\ 7 & \frac{1}{2} \\ 2 & \sqrt{2} \end{bmatrix} \quad B = \begin{bmatrix} -9 & 2 \\ 6.2 & 2 \\ \frac{1}{3} & \sqrt{8} \end{bmatrix} \quad C = \begin{bmatrix} 11 & 4 \\ 1 & -\frac{1}{3} \\ 1 & 6 \end{bmatrix} \quad D = \begin{bmatrix} 2 & 3 & 8 \\ -1 & 6 & \frac{1}{6} \end{bmatrix}$$

27. $A + B$ $\quad \begin{bmatrix} -3 & 1 \\ 13.2 & \frac{5}{2} \\ \frac{7}{3} & 3\sqrt{2} \end{bmatrix}$

28. $A + C$ $\quad \begin{bmatrix} 17 & 3 \\ 8 & \frac{1}{6} \\ 3 & 6+\sqrt{2} \end{bmatrix}$

29. $C - A + B$ $\quad \begin{bmatrix} -4 & 7 \\ 0.2 & \frac{7}{6} \\ -\frac{2}{3} & 6+\sqrt{2} \end{bmatrix}$

30. $B - A - C$ $\quad \begin{bmatrix} -26 & -1 \\ -1.8 & \frac{11}{6} \\ -\frac{8}{3} & \sqrt{2}-6 \end{bmatrix}$

31. $B + D$ Not possible

32. $C + D$ Not possible

Objective 3: Multiply a Matrix by a Scalar

33. Explain how to multiply a matrix by a scalar. Multiply each element in the matrix by the scalar.

34. Given matrix A, explain how to find its additive inverse $-A$. Multiply each element of A by -1.

For Exercises 35–42, use $A = \begin{bmatrix} 2 & 4 & -9 \\ 1 & \sqrt{3} & \frac{1}{2} \end{bmatrix}$ **and** $B = \begin{bmatrix} -1 & 0 & 4 \\ 2 & 9 & \frac{2}{3} \end{bmatrix}$. **(See Example 4)**

35. $3A$ $\begin{bmatrix} 6 & 12 & -27 \\ 3 & 3\sqrt{3} & \frac{3}{2} \end{bmatrix}$

36. $-6B$ $\begin{bmatrix} 6 & 0 & -24 \\ -12 & -54 & -4 \end{bmatrix}$

37. $-2A - 7B$ $\begin{bmatrix} 3 & -8 & -10 \\ -16 & -2\sqrt{3}-63 & -\frac{17}{3} \end{bmatrix}$

38. $4A - 3B$ $\begin{bmatrix} 11 & 16 & -48 \\ -2 & 4\sqrt{3}-27 & 0 \end{bmatrix}$

39. $-4(A + B)$ $\begin{bmatrix} -4 & -16 & 20 \\ -12 & -4\sqrt{3}-36 & -\frac{14}{3} \end{bmatrix}$

40. $2(A - B)$ $\begin{bmatrix} 6 & 8 & -26 \\ -2 & 2\sqrt{3}-18 & -\frac{1}{3} \end{bmatrix}$

41. $-3A + 5(A - B)$

42. $-8A - 2(A + B)$

For Exercises 43–48, use $A = \begin{bmatrix} 1 & 6 \\ 4 & -2 \end{bmatrix}$ **and** $B = \begin{bmatrix} 2 & -4 \\ 6 & 9 \end{bmatrix}$ **and solve for** X. **(See Example 5)**

43. $2X - B = A$ $\quad X = \begin{bmatrix} \frac{3}{2} & 1 \\ 5 & \frac{7}{2} \end{bmatrix}$

44. $3X + A = B$ $\quad X = \begin{bmatrix} \frac{1}{3} & -\frac{10}{3} \\ \frac{2}{3} & \frac{11}{3} \end{bmatrix}$

45. $A + 5X = B$ $\quad X = \begin{bmatrix} \frac{1}{5} & -2 \\ \frac{2}{5} & \frac{11}{5} \end{bmatrix}$

46. $B - 4X = A$ $\quad X = \begin{bmatrix} \frac{1}{4} & -\frac{5}{2} \\ \frac{1}{2} & \frac{11}{4} \end{bmatrix}$

47. $2A - B = 10X$ $\quad X = \begin{bmatrix} 0 & \frac{8}{5} \\ \frac{1}{5} & -\frac{13}{10} \end{bmatrix}$

48. $3B - A = -2X$ $\quad X = \begin{bmatrix} -\frac{5}{2} & 9 \\ -7 & -\frac{29}{2} \end{bmatrix}$

Objective 4: Multiply Matrices

49. Given that A is a 4×2 matrix and B is a 2×1 matrix,
a. Is AB defined? If so, what is the order of AB? Yes; 4×1
b. Is BA defined? If so, what is the order of BA? No

50. Given that C is a 3×7 matrix and D is a 7×2 matrix,
a. Is CD defined? If so, what is the order of CD? Yes; 3×2
b. Is DC defined? If so, what is the order of DC? No

51. Given that E is a 5×1 matrix and F is a 1×5 matrix,
a. Is EF defined? If so, what is the order of EF? Yes; 5×5
b. Is FE defined? If so, what is the order of FE? Yes; 1×1

52. Given that G is a 1×6 matrix and H is a 6×1 matrix,
a. Is GH defined? If so, what is the order of GH? Yes; 1×1
b. Is HG defined? If so, what is the order of HG? Yes; 6×6

For Exercises 53–64, (See Examples 6–7)
a. Find AB if possible.
b. Find BA if possible.
c. Find A^2 if possible. (*Hint:* $A^2 = A \cdot A$.)

53. $A = \begin{bmatrix} 2 & 3 \\ 5 & 7 \end{bmatrix}$ and $B = \begin{bmatrix} 1 & 4 \\ -1 & 3 \end{bmatrix}$

54. $A = \begin{bmatrix} 1 & -6 \\ 5 & 10 \end{bmatrix}$ and $B = \begin{bmatrix} -2 & 3 \\ 7 & -1 \end{bmatrix}$

55. $A = \begin{bmatrix} 2 & 4 \\ -6 & 3 \\ 1 & 7 \end{bmatrix}$ and $B = \begin{bmatrix} 1 & 4 & -1 \\ -2 & 0 & 10 \end{bmatrix}$

56. $A = \begin{bmatrix} 1 & 3 & 4 \\ -2 & 5 & 6 \end{bmatrix}$ and $B = \begin{bmatrix} 1 & 3 \\ 9 & 2 \\ 0 & 4 \end{bmatrix}$

57. $A = \begin{bmatrix} -9 & 2 & 3 \\ -1 & 5 & 4 \\ 0 & 1 & 7 \end{bmatrix}$ and $B = \begin{bmatrix} -1 \\ 5 \\ 0 \end{bmatrix}$

58. $A = \begin{bmatrix} 2 \\ -1 \\ 3 \end{bmatrix}$ and $B = \begin{bmatrix} 2 & 7 & -1 \\ 0 & 4 & 1 \\ 0 & 3 & -6 \end{bmatrix}$

59. $A = \begin{bmatrix} 1 & \frac{1}{2} \end{bmatrix}$ and $B = \begin{bmatrix} -\frac{1}{3} \\ 2 \end{bmatrix}$

60. $A = \begin{bmatrix} 4 \\ \frac{3}{4} \\ 1 \end{bmatrix}$ and $B = \begin{bmatrix} -5 & \frac{1}{2} & \frac{1}{3} \end{bmatrix}$

61. $A = \begin{bmatrix} 4 \\ -6 \end{bmatrix}$ and $B = \begin{bmatrix} 1 & 2 & 5 & 6 \end{bmatrix}$

62. $A = [4]$ and $B = \begin{bmatrix} -3 & 4 & 1 \end{bmatrix}$ a. $\begin{bmatrix} -12 & 16 & 4 \end{bmatrix}$ b. Not possible c. $[16]$

63. $A = [-5]$ and $B = [5]$ a. $[-25]$ b. $[-25]$ c. $[25]$

64. $A = [6]$ and $B = [-2]$ a. $[-12]$ b. $[-12]$ c. $[36]$

For Exercises 65–68, find AB and BA.

65. $A = \begin{bmatrix} 3.1 & -2.3 \\ 1.1 & 6.5 \end{bmatrix}$ and $B = \begin{bmatrix} 1 & 0 \\ 0 & 1 \end{bmatrix} \begin{bmatrix} 3.1 & -2.3 \\ 1.1 & 6.5 \end{bmatrix}$

66. $A = \begin{bmatrix} 1 & 0 \\ 0 & 1 \end{bmatrix}$ and $B = \begin{bmatrix} 0.05 & -0.07 \\ 0.16 & 0.09 \end{bmatrix} \begin{bmatrix} 0.05 & -0.07 \\ 0.16 & 0.09 \end{bmatrix}$

67. $A = \begin{bmatrix} 1 & 0 & 0 \\ 0 & 1 & 0 \\ 0 & 0 & 1 \end{bmatrix}$ and $B = \begin{bmatrix} \frac{9}{5} & -3 & \sqrt{6} \\ 5 & \frac{1}{2} & 2 \\ 3 & 0 & 1 \end{bmatrix} \begin{bmatrix} \frac{9}{5} & -3 & \sqrt{6} \\ 5 & \frac{1}{2} & 2 \\ 3 & 0 & 1 \end{bmatrix}$

68. $A = \begin{bmatrix} \frac{2}{3} & 5 & \sqrt{2} \\ 1 & 0 & 3 \\ -\frac{1}{4} & 0 & 8 \end{bmatrix}$ and $B = \begin{bmatrix} 1 & 0 & 0 \\ 0 & 1 & 0 \\ 0 & 0 & 1 \end{bmatrix} \begin{bmatrix} \frac{2}{3} & 5 & \sqrt{2} \\ 1 & 0 & 3 \\ -\frac{1}{4} & 0 & 8 \end{bmatrix}$

Objective 5: Apply Operations on Matrices

69. Matrix D gives the dealer invoice prices for sedan and hatchback models of a car with manual transmission or automatic transmission. Matrix M gives the MSRP (manufacturer's suggested retail price) for the cars.

	Sedan	Hatchback	
$D =$	$29,000	$27,500	Manual
	$28,500	$26,900	Automatic

	Sedan	Hatchback	
$M =$	$32,600	$29,900	Manual
	$31,900	$28,900	Automatic

a. Compute $M - D$ and interpret the result.

b. A buyer thinks that a fair price is 6% above dealer invoice. Use scalar multiplication to determine a matrix F that gives the fair price for these cars for each type of transmission.

70. In matrix C, a coffee shop records the cost to produce a cup of standard Columbian coffee and the cost to produce a cup of hot chocolate. Matrix P contains the selling prices to the customer.

	Coffee	Chocolate	
$C =$	$0.90	$0.84	Small
	$1.26	$1.15	Medium
	$1.64	$1.50	Large

	Coffee	Chocolate	
$P =$	$3.05	$2.25	Small
	$3.65	$3.05	Medium
	$4.15	$3.65	Large

a. Compute $P - C$ and interpret its meaning.

b. If the tax rate in a certain city is 7%, use scalar multiplication to find a matrix F that gives the final price to the customer (including sales tax) for both beverages for each size. Round each entry to the nearest cent.

71. A street vendor at a parade sells fresh lemonade, soda, bottled water, and iced-tea, and the unit price for each item is given in matrix P. The number of units sold of each item is given in matrix N. Compute NP and interpret the result.

$$N = \begin{bmatrix} \text{Lemonade} & \text{Soda} & \text{Water} & \text{Tea} \\ 150 & 270 & 440 & 80 \end{bmatrix},$$

$$P = \begin{bmatrix} \$2.50 \\ \$1.50 \\ \$2.00 \\ \$2.00 \end{bmatrix} \begin{matrix} \text{Lemonade} \\ \text{Soda} \\ \text{Water} \\ \text{Tea} \end{matrix}$$

[$1820]; The value $1820 represents the total revenue from the sale of these four items.

72. A math course has 4 exams weighted 20%, 25%, 25%, and 30%, respectively, toward the final grade. Suppose that a student earns grades of 75, 84, 92, and 86, respectively, on the four exams. The weights are given in matrix W and the test grades are given in matrix G. Compute WG and interpret the result.

$$W = \begin{bmatrix} \text{Test 1} & \text{Test 2} & \text{Test 3} & \text{Test 4} \\ 0.20 & 0.25 & 0.25 & 0.30 \end{bmatrix}, \quad G = \begin{bmatrix} 75 \\ 84 \\ 92 \\ 86 \end{bmatrix} \begin{matrix} \text{Test 1} \\ \text{Test 2} \\ \text{Test 3} \\ \text{Test 4} \end{matrix}$$

[84.8]; The value 84.8 is the student's overall course grade.

73. An electronics store sells three models of tablets. The number of each model sold during "Black Friday" weekend is given in matrix A. The selling price and profit for each model are given in matrix B. (**See Example 8**)

	Model A	B	C	
$A =$	84	70	32	Friday
	62	48	16	Saturday
	70	40	12	Sunday

	Selling Price	Profit	
$B =$	$499	$200	A
	$599	$240	B Model
	$629	$280	C

a. Compute AB and interpret the result.

b. Determine the total revenue for Sunday.

c. Determine the total profit for the 3-day period for these three models.

74. A gas station manager records the number of gallons of Regular, Plus, and Premium gasoline sold during the week (Monday–Friday) and on the weekends (Saturday–Sunday) in matrix A. The selling price and profit for 1 gal of each type of gasoline is given in matrix B.

	Regular	Plus	Premium	
$A =$	4600	1850	720	Weekdays
	2300	620	480	Weekend

	Selling Price	Profit	
$B =$	$3.59	$0.21	Regular
	$3.79	$0.24	Plus
	$4.19	$0.18	Premium

a. Compute AB and interpret the result.

b. Determine the profit for the weekend.

c. Determine the revenue for the entire week.

75. The labor costs per hour for an electrician, plumber, and air-conditioning/heating expert are given in matrix L. The time required from each specialist for three new model homes is given in matrix T.

$$
\begin{array}{l}
\text{Cost/hr} \\
L = \begin{bmatrix} \$45 \\ \$38 \\ \$35 \end{bmatrix} \begin{array}{l} \textbf{Electrician} \\ \textbf{Plumber} \\ \textbf{AC/heating} \end{array}
\end{array}
$$

$$
\begin{array}{l}
\qquad\quad\text{Time (hr)} \\
\;\;\textbf{Electrician}\;\;\textbf{Plumber}\;\;\textbf{AC/heating} \\
T = \begin{bmatrix} 22 & 16 & 14 \\ 28 & 21 & 18 \\ 18 & 14 & 9 \end{bmatrix} \begin{array}{l} \textbf{Model 1} \\ \textbf{Model 2} \\ \textbf{Model 3} \end{array}
\end{array}
$$

a. Which product LT or TL gives the total cost for these three services for each model?
TL (The product LT is not possible.)

b. Find a matrix that gives the total cost for these three services for each model.

77. A student researches the cost for three cell phone plans. Matrix C contains the cost per text message and the cost per minute over the maximum number of minutes allowed in each plan. Matrix N_1 contains the number of text messages and the number of minutes over the maximum incurred for 1 month. Matrix N_3 represents the number of text messages and number of minutes over the maximum for 3 months.

$$
\begin{array}{l}
\;\textbf{Cost/text}\quad\textbf{Cost/min} \\
C = \begin{bmatrix} \$0.25 & \$0.40 \\ \$0 & \$0.40 \\ \$0.10 & \$0 \end{bmatrix} \begin{array}{l} \textbf{Plan A} \\ \textbf{Plan B} \\ \textbf{Plan C} \end{array}
\end{array}
$$

$$
N_1 = \begin{bmatrix} 24 \\ 100 \end{bmatrix} \begin{array}{l} \textbf{Number of texts} \\ \textbf{Minutes over} \end{array}
$$

$$
\begin{array}{l}
\textbf{Month}\quad\textbf{Month}\quad\textbf{Month} \\
\;\;\;\textbf{1}\qquad\;\;\textbf{2}\qquad\;\;\textbf{3} \\
N_3 = \begin{bmatrix} 24 & 56 & 30 \\ 100 & 24 & 0 \end{bmatrix} \begin{array}{l} \textbf{Number of texts} \\ \textbf{Minutes over} \end{array}
\end{array}
$$

a. Find the product CN_1 and interpret its meaning.

b. Find the product CN_3 and interpret its meaning.

a. $CN_1 = \begin{bmatrix} \$46 \\ \$40 \\ \$2.40 \end{bmatrix}$; The matrix CN_1 represents the additional cost for

24 text messages and 100 extra minutes for each of the cell phone plans.

b. $CN_3 = \begin{bmatrix} \$46 & \$23.60 & \$7.50 \\ \$40 & \$9.60 & \$0 \\ \$2.40 & \$5.60 & \$3 \end{bmatrix}$; The matrix CN_3 represents the

additional cost per month for each plan. For example, row 1 represents the cost for plan A for months 1, 2, and 3, respectively.

76. The number of calories burned per hour for three activities is given in matrix N for a 140-lb woman training for a triathlon. The time spent on each activity for two different training days is given in matrix T.

$$
\begin{array}{l}
\text{Calories/hr} \\
N = \begin{bmatrix} 540 \\ 400 \\ 360 \end{bmatrix} \begin{array}{l} \textbf{Running} \\ \textbf{Bicycling} \\ \textbf{Swimming} \end{array}
\end{array}
$$

$$
\begin{array}{l}
\qquad\qquad\quad\text{Time} \\
\;\textbf{Running}\;\;\textbf{Bicycling}\;\;\textbf{Swimming} \\
T = \begin{bmatrix} 45\text{ min} & 1\text{ hr} & 30\text{ min} \\ 1\text{ hr} & 1\text{ hr }30\text{ min} & 45\text{ min} \end{bmatrix} \begin{array}{l} \textbf{Day 1} \\ \textbf{Day 2} \end{array}
\end{array}
$$

a. Which product NT or TN gives the total number of calories burned from these activities for each day?
TN (The product NT is not possible.)

b. Find a matrix that gives the total number of calories burned from these activities for each day. $TN = \begin{bmatrix} 985 \\ 1410 \end{bmatrix}$

78. Refer to Exercise 77. Suppose that matrix B represents the base cost for each cell phone plan and matrix T represents the tax for each plan.

$$
B = \begin{bmatrix} \$39.99 \\ \$49.99 \\ \$59.99 \end{bmatrix} \begin{array}{l} \textbf{Plan A} \\ \textbf{Plan B} \\ \textbf{Plan C} \end{array}
$$

$$
T = \begin{bmatrix} \$11.96 \\ \$13.04 \\ \$14.91 \end{bmatrix} \begin{array}{l} \textbf{Plan A} \\ \textbf{Plan B} \\ \textbf{Plan C} \end{array}
$$

a. Compute $B + CN_1 + T$ and interpret its meaning.

b. Which cell phone plan is the least expensive if the student has 60 text messages and talks 20 min more than the maximum?

a. $B + CN_1 + T = \begin{bmatrix} \$97.95 \\ \$103.03 \\ \$77.30 \end{bmatrix}$; This matrix represents the

total cost (including the base cost and tax) for each plan for 24 text messages and 100 additional minutes.

b. Plan B

79. a. Write a matrix A that represents the coordinates of the vertices of the triangle. Place the x-coordinate of each point in the first row of A and the corresponding y-coordinate in the second row of A.
(**See Example 9**)

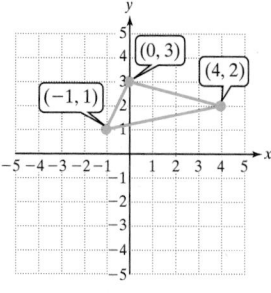

b. Use addition of matrices to shift the triangle 2 units to the right and 4 units downward.

c. Find the product $\begin{bmatrix} -1 & 0 \\ 0 & 1 \end{bmatrix} \cdot A$ and explain the effect on the graph of the triangle.

d. Find the product $\begin{bmatrix} 1 & 0 \\ 0 & -1 \end{bmatrix} \cdot A$ and explain the effect on the graph of the triangle.

e. Find $\begin{bmatrix} 1 & 0 \\ 0 & -1 \end{bmatrix} \cdot A + \begin{bmatrix} -1 & -1 & -1 \\ 2 & 2 & 2 \end{bmatrix}$ and explain the effect on the graph of the triangle.

80. a. Write a matrix A that represents the coordinates of the vertices of the triangle. Place the x-coordinate of each point in the first row of A and the corresponding y-coordinate in the second row of A.

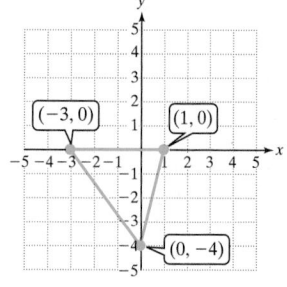

b. Use addition of matrices to shift the triangle 1 unit to the left and 3 units upward.

c. Find the product $\begin{bmatrix} -1 & 0 \\ 0 & 1 \end{bmatrix} \cdot A$ and explain the effect on the graph of the triangle.

d. Find the product $\begin{bmatrix} 1 & 0 \\ 0 & -1 \end{bmatrix} \cdot A$ and explain the effect on the graph of the triangle.

e. Find $\begin{bmatrix} -1 & 0 \\ 0 & 1 \end{bmatrix} \cdot A + \begin{bmatrix} 2 & 2 & 2 \\ -5 & -5 & -5 \end{bmatrix}$ and explain the effect on the graph of the triangle.

81. a. Write a matrix A that represents the coordinates of the vertices of the quadrilateral.

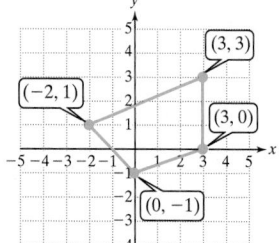

b. What operation on A will shift the graph of the quadrilateral 3 units downward?

c. What operation on A will shift the graph 4 units to the left?

d. Use matrix multiplication to reflect the graph across the x-axis.

e. Use matrix multiplication to reflect the graph across the y-axis.

82. a. Write a matrix A that represents the coordinates of the vertices of the quadrilateral.

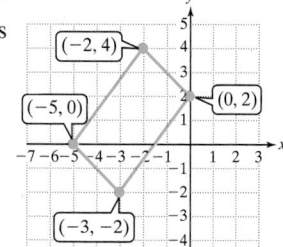

b. What operation on A will shift the graph of the quadrilateral 6 units upward?

c. What operation on A will shift the graph 2 units to the right?

d. Use matrix multiplication to reflect the graph across the x-axis.

e. Use matrix multiplication to reflect the graph across the y-axis.

83. a. Write a matrix A that represents the coordinates of the vertices of the triangle.

b. Multiply $\begin{bmatrix} \frac{\sqrt{3}}{2} & -\frac{1}{2} \\ \frac{1}{2} & \frac{\sqrt{3}}{2} \end{bmatrix} \cdot A$ and round each entry to 1 decimal place.

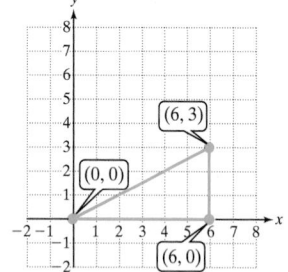

c. Graph the figure represented by the matrix from part (b). What effect does this product have on the graph of the triangle?

84. a. Write a matrix A that represents the coordinates of the vertices of the triangle.

b. Multiply $\begin{bmatrix} \frac{1}{2} & -\frac{\sqrt{3}}{2} \\ \frac{\sqrt{3}}{2} & \frac{1}{2} \end{bmatrix} \cdot A$ and round each entry to 1 decimal place.

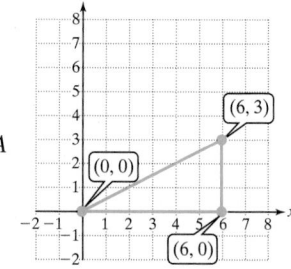

c. Graph the figure represented by the matrix from part (b). What effect does this product have on the graph of the triangle?

For Exercises 85–86, use the following gray scale.

0	1	2	3	4	5	6	7
white	light gray	medium light	gray	medium gray	medium dark	dark gray	black

85. a. Write a 5×3 matrix that represents the letter E in dark gray on a white background.

  b. Use matrix addition to change the pixels so that the letter E is medium dark on a light gray background.

86. a. Write a 5×3 matrix that represents the letter T in medium gray on a medium light background.

  b. Use matrix addition to change the pixels so that the letter T is dark gray on a light gray background.

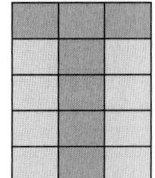

Mixed Exercises

For Exercises 87–92, use matrices A, B, and C to prove the given properties. Assume that the elements within A, B, and C are real numbers.

$$A = \begin{bmatrix} a_1 & a_2 \\ a_3 & a_4 \end{bmatrix} \quad B = \begin{bmatrix} b_1 & b_2 \\ b_3 & b_4 \end{bmatrix} \quad C = \begin{bmatrix} c_1 & c_2 \\ c_3 & c_4 \end{bmatrix}$$

87. Commutative property of matrix addition

$$A + B = B + A$$

88. Associative property of matrix addition

$$A + (B + C) = (A + B) + C$$

89. Inverse property of matrix addition

$$A + (-A) = 0$$

90. Identity property of matrix addition

$$A + 0 = A$$

91. Associative property of scalar multiplication

$$s(tA) = (st)A$$

92. Distributive property of scalar multiplication

$$t(A + B) = tA + tB$$

93. Given $A = \begin{bmatrix} i & 0 \\ 0 & i \end{bmatrix}$, find A^2, A^3, and A^4.

 $\left(Hint\text{: Recall that } i = \sqrt{-1}.\right)$ Discuss the similarities between A^n and i^n, where n is a positive integer.

94. Given $B = \begin{bmatrix} 0 & -i \\ i & 0 \end{bmatrix}$, find B^2. $\begin{bmatrix} 1 & 0 \\ 0 & 1 \end{bmatrix}$

95. a. For real numbers a, b, c, and d, find the product.

$$\begin{bmatrix} a & 0 \\ 0 & b \end{bmatrix} \begin{bmatrix} c & 0 \\ 0 & d \end{bmatrix} \quad \begin{bmatrix} ac & 0 \\ 0 & bd \end{bmatrix}$$

  b. Based on the form of the product of part (a), compute the following product mentally.

$$\begin{bmatrix} 3 & 0 \\ 0 & 7 \end{bmatrix} \begin{bmatrix} 1 & 0 \\ 0 & 2 \end{bmatrix} \quad \begin{bmatrix} 3 & 0 \\ 0 & 14 \end{bmatrix}$$

96. Find the product.

$$\begin{bmatrix} a & a \\ b & b \end{bmatrix} \begin{bmatrix} 1 & -1 \\ -1 & 1 \end{bmatrix} \quad \begin{bmatrix} 0 & 0 \\ 0 & 0 \end{bmatrix}$$

Write About It

97. Explain why the product of a 3×2 matrix by a 3×2 matrix is undefined. The number of columns in the first matrix is not equal to the number of rows in the second matrix.

99. Given a matrix, $A = [a_{ij}]$, explain how to find the additive inverse of A. To find $-A$, take the additive inverse of each individual element of A. That is, $-A = [-a_{ij}]$.

98. Explain how to add or subtract matrices.
To add or subtract matrices first confirm that the matrices have the same order. Then add or subtract their corresponding elements.

100. Given two matrices A and B, how can you determine the order of AB, assuming that the product is defined?
The order of product AB will be the number of rows of A by the number of columns of B.

Technology Connections

For Exercises 101–104, refer to matrices A, B and C and perform the indicated operations on a calculator.

$$A = \begin{bmatrix} 1.05 & 3.9 \\ 4.12 & -9.4 \\ -2.4 & 1.5 \end{bmatrix} \quad B = \begin{bmatrix} -10 & 30 \\ 24 & -36 \\ 18 & -8 \end{bmatrix} \quad C = \begin{bmatrix} 6.2 \\ 4.9 \end{bmatrix}$$

101. $2.5A - 3.6B$

102. $-6.4(A + B)$

103. $-3AC$

104. $7.5BC$

OBJECTIVES

1. Identify Identity and Inverse Matrices
2. Determine the Inverse of a Matrix
3. Solve Systems of Linear Equations Using the Inverse of a Matrix

1. Identify Identity and Inverse Matrices

The identity element under the multiplication of real numbers is 1 because $a \cdot 1 = a$ and $1 \cdot a = a$. We now investigate a similar property for the product of square matrices. The **identity matrix** I_n for matrix multiplication is the $n \times n$ square matrix with 1's along the main diagonal and 0's for all other elements. For example:

The identity matrix of order 2

The identity matrix of order 3

$$I_2 = \begin{bmatrix} 1 & 0 \\ 0 & 1 \end{bmatrix} \quad \text{and} \quad I_3 = \begin{bmatrix} 1 & 0 & 0 \\ 0 & 1 & 0 \\ 0 & 0 & 1 \end{bmatrix}$$

For an $n \times n$ square matrix A, we have that

$$AI_n = A \quad \text{and} \quad I_nA = A \quad \text{(Identity property of matrix multiplication)}$$

In Example 1, we illustrate the identity property of matrix multiplication using a 2×2 matrix.

Classroom Example: p. 883
Exercise 10

EXAMPLE 1   Illustrating the Identity Property of Matrix Multiplication

Given $A = \begin{bmatrix} a & b \\ c & d \end{bmatrix}$ show that

a. $AI_2 = A$ **b.** $I_2A = A$

Solution:

a. $AI_2 = \begin{bmatrix} a & b \\ c & d \end{bmatrix}\begin{bmatrix} 1 & 0 \\ 0 & 1 \end{bmatrix} = \begin{bmatrix} a(1) + b(0) & a(0) + b(1) \\ c(1) + d(0) & c(0) + d(1) \end{bmatrix} = \begin{bmatrix} a & b \\ c & d \end{bmatrix}$ ✓

b. $I_2A = \begin{bmatrix} 1 & 0 \\ 0 & 1 \end{bmatrix}\begin{bmatrix} a & b \\ c & d \end{bmatrix} = \begin{bmatrix} 1(a) + 0(c) & 1(b) + 0(d) \\ 0(a) + 1(c) & 0(b) + 1(d) \end{bmatrix} = \begin{bmatrix} a & b \\ c & d \end{bmatrix}$ ✓

Skill Practice 1   Given $A = \begin{bmatrix} 3 & -4 \\ -2 & 10 \end{bmatrix}$ show that

a. $AI_2 = A$ **b.** $I_2A = A$

For a nonzero real number a, the multiplicative inverse of a is $\frac{1}{a}$ because $a \cdot \frac{1}{a} = 1$ and $\frac{1}{a} \cdot a = 1$. We define the multiplicative inverse of a square matrix in a similar fashion.

Answers

1. a. $\begin{bmatrix} 3 & -4 \\ -2 & 10 \end{bmatrix}\begin{bmatrix} 1 & 0 \\ 0 & 1 \end{bmatrix} = \begin{bmatrix} 3 & -4 \\ -2 & 10 \end{bmatrix}$

b. $\begin{bmatrix} 1 & 0 \\ 0 & 1 \end{bmatrix}\begin{bmatrix} 3 & -4 \\ -2 & 10 \end{bmatrix} = \begin{bmatrix} 3 & -4 \\ -2 & 10 \end{bmatrix}$

Multiplicative Inverse of a Square Matrix
Let A be an $n \times n$ matrix and let I_n be the identity matrix of order n. If there exists an $n \times n$ matrix A^{-1} such that $$AA^{-1} = I_n \quad \text{and} \quad A^{-1}A = I_n$$ then A^{-1} (read as "A inverse") is the **multiplicative inverse** of A.

Classroom Example: p. 883
Exercise 14

EXAMPLE 2   Determining Whether Two Matrices are Inverses

Determine whether $A = \begin{bmatrix} 3 & 2 \\ 7 & 5 \end{bmatrix}$ and $B = \begin{bmatrix} 5 & -2 \\ -7 & 3 \end{bmatrix}$ are inverses.

Solution:

We must show that $AB = I_2$ and $BA = I_2$.

$$AB = \begin{bmatrix} 3 & 2 \\ 7 & 5 \end{bmatrix}\begin{bmatrix} 5 & -2 \\ -7 & 3 \end{bmatrix} = \begin{bmatrix} 3(5) + 2(-7) & 3(-2) + 2(3) \\ 7(5) + 5(-7) & 7(-2) + 5(3) \end{bmatrix} = \begin{bmatrix} 1 & 0 \\ 0 & 1 \end{bmatrix} \checkmark$$

$$BA = \begin{bmatrix} 5 & -2 \\ -7 & 3 \end{bmatrix}\begin{bmatrix} 3 & 2 \\ 7 & 5 \end{bmatrix} = \begin{bmatrix} 5(3) + (-2)(7) & 5(2) + (-2)(5) \\ -7(3) + 3(7) & -7(2) + 3(5) \end{bmatrix} = \begin{bmatrix} 1 & 0 \\ 0 & 1 \end{bmatrix} \checkmark$$

Skill Practice 2

Determine whether $A = \begin{bmatrix} 4 & 3 \\ 13 & 10 \end{bmatrix}$ and $B = \begin{bmatrix} 10 & -3 \\ -13 & 4 \end{bmatrix}$ are inverses.

2. Determine the Inverse of a Matrix

In Example 2, we showed that two given matrices are inverses. Now we look at the task of finding the inverse of a matrix. In Example 3, we will find the inverse of a 2×2 matrix directly from the definition. Then, we develop a general procedure to find the inverse of an $n \times n$ matrix, if the inverse exists.

Classroom Example: p. 883
Exercise 20

EXAMPLE 3   Finding the Inverse of a Matrix

Given $A = \begin{bmatrix} 2 & 9 \\ 1 & 5 \end{bmatrix}$, find A^{-1}.

Solution:

We need to find a matrix $A^{-1} = \begin{bmatrix} x_1 & x_2 \\ x_3 & x_4 \end{bmatrix}$ such that $AA^{-1} = I_2$. That is,

$$\begin{bmatrix} 2 & 9 \\ 1 & 5 \end{bmatrix}\begin{bmatrix} x_1 & x_2 \\ x_3 & x_4 \end{bmatrix} = \begin{bmatrix} 1 & 0 \\ 0 & 1 \end{bmatrix} \longrightarrow \begin{bmatrix} 2x_1 + 9x_3 & 2x_2 + 9x_4 \\ x_1 + 5x_3 & x_2 + 5x_4 \end{bmatrix} = \begin{bmatrix} 1 & 0 \\ 0 & 1 \end{bmatrix}$$

Answer

2. Yes; $AB = I_2$ and $BA = I_2$.

For the matrices to be equal, their corresponding elements must be equal. This results in two systems of equations.

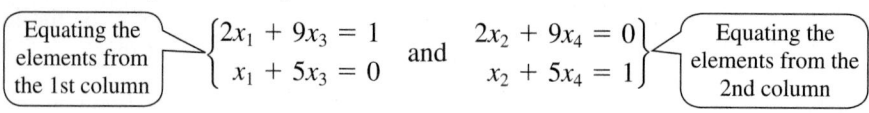

Equating the elements from the 1st column
$$\begin{cases} 2x_1 + 9x_3 = 1 \\ x_1 + 5x_3 = 0 \end{cases} \text{ and } \begin{cases} 2x_2 + 9x_4 = 0 \\ x_2 + 5x_4 = 1 \end{cases}$$
Equating the elements from the 2nd column

The corresponding augmented matrices are

$$\left[\begin{array}{cc|c} 2 & 9 & 1 \\ 1 & 5 & 0 \end{array}\right] \quad \text{and} \quad \left[\begin{array}{cc|c} 2 & 9 & 0 \\ 1 & 5 & 1 \end{array}\right]$$

We can solve both systems simultaneously by placing the identity matrix to the right of the array of coefficients.

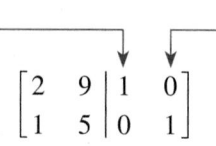

$$\left[\begin{array}{cc|cc} 2 & 9 & 1 & 0 \\ 1 & 5 & 0 & 1 \end{array}\right]$$

Apply Gauss-Jordan elimination.

$$\left[\begin{array}{cc|cc} 2 & 9 & 1 & 0 \\ 1 & 5 & 0 & 1 \end{array}\right] \xrightarrow{R_1 \Leftrightarrow R_2} \left[\begin{array}{cc|cc} 1 & 5 & 0 & 1 \\ 2 & 9 & 1 & 0 \end{array}\right] \xrightarrow{-2R_1 + R_2 \to R_2} \left[\begin{array}{cc|cc} 1 & 5 & 0 & 1 \\ 0 & -1 & 1 & -2 \end{array}\right]$$

$$\xrightarrow{-1R_2 \to R_2} \left[\begin{array}{cc|cc} 1 & 5 & 0 & 1 \\ 0 & 1 & -1 & 2 \end{array}\right] \xrightarrow{-5R_2 + R_1 \to R_1} \left[\begin{array}{cc|cc} 1 & 0 & 5 & -9 \\ 0 & 1 & -1 & 2 \end{array}\right]$$

This result represents the following two matrices and their corresponding systems of equations.

$$\begin{array}{cc} x_1 & x_3 \\ \downarrow & \downarrow \end{array}$$
$$\left[\begin{array}{cc|c} 1 & 0 & 5 \\ 0 & 1 & -1 \end{array}\right] \quad \text{and} \quad \left[\begin{array}{cc|c} 1 & 0 & -9 \\ 0 & 1 & 2 \end{array}\right]$$

$$\begin{array}{cc} x_2 & x_4 \\ \downarrow & \downarrow \end{array}$$

$x_1 = 5, x_3 = -1$ $x_2 = -9, x_4 = 2$

Avoiding Mistakes

The solution to Example 3 can be checked by verifying that $AA^{-1} = I_2$ and $A^{-1}A = I_2$.

Therefore, we have: $A^{-1} = \begin{bmatrix} x_1 & x_2 \\ x_3 & x_4 \end{bmatrix} = \begin{bmatrix} 5 & -9 \\ -1 & 2 \end{bmatrix}$

Skill Practice 3 Given $B = \begin{bmatrix} 9 & 7 \\ 5 & 4 \end{bmatrix}$, find B^{-1}.

In Example 3, we transformed the left side of the augmented matrix into the identity matrix. That is, we performed row operations to make the array on the left of the vertical bar have 1's along the main diagonal, and 0's elsewhere. In the process, we make the following important observation.

- The array of elements to the right of the vertical bar is the inverse of A.

This observation leads to a general procedure to find the multiplicative inverse of a matrix, provided that the inverse exists.

Finding the Multiplicative Inverse of a Square Matrix

Let A be an $n \times n$ matrix for which A^{-1} exists, and let I_n be the $n \times n$ identity matrix. To find A^{-1},

Step 1 Write a matrix of the form $[A \mid I_n]$.

Step 2 Perform row operations to write the matrix in the form $[I_n \mid B]$.

Step 3 The matrix B is A^{-1}.

Answer

3. $B^{-1} = \begin{bmatrix} 4 & -7 \\ -5 & 9 \end{bmatrix}$

element is the determinant of the resulting matrix obtained by deleting the ith row and jth column. For example, consider the matrix:

$$A = [a_{ij}] = \begin{bmatrix} 5 & -1 & 6 \\ 0 & -7 & 1 \\ 4 & 2 & 6 \end{bmatrix}$$

The element a_{11} is 5. The minor of this element is found by deleting the first row and first column and then evaluating the determinant of the remaining 2×2 matrix:

$$\begin{bmatrix} 5 & -1 & 6 \\ 0 & -7 & 1 \\ 4 & 2 & 6 \end{bmatrix} \qquad M_{11} = \begin{vmatrix} -7 & 1 \\ 2 & 6 \end{vmatrix} = (-7)(6) - (1)(2) = -44$$

The element a_{32} is 2. The minor of this element is found by deleting the third row and second column and then evaluating the determinant of the remaining 2×2 matrix:

$$\begin{bmatrix} 5 & -1 & 6 \\ 0 & -7 & 1 \\ 4 & 2 & 6 \end{bmatrix} \qquad M_{32} = \begin{vmatrix} 5 & 6 \\ 0 & 1 \end{vmatrix} = (5)(1) - (6)(0) = 5$$

Classroom Examples: p. 896
Exercises 18, 20

EXAMPLE 2   **Determining the Minor for Elements in a Matrix**

Find the minor for each element in the first column of the matrix.

$$\begin{bmatrix} 3 & 4 & -1 \\ 2 & -4 & 5 \\ 0 & 1 & -6 \end{bmatrix}$$

Solution:

For the element a_{11}:

$$\begin{bmatrix} 3 & 4 & -1 \\ 2 & -4 & 5 \\ 0 & 1 & -6 \end{bmatrix} \qquad M_{11} = \begin{vmatrix} -4 & 5 \\ 1 & -6 \end{vmatrix} = (-4)(-6) - (5)(1) = 19$$

For the element a_{21}:

$$\begin{bmatrix} 3 & 4 & -1 \\ 2 & -4 & 5 \\ 0 & 1 & -6 \end{bmatrix} \qquad M_{21} = \begin{vmatrix} 4 & -1 \\ 1 & -6 \end{vmatrix} = (4)(-6) - (-1)(1) = -23$$

For the element a_{31}:

$$\begin{bmatrix} 3 & 4 & -1 \\ 2 & -4 & 5 \\ 0 & 1 & -6 \end{bmatrix} \qquad M_{31} = \begin{vmatrix} 4 & -1 \\ -4 & 5 \end{vmatrix} = (4)(5) - (-1)(-4) = 16$$

Skill Practice 2   Find the minor for each element in the second row of the matrix from Example 2.

Next, we define the determinant for a 3×3 matrix.

Answer

2. $M_{21} = -23$; $M_{22} = -18$; $M_{23} = 3$

Determinant of a 3 × 3 Matrix

$$\begin{vmatrix} a_1 & b_1 & c_1 \\ a_2 & b_2 & c_2 \\ a_3 & b_3 & c_3 \end{vmatrix} = a_1 \begin{vmatrix} b_2 & c_2 \\ b_3 & c_3 \end{vmatrix} - a_2 \begin{vmatrix} b_1 & c_1 \\ b_3 & c_3 \end{vmatrix} + a_3 \begin{vmatrix} b_1 & c_1 \\ b_2 & c_2 \end{vmatrix} \text{ or equivalently,}$$

$$= a_1(b_2c_3 - c_2b_3) - a_2(b_1c_3 - c_1b_3) + a_3(b_1c_2 - c_1b_2) \quad \text{or}$$

$$= a_1b_2c_3 + b_1c_2a_3 + c_1a_2b_3 - a_3b_2c_1 - b_3c_2a_1 - c_3a_2b_1$$

From this definition, we see that the determinant of the given 3 × 3 matrix can be written as

$$a_1 \cdot (\text{minor of } a_1) - a_2 \cdot (\text{minor of } a_2) + a_3 \cdot (\text{minor of } a_3)$$

Exercise 26

EXAMPLE 3 **Evaluating the Determinant of a 3 × 3 Matrix**

Evaluate the determinant of the matrix. $A = \begin{bmatrix} 2 & 4 & 2 \\ 1 & -3 & 0 \\ -5 & 5 & -1 \end{bmatrix}$

Solution:

$$|A| = \begin{vmatrix} 2 & 4 & 2 \\ 1 & -3 & 0 \\ -5 & 5 & -1 \end{vmatrix} = 2 \cdot \overbrace{\begin{vmatrix} -3 & 0 \\ 5 & -1 \end{vmatrix}}^{\text{minor of 2}} - (1) \overbrace{\begin{vmatrix} 4 & 2 \\ 5 & -1 \end{vmatrix}}^{\text{minor of 1}} + (-5) \cdot \overbrace{\begin{vmatrix} 4 & 2 \\ -3 & 0 \end{vmatrix}}^{\text{minor of } -5}$$

$$= 2\big[(-3)(-1) - (0)(5)\big] - 1\big[(4)(-1) - (2)(5)\big] + (-5)\big[(4)(0) - (2)(-3)\big]$$

$$= 2(3) - 1(-14) - 5(6)$$

$$= -10$$

Skill Practice 3 Evaluate the determinant. $\begin{vmatrix} -2 & 4 & 9 \\ 5 & -1 & 2 \\ 1 & 1 & 6 \end{vmatrix}$

Although we defined the determinant of a 3 × 3 matrix by expanding the minors of the elements in the first column, *any row or column can be used*. However, we must choose the correct sign to apply to the product of factors of each term. The following array of signs is helpful.

row 1, column 1
1 + 1 = 2 (even)

row 1, column 2
1 + 2 = 3 (odd)

$$\begin{bmatrix} + & - & + \\ - & + & - \\ + & - & + \end{bmatrix}$$

Notice that the sign is positive if the sum of the row and column numbers is even. The sign is negative if the sum of the row and column numbers is odd.

Answer

3. -42

To evaluate the determinant of an $n \times n$ matrix, first choose any row or column. For each element a_{ij} in the selected row or column, multiply the minor by 1 or -1 depending on whether the sum of the row and column numbers is even or odd. That is, for the element a_{ij}, we have $(-1)^{i+j}M_{ij}$. This product is called the *cofactor* of the element a_{ij}.

> ### Cofactor of an Element of a Matrix
>
> Given a square matrix $A = [a_{ij}]$, the **cofactor** of a_{ij} is $(-1)^{i+j}M_{ij}$, where M_{ij} is the minor of a_{ij}.

Using the definition of the cofactor of an element of an $n \times n$ matrix, we can now present a generic method to find the determinant of the matrix.

> ### Evaluating the Determinant of an $n \times n$ Matrix by Expanding Cofactors
>
> **Step 1** Choose any row or column.
> **Step 2** Multiply each element in the selected row or column by its cofactor.
> **Step 3** The value of the determinant is the sum of the products from step 2.

It is important to note that this three-step process to evaluate a determinant works for any $n \times n$ matrix, including a 2×2 matrix. For example, given $A = \begin{bmatrix} a & b \\ c & d \end{bmatrix}$, we can evaluate the determinant of A by expanding cofactors about the first row. The cofactor of a is d, and the cofactor of b is $-c$. Therefore,

$$|A| = a(d) + b(-c)$$
$$= ad - bc \text{ as expected.}$$

Classroom Example: p. 896
Exercise 28

EXAMPLE 4   Evaluating the Determinant of a 3 × 3 Matrix

Evaluate the determinant of the matrix from Example 3 by expanding cofactors about the elements in the third row.

$$A = \begin{bmatrix} 2 & 4 & 2 \\ 1 & -3 & 0 \\ -5 & 5 & -1 \end{bmatrix}$$

Solution:

$$\begin{vmatrix} 2 & 4 & 2 \\ 1 & -3 & 0 \\ -5 & 5 & -1 \end{vmatrix} = \overbrace{(-5)(-1)^{3+1} \begin{vmatrix} 4 & 2 \\ -3 & 0 \end{vmatrix}}^{\text{cofactor of } -5} + \overbrace{(5)(-1)^{3+2} \begin{vmatrix} 2 & 2 \\ 1 & 0 \end{vmatrix}}^{\text{cofactor of } 5} + \overbrace{(-1)(-1)^{3+3} \begin{vmatrix} 2 & 4 \\ 1 & -3 \end{vmatrix}}^{\text{cofactor of } -1}$$

$$= (-5)(1)\big[(4)(0) - (2)(-3)\big] + (5)(-1)\big[(2)(0) - (2)(1)\big] + (-1)(1)\big[(2)(-3) - (4)(1)\big]$$
$$= -5(6) - 5(-2) - 1(-10)$$
$$= -30 + 10 + 10$$
$$= -10$$

This is the same result as in Example 3.

Skill Practice 4   Evaluate the determinant by expanding cofactors about the elements in the third column.

$$\begin{vmatrix} -2 & 4 & 9 \\ 5 & -1 & 2 \\ 1 & 1 & 6 \end{vmatrix}$$

TECHNOLOGY CONNECTIONS

Evaluating a Determinant

To find the determinant of a matrix on a graphing calculator, first enter the matrix in the calculator by using the MATRIX menu and EDIT menu. Select a name for the matrix such as [A]. Then access the determinant function det(under the MATRIX and **MATH** menus. Enter det([A]) in the home screen and press **ENTER**.

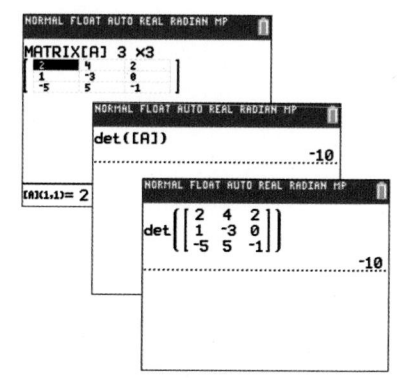

Alternatively, if the calculator is in "MATHPRINT" mode, enter the matrix by selecting **ALPHA** F3.

TIP   The determinant of a 3 × 3 matrix can also be evaluated by using the "method of diagonals."

Step 1: Recopy columns 1 and 2 to the right of the matrix.

Step 2: Multiply the elements on the diagonals labeled d_1 through d_6 (each diagonal has three elements).

Step 3: The value of the determinant is $(d_1 + d_2 + d_3) - (d_4 + d_5 + d_6)$.

The determinant from Example 4 is evaluated as follows:

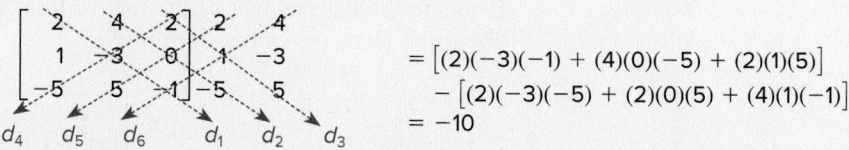

$$= \left[(2)(-3)(-1) + (4)(0)(-5) + (2)(1)(5) \right]$$
$$- \left[(2)(-3)(-5) + (2)(0)(5) + (4)(1)(-1) \right]$$
$$= -10$$

Some students find the method of diagonals to be a faster technique to find the determinant of a 3 × 3 matrix. However, it is critical to note that the method of diagonals only works for the determinant of a 3 × 3 matrix.

Perhaps one of the most important applications of the determinant of a matrix is to determine whether the matrix has an inverse. In Section 9.4, we found the inverse of a 2 × 2 matrix $A = \begin{bmatrix} a & b \\ c & d \end{bmatrix}$ by applying the formula $A^{-1} = \dfrac{1}{ad - bc} \begin{bmatrix} d & -b \\ -c & a \end{bmatrix}$. This is equivalent to $A^{-1} = \dfrac{1}{|A|} \begin{bmatrix} d & -b \\ -c & a \end{bmatrix}$. Furthermore, the inverse exists if $ad - bc \neq 0$ or equivalently, if $|A| \neq 0$. This is true in general for an $n \times n$ matrix.

Using Determinants to Determine if a Matrix Is Invertible

Let A be an $n \times n$ matrix. Then A is invertible if and only if $|A| \neq 0$.

In Example 5, we will use the determinant of A to determine if A has an inverse.

Classroom Example: p. 896
Exercise 30

EXAMPLE 5 **Using a Determinant to Determine if a Matrix Is Invertible**

Use $|A|$ to detrmine if A is invertible.
$$A = \begin{vmatrix} -2 & 0 & 1 & 2 \\ 6 & 2 & -2 & 5 \\ 5 & 3 & -1 & 1 \\ 0 & 4 & 2 & 1 \end{vmatrix}$$

Solution:

To evaluate a determinant, we can simplify the arithmetic by expanding around the row or column that contains the greatest number of 0 elements. In this case, we have chosen row 1.

$$\begin{vmatrix} -2 & 0 & 1 & 2 \\ 6 & 2 & -2 & 5 \\ 5 & 3 & -1 & 1 \\ 0 & 4 & 2 & 1 \end{vmatrix} = -2(-1)^{1+1}\begin{vmatrix} 2 & -2 & 5 \\ 3 & -1 & 1 \\ 4 & 2 & 1 \end{vmatrix} + 0(-1)^{1+2}\begin{vmatrix} 6 & -2 & 5 \\ 5 & -1 & 1 \\ 0 & 2 & 1 \end{vmatrix}$$

$$+ 1(-1)^{1+3}\begin{vmatrix} 6 & 2 & 5 \\ 5 & 3 & 1 \\ 0 & 4 & 1 \end{vmatrix} + 2(-1)^{1+4}\begin{vmatrix} 6 & 2 & -2 \\ 5 & 3 & -1 \\ 0 & 4 & 2 \end{vmatrix}$$

The second term is zero. Evaluating the determinants in the first, third, and fourth terms (shown in red), we have:

$$\begin{vmatrix} -2 & 0 & 1 & 2 \\ 6 & 2 & -2 & 5 \\ 5 & 3 & -1 & 1 \\ 0 & 4 & 2 & 1 \end{vmatrix} = -2(1)(42) + 0 + 1(1)(84) + 2(-1)(0) = 0$$

The determinant of A is zero. Therefore, A is not invertible.

Skill Practice 5 Use $|A|$ to determine if A is invertible. $A = \begin{vmatrix} 3 & 2 & 1 & 0 \\ 4 & 0 & 3 & 1 \\ 2 & 0 & 0 & 5 \\ 3 & -1 & 0 & 9 \end{vmatrix}$

3. Apply Cramer's Rule

We have learned several methods to solve a system of linear equations: the substitution method, the addition method, Gaussian elimination, Gauss-Jordan elimination, and the application of matrix inverses. We now present another method called Cramer's rule.

Cramer's rule involves finding the ratio of several determinants derived from the coefficients of the equations within the system. For example, consider the following system of equations.

$$a_1 x + b_1 y = c_1$$
$$a_2 x + b_2 y = c_2$$

Answer

5. $|A| = -9$; Since $|A| \neq 0$, A is invertible.

Using the addition method to solve for x, we have

$$a_1x + b_1y = c_1 \xrightarrow{\text{Multiply by } b_2.} a_1b_2x + b_1b_2y = c_1b_2$$
$$a_2x + b_2y = c_2 \xrightarrow{\text{Multiply by } -b_1.} \underline{-a_2b_1x - b_1b_2y = -c_2b_1}$$
$$(a_1b_2 - a_2b_1)x \qquad = c_1b_2 - c_2b_1$$

$$x = \frac{c_1b_2 - c_2b_1}{a_1b_2 - a_2b_1} = \frac{\begin{vmatrix} c_1 & b_1 \\ c_2 & b_2 \end{vmatrix}}{\begin{vmatrix} a_1 & b_1 \\ a_2 & b_2 \end{vmatrix}}$$

Using similar logic, we can show that
$$y = \frac{a_1c_2 - a_2c_1}{a_1b_2 - a_2b_1} = \frac{\begin{vmatrix} a_1 & c_1 \\ a_2 & c_2 \end{vmatrix}}{\begin{vmatrix} a_1 & b_1 \\ a_2 & b_2 \end{vmatrix}}$$

These results are summarized as Cramer's rule for a system of linear equations in two variables.

Cramer's Rule for a System of Two Linear Equations in Two Variables

Given the system
$$a_1x + b_1y = c_1$$
$$a_2x + b_2y = c_2$$

let $D = \begin{vmatrix} a_1 & b_1 \\ a_2 & b_2 \end{vmatrix}$, $D_x = \begin{vmatrix} c_1 & b_1 \\ c_2 & b_2 \end{vmatrix}$, and $D_y = \begin{vmatrix} a_1 & c_1 \\ a_2 & c_2 \end{vmatrix}$.

Then if $D \neq 0$, the system has the unique solution: $x = \dfrac{D_x}{D}$ and $y = \dfrac{D_y}{D}$

TIP Here are some memory tips to remember the patterns presented in Cramer's rule.

$$a_1x + b_1y = c_1$$
$$a_2x + b_2y = c_2$$

Coefficients of
x terms   y terms

1. The determinant D is the determinant of the coefficients of x and y. $\qquad D = \begin{vmatrix} a_1 & b_1 \\ a_2 & b_2 \end{vmatrix}$

x-coefficients replaced by c_1 and c_2

2. The determinant D_x has the column of x term coefficients replaced by c_1 and c_2. $\qquad D_x = \begin{vmatrix} c_1 & b_1 \\ c_2 & b_2 \end{vmatrix}$

y-coefficients replaced by c_1 and c_2

3. The determinant D_y has the column of y term coefficients replaced by c_1 and c_2. $\qquad D_y = \begin{vmatrix} a_1 & c_1 \\ a_2 & c_2 \end{vmatrix}$

Classroom Example: p. 896
Exercise 36

EXAMPLE 6   **Solving a 2 × 2 System by Using Cramer's Rule**

Solve the system by using Cramer's rule. $-11x + 6y = 4$
$2x - 5y = -3$

Solution:

For this system, $a_1 = -11$, $b_1 = 6$, $c_1 = 4$, $a_2 = 2$, $b_2 = -5$, and $c_2 = -3$.

$$D = \begin{vmatrix} a_1 & b_1 \\ a_2 & b_2 \end{vmatrix} \longrightarrow D = \begin{vmatrix} -11 & 6 \\ 2 & -5 \end{vmatrix} = (-11)(-5) - (6)(2) = 43$$

$$D_x = \begin{vmatrix} c_1 & b_1 \\ c_2 & b_2 \end{vmatrix} \longrightarrow D_x = \begin{vmatrix} 4 & 6 \\ -3 & -5 \end{vmatrix} = (4)(-5) - (6)(-3) = -2$$

$$D_y = \begin{vmatrix} a_1 & c_1 \\ a_2 & c_2 \end{vmatrix} \longrightarrow D_y = \begin{vmatrix} -11 & 4 \\ 2 & -3 \end{vmatrix} = (-11)(-3) - (4)(2) = 25$$

Therefore, $x = \dfrac{D_x}{D} = \dfrac{-2}{43}$ and $y = \dfrac{D_y}{D} = \dfrac{25}{43}$.

The solution set is $\left\{ \left(-\dfrac{2}{43}, \dfrac{25}{43} \right) \right\}$.

Skill Practice 6   Solve the system by using Cramer's rule. $3x - 4y = 9$
$-5x + 6y = 2$

The patterns associated with Cramer's rule can be generalized to solve a system of n linear equations in n variables.

Cramer's Rule for a System of n Linear Equations in n Variables

Consider the following system of n linear equations in n variables.

$$a_{11}x_1 + a_{12}x_2 + \cdots + a_{1n}x_n = b_1$$
$$a_{21}x_1 + a_{22}x_2 + \cdots + a_{2n}x_n = b_2$$
$$\vdots$$
$$a_{n1}x_1 + a_{n2}x_2 + \cdots + a_{nn}x_n = b_n$$

If the system has a unique solution, then the solution is $(x_1, x_2, \ldots, x_i, \ldots, x_n)$, where

$$x_1 = \frac{D_1}{D}, x_2 = \frac{D_2}{D}, \ldots, x_i = \frac{D_i}{D}, \ldots, x_n = \frac{D_n}{D}.$$

$D \neq 0$ is the determinant of the coefficient matrix, and D_i is the determinant formed by replacing the ith column of the coefficient matrix by the column of constants $b_1, b_2, \ldots, b_n$.

In Example 7, we apply Cramer's rule to a system of equations with three variables.

Answer

6. $\left\{ \left(-31, -\dfrac{51}{2} \right) \right\}$

Classroom Example: p. 896
Exercise 44

EXAMPLE 7   Solving a 3 × 3 System by Using Cramer's Rule

Solve the system by using Cramer's rule.

$$\begin{aligned} 2x - 3y + 5z &= 11 \\ -5x + 7y - 2z &= -6 \\ 9x - 2y + 3z &= 4 \end{aligned}$$

Solution:

Evaluate the determinants D, D_x, D_y, and D_z.

$$D = \begin{vmatrix} 2 & -3 & 5 \\ -5 & 7 & -2 \\ 9 & -2 & 3 \end{vmatrix} = -222 \qquad D_x = \begin{vmatrix} 11 & -3 & 5 \\ -6 & 7 & -2 \\ 4 & -2 & 3 \end{vmatrix} = 77$$

$$D_y = \begin{vmatrix} 2 & 11 & 5 \\ -5 & -6 & -2 \\ 9 & 4 & 3 \end{vmatrix} = 117 \qquad D_z = \begin{vmatrix} 2 & -3 & 11 \\ -5 & 7 & -6 \\ 9 & -2 & 4 \end{vmatrix} = -449$$

Therefore, $x = \dfrac{D_x}{D} = -\dfrac{77}{222}$, $y = \dfrac{D_y}{D} = -\dfrac{117}{222}$ or $-\dfrac{39}{74}$, and $z = \dfrac{D_z}{D} = \dfrac{449}{222}$.

The solution set is $\left\{ \left(-\dfrac{77}{222}, -\dfrac{39}{74}, \dfrac{449}{222} \right) \right\}$.

> **Avoiding Mistakes**
>
> It is important to note that each equation in the system of equations should be written in standard form before applying Cramer's rule.

Skill Practice 7   Solve the system by using Cramer's rule.

$$\begin{aligned} 5x + 3y - 3z &= -14 \\ 3x - 4y + z &= 2 \\ x + 7y + z &= 6 \end{aligned}$$

Although Cramer's rule may seem cumbersome for solving a 3 × 3 system of linear equations, it provides convenient formulas that can be programmed into a computer or calculator to solve the system. However, it is important to remember that Cramer's rule does not apply if $D = 0$. In such a case, the system of equations is either inconsistent (has no solution) or the equations are dependent (the system has infinitely many solutions).

Classroom Example: p. 896
Exercise 38

EXAMPLE 8   Identifying Whether Cramer's Rule Applies

Solve the system by using Cramer's rule if possible. Otherwise, use a different method.

$$\begin{aligned} x + 3y &= 6 \\ -2x - 6y &= -12 \end{aligned}$$

Solution:

Evaluate D. $D = \begin{vmatrix} 1 & 3 \\ -2 & -6 \end{vmatrix} = 1(-6) - 3(-2) = -6 + 6 = 0$

Since $D = 0$, Cramer's rule does not apply. Using Gauss-Jordan elimination, we have

$$\begin{bmatrix} 1 & 3 & | & 6 \\ -2 & -6 & | & -12 \end{bmatrix} \xrightarrow{2R_1 + R_2 \rightarrow R_2} \begin{bmatrix} 1 & 3 & | & 6 \\ 0 & 0 & | & 0 \end{bmatrix}$$

The last row of the augmented matrix represents the equation $0 = 0$, which is true for all values of x and y. The system reduces to the single equation $x + 3y = 6$ and there are infinitely many solutions.

The solution set is $\{(6 - 3y, y) \mid y \text{ is any real number}\}$.

Answer

7. $\left\{ \left(-\dfrac{13}{34}, \dfrac{5}{17}, \dfrac{147}{34} \right) \right\}$

Skill Practice 8 Solve the system by using Cramer's rule if possible. Otherwise, use a different method.

$$x + 4y = 2$$
$$3x + 12y = 4$$

Answer

8. {}

| SECTION 9.5 | Practice Exercises |

Prerequisite Review

For Exercises R.1–R.2, simplify the exponential expression.

R.1. $(-1)^4$ 1

R.2. $(-1)^7$ -1

R.3. Write an equation of the line that passes through the points $(-3, 5)$ and $(-2, 4)$. Write the equation in slope-intercept form. $y = -x + 2$

Concept Connections

1. Associated with every square matrix A is a real number denoted by $|A|$ called the _____determinant_____ of A.

2. For a 2×2 matrix $A = \begin{bmatrix} a & b \\ c & d \end{bmatrix}$, $|A| = \begin{vmatrix} a & b \\ c & d \end{vmatrix} = $ _____$ad - bc$_____.

3. Given $A = [a_{ij}]$, the _____minor_____ of the element a_{ij} is the determinant obtained by deleting the ith row and jth column.

4. Given $A = [a_{ij}]$, then the value $(-1)^{i+j} M_{ij}$ is called the _____cofactor_____ of the element a_{ij}.

5. The determinant of a 3×3 matrix
$$A = \begin{bmatrix} a_1 & b_1 & c_1 \\ a_2 & b_2 & c_2 \\ a_3 & b_3 & c_3 \end{bmatrix} \text{ is given by}$$
$$|A| = a_1 \begin{vmatrix} \Box & \Box \\ \Box & \Box \end{vmatrix} - a_2 \begin{vmatrix} \Box & \Box \\ \Box & \Box \end{vmatrix} + a_3 \begin{vmatrix} \Box & \Box \\ \Box & \Box \end{vmatrix}$$
$$\begin{vmatrix} b_2 & c_2 \\ b_3 & c_3 \end{vmatrix}; \begin{vmatrix} b_1 & c_1 \\ b_3 & c_3 \end{vmatrix}; \begin{vmatrix} b_1 & c_1 \\ b_2 & c_2 \end{vmatrix}$$

6. Suppose that the given system has one solution.
$$a_1x + b_1y = c_1$$
$$a_2x + b_2y = c_2$$
$\dfrac{D_x}{D}, \dfrac{D_y}{D};$

Cramer's rule gives the solution as $x = \dfrac{\Box}{\Box}$ and $y = \dfrac{\Box}{\Box}$,

where $D = \begin{vmatrix} \Box & \Box \\ \Box & \Box \end{vmatrix}$, $D_x = \begin{vmatrix} \Box & \Box \\ \Box & \Box \end{vmatrix}$, and

$D_y = \begin{vmatrix} \Box & \Box \\ \Box & \Box \end{vmatrix}$. $D = \begin{vmatrix} a_1 & b_1 \\ a_2 & b_2 \end{vmatrix}$, $D_x = \begin{vmatrix} c_1 & b_1 \\ c_2 & b_2 \end{vmatrix}$, and $D_y = \begin{vmatrix} a_1 & c_1 \\ a_2 & c_2 \end{vmatrix}$

Objective 1: Evaluate the Determinant of a 2 × 2 Matrix

For Exercises 7–16, evaluate the determinant of the matrix. (See Example 1)

7. $A = \begin{bmatrix} 3 & -2 \\ 6 & 5 \end{bmatrix}$ 27

8. $B = \begin{bmatrix} 7 & 12 \\ -1 & 4 \end{bmatrix}$ 40

9. $C = \begin{bmatrix} \frac{2}{3} & \frac{1}{5} \\ 10 & 12 \end{bmatrix}$ 6

10. $D = \begin{bmatrix} \frac{8}{9} & 4 \\ \frac{5}{2} & 18 \end{bmatrix}$ 6

11. $E = \begin{bmatrix} -3 & 0 \\ 4 & 0 \end{bmatrix}$ 0

12. $F = \begin{bmatrix} 0 & 9 \\ 0 & 4 \end{bmatrix}$ 0

13. $G = \begin{bmatrix} x & 4 \\ 9 & x \end{bmatrix}$ $x^2 - 36$

14. $H = \begin{bmatrix} y & 16 \\ 4 & y \end{bmatrix}$ $y^2 - 64$

15. $T = \begin{bmatrix} e^x & e^{2x} \\ 4 & -e^x \end{bmatrix}$ $-5e^{2x}$

16. $V = \begin{bmatrix} \log x & \log x \\ 2 & 5 \end{bmatrix}$ $3 \log x$

Objective 2: Evaluate the Determinant of an $n \times n$ Matrix

For Exercises 17–22, refer to the matrix $A = [a_{ij}] = \begin{bmatrix} -6 & 11 & 8 \\ 4 & -2 & -5 \\ -3 & 7 & 10 \end{bmatrix}$.

a. Find the minor of the given element. (**See Example 2**)

b. Find the cofactor of the given element.

17. a_{12} a. 25   b. -25

18. a_{23} a. -9 b. 9

19. a_{31} a. -39 b. -39

20. a_{13} a. 22   b. 22

21. a_{22} a. -36 b. -36

22. a_{33} a. -32 b. -32

For Exercises 23–32, evaluate the determinant of the matrix and state whether the matrix is invertible. (**See Examples 3–5**)

23. $A = \begin{bmatrix} 4 & 1 & 3 \\ 0 & -1 & 2 \\ 5 & 8 & 0 \end{bmatrix}$ -39; Yes

24. $B = \begin{bmatrix} 9 & 5 & -1 \\ 2 & 0 & 4 \\ 7 & -2 & 0 \end{bmatrix}$ 216; Yes

25. $C = \begin{bmatrix} 5 & 1 & 6 \\ 2 & 3 & 4 \\ 8 & -1 & 7 \end{bmatrix}$ -13; Yes

26. $D = \begin{bmatrix} -3 & 1 & -2 \\ 10 & 5 & 8 \\ 6 & 7 & -4 \end{bmatrix}$ 236; Yes

27. $E = \begin{bmatrix} 2 & 0 & 1 \\ 1 & -1 & 2 \\ 3 & 1 & 0 \end{bmatrix}$ 0; No

28. $F = \begin{bmatrix} 1 & -3 & 17 \\ 1 & -1 & 7 \\ 2 & -5 & 29 \end{bmatrix}$ 0; No

29. $G = \begin{bmatrix} 5 & 6 & 4 & 1 \\ 2 & 0 & 3 & 0 \\ 0 & 1 & 4 & 0 \\ -1 & 2 & 0 & 0 \end{bmatrix}$ 13; Yes

30. $H = \begin{bmatrix} 8 & 0 & 5 & 1 \\ 0 & 3 & 4 & -2 \\ 2 & 6 & 3 & 0 \\ -1 & 0 & 0 & 0 \end{bmatrix}$ -75; Yes

31. $T = \begin{bmatrix} 3 & 8 & 1 & 4 \\ -2 & 4 & 0 & 5 \\ -1 & 1 & 0 & -1 \\ 0 & 5 & 2 & 3 \end{bmatrix}$ 121; Yes

32. $W = \begin{bmatrix} 2 & 5 & 2 & 4 \\ 0 & 0 & -3 & 1 \\ 4 & 8 & 0 & 1 \\ -1 & 2 & 0 & 5 \end{bmatrix}$ 137; Yes

Objective 3: Apply Cramer's Rule

For Exercise 33–48, solve the system if possible by using Cramer's rule. If Cramer's rule does not apply, solve the system by using another method. (**See Examples 6–8**)

33. $\begin{aligned} 2x + 10y &= 11 \\ 3x - 5y &= 6 \end{aligned}$ $\left\{ \left(\dfrac{23}{8}, \dfrac{21}{40} \right) \right\}$

34. $\begin{aligned} -5x - 8y &= 3 \\ 4x + 7y &= 13 \end{aligned}$ $\left\{ \left(-\dfrac{125}{3}, \dfrac{77}{3} \right) \right\}$

35. $\begin{aligned} -10x + 4y &= 7 \\ 6x &= 7y + 2 \end{aligned}$ $\left\{ \left(-\dfrac{57}{46}, -\dfrac{31}{23} \right) \right\}$

36. $\begin{aligned} 11x + 6y &= 8 \\ 2x &= 9y + 5 \end{aligned}$ $\left\{ \left(\dfrac{34}{37}, -\dfrac{13}{37} \right) \right\}$

37. $\begin{aligned} 3(x - y) &= y + 8 \\ y &= \tfrac{3}{4}x - 2 \end{aligned}$

38. $\begin{aligned} 5(x + y) &= 7x + 4 \\ 4x &= 10y - 8 \end{aligned}$

39. $\begin{aligned} y &= -3x + 7 \\ \tfrac{1}{2}x + \tfrac{1}{6}y &= 1 \end{aligned}$ { }

40. $\begin{aligned} x &= 4y + 5 \\ 3(x - 4) &= 12y \end{aligned}$ { }

41. $\begin{aligned} 11x \qquad\;\; - 3z &= 1 \\ 2y + 9z &= 6 \\ 4x + 5y \qquad &= -9 \end{aligned}$ $\left\{ \left(\dfrac{63}{157}, -\dfrac{333}{157}, \dfrac{536}{471} \right) \right\}$

42. $\begin{aligned} -2x + 6y \qquad\;\; &= 9 \\ 5y + 7z &= 1 \\ 4x \qquad\; - 3z &= -8 \end{aligned}$ $\left\{ \left(-\dfrac{151}{66}, \dfrac{73}{99}, -\dfrac{38}{99} \right) \right\}$

43. $\begin{aligned} 2x - 5y + z &= 11 \\ 3x + 7y - 4z &= 8 \\ x - 9y + 2z &= 4 \end{aligned}$ $\left\{ \left(\dfrac{13}{2}, \dfrac{3}{2}, \dfrac{11}{2} \right) \right\}$

44. $\begin{aligned} -5x - 6y + 8z &= 1 \\ 2x + y - 4z &= 5 \\ 3x - 4y - z &= -2 \end{aligned}$ $\left\{ \left(-\dfrac{239}{57}, -\dfrac{97}{57}, -\dfrac{215}{57} \right) \right\}$

45. $\begin{aligned} 2x - 3y + z &= 6 \\ -4x + 6y - 2z &= -12 \\ 6x - 9y + 3z &= 18 \end{aligned}$

46. $\begin{aligned} -x + y - 3z &= 4 \\ 3x - 3y + 9z &= -12 \\ -2x + 2y - 6z &= 8 \end{aligned}$

47. $\begin{aligned} x - 2y + 3z &= -1 \\ 5x - 7y + 3z &= 1 \\ x \qquad\; - 5z &= 2 \end{aligned}$ { }

48. $\begin{aligned} x + 3y - 5z &= 10 \\ -2x - 4y + 8z &= -14 \\ x + y - 3z &= 5 \end{aligned}$ { }

For Exercises 49–50, use Cramer's rule to solve for the indicated variable.

49. $x_1 + 2x_2 + 3x_3 - 4x_4 = 3$
$\qquad 5x_2 \qquad + x_4 = 9$
$\quad x_1 \qquad + 4x_3 \qquad = -1$
$\qquad\qquad 5x_3 - 2x_4 = 8$ Solve for x_2. $x_2 = \dfrac{113}{60}$

50. $-2x_1 - x_2 + x_3 + 3x_4 = 10$
$\quad x_1 \qquad + 5x_3 \qquad = 4$
$\qquad 2x_2 \qquad + x_4 = -1$
$\qquad\qquad 4x_3 + 2x_4 = 7$ Solve for x_3. $x_3 = \dfrac{21}{16}$

Mixed Exercises

Determinants can be used to determine whether three points are collinear (lie on the same line). Given the ordered pairs (x_1, y_1), (x_2, y_2), and (x_3, y_3), the points are collinear if the determinant to the right equals zero. For Exercises 51–54, determine if the points are collinear.

$$\begin{vmatrix} x_1 & y_1 & 1 \\ x_2 & y_2 & 1 \\ x_3 & y_3 & 1 \end{vmatrix}$$

51. $(3, 6)$, $(6, 10)$, $(-3, -2)$ Yes

52. $(-2, 1)$, $(-4, -4)$, $(4, 16)$ Yes

53. $(4, -3)$, $(5, -7)$, $(8, -14)$ No

54. $(0, 6)$, $(1, 4)$, $(4, -6)$ No

The equation at the right represents an equation of the line passing through the distinct points (x_1, y_1) and (x_2, y_2). For Exercises 55–56,

$$\begin{vmatrix} x & y & 1 \\ x_1 & y_1 & 1 \\ x_2 & y_2 & 1 \end{vmatrix} = 0$$

a. Use the determinant equation to write an equation of the line passing through the given points.

b. Write the equation of the line in slope-intercept form.

55. $(-3, 2)$ and $(-4, 6)$ a. $\begin{vmatrix} x & y & 1 \\ -3 & 2 & 1 \\ -4 & 6 & 1 \end{vmatrix} = 0$ b. $y = -4x - 10$

56. $(-4, 1)$ and $(-5, 4)$ a. $\begin{vmatrix} x & y & 1 \\ -4 & 1 & 1 \\ -5 & 4 & 1 \end{vmatrix} = 0$ b. $y = -3x - 11$

For Exercises 57–58, use the formula at the right to find the area of a triangle with vertices (x_1, y_1), (x_2, y_2), and (x_3, y_3). Choose the $+$ or $-$ sign so that the value of the area is positive.

$$\text{Area} = \pm\frac{1}{2}\begin{vmatrix} x_1 & y_1 & 1 \\ x_2 & y_2 & 1 \\ x_3 & y_3 & 1 \end{vmatrix}$$

57. $(1, 0)$, $(7, -2)$, $(4, -5)$ 12 square units

58. $(-2, 1)$, $(-1, -6)$, $(-8, -5)$ 24 square units

Given a square matrix A, elementary row operations (or column operations) performed on A affect the value of $|A|$ in the following ways:

- Interchanging any two rows (or columns) of A will change the sign of $|A|$.
- Multiplying a row (or column) of A by a constant real number k multiplies $|A|$ by k.
- Adding a multiple of a row (or column) of A to another row (or column) of A does not change the value of $|A|$.

For Exercises 59–64, demonstrate these three properties.

59. Given $A = \begin{bmatrix} 5 & 2 \\ -3 & 6 \end{bmatrix}$ and $B = \begin{bmatrix} -3 & 6 \\ 5 & 2 \end{bmatrix}$,

   a. Evaluate $|A|$.   36

   b. Evaluate $|B|$. -36

   c. How are A and B related and how are $|A|$ and $|B|$ related?

60. Given $A = \begin{bmatrix} 2 & -7 \\ 4 & 10 \end{bmatrix}$ and $B = \begin{bmatrix} 4 & 10 \\ 2 & -7 \end{bmatrix}$,

   a. Evaluate $|A|$.   48

   b. Evaluate $|B|$. -48

   c. How are A and B related and how are $|A|$ and $|B|$ related?

61. Given $A = \begin{bmatrix} 1 & -3 \\ 4 & 1 \end{bmatrix}$ and $B = \begin{bmatrix} 2 & -6 \\ 4 & 1 \end{bmatrix}$,

   a. Evaluate $|A|$.   13

   b. Evaluate $|B|$.   26

   c. How are A and B related and how are $|A|$ and $|B|$ related?
   Row 1 of matrix B is 2 times row 1 of matrix A. The value $|B| = 2|A|$.

62. Given $A = \begin{bmatrix} 2 & 1 \\ 5 & 7 \end{bmatrix}$ and $B = \begin{bmatrix} -6 & -3 \\ 5 & 7 \end{bmatrix}$,

   a. Evaluate $|A|$.   9

   b. Evaluate $|B|$. -27

   c. How are A and B related and how are $|A|$ and $|B|$ related?
   Row 1 of matrix B is -3 times row 1 of matrix A. The value $|B| = -3|A|$.

63. Given $A = \begin{bmatrix} 1 & 2 \\ 3 & 4 \end{bmatrix}$ and $B = \begin{bmatrix} 1 & 2 \\ 6 & 10 \end{bmatrix}$,

   a. Evaluate $|A|$. -2

   b. Evaluate $|B|$. -2

   c. How are A and B related and how are $|A|$ and $|B|$ related?
   Row 2 of matrix B is the same as the sum of 3 times row 1 of A and row 2 of A. The value of $|A|$ equals $|B|$.

64. Given $A = \begin{bmatrix} 1 & 1 \\ 5 & 6 \end{bmatrix}$ and $B = \begin{bmatrix} 1 & 1 \\ 3 & 4 \end{bmatrix}$,

   a. Evaluate $|A|$.   1

   b. Evaluate $|B|$.   1

   c. How are A and B related and how are $|A|$ and $|B|$ related?
   Row 2 of matrix B is the same as the sum of -2 times row 1 of A and row 2 of A. The value of $|A|$ equals $|B|$.

Given a square matrix A, if either of the following conditions are true, then $|A| = 0$.

- A row (or column) of A consists entirely of zeros.
- One row (or column) is a constant multiple of another row (or column).

For Exercises 65–68, demonstrate these two properties.

65. Given $A = \begin{bmatrix} 3 & 5 \\ 6 & 10 \end{bmatrix}$, find $|A|$.   0

66. Given $A = \begin{bmatrix} -2 & 7 \\ -6 & 21 \end{bmatrix}$, find $|A|$.   0

67. Given $A = \begin{bmatrix} 4 & -5 & 0 \\ 3 & -1 & 0 \\ 0 & 1 & 0 \end{bmatrix}$, find $|A|$.   0

68. Given $A = \begin{bmatrix} 5 & 3 & 1 \\ -1 & 0 & 3 \\ 0 & 0 & 0 \end{bmatrix}$, find $|A|$.   0

69. Evaluate $|I_2|$.   1

70. Evaluate $|I_3|$.   1

71. Evaluate $\begin{vmatrix} a & 0 & 0 \\ 0 & b & 0 \\ 0 & 0 & c \end{vmatrix}$.   abc

72. Evaluate $\begin{vmatrix} x & 0 \\ 0 & x \end{vmatrix}$. x^2

73. If A and B are square matrices, then the product property of determinants indicates that $|AB| = |A| \cdot |B|$. Use matrix A and matrix B to demonstrate this property.

$$A = \begin{bmatrix} 4 & -2 \\ 3 & 1 \end{bmatrix} \text{ and } B = \begin{bmatrix} -5 & 1 \\ 3 & 2 \end{bmatrix}$$

$|AB| = -130$; $|A| = 10$ and $|B| = -13$.
So, $|A| \cdot |B| = (10)(-13) = -130$ and, therefore, $|A| \cdot |B| = |AB|$.

74. The **transpose** of a square matrix A, denoted as A^T, is a square matrix that results by writing the rows of A as the columns of A^T.

a. Given $A = \begin{bmatrix} 1 & 2 & 5 \\ 0 & 8 & 4 \\ 3 & 7 & 6 \end{bmatrix}$, find A^T.   a. $A^T = \begin{bmatrix} 1 & 0 & 3 \\ 2 & 8 & 7 \\ 5 & 4 & 6 \end{bmatrix}$

b. Show that $|A| = |A^T|$.   b. $|A| = -76$; $|A^T| = -76$

Write About It

75. What is the difference between the minor of an element a_{ij} and the cofactor of the element?

76. Explain the difference between the notation $\begin{bmatrix} a & b \\ c & d \end{bmatrix}$ and $\begin{vmatrix} a & b \\ c & d \end{vmatrix}$.

77. The determinant of a square matrix can be computed by expanding the cofactors of the elements in any row or column. How would you choose which row or column?

78. Consider the system shown here. Describe the pattern associated with constructing the determinants used with Cramer's rule.

$$a_1x + b_1y = c_1$$
$$a_2x + b_2y = c_2$$

Technology Connections

For Exercises 79–82, use a graphing utility to evaluate the determinant of the matrix. Round to the nearest whole unit.

79. $\begin{bmatrix} \sqrt{3} & e & 1.6 \\ \log 5 & -2\pi & \ln 3 \\ -4 & 8.4 & -\sqrt{6} \end{bmatrix}$ -27

80. $\begin{bmatrix} 8.9 & -2.3 & 3.8 \\ -1.7 & 0.9 & 4.6 \\ 2.7 & 10.1 & 14.9 \end{bmatrix}$ -455

81. $A = \begin{bmatrix} -0.4 & 1.5 & 9 & 11.3 \\ -3.5 & 0.2 & -1.1 & 3 \\ 8 & 9.4 & -5.4 & 2 \\ -1 & 4.6 & 10.8 & -9.7 \end{bmatrix}$ 10,112

82. $B = \begin{bmatrix} -2\pi & e^2 & 9.1 & \log 2 \\ \log 50 & -\sqrt{11} & 4.3 & \pi \\ -4.9 & 0 & e^2 & 8.1 \\ \sqrt{7} & \ln 7 & -9.7 & 0 \end{bmatrix}$ -1993

PROBLEM RECOGNITION EXERCISES

Using Multiple Methods to Solve Systems of Linear Equations

For Exercises 1–4, solve the system of equations using

  a. The substitution method or the addition method (see Sections 8.1 and 8.2).

  b. Gaussian elimination (see Section 9.1).

  c. Gauss-Jordan elimination (see Section 9.1).

  d. The inverse of the coefficient matrix (see Section 9.4).

  e. Cramer's rule (see Section 9.5).

1. $x = -3y - 10$ $\{(2, -4)\}$ **2.** $2x = 2 - 8y$ $\{(5, -1)\}$
$\quad\; -3x - 7y = 22$ $\qquad\qquad\qquad 3x + 10y = 5$

3. $\;\; x + 2y - z = 0$
$\quad\; 2x \quad\;\; + z = 4$ $\{(2, -1, 0)\}$
$\quad\; 2x - y + 2z = 5$

4. $x + 4y + 2z = 10$ $\{(2, 0, 4)\}$
$\qquad\quad\; 2y + z = 4$
$\quad x + y \qquad = 2$

For Exercises 5–8,

  a. Evaluate the determinant of the coefficient matrix.

  b. Based on the value of the determinant from part (a), can an inverse matrix or Cramer's rule be used to solve the system?

  c. Solve the system using an appropriate method.

5. $1.5x - 2y = 3$
$\quad\; -3x + 4y = 12$

6. $5x - 2y = 1$
$\quad\; x - 0.4y = 4$

7. $\;\; x - 3y + 7z = 1$
$\quad\; -2x + 5y - 11z = -3$
$\quad\; x - 5y + 13z = -1$

8. $\;\; x - 2y + 3z = -7$
$\quad\; -2x + y \qquad = -1$
$\quad\; x \qquad - z = 3$

Additional answers can be found in the Instructor Answer Appendix.

CHAPTER 9   KEY CONCEPTS

SECTION 9.1 Solving Systems of Linear Equations Using Matrices	Reference
A system of linear equations can be represented by an **augmented matrix.**	p. 838
Elementary row operations can be used to write a matrix in row-echelon form or reduced row-echelon form. 1. Interchange two rows. 2. Multiply a row by a nonzero constant. 3. Add a nonzero multiple of one row to another row.	p. 839
The method of **Gaussian elimination** uses elementary row operations to write an augmented matrix in row-echelon form so that the system can be solved by back substitution.	p. 841
With the method of **Gauss-Jordan elimination** the augmented matrix is written in *reduced* row-echelon form so that the solution can be found by inspection.	p. 843

SECTION 9.2 Inconsistent Systems and Dependent Equations	Reference
A system of equations that has no solution is called an **inconsistent system.** An inconsistent system is detected algebraically if a contradiction is reached when solving the system.	p. 849
A system of linear equations may have infinitely many solutions. In such a case, the equations are said to be **dependent.** Dependent equations are detected algebraically if an identity is reached when solving the system.	p. 850

SECTION 9.3 Operations on Matrices	Reference
An $m \times n$ matrix has m rows and n columns.	p. 859
Adding and subtracting matrices: If A and B represent two matrices of the same order, then $A + B$ and $A - B$ are found by adding or subtracting the corresponding elements.	p. 861
Scalar multiplication: Let $A = [a_{ij}]$ be an $m \times n$ matrix and let k be a real number. Then, $kA = [ka_{ij}]$.	p. 863
Matrix multiplication: Let A be an $m \times p$ matrix and let B be a $p \times n$ matrix, then the product AB is an $m \times n$ matrix. For the matrix AB, the element in the ith row and jth column is the sum of the products of the corresponding elements in the ith row of A and the jth column of B (the inner product of the ith row and jth column). *Note:* If the number of columns in A does not equal the number of rows in B, then it is not possible to compute the product AB.	p. 865

SECTION 9.4 Inverse Matrices and Matrix Equations	Reference
The **identity matrix** I_n for the multiplication of matrices is the $n \times n$ square matrix with 1's along the main diagonal and 0's for all other elements.	p. 876
Inverse of a square matrix: Let A be an $n \times n$ matrix. If there exists an $n \times n$ matrix A^{-1} such that $$AA^{-1} = I_n \quad \text{and} \quad A^{-1}A = I_n$$ then A^{-1} is the **multiplicative inverse** of A.	p. 877
A matrix that does not have an inverse is called a **singular matrix.** A matrix that does have an inverse is said to be **invertible** or **nonsingular.**	p. 879
Finding the inverse of a nonsingular matrix: Let A be an $n \times n$ matrix for which A^{-1} exists. To find A^{-1}: **Step 1** Write a matrix of the form $[A \mid I_n]$. **Step 2** Perform row operations to write the matrix in the form $[I_n \mid B]$. **Step 3** The matrix B is A^{-1}.	p. 878
The inverse of a nonsingular 2×2 matrix A can also be found as follows. If $A = \begin{bmatrix} a & b \\ c & d \end{bmatrix}$, then $A^{-1} = \dfrac{1}{ad - bc} \cdot \begin{bmatrix} d & -b \\ -c & a \end{bmatrix}$, or equivalently $A^{-1} = \dfrac{1}{|A|}\begin{bmatrix} d & -b \\ -c & a \end{bmatrix}$.	p. 880
Solving a system using inverse matrices: Suppose that $AX = B$ represents a system of n linear equations in n variables with a unique solution. Then, $$X = A^{-1}B$$ where A is the coefficient matrix, B is the matrix of constants, and X is the matrix of variables.	p. 882

SECTION 9.5 Determinants and Cramer's Rule	Reference
Associated with every square matrix is a real number called the determinant of the matrix.	p. 886
Determinant of a 2 × 2 matrix: Given $A = \begin{bmatrix} a & b \\ c & d \end{bmatrix}$, the determinant of A is denoted by $|A|$ or $\begin{vmatrix} a & b \\ c & d \end{vmatrix}$ and the value is $ad - bc$.	p. 886

The minor and cofactor of an element of a matrix:	p. 887
Given an $n \times n$ matrix $A = [a_{ij}]$, the **minor** M_{ij} of an element a_{ij} is the determinant of the resulting matrix obtained by deleting the ith row and jth column.	
The **cofactor** of a_{ij} is $(-1)^{i+j}M_{ij}$ where M_{ij} is the minor of a_{ij}.	p. 889

Determinant of a 3 × 3 matrix:	p. 888
$$\begin{vmatrix} a_1 & b_1 & c_1 \\ a_2 & b_2 & c_2 \\ a_3 & b_3 & c_3 \end{vmatrix} = a_1 \begin{vmatrix} b_2 & c_2 \\ b_3 & c_3 \end{vmatrix} - a_2 \begin{vmatrix} b_1 & c_1 \\ b_3 & c_3 \end{vmatrix} + a_3 \begin{vmatrix} b_1 & c_1 \\ b_2 & c_2 \end{vmatrix}$$	
Find the determinant of an $n \times n$ matrix:	p. 889
Step 1 Choose any row or column.	
Step 2 Multiply each element in the selected row or column by its cofactor.	
Step 3 The value of the determinant is the sum of the products from step 2.	

Cramer's rule for a system of two equations and two variables:	p. 892
Given $\begin{aligned} a_1x + b_1y &= c_1 \\ a_2x + b_2y &= c_2 \end{aligned}$	
if $D \neq 0$, then $x = \dfrac{D_x}{D}$ and $y = \dfrac{D_y}{D}$, where	
$D = \begin{vmatrix} a_1 & b_1 \\ a_2 & b_2 \end{vmatrix}, D_x = \begin{vmatrix} c_1 & b_1 \\ c_2 & b_2 \end{vmatrix}$, and $D_y = \begin{vmatrix} a_1 & c_1 \\ a_2 & c_2 \end{vmatrix}$.	

The patterns associated with Cramer's rule can be generalized to solve a system of n linear equations in n variables.	p. 893

When applying Cramer's rule, if $D = 0$, then the system has either no solution or infinitely many solutions.	p. 894

Expanded Chapter Summary available at www.mhhe.com/millerprecalculus.

CHAPTER 9   Review Exercises

SECTION 9.1

1. Write a system of linear equations represented by the augmented matrix. Then write the solution set.

$$\begin{aligned} x - 2y + 3z &= -1 \\ y + 4z &= -11 \\ z &= -2; \end{aligned} \qquad \begin{bmatrix} 1 & -2 & 3 & | & -1 \\ 0 & 1 & 4 & | & -11 \\ 0 & 0 & 1 & | & -2 \end{bmatrix}$$

Solution set: $\{(-1, -3, -2)\}$

For Exercises 2–4, perform the elementary row operations on the matrix. $\begin{bmatrix} 2 & -3 & | & 1 \\ 5 & 6 & | & -4 \end{bmatrix}$

2. $R_1 \Leftrightarrow R_2$ $\begin{bmatrix} 5 & 6 & | & -4 \\ 2 & -3 & | & 1 \end{bmatrix}$ **3.** $\dfrac{1}{2}R_1 \to R_1$ $\begin{bmatrix} 1 & -\frac{3}{2} & | & \frac{1}{2} \\ 5 & 6 & | & -4 \end{bmatrix}$

4. $-2R_1 + R_2 \to R_2$ $\begin{bmatrix} 2 & -3 & | & 1 \\ 1 & 12 & | & -6 \end{bmatrix}$

For Exercises 5–8, solve the system by using Gaussian elimination or Gauss-Jordan elimination.

5. $\begin{aligned} -2x + y &= -16 \\ x - 2y &= 17 \end{aligned}$ **6.** $\begin{aligned} 2(x - 6y) &= 36 \\ 47y &= 7x - 141 \end{aligned}$

$\{(5, -6)\}$ $\{(0, -3)\}$

7. $\begin{aligned} 2x - 5y + 18z &= 44 \\ x - 3y + 11z &= 27 \\ x - 2y + 11z &= 29 \end{aligned}$ **8.** $\begin{aligned} w + x \quad\quad - 2z &= 3 \\ 2x - 3z &= 3 \\ 2w + y + z &= 3 \\ 4y - z &= 9 \end{aligned}$

$\{(0, 2, 3)\}$ $\{(1, 0, 2, -1)\}$

9. Lily borrowed a total of $10,000. She borrowed part of the money from her friend Sly who did not charge her interest. She borrowed part of the money from a credit union at 5% simple interest, and she borrowed the rest of the money from a bank at 7.5% interest. At the end of 1 yr, she owed $500 in interest. If she borrowed $1000 less from her friend than she did from the bank, determine how much she borrowed from each source.   Lily borrowed $1000 from her friend, $7000 from the credit union, and $2000 from the bank.

SECTION 9.2

For Exercises 10–13, determine the solution set for the system represented by each augmented matrix.

10. $\begin{bmatrix} 1 & 0 & | & 4 \\ 0 & 1 & | & 0 \end{bmatrix}$ $\{(4, 0)\}$ **11.** $\begin{bmatrix} 1 & -2 & | & 6 \\ 0 & 0 & | & 1 \end{bmatrix}$ $\{\}$

12. $\begin{bmatrix} 1 & 4 & | & 0 \\ 0 & 0 & | & 0 \end{bmatrix}$

$\{(-4y, y) \mid y \text{ is any real number}\}$

13. $\begin{bmatrix} 1 & 0 & -3 & | & 0 \\ 0 & 1 & 2 & | & 1 \\ 0 & 0 & 0 & | & 0 \end{bmatrix}$

$\{(3z, -2z + 1, z) \mid z \text{ is any real number}\}$

For Exercises 14–19, solve the system by using Gaussian elimination or Gauss-Jordan elimination.

14. $3x + 6y = -9$
$x + 2y = -3$
$\{(-2y - 3, y) \mid y \text{ is any real number}\}$

15. $-(2x - y) = 8 - y$ $\{\ \}$
$y = x - 6$

16. $x - 2y = 3z - 10$
$x - y = z - 7$ $\{\ \}$
$3x - 7y - 11z = -320$

17. $x \qquad - 3z = 5$
$-2x + y + 10z = -7$
$x + y + z = 8$
$\{(3z + 5, -4z + 3, z) \mid z \text{ is any real number}\}$

18. $2x = 3y - z - 4$
$x - 2y = z + 2y - 2$
$x + y = 2z - 2$
$\{(-2, 0, 0)\}$

19. $5y = x + 2z + 1$
$2(x - 5y) + 4z = -2$
$3(x + 2z) = 15y - 3$

20. The solution set to a system of dependent equations is given. Write three ordered triples that are solutions to the system. Answers may vary.
For example: $(-3, 2, 0), (-1, 3, 1), (1, 4, 2)$
$\{(2z - 3, z + 2, z) \mid z \text{ is any real number}\}$

21. a. Assume that traffic flows freely around the traffic circle. The flow rates given are measured in vehicles per hour. If the flow rate x_3 is 130 vehicles per hour, determine the flow rates x_1 and x_2.

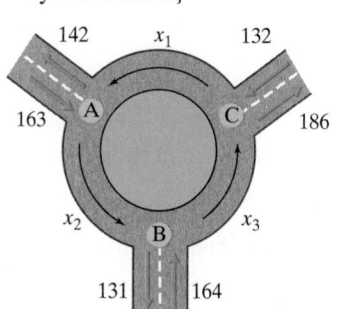

b. If traffic between intersections B and C flows at a rate of between 100 and 150 vehicles per hour, inclusive, find the range of values for x_1 and x_2.

22. a. Assume that traffic flows freely through intersections A, B, C, and D and that all flow rates are measured in vehicles per hour. If the flow rate x_4 is 220 vehicles per hour, find the flow rates x_1, x_2, and x_3.

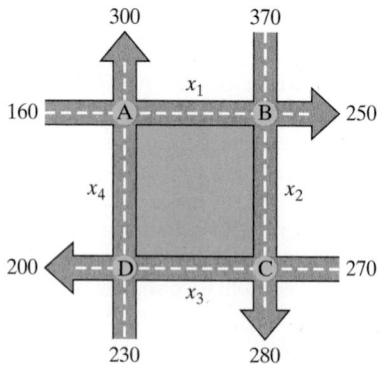

b. If the flow rate x_4 is between 200 and 250 vehicles per hour, inclusive, find the range of values x_1, x_2, and x_3.

SECTION 9.3

For Exercises 23–26,

a. Give the order of the matrix.

b. Classify the matrix as a square matrix, row matrix, column matrix, or none of these.

23. $\begin{bmatrix} 1 & 2 \\ -3 & \pi \\ 4.1 & \sqrt{2} \end{bmatrix}$ a. 3×2
b. None of these

24. $\begin{bmatrix} 8 & 4 & 1 & -6 \end{bmatrix}$
a. 1×4
b. Row matrix

25. $\begin{bmatrix} -3.1 \\ 8.7 \end{bmatrix}$ a. 2×1
b. Column matrix

26. $\begin{bmatrix} -3 & 8 \\ 0 & 0 \end{bmatrix}$ a. 2×2
b. Square matrix

For Exercises 27–28, determine the value of the given element of the matrix $A = [a_{ij}]$.

$$A = \begin{bmatrix} -1 & 8 & -3 \\ 4 & 6 & 9 \end{bmatrix}$$

27. a_{21} 4

28. a_{12} 8

29. For what value of x, y, and z will $A = B$?

$$A = \begin{bmatrix} 3 & -4 \\ x & z \end{bmatrix} \quad \text{and} \quad B = \begin{bmatrix} y & -4 \\ 6 & 8 \end{bmatrix}$$ $x = 6, y = 3, z = 8$

30. Solve the equation $-3X + A = B$ for X, given that

$$A = \begin{bmatrix} 2 & -7 \\ 2 & -5 \end{bmatrix} \text{ and } B = \begin{bmatrix} 5 & 2 \\ -1 & 7 \end{bmatrix}. \quad X = \begin{bmatrix} -1 & -3 \\ 1 & -4 \end{bmatrix}$$

For Exercises 31–40, perform the indicated operations if possible.

$$A = \begin{bmatrix} -4 & 1 \\ 6 & -2 \\ 1 & 3 \end{bmatrix} \quad B = \begin{bmatrix} 2 & 3 & -7 \\ 1 & 5 & -6 \end{bmatrix} \quad C = \begin{bmatrix} \pi & 4 \\ -3 & 1 \\ 0 & 5 \end{bmatrix}$$

31. $3A$

32. $-2B$

33. $A + B$
Not possible

34. $B + C$
Not possible

35. $2A - C$

36. $4A + 3B$
Not possible

37. AB

38. BC

39. AC
Not possible

40. CA
Not possible

For Exercises 41–44, perform the indicated operations if possible.

$$A = \begin{bmatrix} 2 & 6 \\ -1 & 4 \end{bmatrix} \quad B = \begin{bmatrix} 1 \\ -3 \end{bmatrix} \quad C = \begin{bmatrix} 2 & 7 \end{bmatrix}$$

41. A^2

42. AB

43. BC

44. CB
$CB = [-19]$

45. A company owns two movie theaters in town. The number of popcorns and drinks sold for each theater is given in matrix Q. The price per item is given in matrix P. Find the product QP and interpret the result.

	Popcorn (small)	Popcorn (large)	Drinks (small)	Drinks (large)	
$Q =$	386	244	418	216	Theater 1
	450	382	476	262	Theater 2

$$P = \begin{bmatrix} \$8.50 \\ \$6.50 \\ \$5.50 \\ \$3.50 \end{bmatrix} \begin{matrix} \textbf{Popcorn (small)} \\ \textbf{Popcorn (large)} \\ \textbf{Drinks (small)} \\ \textbf{Drinks (large)} \end{matrix}$$

46. Matrix M gives the manufacturer price for four models of dining room tables. Matrix P gives the retail price to the customer.

$$M = \begin{matrix} & \textbf{Wood} & \textbf{Metal} \\ & \begin{bmatrix} \$1050 & \$940 \\ \$890 & \$800 \end{bmatrix} & \begin{matrix} \textbf{Large} \\ \textbf{Small} \end{matrix} \end{matrix}$$

$$P = \begin{matrix} & \textbf{Wood} & \textbf{Metal} \\ & \begin{bmatrix} \$1365 & \$1222 \\ \$1157 & \$1040 \end{bmatrix} & \begin{matrix} \textbf{Large} \\ \textbf{Small} \end{matrix} \end{matrix}$$

a. Compute $P - M$ and interpret its meaning.

b. If the tax rate in a certain city is 6%, use scalar multiplication to find a matrix F that gives the final price (including sales tax) to the customer for each model.

47. a. Write a matrix A that represents the coordinates of the vertices of the triangle. Place the x-coordinate of each point in the first row of A and the corresponding y-coordinate in the second row of A.

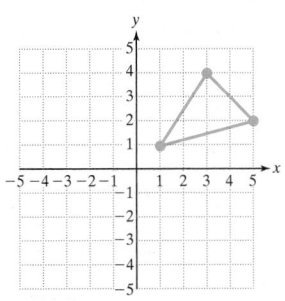

b. Use addition of matrices to shift the triangle 3 units to the left and 1 unit downward.

c. Find the product $\begin{bmatrix} -1 & 0 \\ 0 & 1 \end{bmatrix} \cdot A$ and explain the effect on the graph of the triangle.

d. Find the product $\begin{bmatrix} 1 & 0 \\ 0 & -1 \end{bmatrix} \cdot A$ and explain the effect on the graph of the triangle.

SECTION 9.4

48. Given two $n \times n$ matrices A and B, what are the criteria for the matrices to be inverses? $AB = I_n$ and $BA = I_n$

49. Determine whether A and B are inverses.

$$A = \begin{bmatrix} 4 & 1 \\ 3 & 2 \end{bmatrix} \quad B = \begin{bmatrix} 2 & -3 \\ -1 & 4 \end{bmatrix} \quad \text{No}$$

50. Determine whether A and B are inverses.

$$A = \begin{bmatrix} 1 & 3 \\ 2 & 1 \end{bmatrix} \quad B = \begin{bmatrix} -\frac{1}{5} & \frac{3}{5} \\ \frac{2}{5} & -\frac{1}{5} \end{bmatrix} \quad \text{Yes}$$

For Exercises 51–56, determine the inverse of the given matrix if possible. Otherwise state that the matrix is singular.

51. $A = \begin{bmatrix} 5 & -2 \\ 1 & 2 \end{bmatrix}$

52. $A = \begin{bmatrix} \frac{1}{4} & \frac{3}{8} \\ \frac{1}{8} & -\frac{1}{16} \end{bmatrix}$

53. $A = \begin{bmatrix} 2 & 3 \\ 16 & 24 \end{bmatrix}$
Singular matrix

54. $A = \begin{bmatrix} 1 & 0 & 1 \\ -1 & 5 & -3 \\ 1 & -3 & 2 \end{bmatrix}$

55. $A = \begin{bmatrix} -5 & 4 & 1 \\ 15 & -12 & -4 \\ 4 & -3 & -1 \end{bmatrix}$

56. $A = \begin{bmatrix} 1 & -3 & -17 \\ 1 & -2 & -12 \\ 0 & -2 & -10 \end{bmatrix}$
Singular matrix

57. Write the system of equations as a matrix equation of the form $AX = B$, where A is the coefficient matrix, X is the column matrix of variables, and B is the column matrix of constants.

$$\begin{aligned} -3x + 7y &= 6 \\ 4x \quad\quad + 2z &= -3 \\ 2x - y + 5z &= -13 \end{aligned}$$

For Exercises 58–61, solve the system using the inverse of the coefficient matrix.

58. $\frac{1}{4}x + \frac{3}{8}y = 4$ (See Exercise 52 for A^{-1}.)

$\frac{1}{8}x - \frac{1}{16}y = -2$ $\{(-8, 16)\}$

59. $5x - 2y = 26$ (See Exercise 51 for A^{-1}.)

$x + 2y = -2$ $\{(4, -3)\}$

60. $x \quad\quad + z = 2$ (See Exercise 54 for A^{-1}.)

$-x + 5y - 3z = -6$

$x - 3y + 2z = 3$ $\{(-5, 2, 7)\}$

61. $-5x + 4y + z = 6$ (See Exercise 55 for A^{-1}.)

$15x - 12y - 4z = -21$

$4x - 3y - z = -5$ $\{(1, 2, 3)\}$

SECTION 9.5

For Exercises 62–65, refer to the matrix

$$A = [a_{ij}] = \begin{bmatrix} -5 & 4 & -2 \\ 1 & 0 & 6 \\ 8 & -9 & 0 \end{bmatrix}.$$

a. Find the minor of the given element.

b. Find the cofactor of the given element.

62. a_{13} a. -9 b. -9 **63.** a_{31} a. 24 b. 24 **64.** a_{32} a. -28 b. 28 **65.** a_{23} a. 13 b. -13

For Exercises 66–71, evaluate the determinant of the given matrix.

66. $A = \begin{bmatrix} 9 & -4 \\ 2 & -3 \end{bmatrix}$ -19

67. $B = \begin{bmatrix} 3 & x \\ x & 27 \end{bmatrix}$ $81 - x^2$

68. $C = \begin{bmatrix} 9 & -15 \\ -3 & 5 \end{bmatrix}$ 0

69. $D = \begin{bmatrix} 4 & -1 & 0 \\ 6 & 8 & -2 \\ 1 & 5 & 3 \end{bmatrix}$ 156

70. $E = \begin{bmatrix} 4 & -9 & 0 \\ -3 & 8 & 0 \\ 6 & 1 & 0 \end{bmatrix}$ 0

71. $F = \begin{bmatrix} -2 & 0 & 3 & 1 \\ 1 & 1 & 0 & 5 \\ 4 & 0 & 0 & -2 \\ 0 & -3 & 0 & 6 \end{bmatrix}$ -270

For Exercises 72–76, solve the system by using Cramer's rule if possible. If Cramer's rule does not apply, use another method to solve the system.

72. $3x - 7y = 11$

$4x + 2y = 3$ $\left\{ \left(\frac{43}{34}, -\frac{35}{34} \right) \right\}$

73. $9x = 3y + 5$

$-2(x + 3y) = 4$ $\left\{ \left(\frac{3}{10}, -\frac{23}{30} \right) \right\}$

74. $2x + 5y = 10$

$10y = -4(x - 5)$

$\left\{ \left(\dfrac{10 - 5y}{2}, y \right) \middle| y \text{ is any real number} \right\}$

76. $2x + y - z = 5$

$7x + 7y - 6z = 5$

$3x + 5y - 4z = -1$ $\{\ \}$

75. $3x - 2y + z = 4$

$5x + 3y + 6z = 1$

$-2x + 5z = 7$

$\left\{ \left(-\dfrac{7}{25}, \dfrac{222}{125}, \dfrac{161}{125} \right) \right\}$

For Exercises 77–78, use Cramer's rule to solve for the indicated variable.

77. $-6x + 7y = 8$ \quad Solve for y. \quad $y = -\dfrac{94}{67}$

$2x + 5y + z = -3$

$3x + 2z = 11$

78. $3x_1 + 4x_3 = 6$ \quad Solve for x_4. \quad $x_4 = -\dfrac{43}{5}$

$4x_1 + 2x_3 + x_4 = -7$

$x_2 - 3x_4 = 2$

$5x_3 + x_4 = 1$

CHAPTER 9 Test

For Exercises 1–3, perform the elementary row operations on the matrix $A = \begin{bmatrix} 3 & 1 & 4 & | & -2 \\ 1 & 5 & -3 & | & 1 \\ 0 & 4 & 2 & | & 6 \end{bmatrix}$.

1. $R_1 \Leftrightarrow R_2$ \quad **2.** $-3R_2 + R_1 \to R_1$ \quad **3.** $\frac{1}{4}R_3 \to R_3$

4. Explain why the matrix is not in reduced row-echelon form.

$\begin{bmatrix} 1 & 0 & 6 & | & 4 \\ 0 & 1 & 0 & | & 2 \\ 0 & 0 & 1 & | & -3 \end{bmatrix}$

The elements above the leading 1 in the third column are not all zero. Specifically, the element in row 1, column 3 should be 0.

For Exercises 5–7, write the solution set for the system represented by each augmented matrix.

5. $\begin{bmatrix} 1 & 4 & | & 2 \\ 0 & 1 & | & 3 \end{bmatrix}$ \quad $\{(-10, 3)\}$

6. $\begin{bmatrix} 1 & 0 & 0 & | & 4 \\ 0 & 1 & 0 & | & 2 \\ 0 & 0 & 0 & | & 1 \end{bmatrix}$ \quad $\{\ \}$

7. $\begin{bmatrix} 1 & 0 & -3 & | & 0 \\ 0 & 1 & 2 & | & 5 \\ 0 & 0 & 0 & | & 0 \end{bmatrix}$ \quad $\{(3z, -2z + 5, z) \mid z \text{ is any real number}\}$

For Exercises 8–9, refer to the matrix $A = \begin{bmatrix} -1 & 3 & 0 \\ 2 & 5 & -4 \\ 6 & 9 & -8 \end{bmatrix}$.

a. Find the minor of the given element.

b. Find the cofactor of the given element.

8. a_{12} \quad a. 8 \quad b. -8 \qquad **9.** a_{31} \quad a. -12 \quad b. -12

For Exercises 10–11, evaluate the determinant of the matrix.

10. $A = \begin{bmatrix} 3 & 7 \\ 4 & -1 \end{bmatrix}$ \quad -31

11. $B = \begin{bmatrix} -3 & 4 & 7 \\ 1 & -2 & 3 \\ 6 & 5 & 0 \end{bmatrix}$ \quad 236

12. Use the determinant of A to determine whether A has an inverse.

$A = \begin{bmatrix} 1 & 2 & 3 \\ 1 & 4 & 11 \\ 2 & 5 & 10 \end{bmatrix}$ \quad $|A| = 0$. Therefore, A is singular (does not have an inverse).

For Exercises 13–16, solve the system by using Gaussian elimination or Gauss-Jordan elimination.

13. $-3(x + y) = 3y - 12$

$-3x = 4y - 6$ \quad $\{(-2, 3)\}$

14. $-3x = 11 - 18y$

$x - 6y = 2$ \quad $\{\ \}$

15. $x - 2y = 5z + 4$

$6y + 18z = 2x - 8$

$-3x + 8y + 20z = -18$ \quad $\{(-2, -8, 2)\}$

16. $2x + 6y + 30z = 2$

$x + 2y + 11z = 0$

$-3x - 6y - 33z = 0$ \quad $\{(-3z - 2, -4z + 1, z) \mid z \text{ is any real number}\}$

17. Solve the system by using Cramer's rule.

$3x - 5y = 7$

$11x + 2y = 8$ \quad $\left\{ \left(\dfrac{54}{61}, \dfrac{53}{61} \right) \right\}$

18. Use Cramer's rule to solve for x.

$2x = 3y - 4z + 11$

$9x + y = z - 1$

$3x - 4y = 7$ \quad $x = \dfrac{7}{31}$

19. Solve the equation $A - 4X = B$ for X, given that

$A = \begin{bmatrix} 2 & 5 \\ -2 & -3 \end{bmatrix}$ and $B = \begin{bmatrix} 6 & -3 \\ 14 & 5 \end{bmatrix}$. \quad $X = \begin{bmatrix} -1 & 2 \\ -4 & -2 \end{bmatrix}$

For Exercises 20–23, perform the indicated operations if possible.

$A = \begin{bmatrix} 4 & 1 & -3 \\ 2 & 4 & 6 \end{bmatrix}$ \quad $B = \begin{bmatrix} 1 & 9 \\ 0 & -1 \\ 3 & 5 \end{bmatrix}$ \quad $C = \begin{bmatrix} 0 & 1 & -4 \\ 2 & -1 & 8 \end{bmatrix}$

20. $2A - 3C$ \quad **21.** $A + B$ \quad **22.** AB \quad **23.** BA

Not possible

24. Determine whether A and B are inverses. \quad Yes

$A = \begin{bmatrix} -1 & \frac{2}{3} & -\frac{2}{3} \\ 1 & -\frac{1}{3} & \frac{1}{3} \\ 1 & -\frac{2}{3} & \frac{5}{3} \end{bmatrix}$ and $B = \begin{bmatrix} 1 & 2 & 0 \\ 4 & 3 & 1 \\ 1 & 0 & 1 \end{bmatrix}$

For Exercises 25–27, determine the inverse of the matrix if possible. Otherwise, state that the matrix is singular.

25. $A = \begin{bmatrix} 3 & 2 \\ 5 & 4 \end{bmatrix}$ \quad $A^{-1} = \begin{bmatrix} 2 & -1 \\ -\frac{5}{2} & \frac{3}{2} \end{bmatrix}$

26. $A = \begin{bmatrix} 2 & 5 & 10 \\ 1 & 3 & 7 \\ 1 & 4 & 11 \end{bmatrix}$ \quad Singular matrix

27. $A = \begin{bmatrix} 3 & -1 & -1 \\ 2 & -1 & 1 \\ -5 & 2 & 1 \end{bmatrix}$ $A^{-1} = \begin{bmatrix} 3 & 1 & 2 \\ 7 & 2 & 5 \\ 1 & 1 & 1 \end{bmatrix}$

For Exercises 28–29, solve the system by using the inverse of the coefficient matrix.

28. $3x + 2y = 13$ (See Exercise 25 for A^{-1}.)
$5x + 4y = 25$ $\{(1, 5)\}$

29. $3x - y - z = 8$ (See Exercise 27 for A^{-1}.)
$2x + z = y$
$-5x + 2y + z = -11$ $\{(2, 1, -3)\}$

30. a. Assume that traffic flows freely around the traffic circle. The flow rates given are measured in vehicles per hour. If the flow rate x_3 is 210 vehicles per hour, determine the flow rates x_1 and x_2.

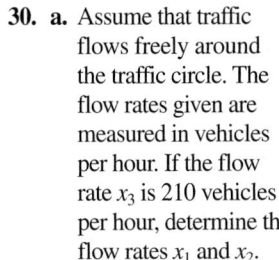

b. If traffic between intersections B and C flows at a rate of between 200 and 250 vehicles per hour, inclusive, find the range of values for x_1 and x_2.

31. Matrix C represents the number of calories burned per hour of exercise riding a bicycle, running, and walking for individuals of two different weights. Matrix N represents the number of hours spent working out with each type of activity for a given week. Find the product CN and interpret its meaning.

$$\begin{array}{ccc} \text{Bike} & \text{Run} & \text{Walk} \end{array}$$
$$C = \begin{bmatrix} 400 & 500 & 320 \\ 550 & 780 & 480 \end{bmatrix} \begin{array}{l} \textbf{120-lb person} \\ \textbf{180-lb person} \end{array}$$

$$N = \begin{bmatrix} 6 \\ 3 \\ 5 \end{bmatrix} \begin{array}{l} \textbf{Bike} \\ \textbf{Run} \\ \textbf{Walk} \end{array}$$

32. Given three points (x_1, y_1), (x_2, y_2), and (x_3, y_3), the points are collinear if the following determinant is equal to zero.

$$\begin{vmatrix} x_1 & y_1 & 1 \\ x_2 & y_2 & 1 \\ x_3 & y_3 & 1 \end{vmatrix}$$

Determine if the points $(4, -11)$, $(-1, -1)$, and $(-5, 7)$ are collinear. Yes

CHAPTER 9 Cumulative Review Exercises

1. Given $f(x) = x^3$, find the difference quotient $\dfrac{f(x + h) - f(x)}{h}$. $3x^2 + 3xh + h^2$

2. Given $f(x) = -2(x - 4)^2 - 5$,

a. Determine the vertex of the graph of the parabola. $(4, -5)$

b. Determine the intervals over which the graph is increasing. $(-\infty, 4)$

c. Determine the intervals over which the graph is decreasing. $(4, \infty)$

d. Write the domain in interval notation. $(-\infty, \infty)$

e. Write the range in interval notation. $(-\infty, -5]$

3. Given $f(x) = \sqrt{2x - 3}$, find the average rate of change on the interval $\left[2, \frac{7}{2}\right]$. $\dfrac{2}{3}$

4. Simplify each expression.

a. i^{57} i

b. i^{22} -1

5. Divide and write the answer in the form $a + bi$. $\dfrac{3 + 4i}{2 - 3i}$

6. Find the distance between the points $(1, -7)$ and $(4, 2)$. $3\sqrt{10}$

7. Determine the center and radius of the circle defined by $\left(x + \frac{5}{3}\right)^2 + y^2 = 11$. Center: $\left(-\frac{5}{3}, 0\right)$; Radius: $\sqrt{11}$

8. Technetium-99m (abbreviated ^{99m}Tc) is a short-lived gamma-ray emitter that is used in nuclear medicine. Suppose that a patient is given a small amount of ^{99m}Tc

for a diagnostic test. Further suppose that the amount of ^{99m}Tc decays exponentially according to the model $Q(t) = Q_0 e^{-kt}$. The value Q_0 is the initial amount administered, and $Q(t)$ represents the amount of ^{99m}Tc remaining after t hours. The value of k is the decay constant.

a. If 94% of the technetium has decayed after 24 hr (that is, 6% remains), determine the value of k. Round to 3 decimal places. $k = 0.117$

b. If 30 mCi is initially given to a patient for blood pool imaging of the heart, determine the amount remaining after 10 hr. Round to 1 decimal place. 9.3 mCi

c. Determine the amount of time required for the amount of ^{99m}Tc to fall below 1% of the original amount given. Round to 1 decimal place. More than 39.4 hr

For Exercises 9–12,

a. Graph the function.

b. Write the domain in interval notation.

c. Write the range in interval notation.

9. $f(x) = e^{x-2} + 1$

10. $g(x) = \ln(x + 4)$

11. $h(x) = \sqrt{-x}$

12. $k(x) = \begin{cases} (x - 1)^2 - 2 & \text{for } x \le 1 \\ x + 1 & \text{for } x > 1 \end{cases}$

For Exercises 13–14, solve the system of equations.

13. $3(x - y) = y - 3$ $\{(7, 6)\}$

$-2x + \dfrac{1}{2}y = -11$

14. $2x - y + 5z = -2$ $\{(-4z, -3z + 2, z) \mid z \text{ is any real number}\}$

$\quad x \quad\;\; + 4z = 0$

$\quad x + y = 2 - 7z$

15. Determine if the graph of the equation $|x| + y^2 = 9$ is symmetric with respect to the x-axis, y-axis, origin, or none of these. x-axis, y-axis, and origin

16. Determine if the function defined by $f(x) = 6x^3 - 4x$ is even, odd, or neither.   odd

For Exercises 17–20, refer to the given matrices and perform the indicated operations.

$$A = \begin{bmatrix} 2 & -3 \\ 1 & 6 \end{bmatrix} \quad B = \begin{bmatrix} 1 & 4 \\ 0 & 7 \end{bmatrix} \quad C = \begin{bmatrix} -1 & 4 & 0 \\ 4 & 2 & 6 \\ 3 & -2 & 0 \end{bmatrix}$$

17. AB **18.** $|C|$ 60   **19.** $4A - B$ **20.** A^{-1}

For Exercises 21–23, find the union or intersection of the given sets. Write the answers in set-builder notation.

$$A = \{x \mid 3 < x\}, B = \{x \mid -2 < x \le 4\}, C = \{x \mid x \ge 5\}$$

21. $A \cup B$ **22.** $A \cap B$ **23.** $A \cap C$

$\{x \mid x > -2\}$ $\{x \mid 3 < x \le 4\}$ $\{x \mid x \ge 5\}$

For Exercises 24–27, solve the equation or inequality. Write the solution set to the inequalities in interval notation if possible.

24. $2x - 3 - 5\sqrt{2x - 3} + 4 = 0$ $\left\{ \dfrac{19}{2}, 2 \right\}$

25. $|-x + 4| = |x - 4|$ $\mathbb{R}$

26. $\dfrac{x + 4}{2x - 1} \ge 1$ $\left(\dfrac{1}{2}, 5 \right]$

27. $\log_2(3x - 4) = \log_2(x + 1) + 1$ $\{6\}$

28. Find all zeros of $f(x) = 2x^3 - x^2 + 18x - 9$.   The zeros are $\frac{1}{2}$, $3i$, and $-3i$.

29. Find the asymptotes. $r(x) = \dfrac{2x - 1}{x^2 - 9}$

Vertical asymptotes: $x = 3$, $x = -3$; Horizontal asymptote: $y = 0$

30. Write the expression as a single logarithm.

$$-3 \log x - \dfrac{1}{2} \log y + 5 \log z \quad \log\!\left(\dfrac{z^5}{x^3 \sqrt{y}} \right)$$

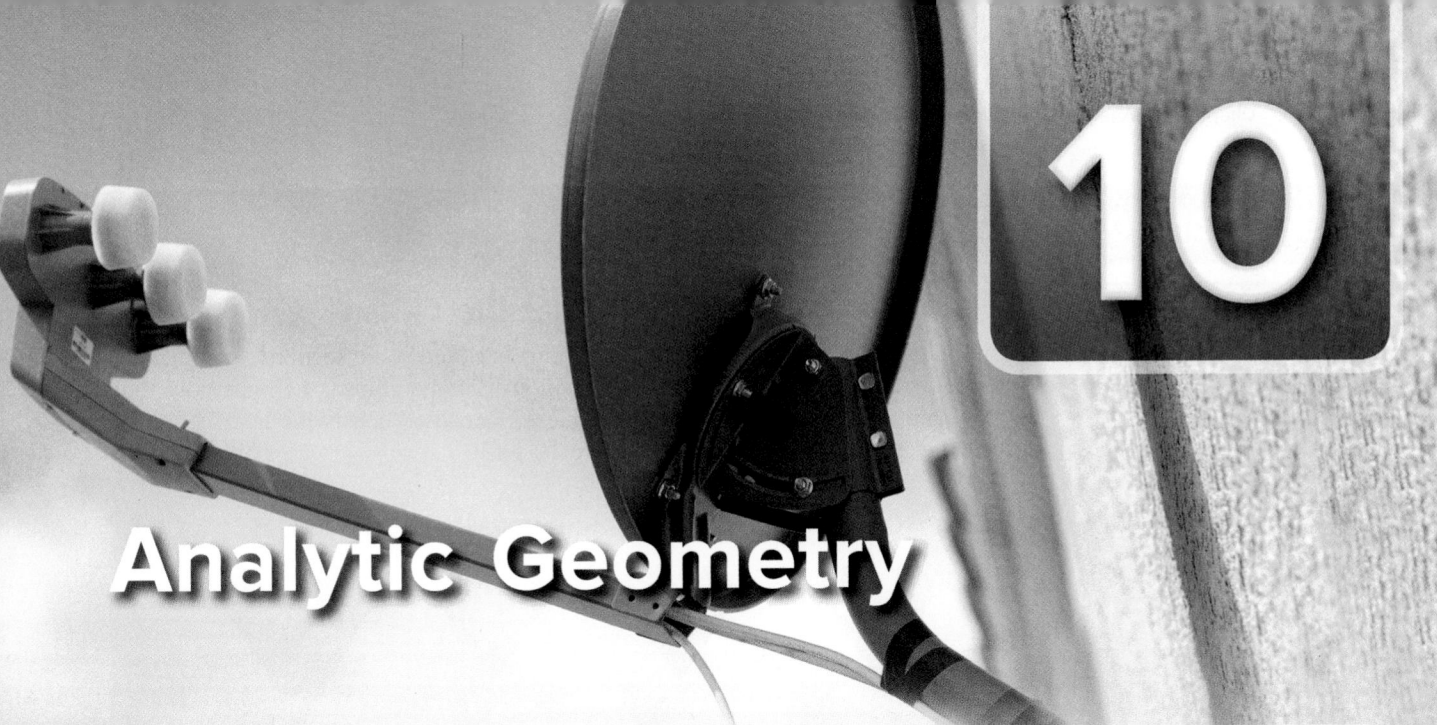

10

Analytic Geometry

Chapter Outline

Many television viewers today rely on a satellite dish to provide a wide variety of programming options. A satellite dish is a parabolic device that receives microwave signals from broadcast centers via communications satellites. The data are then compressed and rendered into a digital form that can be viewed on an electronic device such as a television, computer, or smartphone.

In this chapter, we begin with the study of three types of curves called conic sections: the ellipse, the hyperbola, and the parabola. As we work through the chapter, we will learn how the reflective properties of these curves are used in numerous applications. For example, microwaves received by a satellite dish bounce off the parabolic surface to a common point called the focus. It is at this point that the receiver is placed to gather and interpret the signal that ultimately delivers hundreds of channels to our television set.

Later in the chapter, we represent equations of conics using polar coordinates. This is advantageous in certain applications such as modeling planetary motions. In addition, we will learn how to represent curves in a plane using a third variable called a parameter. In particular, using time as a parameter enables us to determine the time and location of a point traveling along a curve.

SECTION 10.1 The Ellipse

Instructor Note:

Three-dimensional models of the conic sections can help students visualize these intersections more clearly.

1. Graph an Ellipse Centered at the Origin

In this chapter, we investigate four types of curves called conic sections. Specifically, these are the circle (covered in detail in Section 1.2), the ellipse, the hyperbola, and the parabola. Conic sections derive their names because each is the intersection of a plane and a double-napped cone (Figure 10-1).

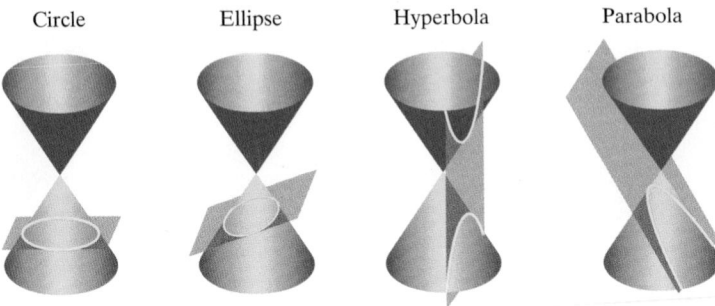

Circle   Ellipse   Hyperbola   Parabola

Figure 10-1

Recall from Section 1.2 that a circle is the set of all points in a plane, r units from a fixed point (the center of the circle), where r is the radius of the circle. We derived the standard form of an equation of a circle from the distance formula by noting that the distance between the center of the circle (h, k) and a point on the circle (x, y) is r units.

Standard Form of an Equation of a Circle

Given a circle centered at (h, k) with radius r, the standard form of an equation of the circle is given by $(x - h)^2 + (y - k)^2 = r^2$ for $r > 0$.

Example

$(x - 3)^2 + (y + 1)^2 = 4$

Standard form

$(x - 3)^2 + [y - (-1)]^2 = 2^2$

Center: $(3, -1)$

Radius: 2

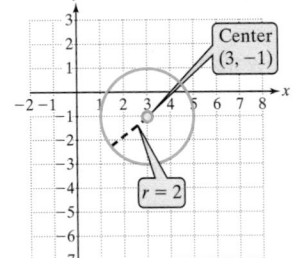

We now study a second type of conic section called an ellipse. An ellipse is an oval-shaped curve that appears in applications in many fields including architecture, acoustics, medicine, and astronomy. To determine an equation of an ellipse, we first need to formalize the definition.

Definition of an Ellipse

An **ellipse** is the set of all points (x, y) in a plane such that the sum of the distances between (x, y) and two fixed points is a constant. The fixed points are called the **foci** (plural of **focus**) of the ellipse.

To visualize the definition of an ellipse, consider the following application. Suppose Sonya wants to cut an elliptical rug from a rectangular rug to avoid a stain made by the family dog. She places two tacks along the center horizontal line. Then she ties the ends of a slack piece of rope to each tack. With the rope pulled tight, she traces out a curve. This curve is an ellipse, and the tacks are located at the foci of the ellipse (Figure 10-2).

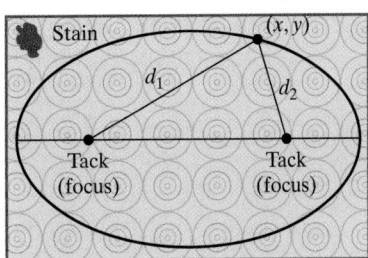

Figure 10-2

An ellipse may be elongated in any direction; however, we will study only those that are elongated horizontally and vertically. Figure 10-3 and Figure 10-4 show ellipses centered at the origin with foci on the x-axis and y-axis, respectively. The line through the foci intersects the ellipse at two points called **vertices.** The line segment with endpoints at the vertices is called the **major axis.** The **center** of the ellipse is the midpoint of the major axis. The line segment perpendicular to the major axis and passing through the center of the ellipse with endpoints on the ellipse is called the **minor axis.**

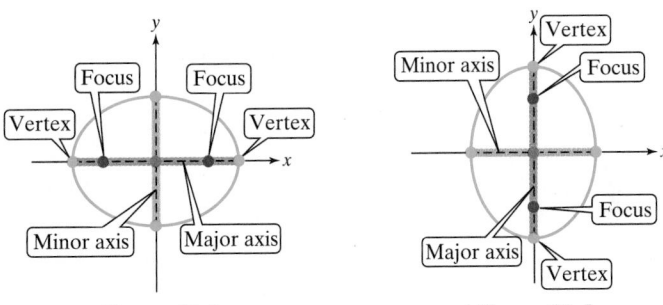

Figure 10-3   **Figure 10-4**

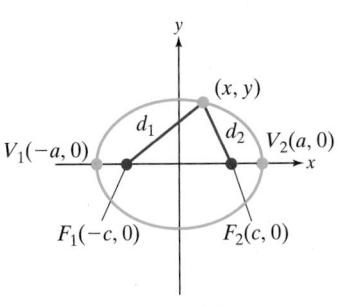

Figure 10-5

We will derive the standard form of an equation of an ellipse centered at the origin with foci on the x-axis. In Figure 10-5, the foci are labeled as $F_1(-c, 0)$ and $F_2(c, 0)$, and the vertices are labeled as $V_1(-a, 0)$ and $V_2(a, 0)$. Let (x, y) be an arbitrary point on the ellipse and let the distances between (x, y) and the two foci be d_1 and d_2.

From Figure 10-5, the distance from F_1 to V_2 is $(a + c)$. The distance from F_2 to V_2 is $(a - c)$. By the definition of an ellipse, the sum of the distances between a point on the ellipse (such as V_2) and the two foci must be constant. Adding $(a + c)$ and $(a - c)$, we see that this constant is $2a$.

$$(a + c) + (a - c) = 2a$$

The sum of the distances between a point on an ellipse and the two foci is equal to the length of the major axis, $2a$.

The sum of the distances between (x, y) and the two foci is equal to $2a$.

$$\underset{d_1}{} \quad + \quad \underset{d_2}{} \quad = 2a$$

$$\sqrt{[(x - (-c)]^2 + (y - 0)^2} + \sqrt{(x - c)^2 + (y - 0)^2} = 2a \quad \text{Apply the distance formula.}$$

$$\sqrt{(x + c)^2 + y^2} + \sqrt{(x - c)^2 + y^2} = 2a \quad \text{Simplify within parentheses.}$$

In Exercise 99, we guide you through a series of algebraic steps to eliminate the radicals. Then collecting the variable terms on one side and the constant terms on the other side, the equation becomes

$$(a^2 - c^2)x^2 + a^2 y^2 = a^2(a^2 - c^2) \qquad (1)$$

From Figure 10-5, we see that the distance between F_1 and F_2 is $2c$. Furthermore, since $d_1 + d_2 = 2a$, we have that $2c < 2a$, which implies that $c < a$ and that $a^2 - c^2 > 0$. Letting $b^2 = a^2 - c^2$, Equation (1) becomes

$$b^2 x^2 + a^2 y^2 = a^2 b^2$$

$$\frac{b^2 x^2}{a^2 b^2} + \frac{a^2 y^2}{a^2 b^2} = \frac{a^2 b^2}{a^2 b^2} \qquad \text{Divide both sides by } a^2 b^2.$$

$$\frac{x^2}{a^2} + \frac{y^2}{b^2} = 1 \qquad \text{Standard form of an equation of an ellipse}$$

This equation represents an ellipse centered at the origin with foci on the x-axis. The length of the major axis is $2a$ and the length of the minor axis is $2b$. The value a is also referred to as the length of the **semimajor axis** (the distance from the center of an ellipse to a vertex). The value b is referred to as the length of the **semiminor axis** (half the length of the minor axis).

In a similar manner, we can develop the standard form of an equation of an ellipse centered at the origin with foci on the y-axis. These two equations are summarized as follows.

Standard Forms of an Equation of an Ellipse Centered at the Origin

The standard forms of an equation of an ellipse centered at the origin are as follows. Assume that

- $a > b > 0$.
- The length of the major axis is $2a$, and the length of the minor axis is $2b$.

	Major Axis: x-axis	Major Axis: y-axis
Equation:	$\dfrac{x^2}{a^2} + \dfrac{y^2}{b^2} = 1$	$\dfrac{x^2}{b^2} + \dfrac{y^2}{a^2} = 1$
Center:	$(0, 0)$	$(0, 0)$
Foci: (*Note*: $c^2 = a^2 - b^2$)	$(c, 0)$ and $(-c, 0)$	$(0, c)$ and $(0, -c)$
Vertices: Endpoints: major axis	$(a, 0)$ and $(-a, 0)$	$(0, a)$ and $(0, -a)$
Endpoints: minor axis	$(0, b)$ and $(0, -b)$	$(b, 0)$ and $(-b, 0)$
Graph:	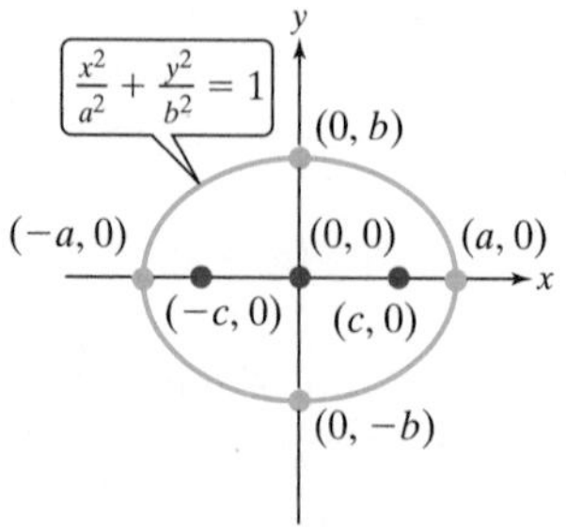	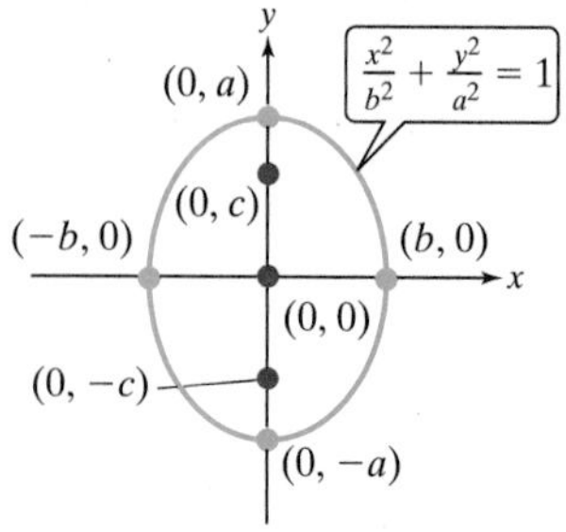

Avoiding Mistakes

It is important to note that the foci are not actually on the graph of an elliptical curve. However, the foci are used to define the curve.

For an ellipse centered at the origin, the x- and y-intercepts are the endpoints of the major and minor axes. For example, for an ellipse centered at (0, 0) with foci on the x-axis, we have:

Standard form:

> *x*-intercept: substitute 0 for *y*.

> *y*-intercept: substitute 0 for *x*.

$$\frac{x^2}{a^2} + \frac{y^2}{b^2} = 1 \longrightarrow \frac{x^2}{a^2} + \frac{(0)^2}{b^2} = 1 \qquad \frac{(0)^2}{a^2} + \frac{y^2}{b^2} = 1$$

$$\frac{x^2}{a^2} = 1 \qquad\qquad \frac{y^2}{b^2} = 1$$

$$x^2 = a^2 \qquad\qquad y^2 = b^2$$

$$x = \pm a \qquad\qquad y = \pm b$$

In Example 1, we graph an ellipse and identify the key features.

Classroom Example: p. 920
Exercise 14

Point of Interest

The National Center for the Performing Arts in Beijing, Peoples Republic of China, is a semielliptical dome made of titanium and glass and surrounded by an artificial lake. On a calm day, the reflection from the lake gives the illusion of a full ellipse.

EXAMPLE 1   Graphing an Ellipse Centered at the Origin

Given $\dfrac{x^2}{16} + \dfrac{y^2}{9} = 1$, graph the ellipse, and identify the center, vertices, and foci.

Solution:

$$\frac{x^2}{16} + \frac{y^2}{9} = 1$$

The equation is in the standard form of an ellipse. Since $16 > 9$, $a^2 = 16$ and $b^2 = 9$. Since the greater number in the denominator is found in the x^2 term, the ellipse is elongated horizontally and the foci are on the x-axis.

$$\frac{x^2}{(4)^2} + \frac{y^2}{(3)^2} = 1$$

$\boxed{a = 4}$ $\boxed{b = 3}$

$a^2 = 16$, and because $a > 0$, we have that $a = 4$.
$b^2 = 9$, and because $b > 0$, we have that $b = 3$.

To graph the ellipse, plot the vertices (endpoints of the major axis) and the endpoints of the minor axis, and sketch the ellipse through the points.

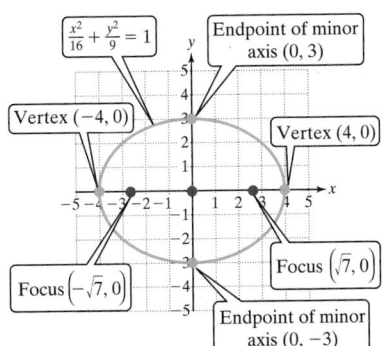

Center: (0, 0)

Vertices: $(a, 0)$ and $(-a, 0)$
(4, 0) and $(-4, 0)$

Minor axis endpoints: $(0, b)$ and $(0, -b)$
(0, 3) and $(0, -3)$

To find the foci, use the relationship
$$c^2 = a^2 - b^2$$
$$c^2 = (4)^2 - (3)^2 = 16 - 9 = 7$$
$$c^2 = 7 \quad \text{(Recall that } c > 0.\text{)}$$
$$c = \sqrt{7} \approx 2.65$$

Foci: $(c, 0)$ and $(-c, 0)$
$(\sqrt{7}, 0)$ and $(-\sqrt{7}, 0)$

It is important to note that the center and foci of the ellipse are not actually part of the curve. For this reason, we graphed the curve in blue and plotted the foci and center in red.

Skill Practice 1   Graph the ellipse, and identify the center, vertices, minor axis endpoints, and foci. $\dfrac{x^2}{9} + \dfrac{y^2}{4} = 1$

Answer

1.

Center: (0, 0)
Vertices: (3, 0) and $(-3, 0)$
Minor axis endpoints: (0, 2) and (0, -2)
Foci: $(\sqrt{5}, 0)$ and $(-\sqrt{5}, 0)$

The graph of an ellipse does not define y as a function of x (the graph fails the vertical line test). However, an ellipse can be defined by the union of two functions:

an upper semiellipse and a lower semiellipse. This is helpful to graph an ellipse on a graphing utility.

From Example 1, we have

$$\frac{x^2}{16} + \frac{y^2}{9} = 1 \longrightarrow y = \pm\sqrt{9 - \frac{9x^2}{16}}$$

The functions represent the top and bottom semiellipses, respectively.

$$y = \sqrt{9 - \frac{9x^2}{16}} \quad \text{top semiellipse}$$

$$y = -\sqrt{9 - \frac{9x^2}{16}} \quad \text{bottom semiellipse}$$

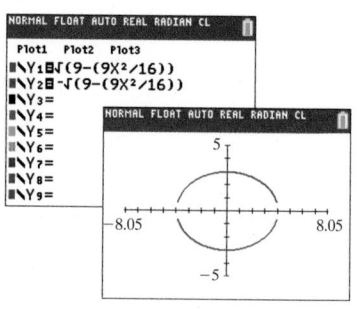

Classroom Example: p. 920
Exercise 20

TIP For an equation of an ellipse written in standard form, the term containing the larger denominator indicates the direction of elongation of the graph. In Example 2, the larger denominator of 25 is located in the y term. Notice that the ellipse is elongated vertically.

EXAMPLE 2   Graphing an Ellipse Centered at the Origin

Given $25x^2 + 9y^2 = 225$, graph the ellipse and identify the center, vertices, minor axis endpoints, and foci.

Solution:

$$25x^2 + 9y^2 = 225$$
For the equation to be in the standard form for an ellipse, we require the constant term to be 1.

$$\frac{25x^2}{225} + \frac{9y^2}{225} = \frac{225}{225}$$
Divide both sides by 225.

$$\frac{x^2}{9} + \frac{y^2}{25} = 1$$
Since $25 > 9$, $a^2 = 25$ and $b^2 = 9$. Since the greater number in the denominator is found in the y^2 term, the ellipse is elongated vertically and the foci are on the y-axis.

$$\frac{x^2}{(3)^2} + \frac{y^2}{(5)^2} = 1$$
$a^2 = 25$, and because $a > 0$, we have that $a = 5$.
$b^2 = 9$, and because $b > 0$, we have that $b = 3$.

$\boxed{b = 3}$ $\boxed{a = 5}$

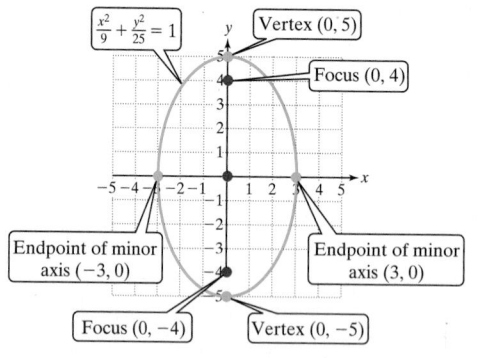

Center: $(0, 0)$

Vertices: $(0, a)$ and $(0, -a)$
$(0, 5)$ and $(0, -5)$

Minor axis endpoints: $(b, 0)$ and $(-b, 0)$
$(3, 0)$ and $(-3, 0)$

To find the foci, use the relationship
$$c^2 = a^2 - b^2$$
$$c^2 = (5)^2 - (3)^2 = 25 - 9 = 16$$
$$c^2 = 16$$
$$c = 4 \quad \text{(Recall that } c > 0.)$$

Foci: $(0, c)$ and $(0, -c)$
$(0, 4)$ and $(0, -4)$

Skill Practice 2   Graph the ellipse and identify the center, vertices, minor axis endpoints, and foci. $25x^2 + 9y^2 = 900$

Answer
2.

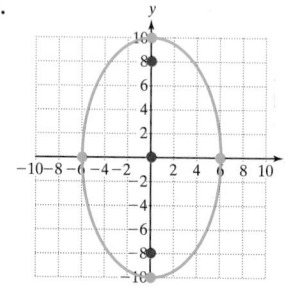

Center: $(0, 0)$
Vertices: $(0, 10)$, $(0, -10)$
Minor axis endpoints: $(6, 0)$, $(-6, 0)$
Foci: $(0, 8)$, $(0, -8)$

2. Graph an Ellipse Centered at (h, k)

In Section 1.2, we learned that the standard form of an equation of a circle centered at the origin with radius r is given by $x^2 + y^2 = r^2$. For a circle centered at (h, k), we have the equation

$$(x - h)^2 + (y - k)^2 = r^2 \quad \text{Circle centered at } (h, k) \text{ with radius } r$$

Replacing x by $x - h$ and y by $y - k$ shifts the graph of $x^2 + y^2 = r^2$ a total of h units horizontally and k units vertically. The same is true for an equation of an ellipse.

Standard Forms of an Equation of an Ellipse Centered at (h, k)

The standard forms of an equation of an ellipse centered at (h, k) are as follows. Assume that

- $a > b > 0$.
- The length of the major axis is $2a$, and the length of the minor axis is $2b$.

	Major Axis Horizontal	**Major Axis Vertical**
Equation:	$\dfrac{(x - h)^2}{a^2} + \dfrac{(y - k)^2}{b^2} = 1$	$\dfrac{(x - h)^2}{b^2} + \dfrac{(y - k)^2}{a^2} = 1$
Center:	(h, k)	(h, k)
Foci: (*Note*: $c^2 = a^2 - b^2$)	($h + c$, k) and ($h - c$, k)	(h, $k + c$) and (h, $k - c$)
Vertices: Endpoints: major axis	($h + a$, k) and ($h - a$, k)	(h, $k + a$) and (h, $k - a$)
Endpoints: minor axis	(h, $k + b$) and (h, $k - b$)	($h + b$, k) and ($h - b$, k)
Graph:		

To graph an ellipse centered at (h, k), it is helpful to locate the center first as well as to identify the orientation of the ellipse (elongated horizontally or elongated vertically). Then use the values of a and b to locate the endpoints of the major and minor axes. This is demonstrated in Example 3.

Classroom Example: p. 920
Exercise 26

EXAMPLE 3   Graphing an Ellipse Centered at (h, k)

Graph the ellipse and identify the center, vertices, endpoints of the minor axis, and foci. $\dfrac{(x - 3)^2}{16} + \dfrac{(y + 2)^2}{25} = 1$

Solution:

$\dfrac{(x - 3)^2}{16} + \dfrac{(y + 2)^2}{25} = 1$

The standard form of an equation of an ellipse is given. The quantity $(y + 2)^2$ is equivalent to $[y - (-2)]^2$. Therefore, the center of the ellipse is $(3, -2)$.

$\dfrac{(x - 3)^2}{(4)^2} + \dfrac{[y - (-2)]^2}{(5)^2} = 1$

$\underbrace{}_{b = 4} \qquad \underbrace{}_{a = 5}$

Since $25 > 16$, $a^2 = 25$ and $b^2 = 16$. $a = 5$ and $b = 4$.

Since the greater number in the denominator is found in the y^2 term, the ellipse is elongated vertically.

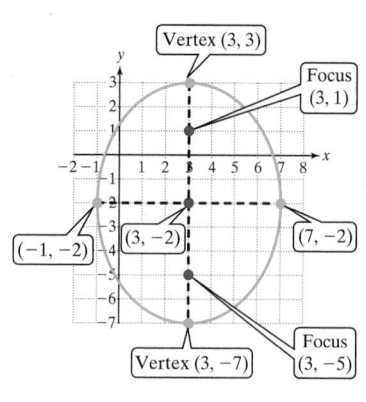

Center: $(3, -2)$
The vertices are a units above and below the center.
Vertices: $(3, -2 + 5)$ and $(3, -2 - 5)$
    $(3, 3)$ and $(3, -7)$
The minor axis endpoints are b units to the right and left of the center.
Minor axis endpoints: $(3 + 4, -2)$ and $(3 - 4, -2)$
    $(7, -2)$ and $(-1, -2)$
To find the foci, use the relationship
$c^2 = a^2 - b^2$
$c^2 = (5)^2 - (4)^2 = 25 - 16 = 9$
    $c = 3$ (Recall that $c > 0$.)
The foci are c units above and below the center. The foci are $(3, 1)$ and $(3, -5)$.

Skill Practice 3 Graph the ellipse and identify the center, vertices, foci, and endpoints of the minor axis. $\dfrac{(x + 1)^2}{25} + \dfrac{(y - 2)^2}{9} = 1$

The equation of the ellipse from Example 3 can be expanded by expanding the binomials and combining like terms.

$$\frac{(x - 3)^2}{16} + \frac{(y + 2)^2}{25} = 1 \qquad \text{Standard form}$$

$$400\left[\frac{(x - 3)^2}{16} + \frac{(y + 2)^2}{25}\right] = 400(1) \qquad \text{Clear fractions.}$$

$$25(x - 3)^2 + 16(y + 2)^2 = 400$$

$$25(x^2 - 6x + 9) + 16(y^2 + 4y + 4) = 400 \qquad \text{Expand the binomials.}$$

$$25x^2 + 16y^2 - 150x + 64y - 111 = 0 \qquad \text{Expanded form}$$

Given an equation of an ellipse in expanded form, we can reverse this process by completing the square to write the equation in standard form. The benefit of writing an equation in standard form is that we can easily identify the center of the ellipse and the values of a and b.

Classroom Example: p. 920
Exercise 34

EXAMPLE 4 Writing an Equation of an Ellipse in Standard Form

Given $3x^2 + 7y^2 + 6x - 28y + 10 = 0$,

a. Write the equation of the ellipse in standard form.

b. Identify the center, vertices, foci, and endpoints of the minor axis.

Answer

3.

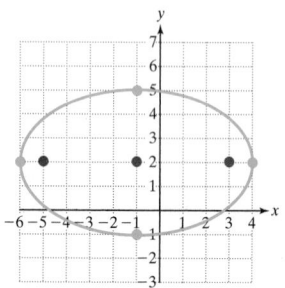

Center: $(-1, 2)$
Vertices: $(4, 2)$, $(-6, 2)$
Minor axis endpoints: $(-1, 5)$, $(-1, -1)$
Foci: $(3, 2)$, $(-5, 2)$

Solution:

a. $3x^2 + 7y^2 + 6x - 28y + 10 = 0$ Group the x terms and group the y terms.

$3x^2 + 6x + 7y^2 - 28y = -10$ Move the constant term to the right side of the equation.

$3(x^2 + 2x + \underline{}) + 7(y^2 - 4y + \underline{}) = -10$ Factor out the leading coefficient from the x and y terms. Leave room to complete the square within parentheses.

3 · 1 is added to   7 · 4 is added to
the left side.   the left side.

$$3(x^2 + 2x + 1) + 7(y^2 - 4y + 4) = -10 + 3 + 28$$

$3 \cdot 1 \quad 7 \cdot 4$

Complete the square.
Note that $\left[\frac{1}{2}(2)\right]^2 = 1$ and $\left[\frac{1}{2}(-4)\right]^2 = 4$.

The values 1 and 4 are added inside parentheses. However, as a result of the factors of 3 and 7 in front of the parentheses, $3 \cdot 1$ and $7 \cdot 4$ are actually added to the left side of the equation.

> To balance the equation add $3 \cdot 1$ and $7 \cdot 4$ to the right side.

$$3(x + 1)^2 + 7(y - 2)^2 = 21$$

To write the equation in standard form, we require that the constant on the right side of the equation be 1.

$$\frac{3(x + 1)^2}{21} + \frac{7(y - 2)^2}{21} = \frac{21}{21}$$

Divide both sides by 21.

$$\frac{(x + 1)^2}{7} + \frac{(y - 2)^2}{3} = 1$$

Standard form

b. $$\frac{[x - (-1)]^2}{7} + \frac{(y - 2)^2}{3} = 1$$

The center is $(-1, 2)$.
Since $7 > 3$, $a^2 = 7$ and $b^2 = 3$. The greater number in the denominator is in the x^2 term. Therefore, the ellipse is elongated horizontally.

$$\frac{[x - (-1)]^2}{(\sqrt{7})^2} + \frac{(y - 2)^2}{(\sqrt{3})^2} = 1$$

$a = \sqrt{7} \qquad b = \sqrt{3}$

$a^2 = 7$. Because $a > 0$, $a = \sqrt{7}$.
$b^2 = 3$. Because $b > 0$, $b = \sqrt{3}$.
$c^2 = a^2 - b^2 = 7 - 3 = 4$
$c^2 = 4$, and because $c > 0$, we have $c = 2$.

The center is $(-1, 2)$.
The vertices are $\left(-1 + \sqrt{7}, 2\right)$ and $\left(-1 - \sqrt{7}, 2\right)$.
The foci are $(1, 2)$ and $(-3, 2)$.
The endpoints of the minor axis are $\left(-1, 2 + \sqrt{3}\right)$ and $\left(-1, 2 - \sqrt{3}\right)$.

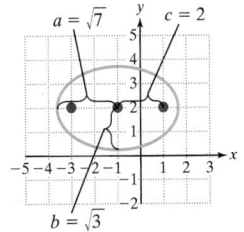

Skill Practice 4 Repeat Example 4 with $2x^2 + 11y^2 - 12x + 44y + 40 = 0$.

Now that we know the standard form of an equation of an ellipse, we can build an equation based on known information about the ellipse. This is demonstrated in Example 5.

EXAMPLE 5 Finding an Equation of an Ellipse from Given Information

Classroom Example: p. 921
Exercise 54

Determine the standard form of an equation of an ellipse with vertices $(4, 2)$ and $(-4, 2)$ and foci $\left(\sqrt{15}, 2\right)$ and $\left(-\sqrt{15}, 2\right)$.

Solution:

Plotting the given points tells us that the ellipse is elongated horizontally. Therefore, the standard form is

$$\frac{(x - h)^2}{a^2} + \frac{(y - k)^2}{b^2} = 1$$

The center is midway between the vertices. Therefore, the center is $(0, 2)$, indicating that $h = 0$ and $k = 2$.

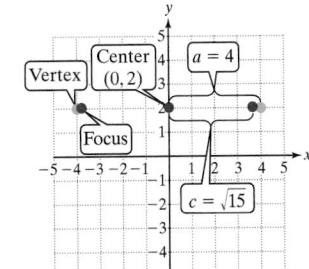

Answers

4. a. $\dfrac{(x - 3)^2}{11} + \dfrac{(y + 2)^2}{2} = 1$

   b. Center: $(3, -2)$; Vertices: $\left(3 + \sqrt{11}, -2\right)$ and $\left(3 - \sqrt{11}, -2\right)$
   Foci: $(0, -2)$ and $(6, -2)$; Minor axis endpoints: $\left(3, -2 + \sqrt{2}\right)$ and $\left(3, -2 - \sqrt{2}\right)$

The vertices are located 4 units away from the center. This means that $a = 4$. The foci are located $\sqrt{15}$ units from the center, indicating that $c = \sqrt{15}$. The only missing piece of information is the value of b. We can find b from the relationship $c^2 = a^2 - b^2$ or equivalently $b^2 = a^2 - c^2$.

$$b^2 = (4)^2 - \left(\sqrt{15}\right)^2$$
$$b^2 = 1$$
$$b = 1 \qquad \text{(Recall that } b > 0.\text{)}$$

$$\frac{(x - 0)^2}{(4)^2} + \frac{(y - 2)^2}{(1)^2} = 1 \qquad \text{Substitute } h = 0, k = 2, a = 4, \text{ and } b = 1 \text{ into the standard form.}$$

$$\frac{x^2}{16} + \frac{(y - 2)^2}{1} = 1 \quad \text{or} \quad \frac{x^2}{16} + (y - 2)^2 = 1 \qquad \text{Standard form}$$

Skill Practice 5   Determine the standard form of an equation of an ellipse with vertices $(3, 3)$ and $(-3, 3)$ and foci $\left(2\sqrt{2}, 3\right)$ and $\left(-2\sqrt{2}, 3\right)$.

3. Use Ellipses in Applications

The ellipse and its three-dimensional counterpart, the ellipsoid (an object in the shape of a blimp), have many useful applications that arise from the reflective property of the ellipse. For example, consider an elliptical-shaped dome. Sound or light waves emitted from one focus are reflected off the surface to the other focus. This principle is applied in Example 6.

Classroom Example: p. 921
Exercise 58

EXAMPLE 6   **Applying the Reflective Property of an Ellipse**

A room in a museum has a semielliptical dome with a major axis of 50 ft and a semiminor axis of 12 ft. A light is placed at one focus of the ellipse and the light that reflects off the ceiling will be directed to the other focus. The museum curator wants to place a sculpture so that the sculpture is at the other focus where it will be illuminated by the light.

a. Choose a coordinate system so that the ellipse is oriented with horizontal major axis with the center placed at (0, 0). Write an equation of the semiellipse.

b. Approximately where should the sculpture be placed?

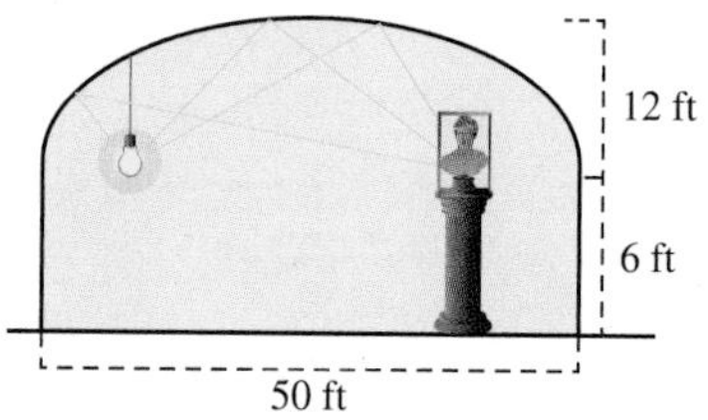

Solution:

a. A graph of the semiellipse is shown in Figure 10-6. The length of the semimajor axis is 25 ft ($a = 25$) and the length of the semiminor axis is 12 ft ($b = 12$). An equation of the full ellipse is

$$\frac{x^2}{25^2} + \frac{y^2}{12^2} = 1 \quad \text{or} \quad \frac{x^2}{625} + \frac{y^2}{144} = 1.$$

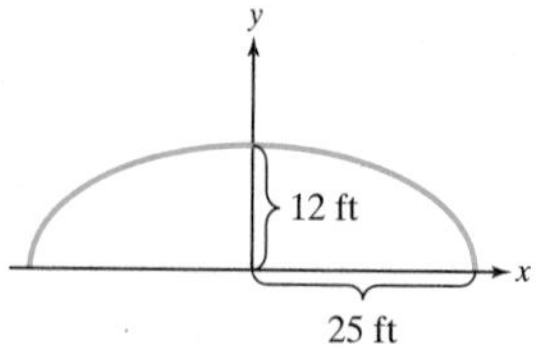

Figure 10-6

We want the top half only, which is given by $\dfrac{x^2}{625} + \dfrac{y^2}{144} = 1$ for $y \geq 0$,

or equivalently $y = 12\sqrt{1 - \dfrac{x^2}{625}}$.

Answer

5. $\dfrac{x^2}{9} + (y - 3)^2 = 1$

b. To find the location of the sculpture, we must find the location of the other focus.

$$c^2 = a^2 - b^2$$
$$c^2 = (25)^2 - (12)^2 = 625 - 144 = 481$$
$$c = \pm\sqrt{481} \approx \pm21.9$$

The sculpture should be located approximately 21.9 ft from the center of the room along the major axis, opposite the light source, and 6 ft up.

Skill Practice 6   A tunnel has vertical sides of 7 ft with a semielliptical top. The width of the tunnel is 10 ft, and the height at the top is 10 ft.

a. Write an equation of the semiellipse. For convenience, place the coordinate system with (0, 0) at the center of the ellipse.

b. To construct the tunnel, an engineer needs to find the location of the foci. How far from the center are the foci?

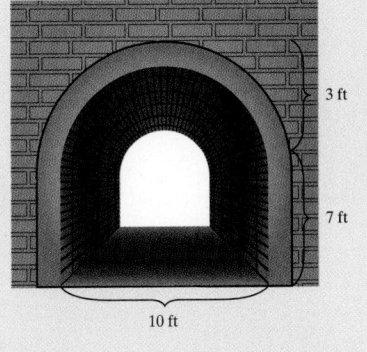

3 ft

7 ft

10 ft

---

4. Determine and Apply the Eccentricity of an Ellipse

The degree of elongation of an ellipse is measured by the eccentricity *e* of the ellipse.

Eccentricity of an Ellipse

For an ellipse defined by $\dfrac{(x-h)^2}{a^2} + \dfrac{(y-k)^2}{b^2} = 1$ or $\dfrac{(x-h)^2}{b^2} + \dfrac{(y-k)^2}{a^2} = 1$,

the **eccentricity** *e* is given by $e = \dfrac{c}{a}$, where $a > b > 0$, $c > 0$, and $c^2 = a^2 - b^2$.

Note: The eccentricity of an ellipse is a number between 0 and 1; that is, $0 < e < 1$.

With this definition, we see that the eccentricity of an ellipse is the ratio of the distance between a focus and the center to the distance between a vertex and the center; that is, $e = \frac{c}{a}$.

- If the foci are near the vertices (that is, *a* is only slightly greater than *c*), then the ellipse is more elongated (Figure 10-7).
- If the foci are near the center, then the ellipse is more circular (Figure 10-8).

Answers

6. a. $\dfrac{x^2}{25} + \dfrac{y^2}{9} = 1;\ y \geq 0$ or

equivalently $y = 3\sqrt{1 - \dfrac{x^2}{25}}$

b. The foci are 4 ft to the right and left of the center of the semiellipse. These are at points 7 ft above the ground and 4 ft left and right of the center of the tunnel.

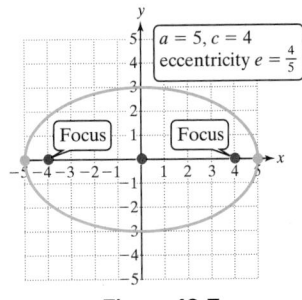

Figure 10-7

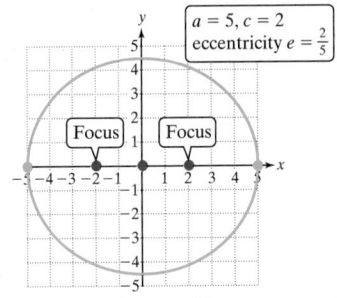

Figure 10-8

The orbits of the planets in our solar system are elliptical with the Sun at one focus. The eccentricities given in Table 10-1 indicate that the orbit of Venus is the most circular and the orbit of Pluto is the most elongated. (*Source:* NASA, www.nasa.gov)

Table 10-1

Planet	Eccentricity
Mercury	0.2056
Venus	0.0067 ←
Earth	0.0167
Mars	0.0935
Jupiter	0.0489
Saturn	0.0565
Uranus	0.0457
Neptune	0.0113
Pluto (dwarf planet)	0.2488 ←

The orbit of Venus is the most circular.

The orbit of Pluto is the most elongated.

Classroom Example: p. 922
Exercise 76

EXAMPLE 7   Using Eccentricity in an Application

The point in a planet's orbit where it is closest to the Sun is called **perihelion** and the point where it is farthest away is called **aphelion** (Figure 10-9). Suppose that Mars is 2.0662×10^8 km from the Sun at perihelion. Use the eccentricity from Table 10-1 to find the distance between Mars and the Sun at aphelion.

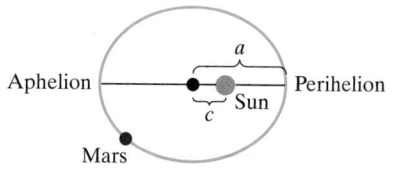

Figure 10-9

Solution:

For convenience, we have labeled the center of the ellipse as point C and the focus at which the Sun resides as point S. Aphelion and perihelion are labeled A and P, respectively (Figure 10-10).

We need to find the distance between Mars and the Sun at aphelion. This is given by $a + c$. Since a and c are two unknown values, we need two independent relationships between a and c.

Figure 10-10

The distance at perihelion and the eccentricity give such information.

(1) $a - c = 2.0662 \times 10^8$ (distance at perihelion)   This a nonlinear system of equations.

(2) $\dfrac{c}{a} = 0.0935 \longrightarrow c = 0.0935a$ (eccentricity)

$$a - 0.0935a = 2.0662 \times 10^8 \qquad \text{Use the substitution method.}$$
$$0.9065a = 2.0662 \times 10^8 \qquad \text{Combine like terms.}$$
$$a \approx 2.2793 \times 10^8 \qquad \text{Divide by 0.9065.}$$
$$c = 0.0935(2.2793 \times 10^8) \qquad \text{Back substitute.}$$
$$c = 2.1311 \times 10^7$$

Thus, $a + c = 2.2793 \times 10^8 + 2.1311 \times 10^7$
$= 2.4924 \times 10^8$

The distance between Mars and the Sun at aphelion (the farthest point from the Sun) is approximately 2.4924×10^8 km.

Skill Practice 7   Pluto is 7.376×10^9 km at aphelion (farthest point from the Sun). Use the eccentricity of Pluto's orbit of 0.2488 to find the closest distance between Pluto and the Sun.

Answer

7. 4.437×10^9 km

Prerequisite Review

For Exercises R.1–R.3, determine the center and radius of the circle.

R.1. $x^2 + (y + 5)^2 = 8$ Center: $(0, -5)$ **R.2.** $x^2 + y^2 = 25$ Center: $(0, 0)$ **R.3.** $(x - 4)^2 + (y + 3)^2 = 81$ Center: $(4, -3)$
Radius: $2\sqrt{2}$ Radius: 5   Radius: 9

For Exercises R.4–R.5, find the value of n so that the expression is a perfect square trinomial. Then factor the trinomial.

R.4. $a^2 - 8a + n$ $n = 16; (a - 4)^2$ **R.5.** $y^2 + 7y + n$ $n = \dfrac{49}{4}; \left(y + \dfrac{7}{2}\right)^2$

R.6. Complete the square and write the equation of the circle in standard form. $x^2 + y^2 + 10x - 10y + 41 = 0$
$(x + 5)^2 + (y - 5)^2 = 9$

Concept Connections

1. The circle, the ellipse, the hyperbola, and the parabola are categories of ___conic___ sections.

2. An ___ellipse___ is a set of points (x, y) in a plane such that the sum of the distances between (x, y) and two fixed points called ___foci___ is a constant.

3. The line through the foci intersects an ellipse at two points called ___vertices___.

4. The line segment with endpoints at the vertices of an ellipse is called the ___major___ axis.

5. The line segment perpendicular to the major axis, with endpoints on the ellipse, and passing through the center of the ellipse, is called the ___minor___ axis.

6. Given $\dfrac{x^2}{a^2} + \dfrac{y^2}{b^2} = 1$, where $a > b > 0$, the ordered pairs representing the vertices are ___$(a, 0)$___ and ___$(-a, 0)$___. The ordered pairs representing the endpoints of the minor axis are ___$(0, b)$___ and ___$(0, -b)$___.

7. Given $\dfrac{(x - h)^2}{b^2} + \dfrac{(y - k)^2}{a^2} = 1$, where $a > b > 0$, the ordered pairs representing the endpoints of the vertices are ___$(h, k + a)$___ and ___$(h, k - a)$___. The ordered pairs representing the endpoints of the minor axis are ___$(h + b, k)$___ and ___$(h - b, k)$___.

8. When referring to the standard form of an equation of an ellipse, the ___eccentricity___ e is defined as $e = \dfrac{\square}{\square}$. $\dfrac{c}{a}$

Objective 1: Graph an Ellipse Centered at the Origin

For Exercises 9–10,

a. Use the distance formula to find the distances d_1, d_2, d_3, and d_4.   **b.** Find the sum $d_1 + d_2$.

c. Find the sum $d_3 + d_4$.   **d.** How do the sums from parts (b) and (c) compare?

e. How do the sums of the distances from parts (b) and (c) relate to the length of the major axis?

9.

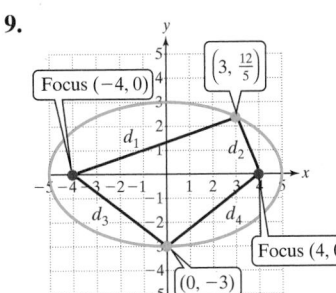

a. $d_1 = \dfrac{37}{5}, d_2 = \dfrac{13}{5},$
   $d_3 = 5, d_4 = 5$
b. 10
c. 10
d. They are the same.
e. The sums equal the length of the major axis.

10.

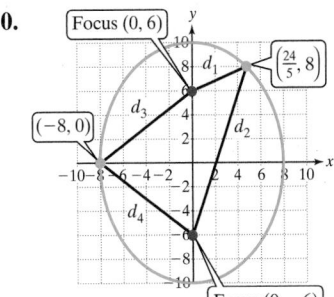

a. $d_1 = \dfrac{26}{5}, d_2 = \dfrac{74}{5},$
   $d_3 = 10, d_4 = 10$
b. 20
c. 20
d. They are the same.
e. The sums equal the length of the major axis.

For Exercises 11–12, from the equation of the ellipse, determine if the major axis is horizontal or vertical.

11. a. $\dfrac{x^2}{2} + \dfrac{y^2}{5} = 1$ **b.** $\dfrac{x^2}{5} + \dfrac{y^2}{2} = 1$ **12. a.** $\dfrac{x^2}{11} + \dfrac{y^2}{10} = 1$ **b.** $\dfrac{x^2}{10} + \dfrac{y^2}{11} = 1$
Vertical   Horizontal   Horizontal   Vertical

For Exercises 13–22,

a. Identify the center of the ellipse.

b. Determine the value of a.

c. Determine the value of b.

d. Identify the vertices.

e. Identify the endpoints of the minor axis.

f. Identify the foci.

g. Determine the length of the major axis.

h. Determine the length of the minor axis.

i. Graph the ellipse. **(See Examples 1–2)**

13. $\dfrac{x^2}{100} + \dfrac{y^2}{25} = 1$

14. $\dfrac{x^2}{64} + \dfrac{y^2}{49} = 1$

15. $\dfrac{x^2}{25} + \dfrac{y^2}{100} = 1$

16. $\dfrac{x^2}{49} + \dfrac{y^2}{64} = 1$

17. $4x^2 + 25y^2 = 100$

18. $9x^2 + 64y^2 = 576$

19. $-36x^2 - 4y^2 = -36$

20. $-64x^2 - 16y^2 = -64$

21. $\dfrac{x^2}{12} + \dfrac{y^2}{5} = 1$

22. $\dfrac{x^2}{18} + \dfrac{y^2}{7} = 1$

Objective 2: Graph an Ellipse Centered at (h, k)

For Exercises 23–32,

a. Identify the center of the ellipse.

b. Identify the vertices.

c. Identify the endpoints of the minor axis.

d. Identify the foci.

e. Graph the ellipse. **(See Example 3)**

23. $\dfrac{(x-1)^2}{25} + \dfrac{(y+6)^2}{16} = 1$

24. $\dfrac{(x-3)^2}{25} + \dfrac{(y+4)^2}{9} = 1$

25. $\dfrac{(x+4)^2}{49} + \dfrac{(y-2)^2}{64} = 1$

26. $\dfrac{(x+1)^2}{36} + \dfrac{(y-5)^2}{81} = 1$

27. $(x-6)^2 + \dfrac{y^2}{9} = 1$

28. $(x-2)^2 + \dfrac{y^2}{4} = 1$

29. $x^2 + 9(y+1)^2 = 81$

30. $x^2 + 4(y+4)^2 = 100$

31. $\dfrac{4(x-3)^2}{25} + \dfrac{16(y-2)^2}{49} = 1$

32. $\dfrac{4(x-4)^2}{81} + \dfrac{16(y-3)^2}{225} = 1$

For Exercises 33–42,

a. Write the equation of the ellipse in standard form. **(See Example 4)**

b. Identify the center, vertices, endpoints of the minor axis, and foci.

33. $3x^2 + 5y^2 + 12x - 60y + 177 = 0$

34. $7x^2 + 11y^2 + 70x - 66y + 197 = 0$

35. $3x^2 + 2y^2 - 30x - 4y + 59 = 0$

36. $5x^2 + 2y^2 - 40x - 12y + 78 = 0$

37. $4x^2 + y^2 + 14y + 45 = 0$

38. $x^2 + 49y^2 - 6x - 40 = 0$

39. $36x^2 + 100y^2 - 180x + 800y + 925 = 0$

40. $4x^2 + 9y^2 - 12x + 18y - 18 = 0$

41. $4x^2 + 9y^2 - 8x + 3 = 0$

42. $25x^2 + 16y^2 + 64y + 63 = 0$

For Exercises 43–56, write the standard form of an equation of an ellipse subject to the given conditions. (See Example 5)

43. Vertices: (4, 0) and (−4, 0);
Foci: (3, 0) and (−3, 0) $\dfrac{x^2}{16} + \dfrac{y^2}{7} = 1$

44. Vertices: (6, 0) and (−6, 0);
Foci: (5, 0) and (−5, 0) $\dfrac{x^2}{36} + \dfrac{y^2}{11} = 1$

45. Endpoints of minor axis: $\left(\sqrt{17}, 0\right)$ and $\left(-\sqrt{17}, 0\right)$;
Foci: (0, 9) and (0, −9) $\dfrac{x^2}{17} + \dfrac{y^2}{98} = 1$

46. Endpoints of minor axis: $\left(\sqrt{21}, 0\right)$ and $\left(-\sqrt{21}, 0\right)$;
Foci: (0, 5) and (0, −5) $\dfrac{x^2}{21} + \dfrac{y^2}{46} = 1$

47. Major axis parallel to the x-axis; Center: (2, 3);
Length of major axis: 14 units;
Length of minor axis: 10 units $\dfrac{(x-2)^2}{49} + \dfrac{(y-3)^2}{25} = 1$

48. Major axis parallel to the y-axis; Center: (1, 5);
Length of major axis: 22 units;
Length of minor axis: 14 units $\dfrac{(x-1)^2}{49} + \dfrac{(y-5)^2}{121} = 1$

49. Foci: (0, 1) and (8, 1);
Length of minor axis: 6 units $\dfrac{(x-4)^2}{25} + \dfrac{(y-1)^2}{9} = 1$

50. Foci: (−3, 3) and (7, 3);
Length of minor axis: 8 units $\dfrac{(x-2)^2}{41} + \dfrac{(y-3)^2}{16} = 1$

51. Vertices: (0, 5) and (0, −5);
Passes through $\left(\frac{16}{5}, 3\right)$ $\dfrac{x^2}{16} + \dfrac{y^2}{25} = 1$

52. Vertices: (0, 13) and (0, −13);
Passes through $\left(\frac{25}{13}, 12\right)$ $\dfrac{x^2}{25} + \dfrac{y^2}{169} = 1$

53. Vertices: (3, 4) and (3, −4); $\dfrac{(x-3)^2}{5} + \dfrac{y^2}{16} = 1$
Foci: $(3, \sqrt{11})$ and $(3, -\sqrt{11})$

54. Vertices: (4, 5) and (−4, 5); $\dfrac{x^2}{16} + \dfrac{(y-5)^2}{10} = 1$
Foci: $(\sqrt{6}, 5)$ and $(-\sqrt{6}, 5)$

55. Vertices: (2, 1) and (−12, 1); $\dfrac{(x+5)^2}{49} + \dfrac{(y-1)^2}{16} = 1$
Foci: $(-5 - \sqrt{33}, 1)$ and $(-5 + \sqrt{33}, 1)$

56. Vertices: (−8, 4) and (2, 4); $\dfrac{(x+3)^2}{25} + \dfrac{(y-4)^2}{4} = 1$
Foci: $(-3 - \sqrt{21}, 4)$ and $(-3 + \sqrt{21}, 4)$

Objective 3: Use Ellipses in Applications

57. The reflective property of an ellipse is used in lithotripsy. Lithotripsy is a technique for treating kidney stones without surgery. Instead, high-energy shock waves are emitted from one focus of an elliptical shell and reflected painlessly to a patient's kidney stone located at the other focus. The vibration from the shock waves shatter the stone into pieces small enough to pass through the patient's urine.

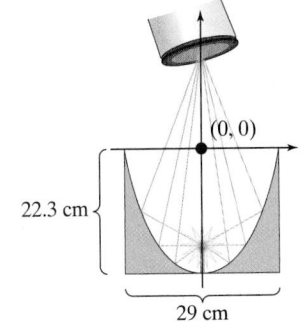

22.3 cm

(0, 0)

29 cm

A vertical cross section of a lithotripter is in the shape of a semiellipse with the dimensions shown. Approximate the distance from the center along the major axis where the patient's kidney stone should be located so that the shock waves will target the stone. Round to 2 decimal places. **(See Example 6)**   16.94 cm

58. The reflective property of an ellipse is the principle behind "whispering galleries." These are rooms with elliptically shaped ceilings such that a person standing at one focus can hear even the slightest whisper spoken by another person standing at the other focus.

Suppose that a dome has a semielliptical ceiling, 96 ft long and 23 ft high. Choose a coordinate system so that the center of the semiellipse is (0, 0) with vertices (−48, 0) and (48, 0) and with the top of the ceiling at (0, 23).

Each person should be approximately 42.1 ft on either side of the center along the major axis to hear the whispering effect.

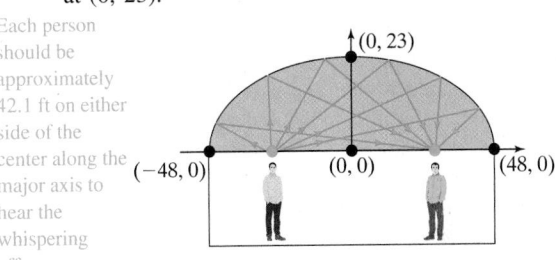

(0, 23)

(−48, 0)   (0, 0)   (48, 0)

Approximately how far from the center along the major axis should each person be standing to hear the "whispering" effect? Round to 1 decimal place.

59. A homeowner wants to make an elliptical rug from a 12-ft by 10-ft rectangular piece of carpeting.

a. What lengths of the major and minor axes would maximize the area of the new rug?
Major axis: 12 ft; Minor axis: 10 ft

b. Write an equation of the ellipse with maximum area. Use a coordinate system with the origin at the center of the rug and horizontal major axis. $\dfrac{x^2}{36} + \dfrac{y^2}{25} = 1$

c. To cut the rectangular piece of carpeting, the homeowner needs to know the location of the foci. Then she will insert tacks at the foci, take a piece of string the length of the major axis, and fasten the ends to the tacks. Drawing the string tight, she'll use a piece of chalk to trace the ellipse. At what coordinates should the tacks be located? Describe the location.   Place the tacks at $(\sqrt{11}, 0)$ and $(-\sqrt{11}, 0)$. These are approximately 3.3 ft to the left and right of the center.

60. Coordinate axes are superimposed on a drawing of an elliptical lightbulb.

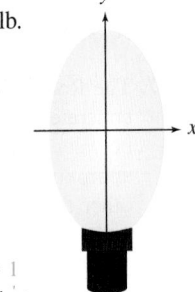

a. If the vertices are 1.5 in. from the center and the width of the bulb is 1 in. at its widest point, write an equation of the elliptical curve along the outside edge of the bulb (do not consider the metal fixture at the bottom as part of the elliptical shape). $\dfrac{x^2}{0.25} + \dfrac{y^2}{2.25} = 1$

b. If the company logo is to be placed at one of the foci, at what coordinates could the logo be placed? Describe the location.
Place the logo at $(0, \sqrt{2})$ or $(0, -\sqrt{2})$. These are approximately 1.4 in. above or below the center.

61. Charles and Bernice ("Ray") Eames were American designers who made major contributions to modern architecture and furniture design. Suppose that a manufacturer wants to make an Eames elliptical coffee table 90 in. long and 30 in. wide out of an 8-ft by 4-ft piece of birch plywood. If the center of a piece of plywood is positioned at (0, 0), determine the distance from the center at which the foci should be located to draw the ellipse.   The foci should be located $\pm 30\sqrt{2}$ in. (approximately 42.4 in.) from the center along the major axis.

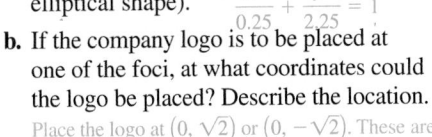

62. A window above a door is to be made in the shape of a semiellipse. If the window is 10 ft at the base and 3 ft high at the center, determine the distance from the center at which the foci are located.
The foci are located 4 ft from the center along the major axis.

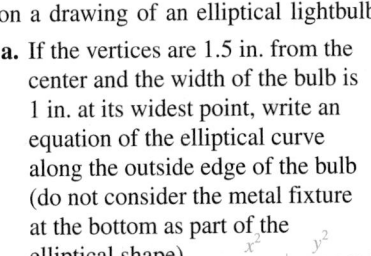

10 ft

3 ft

Objective 4: Determine and Apply the Eccentricity of an Ellipse

63. Choose one: The eccentricity of an ellipse is a number   b
    a. less than 0.
    b. between 0 and 1.
    c. greater than 1.

64. Choose one: An ellipse with eccentricity close to 0 will appear   b
    a. more elongated.
    b. more circular.

65. Choose one: An ellipse with eccentricity close to 1 will appear   a
    a. more elongated.
    b. more circular.

66. Choose one: If the foci of an ellipse are close to the center of the ellipse, then the graph will appear   b
    a. more elongated.
    b. more circular.

For Exercises 67–72, determine the eccentricity of the ellipse.

67. $\dfrac{x^2}{169} + \dfrac{y^2}{25} = 1$ $e = \dfrac{12}{13}$

68. $\dfrac{x^2}{100} + \dfrac{y^2}{64} = 1$ $e = \dfrac{3}{5}$

69. $\dfrac{\left(x + \frac{4}{5}\right)^2}{144} + \dfrac{\left(y - \frac{5}{3}\right)^2}{225} = 1$ $e = \dfrac{3}{5}$

70. $\dfrac{\left(x - \frac{1}{3}\right)^2}{9} + \dfrac{\left(y + \frac{7}{9}\right)^2}{25} = 1$ $e = \dfrac{4}{5}$

71. $\dfrac{x^2}{12} + \dfrac{(y - 3)^2}{6} = 1$ $e = \dfrac{\sqrt{2}}{2}$

72. $\dfrac{(x + 7)^2}{18} + \dfrac{y^2}{12} = 1$ $e = \dfrac{\sqrt{3}}{3}$

73. Halley's Comet and the Earth both orbit the Sun in elliptical paths with the Sun at one focus. The eccentricity of the comet's orbit is 0.967 and the eccentricity of the Earth's orbit is 0.0167. The eccentricity for the Earth is close to zero, whereas the eccentricity for Halley's Comet is close to 1. Based on this information, how do the orbits compare?   The Earth's orbit is more circular, and the orbit for Halley's Comet is very elongated.

74. Halley's Comet and the comet Hale-Bopp both orbit the Sun in elliptical paths with the Sun at one focus. The eccentricity of Halley's Comet's orbit is 0.967 and the eccentricity of comet Hale-Bopp's orbit is 0.995. Which comet has a more elongated orbit?   Comet Hale-Bopp's orbit is more elongated because the eccentricity is greater.

75. The moon's orbit around the Earth is elliptical with the Earth at one focus and with eccentricity 0.0549. If the distance between the moon and Earth at the closest point is 363,300 km, determine the distance at the farthest point. Round to the nearest 100 km. **(See Example 7)**   405,500 km

76. The planet Saturn orbits the Sun in an elliptical path with the Sun at one focus. The eccentricity of the orbit is 0.0565 and the distance between the Sun and Saturn at perihelion (the closest point) is 1.353×10^9 km. Determine the distance at aphelion (the farthest point). Round to the nearest million kilometers. 1.515×10^9 km

77. The Roman Coliseum is an elliptical stone and concrete amphitheater in the center of Rome, built between 70 A.D. and 80 A.D. The Coliseum seated approximately 50,000 spectators and was used among other things for gladiatorial contests.

    a. Using a vertical major axis, write an equation of the ellipse representing the center arena if the maximum length is 287 ft and the maximum width is 180 ft. Place the origin at the center of the arena. $\dfrac{x^2}{90^2} + \dfrac{y^2}{143.5^2} = 1$
    b. Approximate the eccentricity of the center arena. Round to 2 decimal places. $e \approx 0.78$
    c. Find an equation of the outer ellipse if the maximum length is 615 ft and the maximum width is 510 ft. $\dfrac{x^2}{255^2} + \dfrac{y^2}{307.5^2} = 1$
    d. Approximate the eccentricity of the outer ellipse. Round to 2 decimal places. $e \approx 0.56$
    e. Explain how you know that the outer ellipse is more circular than the inner ellipse.   The outer ellipse is more circular than the inner ellipse because the eccentricity is closer to zero.

78. *The Ellipse*, also called *President's Park South*, is a park in Washington, D.C. The lawn area is elliptical with a major axis of 1058 ft and minor axis of 903 ft.
    a. Find an equation of the elliptical boundary. Take the horizontal axes to be the major axis and locate the origin of the coordinate system at the center of the ellipse.
    b. Approximate the eccentricity of the ellipse. Round to 2 decimal places. $e \approx 0.52$

For Exercises 79–82, write the standard form of an equation of the ellipse subject to the following conditions.

79. Center: (0, 0); Eccentricity: $\frac{15}{17}$; $\frac{x^2}{64} + \frac{y^2}{289} = 1$
Major axis vertical of length 34 units

80. Center: (0, 0); Eccentricity: $\frac{40}{41}$; $\frac{x^2}{81} + \frac{y^2}{1681} = 1$
Major axis vertical of length 82 units

81. Foci: (0, −1) and (8, −1); $\frac{(x-4)^2}{25} + \frac{(y+1)^2}{9} = 1$
Eccentricity: $\frac{4}{5}$

82. Foci: (0, −1) and (−6, −1); $\frac{(x+3)^2}{25} + \frac{(y+1)^2}{16} = 1$
Eccentricity: $\frac{3}{5}$

Mixed Exercises

83. A circular vent pipe is placed on a flat roof.

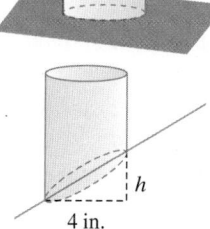
4 in.

a. Write an equation of the circular cross section that the pipe makes with the roof. Assume that the origin is placed at the center of the circle. $x^2 + y^2 = 4$

b. Now suppose that the pipe is placed on a roof with a slope of $\frac{3}{5}$. What shape will the cross section of the pipe form with the plane of the roof?   Elliptical

4 in.

c. Determine the length of the major and minor axes. Find the exact value and approximate to 1 decimal place if necessary.
Major axis: $\frac{4\sqrt{34}}{5}$ in. ≈ 4.7 in.; Minor axis: 4 in.

84. A cylindrical glass of water with diameter 3.5 in. sits on a horizontal counter top.

3.5 in.

a. Write an equation of the circular surface of the water. Assume that the origin is placed at the center of the circle. $x^2 + y^2 = 3.0625$

b. If the glass is tipped 30°, what shape will the surface of the water have?   Elliptical

c. With the glass tipped 30°, the waterline makes a slope of $\frac{1}{2}$ with the coordinate system shown. Determine the length of the major and minor axes. Round to 1 decimal place.
Major axis: ≈ 3.9 in.; Minor axis: 3.5 in.

30° h
3.5 in.

85. The graph of $\dfrac{x^2}{9} + \dfrac{y^2}{4} = 1$ represents an ellipse.

Determine the part of the ellipse represented by the given equation.

a. $y = 2\sqrt{1 - \dfrac{x^2}{9}}$ **b.** $y = -2\sqrt{1 - \dfrac{x^2}{9}}$

c. $x = 3\sqrt{1 - \dfrac{y^2}{4}}$ **d.** $x = -3\sqrt{1 - \dfrac{y^2}{4}}$

86. The graph of $\dfrac{x^2}{16} + \dfrac{y^2}{81} = 1$ represents an ellipse.

Determine the part of the ellipse represented by the given equation.

a. $x = -4\sqrt{1 - \dfrac{y^2}{81}}$ **b.** $x = 4\sqrt{1 - \dfrac{y^2}{81}}$

c. $y = 9\sqrt{1 - \dfrac{x^2}{16}}$ **d.** $y = -9\sqrt{1 - \dfrac{x^2}{16}}$

For Exercises 87–90, solve the system of equations.

87. $\dfrac{x^2}{25} + \dfrac{y^2}{9} = 1$
$3x + 5y = 15$
$\{(0, 3), (5, 0)\}$

88. $13y = 12x - 156$
$\dfrac{x^2}{169} + \dfrac{y^2}{144} = 1$
$\{(0, -12), (13, 0)\}$

89. $\dfrac{x^2}{4} + \dfrac{y^2}{16} = 1$
$y = -x^2 + 4$
$\{(0, 4), (-2, 0), (2, 0)\}$

90. $\dfrac{x^2}{64} + \dfrac{y^2}{289} = 1$
$y = \dfrac{13}{64}x^2 - 13$
$\{(8, 0), (-8, 0)\}$

Given an ellipse with major axis of length 2a and minor axis of length 2b, the area is given by $A = \pi ab$. The perimeter is approximated by $P \approx \pi\sqrt{2(a^2 + b^2)}$. For Exercises 91–92,

a. Determine the area of the ellipse.   **b.** Approximate the perimeter.

91. $\dfrac{x^2}{8} + \dfrac{(y+3)^2}{4} = 1$
a. $A = 4\pi\sqrt{2}$ square units
b. $P \approx 2\pi\sqrt{6}$ units

92. $\dfrac{(x-1)^2}{9} + \dfrac{y^2}{11} = 1$
a. $A = 3\pi\sqrt{11}$ square units
b. $P \approx 2\pi\sqrt{10}$ units

93. A line segment with endpoints on an ellipse, perpendicular to the major axis, and passing through a focus, is called a *latus rectum* of the ellipse. Show that the length of a latus rectum is $\dfrac{2b^2}{a}$ for the ellipse.

$$\frac{x^2}{a^2} + \frac{y^2}{b^2} = 1$$

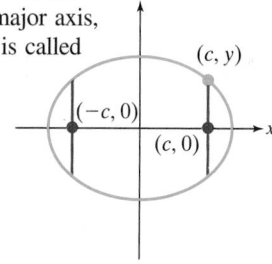

[*Hint*: Substitute (*c*, *y*) into the equation and solve for *y*. Recall that $c^2 = a^2 - b^2$.]

94. A line segment with endpoints on an ellipse and passing through a focus of the ellipse is called a focal chord. Given the ellipse

$$\frac{x^2}{25} + \frac{y^2}{16} = 1$$

a. Show that one focus of the ellipse lies on the line $y = \frac{4}{3}x + 4$.

b. Determine the points of intersection between the ellipse and the line.

c. Approximate the length of the focal chord that lies on the line $y = \frac{4}{3}x + 4$. Round to 2 decimal places.

Write About It

95. An elliptical pool table is in the shape of an ellipse with one pocket located at one focus of the ellipse. If a ball is located at the other focus, explain why a player can strike the ball in any direction to have the ball land in the pocket.

96. Given an equation of an ellipse in standard form, how do you determine whether the major axis is horizontal or vertical?

97. Explain the difference between the graphs of the two equations.

$$4x^2 + 9y^2 = 36 \qquad 4x + 9y = 36$$

98. Discuss the solution set of the equation.

$$\frac{(x-4)^2}{9} + \frac{(x+2)^2}{16} = -1$$

Expanding Your Skills

99. This exercise guides you through the steps to find the standard form of an equation of an ellipse centered at the origin with foci on the x-axis.

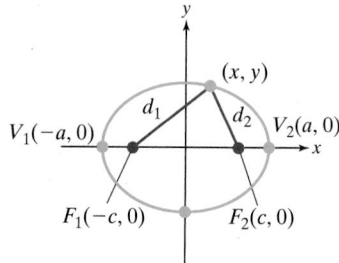

  a. Refer to the figure to verify that the distance from F_1 to V_2 is $(a + c)$ and the distance from F_2 to V_2 is $(a - c)$. Verify that the sum of these distances is $2a$.

  b. Write an expression that represents the sum of the distances from F_1 to (x, y) and from F_2 to (x, y). Then set this expression equal to $2a$.

  c. Given the equation $\sqrt{(x + c)^2 + y^2} + \sqrt{(x - c)^2 + y^2} = 2a$, isolate the leftmost radical and square both sides of the equation. Show that the equation can be written as $a\sqrt{(x - c)^2 + y^2} = a^2 - xc$.

  d. Square both sides of the equation $a\sqrt{(x - c)^2 + y^2} = a^2 - xc$ and show that the equation can be written as $(a^2 - c^2)x^2 + a^2y^2 = a^2(a^2 - c^2)$. (*Hint*: Collect variable terms on the left side of the equation and constant terms on the right side.)

  e. Replace $a^2 - c^2$ by b^2. Then divide both sides of the equation by a^2b^2. Verify that the resulting equation is $\dfrac{x^2}{a^2} + \dfrac{y^2}{b^2} = 1$.

100. Find the points on the ellipse that are twice the distance from one focus to the other.

$$\frac{x^2}{25} + \frac{y^2}{9} = 1$$

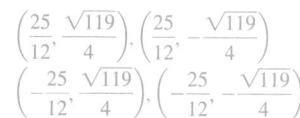

101. An ellipsoid is a three-dimensional surface that resembles the shape of the blimps we see at sporting events. Mathematically, an equation of an ellipsoid centered at the origin of a three-dimensional coordinate system is given by

$$\frac{x^2}{a^2} + \frac{y^2}{b^2} + \frac{z^2}{c^2} = 1.$$

  a. Explain how the formula for an ellipsoid is similar to a two-dimensional formula for an ellipse centered at the origin.

  b. The graph of $\dfrac{x^2}{9} + \dfrac{y^2}{36} + \dfrac{z^2}{4} = 1$ can be generated using computer software (see figure). Write an equation that results if $z = 0$. What does this equation represent?

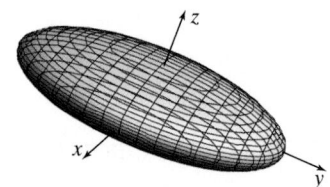

  c. Write the equation that results if $x = 0$. What does this equation represent?

  d. Write the equation that results if $y = 0$. What does this equation represent?

Technology Connections

For Exercises 102–105, graph the ellipse from the given exercise on a square viewing window.

102. Exercise 16   **103.** Exercise 15   **104.** Exercise 36   **105.** Exercise 35

SECTION 10.2 | The Hyperbola

1. Graph a Hyperbola Centered at the Origin

In this section, we will study another type of conic section called the hyperbola (Figure 10-11).

> **Definition of a Hyperbola**
>
> A **hyperbola** is the set of all points (x, y) in a plane such that the difference in distances between (x, y) and two fixed points (foci) is a positive constant.

In Figure 10-12, we present a hyperbola with foci on the x-axis. By definition, the curve is the set of points such that $d_1 - d_2$ is a positive constant.

TIP The definition of a hyperbola differs from the definition of an ellipse in that the curve is the set of points for which the difference (rather than the sum) of d_1 and d_2 is a constant.

Instructor Note:
Students may think that a hyperbola is two parabolas. Emphasize that they are not the same curves. The curvature on the branches of a hyperbola is different than that of a parabola.

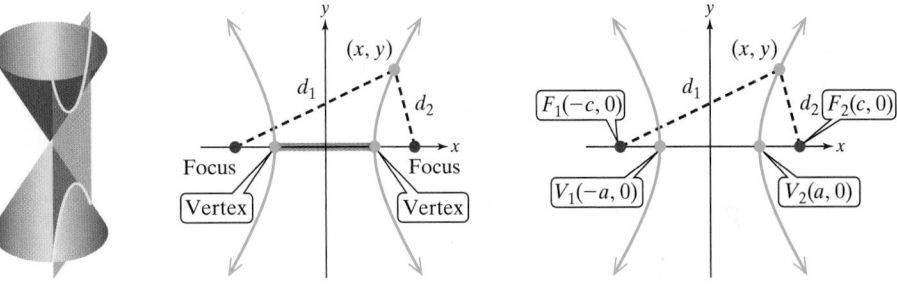

Figure 10-11   **Figure 10-12**   **Figure 10-13**

Notice that the hyperbola consists of two parts called **branches.** The points where the hyperbola intersects the line through the foci are called the **vertices.** The line segment between the vertices (highlighted in red) is called the **transverse axis.** The midpoint of the transverse axis is the **center** of the hyperbola.

To derive the standard form of an equation of a hyperbola, consider a hyperbola centered at the origin with foci on the x-axis (Figure 10-13). The points $F_1(-c, 0)$ and $F_2(c, 0)$ are the foci and the vertices are $V_1(-a, 0)$ and $V_2(a, 0)$.

The distance between F_1 and V_2 is $c + a$. The distance between F_2 and V_2 is $c - a$. By the definition of a hyperbola, the positive difference between a point on the hyperbola (such as V_2) and the two foci is a positive constant. Subtracting $(c + a)$ and $(c - a)$, we have

$$(c + a) - (c - a) = 2a$$

Therefore, the difference of the distances between (x, y) and the two foci is equal to $2a$.

$$d_1 \qquad - \qquad d_2 \qquad = 2a$$

$$\sqrt{[x - (-c)]^2 + (y - 0)^2} - \sqrt{(x - c)^2 + (y - 0)^2} = 2a \qquad \text{Apply the distance formula.}$$

$$\sqrt{(x + c)^2 + y^2} - \sqrt{(x - c)^2 + y^2} = 2a \qquad \text{Simplify within parentheses.}$$

In Exercise 82, we guide you through a series of algebraic steps to eliminate the radicals. Then collecting the variable terms on the left side and the constant terms on the right side, the equation becomes

$$(c^2 - a^2)x^2 - a^2 y^2 = a^2(c^2 - a^2) \qquad (1)$$

From Figure 10-13, we see that $d_1 - d_2 = 2a < 2c$. Therefore, $a < c$ and $c^2 - a^2 > 0$. Substituting $b^2 = c^2 - a^2$ in Equation (1) we have

$$b^2x^2 - a^2y^2 = a^2b^2$$

$$\frac{b^2x^2}{a^2b^2} - \frac{a^2y^2}{a^2b^2} = \frac{a^2b^2}{a^2b^2} \qquad \text{Divide both sides by } a^2b^2.$$

$$\frac{x^2}{a^2} - \frac{y^2}{b^2} = 1 \qquad \text{Standard form}$$

This equation represents a hyperbola centered at the origin with foci on the x-axis. In a similar manner, we can develop the standard equation of a hyperbola centered at the origin with foci on the y-axis. These two equations are summarized as follows.

Avoiding Mistakes

Notice that we do not have the restriction that $a > b$ as with the ellipse.

Standard Forms of an Equation of a Hyperbola Centered at the Origin

The standard forms of an equation of a hyperbola centered at the origin are as follows. Assume that $a > 0$ and $b > 0$.

	Transverse Axis: x-axis	Transverse Axis: y-axis
Equation:	$\frac{x^2}{a^2} - \frac{y^2}{b^2} = 1$	$\frac{y^2}{a^2} - \frac{x^2}{b^2} = 1$
Center:	(0, 0)	(0, 0)
Foci: ($c^2 = a^2 + b^2$)	$(c, 0)$ and $(-c, 0)$	$(0, c)$ and $(0, -c)$
Vertices:	$(a, 0)$ and $(-a, 0)$	$(0, a)$ and $(0, -a)$
Asymptotes:	$y = \frac{b}{a}x$ and $y = -\frac{b}{a}x$	$y = \frac{a}{b}x$ and $y = -\frac{a}{b}x$
Graph:	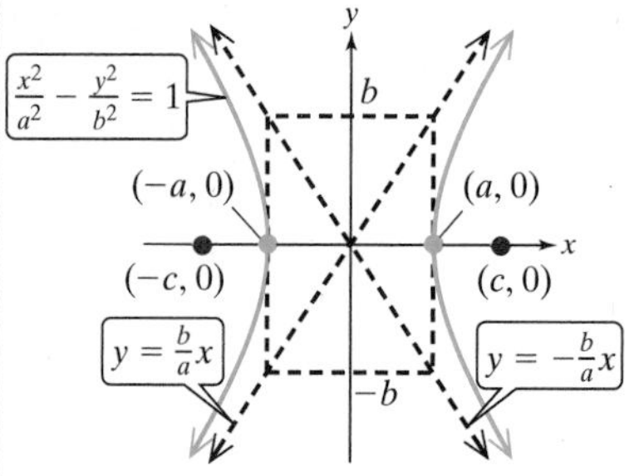 Figure 10-14	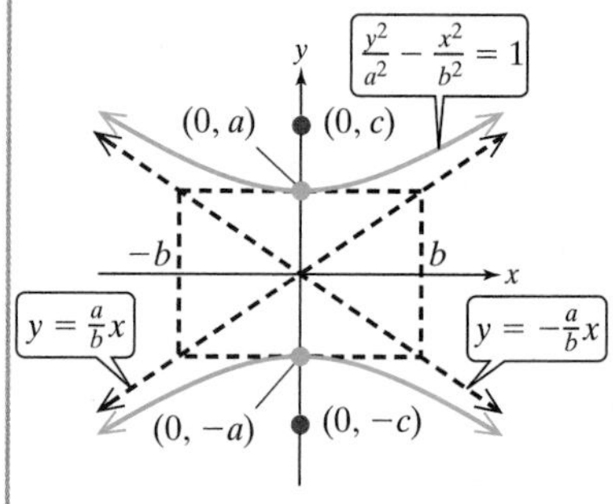 Figure 10-15

TIP Notice that an equation of a hyperbola shows a *difference* between the square terms, whereas an equation of an ellipse shows a sum of the square terms.

From the standard forms of an equation of a hyperbola, notice that the variable term with the positive coefficient always contains a^2 in the denominator. This term also indicates the orientation of the hyperbola.

- If the variable term with the positive coefficient is the x^2 term, then the transverse axis is horizontal and the branches of the hyperbola open horizontally.
- If the variable term with the positive coefficient is the y^2 term, then the transverse axis is vertical and the branches of the hyperbola open vertically.

As x or y increase, the branches of a hyperbola approach a pair of asymptotes. For example, given the hyperbola $\dfrac{x^2}{a^2} - \dfrac{y^2}{b^2} = 1$ (horizontal transverse axis),

$$y = \pm b\sqrt{\dfrac{x^2}{a^2} - 1}$$

$$= \pm b\sqrt{\dfrac{x^2 - a^2}{a^2}}$$

$$= \pm \dfrac{b}{a}\sqrt{x^2 - a^2}$$

As $x \to \infty$, the term a^2 is insignificant (very small relative to x^2), so the radical approaches $\sqrt{x^2} = x$. Thus, for large values of x, the graph of the hyperbola approaches $y = \pm\dfrac{b}{a}x$. Likewise, as $x \to -\infty$, the graph of the hyperbola approaches $y = \pm\dfrac{b}{a}x$.

As with the graph of a rational function, the asymptotes are lines of reference that help graph the curve. To identify the asymptotes, first sketch a **reference rectangle** with dimensions $2a$ by $2b$, centered at the center of the hyperbola (Figures 10-14 and 10-15). The asymptotes are the lines through the opposite corners of the rectangle.

The length of the transverse axis is $2a$. The line segment passing through the center of the hyperbola, perpendicular to the transverse axis, and with endpoints on the reference rectangle is called the **conjugate axis.** The length of the conjugate axis is $2b$.

To graph a hyperbola, we suggest the following guidelines.

Graphing a Hyperbola

Step 1 Identify the center and vertices.

Step 2 Draw the reference rectangle centered at the center of the hyperbola, with dimensions $2a$ and $2b$ as shown in Figures 10-14 and 10-15.

Step 3 Draw the asymptotes through the opposite corners of the rectangle.

Step 4 Sketch each branch of the hyperbola starting at the vertices and approaching the asymptotes.

Classroom Example: p. 936
Exercise 14

EXAMPLE 1   Graphing a Hyperbola Centered at the Origin

Given $\dfrac{x^2}{4} - \dfrac{y^2}{9} = 1$,

a. Graph the hyperbola.

b. Identify the center, vertices, foci, and equations of the asymptotes.

Point of Interest

The James S. McDonnell Planetarium in St. Louis, Missouri, has vertical cross sections in the shape of a hyperbola.

Solution:

a. $\dfrac{x^2}{4} - \dfrac{y^2}{9} = 1$

Step 1: The equation is in the standard form of a hyperbola centered at $(0, 0)$.

The x^2 term has the positive coefficient, indicating that the transverse axis is horizontal. The denominator of the x^2 term is a^2.

$\dfrac{x^2}{(2)^2} - \dfrac{y^2}{(3)^2} = 1$

$\overbrace{a = 2}\quad\overbrace{b = 3}$

$a^2 = 4$, and because $a > 0$, we have that $a = 2$.
$b^2 = 9$, and because $b > 0$, we have that $b = 3$.

Vertices: $(a, 0)$ and $(-a, 0)$
$\qquad\qquad (2, 0)$ and $(-2, 0)$
Note that $c^2 = a^2 + b^2 = 4 + 9 = 13$.
Since $c > 0$, $c = \sqrt{13}$.
Foci: $(c, 0)$ and $(-c, 0)$
$\qquad \left(\sqrt{13}, 0\right)$ and $\left(-\sqrt{13}, 0\right)$

Steps 2–3: Draw the reference rectangle and asymptotes.

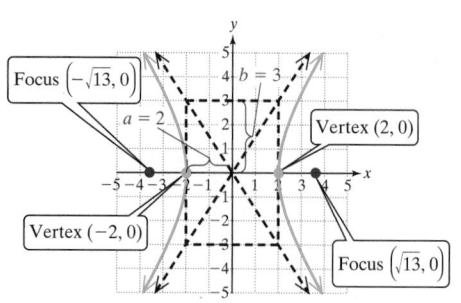

The reference rectangle is centered at the center of the hyperbola, $(0, 0)$.

Draw the rectangle with width $2a$ in the direction of the transverse axis. The height of the rectangle is the length of the conjugate axis $2b$. Draw the asymptotes through the opposite corners of the reference rectangle.

Step 4: Sketch each branch of the hyperbola starting at the vertices and approaching the asymptotes.

b. The center is $(0, 0)$.

The vertices are $(-2, 0)$ and $(2, 0)$.

The foci are $\left(-\sqrt{13}, 0\right)$ and $\left(\sqrt{13}, 0\right)$.

The equations of the asymptotes are $y = \frac{b}{a}x$ and $y = -\frac{b}{a}x$.

Therefore, the asymptotes are $y = \frac{3}{2}x$ and $y = -\frac{3}{2}x$.

TIP If you find it difficult to memorize the equations of the asymptotes, the equations can be found by using the point-slope formula and the points defining the opposite corners of the reference rectangle.

Skill Practice 1 Repeat Example 1 with the equation: $\dfrac{x^2}{9} - \dfrac{y^2}{4} = 1$

In Example 2, we graph a hyperbola in which the transverse axis is vertical and the conjugate axis is horizontal.

Classroom Example: p. 936
Exercise 18

EXAMPLE 2 Graphing a Hyperbola Centered at the Origin

Given $16y^2 - 9x^2 = 144$,

a. Graph the hyperbola.

b. Identify the center, vertices, foci, and equations of the asymptotes.

Solution:

a. $16y^2 - 9x^2 = 144$ For the equation to be in the standard form for a hyperbola, we require the constant term to be 1.

$\dfrac{16y^2}{144} - \dfrac{9x^2}{144} = \dfrac{144}{144}$ Divide both sides by 144.

$\dfrac{y^2}{9} - \dfrac{x^2}{16} = 1$ The y^2 term has the positive coefficient, indicating that the transverse axis is vertical. The denominator of the y^2 term is a^2.

$\dfrac{y^2}{(3)^2} - \dfrac{x^2}{(4)^2} = 1$

$a = 3$ $b = 4$

Step 1: The hyperbola is centered at $(0, 0)$.

$a^2 = 9$, and because $a > 0$, we have that $a = 3$.
$b^2 = 16$, and because $b > 0$, we have that $b = 4$.

Vertices: $(0, a)$ and $(0, -a)$
      $(0, 3)$ and $(0, -3)$

Note that $c^2 = a^2 + b^2 = 9 + 16 = 25$ and $c = 5$.

Foci: $(0, c)$ and $(0, -c)$
   $(0, 5)$ and $(0, -5)$

Answers

1. a.

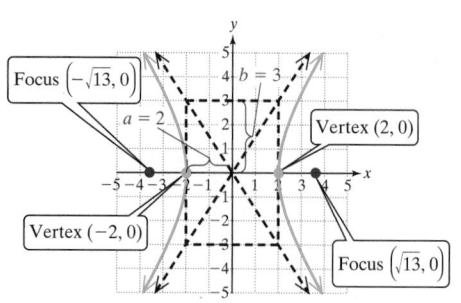

b. Center: $(0, 0)$

Vertices: $(-3, 0)$, $(3, 0)$

Foci: $\left(-\sqrt{13}, 0\right)$, $\left(\sqrt{13}, 0\right)$

Asymptotes: $y = \frac{2}{3}x$ and $y = -\frac{2}{3}x$.

Steps 2–3: Draw the reference rectangle and asymptotes.

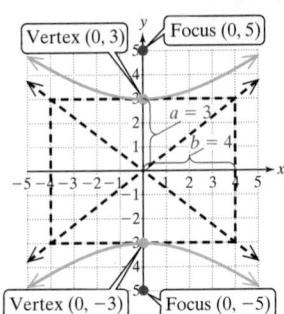

Draw the rectangle centered at the origin and with height $2a$ and width $2b$.

Draw the asymptotes through the opposite corners of the reference rectangle.

Step 4: Sketch each branch of the hyperbola starting at the vertices and approaching the asymptotes.

b. The center is $(0, 0)$.

The vertices are $(0, 3)$ and $(0, -3)$.

The foci are $(0, -5)$ and $(0, 5)$.

The equations of the asymptotes are $y = \frac{a}{b}x$ and $y = -\frac{a}{b}x$.

Therefore, the asymptotes are $y = \frac{3}{4}x$ and $y = -\frac{3}{4}x$.

Skill Practice 2   Repeat Example 2 with the equation: $4y^2 - x^2 = 16$

2. Graph a Hyperbola Centered at (h, k)

In the standard form of a hyperbola, if x and y are replaced by $x - h$ and $y - k$, then the graph of the hyperbola is shifted h units horizontally and k units vertically. The hyperbola is then centered at (h, k).

Standard Forms of an Equation of a Hyperbola Centered at (h, k)

The standard forms of an equation of a hyperbola centered at (h, k) are as follows. Assume that $a > 0$ and $b > 0$.

	Transverse Axis Horizontal	**Transverse Axis Vertical**
Equation:	$\dfrac{(x - h)^2}{a^2} - \dfrac{(y - k)^2}{b^2} = 1$	$\dfrac{(y - k)^2}{a^2} - \dfrac{(x - h)^2}{b^2} = 1$
Center:	(h, k)	(h, k)
Foci: (*Note:* $c^2 = a^2 + b^2$)	$(h + c, k)$ and $(h - c, k)$	$(h, k + c)$ and $(h, k - c)$
Vertices:	$(h + a, k)$ and $(h - a, k)$	$(h, k + a)$ and $(h, k - a)$
Asymptotes:	$y - k = \pm\frac{b}{a}(x - h)$	$y - k = \pm\frac{a}{b}(x - h)$
Graph:		

Answers

2. a.

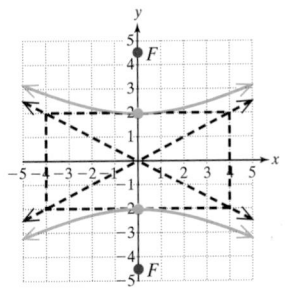

b. Center: $(0, 0)$
Vertices: $(0, 2)$, $(0, -2)$
Foci: $(0, 2\sqrt{5})$, $(0, -2\sqrt{5})$
Asymptotes: $y = \frac{1}{2}x$ and
$y = -\frac{1}{2}x$.

To graph a hyperbola centered at (h, k), first locate the center and identify the orientation of the hyperbola (transverse axis horizontal or vertical). Then use the values of a and b to draw the reference rectangle. Draw the asymptotes through the corners of the reference rectangle. Finally, sketch the hyperbola starting at the vertices and approaching the asymptotes.

Classroom Example: p. 936
Exercise 26

EXAMPLE 3 **Graphing a Hyperbola Centered at (h, k)**

Given $-\dfrac{(x-2)^2}{9} + (y+3)^2 = 1,$

a. Graph the hyperbola.

b. Identify the center, vertices, foci, and equations of the asymptotes.

Solution:

a. $-\dfrac{(x-2)^2}{9} + \dfrac{(y+3)^2}{1} = 1$

$\dfrac{[y-(-3)]^2}{(1)^2} - \dfrac{(x-2)^2}{(3)^2} = 1$

$\boxed{a=1}$ $\boxed{b=3}$

Step 1: Center: $(2, -3)$

The y^2 term has a positive coefficient. Therefore, the transverse axis is vertical. We have $a^2 = 1$ and $b^2 = 9$. Thus, $a = 1$ and $b = 3$.

Steps 2–3: Draw the reference rectangle and asymptotes.

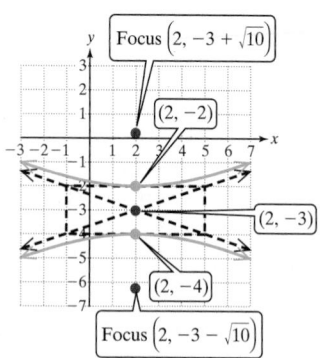

Focus $\left(2, -3 + \sqrt{10}\right)$

$(2, -2)$

$(2, -3)$

$(2, -4)$

Focus $\left(2, -3 - \sqrt{10}\right)$

Draw the rectangle centered at $(2, -3)$ with vertical sides of length $2a$ and horizontal sides of length $2b$.

Draw the asymptotes through the opposite corners of the reference rectangle.

The vertices are a units above and below the center.

Vertices: $(2, -3 + 1)$ and $(2, -3 - 1)$
$(2, -2)$ and $(2, -4)$

Step 4: Sketch the hyperbola.

To find the foci, use the relationship:
$$c^2 = a^2 + b^2$$
$$c^2 = 1 + 9 = 10$$
$$c = \sqrt{10} \quad \text{(Recall that } c > 0.\text{)}$$

The foci are c units above and below the center.

The foci are $\left(2, -3 + \sqrt{10}\right)$ and $\left(2, -3 - \sqrt{10}\right)$.

b. The center is $(2, -3)$.

The vertices are $(2, -2)$ and $(2, -4)$.

The foci are $\left(2, -3 + \sqrt{10}\right)$ and $\left(2, -3 - \sqrt{10}\right)$.

The equations of the asymptotes can be found by using the point-slope formula and the points through the opposite vertices of the reference rectangle. Alternatively, the asymptotes can be found by using $y - k = \pm\dfrac{a}{b}(x - h)$, where h and k are the coordinates of the center.

$$y - k = \dfrac{a}{b}(x - h) \qquad\qquad y - k = -\dfrac{a}{b}(x - h)$$

$$y - (-3) = \dfrac{1}{3}(x - 2) \qquad\qquad y - (-3) = -\dfrac{1}{3}(x - 2)$$

$$y = \dfrac{1}{3}x - \dfrac{11}{3} \qquad\qquad y = -\dfrac{1}{3}x - \dfrac{7}{3}$$

Answers

3. a.

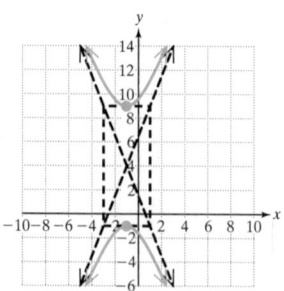

b. Center: $(-1, 4)$
Vertices: $(-1, 9), (-1, -1)$
Foci: $\left(-1, 4 + \sqrt{29}\right),$
$\left(-1, 4 - \sqrt{29}\right)$
Asymptotes: $y = -\dfrac{5}{2}x + \dfrac{3}{2},$
$y = \dfrac{5}{2}x + \dfrac{13}{2}$

Skill Practice 3 Repeat Example 3 with the equation:

$$-\dfrac{(x+1)^2}{4} + \dfrac{(y-4)^2}{25} = 1$$

TECHNOLOGY CONNECTIONS

Graphing a Hyperbola

To graph the hyperbola from Example 3 on a graphing utility, first solve the equation for y. We have

$$-\frac{(x-2)^2}{9} + (y+3)^2 = 1$$

$$y = -3 \pm \sqrt{1 + \frac{(x-2)^2}{9}}$$

The functions represent the top and bottom branches of the hyperbola, respectively.

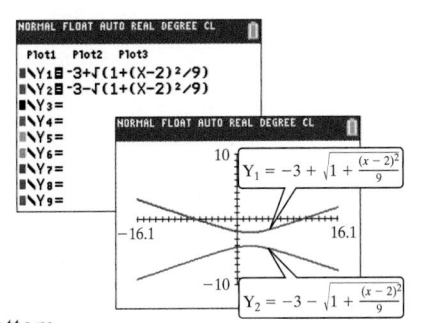

The equation of the hyperbola from Example 3 can be expanded by expanding the binomials and combining like terms.

$$-\frac{(x-2)^2}{9} + (y+3)^2 = 1 \qquad \text{Standard form}$$

$$9 \cdot \left[-\frac{(x-2)^2}{9} + (y+3)^2 \right] = 9 \cdot (1) \qquad \text{Clear fractions.}$$

$$-(x-2)^2 + 9(y+3)^2 = 9$$

$$-x^2 + 4x - 4 + 9y^2 + 54y + 81 = 9 \qquad \text{Expand the binomials.}$$

$$-x^2 + 9y^2 + 4x + 54y + 68 = 0 \qquad \text{Expanded form}$$

Given an equation of a hyperbola in expanded form, we can reverse this process by completing the square to write the equation in standard form. This is shown in Example 4.

Classroom Example: p. 936
Exercise 36

EXAMPLE 4   **Writing an Equation of a Hyperbola in Standard Form**

Write the equation of the hyperbola in standard form. $11x^2 - 2y^2 + 66x - 4y + 75 = 0$

Solution:

$11x^2 - 2y^2 + 66x - 4y + 75 = 0$ Group the x variable terms and group the y variable terms.

$11x^2 + 66x - 2y^2 - 4y = -75$ Move the constant term to the right side of the equation.

$11(x^2 + 6x \quad) - 2(y^2 + 2y \quad) = -75$ Factor out the leading coefficient from the x and y terms. Leave room to complete the square within parentheses.

11 · 9 is added to   −2 · 1 is added to
the left side.   the left side.

$11 \cdot 9 \quad -2 \cdot 1$ Complete the square.
Note that $\left[\frac{1}{2}(6)\right]^2 = 9$ and $\left[\frac{1}{2}(2)\right]^2 = 1$.

$11(x^2 + 6x + 9) - 2(y^2 + 2y + 1) = -75 + 99 - 2$ The values 9 and 1 were added inside parentheses. However, $11 \cdot 9$ and $-2 \cdot 1$ were actually added to the left side as a result of the factors in front of the parentheses.

$11(x+3)^2 - 2(y+1)^2 = 22$ To write the equation in standard form, we require a 1 on the right side of the equation.

$$\frac{11(x+3)^2}{22} - \frac{2(y+1)^2}{22} = \frac{22}{22} \qquad \text{Divide both sides by 22.}$$

$$\frac{(x+3)^2}{2} - \frac{(y+1)^2}{11} = 1 \qquad \text{Standard form}$$

Skill Practice 4 Write the equation of the hyperbola in standard form.
$3x^2 - 7y^2 + 30x + 56y - 58 = 0$

In Example 5, we determine the standard form of an equation of a hyperbola based on known information about the hyperbola.

Classroom Example: p. 937
Exercise 42

EXAMPLE 5　Finding an Equation of a Hyperbola

Determine the standard form of an equation of a hyperbola with vertices $(-4, 0)$ and $(8, 0)$ and foci $(-8, 0)$ and $(12, 0)$.

Solution:

Plotting the given points tells us that the foci are aligned horizontally. Therefore, the transverse axis of the hyperbola is horizontal. The standard form is

$$\frac{(x - h)^2}{a^2} - \frac{(y - k)^2}{b^2} = 1$$

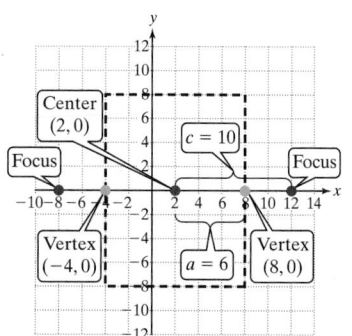

The center is midway between the vertices. Therefore, the center is $(2, 0)$, indicating that $h = 2$ and $k = 0$.

The vertices are located 6 units away from the center. This means that $a = 6$.

The foci are located 10 units from the center, indicating that $c = 10$.

The only missing piece of information is the value of b. We can find b from the relationship $c^2 = a^2 + b^2$ or equivalently $b^2 = c^2 - a^2$.

$b^2 = (10)^2 - (6)^2$

$b^2 = 64$

$b = 8$　　(Recall that $b > 0$.)

$$\frac{(x - 2)^2}{(6)^2} - \frac{(y - 0)^2}{(8)^2} = 1$$　　Substitute $h = 2$, $k = 0$, $a = 6$, and $b = 8$.

$$\frac{(x - 2)^2}{36} - \frac{y^2}{64} = 1$$　　Standard form

Skill Practice 5 Determine the standard form of an equation of a hyperbola with vertices $(-9, 0)$ and $(1, 0)$ and foci $(-17, 0)$ and $(9, 0)$.

3. Determine the Eccentricity of a Hyperbola

The branches of a hyperbola may be wider or narrower depending on the eccentricity of the hyperbola. Like the ellipse, the eccentricity of a hyperbola is given by $e = \frac{c}{a}$.

Answers

4. $\dfrac{(x + 5)^2}{7} - \dfrac{(y - 4)^2}{3} = 1$

5. $\dfrac{(x + 4)^2}{25} - \dfrac{y^2}{144} = 1$

Eccentricity of a Hyperbola

For a hyperbola defined by $\dfrac{(x - h)^2}{a^2} - \dfrac{(y - k)^2}{b^2} = 1$ or

$\dfrac{(y - k)^2}{a^2} - \dfrac{(x - h)^2}{b^2} = 1$, the **eccentricity** e is given by $e = \dfrac{c}{a}$, where $a > 0$, $c > 0$, and $c^2 = a^2 + b^2$.

Note: The eccentricity of a hyperbola is a number greater than 1; that is, $e > 1$.

In Example 6, we compare the graphs of two hyperbolas with different eccentricities.

Classroom Example: p. 937
Exercise 52

EXAMPLE 6   Finding the Eccentricity of a Hyperbola

Determine the eccentricity of the hyperbola.

a. $\dfrac{x^2}{16} - \dfrac{y^2}{9} = 1$ **b.** $\dfrac{x^2}{9} - \dfrac{y^2}{16} = 1$

Solution:

a. $\dfrac{x^2}{16} - \dfrac{y^2}{9} = 1$

$\dfrac{x^2}{(4)^2} - \dfrac{y^2}{(3)^2} = 1$

$\boxed{a = 4}$ $\boxed{b = 3}$

The x^2 term has a positive coefficient. Therefore, $a^2 = 16$.

$a^2 = 16$, and because $a > 0$, we have that $a = 4$.
$b^2 = 9$, and because $b > 0$, we have that $b = 3$.
$c^2 = a^2 + b^2 = (4)^2 + (3)^2 = 25$
$c = 5$ (Recall that $c > 0$.)
Therefore, $e = \dfrac{c}{a} = \dfrac{5}{4}$.

b. $\dfrac{x^2}{9} - \dfrac{y^2}{16} = 1$

$\dfrac{x^2}{(3)^2} - \dfrac{y^2}{(4)^2} = 1$

$\boxed{a = 3}$ $\boxed{b = 4}$

The x^2 term has a positive coefficient. Therefore, $a^2 = 9$.

$a^2 = 9$, and because $a > 0$, we have that $a = 3$.
$b^2 = 16$, and because $b > 0$, we have that $b = 4$.
$c^2 = a^2 + b^2 = (3)^2 + (4)^2 = 25$
$c = 5$ (Recall that $c > 0$.)
Therefore, $e = \dfrac{c}{a} = \dfrac{5}{3}$.

The graphs of the hyperbolas are shown in Figures 10-16 and 10-17. Compare the graphs, and notice that the hyperbola in Figure 10-17 has a greater eccentricity and that the branches open wider.

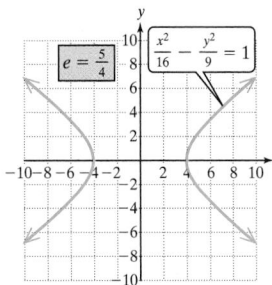

Figure 10-16

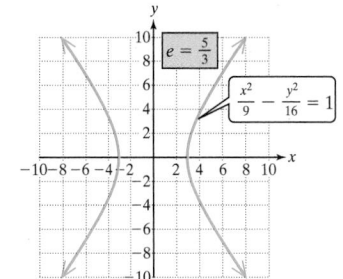

Figure 10-17

Skill Practice 6   Determine the eccentricity of the hyperbola.

a. $\dfrac{y^2}{25} - \dfrac{x^2}{144} = 1$ **b.** $\dfrac{y^2}{144} - \dfrac{x^2}{25} = 1$

Answers

6. a. $e = \dfrac{13}{5}$ **b.** $e = \dfrac{13}{12}$

4. Use Hyperbolas in Applications

Applications of hyperbolas arise in many fields. For example, astronomers know that some comets, such as Halley's comet, follow an elliptical path around the Sun. Other comets have a hyperbolic orbit because they are not captured by the Sun's gravity (Figure 10-18).

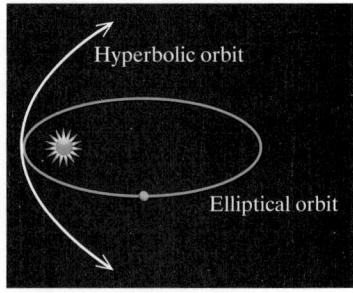

Figure 10-18

Hyperbolas also provide the mathematical basis for LORAN (LOng RAnge Navigation), a technique used for navigation before global positioning systems. A ship using LORAN measures the difference in time between synchronized radio signals from two different transmitters. Using the relationship $d = rt$, the difference in distances between the ship and the two transmitters $(d_1 - d_2)$ can be computed. Thus, the position of the ship is located on the hyperbolic path with the two transmitters at the foci (Figure 10-19).

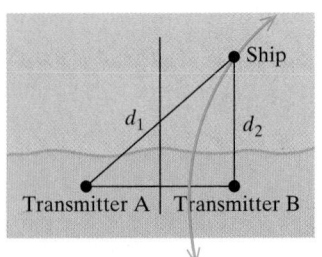

Figure 10-19

The same principle used in LORAN can also be used with sound waves to locate the source of sound heard from two different receivers. This is demonstrated in Example 7.

Classroom Example: p. 937
Exercise 60

EXAMPLE 7   Using a Hyperbola in an Application

Suppose that two people located 3 mi (15,840 ft) apart at points A and B hear a clap of thunder (Figure 10-20). The time difference between the sound heard at point A and the sound heard at point B is 10 sec. If sound travels 1100 ft/sec, find an equation of the hyperbola (with foci at A and B) on which the point of origination of the sound must lie.

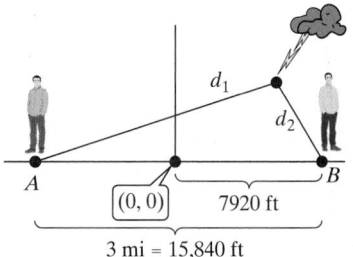

Figure 10-20

Solution:

First, we choose a coordinate system with the origin midway between the two observers. The two observers are at the points (7920, 0) and (−7920, 0). Let d_1 and d_2 represent the distances between each observer and the source of the sound (Figure 10-20).

The difference in distance $d_1 - d_2$ can be computed from the relationship $d = rt$.

$$d_1 - d_2 = (1100 \text{ ft/s})(10 \text{ s}) = 11{,}000 \text{ ft}$$

By the definition of a hyperbola, the source of the sound must be at a point on the hyperbola for which $d_1 - d_2 = 11{,}000$. The center of the hyperbola is (0, 0), and the transverse axis is horizontal with foci (7920, 0) and (−7920, 0). The standard form of the equation of the hyperbola is

$$\frac{x^2}{a^2} - \frac{y^2}{b^2} = 1$$

The value $d_1 - d_2 = 2a$
$$11{,}000 = 2a$$
$$a = 5500 \text{ ft}$$

From the location of the foci, $c = 7920$ ft.
$c^2 = a^2 + b^2$. Taking $b > 0$, we have $b = \sqrt{c^2 - a^2}$.
$$b = \sqrt{(7920)^2 - (5500)^2} \approx 5699 \text{ ft}$$

The source of the sound must be at a point on the curve $\dfrac{x^2}{(5500)^2} - \dfrac{y^2}{(5699)^2} = 1$.

Answer

7. $\dfrac{x^2}{(8800)^2} - \dfrac{y^2}{(5837)^2} = 1$

Skill Practice 7   Repeat Example 7 with two observers 4 mi apart, who hear a clap of thunder 16 sec apart. (*Hint*: 5280 ft = 1 mi.)

The hyperbola, like the ellipse, has an interesting property of reflection. Light directed toward one focus of a hyperbolic mirror is reflected toward the other focus of the hyperbola (Figure 10-21). This property is used in the design of Cassegrain reflecting telescopes. The telescope is a combination of a primary parabolic mirror and a secondary hyperbolic mirror (Figure 10-22). Light from distant stars is reflected off the parabolic mirror toward the focus of the hyperbolic mirror. The light is then reflected again toward an eyepiece located at the other focus of the hyperbola.

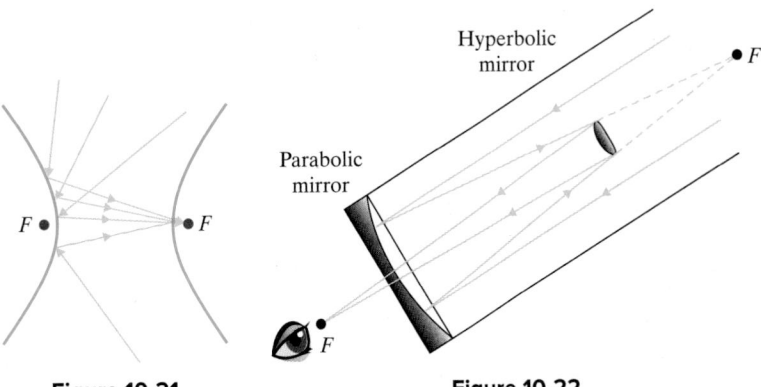

Figure 10-21   **Figure 10-22**

SECTION 10.2   Practice Exercises

Prerequisite Review

R.1. Use the distance formula to find the distance between the points (6, 12) and (−1, 12).   7

For Exercises R.2–R.3, write an equation of the line satisfying the given conditions. Write the answer in slope-intercept form.

R.2. The line passes through the point (−8, 8) and has a slope of −8. $y = -8x - 56$

R.3. The line passes through (−8, 8) and (1, −2).
$$y = -\frac{10}{9}x - \frac{8}{9}$$

Concept Connections

1. A _____hyperbola_____ is the set of points (x, y) in a plane such that the difference in distances between (x, y) and two fixed points (called _____foci_____) is a positive constant.

2. The points where a hyperbola intersects the line through the foci are called the _____vertices_____.

3. The line segment between the vertices of a hyperbola is called the _____transverse_____ axis.

4. The line segment perpendicular to the transverse axis passing through the center of a hyperbola, and with endpoints on the reference rectangle is called the _____conjugate_____ axis.

5. The equation $\dfrac{x^2}{a^2} - \dfrac{y^2}{b^2} = 1$ represents a   horizontal hyperbola with a (horizontal/vertical) transverse axis. The vertices are given by the ordered pairs ____$(a, 0)$____ and ____$(-a, 0)$____. The asymptotes are given by the equations ____$y = \frac{b}{a}x$____ and ____$y = -\frac{b}{a}x$____.

6. The equation $\dfrac{y^2}{a^2} - \dfrac{x^2}{b^2} = 1$ represents a   vertical hyperbola with a (horizontal/vertical) transverse axis. The vertices are given by the ordered pairs ____$(0, a)$____ and ____$(0, -a)$____. The asymptotes are given by the equations ____$y = \frac{a}{b}x$____ and ____$y = -\frac{a}{b}x$____.

Objective 1: Graph a Hyperbola Centered at the Origin

For Exercises 7–8,

a. Use the distance formula to find the distances d_1, d_2, d_3, and d_4.

b. Find the difference of the distances: $d_1 - d_2$.

c. Find the difference of the distances: $d_3 - d_4$.

d. How do the results from parts (b) and (c) compare?

e. How do the results of part (b) and part (c) compare to the length of the transverse axis?

7.

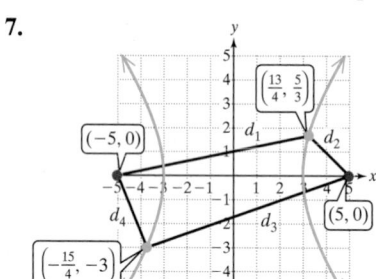

a. $d_1 = \dfrac{101}{12}$, $d_2 = \dfrac{29}{12}$, $d_3 = \dfrac{37}{4}$, $d_4 = \dfrac{13}{4}$

b. 6

c. 6

d. They are the same.

e. The difference in distances equals the length of the transverse axis.

8.

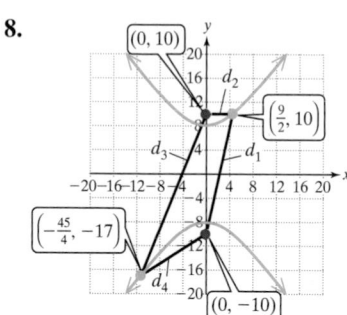

a. $d_1 = \dfrac{41}{2}$, $d_2 = \dfrac{9}{2}$, $d_3 = \dfrac{117}{4}$, $d_4 = \dfrac{53}{4}$

b. 16

c. 16

d. They are the same.

e. The difference in distances equals the length of the transverse axis.

For Exercises 9–12, determine whether the transverse axis and foci of the hyperbola are on the x-axis or the y-axis.

9. $\dfrac{x^2}{15} - \dfrac{y^2}{20} = 1$ *x*-axis

10. $\dfrac{y^2}{12} - \dfrac{x^2}{18} = 1$ *y*-axis

11. $-x^2 + \dfrac{y^2}{3} = 1$ *y*-axis

12. $-\dfrac{y^2}{16} + x^2 = 1$ *x*-axis

For Exercises 13–22,

a. Identify the center.   **b.** Identify the vertices.

c. Identify the foci.   **d.** Write equations for the asymptotes.

e. Graph the hyperbola. **(See Examples 1–2)**

13. $\dfrac{x^2}{16} - \dfrac{y^2}{25} = 1$

14. $\dfrac{x^2}{25} - \dfrac{y^2}{36} = 1$

15. $\dfrac{y^2}{4} - \dfrac{x^2}{36} = 1$

16. $\dfrac{y^2}{9} - \dfrac{x^2}{49} = 1$

17. $25y^2 - 81x^2 = 2025$

18. $49y^2 - 16x^2 = 784$

19. $-5x^2 + 7y^2 = -35$

20. $-7x^2 + 11y^2 = -77$

21. $\dfrac{4x^2}{25} - \dfrac{16y^2}{49} = 1$

22. $\dfrac{4x^2}{81} - \dfrac{16y^2}{225} = 1$

Objective 2: Graph a Hyperbola Centered at (h, k)

For Exercises 23–32,

a. Identify the center.   **b.** Identify the vertices.

c. Identify the foci.   **d.** Write equations for the asymptotes.

e. Graph the hyperbola. **(See Example 3)**

23. $\dfrac{(x - 4)^2}{9} - \dfrac{(y + 2)^2}{16} = 1$

24. $\dfrac{(x - 3)^2}{36} - \dfrac{(y + 1)^2}{64} = 1$

25. $\dfrac{(y - 5)^2}{49} - \dfrac{(x + 3)^2}{25} = 1$

26. $\dfrac{(y - 4)^2}{36} - \dfrac{(x + 5)^2}{16} = 1$

27. $100(y - 7)^2 - 81(x + 4)^2 = -8100$

28. $49(y - 3)^2 - 100(x + 6)^2 = -4900$

29. $y^2 - \dfrac{(x - 3)^2}{12} = 1$

30. $y^2 - \dfrac{(x - 2)^2}{18} = 1$

31. $x^2 - \dfrac{(y - 4)^2}{8} = 1$

32. $x^2 - \dfrac{(y - 6)^2}{24} = 1$

For Exercises 33–40,

a. Write the equation of the hyperbola in standard form. **(See Example 4)**

b. Identify the center, vertices, and foci.

33. $7x^2 - 5y^2 + 42x + 10y + 23 = 0$

34. $5x^2 - 3y^2 + 10x + 24y - 73 = 0$

35. $-5x^2 + 9y^2 + 20x - 72y + 79 = 0$

36. $-7x^2 + 16y^2 - 70x + 96y - 143 = 0$

37. $4x^2 - y^2 - 10y - 29 = 0$

38. $9x^2 - y^2 - 14y - 58 = 0$

39. $-36x^2 + 64y^2 + 108x + 256y - 401 = 0$

40. $-144x^2 + 25y^2 + 720x - 50y - 4475 = 0$

For Exercises 41–50, write the standard form of the equation of the hyperbola subject to the given conditions. (See Example 5)

41. Vertices: $(12, 0)$, $(-12, 0)$; Foci: $(13, 0)$, $(-13, 0)$

42. Vertices: $(40, 0)$, $(-40, 0)$; Foci: $(41, 0)$, $(-41, 0)$

43. Vertices: $(0, 12)$, $(0, -12)$; Asymptotes: $y = \pm\frac{4}{3}x$

44. Vertices: $(0, 15)$, $(0, -15)$; Asymptotes: $y = \pm\frac{15}{8}x$

45. Vertices: $(-3, -3)$, $(7, -3)$; $\dfrac{(x-2)^2}{25} - \dfrac{(y+3)^2}{49} = 1$
Slope of the asymptotes: $\pm\frac{7}{5}$

46. Vertices: $(2, 1)$, $(-10, 1)$; $\dfrac{(x+4)^2}{36} - \dfrac{(y-1)^2}{25} = 1$
Slope of the asymptotes: $\pm\frac{5}{6}$

47. Vertices: $(-3, 0)$, $(-3, 8)$; $\dfrac{(y-4)^2}{16} - \dfrac{(x+3)^2}{5} = 1$
Foci: $\left(-3, 4 + \sqrt{21}\right)$, $\left(-3, 4 - \sqrt{21}\right)$

48. Vertices: $(2, -1)$, $(2, -11)$;
Foci: $\left(2, -6 + \sqrt{31}\right)$, $\left(2, -6 - \sqrt{31}\right)$

49. Corners of the reference rectangle: $(8, 7)$, $(-6, 7)$, $(8, -3)$, $(-6, -3)$; Horizontal transverse axis
$\dfrac{(x-1)^2}{49} - \dfrac{(y-2)^2}{25} = 1$

50. Corners of the reference rectangle: $(7, 6)$, $(-1, 6)$, $(7, 0)$, $(-1, 0)$; Horizontal transverse axis
$\dfrac{(x-3)^2}{16} - \dfrac{(y-3)^2}{9} = 1$

Objective 3: Determine the Eccentricity of a Hyperbola

For Exercises 51–52,

a. Determine the eccentricity of each hyperbola. **(See Example 6)**

b. Based on the eccentricity, match the equation with its graph. The scaling is the same for both graphs.

51. Equation 1: $\dfrac{x^2}{144} - \dfrac{y^2}{81} = 1$  a. Equation 1: $e = \frac{5}{4}$,
Equation 2: $e = \frac{5}{3}$
b. Graph B represents Equation 1.
Graph A represents Equation 2.

Equation 2: $\dfrac{x^2}{81} - \dfrac{y^2}{144} = 1$

52. Equation 1: $\dfrac{x^2}{225} - \dfrac{y^2}{64} = 1$ a. Equation 1: $e = \frac{17}{15}$,
Equation 2: $e = \frac{17}{8}$
b. Graph A represents Equation 1.
Graph B represents Equation 2.

Equation 2: $\dfrac{x^2}{64} - \dfrac{y^2}{225} = 1$

A. 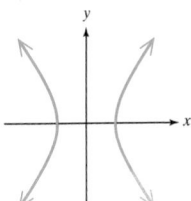   **B.**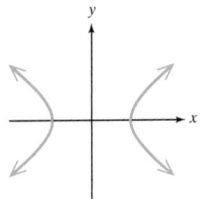

A. 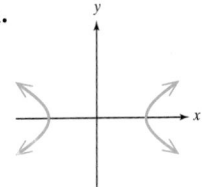   **B.**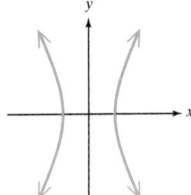

For Exercises 53–54, determine the eccentricity.

53. $\dfrac{\left(y - \frac{2}{3}\right)^2}{1600} - \dfrac{\left(x + \frac{7}{4}\right)^2}{81} = 1$ $e = \dfrac{41}{40}$

54. $\dfrac{(y - 3.8)^2}{49} - \dfrac{(x - 2.7)^2}{576} = 1$ $e = \dfrac{25}{7}$

55. Determine the eccentricity of a hyperbola with a horizontal transverse axis of length 24 units and asymptotes $y = \pm\frac{3}{4}x$. $e = \dfrac{5}{4}$

56. Determine the eccentricity of a hyperbola with a vertical transverse axis of length 48 units and asymptotes $y = \pm\frac{12}{5}x$. $e = \dfrac{13}{12}$

57. Determine the standard form of an equation of a hyperbola with eccentricity $\frac{5}{4}$ and vertices $(-1, -1)$ and $(7, -1)$. $\dfrac{(x-3)^2}{16} - \dfrac{(y+1)^2}{9} = 1$

58. Determine the standard form of an equation of a hyperbola with eccentricity $\frac{13}{12}$ and vertices $(-2, 8)$ and $(-2, -16)$. $\dfrac{(y+4)^2}{144} - \dfrac{(x+2)^2}{25} = 1$

Objective 4: Use Hyperbolas in Applications

59. Two radio transmitters are 1000 mi apart at points A and B along a coastline. Using LORAN on the ship, the time difference between the radio signals is 4 milliseconds (0.004 sec). If radio signals travel 186 mi/millisecond, find an equation of the hyperbola with foci at A and B, on which the ship is located. **(See Example 7)**

60. Suppose that two microphones 1500 m apart at points A and B detect the sound of a rifle shot. The time difference between the sound detected at point A and the sound detected at point B is 4 sec. If sound travels at approximately 330 m/sec, find an equation of the hyperbola with foci at A and B defining the points where the shooter may be located.

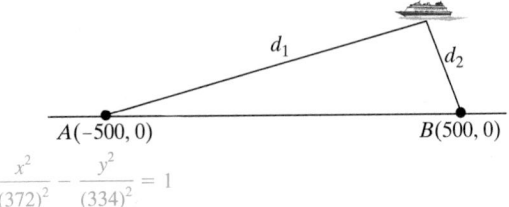

$\dfrac{x^2}{(372)^2} - \dfrac{y^2}{(334)^2} = 1$

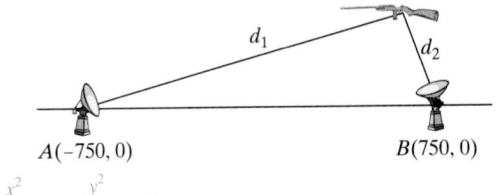

$\dfrac{x^2}{(660)^2} - \dfrac{y^2}{(356)^2} = 1$

61. In some designs of eyeglasses, the surface is "aspheric," meaning that the contour varies slightly from spherical. An aspheric lens is often used to correct for spherical aberration—a distortion due to increased refraction of light rays when they strike the lens near its edge. Aspheric lenses are often designed with hyperbolic cross sections.

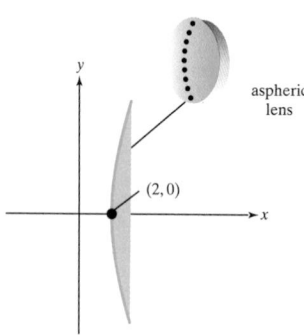

aspheric lens

Write an equation of the cross section of the hyperbolic lens shown if the center is (0, 0), one vertex is (2, 0), and the focal length (distance between center and foci) is $\sqrt{85}$. Assume that all units are in millimeters. $\frac{x^2}{4} - \frac{y^2}{81} = 1; x \geq 0$ or $x = 2\sqrt{1 + \frac{y^2}{81}}$

62. In 1911, Ernest Rutherford discovered the nucleus of the atom. Experiments leading to this discovery involved the scattering of alpha particles by the heavy nuclei in gold foil. When alpha particles are thrust toward the gold nuclei, the particles are deflected and follow a hyperbolic path.

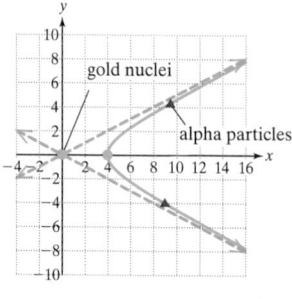

Suppose that the minimum distance that the alpha particles get to the gold nuclei is 4 microns (1 micron is one-millionth of a meter) and that the hyperbolic path has asymptotes of $y = \pm\frac{1}{2}x$. Determine an equation of the path of the particles shown. $\frac{x^2}{16} - \frac{y^2}{4} = 1; x \geq 0$ or $x = 4\sqrt{1 + \frac{y^2}{4}}$

63. Atomic particles with like charges tend to repel one another. Suppose that two beams of like-charged particles are hurled toward each other from two parallel atomic accelerators. The path defined by the particles is $x^2 - 4y^2 = 36$, where x and y are measured in microns. What is the minimum distance between the particles?   12 microns

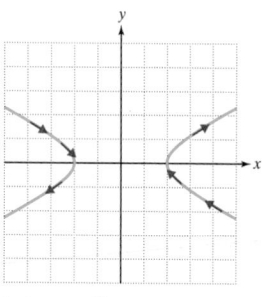

64. A returning boomerang is a V-shaped throwing device made from two wings that are set at a slight tilt and that have an airfoil design. One side is rounded and the other side is flat, similar to an airplane propeller. When thrown properly, a boomerang follows a circular flight path and should theoretically return close to the point of release.

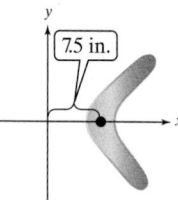

7.5 in.

The boomerang pictured is approximately in the shape of one branch of a hyperbola (although the two wings are in slightly different planes). To construct the hyperbola, an engineer needs to know the location of the foci. Determine the location of the focus to the right of the center if the vertex is 7.5 in. from the center and the equations of the asymptotes are $y = \pm\frac{4}{5}x$. Round the coordinates to the nearest tenth of an inch.   (9.6, 0)

65. In September 2009, Australian astronomer Robert H. McNaught discovered comet C/2009 R_1 (McNaught). The orbit of this comet is hyperbolic with the Sun at one focus. Because the orbit is not elliptical, the comet will not be captured by the Sun's gravitational pull and instead will pass by the Sun only once. The comet reached perihelion on July 2, 2010. (*Source:* Minor Planet Center, http://minorplanetcenter.net/)

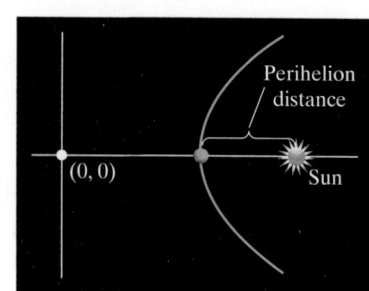

The path of the comet can be modeled by the equation

$$\frac{x^2}{(1191.2)^2} - \frac{y^2}{(30.9)^2} = 1$$

where x and y are measured in AU (astronomical units).

a. Determine the distance (in AU) at perihelion. Round to 1 decimal place.   0.4 AU

b. Using the rounded value from part (a), if 1 AU $\approx$ 93,000,000 mi, find the distance in miles.   37,200,000 mi

66. The cross section of a cooling tower of a nuclear power plant is in the shape of a hyperbola, and can be modeled by the equation

$$\frac{x^2}{625} - \frac{(y - 80)^2}{2500} = 1$$

where x and y are measured in meters. The top of the tower is 120 m above the base.

a. Determine the diameter of the tower at the base. Round to the nearest meter.   94 m

b. Determine the diameter of the tower at the top. Round to the nearest meter.   64 m

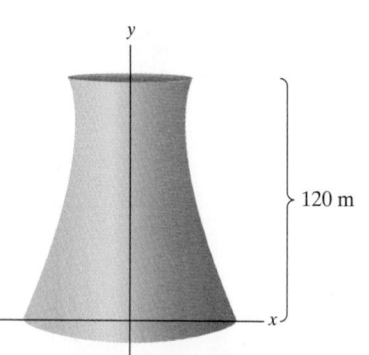

120 m

Mixed Exercises

For Exercises 67–70, identify the equation as representing an ellipse or a hyperbola, and match the equation with the graph.

67. $\dfrac{(x-5)^2}{49} + \dfrac{(y+2)^2}{36} = 1$ Ellipse; B

68. $\dfrac{(x-5)^2}{36} + \dfrac{(y+2)^2}{49} = 1$ Ellipse; C

69. $\dfrac{(x-5)^2}{49} - \dfrac{(y+2)^2}{36} = 1$ Hyperbola; D

70. $-\dfrac{(x-5)^2}{49} + \dfrac{(y+2)^2}{36} = 1$ Hyperbola; A

A.
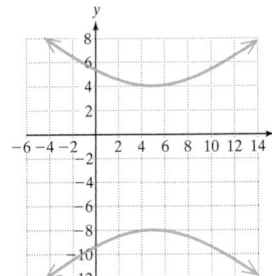

B.
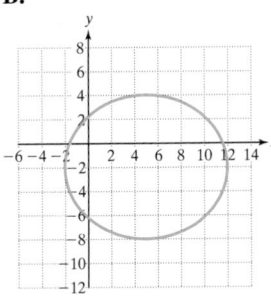

C.
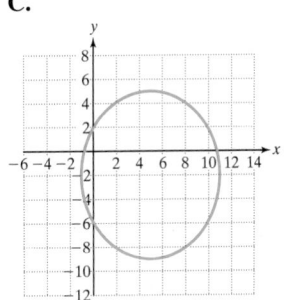

D.
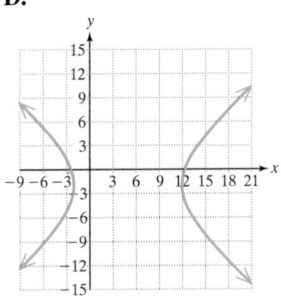

For Exercises 71–74, find the standard form of the equation of the ellipse or hyperbola shown.

71.
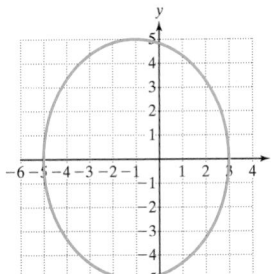
$\dfrac{(x+1)^2}{16} + \dfrac{y^2}{25} = 1$

72.
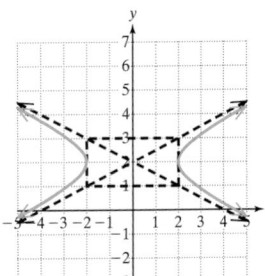
$\dfrac{x^2}{4} - (y-2)^2 = 1$

73.
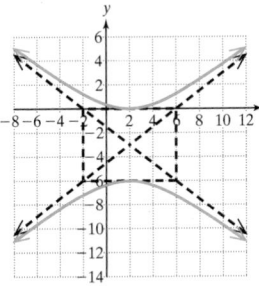
$\dfrac{(y+3)^2}{9} - \dfrac{(x-2)^2}{16} = 1$

74.
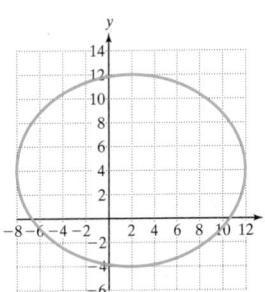
$\dfrac{(x-2)^2}{100} + \dfrac{(y-4)^2}{64} = 1$

For Exercises 75–76, solve the system of equations.

75. $\quad 9x^2 - 4y^2 = 36$
$\quad -13x^2 + 8y^2 = 8$
$\{(4, 3\sqrt{3}), (4, -3\sqrt{3}), (-4, 3\sqrt{3}), (-4, -3\sqrt{3})\}$

76. $x^2 + 4y^2 = 36$
$\quad 4x^2 - y^2 = 8$
$\{(2, 2\sqrt{2}), (2, -2\sqrt{2}), (-2, 2\sqrt{2}), (-2, -2\sqrt{2})\}$

Write About It

77. Given an equation of a hyperbola in standard form, how do you determine whether the transverse axis is horizontal or vertical?

78. Discuss the solution set of the equation.
$$\dfrac{x^2}{4} - \dfrac{y^2}{9} = 0$$

Expanding Your Skills

79. What is the eccentricity of a hyperbola if the asymptotes are perpendicular? $\sqrt{2}$

80. a. Describe the graph of $y = 4\sqrt{1 + \dfrac{x^2}{9}}$.

b. Does the equation $y = 4\sqrt{1 + \dfrac{x^2}{9}}$ define y as a function of x?

c. Sketch $y = 4\sqrt{1 + \dfrac{x^2}{9}}$.

d. Determine the interval(s) over which the function is increasing.

e. Determine the interval(s) over which the function is decreasing.

81. A line segment with endpoints on a hyperbola, perpendicular to the transverse axis, and passing through a focus is called a *latus rectum* of the hyperbola (shown in red). Show that the length of a latus rectum is $\dfrac{2b^2}{a}$ for the hyperbola

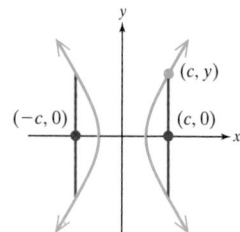

$$\frac{x^2}{a^2} - \frac{y^2}{b^2} = 1$$

[*Hint*: Substitute (c, y) into the equation and solve for y. Recall that $c^2 = a^2 + b^2$.]

82. This exercise guides you through the steps to find the standard form of an equation of a hyperbola centered at the origin with foci on the x-axis.

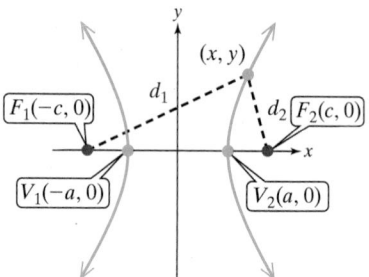

a. Refer to the figure to verify that the distance between F_1 and V_2 is $(c + a)$ and the distance between F_2 and V_2 is $(c - a)$. Verify that the *difference* between these distances is $2a$.

b. Write an expression that represents the difference of the distances from F_1 to (x, y) and from F_2 to (x, y). Then set this expression equal to $2a$.

c. Given the equation $\sqrt{(x + c)^2 + y^2} - \sqrt{(x - c)^2 + y^2} = 2a$, isolate the leftmost radical and square both sides of the equation. Show that the equation can be written as $cx - a^2 = a\sqrt{(x - c)^2 + y^2}$.

d. Square both sides of the equation $cx - a^2 = a\sqrt{(x - c)^2 + y^2}$ and show that the equation can be written as $(c^2 - a^2)x^2 - a^2y^2 = a^2(c^2 - a^2)$. (*Hint*: Collect variable terms on the left side of the equation and constant terms on the right side.)

e. Replace $c^2 - a^2$ by b^2. Then divide both sides of the equation by a^2b^2. Verify that the resulting equation is $\dfrac{x^2}{a^2} - \dfrac{y^2}{b^2} = 1$.

83. A hyperboloid of one sheet is a three-dimensional surface generated by an equation of the form $\dfrac{x^2}{a^2} + \dfrac{y^2}{b^2} - \dfrac{z^2}{c^2} = 1$. The surface has hyperbolic cross sections and either circular cross sections or elliptical cross sections.

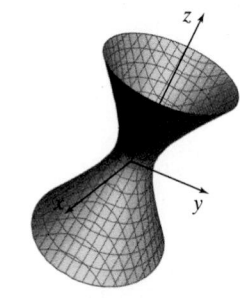

a. Write the equation with $z = 0$. What type of curve is represented by this equation?

b. Write the equation with $x = 0$. What type of curve is represented by this equation?

c. Write the equation with $y = 0$. What type of curve is represented by this equation?

84. a. Radio signals emitted from points $(8, 0)$ and $(-8, 0)$ indicate that a plane is 8 mi closer to $(8, 0)$ than to $(-8, 0)$. Find an equation of the hyperbola that passes through the plane's location and with foci $(8, 0)$ and $(-8, 0)$. All units are in miles.

b. At the same time, radio signals emitted from points $(0, 8)$ and $(0, -8)$ indicate that the plane is 4 mi farther from $(0, 8)$ than from $(0, -8)$. Find an equation of the hyperbola that passes through the plane's location and with foci $(0, 8)$ and $(0, -8)$.

c. From the figure, the plane is located in the fourth quadrant of the coordinate system. Solve the system of equations defining the two hyperbolas for the point of intersection in the fourth quadrant. This is the location of the plane. Then round the coordinates to the nearest tenth of a mile.

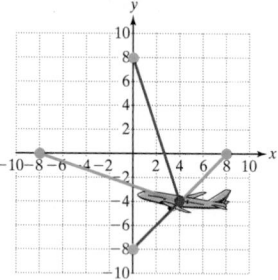

Technology Connections

For Exercises 85–88, graph the hyperbola from the given exercise.

85. Exercise 15   **86.** Exercise 16   **87.** Exercise 37   **88.** Exercise 38

1. Identify the Focus and Directrix of a Parabola

A parabola is another type of conic section generated by the cross section of a cone intersected by a plane (Figure 10-23). An equation of the form $y = ax^2 + bx + c$ ($a \neq 0$) is a parabola opening upward if $a > 0$ and opening downward if $a < 0$. We now extend our study of parabolas to include those that open to the left and right. To do so, we will define a parabola geometrically.

> ### Definition of a Parabola
>
> A **parabola** is the set of all points in a plane that are equidistant from a fixed line (called the **directrix**) and a fixed point (called the **focus**).

Figure 10-23

Figure 10-24 shows a parabola opening upward. The line perpendicular to the directrix and passing through the focus is called the **axis of symmetry.** The **vertex** of the parabola is the point of intersection of the parabola and the axis of symmetry.

A point $P(x, y)$ is on the parabola if $d_1 = d_2$, where d_1 is the distance between the focus and P, and d_2 is the perpendicular distance between P and the directrix.

To develop an algebraic equation that represents a parabola, we will consider a parabola with vertex at the origin and with a vertical line of symmetry (that is, the parabola opens upward or downward). See Figure 10-25. In the discussion that follows, the same logic can be applied to a parabola with a horizontal line of symmetry (the parabola opens to the left or to the right).

The distance between the vertex and the focus of a parabola is called the **focal length** and is often represented by $|p|$. Suppose that we choose a coordinate system with the vertex at the origin and focus $F(0, p)$. Because the distance between any point on the parabola and the focus must equal the distance between the point and the directrix, the directrix is a horizontal line $|p|$ units from the vertex. The equation of the directrix is $y = -p$.

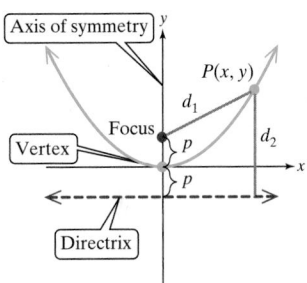

Figure 10-24

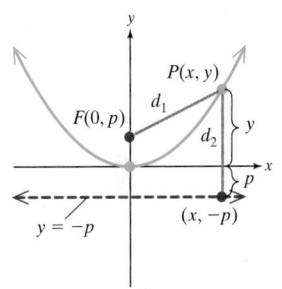

Figure 10-25

Let $P(x, y)$ represent any point on the parabola. Then the distance between $F(0, p)$ and $P(x, y)$ can be found by using the distance formula.

$$d_1 = \sqrt{(x - 0)^2 + (y - p)^2}$$

The distance from $P(x, y)$ to the directrix is the vertical distance between $P(x, y)$ and the point $(x, -p)$ on the directrix.

$$d_2 = y + p$$

Equating d_1 and d_2, we have

$$\sqrt{x^2 + (y - p)^2} = y + p$$

$$x^2 + (y - p)^2 = (y + p)^2 \qquad \text{Square both sides.}$$

$$x^2 + y^2 - 2py + p^2 = y^2 + 2py + p^2 \qquad \text{Expand the binomials.}$$

$$x^2 = 4py$$

> **Standard form**
> If $p > 0$, the parabola opens upward.
> If $p < 0$, the parabola opens downward.

> **TIP** The shortest distance between P and the directrix is the *perpendicular distance*. This is the length of the line segment perpendicular to the directrix with one endpoint at P and the other endpoint on the directrix.

The equation $x^2 = 4py$ represents a parabola with vertex at the origin and focus on the y-axis. Using similar logic, we can derive the standard form of an equation of a parabola with focus on the x-axis.

Standard Forms of an Equation of a Parabola with Vertex at the Origin		
	Axis of Symmetry: y-axis	**Axis of Symmetry: x-axis**
Equation:	$x^2 = 4py$	$y^2 = 4px$
Vertex:	$(0, 0)$	$(0, 0)$
Focus:	$(0, p)$	$(p, 0)$
Directrix:	$y = -p$	$x = -p$
Axis of symmetry:	$x = 0$	$y = 0$
Graph: $(p > 0)$		
Graph: $(p < 0)$		

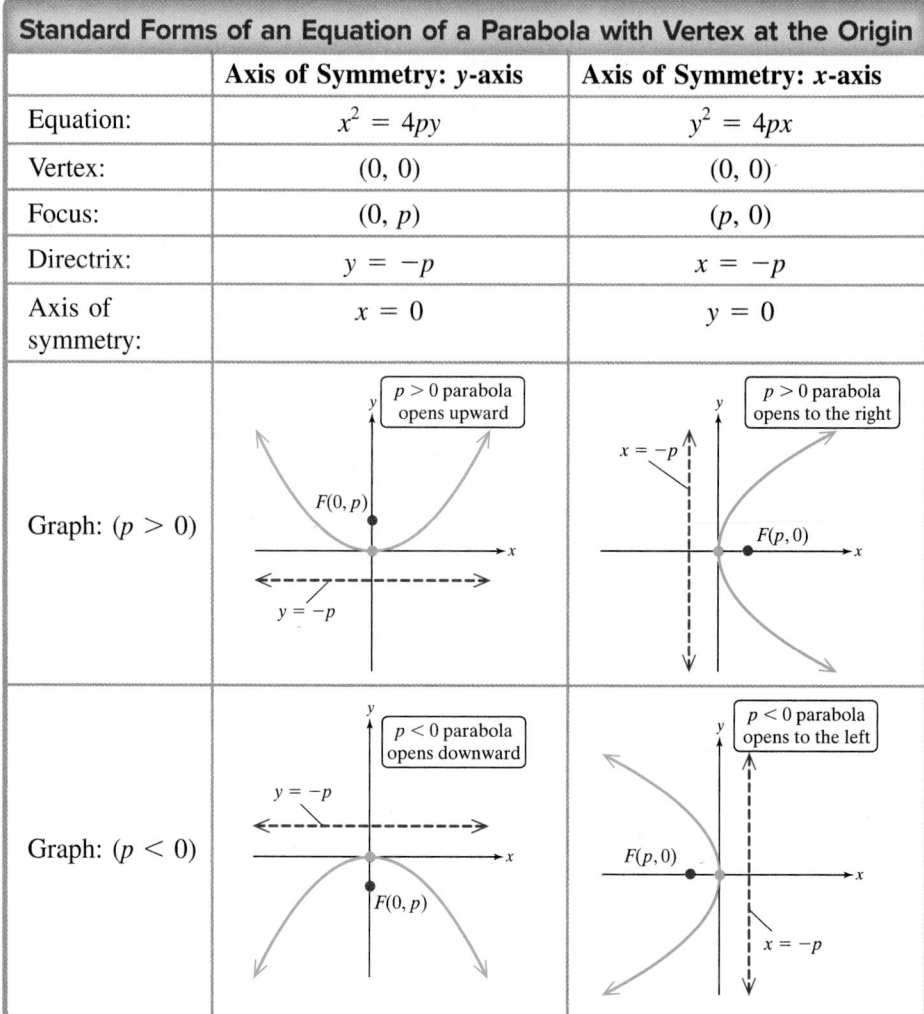

From Section 2.1, we became accustomed to writing an equation of a quadratic function with vertex at the origin in the form $y = ax^2$. The benefit of writing the equation as $x^2 = 4py$ is that we can identify the value of p, which gives us the distance $|p|$ between the vertex and the focus.

Finding the location of the focus is particularly important in applications. For example, a flashlight has a mirror with cross section in the shape of a parabola (Figure 10-26). The bulb is located at the focus. The light emitted from the bulb will reflect off the mirror in lines parallel to the axis of symmetry to form a beam of light. This is a result of the reflective property of a parabola.

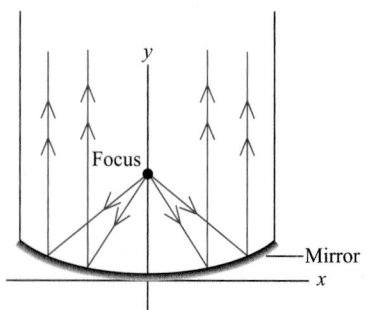

Figure 10-26

EXAMPLE 1   Determining the Focus and Directrix of a Parabola

An engineer designs a flashlight with a parabolic mirror. A coordinate system is chosen so that the vertex of a cross section through the center of the mirror is located at $(0, 0)$. The equation of the parabola is modeled by $x^2 = 8y$, where x and y are measured in centimeters.

  a. Where should the bulb be placed to make use of the reflective property of the parabola? That is, find the location of the focus.

  b. Write an equation of the directrix.

Solution:

  a. $x^2 = 8y$ is in the form $x^2 = 4py$ with $4p = 8$.

$$4p = 8$$
$$p = 2$$

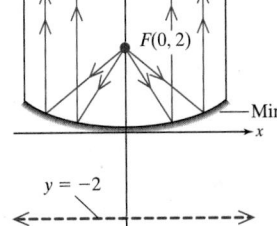

Since the equation is in the form $x^2 = 4py$, the focus is on the y-axis. Furthermore, since $p > 0$, the parabola opens upward, and the focus is $|p|$ units above the vertex.

The focus is $(0, 2)$.
The bulb should be placed 2 cm above the vertex.

  b. The directrix is the horizontal line $|p|$ units below the vertex.

The equation of the directrix is $y = -2$.

Skill Practice 1   Given a parabola defined by $x^2 = 4y$, find the focus and an equation of the directrix.

2. Graph a Parabola with Vertex at the Origin

In Examples 2 and 3, we practice graphing parabolas with vertex at the origin. To sketch the graph, we will identify the location of the vertex and the orientation of the parabola (up or down versus left or right). To determine how "wide" to graph the parabola, we will determine the length of the line segment passing through the focus, perpendicular to the axis of symmetry with endpoints on the parabola. This line segment is called the **latus rectum.**

For the parabola defined by $x^2 = 4py$, the endpoints of the latus rectum will have the same y-coordinate as the focus. Therefore, to find the endpoints of the latus rectum, we can substitute p for y in the equation (Figure 10-27).

$$x^2 = 4py$$
$$x^2 = 4p(p) \quad \text{Substitute } p \text{ for } y.$$
$$x^2 = 4p^2$$
$$x = \pm 2|p|$$

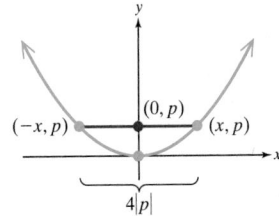

Figure 10-27

Thus, the endpoints of the latus rectum are located $2|p|$ units to the left and right of the focus at $(-2|p|, p)$ and $(2|p|, p)$. The length of the latus rectum is called the **focal diameter** and is equal to $4|p|$.

    In Example 2, we graph a parabola by using the vertex, orientation of the parabola, and focal diameter.

EXAMPLE 2   Graphing a Parabola with Vertex at the Origin

Given $x^2 = -12y$,

a. Identify the vertex, focus, and focal diameter.

b. Identify the endpoints of the latus rectum and graph the parabola.

c. Write equations for the directrix and axis of symmetry.

Solution:

a. $x^2 = -12y$ The equation is in the standard form $x^2 = 4py$. From the standard form, we see that the focus is on the y-axis and that the parabola will open upward or downward. The vertex is $(0, 0)$, with $4p = -12$.

$4p = -12$

$p = -3$ Because $p < 0$, the parabola will open downward.

The vertex is $(0, 0)$.

The focus is $(0, p)$, which is $(0, -3)$.

The focal diameter is $4|p|$, which is $4|(-3)| = 12$.

TIP The focal diameter gives the length of the latus rectum. This is helpful for graphing a parabola because it tells us the "width" of the parabola at the focus.

b. Plot the vertex and locate the focus. Note that the focus is shown in red because it is not actually part of the graph. The focal diameter is 12 units. Therefore, plot points 6 units to the left and right of the focus. These are the endpoints of the latus rectum: $(6, -3)$ and $(-6, -3)$.

c. Equation of the directrix: $y = 3$

Equation of the axis of symmetry: $x = 0$

Skill Practice 2   Repeat Example 2 with the equation: $x^2 = -8y$

In Example 3, we graph a parabola opening horizontally.

EXAMPLE 3   Graphing a Parabola with Vertex at the Origin

Given $2y^2 = 32x$,

a. Identify the vertex, focus, and focal diameter.

b. Identify the endpoints of the latus rectum and graph the parabola.

c. Write equations for the directrix and axis of symmetry.

Solution:

a. $2y^2 = 32x$ Divide both sides by 2 to obtain a coefficient of 1 on the square term.

$y^2 = 16x$ The equation is in the standard form $y^2 = 4px$. From the standard form, we see that the focus is on the x-axis, and that the parabola will open to the left or right. The vertex is $(0, 0)$ and $4p = 16$.

$4p = 16$

$p = 4$ Because $p > 0$, the parabola will open to the right.

The vertex is $(0, 0)$.

The focus is $(p, 0)$, which is $(4, 0)$.

The focal diameter is $4|p|$, which is $4|(4)| = 16$.

Answers

2. a. Vertex: $(0, 0)$; Focus: $(0, -2)$; Focal diameter: 8
b. $(-4, -2), (4, -2)$
c. Directrix: $y = 2$; Axis of symmetry: $x = 0$

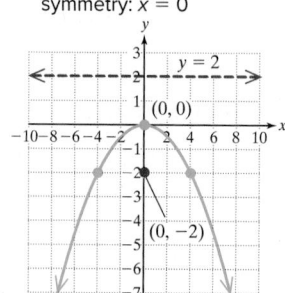

b. Plot the vertex and locate the focus. The focal diameter is 16 units. Therefore, plot points 8 units above and 8 units below the focus. These are the endpoints of the latus rectum, $(4, 8)$ and $(4, -8)$.

c. Equation of the directrix: $x = -4$

Equation of the axis of symmetry: $y = 0$

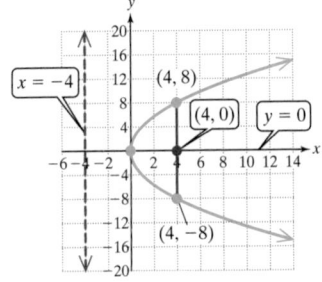

Skill Practice 3   Repeat Example 3 with the equation: $y^2 = -12x$

3. Graph a Parabola with Vertex (*h, k*)

In the standard form of a parabola, if x and y are replaced by $x - h$ and $y - k$, then the graph of the parabola is shifted h units horizontally and k units vertically. The vertex is (h, k).

Standard Forms of an Equation of a Parabola with Vertex (*h, k*)		
	Vertical Axis of Symmetry	**Horizontal Axis of Symmetry**
Equation:	$(x - h)^2 = 4p(y - k)$	$(y - k)^2 = 4p(x - h)$
Vertex:	(h, k)	(h, k)
Focus:	$(h, k + p)$	$(h + p, k)$
Directrix:	$y = k - p$	$x = h - p$
Axis of symmetry:	$x = h$	$y = k$
Graph:	$x = h$ $(h, k + p)$ $y = k - p$ (h, k)	$x = h - p$ (h, k) $(h + p, k)$ $y = k$
Note:	• If $p > 0$, the parabola opens upward as shown. • If $p < 0$, the parabola opens downward.	• If $p > 0$, the parabola opens to the right as shown. • If $p < 0$, the parabola opens to the left.

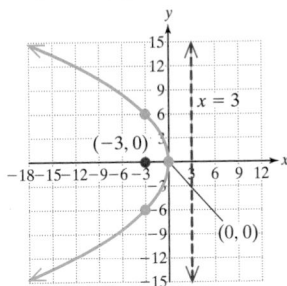

To graph a parabola with vertex (h, k), first locate the vertex and identify the orientation of the parabola (vertical axis of symmetry or horizontal axis of symmetry). Then identify the focus, directrix, and endpoints of the latus rectum. Finally, sketch the parabola through the vertex and endpoints of the latus rectum.

EXAMPLE 4   **Graphing a Parabola with Vertex (h, k)**

Given $(y + 3)^2 = -4(x - 2)$,

    a. Identify the vertex, focus, and focal diameter.

    b. Identify the endpoints of the latus rectum and graph the parabola.

    c. Write equations for the directrix and axis of symmetry.

Solution:

    a. $(y + 3)^2 = -4(x - 2)$ The equation is in the form $(y - k)^2 = 4p(x - h)$, where $h = 2$ and $k = -3$. The vertex is $(2, -3)$, and $4p = -4$. This implies that $p = -1$.

         $4p = -4$

         $p = -1$ The x term is linear. Therefore, the parabola will open in the x direction (left or right). Furthermore, since $p < 0$, the parabola will open to the left.

    The vertex is $(2, -3)$.

    The focus is $(1, -3)$.   Since $p = -1$, the focus is 1 unit to the left of the vertex.

    The focal diameter is 4.   The focal diameter is $4|p| = 4|(-1)| = 4$.

    b. Plot the vertex and locate the focus. The focal diameter is 4 units. Therefore, plot points 2 units above and below the focus. These are the endpoints of the latus rectum: $(1, -1)$ and $(1, -5)$.

       Sketch the curve through the vertex and endpoints of the latus rectum.

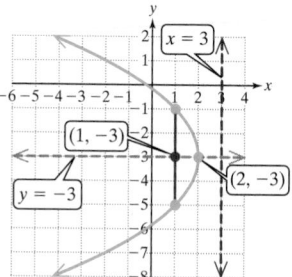

    c. Equation of the directrix: $x = 3$

       Equation of the axis of symmetry: $y = -3$

Skill Practice 4   Repeat Example 4 with the equation: $(y - 2)^2 = -8(x - 4)$

TECHNOLOGY CONNECTIONS

Graphing a Parabola

To graph the parabola from Example 4 on a graphing utility, first solve the equation for y.

$$(y + 3)^2 = -4(x - 2)$$

$$y = -3 \pm \sqrt{-4(x - 2)} \qquad \text{Solve for } y.$$

The functions represent the top and bottom branches of the parabola, respectively.

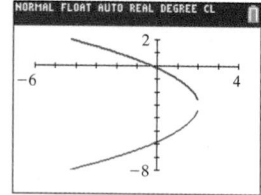

Answers

4. a. Vertex: $(4, 2)$; Focus: $(2, 2)$; Focal diameter: 8

    b. $(2, 6)$, $(2, -2)$

    c. Directrix: $x = 6$; Axis of symmetry: $y = 2$

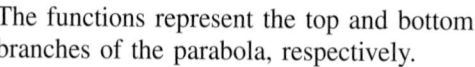

Sometimes an equation of a parabola is given in expanded form. We can complete the square to write the equation in standard form. The process involves completing the square on either the x or y term, depending on which variable is squared in the equation. This is demonstrated in Example 5.

EXAMPLE 5   Writing an Equation of a Parabola
in Standard Form

Given the equation of the parabola $4x^2 - 20x - 8y + 57 = 0$,

 a. Write the equation in standard form.

 b. Identify the vertex, focus, and focal diameter.

 c. Graph the parabola.

Solution:

 a. $4x^2 - 20x - 8y + 57 = 0$

   $4(x^2 - 5x \quad\quad) = 8y - 57$

The x term is squared. Therefore, we complete the square on x on the left side. Move the y term and constant term to the right.

$4 \cdot \frac{25}{4}$ is added to the left side. $4 \cdot \frac{25}{4}$

$4\left(x^2 - 5x + \frac{25}{4}\right) = 8y - 57 + 25$

Complete the square.
Note that $\left[\frac{1}{2} \cdot (-5)\right]^2 = \left[-\frac{5}{2}\right]^2 = \frac{25}{4}$.

$4\left(x - \frac{5}{2}\right)^2 = 8y - 32$

The value $\frac{25}{4}$ was added inside the parentheses. However, $4 \cdot \frac{25}{4}$ was actually added to the left side as a result of the factor 4 in front of the parentheses.

$\dfrac{4\left(x - \frac{5}{2}\right)^2}{4} = \dfrac{8y - 32}{4}$

To write the equation in standard form, we require the coefficient of the square term to be 1. Divide both sides by 4.

$\left(x - \frac{5}{2}\right)^2 = 2y - 8$

$\left(x - \frac{5}{2}\right)^2 = 2(y - 4)$

Factor out the leading coefficient on the right side of the equation. The equation is in the form $(x - h)^2 = 4p(y - k)$.

 b. $4p = 2$

   $p = \frac{1}{2}$

The y term is linear. Therefore, the parabola will open in the y direction (upward or downward). Furthermore, since $p > 0$, the parabola will open upward.

The vertex is $\left(\frac{5}{2}, 4\right)$.

The focus is $\left(\frac{5}{2}, 4 + \frac{1}{2}\right) = \left(\frac{5}{2}, \frac{9}{2}\right)$.   Since $p = \frac{1}{2}$, the focus is $\frac{1}{2}$ unit above the vertex.

The focal diameter is 2.   The focal diameter is $4|p| = 4\left|\frac{1}{2}\right| = 2$.

 c. Plot the vertex and locate the focus. The focal diameter is 2 units. Therefore, plot points 1 unit to the left and right of the focus. These are the points $\left(\frac{3}{2}, \frac{9}{2}\right)$ and $\left(\frac{7}{2}, \frac{9}{2}\right)$.

Sketch the curve.

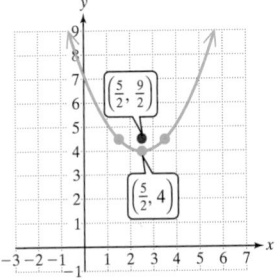

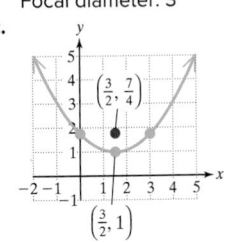

Skill Practice 5   Repeat Example 5 with the equation:

$4x^2 - 12x - 12y + 21 = 0$

In Example 6, we determine the standard form of an equation of a parabola based on known information about the parabola.

Classroom Example: p. 951
Exercise 66

EXAMPLE 6   Writing an Equation of a Parabola

Determine the standard form of an equation of the parabola with focus (1, 2) and directrix $x = 7$.

Solution:

The directrix is a vertical line $x = 7$ and the focus is 6 units to the left of the directrix (Figure 10-28). The vertex is halfway between the directrix and focus. Therefore, the vertex is 3 units to the right of the focus at (4, 2). This indicates that $h = 4$ and $k = 2$.

The parabola must open away from the directrix and toward the focus which in this case is to the left. The standard equation is

$$(y - k)^2 = 4p(x - h)$$

The distance between the focus and vertex is 3. Therefore, $|p| = 3$. Because the parabola opens to the left, $p < 0$, indicating that $p = -3$.

The equation is: $(y - k)^2 = 4p(x - h)$
$(y - 2)^2 = 4(-3)(x - 4)$
$(y - 2)^2 = -12(x - 4)$

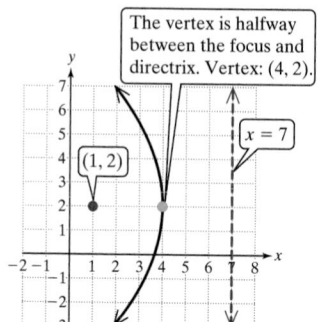

The vertex is halfway between the focus and directrix. Vertex: (4, 2).

Figure 10-28

Skill Practice 6   Determine the standard form of an equation of a parabola with focus (−2, 3) and directrix $y = 7$.

4. Use Parabolas in Applications

We complete this section by determining an equation of a parabola used to construct a satellite dish. A parabolic satellite dish uses the reflective property of a parabola to gather radio waves and direct the signal to a common focus to make the signal stronger.

Classroom Example: p. 952
Exercise 70

EXAMPLE 7   Applying an Equation of a Parabola

A satellite dish is in the shape of a paraboloid. Cross sections taken parallel to the direction the dish opens are parabolic. Cross sections taken perpendicular to the direction the dish opens are circular. The diameter of the dish is 30 in. and the depth is 3 in.

a. Find an equation of a parabolic cross section through the vertex of the dish.

b. Where should the receiver be placed?

Solution:

a. Orient the dish with the vertex at the origin of a rectangular coordinate system with the dish opening along the positive x-axis (Figure 10-29). An equation of a parabolic cross section through the origin is of the form $y^2 = 4px$. To solve for the value of p, we can

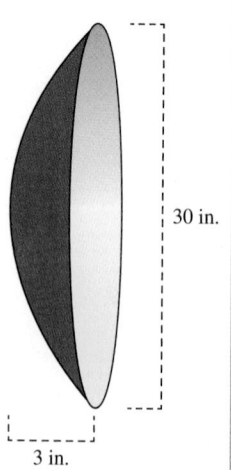

30 in.

3 in.

Answer
6. $(x + 2)^2 = -8(y - 5)$

use a known point on the parabola such as (3, 15) or
(3, −15). Using (3, 15), we have

$$y^2 = 4px$$
$$(15)^2 = 4p(3)$$
$$225 = 12p$$
$$p = 18.75$$

Therefore, an equation of a parabolic cross section through
the origin is

$$y^2 = 4(18.75)x$$
$$y^2 = 75x$$

b. The receiver should be placed at the focus of the parabola. Since the focal
length is $|p| = |18.75| = 18.75$, the receiver should be placed at (18.75, 0).

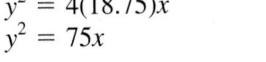

Figure 10-29

Answers

7. a. $y^2 = 87.5x$

  b. (21.875, 0)

Skill Practice 7   Repeat Example 7 with a radio telescope of diameter 70 m
and depth 14 m.

SECTION 10.3   Practice Exercises

Prerequisite Review

R.1. Use the distance formula to find the distance between the points (−1, −2) and (−2, −9). $5\sqrt{2}$

For Exercises R.2–R.3, write an equation of the line satisfying the given conditions.

R.2. The line is parallel to the line $y = 4$ and passes through (−4, 3). $y = 3$

R.3. The line is perpendicular to the x-axis and passes through (2, 1). $x = 2$

Concept Connections

1. A ___parabola___ is the set of all points in a plane that are
equidistant from a fixed line (called the ___directrix___)
and a fixed point (called the ___focus___).

2. The line perpendicular to the directrix and passing
through the focus of a parabola is called the axis
of ___symmetry___.

3. The ___vertex___ of a parabola is the point of
intersection of the parabola and the axis of symmetry.

4. The distance between the vertex and the focus of a
parabola is called the ___focal___ length and is often
represented by $|p|$.

5. Given $y^2 = 4px$ with $p < 0$, the parabola opens
(upward/downward/left/right).   left

6. Given $x^2 = 4py$ with $p > 0$, the parabola opens
(upward/downward/left/right).   upward

7. The line segment perpendicular to the axis of symmetry,
passing through the focus and with endpoints on the
parabola is called the ___latus___ ___rectum___.

8. The length of the latus rectum is called the ___focal___
diameter.

9. Given $(x - h)^2 = 4p(y - k)$, the ordered pairs
representing the vertex and focus are ___(h, k)___ and
___(h, k + p)___, respectively. The directrix is the line
defined by the equation ___$y = k - p$___.

10. Given $(y - k)^2 = 4p(x - h)$, the ordered pairs
representing the vertex and focus are ___(h, k)___ and
___(h + p, k)___, respectively. The directrix is the line
defined by the equation ___$x = h - p$___.

11. If the directrix is horizontal, then the parabola opens
(horizontally/vertically).   vertically

12. If the line of symmetry is horizontal, then the parabola
opens (horizontally/vertically).   horizontally

For Exercises 13–14, the graph of a parabola is given.

a. Determine the distances d_1, d_2, d_3, and d_4.

b. Compare d_1 and d_2.

c. Compare d_3 and d_4.

13.

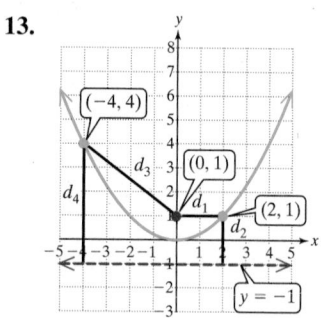

a. $d_1 = 2$, $d_2 = 2$, $d_3 = 5$, $d_4 = 5$
b. They are the same.
c. They are the same.

14.

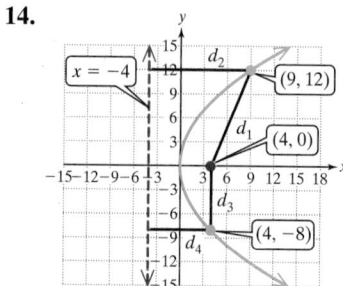

a. $d_1 = 13$, $d_2 = 13$,
   $d_3 = 8$, $d_4 = 8$
b. They are the same.
c. They are the same.

Objective 1: Identify the Focus and Directrix of a Parabola

For Exercises 15–22, a model of the form $x^2 = 4py$ or $y^2 = 4px$ is given.

a. Determine the value of p.

b. Identify the focus of the parabola.

c. Write an equation for the directrix. (**See Example 1**)

15. $x^2 = 24y$
a. $p = 6$ b. $(0, 6)$ c. $y = -6$

16. $x^2 = 12y$ a. $p = 3$ b. $(0, 3)$ c. $y = -3$

17. $y^2 = 36x$
a. $p = 9$ b. $(9, 0)$ c. $x = -9$

18. $y^2 = 48x$
a. $p = 12$ b. $(12, 0)$ c. $x = -12$

19. $x^2 = -5y$
a. $p = -\frac{5}{4}$ b. $(0, -\frac{5}{4})$ c. $y = \frac{5}{4}$

20. $x^2 = -11y$
a. $p = -\frac{11}{4}$ b. $(0, -\frac{11}{4})$ c. $y = \frac{11}{4}$

21. $-x = y^2$
a. $p = -\frac{1}{4}$ b. $(-\frac{1}{4}, 0)$ c. $x = \frac{1}{4}$

22. $-3x = y^2$ a. $p = -\frac{3}{4}$ b. $(-\frac{3}{4}, 0)$ c. $x = \frac{3}{4}$

23. A 20-in. satellite dish for a television has parabolic cross sections. A coordinate system is chosen so that the vertex of a cross section through the center of the dish is located at $(0, 0)$. The equation of the parabola is modeled by $x^2 = 25.2y$, where x and y are measured in inches.

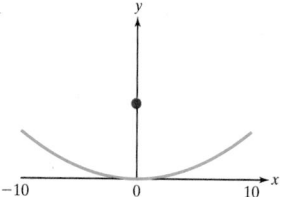

a. Where should the receiver be placed to maximize signal strength? That is, where is the focus?
   (**See Example 1**)   Focus: $(0, 6.3)$; Place the receiver 6.3 in. above the center of the dish.
b. Determine the equation of the directrix. $y = -6.3$

24. Solar cookers provide an alternative form of cooking in regions of the world where consistent sources of fuel are not readily available. Suppose that a 36-in. solar cooker has parabolic cross sections. A coordinate system is chosen with the origin placed at the vertex of a cross section through the center of the mirror. The equation of the parabola is modeled by $x^2 = 82y$, where x and y are measured in inches.

a. Where should a pot be placed to maximize heat? That is, where is the focus?   Focus: $(0, 20.5)$; The pot should be placed 20.5 in. above the center.
b. Determine the equation of the directrix. $y = -20.5$

25. If a cross section of the parabolic mirror in a flashlight has an equation $y^2 = 2x$, where should the bulb be placed? $\frac{1}{2}$ unit to the right of the vertex

26. A cross section of the parabolic mirror in a car headlight is modeled by $y^2 = 12x$. Where should the bulb be placed?   3 units to the right of the vertex

Objective 2: Graph a Parabola with Vertex at the Origin

For Exercises 27–34, an equation of a parabola $x^2 = 4py$ or $y^2 = 4px$ is given.

a. Identify the vertex, value of p, focus, and focal diameter of the parabola.

b. Identify the endpoints of the latus rectum.

c. Graph the parabola.

d. Write equations for the directrix and axis of symmetry. (**See Examples 2–3**)

27. $x^2 = -4y$ **28.** $x^2 = -20y$ **29.** $10y^2 = 80x$ **30.** $3y^2 = 12x$

31. $4x^2 = 40y$ **32.** $2x^2 = 14y$ **33.** $y^2 = -x$ **34.** $y^2 = -2x$

Objective 3: Graph a Parabola with Vertex (h, k)

For Exercises 35–44, an equation of a parabola $(x - h)^2 = 4p(y - k)$ or $(y - k)^2 = 4p(x - h)$ is given.

a. Identify the vertex, value of p, focus, and focal diameter of the parabola.

b. Identify the endpoints of the latus rectum.

c. Graph the parabola.

d. Write equations for the directrix and axis of symmetry. (**See Example 4**)

35. $(y + 1)^2 = -12(x - 4)$ **36.** $(y + 4)^2 = -16(x - 2)$ **37.** $(x - 1)^2 = -4(y + 5)$

38. $(x - 5)^2 = -8(y + 2)$ **39.** $(x + 3)^2 = 2\left(y - \frac{3}{2}\right)$ **40.** $(x + 2)^2 = \left(y - \frac{7}{4}\right)$

41. $-(y - 3)^2 = 7\left(x - \frac{1}{4}\right)$ **42.** $-(y - 6)^2 = 9\left(x - \frac{3}{4}\right)$ **43.** $2(y - 3) = \frac{1}{10}(x + 6)^2$

44. $8(y - 2) = \frac{1}{2}(x + 3)^2$

For Exercises 45–52, an equation of a parabola is given.

a. Write the equation of the parabola in standard form.

b. Identify the vertex, focus, and focal diameter. (**See Example 5**)

45. $x^2 - 6x - 4y + 5 = 0$ **46.** $x^2 - 4x - 8y - 20 = 0$ **47.** $y^2 + 4y + 8x + 52 = 0$

48. $y^2 + 8y + 4x + 36 = 0$ **49.** $4x^2 - 28x + 24y + 73 = 0$ **50.** $4x^2 + 36x + 40y + 1 = 0$

51. $16y^2 + 24y - 16x + 57 = 0$ **52.** $16y^2 - 56y - 16x + 81 = 0$

For Exercises 53–58, fill in the blanks. Let $|p|$ represent the focal length (**distance between the vertex and focus**).

53. If the directrix of a parabola is given by $y = 4$ and the focus is (1, 10), then the vertex is given by the ordered pair ___(1, 7)___ and the value of p is ___3___.

54. If the directrix of a parabola is given by $x = 6$ and the focus is (2, 1), then the vertex is given by the ordered pair ___(4, 1)___ and the value of p is ___-2___.

55. If the vertex of a parabola is (−3, 0) and the focus is (−7, 0), then the directrix is given by the equation ___$x = 1$___, and the value of p is ___-4___.

56. If the vertex of a parabola is (4, −2) and the focus is (4, −7), then the directrix is given by the equation ___$y = 3$___, and the value of p is ___-5___.

57. If the focal length of a parabola is 3 units, then the length of the latus rectum is ___12___ units.

58. If the focal diameter of a parabola is 8 units, then the focal length is ___2___ units.

59. If the focal length of a parabola is 6 units and the vertex is (3, 2), is it possible to determine whether the parabola opens upward, downward, to the right, or to the left? Explain. No; Focal length is a distance and gives no information regarding the orientation of a parabola.

60. If the axis of symmetry of a parabola is given by $x = 4$, is it possible for the directrix to have the equation $x = 3$? Explain. No; The directrix must be perpendicular to the axis of symmetry.

For Exercises 61–68, determine the standard form of an equation of the parabola subject to the given conditions. (**See Example 6**)

61. Vertex: (0, 0); Directrix: $x = 4$ $y^2 = -16x$ **62.** Vertex: (0, 0); Directrix: $y = -2$ $x^2 = 8y$

63. Focus: (2, 4); Vertex: (2, 1) $(x - 2)^2 = 12(y - 1)$ **64.** Focus: (5, 3): Vertex: (3, 3) $(y - 3)^2 = 8(x - 3)$

65. Focus: (−6, −2); Directrix: $y = 0$ $(x + 6)^2 = -4(y + 1)$ **66.** Focus: (−4, 5); Directrix: $x = 0$ $(y - 5)^2 = -8(x + 2)$

67. Vertex: (2, 3); Parabola passes through (6, 5). (*Hint*: There are two possible answers.) $(x - 2)^2 = 8(y - 3)$ or $(y - 3)^2 = (x - 2)$

68. Vertex: (−4, 2); Parabola passes through (8, 14) (*Hint*: There are two possible answers.) $(x + 4)^2 = 12(y - 2)$ or $(y - 2)^2 = 12(x + 4)$

Objective 4: Use Parabolas in Applications

69. Suppose that a solar cooker has a parabolic mirror (see figure).

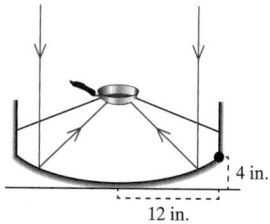

12 in.

4 in.

a. Use a coordinate system with origin at the vertex of the mirror and write an equation of a parabolic cross section of the mirror. $x^2 = 36y$ for $-12 \le x \le 12$

b. Where should a pot be placed so that it receives maximum heat? (**See Example 7**)

(0, 9); The pot should be placed 9 in. above the vertex.

70. A solar water heater is made from a long sheet of metal bent so that the cross sections are parabolic. A long tube of water is placed inside the curved surface so that the height of the tube is equal to the focal length of the parabolic cross section. In this way, water in the tube is exposed to maximum heat.

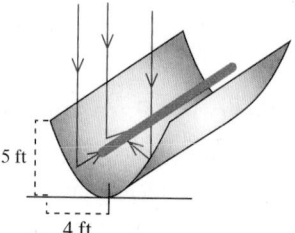

5 ft

4 ft

a. Determine the focal length of the parabolic cross sections so that the engineer knows where to place the tube.   0.8 ft

b. Use a coordinate system with origin at the vertex of a parabolic cross section and write an equation of the parabola. $x^2 = 3.2y$ for $-4 \le x \le 4$

71. The Hubble Space Telescope was launched into space in 1990 and now orbits the Earth at 5 mi/sec at a distance of 353 mi above the Earth. From its location in space, the Hubble is free from the distortion of the Earth's atmosphere, enabling it to return magnificent images from distant stars and galaxies. The quality of the Hubble's images result from a large parabolic mirror, 2.4 m (7.9 ft) in diameter, that collects light from space.

Suppose that a coordinate system is chosen so that the vertex of a cross section through the center of the mirror is located at (0, 0). Furthermore, the focal length is 57.6 m.

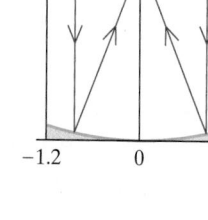

a. Assume that x and y are measured in meters. Write an equation of the parabolic cross section of the mirror for $-1.2 \le x \le 1.2$. See figure.
$x^2 = 230.4y$ for $-1.2 \le x \le 1.2$

b. How thick is the mirror at the edge? That is, what is the y value for $x = 1.2$?   0.00625 m or 6.25 mm

72. The James Webb Space Telescope (JWST) is a new space telescope currently under construction. The JWST will orbit the Sun in an orbit roughly 1.5 million km from the earth, in a position with the Earth between the telescope and the Sun. The primary mirror of the JWST will consist of 18 hexagonally shaped segments that when fitted together will be 6.5 m in diameter (over 2.5 times the diameter of the Hubble). With a larger mirror, the JWST will be able to collect more light to "see" deeper into space.

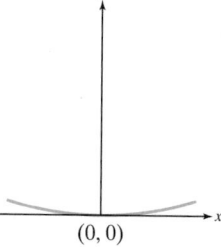

The primary mirror when pieced together will function as a parabolic mirror with a focal length of 131.4 m. Suppose that a coordinate system is chosen with (0, 0) at the center of a cross section of the primary mirror. Write an equation of the parabolic cross section of the mirror for $-3.25 \le x \le 3.25$. Assume that x and y are measured in meters.

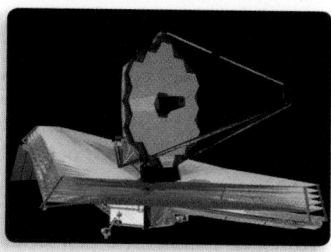

(0, 0)

$x^2 = 525.6y$ for $-3.25 \le x \le 3.25$

73. The Subaru telescope is a large optical-infrared telescope at the summit of Mauna Kea, Hawaii. The telescope has a parabolic mirror 8.2 m in diameter with a focal length of 15 m.

  a. Suppose that a cross section of the mirror is taken through the vertex, and that a coordinate system is set up with (0, 0) placed at the vertex. If the focus is (0, 15), find an equation representing the curve. $x^2 = 60y$ for $-4.1 \le x \le 4.1$

  b. Determine the vertical displacement of the mirror relative to horizontal at the edge of the mirror. That is, find the y value at a point 4.1 m to the left or right of the vertex. $\dfrac{1681}{6000}$ m ≈ 0.28 m

  c. What is the average slope between the vertex of the parabola and the point on the curve at the right edge? $m = \dfrac{41}{600} \approx 0.068$

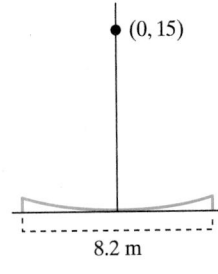

(0, 15)

8.2 m

75. A parabolic arch forms a structure that allows equal vertical loading along its length.

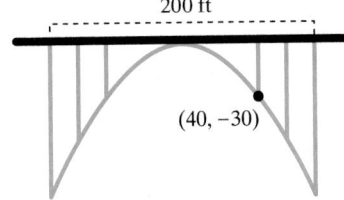

200 ft

(40, −30)

  a. Take the origin at the vertex of the parabolic arch, and write an equation of the parabola. $x^2 = -\dfrac{160}{3}y$

  b. Determine the focal length. $\dfrac{40}{3}$ ft

  c. Determine the length of the vertical support 60 ft from the center.   67.5 ft

74. A parabolic mirror on a telescope has a focal length of 16 cm.

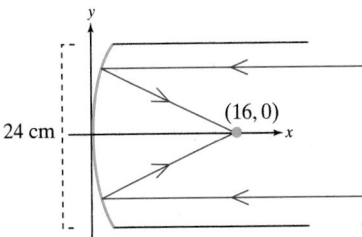

24 cm   (16, 0)

  a. For the coordinate system shown, write an equation of the parabolic cross section of the mirror. $y^2 = 64x$ for $-12 \le y \le 12$

  b. Determine the displacement of the mirror relative to y-axis at the edge of the mirror. That is, find the x value at a point 12 cm above or below the vertex.   2.25 cm

76. A cable hanging freely between two vertical support beams forms a curve called a catenary. The shape of a catenary resembles a parabola but mathematically the two functions are quite different.

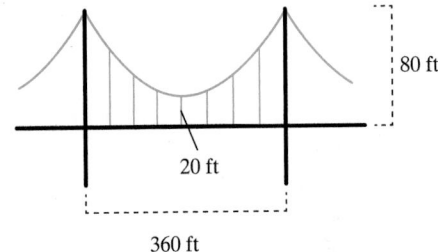

80 ft

20 ft

360 ft

  a. On a graphing utility, graph a catenary defined by $y = \dfrac{1}{2}(e^x + e^{-x})$ and graph the parabola defined by $y = x^2 + 1$.

  b. A catenary and a parabola are so similar in shape that we can often use a parabolic curve to approximate the shape of a catenary. For example, a bridge has cables suspended from a larger approximately parabolic cable. Take the origin at a point on the road directly below the vertex and write an equation of the parabolic cable.

  c. Determine the focal length of the parabolic cable.

  d. Determine the length of the vertical support cable 100 ft from the vertex. Round to the nearest tenth of a foot.

Mixed Exercises

For Exercises 77–78, solve the system of nonlinear equations.

77. $(y + 3)^2 = 4(x - 4)$ {(5, −1), (8, −7)}

   $2x + y = 9$

78. $(y - 2)^2 = -8(x - 1)$ {(−1, 6), (−7, −6)}

   $2x - y = -8$

Write About It

79. Given an equation of a parabola $(x - h)^2 = 4p(y - k)$ or $(y - k)^2 = 4p(x - h)$, how can you determine whether the parabola opens vertically or horizontally?

If the y term is linear, then the parabola opens vertically. If the x term is linear, then the parabola opens horizontally.

80. Give an example of the reflective property of a parabola.

Light reflecting off a parabolic mirror (such as in a telescope) will be directed to the focus. Light emitted at the focus of a parabolic mirror (such as in a flashlight) will reflect off the parabolic mirror to form a beam of light parallel to the axis of symmetry.

Expanding Your Skills

81. The surface defined by the equation $z = 4x^2 + y^2$ is called an elliptical paraboloid.

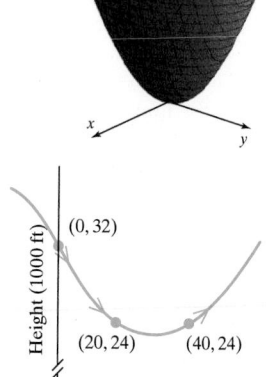

  a. Write the equation with $x = 0$. What type of curve is represented by this equation?

  b. Write the equation with $y = 0$. What type of curve is represented by this equation?

  c. Write the equation with $z = 0$. What type of curve is represented by this equation?

Technology Connections

82. A jet flies in a parabolic arc to simulate partial weightlessness. The curve shown in the figure represents the plane's height y (in 1000 ft) versus the time t (in sec).

  a. For each ordered pair, substitute the t and y values into the model $y = at^2 + bt + c$ to form a linear equation with three unknowns a, b, and c. Together, these form a system of three linear equations with three unknowns.

  b. Use a graphing utility to solve for a, b, and c.

  c. Substitute the known values of a, b, and c into the model $y = at^2 + bt + c$.

  d. Determine the vertex of the parabola.

  e. Determine the focal length of the parabola.

Height (1000 ft) vs. Time (sec): $(0, 32)$, $(20, 24)$, $(40, 24)$

For Exercises 83–86, graph the parabola from the given exercise.

83. Exercise 29   **84.** Exercise 30   **85.** Exercise 41   **86.** Exercise 42

PROBLEM RECOGNITION EXERCISES

Comparing Equations of Conic Sections and the General Equation

For Exercises 1–8, identify each equation as representing a circle, an ellipse, a hyperbola, or a parabola.

- If the equation represents a circle, identify the center and radius.
- If the equation represents an ellipse, identify the center, vertices, endpoints of the minor axis, foci, and eccentricity.
- If the equation represents a hyperbola, identify the center, vertices, foci, equations of the asymptotes, and eccentricity.
- If the equation represents a parabola, identify the vertex, focus, equation of the directrix, and equation of the axis of symmetry.

1. $\dfrac{(x - 2)^2}{16} - \dfrac{(y + 2)^2}{9} = 1$

2. $9x^2 + 25(y - 2)^2 = 225$

3. $(y - 5)^2 = -(x + 2)$

4. $(x - 3)^2 + (y + 7)^2 = 25$

5. $16(x + 1)^2 + y^2 = 16$

6. $(x - 1)^2 = 10(y + 3)$

7. $-(x + 1)^2 - (y + 6)^2 = -16$

8. $-\dfrac{(x + 4)^2}{25} + \dfrac{(y - 1)^2}{144} = 1$

Suppose that the binomials within an equation of a conic section are expanded and all terms are collected on the left side of the equation. The result will be an equation of the conic section written in **general form:**

$$Ax^2 + Cy^2 + Dx + Ey + F = 0, \text{ where } A \text{ and } C \text{ are not both zero.}$$

Furthermore, given an equation of a conic section in general form, we can complete the square to identify the type of conic that the equation represents. In some cases, however, the graph of an equation in general form is a **degenerate conic section.** Geometrically, a degenerate case occurs when a plane intersects a double-napped cone in a point, a line, or a pair of intersecting lines.

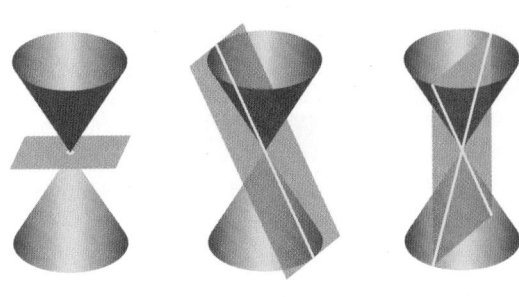

General Equation of a Conic Section

The graph of the equation $Ax^2 + Cy^2 + Dx + Ey + F = 0$, where A and C are not both zero, is a conic section or a degenerate conic. If the graph represents a conic section, then the graph is

1. A circle if $A = C$.

2. An ellipse if $AC > 0$ (A and C have the same sign).

3. A hyperbola if $AC < 0$ (A and C have opposite signs).

4. A parabola if $AC = 0$ (A or C equals zero).

TIP The general form of an equation of a conic section also has a term Bxy, which rotates the curve. Some courses in trigonometry address rotated conics.

For Exercises 9–16, determine whether the equation represents a circle, an ellipse, a hyperbola, or a parabola. Write the equation in standard form.

9. $9x^2 - 16y^2 - 36x - 64y - 172 = 0$ Hyperbola; $\dfrac{(x-2)^2}{16} - \dfrac{(y+2)^2}{9} = 1$

10. $9x^2 + 25y^2 - 100y - 125 = 0$ Ellipse; $\dfrac{x^2}{25} + \dfrac{(y-2)^2}{9} = 1$

11. $y^2 - 10y + x + 27 = 0$
Parabola; $(y-5)^2 = -(x+2)$

12. $x^2 + y^2 - 6x + 14y + 33 = 0$
Circle; $(x-3)^2 + (y+7)^2 = 25$

13. $16x^2 + y^2 + 32x = 0$ Ellipse; $(x+1)^2 + \dfrac{y^2}{16} = 1$

14. $x^2 - 2x - 10y - 29 = 0$ Parabola; $(x-1)^2 = 10(y+3)$

15. $x^2 + y^2 + 2x + 12y + 21 = 0$
Circle; $(x+1)^2 + (y+6)^2 = 16$

16. $-144x^2 + 25y^2 - 1152x - 50y - 5879 = 0$
Hyperbola; $\dfrac{(y-1)^2}{144} - \dfrac{(x+4)^2}{25} = 1$

For Exercises 17–22, an equation of a degenerate conic section is given. Complete the square and describe the graph of each equation.

17. $x^2 + y^2 - 4x + 2y + 5 = 0$

18. $9x^2 + 4y^2 - 24y + 36 = 0$

19. $4x^2 - y^2 - 32x - 4y + 60 = 0$

20. $9x^2 - 4y^2 - 90x - 8y + 221 = 0$

21. $9x^2 + 4y^2 - 16y + 52 = 0$

22. $x^2 + y^2 + 12x + 37 = 0$

23. Given $Ax^2 + Cy^2 + Dx + Ey + F = 0$, show that the equation represents a circle if $A = C \neq 0$ and $D^2 + E^2 - 4AF > 0$.

24. Consider $Ax^2 + Cy^2 + Dx + Ey + F = 0$, for $A \neq 0$ and $C \neq 0$.

   a. For A and C of the same sign, show that the equation represents an ellipse if $\dfrac{D^2}{4A} + \dfrac{E^2}{4C} - F$ has the same sign as A.

   b. For A and C of opposite signs, show that the equation represents a hyperbola if $\dfrac{D^2}{4A} + \dfrac{E^2}{4C} - F \neq 0$.

25. Given $Ax^2 + Cy^2 + Dx + Ey + F = 0$, show that the equation represents a parabola if $A = 0$ but $C \neq 0$ and $D \neq 0$.

SECTION 10.4 Rotation of Axes

Instructor Note:
Consider assigning the Problem Recognition Exercises on pages 954–955 before starting this section.

Classroom Example: p. 966
Exercise 6

1. Identify Conic Sections with No Rotation

In Sections 10.1–10.3, we analyzed equations of the form

$$Ax^2 + Cy^2 + Dx + Ey + F = 0 \qquad (1)$$

Equation (1) represents a conic section or a degenerate case. For nondegenerate cases, we can complete the square on the quadratic expressions in x and y to determine the type of conic section represented by the equation. Alternatively we can identify the type of conic section directly from the equation based on the values of A and C.

General Equation of a Conic with No Rotation

The graph of the equation $Ax^2 + Cy^2 + Dx + Ey + F = 0$, where A and C are not both zero, is a conic section or a degenerate case. If the graph represents a conic section, then the graph is

1. A circle if $A = C$.
2. An ellipse if $AC > 0$ (A and C have the same sign).
3. A hyperbola if $AC < 0$ (A and C have opposite signs).
4. A parabola if $AC = 0$ (A or C equals zero).

EXAMPLE 1   Identifying Conic Sections with No Rotation

The equations represent nondegenerate conic sections. Identify the type of conic section.

a. $x^2 - y^2 - 4x = 0$ **b.** $y^2 - 12x - 6y + 23 = 0$
c. $36x^2 + 36y^2 - 18x + 24y + 5 = 0$ **d.** $4x^2 + y^2 + 24x - 2y + 3 = 0$

Solution:

a. $x^2 - y^2 - 4x = 0$

The coefficients of the square terms are $A = 1$ and $C = -1$. Since A and C have opposite signs ($AC < 0$), the graph is a hyperbola.

b. $y^2 - 12x - 6y + 23 = 0$

The equation does not have an x^2 term. Since $A = 0$, the graph is a parabola.

c. $36x^2 + 36y^2 - 18x + 24y + 5 = 0$

Since A and C are equal ($A = C = 36$), the graph is a circle.

d. $4x^2 + y^2 + 24x - 2y + 3 = 0$

The coefficients of the square terms are $A = 4$ and $C = 1$. Since A and C have the same sign ($AC > 0$), the graph is an ellipse.

Skill Practice 1 The equations represent nondegenerate conic sections. Identify the type of conic section.

a. $x^2 + y^2 - 4x - 14 = 0$ **b.** $4x^2 - 9y^2 + 54y - 117 = 0$
c. $4x^2 - 16x - y + 15 = 0$ **d.** $25x^2 + 4y^2 + 250x - 16y + 541 = 0$

Answers
1. **a.** Circle   **b.** Hyperbola
    c. Parabola   **d.** Ellipse

2. Use the Rotation of Axes Formulas

We now consider the most general second-degree polynomial equation in two variables.

$$Ax^2 + Bxy + Cy^2 + Dx + Ey + F = 0 \qquad (2)$$

Equation (2) differs from equation (1) $Ax^2 + Cy^2 + Dx + Ey + F = 0$ because of the existence of the Bxy term. We will see that the graph of (2) is also a conic section (or degenerate case). Excluding degenerate cases, the term Bxy introduces a rotation of the graph through an angle θ (Figure 10-30). To analyze the graph, we can rotate the x- and y-axes about the origin through an angle θ to form a new coordinate system with axes x' and y' (Figure 10-31). Relative to the x'- and y'-axes, the graph has no rotation.

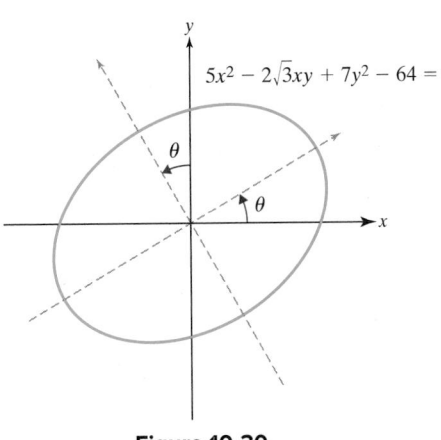

y
$5x^2 - 2\sqrt{3}xy + 7y^2 - 64 = 0$
θ
θ
x

Figure 10-30

y'
$x'^2 + 2y'^2 - 16 = 0$
x'
$\frac{x'^2}{16} + \frac{y'^2}{8} = 1$

Figure 10-31

The goal in this section is to transform the original equation (2) in the variables x and y to a new equation in x' and y' with $B = 0$.

Suppose that a point $P(x, y)$ relative to the xy-plane has coordinate $P(x', y')$ relative to the $x'y'$-plane (Figure 10-32). Let r represent the distance between P and the origin and let α be the angle made by line segment $\overline{OP}$ and the positive x'-axis.

P(x, y)
P(x', y')
y
y'
x'
y'
y
r
α
θ
x'
O
x
x

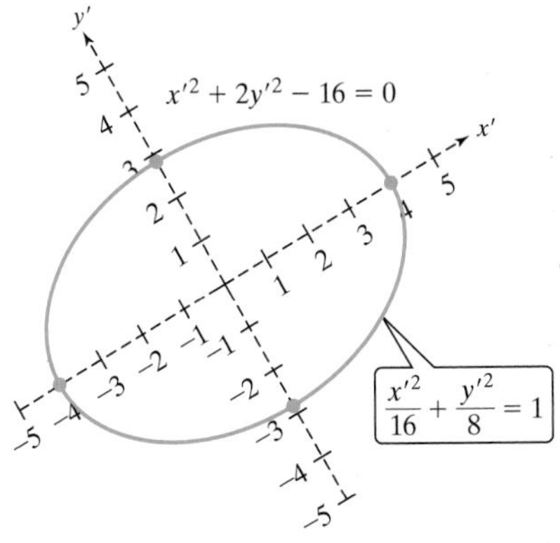

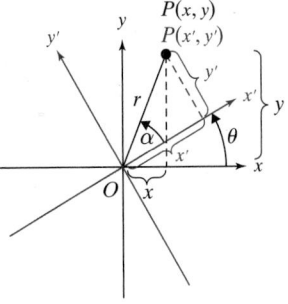

Figure 10-32

From Figure 10-32, we have

(3) $x' = r\cos\alpha$ (5) $x = r\cos(\theta + \alpha)$

(4) $y' = r\sin\alpha$ (6) $y = r\sin(\theta + \alpha)$

Applying the sum formula for cosine in equation (5), we have

$$\begin{aligned} x &= r\cos(\theta + \alpha) \\ &= r(\cos\theta\cos\alpha - \sin\theta\sin\alpha) \\ &= (r\cos\alpha)\cos\theta - (r\sin\alpha)\sin\theta \\ &= x'\cos\theta - y'\sin\theta \qquad \text{By equations (3) and (4)} \end{aligned}$$

Applying the sum formula for sine in equation (6), we have

$$\begin{aligned} y &= r\sin(\theta + \alpha) \\ &= r(\sin\theta\cos\alpha + \cos\theta\sin\alpha) \\ &= (r\cos\alpha)\sin\theta + (r\sin\alpha)\cos\theta \\ &= x'\sin\theta + y'\cos\theta \qquad \text{By equations (3) and (4)} \end{aligned}$$

Treating the equations $x = x'\cos\theta - y'\sin\theta$ and $y = x'\sin\theta + y'\cos\theta$ as a system of equations we can also solve for x' and y' in terms of x and y (see Exercise 51).

Rotation of Axes

Suppose that the x- and y-axes in a rectangular coordinate system are rotated about the origin through an angle θ to form the x'- and y'-axes. Then the coordinates of $P(x, y)$ in the xy-plane are related to $P(x', y')$ in the $x'y'$-plane as follows.

x and y in terms of x' and y'	x' and y' in terms of x and y
(7) $x = x'\cos\theta - y'\sin\theta$	(9) $x' = x\cos\theta + y\sin\theta$
(8) $y = x'\sin\theta + y'\cos\theta$	(10) $y' = -x\sin\theta + y\cos\theta$

Classroom Example: p. 966
Exercise 10

EXAMPLE 2 Identifying Coordinates of a Point after Rotating the Coordinate Axes

Suppose that the x- and y-axes are rotated $60°$ about the origin to form the x'- and y'-axes. Given $P(6, -2)$ in the xy-plane, find the coordinates of P in the $x'y'$-plane.

Solution:

$P(6, -2)$ corresponds to $x = 6$ and $y = -2$.

Using rotation equations (9) and (10),

TIP In Figure 10-33, the coordinates of point P are given relative to the xy-plane and the $x'y'$-plane.

we have $x' = x\cos\theta + y\sin\theta$

$\qquad = 6\cos 60° + (-2)\sin 60°$

$\qquad = 6\left(\dfrac{1}{2}\right) + (-2)\left(\dfrac{\sqrt{3}}{2}\right) = 3 - \sqrt{3}$

$y' = -x\sin\theta + y\cos\theta$

$\qquad = -6\sin 60° - 2\cos 60°$

$\qquad = -6\left(\dfrac{\sqrt{3}}{2}\right) + (-2)\left(\dfrac{1}{2}\right) = -3\sqrt{3} - 1$

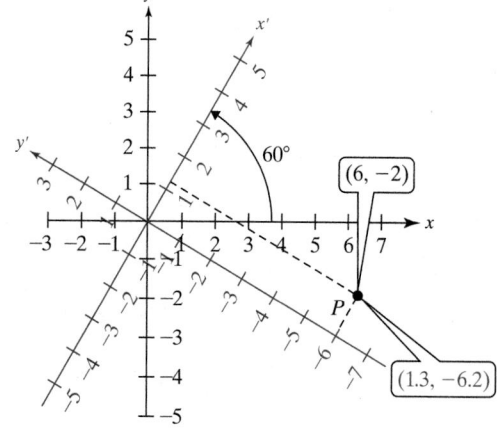

Figure 10-33

$P(x', y')$ is $\left(3 - \sqrt{3},\ -3\sqrt{3} - 1\right) \approx (1.3, -6.2)$.

Skill Practice 2 Suppose that the x- and y-axes are rotated $45°$ about the origin to form the x'- and y'-axes. Given $P(-4, 2)$ in the xy-plane, find the coordinate of P in the $x'y'$-plane.

The equation $y = \dfrac{1}{x}$ was first graphed in Section 1.6 (Figure 10-34). The graph looks like a hyperbola rotated through some angle about the origin. This is indeed the case. The equation can be written as $xy = 1$ and the left side of the equation is the Bxy term in a general second-degree polynomial equation. In Example 3, we show that the graph of $xy = 1$ is a hyperbola rotated $45°$ from the x-axis.

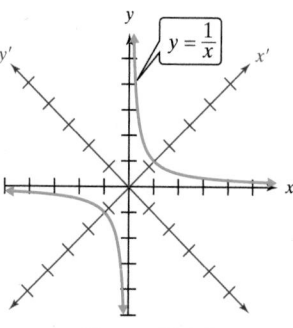

Figure 10-34

Answer

2. $\left(-\sqrt{2}, 3\sqrt{2}\right)$

EXAMPLE 3   Rotating Axes and Transforming an Equation

Assume that the x- and y-axes are rotated through an angle of $45°$ about the origin to form the x'- and y'-axes. Write the equation $xy = 1$ in $x'y'$-coordinates. Discuss the result.

Solution:

Use the rotation formulas (7) and (8) and replace x by $x' \cos\theta - y' \sin\theta$ and y by $x' \sin\theta + y' \cos\theta$.

$$xy = 1$$

$$(x' \cos 45° - y' \sin 45°)(x' \sin 45° + y' \cos 45°) = 1$$

$$\left[x'\left(\frac{\sqrt{2}}{2}\right) - y'\left(\frac{\sqrt{2}}{2}\right) \right]\left[x'\left(\frac{\sqrt{2}}{2}\right) + y'\left(\frac{\sqrt{2}}{2}\right) \right] = 1$$

$$\left(\frac{\sqrt{2}}{2}\right)\left(\frac{\sqrt{2}}{2}\right)(x' - y')(x' + y') = 1 \qquad \text{Factor } \frac{\sqrt{2}}{2} \text{ from each binomial factor.}$$

$$\frac{1}{2}(x'^2 - y'^2) = 1 \qquad \text{Multiply conjugates.}$$

$$x'^2 - y'^2 - 2 = 0 \qquad \text{Multiply both sides by 2 and collect terms on the left.}$$

The equation is of the form $A'x'^2 + C'y'^2 + D'x' + E'y' + F' = 0$. Notice that relative to the $x'y'$-coordinate system, there is no rotation term $B'x'y'$. Furthermore, $A' = 1$ and $C' = -1$ so the product $A'C' = -1 < 0$ indicating that the graph is a hyperbola.

We can write the equation in the standard form of a hyperbola, $\frac{x'^2}{a^2} - \frac{y'^2}{b^2} = 1$.

$$\frac{x'^2}{2} - \frac{y'^2}{2} = 1 \quad \text{where } a^2 = 2 \text{ and } b^2 = 2.$$

The center of the hyperbola is at the origin and the vertices are on the x'-axis at $(-\sqrt{2}, 0)$ and $(\sqrt{2}, 0)$. The equations of the asymptotes are given by $y' = \pm\frac{b}{a}x'$ which is $y' = \pm\frac{\sqrt{2}}{\sqrt{2}}x'$ or simply $y' = \pm x'$. These are the x- and y-axes in the original coordinate system (Figure 10-35).

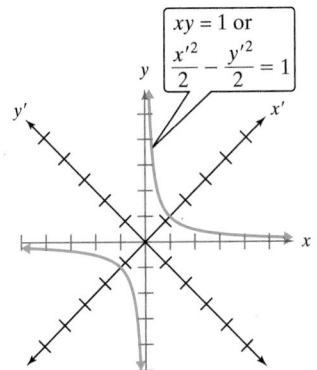

$xy = 1$ or
$\frac{x'^2}{2} - \frac{y'^2}{2} = 1$

Figure 10-35

Skill Practice 3   Write the equation $xy = 8$ in terms of x' and y' given that the angle of rotation of the x'- and y'-axes is $45°$ from the x- and y-axes.

3. Transform Equations by Rotation of Axes

The technique used in Example 3 can be used to transform any equation of the form

$$Ax^2 + Bxy + Cy^2 + Dx + Ey + F = 0 \qquad (2)$$

to an equation in x' and y' with no $x'y'$ term by choosing an appropriate angle of rotation, θ. To find θ, substitute $x = x' \cos\theta - y' \sin\theta$ and $y = x' \sin\theta + y' \cos\theta$ into (2).

$$Ax^2 + Bxy + Cy^2 + Dx + Ey + F = 0 \qquad (2)$$

$$A(x' \cos\theta - y' \sin\theta)^2 + B(x' \cos\theta - y' \sin\theta)(x' \sin\theta + y' \cos\theta)$$

$$+ C(x' \sin\theta + y' \cos\theta)^2 + D(x' \cos\theta - y' \sin\theta)$$

$$+ E(x' \sin\theta + y' \cos\theta) + F = 0$$

Answer

3. $x'^2 - y'^2 - 16 = 0$ or

$\dfrac{x'^2}{16} - \dfrac{y'^2}{16} = 1$

Expanding binomial factors and collecting like terms results in an equation of the form

$$A'x'^2 + B'x'y' + C'y'^2 + D'x' + E'y' + F' = 0$$

where
$$A' = A\cos^2\theta + B\sin\theta\cos\theta + C\sin^2\theta$$
$$B' = 2(C - A)\sin\theta\cos\theta + B(\cos^2\theta - \sin^2\theta)$$
$$C' = A\sin^2\theta - B\sin\theta\cos\theta + C\cos^2\theta$$
$$D' = D\cos\theta + E\sin\theta$$
$$E' = -D\sin\theta + E\cos\theta$$
$$F' = F$$

To eliminate the $x'y'$ term, we require $B' = 0$. Thus,

$$2(C - A)\sin\theta\cos\theta + B(\cos^2\theta - \sin^2\theta) = 0$$

> Recall that
> $2\sin\theta\cos\theta = \sin 2\theta$ and
> $\cos^2\theta - \sin^2\theta = \cos 2\theta$.

$$(C - A)\sin 2\theta + B\cos 2\theta = 0$$
$$B\cos 2\theta = (A - C)\sin 2\theta$$
$$\cot 2\theta = \frac{A - C}{B}$$

> Divide both sides by $B\sin 2\theta$.

There are an infinite number of solutions to the equation $\cot 2\theta = \dfrac{A - C}{B}$, however, we choose θ to be an acute angle. The cotangent function is used rather than the tangent function because $\cot^{-1}\left(\dfrac{A - C}{B}\right)$ returns $0 < 2\theta < \pi$ and thus, $0 < \theta < \dfrac{\pi}{2}$.

Formula for Determining an Angle of Rotation

To write the general second-degree polynomial equation

$$Ax^2 + Bxy + Cy^2 + Dx + Ey + F = 0 \qquad (2)$$

with no xy term, rotate the axes through an acute angle θ given by

$$\cot 2\theta = \frac{A - C}{B}$$

The resulting equation is $A'x'^2 + C'y'^2 + D'x' + E'y' + F' = 0$, where the coefficients are found by substituting $x'\cos\theta - y'\sin\theta$ for x and $x'\sin\theta + y'\cos\theta$ for y into (2).

In Example 4, we will eliminate the xy-term in the equation $5x^2 - 2\sqrt{3}xy + 7y^2 - 64 = 0$ and verify that it represents the graph of an ellipse as shown in Figures 10-30 and 10-31.

Classroom Example: p. 967
Exercise 24

EXAMPLE 4   **Eliminating the *xy* Term in an Equation**

Given $5x^2 - 2\sqrt{3}xy + 7y^2 - 64 = 0$,

a. Determine an acute angle of rotation to eliminate the xy-term.

b. Use a rotation of axes to eliminate the xy-term in the equation.

c. Identify the type of curve represented by the equation.

d. Sketch the graph.

Solution:

a. To eliminate the xy-term in the equation $5x^2 - 2\sqrt{3}xy + 7y^2 - 64 = 0$, we need to rotate the x- and y-axes through an angle θ that satisfies

$$\cot 2\theta = \frac{A - C}{B} \Rightarrow \cot 2\theta = \frac{5 - 7}{-2\sqrt{3}} = \frac{1}{\sqrt{3}} \Rightarrow 2\theta = 60° \Rightarrow \theta = 30°$$

b. Using rotation formulas (7) and (8), we have

$$x = x'\cos\theta - y'\sin\theta = x'\cos 30° - y'\sin 30° = \frac{\sqrt{3}}{2}x' - \frac{1}{2}y'$$

$$y = x'\sin\theta + y'\cos\theta = x'\sin 30° + y'\cos 30° = \frac{1}{2}x' + \frac{\sqrt{3}}{2}y'$$

Substitute $x = \dfrac{\sqrt{3}}{2}x' - \dfrac{1}{2}y'$ and $y = \dfrac{1}{2}x' + \dfrac{\sqrt{3}}{2}y'$ into the original equation.

$$\boxed{A} \quad 5\left(\frac{\sqrt{3}}{2}x' - \frac{1}{2}y'\right)^2 - 2\sqrt{3}\left(\frac{\sqrt{3}}{2}x' - \frac{1}{2}y'\right)\left(\frac{1}{2}x' + \frac{\sqrt{3}}{2}y'\right)$$

$$+ 7\left(\frac{1}{2}x' + \frac{\sqrt{3}}{2}y'\right)^2 - 64 = 0$$

To facilitate the "bookkeeping" consider simplifying the first three terms on the left side of equation $\boxed{A}$ separately.

$$5\left(\frac{\sqrt{3}}{2}x' - \frac{1}{2}y'\right)^2 = 5\left(\frac{3}{4}x'^2 - \frac{\sqrt{3}}{2}x'y' + \frac{1}{4}y'^2\right)$$

$$= \frac{15}{4}x'^2 - \frac{5\sqrt{3}}{2}x'y' + \frac{5}{4}y'^2$$

$$-2\sqrt{3}\left(\frac{\sqrt{3}}{2}x' - \frac{1}{2}y'\right)\left(\frac{1}{2}x' + \frac{\sqrt{3}}{2}y'\right) = -2\sqrt{3}\left(\frac{\sqrt{3}}{4}x'^2 + \frac{3}{4}x'y' - \frac{1}{4}x'y' - \frac{\sqrt{3}}{4}y'^2\right)$$

$$= -2\sqrt{3}\left(\frac{\sqrt{3}}{4}x'^2 + \frac{1}{2}x'y' - \frac{\sqrt{3}}{4}y'^2\right)$$

$$= -\frac{3}{2}x'^2 - \sqrt{3}x'y' + \frac{3}{2}y'^2$$

$$7\left(\frac{1}{2}x' + \frac{\sqrt{3}}{2}y'\right)^2 = 7\left(\frac{1}{4}x'^2 + \frac{\sqrt{3}}{2}x'y' + \frac{3}{4}y'^2\right)$$

$$= \frac{7}{4}x'^2 + \frac{7\sqrt{3}}{2}x'y' + \frac{21}{4}y'^2$$

Equation $\boxed{A}$ becomes

$$\frac{15}{4}x'^2 - \frac{5\sqrt{3}}{2}x'y' + \frac{5}{4}y'^2 - \frac{3}{2}x'^2 - \sqrt{3}x'y' + \frac{3}{2}y'^2 + \frac{7}{4}x'^2 + \frac{7\sqrt{3}}{2}x'y' + \frac{21}{4}y'^2 - 64 = 0$$

$$4x'^2 + 8y'^2 - 64 = 0$$

$$x'^2 + 2y'^2 - 16 = 0$$

c. The equation $x'^2 + 2y'^2 - 16 = 0$ is of the form $A'x'^2 + C'y'^2 + D'x' + E'y' + F' = 0$ with $A' = 1$ and $C' = 2$. Since A' and C' have the same sign ($A'C' > 0$), the equation is an ellipse.

d. To graph the equation, divide both sides of the equation by 16 and write the equation in standard form.

$$\frac{x'^2}{16} + \frac{y'^2}{8} = 1$$

Relative to the $x'y'$-plane, the equation is an ellipse with center $(0, 0)$.

$a = \sqrt{16} = 4$ and $b = \sqrt{8} = 2\sqrt{2}$

The vertices are $(4, 0)$ and $(-4, 0)$ in the $x'y'$-plane.

The endpoints of the minor axis are $\left(0, 2\sqrt{2}\right)$ and $\left(0, -2\sqrt{2}\right)$ in the $x'y'$-plane.

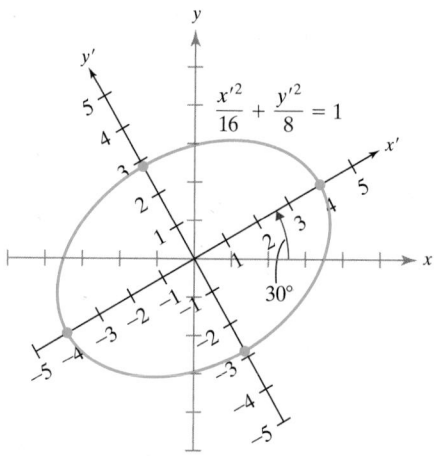

Skill Practice 4   Given $5x^2 + 2\sqrt{3}xy + 7y^2 - 32 = 0$,

a. Determine an acute angle of rotation to eliminate the xy-term.
b. Use a rotation of axes to eliminate the xy-term in the equation.
c. Identify the type of curve represented by the equation.
d. Sketch the graph.

---

4. Use the Discriminant to Identify Conic Sections Containing a *Bxy* Term

In Example 4, we have an equation of a conic section of the form

$$Ax^2 + Bxy + Cy^2 + Dx + Ey + F = 0 \qquad (2)$$

To eliminate the rotation term Bxy, we rotated the x- and y-axes through a suitable angle θ to form the x'- and y'-axes and an equation of the form $A'x'^2 + C'y'^2 + D'x' + E'y' + F' = 0$. Since the new equation has no rotation term, we then used the values of A' and C' to identify the type of conic section represented by the equation.

This method to identify a conic section involves tedious algebra. Fortunately, we now present a means to identify the type of conic section represented by equation (2) by using the coefficients A, B, and C.

Suppose that the x- and y-axes are rotated about the origin through an angle θ to form the x'- and y'-axes. As a result, the equation $Ax^2 + Bxy + Cy^2 + Dx + Ey + F = 0$ is changed into the equation $A'x'^2 + B'x'y' + C'y'^2 + D'x' + E'y' + F' = 0$. From the equations on page 960 relating A, B, and C to A', B', and C', we can show that

$$B^2 - 4AC = B'^2 - 4A'C'.$$

When θ is chosen such that $B' = 0$, we have

$$B^2 - 4AC = -4A'C'.$$

Excluding degenerate cases, the type of conic section represented by the equation $A'x'^2 + C'y'^2 + D'x' + E'y' + F' = 0$ can be determined by the product $A'C'$.

The equation $A'x'^2 + C'y'^2 + D'x' + E'y' + F' = 0$ represents
    An ellipse (or a circle) if $A'C' > 0$.
    A hyperbola if $A'C' < 0$.
    A parabola if $A'C' = 0$.

Since $B^2 - 4AC = -4A'C'$, nonzero values of $B^2 - 4AC$ have the *opposite* sign of $A'C'$. Therefore, we can make the following statements about the equation $Ax^2 + Bxy + Cy^2 + Dx + Ey + F = 0$.

Answers
4. a. $60°$ **b.** $2x'^2 + y'^2 = 8$
c. Ellipse   **d.** $\dfrac{x'^2}{4} + \dfrac{y'^2}{8} = 1$

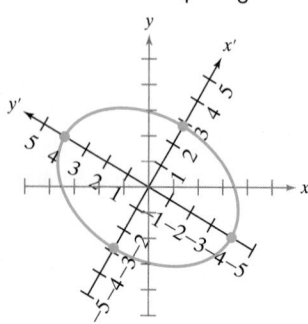

TIP The discriminant
$B^2 - 4AC = B'^2 - 4A'C'$ is
unchanged under rotation.
That is, the value of
$B^2 - 4AC$ and $B'^2 - 4A'C'$
are the same regardless of
the angle of rotation. For this
reason, we say that the
discriminant is **invariant**
under rotation.

Identify Conic by the Discriminant

Excluding degenerate cases, the equation $Ax^2 + Bxy + Cy^2 + Dx + Ey + F = 0$ represents

An ellipse (or a circle) if $B^2 - 4AC < 0$.
A hyperbola if $B^2 - 4AC > 0$.
A parabola if $B^2 - 4AC = 0$.

The quantity $B^2 - 4AC$ is called the **discriminant** of the equation.

In Examples 5 and 6, we use the discriminant to identify the type of conic section represented by a general second-degree polynomial equation. Then we graph the equation.

Classroom Example: p. 967
Exercise 40

EXAMPLE 5   Graphing a Rotated Conic Section

The equation $x^2 + 6xy + 9y^2 - 7\sqrt{10}x - 11\sqrt{10}y + 30 = 0$ is a conic section (nondegenerate case).

a. Use the discriminant to determine the type of conic section represented.
b. Use a suitable rotation to eliminate the xy term.
c. Sketch the graph.

Solution:

a. $x^2 + 6xy + 9y^2 - 7\sqrt{10}x - 11\sqrt{10}y + 30 = 0$
Since $B^2 - 4AC = (6)^2 - 4(1)(9) = 0$, the
equation represents a parabola. $A = 1, B = 6,$ and $C = 9$

b. We need to rotate the x- and y-axes through an
angle θ that satisfies

$$\cot 2\theta = \frac{A - C}{B} \Rightarrow \cot 2\theta = \frac{1 - 9}{6} = -\frac{4}{3}.$$

In order to apply rotation formulas (7) and (8),
we need to know the value of $\sin\theta$ and $\cos\theta$.
Since θ is an acute angle and $\cot 2\theta < 0$, angle
2θ is a second quadrant angle (Figure 10-36). Thus,

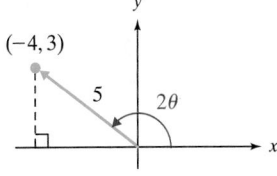

Figure 10-36

$$\cos 2\theta = -\frac{4}{5}, \text{ and using the half-angle formulas, we have}$$

$$\cos\theta = \sqrt{\frac{1 + \cos 2\theta}{2}} = \sqrt{\frac{1 + \left(-\frac{4}{5}\right)}{2}} = \sqrt{\frac{1}{10}} = \frac{\sqrt{10}}{10}$$

$$\sin\theta = \sqrt{\frac{1 - \cos 2\theta}{2}} = \sqrt{\frac{1 - \left(-\frac{4}{5}\right)}{2}} = \sqrt{\frac{9}{10}} = \frac{3\sqrt{10}}{10}$$

From the equation $\cos\theta = \dfrac{\sqrt{10}}{10}$, we have $\theta \approx 71.6°$.

We now apply rotation formulas (7) and (8).

$$x = x'\cos\theta - y'\sin\theta = \frac{\sqrt{10}}{10}x' - \frac{3\sqrt{10}}{10}y' = \frac{\sqrt{10}}{10}(x' - 3y')$$

$$y = x'\sin\theta + y'\cos\theta = \frac{3\sqrt{10}}{10}x' + \frac{\sqrt{10}}{10}y' = \frac{\sqrt{10}}{10}(3x' + y')$$

Before substituting formulas (7) and (8) into the original equation, we will write expressions for x^2, y^2, and xy in terms of x' and y'.

$$x^2 = \left[\frac{\sqrt{10}}{10}(x' - 3y')\right]^2 = \frac{1}{10}(x'^2 - 6x'y' + 9y'^2)$$

$$y^2 = \left[\frac{\sqrt{10}}{10}(3x' + y')\right]^2 = \frac{1}{10}(9x'^2 + 6x'y' + y'^2)$$

$$xy = \left[\frac{\sqrt{10}}{10}(x' - 3y')\right]\left[\frac{\sqrt{10}}{10}(3x' + y')\right] = \frac{1}{10}(3x'^2 - 8x'y' - 3y'^2)$$

The equation $x^2 + 6xy + 9y^2 - 7\sqrt{10}x - 11\sqrt{10}y + 30 = 0$ becomes

$$\frac{1}{10}(x'^2 - 6x'y' + 9y'^2) + 6 \cdot \frac{1}{10}(3x'^2 - 8x'y' - 3y'^2)$$

$$+ 9 \cdot \frac{1}{10}(9x'^2 + 6x'y' + y'^2)$$

$$-7\sqrt{10} \cdot \frac{\sqrt{10}}{10}(x' - 3y') - 11\sqrt{10} \cdot \frac{\sqrt{10}}{10}(3x' + y') + 30 = 0$$

Multiplying both sides by 10 gives

$$(x'^2 - 6x'y' + 9y'^2) + 6(3x'^2 - 8x'y' - 3y'^2) + 9(9x'^2 + 6x'y' + y'^2)$$
$$-70(x' - 3y') - 110(3x' + y') + 300 = 0$$

Applying the distributive property and combining like terms gives

$$100x'^2 - 400x' + 100y' + 300 = 0$$

Dividing both sides by 100 results in $x'^2 - 4x' + y' + 3 = 0$.

c. We can find the vertex of the parabola by completing the square.

$$x'^2 - 4x' + 4 = -y' - 3 + 4$$
$$(x' - 2)^2 = -(y' - 1)$$

Draw the x'- and y'-axes 71.6° counterclockwise from the x- and y-axes. In the $x'y'$-plane the vertex is (2, 1) and the axis of symmetry is parallel to the y'-axis.

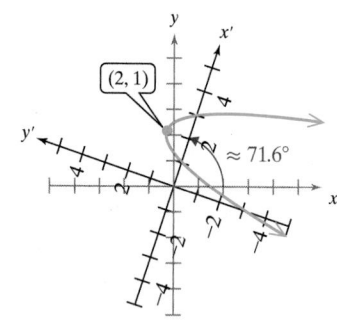

Skill Practice 5 The equation $x^2 + 4xy + 4y^2 + 6\sqrt{5}x + 7\sqrt{5}y + 20 = 0$ is a conic section (nondegenerate case).

   a. Determine the type of conic section represented by the equation.

   b. Use a suitable rotation to eliminate the xy term.

   c. Sketch the graph.

EXAMPLE 6 Graphing a Rotated Conic Section with a Graphing Utility

The equation $27x^2 - 120xy - 92y^2 - 96x + 40y + 520 = 0$ represents a conic section (nondegenerate case).

a. Identify the type of conic section.

b. Graph the equation on a graphing utility.

Solution:

a. $27x^2 - 120xy - 92y^2 - 96x + 40y + 520 = 0$

Since $B^2 - 4AC = (-120)^2 - 4(27)(-92) = 24{,}336 > 0$, the equation represents a hyperbola.

b. The equation is quadratic in the y variable and we can write it in the form $ay^2 + by + c = 0$.

$-92y^2 + (-120x + 40)y + (27x^2 - 96x + 520) = 0$

Apply the quadratic formula with $a = -92$, $b = -120x + 40$, and $c = 27x^2 - 96x + 520$.

$$y = \frac{-(-120x + 40) \pm \sqrt{(-120x + 40)^2 - 4(-92)(27x^2 - 96x + 520)}}{2(-92)}$$

TIP For the purpose of entering the equation from Example 6 into a calculator, we do not need to simplify the radical.

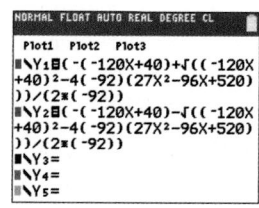 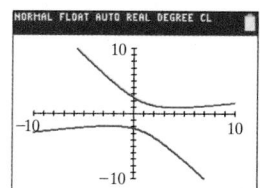

Skill Practice 6   The equation $44x^2 + 600xy - 551y^2 - 624x - 260y - 2028 = 0$ represents a conic section (nondegenerate case).

a. Identify the type of conic section.

b. Graph the equation on a graphing utility.

Answers

6. a. Hyperbola

b.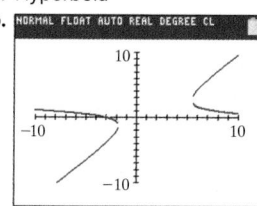

SECTION 10.4   Practice Exercises

Prerequisite Review

R.1. Given $\cos\alpha = \dfrac{12}{13}$ for $\dfrac{3\pi}{2} < \alpha < 2\pi$, find the exact values of the following.

a. $\sin\dfrac{\alpha}{2}$ $\dfrac{\sqrt{26}}{26}$

b. $\cos\dfrac{\alpha}{2}$ $-\dfrac{5\sqrt{26}}{26}$

R.2. Use a calculator to approximate the degree measure (to 1 decimal place) of θ subject to the given conditions.

$\sin\theta = \dfrac{12}{19}$ and $90° \le \theta \le 180°$ $140.8°$

R.3. Find the exact value or state that the expression is undefined.

$\cos^{-1}\left(-\dfrac{\sqrt{3}}{2}\right)$ $\dfrac{5\pi}{6}$

R.4. Solve by using the quadratic formula.

$w^2 - 3w - 5 = 0$

$\left\{\dfrac{3 \pm \sqrt{29}}{2}\right\}$

R.5. Solve for the indicated variable.

$v = 2\pi rq + \pi r^2 q$ for r. $r = \dfrac{-\pi q \pm \sqrt{\pi^2 q^2 + \pi q v}}{\pi q}$

R.6. Determine the value of n that makes the polynomial a perfect square trinomial. Then factor as the square of a binomial.

$v^2 + 30v + n$ $n = 225$; $(v + 15)^2$

Concept Connections

1. Suppose that the equation $Ax^2 + Cy^2 + Dx + Ey + F = 0$ represents a conic section (nondegenerative case). In terms of A and C, the equation represents a circle if _____ $A = C$ _____, an ellipse if _____ $AC > 0$ _____, a hyperbola if _____ $AC < 0$ _____, and a parabola if _____ $AC = 0$ _____.

2. Suppose that the equation $Ax^2 + Bxy + Cy^2 + Dx + Ey + F = 0$ represents a conic section (nondegenerative case). In terms of the discriminant, $B^2 - 4AC$, the equation represents an ellipse if _____ $B^2 - 4AC < 0$ _____, a hyperbola if _____ $B^2 - 4AC > 0$ _____, and a parabola if _____ $B^2 - 4AC = 0$ _____.

3. Suppose that the equation $Ax^2 + Bxy + Cy^2 + Dx + Ey + F = 0$ represents a conic section (nondegenerative case). Which term determines whether the conic section is rotated? Bxy

4. The conic section defined by $Ax^2 + Bxy + Cy^2 + Dx + Ey + F = 0$ will have no rotation relative to the new coordinate axes x' and y' if the x- and y-axes are rotated through an angle θ satisfying the relationship $\cot 2\theta =$ _____ $\dfrac{A - C}{B}$ (answer in terms of A, B, and C).

Objective 1: Identify Conic Sections with No Rotation

For Exercises 5–8, the given equation represents a conic section (nondegenerative case). Identify the type of conic section. (See Example 1)

5. a. $8x^2 - 2y^2 - 3x + 2y - 6 = 0$ Hyperbola

b. $-6y^2 + 4x - 12y - 24 = 0$ Parabola

c. $-9x^2 - 4y^2 - 18x + 12y = 0$ Ellipse

7. a. $0.1x^2 + 0.6x - 1.6 = 0.2y - 0.1y^2$ Circle

b. $2x^2 - 7xy = -y^2 + 4x - 2y - 1$ Ellipse

c. $8x + 2y = y^2 + 4$ Parabola

6. a. $-5x^2 + 8y + 4 = 0$ Parabola

b. $3x^2 + 3y^2 - 4x + 2y - 8 = 0$ Circle

c. $-2x^2 + y^2 - 4y + 1 = 0$ Hyperbola

8. a. $6x^2 + 3x + 10y = 10y^2 + 8$ Hyperbola

b. $3x^2 + 18xy = 5x + 2y + 9$ Hyperbola

c. $4x^2 + 8x - 5 = -y^2 + 6y + 3$ Ellipse

Objective 2: Use the Rotation of Axes Formulas

For Exercises 9–14, assume that the x- and y-axes are rotated through angle θ about the origin to form the x'- and y'-axes. (See Example 2)

9. Given $\theta = 60°$ and $P(2, -10)$ in the xy-plane, find the coordinates of P in the $x'y'$-plane. $(1 - 5\sqrt{3}, -5 - \sqrt{3})$

10. Given $\theta = 30°$ and $P(-4, 8)$ in the xy-plane, find the coordinates of P in the $x'y'$-plane. $(4 - 2\sqrt{3}, 2 + 4\sqrt{3})$

11. Given $\theta = 45°$ and $P(-3, -6)$ in the $x'y'$-plane, find the coordinates of P in the xy-plane. $\left(\dfrac{3\sqrt{2}}{2}, -\dfrac{9\sqrt{2}}{2}\right)$

12. Given $\theta = 45°$ and $P(-\sqrt{2}, 5\sqrt{2})$ in the $x'y'$-plane, find the coordinates of P in the xy-plane. $(-6, 4)$

13. Given $\cos\theta = \dfrac{4}{5}$ and $\sin\theta = \dfrac{3}{5}$ and $P(-5, -15)$ in the xy-plane, find the coordinates of P in the $x'y'$-plane. $(-13, -9)$

14. Given $\cos\theta = \dfrac{5}{13}$ and $\sin\theta = \dfrac{12}{13}$ and $P(-39, 26)$ in the xy-plane, find the coordinates of P in the $x'y'$-plane. $(9, 46)$

Objective 3: Transform Equations by Rotation of Axes

For Exercises 15–18, assume that the x- and y-axes are rotated through angle θ about the origin to form the x'- and y'-axes. Write the equation in $x'y'$-coordinates. (See Example 3)

15. $xy + 2 = 0$, $\theta = 45°$ $x'^2 - y'^2 + 4 = 0$

16. $xy - 4 = 0$, $\theta = 45°$ $x'^2 - y'^2 - 8 = 0$

17. $7x^2 - 2\sqrt{3}xy + 5y^2 - 48 = 0$, $\theta = 60°$ $x'^2 + 2y'^2 - 12 = 0$

18. $13x^2 + 6\sqrt{3}xy + 7y^2 - 32 = 0$, $\theta = 30°$ $4x'^2 + y'^2 - 8 = 0$

For Exercises 19–22, assume that $0 < 2\theta < \pi$. Find the exact values of $\cos\theta$ and $\sin\theta$. Then approximate the value of θ to the nearest tenth of a degree if necessary.

19. $\cot 2\theta = -\dfrac{\sqrt{3}}{3}$

20. $\cot 2\theta = \dfrac{\sqrt{3}}{3}$

21. $\cot 2\theta = \dfrac{24}{7}$

22. $\cot 2\theta = -\dfrac{8}{15}$

For Exercises 23–28,

a. Determine an acute angle of rotation to eliminate the xy-term. (See Example 4)

b. Use a rotation of axes to eliminate the xy-term to write the equation in the form $A'x'^2 + C'y'^2 + D'x' + E'y' + F' = 0$.

c. Identify the type of curve represented by the equation.

d. Write the equation in standard form and sketch the graph.

23. $3x^2 - 2xy + 3y^2 - 16 = 0$

24. $2x^2 + 2xy + 2y^2 - 9 = 0$

25. $11x^2 - 10\sqrt{3}xy + y^2 - 16 = 0$

26. $3x^2 - 10\sqrt{3}xy + 13y^2 + 18 = 0$

27. $x^2 - 2\sqrt{3}xy + 3y^2 - 16\sqrt{3}x - 16y = 0$

28. $x^2 + 2\sqrt{3}xy + 3y^2 - 8\sqrt{3}x + 8y = 0$

Objective 4: Use the Discriminant to Identify Conic Sections Containing a *Bxy* Term

29. The equation $3x^2 + 6xy + 3y^2 - 9x + 12y - 4 = 0$ has coefficients $A = 3$ and $C = 3$. Although $A = C$, the graph of the equation is not a circle, but rather a parabola. Why?

30. The equation $9x^2 + 10xy + y^2 - 3x + 2y - 4 = 0$ has coefficients $A = 9$ and $C = 1$. Although $AC > 0$ (A and C have the same sign), the graph of the equation is not an ellipse, but rather a hyperbola. Why?

For Exercises 31–34, the given equation represents a conic section (nondegenerative case). Identify the type of conic section. (See Examples 5 and 6)

31. a. $4x^2 - 12xy + 6y^2 + 2x - 3y - 8 = 0$ Hyperbola

  b. $4x^2 - 12xy + 9y^2 + 2x - 3y - 8 = 0$ Parabola

32. a. $-3x^2 + xy - 4y^2 + 2x - 4y - 4 = 0$ Ellipse

  b. $-3x^2 + 10xy - 4y^2 + 2x - 4y - 4 = 0$ Hyperbola

33. a. $x^2 + 6xy + 4x = 3y^2 - 2y + 10$ Hyperbola

  b. $6y^2 - 3y = 8 - 2x - 4x^2$ Ellipse

34. a. $x^2 - 3xy + 12x = -6y^2 - 12y + 1$ Ellipse

  b. $-3x^2 + 12x = 4y + 4$ Parabola

For Exercises 35–42, the equation represents a conic section (nondegenerative case).

  a. Use the discriminant to determine the type of conic section represented. (See Example 5)

  b. Use a rotation of axes to eliminate the *xy*-term to write the equation in the form
  $A'x'^2 + C'y'^2 + D'x' + E'y' + F' = 0$.

  c. Write the equation in standard form and sketch the graph.

35. $3x^2 + 26\sqrt{3}xy - 23y^2 - 144 = 0$

36. $13x^2 + 6\sqrt{3}xy + 7y^2 - 64 = 0$

37. $16x^2 - 24xy + 9y^2 + 140x - 230y + 625 = 0$

38. $4x^2 - 6xy - 4y^2 + \sqrt{10}x + 3\sqrt{10}y - 25 = 0$

39. $40x^2 + 20xy + 25y^2 + 24\sqrt{5}x - 48\sqrt{5}y = 0$

40. $x^2 + 6xy + 9y^2 - 9\sqrt{10}x - 17\sqrt{10}y + 90 = 0$

41. $5x^2 + 6xy + 5y^2 - 12\sqrt{2}x - 4\sqrt{2}y - 16 = 0$

42. $7x^2 + 50xy + 7y^2 + 78\sqrt{2}x + 114\sqrt{2}y - 18 = 0$

For Exercises 43–48, the equation represents a conic section (nondegenerative case).

  a. Identify the type of conic section. (See Example 6)

  b. Graph the equation on a graphing utility.

43. $4x^2 - 4xy + 5y^2 - 20 = 0$ a. Ellipse

44. $6x^2 + 4\sqrt{3}xy + 2y^2 - 18x + 18\sqrt{3}y - 72 = 0$
  a. Parabola

45. $2x^2 - 6xy + 3y^2 - 4x + 12y - 9 = 0$ a. Hyperbola

46. $5x^2 - 3xy + 2y^2 - 6 = 0$ a. Ellipse

47. $4x^2 + 8xy + 4y^2 - 2x - 5y - 2 = 0$ a. Parabola

48. $4x^2 + 8\sqrt{3}xy + 3y^2 + 2x - 12y - 6 = 0$ a. Hyperbola

Mixed Exercises

49. Given $y = \dfrac{\sqrt{3}}{3}x$

  a. What angle does the line $y = \dfrac{\sqrt{3}}{3}x$ make with the positive *x*-axis?   30°

  b. Suppose that the *x*- and *y*-axes are rotated through an angle of 30° to form the x'- and y'-axes. Write the equation $y = \dfrac{\sqrt{3}}{3}x$ in terms of x' and y'.
  $y' = 0$ (The new equation is the x'-axis.)

  c. Suppose that the *x*- and *y*-axes are rotated through an angle of 60° to form the x'- and y'-axes. Write the equation $y = \dfrac{\sqrt{3}}{3}x$ in terms of x' and y'.
  $y' = -\dfrac{\sqrt{3}}{3}x'$

50. Given $y = \dfrac{\sqrt{2}}{2}x$

  a. What angle does the line $y = \dfrac{\sqrt{2}}{2}x$ make with the positive *x*-axis?   45°

  b. Suppose that the *x*- and *y*-axes are rotated through an angle of 45° to form the x'- and y'-axes. Write the equation $y = \dfrac{\sqrt{2}}{2}x$ in terms of x' and y'.
  $y' = 0$ (The new equation is the x'-axis.)

  c. Suppose that the *x*- and *y*-axes are rotated through an angle of 90° to form the x'- and y'-axes. Write the equation $y = \dfrac{\sqrt{2}}{2}x$ in terms of x' and y'. $y' = -\dfrac{\sqrt{2}}{2}x'$

51. Solve the system for x' and y'. (*Hint*: Multiply the first equation by $\cos\theta$ and the second equation by $\sin\theta$ to eliminate the y' term.)

$x = x'\cos\theta - y'\sin\theta$

$y = x'\sin\theta + y'\cos\theta$ $\{(x\cos\theta + y\sin\theta, -x\sin\theta + y\cos\theta)\}$

53. Discuss the graph of $x^2 - 2xy + y^2 - 1 = 0$.

52. Consider the parabola with axis of symmetry $y = x$ and passing through the points $(1, 0)$, $(0, 1)$, and $(1, 1)$ in the xy-plane. Find an equation of the parabola

a. In the $x'y'$-plane, where the x'-axis is aligned with the line $y = x$.

b. In the xy-plane. $x^2 - 2xy + y^2 + x + y - 2 = 0$

54. Discuss the graph of $x^2 - 8xy + y^2 = 0$.

Write About It

55. Given $Ax^2 + Bxy + Cy^2 + Dx + Ey + F = 0$, what affect does the term Bxy have on the graph of the equation?

56. Given $Ax^2 + Bxy + Cy^2 + Dx + Ey + F = 0$, explain how to find the sine and cosine of the angle θ by which the conic section is rotated.

57. Suppose that the x'- and y'-axes are formed by a $30°$-counterclockwise rotation from the x- and y-axes. If a point $P(2, 1)$ is located in the xy-plane, in which quadrant in the $x'y'$-plane will P be located? Why?

58. Given an equation in x and y, many graphing utilities require that the y variable be isolated. Explain how to isolate the y variable in the equation $2x^2 + 3xy + 5y^2 - 4x + 2y - 6 = 0$.

Expanding Your Skills

59. Use the relationships between A, B, and C and A', B', and C' on page 960 to show that $A + C$ is invariant under rotation. That is, show that $A + C = A' + C'$.

61. For a positive real number r, explain why the equation $x^2 + y^2 = r^2$ is or is not invariant under rotation.

63. Recall from Section 9.3 that a point $P(x, y)$ can be represented by the matrix $\begin{bmatrix} x \\ y \end{bmatrix}$. By applying matrix addition, subtraction, or multiplication we can translate the point or rotate the point to a new location. Given $P(x, y)$, the product $\begin{bmatrix} \cos\theta & -\sin\theta \\ \sin\theta & \cos\theta \end{bmatrix}\begin{bmatrix} x \\ y \end{bmatrix}$ gives the coordinates of the point rotated by an angle θ about the origin.

a. Write a product of matrices that rotates the point $(6, 2)$ counterclockwise $60°$ about the origin.

b. Compute the product in part (a).

c. Graph the point $(6, 2)$ and the point found in part (b) to illustrate the rotation. Round the coordinates of the rotated point to 1 decimal place.

60. Show that the distance between two points $P_1(x_1, y_1)$ and $P_2(x_2, y_2)$ is invariant under rotation. That is, show that the distance between $P_1(x_1, y_1)$ and $P_2(x_2, y_2)$ equals the distance between $P_1(x_1', y_1')$ and $P_2(x_2', y_2')$.

62. Show that $x^2 + y^2 = r^2$ is invariant under rotation.

64. The set of points (x_1, y_1), (x_2, y_2), ... (x_n, y_n) can be represented as the matrix $\begin{bmatrix} x_1 & x_2 & \cdots & y_n \\ y_1 & y_2 & \cdots & y_n \end{bmatrix}$.

a. Write a matrix to represent the points defining the given figure.

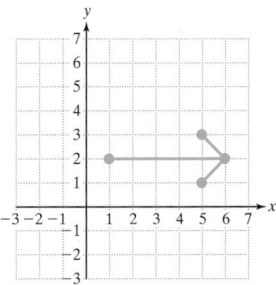

b. Suppose a computer programmer wants to rotate the matrix $90°$ counterclockwise. Write a product of matrices that will perform this rotation and compute the result.

c. Plot the original points and arrow along with the rotated points and arrow.

SECTION 10.5   Polar Equations of Conics

1. Identify and Graph Conics in Polar Form

In Section 10.3, we defined a parabola as the set of points in a plane equidistant from a fixed point (focus) and a fixed line (directrix). However, the ellipse and hyperbola were defined in terms of two points (their foci). We now give a unified approach in which all three categories of conics are defined in terms of a fixed point and fixed line.

Alternative Definitions of Conics

Let l be a fixed line (**directrix**) in a plane and let F be a fixed point (**focus**) in the plane, not on the line. A **conic** is the set of points P in the plane such that the ratio of the distance between P and F to the distance from P to l is a constant positive real number e (the **eccentricity**). That is,

$$\underbrace{d(P, F)}_{\text{Distance between } P \text{ and } F} \over \underbrace{d(P, l)}_{\text{Distance between } P \text{ and } l}} = e \qquad (1)$$

Furthermore,

*1. If $e = 1$, the conic is a parabola.
  2. If $e < 1$, the conic is an ellipse.
  3. If $e > 1$, the conic is a hyperbola.

*We show that these statements are true in online appendix, Section A.2.

A polar coordinate system with the pole placed at the focus F is well suited to illustrate these alternative definitions of conics. For example, in Figures 10-37 through 10-39 we show a parabola, ellipse, and hyperbola with a vertical directrix. The point Q is the point on line l closest to P (that is, line segment $\overline{QP}$ is the perpendicular distance from P to l).

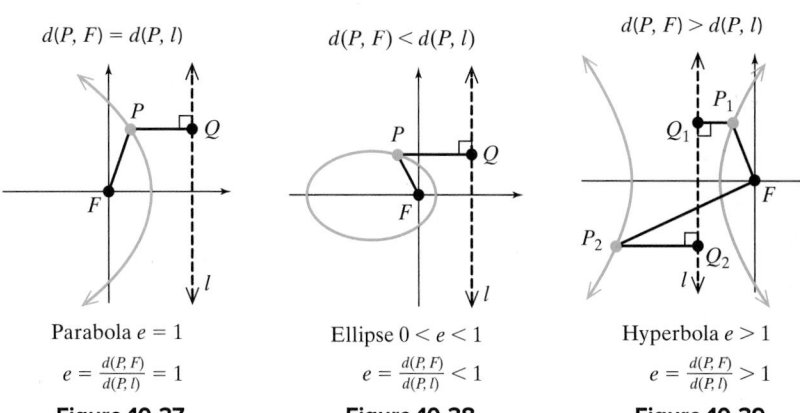

$d(P, F) = d(P, l)$	$d(P, F) < d(P, l)$	$d(P, F) > d(P, l)$
Parabola $e = 1$	Ellipse $0 < e < 1$	Hyperbola $e > 1$
$e = \dfrac{d(P, F)}{d(P, l)} = 1$	$e = \dfrac{d(P, F)}{d(P, l)} < 1$	$e = \dfrac{d(P, F)}{d(P, l)} > 1$
Figure 10-37	**Figure 10-38**	**Figure 10-39**

To develop polar equations of conics for the case of a vertical directrix, begin by placing the origin (pole) at the focus F. Then place the directrix l parallel to the y-axis $\left(\text{line } \theta = \dfrac{\pi}{2}\right)$ at a distance of $d > 0$ units to the right of the y-axis (Figure 10-40). Let Q represent the point on l closest to P.

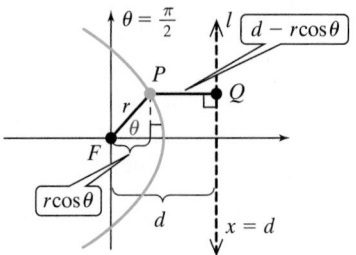

Figure 10-40

The directrix is perpendicular to the polar axis and is represented in rectangular coordinates as $x = d$. Suppose that point P has rectangular coordinates (x, y) and polar coordinates (r, θ). Then from Figure 10-40, $d(P, F) = r$ and $d(P, Q) = d - r\cos\theta$.

The condition $\dfrac{d(P, F)}{d(P, Q)} = e$ can be written as

$$d(P, F) = e \cdot d(P, Q) \text{ or as } r = e(d - r\cos\theta).$$

Solving for r gives $r = \dfrac{ed}{1 + e\cos\theta}.$

- If the directrix is a vertical line $d > 0$ units to the *left* of the pole, then an equation of the directrix is $x = -d$ and the equation of the conic is
$$r = \dfrac{ed}{1 - e\cos\theta}.$$

- If the directrix is horizontal (parallel to the polar axis) $d > 0$ units above or below the pole, then $y = d$ and $y = -d$ are equations of the directrix and the equation of the conic is
$$r = \dfrac{ed}{1 + e\sin\theta} \text{ or } r = \dfrac{ed}{1 - e\sin\theta}.$$

Polar Equations of Conics

A polar equation of the form

$$(2)\ r = \dfrac{ed}{1 \pm e\cos\theta} \qquad \text{or} \qquad (3)\ r = \dfrac{ed}{1 \pm e\sin\theta}$$

is a conic with focus at the pole and eccentricity $e > 0$. The value $|d|$ is the distance between the focus at the pole and the directrix. The conic is

1. A parabola if $e = 1$ (axis of symmetry perpendicular to the directrix).
2. An ellipse if $0 < e < 1$ (major axis perpendicular to the directrix).
3. A hyperbola if $e > 1$ (transverse axis perpendicular to the directrix).

Given a polar equation of a conic, the location of the directrix and orientation of the curve for $d > 0$ are summarized in Figure 10-41. If $d < 0$, the curve is reflected across the pole.

TIP From Example 3,

$r = \dfrac{\frac{9}{5}}{1 + \frac{4}{5}\sin\theta}$ is an ellipse with horizontal directrix above the pole.

We know that $e = \frac{4}{5}$ and $ed = \frac{9}{5}$. Thus, $\frac{4}{5}d = \frac{9}{5}$ or $d = \frac{9}{4}$.

An equation of the directrix in rectangular coordinates is $y = d$, which is $y = \frac{9}{4}$.

A sketch of the ellipse is given in Figure 10-45. The distance between the vertices $\left(1, \dfrac{\pi}{2}\right)$ and $\left(-9, \dfrac{\pi}{2}\right)$ is the length of the major axis. Thus, $2a = 1 - (-9) = 10$, which means that $a = 5$. Since $e = \dfrac{c}{a}$, then

$c = ae = 5 \cdot \dfrac{4}{5} = 4$.

For an ellipse, $c^2 = a^2 - b^2$, so
$b = \sqrt{a^2 - c^2} = \sqrt{5^2 - 4^2} = 3$.

Therefore, the length of the minor axis is $2b = 2(3) = 6$.

Figure 10-45

Skill Practice 3   Graph the equation. $r = \dfrac{16}{5 - 3\cos\theta}$

The polar equation of the ellipse from Example 3 can be converted to rectangular coordinates to verify the results of Example 3 (see Exercise 33).

2. Rotate Conics in Polar Form

The rotation of a conic about the origin is much more convenient when the conic is expressed in polar form versus rectangular form. Consider the graph of $r = f(\theta)$. Replacing θ by $(\theta - \alpha)$ results in

- A counterclockwise rotation (rotation through a positive angle) if $\alpha > 0$.
- A clockwise rotation (rotation through a negative angle) if $\alpha < 0$.

In Example 4, we rotate the ellipse from Example 3.

Classroom Examples: p. 977
Exercises 28, 30

TIP Notice the similarity in the process to rotate a polar graph about the origin to the process to shift a graph horizontally or vertically in a rectangular coordinate system.

For example, consider an equation of an ellipse in rectangular form,

$\dfrac{(x - h)^2}{a^2} + \dfrac{(y - k)^2}{b^2} = 1$

If $h > 0$, the ellipse is shifted horizontally in the positive x direction.

If $h < 0$, the ellipse is shifted horizontally in the negative x direction.

EXAMPLE 4   Rotating a Conic in Polar Form

Write an equation of the ellipse $r = \dfrac{9}{5 + 4\sin\theta}$ rotated about the origin through the given angle. Then graph the result on a graphing utility.

a. Counterclockwise $\dfrac{\pi}{4}$

b. Clockwise $\dfrac{\pi}{3}$

Solution:

a. Substitute $\theta - \dfrac{\pi}{4}$ for θ.

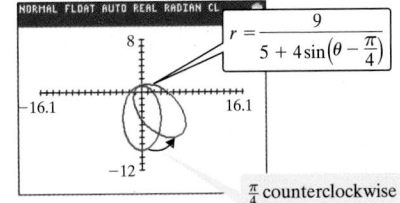

$\frac{\pi}{4}$ counterclockwise

b. Substitute $\theta + \dfrac{\pi}{3}$ for θ.

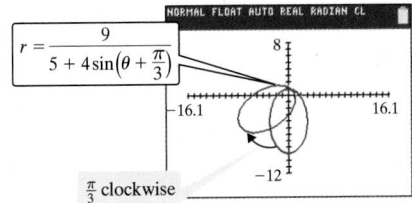

$\frac{\pi}{3}$ clockwise

Skill Practice 4   Rotate the ellipse $r = \dfrac{16}{5 - 3\cos\theta}$ and graph the result on a graphing utility.

a. Counterclockwise $\dfrac{2\pi}{3}$

b. Clockwise $\dfrac{\pi}{4}$

Answers

3. See page SA-72.
4. See page SA-72.

SECTION 10.5   Practice Exercises

Prerequisite Review

R.1. Given $5x^2 + 11y^2 + 40x - 132y + 421 = 0$,

    a. Write the equation of the ellipse in standard form.

    b. Identify the center, vertices, endpoints of the minor axis, and foci.

R.2. Given $11x^2 - 5y^2 + 22x + 20y - 64 = 0$,

    a. Write the equation of the hyperbola in standard form.

    b. Identify the center, vertices, and foci.

R.3. Given $y^2 - 2y + 4x - 15 = 0$,

    a. Write the equation of the parabola in standard form.

    b. Identify the vertex, focus, and directrix.

For Exercises R.4–R.5, convert the polar equation to rectangular form and identify the type of curve represented.

R.4. $r = -6\cos\theta$

    $(x + 3)^2 + y^2 = 9$; A circle centered at $(-3, 0)$ with radius 3.

R.5. $r = \dfrac{4}{\sin\theta}$

    $y = 4$; A horizontal line passing through the y-axis at 4.

Concept Connections

1. Consider a conic defined by the set of points P satisfying the ratio $\dfrac{d(P, F)}{d(P, l)} = e$, where F is the focus at the pole and l is the directrix. The conic is a parabola if ____$e = 1$____, an ellipse if ____$0 < e < 1$____, and a hyperbola if ____$e > 1$____.

2. For a conic defined by $r = \dfrac{ed}{1 - e\cos\theta}$, the directrix is (horizontal/vertical) with equation ____vertical; $x = -d$____.

3. For a conic defined by $r = \dfrac{ed}{1 + e\sin\theta}$, the directrix is (horizontal/vertical) with equation ____horizontal; $y = d$____.

4. Given an equation of a conic in polar form $r = f(\theta)$, substituting $(\theta - \alpha)$ results in a (clockwise/counterclockwise) rotation about the origin if $\alpha > 0$. The rotation is (clockwise/counterclockwise) if $\alpha < 0$.   counterclockwise; clockwise

Objective 1: Identify and Graph Conics in Polar Form

For Exercises 5–10,

    a. Determine the value of e and d.

    b. Identify the type of conic represented by the equation.

    c. Write an equation for the directrix in rectangular coordinates.

    d. Match the equation with its graph. Choose from A–F.

5. $r = \dfrac{10}{5 + 5\sin\theta}$

6. $r = \dfrac{10}{2 - 5\sin\theta}$

7. $r = \dfrac{10}{5 - 2\cos\theta}$

8. $r = \dfrac{5}{2 + 2\cos\theta}$

9. $r = \dfrac{5}{2 + 4\cos\theta}$

10. $r = \dfrac{5}{4 + 2\cos\theta}$

A.

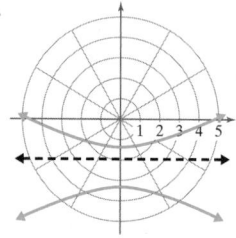

B.

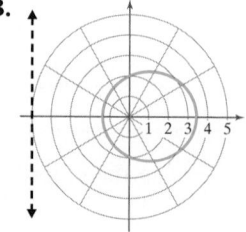

C.

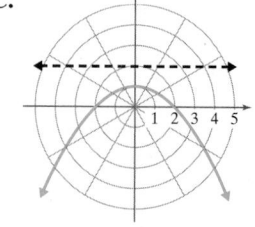

D.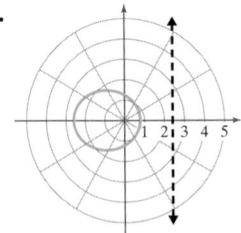

E.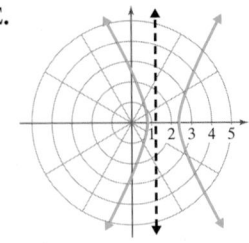

F.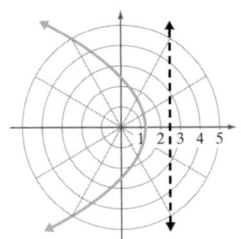

For Exercises 11–22, graph the conic. (See Examples 1–3)

11. $r = \dfrac{5}{1 + \cos \theta}$

12. $r = \dfrac{3}{1 + \sin \theta}$

13. $r = \dfrac{5}{2 + 3\sin \theta}$

14. $r = \dfrac{8}{3 - 5\cos \theta}$

15. $r = \dfrac{6}{2 + \sin \theta}$

16. $r = \dfrac{9}{3 + \cos \theta}$

17. $r = \dfrac{-6}{-2 + 2\sin \theta}$

18. $r = \dfrac{-15}{5\cos \theta - 5}$

19. $r = \dfrac{10\sec \theta}{5\sec \theta - 3}$

20. $r = \dfrac{10\csc \theta}{4\csc \theta - 3}$

21. $5r\sin \theta + 3r = 16$

22. $4r - 5r\cos \theta = 9$

For Exercises 23–24, refer to the polar equation of the hyperbola.

   a. Determine the center in rectangular coordinates.

   b. Determine equations of the asymptotes in rectangular coordinates. **(See Example 2)**

23. $r = \dfrac{5}{2 + 3\sin \theta}$ (Exercise 13)

24. $r = \dfrac{8}{3 - 5\cos \theta}$ (Exercise 14)

For Exercises 25–26, refer to the polar equation of the ellipse.

   a. Determine the center in rectangular coordinates.

   b. Determine the length of the major axis.

   c. Determine the length of the minor axis.

25. $r = \dfrac{10\sec \theta}{5\sec \theta - 3}$ (Exercise 19)

26. $r = \dfrac{10\csc \theta}{4\csc \theta - 3}$ (Exercise 20)

Objective 2: Rotate Conics in Polar Form

For Exercises 27–30, write an equation of the conic rotated about the origin through the given angle. Then graph the result on a graphing utility. (See Example 4)

27. $r = \dfrac{1}{1 - \sin \theta}$; Clockwise $\dfrac{\pi}{3}$

28. $r = \dfrac{8}{5 + 6\cos \theta}$; Clockwise $\dfrac{\pi}{4}$

29. $r = \dfrac{12}{4 - 3\cos \theta}$; Counterclockwise $\dfrac{7\pi}{6}$

30. $r = \dfrac{5}{2 + 2\sin \theta}$; Counterclockwise $\dfrac{4\pi}{3}$

31. a. Write a polar equation to represent the vertical line defined by $x = 2$ in rectangular form.

   b. Use the polar equation from part (a) to write a polar equation of the line rotated about the origin 20° clockwise.

   c. Graph the equations from parts (a) and (b).

32. a. Write a polar equation to represent the horizontal line defined by $y = 4$ in rectangular form.

   b. Use the polar equation from part (a) to write a polar equation of the line rotated about the origin 15° counterclockwise.

   c. Graph the equations from parts (a) and (b).

Mixed Exercises

For Exercises 33–36, convert the equation to rectangular coordinates. Compare the result to the indicated example or exercise.

33. $r = \dfrac{9}{5 + 4\sin \theta}$; Example 3

34. $r = \dfrac{12}{3 - 6\cos \theta}$; Example 2

35. $r = \dfrac{5}{1 + \cos \theta}$; Exercise 11

36. $r = \dfrac{6}{2 + \sin \theta}$; Exercise 15

For Exercises 37–40, write a polar equation of a conic subject to the given conditions.

37. $e = 1$, Directrix $y = 2$ $r = \dfrac{2}{1 + \sin\theta}$ **38.** $e = \dfrac{5}{4}$, Directrix $x = -2$ $r = \dfrac{\frac{5}{2}}{1 - \frac{5}{4}\cos\theta}$

39. $e = \dfrac{3}{5}$, Directrix is perpendicular to the polar axis and 10 units to the right of the pole. $r = \dfrac{6}{1 + \frac{3}{5}\cos\theta}$

40. $e = 1$, Directrix is parallel to the polar axis and 3 units below the pole. $r = \dfrac{3}{1 - \sin\theta}$

Write About It

41. The equations $r_1 = \dfrac{6}{2 - 3\cos\theta}$ and $r_2 = \dfrac{6}{3 - 2\cos\theta}$ look similar. Discuss the difference in their graphs.

42. Discuss the difference between the graphs of $r_1 = \dfrac{8}{4 - \cos\theta}$ and $r_2 = \dfrac{8}{4 - \sin\theta}$.

Expanding Your Skills

43. Recall from Section 10.1 that the planets travel in elliptical orbits with the Sun at one focus. Assume that the Sun is located at the pole and the major axis of the ellipse is aligned with the polar axis as shown in the figure. We can represent the ellipse in polar form in terms of the eccentricity e and length of the semimajor axis a.

Let c represent the distance between the center C and either focus, F_1 and F_2. Let r and r_1 represent the distances between the foci and point P on the ellipse.

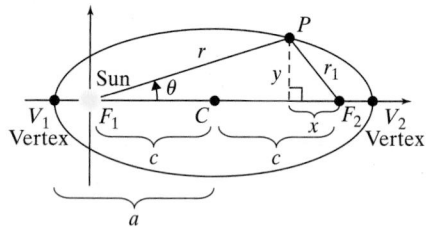

a. Show that $c = ea$.

b. Show that $r_1 = \sqrt{4e^2a^2 - 4ear\cos\theta + r^2}$.

c. From the definition of an ellipse, $r + r_1 = 2a$. Use this relationship to show that the ellipse is represented in polar form by $r = \dfrac{a(1 - e^2)}{1 - e\cos\theta}$.

44. Expressing a planet's elliptical orbit in polar form with the Sun at one focus is particularly convenient for finding the planet's distance at perihelion and aphelion (closest and farthest points between the planet and the Sun). Refer to the figure in Exercise 43. Write expressions in terms of a and e representing the distances R_p and R_a between the planet and Sun at perihelion and aphelion, respectively. $R_p = a(1 - e)$ and $R_a = a(1 + e)$

For Exercises 45–46, use the results of Exercises 43–44 to

a. Find a polar equation of the planet's orbit.

b. Find the distances R_p and R_a between the planet and Sun at perihelion and aphelion, respectively. Round to the nearest million kilometers.

45. Mercury: $a = 5.790 \times 10^7$ km, $e = 0.2056$ **46.** Mars: $a = 2.279 \times 10^8$ km, $e = 0.0935$

Technology Connections

47. Graph the equations $r_1 = \dfrac{2}{1 - 0.6\cos\theta}$ and $r_2 = \dfrac{-2}{1 - 0.6\cos\theta}$ on the same viewing window. How does the negative value of d affect the graph of the second equation?

48. Graph the equations $r = \dfrac{1}{1 - e\sin\theta}$ for $e = 0.2, 0.4, 0.6,$ and 0.8 on the same viewing window. How does the value of e affect the ellipse?

49. Graph the equations $r = \dfrac{1}{1 - e\sin\theta}$ for $e = 0.8, 1,$ and 1.2 on a viewing window with $-16.1 \le x \le 16.1$ and $-10 \le y \le 10$. Comment on the effect of e on the graph.

SECTION 10.6 **Plane Curves and Parametric Equations**

OBJECTIVES

1. Graph a Plane Curve by Plotting Points
2. Eliminate the Parameter
3. Write Parametric Equations to Represent a Curve
4. Apply Parametric Equations

1. Graph a Plane Curve by Plotting Points

Suppose that a target in a video game starts at the lower left corner of the screen and moves across the screen along the path shown (Figure 10-46).

Placing the origin at the lower left corner of the screen, the path of the target is given by $y = 2x$. However, a player who wants to shoot the target must also know *when* the object is at a given point (x, y). That is, to adequately model the motion of the target, we must know its x- and y-coordinates at a given time t. The variable t is called a *parameter*. From the information given in Figure 10-46, we can express x and y (in pixels) in terms of $t \geq 0$ (in seconds) as follows:

$$x = 60t \text{ and } y = 120t$$

These equations are called parametric equations because x and y are defined in terms of a parameter. Furthermore, the collection of points $(x(t), y(t))$ defined by these equations is called a plane curve.

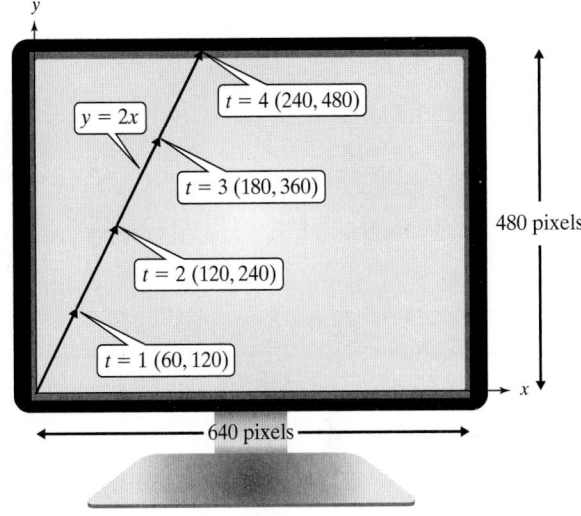

Figure 10-46

TIP In many applications of parametric equations, the parameter often represents the time t during which an object is in motion or an angle θ through which an object moves.

Plane Curves and Parametric Equations

Let f and g be functions of t over an interval I. Then the set of points defined by $(x, y) = (f(t), g(t))$ is called a **plane curve**. The equations

$$x = f(t) \text{ and } y = g(t)$$

are called **parametric equations** of the curve, and t is called the **parameter.**

One method to graph a plane curve defined by parametric equations is to create a table of points. Choose arbitrary values of the parameter t and compute the corresponding values of x and y. Plot the points (x, y) in the order of increasing t and connect the points with a curve. Then use arrows to indicate the **orientation** of the curve. This is the direction of the path for increasing values of t.

Classroom Example: p. 988
Exercise 8

EXAMPLE 1   Graphing a Plane Curve

Sketch the plane curve defined by the parametric equations over the given interval.

$$x = t^2 - 2t \text{ and } y = t - 2 \text{ over the interval } -2 \leq t < 3$$

Solution:

In Table 10-2, we have selected arbitrary values of t on the interval $[-2, 3)$ and have computed the corresponding values of x and y. After plotting the points, we draw a curve through the points and place arrows indicating the direction of increasing values of t (Figure 10-47).

Although the value $t = 3$ is not on the interval defining the curve, we evaluate x and y at $t = 3$ to determine the location of the open dot at the upper endpoint.

t	x	y
-2	8	-4
-1	3	-3
0	0	-2
1	-1	-1
2	0	0
3	3	1

Table 10-2

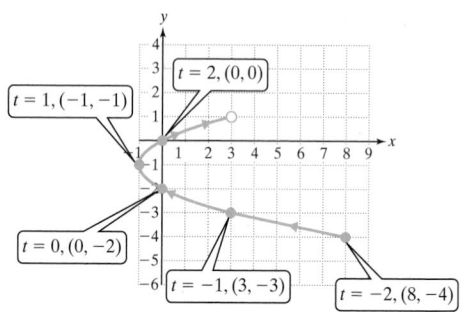

Figure 10-47

> **Skill Practice 1** Sketch the plane curve defined by the parametric
>
> equations $x = |t| - 2$ and $y = \dfrac{t}{2}$ over the interval $-4 < t \le 4$.

In Example 1, the interval for t was restricted to $[-2, 3)$. If no interval is given, we will assume that t can be any real number for which x and y are defined. In Example 1, if the restriction of t were lifted, the parametric equations would define an entire parabola rather than just the part shown.

2. Eliminate the Parameter

Sometimes we can convert parametric equations to an equation in rectangular coordinates by a process called eliminating the parameter. The purpose of eliminating the parameter is to make use of our familiarity with the graphs of many equations and functions expressed in rectangular coordinates.

One way to eliminate a parameter t is to solve for t or an expression involving t in one equation. Then substitute the result in the other equation. However, we must also note restrictions on the domain of the rectangular equation so that the resulting graph is consistent with the restrictions on x and y in the parametric equations. This is illustrated in Example 2.

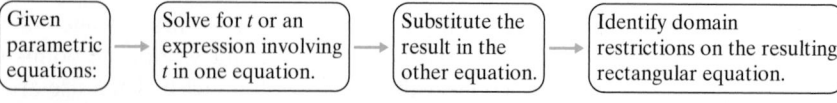

Classroom Example: p. 989
Exercise 12

Answer

1.

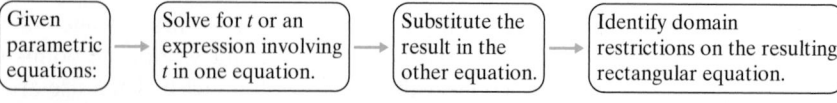

EXAMPLE 2   Eliminating a Parameter

Given $x = \sqrt{t}$ and $y = -t + 4$,

a. Eliminate the parameter and write an equation in rectangular coordinates.

b. Sketch the curve and indicate its orientation.

Solution:

First note that $x = \sqrt{t}$ has the implied restriction that $t \geq 0$. Furthermore, x is equated to the principle square root of t, which implies that $x \geq 0$. Given the equation $y = -t + 4$, for $t \geq 0$, we have the restriction to $y \leq 4$.

a. From the equation $x = \sqrt{t}$, we have $t = x^2$ for $x \geq 0$. Substituting $t = x^2$ into the equation $y = -t + 4$ gives

$$y = -x^2 + 4 \text{ for } x \geq 0$$

b. The equation $y = -x^2 + 4$ is a parabola opening downward with vertex at $(0, 4)$. Since $x \geq 0$, we take only the right branch of the parabola.

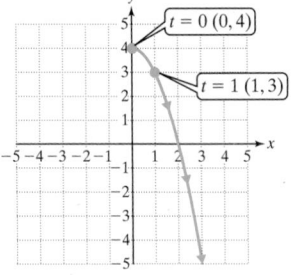

To determine the orientation of the curve, substitute two values of t into the parametric equations. Plot the corresponding points and draw arrows indicating the direction of increasing values of t.

When $t = 0$, $x = \sqrt{0} = 0$, $y = -(0) + 4 = 4$. The point is $(0, 4)$.
When $t = 1$, $x = \sqrt{1} = 1$, $y = -(1) + 4 = 3$. The point is $(1, 3)$.

The curve is oriented downward for increasing values of t.

Skill Practice 2   Repeat Example 2 with the equations $x = \sqrt{t - 1}$ and $y = t - 1$.

Another method to eliminate a parameter is to combine the equations by using an identity such as $\sin^2 \theta + \cos^2 \theta = 1$.

Classroom Example: p. 989
Exercise 24

EXAMPLE 3   Eliminating a Parameter

Given $x = -2 \sin \theta$ and $y = 2 \cos \theta$,

a. Eliminate the parameter and write an equation in rectangular coordinates.

b. Sketch the curve and indicate its orientation.

Solution:

The sine and cosine functions are defined for all real numbers θ. Since $-1 \leq \sin \theta \leq 1$ and $-1 \leq \cos \theta \leq 1$, the values of x and y are between -2 and 2.

a. Solving for θ in either equation would require an inverse trigonometric function. Instead, we make use of the fact that $\sin^2 \theta + \cos^2 \theta = 1$. Squaring both sides of the two equations and taking the sum $x^2 + y^2$ gives

$$\begin{aligned} x^2 + y^2 &= (-2 \sin \theta)^2 + (2 \cos \theta)^2 \\ &= 4 \sin^2 \theta + 4 \cos^2 \theta \\ &= 4(\sin^2 \theta + \cos^2 \theta) \\ &= 4(1) \\ &= 4 \end{aligned}$$

The curve represented in rectangular coordinates is $x^2 + y^2 = 4$.

Answers

2. a. $y = x^2$ for $x \geq 0$

b.

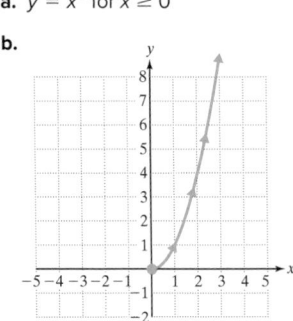

b. The graph of $x^2 + y^2 = 4$ is a circle centered at the origin with radius 2.

To determine the orientation of the curve, substitute two values of θ into the parametric equations. Plot the corresponding points and draw arrows indicating the direction of increasing values of θ.

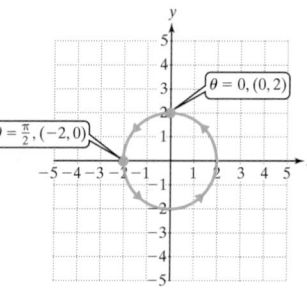

When $\theta = 0$, $x = -2\sin(0) = 0$,
$$y = 2\cos(0) = 2.$$

When $\theta = \dfrac{\pi}{2}$, $x = -2\sin\left(\dfrac{\pi}{2}\right) = -2$, $y = 2\cos\left(\dfrac{\pi}{2}\right) = 0$.

The orientation of the curve is counterclockwise.

Skill Practice 3 Repeat Example 3 with the equations $x = 4\cos\theta$ and $y = -4\sin\theta$.

In Example 3, the implied domain of the parametric equations is $(-\infty, \infty)$. Thus, the circle is traced repeatedly for θ values on intervals of 2π. For example:

- For θ on the interval $[0, 2\pi]$, the curve starts at $(0, 2)$ and is traced one time counterclockwise.

- For θ on the interval $[0, 4\pi]$, the curve starts at $(0, 2)$ and is traced two times counterclockwise.

- For θ on the interval $\left[\dfrac{\pi}{2}, \dfrac{13\pi}{2}\right]$, the curve starts at $(-2, 0)$ and is traced three times counterclockwise.

It is also important to note that the orientation of a curve defined by parametric equations may change over the interval defining the parameter. In Exercises 21–22, we show a situation where a curve is traversed back and forth in an oscillating manner.

3. Write Parametric Equations to Represent a Curve

An equation in x and y in rectangular coordinates defines a set of points (x, y). Suppose that a parameter t is introduced to represent time. Different parametric equations (that is, different parameterizations) can be used to traverse the curve at different speeds and in different directions. This is demonstrated in Example 4.

Classroom Example: p. 989
Exercise 32

Answers

3. a. $x^2 + y^2 = 16$

b.

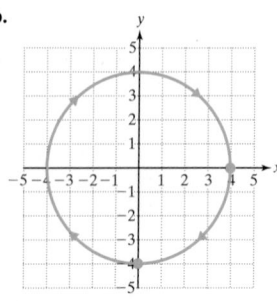

EXAMPLE 4 Writing Parametric Equations to Represent an Equation in Rectangular Coordinates

Write parametric equations for the curve defined by $y = x + 2$ with the given definition for x.

    a. $x = t$ **b.** $x = 3t$ **c.** $x = -\dfrac{t}{2}$

Solution:

  a. Substituting $x = t$ into the equation $y = x + 2$ yields $y = t + 2$.
     The parametric equations are $x = t$ and $y = t + 2$.

b. Substituting $x = 3t$ into the equation $y = x + 2$ yields $y = 3t + 2$.

The parametric equations are $x = 3t$ and $y = 3t + 2$.

c. Substituting $x = -\dfrac{t}{2}$ into the equation $y = x + 2$ yields $y = -\dfrac{t}{2} + 2$.

The parametric equations are $x = -\dfrac{t}{2}$ and $y = -\dfrac{t}{2} + 2$.

Skill Practice 4   Write parametric equations for the curve defined by $y = 2x - 3$ with the given definition for x.

a. $x = t$ **b.** $x = -4t$ **c.** $x = \dfrac{t}{5}$

The graphs of the three parametric representations of the line $y = x + 2$ are shown in Figures 10-48 through 10-50.

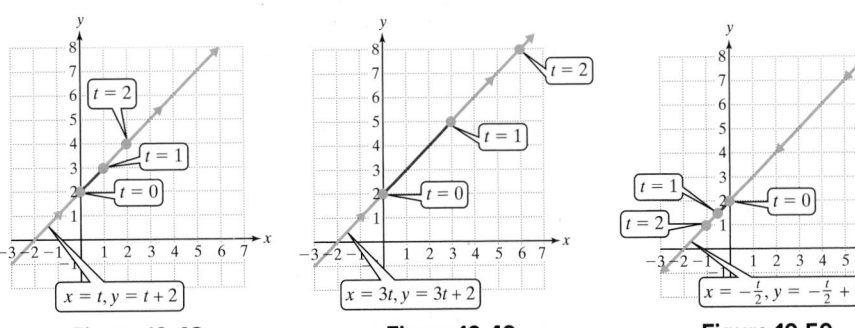

Figure 10-48   **Figure 10-49**   **Figure 10-50**

For a given interval of t, the different parameterizations traverse different parts of the curve. Notice that for the interval $0 \leq t \leq 1$, different portions of the line (shown in red) are traversed for each parameterization. Furthermore, if t represents time, then the curve in Figure 10-49 is traversed most quickly because over a 1-sec interval, a longer portion of the curve is represented. Also notice that the curve defined by the parameterization in Figure 10-50 is traversed in the opposite direction than the other two curves.

It is also important to note that a parameterization must be chosen carefully to represent the entire curve defined in rectangular coordinates. For example, the parameterization $x = \sqrt{t}$ and $y = \sqrt{t} + 2$ would not represent the entire line $y = x + 2$. This parameterization would only define the portion of the line from the point $(0, 2)$ to the right.

4. Apply Parametric Equations

Parametric equations are particularly well suited for describing objects in motion because curves can be expressed using time as the parameter. Two common applications and the associated parametric equations are summarized as follows.

Answers

4. a. $x = t$ and $y = 2t - 3$
  b. $x = -4t$ and $y = -8t - 3$
  c. $x = \dfrac{t}{5}$ and $y = \dfrac{2t}{5} - 3$

Modeling Motion Using Parametric Equations

Let $t \geq 0$ represent the time under which an object is in motion along a plane curve, and let (x_0, y_0) represent the initial position of the object at $t = 0$.

Uniform linear motion $x = x_0 + at$ $y = y_0 + bt$	The values of a and b are the horizontal and vertical components of velocity, respectively.
Projectile motion* $x = (v_0 \cos \theta)t + x_0$ $y = -\dfrac{1}{2}gt^2 + (v_0 \sin \theta)t + y_0$ **Assuming negligible air resistance.*	θ is the angle of launch relative to the horizontal, and v_0 is the initial speed (magnitude of the initial velocity). The value of g is the acceleration due to gravity, where $g = 32$ ft/sec^2 or $g = 9.8$ m/sec^2 on the surface of the Earth.

In Example 5, we investigate an application of motion along a straight path.

Classroom Example: p. 990
Exercise 50

EXAMPLE 5   An Application of Motion Along a Straight Path

An ant walks across a 72-in. by 36-in. picnic table along a straight path as shown in Figure 10-51. The origin of a rectangular coordinate system is placed at the lower left corner of the table.

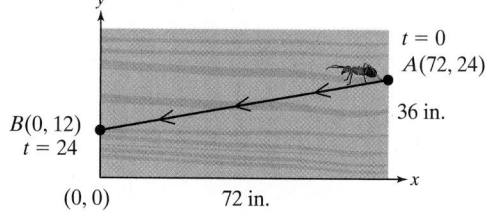

Figure 10-51

a. If the ant walks from point $A(72, 24)$ to point $B(0, 12)$ in 24 sec, write parametric equations to represent the ant's path, as a function of the time t (in sec). Assume that the distances in the figure are in inches.

b. Determine the ant's location 8 sec after leaving point A.

Solution:

a. For a straight path we use the equations $x = x_0 + at$ and $y = y_0 + bt$. The initial position (x_0, y_0) of the ant is $(72, 24)$. Thus, $x_0 = 72$ and $y_0 = 24$.

The ant moves from $x = 72$ to $x = 0$ in 24 sec. Therefore, the ant's horizontal component of velocity is

$$a = \frac{0 - 72}{24} = -3 \text{ in./sec}$$

> **TIP** The horizontal component of velocity is negative because the ant is moving from right to left.

The ant moves from $y = 24$ to $y = 12$ in 24 sec. Thus, the vertical component of velocity is

$$b = \frac{12 - 24}{24} = -\frac{1}{2} \text{ in./sec}$$

> **TIP** The vertical component of velocity is negative because the ant is moving downward.

> **TIP** For an object moving under uniform linear motion from point (x_1, y_1) to point (x_2, y_2) in time Δt, the horizontal and vertical components of velocity are given by
>
> $$a = \frac{x_2 - x_1}{\Delta t} \text{ and}$$
>
> $$b = \frac{y_2 - y_1}{\Delta t}$$
>
> In each case, notice that the pattern is
>
> $$\frac{\left(\begin{array}{c}\text{final}\\\text{position}\end{array}\right) - \left(\begin{array}{c}\text{initial}\\\text{position}\end{array}\right)}{\Delta t}$$

The ant's path is given by

$$x = x_0 + at = 72 + (-3)t \text{ and } y = y_0 + bt = 24 + \left(-\frac{1}{2}\right)t. \text{ Thus,}$$

$$x = 72 - 3t \text{ and } y = 24 - \frac{1}{2}t.$$

b. The ant's location at $t = 8$ sec is given by

$$x = 72 - 3(8) = 48 \text{ and } y = 24 - \frac{1}{2}(8) = 20.$$

The location is (48, 20).

> **Skill Practice 5**   Repeat Example 5 with an ant that begins at $A(72, 10)$ and walks to $B(20, 36)$ in 13 sec.

In Example 6, we use parametric equations to model the path of a projectile.

Classroom Example: p. 990
Exercise 52

EXAMPLE 6   An Application of Projectile Motion

The shot put is a track-and-field event in the Olympics in which competitors throw an iron ball (the shot) as far as they can. The weight of the shot is 7.26 kg (16 lb) for men and 4 kg (8.8 lb) for women. Suppose that a male competitor throws the shot from an initial height of 2 m, at an angle of 38° to the horizontal, with an initial speed of 12 m/sec (Figure 10-52).

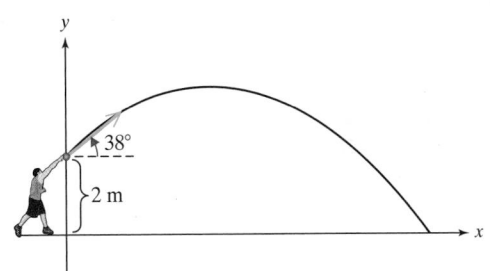

Figure 10-52

a. Write parametric equations to model the path of the shot as a function of the time t (in sec).

b. For how many seconds is the shot in the air? Round to the nearest hundredth of a second.

c. Determine the horizontal distance traveled by the shot. Round to the nearest hundredth of a foot.

d. When does the shot reach its maximum height? Find the exact value and round to the nearest hundredth of a second.

e. Determine the maximum height. Round to the nearest hundredth of a foot.

Solution:

a. To begin, we choose a coordinate system with the origin placed at ground level, directly below the point of release. Thus, $x_0 = 0$ m, $y_0 = 2$ m, $\theta = 38°$, $v_0 = 12$ m/sec, and $g = 9.8$ m/sec^2.

Using the model for projectile motion, we have

$$x = (v_0 \cos \theta)t + x_0 = (12 \cos 38°)t + 0$$

$$y = -\frac{1}{2}gt^2 + (v_0 \sin \theta)t + y_0 = -\frac{1}{2}(9.8)t^2 + (12 \sin 38°)t + 2.$$

Simplifying, we have $x = 12 \cos 38°t$ and $y = -4.9t^2 + 12 \sin 38°t + 2$, where x and y are in meters and t is in seconds.

Point of Interest

In Section 5.3, we learned that launching a projectile at a 45° angle results in the maximum horizontal distance. However, experiments performed on world-class shot putters show that the initial speed that an athlete can generate throwing the shot decreases with increasing projection angle. Based on empirical data, the optimum projection angle for the shot put is approximately 35°. (*Source: Journal of Sports Sciences*)

Answers

5. a. $x = 72 - 4t$ and $y = 10 + 2t$.
   b. At $t = 8$ sec, the ant's location is (40, 26).

b. The shot will hit the ground when $y = 0$.

$$-4.9t^2 + 12\sin 38°t + 2 = 0$$

$$t = \frac{-12\sin 38° \pm \sqrt{(12\sin 38°)^2 - 4(-4.9)(2)}}{2(-4.9)}$$

$t \approx 1.74$ sec and $t \approx -0.23$ sec. — Reject the negative solution.

The shot will hit the ground in approximately 1.74 sec.

c. The horizontal distance traveled by the shot is the x value at the time the shot hits the ground. Compute x at $t \approx 1.74$.

$$x \approx 12\cos 38°(1.74) \approx 16.45 \text{ m}$$

The shot travels approximately 16.45 m. This is the length of the throw and score for the competitor.

d. The path of the shot is parabolic. The maximum height is reached at the vertex. To find the time t at which the shot reaches its maximum height, use the vertex formula with $y = -4.9t^2 + 12\sin 38°t + 2$.

$$t = \frac{-b}{2a} = \frac{-12\sin 38°}{2(-4.9)} \approx 0.75$$

The shot will reach its maximum height approximately 0.75 sec after release.

e. The maximum height is the y value at the vertex. Compute y at $t \approx 0.75$.

$$y = -4.9(0.75)^2 + 12\sin 38°(0.75) + 2 \approx 4.78$$

The maximum height is approximately 4.78 m.

Skill Practice 6 Repeat Example 6 with a cannon ball launched from a cannon at an initial height of 1 m, at an angle of 30° to the horizontal, with an initial speed of 200 m/sec. Round values to 1 decimal place.

TIP The value of Tstep in the WINDOW menu affects the speed at which the calculator graphs the curve. Small values of Tstep result in more points being plotted and the curve graphed more slowly. This can be a helpful feature when you want to identify the orientation of a curve.

TECHNOLOGY CONNECTIONS

Graphing Parametric Equations

We can graph the parametric equations from Example 6 on a graphing utility. First select the "PARAMETRIC" option in the MODE menu. Then enter the pair of parametric equations in the equation editor, set the window parameters, and select the GRAPH button.

Answers

6. a. $x(t) = 100\sqrt{3}t$ and
$y(t) = -4.9t^2 + 100t + 1$
b. 20.4 sec **c.** 3533.4 m
d. 10.2 sec **e.** 511.2 m

We conclude this section with a discussion of a special curve with interesting physical properties. Suppose that a circle rolls along a straight path in a plane. The curve traced out by a fixed point P on the circumference of the circle is called a **cycloid** (Figure 10-53).

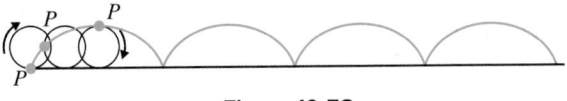

Figure 10-53

An equation of a cycloid in rectangular coordinates is quite cumbersome. Fortunately, the curve is defined nicely with parametric equations using the angle of rotation as the parameter.

Consider a circle with center C and radius r that rolls along the x-axis. Choose the origin to be one of the points where the fixed point $P(x, y)$ on the circumference of the circle touches the x-axis. In Figure 10-54, we show the position of the circle after it has rolled to the right through an angle of $0 < \theta < \dfrac{\pi}{2}$ radians.

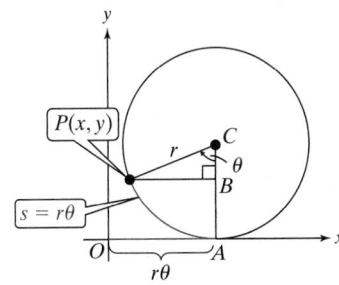

Figure 10-54

The distance $d(O, A)$ that the circle rolls along the x-axis equals the arc length s between A and P. Furthermore, recall that arc length is computed as $s = r\theta$ for θ in radians.

Thus, $d(O, A) = r\theta$. The x-coordinate of point P is given by

$$x = d(O, A) - d(P, B). \text{ Thus, } x = r\theta - r\sin\theta.$$

The y-coordinate of P is given by

$$y = d(C, A) - d(C, B). \text{ Thus, } y = r - r\cos\theta.$$

In summary, the cycloid is represented by $x = r(\theta - \sin\theta)$ and $y = r(1 - \cos\theta)$.

An inverted cycloid is generated by the same equations with a negative value of r (Figure 10-55). This curve has the interesting physical property that it is a solution to the *brachistochrone* problem posed in 1696 by Swiss mathematician Johann Bernoulli. Suppose that two points A and B lie in a vertical plane with A higher than B, but not directly above B (Figure 10-56). The brachistochrone problem asks what curve would minimize the time required for an object to slide from A to B with only the force of gravity to move it? The answer is a half-arch of an inverted cycloid. In Figure 10-56, we show several paths from A to B. Note that although a line is the shortest *distance* between points A and B, the shortest *time* to move from A to B under the influence of gravity occurs along the path of an inverted cycloid.

TIP The word "brachistochrone" is derived from the Greek (*brachistos*) meaning "shortest" and (*chronos*) meaning "time."

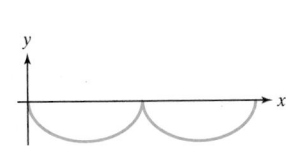

Figure 10-55   Inverted cycloid

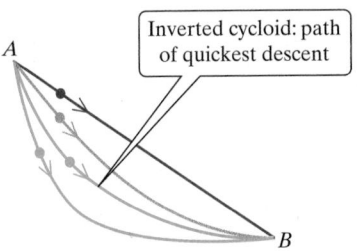

Inverted cycloid: path of quickest descent

Figure 10-56

TIP The word "tauto-chrone" is derived from the Greek (*tauto*) meaning "same" and (*chronos*) meaning "time."

The inverted cycloid also satisfies the *tautochrone* property (property of equal time). Let Q be the lowest point on one arch of an inverted cycloid. If several objects are placed at different points along the cycloid and allowed to slide down the curve under the influence of gravity, they will all reach point Q at the same time (Figure 10-57). This discovery intrigued seventeenth-century clockmakers.

Objects A, B, and C will reach point Q at the same time.

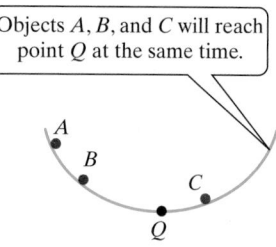

Figure 10-57

A clock made from a simple pendulum (a bob hanging from a thread swinging back and forth) has slight variations in the period with increasing amplitude of the swing. To address this problem, Dutch mathematician Christiaan Huygens constrained the swing of the pendulum between two metal plates in the shape of an inverted cycloid. As the pendulum swings, the thread wraps and unwraps around the cycloid, causing the bob to follow the path of a cycloid (Figure 10-58). Theoretically, in the absence of friction, this design would ensure equal times for each back-and-forth swing.

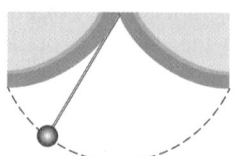

Figure 10-58

SECTION 10.6   Practice Exercises

Prerequisite Review

R.1. Solve the equation.

$$5w^2 + 2w - 1 = 0 \qquad \left\{\frac{-1 \pm \sqrt{6}}{5}\right\}$$

R.2. Find the vertex by using the vertex formula.

$$T(x) = x^2 - 8x + 7 \quad (4, -9)$$

Concept Connections

1. If f and g are continuous functions of t over an interval I, then the set of points $(f(t), g(t))$ is called a ____plane____ curve. The equations $x = f(t)$ and $y = g(t)$ are called ____parametric____ equations and t is called the ____parameter____.

2. The parametric equations $x = 5\cos\theta$ and $y = -5\sin\theta$ define a circle. In which direction is the curve oriented? (clockwise or counterclockwise)   clockwise

3. Identify the endpoints of the curve defined by $x = 3 - t$ and $y = 2t + 1$ on the interval $-2 \le t \le 1$. $(5, -3)$ and $(2, 3)$

4. Suppose that an object travels along a straight path in a rectangular coordinate system with units of distance in feet. If the object moves from the point $(12, 18)$ to $(36, 16)$ in 4 sec, the horizontal component of velocity is ____6 ft/sec____ and the vertical component of velocity is ____-0.5 ft/sec____.

Objective 1: Graph a Plane Curve by Plotting Points

For Exercises 5–6,

a. Use the parametric equations to compute x and y for the given values of t.

b. Plot the points, sketch the curve, and indicate the orientation of the curve.

5. $x = 2t$ and $y = t + 1$ for $-3 \le t \le 1$

t	-3	-2	-1	0	1
x	-6	-4	-2	0	2
y	-2	-1	0	1	2

6. $x = -t$ and $y = 3t - 1$ for $0 \le t \le 4$

t	0	1	2	3	4
x	0	-1	-2	-3	-4
y	-1	2	5	8	11

For Exercises 7–10, sketch the plane curve by plotting points. Indicate the orientation of the curve. (See Example 1)

7. $x = -t^2$ and $y = t + 3$ for $-1 < t \le 3$

8. $x = t^2 - 4$ and $y = t - 1$ for $-2 \le t < 2$

9. $x = t - 4$ and $y = \sqrt{t}$

10. $x = -t - 3$ and $y = |t|$

Objective 2: Eliminate the Parameter

For Exercises 11–26,

a. Eliminate the parameter and write an equation in rectangular coordinates.

b. Sketch the curve and indicate its orientation. (**See Examples 2–3**)

11. $x = t + 2$ and $y = t^2 + t$ a. $y = x^2 - 3x + 2$

12. $x = t - 1$ and $y = t^2 - 2t$ a. $y = x^2 - 1$

13. $x = t - 1$ and $y = \dfrac{t}{t-1}$ a. $y = \dfrac{x+1}{x}$ or $y = 1 + \dfrac{1}{x}, x \neq 0$

14. $x = t + 2$ and $y = \dfrac{t}{t+2}$ a. $y = \dfrac{x-2}{x}$ or $y = 1 - \dfrac{2}{x}, x \neq 0$

15. $x = |t + 3|$ and $y = 1 + t$ a. $x = |y + 2|$

16. $x = |t - 3|$ and $y = t - 2$ a. $x = |y - 1|$

17. $x = e^{-t}$ and $y = e^t$ a. $y = \dfrac{1}{x}, x > 0$

18. $x = 2e^t$ and $y = e^t - 2$ a. $y = \dfrac{x}{2} - 2, x > 0$

19. $x = \sqrt{t}$ and $y = \sqrt{16 - t}$

20. $x = -\sqrt{9 - t}$ and $y = \sqrt{t}$

21. $x = \cos t$ and $y = \sin^2 t - 4$
   (*Hint:* Use the identity $\sin^2 t + \cos^2 t = 1$.)
   a. $y = -x^2 - 3, -1 \leq x \leq 1, -4 \leq y \leq -3$

22. $x = -\sin t$ and $y = 1 + \cos^2 t$
   a. $y = 2 - x^2, -1 \leq x \leq 1, 1 \leq y \leq 2$

23. $x = 5\cos\theta$ and $y = -5\sin\theta$ a. $x^2 + y^2 = 25$

24. $x = 4\sin\theta$ and $y = -4\cos\theta$ a. $x^2 + y^2 = 16$

25. $x = 1 + 2\cos\theta$ and $y = -2 + 3\sin\theta$
   (*Hint:* Solve for $\sin\theta$ and $\cos\theta$. Square the results and apply the identity $\sin^2\theta + \cos^2\theta = 1$.)
   a. $\dfrac{(x-1)^2}{4} + \dfrac{(y+2)^2}{9} = 1$

26. $x = 5 + 4\sin\theta$ and $y = -3 + 2\cos\theta$
   a. $\dfrac{(x-5)^2}{16} + \dfrac{(y+3)^2}{4} = 1$

For Exercises 27–30, eliminate the parameter and write an equation in rectangular coordinates to represent the given curve.

27. Circle: $x = h + r\cos\theta$ and $y = k + r\sin\theta$ (*Hint:* Solve for $\cos\theta$ and $\sin\theta$. Then square both equations and use a Pythagorean identity.) $(x - h)^2 + (y - k)^2 = r^2$

28. Line through the points (x_1, y_1) and (x_2, y_2):
   $x = x_1 + t(x_2 - x_1)$ and $y = y_1 + t(y_2 - y_1)$
   $y - y_1 = \dfrac{y_2 - y_1}{x_2 - x_1}(x - x_1)$

29. Ellipse: $x = h + a\cos\theta$ and $y = k + b\sin\theta$

30. Hyperbola: $x = h + a\sec\theta$ and $y = k + b\tan\theta$

Objective 3: Write Parametric Equations to Represent a Curve

For Exercises 31–34, write parametric equations for the given curve for the given definitions of x. (See Example 4)

31. $y = 2x - 4$
   a. $x = t$ **b.** $x = \dfrac{t}{3}$ **c.** $x = -3t$

32. $y = -4x + 1$
   a. $x = t$ **b.** $x = \dfrac{t}{2}$ **c.** $x = -4t$

33. $y = x + 4, x > 0$
   a. $x = t, t > 0$ **b.** $x = e^t$

34. $y = -2x, x \geq 0$
   a. $x = t, t \geq 0$ **b.** $x = t^2$

For Exercises 35–42, use the results of Exercises 27–30 and use the parameter t to write parametric equations representing the given curve. Answers may vary.

35. Line passing through $(0, 2)$ and $(3, 1)$
   $x = 3t$ and $y = 2 - t$ or $x = 3 - 3t$ and $y = 1 + t$

36. Line passing through $(0, -1)$ and $(4, 2)$
   $x = 4t$ and $y = -1 + 3t$ or $x = 4 - 4t$ and $y = 2 - 3t$

37. Circle with center $(4, -3)$ and radius 5
   $x = 4 + 5\cos t$ and $y = -3 + 5\sin t$

38. Circle with center $\left(-\dfrac{1}{2}, 1\right)$ and radius $\dfrac{1}{2}$

39. Hyperbola with center $(0, 0)$, vertices $(0, \pm 2)$, and asymptotes $y = \pm\dfrac{2}{3}x$ $x = 3\tan t$ and $y = 2\sec t$

40. Hyperbola with center $(1, 3)$, vertices $(-2, 3)$ and $(4, 3)$, and foci $(-4, 3)$ and $(6, 3)$
   $x = 1 + 3\sec t$ and $y = 3 + 4\tan t$

41. Ellipse with center $(-2, 1)$, vertices $(1, 1)$ and $(-5, 1)$, and endpoints of the minor axis $(-2, 3)$ and $(-2, -1)$
   $x = -2 + 3\cos t$ and $y = 1 + 2\sin t$

42. Ellipse with center $(0, 0)$, vertices $(0, \pm 5)$, and foci $(0, \pm 4)$ $x = 3\cos t$ and $y = 5\sin t$

For Exercises 43–46, write parametric equations on a restricted interval for *t* to define the given graph.

43.

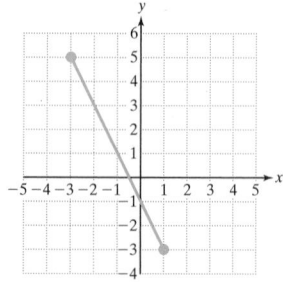

$x = -3 + 4t$ and $y = 5 - 8t$ or
$x = 1 - 4t$ and $y = -3 + 8t$
both for $0 \le t \le 1$

44.

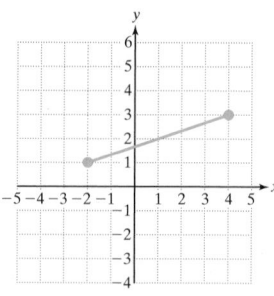

$x = -2 + 6t$ and $y = 1 + 2t$ or
$x = 4 - 6t$ and $y = 3 - 2t$
both for $0 \le t \le 1$

45. Right half of an ellipse

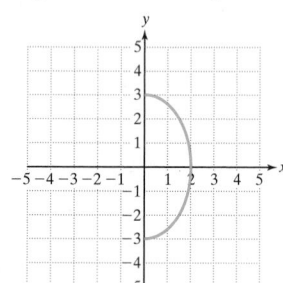

$x = 2\cos t$ and $y = 3\sin t$,
$-\dfrac{\pi}{2} \le t \le \dfrac{\pi}{2}$

46. Top branch of a hyperbola

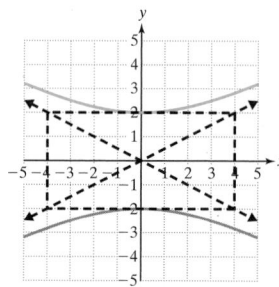

$x = 4\tan t$ and $y = 2\sec t$,
$-\dfrac{\pi}{2} < t < \dfrac{\pi}{2}$

Objective 4: Apply Parametric Equations

For Exercises 47–48, an object undergoes uniform linear motion on a straight path from point *A* to point *B* in Δ*t* sec. Write parametric equations over an interval *I* that describe the motion along the path.

47. $A = (1, 5)$, $B = (7, 3)$, and $\Delta t = 2$ sec
$x = 1 + 3t$ and $y = 5 - t$ for $0 \le t \le 2$

48. $A = (10, 7)$, $B = (-14, 1)$, and $\Delta t = 6$ sec
$x = 10 - 4t$ and $y = 7 - t$ for $0 \le t \le 6$

49. Suppose that the origin on a computer screen is in the lower left corner of the screen. A target moves along a straight line at a constant speed from point $A(670, 450)$ to point $B(250, 210)$ in 4 sec. **(See Example 5)**

   a. Write parametric equations to represent the target's path as a function of the time *t* (in sec) after the target leaves its initial position.
$x = 670 - 105t$ and $y = 450 - 60t$ for $0 \le t \le 4$

   b. Where is the target located 1.4 sec after motion starts?
(523, 366)

50. A hospital is located 3 mi west and 4 mi north of the center of town. Suppose that a Medi-Vac helicopter flies at a constant speed from the hospital to the location of an accident 15 mi east and 20 mi south of the center of town in 15 min $\left(\frac{1}{4}\text{ hr}\right)$. Choose a coordinate system with the origin at the center of town.

   a. Write parametric equations to represent the path of the helicopter as a function of the time *t* (in hr) after the helicopter leaves the hospital.
$x = -3 + 72t$ and $y = 4 - 96t$ for $0 \le t \le 0.25$

   b. Where is the helicopter located 10 min after leaving the hospital?   9 mi east and 12 mi south of the center of town

51. An artillery specialist practicing at a firing range on a flat stretch of ground fires a shell from 1 m above the ground, with a muzzle velocity of 96 m/sec at an angle of 16° from the horizontal. Choose a coordinate system with the origin at ground level directly below the launch position. **(See Example 6)**

   a. Write parametric equations that model the path of the shell as a function of time *t* (in sec) after launch.
$x = (96\cos 16°)t$ and $y = -4.9t^2 + (96\sin 16°)t + 1$

   b. Approximate the time required for the shell to hit the ground. Round to the nearest hundredth of a second.
$t \approx 5.44$ sec

   c. Approximate the horizontal distance that the shell travels before it hits the ground. Round to the nearest meter.   502 m

   d. When is the shell at its maximum height? Find the exact value and an approximation to the nearest hundredth of a second. $t = \dfrac{96\sin 16°}{9.8} \approx 2.7$ sec

   e. Determine the maximum height. Round to the nearest meter.   37 m

52. A golfer tees off on level ground, and hits the ball with an initial speed of 52 m/sec at an angle of 34° above the horizontal. Choose a coordinate system with the origin at the point where the ball is struck.

   a. Write parametric equations that model the path of the ball as a function of time *t* (in sec).
$x = (52\cos 34°)t$ and $y = -4.9t^2 + (52\sin 34°)t$

   b. For how long will the ball be in the air before it hits the ground? Round to the nearest hundredth of a second. $t \approx 5.93$ sec

   c. Approximate the horizontal distance that the ball travels before it hits the ground. Round to the nearest foot.   256 m

   d. When is the ball at its maximum height? Find the exact value and an approximation to the nearest hundredth of a second. $t = \dfrac{52\sin 34°}{9.8} \approx 2.97$ sec

   e. What is the maximum height? Round to the nearest foot.   43 m

53. Anna hits a softball at a height of 3 ft from the ground. The softball leaves her bat traveling with an initial speed of 80 ft/sec, at an angle of 30° from the horizontal. Choose a coordinate system with the origin at ground level directly under the point where the ball is struck.

   a. Write parametric equations that model the path of the ball as a function of time t (in sec). $x = 40\sqrt{3}t$ and $y = -16t^2 + 40t + 3$

   b. When is the ball at its maximum height?   1.25 sec

   c. What is the maximum height? Round to the nearest foot.   28 ft

   d. If an outfielder catches the ball at a height of 5 ft, how long was the ball in the air after being struck? Give the exact answer and the answer rounded to the nearest hundredth of a second. $t = \dfrac{5 + \sqrt{23}}{4} \approx 2.45$ sec

   e. How far is the outfielder from home plate when she catches the ball? Round to the nearest foot.   170 ft

54. Tony hits a baseball at a height of 3 ft from the ground. The ball leaves his bat traveling with an initial speed of 120 ft/sec at an angle of 45° from the horizontal. Choose a coordinate system with the origin at ground level directly under the point where the ball is struck.

   a. Write parametric equations that model the path of the ball as a function of time t (in sec). $x = 60\sqrt{2}t$ and $y = -16t^2 + 60\sqrt{2}t + 3$

   b. When is the ball at its maximum height? Give the exact value and round to the nearest hundredth of a second. $t = \dfrac{15\sqrt{2}}{8} \approx 2.65$ sec

   c. What is the maximum height?   115.5 ft

   d. If an outfielder catches the ball at a height of 6 ft, for how long was the ball in the air after being struck? Give the exact answer and the answer rounded to the nearest hundredth of a second.

   e. How far is the outfielder from home plate when he catches the ball? Round to the nearest foot.   447 ft

Mixed Exercises

55. Two cars approach an intersection. Car A is 25 mi west of the intersection traveling 40 mph. Car B is 30 mi north of the intersection traveling 50 mph. Place the origin of a rectangular coordinate system at the intersection.

   a. Write parametric equations that model the path of each car as a function of the time $t \geq 0$ (in hr).

   b. Determine the times required for each car to reach the intersection. Based on these results, will the cars crash?

   c. Write the distance between the cars as a function of the time t. $d(t) = \sqrt{4100t^2 - 5000t + 1525}$

   d. Determine the time at which the two cars are at their closest point. [*Hint*: The function from part (c) is minimized when the radicand is minimized.] $\dfrac{25}{41}$ hr

   e. How close are the cars at their closest point? Round to the nearest hundredth of a mile.   0.78 mi

56. Two planes flying at the same altitude are on a course to fly over a control tower. Plane A is 50 mi east of the tower flying 125 mph. Plane B is 90 mi south of the tower flying 200 mph. Place the origin of a rectangular coordinate system at the intersection.

   a. Write parametric equations that model the path of each plane as a function of the time $t \geq 0$ (in hr).

   b. Determine the times required for each plane to reach a point directly above the tower. Based on these results, will the planes crash?

   c. Write the distance between the planes as a function of the time t. $d(t) = \sqrt{55,625t^2 - 48,500t + 10,600}$

   d. How close do the planes pass? Round to the nearest tenth of a mile.   5.3 mi

57. Suppose that a mortar is fired from ground level at an angle of 45° with an initial speed of 200 ft/sec. Choose a coordinate system with the origin at the point of launch.

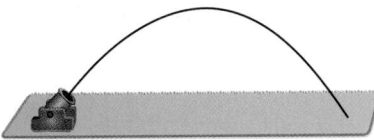

   a. Write parametric equations to define the path of the mortar as a function of the time t (in sec). $x(t) = 100\sqrt{2}t$ and $y(t) = -16t^2 + 100\sqrt{2}t$

   b. What is the range of the mortar? That is, what is the horizontal distance traveled from the point of launch to the point where the mortar lands?   1250 ft

   c. What are the coordinates of the mortar at its maximum height? $\left(625, \dfrac{625}{2}\right)$

   d. Eliminate the parameter and write an equation in rectangular coordinates to represent the path.

   e. If a drone will be at position (848.5, 272.5) at a time 6 sec after the mortar is launched, will the mortar hit the drone? Assume a 1-ft margin of error.   Yes

58. A daredevil wants to jump over a canyon on his motorcycle. He travels approximately 88 ft/sec (60 mph) at an angle of 30° when the motorcycle leaves the ramp at the edge of the canyon. Choose a coordinate system with the origin at the end of the take-off ramp and assume that he lands at the same height.

   a. Write parametric equations to define the path of the motorcycle as a function of the time t (in sec) after leaving the ramp. $x = 44\sqrt{3}t$ and $y = -16t^2 + 44t$

   b. If a bird is at a position (90, 26) at a time 1.2 sec after the motorcycle leaves the ramp, will the motorcycle hit the bird? Assume a 1-ft margin of error.   No

   c. What is the horizontal distance across the canyon from the point of take-off to the point of landing? Round to the nearest foot.   210 ft

   d. What are the coordinates (to the nearest foot) of the motorcycle at its maximum height?   (105, 30)

   e. Eliminate the parameter and write an equation in rectangular coordinates to represent the path.

$$y = -\dfrac{1}{363}x^2 + \dfrac{\sqrt{3}}{3}x$$

Recall from Section 7.1 that a point (x, y) in rectangular coordinates and a point (r, θ) in polar coordinates are related by the equations $x = r\cos\theta$ and $y = r\sin\theta$. Thus, the graph of the polar equation $r = f(\theta)$ is the same as the graph of the parametric equations $x = f(\theta)\cos\theta$ and $y = f(\theta)\sin\theta$. For Exercises 59–60, a polar equation is given.

a. Use a graphing utility in "polar" mode to graph the polar equation.

b. Write parametric equations to represent the graph of the polar equation.

c. Use a graphing utility in "parametric" mode to graph the parametric equations from part (b).

59. $r = 4\sin 2\theta$ for $0 \le \theta \le 2\pi$ **60.** $r = 3 - 4\cos\theta$ for $0 \le \theta \le 2\pi$

Write About It

For Exercises 61–62, describe the differences in the graphs of C_1 through C_3.

61. C_1: $x = t$ and $y = t^2 - 1$

     C_2: $x = e^t$ and $y = e^{2t} - 1$

     C_3: $x = \cos t$ and $y = \cos^2 t - 1$

62. C_1: $x = t$ and $y = \sqrt{25 - t^2}$

     C_2: $x = e^t$ and $y = \sqrt{25 - e^{2t}}$

     C_3: $x = \cos t$ and $y = \sqrt{25 - \cos^2 t}$

63. Do the parametric equations $x = \sqrt{t}$ and $y = t + 5$ define a parabola? Why or why not?

64. Is the point $(16, 12)$ on the curve defined by $x = 2t$ and $y = t + 1$? Why or why not?

Expanding Your Skills

65. For a short interval of time, a military supply plane flies on a hyperbolic path given by the equation

$$\frac{y^2}{122.5^2} - \frac{x^2}{100^2} = 1; \; y \ge 0,$$

where x and y are measured in meters.

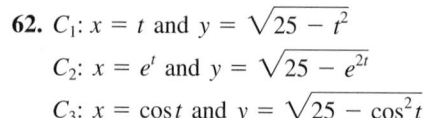

     a. What are the coordinates of the point on the flight path closest to the ground?   (0, 122.5)

     b. When the plane reaches its closest point to the ground, it drops a bag of supplies to people on the ground. Assuming that the plane is traveling 200 m/sec due east at the time of the drop, write parametric equations representing the path of the bag as a function of the time t (in sec) after the drop. $x(t) = 200t$ and $y(t) = -4.9t^2 + 122.5$

     c. Determine the coordinates of the point where the bag hits the ground.   (1000, 0)

66. Suppose that the position of one particle at time t is given by

$$x_1 = 3 + 3\cos t \text{ and } y_1 = 5\sin t \text{ for } 0 \le t \le 2\pi$$

Suppose that the position of a second particle is given by

$$x_2 = \frac{6}{\pi}t \text{ and } y_2 = \frac{10}{\pi}t \text{ for } 0 \le t \le 2\pi$$

     a. Use a graphing utility to graph the paths of both particles simultaneously.

     b. In how many points do the graphs intersect?   two

     c. Will the particles ever collide? If so, where and at what time?

         Yes. The particles collide at $t = \frac{\pi}{2}$ sec at the point $(3, 5)$.

67. Suppose that an imaginary string of unlimited length is attached to a point on a circle and pulled taut so that it is tangent to the circle. Keeping the string pulled tight, wind (or unwind) the string around the circle. The path traced out by the end of the string as it moves is called an *involute* of the circle (shown in blue).

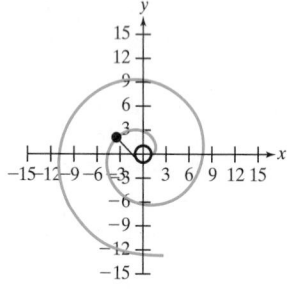
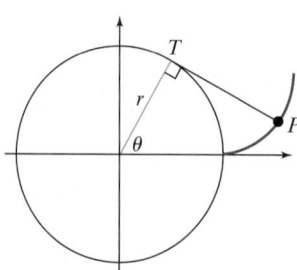

Given a circle of radius r with parametric equations $x = r\cos\theta$, $y = r\sin\theta$, show that the parametric equations of the involute of the circle are $x = r(\cos\theta + \theta\sin\theta)$ and $y = r(\sin\theta - \theta\cos\theta)$.

Technology Connections

For Exercises 68–70, use a graphing utility to graph the given curve on the recommended viewing window.

68. Witch of Agnesi:

$x = 2\cot t$ and $y = 2\sin^2 t$

t: $[0, 2\pi, 0.1]$, x: $[-6.4, 6.4, 1]$, y: $[-4, 4, 1]$

69. Lissajous curve:

$x = 3\cos 3\theta$ and $y = 3\sin 4\theta$

t: $[0, 2\pi, 0.1]$, x: $[-6.4, 6.4, 1]$, y: $[-4, 4, 1]$

70. Involute of a circle:

$x = \cos t + t\sin t$ and $y = \sin t - t\cos t$

t: $[0, 6\pi, 0.1]$, x: $[-32, 32, 4]$, y: $[-20, 20, 4]$

A cycloid is a curve created by rolling a circle along a line. We can also create families of curves by rolling a circle around another *circle*. If a circle of radius r rolls around the *interior* of a larger circle of radius R, a fixed point on the smaller circle traces out a curve called a **hypocycloid.** For Exercises 71–72, use the following parametric equations for a hypocycloid.

$$x = (R - r)\cos t + r\cos\left[\left(\frac{R - r}{r}\right)t\right] \text{ and } y = (R - r)\sin t - r\sin\left[\left(\frac{R - r}{r}\right)t\right]$$

71. a. Write parametric equations for a hypocycloid with $R = 3$ and $r = 1$. The curve defined by these equations is called a *deltoid.* $x = 2\cos t + \cos 2t$ and $y = 2\sin t - \sin 2t$

b. Graph the exterior circle given by $x = 3\cos t$ and $y = 3\sin t$ and the deltoid from part (a). Use $0 \le t \le 2\pi$ and a viewing window of $[-4.8, 4.8, 1]$ by $[-3, 3, 1]$.

c. For what values of t on the interval $[0, 2\pi)$ do the deltoid curve and circle intersect? $t = 0, t = \frac{2\pi}{3}, t = \frac{4\pi}{3}$

72. a. Write parametric equations for a hypocycloid with $R = 4$ and $r = 1$. The curve defined by these equations is called an *astroid* (not to be confused with asteroid). $x = 3\cos t + \cos 3t$ and $y = 3\sin t - \sin 3t$

b. Graph the exterior circle given by $x = 4\cos t$ and $y = 4\sin t$ and the astroid from part (a). Use $0 \le t \le 2\pi$ and a viewing window of $[-6.4, 6.4, 1]$ by $[-4, 4, 1]$.

c. For what values of t on the interval $[0, 2\pi)$ do the astroid curve and circle intersect? $t = 0, t = \frac{\pi}{2}, \pi, t = \frac{3\pi}{2}$

If a circle of radius r rolls around the *exterior* of a circle of radius $R \ge r$, a fixed point on the outer circle traces out a curve called an **epicycloid.** For Exercises 73–74, use the following parametric equations for an epicycloid.

$$x = (R + r)\cos t - r\cos\left[\left(\frac{R + r}{r}\right)t\right] \text{ and } y = (R + r)\sin t - r\sin\left[\left(\frac{R + r}{r}\right)t\right]$$

73. a. Write parametric equations for a epicycloid with $R = 1$ and $r = 1$. The curve defined by these equations is called a *cardioid.* $x = 2\cos t - \cos 2t$ and $y = 2\sin t - \sin 2t$

b. Graph the circle given by $x = \cos t$ and $y = \sin t$ and the *cardioid* from part (a). Use $0 \le t \le 2\pi$ and a viewing window of $[-6.4, 6.4, 1]$ by $[-4, 4, 1]$.

74. a. Write parametric equations for a epicycloid with $R = 2$ and $r = 1$. The curve defined by these equations is called a *nephroid* meaning "kidney-shaped." $x = 3\cos t - \cos 3t$ and $y = 3\sin t - \sin 3t$

b. Graph the circle given by $x = 2\cos t$ and $y = 2\sin t$ and the *nephroid* from part (a). Use $0 \le t \le 2\pi$ and a viewing window of $[-6.4, 6.4, 1]$ by $[-4, 4, 1]$.

CHAPTER 10 KEY CONCEPTS

SECTION 10.1 The Ellipse	Reference
An **ellipse** is the set of all points (x, y) in a plane such that the sum of the distances between (x, y) and two fixed points is a constant. The fixed points are called the **foci** of the ellipse.	p. 908

	Reference
Standard forms of an equation of an ellipse:	p. 913

• Major axis horizontal:

Equation: $\dfrac{(x-h)^2}{a^2} + \dfrac{(y-k)^2}{b^2} = 1 \qquad a > b > 0; c^2 = a^2 - b^2 \ (c > 0)$

Center: (h, k)

Vertices: $(h + a, k), (h - a, k)$

Foci: $(h + c, k), (h - c, k)$

Minor axis endpoints: $(h, k + b), (h, k - b)$

• Major axis vertical:

Equation: $\dfrac{(x-h)^2}{b^2} + \dfrac{(y-k)^2}{a^2} = 1 \qquad a > b > 0; c^2 = a^2 - b^2 \ (c > 0)$

Center: (h, k)

Vertices: $(h, k + a), (h, k - a)$

Foci: $(h, k + c), (h, k - c)$

Minor axis endpoints: $(h + b, k), (h - b, k)$

	Reference
$2a$ is the length of the major axis. $2b$ is the length of the minor axis.	p. 910
The **eccentricity** $e = \frac{c}{a}$ of an ellipse measures the degree of elongation. Note that for an ellipse, $0 < e < 1$.	p. 917

SECTION 10.2 The Hyperbola	Reference
A **hyperbola** is the set of all points (x, y) in a plane such that the difference in distances between (x, y) and two fixed points (foci) is a positive constant.	p. 925

	Reference
Standard forms of an equation of a hyperbola:	p. 929

• Transverse axis horizontal:

Equation: $\dfrac{(x-h)^2}{a^2} - \dfrac{(y-k)^2}{b^2} = 1 \qquad a > 0, b > 0; c^2 = a^2 + b^2 \ (c > 0)$

Center: (h, k)

Vertices: $(h + a, k), (h - a, k)$

Foci: $(h + c, k), (h - c, k)$

Asymptotes: $y - k = \pm\frac{b}{a}(x - h)$

• Transverse axis vertical:

Equation: $\dfrac{(y-k)^2}{a^2} - \dfrac{(x-h)^2}{b^2} = 1 \qquad a > 0, b > 0; c^2 = a^2 + b^2 \ (c > 0)$

Center: (h, k)

Vertices: $(h, k + a), (h, k - a)$

Foci: $(h, k + c), (h, k - c)$

Asymptotes: $y - k = \pm\frac{a}{b}(x - h)$

	Reference
$2a$ is the length of the transverse axis. $2b$ is the length of the conjugate axis.	p. 925
The **eccentricity** of a hyperbola is given by $e = \frac{c}{a}$. Note that for a hyperbola, $e > 1$.	p. 933

	Reference
SECTION 10.3 The Parabola	
A **parabola** is the set of all points in a plane that are equidistant from a fixed line (called the **directrix**) and a fixed point (called the **focus**).	p. 941
Standard forms of an equation of a parabola:	p. 942

- Vertical axis of symmetry:
  Equation: $(x - h)^2 = 4p(y - k)$
  Vertex: (h, k)
  Focus: $(h, k + p)$
  Directrix: $y = k - p$
  Axis of symmetry: $x = h$

- Horizontal axis of symmetry:
  Equation: $(y - k)^2 = 4p(x - h)$
  Vertex: (h, k)
  Focus: $(h + p, k)$
  Directrix: $x = h - p$
  Axis of symmetry: $y = k$

- If $p > 0$, the parabola opens upward as shown.
- If $p < 0$, the parabola opens downward.

- If $p > 0$, the parabola opens to the right as shown.
- If $p < 0$, the parabola opens to the left.

	Reference
The value of $|p|$ is the focal length (distance between the vertex and focus).	p. 942
The **latus rectum** is the line segment perpendicular to the axis of symmetry, through the focus, with endpoints on the parabola.	p. 943
The value of $4|p|$ is the focal diameter (length of the latus rectum).	p. 943

	Reference
SECTION 10.4 Rotation of Axes	
Rotation of axes:	p. 958

Suppose that the x- and y-axes in a rectangular coordinate system are rotated about the origin through an angle θ to form the x'- and y'-axes.

x and y in terms of x' and y'	x' and y' in terms of x and y
$x = x'\cos\theta - y'\sin\theta$	$x' = x\cos\theta + y\sin\theta$
$y = x'\sin\theta + y'\cos\theta$	$y' = -x\sin\theta + y\cos\theta$

	Reference
Determine the angle of rotation:	p. 960

To write the general second-degree polynomial equation

$$Ax^2 + Bxy + Cy^2 + Dx + Ey + F = 0$$

with no xy term, rotate the axes through an acute angle θ given by

$$\cot 2\theta = \frac{A - C}{B}$$

	Reference
Graphing a rotated conic:	p. 961

1. Determine the angle of rotation.
2. Substitute $x = x'\cos\theta - y'\sin\theta$ and $y = x'\sin\theta + y'\cos\theta$ into the equation $Ax^2 + Bxy + Cy^2 + Dx + Ey + F = 0$. The resulting equation will have no xy term.
3. Write the equation from Step 2 in standard form and graph the conic relative to the x'- and y'-axes.

	Reference
Identify conics by the discriminant of the equation, $B^2 - 4AC$:	p. 963

Excluding degenerate cases, the equation $Ax^2 + Bxy + Cy^2 + Dx + Ey + F = 0$ represents

An ellipse (or a circle) if $B^2 - 4AC < 0$.

A hyperbola if $B^2 - 4AC > 0$.

A parabola if $B^2 - 4AC = 0$.

SECTION 10.5 Polar Equations of Conics	Reference
Alternative definition of conics: Let l be a fixed line (**directrix**) in a plane and let F be a fixed point (**focus**) in the plane, not on the line. A **conic** is the set of points P in the plane such that the ratio of the distance between P and F to the distance from P to l is a constant positive real number e (the **eccentricity**). That is, $\dfrac{d(P, F)}{d(P, l)} = e.$	p. 969
Polar equations of conics: A polar equation of the form $r = \dfrac{ed}{1 \pm e\cos\theta}$ or $r = \dfrac{ed}{1 \pm e\sin\theta}$ is a conic with focus at the pole and eccentricity $e > 0$. The value $|d|$ is the distance between the focus at the pole and the directrix. The conic is 1. A parabola if $e = 1$ (axis of symmetry perpendicular to the directrix). 2. An ellipse if $0 < e < 1$ (major axis perpendicular to the directrix). 3. A hyperbola if $e > 1$ (transverse axis perpendicular to the directrix). The location of the directrix and orientation of the curve for $d > 0$ are summarized in Figure 10-41 on page 971.	p. 970
Guidelines to graph a conic in polar form: 1. Write the polar equation in the form $r = \dfrac{ed}{1 \pm e\cos\theta}$ or $r = \dfrac{ed}{1 \pm e\sin\theta}.$ 2. Determine the value of the eccentricity e and identify the type of conic. 3. Determine the orientation of the conic. 4. Plot points at $\theta = 0, \dfrac{\pi}{2}, \pi,$ and $\dfrac{3\pi}{2}.$ 5. Identify key characteristics of the conic as desired (center, vertices, length of major/minor/transverse/conjugate axes, directrix).	p. 971
Rotation of conics in polar form: The rotation of a conic about the origin is much more convenient when the conic is expressed in polar versus rectangular form. Consider the graph of $r = f(\theta)$. Replacing θ by $(\theta - \alpha)$ results in • A counterclockwise rotation (rotation through a positive angle) if $\alpha > 0$. • A clockwise rotation (rotation through a negative angle) if $\alpha < 0$.	p. 975

SECTION 10.6 Plane Curves and Parametric Equations	Reference
Plane curves: Let f and g be functions of t over an interval I. Then the set of points defined by $(x, y) = (f(t), g(t))$ is called a plane curve. The equations $$x = f(t) \text{ and } y = g(t)$$ are called parametric equations of the curve, and t is called the parameter.	p. 979
Graphing a plane curve: • One method to graph a plane curve defined by parametric equations is to create a table of points. • Another method to graph a plane curve is to eliminate the parameter t, by solving for t or an expression involving t in one equation. Then substitute the result in the other equation. Note any restrictions on the domain of the rectangular equation so that the resulting graph is consistent with the restrictions on x and y in the parametric equations.	pp. 979–982

Applications of parametric equations:	p. 984

Parametric equations are particularly well suited for describing objects in motion. Let $t \geq 0$ represent the time under which an object is in motion along a plane curve, and let (x_0, y_0) represent the initial position of the object at $t = 0$.

Uniform linear motion

$x = x_0 + at$
$y = y_0 + bt$

The values of a and b are the horizontal and vertical components of velocity, respectively.

Projectile motion*

$x = (v_0 \cos \theta)t + x_0$

$y = -\dfrac{1}{2}gt^2 + (v_0 \sin \theta)t + y_0$

*Assuming negligible air resistance.

θ is the angle of launch relative to the horizontal, and v_0 is the initial speed (magnitude of the initial velocity). The value of g is the acceleration due to gravity, where $g = 32$ ft/sec² or $g = 9.8$ m/sec² on the surface of the Earth.

Expanded Chapter Summary available at www.mhhe.com/millerprecalculus.

CHAPTER 10   Review Exercises

SECTION 10.1

1. How are the graphs of these ellipses similar and how are they different?

$$\dfrac{x^2}{100} + \dfrac{y^2}{225} = 1 \qquad \dfrac{(x-1)^2}{100} + \dfrac{(y+7)^2}{225} = 1$$

For Exercises 2–5, an equation of an ellipse is given.

a. Identify the center.
b. Identify the vertices.
c. Identify the endpoints of the minor axis.
d. Identify the foci.
e. Determine the eccentricity.
f. Graph the ellipse.

2. $\dfrac{x^2}{289} + \dfrac{y^2}{64} = 1$ **3.** $15x^2 + 9y^2 = 135$

4. $\dfrac{(x+1)^2}{9} + \dfrac{(y-4)^2}{25} = 1$ **5.** $\dfrac{(x-1)^2}{16} + \dfrac{(y-2)^2}{9} = 1$

For Exercises 6–7,

a. Write the equation of the ellipse in standard form.
b. Identify the center, vertices, endpoints of the minor axis, and foci.

6. $5x^2 + 8y^2 + 40x - 16y + 48 = 0$

7. $100x^2 + 64y^2 - 100x - 1575 = 0$

For Exercises 8–10, write the standard form of an equation of the ellipse subject to the given conditions.

8. Vertices: $(4, 0)$ and $(-4, 0)$;
Foci: $\left(\sqrt{11}, 0\right)$ and $\left(-\sqrt{11}, 0\right)$ $\dfrac{x^2}{16} + \dfrac{y^2}{5} = 1$

9. Endpoints of minor axis: $(0, 1)$ and $(6, 1)$;
Foci: $(3, 5)$ and $(3, -3)$ $\dfrac{(x-3)^2}{9} + \dfrac{(y-1)^2}{25} = 1$

10. Minor axis parallel to x-axis; Center: $(2, -4)$;
Length of major axis: 20 units;
Length of minor axis: 12 units $\dfrac{(x-2)^2}{36} + \dfrac{(y+4)^2}{100} = 1$

11. Suppose that one ellipse has an eccentricity of $\frac{4}{7}$ and a second ellipse has an eccentricity of $\frac{5}{7}$. Which ellipse is more elongated?   The second ellipse with eccentricity $\frac{5}{7}$ is more elongated.

12. Jupiter orbits the Sun in an elliptical path with the Sun at one focus. At perihelion, Jupiter is closest to the Sun at 7.4052×10^8 km. If the eccentricity of the orbit is 0.0489, determine the distance at aphelion (farthest point between Jupiter and the Sun). Round to the nearest million km. 8.17×10^8 km

13. A bridge over a gorge is supported by an arch in the shape of a semiellipse. The length of the bridge is 400 ft and the maximum height is 100 ft. Find the height of the arch 50 ft from the center. Round to the nearest foot.   The height 50 ft from the center is approximately 97 ft.

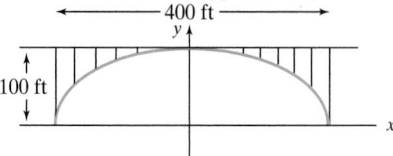

14. Elliptical concrete pipes have a greater capacity for shallow flow than circular pipes. For this reason, elliptical pipes are often used in storm drains, culverts, and sewers.

An engineer wants to design an elliptical concrete pipe with a maximum horizontal opening of 4 ft and a maximum vertical opening of 3 ft.

a. Write an equation for an elliptical cross section of the pipe. For convenience, place the coordinate system with (0, 0) at the center of the pipe. $\frac{x^2}{4} + \frac{y^2}{2.25} = 1$

b. To construct the pipe, the engineer needs to know the location of the foci. How far from the center are the foci? $\frac{\sqrt{7}}{2}$ ft

SECTION 10.2

For Exercises 15–16, an equation of a hyperbola is given. Determine whether the transverse axis is horizontal or vertical.

15. $\frac{x^2}{11} - \frac{y^2}{16} = 1$ Horizontal **16.** $-\frac{x^2}{11} + \frac{y^2}{16} = 1$ Vertical

For Exercises 17–20, an equation of a hyperbola is given.

a. Identify the center.

b. Identify the vertices.

c. Identify the foci.

d. Write equations for the asymptotes.

e. Graph the hyperbola.

17. $\frac{x^2}{9} - \frac{y^2}{4} = 1$ **18.** $-3x^2 + 2y^2 = 18$

19. $-\frac{(x+3)^2}{16} + \frac{(y-2)^2}{9} = 1$

20. $\frac{(x-1)^2}{4} - (y+5)^2 = 1$

For Exercises 21–22,

a. Write the equation of the hyperbola in standard form.

b. Identify the center, vertices, and foci.

21. $11x^2 - 7y^2 + 44x - 56y - 145 = 0$

22. $-9x^2 + y^2 + 2y - 8 = 0$

For Exercises 23–25, write the standard form of an equation of a hyperbola subject to the given conditions.

23. Vertices: (4, 0), (−4, 0); $\frac{x^2}{16} - \frac{y^2}{9} = 1$
Foci: (5, 0), (−5, 0)

24. Vertices (0, 13), (0, −11); $\frac{(y-1)^2}{144} - \frac{x^2}{25} = 1$
Slope of asymptotes: $\pm\frac{12}{5}$

25. Corners of reference rectangle: (19, 15), (19, −1), (−15, −1), (−15, 15); Horizontal transverse axis

26. Suppose that one hyperbola has an eccentricity of $\frac{41}{40}$ and a second hyperbola has an eccentricity of $\frac{41}{9}$. For which hyperbola do the branches open "wider"?
The second hyperbola with eccentricity $\frac{41}{9}$ will have branches opening wider.

27. Solve the system of equations.

$$6x^2 - 2y^2 = -12$$
$$2x^2 + y^2 = 11$$
$\{(1, 3), (1, -3), (-1, 3), (-1, -3)\}$

28. Suppose that two people standing 2 mi (10,560 ft) apart at points A and B hear a car backfire. The time difference between the sound heard at point A and the sound heard at point B is 8 sec. If sound travels 1100 ft/sec, find an equation of the hyperbola (with foci at A and B) on which the point of origination of the sound must lie.
$\frac{x^2}{(4400)^2} - \frac{y^2}{(2919)^2} = 1$

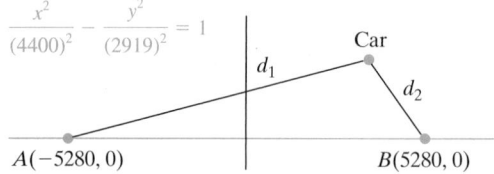

29. A 120-ft flight control tower in the shape of a hyperboloid has hyperbolic cross sections perpendicular to the ground. Placing the origin at the bottom and center of the tower, the center of a hyperbolic cross section is (0, 30) with one focus at $(15\sqrt{10}, 30)$ and one vertex at (15, 30). All units are in feet.

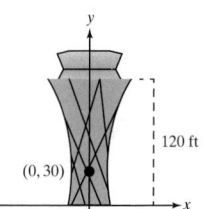

a. Write an equation of a hyperbolic cross section through the origin. Assume that there are no restrictions on x or y. $\frac{x^2}{15^2} - \frac{(y-30)^2}{45^2} = 1$

b. Determine the diameter of the tower at the base. Round to the nearest foot.   36 ft

c. Determine the diameter of the tower at the top. Round to the nearest foot.   67 ft

SECTION 10.3

For Exercises 30–33, an equation of a parabola is given in standard form.

a. Determine the value of p.

b. Identify the vertex.

c. Identify the focus.

d. Determine the focal diameter.

e. Determine the endpoints of the latus rectum.

f. Write an equation of the directrix.

g. Write an equation for the axis of symmetry.

h. Graph the parabola.

30. $x^2 = -2y$ **31.** $\frac{1}{10}y^2 = 2x$

32. $(y+2)^2 = -4(x-3)$ **33.** $\left(x - \frac{1}{2}\right)^2 = 6(y-3)$

For Exercises 34–35, an equation of a parabola is given.

a. Write the equation of the parabola in standard form.

b. Identify the vertex, focus, and directrix.

34. $x^2 - 10x - 4y + 17 = 0$

35. $4y^2 - 12y + 56x + 177 = 0$

For Exercises 36–37, fill in the blank. Let $|p|$ represent the focal length (distance between the vertex and focus).

36. If the directrix of a parabola is given by $y = 3$ and the focus is (4, 1), then the vertex is given by the ordered pair ____(4, 2)____, and the value of p is ____−1____.

37. If the vertex of a parabola is (10, −4) and the focus is (2, −4), then the directrix is given by the equation ____$x = 18$____, and the value of p is ____−8____.

For Exercises 38–40, determine the standard form of the parabola subject to the given conditions.

38. Vertex: (3, 2); Directrix: $x = 5$ $(y - 2)^2 = -8(x - 3)$

39. Vertex: (−4, 7); Focus (−4, −1) $(x + 4)^2 = -32(y - 7)$

40. Focus: (0, −3); Directrix: $y = -6$ $x^2 = 6\left(y + \frac{9}{2}\right)$

41. Solve the system of equations.

$$(y + 2)^2 = 3(x - 5)$$ $\{(8, 1), (5, -2)\}$
$$x - y = 7$$

42. The longest span of the Brooklyn Bridge is suspended from cables supported by two towers approximately 1596 ft apart. The top of the towers are 142 ft above the road. Assuming that the cables are parabolic and are at road level halfway between the towers, find the height of the cables above the roadway 200 ft from either end. Round to the nearest foot.   80 ft

43. Two large airship hangars were built in Orly, France, in the early 1900's but were destroyed in World War II by American aircraft. The hangars were 175 m in length, 90 m wide, and 60 m high and were formed by a series of parabolic arches.

a. Set up a coordinate system with (0, 0) at the vertex of one of the arches and write an equation of the parabola. $x^2 = -\frac{135}{4}y$ or $y = -\frac{4}{135}x^2$ for $-45 \le x \le 45$

b. What is the focal length of an arch? $\frac{135}{16}$ ft

SECTION 10.4

For Exercises 44–45, the given equation represents a conic section (nondegenerative case). Identify the type of conic section.

44. a. $3x^2 + 6xy + 2y^2 - 4x + y - 4 = 0$ Hyperbola

b. $3x^2 + 2y^2 - 4x + y - 4 = 0$ Ellipse

c. $9x^2 - 12xy = -4y^2 - 4y + 8$ Parabola

45. a. $2x^2 + xy + 6y^2 + 4x - 12y - 10 = 0$ Ellipse

b. $2x^2 + 10x = y^2 + 8y - 6$ Hyperbola

c. $x^2 + 4y^2 = 4xy - x + 6y + 3$ Parabola

For Exercises 46–48, assume that the x- and y-axes are rotated through angle θ about the origin to form the x'- and y'-axes.

46. Given $\theta = 60°$ and $P(-8, -10)$ in the xy-plane, find the coordinates of P in the $x'y'$-plane. $(-4 - 5\sqrt{3}, 4\sqrt{3} - 5)$

47. Given $\theta = 45°$ and $P(8\sqrt{2}, -4\sqrt{2})$ in the $x'y'$-plane, find the coordinates of P in the xy-plane.   (12, 4)

48. Given $\theta = 45°$, write the equation $xy + 10 = 0$ in $x'y'$-coordinates. $x'^2 - y'^2 + 20 = 0$

For Exercises 49–51, an equation of a conic section (nondegenerative case) is given.

a. Identify the type of conic section.

b. Determine an acute angle of rotation to eliminate the xy term.

c. Use a rotation of axes to eliminate the xy term in the equation.

d. Sketch the graph.

49. $47x^2 + 34\sqrt{3}xy + 13y^2 - 64 = 0$

50. $5x^2 - 6xy + 5y^2 - 34\sqrt{2}x + 30\sqrt{2}y + 122 = 0$

51. $16x^2 - 24xy + 9y^2 + 60x + 80y = 0$

For Exercises 52–53, the equation represents a conic section (nondegenerative case).

a. Identify the type of conic section.

b. Graph the equation on a graphing utility.

52. $4x^2 + 3xy - 2y^2 + 8x - 12y - 10 = 0$

53. $2x^2 - 2xy + 3y^2 + 8x - 6y - 4 = 0$

SECTION 10.5

For Exercises 54–59,

a. Identify the type of conic represented by the equation.

b. Give the eccentricity and write an equation for the directrix in rectangular coordinates.

c. Give the coordinates of the vertex or vertices in polar coordinates.

d. Sketch the graph.

54. $r = \dfrac{3}{2 - 2\sin\theta}$

55. $r = \dfrac{2}{3 - 2\sin\theta}$

56. $r = \dfrac{2}{2 - 3\sin\theta}$

57. $r = \dfrac{1}{1 + 2\cos\theta}$

58. $r = \dfrac{1}{2 + \cos\theta}$

59. $r = \dfrac{2}{1 + \cos\theta}$

60. Write an equation of the parabola $r = \dfrac{10}{4 + 4\sin\theta}$ rotated about the origin through the given angle.

a. Counterclockwise $\dfrac{2\pi}{3}$ **b.** Clockwise $\dfrac{3\pi}{4}$

61. Write an equation of the hyperbola $r = \dfrac{8}{4 - 5\cos\theta}$ rotated about the origin through the given angle.

a. Counterclockwise $\dfrac{\pi}{3}$ b. Clockwise $\dfrac{\pi}{2}$

For Exercises 62–63, write a polar equation of a conic subject to the given conditions.

62. $e = \dfrac{3}{4}$, Directrix $x = 4$ $r = \dfrac{3}{1 + \frac{3}{4}\cos\theta}$

63. $e = 1$, Directrix is parallel to the polar axis and 3 units below the pole. $r = \dfrac{3}{1 - \sin\theta}$

SECTION 10.6

64. $x = |4 - t|$ and $y = t - 2$ for $1 \le t < 5$

a. Use the parametric equations to compute x and y for the given values of t.

t	1	2	3	4	5
x	3	2	1	0	1
y	-1	0	1	2	3

b. Plot the points, sketch the curve, and indicate the orientation of the curve.

65. $x = 4 - t^2$ and $y = -2t$ for $-2 \le t < 2$

Sketch the plane curve by plotting points. Indicate the orientation of the curve.

For Exercises 66–70,

a. Eliminate the parameter and write an equation in rectangular coordinates.

b. Sketch the curve and indicate its orientation.

66. $x = t^4 + 1$ and $y = t^2$ a. $y = \sqrt{x - 1}$

67. $x = 2\sqrt{t} - 4$ and $y = \sqrt{t} + 1$ a. $y = \frac{1}{2}x + 3, x \ge -4$

68. $x = -7\sin\theta$ and $y = 7\cos\theta$ a. $x^2 + y^2 = 49$

69. $x = 2 + 3\cos\theta$ and $y = -1 + 4\sin\theta$

70. $x = 4\cos^2 t$ and $y = 4\sin^2 t$

71. For each given curve, the parameter t represents the time over which an object traverses the curve. Describe the curves and discuss the motion along the curves.

C_1: $x = 2\cos\theta$ and $y = 4\sin\theta, 0 \le \theta \le \pi$

C_2: $x = -2\sin\theta$ and $y = -4\cos\theta, 0 \le \theta \le \pi$

72. An object undergoes uniform linear motion on a path from point $A(10, -2)$ to point $B(-6, 6)$ in 4 sec. Write parametric equations over an interval I that describe the motion along the path.
$x = 10 - 4t$ and $y = -2 + 2t, 0 \le t \le 4$

73. Suppose that the origin on a computer screen is at the lower left corner of the screen. A target moves along a straight line at a constant speed from point $A(52, 410)$ to point $B(412, 140)$ in 3 sec.

a. Write parametric equations to represent the target's path as a function of the time t (in sec) after the target leaves its initial position. All distances are in pixels.
$x = 52 + 120t$ and $y = 410 - 90t, 0 \le t \le 3$

b. Where is the target located 0.8 sec after motion starts?
(148, 338)

c. Suppose that a bullet is fired from a gun at a position of (212, 150). Suppose the bullet travels with uniform speed where both the horizontal component of velocity and vertical component of velocity is 40 pixels per second in the positive direction. Write parametric equations to represent the path of the bullet.
$x = 212 + 40t$ and $y = 150 + 40t$

d. If the bullet leaves the gun at the same time that the target begins its motion, will the bullet hit the target? If so, when and where?   Yes. The bullet will hit the target after 2 sec at a position of (292, 230).

74. A stunt man drives a car at a speed of 25 m/sec off a 10-m cliff. The road leading to the edge of the cliff is inclined upward at an angle of 16°. Choose a coordinate system with the origin at the base of the cliff directly under the point where the car leaves the edge.

a. Write parametric equations defining the path of the car.
$x = (25\cos 16°)t$ and $y = -4.9t^2 + (25\sin 16°)t + 10$

b. How long is the car in the air? Round to the nearest tenth of a second.   2.3 sec

c. How far from the base of the cliff will the car land? Round to the nearest foot.   55 ft

75. Suppose that a baseball is thrown at an angle of 30° with an initial speed of 88 ft/sec from an initial height of 3 ft. Choose a coordinate system with the origin at ground level directly under the point of release.

a. Write parametric equations defining the path of the ball. $x = 44\sqrt{3}t$ and $y = -16t^2 + 44t + 3$

b. When will the ball reach its highest point?
1.375 sec after release

c. Determine the coordinates of the ball at its highest point. Give the exact values and the coordinates rounded to the nearest tenth of a foot.
$(60.5\sqrt{3}, 33.25) \approx (104.8, 33.25)$

d. If another player catches the ball at a height of 4 ft on the way down, how long was the ball in the air? Round to the nearest hundredth of a second.   2.73 sec

e. How far apart are the two players? Round to the nearest foot.   208 ft

f. Eliminate the parameter and write an equation in rectangular coordinates to represent the path of the ball. $y = -\dfrac{1}{363}x^2 + \dfrac{\sqrt{3}}{3}x + 3$

CHAPTER 10   Test

1. How are the graphs of these ellipses similar, and how are they different?

$$\frac{x^2}{36} + \frac{y^2}{144} = 1 \qquad \frac{x^2}{144} + \frac{y^2}{36} = 1$$

2. Write the standard form of an equation of a circle with center (h, k) and radius r. $(x - h)^2 + (y - k)^2 = r^2$

3. Write the standard form of an equation of an ellipse with center (h, k) and major axis horizontal. $\frac{(x - h)^2}{a^2} + \frac{(y - k)^2}{b^2} = 1$

4. Write the standard form of an equation of an ellipse with center (h, k) and major axis vertical. $\frac{(x - h)^2}{b^2} + \frac{(y - k)^2}{a^2} = 1$

5. Write the standard form of an equation of a hyperbola with center (h, k) and transverse axis vertical.

6. Write the standard form of an equation of a hyperbola with center (h, k) and transverse axis horizontal.

7. Write the standard form of an equation of a parabola with vertex (h, k) and vertical axis of symmetry.
   $(x - h)^2 = 4p(y - k)$

8. Write the standard form of an equation of a parabola with vertex (h, k) and horizontal axis of symmetry.
   $(y - k)^2 = 4p(x - h)$

For Exercises 9–16,

a. Identify the equation as representing a circle, an ellipse, a hyperbola, or a parabola.

b. Graph the curve.

c. Identify key features of the graph. That is,

- If the equation represents a circle, identify the center and radius.
- If the equation represents an ellipse, identify the center, vertices, endpoints of the minor axis, foci, and eccentricity.
- If the equation represents a hyperbola, identify the center, vertices, foci, equations of the asymptotes, and eccentricity.
- If the equation represents a parabola, identify the vertex, focus, endpoints of the latus rectum, equation of the directrix, and equation of the axis of symmetry.

9. $\frac{(x - 2)^2}{9} + \frac{y^2}{16} = 1$

10. $\frac{(x - 2)^2}{9} - \frac{y^2}{16} = 1$

11. $\frac{(x - 2)^2}{16} + y = 0$

12. $x^2 + y^2 - 4x - 6y + 1 = 0$

13. $-9x^2 + 16y^2 + 64y - 512 = 0$

14. $9x^2 + 25y^2 + 72x - 50y - 731 = 0$

15. $\frac{(x + 4)^2}{4} + \frac{(y - 1)^2}{4} = 1$ 16. $y^2 - 8y - 8x + 40 = 0$

17. The entrance to a tunnel is in the shape of a semiellipse over a 24-ft by 8-ft rectangular opening. The height at the center of the opening is 14 ft.

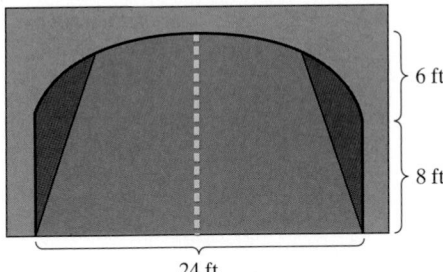

a. Determine the height of the opening at a point 6 ft from the edge of the tunnel. Round to 1 decimal place.
   13.2 ft

b. Can a 10-ft-high truck pass through the opening if the truck's outer wheel passes at a point 3 ft from the edge of the tunnel?   Yes; The height of the opening is approximately 12 ft at a point 3 ft from the edge.

c. Set up a coordinate system with the origin placed on the ground in the center of the roadway. Write a function that represents the height of the opening $h(x)$ (in ft) as a function of the horizontal distance x (in ft) from the center. $h(x) = 6\sqrt{1 - \frac{x^2}{144}} + 8$

18. Neptune orbits the Sun in an elliptical path with the Sun at one focus. At aphelion, Neptune is farthest from the Sun at 4.546×10^9 km. If the eccentricity of the orbit is 0.0113, determine the distance at perihelion (closest point between Neptune and the Sun). Round to the nearest million km. 4.444×10^9 km

19. Suppose that two people standing 1.5 mi (7920 ft) apart at points A and B hear a car accident. The time difference between the sound heard at point A and the sound heard at point B is 6 sec. If sound travels 1100 ft/sec, find an equation of the hyperbola (with foci at A and B) at which the accident occurred. $\frac{x^2}{(3300)^2} - \frac{y^2}{(2189)^2} = 1$

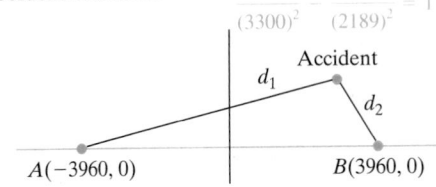

20. A building in the shape of a hyperboloid has hyperbolic cross sections perpendicular to the ground. Placing the origin at the bottom and center of the building, the center of a hyperbolic cross section is (0, 120) with one vertex at (12, 120) and one focus at $(12\sqrt{10}, 90)$. All units are in feet.

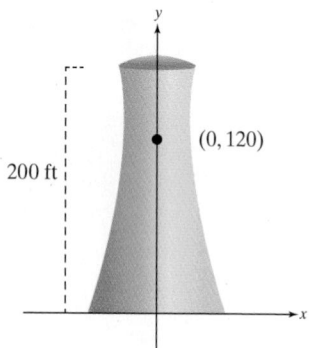

a. Write an equation of a hyperbolic cross section through the origin. Assume that there are no restrictions on x or y. $\frac{x^2}{144} - \frac{(y-120)^2}{1296} = 1$

b. Determine the diameter of the building at the base. Round to the nearest tenth of a foot. 83.5 ft

c. Determine the diameter of the building at the top. Round to the nearest tenth of a foot. 58.5 ft

21. The middle span of the Golden Gate Bridge in San Francisco is supported by cables from two towers 4200 ft apart. The tops of the towers are 500 ft above the roadway. Assuming that the cables are parabolic and are at road level halfway between the towers, find the height of the cables 1000 ft from the center of the bridge. Round to the nearest foot. 113 ft

22. Write the standard form of an equation of a parabola with focus (1, 6) and directrix $y = -2$. $(x-1)^2 = 16(y-2)$

23. Write the standard form of an equation of an ellipse with vertices (2, −3) and (6, −3) and foci (3, −3) and (5, −3).

24. Write the standard form of an equation of a hyperbola with vertices (4, −1) and (4, 7) and foci (4, −3) and (4, 9).

25. Write the standard form of an equation of a hyperbola with vertices (−1, −3) and (5, −3) and slope of one asymptote $\frac{4}{3}$. $\frac{(x-2)^2}{9} - \frac{(y+3)^2}{16} = 1$

26. If the major axis of an ellipse is 26 units in length and the minor axis is 10 units in length, determine the eccentricity of the ellipse. $\frac{12}{13}$

27. If the transverse axis of a hyperbola is 20 units and the conjugate axis is 12 units, determine the eccentricity of the hyperbola. $\frac{\sqrt{34}}{5}$

28. If the directrix of a parabola is $x = 2$, and the focus is (−4, 0), determine the focal length and the focal diameter. Focal length: 3; Focal diameter: 12

29. If the asymptotes of a hyperbola are $y = \frac{3}{5}x + 1$ and $y = -\frac{3}{5}x + 1$, identify the center of the hyperbola. Center: (0, 1)

30. Solve the system of equations.

$$3x^2 - 4y^2 = -13$$
$$5x^2 + 2y^2 = 13$$

$\{(1, 2), (1, -2), (-1, 2), (-1, -2)\}$

31. Describe the graph of the equation.

$$x = -4\sqrt{1 - \frac{y^2}{9}}$$

32. Each equation represents a conic section (nondegenerative case). Identify the type of conic section.

a. $50x^2 - 20xy + 2y^2 - 100x + 8y - 9 = 0$ Parabola

b. $4x^2 - 16x = -9y^2 + 18y + 4$ Ellipse

c. $x^2 + 4xy = -2y^2 + 2x - 16y + 4$ Hyperbola

33. Identify the conic each equation represents and give the eccentricity.

a. $r = \dfrac{6}{2 + 3\sin\theta}$ Hyperbola; $e = \dfrac{3}{2}$

b. $r = \dfrac{10}{3 - 3\sin\theta}$ Parabola; $e = 1$

c. $r = \dfrac{12}{6 + 3\cos\theta}$ Ellipse; $e = \dfrac{1}{2}$

For Exercises 34–35, assume that the x- and y-axes are rotated through angle θ about the origin to form the x'- and y'-axes.

34. Given $\theta = 30°$ and $P(8\sqrt{3}, -2\sqrt{3})$ in the xy-plane, find the coordinates of P in the x'y'-plane. $(12 - \sqrt{3}, -4\sqrt{3} - 3)$

35. Given $\theta = 45°$, write the equation $x^2 + 2xy + y^2 + 3\sqrt{2}x + \sqrt{2}y + 2 = 0$ in x'y'-coordinates. $x'^2 + 2x' - y' + 1 = 0$

For Exercises 36–37, an equation of a conic (nondegenerative case) is given.

a. Identify the type of conic.

b. Determine an acute angle of rotation to eliminate the xy term. Round to the nearest tenth of a degree if necessary.

c. Use a rotation of axes to eliminate the xy term in the equation.

d. Sketch the graph.

36. $13x^2 - 10xy + 13y^2 - 24\sqrt{2}x - 24\sqrt{2}y = 0$

37. $8x^2 + 6xy + 9 = 0$

For Exercises 38–39, an equation of a conic is given.

a. Identify the type of conic represented by the equation.

b. Give the eccentricity and write an equation of the directrix in rectangular coordinates.

c. Give the coordinates of the vertex or vertices in polar coordinates.

d. Sketch the graph.

38. $r = \dfrac{8}{4 + 6\sin\theta}$

39. $r = \dfrac{14}{6 - 6\cos\theta}$

40. Write a polar equation of a conic with $e = 2$ directrix $y = 3$. $r = \dfrac{6}{1 + 2\sin\theta}$

For Exercises 41–42,

a. Eliminate the parameter.

b. Sketch the curve and show the orientation of the curve.

41. $x = 2 - t$ and $y = -2 + 3t$ for $-1 \leq t < 2$
  a. $y = -3x + 4, 0 < x \leq 3$

42. $x = 2 - 3\sin\theta$ and $y = 3\cos\theta$ a. $(x - 2)^2 + y^2 = 9$

43. Write parametric equations describing the motion of a particle moving under uniform linear motion from point $A(3, 4)$ to point $B(12, -2)$ in 3 sec.
  $x = 3 + 3t$ and $y = 4 - 2t$, for $0 \leq t \leq 3$

44. A pyrotechnic rocket is fired from a platform 2 ft high at an angle of $60°$ from the horizontal with an initial speed of 72 ft/sec. Choose a coordinate system with the origin at ground level directly below the launch position.

  a. Write parametric equations that model the path of the shell as a function of the time t (in sec) after launch.
    $x = 36t$ and $y = -16t^2 + 36\sqrt{3}t + 2$

  b. Approximate the time required for the shell to hit the ground. Round to the nearest hundredth of a second.
    3.93 sec

  c. Approximate the horizontal distance that the shell travels before it hits the ground. Round to the nearest foot.   141 ft

d. When is the shell at its maximum height? Find the exact value and an approximation to the nearest hundredth of a second. $t = \dfrac{9\sqrt{3}}{8} \approx 1.95$ sec

e. Determine the maximum height.   62.75 ft

45. The equation $5x^2 - 10xy + 5y^2 + 10x - 15y - 9 = 0$ represents a conic section (nondegenerative case).

  a. Identify the type of conic section.   Parabola

  b. Graph the equation on a graphing utility.

46. The equation $r = \dfrac{8}{4 - 2\cos\theta}$ represents a conic.

  a. Identify the type of conic.   Ellipse

  b. Give a new equation if the graph of the conic is rotated clockwise about the origin through the angle $\theta = \dfrac{5\pi}{6}$.

  c. Graph both equations on a graphing utility.

47. Use the parametric mode to graph the curve on a graphing utility.
    $x = \sin 4t$ and $y = 3\cos t, 0 \leq t \leq 2\pi$

CHAPTER 10   Cumulative Review Exercises

1. Use long division to divide.
  $(3x^4 - 2x^3 - 5x + 1) \div (x^2 + 3)$ $3x^2 - 2x - 9 + \dfrac{x + 28}{x^2 + 3}$

2. Is -4 a zero of the polynomial
  $f(x) = 3x^4 - 6x^3 + 5x - 12$?   No

3. Is $(x - 3)$ a factor of $2x^4 - 7x^3 + 8x^2 - 17x + 6$?   Yes

For Exercises 4–5, simplify completely.

4. $\dfrac{\sqrt[3]{54x^4}}{\sqrt[3]{2x^5}}$ $\dfrac{3\sqrt[3]{x^2}}{x}$

5. $\left(\dfrac{-2x^2y^{-1}}{z^3}\right)^{-4}\left(\dfrac{4x^{-5}}{y^7}\right)^2$ $\dfrac{z^{12}}{x^{18}y^{10}}$

6. Find the midpoint of the line segment with endpoints $(-362, 147)$ and $(118, 24)$. $\left(-122, \dfrac{171}{2}\right)$

7. Given $f(x) = \sqrt[3]{x} - 7$, write an equation defining $f^{-1}(x)$.
  $f^{-1}(x) = x^3 + 7$

8. A boat travels 10 mph at a bearing of N30°E in still water. A 2-mph current is directed due west. Determine the actual speed of the boat. Write the exact value and the speed rounded to the nearest tenth of a mph.
  $2\sqrt{21} \approx 9.2$ mph

9. Determine the asymptotes of the graph of
  $y = \csc\left(2x - \dfrac{\pi}{2}\right)$. $x = \left(\dfrac{2n + 1}{4}\right)\pi$

For Exercises 10–15, solve the equation or inequality. Write the solution set to the inequalities in interval notation if possible.

10. $e^{3x+2} = 22$ $\left\{\dfrac{-2 + \ln 22}{3}\right\}$

11. $2 - 3\{4 - [x + 2(x - 1) - 3] + 4\} > 0$ $\left(\dfrac{37}{9}, \infty\right)$

12. $21 \leq 9 - 2|x + 1|$ $\{\ \}$ **13.** $2x^2 - 8x + 3 \geq 0$

14. $\log_3 x + \log_3(x - 6) = 3$ **15.** $e^{2x} - 8e^x + 15 = 0$
  $\{9\}$; The value -3 does not check. $\{\ln 5, \ln 3\}$

16. Graph the solution set.
  $x^2 + (y - 1)^2 \leq 9$

17. Graph $y = 2\sin\left(x - \dfrac{\pi}{4}\right)$ over one completed period.

18. Find the partial fraction decomposition for
  $\dfrac{8x - 27}{x^2 - 7x + 12}$ $\dfrac{3}{x - 3} + \dfrac{5}{x - 4}$

19. Given $f(x) = \dfrac{1}{x - 2}$ and $g(x) = \sqrt{x - 1}$,

  a. Find $(f \circ g)(x)$. $(f \circ g)(x) = \dfrac{1}{\sqrt{x - 1} - 2}$

  b. Write the domain of $(f \circ g)(x)$ in interval notation.
    $[1, 5) \cup (5, \infty)$

20. Given $f(x) = \dfrac{x^2 - 4x}{x + 3}$,

     a. Determine the x-intercept(s) of the graph of f.   (0, 0), (4, 0)

     b. Determine the y-intercept.   (0, 0)

     c. Find the vertical asymptote(s). $x = -3$

     d. Find the horizontal asymptote.   None

     e. Find the slant asymptote. $y = x - 7$

For Exercises 21–24, graph the equations and functions.

21. $\dfrac{x^2}{4} + \dfrac{(y - 1)^2}{9} = 1$ **22.** $\dfrac{x^2}{4} - \dfrac{(y - 1)^2}{9} = 1$

23. $x^2 = 4(y - 2)$ **24.** $f(x) = \left(\dfrac{1}{2}\right)^x + 2$

For Exercises 25–26, solve the system.

25. $3x - 4y + 2z = 5$

           $5y - 3z = -12$

     $7x \qquad + 2z = 1$

     $\{(-1, 0, 4)\}$

26. $x^2 - 5y^2 = 4$

     $4x^2 + y^2 = 37$

     $\{(3, 1), (3, -1), (-3, 1), (-3, -1)\}$

For Exercises 27–29, refer to the matrices A and B. Simplify each expression.

$$A = \begin{bmatrix} 2 & 4 \\ -1 & 6 \end{bmatrix} \quad B = \begin{bmatrix} -3 & 8 \\ 0 & 4 \end{bmatrix}$$

27. $2AB$ $\begin{bmatrix} -12 & 64 \\ 6 & 32 \end{bmatrix}$

28. A^{-1} $\begin{bmatrix} \frac{3}{8} & -\frac{1}{4} \\ \frac{1}{16} & \frac{1}{8} \end{bmatrix}$

29. $|B|$ -12

30. Use Cramer's rule to solve for z.

     $3x - 4y + z = 3$

     $2x + 3y \qquad = 8$

           $5y + 8z = 11$

     $z = \dfrac{97}{146}$

Sequences, Series, Induction, and Probability

Chapter Outline

Have you ever wondered why one individual pays more than another for car insurance? Insurance rates are based on a number of variables including age, gender, Zip code, past driving and claims history, and even grades in school. Insurance companies hire actuaries, special statisticians who analyze data to estimate the probability and cost of accidents for people in various demographic groups. In turn, premiums are established to offset the theoretical cost to cover a customer based on the customer's risk factors.

In this chapter, we study sequences and series, the binomial theorem, and a technique of proof called mathematical induction. We complete the chapter by presenting several methods of counting that will ultimately enable us to calculate simple probabilities. The study of probability is the first step toward unraveling the mystery of car insurance premiums.

Sequences and Series

1. Write Terms of a Sequence from the *n*th Term

In day-to-day life, we think of a **sequence** as a progression of elements with some order or pattern. For example, suppose that a new college graduate has to choose between two job offers. Job 1 offers $75,000 the first year with a $4000 raise each year thereafter. Job 2 offers $75,000 with a 5% raise each year. The sequences in Table 11-1 represent the salaries for each job as a function of the year number.

Table 11-1

	Year 1	Year 2	Year 3	Year 4	Year 5	Year 6
Job 1	$75,000	$79,000	$83,000	$87,000	$91,000	$95,000
Job 2	$75,000	$78,750	$82,687.50	$86,821.86	$91,162.97	$95,721.12

By studying the two sequences, we see that by year 5, job 2 has a better salary. If we take the *sum* of the first *n* terms of a sequence, we have a **finite series.** The total salary through the first 5 yr favors job 1 ($415,000 for job 1 versus $414,422.33 for job 2). However, by the end of year 6, job 2 not only pays more each year, but has a greater total salary for the 6-yr period ($510,143.45 for job 2 versus $510,000 for job 1).

From Table 11-1, we can think of a sequence as a function pairing the year number (a positive integer) with the salary. For this reason, it makes sense to define a sequence mathematically as a function whose domain is the set of positive integers.

> ### Finite and Infinite Sequences
>
> An **infinite sequence** is a function whose domain is the set of positive integers. A **finite sequence** is a function whose domain is the set of the first *n* positive integers.

> **TIP** In this text, we address sequences with real-valued terms only.

> **TIP** Lowercase letters such as *a*, *b*, *c*, and so on are often used to represent the terms of a sequence.

Although a sequence is defined as a function, we typically do not use function notation to denote the terms of a sequence. Instead, the terms of the sequence are denoted by the letter *a* with a subscript representing the term number. The sequence of salaries for job 1 would be denoted by

$$\underset{\text{term 1}}{a_1 = 75{,}000} \qquad \underset{\text{term 2}}{a_2 = 79{,}000} \qquad \underset{\text{term 3}}{a_3 = 83{,}000} \qquad \text{and so on.}$$

The notation $a_1 = 75{,}000$ is used in place of function notation $f(1) = 75{,}000$. Likewise $a_2 = 79{,}000$ is used instead of $f(2) = 79{,}000$ and so on. For a positive integer n, the value a_n is called the ***n*th term** or **general term** of the sequence. The notation $\{a_n\}$ represents the entire sequence.

$$\{a_n\} = a_1, a_2, a_3, \ldots, a_n, \ldots$$

Classroom Examples: p. 1014
Exercises 8, 10

EXAMPLE 1 Writing Several Terms of a Sequence

Write the first four terms of the sequence defined by the nth term.

a. $a_n = 3n^2 - 4$ **b.** $c_n = 9\left(-\frac{1}{3}\right)^n$

Solution:

a. $a_n = 3n^2 - 4$ To find the first four terms, substitute $n = 1, 2, 3,$ and 4.

$a_1 = 3(1)^2 - 4 = -1$

$a_2 = 3(2)^2 - 4 = 8$

$a_3 = 3(3)^2 - 4 = 23$

$a_4 = 3(4)^2 - 4 = 44$ The first four terms are $-1, 8, 23, 44$.

b. $c_n = 9\left(-\frac{1}{3}\right)^n$ To find the first four terms, substitute $n = 1, 2, 3,$ and 4.

$c_1 = 9\left(-\frac{1}{3}\right)^1 = -3$

$c_2 = 9\left(-\frac{1}{3}\right)^2 = 1$

$c_3 = 9\left(-\frac{1}{3}\right)^3 = -\frac{1}{3}$

$c_4 = 9\left(-\frac{1}{3}\right)^4 = \frac{1}{9}$ The first four terms are $-3, 1, -\frac{1}{3}, \frac{1}{9}$.

Skill Practice 1   Write the first four terms of the sequence defined by the nth term.

a. $b_n = 2n + 3$ **b.** $d_n = (-1)^n \cdot n^2$

TECHNOLOGY CONNECTIONS

Displaying Terms of a Sequence

If the nth term of a sequence is known, a seq function on a graphing utility can display a list of terms. Finding the first four terms of the sequence from Example 1(b) is outlined here.

- Select the **MODE** key and highlight SEQ.
- Then access the seq function from the catalog or from the LIST menu (OPS submenu) and enter the parameters as prompted.
- Paste the seq command on the home screen and select **ENTER**.

$\text{seq}(9*(-1/3)\wedge n,\ n,\ 1,\ 4,\ 1) > \text{Frac}$

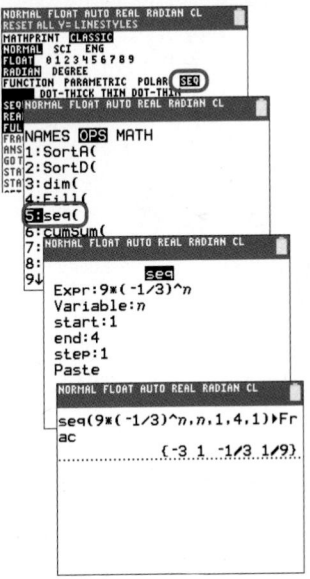

Notice that the terms of the sequence given in Example 1(b) alternate in signs. Such a sequence is called an **alternating sequence**. The alternation in signs can be represented in the nth term by a factor of -1 raised to a variable power that alternates between an even and odd integer. For example:

$a_n = (-1)^n$ odd-numbered terms, negative; even-numbered terms, positive

$b_n = (-1)^{n+1}$ odd-numbered terms, positive; even-numbered terms, negative

Answers

1. a. 5, 7, 9, 11   **b.** $-1, 4, -9, 16$

Because a sequence is defined only for positive integers, its graph is a discrete set of points (the points are not connected). Compare the graphs of the sequence defined by $a_n = 8\left(\frac{1}{2}\right)^n$ (see Figure 11-1) versus the function defined by $f(x) = 8\left(\frac{1}{2}\right)^x$ (see Figure 11-2).

Classroom Example: p. 1014
Exercise 28

> **TIP** In some cases, we may define a sequence with a domain beginning at zero or some other whole number.
>
> For example, $a_n = \dfrac{4}{n-1}$
>
> is not defined for $n = 1$. Therefore, we might restrict the domain to the set of integers greater than or equal to 2.

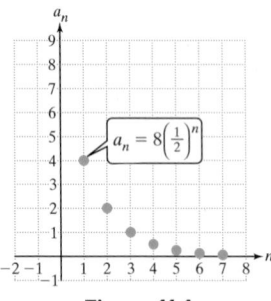

Figure 11-1

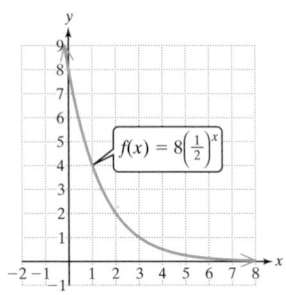

Figure 11-2

The graph of $\{a_n\}$ consists of a discrete set of points that correspond to points on the graph of $f(x) = 8\left(\frac{1}{2}\right)^x$ for positive integers x.

2. Write Terms of a Sequence Defined Recursively

The sequences in Example 1 were defined as a function of the nth term. Now we look at sequences defined *recursively*, using a recursive formula. A **recursive formula** defines the nth term of a sequence as a function of one or more terms preceding it. This is demonstrated in Example 2.

EXAMPLE 2	**Writing Terms of a Sequence Defined Recursively**

Write the first five terms of the sequence defined by $a_1 = 4$ and $a_n = 2a_{n-1} + 1$ for $n > 1$.

Solution:

The first term is given: $a_1 = 4$.

Every term thereafter is defined by $a_n = 2a_{n-1} + 1$

> Value of the nth term is

> two times the value of the preceding term, plus 1.

$a_1 = 4$

$a_2 = 2a_1 + 1 = 2(4) + 1 = 9$ Substitute $a_1 = 4$.

$a_3 = 2a_2 + 1 = 2(9) + 1 = 19$ Substitute $a_2 = 9$.

$a_4 = 2a_3 + 1 = 2(19) + 1 = 39$ Substitute $a_3 = 19$.

$a_5 = 2a_4 + 1 = 2(39) + 1 = 79$ The first five terms are 4, 9, 19, 39, 79.

Skill Practice 2   Write the first five terms of the sequence defined by:

$c_1 = 5, \quad c_n = 3c_{n-1} - 4$

Instructor Note:
Consider showing an example with a sequence defined by the nth term and defined recursively. For example:

$a_n = 7 + (n - 1)(2)$

and

$a_1 = 7, a_n = a_{n-1} + 2$

Both represent the same sequence.

Perhaps the most famous sequence is the Fibonacci sequence, named after the twelfth-century Italian mathematician Leonardo of Pisa (known as Fibonacci). The Fibonacci sequence is defined recursively as

$$a_1 = 1, \quad a_2 = 1, \quad a_n = a_{n-2} + a_{n-1}$$

Answer
2. 5, 11, 29, 83, 245

The first two terms of the Fibonacci sequence are 1, and each term thereafter is the sum of its two predecessors. We have:

Fibonacci sequence: 1, 1, 2, 3, 5, 8, 13, 21, 34, 55, 89, …

Point of Interest

The Fibonacci sequence is of particular interest because it is often observed in nature. For example, male honey bees hatch from eggs that have not been fertilized, so each male honey bee has only 1 parent, a female. A female honey bee hatches from a fertilized egg, so she has two parents: one male and one female. Notice that the number of ancestors for n generations for a male honey bee follows the Fibonacci sequence (Figure 11-3).

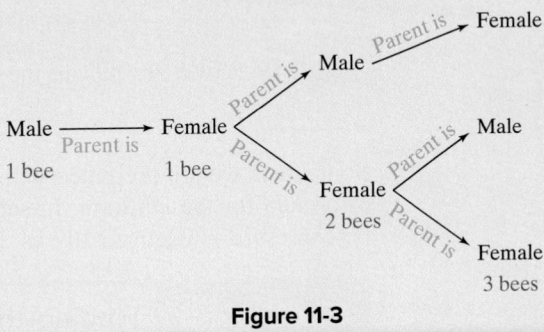

Figure 11-3

3. Use Factorial Notation

We now introduce factorial notation to denote the product of the first n positive integers.

Definition of $n!$

- Let n be a positive integer. The expression $n!$ (read as "n factorial") is defined as $n! = (n)(n-1)(n-2) \cdots (2)(1)$
- Zero factorial is defined as $0! = 1$.

Examples:

$0! = 1$

$1! = 1$

$2! = 2 \cdot 1 = 2$

$3! = 3 \cdot 2 \cdot 1 = 6$

$4! = 4 \cdot 3 \cdot 2 \cdot 1 = 24$

$5! = 5 \cdot 4 \cdot 3 \cdot 2 \cdot 1 = 120$

$6! = 6 \cdot 5 \cdot 4 \cdot 3 \cdot 2 \cdot 1 = 720$

$7! = 7 \cdot 6 \cdot 5 \cdot 4 \cdot 3 \cdot 2 \cdot 1 = 5040$

Classroom Examples: p. 1015
Exercises 40, 42

EXAMPLE 3   Evaluating Expressions with Factorial Notation

Evaluate.

a. $\dfrac{8!}{3! \cdot 5!}$

b. $\dfrac{n!}{(n+2)!}$

Solution:

a. $\dfrac{8!}{3! \cdot 5!} = \dfrac{8 \cdot 7 \cdot \overset{1}{\cancel{6}} \cdot \overset{1}{\cancel{5!}}}{(\cancel{3} \cdot 2 \cdot 1)(\cancel{5!})} = 56$

b. $\dfrac{n!}{(n+2)!} = \dfrac{\overset{1}{\cancel{n!}}}{(n+2)(n+1)(\cancel{n!})} = \dfrac{1}{(n+2)(n+1)}$

Skill Practice 3   Evaluate.

a. $\dfrac{9!}{2! \cdot 7!}$

b. $\dfrac{(n-1)!}{n!}$

Answers

3. a. 36   **b.** $\dfrac{1}{n}$

In Example 4, we find a specific term of a sequence in which the expression for the *n*th term contains factorial notation.

Classroom Example: p. 1015
Exercise 46

EXAMPLE 4   Finding a Specific Term of a Sequence

Given the sequence defined by $b_n = \dfrac{n^2}{(n+1)!}$, find b_6.

Solution:

$$b_n = \frac{n^2}{(n+1)!} \xrightarrow[\text{6 for } n.]{\text{Substitute}} b_6 = \frac{(6)^2}{(6+1)!} = \frac{36}{7!} = \frac{36}{7 \cdot 6 \cdot 5 \cdot 4 \cdot 3 \cdot 2 \cdot 1} = \frac{1}{140}$$

Skill Practice 4   Given the sequence defined by $c_n = \dfrac{2^n}{(n+1)!}$, find c_4.

Sometimes we are presented with several terms of a sequence and are asked to find a formula for the *n*th term. In such a case, the goal is to look for a pattern that can be expressed mathematically as a function of the term number.

Classroom Example: p. 1015
Exercise 52

EXAMPLE 5   Finding the *n*th Term of a Sequence

Find the *n*th term a_n of a sequence whose first four terms are given.

a. $\dfrac{2}{3}, \dfrac{3}{4}, \dfrac{4}{5}, \dfrac{5}{6}, \ldots$ **b.** $\dfrac{1}{5}, -\dfrac{4}{25}, \dfrac{9}{125}, -\dfrac{16}{625}, \ldots$

Solution:

a. Term number: $n = 1\ \ 2\ \ 3\ \ 4$

$\dfrac{2}{3}, \dfrac{3}{4}, \dfrac{4}{5}, \dfrac{5}{6}, \ldots$

$a_n = \dfrac{n+1}{n+2}$

The numerator is 1 more than the term number: $n + 1$

The denominator is 2 more than the term number: $n + 2$

b. Term number: $n = 1\ \ \ \ 2\ \ \ \ 3\ \ \ \ 4$

$\dfrac{1}{5}, -\dfrac{4}{25}, \dfrac{9}{125}, -\dfrac{16}{625}, \ldots$

$a_n = (-1)^{n+1} \cdot \dfrac{n^2}{5^n}$

Terms in the numerator are perfect squares: $1^2, 2^2, 3^2, 4^2, \ldots, n^2$

Terms in the denominator are powers of 5: $5^1, 5^2, 5^3, 5^4, \ldots, 5^n$

The signs are positive for odd-numbered terms. Therefore, we want a factor of -1 raised to an even exponent on odd-numbered terms and an odd exponent for even-numbered terms: $(-1)^{n+1}$

Skill Practice 5   Find the *n*th term a_n of a sequence whose first four terms are given.

a. $-5, 6, -7, 8, \ldots$ **b.** $\dfrac{1}{3}, \dfrac{1 \cdot 2}{9}, \dfrac{1 \cdot 2 \cdot 3}{27}, \dfrac{1 \cdot 2 \cdot 3 \cdot 4}{81}, \ldots$

Answers

4. $\dfrac{2}{15}$

5. a. $a_n = (-1)^n (n+4)$

b. $a_n = \dfrac{n!}{3^n}$

4. Use Summation Notation

Consider an infinite sequence $\{a_n\} = a_1, a_2, a_3, \ldots$. The sum S of such a sequence is called an **infinite series** and is represented by $S = a_1 + a_2 + a_3 + \cdots$. The sum of the first *n* terms of the sequence is called the *n*th **partial sum** of the sequence and is denoted by S_n. The *n*th partial sum of a sequence is a **finite series.**

In mathematics, the Greek letter Σ (sigma) is used to represent summations.

> ### Summation Notation
>
> Given a sequence $a_1, a_2, a_3, \ldots$, the nth partial sum S_n is a finite series and is represented by
>
> n is the upper limit of summation.
>
> $$S_n = \sum_{i=1}^{n} a_i = a_1 + a_2 + a_3 + \cdots + a_n$$
>
> i is the index of summation.   1 is the lower limit of summation.
>
> The sum of *all* terms in the sequence is an infinite series and is given by
>
> $$\sum_{i=1}^{\infty} a_i = a_1 + a_2 + a_3 + \cdots$$

TIP The summation $\sum_{i=1}^{n} a_i$ is read as the sum of a_i for i equals 1 to n.

Any letter such as i, j, k, and n may be used for the index of summation. However, if i is used for the index of summation, do not confuse it with the imaginary number $i = \sqrt{-1}$.

Classroom Examples: p. 1015
Exercises 58, 62, 64

EXAMPLE 6   Using Summation Notation

Write the terms for each series and evaluate the sum.

a. $\displaystyle\sum_{i=1}^{5} (2i + 3)$ **b.** $\displaystyle\sum_{k=3}^{6} (-1)^k \left(\frac{1}{k}\right)$ **c.** $\displaystyle\sum_{n=1}^{6} 3$

Solution:

a.
$$\sum_{i=1}^{5} (2i + 3) = \overset{i=1}{[2(1)+3]} + \overset{i=2}{[2(2)+3]} + \overset{i=3}{[2(3)+3]} + \overset{i=4}{[2(4)+3]} + \overset{i=5}{[2(5)+3]}$$
$$= 5 + 7 + 9 + 11 + 13$$
$$= 45$$

b.
$$\sum_{k=3}^{6} (-1)^k \left(\frac{1}{k}\right) = \left[(-1)^3\left(\frac{1}{3}\right)\right] + \left[(-1)^4\left(\frac{1}{4}\right)\right] + \left[(-1)^5\left(\frac{1}{5}\right)\right] + \left[(-1)^6\left(\frac{1}{6}\right)\right]$$
$$= -\frac{1}{3} + \frac{1}{4} - \frac{1}{5} + \frac{1}{6}$$
$$= -\frac{7}{60}$$

c.
$$\sum_{n=1}^{6} 3 = \underbrace{3 + 3 + 3 + 3 + 3 + 3}_{6 \text{ terms}} = 18$$

> **Skill Practice 6**   Evaluate the sum.
>
> **a.** $\displaystyle\sum_{i=1}^{5} (3i + 1)$ **b.** $\displaystyle\sum_{n=2}^{5} (-1)^{n+1} \frac{1}{2^n}$ **c.** $\displaystyle\sum_{k=1}^{5} 4$

Answers
6. a. 50   **b.** $-\dfrac{5}{32}$ **c.** 20

TECHNOLOGY CONNECTIONS

Evaluating a Finite Series

A graphing utility can be used to evaluate a finite series if the *n*th term of the corresponding sequence is known. The Σ symbol can be accessed by selecting (ALPHA) F2.

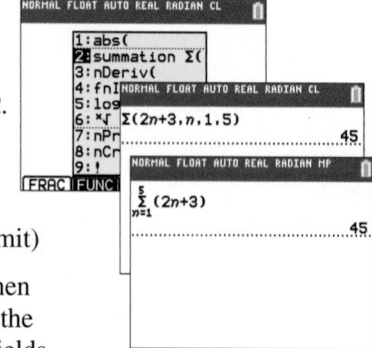

- If the calculator is in "Classic" mode, the sum can be evaluated by entering

  Σ (*n*th term, variable, lower limit, upper limit)

- If the calculator is in "Mathprint" mode, then the user is prompted to enter the *n*th term, the variable, and limits of summation in text fields.

It is important to note that the lower limit of summation need not be 1. In Example 6(b) for instance, the index of summation ranges from 3 to 6. In fact, sometimes we adjust the limits of summation and the expression being summed to write the summation in a more convenient form. For example, the following three series are all equivalent.

$$\sum_{i=1}^{5} 3i = 3(1) + 3(2) + 3(3) + 3(4) + 3(5)$$
$$= 3 + 6 + 9 + 12 + 15 = 45$$

$$\sum_{j=0}^{4} 3(j + 1) = 3(0 + 1) + 3(1 + 1) + 3(2 + 1) + 3(3 + 1) + 3(4 + 1)$$
$$= 3 + 6 + 9 + 12 + 15 = 45$$

$$\sum_{k=2}^{6} 3(k - 1) = 3(2 - 1) + 3(3 - 1) + 3(4 - 1) + 3(5 - 1) + 3(6 - 1)$$
$$= 3 + 6 + 9 + 12 + 15 = 45$$

Classroom Examples: pp. 1015–1016
Exercises 76, 86

EXAMPLE 7   Write a Series Using Summation Notation

Write each sum using summation notation.

a. $\dfrac{\sqrt{3}}{1!} + \dfrac{\sqrt{4}}{2!} + \dfrac{\sqrt{5}}{3!} + \cdots + \dfrac{\sqrt{n + 2}}{n!}$ **b.** $\dfrac{2}{1} - \dfrac{3}{4} + \dfrac{4}{9} - \dfrac{5}{16} + \dfrac{6}{25}$

Solution:

a. $\dfrac{\sqrt{3}}{1!} + \dfrac{\sqrt{4}}{2!} + \dfrac{\sqrt{5}}{3!} + \cdots + \dfrac{\sqrt{n + 2}}{n!}$ The series consists of *n* terms where a formula for the *n*th term is given.

$= \displaystyle\sum_{i=1}^{n} \dfrac{\sqrt{i + 2}}{i!}$ We have used *n* as the upper limit of summation. Therefore, use a different variable for the index of summation. We have chosen *i*.

b.   Look for a relationship between the term number and the values in the numerator and denominator. Choose a variable such as *i* for the index of summation.

Term: 1   2   3   4   5

$\dfrac{2}{1} \quad - \dfrac{3}{4} \quad + \dfrac{4}{9} \quad - \dfrac{5}{16} \quad + \dfrac{6}{25}$ ← The numerator is 1 more than the term number: $(i + 1)$
    ← The denominator is the square of the term number: i^2

$\displaystyle\sum_{i=1}^{5} (-1)^{i+1} \cdot \dfrac{i + 1}{i^2}$ The odd-numbered terms are positive. This can be generated by the factor $(-1)^{i+1}$.

Skill Practice 7   Write each sum using summation notation.

a. $\dfrac{2!}{7} + \dfrac{3!}{14} + \dfrac{4!}{21} + \cdots + \dfrac{(n+1)!}{7n}$

b. $-\dfrac{1}{5} + \dfrac{8}{25} - \dfrac{27}{125} + \dfrac{64}{625}$

Several important properties of summation are given in Table 11-2.

Table 11-2   Properties of Summation

If $\{a_n\}$ and $\{b_n\}$ are sequences, and c is a real number, then:

1. $\displaystyle\sum_{i=1}^{n} c = cn$ Adding a constant c a total of n times equals cn.

2. $\displaystyle\sum_{i=1}^{n} ca_i = c\sum_{i=1}^{n} a_i$ A constant factor can be factored out of a summation.

3. $\displaystyle\sum_{i=1}^{n} (a_i \pm b_i) = \sum_{i=1}^{n} a_i \pm \sum_{i=1}^{n} b_i$ A single sum or difference can be regrouped as two sums or differences.

The proof of Property 2 follows from the distributive property of real numbers.

$$\sum_{i=1}^{n} ca_i = ca_1 + ca_2 + ca_3 + \cdots + ca_n \qquad \text{Expand the terms in the series.}$$

$$= c(a_1 + a_2 + a_3 + \cdots + a_n) \qquad \text{Apply the distributive property.}$$

$$= c\sum_{i=1}^{n} a_i \qquad \text{Write the sum using summation notation.}$$

The proofs of properties 1 and 3 are examined in Exercises 89 and 90.

Answers

7. a. $\displaystyle\sum_{i=1}^{n} \dfrac{(i+1)!}{7i}$ **b.** $\displaystyle\sum_{i=1}^{4} (-1)^i \dfrac{i^3}{5^i}$

SECTION 11.1   Practice Exercises

Prerequisite Review

For Exercises R.1–R.3, evaluate the function for the given value.

R.1. $g(x) = \dfrac{5}{x}$; $g(15)$ $\dfrac{1}{3}$

R.2. $f(x) = 5$; $f(4)$ 5

R.3. $h(x) = x^2 + 3x$; $h(3)$ 18

For Exercises R.4–R.5, simplify the rational expression.

R.4. $\dfrac{(r-2)^2(2r-1)^4}{(r-2)^4(2r-1)}$ $\dfrac{(2r-1)^3}{(r-2)^2}$

R.5. $\dfrac{p^2(p-6)^5}{p^5(p-6)^2}$ $\dfrac{(p-6)^3}{p^3}$

Concept Connections

1. An infinite ___sequence___ is a function whose domain is the set of positive integers. A ___finite___ sequence is a function whose domain is the set of the first n positive integers.

2. An ___alternating___ sequence is a sequence in which consecutive terms alternate in sign.

3. A ___recursive___ formula defines the nth term of a sequence as a function of one or more terms preceding it.

4. For $n \geq 1$, the expression ___$n!$___ represents the product of the first n positive integers $n(n-1)(n-2)\cdots(2)(1)$. Furthermore, by definition, $0! =$ ___1___.

5. Given $\sum\limits_{i=1}^{n} a_i$, the variable i is called the ___index___ of ___summation___. The value 1 is called the ___lower___ limit of summation. The value n is called the upper ___limit___ of summation.

6. One property of summation indicates that $\sum\limits_{i=1}^{n} c = $ ___cn___ .

Objective 1: Write Terms of a Sequence from the nth Term

For Exercises 7–14, the nth term of a sequence is given. Write the first four terms of the sequence. (See Example 1)

7. $a_n = 2n^2 + 3$ 5, 11, 21, 35

8. $b_n = -n^3 + 5$ 4, −3, −22, −59

9. $c_n = 12\left(-\frac{1}{2}\right)^n$ −6, 3, −$\frac{3}{2}$, $\frac{3}{4}$

10. $d_n = 64\left(-\frac{1}{4}\right)^n$ −16, 4, −1, $\frac{1}{4}$

11. $a_n = \sqrt{n+3}$ 2, $\sqrt{5}$, $\sqrt{6}$, $\sqrt{7}$

12. $b_n = \sqrt{n-1}$ 0, 1, $\sqrt{2}$, $\sqrt{3}$

13. $a_n = e^{3\ln n}$ 1, 8, 27, 64

14. $c_n = \ln e^{2n}$ 2, 4, 6, 8

15. If the nth term of a sequence is $(-1)^n n^2$, which terms are positive and which are negative?
The odd-numbered terms are negative, and the even-numbered terms are positive.

16. If the nth term of a sequence is $(-1)^{n+1} \dfrac{1}{n}$, which terms are positive and which are negative?
The odd-numbered terms are positive, and the even-numbered terms are negative.

For Exercises 17–20, the nth term of a sequence is given. Find the indicated term.

17. $a_n = 2^n - 1$; find a_{10} 1023

18. $b_n = \dfrac{5}{n} - 1$; find b_{20} −$\frac{3}{4}$

19. $c_n = 5n - 4$; find c_{157} 781

20. $d_n = 6n + 7$; find d_{204} 1231

For Exercises 21–24, match the sequence or function with its graph.

21. $a_n = \sqrt{n}$ c

22. $f(x) = \sqrt{x}$ b

23. $f(x) = \dfrac{3}{x}$ a

24. $a_n = \dfrac{3}{n}$ d

a.
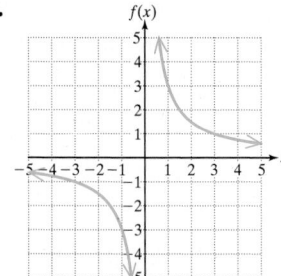

b.
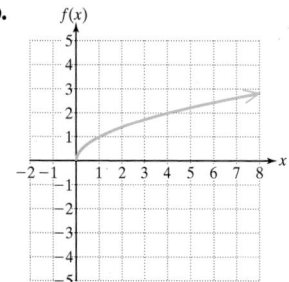

c.
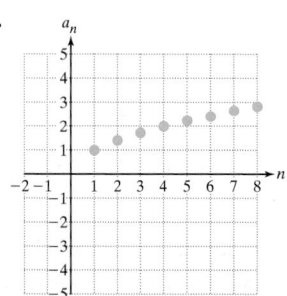

d.
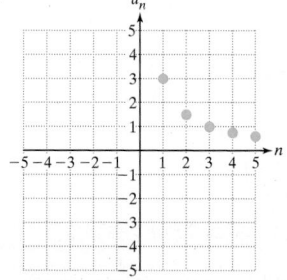

Objective 2: Write Terms of a Sequence Defined Recursively

For Exercises 25–30, write the first five terms of the sequence defined recursively. (See Example 2)

25. $b_1 = -3$; $b_n = 3b_{n-1} + 4$
−3, −5, −11, −29, −83

26. $a_1 = 11$; $a_n = 4a_{n-1} + 3$
11, 47, 191, 767, 3071

27. $c_1 = 4$; $c_n = \frac{1}{2}c_{n-1} + 4$
4, 6, 7, $\frac{15}{2}$, $\frac{31}{4}$

28. $d_1 = 30$; $d_n = \frac{1}{3}d_{n-1} - 1$
30, 9, 2, −$\frac{1}{3}$, −$\frac{10}{9}$

29. $a_1 = 5$; $a_n = \dfrac{1}{a_{n-1}}$
5, $\frac{1}{5}$, 5, $\frac{1}{5}$, 5

30. $b_1 = 10$; $b_n = -\dfrac{1}{b_{n-1}}$
10, −$\frac{1}{10}$, 10, −$\frac{1}{10}$, 10

31. a. Write the first 10 terms of the Fibonacci sequence.
1, 1, 2, 3, 5, 8, 13, 21, 34, 55

  b. The formula $F_n = \dfrac{1}{\sqrt{5}}\left(\dfrac{1+\sqrt{5}}{2}\right)^n - \dfrac{1}{\sqrt{5}}\left(\dfrac{1-\sqrt{5}}{2}\right)^n$ gives the nth term of the Fibonacci sequence. Use a calculator to verify this statement for $n = 1$, $n = 2$, and $n = 3$. $F_1 = 1$, $F_2 = 1$, $F_3 = 2$

32. The numbers in the sequence defined by $a_1 = 1$, $a_2 = 3$, and $a_n = a_{n-1} + a_{n-2}$ for $n \geq 3$ are referred to as Lucas numbers in honor of French mathematician Edouard Lucas (1842–1891).

  a. Find the first eight Lucas numbers. 1, 3, 4, 7, 11, 18, 29, 47

  b. The formula $L_n = \left(\dfrac{1+\sqrt{5}}{2}\right)^n + \left(\dfrac{1-\sqrt{5}}{2}\right)^n$ gives the nth Lucas number. Use a calculator to verify this statement for $n = 1$, $n = 2$, and $n = 3$.
  $L_1 = 1$, $L_2 = 3$, $L_3 = 4$

Objective 3: Use Factorial Notation

For Exercises 33–44, evaluate the expression. (See Example 3)

33. $7!$ 5040

34. $8!$ 40,320

35. $0!$ 1

36. $1!$ 1

37. $\dfrac{8!}{6!}$ 56

38. $\dfrac{12!}{9!}$ 1320

39. $\dfrac{9!}{5! \cdot 4!}$ 126

40. $\dfrac{10!}{6! \cdot 4!}$ 210

41. $\dfrac{(n-1)!}{(n+1)!}$ $\dfrac{1}{n(n+1)}$

42. $\dfrac{(n-2)!}{n!}$ $\dfrac{1}{n(n-1)}$

43. $\dfrac{(2n)!}{(2n+1)!}$ $\dfrac{1}{2n+1}$

44. $\dfrac{(2n-1)!}{(2n)!}$ $\dfrac{1}{2n}$

For Exercises 45–48, the nth term of a sequence is given. Find the indicated term. (See Example 4)

45. $a_n = \dfrac{2^n}{(n+2)!}$; find a_5 $a_5 = \dfrac{2}{315}$

46. $b_n = \dfrac{4n}{(n+1)!}$; find b_3 $b_3 = \dfrac{1}{2}$

47. $c_n = \dfrac{(3n)!}{5n}$; find c_3 $c_3 = 24{,}192$

48. $d_n = \dfrac{(n+3)!}{3^n}$; find d_3 $d_3 = \dfrac{80}{3}$

For Exercises 49–56, find the nth term a_n of a sequence whose first four terms are given. (See Example 5)

49. $2, 4, 8, 16, \ldots$ $a_n = 2^n$

50. $3, 9, 27, 81, \ldots$ $a_n = 3^n$

51. $-\dfrac{2}{3}, -\dfrac{3}{6}, -\dfrac{4}{9}, -\dfrac{5}{12}, \ldots$ $a_n = -\dfrac{n+1}{3n}$

52. $-\dfrac{8}{11}, -\dfrac{9}{22}, -\dfrac{10}{33}, -\dfrac{11}{44}, \ldots$ $a_n = -\dfrac{n+7}{11n}$

53. $-1, 4, -9, 16, \ldots$ $a_n = (-1)^n(n^2)$

54. $1, -8, 27, -64, \ldots$ $a_n = (-1)^{n+1}(n^3)$

55. $\dfrac{1\cdot 2}{2}, \dfrac{1\cdot 2\cdot 3}{4}, \dfrac{1\cdot 2\cdot 3\cdot 4}{6}, \dfrac{1\cdot 2\cdot 3\cdot 4\cdot 5}{8}, \ldots$ $a_n = \dfrac{(n+1)!}{2n}$

56. $\dfrac{5}{4}, \dfrac{10}{9}, \dfrac{15}{16}, \dfrac{20}{25}, \ldots$ $a_n = \dfrac{5n}{(n+1)^2}$

Objective 4: Use Summation Notation

For Exercises 57–70, find the sum. (See Example 6)

57. $\displaystyle\sum_{i=1}^{6}(3i-4)$ 39

58. $\displaystyle\sum_{i=1}^{5}(2i+7)$ 65

59. $\displaystyle\sum_{j=1}^{5}(-2j^2)$ −110

60. $\displaystyle\sum_{k=1}^{3}(-3k^3)$ −108

61. $\displaystyle\sum_{j=2}^{5}\left(\dfrac{1}{2}\right)^j$ $\dfrac{15}{32}$

62. $\displaystyle\sum_{n=2}^{4}\left(\dfrac{1}{3}\right)^n$ $\dfrac{13}{81}$

63. $\displaystyle\sum_{j=1}^{20}6$ 120

64. $\displaystyle\sum_{i=1}^{50}4$ 200

65. $\displaystyle\sum_{k=3}^{7}(-1)^k(6k)$ −30

66. $\displaystyle\sum_{n=3}^{8}(-1)^{n-1}(2n)$ −6

67. $\displaystyle\sum_{m=1}^{3}\dfrac{m+1}{m}$ $\dfrac{29}{6}$

68. $\displaystyle\sum_{n=1}^{4}\dfrac{n-1}{n}$ $\dfrac{23}{12}$

69. $\displaystyle\sum_{k=1}^{3}(k+2)(k+3)$ 62

70. $\displaystyle\sum_{j=1}^{4}(j+1)(j-1)$ 26

71. a. Evaluate $\displaystyle\sum_{i=1}^{n}(-1)^i$ if n is even.   0

   b. Evaluate $\displaystyle\sum_{i=1}^{n}(-1)^i$ if n is odd.   −1

72. a. Evaluate $\displaystyle\sum_{i=1}^{n}(-1)^{i+1}$ if n is even.   0

   b. Evaluate $\displaystyle\sum_{i=1}^{n}(-1)^{i+1}$ if n is odd.   1

73. Given the sequence defined by $a_n = n^2 - 2n$, find the fifth partial sum.   25

74. Given the sequence defined by $b_n = n^3 - 3n^2$, find the fourth partial sum.   10

For Exercises 75–86, write the sum using summation notation. There may be multiple representations. Use i as the index of summation. (See Example 7)

75. $\dfrac{1}{2} + \dfrac{4}{3} + \dfrac{9}{4} + \dfrac{16}{5} + \cdots + \dfrac{n^2}{n+1}$

76. $3 + \dfrac{1}{2} + \dfrac{5}{27} + \dfrac{3}{32} + \cdots + \dfrac{n+2}{n^3}$ $\displaystyle\sum_{i=1}^{n}\dfrac{i+2}{i^3}$

77. $1 + 2 + 3 + 4 + 5$ $\displaystyle\sum_{i=1}^{5}i$

78. $2 + 4 + 6 + 8 + 10 + 12$ $\displaystyle\sum_{i=1}^{6}(2i)$

79. $8 + 8 + 8 + 8$ $\displaystyle\sum_{i=1}^{4}8$

80. $11 + 11 + 11 + 11 + 11$ $\displaystyle\sum_{i=1}^{5}11$

81. $\dfrac{1}{3} - \dfrac{1}{9} + \dfrac{1}{27} - \dfrac{1}{81}$ $\displaystyle\sum_{i=1}^{4} (-1)^{i+1} \dfrac{1}{3^i}$

82. $-\dfrac{1}{2} + \dfrac{1}{4} - \dfrac{1}{8} + \dfrac{1}{16} - \dfrac{1}{32}$ $\displaystyle\sum_{i=1}^{5} (-1)^{i} \dfrac{1}{2^i}$

83. $c^3 + c^4 + c^5 + \cdots + c^{20}$ $\displaystyle\sum_{i=1}^{18} c^{i+2}$

84. $a + ar + ar^2 + \cdots + ar^{12}$ $\displaystyle\sum_{i=1}^{13} ar^{j-1}$

85. $\dfrac{x}{1} + \dfrac{x^2}{2} + \dfrac{x^3}{6} + \dfrac{x^4}{24}$ (*Hint*: Note the pattern produced by $n!$ for $n = 1$, $n = 2$, and so on.) $\displaystyle\sum_{i=1}^{4} \dfrac{x^i}{i!}$

86. $\dfrac{1}{x+1} + \dfrac{2}{x+2} + \dfrac{6}{x+3} + \dfrac{24}{x+4} + \dfrac{120}{x+5}$ $\displaystyle\sum_{i=1}^{5} \dfrac{i!}{x+i}$

For Exercises 87–88, rewrite each series as an equivalent series with the new index of summation.

87. $\displaystyle\sum_{i=1}^{8} i^2 = \sum_{j=0}^{\square} \square = \sum_{k=2}^{\square} \square$ $\displaystyle\sum_{i=1}^{8} i^2 = \sum_{j=0}^{7} (j+1)^2 = \sum_{k=2}^{9} (k-1)^2$

88. $\displaystyle\sum_{i=1}^{5} (4i) = \sum_{j=0}^{\square} \square = \sum_{k=2}^{\square} \square$
$\displaystyle\sum_{i=1}^{5} (4i) = \sum_{j=0}^{4} [4(j+1)] = \sum_{k=2}^{6} [4(k-1)]$

Mixed Exercises

89. Prove that $\displaystyle\sum_{i=1}^{n} c = cn$.

90. Prove that $\displaystyle\sum_{i=1}^{n} (a_i + b_i) = \sum_{i=1}^{n} a_i + \sum_{i=1}^{n} b_i$.

For Exercises 91–94, use the sums $\displaystyle\sum_{i=1}^{50} i^2 = 42{,}925$ **and** $\displaystyle\sum_{i=1}^{50} i = 1275$ **and the properties of summation given on page 1013 to evaluate the given expression.**

91. $\displaystyle\sum_{i=1}^{50} (i^2 + 3i)$ 46,750

92. $\displaystyle\sum_{i=1}^{50} (2i^2 - i)$ 84,575

93. $\displaystyle\sum_{i=1}^{50} (5i + 4)$ 6575

94. $\displaystyle\sum_{i=1}^{50} (6i - 7)$ 7300

For Exercises 95–98, determine whether the statement is true or false. If a statement is false, explain why.

95. $\displaystyle\sum_{i=1}^{n} (3i + 7) = 3 \sum_{i=1}^{n} i + 7n$ True

96. $\displaystyle\sum_{i=1}^{n} (i^2 - 4i + 5) = \sum_{i=1}^{n} i^2 - 4 \sum_{i=1}^{n} i + 5n$ True

97. $\displaystyle\sum_{i=1}^{n} a_i b_i = \sum_{i=1}^{n} a_i \sum_{i=1}^{n} b_i$ False; $\displaystyle\sum_{i=1}^{n} a_i b_i = a_1 b_1 + a_2 b_2 + \cdots + a_n b_n \neq (a_1 + a_2 + \cdots + a_n)(b_1 + b_2 + \cdots + b_n)$

98. $\displaystyle\sum_{i=1}^{n} \dfrac{a_i}{b_i} = \dfrac{\displaystyle\sum_{i=1}^{n} a_i}{\displaystyle\sum_{i=1}^{n} b_i}$ False; $\displaystyle\sum_{i=1}^{n} \dfrac{a_i}{b_i} = \dfrac{a_1}{b_1} + \dfrac{a_2}{b_2} + \cdots + \dfrac{a_n}{b_n} \neq \dfrac{(a_1 + a_2 + \cdots + a_n)}{(b_1 + b_2 + \cdots + b_n)}$

99. Expenses for a company for year 1 are $24,000. Every year thereafter, expenses increase by $1000 plus 3% of the cost of the prior year. Let a_1 represent the original cost for year 1; that is, $a_1 = 24{,}000$. Use a recursive formula to find the cost a_n in terms of a_{n-1} for each subsequent year, $n \geq 2$. $a_n = 1.03 a_{n-1} + 1000;\ n \geq 2$

100. A retirement account initially has $500,000 and grows by 5% per year. Furthermore, the account owner adds $12,000 to the account each year after the first. Let a_1 represent the original amount in the account; that is, $a_1 = 500{,}000$. Use a recursive formula to find the amount in the account a_n in terms of a_{n-1} for each subsequent year, $n \geq 2$. $a_n = 1.05 a_{n-1} + 12{,}000;\ n \geq 2$

101. In a business meeting, every person at the meeting shakes every other person's hand exactly one time. The total number of handshakes for n people at the meeting is given by $a_n = \frac{1}{2} n(n-1)$. Evaluate a_{12} and interpret its meaning in the context of this problem.   66; If 12 people are present at the meeting, then there will be 66 handshakes.

102. Given a polygon of $n \geq 3$ sides, the sum of the interior angles within the polygon is given by $s_n = 180(n - 2)$. Evaluate s_{10} and interpret its meaning in the context of this problem.   1440; The sum of the angles within a polygon of 10 sides (a decagon) is 1440°.

Given a sequence $a_1, a_2, a_3, \dots, a_n$, the arithmetic mean $\bar{a}$ is given by $\bar{a} = \dfrac{1}{n} \displaystyle\sum_{i=1}^{n} a_i$. Use the arithmetic mean for Exercises 103–104.

103. Consider the sequence defined by $\{a_n\} = 18, 32, 44, 20, 36, 28, 32, 38$. Evaluate $\displaystyle\sum_{i=1}^{8} (a_i - \bar{a})^2$.   544

104. Show that $\displaystyle\sum_{i=1}^{n} (a_i - \bar{a}) = 0$.

$\displaystyle\sum_{i=1}^{n} (a_i - \bar{a}) = \sum_{i=1}^{n} a_i - \sum_{i=1}^{n} \bar{a} = \sum_{i=1}^{n} a_i - n\bar{a} = \sum_{i=1}^{n} a_i - n \cdot \dfrac{1}{n} \sum_{i=1}^{n} a_i = 0$

Write About It

105. Explain the difference between the graph of $a_n = n^2$ and $f(x) = x^2$.

106. What is the difference between a sequence and a series?   A sequence is an ordered list of terms. A series is the sum of the terms of a sequence.

107. Given the sequence defined by $a_n = \dfrac{n}{n-1}$, explain why the domain must be restricted to positive integers $n \geq 2$.   The sequence defined by $a_n = \dfrac{n}{n-1}$ is not defined for $n = 1$ because the denominator would be zero.

108. The value 1206! is too large to evaluate on most calculators. Explain how you would evaluate $\dfrac{1206!}{1204!}$ on a calculator.   Write 1206! as (1206)(1205)(1204!) and then simplify the fraction before multiplying. The result is the product $(1206)(1205) = 1{,}453{,}230$.

Expanding Your Skills

109. For $i = \sqrt{-1}$, find the first eight terms of the sequence defined by $a_n = i^n$. $i, -1, -i, 1, i, -1, -i, 1$

110. For $i = \sqrt{-1}$, find the first eight terms of the sequence defined by $a_n = 1 - i^n$. $1 - i, 2, 1 + i, 0, 1 - i, 2, 1 + i, 0$

111. The terms of the sequence defined by $a_1 = x$ and $a_n = \frac{1}{2}\left(a_{n-1} + \frac{x}{a_{n-1}}\right)$ for $n > 1$ give successively better approximations of $\sqrt{x}$ for $x > 1$. Approximate $\sqrt{2}$ by substituting 2 for x and finding the first four terms of the sequence. Round to 4 decimal places if necessary. $2, 1.5, 1.4167, 1.4142$

112. Find a formula for the nth term of the sequence. $\sqrt{3}, \sqrt{\sqrt{3}}, \sqrt{\sqrt{\sqrt{3}}}, \dots$ $a_n = 3^{1/(2^n)}$

Technology Connections

For Exercises 113–114, use a graphing utility to find the first four terms of the sequence.

113. $a_n = 12\left(-\frac{1}{2}\right)^n$ (Exercise 9)

114. $a_n = 64\left(-\frac{1}{4}\right)^n$ (Exercise 10)

For Exercises 115–116, use a graphing utility to find the sum.

115. $\sum_{k=3}^{7}(-1)^k(6k)$ (Exercise 65)

116. $\sum_{n=3}^{8}(-1)^{n-1}(2n)$ (Exercise 66)

117. Using calculus, we can show that the series $\sum_{k=0}^{n}\frac{1}{k!}$ approaches e as n approaches infinity. Investigate this statement by evaluating the sum for $n = 10$ and $n = 50$.

118. Using calculus, we can show that the series $\sum_{k=1}^{n}(-1)^{k-1}\frac{(0.5)^k}{k}$ approaches $\ln 1.5$ as n approaches infinity. Investigate this statement by evaluating the sum for $n = 10$ and $n = 50$.

Additional answers can be found in the Instructor Answer Appendix.

SECTION 11.2   Arithmetic Sequences and Series

OBJECTIVES

1. Identify Specific and General Terms of an Arithmetic Sequence
2. Evaluate a Finite Arithmetic Series
3. Apply Arithmetic Sequences and Series

1. Identify Specific and General Terms of an Arithmetic Sequence

In this section and Section 11.3, we study two special types of sequences. The first is called an arithmetic sequence. For example, consider the salary plan for a job that pays $75,000 the first year with a $4000 raise each year thereafter. The sequence of salaries for the first 5 yr is

Year 1	Year 2	Year 3	Year 4	Year 5
$75,000	$79,000	$83,000	$87,000	$91,000

Notice that each term after the first results from adding a fixed constant ($4000) to its predecessor. This is the characteristic that makes this sequence arithmetic.

> **Arithmetic Sequence**
>
> An **arithmetic sequence** $\{a_n\}$ is a sequence of the form
>
> $$a_1, a_1 + d, a_1 + 2d, a_1 + 3d, a_1 + 4d, \dots$$
>
> - The value a_1 is the first term, and d is called the **common difference** of the sequence.
> - The value of d is the difference of any term after the first and its predecessor. $d = a_{n+1} - a_n$
> - The nth term of the sequence is given by $a_n = a_1 + (n - 1)d$.
> - The sequence can be defined recursively as $a_1, a_n = a_{n-1} + d$ for $n \geq 2$.

Classroom Examples: p. 1025
Exercises 10, 12

EXAMPLE 1   Identifying an Arithmetic Sequence and the Common Difference

Determine whether the sequence is arithmetic. If so, identify the common difference.

a. 35, 25, 15, 5, −5, …   **b.** 1, 5, 10, 16, 23, …

Solution:

a. 35, 25, 15, 5, −5, …

$$a_2 - a_1 = 25 - 35 = -10$$ The sequence is arithmetic because the difference
$$a_3 - a_2 = 15 - 25 = -10$$ between each term and its predecessor is the same constant.

$$a_4 - a_3 = 5 - 15 = -10$$ The common difference is $d = -10$.
$$a_5 - a_4 = -5 - 5 = -10$$

b. 1, 5, 10, 16, 23, …

$$a_2 - a_1 = 5 - 1 = 4$$ The sequence is *not* arithmetic because the
$$a_3 - a_2 = 10 - 5 = 5$$ difference between a_2 and a_1 is different than the
                               difference between a_3 and a_2. That is, the difference
                               between consecutive terms is not the same.

Skill Practice 1   Determine whether the sequence is arithmetic. If so, identify the common difference.

a. 12, 5, −2, −9, −16, …   **b.** 1, 4, 9, 16, 25, …

Classroom Example: p. 1025
Exercise 16

EXAMPLE 2   Writing the Terms of a Sequence

a. Write the first five terms of an arithmetic sequence with first term −5 and common difference 4.

b. Write a recursive formula to define the sequence.

Solution:

a. −5, −1, 3, 7, 11   Because the common difference is 4, each term after the first
$$+4 \quad +4 \quad +4 \quad +4$$ must be 4 *more than* its predecessor.

b. The recursive formula for an arithmetic sequence is a_1, $a_n = a_{n-1} + d$ for $n \geq 2$.

The sequence is defined by $a_1 = -5$ and $a_n = a_{n-1} + 4$ for $n \geq 2$.

Skill Practice 2

a. Write the first four terms of an arithmetic sequence with first term 8 and common difference −3.

b. Write a recursive formula to define the sequence.

Classroom Example: p. 1025
Exercise 28

EXAMPLE 3   Applying an Arithmetic Sequence

A park ranger in Everglades National Park measures the water level in one region of the park for a 5-day period during a drought.

Day number	1	2	3	4	5
Water level (in.)	54.0	53.2	52.4	51.6	50.8

Answers

1. a. Arithmetic; $d = -7$
   b. Not arithmetic

2. a. 8, 5, 2, −1
   b. $a_1 = 8$ and
   $a_n = a_{n-1} - 3$ for $n \geq 2$.

a. Based on the given data, does the water level follow an arithmetic progression?

b. Write an expression for the *n*th term of the sequence where *n* represents the day number.

c. Predict the water level on day 30 if this trend continues.

Solution:

a. $a_2 - a_1 = 53.2 - 54.0 = -0.8$ The sequence is arithmetic because the
     $a_3 - a_2 = 52.4 - 53.2 = -0.8$ difference between each term and its
     $a_4 - a_3 = 51.6 - 52.4 = -0.8$ predecessor is the same constant.
     $a_5 - a_4 = 50.8 - 51.6 = -0.8$ The common difference is $d = -0.8$ in.

b. $a_n = a_1 + (n - 1)d$ Formula for the *n*th term of an arithmetic
                          sequence

     $a_n = 54 + (n - 1)(-0.8)$ Substitute 54 for a_1 and -0.8 for d.

c. $a_{30} = 54 + (30 - 1)(-0.8)$ To find a_{30}, substitute 30 for *n*.
        $= 30.8$

The water level on day 30 will be 30.8 in. if this trend continues.

Skill Practice 3   A homeowner has kept records of the average monthly electric bill for 4 yr.

Year	1	2	3	4
Amount ($)	102.60	108.00	113.40	118.80

a. Do the average monthly electric bills follow an arithmetic progression?

b. Write an expression for the *n*th term of the sequence.

c. Predict the average monthly bill for year 6 if this trend continues.

The *n*th term of the sequence from Example 3 can be written in several equivalent algebraic forms.

$$a_n = 54 + (n - 1)(-0.8)$$
$$a_n = 54 - 0.8n + 0.8 \qquad \text{Apply the distributive property.}$$
$$a_n = -0.8n + 54.8$$

The expression $a_n = -0.8n + 54.8$ resembles the slope-intercept form of a linear function $f(x) = mx + b$. In fact, an arithmetic sequence is a linear function whose domain is the set of positive integers. The graph of the sequence from Example 3 is shown in Figure 11-4. The value of the common difference is negative ($d = -0.8$), indicating that the progression of points slopes downward.

In Example 4, we find specific terms of an arithmetic sequence given information about the sequence.

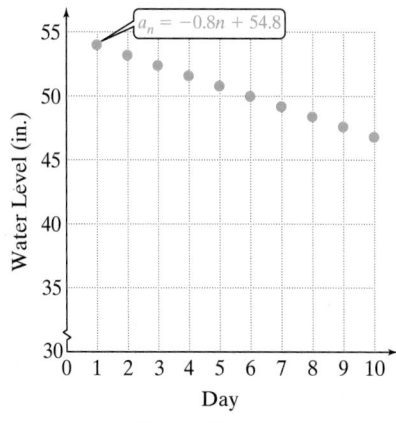

Figure 11-4

Classroom Example: p. 1025
Exercise 34

EXAMPLE 4 Finding a Specific Term of an Arithmetic Sequence

Find the ninth term of the arithmetic sequence in which $a_1 = -4$ and $a_{22} = 164$.

Solution:

To find the value of a_9, we need to determine the common difference d. To find d, substitute $a_1 = -4$, $n = 22$, and $a_{22} = 164$ into the formula for the nth term.

$$a_n = a_1 + (n - 1)d$$
$$164 = -4 + (22 - 1)d$$
$$164 = -4 + 21d$$
$$d = 8$$

Therefore, $a_n = -4 + (n - 1)(8)$.   With the values of a_1 and d known, the nth term is represented by $a_n = -4 + (n - 1)(8)$.

$$a_n = -4 + 8n - 8$$ Simplify.
$$a_n = 8n - 12$$
$$a_9 = 8(9) - 12$$ To find a_9 substitute $n = 9$.
$$= 60$$

Skill Practice 4 Find the tenth term of the arithmetic sequence in which $a_1 = 12$ and $a_{30} = 128$.

In Example 5, we use the general formula for the nth term of an arithmetic sequence to find the number of terms in an arithmetic sequence.

Classroom Example: p. 1026
Exercise 36

EXAMPLE 5 Finding the Number of Terms in an Arithmetic Sequence

Find the number of terms of the finite arithmetic sequence 7, 3, -1, -5, ... , -113.

Solution:

$$a_n = a_1 + (n - 1)d$$ The first term of the sequence is 7. The nth term is -113. The common difference is $d = -4$.

$$-113 = 7 + (n - 1)(-4)$$ To find the number of terms n, substitute $a_1 = 7$, $d = -4$, and $a_n = -113$ into the formula for the nth term.
$$-113 = 7 - 4n + 4$$
$$-113 = 11 - 4n$$
$$4n = 124$$
$$n = 31$$ There are 31 terms.

Skill Practice 5 Find the number of terms of the finite arithmetic sequence 16, 11, 6, 1, ... , -239.

Answers
4. $a_{10} = 48$
5. 52 terms

Classroom Example: p. 1026
Exercise 44

> **EXAMPLE 6**   Finding a Specific Term of an Arithmetic Sequence

For an arithmetic sequence, $a_{15} = 49$ and $a_{27} = 85$. Find the 500th term.

Solution:

Substituting $a_{15} = 49$ and $a_{27} = 85$ into the formula $a_n = a_1 + (n - 1)d$, we can set up a system of linear equations to solve for a_1 and d.

$a_n = a_1 + (n - 1)d$	$a_n = a_1 + (n - 1)d$	Substitute $a_{15} = 49$ and
$49 = a_1 + (15 - 1)d$	$85 = a_1 + (27 - 1)d$	$a_{27} = 85$ into the formula
$49 = a_1 + 14d$	$85 = a_1 + 26d$	for the nth term.

The two equations form a system of linear equations in two variables.

$$49 = a_1 + 14d \xrightarrow{\text{Multiply by } -1} \quad -49 = -a_1 - 14d \qquad 49 = a_1 + 14d$$
$$85 = a_1 + 26d \qquad \qquad \underline{85 = a_1 + 26d} \qquad 49 = a_1 + 14(3)$$
$$36 = 12d \qquad 7 = a_1$$
$$3 = d$$

With $d = 3$ and $a_1 = 7$, we have an nth term of $a_n = 7 + (n - 1)3$. Thus, the 500th term is given by $a_{500} = 7 + (500 - 1)3 = 1504$.

> **Skill Practice 6**   For an arithmetic sequence, $a_{11} = 65$ and $a_{25} = 149$. Find the 400th term.

TIP In Example 6, we can also find the value of d by dividing the difference of a_{27} and a_{15} by the number of terms between a_{27} and a_{15}. That is,

$$d = \frac{85 - 49}{27 - 15} = 3$$

2. Evaluate a Finite Arithmetic Series

Consider the finite arithmetic sequence 1, 4, 7, 10, 13. The sum of the first n terms of the sequence is called the **nth partial sum** and is denoted by S_n. In this case, we have

$$
\begin{aligned}
S_1 &= 1 &&= 1 \\
S_2 &= 1 + 4 &&= 5 \\
S_3 &= 1 + 4 + 7 &&= 12 \\
S_4 &= 1 + 4 + 7 + 10 &&= 22 \\
S_5 &= 1 + 4 + 7 + 10 + 13 &&= 35
\end{aligned}
$$

For a large number of terms, adding the terms individually would be a cumbersome process, so instead, we observe the following pattern. We write the sum of the first five terms of the sequence in both ascending and descending order.

$$
\begin{aligned}
\text{Ascending order:} \quad S_5 &= 1 + 4 + 7 + 10 + 13 \\
\text{Descending order:} \quad S_5 &= 13 + 10 + 7 + 4 + 1 \\
\hline
2S_5 &= 14 + 14 + 14 + 14 + 14 \qquad \text{Now add to get } 2S_5.
\end{aligned}
$$

$$2S_5 = 5(14)$$
$$\frac{2S_5}{2} = \frac{5(14)}{2}$$
$$S_5 = 35 \checkmark$$

TIP The sum of the first n terms of an arithmetic sequence is called a **finite arithmetic series.**

Answer

6. $a_{400} = 2399$

By adding the terms in ascending and descending order, we double the sum but create a pattern that is easily added. We can use a similar process to find the sum S_n of the first n terms of an arithmetic sequence: $a_1 + a_2 + a_3 + \cdots + a_{n-1} + a_n$.

$$S_n = a_1 \qquad\quad + (a_1 + d) + (a_1 + 2d) + \cdots + a_n \qquad \text{Ascending order}$$

$$\underline{S_n = a_n \qquad\quad + (a_n - d) + (a_n - 2d) + \cdots + a_1} \qquad \text{Descending order}$$

$$2S_n = (a_1 + a_n) + (a_1 + a_n) + (a_1 + a_n) + \cdots + (a_1 + a_n)$$

$$2S_n = n(a_1 + a_n)$$

$$S_n = \frac{n}{2}(a_1 + a_n)$$

---

*n*th Partial Sum of an Arithmetic Sequence

The sum S_n of the first n terms of an arithmetic sequence is given by

$$S_n = \frac{n}{2}(a_1 + a_n)$$

where a_1 is the first term of the sequence, and a_n is the nth term of the sequence.

In Example 7, we use the formula for the nth partial sum of an arithmetic sequence to find the sum of the first 50 positive even integers.

Classroom Example: p. 1026
Exercise 52

EXAMPLE 7   Finding an *n*th Partial Sum of an Arithmetic Sequence

Find the sum of the first 50 terms of the sequence. 2, 4, 6, 8, 10, ...

Solution:

The sequence is arithmetic because each term is 2 more than its predecessor. The common difference is 2. To find the sum of the first 50 terms, we need to know the values of a_1 and a_{50}. To find a_{50}, we write a formula for the nth term and then evaluate the expression for $n = 50$.

$$a_n = a_1 + (n - 1)d \qquad \text{General expression for the } n\text{th term}$$
$$a_n = 2 + (n - 1)(2) \qquad \text{Substitute } a_1 = 2 \text{ and } d = 2.$$
$$a_n = 2n$$

$$a_{50} = 2(50) \qquad \text{To find } a_{50}, \text{ substitute 50 for } n.$$
$$a_{50} = 100$$

$$S_n = \frac{n}{2}(a_1 + a_n) \qquad \text{Now find the sum of the first 50 terms.}$$

$$S_{50} = \frac{50}{2}(2 + 100) \qquad \text{Substitute } n = 50, a_1 = 2, \text{ and } a_{50} = 100.$$

$$S_{50} = 2550 \qquad \text{The sum of the first 50 positive even integers is 2550.}$$

Instructor Note:
Consider showing students that the formula for the sum of a finite arithmetic series is the average of the first and last terms times the number of terms.

Skill Practice 7   Find the sum of the first 50 terms of the sequence. 1, 3, 5, 7, 9, ...

Answer

7. 2500

Point of Interest

An old story suggests that at the age of 7, Carl Friedrich Gauss (see page 287) amazed his teachers by quickly computing the sum of the first 100 positive integers. Gauss realized that the sum of the integers from 1 to 100 was the same as 50 pairs of numbers each summing to 101. As the story goes, Gauss returned the correct sum of 5050 within minutes by taking the product of 50(101).

Sum of the first 100 positive integers: $1 + 2 + 3 + \cdots + 98 + 99 + 100$
50 pairs, each summing to 101.

sum 101

In Example 8, we evaluate an arithmetic series written in summation notation.

Classroom Example: p. 1026
Exercise 58

EXAMPLE 8 Evaluating a Finite Arithmetic Series

Find the sum. $\displaystyle\sum_{i=1}^{60}(3i + 5)$

Solution:

The expression $3i + 5$ is linear in the variable i. This tells us that the terms of the series form an arithmetic progression. To verify, we can write several terms of the sum.

$$\sum_{i=1}^{60}(3i + 5) = [3(1) + 5] + [3(2) + 5] + [3(3) + 5] + \cdots + [3(60) + 5]$$
$$= 8 + 11 + 14 + \cdots + 185$$

The individual terms in the series: 8, 11, 14, … , 185 form an arithmetic sequence $\{a_n\}$ with a common difference of 3. The first term is $a_1 = 8$, and the 60th term is $a_{60} = 185$. The value of the series is equal to the 60th partial sum of the sequence of terms.

$$S_n = \frac{n}{2}(a_1 + a_n)$$
$$S_{60} = \frac{60}{2}(8 + 185) \qquad \text{Substitute } n = 60, a_1 = 8, \text{ and } a_{60} = 185.$$
$$S_{60} = 5790$$

Skill Practice 8   Find the sum. $\displaystyle\sum_{i=1}^{80}(4i + 3)$

Avoiding Mistakes

When we apply the formula $S_n = \frac{n}{2}(a_1 + a_n)$ to find the sum $\sum_{i=1}^{n} a_n$, the index of summation, i, must begin at 1.

3. Apply Arithmetic Sequences and Series

In Example 9, we use an arithmetic sequence and series in an application.

Classroom Example: p. 1026
Exercise 66

EXAMPLE 9 Applying an Arithmetic Sequence and Series

Suppose that a job offers a starting salary of $75,000 with a raise of $4000 every year thereafter.

a. Write an expression for the nth term of an arithmetic sequence that represents the salary as a function of the number of years of employment, n.

b. Find the total income for an employee who works at the job for 20 yr.

Answer
8. 13,200

Solution:

a. $a_n = a_1 + (n - 1)d$

$a_n = 75{,}000 + (n - 1)(4000)$ Substitute $a_1 = 75{,}000$ and $d = 4000$.

$a_n = 75{,}000 + 4000n - 4000$ Simplify.

$a_n = 4000n + 71{,}000$

b. $a_{20} = 75{,}000 + (20 - 1)(4000)$ To find the total income over 20 yr, we need to

$a_{20} = 151{,}000$ know a_1 and a_{20}. Use the nth term to find a_{20}.

$S_n = \dfrac{n}{2}(a_1 + a_n)$

$S_{20} = \dfrac{20}{2}(75{,}000 + 151{,}000)$ Substitute $n = 20$, $a_1 = 75{,}000$, and

$a_{20} = 151{,}000$.

$S_{20} = 2{,}260{,}000$

Skill Practice 9   A teaching position has a starting salary of \$60,000 with a raise of \$3000 every year thereafter.

   a. Write an expression for the nth term of an arithmetic sequence that represents the salary as a function of the number of years of employment, n.

   b. Find the total income for an employee who works at the job for 30 yr.

Answers

9 a. $a_n = 3000n + 57{,}000$

   b. \$3,105,000

SECTION 11.2   Practice Exercises

Prerequisite Review

R.1. Given $f(x) = 7x - 9$, evaluate $f(2)$. $f(2) = 5$

R.2. Determine the slope and y-intercept of the line $y = -\dfrac{1}{4}x + 8$.   Slope: $-\dfrac{1}{4}$; y-intercept: $(0, 8)$

For Exercises R.3–R.4, solve the system.

R.3. $6x - 2y = 10$ $\{(1, -2)\}$
   $2x - 10y = 22$

R.4. $0.3x - 0.4y = -1.6$ $\{(-4, 1)\}$
   $0.9x + 0.1y = -3.5$

Concept Connections

1. A(n) ____arithmetic____ sequence is a sequence in which each term after the first is found by adding a fixed constant to its predecessor.

2. The difference between any term after the first and its predecessor in an arithmetic sequence is called the ____common____ ____difference____ and is denoted by d.

3. Given an arithmetic sequence with first term a_1 and common difference d, the nth term is represented by the formula $a_n = \underline{\ a_1 + (n-1)d\ }$ or by the recursive formula $a_n = \underline{\ a_{n-1} + d\ }$ for $n \geq 2$.

4. An arithmetic sequence is a linear function whose domain is the set of ____positive____ integers.

5. The sum of the first n terms of a sequence is called the nth ____partial____ sum and is denoted by S_n.

6. Given an arithmetic sequence with first term a_1 and nth term a_n, the nth partial sum is given by the formula $S_n = \underline{\hspace{2cm}}$. $\dfrac{n}{2}(a_1 + a_n)$

Objective 1: Identify Specific and General Terms of an Arithmetic Sequence

For Exercises 7–14, determine whether the sequence is arithmetic. If so, find the common difference. (See Example 1)

7. 15, 19, 23, 27, ...
Yes; $d = 4$

8. 256, 268, 280, 292, ...
Yes; $d = 12$

9. 9, -2, -13, -24, ...
Yes; $d = -11$

10. 8, 0, -8, -16, ...
Yes; $d = -8$

11. 18, 22, 27, 33, ...   No

12. 2, 4, 8, 16, ...   No

13. 4, $\dfrac{14}{3}$, $\dfrac{16}{3}$, 6, ...
Yes; $d = \dfrac{2}{3}$

14. 3, $\dfrac{15}{4}$, $\dfrac{9}{2}$, $\dfrac{21}{4}$, ...
Yes; $d = \dfrac{3}{4}$

For Exercises 15–18,

a. Write the first five terms of an arithmetic sequence with the given first term and common difference.

b. Write a recursive formula to define the sequence. (**See Example 2**)

15. $a_1 = 3$, $d = 10$

16. $a_1 = 6$, $d = 5$

17. $a_1 = 4$, $d = -2$

18. $a_1 = 5$, $d = -3$

For Exercises 19–24,

a. Write a nonrecursive formula for the nth term of the arithmetic sequence $\{a_n\}$ based on the given information.

b. Find the indicated term.

19. a. $a_1 = 7$, $d = 10$ $a_n = 10n - 3$

b. Find a_{22}. $a_{22} = 217$

20. a. $a_1 = 102$, $d = 4$ $a_n = 4n + 98$

b. Find a_{43}. $a_{43} = 270$

21. a. $a_1 = -12$, $d = 5$ $a_n = 5n - 17$

b. Find a_{20}. $a_{20} = 83$

22. a. $a_1 = -4$, $d = 6$ $a_n = 6n - 10$

b. Find a_{18}. $a_{18} = 98$

23. a. $a_1 = \dfrac{1}{2}$, $d = \dfrac{1}{3}$ $a_n = \dfrac{1}{3}n + \dfrac{1}{6}$

b. Find a_{10}. $a_{10} = \dfrac{7}{2}$

24. a. $a_1 = \dfrac{2}{3}$, $d = \dfrac{1}{2}$ $a_n = \dfrac{1}{2}n + \dfrac{1}{6}$

b. Find a_8. $a_8 = \dfrac{25}{6}$

25. Jim has 8 unread emails in his inbox before going on vacation. While on vacation, Jim does not read email. If he receives an average of 22 emails each day, write the nth term of a sequence defining the number of unread emails in his box at the end of day n of his vacation. $a_n = 22n + 8$

26. Sandy has a personal trainer who encourages her to get plenty of cardiovascular exercise. In her first week of training, Sandy walks for 10 min on a treadmill every day. Each week thereafter, she increases the time on the treadmill by 5 min. Write the nth term of a sequence defining the number of minutes that Sandy spends on the treadmill per day for her nth week at the gym.
$a_n = 5n + 5$

27. A new drug and alcohol rehabilitation program performs outreach for members of the community. The number of participants for a 4-week period is given in the table. (**See Example 3**)

Week number	1	2	3	4
Number of participants	34	50	66	82

a. Based on the data given, does the number of participants follow an arithmetic progression?   Yes

b. Write an expression for the nth term of the sequence representing the number of participants, where n represents the week number. $a_n = 16n + 18$

c. Predict the number of participants in week 10 if this trend continues.   178

28. A student studying to be a veterinarian's assistant keeps track of a kitten's weight each week for a 5-week period after birth.

Week number	1	2	3	4	5
Weight (lb)	0.6	0.88	1.16	1.44	1.72

a. Based on the data given, does the weight of the kitten follow an arithmetic progression?   Yes

b. Write an expression for the nth term of the sequence representing the kitten's weight, n weeks after birth.
$a_n = 0.28n + 0.32$

c. If the weight of the kitten continues to increase linearly for 3 months, predict the kitten's weight 12 weeks after birth.   3.68 lb

29. Suppose that an object starts with an initial velocity of v_0 (in ft/sec) and moves under a constant acceleration a (in ft/sec²). Then the velocity v_n (in ft/sec) after n seconds is given by $v_n = v_0 + an$. Show that this sequence is arithmetic. (*Hint*: Show that $v_{n+1} - v_n$ is constant.)
$v_{n+1} - v_n = [v_0 + a(n + 1)] - (v_0 + an)$
$= v_0 + an + a - v_0 - an$
$= a$
The difference between two consecutive terms is the constant a. Therefore, the sequence is arithmetic.

30. Suppose that an object is dropped from rest from an airplane. Assuming negligible air resistance, the vertical acceleration is 32 ft/sec². Using the formula for v_n given in Exercise 29,

a. Write the nth term v_n of the sequence representing the velocity in the downward direction after n seconds. $v_n = 32n$

b. Determine the vertical velocity 1 sec after release and 5 sec after release. $v_1 = 32$ ft/sec; $v_5 = 160$ ft/sec

31. Find the 8th term of an arithmetic sequence with $a_1 = -2$ and $a_{15} = 68$. (**See Example 4**) $a_8 = 33$

32. Find the 19th term of an arithmetic sequence with $a_1 = -11$ and $a_{30} = 163$. $a_{19} = 97$

33. Find the 35th term of an arithmetic sequence with $a_1 = 50$ and $a_{22} = -265$. $a_{35} = -460$

34. Find the 46th term of an arithmetic sequence with $a_1 = 210$ and $a_{60} = -262$. $a_{46} = -150$

For Exercises 35–38, find the number of terms of the finite arithmetic sequence. (See Example 5)

35. 8, 14, 20, 26, ... , 320 $n = 53$

36. 7, 16, 25, 34, ... , 574 $n = 64$

37. 11, 10.7, 10.4, 10.1, ... , −3.4 $n = 49$

38. 9, 8.4, 7.8, 7.2, ... , −39 $n = 81$

39. Given an arithmetic sequence with $a_{14} = 148$ and $a_{35} = 316$, find a_1 and d. $a_1 = 44; d = 8$

40. Given an arithmetic sequence with $a_{12} = -76$ and $a_{51} = -193$, find a_1 and d. $a_1 = -43; d = -3$

For Exercises 41–46, two terms of an arithmetic sequence are given. Find the indicated term. (See Example 6)

41. $a_{15} = 86$, $a_{34} = 200$; Find a_{150}. $a_{150} = 896$

42. $a_{12} = 52$, $a_{51} = 208$; Find a_{172}. $a_{172} = 692$

43. $b_{32} = -303$, $b_{54} = -567$; Find b_{214}. $b_{214} = -2487$

44. $b_{64} = -456$, $b_{81} = -575$; Find b_{105}. $b_{105} = -743$

45. $c_{14} = 7.5$, $c_{101} = 29.25$; Find c_{400}. $c_{400} = 104$

46. $c_{21} = -11.16$, $c_{116} = -7.36$; Find c_{505}. $c_{505} = 8.2$

47. If the third and fourth terms of an arithmetic sequence are −6 and −9, what are the first and second terms? $a_1 = 0, a_2 = -3$

48. If the third and fourth terms of an arithmetic sequence are 12 and 16, what are the first and second terms? $a_1 = 4, a_2 = 8$

Objective 2: Evaluate a Finite Arithmetic Series

For Exercises 49–50, find the sum.

49. 4 + 7 + 10 + 13 + 16 + 19 + 22 + 25 + 28 + 31 + 34 + 37   246

50. 8 + 12 + 16 + 20 + 24 + 28 + 32 + 36 + 40 + 44   260

51. Find the sum of the first 40 terms of the sequence. {1, 6, 11, 16, ...} **(See Example 7)**   3940

52. Find the sum of the first 60 terms of the sequence. {2, 10, 18, 26, ...}   14,280

53. Find the sum. 5 + 4.5 + 4 + 3.5 + ⋯ + (−30.5)   −918

54. Find the sum. 8 + 7.2 + 6.4 + 5.6 + ⋯ + (−45.6)   −1278.4

For Exercises 55–64, find the sum. (See Example 8)

55. $\sum_{j=1}^{18} (j + 6)$ 279

56. $\sum_{j=1}^{15} (j - 3)$ 75

57. $\sum_{i=1}^{50} (2i + 6)$ 2850

58. $\sum_{i=1}^{40} (3i - 7)$ 2180

59. $\sum_{k=1}^{162} \left(3 - \tfrac{1}{2}k\right)$ −6115.5

60. $\sum_{k=1}^{141} \left(4 - \tfrac{1}{4}k\right)$ −1938.75

61. −1 + 4 + 9 + ⋯ + 49   264

62. 12 + 16 + 20 + ⋯ + 84   912

63. −7 + (−11) + (−15) + ⋯ + (−39)   −207

64. −18 + (−23) + (−28) + ⋯ + (−183)   −3417

Objective 3: Apply Arithmetic Sequences and Series

65. Suzanne must choose between two job offers. The first job offers $64,000 for the first year. Each year thereafter, she would receive a $3200 raise. The second job offers $60,000 in the first year. Each year thereafter, she would receive a $5000 raise. **(See Example 9)**

   a. If she anticipates working for the company for 5 yr, find the total amount she would earn from each job.   Job 1 for 5 yr: $352,000; Job 2 for 5 yr: $350,000

   b. If she anticipates working for the company for 10 yr, find the total amount she would earn from each job.   Job 1 for 10 yr: $784,000; Job 2 for 10 yr: $825,000

66. José must choose between two job offers. The first job pays $50,000 the first year. Each year thereafter, he would receive a raise of $2400. A second job offers $54,000 the first year with a raise of $2000 each year thereafter. However, with the second job, José would have to pay $100 per month out of his paycheck for health insurance.

   a. If José anticipates working for the company for 6 yr, find the total amount he would earn from each job.   Job 1 for 6 yr: $336,000; Job 2 for 6 yr: $346,800

   b. If he anticipates working for the company for 12 yr, find the total amount he would earn from each job.   Job 1 for 12 yr: $758,400; Job 2 for 12 yr: $765,600

67. An object in free fall is dropped from a tall cliff. It falls 16 ft in the first second, 48 ft in the second second, 80 ft in the third second, and so on.

   a. Write a formula for the nth term of an arithmetic sequence that represents the distance d_n (in ft) that the object will fall in the nth second. $d_n = 32n - 16$

   b. How far will the object fall in the 8th second?   240 ft

   c. What is the total distance that the object will fall in 8 sec?   1024 ft

68. A ball rolling down an inclined plane rolls 4 in. in the first second, 8 in. in the second second, 12 inches in the third second, and so on.

   a. Write a formula for the nth term of an arithmetic sequence that represents the distance d_n (in inches) that the ball will roll in the nth second. $d_n = 4n$

   b. How far will the ball roll in the 10th second?   40 in.

   c. What is the total distance that the ball will travel in 10 sec?   220 in.

69. The students at Prairiewood Elementary plan to make a pyramid out of plastic cups. The bottom row has 15 cups. Moving up the pyramid, the number of cups in each row decreases by 1.

    a. If the students build the pyramid so that there are 12 rows, how many cups will be at the top?　4 cups

    b. How many total cups will be required?　114 cups

70. A theater has 32 rows. The first row has 18 seats, and each row that follows has three more seats than the row in front.

    a. Determine the number of seats in row 32.　111 seats

    b. Determine the total number of seats in the theater.　2064 seats

Mixed Exercises

71. Refer to the graph of the sequence $\{b_n\}$.

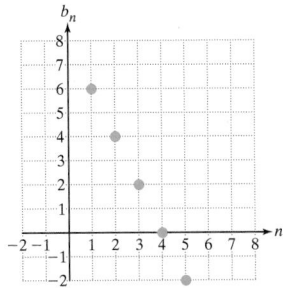

    a. Estimate the first four terms of the sequence.　6, 4, 2, 0

    b. Write a formula for the nth term of the sequence.
$b_n = -2n + 8$

    c. Find the 30th term.　$b_{30} = -52$

    d. Find the sum of the first 30 terms.　-690

    e. If $b_n = -180$, what is n?　$n = 94$

    f. Find the difference between b_{88} and b_{20}.　-136

72. Refer to the graph of the sequence $\{a_n\}$.

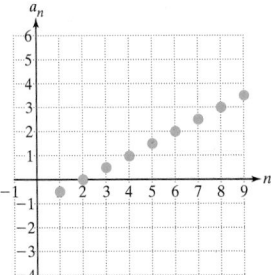

    a. Estimate the first four terms of the sequence.
$-0.5, 0, 0.5, 1$

    b. Write a formula for the nth term of the sequence.
$a_n = 0.5n - 1$

    c. Find the 20th term.　$a_{20} = 9$

    d. Find the sum of the first 20 terms.　85

    e. If $a_n = 256$, what is n?　$n = 514$

    f. Find the difference between a_{50} and a_{32}.　9

73. Find the sum of the integers from -20 to 256.　32,686

74. Find the sum of the integers from -102 to 57.　-3600

75. Compute the sum of the first 50 positive integers that are exactly divisible by 5.　6375

76. Compute the sum of the first 60 positive integers that are exactly divisible by 4.　7320

77. Compute the sum of all integers between 40 and 100 that are exactly divisible by 3.　1410

78. Compute the sum of all integers between 60 and 150 that are exactly divisible by 8.　1144

79. a. Use the formula $S_n = \frac{n}{2}(a_1 + a_n)$ to show that the sum of the first n positive integers is $S_n = \frac{n}{2}(1 + n)$.
Substitute $a_1 = 1$ and $a_n = n$. $S_n = \frac{n}{2}(a_1 + a_n) = \frac{n}{2}(1 + n)$

    b. Find the sum of the first 100 positive integers.
5050

    c. Find the sum of the first 1000 positive integers.　500,500

80. a. Use the formula $S_n = \frac{n}{2}(a_1 + a_n)$ to show that the sum 1, 3, 5, ... $(2n - 1) = n^2$.

    b. Find the sum of the first 100 positive odd integers.　10,000

The arithmetic mean (average) of two numbers c and d is given by $\bar{x} = \dfrac{c + d}{2}$. The value $\bar{x}$ is equidistant between c and d, so the sequence $c, \bar{x}, d$ is an arithmetic sequence. Inserting k equally spaced values between c and d, yields the arithmetic sequence $c, \bar{x}_1, \bar{x}_2, \bar{x}_3, \bar{x}_4, \ldots, \bar{x}_k, d$. Use this information for Exercises 81–82.

81. Insert three arithmetic means between 4 and 28. (*Hint:* There will be a total of five terms. Write the nth term of an arithmetic sequence with $a_1 = 4$ and $a_5 = 28$, and then find a_2, a_3, and a_4.)
The three arithmetic means between 4 and 28 are 10, 16, and 22.

82. Insert four arithmetic means between 19 and 64.
The four arithmetic means between 19 and 64 are 28, 37, 46, and 55.

Write About It

83. Suppose that $\{a_1, a_2, a_3, \ldots\}$ is an arithmetic sequence with common difference d. Explain why $\{a_1, a_3, a_5, \ldots\}$ is also an arithmetic sequence.　The terms in the given sequence $a_1, a_2, a_3, \ldots$ differ by d units. Therefore, every other term would differ by $2d$ units. Thus, $a_1, a_3, a_5, \ldots$ is an arithmetic sequence with common difference $2d$.

84. Suppose you are helping a friend with the homework for this section. Explain how to construct an arithmetic sequence.　Select any number for a_1 and any number for d. To find each term after a_1, add d to the preceding term.

Expanding Your Skills

85. Determine the *n*th term of the arithmetic sequence whose *n*th partial sum is $n^2 + 2n$. [*Hint:* The *n*th term of the sequence is the difference between the sum of the first *n* terms and the first $(n - 1)$ terms.]
$a_n = 2n + 1$; The sequence is $\{3, 5, 7, \ldots, (2n + 1), \ldots\}$.

86. Determine the *n*th term of the arithmetic sequence whose *n*th partial sum is $2n^2$.
$a_n = 4n - 2$; The sequence is $\{2, 6, 10, 14, \ldots, (4n - 2), \ldots\}$.

SECTION 11.3 Geometric Sequences and Series

1. Identify Specific and General Terms of a Geometric Sequence

The sequence 2, 4, 8, 16, 32, … is not an arithmetic sequence because the difference between consecutive terms is not the same constant. However, a different pattern exists. Notice that each term after the first is 2 times the preceding term. This sequence is called a geometric sequence.

> **Geometric Sequence**
>
> A **geometric sequence** $\{a_n\}$ is a sequence of the form
>
> $$a_1, a_1r, a_1r^2, a_1r^3, a_1r^4, \ldots$$
>
> - The value a_1 is the first term, and r is called the **common ratio** of the sequence.
> - The value of r is the quotient of any term after the first and its predecessor.
>
> $$r = \frac{a_{n+1}}{a_n}$$
>
> - The *n*th term of the sequence is given by $a_n = a_1r^{n-1}$.
> - The sequence can be defined recursively as $a_1, a_n = a_{n-1}r$ for $n \geq 2$.

A geometric sequence is recognized by dividing any term after the first by its predecessor. In all cases the quotient is the same ratio r.

Classroom Examples: p. 1037
Exercises 12, 18

EXAMPLE 1 Identifying a Geometric Sequence and the Common Ratio

Determine whether the sequence is geometric. If so, identify the common ratio.

a. $18, 6, 2, \dfrac{2}{3}, \ldots$ 　　　　**b.** $3, 12, 48, 240, \ldots$

Solution:

a. $18, 6, 2, \dfrac{2}{3}, \ldots$

$\dfrac{a_2}{a_1} = \dfrac{6}{18} = \dfrac{1}{3}$,　$\dfrac{a_3}{a_2} = \dfrac{2}{6} = \dfrac{1}{3}$,　　The sequence is geometric because the ratio between each term and its predecessor is the same constant.

$\dfrac{a_4}{a_3} = \dfrac{\frac{2}{3}}{2} = \dfrac{2}{3} \cdot \dfrac{1}{2} = \dfrac{1}{3}$　　The common ratio is $r = \frac{1}{3}$.

b. $3, 12, 48, 240, \ldots$

$\dfrac{a_2}{a_1} = \dfrac{12}{3} = 4$,　$\dfrac{a_3}{a_2} = \dfrac{48}{12} = 4$,　　The sequence is *not* geometric because the ratio between a_4 and a_3 is different than the ratio of other pairs of consecutive terms.

$\dfrac{a_4}{a_3} = \dfrac{240}{48} = 5$

> **Skill Practice 1** Determine whether the sequence is geometric. If so, identify the common ratio.
>
> **a.** $80, 20, 5, \dfrac{5}{4}, \ldots$ **b.** $1, 8, 64, 640, \ldots$

Example 2 illustrates the fundamental characteristic of a geometric sequence; that is, each term of a geometric sequence is a constant multiple of the preceding term.

Classroom Example: p. 1037
Exercise 22

EXAMPLE 2   Writing Several Terms of a Geometric Sequence

Write the first five terms of a geometric sequence with $a_1 = 5$ and $r = -2$.

Solution:

By definition a geometric sequence follows the pattern $a_1, a_1r, a_1r^2, a_1r^3, \ldots$. The first five terms are

$a_1 = 5$

$a_2 = 5(-2) = -10$

$a_3 = 5(-2)^2 = 20$

$a_4 = 5(-2)^3 = -40$

$a_5 = 5(-2)^4 = 80$

> **TIP** Alternatively, we can use a recursive formula: $a_1 = 5$ and $a_n = a_{n-1} \cdot r$ for $n \geq 2$.
>
> $a_2 = a_1 \cdot (-2) = 5(-2) = -10$
>
> $a_3 = a_2 \cdot (-2) = -10(-2) = 20$
>
> $a_4 = a_3 \cdot (-2) = 20(-2) = -40$
>
> $a_5 = a_4 \cdot (-2) = -40(-2) = 80$

> **Skill Practice 2** Write the first five terms of the geometric sequence with $a_1 = 2$ and $r = -3$.

In Example 3, we write an expression for the nth term of a geometric sequence given the first four terms.

Classroom Example: p. 1037
Exercise 30

EXAMPLE 3   Writing the nth Term of a Geometric Sequence

Write a formula for the nth term of the geometric sequence. $2, 3, \dfrac{9}{2}, \dfrac{27}{4}, \ldots$

Solution:

$r = \dfrac{a_2}{a_1} = \dfrac{3}{2}$ Dividing any term by its predecessor, we have a common ratio of $\frac{3}{2}$.

$a_n = a_1 r^{n-1}$ Begin with the formula for the nth term of a geometric sequence.

$a_n = 2\left(\dfrac{3}{2}\right)^{n-1}$ Substitute $a_1 = 2$ and $r = \frac{3}{2}$.

> **Skill Practice 3** Write the nth term of the geometric sequence.
>
> $4, 5, \dfrac{25}{4}, \dfrac{125}{16}, \ldots$

A graph of several terms of the geometric sequence from Example 3 is shown in Figure 11-5. The points representing the sequence coincide with points on the graph of the exponential function $f(x) = 2\left(\frac{3}{2}\right)^{x-1}$ for positive integer values of x (Figure 11-6).

Answers

1. a. Geometric; $r = \dfrac{1}{4}$

  b. Not geometric

2. $2, -6, 18, -54, 162$

3. $a_n = 4\left(\dfrac{5}{4}\right)^{n-1}$

In fact, a geometric sequence with $r > 0$ and $r \neq 1$ is an exponential function whose domain is restricted to the set of positive integers.

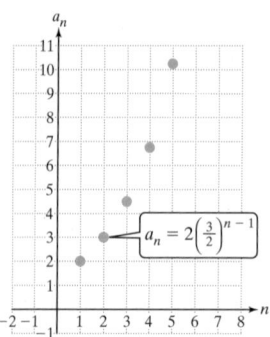

Figure 11-5

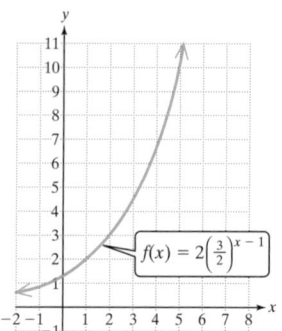

Figure 11-6

Classroom Example: p. 1037
Exercise 36

EXAMPLE 4 Finding a Specified Term of a Geometric Sequence

Find the fifth term of a geometric sequence $\{a_n\}$ given that $a_1 = 15$ and $a_2 = -9$.

Solution:

$$r = \frac{a_2}{a_1} = \frac{-9}{15} = -\frac{3}{5} \qquad \text{Divide } a_2 \text{ by } a_1 \text{ to obtain the common ratio } -\tfrac{3}{5}.$$

$$a_n = a_1 r^{n-1} \qquad \text{Use the formula for the } n\text{th term of a geometric sequence.}$$

$$a_n = 15\left(-\frac{3}{5}\right)^{n-1} \qquad \text{Substitute } a_1 = 15 \text{ and } r = -\tfrac{3}{5} \text{ to get the } n\text{th term for this sequence.}$$

$$a_5 = 15\left(-\frac{3}{5}\right)^{5-1} \qquad \text{To find } a_5, \text{ substitute 5 for } n.$$

$$a_5 = \frac{243}{125}$$

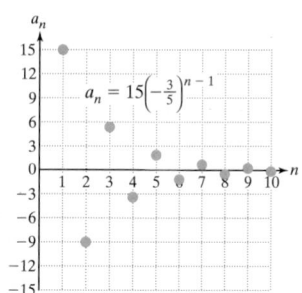

Figure 11-7

Skill Practice 4 Find the sixth term of a geometric sequence $\{a_n\}$ given that $a_1 = 64$ and $a_2 = -16$.

Example 4 illustrates that a geometric sequence with a negative common ratio is an alternating sequence. The graph of the first 10 terms of the sequence is shown in Figure 11-7.

Classroom Example: p. 1037
Exercise 48

EXAMPLE 5 Writing the nth Term of a Geometric Sequence

Given the terms $a_2 = 80$ and $a_5 = 40.96$ of a geometric sequence, find r, a_1, and a_n.

Solution:

In this example, we need to find two unknown quantities. One strategy is to begin with the formula for the nth term of the geometric sequence and substitute the known values for terms 2 and 5 of the sequence. Then solve the resulting system of nonlinear equations.

$$a_n = a_1 r^{n-1}$$
$$80 = a_1 r^{(2-1)} \longrightarrow 80 = a_1 r \qquad \text{Substitute } a_2 = 80.$$
$$40.96 = a_1 r^{(5-1)} \longrightarrow 40.96 = a_1 r^4 \qquad \text{Substitute } a_5 = 40.96.$$

Answer

4. $a_6 = -\dfrac{1}{16}$

Substitute $a_1 = \dfrac{80}{r}$ from the first equation into the second equation: $40.96 = a_1 r^4$.

$$40.96 = \frac{80}{r} r^4$$

$$40.96 = 80r^3$$

$$0.512 = r^3 \qquad \text{Divide by 80.}$$

$$r = 0.8 \qquad \text{Take the cube root of both sides.}$$

$$a_1 = \frac{80}{r} = \frac{80}{0.8} = 100 \qquad \text{To solve for } a_1, \text{ substitute } r = 0.8 \text{ into the equation } a_1 = \frac{80}{r}.$$

The value of r is 0.8, the value of a_1 is 100, and the nth term is $a_n = 100(0.8)^{n-1}$.

> **Skill Practice 5**   Given the terms $a_2 = 54$ and $a_5 = 182.25$ of a geometric sequence, find r, a_1, and a_n.

2. Evaluate Finite Geometric Series

The nth partial sum S_n of the first n terms of a geometric sequence is a **finite geometric series.** Consider the geometric series:

$$S_n = a_1 + a_1 r + a_1 r^2 + a_1 r^3 + \cdots + a_1 r^{n-1}$$

Suppose that we subtract rS_n from S_n.

$$
\begin{array}{llll}
S_n = a_1 & + a_1 r & + a_1 r^2 & + \cdots + a_1 r^{n-1} \\
rS_n = a_1 r & + a_1 r^2 & + a_1 r^3 & + \cdots + a_1 r^n \\
\hline
\end{array}
$$

$$S_n - rS_n = (a_1 - a_1 r) + (a_1 r - a_1 r^2) + (a_1 r^2 - a_1 r^3) + \cdots + (a_1 r^{n-1} - a_1 r^n)$$

$$S_n - rS_n = a_1 - a_1 r^n \qquad \text{The terms in red form a sum of zero.}$$

$$S_n(1 - r) = a_1(1 - r^n) \qquad \text{Factor each side of the equation.}$$

$$S_n = \frac{a_1(1 - r^n)}{1 - r} \qquad \text{Divide by } (1 - r).$$

> **nth Partial Sum of a Geometric Sequence**
>
> The sum S_n of the first n terms of a geometric sequence is given by
>
> $$S_n = \frac{a_1(1 - r^n)}{1 - r}$$
>
> where a_1 is the first term of the sequence and r is the common ratio, $r \neq 1$.

Classroom Example: p. 1038
Exercise 52

> **EXAMPLE 6**   **Evaluating a Finite Geometric Series**
>
> Find the sum. $\displaystyle\sum_{i=1}^{6} 4\left(\frac{1}{2}\right)^{i-1}$
>
> **Solution:**
>
> $$\sum_{i=1}^{6} 4\left(\frac{1}{2}\right)^{i-1} = 4 + 2 + 1 + \frac{1}{2} + \frac{1}{4} + \frac{1}{8}$$
>
> The individual terms in the series form a geometric sequence with $a_1 = 4$ and $r = \frac{1}{2}$. The value of the series is the sixth partial sum of the sequence of terms.
>
> $$S_n = \frac{a_1(1 - r^n)}{1 - r} = \frac{4\left[1 - \left(\frac{1}{2}\right)^6\right]}{1 - \frac{1}{2}} = \frac{4\left(1 - \frac{1}{64}\right)}{\frac{1}{2}} = \frac{63}{8} \qquad \text{Apply the formula for the } n\text{th partial sum with } n = 6.$$

Answers

5. $r = 1.5$; $a_1 = 36$; $a_n = 36(1.5)^{n-1}$

6. $\dfrac{1093}{81}$

> **Skill Practice 6**   Find the sum. $\displaystyle\sum_{i=1}^{7} 9\left(\frac{1}{3}\right)^{i-1}$

Classroom Example: p. 1038
Exercise 56

EXAMPLE 7 Evaluating a Finite Geometric Series

Find the sum of the finite geometric series. $\quad 5 + 10 + 20 + \cdots + 5120$

Solution:

The common ratio is 2 and $a_1 = 5$. The nth term of the sequence of terms can be written as $a_n = 5(2)^{n-1}$. To find the number of terms n, substitute 5120 for a_n.

$$5120 = 5(2)^{n-1}$$
$$1024 = 2^{n-1} \qquad \text{Divide both sides by 5.}$$
$$2^{10} = 2^{n-1} \qquad \text{Apply the equivalence property of exponential expressions.}$$
$$10 = n - 1$$
$$n = 11 \qquad \text{The sequence has 11 terms.}$$

> **TIP** The equivalence property of exponential expressions indicates that if $b^n = b^m$, then $n = m$. That is, if two exponential expressions with the same base are equal, then their exponents are equal.

With $a_1 = 5$, $r = 2$, and $n = 11$, we have

$$S_n = \frac{a_1(1 - r^n)}{1 - r} = \frac{5(1 - 2^{11})}{1 - 2} = 10{,}235$$

Skill Practice 7 Find the sum of the finite geometric series.

$$3 + 6 + 12 + \cdots + 768$$

3. Evaluate Infinite Geometric Series

For a positive integer n, the series $\sum_{i=1}^{n} a_i = a_1 + a_2 + a_3 + \cdots + a_n$ is called a **finite series** because there are a finite number of terms. The series $\sum_{i=1}^{\infty} a_i = a_1 + a_2 + a_3 + \cdots$ is called an **infinite series** because there are an infinite number of terms. Although it is impossible to add an infinite number of terms on a term-by-term basis, we might ask if the sum approaches a limiting value. If the nth partial sum S_n of an infinite series approaches a number L as $n \to \infty$, we say that the series **converges** and we call L the sum of the series. If a series does not converge, we say that the series **diverges** (the sum does not exist).

Consider the geometric sequence $\frac{1}{2}, \frac{1}{4}, \frac{1}{8}, \frac{1}{16}, \ldots, \left(\frac{1}{2}\right)^n, \ldots$ and the corresponding infinite geometric series $\frac{1}{2} + \frac{1}{4} + \frac{1}{8} + \frac{1}{16} + \cdots + \left(\frac{1}{2}\right)^n + \cdots$. To visualize the infinite series, suppose that we add $\frac{1}{2}$ of a pie plus $\frac{1}{4}$ of a pie plus $\frac{1}{8}$ of a pie and so on. Will we approach some finite amount of pie? From Figure 11-8, it appears that the sum will be 1 whole unit.

For an infinite geometric sequence with $-1 < r < 1$ (equivalently $|r| < 1$), the value of r^n will become smaller as n gets larger. For example:

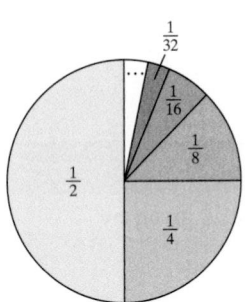

Figure 11-8

$$\approx 0.0009766 \qquad \approx 0.0000305$$

$$\frac{1}{2}, \frac{1}{4}, \frac{1}{8}, \frac{1}{16}, \ldots, \left(\frac{1}{2}\right)^{10}, \ldots, \left(\frac{1}{2}\right)^{15}, \ldots$$

In fact, for $|r| < 1$, as $n \to \infty$, $r^n \to 0$. Therefore, as n approaches infinity

$$S_n = \frac{a_1(1 - r^n)}{1 - r} \quad \text{approaches} \quad \frac{a_1(1 - 0)}{1 - r} = \frac{a_1}{1 - r}.$$

Answer

7. 1533

> ## Sum of an Infinite Geometric Series
>
> Given an infinite geometric series $a_1 + a_1r + a_1r^2 + a_1r^3 + \cdots$, with $|r| < 1$, the sum S of all terms in the series is given by
>
> $$S = \frac{a_1}{1 - r}$$
>
> *Note:* If $|r| \geq 1$, then the sum does not exist.

Classroom Examples: p. 1038
Exercises 64, 66

EXAMPLE 8   Evaluating an Infinite Geometric Series

Find the sum if possible.   **a.** $\displaystyle\sum_{i=1}^{\infty} 5\left(-\frac{1}{3}\right)^{i-1}$ **b.** $2 + \dfrac{8}{3} + \dfrac{32}{9} + \dfrac{128}{27} + \cdots$

Solution:

a. $\displaystyle\sum_{i=1}^{\infty} 5\left(-\frac{1}{3}\right)^{i-1} = 5 - \dfrac{5}{3} + \dfrac{5}{9} - \dfrac{5}{27} + \dfrac{5}{81} + \cdots$

The sum is an infinite geometric series with $a_1 = 5$ and $r = -\frac{1}{3}$.

Because $|r| = \left|-\frac{1}{3}\right| < 1$, we have $S = \dfrac{a_1}{1 - r} = \dfrac{5}{1 - \left(-\frac{1}{3}\right)} = \dfrac{15}{4}$.

The sum is $\dfrac{15}{4}$.

b. $2 + \dfrac{8}{3} + \dfrac{32}{9} + \dfrac{128}{27} + \cdots$ is a geometric series with $a_1 = 2$ and $r = \frac{4}{3}$.

Because $|r| = \left|\frac{4}{3}\right| \geq 1$, the sum does not exist.

Skill Practice 8   Find the sum if possible.

a. $\displaystyle\sum_{i=1}^{\infty} 4\left(\frac{3}{4}\right)^{i-1}$ **b.** $2 + \dfrac{5}{2} + \dfrac{25}{8} + \dfrac{125}{32} + \cdots$

In Example 9, we use the formula for the sum of an infinite geometric series to write a repeating decimal as a fraction.

Classroom Examples: p. 1038
Exercises 74, 76

EXAMPLE 9   Writing a Repeating Decimal as a Fraction

a. Write $0.\overline{5}$ as a fraction.   **b.** Write $1.9\overline{75}$ as a fraction.

Solution:

a. $0.\overline{5} = 0.5555\ldots$

$$= \underbrace{\frac{5}{10} + \frac{5}{100} + \frac{5}{1000} + \frac{5}{10,000} + \cdots}_{\substack{\text{Infinite geometric series} \\ \text{with } a_1 = \frac{5}{10} \text{ and } r = \frac{1}{10}}}$$

$\longrightarrow S = \dfrac{a_1}{1 - r} = \dfrac{\frac{5}{10}}{1 - \frac{1}{10}} = \dfrac{\frac{5}{10}}{\frac{9}{10}} = \dfrac{5}{9}$

$0.\overline{5} = \dfrac{5}{9}$

Answers

8. a. 16

　b. $r = \frac{5}{4} \geq 1$, so the sum does not exist.

b. $1.9\overline{75} = 1.9757575\ldots$

$$= \frac{19}{10} + \left[\frac{75}{1000} + \frac{75}{100{,}000} + \cdots \right]$$

Infinite geometric series with $a_1 = \frac{75}{1000}$ and $r = \frac{1}{100}$ $\longrightarrow$ $S = \dfrac{a_1}{1-r} = \dfrac{\frac{75}{1000}}{1-\frac{1}{100}} = \dfrac{\frac{75}{1000}}{\frac{99}{100}} = \dfrac{75}{990} = \dfrac{5}{66}$

$$1.9\overline{75} = \frac{19}{10} + \frac{5}{66} = \frac{326}{165}$$

Skill Practice 9 **a.** Write $0.\overline{7}$ as a fraction. **b.** Write $0.3\overline{4}$ as a fraction.

When money is infused into the economy, a percentage of the money received by individuals or businesses is often respent over and over again. Economists call this the **multiplier effect.**

Classroom Example: p. 1038
Exercise 82

EXAMPLE 10 Investigating an Application of an Infinite Geometric Series—The Multiplier Effect

Suppose that \$200 million is spent annually by tourists in a certain state. Further suppose that 75% of the money is respent in the state and then respent over and over again, each time at a rate of 75%. Determine the theoretical total amount spent from the initial \$200 million if the money can be respent an infinite number of times.

Solution:

The total amount spent can be represented by the infinite geometric series where all values are in \$ millions.

75% of 200   75% of 150

$$200 + 150 + 112.5 + \cdots$$

We have $a_1 = 200$ and $r = 0.75$. Because $|r| = |0.75| < 1$, we have

$$S = \frac{a_1}{1-r} = \frac{200}{1-0.75} = 800.$$

Theoretically, the total amount spent is \$800 million.

Skill Practice 10 Suppose that after a tax rebate an individual spends \$210. The money is then respent over and over again, each time at a rate of 70%. Determine the total amount spent. Assume that the money can be respent an infinite number of times.

4. Find the Value of an Annuity

In Section 3.2, we studied applications of exponential functions involving compound interest. We learned that if P dollars is invested in an account at interest rate r, compounded annually for t years, then the amount A in the account is given by

$$A = P(1 + r)^t$$

However, rather than making a one-time lump sum payment of P dollars, many individuals will invest smaller amounts at regular and more frequent intervals. Such a sequence of fixed payments made (or received) by an individual over a fixed period of time is called an **annuity.**

Answers

9. a. $\dfrac{7}{9}$ **b.** $\dfrac{31}{90}$

10. \$700

Suppose that an individual invests P dollars at the *end* of each year for 4 yr at interest rate r. The money deposited at the end of the first year will have 3 yr in which to earn interest. The money invested at the end of the second year will earn interest for 2 yr, and the money invested at the end of the third year will earn interest for 1 yr. However, the money invested at the end of the fourth year will not earn any interest.

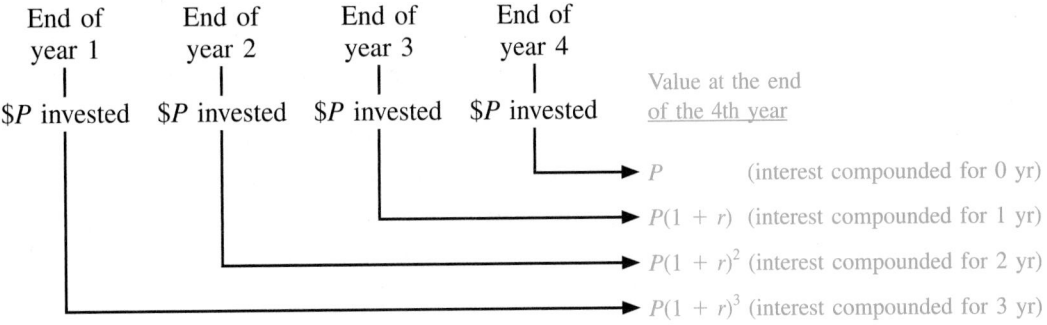

The total amount A, also called the **future value** of the annuity, is given by

$$A = P + P(1 + r) + P(1 + r)^2 + P(1 + r)^3$$

This is a finite geometric series with $a_1 = P$ and common ratio of $(1 + r)$. Using the relationship $S_n = \dfrac{a_1(1 - r^n)}{1 - r}$, the total amount invested for t years is given by

$$A = \frac{P[1 - (1 + r)^t]}{1 - (1 + r)} = \frac{P[1 - (1 + r)^t]}{-r} \text{ or simply } A = \frac{P[(1 + r)^t - 1]}{r}$$

If deposits are made n times per year, then the interest rate per compounding period is $\frac{r}{n}$, and the total number of times the money is compounded is nt.

Instructor Note:
Encourage students to be careful with the order of operations when applying the formula for the future value of an annuity.

TIP In this section we study **ordinary annuities.** These are annuities in which money is invested at the *end* of the compounding periods. If the money were invested at the beginning of the compounding periods, such an investment is called an **annuity due** and the future value is computed with a different formula.

Future Value of an Ordinary Annuity

Suppose that P dollars is invested at the end of each compounding period n times per year at interest rate r. Then the value A (in \$) of the annuity after t years is given by

$$A = \frac{P\left[\left(1 + \frac{r}{n}\right)^{nt} - 1\right]}{\frac{r}{n}}$$

Classroom Example: p. 1039
Exercise 88

EXAMPLE 11 Finding the Value of an Annuity

Suppose that an employee working for a state college puts aside \$150 at the end of each month in a tax-sheltered annuity. The annuity pays 6% annual interest compounded monthly. Suppose that the employee begins contributing at age 28.

  a. How much will the annuity be worth by the time the employee reaches age 62?

  b. How much interest will have been earned?

Point of Interest

The contributions made to the investment in Example 11 are **tax deferred.** This means that the principal invested each month has not yet been taxed. Because the principal is invested pretax, the individual potentially has more money available to invest. Furthermore, no taxes are paid on either the principal or interest until the money is withdrawn. This type of plan is meant as a long-term investment, and withdrawals are typically taken after age 59½ without penalty.

Answers

11. a. $81,007.17   **b.** $51,007.17

Solution:

a. $A = \dfrac{P\left[\left(1 + \frac{r}{n}\right)^{nt} - 1\right]}{\frac{r}{n}}$

$A = \dfrac{150\left[\left(1 + \frac{0.06}{12}\right)^{(12)(34)} - 1\right]}{\frac{0.06}{12}}$

$\begin{aligned} P &= \$150 \\ r &= 0.06 \\ n &= 12 \text{ (the money is invested monthly)} \\ t &= 62 - 28 = 34 \text{ yr} \end{aligned}$

$A = 199{,}548.50$

The annuity will be worth $199,548.50 when the employee reaches age 62.

b. $150 is invested 12 times per year for 34 yr.

Principal invested: ($150)(12)(34) = $61,200

The amount of interest is $199,548.50 − $61,200 = $138,348.50.

Skill Practice 11   Suppose that an employee contributes $100 to an annuity at the end of each month for 25 yr. If the annuity earns 7%,

a. Determine the value of the annuity at the end of the 25-yr period.

b. How much interest will be earned?

SECTION 11.3   Practice Exercises

Prerequisite Review

R.1. Simplify. $-\dfrac{5}{6} \div \dfrac{2}{3} \quad -\dfrac{5}{4}$

R.2. Solve for x. $3^x = 81$ $\{4\}$

R.3. For $g(x) = \left(\dfrac{1}{4}\right)^x$ find $g(0)$, $g(1)$, $g(2)$, $g(-1)$, and $g(-2)$. $g(0) = 1, g(1) = \dfrac{1}{4}, g(2) = \dfrac{1}{16}, g(-1) = 4, g(-2) = 16$

R.4. For $h(x) = 16^x$ find $h(0)$, $h(1)$, and $h(-1)$. $h(0) = 1, h(1) = 16, h(-1) = \dfrac{1}{16}$

R.5. Suppose that an investor deposits $14,000 in a savings account for 10 yr at 2% interest. Use the model $A(t) = P\left(1 + \dfrac{r}{n}\right)^{nt}$ for P dollars in principal invested at an interest rate r compounded n times per year for t years for the following compounding options. Round to the nearest dollar.

a. Interest compounded annually $17,066

b. Interest compounded quarterly $17,091

c. Interest compounded monthly $17,097

Concept Connections

1. A ___geometric___ sequence is a sequence in which each term after the first is the product of the preceding term and a fixed nonzero real number, called the common ___ratio___, r.

2. The nth term of a geometric sequence with first term a_1 and common ratio r is given by $a_n = $ ___$a_1 r^{n-1}$___.

3. The nth partial sum S_n of the first n terms of a geometric sequence is a (finite/infinite) geometric series.   finite

4. The sum S_n of the first n terms of a geometric sequence with first term a_1 and common ratio r is given by the formula $S_n = $ ___$\dfrac{a_1(1 - r^n)}{1 - r}$___.

5. Given a geometric sequence with $|r| < 1$, the value of $r^n \rightarrow$ ___0___ as $n \rightarrow \infty$.

6. Given an infinite geometric series with first term a_1 and common ratio r, if $|r| < 1$, then the sum S is given by the formula $S = $ ___$\dfrac{a_1}{1 - r}$___. If $|r| \geq 1$, then the sum (does/does not) exist.   does not

7. A sequence of payments made at equal intervals over a fixed period of time is called an ___annuity___.

8. Suppose that an infinite series $a_1 + a_2 + a_3 + \cdots + a_n$ approaches a value L as $n \rightarrow \infty$. Then the series ___converges___. Otherwise, the series ___diverges___.

Objective 1: Identify Specific and General Terms of a Geometric Sequence

For Exercises 9–18, determine whether the sequence is geometric. If so, find the value of r. (See Example 1)

9. 6, 18, 54, 162, …
Yes; $r = 3$

10. 4, 20, 100, 500, …
Yes; $r = 5$

11. $-7, \dfrac{7}{2}, -\dfrac{7}{4}, \dfrac{7}{8}, \ldots$ Yes; $r = -\dfrac{1}{2}$

12. $5, -\dfrac{5}{3}, \dfrac{5}{9}, -\dfrac{5}{27}, \ldots$ Yes; $r = -\dfrac{1}{3}$

13. 3, 12, 60, 360, … No

14. 7, 14, 42, 88, … No

15. $\sqrt{5}, 5, 5\sqrt{5}, 25, \ldots$
Yes; $r = \sqrt{5}$

16. $\sqrt[3]{7}, \sqrt[3]{49}, 7, 7\sqrt[3]{7}, \ldots$
Yes; $r = \sqrt[3]{7}$

17. $2, \dfrac{4}{t}, \dfrac{8}{t^2}, \dfrac{16}{t^3}, \ldots$ Yes; $r = \dfrac{2}{t}$

18. $\dfrac{5}{a^2}, \dfrac{15}{a^4}, \dfrac{45}{a^6}, \dfrac{135}{a^8}, \ldots$ Yes; $r = \dfrac{3}{a^2}$

For Exercises 19–24, write the first five terms of a geometric sequence $\{a_n\}$ based on the given information about the sequence. (See Example 2)

19. $a_1 = 7$ and $r = 2$ 7, 14, 28, 56, 112

20. $a_1 = 2$ and $r = 3$ 2, 6, 18, 54, 162

21. $a_1 = 24$ and $r = -\dfrac{2}{3}$ $24, -16, \dfrac{32}{3}, -\dfrac{64}{9}, \dfrac{128}{27}$

22. $a_1 = 80$ and $r = -\dfrac{4}{5}$ $80, -64, \dfrac{256}{5}, -\dfrac{1024}{25}, \dfrac{4096}{125}$

23. $a_1 = 36$ and $a_n = \dfrac{1}{2}a_{n-1}$ for $n \geq 2$ $36, 18, 9, \dfrac{9}{2}, \dfrac{9}{4}$

24. $a_1 = 27, a_n = \dfrac{1}{3}a_{n-1}$ for $n \geq 2$ $27, 9, 3, 1, \dfrac{1}{3}$

For Exercises 25–30, write a formula for the nth term of the geometric sequence. (See Example 3)

25. 5, 10, 20, 40, … $a_n = 5(2)^{n-1}$

26. 3, 6, 12, 24, … $a_n = 3(2)^{n-1}$

27. $-2, -1, -\dfrac{1}{2}, -\dfrac{1}{4}, \ldots$ $a_n = -2\left(\dfrac{1}{2}\right)^{n-1}$

28. $-8, -2, -\dfrac{1}{2}, -\dfrac{1}{8}, \ldots$ $a_n = -8\left(\dfrac{1}{4}\right)^{n-1}$

29. $\dfrac{16}{3}, -4, 3, -\dfrac{9}{4}, \ldots$ $a_n = \dfrac{16}{3}\left(-\dfrac{3}{4}\right)^{n-1}$

30. $\dfrac{18}{5}, -\dfrac{6}{5}, \dfrac{2}{5}, -\dfrac{2}{15}, \ldots$ $a_n = \dfrac{18}{5}\left(-\dfrac{1}{3}\right)^{n-1}$

31. A farmer depreciates a \$100,000 tractor. He estimates that the resale value of the tractor n years after purchase is 85% of its value from the previous year.

   a. Write a formula for the nth term of a sequence that represents the resale value of the tractor n years after purchase. $a_n = 100,000(0.85)^n$ or $85,000(0.85)^{n-1}$

   b. What will the resale value be 5 yr after purchase? Round to the nearest \$1000.   \$44,000

32. A Coulter Counter is a device used to count the number of microscopic particles in a fluid, most notably, cells in blood. A hospital depreciates a \$9000 Coulter Counter at a rate of 75% per year after purchase.

   a. Write a formula for the nth term of a sequence that represents the resale value of the device n years after purchase. $a_n = 9000(0.75)^n$ or $6750(0.75)^{n-1}$

   b. What will the resale value be 4 yr after purchase? Round to the nearest \$100.   \$2800

33. Doctors in a certain city report 24 confirmed cases of the flu to the health department. At that time, the health department declares a flu epidemic. If the number of reported cases increases by roughly 30% each week thereafter, find the number of cases 10 weeks after the initial report. Round to the nearest whole unit.   331 cases

34. After a 5-yr slump in the real estate market, housing prices stabilize and even begin to appreciate in value. One homeowner buys a house for \$140,000 and finds that the value of the property increases by 3% per year thereafter. Assuming that the trend continues, find the value of the home 15 yr later. Round to the nearest \$1000.   \$218,000

For Exercises 35–42, find the indicated term of a geometric sequence from the given information. (See Example 4)

35. $a_1 = 12$ and $a_2 = -8$. Find the sixth term. $a_6 = -\dfrac{128}{81}$

36. $a_1 = 16$ and $a_2 = -12$. Find the fifth term. $a_5 = \dfrac{81}{16}$

37. $a_1 = 2$ and $a_4 = 16$. Find a_{12}. $a_{12} = 4096$

38. $a_1 = 4$ and $a_4 = 108$. Find a_{10}. $a_{10} = 78{,}732$

39. $a_2 = -6$ and $r = \dfrac{1}{2}$. Find a_7. $a_7 = -\dfrac{3}{16}$

40. $a_2 = -15$ and $r = \dfrac{1}{3}$. Find a_8. $a_8 = -\dfrac{5}{243}$

41. $a_5 = -\dfrac{16}{9}$ and $r = -\dfrac{2}{3}$. Find a_1. $a_1 = -9$

42. $a_6 = \dfrac{5}{16}$ and $r = -\dfrac{1}{2}$. Find a_1. $a_1 = -10$

43. If the second and third terms of a geometric sequence are 15 and 75, what is the first term? $a_1 = 3$

44. If the second and third terms of a geometric sequence are 4 and 1, what is the first term? $a_1 = 16$

For Exercises 45–48, find a_1 and r for a geometric sequence $\{a_n\}$ from the given information. (See Example 5)

45. $a_2 = 18$ and $a_5 = 144$ $a_1 = 9; r = 2$

46. $a_2 = 21$ and $a_7 = 5103$ $a_1 = 7; r = 3$

47. $a_3 = 72$ and $a_6 = -\dfrac{243}{8}$ $a_1 = 128; r = -\dfrac{3}{4}$

48. $a_3 = 45$ and $a_6 = -\dfrac{243}{25}$ $a_1 = 125; r = -\dfrac{3}{5}$

Objectives 2–3: Evaluate Finite and Infinite Geometric Series

For Exercises 49–72, find the sum of the geometric series, if possible. (See Examples 6–8)

49. $\displaystyle\sum_{n=1}^{10} 3(2)^{n-1}$ 3069

50. $\displaystyle\sum_{n=1}^{8} 4(3)^{n-1}$ 13,120

51. $\displaystyle\sum_{k=1}^{7} 6\left(\frac{2}{3}\right)^{k-1}$ $\frac{4118}{243}$

52. $\displaystyle\sum_{j=1}^{7} 2\left(\frac{3}{4}\right)^{j-1}$ $\frac{14,197}{2048}$

53. $15 + 5 + \dfrac{5}{3} + \dfrac{5}{9} + \dfrac{5}{27} + \dfrac{5}{81}$ $\frac{1820}{81}$

54. $50 + 10 + 2 + \dfrac{2}{5} + \dfrac{2}{25} + \dfrac{2}{125}$ $\frac{7812}{125}$

55. $2 + 6 + 18 + \cdots + 13,122$ 19,682

56. $4 + 12 + 36 + \cdots + 78,732$ 118,096

57. $1 + \dfrac{2}{3} + \dfrac{4}{9} + \cdots + \dfrac{32}{243}$ $\frac{665}{243}$

58. $\dfrac{8}{3} + 2 + \dfrac{3}{2} + \cdots + \dfrac{243}{512}$ $\frac{14,197}{1536}$

59. $1 + \dfrac{1}{5} + \dfrac{1}{25} + \dfrac{1}{125} + \cdots$ $\frac{5}{4}$

60. $1 + \dfrac{1}{6} + \dfrac{1}{36} + \dfrac{1}{216} + \cdots$ $\frac{6}{5}$

61. $-2 - \dfrac{1}{2} - \dfrac{1}{8} - \dfrac{1}{32} - \cdots$ $-\frac{8}{3}$

62. $-5 - 1 - \dfrac{1}{5} - \dfrac{1}{25} - \cdots$ $-\frac{25}{4}$

63. $2 + 8 + 32 + 128 + \cdots$ Sum does not exist

64. $1 + 6 + 36 + 216 + \cdots$ Sum does not exist

65. $\displaystyle\sum_{j=1}^{\infty}\left(\frac{2}{3}\right)^{j-1}$ 3

66. $\displaystyle\sum_{i=1}^{\infty}\left(\frac{3}{4}\right)^{i-1}$ 4

67. $\displaystyle\sum_{k=1}^{\infty}\left(\frac{3}{2}\right)^{k-1}$ Sum does not exist

68. $\displaystyle\sum_{n=1}^{\infty}\left(\frac{4}{3}\right)^{n-1}$ Sum does not exist

69. $\displaystyle\sum_{i=1}^{12} 4(2)^{i}$ [*Hint*: Rewrite the expression within the summation so that the base of 2 appears to the $(i-1)$ power.] 32,760

70. $\displaystyle\sum_{j=1}^{10} 5(2)^{j}$ 10,230

71. $\displaystyle\sum_{n=3}^{\infty} 4\left(\frac{1}{2}\right)^{n-1}$ 2
(*Hint*: Evaluate the infinite sum from $n = 1$ to infinity. Then subtract the terms corresponding to $n = 1$ and $n = 2$.)

72. $\displaystyle\sum_{n=4}^{\infty} 18\left(\frac{1}{3}\right)^{n-1}$ 1

For Exercises 73–80, write the repeating decimal as a fraction. (See Example 9)

73. $0.\overline{8}$ $\frac{8}{9}$

74. $0.\overline{2}$ $\frac{2}{9}$

75. $0.\overline{64}$ $\frac{29}{45}$

76. $0.7\overline{8}$ $\frac{71}{90}$

77. $0.\overline{81}$ $\frac{9}{11}$

78. $0.\overline{72}$ $\frac{8}{11}$

79. $3.4\overline{25}$ $\frac{3391}{990}$

80. $4.1\overline{62}$ $\frac{4121}{990}$

81. Bike Week in Daytona Beach brings an estimated 500,000 people to the town. Suppose that each person spends an average of $300.

  a. How much money is infused into the local economy during Bike Week? $150,000,000

  b. If the money is respent in the community over and over again at a rate of 68%, determine the total amount spent. Assume that the money is respent an infinite number of times. (**See Example 10**) $468,750,000

82. An individual with questionable integrity prints and spends $12,000 in counterfeit money. If the "money" is respent over and over again each time at a rate of 76%, determine the total amount spent. Assume that the "money" is respent an infinite number of times without being detected. $50,000

83. Rafael received an inheritance of $18,000. He saves $6480 and then spends $11,520 of the money on college tuition, books, and living expenses for school. If the money is respent over and over again in the community an infinite number of times, at a rate of 64%, determine the total amount spent. $32,000

84. A tax rebate returns $100 million to individuals in the community. Suppose that $25,000,000 is put into savings, and that $75,000,000 is spent. If the money is spent over and over again an infinite number of times, each time at a rate of 75%, determine the total amount spent. $300,000,000

Objective 4: Find the Value of an Annuity

For Exercises 85–86, find the value of an ordinary annuity in which regular payments of P dollars are made at the end of each compounding period, n times per year, at an interest rate r for t years. (See Example 11)

  85. $P = \$200$, $n = 12$, $r = 5\%$, $t = 30$ yr $166,451.73

  86. $P = \$100$, $n = 24$, $r = 5.5\%$, $t = 28$ yr $159,551.08

87. a. An employee invests $100 per month in an ordinary annuity. If the interest rate is 6%, find the value of the annuity after 20 yr. (**See Example 11**) $46,204.09

b. If the employee invests $200 instead of $100 at 6%, find the value of the annuity after 20 yr. Compare the result to part (a). $92,408.18; The value of the annuity doubles.

c. If the employee invests $100 per month in the annuity at 6% interest, find the value after 40 yr. Compare the result to part (a).
$199,149.07; The value of the annuity more than doubles.

88. a. An employee invests $500 per month in an ordinary annuity. If the interest rate is 5%, find the value of the annuity after 18 yr. $174,601.01

b. If the employee invests $1000 per month in the annuity instead of $500 at 5% interest, find the value of the annuity after 18 yr. Compare the result to part (a).
$349,202.02; The value of the annuity doubles.

c. If the employee invests $500 per month in the annuity at 5% interest, find the value of the annuity after 36 yr. Compare the result to part (a).
$603,247.96; The value of the annuity more than doubles.

Mixed Exercises

89. a. Given a geometric sequence whose nth term is $a_n = 6(0.4)^n$, are the terms of this sequence increasing or decreasing? Decreasing

b. Given a geometric sequence whose nth term is $a_n = 3(1.4)^n$, are the terms of this sequence increasing or decreasing? Increasing

90. a. Given a geometric sequence whose nth term is $a_n = -8(0.2)^n$, are the terms of this sequence increasing or decreasing? Increasing

b. Given a geometric sequence whose nth term is $a_n = -5(1.6)^n$, are the terms of this sequence increasing or decreasing? Decreasing

For Exercises 91–94, match the sequence with its graph.

91. $a_n = 5(0.8)^n$ Graph b **92.** $a_n = 2(1.2)^n$ Graph c **93.** $a_n = 5(-0.8)^n$ Graph d **94.** $a_n = 2(-1.2)^n$ Graph a

a. 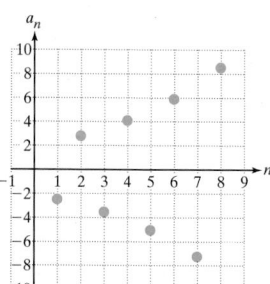 **b.** 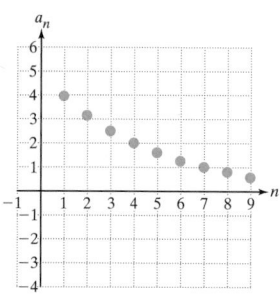 **c.** **d.**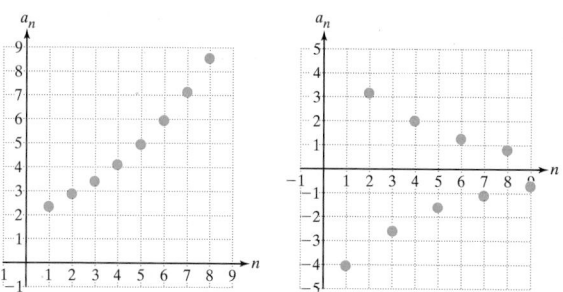

95. The initial swing (one way) of a pendulum makes an arc of 24 in. Each swing (one way) thereafter makes an arc of 98% of the length of the previous swing. What is the total arc length that the pendulum travels? 1200 in. or 100 ft.

96. The initial swing (one way) of a pendulum makes an arc of 4 ft. Each swing (one way) thereafter makes an arc of 90% of the length of the previous swing. What is the total arc length that the pendulum travels? 40 ft

97. A child drops a ball from a height of 4 ft. With each bounce, the ball rebounds to 50% of its original height. After the ball falls from its initial height of 4 ft, the vertical distance traveled for every bounce thereafter is doubled (the ball travels up and down).

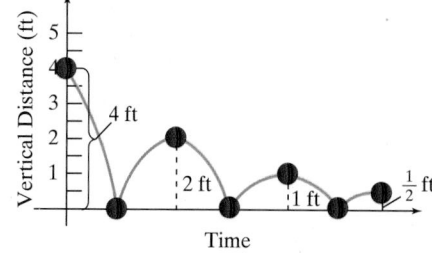

The total vertical distance traveled by the ball is given by

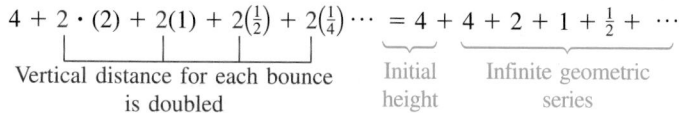

$$4 + 2 \cdot (2) + 2(1) + 2\left(\tfrac{1}{2}\right) + 2\left(\tfrac{1}{4}\right) \cdots = 4 + 4 + 2 + 1 + \tfrac{1}{2} + \cdots$$

Vertical distance for each bounce is doubled · Initial height · Infinite geometric series

Determine the total vertical distance traveled by the ball. 12 ft

98. A ball is dropped from a height of 12 ft. With each bounce, the ball rebounds to $\frac{3}{4}$ of its height. Determine the total vertical distance traveled by the ball. 84 ft

99. Suppose that an individual is paid $0.01 on day 1 and every day thereafter, the payment is doubled.

a. Write a formula for the nth term of a sequence that gives the payment (in $) on day n.
$a_n = 0.01(2)^{n-1}$ (dollars)

b. How much will the individual earn on day 10? day 20? and day 30? Day 10: $5.12; Day 20: $5242.88; Day 30: $5,368,709.12

c. What is the total amount earned in 30 days? $10,737,418.23

100. The vibration of sound is measured in cycles per second, also called hertz (Hz). The frequency for middle C on a piano is 256 Hz. The C above middle C (one octave above) is 512 Hz. The frequencies of musical notes follow a geometric progression.

a. Find the frequency for C two octaves above middle C. 1024 Hz

b. Find the frequency for C one octave *below* middle C. 128 Hz

101. The yearly salary for job A is $60,000 initially with an annual raise of $3000 every year thereafter. The yearly salary for job B is $56,000 for year 1 with an annual raise of 6%.

   a. Consider a sequence representing the salary for job A for year *n*. Is this an arithmetic or geometric sequence? Find the total earnings for job A over 20 yr. Arithmetic; $1,770,000

   b. Consider a sequence representing the salary for job B for year *n*. Is this an arithmetic or geometric sequence? Find the total earnings for job B over 20 yr. Round to the nearest dollar. Geometric; $2,059,993

   c. What is the difference in total salary between the two jobs over 20 yr? $289,993

102. a. Jacob has a job that pays $48,000 the first year. He receives a 4% raise each year thereafter. Find the sum of his yearly salaries over a 20-yr period. Round to the nearest dollar. $1,429,348

   b. Cherise has a job that pays $48,000 the first year. She receives a 4.5% raise each year thereafter. Find the sum of her yearly salaries over a 20-yr period. Round to the nearest dollar. $1,505,828

   c. How much more will Cherise earn than Jacob over the 20-yr period? $76,480

103. If a fair coin is flipped *n* times, the number of head/tail arrangements follows a geometric sequence. In the figure, if the coin is flipped 1 time, there are two possible outcomes, H or T. If the coin is flipped 2 times, then there are four possible outcomes: HH, HT, TH, and TT.

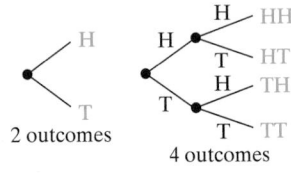

2 outcomes   4 outcomes

   a. Write a formula for the *n*th term of a sequence representing the number of outcomes if a fair coin is flipped *n* times. $a_n = 2^n$

   b. How many outcomes are there if a fair coin is flipped 10 times? 1024

104. An ancestor is a person from whom an individual is descended (a parent, a grandparent, a great-grandparent, and so on). Every individual has two biological parents, a mother and a father. The mother and father in turn each have two parents and so on.

1st Generation	2nd Generation	3rd Generation
2 parents	4 grandparents	8 great-grandparents

The terms of the sequence 2, 4, 8, 16, ... give the additional number of ancestors for each generation back in time. Determine the total number of ancestors that an individual has going back 12 generations. 8190

Write About It

105. Explain the difference between an arithmetic sequence and a geometric sequence. In an arithmetic sequence, the difference between a term and its predecessor is a fixed constant. In a geometric sequence, the ratio between a term and its predecessor is a fixed constant.

106. Explain why a finite number of terms is not sufficient to determine whether an infinite sequence is arithmetic or geometric. For example, explain why 4, 16, ... can be arithmetic, geometric, or neither.

107. A geometric sequence $a_1, a_2, a_3, \ldots$ has a common ratio *r*. Explain why the sequence $a_1, a_3, a_5, \ldots$ is also geometric and determine the common ratio. Given the sequence $a_1, a_2, a_3, \ldots$, each term after the first is obtained by multiplying the preceding term by *r*. Therefore, every other term is obtained by multiplying by r^2. Therefore, the sequence $a_1, a_3, a_5, \ldots$ is geometric with common ratio r^2.

Expanding Your Skills

108. Show that $x, x + 2, x + 4, x + 6, \ldots$ is *not* a geometric sequence.

109. If $a_1, a_2, a_3, a_4, \ldots$ is a geometric sequence with common ratio *r*, show that $\frac{1}{a_1}, \frac{1}{a_2}, \frac{1}{a_3}, \frac{1}{a_4} \ldots$ is also a geometric sequence and determine the value of *r*.

110. Suppose that $a_1, a_2, a_3, \ldots$ is an arithmetic sequence with common difference *d*. Show that $10^{a_1}, 10^{a_2}, 10^{a_3}, 10^{a_4}, \ldots$ is a geometric sequence and find the common ratio *r*.

111. Suppose that $a_1, a_2, a_3, \ldots$ is a geometric sequence with $r > 0$ and $a_1 > 0$. Show that the sequence $\log a_1$, $\log a_2, \log a_3, \ldots$ is arithmetic and find the common difference *d*.

112. Given the series $\frac{1}{2} + \frac{1}{4} + \frac{1}{8} + \frac{1}{16} + \cdots$,

   a. Find the sum. 1

   b. How many terms must be taken so that the *n*th partial sum is within $\frac{1}{100}$ of the actual sum? 7 or more terms must be taken.

113. Determine whether the sequence ln 1, ln 2, ln 4, ln 8, ... is arithmetic or geometric. If the sequence is arithmetic, find *d*. If the sequence is geometric, find *r*. Arithmetic; $d = \ln 2$

PROBLEM RECOGNITION EXERCISES

Comparing Arithmetic and Geometric Sequences and Series

For Exercises 1–10, determine if the nth term of the sequence defines an arithmetic sequence, a geometric sequence, or neither. If the sequence is arithmetic, find the common difference d. If the sequence is geometric, find the common ratio r.

1. $a_n = \dfrac{2n}{3} + \dfrac{1}{4}$ Arithmetic; $d = \dfrac{2}{3}$

2. $a_n = -\dfrac{3}{5}n - \dfrac{1}{3}$ Arithmetic; $d = -\dfrac{3}{5}$

3. $a_n = (-1)^{n-1}$ Geometric; $r = -1$

4. $a_n = (-2)^n$ Geometric; $r = -2$

5. $a_n = \dfrac{n-1}{n+1}$ Neither

6. $a_n = \dfrac{3n-4}{2n+1}$ Neither

7. $a_n = 5(\sqrt{5})^{n-1}$ Geometric; $r = \sqrt{5}$

8. $a_n = 6(\sqrt{2})^{n+1}$ Geometric; $r = \sqrt{2}$

9. $a_n = 5 + \sqrt{2}n$ Arithmetic; $d = \sqrt{2}$

10. $a_n = \dfrac{\sqrt{3}}{2}n + 4$ Arithmetic; $d = \dfrac{\sqrt{3}}{2}$

For Exercises 11–28, evaluate the sum if possible.

11. $\displaystyle\sum_{i=1}^{1506} 5$ 7530

12. $\displaystyle\sum_{i=1}^{417} (-3)$ -1251

13. $\displaystyle\sum_{i=1}^{716} (-1)^{i+1}$ 0

14. $\displaystyle\sum_{j=1}^{2651} (-1)^{j}$ -1

15. $\displaystyle\sum_{n=1}^{3} \dfrac{n-2}{n+1}$ $-\dfrac{1}{4}$

16. $\displaystyle\sum_{j=1}^{4} (j^4 - 2j^2)$ 294

17. $\displaystyle\sum_{n=1}^{4} -3\left(\dfrac{1}{3}\right)^{n-1}$ $-\dfrac{40}{9}$

18. $\displaystyle\sum_{n=1}^{5} -2\left(\dfrac{3}{4}\right)^{n-1}$ $-\dfrac{781}{128}$

19. $\displaystyle\sum_{n=1}^{\infty} 6\left(\dfrac{5}{3}\right)^{n-1}$ The sum does not exist.

20. $\displaystyle\sum_{i=1}^{\infty} 7\left(\dfrac{4}{3}\right)^{i-1}$ The sum does not exist.

21. $\displaystyle\sum_{i=1}^{27} (-6i - 4)$ -2376

22. $\displaystyle\sum_{i=1}^{41} (-5i - 3)$ -4428

23. $\displaystyle\sum_{n=1}^{\infty} 8\left(\dfrac{1}{2}\right)^{n}$ 8

24. $\displaystyle\sum_{n=1}^{\infty} 12\left(\dfrac{1}{3}\right)^{n}$ 6

25. $36 + 30 + 25 + \dfrac{125}{6} + \cdots$ 216

26. $54 + 36 + 24 + 16 + \cdots$ 162

27. $3 + 11 + 19 + 27 + \cdots + 363$ 8418

28. $36 + 29 + 22 + 15 + \cdots + (-419)$ $-12{,}639$

SECTION 11.4   Mathematical Induction

OBJECTIVES

1. Prove a Statement Using Mathematical Induction
2. Prove a Statement Using the Extended Principle of Mathematical Induction

1. Prove a Statement Using Mathematical Induction

In this section, we present a technique of mathematical proof that enables us to prove the validity of a statement that is true over the set of positive integers. This technique is called mathematical induction.

Consider the following sum:

$$\frac{1}{1 \cdot 2} + \frac{1}{2 \cdot 3} + \frac{1}{3 \cdot 4} + \frac{1}{4 \cdot 5} + \cdots + \frac{1}{n(n+1)}$$

The sequence of terms that make up the sum is neither arithmetic nor geometric. Therefore, we have no formula readily available to evaluate the sum of the first n terms.

We might consider evaluating the nth partial sum for several positive integers n to determine if a pattern exists. For each value of n, we have a corresponding statement P_n involving the value of the partial sum.

$n = 1$ $\qquad \dfrac{1}{1 \cdot 2} = \dfrac{1}{2}$ $\qquad$ Call this statement P_1.

$n = 2$ $\qquad \dfrac{1}{1 \cdot 2} + \dfrac{1}{2 \cdot 3} = \dfrac{1}{2} + \dfrac{1}{6} = \dfrac{2}{3}$ $\qquad$ Call this statement P_2.

$n = 3$ $\qquad \dfrac{1}{1 \cdot 2} + \dfrac{1}{2 \cdot 3} + \dfrac{1}{3 \cdot 4} = \dfrac{1}{2} + \dfrac{1}{6} + \dfrac{1}{12} = \dfrac{3}{4}$ $\qquad$ Call this statement P_3.

$n = 4$ $\qquad \dfrac{1}{1 \cdot 2} + \dfrac{1}{2 \cdot 3} + \dfrac{1}{3 \cdot 4} + \dfrac{1}{4 \cdot 5} = \dfrac{1}{2} + \dfrac{1}{6} + \dfrac{1}{12} + \dfrac{1}{20} = \dfrac{4}{5}$ $\qquad$ Call this statement P_4.

> **TIP** The proof of the statement
>
> $$\dfrac{1}{1 \cdot 2} + \dfrac{1}{2 \cdot 3} + \cdots + \dfrac{1}{n(n+1)}$$
> $$= \dfrac{n}{n+1}$$
>
> is given in Example 2.

For the first four terms, the numerator is the same as the term number n and the denominator is one more than the term number. From this observation, we might make the following hypothesis, which we call P_n.

$$P_n: \quad \dfrac{1}{1 \cdot 2} + \dfrac{1}{2 \cdot 3} + \dfrac{1}{3 \cdot 4} + \dfrac{1}{4 \cdot 5} + \cdots + \dfrac{1}{n(n+1)} = \dfrac{n}{n+1}$$

In words, the hypothesis P_n suggests that the sum of the first n terms of the sequence

$$\left\{ \dfrac{1}{n(n+1)} \right\} \text{ is } \dfrac{n}{n+1}. \text{ Equivalently, } \sum_{i=1}^{n} \dfrac{1}{i(i+1)} = \dfrac{n}{n+1}.$$

We have shown that the statement P_n is true for $n = 1, 2, 3,$ and 4, but what about values of n thereafter? Because there are infinitely many positive integers, there are actually infinitely many statements to prove and we cannot approach them on a case-by-case basis. Instead we will use the principle of mathematical induction.

> ### Principle of Mathematical Induction
>
> Let P_n be a statement involving the positive integer n, and let k be an arbitrary positive integer. Then P_n is true for all positive integers n if
>
> **1.** P_1 is true, and
> **2.** The truth of P_k implies the truth of P_{k+1}.

P_n represents the statements, P_1, P_2, P_3, and so on. Mathematical induction is a two-part process to prove all the statements in the sequence. We first need to show that the statement is true for $n = 1$. That is, prove that P_1 is true. The second part involves proving that if any statement P_k in the sequence is true, then the statement that follows is also true (P_k implies P_{k+1}).

Intuitively, mathematical induction is similar to the sequential effect of falling dominos. Pushing the first domino down is analogous to proving P_1. Then showing that any falling domino in the sequence will cause the next domino to fall is the idea behind part 2 in mathematical induction. That is, if domino 1 falls, and any falling domino makes the next domino in line fall, then all the dominos will fall.

In Examples 1 and 2, we use mathematical induction to prove a statement involving a summation.

Classroom Example: p. 1047
Exercise 4

EXAMPLE 1   **Using Mathematical Induction**

Use mathematical induction to prove that $1 + 3 + 5 + 7 + \cdots + (2n - 1) = n^2$.

Solution:

Let P_n denote the statement $1 + 3 + 5 + 7 + \cdots + (2n - 1) = n^2$.

  1. Show that P_1 is true.

    For $n = 1$, the sum is 1 which equals $(1)^2$.

    Therefore, P_1 is true.

> We need to show that P_1 is true. That is, we need to show that the left and right sides of the statement are equal for $n = 1$.

  2. Next, assume that P_k is true and show that P_{k+1} is true.

    Assume that P_k is true.

    Assume that $1 + 3 + 5 + 7 + \cdots + (2k - 1) = k^2$.

    We must show that P_{k+1} is true.

> The statement that P_k is true is called the **inductive hypothesis.**

    We must show that $1 + 3 + 5 + \cdots + (2k - 1) + [2(k + 1) - 1] = (k + 1)^2$.   (1)

Avoiding Mistakes

It is important to use parentheses when substituting $(k + 1)$ into a statement for n.

The left side of statement P_{k+1} can be written as

$$\overbrace{[1 + 3 + 5 + 7 + \cdots + (2k - 1)]}^{\text{Sum of first } k \text{ terms}} + \overbrace{[2(k + 1) - 1]}^{(k+1)\text{th term}}$$

$$= \qquad\qquad k^2 \qquad\qquad + [2(k + 1) - 1]$$

> By the inductive hypothesis, we can replace the sum of the first k terms by k^2.

$$= k^2 + (2k + 2 - 1)$$
$$= k^2 + 2k + 1$$
$$= (k + 1)^2 \text{ as desired. This matches equation (1).}$$

We have shown that P_1 is true, and that if P_k is true, then P_{k+1} is true. By mathematical induction, we conclude that the statement is true for all positive integers.

> **Skill Practice 1**   Use mathematical induction to prove
> P_n: $2 + 4 + 6 + 8 + \cdots + 2n = n(n + 1)$.

In Example 2, we prove the statement introduced at the beginning of this section.

Classroom Example: p. 1047
Exercise 14

EXAMPLE 2   **Using Mathematical Induction**

Use mathematical induction to prove that

$$\frac{1}{1 \cdot 2} + \frac{1}{2 \cdot 3} + \frac{1}{3 \cdot 4} + \cdots + \frac{1}{n(n + 1)} = \frac{n}{n + 1}.$$

Solution:

Let P_n denote the statement $\dfrac{1}{1 \cdot 2} + \dfrac{1}{2 \cdot 3} + \dfrac{1}{3 \cdot 4} + \cdots + \dfrac{1}{n(n + 1)} = \dfrac{n}{n + 1}$.

  1. Show that P_1 is true.

    For $n = 1$, the sum equals $\underbrace{\dfrac{1}{1 \cdot 2} = \dfrac{1}{2}}_{\text{left side of statement}}$ which is the same as $\underbrace{\dfrac{n}{n + 1} = \dfrac{1}{1 + 1} = \dfrac{1}{2}}_{\text{right side of statement}}$.

Answer

1. See page SA-78.

2. Next, assume that P_k is true.

Assume that $\dfrac{1}{1 \cdot 2} + \dfrac{1}{2 \cdot 3} + \cdots + \dfrac{1}{k(k+1)} = \dfrac{k}{k+1}$. Assume that P_k is true. This is the inductive hypothesis.

We must show that P_{k+1} is true.

We must show that

$$\dfrac{1}{1 \cdot 2} + \dfrac{1}{2 \cdot 3} + \cdots + \dfrac{1}{k(k+1)} + \dfrac{1}{(k+1)[(k+1)+1]} = \dfrac{k+1}{(k+1)+1} = \dfrac{k+1}{k+2}. \quad (1)$$

The left side of statement P_{k+1} can be written as

$$\overbrace{\dfrac{1}{1 \cdot 2} + \dfrac{1}{2 \cdot 3} + \cdots + \dfrac{1}{k(k+1)}}^{\text{Sum of first } k \text{ terms}} + \overbrace{\dfrac{1}{(k+1)[(k+1)+1]}}^{(k+1)\text{th term}}$$

$$= \dfrac{k}{k+1} + \dfrac{1}{(k+1)[(k+1)+1]}$$ By the inductive hypothesis, replace the sum of the first k terms by $\frac{k}{k+1}$.

$$= \dfrac{k}{k+1} + \dfrac{1}{(k+1)(k+2)}$$ Simplify.

$$= \dfrac{k}{(k+1)} \cdot \dfrac{(k+2)}{(k+2)} + \dfrac{1}{(k+1)(k+2)}$$ Write terms with a common denominator.

$$= \dfrac{k^2 + 2k + 1}{(k+1)(k+2)}$$ Add the fractions.

$$= \dfrac{(k+1)^2}{(k+1)(k+2)}$$ Factor the numerator.

$$= \dfrac{k+1}{k+2}$$ as desired. This matches equation (1).

We have shown that P_1 is true, and that if P_k is true, then P_{k+1} is true. By mathematical induction, we conclude that the statement is true for all positive integers.

Skill Practice 2 Use mathematical induction to prove that

$$P_n\colon \dfrac{1}{1 \cdot 3} + \dfrac{1}{3 \cdot 5} + \dfrac{1}{5 \cdot 7} + \cdots + \dfrac{1}{(2n-1)(2n+1)} = \dfrac{n}{2n+1}.$$

Mathematical induction can be used to prove the following useful summation formulas involving powers of the first n positive integers (Table 11-3). The proofs of these formulas are addressed in Exercises 17–20.

Table 11-3 Sums of Powers

Let n represent a positive integer. Then,

1. $\displaystyle\sum_{i=1}^{n} 1 = n$ 2. $\displaystyle\sum_{i=1}^{n} i = \dfrac{n(n+1)}{2}$

3. $\displaystyle\sum_{i=1}^{n} i^2 = \dfrac{n(n+1)(2n+1)}{6}$ 4. $\displaystyle\sum_{i=1}^{n} i^3 = \dfrac{n^2(n+1)^2}{4}$

Answer

2. See page SA-78.

In Examples 3 and 4, we use mathematical induction to prove statements that do not involve a sum.

Classroom Example: p. 1047
Exercise 22

EXAMPLE 3   Using Mathematical Induction

Use mathematical induction to prove that 4 is a factor of $9^n - 1$.

Solution:

Let P_n be the statement: 4 is a factor of $9^n - 1$.

To show that 4 is a factor of any expression, we need to write the expression as the product of 4 and some positive integer that we call a.

1. Show that P_n is true for $n = 1$.

   P_1 reads: 4 is a factor of $9^1 - 1 = 8$.

   P_1 is true because $8 = 4 \cdot 2$, indicating that 4 is indeed a factor of 8.

2. Next, assume that P_k is true and show that P_{k+1} is true:

   Assume that 4 is a factor of $9^k - 1$.

   This implies that $9^k - 1 = 4a$ or equivalently $9^k = 4a + 1$.   Inductive hypothesis

   We must show that 4 is a factor of $9^{k+1} - 1$.

$$9^{k+1} - 1 = 9 \cdot 9^k - 1$$
$$= 9 \cdot (4a + 1) - 1 \qquad \text{From the inductive hypothesis, replace } 9^k \text{ by } 4a + 1.$$
$$= 36a + 9 - 1 \qquad \text{Apply the distributive property.}$$
$$= 36a + 8$$
$$= 4(9a + 2) \qquad \text{Factor. The value 4 is a factor of the expression } 4(9a + 2). \text{ Therefore, 4 is a factor of } 9^{k+1} - 1.$$

We have shown that P_1 is true, and that if P_k is true, then P_{k+1} is true. By mathematical induction, we conclude that the statement is true for all positive integers.

Skill Practice 3   Use mathematical induction to prove that 2 is a factor of $5^n + 1$.

2. Prove a Statement Using the Extended Principle of Mathematical Induction

Mathematical induction can be extended to prove statements that might hold true only for integers greater than or equal to some positive integer j. In such a case, we use the extended principle of mathematical induction.

Extended Principle of Mathematical Induction

Let P_n be a statement involving the positive integer n. Then P_n is true for all positive integers $n \geq j$ if

1. P_j is true, and
2. For an integer k, if $k \geq j$, the truth of P_k implies the truth of P_{k+1}.

Answer
3. See page SA-78.

Classroom Example: p. 1047
Exercise 30

| EXAMPLE 4 | Using the Extended Principle of Mathematical Induction |

Use mathematical induction to prove that $n! > 2^n$ for positive integers $n \geq 4$.

Solution:

Let P_n be the statement: $n! > 2^n$.

1. First show that P_n is true for $n = 4$.

   To prove that P_4 is true, show that $4! > 2^4$.
   $4! = 4 \cdot 3 \cdot 2 \cdot 1 = 24$ and $2^4 = 16$. Therefore, $4! > 2^4$, indicating that P_4 is true.

2. Next, assume that P_k is true for $k \geq 4$, and show that P_{k+1} is true.

   Assume that $k! > 2^k$ and show that $(k + 1)! > 2^{k+1}$ where $k \geq 4$.

   $$\begin{aligned}
   (k + 1)! &= (k + 1) \cdot k! \\
   &> (k + 1)(2^k) \quad &&\text{By the inductive hypothesis, } k! > 2^k. \text{ Therefore, we can} \\
   & &&\text{replace } k! \text{ by } 2^k \text{ and replace the } = \text{sign with } >. \\
   &> 2 \cdot 2^k \quad &&\text{For } k \geq 4, \text{ the expression } (k + 1) > 2. \text{ Therefore, we can} \\
   & &&\text{replace } (k + 1) \text{ by 2 and maintain the } > \text{symbol.} \\
   &= 2^{k+1}
   \end{aligned}$$

   From the string of inequalities we have shown that $(k + 1)! > 2^{k+1}$.

We have shown that P_4 is true, and that if P_k is true, then P_{k+1} is true. By mathematical induction, we conclude that $n! > 2^n$ is true for all integers, $n \geq 4$.

> **TIP** In part 2 of Example 4, we manipulate the expression $(k + 1)!$ on the left side to show that it is greater than the expression on the right side for $k \geq 4$.

Skill Practice 4 Use mathematical induction to prove that $\left(\dfrac{3}{2}\right)^n > 2n$ for $n \geq 7$.

Point of Interest

Mathematical induction is one type of mathematical proof, but there are many other techniques, including proof by contradiction. This technique uses the premise that a statement's being false would imply a contradiction. This approach was used in the complex proof of Fermat's last theorem.

The equation $x^2 + y^2 = z^2$ has infinitely many positive integer solutions for a, b, and c. Such solutions are called Pythagorean triples, such as $a = 5$, $b = 12$, and $c = 13$. However, French mathematician Pierre de Fermat (1601–1665) posed the statement that the equation $x^n + y^n = z^n$ has *no* such solutions for positive integers $n > 2$. Fermat made this claim, now famously called Fermat's last theorem, while annotating a copy of Diophantus' *Arithmetika*. Fermat included a comment that he had "found a remarkable proof of this fact, but there is not enough space in the margin to write it."

Proof of this theorem eluded mathematicians for over 350 yr until Andrew Wiles of Princeton University announced in June of 1993 that he had solved the problem. Wiles's proof is extremely complex and was revised to correct a slight but critical flaw shortly after publication. The proof involves the use of mathematics that did not exist in Fermat's time, and for this reason mathematicians believe that Fermat did not have a simpler proof.

Answer

4. See page SA-78.

SECTION 11.4   Practice Exercises

Prerequisite Review

For Exercises R.1–R.4, factor completely.

R.1. $x^4 + x^2$ $x^2(x^2 + 1)$

R.2. $-15r^4 - 60r^3$ $-15r^3(r + 4)$

R.3. $w^2 + 4w + 4$ $(w + 2)^2$

R.4. $v^2 - 9v + 18$ $(v - 3)(v - 6)$

For Exercises R.5–R.6, simplify.

R.5. $t - 4 + \dfrac{1}{t + 4}$ $\dfrac{t^2 - 15}{t + 4}$

R.6. $\dfrac{9}{a(a - 3)} + \dfrac{3}{a}$ $\dfrac{3}{a - 3}$

Concept Connections

1. Let P_n be a statement involving the positive integer n, and let k be an arbitrary positive integer. Proof by mathematical __induction__ indicates that P_n is true for all positive integers n if (1) ____P_1____ is true, and (2) the truth of P_k implies the truth of ____P_{k+1}____.

2. The statement that P_k is true is called the ____inductive____ hypothesis.

Objective 1: Prove a Statement Using Mathematical Induction

For Exercises 3–16, use mathematical induction to prove the given statement for all positive integers n. (See Examples 1–2)

3. $2 + 6 + 10 + \cdots + (4n - 2) = 2n^2$

4. $2 + 8 + 14 + \cdots + (6n - 4) = n(3n - 1)$

5. $5 + 8 + 11 + \cdots + (3n + 2) = \dfrac{n}{2}(3n + 7)$

6. $4 + 9 + 14 + \cdots + (5n - 1) = \dfrac{n}{2}(5n + 3)$

7. $8 + 4 + 0 + \cdots + (-4n + 12) = -2n(n - 5)$

8. $12 + 6 + 0 + \cdots + (-6n + 18) = -3n(n - 5)$

9. $1 + 2 + 2^2 + 2^3 + \cdots + 2^{n-1} = 2^n - 1$

10. $1 + 3 + 3^2 + 3^3 + \cdots + 3^{n-1} = \dfrac{1}{2}(3^n - 1)$

11. $\dfrac{3}{4} + \dfrac{3}{16} + \dfrac{3}{64} + \cdots + \dfrac{3}{4^n} = 1 - \left(\dfrac{1}{4}\right)^n$

12. $\dfrac{1}{2} + \dfrac{1}{4} + \dfrac{1}{8} + \cdots + \dfrac{1}{2^n} = 1 - \left(\dfrac{1}{2}\right)^n$

13. $1 \cdot 2 + 2 \cdot 3 + 3 \cdot 4 + \cdots + n(n + 1)$
$= \dfrac{n(n + 1)(n + 2)}{3}$

14. $1 \cdot 3 + 2 \cdot 5 + 3 \cdot 7 + \cdots + n(2n + 1)$
$= \dfrac{n(n + 1)(4n + 5)}{6}$

15. $\left(1 - \dfrac{1}{2}\right)\left(1 - \dfrac{1}{3}\right)\left(1 - \dfrac{1}{4}\right)\cdots\left(1 - \dfrac{1}{n + 1}\right)$
$= \dfrac{1}{n + 1}$

16. $\left(1 - \dfrac{1}{2^2}\right)\left(1 - \dfrac{1}{3^2}\right)\left(1 - \dfrac{1}{4^2}\right)\cdots\left(1 - \dfrac{1}{(n + 1)^2}\right)$
$= \dfrac{n + 2}{2(n + 1)}$

For Exercises 17–24, use mathematical induction to prove the given statement for all positive integers n. (See Example 3)

17. $\displaystyle\sum_{i=1}^{n} 1 = n$

18. $\displaystyle\sum_{i=1}^{n} i = \dfrac{n(n + 1)}{2}$

19. $\displaystyle\sum_{i=1}^{n} i^2 = \dfrac{n(n + 1)(2n + 1)}{6}$

20. $\displaystyle\sum_{i=1}^{n} i^3 = \dfrac{n^2(n + 1)^2}{4}$

21. 2 is a factor of $5^n - 3$.

22. 2 is a factor of $7^n - 3$.

23. $4^n - 1$ is divisible by 3.

24. $5^n - 1$ is divisible by 4.

Objective 2: Prove a Statement Using the Extended Principle of Mathematical Induction

For Exercises 25–28, use trial-and-error to determine the smallest positive integer n for which the given statement is true.

25. $n! > 3^n$ $n = 7$

26. $(n + 1)! > 4^n$ $n = 6$

27. $3n < 2^n$ $n = 4$

28. $5n < 3^n$ $n = 3$

For Exercises 29–32, use mathematical induction to prove the given statement. (See Example 4)

29. $n! > 3^n$ for positive integers $n \geq 7$.

30. $(n + 1)! > 4^n$ for positive integers $n \geq 6$.

31. $3n < 2^n$ for positive integers $n \geq 4$.

32. $5n < 3^n$ for positive integers $n \geq 3$.

Mixed Exercises

For Exercises 33–36, use mathematical induction to prove the given statement for all positive integers n and real numbers x and y.

33. $(xy)^n = x^n y^n$

34. $\left(\dfrac{x}{y}\right)^n = \dfrac{x^n}{y^n}$ provided that $y \neq 0$.

35. If $x > 1$, then $x^n > x^{n-1}$.

36. If $0 < x < 1$, then $x^n < x^{n-1}$.

We have used mathematical induction to prove that a statement is true for all positive integers n. To show that a statement is *not* true, all we need is one case in which the statement is false. This is called a counterexample. For Exercises 37–38, find a counterexample to show that the given statement is *not* true.

37. The expression $n^2 - n + 11$ is prime for all positive integers n. The statement is false for $n = 11$.

38. The inequality $5^n > n!$ is true for all positive integers n. The statement is false for $n = 12$.

Write About It

39. Explain the difference between the principle of mathematical induction and the extended principle of mathematical induction.

40. Suppose that a fellow student showed that the expression $n^2 + n + 1$ is prime for $n = 1, 2$, and 3. Explain why this is not a sufficient proof that the expression is prime for all positive integers n.

Expanding Your Skills

41. Show that $n^2 - n$ is even for all positive integers n.

42. Show that $n^2 - n + 1$ is odd for all positive integers n.

For Exercises 43–44, use the Fibonacci sequence $\{F_n\} = \{1, 1, 2, 3, 5, 8, 13, \ldots\}$. Recall that the Fibonacci sequence can be defined recursively as $F_1 = 1$, $F_2 = 1$, and $F_n = F_{n-1} + F_{n-2}$ for $n \geq 3$.

43. Prove that $F_1 + F_2 + F_3 + \cdots + F_n = F_{n+2} - 1$ for positive integers $n \geq 3$.

44. Prove that $F_1 + F_3 + F_5 + \cdots + F_{2n-1} = F_{2n}$ for all positive integers n.

Additional answers can be found in the Instructor Answer Appendix.

SECTION 11.5 The Binomial Theorem

OBJECTIVES

1. Determine Binomial Coefficients
2. Apply the Binomial Theorem
3. Find a Specific Term in a Binomial Expansion

1. Determine Binomial Coefficients

In Section R.3 we learned how to square a binomial.

$$(a + b)^2 = a^2 + 2ab + b^2$$

The expression $a^2 + 2ab + b^2$ is called the **binomial expansion** of $(a + b)^2$. To expand $(a + b)^3$, we can find the product $(a + b)(a + b)^2$.

$$(a + b)(a + b)^2 = (a + b)(a^2 + 2ab + b^2)$$

$$= a^3 + 2a^2b + ab^2 + a^2b + 2ab^2 + b^3$$

$$= a^3 + 3a^2b + 3ab^2 + b^3$$

Similarly, to expand $(a + b)^4$, we can multiply $(a + b)$ by $(a + b)^3$. Using this method, we can expand several powers of $(a + b)$.

$$(a + b)^0 = 1$$

$$(a + b)^1 = 1a + 1b$$

$$(a + b)^2 = 1a^2 + 2ab + 1b^2$$

$$(a + b)^3 = 1a^3 + 3a^2b + 3ab^2 + 1b^3$$

$$(a + b)^4 = 1a^4 + 4a^3b + 6a^2b^2 + 4ab^3 + 1b^4$$

$$(a + b)^5 = 1a^5 + 5a^4b + 10a^3b^2 + 10a^2b^3 + 5ab^4 + 1b^5$$

1
1 1
1 2 1
1 3 3 1
1 4 6 4 1
1 5 10 10 5 1

Figure 11-9

From the expansion of $(a + b)^n$, we note the following patterns.

- The exponent on a decreases from n to 0 on sequential terms from left to right.
- The exponent on b increases from 0 to n on sequential terms from left to right.
- The sum of the exponents on each term (that is, the degree of each term) is n.
- The number of terms in the expansion is $(n + 1)$. For example, the expansion of $(a + b)^4$ has five terms.

The coefficients for the expansion of $(a + b)^n$ follow a triangular array of numbers called **Pascal's triangle** (Figure 11-9), named after the French mathematician Blaise Pascal (1623–1662). Each row begins and ends with a 1, and each entry in between is the sum of the two diagonal entries from the row above. For example, in the last row of Figure 11-9, we have

$$1, \quad 1 + 4 = 5, \quad 4 + 6 = 10, \quad 6 + 4 = 10, \quad 4 + 1 = 5, \text{ and } 1$$

Classroom Example: p. 1053
Exercise 8

EXAMPLE 1   Expanding a Binomial

Expand. $(a + b)^6$

Solution:

The expansion of $(a + b)^6$ will have 7 terms (one more than the exponent of 6). Using the entries of the seventh row of Pascal's triangle as the coefficients, the expansion of $(a + b)^6$ is:

```
            1
          1   1
        1   2   1
      1   3   3   1
    1   4   6   4   1
  1   5  10  10   5   1
1   6  15  20  15   6   1
```

$$(a + b)^6 = 1a^6 + 6a^5b + 15a^4b^2 + 20a^3b^3 + 15a^2b^4 + 6ab^5 + 1b^6$$

Skill Practice 1   Expand. $(a + b)^7$

Determining the coefficients of a binomial expansion using Pascal's triangle would be cumbersome for expansions of higher degree. Instead we can compute the coefficients of a binomial expansion using the following formula.

TIP The expression $\binom{n}{r}$ will also be used in Section 11.6 when we study counting principles. $\binom{n}{r}$ represents the number of ways we can *choose* a group of r items in any order from a group of n items.

Coefficients of a Binomial Expansion

Let n and r be nonnegative integers with $n \geq r$. The expression $\binom{n}{r}$ (read as "n choose r") is called a **binomial coefficient** and is defined by

$$\binom{n}{r} = \frac{n!}{r!(n - r)!}$$

Note: The notation $_nC_r$ is often used in place of $\binom{n}{r}$.

Classroom Example: p. 1053
Exercise 10

EXAMPLE 2   Computing Binomial Coefficients

Evaluate.

a. $\binom{6}{0}$ **b.** $\binom{6}{1}$ **c.** $\binom{6}{2}$ **d.** $\binom{6}{6}$

Answer

1. $a^7 + 7a^6b + 21a^5b^2 + 35a^4b^3 + 35a^3b^4 + 21a^2b^5 + 7ab^6 + b^7$

Solution:

For parts (a)–(d), apply $\dbinom{n}{r} = \dfrac{n!}{r!(n-r)!}$

a. $\dbinom{6}{0} = \dfrac{6!}{0! \cdot (6-0)!} = \dfrac{6!}{0! \cdot 6!} = \dfrac{6!}{1 \cdot 6!} = 1$ $n = 6$ and $r = 0$.
Recall that $0! = 1$.

b. $\dbinom{6}{1} = \dfrac{6!}{1! \cdot (6-1)!} = \dfrac{6!}{1! \cdot 5!} = \dfrac{6 \cdot 5!}{1 \cdot 5!} = 6$ $n = 6$ and $r = 1$.

c. $\dbinom{6}{2} = \dfrac{6!}{2! \cdot (6-2)!} = \dfrac{6!}{2! \cdot 4!} = \dfrac{6 \cdot 5 \cdot 4!}{2 \cdot 1 \cdot 4!} = 15$ $n = 6$ and $r = 2$.

d. $\dbinom{6}{6} = \dfrac{6!}{6! \cdot (6-6)!} = \dfrac{6!}{6! \cdot 0!} = \dfrac{6!}{6! \cdot 1} = 1$ $n = 6$ and $r = 6$.

Skill Practice 2   Evaluate.

a. $\dbinom{5}{0}$ **b.** $\dbinom{5}{1}$ **c.** $\dbinom{5}{2}$ **d.** $\dbinom{5}{5}$

TECHNOLOGY CONNECTIONS

Evaluating Binomial Coefficients

To evaluate binomial coefficients on a graphing utility, use the $_nC_r$ function found in the **MATH** menu under PRB.

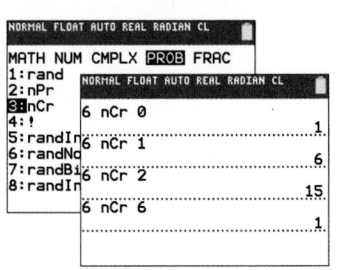

---

2. Apply the Binomial Theorem

From Example 2, notice that $\dbinom{6}{0} = 1$, $\dbinom{6}{1} = 6$, and $\dbinom{6}{2} = 15$ are the coefficients for the first three terms of the expansion of $(a+b)^6$. The value $\dbinom{6}{6} = 1$ is the coefficient of the last term of the expansion. Using the values of $\dbinom{n}{r}$ for the coefficients, we have a formula for the expansion of the binomial $(a+b)^n$. This is called the **binomial theorem.**

> **TIP** The binomial theorem can be proved by using mathematical induction. See the proof in the online appendix at www.mhhe.com/millercollegealgebra.

The Binomial Theorem

Let n be a positive integer. The expansion of $(a+b)^n$ is given by

$$(a+b)^n = \binom{n}{0}a^n + \binom{n}{1}a^{n-1}b + \binom{n}{2}a^{n-2}b^2 + \cdots + \binom{n}{n-1}ab^{n-1} + \binom{n}{n}b^n$$

$$= \sum_{r=0}^{n} \binom{n}{r}a^{n-r}b^r$$

Answers
2. a. 1   **b.** 5   **c.** 10   **d.** 1

Classroom Example: p. 1053
Exercise 16

EXAMPLE 3   Applying the Binomial Theorem

Expand by using the binomial theorem. $(2x + 3)^4$

Solution:

The expression $(2x + 3)^4$ is in the form $(a + b)^4$ with $a = 2x$ and $b = 3$.

$$(a + b)^4 = \binom{4}{0}a^4 + \binom{4}{1}a^3b + \binom{4}{2}a^2b^2 + \binom{4}{3}ab^3 + \binom{4}{4}b^4$$

$$(2x + 3)^4 = \binom{4}{0}(2x)^4 + \binom{4}{1}(2x)^3(3) + \binom{4}{2}(2x)^2(3)^2 + \binom{4}{3}(2x)(3)^3 + \binom{4}{4}(3)^4$$

$$= 1(16x^4) + 4(8x^3)(3) + 6(4x^2)(9) + 4(2x)(27) + 1(81)$$

$$= 16x^4 + 96x^3 + 216x^2 + 216x + 81$$

Skill Practice 3   Expand by using the binomial theorem. $(4y + 5)^4$

In Example 4, we expand a difference of terms.

Classroom Example: p. 1053
Exercise 22

EXAMPLE 4   Applying the Binomial Theorem

Expand by using the binomial theorem. $(3x^2 - 5y)^3$

Solution:

The expression $(3x^2 - 5y)^3 = [3x^2 + (-5y)]^3$ is in the form $(a + b)^3$ with $a = 3x^2$ and $b = -5y$.

$$(a + b)^3 = \binom{3}{0}a^3 + \binom{3}{1}a^2b + \binom{3}{2}ab^2 + \binom{3}{3}b^3$$

$$[3x^2 + (-5y)]^3 = \binom{3}{0}(3x^2)^3 + \binom{3}{1}(3x^2)^2(-5y) + \binom{3}{2}(3x^2)(-5y)^2 + \binom{3}{3}(-5y)^3$$

$$= 1(27x^6) + 3(9x^4)(-5y) + 3(3x^2)(25y^2) + 1(-125y^3)$$

$$= 27x^6 - 135x^4y + 225x^2y^2 - 125y^3$$

Skill Practice 4   Expand by using the binomial theorem. $(2t^4 - 3v)^3$

3. Find a Specific Term in a Binomial Expansion

Consider the first four terms of the binomial expansion of $(a + b)^n$.

The exponent on b is one less than the term number.

$$\binom{n}{0}a^nb^0 + \binom{n}{1}a^{n-1}b^1 + \binom{n}{2}a^{n-2}b^2 + \binom{n}{3}a^{n-3}b^3$$

These values match the exponent on b.

From the observed pattern, the kth term has a factor of b raised to the $k - 1$ power. That is, the exponent on b is one less than the term number. Furthermore, the sum of the exponents must equal n. Therefore, the factor a is raised to the $n - (k - 1)$ power. That is, the exponent on a is n minus the exponent on b.

Answers

3. $256y^4 + 1280y^3 + 2400y^2$
$\quad + 2000y + 625$
4. $8t^{12} - 36t^8 v + 54t^4 v^2 - 27v^3$

> ### *k*th Term of a Binomial Expansion
>
> Let n and k be positive integers with $k \leq n + 1$. The kth term of $(a + b)^n$ is
>
> $$\binom{n}{k-1}a^{n-(k-1)}b^{k-1}$$

Classroom Example: p. 1053
Exercise 34

EXAMPLE 5   Finding a Specific Term of a Binomial Expansion

Find the eighth term of $(2x + y^4)^{10}$.

Solution:

To find the eighth term of $(a + b)^{10}$, we require the exponent on b to be 7 (one less than the term number). Therefore, the exponent on a must be 3 so that the sum of the exponents is 10.

Therefore, the eighth term of $(a + b)^{10}$ is $\binom{10}{7}a^3b^7$.

> Sum of exponents must be 10.

> 1 less than the term number

> same as exponent on b

The eighth term of $(2x + y^4)^{10}$ is $\binom{10}{7}(2x)^3(y^4)^7 = 120(8x^3)(y^{28}) = 960x^3y^{28}$.

Alternatively, we can apply the formula for the kth term of a binomial expansion with $n = 10$, $k = 8$, $a = 2x$, and $b = y^4$.

$$\binom{n}{k-1}a^{n-(k-1)}b^{k-1} = \binom{10}{8-1}(2x)^{10-(8-1)}(y^4)^{8-1} = \binom{10}{7}(2x)^3(y^4)^7 = 960x^3y^{28}$$

The eighth term of the expansion is $960x^3y^{28}$.

Skill Practice 5   Find the seventh term of $(3c + d^5)^9$.

Answer

5. $2268c^3d^{30}$

SECTION 11.5   Practice Exercises

Prerequisite Review

For Exercises R.1–R.5, multiply and simplify.

R.1. $(5v + 2)^2$ $25v^2 + 20v + 4$ **R.2.** $(8a^2 - 9b^3)^2$ **R.3.** $(4x + \sqrt{15})^2$ $16x^2 + 8\sqrt{15}x + 15$

                                           $64a^4 - 144a^2b^3 + 81b^6$

R.4. $\left(\dfrac{1}{5}m + 2\right)^2$ $\dfrac{1}{25}m^2 + \dfrac{4}{5}m + 4$ **R.5.** $(4a - 3b)^3$ $64a^3 - 144a^2b + 108ab^2 - 27b^3$

Concept Connections

1. The expression $a^3 + 3a^2b + 3ab^2 + b^3$ is called the _____binomial_____ expansion of $(a + b)^3$.

2. Consider $(a + b)^n$, where n is a whole number. How many terms are in the binomial expansion? $n + 1$

3. Consider $(a + b)^n$, where n is a whole number. What is the degree of each term in the expansion? n

4. Consider $(a + b)^n$, where n is a whole number. The coefficients of the terms in the expansion can be found by using _____Pascal's_____ triangle or by using $\binom{n}{r}$.

5. For positive integers n and k ($k \leq n + 1$), the kth term of $(a + b)^n$ is given by $\left(\quad \right) a^{\square} b^{\square}$. $\binom{n}{k-1} a^{n-(k-1)} b^{k-1}$

6. Given $(a + b)^{17}$, the 12th term is given by $\left(\quad \right) a^{\square} b^{\square}$. $\binom{17}{11} a^6 b^{11}$

Objective 1: Determine Binomial Coefficients

7. a. Write the first five rows of Pascal's triangle.
See Figure 11-9 on page 1048.
b. Write the expansion of $(x - y)^4$. (**See Example 1**)
$x^4 - 4x^3 y + 6x^2 y^2 - 4xy^3 + y^4$

8. a. Write the first four rows of Pascal's triangle.
See Figure 11-9 on page 1048.
b. Write the expansion of $(c - d)^3$. $c^3 - 3c^2 d + 3cd^2 - d^3$

For Exercises 9–10, evaluate the given expressions. Compare the results to the fifth and sixth rows of Pascal's triangle. (See Example 2)

9. a. $\binom{4}{0}$ 1   **b.** $\binom{4}{1}$ 4   **c.** $\binom{4}{2}$ 6   **10. a.** $\binom{5}{0}$ 1   **b.** $\binom{5}{1}$ 5   **c.** $\binom{5}{2}$ 10

d. $\binom{4}{3}$ 4   **e.** $\binom{4}{4}$ 1   **d.** $\binom{5}{3}$ 10   **e.** $\binom{5}{4}$ 5   **f.** $\binom{5}{5}$ 1

For Exercises 11–14, evaluate the expression.

11. $\binom{13}{3}$ 286   **12.** $\binom{17}{15}$ 136   **13.** $\binom{11}{5}$ 462   **14.** $\binom{9}{4}$ 126

Objective 2: Apply the Binomial Theorem

For Exercises 15–28, expand the binomial by using the binomial theorem. (See Examples 3–4)

15. $(3x + 1)^5$ **16.** $(5x + 3)^5$ **17.** $(7x + 3)^3$ $343x^3 + 441x^2 + 189x + 27$ **18.** $(6x + 5)^3$ $216x^3 + 540x^2 + 450x + 125$

19. $(2x - 5)^4$ **20.** $(4x - 1)^4$ $256x^4 - 256x^3 + 96x^2 - 16x + 1$ **21.** $(2x^3 - y)^5$ **22.** $(3y^2 - z)^5$

23. $(p^2 - w^4)^6$ **24.** $(t^3 - v^5)^6$ **25.** $(0.2 + 0.1k)^4$ **26.** $(0.1 + 0.3m)^4$

27. $\left(\dfrac{c}{2} - d \right)^3$ **28.** $\left(\dfrac{x}{3} - n \right)^3$ $\dfrac{1}{27}x^3 - \dfrac{1}{3}x^2 n + xn^2 - n^3$

Objective 3: Find a Specific Term in a Binomial Expansion

For Exercises 29–40, find the indicated term of the binomial expansion. (See Example 5)

29. $(m + n)^{10}$; seventh term $210m^4 n^6$ **30.** $(p + q)^{11}$; ninth term $165p^3 q^8$ **31.** $(c - d)^8$; fourth term $-56c^5 d^3$

32. $(x - y)^9$; sixth term $-126x^4 y^5$ **33.** $(u^2 + 2v^4)^{15}$; tenth term $2{,}562{,}560u^{12} v^{36}$ **34.** $(y^3 + 2z^2)^{14}$; tenth term $1{,}025{,}024y^{15} z^{18}$

35. $(\sqrt{3}x^2 + y^3)^9$; fourth term $2268x^{12} y^9$ **36.** $(\sqrt{2}u^3 + v^4)^8$; third term $224u^{18} v^8$ **37.** $(h^4 - 1)^{12}$; middle term $924h^{24}$

38. $(k^3 - 1)^{10}$; middle term $-252k^{15}$ **39.** $(p^4 + 3q)^8$; term containing p^{12}. $13{,}608\, p^{12} q^5$ **40.** $(a^3 + 4b)^6$; term containing a^9. $1280a^9 b^3$

Mixed Exercises

41. Expand $(e^x - e^{-x})^4$. $e^{4x} - 4e^{2x} + 6 - \dfrac{4}{e^{2x}} + \dfrac{1}{e^{4x}}$

42. Expand $(e^x + e^{-x})^3$. $e^{3x} + 3e^x + \dfrac{3}{e^x} + \dfrac{1}{e^{3x}}$

43. Use the binomial theorem to expand $(x + y - z)^3$.
$x^3 + 3x^2 y + 3xy^2 + y^3 - 3x^2 z - 6xyz - 3y^2 z + 3xz^2 + 3yz^2 - z^3$

44. Use the binomial theorem to expand $(a + b - 2)^3$.
$a^3 + 3a^2 b + 3ab^2 + b^3 - 6a^2 - 12ab - 6b^2 + 12a + 12b - 8$

45. Use the binomial theorem to find $(1.01)^4$. [*Hint*: Write the expression as $(1 + 0.01)^4$.] 1.04060401

46. Use the binomial theorem to find $(1.1)^5$. 1.61051

47. Simplify. $(x + y)^3 - (x - y)^3$ $6x^2 y + 2y^3$

48. Simplify. $(x + 1)^3 - (x - 1)^3$ $6x^2 + 2$

For Exercises 49–52, simplify the difference quotient: $\dfrac{f(x + h) - f(x)}{h}$

49. $f(x) = 2x^3 + 4$ $6x^2 + 6xh + 2h^2$ **50.** $f(x) = 4x^3 + 3$ $12x^2 + 12xh + 4h^2$ **51.** $f(x) = x^4 - 5x^2 + 1$ $4x^3 + 6x^2 h + 4xh^2 + h^3 - 10x - 5h$ **52.** $f(x) = x^4 - 6x^2 - 4$ $4x^3 + 6x^2 h + 4xh^2 + h^3 - 12x - 6h$

For Exercises 53–56, use the binomial theorem to find the value of the complex number raised to the given power. Recall that $i = \sqrt{-1}$.

53. $(2 + i)^3$ $2 + 11i$ **54.** $(3 + i)^3$ $18 + 26i$ **55.** $(5 - 2i)^4$ $41 - 840i$ **56.** $(6 - 5i)^4$ $-3479 - 1320i$

57. Show that $\binom{n}{r}$ and $\binom{n}{n - r}$ are equivalent.

58. Show that $\binom{n}{0} = 1$ and $\binom{n}{n} = 1$.

59. a. Write the sequence corresponding to the sum of the numbers in each row of Pascal's triangle for the first nine rows. 1, 2, 4, 8, 16, 32, 64, 128, 256

   b. Let n represent the row number in Pascal's triangle. Write a formula for the nth term of the sequence representing the sum of the numbers in row n. $a_n = 2^{n-1}$

Write About It

60. Explain why $\begin{pmatrix} -13 \\ 3 \end{pmatrix}$ is undefined. In the expression $\begin{pmatrix} -13 \\ 3 \end{pmatrix}$, $n = -13$. However, $n! = (-13)!$ is undefined.

61. Explain why $\begin{pmatrix} 3 \\ 5 \end{pmatrix}$ is undefined. In this expression $n = 3$ and $r = 5$. The value $(n - r)!$ is $(3 - 5)! = (-2)!$, which is undefined.

62. Explain why many graphing utilities give an error message for the expression $\dfrac{120!}{118!}$. How can this expression be evaluated by hand? The numerator and denominator are each too large to represent on a calculator. However, to simplify this expression by hand, we can reduce common factors from the numerator and denominator before multiplying; that is, $\dfrac{120!}{118!} = \dfrac{120 \cdot 119 \cdot 118!}{118!} = 120 \cdot 119 = 14{,}280.$

Expanding Your Skills

Stirling's formula (named after Scottish mathematician, James Stirling: 1692–1770) is used to approximate large values of $n!$.
Stirling's formula is $n! \approx \sqrt{2\pi n}\left(\dfrac{n}{e}\right)^n$. For Exercises 63–64,

a. Use Stirling's formula to approximate the given expression. Round to the nearest whole unit.

b. Compute the actual value of the expression.

c. Determine the percent difference between the approximate value and the actual value. Round to the nearest tenth of a percent.

63. $8!$ **a.** 39,902   **b.** 40,320   **c.** 1.0% error

64. $12!$ **a.** 475,687,487   **b.** 479,001,600   **c.** 0.7% error

Using techniques from calculus, we can show that $(1 + x)^n = 1 + nx + \dfrac{n(n - 1)x^2}{2!} + \dfrac{n(n - 1)(n - 2)}{3!}x^3 + \cdots$, for $|x| < 1$.

This formula can be used to evaluate binomial expressions raised to noninteger exponents. For Exercises 65–66, use the first four terms of this infinite series to approximate the given expression. Round to 3 decimal places if necessary.

65. $(1.4)^{3/2}$ [*Hint*: $(1.4)^{3/2} = (1 + 0.4)^{3/2}$]   1.656

66. $\sqrt[4]{(1.3)^3}$ 1.218

SECTION 11.6   Principles of Counting

OBJECTIVES

1. Apply the Fundamental Principle of Counting
2. Count Permutations
3. Count Combinations

1. Apply the Fundamental Principle of Counting

Many states have lotteries as a means to raise income. The game "Florida Lotto" for example, pays a grand prize to players who choose six distinct numbers from 1 to 53 (in any order) that match the same group of six numbers in the drawing. The number of such six-number combinations is 22,957,480 (see Example 8). If a player chooses one group of six numbers, then the probability (likelihood) of winning the grand prize is $\frac{1}{22{,}957{,}480}$. This value (roughly 1 in 23 million) means that it is highly unlikely to win the grand prize. To put this in perspective, the following events are more likely to happen than winning the grand prize in "Florida Lotto."

- A pedestrian being killed by a motor vehicle: Probability $\approx \frac{1}{1{,}000{,}000}$
- Flipping a coin 24 times and getting all heads: Probability $= \frac{1}{16{,}777{,}216}$
- Dying from a venomous insect or animal bite: Probability $\approx \frac{1}{100{,}000}$

In this section and in Section 11.7, we present basic principles of counting and how these principles apply to probability. We begin in Example 1, by using a figure called a tree diagram to organize different outcomes of a sequence of events.

Classroom Example: p. 1063
Exercise 16

EXAMPLE 1   Counting the Outcomes of a
Sequence of Events

Suppose that a frozen custard shop offers 4 flavors of custard, 2 types of syrup,
and 2 toppings. If a customer can select 1 item from each group for a sundae,
how many different sundaes can be made?

Custard Flavors	Syrups	Toppings
Vanilla (V)	Hot fudge (H)	Nuts (N)
Chocolate (C)	Butterscotch (B)	Granola (G)
Mint chip (M)		
Peanut butter (P)		

Solution:

We can depict the 4 flavors of custard
by the left-most branches in Figure 11-10.
Then for each branch of custard, there
are 2 branches representing the type of
syrup. Finally, for each custard-syrup
arrangement, there are 2 toppings.

The tree diagram shows 16 different
sundaes:

VHN, VHG, VBN, VBG, CHN, CHG,
CBN, CBG, MHN, MHG, MBN, MBG,
PHN, PHG, PBN, and PBG.

Notice that the number of outcomes is
easily found by multiplying the
number of choices from each group:

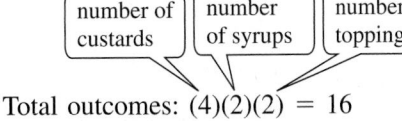

Total outcomes: $(4)(2)(2) = 16$

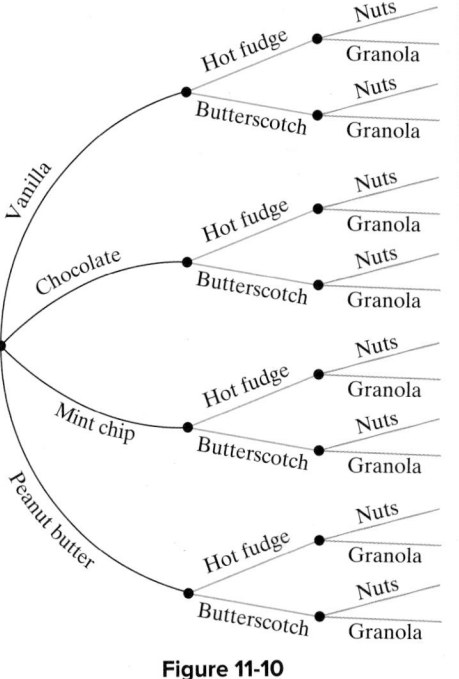

Figure 11-10

Skill Practice 1   A pizza can be made from either thick or thin crust, and
a choice of 6 toppings. How many different types of pizza can be made?

The result from Example 1 suggests that the number of outcomes for the sequence
of events is the product of the number of outcomes for each individual event. This
is generalized as the **fundamental principle of counting.**

Fundamental Principle of Counting

If one event can occur in m different ways and a second event can occur in n
different ways, then the number of ways that the two events can occur in
sequence is $m \cdot n$.

In Example 2, we demonstrate that the fundamental principle of counting can be
applied for a sequence of many events.

Answer

1. 12

Classroom Example: p. 1063
Exercise 20

EXAMPLE 2 Applying the Fundamental Principle of Counting

A computer password must have three letters, followed by two digits. How many passwords can be made if

a. There are no restrictions on the letters or digits?

b. No letter or digit may be used more than once?

Solution:

a. The computer password is a sequence of five characters. The number of arrangements of letters and digits is given by the product of the number of outcomes for each character. There are 26 letters of the alphabet from which to choose. The digits in the password may be selected from 0, 1, 2, 3, 4, 5, 6, 7, 8, 9. Therefore, the total number of passwords is given by

There are 26 choices for each letter.   There are 10 choices for each digit.

$$26 \cdot 26 \cdot 26 \cdot 10 \cdot 10 = 1{,}757{,}600$$

b. If no letter may be repeated, then the number of letters from which to choose must be decreased by one when the second letter is selected. Likewise, the number of letters must be decreased by one again when the third letter is selected. The same logic is used when selecting the digits. Once the first digit is selected in the password, there are only nine remaining digits available for the second choice.

The number of letters is decreased by one for each letter in the sequence.   There are 10 choices for the first digit, but only 9 remaining for the second.

$$26 \cdot 25 \cdot 24 \cdot 10 \cdot 9 = 1{,}404{,}000$$

Skill Practice 2 A code for an alarm system is made up of two letters, followed by four digits. How many codes can be made if

a. There are no restrictions on the letters or digits?

b. No letter or digit may be used more than once?

Classroom Example: p. 1063
Exercise 22

EXAMPLE 3 Applying the Fundamental Principle of Counting

A quiz has five true/false questions and five multiple-choice questions. The multiple-choice questions each have four possible answers of which only one is correct. If a student forgets to study and must guess on each question, in how many ways can the student answer the questions on the test?

Solution:

The quiz questions form a 10-stage event. Each true/false question has two possible choices (true or false). Each multiple-choice question has four possible choices (for example: a, b, c, or d). Therefore, the number of ways that the questions can be answered on the test is given by

Each true/false question has 2 choices   Each multiple-choice question has 4 choices

$$2 \cdot 2 \cdot 2 \cdot 2 \cdot 2 \cdot 4 \cdot 4 \cdot 4 \cdot 4 \cdot 4 = 32{,}768$$

Answers

2. **a.** $26 \cdot 26 \cdot 10 \cdot 10 \cdot 10 \cdot 10 = 6{,}760{,}000$

   b. $26 \cdot 25 \cdot 10 \cdot 9 \cdot 8 \cdot 7 = 3{,}276{,}000$

> **Skill Practice 3**   A quiz has four true/false questions and six multiple-choice questions. The multiple-choice questions each have five possible answers of which only one is correct. In how many ways can the questions on the test be answered?

2. Count Permutations

We now look at a situation in which n items are to be arranged in order. Such an arrangement is called a **permutation.** For example, consider the different arrangements of the letters in the word FIVE.

FIVE	FIEV	FVIE	FVEI	FEIV	FEVI
IFVE	IFEV	IVEF	IVFE	IEVF	IEFV
VFIE	VFEI	VEIF	VEFI	VIEF	VIFE
EFIV	EFVI	EIFV	EIVF	EVIF	EVFI

} 24 arrangements

There are 24 such arrangements. We can arrive at the same conclusion by applying the fundamental principle of counting. There are four letters available for the first choice in the arrangement. That leaves 3 letters remaining for the second choice, followed by 2 letters for the third choice, and only 1 letter for the last choice. Therefore, the number of permutations of 4 distinct letters is given by

$$4 \cdot 3 \cdot 2 \cdot 1 = 24$$

We generalize this as follows.

> **Number of Permutations of n Distinguishable Elements**
>
> The number of permutations of n distinct elements is $n!$.
>
> *Note*: This means that there are $n!$ ways to arrange n distinguishable items in various orders.

Now suppose that we wanted to arrange the letters in the word NINE. Notice that if we were to switch the first and third letters (both "N's") we would get the same arrangement. This is because the two N's are indistinguishable. This means that when counting the number of permutations we must "divide out" the number of ways that the two N's can be arranged. The two N's can be arranged in 2! different ways. This leads to the following result.

> **Number of Permutations of n Elements, Some Indistinguishable**
>
> Consider a set of n elements with r_1 duplicates of one kind, r_2 duplicates of a second kind, ... , r_k duplicates of a kth kind. Then the number of distinguishable permutations of the n elements of the set is
>
> $$\frac{n!}{r_1! \cdot r_2! \cdot \cdots \cdot r_k!}$$
>
> *Note*: The factors of $r_1!$, $r_2!$, ... , $r_k!$ in the denominator remove the repetition of arrangements that arise from the indistinguishable elements.

Answer

3. 250,000

Classroom Example: p. 1063
Exercise 26

EXAMPLE 4 **Counting Permutations**

a. Determine the number of ways that 6 people can be arranged in line at a ticket counter.

b. Determine the number of ways that the letters in the word MICROSCOPIC can be arranged.

Solution:

a. No two people can be the same. Therefore, we are arranging 6 different (distinguishable) items in various orders. The number of such permutations is given by

$6! = 720$ There are 720 ways in which 6 people can be arranged in line.

b. The word MICROSCOPIC has 11 letters. However, the letter I appears twice, the letter O appears twice, and the letter C appears 3 times. The number of unique arrangements of these letters is

$$\frac{11!}{2! \cdot 2! \cdot 3!} = 1{,}663{,}200$$

Skill Practice 4

a. In how many ways can 7 different books be arranged on a bookshelf?

b. Determine the number of ways that the letters in the word RIFFRAFF can be arranged.

In Example 5, we determine the number of permutations from a group of n items in which fewer than n items are selected at one time.

Classroom Example: p. 1064
Exercise 48

EXAMPLE 5 **Finding Permutations of n Items Taken r at a Time**

Suppose that 5 students (Alberto, Beth, Carol, Dennis, and Erik) submit applications for scholarships. There are two scholarships available. The first is for $1000 and the second is for $500. In how many ways can 2 students from the 5 be selected to receive the scholarships?

Solution:

We begin by labeling the students as A, B, C, D, and E. We can construct a list of the possible outcomes.

AB	AC	AD	AE	BC	BD	BE	CD	CE	DE
BA	CA	DA	EA	CB	DB	EB	DC	EC	ED

Each outcome in the second row involves the same 2 people as the outcome directly above in the first row, but in the reverse order. The order of selection is important. For example:

- AB means that Alberto gets $1000 and Beth gets $500, whereas
- BA means that Beth gets $1000 and Alberto gets $500.

There are 20 possible ways to select 2 people from 5 people in which the order of selection is relevant. That is, there are 20 permutations of 5 people taken 2 at a time.

Answers

4. a. $7! = 5040$ **b.** $\dfrac{8!}{2! \cdot 4!} = 840$

5. 12

Skill Practice 5   Determine the number of ways that 2 students from a group of 4 students can be selected to hold the positions of president and vice president of student government.

The number of permutations of 5 people taken 2 at a time can also be computed by using the fundamental principle of counting. There are 5 candidates for the $1000 scholarship, which leaves 4 students left over for the $500 scholarship. The product is $5 \cdot 4 = 20$.

This result can also be obtained by using factorial notation:

Number of people in the group (5).

$$\frac{5!}{(5-2)!} = \frac{5!}{3!} = \frac{5 \cdot 4 \cdot (3 \cdot 2 \cdot 1)}{(3 \cdot 2 \cdot 1)} = 20$$

Number of people to be selected (2).

Suppose that n represents the number of distinct elements in a group from which r elements will be chosen in a particular order. We call each arrangement a **permutation** of n elements taken r at a time and denote the number of such permutations as $_nP_r$.

Number of Permutations of n Elements Taken r at a Time

The number of permutations of n elements taken r at a time is given by

r factors

$$_nP_r = \frac{n!}{(n-r)!}, \text{ or equivalently, } _nP_r = \overbrace{n(n-1)(n-2)\cdots(n-r+1)}$$

Note: $_nP_r$ counts the number of permutations of n items taken r at a time under the assumption that no item can be selected more than once, and that each of the n items is distinguishable.

Classroom Example: p. 1064
Exercise 50

EXAMPLE 6   Counting Permutations in an Application

If 8 horses enter a race, in how many ways can the horses finish first, second, and third?

Solution:

We must find the number of ways that 3 horses can be selected from 8 horses in a prescribed order (first, second, third). This is given by $_8P_3$.

$$_8P_3 = \frac{8!}{(8-3)!} = \frac{8!}{5!} = \frac{8 \cdot 7 \cdot 6 \cdot 5!}{5!} = 8 \cdot 7 \cdot 6 = 336$$

$n = 8$

Alternatively: $_8P_3 = \underbrace{8 \cdot 7 \cdot 6}_{3 \text{ factors}} = 336$

TIP The alternative formula for $_nP_r$ indicates that we multiply n times the consecutive integers less than n until a total of r factors is reached.

There are 336 possible first-, second-, and third-place arrangements.

Skill Practice 6   A judge at the County Fair must give blue, red, and white ribbons for first-, second-, and third-place entries in a poetry contest. If there are 12 contestants, in how many ways can the judge award the ribbons?

Answer
6. 1320

TECHNOLOGY CONNECTIONS

Evaluating a Number of Permutations, $_nP_r$

Most graphing utilities can evaluate the number of permutations of n elements taken r at a time. For example, use the $_nP_r$ function found in the **MATH** menu under PRB.

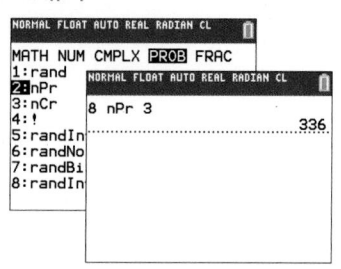

---

3. Count Combinations

Consider the situation in Example 5 in which 5 students are selected for 2 scholarships. If the scholarships were for *equal* amounts (say for $500 each), then the order in which the 2 students are selected does not matter. This is because the "prizes" are indistinguishable; that is, the outcomes AB and BA are the same because student A and student B would each receive $500.

In this scenario, we want to select a set of 2 people from a set of 5 people without regard to order. In such a case, the outcomes are called *combinations* (rather than permutations). From the list of permutations given in Example 5, we can strike through the redundant cases that arise from the order of the two individuals in the group. As a result, there are 10 combinations of 5 people taken 2 at a time.

AB	AC	AD	AE	BC	BD	BE	CD	CE	DE
~~BA~~	~~CA~~	~~DA~~	~~EA~~	~~CB~~	~~DB~~	~~EB~~	~~DC~~	~~EC~~	~~ED~~

To find the number of combinations, we divide the number of permutations (in this case 20) by 2! because there are 2! ways in which two letters can be arranged in order.

$$\text{Number of combinations} = \frac{_5P_2}{2!} = \frac{20}{2} = 10$$

To generalize, suppose that n represents the number of distinct elements in a group from which r elements will be chosen in *no particular order*. We call each selection a **combination** of n elements taken r at a time and denote the number of such combinations as $_nC_r$.

> **TIP** The number of combinations $_nC_r$ is equal to the number of permutations $_nP_r$ divided by $r!$. The value $r!$ in the denominator "divides out" the redundant cases involving different arrangements of r elements within the same group.

Number of Combinations of n Elements Taken r at a Time

The number of combinations of n elements taken r at a time is given by

$$_nC_r = \frac{n!}{r! \cdot (n-r)!}, \text{ or equivalently, } _nC_r = \frac{_nP_r}{r!}$$

Classroom Example: p. 1064
Exercise 52

EXAMPLE 7 **Comparing Combinations and Permutations**

a. In how many ways can 3 students from a group of 15 students be selected to serve on a committee?

b. In how many ways can 3 students from a group of 15 students be selected to serve on a committee, if the students will hold the offices of president, vice president, and treasurer?

Solution:

a. In this situation, the students will serve indistinguishable roles on the committee. Therefore, the order in which the 3 students are selected is not relevant. The number of such committees is the number of *combinations* of $n = 15$ students taken $r = 3$ at a time.

$$_{15}C_3 = \frac{15!}{3! \cdot (15 - 3)!} = \frac{15!}{3! \cdot 12!} = \frac{\overset{5}{\cancel{15}} \cdot \overset{7}{\cancel{14}} \cdot 13 \cdot \overset{1}{\cancel{12!}}}{3 \cdot 2 \cdot 1 \cdot \cancel{12!}} = 455$$

b. Now the students will be selected to a committee and assigned different roles (president, vice president, and treasurer). In this case, the order in which the students are selected matters. We need to count the number of *permutations* of $n = 15$ students taken $r = 3$ at a time.

$$_{15}P_3 = \frac{15!}{(15 - 3)!} = \frac{15!}{12!} = \frac{15 \cdot 14 \cdot 13 \cdot \overset{1}{\cancel{12!}}}{\cancel{12!}} = 2730$$

Skill Practice 7   Suppose that 20 people enter a raffle.

    a. In how many ways can four different people among the 20 be selected to receive prizes of $50 each?

    b. In how many ways can four different people among the 20 be selected to receive prizes of $50, $25, $10, and $5?

In Example 7, notice that there are $3! = 6$ times as many permutations as combinations. The reason is that for each combination of 3 items, such as A, B, and C, there are $3! = 6$ permutations: ABC, ACB, BAC, BCA, CAB, CBA.

Classroom Example: p. 1064
Exercise 56

EXAMPLE 8   Counting Combinations in an Application

In the game "Florida Lotto," a player must select a group of 6 numbers (without regard to order) from the numbers 1 to 53. If the 6 numbers match the numbers in the drawing, then the player wins the grand prize. How many groups of 6 numbers are possible?

Solution:

In this situation, the order in which the group of 6 numbers is selected does not matter. Therefore, we need to compute the number of combinations of $n = 53$ numbers taken $r = 6$ at a time.

$$_{53}C_6 = \frac{53!}{6! \cdot (53 - 6)!} = \frac{53!}{6! \cdot 47!} = 22{,}957{,}480$$

There are 22,957,480 different groups of 6 numbers taken from 53 numbers.

Skill Practice 8   The California lottery game "Fantasy 5" offers a grand prize to a player who selects the correct group of 5 numbers (in any order) from the numbers 1 to 39. How many groups of 5 numbers are possible?

Point of Interest

As mentioned in the section opener, the probability of winning the grand prize in "Florida Lotto" for a player who plays 1 combination of 6 numbers is $\frac{1}{22,957,480}$.

Answers

7. a. 4845   **b.** 116,280

8. 575,757

TECHNOLOGY CONNECTIONS

Evaluating a Number of Combinations, $_nC_r$

Most graphing utilities can evaluate the number of combinations of n elements taken r at a time. For example, use the $_nC_r$ function found in the **MATH** menu under PRB.

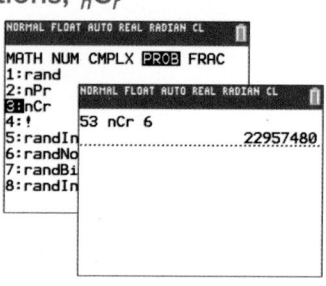

In Example 9 we encounter a situation in which more than one technique of counting must be used.

Classroom Example: p. 1064
Exercise 58

EXAMPLE 9   Applying Multiple Principles of Counting

Suppose that a committee of 3 women and 2 men must be formed from a group of 8 women and 7 men. In how many ways can such a committee be formed?

Solution:

This situation can be thought of as a sequence of two events in which we apply the fundamental principle of counting.

$$\begin{pmatrix} \text{Total number} \\ \text{of committees} \end{pmatrix} = \begin{pmatrix} \text{Number of ways to select} \\ \text{3 women from 8 women} \\ \text{in any order} \end{pmatrix} \cdot \begin{pmatrix} \text{Number of ways to select} \\ \text{2 men from 7 men} \\ \text{in any order} \end{pmatrix}$$

$$= {}_8C_3 \cdot {}_7C_2$$
$$= 56 \cdot 21$$
$$= 1176$$

Avoiding Mistakes

The roles of the men and women on the committee are indistinguishable. Therefore, count the number of combinations rather than permutations.

Skill Practice 9   The coach of a co-ed softball team must select 4 women and 5 men from a group of 7 women and 10 men to play in a game. In how many ways can such a team be formed?

Answer
9. 8820

SECTION 11.6   Practice Exercises

Prerequisite Review

For Exercises R.1–R.4, evaluate the expression.

R.1. 9!   362,880   **R.2.** 0!   1   **R.3.** $\dfrac{7!}{5!}$ 42   **R.4.** $\dfrac{10!}{4!\,6!}$ 210

Concept Connections

1. The fundamental __principle__ of __counting__ indicates that if one event can occur in m different ways, and a second event can occur in n different ways, then the two events can occur in sequence in ___$m \cdot n$___ different ways.

2. If n items are arranged in order, then each arrangement is called a __permutation__ of n items.

3. The number of ways that n distinguishable items can be arranged in various orders is ___$n!$___.

4. Consider a set of n elements of which one element is repeated r times. Then the number of permutations of the elements of the set is given by $\dfrac{\square}{\square}$. $\dfrac{n!}{r!}$

5. Suppose that n represents the number of distinct elements in a group from which r elements will be chosen in a particular order. Then each arrangement is called a permutation of n items taken ___r___ at a time.

6. Suppose that n represents the number of elements in a group from which r elements will be selected in no particular order. Then each group selected is called a combination of n elements taken ___r___ at a time.

7. The number of permutations of n elements taken r at a time is denoted by ${}_nP_r$ and is computed by

$${}_nP_r = \frac{\square}{\square} \quad \text{or} \quad {}_nP_r = n(n-1)(n-2)\cdots(n-r+1). \qquad \frac{n!}{(n-r)!}$$

8. The number of combinations of n elements taken r at a time is denoted by ${}_nC_r$ and is computed by _____. $\frac{n!}{r!\cdot(n-r)!}$ or $\frac{{}_nP_r}{r!}$

Objective 1: Apply the Fundamental Principle of Counting

For Exercises 9–14, consider the set of integers from 1 to 20, inclusive. If one number is selected, in how many ways can we obtain

9. an even number?   10

10. an odd number?   10

11. a prime number?   8

12. a number that is a multiple of 5?   4

13. a number that is a multiple of 10?   2

14. a number that is divisible by 4?   5

15. At a hospital, the dinner menu consists of 4 choices of entrée, 3 choices of salad, 6 choices of beverage, and 4 choices of dessert. How many different meals can be formed if a patient chooses one item from each category? (**See Example 1**) $4\cdot3\cdot6\cdot4 = 288$

16. Debbie travels several times a year for her job. She takes 2 pairs of slacks, 6 blouses, and 4 scarves, all of different colors. Assuming that the items all match together well, how many different outfits does Debbie have if she selects one item from each category? $2\cdot6\cdot4 = 48$

17. A license plate has 3 letters followed by 3 digits.

   a. How many license plates can be made if there are no restrictions on the letters or digits? (**See Example 2**) $26^3\cdot10^3 = 17{,}576{,}000$

   b. How many license plates can be made if no digit or letter may be repeated? $26\cdot25\cdot24\cdot10\cdot9\cdot8 = 11{,}232{,}000$

18. A security company requires its employees to have a 7-character computer password that must consist of 5 letters and 2 digits.

   a. How many passwords can be made if there are no restrictions on the letters or digits? $26^5\cdot10^2 = 1{,}188{,}137{,}600$

   b. How many passwords can be made if no digit or letter may be repeated? $26\cdot25\cdot24\cdot23\cdot22\cdot10\cdot9 = 710{,}424{,}000$

19. The call letters for a radio station must begin with either K or W.

   a. How many 4-letter arrangements are possible assuming that letters may be repeated? $2\cdot26\cdot26\cdot26 = 35{,}152$

   b. How many 4-letter arrangements are possible assuming that the letters may not be repeated? $2\cdot25\cdot24\cdot23 = 27{,}600$

20. An employee identification code for a hospital consists of 2 letters from the set {A, B, C, D} followed by 4 digits.

   a. How many identification codes are possible if both letters and digits may be repeated? $4\cdot4\cdot10\cdot10\cdot10\cdot10 = 160{,}000$

   b. How many identification codes are possible if letters and digits may not be repeated? $4\cdot3\cdot10\cdot9\cdot8\cdot7 = 60{,}480$

21. An online survey is used to monitor customer service. The survey consists of 14 questions. Ten questions have 5 possible responses (strongly agree, agree, neutral, disagree, and strongly disagree). The remaining 4 questions are yes/no questions. In how many different ways can the survey be filled in? (**See Example 3**) $5^{10}\cdot2^4 = 156{,}250{,}000$

22. A test consists of 3 multiple-choice questions, each with four possible responses, and 7 true/false questions. In how many ways can a student answer the questions on the test? $4^3\cdot2^7 = 8192$

23. On a computer, 1 *bit* is a single binary digit and has two possible outcomes: either 1 or 0. One *byte* is 8 bits. That is, a byte is a sequence of 8 binary digits. How many arrangements of 1's and 0's can be made with one byte? $2^8 = 256$

24. Older models of garage door remote controls have a sequence of 10 switches that are individually placed in an up or down position. The remote control can "talk to" the overhead door unit if the 10 corresponding switches in the unit are in the same up/down sequence. How many up/down sequences are possible in an arrangement of 10 switches? $2^{10} = 1024$

Objectives 2–3: Count Permutations and Combinations

25. **a.** In how many ways can the letters in the word FLORIDA be arranged? (**See Example 4**) $7! = 5040$

   b. In how many ways can the letters in the word MISSISSIPPI be arranged? $\frac{11!}{4!\cdot4!\cdot2!} = 34{,}650$

27. In how many ways can the word WRONG be misspelled? $5! - 1 = 119$

26. **a.** In how many ways can the letters in the word XRAY be arranged? $4! = 24$

   b. In how many ways can the letters in the word MAMMOGRAM be arranged? $\frac{9!}{4!\cdot2!} = 7560$

28. In how many ways can the word EXACTLY be misspelled? $7! - 1 = 5039$

29. A delivery truck must make 4 stops at locations A, B, C, and D. Over several weeks, management asks the driver to drive each possible route and record the time required to complete the route. This is to determine the most time-efficient route. How many possible routes are there? $4! = 24$

30. In how many ways can 6 people in a family be lined up for a photograph? $6! = 720$

For Exercises 31–36, evaluate $_nP_r$.

31. $_6P_4$ 360

32. $_8P_5$ 6720

33. $_{12}P_2$ 132

34. $_{11}P_3$ · 990

35. $_9P_9$ 362,880

36. $_5P_5$ 120

37. Evaluate $_{20}P_3$ and interpret its meaning.   6840; There are 6840 ways to select 3 distinct items in a specific order from a group of 20 items.

38. Evaluate $_{15}P_6$ and interpret its meaning.   3,603,600; There are 3,603,600 ways to select 6 distinct items in a specific order from a group of 15 items.

For Exercises 39–44, evaluate $_nC_r$.

39. $_6C_4$ 15

40. $_8C_5$ 56

41. $_{12}C_2$ 66

42. $_{11}C_3$ 165

43. $_9C_9$ 1

44. $_5C_5$ 1

45. Evaluate $_{20}C_3$ and interpret its meaning. 1140; There are 1140 ways to select 3 distinct items in *no specific order* from a group of 20 items.

46. Evaluate $_{15}C_6$ and interpret its meaning. 5005; There are 5005 ways to select 6 distinct items in *no specific order* from a group of 15 items.

47. Given {A, B, C},

    a. List all the permutations of two elements from the set. **(See Example 5)**   AB, BA, AC, CA, BC, CB

    b. List all the combinations of two elements from the set.

48. Given {W, X, Y, Z},

    a. List all the permutations of three elements from the set.

    b. List all the combinations of three elements from the set.   WXY, WXZ, WYZ, XYZ

49. Suppose that 9 horses run a race. How many first-, second-, and third-place finishes are possible? **(See Example 6)**
$_9P_3 = 9 \cdot 8 \cdot 7 = 504$

50. In how many ways can a judge award blue, red, and yellow ribbons to 3 films at a film festival if there are 10 films entered? $_{10}P_3 = 10 \cdot 9 \cdot 8 = 720$

51. a. In a drama class, 5 students are to be selected from 24 students to perform a synchronized dance. In how many ways can 5 students be selected from the 24? **(See Example 7)** $_{24}C_5 = 42{,}504$

    b. Determine the number of ways that 5 students can be selected from the class of 24 to play 5 different roles in a short play. $_{24}P_5 = 5{,}100{,}480$

52. a. There are 100 members of the U.S. Senate. Suppose that 4 senators currently serve on a committee. In how many ways can 4 more senators be selected to serve on the committee? $_{96}C_4 = 3{,}321{,}960$

    b. In how many ways can a group of 3 U.S. senators be selected from a group of 7 senators to fill the positions of chair, vice-chair, and secretary for the Ethics Committee? $_7P_3 = 210$

53. A chess tournament has 16 players. Every player plays every other player exactly once. How many chess matches will be played? $_{16}C_2 = 120$

54. Suppose that a tennis tournament has 64 players. In how many ways can the 64 players be paired to play in the first round? Assume that each player can play any other player without regard to seeding. $_{64}C_2 = 2016$

55. In the Minnesota Lotto game "Gopher 5" a player wins the grand prize by choosing the same group of 5 numbers from 1 through 47 as is chosen by the computer. How many 5-number groups are possible? **(See Example 8)** $_{47}C_5 = 1{,}533{,}939$

56. In the New York state lottery game "Lotto" a player wins the grand prize by choosing the same group of 6 numbers from 1 through 59 as is chosen by the computer. How many 6-number groups are possible? $_{59}C_6 = 45{,}057{,}474$

57. At a ballroom dance lesson, the instructor chooses 3 men and 3 women to demonstrate a new pattern. If there are 9 women and 7 men in the class, in how many ways can the instructor choose 3 men and 3 women? **(See Example 9)** $(_9C_3) \cdot (_7C_3) = 2940$

58. A committee of 4 men and 4 women is to be made from a group of 12 men and 9 women. In how many ways can such a committee be formed?
$(_{12}C_4) \cdot (_9C_4) = 62{,}370$

Mixed Exercises

59. In a "Pick-4" game, a player wins a prize for matching a 4-digit number from 0000 to 9999 with the number randomly selected during the drawing. How many 4-digit numbers can a player choose? Assume that a number can start with a zero or zeros such as 0001.
$10^4 = 10{,}000$

60. In a "Numbers" game, a player wins a prize for matching a 3-digit number from 000 to 999 with the number randomly selected during the drawing. How many 3-digit numbers can a player choose? Assume that a number can start with a zero or zeros such as 0001.
$10^3 = 1000$

61. Liza is a basketball coach and must select 5 players out of 12 players to start a game. In how many ways can she select the 5 players if each player is equally qualified to play each position? $_{12}C_5 = 792$

62. Jean has a list of 8 books that she knows she must read for a class in the upcoming fall semester of school. She wants to get a head start by reading several of the books during the summer. If she has time in the summer to read 5 of the 8 books, in how many ways can she select 5 books from 8 books? $_8C_5 = 56$

63. In how many ways can a manager assign 5 employees at a coffee shop to 5 different tasks? Assume that each employee is assigned to exactly one task. $_5P_5 = 5! = 120$

64. In how many ways can a class of 12 kindergarten children line up at the cafeteria? $_{12}P_{12} = 12! = 479{,}001{,}600$

65. Twenty batteries have been sitting in a drawer for 2 yr. There are 4 dead batteries among the 20. If three batteries are selected at random, determine the number of ways in which

a. 3 dead batteries can be selected. $_4C_3 = 4$

b. 3 good batteries can be selected. $_{16}C_3 = 560$

c. 2 good batteries and 1 dead battery can be selected. $(_{16}C_2) \cdot (_4C_1) = 480$

66. There are 30 seeds in a package. Five seeds are defective (will not germinate). If four seeds are selected at random, determine the number of ways in which

a. 4 defective seeds can be selected. $_5C_4 = 5$

b. 4 good seeds can be selected. $_{25}C_4 = 12{,}650$

c. 2 good seeds and 2 defective seeds can be selected. $(_{25}C_2) \cdot (_5C_2) = 3000$

67. A "combination" lock is opened by correctly "dialing" 3 numbers from 0 to 39, inclusive. The user who knows the code turns the dial to the right to the first number in the code, then to the left to find the second number in the code, and then back to the right for the third number in the code. If someone does not know the code and tries to guess, how many guesses are possible? $40^3 = 64{,}000$

68. A palindrome is an arrangement of letters that reads the same way forwards and backwards. For example, one five-letter palindrome is: ABCBA.

a. How many 5-letter palindromes are possible from a 26-letter alphabet? $26 \cdot 26 \cdot 26 \cdot 1 \cdot 1 = 17{,}576$

b. How many 4-letter palindromes are possible from a 26-letter alphabet? $26 \cdot 26 \cdot 1 \cdot 1 = 676$

A line segment connecting any two nonadjacent vertices of a polygon is called a *diagonal* of the polygon. For Exercises 69–72, determine the number of diagonals for the given polygon.

69. quadrilateral (4 sides) 2

70. pentagon (5 sides) 5

71. hexagon (6 sides) 9

72. octagon (8 sides) 20

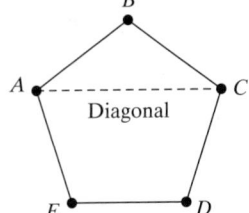

For Exercises 73–78, consider the set of numbers {0, 1, 2, 3, 4, 5}. How many 3-digit codes can be formed with the given restrictions?

73. The code has no restrictions. $6^3 = 216$

74. The code may not contain repeated digits. $_6P_3$ or $6 \cdot 5 \cdot 4 = 120$

75. The code must represent a 3-digit number. (*Hint*: This means that the first digit cannot be zero.) $5 \cdot 6 \cdot 6 = 180$

76. The code must represent a 3-digit number that is a multiple of 5. $5 \cdot 6 \cdot 2 = 60$

77. The code must represent an even 3-digit number. $5 \cdot 6 \cdot 3 = 90$

78. All three digits in the code must be the same. 6

79. In Exercise 103 from Section 11.3, we learned that if a fair coin is flipped n times, the number of head/tail arrangements follows a geometric sequence.

a. Determine the number of head/tail arrangements if a fair coin is flipped 3 times, 4 times, and 5 times. 8; 16; 32

b. If a couple has 3 children, how many boy/girl sequences are possible for the three births? List the outcomes using "B" for boy and "G" for girl. 8; BBB, BBG, BGB, BGG, GBB, GBG, GGB, GGG

c. If a couple has 4 children, how many boy/girl sequences are possible for the four births? 16

80. Social Security numbers assigned in the United States are comprised of 9 digits of the form _ _ _-_ _-_ _ _ _. Would there be enough Social Security numbers for the population of China if the Chinese used the same system? (*Hint*: The population of China is approximately 1.5 billion.) No; the number of possible 9-digit Social Security numbers is $10^9 = 1{,}000{,}000{,}000$

81. Airlines often oversell seats on an airplane. This is done so that the seats for the few passengers that are no-shows will still have been sold. Sometimes, however, all passengers show up and there are more ticketed passengers than seats. Suppose that one flight has 160 passengers and only 156 seats. Determine the number of ways that the airline can select 4 people at random to place on a different flight. $_{160}C_4 = 26{,}294{,}360$

82. In how many ways can a platoon leader select 4 soldiers among 15 soldiers to secure a building? $_{15}C_4 = 1365$

83. A car can comfortably hold a family of five. If two people among the five can drive, how many different seating arrangements are possible? $2 \cdot 4 \cdot 3 \cdot 2 \cdot 1 = 48$

84. A television station must play twelve 30-sec commercials during a half-hour show. In how many ways can the commercials be aired? $12! = 479{,}001{,}600$

Write About It

85. Consider a horse race with 8 horses. Explain how the fundamental principle of counting or the permutation rule can be used to determine the number of first-, second-, and third-place arrangements.

86. Explain why the number of combinations of n items taken r at a time can be computed by $\dfrac{{}_nP_r}{r!}$.

Expanding Your Skills

87. Three biology books, 4 math books, and 2 physics books are to be placed on a book shelf where the books in each discipline are grouped together. In how many ways can the books be arranged on the book shelf?
$[({}_3P_3) \cdot ({}_4P_4) \cdot ({}_2P_2)] \cdot ({}_3P_3) = 1728$

88. A softball team has 9 players consisting of 3 women and 6 men. In how many ways can the coach arrange the batting order if the men must bat consecutively and the women must bat consecutively?
$[({}_3P_3) \cdot ({}_6P_6)] \cdot ({}_2P_2) = 8640$

Additional answers can be found in the Instructor Answer Appendix.

SECTION 11.7   Introduction to Probability

OBJECTIVES

1. Determine Theoretical Probabilities
2. Determine Empirical Probabilities
3. Find the Probability of the Union of Two Events
4. Find the Probability of Sequential Independent Events

1. Determine Theoretical Probabilities

The Centers for Disease Control publishes National Vital Statistics Reports every year that provide data for birth rates and mortality rates based on gender, race, age, and other factors. The data in Table 11-4 give the 1-yr survival rates for people in the United States for selected ages.

From the table, the probability that a 20-yr-old will live to age 21 is 0.9991. Put another way, this means that approximately 99.91% of 20-yr-olds will live to age 21 (see Example 5).

A **probability** is a value assigned to an event that quantifies the likelihood of the event to occur. To compute the probability of an event, we first need to define several key terms. An **experiment** is a test with an uncertain outcome. The set of all possible outcomes of an experiment is called the **sample space** of the experiment. For example:

Table 11-4

Age (yr)	Probability of 1-yr Survival
10	0.9999
20	0.9991
30	0.9990
40	0.9981
50	0.9959
60	0.9907
70	0.9791
80	0.9434
90	0.8543

Experiment:	**Sample Space:**
Flip a coin	{head, tail}
Roll a single 6-sided die	{1, 2, 3, 4, 5, 6}

An **event** is a subset of the sample space and is often denoted by E or some other capital letter.

Description of Event:	**Set Representing the Event:**
Flip a coin and the outcome is "head"	{head}
Roll a six-sided die and an even number comes up	{2, 4, 6}

The number of elements in an event E is often denoted by $n(E)$. For example, given the event $E = \{2, 4, 6\}$, $n(E) = 3$.

If we assume that all elements in a sample space are equally likely to occur, then we define the theoretical probability of an event as follows.

TIP The word "die" is the singular of the word "dice." That is, we roll one die but we roll two dice.

Theoretical Probability of Event *E*

Let *S* represent a sample space with equally likely outcomes, and let *E* be a subset of *S*. Then the probability of event *E*, denoted by $P(E)$, is given by

$$P(E) = \frac{n(E)}{n(S)} \quad \begin{array}{l}\longleftarrow \text{Number of elements in the event} \\ \longleftarrow \text{Number of elements in the sample space}\end{array}$$

The probability of an event is the relative frequency of the event compared to the sample space. For example, if *E* is the event of rolling an even number on a die, then $E = \{2, 4, 6\}$ and $S = \{1, 2, 3, 4, 5, 6\}$. Then,

$P(E) = \frac{n(E)}{n(S)} = \frac{3}{6} = \frac{1}{2}$

The value of a probability can be written as a fraction, as a decimal, or as a percent. Therefore, $P(E) = \frac{1}{2}$, or 0.5, or 50%.

> **Avoiding Mistakes**
>
> The probability $\frac{1}{2}$ does not mean that exactly 1 out of every 2 rolls of a die will result in an even number. Instead, it is a ratio that we expect experimental observations to approach after the experiment is performed a large number of times.

Classroom Example: p. 1076
Exercise 26

EXAMPLE 1   Finding the Probability of an Event

Suppose that a box contains two red, three blue, and five green marbles. If one marble is selected at random, find the probability of the event.

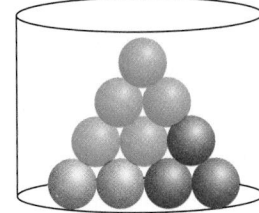

a. E_1:   A red marble is selected.

b. E_2:   A blue marble is selected.

c. E_3:   A white marble is selected.

Solution:

Each of the 10 marbles in the box is equally likely to be selected. Denote the sample space as $S = \{R_1, R_2, B_1, B_2, B_3, G_1, G_2, G_3, G_4, G_5\}$.

a. $E_1 = \{R_1, R_2\}$

Therefore, $P(E_1) = \frac{n(E_1)}{n(S)} = \frac{2}{10} = \frac{1}{5}$.

The number of red marbles is 2.

The probability of selecting a red marble is $\frac{1}{5}$.

b. $E_2 = \{B_1, B_2, B_3\}$

Therefore, $P(E_2) = \frac{n(E_2)}{n(S)} = \frac{3}{10}$.

The number of blue marbles is 3.

The probability of selecting a blue marble is $\frac{3}{10}$.

c. $E_3 = \{\ \}$

Therefore, $P(E_3) = \frac{n(E_3)}{n(S)} = \frac{0}{10} = 0$.

The number of white marbles is 0.

It is intuitively obvious that a white marble cannot be selected. Mathematically, a probability of 0 means that the event is impossible.

> **Skill Practice 1**   A sock drawer has 6 blue socks, 2 white socks, and 12 black socks. If one sock is selected at random, find the probability of the event.
>
> **a.** E_1:   A blue sock is selected.   **b.** E_2:   A black sock is selected.
>
> **c.** E_3:   A brown sock is selected.

Answers

1. a. $\frac{3}{10}$ **b.** $\frac{3}{5}$ **c.** 0

The number of outcomes in an event is a nonnegative number less than or equal to the number of outcomes in the sample space. Therefore, the value of a probability is a number between 0 and 1, inclusive. That is, $0 \leq P(E) \leq 1$. Furthermore,

- If $P(E) = 0$, then the event E is called an **impossible event.**
- If $P(E) = 1$, then the event E is called a **certain event.**

In Example 1(c), it is impossible to draw a white marble from the box. However, we would be 100% certain of drawing a red, blue, or green marble from the box and the probability would be 1.

Classroom Example: p. 1076
Exercise 28

EXAMPLE 2   Finding the Probability of an Event

An American roulette wheel has 38 slots, numbered 1 through 36, 0, and 00. Eighteen slots are red, 18 are black, and 2 are green. The dealer spins the wheel in one direction and rolls a small ball in the opposite direction until both come to rest. The ball is equally likely to fall in any one of the 38 slots. On a given spin of an American roulette wheel, find the probability of the event.

  a. E_1:   The ball lands on an odd number.
  b. E_2:   The ball lands on a green slot.
  c. E_3:   The ball does not land on a green slot.

Solution:

The sample space has 38 outcomes. That is $n(S) = 38$.

  a. There are 18 odd numbers. Therefore, $n(E_1) = 18$ and
$$P(E_1) = \frac{n(E_1)}{n(S)} = \frac{18}{38} = \frac{9}{19} \approx 0.4737.$$

  b. There are 2 green slots on the wheel. Therefore, $n(E_2) = 2$ and
$$P(E_2) = \frac{n(E_2)}{n(S)} = \frac{2}{38} = \frac{1}{19} \approx 0.0526.$$

  c. There are 36 slots that are not green. Therefore, $n(E_3) = 36$ and
$$P(E_3) = \frac{n(E_3)}{n(S)} = \frac{36}{38} = \frac{18}{19} \approx 0.9474.$$

Skill Practice 2   On a given spin of an American roulette wheel, find the probability of the event.

  a. E_1:   The ball lands on the number 7.
  b. E_2:   The ball lands on a red slot.
  c. E_3:   The ball does not land on a red slot.

From Example 2(b) and 2(c), notice that event E_3 consists of all elements in the sample space not in event E_2. Therefore, the union of the two events make up the entire sample space, and $P(E_2) + P(E_3) = \frac{1}{19} + \frac{18}{19} = 1$. In this case, we say that events E_2 and E_3 are *complementary events.*

Answers

2. a. $\frac{1}{38}$ **b.** $\frac{9}{19}$ **c.** $\frac{10}{19}$

Definition of Complementary Events

Let E be an event relative to sample space S. The **complement of E,** denoted by $\overline{E}$ (or sometimes by $\sim E$ or E') is the set of outcomes in the sample space but not in event E. It follows that

$$P(E) + P(\overline{E}) = 1.$$

Classroom Example: p. 1076 Exercise 34

EXAMPLE 3 Finding the Probabilities of Complementary Events

Suppose that 2 dice are rolled. Determine the probability that

a. The sum of the numbers showing on the dice is 7.

b. The sum of the numbers showing on the dice is *not* 7.

Solution:

Each individual die has 6 equally likely outcomes. Therefore, by the fundamental principle of counting, the number of ways the pair of dice can fall is $6 \cdot 6 = 36$ (Figure 11-11).

First Die

Second Die	1	2	3	4	5	6
1	(1, 1)	(1, 2)	(1, 3)	(1, 4)	(1, 5)	(1, 6)
2	(2, 1)	(2, 2)	(2, 3)	(2, 4)	(2, 5)	(2, 6)
3	(3, 1)	(3, 2)	(3, 3)	(3, 4)	(3, 5)	(3, 6)
4	(4, 1)	(4, 2)	(4, 3)	(4, 4)	(4, 5)	(4, 6)
5	(5, 1)	(5, 2)	(5, 3)	(5, 4)	(5, 5)	(5, 6)
6	(6, 1)	(6, 2)	(6, 3)	(6, 4)	(6, 5)	(6, 6)

Figure 11-11

a. Let E represent the event that the numbers showing on the dice have a sum of 7. A sum of 7 will occur if the dice land on one of the following 6 outcomes.

$E = \{(1, 6), (2, 5), (3, 4), (4, 3), (5, 2), (6, 1)\}$. Therefore, $n(E) = 6$ and

$$P(E) = \frac{n(E)}{n(S)} = \frac{6}{36} = \frac{1}{6}.$$ The probability of rolling a sum of 7 is $\frac{1}{6}$.

b. The event that the numbers on the dice do not present a sum of 7 is the complement of event E. Therefore, we have the option of using the formula $P(E) + P(\overline{E}) = 1$.

$$P(\overline{E}) = 1 - P(E) = 1 - \frac{1}{6} = \frac{5}{6}$$ The probability of rolling a sum other than 7 is $\frac{5}{6}$.

Skill Practice 3 Suppose that two dice are rolled. Determine the probability that

a. The sum of the numbers showing on the dice is 8.

b. The sum of the numbers showing on the dice is *not* 8.

In Example 4, we use the techniques of counting learned in Section 11.6 to determine the probability of an event.

Classroom Example: p. 1076 Exercise 40

EXAMPLE 4 Finding the Probability of an Event by Using Counting Techniques

Suppose that 5 women and 3 men apply for 2 job openings. If the applicants are equally qualified, find the probability that both positions are filled by women.

Answers

3. a. $\frac{5}{36}$ b. $\frac{31}{36}$

Solution:

Let E represent the event that two women are chosen. The number of elements in E is the number of possible ways to select 2 women from a group of 5 women without regard to order. That is,

$$n(E) = {_5}C_2 = 10$$

The sample space S consists of all possible ways in which 2 people can be selected from 8 people without regard to order. That is,

$$n(S) = {_8}C_2 = 28$$

The probability that both positions will be filled by women is

$$P(E) = \frac{n(E)}{n(S)} = \frac{{_5}C_2}{{_8}C_2} = \frac{10}{28} = \frac{5}{14} \approx 0.3571$$

> **Skill Practice 4** Suppose that a committee of three people is to be formed from a group of 8 men and 6 women. Find the probability that the committee will consist of all men.

2. Determine Empirical Probabilities

Suppose that the Centers for Disease Control wants to measure the 1-yr survival rates for Americans for specific ages. There is no way that this can be derived theoretically. Instead, the probability can be approximated through observations. A probability computed in this way is called an **empirical probability.**

> **Computing Empirical Probability**
>
> The empirical probability of an event E is given by
>
> $$P(E) = \frac{\text{Number of times the event } E \text{ occurs}}{\text{Number of times the experiment is perfomed}}$$

Classroom Example: p. 1077
Exercise 46

EXAMPLE 5 Computing an Empirical Probability

Data from the Centers for Disease Control for a recent year indicate that there were approximately 4,295,000 individuals of age 20 in the United States. One year later 3887 of these individuals had died and 4,291,113 had lived. Determine the probability that a 20-yr-old will survive to age 21. (*Source*: www.cdc.gov)

Solution:

Let E represent the event that a 20-yr-old lives to age 21.

The experiment involves 4,295,000 individuals of age 20. The number of observed outcomes where an individual lives to the age of 21 is 4,291,113. Therefore, the empirical probability is given by

$$P(E) = \frac{4,291,113}{4,295,000} \approx 0.9991$$

> **Skill Practice 5** Suppose that of approximately 4,224,100 individuals of age 40 in the United States, 8038 die before the age of 41 and 4,216,062 survive. Determine the probability that a 40-yr-old will live to age 41.

Answers

4. $\frac{2}{13}$ 5. 0.9981

3. Find the Probability of the Union of Two Events

We now study the probability of the union of two events. Suppose that one card is drawn from a standard deck of cards. The sample space has the following properties (Figure 11-12).

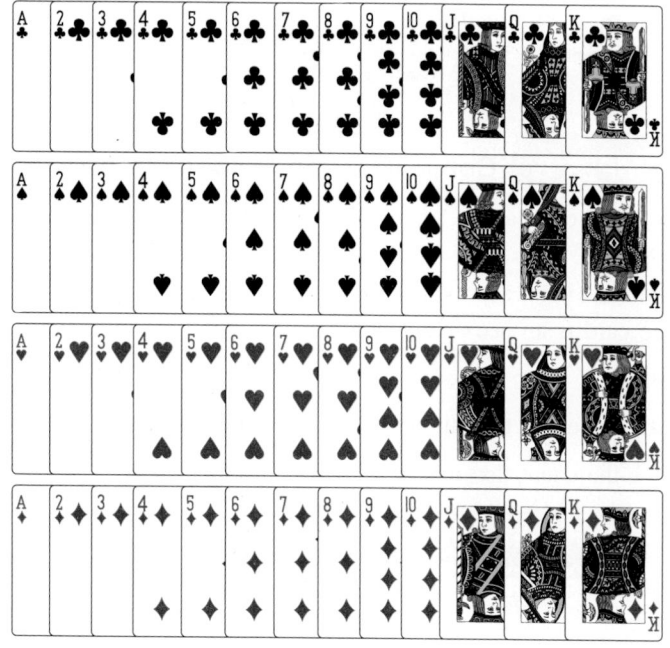

- The deck consists of 52 cards.

- The cards are divided into four suits (or categories) called spades (♠), clubs (♣), hearts (♥), and diamonds (♦).

- Each suit consists of 13 cards labeled: Ace (A), 2, 3, 4, 5, 6, 7, 8, 9, 10, Jack (J), Queen (Q), and King (K).

- There are 26 black cards (spades and clubs) and 26 red cards (hearts and diamonds).

Figure 11-12

Let A be the event that an ace is drawn: {A♠, A♣, A♥, A♦}.

Let K be the event that a king is drawn: {K♠, K♣, K♥, K♦}.

Events A and K have no elements in common (a single card cannot be both an ace and a king). Therefore, the intersection of A and K is the empty set, $A \cap K = \{ \}$, and we say that events A and K are **mutually exclusive** (do not overlap).

Now let event A be the event that an ace is drawn, and let S be the event that a spade is drawn (Figure 11-13).

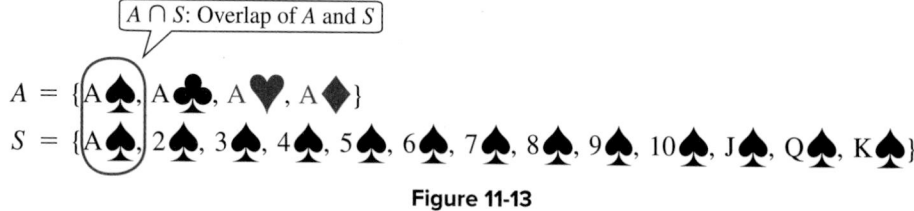

Figure 11-13

Notice that events A and S share the comment element of the ace of spades, A♠. Therefore, events A and S are *not* mutually exclusive, meaning that they overlap.

We are now ready to generalize. To find the probability of the event (A or B), denoted $P(A$ or $B)$, find the probability of the *union* of A and B.

Classroom Examples: p. 1077
Exercises 50, 52

Probability of (A or B)

Given events A and B in the same sample space, $P(A$ or $B)$ is given by

$$P(A \cup B) = P(A) + P(B) - P(A \cap B)$$

Note: If A and B are mutually exclusive, then $P(A \cap B) = 0$ and we have $P(A \cup B) = P(A) + P(B)$.

EXAMPLE 6   Computing the Probability of *A* or *B*

Suppose that one card is selected at random from a standard deck of cards. Find the probability that

a. The card is an ace or a king.

b. The card is an ace or a spade.

Solution:

Let A be the event that an ace is drawn: {A♠, A♣, A♥, A♦}.

Let K be the event that a king is drawn: {K♠, K♣, K♥, K♦}.

Let S be the event that a spade is drawn:

$$S = \{A♠, 2♠, 3♠, 4♠, 5♠, 6♠, 7♠, 8♠, 9♠, 10♠, J♠,$$
$$Q♠, K♠\}.$$

a. Events A and K are mutually exclusive. Therefore, to find $P(A$ or $K)$, we have

> **TIP** In Example 6(a), since A and K are mutually exclusive, then $P(A \cap K) = 0$.

$$P(A \cup K) = P(A) + P(K)$$
$$= \frac{4}{52} + \frac{4}{52} \qquad \text{There are 4 aces in the deck out of 52 cards. } P(A) = \frac{4}{52}.$$
$$\qquad\qquad\qquad \text{There are 4 kings in the deck out of 52 cards. } P(K) = \frac{4}{52}.$$
$$= \frac{8}{52} \text{ or } \frac{2}{13}$$

b. From Figure 11-13, events A and S are *not* mutually exclusive. Therefore, to find $P(A$ or $S)$, we have:

> **TIP** As an alternative to computing $P(A) + P(S) - P(A \cap S)$, count the number of elements in the event $(A \cup S)$, being careful not to count the events common to A and S twice. That is, count 13 spades plus the 3 aces that are not spades. This gives a total of 16 elements in event $(A \cup S)$. Thus,
>
> $$P(A \cup S) = \frac{16}{52} = \frac{4}{13}$$

$$P(A \cup S) = P(A) + P(S) - P(A \cap S)$$
$$= \frac{4}{52} + \frac{13}{52} - \frac{1}{52} \boxed{\text{ace of spades}}$$
$$= \frac{16}{52} \text{ or } \frac{4}{13}$$

There are 4 aces in the deck out of 52 cards. $P(A) = \frac{4}{52}$.

There are 13 spades in the deck out of 52 cards. $P(S) = \frac{13}{52}$.

There is 1 spade that is an ace in the deck. Therefore, $P(A \cap S) = \frac{1}{52}$.

Skill Practice 6   Suppose that one card is drawn at random from a standard deck. Find the probability that

a. The card is a 2 or a 10.

b. The card is a 2 or a red card.

Answers

6. a. $\dfrac{2}{13}$ **b.** $\dfrac{7}{13}$

Classroom Examples: p. 1078
Exercises 62, 64

EXAMPLE 7   Computing an Empirical Probability of (*A* or *B*)

The safety and security department at a college asked a sample of 265 students to respond to the following question.

"Do you think that the college has adequate lighting on campus at night?"

The table gives the results of the survey based on gender and response.

	Yes	No	No Opinion	Total
Male	92	7	4	103
Female	36	102	24	162
Total	128	109	28	265

If one student is selected at random from the group, find the probability that

a. The student answered "Yes" or had "No Opinion."

b. The student answered "No" or was female.

Solution:

a. Let *Y* be the event that the student answered "Yes."

Let *O* be the event that the student had "No Opinion."

Y and *O* are mutually exclusive events (do not overlap).

	Yes	No	No Opinion	Total
Male	92	7	4	103
Female	36	102	24	162
Total	128	109	28	265

$P(Y \text{ or } O) = P(Y) + P(O)$

$= \dfrac{128}{265} + \dfrac{28}{265}$ $P(Y) = \frac{128}{265}$ and $P(O) = \frac{28}{265}$.

$= \dfrac{156}{265}$

TIP As an alternative to computing $P(N) + P(F) - P(N \cap F)$, count the number of elements in the event $(N \cup F)$, being careful not to count the elements common to *N* and *F* twice; that is, count 162 females plus the 7 males who answered "No" for a total of 169 elements in event $(N \cup F)$. Thus, $P(N \cup F) = \frac{169}{265}$.

b. Let *N* be the event that the student answered "No."

Let *F* be the event that the student is female.

N and *F* are *not* mutually exclusive (the events overlap).

	Yes	No	No Opinion	Total
Male	92	7	4	103
Female	36	102	24	162
Total	128	109	28	265

$P(N \text{ or } F) = P(N) + P(F) - P(N \cap F)$

$= \dfrac{109}{265} + \dfrac{162}{265} - \dfrac{102}{265}$ $P(N) = \frac{109}{265}, P(F) = \frac{162}{265}$, and $P(N \cap F) = \frac{102}{265}$

$= \dfrac{169}{265}$ The 102 females who answered "No" are the elements in the intersection of *N* and *F*.

Skill Practice 7   Refer to the data given in Example 7. If one student is selected at random from the group, find the probability that

a. The student answered "Yes" or "No."

b. The student is male or had no opinion.

Answers

7. a. $\dfrac{237}{265}$ **b.** $\dfrac{127}{265}$

4. Find the Probability of Sequential Independent Events

When a coin is tossed, the probability that it lands heads up is $\frac{1}{2}$. Now suppose that the coin is flipped two times in succession. What is the probability that it lands heads up twice in a row? Our intuition tells us that the probability should be less than $\frac{1}{2}$ because it is more unlikely that the desired outcome will happen twice in a row than one time. We also know that the outcome of the first event (heads up) does not affect the outcome of the second event (also heads up). For this reason, the two events are called **independent events.**

Probability of a Sequence of Independent Events

If events A and B are independent events, then the probability that both A and B will occur is

$$P(A \text{ and } B) = P(A) \cdot P(B)$$

For example,

Let A represent the event that a coin lands heads up on the first toss.

Let B represent the event that a coin lands heads up on the second toss.

Then the probability that the coin will land heads up on both tosses is

$$P(A \text{ and } B) = P(A) \cdot P(B) = \frac{1}{2} \cdot \frac{1}{2} = \frac{1}{4}$$

Classroom Example: p. 1078
Exercise 68

EXAMPLE 8   Finding the Probability of Independent Events

Suppose that a family plans to have three children. Find the probability that all three children will be boys.

Solution:

Each birth is independent of the birth that precedes it. For each birth, the probability that the child is born a boy is $\frac{1}{2}$.

$P(\text{all boys}) = P(\text{boy on 1st}) \cdot P(\text{boy on 2nd}) \cdot P(\text{boy on 3rd})$

$$= \frac{1}{2} \cdot \frac{1}{2} \cdot \frac{1}{2} = \frac{1}{8}$$

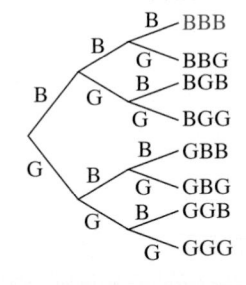

Outcomes

> **TIP** In Example 8 there are 8 elements in the sample space:
>
> BBB, BBG, BGB, BGG,
> GBB, GBG, GGB, GGG
>
> The event BBB occurs 1 time. Thus, the probability of all boys is $\frac{1}{8}$.

Skill Practice 8   Suppose that a family plans to have five children. Find the probability that all five children will be girls.

Classroom Example: p. 1078
Exercise 70

EXAMPLE 9   Finding the Probability of Independent Events

In baseball, a player's batting "average" is the quotient of the number of hits to the number of "at bats." It can also be interpreted as the probability that the player will get a hit on a given time at bat. If Albert Pujols had a batting average of 0.279 for a recent season, determine the probability that he would get a hit on four consecutive times at bat.

Answer

8. $\dfrac{1}{32}$

Solution:

Let H represent the event that Albert Pujols gets a hit on a given time at bat. Furthermore, each time at bat is independent of the time before.

We have that $P(H) = 0.279$. Therefore,

$$P(4 \text{ consecutive hits}) = (0.279)(0.279)(0.279)(0.279)$$
$$\approx 0.006$$

If Albert Pujols has four times at bat, there is less than a 1% chance that he will get a hit all four times.

> **Skill Practice 9**   Suppose the probability that a person will catch a winter "cold" is 0.16. What is the probability that four unrelated people will all catch winter "colds"?

Answer

9. $(0.16)^4 \approx 0.000655$

SECTION 11.7   Practice Exercises

Prerequisite Review

R.1. Given: $P = \{a, b, c, d, e, f, g, h, i\}$ and $Q = \{c, f, h, o, u\}$
List the elements of the following sets.
  a. $P \cap Q$ $\{c, f, h\}$ **b.** $P \cup Q$ $\{a, b, c, d, e, f, g, h, i, o, u\}$

R.2. Given: $M = \{-4, -2, 1, 3, 5\}$ and $N = \{-5, -4, -3, -2, -1\}$
List the elements of the following sets.
  a. $M \cap N$ $\{-4, -2\}$ **b.** $M \cup N$ $\{-5, -4, -3, -2, -1, 1, 3, 5\}$

Concept Connections

1. The ___sample___ ___space___ of an experiment is the set of all possible outcomes.

2. An ___event___ is a subset of the sample space of an experiment.

3. If $P(E) = 0$, then E is called an ___impossible___ event. If $P(E) = 1$, then E is called a ___certain___ event.

4. The notation $\overline{E}$ represents the ___complement___ of event E. Furthermore, $P(E) + P(\overline{E}) = $ ___1___.

5. Two events in a sample space are ___mutually___ ___exclusive___ if they do not share any common elements. That is, the two events do not overlap.

6. If two events A and B are not mutually exclusive, then $P(A \cup B)$ can be computed by the formula $P(A \cup B) = $ $P(A) + P(B) - P(A \cap B)$

7. If two events A and B are mutually exclusive, then $P(A \cap B) = $ ___0___. As a result, $P(A \cup B)$ can be computed from the fomula $P(A \cup B) = $ $P(A) + P(B)$.

8. For two independent events A and B, $P(A \text{ and } B) = $ $P(A) \cdot P(B)$.

Objective 1: Determine Theoretical Probabilities

9. Which of the values can represent the probability of an event?
  a. 1.84   **b.** $-\frac{3}{7}$ **c.** 0.00   **d.** 1.00
  e. 250%   **f.** 6.1   **g.** 0.61   **h.** 6.1%
  c, d, g, h

10. Which of the values can represent the probability of an event?
  a. 2.32   **b.** 0.231   **c.** 2.31%   **d.** −1
  e. $\frac{3}{8}$ **f.** $-\frac{1}{4}$ **g.** $\sqrt{2}$ **h.** 125%
  b, c, e

For Exercises 11–16, match the probability with a statement a, b, c, d, e, or f.

a. Event E is certain to happen.
b. Event E cannot happen.
c. Event E is very likely to happen.
d. Event E is very unlikely to happen.
e. Event E is somewhat likely to happen.
f. Event E is as likely to happen as not to happen.

11. $P(E) = 0.994$ c

12. $P(E) = 1$ a

13. $P(E) = 0.75$ e

14. $P(E) = 0.003$ d

15. $P(E) = 0.5$ f

16. $P(E) = 0$ b

For Exercises 17–24, consider an experiment where a single 10-sided die is rolled with the outcomes 1, 2, 3, 4, 5, 6, 7, 8, 9, 10. Determine the probability of each event.

17. A number less than 5 is rolled. $\dfrac{4}{10} = \dfrac{2}{5}$

18. A number less than 3 is rolled. $\dfrac{2}{10} = \dfrac{1}{5}$

19. A number between 4 and 10, inclusive, is rolled. $\dfrac{7}{10}$

20. A number between 2 and 7, inclusive, is rolled. $\dfrac{6}{10} = \dfrac{3}{5}$

21. A number greater than 10 is rolled.   0

22. A number less than 1 is rolled.   0

23. A number greater than or equal to 1 is rolled.   1

24. A number less than or equal to 10 is rolled.   1

25. A course in early civilization has 6 freshmen, 8 sophomores, and 16 juniors. If one student is selected at random, find the probability of the following events. **(See Example 1)**
    a. A junior is selected. $\dfrac{16}{30} = \dfrac{8}{15}$
    b. A freshman is selected. $\dfrac{6}{30} = \dfrac{1}{5}$
    c. A senior is selected.   0

26. Suppose that a box contains 4 chocolate chip cookies, 8 molasses cookies, and 12 raisin cookies. If one cookie is selected at random, find the probability of the following events.
    a. A chocolate chip cookie is selected. $\dfrac{4}{24} = \dfrac{1}{6}$
    b. A molasses cookie is selected. $\dfrac{8}{24} = \dfrac{1}{3}$
    c. A ginger cookie is selected.   0

For Exercises 27–28, consider an American roulette wheel. (See Example 2) For a given spin of the wheel, find the probability of the following events.

27. a. The ball lands on an even number (do not include 0 and 00). $\dfrac{18}{38} = \dfrac{9}{19}$
    b. The ball lands on a number that is a multiple of 5 (do not include 0 and 00). $\dfrac{7}{38}$
    c. The ball does not land on the number 8. $\dfrac{37}{38}$

28. a. The ball lands on a black slot. $\dfrac{18}{38} = \dfrac{9}{19}$
    b. The ball lands on a number that is a multiple of 6 (do not include 0 and 00). $\dfrac{6}{38} = \dfrac{3}{19}$
    c. The ball does not land on the number 12. $\dfrac{37}{38}$

29. If $P(E) = 0.842$, what is the value of $P(\overline{E})$?   0.158

30. If $P(A) = 0.431$, what is the value of $P(\overline{A})$?   0.569

31. According to the Centers for Disease Control, the probability that a live birth will be of twins in the United States is 0.016. What is the probability that a live birth will *not* be of twins?   0.984

32. A baseball player with a batting average of 0.291 has a probability of 0.291 of getting a hit for a given time at bat. What is the probability that the player will *not* get a hit for a given time at bat?   0.709

For Exercises 33–36, consider the sample space when two fair dice are rolled. (See Example 3) Determine the probabilities for the following events.

33. a. The sum of the numbers on the dice is 4.
    b. The sum of the numbers on the dice is *not* 4.

34. a. The sum of the numbers on the dice is 12.
    b. The sum of the numbers on the dice is *not* 12.

35. a. The sum of the numbers on the dice is greater than 9.
    b. The sum of the numbers on the dice is less than 4.

36. a. The sum of the numbers on the dice is greater than or equal to 8.
    b. The sum of the numbers on the dice is less than or equal to 5.

37. After a nationally televised trial, a poll of viewers indicated that 68% thought the defendant was guilty, 22% thought the defendant was not guilty, and the rest were undecided. What is the probability that a person selected from the viewing audience was undecided?   0.10

38. If a couple plans to have three children, the probability that all three will be boys is 0.125. What is the probability that the couple will have at least one girl?   0.875

39. Suppose that a jury pool consists of 18 women and 16 men.
    a. What is the probability that a jury of 9 people taken at random from the pool will consist only of women? **(See Example 4)**
    b. What is the probability that the jury will consist only of men?
    c. Why do the probabilities from parts (a) and (b) not add up to 1?

40. Suppose that 20 good batteries and 6 defective batteries are in a drawer.
    a. If 4 batteries are drawn at random, what is the probability that all four will be defective?
    b. What is the probability that all four will be good?
    c. Why do the probabilities from parts (a) and (b) not add up to 1?

41. In the Illinois state lottery game "Little Lotto," a player wins the grand prize by choosing the same group of five numbers from 1 through 39 as is chosen by the computer. What is the probability that a player will win the grand prize by playing 1 ticket?

42. In the New York state lottery game "Lotto," a player wins the grand prize by choosing the same group of 6 numbers from 1 through 59 as is chosen by the computer. What is the probability that a player will win the grand prize by playing 5 different tickets?

Scientist Gregor Mendel (1822–1884) is often called the "father of modern genetics" and is famous for his work involving the inheritance of certain traits in pea plants. Suppose that the genes controlling the color of peas are Y for yellow and y for green. Each plant has two genes, one from the female (seed) and one from the male (pollen). The Y gene is dominant, and therefore a plant with genes YY will have yellow peas, a plant with genes Yy or yY will have yellow peas, and a plant with genes yy will have green peas.

If a plant with two yellow genes (YY) is crossed with a plant with two green genes (yy), the result is four hybrid offspring with genotypes Yy. The offspring will be yellow, but will carry the recessive green gene.

		Parent 2	
		y	y
Parent 1	Y	Yy	Yy
	Y	Yy	Yy

43. Suppose that both parent pea plants are hybrids with genotype Yy.

   a. Make a chart showing the possible genotypes of the offspring.

		Parent 2	
		Y	y
Parent 1	Y	YY	Yy
	y	yY	yy

   b. What is the probability that a given offspring will have green peas? $\frac{1}{4}$

   c. What is the probability that a given offspring will have yellow peas? $\frac{3}{4}$

44. Suppose that one parent pea plant has genotype YY and the other has genotype Yy.

   a. Make a chart showing the possible genotypes of the offspring.

		Parent 2	
		Y	y
Parent 1	Y	YY	Yy
	Y	YY	Yy

   b. What is the probability that a given offspring will have green peas? 0

   c. What is the probability that a given offspring will have yellow peas? 1

Objective 2: Determine Empirical Probabilities

45. At a hospital specializing in treating heart disease, it was found that 222 out of 4624 patients undergoing open heart mitral valve surgery died during surgery or within 30 days after surgery. Determine the probability that a patient will not survive the surgery or 30 days after the surgery. This is called the mortality rate. Round to 3 decimal places. (**See Example 5**) 0.048

46. China has the largest population of any country with approximately 1.5 billion people. In a recent year, census results indicated that 199,500,000 Chinese were over the age of 60. If a person is selected at random from the population of China, what is the probability that the person is over 60 years old? Round to 3 decimal places. 0.133

47. For a certain district, a random sample of registered voters results in the distribution by political party given in the graph. Based on these results, what is the probability of selecting a voter at random from the district and getting

   a. A Democrat? $\frac{103}{250}$ or 0.412

   b. A voter that is neither Democrat, Republican, nor Independent? $\frac{9}{250}$ or 0.036

Party Distribution

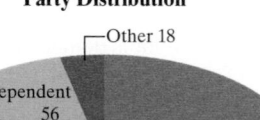
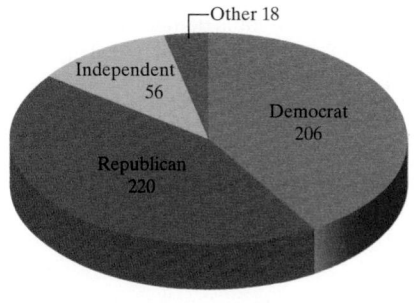

48. The final exam grades for a sample of students in a Freshmen English class at a large university result in the following grade distribution. Based on these results, what is the probability of selecting a student at random taking Freshmen English and getting a student who received

   a. An "A." $\frac{8}{59} \approx 0.1356$ **b.** A "C." $\frac{21}{59} \approx 0.3559$

Distribution of Grades

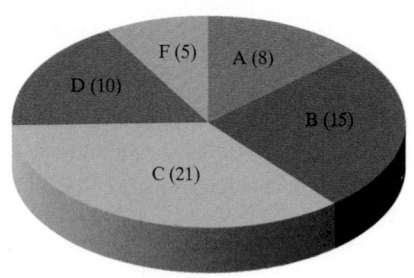

Objective 3: Find the Probability of the Union of Two Events

For Exercises 49–58, consider the sample space for a single card drawn from a standard deck. (**See Example 6 and Figure 11-12**) Find the probability that the card drawn is

49. A jack or a queen. $\frac{8}{52} = \frac{2}{13}$

50. An ace or a 2. $\frac{8}{52} = \frac{2}{13}$

51. A jack or a diamond. $\frac{16}{52} = \frac{4}{13}$

52. A 5 or a heart. $\frac{16}{52} = \frac{4}{13}$

53. A face card (jack, queen, or king). $\frac{12}{52} = \frac{3}{13}$

54. A card numbered between 5 and 10, inclusive. $\frac{24}{52} = \frac{6}{13}$

55. A face card or a red card. $\dfrac{32}{52} = \dfrac{8}{13}$

56. A card numbered between 5 and 10, inclusive, or a black card. $\dfrac{38}{52} = \dfrac{19}{26}$

57. A heart, club, or spade. $\dfrac{39}{52} = \dfrac{3}{4}$

58. An ace, 2, or 3. $\dfrac{12}{52} = \dfrac{3}{13}$

For Exercises 59–66, use the data in the table categorizing cholesterol levels by the ages of the individuals in a study. If one person from the study is chosen at random, find the probability of the given event. (See Example 7)

	Normal Cholesterol	Elevated Cholesterol	Total
30 and under	14	4	18
31–60	52	28	80
61 or older	22	80	102
Total	88	112	200

59. The person has elevated cholesterol. $\dfrac{112}{200} = \dfrac{14}{25}$

60. The person is 61 or older. $\dfrac{102}{200} = \dfrac{51}{100}$

61. The person is 60 or under. $\dfrac{98}{200} = \dfrac{49}{100}$

62. The person is 31 or older. $\dfrac{182}{200} = \dfrac{91}{100}$

63. The person has normal cholesterol or is 61 or older. $\dfrac{168}{200} = \dfrac{21}{25}$

64. The person has elevated cholesterol or is 30 or under. $\dfrac{126}{200} = \dfrac{63}{100}$

65. The person is between 31 and 60, inclusive, or has elevated cholesterol. $\dfrac{164}{200} = \dfrac{41}{50}$

66. The person is 61 or older or has elevated cholesterol. $\dfrac{134}{200} = \dfrac{67}{100}$

Objective 4: Find the Probability of Sequential Independent Events

67. Suppose that a die is rolled followed by the flip of a coin. Find the probability that the outcome is a 5 on the die followed by the coin turning up heads. (**See Example 8**) $\dfrac{1}{6} \cdot \dfrac{1}{2} = \dfrac{1}{12}$

68. If a code for an alarm is a 4-digit sequence, determine the probability that someone guesses each digit correctly. $\left(\dfrac{1}{10}\right)^4 = \dfrac{1}{10{,}000}$

69. The 5-yr survival rate for stage I breast cancer is 88%. (*Source*: American Cancer Society, www.cancer.org) If two women with stage I breast cancer are selected at random, what is the probability that they both survive 5 yr? (**See Example 9**) $(0.88)(0.88) = 0.7744$

70. Basketball player Lebron James makes approximately 76% of free throws. If he plays in a game in which he shoots 6 free throws, what is the probability that he will make all 6? $(0.76)^6 \approx 0.1927$

71. In the 2010 Wimbledon Championships, John Isner from the United States and Nicolas Mahut from France played a first-round tennis match that became the longest match in tennis history. (The match stretched over a 3-day period with Isner winning 70-68 in the fifth set.) In 2011, after a random draw, the two men met again in the first round of Wimbledon. This is highly improbable. If there are 128 men in the tournament, estimate the probability that

a. Isner and Mahut would meet in the first round at Wimbledon in any given year. Assume that any player can play any other player in the first round (that is, disregard the fact that seeded players do not play one another in the first round). $\dfrac{1}{127} \approx 0.007874$

b. Isner and Mahut would meet in the first round 2 yr in a row. $\dfrac{1}{127} \cdot \dfrac{1}{127} \approx 0.000062$

72. A traffic light at an intersection has a 120-sec cycle. The light is green for 80 sec, yellow for 5 sec, and red for 35 sec.

a. When a motorist approaches the intersection, find the probability that the light will be red. (Assume that the color of the light is defined as the color when the car is 100 ft from the intersection. This is the approximate distance at which the driver makes a decision to stop or go.) $\dfrac{35}{120} = \dfrac{7}{24} \approx 0.292$

b. If a motorist approaches the intersection twice during the day, find the probability that the light will be red both times. $\dfrac{35}{120} \cdot \dfrac{35}{120} = \dfrac{49}{576} \approx 0.085$

73. A test has 10 questions. Five questions are true/false and five questions are multiple-choice. Each multiple-choice question has 4 possible responses of which exactly one is correct. Find the probability that a student guesses on each question and gets a perfect score. $\left(\dfrac{1}{2}\right)^5 \cdot \left(\dfrac{1}{4}\right)^5 = \dfrac{1}{32{,}768} \approx 0.0000305$

74. A quiz has 6 multiple-choice questions and each question has 5 possible responses of which exactly one is correct. Find the probability that a student answers each question *incorrectly*. $\left(\dfrac{4}{5}\right)^6 = \dfrac{4096}{15{,}625} = 0.262144$

Mixed Exercises

The blood type of an individual is classified according to the presence of certain antigens, substances that cause the immune system to produce antibodies. These antigens are denoted by A, B, and Rh. If an individual's blood contains either the A or B antigen, these letters are listed in the blood type. If neither A nor B is present, then the letter O is used. If the Rh antigen is present, the blood is said to be Rh positive (Rh^+); otherwise, the blood is Rh negative (Rh^-). Under this system, a person with AB^+ blood has all three antigens, and group O^- is absent all three antigens.

Type	Probability
O^+	0.374
O^-	0.066
A^+	0.357
A^-	0.063
B^+	0.085
B^-	0.015
AB^+	0.034
AB^-	0.006

The distribution of blood types for people living in the United States is given in the table. (*Source*: Stanford School of Medicine, http://bloodcenter.stanford.edu) Refer to the table for Exercises 75–82. Round to 3 decimal places when necessary.

75. a. If an individual is randomly selected from the population, find the probability that the individual will have the Rh factor.   0.85

   b. If three people are selected at random, find the probability that they all have the Rh factor. $(0.85)^3 \approx 0.614$

77. a. Which blood type is most common ? O^+

   b. Which blood type is most rare? AB^-

76. a. If a person is selected at random from the population, find the probability that the individual has the A antigen.   0.46

   b. If four people are randomly selected, find the probability that all four have the A antigen. $(0.46)^4 \approx 0.045$

78. Across all blood types, is a person more likely to have the Rh factor or less likely to have the Rh factor?   more likely

Doctors know that certain restrictions apply when considering the administration of blood to a patient. The antigens of the blood donor and recipient must be compatible. If an antigen is absent from the recipient's blood, then the recipient cannot receive blood from a person who has that antigen. For example, a person with B^+ blood cannot receive AB^+ blood because of the presence of the A antigen from the donor. Use this information and the table with Exercises 75–78 for Exercises 79–82.

79. Suppose that a person is randomly selected from the population. What is the probability that this individual's blood can be used for a transfusion for a person with type B^+ blood?   0.54

81. a. What is the probability that an individual from the population can donate blood to a person with type O^- blood.   0.066

   b. Explain why a person with type O^- blood is called a universal donor. O^- blood is absent all three antigens and will not introduce a new antigen to the recipient's blood.

83. A slot machine in a casino has three wheels that all spin independently. Each wheel has 11 stops, denoted by 0 through 9, and bar. What is the probability that a given outcome is bar-bar-bar? $\left(\dfrac{1}{11}\right)^3 = \dfrac{1}{1331}$

80. Suppose that a person is randomly selected from the population. What is the probability that this individual's blood can be used for a transfusion for a person with type A^+ blood?   0.86

82. a. What is the probability that an individual from the population can donate blood to a person with AB^+ blood?   1

   b. Explain why a person with AB^+ blood can receive blood from anyone. AB^+ blood has all three antigens. Therefore, blood from a donor will not introduce a new antigen to the recipient's blood.

84. Airlines often overbook flights because a small percentage of passengers do not show up (perhaps due to missed connections). Past history indicates that for a certain route, the probability that an individual passenger will not show up is 0.04. Suppose that 61 people bought tickets for a flight that has 60 seats. Determine the probability that there will not be enough seats. Round to 3 decimal places. $(0.96)^{61} \approx 0.083$

85. South Florida humorist Dave Barry often wrote about his dog, Zippy. Suppose that the Barry home has 3200 ft^2 of living area with tile floor. Further suppose that an expensive 8 ft by 10 ft oriental rug is placed on the floor. If Zippy had an "accident" in the house, what is the probability that it would happen on the expensive rug?   0.025

For Exercises 86–88,

a. Shade the area bounded by the given inequalities on a coordinate grid showing $-5 \le x \le 5$ and $-5 \le y \le 5$.

b. Suppose that an enthusiastic mathematics student makes a square dart board out of the portion of the rectangular coordinate system defined by $-5 \le x \le 5$ and $-5 \le y \le 5$. Find the probability that a dart thrown at the target will land in the shaded region.

86. $y \ge |x|$ and $y \le 4$ **87.** $|y| \le 3$ and $|x| \le 2$ **88.** $x^2 + y^2 \le 9$

Write About It

89. In a carnival game, players win a prize if they can toss a ring around the neck of a bowling pin. How would you go about estimating the probability of winning the game?

91. Give an example of two events that are not mutually exclusive.   Answers will vary.

90. Give an example of two events that are mutually exclusive.   Answers will vary.

92. Explain why a probability of $\frac{5}{4}$ is impossible.

Expanding Your Skills

93. Suppose that a box of DVDs contains 10 action movies and 5 comedies.

   a. If two DVDs are selected from the box with replacement, determine the probability that both are comedies. $\frac{5}{15} \cdot \frac{5}{15} = \frac{1}{9}$

   b. It probably seems more reasonable that someone would select two *different* DVDs from the box. That is, the first DVD would not be replaced before the second DVD is selected. In such a case, are the events of selecting comedies on the first and second picks independent events?   No

   c. If two DVDs are selected from the box *without* replacement, determine the probability that both are comedies. $\frac{5}{15} \cdot \frac{4}{14} = \frac{2}{21}$

95. If five cards are dealt from a standard deck of 52 cards, find the probability that

   a. The cards are all hearts.

   b. The cards are all of the same suit.

94. Suppose that 12 students (5 freshmen and 7 sophomores) are being considered for two different scholarships. One scholarship is for $500 and the other is for $250.

   a. Two students are selected at random from the group of 12 to receive the scholarships. If a student may receive both scholarships, determine the probability that both students are freshmen. $\frac{5}{12} \cdot \frac{5}{12} = \frac{25}{144}$

   b. Now suppose that an individual student may not receive both scholarships. Determine the probability that both students chosen are freshmen. $\frac{5}{12} \cdot \frac{4}{11} = \frac{5}{33}$

96. If five cards are dealt from a standard deck of 52 cards, find the probability that

   a. The cards consist of four aces.

   b. The cards are four of a kind (four cards with the same face value).

CHAPTER 11   KEY CONCEPTS

SECTION 11.1 Sequences and Series	Reference
An **infinite sequence** is a function whose domain is the set of positive integers. A **finite sequence** is a function whose domain is the set of the first n positive integers.	p. 1006
A sequence in which consecutive terms alternate in sign is an **alternating sequence.**	p. 1007
A **recursive formula** defines the nth term of a sequence as a function of one or more terms preceding it.	p. 1008
For a positive integer n, the quantity $\boldsymbol{n!}$ ("n factorial") is defined as $n! = (n)(n-1)(n-2) \cdots (2)(1)$ By definition, $0! = 1$.	p. 1009
The $\boldsymbol{n}$**th partial sum** of a sequence $\{a_n\}$ is a finite series and is given by $S_n = \displaystyle\sum_{i=1}^{n} a_i = a_1 + a_2 + a_3 + \cdots + a_n$	p. 1010
The sum of *all* terms in an infinite sequence is an **infinite series**: $\displaystyle\sum_{i=1}^{\infty} a_i = a_1 + a_2 + a_3 + \cdots$	p. 1011
Properties of summation: If $\{a_n\}$ and $\{b_n\}$ are sequences, and c is a real number, then 1. $\displaystyle\sum_{i=1}^{n} c = cn$ 2. $\displaystyle\sum_{i=1}^{n} ca_i = c\sum_{i=1}^{n} a_i$ 3. $\displaystyle\sum_{i=1}^{n} (a_i \pm b_i) = \sum_{i=1}^{n} a_i \pm \sum_{i=1}^{n} b_i$	p. 1013

SECTION 11.2 Arithmetic Sequences and Series	Reference
An **arithmetic sequence** is a sequence in which each term after the first differs from its predecessor by a common difference d.	p. 1017
The nth term of an arithmetic sequence: $$a_n = a_1 + (n-1)d$$ where a_1 is the first term of the sequence and d is the common difference.	p. 1017
Sum of the first n terms of an arithmetic sequence (nth partial sum): $$S_n = \frac{n}{2}(a_1 + a_n)$$ where a_1 is the first term of the sequence and a_n is the nth term of the sequence.	p. 1022

SECTION 11.3 Geometric Sequences and Series	Reference
A **geometric sequence** is a sequence in which each term after the first is the product of the preceding term and a fixed nonzero real number, called the **common ratio.** If the first term of a geometric sequence is a_1, and the common ratio is r, then the nth term of a geometric sequence is given by $a_n = a_1 r^{n-1}$.	p. 1028
nth partial sum of a geometric sequence: The sum S_n of the first n terms of a geometric sequence is given by $S_n = \dfrac{a_1(1 - r^n)}{1 - r}$ where a_1 is the first term of the sequence and r is the common ratio, $r \neq 1$.	p. 1031
Sum of an infinite geometric series: Given an infinite geometric series $a_1 + a_1 r + a_1 r^2 + a_1 r^3 + \cdots$ with $|r| < 1$, the sum S of all terms in the series is given by $$S = \frac{a_1}{1 - r}$$ *Note:* If $|r| \geq 1$, then the sum does not exist.	p. 1033
Future value of an ordinary annuity: Suppose that P dollars is invested at the end of each compounding period n times per year at interest rate r. Then the value A (in \$) of the annuity after t years is given by $$A = \frac{P\left[\left(1 + \frac{r}{n}\right)^{nt} - 1\right]}{\frac{r}{n}}$$	p. 1035

SECTION 11.4 Mathematical Induction	Reference
Mathematical induction is a technique of mathematical proof that is often used to prove that a statement is true for all positive integers.	p. 1042
The principle of mathematical induction: Let P_n be a statement involving the positive integer n, and let k be an arbitrary positive integer. Then P_n is true for all positive integers n if 1. P_1 is true, and 2. The truth of P_k implies the truth of P_{k+1}.	p. 1042
The assumption that P_k is true is called the **inductive hypothesis.**	p. 1043
Extended principle of induction: Mathematical induction can be extended to prove statements that might hold true only for integers greater than or equal to some positive integer j. In such a case, • First prove the statement for $n = j$. • Then show that the truth of the statement for an integer greater than or equal to j implies the truth of the statement for the integer that follows.	p. 1045

SECTION 11.5 The Binomial Theorem	Reference
Binomial theorem: Let n be a positive integer. The expansion of $(a + b)^n$ is given by $$(a + b)^n = \binom{n}{0}a^n + \binom{n}{1}a^{n-1}b + \binom{n}{2}a^{n-2}b^2 + \cdots + \binom{n}{n-1}ab^{n-1} + \binom{n}{n}b^n$$ $$= \sum_{r=0}^{n} \binom{n}{r}a^{n-r}b^r$$ The values $\binom{n}{0}, \binom{n}{1}, \binom{n}{2}, \ldots, \binom{n}{n}$ are called the **binomial coefficients** of the expansion of $(a + b)^n$. The binomial coefficients can also be found by using **Pascal's triangle.** 1 1 1 1 2 1 1 3 3 1 1 4 6 4 1 1 ...	pp. 1148–1150
Finding the kth term of a binomial expansion: Let n and k be positive integers with $k \le n + 1$. The kth term of $(a + b)^n$ is $\binom{n}{k-1}a^{n-(k-1)}b^{k-1}$	p. 1052

SECTION 11.6 Principles of Counting	Reference
Fundamental principle of counting: If one event can occur in m different ways and a second event can occur in n different ways, then the number of ways that the two events can occur in sequence is $m \cdot n$.	p. 1055
A **permutation** is an ordered arrangement of distinct items.	p. 1057
• The number of permutations of n distinct elements is $n!$.	
• Consider a set of n elements with r_1 duplicates of one kind, r_2 duplicates of a second kind, ... , r_k duplicates of a kth kind. Then the number of distinguishable permutations of the n elements of the set is: $$\frac{n!}{r_1! \cdot r_2! \cdot \cdots \cdot r_k!}$$	p. 1057
• The number of permutations of n elements taken r at a time is given by $$_nP_r = \frac{n!}{(n-r)!} \text{ or equivalently, } _nP_r = n(n-1)(n-2)\cdots(n-r+1)$$	p. 1059
A **combination** is a collection of distinct items taken without regard to order. The number of combinations of n elements taken r at a time is given by $_nC_r = \dfrac{n!}{r! \cdot (n-r)!}$.	p. 1060

SECTION 11.7 Introduction to Probability	Reference
Theoretical probability of an event: Let S represent a sample space with equally likely outcomes, and let E be a subset of S. Then the probability of event E, denoted by $P(E)$, is given by $$P(E) = \frac{n(E)}{n(S)}$$	p. 1067
Let E be an event relative to sample space S. The **complement of E,** denoted by $\bar{E}$, is the set of outcomes in the sample space but not in event E. It follows that $P(E) + P(\bar{E}) = 1$.	p. 1069
Empirical probability is computed based on observed outcomes of the relative frequency of an event to the number of times an experiment is performed.	p. 1070

Two events are **mutually exclusive** if they share no common elements.	p. 1072
The probability of *A* or *B*: Given events *A* and *B* in a same sample space, $P(A \text{ or } B)$ is given by $P(A \cup B) = P(A) + P(B) - P(A \cap B)$. If *A* and *B* are mutually exclusive, then $P(A \cup B) = P(A) + P(B)$.	
Probability of a sequence of independent events: Two events are **independent** if the probability of one event does not affect the probability of the second event. If events *A* and *B* are independent events, then the probability that both *A* and *B* will occur is $$P(A \text{ and } B) = P(A) \cdot P(B).$$	p. 1074

Expanded Chapter Summary available at www.mhhe.com/millerprecalculus.

CHAPTER 11   Review Exercises

SECTION 11.1

For Exercises 1–2, simplify the expression.

1. $\dfrac{8!}{3! \cdot 5!}$ 56

2. $\dfrac{(2n + 1)!}{(2n - 3)!}$
$(2n + 1)(2n)(2n - 1)(2n - 2)$

For Exercises 3–4, write the first five terms of the sequence.

3. $a_n = \dfrac{(n + 1)!}{n!}$
2, 3, 4, 5, 6

4. $c_1 = 5; c_n = -2c_{n-1} + 1$
5, −9, 19, −37, 75

5. Given the sequence defined by $b_n = (-1)^{n-1} \cdot n$, which terms are positive and which are negative? The odd-numbered terms are positive, and the even-numbered terms are negative.

6. Given $b_n = \dfrac{n + 4}{3n}$, find b_{45}. $\dfrac{49}{135}$

7. a. Write the first five terms of the sequence defined by $a_n = 2n + 3$.   5, 7, 9, 11, 13

b. Evaluate $\displaystyle\sum_{i=1}^{5} (2i + 3)$.   45

c. Use the formula $S_n = \dfrac{n}{2}(a_1 + a_n)$ to verify the fifth partial sum of the arithmetic sequence from part (a).
$S_5 = 45$

8. Write the sum $4 - \dfrac{4}{3} + \dfrac{4}{5} - \dfrac{4}{7} + \cdots$ using summation notation with *n* as the index of summation. (Using techniques from calculus, we can show that this sum converges to π.) $\displaystyle\sum_{n=1}^{\infty} (-1)^{n+1} \dfrac{4}{2n - 1}$

For Exercises 9–11, find the sum.

9. $\displaystyle\sum_{i=3}^{8} 2i$ 66

10. $\displaystyle\sum_{i=1}^{34} 10$ 340

11. $\displaystyle\sum_{i=1}^{60} (-1)^{i+1}$ 0

12. Write an expression for the apparent *n*th term a_n for the sequence. 10, −30, 90, −270, … $a_n = 10(-3)^{n-1}$

For Exercises 13–14, write the sum using summation notation. Use *i* as the index of summation.

13. $\dfrac{3}{2} + \dfrac{4}{4} + \dfrac{5}{8} + \dfrac{6}{16} + \dfrac{7}{32}$ $\displaystyle\sum_{i=1}^{5} \dfrac{i + 2}{2^i}$

14. $x^2 + x^2 y + x^2 y^2 + x^2 y^3$ $\displaystyle\sum_{i=1}^{4} x^2 y^{i-1}$

15. Rewrite the series as an equivalent series with the new index of summation.

$$\sum_{i=1}^{10} i^3 = \sum_{j=0}^{\square} \square = \sum_{k=2}^{\square} \square$$

16. Determine if the statement is true or false.   True

$$\sum_{i=1}^{100} (4i - 2) = -200 + 4 \sum_{i=1}^{100} i$$

17. Suppose that a single cell of bacteria divides every 20 min for 4 hr. Write a formula for the sequence $\{a_n\}$ representing the number of cells after the *n*th cell division. $a_n = 2^n; 1 \le n \le 12$

SECTION 11.2

For Exercises 18–20, determine whether the sequence is arithmetic. If so, identify the common difference *d*.

18. 11, 10.2, 9.4, 8.6, 7.8, …   Yes; $d = -0.8$

19. $4, \dfrac{21}{4}, \dfrac{13}{2}, \dfrac{31}{4}, \ldots$ Yes; $d = \dfrac{5}{4}$

20. 9, 12, 16, 21, 27, …   No

21. Determine the first five terms of the arithmetic sequence $\{a_n\}$ with $a_1 = 4$, and $d = 8$.   4, 12, 20, 28, 36

22. a. Write an expression for the *n*th term of the arithmetic sequence $\{a_n\}$ with $a_1 = -19$, and $d = 5$. $a_n = 5n - 24$

b. Find a_{36}. $a_{36} = 156$

23. Find the 23rd term of an arithmetic sequence with $a_1 = 15$ and $a_{57} = 239$. $a_{23} = 103$

24. Given an arithmetic sequence with $a_{15} = 86$ and $a_{37} = 240$, find the 104th term. $a_{104} = 709$

25. Find the number of terms of the arithmetic sequence. 11, 14, 17, 20, 23, ... , 122 $n = 38$

26. A sales person working for a heating and air-conditioning company earns an annual base salary of $30,000 plus $500 on every new system he sells. Suppose that $\{a_n\}$ is a sequence representing the sales person's total yearly income based on the number of units sold n.

   a. Write a formula for the nth term of a sequence that represents the sales person's total income for n units sold. In this case, define $\{a_n\}$ with domain $n \geq 0$ to allow for the possibility of 0 units sold.
   $a_n = 30,000 + 500n$

   b. How much will the sales person earn in a year for selling 42 new units? $51,000

27. Find the sum of the first 35 terms of the arithmetic sequence $-1, -9, -17, -25, \dots$. -4795

For Exercises 28–30, find the sum.

28. $3 + 10 + 17 + 24 + \cdots + 437$
   13,860

29. $\sum_{n=1}^{36} (2 - 5n)$ -3258 **30.** $\sum_{i=1}^{68} \left(5 + \tfrac{1}{2}i\right)$ 1513

31. How long will it take to pay off a debt of $3960 if $50 is paid off the first month, $60 is paid off the second month, $70 is paid off the third month, and so on? 24 months

SECTION 11.3

For Exercises 32–34, determine whether the sequence is geometric. If so, find the value of r.

32. $\dfrac{3}{10}, \dfrac{3}{1000}, \dfrac{3}{100,000}, \cdots$ Yes; $r = \tfrac{1}{100}$ **33.** 3, 9, 36, 180, ... No

34. $5p^3, -5p^5, 5p^7, -5p^9, \dots$ Yes; $r = -p^2$

35. Write the first five terms of a geometric sequence with $a_1 = 120$, and $r = \tfrac{2}{3}$. 120, 80, $\dfrac{160}{3}, \dfrac{320}{9}, \dfrac{640}{27}$

36. Write a formula for the nth term of the geometric sequence.
   $-40, 20, -10, 5, \dots$ $a_n = -40\left(-\dfrac{1}{2}\right)^{n-1}$

For Exercises 37–39, find the indicated term of a geometric sequence from the given information.

37. $a_1 = 4$ and $a_2 = 12$. Find a_6. $a_6 = 972$

38. $a_1 = -15$ and $a_4 = -\dfrac{5}{9}$. Find a_7. $a_7 = -\dfrac{5}{243}$

39. $a_7 = \dfrac{1}{64}$ and $r = -\dfrac{1}{4}$. Find a_1. $a_1 = 64$

40. Find a_1 and r for a geometric sequence given that $a_3 = 18$ and $a_6 = 486$. $a_1 = 2$ and $r = 3$

For Exercises 41–46, find the sum of the geometric series, if possible.

41. $\sum_{n=1}^{7} 5(3)^{n-1}$ 5465 **42.** $\sum_{i=1}^{6} 12\left(-\dfrac{2}{3}\right)^{i-1}$ $\dfrac{532}{81}$

43. $\sum_{k=1}^{\infty} 5\left(\dfrac{5}{6}\right)^{k-1}$ 30 **44.** $\sum_{i=1}^{\infty} \dfrac{2}{3}(4)^{i-1}$ The sum is undefined.

45. $\sum_{n=3}^{\infty} 6\left(\dfrac{1}{2}\right)^{n-1}$ 3 **46.** $36 - 12 + 4 - \dfrac{4}{3} + \cdots$ 27

47. Write $0.8\overline{7}$ as a fraction. $\dfrac{79}{90}$

48. An estimated 150,000 people attended the Coconut Grove art festival over a 3-day period. Admission to the event is $10 per person. In addition, suppose that each person spends an average of $100 on art, drinks, and food.

   a. How much money is initially infused into the local economy during the festival for admission, art, drinks, and food. $16,500,000

   b. If the money is later respent in the community over and over again at a rate of 70%, determine the total amount spent. Assume that the money is respent an infinite number of times. $55,000,000

For Exercise 49–50, find the value of an ordinary annuity in which regular payments of P dollars are made at the end of each compounding period, n times per year, at an interest rate r for t years.

49. $P = \$150$, $n = 12$, $r = 4\%$, $t = 16$ yr $40,250.86

50. $P = \$300$, $n = 12$, $r = 4\%$, $t = 32$ yr $233,009.28

51. a. At age 28, an employee begins investing $100 each pay period (twice per month) in an ordinary annuity. If the interest rate is 5.5%, find the value of the annuity when the employee retires at age 62. $238,884.21

   b. Determine the value of the annuity if the employee waits to retire at age 65. $289,503.57

52. The initial swing (one way) of a pendulum makes an arc of 2 ft. Each swing (one way) thereafter makes an arc of 85% of the length of the previous swing. What is the total arc length that the pendulum travels? $\dfrac{40}{3}$ ft or $13\dfrac{1}{3}$ ft

SECTION 11.4

For Exercises 53–56, use mathematical induction to prove the given statement for all positive integers n.

53. $3 + 7 + 11 + \cdots + (4n - 1) = n(2n + 1)$

54. $-5 + 2 + 9 + \cdots + (7n - 12) = \dfrac{n}{2}(7n - 17)$

55. $1 + 4 + 16 + \cdots + 4^{n-1} = \dfrac{1}{3}(4^n - 1)$

56. $10^n - 1$ is divisible by 3.

57. Use mathematical induction to show that $4^n < (n + 2)!$ for integers $n \geq 2$.

SECTION 11.5

58. a. Write the first four rows of Pascal's triangle.
See Figure 11-9 on page 1048.
   b. Expand $(a + b)^3$. $a^3 + 3a^2b + 3ab^2 + b^3$

59. Evaluate the given expression. Compare the results to the result of Exercise 58.

   a. $\begin{pmatrix} 3 \\ 0 \end{pmatrix}$ 1   **b.** $\begin{pmatrix} 3 \\ 1 \end{pmatrix}$ 3   **c.** $\begin{pmatrix} 3 \\ 2 \end{pmatrix}$ 3   **d.** $\begin{pmatrix} 3 \\ 3 \end{pmatrix}$ 1

The values of $\begin{pmatrix} 3 \\ 0 \end{pmatrix}, \begin{pmatrix} 3 \\ 1 \end{pmatrix}, \begin{pmatrix} 3 \\ 2 \end{pmatrix}$, and $\begin{pmatrix} 3 \\ 3 \end{pmatrix}$ match the entries in the fourth row of Pascal's triangle.

For Exercises 60–64, expand the binomial by using the binomial theorem.

60. $(4y + 3)^4$ **61.** $(2x - 3)^5$ **62.** $(5c^3 - d^2)^4$

63. $(t^5 + u^3)^6$ **64.** $\left(\dfrac{x}{2} + 2y \right)^3$

For Exercises 65–67, find the indicated term of the binomial expansion.

65. $(5x + 4y)^6$; fifth term $96{,}000x^2y^4$

66. $(3x^4 - 2)^8$; middle term $90{,}720x^{16}$

67. $(2c^2 - d^5)^9$; Term containing d^{25} $-2016c^8d^{25}$

68. Given $f(x) = 4x^3 + 2x$, find the difference quotient.
$12x^2 + 12xh + 4h^2 + 2$

69. Use the binomial theorem to find the value of $(3 + 2i)^4$ where i is the imaginary unit. $-119 + 120i$

SECTION 11.6

70. Gaynelle can travel one of 3 roads from her home to school. From school to work there are 4 different routes. How many different routes are available for Gaynelle to travel from home to school to work? $4 \cdot 3 = 12$

71. A disc jockey has 7 songs that he must play in a half-hour period. In how many different ways can he arrange the 7 songs? $7! = 5040$

72. In how many ways can the letters in the word SHUFFLE be arranged? $\dfrac{7!}{2!} = 2520$

73. In how many ways can the word SPACE be misspelled? $5! - 1 = 119$

74. A quiz consists of 6 true/false questions and 4 multiple-choice questions. The multiple-choice questions each have 5 possible responses (a, b, c, d, e) of which only one is correct. In how many different ways can a student fill out the answers to the quiz? $2^6 \cdot 5^4 = 40{,}000$

75. A 3-digit code is to be made from the set of digits $\{4, 5, 6, 7, 8\}$.

   a. How many codes can be formed if there are no restrictions? $5^3 = 125$

   b. How many codes can be formed if the corresponding 3-digit number is to be an even number? $5 \cdot 5 \cdot 3 = 75$

   c. How many codes can be formed if the corresponding 3-digit number is to be a multiple of 5 and there can be no repetition of digits? $4 \cdot 3 \cdot 1 = 12$

76. Evaluate $_{21}P_4$ and interpret its meaning.

77. Evaluate $_{21}C_4$ and interpret its meaning.

78. Evaluate $_{10}P_3$ and $_{10}C_3$ and compare the results.

79. In how many ways can a statistician select a sample of 15 people from a population of 90 people?
$_{90}C_{15} \approx 4.58 \times 10^{16}$

80. How many triangles can be made if the vertices are from three of the six points on the circle? One such triangle is shown in the figure. $_6C_3 = 20$

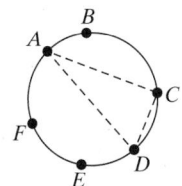

81. The Daytona 500 auto race has 40 cars that initially start the race. How many first-, second-, and third-place finishes can occur? $_{40}P_3$ or $40 \cdot 39 \cdot 38 = 59{,}280$

82. Suppose that a drama class has 22 students.

   a. In how many ways can four students be selected to take part in a survey? $_{22}C_4 = 7315$

   b. In how many ways can four students be selected to act out a scene from a play involving 4 different parts?
   $_{22}P_4 = 175{,}560$

83. To meet the graduation requirements, a student must take 2 English classes out of 10 available, 3 math classes out of 6 available, and 2 history classes out of 7 available. Assuming that the student has met the prerequisites of each course and that there are no scheduling conflicts, determine the number of ways in which the student can select these courses. $(_{10}C_2) \cdot (_6C_3) \cdot (_7C_2) = 18{,}900$

SECTION 11.7

84. Which of the following can represent the probability of an event?   a, b, d, e, h

   a. 0   **b.** 1   **c.** 1.2   **d.** 0.12

   e. 1.2%   **f.** 120%   **g.** -0.12 **h.** $\frac{4}{5}$

85. If the probability of an event is 0.0042, is the event likely to occur or not likely to occur?   Not likely

86. If the probability of an event is $\frac{87}{90}$, is the event likely to occur or not likely to occur?   Likely

87. If $P(E) = 0.73$, what is the probability of $P(\overline{E})$?   0.27

88. Suppose that a box containing music CDs has 4 with country music, 6 with rock music, 3 with jazz music, and 7 with rap music. If one CD is selected at random from the box, determine the probability that

   a. The CD has rap music. $\dfrac{7}{20}$

   b. The CD has jazz music. $\dfrac{3}{20}$

   c. The CD does not have jazz music. $\dfrac{17}{20}$

   d. The CD has classical music.   0

89. Suppose that two fair dice are rolled. Determine the probability that

a. The sum of the numbers on the dice is 6. $\dfrac{5}{36}$

b. The sum of the numbers on the dice is greater than 9. $\dfrac{6}{36} = \dfrac{1}{6}$

c. The numbers on the dice form a sum that is a multiple of 5. $\dfrac{7}{36}$

90. For a recent season, the batting average for baseball player Jose Iglesias was 0.306. (This means that the probability that Iglesias will get a hit on a given time at bat is 0.306.)

a. Determine the probability that Iglesias will not get a hit on a given time at bat. 0.694

b. If Iglesias is at bat three times in a game, what is the probability that he will get a hit all three times? $(0.306)^3 \approx 0.029$

c. If Iglesias is at bat three times in a game, what is the probability that he will not get a hit on any of the three times at bat? $(0.694)^3 \approx 0.334$

91. Suppose that 15 lightbulbs are in a cabinet and that 4 are defective. If three bulbs are chosen at random,

a. What is the probability that all three will be defective?

b. What is the probability that all three will be good?

c. Why do the probabilities from parts (a) and (b) not add up to 1?

92. Suppose that a lottery game has the player select 5 distinct numbers from 1 to 30, inclusive. The player wins by choosing numbers that match those randomly selected in a drawing.

a. What is the probability that a player will win if the numbers do not have to be selected in any particular order? $\dfrac{1}{{}_{30}C_5} = \dfrac{1}{142,506}$

b. If the player has to pick the correct 5 numbers in a specific order, what is the probability that the player will win? $\dfrac{1}{{}_{30}P_5} = \dfrac{1}{17,100,720}$

93. For a recent year, the Centers for Disease Control reported that the probability that a 50-yr-old will live to age 51 is 0.9959. In a group of ten 50-yr-olds, what is the probability that all ten survive to the age of 51? $(0.9959)^{10} \approx 0.9597$

For Exercises 94–99, use the data in the table categorizing smokers and nonsmokers according to their blood pressure (BP) levels. If one person is chosen at random, find the probability of the given event.

	Normal BP	Elevated BP	Total
Smokers	42	28	70
Nonsmokers	80	10	90
Total	122	38	160

94. The person has elevated blood pressure. $\dfrac{38}{160} = \dfrac{19}{80}$

95. The person is a nonsmoker. $\dfrac{90}{160} = \dfrac{9}{16}$

96. The person has elevated blood pressure or is a nonsmoker. $\dfrac{118}{160} = \dfrac{59}{80}$

97. The person has normal blood pressure or is a smoker. $\dfrac{150}{160} = \dfrac{15}{16}$

98. The person is a smoker or has elevated blood pressure. $\dfrac{80}{160} = \dfrac{1}{2}$

99. The person is a nonsmoker or has normal blood pressure. $\dfrac{132}{160} = \dfrac{33}{40}$

For Exercises 100–103, refer to the sample space for a card drawn from a standard deck. See page 1071.

100. If one card is drawn at random from a standard deck, what is the probability that it is an 8 or a club? $\dfrac{16}{52} = \dfrac{4}{13}$

101. If one card is drawn at random from a standard deck, what is the probability that it is a red card or a 5? $\dfrac{28}{52} = \dfrac{7}{13}$

102. If *two* cards are drawn at random with replacement from a standard deck, what is the probability that both are hearts? $\dfrac{13}{52} \cdot \dfrac{13}{52} = \dfrac{1}{16}$

103. If *two* cards are drawn at random with replacement from a standard deck, what is the probability that both are kings? $\dfrac{4}{52} \cdot \dfrac{4}{52} = \dfrac{1}{169}$

CHAPTER 11 Test

1. Write the first six terms of the sequence defined by $a_1 = -2$, $a_2 = 3$, $a_n = a_{n-2} + a_{n-1}$ for $n \geq 3$. $-2, 3, 1, 4, 5, 9$

2. Simplify. $\dfrac{(3n + 1)!}{(3n - 1)!}$ $(3n + 1)(3n)$

For Exercises 3–5,

a. Determine whether the sequence is arithmetic, geometric, or neither.

b. If the sequence is arithmetic, determine d. If the sequence is geometric, determine r.

c. Write an expression a_n for the apparent nth term of the sequence.

3. $0.139, 0.00139, 0.0000139, \ldots$
a. Geometric   **b.** $r = 0.01$ **c.** $a_n = 0.139(0.01)^{n-1}$

4. $0.52, 0.68, 0.84, 1.00, \ldots$
a. Arithmetic   **b.** $d = 0.16$ **c.** $a_n = 0.16n + 0.36$

5. $\dfrac{3}{5}, \dfrac{6}{25}, \dfrac{9}{125}, \dfrac{12}{625}, \ldots$
a. Neither   **b.** Not applicable   **c.** $a_n = \dfrac{3n}{5^n}$

6. Write the first four terms of the geometric sequence with $a_1 = 4$ and $r = \dfrac{3}{2}$. $4, 6, 9, \dfrac{27}{2}$

7. Write the first five terms of the arithmetic sequence with $a_1 = 10$ and $a_{20} = 67$. 10, 13, 16, 19, 22

8. Find the 78th term of an arithmetic sequence with $a_1 = 64$ and $d = -11$. $a_{78} = -783$

9. Find the 6th term of a geometric sequence with $a_1 = -3$ and $r = 2$. $a_6 = -96$

10. Find the number of terms in the arithmetic sequence.
$$-15, -19, -23, -27, \ldots, -679 \quad 167$$

11. Find the number of terms in the geometric sequence. $\quad 9$
$$16, 8, 4, 2, \ldots, \frac{1}{16}$$

12. Given an arithmetic sequence with $a_{12} = -21$ and $a_{50} = -97$, find a_1 and d. $\quad a_1 = 1, d = -2$

13. Given a geometric sequence with $a_3 = 20$ and $a_8 = 640$, find a_1 and r. $\quad a_1 = 5, r = 2$

For Exercises 14–18, evaluate the sum if possible.

14. $\dfrac{3}{2} + \dfrac{3}{4} + \dfrac{3}{8} + \dfrac{3}{16} + \cdots \quad 3$

15. $\displaystyle\sum_{k=1}^{54} (3k + 7) \quad 4833$

16. $\displaystyle\sum_{i=1}^{6} 4\left(\dfrac{3}{2}\right)^{i-1} \quad \dfrac{665}{8}$

17. $\displaystyle\sum_{k=1}^{4} k! \quad 33$

18. $\displaystyle\sum_{i=1}^{\infty} \dfrac{1}{2}\left(\dfrac{6}{5}\right)^{i-1} \quad$ The sum does not exist.

19. Suppose that a county fair has an estimated 50,000 people attend over a 2-week period. Admission to the fair is $5.00. In addition, suppose that each person spends an average of $10 on food, drinks, and rides.

   a. How much money is initially infused into the local community for admission, food, drinks, and rides? $750,000

   b. If the money is respent in the local community over and over again at a rate of 65%, determine the total amount spent. Assume that the money is respent an infinite number of times. Round to the nearest dollar. $2,142,857

20. An employee invests $400 per month in an ordinary annuity. If the interest rate is 5.2%, find the value of the annuity after 25 yr. $245,446.68

21. Lakeisha wants to put down new tile in her home. The tile costs $3.50/ft^2 and labor is $4.00/ft^2. Write the nth term of a sequence representing the cost to tile an n by n square foot area where n is an integer and $n \geq 1$ ft. $a_n = 7.50n^2$

22. How many days will it take Johan to read a 920-page book if he reads 20 pages the first day, 25 pages the second day, 30 pages the third day, and so on? $\quad$ 16 days

For Exercises 23–25, use mathematical induction to prove the given statement for all positive integers n.

23. $6 + 10 + 14 + \cdots + (4n + 2) = n(2n + 4)$

24. $1 + 5 + 25 + \cdots + 5^{n-1} = \frac{1}{4}(5^n - 1)$

25. 2 is a factor of $7^n - 5$.

For Exercises 26–27, expand the binomial by using the binomial theorem.

26. $\left(\dfrac{x}{2} + 3\right)^4$

27. $(4c^2 - t^4)^5$

For Exercises 28–29, find the indicated term of the binomial expansion.

28. $(-2t + v^2)^{10}$; eighth term $\quad -960t^3v^{14}$

29. $(3x + y^2)^7$; Term containing y^6. $\quad 2835x^4y^6$

30. Evaluate $_{13}P_5$ and $_{13}C_5$. $\quad _{13}P_5 = 154{,}440$ and $_{13}C_5 = 1287$

31. Explain the difference between a permutation and a combination of n items taken r at a time.

32. A musician plans to perform 9 selections. In how many ways can she arrange the musical selections?
$9! = 362{,}880$

33. How many outfits can be made from 4 pairs of slacks, 5 shirts, and 3 ties, if one selection from each category is made? Assume that all items fashionably match.
$4 \cdot 5 \cdot 3 = 60$

34. In how many ways can the word HIPPOPOTAMUS be misspelled? $\quad \dfrac{12!}{3! \cdot 2!} - 1 = 39{,}916{,}799$

35. Suppose that a jury pool consists of 30 women and 26 men.

   a. In how many ways can a jury of 12 people be selected? $\quad _{56}C_{12} \approx 5.584 \times 10^{11}$

   b. In how many ways can a jury of 6 women and 6 men be selected? $\quad (_{30}C_6) \cdot (_{26}C_6) \approx 1.367 \times 10^{11}$

   c. What is the probability of randomly selecting a jury of 6 women and 6 men? $\dfrac{(_{30}C_6) \cdot (_{26}C_6)}{_{56}C_{12}} \approx 0.245$

   d. What is the probability of randomly selecting a jury of all men? $\dfrac{_{26}C_{12}}{_{56}C_{12}} \approx 0.0000173$

36. A review sheet for a history test has 10 essay questions. Suppose that the professor picks 3 questions from the review sheet to put on the test. In how many ways can the professor choose 3 questions from 10 questions?
$_{10}C_3 = 120$

37. After a service call by a plumber, the company follows up with a survey to rate the service and professionalism of the technician. The survey has 6 yes/no questions and 4 multiple-choice questions each with 3 possible responses. In how many different ways can a customer fill out the survey?
$2^6 \cdot 3^4 = 5184$

38. Suppose that 50 people buy raffle tickets.

   a. In how many ways can 4 people who bought tickets be selected if each is to receive a $20 gift certificate to a restaurant? $\quad _{50}C_4 = 230{,}300$

   b. In how many ways can 4 people who bought tickets be selected if the first person wins a $10 gift certificate, the second person wins a $25 gift certificate, the third person wins a $50 gift certificate, and the fourth person wins a $200 camera? $\quad _{50}P_4 = 5{,}527{,}200$

39. For a recent year, approximately 36,000 people were killed in the United States in motor vehicle accidents. If the population had 300,000,000 people at that time, estimate the probability of being killed in a motor vehicle crash. $\dfrac{36{,}000}{300{,}000{,}000} = 0.00012$

40. A cable company advertises short wait times for customer service calls. The graph shows the wait times (in seconds) for a sample of customers. Based on the data given, if a customer is selected at random, find the probability that

Number of Customers vs. Wait Time, x

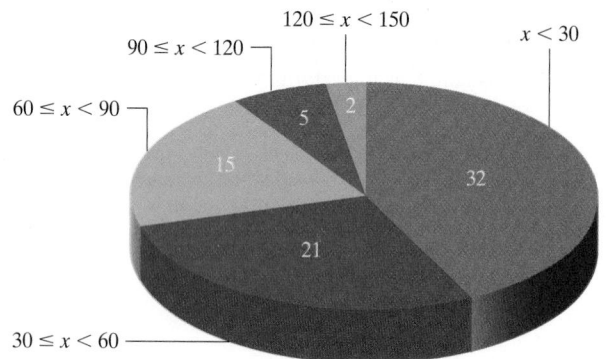

a. The customer will wait less than 30 sec. $\dfrac{32}{75}$

b. The customer will wait at least 30 sec. $\dfrac{43}{75}$

c. The customer will wait at least 90 sec but less than 120 sec. $\dfrac{5}{75} = \dfrac{1}{15}$

d. The customer will wait more than 150 sec. 0

41. If two fair dice are thrown, find the probability that the sum is between 6 and 8, inclusive. $\dfrac{16}{36} = \dfrac{4}{9}$

42. Suppose that two cards are drawn from a standard deck with replacement. What is the probability that an ace is selected, followed by a heart? $\dfrac{4}{52} \cdot \dfrac{13}{52} = \dfrac{1}{52}$

43. For a recent year, 31.9% of Americans living below the poverty level were not covered by health insurance. (*Source:* U.S. Census Bureau, www.census.gov) If three people were selected at random from this population, what is the probability that all three would not have coverage? $(0.319)^3 \approx 0.0325$

For Exercises 44–47, use the data in the table categorizing the type of payment used at a grocery store according to the gender of the customer. If one person is chosen at random, find the probability of the given event.

	Cash	Credit Card	Check	Total
Male	19	24	13	56
Female	14	30	20	64
Total	33	54	33	120

44. The customer is female. $\dfrac{64}{120} = \dfrac{8}{15}$

45. The customer is male or paid by check. $\dfrac{76}{120} = \dfrac{19}{30}$

46. The customer paid by credit card or by cash. $\dfrac{87}{120} = \dfrac{29}{40}$

47. The customer paid cash or was female. $\dfrac{83}{120}$

CHAPTER 11   Cumulative Review Exercises

For Exercises 1–4, consider sets A and B and determine if the statement is true or false.

$$A = \{x \mid 4 > x\}, B = \{x \mid x \le 1\}$$

1. $4 \in A$ False

2. $1 \in B$ True

3. $A \cap B = B$ True

4. $A \cup B = B$ False

5. a. Write an expression for the distance between t and 5 on the number line. $|t - 5|$ or $|5 - t|$

b. Simplify the expression from part (a) for $t < 5$. $5 - t$

6. Simplify without using a calculator. 2.7×10^{28}

$$\frac{(6.0 \times 10^{13})(9.0 \times 10^8)}{2.0 \times 10^{-6}}$$

For Exercises 7–10, simplify the expression.

7. $-27^{-4/3}$ $-\dfrac{1}{81}$

8. $(\sqrt{x} + 5\sqrt{2})(3\sqrt{x} - \sqrt{2})$ $3x + 14\sqrt{2x} - 10$

9. $\dfrac{4 - x^{-2}}{6 - 3x^{-1}}$ $\dfrac{2x + 1}{3x}$

10. i^{127} $-i$

For Exercises 11–18, solve the equation or inequality. Write the solution set to the inequalities in interval notation.

11. $\sqrt{2x^2 - x - 14} + 4 = x$ $\{ \}$; The values 3 and -10 do not check.

12. $(x^2 + x)^2 - 14(x^2 + x) + 24 = 0$ $\{-4, 3, -2, 1\}$

13. $|3x - 5| = |2x + 1|$ $\left\{ 6, \dfrac{4}{5} \right\}$

14. $2 \tan x - 5 = 0$ for $0 \le x < 2\pi$ $\left\{ \tan^{-1}\dfrac{5}{2}, \pi + \tan^{-1}\dfrac{5}{2} \right\}$

15. $0 \le |x + 7| - 6$ $(-\infty, -13] \cup [-1, \infty)$

16. $\dfrac{2x + 1}{x + 4} \ge 1$ $(-\infty, -4) \cup [3, \infty)$

17. $\log_4(2x + 7) = 2 + \log_4 x$ $\left\{ \dfrac{1}{2} \right\}$

18. $5e^{x+1} - 100 = 0$ $\{-1 + \ln 20\}$

19. Given $f(x) = 2x^3 - 5x^2 - 28x + 15$,

a. Find all the zeros of $f(x)$. $-3, 5, \dfrac{1}{2}$

b. Identify the x-intercepts of the graph of f. $(-3, 0), (5, 0), \left(\dfrac{1}{2}, 0\right)$

c. Determine the y-intercept of the graph of f. $(0, 15)$

d. Graph $y = f(x)$.

e. Solve the inequality $2x^3 - 5x^2 - 28x + 15 < 0$. $(-\infty, -3) \cup \left(\dfrac{1}{2}, 5\right)$

20. An object is launched from a height of 4 ft at an angle of 60° from the horizontal with an initial speed of 112 ft/sec. Define a coordinate system with the origin at ground level directly below the point of launch. Use an acceleration due to gravity of -32 ft/sec².

a. Write parametric equations to model the path of the object t seconds after launch. $x = 56t$ and $y = -16t^2 + 56\sqrt{3}t + 4$

b. Determine the maximum height of the object.   151 ft

c. How far from the starting point will the object hit the ground? Round to the nearest foot.   342 ft

21. Solve the triangle with sides of length $a = 5$ cm, $b = 7$ cm, and $c = 10$ cm. Round the measures of the angles to the nearest tenth of a degree. $A = 27.7°, B = 40.5°, C = 111.8°$

For Exercises 22–25, graph the equation.

22. $\dfrac{y^2}{9} - \dfrac{x^2}{16} = 1$

23. $\dfrac{(x + 1)^2}{4} + \dfrac{(y - 2)^2}{9} = 1$

24. $y = \log_2(x + 3)$

25. $y = \begin{cases} |x| + 1 & \text{for } x \le 1 \\ -x + 1 & \text{for } x > 1 \end{cases}$

26. Explain how the graph of $f(x) = 2\sqrt{x - 1} + 3$ is related to the graph of $y = \sqrt{x}$.

27. Given $f(x) = 4x^3 - 3x$,

   a. Find the difference quotient.
     $12x^2 + 12xh + 4h^2 - 3$
   b. Find the average rate of change on the interval $[1, 3]$.
     49
   c. Determine if the function is even, odd, or neither.
     Odd

28. Given $g(x) = 5x - 1$,

   a. Is g a one-to-one function?   Yes
   b. Write an equation for $g^{-1}(x)$. $g^{-1}(x) = \dfrac{x + 1}{5}$

29. Given $f(x) = x^3 - x^2 - 7x + 15$,

   a. Is $2 + i$ a zero of $f(x)$?   Yes
   b. Is $(x + 3)$ a factor of $f(x)$?   Yes

30. Given $r(x) = \dfrac{3x^2 + 2}{x^2 - 5x - 14}$,

   a. Determine the vertical asymptote(s).
     $x = 7, x = -2$
   b. Determine the horizontal or slant asymptote if either exist.   Horizontal asymptote: $y = 3$

31. The population of a city was 320,000 in the year 2000. By 2012, the population reached 360,800.

   a. Write a model of the form $P(t) = P_0 e^{kt}$ to represent the population $P(t)$ for a time t years since 2000.
     $P(t) = 320{,}000e^{0.01t}$
   b. Approximate the population in the year 2015 assuming that this trend continued. Round to the nearest 100 people.   371,800

   c. Determine the amount of time for the population to reach 400,000 if this trend continues. Round to the nearest tenth of a year.   22.3 yr

32. Write the expression in terms of $\log x$, $\log y$, and $\log z$.
   $3 \log x - 5 \log y - \tfrac{1}{2} \log z$

$$\log\left(\dfrac{x^3}{y^5\sqrt{z}}\right)$$

33. Simplify. $\log_2 \dfrac{1}{16}$ -4

34. Given $(-4\sqrt{3}, -4)$ in rectangular coordinates, find two representations in polar coordinates: one with $r > 0$ and one with $r < 0$. $\left(8, \dfrac{7\pi}{6}\right)$ and $\left(-8, \dfrac{\pi}{6}\right)$

For Exercises 35–37, solve the system of equations.

35. $2x - 5y = 13$
    $-3x + 2y = -3$
    $\{(-1, -3)\}$

36. $4x \quad\quad - 2z = -4$
    $6y + 5z = 8$
    $7x - 3y \quad\quad = 13$
    $\{(1, -2, 4)\}$

37. $2x - y = 6$
    $x^2 + y = 9$ $\{(-5, -16), (3, 0)\}$

38. Use Gaussian elimination or Gauss-Jordan elimination to solve the system of equations.

$$-2x - 9y + 16z = -15$$
$$x + \ y - \ z = 4$$
$$x - 2y + \ 5z = 1$$

   $\{(-z + 3, 2z + 1, z) \mid z \text{ is a real number}\}$

39. Use Cramer's rule to solve the system.

$$13x - 2y = 11$$
$$5x + 3y = 6$$

   $\left\{\left(\dfrac{45}{49}, \dfrac{23}{49}\right)\right\}$

40. Graph the solution set.

$$x^2 + y^2 \le 16$$
$$y > -x^2 + 4$$

For Exercises 41–46, refer to the matrices given and perform the indicated operations, if possible.

$$A = \begin{bmatrix} 4 & -3 \\ 5 & 9 \end{bmatrix} \qquad B = \begin{bmatrix} -1 & 6 \\ 3 & 7 \end{bmatrix}$$

$$C = \begin{bmatrix} -1 & 4 & 0 \\ 3 & 6 & 1 \\ 2 & -5 & 7 \end{bmatrix} \qquad D = \begin{bmatrix} -5 & 1 & 0 \\ 8 & 4 & 6 \end{bmatrix}$$

41. $-5A + 2B$ $\begin{bmatrix} -22 & 27 \\ -19 & -31 \end{bmatrix}$

42. $3C - A$ Not possible

43. $|A|$ 51

44. $|C|$ -123

45. CD Not possible

46. AD $\begin{bmatrix} -44 & -8 & -18 \\ 47 & 41 & 54 \end{bmatrix}$

47. Given $A = \begin{bmatrix} 1 & 2 & -1 \\ 0 & 1 & 3 \\ 1 & 0 & 2 \end{bmatrix}$, find A^{-1}. $A^{-1} = \begin{bmatrix} \frac{2}{9} & -\frac{4}{9} & \frac{7}{9} \\ \frac{1}{3} & \frac{1}{3} & -\frac{1}{3} \\ -\frac{1}{9} & \frac{2}{9} & \frac{1}{9} \end{bmatrix}$

48. Use the inverse matrix from Exercise 47 to solve the system of equations. $\{(-3, 1, 4)\}$

$$x + 2y - \ z = -5$$
$$y + 3z = 13$$
$$x \quad\quad + 2z = 5$$

49. Given $(y + 2)^2 = -8(x - 1)$,

   a. Determine the vertex of the graph of the parabola.
     $(1, -2)$
   b. Identify the focus. $(-1, -2)$

   c. Write an equation for the directrix. $x = 3$

50. Determine whether the sequence is arithmetic or geometric and find the value of the common difference or the common ratio.   Geometric; $r = -a^3$

$$-a^2, a^5, -a^8, a^{11}, \dots$$

51. Find the 500th term of an arithmetic sequence with $a_1 = 6.9$ and $d = 0.3$. $a_{500} = 156.6$

For Exercises 52–53, find the sum.

52. $\displaystyle\sum_{k=1}^{74}(5k+3)$ 14,097

53. $\displaystyle\sum_{i=1}^{8}6(2)^{i-1}$ 1530

54. Find the sum if possible. $6+2+\dfrac{2}{3}+\dfrac{2}{9}+\cdots$ 9

55. Evaluate the expressions.

a. $8!$ 40,320   b. $\dbinom{12}{11}$ 12   c. $_{15}P_3$ 2730

56. Expand the binomial. $(3x-y^2)^5$
$243x^5-405x^4y^2+270x^3y^4-90x^2y^6+15xy^8-y^{10}$

57. Simplify. $\dfrac{(3n-1)!}{2!(3n+1)!}$ $\dfrac{1}{2(3n+1)(3n)}$

58. In how many ways can 5 children be arranged in a line for a photograph? $5!=120$

59. Suppose that four people are to be randomly selected from a group of 8 women and 5 men. What is the probability of selecting 2 women and 2 men? $\dfrac{_8C_2\cdot\,_5C_2}{_{13}C_4}\approx0.3916$

60. If one card is selected from a standard deck of cards, what is the probability that the card selected is a diamond or an ace? (*Hint*: Refer to Figure 11-12 from page 1071.) $\dfrac{16}{52}=\dfrac{4}{13}$

Instructor Answer Appendix

CHAPTER R

Section R.1 Practice Exercises, pp. 11–15

25. 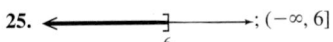 ; $(-\infty, 6]$

26. ——————; $(-\infty, -4)$ at -4

27. ——————; $\left(-\dfrac{7}{6}, \dfrac{1}{3}\right]$ at $-\dfrac{7}{6}$, $\dfrac{1}{3}$

28. ——————; $\left[-\dfrac{4}{3}, \dfrac{7}{4}\right)$ at $-\dfrac{4}{3}$, $\dfrac{7}{4}$

29. ——————; $(4, \infty)$ at 4

30. ——————; $[-3, \infty)$ at -3

31. ——————; $\{x \mid -3 < x \le 7\}$ at -3, 7

32. ——————; $\{x \mid -4 \le x < -1\}$ at -4, -1

33. ——————; $\{x \mid x \le 6.7\}$ at 6.7

34. ——————; $\{x \mid x < -3.2\}$ at -3.2

35. ——————; $\left\{x \mid x \ge -\dfrac{3}{5}\right\}$ at $-\dfrac{3}{5}$

36. ——————; $\left\{x \mid x > \dfrac{7}{8}\right\}$ at $\dfrac{7}{8}$

119. The commutative property of addition indicates that the order in which two quantities are added does not affect the sum. The associative property of addition indicates that the manner in which quantities are grouped under addition does not affect the sum.

Section R.2 Practice Exercises, pp. 25–29

127. In the expression $6x^0$, the exponent 0 applies to x only. In the expression $(6x)^0$, the exponent 0 applies to a base of $(6x)$. The first expression simplifies to 6, and the second expression simplifies to 1.

128. Scientific notation is used to represent very large and very small numbers to eliminate the need to write numerous zeros. It is also helpful to keep track of the location of the decimal point when performing calculations.

Section R.3 Practice Exercises, pp. 39–42

145. After exhausting all possible combinations of factors, no combination results in $a^2 + b^2$. That is, $(a - b)(a + b) \ne a^2 + b^2$.
$(a - b)(a - b) \ne a^2 + b^2$. $(a + b)(a + b) \ne a^2 + b^2$.

146. In each case, factor out x to the smallest exponent to which it appears in both terms. That is, $5x^4 + 4x^3 = x^3(5x + 4)$ and $5x^{-4} + 4x^{-3} = x^{-4}(5 + 4x)$

153. False; For example the sum $(x^5 + 2x) + (-x^5 + 4x)$ equals $6x$ which is a first degree polynomial.

156. False; The product of two fourth degree polynomials $\left(a_4x^4 + a_3x^3 + a_2x^2 + a_1x + a_0\right)\left(b_4x^4 + b_3x^3 + b_2x^2 + b_1x + b_0\right)$ will have leading term $\left(a_4x^4\right)\left(b_4x^4\right) = a_4b_4x^8$ which is an 8th degree polynomial.

Section R.4 Practice Exercises, pp. 51–54

87. **a.** At 1 hr: 4.8 ng/mL; At 12 hr: 3.9 ng/mL; At 24 hr: 1.0 ng/mL; At 48 hr: 0.3 ng/mL

88. **a.** At 2 hr: 17.5°C; At 4 hr: 9.2°C; At 12 hr: 1.8°C; At 24 hr: 0.5°C

Section R.5 Practice Exercises, pp. 68–71

131. $d = \dfrac{2S}{n} - a$ or $d = \dfrac{2S - an}{n}$

132. $a = \dfrac{2S - n^2d + nd}{2n}$

135. $x = \dfrac{5 - ay}{6 - b}$ or $x = \dfrac{ay - 5}{b - 6}$

136. $x = \dfrac{d - 2y}{3 - c}$ or $x = \dfrac{2y - d}{c - 3}$

137. $r = \sqrt{\dfrac{A}{\pi}}$ or $r = \dfrac{\sqrt{A\pi}}{\pi}$

138. $r = \sqrt{\dfrac{V}{\pi h}}$ or $r = \dfrac{\sqrt{V\pi h}}{\pi h}$

141. $w = \dfrac{c \pm \sqrt{c^2 + 4kr}}{2k}$

142. $y = \dfrac{-m \pm \sqrt{m^2 + 4dp}}{2d}$

143. $t = \dfrac{-v_0 \pm \sqrt{v_0^2 + 2as}}{a}$

144. $r = \dfrac{-\pi h \pm \sqrt{\pi^2 h^2 + \pi h S}}{\pi h}$

146. $R_3 = \dfrac{RR_1R_2}{R_1R_2 - RR_2 - RR_1}$

157. The value 5 is not defined within two of the expressions in the equation. Substituting 5 into the equation would result in division by 0.

158. The equation is an identity. The solution set is all real numbers.

Section R.6 Practice Exercises, pp. 80–82

105. $(a + bi)(c + di)$
$= ac + adi + bci + bdi^2$
$= ac + (ad + bc)i + bd(-1)$
$= (ac - bd) + (ad + bc)i$

106. $(a + bi)^2$
$= (a)^2 + 2(a)(bi) + (bi)^2$
$= a^2 + (2ab)i + b^2i^2$
$= a^2 + (2ab)i + b^2(-1)$
$= (a^2 - b^2) + (2ab)i$

109. The second step does not follow because the multiplication property of radicals can be applied only if the individual radicals are real numbers. Because $\sqrt{-9}$ and $\sqrt{-4}$ are imaginary numbers, the correct logic for simplification would be
$\sqrt{-9} \cdot \sqrt{-4} = i\sqrt{9} \cdot i\sqrt{4} = i^2\sqrt{36} = -1 \cdot 6 = -6$

110. The product $(a + b)(a - b)$ simplifies to $a^2 - b^2$. The product $(a + bi)(a - bi)$ simplifies to $a^2 - (bi)^2$, which simplifies to $a^2 + b^2$.

111. Any real number. For example: 5.

112. Any complex number and its conjugate. For example: $2 + 5i$ and $2 - 5i$. In general, for real numbers, a and b, $(a + bi)(a - bi) = a^2 + b^2$, which is a real number.

Section R.8 Practice Exercises, pp. 102–105

9. $\{x \mid x < -11\}$; $(-\infty, -11)$; ——————) at -11

10. $\{t \mid t > -2\}$; $(-2, \infty)$; ——————(at -2

11. $\{w \mid w \le 3\}$; $(-\infty, 3]$; ——————] at 3

12. $\{y \mid y \ge -4\}$; $[-4, \infty)$; ——————[at -4

13. $\{a \mid a \le 8.5\}$; $(-\infty, 8.5]$; ——————] at 8.5

14. $\{x \mid x \ge -0.5\}$; $[-0.5, \infty)$; ——————[at -0.5

15. $\{c \mid c < 2\}$; $(-\infty, 2)$; ——————) at 2

16. $\{m \mid m > 0\}$; $(0, \infty)$; ——————(at 0

17. $\left\{x \mid x < -\frac{13}{2}\right\}$; $\left(-\infty, -\frac{13}{2}\right)$; ⟵――――→
$-\frac{13}{2}$

18. $\left\{y \mid y < \frac{8}{3}\right\}$; $\left(-\infty, \frac{8}{3}\right)$; ⟵――――→
$\frac{8}{3}$

19. $\left\{x \mid x \leq \frac{17}{6}\right\}$; $\left(-\infty, \frac{17}{6}\right]$; ⟵――――→
$\frac{17}{6}$

20. $\{t \mid t \leq -44\}$; $(-\infty, -44]$; ⟵――――→
-44

23. $\left\{x \mid x \geq -\frac{5}{6}\right\}$; $\left[-\frac{5}{6}, \infty\right)$; ――――――→
$-\frac{5}{6}$

24. $\left\{x \mid x \geq \frac{11}{2}\right\}$; $\left[\frac{11}{2}, \infty\right)$; ――――――→
$\frac{11}{2}$

25. $\mathbb{R}$; $(-\infty, \infty)$; ⟵――――→
26. $\mathbb{R}$; $(-\infty, \infty)$; ⟵――――→
27. a. $[-2, 4)$
$-2 \quad 4$

b. $(-\infty, \infty)$ ⟵――――→

28. a. $(-5, -2]$
$-5 \quad -2$

b. $(-\infty, \infty)$ ⟵――――→

29. a. $(-\infty, 5]$
5

b. $(-\infty, -6)$ ⟵――――→
-6

30. a. $(7, \infty)$ ――――――→
7

b. $[8, \infty)$ ――――――→
8

31. a. $(-\infty, 3.2]$ ⟵――――→
3.2

b. $(-\infty, 18)$ ⟵――――→
18

32. a. $[2.5, \infty)$ ――――――→
2.5

b. $(-10, \infty)$ ――――――→
-10

33. a. $(-\infty, -2] \cup \left(-\frac{1}{3}, \infty\right)$
$-2 \quad -\frac{1}{3}$ **b.** $\{\ \}$

34. a. $(-\infty, 8) \cup (9, \infty)$
$8 \quad 9$ **b.** $\{\ \}$

37. $[-4, 2)$
$-4 \quad 2$

38. $(3, 5]$
$3 \quad 5$

39. $\left[\frac{6}{5}, 2\right)$
$\frac{6}{5} \quad 2$

40. $\left[-\frac{5}{4}, 4\right]$
$-\frac{5}{4} \quad 4$

41. $\left[-\frac{5}{2}, \frac{13}{2}\right]$
$-\frac{5}{2} \quad \frac{13}{2}$

42. $\left(-2, \frac{6}{5}\right)$
$-2 \quad \frac{6}{5}$ **57.** $\left(-\infty, -\frac{12}{5}\right) \cup \left(\frac{4}{5}, \infty\right)$

75. An average score in league play between 140 and 220, inclusive, would produce a handicap of 72 or less.

86. b. $[31, 37]$; The motorist was traveling between 31 mph and 37 mph, inclusive. The motorist should receive a ticket.

97. The steps are the same with the following exception. If both sides of an inequality are multiplied or divided by a negative real number, then the direction of the inequality sign must be reversed.

98. The statement $8 < x < 2$ is equivalent to $8 < x$ and $x < 2$. No real number is greater than 8 and simultaneously less than 2.

99. The inequality $|x - 3| \leq 0$ will be true only for values of x for which $x - 3 = 0$ (the absolute value will never be less than 0). The solution set is $\{3\}$. The inequality $|x - 3| > 0$ is true for all values of x excluding 3. The solution set is $\{x \mid x < 3 \text{ or } x > 3\}$.

100. Taking the square root of both sides of the equation $x^2 = 4$ results in $\sqrt{x^2} = \sqrt{4}$ or equivalently $|x| = 2$. The solution set for each equation is $\{2, -2\}$, indicating that the equations are equivalent.

Chapter R Review Exercises, pp. 112–115

2. a. $\{10, 11, 12, 13, 14, 16\}$
b. $\{10, 12\}$
c. $\{7, 8, 9, 10, 11, 12, 13\}$
d. $\{10, 11\}$
e. $\{7, 8, 9, 10, 11, 12, 14, 16\}$
f. $\{10\}$

3. a. $\mathbb{R}$
b. $\{x \mid -2 \leq x < 7\}$
c. $\{x \mid x < 7\}$
d. $\{x \mid x < -3\}$
e. $\{x \mid x < -3 \text{ or } x \geq -2\}$
f. $\{\ \}$

82. $\left\{\dfrac{5 \pm 3\sqrt{3}}{2}\right\}$ **99.** $\dfrac{1}{2} + \dfrac{1}{5}i$

122. $\{p \mid p \leq 27\}$; $(-\infty, 27]$; ⟵――――→
27

123. $\{y \mid y > 11\}$; $(11, \infty)$; ――――――→
11

124. a. $\{t \mid t \leq 6\}$; $(-\infty, 6]$ ⟵――――→
6

b. $\{t \mid t < -12\}$; $(-\infty, -12)$ ⟵――――→
-12

125. a. $\{x \mid x \leq 6 \text{ or } x > 1\}$; $(-\infty, -6] \cup (1, \infty)$ **b.** $\{\ \}$
$-6 \quad 1$

126. $\left\{x \mid -2 \leq x \leq \frac{5}{2}\right\}$; $\left[-2, \frac{5}{2}\right]$
$-2 \quad \frac{5}{2}$

127. $\{x \mid 3 < x < 11\}$; $(3, 11)$
$3 \quad 11$

Chapter R Test, pp. 115–117

52. $t = \dfrac{-v_0 \pm \sqrt{v_0^2 + 128}}{-32}$ or $t = \dfrac{v_0 \pm \sqrt{v_0^2 + 128}}{32}$

CHAPTER 1

Section 1.1 Practice Exercises, pp. 127–131

9.
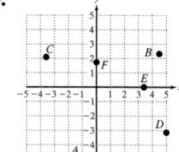

10.
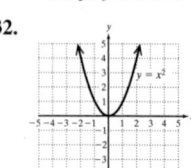

31.
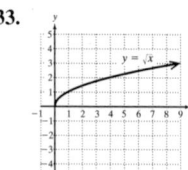

32.
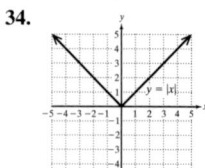

33.

34.

35.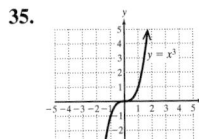

36.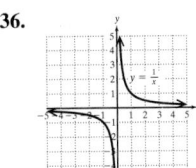

37.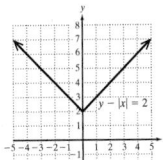

38.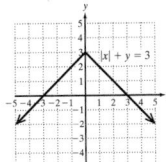

39.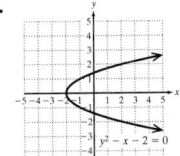

40.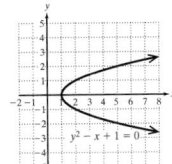

41.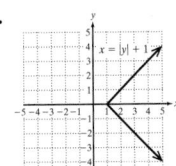

42.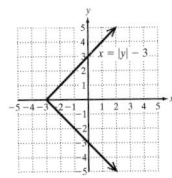

43.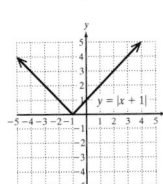

44.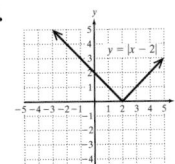

45. x-intercepts: $(-1, 0)$, $(9, 0)$; y-intercepts: $(0, -3)$, $(0, 3)$

46. x-intercepts: $(-16, 0)$, $(4, 0)$; y-intercepts: $(0, -8)$, $(0, 8)$

47. x-intercept: $(-2, 0)$; y-intercept: None   **48.** x-intercept: None; y-intercept $(0, 1)$ **49.** x-intercept: $(0, 0)$; y-intercept $(0, 0)$

50. x-intercepts: $(0, 0)$, $(6, 0)$; y-intercept $(0, 0)$

79. The points (x_1, y_1) and (x_2, y_2) define the endpoints of the hypotenuse d of a right triangle. The lengths of the legs of the triangle are $|x_2 - x_1|$ and $|y_2 - y_1|$. Applying the Pythagorean theorem produces $d^2 = |x_2 - x_1|^2 + |y_2 - y_1|^2$, or equivalently $d = \sqrt{(x_2 - x_1)^2 + (y_2 - y_1)^2}$ for $d \geq 0$.

80. The midpoint formula results in an ordered pair. The x-coordinate of the midpoint is the average of the x-coordinates of the endpoints. The y-coordinate of the midpoint is the average of the y-coordinates of the endpoints.   **81.** To find the x-intercept(s), substitute 0 for y and solve for x. To find the y-intercept(s), substitute 0 for x and solve for y.

82. The graph of the equation represents the set of all solutions to the equation graphed in a rectangular coordinate system.

89.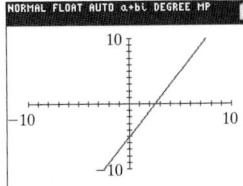

90.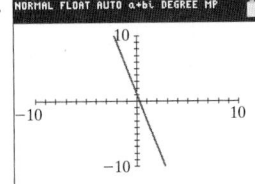

91.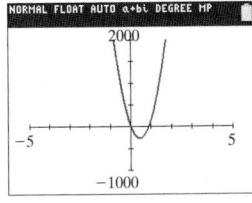

92.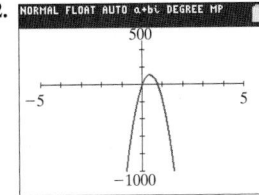

93.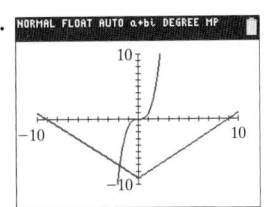

94.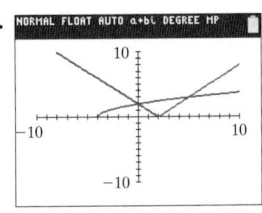

Section 1.2 Practice Exercises, pp. 135–137

17. a. $(x + 2)^2 + (y - 5)^2 = 1$ **18. a.** $(x + 3)^2 + (y - 2)^2 = 16$

b. 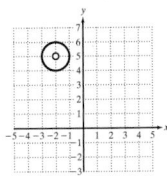   **b.**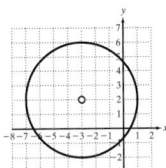

19. a. $(x + 4)^2 + (y - 1)^2 = 9$ **20. a.** $(x - 6)^2 + (y + 2)^2 = 36$

b. 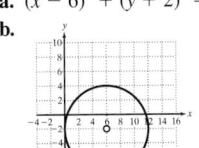   **b.**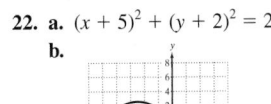

21. a. $(x + 4)^2 + (y + 3)^2 = 11$ **22. a.** $(x + 5)^2 + (y + 2)^2 = 21$

b. 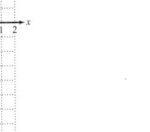   **b.**

23. a. $x^2 + y^2 = 6.76$ **24. a.** $x^2 + y^2 = 17.64$

b. 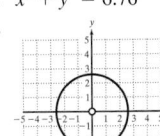   **b.**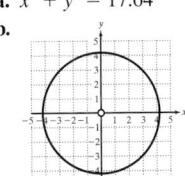

25. a. $(x - 2)^2 + (y - 1)^2 = 25$ **26. a.** $(x - 6)^2 + (y - 1)^2 = 5$

b. 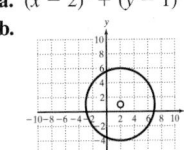   **b.**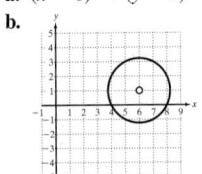

27. a. $(x + 2)^2 + (y + 1)^2 = 100$ **28. a.** $(x - 3)^2 + (y - 1)^2 = 25$

b. 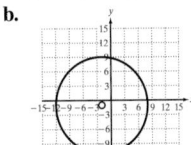   **b.**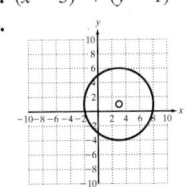

29. a. $(x - 4)^2 + (y - 6)^2 = 16$ **30. a.** $(x + 2)^2 + (y + 4)^2 = 16$

b. 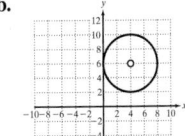   **b.**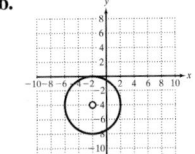

31. a. $(x - 5)^2 + (y + 5)^2 = 25$ **32. a.** $(x + 3)^2 + (y - 3)^2 = 9$
b.
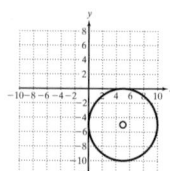
b.
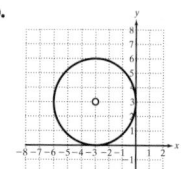

c.
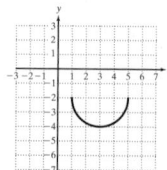
d.
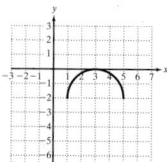

57.
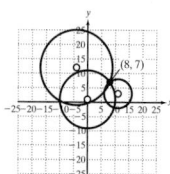
58.
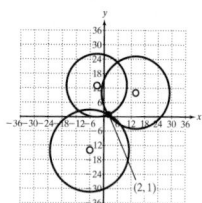

71.
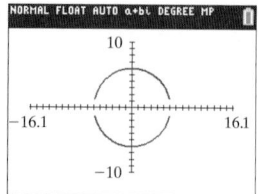
72.
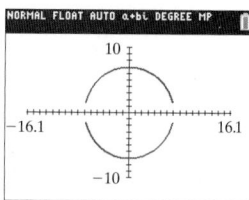

73.
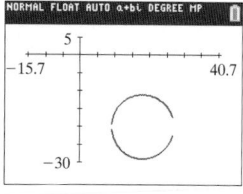
74.
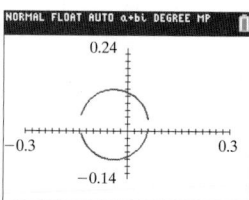

65. a.
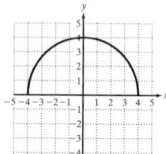
b.
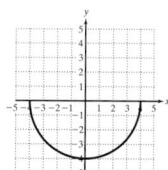

Section 1.3 Practice Exercises, pp. 145–151

123. If two points in a set of ordered pairs are aligned vertically in a graph, then they have the same x-coordinate but different y-coordinates. This contradicts the definition of a function. Therefore, the points do not define y as a function of x.

c.
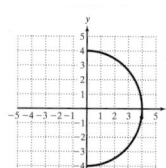
d.
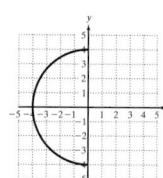

125. c. $A(P) = \left(\dfrac{P}{4}\right)^2$ or $A(P) = \dfrac{P^2}{16}$

Section 1.4 Practice Exercises, pp. 161–166

9. x-intercept: $(-4, 0)$;   **10.** x-intercept: $(-2, 0)$;
  y-intercept: $(0, 3)$ y-intercept: $(0, 4)$

66. a.
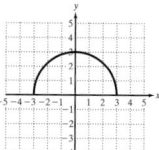
b.
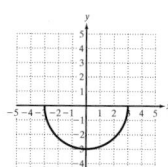

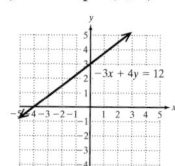
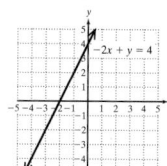

c.
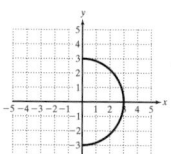
d.
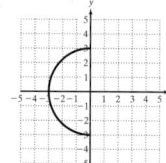

11. x-intercept: $\left(\dfrac{2}{5}, 0\right)$;   **12.** x-intercept: $\left(\dfrac{3}{2}, 0\right)$;
  y-intercept: $(0, 1)$ y-intercept: $(0, 2)$

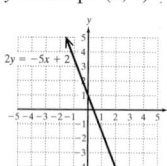
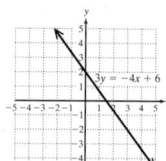

67. a.
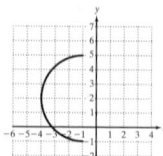
b.
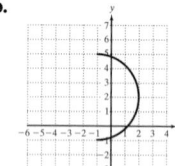

13. x-intercept: $(-6, 0)$;   **14.** x-intercept: None;
  y-intercept: None y-intercept: $(0, 4)$

c.
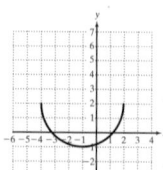
d.
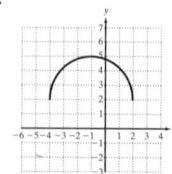

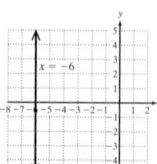
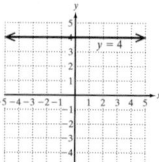

68. a.
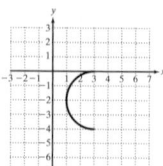
b.
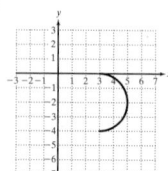

15. x-intercept: None;   **16.** x-intercept: $(2, 0)$;
  y-intercept: $(0, 2)$ y-intercept: None

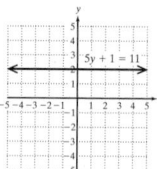
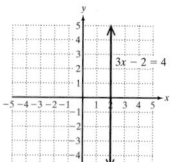

17. x-intercept: $(5, 0)$;
   y-intercept: $(0, 2)$

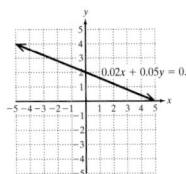

18. x-intercept: $(7, 0)$;
   y-intercept: $(0, 3)$

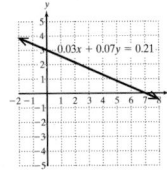

19. x-intercept: $(0, 0)$;
   y-intercept: $(0, 0)$

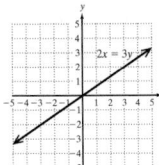

20. x-intercept: $(0, 0)$;
   y-intercept: $(0, 0)$

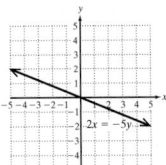

51. a. $y = \frac{1}{2}x - 2$; $m = \frac{1}{2}$;
   y-intercept: $(0, -2)$

   b.

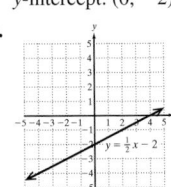

52. a. $y = 3x - 6$; $m = 3$,
   y-intercept: $(0, -6)$

   b.

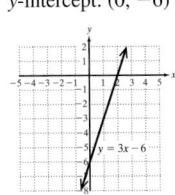

53. a. $y = \frac{3}{2}x + 2$; $m = \frac{3}{2}$;
   y-intercept: $(0, 2)$

   b.

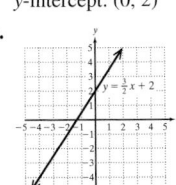

54. a. $y = \frac{5}{3}x + 2$; $m = \frac{5}{3}$;
   y-intercept: $(0, 2)$

   b.

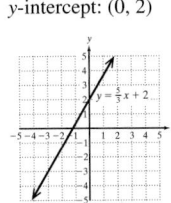

55. a. $y = \frac{3}{4}x$; $m = \frac{3}{4}$;
   y-intercept: $(0, 0)$

   b.

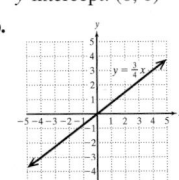

56. a. $y = -\frac{2}{3}x$; $m = -\frac{2}{3}$;
   y-intercept: $(0, 0)$

   b.

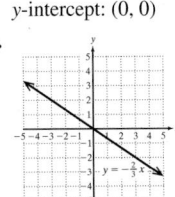

57. a. $y = 7$; $m = 0$;
   y-intercept: $(0, 7)$

   b.

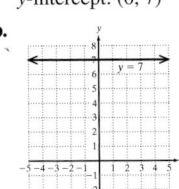

58. a. $y = -1$; $m = 0$;
   y-intercept: $(0, -1)$

   b.

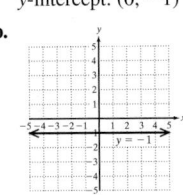

59. a. $y = -\frac{1}{3}x + 1$; $m = -\frac{1}{3}$;
   y-intercept: $(0, 1)$

   b.

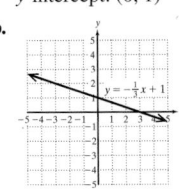

60. a. $y = -\frac{3}{4}x + 3$; $m = -\frac{3}{4}$;
   y-intercept: $(0, 3)$

   b.

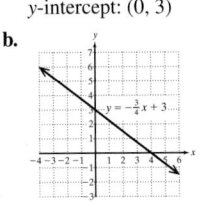

61. a. $y = -\frac{7}{4}x + 7$; $m = -\frac{7}{4}$;
   y-intercept: $(0, 7)$

   b.

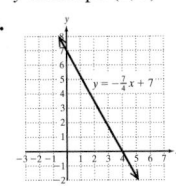

62. a. $y = -\frac{4}{3}x + 4$; $m = -\frac{4}{3}$;
   y-intercept: $(0, 4)$

   b.

75. a. $y = 2x - 6$ **b.** $f(x) = 2x - 6$

76. a. $y = -\frac{1}{2}x - 3$ **b.** $f(x) = -\frac{1}{2}x - 3$

77. a. $y = -\frac{4}{3}x + \frac{19}{3}$ **b.** $f(x) = -\frac{4}{3}x + \frac{19}{3}$

78. a. $y = -\frac{7}{3}x + \frac{2}{3}$ **b.** $f(x) = -\frac{7}{3}x + \frac{2}{3}$

83. a. -262; The number of new flu cases dropped by 262 per month during this time interval.

   b. Between months 4 and 6: -683 cases/month; Between months 10 and 12: -110/month

   c.

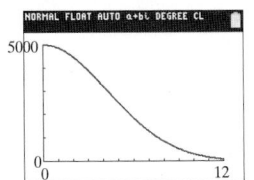

   The number of new flu cases dropped slowly during the first two months. Then the rate of new cases dropped more rapidly between months 4 and 6 (perhaps as health department officials managed the outbreak). Finally, the rate of new cases dropped more slowly toward the end of the outbreak.

84. a. 0.4 m/sec   **b.** 0.24 m/sec

   c.

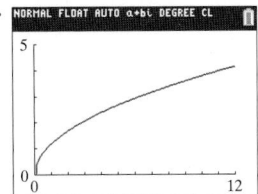

   From the graph, the longer the wavelength, the faster the wave. However, the *rate* at which a wave gains speed decreases as wavelength increases.

99. The line will be slanted if both A and B are nonzero. If A is zero and B is not zero, then the equation can be written in the form $y = k$ and the graph is a horizontal line. If B is zero and A is not zero, then the equation can be written in the form $x = k$, and the graph is a vertical line.

109. a. $\{-1.5\}$ **b.** $(-\infty, -1.5)$ **c.** $(-1.5, \infty)$

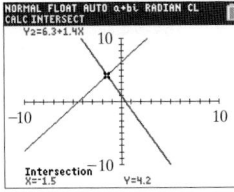

110. a. $\{0.5625\}$ **b.** $(-\infty, 0.5625)$ **c.** $(0.5625, \infty)$

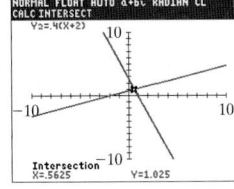

111. a. $\{-4, 7.8\}$

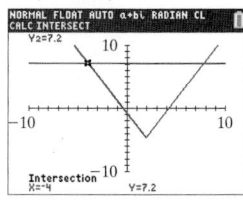 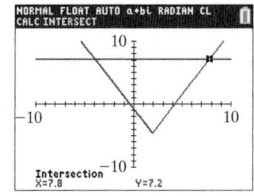

b. $(-\infty, -4] \cup [7.8, \infty)$ **c.** $[-4, 7.8]$

112. a. $\{-4.5, 7.9\}$

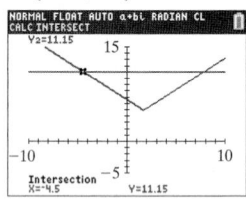 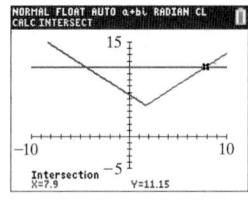

b. $(-\infty, -4.5] \cup [7.9, \infty)$ **c.** $[-4.5, 7.9]$

113. The lines are not exactly the same. The slopes are different.

114. The lines are not exactly the same. The y-intercepts are different.

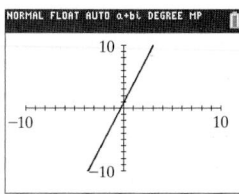 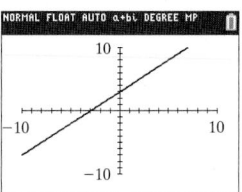

Section 1.5 Practice Exercises, pp. 176–182

65. a.

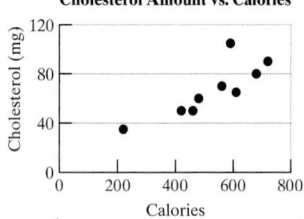

Cholesterol Amount vs. Calories

66. a.

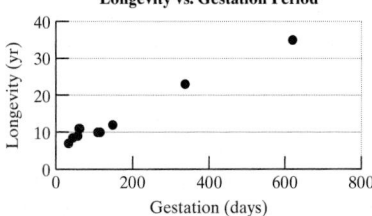

Longevity vs. Gestation Period

71. b.

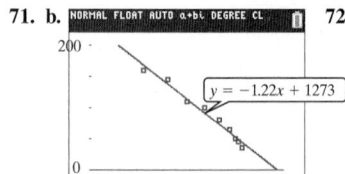

$y = -1.22x + 1273$

72. b.

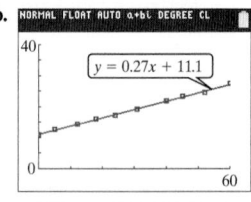

$y = 0.27x + 11.1$

73. b.

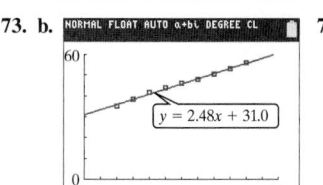

$y = 2.48x + 31.0$

74. b.

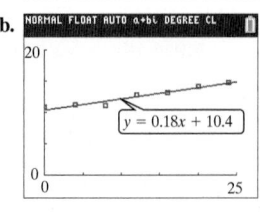

$y = 0.18x + 10.4$

75. b.

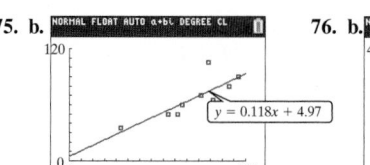

$y = 0.118x + 4.97$

76. b.

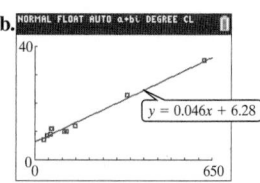

$y = 0.046x + 6.28$

83. If the slopes of the two lines are the same and the y-intercepts are different, then the lines are parallel. If the slope of one line is the opposite of the reciprocal of the slope of the other line, then the lines are perpendicular.

Problem Recognition Exercises, p. 182

1.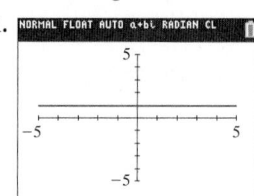

2.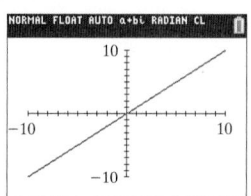

3.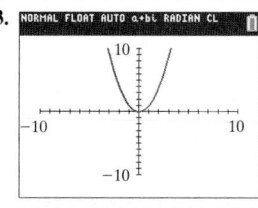

4.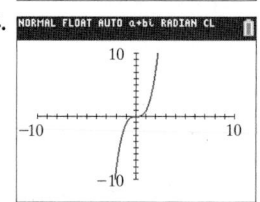

5.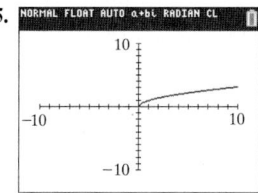

6.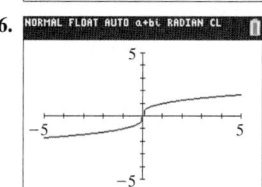

7.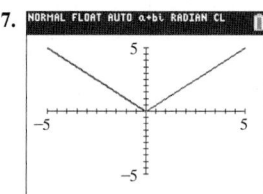

8.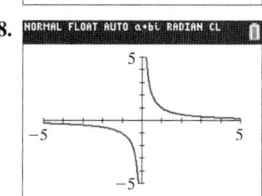

9. The graphs have the shape of $y = x^2$ with a vertical shift.

10. The graphs have the shape of $y = |x|$ with a vertical shift.

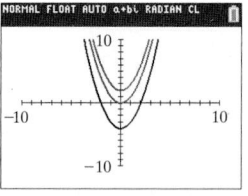 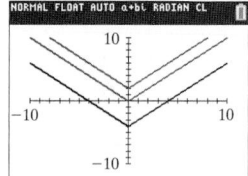

11. The graphs have the shape of $y = \sqrt{x}$ with a horizontal shift.

12. The graphs have the shape of $y = x^2$ with a horizontal shift.

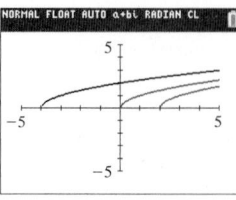 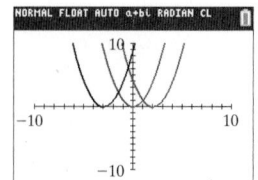

13. The graph of $g(x) = -|x|$ has the shape of the graph of $y = |x|$ but is reflected across the x-axis.

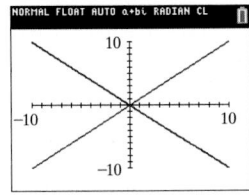

14. The graph of $g(x) = -\sqrt{x}$ has the shape of the graph of $y = \sqrt{x}$ but is reflected across the x-axis.

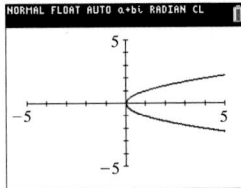

15. The graphs have the shape of $y = x^2$ but show a vertical shrink or stretch.

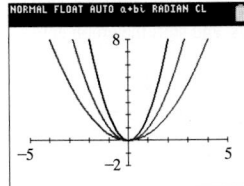

16. The graphs have the shape of $y = |x|$ but show a vertical shrink or stretch.

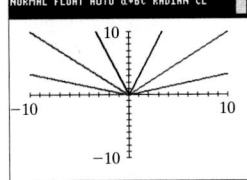

17. The graph of $g(x) = \sqrt{-x}$ has the shape of the graph of $y = \sqrt{x}$ but is reflected across the y-axis.

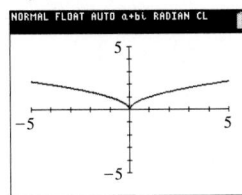

18. The graph of $g(x) = \sqrt[3]{-x}$ has the shape of the graph of $y = \sqrt[3]{x}$ but is reflected across the y-axis.

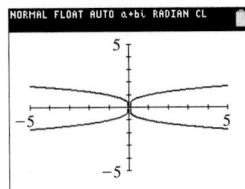

Section 1.6 Practice Exercises, pp. 193–197

R.1.

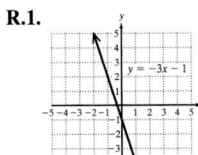

R.2.

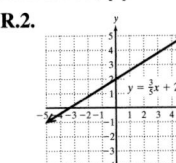

R.3.

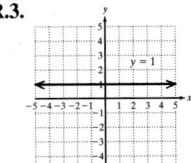

15.

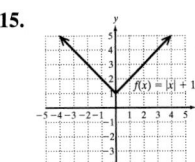

16.

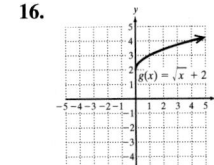

17.

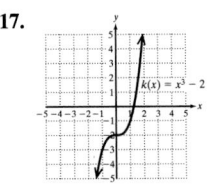

18.

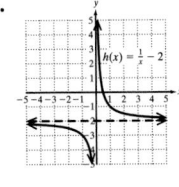

19.

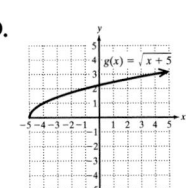

20.

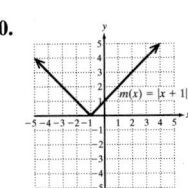

21.

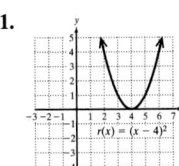

22.

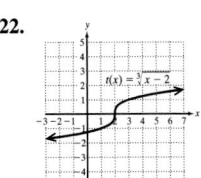

23.

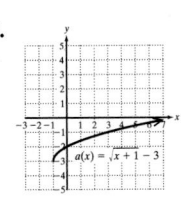

24.

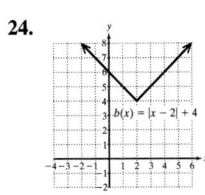

25.

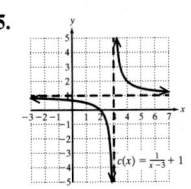

26.

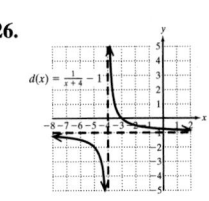

27.

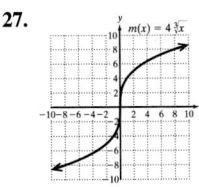

28.

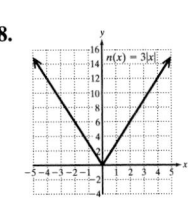

29.

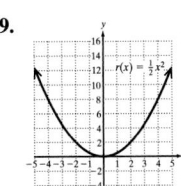

30.

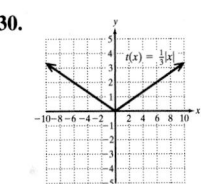

31.

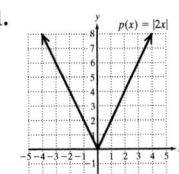

32.

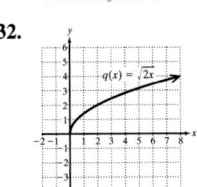

33.

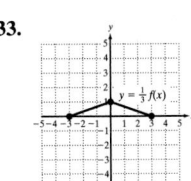

34.

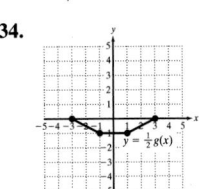

35.

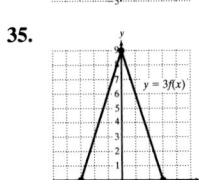

36.

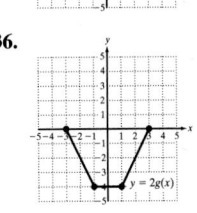

37.

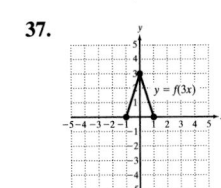

38.

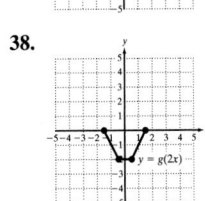

39.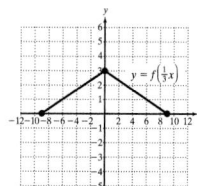

40.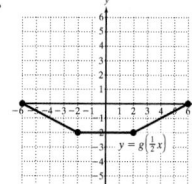

57. Parent function: $f(x) = x^2$. Shift the graph of f to the right 2.1 units, shrink the graph vertically by a factor of $\frac{1}{3}$, and shift the graph upward 7.9 units.

58. Parent function: $f(x) = \sqrt{x}$. Shift the graph of f to the left 4.3 units, shrink the graph vertically by a factor of $\frac{1}{2}$, and shift the graph downward 8.4 units.

59. Parent function: $f(x) = \sqrt{x}$. Shift the graph of f to the left 5 units, shrink the graph horizontally by a factor of $\frac{1}{2}$, stretch the graph vertically by a factor of 2, and reflect the graph across the y-axis.

41.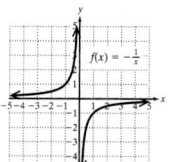

42.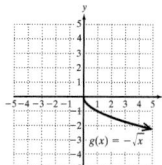

60. Parent function: $f(x) = |x|$. Shift the graph of f to the right 4 units, stretch the graph horizontally by a factor of 2, stretch the graph vertically by a factor of 3, and reflect the graph over the y-axis.

61. Parent function: $f(x) = \sqrt{x}$. Stretch the graph of f horizontally by a factor of 3, reflect across the x-axis, and shift the graph downward 6 units.

43.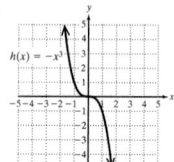

44.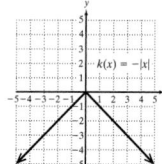

62. Parent function: $f(x) = |x|$. Shrink the graph of f horizontally by a factor of $\frac{1}{2}$, reflect the over the x-axis, and shift the graph upward 8 units.

63.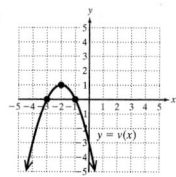

64.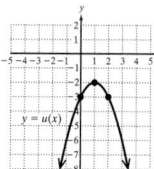

45.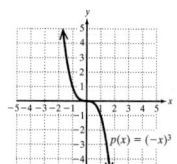

46.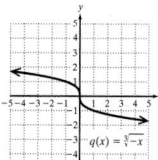

65.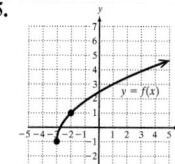

66.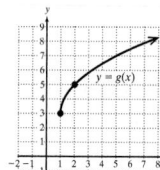

47.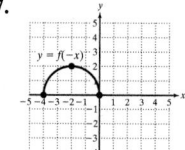

48.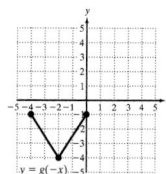

67.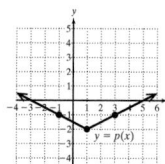

68.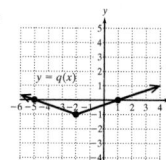

49.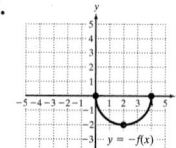

50.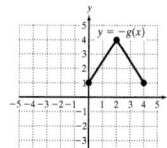

69.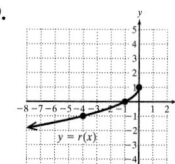

70.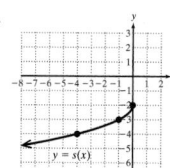

51.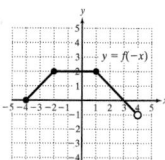

52.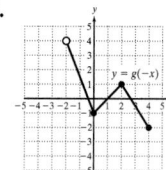

71.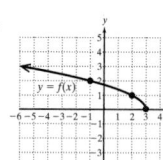

72.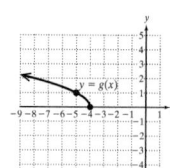

53.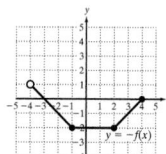

54.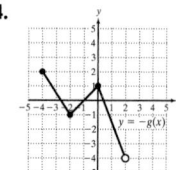

73.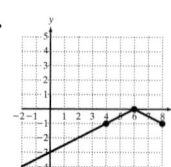

74.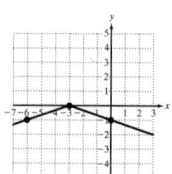

55. Parent function: $f(x) = \dfrac{1}{x}$; Shift the graph of f to the left 1 unit, stretch the graph vertically by a factor of 3, and shift the graph downward by 2 units.

56. Parent function: $f(x) = \dfrac{1}{x}$; Shift the graph of f to the right 4 units, stretch the graph vertically by a factor of 5, and shift the graph upward by 1 unit.

75.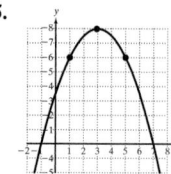

76.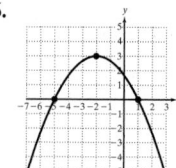

77.
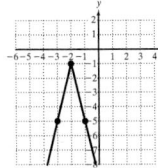

78.
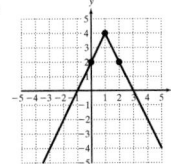

79.
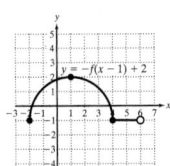

80.
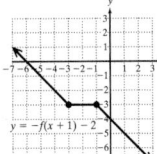

81.
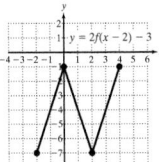

82.
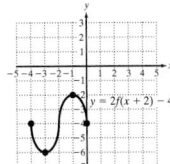

83.
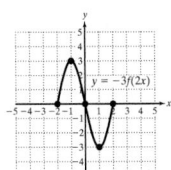

84.
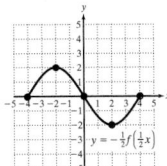

85.
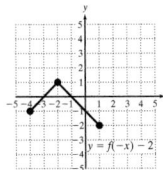

86.
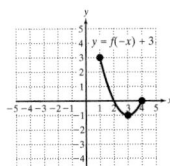

91. As written, $g(x) = |2x|$ is in the form $g(x) = f(ax)$ with $a > 1$. This indicates a horizontal shrink. However, $g(x)$ can also be written as $g(x) = |2| \cdot |x| = 2|x|$. This is written in the form $g(x) = af(x)$ with $a > 1$. This represents a vertical stretch.

92. As written, $h(x) = \sqrt{\frac{1}{2}x}$ is in the form $h(x) = f(ax)$ with $0 < a < 1$. This indicates a horizontal stretch. However, $h(x)$ can also be written as $h(x) = \sqrt{\frac{1}{2}} \cdot \sqrt{x}$. This is written in the form $h(x) = af(x)$ with $0 < a < 1$. This represents a vertical shrink.

94. The equation $g(x) = \dfrac{1}{-x+1}$ can also be written as $g(x) = \dfrac{1}{-(x-1)} = -\dfrac{1}{x-1}$. Written in this alternative form, we see that shifting f one unit to the right and reflecting over the x-axis produces the same graph.

95. $f(x) = (x-2)^2 - 3$ **96.** $f(x) = |x+3| - 4$

97. $f(x) = \dfrac{1}{x+3}$ **98.** $f(x) = \sqrt{x} - 2$

99. $f(x) = -x^3 + 1$ **100.** $f(x) = -(x+2)^2$

103. a.
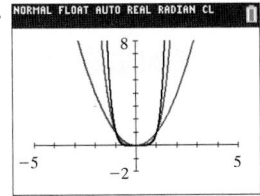

b.
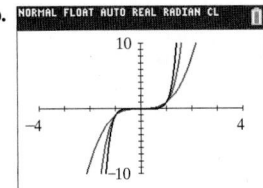

c. The general shape of $y = x^n$ is similar to the graph of $y = x^2$ for even values of n greater than 1.

d. The general shape of $y = x^n$ is similar to the graph of $y = x^3$ for odd values of n greater than 1.

Section 1.7 Practice Exercises, pp. 209–215

R.2. Interval notation: $(-\infty, 8)$
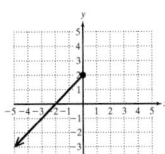

R.3. Interval notation: $[-2.4, 5.8)$
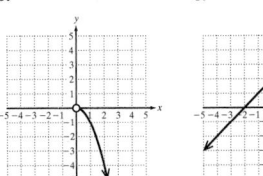

R.4. Interval notation: $\left[-\dfrac{9}{2}, \infty\right)$
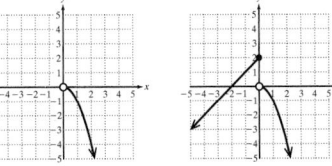

57. a.   **b.**   **c.**
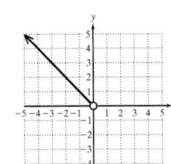 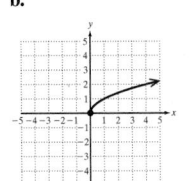 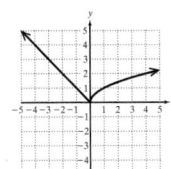

58. a.   **b.**   **c.**
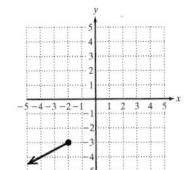 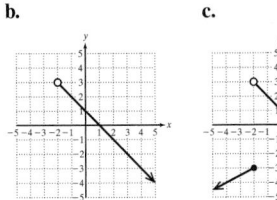

59. a.   **b.**   **c.**
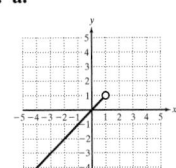 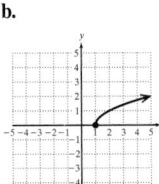 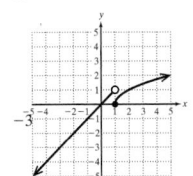

60. a.   **b.**   **c.**
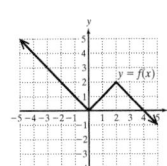 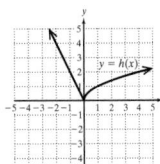

61.   **62.**
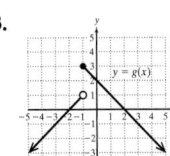 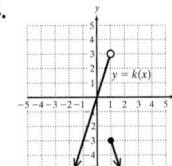

63.   **64.**
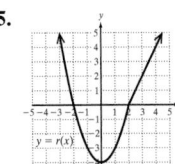 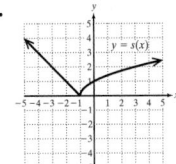

65.   **66.**

67.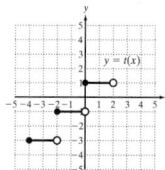

68.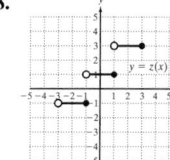

69.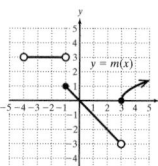

70.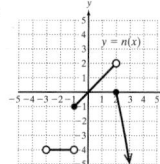

71. a. 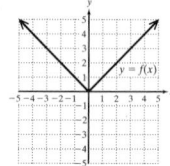   **b.** $y = |x|$

81.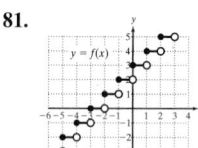

82.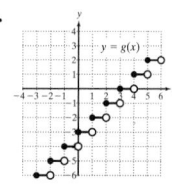

83.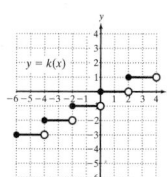

84.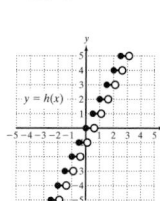

85. $C(x) = \begin{cases} 0.44 & \text{for } 0 < x \leq 1 \\ 0.61 & \text{for } 1 < x \leq 2 \\ 0.78 & \text{for } 2 < x \leq 3 \\ 0.95 & \text{for } 3 < x \leq 3.5 \end{cases}$

87. $S(x) = \begin{cases} 2000 & \text{for } 0 \leq x < 40{,}000 \\ 2000 + 0.05(x - 40{,}000) & \text{for } x \geq 40{,}000 \end{cases}$

99. At $x = -2$, the function has a relative minimum of 0. At $x = 0$, the function has a relative maximum of 2. At $x = 2$, the function has a relative minimum of 0.

100. At $x = -2$, the function has a relative maximum of 3. At $x = 0$, the function has a relative minimum of 1. At $x = 2$, the function has a relative maximum of 3.

101. At $x = -2$, the function has a relative minimum of -4. At $x = 0$, the function has a relative maximum of 0. At $x = 2$, the function has a relative minimum of -4.

102. At $x = -3$, the function has a relative maximum of 5. At $x = 0$, the function has a relative minimum of 0. At $x = 3$, the function has a relative maximum of 5.

103. c. The function has relative minima of 3 ft and 3.5 ft at approximately 8 days and 18 days after recording began. The function has a relative maximum of 4.5 ft at a time 12 days after recording began.   **d.** The weather was dry on the intervals of decreasing depth, and water from the pond evaporated. The weather was rainy during intervals of increasing depth.

105. $f(x) = \begin{cases} -2 & \text{for } x < 1 \\ 3 & \text{for } x \geq 1 \end{cases}$ **106.** $f(x) = \begin{cases} 4 & \text{for } x \leq -2 \\ 1 & \text{for } x > -2 \end{cases}$

107. $f(x) = \begin{cases} -|x| & \text{for } x < 2 \\ -2 & \text{for } x \geq 2 \end{cases}$ **108.** $f(x) = \begin{cases} x^2 & \text{for } x < 2 \\ 4 & \text{for } x \geq 2 \end{cases}$

109. $f(x) = \begin{cases} \dfrac{1}{x} & \text{for } x < 0 \\ x & \text{for } x > 0 \end{cases}$ **110.** $f(x) = \begin{cases} -x & \text{for } x < 0 \\ \sqrt{x} + 1 & \text{for } x \geq 0 \end{cases}$

111. a.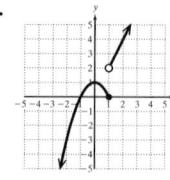
b. $(-\infty, \infty)$ **c.** $(-\infty, 1] \cup (2, \infty)$
d. $f(-1) = 0, f(1) = 0,$ and $f(2) = 4$
e. $x = 3$ **f.** $x = -2$
g. Increasing: $(-\infty, 0) \cup (1, \infty)$;
Decreasing: $(0, 1)$;
Never constant

112. a.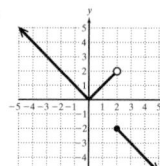
b. $(-\infty, \infty)$ **c.** $(-\infty, -2] \cup [0, \infty)$
d. $f(-1) = 1, f(1) = 1,$ and $f(2) = -2$
e. $x = -6$ **f.** $x = 3$
g. Increasing: $(0, 2)$;
Decreasing: $(-\infty, 0) \cup (2, \infty)$;
Never constant

115. If replacing y by $-y$ in the equation results in an equivalent equation, then the graph is symmetric to the x-axis. If replacing x by $-x$ in the equation results in an equivalent equation, then the graph is symmetric to the y-axis. If replacing both x by $-x$ and y by $-y$ results in an equivalent equation, then the graph is symmetric to the origin.

127. $f(x) = \begin{cases} 0.1x & \text{if } 0 < x \leq 8925 \\ 892.50 + 0.15(x - 8925) & \text{if } 8925 < x \leq 36{,}250 \\ 4991.25 + 0.25(x - 36{,}250) & \text{if } 36{,}250 < x \leq 87{,}850 \end{cases}$

or

$f(x) = \begin{cases} 0.1x & \text{if } 0 < x \leq 8925 \\ 0.15x - 446.25 & \text{if } 8925 < x \leq 36{,}250 \\ 0.25x - 4071.25 & \text{if } 36{,}250 < x \leq 87{,}850 \end{cases}$

128.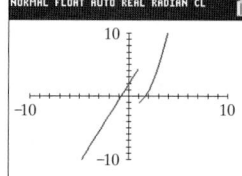

129.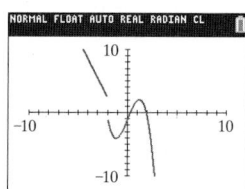

130.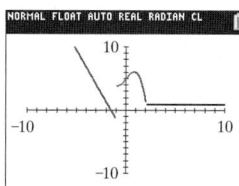

131.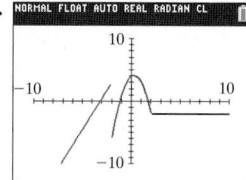

132. a. Relative maximum of 4.667 at $x = 1.667$
b. Increasing on $(-\infty, 1.667)$; Decreasing on $(1.667, \infty)$

133. a. Relative minimum of -7.825 at $x = 3.750$
b. Increasing on $(3.750, \infty)$; Decreasing on $(-\infty, 3.750)$

134. a. Relative maximum of 7.824 at $x = -3.390$;
Relative minimum of -7.936 at $x = 0.590$

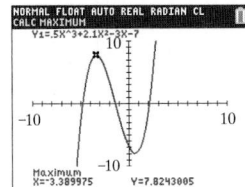 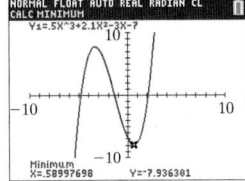

b. Increasing on $(-\infty, -3.390) \cup (0.590, \infty)$
Decreasing on $(-3.390, 0.590)$

135. a. Relative maximum of 3.726 at $x = 0.667$;
Relative minimum of -2.625 at $x = -2.500$

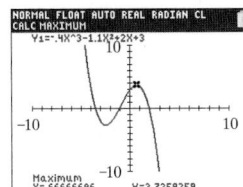 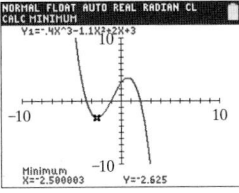

b. Increasing on $(-2.500, 0.667)$;
Decreasing on $(-\infty, -2.500) \cup (0.667, \infty)$

Section 1.8 Practice Exercises, pp. 224–229

19. $(r - p)(x) = -x^2 - 6x; (-\infty, \infty)$

20. $(p - r)(x) = x^2 + 6x; (-\infty, \infty)$

21. $(p \cdot q)(x) = (x^2 + 3x)\sqrt{1 - x}; (-\infty, 1]$

22. $(r \cdot q)(x) = -3x\sqrt{1 - x}; (-\infty, 1]$

23. $\left(\dfrac{q}{p}\right)(x) = \dfrac{\sqrt{1 - x}}{x^2 + 3x}; (-\infty, -3) \cup (-3, 0) \cup (0, 1]$

24. $\left(\dfrac{q}{r}\right)(x) = \dfrac{\sqrt{1 - x}}{-3x}; (-\infty, 0) \cup (0, 1]$

25. $\left(\dfrac{p}{q}\right)(x) = \dfrac{x^2 + 3x}{\sqrt{1 - x}}; (-\infty, 1)$ **26.** $\left(\dfrac{r}{q}\right)(x) = \dfrac{-3x}{\sqrt{1 - x}}; (-\infty, 1)$

27. $(s \cdot t)(x) = \dfrac{-1}{x + 3}$;
Domain: $(-\infty, -3) \cup (-3, 2) \cup (2, 3) \cup (3, \infty)$

28. $\left(\dfrac{s}{t}\right)(x) = -\dfrac{(x - 2)^2}{(x - 3)^2(x + 3)}$;
Domain: $(-\infty, -3) \cup (-3, 2) \cup (2, 3) \cup (3, \infty)$

29. $(s + t)(x) = -\dfrac{x^3 - 4x^2 - 5x + 23}{(x + 3)(x - 3)(x - 2)}$;
Domain: $(-\infty, -3) \cup (-3, 2) \cup (2, 3) \cup (3, \infty)$

30. $(s - t)(x) = \dfrac{x^3 - 2x^2 - 13x + 31}{(x + 3)(x - 3)(x - 2)}$;
Domain: $(-\infty, -3) \cup (-3, 2) \cup (2, 3) \cup (3, \infty)$

31. $(s \cdot v)(x) = \dfrac{(x - 2)\sqrt{x + 3}}{(x - 3)(x + 3)}$;
Domain: $(-3, 3) \cup (3, \infty)$

32. $\left(\dfrac{v}{s}\right)(x) = \dfrac{\sqrt{x + 3}(x + 3)(x - 3)}{x - 2}$;
Domain: $(-3, 2) \cup (2, 3) \cup (3, \infty)$

33. a. $5x + 5h + 9$ **b.** 5   **34. a.** $8x + 8h + 4$ **b.** 8
35. a. $x^2 + 2xh + h^2 + 4x + 4h$ **b.** $2x + h + 4$
36. a. $x^2 + 2xh + h^2 - 3x - 3h$ **b.** $2x + h - 3$

65. $(n \circ p)(x) = x^2 - 9x - 5; (-\infty, \infty)$

66. $(p \circ n)(x) = x^2 - 19x + 70; (-\infty, \infty)$

67. $(m \circ n)(x) = \sqrt{x + 3}; [-3, \infty)$

68. $(n \circ m)(x) = \sqrt{x + 8} - 5; [-8, \infty)$

69. $(q \circ n)(x) = \dfrac{1}{x - 15}; (-\infty, 15) \cup (15, \infty)$

70. $(q \circ p)(x) = \dfrac{1}{x^2 - 9x - 10}; (-\infty, -1) \cup (-1, 10) \cup (10, \infty)$

71. $(q \circ r)(x) = \dfrac{1}{|2x + 3| - 10}; \left(-\infty, -\dfrac{13}{2}\right) \cup \left(-\dfrac{13}{2}, \dfrac{7}{2}\right) \cup \left(\dfrac{7}{2}, \infty\right)$

72. $(q \circ m)(x) = \dfrac{1}{\sqrt{x + 8} - 10}; [-8, 92) \cup (92, \infty)$

73. $(n \circ r)(x) = |2x + 3| - 5; (-\infty, \infty)$

74. $(r \circ n)(x) = |2x - 7|; (-\infty, \infty)$

75. $(q \circ q)(x) = -\dfrac{x - 10}{10x - 101}; (-\infty, 10) \cup \left(10, \dfrac{101}{10}\right) \cup \left(\dfrac{101}{10}, \infty\right)$

76. $(p \circ p)(x) = x^4 - 18x^3 + 72x^2 + 81x; (-\infty, \infty)$

77. $(f \circ g)(x) = -\dfrac{3}{x + 14}; (-\infty, -14) \cup (-14, 2]$

78. $(f \circ g)(x) = -\dfrac{4}{x + 6}; (-\infty, -6) \cup (-6, 3]$

79. $(f \circ g)(x) = \dfrac{9}{25 - x^2}; (-\infty, -5) \cup (-5, -4) \cup (-4, 4) \cup (4, 5) \cup (5, \infty)$

80. $(f \circ g)(x) = \dfrac{3}{4x^2 - 1}; (-\infty, -1) \cup \left(-1, -\dfrac{1}{2}\right) \cup \left(-\dfrac{1}{2}, \dfrac{1}{2}\right) \cup \left(\dfrac{1}{2}, 1\right) \cup (1, \infty)$

81. $(f \circ f)(x) = \dfrac{x - 2}{-2x + 5}; (-\infty, 2) \cup \left(2, \dfrac{5}{2}\right) \cup \left(\dfrac{5}{2}, \infty\right)$

113. $(A \circ A)(x) = (1.045)^2 x$ represents the amount of money in the account after 2 yr compounded annually.

114. $(P \circ P)(x) = (0.98)^2 x$ represents the population 2 yr later.

116. a. $A_1(x) = \pi(x + 5)^2$ represents the area of the outer circle in terms of the radius of the inner circle, x.   **b.** $A_2(x) = \pi x^2$ represents the area of the inner circle based on its radius, x.
c. $(A_1 - A_2)(x) = 10\pi x + 25\pi$ represents the area of the region outside the inner circle and inside the outer circle.

118. The domain of $\left(\dfrac{f}{g}\right)(x)$ is the intersection of the domains of f and g with the further restriction to exclude values of x for which $g(x) = 0$.

119. The domain of $(f \circ g)(x)$ is the set of real numbers x in the domain of g such that $g(x)$ is in the domain of f.

120. For a positive real number h, the difference quotient represents the average rate of change of a function f between points $(x, f(x))$ and $(x + h, f(x + h))$. This is also interpreted as the slope of the secant line between the points $(x, f(x))$ and $(x + h, f(x + h))$.

Chapter 1 Review Exercises, pp. 231–235

7.

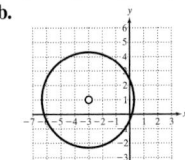

11. a. $(x + 3)^2 + (y - 1)^2 = 11$ **12. a.** $x^2 + y^2 = 10.24$
b.   **b.**

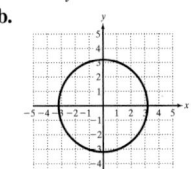

13. a. $(x - 4)^2 + (y - 1)^2 = 25$ **14. a.** $(x - 4)^2 + (y + 4)^2 = 16$

b.

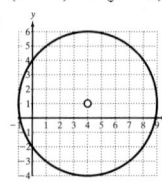

b.

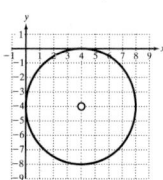

26. $P(189) = 265.70$ means that a drill that costs \$189 from the manufacturer will cost the customer \$265.70 after the department store markup and sales tax.

32. $(-\infty, -3) \cup (-3, 3) \cup (3, \infty)$

37. x-intercept: $(-4, 0)$;   **38.** x-intercept: $(0, 0)$;
y-intercept: $(0, 2)$ y-intercept: $(0, 0)$

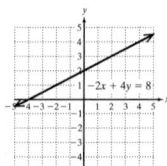 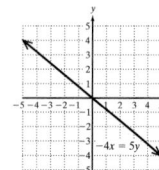

39. x-intercept: None;   **40.** x-intercept: $\left(\frac{5}{3}, 0\right)$;
y-intercept: $(0, 2)$ y-intercept: None

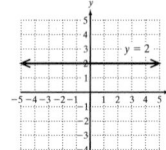 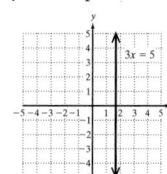

66. a.

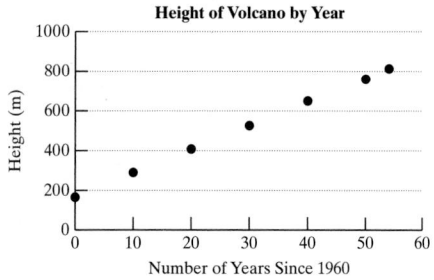

67. b.

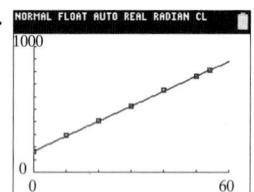

69. 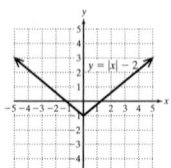   **70.**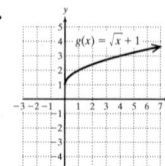

71. 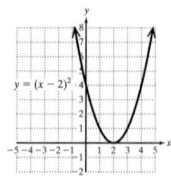   **72.**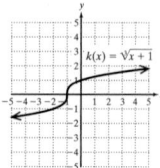

73. 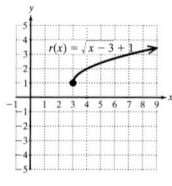   **74.**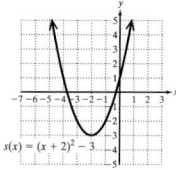

75. 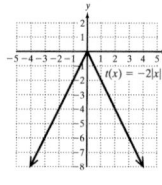   **76.**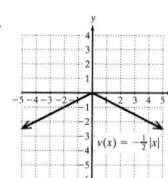

77. 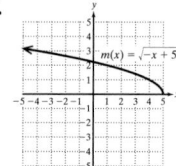   **78.**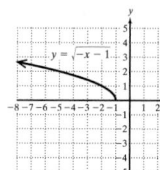

79. 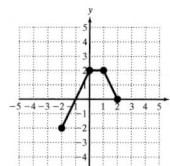   **80.**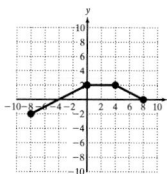

81. 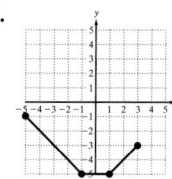   **82.**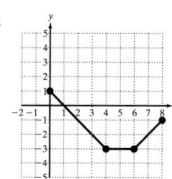

83. 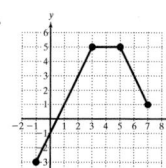   **84.**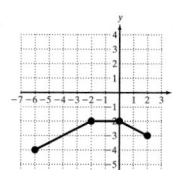

96. 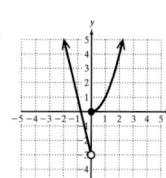   **97.**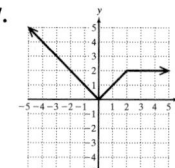

98.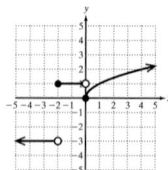

102. At $x = -2$ the function has a relative minimum of -2. At $x = 4$, the function has a relative minimum of -1. At $x = 2$, the function has a relative maximum of 1.

103. At $x = -2$, the function has a relative maximum of 4.

111. $(n - m)(x) = x^2$; Domain: $(-\infty, \infty)$

112. $\left(\dfrac{p}{n}\right)(x) = \dfrac{\sqrt{x - 2}}{x^2 - 4x}$; Domain: $[2, 4) \cup (4, \infty)$

113. $\left(\dfrac{n}{p}\right)(x) = \dfrac{x^2 - 4x}{\sqrt{x - 2}}$; Domain: $(2, \infty)$

114. $(m \cdot p)(x) = -4x\sqrt{x - 2}$; Domain: $[2, \infty)$

115. $(q \circ n)(x) = \dfrac{1}{x^2 - 4x - 5}$; Domain: $(-\infty, -1) \cup (-1, 5) \cup (5, \infty)$

116. $(q \circ p)(x) = \dfrac{1}{\sqrt{x - 2} - 5}$; Domain: $[2, 27) \cup (27, \infty)$

121. c. $(n \circ d)(t) = \dfrac{15t}{7}$ represents the number of gallons of gasoline used in t hours.

Chapter 1 Test, pp. 236–237

10. c.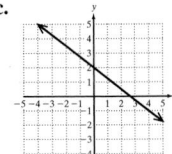

13.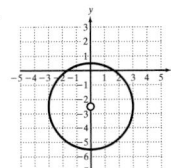

14.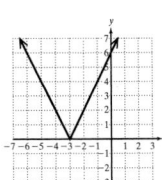

15.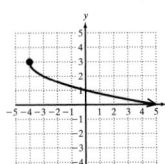

16.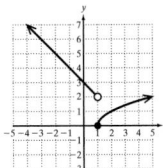

25. $\left(\dfrac{g}{f}\right)(x) = \dfrac{1}{(x - 3)(x - 4)}$; Domain: $(-\infty, 3) \cup (3, 4) \cup (4, \infty)$

26. $(g \circ h)(x) = \dfrac{1}{\sqrt{x - 5} - 3}$; Domain: $[5, 14) \cup (14, \infty)$

29. a.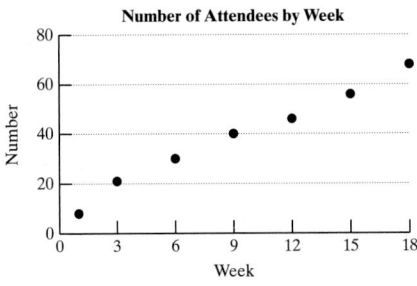

30. b.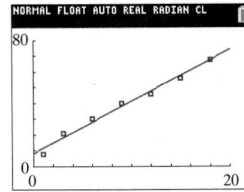

Chapter 1 Cumulative Review Exercises, pp. 237–238

3. $(g \circ f)(x) = \dfrac{1}{-x^2 + 3x}$; Domain: $(-\infty, 0) \cup (0, 3) \cup (3, \infty)$

4. $(g \cdot h)(x) = \dfrac{\sqrt{x + 2}}{x}$; Domain: $[-2, 0) \cup (0, \infty)$

8.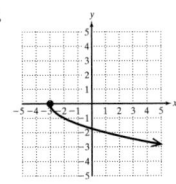

9.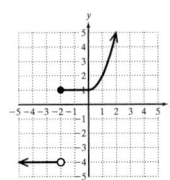

CHAPTER 2

Section 2.1 Practice Exercises, pp. 248–253

7. a. Downward **b.** $(4, 1)$ **c.** $(3, 0)$ and $(5, 0)$ **d.** $(0, -15)$
e. 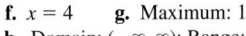 **f.** $x = 4$ **g.** Maximum: 1
h. Domain: $(-\infty, \infty)$; Range: $(-\infty, 1]$

8. a. Downward **b.** $(-2, 4)$ **c.** $(-4, 0)$ and $(0, 0)$ **d.** $(0, 0)$
e. 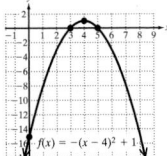 **f.** $x = -2$ **g.** Maximum: 4
h. Domain: $(-\infty, \infty)$; Range: $(-\infty, 4]$

9. a. Upward **b.** $(-1, -8)$ **c.** $(-3, 0)$ and $(1, 0)$ **d.** $(0, -6)$
e. 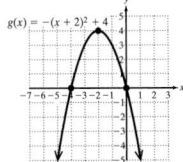 **f.** $x = -1$ **g.** Minimum: -8
h. Domain: $(-\infty, \infty)$; Range: $[-8, \infty)$

10. a. Upward **b.** $(3, -2)$ **c.** $(2, 0)$ and $(4, 0)$ **d.** $(0, 16)$
e. 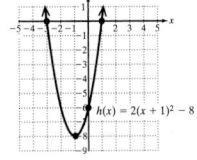 **f.** $x = 3$ **g.** Minimum: -2
h. Domain: $(-\infty, \infty)$; Range: $[-2, \infty)$

11. a. Upward **b.** $(1, 0)$ **c.** $(1, 0)$ **d.** $(0, 3)$
e. **f.** $x = 1$ **g.** Minimum: 0
h. Domain: $(-\infty, \infty)$; Range: $[0, \infty)$

12. a. Upward **b.** $(-2, 0)$ **c.** $(-2, 0)$ **d.** $(0, 2)$
e. **f.** $x = -2$ **g.** Minimum: 0
h. Domain: $(-\infty, \infty)$; Range: $[0, \infty)$

13. a. Downward   **b.** $(-4, 1)$ **c.** $\left(-4 + \sqrt{5}, 0\right)$ and $\left(-4 - \sqrt{5}, 0\right)$
d. $\left(0, -\dfrac{11}{5}\right)$
e.

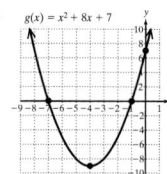

f. $x = -4$ **g.** Maximum: 1
h. Domain: $(-\infty, \infty)$; Range: $(-\infty, 1]$

$p(x) = -\frac{1}{5}(x + 4)^2 + 1$

14. a. Downward   **b.** $(1, 1)$ **c.** $\left(1 + \sqrt{3}, 0\right)$ and $\left(1 - \sqrt{3}, 0\right)$
d. $\left(0, \dfrac{2}{3}\right)$
e.

$q(x) = -\frac{1}{3}(x - 1)^2 + 1$

f. $x = 1$ **g.** Maximum: 1
h. Domain: $(-\infty, \infty)$; Range: $(-\infty, 1]$

15. a. $f(x) = (x + 3)^2 - 4$ **b.** $(-3, -4)$ **c.** $(-1, 0)$ and $(-5, 0)$
d. $(0, 5)$
e.

$f(x) = x^2 + 6x + 5$

f. $x = -3$ **g.** Minimum: -4
h. Domain: $(-\infty, \infty)$; Range: $[-4, \infty)$

16. a. $g(x) = (x + 4)^2 - 9$ **b.** $(-4, -9)$ **c.** $(-7, 0)$ and $(-1, 0)$
d. $(0, 7)$
e.

$g(x) = x^2 + 8x + 7$

f. $x = -4$ **g.** Minimum: -9
h. Domain: $(-\infty, \infty)$; Range: $[-9, \infty)$

17. a. $p(x) = 3(x - 2)^2 - 19$ **b.** $(2, -19)$
c. $\left(\dfrac{6 + \sqrt{57}}{3}, 0\right)$ and $\left(\dfrac{6 - \sqrt{57}}{3}, 0\right)$ **d.** $(0, -7)$
e.

$p(x) = 3x^2 - 12x - 7$

f. $x = 2$ **g.** Minimum: -19
h. Domain: $(-\infty, \infty)$; Range: $[-19, \infty)$

18. a. $q(x) = 2(x - 1)^2 - 5$ **b.** $(1, -5)$
c. $\left(\dfrac{2 + \sqrt{10}}{2}, 0\right)$ and $\left(\dfrac{2 - \sqrt{10}}{2}, 0\right)$ **d.** $(0, -3)$
e.

$q(x) = 2x^2 - 4x - 3$

f. $x = 1$ **g.** Minimum: -5
h. Domain: $(-\infty, \infty)$; Range: $[-5, \infty)$

19. a. $c(x) = -2\left(x + \dfrac{5}{2}\right)^2 + \dfrac{33}{2}$ **b.** $\left(-\dfrac{5}{2}, \dfrac{33}{2}\right)$
c. $\left(\dfrac{-5 + \sqrt{33}}{2}, 0\right)$ and $\left(\dfrac{-5 - \sqrt{33}}{2}, 0\right)$ **d.** $(0, 4)$

e.

$c(x) = -2x^2 - 10x + 4$

f. $x = -\dfrac{5}{2}$ **g.** Maximum: $\dfrac{33}{2}$
h. Domain: $(-\infty, \infty)$;
Range: $\left(-\infty, \dfrac{33}{2}\right]$

20. a. $d(x) = -3\left(x + \dfrac{3}{2}\right)^2 + \dfrac{59}{4}$ **b.** $\left(-\dfrac{3}{2}, \dfrac{59}{4}\right)$
c. $\left(\dfrac{-9 + \sqrt{177}}{6}, 0\right)$ and $\left(\dfrac{-9 - \sqrt{177}}{6}, 0\right)$ **d.** $(0, 8)$
e.

$d(x) = -3x^2 - 9x + 8$

f. $x = -\dfrac{3}{2}$ **g.** Maximum: $\dfrac{59}{4}$
h. Domain: $(-\infty, \infty)$;
Range: $\left(-\infty, \dfrac{59}{4}\right]$

21. a. $h(x) = -2\left(x - \dfrac{7}{4}\right)^2 + \dfrac{49}{8}$ **b.** $\left(\dfrac{7}{4}, \dfrac{49}{8}\right)$
c. $(0, 0)$ and $\left(\dfrac{7}{2}, 0\right)$ **d.** $(0, 0)$
e.

f. $x = \dfrac{7}{4}$ **g.** Maximum: $\dfrac{49}{8}$
h. Domain: $(-\infty, \infty)$;
Range: $\left(-\infty, \dfrac{49}{8}\right]$

22. a. $k(x) = 3\left(x - \dfrac{4}{3}\right)^2 - \dfrac{16}{3}$ **b.** $\left(\dfrac{4}{3}, -\dfrac{16}{3}\right)$
c. $(0, 0)$ and $\left(\dfrac{8}{3}, 0\right)$ **d.** $(0, 0)$
e.

f. $x = \dfrac{4}{3}$ **g.** Minimum: $-\dfrac{16}{3}$
h. Domain: $(-\infty, \infty)$;
Range: $\left[-\dfrac{16}{3}, \infty\right)$

23. a. $p(x) = \left(x + \dfrac{9}{2}\right)^2 - \dfrac{13}{4}$ **b.** $\left(-\dfrac{9}{2}, -\dfrac{13}{4}\right)$
c. $\left(\dfrac{-9 + \sqrt{13}}{2}, 0\right)$ and $\left(\dfrac{-9 - \sqrt{13}}{2}, 0\right)$ **d.** $(0, 17)$
e.

f. $x = -\dfrac{9}{2}$ **g.** Minimum: $-\dfrac{13}{4}$
h. Domain: $(-\infty, \infty)$;
Range: $\left[-\dfrac{13}{4}, \infty\right)$

24. a. $q(x) = \left(x + \dfrac{11}{2}\right)^2 - \dfrac{17}{4}$ **b.** $\left(-\dfrac{11}{2}, -\dfrac{17}{4}\right)$
c. $\left(\dfrac{-11 + \sqrt{17}}{2}, 0\right)$ and $\left(\dfrac{-11 - \sqrt{17}}{2}, 0\right)$ **d.** $(0, 26)$
e.

f. $x = -\dfrac{11}{2}$ **g.** Minimum: $-\dfrac{17}{4}$
h. Domain: $(-\infty, \infty)$;
Range: $\left[-\dfrac{17}{4}, \infty\right)$

33. a. Downward **b.** $(1, -3)$ **c.** None **d.** $(0, -4)$
e.
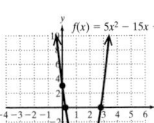
  f. $x = 1$ **g.** Maximum: -3
  h. Domain: $(-\infty, \infty)$;
  Range: $(-\infty, -3]$

34. a. Downward **b.** $(-3, -1)$ **c.** None **d.** $(0, -10)$
e.
  f. $x = -3$ **g.** Maximum: -1
  h. Domain: $(-\infty, \infty)$;
  Range: $(-\infty, -1]$

35. a. Upward **b.** $\left(\dfrac{3}{2}, -\dfrac{33}{4}\right)$
c. $\left(\dfrac{15 + \sqrt{165}}{10}, 0\right)$ and $\left(\dfrac{15 - \sqrt{165}}{10}, 0\right)$ **d.** $(0, 3)$
e.
  f. $x = \dfrac{3}{2}$ **g.** Minimum: $-\dfrac{33}{4}$
  h. Domain: $(-\infty, \infty)$;
  Range: $\left[-\dfrac{33}{4}, \infty\right)$

36. a. Upward **b.** $\left(\dfrac{5}{2}, -\dfrac{35}{2}\right)$
c. $\left(\dfrac{5 + \sqrt{35}}{2}, 0\right)$ and $\left(\dfrac{5 - \sqrt{35}}{2}, 0\right)$ **d.** $(0, -5)$
e.
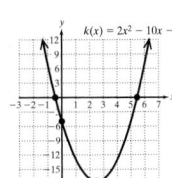
  f. $x = \dfrac{5}{2}$ **g.** Minimum: $-\dfrac{35}{2}$
  h. Domain: $(-\infty, \infty)$;
  Range: $\left[-\dfrac{35}{2}, \infty\right)$

37. a. Upward **b.** $(0, 3)$ **c.** None **d.** $(0, 3)$
e.
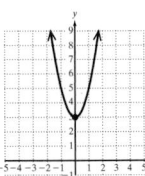
  f. $x = 0$ **g.** Minimum: 3
  h. Domain: $(-\infty, \infty)$; Range: $[3, \infty)$

38. a. Downward **b.** $(0, -1)$ **c.** None **d.** $(0, -1)$
e.
  f. $x = 0$ **g.** Maximum: -1
  h. Domain: $(-\infty, \infty)$; Range: $(-\infty, -1]$

39. a. Downward **b.** $(-5, 0)$ **c.** $(-5, 0)$ **d.** $(0, -50)$
e.
  f. $x = -5$ **g.** Maximum: 0
  h. Domain: $(-\infty, \infty)$; Range: $(-\infty, 0]$

40. a. Upward **b.** $(2, 0)$ **c.** $(2, 0)$ **d.** $(0, 8)$
e.
  f. $x = 2$ **g.** Minimum: 0
  h. Domain: $(-\infty, \infty)$; Range: $[0, \infty)$

41. a. Upward **b.** $\left(\dfrac{1}{2}, \dfrac{11}{4}\right)$ **c.** None **d.** $(0, 3)$
e.
  f. $x = \dfrac{1}{2}$ **g.** Minimum: $\dfrac{11}{4}$
  h. Domain: $(-\infty, \infty)$;
  Range: $\left[\dfrac{11}{4}, \infty\right)$

42. a. Upward **b.** $\left(\dfrac{5}{2}, \dfrac{3}{4}\right)$ **c.** None **d.** $(0, 7)$
e.
  f. $x = \dfrac{5}{2}$ **g.** Minimum: $\dfrac{3}{4}$
  h. Domain: $(-\infty, \infty)$;
  Range: $\left[\dfrac{3}{4}, \infty\right)$

79. For a parabola opening upward, such as the graph of $f(x) = x^2$, the minimum value is the y-coordinate of the vertex. There is no maximum value because the y values of the function become arbitrarily large for large values of $|x|$.

80. The vertex $(4, -3)$ is below the x-axis. If the parabola opens downward, then the vertex is the maximum point. If the maximum point is below the x-axis, then no points on the curve will intersect the x-axis.

81. No function defined by $y = f(x)$ can have two y-intercepts because the graph would fail the vertical line test.

82. The x-intercepts are the real solutions of the equation $ax^2 + bx + c = 0$. The discriminant $b^2 - 4ac$ determines the number and type of solutions to the equation and thus the number of x-intercepts. If $b^2 - 4ac < 0$, then the solutions to the equation are not real and the function will have no x-intercepts. If $b^2 - 4ac = 0$, then the equation has one real solution and the function has one x-intercept. If $b^2 - 4ac > 0$, then the equation has two real solutions and the function has two x-intercepts.

83. Because a parabola is symmetric with respect to the vertical line through the vertex, the x-coordinate of the vertex must be equidistant from the x-intercepts. Therefore, given $y = f(x)$, the x-coordinate of the vertex is 4 because 4 is midway between 2 and 6. The y-coordinate of the vertex is $f(4)$.

84. Given an equation of a parabola in the form $y = af(x - h)^2 + k$, the orientation is determined by a and the vertex is (h, k). If the vertex is above the x-axis and the parabola opens upward ($a > 0$), then the graph has no x-intercepts. Likewise, if the vertex is below the x-axis and the parabola opens downward ($a < 0$), then the graph has no x-intercepts.

Section 2.2 Practice Exercises, pp. 265–269

R.3.   **R.4.**   **R.5.**
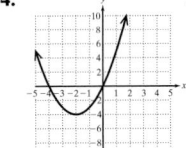

13. Down left and down right. As $x \to -\infty$, $f(x) \to -\infty$, and as $x \to \infty$, $f(x) \to -\infty$.

14. Down left and down right. As $x \to -\infty$, $f(x) \to -\infty$, and as $x \to \infty$, $f(x) \to -\infty$.

15. Down left and up right. As $x \to -\infty$, $f(x) \to -\infty$, and as $x \to \infty$, $f(x) \to \infty$.

16. Down left and up right. As $x \to -\infty$, $f(x) \to -\infty$, and as $x \to \infty$, $f(x) \to \infty$.

17. Up left and down right. As $x \to -\infty$, $f(x) \to \infty$, and as $x \to \infty$, $f(x) \to -\infty$.

18. Up left and down right. As $x \to -\infty$, $f(x) \to \infty$, and as $x \to \infty$, $f(x) \to -\infty$.

19. Up left and up right. As $x \to -\infty$, $f(x) \to \infty$, and as $x \to \infty$, $f(x) \to \infty$.

20. Up left and up right. As $x \to -\infty$, $f(x) \to \infty$, and as $x \to \infty$, $f(x) \to \infty$.

23. $-2, 5, -5$; each of multiplicity 1

24. $-5, 1, -1$; each of multiplicity 1

25. $0, -4, \dfrac{5}{2}$; each of multiplicity 1   **26.** $0, 2, \dfrac{7}{3}$; each of multiplicity 1

27. 0 (multiplicity 3), 5 (multiplicity 2)

28. 0 (multiplicity 4), -2 (multiplicity 2)

29. 0 (multiplicity 1), -2 (multiplicity 3), -4 (multiplicity 1)

30. 0 (multiplicity 4), -1 (multiplicity 3), 2 (multiplicity 2)

31. $0, \dfrac{5}{3}, -\dfrac{9}{2}, \pm\sqrt{3}$; each of multiplicity 1

32. $0, \dfrac{1}{5}, -\dfrac{8}{3}, \pm\sqrt{5}$; each of multiplicity 1

33. $3 \pm \sqrt{5}$; each of multiplicity 1
34. $2 \pm \sqrt{11}$; each of multiplicity 1

35. $-3, -\dfrac{1}{2}, \dfrac{1}{2}, 3$; each of multiplicity 1

36. $-4, -\dfrac{1}{2}, \dfrac{1}{2}, 4$; each of multiplicity 1

37. $-\sqrt{7}$ (multiplicity 1), 0 (multiplicity 4), $\sqrt{7}$ (multiplicity 1)
38. $-\sqrt{5}$ (multiplicity 1), 0 (multiplicity 3), $\sqrt{5}$ (multiplicity 1)

45. Not a polynomial function. The graph is not smooth.
46. Not a polynomial function. The graph is not smooth.

47. Polynomial function   **a.** Minimum degree 3
  b. Leading coefficient positive; degree odd
  c. -4 (odd multiplicity), 1 (odd multiplicity), 3 (odd multiplicity)

48. Polynomial function   **a.** Minimum degree 4
  b. Leading coefficient positive; degree even
  c. -4 (odd multiplicity), -1 (odd multiplicity), 3 (even multiplicity)

49. Polynomial function   **a.** Minimum degree 6
  b. Leading coefficient negative; degree even
  c. -4 (odd multiplicity), -3 (odd multiplicity), -1 (even multiplicity), 2 (odd multiplicity), $\frac{7}{2}$ (odd multiplicity)

50. Polynomial function   **a.** Minimum degree 3
  b. Leading coefficient negative; degree odd
  c. -4 (odd multiplicity), -1 (odd multiplicity), 2 (odd multiplicity)

51. Not a polynomial function. The graph is not continuous.
52. Not a polynomial function. The graph is not continuous.

53. a. $y = x^6$
  b. Shrink $y = x^6$ vertically by a factor of $\dfrac{1}{3}$. Reflect across the x-axis. Shift downward 2 units.
  c. Graph iii

54. a. $y = x^4$
  b. Shift $y = x^4$ to the right 3 units. Shrink vertically by a factor of $\dfrac{1}{2}$. Reflect across the x-axis.
  c. Graph v

55. a. $y = x^3$
  b. Shift $y = x^3$ to the left 2 units. Reflect across the x-axis. Shift upward 3 units.
  c. Graph i

56. a. $y = x^3$
  b. Shift $y = x^3$ to the left 4 units. Stretch vertically by a factor of 2. Shift downward 3 units.
  c. Graph vi

57. a. $y = x^5$
  b. Shift $y = x^5$ to the right 3 units. Reflect across the y-axis. Shift upward 1 unit.
  c. Graph iv

58. a. $y = x^4$
  b. Shift $y = x^4$ to the left 3 units. Reflect across the y-axis. Shift downward 1 unit.
  c. Graph ii

59.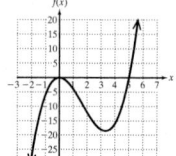

60.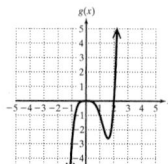

61.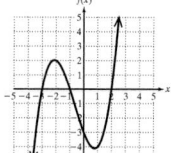

62.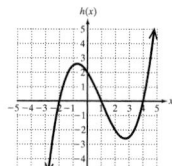

63.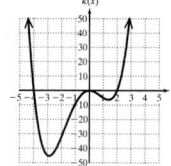

64.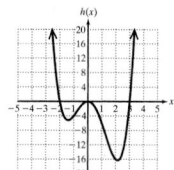

65.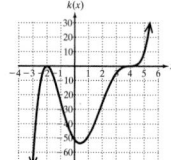

66.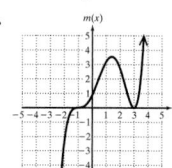

67.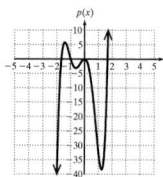

68.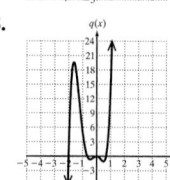

69.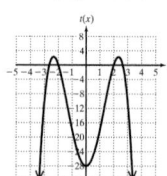

70.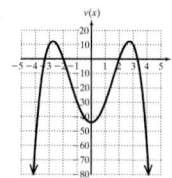

71.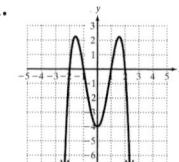

72.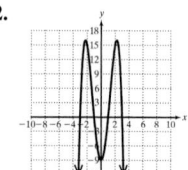

73.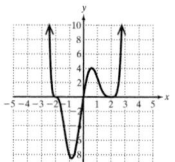

74.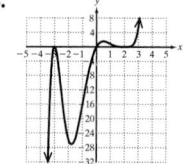

75. 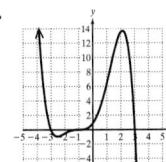   **76.**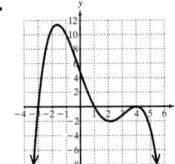

79. False. An nth-degree polynomial has at most $n - 1$ turning points. Therefore, a third-degree polynomial has at most 2 turning points.

80. False. A third-degree polynomial may have *at most* 2 turning points, but may have fewer. For example the graph of $y = x^3$ has no turning points.

82. False. There are infinitely many such polynomials. For example, $f(x) = (x - 2)(x - 4)(x - 6)$ and $g(x) = 2(x - 2)(x - 4)(x - 6)$ are two polynomials with the required zeros.

83. False. If the leading coefficient is negative, the graph will be down to the far left and down to the far right.

91. The x-intercepts are the real solutions to the equation $f(x) = 0$.

92. An x-intercept is a cross point if the corresponding zero of the polynomial has an odd multiplicity. An x-intercept is a touch point if the corresponding zero of the polynomial has an even multiplicity.

93. A function is continuous if its graph can be drawn without lifting the pencil from the paper.

94. All polynomial functions $y = f(x)$ are continuous. Therefore, for $a < b$, if $f(a)$ and $f(b)$ have different signs, then at some point on the interval $[a, b]$, $f(x)$ must be 0. That is, for the function to change sign from positive to negative or from negative to positive, the function must have a y value of 0 somewhere in between.

95. a. $f(3) = 2; f(4) = 6$
   b. By the intermediate value theorem, because $f(3) = 2$ and $f(4) = 6$, then f must take on every value between 2 and 6 on the interval $[3, 4]$.
   c. $x = \dfrac{3 + \sqrt{17}}{2} \approx 3.56$

96. a. $f(-4) = 3; f(-3) = 6$
   b. By the intermediate value theorem, because $f(-4) = 3$ and $f(-3) = 6$, then f must take on every value between 3 and 6 on the interval $[-4, -3]$.   **c.** $x = -2 - \sqrt{2} \approx -3.41$

97. $V(t) = -0.0406t^3 + 0.154t^2 + 0.173t - 0.0024$

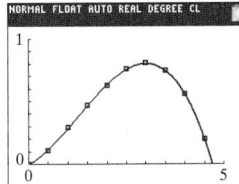

98. $T(x) = 0.76x^3 - 17.7x^2 + 71.2x + 111$

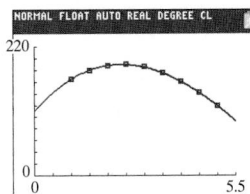

99. a. $V = lwh$
   $= (30 - 2x)(24 - 2x)(x)$
   $= 4x^3 - 108x^2 + 720x$
   The domain is restricted to $0 < x < 12$ because the width of the rectangular sheet is 24 in. The maximum amount that can be removed from each end would be half of 24 in.

b.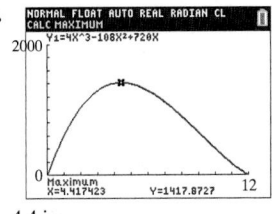
   4.4 in.

100. Window b is better.   **101.** Window b is better.

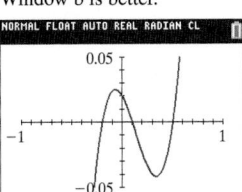   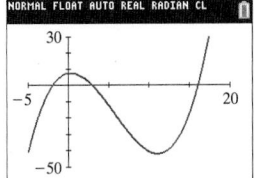

102. 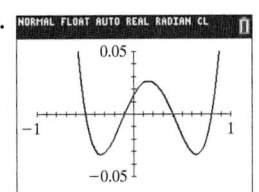   **103.**

Section 2.3 Practice Exercises, pp. 279–282

R.2. a. $(6 - i\sqrt{3})^2 - 12(6 - i\sqrt{3}) + 39 = 0$ ✓
   b. $(6 + i\sqrt{3})^2 - 12(6 + i\sqrt{3}) + 39 = 0$ ✓

7. a. $3x + 12 + \dfrac{65}{2x - 5}$ **b.** Dividend: $6x^2 + 9x + 5$;
   Divisor: $2x - 5$; Quotient: $3x + 12$; Remainder: 65
   c. $(2x - 5)(3x + 12) + 65 = 6x^2 + 9x + 5$ ✓

8. a. $4x - 2 + \dfrac{11}{3x + 4}$ **b.** Dividend: $12x^2 + 10x + 3$;
   Divisor: $3x + 4$; Quotient: $4x - 2$; Remainder: 11
   c. $(3x + 4)(4x - 2) + 11 = 12x^2 + 10x + 3$ ✓

73. $f(x) = x^4 - \dfrac{5}{2}x^3 + \dfrac{3}{2}x^2$ or $f(x) = 2x^4 - 5x^3 + 3x^2$

74. $f(x) = x^5 - \dfrac{9}{2}x^4 + 5x^3$ or $f(x) = 2x^5 - 9x^4 + 10x^3$

Section 2.4 Practice Exercises, pp. 294–298

9. $\pm 1, \pm 2, \pm 3, \pm 6, \pm\dfrac{1}{2}, \pm\dfrac{3}{2}, \pm\dfrac{1}{4}, \pm\dfrac{3}{4}$

10. $\pm 1, \pm 2, \pm 5, \pm 10, \pm\dfrac{1}{5}, \pm\dfrac{2}{5}, \pm\dfrac{1}{25}, \pm\dfrac{2}{25}$

11. $\pm 1, \pm 2, \pm 4, \pm 8, \pm\dfrac{1}{2}, \pm\dfrac{1}{3}, \pm\dfrac{2}{3}, \pm\dfrac{4}{3}, \pm\dfrac{8}{3}, \pm\dfrac{1}{4}, \pm\dfrac{1}{6}, \pm\dfrac{1}{12}$

12. $\pm 1, \pm 2, \pm 3, \pm 6, \pm\dfrac{1}{2}, \pm\dfrac{3}{2}, \pm\dfrac{1}{4}, \pm\dfrac{3}{4}, \pm\dfrac{1}{8}, \pm\dfrac{3}{8}, \pm\dfrac{1}{16}, \pm\dfrac{3}{16}$

17. $-3, \dfrac{1}{2}, 2, 4$ **18.** $-2, -1, \dfrac{2}{3}, 3$

23. 2 (multiplicity 2), $\dfrac{1}{3}, -4$ **24.** -3 (multiplicity 2), $\dfrac{5}{2}, -1$

33. a. $2 \pm 5i, \pm\sqrt{7}$
   b. $[x - (2 + 5i)][x - (2 - 5i)](x - \sqrt{7})(x + \sqrt{7})$
   c. $\{2 \pm 5i, \pm\sqrt{7}\}$ **34. a.** $3 \pm i, \pm\sqrt{5}$
   b. $[x - (3 + i)][x - (3 - i)](x - \sqrt{5})(x + \sqrt{5})$
   c. $\{3 \pm i, \pm\sqrt{5}\}$ **35. a.** $4 \pm i, \dfrac{4}{3}$
   b. $[x - (4 + i)][x - (4 - i)](3x - 4)$ **c.** $\left\{4 \pm i, \dfrac{4}{3}\right\}$

36. a. $5 \pm i, \dfrac{4}{5}$ **b.** $[x - (5 + i)][x - (5 - i)](5x - 4)$

c. $\left\{5 \pm i, \dfrac{4}{5}\right\}$ **37. a.** $-3 \pm 2i, -\dfrac{1}{4}, 1, -4$

b. $[x - (-3 + 2i)][x - (-3 - 2i)](4x + 1)(x - 1)(x + 4)$

c. $\left\{-3 \pm 2i, -\dfrac{1}{4}, 1, -4\right\}$ **38. a.** $-1 \pm 2i, \dfrac{5}{2}, -1, 3$

b. $[x - (-1 + 2i)][x - (-1 - 2i)](2x - 5)(x + 1)(x - 3)$

c. $\left\{-1 \pm 2i, \dfrac{5}{2}, -1, 3\right\}$

73. -3 (multiplicity 2) and $\dfrac{1}{2}$ (multiplicity 2)

74. -2 (multiplicity 2) and $\dfrac{1}{3}$ (multiplicity 2)

75. $\dfrac{1}{2}, 3, 1 \pm 2i$ (each with multiplicity 1)

76. $\dfrac{2}{3}, 2, 2 \pm i$ (each with multiplicity 1)

77. $\dfrac{5}{2}$ (multiplicity 2), $1 \pm i$ (each with multiplicity 1)

78. $\dfrac{1}{3}$ (multiplicity 2), $2 \pm i$ (each with multiplicity 1)

85. False. For example, the graph of $f(x) = x^4 + 1$ has no x-intercepts. Thus, $x^4 + 1$ has no real zeros.

86. False. This is true only if the polynomial has real coefficients.

87. False. For example, the graph of $f(x) = x^{10} + 1$ has no x-intercepts.

88. False. The graph of $y = f(x)$ may touch or cross the x-axis between a and b. Thus, $f(x)$ may have one or more zeros between a and b.

90. False. This might be true in special cases (for example, if the graph of the polynomial were symmetric with respect to the y-axis), but the statement is false in general.

93. a. $f(2) = -2$ and $f(3) = 1$. Since $f(2)$ and $f(3)$ have opposite signs, the intermediate value theorem guarantees that f has at least one real zero between 2 and 3.
b. $\dfrac{7 \pm \sqrt{17}}{4}$; Furthermore, $\dfrac{7 + \sqrt{17}}{4} \approx 2.78$ is on the interval $[2, 3]$.

95. If a polynomial has real coefficients, then all nonreal zeros must come in conjugate pairs. This means that if the polynomial has nonreal zeros, there would be an even number of them. A third-degree polynomial has 3 zeros (including multiplicities). Therefore, it would have either 2 or 0 nonreal zeros, leaving room for either 1 or 3 real zeros.

113. The number $\sqrt{5}$ is a real solution to the equation $x^2 - 5 = 0$ and a zero of the polynomial $f(x) = x^2 - 5$. However, by the rational zeros theorem, the only possible rational zeros of $f(x)$ are ± 1 and ± 5. This means that $\sqrt{5}$ is irrational.

Section 2.5 Practice Exercises, pp. 315–321

37. a. $\dfrac{1 + \dfrac{3}{x} + \dfrac{1}{x^2}}{2 + \dfrac{5}{x^2}}$ **38. a.** $\dfrac{3 - \dfrac{2}{x} + \dfrac{7}{x^2}}{5 + \dfrac{1}{x^3}}$

39. Vertical asymptote: $x = 0$; Slant asymptote: $y = 2x$
40. Vertical asymptote: $x = 0$; Slant asymptote: $y = 3x$
41. Vertical asymptote: $x = -6$; Slant asymptote: $y = -3x + 22$
42. Vertical asymptote: $x = -3$; Slant asymptote: $y = -2x + 3$

49.

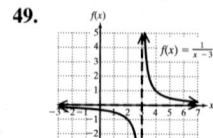

50.

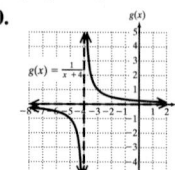

51.

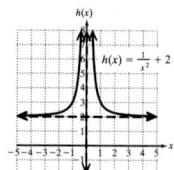

52.

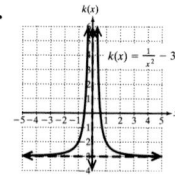

53.

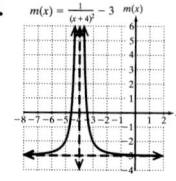

54.

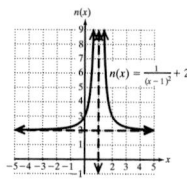

55.

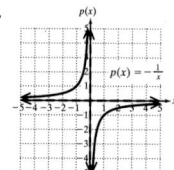

56.

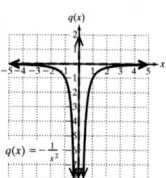

57. a. $(-3, 0)$ and $\left(\dfrac{7}{2}, 0\right)$ **b.** $x = -2$ and $x = -\dfrac{1}{4}$

c. Horizontal asymptote: $y = \dfrac{1}{2}$ **d.** $\left(0, -\dfrac{21}{2}\right)$

58. a. $\left(\dfrac{4}{3}, 0\right)$ and $(6, 0)$ **b.** $x = \dfrac{3}{2}$ and $x = -5$

c. Horizontal asymptote: $y = \dfrac{3}{2}$ **d.** $\left(0, -\dfrac{8}{5}\right)$

59. a. $\left(\dfrac{9}{4}, 0\right)$ **b.** $x = 3$ and $x = -3$

c. Horizontal asymptote: $y = 0$ **d.** $(0, 1)$

60. a. $\left(\dfrac{8}{5}, 0\right)$ **b.** $x = 2$ and $x = -2$

c. Horizontal asymptote: $y = 0$ **d.** $(0, 2)$

61. a. $\left(\dfrac{1}{5}, 0\right)$ and $(-3, 0)$ **b.** $x = -2$

c. Slant asymptote: $y = 5x + 4$ **d.** $\left(0, -\dfrac{3}{2}\right)$

62. a. $\left(-\dfrac{3}{4}, 0\right)$ and $(-2, 0)$ **b.** $x = -3$

c. Slant asymptote: $y = 4x - 1$ **d.** $(0, 2)$

63.

64.

65.

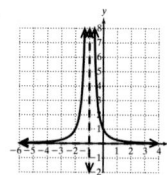

66.

67.

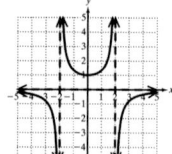

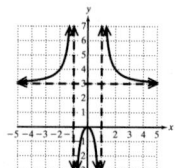

68.

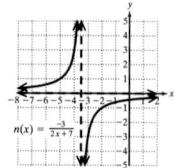

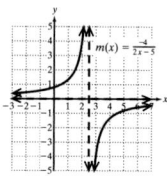

69. 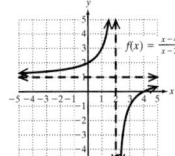 $f(x) = \frac{x-4}{x-2}$

70. 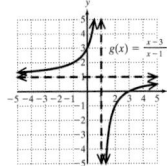 $g(x) = \frac{x-3}{x-1}$

71. 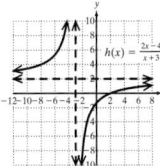 $h(x) = \frac{2x-4}{x+3}$

72. 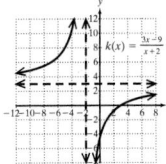 $k(x) = \frac{3x-9}{x+2}$

73. 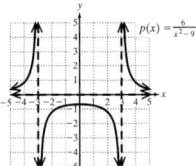 $p(x) = \frac{6}{x^2-9}$

74. 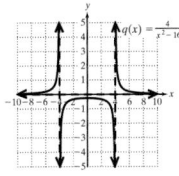 $q(x) = \frac{4}{x^2-16}$

75. 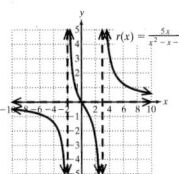 $r(x) = \frac{5x}{x^2-x-6}$

76. 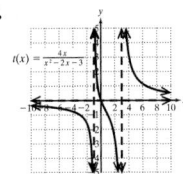 $t(x) = \frac{4x}{x^2-2x-3}$

77. 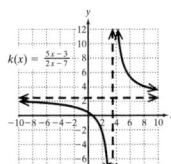 $k(x) = \frac{5x-3}{2x-7}$

78. 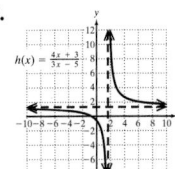 $h(x) = \frac{4x+3}{3x-5}$

79. 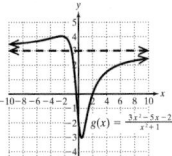 $g(x) = \frac{3x^2-5x-2}{x^2+1}$

80. 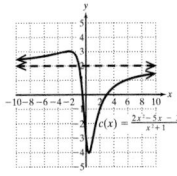 $c(x) = \frac{2x^2-5x-3}{x^2+1}$

81. 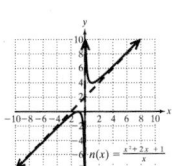 $n(x) = \frac{x^2+2x+1}{x}$

82. 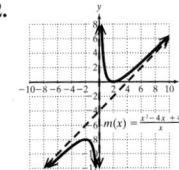 $m(x) = \frac{x^2-4x+4}{x}$

83. 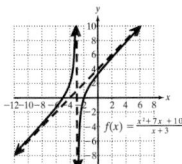 $f(x) = \frac{x^2+7x+10}{x+3}$

84. 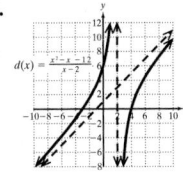 $d(x) = \frac{x^2-x-12}{x-2}$

85. 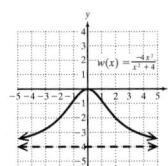 $w(x) = \frac{-4x^2}{x^2+4}$

86. 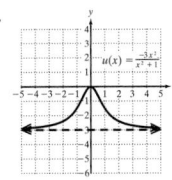 $u(x) = \frac{-3x^2}{x^2+1}$

87. 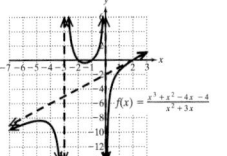 $f(x) = \frac{x^3+x^2-4x-4}{x^2+3x}$

88. 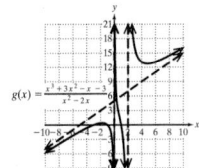 $g(x) = \frac{x^3+3x^2-x-3}{x^2-2x}$

89. 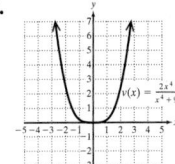 $v(x) = \frac{2x^4}{x^4+9}$

90. 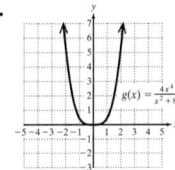 $g(x) = \frac{4x^4}{x^4+8}$

91. a. $C(x) = 109.94 + 20x$ **b.** $\overline{C}(x) = \dfrac{109.94 + 20x}{x}$

   c. $\overline{C}(5) = 41.99$; $\overline{C}(30) = 23.67$; $\overline{C}(120) = 20.92$

   d. The average cost would approach \$20 per session. This is the same as the fee paid to the gym in the absence of fixed costs.

92. a. $C(x) = 1920 + 40x$ **b.** $\overline{C}(x) = \dfrac{1920 + 40x}{x}$

   c. $\overline{C}(20) = 136$; $\overline{C}(50) = 78.4$; $\overline{C}(100) = 59.2$; $\overline{C}(200) = 49.6$

   d. $\overline{C}(200) = 49.6$ means that if 200,000 pages are printed, then the average cost per thousand pages is \$49.60 (or equivalently \$0.0496 per page).

   e. The average cost would approach \$40 per thousand pages or equivalently \$0.04 per page. This is the cost per page in the absence of fixed costs.

93. a. $R(x) = \dfrac{6x}{x+6}$

   b.

x	6	12	18	30
$R(x)$	3	4	4.5	5

   c. 6 Ω; Even for large values of x, the total resistance will always be less than 6 Ω. This is consistent with the statement that the total resistance is always less than the resistance in any individual branch of the circuit.

95. a.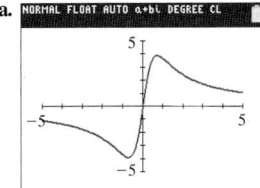

96. a.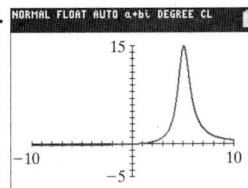

99. a. $F(v) = 560\left(\dfrac{772.4}{772.4 - v}\right)$ **b.**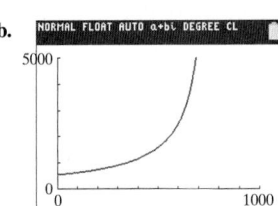

100. a. $F(v) = 600\left(\dfrac{772.4}{772.4 - v}\right)$

103. a. $f(x) = 2 + \dfrac{1}{x+3}$

   b. 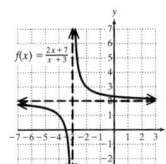 $f(x) = \frac{2x+7}{x+3}$

104. a. $f(x) = 5 + \dfrac{1}{x+2}$

   b. 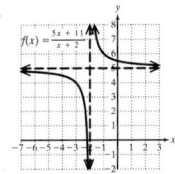 $f(x) = \frac{5x+11}{x+2}$

111. a. $(-\infty, 2) \cup (2, \infty)$ **b.** $f(x) = x + 3$ where $x \neq 2$ **c.** None
   d. $x = 2$ **e.** Graph iii

112. a. $(-\infty, -1) \cup (-1, \infty)$ **b.** $f(x) = -x + 3$ where $x \neq -1$
   c. None   **d.** $x = -1$ **e.** Graph ii

113. a. $(-\infty, -5) \cup (-5, -4) \cup (-4, \infty)$ **b.** $f(x) = \dfrac{2}{x + 4}$ where $x \neq -5$
   c. $x = -4$ **d.** $x = -5$ **e.** Graph iv

114. a. $(-\infty, -3) \cup (-3, 1) \cup (1, \infty)$ **b.** $f(x) = \dfrac{2}{x + 3}$ where $x \neq 1$
   c. $x = -3$ **d.** $x = 1$ **e.** Graph i

Problem Recognition Exercises, p. 322

8. $\left(\dfrac{1 + \sqrt{281}}{10}, 1\right) \approx (1.78, 1)$ and $\left(\dfrac{1 - \sqrt{281}}{10}, 1\right) \approx (-1.58, 1)$

16. $\left(\dfrac{-1 + \sqrt{85}}{7}, 1\right) \approx (1.17, 1)$ and $\left(\dfrac{-1 - \sqrt{85}}{7}, 1\right) \approx (-1.46, 1)$

Section 2.6 Practice Exercises, pp. 331–336

15. a. $\left\{\dfrac{3}{5}, 5\right\}$ **b.** $\left(\dfrac{3}{5}, 5\right)$ **c.** $\left[\dfrac{3}{5}, 5\right]$ **d.** $\left(-\infty, \dfrac{3}{5}\right) \cup (5, \infty)$
   e. $\left(-\infty, \dfrac{3}{5}\right] \cup [5, \infty)$ **16. a.** $\left\{-\dfrac{7}{3}, 2\right\}$ **b.** $\left(-\dfrac{7}{3}, 2\right)$
   c. $\left[-\dfrac{7}{3}, 2\right]$ **d.** $\left(-\infty, -\dfrac{7}{3}\right) \cup (2, \infty)$ **e.** $\left(-\infty, -\dfrac{7}{3}\right] \cup [2, \infty)$

18. b. $(-\infty, -9) \cup (-1, \infty)$ **c.** $(-\infty, -9] \cup [-1, \infty)$
19. d. $(-\infty, -6) \cup (-6, \infty)$ **20. d.** $(-\infty, 7) \cup (7, \infty)$
27. $\left(\dfrac{-3 - \sqrt{59}}{5}, \dfrac{-3 + \sqrt{59}}{5}\right)$ **28.** $\left(\dfrac{-2 - \sqrt{22}}{3}, \dfrac{-2 + \sqrt{22}}{3}\right)$
39. $\left(-\infty, -\dfrac{5}{2}\right) \cup (-2, 2)$ **40.** $(-\infty, -1) \cup \left(1, \dfrac{4}{3}\right)$
43. $\left(-\infty, \dfrac{3}{5}\right] \cup [5, \infty)$ **45.** $\left(-\infty, -\dfrac{1}{3}\right) \cup \left(0, \dfrac{5}{2}\right) \cup \left(\dfrac{5}{2}, 4\right)$
46. $\left(-\dfrac{1}{4}, 0\right)$ **51.** $\left(-\infty, \dfrac{3}{4}\right) \cup \left(\dfrac{3}{4}, \infty\right)$

103. a.

Sign of $(x - a)^2$:	+	+	+	+
Sign of $(b - x)$:	+	+	−	−
Sign of $(x - c)^3$:	−	−	−	+
Sign of $(x - a)^2(b - x)(x - c)^3$:	−	−	+	−

   a b c

   b. (b, c) **c.** $(-\infty, a) \cup (a, b) \cup (c, \infty)$
104. a.

Sign of $(a - x)$:	+	−	−	−
Sign of $(x - b)^2$:	+	+	+	+
Sign of $(c - x)^5$:	+	+	+	−
Sign of $\dfrac{(a - x)(x - b)^2}{(c - x)^5}$:	+	−	−	+

   a b c

   b. $(-\infty, a) \cup (c, \infty)$ **c.** $(a, b) \cup (b, c)$
105. The solution set to the inequality $f(x) < 0$ corresponds to the values of x for which the graph of $y = f(x)$ is below the x-axis.
106. The solution set to the inequality $f(x) \geq 0$ corresponds to the values of x for which the graph of $y = f(x)$ is on or above the x-axis.
107. Both the numerator and denominator of the rational expression are positive for all real numbers x. Therefore, the expression cannot be negative for any real number.
108. The rational expression is not defined for $x = 1$. Therefore, $x = 1$ is not included in the solution set. At $x = 3$, the expression equals zero, indicating that 3 is in the solution set. The solution set is $(1, 3]$.
121. a. $0.552x^3 + 4.13x^2 - 1.84x - 10.2 < 0$

b.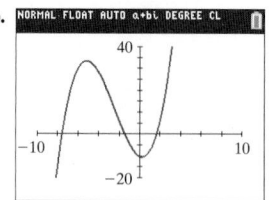

c. The real zeros are approximately -7.6, -1.5, and 1.6.
d. $(-\infty, -7.6) \cup (-1.5, 1.6)$
122. a. $0.24x^4 + 1.8x^3 + 3.3x^2 + 2.84x - 6.3 < 0$

b.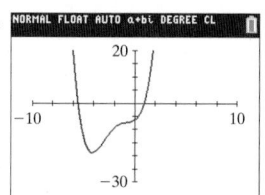

c. The real zeros are approximately -5.6 and 0.9.
d. $(-\infty, -5.6) \cup (0.9, \infty)$
123. a.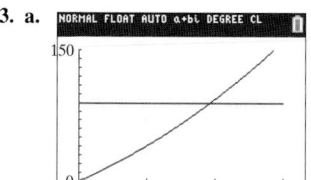

b. $(1.9, 90)$
c. The radius should be no more than 1.9 in. to keep the amount of aluminum to at most 90 in.^{2}.
124. a.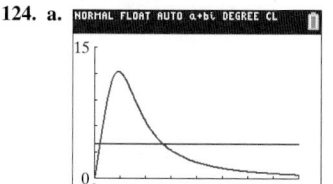

b. $(0.8, 4)$ and $(10.8, 4)$
c. It is safe to give a second dose approximately 10.8 hr after the first dose.

Chapter 2 Review Exercises, pp. 348–351

2. a. $f(x) = (x - 4)^2 - 1$
   b. Upward
   c. $(4, -1)$
   d. $(3, 0)$ and $(5, 0)$
   e. $(0, 15)$
   g. $x = 4$
   h. Minimum value: -1
   i. Domain: $(-\infty, \infty)$; Range: $[-1, \infty)$
   f.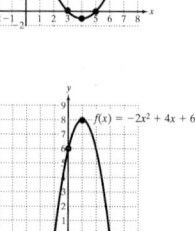

3. a. $f(x) = -2(x - 1)^2 + 8$
   b. Downward
   c. $(1, 8)$
   d. $(-1, 0)$ and $(3, 0)$
   e. $(0, 6)$ **g.** $x = 1$
   h. Maximum value: 8
   i. Domain: $(-\infty, \infty)$; Range: $(-\infty, 8]$
   f.

8. a. Up to the left, down to the right;
   As $x \to -\infty, f(x) \to \infty$, and as $x \to \infty, f(x) \to -\infty$.
   b. Zeros: $\dfrac{5}{2}, -\dfrac{5}{2}, 4$ (each with multiplicity 1)

c. $\left(\dfrac{5}{2}, 0\right), \left(-\dfrac{5}{2}, 0\right), (4, 0)$ **f.**

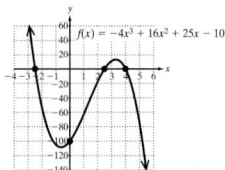

d. $(0, -100)$

e. Neither even nor odd

9. a. Up to the left, up to the right;
   As $x \to -\infty, f(x) \to \infty$, and as $x \to \infty, f(x) \to \infty$.

b. Zeros: 3, −3, 1, −1 (each with   **f.**
   multiplicity 1)

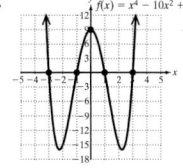

c. $(3, 0), (-3, 0), (1, 0), (-1, 0)$

d. $(0, 9)$ **e.** Even function

10. a. Up to the left, up to the right;
   As $x \to -\infty, f(x) \to \infty$, and as $x \to \infty, f(x) \to \infty$.

b. Zeros: 2, −3 (each with multiplicity 1)   **f.**
   and −1 (multiplicity 2)   **c.** $(2, 0)$,
   $(-3, 0), (-1, 0)$

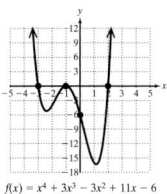

d. $(0, -6)$ **e.** Neither even nor odd

11. a. Down to the left, up to the right;
   As $x \to -\infty, f(x) \to -\infty$, and as $x \to \infty, f(x) \to \infty$.

b. Zeros: 0 (with multiplicity 3) and   **f.**
   $4 \pm \sqrt{3}$ (each with multiplicity 1)

c. $(0, 0), \left(4 + \sqrt{3}, 0\right), \left(4 - \sqrt{3}, 0\right)$

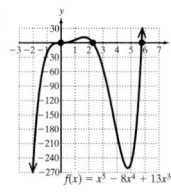

d. $(0, 0)$

e. Neither even nor odd

15. False. There are infinitely many such polynomials. For example, any polynomial of the form $f(x) = a(x - 2)(x - 3)(x - 4)$ has the required zeros.

17. a. $-2x^2 + 3x - 9 + \dfrac{22x - 28}{x^2 + x - 3}$

b. Dividend: $-2x^4 + x^3 + 4x - 1$; Divisor: $x^2 + x - 3$;
   Quotient: $-2x^2 + 3x - 9$; Remainder: $22x - 28$

18. a. $x^3 - 5x + 4$ **b.** Dividend: $3x^4 - 2x^3 - 15x^2 + 22x - 8$;
   Divisor: $3x - 2$; Quotient: $x^3 - 5x + 4$; Remainder: 0

48.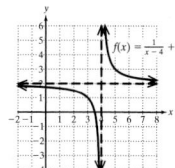

49.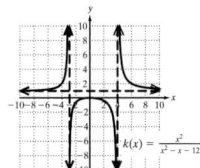

50.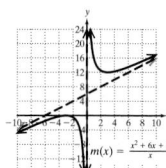

51.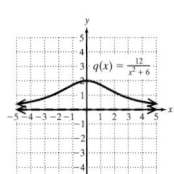

Chapter 2 Test, pp. 351–352

1. f.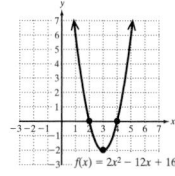

2. g.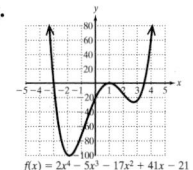

9. e. $\pm 1, \pm\dfrac{1}{3}, \pm\dfrac{2}{3}, \pm\dfrac{4}{3}, -2, -3$;   **h.**
   From part (c), the value 2 itself
   is not a zero of $f(x)$. Likewise,
   from part (d), the value −4
   itself is not a zero. Therefore,
   2 and −4 are also eliminated
   from the list of possible
   rational zeros.

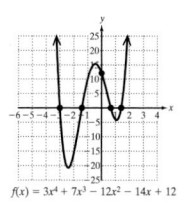

12. Vertical asymptote: $x = 7$; Slant asymptote: $y = 2x + 11$

13. Vertical asymptotes: $x = \dfrac{1}{2}, x = -\dfrac{1}{2}$; Horizontal asymptote: $y = 0$

15.   **16.**   **17.**

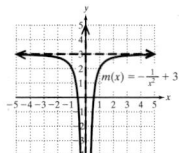 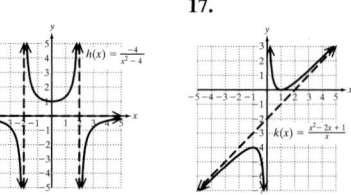

31. a. $y(20) = 140.3$ means that with 20,000 plants per acre, the yield
   will be 140.3 bushels per acre; $y(30) = 172$ means that with
   30,000 plants per acre, the yield will be 172 bushels per acre;
   $y(60) = 143.5$ means that with 60,000 plants per acre, the yield
   will be 143.5 bushels per acre.

Chapter 2 Cumulative Review Exercises, p. 353

3. a. Down to the left and up to the right; As $x \to -\infty, f(x) \to -\infty$,
   and as $x \to \infty, f(x) \to \infty$.

e.   **8.**

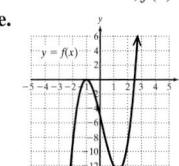 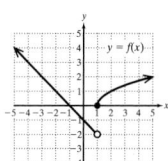

CHAPTER 3

Section 3.1 Practice Exercises, pp. 363–368

23. Yes; If $f(a) = f(b)$, then $4a - 7 = 4b - 7$, which implies that $a = b$.

24. Yes; If $h(a) = h(b)$, then $-3a + 2 = -3b + 2$, which implies that $a = b$.

25. Yes; If $g(a) = g(b)$, then $a^3 + 8 = b^3 + 8$, which implies that $a = b$.

26. Yes; If $k(a) = k(b)$, then $a^3 - 27 = b^3 - 27$, which implies that $a = b$.

39. a. If $f(a) = f(b)$, then $2a - 3 = 2b - 3$, which implies that $a = b$.
   The function is one-to-one.

b. $f^{-1}(x) = \dfrac{x + 3}{2}$

c.

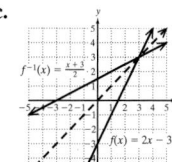

40. a. If $f(a) = f(b)$, then $4a + 4 = 4b + 4$, which implies that $a = b$.
   The function is one-to-one.

b. $f^{-1}(x) = \dfrac{x - 4}{4}$

c.

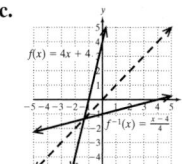

53. a.

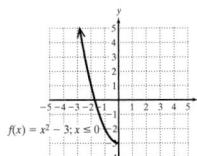

f.

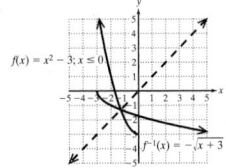

54. a.

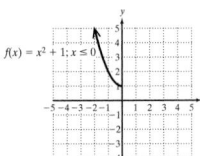

f.

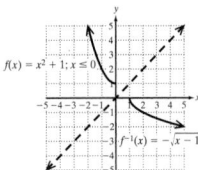

55. a.

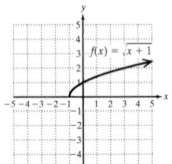

g.

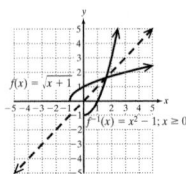

56. a.

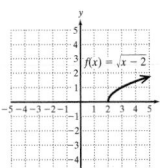

g.

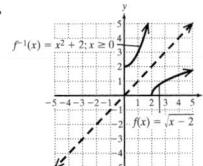

71.

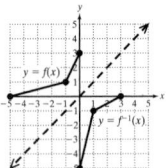

72.

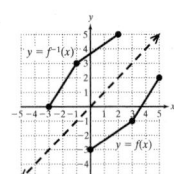

73.

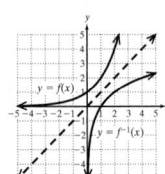

74.

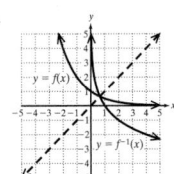

81. b. $w^{-1}(x) = \dfrac{x - 1220}{-1.17}$; The inverse gives the barometric pressure

$w^{-1}(x)$ for a given wind speed x.

82. b. $F^{-1}(x) = \dfrac{5}{9}(x - 32)$; The inverse gives the temperature in Celsius

$F^{-1}(x)$ for a given temperature in Fahrenheit x.

83. b. $T^{-1}(x) = \dfrac{x}{6.33}$ **c.** $T^{-1}(x)$ represents the mass of a mammal

based on the amount of air inhaled per breath, x.

d. $T^{-1}(170) = 27$ means that a mammal that inhales 170 mL of air per breath during normal respiration is approximately 27 kg (this is approximately 60 lb—the size of a Labrador retriever).

Section 3.2 Practice Exercises, pp. 376–381

15. Domain: $(-\infty, \infty)$;
Range: $(0, \infty)$

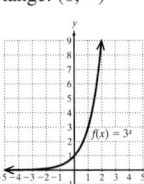

16. Domain: $(-\infty, \infty)$;
Range: $(0, \infty)$

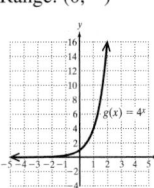

17. Domain: $(-\infty, \infty)$;
Range: $(0, \infty)$

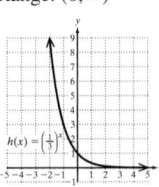

18. Domain: $(-\infty, \infty)$;
Range: $(0, \infty)$

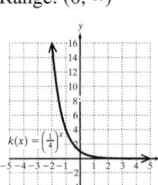

19. Domain: $(-\infty, \infty)$;
Range: $(0, \infty)$

20. Domain: $(-\infty, \infty)$;
Range: $(0, \infty)$

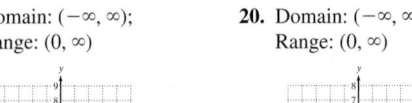

21. Domain: $(-\infty, \infty)$;
Range: $(0, \infty)$

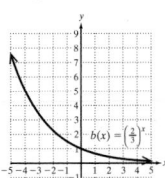

22. Domain: $(-\infty, \infty)$;
Range: $(0, \infty)$

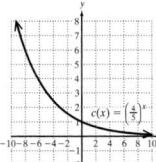

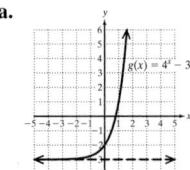

23. a.

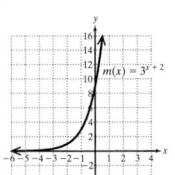

24. a.

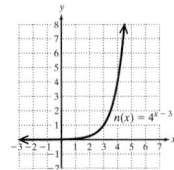

b. Domain: $(-\infty, \infty)$;
Range: $(2, \infty)$
c. $y = 2$

b. Domain: $(-\infty, \infty)$;
Range: $(-3, \infty)$
c. $y = -3$

25. a.

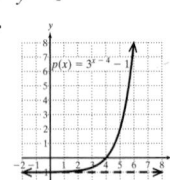

26. a.

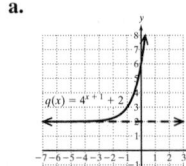

b. Domain: $(-\infty, \infty)$;
Range: $(0, \infty)$
c. $y = 0$

b. Domain: $(-\infty, \infty)$;
Range: $(0, \infty)$
c. $y = 0$

27. a.

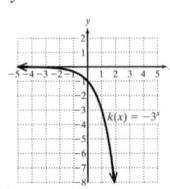

28. a.

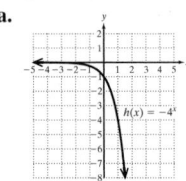

b. Domain: $(-\infty, \infty)$;
Range: $(-1, \infty)$
c. $y = -1$

b. Domain: $(-\infty, \infty)$;
Range: $(2, \infty)$
c. $y = 2$

29. a.

30. a.

b. Domain: $(-\infty, \infty)$;
Range: $(-\infty, 0)$
c. $y = 0$

b. Domain: $(-\infty, \infty)$;
Range: $(-\infty, 0)$
c. $y = 0$

31. a.

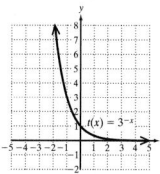

b. Domain: $(-\infty, \infty)$;
Range: $(0, \infty)$
c. $y = 0$

33. a.

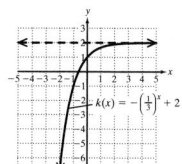

b. Domain: $(-\infty, \infty)$;
Range: $(-3, \infty)$
c. $y = -3$

35. a.

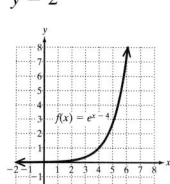

b. Domain: $(-\infty, \infty)$;
Range: $(-\infty, 2)$
c. $y = 2$

39. a.

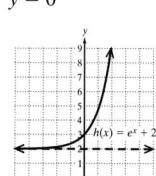

b. Domain: $(-\infty, \infty)$;
Range: $(0, \infty)$
c. $y = 0$

41. a.

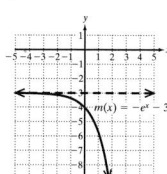

b. Domain: $(-\infty, \infty)$;
Range: $(2, \infty)$
c. $y = 2$

43. a.

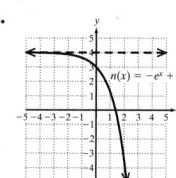

b. Domain: $(-\infty, \infty)$;
Range: $(-\infty, -3)$
c. $y = -3$

32. a.

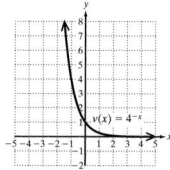

b. Domain: $(-\infty, \infty)$;
Range: $(0, \infty)$
c. $y = 0$

34. a.

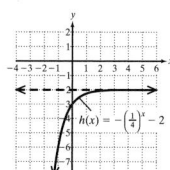

b. Domain: $(-\infty, \infty)$;
Range: $(1, \infty)$
c. $y = 1$

36. a.

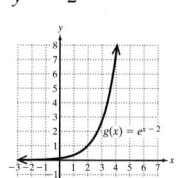

b. Domain: $(-\infty, \infty)$;
Range: $(-\infty, -2)$
c. $y = -2$

40. a.

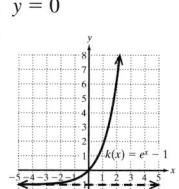

b. Domain: $(-\infty, \infty)$;
Range: $(0, \infty)$
c. $y = 0$

42. a.

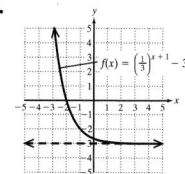

b. Domain: $(-\infty, \infty)$;
Range: $(-1, \infty)$
c. $y = -1$

44. a.

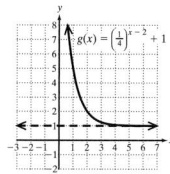

b. Domain: $(-\infty, \infty)$;
Range: $(-\infty, 4)$
c. $y = 4$

53. a. $A(28.9) = 5$ means that after 28.9 yr, the amount of ^{90}Sr remaining is 5 μg. After one half-life, the amount of substance has been halved.   **b.** $A(57.8) = 2.5$ means that after 57.8 yr, the amount of ^{90}Sr remaining is 2.5 μg. After two half-lives, the amount of substance has been halved, twice.   **c.** $A(100) = 0.909$ means that after 100 yr, the amount of ^{90}Sr remaining is approximately 0.909 μg.

54. a. $A(138.4) = 0.05$ means that after 138.4 yr, the amount of ^{210}Po remaining is 0.05 mg. After one half-life, the amount of substance has been halved.   **b.** $A(276.8) = 0.025$ means that after 276.8 yr, the amount of ^{210}Po remaining is 0.025 mg. After two half-lives, the amount of substance has been halved, twice.   **c.** $A(500) = 0.008$ means that after 500 yr, the amount of ^{210}Po remaining is approximately 0.008 mg.

55. b. $P(0) = 310$ means that in the year 2010, the U.S. population was approximately 310 million. This is the initial population in 2010.
c. $P(10) = 341$ means that in the year 2020, the U.S. population will be approximately 341 million if this trend continues.
e. $P(200) = 2137$; In the year 2210 the U.S population will be approximately 2.137 billion. The model cannot continue indefinitely because the population will become too large to be sustained from the available resources.

56. b. $P(0) = 34$ means that in the year 2010, the Canadian population was approximately 34 million. This is the initial population in 2010.
c. $P(5) = 35$ means that in the year 2015, the population of Canada was approximately 35 million.

58. a. $A(15) = 161.12$ means that in the year 2025, approximately $161.12 will be needed to buy something that cost $100 in 2010.

65. a. and **d.**   **66. a.** and **d.**

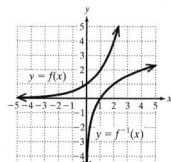   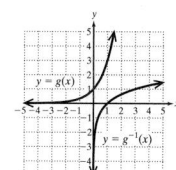

69. The range of an exponential function is the set of positive real numbers; that is, 2^x is nonnegative for all values of x in the domain.

77. $\left(\dfrac{e^x + e^{-x}}{2}\right)^2 - \left(\dfrac{e^x - e^{-x}}{2}\right)^2$

$= \dfrac{1}{4}[(e^{2x} + 2 + e^{-2x}) - (e^{2x} - 2 + e^{-2x})]$

$= \dfrac{1}{4}(4) = 1$

78. $2\left(\dfrac{e^x - e^{-x}}{2}\right)\left(\dfrac{e^x + e^{-x}}{2}\right)$

$= 2 \cdot \dfrac{1}{4}[(e^x - e^{-x})(e^x + e^{-x})]$

$= \dfrac{1}{2}(e^{2x} - e^{-2x}) = \dfrac{e^{2x} - e^{-2x}}{2}$

81. The graphs of Y_2 and Y_3 are close approximations of $Y_1 = e^x$ near $x = 0$.

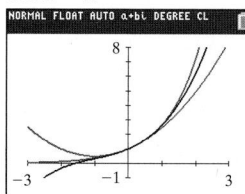

Section 3.3 Practice Exercises, pp. 391–395

65.

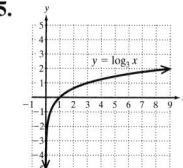

66.

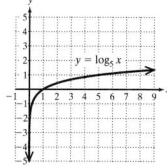

67.

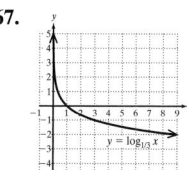

68.

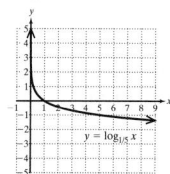

69.
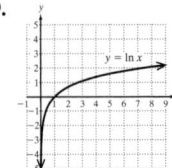

70.
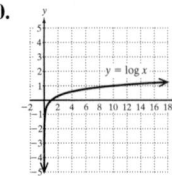

71. a.
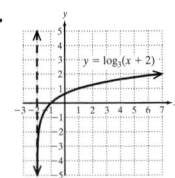

b. Domain: $(-2, \infty)$;
Range: $(-\infty, \infty)$
c. $x = -2$

72. a.
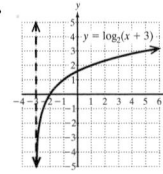

b. Domain: $(-3, \infty)$;
Range: $(-\infty, \infty)$
c. $x = -3$

73. a.
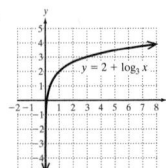

b. Domain: $(0, \infty)$;
Range: $(-\infty, \infty)$
c. $x = 0$

74. a.
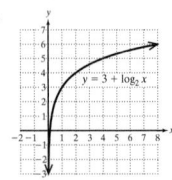

b. Domain: $(0, \infty)$;
Range: $(-\infty, \infty)$
c. $x = 0$

75. a.
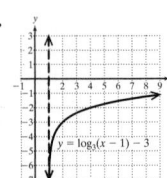

b. Domain: $(1, \infty)$;
Range: $(-\infty, \infty)$
c. $x = 1$

76. a.
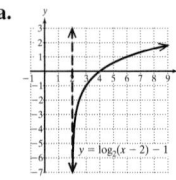

b. Domain: $(2, \infty)$;
Range: $(-\infty, \infty)$
c. $x = 2$

77. a.
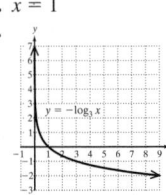

b. Domain: $(0, \infty)$;
Range: $(-\infty, \infty)$
c. $x = 0$

78. a.
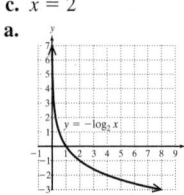

b. Domain: $(0, \infty)$;
Range: $(-\infty, \infty)$
c. $x = 0$

123. a. The graphs match closely on the interval $(0, 2)$.

124. It appears that the functions are inverses.

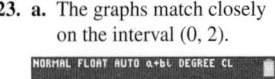

Problem Recognition Exercises, p. 396

1. a. $(-\infty, \infty)$ **b.** $\{3\}$ **c.** No x-intercept   **d.** $(0, 3)$
e. No asymptotes   **f.** Never increasing   **g.** Never decreasing
h. Graph E   **2. a.** $(-\infty, \infty)$ **b.** $(-\infty, \infty)$ **c.** $\left(\frac{3}{2}, 0\right)$
d. $(0, -3)$ **e.** No asymptotes   **f.** $(-\infty, \infty)$
g. Never decreasing   **h.** Graph G
3. a. $(-\infty, \infty)$ **b.** $[-4, \infty)$ **c.** $(1, 0)$ and $(5, 0)$ **d.** $(0, 5)$
e. No asymptotes   **f.** $(3, \infty)$ **g.** $(-\infty, 3)$ **h.** Graph N

4. a. $(-\infty, \infty)$ **b.** $(-\infty, \infty)$ **c.** $(2, 0)$ **d.** $\left(0, -\sqrt[3]{2}\right)$
e. No asymptotes   **f.** $(-\infty, \infty)$ **g.** Never decreasing
h. Graph B
5. a. $(-\infty, 1) \cup (1, \infty)$ **b.** $(-\infty, 0) \cup (0, \infty)$ **c.** None
d. $(0, -2)$ **e.** Vertical asymptote: $x = 1$; Horizontal asymptote:
$y = 0$ **f.** Never increasing   **g.** $(-\infty, 1) \cup (1, \infty)$
h. Graph L
6. a. $(-\infty, -2) \cup (-2, \infty)$ **b.** $(-\infty, 3) \cup (3, \infty)$ **c.** $(0, 0)$
d. $(0, 0)$ **e.** Vertical asymptote: $x = -2$; Horizontal asymptote:
$y = 3$ **f.** $(-\infty, -2) \cup (-2, \infty)$ **g.** Never decreasing
h. Graph A
7. a. $(-\infty, \infty)$ **b.** $(0, \infty)$ **c.** No x-intercept   **d.** $(0, 1)$
e. Horizontal asymptote: $y = 0$ **f.** $(-\infty, \infty)$
g. Never decreasing   **h.** Graph M
8. a. $(-\infty, \infty)$ **b.** $(-\infty, 0]$ **c.** $(-3, 0)$ **d.** $(0, -9)$
e. No asymptotes   **f.** $(-\infty, -3)$ **g.** $(-3, \infty)$ **h.** Graph C
9. a. $(-\infty, \infty)$ **b.** $[-1, \infty)$ **c.** $(3, 0)$ and $(5, 0)$ **d.** $(0, 3)$
e. No asymptotes   **f.** $(4, \infty)$ **g.** $(-\infty, 4)$ **h.** Graph I
10. a. $(-\infty, \infty)$ **b.** $(-\infty, 3]$ **c.** $(-3, 0)$ and $(3, 0)$ **d.** $(0, 3)$
e. No asymptotes   **f.** $(-\infty, 0)$ **g.** $(0, \infty)$ **h.** Graph D
11. a. $(-\infty, 3]$ **b.** $[0, \infty)$ **c.** $(3, 0)$ **d.** $\left(0, \sqrt{3}\right)$
e. No asymptotes   **f.** Never increasing   **g.** $(-\infty, 3)$ **h.** Graph F
12. a. $[3, \infty)$ **b.** $[0, \infty)$ **c.** $(3, 0)$ **d.** No y-intercept
e. No asymptotes   **f.** $(3, \infty)$ **g.** Never decreasing   **h.** Graph K
13. a. $(-\infty, \infty)$ **b.** $(2, \infty)$ **c.** No x-intercept   **d.** $(0, 3)$
e. Horizontal asymptote: $y = 2$ **f.** $(-\infty, \infty)$ **g.** Never decreasing
h. Graph H
14. a. $(-2, \infty)$ **b.** $(-\infty, \infty)$ **c.** $(-1, 0)$ **d.** $(0, \ln 2)$
e. Vertical asymptote: $x = -2$ **f.** $(-2, \infty)$ **g.** Never decreasing
h. Graph J

Section 3.4 Practice Exercises, pp. 403–406

39. $\log 2 + \log x + 8 \log(x^2 + 3) - \frac{1}{2}\log(4 - 3x)$

40. $\log 5 + \log y + 7 \log(4x + 1) - \frac{1}{3}\log(2 - 7x)$

43. $2 + 2\log_2 a + \frac{1}{2}\log_2 (3 - b) - \log_2 c - 2 \log_2 (b + 4)$

44. $3 + 3\log_3 x + \frac{1}{2}\log_3 (y^2 - 1) - \log_3 y - 2 \log_3 (x - 1)$

55. $\log_8\left(\dfrac{m^4}{n^3 p^2}\right)$ **56.** $\log_3\left(\dfrac{x^8}{z^2 y^7}\right)$ **57.** $\ln\left(\dfrac{x}{x^2 - 9}\right)^3$

91. False; $\log_5\left(\dfrac{1}{125}\right) \neq \dfrac{1}{\log_5 125}$ (The left side is -3 and the right
side is $\frac{1}{3}$.)

92. False; $\log_6\left(\dfrac{1}{36}\right) \neq \dfrac{1}{\log_6 36}$ (The left side is -2 and the right side is $\frac{1}{2}$.)

95. False; $\log(10 \cdot 10) \neq (\log 10)(\log 10)$ (The left side is 2 and the right
side is 1.)

96. False; $\log\left(\dfrac{100}{10}\right) \neq \dfrac{\log 100}{\log 10}$ (The left side is 1 and the right side is 2.)

101. a. $\dfrac{\ln(x + h) - \ln x}{h}$ **b.** $\dfrac{1}{h}[\ln(x + h) - \ln x] = \dfrac{1}{h}\ln\left(\dfrac{x + h}{x}\right)$

$= \ln\left(\dfrac{x + h}{x}\right)^{1/h}$

102. $-\ln(x - \sqrt{x^2 - 1})$

$= \ln(x - \sqrt{x^2 - 1})^{-1}$

$= \ln\left(\dfrac{1}{x - \sqrt{x^2 - 1}}\right)$

$= \ln\left(\dfrac{1}{x - \sqrt{x^2 - 1}} \cdot \dfrac{x + \sqrt{x^2 - 1}}{x + \sqrt{x^2 - 1}}\right)$

$$= \ln\left[\frac{x + \sqrt{x^2 - 1}}{x^2 - (x^2 - 1)}\right]$$

$$= \ln\left(x + \sqrt{x^2 - 1}\right)$$

103. $\log\left(\dfrac{-b + \sqrt{b^2 - 4ac}}{2a}\right) + \log\left(\dfrac{-b - \sqrt{b^2 - 4ac}}{2a}\right)$

$$= \log\left(\frac{-b + \sqrt{b^2 - 4ac}}{2a} \cdot \frac{-b - \sqrt{b^2 - 4ac}}{2a}\right)$$

$$= \log\left[\frac{b^2 - (b^2 - 4ac)}{4a^2}\right]$$

$$= \log\left(\frac{4ac}{4a^2}\right)$$

$$= \log\left(\frac{c}{a}\right) = \log c - \log a$$

104. $\ln\left(\dfrac{c + \sqrt{c^2 - x^2}}{c - \sqrt{c^2 - x^2}}\right)$

$$= \ln\left(\frac{c + \sqrt{c^2 - x^2}}{c - \sqrt{c^2 - x^2}} \cdot \frac{c + \sqrt{c^2 - x^2}}{c + \sqrt{c^2 - x^2}}\right)$$

$$= \ln\left[\frac{(c + \sqrt{c^2 - x^2})^2}{c^2 - (c^2 - x^2)}\right]$$

$$= \ln\left[\frac{(c + \sqrt{c^2 - x^2})^2}{x^2}\right]$$

$$= \ln\left(c + \sqrt{c^2 - x^2}\right)^2 - \ln x^2$$

$$= 2\ln\left(c + \sqrt{c^2 - x^2}\right) - 2\ln x$$

107. Let $M = \log_b x$ and $N = \log_b y$, which implies that $b^M = x$ and $b^N = y$.
Then $\dfrac{x}{y} = \dfrac{b^M}{b^N} = b^{M-N}$. Writing the expression $\dfrac{x}{y} = b^{M-N}$ in
logarithmic form, we have $\log_b\left(\dfrac{x}{y}\right) = M - N$, or equivalently,
$\log_b\left(\dfrac{x}{y}\right) = \log_b x - \log_b y$ as desired.

109. 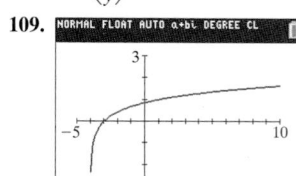 **110.**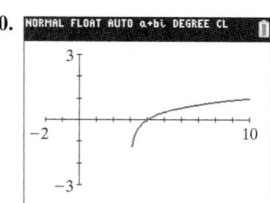

111. 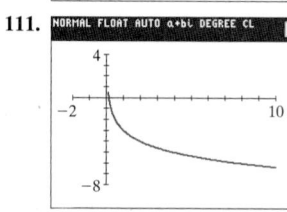 **112.**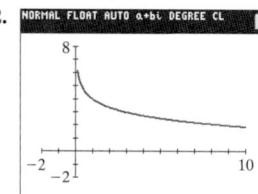

113. a. The graphs are the same.

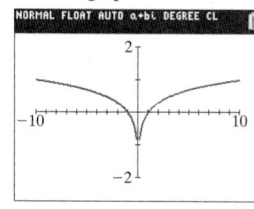

b. $\dfrac{1}{2}\log x^2 = \log(x^2)^{1/2} = \log\sqrt{x^2} = \log|x|$

114. The graphs differ by a vertical shift. This can be shown algebraically
by using the product property of logarithms. For example, $\ln(2x) = \ln 2 + \ln x$. Therefore, the graph of $y = \ln x$ is shifted upward by $\ln 2$ units.

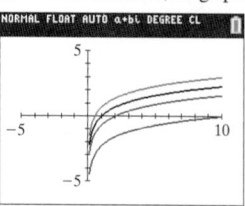

Section 3.5 Practice Exercises, pp. 416–420

21. $\left\{\dfrac{\log 128{,}100 - 3}{4}\right\}; x \approx 0.5269$ **22.** $\left\{\dfrac{\log 79{,}800 - 5}{8}\right\};$
$x \approx -0.0122$ **23.** $\left\{\dfrac{\ln 3}{0.2}\right\}$ or $\{5\ln 3\}; t \approx 5.4931$

24. $\left\{\dfrac{\ln 4}{0.5}\right\}$ or $\{2\ln 4\}; t \approx 2.7726$ **26.** $\left\{\dfrac{3 + \ln 4}{4}\right\}; m \approx 1.0966$

27. $\left\{\dfrac{5\ln 3}{2\ln 5 - 6\ln 3}\right\}; x \approx -1.6286$

28. $\left\{\dfrac{\ln 7}{4\ln 7 - 5\ln 3}\right\}; x \approx 0.8495$

29. $\left\{\dfrac{\ln 2 - 4\ln 7}{3\ln 7 + 6\ln 2}\right\}; x \approx -0.7093$

30. $\left\{\dfrac{\ln 11 - 3\ln 9}{2\ln 9 + 8\ln 11}\right\}; x \approx -0.1779$

47. $\left\{\dfrac{4 - e^3}{3}\right\}; t \approx -5.3618$ **48.** $\left\{\dfrac{6 - e^5}{5}\right\}; t \approx -28.4826$

107. $\left\{e^4, \dfrac{1}{e^4}\right\}; x \approx 54.5982, x \approx 0.0183$

108. $\left\{e^2, \dfrac{1}{e^2}\right\}; x \approx 7.3891, x \approx 0.1353$

113. Take a logarithm of any base b on each side of the equation. Then
apply the power property of logarithms to write the product of x and
the $\log_b 4$. Finally divide both sides by $\log_b 4$.

114. Use the product property of logarithms to write a single logarithm on the
left side of the equation. Then write the logarithmic equation in its
equivalent exponential form. The resulting equation is linear which is
solved by applying the addition and multiplication properties of equality.

118. $\left\{\dfrac{1}{e^2}, \dfrac{1}{e}\right\}; x \approx 0.1353, x \approx 0.3679$

Section 3.6 Practice Exercises, pp. 430–436

5. $k = -\dfrac{\ln\left(\dfrac{Q}{Q_0}\right)}{t}$ or $\dfrac{\ln Q_0 - \ln Q}{t}$ **6.** $t = -\dfrac{\ln\left(\dfrac{N}{N_0}\right)}{0.025}$ or $\dfrac{\ln N_0 - \ln N}{0.025}$

11. $t = \dfrac{\ln\left(\dfrac{A}{P}\right)}{\ln(1 + r)}$ or $\dfrac{\ln A - \ln P}{\ln(1 + r)}$ **12.** $r = \dfrac{\ln\left(\dfrac{A}{P}\right)}{t}$ or $\dfrac{\ln A - \ln P}{t}$

39. a. exponential   **40. a.** exponential

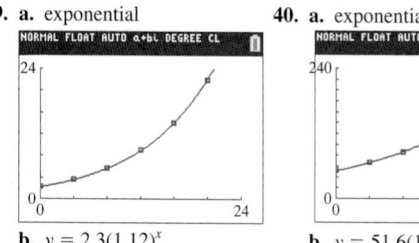

b. $y = 2.3(1.12)^x$ **b.** $y = 51.6(1.3)^x$

41. a. linear

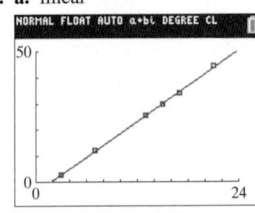

b. $y = 2.28x - 4.08$

43. a. logarithmic

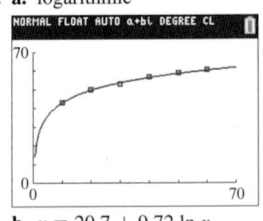

b. $y = 20.7 + 9.72 \ln x$

45. a. logistic

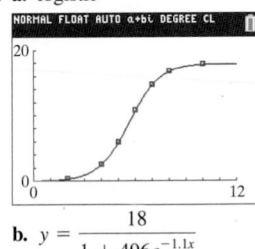

b. $y = \dfrac{18}{1 + 496e^{-1.1x}}$

42. a. linear

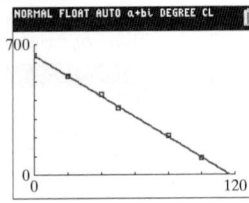

b. $y = -5.47x + 641$

44. a. logarithmic

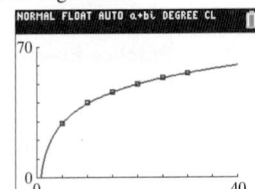

b. $y = 5.08 + 15 \ln x$

46. a. logistic

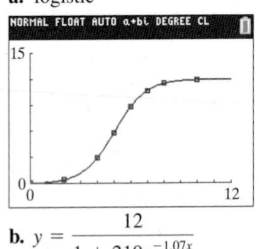

b. $y = \dfrac{12}{1 + 219e^{-1.07x}}$

53. A visual representation of the data can be helpful in determining the type of equation or function that best models the data.

54. The average rate of change is constant for a linear function for all intervals $[a, b]$ in the domain of the function. For an exponential function defined by $f(x) = b^x$ $(b > 1)$, the average rate of change increases for increasing values of x.

55. An exponential growth model has unbounded growth, whereas a logistic growth model imposes a limiting value on the dependent variable. That is, a logistic growth model has an upper bound restricting the amount of growth.

56. The corresponding expression with base e is given by $e^{(\ln b)t}$.

57. a. $t = -\dfrac{\ln\left(1 - \dfrac{Ar}{12P}\right)}{12 \ln\left(1 + \dfrac{r}{12}\right)}$

b. This represents the amount of time (in yr) required to completely pay off a loan of A dollars at interest rate r, by paying P dollars per month.

58. a. $t = \dfrac{1}{k} \cdot \ln\left[\dfrac{P(N - P_0)}{P_0(N - P)}\right]$

b. This represents the amount of time required for the population to reach a value P, given an initial population P_0 and a limiting value N.

Chapter 3 Review Exercises, pp. 439–441

3. Yes; If $f(a) = f(b)$, then $a^3 - 1 = b^3 - 1$, which implies that $a = b$.

4. No; For example, $f(-2) = 3$ and $f(2) = 3$. The points $(2, 3)$ and $(-2, 3)$ have the same y value but different x values. Therefore, $f(a) = f(b) = 3$ does not imply that $a = b$.

9. a.

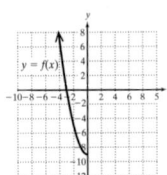

f.

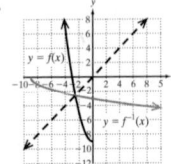

10. a.

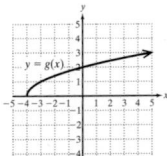

f.

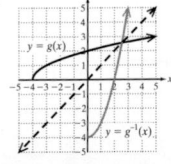

13. a.

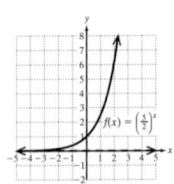

b. $(-\infty, \infty)$ **c.** $(0, \infty)$
d. $y = 0$

14. a.

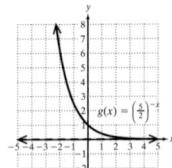

b. $(-\infty, \infty)$ **c.** $(0, \infty)$
d. $y = 0$

15. a.

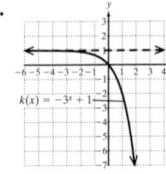

b. $(-\infty, \infty)$ **c.** $(-\infty, 1)$
d. $y = 1$

16. a.

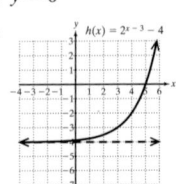

b. $(-\infty, \infty)$ **c.** $(-4, \infty)$
d. $y = -4$

20. b. $R(4.2) = 64$ means that after 4.2 days (1 biological half-life), the radioactivity level is 64 mCi (half the original amount).

38. a.

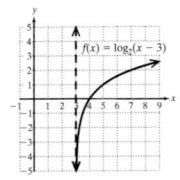

b. $(3, \infty)$ **c.** $(-\infty, \infty)$
d. $x = 3$

39. a.

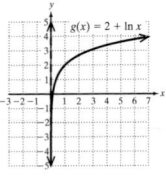

b. $(0, \infty)$ **c.** $(-\infty, \infty)$
d. $x = 0$

49. $2 - \dfrac{1}{2}\log(c^2 + 10)$ **50.** $-3 + 2\log_2 a + \log_2 b$

51. $\dfrac{1}{3}\ln a + \dfrac{2}{3}\ln b - \ln c - 5 \ln d$

52. $2 \log x + 5 \log(2x + 1) - \dfrac{1}{2}\log(1 - x)$

63. $\left\{\dfrac{\ln 51}{\ln 7}\right\}; x \approx 2.0206$ **64.** $\left\{\dfrac{\ln 537}{\ln 11}\right\}; w \approx 2.6215$

65. $\left\{\dfrac{\ln 3}{3 \ln 4 - 2 \ln 3}\right\}; x \approx 0.5600$

66. $\left\{\dfrac{3 \ln 2 - 5 \ln 7}{2 \ln 7 - \ln 2}\right\}; c \approx -2.3917$

67. $\left\{\dfrac{\ln\left(\frac{2.989}{400}\right)}{-2}\right\}$ or $\left\{\dfrac{\ln 400 - \ln 2.989}{2}\right\}; t \approx 2.4483$

68. $\left\{\dfrac{\log 29}{1.2}\right\}; t \approx 1.2187$

91. b.

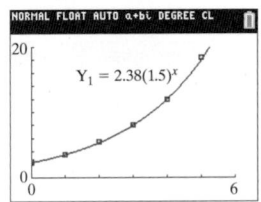

Chapter 3 Test, pp. 441–442

1. b. $(f \circ f^{-1})(x) = 4\left(\sqrt[3]{\dfrac{x+1}{4}}\right)^3 - 1 = 4\left(\dfrac{x+1}{4}\right) - 1 = x + 1 - 1 = x$

$(f^{-1} \circ f)(x) = \sqrt[3]{\dfrac{4x^3 - 1 + 1}{4}} = \sqrt[3]{\dfrac{4x^3}{4}} = \sqrt[3]{x^3} = x$

2. b.

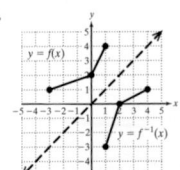

4. a. Domain: $(-\infty, 0]$; Range: $(-\infty, 1]$
b. $f^{-1}(x) = -\sqrt{1 - x}$
c. Domain: $(-\infty, 1]$; Range: $(-\infty, 0]$

5. a. Domain: $(0, \infty)$; Range: $(-\infty, \infty)$
  b. $f^{-1}(x) = 10^x$ **c.** Domain: $(-\infty, \infty)$; Range: $(0, \infty)$
6. a. Domain: $(-\infty, \infty)$; Range: $(1, \infty)$ **b.** $f^{-1}(x) = \log_3(x - 1)$
  c. Domain: $(1, \infty)$; Range: $(-\infty, \infty)$
7. a. Domain: $[-5, \infty)$; Range $[0, \infty)$ **b.** $f^{-1}(x) = x^2 - 5; x \geq 0$
  c. Domain: $[0, \infty)$; Range: $[-5, \infty)$

8. a.
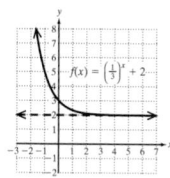
  b. $(-\infty, \infty)$ **c.** $(2, \infty)$
  d. $y = 2$

9. a.
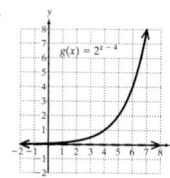
  b. $(-\infty, \infty)$ **c.** $(0, \infty)$
  d. $y = 0$

10. a.
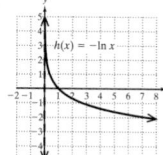
  b. $(0, \infty)$ **c.** $(-\infty, \infty)$
  d. $x = 0$

11. a.
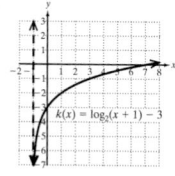
  b. $(-1, \infty)$ **c.** $(-\infty, \infty)$
  d. $x = -1$

28. $\left\{ \dfrac{\ln 53}{\ln 5} - 3 \right\}; x \approx -0.5331$ **29.** $\left\{ \dfrac{7 \ln 2 - 3 \ln 3}{2 \ln 3 - \ln 2} \right\}; c \approx 1.0346$

30. $\left\{ \dfrac{\ln 2}{4} \right\}; x \approx 0.1733$

42. b.
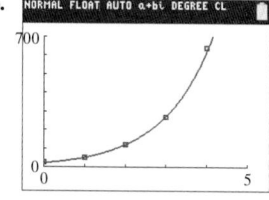

Chapter 3 Cumulative Review Exercises, p. 443

16.
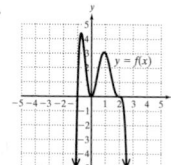

17. c.
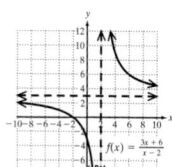

CHAPTER 4

Section 4.1 Practice Exercises, pp. 456–462

25.
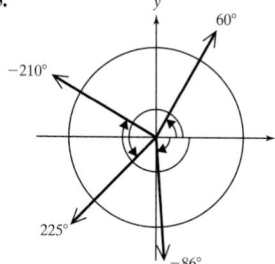

26.
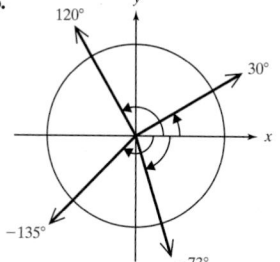

35.
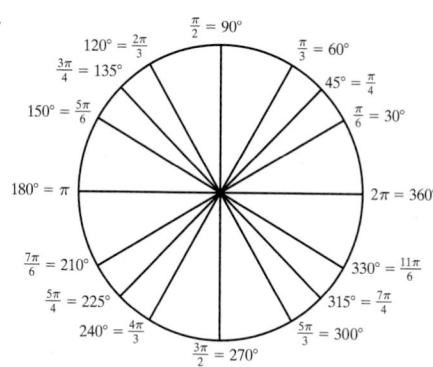

121. The four quadrants are divided by 0°, 90°, 180°, 270°, and 360° angles (likewise 0, $\frac{\pi}{2}$, π, $\frac{3\pi}{2}$, and 2π radians). Compare the measure of the given angle to these benchmarks. It may first be necessary to find an angle between 0° and 360° (or 0 and 2π radians) coterminal to the given angle.

122. Points A and B travel in circular paths around the center of the fan and have the same angular speed. The radius for the path of B is greater than the radius for the path of A. Therefore, point B travels a greater distance per unit time, and has the greater linear speed.

129.
```
NORMAL FLOAT AUTO REAL RADIAN CL
-216.479▶DMS
            -216°28'44.4"
42°13'5.9"
            42.21830556
```

130.
```
NORMAL FLOAT AUTO REAL RADIAN CL
-14.908▶DMS
            -14°54'28.8"
71°19'4.7"
            71.31797222
```

131. Calculator in radian mode.

```
NORMAL FLOAT AUTO REAL RADIAN CL
(147°26'9")ᴿ
            2.573240727
(-228.459)ᴿ
            -3.987361756
```

132. Calculator in radian mode.

```
NORMAL FLOAT AUTO REAL RADIAN CL
(36°4'47")ᴿ
            .629709946
(-25.716)ᴿ
            -.4488288704
```

133. Calculator in degree mode.

```
NORMAL FLOAT AUTO REAL DEGREE CL
(4π/9)ᴿᵒ
            80
(-5.718)ᴿᵒ
            -327.6172673
```

134. Calculator in degree mode.

```
NORMAL FLOAT AUTO REAL DEGREE CL
(11π/18)ᴿᵒ
            110
(-1.356)ᴿᵒ
            -77.69307702
```

Section 4.2 Practice Exercises, pp. 476–481

15. $\sin t = \dfrac{\sqrt{7}}{4}, \cos t = -\dfrac{3}{4}, \tan t = -\dfrac{\sqrt{7}}{3},$
$\csc t = \dfrac{4\sqrt{7}}{7}, \sec t = -\dfrac{4}{3}, \cot t = -\dfrac{3\sqrt{7}}{7}$

16. $\sin t = \dfrac{7}{25}, \cos t = -\dfrac{24}{25}, \tan t = -\dfrac{7}{24},$
$\csc t = \dfrac{25}{7}, \sec t = -\dfrac{25}{24}, \cot t = -\dfrac{24}{7}$

17. $\sin t = -\dfrac{15}{17}, \cos t = \dfrac{8}{17}, \tan t = -\dfrac{15}{8},$
$\csc t = -\dfrac{17}{15}, \sec t = \dfrac{17}{8}, \cot t = -\dfrac{8}{15}$

18. $\sin t = -\dfrac{3\sqrt{38}}{19}$, $\cos t = -\dfrac{\sqrt{19}}{19}$, $\tan t = 3\sqrt{2}$,

$\csc t = -\dfrac{\sqrt{38}}{6}$, $\sec t = -\sqrt{19}$, $\cot t = \dfrac{\sqrt{2}}{6}$

20. $P\left(-\dfrac{\sqrt{3}}{2}, \dfrac{1}{2}\right)$;

$\sin t = \dfrac{1}{2}$, $\cos t = -\dfrac{\sqrt{3}}{2}$, $\tan t = -\dfrac{\sqrt{3}}{3}$,

$\csc\theta = 2$, $\sec\theta = -\dfrac{2\sqrt{3}}{3}$, $\cot\theta = -\sqrt{3}$

21. $P\left(-\dfrac{1}{2}, -\dfrac{\sqrt{3}}{2}\right)$

$\sin t = -\dfrac{\sqrt{3}}{2}$, $\cos t = -\dfrac{1}{2}$, $\tan t = \sqrt{3}$,

$\csc t = -\dfrac{2\sqrt{3}}{3}$, $\sec t = -2$, $\cot t = \dfrac{\sqrt{3}}{3}$

22. $P\left(-\dfrac{\sqrt{2}}{2}, -\dfrac{\sqrt{2}}{2}\right)$

$\sin t = -\dfrac{\sqrt{2}}{2}$, $\cos t = -\dfrac{\sqrt{2}}{2}$, $\tan t = 1$,

$\csc t = -\sqrt{2}$, $\sec t = -\sqrt{2}$, $\cot t = 1$

23. $P\left(\dfrac{\sqrt{2}}{2}, \dfrac{\sqrt{2}}{2}\right)$

$\sin t = \dfrac{\sqrt{2}}{2}$, $\cos t = \dfrac{\sqrt{2}}{2}$, $\tan t = 1$,

$\csc t = \sqrt{2}$, $\sec t = \sqrt{2}$, $\cot t = 1$

43. $\tan t = \dfrac{\sqrt{5}}{2}$, $\csc t = \dfrac{3\sqrt{5}}{5}$, $\sec t = \dfrac{3}{2}$, $\cot t = \dfrac{2\sqrt{5}}{5}$

44. $\tan t = \dfrac{3\sqrt{7}}{7}$, $\csc t = \dfrac{4}{3}$, $\sec t = \dfrac{4\sqrt{7}}{7}$, $\cot t = \dfrac{\sqrt{7}}{3}$

45. $\tan t = \dfrac{\sqrt{39}}{5}$, $\sec t = -\dfrac{8}{5}$, $\csc t = -\dfrac{8\sqrt{39}}{39}$, $\cot t = \dfrac{5\sqrt{39}}{39}$

46. $\tan t = -\dfrac{28}{45}$, $\sec t = -\dfrac{53}{45}$, $\csc t = \dfrac{53}{28}$, $\cot t = -\dfrac{45}{28}$

47. Divide both sides of $\sin^2 t + \cos^2 t = 1$ by $\cos^2 t$.

$\dfrac{\sin^2 t}{\cos^2 t} + \dfrac{\cos^2 t}{\cos^2 t} = \dfrac{1}{\cos^2 t}$

$\tan^2 t + 1 = \sec^2 t$

48. Divide both sides of $\sin^2 t + \cos^2 t = 1$ by $\sin^2\theta$.

$\dfrac{\sin^2 t}{\sin^2 t} + \dfrac{\cos^2 t}{\sin^2 t} = \dfrac{1}{\sin^2 t}$

$1 + \cot^2 t = \csc^2 t$

79. a. $\sin t = \dfrac{\sqrt{3}}{2}$, $\cos t = -\dfrac{1}{2}$, $\tan t = -\sqrt{3}$

b. $\sin t = \dfrac{\sqrt{3}}{2}$, $\cos t = -\dfrac{1}{2}$, $\tan t = -\sqrt{3}$

80. a. $\sin t = \dfrac{1}{2}$, $\cos t = \dfrac{\sqrt{3}}{2}$, $\tan t = \dfrac{\sqrt{3}}{3}$

b. $\sin t = \dfrac{1}{2}$, $\cos t = \dfrac{\sqrt{3}}{2}$, $\tan t = \dfrac{\sqrt{3}}{3}$

105. As t increases from 0 to $\dfrac{\pi}{2}$, the value of $\cos t$ decreases from 1 to 0 and the value of $\sin t$ increases from 0 to 1.

106. As t increases from 0 to $\dfrac{\pi}{2}$, the value of $\cos t$ decreases from 1 to 0. The value of $\sec t$ increases from 1 when $t = 0$ and becomes arbitrarily large as t approaches $\dfrac{\pi}{2}$. The value of $\sec t$ is undefined at $\dfrac{\pi}{2}$.

107. $\cos t = x$ and $\sin t = y$ for (x, y) satisfying the equation $x^2 + y^2 = 1$. It follows that $|x| = \sqrt{1 - y^2}$ and $|y| = \sqrt{1 - x^2}$ which implies that $-1 \le x \le 1$ and $-1 \le y \le 1$.

108. Yes; each value of the independent variable corresponds to one and only one location on the unit circle, which is then used to evaluate the function.

119. Let the real number t correspond to the point $P(x, y)$ on the unit circle. Then $t + \pi$ corresponds to the point $P(-x, -y)$ on the unit circle. By definition, $\tan t = \dfrac{y}{x}$ and $\tan(t + \pi) = \dfrac{-y}{-x} = \dfrac{y}{x}$. So we have shown that $\tan(t + \pi) = \tan t$.

120. Let the real number t correspond to the point $P(x, y)$ on the unit circle. Then $t + \pi$ corresponds to the point $P(-x, -y)$ on the unit circle. By definition, $\cot t = \dfrac{x}{y}$ and $\cot(t + \pi) = \dfrac{-x}{-y} = \dfrac{x}{y}$. So we have shown that $\cot(t + \pi) = \cot t$.

121. Let the real number t correspond to the point $P(x, y)$ on the unit circle. Since the circumference of the unit circle is 2π, adding 2π to t results in the same terminal point (x, y). Consequently, $\sin(t + 2\pi) = \sin t$ for all t.

To show that 2π is the *smallest* positive value p for which $\sin(t + p) = \sin t$, we will do a proof by contradiction. Assume the contrary; that is, assume that there exists a value p on the interval $0 < p < 2\pi$ such that $\sin(t + p) = \sin t$. Substituting $t = 0$ results in $\sin(0 + p) = \sin 0$, or equivalently $\sin p = 0$. The only solution on the interval $0 < p < 2\pi$ is $p = \pi$. Now substitute $t = \dfrac{\pi}{2}$ into the equation $\sin(t + p) = \sin t$. We have $\sin\left(\dfrac{\pi}{2} + \pi\right) = \sin\dfrac{\pi}{2}$. This is a contradiction because $\sin\dfrac{3\pi}{2} = -1$ but $\sin\dfrac{\pi}{2} = 1$. Therefore, our original assumption that there exists a value p on the interval $0 < p < 2\pi$ such that $\sin(t + p) = \sin t$ is incorrect. The smallest possible value of p is 2π.

122. Let the real number t correspond to the point $P(x, y)$ on the unit circle. Since the circumference of the unit circle is 2π, adding 2π to t results in the same terminal point (x, y). Consequently, $\cos(t + 2\pi) = \cos t$ for all t.

To show that 2π is the *smallest* positive value p for which $\cos(t + p) = \cos t$, we will do a proof by contradiction. Assume the contrary. That is, assume that there exists a value p on the interval $0 < p < 2\pi$ such that $\cos(t + p) = \cos t$. Substituting $t = 0$ results in $\cos(0 + p) = \cos 0$ or equivalently $\cos p = 1$. This is a contradiction because there is no solution to this equation on the interval $0 < p < 2\pi$. (The values of p that would satisfy this equation are multiples of 2π which do not fall strictly between 0 and 2π.) Therefore, our original assumption that there exists a value p on the interval $0 < p < 2\pi$ such that $\cos(t + 2\pi) = \cos t$ is incorrect. The smallest possible value of p is 2π.

Section 4.3 Practice Exercises, pp. 491–496

15. Hypotenuse: 13;

$\sin\theta = \dfrac{5}{13}$, $\cos\theta = \dfrac{12}{13}$, $\tan\theta = \dfrac{5}{12}$,

$\csc\theta = \dfrac{13}{5}$, $\sec\theta = \dfrac{13}{12}$, $\cot\theta = \dfrac{12}{5}$

16. Hypotenuse: 17;

$\sin\theta = \dfrac{8}{17}$, $\cos\theta = \dfrac{15}{17}$, $\tan\theta = \dfrac{8}{15}$,

$\csc\theta = \dfrac{17}{8}$, $\sec\theta = \dfrac{17}{15}$, $\cot\theta = \dfrac{15}{8}$

17. Hypotenuse: $\sqrt{61}$;

$\sin\theta = \dfrac{5\sqrt{61}}{61}$, $\cos\theta = \dfrac{6\sqrt{61}}{61}$, $\tan\theta = \dfrac{5}{6}$,

$\csc\theta = \dfrac{\sqrt{61}}{5}$, $\sec\theta = \dfrac{\sqrt{61}}{6}$, $\cot\theta = \dfrac{6}{5}$

18. Hypotenuse: $\sqrt{13}$;

$\sin\theta = \dfrac{2\sqrt{13}}{13}$, $\cos\theta = \dfrac{3\sqrt{13}}{13}$, $\tan\theta = \dfrac{2}{3}$,

$\csc\theta = \dfrac{\sqrt{13}}{2}$, $\sec\theta = \dfrac{\sqrt{13}}{3}$, $\cot\theta = \dfrac{3}{2}$

19. Hypotenuse $= 2\sqrt{10}$ cm;

$\sin\theta = \dfrac{\sqrt{10}}{10}$, $\cos\theta = \dfrac{3\sqrt{10}}{10}$, $\tan\theta = \dfrac{1}{3}$,

$\csc\theta = \sqrt{10}$, $\sec\theta = \dfrac{\sqrt{10}}{3}$, $\cot\theta = 3$

20. Hypotenuse $= 3\sqrt{26}$ cm;

$\sin\theta = \dfrac{\sqrt{26}}{26}$, $\cos\theta = \dfrac{5\sqrt{26}}{26}$, $\tan\theta = \dfrac{1}{5}$,

$\csc\theta = \sqrt{26}$, $\sec\theta = \dfrac{\sqrt{26}}{5}$, $\cot\theta = 5$

21. Leg $= 4$ m;

$\sin\theta = \dfrac{2\sqrt{11}}{11}$, $\cos\theta = \dfrac{\sqrt{77}}{11}$, $\tan\theta = \dfrac{2\sqrt{7}}{7}$,

$\csc\theta = \dfrac{\sqrt{11}}{2}$, $\sec\theta = \dfrac{\sqrt{77}}{7}$, $\cot\theta = \dfrac{\sqrt{7}}{2}$

22. Leg $= 3$ in.;

$\sin\theta = \dfrac{\sqrt{21}}{14}$, $\cos\theta = \dfrac{5\sqrt{7}}{14}$, $\tan\theta = \dfrac{\sqrt{3}}{5}$,

$\csc\theta = \dfrac{2\sqrt{21}}{3}$, $\sec\theta = \dfrac{2\sqrt{7}}{5}$, $\cot\theta = \dfrac{5\sqrt{3}}{3}$

23. $\sin\theta = \dfrac{4\sqrt{65}}{65}$, $\cos\theta = \dfrac{7\sqrt{65}}{65}$,

$\csc\theta = \dfrac{\sqrt{65}}{4}$, $\sec\theta = \dfrac{\sqrt{65}}{7}$, $\cot\theta = \dfrac{7}{4}$

24. $\sin\theta = \dfrac{3\sqrt{10}}{10}$, $\tan\theta = 3$, $\csc\theta = \dfrac{\sqrt{10}}{3}$, $\sec\theta = \sqrt{10}$, $\cot\theta = \dfrac{1}{3}$

39. False; For example, $\sin\dfrac{\pi}{4}\tan\dfrac{\pi}{4} = \dfrac{\sqrt{2}}{2}\cdot 1 = \dfrac{\sqrt{2}}{2} \neq 1$

42. False; For example

$\csc\dfrac{\pi}{3}\cdot\cot\dfrac{\pi}{3} = \dfrac{2\sqrt{3}}{3}\cdot\dfrac{\sqrt{3}}{3} = \dfrac{2}{3}$

$\sec\dfrac{\pi}{3} = 2$

$\dfrac{2}{3} \neq 2$

44. False; For example,

$\sin\dfrac{\pi}{6}\cos\dfrac{\pi}{6}\tan\dfrac{\pi}{6} + 1 = \dfrac{1}{2}\cdot\dfrac{\sqrt{3}}{2}\cdot\dfrac{\sqrt{3}}{3} + 1 = \dfrac{5}{4}$

$\cos^2\left(\dfrac{\pi}{6}\right) = \left(\cos\dfrac{\pi}{6}\right)^2 = \left(\dfrac{\sqrt{3}}{2}\right)^2 = \dfrac{3}{4}$

$\dfrac{5}{4} \neq \dfrac{3}{4}$

61. Yes. The tops of the goalposts are approximately 41 ft high (rounded to the nearest foot). This means that the vertical posts are approximately 31 ft.

74. Relative to angle θ, side b is the opposite side and side a is the adjacent side.

So $\tan\theta = \dfrac{\text{opp}}{\text{adj}} = \dfrac{b}{a}$.

Relative to angle $(90° - \theta)$, side b is the adjacent side and side a is the opposite side.

So $\cot(90° - \theta) = \dfrac{\text{adj}}{\text{opp}} = \dfrac{b}{a}$.

This shows that $\tan\theta = \cot(90° - \theta)$.

79. $\cot A = \dfrac{a+b}{h}$ and $\cot B = \dfrac{b}{h}$.

Solving each equation for b yields, $b = h\cot A - a$ and $b = h\cot B$.
Equating the expressions for b, gives $h\cot A - a = h\cot B$, which implies that $h\cot A - h\cot B = a$.

80. Let h represent the height of the perpendicular line segment from vertex C to side c.

Then $\sin A = \dfrac{h}{b}$, which implies that $h = b\sin A$.

The area of the triangle is

$A = \dfrac{1}{2}(\text{base})(\text{height}) = \dfrac{1}{2}c(b\sin A) = \dfrac{1}{2}bc\sin A$.

Section 4.4 Practice Exercises, pp. 504–507

15. $\sin\theta = -\dfrac{12}{13}$, $\cos\theta = \dfrac{5}{13}$, $\tan\theta = -\dfrac{12}{5}$,

$\csc\theta = -\dfrac{13}{12}$, $\sec\theta = \dfrac{13}{5}$, $\cot\theta = -\dfrac{5}{12}$

16. $\sin\theta = \dfrac{15}{17}$, $\cos\theta = -\dfrac{8}{17}$, $\tan\theta = -\dfrac{15}{8}$,

$\csc\theta = \dfrac{17}{15}$, $\sec\theta = -\dfrac{17}{8}$, $\cot\theta = -\dfrac{8}{15}$

17. $\sin\theta = -\dfrac{5\sqrt{34}}{34}$, $\cos\theta = -\dfrac{3\sqrt{34}}{34}$, $\tan\theta = \dfrac{5}{3}$,

$\csc\theta = -\dfrac{\sqrt{34}}{5}$, $\sec\theta = -\dfrac{\sqrt{34}}{3}$, $\cot\theta = \dfrac{3}{5}$

18. $\sin\theta = -\dfrac{3\sqrt{13}}{13}$, $\cos\theta = \dfrac{2\sqrt{13}}{13}$, $\tan\theta = -\dfrac{3}{2}$,

$\csc\theta = -\dfrac{\sqrt{13}}{3}$, $\sec\theta = \dfrac{\sqrt{13}}{2}$, $\cot\theta = -\dfrac{2}{3}$

19. $\sin\theta = \dfrac{4}{5}$, $\cos\theta = -\dfrac{3}{5}$, $\tan\theta = -\dfrac{4}{3}$,

$\csc\theta = \dfrac{5}{4}$, $\sec\theta = -\dfrac{5}{3}$, $\cot\theta = -\dfrac{3}{4}$

20. $\sin\theta = -\dfrac{8}{17}$, $\cos\theta = \dfrac{15}{17}$, $\tan\theta = -\dfrac{8}{15}$,

$\csc\theta = -\dfrac{17}{8}$, $\sec\theta = \dfrac{17}{15}$, $\cot\theta = -\dfrac{15}{8}$

91. $\sin(240° - 120°) = \sin 120° = \dfrac{\sqrt{3}}{2}$;

$\sin 240°\cos 120° - \cos 240°\sin 120°$

$= \left(-\dfrac{\sqrt{3}}{2}\right)\left(-\dfrac{1}{2}\right) - \left(-\dfrac{1}{2}\right)\left(\dfrac{\sqrt{3}}{2}\right) = \dfrac{\sqrt{3}}{2}$

92. $\cos(120° - 330°) = \cos(-210°) = -\dfrac{\sqrt{3}}{2}$;

$\cos 120°\cos 330° + \sin 120°\sin 330°$

$= \left(-\dfrac{1}{2}\right)\left(\dfrac{\sqrt{3}}{2}\right) + \left(\dfrac{\sqrt{3}}{2}\right)\left(-\dfrac{1}{2}\right) = -\dfrac{\sqrt{3}}{2}$

93. $\tan(210° + 120°) = \tan 330° = -\dfrac{\sqrt{3}}{3}$; $\dfrac{\tan 210° + \tan 120°}{1 - \tan 210°\tan 120°}$

$= \dfrac{\dfrac{\sqrt{3}}{3} - \sqrt{3}}{1 - \left(\dfrac{\sqrt{3}}{3}\right)(-\sqrt{3})} = -\dfrac{\sqrt{3}}{3}$

94. $\cot(300° + 150°) = \cot 450° = 0$; $\dfrac{\cot 300°\cot 150° - 1}{\cot 300° + \cot 150°}$

$= \dfrac{\left(-\dfrac{\sqrt{3}}{3}\right)(-\sqrt{3}) - 1}{-\dfrac{\sqrt{3}}{3} + (-\sqrt{3})} = 0$

97. Let $P(x, y)$ be a point on the terminal side of θ in standard position, and let $r = \sqrt{x^2 + y^2}$ be the distance between P and the origin.

We know that $r = \sqrt{x^2 + y^2} \geq y$.

Therefore, $\sin\theta = \dfrac{y}{r} \leq 1$.

Likewise, we know that $r = \sqrt{x^2 + y^2} \geq x$. Therefore,

$\cos\theta = \dfrac{x}{r} \leq 1$.

99. a.–b.

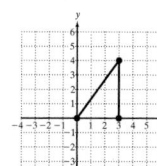

$$\text{Slope} = \frac{4}{3}$$

$$\text{Tangent of the angle} = \frac{4}{3}$$

c. The results are the same.

101. a. Using the right triangle formed by point P and the center of circle A, $\sin\theta = \dfrac{\text{opp}}{\text{hyp}} = \dfrac{a}{x+a}$.

Using the right triangle formed by point P and the center of circle B,

$$\sin\theta = \frac{\text{opp}}{\text{hyp}} = \frac{b}{x+2a+b}.$$

b.
$$\frac{a}{x+a} = \frac{b}{x+2a+b}$$
$$a(x+2a+b) = b(x+a)$$
$$ax + 2a^2 + ab = bx + ba$$
$$2a^2 = bx - ax$$
$$2a^2 = x(b-a)$$
$$\frac{2a^2}{b-a} = x$$

$$\sin\theta = \frac{a}{x+a} = \frac{a}{\left(\dfrac{2a^2}{b-a}+a\right)} \cdot \frac{(b-a)}{(b-a)}$$

$$= \frac{a(b-a)}{2a^2+ab-a^2} = \frac{a(b-a)}{a^2+ab}$$

$$= \frac{a(b-a)}{a(a+b)} = \frac{b-a}{a+b}$$

Section 4.5 Practice Exercises, pp. 520–526

R.1.

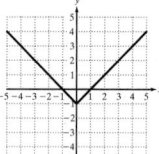

R.2.

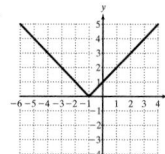

R.3.

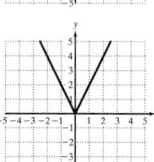

R.4.

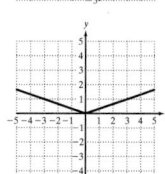

R.5.

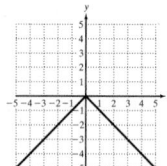

9.

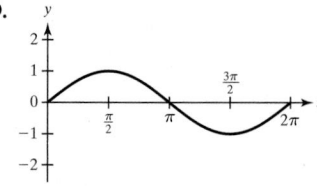

10.

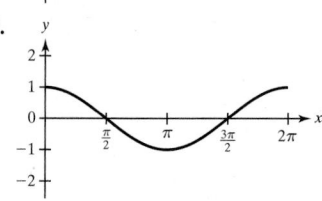

19.

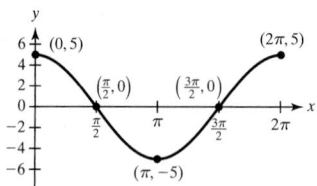

20.

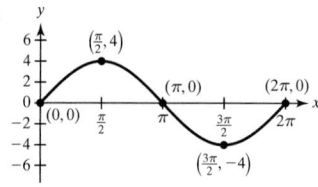

21.

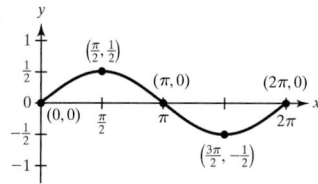

22.

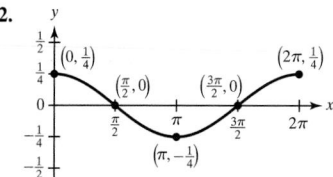

23.

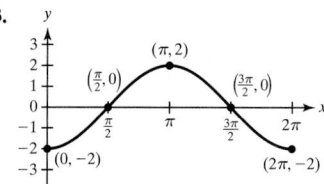

24.

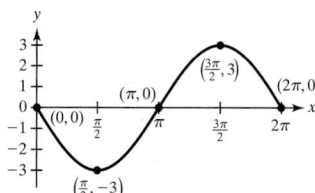

27. a. Amplitude $= 2$, Period $= \dfrac{2\pi}{3}$

b.

28. a. Amplitude $= 6$, Period $= \dfrac{\pi}{2}$

b.

29. a. Amplitude = 4, Period = 6

b.

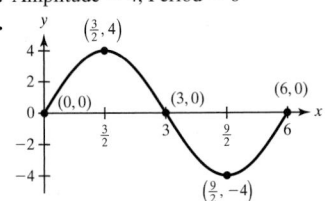

30. a. Amplitude = 5, Period = 12

b.

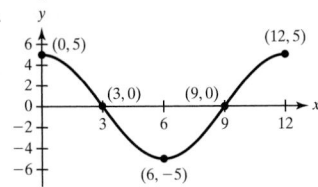

31. a. Amplitude = 1, Period = 6π

b.

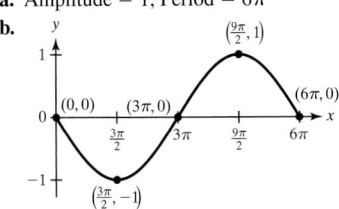

32. a. Amplitude = 1, Period = 4π

b.

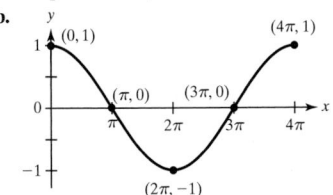

36. c. Inhalation: (0, 3.5);
Exhalation: (3.5, 7)

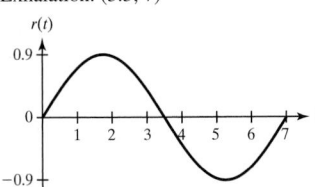

39. a. Amplitude = 2, Period = 2π, Phase shift = $-\pi$

b.

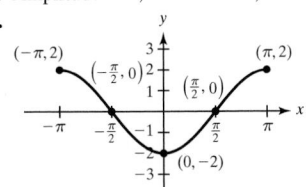

40. a. Amplitude = 4, Period = 2π, Phase shift = $-\dfrac{\pi}{2}$

b.

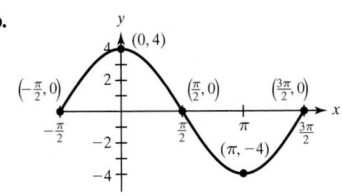

41. a. Amplitude = 1, Period = π, Phase shift = $\dfrac{\pi}{6}$

b.

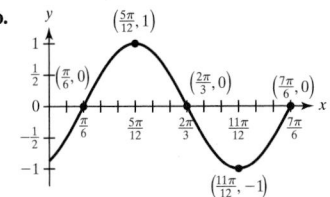

42. a. Amplitude = 1, Period = $\dfrac{2\pi}{3}$, Phase shift = $\dfrac{\pi}{12}$

b.

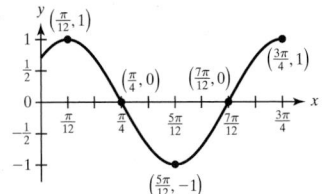

43. a. Amplitude = 6, Period = 4π, Phase shift = $-\dfrac{\pi}{2}$

b.

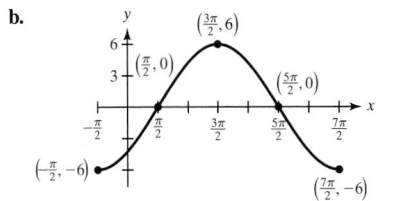

44. a. Amplitude = 5, Period = 6π, Phase shift = $-\dfrac{\pi}{2}$

b.

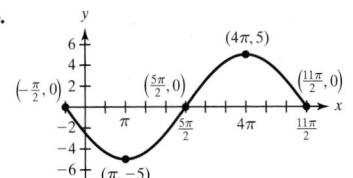

55. a. Amplitude = 3, Period = $\dfrac{\pi}{2}$, Phase shift = $\dfrac{\pi}{4}$, Vertical shift = 5

b.

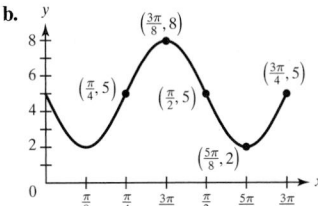

56. a. Amplitude = 2, Period = $\dfrac{2\pi}{3}$, Phase shift = $\dfrac{\pi}{3}$, Vertical shift = -4

b.

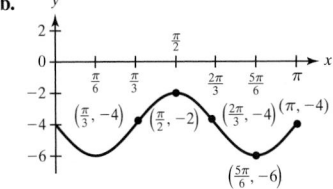

57. a. Amplitude $= 4$, Period $= \dfrac{2\pi}{3}$, Phase shift $= \dfrac{\pi}{6}$, Vertical shift $= -1$

b.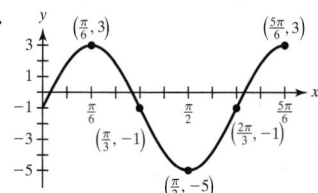

58. a. Amplitude $= 5$, Period $= \pi$, Phase shift $= \dfrac{\pi}{4}$, Vertical shift $= 1$

b.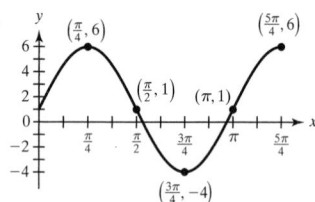

59. a. Amplitude $= \dfrac{1}{2}$, Period $= 6\pi$, Phase shift $= 0$, Vertical shift $= 0$

b.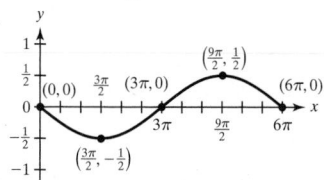

60. a. Amplitude $= \dfrac{2}{3}$, Period $= 4\pi$, Phase shift $= 0$, Vertical shift $= 0$

b.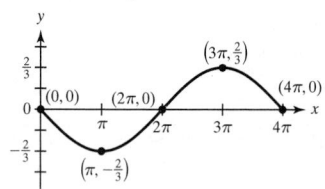

61. a. Amplitude $= 1.6$, Period $= 2$, Phase shift $= 0$, Vertical shift $= 0$

b.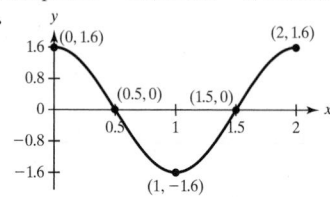

62. a. Amplitude $= 2.4$, Period $= 0.5$, Phase shift $= 0$, Vertical shift $= 0$

b.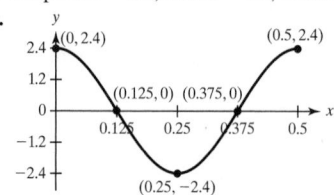

63. a. Amplitude $= 2$, Period $= \pi$, Phase shift $= -\dfrac{\pi}{2}$, Vertical shift $= 5$

b.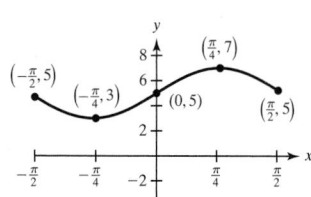

64. a. Amplitude $= 3$, Period $= \dfrac{\pi}{2}$, Phase shift $= -\dfrac{\pi}{4}$, Vertical shift $= -7$

b.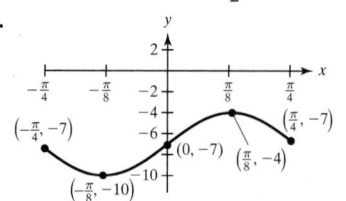

65. a. Amplitude $= 1$, Period $= 4$, Phase shift $= -2$, Vertical shift $= 2$

b.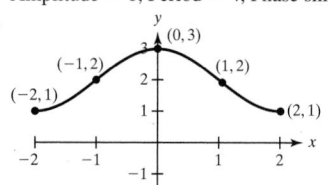

66. a. Amplitude $= 1$, Period $= 6$, Phase shift $= -3$, Vertical shift $= -2$

b.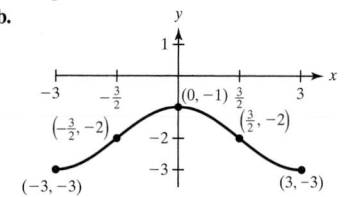

67. a. Amplitude $= 1$, Period $= 8$, Phase shift $= -2$, Vertical shift $= -3$

b.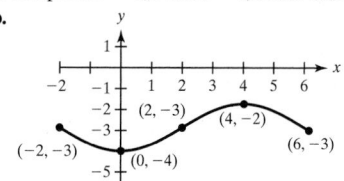

68. a. Amplitude $= 1$, Period $= 6$, Phase shift $= -\dfrac{3}{2}$, Vertical shift $= 4$

b.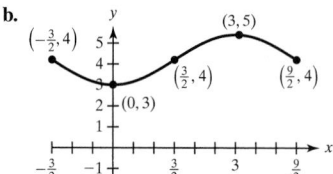

70. a. The two graphs are symmetric with respect to the line $y = 12$.

b. $n(t) = -2.65 \sin(0.51t - 1.32) + 12$

c. The points of intersection represent the numbers of the months since January 1 when the duration of daylight and darkness are each 12 hr. In the northern hemisphere, these correspond to the first day of spring (March equinox) and first day of fall (September equinox).

d. $d(t)$ has a relative maximum and $n(t)$ has a relative minimum on the first day of summer (the longest day). $d(t)$ has a relative minimum and $n(t)$ has a relative maximum on the first day of winter (the shortest day).

e. The value of $T(t)$ is the sum of the number of hours of daylight and the number of hours of darkness. This is the constant function $T(t) = 24$.

71. a. $P(t) = 0.15 \cos\left(\dfrac{\pi}{6}t - \dfrac{\pi}{6}\right) + 0.19$

b.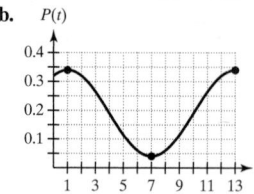

72. a. $H(t) = -11\cos\left(\dfrac{\pi}{6}t\right) + 75$

b.

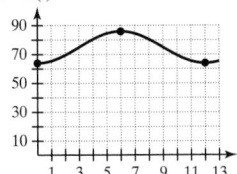

75. b.

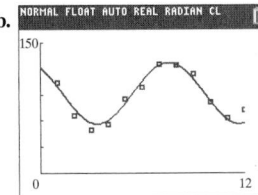

76. b.

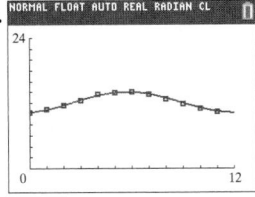

81. $y = 4\sin(6x - 3\pi) + 5$ or $y = -4\sin(6x - 3\pi) + 5$

82. $y = 2\cos\left(8x + \dfrac{8\pi}{3}\right) + 7$ or $y = -2\cos\left(8x + \dfrac{8\pi}{3}\right) + 7$

85. a. $y = 2\cos\left(x - \dfrac{\pi}{4}\right) + 3$ **b.** $y = 2\sin\left(x + \dfrac{\pi}{4}\right) + 3$

86. a. $y = 3\cos\left(2x - \dfrac{\pi}{2}\right) + 1$ **b.** $y = 3\sin 2x + 1$

87. a. The graph of $y = \sin x$ is compressed horizontally by a factor of 2.
   b. The graph of $y = \sin x$ is vertically stretched by a factor of 2.

88. a. The graph of $y = \cos x$ is vertically compressed by a factor of $\dfrac{1}{3}$.
   b. The graph of $y = \cos x$ is stretched horizontally by a factor of 3.

89. a. The graph of $y = \sin x$ is shifted 2 units to the left.
   b. The graph of $y = \sin x$ is shifted vertically upward 2 units.

90. a. The graph of $y = \cos x$ is shifted vertically downward 4 units.
   b. The graph of $y = \cos x$ is shifted 4 units to the right.

91. No. If f were one-to-one, then $f(a) = f(b)$ would imply that $a = b$.
   However, as a counterexample, $f\left(\dfrac{\pi}{4}\right) = f\left(\dfrac{3\pi}{4}\right) = \dfrac{\sqrt{2}}{2}$ but $\dfrac{\pi}{4} \neq \dfrac{3\pi}{4}$.

92. No. If f were one-to-one, then $f(a) = f(b)$ would imply that $a = b$.
   However, as a counterexample, $f\left(\dfrac{\pi}{3}\right) = f\left(\dfrac{5\pi}{3}\right) = \dfrac{1}{2}$ but $\dfrac{\pi}{3} \neq \dfrac{5\pi}{3}$.

93. Yes. $(f + g)(x) = f(x) + g(x) = f(x + P) + g(x + P) = (f + g)(x + P)$ for all x in the domain of f and g.

94. If a function f is increasing over its entire domain, then for any interval $[a, b]$ in the domain of f, $f(a) < f(b)$. However, if f were periodic, then there would exist a positive number P such that $f(a + P) = f(a)$. This contradicts the fact that f is increasing.

95. a. $\dfrac{3}{\pi} \approx 0.9549$ **b.** $\dfrac{3(\sqrt{3} - 1)}{\pi} \approx 0.6991$

   c. $\dfrac{3(2 - \sqrt{3})}{\pi} \approx 0.2559$

   The average rate of change is positive and decreasing for the intervals given here. Likewise, the slopes of the lines are positive and decreasing.

96. a. $\dfrac{3(\sqrt{3} - 2)}{\pi} \approx -0.2559$ **b.** $\dfrac{3(1 - \sqrt{3})}{\pi} \approx -0.6991$

   c. $-\dfrac{3}{\pi} \approx -0.9549$

   The average rate of change is negative and decreasing for the intervals given here. Likewise, the slopes of the lines are negative and decreasing.

101.

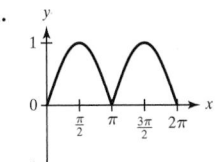

102.

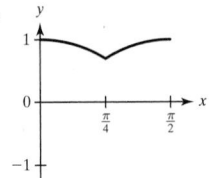

103. a. The functions have different amplitudes.

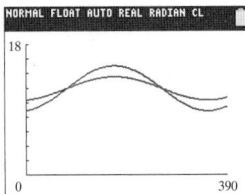

   b. The constant term of 12 is the vertical shift of the parent sine function. It represents the "equilibrium" value for duration of sunlight from which the amplitude is measured.

   c. The factors 3.1 and 1.6, represent the amplitude of the functions a and m, respectively. The amplitude represents the greatest amount of deviation above and below the line $y = 12$.

   d. The period of each function is 365 days.

   e. The horizontal shift represents the number of days from January 1 to the point where the duration of light and darkness is the same (12 hr each). This is the approximate number of days to the March equinox (first day of spring).

   f. (80, 12) and (262.5, 12)

   g. The points of intersection represent the points where the duration of light and darkness is the same. They represent the March and September equinoxes where the durations of light and darkness are each 12 hr.

104. Each polynomial function gives a better approximation for values of x close to zero than for values farther from zero. Furthermore, the greater the degree of the polynomial, the better the approximation of $\cos x$ for each value of x.

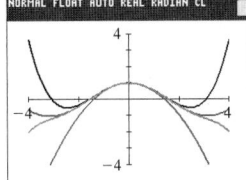

105. Each polynomial function gives a better approximation for values of x close to zero than for values farther from zero. Furthermore, the greater the degree of the polynomial, the better the approximation of $\sin x$ for each value of x.

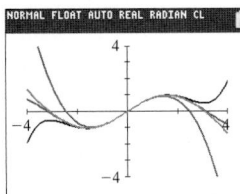

106. $x = 0, x = \dfrac{\pi}{3}, x = \pi$

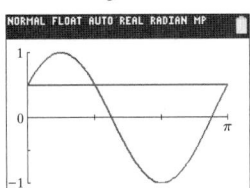

107. $x = \dfrac{\pi}{8}, x = \dfrac{9\pi}{8}$

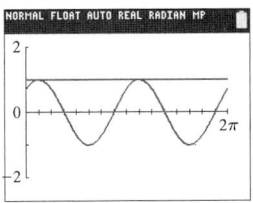

108. $x \approx -2.25,\ x \approx -0.19$ **109.** $x \approx 1.04,\ x \approx 4.44$

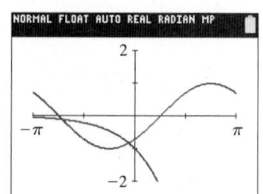   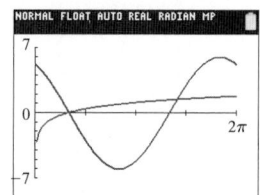

10.

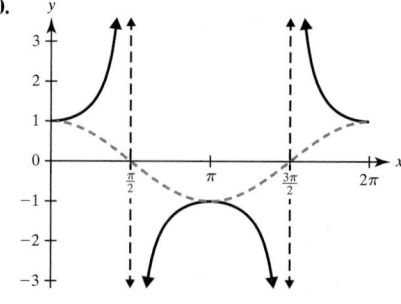

110. a. 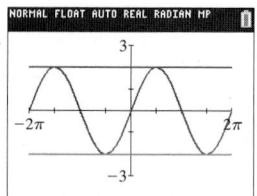 **b.**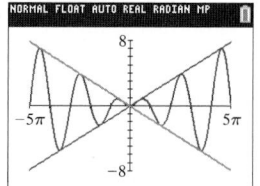

c.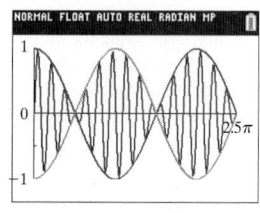

17.

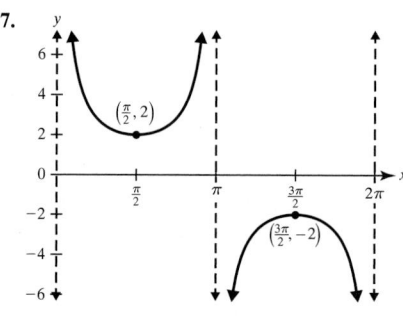

Section 4.6 Skill Practice

Answer 1

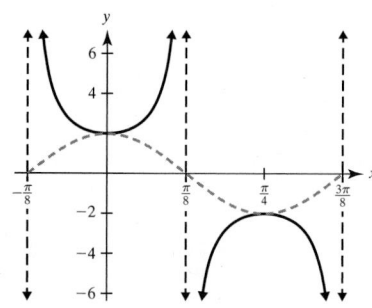

18.

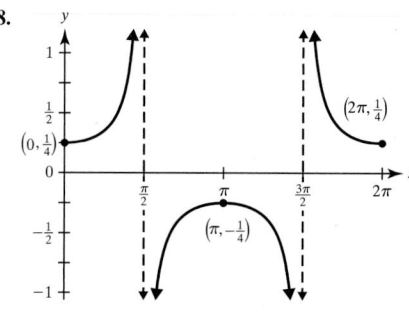

Answer 4

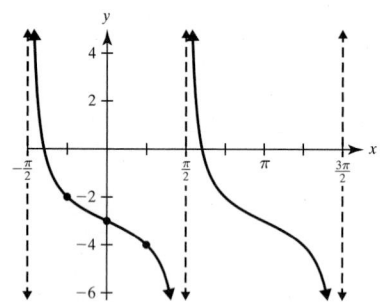

19.

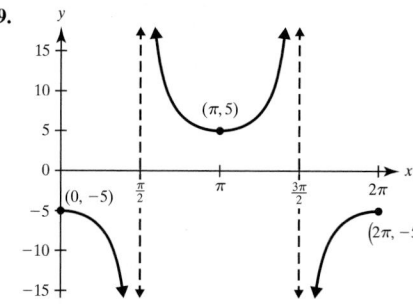

Section 4.6 Practice Exercises, pp. 535–539

9.

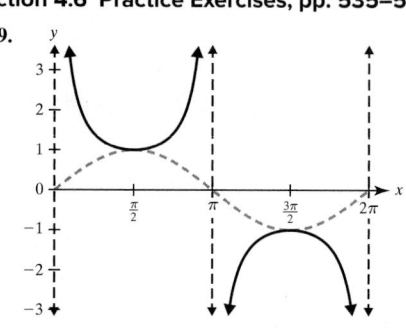

20.

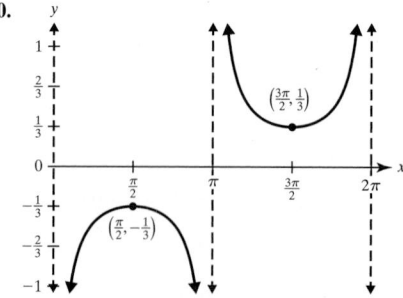

21.

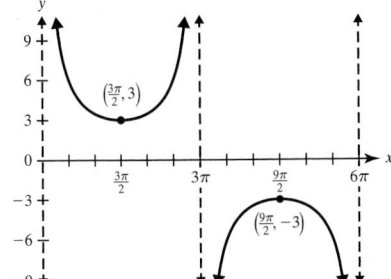

22.

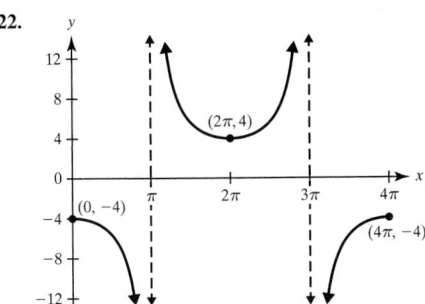

23.

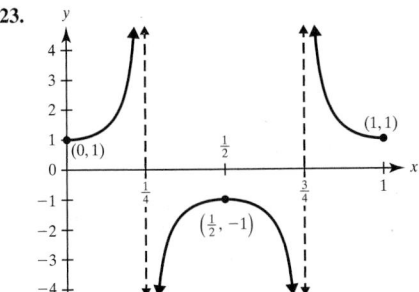

24.

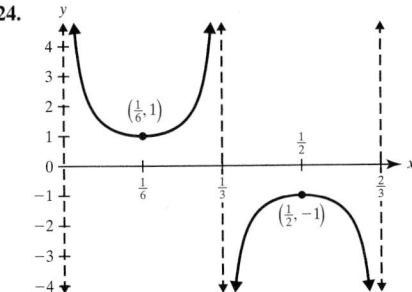

25.

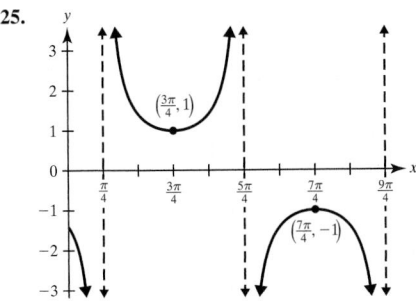

26.

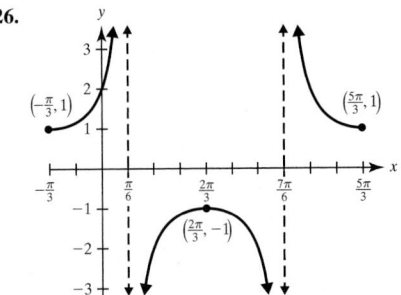

27.

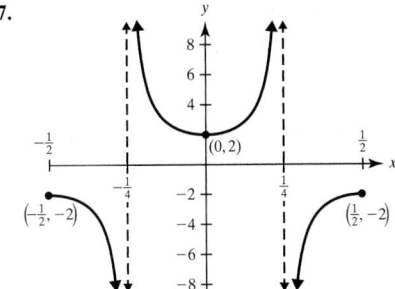

28.

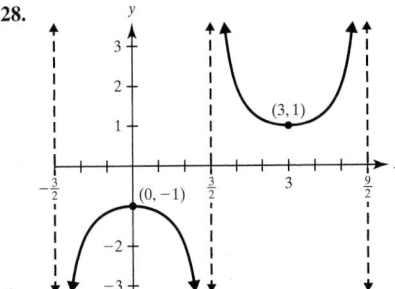

29.

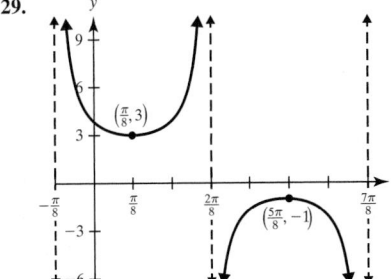

30.

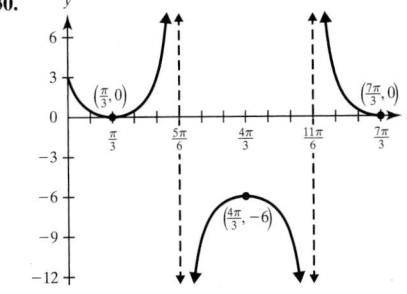

31.

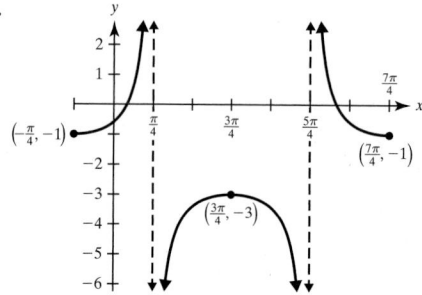

b.

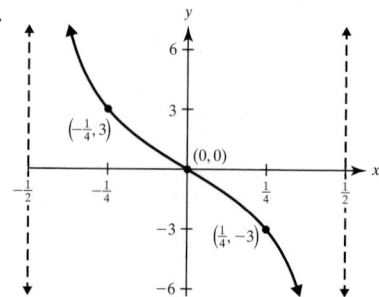

32.

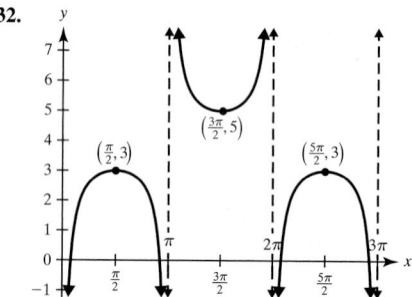

42. a.

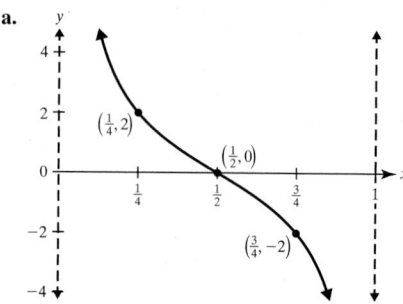

39. a.

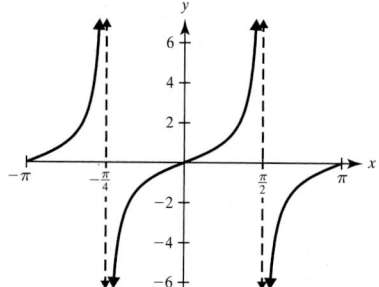

b.

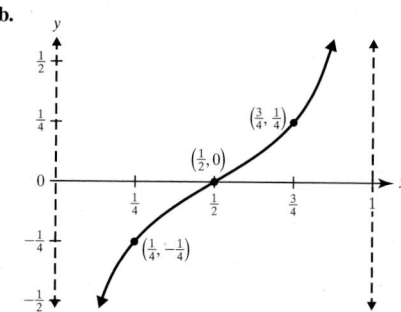

40. a.

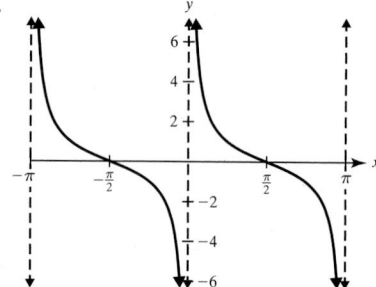

43.

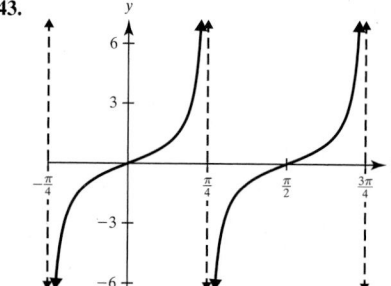

41. a.

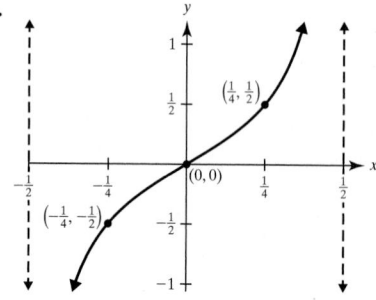

44.

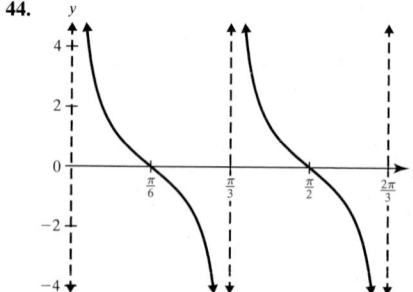

45.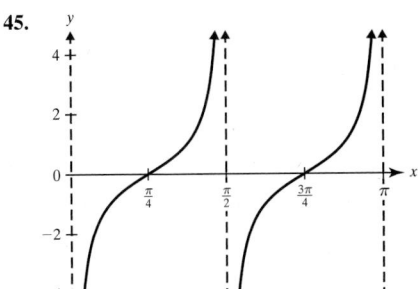

46.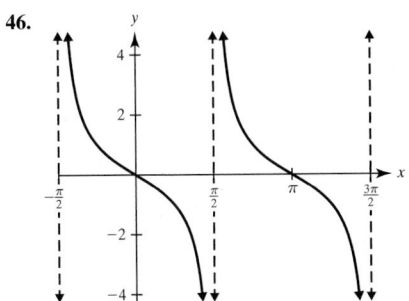

47.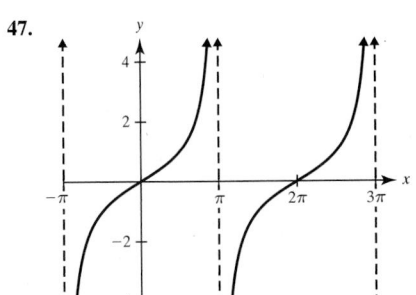

48.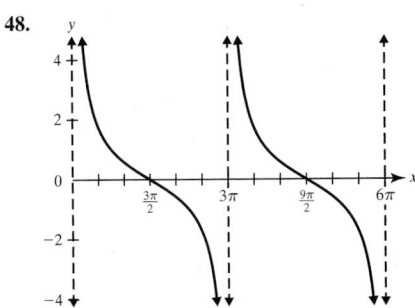

49.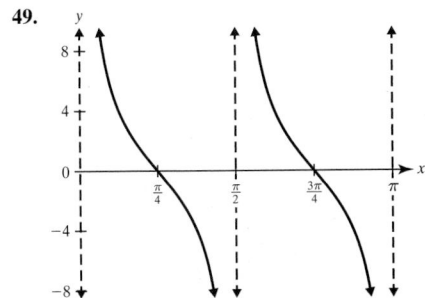

50.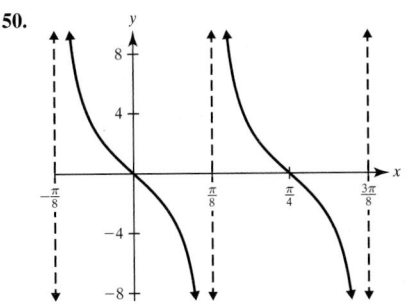

51.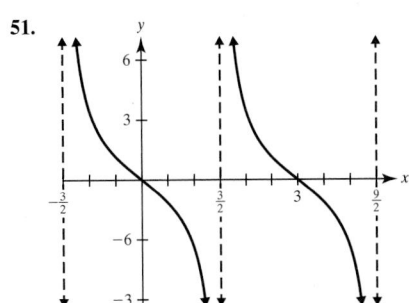

52.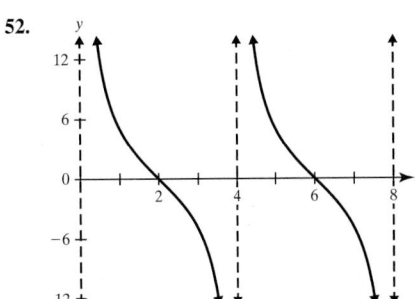

53.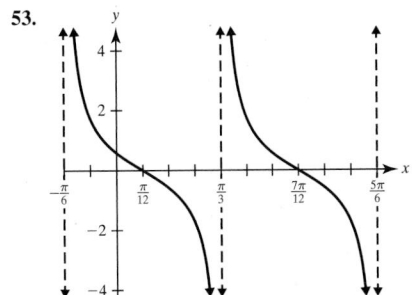

54.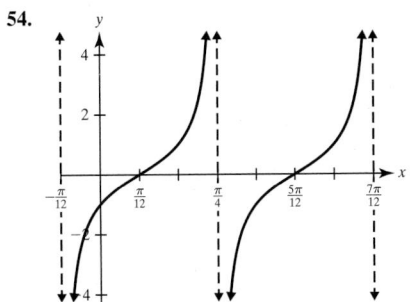

55.

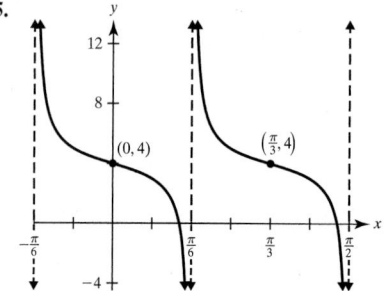

56.

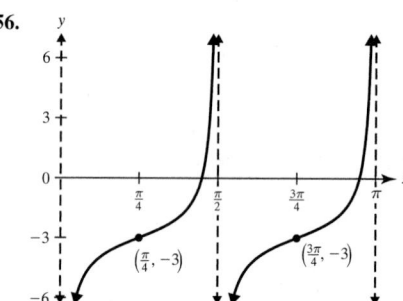

57.

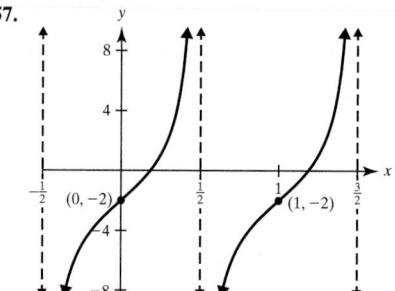

58.

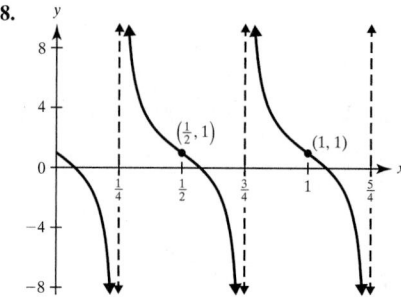

61. a. $(f \circ g)(x) = \tan\dfrac{x}{4}$

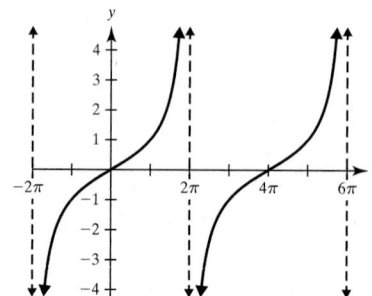

b. $(g \circ f)(x) = \dfrac{1}{4}\tan x$

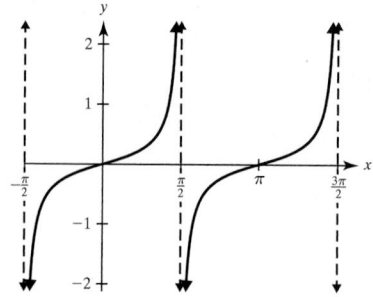

62. a. $(f \circ g)(x) = -2\cot x$

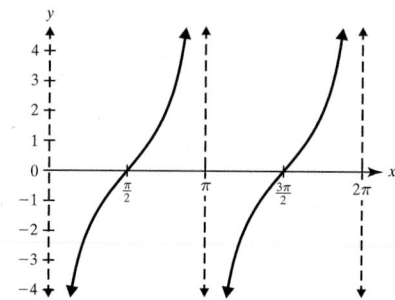

b. $(g \circ f)(x) = \cot(-2x)$

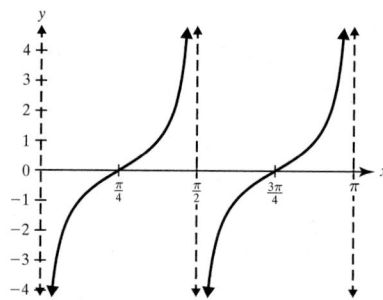

63. a. $(f \circ g)(x) = -3\csc x$

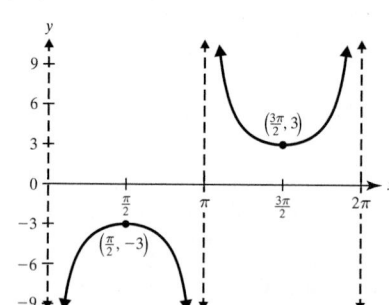

b. $(g \circ f)(x) = \csc(-3x)$

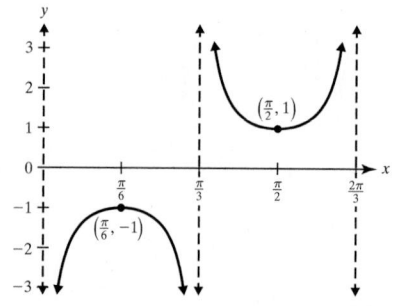

64. a. $(f \circ g)(x) = \frac{1}{3}\sec x$

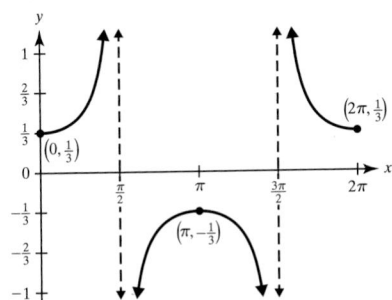

b. $(g \circ f)(x) = \sec\frac{x}{3}$

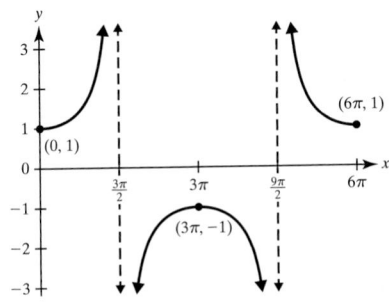

69. Solve the inequality $-\frac{\pi}{2} < Bx - C < \frac{\pi}{2}$. The endpoints of the resulting interval give the location of two consecutive vertical asymptotes: $x = \frac{1}{B}\left(C - \frac{\pi}{2}\right)$ and $x = \frac{1}{B}\left(C + \frac{\pi}{2}\right)$.

70. Solve the inequality $0 < Bx - C < \pi$. The endpoints of the resulting interval give the location of two consecutive vertical asymptotes: $x = \frac{C}{B}$ and $x = \frac{C + \pi}{B}$.

71. First graph the related reciprocal function, $y = A\cos(Bx - C)$, without the vertical shift. Then graph $y = A\sec(Bx - C)$ by first graphing the vertical asymptotes. The vertical asymptotes occur where $y = A\cos(Bx - C) = 0$.

 Then plot the relative maximum and minimum points for $y = A\sec(Bx - C)$. These correspond to the relative minimum and maximum points of $y = A\cos(Bx - C)$. Sketch the general shape of the "parent" function $y = \sec x$ between the asymptotes.

 Then shift the graph of $y = A\sec(Bx - C)$ vertically D units.

72. First graph the related reciprocal function, $y = A\sin(Bx - C)$, without the vertical shift. Then graph $y = A\csc(Bx - C)$ by first graphing the vertical asymptotes. The vertical asymptotes occur where $y = A\sin(Bx - C) = 0$.

 Then plot the relative maximum and minimum points for $y = A\csc(Bx - C)$. These correspond to the relative minimum and maximum points of $y = A\sin(Bx - C)$. Sketch the general shape of the "parent" function $y = \csc x$ between the asymptotes.

 Then shift the graph of $y = A\csc(Bx - C)$ vertically D units.

77. Let x and y represent the hypotenuse of the upper and lower right triangles, respectively.

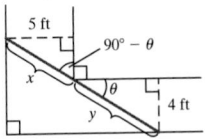

$\sin(90° - \theta) = \frac{5}{x} = \cos\theta$ and $\sin\theta = \frac{4}{y}$.

Therefore, $x = \frac{5}{\cos\theta} = 5\sec\theta$ and $y = \frac{4}{\sin\theta} = 4\csc\theta$.

$L = x + y = 5\sec\theta + 4\csc\theta$

78. The function values are close for x taken near 0.

79. The function values are close for x taken near 0.

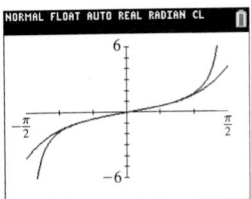 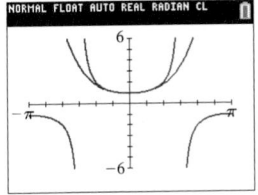

80. b.

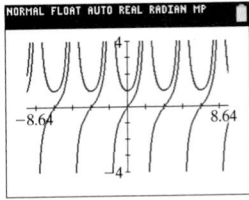

81. b.

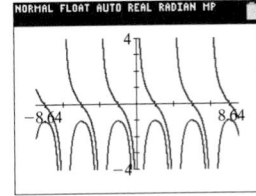

Section 4.7 Practice Exercises, pp. 551–556

R.4.

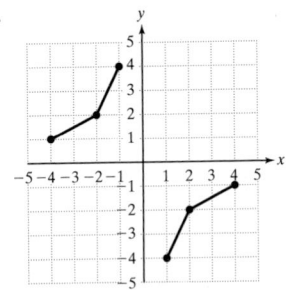

33. In radians:

NORMAL FLOAT AUTO REAL RADIAN MP
$\cos^{-1}(3/8)$
1.186399552
$\tan^{-1}(25)$
1.53081764
$\sin^{-1}(0.05)$
.0500208568

In degrees:

NORMAL FLOAT AUTO REAL DEGREE MP
$\cos^{-1}(3/8)$
67.97568716
$\tan^{-1}(25)$
87.70938996
$\sin^{-1}(0.05)$
2.865983983

34. In radians:

NORMAL FLOAT AUTO REAL RADIAN MP
$\sin^{-1}(0.93)$
1.194412844
$\cos^{-1}(0.17)$
1.399966658
$\tan^{-1}(7/4)$
1.051650213

In degrees:

NORMAL FLOAT AUTO REAL DEGREE MP
$\sin^{-1}(0.93)$
68.43481498
$\cos^{-1}(0.17)$
80.21218094
$\tan^{-1}(7/4)$
60.2551187

35. In radians:

NORMAL FLOAT AUTO REAL RADIAN MP
$\tan^{-1}(-28)$
-1.535097214
$\cos^{-1}(\sqrt{3}/5)$
1.217054721
$\sin^{-1}(-0.14)$
-.1404614147

In degrees:

NORMAL FLOAT AUTO REAL DEGREE MP
$\tan^{-1}(-28)$
-87.95459151
$\cos^{-1}(\sqrt{3}/5)$
69.73209894
$\sin^{-1}(-0.14)$
-8.047846247

36. In radians:

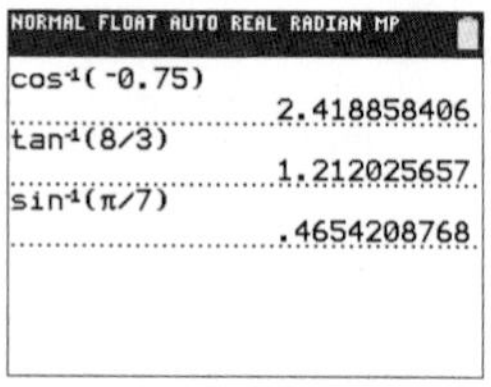
NORMAL FLOAT AUTO REAL RADIAN MP
cos⁻¹(-0.75)
2.418858406
tan⁻¹(8/3)
1.212025657
sin⁻¹(π/7)
.4654208768

In degrees:

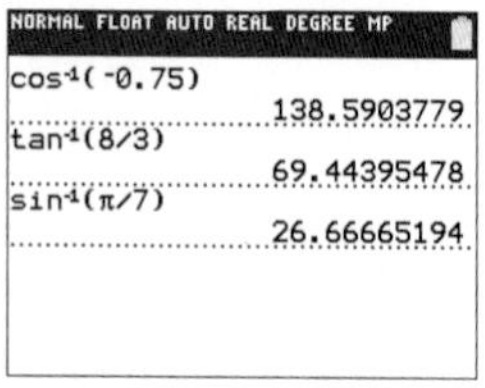
NORMAL FLOAT AUTO REAL DEGREE MP
cos⁻¹(-0.75)
138.5903779
tan⁻¹(8/3)
69.44395478
sin⁻¹(π/7)
26.66665194

85. Let $\theta = \sec^{-1}x$ for $x \geq 1$.
Then, $\sec\theta = x$

$$\frac{1}{\cos\theta} = x$$
$$\cos\theta = \frac{1}{x}$$
$$\theta = \cos^{-1}\frac{1}{x}$$

Hence, $\sec^{-1}x = \cos^{-1}\frac{1}{x}$.

86. Let $\theta = \csc^{-1}x$ for $x \geq 1$.
Then, $\csc\theta = x$

$$\frac{1}{\sin\theta} = x$$
$$\sin\theta = \frac{1}{x}$$
$$\theta = \sin^{-1}\frac{1}{x}$$

Hence, $\csc^{-1}x = \sin^{-1}\frac{1}{x}$.

87. Let $\alpha = \sec^{-1}x$ for $x \geq 1$, which implies that $\sec\alpha = x$ and

$$\csc\left(\frac{\pi}{2} - \alpha\right) = x$$
$$\frac{\pi}{2} - \alpha = \csc^{-1}x$$

Substituting $\alpha = \sec^{-1}x$ yields

$$\frac{\pi}{2} - \sec^{-1}x = \csc^{-1}x$$
$$\frac{\pi}{2} = \sec^{-1}x + \csc^{-1}x$$

88.

Inverse Function	Domain	Range
$y = \sin^{-1}x$	$[-1, 1]$	$\left[-\frac{\pi}{2}, \frac{\pi}{2}\right]$
$y = \csc^{-1}x$	$(-\infty, -1] \cup [1, \infty)$	$\left[-\frac{\pi}{2}, 0\right) \cup \left(0, \frac{\pi}{2}\right]$
$y = \cos^{-1}x$	$[-1, 1]$	$[0, \pi]$
$y = \sec^{-1}x$	$(-\infty, -1] \cup [1, \infty)$	$\left[0, \frac{\pi}{2}\right) \cup \left(\frac{\pi}{2}, \pi\right]$
$y = \tan^{-1}x$	$(-\infty, \infty)$	$\left(-\frac{\pi}{2}, \frac{\pi}{2}\right)$
$y = \cot^{-1}x$	$(-\infty, \infty)$	$(0, \pi)$

105. $f^{-1}(x) = \sin^{-1}\left(\frac{x-2}{3}\right)$;
Domain: $\{x \mid -1 \leq x \leq 5\}$;
Range: $\left\{y \,\middle|\, -\frac{\pi}{2} \leq y \leq \frac{\pi}{2}\right\}$

106. $g^{-1}(x) = \cos^{-1}\left(\frac{x+4}{6}\right)$;
Domain: $\{x \mid -10 \leq x \leq 2\}$;
Range: $\{y \mid 0 \leq y \leq \pi\}$

107. $h^{-1}(x) = \tan^{-1}\left(x - \frac{\pi}{4}\right)$;
Domain: $\mathbb{R}$;
Range: $\left\{y \,\middle|\, -\frac{\pi}{2} < y < \frac{\pi}{2}\right\}$

108. $k^{-1}(x) = \sin^{-1}(x - \pi)$;
Domain:
$\{x \mid -1 + \pi \leq x \leq 1 + \pi\}$;
Range: $\left\{y \,\middle|\, -\frac{\pi}{2} \leq y \leq \frac{\pi}{2}\right\}$

110. a.
$$\tan\frac{\alpha}{2} = \frac{\frac{1}{2}d}{f}$$
$$\frac{\alpha}{2} = \arctan\frac{d}{2f}$$
$$\alpha = 2\arctan\frac{d}{2f}$$

111. $\theta = \cos^{-1}\frac{\sqrt{25 - x^2}}{5}$ or $\theta = \sin^{-1}\frac{x}{5}$ or $\theta = \tan^{-1}\frac{x}{\sqrt{25 - x^2}}$

112. $\theta = \tan^{-1}\frac{\sqrt{x^2 - 4}}{2}$ or $\theta = \sin^{-1}\frac{\sqrt{x^2 - 4}}{x}$ or $\theta = \cos^{-1}\frac{2}{x}$

113. $\theta = \sin^{-1}\frac{x}{6}$ or $\theta = \cos^{-1}\frac{\sqrt{36 - x^2}}{6}$ or $\theta = \tan^{-1}\frac{x}{\sqrt{36 - x^2}}$

114. $\theta = \tan^{-1}\frac{3}{x}$ or $\theta = \cos^{-1}\frac{x}{\sqrt{x^2 + 9}}$ or $\theta = \sin^{-1}\frac{3}{\sqrt{x^2 + 9}}$

125. c.
$$A_{\text{trapezoid}} = \frac{1}{2}(b_1 + b_2)h$$
$$= \frac{1}{2}[f(0) + f(1)](1 - 0)$$
$$= \frac{1}{2}\left(1 + \frac{1}{2}\right)(1) = \frac{3}{4};$$

The answer from part (a) is $\frac{\pi}{4} \approx 0.785$, and the approximated value is $\frac{3}{4} = 0.75$. These are very close.

126. c.
$$A_{\text{trapezoid}} = \frac{1}{2}(b_1 + b_2)h$$
$$= \frac{1}{2}[f(0) + f(0.5)](0.5 - 0)$$
$$= \frac{1}{2}\left(1 + \frac{2\sqrt{3}}{3}\right)(0.5) = \frac{3 + 2\sqrt{3}}{12};$$

The answer from part (a) is $\frac{\pi}{6} \approx 0.524$, and the approximated value is $\frac{3 + 2\sqrt{3}}{12} \approx 0.539$. These are very close.

127. c.
$$\tan(\theta + \alpha) = \frac{63}{x}$$
$$\theta + \alpha = \tan^{-1}\frac{63}{x}$$
$$\theta = \tan^{-1}\frac{63}{x} - \alpha$$
$$\theta = \tan^{-1}\frac{63}{x} - \tan^{-1}\frac{10}{x}$$

128. b.

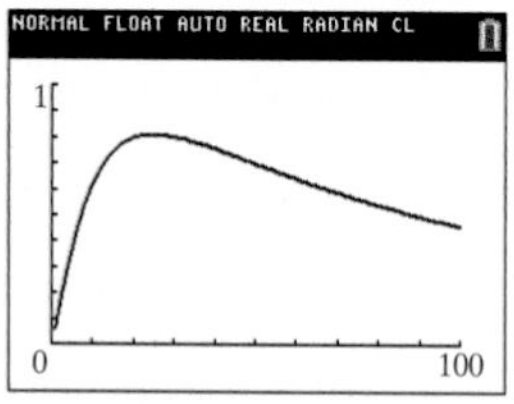
NORMAL FLOAT AUTO REAL RADIAN CL
1
0
100

c. 25.1 ft

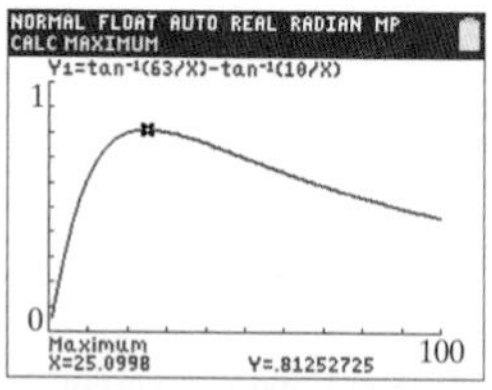
NORMAL FLOAT AUTO REAL RADIAN MP
CALC MAXIMUM
Y₁=tan⁻¹(63/X)-tan⁻¹(10/X)
1
0
Maximum
X=25.0998
Y=.81252725
100

129. a.

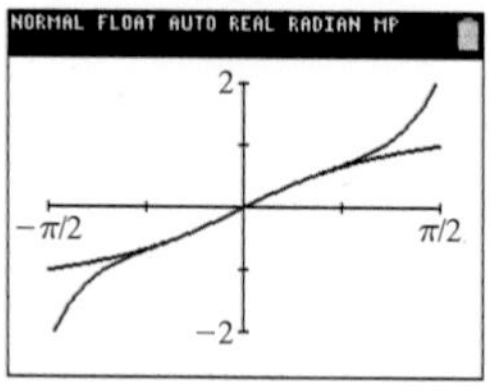
NORMAL FLOAT AUTO REAL RADIAN MP
2
−π/2
π/2
−2

b.

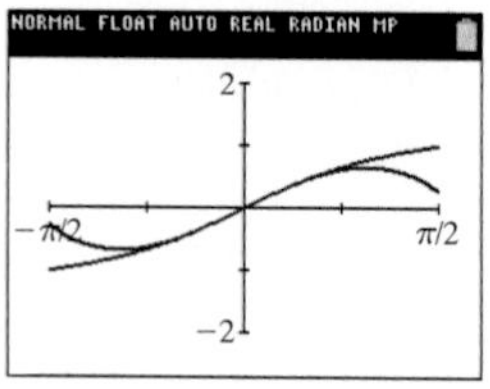
NORMAL FLOAT AUTO REAL RADIAN MP
2
−π/2
π/2
−2

Chapter 4 Review Exercises, pp. 561–564

21. a. $\dfrac{2}{7}$ **b.** $-\dfrac{3\sqrt{5}}{7}$ **c.** $-\dfrac{2\sqrt{5}}{15}$ **d.** $\dfrac{7}{2}$ **e.** $-\dfrac{7\sqrt{5}}{15}$ **f.** $-\dfrac{3\sqrt{5}}{2}$

22. a. $\left(-\dfrac{1}{2}, \dfrac{\sqrt{3}}{2}\right)$ **b.** $\left(-\dfrac{\sqrt{2}}{2}, -\dfrac{\sqrt{2}}{2}\right)$ **c.** $\left(\dfrac{\sqrt{3}}{2}, -\dfrac{1}{2}\right)$

34. $\sin\theta = \dfrac{5\sqrt{3}}{9}$, $\cos\theta = \dfrac{\sqrt{6}}{9}$, $\tan\theta = \dfrac{5\sqrt{2}}{2}$,

$\csc\theta = \dfrac{3\sqrt{3}}{5}$, $\sec\theta = \dfrac{3\sqrt{6}}{2}$, $\cot\theta = \dfrac{\sqrt{2}}{5}$

35. $\sin\theta = \dfrac{33}{65}$, $\cos\theta = \dfrac{56}{65}$, $\tan\theta = \dfrac{33}{56}$,

$\csc\theta = \dfrac{65}{33}$, $\sec\theta = \dfrac{65}{56}$, $\cot\theta = \dfrac{56}{33}$

36. $\csc\theta = \dfrac{7\sqrt{6}}{12}$, $\tan\theta = \dfrac{2\sqrt{6}}{5}$ **37.** $\sin\theta = \dfrac{3\sqrt{10}}{10}$, $\sec\theta = \sqrt{10}$

40. $\tan\theta = \dfrac{99}{20}$, $\csc\theta = \dfrac{101}{99}$, $\sec\theta = \dfrac{101}{20}$, $\cot\theta = \dfrac{20}{99}$

49. $\sin\theta = -\dfrac{5\sqrt{34}}{34}$, $\cos\theta = \dfrac{3\sqrt{34}}{34}$, $\tan\theta = -\dfrac{5}{3}$,

$\csc\theta = -\dfrac{\sqrt{34}}{5}$, $\sec\theta = \dfrac{\sqrt{34}}{3}$, $\cot\theta = -\dfrac{3}{5}$

70.

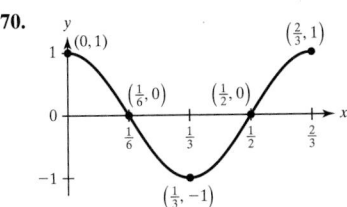

71.

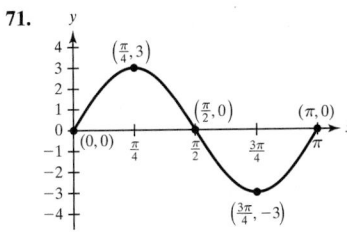

72.

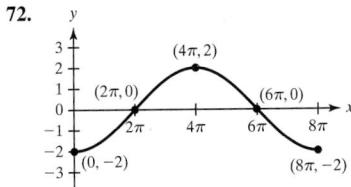

73. a. Amplitude $= \dfrac{1}{3}$, Period $= \pi$,   **b.** Amplitude $= 5$, Period $= 1$,

Phase shift $= -\dfrac{\pi}{2}$ Phase shift $= \dfrac{1}{2}$

74.

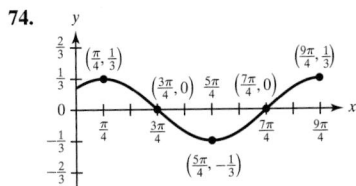

75.

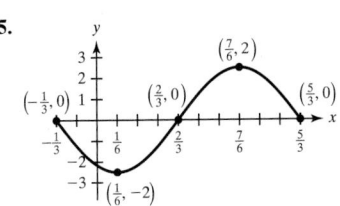

76. Amplitude $= 4$, Period $= \dfrac{2\pi}{3}$,

Phase shift $= -\dfrac{\pi}{3}$, Vertical shift $= -2$

77. Amplitude $= \dfrac{1}{4}$, Period $= 6$, Phase shift $= \dfrac{3}{2}$, Vertical shift $= 5$

78.

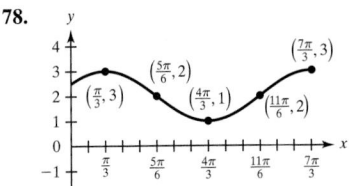

79.

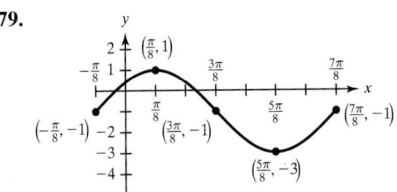

81. b.

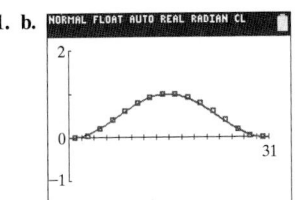

82. Period: $\dfrac{2\pi}{3}$; Asymptotes: $x = 0$, $x = \dfrac{\pi}{3}$

83. Period: $\dfrac{\pi}{3}$; Asymptotes: $x = -\dfrac{\pi}{6}$, $x = \dfrac{\pi}{6}$

84. Period: 1; Asymptotes: $x = 0$, $x = 1$

85. Period: 2π; Asymptotes: $x = \dfrac{\pi}{2}$, $x = \dfrac{3\pi}{2}$

86.

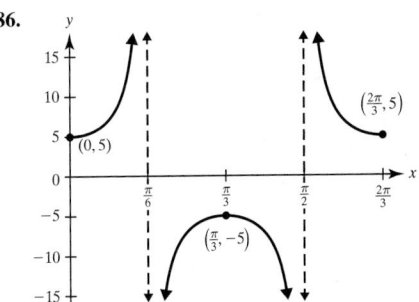

87.

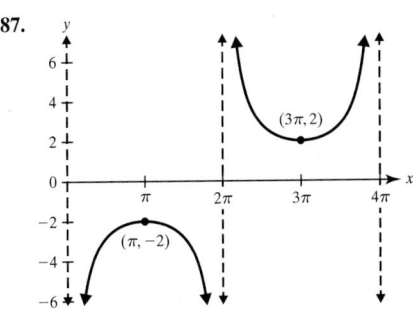

88.

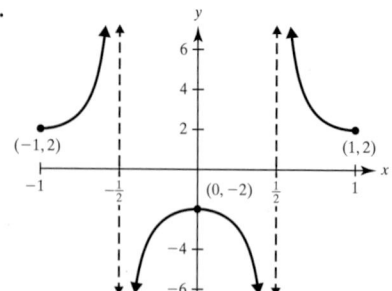

89.

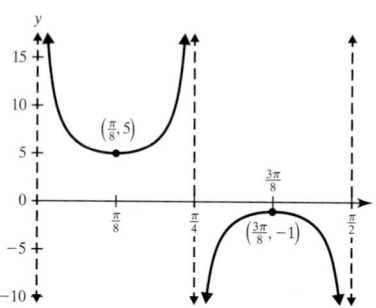

90.

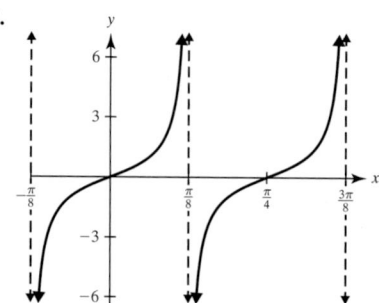

91.

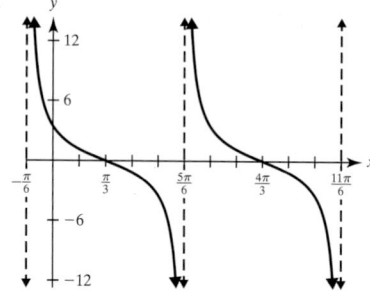

92.

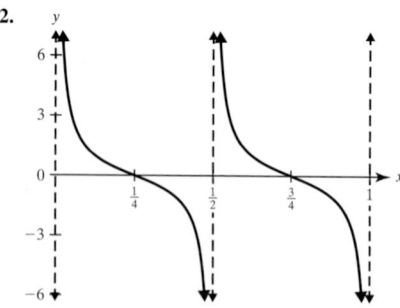

93.

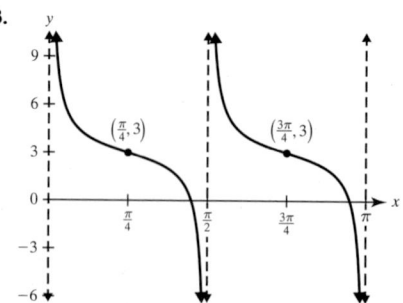

110. a.–c.

In radians: In degrees:

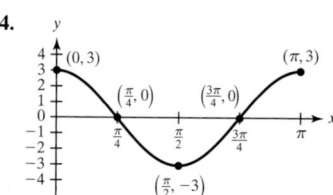
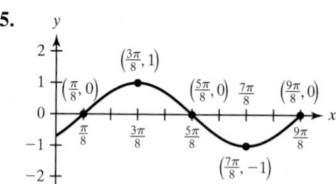
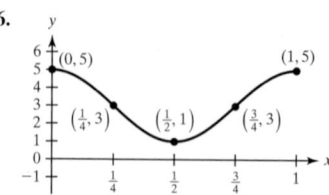

Chapter 4 Test, pp. 564–565

7. $\cos\theta = \dfrac{\sqrt{11}}{6}$, $\tan\theta = \dfrac{5\sqrt{11}}{11}$

31. Amplitude $= \dfrac{3}{4}$, Period $= 1$, Phase shift $= \dfrac{1}{12}$, Vertical shift $= 0$

32. Amplitude $= 5$, Period $= \dfrac{2\pi}{5}$, Phase shift $= -\dfrac{\pi}{5}$, Vertical shift $= 7$

33.

34.

35.

36.

39.

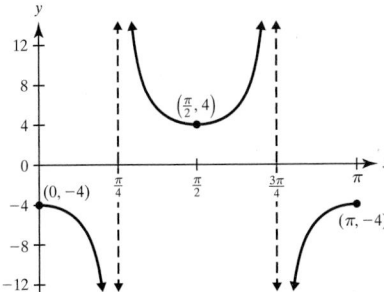

40.

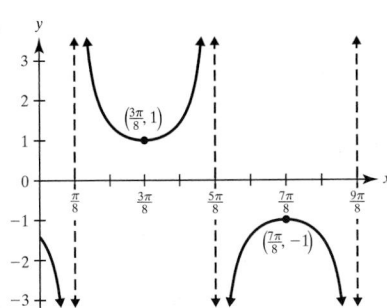

41.

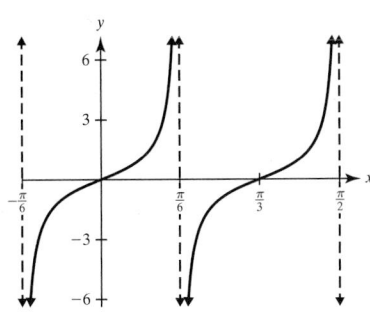

42.

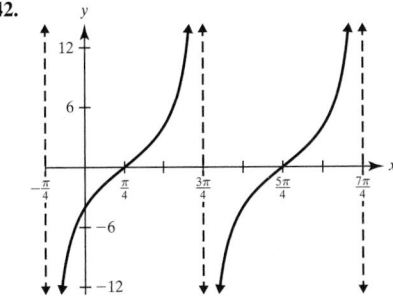

49. a. $\theta = \tan^{-1}\dfrac{6}{x}$

b. θ is greater than $45°$ because the ratio $\dfrac{6}{3.2}$ is greater than 1.

Chapter 4 Cumulative Review Exercises, p. 566

13. a.

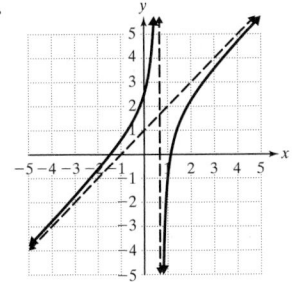

CHAPTER 5

Section 5.1 Skill Practice
Answer 5

$$\frac{1}{1 - \sin x} - \frac{1}{1 + \sin x} = \frac{1}{(1 + \sin x)} \cdot \frac{1}{1 - \sin x} - \frac{1}{1 + \sin x} \cdot \frac{(1 - \sin x)}{(1 - \sin x)}$$

$$= \frac{1 + \sin x - 1 + \sin x}{(1 + \sin x)(1 - \sin x)} = \frac{2\sin x}{1 - \sin^2 x} = \frac{2\sin x}{\cos^2 x}$$

$$= 2 \cdot \frac{\sin x}{\cos x} \cdot \frac{1}{\cos x} = 2\tan x \sec x$$

Answer 6

$$1 - \frac{\sin^2 t}{1 + \cos t} = 1 - \frac{\sin^2 t}{(1 + \cos t)} \cdot \frac{(1 - \cos t)}{(1 - \cos t)} = 1 - \frac{\sin^2 t(1 - \cos t)}{1 - \cos^2 t}$$

$$= 1 - \frac{\sin^2 t(1 - \cos t)}{\sin^2 t} = 1 - (1 - \cos t) = \cos t$$

Answer 7

$$\ln|\cot x| + \ln|\sin x| = \ln\left|\frac{\cos x}{\sin x}\right| + \ln|\sin x|$$

$$= \ln\left(\left|\frac{\cos x}{\sin x}\right| \cdot |\sin x|\right) = \ln|\cos x|$$

Section 5.1 Practice Exercises, pp. 575–578

7. a. $ab;\ \dfrac{1}{a} = \dfrac{b}{ab},\ \dfrac{1}{b} = \dfrac{a}{ab}$

b. $\cos x \sin x;\ \dfrac{1}{\cos x} = \dfrac{\sin x}{\cos x \sin x},\ \dfrac{1}{\sin x} = \dfrac{\cos x}{\cos x \sin x}$

8. a. $b;\ \dfrac{a}{b} = \dfrac{a}{b},\ a = \dfrac{ab}{b}$ **b.** $\cos x;\ \dfrac{\sin x}{\cos x} = \dfrac{\sin x}{\cos x},\ \sin x = \dfrac{\sin x \cos x}{\cos x}$

9. a. $b(1 + a);\ \dfrac{1}{1 + a} = \dfrac{b}{b(1 + a)},\ \dfrac{a}{b} = \dfrac{a + a^2}{b(1 + a)}$

b. $\cos x(1 + \sin x);$

$$\frac{1}{1 + \sin x} = \frac{\cos x}{\cos x(1 + \sin x)},\ \frac{\sin x}{\cos x} = \frac{\sin x + \sin^2 x}{\cos x(1 + \sin x)}$$

10. a. $a(1 - a);\ \dfrac{1}{a} = \dfrac{1 - a}{a(1 - a)},\ \dfrac{a}{1 - a} = \dfrac{a^2}{a(1 - a)}$

b. $\cos x(1 - \cos x);\ \dfrac{1}{\cos x} = \dfrac{1 - \cos x}{\cos x(1 - \cos x)},\ \dfrac{\cos x}{1 - \cos x} = \dfrac{\cos^2 x}{\cos x(1 - \cos x)}$

13. a. $(a + b)(a - b)$ **b.** $(\sin x + \cos x)(\sin x - \cos x)$

14. a. Not factorable over the real numbers
b. Not factorable over the real numbers

15. a. $(a + b)(a^2 - ab + b^2)$
b. $(\sin x + \sec x)(\sin^2 x - \sin x \sec x + \sec^2 x)$

16. a. $(a - b)(a^2 + ab + b^2)$
b. $(\cos x - \cot x)(\cos^2 x + \cos x \cot x + \cot^2 x)$

17. a. $(a + b)^2$ **b.** $(\cos x + \csc x)^2$ **18. a.** $(a - b)^2$ **b.** $(\sin x - \tan x)^2$

19. a. $(3a + 2b)(a - 2b)$ **b.** $(3\sin x + 2\cos x)(\sin x - 2\cos x)$

20. a. $(2a - b)(2a - 5b)$ **b.** $(2\cos x - \sin x)(2\cos x - 5\sin x)$

37. $\sin(-x) + \csc x = -\sin x + \dfrac{1}{\sin x} = -\sin x \cdot \dfrac{\sin x}{\sin x} + \dfrac{1}{\sin x}$

$$= \frac{-\sin^2 x + 1}{\sin x} = \frac{\cos^2 x}{\sin x} = \frac{\cos x}{\sin x} \cdot \cos x = \cot x \cos x$$

38. $\sin(-x) + \cot(-x)\cos(-x) = -\sin x - \dfrac{\cos x}{\sin x} \cdot \cos x$

$$= -\sin x \cdot \frac{\sin x}{\sin x} - \frac{\cos^2 x}{\sin x} = \frac{-\sin^2 x - \cos^2 x}{\sin x}$$

$$= -\frac{1}{\sin x} = -\csc x$$

39. $\cot x[\cot(-x) + \tan(-x)] = \cot x(-\cot x - \tan x)$

$$= -\cot^2 x - \cot x \tan x = -\cot x^2 - 1 = -\csc^2 x$$

40. $\dfrac{\csc^2(-x)\tan x}{\cot(-x)} = \dfrac{(-\csc x)^2\tan x}{-\cot x} = -\csc^2 x\tan^2 x$

$= -\dfrac{1}{\sin^2 x}\cdot\dfrac{\sin^2 x}{\cos^2 x} = -\dfrac{1}{\cos^2 x} = -\sec^2 x$

41. $\dfrac{\sec x}{\tan x} - \dfrac{\tan x}{\sec x} = \dfrac{\sec x}{\tan x}\cdot\dfrac{\sec x}{\sec x} - \dfrac{\tan x}{\sec x}\cdot\dfrac{\tan x}{\tan x} = \dfrac{\sec^2 x - \tan^2 x}{\sec x\tan x}$

$= \dfrac{1}{\sec x\tan x} = \cos x\cot x$

42. $\dfrac{\cot x}{\csc x} - \dfrac{\csc x}{\cot x} = \dfrac{\cot x}{\csc x}\cdot\dfrac{\cot x}{\cot x} - \dfrac{\csc x}{\cot x}\cdot\dfrac{\csc x}{\csc x} = \dfrac{\cot^2 x - \csc^2 x}{\csc x\cot x}$

$= \dfrac{-1}{\csc x\cot x} = -\sin x\tan x$

43. $\dfrac{1}{1+\sin t} + \dfrac{1}{1-\sin t} = \dfrac{1}{(1+\sin t)}\cdot\dfrac{(1-\sin t)}{(1-\sin t)} + \dfrac{1}{(1-\sin t)}\cdot\dfrac{(1+\sin t)}{(1+\sin t)}$

$= \dfrac{1-\sin t}{1-\sin^2 t} + \dfrac{1+\sin t}{1-\sin^2 t} = \dfrac{1-\sin t + 1+\sin t}{1-\sin^2 t} = \dfrac{2}{\cos^2 t} = 2\sec^2 t$

44. $\dfrac{1}{\sec t - 1} - \dfrac{1}{\sec t + 1} = \dfrac{1}{(\sec t - 1)}\cdot\dfrac{(\sec t + 1)}{(\sec t + 1)} - \dfrac{1}{(\sec t + 1)}\cdot\dfrac{(\sec t - 1)}{(\sec t - 1)}$

$= \dfrac{\sec t + 1}{\sec^2 t - 1} - \dfrac{\sec t - 1}{\sec^2 t - 1} = \dfrac{\sec t + 1 - \sec t + 1}{\sec^2 t - 1} = \dfrac{2}{\tan^2 t} = 2\cot^2 t$

45. $\dfrac{1}{1+\sec x} - \dfrac{1}{1-\sec x} = \dfrac{1}{(1+\sec x)}\cdot\dfrac{(1-\sec x)}{(1-\sec x)}$

$- \dfrac{1}{(1-\sec x)}\cdot\dfrac{(1+\sec x)}{(1+\sec x)} = \dfrac{1-\sec x - (1+\sec x)}{(1+\sec x)(1-\sec x)}$

$= \dfrac{-2\sec x}{1-\sec^2 x} = \dfrac{-2\sec x}{-\tan^2 x} = \dfrac{\frac{2}{\cos x}}{\frac{\sin^2 x}{\cos^2 x}} = \dfrac{2}{\cos x}\cdot\dfrac{\cos^2 x}{\sin^2 x}$

$= \dfrac{2\cos x}{\sin^2 x} = 2\cdot\dfrac{\cos x}{\sin x}\cdot\dfrac{1}{\sin x} = 2\cot x\csc x$

46. $\dfrac{1}{1-\csc x} - \dfrac{1}{1+\csc x} = \dfrac{1}{(1-\csc x)}\cdot\dfrac{(1+\csc x)}{(1+\csc x)}$

$- \dfrac{1}{(1+\csc x)}\cdot\dfrac{(1-\csc x)}{(1-\csc x)} = \dfrac{1+\csc x - (1-\csc x)}{(1-\csc x)(1+\csc x)}$

$= \dfrac{2\csc x}{1-\csc^2 x} = \dfrac{2\csc x}{-\cot^2 x} = -\dfrac{\frac{2}{\sin x}}{\frac{\cos^2 x}{\sin^2 x}} = -\dfrac{2}{\sin x}\cdot\dfrac{\sin^2 x}{\cos^2 x}$

$= -\dfrac{2\sin x}{\cos^2 x} = -2\cdot\dfrac{\sin x}{\cos x}\cdot\dfrac{1}{\cos x} = -2\tan x\sec x$

47. $\dfrac{\sin\theta}{\csc\theta - \cot\theta} = \dfrac{\sin\theta}{(\csc\theta - \cot\theta)}\cdot\dfrac{(\csc\theta + \cot\theta)}{(\csc\theta + \cot\theta)}$

$= \dfrac{\sin\theta\csc\theta + \sin\theta\cot\theta}{\csc^2\theta - \cot^2\theta} = \dfrac{\sin\theta\cdot\frac{1}{\sin\theta} + \sin\theta\cdot\frac{\cos\theta}{\sin\theta}}{1} = 1+\cos\theta$

48. $\dfrac{\cos\theta}{\tan\theta + \sec\theta} = \dfrac{\cos\theta}{(\tan\theta + \sec\theta)}\cdot\dfrac{(\tan\theta - \sec\theta)}{(\tan\theta - \sec\theta)}$

$= \dfrac{\cos\theta\tan\theta - \cos\theta\sec\theta}{\tan^2\theta - \sec^2\theta} = \dfrac{\cos\theta\cdot\frac{\sin\theta}{\cos\theta} - \cos\theta\cdot\frac{1}{\cos\theta}}{-1} = 1-\sin\theta$

49. $\dfrac{1}{\cos x + \sin x\cos x} = \dfrac{1}{\cos x(1+\sin x)}\cdot\dfrac{(1-\sin x)}{(1-\sin x)}$

$= \dfrac{1-\sin x}{\cos x(1-\sin^2 x)} = \dfrac{1-\sin x}{\cos x(\cos^2 x)} = \dfrac{1-\sin x}{\cos^3 x}$

50. $\dfrac{1}{\sin x - \sin^2 x} = \dfrac{1}{\sin x(1-\sin x)}\cdot\dfrac{(1+\sin x)}{(1+\sin x)}$

$= \dfrac{1+\sin x}{\sin x(1-\sin^2 x)} = \dfrac{1+\sin x}{\sin x\cos^2 x}$

51. $\dfrac{\sec x + 1}{\sec x - 1} = \dfrac{(\sec x + 1)}{(\sec x - 1)}\cdot\dfrac{(\sec x + 1)}{(\sec x + 1)} = \dfrac{(\sec x + 1)^2}{\sec^2 x - 1} = \dfrac{(\sec x + 1)^2}{\tan^2 x}$

$= \left(\dfrac{\sec x + 1}{\tan x}\right)^2 = \left(\dfrac{\sec x}{\tan x} + \dfrac{1}{\tan x}\right)^2 = \left(\dfrac{\frac{1}{\cos x}}{\frac{\sin x}{\cos x}} + \cot x\right)^2$

$= \left(\dfrac{1}{\cos x}\cdot\dfrac{\cos x}{\sin x} + \cot x\right)^2 = (\csc x + \cot x)^2$

52. $\dfrac{\sin x + \cos x}{\sin x - \cos x} = \dfrac{(\sin x + \cos x)}{(\sin x - \cos x)}\cdot\dfrac{(\sin x + \cos x)}{(\sin x + \cos x)}$

$= \dfrac{(\sin x + \cos x)^2}{\sin^2 x - \cos^2 x} = \dfrac{\sin^2 x + 2\sin x\cos x + \cos^2 x}{1-\cos^2 x - \cos^2 x}$

$= \dfrac{1+2\sin x\cos x}{1-2\cos^2 x}$

53. $\ln|\cos t| - \ln|\cot t| = \ln|\cos t| - \ln\dfrac{|\cos t|}{|\sin t|}$

$= \ln|\cos t| - (\ln|\cos t| - \ln|\sin t|)$

$= \ln|\cos t| - \ln|\cos t| + \ln|\sin t| = \ln|\sin t|$

54. $\ln|\cot t| - \ln|\tan t| = \ln\left|\dfrac{\cot t}{\tan t}\right| = \ln|\cot^2 t| = 2\ln|\cot t|$

55. $\ln|\sec\theta + \tan\theta| = \ln\left|(\sec\theta + \tan\theta)\cdot\dfrac{(\sec\theta - \tan\theta)}{(\sec\theta - \tan\theta)}\right|$

$= \ln\left|\dfrac{\sec^2\theta - \tan^2\theta}{\sec\theta - \tan\theta}\right| = \ln\left|\dfrac{1}{\sec\theta - \tan\theta}\right|$

$= \ln 1 - \ln|\sec\theta - \tan\theta| = -\ln|\sec\theta - \tan\theta|$

56. $\ln|\csc\theta + \cot\theta| = \ln\left|(\csc\theta + \cot\theta)\cdot\dfrac{(\csc\theta - \cot\theta)}{(\csc\theta - \cot\theta)}\right|$

$= \ln\left|\dfrac{\csc^2\theta - \cot^2\theta}{\csc\theta - \cot\theta}\right| = \ln\left|\dfrac{1}{\csc\theta - \cot\theta}\right|$

$= \ln 1 - \ln|\csc\theta - \cot\theta| = -\ln|\csc\theta - \cot\theta|$

57. LHS: $\dfrac{\cos x}{1-\sin x} = \dfrac{\cos x}{(1-\sin x)}\cdot\dfrac{(1+\sin x)}{(1+\sin x)} = \dfrac{\cos x(1+\sin x)}{1-\sin^2 x}$

$= \dfrac{\cos x(1+\sin x)}{\cos^2 x} = \dfrac{1+\sin x}{\cos x}$

RHS: $\sec x + \tan x = \dfrac{1}{\cos x} + \dfrac{\sin x}{\cos x} = \dfrac{1+\sin x}{\cos x}$

58. LHS: $\dfrac{1+\tan^2 x}{\tan x} = \dfrac{\sec^2 x}{\tan x} = \dfrac{\frac{1}{\cos^2 x}}{\frac{\sin x}{\cos x}} = \dfrac{1}{\cos^2 x}\cdot\dfrac{\cos x}{\sin x} = \dfrac{1}{\cos x\sin x}$

RHS: $\dfrac{\cos x}{\sin x - \sin^3 x} = \dfrac{\cos x}{\sin x(1-\sin^2 x)} = \dfrac{\cos x}{\sin x\cos^2 x} = \dfrac{1}{\cos x\sin x}$

59. LHS: $\cot^2 t - \cos^2 t = \dfrac{\cos^2 t}{\sin^2 t} - \cos^2 t\cdot\dfrac{\sin^2 t}{\sin^2 t}$

$= \dfrac{\cos^2 t - \cos^2 t\sin^2 t}{\sin^2 t} = \dfrac{\cos^2 t(1-\sin^2 t)}{\sin^2 t}$

$= \dfrac{\cos^2 t}{\sin^2 t}\cdot\cos^2 t = \cot^2 t\cos^2 t$

RHS: $\csc^2 t\cos^2 t - \cos^2 t = \cos^2 t(\csc^2 t - 1) = \cos^2 t\cot^2 t$

60. LHS: $\tan^2 t + \sin^2 t = \dfrac{\sin^2 t}{\cos^2 t} + \sin^2 t\cdot\dfrac{\cos^2 t}{\cos^2 t}$

$= \dfrac{\sin^2 t + \sin^2 t\cos^2 t}{\cos^2 t} = \dfrac{\sin^2 t(1+\cos^2 t)}{\cos^2 t} = \dfrac{(1-\cos^2 t)(1+\cos^2 t)}{\cos^2 t}$

$= \dfrac{1-\cos^4 t}{\cos^2 t}$

RHS: $\sec^2 t - \cos^2 t = \dfrac{1}{\cos^2 t} - \cos^2 t\cdot\dfrac{\cos^2 t}{\cos^2 t} = \dfrac{1-\cos^4 t}{\cos^2 t}$

73. $(\cos^2\theta - 1)(\csc^2\theta - 1) = -\sin^2\theta \cdot \cot^2\theta$

$$= -\sin^2\theta \cdot \frac{\cos^2\theta}{\sin^2\theta} = -\cos^2\theta$$

74. $(1 - \sin^2\theta)(1 + \tan^2\theta) = \cos^2\theta \cdot \sec^2\theta = 1$

75. $\sec x \cdot \frac{\tan x}{\sin x} = \frac{1}{\cos x} \cdot \frac{\frac{\sin x}{\cos x}}{\sin x} = \frac{1}{\cos x} \cdot \frac{\sin x}{\cos x} \cdot \frac{1}{\sin x} = \frac{1}{\cos^2 x} = \sec^2 x$

76. $\sec x \cdot \frac{\cot x}{\sin x} = \frac{1}{\cos x} \cdot \frac{\frac{\cos x}{\sin x}}{\sin x} = \frac{1}{\cos x} \cdot \frac{\cos x}{\sin x} \cdot \frac{1}{\sin x} = \frac{1}{\sin^2 x} = \csc^2 x$

77. $\frac{\sec x \sin x}{\csc x \cos x} = \frac{\frac{1}{\cos x} \cdot \sin x}{\frac{1}{\sin x} \cdot \cos x} = \frac{\frac{\sin x}{\cos x}}{\frac{\cos x}{\sin x}} = \frac{\sin x}{\cos x} \cdot \frac{\sin x}{\cos x} = \frac{\sin^2 x}{\cos^2 x} = \tan^2 x$

78. $\frac{\sin x \cot x}{\cos x \tan x} = \frac{\sin x \cdot \frac{\cos x}{\sin x}}{\cos x \cdot \frac{\sin x}{\cos x}} = \frac{\cos x}{\sin x} = \cot x$

79. $\frac{2 + \sin^2 t - 3\sin^4 t}{\cos^2 t} = \frac{(1 - \sin^2 t)(2 + 3\sin^2 t)}{\cos^2 t}$

$$= \frac{\cos^2 t (2 + 3\sin^2 t)}{\cos^2 t} = 2 + 3\sin^2 t$$

80. $\frac{2\tan^4 x + 7\tan^2 x + 5}{\sec^2 x} = \frac{(\tan^2 x + 1)(2\tan^2 x + 5)}{\sec^2 x}$

$$= \frac{\sec^2 x (2\tan^2 x + 5)}{\sec^2 x} = 2\tan^2 x + 5$$

81. $\frac{\tan x \sin x}{\tan x - \sin x} = \frac{\left(\frac{\sin x}{\cos x} \cdot \sin x\right)}{\left(\frac{\sin x}{\cos x} - \sin x\right)} \cdot \frac{\cos x}{\cos x} = \frac{\sin^2 x}{\sin x - \sin x \cos x}$

$$= \frac{\sin^2 x}{\sin x(1 - \cos x)} = \frac{\sin x}{1 - \cos x}$$

82. $\frac{\csc x \cos x}{\csc x - \sin x} = \frac{\left(\frac{1}{\sin x} \cdot \cos x\right)}{\left(\frac{1}{\sin x} - \sin x\right)} \cdot \frac{\sin x}{\sin x} = \frac{\cos x}{1 - \sin^2 x}$

$$= \frac{\cos x}{\cos^2 x} = \frac{1}{\cos x} = \sec x$$

83. $(\sec\theta + \tan\theta)^2 = \sec^2\theta + 2\sec\theta\tan\theta + \tan^2\theta$

$$= \frac{1}{\cos^2\theta} + 2 \cdot \frac{1}{\cos\theta} \cdot \frac{\sin\theta}{\cos\theta} + \frac{\sin^2\theta}{\cos^2\theta} = \frac{1 + 2\sin\theta + \sin^2\theta}{\cos^2\theta}$$

$$= \frac{(1 + \sin\theta)^2}{1 - \sin^2\theta} = \frac{(1 + \sin\theta)^2}{(1 + \sin\theta)(1 - \sin\theta)} = \frac{1 + \sin\theta}{1 - \sin\theta}$$

84. $(\csc t + \cot t)^2 = \csc^2 t + 2\csc t\cot t + \cot^2 t$

$$= \frac{1}{\sin^2 t} + 2 \cdot \frac{1}{\sin t} \cdot \frac{\cos t}{\sin t} + \frac{\cos^2 t}{\sin^2 t} = \frac{1 + 2\cos t + \cos^2 t}{\sin^2 t}$$

$$= \frac{(1 + \cos t)^2}{1 - \cos^2 t} = \frac{(1 + \cos t)^2}{(1 + \cos t)(1 - \cos t)} = \frac{1 + \cos t}{1 - \cos t}$$

85. $(1 + \sin x)[1 + \sin(-x)] = (1 + \sin x)(1 - \sin x)$

$$= 1 - \sin^2 x = \cos^2 x$$

86. $[1 + \csc(-x)](1 + \csc x) = (1 - \csc x)(1 + \csc x)$

$$= 1 - \csc^2 x = -\cot^2 x$$

87. $\frac{\sin x}{\cos x + 1} = \frac{\sin x}{\cos x + 1} \cdot \frac{\cos x - 1}{\cos x - 1} = \frac{\sin x(\cos x - 1)}{\cos^2 x - 1}$

$$= \frac{\sin x(\cos x - 1)}{-\sin^2 x} = \frac{\cos x - 1}{-\sin x} = \frac{1 - \cos x}{\sin x}$$

88. $\frac{\tan x}{\sec x - 1} = \frac{\tan x}{\sec x - 1} \cdot \frac{\sec x + 1}{\sec x + 1} = \frac{\tan x(\sec x + 1)}{\sec^2 x - 1}$

$$= \frac{\tan x(\sec x + 1)}{\tan^2 x} = \frac{\sec x + 1}{\tan x}$$

89. $\frac{1 + \sin\theta}{\tan\theta} = \frac{\frac{1}{\sin\theta} \cdot (1 + \sin\theta)}{\frac{1}{\sin\theta} \cdot \left(\frac{\sin\theta}{\cos\theta}\right)} = \frac{\frac{1}{\sin\theta} + 1}{\frac{1}{\cos\theta}} = \frac{\csc\theta + 1}{\sec\theta}$

90. $\frac{\sec\theta + 1}{\csc\theta} = \frac{\cos\theta\left(\frac{1}{\cos\theta} + 1\right)}{\cos\theta\left(\frac{1}{\sin\theta}\right)} = \frac{1 + \cos\theta}{\cot\theta}$

91. $\frac{\sin^3 t - \cos^3 t}{\cos t \sin t - \cos^2 t} = \frac{(\sin t - \cos t)(\sin^2 t + \sin t\cos t + \cos^2 t)}{\cos t(\sin t - \cos t)}$

$$= \frac{1 + \sin t\cos t}{\cos t} = \frac{1}{\cos t} + \frac{\sin t\cos t}{\cos t} = \sec t + \sin t$$

92. $\frac{\cos^3 t + \sin^3 t}{\cos t \sin t + \sin^2 t} = \frac{(\cos t + \sin t)(\cos^2 t - \sin t\cos t + \sin^2 t)}{\sin t(\cos t + \sin t)}$

$$= \frac{1 - \sin t\cos t}{\sin t} = \frac{1}{\sin t} - \frac{\sin t\cos t}{\sin t} = \csc t - \cos t$$

93. $\frac{\tan^4\theta - 4}{\sec^2\theta + 1} = \frac{(\tan^2\theta + 2)(\tan^2\theta - 2)}{\sec^2\theta + 1} = \frac{(\sec^2\theta - 1 + 2)(\sec^2\theta - 1 - 2)}{\sec^2\theta + 1}$

$$= \frac{(\sec^2\theta + 1)(\sec^2\theta - 3)}{\sec^2\theta + 1} = \sec^2\theta - 3$$

94. $\frac{\sin^4\theta - 9}{\cos^2\theta + 2} = \frac{(\sin^2\theta - 3)(\sin^2\theta + 3)}{\cos^2\theta + 2} = \frac{(1 - \cos^2\theta - 3)(1 - \cos^2\theta + 3)}{\cos^2\theta + 2}$

$$= \frac{(-\cos^2\theta - 2)(4 - \cos^2\theta)}{\cos^2\theta + 2} = -(4 - \cos^2\theta) = \cos^2\theta - 4$$

95. $\frac{\cot^3 z - 1}{\cot z - 1} - \cot z = \frac{(\cot z - 1)(\cot^2 z + \cot z + 1)}{\cot z - 1} - \cot z$

$$= \cot^2 z + \cot z + 1 - \cot z = \cot^2 z + 1 = \csc^2 z$$

96. $\frac{\tan^3 z + 1}{\tan z + 1} + \tan z = \frac{(\tan z + 1)(\tan^2 z - \tan z + 1)}{\tan z + 1} + \tan z$

$$= \tan^2 z - \tan z + 1 + \tan z = \tan^2 z + 1 = \sec^2 z$$

97. $\cos x\tan x - \sec x\cot x = \cos x \cdot \frac{\sin x}{\cos x} - \frac{1}{\cos x} \cdot \frac{\cos x}{\sin x}$

$$= \sin x - \frac{1}{\sin x} = \frac{\sin^2 x - 1}{\sin x} = -\frac{\cos^2 x}{\sin x} = -\frac{\cos x}{\sin x} \cdot \cos x$$

$$= -\cot x\cos x$$

98. $\sin x\cot x + \cos x\tan^2 x = \sin x \cdot \frac{\cos x}{\sin x} + \cos x \cdot \frac{\sin^2 x}{\cos^2 x}$

$$= \cos x + \frac{\sin^2 x}{\cos x} = \frac{\cos^2 x + \sin^2 x}{\cos x} = \frac{1}{\cos x} = \sec x$$

99. $\log|\cos x| - \log|\sin x| + \log|\sec x| = \log\left|\frac{\cos x\sec x}{\sin x}\right|$

$$= \log\left|\frac{1}{\sin x}\right| = \log|\csc x|$$

100. $\ln|\cot x| + \ln|\sec x| + \ln|\sin x| = \ln|\cot x \cdot \sec x \cdot \sin x|$

$$= \ln\left|\frac{\cos x}{\sin x} \cdot \frac{1}{\cos x} \cdot \sin x\right| = \ln 1 = 0$$

101. $e^{\ln(\sin^2 x + \cos^2 x)} = \sin^2 x + \cos^2 x = 1$

102. $\log 100^{\sin x} = \sin x\log 100 = 2\sin x$

103. $\frac{1}{\sin x + \cos x} + \frac{1}{\sin x - \cos x} = \frac{\sin x - \cos x + \sin x + \cos x}{\sin^2 x - \cos^2 x}$

$$= \frac{2\sin x}{1 - \cos^2 x - \cos^2 x} = \frac{2\sin x}{1 - 2\cos^2 x}$$

104. $\frac{1}{\sin x + \cos x} - \frac{\sin x}{(\sin x + \cos x)^2} = \frac{\sin x + \cos x - \sin x}{(\sin x + \cos x)^2}$

$$= \frac{\cos x}{\sin^2 x + 2\sin x\cos x + \cos^2 x} = \frac{\cos x}{1 + 2\sin x\cos x}$$

105. $(f \circ g)(x) = \sqrt{1 - \cos^2 x} = \sqrt{\sin^2 x} = |\sin x|$

106. $(f \circ g)(x) = \dfrac{1}{\sqrt{\tan^2 x + 1}} = \dfrac{1}{\sqrt{\sec^2 x}} = \dfrac{1}{|\sec x|} = |\cos x|$

107. The logic is incorrect because it shows that $\sin x \sec x = 1$ only for one value of x. For this statement to be true, it must be true for *all* values of x for which $\sin x$ and $\sec x$ are defined. The counterexample $\sin\dfrac{\pi}{6}\sec\dfrac{\pi}{6} = \dfrac{1}{2} \cdot \dfrac{2\sqrt{3}}{3} = \dfrac{\sqrt{3}}{3} \neq 1$ shows that $\sin x \sec x = 1$ is not true in general.

108. The procedure is incorrect. We are trying to show that the expression on one side equals the expression on the other. That is, we cannot *assume* the equality of the statement and multiply both sides by the common denominator. We must *prove* the equality of the statement first.

109. One method is to use the tools of algebra and the fundamental trigonometric identities to manipulate one side of the equation until we get the expression on the other side of the equation.

Alternatively, we can manipulate each side separately until we reach a common, equivalent expression.

110. $\sqrt{\dfrac{1 + \cos x}{1 - \cos x}} = \sqrt{\dfrac{(1 + \cos x)}{(1 - \cos x)} \cdot \dfrac{(1 + \cos x)}{(1 + \cos x)}} = \sqrt{\dfrac{(1 + \cos x)^2}{1 - \cos^2 x}}$

$= \sqrt{\dfrac{(1 + \cos x)^2}{\sin^2 x}} = \dfrac{1 + \cos x}{\sin x} = \dfrac{1}{\sin x} + \dfrac{\cos x}{\sin x} = \csc x + \cot x$

111. $\sqrt{\dfrac{\csc x - \cot x}{\csc x + \cot x}} = \sqrt{\dfrac{(\csc x - \cot x)}{(\csc x + \cot x)} \cdot \dfrac{(\csc x - \cot x)}{(\csc x - \cot x)}}$

$= \sqrt{\dfrac{(\csc x - \cot x)^2}{\csc^2 x - \cot^2 x}} = \sqrt{\dfrac{(\csc x - \cot x)^2}{1}} = \csc x - \cot x$

$= \dfrac{1}{\sin x} - \dfrac{\cos x}{\sin x} = \dfrac{1 - \cos x}{\sin x}$

112. LHS $= \sin^2 x + 2\sin x\cos y + \cos^2 y + \sin^2 x - \cos^2 y$
$= 2\sin^2 x + 2\sin x\cos y = 2\sin x(\sin x + \cos y)$

113. LHS $= \tan^2 x - \tan^2 y + 2\tan^2 y - 2\tan x\tan y$
$= \tan^2 x - 2\tan x\tan y + \tan^2 y = (\tan x - \tan y)^2$

114. LHS $= A^2\cot^2\theta - B^2 - B^2\cot^2\theta + A^2$
$= A^2\cot^2\theta - B^2\cot^2\theta + A^2 - B^2 = (A^2 - B^2)\cot^2\theta + (A^2 - B^2)$
$= (A^2 - B^2)(\cot^2\theta + 1) = (A^2 - B^2)\csc^2\theta$

115. LHS $= \dfrac{\tan x(\sec^2 x - 25)}{\sec^2 x(\sec x + 5) - (\sec x + 5)} = \dfrac{\tan x(\sec x - 5)(\sec x + 5)}{(\sec x + 5)(\sec^2 x - 1)}$

$= \dfrac{\tan x(\sec x - 5)}{\tan^2 x} = \dfrac{\sec x - 5}{\tan x} = \dfrac{\sec x}{\tan x} - \dfrac{5}{\tan x}$

$= \dfrac{1}{\cos x} \cdot \dfrac{\cos x}{\sin x} - 5\cot x = \csc x - 5\cot x$

116. LHS $= (\sin^3 x - \cos^3 x)(\sin^3 x + \cos^3 x)$
$= (\sin x - \cos x)(\sin^2 x + \sin x\cos x + \cos^2 x)(\sin x + \cos x)(\sin^2 x - \sin x\cos x + \cos^2 x)$
$= (\sin^2 x - \cos^2 x)(1 + \sin x\cos x)(1 - \sin x\cos x)$
$= (\sin^2 x - \cos^2 x)(1 - \sin^2 x\cos^2 x)$

117. a. Conditional equation;

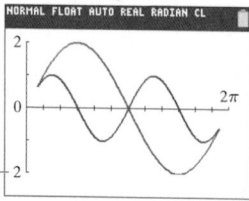

b. Identity;

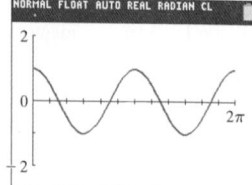

c. Contradiction;

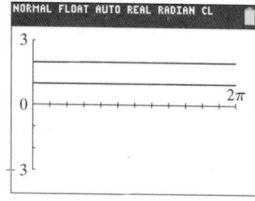

118. $y = 5$

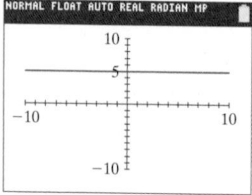

$(2\cos x - \sin x)^2 + (\cos x + 2\sin x)^2 = 4\cos^2 x - 4\cos x\sin x$
$+ \sin^2 x + \cos^2 x + 4\cos x\sin x + 4\sin^2 x$
$= 5\cos^2 x + 5\sin^2 x = 5(\cos^2 x + \sin^2 x) = 5$

Section 5.2 Practice Exercises, pp. 586–590

41. $\cos\left(\dfrac{\pi}{2} + x\right) = \cos\dfrac{\pi}{2}\cos x - \sin\dfrac{\pi}{2}\sin x = (0)\cos x - (1)\sin x$
$= -\sin x$

42. $\sin\left(\dfrac{\pi}{2} + x\right) = \sin\dfrac{\pi}{2}\cos x + \cos\dfrac{\pi}{2}\sin x = (1)\cos x + (0)\sin x = \cos x$

43. $\tan(\pi + x) = \dfrac{\tan\pi + \tan x}{1 - \tan\pi\tan x} = \dfrac{0 + \tan x}{1 - (0)\tan x} = \tan x$

44. $\cos(x - \pi) = \cos x\cos\pi + \sin x\sin\pi = \cos x(-1) + \sin x(0) = -\cos x$

45. $\sin\left(\dfrac{3\pi}{2} + x\right) = \sin\dfrac{3\pi}{2}\cos x + \cos\dfrac{3\pi}{2}\sin x = (-1)\cos x + (0)\sin x$
$= -\cos x$

46. $\tan(\pi - x) = \dfrac{\tan\pi - \tan x}{1 + \tan\pi\tan x} = \dfrac{0 - \tan x}{1 + (0)\tan x} = -\tan x$

47. $\sin(x + y) + \sin(x - y) = \sin x\cos y + \cos x\sin y + \sin x\cos y - \cos x\sin y$
$= 2\sin x\cos y$

48. $\cos(x - y) - \cos(x + y) = \cos x\cos y + \sin x\sin y - (\cos x\cos y - \sin x\sin y)$
$= 2\sin x\sin y$

49. $\dfrac{\cos(x + y)}{\sin x\cos y} = \dfrac{\cos x\cos y - \sin x\sin y}{\sin x\cos y}$

$= \dfrac{\cos x\cos y}{\sin x\cos y} - \dfrac{\sin x\sin y}{\sin x\cos y} = \cot x - \tan y$

50. $\dfrac{\sin(x + y)}{\cos x\sin y} = \dfrac{\sin x\cos y + \cos x\sin y}{\cos x\sin y}$

$= \dfrac{\sin x\cos y}{\cos x\sin y} + \dfrac{\cos x\sin y}{\cos x\sin y} = \tan x\cot y + 1$

51. $\sin\left(x + \dfrac{\pi}{4}\right) + \sin\left(x - \dfrac{\pi}{4}\right)$
$= \sin x\cos\dfrac{\pi}{4} + \cos x\sin\dfrac{\pi}{4} + \sin x\cos\dfrac{\pi}{4} - \cos x\sin\dfrac{\pi}{4}$
$= \sin x\left(\dfrac{\sqrt{2}}{2}\right) + \sin x\left(\dfrac{\sqrt{2}}{2}\right) = 2\sin x\left(\dfrac{\sqrt{2}}{2}\right) = \sqrt{2}\sin x$

52. $\cos\left(x + \dfrac{\pi}{4}\right) - \cos\left(x - \dfrac{\pi}{4}\right)$
$= \cos x\cos\dfrac{\pi}{4} - \sin x\sin\dfrac{\pi}{4} - \left(\cos x\cos\dfrac{\pi}{4} + \sin x\sin\dfrac{\pi}{4}\right)$
$= \cos x\cos\dfrac{\pi}{4} - \sin x\sin\dfrac{\pi}{4} - \cos x\cos\dfrac{\pi}{4} - \sin x\sin\dfrac{\pi}{4}$
$= -\sin x\left(\dfrac{\sqrt{2}}{2}\right) - \sin x\left(\dfrac{\sqrt{2}}{2}\right)$
$= -2\sin x\left(\dfrac{\sqrt{2}}{2}\right) = -\sqrt{2}\sin x$

53. $\tan\left(x - \dfrac{\pi}{4}\right) = \dfrac{\tan x - \tan\dfrac{\pi}{4}}{1 + \tan x\tan\dfrac{\pi}{4}} = \dfrac{\tan x - 1}{1 + \tan x} = \dfrac{\left(\dfrac{\sin x}{\cos x} - 1\right)}{\left(1 + \dfrac{\sin x}{\cos x}\right)} \cdot \dfrac{\cos x}{\cos x}$

$= \dfrac{\sin x - \cos x}{\cos x + \sin x}$

54. $\tan\left(x + \dfrac{\pi}{4}\right) = \dfrac{\tan x + \tan\dfrac{\pi}{4}}{1 - \tan x\tan\dfrac{\pi}{4}} = \dfrac{\tan x + 1}{1 - \tan x} = \dfrac{\left(\dfrac{\sin x}{\cos x} + 1\right)}{\left(1 - \dfrac{\sin x}{\cos x}\right)} \cdot \dfrac{\cos x}{\cos x}$

$= \dfrac{\cos x + \sin x}{\cos x - \sin x}$

55. $\dfrac{\cos(\alpha - \beta)}{\cos(\alpha + \beta)} = \dfrac{\cos\alpha\cos\beta + \sin\alpha\sin\beta}{\cos\alpha\cos\beta - \sin\alpha\sin\beta} \cdot \dfrac{\left(\dfrac{1}{\sin\alpha\sin\beta}\right)}{\left(\dfrac{1}{\sin\alpha\sin\beta}\right)}$

$= \dfrac{\dfrac{\cos\alpha\cos\beta}{\sin\alpha\sin\beta} + \dfrac{\sin\alpha\sin\beta}{\sin\alpha\sin\beta}}{\dfrac{\cos\alpha\cos\beta}{\sin\alpha\sin\beta} - \dfrac{\sin\alpha\sin\beta}{\sin\alpha\sin\beta}} = \dfrac{\cot\alpha\cot\beta + 1}{\cot\alpha\cot\beta - 1}$

56. $\dfrac{\sin(\alpha - \beta)}{\sin(\alpha + \beta)} = \dfrac{\sin\alpha\cos\beta - \cos\alpha\sin\beta}{\sin\alpha\cos\beta + \cos\alpha\sin\beta} \cdot \dfrac{\left(\dfrac{1}{\cos\alpha\cos\beta}\right)}{\left(\dfrac{1}{\cos\alpha\cos\beta}\right)}$

$= \dfrac{\dfrac{\sin\alpha\cos\beta}{\cos\alpha\cos\beta} - \dfrac{\cos\alpha\sin\beta}{\cos\alpha\cos\beta}}{\dfrac{\sin\alpha\cos\beta}{\cos\alpha\cos\beta} + \dfrac{\cos\alpha\sin\beta}{\cos\alpha\cos\beta}} = \dfrac{\tan\alpha - \tan\beta}{\tan\alpha + \tan\beta}$

57. $\csc(x + y) = \dfrac{1}{\sin(x + y)} = \dfrac{1}{\sin x\cos y + \cos x\sin y} \cdot \dfrac{\left(\dfrac{1}{\cos x\cos y}\right)}{\left(\dfrac{1}{\cos x\cos y}\right)}$

$= \dfrac{\dfrac{1}{\cos x\cos y}}{\dfrac{\sin x\cos y}{\cos x\cos y} + \dfrac{\cos x\sin y}{\cos x\cos y}} = \dfrac{\sec x\sec y}{\tan x + \tan y}$

58. $\sec(x - y) = \dfrac{1}{\cos(x - y)} = \dfrac{1}{\cos x\cos y + \sin x\sin y} \cdot \dfrac{\left(\dfrac{1}{\sin x\cos y}\right)}{\left(\dfrac{1}{\sin x\cos y}\right)}$

$= \dfrac{\dfrac{1}{\sin x\cos y}}{\dfrac{\cos x\cos y}{\sin x\cos y} + \dfrac{\sin x\sin y}{\sin x\cos y}} = \dfrac{\csc x\sec y}{\cot x + \tan y}$

59. $\cos(A + B)\cos(A - B) = (\cos A\cos B - \sin A\sin B)(\cos A\cos B + \sin A\sin B)$
$= \cos^2 A\cos^2 B - \sin^2 A\sin^2 B = \cos^2 A\cos^2 B - (1 - \cos^2 A)(1 - \cos^2 B)$
$= \cos^2 A\cos^2 B - 1 + \cos^2 A + \cos^2 B - \cos^2 A\cos^2 B = \cos^2 A + \cos^2 B - 1$

60. $\sin(A + B)\sin(A - B) = (\sin A\cos B + \cos A\sin B)(\sin A\cos B - \cos A\sin B)$
$= \sin^2 A\cos^2 B - \cos^2 A\sin^2 B = \sin^2 A(1 - \sin^2 B) - (1 - \sin^2 A)\sin^2 B$
$= \sin^2 A - \sin^2 A\sin^2 B - \sin^2 B + \sin^2 A\sin^2 B = \sin^2 A - \sin^2 B$

61. $\tan(x + y) - \tan(x - y) = \dfrac{\tan x + \tan y}{1 - \tan x\tan y} - \dfrac{\tan x - \tan y}{1 + \tan x\tan y}$

$= \dfrac{(\tan x + \tan y)(1 + \tan x\tan y) - (\tan x - \tan y)(1 - \tan x\tan y)}{(1 - \tan x\tan y)(1 + \tan x\tan y)}$

$= \dfrac{\tan x + \tan^2 x\tan y + \tan y + \tan x\tan^2 y - \tan x + \tan^2 x\tan y + \tan y - \tan x\tan^2 y}{1 - \tan^2 x\tan^2 y}$

$= \dfrac{2\tan^2 x\tan y + 2\tan y}{1 - \tan^2 x\tan^2 y} = \dfrac{2\tan y(\tan^2 x + 1)}{1 - \tan^2 x\tan^2 y} = \dfrac{2\tan y\sec^2 x}{1 - \tan^2 x\tan^2 y}$

62. $\tan(x + y) + \tan(x - y) = \dfrac{\tan x + \tan y}{1 - \tan x\tan y} + \dfrac{\tan x - \tan y}{1 + \tan x\tan y}$

$= \dfrac{(\tan x + \tan y)(1 + \tan x\tan y) + (\tan x - \tan y)(1 - \tan x\tan y)}{(1 - \tan x\tan y)(1 + \tan x\tan y)}$

$= \dfrac{\tan x + \tan^2 x\tan y + \tan y + \tan x\tan^2 y + \tan x - \tan^2 x\tan y - \tan y + \tan x\tan^2 y}{1 - \tan^2 x\tan^2 y}$

$= \dfrac{2\tan x + 2\tan x\tan^2 y}{1 - \tan^2 x\tan^2 y} = \dfrac{2\tan x(1 + \tan^2 y)}{1 - \tan^2 x\tan^2 y} = \dfrac{2\tan x\sec^2 y}{1 - \tan^2 x\tan^2 y}$

63. a. $10\sin(x + 5.640)$
   b. $10\sin(x + 5.640) = 10(\sin x\cos 5.640 + \cos x\sin 5.640)$
   $= 10\cos 5.640 \cdot \sin x + 10\sin 5.640 \cdot \cos x \approx 8\sin x - 6\cos x$

64. a. $17\sin(x + 2.652)$
   b. $17\sin(x + 2.652) = 17(\sin x\cos 2.652 + \cos x\sin 2.652)$
   $= 17\cos 2.652 \cdot \sin x + 17\sin 2.652 \cdot \cos x$
   $\approx -15\sin x + 8\cos x$

65. a. $\sqrt{10}\sin(x + 4.391)$
   b. $\sqrt{10}\sin(x + 4.391) = \sqrt{10}(\sin x\cos 4.391 + \cos x\sin 4.391)$
   $= \sqrt{10}\cos 4.391 \cdot \sin x + \sqrt{10}\sin 4.391 \cdot \cos x$
   $\approx -\sin x - 3\cos x$

66. a. $\sqrt{85}\sin(x + 1.352)$
   b. $\sqrt{85}\sin(x + 1.352) = \sqrt{85}(\sin x\cos 1.352 + \cos x\sin 1.352)$
   $= \sqrt{85}\cos 1.352 \cdot \sin x + \sqrt{85}\sin 1.352 \cdot \cos x$
   $\approx 2\sin x + 9\cos x$

67. a. $y = 2\sin\left(x + \dfrac{\pi}{6}\right)$ **b.**

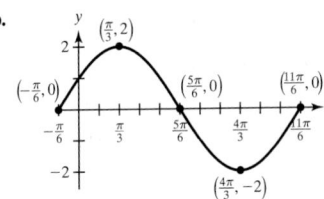

68. a. $y = 2\sin\left(x + \dfrac{7\pi}{4}\right)$ or $y = 2\sin\left(x - \dfrac{\pi}{4}\right)$
   b.

69. a. $y = \sin\left(x + \dfrac{2\pi}{3}\right)$ **b.**

70. a. $y = \sin\left(x + \dfrac{7\pi}{6}\right)$ or $y = \sin\left(x - \dfrac{5\pi}{6}\right)$
   b.

73. $\cos(u + v) = \cos[u - (-v)] = \cos u\cos(-v) + \sin u\sin(-v)$
   $= \cos u\cos v - \sin u\sin v$

74. $\tan(u - v) = \tan[u + (-v)] = \dfrac{\tan u + \tan(-v)}{1 - \tan u\tan(-v)} = \dfrac{\tan u - \tan v}{1 + \tan u\tan v}$

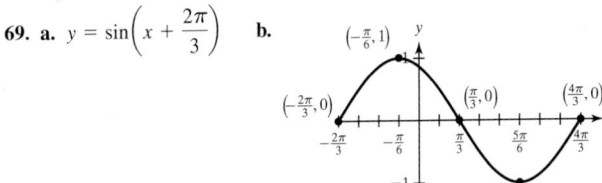

75. $\sin(u + v) = \cos\left[\dfrac{\pi}{2} - (u + v)\right] = \cos\left[\left(\dfrac{\pi}{2} - u\right) - v\right]$

$\qquad = \cos\left(\dfrac{\pi}{2} - u\right)\cos v + \sin\left(\dfrac{\pi}{2} - u\right)\sin v$

$\qquad = \sin u \cos v + \cos u \sin v$

76. $\sin(u - v) = \cos\left[\dfrac{\pi}{2} - (u - v)\right] = \cos\left[\left(\dfrac{\pi}{2} - u\right) + v\right]$

$\qquad = \cos\left(\dfrac{\pi}{2} - u\right)\cos v - \sin\left(\dfrac{\pi}{2} - u\right)\sin v = \sin u \cos v - \cos u \sin v$

77. $\dfrac{f(x + h) - f(x)}{h} = \dfrac{\sin(x + h) - \sin x}{h}$

$\qquad = \dfrac{\sin x \cos h + \cos x \sin h - \sin x}{h}$

$\qquad = \dfrac{\sin x \cos h - \sin x}{h} + \dfrac{\cos x \sin h}{h}$

$\qquad = \sin x\left(\dfrac{\cos h - 1}{h}\right) + \cos x\left(\dfrac{\sin h}{h}\right)$

$\qquad = \sin x\left[-\left(\dfrac{1 - \cos h}{h}\right)\right] + \cos x\left(\dfrac{\sin h}{h}\right)$

$\qquad = -\sin x\left(\dfrac{1 - \cos h}{h}\right) + \cos x\left(\dfrac{\sin h}{h}\right)$

78. $\dfrac{f(x + h) - f(x)}{h} = \dfrac{\cos(x + h) - \cos x}{h}$

$\qquad = \dfrac{\cos x \cos h - \sin x \sin h - \cos x}{h}$

$\qquad = \dfrac{\cos x \cos h - \cos x}{h} - \dfrac{\sin x \sin h}{h}$

$\qquad = \cos x\left(\dfrac{\cos h - 1}{h}\right) - \sin x\left(\dfrac{\sin h}{h}\right)$

$\qquad = \cos x\left[-\left(\dfrac{1 - \cos h}{h}\right)\right] - \sin x\left(\dfrac{\sin h}{h}\right)$

$\qquad = -\cos x\left(\dfrac{1 - \cos h}{h}\right) - \sin x\left(\dfrac{\sin h}{h}\right)$

91. $\cos(a + b + c) = \cos(a + b)\cos c - \sin(a + b)\sin c$

$\qquad = (\cos a \cos b - \sin a \sin b)\cos c - (\sin a \cos b + \cos a \sin b)\sin c$

$\qquad = \cos a \cos b \cos c - \sin a \sin b \cos c - \sin a \cos b \sin c$

$\qquad\quad - \cos a \sin b \sin c$

92. $\sin(a + b + c) = \sin(a + b)\cos c + \cos(a + b)\sin c$

$\qquad = (\sin a \cos b + \cos a \sin b)\cos c + (\cos a \cos b - \sin a \sin b)\sin c$

$\qquad = \sin a \cos b \cos c + \cos a \sin b \cos c + \cos a \cos b \sin c$

$\qquad\quad - \sin a \sin b \sin c$

93. $\text{LHS} = (\cos x \cos y - \sin x \sin y)\cos y + (\sin x \cos y + \cos x \sin y)\sin y$

$\qquad = \cos x \cos^2 y - \sin x \sin y \cos y + \sin x \cos y \sin y + \cos x \sin^2 y$

$\qquad = \cos x \cos^2 y + \cos x \sin^2 y = \cos x(\cos^2 y + \sin^2 y) = \cos x$

94. $\text{LHS} = (\cos x \cos y + \sin x \sin y)\sin y + (\sin x \cos y - \cos x \sin y)\cos y$

$\qquad = \cos x \cos y \sin y + \sin x \sin^2 y + \sin x \cos^2 y - \cos x \sin y \cos y$

$\qquad = \sin x \sin^2 y + \sin x \cos^2 y = \sin x(\sin^2 y + \cos^2 y) = \sin x$

95. $\text{LHS} = \dfrac{1}{\cos(x + y)} = \dfrac{1}{(\cos x \cos y - \sin x \sin y)}$

$\qquad \cdot \dfrac{(\cos x \cos y + \sin x \sin y)}{(\cos x \cos y + \sin x \sin y)} = \dfrac{\cos x \cos y + \sin x \sin y}{\cos^2 x \cos^2 y - \sin^2 x \sin^2 y}$

$\qquad = \dfrac{\cos(x - y)}{\cos^2 x(1 - \sin^2 y) - (1 - \cos^2 x)\sin^2 y}$

$\qquad = \dfrac{\cos(x - y)}{\cos^2 x - \cos^2 x \sin^2 y - \sin^2 y + \cos^2 x \sin^2 y} = \dfrac{\cos(x - y)}{\cos^2 x - \sin^2 y}$

96. $\text{LHS} = \dfrac{1}{\cos(x - y)} = \dfrac{1}{(\cos x \cos y + \sin x \sin y)}$

$\qquad \cdot \dfrac{(\cos x \cos y - \sin x \sin y)}{(\cos x \cos y - \sin x \sin y)} = \dfrac{\cos x \cos y - \sin x \sin y}{\cos^2 x \cos^2 y - \sin^2 x \sin^2 y}$

$\qquad = \dfrac{\cos(x + y)}{\cos^2 x(1 - \sin^2 y) - (1 - \cos^2 x)\sin^2 y}$

$\qquad = \dfrac{\cos(x + y)}{\cos^2 x - \cos^2 x \sin^2 y - \sin^2 y + \cos^2 x \sin^2 y} = \dfrac{\cos(x + y)}{\cos^2 x - \sin^2 y}$

97. $\sin(A - B) = \sin A \cos B - \cos A \sin B$

$\qquad = \dfrac{a}{c} \cdot \dfrac{a}{c} - \dfrac{b}{c} \cdot \dfrac{b}{c} = \dfrac{a^2 - b^2}{c^2}$

$\quad \cos(A - B) = \cos A \cos B + \sin A \sin B$

$\qquad = \dfrac{b}{c} \cdot \dfrac{a}{c} + \dfrac{a}{c} \cdot \dfrac{b}{c} = \dfrac{2ab}{c^2}$

98. Since none of the angles is a right angle, $C \neq 90°$, and $A + B \neq 90°$.

$\qquad A + B + C = 180°$

$\qquad\quad A + B = 180° - C$

$\qquad \tan(A + B) = \tan(180° - C)$

$\qquad \dfrac{\tan A + \tan B}{1 - \tan A \tan B} = \dfrac{\tan 180° - \tan C}{1 + \tan 180° \tan C}$

$\qquad \dfrac{\tan A + \tan B}{1 - \tan A \tan B} = -\tan C$

$\qquad \tan A + \tan B = -\tan C(1 - \tan A \tan B)$

$\qquad \tan A + \tan B = -\tan C + \tan C \tan A \tan B$

$\quad \tan A + \tan B + \tan C = (\tan A)(\tan B)(\tan C)$

99. $\tan\theta = \dfrac{(mx_2 + b) - (mx_1 + b)}{x_2 - x_1}$

$\qquad = \dfrac{mx_2 - mx_1}{x_2 - x_1} = \dfrac{m(x_2 - x_1)}{x_2 - x_1} = m$

100. a. Since $m_2 > m_1$, $\theta_2 > \theta_1$ and $\alpha = \theta_2 - \theta_1$.

$\qquad \tan\alpha = \tan(\theta_2 - \theta_1)$

$\qquad\quad = \dfrac{\tan\theta_2 - \tan\theta_1}{1 + \tan\theta_2 \tan\theta_1}$

$\qquad$ From Exercise 99, $\tan\theta_1 = m_1$ and $\tan\theta_2 = m_2$. Therefore,

$\qquad \dfrac{\tan\theta_2 - \tan\theta_1}{1 + \tan\theta_2 \tan\theta_1} = \dfrac{m_2 - m_1}{1 + m_2 m_1}$.

$\quad$ **b.** $26.6°$

101. a. $y = -2\sin x$

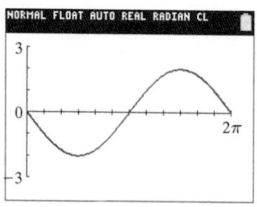

$\quad$ **b.** $\cos\left(x + \dfrac{\pi}{2}\right) - \cos\left(x - \dfrac{\pi}{2}\right)$

$\qquad = \left(\cos x \cos\dfrac{\pi}{2} - \sin x \sin\dfrac{\pi}{2}\right) - \left(\cos x \cos\dfrac{\pi}{2} + \sin x \sin\dfrac{\pi}{2}\right)$

$\qquad = \cos x \cos\dfrac{\pi}{2} - \sin x \sin\dfrac{\pi}{2} - \cos x \cos\dfrac{\pi}{2} - \sin x \sin\dfrac{\pi}{2}$

$\qquad = -2\sin x \sin\dfrac{\pi}{2} = -2\sin x(1) = -2\sin x$

102. a. $y = \sin x$

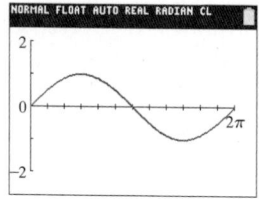

$\quad$ **b.** $\sin 3x \cos 2x - \cos 3x \sin 2x$

$\qquad = \sin(3x - 2x) = \sin x$

Section 5.3 Practice Exercises, pp. 597–600

7. a. $-\dfrac{120}{169}$ **b.** $-\dfrac{119}{169}$ **c.** $\dfrac{120}{119}$

8. a. $\dfrac{8\sqrt{33}}{49}$ **b.** $\dfrac{17}{49}$ **c.** $\dfrac{8\sqrt{33}}{17}$

9. a. $-\dfrac{4}{5}$ **b.** $-\dfrac{3}{5}$ **c.** $\dfrac{4}{3}$

10. a. $\dfrac{5\sqrt{39}}{32}$ **b.** $-\dfrac{7}{32}$ **c.** $-\dfrac{5\sqrt{39}}{7}$

11. a. $-\dfrac{20}{29}$ **b.** $-\dfrac{21}{29}$ **c.** $\dfrac{20}{21}$

12. a. $\dfrac{6\sqrt{2}}{19}$ **b.** $-\dfrac{17}{19}$ **c.** $-\dfrac{6\sqrt{2}}{17}$

13. a. $-\dfrac{840}{1369}$ **b.** $-\dfrac{1081}{1369}$ **c.** $\dfrac{840}{1081}$

14. a. $\dfrac{336}{625}$ **b.** $-\dfrac{527}{625}$ **c.** $-\dfrac{336}{527}$

21. $\dfrac{\sin 2\theta}{1-\cos 2\theta} = \dfrac{2\sin\theta\cos\theta}{1-(\cos^2\theta-\sin^2\theta)} = \dfrac{2\sin\theta\cos\theta}{1-\cos^2\theta+\sin^2\theta}$

$= \dfrac{2\sin\theta\cos\theta}{\sin^2\theta+\sin^2\theta} = \dfrac{2\sin\theta\cos\theta}{2\sin^2\theta} = \cot\theta$

22. $\dfrac{\sin 2\theta}{\sin 4\theta} = \dfrac{2\sin\theta\cos\theta}{2\sin 2\theta\cos 2\theta} = \dfrac{2\sin\theta\cos\theta}{2\cdot(2\sin\theta\cos\theta)\cos 2\theta} = \dfrac{1}{2}\sec 2\theta$

23. $\cos^4 x - \sin^4 x = (\cos^2 x - \sin^2 x)(\cos^2 x + \sin^2 x) = \cos 2x$

24. $(\sin x - \cos x)^2 = \sin^2 x - 2\sin x\cos x + \cos^2 x = 1 - \sin 2x$

25. $\dfrac{\cos 3x}{\sin 2x} = \dfrac{\cos(2x+x)}{2\sin x\cos x} = \dfrac{\cos 2x\cos x - \sin 2x\sin x}{2\sin x\cos x}$

$= \dfrac{(1-2\sin^2 x)\cos x - 2\sin x\cos x\cdot\sin x}{2\sin x\cos x}$

$= \dfrac{\cos x - 2\sin^2 x\cos x - 2\sin^2 x\cos x}{2\sin x\cos x} = \dfrac{\cos x - 4\sin^2 x\cos x}{2\sin x\cos x}$

$= \dfrac{1}{2\sin x} - 2\sin x = \dfrac{1}{2}\csc x - 2\sin x$

26. $\dfrac{\sin 3x}{\sin 2x} = \dfrac{\sin(2x+x)}{2\sin x\cos x} = \dfrac{\sin 2x\cos x + \cos 2x\sin x}{2\sin x\cos x}$

$= \dfrac{2\sin x\cos x\cdot\cos x + (2\cos^2 x - 1)\sin x}{2\sin x\cos x}$

$= \dfrac{2\cos^2 x\sin x + 2\cos^2 x\sin x - \sin x}{2\sin x\cos x} = \dfrac{4\cos^2 x\sin x - \sin x}{2\sin x\cos x}$

$= 2\cos x - \dfrac{1}{2\cos x} = 2\cos x - \dfrac{1}{2}\sec x$

27. $\sin 2A - \tan A = 2\sin A\cos A - \dfrac{\sin A}{\cos A} = \dfrac{2\sin A\cos^2 A - \sin A}{\cos A}$

$= \dfrac{\sin A(2\cos^2 A - 1)}{\cos A} = \tan A\cos 2A$

28. $\tan 2A - \tan A = \dfrac{\sin 2A}{\cos 2A} - \dfrac{\sin A}{\cos A} = \dfrac{2\sin A\cos A}{\cos^2 A - \sin^2 A} - \dfrac{\sin A}{\cos A}$

$= \dfrac{2\sin A\cos^2 A - \sin A(\cos^2 A - \sin^2 A)}{\cos A(\cos^2 A - \sin^2 A)}$

$= \dfrac{\sin A\cos^2 A + \sin^3 A}{\cos A(\cos^2 A - \sin^2 A)} = \dfrac{\sin A(\cos^2 A + \sin^2 A)}{\cos A(\cos^2 A - \sin^2 A)}$

$= \dfrac{\sin A}{\cos A}\cdot\dfrac{1}{\cos^2 A - \sin^2 A} = \tan A\cdot\dfrac{1}{\cos 2A} = \tan A\sec 2A$

29. $\tan 2x = \dfrac{2\tan x}{1-\tan^2 x} = \dfrac{2\tan x}{1-(\sec^2 x - 1)} = \dfrac{2\tan x}{2-\sec^2 x}$

30. $2\cot 2x = 2\cdot\dfrac{\cos 2x}{\sin 2x} = 2\cdot\dfrac{\cos^2 x - \sin^2 x}{2\sin x\cos x}$

$= \dfrac{\cos^2 x}{\sin x\cos x} - \dfrac{\sin^2 x}{\sin x\cos x} = \cot x - \tan x$

31. $\dfrac{\cos 2x}{\sin x} - \dfrac{\sin 2x}{\cos x} = \dfrac{1-2\sin^2 x}{\sin x} - \dfrac{2\sin x\cos x}{\cos x}$

$= \dfrac{\cos x(1-2\sin^2 x) - 2\sin^2 x\cos x}{\sin x\cos x} = \dfrac{\cos x - 4\sin^2 x\cos x}{\sin x\cos x}$

$= \dfrac{\cos x}{\sin x\cos x} - \dfrac{4\sin^2 x\cos x}{\sin x\cos x} = \csc x - 4\sin x$

32. $\dfrac{\sin 2x}{\sin x} - \dfrac{\cos 2x}{\cos x} = \dfrac{2\sin x\cos x}{\sin x} - \dfrac{\cos^2 x - \sin^2 x}{\cos x}$

$= \dfrac{\cos x(2\sin x\cos x) - (\cos^2 x - \sin^2 x)\sin x}{\sin x\cos x}$

$= \dfrac{2\cos^2 x\sin x - \cos^2 x\sin x + \sin^3 x}{\sin x\cos x}$

$= \dfrac{\cos^2 x\sin x + \sin^3 x}{\sin x\cos x} = \dfrac{\sin x(\cos^2 x + \sin^2 x)}{\sin x\cos x} = \sec x$

33. $\sin 4\theta = \sin[2(2\theta)] = 2\sin 2\theta\cos 2\theta$

$= 2\cdot(2\sin\theta\cos\theta)(\cos^2\theta - \sin^2\theta)$

$= 4\cos^3\theta\sin\theta - 4\sin^3\theta\cos\theta$

34. $\cos 4\theta = \cos[2(2\theta)] = \cos^2 2\theta - \sin^2 2\theta$

$= (\cos^2\theta - \sin^2\theta)^2 - (2\sin\theta\cos\theta)^2$

$= \cos^4\theta - 2\cos^2\theta\sin^2\theta + \sin^4\theta - 4\sin^2\theta\cos^2\theta$

$= \sin^4\theta - 6\sin^2\theta\cos^2\theta + \cos^4\theta$

45. The signs of $\sin\alpha$ and $\tan\dfrac{\alpha}{2}$ are the same for α taken in any quadrant.

Quadrant for α	I	II	III	IV
Quadrant for $\dfrac{\alpha}{2}$	I	I	II	II
Sign of $\sin\alpha$	+	+	−	−
Sign of $\tan\dfrac{\alpha}{2}$	+	+	−	−

47. a. $\dfrac{\sqrt{122}}{122}$ **b.** $\dfrac{11\sqrt{122}}{122}$ **c.** $\dfrac{1}{11}$ **48. a.** $\dfrac{\sqrt{26}}{26}$ **b.** $-\dfrac{5\sqrt{26}}{26}$

c. $-\dfrac{1}{5}$ **49. a.** $\dfrac{7\sqrt{74}}{74}$ **b.** $-\dfrac{5\sqrt{74}}{74}$ **c.** $-\dfrac{7}{5}$

50. a. $\dfrac{11\sqrt{130}}{130}$ **b.** $\dfrac{3\sqrt{130}}{130}$ **c.** $\dfrac{11}{3}$

51. $\sin^2\dfrac{\theta}{2} = \dfrac{1-\cos\theta}{2}\cdot\dfrac{\frac{1}{\sin\theta}}{\frac{1}{\sin\theta}} = \dfrac{\frac{1}{\sin\theta}-\frac{\cos\theta}{\sin\theta}}{\frac{2}{\sin\theta}} = \dfrac{\csc\theta - \cot\theta}{2\csc\theta}$

52. $\cos^2\dfrac{\theta}{2} = \dfrac{1+\cos\theta}{2}\cdot\dfrac{\frac{1}{\cos\theta}}{\frac{1}{\cos\theta}} = \dfrac{\frac{1}{\cos\theta}+\frac{\cos\theta}{\cos\theta}}{\frac{2}{\cos\theta}} = \dfrac{\sec\theta + 1}{2\sec\theta}$

53. $\tan\dfrac{x}{2} + \cot\dfrac{x}{2} = \dfrac{1-\cos x}{\sin x} + \dfrac{1+\cos x}{\sin x} = \dfrac{1-\cos x + 1 + \cos x}{\sin x}$

$= \dfrac{2}{\sin x} = 2\csc x$

54. $\csc x - \cot x = \dfrac{1}{\sin x} - \dfrac{\cos x}{\sin x} = \dfrac{1-\cos x}{\sin x} = \tan\dfrac{x}{2}$

55. $2\sin^2\dfrac{x}{2} + 4\cos^2\dfrac{x}{2} - 3 = 2\left(\dfrac{1-\cos x}{2}\right) + 4\left(\dfrac{1+\cos x}{2}\right) - 3$

$= 1 - \cos x + 2 + 2\cos x - 3 = \cos x$

56. $\cos^2\dfrac{x}{2} - 3\sin^2\dfrac{x}{2} + 1 = \dfrac{1+\cos x}{2} - 3\left(\dfrac{1-\cos x}{2}\right) + 1$

$= \dfrac{1}{2} + \dfrac{1}{2}\cos x - \dfrac{3}{2} + \dfrac{3}{2}\cos x + 1 = 2\cos x$

57. $\dfrac{1+\cos 2\theta}{2} = \dfrac{1+(2\cos^2\theta - 1)}{2} = \cos^2\theta$

58. $\tan^2\theta = \dfrac{\sin^2\theta}{\cos^2\theta} = \dfrac{\dfrac{1-\cos 2\theta}{2}}{\dfrac{1+\cos 2\theta}{2}} = \dfrac{1-\cos 2\theta}{2} \cdot \dfrac{2}{1+\cos 2\theta} = \dfrac{1-\cos 2\theta}{1+\cos 2\theta}$

71. b.

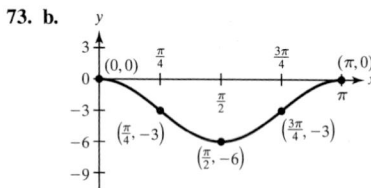

72. b.

73. b.

74. b.

83. a. $A = \cos\theta\sin\theta + \sin\theta$

b. $A = \cos\theta\sin\theta + \sin\theta$

$= \dfrac{1}{2}(2\cos\theta\sin\theta) + \sin\theta$

$= \dfrac{1}{2}\sin 2\theta + \sin\theta$

85. The ball is kicked from ground level, so $h_0 = 0$, and we want to know the distance x that it travels when the ball hits the ground ($y = 0$).

Solve for x: $0 = -\dfrac{g}{2v_0^2\cos^2\theta}x^2 + (\tan\theta)x$

$0 = x\left(-\dfrac{g}{2v_0^2\cos^2\theta}x + \tan\theta\right)$

Discard the solution $x = 0$.

$-\dfrac{g}{2v_0^2\cos^2\theta}x + \tan\theta = 0$

$x = \dfrac{2v_0^2\cos^2\theta\tan\theta}{g} = \dfrac{2v_0^2\cos\theta\sin\theta}{g}$

$= \dfrac{2v_0^2\dfrac{1}{2}(2\cos\theta\sin\theta)}{g} = \dfrac{v_0^2\sin 2\theta}{g}$

91. The sign is determined by the location of angle $\dfrac{\alpha}{2}$ and the function being evaluated. For example if $\dfrac{\alpha}{2}$ is in Quadrant II, then $\sin\dfrac{\alpha}{2}$ will be positive and $\cos\dfrac{\alpha}{2}$ and $\tan\dfrac{\alpha}{2}$ will be negative.

92. Write $\sin 2\theta$ as $\sin(\theta + \theta)$ and write $\cos 2\theta$ as $\cos(\theta + \theta)$. Then simplify the results. To derive the "alternative" formulas for double-angle cosine, use the Pythagorean identity $\sin^2\theta + \cos^2\theta = 1$ and replace $\sin^2\theta$ by $1 - \cos^2\theta$ and $\cos^2\theta$ by $1 - \sin^2\theta$.

93. Begin with the double-angle cosine formulas $\cos 2\theta = 1 - 2\sin^2\theta$ and $\cos 2\theta = 2\cos^2\theta - 1$. Then solve each equation for $\sin^2\theta$ and $\cos^2\theta$, respectively.

94. Begin with the power-reducing formulas, $\sin^2\theta = \dfrac{1 - \cos 2\theta}{2}$ and $\cos^2\theta = \dfrac{1 + \cos 2\theta}{2}$. Then substitute $\dfrac{\alpha}{2}$ for θ, and apply the square root property to solve for $\sin\dfrac{\alpha}{2}$ and $\cos\dfrac{\alpha}{2}$.

95. $\sin 2A = 2\sin A\cos A = 2 \cdot \dfrac{a}{c} \cdot \dfrac{b}{c} = 2\cos B\sin B = \sin 2B$

96. $\cos 2A = 2\cos^2 A - 1 = 2\left(\dfrac{b}{c}\right)^2 - 1 = 2\sin^2 B - 1$

$= -(1 - 2\sin^2 B) = -\cos 2B$

97. $\tan 2A = \dfrac{2\tan A}{1 - \tan^2 A} = \dfrac{\left(2 \cdot \dfrac{a}{b}\right) \cdot b^2}{\left(1 - \dfrac{a^2}{b^2}\right) \cdot b^2} = \dfrac{2ab}{b^2 - a^2}$

98. $\cos 2A = \cos^2 A - \sin^2 A = \left(\dfrac{b}{c}\right)^2 - \left(\dfrac{a}{c}\right)^2 = \dfrac{b^2 - a^2}{c^2}$

99. $\cos\dfrac{A}{2} = \sqrt{\dfrac{1 + \cos A}{2}} = \sqrt{\dfrac{\left(1 + \dfrac{b}{c}\right) \cdot c}{(2) \cdot c}} = \sqrt{\dfrac{c + b}{2c}}$

100. $\tan\dfrac{A}{2} = \dfrac{1 - \cos A}{\sin A} = \dfrac{\left(1 - \dfrac{b}{c}\right) \cdot c}{\left(\dfrac{a}{c}\right) \cdot c} = \dfrac{c - b}{a}$

Section 5.4 Practice Exercises, pp. 604–607

15. LHS: $\dfrac{1}{2}(\cos[(x + y) - (x - y)] - \cos[(x + y) + (x - y)])$

$= \dfrac{1}{2}(\cos 2y - \cos 2x) = \dfrac{1}{2}[1 - 2\sin^2 y - (1 - 2\sin^2 x)]$

$= \dfrac{1}{2}(-2\sin^2 y + 2\sin^2 x) = \sin^2 x - \sin^2 y$

16. LHS: $\dfrac{1}{2}(\cos[(x + y) - (x - y)] + \cos[(x + y) + (x - y)])$

$= \dfrac{1}{2}(\cos 2y + \cos 2x) = \dfrac{1}{2}[1 - 2\sin^2 y + 2\cos^2 x - 1]$

$= \dfrac{1}{2}(2\cos^2 x - 2\sin^2 y) = \cos^2 x - \sin^2 y$

17. LHS $= 2 \cdot \dfrac{1}{2}\left(\sin\left[\left(\dfrac{\pi}{4} + \dfrac{x - y}{2}\right) + \left(\dfrac{\pi}{4} - \dfrac{x + y}{2}\right)\right]\right.$

$\left. + \sin\left[\left(\dfrac{\pi}{4} + \dfrac{x - y}{2}\right) - \left(\dfrac{\pi}{4} - \dfrac{x + y}{2}\right)\right]\right)$

$= \sin\left(\dfrac{\pi}{4} + \dfrac{\pi}{4} + \dfrac{x - y}{2} - \dfrac{x + y}{2}\right)$

$+ \sin\left(\dfrac{\pi}{4} - \dfrac{\pi}{4} + \dfrac{x - y}{2} + \dfrac{x + y}{2}\right)$

$= \sin\left(\dfrac{\pi}{2} - y\right) + \sin x = \cos y + \sin x$

18. LHS: $2 \cdot \dfrac{1}{2}\left(\sin\left[\left(\dfrac{\pi}{4} + \dfrac{x - y}{2}\right) + \left(\dfrac{\pi}{4} + \dfrac{x + y}{2}\right)\right]\right.$

$\left. + \sin\left[\left(\dfrac{\pi}{4} + \dfrac{x - y}{2}\right) - \left(\dfrac{\pi}{4} + \dfrac{x + y}{2}\right)\right]\right)$

$= \sin\left(\dfrac{\pi}{4} + \dfrac{\pi}{4} + \dfrac{x - y}{2} + \dfrac{x + y}{2}\right)$

$+ \sin\left(\dfrac{\pi}{4} - \dfrac{\pi}{4} + \dfrac{x - y}{2} - \dfrac{x + y}{2}\right)$

$= \sin\left(\dfrac{\pi}{2} + x\right) + \sin(-y) = \cos x - \sin y$

27. $\dfrac{\cos 3x - \cos x}{\cos 3x + \cos x} = \dfrac{-2\sin 2x \sin x}{2\cos 2x \cos x} = -\tan 2x \tan x$

28. $\dfrac{\cos x - \cos 3x}{\sin x - \sin 3x} = \dfrac{-2\sin 2x \sin(-x)}{2\cos 2x \sin(-x)} = -\tan 2x$

29. $\sin 2x + \sin 4x + \sin 6x$

$= 2\sin 3x \cos(-x) + \sin[2(3x)]$

$= 2\sin 3x \cos x + 2\sin 3x \cos 3x$

$= 2\sin 3x (\cos x + \cos 3x)$

$= 2\sin 3x[2\cos 2x \cos(-x)]$

$= 4\cos x \cos 2x \sin 3x$

30. $\cos 2x + \cos 4x + \cos 6x$

$= 2\cos 3x \cos(-x) + \cos[2(3x)]$

$= 2\cos 3x \cos x + 2\cos^2 3x - 1$

$= 2\cos 3x (\cos x + \cos 3x) - 1$

$= 2\cos 3x[2\cos 2x \cos(-x)] - 1$

$= 4\cos x \cos 2x \cos 3x - 1$

31. $\sin 5x - \sin 3x + 2\sin x \cos 2x$

$= 2\cos 4x \sin x + 2\sin x \cos 2x$

$= 2\sin x(\cos 4x + \cos 2x)$

$= 2\sin x[2\cos 3x \cos x]$

$= 2\sin x \cos x \cdot 2\cos 3x$

$= \sin 2x \cdot 2\cos 3x$

$= 2\sin 2x \cos 3x$

32. $\sin t + \sin 3t + \sin 5t + \sin 7t$

$= 2\sin 2t \cos(-t) + 2\sin 6t \cos(-t)$

$= 2\sin 2t \cos t + 2\sin 6t \cos t$

$= 2\cos t(\sin 2t + \sin 6t)$

$= 2\cos t[2\sin 4t \cos(-2t)]$

$= 4\cos t \cos 2t \sin 4t$

33. $\dfrac{\cos 3x - \cos 2x + \cos x}{\sin 3x - \sin 2x + \sin x}$

$= \dfrac{\cos 3x + \cos x - \cos 2x}{\sin 3x + \sin x - \sin 2x} = \dfrac{2\cos 2x \cos x - \cos 2x}{2\sin 2x \cos x - \sin 2x}$

$= \dfrac{\cos 2x(2\cos x - 1)}{\sin 2x(2\cos x - 1)} = \cot 2x$

34. $\dfrac{\sin 4t + \sin 3t + \sin 2t}{\cos 4t + \cos 3t + \cos 2t}$

$= \dfrac{\sin 4t + \sin 2t + \sin 3t}{\cos 4t + \cos 2t + \cos 3t} = \dfrac{2\sin 3t \cos t + \sin 3t}{2\cos 3t \cos t + \cos 3t}$

$= \dfrac{\sin 3t(2\cos t + 1)}{\cos 3t(2\cos t + 1)} = \tan 3t$

37. $\dfrac{1}{2}[\cos(u - v) + \cos(u + v)]$

$= \dfrac{1}{2}(\cos u \cos v + \sin u \sin v + \cos u \cos v - \sin u \sin v)$

$= \dfrac{1}{2}(2\cos u \cos v) = \cos u \cos v$

38. $\dfrac{1}{2}[\sin(u + v) + \sin(u - v)]$

$= \dfrac{1}{2}(\sin u \cos v + \cos u \sin v + \sin u \cos v - \cos u \sin v)$

$= \dfrac{1}{2}(2\sin u \cos v) = \sin u \cos v$

39. In the product-to-sum formula

$\cos u \cos v = \dfrac{1}{2}[\cos(u - v) + \cos(u + v)]$, let $u = \dfrac{x + y}{2}$ and $v = \dfrac{x - y}{2}$.

Then, $\cos\dfrac{x + y}{2}\cos\dfrac{x - y}{2}$

$= \dfrac{1}{2}\left[\cos\left(\dfrac{x + y}{2} - \dfrac{x - y}{2}\right) + \cos\left(\dfrac{x + y}{2} + \dfrac{x - y}{2}\right)\right]$

$= \dfrac{1}{2}(\cos y + \cos x)$

Therefore, $\cos x + \cos y = 2\cos\dfrac{x + y}{2}\cos\dfrac{x - y}{2}$.

40. In the product-to-sum formula

$\sin u \cos v = \dfrac{1}{2}[\sin(u + v) + \sin(u - v)]$,

let $u = \dfrac{x + y}{2}$ and $v = \dfrac{x - y}{2}$. Then, $\sin\dfrac{x + y}{2}\cos\dfrac{x - y}{2}$

$= \dfrac{1}{2}\left[\sin\left(\dfrac{x + y}{2} + \dfrac{x - y}{2}\right) + \sin\left(\dfrac{x + y}{2} - \dfrac{x - y}{2}\right)\right]$

$= \dfrac{1}{2}(\sin x + \sin y)$

Therefore, $\sin x + \sin y = 2\sin\dfrac{x + y}{2}\cos\dfrac{x - y}{2}$.

41. c.

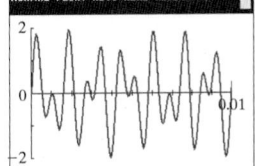

42. c.

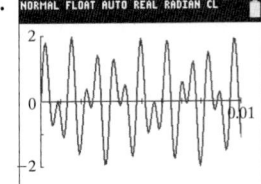

43. Formula (2) converts a product of cosines to a sum of cosines, whereas formula (6) converts a sum of cosines to a product of cosines. The two formulas work in reverse.

44. The sum-to-product formulas involve a sum or difference of sines or a sum or difference of cosines of the same or different argument. Furthermore, the sum-to-product formulas require a coefficient of 1 on the sine and cosine terms. The reduction formula reduces a sum of $\sin x$ and $\cos x$ (each has the same argument) to a single term involving the sine function with a phase shift.

45. LHS: $\dfrac{\sin[A + (B - C)]}{2} - \dfrac{\sin[A - (B - C)]}{2}$

$+ \dfrac{\sin[B + (C - A)]}{2} - \dfrac{\sin[B - (C - A)]}{2}$

$+ \dfrac{\sin[C + (A - B)]}{2} - \dfrac{\sin[C - (A - B)]}{2}$

$= \dfrac{\sin(A + B - C)}{2} - \dfrac{\sin(A - B + C)}{2} + \dfrac{\sin(B + C - A)}{2}$

$- \dfrac{\sin(B - C + A)}{2} + \dfrac{\sin(C + A - B)}{2} - \dfrac{\sin(C - A + B)}{2} = 0$

46. LHS: $\dfrac{\sin[A + (B - C)]}{2} - \dfrac{\sin[A - (B - C)]}{2}$

$\qquad - \dfrac{\sin[B + (C - A)]}{2} + \dfrac{\sin[B - (C - A)]}{2} - \dfrac{\sin[C + (A - B)]}{2}$

$\qquad + \dfrac{\sin[C - (A - B)]}{2}$

$= \dfrac{\sin(A + B - C)}{2} - \dfrac{\sin(A - B + C)}{2} - \dfrac{\sin(B + C - A)}{2}$

$\qquad + \dfrac{\sin(B - C + A)}{2} - \dfrac{\sin(C + A - B)}{2} + \dfrac{\sin(C - A + B)}{2}$

$= 2 \cdot \dfrac{\sin(A + B - C)}{2} - 2 \cdot \dfrac{\sin(A - B + C)}{2}$

$= \sin(A + [B - C]) - \sin(A - [B - C]) = 2\cos A \sin(B - C)$

47. $\sin A + \sin B = 2\sin\dfrac{A + B}{2}\cos\dfrac{A - B}{2}$ and $\sin C = 2\sin\dfrac{C}{2}\cos\dfrac{C}{2}$.

Therefore, $\sin A + \sin B + \sin C$

$= 2\sin\dfrac{A + B}{2}\cos\dfrac{A - B}{2} + 2\sin\dfrac{C}{2}\cos\dfrac{C}{2}$

$= 2\sin\dfrac{\pi - C}{2}\cos\dfrac{A - B}{2} + 2\sin\dfrac{C}{2}\cos\dfrac{C}{2}$

$= 2\sin\left(\dfrac{\pi}{2} - \dfrac{C}{2}\right)\cos\dfrac{A - B}{2} + 2\sin\dfrac{C}{2}\cos\dfrac{C}{2}$

$= 2\cos\dfrac{C}{2}\cos\dfrac{A - B}{2} + 2\sin\dfrac{C}{2}\cos\dfrac{C}{2}$

$= 2\cos\dfrac{C}{2}\left(\cos\dfrac{A - B}{2} + \sin\dfrac{C}{2}\right)$

$= 2\cos\dfrac{C}{2}\left[\cos\dfrac{A - B}{2} + \cos\left(\dfrac{\pi}{2} - \dfrac{C}{2}\right)\right]$

$= 2\cos\dfrac{C}{2}\left(\cos\dfrac{A - B}{2} + \cos\dfrac{\pi - C}{2}\right)$

$= 2\cos\dfrac{C}{2}\left(\cos\dfrac{A - B}{2} + \cos\dfrac{A + B}{2}\right)$

$= 2\cos\dfrac{C}{2}\left[2\cos\dfrac{(A - B) + (A + B)}{4}\cos\dfrac{(A - B) - (A + B)}{4}\right]$

$= 4\cos\dfrac{C}{2}\cos\dfrac{A}{2}\cos\left(-\dfrac{B}{2}\right) = 4\cos\dfrac{A}{2}\cos\dfrac{B}{2}\cos\dfrac{C}{2}$

48. $\cos A + \cos B + \cos C = 2\cos\dfrac{A + B}{2}\cos\dfrac{A - B}{2} + 1 - 2\sin^2\dfrac{C}{2}$

$= 2\cos\dfrac{\pi - C}{2}\cos\dfrac{A - B}{2} + 1 - 2\sin^2\dfrac{C}{2}$

$= 2\cos\left(\dfrac{\pi}{2} - \dfrac{C}{2}\right)\cos\dfrac{A - B}{2} + 1 - 2\sin^2\dfrac{C}{2}$

$= 2\sin\dfrac{C}{2}\cos\dfrac{A - B}{2} + 1 - 2\sin^2\dfrac{C}{2}$

$= 2\sin\dfrac{C}{2}\left(\cos\dfrac{A - B}{2} - \sin\dfrac{C}{2}\right) + 1$

$= 2\sin\dfrac{C}{2}\left[\cos\dfrac{A - B}{2} - \cos\left(\dfrac{\pi}{2} - \dfrac{C}{2}\right)\right] + 1$

$= 2\sin\dfrac{C}{2}\left(\cos\dfrac{A - B}{2} - \cos\dfrac{\pi - C}{2}\right) + 1$

$= 2\sin\dfrac{C}{2}\left(\cos\dfrac{A - B}{2} - \cos\dfrac{A + B}{2}\right) + 1$

$= 2\sin\dfrac{C}{2}\left[-2\sin\dfrac{(A - B) + (A + B)}{4}\sin\dfrac{(A - B) - (A + B)}{4}\right] + 1$

$= 2\sin\dfrac{C}{2}\left[-2\sin\dfrac{A}{2}\sin\left(-\dfrac{B}{2}\right)\right] + 1 = 4\sin\dfrac{A}{2}\sin\dfrac{B}{2}\sin\dfrac{C}{2} + 1$

49. a. 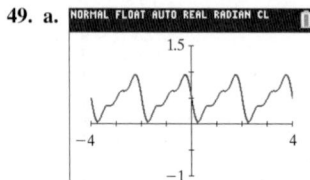   **b.**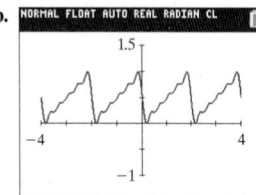

50. a. 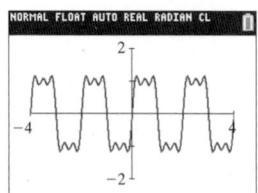   **b.**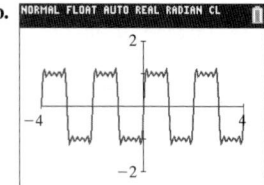

51. b.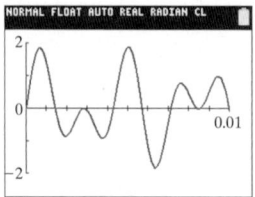

Section 5.5 Practice Exercises, pp. 617–622

2. $\dfrac{\pi}{4}, \dfrac{3\pi}{4}, \dfrac{5\pi}{4}, \dfrac{7\pi}{4}; \dfrac{\pi}{4} + \dfrac{n\pi}{2}$

13. a. $\left\{\dfrac{\pi}{6}, \dfrac{5\pi}{6}\right\}$ **b.** $\left\{x \,\middle|\, x = \dfrac{\pi}{6} + 2n\pi, x = \dfrac{5\pi}{6} + 2n\pi\right\}$

14. a. $\left\{\dfrac{3\pi}{4}, \dfrac{5\pi}{4}\right\}$ **b.** $\left\{x \,\middle|\, x = \dfrac{3\pi}{4} + 2n\pi, x = \dfrac{5\pi}{4} + 2n\pi\right\}$

15. a. $\left\{\dfrac{\pi}{3}, \dfrac{5\pi}{3}\right\}$ **b.** $\left\{x \,\middle|\, x = \dfrac{\pi}{3} + 2n\pi, x = \dfrac{5\pi}{3} + 2n\pi\right\}$

16. a. $\left\{\dfrac{\pi}{3}, \dfrac{2\pi}{3}\right\}$ **b.** $\left\{x \,\middle|\, x = \dfrac{\pi}{3} + 2n\pi, x = \dfrac{2\pi}{3} + 2n\pi\right\}$

17. a. $\left\{\dfrac{3\pi}{4}, \dfrac{7\pi}{4}\right\}$ **b.** $\left\{x \,\middle|\, x = \dfrac{3\pi}{4} + n\pi\right\}$

18. a. $\left\{\dfrac{5\pi}{6}, \dfrac{11\pi}{6}\right\}$ **b.** $\left\{x \,\middle|\, x = \dfrac{5\pi}{6} + n\pi\right\}$

23. a. $\left\{x \,\middle|\, x = \dfrac{\pi}{12} + n\pi, x = \dfrac{11\pi}{12} + n\pi\right\}$ **b.** $\left\{\dfrac{\pi}{12}, \dfrac{11\pi}{12}, \dfrac{13\pi}{12}, \dfrac{23\pi}{12}\right\}$

24. a. $\left\{x \,\middle|\, x = \dfrac{\pi}{2} + \dfrac{2n\pi}{3}\right\}$ **b.** $\left\{\dfrac{\pi}{2}, \dfrac{7\pi}{6}, \dfrac{11\pi}{6}\right\}$

25. a. $\left\{x \,\middle|\, x = \dfrac{2\pi}{9} + \dfrac{n\pi}{3}\right\}$ **b.** $\left\{\dfrac{2\pi}{9}, \dfrac{5\pi}{9}, \dfrac{8\pi}{9}, \dfrac{11\pi}{9}, \dfrac{14\pi}{9}, \dfrac{17\pi}{9}\right\}$

26. a. $\left\{x \,\middle|\, x = \dfrac{3\pi}{8} + n\pi, x = \dfrac{5\pi}{8} + n\pi\right\}$ **b.** $\left\{\dfrac{3\pi}{8}, \dfrac{5\pi}{8}, \dfrac{11\pi}{8}, \dfrac{13\pi}{8}\right\}$

27. a. $\left\{x \,\middle|\, x = \dfrac{\pi}{2} + 4n\pi, x = \dfrac{3\pi}{2} + 4n\pi\right\}$ **b.** $\left\{\dfrac{\pi}{2}, \dfrac{3\pi}{2}\right\}$

28. a. $\left\{x \,\middle|\, x = \dfrac{\pi}{2} + 2n\pi\right\}$ **b.** $\left\{\dfrac{\pi}{2}\right\}$

29. a. $\left\{x \,\middle|\, x = \dfrac{\pi}{2} + 2n\pi, x = \dfrac{7\pi}{6} + 2n\pi\right\}$ **b.** $\left\{\dfrac{\pi}{2}, \dfrac{7\pi}{6}\right\}$

30. a. $\left\{x \,\middle|\, x = \dfrac{\pi}{12} + 2n\pi, x = \dfrac{17\pi}{12} + 2n\pi\right\}$ **b.** $\left\{\dfrac{\pi}{12}, \dfrac{17\pi}{12}\right\}$

31. a. $\left\{x \,\middle|\, x = \dfrac{\pi}{4} + n\pi\right\}$ **b.** $\left\{\dfrac{\pi}{4}, \dfrac{5\pi}{4}\right\}$

32. a. $\left\{x \,\middle|\, x = \dfrac{\pi}{3} + n\pi\right\}$ **b.** $\left\{\dfrac{\pi}{3}, \dfrac{4\pi}{3}\right\}$ **33.** $\left\{\dfrac{\pi}{3}, \dfrac{3\pi}{4}, \dfrac{5\pi}{3}, \dfrac{7\pi}{4}\right\}$

34. $\left\{\dfrac{\pi}{4}, \dfrac{7\pi}{6}, \dfrac{5\pi}{4}, \dfrac{11\pi}{6}\right\}$ **38.** $\left\{\dfrac{\pi}{6}, \dfrac{5\pi}{6}\right\}$ **39.** $\left\{\dfrac{\pi}{6}, \dfrac{5\pi}{6}, \dfrac{3\pi}{2}\right\}$

41. $\left\{\dfrac{\pi}{3}, \dfrac{\pi}{2}, \dfrac{2\pi}{3}, \dfrac{4\pi}{3}, \dfrac{3\pi}{2}, \dfrac{5\pi}{3}\right\}$ **43.** $\left\{\dfrac{\pi}{6}, \dfrac{5\pi}{6}\right\}$

49. $\left\{0, \dfrac{\pi}{3}, \dfrac{2\pi}{3}, \pi, \dfrac{4\pi}{3}, \dfrac{5\pi}{3}\right\}$ **50.** $\left\{0, \dfrac{\pi}{4}, \dfrac{3\pi}{4}, \pi, \dfrac{5\pi}{4}, \dfrac{7\pi}{4}\right\}$

51. $\left\{\dfrac{\pi}{12}, \dfrac{5\pi}{12}, \dfrac{\pi}{2}, \dfrac{13\pi}{12}, \dfrac{17\pi}{12}, \dfrac{3\pi}{2}\right\}$ **52.** $\left\{0, \dfrac{\pi}{2}, \pi, \dfrac{7\pi}{6}, \dfrac{3\pi}{2}, \dfrac{11\pi}{6}\right\}$

57. $\left\{\cos^{-1}\dfrac{3}{11},\ 2\pi - \cos^{-1}\dfrac{3}{11}\right\}$; $\{74.2°,\ 285.8°\}$

58. $\{\pi - \tan^{-1}3,\ 2\pi - \tan^{-1}3\}$; $\{108.4°,\ 288.4°\}$

59. $\left\{\pi - \tan^{-1}\dfrac{7}{4},\ 2\pi - \tan^{-1}\dfrac{7}{4}\right\}$; $\{119.7°,\ 299.7°\}$

60. $\left\{\sin^{-1}\dfrac{1}{6},\ \pi - \sin^{-1}\dfrac{1}{6}\right\}$; $\{9.6°,\ 170.4°\}$

61. $\left\{\dfrac{\pi}{2},\ \sin^{-1}\dfrac{3}{4},\ \pi - \sin^{-1}\dfrac{3}{4}\right\}$; $\{90.0°,\ 48.6°,\ 131.4°\}$

62. $\left\{0,\ \cos^{-1}\left(-\dfrac{2}{3}\right),\ 2\pi - \cos^{-1}\left(-\dfrac{2}{3}\right)\right\}$; $\{0.0°,\ 131.8°,\ 228.2°\}$

71. $\left\{\dfrac{\pi}{18},\ \dfrac{5\pi}{18},\ \dfrac{13\pi}{18},\ \dfrac{17\pi}{18},\ \dfrac{25\pi}{18},\ \dfrac{29\pi}{18}\right\}$ **72.** $\left\{\dfrac{\pi}{3},\ \dfrac{2\pi}{3},\ \dfrac{4\pi}{3},\ \dfrac{5\pi}{3}\right\}$

77. $\left\{0,\ \dfrac{\pi}{6},\ \dfrac{\pi}{3},\ \dfrac{2\pi}{3},\ \dfrac{5\pi}{6},\ \pi,\ \dfrac{4\pi}{3},\ \dfrac{3\pi}{2},\ \dfrac{5\pi}{3}\right\}$

78. $\left\{\dfrac{\pi}{6},\ \dfrac{\pi}{2},\ \dfrac{5\pi}{6},\ \dfrac{7\pi}{6},\ \dfrac{3\pi}{2},\ \dfrac{11\pi}{6}\right\}$ **79.** $\left\{0,\ \dfrac{\pi}{5},\ \dfrac{3\pi}{5},\ \pi,\ \dfrac{7\pi}{5},\ \dfrac{9\pi}{5}\right\}$

80. $\left\{\dfrac{\pi}{6},\ \dfrac{5\pi}{6},\ \dfrac{7\pi}{6},\ \dfrac{11\pi}{6}\right\}$

81. $\left\{\cos^{-1}\dfrac{2}{3},\ 2\pi - \cos^{-1}\dfrac{2}{3},\ \cos^{-1}\left(-\dfrac{1}{5}\right),\ 2\pi - \cos^{-1}\left(-\dfrac{1}{5}\right)\right\}$

82. $\left\{\sin^{-1}\dfrac{3}{5},\ \pi - \sin^{-1}\dfrac{3}{5},\ \pi + \sin^{-1}\dfrac{1}{4},\ 2\pi - \sin^{-1}\dfrac{1}{4}\right\}$

83. $\left\{\cos^{-1}\left(\dfrac{1 + \sqrt{2}}{4}\right),\ 2\pi - \cos^{-1}\left(\dfrac{1 + \sqrt{2}}{4}\right),\right.$

$\left. \cos^{-1}\left(\dfrac{1 - \sqrt{2}}{4}\right),\ 2\pi - \cos^{-1}\left(\dfrac{1 - \sqrt{2}}{4}\right)\right\}$

84. $\left\{\sin^{-1}\left(\dfrac{3 + \sqrt{5}}{8}\right),\ \pi - \sin^{-1}\left(\dfrac{3 + \sqrt{5}}{8}\right),\ \sin^{-1}\left(\dfrac{3 - \sqrt{5}}{8}\right),\right.$

$\left. \pi - \sin^{-1}\left(\dfrac{3 - \sqrt{5}}{8}\right)\right\}$

85. $\left\{\sin^{-1}(2 - \sqrt{3}),\ \pi - \sin^{-1}(2 - \sqrt{3})\right\}$

92. a. The period is $\dfrac{1}{70}$ min and represents the time for one complete heart beat. This implies that the heart rate is 70 beats per minute.

b. Maximum: 110 mmHg; Minimum 70 mmHg; The blood pressure *is* "110 over 70."

c. $t = \left(\dfrac{1}{280} + \dfrac{n}{70}\right)$ min for integers $n \geq 0$. The blood pressure is maximized $\dfrac{1}{280}$ min after recording begins and every $\dfrac{1}{70}$ min thereafter.

95. After factoring, we have $(\cos x - 4)(\cos x + 3) = 0$. Therefore, $\cos x = 4$ or $\cos x = -3$. Both 4 and -3 are outside the range of the cosine function. Therefore, there are no values of x for which $\cos x = 4$ or $\cos x = -3$.

97. The general solution consists of *all* solutions to the equation over the set of real numbers. The solutions taken over the interval $[0, 2\pi)$ are only those solutions that fall within this interval.

98. Set Y_1 and Y_2 equal to the left and right sides of the equation and graph the functions. Then use the ISECT feature to find the point(s) of intersection. The solution(s) are the x-coordinates of the point(s) of intersection. Alternatively, rewrite the equation as $\sin 3x + x - 1 = 0$ and set Y_1 equal to the expression on the left side of the equation. Use the ZERO command to find the zeros of the function.

101. $\left\{\dfrac{\pi}{4},\ \dfrac{2\pi}{3},\ \dfrac{3\pi}{4},\ \dfrac{5\pi}{4},\ \dfrac{4\pi}{3},\ \dfrac{7\pi}{4}\right\}$

104. $\left\{\dfrac{\pi}{4},\ \dfrac{\pi}{3},\ \dfrac{2\pi}{3},\ \dfrac{3\pi}{4},\ \dfrac{5\pi}{4},\ \dfrac{4\pi}{3},\ \dfrac{5\pi}{3},\ \dfrac{7\pi}{4}\right\}$ **105.** $\left\{\dfrac{\pi}{6},\ \dfrac{5\pi}{6},\ \dfrac{3\pi}{2}\right\}$

113. a. $\sqrt{2}\sin\left(x + \dfrac{7\pi}{4}\right) = -1$ **b.** $\left\{0,\ \dfrac{3\pi}{2}\right\}$

122. a. $A(x) = 2x\cos x$

b.

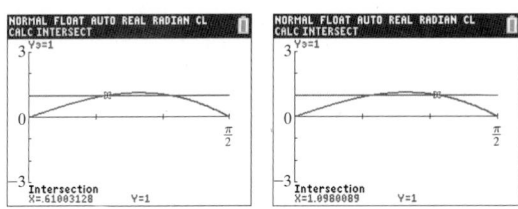

c.

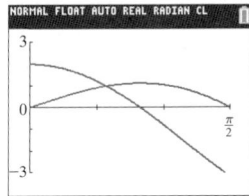

The area is 1 square unit when $x \approx 0.61$ and $x \approx 1.10$.

d. The maximum value of $A(x)$ occurs where $y = 2\cos x - 2x\sin x$ equals zero.

Problem Recognition Exercises, p. 622

1. a. $\sin^4 x - \cos^4 x$

$= (\sin^2 x + \cos^2 x)(\sin^2 x - \cos^2 x)$

$= (1)(\sin^2 x - \cos^2 x)$

$= -(\cos^2 x - \sin^2 x)$

$= -\cos 2x$

b. $\left\{\dfrac{\pi}{2},\ \dfrac{3\pi}{2}\right\}$

2. a. $\tan x\sin x + \cos x = \dfrac{\sin x}{\cos x} \cdot \sin x + \cos x$

$= \dfrac{\sin^2 x}{\cos x} + \dfrac{\cos^2 x}{\cos x} = \dfrac{1}{\cos x} = \sec x$

b. $\left\{\dfrac{\pi}{3},\ \dfrac{5\pi}{3}\right\}$

3. a. $\cot x - \sec x\csc x\cos 2x$

$= \dfrac{\cos x}{\sin x} - \dfrac{1}{\cos x} \cdot \dfrac{1}{\sin x}(1 - 2\sin^2 x)$

$= \dfrac{\cos^2 x - (1 - 2\sin^2 x)}{\sin x\cos x}$

$= \dfrac{1 - \sin^2 x - 1 + 2\sin^2 x}{\sin x\cos x}$

$= \dfrac{\sin^2 x}{\sin x\cos x} = \dfrac{\sin x}{\cos x} = \tan x$

b. $\left\{\dfrac{\pi}{4},\ \dfrac{5\pi}{4}\right\}$

4. a. $\dfrac{\sec x + \tan x}{\cos x + \cot x} = \dfrac{\dfrac{1}{\cos x} + \dfrac{\sin x}{\cos x}}{\cos x + \dfrac{\cos x}{\sin x}}$ **b.** $\{0, \pi\}$

$= \dfrac{\left(\dfrac{1}{\cos x} + \dfrac{\sin x}{\cos x}\right)\sin x \cos x}{\left(\cos x + \dfrac{\cos x}{\sin x}\right)\sin x \cos x}$

$= \dfrac{\sin x + \sin^2 x}{\sin x \cos^2 x + \cos^2 x}$

$= \dfrac{\sin x(1 + \sin x)}{\cos^2 x(\sin x + 1)} = \dfrac{\sin x}{\cos x \cdot \cos x}$

$= \dfrac{1}{\cos x} \cdot \dfrac{\sin x}{\cos x} = \sec x \tan x$

5. a. $\dfrac{2\cot x}{1 + \cot^2 x} = \dfrac{2 \cdot \dfrac{\cos x}{\sin x}}{1 + \left(\dfrac{\cos x}{\sin x}\right)^2} = \dfrac{\dfrac{2\cos x}{\sin x}}{\dfrac{\sin^2 x + \cos^2 x}{\sin^2 x}}$

$= \dfrac{\dfrac{2\cos x}{\sin x}}{\dfrac{1}{\sin^2 x}} = \dfrac{2\cos x}{\sin x} \cdot \sin^2 x$

$= 2\cos x \sin x = \sin 2x$

b. $\left\{0, \dfrac{\pi}{2}, \pi, \dfrac{3\pi}{2}\right\}$

6. a. $\cos 2x - \cos 4x = -2\sin 3x \sin(-x)$

$= -2\sin 3x(-\sin x) = 2\sin 3x \sin x$

b. $\left\{0, \dfrac{\pi}{3}, \dfrac{2\pi}{3}, \pi, \dfrac{4\pi}{3}, \dfrac{5\pi}{3}\right\}$

7. a. $\dfrac{\sqrt{2} + \sqrt{6}}{4}$ **b.** $\sqrt{\dfrac{1 + \dfrac{\sqrt{3}}{2}}{2}}$

c. $\sqrt{\dfrac{1 + \dfrac{\sqrt{3}}{2}}{2}} = \sqrt{\dfrac{2 + \sqrt{3}}{4}} = \sqrt{\dfrac{2 + \sqrt{3}}{4} \cdot \dfrac{2}{2}}$

$= \sqrt{\dfrac{4 + 2\sqrt{3}}{8}} = \sqrt{\dfrac{1 + 2\sqrt{3} + (\sqrt{3})^2}{8}}$

$= \sqrt{\dfrac{(1 + \sqrt{3})^2}{8}} = \dfrac{1 + \sqrt{3}}{2\sqrt{2}} = \dfrac{1 + \sqrt{3}}{2\sqrt{2}} \cdot \dfrac{\sqrt{2}}{\sqrt{2}}$

$= \dfrac{\sqrt{2} + \sqrt{6}}{4}$

8. a. $\dfrac{\sqrt{6} + \sqrt{2}}{4}$ **b.** $\sin\dfrac{\pi}{24} = \dfrac{\sqrt{8 - 2\sqrt{6} - 2\sqrt{2}}}{4}$;

$\cos\dfrac{\pi}{24} = \dfrac{\sqrt{8 + 2\sqrt{6} + 2\sqrt{2}}}{4}$

Chapter 5 Review Exercises, pp. 625–627

7. $\dfrac{\sin x \tan^2 x - \sin x}{\cos x \tan x + \cos x} = \dfrac{\sin x(\tan^2 x - 1)}{\cos x(\tan x + 1)}$

$= \tan x \dfrac{(\tan x + 1)(\tan x - 1)}{\tan x + 1} = \tan x(\tan x - 1)$

8. $\dfrac{3\sin^2 x - \sin x - 2}{\cos x \sin x - \cos x} = \dfrac{(3\sin x + 2)(\sin x - 1)}{\cos x(\sin x - 1)}$

$= \dfrac{3\sin x + 2}{\cos x} = 3 \cdot \dfrac{\sin x}{\cos x} + 2 \cdot \dfrac{1}{\cos x} = 3\tan x + 2\sec x$

9. $\dfrac{1}{\csc x - 1} + \dfrac{1}{\csc x + 1} = \dfrac{(\csc x + 1) + (\csc x - 1)}{(\csc x - 1)(\csc x + 1)} = \dfrac{2\csc x}{\csc^2 x - 1}$

$= \dfrac{2\csc x}{\cot^2 x} = 2 \cdot \dfrac{1}{\sin x} \cdot \dfrac{\sin^2 x}{\cos^2 x} = \dfrac{2\sin x}{\cos^2 x}$

$= 2 \cdot \dfrac{\sin x}{\cos x} \cdot \dfrac{1}{\cos x} = 2\tan x \sec x$

10. $\dfrac{1 - \sin x}{\cos x} = \dfrac{(1 - \sin x)}{\cos x} \cdot \dfrac{(1 + \sin x)}{(1 + \sin x)} = \dfrac{1 - \sin^2 x}{\cos x(1 + \sin x)}$

$= \dfrac{\cos^2 x}{\cos x(1 + \sin x)} = \dfrac{\cos x}{1 + \sin x}$

11. $\ln|\csc x| + \ln|\tan x| = \ln|\csc x \cdot \tan x| = \ln\left|\dfrac{1}{\sin x} \cdot \dfrac{\sin x}{\cos x}\right|$

$= \ln\left|\dfrac{1}{\cos x}\right| = \ln|\sec x|$

12. LHS $= \ln\left|\dfrac{\tan x \cdot \cos x}{\sin x}\right| = \ln\left|\dfrac{\dfrac{\sin x}{\cos x} \cdot \cos x}{\sin x}\right|$

$= \ln\left|\dfrac{\sin x}{\sin x}\right| = \ln 1 = 0$

13. LHS $= \cos x + (-\sin x)(-\tan x) = \cos x + \sin x \tan x$

$= \cos x + \sin x \cdot \dfrac{\sin x}{\cos x} = \dfrac{\cos^2 x + \sin^2 x}{\cos x} = \dfrac{1}{\cos x} = \sec x$

14. LHS $= (-\csc x + 1)(\csc x + 1) = (1 - \csc x)(1 + \csc x)$

$= 1 - \csc^2 x = -\cot^2 x$

27. $\dfrac{48 - 15\sqrt{13}}{119}$

29. $\dfrac{\cos(x - y)}{\cos x \sin y} = \dfrac{\cos x \cos y + \sin x \sin y}{\cos x \sin y} = \dfrac{\cos x \cos y}{\cos x \sin y} + \dfrac{\sin x \sin y}{\cos x \sin y}$

$= \cot y + \tan x$

30. $\dfrac{\sin(x - y)}{\cos x \cos y} = \dfrac{\sin x \cos y - \cos x \sin y}{\cos x \cos y} = \dfrac{\sin x \cos y}{\cos x \cos y} - \dfrac{\cos x \sin y}{\cos x \cos y}$

$= \tan x - \tan y$

31. $\dfrac{\tan\left(x - \dfrac{3\pi}{4}\right)}{\tan\left(x + \dfrac{\pi}{4}\right)} = \tan\left(x - \dfrac{3\pi}{4}\right) \cdot \dfrac{1}{\tan\left(x + \dfrac{\pi}{4}\right)}$

$= \dfrac{\tan x - \tan\dfrac{3\pi}{4}}{1 + \tan x \tan\dfrac{3\pi}{4}} \cdot \dfrac{1 - \tan x \tan\dfrac{\pi}{4}}{\tan x + \tan\dfrac{\pi}{4}}$

$= \dfrac{\tan x - (-1)}{1 + \tan x(-1)} \cdot \dfrac{1 - \tan x(1)}{\tan x + (1)} = \dfrac{(\tan x + 1)(1 - \tan x)}{(1 - \tan x)(\tan x + 1)} = 1$

32. LHS $= \sin x \cos\dfrac{\pi}{4} + \cos x \sin\dfrac{\pi}{4} - \left(\cos x \cos\dfrac{\pi}{4} - \sin x \sin\dfrac{\pi}{4}\right)$

$= \dfrac{\sqrt{2}}{2}\sin x + \dfrac{\sqrt{2}}{2}\cos x - \dfrac{\sqrt{2}}{2}\cos x + \dfrac{\sqrt{2}}{2}\sin x$

$= \sqrt{2}\sin x$

35. $\dfrac{2\tan x}{1 + \tan^2 x} = \dfrac{2\tan x}{\sec^2 x} = 2 \cdot \dfrac{\sin x}{\cos x} \cdot \cos^2 x$

$= 2\sin x \cos x = \sin 2x$

36. $\dfrac{\cos 2x}{\sin x} = \dfrac{1 - 2\sin^2 x}{\sin x} = \dfrac{1}{\sin x} - \dfrac{2\sin^2 x}{\sin x} = \csc x - 2\sin x$

37. $\sin^2\dfrac{x}{2} - \cos^2\dfrac{x}{2} = \dfrac{1 - \cos x}{2} - \dfrac{1 + \cos x}{2} = \dfrac{-2\cos x}{2} = -\cos x$

38. $\tan x \tan\dfrac{x}{2} = \dfrac{\sin x}{\cos x} \cdot \dfrac{1 - \cos x}{\sin x} = \dfrac{1 - \cos x}{\cos x} = \dfrac{1}{\cos x} - 1 = \sec x - 1$

41. a. $-\dfrac{120}{169}$ **b.** $\dfrac{119}{169}$ **c.** $\dfrac{5\sqrt{26}}{26}$ **d.** $\dfrac{\sqrt{26}}{26}$

42. a. $\dfrac{24}{25}$ **b.** $\dfrac{7}{25}$ **c.** $\dfrac{3\sqrt{10}}{10}$ **d.** $-\dfrac{\sqrt{10}}{10}$

43. a. $-\dfrac{4}{5}$ **b.** $-\dfrac{3}{5}$ **c.** $\sqrt{\dfrac{5+\sqrt{5}}{10}}$ **d.** $\sqrt{\dfrac{5-\sqrt{5}}{10}}$

44. a. $-\dfrac{3}{5}$ **b.** $-\dfrac{4}{5}$ **c.** $\sqrt{\dfrac{10-\sqrt{10}}{20}}$ **d.** $-\sqrt{\dfrac{10+\sqrt{10}}{20}}$

47. $\dfrac{1}{2}\cos 2x - \dfrac{1}{2}\cos 10x$ **48.** $\dfrac{1}{2}\sin 4x + \dfrac{1}{2}\sin 8x$

49. $\dfrac{1}{2}\sin 3x - \dfrac{1}{2}\sin 9x$ **50.** $\dfrac{1}{2}\cos 5x + \dfrac{1}{2}\cos 15x$

59. $\cos\left(\dfrac{\pi}{4}+t\right) - \cos\left(\dfrac{\pi}{4}-t\right)$

$= -2\sin\dfrac{\left(\frac{\pi}{4}+t\right)+\left(\frac{\pi}{4}-t\right)}{2}\sin\dfrac{\left(\frac{\pi}{4}+t\right)-\left(\frac{\pi}{4}-t\right)}{2}$

$= -2\sin\dfrac{\pi}{4}\sin t = -2\cdot\dfrac{\sqrt{2}}{2}\sin t = -\sqrt{2}\sin t$

60. $\sin\left(\theta-\dfrac{\pi}{4}\right) + \sin\left(\theta+\dfrac{\pi}{4}\right)$

$= 2\sin\dfrac{\left(\theta-\frac{\pi}{4}\right)+\left(\theta+\frac{\pi}{4}\right)}{2}\cos\dfrac{\left(\theta-\frac{\pi}{4}\right)-\left(\theta+\frac{\pi}{4}\right)}{2}$

$= 2\sin\theta\cos\left(-\dfrac{\pi}{4}\right) = 2\sin\theta\cdot\dfrac{\sqrt{2}}{2} = \sqrt{2}\sin\theta$

61. $\dfrac{\sin 4x - \sin 2x}{\cos 4x - \cos 2x} = \dfrac{2\cos 3x\sin x}{-2\sin 3x\sin x} = -\cot 3x$

62. $\sin 3x + \sin 5x + \sin 8x = 2\sin 4x\cos(-x) + \sin[2(4x)]$

$= 2\sin 4x\cos x + 2\sin 4x\cos 4x$

$= 2\sin 4x(\cos x + \cos 4x) = 2\sin 4x\left[2\cos\dfrac{5x}{2}\cos\left(\dfrac{-3x}{2}\right)\right]$

$= 4\sin 4x\cos\dfrac{5x}{2}\cos\dfrac{3x}{2}$

63. a. $\left\{x\,\middle|\,x=\dfrac{4\pi}{3}+2n\pi,\ x=\dfrac{5\pi}{3}+2n\pi\right\}$ **b.** $\left\{\dfrac{4\pi}{3},\dfrac{5\pi}{3}\right\}$

64. a. $\left\{x\,\middle|\,x=\dfrac{\pi}{4}+2n\pi,\ x=\dfrac{7\pi}{4}+2n\pi\right\}$ **b.** $\left\{\dfrac{\pi}{4},\dfrac{7\pi}{4}\right\}$

65. a. $\left\{x\,\middle|\,x=\dfrac{\pi}{6}+n\pi,\ x=\dfrac{5\pi}{6}+n\pi\right\}$ **b.** $\left\{\dfrac{\pi}{6},\dfrac{5\pi}{6},\dfrac{7\pi}{6},\dfrac{11\pi}{6}\right\}$

66. a. $\left\{x\,\middle|\,x=\dfrac{\pi}{3}+n\pi,\ x=\dfrac{2\pi}{3}+n\pi\right\}$ **b.** $\left\{\dfrac{\pi}{3},\dfrac{2\pi}{3},\dfrac{4\pi}{3},\dfrac{5\pi}{3}\right\}$

67. a. $\left\{x\,\middle|\,x=\dfrac{3\pi}{4}+n\pi\right\}$ **b.** $\left\{\dfrac{3\pi}{4},\dfrac{7\pi}{4}\right\}$

68. a. $\left\{x\,\middle|\,x=\dfrac{2\pi}{9}+\dfrac{2n\pi}{3},\ x=\dfrac{4\pi}{9}+\dfrac{2n\pi}{3}\right\}$

   b. $\left\{\dfrac{2\pi}{9},\dfrac{4\pi}{9},\dfrac{8\pi}{9},\dfrac{10\pi}{9},\dfrac{14\pi}{9},\dfrac{16\pi}{9}\right\}$

69. a. $\left\{x\,\middle|\,x=\dfrac{\pi}{2}+2n\pi\right\}$ **b.** $\left\{\dfrac{\pi}{2}\right\}$

70. a. $\left\{x\,\middle|\,x=\dfrac{2\pi}{3}+4n\pi,\ x=\dfrac{4\pi}{3}+4n\pi\right\}$ **b.** $\left\{\dfrac{2\pi}{3},\dfrac{4\pi}{3}\right\}$

71. $\left\{\dfrac{3\pi}{4}\right\}$ **72.** $\left\{\dfrac{2\pi}{3},\dfrac{5\pi}{3}\right\}$ **73.** $\left\{\sin^{-1}\dfrac{1}{5},\ \pi-\sin^{-1}\dfrac{1}{5}\right\}$

74. $\left\{\cos^{-1}\dfrac{3}{4},\ 2\pi-\cos^{-1}\dfrac{3}{4}\right\}$ **75.** $\left\{\pi-\tan^{-1}\dfrac{3}{8},\ 2\pi-\tan^{-1}\dfrac{3}{8}\right\}$

77. $\left\{\dfrac{\pi}{3},\dfrac{5\pi}{3}\right\}$ **78.** $\left\{\dfrac{7\pi}{6},\dfrac{11\pi}{6}\right\}$

79. $\left\{\cos^{-1}\left(\dfrac{-2+\sqrt{21}}{17}\right),\ 2\pi-\cos^{-1}\left(\dfrac{-2+\sqrt{21}}{17}\right),\right.$

   $\left.\cos^{-1}\left(\dfrac{-2-\sqrt{21}}{17}\right),\ 2\pi-\cos^{-1}\left(\dfrac{-2+\sqrt{21}}{17}\right)\right\}$

80. $\left\{\sin^{-1}\dfrac{4}{5},\ \pi-\sin^{-1}\dfrac{4}{5},\dfrac{7\pi}{6},\dfrac{11\pi}{6}\right\}$ **81.** $\left\{\dfrac{\pi}{6},\dfrac{5\pi}{6}\right\}$

82. $\left\{\dfrac{\pi}{3},\dfrac{5\pi}{3}\right\}$ **83.** $\left\{\dfrac{\pi}{2},\dfrac{7\pi}{6},\dfrac{3\pi}{2},\dfrac{11\pi}{6}\right\}$ **84.** $\left\{\dfrac{\pi}{6},\dfrac{5\pi}{6},\dfrac{3\pi}{2}\right\}$

85. $\left\{0,\dfrac{\pi}{2},\pi,\dfrac{3\pi}{2}\right\}$ **86.** $\left\{0,\dfrac{\pi}{3},\dfrac{\pi}{2},\dfrac{2\pi}{3},\pi,\dfrac{4\pi}{3},\dfrac{3\pi}{2},\dfrac{5\pi}{3}\right\}$

87. $\left\{\dfrac{\pi}{2},\pi\right\}$ **88.** $\left\{\dfrac{3\pi}{2}\right\}$

Chapter 5 Test, pp. 627–628

3. LHS $= \sin\theta\sec\theta + \sin\theta\csc\theta - \cos\theta\sec\theta - \cos\theta\csc\theta$

$= \sin\theta\cdot\dfrac{1}{\cos\theta} + 1 - 1 - \cos\theta\cdot\dfrac{1}{\sin\theta} = \tan\theta - \cot\theta$

4. $\dfrac{\cot x - \tan x}{\cot x + \tan x} = \dfrac{\left(\dfrac{\cos x}{\sin x} - \dfrac{\sin x}{\cos x}\right)}{\left(\dfrac{\cos x}{\sin x} + \dfrac{\sin x}{\cos x}\right)}\cdot\dfrac{\sin x\cos x}{\sin x\cos x} = \dfrac{\cos^2 x - \sin^2 x}{\cos^2 x + \sin^2 x}$

$= \cos 2x$

5. $\dfrac{\sin 7x - \sin 5x}{\cos 7x + \cos 5x} = \dfrac{2\cos\left(\dfrac{7x+5x}{2}\right)\sin\left(\dfrac{7x-5x}{2}\right)}{2\cos\left(\dfrac{7x+5x}{2}\right)\cos\left(\dfrac{7x-5x}{2}\right)}$

$= \dfrac{2\cos 6x\sin x}{2\cos 6x\cos x} = \tan x$

6. $\dfrac{\sin^3 x - \cos^3 x}{\sin x - \cos x} = \dfrac{(\sin x - \cos x)(\sin^2 x + \sin x\cos x + \cos^2 x)}{\sin x - \cos x}$

$= \sin x\cos x + 1 = \dfrac{1}{2}(2\sin x\cos x) + 1 = \dfrac{1}{2}\sin 2x + 1$

7. LHS $= 4\cdot\dfrac{1+\cos x}{2} + 2\cdot\dfrac{1-\cos x}{2} - 3$

$= 2 + 2\cos x + 1 - \cos x - 3 = \cos x$

8. $\dfrac{(1-\cos x)}{(1+\cos x)}\cdot\dfrac{(1-\cos x)}{(1-\cos x)} = \dfrac{1 - 2\cos x + \cos^2 x}{1-\cos^2 x}$

$= \dfrac{1-2\cos x + \cos^2 x}{\sin^2 x} = \dfrac{1}{\sin^2 x} - 2\cdot\dfrac{1}{\sin x}\cdot\dfrac{\cos x}{\sin x} + \dfrac{\cos^2 x}{\sin^2 x}$

$= \csc^2 x - 2\csc x\cot x + \cot^2 x = (\csc x - \cot x)^2$

11. $-\cos 2x + \dfrac{3}{8}\cos 4x + \dfrac{13}{8}$

21. a. $\left\{x\,\middle|\,x=\dfrac{\pi}{6}+n\pi\right\}$ **b.** $\left\{\dfrac{\pi}{6},\dfrac{7\pi}{6}\right\}$

22. a. $\left\{x\,\middle|\,x=\dfrac{\pi}{8}+n\pi,\ x=\dfrac{7\pi}{8}+n\pi\right\}$ **b.** $\left\{\dfrac{\pi}{8},\dfrac{7\pi}{8},\dfrac{9\pi}{8},\dfrac{15\pi}{8}\right\}$

23. $\left\{\dfrac{\pi}{4},\dfrac{3\pi}{4},\dfrac{\pi}{2},\dfrac{5\pi}{4},\dfrac{3\pi}{2},\dfrac{7\pi}{4}\right\}$ **26.** $\left\{\dfrac{\pi}{4},\dfrac{7\pi}{12},\dfrac{11\pi}{12},\dfrac{5\pi}{4},\dfrac{19\pi}{12},\dfrac{23\pi}{12}\right\}$

27. $\left\{\cos^{-1}\left(\dfrac{-1+\sqrt{33}}{12}\right),\ 2\pi-\cos^{-1}\left(\dfrac{-1+\sqrt{33}}{12}\right),\right.$

   $\left.\cos^{-1}\left(\dfrac{-1-\sqrt{33}}{12}\right),\ 2\pi-\cos^{-1}\left(\dfrac{-1-\sqrt{33}}{12}\right)\right\}$

29. $\left\{\pi + \sin^{-1}\dfrac{5}{8},\ 2\pi-\sin^{-1}\dfrac{5}{8}\right\}$ **30.** $\left\{0,\dfrac{\pi}{3},\dfrac{\pi}{2},\pi,\dfrac{3\pi}{2},\dfrac{5\pi}{3}\right\}$

31. $\left\{-0.6891,\,0,\,0.6891\right\}$

Chapter 5 Cumulative Review Exercises, p. 628

1. $\left\{\dfrac{1}{3},\dfrac{2}{3}\right\}$ **2.** $\left\{-\dfrac{1}{2},\dfrac{1}{2},-i,i\right\}$ **3.** $\left\{\dfrac{\pi}{6},\dfrac{5\pi}{6},\dfrac{7\pi}{6},\dfrac{11\pi}{6}\right\}$

4. $(-\infty,-1]\cup\left[\dfrac{1}{4},\infty\right)$ **5.** $\left(-1,\dfrac{1}{4}\right)$ **6.** $\left\{\dfrac{1}{4}\right\}$ **8.** $\left\{\dfrac{1}{4}\right\}$

9. a.

$\begin{array}{r|rrrr} 2i & 4 & 3 & 15 & 12 & -4 \\ & & 8i & -16+6i & -12-2i & 4 \\ \hline & 4 & 3+8i & -1+6i & -2i & 0 \end{array}$

12. Domain: $(-\infty,\infty)$; Range: $[0,\infty)$

13. Domain: $(-\infty,\infty)$; Range: $[-2,2]$

14. Domain: $[2,\infty)$; Range: $[3,\infty)$

15. Domain: $[-1, 1]$; Range: $\left[-\dfrac{\pi}{2}, \dfrac{\pi}{2}\right]$

18. $\cot^2\theta - \cos^2\theta = \dfrac{\cos^2\theta}{\sin^2\theta} - \cos^2\theta = \dfrac{\cos^2\theta - \cos^2\theta\sin^2\theta}{\sin^2\theta}$

$= \dfrac{\cos^2\theta(1 - \sin^2\theta)}{\sin^2\theta} = \cos^2\theta \cdot \dfrac{\cos^2\theta}{\sin^2\theta} = \cot^2\theta\cos^2\theta$

19. LHS: $(2\sin x\cos x)\cos x - (2\cos^2 x - 1)\sin x$

$= 2\cos^2 x\sin x - 2\cos^2 x\sin x + \sin x = \sin x$

CHAPTER 6

Section 6.1 Practice Exercises, pp. 635–641

39.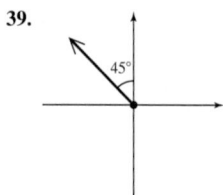

40.

41.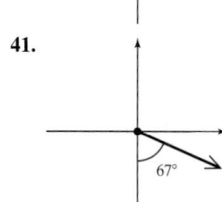

42.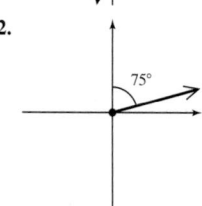

73. Determine the acute angle α from P to Q measured from a north-south line. Then write the bearing as NαE, NαW, SαE, or SαW, depending on whether Q is to the northeast, northwest, southeast, or southwest of P, respectively.

74. Solving a triangle means that we find the measures of the angles and sides based on given information about the triangle.

79. b.

80. b.

Section 6.2 Practice Exercises, pp. 650–655

7. $C = 35°$,

$a = \dfrac{42\sin 43°}{\sin 102°} \approx 29.3$,

$c = \dfrac{42\sin 35°}{\sin 102°} \approx 24.6$

8. $A = 50°$,

$a = \dfrac{60\sin 50°}{\sin 98°} \approx 46.4$,

$b = \dfrac{60\sin 32°}{\sin 98°} \approx 32.1$

9. $B = 66°$,

$a = \dfrac{73\sin 17.8°}{\sin 66°} \approx 24.4$,

$c = \dfrac{73\sin 96.2°}{\sin 66°} \approx 79.4$

10. $A = 45.5°$,

$b = \dfrac{19\sin 110.4°}{\sin 45.5°} \approx 25.0$,

$c = \dfrac{19\sin 24.1°}{\sin 45.5°} \approx 10.9$

11. $Q = 113°$,

$q = \dfrac{18.9\sin 113°}{\sin 19°} \approx 53.4$,

$r = \dfrac{18.9\sin 48°}{\sin 19°} \approx 43.1$

12. $Q = 51°$,

$p = \dfrac{200.3\sin 67°}{\sin 51°} \approx 237.2$,

$r = \dfrac{200.3\sin 62°}{\sin 51°} \approx 227.6$

38. a. The distance from A to the helicopter is 474 ft and the distance from B to the helicopter is 637 ft.

70. a. $a = \dfrac{5\sqrt{2}}{\sin 65°}$; $b = \dfrac{5\sqrt{2}\sin 25°}{\sin 65°\sin 40°}$; $c = \dfrac{5\sin 25°}{\sin 65°\sin 40°\sin 70°}$

b. $a \approx 7.8$ cm; $b \approx 5.1$ cm; $c \approx 3.9$ cm

71. a. $a = \dfrac{20\sin 40°}{\sin 95°}$; $b = \dfrac{10\sqrt{3}\sin 40°}{\sin 95°\sin 35°}$; $c = \dfrac{10\sqrt{3}\sin 40°}{\sin 35°\sin 50°}$

b. $a \approx 12.9$ in.; $b \approx 19.5$ in.; $c \approx 25.3$ in.

74. Compare the side opposite the given angle to the altitude of the triangle.
1. If the length of the side opposite the given angle is shorter than the altitude, then no triangle is possible.
2. If the length of the side opposite the given angle equals the altitude, then one right triangle is formed.
3. If the length of the side opposite the given angle is greater than the length of the known side adjacent to the given angle then one triangle is possible.
4. If the length of the side opposite the given angle is greater than the altitude but less than the length of the known side adjacent to the given angle, then two triangles are possible.

75. Confirm that the longest side is opposite the largest angle, and that the shortest side is opposite the shortest angle. Verify that the three ratios $\dfrac{a}{\sin A}$, $\dfrac{b}{\sin B}$, and $\dfrac{c}{\sin C}$ are the same. Verify that the sum of the measures of angles A, B, and C equals $180°$.

77. Let $y = AC$, $\gamma = \angle BCA$, and $\beta = \angle ABC$.

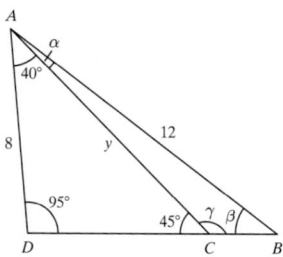

Then,

(1) $\gamma = 180° - 45° = 135°$

(2) $\angle D = 180° - (40° + 45°) = 95°$

(3) $\dfrac{8}{\sin 45°} = \dfrac{y}{\sin 95°} \Rightarrow y = \dfrac{8\sin 95°}{\sin 45°}$

(4) $\dfrac{y}{\sin\beta} = \dfrac{12}{\sin 135°} \Rightarrow \sin\beta = \dfrac{y\sin 135°}{12}$

and from (3), $\sin\beta = \dfrac{8\sin 95°\sin 135°}{12\sin 45°}$.

Since $\sin 135° = \sin 45°$ and $\sin 95° = \sin 85°$, it follows that

$\sin\beta = \dfrac{2}{3}\sin 85°$.

$\alpha = 180 - \gamma - \beta$. Therefore, $\alpha = 180° - 135° - \sin^{-1}\left(\dfrac{2}{3}\sin 85°\right)$.

$\alpha = 45° - \sin^{-1}\left(\dfrac{2}{3}\sin 85°\right)$

$45° - \alpha = \sin^{-1}\left(\dfrac{2}{3}\sin 85°\right)$

$\sin(45° - \alpha) = \dfrac{2}{3}\sin 85°$

78. From the law of sines,

$$\dfrac{\sin A}{\sin B} = \dfrac{a}{b} \qquad (1)$$

Subtract 1 from both sides of equation (1).

$\dfrac{\sin A}{\sin B} - 1 = \dfrac{a}{b} - 1$

$$\dfrac{\sin A - \sin B}{\sin B} = \dfrac{a - b}{b} \qquad (2)$$

Add 1 to both sides of equation (1).

$\dfrac{\sin A}{\sin B} + 1 = \dfrac{a}{b} + 1$

$$\frac{\sin A + \sin B}{\sin B} = \frac{a + b}{b} \qquad (3)$$

Divide (2) by (3).

$$\frac{\dfrac{\sin A - \sin B}{\sin B}}{\dfrac{\sin A + \sin B}{\sin B}} = \frac{\dfrac{a - b}{b}}{\dfrac{a + b}{b}} \quad \text{which is equivalent to}$$

$$\frac{a - b}{a + b} = \frac{\sin A - \sin B}{\sin A + \sin B}$$

Apply sum-to-product formulas.

$$\frac{a - b}{a + b} = \frac{\sin A - \sin B}{\sin A + \sin B} = \frac{2\cos\left(\frac{A + B}{2}\right)\sin\left(\frac{A - B}{2}\right)}{2\sin\left(\frac{A + B}{2}\right)\cos\left(\frac{A - B}{2}\right)}$$

$$= \cot\left(\frac{A + B}{2}\right)\tan\left(\frac{A - B}{2}\right) = \frac{\tan\left(\frac{A - B}{2}\right)}{\tan\left(\frac{A + B}{2}\right)}$$

Thus, $\dfrac{a - b}{a + b} = \dfrac{\tan\left(\frac{A - B}{2}\right)}{\tan\left(\frac{A + B}{2}\right)}$.

79. Given, $\dfrac{a}{c} = \dfrac{\sin A}{\sin C}$ and $\dfrac{b}{c} = \dfrac{\sin B}{\sin C}$,

we have $\dfrac{a}{c} + \dfrac{b}{c} = \dfrac{\sin A}{\sin C} + \dfrac{\sin B}{\sin C}$

or equivalently $\dfrac{a + b}{c} = \dfrac{\sin A + \sin B}{\sin C}$.

Apply a sum-to-product formula in the numerator and the double angle formula for sine in the denominator,

$$\frac{a + b}{c} = \frac{2\sin\frac{A + B}{2}\cos\frac{A - B}{2}}{2\sin\frac{C}{2}\cos\frac{C}{2}} \qquad (1)$$

Since $A + B = 180° - C$,

$\sin\frac{A + B}{2} = \sin\frac{180° - C}{2} = \sin\left(90° - \frac{C}{2}\right) = \cos\frac{C}{2}$.

Equation (1) becomes

$$\frac{a + b}{c} = \frac{2\cos\frac{C}{2}\cos\frac{A - B}{2}}{2\sin\frac{C}{2}\cos\frac{C}{2}} = \frac{\cos\frac{A - B}{2}}{\sin\frac{C}{2}}.$$

80.

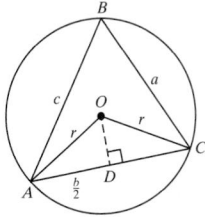

From the right the triangle $\triangle AOD$, we have

$$\sin(\angle AOD) = \frac{\frac{b}{2}}{r} = \frac{b}{2r} = \frac{b}{d}.$$

This implies that $d = \dfrac{b}{\sin(\angle AOD)}$.

From geometry, we know that $\angle B = \frac{1}{2}\angle AOC = \angle AOD$.

Thus, $d = \dfrac{b}{\sin B}$.

Section 6.3 Practice Exercises, pp. 662–666

52. The angle opposite the longest side of a triangle may potentially be greater than 90°. However, the angles opposite the two shorter sides of a triangle are guaranteed to be acute angles. Using the law of sines to find the measure of an angle involves evaluating the inverse sine function with a positive argument. In such a case, the inverse sine function will return an acute angle. Therefore, we should only use the law of sines to find the measures of the angles opposite the two shorter sides.

53. The only angle in a triangle that can potentially be greater than 90° is the angle opposite the longest side. Using the law of cosines to find the measure of an angle involves evaluating the inverse cosine function. If the argument is negative, then the inverse cosine will return an obtuse angle. If the argument is positive or zero, the inverse

cosine will return an acute angle or 90° angle, respectively. Once the measure of the largest angle is determined, the law of sines can be used to find the remaining acute angles without the threat of the ambiguous case. Alternatively, the law of cosines can be used to find the measures of the remaining angles as well.

54. When applying the law of sines or cosines, do no intermediate rounding. Round the final answers only after all calculations have been performed. If possible, use the lengths of the given sides or measures of the given angles rather than rounded values from a prior calculation.

55. $c^2 = a^2 + b^2$; The Pythagorean theorem is a special case of the law of cosines (the case where one angle in the triangle is a right angle).

56. No. There are infinitely many triangles that can be constructed with the same three angles, so the lengths of the sides cannot be uniquely determined. The corresponding sides of two triangles with the same angles are proportional. You have already encountered such a situation when you studied similar triangles.

57. a. Let $y = CP = x + r$. From the law of cosines,

$$L^2 = r^2 + y^2 - 2ry\cos\alpha.$$

The equation is quadratic in y:

$$y^2 + (-2r\cos\alpha)y + r^2 - L^2 = 0$$

Choosing the positive solution from the quadratic formula,

$$y = \frac{2r\cos\alpha + \sqrt{4r^2\cos^2\alpha - 4(r^2 - L^2)}}{2} \quad \text{and therefore,}$$

$$y = r\cos\alpha + \sqrt{r^2\cos^2\alpha - r^2 + L^2}.$$

Since $y = x + r$, then $x = y - r$ and we have

$$x = r\cos\alpha + \sqrt{r^2\cos^2\alpha - r^2 + L^2} - r.$$

c. Maximum distance: 5.9 in.;
Minimum distance 2.5 in.

58. From the law of cosines,

(1) $\cos A = \dfrac{b^2 + c^2 - a^2}{2bc}$ (2) $\cos B = \dfrac{a^2 + c^2 - b^2}{2ac}$

(3) $\cos C = \dfrac{a^2 + b^2 - c^2}{2ab}$

Therefore, $\dfrac{\cos A}{a} + \dfrac{\cos B}{b} + \dfrac{\cos C}{c}$

$$= \frac{b^2 + c^2 - a^2}{2abc} + \frac{a^2 + c^2 - b^2}{2abc} + \frac{a^2 + b^2 - c^2}{2abc} = \frac{a^2 + b^2 + c^2}{2abc}$$

59. From the law of cosines:

(1) $\cos B = \dfrac{a^2 + c^2 - b^2}{2ac}$ (2) $\cos C = \dfrac{a^2 + b^2 - c^2}{2ab}$

Therefore,

$2ac\cos B = a^2 + c^2 - b^2$ and $2ab\cos C = a^2 + b^2 - c^2$ and

$2ac\cos B + 2ab\cos C = a^2 + c^2 - b^2 + a^2 + b^2 - c^2$

$2ac\cos B + 2ab\cos C = 2a^2$. Dividing both sides by $2a$, we obtain

$a = b\cos C + c\cos B$.

60. a. $(\text{Area})^2 = \dfrac{1}{4}b^2c^2\sin^2 A$

b. $(\text{Area})^2 = \dfrac{1}{4}b^2c^2(1 - \cos^2 A)$

c. $(\text{Area})^2 = \dfrac{1}{4}b^2c^2(1 + \cos A)(1 - \cos A)$

d. Area $= \sqrt{\dfrac{1}{4}b^2c^2(1 + \cos A)(1 - \cos A)}$

$$= \sqrt{\left[\frac{1}{2}bc(1 + \cos A)\right]\left[\frac{1}{2}bc(1 - \cos A)\right]}$$

e. $\dfrac{1}{2}bc(1 + \cos A) = \dfrac{1}{2}bc\left(1 + \dfrac{b^2 + c^2 - a^2}{2bc}\right)$

$$= \frac{bc}{2} + \frac{b^2 + c^2 - a^2}{4} = \frac{2bc + b^2 + c^2 - a^2}{4}$$

$$= \frac{b^2 + 2bc + c^2 - a^2}{4} = \frac{(b + c)^2 - a^2}{4}$$

$$= \frac{[(b + c) + a][(b + c) - a]}{4} = \frac{a + b + c}{2} \cdot \frac{-a + b + c}{2}$$

f. Adding and subtracting a in the numerator of the expression

$$\frac{-a + b + c}{2} \text{ gives } \frac{-a + b + c + a - a}{2} = \frac{a + b + c + -2a}{2}$$

$$= \frac{a + b + c}{2} - \frac{2a}{2} = s - a.$$

Therefore, $\frac{1}{2}bc(1 + \cos A) = \frac{a + b + c}{2} \cdot \frac{-a + b + c}{2} = s(s - a)$.

g. Substitute $\frac{1}{2}bc(1 + \cos A) = s(s - a)$ and

$\frac{1}{2}bc(1 - \cos A) = (s - b)(s - c)$ into

$$\text{Area} = \sqrt{\left[\frac{1}{2}bc(1 + \cos A)\right]\left[\frac{1}{2}bc(1 - \cos A)\right]} \text{ to get}$$

$$\text{Area} = \sqrt{s(s - a)(s - b)(s - c)}.$$

Problem Recognition Exercises, p. 666

7. Two possible triangles:
$c \approx 2.62, B \approx 126.09°, C \approx 13.91°$ or
$c \approx 10.86, B \approx 53.91°, C \approx 86.09°$

13. a. $BC \approx 17.83$ cm, $AC \approx 16.11$ cm, $DB \approx 14.48$ cm,
$DC \approx 11.74$ cm, $ED \approx 25.98$ cm, $EC \approx 17.52$ cm,
$EA \approx 27.56$ cm   **b.** $\angle AEC = 33.32°, \angle CAE \approx 36.68°$

14. From the figure

$$\cos B = \frac{BP}{c} \Rightarrow BP = c \cos B \text{ and } \cos C = \frac{PC}{b} \Rightarrow PC = b \cos C.$$

Therefore, $a = BP + PC = c \cos B + b \cos C$.

15. a. Eddie: (900, 0), Donna: (1705, 1045)

Section 6.4 Practice Exercises, pp. 672–677

7. a. 6 cm   **b.** $\frac{1}{40}$ sec   **c.** 40 Hz   **d.** 0 sec   **e.** $\frac{1}{160}$ sec

8. a. 2 cm   **b.** $\frac{1}{60}$ sec   **c.** 60 Hz   **d.** 0 sec   **e.** $\frac{1}{240}$ sec

9. a. 0.25 cm   **b.** 8 sec   **c.** $\frac{1}{8}$ Hz   **d.** 0 sec   **e.** 4 sec

10. a. 0.125 cm   **b.** 4 sec   **c.** $\frac{1}{4}$ Hz   **d.** 0 sec   **f.** 2 sec

11. a. 1 cm   **b.** 2π sec   **c.** $\frac{1}{2\pi}$ Hz   **d.** $\frac{\pi}{6}$ sec   **e.** $\frac{2\pi}{3}$ sec

12. a. 1 cm   **b.** 2π sec   **c.** $\frac{1}{2\pi}$ Hz   **d.** $-\frac{\pi}{4}$ sec   **e.** $\frac{\pi}{4}$ sec

13. a. $\frac{1}{4}$ cm   **b.** 4π sec   **c.** $\frac{1}{4\pi}$ Hz   **d.** -2 sec   **e.** $(2\pi - 2)$ sec

14. a. $\frac{1}{3}$ cm   **b.** 8π sec   **c.** $\frac{1}{8\pi}$ Hz   **d.** -2 sec   **e.** $(4\pi - 2)$ sec

27. c. $(-45\sqrt{3}, 47.5) \approx (-77.9, 47.5)$; The children are approximately 77.9 ft to the left of the center of the wheel and 47.5 ft above the ground on the way down.

28. c. $\left(\frac{15\sqrt{3}}{2}, \frac{45}{2}\right) \approx (13.0, 22.5)$; The point is approximately 13.0 ft to the right of the center of the wheel and 22.5 ft above the water.

41. a.

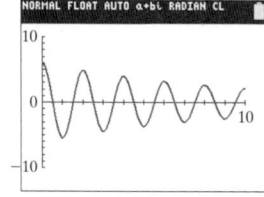

b.

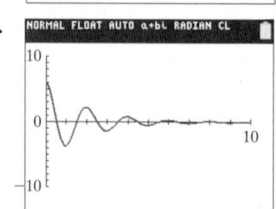

c.

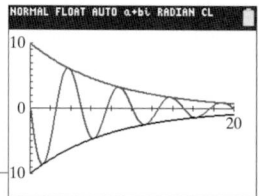

The maximum displacement for each cycle decreases more rapidly with time for greater values of c. That is, the curve converges to the t-axis more rapidly for greater values of c.

42. The distance from the equilibrium position is given by the absolute value of the displacement. Using the properties of absolute value and that $|\cos \omega t| \le 1$, we have
$|d| = |ae^{-ct} \cos \omega t| = |ae^{-ct}| \cdot |\cos \omega t| \le |ae^{-ct}| \cdot (1)$
Therefore, $|d| \le |ae^{-ct}|$. Since $a > 0$, we know that $ae^{-ct} > 0$ and we can drop the absolute value bars, $|d| \le ae^{-ct}$. Solving the inequality gives $-ae^{-ct} \le d \le ae^{-ct}$.

43. Bounding curves: $y = 10e^{-0.125t}$ and $y = -10e^{-0.125t}$

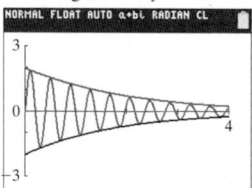

44. Bounding curves: $y = 2e^{-0.55t}$ and $y = -2e^{-0.55t}$

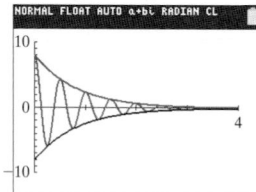

45. Bounding curves: $y = 8e^{-1.2t}$ and $y = -8e^{-1.2t}$

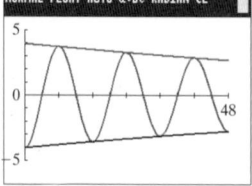

46. Bounding curves: $y = 4e^{-0.008t}$ and $y = -4e^{-0.008t}$

49. If the displacement of the object is at its equilibrium position at $t = 0$, then the sine function is preferred. If the object is at a displacement of a or $-a$ at $t = 0$, then the cosine model is preferred.

50. With simple harmonic motion, an object moves in a sinusoidal path indefinitely with constant amplitude. With damped harmonic motion, damping forces dissipate kinetic energy (energy associated with movement) as heat causing the amplitude of oscillation to decrease with time.

51. b. The graph will follow the curve $y = x$ with small oscillations above and below the line.

c.
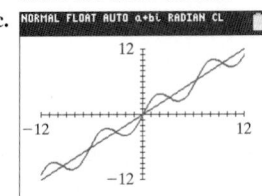

52. b. The graph will follow the curve $y = e^x$ with small oscillations above and below. However, the oscillations contributed by the term $\cos x$ will be greater for the portion of the curve with negative values of x. As the exponential function grows for positive values of x, the contribution by $\cos x$ will be very small (almost unnoticeable) relative to the contribution of the term e^x.

c.
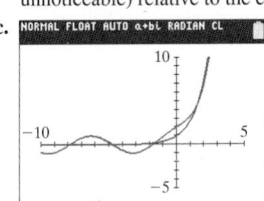

55. a. $d = 0.8e^{-0.03t}\cos\left(t\sqrt{\dfrac{\pi^2}{9} - 0.0009}\right)$

b.
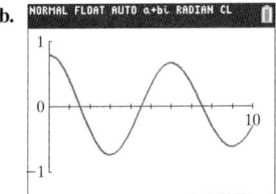

56. a. $d = 0.6e^{-0.05t}\cos\left(t\sqrt{\pi^2 - 0.0025}\right)$

b.
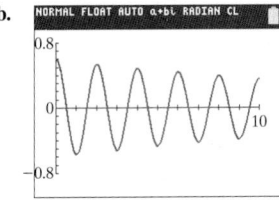

57. b.
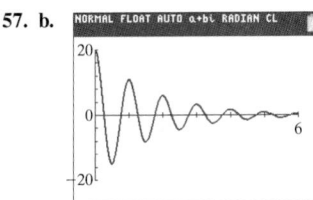

58. b.
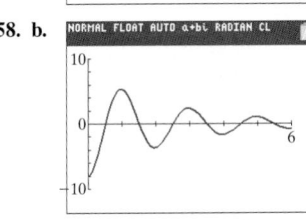

59. b. $A = 57.2 - 23.5\cos\left(\dfrac{2\pi}{365}t + \dfrac{4\pi}{73}\right)$

c.
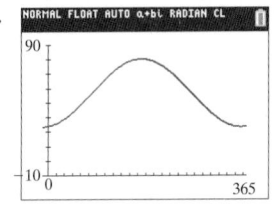

60. b. $A = 50.3 - 23.5\cos\left(\dfrac{2\pi}{365}t + \dfrac{4\pi}{73}\right)$

c.
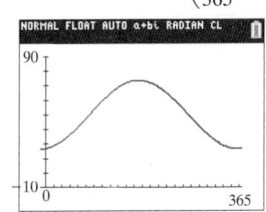

Chapter 6 Review Exercises, pp. 679–681

28. a. $AD = \dfrac{64\sin 80° \sin 72.5°}{\sin 58° \sin 69.3°}$

34. $\angle A \approx 48.4°,\ \angle B \approx 23.2°,\ \angle C \approx 108.4°$

43. a. 8 cm **b.** 12 sec **c.** $\dfrac{1}{12}\text{ Hz}$ **d.** 0 sec

44. a. 5 cm **b.** $\dfrac{1}{15}\text{ sec}$ **c.** 15 Hz **d.** 0 sec

45. a. 2 cm **b.** $2\pi\text{ sec}$ **c.** $\dfrac{1}{2\pi}\text{ Hz}$ **d.** $\dfrac{\pi}{3}\text{ sec}$

46. a. $\dfrac{1}{2}\text{ cm}$ **b.** $6\pi\text{ sec}$ **c.** $\dfrac{1}{6\pi}\text{ Hz}$ **d.** $-\dfrac{1}{2}\text{ sec}$

50. b. $y = 103 - 100\cos\dfrac{2\pi}{3}t$

c. $(-50\sqrt{3}, 153) \approx (-86.6, 153)$; The couple is approximately 86.6 ft to the left of the center of the wheel and 153 ft above the ground on the way down.

51. Bounding curves:
$y = 4e^{-0.03t}$ and $y = -4e^{-0.03t}$

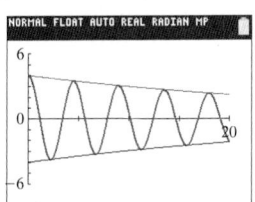

52. Bounding curves:
$y = 6e^{-1.3t}$ and $y = -6e^{-1.3t}$

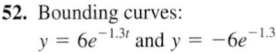
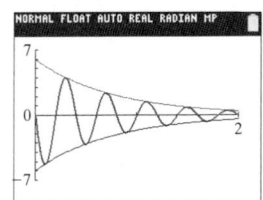

Chapter 6 Test, pp. 681–683

20. a. $a = \dfrac{12\sin 33°}{\sin 108°};\ b = \dfrac{12\sin 33° \sin 27°}{\sin 108° \sin 75°}$

21. The ratio of the length of a side of a triangle to the sine of the angle opposite it is the same for all three side-angle combinations.

23. Period $= \pi$, Phase shift $= \dfrac{\pi}{8}$

27. Bounding curves: $y = 4e^{-0.25t}$ and $y = -4e^{-0.25t}$

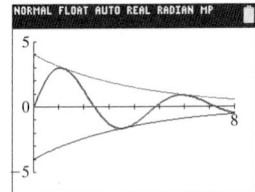

Chapter 6 Cumulative Review Exercises, pp. 683–684

1. $y = \dfrac{2}{3}x + \dfrac{5}{2}$

6. Vertical asymptotes: $x = -\dfrac{\pi}{4},\ x = \dfrac{\pi}{4}$

8. $y = 6\sin\left(2x - \dfrac{\pi}{3}\right)$ or $y = -6\sin\left(2x - \dfrac{\pi}{3}\right)$

11. a. $Z = \dfrac{kA^3}{\sqrt{B}}$

13. $S(x) = \begin{cases} 2500 & \text{for } 0 \le x \le 30{,}000 \\ 2500 + 0.05(x - 30{,}000) & \text{for } x > 30{,}000 \end{cases}$

CHAPTER 7

Section 7.1 Practice Exercises, pp. 693–696

15.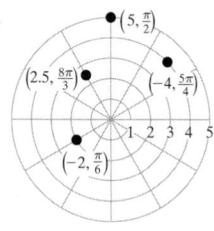

16.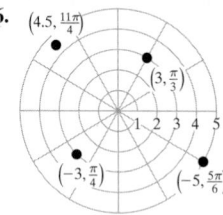

17.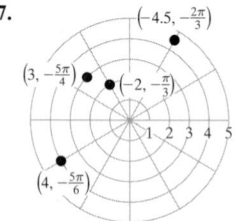

18.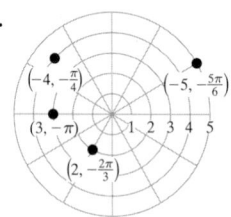

19. a. $\left(7, \dfrac{17\pi}{6}\right), \left(7, -\dfrac{7\pi}{6}\right)$ **b.** $\left(-7, \dfrac{11\pi}{6}\right), \left(-7, -\dfrac{\pi}{6}\right)$

20. a. $\left(3.6, \dfrac{5\pi}{2}\right), \left(3.6, -\dfrac{3\pi}{2}\right)$ **b.** $\left(-3.6, \dfrac{3\pi}{2}\right), \left(-3.6, -\dfrac{\pi}{2}\right)$

21. a. $\left(8, \dfrac{3\pi}{4}\right), \left(8, \dfrac{11\pi}{4}\right)$ **b.** $\left(-8, \dfrac{15\pi}{4}\right), \left(-8, -\dfrac{\pi}{4}\right)$

22. a. $\left(2, \dfrac{\pi}{3}\right), \left(2, \dfrac{7\pi}{3}\right)$ **b.** $\left(-2, -\dfrac{2\pi}{3}\right), \left(-2, \dfrac{10\pi}{3}\right)$

57. $x^2 + y^2 = 9$; A circle centered at the origin with radius 3

58. $x^2 + y^2 = 49$; A circle centered at the origin with radius 7

59. $y = 2$; A horizontal line crossing the y-axis at 2

60. $x = 1$; A vertical line crossing the x-axis at 1

61. $(x - 2)^2 + y^2 = 4$; A circle centered at $(2, 0)$ with radius 2

62. $x^2 + (y + 1)^2 = 1$; A circle centered at $(0, -1)$ with radius 1

63. $y = -x$; A line through the origin with a slope of -1

64. $y = \dfrac{\sqrt{3}}{3}x$; A line through the origin with slope $\dfrac{\sqrt{3}}{3}$

65. $x^2 + \left(y - \dfrac{a}{2}\right)^2 = \dfrac{a^2}{4}$; Circle centered at $\left(0, \dfrac{a}{2}\right)$ with radius $\dfrac{a}{2}$

66. $\left(x - \dfrac{a}{2}\right)^2 + y^2 = \dfrac{a^2}{4}$; Circle centered at $\left(\dfrac{a}{2}, 0\right)$ with radius $\dfrac{a}{2}$

67. $y = a$; Horizontal line passing through the y-axis at a

68. $x = a$; Vertical line passing through the x-axis at a

69. $(x - h)^2 + (y - k)^2 = h^2 + k^2$; Circle centered at (h, k) in rectangular coordinates with radius $\sqrt{h^2 + k^2}$

70. $\left(x - \dfrac{h}{2}\right)^2 + \left(y - \dfrac{k}{2}\right)^2 = \dfrac{1}{4}(h^2 + k^2)$; Circle centered at $\left(\dfrac{h}{2}, \dfrac{k}{2}\right)$ in rectangular coordinates with radius $\dfrac{1}{2}\sqrt{h^2 + k^2}$

73. No; the argument in the second ordered pair would have to be θ plus or minus an odd multiple of π: $(-r, \theta \pm (2n + 1)\pi)$.

74. No; the argument in the second ordered pair would have to be θ plus or minus an odd multiple of π: $(r, \theta \pm (2n + 1)\pi)$.

75. No; the ordered pair $(r, \pi + \theta)$ would represent the same point as $(-r, \theta)$.

83. The graph is the set of points inside the circle centered at the pole with radius 8.

84. The graph is the set of points on and outside the circle centered at the pole with radius 5.

85. The graph is the set of points on and between the circles centered at the pole with radii 2 and 4.

86. The graph is the set of points between the circles centered at the pole with radii 1 and 2.

87. The graph is the set of points between the line $\theta = 0$ (x-axis) and the line $\theta = \dfrac{\pi}{4}$ (line $y = x$).

88. The graph is the set of points in the second and fourth quadrants of a rectangular coordinate system.

89. a. 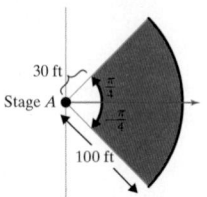   **90. a.**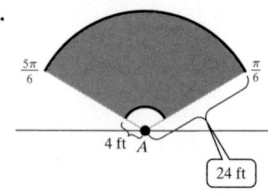

93. If $r < 0$, the point (r, θ) is $|r|$ units from the pole in the *opposite* direction of θ. That is, a negative value of r tells us to plot a point $|r|$ units in the direction $\theta + \pi$.

94. For any ordered pair (r, θ), the ordered pair $(r, \theta + 2n\pi)$ represents the same point for all integers n because the arguments θ and $\theta + 2n\pi$ are coterminal. Likewise, the ordered pairs (r, θ) and $(-r, \theta + (2n + 1)\pi)$ represent the same point for all integers n.

95. a. $P(r_1\cos\theta_1, r_1\sin\theta_1)$ and $Q(r_2\cos\theta_2, r_2\sin\theta_2)$

b. $d^2 = (r_2\cos\theta_2 - r_1\cos\theta_1)^2 + (r_2\sin\theta_2 - r_1\sin\theta_1)^2$

$= r_2^2\cos^2\theta_2 - 2r_1r_2\cos\theta_2\cos\theta_1 + r_1^2\cos^2\theta_1$

$\quad + r_2^2\sin^2\theta_2 - 2r_1r_2\sin\theta_2\sin\theta_1 + r_1^2\sin^2\theta_1$

$= r_2^2(\cos^2\theta_2 + \sin^2\theta_2) + r_1^2(\cos^2\theta_1 + \sin^2\theta_1)$

$\quad -2r_1r_2\cos\theta_2\cos\theta_1 - 2r_1r_2\sin\theta_2\sin\theta_1$

$= r_1^2 + r_2^2 - 2r_1r_2(\cos\theta_2\cos\theta_1 + \sin\theta_2\sin\theta_1)$

$= r_1^2 + r_2^2 - 2r_1r_2\cos(\theta_2 - \theta_1)$

Therefore, $d = \sqrt{r_1^2 + r_2^2 - 2r_1r_2\cos(\theta_2 - \theta_1)}$.

96. a. The distance is $|r_1 - r_2|$. The two points fall on the same line through the pole. Recall that for two points a and b on a number line, the distance between a and b is given by $|a - b|$ or $|b - a|$.

b. The distance is $\sqrt{r_1^2 + r_2^2}$. Since the arguments of the points differ by $\dfrac{\pi}{2}$, the two points and the pole form a right triangle. The distance between the points is given by the Pythagorean theorem.

97. Since $\theta_1 \neq \theta_2$ points P, Q, and the pole O are not collinear and thus form the vertices of a triangle, $\triangle POQ$.

Side $OP = r_1$ and side $OQ = r_2$, and the angle between them is $\angle POQ = |\theta_2 - \theta_1|$.

From the law of cosines, we have $PQ^2 = r_1^2 + r_2^2 - 2r_1r_2\cos|\theta_2 - \theta_1|$.

Noting that the cosine function is even, we have $\cos(\theta_2 - \theta_1) = \cos[-(\theta_2 - \theta_1)] = \cos(\theta_1 - \theta_2)$, so we can drop the absolute value bars.

Therefore, the distance between P and Q is $PQ = \sqrt{r_1^2 + r_2^2 - 2r_1r_2\cos(\theta_2 - \theta_1)}$.

99. 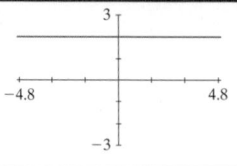   **100.**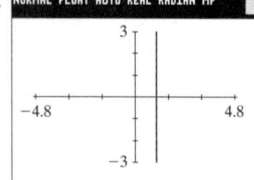

101. 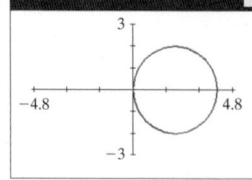   **102.**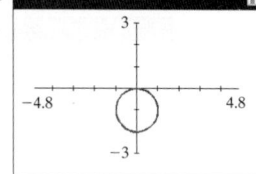

Section 7.2 Practice Exercises, pp. 707–711

7.
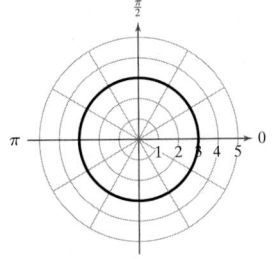

8.
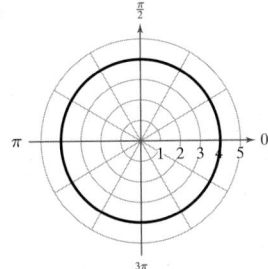

9.
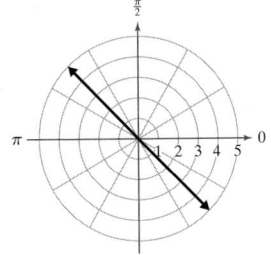

10.
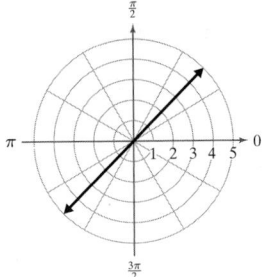

11.
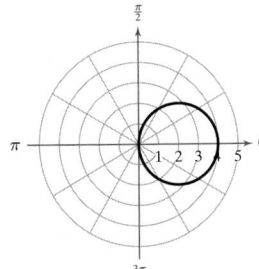

12.
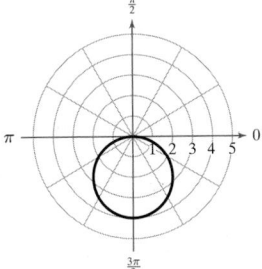

13.
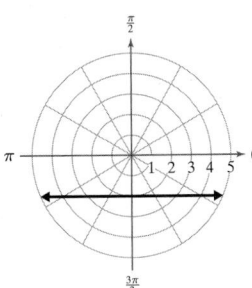

14.
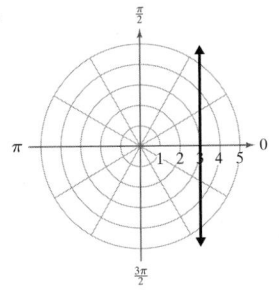

18. a. Yes **b.** Inconclusive **c.** Inconclusive
19. a. Inconclusive **b.** Yes **c.** Inconclusive
20. a. Inconclusive **b.** Yes **c.** Inconclusive
21. a. Inconclusive **b.** Inconclusive **c.** Inconclusive
22. a. Inconclusive **b.** Inconclusive **c.** Yes
23. Cardioid   **24.** Cardioid

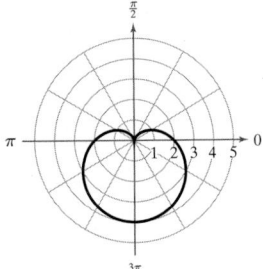

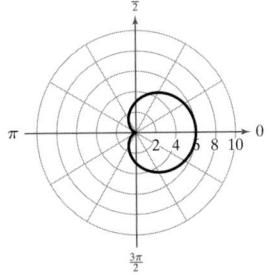

25. Limaçon with a loop
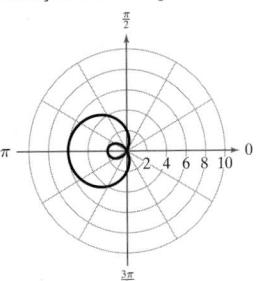

26. Limaçon with a loop
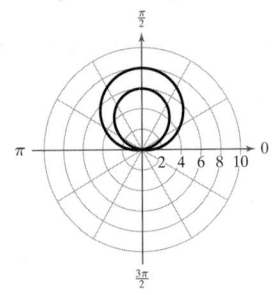

27. Rose (5 petals)
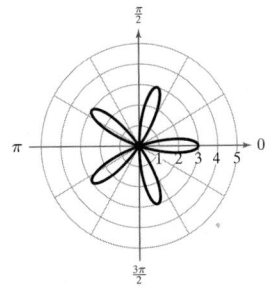

28. Rose (8 petals)
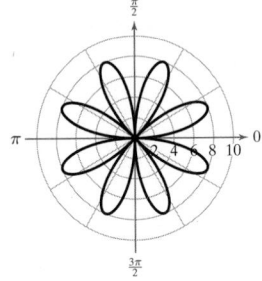

29. Dimpled limaçon
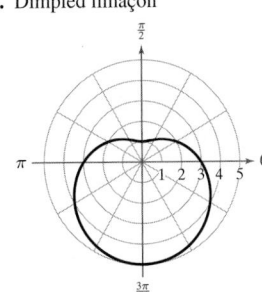

30. Dimpled limaçon
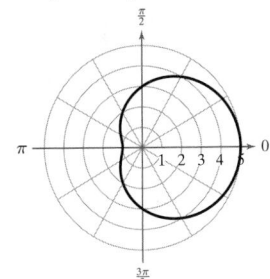

31. a. Rose (4 petals)   **b.** $r = 5\cos 2\theta$
32. a. Rose (3 petals)   **b.** $r = 4\sin 3\theta$
33. a. Circle (center on line $\theta = \frac{\pi}{2}$)   **b.** $r = -4\sin\theta$
34. a. Circle (center on polar axis)   **b.** $r = -3\cos\theta$
35. a. Lemniscate   **b.** $r^2 = 4\sin 2\theta$
36. a. Lemniscate   **b.** $r^2 = 9\cos 2\theta$
37. a. Limaçon (cardioid)   **b.** $r = 2 - 2\cos\theta$
38. a. Limaçon (cardioid)   **b.** $r = 2 + 2\sin\theta$
39. a. Horizontal line   **b.** $r = \dfrac{2}{\sin\theta}$ or $r = 2\csc\theta$

40. a. Vertical line   **b.** $r = \dfrac{-1}{\cos\theta}$ or $r = -\sec\theta$

43.
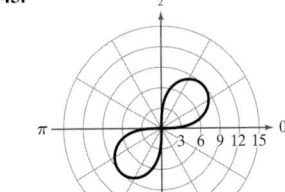

44.
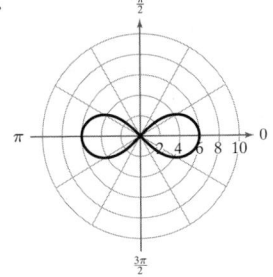

45. 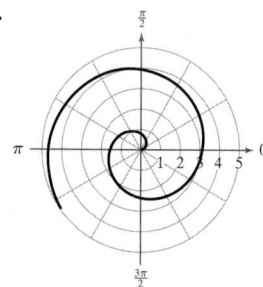   **46.**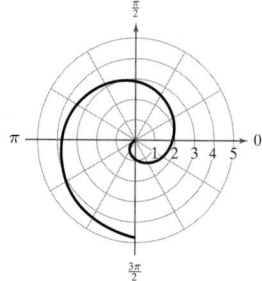

59. a. $\left\{ 0, \dfrac{\pi}{4}, \dfrac{\pi}{2}, \dfrac{3\pi}{4}, \pi, \dfrac{5\pi}{4}, \dfrac{3\pi}{2}, \dfrac{7\pi}{4} \right\}$

b.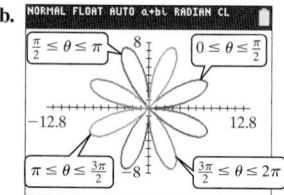

60. a. $\left\{ 0, \dfrac{\pi}{3}, \dfrac{2\pi}{3} \right\}$ **b.**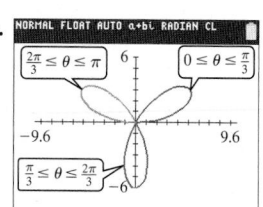

61. Since $\sin \theta = \sin(\pi - \theta)$, replacing θ by $\pi - \theta$ in a polar equation results in an equivalent equation. Therefore, the graph is symmetric with respect to the line $\theta = \dfrac{\pi}{2}$.

62. Since $\cos \theta = \cos(-\theta)$, replacing θ by $-\theta$ in a polar equation results in an equivalent equation. Therefore, the graph is symmetric with respect to the polar axis.

67. a. $\theta = \dfrac{\pi}{2}$ and $\theta = \dfrac{3\pi}{2}$ **b.**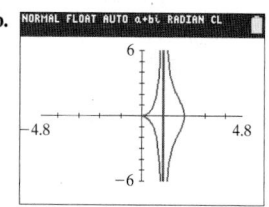

c. The graph appears to have a vertical asymptote.
For example, when θ approaches $\dfrac{\pi}{2}$, the value of $|\sec \theta|$ (and subsequently $|r|$) becomes infinitely large. This means that the constant term on the right side of the equation does not contribute much to the value of r. Thus, for values of θ close to $\dfrac{\pi}{2}$, $r \approx \sec \theta$.
Converting to rectangular coordinates, $r \approx \dfrac{1}{\cos \theta} \Rightarrow r\cos \theta = 1$ or equivalently, $x = 1$.

68. b.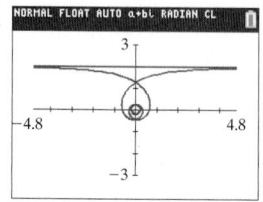

69. a.–b.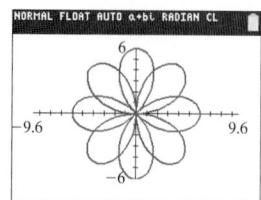

c. The graph of $r = 6\sin\left[2\left(\theta + \dfrac{\pi}{4} \right) \right]$ is the same as the graph of $r = 6\sin 2\theta$, but rotated $-\dfrac{\pi}{4}$ radians.

70. a.–c.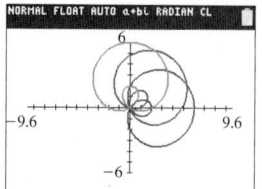

d. The graph of $r = f(\theta - \alpha)$ appears to be the same as the graph of $r = f(\theta)$ but rotated by α units.

71. a.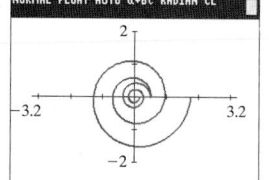

b. For $b > 0$, the spiral gets larger for increasing values of θ.
For $b < 0$, the spiral gets smaller for increasing values of θ.

72. a.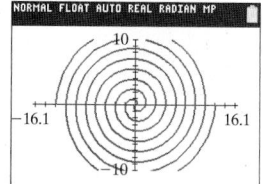

73.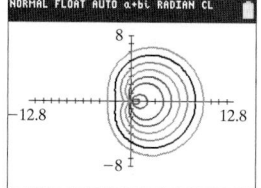

When $\dfrac{a}{b} < 1$ the graph has an inner loop and the inner loop shrinks as $\dfrac{a}{b}$ approaches 1.

When $\dfrac{a}{b} = 1$, the inner loop converges to a sharp point.

For $1 < \dfrac{a}{b} < 2$ the sharp point becomes a smooth dimple and the degree of curvature of the dimple decreases as $\dfrac{a}{b}$ approaches 2.

When $\dfrac{a}{b} = 2$, the dimple disappears and the "left" side of the curve is rounded outwards.

74.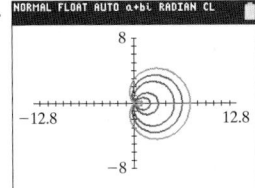

As $\dfrac{a}{b}$ approaches 1, the size of the outer loop increases and the size of the inner loop decreases.

77. d.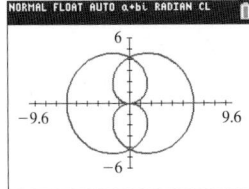

78. c. The graph is also symmetric with respect to the line $\theta = \dfrac{\pi}{2}$ and to the pole.

d.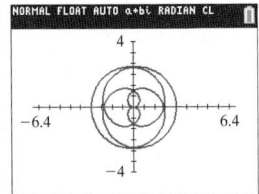

Problem Recognition Exercises, pp. 711–712

9. The graph is a parabola opening upward with vertex at $\left(0, -\dfrac{3}{2}\right)$.

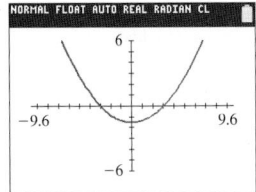

10. The graph is a parabola opening to the left with vertex at $(1, 0)$.

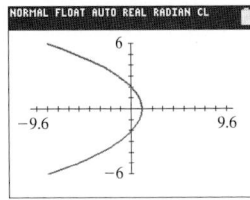

Section 7.3 Practice Exercises, pp. 722–725

7.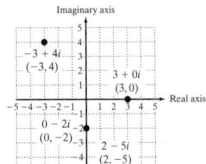

8.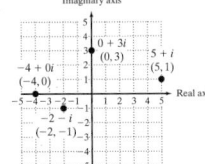

15. a. $3\sqrt{2}\left(\cos\dfrac{\pi}{4} + i\sin\dfrac{\pi}{4}\right)$ **b.** $3\sqrt{2}\left(\cos\dfrac{5\pi}{4} + i\sin\dfrac{5\pi}{4}\right)$

16. a. $4\left(\cos\dfrac{5\pi}{3} + i\sin\dfrac{5\pi}{3}\right)$ **b.** $4\left(\cos\dfrac{2\pi}{3} + i\sin\dfrac{2\pi}{3}\right)$

17. a. $8\left(\cos\dfrac{5\pi}{6} + i\sin\dfrac{5\pi}{6}\right)$ **b.** $20\left(\cos\dfrac{11\pi}{6} + i\sin\dfrac{11\pi}{6}\right)$

18. a. $6\left(\cos\dfrac{7\pi}{4} + i\sin\dfrac{7\pi}{4}\right)$ **b.** $2\left(\cos\dfrac{3\pi}{4} + i\sin\dfrac{3\pi}{4}\right)$

19. a. $25\left[\cos\left(\tan^{-1}\dfrac{7}{24}\right) + i\sin\left(\tan^{-1}\dfrac{7}{24}\right)\right]$

b. $29\left[\cos\left(\pi + \tan^{-1}\left(-\dfrac{21}{20}\right)\right) + i\sin\left(\pi + \tan^{-1}\left(-\dfrac{21}{20}\right)\right)\right]$

20. a. $37\left[\cos\left(2\pi + \tan^{-1}\left(-\dfrac{12}{35}\right)\right) + i\sin\left(2\pi + \tan^{-1}\left(-\dfrac{12}{35}\right)\right)\right]$

b. $53\left[\cos\left(\pi + \tan^{-1}\dfrac{45}{28}\right) + i\sin\left(\pi + \tan^{-1}\dfrac{45}{28}\right)\right]$

21. a. $17(\cos 0 + i\sin 0)$ **b.** $4\left(\cos\dfrac{3\pi}{2} + i\sin\dfrac{3\pi}{2}\right)$

22. a. $10\left(\cos\dfrac{\pi}{2} + i\sin\dfrac{\pi}{2}\right)$ **b.** $12(\cos\pi + i\sin\pi)$

31. a. $800(\cos 45° + i\sin 45°)$ **b.** $\dfrac{1}{2}(\cos 17° + i\sin 17°)$

32. a. $243(\cos 120° + i\sin 120°)$ **b.** $3(\cos 14° + i\sin 14°)$

33. a. $18\left(\cos\dfrac{\pi}{3} + i\sin\dfrac{\pi}{3}\right)$ **b.** $\dfrac{1}{2}\left(\cos\dfrac{7\pi}{6} + i\sin\dfrac{7\pi}{6}\right)$

34. a. $20\left(\cos\dfrac{\pi}{6} + i\sin\dfrac{\pi}{6}\right)$ **b.** $5\left(\cos\dfrac{5\pi}{3} + i\sin\dfrac{5\pi}{3}\right)$

35. a. $\cos 160° + i\sin 160°$ **b.** $\cos 280° + i\sin 280°$
36. a. $\cos 90° + i\sin 90°$ **b.** $\cos 304° + i\sin 304°$

37. a. $\dfrac{1}{16}\left(\cos\dfrac{\pi}{2} + i\sin\dfrac{\pi}{2}\right)$ **b.** $9\left(\cos\dfrac{5\pi}{3} + i\sin\dfrac{5\pi}{3}\right)$

38. a. $25(\cos\pi + i\sin\pi)$ **b.** $\dfrac{1}{36}\left(\cos\dfrac{3\pi}{2} + i\sin\dfrac{3\pi}{2}\right)$

39. a. $12(\cos 0 + i\sin 0)$ **b.** $\dfrac{2}{3}\left(\cos\dfrac{3\pi}{2} + i\sin\dfrac{3\pi}{2}\right)$

40. a. $100\left(\cos\dfrac{3\pi}{2} + i\sin\dfrac{3\pi}{2}\right)$ **b.** $\dfrac{1}{2}(\cos\pi + i\sin\pi)$

41. a. $288\left(\cos\dfrac{11\pi}{6} + i\sin\dfrac{11\pi}{6}\right)$ **b.** $2\left(\cos\dfrac{\pi}{2} + i\sin\dfrac{\pi}{2}\right)$

42. a. $384\left(\cos\dfrac{\pi}{6} + i\sin\dfrac{\pi}{6}\right)$ **b.** $6\left(\cos\dfrac{\pi}{2} + i\sin\dfrac{\pi}{2}\right)$

55. $-3, 3, -\dfrac{3}{2} - \dfrac{3\sqrt{3}}{2}i, \dfrac{3}{2} - \dfrac{3\sqrt{3}}{2}i, -\dfrac{3}{2} + \dfrac{3\sqrt{3}}{2}i, \dfrac{3}{2} + \dfrac{3\sqrt{3}}{2}i$

56. $5, -5, 5i, -5i$

57. $-3i, -\dfrac{3\sqrt{3}}{2} + \dfrac{3}{2}i, \dfrac{3\sqrt{3}}{2} + \dfrac{3}{2}i$

58. $5i, -\dfrac{5\sqrt{3}}{2} - \dfrac{5}{2}i, \dfrac{5\sqrt{3}}{2} - \dfrac{5}{2}i$

59. $2 + 2\sqrt{3}i, -2 - 2\sqrt{3}i$ **60.** $-5\sqrt{3} + 5i, 5\sqrt{3} - 5i$

61. $4\left(\cos\dfrac{3\pi}{8} + i\sin\dfrac{3\pi}{8}\right), 4\left(\cos\dfrac{11\pi}{8} + i\sin\dfrac{11\pi}{8}\right)$

62. $7\left(\cos\dfrac{11\pi}{12} + i\sin\dfrac{11\pi}{12}\right), 7\left(\cos\dfrac{23\pi}{12} + i\sin\dfrac{23\pi}{12}\right)$

63. $2\left(\cos\dfrac{\pi}{30} + i\sin\dfrac{\pi}{30}\right), 2\left(\cos\dfrac{13\pi}{30} + i\sin\dfrac{13\pi}{30}\right),$
$2\left(\cos\dfrac{5\pi}{6} + i\sin\dfrac{5\pi}{6}\right), 2\left(\cos\dfrac{37\pi}{30} + i\sin\dfrac{37\pi}{30}\right),$
$2\left(\cos\dfrac{49\pi}{30} + i\sin\dfrac{49\pi}{30}\right)$

64. $\cos\dfrac{7\pi}{24} + i\sin\dfrac{7\pi}{24}, \cos\dfrac{5\pi}{8} + i\sin\dfrac{5\pi}{8},$
$\cos\dfrac{23\pi}{24} + i\sin\dfrac{23\pi}{24}, \cos\dfrac{31\pi}{24} + i\sin\dfrac{31\pi}{24},$
$\cos\dfrac{13\pi}{8} + i\sin\dfrac{13\pi}{8}, \cos\dfrac{47\pi}{24} + i\sin\dfrac{47\pi}{24}$

65. $3(\cos 38° + i\sin 38°), 3(\cos 158° + i\sin 158°),$
$3(\cos 278° + i\sin 278°)$

66. $4(\cos 57° + i\sin 57°), 4(\cos 147° + i\sin 147°),$
$4(\cos 237° + i\sin 237°), 4(\cos 327° + i\sin 327°)$

67. $9(\cos 157.5° + i\sin 157.5°), 9(\cos 337.5° + i\sin 337.5°)$

68. $\sqrt{3}(\cos 75° + i\sin 75°), \sqrt{3}(\cos 255° + i\sin 255°)$

69. $\sqrt[6]{4}(\cos 40° + i\sin 40°), \sqrt[6]{4}(\cos 100° + i\sin 100°),$
$\sqrt[6]{4}(\cos 160° + i\sin 160°), \sqrt[6]{4}(\cos 220° + i\sin 220°),$
$\sqrt[6]{4}(\cos 280° + i\sin 280°), \sqrt[6]{4}(\cos 340° + i\sin 340°)$

70. $\sqrt[5]{10}(\cos 66° + i\sin 66°), \sqrt[5]{10}(\cos 138° + i\sin 138°),$
$\sqrt[5]{10}(\cos 210° + i\sin 210°), \sqrt[5]{10}(\cos 282° + i\sin 282°),$
$\sqrt[5]{10}(\cos 354° + i\sin 354°)$

71. The value 1 is an nth root of 1 and can be represented as $1(\cos 0° + i\sin 0°)$ or $1(\cos 0 + i\sin 0)$. The moduli of the remaining roots are also 1, and the arguments differ by $\dfrac{360}{n}$ degrees, or $\dfrac{2\pi}{n}$ radians.

72. $1(\cos 0 + i\sin 0), 1\left(\cos\dfrac{2\pi}{5} + i\sin\dfrac{2\pi}{5}\right), 1\left(\cos\dfrac{4\pi}{5} + i\sin\dfrac{4\pi}{5}\right),$
$1\left(\cos\dfrac{6\pi}{5} + i\sin\dfrac{6\pi}{5}\right), 1\left(\cos\dfrac{8\pi}{5} + i\sin\dfrac{8\pi}{5}\right)$

73. $\left[2\left(\cos\dfrac{2\pi}{9} + i\sin\dfrac{2\pi}{9}\right)\right]^3 = 2^3\left[\cos\left(3\cdot\dfrac{2\pi}{9}\right) + i\sin\left(3\cdot\dfrac{2\pi}{9}\right)\right]$
$= 8\left(\cos\dfrac{2\pi}{3} + i\sin\dfrac{2\pi}{3}\right)$

74. $(1-i)^3 = \left[\sqrt{2}\left(\cos\dfrac{7\pi}{4} + i\sin\dfrac{7\pi}{4}\right)\right]^3$
$= (\sqrt{2})^3\left[\cos\left(3\cdot\dfrac{7\pi}{4}\right) + i\sin\left(3\cdot\dfrac{7\pi}{4}\right)\right]$
$= 2\sqrt{2}\left(\cos\dfrac{21\pi}{4} + i\sin\dfrac{21\pi}{4}\right) = 2\sqrt{2}\left(\cos\dfrac{5\pi}{4} + i\sin\dfrac{5\pi}{4}\right)$

75.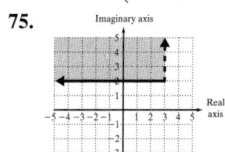

76.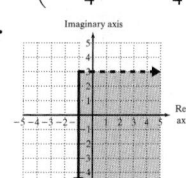

77.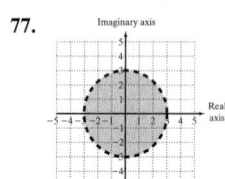

78.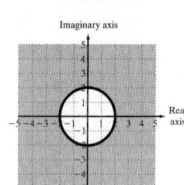

79. $-\sqrt{6} - \sqrt{2} + (-\sqrt{6} + \sqrt{2})i$

80. $-9\sqrt{6} + 9\sqrt{2} + (9\sqrt{6} + 9\sqrt{2})i$

81. $\dfrac{\sqrt{6} + \sqrt{2}}{4} + \left(\dfrac{\sqrt{6} - \sqrt{2}}{4}\right)i$

82. $\dfrac{\sqrt{6} - \sqrt{2}}{4} + \left(\dfrac{-\sqrt{6} - \sqrt{2}}{4}\right)i$

83. $z_1 = -2 - 2\sqrt{3}i$ and $z_2 = 2i$
$z_1 z_2 = (-2 - 2\sqrt{3}i)(2i) = -4i - 4\sqrt{3}i^2$
$= 4\sqrt{3} - 4i = 8\left(\cos\dfrac{11\pi}{6} + i\sin\dfrac{11\pi}{6}\right)$

84. a. $\dfrac{z_1}{z_2} = \dfrac{(2+2i)}{(3i)}\cdot\dfrac{(-3i)}{(-3i)} = \dfrac{6 - 6i}{9} = \dfrac{2}{3} - \dfrac{2}{3}i$
$= \dfrac{2\sqrt{2}}{3}\left(\cos\dfrac{7\pi}{4} + i\sin\dfrac{7\pi}{4}\right)$

b. $\dfrac{2\sqrt{2}\left(\cos\dfrac{\pi}{4} + i\sin\dfrac{\pi}{4}\right)}{3\left(\cos\dfrac{\pi}{2} + i\sin\dfrac{\pi}{2}\right)}$
$= \dfrac{2\sqrt{2}}{3}\left(\cos\dfrac{7\pi}{4} + i\sin\dfrac{7\pi}{4}\right)$

85. $\dfrac{z_1}{z_2} = \dfrac{r_1(\cos\theta_1 + i\sin\theta_1)}{r_2(\cos\theta_2 + i\sin\theta_2)}$
$= \dfrac{r_1(\cos\theta_1 + i\sin\theta_1)\cdot(\cos\theta_2 - i\sin\theta_2)}{r_2(\cos\theta_2 + i\sin\theta_2)\cdot(\cos\theta_2 - i\sin\theta_2)}$
$= \dfrac{r_1}{r_2}\cdot\dfrac{\cos\theta_1\cos\theta_2 - i\cos\theta_1\sin\theta_2 + i\sin\theta_1\cos\theta_2 - i^2\sin\theta_1\sin\theta_2}{\cos^2\theta_2 - i^2\sin^2\theta_2}$
$= \dfrac{r_1}{r_2}\cdot\dfrac{\cos\theta_1\cos\theta_2 + \sin\theta_1\sin\theta_2 + i(\sin\theta_1\cos\theta_2 - \cos\theta_1\sin\theta_2)}{\cos^2\theta_2 + \sin^2\theta_2}$
$= \dfrac{r_1}{r_2}[\cos(\theta_1 - \theta_2) + i\sin(\theta_1 - \theta_2)]$

86. Let $z_1 = a + bi$ in rectangular form. We must show that $z_2 = a - bi$. Because the cosine function is even and the sine function is odd, we have $z_2 = r[\cos(-\theta) + i\sin(-\theta)] = r(\cos\theta - i\sin\theta) = a - bi$

87. a. $z^{-1} = \dfrac{1}{z} = \dfrac{1}{r(\cos\theta + i\sin\theta)}\cdot\dfrac{(\cos\theta - i\sin\theta)}{(\cos\theta - i\sin\theta)}$
$= \dfrac{1}{r}\cdot\dfrac{\cos\theta - i\sin\theta}{\cos^2\theta + \sin^2\theta} = r^{-1}[\cos(-\theta) + i\sin(-\theta)]$

b. $z^{-2} = \dfrac{1}{z^2} = \dfrac{1}{r^2(\cos 2\theta + i\sin 2\theta)}\cdot\dfrac{(\cos 2\theta - i\sin 2\theta)}{(\cos 2\theta - i\sin 2\theta)}$
$= \dfrac{1}{r^2}\cdot\dfrac{\cos 2\theta - i\sin 2\theta}{\cos^2 2\theta + \sin^2 2\theta} = r^{-2}[\cos(-2\theta) + i\sin(-2\theta)]$

93. Setting one side equal to zero and factoring yields $(x + 2)(x^2 - 2x + 4)(x - 2)(x^2 + 2x + 4) = 0$. Solving for x gives $2, 1 + \sqrt{3}i, -1 + \sqrt{3}i, -2, -1 - \sqrt{3}i, 1 - \sqrt{3}i$ as expected.

94. Setting one side equal to zero and factoring yields $(x + 4)(x^2 - 4x + 16) = 0$. Solving for x gives $-4, 2 + 2\sqrt{3}i$, and $2 - 2\sqrt{3}i$ as expected.

95. $2\left(\cos\dfrac{5\pi}{16} + i\sin\dfrac{5\pi}{16}\right), 2\left(\cos\dfrac{13\pi}{16} + i\sin\dfrac{13\pi}{16}\right),$
$2\left(\cos\dfrac{21\pi}{16} + i\sin\dfrac{21\pi}{16}\right), 2\left(\cos\dfrac{29\pi}{16} + i\sin\dfrac{29\pi}{16}\right)$

96. $2\left(\cos\dfrac{\pi}{20} + i\sin\dfrac{\pi}{20}\right), 2\left(\cos\dfrac{9\pi}{20} + i\sin\dfrac{9\pi}{20}\right),$
$2\left(\cos\dfrac{17\pi}{20} + i\sin\dfrac{17\pi}{20}\right), 2\left(\cos\dfrac{5\pi}{4} + i\sin\dfrac{5\pi}{4}\right),$
$2\left(\cos\dfrac{33\pi}{20} + i\sin\dfrac{33\pi}{20}\right)$

97. $\cos\dfrac{5\pi}{18} + i\sin\dfrac{5\pi}{18}, \cos\dfrac{17\pi}{18} + i\sin\dfrac{17\pi}{18}, \cos\dfrac{29\pi}{18} + i\sin\dfrac{29\pi}{18}$

98. $2\left(\cos\dfrac{\pi}{6} + i\sin\dfrac{\pi}{6}\right), 2\left(\cos\dfrac{5\pi}{6} + i\sin\dfrac{5\pi}{6}\right), 2\left(\cos\dfrac{3\pi}{2} + i\sin\dfrac{3\pi}{2}\right)$

99. The absolute value of z is the distance between (a, b) and the origin $(0, 0)$ in the complex plane. The absolute value of a real number x is the distance between x and 0 on the number line.

100. No, the value $z = r(\cos\theta + i\sin\theta)$ has infinitely many representations $z = r[\cos(\theta + 2n\pi) + i\sin(\theta + 2n\pi)]$, where n is any integer. However, θ is normally taken on the interval $0 \le \theta < 2\pi$.

101. a. $\cos\dfrac{\pi}{4} + i\sin\dfrac{\pi}{4}$ or $\dfrac{\sqrt{2}}{2} + \dfrac{\sqrt{2}}{2}i$

b. $e^{\pi i} + 1 = (\cos\pi + i\sin\pi) + 1$
$= (-1 + 0) + 1 = 0$

102. $e^{ix} = 1 + \dfrac{ix}{1} + \dfrac{(ix)^2}{2 \cdot 1} + \dfrac{(ix)^3}{3 \cdot 2 \cdot 1} + \dfrac{(ix)^4}{4 \cdot 3 \cdot 2 \cdot 1}$

$\qquad + \dfrac{(ix)^5}{5 \cdot 4 \cdot 3 \cdot 2 \cdot 1} + \dfrac{(ix)^6}{6 \cdot 5 \cdot 4 \cdot 3 \cdot 2 \cdot 1}$

$\qquad + \dfrac{(ix)^7}{7 \cdot 6 \cdot 5 \cdot 4 \cdot 3 \cdot 2 \cdot 1} + \cdots$

$\quad = 1 + \dfrac{ix}{1} - \dfrac{x^2}{2 \cdot 1} - \dfrac{ix^3}{3 \cdot 2 \cdot 1} + \dfrac{x^4}{4 \cdot 3 \cdot 2 \cdot 1} + \dfrac{ix^5}{5 \cdot 4 \cdot 3 \cdot 2 \cdot 1}$

$\qquad - \dfrac{x^6}{6 \cdot 5 \cdot 4 \cdot 3 \cdot 2 \cdot 1} - \dfrac{ix^7}{7 \cdot 6 \cdot 5 \cdot 4 \cdot 3 \cdot 2 \cdot 1} + \cdots$

$\quad = \left(1 - \dfrac{x^2}{2 \cdot 1} + \dfrac{x^4}{4 \cdot 3 \cdot 2 \cdot 1} - \dfrac{x^6}{6 \cdot 5 \cdot 4 \cdot 3 \cdot 2 \cdot 1} + \cdots \right)$

$\qquad + i\left(x - \dfrac{x^3}{3 \cdot 2 \cdot 1} + \dfrac{x^5}{5 \cdot 4 \cdot 3 \cdot 2 \cdot 1} - \dfrac{x^7}{7 \cdot 6 \cdot 5 \cdot 4 \cdot 3 \cdot 2 \cdot 1} \cdots \right)$

$\quad = \cos x + i \sin x$

103.

$$\left(3e^{\pi i/12}\right)^8 - 81\left(3e^{\pi i/12}\right)^4 + 6561 \overset{?}{=} 0$$

$$3^8 e^{8\pi i/12} - 3^4 \cdot 3^4 e^{4\pi i/12} + 3^8 \overset{?}{=} 0$$

$$3^8\left(e^{2\pi i/3} - e^{\pi i/3} + 1\right) \overset{?}{=} 0$$

$$3^8\left[\cos\dfrac{2\pi}{3} + i\sin\dfrac{2\pi}{3} - \left(\cos\dfrac{\pi}{3} + i\sin\dfrac{\pi}{3}\right) + 1\right] \overset{?}{=} 0$$

$$3^8\left(-\dfrac{1}{2} + \dfrac{\sqrt{3}}{2}i - \dfrac{1}{2} - \dfrac{\sqrt{3}}{2}i + 1\right) \overset{?}{=} 0$$

$$3^8(0) = 0 \checkmark$$

104.

$$\left(e^{7\pi i/12}\right)^8 - \left(e^{7\pi i/12}\right)^4 + 1 \overset{?}{=} 0$$

$$e^{56\pi i/12} - e^{28\pi i/12} + 1 \overset{?}{=} 0$$

$$e^{14\pi i/3} - e^{7\pi i/3} + 1 \overset{?}{=} 0$$

$$\cos\dfrac{14\pi}{3} + i\sin\dfrac{14\pi}{3} - \left(\cos\dfrac{7\pi}{3} + i\sin\dfrac{7\pi}{3}\right) + 1 \overset{?}{=} 0$$

$$-\dfrac{1}{2} + \dfrac{\sqrt{3}}{2}i - \dfrac{1}{2} - \dfrac{\sqrt{3}}{2}i + 1 \overset{?}{=} 0 \checkmark$$

Section 7.4 Practice Exercises, pp. 738–743

R.3.

R.4.

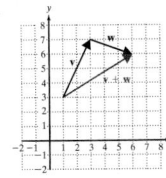

11. a. $\|\mathbf{v}\| = \sqrt{10}$ **b.** $\|\mathbf{w}\| = \sqrt{10}$ **c.** $\mathbf{v} = \mathbf{w}$; The vectors have the same magnitude and direction.

12. a. $\|\mathbf{v}\| = \sqrt{13}$ **b.** $\|\mathbf{w}\| = \sqrt{13}$ **c.** $\mathbf{v} = \mathbf{w}$; The vectors have the same magnitude and direction.

13. a. $\|\mathbf{v}\| = \sqrt{34}$ **b.** $\|\mathbf{w}\| = \sqrt{34}$ **c.** $\mathbf{v} \neq \mathbf{w}$; The vectors have the same magnitude but different directions.

14. a. $\|\mathbf{v}\| = \sqrt{865}$ **b.** $\|\mathbf{w}\| = \sqrt{865}$ **c.** $\mathbf{v} \neq \mathbf{w}$; The vectors have the same magnitude but different directions.

15. a. $\|\mathbf{v}\| = 5$ **b.** $\|\mathbf{w}\| = 10$ **c.** $\mathbf{v} \neq \mathbf{w}$; The vectors have the same direction, but different magnitudes.

16. a. $\|\mathbf{v}\| = 34$ **b.** $\|\mathbf{w}\| = 51$ **c.** $\mathbf{v} \neq \mathbf{w}$; The vectors have the same direction, but different magnitudes.

17. a. 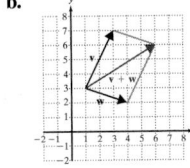 **b.**

18. a. 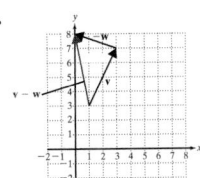 **b.**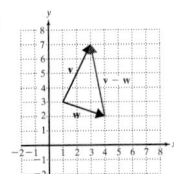

19. 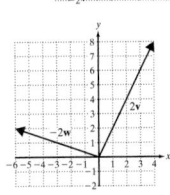 **20.**

21. b. **22. b.**

47. $(\mathbf{v} + \mathbf{w}) + \mathbf{u} = (\langle a_1, b_1 \rangle + \langle a_2, b_2 \rangle) + \langle a_3, b_3 \rangle$

$\qquad = \langle a_1 + a_2, b_1 + b_2 \rangle + \langle a_3, b_3 \rangle = \langle a_1 + a_2 + a_3, b_1 + b_2 + b_3 \rangle$

$\qquad = \langle a_1 + (a_2 + a_3), b_1 + (b_2 + b_3) \rangle$

$\qquad = \langle a_1, b_1 \rangle + \langle a_2 + a_3, b_2 + b_3 \rangle = \langle a_1, b_1 \rangle + (\langle a_2, b_2 \rangle + \langle a_3, b_3 \rangle)$

$\qquad = \mathbf{v} + (\mathbf{w} + \mathbf{u})$

48. $\mathbf{v} + \mathbf{0} = \langle a_1, b_1 \rangle + \langle 0, 0 \rangle = \langle a_1 + 0, b_1 + 0 \rangle$

$\qquad = \langle 0 + a_1, 0 + b_1 \rangle = \langle 0, 0 \rangle + \langle a_1, b_1 \rangle = \mathbf{0} + \mathbf{v}$

49. $\mathbf{v} + (-\mathbf{v}) = \langle a_1, b_1 \rangle + \langle -a_1, -b_1 \rangle = \langle a_1 + (-a_1), b_1 + (-b_1) \rangle$

$\qquad = \langle 0, 0 \rangle = \mathbf{0}$

50. $c(\mathbf{v} + \mathbf{w}) = c(\langle a_1, b_1 \rangle + \langle a_2, b_2 \rangle) = c\langle a_1 + a_2, b_1 + b_2 \rangle$

$\qquad = \langle c(a_1 + a_2), c(b_1 + b_2) \rangle = \langle ca_1 + ca_2, cb_1 + cb_2 \rangle$

$\qquad = \langle ca_1, cb_1 \rangle + \langle ca_2, cb_2 \rangle = c\langle a_1, b_1 \rangle + c\langle a_2, b_2 \rangle = c\mathbf{v} + c\mathbf{w}$

51. $(c + d)\mathbf{v} = (c + d)\langle a_1, b_1 \rangle = \langle (c + d)a_1, (c + d)b_1 \rangle$

$\qquad = \langle ca_1 + da_1, cb_1 + db_1 \rangle = \langle ca_1, cb_1 \rangle + \langle da_1, db_1 \rangle$

$\qquad = c\langle a_1, b_1 \rangle + d\langle a_1, b_1 \rangle = c\mathbf{v} + d\mathbf{v}$

52. $(cd)\mathbf{v} = (cd)\langle a_1, b_1 \rangle = \langle (cd)a_1, (cd)b_1 \rangle = \langle c(da_1), c(db_1) \rangle$

$\qquad = c\langle da_1, db_1 \rangle = c(d\mathbf{v})$

53. $0\langle a_1, b_1 \rangle = \langle 0 \cdot a_1, 0 \cdot b_1 \rangle = \langle 0, 0 \rangle = \mathbf{0}$

54. $\|c\mathbf{v}\| = \|c\langle a_1, b_1 \rangle\| = \|\langle ca_1, cb_1 \rangle\| = \sqrt{(ca_1)^2 + (cb_1)^2}$

$\qquad = \sqrt{c^2 a_1^2 + c^2 b_1^2} = \sqrt{c^2(a_1^2 + b_1^2)}$

$\qquad = |c|\sqrt{a_1^2 + b_1^2} = |c|\|v\|$

105. a.

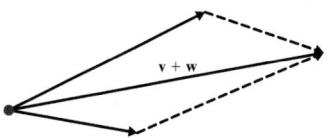

106. b.

107. A vector in standard position has initial point at the origin. The terminal point is a units horizontally and b units vertically from the origin. Therefore, the terminal point is (a, b).

108. The initial point (tail) of $\mathbf{w}$ is placed at the terminal point (tip) of $\mathbf{v}$. The sum is the vector from the initial point of $\mathbf{v}$ to the terminal point of $\mathbf{w}$. Alternatively, $\mathbf{v}$ and $\mathbf{w}$ can be placed with the same initial point. The sum $\mathbf{v} + \mathbf{w}$ is the diagonal of the parallelogram extending from the common initial point to the opposite vertex.

109. If k is positive, the vector $k\mathbf{v}$ is in the same direction of $\mathbf{v}$. If k is negative, the vector $k\mathbf{v}$ is in the opposite direction of $\mathbf{v}$. If $k = 0$, the product is the zero vector. The value of k either stretches or shrinks the length (magnitude) of $\mathbf{v}$ by a factor of $|k|$.

110. To add vectors in component form, add the corresponding components. Likewise, to subtract vectors, subtract the corresponding components.

Section 7.5 Practice Exercises, pp. 752–756

29. $(c\mathbf{v}) \cdot \mathbf{w} = \left[c\langle a_1, b_1\rangle\right] \cdot \langle a_2, b_2\rangle = \langle ca_1, cb_1\rangle \cdot \langle a_2, b_2\rangle$
$= ca_1a_2 + cb_1b_2 = c(a_1a_2 + b_1b_2) = c(\mathbf{v} \cdot \mathbf{w})$

31. $c(\mathbf{v} \cdot \mathbf{w}) = c\left[\langle a_1, b_1\rangle \cdot \langle a_2, b_2\rangle\right] = c(a_1a_2 + b_1b_2)$
$= ca_1a_2 + cb_1b_2 = (ca_1)a_2 + (cb_1)b_2$
$= \langle ca_1, cb_1\rangle \cdot \langle a_2, b_2\rangle = c\mathbf{v} \cdot \mathbf{w}$

32. $\mathbf{v} \cdot (\mathbf{w} + \mathbf{u}) = \langle a_1, b_1\rangle \cdot (\langle a_2, b_2\rangle + \langle a_3, b_3\rangle)$
$= \langle a_1, b_1\rangle \cdot \langle a_2 + a_3, b_2 + b_3\rangle = a_1(a_2 + a_3) + b_1(b_2 + b_3)$
$= a_1a_2 + a_1a_3 + b_1b_2 + b_1b_3 = a_1a_2 + b_1b_2 + a_1a_3 + b_1b_3$
$= \langle a_1, b_1\rangle \cdot \langle a_2, b_2\rangle + \langle a_1, b_1\rangle \cdot \langle a_3, b_3\rangle = \mathbf{v} \cdot \mathbf{w} + \mathbf{v} \cdot \mathbf{u}$

51. a. $\text{proj}_{\mathbf{w}}\mathbf{v} = \left\langle \dfrac{3}{2}, \dfrac{9}{2}\right\rangle$ **b.** $\mathbf{v}_1 = \left\langle \dfrac{3}{2}, \dfrac{9}{2}\right\rangle, \mathbf{v}_2 = \left\langle -\dfrac{9}{2}, \dfrac{3}{2}\right\rangle$

c. $c = \dfrac{3}{2}; \mathbf{v}_1 = \dfrac{3}{2}\mathbf{w}$ **d.** $\mathbf{v}_2 \cdot \mathbf{w} = \left\langle -\dfrac{9}{2}, \dfrac{3}{2}\right\rangle \cdot \langle 1, 3\rangle = -\dfrac{9}{2} + \dfrac{9}{2} = 0$;
Thus, $\mathbf{v}_2$ and $\mathbf{w}$ are orthogonal.

e. $\mathbf{v}_1 + \mathbf{v}_2 = \left\langle \dfrac{3}{2}, \dfrac{9}{2}\right\rangle + \left\langle -\dfrac{9}{2}, \dfrac{3}{2}\right\rangle = \left\langle -\dfrac{6}{2}, \dfrac{12}{2}\right\rangle = \langle -3, 6\rangle$

52. a. $\text{proj}_{\mathbf{w}}\mathbf{v} = \left\langle \dfrac{7}{2}, -\dfrac{7}{2}\right\rangle$ **b.** $\mathbf{v}_1 = \left\langle \dfrac{7}{2}, -\dfrac{7}{2}\right\rangle, \mathbf{v}_2 = \left\langle \dfrac{1}{2}, \dfrac{1}{2}\right\rangle$

c. $c = \dfrac{7}{4}; \mathbf{v}_1 = \dfrac{7}{4}\mathbf{w}$ **d.** $\mathbf{v}_2 \cdot \mathbf{w} = \left\langle \dfrac{1}{2}, \dfrac{1}{2}\right\rangle \cdot \langle 2, -2\rangle = 1 - 1 = 0$;
Thus, $\mathbf{v}_2$ and $\mathbf{w}$ are orthogonal.

e. $\mathbf{v}_1 + \mathbf{v}_2 = \left\langle \dfrac{7}{2}, -\dfrac{7}{2}\right\rangle + \left\langle \dfrac{1}{2}, \dfrac{1}{2}\right\rangle = \left\langle \dfrac{8}{2}, -\dfrac{6}{2}\right\rangle = \langle 4, -3\rangle$

53. a. $\text{proj}_{\mathbf{w}}\mathbf{v} = -\dfrac{2}{5}\mathbf{i} + \dfrac{1}{5}\mathbf{j}$ **b.** $\mathbf{v}_1 = -\dfrac{2}{5}\mathbf{i} + \dfrac{1}{5}\mathbf{j}, \mathbf{v}_2 = \dfrac{22}{5}\mathbf{i} + \dfrac{44}{5}\mathbf{j}$

54. a. $\text{proj}_{\mathbf{w}}\mathbf{v} = -\dfrac{48}{13}\mathbf{i} + \dfrac{32}{13}\mathbf{j}$

b. $\mathbf{v}_1 = -\dfrac{48}{13}\mathbf{i} + \dfrac{32}{13}\mathbf{j}, \mathbf{v}_2 = -\dfrac{56}{13}\mathbf{i} - \dfrac{84}{13}\mathbf{j}$

55. a. $\mathbf{F} = -225\mathbf{j}$ **b.** $\mathbf{F}_1 = -66.1\mathbf{i} - 21.5\mathbf{j}$ **c.** 70 lb
56. a. $\mathbf{F} = -200\mathbf{j}$ **b.** $\mathbf{F}_1 = 46.9\mathbf{i} - 11.7\mathbf{j}$ **c.** 48 lb
57. a. $\mathbf{F}_2 = 66.1\mathbf{i} - 203.5\mathbf{j}$ **b.** 214 lb
58. a. $\mathbf{F}_2 = -46.9\mathbf{i} - 188.3\mathbf{j}$ **b.** 194 lb

93. The dot product of two nonzero vectors $\mathbf{v}$ and $\mathbf{w}$ is given by $\mathbf{v} \cdot \mathbf{w} = \|\mathbf{v}\|\|\mathbf{w}\|\cos\theta$. Since the magnitudes of $\mathbf{v}$ and $\mathbf{w}$ are positive, the sign of the dot product is determined by the factor $\cos\theta$. If $\mathbf{v} \cdot \mathbf{w} > 0$, then $\cos\theta > 0$ and $0° < \theta < 90°$. If $\mathbf{v} \cdot \mathbf{w} < 0$, then $\cos\theta < 0$ and $90° < \theta < 180°$.

95. $\|\mathbf{v} - \mathbf{w}\|^2 = (\mathbf{v} - \mathbf{w}) \cdot (\mathbf{v} - \mathbf{w})$
$= \mathbf{v} \cdot (\mathbf{v} - \mathbf{w}) - \mathbf{w} \cdot (\mathbf{v} - \mathbf{w}) = \mathbf{v} \cdot \mathbf{v} - \mathbf{v} \cdot \mathbf{w} - \mathbf{w} \cdot \mathbf{v} + \mathbf{w} \cdot \mathbf{w}$
$= \|\mathbf{v}\|^2 - 2(\mathbf{v} \cdot \mathbf{w}) + \|\mathbf{w}\|^2 = \|\mathbf{v}\|^2 + \|\mathbf{w}\|^2 - 2\mathbf{v} \cdot \mathbf{w}$

96. $\|\mathbf{v} + \mathbf{w}\|^2 - \|\mathbf{v} - \mathbf{w}\|^2$
$= (\mathbf{v} + \mathbf{w}) \cdot (\mathbf{v} + \mathbf{w}) - (\mathbf{v} - \mathbf{w}) \cdot (\mathbf{v} - \mathbf{w})$
$= \mathbf{v} \cdot \mathbf{v} + 2(\mathbf{v} \cdot \mathbf{w}) + \mathbf{w} \cdot \mathbf{w} - [\mathbf{v} \cdot \mathbf{v} - 2(\mathbf{v} \cdot \mathbf{w}) + \mathbf{w} \cdot \mathbf{w}]$
$= \mathbf{v} \cdot \mathbf{v} + 2(\mathbf{v} \cdot \mathbf{w}) + \mathbf{w} \cdot \mathbf{w} - \mathbf{v} \cdot \mathbf{v} + 2(\mathbf{v} \cdot \mathbf{w}) - \mathbf{w} \cdot \mathbf{w}$
$= 4(\mathbf{v} \cdot \mathbf{w})$

97. $\|\mathbf{v} + \mathbf{w}\|^2 + \|\mathbf{v} - \mathbf{w}\|^2$
$= (\mathbf{v} + \mathbf{w}) \cdot (\mathbf{v} + \mathbf{w}) + (\mathbf{v} - \mathbf{w}) \cdot (\mathbf{v} - \mathbf{w})$
$= \mathbf{v} \cdot \mathbf{v} + 2(\mathbf{v} \cdot \mathbf{w}) + \mathbf{w} \cdot \mathbf{w} + \mathbf{v} \cdot \mathbf{v} - 2(\mathbf{v} \cdot \mathbf{w}) + \mathbf{w} \cdot \mathbf{w}$
$= 2\|\mathbf{v}\|^2 + 2\|\mathbf{w}\|^2 = 2(\|\mathbf{v}\|^2 + \|\mathbf{w}\|^2)$

98. Assume that $\|\mathbf{v}\| = \|\mathbf{w}\|$.
To show that $\mathbf{v} + \mathbf{w}$ and $\mathbf{v} - \mathbf{w}$ are orthogonal, we must show that their dot product is zero.
We have $(\mathbf{v} + \mathbf{w}) \cdot (\mathbf{v} - \mathbf{w}) = \mathbf{v} \cdot \mathbf{v} - \mathbf{v} \cdot \mathbf{w} + \mathbf{w} \cdot \mathbf{v} - \mathbf{w} \cdot \mathbf{w}$
$= \|\mathbf{v}\|^2 - \|\mathbf{w}\|^2 = 0.$

Chapter 7 Review Exercises, pp. 759–761

1. a. $\left(4, \dfrac{8\pi}{3}\right), \left(4, -\dfrac{4\pi}{3}\right)$ **b.** $\left(-4, -\dfrac{\pi}{3}\right), \left(-4, \dfrac{5\pi}{3}\right)$

2. a. $\left(5, \dfrac{7\pi}{6}\right), \left(5, -\dfrac{5\pi}{6}\right)$ **b.** $\left(-5, \dfrac{13\pi}{6}\right), \left(-5, -\dfrac{11\pi}{6}\right)$

7. $r = \dfrac{12}{\cos\theta + 5\sin\theta}$

13. $x^2 + y^2 = 256$; A circle centered at the origin with radius 16.
14. $(x - 8)^2 + y^2 = 64$; A circle centered at $(8, 0)$ with radius 8.
15. $y = -\sqrt{3}x$; A line through the origin with slope $-\sqrt{3}$.
16. $y = -10$; A horizontal line passing through the y-axis at -10.
17. Line through the origin with slope 1   **18.** Vertical line

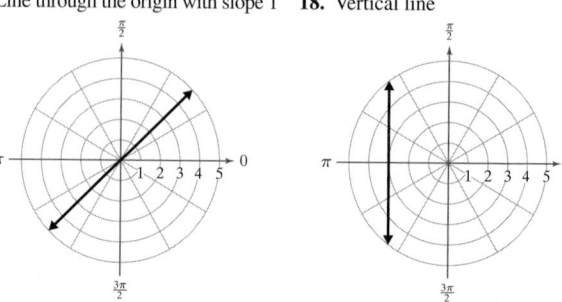

19. Cardioid   **20.** Cardioid

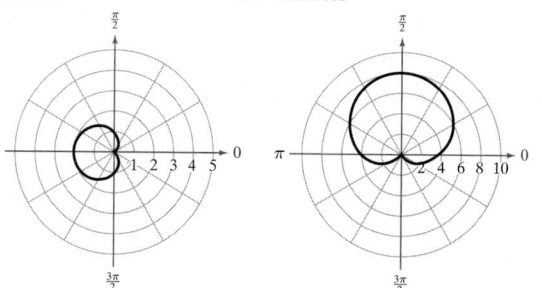

21. Dimpled limaçon   **22.** Limaçon with a loop

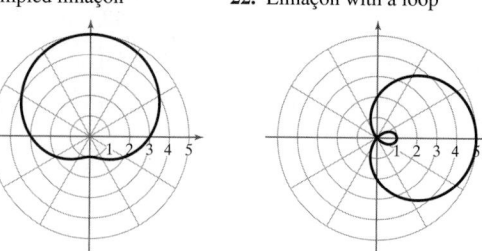

23. Rose (3 petals)   **24.** Rose (7 petals)

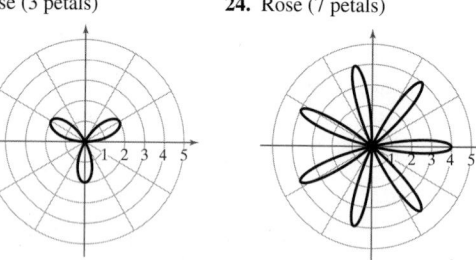

25. Lemniscate   **26.** Spiral

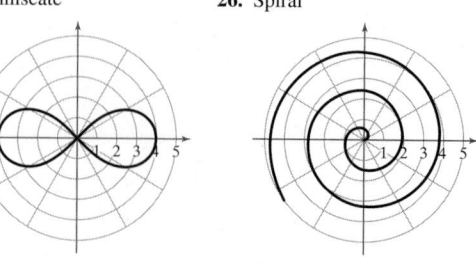

27. a. Yes **b.** Yes **c.** Inconclusive **28. a.** Yes **b.** Yes **c.** Inconclusive **29. a.** Yes **b.** Inconclusive **c.** Inconclusive
30. a. Inconclusive **b.** Yes **c.** Inconclusive

31. Modulus $= 12$; $12\left(\cos\dfrac{5\pi}{3} + i\sin\dfrac{5\pi}{3}\right)$

32. Modulus $= 7\sqrt{2}$; $7\sqrt{2}\left(\cos\dfrac{5\pi}{4} + i\sin\dfrac{5\pi}{4}\right)$

35. a. $32(\cos 200° + i\sin 200°)$ **b.** $2(\cos 280° + i\sin 280°)$

36. a. $150\left(\cos\dfrac{2\pi}{3} + i\sin\dfrac{2\pi}{3}\right)$ **b.** $\dfrac{2}{3}\left(\cos\dfrac{\pi}{6} + i\sin\dfrac{\pi}{6}\right)$

39. $-7, 7, -7i, 7i$
40. $-4, 4, -2 - 2\sqrt{3}i, 2 - 2\sqrt{3}i, -2 + 2\sqrt{3}i, 2 + 2\sqrt{3}i$
41. $5(\cos 60° + i\sin 60°), 5(\cos 240° + i\sin 240°)$
42. $4(\cos 100° + i\sin 100°), 4(\cos 220° + i\sin 220°),$
$4(\cos 340° + i\sin 340°)$

47.
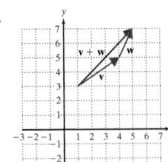

48.
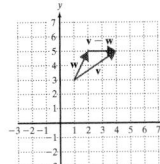

55. $\mathbf{u_v} = -\dfrac{8}{17}\mathbf{i} + \dfrac{15}{17}\mathbf{j}$ **56.** $\mathbf{u_v} = -\dfrac{9}{41}\mathbf{i} - \dfrac{40}{41}\mathbf{j}$
61. a. $\mathbf{R} = \langle -3, -7\rangle$ **b.** $\mathbf{F} = \langle 3, 7\rangle$
62. a. $\mathbf{R} = 6\mathbf{i} + 7\mathbf{j}$ **b.** $\mathbf{F} = -6\mathbf{i} - 7\mathbf{j}$
63. $\mathbf{v} = 210\sqrt{2}\mathbf{i} - 210\sqrt{2}\mathbf{j}$
65. a. $\mathbf{v} \approx 77.1\mathbf{i} + 91.9\mathbf{j}$ **b.** $\mathbf{v_w} \approx -14.1\mathbf{i} + 14.1\mathbf{j}$
 c. $\mathbf{v_T} \approx 63.0\mathbf{i} + 106.0\mathbf{j}$; True speed ≈ 123 mph
74. b. $\mathbf{v_1} = \dfrac{24}{5}\mathbf{i} + \dfrac{32}{5}\mathbf{j}, \mathbf{v_2} = -\dfrac{24}{5}\mathbf{i} + \dfrac{18}{5}\mathbf{j}$

Chapter 7 Test, p. 762

7. Rose (3 petals)
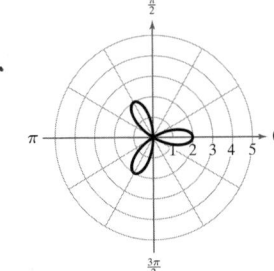

8. Limaçon with a loop
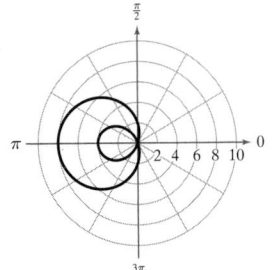

9. Cardioid
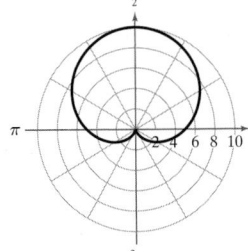

10. Rose (4 petals)
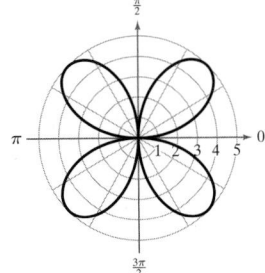

18. $2\left(\cos\dfrac{5\pi}{18} + i\sin\dfrac{5\pi}{18}\right), 2\left(\cos\dfrac{17\pi}{18} + i\sin\dfrac{17\pi}{18}\right),$
$2\left(\cos\dfrac{29\pi}{18} + i\sin\dfrac{29\pi}{18}\right)$

26. $\mathbf{u_v} = -\dfrac{4}{5}\mathbf{i} + \dfrac{3}{5}\mathbf{j}$ **32.** $\text{proj}_w\mathbf{v} = \dfrac{25}{61}\mathbf{i} + \dfrac{30}{61}\mathbf{j}$

33. a. $\mathbf{v_p} = 134.4\mathbf{i} - 350.1\mathbf{j}$; $\mathbf{v_w} = 28\mathbf{i}$ **b.** $\mathbf{v_g} = 162.4\mathbf{i} - 350.1\mathbf{j}$;
Speed ≈ 386 mph **c.** S24.9°E

Chapter 7 Cumulative Review Exercises, p. 763

1. $y = \dfrac{2A - x^2}{x}$ or $y = \dfrac{2A}{x} - x$ **2.** $y = \pm\sqrt{(10 - z)^2 - x^2}$

3. $y = \dfrac{-t \pm \sqrt{t^2 + rs}}{s}$ **4.** $y = \dfrac{\ln A_0 - \ln A}{k}$

5. $y = y_0 \cdot 10^{(R/10)}$ **6.** $(-\infty, -2) \cup [-1, 4]$
16. Domain: $(-\infty, 0) \cup (0, \infty)$, Range: $(-\infty, 0) \cup (0, \infty)$

17. a. Period $= \dfrac{\pi}{2}$, Phase shift $= -\dfrac{\pi}{6}$ **b.** Period $= \pi$, Phase shift $= -\dfrac{\pi}{6}$

19. $\text{proj}_w\mathbf{v} = \left\langle\dfrac{81}{10}, -\dfrac{27}{10}\right\rangle$

CHAPTER 8

Section 8.1 Practice Exercises, pp. 775–779

R.5.
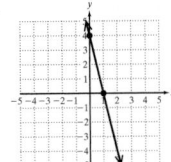

R.6.
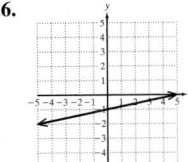

x-intercept: (1, 0); slope: $m = \dfrac{1}{5}$; y-intercept: (0, −1)
y-intercept: (0, 4)

31. $\{(x, -3x + 6)\,|\,x \text{ is any real number}\}$
or $\left\{\left(\dfrac{6 - y}{3}, y\right)\,\middle|\,y \text{ is any real number}\right\}$;
The equations are dependent.
32. $\{(x, 2x - 8)\,|\,x \text{ is any real number}\}$
or $\left\{\left(\dfrac{y + 8}{2}, y\right)\,\middle|\,y \text{ is any real number}\right\}$;
The equations are dependent.
45. $\{(x, 4x - 2)\,|\,x \text{ is any real number}\}$
or $\left\{\left(\dfrac{y + 2}{4}, y\right)\,\middle|\,y \text{ is any real number}\right\}$
46. $\{(x, -5x + 3)\,|\,x \text{ is any real number}\}$
or $\left\{\left(\dfrac{3 - y}{5}, y\right)\,\middle|\,y \text{ is any real number}\right\}$

71. a.
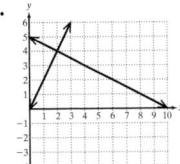

72. a.
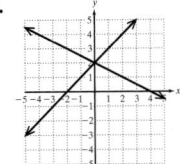

91. If the system represents two intersecting lines, then the lines intersect in exactly one point. The solution set consists of the ordered pair representing that point. If the lines in the system are parallel, then the lines do not intersect and the system has no solution. If the equations in a system of linear equations represent the same line, then the solution set is the set of points on the line.
92. If the system of equations reduces to an identity such as $0 = 0$, then the equations are dependent.
93. If the system of equations reduces to a contradiction such as $0 = 1$, then the system has no solution and is said to be inconsistent.
94. The solutions to the system are ordered pairs generated by the equation $y = x + 2$. In each ordered pair, y depends on x (y is 2 more than x). Likewise, x depends on y (x is 2 less than y).

96. $\{(2.470, -2.253)\}$

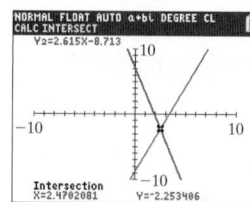

97. $\{(2.017, -0.015)\}$

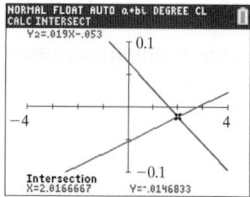

98. $\{(-0.529, -13.806)\}$

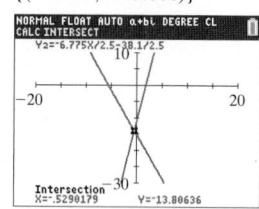

99. $\{(1.028, 15.772)\}$

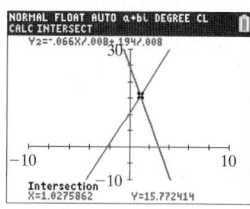

Section 8.2 Practice Exercises, pp. 788–791

33. $\left\{\left(x, \dfrac{-3x-11}{2}, \dfrac{x+7}{2}\right)\Big| x \text{ is any real number}\right\}$ or
$\left\{\left(\dfrac{-2y-11}{3}, y, \dfrac{5-y}{3}\right)\Big| y \text{ is any real number}\right\}$ or
$\{(2z-7, -3z+5, z)|z \text{ is any real number}\}$

34. $\left\{\left(x, \dfrac{26-3x}{17}, \dfrac{10-9x}{17}\right)\Big| x \text{ is any real number}\right\}$ or
$\left\{\left(\dfrac{26-17y}{3}, y, 3y-4\right)\Big| y \text{ is any real number}\right\}$ or
$\left\{\left(\dfrac{10-17z}{9}, \dfrac{z+4}{3}, z\right)\Big| z \text{ is any real number}\right\}$

35. $\left\{\left(x, \dfrac{1}{3}x, \dfrac{12-5x}{3}\right)\Big| x \text{ is any real number}\right\}$ or
$\{(3y, y, 4-5y)|y \text{ is any real number}\}$ or
$\left\{\left(\dfrac{12-3z}{5}, \dfrac{4-z}{5}, z\right)\Big| z \text{ is any real number}\right\}$

36. $\left\{\left(x, \dfrac{-2x-2}{7}, \dfrac{x-13}{7}\right)\Big| x \text{ is any real number}\right\}$ or
$\left\{\left(\dfrac{-7y-2}{2}, y, \dfrac{-y-4}{2}\right)\Big| y \text{ is any real number}\right\}$ or
$\{(7z+13, -2z-4, z)|z \text{ is any real number}\}$

49. c.

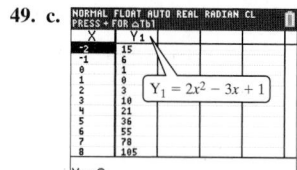

50. c.

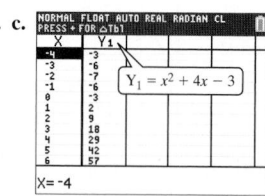

59. Pair up two equations in the system and eliminate a variable. Choose a different pair of two equations from the system and eliminate the same variable. The result should be a system of two linear equations in two variables. Solve this system using either the substitution or addition method. Then back substitute to find the third variable.

Section 8.3 Practice Exercises, pp. 798–800

48. For a repeated linear factor that occurs three times, one fraction must be given with a denominator of $(ax+b)$, one fraction must have denominator $(ax+b)^2$, and one denominator must have $(ax+b)^3$. The numerators in each fraction must be constants, which we might call A, B, and C.

51. b. $\left(\dfrac{1}{1}-\dfrac{1}{3}\right)+\left(\dfrac{1}{2}-\dfrac{1}{4}\right)+\left(\dfrac{1}{3}-\dfrac{1}{5}\right)+\left(\dfrac{1}{4}-\dfrac{1}{6}\right)+\left(\dfrac{1}{5}-\dfrac{1}{7}\right)+\cdots$

52. b. $\left(\dfrac{1}{1}-\dfrac{1}{4}\right)+\left(\dfrac{1}{2}-\dfrac{1}{5}\right)+\left(\dfrac{1}{3}-\dfrac{1}{6}\right)+\left(\dfrac{1}{4}-\dfrac{1}{7}\right)+\left(\dfrac{1}{5}-\dfrac{1}{8}\right)+\cdots$

Section 8.4 Practice Exercises, pp. 806–809

R.1.

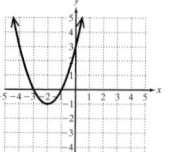

R.2.

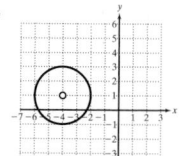

R.3.

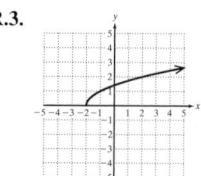

3. a.

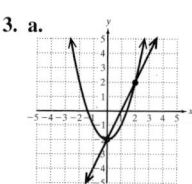

4. a.

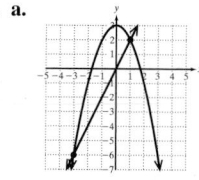

b. $\{(2, 2), (0, -2)\}$

b. $\{(1, 2), (-3, -6)\}$

5. a.

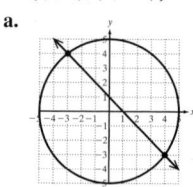

6. a.

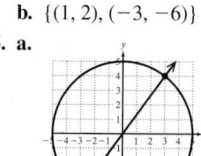

b. $\{(-3, 4), (4, -3)\}$

b. $\{(3, 4), (-3, -4)\}$

7. a.

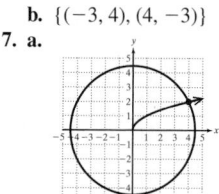

8. a.

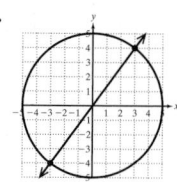

b. $\{(4, 2)\}$

b. $\{(3, 1)\}$

9. a.

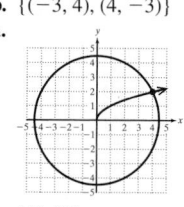

10. a.

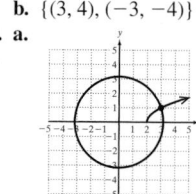

b. $\{ \}$

b. $\{ \}$

11. a.

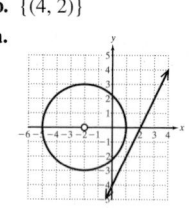

12. a.

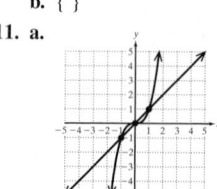

b. $\{(-1, -1), (0, 0), (1, 1)\}$

b. $\{(-1, -1), (0, 0), (1, 1)\}$

13. a.

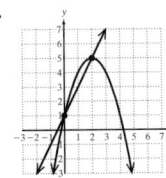

b. $\{(0, 1), (2, 5)\}$

14. a.

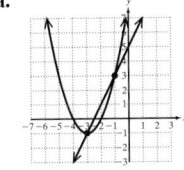

b. $\{(-1, 3), (-3, -1)\}$

Section 8.5 Practice Exercises, pp. 818–820

11. a.

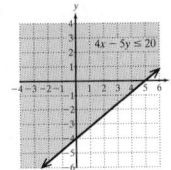

b. The bounding line would be drawn as a dashed line.

c. The bounding line would be dashed and the graph would be shaded strictly below the line.

12. a.

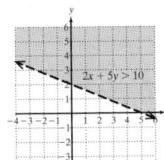

b. The bounding line would be drawn as a solid line.

c. The bounding line would be dashed and the graph would be shaded strictly below the line.

13.

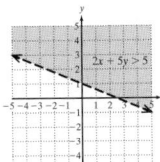

14.

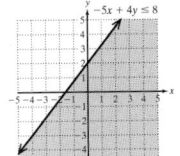

15.

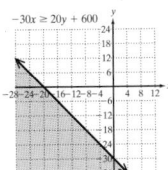

16.

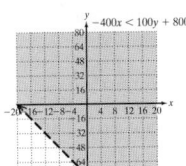

17.

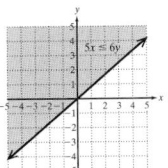

18.

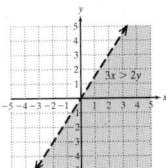

19.

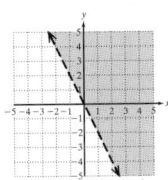

20.

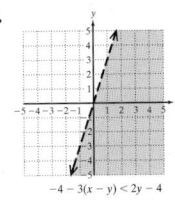

21.

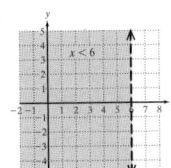

22.

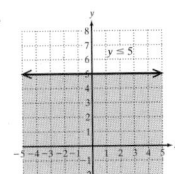

23.

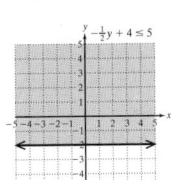

24.

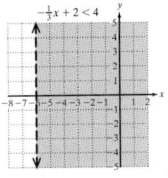

25. a.

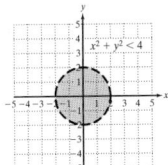

26. a.

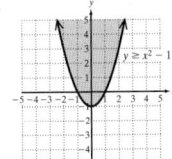

27.

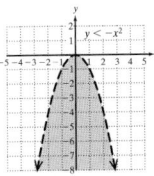

28.

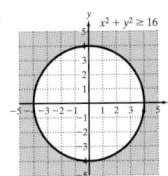

29.

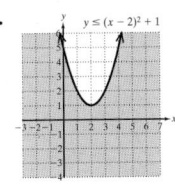

30.

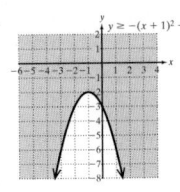

31.

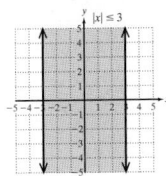

32.

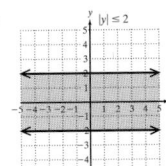

33.

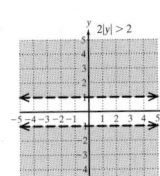

34.

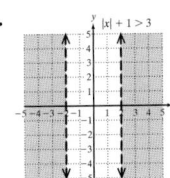

35.

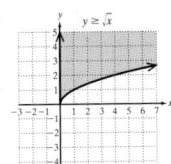

36.

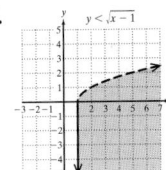

41.

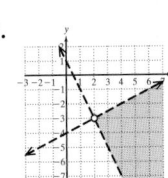

42.

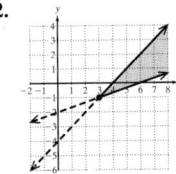

43.

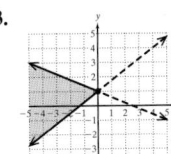

44.

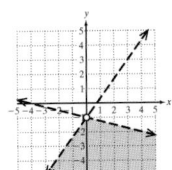

45.

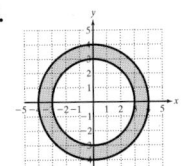

46.

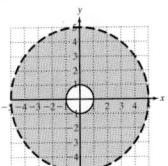

49.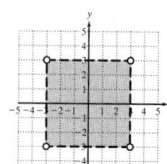

50.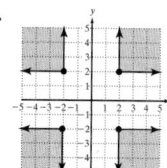

51.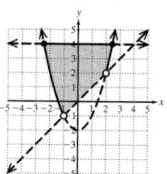

52.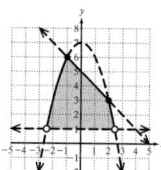

53.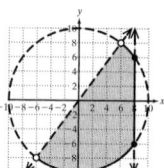

54.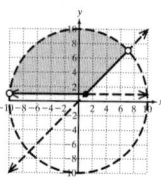

55.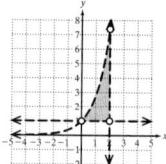

56.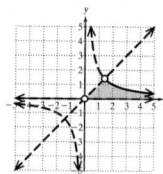

65. f.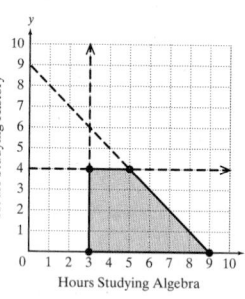

66. f.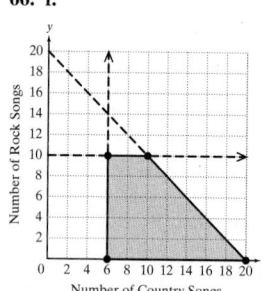

67. e.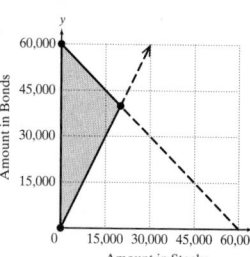

68. e.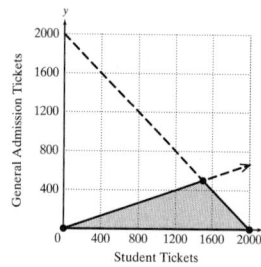

71. $y > x - 1$
$y > -4x - 16$
$y < -\frac{1}{4}x + \frac{11}{4}$

72. $y < x$
$y < -\frac{1}{2}x + \frac{3}{2}$
$y > \frac{1}{3}x - \frac{8}{3}$

75. If the inequality is strict—that is, posed with $<$ or $>$—then the bounding line or curve should be dashed.

76. After graphing the bounding line or curve to the solution set, select a test point from a given region of the plane. If the ordered pair is a solution to the inequality, then all other points in that region are also solutions.

77. Find the solution set to each individual inequality in the system. Then to find the solution set for the system of inequalities, take the intersection of the solution sets to the individual inequalities.

79.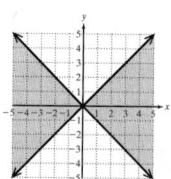

80.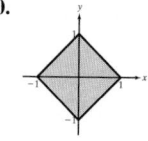

81.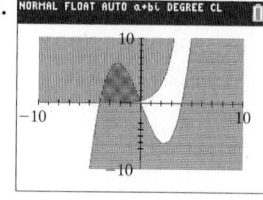

82.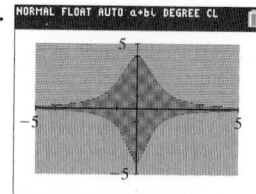

Problem Recognition Exercises, p. 821

1. a. 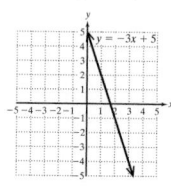 **b.** 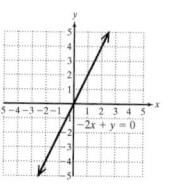 **c.** $\{(1, 2)\}$

d. 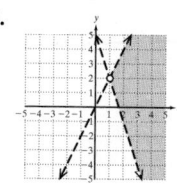 **e.**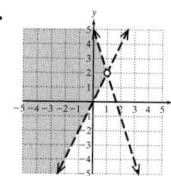

2. a. 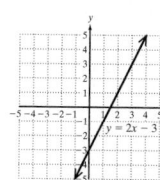 **b.** 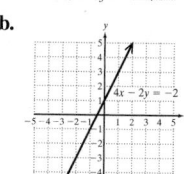 **c.** $\{\ \}$

d. 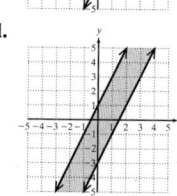 **e.** $\{\ \}$

3. a. $\{(0, 0)\}$

b. 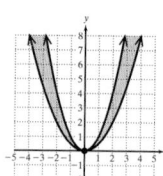 **c.**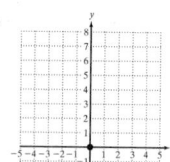

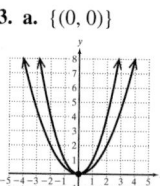

4. a. $\{(2, 1), (5, 4)\}$

b. 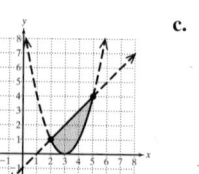 **c.**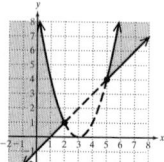

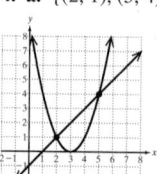

Section 8.6 Practice Exercises, pp. 826–830

13. a. Vertices:
$(0, 0), (20, 40), (60, 0)$

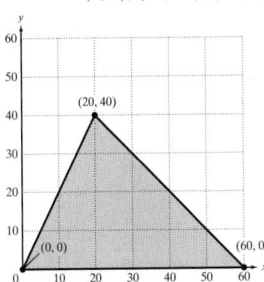

b. $x = 60$, $y = 0$
c. Maximum: 15,000

15. a. Vertices:
$(0, 50), (10, 20), (20, 0)$

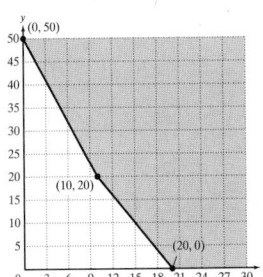

b. $x = 20$, $y = 0$
c. Minimum: 60

17. a. Vertices:
$(0, 0), (0, 40), (8, 40),$
$(36, 12), (36, 0)$

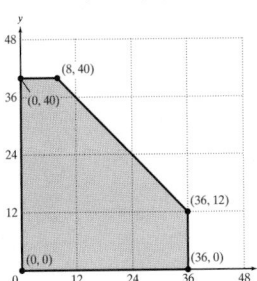

b. $x = 36$, $y = 12$
c. Maximum: 6480

21. c.

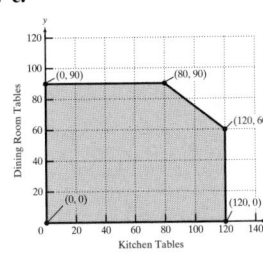

14. a. Vertices:
$(0, 0), (0, 25), (10, 20), (20, 0)$

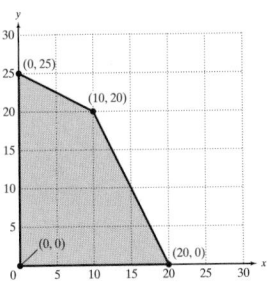

b. $x = 10$, $y = 20$
c. Maximum: 254

16. a. Vertices:
$(0, 20), (12, 4), (18, 0)$

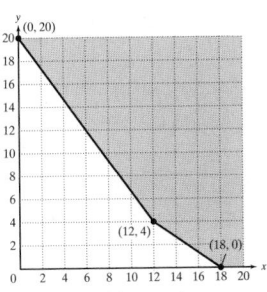

b. $x = 12$, $y = 4$
c. Minimum: 78

18. a. Vertices:
$(0, 0), (0, 8), (4, 8),$
$(10, 2), (10, 0)$

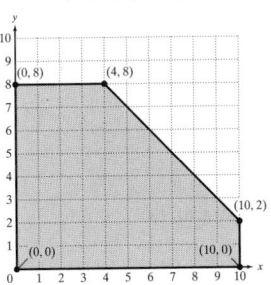

b. $x = 4$, $y = 8$
c. Maximum: 760

e. Profit at $(0, 0)$: $z = 0$
Profit at $(0, 90)$: $z = 21{,}600$
Profit at $(80, 90)$: $z = 34{,}400$
Profit at $(120, 60)$: $z = 33{,}600$
Profit at $(120, 0)$: $z = 19{,}200$

f. The greatest profit is realized when 80 kitchen tables and 90 dining room tables are produced.

22. b. $x \geq 0$
$y \geq 0$
$x + y \leq 18$
$y \geq 4$
$y \geq 2x$

e. Income at $(0, 18)$: $z = \$360$
Income at $(6, 12)$: $z = \$384$
Income at $(2, 4)$: $z = \$128$
Income at $(0, 4)$: $z = \$80$

f. Guyton should tutor 6 hr of chemistry and 12 hr of math to maximize his income.

h. Guyton will maximize his income when he tutors the maximum number of hours that he can work into his schedule which is 18 hr.

c.

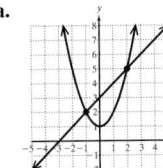

31. Linear programming is a technique that enables us to maximize or minimize a function under specific constraints.

32. An objective function is a function that is to be optimized, such as a function representing cost or profit.

33. The feasible region for a linear programming application is found by first identifying the constraints on the relevant variables. Then the regions defined by the individual constraints are graphed. The intersection of the constraints defines the feasible region.

34. The vertices are found by finding the points of intersection of the bounding lines representing the constraints.

Chapter 8 Review Exercises, pp. 832–834

8. $\left\{ \left(x, \dfrac{3}{4}x \right) \middle| x \text{ is any real number} \right\}$
or $\left\{ \left(\dfrac{4}{3}y, y \right) \middle| y \text{ is any real number} \right\}$;
The equations are dependent.

15. Infinitely many solutions; The equations are dependent.

16. No solution; The system is inconsistent.

17. $\left\{ \left(x, -\dfrac{1}{5}x, -\dfrac{23}{5}x \right) \middle| x \text{ is any real number} \right\}$ or $\left\{ (-5y, y, 23y) \middle| y \text{ is any real number} \right\}$ or $\left\{ \left(-\dfrac{5}{23}z, \dfrac{1}{23}z, z \right) \middle| z \text{ is any real number} \right\}$

22. $\dfrac{A}{x + 4} + \dfrac{B}{(x + 4)^2}$ **23.** $\dfrac{A}{x + 2} + \dfrac{B}{x - 1}$

24. $\dfrac{A}{x} + \dfrac{Bx + C}{x^2 + 5}$ **25.** $\dfrac{A}{x} + \dfrac{B}{x^2} + \dfrac{C}{2x + 3}$

26. $\dfrac{A}{x} + \dfrac{B}{2x + 5} + \dfrac{C}{(2x + 5)^2} + \dfrac{D}{(2x + 5)^3} + \dfrac{Ex + F}{x^2 + 2} + \dfrac{Gx + H}{(x^2 + 2)^2}$

27. $\dfrac{Ax + B}{x^2 + 1} + \dfrac{Cx + D}{x^2 + 4}$ **28.** $\dfrac{3}{x + 2} + \dfrac{-4}{x - 1}$

29. $\dfrac{5}{x + 4} + \dfrac{2}{(x + 4)^2}$ **30.** $x + 2 + \dfrac{3}{x} + \dfrac{5}{x^2} + \dfrac{1}{2x + 3}$

31. $\dfrac{-2}{x} + \dfrac{4x + 1}{x^2 + 5}$ **32.** $\dfrac{2x - 5}{x^2 + 1} + \dfrac{4}{x^2 + 4}$

33. a.

34. a.

b. $\{(-1, 2), (2, 5)\}$ **b.** $\{(2, 1)\}$

42. a. 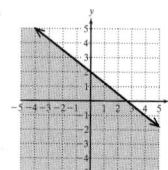 **b.**

43. a. **b.**

44. **45.**

46. **47.**

48. **50.**

51. **52.**

54. e. 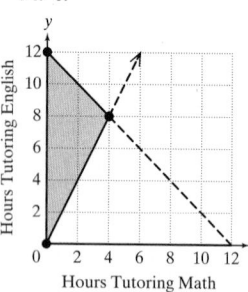 **57. a.**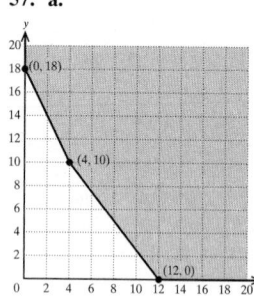

Chapter 8 Test, pp. 834–835

6. $\left\{\left(x, \dfrac{10x - 3}{4}\right)\middle|\, x \text{ is any real number}\right\}$

or $\left\{\left(\dfrac{4y + 3}{10}, y\right)\middle|\, y \text{ is any real number}\right\}$;

The equations are dependent.

15. $\left\{\left(x, \dfrac{22 - 3x}{2}, \dfrac{x - 6}{2}\right)\middle|\, x \text{ is any real number}\right\}$ or

$\left\{\left(\dfrac{22 - 2y}{3}, y, \dfrac{2 - y}{3}\right)\middle|\, y \text{ is any real number}\right\}$ or

$\left\{(2z + 6,\, 2 - 3z,\, z)\,\middle|\, z \text{ is any real number}\right\}$

16. The manager should mix 4 lb of peanuts with 16 lb of the 45% mixture.

18. Dylan invested $3000 in the risky stock, $7000 in the second stock, and $5000 in the third stock.

22. $\dfrac{A}{x + 1} + \dfrac{B}{3x - 2}$

23. $\dfrac{A}{x} + \dfrac{B}{x^2} + \dfrac{C}{x^3} + \dfrac{D}{x - 3} + \dfrac{Ex + F}{x^2 + 5x + 1} + \dfrac{Gx + H}{(x^2 + 5x + 1)^2}$

24. $\dfrac{-7}{x + 3} + \dfrac{2}{2x + 5}$ **25.** $\dfrac{6}{x + 2} + \dfrac{-4}{(x + 2)^2}$

27. $\dfrac{-3}{x} + \dfrac{4x - 2}{x^2 + 7}$ **28.** $\dfrac{7x + 1}{x^2 + 2} + \dfrac{3}{x^2 + 9}$

29. 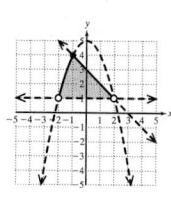 **30.**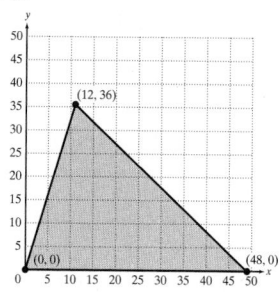

31. **32.**

33. **36. a.**

CHAPTER 9

Section 9.1 Practice Exercises, pp. 845–848

9. $\begin{bmatrix} -3 & 2 & -1 & | & 4 \\ 8 & 0 & 4 & | & 12 \\ 0 & 2 & -5 & | & 1 \end{bmatrix}$ **10.** $\begin{bmatrix} -4 & -1 & 1 & | & 8 \\ 2 & 0 & 5 & | & 11 \\ 0 & 1 & -7 & | & -6 \end{bmatrix}$

27. $\begin{bmatrix} 1 & 5 & 6 & | & 2 \\ 4 & -2 & -3 & | & 10 \\ 2 & 1 & 5 & | & 1 \end{bmatrix}$ **28.** $\begin{bmatrix} 2 & 1 & 5 & | & 1 \\ 1 & 5 & 6 & | & 2 \\ 4 & -2 & -3 & | & 10 \end{bmatrix}$

29. $\begin{bmatrix} 1 & 5 & 6 & | & 2 \\ 2 & 1 & 5 & | & 1 \\ 1 & -\frac{1}{2} & -\frac{3}{4} & | & \frac{5}{2} \end{bmatrix}$ **30.** $\begin{bmatrix} 1 & 5 & 6 & | & 2 \\ 1 & \frac{1}{2} & \frac{5}{2} & | & \frac{1}{2} \\ 4 & -2 & -3 & | & 10 \end{bmatrix}$

31. $\begin{bmatrix} 1 & 5 & 6 & | & 2 \\ 0 & -9 & -7 & | & -3 \\ 4 & -2 & -3 & | & 10 \end{bmatrix}$ **32.** $\begin{bmatrix} 1 & 5 & 6 & | & 2 \\ 2 & 1 & 5 & | & 1 \\ 0 & -22 & -27 & | & 2 \end{bmatrix}$

37. No; The elements on the main diagonal are not 1 with zeros above and below. **38.** No; The element in the first row, third column should be 0. **61.** He borrowed $10,000 from his parents, $8000 from the credit union, and $2000 from the bank. **62.** She invested $10,000 in the CD, $20,000 in the stock, and $10,000 in the bond fund. **67.** Interchanging two rows in an augmented matrix represents interchanging two equations in a system of equations. This operation does not affect the solution set of the system.

68. Multiplying a row of an augmented matrix by a nonzero constant represents multiplying an equation within a system of equations by a nonzero constant. By the multiplication property of equality, this operation results in an equivalent equation.

71. $\{(9.32, -17.48, 12.93)\}$ **72.** $\{(-1.09, 7.86, 6.76)\}$

NORMAL FLOAT AUTO REAL RADIAN MP
$$\text{rref}\begin{bmatrix} .52 & -3.71 & -4.68 & 9.1 \\ .02 & .06 & .11 & .5\blacktriangleright \\ .972 & .816 & .417 & .18 \end{bmatrix}$$
$$\begin{bmatrix} 1 & 0 & 0 & 9.3175985 \\ 0 & 1 & 0 & -17.4823735 \\ 0 & 0 & 1 & 12.93264036 \end{bmatrix}$$

NORMAL FLOAT AUTO REAL RADIAN MP
$$\text{rref}\begin{bmatrix} -3.61 & 8.17 & -5.62 & 30 \\ 8.04 & -3.16 & 9.18 & 2\blacktriangleright \\ -.16 & .09 & .55 & 4 \end{bmatrix}$$
$$\begin{bmatrix} 1 & 0 & 0 & -1.094653428 \\ 0 & 1 & 0 & 7.861951324 \\ 0 & 0 & 1 & 6.758690604 \end{bmatrix}$$

73. a. $2400a + 800b + c = 36{,}000$ **b.**
$2000a + 500b + c = 30{,}000$
$3000a + 1000b + c = 44{,}000$

NORMAL FLOAT AUTO REAL RADIAN MP
$$\text{rref}\begin{bmatrix} 2400 & 800 & 1 & 36000 \\ 2000 & 500 & 1 & 30000 \\ 3000 & 1000 & 1 & 44000 \end{bmatrix}$$
$$\begin{bmatrix} 1 & 0 & 0 & 12 \\ 0 & 1 & 0 & 4 \\ 0 & 0 & 1 & 4000 \end{bmatrix}$$

74. a. $80a + 21b + c = 180$ **b.**
$150a + 28b + c = 250$
$75a + 18b + c = 160$

NORMAL FLOAT AUTO REAL RADIAN MP
$$\text{rref}\begin{bmatrix} 80 & 21 & 1 & 180 \\ 150 & 28 & 1 & 250 \\ 75 & 18 & 1 & 160 \end{bmatrix}$$
$$\begin{bmatrix} 1 & 0 & 0 & .4 \\ 0 & 1 & 0 & 6 \\ 0 & 0 & 1 & 22 \end{bmatrix}$$

75. b.
NORMAL FLOAT AUTO REAL RADIAN MP
$$\text{rref}\begin{bmatrix} 9 & -3 & 1 & -7.28 \\ 1 & -1 & 1 & 3.68 \\ 100 & 10 & 1 & 18.2 \end{bmatrix}$$
$$\begin{bmatrix} 1 & 0 & 0 & -.32 \\ 0 & 1 & 0 & 4.2 \\ 0 & 0 & 1 & 8.2 \end{bmatrix}$$

76. b.
NORMAL FLOAT AUTO REAL RADIAN MP
$$\text{rref}\begin{bmatrix} 9 & 3 & 1 & 6.95 \\ 4 & -2 & 1 & 20.2 \\ 144 & 12 & 1 & 39.8 \end{bmatrix}$$
$$\begin{bmatrix} 1 & 0 & 0 & .45 \\ 0 & 1 & 0 & -3.1 \\ 0 & 0 & 1 & 12.2 \end{bmatrix}$$

Section 9.2 Practice Exercises, pp. 856–859

21. $\left\{ \left(\dfrac{10 - 7y}{2}, y \right) \,\middle|\, y \text{ is any real number} \right\}$

31. $\left\{ \left(\dfrac{12 - 3y - 4z}{2}, y, z \right) \,\middle|\, y \text{ and } z \text{ are any real numbers} \right\}$ or
$\{(x, y, z) \,|\, 2x + 3y + 4z = 12\}$

32. $\left\{ \left(2y + 7z - 14, y, z \right) \,\middle|\, y \text{ and } z \text{ are any real numbers} \right\}$ or
$\{(x, y, z) \,|\, x - 2y - 7z = -14\}$

45. d. $\begin{aligned} x_1 \quad\;\;\; + x_3 &= 370 \\ x_1 + x_2 \quad\;\; &= 400 \\ x_2 - x_3 &= 30 \end{aligned}$ **e.** $\begin{bmatrix} 1 & 0 & 1 & 370 \\ 0 & 1 & -1 & 30 \\ 0 & 0 & 0 & 0 \end{bmatrix}$

  f. $x_1 = 250$ vehicles per hour; **g.** $220 \le x_1 \le 270$ vehicles per hour;
  $x_2 = 150$ vehicles per hour $130 \le x_2 \le 180$ vehicles per hour

46. d. $\begin{aligned} x_1 \quad\;\;\; + x_3 &= 391 \\ x_1 + x_2 \quad\;\; &= 426 \\ x_2 - x_3 &= 35 \end{aligned}$ **e.** $\begin{bmatrix} 1 & 0 & 1 & 391 \\ 0 & 1 & -1 & 35 \\ 0 & 0 & 0 & 0 \end{bmatrix}$

  f. $x_1 = 235$ vehicles per hour; **g.** $191 \le x_1 \le 291$ vehicles per hour;
  $x_2 = 191$ vehicles per hour $135 \le x_2 \le 235$ vehicles per hour

47. a. $x_1 = 60$ vehicles per hour; **b.** $40 \le x_1 \le 90$ vehicles per hour;
  $x_2 = 210$ vehicles per hour; $190 \le x_2 \le 240$ vehicles per hour;
  $x_3 = 170$ vehicles per hour $150 \le x_3 \le 200$ vehicles per hour

48. a. $x_1 = 75$ vehicles per hour; **b.** $45 \le x_1 \le 95$ vehicles per hour;
  $x_2 = 187$ vehicles per hour; $157 \le x_2 \le 207$ vehicles per hour;
  $x_3 = 215$ vehicles per hour $185 \le x_3 \le 235$ vehicles per hour

49. a. $80x + 400y + 480z = 9280$
  $50x + 350y + 400z = 7800$
  $75x + 525y + 600z = 10{,}500$

  b. $\{\ \}$ **c.** The system of equations reduces to a contradiction. There are no values for x, y, and z that can simultaneously meet the conditions of this problem.

50. a. $\begin{aligned} x + y + 5z &= 11 \\ y + 2z &= 5 \\ 3x + y + 11z &= 22 \end{aligned}$

  b. $\{\ \}$ **c.** The system of equations reduces to a contradiction. There are no values for x, y, and z that can simultaneously meet the conditions of this problem.

51. $5(-2z + 16) + 7(3z - 5) - 11z = 45$ ✓
  $3(-2z + 16) + 5(3z - 5) - 9z = 23$ ✓
  $(-2z + 16) + (3z - 5) - z = 11$

52. $(3z + 6) + 3(-4z + 2) + 9z = 12$ ✓
  $2(3z + 6) + 7(-4z + 2) + 22z = 26$ ✓
  $-5(3z + 6) - 17(-4z + 2) - 53z = -64$ ✓

57. If a row of the reduced row-echelon form results in a contradiction (that is, zeros to the left of the vertical bar and a nonzero element to the right), then the system is inconsistent.   **58.** If the reduced row-echelon form contains a row entirely of zeros, then the system consists of dependent equations.   **59.** The equations are equivalent, meaning that they all have the same solution set. The points in the solution set represent a common plane in space.

61. $\{\ \}$ **62.** $\{\ \}$

NORMAL FLOAT AUTO REAL RADIAN MP
$$\text{rref}\begin{bmatrix} 1 & -3 & 14 & -9 \\ -2 & 7 & -31 & 21 \\ 1 & -5 & 20 & -14 \end{bmatrix}$$
$$\begin{bmatrix} 1 & 0 & 5 & 0 \\ 0 & 1 & -3 & 0 \\ 0 & 0 & 0 & 1 \end{bmatrix}$$

NORMAL FLOAT AUTO REAL RADIAN MP
$$\text{rref}\begin{bmatrix} 1 & -3 & 17 & 1 \\ 1 & -1 & 7 & 2 \\ 2 & -5 & 29 & 5 \end{bmatrix}$$
$$\begin{bmatrix} 1 & 0 & 2 & 0 \\ 0 & 1 & -5 & 0 \\ 0 & 0 & 0 & 1 \end{bmatrix}$$

63. $\{(-2z + 16, 3z - 5, z) \,|\, z$ is any real number$\}$ **64.** $\{(3z + 6, 2 - 4z, z) \,|\, z$ is any real number$\}$

NORMAL FLOAT AUTO REAL RADIAN MP
$$\text{rref}\begin{bmatrix} 5 & 7 & -11 & 45 \\ 3 & 5 & -9 & 23 \\ 1 & 1 & -1 & 11 \end{bmatrix}$$
$$\begin{bmatrix} 1 & 0 & 2 & 16 \\ 0 & 1 & -3 & -5 \\ 0 & 0 & 0 & 0 \end{bmatrix}$$

NORMAL FLOAT AUTO REAL RADIAN MP
$$\text{rref}\begin{bmatrix} 1 & 3 & 9 & 12 \\ 2 & 7 & 22 & 26 \\ -5 & -17 & -53 & -64 \end{bmatrix}$$
$$\begin{bmatrix} 1 & 0 & -3 & 6 \\ 0 & 1 & 4 & 2 \\ 0 & 0 & 0 & 0 \end{bmatrix}$$

65. $\left\{ \left(\dfrac{12 - 3y - 4z}{2}, y, z \right) \,\middle|\, y \text{ and } z \text{ are any real numbers} \right\}$ or
$\{(x, y, z) \,|\, 2x + 3y + 4z = 12\}$

NORMAL FLOAT AUTO REAL RADIAN MP
$$\text{rref}\begin{bmatrix} 2 & 3 & 4 & 12 \\ -4 & -6 & -8 & -24 \\ 1 & 1.5 & 2 & 6 \end{bmatrix}\blacktriangleright\text{Frac}$$
$$\begin{bmatrix} 1 & \frac{3}{2} & 2 & 6 \\ 0 & 0 & 0 & 0 \\ 0 & 0 & 0 & 0 \end{bmatrix}$$

66. $\{(2y + 7z - 14, y, z) \,|\, y \text{ and } z$ are any real numbers$\}$ or
$\{(x, y, z) \,|\, x - 2y - 7z = -14\}$

NORMAL FLOAT AUTO REAL RADIAN MP
$$\text{rref}\begin{bmatrix} -1 & 2 & 7 & 14 \\ 10 & -20 & -70 & -140 \\ -\frac{1}{7} & \frac{2}{7} & 1 & 2 \end{bmatrix}$$
$$\begin{bmatrix} 1 & -2 & -7 & -14 \\ 0 & 0 & 0 & 0 \\ 0 & 0 & 0 & 0 \end{bmatrix}$$

Section 9.3 Practice Exercises, pp. 870–875

41. $\begin{bmatrix} 9 & 8 & -38 \\ -8 & 2\sqrt{3} - 45 & -\frac{7}{3} \end{bmatrix}$ **42.** $\begin{bmatrix} -18 & -40 & 82 \\ -14 & -10\sqrt{3} - 18 & -\frac{19}{3} \end{bmatrix}$

53. a. $\begin{bmatrix} -1 & 17 \\ -2 & 41 \end{bmatrix}$ **b.** $\begin{bmatrix} 22 & 31 \\ 13 & 18 \end{bmatrix}$ **c.** $\begin{bmatrix} 19 & 27 \\ 45 & 64 \end{bmatrix}$

54. a. $\begin{bmatrix} -44 & 9 \\ 60 & 5 \end{bmatrix}$ **b.** $\begin{bmatrix} 13 & 42 \\ 2 & -52 \end{bmatrix}$ **c.** $\begin{bmatrix} -29 & -66 \\ 55 & 70 \end{bmatrix}$

55. a. $\begin{bmatrix} -6 & 8 & 38 \\ -12 & -24 & 36 \\ -13 & 4 & 69 \end{bmatrix}$ **b.** $\begin{bmatrix} -23 & 9 \\ 6 & 62 \end{bmatrix}$ **c.** Not possible

56. a. $\begin{bmatrix} 28 & 25 \\ 43 & 28 \end{bmatrix}$ **b.** $\begin{bmatrix} -5 & 18 & 22 \\ 5 & 37 & 48 \\ -8 & 20 & 24 \end{bmatrix}$ **c.** Not possible

57. a. $\begin{bmatrix} 19 \\ 26 \\ 5 \end{bmatrix}$ **b.** Not possible **c.** $\begin{bmatrix} 79 & -5 & 2 \\ 4 & 27 & 45 \\ -1 & 12 & 53 \end{bmatrix}$

58. a. Not possible **b.** $\begin{bmatrix} -6 \\ -1 \\ -21 \end{bmatrix}$ **c.** Not possible

59. a. $\left[\frac{2}{3}\right]$ **b.** $\begin{bmatrix} -\frac{1}{3} & -\frac{1}{6} \\ 2 & 1 \end{bmatrix}$ **c.** Not possible

60. a. $\begin{bmatrix} -20 & 2 & \frac{4}{3} \\ -\frac{15}{4} & \frac{3}{8} & \frac{1}{4} \\ -5 & \frac{1}{2} & \frac{1}{3} \end{bmatrix}$ **b.** $\left[-\frac{463}{24}\right]$ **c.** Not possible

61. a. $\begin{bmatrix} 4 & 8 & 20 & 24 \\ -6 & -12 & -30 & -36 \end{bmatrix}$ **b.** Not possible

c. Not possible

69. a. $M - D = \begin{bmatrix} \$3600 & \$2400 \\ \$3400 & \$2000 \end{bmatrix}$;

This represents the profit that the dealer clears for each model.

b. $F = 1.06D = \begin{bmatrix} \$30,740 & \$29,150 \\ \$30,210 & \$28,514 \end{bmatrix}$

70. a. $P - C = \begin{bmatrix} \$2.15 & \$1.41 \\ \$2.39 & \$1.90 \\ \$2.51 & \$2.15 \end{bmatrix}$; This represents the profit that

the store clears for both beverages for each size.

b. $F = 1.07P = \begin{bmatrix} \$3.26 & \$2.41 \\ \$3.91 & \$3.26 \\ \$4.44 & \$3.91 \end{bmatrix}$

73. a. $\begin{bmatrix} \$103,974 & \$42,560 \\ \$69,754 & \$28,400 \\ \$66,438 & \$26,960 \end{bmatrix}$; The first column gives the total revenue

for Friday, Saturday, and Sunday, respectively. The second column gives the profit for Friday, Saturday, and Sunday, respectively.

b. \$66,438 **c.** \$97,920

74. a. $\begin{bmatrix} \$26,542.30 & \$1539.60 \\ \$12,618.00 & \$718.20 \end{bmatrix}$; The first column gives the total

revenue for weekdays and weekends, respectively. The second column gives the profit for weekdays and weekends, respectively.

b. \$718.20 **c.** \$39,160.30

75. b. $TL = \begin{bmatrix} \$2088 \\ \$2688 \\ \$1657 \end{bmatrix}$

79. a. $A = \begin{bmatrix} -1 & 0 & 4 \\ 1 & 3 & 2 \end{bmatrix}$ **b.** $\begin{bmatrix} -1 & 0 & 4 \\ 1 & 3 & 2 \end{bmatrix} + \begin{bmatrix} 2 & 2 & 2 \\ -4 & -4 & -4 \end{bmatrix}$

$= \begin{bmatrix} 1 & 2 & 6 \\ -3 & -1 & -2 \end{bmatrix}$

c. $\begin{bmatrix} 1 & 0 & -4 \\ 1 & 3 & 2 \end{bmatrix}$; This matrix represents the reflection of the

triangle across the y-axis. **d.** $\begin{bmatrix} -1 & 0 & 4 \\ -1 & -3 & -2 \end{bmatrix}$; This matrix

represents the reflection of the triangle across the x-axis.

e. $\begin{bmatrix} -2 & -1 & 3 \\ 1 & -1 & 0 \end{bmatrix}$; This matrix represents the reflection of the

triangle across the x-axis, followed by a shift to the left 1 unit and a shift upward 2 units.

80. a. $A = \begin{bmatrix} -3 & 1 & 0 \\ 0 & 0 & -4 \end{bmatrix}$ **b.** $\begin{bmatrix} -3 & 1 & 0 \\ 0 & 0 & -4 \end{bmatrix} + \begin{bmatrix} -1 & -1 & -1 \\ 3 & 3 & 3 \end{bmatrix}$

$= \begin{bmatrix} -4 & 0 & -1 \\ 3 & 3 & -1 \end{bmatrix}$

c. $\begin{bmatrix} 3 & -1 & 0 \\ 0 & 0 & -4 \end{bmatrix}$; This matrix represents the reflection of the

triangle across the y-axis. **d.** $\begin{bmatrix} -3 & 1 & 0 \\ 0 & 0 & 4 \end{bmatrix}$; This matrix

represents the reflection of the triangle across the x-axis.

e. $\begin{bmatrix} 5 & 1 & 2 \\ -5 & -5 & -9 \end{bmatrix}$; This matrix represents the reflection of the

triangle across the y-axis, followed by a shift to the right 2 units and a shift downward 5 units.

81. a. $A = \begin{bmatrix} -2 & 3 & 3 & 0 \\ 1 & 3 & 0 & -1 \end{bmatrix}$ **b.** $A + \begin{bmatrix} 0 & 0 & 0 & 0 \\ -3 & -3 & -3 & -3 \end{bmatrix}$

c. $A + \begin{bmatrix} -4 & -4 & -4 & -4 \\ 0 & 0 & 0 & 0 \end{bmatrix}$

d. $\begin{bmatrix} 1 & 0 \\ 0 & -1 \end{bmatrix} \cdot A = \begin{bmatrix} -2 & 3 & 3 & 0 \\ -1 & -3 & 0 & 1 \end{bmatrix}$

e. $\begin{bmatrix} -1 & 0 \\ 0 & 1 \end{bmatrix} \cdot A = \begin{bmatrix} 2 & -3 & -3 & 0 \\ 1 & 3 & 0 & -1 \end{bmatrix}$

82. a. $A = \begin{bmatrix} -2 & 0 & -3 & -5 \\ 4 & 2 & -2 & 0 \end{bmatrix}$ **b.** $A + \begin{bmatrix} 0 & 0 & 0 & 0 \\ 6 & 6 & 6 & 6 \end{bmatrix}$

c. $A + \begin{bmatrix} 2 & 2 & 2 & 2 \\ 0 & 0 & 0 & 0 \end{bmatrix}$ **d.** $\begin{bmatrix} 1 & 0 \\ 0 & -1 \end{bmatrix} \cdot A = \begin{bmatrix} -2 & 0 & -3 & -5 \\ -4 & -2 & 2 & 0 \end{bmatrix}$

e. $\begin{bmatrix} -1 & 0 \\ 0 & 1 \end{bmatrix} \cdot A = \begin{bmatrix} 2 & 0 & 3 & 5 \\ 4 & 2 & -2 & 0 \end{bmatrix}$

83. a. $\begin{bmatrix} 0 & 6 & 6 \\ 0 & 3 & 0 \end{bmatrix}$ **b.** $\begin{bmatrix} 0 & 3.7 & 5.2 \\ 0 & 5.6 & 3 \end{bmatrix}$

c. It appears that the triangle was rotated approximately 30° counterclockwise.

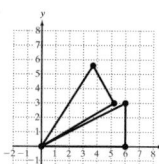

84. a. $\begin{bmatrix} 0 & 6 & 6 \\ 0 & 3 & 0 \end{bmatrix}$ **b.** $\begin{bmatrix} 0 & 0.4 & 3 \\ 0 & 6.7 & 5.2 \end{bmatrix}$

c. It appears that the triangle was rotated approximately 60° counterclockwise.

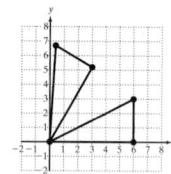

85. a. $\begin{bmatrix} 6 & 6 & 6 \\ 6 & 0 & 0 \\ 6 & 6 & 6 \\ 6 & 0 & 0 \\ 6 & 6 & 6 \end{bmatrix}$

b. $\begin{bmatrix} 6 & 6 & 6 \\ 6 & 0 & 0 \\ 6 & 6 & 6 \\ 6 & 0 & 0 \\ 6 & 6 & 6 \end{bmatrix} + \begin{bmatrix} -1 & -1 & -1 \\ -1 & 1 & 1 \\ -1 & -1 & -1 \\ -1 & 1 & 1 \\ -1 & -1 & -1 \end{bmatrix} = \begin{bmatrix} 5 & 5 & 5 \\ 5 & 1 & 1 \\ 5 & 5 & 5 \\ 5 & 1 & 1 \\ 5 & 5 & 5 \end{bmatrix}$

86. a. $\begin{bmatrix} 4 & 4 & 4 \\ 2 & 4 & 2 \\ 2 & 4 & 2 \\ 2 & 4 & 2 \\ 2 & 4 & 2 \end{bmatrix}$

b. $\begin{bmatrix} 4 & 4 & 4 \\ 2 & 4 & 2 \\ 2 & 4 & 2 \\ 2 & 4 & 2 \\ 2 & 4 & 2 \end{bmatrix} + \begin{bmatrix} 2 & 2 & 2 \\ -1 & 2 & -1 \\ -1 & 2 & -1 \\ -1 & 2 & -1 \\ -1 & 2 & -1 \end{bmatrix} = \begin{bmatrix} 6 & 6 & 6 \\ 1 & 6 & 1 \\ 1 & 6 & 1 \\ 1 & 6 & 1 \\ 1 & 6 & 1 \end{bmatrix}$

87. $A + B = \begin{bmatrix} a_1 & a_2 \\ a_3 & a_4 \end{bmatrix} + \begin{bmatrix} b_1 & b_2 \\ b_3 & b_4 \end{bmatrix}$

$= \begin{bmatrix} a_1 + b_1 & a_2 + b_2 \\ a_3 + b_3 & a_4 + b_4 \end{bmatrix} = \begin{bmatrix} b_1 + a_1 & b_2 + a_2 \\ b_3 + a_3 & b_4 + a_4 \end{bmatrix}$

$= B + A$

88. $A + (B + C) = \begin{bmatrix} a_1 & a_2 \\ a_3 & a_4 \end{bmatrix} + \left(\begin{bmatrix} b_1 & b_2 \\ b_3 & b_4 \end{bmatrix} + \begin{bmatrix} c_1 & c_2 \\ c_3 & b_4 \end{bmatrix} \right)$

$= \begin{bmatrix} a_1 & a_2 \\ a_3 & a_4 \end{bmatrix} + \begin{bmatrix} b_1 + c_1 & b_2 + c_2 \\ b_3 + c_3 & b_4 + c_4 \end{bmatrix}$

$= \begin{bmatrix} a_1 + b_1 + c_1 & a_2 + b_2 + c_2 \\ a_3 + b_3 + c_3 & a_4 + b_4 + c_4 \end{bmatrix}$

$= \begin{bmatrix} (a_1 + b_1) + c_1 & (a_2 + b_2) + c_2 \\ (a_3 + b_3) + c_3 & (a_4 + b_4) + c_4 \end{bmatrix}$

$= \begin{bmatrix} a_1 + b_1 & a_2 + b_2 \\ a_3 + b_3 & a_4 + b_4 \end{bmatrix} + \begin{bmatrix} c_1 & c_2 \\ c_3 & c_4 \end{bmatrix} = (A + B) + C$

89. $A + (-A) = \begin{bmatrix} a_1 & a_2 \\ a_3 & a_4 \end{bmatrix} + \begin{bmatrix} -a_1 & -a_2 \\ -a_3 & -a_4 \end{bmatrix}$

$= \begin{bmatrix} a_1 + (-a_1) & a_2 + (-a_2) \\ a_3 + (-a_3) & a_4 + (-a_4) \end{bmatrix} = \begin{bmatrix} 0 & 0 \\ 0 & 0 \end{bmatrix} = 0$

90. $A + 0 = \begin{bmatrix} a_1 & a_2 \\ a_3 & a_4 \end{bmatrix} + \begin{bmatrix} 0 & 0 \\ 0 & 0 \end{bmatrix}$

$= \begin{bmatrix} a_1 + 0 & a_2 + 0 \\ a_3 + 0 & a_4 + 0 \end{bmatrix} = \begin{bmatrix} a_1 & a_2 \\ a_3 & a_4 \end{bmatrix} = A$

91. $s(tA) = s \cdot \left(t \begin{bmatrix} a_1 & a_2 \\ a_3 & a_4 \end{bmatrix} \right) = s \cdot \begin{bmatrix} ta_1 & ta_2 \\ ta_3 & ta_4 \end{bmatrix}$

$= \begin{bmatrix} sta_1 & sta_2 \\ sta_3 & sta_4 \end{bmatrix} = (st) \begin{bmatrix} a_1 & a_2 \\ a_3 & a_4 \end{bmatrix} = (st)A$

92. $t(A + B) = t \left(\begin{bmatrix} a_1 & a_2 \\ a_3 & a_4 \end{bmatrix} + \begin{bmatrix} b_1 & b_2 \\ b_3 & b_4 \end{bmatrix} \right)$

$= t \begin{bmatrix} a_1 + b_1 & a_2 + b_2 \\ a_3 + b_3 & a_4 + b_4 \end{bmatrix} = \begin{bmatrix} t(a_1 + b_1) & t(a_2 + b_2) \\ t(a_3 + b_3) & t(a_4 + b_4) \end{bmatrix}$

$= \begin{bmatrix} ta_1 + tb_1 & ta_2 + tb_2 \\ ta_3 + tb_3 & ta_4 + tb_4 \end{bmatrix} = \begin{bmatrix} ta_1 & ta_2 \\ ta_3 & ta_4 \end{bmatrix} + \begin{bmatrix} tb_1 & tb_2 \\ tb_3 & tb_4 \end{bmatrix} = tA + tB$

93. $A^2 = \begin{bmatrix} -1 & 0 \\ 0 & -1 \end{bmatrix}$, $A^3 = \begin{bmatrix} -i & 0 \\ 0 & -i \end{bmatrix}$, $A^4 = \begin{bmatrix} 1 & 0 \\ 0 & 1 \end{bmatrix}$; The entries along the main diagonal in matrix A^n are the same as the value of i^n.

101.
```
NORMAL FLOAT AUTO REAL RADIAN MP
2.5[A]-3.6[B]
     [38.625  -98.25]
     [-76.1   106.1 ]
     [-70.8    32.55]
```

102.
```
NORMAL FLOAT AUTO REAL RADIAN MP
-6.4([A]+[B])
     [57.28    -216.96]
     [-179.968  290.56]
     [-99.84     41.6 ]
```

103.
```
NORMAL FLOAT AUTO REAL RADIAN MP
-3[A][C]
        [-76.86]
        [61.548]
        [22.59 ]
```

104.
```
NORMAL FLOAT AUTO REAL RADIAN MP
7.5[B][C]
        [637.5]
        [-207 ]
        [543  ]
```

Section 9.4 Practice Exercises, pp. 883–885

9. a. $AI_2 = \begin{bmatrix} -\frac{7}{8} & \sqrt{5} \\ 5.1 & 8 \end{bmatrix} \begin{bmatrix} 1 & 0 \\ 0 & 1 \end{bmatrix} = \begin{bmatrix} -\frac{7}{8}(1) + \sqrt{5}(0) & -\frac{7}{8}(0) + \sqrt{5}(1) \\ 5.1(1) + 8(0) & 5.1(0) + 8(1) \end{bmatrix}$

$= \begin{bmatrix} -\frac{7}{8} & \sqrt{5} \\ 5.1 & 8 \end{bmatrix}$ ✓

b. $I_2A = \begin{bmatrix} 1 & 0 \\ 0 & 1 \end{bmatrix} \begin{bmatrix} -\frac{7}{8} & \sqrt{5} \\ 5.1 & 8 \end{bmatrix} = \begin{bmatrix} 1(-\frac{7}{8}) + 0(5.1) & 1(\sqrt{5}) + 0(8) \\ 0(-\frac{7}{8}) + 1(5.1) & 0(\sqrt{5}) + 1(8) \end{bmatrix}$

$= \begin{bmatrix} -\frac{7}{8} & \sqrt{5} \\ 5.1 & 8 \end{bmatrix}$ ✓

10. a. $AI_2 = \begin{bmatrix} \sqrt{3} & 1 \\ \pi & 4 \end{bmatrix} \begin{bmatrix} 1 & 0 \\ 0 & 1 \end{bmatrix} = \begin{bmatrix} \sqrt{3}(1) + 1(0) & \sqrt{3}(0) + 1(1) \\ \pi(1) + 4(0) & \pi(0) + 4(1) \end{bmatrix}$

$= \begin{bmatrix} \sqrt{3} & 1 \\ \pi & 4 \end{bmatrix}$ ✓

b. $I_2A = \begin{bmatrix} 1 & 0 \\ 0 & 1 \end{bmatrix} \begin{bmatrix} \sqrt{3} & 1 \\ \pi & 4 \end{bmatrix} = \begin{bmatrix} 1(\sqrt{3}) + 0(\pi) & 1(1) + 0(4) \\ 0(\sqrt{3}) + 1(\pi) & 0(1) + 1(4) \end{bmatrix}$

$= \begin{bmatrix} \sqrt{3} & 1 \\ \pi & 4 \end{bmatrix}$ ✓

11. a. $AI_3 = \begin{bmatrix} 1 & -3 & 4 \\ 9 & 5 & 3 \\ 11 & -6 & -4 \end{bmatrix} \begin{bmatrix} 1 & 0 & 0 \\ 0 & 1 & 0 \\ 0 & 0 & 1 \end{bmatrix}$

$= \begin{bmatrix} 1(1) + -3(0) + 4(0) & 1(0) + -3(1) + 4(0) & 1(0) - 3(0) + 4(1) \\ 9(1) + 5(0) + 3(0) & 9(0) + 5(1) + 3(0) & 9(0) + 5(0) + 3(1) \\ 11(1) - 6(0) - 4(0) & 11(0) - 6(1) - 4(0) & 11(0) - 6(0) - 4(1) \end{bmatrix}$

$= \begin{bmatrix} 1 & -3 & 4 \\ 9 & 5 & 3 \\ 11 & -6 & -4 \end{bmatrix}$ ✓

b. $I_3A = \begin{bmatrix} 1 & 0 & 0 \\ 0 & 1 & 0 \\ 0 & 0 & 1 \end{bmatrix} \begin{bmatrix} 1 & -3 & 4 \\ 9 & 5 & 3 \\ 11 & -6 & -4 \end{bmatrix}$

$= \begin{bmatrix} 1(1) + 0(9) + 0(11) & 1(-3) + 0(5) + 0(-6) & 1(4) + 0(3) + 0(-4) \\ 0(1) + 1(9) + 0(11) & 0(-3) + 1(5) + 0(-6) & 0(4) + 1(3) + 0(-4) \\ 0(1) + 0(9) + 1(11) & 0(-3) + 0(5) + 1(-6) & 0(4) + 0(3) + 1(-4) \end{bmatrix}$

$= \begin{bmatrix} 1 & -3 & 4 \\ 9 & 5 & 3 \\ 11 & -6 & -4 \end{bmatrix}$ ✓

12. a. $AI_3 = \begin{bmatrix} -3 & 9 & 1 \\ 0 & 4 & -1 \\ 5 & 0 & 3 \end{bmatrix} \begin{bmatrix} 1 & 0 & 0 \\ 0 & 1 & 0 \\ 0 & 0 & 1 \end{bmatrix}$

$= \begin{bmatrix} -3(1) + 9(0) + 1(0) & -3(0) + 9(1) + 1(0) & -3(0) + 9(0) + 1(1) \\ 0(1) + 4(0) - 1(0) & 0(0) + 4(1) - 1(0) & 0(0) + 4(0) - 1(1) \\ 5(1) + 0(0) + 3(0) & 5(0) + 0(1) + 3(0) & 5(0) + 0(0) + 3(1) \end{bmatrix}$

$= \begin{bmatrix} -3 & 9 & 1 \\ 0 & 4 & -1 \\ 5 & 0 & 3 \end{bmatrix}$ ✓

b. $I_3A = \begin{bmatrix} 1 & 0 & 0 \\ 0 & 1 & 0 \\ 0 & 0 & 1 \end{bmatrix} \begin{bmatrix} -3 & 9 & 1 \\ 0 & 4 & -1 \\ 5 & 0 & 3 \end{bmatrix}$

$= \begin{bmatrix} 1(-3) + 0(0) + 0(5) & 1(9) + 0(4) + 0(0) & 1(1) + 0(-1) + 0(3) \\ 0(-3) + 1(0) + 0(5) & 0(9) + 1(4) + 0(0) & 0(1) + 1(-1) + 0(3) \\ 0(-3) + 0(0) + 1(5) & 0(9) + 0(4) + 1(0) & 0(1) + 0(-1) + 1(3) \end{bmatrix}$

$= \begin{bmatrix} -3 & 9 & 1 \\ 0 & 4 & -1 \\ 5 & 0 & 3 \end{bmatrix}$ ✓

25. $A^{-1} = \begin{bmatrix} 3 & -2 & -3 \\ 3 & -4 & -2 \\ 2 & -3 & -1 \end{bmatrix}$

26. $A^{-1} = \begin{bmatrix} 1 & 1 & 1 \\ 1 & 2 & 3 \\ 1 & 1 & 2 \end{bmatrix}$

27. $A^{-1} = \begin{bmatrix} 1 & 0 & 2 \\ 1 & -\frac{1}{2} & \frac{3}{2} \\ 1 & -\frac{1}{2} & \frac{1}{2} \end{bmatrix}$

28. $A^{-1} = \begin{bmatrix} \frac{13}{2} & \frac{25}{2} & \frac{3}{2} \\ 4 & 7 & 1 \\ \frac{3}{2} & \frac{5}{2} & \frac{1}{2} \end{bmatrix}$

31. $A^{-1} = \begin{bmatrix} 3 & -3 & -6 \\ -1 & 3 & 5 \\ 2 & -3 & -5 \end{bmatrix}$ **33.** $A^{-1} = \begin{bmatrix} 1 & -3 & -1 & 0 \\ 0 & 1 & 2 & 0 \\ 0 & 1 & 1 & 0 \\ -2 & 0 & -3 & 1 \end{bmatrix}$

35. $\begin{bmatrix} 3 & -4 \\ 2 & 1 \end{bmatrix}\begin{bmatrix} x \\ y \end{bmatrix} = \begin{bmatrix} -1 \\ 14 \end{bmatrix}$ **36.** $\begin{bmatrix} -6 & -1 \\ 2 & 3 \end{bmatrix}\begin{bmatrix} x \\ y \end{bmatrix} = \begin{bmatrix} 1 \\ 13 \end{bmatrix}$

37. $\begin{bmatrix} 9 & -6 & 4 \\ 4 & 0 & -1 \\ 0 & 3 & 1 \end{bmatrix}\begin{bmatrix} x \\ y \\ z \end{bmatrix} = \begin{bmatrix} 27 \\ 1 \\ 0 \end{bmatrix}$

38. $\begin{bmatrix} -3 & 0 & 8 \\ 0 & 6 & -1 \\ 2 & -1 & 6 \end{bmatrix}\begin{bmatrix} x \\ y \\ z \end{bmatrix} = \begin{bmatrix} 4 \\ 7 \\ -15 \end{bmatrix}$

61. a. $A^{-1} = \begin{bmatrix} \frac{3}{4} & -\frac{1}{4} \\ -\frac{5}{8} & \frac{3}{8} \end{bmatrix}$ **b.** $(A^{-1})^{-1} = A = \begin{bmatrix} 3 & 2 \\ 5 & 6 \end{bmatrix}$

62. a. $B^{-1} = \begin{bmatrix} -2 & 1 \\ -\frac{5}{2} & \frac{3}{2} \end{bmatrix}$ **b.** $(B^{-1})^{-1} = B = \begin{bmatrix} -3 & 2 \\ -5 & 4 \end{bmatrix}$

67. $\begin{bmatrix} a & b & | & 1 & 0 \\ c & d & | & 0 & 1 \end{bmatrix} \overset{\frac{1}{a}R_1 \to R_1}{=} \begin{bmatrix} 1 & \frac{b}{a} & | & \frac{1}{a} & 0 \\ c & d & | & 0 & 1 \end{bmatrix}$

$\overset{-c \cdot R_1 + R_2 \to R_2}{=} \begin{bmatrix} 1 & \frac{b}{a} & | & \frac{1}{a} & 0 \\ 0 & d - \frac{cb}{a} & | & -\frac{c}{a} & 1 \end{bmatrix}$

$= \begin{bmatrix} 1 & \frac{b}{a} & | & \frac{1}{a} & 0 \\ 0 & \frac{ad - bc}{a} & | & -\frac{c}{a} & 1 \end{bmatrix}$

$\overset{\frac{a}{ad-bc} \cdot R_2 \to R_2}{=} \begin{bmatrix} 1 & \frac{b}{a} & | & \frac{1}{a} & 0 \\ 0 & 1 & | & -\frac{c}{ad-bc} & \frac{a}{ad-bc} \end{bmatrix}$

$\overset{-\frac{b}{a} \cdot R_2 + R_1 \to R_1}{=} \begin{bmatrix} 1 & 0 & | & \frac{d}{ad-bc} & -\frac{b}{ad-bc} \\ 0 & 1 & | & -\frac{c}{ad-bc} & \frac{a}{ad-bc} \end{bmatrix}$

Therefore, $A^{-1} = \dfrac{1}{ad - bc}\begin{bmatrix} d & -b \\ -c & a \end{bmatrix}$, provided $ad - bc \neq 0$.

68. For example $A = \begin{bmatrix} 0 & 2 \\ 3 & 7 \end{bmatrix}$ and $B = \begin{bmatrix} 1 & 2 \\ 4 & 6 \end{bmatrix}$;

$(AB)^{-1} = \begin{bmatrix} 4 & -1 \\ -\frac{31}{12} & \frac{2}{3} \end{bmatrix}$, whereas $A^{-1}B^{-1} = \begin{bmatrix} \frac{25}{6} & -\frac{4}{3} \\ -\frac{3}{2} & \frac{1}{2} \end{bmatrix}$.

70. $A^{-1} = \begin{bmatrix} -2.15 & 3.77 & 3.86 & 0.90 \\ -1.27 & 1.13 & 1.41 & -0.48 \\ -7.47 & 2.28 & -2.49 & -0.30 \\ -1.52 & 0.27 & 1.25 & -0.01 \end{bmatrix}$

71. $A^{-1} = \begin{bmatrix} 0.29 & -0.10 & 0.08 & -0.27 \\ -0.09 & 0.12 & 0.00 & 0.27 \\ -0.12 & 0.00 & -0.07 & 0.25 \\ 0.15 & 0.03 & -0.06 & -0.01 \end{bmatrix}$

Section 9.5 Practice Exercises, pp. 895–898

37. $\left\{ \left(\dfrac{4y + 8}{3}, y \right) \,\middle|\, y \text{ is any real number} \right\}$

38. $\left\{ \left(\dfrac{5y - 4}{2}, y \right) \,\middle|\, y \text{ is any real number} \right\}$

45. $\left\{ \left(\dfrac{3y - z + 6}{2}, y, z \right) \,\middle|\, y \text{ and } z \text{ are any real numbers} \right\}$ or
$\{(x, y, z) \mid 2x - 3y + z = 6\}$

46. $\{(y - 3z - 4, y, z) \mid y \text{ and } z \text{ are any real numbers}\}$ or
$\{(x, y, z) \mid x - y + 3z = -4\}$

59. c. Rows 1 and 2 are interchanged between matrix A and matrix B. The determinants are opposite in sign.

60. c. Rows 1 and 2 are interchanged between matrix A and matrix B. The determinants are opposite in sign.

75. The minor is the determinant of the matrix obtained by deleting the ith row and jth column of the original matrix. The cofactor is the product of the minor and the factor $(-1)^{i+j}$.

76. The notation $\begin{bmatrix} a & b \\ c & d \end{bmatrix}$ represents a 2×2 matrix, whereas $\begin{vmatrix} a & b \\ c & d \end{vmatrix}$
represents the determinant of the matrix.   **77.** Choose the row or column with the greatest number of zero elements.

78. The determinant D consists of the coefficients of x and y. The determinant D_x has the column of x term coefficients replaced by the constants c_1 and c_2. The determinant D_y has the column of y term coefficients replaced by the constants c_1 and c_2.

Problem Recognition Exercises, p. 899

5. a. $\begin{vmatrix} 1.5 & -2 \\ -3 & 4 \end{vmatrix} = 0$ **b.** No   **c.** { }

6. a. $\begin{vmatrix} 5 & -2 \\ 1 & -0.4 \end{vmatrix} = 0$ **b.** No   **c.** { }

7. a. $\begin{vmatrix} 1 & -3 & 7 \\ -2 & 5 & -11 \\ 1 & -5 & 13 \end{vmatrix} = 0$ **b.** No

c. $\{(2z + 4, 3z + 1, z) \mid z \text{ is any real number}\}$

8. a. $\begin{vmatrix} 1 & -2 & 3 \\ -2 & 1 & 0 \\ 1 & 0 & -1 \end{vmatrix} = 0$ **b.** No

c. $\{(z + 3, 2z + 5, z) \mid z \text{ is any real number}\}$

Chapter 9 Review Exercises, pp. 901–904

19. $\{(5y - 2z - 1, y, z) \mid y \text{ and } z \text{ are any real numbers}\}$ or
$\{(x, y, z) \mid x - 5y + 2z = -1\}$

21. a. $x_1 = 76$ vehicles per hour; $x_2 = 97$ vehicles per hour
b. $46 \le x_1 \le 96$ vehicles per hour; $67 \le x_2 \le 117$ vehicles per hour

22. a. $x_1 = 80$ vehicles per hour; $x_2 = 200$ vehicles per hour; $x_3 = 190$ vehicles per hour   **b.** $60 \le x_1 \le 110$ vehicles per hour; $180 \le x_2 \le 230$ vehicles per hour; $170 \le x_3 \le 220$ vehicles per hour

31. $3A = \begin{bmatrix} -12 & 3 \\ 18 & -6 \\ 3 & 9 \end{bmatrix}$ **32.** $-2B = \begin{bmatrix} -4 & -6 & 14 \\ -2 & -10 & 12 \end{bmatrix}$

35. $2A - C = \begin{bmatrix} -8 - \pi & -2 \\ 15 & -5 \\ 2 & 1 \end{bmatrix}$ **37.** $AB = \begin{bmatrix} -7 & -7 & 22 \\ 10 & 8 & -30 \\ 5 & 18 & -25 \end{bmatrix}$

38. $BC = \begin{bmatrix} 2\pi - 9 & -24 \\ \pi - 15 & -21 \end{bmatrix}$ **41.** $A^2 = \begin{bmatrix} -2 & 36 \\ -6 & 10 \end{bmatrix}$

42. $AB = \begin{bmatrix} -16 \\ -13 \end{bmatrix}$ **43.** $BC = \begin{bmatrix} 2 & 7 \\ -6 & -21 \end{bmatrix}$

45. $QP = \begin{bmatrix} \$7922 \\ \$9843 \end{bmatrix}$; QP is the matrix representing the total revenue from these four items for each theater.

46. a. $P - M = \begin{bmatrix} \$315 & \$282 \\ \$267 & \$240 \end{bmatrix}$; This represents the profit that the store clears for each model.
b. $F = 1.06P = \begin{bmatrix} \$1446.90 & \$1295.32 \\ \$1226.42 & \$1102.40 \end{bmatrix}$

47. a. $A = \begin{bmatrix} 1 & 3 & 5 \\ 1 & 4 & 2 \end{bmatrix}$ **b.** $\begin{bmatrix} 1 & 3 & 5 \\ 1 & 4 & 2 \end{bmatrix} + \begin{bmatrix} -3 & -3 & -3 \\ -1 & -1 & -1 \end{bmatrix}$
$= \begin{bmatrix} -2 & 0 & 2 \\ 0 & 3 & 1 \end{bmatrix}$

c. $\begin{bmatrix} -1 & -3 & -5 \\ 1 & 4 & 2 \end{bmatrix}$; This matrix represents the reflection of the triangle across the y-axis.   **d.** $\begin{bmatrix} 1 & 3 & 5 \\ -1 & -4 & -2 \end{bmatrix}$; This matrix represents the reflection of the triangle across the x-axis.

51. $A^{-1} = \begin{bmatrix} \frac{1}{6} & \frac{1}{6} \\ -\frac{1}{12} & \frac{5}{12} \end{bmatrix}$ **52.** $A = \begin{bmatrix} 1 & 6 \\ 2 & -4 \end{bmatrix}$

54. $A^{-1} = \begin{bmatrix} -1 & 3 & 5 \\ 1 & -1 & -2 \\ 2 & -3 & -5 \end{bmatrix}$ **55.** $A^{-1} = \begin{bmatrix} 0 & -1 & 4 \\ 1 & -1 & 5 \\ -3 & -1 & 0 \end{bmatrix}$

57. $\begin{bmatrix} -3 & 7 & 0 \\ 4 & 0 & 2 \\ 2 & -1 & 5 \end{bmatrix} \begin{bmatrix} x \\ y \\ z \end{bmatrix} = \begin{bmatrix} 6 \\ -3 \\ -13 \end{bmatrix}$

Chapter 9 Test, pp. 904–905

1. $\left[\begin{array}{ccc|c} 1 & 5 & -3 & 1 \\ 3 & 1 & 4 & -2 \\ 0 & 4 & 2 & 6 \end{array}\right]$ **2.** $\left[\begin{array}{ccc|c} 0 & -14 & 13 & -5 \\ 1 & 5 & -3 & 1 \\ 0 & 4 & 2 & 6 \end{array}\right]$

3. $\left[\begin{array}{ccc|c} 3 & 1 & 4 & -2 \\ 1 & 5 & -3 & 1 \\ 0 & 1 & \frac{1}{2} & \frac{3}{2} \end{array}\right]$ **20.** $2A - 3C = \begin{bmatrix} 8 & -1 & 6 \\ -2 & 11 & -12 \end{bmatrix}$

22. $AB = \begin{bmatrix} -5 & 20 \\ 20 & 44 \end{bmatrix}$ **23.** $BA = \begin{bmatrix} 22 & 37 & 51 \\ -2 & -4 & -6 \\ 22 & 23 & 21 \end{bmatrix}$

30. a. $x_1 = 128$ vehicles per hour; $x_2 = 173$ vehicles per hour
    b. $118 \le x_1 \le 168$ vehicles per hour; $163 \le x_2 \le 213$ vehicles per hour

31. $CN = \begin{bmatrix} 5500 \\ 8040 \end{bmatrix}$; This represents the total number of calories burned by two individuals with different weights after biking 6 hr, running 3 hr, and walking 5 hr. For example, the element 5500 in the first row tells us that 5500 cal would be burned by a 120-lb individual who biked 6 hr, ran 3 hr, and walked 5 hr in a given week.

Chapter 9 Cumulative Review Exercises, pp. 905–906

5. $-\dfrac{6}{13} + \dfrac{17}{13}i$

9. a.
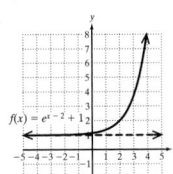
   10. a.
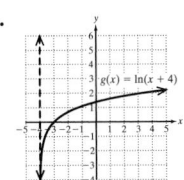

    b. $(-\infty, \infty)$ **c.** $(1, \infty)$ **b.** $(-4, \infty)$ **c.** $(-\infty, \infty)$

11. a.
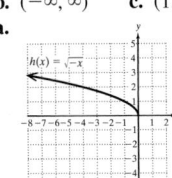
   12. a.
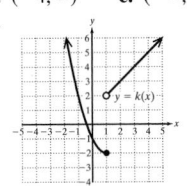

    b. $(-\infty, 0]$ **c.** $[0, \infty)$ **b.** $(-\infty, \infty)$ **c.** $[-2, \infty)$

17. $AB = \begin{bmatrix} 2 & -13 \\ 1 & 46 \end{bmatrix}$ **19.** $4A - B = \begin{bmatrix} 7 & -16 \\ 4 & 17 \end{bmatrix}$

20. $A^{-1} = \begin{bmatrix} \frac{2}{5} & \frac{1}{5} \\ -\frac{1}{15} & \frac{2}{15} \end{bmatrix}$

CHAPTER 10

Section 10.1 Practice Exercises, pp. 919–924

13. a. Center: $(0, 0)$
    b. $a = 10$ **c.** $b = 5$
    d. Vertices: $(10, 0), (-10, 0)$
    e. Endpoints of minor axis: $(0, 5), (0, -5)$
    f. Foci: $(5\sqrt{3}, 0), (-5\sqrt{3}, 0)$
    g. 20   **h.** 10
    i.
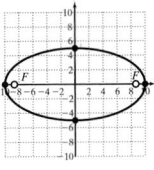

14. a. Center: $(0, 0)$
    b. $a = 8$ **c.** $b = 7$
    d. Vertices: $(8, 0), (-8, 0)$
    e. Endpoints of minor axis: $(0, 7), (0, -7)$
    f. Foci: $(\sqrt{15}, 0), (-\sqrt{15}, 0)$
    g. 16   **h.** 14
    i.
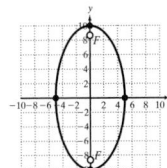

15. a. Center: $(0, 0)$
    b. $a = 10$ **c.** $b = 5$
    d. Vertices: $(0, 10), (0, -10)$
    e. Endpoints of minor axis: $(5, 0), (-5, 0)$
    f. Foci: $(0, 5\sqrt{3}), (0, -5\sqrt{3})$
    g. 20   **h.** 10
    i.
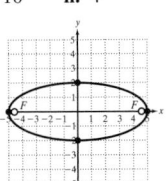

16. a. Center: $(0, 0)$
    b. $a = 8$ **c.** $b = 7$
    d. Vertices: $(0, 8), (0, -8)$
    e. Endpoints of minor axis: $(7, 0), (-7, 0)$
    f. Foci: $(0, \sqrt{15}), (0, -\sqrt{15})$
    g. 16   **h.** 14
    i.
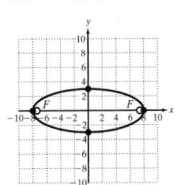

17. a. Center: $(0, 0)$
    b. $a = 5$ **c.** $b = 2$
    d. Vertices: $(5, 0), (-5, 0)$
    e. Endpoints of minor axis: $(0, 2), (0, -2)$
    f. Foci: $(\sqrt{21}, 0), (-\sqrt{21}, 0)$
    g. 10   **h.** 4
    i.
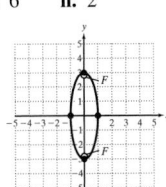

18. a. Center: $(0, 0)$
    b. $a = 8$ **c.** $b = 3$
    d. Vertices: $(8, 0), (-8, 0)$
    e. Endpoints of minor axis: $(0, 3), (0, -3)$
    f. Foci: $(\sqrt{55}, 0), (-\sqrt{55}, 0)$
    g. 16   **h.** 6
    i.
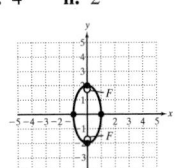

19. a. Center: $(0, 0)$
    b. $a = 3$ **c.** $b = 1$
    d. Vertices: $(0, 3), (0, -3)$
    e. Endpoint of minor axis: $(1, 0), (-1, 0)$
    f. Foci: $(0, 2\sqrt{2}), (0, -2\sqrt{2})$
    g. 6   **h.** 2
    i.
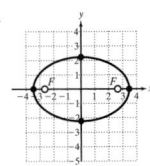

20. a. Center: $(0, 0)$
    b. $a = 2$ **c.** $b = 1$
    d. Vertices: $(0, 2), (0, -2)$
    e. Endpoints of minor axis: $(1, 0), (-1, 0)$
    f. Foci: $(0, \sqrt{3}), (0, -\sqrt{3})$
    g. 4   **h.** 2
    i.
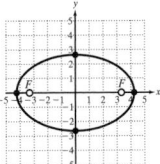

21. a. Center: $(0, 0)$
    b. $a = 2\sqrt{3}$ **c.** $b = \sqrt{5}$
    d. Vertices: $(2\sqrt{3}, 0), (-2\sqrt{3}, 0)$
    e. Endpoints of minor axis: $(0, \sqrt{5}), (0, -\sqrt{5})$
    f. Foci: $(\sqrt{7}, 0), (-\sqrt{7}, 0)$
    g. $4\sqrt{3}$ **h.** $2\sqrt{5}$
    i.

22. a. Center: $(0, 0)$
    b. $a = 3\sqrt{2}$ **c.** $b = \sqrt{7}$
    d. Vertices: $(3\sqrt{2}, 0), (-3\sqrt{2}, 0)$
    e. Endpoints of minor axis: $(0, \sqrt{7}), (0, -\sqrt{7})$
    f. Foci: $(\sqrt{11}, 0), (-\sqrt{11}, 0)$
    g. $6\sqrt{2}$ **h.** $2\sqrt{7}$
    i.

23. a. Center: $(1, -6)$
   b. Vertices: $(6, -6), (-4, -6)$
   c. Endpoints of minor axis:
      $(1, -2), (1, -10)$
   d. Foci: $(4, -6), (-2, -6)$
   e.

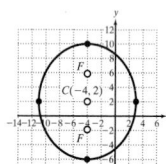

24. a. Center: $(3, -4)$
   b. Vertices: $(8, -4), (-2, -4)$
   c. Endpoints of minor axis:
      $(3, -1), (3, -7)$
   d. Foci: $(-1, -4), (7, -4)$
   e.

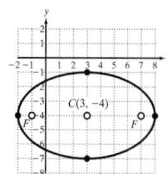

25. a. Center: $(-4, 2)$
   b. Vertices: $(-4, 10), (-4, -6)$
   c. Endpoints of minor axis:
      $(-11, 2), (3, 2)$
   d. Foci: $\left(-4, 2 + \sqrt{15}\right),$
      $\left(-4, 2 - \sqrt{15}\right)$
   e.

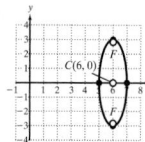

26. a. Center: $(-1, 5)$
   b. Vertices: $(-1, 14), (-1, -4)$
   c. Endpoints of minor axis:
      $(-7, 5), (5, 5)$
   d. Foci: $\left(-1, 5 + 3\sqrt{5}\right),$
      $\left(-1, 5 - 3\sqrt{5}\right)$
   e.

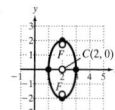

27. a. Center: $(6, 0)$
   b. Vertices: $(6, 3), (6, -3)$
   c. Endpoints of minor axis:
      $(5, 0), (7, 0)$
   d. Foci: $\left(6, 2\sqrt{2}\right), \left(6, -2\sqrt{2}\right)$
   e.

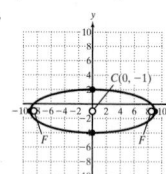

28. a. Center: $(2, 0)$
   b. Vertices: $(2, 2), (2, -2)$
   c. Endpoints of minor axis:
      $(1, 0), (3, 0)$
   d. Foci: $\left(2, \sqrt{3}\right), \left(2, -\sqrt{3}\right)$
   e.

29. a. Center: $(0, -1)$
   b. Vertices: $(9, -1), (-9, -1)$
   c. Endpoints of minor axis:
      $(0, 2), (0, -4)$
   d. Foci: $\left(6\sqrt{2}, -1\right),$
      $\left(-6\sqrt{2}, -1\right)$
   e.

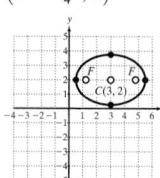

30. a. Center: $(0, -4)$
   b. Vertices: $(10, -4), (-10, -4)$
   c. Endpoints of minor axis:
      $(0, 1), (0, -9)$
   d. Foci: $\left(5\sqrt{3}, -4\right),$
      $\left(-5\sqrt{3}, -4\right)$
   e.

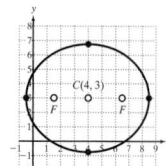

31. a. Center: $(3, 2)$
   b. Vertices: $\left(\frac{1}{2}, 2\right), \left(\frac{11}{2}, 2\right)$
   c. Endpoints of minor axis:
      $\left(3, \frac{1}{4}\right), \left(3, \frac{15}{4}\right)$
   d. Foci: $\left(3 + \frac{\sqrt{51}}{4}, 2\right),$
      $\left(3 - \frac{\sqrt{51}}{4}, 2\right)$
   e.

32. a. Center: $(4, 3)$
   b. Vertices: $\left(-\frac{1}{2}, 3\right), \left(\frac{17}{2}, 3\right)$
   c. Endpoints of minor axis:
      $\left(4, -\frac{3}{4}\right), \left(4, \frac{27}{4}\right)$
   d. Foci: $\left(4 + \frac{3\sqrt{11}}{4}, 3\right),$
      $\left(4 - \frac{3\sqrt{11}}{4}, 3\right)$
   e.

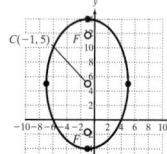

33. a. $\dfrac{(x + 2)^2}{5} + \dfrac{(y - 6)^2}{3} = 1$
   b. Center: $(-2, 6)$;
      Vertices: $\left(-2 - \sqrt{5}, 6\right); \left(-2 + \sqrt{5}, 6\right)$
      Endpoints of minor axis: $\left(-2, 6 + \sqrt{3}\right), \left(-2, 6 - \sqrt{3}\right)$
      Foci: $\left(-2 - \sqrt{2}, 6\right), \left(-2 + \sqrt{2}, 6\right)$

34. a. $\dfrac{(x + 5)^2}{11} + \dfrac{(y - 3)^2}{7} = 1$
   b. Center: $(-5, 3)$;
      Vertices: $\left(-5 - \sqrt{11}, 3\right), \left(-5 + \sqrt{11}, 3\right)$
      Endpoints of minor axis: $\left(-5, 3 + \sqrt{7}\right), \left(-5, 3 - \sqrt{7}\right)$
      Foci: $(-7, 3), (-3, 3)$

35. a. $\dfrac{(x - 5)^2}{6} + \dfrac{(y - 1)^2}{9} = 1$
   b. Center: $(5, 1)$
      Vertices: $(5, 4), (5, -2)$
      Endpoints of minor axis: $\left(5 + \sqrt{6}, 1\right), \left(5 - \sqrt{6}, 1\right)$
      Foci: $\left(5, 1 + \sqrt{3}\right), \left(5, 1 - \sqrt{3}\right)$

36. a. $\dfrac{(x - 4)^2}{4} + \dfrac{(y - 3)^2}{10} = 1$
   b. Center: $(4, 3)$
      Vertices: $\left(4, 3 + \sqrt{10}\right), \left(4, 3 - \sqrt{10}\right)$
      Endpoints of minor axis: $(6, 3), (2, 3)$
      Foci: $\left(4, 3 + \sqrt{6}\right), \left(4, 3 - \sqrt{6}\right)$

37. a. $x^2 + \dfrac{(y + 7)^2}{4} = 1$
   b. Center: $(0, -7)$
      Vertices: $(0, -5), (0, -9)$
      Endpoints of minor axis: $(-1, -7), (1, -7)$
      Foci: $\left(0, -7 + \sqrt{3}\right), \left(0, -7 - \sqrt{3}\right)$

38. a. $\dfrac{(x - 3)^2}{49} + y^2 = 1$
   b. Center: $(3, 0)$
      Vertices: $(-4, 0), (10, 0)$
      Endpoints of minor axis: $(3, 1), (3, -1)$
      Foci: $\left(3 + 4\sqrt{3}, 0\right), \left(3 - 4\sqrt{3}, 0\right)$

39. a. $\dfrac{\left(x - \frac{5}{2}\right)^2}{25} + \dfrac{(y + 4)^2}{9} = 1$
   b. Center: $\left(\frac{5}{2}, -4\right)$
      Vertices: $\left(\frac{15}{2}, -4\right), \left(-\frac{5}{2}, -4\right)$
      Endpoints of minor axis: $\left(\frac{5}{2}, -1\right), \left(\frac{5}{2}, -7\right)$
      Foci: $\left(\frac{13}{2}, -4\right), \left(-\frac{3}{2}, -4\right)$

40. a. $\dfrac{\left(x - \frac{3}{2}\right)^2}{9} + \dfrac{(y + 1)^2}{4} = 1$
   b. Center: $\left(\frac{3}{2}, -1\right)$
      Vertices: $\left(\frac{9}{2}, -1\right), \left(-\frac{3}{2}, -1\right)$
      Endpoints of minor axis: $\left(\frac{3}{2}, 1\right), \left(\frac{3}{2}, -3\right)$
      Foci: $\left(\frac{3}{2} + \sqrt{5}, -1\right), \left(\frac{3}{2} - \sqrt{5}, -1\right)$

41. a. $\dfrac{(x - 1)^2}{\frac{1}{4}} + \dfrac{y^2}{\frac{1}{9}} = 1$
   b. Center: $(1, 0)$; Vertices: $\left(\frac{1}{2}, 0\right), \left(\frac{3}{2}, 0\right)$;
      Endpoints of minor axis: $\left(1, \frac{1}{3}\right), \left(1, -\frac{1}{3}\right)$;
      Foci: $\left(1 - \frac{\sqrt{5}}{6}, 0\right), \left(1 + \frac{\sqrt{5}}{6}, 0\right)$

42. a. $\dfrac{x^2}{\frac{1}{25}} + \dfrac{(y + 2)^2}{\frac{1}{16}} = 1$
   b. Center: $(0, -2)$; Vertices: $\left(0, -\frac{7}{4}\right), \left(0, -\frac{9}{4}\right)$;
      Endpoints of minor axis: $\left(-\frac{1}{5}, -2\right), \left(\frac{1}{5}, -2\right)$;
      Foci: $\left(0, -\frac{37}{20}\right), \left(0, -\frac{43}{20}\right)$

78. a. $\dfrac{x^2}{529^2} + \dfrac{y^2}{451.5^2} = 1$

85. a. Top semiellipse   **b.** Bottom semiellipse
   c. Right semiellipse   **d.** Left semiellipse
86. a. Left semiellipse   **b.** Right semiellipse
   c. Top semiellipse   **d.** Bottom semiellipse

93. $\dfrac{x^2}{a^2} + \dfrac{y^2}{b^2} = 1$ $\dfrac{c^2}{a^2} + \dfrac{y^2}{b^2} = 1$ $y = b\sqrt{1 - \dfrac{c^2}{a^2}}$

$y = b\sqrt{\dfrac{a^2 - c^2}{a^2}}$ $y = \dfrac{b}{a}\sqrt{a^2 - c^2}$

Recall that $c^2 = a^2 - b^2$, or equivalently $b^2 = a^2 - c^2$ and $b > 0$.

Therefore, $y = \dfrac{b}{a}\sqrt{b^2}$ or $y = \dfrac{b^2}{a}$. The length of a latus rectum is

$2y = \dfrac{2b^2}{a}$.

94. a. The foci of the ellipse are $(-3, 0)$ and $(3, 0)$. The point $(-3, 0)$ lies on the line $y = \frac{4}{3}x + 4$ because $0 = \frac{4}{3}(-3) + 4$.

b. $\left(-\dfrac{75}{17}, -\dfrac{32}{17}\right)$ and $(0, 4)$ **c.** Approximately 7.35 units

95. By the reflective property of an ellipse, any shot passing through one focus is reflected through the other focus.

96. If the denominator is greater in the coefficient of the x^2 term, then the ellipse is elongated horizontally and the major axis is horizontal. If the denominator is greater in the coefficient of the y^2 term, then the ellipse is elongated vertically and the major axis is vertical.

97. The first equation represents an ellipse centered at the origin, whereas the second equation represents a line with slope $-\frac{4}{9}$ and y-intercept $(0, 4)$.

98. There are no real numbers x and y that will make the sum of two squares equal to a negative number. Therefore, the solution set is { }. This equation is a degenerate form of an ellipse. (*Note*: Compare this to the degenerate forms of a circle introduced in Section 1.2.)

99. a. $(a + c) + (a - c) = 2a$

b. $\sqrt{(x + c)^2 + y^2} + \sqrt{(x - c)^2 + y^2} = 2a$

c. $\sqrt{(x + c)^2 + y^2} = 2a - \sqrt{(x - c)^2 + y^2}$

$(x + c)^2 + y^2 = 4a^2 - 4a\sqrt{(x - c)^2 + y^2} + (x - c)^2 + y^2$

$4a\sqrt{(x - c)^2 + y^2} = 4a^2 + (x - c)^2 - (x + c)^2$

$4a\sqrt{(x - c)^2 + y^2} = 4a^2 - 4xc$

$a\sqrt{(x - c)^2 + y^2} = a^2 - xc$

d. $\left(a\sqrt{(x - c)^2 + y^2}\right)^2 = (a^2 - xc)^2$

$a^2\left[(x - c)^2 + y^2\right] = a^4 - 2a^2xc + c^2x^2$

$a^2\left[x^2 - 2xc + c^2 + y^2\right] = a^4 - 2a^2xc + c^2x^2$

$a^2x^2 - 2a^2xc + a^2c^2 + a^2y^2 = a^4 - 2a^2xc + c^2x^2$

$a^2x^2 - c^2x^2 + a^2y^2 = a^4 - a^2c^2$

$(a^2 - c^2)x^2 + a^2y^2 = a^2(a^2 - c^2)$

e. $b^2x^2 + a^2y^2 = a^2b^2$

$\dfrac{b^2x^2}{a^2b^2} + \dfrac{a^2y^2}{a^2b^2} = \dfrac{a^2b^2}{a^2b^2}$

$\dfrac{x^2}{a^2} + \dfrac{y^2}{b^2} = 1$

101. a. Both equations have a sum of terms with a variable squared in the numerator and a real number squared in the denominator, all equal to 1.

b. $\dfrac{x^2}{9} + \dfrac{y^2}{36} = 1$; This represents the graph of an ellipse in the xy-plane.

c. $\dfrac{y^2}{36} + \dfrac{z^2}{4} = 1$; This represents the graph of an ellipse in the yz-plane.

d. $\dfrac{x^2}{9} + \dfrac{z^2}{4} = 1$; This represents the graph of an ellipse in the xz-plane.

102. 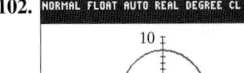 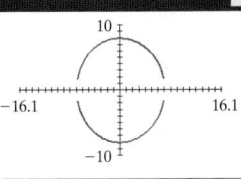   **103.**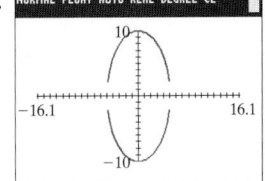

104. 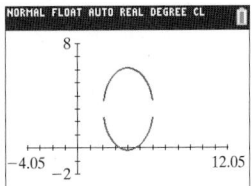   **105.**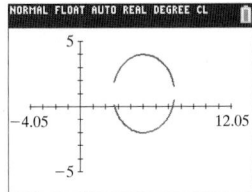

Section 10.2 Practice Exercises, pp. 935–940

13. a. Center: $(0, 0)$
b. Vertices: $(4, 0), (-4, 0)$
c. Foci: $\left(\sqrt{41}, 0\right), \left(-\sqrt{41}, 0\right)$
d. $y = \frac{5}{4}x$ and $y = -\frac{5}{4}x$
e.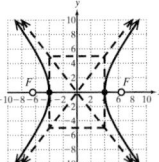

14. a. Center: $(0, 0)$
b. Vertices: $(5, 0), (-5, 0)$
c. Foci: $\left(\sqrt{61}, 0\right), \left(-\sqrt{61}, 0\right)$
d. $y = \frac{6}{5}x$ and $y = -\frac{6}{5}x$
e.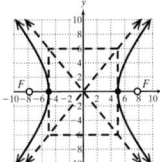

15. a. Center: $(0, 0)$
b. Vertices: $(0, 2), (0, -2)$
c. Foci: $\left(0, 2\sqrt{10}\right), \left(0, -2\sqrt{10}\right)$
d. $y = \frac{1}{3}x$ and $y = -\frac{1}{3}x$
e.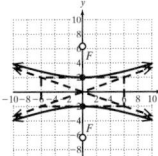

16. a. Center: $(0, 0)$
b. Vertices: $(0, 3), (0, -3)$
c. Foci: $\left(0, \sqrt{58}\right), \left(0, -\sqrt{58}\right)$
d. $y = \frac{3}{7}x$ and $y = -\frac{3}{7}x$
e.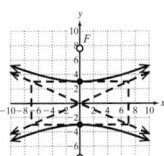

17. a. Center: $(0, 0)$
b. Vertices: $(0, 9), (0, -9)$
c. Foci: $\left(0, \sqrt{106}\right), \left(0, -\sqrt{106}\right)$
d. $y = \frac{9}{5}x$ and $y = -\frac{9}{5}x$
e.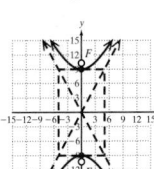

18. a. Center: $(0, 0)$
b. Vertices: $(0, 4), (0, -4)$
c. Foci: $\left(0, \sqrt{65}\right), \left(0, -\sqrt{65}\right)$
d. $y = \frac{4}{7}x$ and $y = -\frac{4}{7}x$
e.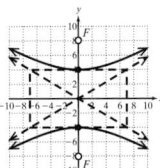

19. a. Center: $(0, 0)$
b. Vertices: $\left(\sqrt{7}, 0\right), \left(-\sqrt{7}, 0\right)$
c. Foci: $\left(2\sqrt{3}, 0\right), \left(-2\sqrt{3}, 0\right)$
d. $y = \frac{\sqrt{35}}{7}x$ and $y = -\frac{\sqrt{35}}{7}x$
e.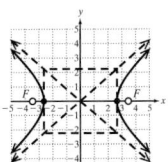

20. a. Center: $(0, 0)$
b. Vertices: $\left(\sqrt{11}, 0\right), \left(-\sqrt{11}, 0\right)$
c. Foci: $\left(3\sqrt{2}, 0\right), \left(-3\sqrt{2}, 0\right)$
d. $y = \frac{\sqrt{77}}{11}x$ and $y = -\frac{\sqrt{77}}{11}x$
e.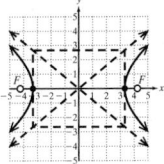

21. a. Center: $(0, 0)$
  b. Vertices: $\left(\frac{5}{2}, 0\right), \left(-\frac{5}{2}, 0\right)$
  c. Foci: $\left(\frac{\sqrt{149}}{4}, 0\right), \left(-\frac{\sqrt{149}}{4}, 0\right)$
  d. $y = \frac{7}{10}x$ and $y = -\frac{7}{10}x$

e.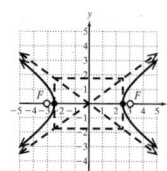

22. a. Center: $(0, 0)$
  b. Vertices: $\left(\frac{9}{2}, 0\right), \left(-\frac{9}{2}, 0\right)$
  c. Foci: $\left(\frac{3\sqrt{61}}{4}, 0\right), \left(-\frac{3\sqrt{61}}{4}, 0\right)$
  d. $y = \frac{5}{6}x$ and $y = -\frac{5}{6}x$

e.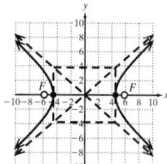

23. a. Center: $(4, -2)$
  b. Vertices: $(1, -2), (7, -2)$
  c. Foci: $(-1, -2), (9, -2)$
  d. $y = \frac{4}{3}x - \frac{22}{3}$ and $y = -\frac{4}{3}x + \frac{10}{3}$

e.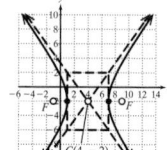

24. a. Center: $(3, -1)$
  b. Vertices: $(-3, -1), (9, -1)$
  c. Foci: $(13, -1), (-7, -1)$
  d. $y = \frac{4}{3}x - 5$ and $y = -\frac{4}{3}x + 3$

e.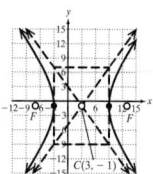

25. a. Center: $(-3, 5)$
  b. Vertices: $(-3, -2), (-3, 12)$
  c. Foci: $\left(-3, 5 + \sqrt{74}\right), \left(-3, 5 - \sqrt{74}\right)$
  d. $y = \frac{7}{5}x + \frac{46}{5}$ and $y = -\frac{7}{5}x + \frac{4}{5}$

e.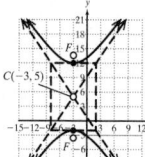

26. a. Center: $(-5, 4)$
  b. Vertices: $(-5, 10), (-5, -2)$
  c. Foci: $\left(-5, 4 + 2\sqrt{13}\right), \left(-5, 4 - 2\sqrt{13}\right)$
  d. $y = \frac{3}{2}x + \frac{23}{2}$ and $y = -\frac{3}{2}x - \frac{7}{2}$

e.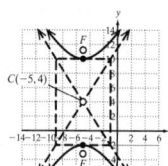

27. a. Center: $(-4, 7)$
  b. Vertices: $(6, 7), (-14, 7)$
  c. Foci: $\left(-4 + \sqrt{181}, 7\right), \left(-4 - \sqrt{181}, 7\right)$
  d. $y = \frac{9}{10}x + \frac{53}{5}$ and $y = -\frac{9}{10}x + \frac{17}{5}$

e.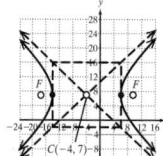

28. a. Center: $(-6, 3)$
  b. Vertices: $(-13, 3), (1, 3)$
  c. Foci: $\left(-6 + \sqrt{149}, 3\right), \left(-6 - \sqrt{149}, 3\right)$
  d. $y = \frac{10}{7}x + \frac{81}{7}$ and $y = -\frac{10}{7}x - \frac{39}{7}$

e.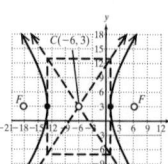

29. a. Center: $(3, 0)$
  b. Vertices: $(3, 1), (3, -1)$
  c. Foci: $\left(3, \sqrt{13}\right), \left(3, -\sqrt{13}\right)$
  d. $y = \frac{\sqrt{3}}{6}x - \frac{\sqrt{3}}{2}$ and $y = -\frac{\sqrt{3}}{6}x + \frac{\sqrt{3}}{2}$

e.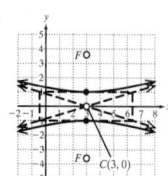

30. a. Center: $(2, 0)$
  b. Vertices: $(2, 1), (2, -1)$
  c. Foci: $\left(2, \sqrt{19}\right), \left(2, -\sqrt{19}\right)$
  d. $y = \frac{\sqrt{2}}{6}x - \frac{\sqrt{2}}{3}$ and $y = -\frac{\sqrt{2}}{6}x + \frac{\sqrt{2}}{3}$

e.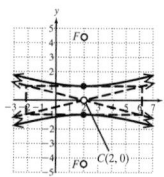

31. a. Center: $(0, 4)$
  b. Vertices: $(1, 4), (-1, 4)$
  c. Foci: $(3, 4), (-3, 4)$
  d. $y = 2\sqrt{2}x + 4$ and $y = -2\sqrt{2}x + 4$

e.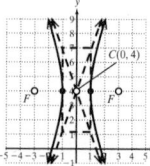

32. a. Center: $(0, 6)$
  b. Vertices: $(1, 6), (-1, 6)$
  c. Foci: $(5, 6), (-5, 6)$
  d. $y = 2\sqrt{6}x + 6$ and $y = -2\sqrt{6}x + 6$

e.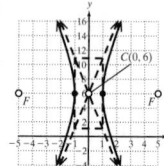

33. a. $\dfrac{(x + 3)^2}{5} - \dfrac{(y - 1)^2}{7} = 1$
  b. Center: $(-3, 1)$; Vertices: $\left(-3 + \sqrt{5}, 1\right), \left(-3 - \sqrt{5}, 1\right)$;
  Foci: $\left(-3 + 2\sqrt{3}, 1\right), \left(-3 - 2\sqrt{3}, 1\right)$

34. a. $\dfrac{(x + 1)^2}{6} - \dfrac{(y - 4)^2}{10} = 1$ **b.** Center: $(-1, 4)$;
  Vertices: $\left(-1 + \sqrt{6}, 4\right), \left(-1 - \sqrt{6}, 4\right)$; Foci: $(3, 4), (-5, 4)$

35. a. $\dfrac{(y - 4)^2}{5} - \dfrac{(x - 2)^2}{9} = 1$
  b. Center: $(2, 4)$; Vertices: $\left(2, 4 + \sqrt{5}\right), \left(2, 4 - \sqrt{5}\right)$;
  Foci: $\left(2, 4 + \sqrt{14}\right), \left(2, 4 - \sqrt{14}\right)$

36. a. $\dfrac{(y + 3)^2}{7} - \dfrac{(x + 5)^2}{16} = 1$
  b. Center: $(-5, -3)$; Vertices: $\left(-5, -3 + \sqrt{7}\right), \left(-5, -3 - \sqrt{7}\right)$;
  Foci: $\left(-5, -3 + \sqrt{23}\right), \left(-5, -3 - \sqrt{23}\right)$

37. a. $x^2 - \dfrac{(y + 5)^2}{4} = 1$ **b.** Center: $(0, -5)$;
  Vertices: $(1, -5), (-1, -5)$; Foci: $\left(\sqrt{5}, -5\right), \left(-\sqrt{5}, -5\right)$

38. a. $x^2 - \dfrac{(y + 7)^2}{9} = 1$ **b.** Center: $(0, -7)$;
  Vertices: $(1, -7), (-1, -7)$; Foci: $\left(\sqrt{10}, -7\right), \left(-\sqrt{10}, -7\right)$

39. a. $\dfrac{(y + 2)^2}{9} - \dfrac{\left(x - \frac{3}{2}\right)^2}{16} = 1$
  b. Center: $\left(\frac{3}{2}, -2\right)$; Vertices: $\left(\frac{3}{2}, 1\right), \left(\frac{3}{2}, -5\right)$; Foci: $\left(\frac{3}{2}, 3\right), \left(\frac{3}{2}, -7\right)$

40. a. $\dfrac{(y - 1)^2}{144} - \dfrac{\left(x - \frac{5}{2}\right)^2}{25} = 1$
  b. Center: $\left(\frac{5}{2}, 1\right)$; Vertices: $\left(\frac{5}{2}, 13\right), \left(\frac{5}{2}, -11\right)$; Foci: $\left(\frac{5}{2}, 14\right), \left(\frac{5}{2}, -12\right)$

41. $\dfrac{x^2}{144} - \dfrac{y^2}{25} = 1$ **42.** $\dfrac{x^2}{1600} - \dfrac{y^2}{81} = 1$ **43.** $\dfrac{y^2}{144} - \dfrac{x^2}{81} = 1$

44. $\dfrac{y^2}{225} - \dfrac{x^2}{64} = 1$ **48.** $\dfrac{(y + 6)^2}{25} - \dfrac{(x - 2)^2}{6} = 1$

77. The transverse axis is horizontal if the coefficient of the x^2 term is positive. The transverse axis is vertical if the coefficient of the y^2 term is positive.

78. The equation is equivalent to $y = \pm\frac{3}{2}x$. The solution set consists of all the points on the pair of lines $y = \frac{3}{2}x$ and $y = -\frac{3}{2}x$. This equation is a degenerate form of a hyperbola.

80. a. The equation represents the upper branch of a hyperbola centered at the origin with a vertical transverse axis.   **b.** Yes
  c.   **d.** $(0, \infty)$ **e.** $(-\infty, 0)$

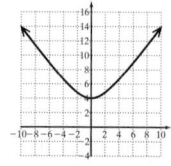

81. $\dfrac{x^2}{a^2} - \dfrac{y^2}{b^2} = 1$ $\dfrac{c^2}{a^2} - \dfrac{y^2}{b^2} = 1$ $y = b\sqrt{\dfrac{c^2}{a^2} - 1}$

$y = b\sqrt{\dfrac{c^2 - a^2}{a^2}}$ $y = \dfrac{b}{a}\sqrt{c^2 - a^2}$

Recall that $c^2 = a^2 + b^2$ or equivalently $b^2 = c^2 - a^2$ and $b > 0$.

Therefore, $y = \dfrac{b}{a}\sqrt{b^2}$ or $y = \dfrac{b^2}{a}$.

The length of a latus rectum is $2y = \dfrac{2b^2}{a}$.

82. a. $(c + a) - (c - a) = 2a$

b. $\sqrt{[x - (-c)]^2 + (y - 0)^2} - \sqrt{(x - c)^2 + (y - 0)^2} = 2a$
$\sqrt{(x + c)^2 + y^2} - \sqrt{(x - c)^2 + y^2} = 2a$

c. $\sqrt{(x + c)^2 + y^2} - \sqrt{(x - c)^2 + y^2} = 2a$
$\sqrt{(x + c)^2 + y^2} = 2a + \sqrt{(x - c)^2 + y^2}$
$(x + c)^2 + y^2 = 4a^2 + 4a\sqrt{(x - c)^2 + y^2} + (x - c)^2 + y^2$
$x^2 + 2cx + c^2 + y^2 = 4a^2 + 4a\sqrt{(x - c)^2 + y^2} + x^2 - 2cx + c^2 + y^2$
$4cx = 4a^2 + 4a\sqrt{(x - c)^2 + y^2}$
$cx - a^2 = a\sqrt{(x - c)^2 + y^2}$

d. $(cx - a^2)^2 = \left(a\sqrt{(x - c)^2 + y^2}\right)^2$
$c^2x^2 - 2a^2cx + a^4 = a^2[(x - c)^2 + y^2]$
$c^2x^2 - 2a^2cx + a^4 = a^2(x^2 - 2cx + c^2 + y^2)$
$c^2x^2 - 2a^2cx + a^4 = a^2x^2 - 2a^2cx + a^2c^2 + a^2y^2$
$c^2x^2 - a^2x^2 - a^2y^2 = a^2c^2 - a^4$
$(c^2 - a^2)x^2 - a^2y^2 = a^2(c^2 - a^2)$

e. Substitute b^2 for $c^2 - a^2$.

$b^2x^2 - a^2y^2 = a^2b^2$ $\dfrac{b^2x^2}{a^2b^2} - \dfrac{a^2y^2}{a^2b^2} = \dfrac{a^2b^2}{a^2b^2}$ $\dfrac{x^2}{a^2} - \dfrac{y^2}{b^2} = 1$

83. a. $\dfrac{x^2}{a^2} + \dfrac{y^2}{b^2} = 1$; This is an equation of an ellipse in the xy-plane.

b. $\dfrac{y^2}{b^2} - \dfrac{z^2}{c^2} = 1$; This is an equation of a hyperbola in the yz-plane
with transverse axis on the y-axis.   **c.** $\dfrac{x^2}{a^2} - \dfrac{z^2}{c^2} = 1$; This is an
equation of a hyperbola in the xz-plane with transverse axis on the
x-axis.

84. a. $\dfrac{x^2}{16} - \dfrac{y^2}{48} = 1$ **b.** $\dfrac{y^2}{4} - \dfrac{x^2}{60} = 1$

c. $\left(\sqrt{\dfrac{195}{11}}, -\sqrt{\dfrac{57}{11}}\right) \approx (4.2, -2.3)$

85. 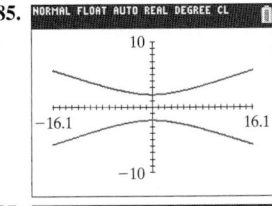   **86.**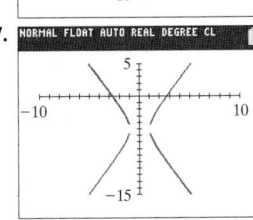

87.   **88.**

Section 10.3 Practice Exercises, pp. 949–954

27. a. Vertex: $(0, 0)$; $p = -1$;
Focus: $(0, -1)$; Focal diameter: 4
b. $(2, -1), (-2, -1)$
d. Directrix: $y = 1$;
Axis of symmetry: $x = 0$

c.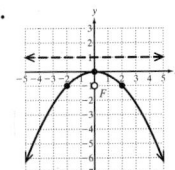

28. a. Vertex: $(0, 0)$; $p = -5$;
Focus: $(0, -5)$; Focal diameter: 20
b. $(10, -5), (-10, -5)$
d. Directrix: $y = 5$;
Axis of symmetry: $x = 0$

c.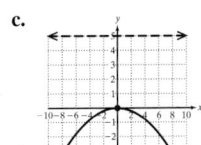

29. a. Vertex: $(0, 0)$; $p = 2$;
Focus: $(2, 0)$; Focal diameter: 8
b. $(2, 4), (2, -4)$
d. Directrix: $x = -2$;
Axis of symmetry: $y = 0$

c.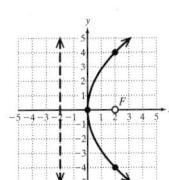

30. a. Vertex: $(0, 0)$; $p = 1$;
Focus: $(1, 0)$; Focal diameter: 4
b. $(1, 2), (1, -2)$
d. Directrix: $x = -1$;
Axis of symmetry: $y = 0$

c.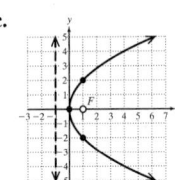

31. a. Vertex: $(0, 0)$; $p = \frac{5}{2}$;
Focus: $\left(0, \frac{5}{2}\right)$; Focal diameter: 10
b. $\left(5, \frac{5}{2}\right), \left(-5, \frac{5}{2}\right)$
d. Directrix: $y = -\frac{5}{2}$;
Axis of symmetry: $x = 0$

c.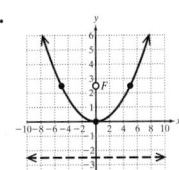

32. a. Vertex: $(0, 0)$; $p = \frac{7}{4}$;
Focus: $\left(0, \frac{7}{4}\right)$; Focal diameter: 7
b. $\left(\frac{7}{2}, \frac{7}{4}\right), \left(-\frac{7}{2}, \frac{7}{4}\right)$
d. Directrix: $y = -\frac{7}{4}$;
Axis of symmetry: $x = 0$

c.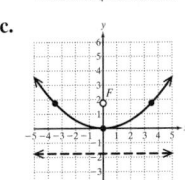

33. a. Vertex: $(0, 0)$; $p = -\frac{1}{4}$;
Focus: $\left(-\frac{1}{4}, 0\right)$; Focal diameter: 1
b. $\left(-\frac{1}{4}, \frac{1}{2}\right), \left(-\frac{1}{4}, -\frac{1}{2}\right)$
d. Directrix: $x = \frac{1}{4}$;
Axis of symmetry: $y = 0$

c.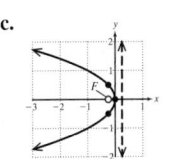

34. a. Vertex: $(0, 0)$; $p = -\frac{1}{2}$;
Focus: $\left(-\frac{1}{2}, 0\right)$; Focal diameter: 2
b. $\left(-\frac{1}{2}, 1\right), \left(-\frac{1}{2}, -1\right)$
d. Directrix: $x = \frac{1}{2}$;
Axis of symmetry: $y = 0$

c.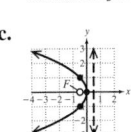

35. a. Vertex: $(4, -1)$; $p = -3$;
Focus: $(1, -1)$; Focal diameter: 12
b. $(1, 5), (1, -7)$
d. Directrix: $x = 7$;
Axis of symmetry: $y = -1$

c.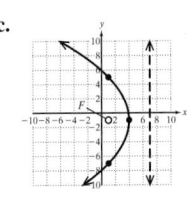

36. a. Vertex: $(2, -4)$; $p = -4$;
Focus: $(-2, -4)$;
Focal diameter: 16
b. $(-2, 4), (-2, -12)$
d. Directrix: $x = 6$;
Axis of symmetry: $y = -4$

c.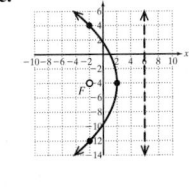

37. a. Vertex: $(1, -5)$; $p = -1$;
Focus: $(1, -6)$; Focal diameter: 4
b. $(3, -6), (-1, -6)$
d. Directrix: $y = -4$;
Axis of symmetry: $x = 1$

c.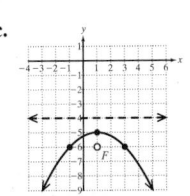

38. a. Vertex: $(5, -2)$; $p = -2$;
   Focus: $(5, -4)$; Focal diameter: 8
b. $(1, -4), (9, -4)$
d. Directrix: $y = 0$;
   Axis of symmetry: $x = 5$

c.
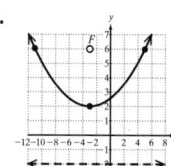

39. a. Vertex: $\left(-3, \frac{3}{2}\right)$; $p = \frac{1}{2}$;
   Focus: $(-3, 2)$; Focal diameter: 2
b. $(-4, 2), (-2, 2)$
d. Directrix: $y = 1$;
   Axis of symmetry: $x = -3$

c.

40. a. Vertex: $\left(-2, \frac{7}{4}\right)$; $p = \frac{1}{4}$;
   Focus: $(-2, 2)$; Focal diameter: 1
b. $\left(-\frac{5}{2}, 2\right), \left(-\frac{3}{2}, 2\right)$
d. Directrix: $y = \frac{3}{2}$
   Axis of symmetry: $x = -2$

c.

41. a. Vertex: $\left(\frac{1}{4}, 3\right)$; $p = -\frac{7}{4}$;
   Focus: $\left(-\frac{3}{2}, 3\right)$;
   Focal diameter: 7
b. $\left(-\frac{3}{2}, \frac{13}{2}\right), \left(-\frac{3}{2}, -\frac{1}{2}\right)$
d. Directrix: $x = 2$
   Axis of symmetry: $y = 3$

c.

42. a. Vertex: $\left(\frac{3}{4}, 6\right)$; $p = -\frac{9}{4}$;
   Focus: $\left(-\frac{3}{2}, 6\right)$; Focal diameter: 9
b. $\left(-\frac{3}{2}, \frac{21}{2}\right), \left(-\frac{3}{2}, \frac{3}{2}\right)$
d. Directrix: $x = 3$
   Axis of symmetry: $y = 6$

c.

43. a. Vertex: $(-6, 3)$; $p = 5$;
   Focus: $(-6, 8)$; Focal diameter: 20
b. $(4, 8), (-16, 8)$
d. Directrix: $y = -2$;
   Axis of symmetry: $x = -6$

c.

44. a. Vertex: $(-3, 2)$; $p = 4$;
   Focus: $(-3, 6)$; Focal diameter: 16
b. $(5, 6), (-11, 6)$
d. Directrix: $y = -2$
   Axis of symmetry: $x = -3$

c.

45. a. $(x - 3)^2 = 4(y + 1)$ **b.** Vertex: $(3, -1)$; Focus: $(3, 0)$;
   Focal diameter: 4   **46. a.** $(x - 2)^2 = 8(y + 3)$
b. Vertex: $(2, -3)$; Focus: $(2, -1)$; Focal diameter: 8
47. a. $(y + 2)^2 = -8(x + 6)$ **b.** Vertex: $(-6, -2)$; Focus: $(-8, -2)$;
   Focal diameter: 8   **48. a.** $(y + 4)^2 = -4(x + 5)$
b. Vertex: $(-5, -4)$; Focus: $(-6, -4)$; Focal diameter: 4
49. a. $\left(x - \frac{7}{2}\right)^2 = -6(y + 1)$ **b.** Vertex: $\left(\frac{7}{2}, -1\right)$; Focus: $\left(\frac{7}{2}, -\frac{5}{2}\right)$;
   Focal diameter: 6   **50. a.** $\left(x + \frac{9}{2}\right)^2 = -10(y - 2)$
b. Vertex: $\left(-\frac{9}{2}, 2\right)$; Focus: $\left(-\frac{9}{2}, -\frac{1}{2}\right)$; Focal diameter: 10
51. a. $\left(y + \frac{3}{4}\right)^2 = (x - 3)$ **b.** Vertex: $\left(3, -\frac{3}{4}\right)$; Focus: $\left(\frac{13}{4}, -\frac{3}{4}\right)$;
   Focal diameter: 1   **52. a.** $\left(y - \frac{7}{4}\right)^2 = (x - 2)$
b. Vertex: $\left(2, \frac{7}{4}\right)$; Focus: $\left(\frac{9}{4}, \frac{7}{4}\right)$; Focal diameter: 1

76. a.
NORMAL FLOAT AUTO REAL DEGREE CL
b. $x^2 = 540(y - 20)$
c. 135 ft   **d.** 38.5 ft

81. a. $z = y^2$; This is an equation of a parabola in the yz-plane.
b. $z = 4x^2$; This is an equation of a parabola in the xz-plane.
c. $0 = 4x^2 + y^2$; This is an equation of a degenerate ellipse.
   The solution set is a single point (the origin).
82. a. $c = 32$ **b.** $a = 0.01, b = -0.6, c = 32$
   $400a + 20b + c = 24$
   $1600a + 40b + c = 24$
c. $y = 0.01t^2 - 0.6t + 32$ or $y = 0.01(t - 30)^2 + 23$
d. $(30, 23)$ **e.** 25,000 ft

83.
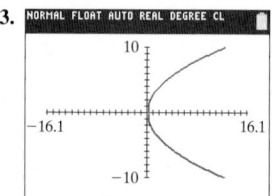
84.
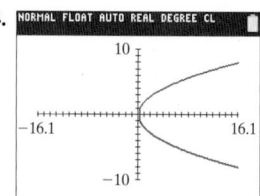

85.
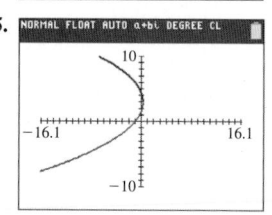
86.
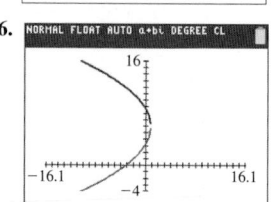

Problem Recognition Exercises, pp. 954–955

1. Hyperbola; Center: $(2, -2)$; Vertices: $(6, -2), (-2, -2)$;
   Foci: $(7, -2), (-3, -2)$; Asymptotes: $y = \frac{3}{4}x - \frac{7}{2}$ and $y = -\frac{3}{4}x - \frac{1}{2}$;
   Eccentricity: $\frac{5}{4}$ **2.** Ellipse; Center: $(0, 2)$; Vertices: $(5, 2), (-5, 2)$;
   Endpoints of minor axis: $(0, 5), (0, -1)$; Foci: $(4, 2), (-4, 2)$;
   Eccentricity: $\frac{4}{5}$ **3.** Parabola; Vertex: $(-2, 5)$; Focus: $\left(-\frac{9}{4}, 5\right)$;
   Directrix: $x = -\frac{7}{4}$; Axis of symmetry: $y = 5$
4. Circle; Center: $(3, -7)$; Radius: 5   **5.** Ellipse; Center: $(-1, 0)$;
   Vertices: $(-1, 4), (-1, -4)$; Endpoints of minor axis: $(-2, 0), (0, 0)$;
   Foci: $\left(-1, \sqrt{15}\right), \left(-1, -\sqrt{15}\right)$; Eccentricity: $\frac{\sqrt{15}}{4}$
6. Parabola; Vertex: $(1, -3)$; Focus: $\left(1, -\frac{1}{2}\right)$;
   Directrix: $y = -\frac{11}{2}$; Axis of symmetry: $x = 1$
7. Circle; Center: $(-1, -6)$; Radius: 4
8. Hyperbola; Center: $(-4, 1)$; Vertices: $(-4, 13), (-4, -11)$;
   Foci: $(-4, 14), (-4, -12)$; Asymptotes: $y = \frac{12}{5}x + \frac{53}{5}$ and
   $y = -\frac{12}{5}x - \frac{43}{5}$; Eccentricity: $\frac{13}{12}$
17. $(x - 2)^2 + (y + 1)^2 = 0$; The graph is a single point: $(2, -1)$.
18. $\frac{x^2}{4} + \frac{(y - 3)^2}{9} = 0$; The graph is a single point: $(0, 3)$.
19. $(x - 4)^2 - \frac{(y + 2)^2}{4} = 0$; The graph is a pair of intersecting lines:
   $y = -2x + 6$ and $y = 2x - 10$.
20. $\frac{(x - 5)^2}{4} - \frac{(y + 1)^2}{9} = 0$; The graph is a pair of intersecting lines:
   $y = \frac{3}{2}x - \frac{17}{2}$ and $y = -\frac{3}{2}x + \frac{13}{2}$.
21. $\frac{x^2}{4} + \frac{(y - 2)^2}{9} = -1$; No solution. There are no real numbers x and
   y that would make the sum of two squares equal to -1.
22. $(x + 6)^2 + y^2 = -1$; No solution. There are no real numbers x and y
   that would make the sum of two squares equal to -1.
23. With $C = A$ and completing the square, the equation
   $Ax^2 + Cy^2 + Dx + Ey + F = 0$ becomes
   $Ax^2 + Ay^2 + Dx + Ey + F = 0$
   $x^2 + y^2 + \frac{D}{A}x + \frac{E}{A}y + \frac{F}{A} = 0$
   $\left(x + \frac{D}{2A}\right)^2 + \left(y + \frac{E}{2A}\right)^2 = \frac{D^2}{4A^2} + \frac{E^2}{4A^2} - \frac{F}{A}$
   $\left(x + \frac{D}{2A}\right)^2 + \left(y + \frac{E}{2A}\right)^2 = \frac{D^2 + E^2 - 4AF}{4A^2}$

This is the standard form of an equation of circle with center $\left(-\dfrac{D}{2A}, -\dfrac{E}{2A}\right)$ and radius $\dfrac{\sqrt{D^2 + E^2 - 4AF}}{2A}$ provided $D^2 + E^2 - 4AF > 0$.

24. Let $k = \dfrac{D^2}{4A} + \dfrac{E^2}{4C} - F$ and assume that $k \neq 0$. Then, by completing the square,

$Ax^2 + Cy^2 + Dx + Ey + F = 0$ becomes

$$A\left(x + \frac{D}{2A}\right)^2 + C\left(y + \frac{E}{2C}\right)^2 = \frac{D^2}{4A} + \frac{E^2}{4C} - F$$

$$A\left(x + \frac{D}{2A}\right)^2 + C\left(y + \frac{E}{2C}\right)^2 = k$$

$$\frac{\left(x + \dfrac{D}{2A}\right)^2}{\dfrac{k}{A}} + \frac{\left(y + \dfrac{E}{2C}\right)^2}{\dfrac{k}{C}} = 1$$

a. The numerator of each term on the left side of the equation is nonnegative. Therefore, the sign of each term is dictated by the denominator. For A and C of the same sign, if k has the same sign as A, it also has the same sign as C and is nonzero. Thus, the denominators $\dfrac{k}{A}$ and $\dfrac{k}{C}$ are both positive. Therefore, the equation is in the standard form of an ellipse with center $\left(-\dfrac{D}{2A}, -\dfrac{E}{2C}\right)$.

b. For A and C of opposite signs and $k \neq 0$, the denominators $\dfrac{k}{A}$ and $\dfrac{k}{C}$ have opposite signs, indicating that the terms on the left side of the equation have opposite signs. Therefore, the equation is a hyperbola with center $\left(-\dfrac{D}{2A}, -\dfrac{E}{2C}\right)$.

25. If $A = 0$, then $Ax^2 + Cy^2 + Dx + Ey + F = 0$ becomes $Cy^2 + Dx + Ey + F = 0$. Completing the square gives

$$C\left(y^2 + \frac{E}{C}y + \frac{E^2}{4C^2}\right) = -Dx - F + \frac{E^2}{4C}$$

$$C\left(y + \frac{E}{2C}\right)^2 = -D\left(x + \frac{F}{D} - \frac{E^2}{4CD}\right)$$

$$\left(y + \frac{E}{2C}\right)^2 = -\frac{D}{C}\left[x - \left(\frac{E^2}{4CD} - \frac{F}{D}\right)\right]$$

Assuming that $C \neq 0$ and $D \neq 0$, this is the standard form of an equation of a parabola opening to the left or right with vertex $\left(\dfrac{E^2}{4CD} - \dfrac{F}{D}, -\dfrac{E}{2C}\right)$.

Section 10.4 Practice Exercises, pp. 965–968

19. $\cos\theta = \dfrac{1}{2}$, $\sin\theta = \dfrac{\sqrt{3}}{2}$, $\theta = 60°$

20. $\cos\theta = \dfrac{\sqrt{3}}{2}$, $\sin\theta = \dfrac{1}{2}$, $\theta = 30°$

21. $\cos\theta = \dfrac{7\sqrt{2}}{10}$, $\sin\theta = \dfrac{\sqrt{2}}{10}$, $\theta \approx 8.1°$

22. $\cos\theta = \dfrac{3\sqrt{34}}{34}$, $\sin\theta = \dfrac{5\sqrt{34}}{34}$, $\theta \approx 59.0°$

23. a. $45°$
   b. $x'^2 + 2y'^2 - 8 = 0$
   c. Ellipse
   d. $\dfrac{x'^2}{8} + \dfrac{y'^2}{4} = 1$

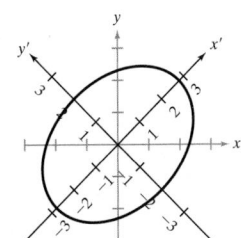

24. a. $45°$
   b. $3x'^2 + y'^2 - 9 = 0$
   c. Ellipse
   d. $\dfrac{x'^2}{3} + \dfrac{y'^2}{9} = 1$

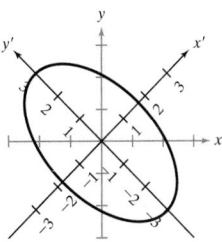

25. a. $60°$
   b. $4y'^2 - x'^2 - 4 = 0$
   c. Hyperbola
   d. $y'^2 - \dfrac{x'^2}{4} = 1$

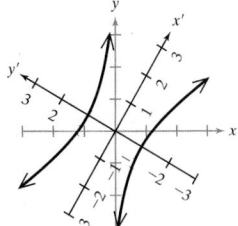

26. a. $30°$
   b. $x'^2 - 9y'^2 - 9 = 0$
   c. Hyperbola
   d. $\dfrac{x'^2}{9} - y'^2 = 1$

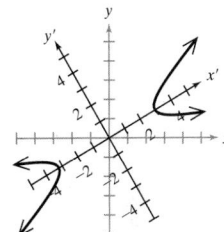

27. a. $30°$
   b. $y'^2 - 8x' = 0$
   c. Parabola
   d. $y'^2 = 8x'$

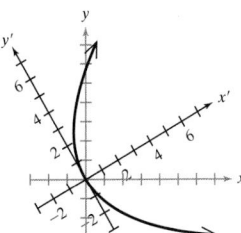

28. a. $60°$
   b. $x'^2 + 4y' = 0$
   c. Parabola
   d. $x'^2 = -4y'$

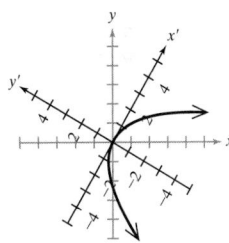

29. The equation has an xy term. Therefore, the discriminant must be used to determine the type of conic section that the equation represents. The value $B^2 - 4AC = (6)^2 - 4(3)(3) = 0$, which means that the equation is a parabola.

30. The equation has an xy term. Therefore, the discriminant must be used to determine the type of conic section that the equation represents. The value $B^2 - 4AC = (10)^2 - 4(9)(1) = 64 > 0$, which means that the equation is a hyperbola.

35. a. Hyperbola
   b. $4x'^2 - 9y'^2 - 36 = 0$
   c. $\dfrac{x'^2}{9} - \dfrac{y'^2}{4} = 1$

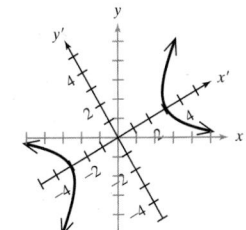

36. a. Ellipse

  b. $4x'^2 + y'^2 - 16 = 0$

  c. $\dfrac{x'^2}{4} + \dfrac{y'^2}{16} = 1$

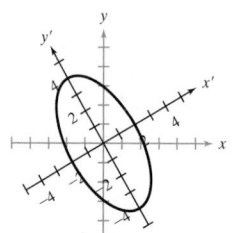

37. a. Parabola

  b. $y'^2 - 4x' - 10y' + 25 = 0$

  c. $(y' - 5)^2 = 4x'$

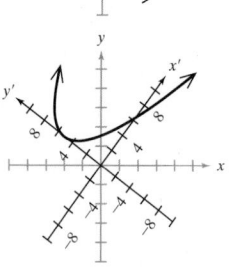

38. a. Hyperbola

  b. $x'^2 - y'^2 - 2x + 5 = 0$

  c. $\dfrac{y'^2}{4} - \dfrac{(x' - 1)^2}{4} = 1$

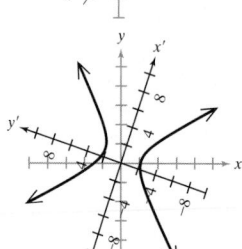

39. a. Ellipse

  b. $9x'^2 + 4y'^2 - 24y' = 0$

  c. $\dfrac{x'^2}{4} + \dfrac{(y' - 3)^2}{9} = 1$

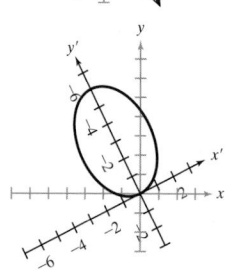

40. a. Parabola

  b. $x'^2 - 6x' + y' + 9 = 0$

  c. $(x' - 3)^2 = -y$

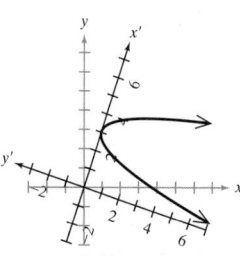

41. a. Ellipse

  b. $4x'^2 + y'^2 - 8x' + 4y' - 8 = 0$

  c. $\dfrac{(x' - 1)^2}{4} + \dfrac{(y' + 2)^2}{16} = 1$

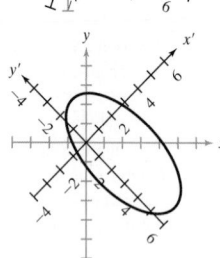

42. a. Hyperbola

  b. $16x'^2 - 9y'^2 + 96x' + 18y' - 9 = 0$

  c. $\dfrac{(x' + 3)^2}{9} - \dfrac{(y' - 1)^2}{16} = 1$

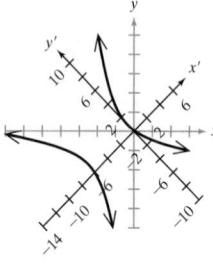

43. b.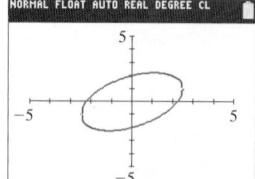

44. b.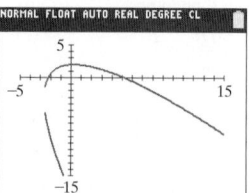

45. b.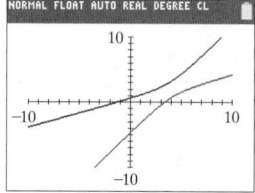

46. b.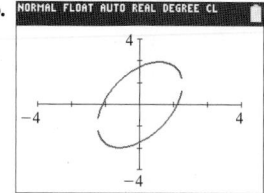

47. b.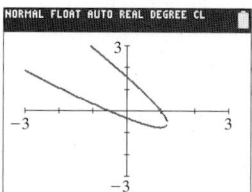

48. b.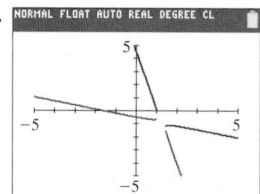

52. a. $2y'^2 + \sqrt{2}x' - 2 = 0$ or $y'^2 = -\dfrac{\sqrt{2}}{2}(x' - \sqrt{2})$

53. The equation can be written as $(x - y)^2 = 1$, which means that $x - y = \pm 1$. The graph is the pair of parallel lines $y = x - 1$ and $y = x + 1$.

54. Applying the quadratic formula to solve for y, we have
$$y = \frac{-(-8x) \pm \sqrt{(-8x)^2 - 4(1)(x^2)}}{2(1)} \text{ or } y = \left(4 \pm \sqrt{15}\right)x.$$
The graph is a pair of intersecting lines through the origin.

55. The Bxy term introduces a rotation.

56. Using the equation $\cot 2\theta = \dfrac{A - C}{B}$, determine the value of $\cos 2\theta$.

Then use the half-angle formulas to find $\cos\theta$ and $\sin\theta$.

57. Since $\tan^{-1}\dfrac{1}{2} \approx 26.6° < 30°$, point P will be below the positive x'-axis. Therefore, P is in Quadrant IV in the $x'y'$-plane.

58. Write the equation in the form $ay^2 + by + c = 0$ and then apply the quadratic formula. In this case, we have $5y^2 + (3x + 2)y + (2x^2 - 4x - 6) = 0$ and

$$y = \frac{-(3x + 2) \pm \sqrt{(3x + 2)^2 - 4(5)(2x^2 - 4x - 6)}}{2(5)}.$$

59. From page 960,

$$A' + C' = (A\cos^2\theta + B\sin\theta\cos\theta + C\sin^2\theta)$$
$$+ (A\sin^2\theta - B\sin\theta\cos\theta + C\cos^2\theta)$$
$$= A(\cos^2\theta + \sin^2\theta) + C(\sin^2\theta + \cos^2\theta)$$
$$= A + C$$

60. Let θ be the angle of rotation between the xy-coordinate system and the $x'y'$-coordinate system. We must show that
$$\sqrt{(x'_2 - x'_1)^2 + (y'_2 - y'_1)^2} = \sqrt{(x_2 - x_1)^2 + (y_2 - y_1)^2}.$$
Using rotation formulas (9) and (10) from page 958, the radicand of $\sqrt{(x'_2 - x'_1)^2 + (y'_2 - y'_1)^2}$ becomes

$$[(x_2\cos\theta + y_2\sin\theta) - (x_1\cos\theta + y_1\sin\theta)]^2$$
$$+ [(-x_2\sin\theta + y_2\cos\theta) - (-x_1\sin\theta + y_1\cos\theta)]^2$$
$$= [(x_2 - x_1)\cos\theta + (y_2 - y_1)\sin\theta]^2$$
$$+ [(x_1 - x_2)\sin\theta + (y_2 - y_1)\cos\theta]^2$$
$$= (x_2 - x_1)^2\cos^2\theta + 2(x_2 - x_1)(y_2 - y_1)\cos\theta\sin\theta$$
$$+ (y_2 - y_1)^2\sin^2\theta + (x_1 - x_2)^2\sin^2\theta$$
$$+ 2(x_1 - x_2)(y_2 - y_1)\sin\theta\cos\theta + (y_2 - y_1)^2\cos^2\theta$$
$$= (x_2 - x_1)^2\cos^2\theta + 2(x_2 - x_1)(y_2 - y_1)\sin\theta\cos\theta$$
$$+ (y_2 - y_1)^2\sin^2\theta + (x_2 - x_1)^2\sin^2\theta$$
$$- 2(x_2 - x_1)(y_2 - y_1)\sin\theta\cos\theta + (y_2 - y_1)^2\cos^2\theta$$
$$= (x_2 - x_1)^2\cos^2\theta + (x_2 - x_1)^2\sin^2\theta + (y_2 - y_1)^2\sin^2\theta$$
$$+ (y_2 - y_1)^2\cos^2\theta$$
$$= (x_2 - x_1)^2(\cos^2\theta + \sin^2\theta) + (y_2 - y_1)^2(\sin^2\theta + \cos^2\theta)$$
$$= (x_2 - x_1)^2 + (y_2 - y_1)^2 \text{ as desired.}$$

61. The graph of $x^2 + y^2 = r^2$ is a circle centered at the origin. Rotating the graph about the origin will result in the same graph. Therefore, the equation $x^2 + y^2 = r^2$ is invariant under rotation.

62. Assume that $x^2 + y^2 = r^2$. We need to show that $x'^2 + y'^2 = x^2 + y^2 = r^2$. Using the rotation formulas (9) and (10) on page 958, we have
$$x'^2 + y'^2 = (x\cos\theta + y\sin\theta)^2 + (-x\sin\theta + y\cos\theta)^2$$
$$= x^2\cos^2\theta + 2xy\cos\theta\sin\theta + y^2\sin^2\theta$$
$$+ x^2\sin^2\theta - 2xy\cos\theta\sin\theta + y^2\cos^2\theta$$
$$= x^2(\cos^2\theta + \sin^2\theta) + y(\sin^2\theta + \cos^2\theta)$$
$$= x^2 + y^2 = r^2$$

63. a. $\begin{bmatrix} \cos 60° & -\sin 60° \\ \sin 60° & \cos 60° \end{bmatrix} \begin{bmatrix} 6 \\ 2 \end{bmatrix}$ or $\begin{bmatrix} \frac{1}{2} & -\frac{\sqrt{3}}{2} \\ \frac{\sqrt{3}}{2} & \frac{1}{2} \end{bmatrix} \begin{bmatrix} 6 \\ 2 \end{bmatrix}$

b. $\begin{bmatrix} 3 - \sqrt{3} \\ 3\sqrt{3} + 1 \end{bmatrix}$

c.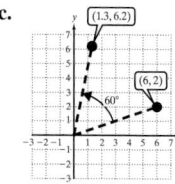

64. a. $\begin{bmatrix} 1 & 6 & 5 & 5 \\ 2 & 2 & 1 & 3 \end{bmatrix}$

b. $\begin{bmatrix} 0 & -1 \\ 1 & 0 \end{bmatrix} \begin{bmatrix} 1 & 6 & 5 & 5 \\ 2 & 2 & 1 & 3 \end{bmatrix} = \begin{bmatrix} -2 & -2 & -1 & -3 \\ 1 & 6 & 5 & 5 \end{bmatrix}$

c.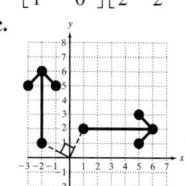

Section 10.5 Skill Practice

Answer 3

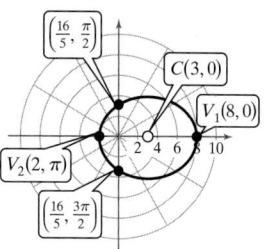

Answer 4

4. a. 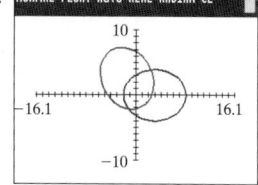 **b.**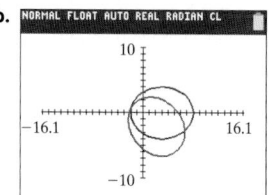

Section 10.5 Practice Exercises, pp. 976–978

R.1. a. $\dfrac{(x + 4)^2}{11} + \dfrac{(y - 6)^2}{5} = 1$

b. Center: $(-4, 6)$; Vertices: $(-4 \pm \sqrt{11}, 6)$; Endpoints of minor axis: $(-4, 6 \pm \sqrt{5})$; Foci: $(-4 \pm \sqrt{6}, 6)$

R.2. a. $\dfrac{(x + 1)^2}{5} - \dfrac{(y - 2)^2}{11} = 1$

b. Center: $(-1, 2)$; Vertices: $(-1 \pm \sqrt{5}, 2)$; Foci: $(3, 2)$ and $(-5, 2)$

R.3. a. $(y - 1)^2 = -4(x - 4)$

b. Vertex: $(4, 1)$; Focus $(3, 1)$; Directrix: $x = 5$

5. a. $e = 1, d = 2$ **b.** Parabola **c.** $y = 2$ **d.** Graph C

6. a. $e = \dfrac{5}{2}, d = 2$ **b.** Hyperbola **c.** $y = -2$ **d.** Graph A

7. a. $e = \dfrac{2}{5}, d = 5$ **b.** Ellipse **c.** $x = -5$ **d.** Graph B

8. a. $e = 1, d = \dfrac{5}{2}$ **b.** Parabola **c.** $x = \dfrac{5}{2}$ **d.** Graph F

9. a. $e = 2, d = \dfrac{5}{4}$ **b.** Hyperbola **c.** $x = \dfrac{5}{4}$ **d.** Graph E

10. a. $e = \dfrac{1}{2}, d = \dfrac{5}{2}$ **b.** Ellipse **c.** $x = \dfrac{5}{2}$ **d.** Graph D

11. 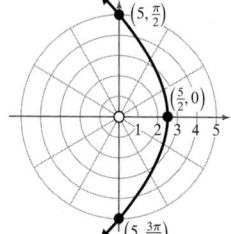 **12.**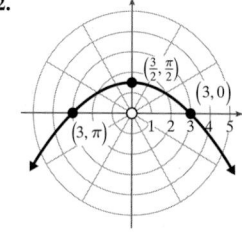

13. 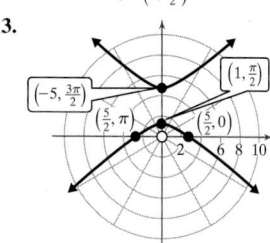 **14.**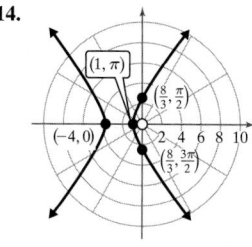

15.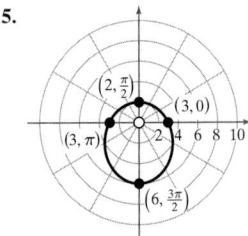

16.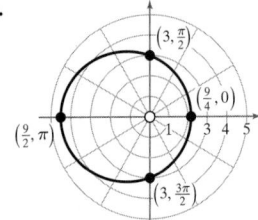

17.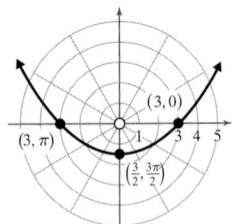

18.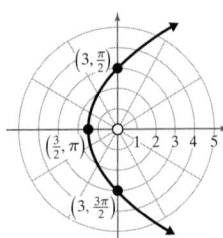

19.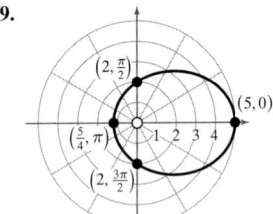

20.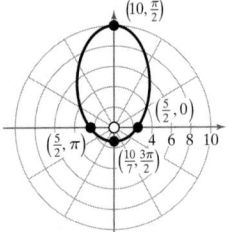

21.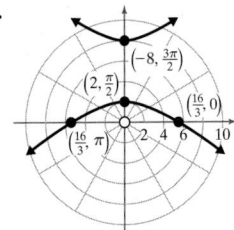

22.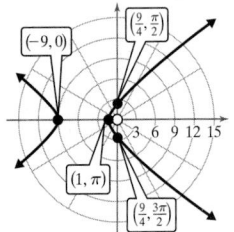

23. a. $(0, 3)$ **b.** $y = \pm\dfrac{2\sqrt{5}}{5}x + 3$

24. a. $\left(-\dfrac{5}{2}, 0\right)$ **b.** $y = \dfrac{4}{3}x + \dfrac{10}{3}$ and $y = -\dfrac{4}{3}x - \dfrac{10}{3}$

25. a. $\left(\dfrac{15}{8}, 0\right)$ **b.** $\dfrac{25}{4}$ units   **c.** 5 units

26. a. $\left(0, \dfrac{30}{7}\right)$ **b.** $\dfrac{80}{7}$ units   **c.** $\dfrac{20\sqrt{7}}{7}$ units

27. $r = \dfrac{1}{1 - \sin\left(\theta + \dfrac{\pi}{3}\right)}$

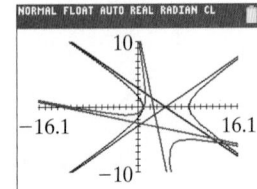

28. $r = \dfrac{8}{5 + 6\cos\left(\theta + \dfrac{\pi}{4}\right)}$

29. $r = \dfrac{12}{4 - 3\cos\left(\theta - \dfrac{7\pi}{6}\right)}$

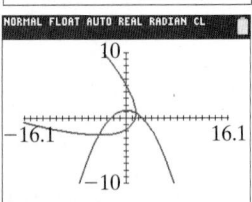

30. $r = \dfrac{5}{2 + 2\sin\left(\theta - \dfrac{4\pi}{3}\right)}$

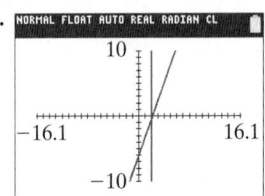

31. a. $r = 2\sec\theta$

  b. $r = 2\sec\left(\theta + \dfrac{\pi}{9}\right)$

  c.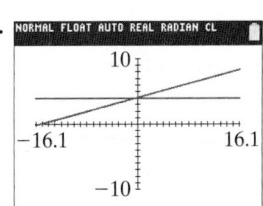

32. a. $r = 4\csc\theta$

  b. $r = 4\csc\left(\theta - \dfrac{\pi}{12}\right)$

  c.

33. $\dfrac{x^2}{9} + \dfrac{(y + 4)^2}{25} = 1$; This is the graph of the same ellipse given in Example 3. The ellipse is centered at $(0, -4)$ with vertices $(0, 1)$ and $(0, -9)$ and minor axis of 6 units.

34. $\dfrac{\left(x + \dfrac{8}{3}\right)^2}{\dfrac{16}{9}} - \dfrac{y^2}{\dfrac{16}{3}} = 1$; This is the graph of the same hyperbola given in Example 2. The hyperbola is centered at $\left(-\dfrac{8}{3}, 0\right)$ with vertices $(-4, 0)$ and $\left(-\dfrac{4}{3}, 0\right)$ and conjugate axis of $\dfrac{8\sqrt{3}}{3} \approx 4.6$ units.

35. $y^2 = -10\left(x - \dfrac{5}{2}\right)$; This is the graph of the same parabola given in Exercise 11: The parabola opens to the left with vertex $\left(\dfrac{5}{2}, 0\right)$, and focal length $\dfrac{5}{2}$ units.

36. $\dfrac{x^2}{12} + \dfrac{(y + 2)^2}{16} = 1$; This is the graph of the same ellipse given in Exercise 15. The ellipse is centered at $(0, -2)$ with vertices $(0, 2)$ and $(0, -6)$ and minor axis of $4\sqrt{3}$ units.

41. The eccentricity of r_1 is $\dfrac{3}{2}$ indicating that the graph is a hyperbola. The eccentricity of r_2 is $\dfrac{2}{3}$ indicating that the graph is an ellipse.

42. Each graph is an ellipse with one focus at the pole. The directrix for r_1 is vertical, indicating that the major axis of the ellipse is horizontal. The directrix for r_2 is horizontal, indicating that the major axis of the ellipse is vertical.

43. a. $e = \dfrac{c}{a} \Rightarrow c = ae$

  b. $x = 2c - r\cos\theta$ or equivalently, $x = 2ea - r\cos\theta$ and $y = r\sin\theta$.
  From the Pythagorean theorem,
  $$r_1^2 = x^2 + y^2 = (2ea - r\cos\theta)^2 + (r\sin\theta)^2$$
  $$= 4e^2a^2 - 4ear\cos\theta + r^2\cos^2\theta + r^2\sin^2\theta$$
  $$= 4e^2a^2 - 4ear\cos\theta + r^2(\cos^2\theta + \sin^2\theta)$$

$= 4e^2a^2 - 4ear\cos\theta + r^2$. Thus,

$r_1 = \sqrt{4e^2a^2 - 4ear\cos\theta + r^2}$.

c. $r + r_1 = 2a$. Thus,

$r + \sqrt{4e^2a^2 - 4ear\cos\theta + r^2} = 2a$

$\sqrt{4e^2a^2 - 4ear\cos\theta + r^2} = 2a - r$

$\left(\sqrt{4e^2a^2 - 4ear\cos\theta + r^2}\right)^2 = (2a - r)^2$

$4e^2a^2 - 4ear\cos\theta + r^2 = 4a^2 - 4ar + r^2$

$4e^2a^2 - 4ear\cos\theta = 4a^2 - 4ar$.

Solving for r, we have

$4ar - 4ear\cos\theta = 4a^2 - 4e^2a^2$

$r(4a - 4ea\cos\theta) = 4a^2 - 4e^2a^2$

$r = \dfrac{4a^2 - 4e^2a^2}{4a - 4ea\cos\theta}$. Dividing by $4a$ gives

$r = \dfrac{a - e^2a}{1 - e\cos\theta}$ or $r = \dfrac{a(1 - e^2)}{1 - e\cos\theta}$.

45. a. $r = \dfrac{5.790 \times 10^7(1 - 0.2056^2)}{1 - 0.2056\cos\theta}$

b. $R_p \approx 46{,}000{,}000$ km and $R_a \approx 70{,}000{,}000$ km

46. a. $r = \dfrac{2.279 \times 10^8(1 - 0.0935^2)}{1 - 0.0935\cos\theta}$

b. $R_p \approx 207{,}000{,}000$ km and $R_a \approx 249{,}000{,}000$ km

47.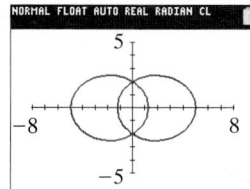

The graph of $r_2 = \dfrac{-2}{1 - 0.6\cos\theta}$ is equivalent to the graph of

$-r_2 = \dfrac{2}{1 - 0.6\cos\theta}$. Replacing r by $-r$ in a polar equation reflects

the graph across the pole. The graph of $r_2 = \dfrac{-2}{1 - 0.6\cos\theta}$ is also

equivalent to the graph of $r_2 = \dfrac{2}{1 + 0.6\cos\theta}$.

48.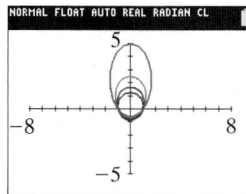

The ellipse becomes more elongated as $e \to 1^-$.

49.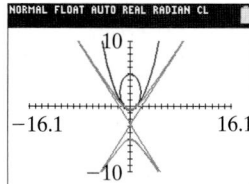

When $0 < e < 1$, the graph is an ellipse. When $e = 1$, the graph is a parabola, and when $e > 1$, the graph is a hyperbola.

Section 10.6 Practice Exercises, pp. 988–993

5. b.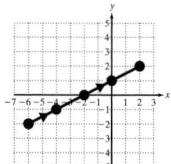

6. b.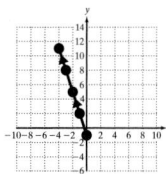

7.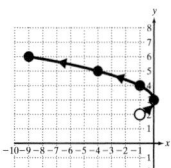

8.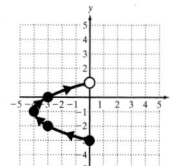

9.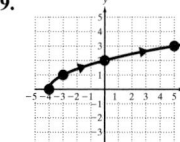

10.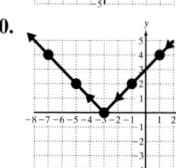

11. b.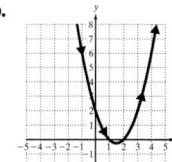

12. b.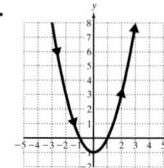

13. b.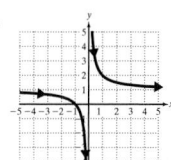

14. b.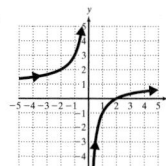

15. b.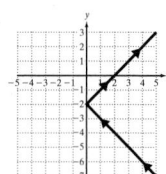

16. b.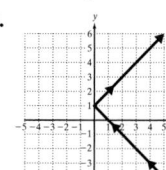

17. b.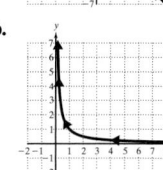

18. b.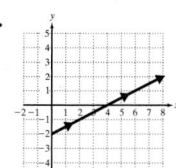

19. a. $x^2 + y^2 = 16$,
$0 \le x \le 4, 0 \le y \le 4$

b.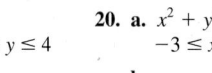

20. a. $x^2 + y^2 = 9$,
$-3 \le x \le 0, 0 \le y \le 3$

b.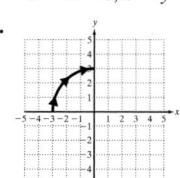

21. b. The curve is traced from right to left for t on the interval $(2n\pi, (2n + 1)\pi)$ and from left to right for t in the interval $((2n + 1)\pi, (2n + 2)\pi)$.

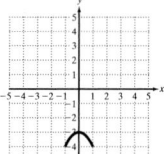

22. b. The curve is traced from left to right for t on the interval $\left(\dfrac{2n - 1}{2}\pi, \dfrac{2n + 1}{2}\pi\right)$ and from right to left for t in the interval $\left(\dfrac{2n + 1}{2}\pi, \dfrac{2n + 3}{2}\pi\right)$.

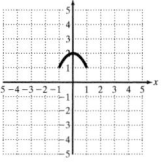

23. b.

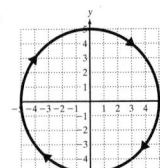

24. b.

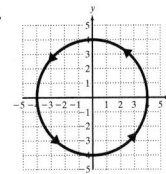

25. b.

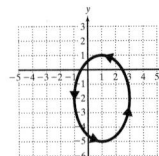

26. b.

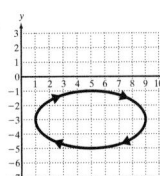

29. $\dfrac{(x-h)^2}{a^2} + \dfrac{(y-k)^2}{b^2} = 1$ **30.** $\dfrac{(x-h)^2}{a^2} - \dfrac{(y-k)^2}{b^2} = 1$

31. a. $x = t$ and $y = 2t - 4$

  b. $x = \dfrac{t}{3}$ and $y = \dfrac{2}{3}t - 4$

  c. $x = -3t$ and $y = -6t - 4$

32. a. $x = t$ and $y = -4t + 1$

  b. $x = \dfrac{t}{2}$ and $y = -2t + 1$

  c. $x = -4t$ and $y = 16t + 1$

33. a. $x = t$ and $y = t + 4, t > 0$

  b. $x = e^t$ and $y = e^t + 4$

34. a. $x = t$ and $y = -2t, t \geq 0$

  b. $x = t^2$ and $y = -2t^2$

38. $x = -\dfrac{1}{2} + \dfrac{1}{2}\cos t$ and $y = 1 + \dfrac{1}{2}\sin t$

54. d. $t = \dfrac{15\sqrt{2} + \sqrt{438}}{8} \approx 5.27$ sec

55. a. Car A: $x = -25 + 40t$ and $y = 0$
    Car B: $x = 0$ and $y = 30 - 50t$
  b. Car A: $t = 0.625$ hr
    Car B: $t = 0.6$ hr
    No. The cars reach the intersection at different times.

56. a. Plane A: $x = 50 - 125t$ and $y = 0$
    Plane B: $x = 0$ and $y = -90 + 200t$
  b. Plane A: $t = 0.4$ hr
    Plane B: $t = 0.45$ hr
    No. The planes reach the tower at different times.

57. d. $y = -\dfrac{x^2}{1250} + x$

59. a.

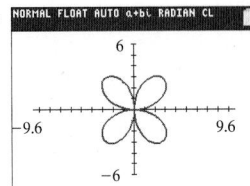

  b. $x = 4\sin 2\theta \cos\theta$ and $y = 4\sin 2\theta \sin\theta$
  c. The graph is the same as part (a).

60. a.

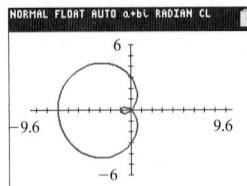

  b. $x = (3 - 4\cos\theta)\cos\theta$ and $y = (3 - 4\cos\theta)\sin\theta$
  c. The graph is the same as part (a).

61. C_1 is the graph of the parabola $y = x^2 - 1$. C_2 is the right branch of the graph of the parabola $y = x^2 - 1$, not including the point $(0, -1)$. C_3 is the graph of the portion of the parabola $y = x^2 - 1$ between $x = -1$ and $x = 1$, inclusive.

62. C_1 is the graph of the upper semicircle of a circle centered at the origin with radius 5. C_2 is the portion of the circle $x^2 + y^2 = 25$ in

the first quadrant. The point $(0, 5)$ is not included and the point $(5, 0)$ is included. C_3 is the graph of the portion of the circle $x^2 + y^2 = 25$ for $-1 \leq x \leq 1$ and $2\sqrt{6} \leq y \leq 5$.

63. No; Eliminating the parameter results in the equation $y = x^2 + 5$. However, the parametric equation $x = \sqrt{t}$ imposes the restriction that $x \geq 0$. Thus, the parametric equations define only the right branch of the parabola $y = x^2 + 5$.

64. No. From the equation $x = 2t$, an x-coordinate of 16 would be produced from a t value of 8. Substituting $t = 8$ into the equation $y = t + 1$ produces a y value of 9 rather than 12.

66. a.

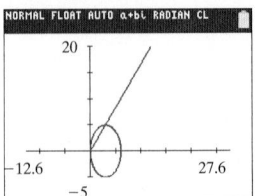

67. We must find the coordinates of point P on the involute curve that begins at the point $(r, 0)$. The distance from T to P is the same as the length of the arc of the circle subtended by angle θ. Thus $d(T, P) = r\theta$. The coordinates of the point T are given by $x = r\cos\theta$ and $y = r\sin\theta$.

To find the coordinates of point P, draw a horizontal line through point P and a vertical line through point T. Point P is represented by $x = r\cos\theta + b$ and $y = a$.

Use right triangle $\triangle TBP$ to find a and b. $\cos\left(\dfrac{\pi}{2} - \theta\right) = \dfrac{b}{r\theta}$ and by the cofunction identity $\sin\theta = \dfrac{b}{r\theta}$, and $b = r\theta\sin\theta$.

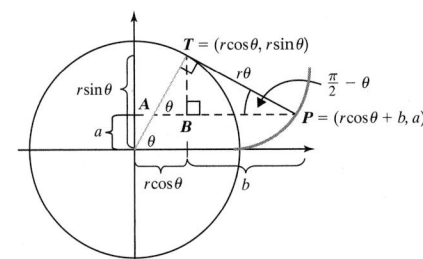

Thus, $x = r\cos\theta + b$
  $= r\cos\theta + r\theta\sin\theta$
  $= r(\cos\theta + \theta\sin\theta)$.

From the same triangle,
$$\sin\left(\dfrac{\pi}{2} - \theta\right) = \dfrac{r\sin\theta - a}{r\theta}$$
$$\cos\theta = \dfrac{r\sin\theta - a}{r\theta}$$
$$r\theta\cos\theta = r\sin\theta - a$$
$$a = r\sin\theta - r\theta\cos\theta$$
Thus, $y = r(\sin\theta - \theta\cos\theta)$.

68.

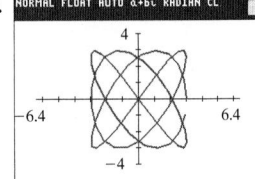

69.

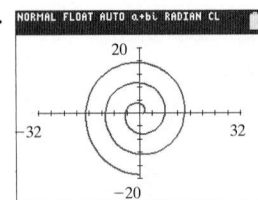

70.

71. b.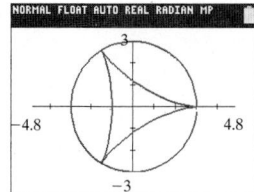

72. b.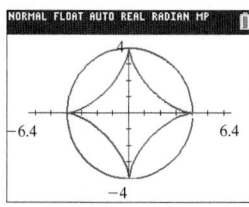

73. b.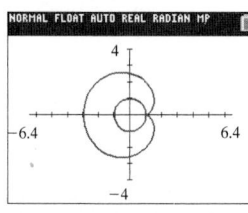

74. b.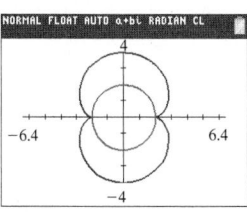

Chapter 10 Review Exercises, pp. 997–1000

1. Each equation represents an ellipse with a vertical major axis of length 30 units and horizontal minor axis of length 20 units. However, the first equation represents an ellipse centered at $(0, 0)$, whereas the second equation represents an ellipse centered at $(1, -7)$.

2. a. $(0, 0)$ **b.** $(-17, 0), (17, 0)$ **f.**
 c. $(0, -8), (0, 8)$
 d. $(-15, 0), (15, 0)$ **e.** $\dfrac{15}{17}$

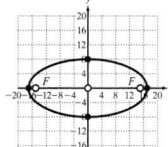

3. a. $(0, 0)$ **b.** $\left(0, -\sqrt{15}\right), \left(0, \sqrt{15}\right)$ **f.**
 c. $(3, 0), (-3, 0)$
 d. $\left(0, -\sqrt{6}\right), \left(0, \sqrt{6}\right)$
 e. $\dfrac{\sqrt{10}}{5}$

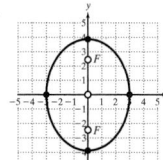

4. a. $(-1, 4)$ **b.** $(-1, -1), (-1, 9)$ **f.**
 c. $(-4, 4), (2, 4)$
 d. $(-1, 0), (-1, 8)$
 e. $\dfrac{4}{5}$

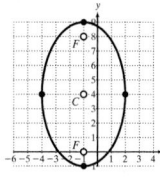

5. a. $(1, 2)$ **b.** $(-3, 2), (5, 2)$ **f.**
 c. $(1, -1), (1, 5)$
 d. $\left(1 - \sqrt{7}, 2\right), \left(1 + \sqrt{7}, 2\right)$
 e. $\dfrac{\sqrt{7}}{4}$

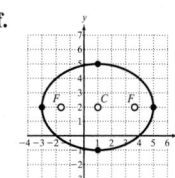

6. a. $\dfrac{(x + 4)^2}{8} + \dfrac{(y - 1)^2}{5} = 1$ **b.** Center: $(-4, 1)$; Vertices: $\left(-4 + 2\sqrt{2}, 1\right), \left(-4 - 2\sqrt{2}, 1\right)$; Endpoints of minor axis: $\left(-4, 1 + \sqrt{5}\right), \left(-4, 1 - \sqrt{5}\right)$; Foci: $\left(-4 + \sqrt{3}, 1\right), \left(-4 - \sqrt{3}, 1\right)$

7. a. $\dfrac{\left(x - \frac{1}{2}\right)^2}{16} + \dfrac{y^2}{25} = 1$ **b.** Center: $\left(\frac{1}{2}, 0\right)$; Vertices: $\left(\frac{1}{2}, 5\right), \left(\frac{1}{2}, -5\right)$; Endpoints of minor axis: $\left(-\frac{7}{2}, 0\right), \left(\frac{9}{2}, 0\right)$; Foci: $\left(\frac{1}{2}, 3\right), \left(\frac{1}{2}, -3\right)$

17. a. $(0, 0)$ **b.** $(3, 0), (-3, 0)$
 c. $\left(\sqrt{13}, 0\right), \left(-\sqrt{13}, 0\right)$
 d. $y = \frac{2}{3}x$ and $y = -\frac{2}{3}x$
 e.

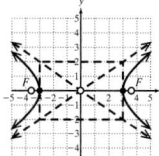

18. a. $(0, 0)$ **b.** $(0, 3), (0, -3)$
 c. $\left(0, \sqrt{15}\right), \left(0, -\sqrt{15}\right)$
 d. $y = \frac{\sqrt{6}}{2}x$ and $y = -\frac{\sqrt{6}}{2}x$
 e.

19. a. $(-3, 2)$ **b.** $(-3, 5), (-3, -1)$ **e.**
 c. $(-3, 7), (-3, -3)$
 d. $y = \frac{3}{4}x + \frac{17}{4}$ and $y = -\frac{3}{4}x - \frac{1}{4}$

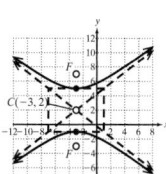

20. a. $(1, -5)$ **b.** $(3, -5), (-1, -5)$ **e.**
 c. $\left(1 + \sqrt{5}, -5\right), \left(1 - \sqrt{5}, -5\right)$
 d. $y = \frac{1}{2}x - \frac{11}{2}$ and $y = -\frac{1}{2}x - \frac{9}{2}$

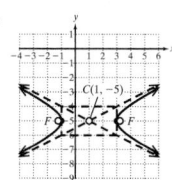

21. a. $\dfrac{(x + 2)^2}{7} - \dfrac{(y + 4)^2}{11} = 1$
 b. Center: $(-2, -4)$; Vertices: $\left(-2 + \sqrt{7}, -4\right), \left(-2 - \sqrt{7}, -4\right)$ Foci: $\left(-2 + 3\sqrt{2}, -4\right), \left(-2 - 3\sqrt{2}, -4\right)$

22. a. $\dfrac{(y + 1)^2}{9} - x^2 = 1$
 b. Center: $(0, -1)$; Vertices: $(0, 2), (0, -4)$ Foci: $\left(0, -1 + \sqrt{10}\right), \left(0, -1 - \sqrt{10}\right)$

25. $\dfrac{(x - 2)^2}{289} - \dfrac{(y - 7)^2}{64} = 1$

30. a. $p = -\frac{1}{2}$ **b.** $(0, 0)$
 c. $\left(0, -\frac{1}{2}\right)$ **d.** 2
 e. $\left(-1, -\frac{1}{2}\right), \left(1, -\frac{1}{2}\right)$
 f. $y = \frac{1}{2}$ **g.** $x = 0$
 h.

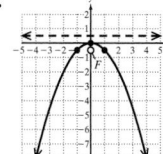

31. a. $p = 5$ **b.** $(0, 0)$
 c. $(5, 0)$ **d.** 20
 e. $(5, 10), (5, -10)$
 f. $x = -5$ **g.** $y = 0$
 h.

32. a. $p = -1$ **b.** $(3, -2)$
 c. $(2, -2)$ **d.** 4
 e. $(2, 0), (2, -4)$
 f. $x = 4$ **g.** $y = -2$
 h.

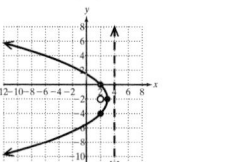

33. a. $p = \frac{3}{2}$ **b.** $\left(\frac{1}{2}, 3\right)$
 c. $\left(\frac{1}{2}, \frac{9}{2}\right)$ **d.** 6
 e. $\left(-\frac{5}{2}, \frac{9}{2}\right), \left(\frac{7}{2}, \frac{9}{2}\right)$
 f. $y = \frac{3}{2}$ **g.** $x = \frac{1}{2}$
 h.

34. a. $(x - 5)^2 = 4(y + 2)$

b. Vertex: $(5, -2)$; Focus: $(5, -1)$; Directrix: $y = -3$

35. a. $\left(y - \frac{3}{2}\right)^2 = -14(x + 3)$

b. Vertex: $\left(-3, \frac{3}{2}\right)$; Focus: $\left(-\frac{13}{2}, \frac{3}{2}\right)$; Directrix: $x = \frac{1}{2}$

49. a. Hyperbola **b.** $30°$ **c.** $16x'^2 - y'^2 - 16 = 0$

d. $x'^2 - \dfrac{y'^2}{16} = 1$

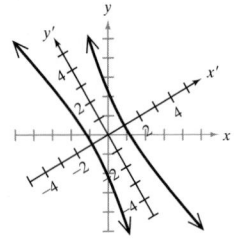

50. a. Ellipse **b.** $45°$ **c.** $x'^2 + 4y'^2 - 2x' + 32y' + 61 = 0$

d. $\dfrac{(x' - 1)^2}{4} + (y' + 4)^2 = 1$

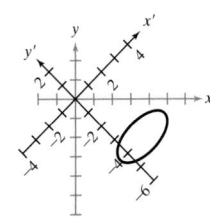

51. a. Parabola **b.** $53.1°$ **c.** $y'^2 + 4x' = 0$

d. $y'^2 = -4x'$

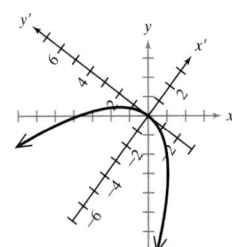

52. a. Hyperbola

b.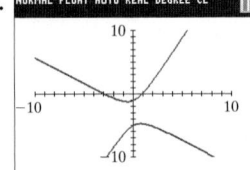

53. a. Ellipse

b.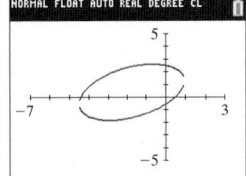

54. a. Parabola **d.**

b. $e = 1$; $y = -\dfrac{3}{2}$

c. Vertex: $\left(\dfrac{3}{4}, \dfrac{3\pi}{2}\right)$

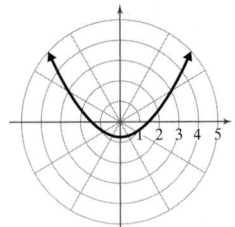

55. a. Ellipse **d.**

b. $e = \dfrac{2}{3}$; $y = -1$

c. Vertices: $\left(2, \dfrac{\pi}{2}\right), \left(\dfrac{2}{5}, \dfrac{3\pi}{2}\right)$

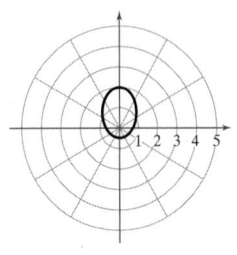

56. a. Hyperbola **d.**

b. $e = \dfrac{3}{2}$; $y = -\dfrac{2}{3}$

c. Vertices: $\left(-2, \dfrac{\pi}{2}\right), \left(\dfrac{2}{5}, \dfrac{3\pi}{2}\right)$

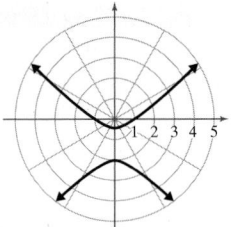

57. a. Hyperbola **d.**

b. $e = 2$; $x = \dfrac{1}{2}$

c. Vertices: $\left(\dfrac{1}{3}, 0\right), (-1, \pi)$

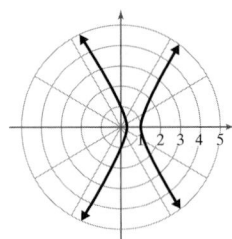

58. a. Ellipse **d.**

b. $e = \dfrac{1}{2}$; $x = 1$

c. Vertices: $\left(\dfrac{1}{3}, 0\right), (1, \pi)$

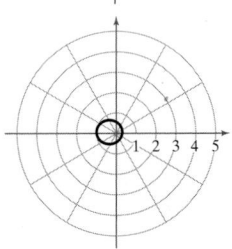

59. a. Parabola **d.**

b. $e = 1$; $x = 2$

c. Vertex: $(1, 0)$

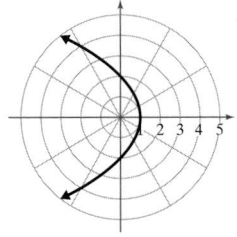

60. a. $r = \dfrac{10}{4 + 4\sin\left(\theta - \dfrac{2\pi}{3}\right)}$ **b.** $r = \dfrac{10}{4 + 4\sin\left(\theta + \dfrac{3\pi}{4}\right)}$

61. a. $r = \dfrac{8}{4 - 5\cos\left(\theta - \dfrac{\pi}{3}\right)}$ **b.** $r = \dfrac{8}{4 - 5\cos\left(\theta + \dfrac{\pi}{2}\right)}$

64. b.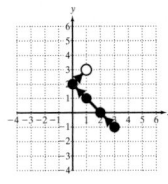

65.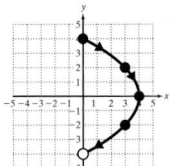

66. b.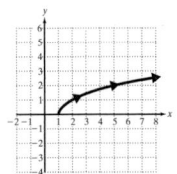

67. b.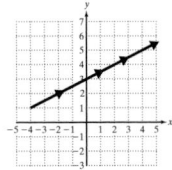

68. b.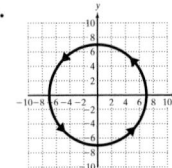

69. a. $\dfrac{(x - 2)^2}{9} + \dfrac{(y + 1)^2}{16} = 1$

b.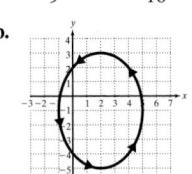

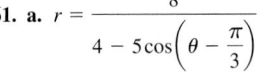

70. a. $y = -x + 4, 0 \le x \le 4$

b. The line segment is traced back and forth on intervals of $\dfrac{\pi}{2}$.

The curve is traced from (4, 0) to (0, 4) on the intervals

$\left(n\pi, \left(\dfrac{2n + 1}{2} \right) \pi \right)$. The curve is traced from (0, 4) to (4, 0)

on the intervals $\left(\left(\dfrac{2n + 1}{2} \right) \pi, (n + 1)\pi \right)$.

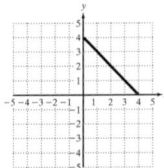

71. C_1 is the top half of the ellipse $\dfrac{x^2}{4} + \dfrac{y^2}{16} = 1$ and C_2 is the left half of

the ellipse. C_1 is traced counterclockwise starting from (2, 0) at $\theta = 0$
and ending at (−2, 0) at $\theta = \pi$. C_2 is traced clockwise starting from
(0, −4) at $\theta = 0$ and ending at (0, 4) at $\theta = \pi$.

Chapter 10 Test, pp. 1001–1003

1. Each equation represents an ellipse centered at the origin with a major
axis of length 24 units and minor axis of length 12 units. However, the
ellipse represented by the first equation has foci on the y-axis, whereas
the second equation represents an ellipse with foci on the x-axis.

5. $\dfrac{(y - k)^2}{a^2} - \dfrac{(x - h)^2}{b^2} = 1$ **6.** $\dfrac{(x - h)^2}{a^2} - \dfrac{(y - k)^2}{b^2} = 1$

9. a. Ellipse   **b.**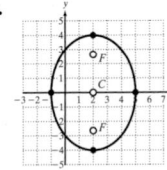

c. Center: (2, 0); Vertices: (2, 4), (2, −4);
Endpoints of minor axis: (−1, 0), (5, 0);
Foci: $(2, \sqrt{7}), (2, -\sqrt{7})$;

Eccentricity: $\dfrac{\sqrt{7}}{4}$

10. a. Hyperbola   **b.**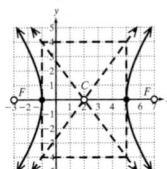

c. Center: (2, 0); Vertices: (−1, 0), (5, 0);
Foci: (−3, 0), (7, 0);
Asymptotes:
$y = \frac{4}{3}x - \frac{8}{3}$ and $y = -\frac{4}{3}x + \frac{8}{3}$;
Eccentricity: $\frac{5}{3}$

11. a. Parabola   **b.**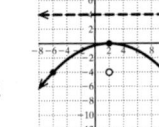

c. Vertex: (2, 0); Focus: (2, −4);
Endpoints of latus rectum:
(10, −4), (−6, −4)
Directrix: $y = 4$; Axis of symmetry: $x = 2$

12. a. Circle   **b.**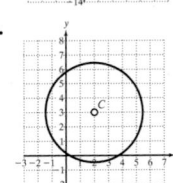

c. Center: (2, 3); Radius: $2\sqrt{3}$

13. a. Hyperbola   **b.**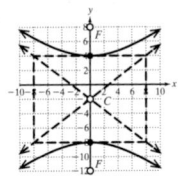

c. Center: (0, −2); Vertices: (0, 4), (0, −8);
Foci: (0, 8), (0, −12);
Asymptotes:
$y = \frac{3}{4}x - 2$ and $y = -\frac{3}{4}x - 2$;
Eccentricity: $\frac{5}{3}$

14. a. Ellipse   **b.**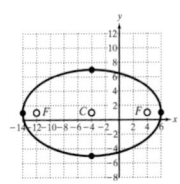

c. Center: (−4, 1);
Vertices: (−14, 1), (6, 1);
Endpoints of minor axis:
(−4, 7), (−4, −5);
Foci: (−12, 1), (4, 1); Eccentricity: $\frac{4}{5}$

15. a. Circle   **b.**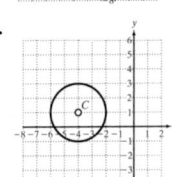

c. Center: (−4, 1); Radius: 2

16. a. Parabola   **b.**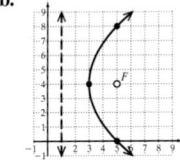

c. Vertex: (3, 4); Focus: (5, 4);
Endpoints of latus rectum:
(5, 0), (5, 8)
Directrix: $x = 1$;
Axis of symmetry: $y = 4$

23. $\dfrac{(x - 4)^2}{4} + \dfrac{(y + 3)^2}{3} = 1$ **24.** $\dfrac{(y - 3)^2}{16} - \dfrac{(x - 4)^2}{20} = 1$

31. The graph is a left semiellipse with center at the origin and major axis
of length 8 units and minor axis of length 6 units.

36. a. Ellipse   **37. a.** Hyperbola

b. 45°   **b.** 18.4°

c. $4x'^2 - 24x' + 9y'^2 = 0$ **c.** $y'^2 - 9x'^2 - 9 = 0$

d. $\dfrac{(x' - 3)^2}{9} + \dfrac{y'^2}{4} = 1$ **d.** $\dfrac{y'^2}{9} - x'^2 = 1$

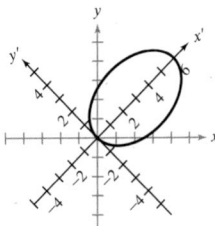   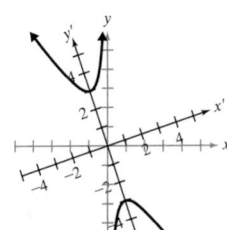

38. a. Hyperbola   **39. a.** Parabola

b. $e = \dfrac{3}{2}; y = \dfrac{4}{3}$ **b.** $e = 1; x = -\dfrac{7}{3}$

c. Vertices: $\left(\dfrac{4}{5}, \dfrac{\pi}{2} \right), \left(-4, \dfrac{3\pi}{2} \right)$ **c.** Vertex: $\left(\dfrac{7}{6}, \pi \right)$

d. 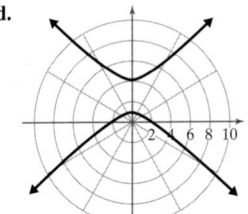   **d.**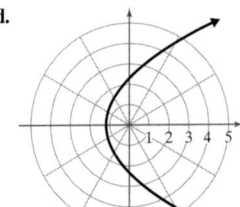

41. b. 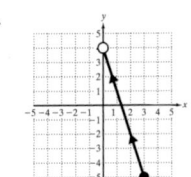   **42. b.**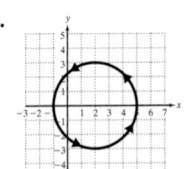

45. b.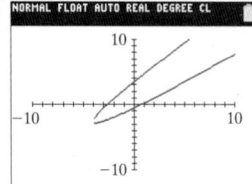

46. b. $r = \dfrac{8}{4 - 2\cos\left(\theta + \dfrac{5\pi}{6}\right)}$

47.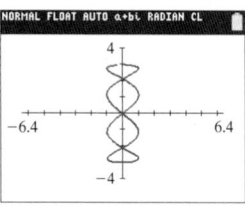

c.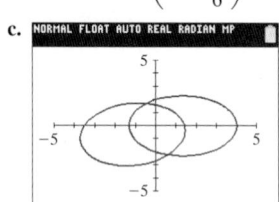

113.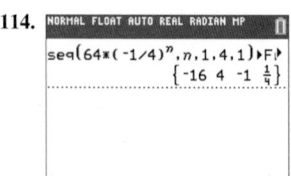

114.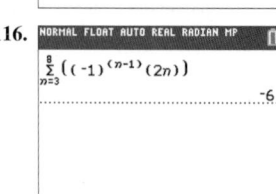

115.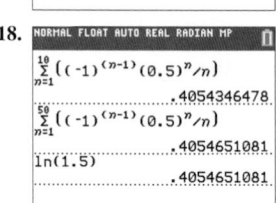

116.

117.

118.

Chapter 10 Cumulative Review Exercises, pp. 1003–1004

13. $\left(-\infty, \dfrac{4 - \sqrt{10}}{2}\right] \cup \left[\dfrac{4 + \sqrt{10}}{2}, \infty\right)$

16.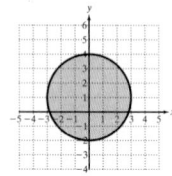

17.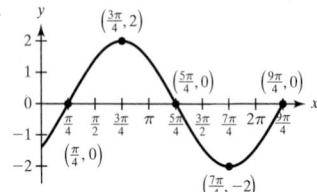

21.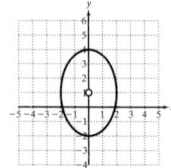

22.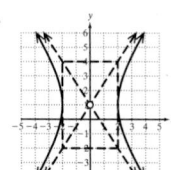

23.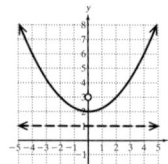

24.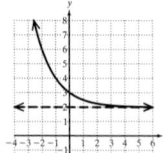

CHAPTER 11

Section 11.1 Practice Exercises, pp. 1013–1017

75. $\displaystyle\sum_{i=1}^{n} \dfrac{i^2}{i + 1}$

89. Expanding the series, there are n terms and each term is c. Therefore, the sum is cn.
$$\sum_{i=1}^{n} c = c + c + c + \cdots + c = cn$$

90. $\displaystyle\sum_{i=1}^{n} (a_i + b_i) = (a_1 + b_1) + (a_2 + b_2) + \cdots + (a_n + b_n)$. Applying the commutative and associative properties of real numbers, the expression equals $(a_1 + a_2 + \cdots + a_n) + (b_1 + b_2 + \cdots + b_n)$
$$= \sum_{i=1}^{n} a_i + \sum_{i=1}^{n} b_i$$

105. The graph of the sequence $a_n = n^2$ is a set of discrete points corresponding to $n = 1, 2, 3, \ldots$, whereas the function $f(x) = x^2$ is a continuous curve over the set of real numbers.

Section 11.2 Practice Exercises, pp. 1024–1028

15. a. 3, 13, 23, 33, 43   **b.** $a_1 = 3$ and $a_n = a_{n-1} + 10$ for $n \ge 2$
16. a. 6, 11, 16, 21, 26   **b.** $a_1 = 6$ and $a_n = a_{n-1} + 5$ for $n \ge 2$
17. a. 4, 2, 0, -2, -4 **b.** $a_1 = 4$ and $a_n = a_{n-1} - 2$ for $n \ge 2$
18. a. 5, 2, -1, -4, -7 **b.** $a_1 = 5$ and $a_n = a_{n-1} - 3$ for $n \ge 2$
80. a. $S_n = \dfrac{n}{2}[1 + (2n - 1)] = \dfrac{n}{2}(2n) = n^2$

Section 11.3 Practice Exercises, pp. 1036–1040

106. Any pattern that may appear for a finite number of terms may change for subsequent terms. The sequence 4, 16, …, for example, may be arithmetic with $d = 12$. It may be geometric with $r = 4$. In other words, we do not know the exact pattern without the nth term given.

108. If the sequence were geometric, the first pair of terms and the second pair of terms would have a common ratio. That is, $\dfrac{a_2}{a_1} = \dfrac{a_3}{a_2}$. However, this is not possible because $\dfrac{x + 2}{x} = \dfrac{x + 4}{x + 2}$ implies that $(x + 2)^2 = x(x + 4)$ and that $x^2 + 4x + 4 = x^2 + 4x$. The last statement is a contradiction.

109. The sequence $\dfrac{1}{a_1}, \dfrac{1}{a_2}, \dfrac{1}{a_3}, \dfrac{1}{a_4}, \ldots$ can be written as $\dfrac{1}{a_1}, \dfrac{1}{a_1 r}, \dfrac{1}{a_1 r^2}, \dfrac{1}{a_1 r^3}, \ldots$ or equivalently as $\dfrac{1}{a_1}, \dfrac{1}{a_1}\left(\dfrac{1}{r}\right), \dfrac{1}{a_1}\left(\dfrac{1}{r}\right)^2, \dfrac{1}{a_1}\left(\dfrac{1}{r}\right)^3, \ldots$. Therefore, the sequence is geometric with common ratio $\dfrac{1}{r}$.

110. The sequence $10^{a_1}, 10^{a_2}, 10^{a_3}, 10^{a_4}, \ldots$ can be written as $10^{a_1}, 10^{a_1 + d}, 10^{a_1 + 2d}, 10^{a_1 + 3d}, \ldots, 10^{a_1 + (n-1)d}, 10^{a_1 + nd}, \ldots$. Dividing the $(n + 1)$th term by the nth term produces a common ratio of 10^d; that is, $\dfrac{10^{a_1 + nd}}{10^{a_1 + (n-1)d}} = 10^{a_1 + nd - [a_1 + (n-1)d]} = 10^d$. The common ratio is 10^d.

111. Each term of the geometric sequence $a_1, a_2, a_3, \ldots$ can be written in the form $a_1 r^{n-1}$. Therefore, $\log a_1, \log a_2, \log a_3, \ldots = \log a_1, \log a_1 r,$ $\log a_1 r^2, \ldots, \log a_1 r^{n-1}, \log a_1 r^n, \ldots$.
$a_{n+1} - a_n = \log a_1 r^n - \log a_1 r^{n-1}$
$\quad = (\log a_1 + \log r^n) - (\log a_1 + \log r^{n-1})$
$\quad = (\log a_1 + n \log r) - [\log a_1 + (n - 1) \log r]$
$\quad = \log r$
The common difference is $\log r$.

Section 11.4 Skill Practice
Answers 1

Let P_n denote the statement that $2 + 4 + \cdots + 2n = n(n + 1)$.
1. P_1 is true because for $n = 1$, the sum is 2 which equals $1(1 + 1)$.
2. Assume that $2 + 4 + \cdots + 2k = k(k + 1)$ (inductive hypothesis). Show that $2 + 4 + \cdots + 2k + 2(k + 1) = (k + 1)[(k + 1) + 1] = (k + 1)(k + 2)$.
By the inductive hypothesis, $[2 + 4 + \cdots + 2k] + 2(k + 1) = [k(k + 1)] + 2k + 2 = k^2 + 3k + 2 = (k + 1)(k + 2)$ as desired.

Answers 2

Let P_n denote the statement $\frac{1}{1 \cdot 3} + \frac{1}{3 \cdot 5} + \cdots + \frac{1}{(2n-1)(2n+1)} = \frac{n}{2n+1}$.

1. P_1 is true because for $n = 1$, the sum is $\frac{1}{1 \cdot 3} = \frac{1}{3}$ which equals $\frac{1}{2(1)+1} = \frac{1}{3}$.

2. Assume that $\frac{1}{1 \cdot 3} + \frac{1}{3 \cdot 5} + \cdots + \frac{1}{(2k-1)(2k+1)} = \frac{k}{2k+1}$ (inductive hypothesis).
   Show that $\frac{1}{1 \cdot 3} + \frac{1}{3 \cdot 5} + \cdots + \frac{1}{(2k-1)(2k+1)} + \frac{1}{[2(k+1)-1][2(k+1)+1]} = \frac{k+1}{2(k+1)+1} = \frac{k+1}{2k+3}$.
   By the inductive hypothesis,
   $\left[\frac{1}{1 \cdot 3} + \frac{1}{3 \cdot 5} + \cdots + \frac{1}{(2k-1)(2k+1)} \right] + \frac{1}{[2(k+1)-1][2(k+1)+1]} = $
   $\frac{k}{2k+1} + \frac{1}{(2k+1)(2k+3)} = \frac{2k^2+3k+1}{(2k+1)(2k+3)} = \frac{(2k+1)(k+1)}{(2k+1)(2k+3)} = \frac{k+1}{2k+3}$ as desired.

Answers 3

Let P_n be the statement that 2 is a factor of $5^n + 1$.

1. P_1 is true because 2 is a factor of $5^1 + 1 = 6$.

2. Assume that 2 is a factor of $5^k + 1$. Then $5^k + 1 = 2a$ for some positive integer a. Equivalently, $5^k = 2a - 1$. Show that 2 is a factor of $5^{k+1} + 1$.
   We have $5^{k+1} + 1 = 5 \cdot 5^k + 1 = 5(2a-1) + 1 = 10a - 4 = 2(5a - 2)$. The value 2 is a factor of $2(5a-2)$, and therefore a factor of $5^{k+1} + 1$ as desired.

Answers 4

Let P_n be the statement that $\left(\frac{3}{2}\right)^n > 2n$ for $n \geq 7$.

1. Show that P_7 is true. That is, show that $\left(\frac{3}{2}\right)^7 > 2(7)$. We have $\left(\frac{3}{2}\right)^7 = \frac{2187}{128} = 17\frac{11}{128}$, which is greater than $2(7) = 14$. Therefore, $\left(\frac{3}{2}\right)^7 > 2(7)$.

2. Assume that $\left(\frac{3}{2}\right)^k > 2k$ for $k \geq 7$. Show that $\left(\frac{3}{2}\right)^{k+1} > 2(k+1)$.
   We have $\left(\frac{3}{2}\right)^{k+1} = \frac{3}{2}\left(\frac{3}{2}\right)^k > \frac{3}{2}(2k) = 3k$ for $k \geq 7$. Furthermore, $3k = 2k + k > 2k + 2$ for $k \geq 7$. Therefore, $\left(\frac{3}{2}\right)^{k+1} > 2(k+1)$ as desired.

Section 11.4 Practice Exercises, pp. 1047–1048

3. Let P_n be the statement $2 + 6 + \cdots + (4n-2) = 2n^2$.
   1. P_1 is true because $2 = 2(1)^2$.
   2. Assume that $2 + 6 + \cdots + (4k-2) = 2k^2$ (Inductive hypothesis).
      Show that $2 + 6 + \cdots + (4k-2) + [4(k+1)-2] = 2(k+1)^2$.
      By the inductive hypothesis,
      $[2 + 6 + \cdots + (4k-2)] + [4(k+1)-2]$
      $= 2k^2 + (4k+2) = 2(k^2+2k+1) = 2(k+1)^2$ as desired.

4. Let P_n be the statement $2 + 8 + \cdots + (6n-4) = n(3n-1)$.
   1. P_1 is true because $2 = 1[3(1)-1]$.
   2. Assume that $2 + 8 + \cdots + (6k-4) = k(3k-1)$ (Inductive hypothesis).
      Show that $2 + 8 + \cdots + (6k-4) + [6(k+1)-4] = (k+1)[3(k+1)-1] = (k+1)(3k+2)$.
      By the inductive hypothesis,
      $[2 + 8 + \cdots + (6k-4)] + [6(k+1)-4]$
      $= k(3k-1) + (6k+2) = 3k^2 + 5k + 2 = (k+1)(3k+2)$ as desired.

5. Let P_n be the statement $5 + 8 + \cdots + (3n+2) = \frac{n}{2}(3n+7)$.
   1. P_1 is true because $5 = \frac{1}{2}[3(1)+7]$.
   2. Assume that $5 + 8 + \cdots + (3k+2) = \frac{k}{2}(3k+7)$ (Inductive hypothesis).
      Show that $5 + 8 + \cdots + (3k+2) + [3(k+1)+2] = \frac{k+1}{2}[3(k+1)+7] = \frac{(k+1)(3k+10)}{2}$.
      By the inductive hypothesis,
      $[5 + 8 + \cdots + (3k+2)] + [3(k+1)+2]$
      $= \frac{k}{2}(3k+7) + 3k + 5 = \frac{3k^2+13k+10}{2}$
      $= \frac{(k+1)(3k+10)}{2}$ as desired.

6. Let P_n be the statement $4 + 9 + \cdots + (5n-1) = \frac{n}{2}(5n+3)$.
   1. P_1 is true because $4 = \frac{1}{2}[5(1)+3]$.
   2. Assume that $4 + 9 + \cdots + (5k-1) = \frac{k}{2}(5k+3)$ (Inductive hypothesis).
      Show that $4 + 9 + \cdots + (5k-1) + [5(k+1)-1]$
      $= \frac{(k+1)}{2}[5(k+1)+3] = \frac{(k+1)(5k+8)}{2}$.
      By the inductive hypothesis,
      $[4 + 9 + 14 + \cdots + (5k-1)] + [5(k+1)-1]$
      $= \frac{k}{2}(5k+3) + (5k+4) = \frac{5k^2+13k+8}{2}$
      $= \frac{(k+1)(5k+8)}{2}$ as desired.

7. Let P_n be the statement $8 + 4 + \cdots + (-4n+12) = -2n(n-5)$.
   1. P_1 is true because $8 = -2(1)(1-5)$.
   2. Assume that $8 + 4 + \cdots + (-4k+12) = -2k(k-5)$ (Inductive hypothesis).
      Show that $8 + 4 + \cdots + (-4k+12) + [-4(k+1)+12]$
      $= -2(k+1)[(k+1)-5] = -2k^2 + 6k + 8$.
      By the inductive hypothesis,
      $[8 + 4 + \cdots + (-4k+12)] + [-4(k+1)+12]$
      $= -2k(k-5) + (-4k+8) = -2k^2 + 6k + 8$ as desired.

8. Let P_n be the statement $12 + 6 + \cdots + (-6n+18) = -3n(n-5)$.
   1. P_1 is true because $12 = -3(1)(1-5)$.
   2. Assume that $12 + 6 + \cdots + (-6k+18) = -3k(k-5)$ (Inductive hypothesis).
      Show that $12 + 6 + \cdots + (-6k+18) + [-6(k+1)+18] = -3(k+1)[(k+1)-5] = -3k^2 + 9k + 12$.
      By the inductive hypothesis,
      $[12 + 6 + \cdots + (-6k+18)] + [-6(k+1)+18]$
      $= -3k(k-5) + (-6k+12) = -3k^2 + 9k + 12$ as desired.

9. Let P_n be the statement $1 + 2 + 2^2 + \cdots + 2^{n-1} = 2^n - 1$.
   1. P_1 is true because $1 = 2^1 - 1$.
   2. Assume that $1 + 2 + 2^2 + \cdots + 2^{k-1} = 2^k - 1$ (Inductive hypothesis).
      Show that $1 + 2 + 2^2 + \cdots + 2^{k-1} + 2^{(k+1)-1} = 2^{k+1} - 1$.
      By the inductive hypothesis,
      $[1 + 2 + 2^2 + \cdots + 2^{k-1}] + 2^{(k+1)-1} = (2^k - 1) + 2^k$
      $= 2 \cdot 2^k - 1 = 2^{k+1} - 1$ as desired.

10. Let P_n be the statement $1 + 3 + 3^2 + \cdots + 3^{n-1} = \frac{1}{2}(3^n - 1)$.
    1. P_1 is true because $1 = \frac{1}{2}(3^1 - 1)$.
    2. Assume that $1 + 3 + 3^2 + \cdots + 3^{k-1} = \frac{1}{2}(3^k - 1)$ (Inductive hypothesis).
       Show that $1 + 3 + 3^2 + \cdots + 3^{k-1} + 3^{(k+1)-1} = \frac{1}{2}(3^{k+1} - 1)$.
       By the inductive hypothesis,
       $[1 + 3 + 3^2 + \cdots + 3^{k-1}] + 3^{(k+1)-1} = \frac{1}{2}(3^k - 1) + 3^k$
       $= \frac{1}{2} \cdot 3^k - \frac{1}{2} + 3^k = \frac{3}{2} \cdot 3^k - \frac{1}{2} = \frac{1}{2}(3 \cdot 3^k - 1)$
       $= \frac{1}{2}(3^{k+1} - 1)$ as desired.

11. Let P_n be the statement $\frac{3}{4} + \frac{3}{16} + \cdots + \frac{3}{4^n} = 1 - \left(\frac{1}{4}\right)^n$.
    1. P_1 is true because $\frac{3}{4} = 1 - \left(\frac{1}{4}\right)^1$.
    2. Assume that $\frac{3}{4} + \frac{3}{16} + \cdots + \frac{3}{4^k} = 1 - \left(\frac{1}{4}\right)^k$ (Inductive hypothesis).
       Show that $\frac{3}{4} + \frac{3}{16} + \cdots + \frac{3}{4^k} + \frac{3}{4^{k+1}} = 1 - \left(\frac{1}{4}\right)^{k+1}$.
       By the inductive hypothesis,
       $\left[\frac{3}{4} + \frac{3}{16} + \cdots + \frac{3}{4^k}\right] + \frac{3}{4^{k+1}} = \left[1 - \left(\frac{1}{4}\right)^k\right] + \frac{3}{4}\left(\frac{1}{4}\right)^k$
       $= 1 - \frac{1}{4}\left(\frac{1}{4}\right)^k = 1 - \left(\frac{1}{4}\right)^{k+1}$ as desired.

12. Let P_n be the statement $\frac{1}{2} + \frac{1}{4} + \cdots + \frac{1}{2^n} = 1 - \left(\frac{1}{2}\right)^n$.

   1. P_1 is true because $\frac{1}{2} = 1 - \left(\frac{1}{2}\right)^1$.

   2. Assume that $\frac{1}{2} + \frac{1}{4} + \cdots + \frac{1}{2^k} = 1 - \left(\frac{1}{2}\right)^k$ (Inductive hypothesis).

   Show that $\frac{1}{2} + \frac{1}{4} + \cdots + \frac{1}{2^k} + \frac{1}{2^{k+1}} = 1 - \left(\frac{1}{2}\right)^{k+1}$.

   By the inductive hypothesis,

$$\left[\frac{1}{2} + \frac{1}{4} + \cdots + \frac{1}{2^k}\right] + \frac{1}{2^{k+1}} = \left[1 - \left(\frac{1}{2}\right)^k\right] + \frac{1}{2}\left(\frac{1}{2}\right)^k$$
$$= 1 - \frac{1}{2}\left(\frac{1}{2}\right)^k = 1 - \left(\frac{1}{2}\right)^{k+1} \text{ as desired.}$$

13. Let P_n be the statement $1 \cdot 2 + 2 \cdot 3 + \cdots + n(n + 1) =$

$$\frac{n(n + 1)(n + 2)}{3}.$$

   1. P_1 is true because $1 \cdot 2 = \dfrac{(1)(1 + 1)(1 + 2)}{3}$.

   2. Assume that $1 \cdot 2 + 2 \cdot 3 + \cdots + k(k + 1) = \dfrac{k(k + 1)(k + 2)}{3}$

   (Inductive hypothesis).

   Show that $[1 \cdot 2 + 2 \cdot 3 + \cdots + k(k + 1)] +$

   $(k + 1)[(k + 1) + 1]$

$$= \frac{(k + 1)[(k + 1) + 1][(k + 1) + 2]}{3} = \frac{k^3 + 6k^2 + 11k + 6}{3}.$$

   By the inductive hypothesis,

   $[1 \cdot 2 + 2 \cdot 3 + \cdots + k(k + 1)] + (k + 1)[(k + 1) + 1]$

$$= \frac{k(k + 1)(k + 2)}{3} + (k + 1)(k + 2)$$
$$= \frac{k^3 + 3k^2 + 2k}{3} + k^2 + 3k + 2$$
$$= \frac{k^3 + 6k^2 + 11k + 6}{3} \text{ as desired.}$$

14. Let P_n be the statement $1 \cdot 3 + 2 \cdot 5 + \cdots + n(2n + 1) =$

$$\frac{n(n + 1)(4n + 5)}{6}.$$

   1. P_1 is true because $1 \cdot 3 = \dfrac{(1)(1 + 1)[4(1) + 5]}{6}$.

   2. Assume that $1 \cdot 3 + 2 \cdot 5 + \cdots + k(2k + 1) =$

   $\dfrac{k(k + 1)(4k + 5)}{6}$ (Inductive hypothesis).

   Show that $1 \cdot 3 + 2 \cdot 5 + \cdots + k(2k + 1) +$

   $(k + 1)[2(k + 1) + 1]$

$$= \frac{(k + 1)[(k + 1) + 1][4(k + 1) + 5]}{6}$$
$$= \frac{4k^3 + 21k^2 + 35k + 18}{6}.$$

   By the inductive hypothesis,

   $[1 \cdot 3 + 2 \cdot 5 + \cdots + k(2k + 1)] + (k + 1)[2(k + 1) + 1]$

$$= \frac{k(k + 1)(4k + 5)}{6} + (k + 1)(2k + 3)$$
$$= \frac{4k^3 + 9k^2 + 5k}{6} + 2k^2 + 5k + 3$$
$$= \frac{4k^3 + 21k^2 + 35k + 18}{6} \text{ as desired.}$$

15. Let P_n be the statement $\left(1 - \dfrac{1}{2}\right)\left(1 - \dfrac{1}{3}\right)\cdots\left(1 - \dfrac{1}{n + 1}\right) = \dfrac{1}{n + 1}$.

   1. P_1 is true because $\left(1 - \dfrac{1}{2}\right) = \dfrac{1}{1 + 1}$.

   2. Assume that $\left(1 - \dfrac{1}{2}\right)\left(1 - \dfrac{1}{3}\right)\cdots\left(1 - \dfrac{1}{k + 1}\right) = \dfrac{1}{k + 1}$

   (Inductive hypothesis).

   Show that $\left(1 - \dfrac{1}{2}\right)\left(1 - \dfrac{1}{3}\right)\cdots\left(1 - \dfrac{1}{k + 1}\right)\left[1 - \dfrac{1}{(k + 1) + 1}\right]$

$$= \frac{1}{(k + 1) + 1} = \frac{1}{k + 2}.$$

By the inductive hypothesis,

$$\left[\left(1 - \frac{1}{2}\right)\left(1 - \frac{1}{3}\right)\cdots\left(1 - \frac{1}{k + 1}\right)\right]\left[1 - \frac{1}{(k + 1) + 1}\right]$$
$$= \frac{1}{k + 1} \cdot \left(1 - \frac{1}{k + 2}\right) = \frac{1}{k + 1} \cdot \frac{k + 1}{k + 2} = \frac{1}{k + 2} \text{ as desired.}$$

16. Let P_n be the statement $\left(1 - \dfrac{1}{2^2}\right)\left(1 - \dfrac{1}{3^2}\right)\cdots\left(1 - \dfrac{1}{(n + 1)^2}\right) =$

$$\frac{n + 2}{2(n + 1)}.$$

   1. P_1 is true because $\left(1 - \dfrac{1}{2^2}\right) = \dfrac{1 + 2}{2(1 + 1)}$.

   2. Assume that $\left(1 - \dfrac{1}{2^2}\right)\left(1 - \dfrac{1}{3^2}\right)\cdots\left(1 - \dfrac{1}{(k + 1)^2}\right)$

   $= \dfrac{k + 2}{2(k + 1)}$ (Inductive hypothesis). Show that

$$\left(1 - \frac{1}{2^2}\right)\left(1 - \frac{1}{3^2}\right)\cdots\left(1 - \frac{1}{(k + 1)^2}\right)\left(1 - \frac{1}{[(k + 1) + 1]^2}\right)$$
$$= \frac{(k + 1) + 2}{2[(k + 1) + 1]} = \frac{k + 3}{2k + 4}.$$

   By the inductive hypothesis,

$$\left[\left(1 - \frac{1}{2^2}\right)\left(1 - \frac{1}{3^2}\right)\cdots\left(1 - \frac{1}{(k + 1)^2}\right)\right]\left(1 - \frac{1}{[(k + 1) + 1]^2}\right)$$
$$= \frac{k + 2}{2(k + 1)}\left[1 - \frac{1}{(k + 2)^2}\right] = \frac{k + 2}{2(k + 1)}\left[\frac{k^2 + 4k + 3}{(k + 2)^2}\right]$$
$$= \frac{k + 2}{2(k + 1)} \cdot \frac{(k + 3)(k + 1)}{(k + 2)^2} = \frac{k + 3}{2k + 4} \text{ as desired.}$$

17. Let P_n be the statement $\sum\limits_{i=1}^{n} 1 = n$.

   1. P_1 is true because $\sum\limits_{i=1}^{1} 1 = 1$.

   2. Assume that $\sum\limits_{i=1}^{k} 1 = k$ (inductive hypothesis).

   Show that $\sum\limits_{i=1}^{k+1} 1 = (k + 1)$. By the inductive hypothesis,

$$\sum_{i=1}^{k+1} 1 = \left(\sum_{i=1}^{k} 1\right) + 1 = k + 1 \text{ as desired.}$$

18. Let P_n be the statement $\sum\limits_{i=1}^{n} i = \dfrac{n(n + 1)}{2}$.

   1. P_1 is true because $\sum\limits_{i=1}^{1} i = 1 = \dfrac{1(1 + 1)}{2}$.

   2. Assume that $\sum\limits_{i=1}^{k} i = \dfrac{k(k + 1)}{2}$ (Inductive hypothesis).

   Show that $\sum\limits_{i=1}^{k+1} i = \dfrac{(k + 1)[(k + 1) + 1]}{2} = \dfrac{(k + 1)(k + 2)}{2}$.

   By the inductive hypothesis,

$$\sum_{i=1}^{k+1} i = \left(\sum_{i=1}^{k} i\right) + (k + 1) = \frac{k(k + 1)}{2} + (k + 1)$$
$$= \frac{k^2 + 3k + 2}{2} = \frac{(k + 1)(k + 2)}{2} \text{ as desired.}$$

19. Let P_n be the statement $\sum\limits_{i=1}^{n} i^2 = \dfrac{n(n + 1)(2n + 1)}{6}$.

   1. P_1 is true because $\sum\limits_{i=1}^{1} i^2 = (1)^2 = \dfrac{1(1 + 1)[2(1) + 1]}{6}$.

2. Assume that $\sum_{i=1}^{k} i^2 = \dfrac{k(k+1)(2k+1)}{6}$ (Inductive hypothesis).

Show that $\sum_{i=1}^{k+1} i^2 = \dfrac{(k+1)[(k+1)+1][2(k+1)+1]}{6} = $

$\dfrac{2k^3 + 9k^2 + 13k + 6}{6}$.

By the inductive hypothesis,

$\sum_{i=1}^{k+1} i^2 = \left(\sum_{i=1}^{k} i^2\right) + (k+1)^2 = \dfrac{k(k+1)(2k+1)}{6} + (k+1)^2$

$= \dfrac{2k^3 + 9k^2 + 13k + 6}{6}$ as desired.

20. Let P_n be the statement $\sum_{i=1}^{n} i^3 = \dfrac{n^2(n+1)^2}{4}$.

  1. P_1 is true because $\sum_{i=1}^{1} i^3 = (1)^3 = \dfrac{(1)^2(1+1)^2}{4}$.

  2. Assume that $\sum_{i=1}^{k} i^3 = \dfrac{k^2(k+1)^2}{4}$ (Inductive hypothesis).

Show that $\sum_{i=1}^{k+1} i^3 = \dfrac{(k+1)^2[(k+1)+1]^2}{4} = $

$\dfrac{k^4 + 6k^3 + 13k^2 + 12k + 4}{4}$.

By the inductive hypothesis,

$\sum_{i=1}^{k+1} i^3 = \left(\sum_{i=1}^{k} i^3\right) + (k+1)^3 = \dfrac{k^2(k+1)^2}{4} + k^3 + 3k^2 + 3k + 1$

$= \dfrac{k^4 + 6k^3 + 13k^2 + 12k + 4}{4}$ as desired.

21. Let P_n be the statement 2 is a factor of $5^n - 3$.
  1. P_1 is true because 2 is a factor of $(5)^1 - 3 = 2$.
  2. Assume that P_k is true; that is, assume that 2 is a factor of $5^k - 3$. This implies that $5^k - 3 = 2a$ and that $5^k = 2a + 3$ for some positive integer a.
Show that P_{k+1} is true; that is, show that 2 is a factor of $5^{k+1} - 3$.
$5^{k+1} - 3 = 5 \cdot 5^k - 3$
          $= 5(2a + 3) - 3$ Replace 5^k by $2a + 3$.
          $= 10a + 12$
          $= 2(5a + 6)$
Therefore, 2 is a factor of $5^{k+1} - 3$ as desired.

22. Let P_n be the statement 2 is a factor of $7^n - 3$.
  1. P_1 is true because 2 is a factor of $(7)^1 - 3 = 4$.
  2. Assume that P_k is true; that is, assume that 2 is a factor of $7^k - 3$. This implies that $7^k - 3 = 2a$ and that $7^k = 2a + 3$ for some positive integer a.
Show that P_{k+1} is true; that is, show that 2 is a factor of $7^{k+1} - 3$.
$7^{k+1} - 3 = 7 \cdot 7^k - 3$
          $= 7 \cdot (2a + 3) - 3$ Replace 7^k by $2a + 3$.
          $= 14a + 18$
          $= 2(7a + 9)$
Therefore, 2 is a factor of $7^{k+1} - 3$ as desired.

23. Let P_n be the statement $4^n - 1$ is divisible by 3. This is equivalent to saying that 3 is a factor of $4^n - 1$.
  1. P_1 is true because 3 is a factor of $(4)^1 - 1 = 3$.
  2. Assume that P_k is true; that is, assume that 3 is a factor of $4^k - 1$. This implies that $4^k - 1 = 3a$ and that $4^k = 3a + 1$ for some positive integer a.
Show that P_{k+1} is true; that is, show that 3 is a factor of $4^{k+1} - 1$.
$4^{k+1} - 1 = 4 \cdot 4^k - 1$
          $= 4 \cdot (3a + 1) - 1$ Replace 4^k by $3a + 1$.
          $= 12a + 3$
          $= 3(4a + 1)$
Therefore, 3 is a factor of $4^{k+1} - 1$ as desired.

24. Let P_n be the statement $5^n - 1$ is divisible by 4. This is equivalent to saying that 4 is a factor of $5^n - 1$.
  1. P_1 is true because 4 is a factor of $(5)^1 - 1 = 4$.
  2. Assume that P_k is true; that is, assume that 4 is a factor of $5^k - 1$. This implies that $5^k - 1 = 4a$ and that $5^k = 4a + 1$ for some positive integer a.
Show that P_{k+1} is true; that is, show that 4 is a factor of $5^{k+1} - 1$.
$5^{k+1} - 1 = 5 \cdot 5^k - 1$
          $= 5 \cdot (4a + 1) - 1$ Replace 5^k by $4a + 1$.
          $= 20a + 4$
          $= 4(5a + 1)$
Therefore, 4 is a factor of $5^{k+1} - 1$ as desired.

29. Let P_n be the statement $n! > 3^n$ for $n \geq 7$.
  1. P_7 is true because $7! = 5040$ and $3^7 = 2187$. Therefore, $7! > 3^7$.
  2. Assume that $k! > 3^k$ for a positive integer $k \geq 7$.
Show that $(k+1)! > 3^{k+1}$.
$(k+1)! = (k+1) \cdot k!$
         $> (k+1)(3^k)$ by the inductive hypothesis.
         $> 3(3^k)$ Since $k \geq 7$, then $(k+1) > 3$.
         $= 3^{k+1}$
Therefore, $(k+1)! > 3^{k+1}$ as desired.

30. Let P_n be the statement $(n+1)! > 4^n$ for $n \geq 6$.
  1. P_6 is true because $(6+1)! = 5040$ and $4^6 = 4096$. Therefore, $(6+1)! > 4^6$.
  2. Assume that $(k+1)! > 4^k$ for a positive integer $k \geq 6$.
Show that $[(k+1)+1]! > 4^{k+1}$, or equivalently, that $(k+2)! > 4^{k+1}$.
$(k+2)! = (k+2) \cdot (k+1)!$
         $> (k+2)(4^k)$ by the inductive hypothesis.
         $> 4(4^k)$ Since $k \geq 6$, $(k+2) > 4$
         $= 4^{k+1}$
Therefore, $(k+2)! > 4^{k+1}$ as desired.

31. Let P_n be the statement $3n < 2^n$ for $n \geq 4$.
  1. P_4 is true because $3(4) = 12$ and $2^4 = 16$.
Therefore, $3(4) < 2^{(4)}$.
  2. Assume that $3k < 2^k$ for a positive integer $k \geq 4$.
Show that $3(k+1) < 2^{k+1}$.
$3(k+1) = 3k + 3$. Furthermore, since $k \geq 4$, $3k > 3$.
         $< 3k + 3k$
         $< 2^k + 2^k$ by the inductive hypothesis.
         $= 2(2^k)$
         $= 2^{k+1}$
Therefore, $3(k+1) < 2^{k+1}$ as desired.

32. Let P_n be the statement $5n < 3^n$ for $n \geq 3$.
  1. P_3 is true because $5(3) = 15$ and $3^3 = 27$. Therefore, $5(3) < 3^3$.
  2. Assume that $5k < 3^k$ for a positive integer $k \geq 3$.
Show that $5(k+1) < 3^{k+1}$.
$5(k+1) = 5k + 5$. Since $k \geq 3$, $10k > 5$.
         $< 5k + 10k$
         $= 15k$
         $= 3(5k)$
         $< 3(3^k)$ by the inductive hypothesis.
         $= 3^{k+1}$
Therefore, $5(k+1) < 3^{k+1}$ as desired.

33. Let P_n be the statement $(xy)^n = x^n y^n$.
  1. P_1 is true because $(xy)^1 = xy = x^1 y^1$.
  2. Assume that $(xy)^k = x^k y^k$ (Inductive hypothesis).
Show that $(xy)^{k+1} = x^{k+1} y^{k+1}$.
Multiplying both sides of the equation $(xy)^k = x^k y^k$ by (xy) gives $(xy)^k(xy) = x^k y^k(xy)$. Using the property $a^n a^m = a^{n+m}$ gives $(xy)^{k+1} = x^{k+1} y^{k+1}$ as desired.

34. Let P_n be the statement $\left(\dfrac{x}{y}\right)^n = \dfrac{x^n}{y^n}$.

    1. P_1 is true because $\left(\dfrac{x}{y}\right)^1 = \dfrac{x}{y} = \dfrac{x^1}{y^1}$.

    2. Assume that $\left(\dfrac{x}{y}\right)^k = \dfrac{x^k}{y^k}$ (Inductive hypothesis).

       Show that $\left(\dfrac{x}{y}\right)^{k+1} = \dfrac{x^{k+1}}{y^{k+1}}$.

       Multiplying both sides of $\left(\dfrac{x}{y}\right)^k = \dfrac{x^k}{y^k}$ by $\left(\dfrac{x}{y}\right)$ gives

       $\left(\dfrac{x}{y}\right)^k\left(\dfrac{x}{y}\right) = \dfrac{x^k}{y^k}\left(\dfrac{x}{y}\right)$. Using the property $a^n a^m = a^{n+m}$ gives

       $\left(\dfrac{x}{y}\right)^{k+1} = \dfrac{x^{k+1}}{y^{k+1}}$ as desired.

35. Let P_n be the statement if $x > 1$, then $x^n > x^{n-1}$.
    1. P_1 is true because $x^1 > 1$ for $x > 1$ and $x^0 = 1$.
       Therefore, $x^1 > x^0$ for $x > 1$.
    2. Assume that $x^k > x^{k-1}$ for $x > 1$.
       Show that $x^{(k+1)} > x^{(k+1)-1}$ or equivalently, $x^{k+1} > x^k$.
       $x^{(k+1)} = x^k \cdot x$
       $> x^{k-1} \cdot x$ by the inductive hypothesis.
       $= x^k x^{-1} \cdot x$
       $= x^k$ as desired.

36. Let P_n be the statement if $0 < x < 1$, then $x^n < x^{n-1}$.
    1. P_1 is true because $x^1 < 1$ for $0 < x < 1$ and $x^0 = 1$.
       Therefore, $x^1 < x^0$ for $0 < x < 1$.
    2. Assume that $x^k < x^{k-1}$ for $0 < x < 1$.
       Show that $x^{(k+1)} < x^{(k+1)-1}$ or equivalently, $x^{k+1} < x^k$
       for $0 < x < 1$.
       $x^{(k+1)} = x^k \cdot x$
       $< x^{k-1} \cdot x$ by the inductive hypothesis.
       $= x^k x^{-1} \cdot x$
       $= x^k$ as desired.

39. The principle of mathematical induction has us test the truth of a statement for $n = 1$. The extended principle of mathematical induction has us test the truth of a statement for the first allowable value of n. In each case, the proof is concluded by showing that the truth of a statement for any other positive integer after the first allowable value of n follows directly from its predecessor.

40. Claiming that the statement is true for all positive integers is actually making an infinite number of claims that cannot all be tested on a case-by-case basis. The statement may be true for $n = 1, 2,$ and 3, but false for other values of n. In fact, the statement is false for $n = 4$.

41. Let P_n be the statement that $n^2 - n$ is even.
    1. P_1 is true because $(1)^2 - (1) = 0$, which is even.
    2. Assume that P_k is true; that is, assume that $k^2 - k$ is even. This implies that $k^2 - k = 2a$ for some positive integer a, and that $k = k^2 - 2a$.
       Show that P_{k+1} is true; that is, show that $(k + 1)^2 - (k + 1)$ is even.
       $(k + 1)^2 - (k + 1)$
       $= k^2 + k$
       $= k^2 + (k^2 - 2a)$ by the inductive hypothesis.
       $= 2k^2 - 2a$
       $= 2(k^2 - a)$, which is an even integer.

42. Let P_n be the statement that $n^2 - n + 1$ is odd.
    1. P_1 is true because $(1)^2 - (1) + 1 = 1$, which is odd.
    2. Assume that P_k is true; that is, assume that $k^2 - k + 1$ is odd. This implies that $k^2 - k + 1 = 2a + 1$ for some positive integer a, and that $k = k^2 - 2a$.
       Show that P_{k+1} is true; that is, show that $(k + 1)^2 - (k + 1) + 1$ is odd.
       $(k + 1)^2 - (k + 1) + 1$
       $= k^2 + k + 1$

       $= k^2 + (k^2 - 2a) + 1$ by the inductive hypothesis.
       $= 2k^2 - 2a + 1$
       $= 2(k^2 - a) + 1$, which is an odd integer.

43. Let P_n be the statement $F_1 + F_2 + \cdots + F_n = F_{n+2} - 1$ for positive integers $n \geq 3$.
    1. P_3 is true because $F_1 + F_2 + F_3 = 1 + 1 + 2 = 4$ and
       $F_{3+2} - 1 = F_5 - 1 = 5 - 1 = 4$.
    2. Assume that P_k is true; that is, assume that
       $F_1 + F_2 + \cdots + F_k = F_{k+2} - 1$.
       Show that P_{k+1} is true; that is, show that
       $F_1 + F_2 + \cdots + F_k + F_{k+1} = F_{[(k+1)+2]} - 1 = F_{k+3} - 1$.
       $F_1 + F_2 + \cdots + F_k + F_{k+1} = (F_1 + F_2 + \cdots + F_k) + F_{k+1}$
       $= (F_{k+2} - 1) + F_{k+1}$ by the inductive hypothesis.
       $= (F_{k+1} + F_{k+2}) - 1$ Replace $F_{k+1} + F_{k+2}$ by F_{k+3}.
       $= F_{k+3} - 1$ as desired.

44. Let P_n be the statement $F_1 + F_3 + \cdots + F_{2n-1} = F_{2n}$ for all positive integers n.
    1. P_1 is true because $F_{2(1)-1} = F_1 = 1$ and $F_{2(1)} = F_2 = 1$.
    2. Assume that P_k is true; that is, assume that
       $F_1 + F_3 + \cdots + F_{2k-1} = F_{2k}$.
       Show that P_{k+1} is true; that is, show that
       $F_1 + F_3 + \cdots + F_{2k-1} + F_{2(k+1)-1} = F_{2(k+1)} = F_{2k+2}$.
       $F_1 + F_3 + \cdots + F_{2k-1} + F_{2(k+1)-1} = (F_1 + F_3 + \cdots + F_{2k-1}) + F_{2k+1}$
       $= (F_{2k}) + F_{2k+1}$ by the inductive hypothesis.
       $= F_{2k+2}$ as desired.

Section 11.5 Practice Exercises, pp. 1052–1054

15. $243x^5 + 405x^4 + 270x^3 + 90x^2 + 15x + 1$
16. $3125x^5 + 9375x^4 + 11{,}250x^3 + 6750x^2 + 2025x + 243$
19. $16x^4 - 160x^3 + 600x^2 - 1000x + 625$
21. $32x^{15} - 80x^{12}y + 80x^9y^2 - 40x^6y^3 + 10x^3y^4 - y^5$
22. $243y^{10} - 405y^8z + 270y^6z^2 - 90y^4z^3 + 15y^2z^4 - z^5$
23. $p^{12} - 6p^{10}w^4 + 15p^8w^8 - 20p^6w^{12} + 15p^4w^{16} - 6p^2w^{20} + w^{24}$
24. $t^{18} - 6t^{15}v^5 + 15t^{12}v^{10} - 20t^9v^{15} + 15t^6v^{20} - 6t^3v^{25} + v^{30}$
25. $0.0016 + 0.0032k + 0.0024k^2 + 0.0008k^3 + 0.0001k^4$
26. $0.0001 + 0.0012m + 0.0054m^2 + 0.0108m^3 + 0.0081m^4$
27. $\dfrac{1}{8}c^3 - \dfrac{3}{4}c^2d + \dfrac{3}{2}cd^2 - d^3$
57. $\dbinom{n}{r} = \dfrac{n!}{r! \cdot (n - r)!}$ and

    $\dbinom{n}{n - r} = \dfrac{n!}{(n - r)! \cdot [n - (n - r)]!} = \dfrac{n!}{(n - r)! \cdot r!}$.

    By the commutative property of multiplication,
    $\dfrac{n!}{r! \cdot (n - r)!} = \dfrac{n!}{(n - r)! \cdot r!}$.

58. $\dbinom{n}{0} = \dfrac{n!}{0! \cdot (n - 0)!} = \dfrac{n!}{0! \cdot n!} = 1$ and

    $\dbinom{n}{n} = \dfrac{n!}{n! \cdot (n - n)!} = \dfrac{n!}{n! \cdot 0!} = 1$.

Section 11.6 Practice Exercises, pp. 1062–1066

47. **b.** AB, AC, BC (*Note:* The order within the individual combinations does not matter. That is, AB or BA represents the same group of two elements.)

48. **a.** WXY, WYX, XWY, XYW, YWX, YXW, WXZ, WZX, XWZ, XZW, ZWX, ZXW, WYZ, WZY, YWZ, YZW, ZWY, ZYW, XYZ, XZY, YXZ, YZX, ZXY, ZYX

85. Using the fundamental principle of counting, we have $8 \cdot 7 \cdot 6 = 336$. There are 8 horses that can cross the finish line first. Once the first horse finishes, there are 7 horses remaining that can come in second. Then there are 6 horses that are available for third place. Alternatively, the number of first-, second-, and third-place ordered arrangements can be found by taking the number of permutations of 8 horses taken 3 at a time, $_8P_3$.

86. The number of combinations of n items taken r at a time does not count specific order among the r items. When counting permutations, we *do* include the order among the r items. By the fundamental principle of counting, r items can be arranged in $r!$ different orders. Therefore, taking the number of permutations and dividing by $r!$ "divides out" the redundancy within the same combination of items due to order.

Section 11.7 Practice Exercises, pp. 1075–1080

33. a. $\dfrac{3}{36} = \dfrac{1}{12}$ **b.** $\dfrac{33}{36} = \dfrac{11}{12}$ **34. a.** $\dfrac{1}{36}$ **b.** $\dfrac{35}{36}$

35. a. $\dfrac{6}{36} = \dfrac{1}{6}$ **b.** $\dfrac{3}{36} = \dfrac{1}{12}$ **36. a.** $\dfrac{15}{36} = \dfrac{5}{12}$ **b.** $\dfrac{10}{36} = \dfrac{5}{18}$

39. a. $\dfrac{{}_{18}C_9}{{}_{34}C_9} = \dfrac{48,620}{52,451,256} \approx 0.00093$

  b. $\dfrac{{}_{16}C_9}{{}_{34}C_9} = \dfrac{11,440}{52,451,256} \approx 0.00022$

  c. The events from parts (a) and (b) are not complementary events. There are many other cases to consider regarding the number of male and female jurors: for example, 4 male, 5 female, etc.

40. a. $\dfrac{{}_6C_4}{{}_{26}C_4} = \dfrac{15}{14,950} \approx 0.001$ **b.** $\dfrac{{}_{20}C_4}{{}_{26}C_4} = \dfrac{4845}{14,950} \approx 0.324$

  c. The events from parts (a) and (b) are not complementary events. There are many other cases to consider regarding the number of defective and good batteries: for example, 2 good, 2 defective, etc.

41. $\dfrac{1}{{}_{39}C_5} = \dfrac{1}{575,757}$ **42.** $\dfrac{5}{{}_{59}C_6} = \dfrac{5}{45,057,474}$

86. a.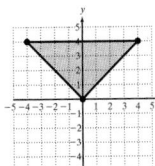

87. a.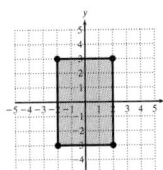

  b. 0.16   **b.** 0.24

88. a.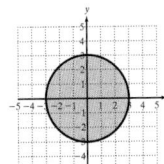

  b. $\dfrac{9\pi}{100} \approx 0.2827$

89. Observe the game being played by other players. Approximate the probability by dividing the number of times a player wins to the number of games played. **92.** The value of a probability is the ratio of the number of elements in an event to the number of elements in the sample space. An event is a subset of the sample space and cannot have more elements than the sample space. Therefore, the value of a probability is always less than or equal to 1. Thus, $\frac{5}{4}$ is impossible for a probability.

95. a. $\dfrac{{}_{13}C_5}{{}_{52}C_5} = \dfrac{1287}{2,598,960} \approx 0.000495$

  b. $(4)\dfrac{{}_{13}C_5}{{}_{52}C_5} = (4)\dfrac{1287}{2,598,960} \approx 0.001981$

96. a. $\dfrac{({}_4C_4)({}_{48}C_1)}{{}_{52}C_5} = \dfrac{48}{2,598,960} \approx 0.0000185$

  b. $13 \cdot \dfrac{({}_4C_4)({}_{48}C_1)}{{}_{52}C_5} = \dfrac{624}{2,598,960} \approx 0.0002401$

Chapter 11 Review Exercises, pp. 1083–1086

15. $\displaystyle\sum_{i=1}^{10} i^3 = \sum_{j=0}^{9} (j+1)^3 = \sum_{k=2}^{11} (k-1)^3$

53. Let P_n be the statement $3 + 7 + \cdots + (4n - 1) = n(2n + 1)$.
  1. P_1 is true because $3 = 1[2(1) + 1]$.
  2. Assume that $3 + 7 + \cdots + (4k - 1) = k(2k + 1)$
    (Inductive hypothesis).
    Show that $3 + 7 + \cdots + (4k - 1) + [4(k + 1) - 1]$
    $= (k + 1)[2(k + 1) + 1] = (k + 1)(2k + 3)$.
    By the inductive hypothesis,
    $[3 + 7 + \cdots + (4k - 1)] + [4(k + 1) - 1] = k(2k + 1) + (4k + 3)$
    $= 2k^2 + 5k + 3$
    $= (k + 1)(2k + 3)$ as desired.

54. Let P_n be the statement $-5 + 2 + \cdots + (7n - 12) = \dfrac{n}{2}(7n - 17)$.

  1. P_1 is true because $-5 = \dfrac{1}{2}[7(1) - 17]$.

  2. Assume that $-5 + 2 + \cdots + (7k - 12) = \dfrac{k}{2}(7k - 17)$
    (Inductive hypothesis).
    Show that $-5 + 2 + \cdots + (7k - 12) + [7(k + 1) - 12]$
    $= \dfrac{k + 1}{2}[7(k + 1) - 17] = \dfrac{(k + 1)(7k - 10)}{2}$.
    By the inductive hypothesis,
    $[-5 + 2 + \cdots + (7k - 12)] + [7(k + 1) - 12] = \dfrac{k}{2}(7k - 17) + (7k - 5)$
    $= \dfrac{7k^2 - 3k - 10}{2}$
    $= \dfrac{(k + 1)(7k - 10)}{2}$ as desired.

55. Let P_n be the statement $1 + 4 + \cdots + 4^{n-1} = \frac{1}{3}(4^n - 1)$.
  1. P_1 is true because $1 = \frac{1}{3}(4^1 - 1)$.
  2. Assume that $1 + 4 + \cdots + 4^{k-1} = \frac{1}{3}(4^k - 1)$
    (Inductive hypothesis).
    Show that $1 + 4 + \cdots + 4^{k-1} + 4^{(k+1)-1} = \frac{1}{3}(4^{k+1} - 1)$.
    By the inductive hypothesis,
    $[1 + 4 + \cdots + 4^{k-1}] + 4^{(k+1)-1} = \frac{1}{3}(4^k - 1) + 4^k$
        $= \frac{1}{3}4^k - \frac{1}{3} + 4^k$
        $= \frac{4}{3} \cdot 4^k - \frac{1}{3}$
        $= \frac{1}{3}(4 \cdot 4^k - 1)$
        $= \frac{1}{3}(4^{k+1} - 1)$ as desired.

56. Let P_n be the statement $10^n - 1$ is divisible by 3. This is equivalent to saying that 3 is a factor of $10^n - 1$.
  1. P_1 is true because 3 is a factor of $(10)^1 - 1 = 9$.
  2. Assume that P_k is true; that is, assume that 3 is a factor of $10^k - 1$. This implies that $10^k - 1 = 3a$ and that $10^k = 3a + 1$ for some positive integer a.
    Show that P_{k+1} is true; that is, show that 3 is a factor of $10^{k+1} - 1$.
    $10^{k+1} - 1 = 10 \cdot 10^k - 1$
              $= 10 \cdot (3a + 1) - 1$ Replace 10^k by $3a + 1$.
              $= 30a + 9$
              $= 3(10a + 3)$
    Therefore, 3 is a factor of $10^{k+1} - 1$ as desired.

57. Let P_n be the statement $4^n < (n + 2)!$ for $n \geq 2$.
  1. P_2 is true because $4^2 = 16$ and $(2 + 2)! = 24$.
    Therefore, $4^2 < (2 + 2)!$.
  2. Assume that $4^k < (k + 2)!$ for an integer $k \geq 2$. Show that $4^{k+1} < [(k + 1) + 2]!$, or equivalently, that $4^{k+1} < (k + 3)!$.
    $4^{k+1} = 4 \cdot 4^k < 4 \cdot (k + 2)!$ by the inductive hypothesis.
    If $k \geq 2$, then $(k + 3) > 4$. Therefore,
    $4^{k+1} < 4(k + 2)! < (k + 3)(k + 2)! = (k + 3)!$ as desired.

60. $256y^4 + 768y^3 + 864y^2 + 432y + 81$
61. $32x^5 - 240x^4 + 720x^3 - 1080x^2 + 810x - 243$
62. $625c^{12} - 500c^9d^2 + 150c^6d^4 - 20c^3d^6 + d^8$
63. $t^{30} + 6t^{25}u^3 + 15t^{20}u^6 + 20t^{15}u^9 + 15t^{10}u^{12} + 6t^5u^{15} + u^{18}$

64. $\frac{1}{8}x^3 + \frac{3}{2}x^2y + 6xy^2 + 8y^3$

76. 143,640; There are 143,640 ways to select 4 distinct items in a specific order from a group of 21 items.   **77.** 5985; There are 5985 ways to select 4 distinct items in no specific order from a group of 21 items.

78. The value $_{10}P_3 = 720$ and $_{10}C_3 = 120$. There are 6 times as many permutations as combinations. This is because each permutation of 3 items can be arranged in $3! = 6$ different orders.

91. a. $\dfrac{_4C_3}{_{15}C_3} = \dfrac{4}{455} \approx 0.00879$ **b.** $\dfrac{_{11}C_3}{_{15}C_3} = \dfrac{165}{455} \approx 0.36264$

   c. The events from parts (a) and (b) are not complementary events. There are many other cases to consider regarding the number of defective and good lightbulbs: for example, 2 good, 1 defective, etc.

Chapter 11 Test, pp. 1086–1088

23. Let P_n be the statement $6 + 10 + \cdots + (4n + 2) = n(2n + 4)$.

   1. P_1 is true because $6 = 1[2(1) + 4]$.

   2. Assume that $6 + 10 + \cdots + (4k + 2) = k(2k + 4)$ (Inductive hypothesis).

   Show that $6 + 10 + \cdots + (4k + 2) + [4(k + 1) + 2]$
   $= (k + 1)[2(k + 1) + 4] = (k + 1)(2k + 6)$.

   By the inductive hypothesis,
   $[6 + 10 + \cdots + (4k + 2)] + [4(k + 1) + 2] = k(2k + 4) + (4k + 6)$
   $= 2k^2 + 8k + 6$
   $= (k + 1)(2k + 6)$ as desired.

24. Let P_n be the statement $1 + 5 + \cdots + 5^{n-1} = \frac{1}{4}(5^n - 1)$.

   1. P_1 is true because $1 = \frac{1}{4}(5^1 - 1)$.

   2. Assume that $1 + 5 + \cdots + 5^{k-1} = \frac{1}{4}(5^k - 1)$ (Inductive hypothesis).

   Show that $1 + 5 + \cdots + 5^{k-1} + 5^{(k+1)-1} = \frac{1}{4}(5^{k+1} - 1)$.

   By the inductive hypothesis,
   $[1 + 5 + \cdots + 5^{k-1}] + 5^{(k+1)-1} = \frac{1}{4}(5^k - 1) + 5^k$
   $= \frac{1}{4}5^k - \frac{1}{4} + 5^k$
   $= \frac{5}{4} \cdot 5^k - \frac{1}{4}$
   $= \frac{1}{4}(5 \cdot 5^k - 1)$
   $= \frac{1}{4}(5^{k+1} - 1)$ as desired.

25. Let P_n be the statement that 2 is a factor of $7^n - 5$.

   1. P_1 is true because 2 is a factor of $(7)^1 - 5 = 2$.

   2. Assume that P_k is true; that is, assume that 2 is a factor of $7^k - 5$. This implies that $7^k - 5 = 2a$ and that $7^k = 2a + 5$ for some positive integer a.

   Show that P_{k+1} is true; that is, show that 2 is a factor of $7^{k+1} - 5$.
   $7^{k+1} - 5 = 7 \cdot 7^k - 5$
   $= 7 \cdot (2a + 5) - 5$ Replace 7^k by $2a + 5$.
   $= 14a + 30$
   $= 2(7a + 15)$

   Therefore, 2 is a factor of $7^{k+1} - 5$ as desired.

26. $\dfrac{x^4}{16} + \dfrac{3}{2}x^3 + \dfrac{27}{2}x^2 + 54x + 81$

27. $1024c^{10} - 1280c^8t^4 + 640c^6t^8 - 160c^4t^{12} + 20c^2t^{16} - t^{20}$

31. A permutation of n items taken r at a time is an arrangement of r items taken from a group of n items in a specific order. A combination of n items taken r at a time is a group of r items taken from a group of n items in no particular order.

Chapter 11 Cumulative Review Exercises, pp. 1088–1090

19. d.   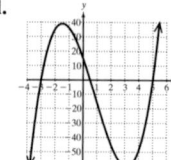   **22.**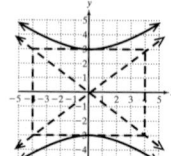

23.   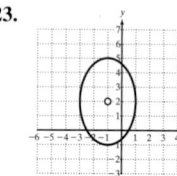   **24.**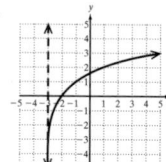

25.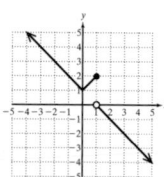

26. The graph of f is the graph of $y = \sqrt{x}$ shifted to the right 1 unit, stretched vertically by a factor of 2, and shifted upwards 3 units.

40.

Photo Credits

Subject Index

Perimeter and Circumference

Rectangle
$P = 2l + 2w$

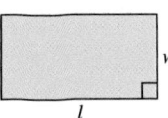

Square
$P = 4s$

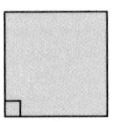

Triangle
$P = a + b + c$

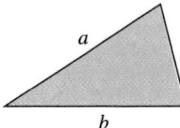

Circle
Circumference: $C = 2\pi r$

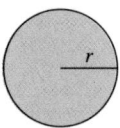

Area

Rectangle
$A = lw$

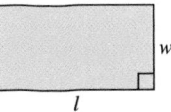

Square
$A = s^2$

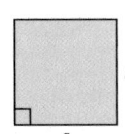

Parallelogram
$A = bh$

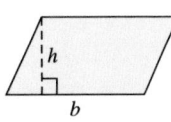

Triangle
$A = \frac{1}{2}bh$

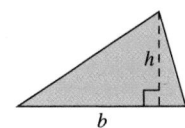

Trapezoid
$A = \frac{1}{2}(b_1 + b_2)h$

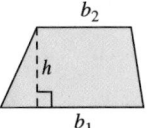

Circle
$A = \pi r^2$

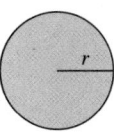

Volume

Rectangular Solid
$V = lwh$

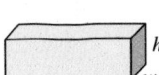

Right Circular Cylinder
$V = \pi r^2 h$

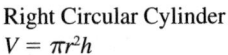
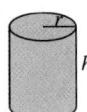

Right Circular Cone
$V = \frac{1}{3}\pi r^2 h$

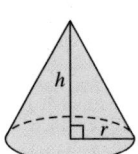

Sphere
$V = \frac{4}{3}\pi r^3$

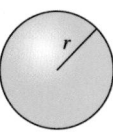

Angles

- Two angles are complementary if the sum of their measures is 90°.

$x + y = 90°$

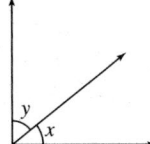

- Two angles are supplementary if the sum of their measures is 180°.

$x + y = 180°$

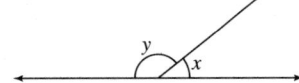

Triangles

- The sum of the measures of the angles of a triangle is 180°.

$x + y + z = 180°$

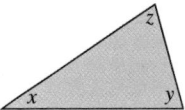

- Given a right triangle with legs of length a and b, and hypotenuse of length c, the Pythagorean theorem indicates that

$a^2 + b^2 = c^2$

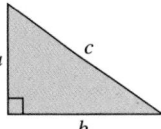

Trigonometric Functions

$\sin t = y$ \qquad $\csc t = \dfrac{1}{y}(y \neq 0)$

$\cos t = x$ \qquad $\sec t = \dfrac{1}{x}(x \neq 0)$

$\tan t = \dfrac{y}{x}(x \neq 0)$ \qquad $\cot t = \dfrac{x}{y}(y \neq 0)$

Right Triangle Trigonometry

$\sin \theta = \dfrac{\text{opp}}{\text{hyp}}$ \qquad $\csc \theta = \dfrac{\text{hyp}}{\text{opp}}$

$\cos \theta = \dfrac{\text{adj}}{\text{hyp}}$ \qquad $\sec \theta = \dfrac{\text{hyp}}{\text{adj}}$

$\tan \theta = \dfrac{\text{opp}}{\text{adj}}$ \qquad $\cot \theta = \dfrac{\text{adj}}{\text{opp}}$

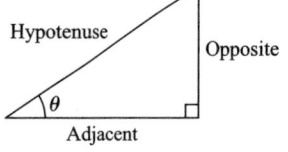

Unit Circle

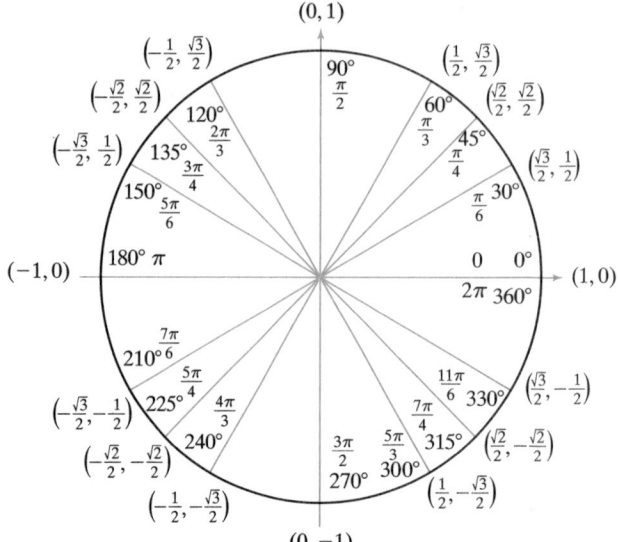

Trigonometric Functions of Any Angle

$\sin \theta = \dfrac{y}{r}$ \qquad $\csc \theta = \dfrac{r}{y}(y \neq 0)$

$\cos \theta = \dfrac{x}{r}$ \qquad $\sec \theta = \dfrac{r}{x}(x \neq 0)$

$\tan \theta = \dfrac{y}{x}(x \neq 0)$

$\cot \theta = \dfrac{x}{y}(y \neq 0)$

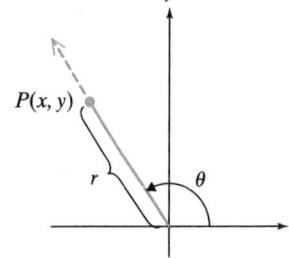

Inverse Trigonometric Functions

$y = \sin^{-1}x \quad \Leftrightarrow \quad \sin y = x$

$-1 \leq x \leq 1 \text{ and } -\dfrac{\pi}{2} \leq y \leq \dfrac{\pi}{2}$

$y = \cos^{-1}x \quad \Leftrightarrow \quad \cos y = x$

$-1 \leq x \leq 1 \text{ and } 0 \leq y \leq \pi$

$y = \tan^{-1}x \quad \Leftrightarrow \quad \tan y = x$

$x \in \mathbb{R} \text{ and } -\dfrac{\pi}{2} < y < \dfrac{\pi}{2}$

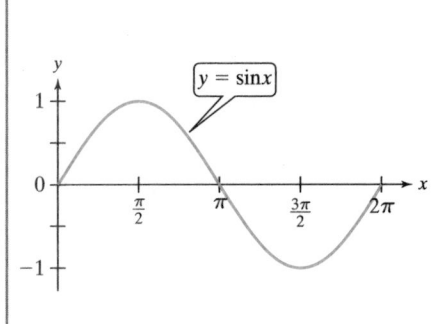

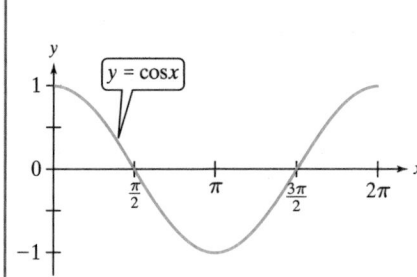

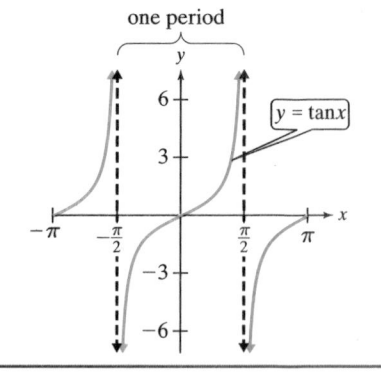

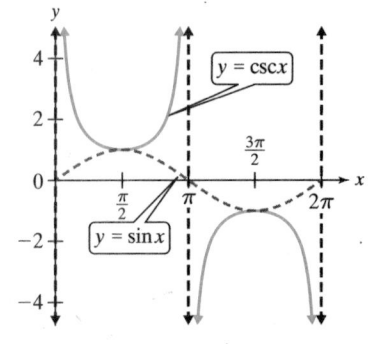

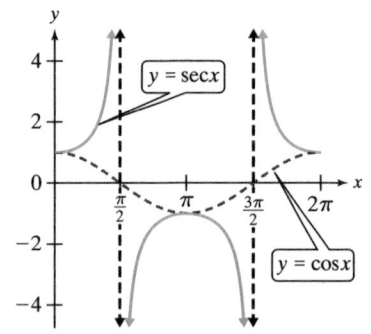

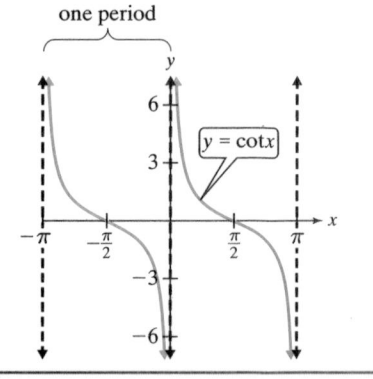